Phagocyte-Pathogen Interactions

Macrophages and the Host Response
to Infection

Phagocyte-Pathogen Interactions

Macrophages and the Host Response to Infection

Edited by

David G. Russell
College of Veterinary Medicine, Cornell University, Ithaca, New York

Siamon Gordon
Sir William Dunn School of Pathology, University of Oxford, Oxford, United Kingdom

Washington, D.C.

Cover: *Leishmania mexicana* in macrophages (courtesy David Russell)

Copyright © 2009 ASM Press
American Society for Microbiology
1752 N Street, N.W.
Washington, DC 20036-2904

Library of Congress Cataloging-in-Publication Data

Phagocyte-pathogen interactions : macrophages and the host response to infection / edited by David G. Russell, Siamon Gordon.
 p. ; cm.
 Includes bibliographical references and index.
 ISBN 978-1-55581-401-4
 1. Phagocytes. 2. Macrophages. 3. Host-parasite relationships. I. Russell, David G., 1956– II. Gordon, Siamon. III. American Society for Microbiology.
 [DNLM: 1. Phagocytes—pathology. 2. Host-Pathogen Interactions—immunology. 3. Macrophages—immunology. QW 690 P5317 2009]

QR185.8.P45P473 2009
616.07'99—dc22

2009016479

All Rights Reserved
Printed in the United States of America

10 9 8 7 6 5 4 3 2 1

Address editorial correspondence to: ASM Press, 1752 N St., N.W., Washington, DC 20036-2904, U.S.A.

Send orders to: ASM Press, P.O. Box 605, Herndon, VA 20172, U.S.A.
Phone: 800-546-2416; 703-661-1593
Fax: 703-661-1501
Email: Books@asmusa.org
Online: estore.asm.org

Contents

Contributors / vii
Foreword: What Would Darwin Have Said? / xi
  Samuel C. Silverstein
Preface / xiii

SECTION I
BIOLOGY OF THE PROFESSIONAL PHAGOCYTE / 1

1 Neutrophils Forever... / 3
TACO W. KUIJPERS, TIMO K. VAN DEN BERG, AND DIRK ROOS

2 Macrophages: Microbial Recognition and Response / 27
ANNETTE PLÜDDEMANN AND SIAMON GORDON

3 Dendritic Cells and Their Tissue Microenvironment during Exposure to Pathogens / 51
A. MORTELLARO, F. GRANUCCI, M. FOTI, AND P. RICCIARDI-CASTAGNOLI

SECTION II
MEMBRANE RECEPTORS: RECOGNITION, ADHESION, PHAGOCYTOSIS, CHEMOTAXIS, AND MIGRATION / 69

4 Fc Receptors and Phagocytosis / 71
STEVEN GREENBERG AND BENJAMIN M. DALE

5 Chemokines and Phagocyte Trafficking / 93
TIMOTHY J. WILLIAMS AND SARA M. RANKIN

6 Toll-Like Receptors / 107
RICARDO T. GAZZINELLI, KATE FITZGERALD, AND DOUGLAS T. GOLENBOCK

7 C-Type Lectins: Multifaceted Receptors in Phagocyte Biology / 123
ALESSANDRA CAMBI AND CARL G. FIGDOR

8 Integrins on Phagocytes / 137
WOUTER L. W. HAZENBOS AND ERIC J. BROWN

9 Intracytosolic Sensing of Pathogens: Nucleic Acid Receptors, NLRs, and the Associated Responses during Infections and Autoinflammatory Diseases / 153
THOMAS HENRY AND DENISE M. MONACK

10 Phagocytes Are a Source of the Fluid-Phase Pattern Recognition Receptor PTX3: Interplay between Cellular and Humoral Innate Immunity / 171
ALBERTO MANTOVANI, BARBARA BOTTAZZI, ANDREA DONI, GIOVANNI SALVATORI, PASCALE JEANNIN, AND CECILIA GARLANDA

11 Leukocyte Chemotaxis / 183
ANN P. WHEELER AND ANNE J. RIDLEY

SECTION III
PHAGOCYTOSIS: SIGNALING, CYTOSKELETON, AND THE PHAGOSOME / 193

12 Signaling for Phagocytosis / 195
JOEL A. SWANSON

13 Membrane Trafficking during Phagosome Formation and Maturation / 209
GREGORY D. FAIRN, ELENA GERSHENZON, AND SERGIO GRINSTEIN

14 Acidification of Endosomes and Phagosomes / 225
SUSHMITA MUKHERJEE AND FREDERICK R. MAXFIELD

15 Actin-Based Motility in Professional Phagocytes / 235
FREDERICK S. SOUTHWICK

16 Functional Analysis of the Intraphagosomal Environment of the Macrophage: Fluorogenic Reporters and the Transcriptional Responses of *Salmonella* and *Mycobacterium* spp. / 249
ROBIN M. YATES, KYLE H. ROHDE, ROBERT B. ABRAMOVITCH, AND DAVID G. RUSSELL

SECTION IV
BRIDGING THE GAP BETWEEN THE INNATE AND ACQUIRED IMMUNE RESPONSE / 265

17 Novel Anti-Inflammatory and Proresolution Lipid Mediators in Induction and Modulation of Phagocyte Function / 267
CHARLES N. SERHAN AND JULIO ALIBERTI

18 Cytokines and Macrophages and Dendritic Cells: Key Modulators of Immune Responses / 281
FRANK KAISER AND ANNE O'GARRA

19 Macrophage Classical Activation / 301
DONALD C. VINH AND STEVEN M. HOLLAND

20 The Functional Heterogeneity of Activated Macrophages / 325
XIA ZHANG AND DAVID M. MOSSER

21 Recognition and Removal of Apoptotic Cells / 341
PETER M. HENSON AND DONNA L. BRATTON

22 Regulation and Antimicrobial Function of Inducible Nitric Oxide Synthase in Phagocytes / 367
CHRISTIAN BOGDAN

SECTION V
PATHOGENS OF THE PROFESSIONAL PHAGOCYTE / 379

23 The Multiple Interactions between *Salmonella* and Phagocytes / 381
JESSICA A. THOMPSON AND DAVID W. HOLDEN

24 *Legionella pneumophila*, a Pathogen of Amoebae and Macrophages / 393
MICHELE S. SWANSON AND ANDREW BRYAN

25 The Role of Phagocytic Cells during *Shigella* Invasion of the Colonic Mucosa / 405
GUY TRAN VAN NHIEU AND PHILIPPE SANSONETTI

26 Autophagy: a Fundamental Cytoplasmic Sanitation Process Operational in All Cell Types Including Macrophages / 419
VOJO DERETIC

27 *Brucella*, a Perfect Trojan Horse in Phagocytes / 427
SUZANA P. SALCEDO AND JEAN-PIERRE GORVEL

28 Interaction of *Candida albicans* with Phagocytes / 437
INÊS FARO-TRINDADE AND GORDON D. BROWN

29 The Parasite Point of View: Insights and Questions on the Cell Biology of *Trypanosoma* and *Leishmania* Parasite-Phagocyte Interactions / 453
KEITH GULL

30 Phagocyte Interactions with the Intracellular Protozoan *Toxoplasma gondii* / 463
ERIC Y. DENKERS

31 Macrophages in Helminth Infection: Effectors, Regulators, and Wound Healers / 477
JUDITH E. ALLEN AND THOMAS A. WYNN

SECTION VI
MODELS OF HOST-PATHOGEN INTERACTIONS / 491

32 *Dictyostelium discoideum*: a Model Phagocyte and a Model for Host-Pathogen Interactions / 493
ZHIRU LI AND RALPH R. ISBERG

33 Phagocytosis in *Drosophila melanogaster* Immune Response / 513
VINCENT LECLERC, ISABELLE CALDELARI, NATALIA VERESHCHAGHINA, AND JEAN-MARC REICHHART

34 The Zebrafish as a Model of Host-Pathogen Interactions / 523
J. MUSE DAVIS AND LALITA RAMAKRISHNAN

35 Whole Genome Screens in Macrophages / 537
BABAK JAVID AND ERIC J. RUBIN

Index / 545

Contributors

ROBERT B. ABRAMOVITCH
Microbiology and Immunology, College of Veterinary Medicine, Cornell University, Ithaca, NY 14853

JULIO ALIBERTI
Divisions of Molecular Immunology and Pulmonary Medicine, Cincinnati Children's Hospital Medical Center, Cincinnati, OH 45229

JUDITH E. ALLEN
Institutes of Evolution, Immunology and Infection Research, University of Edinburgh, Edinburgh EH9 3JT, United Kingdom

CHRISTIAN BOGDAN
Institute of Clinical Microbiology, Immunology and Hygiene, University Clinic of Erlangen, Erlangen, Germany

BARBARA BOTTAZZI
Istituto Clinico Humanitas, IRCCS, via Manzoni 56, 20089, Rozzano, Milan, Italy

DONNA L. BRATTON
Cell Biology and Allergy Divisions, Department of Pediatrics, National Jewish Health, 1400 Jackson Street, Denver, CO 80206

ERIC J. BROWN
Department of Microbial Pathogenesis, Genentech, Inc., 1 DNA Way, M.S. 33, South San Francisco, CA 94080

GORDON D. BROWN
Institute of Infectious Disease and Molecular Medicine, Division of Immunology, University of Cape Town, Observatory, 7925, Cape Town, South Africa

ANDREW BRYAN
Department of Microbiology and Immunology, University of Michigan Medical School, Ann Arbor, MI 48109-5620

ISABELLE CALDELARI
UPR 9022 Institut de Biologie Moléculaire et Cellulaire, 67084 Strasbourg, France

ALESSANDRA CAMBI
Department of Tumor Immunology, Nijmegen Centre for Molecular Life Sciences, Radboud University Nijmegen Medical Centre, Nijmegen, The Netherlands

BENJAMIN M. DALE
Division of Infectious Diseases, Department of Medicine, Immunology Institute, Mount Sinai School of Medicine, New York, NY 10029

J. MUSE DAVIS
Immunology and Molecular Pathogenesis Program, Emory University, 954 Gatewood Drive, Atlanta, GA 30322

ERIC Y. DENKERS
Department of Microbiology and Immunology, College of Veterinary Medicine, Cornell University, Ithaca, NY 14853-6401

VOJO DERETIC
Department of Molecular Genetics and Microbiology, University of New Mexico School of Medicine, Albuquerque, NM 87131

ANDREA DONI
Istituto Clinico Humanitas, IRCCS, via Manzoni 56, 20089, Rozzano, Milan, Italy

GREGORY D. FAIRN
Programme in Cell Biology, Hospital for Sick Children, University of Toronto, Toronto, Ontario, Canada

INÊS FARO-TRINDADE
Institute of Infectious Disease and Molecular Medicine, Division of Immunology, University of Cape Town, Observatory, 7925, Cape Town, South Africa

CARL G. FIGDOR
Department of Tumor Immunology, Nijmegen Centre for Molecular Life Sciences, Radboud University Nijmegen Medical Centre, Nijmegen, The Netherlands

KATE FITZGERALD
Division of Infectious Diseases and Immunology, Dept. of Medicine, University of Massachusetts Medical School, Worcester, MA 01605

MARIA FOTI
Department of Biotechnology and Bioscience, University of Milano-Bicocca, Milano, Italy

CECILIA GARLANDA
Istituto Clinico Humanitas, IRCCS, via Manzoni 56, 20089, Rozzano, Milan, Italy

CONTRIBUTORS

RICARDO T. GAZZINELLI
Div. of Infectious Diseases and Immunology, Dept. of Medicine, University of Massachusetts Medical School, Worcester, MA 01605; Rene Rachou Institute, Oswaldo Cruz Foundation, Institute of Biological Sciences, and Federal University of Minas Gerais, Belo Horizonte, MG, Brazil

ELENA GERSHENZON
Programme in Cell Biology, Hospital for Sick Children and Department of Biochemistry, University of Toronto, Toronto, Ontario, Canada

DOUGLAS T. GOLENBOCK
Div. of Infectious Diseases and Immunology, Dept. of Medicine, University of Massachusetts Medical School, Worcester, MA 01605; Rene Rachou Institute, Oswaldo Cruz Foundation, and Institute of Biological Sciences, Belo Horizonte, MG, Brazil

SIAMON GORDON
Sir William Dunn School of Pathology, University of Oxford, South Parks Road, Oxford OX1 3RE, United Kingdom, and National Cancer Institute, National Institutes of Health, Frederick, MD 21702-1201

JEAN-PIERRE GORVEL
Centre d'Immunologie de Marseille-Luminy, Aix Marseille Université, Faculté de Sciences de Luminy, Case 906, Marseille, 13288 Cedex 9, France; INSERM, U631, Marseille, 13288, France; and CNRS, UMR6102, Marseille, 13288, France

FRANCESCA GRANUCCI
Department of Biotechnology and Bioscience, University of Milano-Bicocca, Milano, Italy

STEVEN GREENBERG
Merck Research Labs, 126 East Lincoln Ave., RY34B-348, Rahway, NJ 07065-0900

SERGIO GRINSTEIN
Programme in Cell Biology, Hospital for Sick Children and Department of Biochemistry, University of Toronto, Toronto, Ontario, Canada

KEITH GULL
Sir William Dunn School of Pathology, University of Oxford, South Parks Road, Oxford OX1 3RE, United Kingdom

WOUTER L. W. HAZENBOS
Department of Microbial Pathogenesis, Genentech, Inc., 1 DNA Way, M.S. 33, South San Francisco, CA 94080

THOMAS HENRY
Stanford University, School of Medicine, Department of Microbiology and Immunology, Fairchild Building, 299 Campus Drive, Stanford, CA 94305-5124

PETER M. HENSON
Cell Biology Division, Department of Pediatrics, National Jewish Health, 1400 Jackson Street, Denver, CO 80206

DAVID W. HOLDEN
Department of Microbiology, Imperial College London, London, United Kingdom

STEVEN M. HOLLAND
Immunopathogenesis Section, Laboratory of Clinical Infectious Diseases, National Institute of Allergy and Infectious Diseases, National Institutes of Health, Bethesda, MD 20892-1684

RALPH R. ISBERG
Howard Hughes Medical Institute and Department of Molecular Biology and Microbiology, Tufts University School of Medicine, 150 Harrison Avenue, Boston, MA 02111

BABAK JAVID
Harvard School of Public Health, 200 Longwood Avenue, Boston, MA 02115

PASCALE JEANNIN
INSERM, U564, Equipe Avenir, 4 rue Larrey, and University Hospital of Angers, Service d'Immunologie, Angers, F-49933 France

FRANK KAISER
National Institute for Medical Research, Division of Immunoregulation, The Ridgeway, Mill Hill, London NW7 1AA, United Kingdom

TACO W. KUIJPERS
Emma Children's Hospital, Academic Medical Centre, University of Amsterdam, Amsterdam, The Netherlands

VINCENT LECLERC
UPR 9022, CNRS, Institut de Biologie Moléculaire et Cellulaire, 67084 Strasbourg, France

ZHIRU LI
Department of Molecular Biology and Microbiology, Tufts University School of Medicine, 150 Harrison Avenue, Boston, MA 02111

ALBERTO MANTOVANI
Istituto Clinico Humanitas, IRCCS, via Manzoni 56, 20089, Rozzano, Milan, Italy, and Department of Translational Medicine, University of Milan, Milan, 20100, Italy

FREDERICK R. MAXFIELD
Department of Biochemistry, Weill Medical College of Cornell University, New York, NY 10065

DENISE M. MONACK
Stanford University, School of Medicine, Department of Microbiology and Immunology, Fairchild Building, 299 Campus Drive, Stanford, CA 94305-5124

ALESSANDRA MORTELLARO
Singapore Immunology Network (SIgN), Agency for Science, Technology and Research (A*STAR), Singapore, and Department of Biotechnology and Bioscience, University of Milano-Bicocca, Milano, Italy

DAVID M. MOSSER
Department of Cell Biology and Molecular Genetics and the Maryland Pathogen Research Institute, University of Maryland, College Park, MD 20742

SUSHMITA MUKHERJEE
Department of Biochemistry, Weill Medical College of Cornell University, New York, NY 10065

ANNE O'GARRA
National Institute for Medical Research, Division of Immunoregulation, The Ridgeway, Mill Hill, London NW7 1AA, United Kingdom

ANNETTE PLÜDDEMANN
Sir William Dunn School of Pathology, University of Oxford, South Parks Road, Oxford OX1 3RE, United Kingdom

LALITA RAMAKRISHNAN
Department of Microbiology, University of Washington, 1959 NE Pacific Street, Seattle, WA 98195

SARA M. RANKIN
Leukocyte Biology Section, NHLI Division, Faculty of Medicine, Imperial College London, South Kensington Campus, Exhibition Road, London SW7 2AZ, United Kingdom

JEAN-MARC REICHHART
UPR 9022 Institut de Biologie Moléculaire et Cellulaire, 67084 Strasbourg, France

PAOLA RICCIARDI-CASTAGNOLI
Singapore Immunology Network (SIgN), Agency for Science, Technology and Research (A*STAR), Singapore, and Department of Biotechnology and Bioscience, University of Milano-Bicocca, Milano, Italy

ANNE J. RIDLEY
King's College London, Randall Division of Cell and Molecular Biophysics, New Hunt's House, Guy's Campus, London SE1 1UL, United Kingdom

KYLE H. ROHDE
Microbiology and Immunology, College of Veterinary Medicine, Cornell University, Ithaca, NY 14853

DIRK ROOS
Sanquin Research at CLB, and Landsteiner Laboratory, Academic Medical Center, University of Amsterdam, Amsterdam, The Netherlands

ERIC J. RUBIN
Harvard School of Public Health, 200 Longwood Avenue, Boston, MA 02115

DAVID G. RUSSELL
Microbiology and Immunology, College of Veterinary Medicine, Cornell University, Ithaca, NY 14853

SUZANA P. SALCEDO
Centre d'Immunologie de Marseille-Luminy, Aix Marseille Université, Faculté de Sciences de Luminy, Case 906, Marseille, 13288 Cedex 9, France; INSERM, U631, Marseille, 13288, France; and CNRS, UMR6102, Marseille, 13288, France

GIOVANNI SALVATORI
Sigma-Tau R&D, via Pontina Km 30.400, 00040, Pomezia, Rome, Italy

PHILIPPE SANSONETTI
Unité de Pathogénie Microbienne Moléculaire, Institut Pasteur, 28, rue du Dr. Roux, 75724 Paris Cedex 15, France

CHARLES N. SERHAN
Center for Experimental Therapeutics and Reperfusion Injury, Department of Anesthesiology, Perioperative and Pain Medicine, Brigham and Women's Hospital and Harvard Medical School, Boston, MA 02115

SAMUEL C. SILVERSTEIN
Department of Physiology and Cellular Biophysics, Columbia University, New York, NY 10032

FREDERICK S. SOUTHWICK
Gainesville, FL 32610-0275

JOEL A. SWANSON
Department of Microbiology and Immunology, University of Michigan Medical School, Ann Arbor, MI 48109-0620

MICHELE S. SWANSON
Department of Microbiology and Immunology, University of Michigan Medical School, Ann Arbor, MI 48109-5620

JESSICA A. THOMPSON
Department of Microbiology, Imperial College London, London, United Kingdom

GUY TRAN VAN NHIEU
Unité de Communication Intercellulaire et Infections Microbiennes, Inserm U971, Collège de France, 11, Place Marcelin Berthelot, 75005 Paris, France

TIMO K. VAN DEN BERG
Sanquin Research at CLB, and Landsteiner Laboratory, Academic Medical Center, University of Amsterdam, Amsterdam, The Netherlands

NATALIA VERESHCHAGHINA
Neuroscience and Behavioral Disorders, Laboratory of Neural Stem Cell, Duke-NUS (National University of Singapore) Graduate Medical School, 2 Julan Bukit Merah, Singapore 169547

DONALD C. VINH
Immunopathogenesis Section, Laboratory of Clinical Infectious Diseases, National Institute of Allergy and Infectious Diseases, National Institutes of Health, Bethesda, MD 20892-1684

ANN P. WHEELER
Queen Mary University of London, Institute of Cell and Molecular Science, Barts, and The London School of Medicine and Dentistry, Blizard Building, 4 Newark St., London E1 2AT, United Kingdom

TIMOTHY J. WILLIAMS
Leukocyte Biology Section, NHLI Division, Faculty of Medicine, Imperial College London, South Kensington Campus, Exhibition Road, London SW7 2AZ, United Kingdom

THOMAS A. WYNN
Immunopathogenesis Section, National Institute of Allergy and Infectious Diseases, National Institutes of Health, Bethesda, MD 20892-8003

ROBIN M. YATES
Microbiology and Immunology, College of Veterinary Medicine, Cornell University, Ithaca, NY 14853

XIA ZHANG
Department of Cell Biology and Molecular Genetics and the Maryland Pathogen Research Institute, University of Maryland, College Park, MD 20742

FOREWORD

What Would Darwin Have Said?

SAMUEL C. SILVERSTEIN

The eminent immunologists, microbial geneticists, and cell and molecular biologists whose reviews appear in this volume have contributed significantly to the spectacular advances in understanding of the molecular pathogenesis of bacterial and protozoan diseases that have characterized the last quarter-century. They are leading representatives of the small army of investigators who have identified the cellular pathways and signaling mechanisms that viral, bacterial, and protozoan pathogens have evolved to cross mucous membranes; evade the innate and acquired defense mechanisms of host epithelial and myeloid cells; counter the toxic effects of defensins, perforins, complement, acid, and reactive nitrogen and oxygen intermediates; and set up housekeeping in a protective and supportive extracellular environment (e.g., biofilms) or intracellular compartment.

We have long recognized that viruses, bacteria, and protozoa are superb probes for various intracellular processes and pathways, e.g., entry into and exit from the endocytic pathway (Horwitz and Silverstein, 1980; Katz et al., 1977; Portnoy et al., 2002; Sibley and Andrews, 2000; Silverstein, 1970; Silverstein and Marcus, 1964); uncovering sites and mechanisms of membrane protein synthesis, transport, glycosylation, and secretion (Margulis and Fester, 1991); and identifying the roles of antibodies in inhibiting infection and promoting destruction of viruses (Silverstein et al., 1976; Silverstein et al., 1977). Microbial pathogens also have proved extraordinarily helpful in unlocking the secrets of the vertebrate immune system, as Zinkernagel and Doherty's seminal papers on major histocompatibility complex (MHC)-restricted killing of lymphocytic choriomeningitis virus- and influenza virus-infected mouse cells demonstrated (Ada, 1994). Similarly, the discoveries that viable *Listeria* cells secrete a hemolysin enabling them to enter their host cell's cytoplasmic matrix, present peptide antigens via the class I MHC, and elicit $CD8^+$ cytotoxic lymphocytes that elicit sterilizing immunity, while heat-killed *Listeria* cells are sequestered in host cell lysosomes and present peptide antigens via class II MHC that promote formation of anti-*Listeria* antibody but do not block progression of *Listeria* infection (Sibley and Andrews, 2000), stimulated a still ongoing avalanche of research on antigen-presenting cells and the mechanisms they utilize to process and present antigens.

Roles for actin and myosin in motility and membrane movement of phagocytes were identified soon after the discovery that all eukaryotic cells contain these muscle proteins (Specter, 2007). But it took more than a decade for investigators to recognize that *Listeria* and *Shigella* express proteins that promote actin filament assembly, and that these filaments propel *Listeria* and *Shigella* through the host cell's cytoplasm and enable them to spread to neighboring cells without exposure to the extracellular environment (Pizarro-Cerda and Cossart, 2006; Portnoy et al., 2002; Sibley and Andrews, 2000). An even greater time interval separates the discovery by Monod and Jacob that galactose activates Lac operon expression, and the later discovery that signals initiated by interaction of enteroinvasive bacteria with intestinal cells activate both bacterial and intestinal cell genes that facilitate penetration of mucosal surfaces and entry of these bacteria into the cytoplasm of macrophages and intestinal epithelial cells (Pedron et al., 2003).

Why did it take investigators so long to search for and to uncover these phenomena? Could it be that, like the British prelate's wife (who is reported to have said of Darwin's theory of evolution, "Let's hope it is not true, and if it is, that not too many people learn of it"), cell biologists and cellular immunologists have resisted the lessons of evolution?

The availability of chip technologies assures that such resistance notwithstanding, the genes and gene products induced in bacteria, protozoa, and host eukaryotic cells by their interactions with one another soon will be identified. Hopefully, such experiments will tell us whether interactions between microbial pathogens and their metazoan hosts stimulate expression of host cell proteins involved in nutrient transport. If so, I for one hope that such findings will stimulate greater interest in the nutrients, nutritional cofactors, and microenvironments to which bacteria and protozoa gain access by taking up residence within eukaryotic cells. Does *Legionella* seek out cisternae of the endoplasmic reticulum because this organelle provides an en-

---

Samuel C. Silverstein, Department of Physiology and Cellular Biophysics, Columbia University, New York, NY 10032.

vironment enriched in iron, sulfur-containing amino acids, and oxidized glutathione? Do the membranes of the vacuoles within which M. *tuberculosis* grows express transporters that facilitate transport of fatty acids from the cytoplasm to the bacterium? Do *Listeria* and *Trypanosoma cruzi* find nutrients in the host cell's cytoplasm that are unavailable or in short supply in extracellular tissue spaces? How and why do prosthetic devices signal biofilm production by bacteria? The answers to these questions will likely provide insights into the physiology of both microbial pathogens and their eukaryotic cell hosts, and may uncover novel methods for inhibiting growth of these pathogens.

Given that archea and eubacteria combined to form the first eukaryotic cells (Pedron et al., 2003), and that free-living unicellular eukaryotes spent at least a billion years preying on prokaryotes and one another and likely exchanging genes with the viruses, bacteria, fungi, and protozoa that evolved in association with them, we should not be surprised that both prokaryotes and eukaryotes retain the ability to signal one another. Indeed, the coevolution of pro- and eukaryotic cells may have exerted some constraints on the signaling and metabolic pathways they employ to accomplish specific tasks. Would a better understanding of these constraints simplify the task of dissecting host-parasite relationships?

The spectacular expansion in knowledge catalyzed by the past two decades of genome sequencing, structural analysis of proteins, and biochemical and genetic dissection of organelle biogenesis and membrane trafficking pathways makes clear that the discovery of each design principle simplifies the task of finding the next one. The principles already in hand enable us to make intelligent guesses about the molecular and cellular mechanisms that govern various pathways in health and disease. No doubt many more of these principles will become evident with the development of ever more powerful tools.

I began this essay with the question, "What would Darwin have said?" It is my guess that he would have been rendered speechless by the extent to which advances in the physical and biological sciences have deepened and extended his evolutionary concepts. Hopefully, the extent of progress, the robustness of contemporary life science research, and the capacity of contemporary computers would have disabused him of his gloomy prediction that the human mind could not "possibly.... add up.... the full effects of many slight variations, accumulated over an almost infinite number of generations" (Darwin, 1979). He might have become persuaded that science itself has evolved to the point at which analyses and experiments he had thought to be beyond human capacity are now within reach. Indeed, these tools have made it possible to identify the viral genes and proteins responsible for the virulence of the influenza virus strain that caused the 1918 pandemic (Tumpey et al., 2005). Similar analyses of retroviruses buried in the genomes of humans and other animals may provide insights into the organization of cells of extinct organisms (Tumpey et al., 2005).

For students of host-pathogen interactions, the take-home message of these advances in molecular paleontology is clear. The past is indeed prologue. As the globe warms and population density increases, so does the likelihood of resurgence of known infectious agents and the appearance of new ones. Each new ecological niche creates new opportunities for transmission to humans and other metazoans of microorganisms previously thought to be harmless members of an insect's, plant's, or animal's endogenous flora. This is the lesson of AIDS, the severe acute respiratory syndrome, *Legionella*, and biofilms. The study of host-microbial pathogen interactions is a timeless endeavor. Evolution assures that investigators in this field will never run out of problems to explore. I expect this conclusion would have brought a smile to Darwin's reserved countenance and perhaps a comment, "How could it have been otherwise?"

REFERENCES

Ada, G. 1994. Twenty years into the saga of MHC-restriction. *Immunol. Cell Biol.* **72:**447–454.

Darwin, C. 1979. *Origin of Species*, p. 453. Gramercy Books, New York, NY.

Helenius, A., J. Kartenbeck, K. Simons, and E. Fries. 1980. On the entry of Semliki forest virus into BHK-21 cells. *J. Cell Biol.* **84:**404–420.

Horwitz, M. A., and S. C. Silverstein. 1980. Legionnaires' disease bacterium (*Legionella pneumophila*) multiples intracellularly in human monocytes. *J. Clin. Investig.* **66:**441–450.

Katz, F. N., J. E. Rothman, V. R. Lingappa, G. Blobel, and H. F. Lodish. 1977. Membrane assembly in vitro: synthesis, glycosylation, and asymmetric insertion of a transmembrane protein. *Proc. Natl. Acad. Sci. USA* **74:**3278–3282.

Margulis, L., and R. Fester. 1991. Bellagio conference and book. *Symbiosis as Source of Evolutionary Innovation: Speciation and Morphogenesis*. Conference—June 25–30, 1989, Bellagio Conference Center, Italy. *Symbiosis* **11:**93–101.

Pedron, T., C. Thibault, and P. J. Sansonetti. 2003. The invasive phenotype of *Shigella flexneri* directs a distinct gene expression pattern in the human intestinal epithelial cell line Caco-2. *J. Biol. Chem.* **278:**33878–33886.

Pizarro-Cerda, J., and P. Cossart. 2006. Bacterial adhesion and entry into host cells. *Cell* **124:**715–727.

Portnoy, D. A., V. Auerbuch, and I. J. Glomski. 2002. The cell biology of *Listeria monocytogenes* infection: the intersection of bacterial pathogenesis and cell-mediated immunity. *J. Cell Biol.* **158:**409–414.

Sibley, L. D., and N. W. Andrews. 2000. Cell invasion by un-palatable parasites. *Traffic* **1:**100–106.

Silverstein, S. 1970. Macrophages and viral immunity. *Semin. Hematol.* **7:**185–214.

Silverstein, S. C., J. K. Christman, and G. Acs. 1976. The reovirus replicative cycle. *Annu. Rev. Biochem.* **45:**375–408.

Silverstein, S. C., and P. I. Marcus. 1964. Early stages of Newcastle disease virus-HeLa cell interaction: an electron microscopic study. *Virology* **23:**370–380.

Silverstein, S. C., R. M. Steinman, and Z. A. Cohn. 1977. Endocytosis. *Annu. Rev. Biochem.* **46:**669–722.

Specter, M. 2007. Darwin's surprise: why are evolutionary biologists bringing back extinct deadly viruses? *New Yorker*, December 3, 2007.

Tumpey, T. M., C. F. Basler, P. V. Aguilar, H. Zeng, A. Solorzano, D. E. Swayne, N. J. Cox, J. M. Katz, J. K. Taubenberger, P. Palese, and A. Garcia-Sastre. 2005. Characterization of the reconstructed 1918 Spanish influenza pandemic virus. *Science* **310:**77–80.

Preface

Our appreciation of the complex biology of the phagocyte has its roots in the last two decades of the 19th century, when Ilya Ilyich Mechnikov made his extraordinary observations on starfish larvae. He watched some highly motile cells in dissected larvae clustering around, and internalizing, particulate material he had added to the cell suspension. From these basic observations he hypothesized that these cells constituted a key arm in the protection of a multicelled organism against infection. At that time the "humoralist" theory of serum-dependent immunity held sway, and the heavyweights of infection biology—Louis Pasteur, Emil Von Behring, and Robert Koch—were extremely skeptical of this cellular theory. Fortunately, however, the seminal nature of Mechnikov's hypothesis was eventually recognized and he was awarded the 1908 Nobel Prize for Medicine, jointly with Paul Ehrlich, a pioneer of humoral immunity. Pasteur recruited him to the Pasteur Institute in Paris, where he remained for the rest of his career.

In multicelled organisms, cells that indulge in phagocytosis have since been recognized to be a specialized cell type, and the term "professional phagocytes" was coined by Michel Rabinovitch to be applied to cells for which phagocytosis is their primary function (Rabinovitch, 1995). To accomplish this task, these cells have a broad range of adhesion receptors capable of recognizing and binding a wide variety of ligands, both self and foreign. Once engaged, these phagocytic receptors trigger rapid rearrangement of the subtending cytoskeleton, which leads to engulfment and internalization of the bound particle. In single-celled organisms such as amoebae, this type of behavior is a means of nutrient acquisition. However, in multicelled animals the roles fulfilled by this behavior are many, and vary with the nature of the professional phagocyte and its tissue environment.

Cells of this phagocyte lineage are derived from hemapoietic stem cells in the bone marrow. These stem cells give rise to a variety of distinct lineages capable of fulfilling a broad range of roles linked to homeostasis as well as defense. The phagocyte lineages that fulfill housekeeping functions include *osteoclasts*, which remodel bone, *Kupffer cells*, which remove damaged erythrocytes from circulation, and *macrophages*, which are responsible for a wide range of homeostatic functions from removing cellular debris from tissues to the regulation of cholesterol levels in the serum.

Cells of the *polymorphonuclear leukocyte* lineage, including *neutrophils*, *granulocytes*, and *eosinophils*, are the advance guard to any foreign invasion. They have a hair-trigger to any microbial insult, are readily recruited to the site of infection, and have a battery of antimicrobial weapons such as a robust superoxide burst and a range of hydrolase-containing secretory granules. Their response is a knee-jerk reaction to infection and can be highly destructive to the surrounding tissue unless downregulated rapidly. The next phagocytes to arrive at the scene are the *blood monocyte-derived macrophages*, which clean up the damaged tissue, and the *dendritic cells*, whose role it is to sample antigen and ferry it back to the nearest draining lymph node to present to T cells. Antigen sampling and presentation to lymphocytes is hard-wired into the developmental cycle of the dendritic cell to maximize the efficiency of the process. The subsequent acquired immune response dampens the neutrophil activity and upregulates the capacity of the macrophage to kill microbes and to present antigen, thus transforming it from a degradative cell to an antimicrobial cell that is now also capable of stimulating lymphocytes.

The different phagocyte lineages act as a team, communicating with one another and developing the tissue response from a relatively nonspecific inflammatory response to one that is antigen specific. These evolving behaviors are orchestrated by a range of receptors and chemical messengers produced by the different phagocytes. With respect to the initial response, recent advances in innate immunity have led to the identification of a range of highly conserved pattern recognition receptors capable of reacting to a wide gamut of molecular patterns that are conserved, or recurring, in the microbial kingdom. These receptors are found both on the cell surface and endocytic pathway (Toll-like receptors [TLRs] and non-TLRs) and within the cell's cytoplasm (nucleotide oligomerization domain [NOD] and retinoic acid-induced gene 1 [RIG-1]-like proteins). Engagement of these receptors initiates a proinflammatory and subsequent anti-inflammatory response through the release of chemokines responsible for the recruitment of mononuclear leukocytes, macrophages, and cytokines, such as tumor necrosis factor-α (which amplifies the response) and interleukin-10 (which downregulates it). This inflammatory reaction continues to build until one starts to see lymphocytes appearing at the tissue site, at which point the inflammation becomes more regulated and the neutrophils are usually supplanted by macrophages. This transition is orchestrated predominantly by cytokines, such as interferon-γ, produced by natural killer cells and T lymphocytes. It is this environment that induces a functional shift in the macrophage, transforming it from a noninflammatory, degradative cell responsible for tissue homeostasis into an antimicrobial cell with the ability to generate reactive oxygen and nitrogen intermediates and to process and present antigen.

This volume contains chapters from recognized experts in phagocyte behavior that explore and explain the complex biology behind the roles fulfilled by these different phagocyte lineages. The chapters discuss the different cell types and how they sense and respond to their environment. They also explore the properties and function of the phagosome itself, which is linked intimately with the divergent roles fulfilled by the professional phagocyte lineages. These chapters include discussion of the link between innate immunity and the development of specific cellular and humoral immune responses. Finally, the volume includes chapters that explore a variety of different infection models and how the phagocytes respond to control the infection and bolster protective immunity. The present volume therefore complements the earlier pathogen-centered volume edited by Cossart et al. (2005).

The volume is not encyclopedic; we have been selective, and the contents are a reflection of the interests of the two editors and their love of the macrophage and its cellular biology. Of all the phagocytes, we feel the macrophage is the most versatile, and yet the least celebrated. Most of the phagocyte lineages are specialized, reminiscent of the fragment attributed to the Greek philosopher Archilocus and popularized by Isaiah Berlin, among others: "The fox knows many things, but the hedgehog knows one big thing" (Berlin, 1953). To our minds, the macrophage is "the fox," and the dendritic cell and other leukocytes are "hedgehogs."

DAVID G. RUSSELL
SIAMON GORDON

REFERENCES

Berlin, I. 1953. *The hedgehog and the fox: an essay on Tolstoy's view of history.* Weidenfeld and Nicolson, London, United Kingdom.

Cossart, P., P. Boquet, S. Normark, and R. Rappuoli (ed.). 2005. *Cellular Biology,* 2nd ed. ASM Press, Washington, DC.

Rabinovitch, M. 1995. Professional and non-professional phagocytes: an introduction. *Trends Cell Biol.* **5:**85–87.

Acknowledgments

We would like to express our gratitude to our assistants Amelia Molloy, Christine Holt, and Sachiko Funaba, who expended considerable effort in the preparation of this volume. We are also very grateful for the editorial support provided by Greg Payne, who worked tirelessly to identify all the editorial issues we missed!

BIOLOGY OF THE PROFESSIONAL PHAGOCYTE

I

1

Neutrophils Forever ...

TACO W. KUIJPERS, TIMO K. VAN DEN BERG, AND DIRK ROOS

Neutrophils constitute the major type of leukocyte in peripheral blood, with counts ranging from 40 to 70% of the leukocytes under normal conditions. Neutrophils are also called polymorphonuclear leukocytes (PMNs) or "granulocytes," but strictly speaking this last name is a designation that includes neutrophilic granulocytes (neutrophils), eosinophilic granulocytes (eosinophils), and basophilic granulocytes (basophils). This nomenclature is derived from the differences in color between the many granular structures in the cytoplasm of these cells, upon standard hematoxylin-eosin staining procedures. Neutrophilic granulocytes protect our bodies against bacterial and fungal infections. For this purpose, neutrophils are equipped with a machinery to sense the site of an infection, to crawl toward the invading microorganisms, and to ingest and kill them. Thus, for proper functioning of this line of defense, sufficient neutrophils must be generated and released from the bone marrow and these cells must be capable of executing a large number of different functions. However, because the neutrophil products are potentially harmful also to the host tissue, many safeguards exist to prevent such unwanted side effects.

DEVELOPMENT

Neutrophils mature in the bone marrow in about 2 weeks, a process in which the myeloid-specific growth factors granulocyte colony-stimulating factor (G-CSF) and granulocyte-monocyte-CSF (GM-CSF) play an important role. In the first half of this period, the neutrophil precursor cells undergo five divisions and differentiate from myeloblasts through promyelocytes to neutrophilic myelocytes. Myeloblasts are cells with a large nucleus relative to the surrounding cytoplasm, which contains no or few granules. In the promyelocyte stage, the so-called azurophil granules are formed, and the specific granules are formed in the myelocytic stage. Because cell divisions still occur after azurophil granule formation has stopped, these granules are distributed over the daughter cells and are complemented with specific granules actively formed during the myelocytic stage. The final ratio of azurophil to specific granules is about 1:2. Later stages of neutrophil differentiation comprise metamyelocytes, band forms, and segmented cells. As the names indicate, these stages are characterized by the typical appearance of the nucleus (Fig. 1A). Divisions do not take place during this period.

Approximately 60% of all nucleated cells in the bone marrow belong to the myeloid series. During myelopoiesis, which includes the development of granulocytic and monocytic lineages, transcription factors from several families are active, including AML1/CBF beta, PU.1, C/EBP isoforms, Ets, c-Myb, HOX, and MZF-1 (Fig. 1B). Few of these factors are expressed exclusively in myeloid cells; instead, it appears that they cooperatively regulate transcription of myeloid-specific genes (Theilgaard-Monch et al., 2005). These factors can be held responsible—in some way or another—for certain phenotypic aberrations when deficient in mice or humans (Anderson et al., 1999; Lekstrom-Hines et al., 1999; Zhang et al., 1998).

The bone marrow comprises a reserve pool of mature neutrophils of approximately 10 times the number of neutrophils in the circulation. Under normal conditions, these cells are released into the blood 2 days after completion of their maturation.

It is not known which factors exactly cause egress of blood cells from the bone marrow under normal conditions. Evidence suggests that protease release by neutrophils in the bone marrow may contribute to hematopoietic progenitor cell (HSC) mobilization. Matrix metalloproteinase-9 (MMP-9), neutrophil elastase (NE), and cathepsin G (CG) accumulate in the bone marrow during G-CSF treatment, where they are thought to degrade key substrates for homing and embedding in the bone marrow environment, including vascular cell adhesion molecule-1 (VCAM-1) and CXCL12 (SDF-1, stromal cell-derived factor-1). Apart from the HSC mobilization, it is believed that both MMP-9 and NE also play an important role in the egress of mature neutrophils. Serine proteases such as NE have been shown to inactivate CXCR4 as well as its ligand CXCL12,

---

Taco W. Kuijpers, Emma Children's Hospital, Academic Medical Centre, University of Amsterdam, Amsterdam, The Netherlands.
Timo K. van den Berg and Dirk Roos, Sanquin Research at CLB, and Landsteiner Laboratory, Academic Medical Center, University of Amsterdam, Amsterdam, The Netherlands.

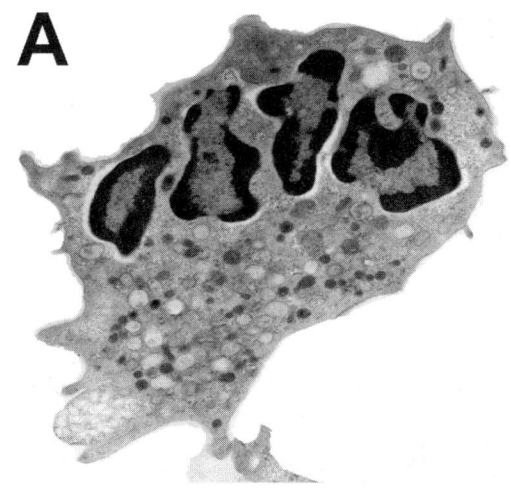

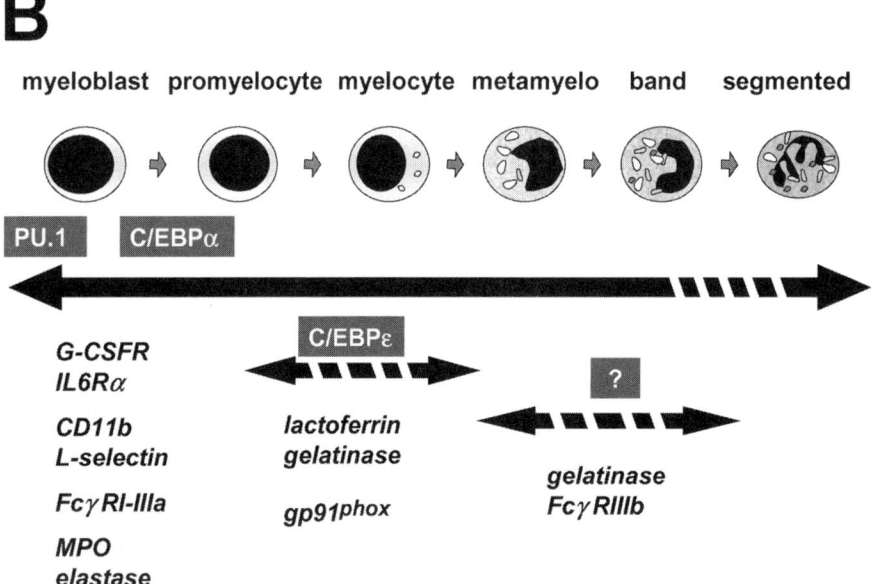

FIGURE 1 (A) Electron microscopy picture showing a human neutrophil with multilobular nucleus and rich granular content. (B) Development of mature neutrophils with transcription factors involved in the differentiation and formation of some important surface proteins and granular components.

accommodating egress of the cells. However, HSC mobilization by G-CSF was found to be normal in MMP-9$^{-/-}$ mice, NE$^{-/-}$ CG$^{-/-}$-double-knockout mice, or mice lacking dipeptidyl peptidase I (CD26), an enzyme required for the functional activation of many hematopoietic serine proteases. Moreover, combined inhibition also had no significant effect on HSC mobilization. VCAM-1 expression on bone marrow stromal cells decreased during G-CSF treatment, but VCAM-1 cleavage has been shown not to be required for HSC mobilization. Furthermore, an as yet unidentified, protease-independent mechanism may contribute to the downregulation of CXCL12 expression (Levesque et al., 2004). Thus, egress depends on both protease-dependent and -independent pathways. Also in humans, CXCR4/CXCL12 interactions have been shown to play a role, as demonstrated by the effects of the CXCR4 antagonist AMD3100 (Broxmeyer et al., 2005). This holds true for neutrophils as well (Liles et al., 2003).

During infections, neutrophil release into the circulation is accelerated, together with band forms (up to 10^{12} per day in adults instead of the usual 10^{11} per day). The appearance of immature neutrophils in the blood is reflected by a decreased chemotactic responsiveness of these cells. In the circulation, about half of the neutrophils are in a marginated pool of cells sequestered in postcapillary venules. This pool is released when epinephrine is generated by exercise or injected intravenously. Neutrophils circulate for only 6 to 10 h; thereafter, they move to the tissues, where they remain active for about 2 to 6 days, depending on the clinical conditions (longer under circumstances of inflammation). Part of the neutrophils will disappear in the gut, but very little is known about other sites of neutrophil apoptosis and destruction by tissue macrophages (see also chapter 27).

Normally, less than 2% of all neutrophils in the body are circulating in the blood. In case the number of neutrophils in the peripheral blood is strongly reduced for unknown reasons (neutropenia) and/or a lack of releasability from the bone marrow by G-CSF or corticosteroids has been found, bone marrow examination and myeloid cell culture may be needed to assess the production, maturation, and differentiation of the bone marrow pool of neutrophils.

NEUTROPHILIC GRANULOCYTES, WHAT IS IN A NAME?

Neutrophil granule constituents are traditionally considered as potent antimicrobial peptides and proteolytic enzymes, specific to the neutrophil. These proteins assist in the killing and digestion of microorganisms but are potentially harmful to the host if released inappropriately. This perception of granule proteins needs to be changed to accommodate the importance of granule membrane proteins for the ability of neutrophils to perceive signals from the environment, to acknowledge the heterogeneity of granules, and to integrate the results obtained by genomics and sensitive proteomics that have greatly expanded the number of proteins known to be localized in neutrophil granules. These modern techniques have revealed that proteins previously believed to be specific to neutrophils are also expressed in a variety of other cell types, for instance, in epithelial cells of skin and mucosal barriers. The important difference is that epithelial cells make these antibacterial peptides and proteins only when an infection or inflammation is established, whereas neutrophils have these antibacterial proteins stored in granules ready for use without any delay when needed.

As mentioned before, the azurophil granule proteins are synthesized at the promyelocytic stage only. Specific granule proteins are synthesized at the myelocyte stage, with gelatinase being formed at an increasing rate at the later metamyelocyte and early-band cell stage, after which granule formation stops (Borregaard et al., 2007). The so-called secretory vesicles are believed to be formed by endocytosis and fuse with the plasma membrane most readily, to some extent already during standard purification steps (Kuijpers, 1991).

The rapidly mobilized secretory vesicles contain, for instance, many of the pattern recognition molecules such as Toll-like receptors (TLRs) as well as the serum lectins Ficolin-1 and pentraxin-3 that have a role in innate immunity (Borregaard et al., 2007). The specific granules seem to support cell movement in the tissues; the azurophil granules are for killing and degrading ingested material. However, not all proteins released from the granules act in a proinflammatory manner. In addition, acute-phase proteins (haptoglobin and orosomucoid) and several protease inhibitors, such as secretory leukocyte protease inhibitor (SLPI), α_1-antitrypsin, cystatin C, and cystatin F, are easily released from the neutrophils. These soluble proteins may help to respond rapidly to the environment and, at the same time, minimize tissue destruction by, for instance, secreting inhibitors of its own proteases. In contrast, the azurophil granules exocytose after considerable cell activation only, which is a fail-safe mechanism preventing unnecessary injury.

The trafficking of molecules and membranes within cells is a prerequisite for all aspects of cellular immune functions, including the delivery and recycling of cell surface proteins, secretion of immune mediators, ingestion of pathogens, and activation of lymphocytes. Soluble N-ethylmaleimide-sensitive factor accessory protein receptor (SNARE) family members mediate membrane fusion during all steps of trafficking, and function in almost all aspects of innate and adaptive immune responses. A SNARE protein on a donor membrane binds to cognate SNAREs on the target membrane (as a trans-SNARE complex) that transiently bridges the two membranes. Members of the SNARE family are characterized by a conserved, central coiled-coil SNARE motif that mediates these protein interactions. The SNARE family is divided into R-SNAREs (usually on vesicles) and Q-SNAREs (usually at the target membrane), depending on whether the central functional residue in their SNARE motif is arginine (R) or glutamine (Q). R-SNAREs (such as the vesicle-associated membrane proteins or VAMPs) are single-transmembrane proteins that contribute one SNARE motif to the trans-SNARE complex. Individual Q-SNARE proteins are subclassified as Qa-, Qb-, Qc- or Qb,c-SNAREs on the basis of the relative position of their SNARE motifs in the assembled trans-SNARE complex. The Q-SNARE functions as a complex that is composed of two or three of these polypeptides (Hong, 2005; Cai et al., 2007).

The density of these fusogenic proteins may dictate the releasability of the different granular compartments. For instance, in neutrophils, VAMP2 is highest in secretory vesicles, decreasing in gelatinase-dense and specific granules (Mollinedo et al., 2006). One of the characteristic exocytotic triggers consists of a rapid and transient elevation in intracellular Ca^{2+} ($[Ca^{2+}]_i$), which will cause exocytosis primarily of the granules that have the highest concentration of SNAREs. That the mRNA of VAMP2 strongly increases during differentiation from promyelocytes to mature neutrophils may explain the differential threshold for release. On the other hand, phosphorylation of Q-SNAREs can also turn them on or off differentially, depending on a single signal or a series of signals from small GTPases of the Rab family or protein kinases for the release of the granular content. Azurophilic granules are mobilized later than specific and gelatinase granules, and instead of fusing with the cell surface, they fuse with phagosomes, delivering their lytic enzymes and bacteriocidal contents to this organelle for destruction of ingested microorganisms. The exocytosis of azurophilic granules is not linked to the exocytosis of specific and gelatinase granules; therefore, a differential distribution and function of SNARE complexes in these two events would be anticipated. VAMP7 can be detected on all granules in neutrophils, but it is prevalent on azurophilic granules, and blocking VAMP7 function specifically inhibits these granules from releasing their contents (Mollinedo et al., 2006; Logan et al., 2006). In most cells, VAMP7 is associated with late endosomes, and its association with neutrophil storage granules, in particular, the azurophilic granules, is reminiscent of the secretory lysosome-like nature of these compartments. VAMP1 and VAMP7 on azurophilic granules couple with the Q-SNARE STX4 on the cell surface (Mollinedo et al., 2006).

Finally, when neutrophils have egressed from the bone marrow and have entered the extravascular tissues during infection, the local conditions initiate another phase of synthesis and secretion of inflammatory chemokines and cytokines, such as interleukin-8 (IL-8), to attract activated monocytes, other neutrophils, and T cells. To support survival and inflammatory responsiveness, other proteins are synthesized, such as the high-avidity immunoglobulin G (IgG) receptor FcγRI (CD64) and factors involved in wound healing (Drewniak et al., 2009).

ELASTASE IS KEY IN CONGENITAL NEUTROPENIA...

In many instances, the inheritance and the exact nature of neutropenia are poorly understood. Yet, in contrast to the sporadic cases of severe congenital neutropenia (SCN), with a permanent form of neutropenia from birth onward and with absolute neutrophil counts of $<200/\mu l$, there is also a disease designated cyclic neutropenia (CN), which is an autosomal dominant disease in which neutrophil production from the bone marrow oscillates with 21-day periodicity. The locus for cyclic neutropenia has been mapped by positional cloning to chromosome 19p13.3, and was identified as the ELA2 gene encoding human NE, the most prominent serine protease of neutrophil and monocyte granules (Dale et al., 2000). Subsequent studies showed that the same ELA2 gene was also somatically mutated in >80% of SCN patients. Both normal and abnormal transcripts are expressed.

Human NE is the major constituent of the azurophil granule and constitutes approximately 1% of the total weight of the neutrophil. NE is synthesized in the myeloblasts as an inactive proenzyme but packaged in the azurophil granules in its active form—associated in a surprisingly specific manner with serglycin (Niemann et al., 2007).

Because mice with heterozygous or homozygous ELA2 deletions do not show neutropenia (Belaaouaj et al., 1998), gain-of-function mutations have been proposed to explain both the neutropenia and the autosomal dominant inheritance (in SCN) (Dale et al., 2000). However, the ELA2 mutations have no consistent effect on proteolysis. Some mutations reduce proteolytic activity of human NE, others retain wild-type levels of activity. Whether enhanced apoptosis in SCN and CN progenitor cells is the result of these mutations, either as a consequence of the leakiness of the granules or of the aberrant routing of the proenzyme, has remained unclear for a long time.

A similar "misrouting" hypothesis has been suggested to explain the effects of defective AP3 in Hermansky-Pudlak type-2 syndrome (HPS 2). AP3 specifically directs posttranslational trafficking of intraluminal proteins from the trans-Golgi network to lysosomes, which in neutrophils consist of the azurophil granules, in which human NE is primarily located. When mutated, NE has been suggested not to react any longer with AP3. AP3 is a 4-subunit complex. Either the β- or μ-subunit of the AP3 adaptor protein complexes can interact with cargo proteins via a dileucine repeat or a tyrosine residue near the carboxyl terminus of the protein. NE lacking the carboxyl terminus interacts with the μ-subunit of AP3 and requires a tyrosine residue (human NE Y199), as expected if NE were indeed an AP3 cargo protein (Benson et al., 2003).

Another link with NE has been found, however, in rare neutropenic patients without ELA2 mutations, in whom heterozygous dominant-negative mutations were found in Gfi1, a zinc finger-containing transcriptional repressor originally recognized for its role in T-cell differentiation (Karsunky et al., 2002; Hock et al., 2003). Patients resembled the Gfi1$^{-/-}$ knockout mouse, including a CD8$^+$ T-cell lymphocytosis and the presence of immature myeloid cells in the peripheral blood, without signs of an overt myelodysplastic syndrome (MDS) or (pre)leukemia (Person et al., 2003). The ELA2 promoter contains a Gfi1-binding motif with repressor activity, and absence of Gfi 1 may indeed lead to ELA2 overexpression in myeloid progenitor cells, thus creating a hostile environment for normal myeloid differentiation.

In sum, mislocalized, misfolded, or overproduced NE somehow drastically reduces the production of neutrophils from the progenitor cells in the bone marrow. As to how mutated elastase and premature apoptosis lead to the clock-like timing of hematopoiesis in CN or a permanent neutropenia is still a complete mystery.

NEUTROPENIA AND THE WRONG KEY

Several causes of neutropenia can be diagnosed (Table 1). In contrast to the more frequently observed neutropenia associated with ELA2 mutations, the classical form of SCN, i.e., Kostmann syndrome, is an autosomal recessive disease. Recently, it was found to be caused by mutations in the HAX1 gene. The role of Hax-1 is unknown. Immunohistochemistry of mouse tissues has determined a prominent Hax-1 expression in epithelial, endothelial, and muscle cells. The protein now seems restricted to the mitochondria and causes a selective neutropenia when defective (Klein et al., 2007). Although a mechanistic link of Hax-1 to the disease is as yet unknown, a link to the intramitochondrial serine protease activity of Omi/HtrA2, with Hax-1 as its substrate, may be the key (Blink et al., 2004; Cilenti et al., 2004).

Defects in neutrophil production or release from the bone marrow are diagnosed more often and more easily than qualitative (functional) phagocytic defects. However, functional defects may also accompany the neutropenia. Such functional defects are seen in SCN as well as in the more complex and syndromal forms of neutropenia that may be more or less isolated hematopoietic abnormalities, such as in Shwachman-Diamond syndrome or cartilage-hair hypoplasia, or part of a developmental bone marrow failure such as in dyskeratosis congenita or Fanconi anemia types. In these syndromes short stature may be part of the clinical complex (Kuijpers, 2002; Kuijpers et al., 2005).

In other syndromes, oculocutaneous abnormalities may be diagnosed, such as in Chediak-Higashi syndrome or in HPS 2 with a concomitant bleeding diathesis. Although there are many genes involved in Hermansky-Pudlak syndrome, the type-2 form is differentiated from other forms of the syndrome because only the type-2 variant is associated with neutropenia. In a novel Mennonite p14 deficiency syndrome the bleeding constitution is not a feature of the syndrome (Bohn et al., 2007). As demonstrated in this latter index family, the endosomal adaptor protein p14 was found to be crucial for the distribution of late endosomes. This p14 was previously characterized only as confining mitogen-activated protein kinase (MAPK) signaling to late endosomes. A homozygous point mutation in the 3′-untranslated region of p14 (also known as MAPBPIP), resulted in decreased protein expression that was important for the function of neutrophils (and that of B cells, cytotoxic T cells, and melanocytes) (Bohn, 2007). These findings may have implications for understanding endosomal membrane dynamics, compartmentalization of cell signal cascades, and their role in immunity.

In Chediak-Higashi syndrome (Karim et al., 2002) and also in HPS-2 (Enders et al., 2006), an "accelerated phase" may suddenly create an often fatal episode when unrecognized or anticipated, because of a virus-associated hemophagocytosis that often proceeds relentlessly unless treated with chemotherapy followed by bone marrow transplantation (BMT). In the accelerated phase, unbridled NK- and T-cell activity and the inadvertent release of cytokines

TABLE 1 Neutropenia: quantitative defects, pathomechanism, and inheritance

Category	Type	Inheritance	Gene defect (chromosome)
Congenital	Reticular dysgenesis	Sporadic/AR	
	Kostmann syndrome	AR	HAX1 (1q21.3)
	G-CSFR defect	AD	GCSFR (1p34.3)
	Cyclic neutropenia	AD	ELA2 (19p13.3)
	Severe chronic neutropenia	Sporadic/AD	ELA2 (19p13.3)
	X-linked neutropenia/myelodysplasia	X-linked	WASP (Xp11.22)
Syndrome-associated	Shwachman-Diamond syndrome	AR	SBDS (7q11)
	Fanconi anemia (groups A–E)	AR	Multiple
	Dyskeratosis congenita	X-linked (AR)	DKC1 (Xq28)
	Cartilage-hair hypoplasia	AR	CHH (9p13)
	Griscelli syndrome, type 2	AR	RAB27A (15q21)
	Chediak-Higashi syndrome	AR	LYST (1q43)
	Hermansky-Pudlak syndrome, type 2	AR	AP3B1 (5q14.1)
	Endosomal p14 deficiency	AR	MAPBPIP (1q22)
	Barth syndrome	X-linked	G4.5 (Xq28)
	Glycogen storage disease, type 1b	AR	G6PT (11q23.3)
	Organic aciduria	AR	Multiple
Immune-mediated	Neonatal alloimmune neutropenia		
	Benign neutropenia of childhood		
	Autoimmune neutropenia (Evans, SLE, JCA, etc.)		Multigenic
	Autoimmune lymphoproliferative syndrome	AD/sporadic	APT1 (10q24)
?	Hyper-IgM syndrome type 1	X-linked	XHIM (Xq26.3)
(Toxicity)	Drug-related neutropenia		

result in the activation of macrophages and, as a consequence, the clinical picture of fever, cachexia, hepatosplenomegaly, aplastic anemia, and sometimes neurological involvement. In these oculocutaneous syndromes, the course of neutropenia is unpredictable, sometimes periodic without a clear cycling pattern, but often moderate and only rarely severe.

Also in the metabolic diseases, such as glycogen storage disease (GSD) type 1b with hypoglycemia and hepatomegaly, and in the mitochondriopathy in Barth syndrome with cardiomyopathy and muscular weakness, the neutropenia is often less prominent than the metabolic manifestations of the disease itself (Kuijpers et al., 2003, 2004).

INFECTIONS, LEUKEMIA RISK, AND RUSSIAN ROULETTE

G-CSF (or GM-CSF) exerts a dual effect: induction of myeloid development through an increase in the number of colony-forming units and a boost in cell survival potential throughout the myeloid stages of differentiation. The effect of G-CSF is transmitted to the myeloid cell through the G-CSF receptor (G-CSFR). G-CSF and G-CSFR knockout mice show a similar phenotype to neutropenia, with decreased numbers of hematopoietic progenitor cells in the bone marrow, reduced expansion and terminal differentiation of these progenitors into mature neutrophils, as well as increased apoptosis of the neutrophils that do mature. A maturational arrest does not occur in these mice (Liu et al., 1996).

To date, a germ line mutation in the G-CSFR has been identified in only one child with congenital neutropenia. Instead, somatic mutations have been regularly observed in patients with SCN. Most of these SCN patients suffer from a neutropenia of <200 neutrophils per μl from birth onward. Mutations in the G-CSFR may be found but are not congenital; instead, these can be acquired over time. A hot spot of such mutations lies in an intracellular domain stretching over 45 amino acids. These mutations result in premature stop codons that cause deletion of the C terminus of the G-CSFR. Although expression levels of the receptor remain unperturbed, the distal maturation signals are now lacking, while leaving the proliferative signaling route via the membrane-proximal part of the cytoplasmic tail of the G-CSFR intact (Dong et al., 1997).

Recent analysis has resulted in a more careful follow-up of subgroups of neutropenia patients. G-CSF is suspected to be a risk factor for developing MDS or acute monoblastic leukemia (AML) in SCN, because the frequency of MDS or AML increases with the dose and duration of G-CSF treatment and because at least one case of AML has occurred in SCN in which the blast count rose and fell directly in response to G-CSF dosing (Jeha et al., 2000). Among SCN patients requiring more than 8 μg/kg/day G-CSF, the incidence of MDS or AML was 40% after 10 years (Rosenberg et al., 2006). Of course, it could be that those with the most severe forms of the disease face the greatest risk of malignant evolution and are unresponsive to all but the largest doses of the cytokine. On the other hand, leukemic transformation had also been documented in SCN before G-CSF therapy was available. Hence, an inherent propensity to malignant transformation within the myeloid lineage may already exist. The syndromal background of deep neutropenia may play an important role herein. For instance, the cumulative annual risk of AML/MDS in SDS is high, being 20% at about 20 years of age, and 36% at about 30 (Donadieu et al., 2005).

Treatment consists of antibiotic prophylaxis to support the patients in reducing colonization with commensals and with particular pathogens, and/or G-CSF administration at a 3-week regimen when infections recur. Final cure of the underlying (isolated) neutropenia can only be reached by hematopoietic stem cell transplantation (HSCT) or BMT, a definite but potentially fatal procedure that will only be performed if benefits from intensified supportive care measures are insufficient.

CHEMOTAXIS AND CELL ACTIVATION

Leukocytes are able to recognize concentration differences in a gradient of chemotaxins and to direct their movement toward the source of these agents, i.e., toward the inflammatory site. This is again a very complicated process, the details of which are largely unknown (Bokoch, 1996; Murdoch and Finn, 2000). Occupation of a threshold difference in the number of chemotaxin receptors on one side of the cell probably induces the cytoskeletal rearrangements needed for movement. As will be described in more detail in the section on neutrophil adhesion and extravasation, adhesion molecules (such as the β_2 integrins on neutrophils) are essential for the connections with the tissue cells or with the extracellular matrix proteins that must be formed at the front of the moving neutrophils and broken at the rear (Kuijpers et al., 1992; Springer, 1994). Moreover, for continued sensing of the chemotaxin gradient, the chemotaxin receptors on the neutrophil must be freed from their ligand for repeated usage. This occurs through internalization of the ligand-receptor complex, intracellular disruption of the connection, and reappearance of the free receptor on the leukocyte surface.

Many of the chemotaxins involved in granulocyte movement are small proteins of approximately 60 to 100 amino acids, very homologous in structure, known as the chemokine superfamily. This family of leukocyte activators consists of over 30 different chemotactic molecules. Most chemokines can be classified into α (CXC) and β (CC) chemokines, distinguished by the presence or absence of a single amino acid between the first two of four conserved cysteines. The γ (single C) and δ (CX3C) classes of chemokines have only recently been coined, each with as yet one member (lymphotactin and fractalkine or neurotactin, respectively). As exemplified by IL-8, the first CXC member described, most CXC chemokines activate neutrophils, whereas the CC chemokines act toward various lymphocyte subsets, monocytes as well as eosinophils and basophils (see also chapter 9). The chemokines are produced by host cells in response to inflammation, injury, hypoxia, or other forms of stress. In addition, chemotactic peptides are released by infecting microorganisms (e.g., formyl-methionyl-leucyl-phenylalanine [fMLP]) and the host complement system (the split product of activated C5 [C5a]). Lipid mediators, such as leukotriene B_4 (LTB_4) and platelet-activating factor, are also strong chemoattractants. Specific receptors on the granulocyte exist for each of these agents. These receptors, as well as those for the chemokines, belong to the seven-span superfamily of integral membrane proteins with seven transmembrane domains. Ligand specificity is created by differences in the extracellular domains. The intracellular domains interact with trimeric guanidine triphosphate (GTP)-binding proteins enabling a link with signal transduction pathways leading to a wide range of functional responses, the complexity of which is still largely unclear.

Small GTPases of the Rho family, in particular, Rac2 in neutrophils (Ambruso et al., 2000), have been shown to be crucial in the signaling leading to cytoskeletal remodeling and cellular motility. Rac GTPases regulate cytoskeletal structure, gene expression, and reactive oxygen species (ROS) production. Rac2-deficient neutrophils cannot chemotax, produce ROS, or degranulate on G-protein-coupled receptor (GPCR) activation. Deficiency in PI3Kγ, an upstream regulator of Rac, causes a similar phenotype. P-Rex1, a guanine-nucleotide exchange factor (GEF) for Rac, is believed to link GPCRs and PI3Kγ to Rac-dependent neutrophil responses as a coincidence detector in both the G$\beta\gamma$ and PtdIns(3,4,5)P3 signaling pathways (Welch et al., 2002). The defects in chemotaxis and superoxide production of Rac2-deficient murine neutrophils are significantly augmented by additional loss of Rac1. In addition, it has been reported that Rac1 deficiency alone results in an inability of neutrophils to detect and to orient in a chemotactic gradient, suggesting that Rac1 is also involved in the chemotactic response of murine neutrophils (Gu et al., 2003; Sun et al., 2004).

Two other GEFs are involved in adhesion and chemotaxis. Vav proteins are specifically required for stable adhesion. The β_2 integrin-induced activation of the small GTPases Cdc42, Rac1, and RhoA is defective in Vav1/3 knockout neutrophils. In contrast, Vav proteins have been shown to be largely dispensable for GPCR-induced signaling events and chemotaxis. Thus, Vav proteins seem to play an essential role in coupling β_2 integrins to Rho GTPases and regulating multiple integrin-induced events that are important in leukocyte adhesion strengthening and phagocytosis (Gakidis et al., 2004). Although DOCK2 does not contain the Dbl homology domain and the pleckstrin homology (PH) domain that are typically found in GEFs, DOCK2 binds to nucleotide-free Rac and catalyzes GTP loading through its Docker (also known as DHR-2) domain. DOCK2-deficient murine neutrophils move toward the chemoattractant source, but they exhibit abnormal migratory behavior with a marked reduction in translocation speed. In DOCK2-deficient neutrophils, chemoattractant-induced activation of both Rac1 and Rac2 are severely impaired, resulting in the loss of polarized accumulation of F-actin and phosphatidylinositol 3,4,5-triphosphate (PIP_3) at the leading edge (besides the impaired ROS production on GPCR stimulation as in Rac2 deficiency) (Kunisaki et al., 2006).

Recent studies have suggested that Rap1 is also involved in neutrophil adhesion. We were the first to describe a syndrome of combined leukocyte adhesion and platelet dysfunction (leukocyte adhesion deficiency [LAD] combined with Glanzmann disease) in the presence of normal integrin expression, designated LAD-1/variant syndrome (Kuijpers et al., 1997). In a boy with a similar clinical syndrome, then named LAD-3, Rap1 activity was found to be absent (Kinashi et al., 2004). Therefore, the Rap1 regulation was suspected to be defective in these clinical pictures. In two subsequent patients with infections and bleeding tendency, LAD-3 was suggested to be caused by a splice site mutation in CalDAG-GEF1 (Pasvolsky et al., 2007). On the other hand, a more extensive family study by our group produced neither evidence for Rap1 activation defects nor any mutations in the gene encoding CalDAG-GEF1 that would result in abnormal mRNA in LAD-1/variant syndrome (Kuijpers et al., 2007). Instead, the syndrome is caused by mutations in a hematopoietic protein, Kindlin-3, encoded by the FERMT3 gene (Kuijpers et al., 2008).

Neutrophils from G-CSFR knockout mice demonstrate a disturbed chemotactic activity toward various chemotactic factors—e.g., chemokines (IL-8 or its analogues) and chemotactic peptides (fMLP or C5a)—although chemotaxin receptors are normally expressed. Also, G-CSF-induced neutrophils of patients with SCN have been suggested to be functionally defective in chemotaxis (Kasper et al., 2000), which may be inherent to the neutrophils generated in these (G-CSF-treated) patients. A defect in chemotaxis by neutrophils with a disturbed GCSFR or ELA2 gene cannot be easily reconciled with the function of either of these molecules per se. One explanation may be their increased susceptibility to cell death signals, perturbing the intracellular machinery for movement (although still morphologically normal upon isolation), as indicated by the rapid decay in functionality (Dransfield et al., 1995; Wolach et al., 2007). Alternatively, the machinery required for motility may be affected because of impaired or incomplete differentiation by either of the aforementioned gene defects, which cannot be fully overcome by G-CSF use in vivo. The latter possibility is strongly suggested by the imperfect maturation of the neutrophils in G-CSF-treated patients with known ELA2 mutations, leading to a deficiency in more granular constituents than only elastase (Donini et al., 2007).

A change in the ratio of expression between small GTPases and their regulatory proteins has been described in SCN, which is a more probable cause of the chemotaxis defect observed in the G-CSF-induced neutrophils (Kasper et al., 2000). Whether this changed ratio relates to the impaired expression of LEF/TCF1 and as a consequence to the reduced regulation of CCND1, BIRC5, MYC, and CEBPA encoding cyclin D1, survivin, c-Myc, and c-EBPα, respectively, the latter being a key transcription factor in neutrophil differentiation, is unknown (Skokowa et al., 2006).

NEUTROPHIL ADHESION AND EXTRAVASATION: ROLLING AND STREAMING

Extravasation is a multistep process involving adhesion molecules and activating agents that act as (pro-) inflammatory mediators. The first step consists of the initial contact between endothelial cells and neutrophils marginated by the fluid flow of the blood. The L-selectin (CD62L) on leukocytes plays a role herein. Within the local environment of an inflammatory tissue reaction, the endothelium begins to express the adhesion molecules E-selectin (CD62E) or P-selectin (CD62P). The low-avidity interaction of these selectins with their ligands on the neutrophils forces the neutrophils to slow down and make a rolling movement along the vessel wall.

Several glycoproteins have been suggested to bind to E-selectin in vitro, such as SLex-carrying CD66 on neutrophils (Kuijpers et al., 1992), but the identification of the predominant and physiological ligands in vivo has remained elusive for some time. More recently, E-selectin ligand-1 (ESL-1), P-selectin glycoprotein ligand-1 (PSGL-1), and CD44 were shown to be involved in the major endothelial-selectin ligand activity on neutrophils. PSGL-1 plays a major role in the initial leukocyte capture, whereas ESL-1 is critical for converting initial tethers into steady slow rolling. CD44 controls rolling velocity and mediates E-selectin-dependent redistribution of PSGL-1 and L-selectin to a major role on slowly rolling leukocytes through p38-MAPK signaling (Hidalgo et al., 2007).

Rolling then enables the neutrophil to make a more stable contact (cell spreading) through the interaction of integrin molecules on the neutrophil surface with accessory molecules such as intercellular adhesion molecule-1 (ICAM-1) on endothelial cells. For this interaction, the integrin molecules must first be activated (see below). The activating signals consist of binding of the prementioned chemotaxins to neutrophil surface receptors and of cross-linking the selectin ligands on the surface of the neutrophils. The chemotaxins may be derived from the endothelial cells or from underlying tissues, and can be exposed on the luminal side of the endothelial cells or diffused from in between the cells (Kuijpers et al., 1992). Once the rolling neutrophils have come to a stop, they can respond to the gradient of chemotaxins, which then guides the firmly adhering cells from the bloodstream between the endothelial cells into the adjacent site of inflammation.

In contrast to the low-avidity binding of neutrophils to selectins, the second and third steps of firm adhesion and subsequent diapedesis depend on changes in binding avidity of the integrin receptors on the leukocytes for adhesion and locomotion. These adhesion molecules consist of heterodimeric integrins of an α-chain covalently associated with a β-chain. One particular β-chain may associate with one of various α-chains. As a consequence, several integrin receptor subfamilies exist. The β_2 integrin receptor subfamily is only expressed on leukocytes and comprises four different heterodimeric proteins, each of which contains a different α-subunit: i.e., $\alpha_L\beta_2$ (LFA-1; CD11a/CD18), $\alpha_M\beta_2$ (CR3; CD11b/CD18), $\alpha_X\beta_2$ (gp150,95; CD11c/CD18), and $\alpha_D\beta_2$ (CD11d/CD18).

Human neutrophils express the β_2 integrins, especially CR3 (CD11b/CD18 or Mac-1), at very high levels and strongly depend on its presence and function, whereas the other leukocytic cell types also express alternative integrin receptors (e.g., β_1 integrins) for adhesion and additional functions related to adhesive properties (Loike et al., 1999). Switching from the low- to high-avidity binding states (and vice versa) of these integrins allows the cell to migrate, a process disturbed in LAD (Table 2). This integrin activation process can take place from within the cells (inside-out signaling) after, e.g., chemotaxin binding to its receptor, but may also occur from the outside (outside-in signaling) after, e.g., binding of ligand to the integrin itself. The first process is disturbed in LAD-1/variant and LAD-3 cells, both in leukocytes (affecting β_2 integrin activation) and in platelets (β_3 integrin activation). As a result, the leukocytes show a deficiency in chemotaxis and adhesion and the platelets show a deficiency in aggregation, leading to a syndrome of recurrent bacterial infections and a bleeding tendency (Kuijpers et al., 1997).

In mice, rolling and adhesion depend in a differential manner on LFA-1 versus Mac-1. From these studies, it was concluded that endothelial ICAM-1 is the main ligand responsible for slow leukocyte rolling mediated by Mac-1, but not LFA-1. Which ligand is used for LFA-1 under these conditions remained unclear (Dunne et al., 2002, 2003). Using similar techniques with a chemokine, equivalent levels of adhesion were observed in wild-type and Mac-1$^{-/-}$ mice, but very little adhesion was seen in LFA-1$^{-/-}$ mice. In contrast, very few Mac-1$^{-/-}$ neutrophils crawled with a 10-fold decrease in displacement. Mac-1$^{-/-}$ neutrophils initiated transmigration closer to the initial site of adhesion, which in turn led to delayed transmigration due to movement through nonoptimal emigration sites (Phillipson et al., 2006). We must be cautious with the translation of these data to human neutrophils. Mouse neutrophils express almost equal amounts of LFA-1 and Mac-1, and these neutrophils highly express the β_1 integrin VLA-4 ($\alpha_4\beta_1$) that may interact with the endothelial counterreceptor

TABLE 2 Neutrophil dysfunction: qualitative defects, pathomechanism, and inheritance

Category	Gene defect	Chromosome	Additional features
Adhesion and motility			
Leukocyte adhesion deficiency type 1	INTG2/CD18	21q22.3	
Leukocyte adhesion deficiency type 2	FUCT1	11p11.2	Retardation, hepatomegaly
Leukocyte adhesion deficiency type 3	CALDAGGEF1?	11q13	Glanzmann thrombasthenia
Leukocyte adhesion deficiency type 1/variant	FERMT3	11q13	Glanzmann thrombasthenia
Rac2 deficiency	RAC2	22q12.2	Multiple functional defects
β-Actin deficiency	ACTB	7p22.1	
NADPH oxidase enzyme activity			
X-linked CGD	CYBB	Xp21.1	
Autosomal-recessive CGD	CYBA	16q24	
	NCF1	7q11.23	
	NCF2	1q25	
Neutrophil glucose-6-phosphate dehydrogenase deficiency	G6PD	Xq28	Hemolytic anemia
TLR-priming and activation defect	IRAK4	12q12	Polysaccharide Ab failure
Myeloperoxidase (MPO) deficiency[a]	MPO	17q23.1	
Proteolytic activity			
Juvenile parodontitis	FPR1	19q13.41	
Papillon-Lefèvre syndrome	CTSC	11q14.2	Hyperkeratosis hand/feet
Specific granule deficiency	CEBPE	14q11.2	
Killing defect of intracellular pathogens			
IL-12p40 deficiency	IL12B	5q31.1-q33.1	
IL-12R deficiency	IL12RB1	5q31.1-q33.1	
IFN-γ receptor 1 deficiency	IFNGR1	6q23-q24	
IFN-γ receptor 2 deficiency	IFNGR2	21q22.1	
Tyk2 deficiency (variant hyper-IgE syndrome)	TYK2	19p13.2	Virus infections (e.g., HSV)
STAT1 deficiency	STAT1	2q32.2	Virus infections (e.g., HSV)
Cytokine activation defect			
Hyper-IgE syndrome	STAT3	17q12	Facies, fractures, dentition
NEMO deficiency	IKBKG	Xq28	Ectodermal dysplasia, conical teeth, sparse hair

[a] No clinical consequence; MPO can be easily picked up by in vitro tests such as dihydrorhodamine (DHR) assay and may result in the misinterpretation of CGD (NADPH oxidase defect) when the investigator is unaware of this pitfall.

VCAM-1. In contrast, VLA-4 expression is extremely low if present at all on circulating human neutrophils.

In contrast to rolling and adhesion of neutrophils, the mechanisms that mediate migration through venular walls are less well understood, although there is now good evidence for the involvement of a number of junctional molecules in the process of migration through endothelial cells. In this context, molecules such as PECAM-1 (CD31), members of the junctional adhesion molecules or JAM family (JAM-A, -B, and -C), CD99, ESAM, ICAM-1, and ICAM-2 have been implicated.

There is much in vivo evidence for the involvement of PECAM-1 in leukocyte transmigration (Vaporciyan et al., 1993; Muller et al., 1993). Endothelial cell PECAM-1 functions as a passive homophilic ligand for neutrophil PECAM-1, which after engagement leads to neutrophil signal transduction, integrin activation, and ultimately transmigration in a stimulus-specific manner. Other surface molecules such as neutrophil CD177 (NB1) may also interact with endothelial PECAM-1. Blocking of both the homophilic CD99 and PECAM-1 interactions together, either on the neutrophil or the endothelial cell, resulted in additive effects on neutrophil transmigration across cultured endothelial cells in vitro, suggesting that these two molecules work at distinct steps in this process (Lou et al., 2007), as was also indicated for ICAM-2 (Huang et al., 2006). ICAM-2 is constitutively expressed on endothelial cells, platelets, and most leukocytes. Its expression on resting endothelial cells is about 10-fold higher than that of ICAM-1 and, in contrast to the rather uniformly expressed ICAM-1, ICAM-2 appears to be more concentrated at endothelial cell junctions promoting the migration of neutrophils (not of lymphocytes) (Huang et al., 2006). Finally, endothelial cell-selective adhesion molecule (ESAM) is strictly limited to endothelial cells and platelets and cannot be found on leukocytes or epithelia. Being localized at endothelial tight junctions, ESAM participates in the transmigration of neutrophils (Wegmann et al., 2006).

The role of the JAM family is confusing. Human JAM-1 is also constitutively expressed on circulating monocytes, neutrophils, erythrocytes, and lymphocyte subsets. The highest vascular expression of JAM-1 has been found in the adult brain, which is known to develop a network of tight junctional strands unique to the vascular system. By analogy with the function of JAM-1 as an organizer of occludin clustering in epithelial cells, one may speculate about a similar function in regulating permeability in endothelial cells. In contrast, endothelial cells of high endothelial venules for lymphocyte traffic through lymph nodes, and lymphatic endothelial cells that have highly specialized cell-cell junctions, both expressed high levels of JAM-2 and low amounts if any of JAM-1. Finally, JAM-3 is found in most endothelial contacts, which would argue for a role for JAM-3 in constitutive functions of vascular endothelial cells. All JAMs may act as a homophilic or heterophilic binding partner, interactions suggested to be strengthened by serving as counterreceptor for LFA-1 (JAM-1), VLA-4 (JAM-2), and CR3 (JAM-3) (Osterman et al., 2002; Aurrand-Lions et al., 2002; Weber et al., 2007), which have not always been confirmed (Sircar et al., 2007; Coroda et al., 2005; Khandoga et al., 2005).

It is unclear whether alternative adhesion molecules exist with a similar propagating role in integrin-dependent migration of human leukocytes into the extravascular space. Other binding molecules and mechanisms play an additional role in the case of epithelial cell linings, such as different types of cadherins being important in opening of tight junctions, as well as CD47, also known as integrin-associated protein (IAP), functioning as an integrator of transepithelial movement (Liu et al., 2002). The precise contribution, involvement, and redundancy of the various adhesion molecules for neutrophil transmigration largely remain to be defined.

SENSING DANGER SIGNALS: PATTERN RECOGNITION

TLRs have emerged as the primary proximal sensory apparatus that enables first-line innate immune surveillance systems to detect the presence of foreign pathogens and to rapidly mount a vigorous immune defense (Akira and Takeda, 2004). TLRs ignite the cytokine response that occurs during infection, and to a large extent, shape the whole of the inflammatory response with all its consequences, both beneficial and harmful. The role of TLRs in normal homeostatic processes and in the pathogenesis of chronic inflammatory diseases, including inflammatory bowel disease, asthma, and atherosclerosis, has recently become apparent.

Both immune-competent cells (such as macrophages, dendritic cells, and neutrophils) and epithelial cells recognize pathogen-associated molecular patterns (PAMPs) on the surface of a wide variety of both pathogens and commensal microorganisms, including gram-positive and gram-negative bacteria, viruses, and fungi (Table 3). Whereas immune cells respond vigorously to invading microorganisms, the epithelial linings have to discriminate more selectively, avoiding a response to the commensal flora in the guts, in the oropharynx, or on our skin. The receptors for PAMPs are called pattern recognition receptors (PRRs), which are expressed on the surface of or within many effector cells of the innate immune system. TLRs are a family of PRRs, and upon binding of their cognate ligands to the leucine-rich repeat (LRR) domain, recruit adaptor molecules to their intracellular signaling domain, leading to the activation of several kinases that ultimately results in the activation of NF-κB and other immune-responsive genes (see chapter 10).

Neutrophils express all TLRs, except TLR3 (Hayashi et al., 2003). However, in comparison with monocytes and macrophages, TLR2, TLR4, and CD14 are expressed at relatively low levels (in addition to the TLR1/TLR2 and TLR6/TLR2 dimers) (Ozinsky et al., 2000; Kurt-Jones et al., 2002). GM-CSF (but not G-CSF) induces increased expression of TLR2 in many, but not in all individuals. TLR1 and TLR4 levels may also be highly variable between donors (Visintin et al., 2001), with estimates of monocyte TLR4 surface expression ranging from 400 to 3,200 molecules per cell, and levels of TLR1 ranging from 0 to 5,400 molecules per cell. The concentration of lipo-

TABLE 3 PAMPs observed on various microorganisms

Organism	Constituent	PRR(s)
Bacteria		
Gram-negative bacteria	LPS	TLR4
Mycoplasma sp.	Diacyl lipopeptides	TLR6/TLR2
Bacteria and mycobacteria	Triacyl lipopeptides	TLR1/TLR2
Group B *Streptococcus*	LTA	TLR6/TLR2
Gram-positive bacteria	Peptidoglycan	TLR2
Neisseria sp.	Porins	TLR2
Mycobacteria	Lipoarabinomannan	TLR2
Flagellated bacteria	Flagellin	TLR5
Bacteria (viruses) and mycobacteria	CpG DNA	TLR9
Uropathogenic bacteria	Not determined	TLR11
Fungi		
Saccharomyces cerevisiae	Zymosan	TLR6/TLR2
Candida albicans	Phospholipomannan	TLR2
Candida albicans	Mannan	TLR4
Cryptococcus neoformans	Glucuronoxylomannan	TLR2 and TLR4
Parasites		
Trypanosoma sp.	Glycoinositolphospholipids	TLR2 and TLR4
Plasmodium sp.	Hemozoin	TLR9
Toxoplasma gondii	Profilin-like molecule	TLR11
Viruses		
Viruses	DNA	TLR9
Viruses	dsRNA	TLR3
RNA viruses	ssRNA	TLR7 and TLR8
RSV, MMTV	Envelope proteins	TLR4
Measles virus	Hemagglutinin protein	TLR2
HCMV, HSV-1	Not determined	TLR2

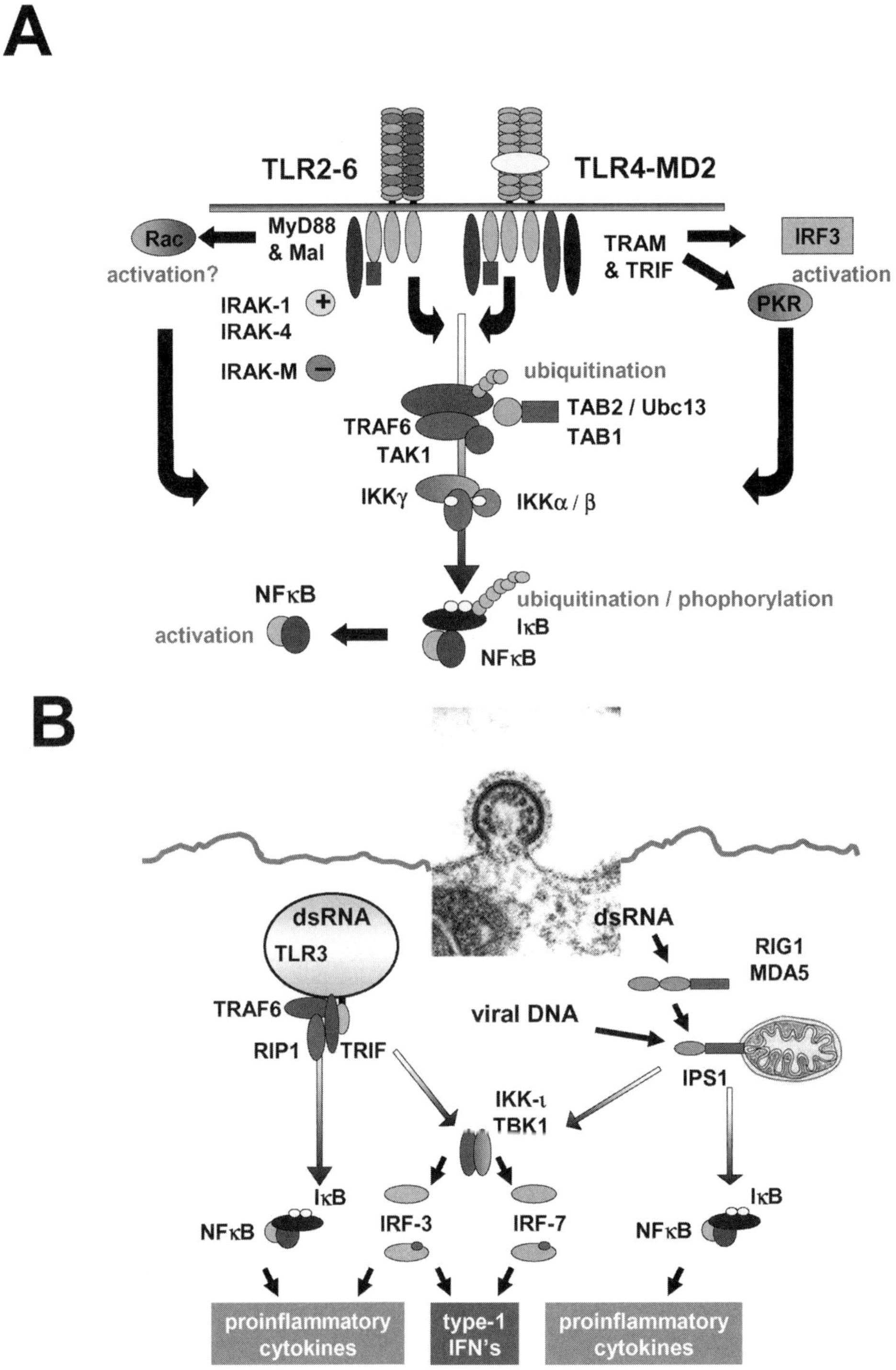

FIGURE 2 (A) TLR activation through TLR2 and TLR4 demonstrates the distinctive mechanisms that can be used by either of these surface proteins. The role of CD14 is limited to the sensing and cooperative activation of TLR4 (not shown). Moreover, TLR4 can activate the cell through MyD88/Mal/IRAK-4 but also via interferon response factor-3 (IRF3) phosphorylation in a TRAM/TRIF-dependent manner through inducible IKK (IKKι) and TANK-binding kinase-1 (TBK1) (see also chapter 10). TLR2 has been suggested to activate small GTPase Rac, but this TLR2-mediated activation pathway remains to be confirmed. Either way, NF-κB activation occurs and is an important step in transcriptional activity and the production of classical proinflammatory proteins. (B) The activation of endosomal TLR3, TLR7, TLR8, and TLR9 are assumed to activate different pathways. This has been indicated by the different signaling cascades. As shown for TLR3, this receptor triggers the TRAM/TRIF pathway to activate NF-κB and the IRF3- and IRF7-mediated gene transcription. *(continued next page)*

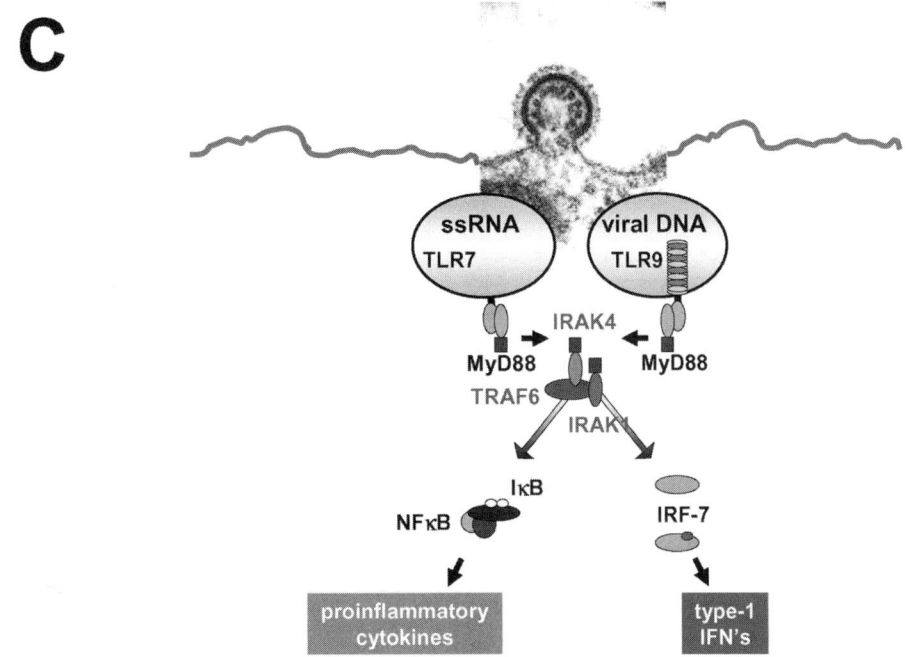

FIGURE 2 *(Continued)* In many cells a similar route of cytoplasmic activation via dsRNA is present which depends on retinoic acidi-inducible glycoprotein-I (RIG-I) or its homologous helicase MDA-5 in a mitochondria-dependent manner. In human neutrophils, TLR3 is not expressed and hence absent; RIG-I expression has not been described in neutrophils. (C) TLR7, -8, and -9 make use of the classical MyD88, which is able to relay their signals in the absence of Mal (in contrast to the TLR2/1, TLR2/6, and TLR4 shown in A). The localization of TLR7, -8, and -9 in human neutrophils is uncertain. IL-8 synthesis in neutrophils completely depends on IRAK-4. However, the immediate induction of adhesion, degranulation, or priming of the NADPH oxidase activity in neutrophils is independent of IRAK-4.

polysaccharide (LPS) required to activate neutrophils is about 100- to 1,000-fold higher than that needed for monocytes. The localization of the neutrophil-expressed TLRs over the various granular compartments may vary (R. van Bruggen and T. W. Kuijpers, unpublished work).

In sum, we can state that cellular responses to recognition patterns in microbial molecules depend on the total repertoire of TLRs and additional sensing receptors displayed on a cell, on necessary cofactors, and on the expression levels of these receptors and cofactors.

TLR AND NLR SIGNALING: WHO IS CALLING?

Most TLR family members have a conserved intracellular signaling motif, the so-called TIR domain. This signaling motif, which is also found in the intracellular domain of the interleukin-1 receptor (IL-1R), is responsible for NF-κB activation/translocation after TLR or IL-1R engagement and is an essential signaling pathway for IL-1β and tumor necrosis factor alpha (TNF-α) secretion (Kawai et al., 1999). All TLRs signal through MyD88 and Mal, which interact with various IL-1 receptor-associated kinases (IRAKs) (see chapter 10). However, both TLR3 and TLR4 can also activate cells through a MyD88-independent pathway in which TRAM and TRIF play a role (Fig. 2A). This can result in phosphorylation and activation of IRF3 or IRF7 that in turn activates a wide range of proinflammatory genes, such as interferon-β (IFN-β) (Fig. 2B and C). These pathways are more active in monocytes and subtypes of dendritic cells. Although mRNA for both TRAM and TRIF are expressed in neutrophils, LPS responsiveness is strictly IRAK-4 dependent (R. Van Bruggen, submitted for publication). In addition, TLR3 is not expressed in neutrophils. The other TLRs are all functionally active but, in contrast to most cell types studied thus far, TLR signaling through TLR7 and TLR9 in neutrophils is largely independent of IRAK-4 signaling via as-yet-unknown mechanisms (Hoarau et al., 2007; R. van Bruggen and T. W. Kuijpers, submitted for publication).

Four homologous IRAKs determine the signaling cascade. Apart from the initially identified IRAK-1 and -2, it is clear now that IRAK-M prevents dissociation of activating IRAK-1 and IRAK-4 from MyD88 and subsequent formation of IRAK-TRAF6 complexes. IRAK-M$^{-/-}$ cells exhibit increased cytokine production upon TLR/IL-1 stimulation, as well as increased inflammatory responses in mice to bacterial infection. Thus, IRAK-M regulates TLR signaling and innate immune homeostasis, in which IRAK-4 has emerged as a dominant and nonredundant signaling candidate. Expression in myeloid cells of IRAK-M seems restricted to macrophages, preventing IRAK-4 signaling (Kobayashi et al., 2002; Suzuki et al., 2002).

Excessive activation of TLR4 by LPS induces endotoxin shock, a serious systemic disorder with a high mortality rate. Therefore, TLR-dependent innate immune responses must be finely regulated, and underlying mechanisms are now being examined extensively (Liew et al., 2005). Several negative regulators of TLR-mediated signaling pathways have been proposed. Cytoplasmic molecules, such as an alternatively spliced short form of MyD88 (MyD88s in rodents), IRAK-M, SOCS1, A20 that deubiquitinates TRAF6 at K63 residues, hence deactivating this molecule, and TRIAD3A that ubiquitinates at K48 residues both TLR4 and TLR9 selectively for breakdown, are all involved in negative regulation of TLR pathways. Membrane-bound SIGIRR, ST2, TRAILR, and RP105 are also implicated in these processes (Diehl et al., 2004; Divanovic et al., 2005; Wald et al., 2003).

TLR-dependent gene induction is also regulated by nuclear IκB proteins, such as IκBζ, Bcl-3, and IκBNS. IκBζ is indispensable for positive regulation of a subset of TLR-dependent genes, such as IL-6 and IL-12p40. In contrast, Bcl-3 and IκBNS seem to be involved in negative regulation (Kuwata et al., 2006). The expression of these "inhibitory" molecules in human neutrophils remains to be elucidated, since only some of these have as yet been excluded (such as ST2 and RP105).

PAMPs are also recognized by a PRR family of cytosolic NOD-like receptors (NLRs; also known as PYPAF or CATERPILLER proteins) (Strober et al., 2006; Ting et al., 2006). NLRs include proteins such as NOD1 (nucleotide-binding oligomerization domain 1), NOD2 (also known as CARD15), NALPs (NACHT-, LRR- and pyrin-domain-containing proteins), IPAF (ICE protease-activating factor or CARD12), and NAIPs (neuronal apoptosis inhibitor proteins). Only some are present in neutrophils (van Bruggen and Kuijpers, unpublished). These cytosolic proteins function as cytoplasmic or intracellular PRRs. For example, peptidoglycan-derived molecules are recognized by NOD proteins within the cytosol of host cells and are involved in activating the NF-κB pathway. The ubiquitously expressed NOD1 senses diaminopimelic acid-containing peptidoglycan present in gram-negative bacteria, whereas epithelial and myeloid NOD2 sense the muramyl dipeptide (MDP) present in most organisms. The signals that IPAF "senses" are again different from the NALP3 signals. Other inflammasomes that contain different members of the NLR protein family are suggested to have their own specificity, which in many instances has not yet been identified.

Neutrophils, dendritic cells, and macrophages express ASC and pro-caspase-1 constitutively. The ligand recognition by NLRs can lead to activation of caspase-1 (previously known as ICE) through the assembly of a cytosolic protein complex that is now known as the "inflammasome" (Fig. 3A). Caspase-1 is required for the processing and subsequent release of active proinflammatory cytokines, such as IL-1β and IL-18. IL-33, a cytokine that is involved in generating a T helper 2 (Th2) cell response, is also cleaved by caspase-1.

NOD1, NOD2, and IPAF have an N-terminal CARD in common with the inflammatory members of the caspase family (i.e., human caspase-1, -4, and -5). Under physiological conditions, NOD1 is expressed by multiple tissues but not by neutrophils (Inohara et al., 1999; Drewniak and Kuijpers, unpublished). NOD2 proteins are predominantly expressed by myeloid cells or cytokine-activated intestinal epithelial cells (Gutierrez et al., 2002). These molecules have recently been linked to a subgroup of familial autoinflammatory disease entities such as Crohn's disease and Blau's syndrome by mutations in NOD2 (Hugot et al., 2001; Miceli-Richard et al., 2001; Ogura et al., 2001).

Some members of the NLR family may also contain a PYRIN domain, designated after the protein pyrin that is genetically mutated in a fever syndrome known as familial Mediterranean fever (FMF) (French FMF Consortium, 1997; International FMF Consortium, 1997). These NLRs were previously known as NALPs or PYPAFs (Fiorentino et al., 2002; Harton et al., 2002; Hlaing et al., 2001; Martinon et al., 2002; Wang et al., 2002) or CATERPILLER proteins (Ting et al., 2006). NALP6 (PYPAF-5, PAN3) and NALP12 (PYPAF-7, PAN6, Monarch1) are highly expressed in neutrophils, whereas others such as NALP1 (DEFCAP, CARD7) or NALP3 (PYPAF-1, cryopyrin, CIAS1) are constitutively expressed in monocytes and macrophages and not in neutrophils. NALP3 as well as IPAF can be induced in neutrophils upon activation. All of these NLRs may have specific microbial ligands, some of which have been identified, and others are still unknown (Fig. 3B). Furthermore, NALP2 (PYPAF-2, PAN1), NALP7 (PYPAF-3), and NALP4 (PYPAF-4, PAN2) may exert an inhibitory role in NF-κB activation, thus preventing the positive effects of some of the other NLR family members (Fiorentino et al., 2002; Bruey et al., 2004; Kinoschita et al., 2005). Similarly, pyrin or a smaller homologue called cellular pyrin-domain-only protein (cPOP) may regulate the inflammasome in a negative way (Dowds et al., 2003; Dorfleutner et al., 2007) (Fig. 3A). Apart from pyrin, which is present in neutrophils, little is known about their neutrophil expression and inflammasome regulation as such.

An alternative molecular mechanism underlying procaspase-1 processing involves interaction between the CARD domains of caspase-1 and a serine/threonine kinase RIP2. The consequence of these CARD interactions is that NOD enhances procaspase-1 oligomerization and processing, thus enhancing caspase-1-induced IL-1β secretion, both in vitro and in vivo (Cartwright et al., 2007). Three CARD-containing inhibitors of caspase-1 have been identified as pseudo-ICE, ICEBERG, and INCA. Their genes are closely linked to the genes for the inflammatory caspases at chromosome 11q22. These regulatory proteins closely resemble caspase-1 in its prodomain, bind to caspase-1, and prevent further active processing (Green and Melino, 2001; Lamkanfi et al., 2004). Some of these proteins are present in neutrophils, although their function clearly needs further study.

RIP2 can interact with NOD1 (and -2) (Fig. 4A). Apart from NOD1, TLRs and surface receptors containing an ITAM motif or associating with an ITAM-containing adaptor protein, such as the common FcR-associated γ-chain of FcγRIIIa or DAP12, have also been demonstrated to signal through another CARD-domain-containing family of proteins. In murine myeloid cells, CARD9 binds to the CARD-containing Bcl-10. In lymphocytes, CARD11, Bcl-10, and MALT1 form a CBM complex that determines signaling via the antigen receptors to activate NF-κB. These CARD-containing complexes activate NF-κB in some cells under certain conditions and p38 MAPK in other cells under similar or alternative circumstances (Gross et al., 2006; Hsu et al., 2007; Harah et al., 2007;

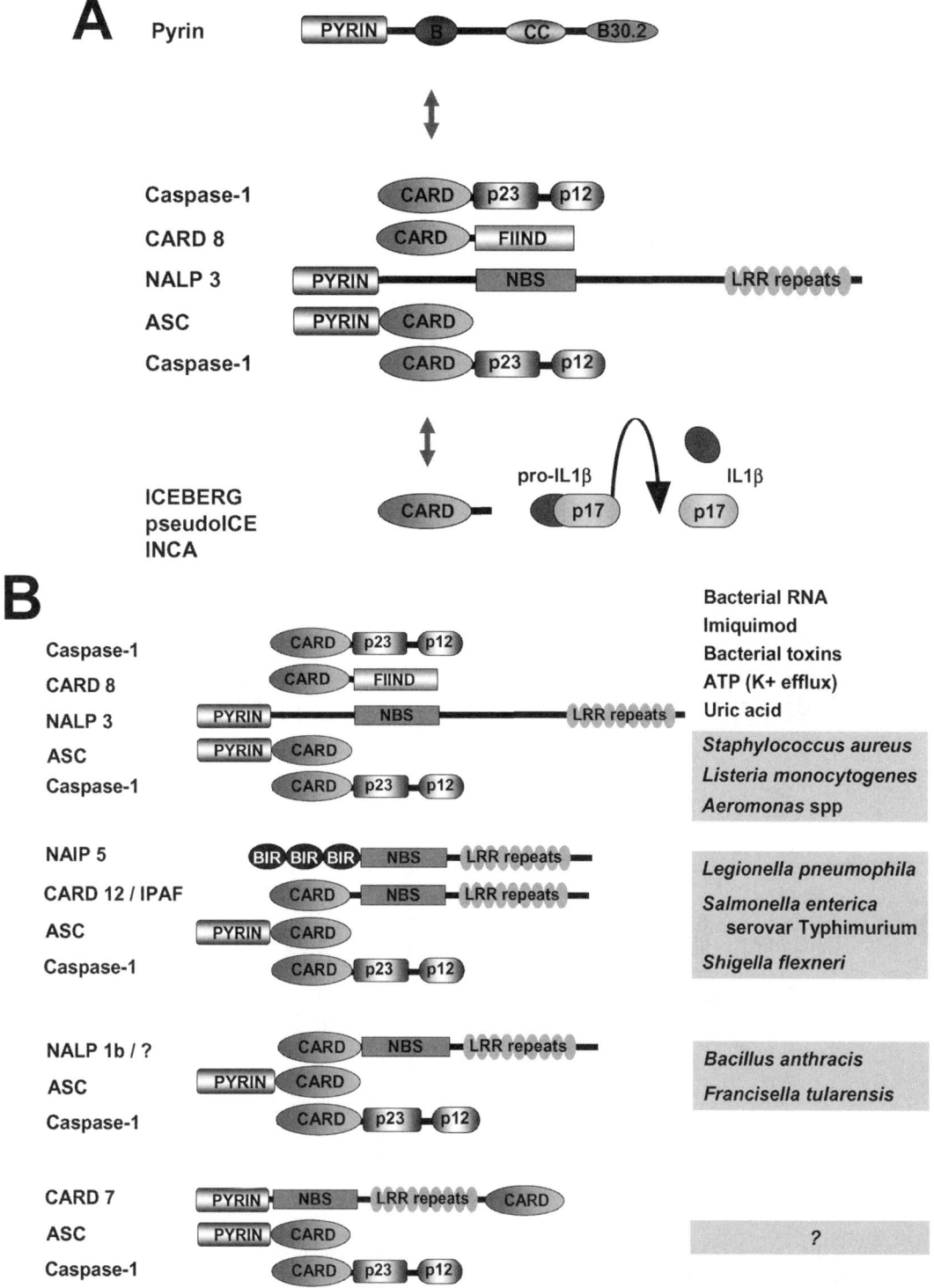

FIGURE 3 (A) NOD-like receptor (NLR) proteins comprise a diverse protein family (over 20 in humans), indicating that NLRs have evolved to acquire specificity to various pathogenic microorganisms, thereby controlling host-pathogen interactions. The NLRs form the backbone of inflammasomes, which are assumed to become active through a process of "close proximity." This means that within these protein complexes of NLRs the associated procaspases cross-activate each other by cleavage into enzymatically active proteases. The NLRs have a series of homologous domains that may interact with each other, such as the caspase recruitment domain (CARD), the nucleotide-binding sequence (NBS), or the leucine-rich repeats (LRRs), which act as the ligand-interacting domains. As indicated here for caspase-1 activity, after the cleavage and activation of procaspase-1, its substrate pro-IL-1β (IL-18 or IL-33) is cleaved into the bioactive IL-1β that is being released by the inflammatory cells. Inhibitory proteins in the cytoplasm such as Pyrin or the CARD-containing proteins INCA, pseudo-ICE, or ICEBERG may prevent the inflammasome from becoming activated. (B) NLR proteins are localized to the cytoplasm and recognize microbial products. Many cells contain different inflammasomes with their own specific ligands that may originate from microbial structures, actively taken up or dissipating from the invading pathogen itself. The inflammasomes identified thus far have been listed with their incriminated microbial agents or pathogen-derived ligands shown at the right. The list is of temporary value because of the rapid evolution of the field of inflammasome research to date.

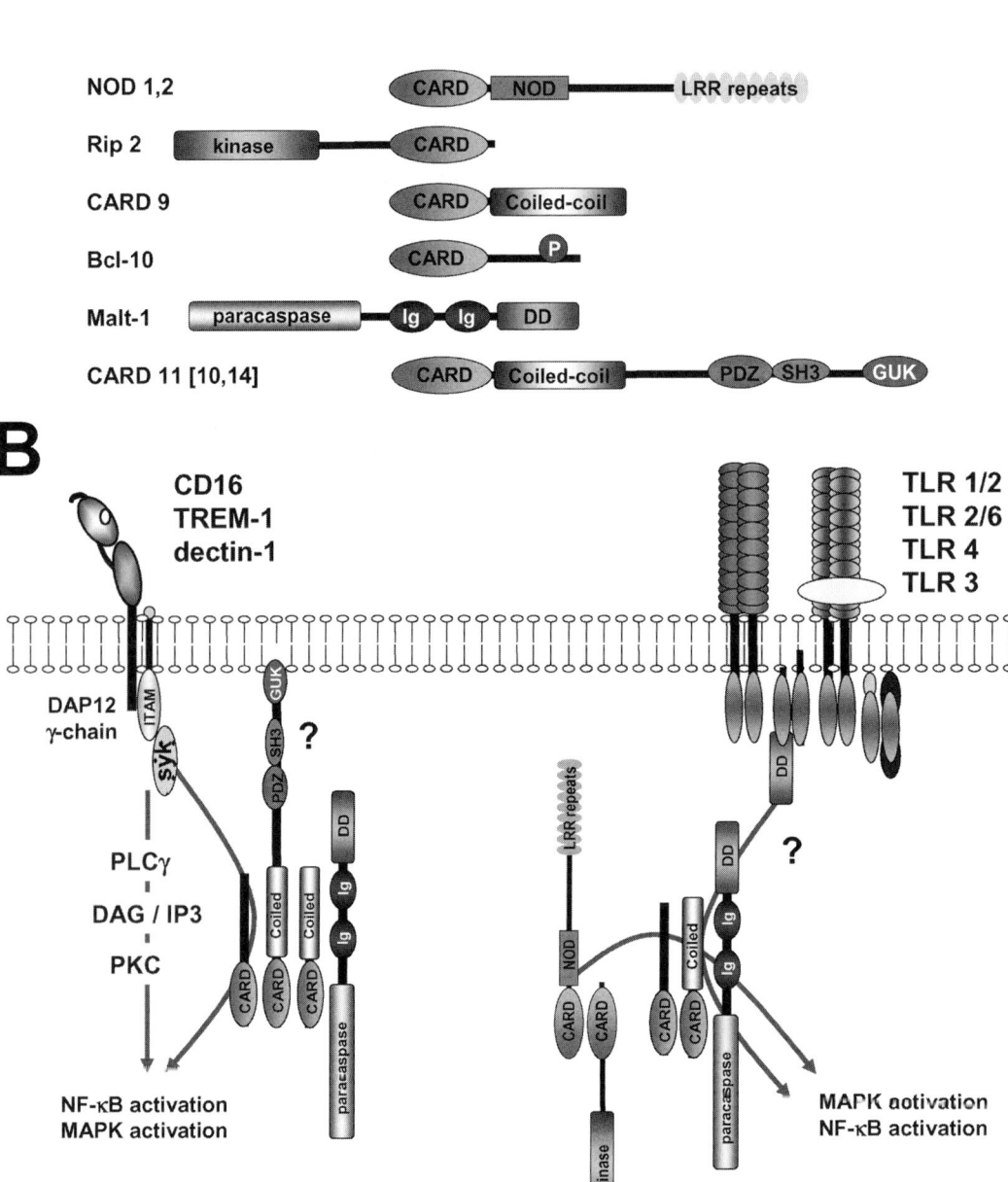

FIGURE 4 (A) Many proteins of the innate immune system contain so-called caspase-recruitment domains (CARDs). Through CARD-CARD interactions the activating protein kinase RIP2 can interact with NOD1 and NOD2. In murine myeloid cells, CARD9 binds to Bcl-10. In both murine and human lymphocytes, additional molecules such as the paracaspase-domain-containing MALT1, and members of the membrane-associated guanylate kinase (MAGUK) family of scaffolding proteins, which coordinate signaling pathways emanating from the plasma membrane. When complexed, these members determine the signaling via the antigen receptors to activate NF-κB. The lymphocyte-specific CARD11 (CARMA1) and proposedly its closest homologues are constitutively oligomerized. This oligomerization of CARD11 via the coiled-coil domains is required for NF-κB activation. IKK triggers Bcl-10 degradation by the ubiquitin-proteasome system through specific phosphorylation of Bcl-10, resulting in inactivation through negative feedback of the NF-κB activation pathway. (B) Apart from the NODs, also TLRs and surface receptors containing an ITAM motif themselves or associating with an ITAM-containing adaptor protein, such as the common FcR-associated γ-chain of FcγRIIIa or DAP12, have been demonstrated to signal through various CARD-containing proteins. Whether such activating modules or platforms also operate in human neutrophils is likely but remains to be formally shown.

Leibundgut-Landmann et al., 2007) (Fig. 4B). Their role in human neutrophils remains to be demonstrated.

The occurrence of a pro- or an anti-inflammatory reaction as well as the switch from a pro- into an anti-inflammatory reaction will not only be dictated by the number of phagocytic cells involved, but also at the molecular level. Activating receptors expressed on neutrophils and monocytes infiltrating into human tissues infected with bacteria (but not inflammatory lesions per se) have been identified as TREM-1 and TREM-2 (Bouchon et al., 2001; Daws et al., 2001). These molecules are upregulated on neutrophils by bacteria or their products (i.e. *Pseudomonas aeruginosa*, *Staphylococcus aureus*, lipoteichoic acid [LTA], LPS) (Bouchon et al., 2001), and probably interact with an as yet unknown ligand, either a changed self-ligand or pathogen-derived products, as another pattern recognition receptor. The TREM genes are localized on chromosome 6p, as are the genes of other activating and sensing receptors with a single Ig-like domain on cells of the innate immune system (e.g. CD83, NKp30, and NKp44) (Ravetch and Lanier, 2000; Young and Uhrberg, 2002; Klesney-Tait et al., 2006).

As a counterregulatory mechanism, neutrophils express a myeloid-specific Ig superfamily member, CD200R, which senses the environment to restrain its tissue-damaging activity via association with the SH2-containing inositol phosphatase (SHIP), consistent with a role in downregulation of myeloid activity (Barclay et al., 2002). For this action, neutrophils have to engage CD200, a glycoprotein exposed by many opposing tissue cells. This scenario is very reminiscent of the signals perceived by cells of the adaptive immune system, as well as natural killer (NK) cells and monocytes/macrophages. Such surface molecules in humans are known as the killer immunoglobulin-like receptors (KIRs), immunoglobulin-like transcripts (ILTs), and signal regulatory proteins (SIRPs).

These receptor families share many properties by generally containing two to three Ig-like domains, and both inhibitory and activating members exist. Their ligands, where known, are self-ligands expressed on the cell surface. Thus, KIRs and at least two ILTs are known to recognize different antigens of the HLA class I of proteins, whereas SIRP1α binds CD47 (integrin-associated protein) on hematopoietic cells (Oldenborg et al., 2000; Young and Uhrberg, 2002). The inhibitory members of these families inhibit cells by virtue of immunoreceptor tyrosine-based inhibitory motifs (ITIMs) in their cytoplasmic tails (or in the tails of associated molecules), as also found in the inhibitory Fcγ receptors FcγRIIb. The tyrosine phosphorylation of these molecules permits binding and activation of tyrosine phosphatases such as SHP1 and -2 (see "Phagocytosis and Microbicidal Activity" below). The activating members of these families stimulate cells by associating to membrane adaptor molecules with cytoplasmic immunoreceptor tyrosine-based activation motifs (ITAMs) as also found in the activating Fcγ receptors. The tyrosine phosphorylation of these molecules permits binding of tyrosine kinases such as Syk and ZAP-70 (Ravetch and Lanier, 2000; Young and Uhrberg, 2002). Studies in CD200 knockout mice have further indicated that an exaggerated innate immune response can result in unforeseen damage, either autoinflammatory or pathogen driven (Barclay et al., 2002). The extent to which many of the responses tested are the result of excessive macrophage or neutrophil reactivity and delayed resolution of the inflammatory response, is as yet unclear. Also, the molecular regulation of expression of CD200 and its in vitro functions are to be described in more detail.

IgG RECEPTORS BRIDGE THE INNATE AND ADAPTIVE IMMUNITY CIRCUITS

Neutrophils operate in concert with antibodies and complement factors, so-called opsonins (see also chapters 4 and 7). Microorganisms covered with these proteins are bound to neutrophils through specific opsonin receptors on the plasma membrane. Antibodies bind with their Fab regions to microbial antigens. In this way, the Fc regions of these antibodies are closely packed together on the microbial surface. This spatial arrangement enhances complement activation, thus leading to binding of C3b and C3bi to the microorganisms and subsequent binding of the microbes to the neutrophil complement receptor type 1 (CR1) and complement receptor type 3 (CR3), respectively. However, the proximity of the Fc regions of the IgG antibodies also promotes direct binding of the opsonized microorganisms to the Fcγ receptors (FcγRs) on the neutrophils.

There are three main classes of FcγRs: FcγRI, FcγRII, and FcγRIII. Each has its characteristic IgG-binding avidity, expression profile among hematopoietic cells, and functional properties. Further complexity arises from the fact that each FcγR is encoded by various genes, and from alternative splicing of these genes. For instance, FcγRII is encoded by three genes (*FCGR2A*, *FCGR2B*, and *FCGR2C*) (Fig. 5A). The FcRs can functionally be divided into activating and inhibitory receptors (Fig. 5B). The activation results from the interaction of FcRs containing an ITAM. The cytoplasmic ITAM motif consists of two copies of the sequence YxxL (Y = tyrosine, L = leucine, x = any amino acid). Within this motif, the tyrosines are phosphorylated after receptor cross-linking, and the integrity of these conserved sequences is required for efficient phagocytosis. The ITAM motif is present in the cytoplasmic tail of FcγRIIa and in the γ-chains associated with FcγRI and FcγRIIIa; the ζ-chain associated with FcγRIIIa contains three copies of this motif. In vitro, the activating pathway initiated by phosphorylation of these receptors can be interrupted by coligation to the FcγRIIb. The FcγRIIb1 and -b2 cytoplasmic regions (65 and 46 amino acids, respectively) are generated by differential mRNA splicing, giving rise to a 19-amino acid insert into the intracellular tail of FcγRIIb1. Both FcγRIIb receptors, FcγRIIb1 and FcγRIIb2, contain only one copy of the cytoplasmic YxxL sequence and, instead of transmitting an activating signal, now transduce inhibitory signals. These FcγRs also contain a so-called inhibitory motif (ITIM) consisting of an ITYSLL sequence in both FcγRIIb isoforms, that can bind to intracellular protein tyrosine phosphatases (i.e., SHP-1, SHP-2) or inositol-5-phosphatase (i.e., SHIP) (Bruhns et al., 2000; Pearse et al., 1999; Ravetch and Bolland, 2001).

Most likely, three types of Fcγ receptors are present on resting neutrophils: the activating FcγRIIa and FcγRIIIb, as well as the inhibitory FcγRIIb2 (van Mirre et al., 2006). In approximately 20% of the Caucasian population, the activating FcγRIIc is expressed by neutrophils as well as monocytes, macrophages, and NK cells (Breunis et al., 2008; W. B. Breunis, unpublished results). Only after activation of neutrophils by interferons or growth factors is the activating FcγRI expressed, which binds monomeric IgG with high affinity. In contrast, the constitutively expressed FcγRIIa, FcγRIIc in some individuals, and FcγRIIIb all bind monomeric IgG only with low avidity,

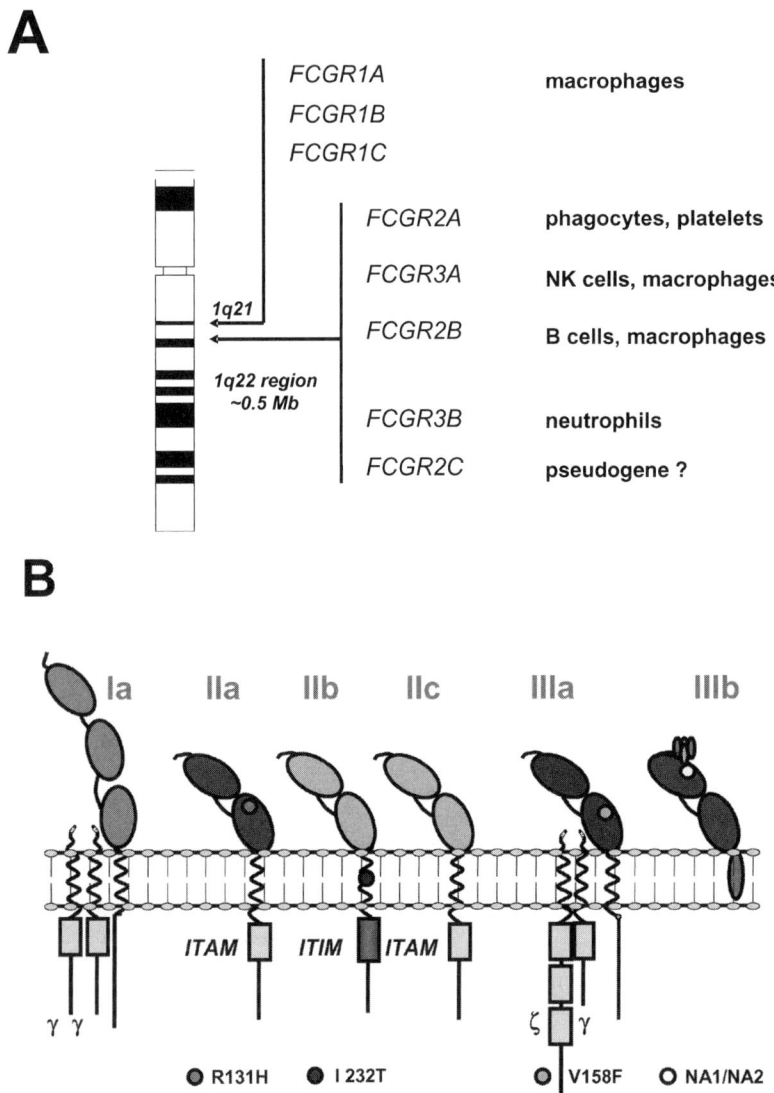

FIGURE 5 (A) The gene loci for FcγRI and FcγRII-III are located on different bands of chromosome 1q. FcγRIIs are encoded by three genes, *FCGR2A*, *FCGR2B*, and *FCGR2C*, which has long been assumed to be a pseudogene. As recently demonstrated most explicitly, approximately 20% of healthy individuals carry an *FCGR2C* gene without a stop codon in exon 3, which creates an open reading frame that results in the functional expression of an activating FcγRIIc. FcγRIIIs are encoded by two genes, one of which is selectively expressed by the neutrophil, FcγRIIIb carrying the allotype NA1 and/or NA2. (B) The FcγRs can functionally be divided into activating and inhibitory receptors. The activation results from the interaction of FcγRs containing an ITAM. The cytoplasmic ITAM motif consists of two copies of the sequence YxxL. Within this motif, the tyrosines are phosphorylated after receptor cross-linking, and the integrity of these conserved sequences is required for efficient phagocytosis. The ITAM motif is present in the cytoplasmic tail of FcγRIIa and in the γ-chains associated with FcγRI and FcγRIIIa; the ζ-chain associated with FcγRIIIa contains three copies of this motif. The single inhibitory IgG receptor, FcγRIIb, contains a so-called ITIM. All FcγRII and FcγRIII isoforms contain amino acid substitutions that are believed to influence their function, as a consequence of genetic variation, i.e., single nucleotide polymorphisms.

but each can efficiently bind immune complexes containing multiple IgG molecules.

A polymorphism in FcγRIIa defines the intrinsic ability to recognize the four IgG subclasses. IgG$_2$ antibodies, often formed against microbial carbohydrate structures, only react with the so-called L131 type of FcγRIIa (with a leucine at amino acid position 131). Individuals with this isotype are better protected against infections with certain microorganisms than individuals with the R131 type of FcγRIIa (arginine at position 131). FcγRIII has two variants, a transmembrane form (FcγRIIIa) expressed on macrophages and NK cells, and a neutrophil-specific form (FcγRIIIb) linked to the plasma membrane by a lipid anchor, which allows very rapid redistribution and early localization of opsonized material. Although there is a polymorphism in the FcγRIIIb (NA1/NA2; depending on differences in glycosylation), its effect on clinical outcome is as yet not as clear as in case of the FcγRIIa. There is also a polymorphic site present in the transmembrane domain of the inhibitory FcγRIIb, of which the meaning in terms of function is starting to become unraveled.

Recently, genomic variation was demonstrated by the presence of copy number variation. We and others have indicated the presence of such variation in cohorts that suffer from systemic forms of autoimmunity, vasculitis, or hematological disease (Aitman et al., 2006; Fanciulli et al., 2007; Breunis et al., 2008). In addition, we have found evidence for the functional expression of FcγRIIc, encoded by FCGR2C, which was previously assumed to be a pseudogene (Ravetch and Bolland, 2001). A single nucleotide change in the stop codon of exon 3 changes the gene and results in expression of the FcγRIIc isoform on the plasma membrane of their phagocytes and NK cells in approximately 20% of the healthy Caucasian population. The remainder is negative, carrying the common FCGR2C pseudogene (Breunis et al., 2008).

Next to the copy number variation in FCGR3B encoding FcγRIIIb (Koene et al., 1998; Aitman et al., 2006), of which the protein is only expressed on neutrophils, we have found recent evidence for similar copy number variations and the gene-dosage effect on the functional expression of FcγRIII, potentially contributing to the susceptibility or course of clinical disease in autoinflammation or infection (Breunis et al., 2008, 2009).

PHAGOCYTOSIS AND MICROBICIDAL ACTIVITY

Binding of opsonized material to neutrophil surface receptors leads to concentration of such receptors around the area of contact. Subsequently, the cell extends pseudopods that engulf the particle. By consecutive receptor binding, these pseudopods fit tightly around the particle and finally fuse with each other to form a closed membrane vesicle (phagosome) around the particle, within the neutrophil. Neutrophils may overeat themselves in infected areas and die of congestion. Macroscopically, this is manifested as pus formation. Apart from phagocytosis, receptor binding also starts two other processes, i.e., the generation of reactive oxygen compounds and the release of granule contents (degranulation). Both reactions are localized events in that they are restricted to the release of microbicidal products into the phagosome. However, the secretion of these products begins before the phagosome is closed, and some of the oxygen compounds and granule enzymes may thus escape into the extracellular environment of the neutrophils (Weiss, 1989). Moreover, neutrophils adhering to opsonized material that is too large to be ingested (e.g., immune complexes deposited along basement membranes) may secrete these products in large quantities into the extracellular space, with serious consequences for the surrounding tissue (Weiss, 1989). The components required for granule traffic and membrane fusion are starting to get unraveled. Although not restricted to the innate immune system, defects in such processes may also result in clinical defects, such as observed in Chediak-Higashi syndrome with giant granules in their granulocytes and—as a consequence of a degranulation defect—serious neutrophil chemotaxis and killing defects.

Degranulation does not occur in resting neutrophils. Only during phagocytosis or adherence of neutrophils to large substrates do intracellular signaling events induce the fusion of granules with the plasma membrane. Neutrophils contain at least two different types of granule. The azurophil granules resemble the lysosomes in other cell types in that they contain acid hydrolases, with a low pH optimum. Moreover, these granules also contain myeloperoxidase (MPO) and a number of serine proteinases. In addition, the azurophil granules also contain large amounts of defensins, small peptides with a broad range of bactericidal activity, and bactericidal permeability-increasing protein (BPI), a very potent antibiotic against gram-negative bacteria. Lysozyme, an enzyme that hydrolyzes certain peptidoglycans of gram-positive bacteria, is present in the azurophil as well as in the specific granules of neutrophils. Proteins exclusively found in the specific granules comprise lactoferrin, an iron-binding and therefore bacteriostatic protein, vitamin B_{12}-binding protein, and the metalloproteinases collagenase and gelatinase. The latter two enzymes help the neutrophil traverse into tissue compartments. Finally, neutrophils also contain so-called secretory vesicles, which actively exchange their membrane-bound receptors and enzymes with the plasma membrane.

Simultaneous with degranulation, a membrane-bound oxidase enzyme complex located in the membrane of secretory vesicles and specific granules is activated to generate reactive oxygen compounds needed in the killing process. This NADPH oxidase complex is composed of several subunits in the plasma membrane (flavocytochrome b_{558} subunits p22-*phox* and gp91-*phox*), and a number of activity-regulating proteins in the cytoplasm (p40-*phox*, p47-*phox*, and p67-*phox*) (Roos et al., 2007). The enzymatic subunit is gp91-*phox*. This protein contains six hydrophobic regions in its N-terminal half that serve as membrane-spanning domains and two heme moieties, each located between two histidines in transmembrane domains of the protein. The C-terminal part of gp91-*phox* is hydrophilic, probably cytosolic, and contains one FAD group and one NADPH-binding domain. The other flavocytochrome b_{558} subunit, p22-*phox*, serves to stabilize gp91-*phox* in the membrane of specific granules and secretory vesicles, and in the phagosomal membrane of activated cells.

The proteins p40-*phox*, p47-*phox*, and p67-*phox* form a complex in the cytosol of resting cells. This complex unfolds during cell activation, initiated by p47-*phox* phosphorylation and stabilized by SH3-polyproline interactions. At the same time, the small GTPase Rac2 (in neutrophils) or Rac1 (in macrophages) is also activated, a process under control of the Rac-GEFs P-Rex1 (for C5a activation) and Vav1 (for fMLP activation). The activated, GTP-containing Rac proteins then bind to a tetratricopeptide (TTP) region in p67-*phox* and to the plasma or phagosomal membrane with their liberated geranyl-geranyl tails. Thus, the result is that the p40/p47/p67-*phox* complex is drawn toward the plasma or phagosomal membrane in activated cells (Fig. 6). This localization is refined by p40- and p47-*phox* interactions of their PX regions with phosphatidyl-phosphoinositides in the activated membranes. Interaction of the cytosolic components with flavocytochrome b_{558} is governed by p47-*phox* SH3 interaction with a polyproline domain in p22-*phox*, by direct interaction of Rac proteins with gp91-*phox* and probably also by charge interactions

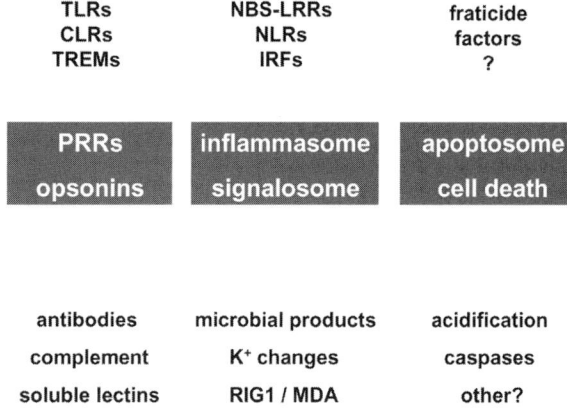

FIGURE 6 The NADPH oxidase complex consists of several proteins that have to assemble into an enzymatically active complex before the generation of superoxide may occur. Cytoplasmic components (the p40/p47/p67-*phox* complex) are drawn toward the plasma or phagosomal membrane where the interactions with flavocytochrome b_{558} are governed by the p47-*phox* SH3 domain interacting with a polyproline domain in p22-*phox*, and by the direct interaction of Rac proteins with gp91-*phox*. Then, the NADPH-binding site in gp91-*phox* is made accessible for NADPH as a substrate and electron donor from the cytosol to start generating superoxide (O_2^-).

between the oxidase components. The result of this whole series of events is that the NADPH-binding site in gp91-*phox* is made accessible for NADPH from the cytosol. Then, and only then, the system starts to generate superoxide (O_2^-). NADPH donates electrons, which are passed on within gp91-*phox* from the NADPH-binding site to FAD, then to the two heme groups and finally to the edge of the protein that faces the phagosome or the extracellular space. At that site, the electrons are attached to molecular oxygen, thus generating superoxide. Superoxide spontaneously dismutates into hydrogen peroxide (H_2O_2), which may then react with chloride ions to form hypochlorous acid (HOCl), in a reaction catalyzed by myeloperoxidase. This enzyme is an abundant constituent of neutrophil azurophil granules, and is released upon cell activation into the phagosome and into the extracellular space (Fig. 6). HOCl is very toxic for a broad range of microorganisms but is rather short-lived. However, it can react with primary and secondary amines, and thus give rise to N-chloramines, some of which are very stable microbicidal agents. Under normal, phagocytosing conditions, neutrophils convert more than 75% of their superoxide into hypochlorous acid and N-chloramines, and thus create a highly toxic environment within the phagosomes and in the cell surroundings (Weiss, 1989).

An interaction between protease activity and NADPH oxidase activity has been indicated by recent studies on the immediate changes after microbial uptake in the phagolysosome (Reeves et al., 2002). The generation of O_2^- in the phagosome not only leads to the influx of protons in the phagosome, for charge compensation, but also to the influx of potassium ions. This last process induces the release of proteases from the negatively charged proteoglycan matrix of the azurophil granules that have fused with the phagosome.

Patients with a defect in the NADPH oxidase system suffer from a syndrome called chronic granulomatous disease (CGD), characterized by recurrent, serious infections with bacteria, yeasts, and fungi. These symptoms manifest in the skin, airways, guts, internal organs, brain, and bones. Genetically, we can distinguish X-linked CGD, with mutations in *CYBB*, the gene that encodes gp91-*phox*, and autosomal recessive CGD, with mutations in *CYBA* (p22-*phox*), *NCF1* (p47-*phox*), or *NCF2* (p67-*phox*). The most common subtype of CGD is X-linked, with approximately 70% of all cases. Mutations in *NCF4* (p40-*phox*) are not known.

NEUTROPHIL APOPTOSIS AS A DEFAULT ROUTE

When released from the bone marrow, neutrophils have a short lifespan and rapidly undergo spontaneous apoptosis within hours in the extravascular tissues. Survival of both the immature myeloid progenitor cells and mature PMNs can be extended by delaying apoptosis through the action of a wide variety of agents, including G-CSF and GM-CSF (Adams and Cory, 1998; Brach et al., 1992; van den Berg et al., 2001). Neutrophils are predisposed to cell death by apoptosis. Apoptosis prevents the cytotoxic contents from the neutrophil granules from being released into the surrounding tissues and facilitates the elimination of cells by tissue macrophages (see also chapter 27). The exact molecular mechanisms underlying apoptosis are unknown, although members of the Bcl-2 protein family and caspases have been shown to be involved in neutrophil apoptosis (Hengartner, 2000; Moulding et al., 1998; Thornberry and Lazebnik, 1998).

Recent work has demonstrated that these two groups of proteins are intimately connected at the level of mitochondria: the Bcl-2 homologues govern the activity of caspases by exerting their effect through the regulation of the mitochondrial function (Maianski et al., 2002). Proapoptotic Bcl-2 proteins, such as Bax, redistribute from the cytosol to the mitochondria to disturb the mitochondrial membrane integrity by forming channels, which facilitates the subsequent release of cytochrome c and the activation of Apaf-1 and downstream caspases. The antiapoptotic protein Bcl-2 is believed to mediate, at least partially, its effect through the inhibition of Bax redistribution and activity. The family of caspase proteases executes the cleavage of specific targets, which, finally, leads to cell disassembly and death. Among these proteases, caspase-3 stands out for the large number of substrates that it destroys, including nuclear proteins, cytoplasmic structures, and cytoskeletal elements. The potential of granulocytes to perform their functional tasks is decreased dramatically by these events.

Regarding the role of Bcl-2 proteins in the process of neutrophil apoptosis, one should realize that mature PMNs were until recently considered to possess no or only a few rudimentary mitochondria, which do not play a role in the active life of the cell. We have shown that freshly isolated neutrophils do contain a large number of mitochondria with a characteristic elongated shape, although their functional capacities are still unclear. As previously reported, in vitro aging of neutrophils is accompanied by a progressive loss of functions, such as adherence, chemotaxis, and phagocytosis, and the generation of reactive oxygen species (Dransfield et al., 1995). Our own findings in this respect have further indicated that the loss of neutrophil function upon aging is not just a consequence of apoptosis but may precede some early hallmarks of this process. When externalization of phosphatidylserine (PS) lipids to the outer leaflet of the plasma membrane was taken as an early event in the execution phase of apoptosis, about one-third of viable aged neutrophils had already lost the ability to migrate, produce H_2O_2, and phagocytize small particles. Even more cells did no longer phagocytize big particles (B. Wolach and T. W. Kuijpers, unpublished results). Thus, functional deterioration is a process independent of cell death and often precedes the earliest events of apoptosis. Neutrophils undergoing spontaneous apoptosis in culture dramatically downregulate their expression of FcγRIIIb, CD62L, and the seven-span receptors fMLPR and CXCR2. In contrast, the reduction in CXCR1 expression is only modest, and CXCR4, which is undetectable on fresh cells, is even upregulated on early apoptotic cells that expose PS (Nagase et al., 2002). Downregulation of chemoattractant receptors on aging and apoptotic neutrophils will likely have dramatic consequences for the trafficking of these cells in vivo. This might reflect mechanisms to prevent the exit of functionally incompetent neutrophils from the circulation and also prevent the exit of these cells from the site of inflammation.

The relevance of CXCR4 de novo expression on apoptotic neutrophils remains to be clarified. Triggering of the receptor may increase intracellular calcium (potentially enhancing apoptosis) but does not seem to induce any chemotaxis (Wolach et al., 2007). In this respect it should be acknowledged that the functional capacity of aging neutrophils can indeed be rescued by both G- and GM-CSF, but at a differential level of efficacy and side effects, rendering G-CSF the growth factor to be preferred for therapeutic usage.

Apart from some rare examples of intramedullary apoptosis of myeloid (progenitor) cells in myelodysplasia or Shwachman syndrome, mature circulating neutrophils do not expose PS. Neither in autoimmune neutropenias, nor in inflammatory conditions of neutropenia as met during septicemia or severe pneumonia, are the neutrophils in the bloodstream engaged in a premature process of cell death. To date, we have observed only in GSD type 1b (GSD1b) that the circulating mature neutrophils are undergoing premature apoptosis (Kuijpers et al., 2003). GSD1b is caused by inherited defects of the glucose-6-phosphate transporter, accompanied by neutropenia and/or neutrophil dysfunction (in chemotaxis, O_2^- generation, and Ca^{2+} mobilization). We now believe that the neutrophil functions in GSD1b are reduced considerably if not completely by the premature aging of the differentiated neutrophils.

The impact of PS exposure can only be speculated on. Enhanced elimination of as yet not fully differentiated myeloid cells by macrophages via the PS receptor most likely occurs in concert with CD14, deposited mannose-binding lectin, and/or complement fragments (Ogura, 2001; Savill and Fadok, 2000). In some of the neutropenic syndromes this is assumed to occur in the environment of the bone marrow prior to neutrophil egress. As demonstrated in our study on GSD1b, the noneliminated neutrophils can be observed in the circulation as the early-apoptotic annexin-V$^+$ neutrophils that outnumber the late-apoptotic neutrophils with clustered mitochondria and condensed nuclei (Kuijpers et al., 2003). These late-apoptotic neutrophilic bodies are likely to be removed more rapidly than the early-apoptotic cells due to auxiliary binding proteins or receptors in addition to PS on the plasma membrane. Splenomegaly in GSD1b may only become apparent when this organ is suddenly overloaded by increased clearance of apoptotic cells during infections and/or exaggerated bone marrow production. In case of severe neutropenia, in vivo G-CSF administration may affect splenic size by extramedullary hematopoiesis, sometimes complicated by clinical hypersplenism requiring dose reduction or splenectomy.

Under normal circumstances, neutrophils will be eliminated mainly in the extravascular space. Macrophages and, most likely, tissue cells such as fibroblasts will cooperatively be involved in this process (see also chapter 27). PS exposure is believed to be insufficient for this process, and additional molecules must be exposed or bound to the surface of apoptotic neutrophils for proper recognition. Because of the large surface area, the disposal of neutrophils in the intestinal tract may also be an important way to both protect the mucosal lining and, at the same time, discard these potentially harmful cells from the body.

CONCLUDING REMARKS

In the first phase of inflammatory responses or other forms of stress, chemotaxins such as C5a, lipid mediators, and so-called chemokines are produced. The secretion of chemokines can be firmly induced by the proinflammatory cytokines derived from tissue macrophages and T cells, such as TNF-α, IL-1β, and IFN-γ, or by bacterial products such as fMLP, LPS, LTA, and peptidoglycans.

The cellular composition and duration of an inflammatory response may differ among the different inflammatory reactions (e.g., bacterial versus parasitic infection, allergy, delayed-type hypersensitivity reaction). The process of recruitment is complex and still incompletely understood. Once recruited, the neutrophils have a wide range of toxic mechanisms to fight any invading microorganism, as described. These mechanisms are strongly regulated and delicately controlled, because an overexcessive or premature induction of the toxic activities may result in the inactivation of protease inhibitors and activation of several cascades of activating substances (e.g., the coagulation, the fibrinolytic, and the complement system). As a consequence, bacteremia may progress to septic shock and disseminated intravascular coagulation, a community-acquired pneumonia may develop into acute or adult respiratory distress syndrome, and hypoxia/reperfusion injury can lead to fatal circulatory collapse. In conclusion, neutrophils are necessary and useful but also very dangerous tools to protect the host from bacterial and fungal infections.

REFERENCES

Adams, M. J., and S. Cory. 1998. The Bcl-2 protein family: arbiters of cell survival. *Science* **281:**1322–1326.

Aitman, T. J., R. Dong, T. J. Vyse, P. J. Norsworthy, M. D. Johnson, J. Smith, J. Mangion, C. Roberton-Lowe, A. J. Marshall, E. Petretto, M. D. Hodges, G. Bhangal, S. G. Patel, K. Sheehan-Rooney, M. Duda, P. R. Cook, D. J. Evans, J. Domin, J. Flint, J. J. Boyle, C. D. Pusey, and H. T. Cook. 2006. Copy number polymorphism in Fcgr3 predisposes to glomerulonephritis in rats and humans. *Nature* **439:**851–855.

Akira, S., and K. Takeda. 2004. Toll-like receptor signalling. *Nat. Rev. Immunol.* **4:**499–511.

Ambruso, D. R., C. Knall, A. N. Abell, J. Panepinto, A. Kurkchubasche, G. Thurman, C. Gonzalez-Aller, A. Hiester, M. de Boer, R. J. Harbeck, R. Oyer, G. L. Johnson, and D. Roos. 2000. Human neutrophil immunodeficiency syndrome is associated with an inhibitory Rac2 mutation. *Proc. Natl. Acad. Sci. USA* **97:**4654–4659.

Anderson, K. L., K. A. Smith, H. Perkin, G. Hermanson, C. G. Anderson, D. J. Jolly, R. A. Maki, and B. E. Torbett. 1999. PU.1 and the granulocyte- and macrophage colony-stimulating factor receptors play distinct roles in late-stage myeloid cell differentiation. *Blood* **94:**2310–2318.

Aurrand-Lions, M., C. Johnson-Leger, and B. Imhof. 2002. The last molecular fortress in leukocyte trans-endothelial migration. *Nat. Immunol.* **3:**116–118.

Barclay, A. N., G. J. Wright, G. Brooke, and M. H. Brown. 2002. CD200 and membrane protein interactions in the control of myeloid cells. *Trends Immunol.* **23:**285–290.

Belaaouaj, A., R. McCarthy, M. Baumann, Z. Gao, T. J. Ley, S. N. Abraham, and S. D. Shapiro. 1998. Mice lacking neutrophil elastase reveal impaired host defense against gram negative bacterial sepsis. *Nat. Med.* **4:**615–618.

Benson, K. F., F. Q. Li, R. E. Person, D. Albani, Z. Duan, J. Wechsler, K. Meade-White, K. Williams, G. M. Acland, G. Niemeyer, C. D. Lothrop, and M. Horwitz. 2003. Mutations associated with neutropenia in dogs and humans disrupt intracellular transport of neutrophil elastase. *Nat. Genet.* **35:**90–96.

Blink, E., N. A. Maianski, E. S. Alnemri, A. S. Zervos, D. Roos, and T. W. Kuijpers. 2004. Intramitochondrial serine protease activity of Omi/HtrA2 is required for caspase-independent cell death of human neutrophils. *Cell Death Differ.* **11:**937–939.

Bohn, G., A. Allroth, G. Brandes, J. Thiel, E. Glocker, A. A. Schäffer, C. Rathinam, N. Taub, D. Teis, C. Zeidler, R. A. Dewey, R. Geffers, J. Buer, L. A. Huber, K. Welte, B. Grimbacher, and C. Klein. 2007. A novel human primary immunodeficiency syndrome caused by deficiency of the endosomal adaptor protein p14. *Nat. Med.* **13:**38–45.

Bokoch, G. M. 1996. Chemoattractant signalling and leukocyte activation. *Blood* **86:**1649–1660.

Borregaard, N., O. E. Sorensen, and K. Theilgaard-Monch. 2007. Neutrophil granules: a library of innate immunity proteins. *Trends Immunol.* **28:**340–345.

Bouchon, A., F. Facchetti, M. A. Welgand, and M. Colonna. 2001. TREM-1 amplifies inflammation and is a crucial mediator of septic shock. *Nature* **410:**1103–1107.

Brach, M. A., S. deVos, H. J. Gruss, and F. Herrmann. 1992. Prolongation of survival of human polymorphonuclear neutrophils by granulocyte-macrophage colony-stimulating factor is caused by inhibition of programmed cell death. *Blood* **80:**2920–2924.

Breunis, W. B., E. van Mirre, M. Bruin, J. Geissler, M. de Boer, M. Peters, D. Roos, M. de Haas, H. R. Koene, and T. W. Kuijpers. 2008. Copy number variation of the activating FCGR2C gene predisposes to idiopathic thrombocytopenia. *Blood* **111:**1029–1038.

Breunis, W. B., E. van Mirre, J. Geissler, N. Laddach, G. J. Wolbink, E. van der Schoot, M. de Haas, M. de Boer, D. Roos, and T. W. Kuijpers. 2009. Copy number variation at the FCGR locus includes FCGR3A, FCGR2C and FCGR3B but not FCGR2A and FCGR2B. *Hum. Mutat.* March 23 [Epub ahead of print].

Broxmeyer H. E., C. M. Orschell, D. W. Clapp, G. Hangoc, S. Cooper, P. A. Plett, W. C. Liles, X. Li, B. Graham-Evans, T. B. Campbell, G. Calandra, G. Bridger, D. C. Dale, and E. F. Srour. 2005. Rapid mobilization of murine and human hematopoietic stem and progenitor cells with AMD3100, a CXCR4 antagonist. *J. Exp. Med.* **201:**1307–1318.

Bruey, J. M., N. Bruey-Sedano, R. Newman, S. Chandler, C. Stehlik, and J. C. Reed. 2004. PAN1/NALP2/PYPAF2, an inducible inflammatory mediator that regulates NF-kappaB and caspase-1 activation in macrophages. *J. Biol. Chem.* **279:**51897–51907.

Bruhns, P., F. Vely, O. Malbec, W. H. Fridman, E. Vivier, and M. Daeron. 2000. Insufficient phosphorylation prevents FcγRIIB from recruiting the SH2 domain-containing protein tyrosine phosphatase SHP-1. *J. Biol. Chem.* **275:**37357–37364.

Cai, H., K. Reinisch, and S. Ferro-Novick. 2007. Coats, tethers, Rabs, and SNAREs work together to mediate the intracellular destination of a transport vesicle. *Dev. Cell.* **12:**671–682.

Cartwright, N., O. Murch, S. K. McMaster, M. J. Paul-Clark, D. A. van Heel, B. Ryffel, V. F. Quesniaux, T. W. Evans, C. Thiemermann, and J. A. Mitchell. 2007. Selective NOD1 agonists cause shock and organ injury/dysfunction in vivo. *Am. J. Respir. Crit. Care Med.* **175:**595–603.

Chamaillard, M., M. Hashimoto, Y. Horie, J. Masumoto, S. Qiu, L. Saab, Y. Ogura, A. Kawasaki, K. Fukase, S. Kusumoto, M. A. Valvano, S. J. Foster, T. W. Mak, G. Nunez, and N. Inohara. 2003. An essential role for NOD1 in host recognition of bacterial peptidoglycan containing diaminopimelic acid. *Nat. Immunol.* **4:**702–707.

Cilenti, L., M. M. Soundarapandian, G. A. Kyriazis, V. Stratico, S. Singh, S. Gupta, J. V. Bonventre, E. S. Alnemri, and A. S. Zervos. 2004. Regulation of HAX-1 anti-apoptotic protein by Omi/HtrA2 protease during cell death. *J. Biol. Chem.* **279:**50295–50301.

Corada, M., S. Chimenti, M. R. Cera, M. Vinci, M. Salio, F. Fiordaliso, N. de Angelis, A. Villa, M. Bossi, L. I. Staszewsky, A. Vecchi, D. Parazzoli, T. Motoike, R. Latini, and E. Dejana. 2005. Junctional adhesion molecule-A-deficient polymorphonuclear cells show reduced diapedesis in peritonitis and heart ischemia-reperfusion injury. *Proc. Natl. Acad. Sci. USA* **102:**10634–10639.

Dale, D. C., R. E. Person, A. A. Boylard, A. G. Aprikyan, C. Bos, M. A. Bonilla, L. A. Boxer, G. Kanourakis, C. Zeidler, K. Welte, K. F. Benson, and M. Horwitz. 2000. Mutations in the gene encoding neutrophil elastase in congenital and cyclic neutropenia. *Blood* **96:**2317–2322.

Daws, M. R., L. L. Lanier, W. R. Seaman, and J. C. Ryan. 2001. Cloning and characterization of a novel mouse myeloid DAP12-associated receptor family. *Eur. J. Immunol.* **31:**783–791.

Diehl, G. E., H. H. Yue, K. Hsieh, A. A. Kuang, M. Ho, L. A. Morici, L. L. Lenz, D. Cado, L. W. Riley, and A. Winoto. 2004. TRAIL-R as a negative regulator of innate immune cell responses. *Immunity* **21:**877–889.

Divanovic, S., A. Trompette, S. F. Atabani, R. Madan, D. T. Golenbock, A. Visintin, R. W. Finberg, A. Tarakhovsky, S. N. Vogel, Y. Belkaid, E. A. Kurt-Jones, and C. L. Karp. 2005. Negative regulation of Toll-like receptor 4 signaling by the Toll-like receptor homolog RP105. *Nat. Immunol.* **6:**571–578.

Donadieu, J., T. Leblanc, B. Bader Meunier, M. Barkaoui, O. Fenneteau, Y. Bertrand, M. Maier-Redelsperger, M. Micheau, J. L. Stephan, N. Phillipe, P. Bordigoni, A. Babin-Boilletot, P. Bensaid, A. M. Manel, E. Vilmer, I. Thuret, S. Blanche, E. Gluckman, A. Fischer, F. Mechinaud, B. Joly, T. Lamy, O. Hermine, B. Cassinat, C. Bellanne-Chantelot, and C. Chomienne; French Severe Chronic Neutropenia Study Group. 2005. Analysis of risk factors for myelodysplasias, leukemias and death from infection among patients with congenital neutropenia. Experience of the French Severe Chronic Neutropenia Study Group. *Haematologica* **90:**45–53.

Dong, F., D. C. Dale, M. A. Bonilla, M. Freedman, A. A. Fasth, H. J. Neijens, J. Palmblad, G. L. Briars, G. Carlsson, A. J. Veerman, K. Welte, B. Lowenberg, and I. P. Touw. 1997. Mutations in the granulocyte colony-stimulating factor receptor gene in patients with severe congenital neutropenia. *Leukemia* **11:**120–125.

Donini, M., S. Fontana, G. Savoldi, W. Vermi, L. Tassone, F. Gentili, E. Zenaro, D. Ferrari, L. D. Notarangelo, F. Porta, F. Facchetti, L. D. Notarangelo, S. Dusi, and R. Badolato. 2007. G-CSF treatment of Severe Congenital Neutropenia reverses neutropenia but does not correct the underlying functional deficiency of the neutrophil in defending against microorganisms. *Blood* **109:**4716–4723.

Dorfleutner, A., N. B. Bryan, S. J. Talbott, K. N. Funya, S. L. Rellick, J. C. Reed, X. Shi, Y. Rojanasakul, D. C. Flynn, and C. Stehlik. 2007. Cellular pyrin domain-only protein 2 is a candidate regulator of inflammasome activation. *Infect. Immun.* **75:**1484–1492.

Dowds, T. A., J. Masumoto, F. F. Chen, Y. Ogura, N. Inohara, and G. Nunez. 2003. Regulation of cryopyrin/Pypaf1 signaling by pyrin, the familial Mediterranean fever gene product. *Biochem. Biophys. Res. Commun.* **302:**575–580.

Dransfield, I., S. C. Stocks, and C. Haslett. 1995. Regulation of cell adhesion molecule expression and function associated with neutrophil apoptosis. *Blood* **85:**3264–3273.

Drewniak, A., B. van Raam, J. Geissler, A. T. J. Tool, O. R. F. Mook, T. K. van den Berg, F. Baas, and T. W. Kuijpers. 2009. Changes in gene expression of granulocytes during in vivo GCSF/dexamethasone mobilization for transfusion purposes. *Blood* April 6 [Epub ahead of print].

Dunne, J. L., C. M. Ballantyne, A. L. Beaudet, and K. Ley. 2002. Control of leukocyte rolling velocity in TNF-alpha-induced inflammation by LFA-1 and Mac-1. *Blood* **99:**336–341.

Dunne, J. L., R. G. Collins, A. L. Beaudet, C. M. Ballantyne, and K. Ley. 2003. Mac-1, but not LFA-1, uses intercellular adhesion molecule-1 to mediate slow leukocyte rolling in TNF-alpha-induced inflammation. *J. Immunol.* **171:**6105–6111.

Enders, A., B. Zieger, K. Schwarz, A. Yoshimi, C. Speckmann, E. M. Knoepfle, U. Kontny, C. Muller, A. Nurden, J. Rohr, M. Henschen, U. Pannicke, C. Niemeyer, P. Nurden, and S. Ehl. 2006. Lethal hemophagocytic lymphohistiocytosis in Hermansky-Pudlak syndrome type II. *Blood* **108:**81–87.

Fanciulli, M., P. J. Norsworthy, E. Petretto, R. Dong, L. Harper, L. Kamesh, J. M. Heward, S. C. Gough, A. de Smith, A. I. Blakemore, P. Froguel, C. J. Owen, S. H. Pearce, L. Teixeira, L. Guillevin, D. S. Graham, C. D. Pusey, H. T. Cook, T. J. Vyse, and T. J. Aitman. 2007. FCGR3B copy number variation is associated with susceptibility to systemic, but not organ-specific, autoimmunity. *Nat. Genet.* **39:**721–723.

Fiorentino, L., C. Stehlik, V. Oliveira, M. E. Ariza, A. Godzik, and J. C. Reed. 2002. A novel PAAD-containing protein that modulates NF-kappa B induction by cytokines tumor necrosis factor-alpha and interleukin-1beta. *J. Biol. Chem.* **277:**35333–35340.

French FMF Consortium. 1997. A candidate gene for familial Mediterranean fever. *Nat. Genet.* **17:**25–31.

Gakidis, M. A., X. Cullere, T. Olson, J. L. Wilsbacher, B. Zhang, S. L. Moores, K. Ley, W. Swat, T. Mayadas, and J. S. Brugge. 2004. Vav GEFs are required for beta2 integrin-dependent functions of neutrophils. *J. Cell Biol.* **166:**273–282.

Green, D. R., and G. Melino. 2001. ICE heats up. *Cell Death Differ.* **8:**549–550.

Gross, O., A. Gewies, K. Finger, M. Schafer, T. Sparwasser, C. Peschel, I. Forster, and J. Ruland. 2006. Card9 controls a non-TLR signalling pathway for innate anti-fungal immunity. *Nature* **442:**651–656.

Gu, Y., M. D. Filippi, J. A. Cancelas, J. E. Siefring, E. P. Williams, A. C. Jasti, C. E. Harris, A. W. Lee, R. Prabhakar, S. J. Atkinson, D. J. Kwiatkowski, and D. A. Williams. 2003. Hematopoietic cell regulation by Rac1 and Rac2 guanosine triphosphatases. *Science* **302:**445–449.

Gutierrez, O., C. Pipaon, N. Inohara, A. Fontalba, Y. Ogura, F. Prosper, G. Nunez, and J. L. Fernandez-Luna. 2002. Induction of Nod2 in myelomonocytic and intestinal epithelial cells via nuclear factor-kappa B activation. *J. Biol. Chem.* **277:**41701–41705.

Harah, H., C. Ishihara, A. Takeuchi, T. Imashnishi, L. Xue, S. W. Morris, M. Inui, T. Takai, A. Shibuya, S. Saijo, Y. Iwakura, N. Ohno, H. Koseki, H. Yoshida, J. M. Penninger, and T. Saito. 2007. The adaptor protein CARD9 is essential for the activation of myeloid cells through ITAM-associated and Toll-like receptors. *Nat. Immunol.* **8:**619–629.

Harton, J. A., M. W. Linhoff, J. Zhang, and J. P. Ting. 2002. CATERPILLER: a large family of mammalian genes containing CARD, pyrin, nucleotide-binding, and leucine-rich repeat domains. *J. Immunol.* **169:**4088–4093.

Hayashi, F., T. K. Means, and A. D. Luster. 2003. Toll-like receptors stimulate human neutrophil function. *Blood* **102:**2660–2669.

Hengartner, M. O. 2000. The biochemistry of apoptosis. *Nature* **407:**770–776.

Hidalgo, A., A. J. Peired, M. K. Wild, D. Vestweber, and P. S. Frenette. 2007. Complete identification of E-selectin ligands on neutrophils reveals distinct functions of PSGL-1, ESL-1, and CD44. *Immunity* **26:**477–489.

Hlaing, T., R. F. Guo, K. A. Dilley, J. M. Loussia, T. A. Morrish, M. M. Shi, C. Vincenz, and P. A. Ward. 2001. Molecular cloning and characterization of DEFCAP-L and -S, two isoforms of a novel member of the mammalian Ced-4 family of apoptosis proteins. *J. Biol. Chem.* **276:**9230–9238.

Hoarau, C., B. Gérard, E. Lescanne, D. Henry, S. François, J. J. Lacapère, J. El Benna, P. M. Dang, B. Grandchamp, Y. Lebranchu, M. A. Gougerot-Pocidalo, and C. Elbim. 2007. TLR9 activation induces normal neutrophil responses in a child with IRAK-4 deficiency: involvement of the direct PI3K pathway. *J. Immunol.* **179:**4754–4765.

Hock, H., M. J. Hamblen, H. M. Rooke, D. Traver, R. T. Bronson, S. Cameron, and S. H. Orkin. 2003. Intrinsic requirement for zinc finger transcription factor Gfi-1 in neutrophil differentiation. *Immunity* **18:**109–120.

Hong, W. 2005. SNAREs and traffic. *Biochim. Biophys. Acta* **1744:**493–517.

Hsu, Y. M., Y. Zhang, Y. You, D. Wang, H. Li, O. Duramad, X. F. Qin, C. Dong, and X. Lin. 2007. The adaptor protein CARD9 is required for innate immune responses to intracellular pathogens. *Nat. Immunol.* **8:**198–205.

Huang, M. T., K. Y. Larbi, C. Scheiermann, A. Woodfin, N. Gerwin, D. O. Haskard, and S. Nourshargh. 2006. ICAM-2 mediates neutrophil transmigration in vivo: evidence for stimulus specificity and a role in PECAM-1-independent transmigration. *Blood* **107:**4721–4727.

Hugot, J. P., M. Chamaillard, H. Zouali, S. Lesage, J. P. Cezard, J. Belaiche, S. Almer, C. Tysk, C. A. O'Morain, M. Gassull, V. Binder, Y. Finkel, A. Cortot, R. Modigliani, P. Laurent-Puig, C. Gower-Rousseau, J. Macry, J. F. Colombel, M. M. Sahbatou, and G. Thomas. 2001. Association of NOD2 leucine-rich repeat variants with susceptibility to Crohn's disease. *Nature* **411:**599–603.

Inohara, N., T. Koseki, L. del Peso, Y. Hu, C. Yee, S. Chen, R. Carrio, J. Merino, D. Liu, J. Ni, and G. Nunez. 1999. Nod1, an Apaf-1-like activator of caspase-9 and nuclear factor kB. *J. Biol. Chem.* **274:**14560–14567.

International FMF Consortium. 1997. Ancient missense mutations in a new member of the RoRet gene family are likely to cause familial Mediterranean fever. *Cell* **90:**797–807.

Jeha, S., K. W. Chan, A. G. Aprikyan, W. K. Hoots, S. Culbert, H. Zietz, D. C. Dale, and M. Albitar. 2000. Spontaneous remission of granulocyte colony-stimulating factor-associated leukemia in a child with severe congenital neutropenia. *Blood* **96:**3647–3649.

Karim, M. A., K. Suzuki, K. Fukai, J. Oh, D. L. Nagle, K. J. Moore, E. Barbosa, T. Falik-Borenstein, A. Filipovich, Y. Ishida, S. Kivrikko, C. Klein, F. Kreuz, A. Levin, H. Miyajima, J. Regueiro, C. Russo, E. Uyama, O. Vierimaa, and R. A. Spritz. 2002. Apparent genotype-phenotype correlation in childhood, adolescent, and adult Chediak-Higashi syndrome. *Am. J. Med. Genet.* **108:**16–22.

Karsunky, H., H. Zeng, T. Schmidt, B. Zevnik, R. Kluge, K. W. Schmid, U. Dührsen, and T. Möröy. Inflammatory reactions and severe neutropenia in mice lacking the transcriptional repressor Gfi1. *Nat. Genet.* **30:**295–300.

Kasper, B., N. Tidow, D. Grothues, and K. Welte. 2000. Differential expression and regulation of GTPases (RhoA and Rac2) and GDIs (LyGDI and RhoGDI) in neutrophils from patients with severe congenital neutropenia. *Blood* **95:**2947–2953.

Kawai, T., O. Adachi, T. Ogawa, K. Takeda, and S. Akira. 1999. Unresponsiveness of MyD88 deficient mice to endotoxin. *Immunity* **11:**115–122.

Khandoga, A., J. S. Kessler, H. Meissner, M. Hanschen, M. Corada, T. Motoike, G. Enders, E. Dejana, and F. Krombach. 2005. Junctional adhesion molecule-A deficiency increases hepatic ischemia-reperfusion injury despite reduction of neutrophil transendothelial migration. *Blood* **106:**725–733.

Kinashi, T., M. Aker, M. Sokolovsky-Eisenberg, V. Grabovsky, C. Tanaka, R. Shamri, S. Feigelson, A. Etzioni, and R. Alon. 2004. LAD-III, a leukocyte adhesion deficiency syndrome associated with defective Rap1 activation and impaired stabilization of integrin bonds. *Blood* **103:**1033–1036.

Kinoshita, T., Y. Wang, M. Hasegawa, R. Imamura, and T. Suda. 2005. PYPAF3, a PYRIN-containing APAF-1-like protein, is a feedback regulator of caspase-1-dependent interleukin-1beta secretion. *J. Biol. Chem.* **280:**21720–21725.

Klein, C., M. Grudzien, G. Appaswamy, M. Germeshausen, I. Sandrock, A. A. Schaffer, C. Rathinam, K. Boztug, B. Schwinzer, N. Rezaei, G. Bohn, M. Melin, G. Carlsson, B. Fadeel, N. Dahl, J. Palmblad, J. I. Henter, C. Zeidler, B. Grimbacher, and K. Welte. 2007. HAX1 deficiency causes autosomal recessive severe congenital neutropenia (Kostmann disease). *Nat. Genet.* **39:**86–92.

Klesney-Tait, J., I. R. Turnbull, and M. Colonna. 2006. The TREM receptor family and signal integration. *Nat. Immunol.* **7:**1266–1273.

Kobayashi, K., L. D. Hernandez, J. E. Galan, C. A. Janeway, Jr, R. Medzhitov, and R. A. Flavell. 2002. IRAK-M is a negative regulator of Toll-like receptor signaling. *Cell* **110:**191–202.

Koene, H. R., M. Kleijer, D. Roos, M. de Haas, and A. E. von dem Borne. 1998. Fc gamma RIIIB gene duplication: evidence for presence and expression of three distinct Fc gamma RIIIB genes in NA(1+,2+)SH(+) individuals. *Blood* **91:**673–679.

Kuijpers, T. W. 2002. Clinical symptoms and neutropenia: the balance of neutrophil development, functional activity, and cell death. *Eur. J. Pediatr.* **161:**S75–S82.

Kuijpers, T. W., M. Alders, A. T. Tool, C. M. Mellink, D. Roos, and R. C. Hennekam. 2005. Hematologic abnormalities in Shwachman Diamond syndrome: lack of genotype-phenotype relationship. *Blood* **106:**356–361.

Kuijpers, T. W., B. C. Hakkert, M. H. Hart, and D. Roos. 1992. Neutrophil migration across monolayers of cytokine-prestimulated endothelial cells: a role for platelet-activating factor and IL-8. *J. Cell Biol.* **117:**565–572.

Kuijpers, T. W., M. A. Maianski, A. T. Tool, K. Becker, B. Plecko, F. Valianpour, R. J. Wanders, R. Pereira, J. Van Hove, A. J. Verhoeven, D. Roos, F. Baas, and P. G. Barth. 2004. Neutrophils in Barth syndrome (BTHS) avidly bind annexin-V in the absence of apoptosis. *Blood* **103:**3915–3923.

Kuijpers, T. W., N. A. Maianski, A. T. Tool, G. P. Smit, J. P. Rake, D. Roos, and G. Visser. 2003. The presence of apoptotic neutrophils in the circulation of patients with Glycogen Storage Disease type 1b (GSD1b). *Blood* **101:**5021–5024.

Kuijpers, T. W., A. T. Tool, C. E. van der Schoot, L. A. Ginsel, J. J. Onderwater, D. Roos, and A. J. Verhoeven. 1991. Membrane surface antigen expression on neutrophils: a reappraisal of the use of surface markers for neutrophil activation. *Blood* **78:**1105–1111.

Kuijpers, T. W., R. van Bruggen, N. Kamerbeek, A. T. Tool, G. Hicsonmez, A. Gurgey, A. Karow, A. J. Verhoeven, K. Seeger, O. Sanal, C. Niemeyer, and D. Roos. 2007. Natural history and early diagnosis of LAD-1/variant syndrome. *Blood* **109:**3529–3537.

Kuijpers, T. W., E. Van De Vijver, M. A. Weterman, M. de Boer, A. T. Tool, T. K. van den Berg, M. Moser, M. E. Jakobs, K. Seeger, O. Sanal, S. Unal, M. Cetin, D. Roos, A. J. Verhoeven, and F. Baas. 2008. LAD-1/variant syndrome is caused by mutations in FERMT3. *Blood* Dec. 8 [Epub ahead of publication].

Kuijpers, T. W., R. A. W. van Lier, D. Hamann, M. de Boer, L. Y. Thung, R. S. Weening, A. J. Verhoeven, and D. Roos. 1997. Leukocyte adhesion deficiency type 1/variant: a novel immunodeficiency syndrome characterized by dysfunctional β2 integrins. *J. Clin. Investig.* **100:**1725–1733.

Kunisaki, Y., A. Nishikimi, Y. Tanaka, R. Takii, M. Noda, A. Inayoshi, K. Watanabe, F. Sanematsu, T. Sasazuki, T. Sasaki, and Y. Fukui. 2006. DOCK2 is a Rac activator that regulates motility and polarity during neutrophil chemotaxis. *J. Cell Biol.* **174:**647–652.

Kurt-Jones, E. A., L. Mandell, C. Whitney, A. Padgett, K. Gosselin, P. E. Newburger, and R. W. Finberg. 2002. Role of toll-like receptor 2 (TLR2) in neutrophil activation: GM-CSF enhances TLR2 expression and TLR2-mediated interleukin 8 responses in neutrophils. *Blood* **100:**1860–1868.

Kuwata, H., M. Matsumoto, K. Atarashi, H. Morishita, T. Hirotani, R. Koga, and K. Takeda. 2006. IkappaBNS inhibits induction of a subset of Toll-like receptor-dependent genes and limits inflammation. *Immunity* **24:**41–51.

Lamkanfi, M., G. Denecker, M. Kalai, K. D'hondt, A. Meeus, W. Declercq, X. Saelens, and P. Vandenabeele. 2004. INCA, a novel human caspase recruitment domain protein that inhibits interleukin-1beta generation. *J. Biol. Chem.* **279:**51729–51738.

Leibundgut-Landmann, S., O. Gross, M. J. Robinson, F. Osorio, E. C. Slack, S. V. Tsoni, V. Schweighoffer, G. D. Brown, J. Ruland, and C. Reiss E Sousa. 2007. Syk- and CARD9-dependent coupling of innate immunity to the induction of T helper cells that produce interleukin 17. *Nat. Immunol.* **8:**630–638.

Lekstrom-Himes, J. A., S. E. Dorman, P. Kopar, S. M. Holland, and J. I. Gallin. 1999. Neutrophil-specific granule deficiency results from a novel mutation with loss of function of the transcription factor CCAAT/enhancer binding protein epsilon. *J. Exp. Med.* **189:**1847–1852.

Levesque, J. P., F. Liu, P. J. Simmons, T. Betsuyaku, R. M. Senior, C. Pham, and D. C. Link. 2004. Characterization of hematopoietic progenitor mobilization in protease-deficient mice. *Blood* **104:**65–72.

Liew, F. Y., D. Xu, E. K. Brint, and L. A. O'Neill. 2005. Negative regulation of Toll-like receptor-mediated immune responses. *Nat. Rev. Immunol.* **5:**446–458.

Liles, W. C., H. E. Broxmeyer, E. Rodger, B. Wood, K. Hübel, S. Cooper, G. Hangoc, G. J. Bridger, G. W. Henson, G. Calandra, and D. C. Dale. 2003. Mobilization of hematopoietic progenitor cells in healthy volunteers by AMD3100, a CXCR4 antagonist. *Blood* **102**:2728–2730.

Liu, F., H. F. Wu, R. Wesselschmidt, T. Kornaga, and D. C. Link. 1996. Impaired production and increased apoptosis of neutrophils in granulocyte colony-stimulating factor receptor-deficient mice. *Immunity* **5**:491–501.

Liu, Y., H. J. Buhring, K. Zen, S. L. Burts, F. J. Schnell, I. R. Williams, and C. A. Parkos. 2002. Signal regulatory protein (SIRPα), a cellular ligand for CD47, regulates neutrophil transmigration. *J. Biol. Chem.* **277**:10028–10036.

Logan, M. R., P. Lacy, S. O. Odemuyiwa, M. Steward, F. Davoine, H. Kita, and R. Moqbel. 2006. A critical role for vesicle-associated membrane protein-7 in exocytosis from human eosinophils and neutrophils. *Allergy* **61**:777–784.

Loike, J. D., L. Cao, S. Budhu, E. E. Marcantonio, J. El Khoury, S. Hoffman, T. A. Yednock, and S. C. Silverstein. 1999. Differential regulation of beta1 integrins by chemoattractants regulates neutrophil migration through fibrin. *J. Cell Biol.* **144**:1047–1056.

Lou, O., P. Alcaide, F. W. Luscinskas, and W. A. Muller. 2007. CD99 is a key mediator of the transendothelial migration of neutrophils. *J. Immunol.* **178**:1136–1143.

Maianski, N. A., F. P. Mul, J. D. van Buul, D. Roos, and T. W. Kuijpers. 2002. Granulocyte Colony-Stimulating factor (G-CSF) inhibits in neutrophils the mitochondria-dependent activation of Caspase-3. *Blood* **99**:672–679.

Martinon, F., K. Burns, and J. Tschopp. 2002. The inflammasome: a molecular platform triggering activation of inflammatory caspases and processing of proIL-beta. *Mol. Cell* **10**:417–426.

Miceli-Richard, C., S. Lesage, M. Rybojad, A. M. Prieur, S. Manouvrier-Hanu, R. Hafner, M. Chamaillard, H. Zouali, G. Thomas, and J. P. Hugot. 2001. CARD15 mutations in Blau syndrome. *Nat. Genet.* **29**:19–20.

Mollinedo, F., J. Calafat, H. Janssen, B. Martin-Martin, J. do Cancha, S. M. Nabokina, and C. Gajate. 2006. Combinatorial SNARE complexes modulate the secretion of cytoplasmic granules in human neutrophils. *J. Immunol.* **177**:2831–2841.

Moulding, D. A., J. A. Quayle, C. A. Hart, and S. W. Edwards. 1998. Mcl-1 expression in human neutrophils: regulation by cytokines and correlation with cell survival. *Blood* **92**:2495–2502.

Muller, W. A., S. A. Weigl, X. Deng, and D. M. Phillips. 1993. PECAM-1 is required for transendothelial migration of leukocytes. *J. Exp. Med.* **178**:449–460.

Murdoch, C., and A. Finn. 2000. Chemokine receptors and their role in inflammation and infectious diseases. *Blood* **95**:3032–3043.

Nagase, H., M. Miyamasu, M. Yamaguchi, M. Imanishi, N. H. Tsuno, K. Matsushima, K. Yamamoto, Y. Morita, and K. Hirai. 2002. Cytokine-mediated regulation of CXCR4 expression in human neutrophils. *J. Leukoc. Biol.* **71**:711–717.

Niemann, C. U., M. Abrink, G. Pejler, R. L. Fischer, E. I. Christensen, S. D. Knight, and N. Borregaard. 2007. Neutrophil elastase depends on serglycin proteoglycan for localization in granules. *Blood* **109**:4478–4486.

Ogura, Y., D. K. Bonen, N. Inohara, D. L. Nicolae, F. F. Chen, R. Ramos, H. Britton, T. Moran, R. Karaliuskas, R. H. Duerr, J. P. Achkar, S. R. Brant, T. M. Bayless, B. S. Kirschner, S. B. Hanauer, G. Nunez, and J. H. Cho. 2001. A frameshift mutation in NOD2 associated with susceptibility to Crohn's disease. *Nature* **411**:603–606.

Oldenborg, P. A., A. Zheleznyak, Y. F. Fang, C. F. Lagenaur, H. D. Gresham, and F. P. Lindberg. 2000. Role of CD47 as a marker of self on red blood cells. *Science* **288**:2051–2054.

Ostermann, G., K. S. Weber, A. Zernecke, A. Schröder, and C. Weber. 2002. JAM-1 is a ligand of the beta(2) integrin LFA-1 involved in transendothelial migration of leukocytes. *Nat. Immunol.* **3**:151–158.

Ozinsky, A., D. M. Underhill, J. D. Fontenot, A. M. Hajjar, K. D. Smith, C. B. Wilson, L. Schroeder, and A. Aderem. 2000. The repertoire for pattern recognition of pathogens by the innate immune system is defined by cooperation between toll-like receptors. *Proc. Natl. Acad. Sci. USA* **97**:13766–13771.

Pasvolsky, R., S. W. Feigelson, S. S. Kilic, A. J. Simon, G. Tal-Lapidot, V. Grabovsky, J. R. Crittenden, N. Amariglio, M. Safran, A. M. Graybiel, G. Rechavi, S. Ben-Dor, A. Etzioni, and R. Alon. 2007. A LAD-III syndrome is associated with defective expression of the Rap-1-activator CalDAG-GEF1 in lymphocytes, neutrophils and platelets. *J. Exp. Med.* **204**:1571–1582.

Pearse, R. N., T. Kawabe, S. Bolland, R. Guinamard, T. Kurosaki, and J. V. Ravetch. 1999. SHIP recruitment attenuates Fc gamma RIIB-induced B cell apoptosis. *Immunity* **10**:753–760.

Person, R. E., F. Q. Li, Z. Duan, J. Wechsler, H. A. Papadaki, G. Eliopoulos, C. Kaufman, S. J. Bertolone, B. Nakamoto, T. Papayannopoulou, H. L. Grimes, and M. Horwitz. 2003. Mutations in proto-oncogene GFI1 cause human neutropenia and target ELA2. *Nat. Genet.* **34**:308–312.

Phillipson, M., B. Heit, P. Colarusso, L. Liu, C. M. Ballantyne, and P. Kubes. 2006. Intraluminal crawling of neutrophils to emigration sites: a molecularly distinct process from adhesion in the recruitment cascade. *J. Exp. Med.* **203**:2569–2575.

Ravetch, J. V., and S. Bolland. 2001. IgG Fc receptors. *Annu. Rev. Immunol.* **19**:275–290.

Ravetch, J. V., and L. L. Lanier. 2000. Immune inhibitory receptors. *Science* **290**:84–89.

Reeves, E. P., H. Lu, H. L. Jacobs, C. G. Messina, S. Bolsover, G. Gabella, E. O. Potma, A. Warley, J. Roes, and A. W. Segal. 2002. Killing activity of neutrophils is mediated through activation of proteases by K+ flux. *Nature* **416**:291–297.

Roos, D., T. W. Kuijpers, and J. T. Curnutte. 2007. Chronic granulomatous disease, p. 525–549. In H. Ochs, A. Fisher, and L. Notarangelo (ed.), *Primary Immunodeficiency Diseases*. Oxford University Press, New York, NY.

Rosenberg, P. S., B. P. Alter, A. A. Bolyard, M. A. Bonilla, L. A. Boxer, B. Cham, C. Fier, M. Freedman, G. Kannourakis, S. Kinsey, B. Schwinzer, C. Zeidler, K. Welte, and D. C. Dale; Severe Chronic Neutropenia International Registry. 2006. The incidence of leukemia and mortality from sepsis in patients with severe congenital neutropenia receiving long-term G-CSF therapy. *Blood* **107**:4628–4635.

Savill, J., and V. Fadok. 2000. Corpse clearance defines the meaning of cell death. *Nature* **407**:784–788.

Shimazu, R., S. Akashi, H. Ogata, Y. Nagai, K. Fukudome, K. Miyake, and M. Kimoto. 1999. MD-2, a molecule that confers lipopolysaccharide responsiveness on Toll-like receptor 4. *J. Exp. Med.* **189**:1777–1782.

Sircar, M., P. F. Bradfield, M. Aurrand-Lions, R. J. Fish, P. Alcaide, L. Yang, G. Newton, D. Lamont, S. Sehrawat, T. Mayadas, T. W. Liang, C. A. Parkos, B. A. Imhof, and F. W. Luscinskas. 2007. Neutrophil transmigration under shear flow conditions in vitro is junctional adhesion molecule-C independent. *J. Immunol.* **178**:5879–5887.

Skokowa, J., G. Cario, M. Uenalan, A. Schambach, M. Germeshausen, K. Battmer, C. Zeidler, U. Lehmann, M. Eder, C. Baum, R. Grosschedl, M. Stanulla, M. Scherr, and K. Welte. 2006. LEF-1 is crucial for neutrophil granulocytopoiesis and its expression is severely reduced in congenital neutropenia. *Nat. Med.* **12**:1191–1197.

Springer, T. A. 1994. Traffic signals for lymphocyte recirculation and leukocyte emigration: the multistep paradigm. *Cell* **76:**301-314.

Strober, W., P. J. Murray, A. Kitani, and T. Watanabe. 2006. Signalling pathways and molecular interactions of NOD1 and NOD2. *Nat. Rev. Immunol.* **6:**9–20.

Sun, C. X., G. P. Downey, F. Zhu, A. L. Koh, H. Thang, and M. Glogauer. 2004. Rac1 is the small GTPase responsible for regulating the neutrophil chemotaxis compass. *Blood* **104:**3758-3765.

Suzuki, N., S. Suzuki, G. S. Duncan, D. G. Millar, T. Wada, C. Mirtsos, H. Takada, A. Wakeham, A. Itie, S. Li, J. M. Penninger, H. Wesche, P. S. Ohashi, T. W. Mak, and W. C. Yeh. 2002. Severe impairment of interleukin-1 and Toll-like receptor signalling in mice lacking IRAK-4. *Nature* **416:** 750–756.

Theilgaard-Monch, K., L. C. Jacobsen, R. Borup, T. Rasmussen, M. D. Bjerregaard, F. C. Nielsen, J. B. Cowland, and N. Borregaard. 2005. The transcriptional program of terminal granulocytic differentiation. *Blood* **105:**1785–1796.

Thornberry, N. A., and Y. Lazebnik. 1998. Caspases: enemies within. *Science* **281:**1312–1316.

Ting, J. P., D. L. Kastner, and H. M. Hoffman. 2006. CATERPILLERs, pyrin and hereditary immunological disorders. *Nat. Rev. Immunol.* **6:**183–195.

van den Berg, J. M., S. Weyer, J. J. Weening, D. Roos, and T. W. Kuijpers. 2001. Divergent effects of tumor necrosis factor alpha on apoptosis of human neutrophils. *J. Leukoc. Biol.* **69:**467–473.

van Mirre, E., W. B. Breunis, J. Geissler, C. E. Hack, M. de Boer, D. Roos, and T. W. Kuijpers. 2006. Neutrophil responsiveness to IgG, as determined by fixed ratios of mRNA levels for activating and inhibitory FcgammaRII (CD32), is stable over time and unaffected by cytokines. *Blood* **108:**584–590.

Vaporciyan, A. A., H. M. DeLisser, H. C. Yan, I. I. Mendiguren, S. R. Thom, M. L. Jones, P. A. Ward, and S. M. Albelda. 1993. Involvement of platelet-endothelial cell adhesion molecule-1 in neutrophil recruitment in vivo. *Science* **262:**1580–1582.

Visintin, A., A. Mazzoni, J. H. Spitzer, D. H. Wyllie, S. K. Dower, and D. M. Segal. 2001. Regulation of toll-like receptors in human monocytes and dendritic cells. *J. Immunol.* **166:**249–255.

Wald, D., J. Qin, Z. Zhao, Y. Qian, M. Naramura, L. Tian, J. Towne, J. E. Sims, G. R. Stark, and X. Li. 2003. SIGIRR, a negative regulator of Toll-like receptor-interleukin 1 receptor signaling. *Nat. Immunol.* **4:**920–927.

Wang, L., G. A. Manji, J. M. Grenier, A. Al-Garawi, S. Merriam, J. M. Lora, B. J. Geddes, M. Briskin, P. S. DiStefano, and J. Bertin. 2002. PYPAF7, a novel PYRIN-containing Apaf1-like protein that regulates activation of NF-kappa B and caspase-1-dependent cytokine processing. *J. Biol. Chem.* **277:**29874–29880.

Weber, C., L. Fraemohs, and E. Dejana. 2007. The role of junctional adhesion molecules in vascular inflammation. *Nat. Rev. Immunol.* **7:**467–477.

Wegmann, F., B. Petri, A. G. Khandoga, C. Moser, A. Khandoga, S. Volkery, H. Li, I. Nasdala, O. Brandau, R. Fässler, S. Butz, F. Krombach, and D. Vestweber. 2006. ESAM supports neutrophil extravasation, activation of Rho, and VEGF-induced vascular permeability. *J. Exp. Med.* **203:** 1671–1677.

Weiss, S. J. 1989. Tissue destruction by neutrophils. *N. Engl. J. Med.* **320:**365–376.

Welch, H. C., W. J. Coadwell, C. D. Ellson, G. J. Ferguson, S. R. Andrews, H. Erdjument-Bromage, P. Tempst, P. T. Hawkins, and L. R. Stephens. 2002. P-Rex1, a PtdIns(3,4,5)P3- and Gbetagamma-regulated guanine-nucleotide exchange factor for Rac. *Cell* **108:**809–821.

Wolach, B., L. J. van der Laan, N. A. Maianski, A. T. Tool, R. van Bruggen, D. Roos, and T. W. Kuijpers. 2007. Growth factors G-CSF and GM-CSF differentially preserve chemotaxis of neutrophils aging in vitro. *Exp. Hematol.* **35:** 541–550.

Young, N. T., and M. Uhrberg. 2002. KIR expression shapes cytotoxic repertoires: a developmental program of survival. *Trends Immunol.* **23:**71–75.

Zhang, P., A. Iwama, M. W. Datta, G. J. Darlington, D. C. Link, and D. G. Tenen. 1998. Upregulation of interleukin 6 and granulocyte colony-stimulating factor receptors by transcription factor CCAAT enhancer binding protein alpha (C/EBP alpha) is critical for granulopoiesis. *J. Exp. Med.* **188:**1173–1184.

2

Macrophages: Microbial Recognition and Response

ANNETTE PLÜDDEMANN AND SIAMON GORDON

INTRODUCTION

The term "macrophage" comes from the Greek meaning "big eaters." In terms of evolution, macrophages seem to be ancient cells, since cells that resemble macrophages are found in the hemolymph of primitive multicellular organisms. In 1882 the Russian microbiologist Ilya Mechnikov recognized the presence of mobile mononuclear phagocytes in the larvae of starfish that have the ability to take up bacteria or organic matter. He postulated that in complex, higher organisms these phagocytes patrol organs and epithelial surfaces to remove non-self-molecules ranging from microbial pathogens to aged or malignant cells. His theory proved to be substantially correct, and he eloquently summarized his findings in his Nobel lecture in 1908 (Mechnikov, 1908):

> Whenever the organism enjoys immunity, the introduction of infectious microbes is followed by the accumulation of mobile cells, of white corpuscles of the blood in particular which absorb the microbes and destroy them. The white corpuscles and the other cells capable of doing this have been designated "phagocytes", i.e. devouring cells, and the whole function that ensures immunity has been given the name of "phagocytosis"...The sum of the very numerous facts established in the archives of science leaves no room to doubt the major part played by the phagocytic system, as the organism's main defence against the danger from infectious agents of all kinds, as well as their poisons. Where natural immunity is concerned, and man enjoys this in respect of a large number of diseases, it is a question of the phagocytes being strong enough to absorb and make the infectious microbes harmless.

Since the discovery of phagocytes, the field of macrophage biology has expanded enormously and, even more than a century after they were first described, we are still discovering new aspects to the underlying mechanisms of these fascinating, multifunctional cells. In particular, they play a crucial role in the recognition and phagocytosis of pathogens, including viruses, bacteria, fungi, and parasites, and initiate the adaptive immune response. Some microorganisms, however, have developed mechanisms to subvert the macrophage killing systems and can survive and replicate inside macrophages, thereby evading the adaptive immune response as well. Macrophages therefore play a central role in the pathogenesis of human disease. While other chapters will discuss many of these aspects in detail, the following provides an overview of macrophage function, placing other chapters in context.

Macrophage Heterogeneity and Recruitment

Macrophages are derived from circulating monocytes and are found in all tissues throughout the body, especially the lung, spleen, liver, and bone marrow. Monocytes originate in the bone marrow from myeloid progenitor cells, the granulocyte-monocyte colony-forming units, which are derived from hematopoietic stem cells and also give rise to neutrophils. In the bone marrow, under the influence of cytokines interleukin-1 (IL-1), IL-3, and IL-6, the monocyte precursor cells first differentiate into proliferating monoblasts, then promonocytes, and finally mature monocytes that are released into the blood and may circulate for several days before migrating to the tissues and differentiating into macrophages (Gordon and Taylor, 2005). Macrophage colony-stimulating factor (M-CSF) (also called CSF-1) is critical for the differentiation into macrophages. The circulating monocytes constitute approximately 5 to 10% of peripheral leukocytes in humans and have a diverse morphology characterized by differential expression of cell surface markers (e.g., CD14 and CD16). Monocyte heterogeneity has been reviewed recently (Gordon and Taylor, 2005; Tacke and Randolph, 2006). Monocyte migration into tissues is mediated by chemokines and chemokine receptors (e.g., fractalkine and its receptor, CX3CR1) that allow the monocytes to adhere to endothelial cells and the tissues where they differentiate into macrophages (Goda et al., 2000; Imhof and Aurrand-Lions, 2004). Macrophages are large cells (approximately 25 to 50 μm in diameter) with an irregular shape and most populations are highly phagocytic. Macrophages play a role in tissue homeostasis, through clearance of senescent cells and remodeling of damaged tissue, and are an important com-

---

Annette Plüddemann and Siamon Gordon, Sir William Dunn School of Pathology, University of Oxford, South Parks Road, Oxford OX1 3RE, United Kingdom.

ponent of the innate immune response to infection. The morphology and specific function of macrophages in tissues is diverse depending on the tissue in which they are found (Hume et al., 1983):

1. *Alveolar macrophages* in the lung express high levels of pattern recognition receptors (e.g., scavenger receptors) and play a role in phagocytosis of airborne pathogens and environmental particles (Arredouani et al., 2005; McCusker and Hoidal, 1989). They require GM-CSFs for their maturation and activity (Stanley et al., 1994).

2. *Microglial cells* are monocyte-derived cells found in the brain parenchyma where they play a role in clearance of apoptotic neurons and in poorly defined homeostasis (Perry and Gordon, 1988). The brain also contains *perivascular macrophages* and macrophage populations in the *choroid plexus* and *meninges*.

3. Macrophages found in the liver are called *Kupffer cells* and they are the largest group of macrophages in the body (Li and Diehl, 2003). They are found in the sinusoids and are in direct contact with the blood.

4. *Langerhans cells* are macrophage-like cells found in the epidermis of the skin and in complex epithelia. They are self-renewing cells generally not replenished by bone marrow-derived cells unless the system is under stress (Merad et al., 2002). Langerhans cells may give rise to myeloid-type dendritic cells that are rich in major histocompatibility complex (MHC) class II molecules and become specialized for antigen presentation. Inflammatory reactions in the skin are partially mediated by these cells. The skin also contains *dermal macrophages* that may give rise to a major proportion of myeloid dendritic cells in draining lymph nodes (Martinez-Pomares et al., 2006; McKenzie et al., 2007).

5. Macrophages in the *lamina propria* of the gut have a high phagocytic and bactericidal capacity, but produce low levels of proinflammatory cytokines (Smith et al., 2005). This phenotype is induced by products of the intestinal stromal cells.

6. *Osteoclasts* are bone marrow-derived multinucleated cells that play a role in bone remodeling (Quinn and Gillespie, 2005). Osteoclast precursors express the receptor for M-CSFs, which are required for their maturation, and signaling through receptor activator of nuclear factor κB (RANK) is essential for the differentiation and activation of osteoclasts (Armstrong et al., 2002; Yoshida et al., 1990).

7. *Peritoneal* and *pleural* macrophages have an important role in bacterial clearance (Lehnert, 1992).

8. *Splenic* macrophages are found in various parts of the spleen (Mebius and Krall, 2005). They are heterogeneous depending on their localization. Marginal metallophils have a distinctive phenotype. In the outer marginal zone they are involved in clearance of pathogens and antigen processing, whereas in the red pulp they clear senescent red blood cells. In the lymphoid regions they interact with B and T lymphocytes. Tingible-body macrophages in the white pulp are involved in clearance of apoptotic lymphocytes.

9. *Thymic* macrophages remove apoptotic lymphocytes during thymocyte differentiation (Surh and Sprent, 1994).

10. *Lymph nodes* are particularly rich in macrophages and contain specialized subpopulations in the subcapsular sinus (Junt et al., 2007). They express high levels of sialoadhesin, a macrophage surface molecule involved in the adhesion and capture of virus particles in afferent lymph (Carrasco and Batista, 2007; Martinez-Pomares et al., 1996; Nakamura et al., 2002; Phan et al., 2007).

11. All zones of the *adrenal gland* contain macrophages, including sinus lining cells and these may be involved in local immune-neuroendocrine interactions (Hume et al., 1984).

12. *Reproductive tissue (uterus, testis, placenta)*. The testis has a large population of resident macrophages that have cytotoxic and phagocytic capacity, but have greatly diminished proinflammatory function (Hedger, 2002). Macrophages in the decidua (the uterine lining that forms the maternal part of the placenta during pregnancy) play a critical role in removal of apoptotic cells which are the result of tissue remodeling of the maternal decidua and invasion of the developing embryo (Abrahams et al., 2004).

The number of macrophages in tissues does not remain constant and they can be recruited via chemotactic signals that are a result of metabolic, immune, and inflammatory stimuli (Van Furth et al., 1973). Macrophages are also activated by either bacterial products (such as lipopolysaccharide [LPS]) or cytokines secreted by T cells (such as interferon-γ [IFN-γ]). Activation results in the upregulation of some receptors, while others are downregulated and secretion of cytokines (e.g., tumor necrosis factor α [TNF-α]) and small molecules, including nitric oxide, is also induced. During inflammation, elicited macrophages acquire distinct phenotypes depending on the nature of the stimulus (Gordon, 2003; Mosser, 2003):

1. *Innate activation* is induced by Toll-like receptor (TLR) ligands such as LPS, lipoteichoic acid (LTA), and peptidoglycan from bacteria and induces increased production of proinflammatory cytokines, inducible nitric oxide synthase (iNOS), reactive oxygen species (ROS), and costimulatory molecules (Mukhopadhyay et al., 2004).

2. *Classical activation* is induced by IFN-γ in synergy with LPS and is characterized by increased production of proinflammatory cytokines, iNOS and ROS, as well as increased expression of MHC class II molecules and CD86. This leads to increased antigen presentation and microbicidal activity, promoting cellular immunity.

3. *Alternative activation* is induced by the Th2 cytokines interleukin 4 (IL-4) or IL-13 and results in increased endocytic activity, as well as upregulation of the expression of mannose receptor, DC-SIGN, and Dectin-1 (these receptors are discussed in detail in "Receptors and Recognition," below). These macrophages also show higher levels of arginase and increased cell growth and parasite killing. Therefore, these macrophages are associated with tissue repair and humoral immunity, as well as modulation of Th1-dependent activation.

4. *Deactivation* of macrophages is induced by the anti-inflammatory cytokines IL-10 and transforming growth factor-β (TGF-β), steroids, or the ligation of the inhibitory receptors CD200R and CD172a by their cognate ligands CD200 and CD47, respectively. Macrophages show increased production of IL-10, TGF-β, and prostaglandin E_2 and reduced expression of MHC class II molecules. These macrophages therefore suppress inflammation and promote tissue repair.

Tissue macrophages are remarkably heterogeneous and can be distinguished based on their expression of differentiation antigens and surface receptors using monoclonal antibodies. Resident and inflammatory macrophages also differ in their expression profile of surface molecules. The phenotypic heterogeneity of monocytes/macrophages during differentiation and activation is summarized by Taylor et al. (2005).

RECEPTORS AND RECOGNITION

A crucial task of macrophages is to distinguish between pathogenic non-self-molecules and innocuous, intact self-molecules, and this is mediated by diverse receptors on the macrophage surface.

Charles Janeway first proposed that cells of the innate immune system express a large repertoire of germ line-encoded pattern recognition receptors (PRRs) that recognize conserved molecular structures on pathogens that are essential and unique to pathogens and not present in the host. He coined the term "pathogen-associated molecular patterns" (PAMPs) for these structures (Janeway and Medzhitov, 2002). However, the structures are not necessarily exclusively associated with pathogens; therefore, this designation may need to be altered. Since Janeway's proposal two decades ago, an increasing number of PRRs have been identified. These include humoral, cell surface, and intracellular receptors. Humoral PRRs (e.g., collectins and pentraxin) generally recognize pathogens from various body fluids and form aggregates that are subsequently cleared by macrophages. Apart from this opsonizing ability, they also have many immunomodulatory properties. Cell surface PRRs function either as phagocytic/endocytic receptors (e.g., scavenger receptor A, mannose receptor) or as sensing molecules (e.g., Toll-like receptors). Phagocytic receptors bind and internalize ligands directly and display temperature-dependent, saturable, and inhibitable ligand binding kinetics of classical receptors. On the other hand, sensors do not bind or internalize ligand directly, but recognize PAMPs and induce a proinflammatory signaling cascade that leads to many antimicrobial effector responses. Macrophages also have intracellular receptors, including the NOD-like receptors and RIG (retinoic acid-inducible protein)-like helicases, that recognize intracellular pathogens and activate macrophages to remove these pathogens. Here, we discuss some of the receptors involved in pathogen recognition.

Macrophage Scavenger Receptors

The macrophage scavenger receptors were discovered by Brown and Goldstein and functionally described for their ability to bind modified low-density lipoproteins (LDLs), such as acetylated LDL (AcLDL) and oxidized LDL (OxLDL), but not native LDL molecules (Goldstein et al., 1979). The family of molecules classified as scavenger receptors has since expanded considerably and consists of unrelated distinct gene products that generally bind modified LDL and selected polyanionic ligands. To date there are eight different classes of scavenger receptors (class A, B, C, D, E, F, G, and H) and this classification is based on their multidomain structure, as proposed by Krieger and coworkers (Krieger, 1997). Another group of scavenger receptors, which contain scavenger receptor cysteine-rich (SRCR) domains, however, do not fall into this classification system as they have not been shown to bind modified LDL or other polyanionic molecules. This includes CD5, CD6, and CD163, which is expressed on macrophages and binds hemoglobin:haptoglobin complexes (Graversen et al., 2002). Although many of the scavenger receptors are restricted to myeloid and certain endothelial cells, some also show a more diverse expression pattern, including epithelial cells. The range of ligands that these receptors bind is extremely diverse and includes many microbial structures as well as an array of endogenous molecules. Many scavenger receptors recognize both gram-positive and gram-negative bacteria and bind proteins, polyribonucleotides, polysaccharides, and lipids (Plüddemann et al., 2006). This chapter focuses on scavenger receptors found on macrophages.

Class A Scavenger Receptor

The first scavenger receptor to be cloned was the class A scavenger receptor (SR-A), which was isolated from bovine lung mRNA (Kodama et al., 1990; Rohrer et al., 1990), and it has since been identified in mice and humans as well (Ashkenas et al., 1993; Matsumoto et al., 1990; Tomokiyo et al., 2002). Macrophages express three naturally occurring isoforms of SR-A, namely SR-AI, SR-AII, and SR-AIII, which are alternative splice variants of the same gene and are collectively referred to as SR-A (Gough et al., 1998; Matsumoto et al., 1990; Rohrer et al., 1990). SR-AI is a trimeric type II transmembrane glycoprotein and consists of a cytoplasmic tail, transmembrane domain, spacer region, α-helical coiled-coil domain, collagenous domain, and C-terminal cysteine-rich domain. The structure of SR-AII resembles that of SR-AI, but it is characterized by a short C terminus, and in SR-AIII this terminus is truncated (Fig. 1). SR-A is found on most myeloid cells, and macrophages express high levels of this receptor (Hughes et al., 1995). However, SR-A is not found on monocytes or neutrophils (Hughes et al., 1995). Mast cells (Brown et al., 2007) and specific subpopulations of bone marrow-derived dendritic cells (DCs) and splenic DCs (Becker et al., 2006) also express SR-A, as do some endothelial cells and smooth muscle cells within atherosclerotic plaques (Naito et al., 1992).

The collagenous region of SR-AI and SR-AII has been shown to be important for ligand binding, but SR-AIII is nonfunctional, remains trapped in the endoplasmic reticulum, and has been shown to have a dominant-negative effect (Gough et al., 1998; Kodama et al., 1990). Ligand binding by SR-A triggers internalization and activates signaling pathways, including protein kinase C and MAP kinase pathways, which results in differential cytokine production depending on the bound ligand (Coller and Paulnock, 2001; Hsu et al., 2001).

SR-A on macrophages has been implicated in a range of functions, including adhesion, phagocytosis, endocytosis, and antigen presentation (Platt and Gordon, 2001). The receptor has been associated with the disease pathologies of atherosclerosis and Alzheimer's disease. SR-A was first discovered for its ability to bind AcLDL, but not native LDL (Goldstein et al., 1979), and was subsequently shown to bind other modified proteins, such as OxLDL (Doi et al., 1993). The structure of SR-A includes a positively charged lysine cluster that, along with other residues spanning the collagenous domain, has been implicated in conformation-dependent ligand binding of these polyanionic molecules (Resnick et al., 1996). Both the apolipoprotein and lipid moieties of modified LDL seem to be recognized by SR-A (Parthsarathy et al., 1987; Terpstra et al., 1998). SR-A-mediated uptake of modified lipoproteins by macrophages leads to the formation of lipid-laden foam cells that occur in atherosclerotic lesions, and this receptor plays a key role in the development of these lesions (Krieger and Herz, 1994; Suzuki et al., 1997). Another function

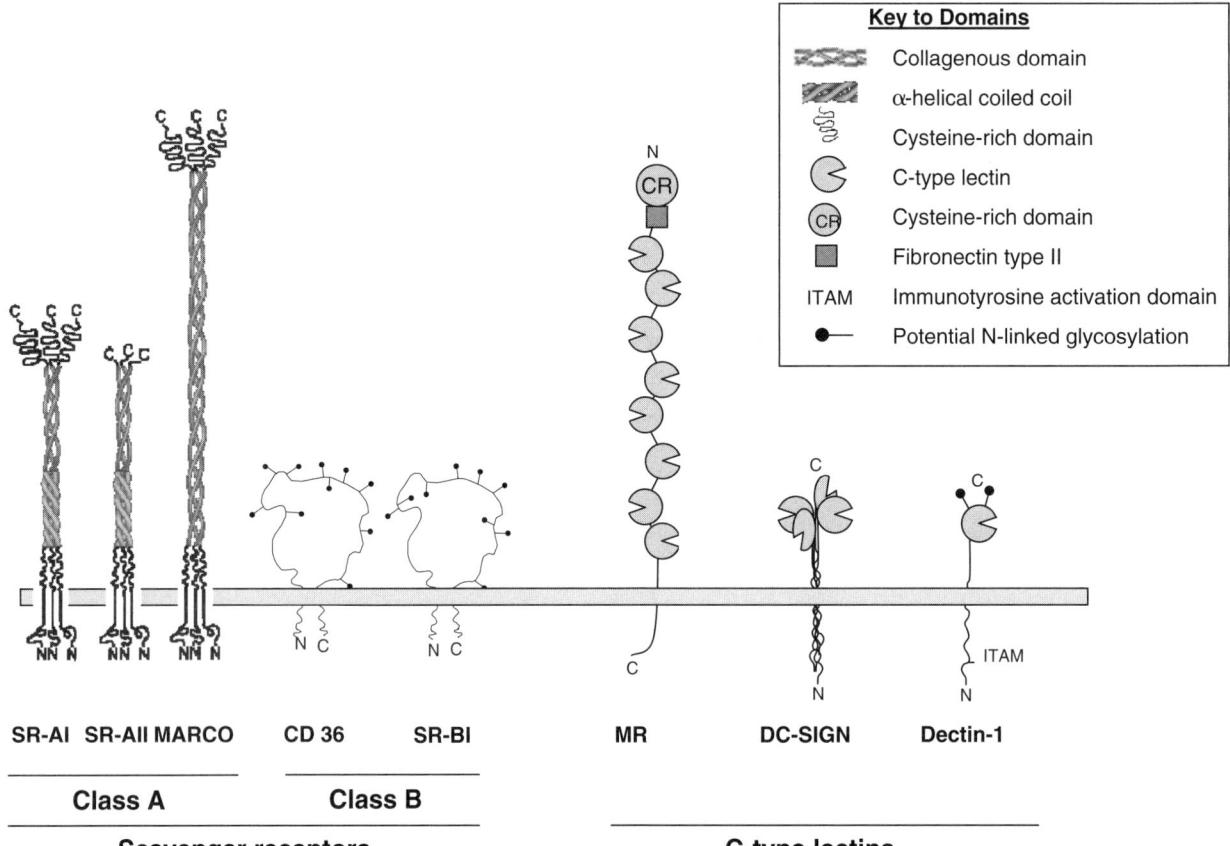

FIGURE 1 Nonopsonic macrophage receptors. The scavenger receptors are grouped together as a result of their ability to bind modified low-density lipoprotein; however, they are structurally diverse. Class A scavenger receptors are structurally closely related. Scavenger receptor A (SR-A) is a trimeric type II transmembrane glycoprotein with distinct cytoplasmic, transmembrane, spacer, α-helical coiled-coil, collagenous, and C-terminal cysteine-rich domains. Macrophage receptor with a collagenous structure (MARCO) lacks the coiled-coil domain and exhibits a longer collagenous domain. The class B scavenger receptors CD36 and SR-BI consist of a large extracellular loop tethered to the membrane by two short transmembrane domains adjacent to the short N and C termini. The C-type lectin receptors all contain one or multiple C-type lectin domain(s). The mannose receptor (MR) is a type I transmembrane receptor with multiple C-type lectin domains (CTLDs), a fibronectin type II domain, and a cysteine rich domain. DC-SIGN is a tetrameric type II transmembrane receptor where each subunit contains one CTLD. Dectin-1 contains one CTLD and has an immunotyrosine activation motif (ITAM) in the cytoplasmic region. See the text for references.

of SR-A on recruited macrophages is adhesion to the tissue extracellular matrix (Fraser et al., 1993), which may be important for retention of macrophages at a site of inflammation. In human smooth muscle cell extracellular matrix, biglycan and decorin (proteoglycans of the extracellular matrix) have been identified as SR-A ligands (Santiago-Garcia et al., 2003). SR-A may therefore contribute to the adhesion of macrophages to the extracellular matrix of atherosclerotic plaques. SR-A on microglia in the brain has been shown to bind β-amyloid fibrils that occur in senile plaques associated with Alzheimer's disease (El Khoury et al., 1996). The binding of these fibrils to SR-A stimulates the production of ROS that have cytotoxic effects on neurons.

Macrophages also have functions in tissue maintenance, and the involvement of SR-A in the clearance of apoptotic cells was demonstrated by a 50% reduction in uptake of apoptotic thymocytes by macrophages in the presence of a monoclonal anti-SR-A antibody (Platt et al., 1996). However, there appears to be redundancy in the receptors involved as there was no significant defect in apoptotic cell removal in SR-A knockout animals (Platt et al., 2000). SR-A may also play a role in antigen presentation, either directly or via internalization of molecular chaperones such as heat shock proteins (Harshyne et al., 2003; Nicoletti et al., 1999; Yokota et al., 1998).

An important role for macrophages is the recognition and removal of pathogens. SR-A recognizes both gram-positive and gram-negative bacteria (see Table 1). Bacterial ligands for SR-A include LTA from gram-positive organisms (Greenberg et al., 1996), LPS from gram-negative bacteria (Hampton et al., 1991), and unmethylated CpG DNA (Zhu et al., 2001). In the absence of opsonins, *Neisseria meningitidis* is almost exclusively recognized by SR-A

TABLE 1 Bacteria that are recognized by scavenger receptor A (SR-A) on macrophages[a]

Bacteria	Gram staining	% binding
Neisseria meningitidis	Gram-negative	90
Neisseria gonorrhoeae	Gram-negative	90
Neisseria lactamica	Gram-negative	90
Neisseria sicca	Gram-negative	90
Neisseria mucosa	Gram-negative	90
Legionella pneumophila	Gram-negative	60
Legionella pneumophila	Gram-negative	20
Sphingomonas capsulata	Gram-negative	40
Bacillus subtilis	Gram-positive	20
Bacillus subtilis spores	Gram-positive	50
Shigella flexneri	Gram-negative	80
Escherichia coli	Gram-negative	60
Haemophilus influenzae	Gram-negative	70
Haemophilus influenzae RD	Gram-negative	0
Streptococcus groups A and B	Gram-positive	20

[a]Bacteria were fluorescently labeled and incubated with bone marrow-derived macrophages from wild-type and SR-A$^{-/-}$ mice at 37°C for 1 h. Macrophages were subsequently washed and binding and uptake of bacteria were measured by flow cytometry. The table shows the percentage reduction in binding and uptake of bacteria by SR-A$^{-/-}$ macrophages when compared with the wild-type macrophages. This indicates the role of SR-A in binding and uptake of the various bacteria. Both gram-positive and gram-negative bacteria are recognized by SR-A. The requirement of SR-A for macrophage binding and uptake varies significantly between bacteria and also depends on growth phase (*L. pneumophila, B. subtilis*) and bacterial strain (*H. influenzae*) (A. Plüddemann and S. Gordon, unpublished observations).

on macrophages, and recent data show that unmodified bacterial surface proteins from this organism are also recognized by SR-A (Peiser et al., 2006). Table 1 summarizes bacterial recognition by SR-A. In vivo data confirm the importance of SR-A in the clearance of bacterial pathogens. SR-A$^{-/-}$ mice are more susceptible to experimental *Listeria monocytogenes* infection as a result of deficient bacterial clearance from liver and spleen (Suzuki et al., 1997). Similarly, another study showed increased susceptibility of SR-A$^{-/-}$ mice to *Staphylococcus aureus* infection (Thomas et al., 2000). A possible anti-inflammatory host-protective role of SR-A was proposed by Haworth et al. (1997), who observed that SR-A$^{-/-}$ mice formed normal granulomas in response to "bacillus of Calmette and Guérin" (BCG) priming. However, these animals were more susceptible to endotoxic shock as a result of increased proinflammatory cytokine secretion in response to additional LPS challenge. Unpublished data from our group further indicate that SR-A$^{-/-}$ mice are less able to clear *N. meningitidis* bacteria, showing higher bacterial levels in both blood and spleen than wild-type mice (A. Plüddemann and J. C. Hoe, unpublished observation).

Macrophage Receptor with a Collagenous Structure

Macrophage receptor with a collagenous structure (MARCO) is another member of the class A scavenger receptors with structural and ligand binding properties broadly similar to those of SR-A. MARCO lacks the coiled-coil domain and exhibits a longer collagenous domain, and the ligand binding region may lie within the cysteine-rich domain (Elomaa et al., 1998) (Fig. 1). MARCO is constitutively expressed by subpopulations of macrophages, in particular, those of the spleen marginal zone, medullary lymph nodes, and resident peritoneal macrophages (Elomaa et al., 1995). However, in most tissue macrophages, MARCO expression can be induced by infectious stimuli (e.g., bacteria) (van der Laan et al., 1997). Like SR-A, this receptor has also been shown to recognize intact gram-positive and gram-negative bacteria and their isolated products, including LPS and CpG DNA (Elomaa et al., 1995; Sankala et al., 2002). MARCO$^{-/-}$ mice display an impaired ability to clear *Streptococcus pneumoniae* infection from the lung, resulting in increased pulmonary inflammation and reduced survival (Arredouani et al., 2004). MARCO also recognizes *N. meningitidis* independent of LPS and can bind to several surface proteins isolated from this pathogen (Mukhopadhyay et al., 2004; S. Mukhopadhyay and S. Gordon, unpublished observation). Endogenous ligands for MARCO include OxLDL and AcLDL, although the latter is a very weak ligand for this receptor (Elomaa et al., 1995; S. Mukhopadhyay and S. Gordon, unpublished observation). Both MARCO and SR-A are important in the organization of the spleen marginal zone, by playing a role in the positioning and differentiation of macrophages (Chen et al., 2005). In the case of MARCO, B lymphocytes of the spleen marginal zone express a ligand for this receptor that seems to be important for retention of B cells in the marginal zone (Karlsson et al., 2003). Bronchial epithelial cells express a uteroglobin-related protein-1 (UGRP-1), which has been described to be a ligand for MARCO, and this interaction may be important in inflammatory processes in the lung (Bin et al., 2003).

Class B Scavenger Receptors

The class B scavenger receptors consist of two receptors, CD36 and SR-BI (CLA-1), both of which are found on macrophages (Terpstra et al., 2000). SR-BI is expressed on monocytes, while CD36 expression is upregulated during monocyte differentiation to macrophages (Huh et al., 1996). CD36 is a type III transmembrane receptor that consists of an extracellular loop with multiple glycosylation sites, anchored to the membrane by two transmembrane domains, and two short intracellular tails (Greenwalt et al., 1992) (see Fig. 1). The central part of the extracellular domain contains the ligand binding region (Puente Navazo et al., 1996). SR-BI (human homologue CLA-1) has a similar loop structure and consists of two alternative splice variants, namely SR-BI and SR-BII (Acton et al., 1994; Webb et al., 1997). Both CD36 and SR-BI can form dimers and multimers. The extensive glycosylation found on these receptors is essential for correct folding, trafficking, and function (Vinals et al., 2003).

CD36 plays a role in fatty acid transport and has a high affinity for long-chain fatty acids (Abumrad et al., 1993). Along with modified LDL, this receptor also binds native very-low-density lipoprotein (VLDL), LDL, and high-density lipoprotein (HDL) (Calvo et al., 1998; Endemann et al., 1993). Cholesterol efflux mechanisms in macrophages are enhanced by ligation of hexarelin (a hexapeptide member of the growth hormone-releasing peptides) to CD36 (Avallone et al., 2006). SR-BI plays a key role in cholesterol homeostasis and functions in HDL metabolism and reverse cholesterol transport. SR-BI binds particularly to native VLDL, LDL, and HDL and mediates selective uptake of HDL cholesteryl ester (Acton et al., 1994; Calvo et al., 1997; Xu et al., 1997). This receptor also recognizes AcLDL and OxLDL, but it binds to AcLDL with higher affinity than does CD36 (Acton et al., 1994). HDL has a protective function in the development of atherosclerosis in that it transports excess unesterified or "free" cholesterol from peripheral tissues (including the artery wall) to the liver, where SR-BI mediates efficient selective HDL-lipid uptake (Trigatti et al., 2003). The binding of HDL is mediated by the HDL apolipoproteins (Bultel-Brienne et al., 2002). SR-BI recognizes apolipoproteins not only in lipid-associated form, but also in lipid-free form (Li et al., 2002). Serum amyloid A (SAA), which can replace HDL apolipoproteins during the acute phase, is a ligand for SR-BI, both in lipid-free and lipid-associated form. SAA presence on HDL does not significantly affect binding of HDL to SR-BI, but impairs selective cholesterol uptake, while promoting cholesterol efflux (Marsche et al., 2007). CD36 and SR-BI recognize hypochlorite-modified LDL, a product of the oxidative burst initiated by macrophages, which has been found in atherosclerotic lesions (Marsche et al., 2003).

Another important function for CD36 is mediation of adhesion between macrophages and activated platelets via thrombospondin, the platelet glycoprotein (Silverstein et al., 1989). Adhesion may play a role in macrophage recruitment to areas of vascular injury. On microglia, both SR-BI and CD36 are able to bind β-amyloid fibrils and mediate release of hydrogen peroxide, therefore implicating these receptors in Alzheimer's disease, along with SR-A (Coraci et al., 2002; Husemann et al., 2001). Increased lipoprotein oxidation and accumulation of lipid peroxidation products have been reported in Alzheimer's disease, and binding of β-amyloid appears to inhibit clearance of modified lipoprotein by CD36 (Kunjathoor et al., 2004).

Clearance of apoptotic cells by macrophages is mediated by CD36 and contributes to peripheral tolerance and prevention of autoimmunity by impairing DC maturation (Puig-Kroger et al., 2006; Ren et al., 1995). Oxidized phosphatidylcholine and oxidized phosphatidylserine displayed on the surface of the plasma membrane of apoptotic cells is recognized by CD36 (Greenberg et al., 2006). Ligation of cell surface CD36 enhances IL-10 secretion and reduces the secretion of TNF-α and IL-1β, which negatively regulates DC maturation (Puig-Kroger et al., 2006). Clearance of apoptotic neutrophils by macrophages has been shown to be mediated by the adhesive protein thrombospondin in cooperation with the vitronectin receptor and CD36 (Savill et al., 1990, 1992).

The class B scavenger receptors on macrophages are important for pathogen recognition. SR-BI and CD36 have been shown to recognize isolated bacterial products such as LPS or diacylated lipopeptide, respectively (Hoebe et al., 2005; Vishnyakova et al., 2003). Gram-negative (*Escherichia coli* and *Salmonella enterica* serovar Typhimurium) and gram-positive (*S. aureus* and *L. monocytogenes*) bacteria have been shown to bind to CD36 and SR-BI (Vishnyakova et al., 2006). CD36$^{-/-}$ animals are unable to clear *S. aureus* infection and develop acute bacteremia (Stuart et al., 2005). It has been suggested that these receptors may play an important role in sepsis by facilitating bacterial adhesion and cytosolic invasion (Vishnyakova et al., 2006). CD36 also recognizes *Plasmodium falciparum* erythrocyte membrane protein-1 (PfEMP-1) on red blood cells containing malaria parasites and has been shown to be important in malaria pathogenesis (Ayi et al., 2005; Serghides et al., 2003).

C-Type Lectins

The C-type lectin receptors recognize carbohydrates on cell surfaces, circulating proteins, and pathogens (McGreal et al., 2004). All classical C-type lectin receptors have a carbohydrate-recognition domain (CRD) that binds carbohydrate molecules in a calcium (Ca^{2+})-dependent manner. Calcium ions are required for lectin activity and play a central role in ligand binding (Drickamer, 1999). However, many members of the C-type lectin family contain CRDs that do not bind Ca^{2+}, and these receptors also interact with noncarbohydrate ligands (Vales-Gomez et al., 2000). Therefore, the term C-type lectin-like domain (CTLD) is used to denote the common fold within this protein family without referring to functional similarities (Weis et al., 1998).

C-type lectin receptors can be divided into three groups based on their CTLD: (i) C-type lectins containing a single CRD, (ii) C-type lectins containing multiple CRDs, and (iii) NK-like C-type lectin-like (NKCL) receptors, which have a single CRD (Taylor et al., 2005). Members of the C-type lectins with a single CRD are type II membrane receptors and include dendritic cell-specific intercellular adhesion molecule-3-grabbing nonintegrin (DC-SIGN) and Dectin-2. C-type lectins containing multiple CRDs are type I membrane receptors and include the mannose receptor, Endo180, and the related DEC-205 and phospholipase A$_2$ receptor (East and Isacke, 2002). The type II membrane receptors termed NKCL receptors include Dectin-1 and CD69. These receptors typically possess a single extracellular CTLD, a stalk region, a transmembrane domain, and a cytoplasmic tail with or without signaling motifs (Taylor et al., 2005).

DC-SIGN

DC-SIGN (CD209), a tetrameric endocytic receptor, is highly expressed by subpopulations of macrophages in the lung and placenta and by mature and immature dendritic cells in the lung, lymph nodes, spleen, dermis, and mucosal tissue (Taylor et al., 2005). It is a C-type lectin with a CTLD, a stalk region, a transmembrane domain, and a cytoplasmic tail containing an internalization motif (Fig. 1). This receptor functions in the adhesion of T cells to DCs via the intercellular adhesion molecule-3 (ICAM-3), explaining the origin of its name (Geijtenbeek et al., 2000). However, it also recognizes pathogens, including viruses (human immunodeficiency virus type 1 [HIV-1], human cytomegalovirus, hepatitis C, dengue, and Ebola), bacteria (*Mycobacterium tuberculosis*, *S. pneumoniae*, *Helicobacter pylori*), fungi (*Candida albicans*), and parasites (*Leishmania mexicana*, *Schistosoma mansoni*) (Appelmelk et al., 2003; Geijtenbeek et al., 2003; Koppel et al., 2005). DC-SIGN generally recognizes N-linked high-mannose structures and fucose-containing glycans on microbes that are not common in endogenous molecules (Gou et al., 2004). In some cases the pathogens modulate the immune response by binding to DC-SIGN; examples of this are HIV-1, dengue virus, and *M. tuberculosis* (Tassaneetrithep et al., 2003; van Kooyk and Geijtenbeek, 2003). HIV-1 exploits the antigen-presenting properties of DC-SIGN on DCs to gain access to T cells that are the primary target of the virus. The binding of mycobacterial lipoarabinomannan to DC-SIGN blocks DC maturation and triggers inhibitory signals that allow this bacterium to evade the immune response (Geijtenbeek et al., 2003). In DCs, DC-SIGN modulates TLR-dependent responses by activating the serine/threonine kinase Raf-1, which leads to acetylation of an activating subunit of NF-κB (p65), resulting in enhanced transcription of the gene encoding the anti-inflammatory cytokine IL-10 (Gringhuis et al., 2007). This occurs only after TLR signaling, thereby modulating the TLR-specific immune response to the pathogen.

Mannose Receptor

The mannose receptor (CD206) is a 180-kDa Ca^{2+}-dependent lectin that functions as an endocytic receptor and contains eight CTLDs, an N-terminal cysteine-rich domain, a domain containing fibronectin type II repeats, a transmembrane domain, and a short cytoplasmic tail (Lee et al., 2003) (Fig. 1). It is expressed by most macrophage populations and specifically binds terminal mannose, fucose, N-acetylglucosamine, or glucose residues. This allows it to distinguish non-self and potentially harmful self from self, as these moieties are commonly found on microorganisms, but not usually in terminal positions on mammalian cell surface oligosaccharides or serum glycoproteins (Mullin et al., 1994). The mannose receptor plays an important role in macrophage recognition of fungi. On alveolar macrophages, for example, it has been identified as a PRR capable of NF-κB activation in response to the fungus *Pneumocystis* (Zhang et al., 2004). The ligand on *Pneumocystis carinii* mediating interaction with the mannose receptor was shown to be the major mannose-rich surface antigen complex termed glycoprotein A (gpA) (O'Riordan et al., 1995). This receptor has been implicated in nonopsonic binding of another pathogenic fungus, *C. albicans*, most likely via mannose residues on the fungal surface (Ezekowitz et al., 1990). Examples of bacteria recognized by the mannose receptor are *Klebsiella pneumoniae* (via LPS) and *S. pneumoniae* (via capsular polysaccharides) (Zamze et al., 2002). The mannose receptor has been shown to mediate phagocytosis of virulent *M. tuberculosis* via the terminal mannosyl residues of the surface lipoglycan lipoarabinomannan (Kang and Schlesinger, 1998; Schlesinger et al., 1994). This receptor-ligand interaction blocks phagosome maturation and fusion with the lysosome, thereby facilitating survival of the bacteria in the phagosome (Kang et al., 2005). In the absence of the mannose receptor or when the receptor is blocked, the inhibition of phagosome maturation is reversed. Another pathogen able to survive inside macrophages that enters in part via the mannose receptor is *Clostridium perfringens*, and it is possible that the ligand is mannose found on the polysaccharide capsule of this organism (O'Brien and Melville, 2003). The mannose receptor has been implicated in the clearance of serum glycoproteins to maintain homeostasis (Lee et al., 2002). The cysteine-rich domain of this receptor has been shown to recognize sulfated glycans, such as those found on glycoprotein hormones produced by the anterior pituitary. Co-localization of cysteine-rich ligands with mannose receptor on macrophages has been observed in the human spleen and, therefore, this receptor may contribute to the filtering capacity of the spleen (Martinez-Pomares et al., 2005). The mannose receptor may play a role in localization of DCs from the marginal zone to the primary lymphoid follicles (Yu et al., 2002).

Dectin-1

Dectin-1 is a small (~28 kDa) type II membrane receptor with a single extracellular CTLD and a cytoplasmic domain with a tyrosine-based activation motif (Adachi et al., 2004; Brown and Gordon, 2001). It is predominantly expressed on myeloid cells (macrophages, DCs, and neutrophils) and is regulated by cytokines and microbial stimuli (Herre et al., 2004; Willment et al., 2003). Carbohydrate recognition is independent of calcium. It recognizes a variety of β-1,3-linked and β-1,6-linked glucans and thus binds and promotes phagocytosis of yeasts such as *Saccharomyces cerevisiae* and *C. albicans* (Brown and Gordon, 2001; Herre et al., 2004). In alveolar macrophages, Dectin-1 has been shown to bind the fungus *P. carinii* (Steele et al., 2003). Dectin-1 can stimulate the oxidative burst in response to β-glucans and is a phagocytic receptor (Herre et al., 2004). A detailed discussion is provided elsewhere in this volume.

"Sensing" Receptors

Macrophages have numerous receptors that are not involved in phagocytosis/endocytosis, but play a role in sensing microbial products and inducing signal transduction. This includes the Toll-like receptors, which have diverse functions and are located on the surface (TLR1, -2, -4, -5, and -6) or in intracellular vesicles (TLR3, -7, -8, and -9) (Ishii et al., 2005; O'Neill, 2006). These receptors differentially recognize various pathogen ligands and induce signaling via adaptor molecules that culminates in the activation of the transcription factor NF-κB, MAP kinase p38 and Jun N-terminal kinase (JNK), and interferon regulatory factors (O'Neill, 2006). This group of receptors has been extensively reviewed recently (Ishii et al., 2005; O'Neill, 2006; Trinchieri and Sher, 2007) and elsewhere in this volume.

Macrophages have a repertoire of cytosolic receptors, namely the NOD-like receptors (NLRs) (Chamaillard et al., 2003; Meylan et al., 2006) and the RIG-like helicases (Seth et al., 2006). The NOD-like receptors are NBS-LRR proteins consisting of a C-terminal leucine-rich repeat (LRR), which has been implicated in ligand binding, a central nucleotide binding site (NBS) or NACHT (for neuronal apoptosis inhibitor protein), and an N-terminal protein-protein interaction domain, which varies between the different molecules in the group and may be a caspase-activating and recruitment domain (CARD), a pyrin domain (PYD), or a baculovirus inhibitor of apoptosis protein repeat (BIR) (Inohara et al., 2005; Inohara and Nunez, 2003). Receptors in this group include nucleotide binding oligomerization domain (NOD), neuronal apoptosis inhibitor (NAIP), and NACHT, LRR, and PYD domains

(NALP) (Inohara et al., 2005). These proteins recognize bacterial products, such as peptidoglycan subunits, in the cytosol (Inohara et al., 2003) and are part of the innate immune response to intracellular bacterial pathogens (Lamkanfi et al., 2007). The RIG-like helicases include retinoic acid-inducible gene I (RIG-I), melanoma differentiation-associated gene-5 (MDA-5), and LGP2, and these molecules have been shown to be intracellular receptors for viral double-stranded RNA (dsRNA). RIG-I has a C-terminal helicase domain responsible for ligand recognition and two tandem CARD domains at the N terminus that activate NF-κB and IRF3 (see "Signal Transduction" below) (Yoneyama et al., 2004). MDA-5 contains two N-terminal CARD domains able to activate the IFN-β promoter, whereas LGP2 lacks the CARD domain and acts as a negative regulator of the RIG-I pathway (Yoneyama et al., 2005).

Humoral Receptors

Humoral PRRs include soluble C-type lectins (collectins), such as lung surfactant proteins SP-A and SP-D and mannose binding lectin (MBL), as well as pentraxins. These molecules are generally not produced by macrophages, but may play a role in bacterial interaction with macrophages by binding to both pathogens and macrophage surface receptors or by modulating macrophage functions. The surfactant proteins bind to macrophages, altering their function by, for example, upregulating the expression of the mannose receptor (Beharka et al., 2005; Kudo et al., 2004) and SR-A (Kuronuma et al., 2004), thereby improving pathogen phagocytosis. The macrophage receptors mediating interaction with surfactant proteins are generally not known; however, SP-A has been shown to mediate the macrophage response to LPS by binding to the macrophage receptor CD14 (Sano et al., 1999). There is preliminary evidence that SP-A may be a ligand for TLR4 (Guillot et al., 2002). In contrast, MBL (which resembles complement protein C1q) activates the classical complement pathway, has been shown to enhance phagocytosis of bacteria, and seems to bind to macrophages via an unknown receptor (Bajtay et al., 2000; Jack et al., 2005; Presanis, 2003). All of the members of the collectin family have been shown to bind to the calreticulin/CD91 receptor complex on alveolar macrophages, although these interactions were described for apoptotic cell clearance (Vandivier et al., 2002). Pentraxins are another group of humoral antimicrobial molecules, some of which, such as PTX3, are produced by several cell types, including mononuclear phagocytes and DCs, in response to primary inflammatory signals and are able to bind complement components and bacteria (Bottazzi et al., 2006; Garlanda et al., 2005). PTX3 therefore plays a role in activation of the classical pathway of complement activation and facilitates recognition of bacteria and other pathogens by macrophages. These molecules are described in detail elsewhere in this volume.

Receptor Interaction

Apart from their individual functions, macrophage receptors also collaborate with each other to either increase the repertoire of innate immune recognition or modulate the biological responses of host cells to pathogens. For example, Dectin-1-mediated production of TNF-α in response to fungal pathogens requires the cytoplasmic tail and immunoreceptor tyrosine activation motif of Dectin-1, as well as TLR2 and the adaptor molecule MyD88 (Brown et al., 2003). Both receptors are required to generate a proinflammatory response that is independent of internalization of the fungal pathogen. Examples of "cross-talk" between macrophage receptors are found among the scavenger receptor family. A mutant mouse strain, "Oblivious," which carries a nonsense mutation in the cd36 gene generated by random germ-line mutagenesis, is insensitive to a subset of TLR2 agonists, such as diacylated bacterial lipopeptide, peptidoglycan, and MALP-2, but is still able to recognize triacylated TLR2 agonists (Hoebe et al., 2005). The "Oblivious" mice are highly susceptible to gram-positive infection and it has been proposed that CD36 may function as a coreceptor for TLR2 in recognizing diacylated TLR2 agonists. Two other scavenger receptors, SREC-I and LOX-1, are responsible for direct binding of the outer membrane protein A (OmpA) of *K. pneumoniae* (Jeannin et al., 2005). However, OmpA-mediated activation of macrophages (and DCs) depends on TLR2. Upon OmpA binding, SREC-I and LOX-1 colocalize with TLR2 and amplify its signaling, which initiates a unique activation program inducing a soluble receptor, the long pentraxin 3 (PTX3). In turn, PTX3 further amplifies TLR2 signaling through a positive feedback loop. These studies show the cooperativity between receptors, and future studies will undoubtedly confirm this as an important paradigm in innate immunity.

PHAGOCYTOSIS

The term "phagocytosis" comes from the Greek words "phagon," which means "eat," and "kytos," which means "cavity." Many cells can ingest small amounts of surrounding medium and particles of up to 100 nm in a process called pinocytosis, and can take up macromolecules, viruses, and small particles by receptor-mediated endocytosis. These processes are generally actin independent and clathrin mediated (Aderem and Underhill, 1999). In contrast, only the professional phagocytes, i.e., mononuclear phagocytes (macrophages, DCs) and polymorphonuclear neutrophils, are capable of receptor-mediated phagocytosis (the uptake of particles larger than $\sim$0.5 μm), which depends on actin. This includes engulfing particles as large as red blood cells that can be 7 to 8 μm in diameter. Macrophages recognize and phagocytose altered or damaged "self" molecules, such as apoptotic cells and tumor cells, as well as infectious "non-self" molecules, such as viruses, fungi, and bacteria. Phagocytosis of pathogens by macrophages initiates the innate immune response that, in turn, activates the adaptive immune response.

The first step of phagocytosis is contact between the particle and the macrophage membrane. This step is mediated by the surface receptors, and phagocytosis is either nonopsonic and mediated by the receptors discussed in "Receptors and Recognition," above, or opsonic, i.e., mediated by circulating molecules called opsonins.

Opsonins

The mechanisms of opsonic phagocytosis have been studied extensively. The most important opsonins are circulating complement proteins and immunoglobulins, such as immunoglobulin G (IgG). IgG can bind either specifically to a pathogen via the F(ab) domain of the antibody or nonspecifically via hydrophobic interaction. Macrophage receptors involved in the recognition of complement-opsonized particles include CR1, CR3, and CR4 (Carroll, 1998). CR1 is a single-chain transmembrane receptor with a complement binding lectin-like domain and binds C3b,

C4b, and C3bi (Brown, 1991). CR3 (CD11b/CD18) and CR4 (Cd11c/CD18) are both integrins and consist of a heterodimer of an α- and a β-chain (CR3, $\alpha_m\beta_2$; CR4, $\alpha_x\beta_2$). These receptors specifically bind C3bi and mediate internalization. Recently, a complement receptor of the immunoglobulin superfamily (CRIg) was identified on Kupffer cells that binds complement fragments C3b and iC3b and is required for efficient binding and phagocytosis of complement-opsonized particles (Helmy et al., 2006). The complement system has been reviewed extensively (Carroll, 1998). In brief, there are two complement activation pathways that consist of a series of reactions where serum proteins are cleaved to form active serine proteases which then cleave the next protein, resulting in a cascade of reactions. The classical complement pathway is made up of 11 proteins, called C1-C9, and is triggered by the presence of IgM and IgG molecules that have been produced against specific pathogens. The alternative complement pathway does not require the presence of specific antibody. This pathway is triggered when circulating C3i protein, which is generated by the spontaneous hydrolysis of C3, binds to the surface of a pathogen by hydrophobic interaction. This interaction, together with other factors, generates a convertase enzyme that, in turn, cleaves more C3, causing an amplification in the reaction. Cleavage of C3 generates two fragments: C3a is a chemotactic factor (also called an anaphylotoxin) that recruits and activates macrophages, and C3b is an opsonin that mediates binding to the complement receptor (CR1) on macrophages. Both pathways also result in the cleavage of C5 to C5a, which is another potent anaphylotoxin. Therefore, complement activation recruits and activates macrophages to deal with the infection and provides opsonins to enhance phagocytosis. In addition to the proteins circulating in the blood, macrophages can also secrete some complement proteins. A pathogen may therefore have both IgG and C3b bound to its surface, and these two opsonins act synergistically to promote phagocytosis.

Phagocytosis Mechanisms

Phagocytosis involves a number of molecules, including macrophage receptors, signal transducers, and the cytoskeleton. Binding of a particle to the membrane receptors triggers actin polymerization, which alters the cell cytoskeleton and results in the formation of pseudopodia around the particle and the formation of a phagocytic cup. Multiple interactions between various receptors on the macrophage surface and the particle aid this process, which is sometimes referred to as the "membrane-zippering effect." The pseudopodia form at sites where the ligand interacts with receptors until they completely engulf the particle and fuse, resulting in a membrane-bound vacuole called the phagosome (Griffin et al., 1975, 1976). Polymerization of F actin, a cytoskeletal protein, is essential for phagocytosis and is found abundantly in pseudopodia along with other actin molecules (Lowry et al., 1998). Fc receptor (FcR)-mediated phagocytosis (discussed further elsewhere in this volume) occurs via this classical zipper model. Vinculin-, paxillin-, α-actinin-, and phosphotyrosine-containing proteins are enriched around the whole particle and internalization requires tyrosine kinases (Allen and Aderem, 1996). In contrast, when particles are opsonized by complement proteins and are internalized via the CR3 receptor, they appear to sink into the cell. In this mechanism of phagocytosis the phagosome membrane is not as tightly apposed to the particle surface and displays distinct points of contact with the particle called foci. These foci have been shown to be made up of vinculin-, paxillin-, α-actinin-, and phosphotyrosine-containing proteins. Internalization does not require tyrosine kinases but depends on microtubules. FcR- and CR3-mediated phagocytosis also differs in the requirement for members of the Rho family of small guanosine triphosphatases (GTPases). Rho activation is required for CR3-mediated internalization, while activation of Rac and Cdc42 is required for phagocytosis mediated by FcR (Caron and Hall, 1998). Signaling is discussed further in the next section. Another key difference between complement receptor- and Fc receptor-mediated phagocytosis is the resulting release of inflammatory mediators. Fc receptor-mediated phagocytosis results in the release of reactive oxygen intermediates and the production of arachidonic acid metabolites, whereas these inflammatory mediators are not produced in complement receptor-mediated phagocytosis (Aderem et al., 1985; Wright and Silverstein, 1983). Thus, FcR ligation is a proinflammatory process, while CR ligation is not. Some evidence also indicates a role for the endoplasmic reticulum in some conditions of phagosome formation (Gagnon et al., 2002; Muller-Taubenberger et al., 2001). This model proposes that the endoplasmic reticulum membrane fuses with plasma membrane underneath the phagocytic cup to provide part of the membrane needed for the formation of nascent organelle (reviewed by Desjardins et al. [2005]).

After internalization, the actin is shed rapidly and the phagosome undergoes maturation, which involves small GTPases such as Rab5 and Rab7. The phagosome fuses with vesicles of the endocytic pathway, resulting in acidification and the production of reactive oxygen and nitrogen species. The phagosome acquires proteolytic enzymes by fusing with the lysosome, and all of these mechanisms play a role in the destruction of the engulfed pathogen. Furthermore, the acquired immune system can be activated either by presentation of an antigen from the pathogen on the cell surface or by induction of cytokines and other costimulatory molecules. Phagocytic mechanisms and signaling are discussed in detail elsewhere in this volume.

SIGNAL TRANSDUCTION

Macrophage receptor ligation results in complex signaling cascades, implicating numerous signaling molecules with varying outcomes (Fig. 2). These include phagocytosis, endocytosis, and/or production of inflammatory mediators. Mammalian cells have three main mitogen-activated protein kinase (MAPK) pathways that mediate changes in gene expression in response to extracellular stimuli, namely the extracellular signal-regulated kinase (ERK) pathway, the cJun N-terminal kinase (JNK) pathway, and the p38 pathway (Clark et al., 2003). Generally, the JNK and p38 pathways are activated by proinflammatory stimuli. These pathways consist of signaling cascades that function by phosphorylation and dephosphorylation reactions, and activated MAPKs can exert their effects directly by phosphorylating transcription factors.

Phagocytic Receptor Signaling

Fc receptors have an immunoreceptor tyrosine-based activation motif (ITAM) in their cytoplasmic tail. Ligation of the FcR by IgG-opsonized particles results in tyrosine phosphorylation of the ITAM domain by members of the Src family of tyrosine kinases (Swanson and Hoppe, 2004). This is followed by the recruitment of Syk tyrosine kinases

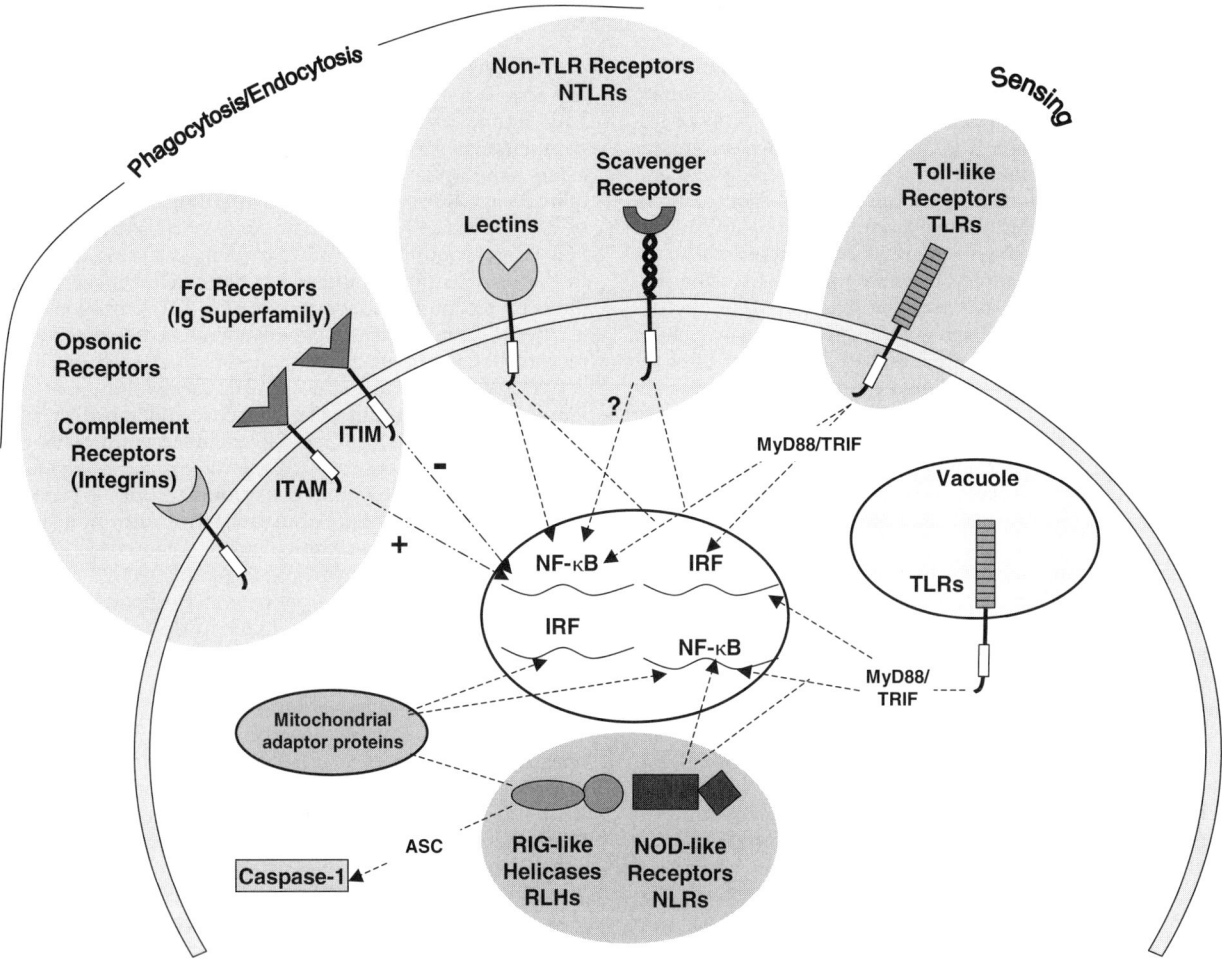

FIGURE 2 Macrophages have many receptors that mediate their diverse functions. The receptors are located on the surface as well as in vacuolar compartments and the cytosol, thereby mediating recognition of extracellular and intracellular pathogens. The opsonic receptors include complement receptors (integrins) and Fc receptors (Ig superfamily) (discussed further elsewhere in this volume). They function in phagocytosis and endocytosis of complement- or antibody-opsonized particles, respectively (Hawlisch and Kohl, 2006; Nimmerjahn and Ravetch, 2006). Fc receptors have either an inhibitory (contain an immunoreceptor tyrosine-based inhibition motif [ITIM]) or activatory (contain an immunoreceptor tyrosine-based activation motif [ITAM]) effect on NF-κB induction (Nimmerjahn and Ravetch, 2006). NF-κB is a family of nuclear transcription factors that regulate production of proinflammatory mediators. Another group of phagocytic/endocytic surface receptors are the non-Toll-like receptors (NTLRs), which include the family of scavenger receptors and the C-type lectins (Brown et al., 2003; Plüddemann et al., 2006). Non-opsonic surface receptors that do not mediate phagocytosis/endocytosis but are important sensors of bacteria, fungi, and viruses are the Toll-like receptors (TLRs) (O'Neill, 2006). Scavenger receptors have been shown to collaborate with TLRs to induce NF-κB and may also directly mediate NF-κB induction upon interaction with ligand. Ligand recognition by lectins induces NF-κB, both directly and in collaboration with TLRs. TLRs can induce both NF-κB and IRFs via a signaling cascade mediated by the adaptor molecules MyD88 and TRIF. Some TLRs are located within vacuoles and play a role in recognition of intracellular pathogens. Cytosolic viruses and bacterial products are recognized by the NOD-like receptors (NLRs) and RIG-like helicases (RLHs) (Creagh and O'Neill, 2006). NLRs induce NF-κB either directly or in collaboration with TLRs. RLHs either induce NF-κB and IRF via mediators that are located on the outer membrane of the mitochondria or induce caspase-1-mediated apoptosis via the adaptor molecule ASC. In addition to NF-κB and IRF induction, there are a multitude of other signaling pathways within macrophages that have been omitted for clarity. Abbreviations: ASC, apoptosis-associated speck-like protein containing a caspase recruitment domain; MyD88, myeloid differentiation primary response gene (Greenberg et al., 1996); NOD, nucleotide binding oligomerization domain; TRIF, TIR domain-containing adaptor-inducing interferon-β. (Drawing, A. Plüddemann.)

to the ITAM domain. Syk then activates a signaling cascade that includes phospholipase C (PLC), phosphatidylinositol 3-kinase (PI3K), SH2 domain-containing inositol phosphatase 1 (SHIP-1) and Shc. The activated PLC cleaves phosphatidylinositol 4,5-bisphosphate (PI(4,5)P2), resulting in diacylglycerol formation. Subsequently, protein kinase C (PKC) and inositol triphosphate (IP3) are activated, triggering the release of Ca^{2+}. In the absence of Syk kinase, IgG-opsonized molecules cannot be phagocytosed. The small GTPases play a role in FcR-mediated phagocytosis. Rho is required for receptor clustering, which is essential for efficient receptor binding and internalization (Hackam et al., 1997). Rac and Cdc42 do not appear to be involved in receptor binding, but are essential for internalization by mediating the accumulation of F actin in the phagocytic cup (Cox et al., 1997). The activity of the NADPH oxidase complex is regulated by Rac (reviewed by Bokoch and Diebold [2002]). Contractile proteins (e.g., myosins, RhoA) and proteins that regulate membrane fusion or fission (e.g., dynamin, Rab5, phospholipase A_2) are also recruited to assist or regulate phagocytosis (Swanson and Hoppe, 2004).

Signaling cascades induced by complement receptor ligation require Syk. This protein tyrosine kinase plays a crucial role in the process from phagosome formation to engulfment, by controlling the accumulation and disassembly of polymerized actin and Rho-GTPase activation (Shi et al., 2006). CR3 receptor ligation results in the translocation of PKC-α to foci on the forming phagosome, which leads to serine phosphorylation of the CD18 subunit of CR3 and the subsequent activation of the receptor and recruitment of α-actinin, vinculin, actin, and paxillin to these foci. Signaling induced by the nonopsonic phagocytic receptors, such as the scavenger receptors and lectins, is discussed in "Receptors and Recognition," above, and elsewhere in this volume.

"Sensing" Receptor Signaling

The sensing receptors on the macrophage surface interact with numerous signaling pathways. The signaling pathways triggered by the TLRs have been studied extensively and have been reviewed (Akira et al., 2006; O'Neill, 2006; O'Neill and Bowie, 2007). TLR ligation by various microbial effectors triggers signaling resulting in both the induction of NF-κB and activation of IRF3 (Kawai et al., 2001). NF-κB is a family of transcription factors consisting of five members (p50, p53, RelA, RelB, and c-Rel). They form homo- and heterodimers and regulate the expression of genes involved in inflammation and immunity. IRF3 is a member of the IRF family that consists of nine proteins containing an N-terminal DNA binding domain and a tryptophan repeat. IRF3 activates the IFN-β promoter upon viral infection. In brief, the TLRs utilize two pathways, one that requires the adaptor MyD88 and another that is independent of this adaptor. The TIR domain in the cytoplasmic tail of the TLRs interacts directly with MyD88, resulting in recruitment of two other adaptor proteins, IRAK (IL1-R-associated kinase) and TRAF6 (tumor necrosis factor receptor-associated factor 6). Subsequently, two major signaling pathways are activated, namely the MAPK cascade and NF-κB, resulting in cytokine induction. In addition to MyD88, another adaptor protein, TIRAP/MAL (TIR domain-containing adaptor protein/MyD88-adaptor-like) is essential for NF-κB activation downstream of TLR2 and TLR4 (Fitzgerald et al., 2003b).

Activation of IRF3 by TLR3 and TLR4 is independent of MyD88 and TIRAP/MAL and depends on an adaptor called TRIF (TIR domain-containing adaptor inducing IFN-β) (Fitzgerald et al., 2003a). However, signaling from TLR4 requires a fourth adaptor, TRIF-related adaptor molecule (TRAM), which is not required for TLR3 signaling. The intracellular TLR9 can induce both inflammatory cytokines and IFN-β in a MyD88-dependent manner. Thus, TLRs use distinct combinations of the different adaptor proteins to effect signaling.

The intracellular NOD-like receptors recognize bacterial products and trigger NF-κB via recruitment of a serine/threonine kinase or induce caspase-1 via a complex of proteins called the inflammasome (Martinon et al., 2002). The inflammasome consists of caspase-1, caspase-5, apoptosis-associated speck-like proteins containing a CARD (ASC), and NALP family proteins. The NALPs recruit ASC through a homotypic interaction between the PYRIN domains, and ASC in turn recruits caspase-1 via its CARD, leading to the activation of IL-1β and IL-18. The RIG-like helicases recognize viral nucleotides via their helicase domain and trigger type I IFN production (Yoneyama et al., 2004). RIG-like helicases interact with an adaptor protein called IPS-1 which is located on the mitochondrial membrane and is essential for the triggering of type I IFN production and induction of NF-κB (Kawai et al., 2005).

Cytokine Signaling

Each cytokine interacts with its particular receptor, initiating a distinct response. Many cytokine receptors interact with the cytoplasmic kinases of the Janus kinase (JAK) family, which initiate a response through the signal transducer and activator of transcription (STAT) molecules (O'Shea et al., 2002). Activation of the cytokine receptors induces phosphorylation and activation of JAK, and this activation is required for most receptor functions. STAT dimers can directly translocate to the nucleus and bind to DNA, thereby controlling gene transcription. Different cytokines interact with different JAK molecules, and this mediates the differential response; for example, IL-4 activates JAK1 and JAK3, IL-10 activates JAK1 and Tyk2, and IL-12 activates JAK2 and Tyk2. Cytokines also stimulate cell proliferation via the MAP kinase pathway. TNF-α and IL-1β activate the MAP kinase family members ERK, JNK, and p38, resulting in increased binding of the transcription factors AP-1, NF-κB, and NFIL-6 to DNA. Macrophages also have a family of proteins called suppressor of cytokine signaling (SOCS) that regulate the signal transduction pathway responses to cytokines (reviewed by Naka et al. [2005] and Yoshimura et al. [2007]).

ANTIMICROBIAL MECHANISMS

The antimicrobial arsenal of macrophages includes the acidification of the phagolysosome; the production of reactive oxygen (NADPH oxidase) and nitrogen intermediates (iNOS, NOS2), lysosomal enzymes, and antimicrobial peptides and proteins (e.g., defensins and bactericidal proteins); and the production of interferon-inducible GTPases. Other mechanisms include the depletion of amino acids essential for bacterial growth, such as arginine (by arginase) and tryptophan (by indoleamine-2,3-dioxygenase [IDO]), as well as the depletion of iron.

The production of reactive oxygen intermediates (ROIs) results from the activation of the NADPH oxidase, which is elicited by IgG binding to the Fc receptor, cytokines (IFN-γ, IL-8), or microbial products, such as bacterial LPS or fungal cell wall components. NADPH oxidase catalysis results in the production of superoxide (O_2^-), which is converted to hydrogen peroxide (H_2O_2) by superoxide dismutase, hydroxyl radicals, and anions (OH^-) by a reaction called the Haber-Weiss reaction and highly reactive singlet oxygen ($^1O_2$) (Bogdan, 2004).

Reactive nitrogen intermediates (RNIs) are generated by NOSs, which ultimately convert L-arginine and molecular oxygen to nitric oxide (NO) and citrulline with the aid of various cofactors (e.g., NADPH, FAD, Ca^{2+}/calmodulin) (24). Macrophages produce iNOS, which is absent in resting cells, but strongly induced by microbial products (e.g., LPS) and the cytokines IFN-γ, IFN-α/β, and TNF. NO can then react with sulfhydryl (SH) groups on amino acids, peptides, and proteins, generating S-nitrosothiols and other toxic RNIs. These RNIs target microbial proteins containing heme groups, iron-sulfur clusters, SH- or amino-groups (e.g., mitochondrial proteins), DNA, and lipids. Therefore they are effective agents against viruses, bacteria, fungi, and protozoa.

For many bacterial and protozoan pathogens, tryptophan and arginine are essential amino acids. In human macrophages, IFN-γ induces IDO and arginase, which deplete tryptophan and arginine, respectively (Bogdan, 2004; Mellor and Munn, 2004). In addition, all microbial pathogens depend on a supply of iron. Phagocytes can produce iron binding proteins (lactoferrin), which compete for the available iron, or downregulate the transferrin receptor, thereby reducing the influx of iron to the cell (Schaible and Kaufmann, 2004). Another mechanism involves hepcidin, a peptide regulator of iron mobilization, which is induced during infection and inflammation and inhibits iron efflux from macrophages (and other cells), reducing the available circulating iron (Ganz, 2006).

Antimicrobial peptides and proteins are constitutively expressed in the lysosomes of macrophages. They act by hydrolyzing microbial proteins (lysosomal proteases), forming pores in the microbial cell wall (defensins), or degrading the cell wall completely (lysozyme) (Levy, 2004).

The IFN-γ-mediated induction of several families of GTPases is another important component of antimicrobial immunity, in particular, for intracellular pathogens (MacMicking, 2004). The three main families of IFN-γ-inducible GTPases are the Mx GTPases, the p65 guanylate binding proteins (GBPs), and the p47 GTPases (IRG proteins) (Martens and Howard, 2006). The functions of these proteins are still not fully understood.

Autophagy is a mechanism by which cells can capture cytoplasmic components for degradation in the lysosome (Schmid et al., 2006; Swanson, 2006). During autophagy, cytoplasmic material is sequestered within double-membraned cisternae, which then fuse to form isolation vacuoles. This process is a source of nutrients during starvation, but presents a host defense mechanism against intracellular pathogens, allowing cells to deliver cytosolic antigens to the MHC class II pathway. For example, bacteria, such as L. monocytogenes, that evade degradation by escaping the phagosome into the cytosol, then become targets for autophagy and delivery to the lysosome (Rich et al., 2003).

ANTIGEN PRESENTATION

Degradation of pathogens in the phagolysosome generates peptides that can be presented to other cells of the immune system via the MHC. Phagosomes can directly mediate the formation of peptide-MHC-II complexes, which are then transported to the surface and presented to T lymphocytes (Ramachandra et al., 1999). Antigen presentation by macrophages generally induces a cellular immunity and suppresses humoral immunity (Desmedt et al., 1998; Uanue and Allen, 1987). DCs (discussed elsewhere in this volume) are the "professional" antigen-presenting cells, and the T-cell-stimulatory capacity of macrophages is generally inferior to that of mature DCs. Macrophages express low levels of MHC class II, but the expression can be upregulated by microbial products and macrophages can therefore present the antigens from the microbe to T lymphocytes (Hsieh et al., 1993). Specifically, production of IL-12 by antigen-presenting macrophages induces the IFN-γ-producing, $CD4^+$ MHC-II-restricted type 1 T-helper cells. This results in the expansion of the T-cell population and production of signals (IFN-γ) by the T cells that activate the macrophage and increase its microbicidal activity. Intracellular pathogens can only be eliminated when the macrophage is activated by the T cell.

In general, endogenous proteins from the cytoplasm are loaded onto MHC-I molecules in the endoplasmic reticulum; however, antigens from pathogens such as *Mycobacteria*, *Leishmania*, *Salmonella*, and *Brucella* can also elicit an MHC-I-dependent response that results in the proliferation of $CD8^+$ cytotoxic T lymphocytes (Ackerman and Cresswell, 2004). This process is referred to as cross-presentation. It has been proposed that peptides generated in phagosomes are translocated to the cytoplasm where they are processed by proteasomes and enter the MHC-I pathway (Kovacsovics-Bankowski and Rock, 1995). Lipid antigens can also be presented to T lymphocytes, and this occurs via CD1 molecules that are found in endosomal compartments (Scotton et al., 2005; van den Elzen et al., 2005).

CYTOKINE SECRETION

Macrophages produce a wide range of cytokines, which include interleukin molecules (IL), tumor necrosis factors (TNF), and chemokines (CC and CXC). These are small proteins that can act either on the cell that produced them or the surrounding cells. Their production is tightly regulated and they affect the function of the cell by interacting with a surface receptor that triggers a signaling cascade, resulting in the induction or inhibition of cytokine-regulated genes. The production of cytokines by macrophages can be triggered by microbial products, by interaction with type 1 T-helper cells, or by soluble factors including other cytokines.

Cytokines can be divided into two main groups, as follows. (i) *Proinflammatory cytokines*. This includes cytokines and chemokines that play a role in recruiting other inflammatory cells, e.g., IL-6, TNF, CC and CXC chemokines; cytokines mediating T-cell and natural killer (NK) cell activation, e.g., IL-1, IL-12, IL-18, and IL-23; and cytokines that activate the macrophage itself, e.g., IL-1, TNF-α, IL-12, IL-18, and IFN-α/β. (ii) *Anti-inflammatory cytokines*. These molecules downregulate the macrophages and reduce inflammation, e.g., IL-4, IL-10, IL-13, and TGF-β (Bogdan and Nathan, 1993). Cytokines affect many different cellular processes, including proliferation, differentiation, inhibition of growth, apoptosis, chemokinesis, and adhesion (Bonecchi et al., 1998; Pilette et al., 2002). Gene expression profiling shows the diverse macrophage responses initiated by cytokines (e.g., IL-4 [Raes et al., 2002; Stein et al., 1992; Kzhyshkowska et al., 2006; Martinez et al., 2006], IL-10 [Park-Min et al., 2005], IFN [Hissong et al., 1995; Modolell et al., 1995], IL-13 [de Waal Malefyt et al., 1993]). The proinflammatory cytokines TNF-α, IL-6, and IL-1 induce acute-phase proteins that opsonize pathogens, and they are pyrogens (induce fever), which may increase leukocyte activity by increasing the body temperature. Ligation of chemokines (e.g., RANTES and macrophage inflammatory protein-1α) to their G-protein receptors results in cell migration by reorganization of the cytoskeleton, formation of focal adhesions, and pseudopodal extensions and retractions. This facilitates migration of leukocytes from the blood to the site of infection in tissue. IFNs inhibit viral replication and cell proliferation and upregulate MHC class I and downregulate MHC class II expression. Cytokines also activate T and B cells, but the T-cell stimulatory capacity of macrophages is less than that of mature DCs. Inflammatory macrophages have been shown to downregulate T-cell activation and proliferation, for example, by the release of T-cell-inhibitory compounds such as nitric oxide and reactive oxygen intermediates (Bogdan, 2001) or depletion of tryptophan (Mellor and Munn, 2004).

INFECTION

Macrophages play a critical role in the recognition and removal of pathogens. As discussed above, pathogens are phagocytosed and phagosome maturation starts with the early endosome stage, which is regulated by the Rab5 GTPase and acquires the early endosomal antigen EEA1, followed by a late endosome stage controlled by Rab7 and characterized by the presence of Lamp1 and Lamp2. The late endosome then acidifies by the proton ATPase pump, fuses to the lysosome, and becomes the phagolysosome that degrades the pathogen. However, some pathogens have evolved mechanisms to evade the killing mechanisms of the macrophage (Fig. 3) (Pizarro-Cerda and Cossart, 2006):

1. The microorganism *escapes into the cytoplasm* by producing a factor that destroys the phagosome membrane (e.g., *L. monocytogenes*, *Francisella tularensis*, and *Shigella flexneri*).
2. The microorganism *changes the development of the phagosome* (e.g., prevents fusion with the lysosome) and resides in a phagosomal compartment (e.g., *M. tuberculosis* and *Legionella pneumophila*).
3. The microorganism *has adapted to survive in the harsh environment of the phagolysosome* and/or requires its otherwise lethal conditions (e.g., *Coxiella burnetti* and *L. mexicana*).
4. The microorganism subverts the signaling mechanisms and abrogates cytokine induction or apoptosis (e.g., viruses).

Here, we discuss in brief a few selected examples of pathogens that are able to survive and even proliferate within macrophages by various mechanisms. Further details are provided elsewhere in this volume.

Mycobacterium tuberculosis

The mycobacterium, *M. tuberculosis*, is one of the most successful intracellular pathogens. This organism is able to survive and replicate inside macrophages and reside within phagosomes that do not undergo fusion with the lysosome. Various macrophage receptors have been implicated in the uptake of *M. tuberculosis*, including the complement receptor CR3 (Velasco-Velazquez et al., 2003) and mannose receptor (Diaz-Silvestre et al., 2005; Kang et al., 2005). The organism has developed multiple mechanisms that interfere with the phagocytic pathway and immune response. *M. tuberculosis* blocks the maturation of phagosomes and reduces the ability of infected cells to present exogenous soluble protein antigens (Ramachandra et al., 1999). The *M. tuberculosis* phagosome acquires the early endosome marker Rab5, but late endosomal markers Rab7 and Lamp1 are excluded, thereby preventing fusion with the lysosome (Hasan et al., 1997). Mycobacteria produce various molecules that block fusion with the lysosome. One example is the cell wall component, phosphatidylinositol lipoarabinomannan (ManLAM), which inhibits the rise in Ca^{2+} concentration in the cytosol that is required for phagosome maturation (Vergne et al., 2003). The *Mycobacterium* phagosome also fails to acidify due to a lack of the proton-ATPase responsible for phagosomal acidification; however, the precise mechanism of this remains unclear (Sturgill-Koszycki et al., 1994). In addition, mycobacterial products interact with the intracellular MAP kinase signaling pathways (Jo et al., 2007); for example, some strains of *M. tuberculosis* produce a phenolic glycolipid that inhibits the release of proinflammatory cytokines TNF-α, IL-6, IL-12, and monocyte chemotactic protein-1 (MCP-1) (Reed et al., 2004). These bacteria have therefore evolved many strategies to ensure their survival within macrophages.

Legionella pneumophila

L. pneumophila is a gram-negative bacillus that causes an acute pneumonia called Legionnaires' disease. It is transmitted by inhalation and is an intracellular pathogen that can survive and replicate within the alveolar macrophages. Macrophages phagocytose the bacteria either by conventional phagocytosis or by a unique coiling mechanism (Elliott and Winn, 1986). One of the receptors shown to mediate uptake of *L. pneumophila* is the complement receptor (Payne and Horwitz, 1987), but nonopsonic receptors such as SR-A also seem to play a role (Pierce et al., 1996; A. Plüddemann and S. Gordon, unpublished observation). The *L. pneumophila* phagosome is rapidly surrounded by organelles, including mitochondria and the rough endoplasmic reticulum (Horwitz, 1983; Tilney et al., 2001), and ribosomes are incorporated into the phagosome membrane. The formation of this phagosome is directed by the bacterial type IV secretion system encoded by the defect in organelle trafficking (*dot*)/intracellular multiplication (*icm*) genes. This secretion system facilitates the transport into the cell of effector molecules that modulate the phagosome and inhibit phagosome/lysosome fusion. Bacteria are able to replicate in this phagosome, and in their postexponential phase they disrupt the phagosomal membrane and exit into the cytoplasm. Finally, *Legionella* induces macrophage apoptosis via activation of caspase-3, and this depends on the type IV secretion system (Gao and Abu Kwaik, 1999a, 1999b). The cytoplasmic receptor Naip5 seems to play a role in resistance to *L. pneumophila* infection, possibly by either blocking caspase activation or by activating NF-κB; however, this requires further investigation (Coers et al., 2007).

Listeria monocytogenes

L. monocytogenes is a gram-positive intracellular pathogen causing the disease listeriosis and is commonly transmitted via contaminated food products. Macrophages phagocytose and kill ~90% of bacteria; however, some are able to sur-

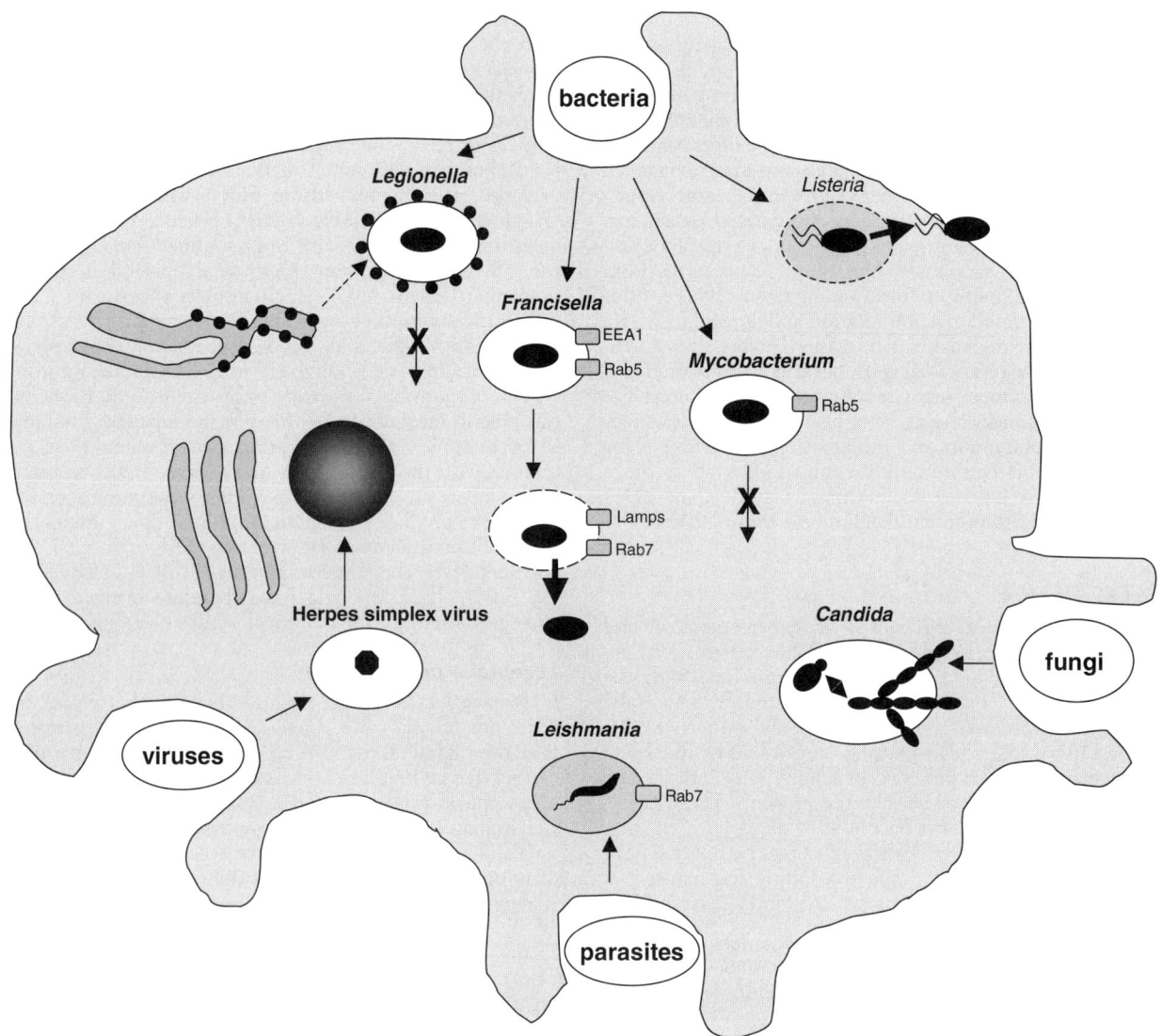

FIGURE 3 Pathogens have developed several mechanisms to survive inside macrophages. *Legionella pneumophila* resides and multiplies in a vacuole studded with ribosomes due to interaction with the rough endoplasmic reticulum (RER). The organism secretes effector molecules via its type IV secretion system into the cell that inhibit phagosome/lysosome fusion. The *Francisella tularensis* phagosome acquires the early endosome markers EEA1 and Rab5 and then matures into a late endosome defined by the presence of the markers Lamp1, Lamp2, and Rab7. The late endosome does not acidify and the phagosomal membrane is disrupted, releasing the bacteria into the cytosol. The *Mycobacterium tuberculosis* phagosome acquires the early endosome marker Rab5 but excludes the late endosomal Lamps and Rab7. This organism also produces molecules that block fusion with the lysosome and resides and replicates in this early endosome. Acidification of the *Listeria monocytogenes* phagosome is essential for the perforation of the phagosomal membrane and escape of the bacteria into the cytosol. Here, they mobilize the actin polymerization machinery to move within the cell and then from cell to cell. *Candida albicans* undergoes a conversion from a unicellular form to a multicellular hyphal form, which allows this fungus to escape the macrophage. The *Leishmania mexicana* phagosome develops into an acidic phagolysosome containing Rab7 where the parasite is able to survive and replicate. Viruses such as the herpes simplex virus are able to inhibit the activation of antiviral mechanisms, such as the activation of interferon regulatory function (IRF) proteins that induce interferon production upon viral infection. See the text for references. (Drawing, A. Plüddemann.)

vive and replicate within macrophages (Campbell, 1994; Ireton and Cossart, 1997). They can rapidly spread to neighboring cells. When macrophages phagocytose *L. monocytogenes* the bacteria rapidly escape the phagosome by using the pore-forming protein listeriolysin O (LLO) (Vazquez-Boland et al., 2001). The phagosome acquires maturation markers and acidifies, which seems to be essential for perforation of the membrane and bacterial escape (Glumski et al., 2002). The exact mechanism by which *Listeria* escapes the vacuole remains unclear, but it appears that both bacterial and host proteins may play a role (Marquis et al., 1997; Vazquez-Boland et al., 2001). *L. monocytogenes* is then able to replicate in the cytosol with the aid of a hexose phosphate transporter and induces polymerization of the host actin filaments (Chico-Calero et al., 2002). The bacterium uses the force generated by this actin polymerization to move within the cell and then from cell to cell. This process is facilitated by a single bacterial protein, ActA, which acts as a scaffold for the actin polymerization machinery by providing multiple binding sites for components of the cytoskeleton (Cameron et al., 2000). Mechanisms of *Listeria* motility are reviewed in detail by Portnoy et al. (2002). Movement of bacteria from cell to cell allows them to avoid cytotoxic T cells and phagocytes.

Francisella tularensis

F. tularensis is a gram-negative intracellular bacterium that causes fatal tularemia (Ellis et al., 2002). It was first isolated in 1911 in Tulare County (California) from rodents that had a plague-like disease. Transmission occurs through inhalation, ingestion, or insect bite and spreads to the lymph nodes, liver, and spleen. Macrophages phagocytose *F. tularensis*, forming asymmetric spacious pseudopod loops not observed with other bacteria (Clemens et al., 2005). The mannose receptor and complement receptor (CR3) have been shown to be involved in uptake of this pathogen (Schulert and Allen, 2006). The spacious phagosome is rapidly remodeled to form a tight phagosome around the bacterium which acquires EEA1 (Clemens et al., 2004; Santic et al., 2005). The phagosome subsequently acquires the late endosomal markers Lamp1 and Lamp2, but appears not to acidify and does not contain acid hydrolases. The mechanisms by which the bacteria facilitate this are as yet unknown. The phagosomal membrane is then gradually disrupted and bacteria are released into the cytosol approximately 8 to 12 h after infection, where they multiply rapidly. After 24 to 48 h macrophages undergo caspase-9-mediated apoptosis, thereby allowing the bacteria to escape the host cell without generating an inflammatory response (Lai and Sjostedt, 2003).

Activation of the macrophage by IFN-γ, however, prevents bacterial replication (Elkins et al., 2003). *F. tularensis* contains a group of genes located in an area called the "pathogenicity island" that have been shown to be critical for intracellular infection (Nano et al., 2004). Bacterial proteins essential for modulation of the phagosome and replication include a protein called IglC and its regulator MglA, but since these bacteria do not have a type III or type IV secretion mechanism, their mode of action remains to be elucidated. *F. tularensis* also inhibits TLR-mediated activation of intracellular signaling and secretion of TNF-α and IL-1β, which depends on expression of a bacterial 23-kDa protein (Telepnev et al., 2003).

Candida albicans

C. albicans is an opportunistic fungal pathogen that occurs in a yeast and filamentous fungal form and is able to switch between these two phenotypes. Macrophage uptake of this organism is mediated by the Dectin-1 receptor, which recognizes β-glucan on the fungal surface (Brown and Gordon, 2001). The fungus has evolved various mechanisms to evade phagocytosis, escape destruction, and enable it to survive inside macrophages. Such mechanisms include the synthesis of specific molecules and/or the activation of enzymes with a protective effect and the dimorphic conversion from unicellular yeast forms to mycelial structures (Lorenz et al., 2004), as well as the triggering of macrophage apoptosis (Ibata-Ombetta et al., 2003). On contact with macrophages, *C. albicans* undergoes a dramatic reprogramming of transcription (Lorenz et al., 2004; Martinez-Esparza et al., 2007). In the early phase cells shift to a starvation mode, including gluconeogenic growth, production of trehalose, activation of fatty acid degradation, and downregulation of translation. In a later phase, as hyphal growth enables *C. albicans* to escape from the macrophage, cells quickly resume glycolytic growth. Other responses triggered in the early phase include machinery for DNA damage repair, oxidative stress responses, peptide uptake systems, and arginine biosynthesis. Macrophage apoptosis is subsequently mediated by phospholipomannan (PLM), a unique glycolipid found on the fungal surface, allowing the fungus to escape the macrophage (Ibata-Ombetta et al., 2003).

Leishmania mexicana

L. mexicana is an intracellular protozoan pathogen that has two morphologically distinct stages in its life cycle: the promastigote and amastigote. The promastigote is extracellular and resides in the gut of the sandfly vector, whereas the amastigotes reside in the phagolysosome of mammalian macrophages (Cohen-Freue et al., 2007). When the promastigotes enter the mammalian host they are rapidly phagocytosed by macrophages and undergo differentiation into the amastigote stage. The phagosome then develops into an acidic phagolysosome where the parasites are able to survive and replicate and may persist for several years (Aebischer, 1994). To survive, *Leishmania* have developed resistance mechanisms to evade sandfly digestion enzymes, host innate immune responses (e.g., complement system), and macrophage defense mechanisms such as NO and ROIs (Peters and Sacks, 2006). *Leishmania* phagocytosis is either opsonic via the complement receptor or nonopsonic via the mannose receptor, and, in both cases, the oxidative burst is not triggered (Wilson and Pearson, 1988; Wright and Silverstein, 1983). They interfere with oxygen metabolism and signal transduction; for example, they avoid triggering the macrophage oxidative burst by inhibition of PKC phosphorylation, which is involved in many activation functions of macrophages (Descoteaux and Turco, 1993; Moore et al., 1993). This seems to be mediated by lipophosphoglycan, the major cell surface glycoconjugate of *L. mexicana*, which can also be released from the surface. Lipophosphoglycan inhibits normal maturation of the phagosome, although the compartment does acidify and contains markers for late endosomes or lysosomes (e.g., Rab7) (Lodge and Descoteaux, 2005). Furthermore, these parasites cause defects in the induction of the acquired immune response, as reviewed by Peters and Sacks (2006).

Viruses

Viruses subvert and evade the host immune system by numerous mechanisms, such as inhibiting host immune responses, preventing antigen presentation, and abrogating induction of apoptosis. Some viruses, such as the vaccinia virus, produce proteins that target the TLR signaling cascade and thereby suppress NF-κB induction (Stack et al., 2005). The hepatitis C virus produces a protease that blocks IRF3 activation by cleaving the TLR3 adaptor TRIF and subverting the activation of type I interferon response (Li et al., 2005). The RIG-like helicase signaling pathway is a target for viruses as it is critical for host defense against RNA viruses, and numerous viral proteins have been shown to be antagonists of this pathway (Andrejeva et al., 2004; Meylan et al., 2005). Proteins from DNA viruses, such as herpes simplex virus type 1 and Kaposi's sarcoma-associated herpesvirus, have been shown to inhibit IRF activation (Lin et al., 2004; Yu et al., 2005). This shows that both RNA and DNA viruses can hamper host type I IFN production and suggest that controlling the IFN response is essential for the survival of a broad range of viruses.

CONCLUSION

Macrophages play a central role in the innate immune system. They are found in tissues throughout the body and are heterogeneous in nature, serving multiple functions such as wound healing, tissue maintenance, clearance of apoptotic cells, and recognition of pathogens. An array of receptors is located both on the cell surface and intracellularly in the cytosol or in vacuoles to enable these cells to recognize their targets and perform their functions. These include opsonic receptors that recognize molecules that have been opsonized by either complement proteins or immunoglobulins, as well as nonopsonic receptors that mediate direct recognition of pathogens and other molecules. Receptor ligation mediates many processes including phagocytosis, production of cytokines and chemokines, and ultimately activation of the humoral immune response to invading pathogens. Macrophages have an array of antimicrobial mechanisms; however, in some cases these mechanisms are subverted by the invading microorganisms, allowing them to survive and replicate inside macrophages and so remain undetected by the humoral immune system. Although extensive progress has been made in understanding the complex macrophage and its functions, one of the key questions that still remains to be fully elucidated is how macrophages distinguish between harmful pathogens and innocuous commensal organisms that reside normally within the body. Currently, extensive studies are directed at elucidating the interaction and requirement of the phagocytic and sensing receptors (Blander and Medzhitov, 2006; Russell and Yates, 2007). Another important question that remains is the role of macrophages in autoimmune disease. Although they contribute to tissue repair and resolution of inflammation, they produce toxic compounds that may play a role in autoimmune disease pathology. A clear understanding of the biology of the macrophage is therefore crucial for the development of therapies against these pathogens and diseases.

REFERENCES

Abrahams, V. M., Y. M. Kim, S. L. Straszewski, R. Romero, and G. Mor. 2004. Macrophages and apoptotic cell clearance during pregnancy. *Am. J. Reprod. Immunol.* **51**:275–282.

Abumrad, N. A., M. R. el-Maghrabi, E. Z. Amri, E. Lopez, and P. A. Grimaldi. 1993. Cloning of a rat adipocyte membrane protein implicated in binding or transport of long-chain fatty acids that is induced during preadipocyte differentiation. Homology with human CD36. *J. Biol. Chem.* **268**:17665–17668.

Ackerman, A. L., and P. Cresswell. 2004. Cellular mechanisms governing cross-presentation of exogenous antigens. *Nat. Immunol.* **5**:678–684.

Acton, S. L., P. E. Scherer, H. F. Lodish, and M. Krieger. 1994. Expression cloning of SR-BI, a CD36-related class B scavenger receptor. *J. Biol. Chem.* **269**:21003–21009.

Adachi, Y., T. Ishii, Y. Ikeda, A. Hoshino, H. Tamura, J. Aketagawa, S. Tanaka, and N. Ohno. 2004. Characterization of beta-glucan recognition site on C-type lectin, dectin 1. *Infect. Immun.* **72**:4159–4171.

Aderem, A., and D. M. Underhill. 1999. Mechanisms of phagocytosis in macrophages. *Annu. Rev. Immunol.* **17**:593–623.

Aderem, A. A., S. D. Wright, S. C. Silverstein, and Z. A. Cohn. 1985. Ligated complement receptors do not activate the arachidonic acid cascade in resident peritoneal macrophages. *J. Exp. Med.* **161**:617–622.

Aebischer, T. 1994. Recurrent cutaneous leishmaniasis: a role for persistent parasites? *Parasitol. Today* **10**:25–28.

Akira, S., S. Uematsu, and O. Takeuchi. 2006. Pathogen recognition and innate immunity. *Cell* **124**:783-801.

Allen, L. A., and A. Aderem. 1996. Molecular definition of distinct cytoskeletal structures involved in complement- and Fc receptor-mediated phagocytosis in macrophages. *J. Exp. Med.* **184**:627–637.

Andrejeva, J., K. S. Childs, D. F. Young, T. S. Carlos, N. Stock, S. Goodbourn, and R. E. Randall. 2004. The V proteins of paramyxoviruses bind the IFN-inducible RNA helicase, mda-5, and inhibit its activation of the IFN-beta promoter. *Proc. Natl. Acad. Sci. USA* **101**:17264–17269.

Appelmelk, B. J., I. van Die, S. J. van Vliet, C. M. J. E. Vandenbroucke-Grauls, T. B. H. Geijtenbeek, and Y. van Kooyk. 2003. Cutting edge: carbohydrate profiling identifies new pathogens that interact with dendritic cell-specific ICAM-3-grabbing nonintegrin on dendritic cells. *J. Immunol.* **170**:1635–1639.

Armstrong, A. P., M. E. Tometsko, M. Glaccum, C. L. Sutherland, D. Cosman, and W. C. Dougall. 2002. A RANK/TRAF6-dependent signal transduction pathway is essential for osteoclast cytoskeletal organization and resorptive function. *J. Biol. Chem.* **277**:44347–44356.

Arredouani, M., Z. Yang, Y. Ning, G. Qin, R. Soininen, K. Tryggvason, and L. Kobzik. 2004. The scavenger receptor MARCO is required for lung defense against pneumococcal pneumonia and inhaled particles. *J. Exp. Med.* **200**:267–272.

Arredouani, M. S., A. Palecanda, H. Koziel, Y. C. Huang, A. Imrich, T. H. Sulahian, Y. Y. Ning, Z. Yang, T. Pikkarainen, M. Sankala, S. O. Vargas, M. Takeya, K. Tryggvason, and L. Kobzik. 2005. MARCO is the major binding receptor for unopsonized particles and bacteria on human alveolar macrophages. *J. Immunol.* **175**:6058–6064.

Ashkenas, J., M. Penman, E. Vasile, S. Acton, M. Freeman, and M. Krieger. 1993. Structures and high and low affinity ligand binding properties of murine type I and type II macrophage scavenger receptors. *J. Lipid Res.* **34**:983–1000.

Avallone, R., A. Demers, A. Rodrigue-Way, K. Bujold, D. Harb, S. Anghel, W. Wahli, S. Marleau, H. Ong, and A. Tremblay. 2006. A growth hormone-releasing peptide that binds scavenger receptor CD36 and ghrelin receptor up-regulates sterol transporters and cholesterol efflux in macrophages through a peroxisome proliferator-activated receptor gamma-dependent pathway. *Mol. Endocrinol.* **20**:3165–3178.

Ayi, K., S. N. Patel, L. Serghides, T. G. Smith, and K. C. Kain. 2005. Nonopsonic phagocytosis of erythrocytes infected with ring-stage Plasmodium falciparum. *Infect. Immun.* **73**:2559–2563.

Bajtay, Z., M. Jozsi, Z. Banki, S. Thiel, N. Thielens, and A. Erdei. 2000. Mannan-binding lectin and C1q bind to distinct structures and exert differential effects on macrophages. *Eur. J. Immunol.* **30:**1706–1713.

Becker, M., A. Cotena, S. Gordon, and N. Platt. 2006. Expression of the class A macrophage scavenger receptor on specific subpopulations of murine dendritic cells limits their endotoxin response. *Eur. J. Immunol.* **36:**950–960.

Beharka, A. A., J. E. Crowther, F. X. McCormack, G. M. Denning, J. Lees, E. Tibesar, and L. S. Schlesinger. 2005. Pulmonary surfactant protein A activates a phosphatidylinositol 3-kinase/calcium signal transduction pathway in human macrophages: participation in the up-regulation of mannose receptor activity. *J. Immunol.* **175:**2227–2236.

Bin, L. H., L. D. Nielson, X. Liu, R. J. Mason, and H. B. Shu. 2003. Identification of uteroglobin-related protein 1 and macrophage scavenger receptor with collagenous structure as a lung-specific ligand-receptor pair. *J. Immunol.* **171:**924–930.

Blander, J. M., and R. Medzhitov. 2006. On regulation of phagosome maturation and antigen presentation. *Nat. Immunol.* **7:**1029–1035.

Bogdan, C. 2001. Nitric oxide and the immune response. *Nat. Immunol.* **2:**907–916.

Bogdan, C. 2004. Reactive oxygen and reactive nitrogen metabolites as effector molecules against infectious pathogens, p. 357–396. *In* S. H. E. Kaufman, R. Medzhitov, and S. Gordon (ed.), *The Innate Immune Response to Infection.* ASM Press, Washington, DC.

Bogdan, C., and C. Nathan. 1993. Modulation of macrophage function by transforming growth factor beta, interleukin-4, and interleukin-10. *Ann. N. Y. Acad. Sci.* **685:**713–739.

Bokoch, G. M., and B. A. Diebold. 2002. Current molecular models for NADPH oxidase regulation by Rac GTPase. *Blood* **100:**2692–2696.

Bonecchi, R., S. Sozzani, J. T. Stine, W. Luini, G. D'Amico, P. Allavena, D. Chantry, and A. Mantovani. 1998. Divergent effects of interleukin-4 and interferon-gamma on macrophage-derived chemokine production: an amplification circuit of polarized T helper 2 responses. *Blood* **92:**2668–2671.

Bottazzi, B., C. Garlanda, G. Salvatori, P. Jeannin, A. Manfredi, and A. Mantovani. 2006. Pentraxins as a key component of innate immunity. *Curr. Opin. Immunol.* **18:**10–15.

Brown, E. J. 1991. Complement receptors and phagocytosis. *Curr. Opin. Immunol.* **3:**76–82.

Brown, G. D., and S. Gordon. 2001. Immune recognition: a new receptor for beta-glucans. *Nature* **413:**36–37.

Brown, G. D., J. Herre, D. L. Williams, J. A. Willment, A. S. Marshall, and S. Gordon. 2003. Dectin-1 mediates the biological effects of beta-glucans. *J. Exp. Med.* **197:**1119–1124.

Brown, J. M., E. J. Swindle, N. M. Kushnir-Sukhov, A. Holian, and D. D. Metcalfe. 2007. Silica-directed mast cell activation is enhanced by scavenger receptors. *Am. J. Respir. Cell Mol. Biol.* **36:**43–52.

Bultel-Brienne, S., S. Lestavel, A. Pilon, I. Laffont, A. Tailleux, J. C. Fruchart, G. Siest, and V. Clavey. 2002. Lipid free apolipoprotein E binds to the class B Type I scavenger receptor I (SR-BI) and enhances cholesteryl ester uptake from lipoproteins. *J. Biol. Chem.* **277:**36092–36099.

Calvo, D., D. Gomez-Coronado, M. A. Lasuncion, and M. A. Vega. 1997. CLA-1 is an 85-kD plasma membrane glycoprotein that acts as a high-affinity receptor for both native (HDL, LDL, and VLDL) and modified (OxLDL and AcLDL) lipoproteins. *Arterioscler. Thromb. Vasc. Biol.* **17:**2341–2349.

Calvo, D., D. Gomez-Coronado, Y. Suarez, M. A. Lasuncion, and M. A. Vega. 1998. Human CD36 is a high affinity receptor for the native lipoproteins HDL, LDL, and VLDL. *J. Lipid Res.* **39:**777–788.

Cameron, L. A., P. A. Giardini, F. S. Soo, and J. A. Theriot. 2000. Secrets of actin-based motility revealed by a bacterial pathogen. *Nat. Rev. Mol. Cell Biol.* **1:**110–119.

Campbell, P. A. 1994. Macrophage-Listeria interactions. *Immunol. Ser.* **60:**313–328.

Caron, E., and A. Hall. 1998. Identification of two distinct mechanisms of phagocytosis controlled by different Rho GTPases. *Science* **282:**1717–1721.

Carrasco, Y. R., and F. D. Batista. 2007. B cells acquire particulate antigen in a macrophage-rich area at the boundary between the follicle and the subcapsular sinus of the lymph node. *Immunity* **27:**160–171.

Carroll, M. C. 1998. The role of complement and complement receptors in induction and regulation of immunity. *Annu. Rev. Immunol.* **16:**545–568.

Chamaillard, M., S. E. Girardin, J. Viala, and D. J. Philpott. 2003. Nods, Nalps and Naip: intracellular regulators of bacterial-induced inflammation. *Cell Microbiol.* **5:**581–592.

Chen, Y., T. Pikkarainen, O. Elomaa, R. Soininen, T. Kodama, G. Kraal, and K. Tryggvason. 2005. Defective microarchitecture of the spleen marginal zone and impaired response to a thymus-independent type 2 antigen in mice lacking scavenger receptors MARCO and SR-A. *J. Immunol.* **175:**8173–8180.

Chico-Calero, I., M. Suarez, B. Gonzalez-Zorn, M. Scortti, J. Slaghuis, W. Goebel, and J. A. Vazquez-Boland. 2002. Hpt, a bacterial homolog of the microsomal glucose-6-phosphate translocase, mediates rapid intracellular proliferation in Listeria. *Proc. Natl. Acad. Sci. USA* **99:**431–436.

Clark, A. R., J. L. Dean, and J. Saklatvala. 2003. Post-transcriptional regulation of gene expression by mitogen-activated protein kinase p38. *FEBS Lett.* **546:**37–44.

Clemens, D. L., B. Y. Lee, and M. A. Horwitz. 2005. Francisella tularensis enters macrophages via a novel process involving pseudopod loops. *Infect. Immun.* **73:**5892–5902.

Clemens, D. L., B. Y. Lee, and M. A. Horwitz. 2004. Virulent and avirulent strains of Francisella tularensis prevent acidification and maturation of their phagosomes and escape into the cytoplasm in human macrophages. *Infect. Immun.* **72:**3204–3217.

Coers, J., R. E. Vance, M. F. Fontana, and W. F. Dietrich. 2007. Restriction of Legionella pneumophila growth in macrophages requires the concerted action of cytokine and Naip5/Ipaf signalling pathways. *Cell Microbiol.* **9:**2344–2357.

Cohen-Freue, G., T. R. Holzer, J. D. Forney, and W. R. McMaster. 2007. Global gene expression in Leishmania. *Int. J. Parasitol.* **37:**1077–1086.

Coller, S. P., and D. M. Paulnock. 2001. Signaling pathways initiated in macrophages after engagement of type A scavenger receptors. *J. Leukoc. Biol.* **70:**142–148.

Coraci, I. S., J. Husemann, J. W. Berman, C. Hulette, J. H. Dufour, G. K. Campanella, A. D. Luster, S. C. Silverstein, and J. B. El-Khoury. 2002. CD36, a class B scavenger receptor, is expressed on microglia in Alzheimer's disease brains and can mediate production of reactive oxygen species in response to beta-amyloid fibrils. *Am. J. Pathol.* **160:**101–112.

Cox, D., P. Chang, Q. Zhang, P. G. Reddy, G. M. Bokoch, and S. Greenberg. 1997. Requirements for both Rac1 and Cdc42 in membrane ruffling and phagocytosis in leukocytes. *J. Exp. Med.* **186:**1487–1494.

Creagh, E. M., and L. A. O'Neill. 2006. TLRs, NLRs and RLRs: a trinity of pathogen sensors that co-operate in innate immunity. *Trends Immunol.* **27:**352–357.

Descoteaux, A., and S. J. Turco. 1993. The lipophosphoglycan of Leishmania and macrophage protein kinase C. *Parasitol. Today* **9:**468–471.

Desjardins, M., M. Houde, and E. Gagnon. 2005. Phagocytosis: the convoluted way from nutrition to adaptive immunity. *Immunol. Rev.* **207**:158–165.

Desmedt, M., P. Rottiers, H. Dooms, W. Fiers, and J. Grooten. 1998. Macrophages induce cellular immunity by activating Th1 cell responses and suppressing Th2 cell responses. *J. Immunol.* **160**:5300–5308.

de Waal Malefyt, R., C. G. Figdor, R. Huijbens, S. Mohan-Peterson, B. Bennett, J. Culpepper, W. Dang, G. Zurawski, and J. E. de Vries. 1993. Effects of IL-13 on phenotype, cytokine production, and cytotoxic function of human monocytes. Comparison with IL-4 and modulation by IFN-gamma or IL-10. *J. Immunol.* **151**:6370–6381.

Diaz-Silvestre, H., P. Espinosa-Cueto, A. Sanchez-Gonzalez, M. A. Esparza-Ceron, A. L. Pereira-Suarez, G. Bernal-Fernandez, C. Espitia, and R. Mancilla. 2005. The 19-kDa antigen of Mycobacterium tuberculosis is a major adhesin that binds the mannose receptor of THP-1 monocytic cells and promotes phagocytosis of mycobacteria. *Microb. Pathog.* **39**:97.

Doi, T., K. Higashino, Y. Kurihara, Y. Wada, T. Miyazaki, H. Nakamura, S. Uesugi, T. Imanishi, Y. Kawabe, H. Itakura, et al. 1993. Charged collagen structure mediates the recognition of negatively charged macromolecules by macrophage scavenger receptors. *J. Biol. Chem.* **268**:2126–2133.

Drickamer, K. 1999. C-type lectin-like domains. *Curr. Opin. Struct. Biol.* **9**:585–590.

East, L., and C. M. Isacke. 2002. The mannose receptor family. *Biochim. Biophys. Acta* **1572**:364.

El Khoury, J., S. E. Hickman, C. A. Thomas, L. Cao, S. C. Silverstein, and J. D. Loike. 1996. Scavenger receptor-mediated adhesion of microglia to beta-amyloid fibrils. *Nature* **382**:716–719.

Elkins, K. L., S. C. Cowley, and C. M. Bosio. 2003. Innate and adaptive immune responses to an intracellular bacterium, Francisella tularensis live vaccine strain. *Microbes Infect.* **5**:135–142.

Elliott, J. A., and W. C. Winn, Jr. 1986. Treatment of alveolar macrophages with cytochalasin D inhibits uptake and subsequent growth of Legionella pneumophila. *Infect. Immun.* **51**:31–36.

Ellis, J., P. C. Oyston, M. Green, and R. W. Titball. 2002. Tularemia. *Clin. Microbiol. Rev.* **15**:631–646.

Elomaa, O., M. Kangas, C. Sahlberg, J. Tuukkanen, R. Sormunen, A. Liakka, I. Thesleff, G. Kraal, and K. Tryggvason. 1995. Cloning of a novel bacteria-binding receptor structurally related to scavenger receptors and expressed in a subset of macrophages. *Cell* **80**:603–609.

Elomaa, O., M. Sankala, T. Pikkarainen, U. Bergmann, A. Tuuttila, A. Raatikainen-Ahokas, H. Sariola, and K. Tryggvason. 1998. Structure of the human macrophage MARCO receptor and characterization of its bacteria-binding region. *J. Biol. Chem.* **273**:4530–4538.

Endemann, G., L. W. Stanton, K. S. Madden, C. M. Bryant, R. T. White, and A. A. Protter. 1993. CD36 is a receptor for oxidized low density lipoprotein. *J. Biol. Chem.* **268**:11811–11816.

Ezekowitz, R. A., K. Sastry, P. Bailly, and A. Warner. 1990. Molecular characterization of the human macrophage mannose receptor: demonstration of multiple carbohydrate recognition-like domains and phagocytosis of yeasts in Cos-1 cells. *J. Exp. Med.* **172**:1785–1794.

Fitzgerald, K. A., S. M. McWhirter, K. L. Faia, D. C. Rowe, E. Latz, D. T. Golenbock, A. J. Coyle, S. M. Liao, and T. Maniatis. 2003a. IKKepsilon and TBK1 are essential components of the IRF3 signaling pathway. *Nat. Immunol.* **4**:491–496.

Fitzgerald, K. A., D. C. Rowe, B. J. Barnes, D. R. Caffrey, A. Visintin, E. Latz, B. Monks, P. M. Pitha, and D. T. Golenbock. 2003b. LPS-TLR4 signaling to IRF-3/7 and NF-kappaB involves the toll adapters TRAM and TRIF. *J. Exp. Med.* **198**:1043–1055.

Fraser, I., D. Hughes, and S. Gordon. 1993. Divalent cation-independent macrophage adhesion inhibited by monoclonal antibody to murine scavenger receptor. *Nature* **364**:343–346.

Gagnon, E., S. Duclos, C. Rondeau, E. Chevet, P. H. Cameron, O. Steele-Mortimer, J. Paiement, J. J. Bergeron, and M. Desjardins. 2002. Endoplasmic reticulum-mediated phagocytosis is a mechanism of entry into macrophages. *Cell* **110**:119–131.

Ganz, T. 2006. Hepcidin—a peptide hormone at the interface of innate immunity and iron metabolism. *Curr. Top. Microbiol. Immunol.* **306**:183–898.

Gao, L. Y., and Y. Abu Kwaik. 1999a. Activation of caspase 3 during Legionella pneumophila-induced apoptosis. *Infect. Immun.* **67**:4886–4894.

Gao, L. Y., and Y. Abu Kwaik. 1999b. Apoptosis in macrophages and alveolar epithelial cells during early stages of infection by Legionella pneumophila and its role in cytopathogenicity. *Infect. Immun.* **67**:862–870.

Garlanda, C., B. Bottazzi, A. Bastone, and A. Mantovani. 2005. Pentraxins at the crossroads between innate immunity, inflammation, matrix deposition, and female fertility. *Annu. Rev. Immunol.* **23**:337–366.

Geijtenbeek, T. B. H., R. Torensma, S. J. van Vliet, G. C. F. van Duijnhoven, G. J. Adema, Y. van Kooyk, and C. G. Figdor. 2000. Identification of DC-SIGN, a novel dendritic cell-specific ICAM-3 receptor that supports primary immune responses. *Cell* **100**:575–585.

Geijtenbeek, T. B. H., S. J. van Vliet, E. A. Koppel, M. Sanchez-Hernandez, C. M. J. E. Vandenbroucke-Grauls, B. Appelmelk, and Y. van Kooyk. 2003. Mycobacteria target DC-SIGN to suppress dendritic cell function. *J. Exp. Med.* **197**:7–17.

Glomski, I. J., M. M. Gedde, A. W. Tsang, J. A. Swanson, and D. A. Portnoy. 2002. The Listeria monocytogenes hemolysin has an acidic pH optimum to compartmentalize activity and prevent damage to infected host cells. *J. Cell Biol.* **156**:1029–1038.

Goda, S., T. Imai, O. Yoshie, O. Yoneda, H. Inoue, Y. Nagano, T. Okazaki, H. Imai, E. T. Bloom, N. Domae, and H. Umehara. 2000. CX3C-chemokine, fractalkine-enhanced adhesion of THP-1 cells to endothelial cells through integrin-dependent and -independent mechanisms. *J. Immunol.* **164**:1313–1320.

Goldstein, J. L., Y. K. Ho, S. K. Basu, and M. S. Brown. 1979. Binding site on macrophages that mediates uptake and degradation of acetylated low density lipoprotein, producing massive cholesterol deposition. *Proc. Natl. Acad. Sci. USA* **76**:333–337.

Gordon, S. 2003. Alternative activation of macrophages. *Nat. Rev. Immunol.* **3**:23–35.

Gordon, S., and P. R. Taylor. 2005. Monocyte and macrophage heterogeneity. *Nat. Rev. Immunol.* **5**:953–564.

Gou, Y., H. Feinberg, E. Conroy, D. A. Mitchell, R. Alvarez, O. Blixt, M. E. Taylor, W. I. Weis, and K. Drickamer. 2004. Structural basis for distinct ligand-binding and targeting properties of the receptors DC-SIGN and DC-SIGNR. *Nat. Struct. Mol. Biol.* **11**:591–598.

Gough, P. J., D. R. Greaves, and S. Gordon. 1998. A naturally occurring isoform of the human macrophage scavenger receptor (SR-A) gene generated by alternative splicing blocks modified LDL uptake. *J. Lipid Res.* **39**:531–543.

Graversen, J. H., M. Madsen, and S. K. Moestrup. 2002. CD163: a signal receptor scavenging haptoglobin-hemoglobin complexes from plasma. *Int. J. Biochem. Cell Biol.* **34**:309–314.

Greenberg, J. W., W. Fischer, and K. A. Joiner. 1996. Influence of lipoteichoic acid structure on recognition by the macrophage scavenger receptor. *Infect. Immun.* **64:**3318–3325.

Greenberg, M. E., M. Sun, R. Zhang, M. Febbraio, R. Silverstein, and S. L. Hazen. 2006. Oxidized phosphatidylserine-CD36 interactions play an essential role in macrophage-dependent phagocytosis of apoptotic cells. *J. Exp. Med.* **203:**2613–2625.

Greenwalt, D. E., R. H. Lipsky, C. F. Ockenhouse, H. Ikeda, N. N. Tandon, and G. A. Jamieson. 1992. Membrane glycoprotein CD36: a review of its roles in adherence, signal transduction, and transfusion medicine. *Blood* **80:**1105–1115.

Griffin, F. M., Jr., J. A. Griffin, J. E. Leider, and S. C. Silverstein. 1975. Studies on the mechanism of phagocytosis. I. Requirements for circumferential attachment of particle-bound ligands to specific receptors on the macrophage plasma membrane. *J. Exp. Med.* **142:**1263–1282.

Griffin, F. M., Jr., J. A. Griffin, and S. C. Silverstein. 1976. Studies on the mechanism of phagocytosis. II. The interaction of macrophages with anti-immunoglobulin IgG-coated bone marrow-derived lymphocytes. *J. Exp. Med.* **144:**788–809.

Gringhuis, S. I., J. den Dunnen, M. Litjens, B. van Het Hof, Y. van Kooyk, and T. B. Geijtenbeek. 2007. C-type lectin DC-SIGN modulates Toll-like receptor signaling via Raf-1 kinase-dependent acetylation of transcription factor NF-kappaB. *Immunity* **26:**605–616.

Guillot, L., V. Balloy, F. X. McCormack, D. T. Golenbock, M. Chignard, and M. Si-Tahar. 2002. Cutting edge: the immunostimulatory activity of the lung surfactant protein-A involves Toll-like receptor 4. *J. Immunol.* **168:**5989–5992.

Hackam, D. J., O. D. Rotstein, A. Schreiber, W. Zhang, and S. Grinstein. 1997. Rho is required for the initiation of calcium signaling and phagocytosis by Fcgamma receptors in macrophages. *J. Exp. Med.* **186:**955–966.

Hampton, R. Y., D. T. Golenbock, M. Penman, M. Krieger, and C. R. Raetz. 1991. Recognition and plasma clearance of endotoxin by scavenger receptors. *Nature* **352:**342–344.

Harshyne, L. A., M. I. Zimmer, S. C. Watkins, and S. M. Barratt-Boyes. 2003. A role for class A scavenger receptor in dendritic cell nibbling from live cells. *J. Immunol.* **170:**2302–2309.

Hasan, Z., C. Schlax, L. Kuhn, I. Lefkovits, D. Young, J. Thole, and J. Pieters. 1997. Isolation and characterization of the mycobacterial phagosome: segregation from the endosomal/lysosomal pathway. *Mol. Microbiol.* **24:**545–553.

Hawlisch, H., and J. Kohl. 2006. Complement and Toll-like receptors: key regulators of adaptive immune responses. *Mol. Immunol.* **43:**13–21.

Haworth, R., N. Platt, S. Keshav, D. Hughes, E. Darley, H. Suzuki, Y. Kurihara, T. Kodama, and S. Gordon. 1997. The macrophage Scavenger Receptor Type A is expressed by activated macrophages and protects the host against lethal endotoxic shock. *J. Exp. Med.* **186:**1431–1439.

Hedger, M. P. 2002. Macrophages and the immune responsiveness of the testis. *J. Reprod. Immunol.* **57:**19–34.

Helmy, K. Y., K. J. Katschke, Jr., N. N. Gorgani, N. M. Kljavin, J. M. Elliott, L. Diehl, S. J. Scales, N. Ghilardi, and M. van Lookeren Campagne. 2006. CRIg: a macrophage complement receptor required for phagocytosis of circulating pathogens. *Cell* **124:**915–927.

Herre, J., S. Gordon, and G. D. Brown. 2004. Dectin-1 and its role in the recognition of [beta]-glucans by macrophages. *Mol. Immunol.* **40:**869–876.

Hissong, B. D., G. I. Byrne, M. L. Padilla, and J. M. Carlin. 1995. Upregulation of interferon-induced indoleamine 2,3-dioxygenase in human macrophage cultures by lipopolysaccharide, muramyl tripeptide, and interleukin-1. *Cell. Immunol.* **160:**264–269.

Hoebe, K., P. Georgel, S. Rutschmann, X. Du, S. Mudd, K. Crozat, S. Sovath, L. Shamel, T. Hartung, U. Zahringer, and B. Beutler. 2005. CD36 is a sensor of diacylglycerides. *Nature* **433:**523–527.

Horwitz, M. A. 1983. The Legionnaires' disease bacterium (Legionella pneumophila) inhibits phagosome-lysosome fusion in human monocytes. *J. Exp. Med.* **158:**2108–2126.

Hsieh, C. S., S. E. Macatonia, C. S. Tripp, S. F. Wolf, A. O'Garra, and K. M. Murphy. 1993. Development of TH1 CD4+ T cells through IL-12 produced by Listeria-induced macrophages. *Science* **260:**547–549.

Hsu, H. Y., S. L. Chiu, M. H. Wen, K. Y. Chen, and K. F. Hua. 2001. Ligands of macrophage scavenger receptor induce cytokine expression via differential modulation of protein kinase signaling pathways. *J. Biol. Chem.* **276:**28719–28730.

Hughes, D. A., I. P. Fraser, and S. Gordon. 1995. Murine macrophage scavenger receptor: in vivo expression and function as receptor for macrophage adhesion in lymphoid and non-lymphoid organs. *Eur. J. Immunol.* **25:**466–473.

Huh, H. Y., S. F. Pearce, L. M. Yesner, J. L. Schindler, and R. L. Silverstein. 1996. Regulated expression of CD36 during monocyte-to-macrophage differentiation: potential role of CD36 in foam cell formation. *Blood* **87:**2020–2028.

Hume, D. A., D. Halpin, H. Charlton, and S. Gordon. 1984. The mononuclear phagocyte system of the mouse defined by immunohistochemical localization of antigen F4/80: macrophages of endocrine organs. *Proc. Natl. Acad. Sci. USA* **81:**4174–4177.

Hume, D. A., A. P. Robinson, G. G. MacPherson, and S. Gordon. 1983. The mononuclear phagocyte system of the mouse defined by immunohistochemical localization of antigen F4/80. Relationship between macrophages, Langerhans cells, reticular cells, and dendritic cells in lymphoid and hematopoietic organs. *J. Exp. Med.* **158:**1522–1536.

Husemann, J., J. D. Loike, T. Kodama, and S. C. Silverstein. 2001. Scavenger receptor class B type I (SR-BI) mediates adhesion of neonatal murine microglia to fibrillar beta-amyloid. *J. Neuroimmunol.* **114:**142–150.

Ibata-Ombetta, S., T. Idziorek, P. A. Trinel, D. Poulain, and T. Jouault. 2003. Candida albicans phospholipomannan promotes survival of phagocytosed yeasts through modulation of bad phosphorylation and macrophage apoptosis. *J. Biol. Chem.* **278:**13086–13093.

Imhof, B. A., and M. Aurrand-Lions. 2004. Adhesion mechanisms regulating the migration of monocytes. *Nat. Rev. Immunol.* **4:**432–444.

Inohara, N., M. Chamaillard, C. McDonald, and G. Nunez. 2005. NOD-LRR proteins: role in host-microbial interactions and inflammatory disease. *Annu. Rev. Biochem.* **74:**355–383.

Inohara, N., and G. Nunez. 2003. NODs: intracellular proteins involved in inflammation and apoptosis. *Nat. Rev. Immunol.* **3:**371–382.

Inohara, N., Y. Ogura, A. Fontalba, O. Gutierrez, F. Pons, J. Crespo, K. Fukase, S. Inamura, S. Kusumoto, M. Hashimoto, S. J. Foster, A. P. Moran, J. L. Fernandez-Luna, and G. Nunez. 2003. Host recognition of bacterial muramyl dipeptide mediated through NOD2. Implications for Crohn's disease. *J. Biol. Chem.* **278:**5509–5512.

Ireton, K., and P. Cossart. 1997. Host-pathogen interactions during entry and actin-based movement of Listeria monocytogenes. *Annu. Rev. Genet.* **31:**113–138.

Ishii, K. J., C. Coban, and S. Akira. 2005. Manifold mechanisms of toll-like receptor-ligand recognition. *J. Clin. Immunol.* **25:**511–521.

Jack, D. L., M. E. Lee, M. W. Turner, N. J. Klein, and R. C. Read. 2005. Mannose-binding lectin enhances phagocytosis and killing of Neisseria meningitidis by human macrophages. *J. Leukoc. Biol.* **77:**328–336.

Janeway, C. A., and R. Medzhitov. 2002. Innate immune recognition. *Ann. Rev. Immunol.* **20**:197–216.

Jeannin, P., B. Bottazzi, M. Sironi, A. Doni, M. Rusnati, M. Presta, V. Maina, G. Magistrelli, J. F. Haeuw, G. Hoeffel, N. Thieblemont, N. Corvaia, C. Garlanda, Y. Delneste, and A. Mantovani. 2005. Complexity and complementarity of outer membrane protein A recognition by cellular and humoral innate immunity receptors. *Immunity* **22**:551–560.

Jo, E. K., C. S. Yang, C. H. Choi, and C. V. Harding. 2007. Intracellular signalling cascades regulating innate immune responses to Mycobacteria: branching out from Toll-like receptors. *Cell. Microbiol.* **9**:1087–1098.

Junt, T., E. A. Moseman, M. Iannacone, S. Massberg, P. A. Lang, M. Boes, K. Fink, S. E. Henrickson, D. M. Shayakhmetov, N. C. Di Paolo, N. van Rooijen, T. R. Mempel, S. P. Whelan, and U. H. von Andrian. 2007. Subcapsular sinus macrophages in lymph nodes clear lymph-borne viruses and present them to antiviral B cells. *Nature* **450**:110–114.

Kang, B. K., and L. S. Schlesinger. 1998. Characterization of mannose receptor-dependent phagocytosis mediated by *Mycobacterium tuberculosis* lipoarabinomannan. *Infect. Immun.* **66**:2769–2777.

Kang, P. B., A. K. Azad, J. B. Torrelles, T. M. Kaufman, A. Beharka, E. Tibesar, L. E. DesJardin, and L. S. Schlesinger. 2005. The human macrophage mannose receptor directs *Mycobacterium tuberculosis* lipoarabinomannan-mediated phagosome biogenesis. *J. Exp. Med.* **202**:987–999.

Karlsson, M. C., R. Guinamard, S. Bolland, M. Sankala, R. M. Steinman, and J. V. Ravetch. 2003. Macrophages control the retention and trafficking of B lymphocytes in the splenic marginal zone. *J. Exp. Med.* **198**:333–340.

Kawai, T., K. Takahashi, S. Sato, C. Coban, H. Kumar, H. Kato, K. J. Ishii, O. Takeuchi, and S. Akira. 2005. IPS-1, an adaptor triggering RIG-I- and Mda5-mediated type I interferon induction. *Nat. Immunol.* **6**:981–988.

Kawai, T., O. Takeuchi, T. Fujita, J. Inoue, P. F. Muhlradt, S. Sato, K. Hoshino, and S. Akira. 2001. Lipopolysaccharide stimulates the MyD88-independent pathway and results in activation of IFN-regulatory factor 3 and the expression of a subset of lipopolysaccharide-inducible genes. *J. Immunol.* **167**:5887–5894.

Kodama, T., M. Freeman, L. Rohrer, J. Zabrecky, P. Matsudaira, and M. Krieger. 1990. Type I macrophage scavenger receptor contains alpha-helical and collagen-like coiled coils. *Nature* **343**:531–535.

Koppel, E. A., E. Saeland, D. J. de Cooker, Y. van Kooyk, and T. B. Geijtenbeek. 2005. DC-SIGN specifically recognizes *Streptococcus pneumoniae* serotypes 3 and 14. *Immunobiology* **210**:203–210.

Kovacsovics-Bankowski, M., and K. L. Rock. 1995. A phagosome-to-cytosol pathway for exogenous antigens presented on MHC class I molecules. *Science* **267**:243–246.

Krieger, M. 1997. The other side of scavenger receptors: pattern recognition for host defense. *Curr. Opin. Lipidol.* **8**:275–280.

Krieger, M., and J. Herz. 1994. Structures and functions of multiligand lipoprotein receptors: macrophage scavenger receptors and LDL receptor-related protein (LRP). *Annu. Rev. Biochem.* **63**:601–637.

Kudo, K., H. Sano, H. Takahashi, K. Kuronuma, S.-I. Yokota, N. Fujii, K.-I. Shimada, I. Yano, Y. Kumazawa, D. R. Voelker, S. Abe, and Y. Kuroki. 2004. Pulmonary collectins enhance phagocytosis of Mycobacterium avium through increased activity of mannose receptor. *J. Immunol.* **172**:7592–7602.

Kunjathoor, V. V., A. A. Tseng, L. A. Medeiros, T. Khan, and K. J. Moore. 2004. beta-Amyloid promotes accumulation of lipid peroxides by inhibiting CD36-mediated clearance of oxidized lipoproteins. *J. Neuroinflammation* **1**:23.

Kuronuma, K., H. Sano, K. Kato, K. Kudo, N. Hyakushima, S. Yokota, H. Takahashi, N. Fujii, H. Suzuki, T. Kodama, S. Abe, and Y. Kuroki. 2004. Pulmonary surfactant protein A augments the phagocytosis of *Streptococcus pneumoniae* by alveolar macrophages through a casein kinase 2-dependent increase of cell surface localization of scavenger receptor A. *J. Biol. Chem.* **279**:21421–21430.

Kzhyshkowska, J., G. Workman, M. Cardo-Vila, W. Arap, R. Pasqualini, A. Gratchev, L. Krusell, S. Goerdt, and E. H. Sage. 2006. Novel function of alternatively activated macrophages: stabilin-1-mediated clearance of SPARC. *J. Immunol.* **176**:5825–5832.

Lai, X. H., and A. Sjostedt. 2003. Delineation of the molecular mechanisms of Francisella tularensis-induced apoptosis in murine macrophages. *Infect. Immun.* **71**:4642–4646.

Lamkanfi, M., A. Amer, T. D. Kanneganti, R. Munoz-Planillo, G. Chen, P. Vandenabeele, A. Fortier, P. Gros, and G. Nunez. 2007. The Nod-like receptor family member Naip5/Birc1e restricts Legionella pneumophila growth independently of caspase-1 activation. *J. Immunol.* **178**:8022–8027.

Lee, S. J., S. Evers, D. Roeder, A. F. Parlow, J. Risteli, L. Risteli, Y. C. Lee, T. Feizi, H. Langen, and M. C. Nussenzweig. 2002. Mannose receptor-mediated regulation of serum glycoprotein homeostasis. *Science* **295**:1898–1901.

Lee, S. J., N.-Y. Zheng, M. Clavijo, and M. C. Nussenzweig. 2003. Normal host defense during systemic candidiasis in mannose receptor-deficient mice. *Infect. Immun.* **71**:437–445.

Lehnert, B. E. 1992. Pulmonary and thoracic macrophage subpopulations and clearance of particles from the lung. *Environ. Health Perspect.* **97**:17–46.

Levy, O. 2004. Antimicrobial proteins and peptides: anti-infective molecules of mammalian leukocytes. *J. Leukoc. Biol.* **76**:909–925.

Li, K., E. Foy, J. C. Ferreon, M. Nakamura, A. C. Ferreon, M. Ikeda, S. C. Ray, M. Gale, Jr., and S. M. Lemon. 2005. Immune evasion by hepatitis C virus NS3/4A protease-mediated cleavage of the Toll-like receptor 3 adaptor protein TRIF. *Proc. Natl. Acad. Sci. USA* **102**:2992–2997.

Li, X., H. Y. Kan, S. Lavrentiadou, M. Krieger, and V. Zannis. 2002. Reconstituted discoidal ApoE-phospholipid particles are ligands for the scavenger receptor BI. The amino-terminal 1-165 domain of ApoE suffices for receptor binding. *J. Biol. Chem.* **277**:21149–21157.

Li, Z., and A. M. Diehl. 2003. Innate immunity in the liver. *Curr. Opin. Gastroenterol.* **19**:565–571.

Lin, R., R. S. Noyce, S. E. Collins, R. D. Everett, and K. L. Mossman. 2004. The herpes simplex virus ICP0 RING finger domain inhibits IRF3- and IRF7-mediated activation of interferon-stimulated genes. *J. Virol.* **78**:1675–1684.

Lodge, R., and A. Descoteaux. 2005. Modulation of phagolysosome biogenesis by the lipophosphoglycan of Leishmania. *Clin. Immunol.* **114**:256–265.

Lorenz, M. C., J. A. Bender, and G. R. Fink. 2004. Transcriptional response of Candida albicans upon internalization by macrophages. *Eukaryot. Cell* **3**:1076–1087.

Lowry, M. B., A. M. Duchemin, J. M. Robinson, and C. L. Anderson. 1998. Functional separation of pseudopod extension and particle internalization during Fc gamma receptor-mediated phagocytosis. *J. Exp. Med.* **187**:161–176.

MacMicking, J. D. 2004. IFN-inducible GTPases and immunity to intracellular pathogens. *Trends Immunol.* **25**:601–609.

Marquis, H., H. Goldfine, and D. A. Portnoy. 1997. Proteolytic pathways of activation and degradation of a bacterial phospholipase C during intracellular infection by Listeria monocytogenes. *J. Cell Biol.* **137**:1381–1392.

Marsche, G., S. Frank, J. G. Raynes, K. F. Kozarsky, W. Sattler, and E. Malle. 2007. The lipidation status of acute-phase protein serum amyloid A determines cholesterol mobilization via scavenger receptor class B, type I. *Biochem. J.* **402:**117–124.

Marsche, G., R. Zimmermann, S. Horiuchi, N. N. Tandon, W. Sattler, and E. Malle. 2003. Class B scavenger receptors CD36 and SR-BI are receptors for hypochlorite-modified low density lipoprotein. *J. Biol. Chem.* **278:**47562–47570.

Martens, S., and J. Howard. 2006. The interferon-inducible GTPases. *Annu. Rev. Cell Dev. Biol.* **22:**559–589.

Martinez, F. O., S. Gordon, M. Locati, and A. Mantovani. 2006. Transcriptional profiling of the human monocyte-to-macrophage differentiation and polarization: new molecules and patterns of gene expression. *J. Immunol.* **177:**7303–7311.

Martinez-Esparza, M., A. Aguinaga, P. Gonzalez-Parraga, P. Garcia-Penarrubia, T. Jouault, and J. C. Arguelles. 2007. Role of trehalose in resistance to macrophage killing: study with a tps1/tps1 trehalose-deficient mutant of Candida albicans. *Clin. Microbiol. Infect.* **13:**384–394.

Martinez-Pomares, L., L. G. Hanitsch, R. Stillion, S. Keshav, and S. Gordon. 2005. Expression of mannose receptor and ligands for its cysteine-rich domain in venous sinuses of human spleen. *Lab. Invest.* **85:**1238–1249.

Martinez-Pomares, L., M. Kosco-Vilbois, E. Darley, P. Tree, S. Herren, J. Y. Bonnefoy, and S. Gordon. 1996. Fc chimeric protein containing the cysteine-rich domain of the murine mannose receptor binds to macrophages from splenic marginal zone and lymph node subcapsular sinus and to germinal centers. *J. Exp. Med.* **184:**1927–1937.

Martinez-Pomares, L., D. Wienke, R. Stillion, E. J. McKenzie, J. N. Arnold, J. Harris, E. McGreal, R. B. Sim, C. M. Isacke, and S. Gordon. 2006. Carbohydrate-independent recognition of collagens by the macrophage mannose receptor. *Eur. J. Immunol.* **36:**1074–1082.

Martinon, F., K. Burns, and J. Tschopp. 2002. The inflammasome: a molecular platform triggering activation of inflammatory caspases and processing of proIL-beta. *Mol. Cell* **10:**417–426.

Matsumoto, A., M. Naito, H. Itakura, S. Ikemoto, H. Asaoka, I. Hayakawa, H. Kanamori, H. Aburatani, F. Takaku, and H. Suzuki. 1990. Human macrophage scavenger receptors: primary structure, expression, and localization in atherosclerotic lesions. *Proc. Natl. Acad. Sci. USA* **87:**9133–9137.

McCusker, K., and J. Hoidal. 1989. Characterization of scavenger receptor activity in resident human lung macrophages. *Exp. Lung Res.* **15:**651–661.

McGreal, E. P., L. Martinez-Pomares, and S. Gordon. 2004. Divergent roles for C-type lectins expressed by cells of the innate immune system. *Mol. Immunol.* **41:**1109–1121.

McKenzie, E. J., P. R. Taylor, R. J. Stillion, A. D. Lucas, J. Harris, S. Gordon, and L. Martinez-Pomares. 2007. Mannose receptor expression and function define a new population of murine dendritic cells. *J. Immunol.* **178:**4975–4983.

Mebius, R. E., and G. Kraal. 2005. Structure and function of the spleen. *Nat. Rev. Immunol.* **5:**606–616.

Mechnikov, I. I. 1908. On the present state of the question of immunity in infectious diseases. *Nobel Lecture.* http://nobelprize.org/nobel_prizes/medicine/laureates/1908/mechnikov-lecture.html.

Mellor, A. L., and D. H. Munn. 2004. IDO expression by dendritic cells: tolerance and tryptophan catabolism. *Nat. Rev. Immunol.* **4:**762–774.

Merad, M., M. G. Manz, H. Karsunky, A. Wagers, W. Peters, I. Charo, I. L. Weissman, J. G. Cyster, and E. G. Engleman. 2002. Langerhans cells renew in the skin throughout life under steady-state conditions. *Nat. Immunol.* **3:**1135–1141.

Meylan, E., J. Curran, K. Hofmann, D. Moradpour, M. Binder, R. Bartenschlager, and J. Tschopp. 2005. Cardif is an adaptor protein in the RIG-I antiviral pathway and is targeted by hepatitis C virus. *Nature* **437:**1167–1172.

Meylan, E., J. Tschopp, and M. Karin. 2006. Intracellular pattern recognition receptors in the host response. *Nature* **442:**39–44.

Modolell, M., I. M. Corraliza, F. Link, G. Soler, and K. Eichmann. 1995. Reciprocal regulation of the nitric oxide synthase/arginase balance in mouse bone marrow-derived macrophages by TH1 and TH2 cytokines. *Eur. J. Immunol.* **25:**1101–1104.

Moore, K. J., S. Labrecque, and G. Matlashewski. 1993. Alteration of Leishmania donovani infection levels by selective impairment of macrophage signal transduction. *J. Immunol.* **150:**4457–4465.

Mosser, D. M. 2003. The many faces of macrophage activation. *J. Leukoc. Biol.* **73:**209–212.

Mukhopadhyay, S., Y. Chen, M. Sankala, L. Peiser, T. Pikkarainen, G. Kraal, K. Tryggvason, and S. Gordon. 2006. MARCO, an innate activation marker of macrophages, is a class A scavenger receptor for Neisseria meningitidis. *Eur. J. Immunol.* **36:**940–949.

Mukhopadhyay, S., L. Peiser, and S. Gordon. 2004. Activation of murine macrophages by Neisseria meningitidis and IFN-gamma in vitro: distinct roles of class A scavenger and Toll-like pattern recognition receptors in selective modulation of surface phenotype. *J. Leukoc. Biol.* **76:**577–584.

Muller-Taubenberger, A., A. N. Lupas, H. Li, M. Ecke, E. Simmeth, and G. Gerisch. 2001. Calreticulin and calnexin in the endoplasmic reticulum are important for phagocytosis. *EMBO J.* **20:**6772–6782.

Mullin, N., K. Hall, and M. Taylor. 1994. Characterization of ligand binding to a carbohydrate-recognition domain of the macrophage mannose receptor. *J. Biol. Chem.* **269:**28405–28413.

Naito, M., H. Suzuki, T. Mori, A. Matsumoto, T. Kodama, and K. Takahashi. 1992. Coexpression of type I and type II human macrophage scavenger receptors in macrophages of various organs and foam cells in atherosclerotic lesions. *Am. J. Pathol.* **141:**591–599.

Naka, T., M. Fujimoto, H. Tsutsui, and A. Yoshimura. 2005. Negative regulation of cytokine and TLR signalings by SOCS and others. *Adv. Immunol.* **87:**61–122.

Nakamura, K., T. Yamaji, P. R. Crocker, A. Suzuki, and Y. Hashimoto. 2002. Lymph node macrophages, but not spleen macrophages, express high levels of unmasked sialoadhesin: implication for the adhesive properties of macrophages in vivo. *Glycobiology* **12:**209–216.

Nano, F. E., N. Zhang, S. C. Cowley, K. E. Klose, K. K. Cheung, M. J. Roberts, J. S. Ludu, G. W. Letendre, A. I. Meierovics, G. Stephens, and K. L. Elkins. 2004. A Francisella tularensis pathogenicity island required for intramacrophage growth. *J. Bacteriol.* **186:**6430–6436.

Nicoletti, A., G. Caligiuri, I. Tornberg, T. Kodama, S. Stemme, and G. K. Hansson. 1999. The macrophage scavenger receptor type A directs modified proteins to antigen presentation. *Eur. J. Immunol.* **29:**512–521.

Nimmerjahn, F., and J. V. Ravetch. 2006. Fcgamma receptors: old friends and new family members. *Immunity* **24:**19–28.

O'Brien, D. K., and S. B. Melville. 2003. Multiple effects on *Clostridium perfringens* binding, uptake and trafficking to lysosomes by inhibitors of macrophage phagocytosis receptors. *Microbiology* **149:**1377–1386.

O'Neill, L. A. 2006. How Toll-like receptors signal: what we know and what we don't know. *Curr. Opin. Immunol.* **18:**3–9.

O'Neill, L. A., and A. G. Bowie. 2007. The family of five: TIR-domain-containing adaptors in Toll-like receptor signalling. *Nat. Rev. Immunol.* **7:**353–364.

O'Riordan, D., J. Standing, and A. Limper. 1995. Pneumocystis carinii glycoprotein A binds macrophage mannose receptors. *Infect. Immun.* **63:**779–784.

O'Shea, J. J., M. Gadina, and R. D. Schreiber. 2002. Cytokine signaling in 2002: new surprises in the Jak/Stat pathway. *Cell* **109**(Suppl.):S121–S131.

Park-Min, K. H., T. T. Antoniv, and L. B. Ivashkiv. 2005. Regulation of macrophage phenotype by long-term exposure to IL-10. *Immunobiology* **210**:77–86.

Parthasarathy, S., L. G. Fong, D. Otero, and D. Steinberg. 1987. Recognition of solubilized apoproteins from delipidated, oxidized low density lipoprotein (LDL) by the acetyl-LDL receptor. *Proc. Natl. Acad. Sci. USA* **84**:537–540.

Payne, N. R., and M. A. Horwitz. 1987. Phagocytosis of *Legionella pneumophila* is mediated by human monocyte complement receptors. *J. Exp. Med.* **166**:1377–1389.

Peiser, L., K. Makepeace, A. Plüddemann, S. Savino, J. C. Wright, M. Pizza, R. Rappuoli, E. R. Moxon, and S. Gordon. 2006. Identification of Neisseria meningitidis nonlipopolysaccharide ligands for class A macrophage scavenger receptor by using a novel assay. *Infect. Immun.* **74**:5191–5199.

Perry, V. H., and S. Gordon. 1988. Macrophages and microglia in the nervous system. *Trends Neurosci.* **11**:273–277.

Peters, N., and D. Sacks. 2006. Immune privilege in sites of chronic infection: Leishmania and regulatory T cells. *Immunol. Rev.* **213**:159–179.

Phan, T. G., I. Grigorova, T. Okada, and J. G. Cyster. 2007. Subcapsular encounter and complement-dependent transport of immune complexes by lymph node B cells. *Nat. Immunol.* **8**:992–1000.

Pierce, M. M., R. E. Gibson, and F. G. Rodgers. 1996. Opsonin-independent adherence and phagocytosis of Listeria monocytogenes by murine peritoneal macrophages. *J. Med. Microbiol.* **45**:258–262.

Pilette, C., Y. Ouadrhiri, J. Van Snick, J. C. Renauld, P. Staquet, J. P. Vaerman, and Y. Sibille. 2002. IL-9 inhibits oxidative burst and TNF-alpha release in lipopolysaccharide-stimulated human monocytes through TGF-beta. *J. Immunol.* **168**:4103–4111.

Pizarro-Cerda, J., and P. Cossart. 2006. Bacterial adhesion and entry into host cells. *Cell* **124**:715–727.

Platt, N., and S. Gordon. 2001. Is the class A macrophage scavenger receptor (SR-A) multifunctional?—The mouse's tale. *J. Clin. Invest.* **108**:649–654.

Platt, N., H. Suzuki, T. Kodama, and S. Gordon. 2000. Apoptotic thymocyte clearance in scavenger receptor class A-deficient mice is apparently normal. *J. Immunol.* **164**:4861–4867.

Platt, N., H. Suzuki, Y. Kurihara, T. Kodama, and S. Gordon. 1996. Role for the class A macrophage scavenger receptor in the phagocytosis of apoptotic thymocytes in vitro. *Proc. Natl. Acad. Sci. USA* **93**:12456–12460.

Plüddemann, A., S. Mukhopadhyay, and S. Gordon. 2006. The interaction of macrophage receptors with bacterial ligands. *Expert Rev. Mol. Med.* **8**:1–25.

Portnoy, D. A., V. Auerbuch, and I. J. Glomski. 2002. The cell biology of Listeria monocytogenes infection: the intersection of bacterial pathogenesis and cell-mediated immunity. *J. Cell Biol.* **158**:409–414.

Presanis, J. S., M. Kojirna, and R. B. Sim. 2003. Biochemistry and genetics of mannan-binding lectin (MBL). *Biochem. Soc. Trans.* **31**:748–752.

Puente Navazo, M. D., L. Daviet, E. Ninio, and J. L. McGregor. 1996. Identification on human CD36 of a domain (155-183) implicated in binding oxidized low-density lipoproteins (Ox-LDL). *Arterioscler. Thromb. Vasc. Biol.* **16**:1033–1039.

Puig-Kroger, A., A. Dominguez-Soto, L. Martinez-Munoz, D. Serrano-Gomez, M. Lopez-Bravo, E. Sierra-Filardi, E. Fernandez-Ruiz, N. Ruiz-Velasco, C. Ardavin, Y. Groner, N. Tandon, A. L. Corbi, and M. A. Vega. 2006. RUNX3 negatively regulates CD36 expression in myeloid cell lines. *J. Immunol.* **177**:2107–2114.

Quinn, J. M., and M. T. Gillespie. 2005. Modulation of osteoclast formation. *Biochem. Biophys. Res. Commun.* **328**:739–745.

Raes, G., P. De Baetselier, W. Noel, A. Beschin, F. Brombacher, and G. Hassanzadeh Gh. 2002. Differential expression of FIZZ1 and Ym1 in alternatively versus classically activated macrophages. *J. Leukoc. Biol.* **71**:597–602.

Ramachandra, L., E. Noss, W. H. Boom, and C. V. Harding. 1999. Phagocytic processing of antigens for presentation by class II major histocompatibility complex molecules. *Cell. Microbiol.* **1**:205–214.

Reed, M. B., P. Domenech, C. Manca, H. Su, A. K. Barczak, B. N. Kreiswirth, G. Kaplan, and C. E. Barry III. 2004. A glycolipid of hypervirulent tuberculosis strains that inhibits the innate immune response. *Nature* **431**:84–87.

Ren, Y., R. L. Silverstein, J. Allen, and J. Savill. 1995. CD36 gene transfer confers capacity for phagocytosis of cells undergoing apoptosis. *J. Exp. Med.* **181**:1857–1862.

Resnick, D., J. E. Chatterton, K. Schwartz, H. Slayter, and M. Krieger. 1996. Structures of class A macrophage scavenger receptors. Electron microscopic study of flexible, multidomain, fibrous proteins and determination of the disulfide bond pattern of the scavenger receptor cysteine-rich domain. *J. Biol. Chem.* **271**:26924–26930.

Rich, K. A., C. Burkett, and P. Webster. 2003. Cytoplasmic bacteria can be targets for autophagy. *Cell. Microbiol.* **5**:455–468.

Rohrer, L., M. Freeman, T. Kodama, M. Penman, and M. Krieger. 1990. Coiled-coil fibrous domains mediate ligand binding by macrophage scavenger receptor type II. *Nature* **343**:570–572.

Russell, D. G., and R. M. Yates. 2007. TLR signalling and phagosome maturation: an alternative viewpoint. *Cell. Microbiol.* **9**:849–850.

Sankala, M., A. Brannstrom, T. Schulthess, U. Bergmann, E. Morgunova, J. Engel, K. Tryggvason, and T. Pikkarainen. 2002. Characterization of recombinant soluble macrophage scavenger receptor MARCO. *J. Biol. Chem.* **277**:33378–33385.

Sano, H., H. Sohma, T. Muta, S. Nomura, D. R. Voelker, and Y. Kuroki. 1999. Pulmonary surfactant protein A modulates the cellular response to smooth and rough lipopolysaccharides by interaction with CD14. *J. Immunol.* **163**:387–395.

Santiago-Garcia, J., T. Kodama, and R. E. Pitas. 2003. The Class A Scavenger Receptor binds to proteoglycans and mediates adhesion of macrophages to the extracellular matrix. *J. Biol. Chem.* **278**:6942–6946.

Santic, M., M. Molmeret, and Y. Abu Kwaik. 2005. Modulation of biogenesis of the Francisella tularensis subsp. novicida-containing phagosome in quiescent human macrophages and its maturation into a phagolysosome upon activation by IFN-gamma. *Cell. Microbiol.* **7**:957–967.

Savill, J., I. Dransfield, N. Hogg, and C. Haslett. 1990. Vitronectin receptor-mediated phagocytosis of cells undergoing apoptosis. *Nature* **343**:170–173.

Savill, J., N. Hogg, Y. Ren, and C. Haslett. 1992. Thrombospondin cooperates with CD36 and the vitronectin receptor in macrophage recognition of neutrophils undergoing apoptosis. *J. Clin. Invest.* **90**:1513–1522.

Schaible, U. E., and S. H. Kaufmann. 2004. Iron and microbial infection. *Nat. Rev. Microbiol.* **2**:946–953.

Schlesinger, L. S., S. R. Hull, and T. M. Kaufman. 1994. Binding of the terminal mannosyl units of lipoarabinomannan from a virulent strain of *Mycobacterium tuberculosis* to human macrophages. *J. Immunol.* **152**:4070–4079.

Schmid, D., J. Dengjel, O. Schoor, S. Stevanovic, and C. Munz. 2006. Autophagy in innate and adaptive immunity against intracellular pathogens. *J. Mol. Med.* **84**:194–202.

Schulert, G. S., and L. A. Allen. 2006. Differential infection of mononuclear phagocytes by Francisella tularensis: role of the macrophage mannose receptor. *J. Leukoc. Biol.* **80:**563–571.

Scotton, C. J., F. O. Martinez, M. J. Smelt, M. Sironi, M. Locati, A. Mantovani, and S. Sozzani. 2005. Transcriptional profiling reveals complex regulation of the monocyte IL-1 beta system by IL-13. *J. Immunol.* **174:**834–845.

Serghides, L., T. G. Smith, S. N. Patel, and K. C. Kain. 2003. CD36 and malaria: friends or foes? *Trends Parasitol.* **19:**461–469.

Seth, R. B., L. Sun, and Z. J. Chen. 2006. Antiviral innate immunity pathways. *Cell Res.* **16:**141–147.

Shi, Y., Y. Tohyama, T. Kadono, J. He, S. M. Shahjahan Miah, R. Hazama, C. Tanaka, K. Tohyama, and H. Yamamura. 2006. Protein-tyrosine kinase Syk is required for pathogen engulfment in complement-mediated phagocytosis. *Blood* **107:**4554–4562.

Silverstein, R. L., A. S. Asch, and R. L. Nachman. 1989. Glycoprotein IV mediates thrombospondin-dependent platelet-monocyte and platelet-U937 cell adhesion. *J. Clin. Invest.* **84:**546–552.

Smith, P. D., C. Ochsenbauer-Jambor, and L. E. Smythies. 2005. Intestinal macrophages: unique effector cells of the innate immune system. *Immunol. Rev.* **206:**149–159.

Stack, J., I. R. Haga, M. Schroder, N. W. Bartlett, G. Maloney, P. C. Reading, K. A. Fitzgerald, G. L. Smith, and A. G. Bowie. 2005. Vaccinia virus protein A46R targets multiple Toll-like-interleukin-1 receptor adaptors and contributes to virulence. *J. Exp. Med.* **201:**1007–1018.

Stanley, E., G. J. Lieschke, D. Grail, D. Metcalf, G. Hodgson, J. A. Gall, D. W. Maher, J. Cebon, V. Sinickas, and A. R. Dunn. 1994. Granulocyte/macrophage colony-stimulating factor-deficient mice show no major perturbation of hematopoiesis but develop a characteristic pulmonary pathology. *Proc. Natl. Acad. Sci. USA* **91:**5592–5596.

Steele, C., L. Marrero, S. Swain, A. G. Harmsen, M. Zheng, G. D. Brown, S. Gordon, J. E. Shellito, and J. K. Kolls. 2003. Alveolar macrophage-mediated killing of Pneumocystis carinii f. sp. muris involves molecular recognition by the dectin-1 {beta}-glucan receptor. *J. Exp. Med.* **198:**1677–1688.

Stein, M., S. Keshav, N. Harris, and S. Gordon. 1992. Interleukin 4 potently enhances murine macrophage mannose receptor activity: a marker of alternative immunologic macrophage activation. *J. Exp. Med.* **176:**287–292.

Stuart, L. M., J. Deng, J. M. Silver, K. Takahashi, A. A. Tseng, E. J. Hennessy, R. A. Ezekowitz, and K. J. Moore. 2005. Response to Staphylococcus aureus requires CD36-mediated phagocytosis triggered by the COOH-terminal cytoplasmic domain. *J. Cell Biol.* **170:**477–485.

Sturgill-Koszycki, S., P. H. Schlesinger, P. Chakraborty, P. L. Haddix, H. L. Collins, A. K. Fok, R. D. Allen, S. L. Gluck, J. Heuser, and D. G. Russell. 1994. Lack of acidification in Mycobacterium phagosomes produced by exclusion of the vesicular proton-ATPase. *Science* **263:**678–681.

Surh, C. D., and J. Sprent. 1994. T-cell apoptosis detected in situ during positive and negative selection in the thymus. *Nature* **372:**100–103.

Suzuki, H., Y. Kurihara, M. Takeya, N. Kamada, M. Kataoka, K. Jishage, H. Sakaguchi, J. K. Kruijt, T. Higashi, T. Suzuki, T. J. van Berkel, S. Horiuchi, K. Takahashi, Y. Yazaki, and T. Kodama. 1997. The multiple roles of macrophage scavenger receptors (MSR) in vivo: resistance to atherosclerosis and susceptibility to infection in MSR knockout mice. *J. Atheroscler. Thromb.* **4:**1–11.

Suzuki, H., Y. Kurihara, M. Takeya, N. Kamada, M. Kataoka, K. Jishage, O. Ueda, H. Sakaguchi, T. Higashi, T. Suzuki, Y. Takashima, Y. Kawabe, O. Cynshi, Y. Wada, M. Honda, H. Kurihara, H. Aburatani, T. Doi, A. Matsumoto, S. Azuma, T. Noda, Y. Toyoda, H. Itakura, Y. Yazaki, T. Kodama, et al. 1997. A role for macrophage scavenger receptors in atherosclerosis and susceptibility to infection. *Nature* **386:**292–296.

Swanson, J. A., and A. D. Hoppe. 2004. The coordination of signaling during Fc receptor-mediated phagocytosis. *J. Leukoc. Biol.* **76:**1093–1103.

Swanson, M. S. 2006. Autophagy: eating for good health. *J. Immunol.* **177:**4945–4951.

Tacke, F., and G. J. Randolph. 2006. Migratory fate and differentiation of blood monocyte subsets. *Immunobiology* **211:**609–618.

Tassaneetrithep, B., T. H. Burgess, A. Granelli-Piperno, C. Trumpfheller, J. Finke, W. Sun, M. A. Eller, K. Pattanapanyasat, S. Sarasombath, D. L. Birx, R. M. Steinman, S. Schlesinger, and M. A. Marovich. 2003. DC-SIGN (CD209) mediates dengue virus infection of human dendritic cells. *J. Exp. Med.* **197:**823–829.

Taylor, P. R., L. Martinez-Pomares, M. Stacey, H.-H. Lin, G. D. Brown, and S. Gordon. 2005. Macrophage receptors and immune recognition. *Annu. Rev. Immunol.* **23:**901–944.

Telepnev, M., I. Golovliov, T. Grundstrom, A. Tarnvik, and A. Sjostedt. 2003. Francisella tularensis inhibits Toll-like receptor-mediated activation of intracellular signalling and secretion of TNF-alpha and IL-1 from murine macrophages. *Cell. Microbiol.* **5:**41–51.

Terpstra, V., D. A. Bird, and D. Steinberg. 1998. Evidence that the lipid moiety of oxidized low density lipoprotein plays a role in its interaction with macrophage receptors. *Proc. Natl. Acad. Sci. USA* **95:**1806–1811.

Terpstra, V., E. S. van Amersfoort, A. G. van Velzen, J. Kuiper, and T. J. van Berkel. 2000. Hepatic and extrahepatic scavenger receptors: function in relation to disease. *Arterioscler. Thromb. Vasc. Biol.* **20:**1860–1872.

Thomas, C. A., Y. Li, T. Kodama, H. Suzuki, S. C. Silverstein, and J. El Khoury. 2000. Protection from lethal gram-positive infection by macrophage scavenger receptor-dependent phagocytosis. *J. Exp. Med.* **191:**147–156.

Tilney, L. G., O. S. Harb, P. S. Connelly, C. G. Robinson, and C. R. Roy. 2001. How the parasitic bacterium Legionella pneumophila modifies its phagosome and transforms it into rough ER: implications for conversion of plasma membrane to the ER membrane. *J. Cell Sci.* **114:**4637–4650.

Tomokiyo, R., K. Jinnouchi, M. Honda, Y. Wada, N. Hanada, T. Hiraoka, H. Suzuki, T. Kodama, K. Takahashi, and M. Takeya. 2002. Production, characterization, and interspecies reactivities of monoclonal antibodies against human class A macrophage scavenger receptors. *Atherosclerosis* **161:**123–132.

Trigatti, B. L., M. Krieger, and A. Rigotti. 2003. Influence of the HDL receptor SR-BI on lipoprotein metabolism and atherosclerosis. *Arterioscler. Thromb. Vasc. Biol.* **23:**1732–1738.

Trinchieri, G., and A. Sher. 2007. Cooperation of Toll-like receptor signals in innate immune defence. *Nat. Rev. Immunol.* **7:**179–190.

Unanue, E. R., and P. M. Allen. 1987. The basis for the immunoregulatory role of macrophages and other accessory cells. *Science* **236:**551–557.

Vales-Gomez, M., H. Reyburn, and J. Strominger. 2000. Interaction between the human NK receptors and their ligands. *Crit. Rev. Immunol.* **20:**223–244.

van den Elzen, P., S. Garg, L. Leon, M. Brigl, E. A. Leadbetter, J. E. Gumperz, C. C. Dascher, T. Y. Cheng, F. M. Sacks, P. A. Illarionov, G. S. Besra, S. C. Kent, D. B. Moody, and M. B. Brenner. 2005. Apolipoprotein-mediated pathways of lipid antigen presentation. *Nature* **437:**906–910.

van der Laan, L. J., M. Kangas, E. A. Dopp, E. Broug-Holub, O. Elomaa, K. Tryggvason, and G. Kraal. 1997. Macrophage scavenger receptor MARCO: in vitro and in vivo regulation and involvement in the anti-bacterial host defense. *Immunol. Lett.* **57:**203–208.

Vandivier, R. W., C. A. Ogden, V. A. Fadok, P. R. Hoffmann, K. K. Brown, M. Botto, M. J. Walport, J. H. Fisher, P. M. Henson, and K. E. Greene. 2002. Role of surfactant proteins A, D, and C1q in the clearance of apoptotic cells in vivo and in vitro: calreticulin and CD91 as a common collectin receptor complex. *J. Immunol.* **169:**3978–3986.

Van Furth, R., M. C. Diesselhoff-den Dulk, and H. Mattie. 1973. Quantitative study on the production and kinetics of mononuclear phagocytes during an acute inflammatory reaction. *J. Exp. Med.* **138:**1314–1330.

van Kooyk, Y., and T. B. Geijtenbeek. 2003. DC-SIGN: escape mechanism for pathogens. *Nat. Rev. Immunol.* **3:**697–709.

Vazquez-Boland, J. A., M. Kuhn, P. Berche, T. Chakraborty, G. Dominguez-Bernal, W. Goebel, B. Gonzalez-Zorn, J. Wehland, and J. Kreft. 2001. Listeria pathogenesis and molecular virulence determinants. *Clin. Microbiol. Rev.* **14:**584–640.

Velasco-Velazquez, M. A., D. Barrera, A. Gonzalez-Arenas, C. Rosales, and J. Agramonte-Hevia. 2003. Macrophage-*Mycobacterium tuberculosis* interactions: role of complement receptor 3. *Microb. Pathog.* **35:**125–131.

Vergne, I., J. Chua, and V. Deretic. 2003. Mycobacterium tuberculosis phagosome maturation arrest: selective targeting of PI3P-dependent membrane trafficking. *Traffic* **4:**600–606.

Vinals, M., S. Xu, E. Vasile, and M. Krieger. 2003. Identification of the N-linked glycosylation sites on the high density lipoprotein (HDL) receptor SR-BI and assessment of their effects on HDL binding and selective lipid uptake. *J. Biol. Chem.* **278:**5325–5332.

Vishnyakova, T. G., A. V. Bocharov, I. N. Baranova, Z. Chen, A. T. Remaley, G. Csako, T. L. Eggerman, and A. P. Patterson. 2003. Binding and internalization of lipopolysaccharide by Cla-1, a human orthologue of rodent scavenger receptor B1. *J. Biol. Chem.* **278:**22771–22780.

Vishnyakova, T. G., R. Kurlander, A. V. Bocharov, I. N. Baranova, Z. Chen, M. S. Abu-Asab, M. Tsokos, D. Malide, F. Basso, A. Remaley, G. Csako, T. L. Eggerman, and A. P. Patterson. 2006. CLA-1 and its splicing variant CLA-2 mediate bacterial adhesion and cytosolic bacterial invasion in mammalian cells. *Proc. Natl. Acad. Sci. USA* **103:**16888–16893.

Webb, N. R., W. J. de Villiers, P. M. Connell, F. C. de Beer, and D. R. van der Westhuyzen. 1997. Alternative forms of the scavenger receptor BI (SR-BI). *J. Lipid Res.* **38:**1490–1495.

Weis, W. I., M. E. Taylor, and K. Drickamer. 1998. The C-type lectin superfamily in the immune system. *Immunol. Rev.* **163:**19–34.

Willment, J. A., H. H. Lin, D. M. Reid, P. R. Taylor, D. L. Williams, S. Y. Wong, S. Gordon, and G. D. Brown. 2003. Dectin-1 expression and function are enhanced on alternatively activated and GM-CSF-treated macrophages and are negatively regulated by IL-10, dexamethasone, and lipopolysaccharide. *J. Immunol.* **171:**4569–4573.

Wilson, M. E., and R. D. Pearson. 1988. Roles of CR3 and mannose receptors in the attachment and ingestion of Leishmania donovani by human mononuclear phagocytes. *Infect. Immun.* **56:**363–369.

Wright, S. D., and S. C. Silverstein. 1983. Receptors for C3b and C3bi promote phagocytosis but not the release of toxic oxygen from human phagocytes. *J. Exp. Med.* **158:**2016–2023.

Xu, S., M. Laccotripe, X. Huang, A. Rigotti, V. I. Zannis, and M. Krieger. 1997. Apolipoproteins of HDL can directly mediate binding to the scavenger receptor SR-BI, an HDL receptor that mediates selective lipid uptake. *J. Lipid Res.* **38:**1289–1298.

Yokota, T., B. Ehlin-Henriksson, and G. K. Hansson. 1998. Scavenger receptors mediate adhesion of activated B lymphocytes. *Exp. Cell Res.* **239:**16–22.

Yoneyama, M., M. Kikuchi, K. Matsumoto, T. Imaizumi, M. Miyagishi, K. Taira, E. Foy, Y. M. Loo, M. Gale, Jr., S. Akira, S. Yonehara, A. Kato, and T. Fujita. 2005. Shared and unique functions of the DExD/H-box helicases RIG-I, MDA5, and LGP2 in antiviral innate immunity. *J. Immunol.* **175:**2851–2858.

Yoneyama, M., M. Kikuchi, T. Natsukawa, N. Shinobu, T. Imaizumi, M. Miyagishi, K. Taira, S. Akira, and T. Fujita. 2004. The RNA helicase RIG-I has an essential function in double-stranded RNA-induced innate antiviral responses. *Nat. Immunol.* **5:**730–737.

Yoshida, H., S. Hayashi, T. Kunisada, M. Ogawa, S. Nishikawa, H. Okamura, T. Sudo, L. D. Shultz, and S. Nishikawa. 1990. The murine mutation osteopetrosis is in the coding region of the macrophage colony stimulating factor gene. *Nature* **345:**442–444.

Yoshimura, A., T. Naka, and M. Kubo. 2007. SOCS proteins, cytokine signalling and immune regulation. *Nat. Rev. Immunol.* **7:**454–465.

Yu, P., Y. Wang, R. K. Chin, L. Martinez-Pomares, S. Gordon, M. H. Kosco-Vibois, J. Cyster, and Y. X. Fu. 2002. B cells control the migration of a subset of dendritic cells into B cell follicles via CXC chemokine ligand 13 in a lymphotoxin-dependent fashion. *J. Immunol.* **168:**5117–5123.

Yu, Y., S. E. Wang, and G. S. Hayward. 2005. The KSHV immediate-early transcription factor RTA encodes ubiquitin E3 ligase activity that targets IRF7 for proteosome-mediated degradation. *Immunity* **22:**59–70.

Zamze, S., L. Martinez-Pomares, H. Jones, P. R. Taylor, R. J. Stillion, S. Gordon, and S. Y. C. Wong. 2002. Recognition of bacterial capsular polysaccharides and lipopolysaccharides by the macrophage mannose receptor. *J. Biol. Chem.* **277:**41613–41623.

Zhang, J., J. Zhu, A. Imrich, M. Cushion, T. B. Kinane, and H. Koziel. 2004. Pneumocystis activates human alveolar macrophage NF-{kappa}B signaling through mannose receptors. *Infect. Immun.* **72:**3147–3160.

Zhu, F. G., C. F. Reich, and D. S. Pisetsky. 2001. The role of the macrophage scavenger receptor in immune stimulation by bacterial DNA and synthetic oligonucleotides. *Immunology* **103:**226–234.

3

Dendritic Cells and Their Tissue Microenvironment during Exposure to Pathogens

A. MORTELLARO, F. GRANUCCI, M. FOTI, AND P. RICCIARDI-CASTAGNOLI

In 1887, Elie Metchnikoff published the first description of phagocytic cells, which he called mononuclear phagocytes (MPs), in the *Archives of Pathological Anatomy* (Metchnikoff, 1887). More than 80 years later, R. Steinman and Z. Cohn subdivided MPs into macrophages and dendritic cells (DCs) (Steinman and Cohn, 1973, 1974), on the basis of differences in effector functions: antimicrobial and scavenging functions for macrophages and professional antigen presentation for DCs. Both macrophages and DCs have since been studied in great detail, but we still know little about the molecular basis of functional property regulation in these MP cells. This lack of knowledge probably results from the complexity and broadness of the functional roles potentially exerted by these two types of cell. In addition to phagocytosis and antigen presentation, these cells regulate the cytokine and chemokine networks and are involved in inflammation, tissue repair and remodeling, pathogen migration and dissemination, cell recruitment, angiogenesis, the activation of innate (NK, NKT, etc.) and adaptive (T and B lymphocytes) immune cells, and interactions with epithelial, endothelial, and stromal cells. This very broad repertoire of functions probably results from the adaptation of DCs and macrophages to the many organs and tissues within which they reside.

Myeloid DCs are found in all mucosal tissues and sites of pathogen entry, such as the gut and the lung, in particular. However, they are also present in the skin, in internal organs, such as the liver, and in the blood, lymph, and all lymphoid tissues, including bone marrow (BM). Human DCs are easy to isolate from blood. Such blood-derived DCs have been extensively studied as prototype DCs, although it remains unclear whether their functional properties are representative of tissue DCs. In mice, DCs have preferentially been isolated from the spleen and BM, and these cells are not readily comparable with DCs from human blood. The dynamic plasticity of DCs and the heterogeneity of the tissues from which they originate have resulted in a complex terminology of DC subsets frequently defined on the basis of the expression of a few molecular markers rather than on effector functions. DC effector functions are often regulated by tissue microenvironment or by exogenous (microbial) or endogenous (stress) signals, with the acquisition of so-called "mature" phenotypes. However, DCs not only initiate adaptive immunity, they also suppress immunity, by rendering peripheral T cells with autoreactive potentials tolerant. As sentinels, these cells must be sensitive to the amount of antigen present and the persistence of antigens. They use a repertoire of nonclonal receptors to signal downstream to the nucleus, conveying information about what is present in the environment (quality and quantity) and the duration of this signal. This complex activity is revealed by transcriptional responses involving the differential expression of thousands of genes and the integration of a number of signaling pathways. The active transcriptional response leads to the acquisition of diverse DC functional phenotypes that orchestrate the appropriate immune response (Granucci et al., 2001a, 2001b). The molecular mechanisms regulating the wide array of DC functions remain unclear. Tissue architecture, and the stromal component of tissues in particular, may play a key role in the acquisition of DC effector functions. Indeed, DCs derived from different organs, such as the spleen, skin, or mucosal tissues, seem to have different functions. The mucosal route of antigen exposure is known to induce antigen tolerance, whereas systemic antigen exposure on blood cells leads to immunity. Various DC effector functions have been associated with mucosal and systemic sites. However, it remains possible that some tissue DCs, such as the Langerhans cells (LCs) in the epidermis, may be derived from different types of BM progenitors and may undergo specific homing and differentiation processes. This unusually high degree of plasticity and heterogeneity has resulted in considerable confusion in DC terminology. Myeloid DCs have been described in very different ways—conventional, mature, immature, semimature,

---

A. Mortellaro and P. Ricciardi-Castagnoli, Singapore Immunology Network (SIgN), Agency for Science, Technology and Research (A*STAR), Singapore, and Department of Biotechnology and Biosciences, University of Milano-Bicocca, Milano, Italy. F. Granucci and M. Foti, Department of Biotechnology and Biosciences, University of Milano-Bicocca, Milano, Italy.

immunogenic, peripheral, resident, inflammatory, regulatory, and killer—to name just a few (Reis e Sousa, 2006). According to R. Zinkernagel, "In immunology, words instill the illusion of comprehension" (Zinkernagel, 2007). This has often been the case in the field of DC biology. In this review we will refer to either DCs (in the steady-state condition) or to effector DCs (describing those DCs that have been induced by exogenous or endogenous signals to undergo a differentiation process that, as in many other immune cells, leads to terminal differentiation and the acquisition of irreversible effector functions). Other types of terminology will be carefully avoided, to show that we do not need to refer to as many DC subsets and that different DC functions may result largely from different DC tissue microenvironments.

In this review, we will deal with the origin, tissue distribution, and common effector functions of myeloid DCs. Plasmacytoid DCs will be described in a different chapter of this book.

BONE MARROW AND DC PROGENITORS

DCs are derived from hematopoietic progenitors of mesodermal origin in the BM. During embryonic hematopoiesis, multipotent hematopoietic stem cells seed the BM and develop into common myeloid progenitors (CMPs), which respond to the ligand of the receptor tyrosine kinase Flk-2 (also known as FLT3-L) (D'Amico and Wu, 2003). The FLT3 receptor is expressed on hematopoietic stem cells and CMPs (Adolfsson et al., 2001), but is also, surprisingly, expressed on splenic DCs (Karsunky et al., 2003). Like granulocyte-monocyte colony-stimulating factor (GM-CSF), FLT3-L supports DC production from BM-derived progenitors. The origin and renewal of DCs has been widely investigated, leading to suggestions that DCs and macrophages might have different progenitors (D'Amico and Wu, 2003). However, a common clonogenic progenitor giving rise to DCs and macrophages has recently been identified in the BM (Fogg et al., 2006). Mouse blood monocytes have been shown to differentiate into DCs in vivo in response to inflammation (Randolph et al., 1999; Geissmann et al., 2003). The common progenitor in mouse BM, known as MDP—for macrophages and DC progenitor—is $CD117^+$ (c-kit), Lin^-, and $CX3CR1^+$. It gives rise exclusively to monocytes, macrophages, and DCs in vivo and resembles the monoblast described by van Furth in 1975 (Goud et al., 1975). The progeny of the MDP lacks plasmacytoid DCs, indicating that myeloid and plasmacytoid DCs belong to different lineages. Indeed, plasmacytoid DCs, which were originally described as interferon-producing cells, have a very different effector activity based principally on their extremely high capacity to produce type I interferon (IFN) when activated by microbes (Chehimi et al., 1989; Svensson et al., 1996).

In transfer experiments designed to assess the potency of MDPs in vivo, donor-derived cells were detected in the spleen 2 days after injection. These cells increased in number until day 7, and then decreased, consistent with limited self-renewal (Fogg et al., 2006). Seven days after MDP transfer, the spleen contained different populations, based on CD11b and c expression: $CD11c^+$ $CD11b^-$, $CD11c^+$ $CD11b^+$, and $CD11c^-$ $CD11b^+$. Some cells in these populations may retain self-renewal capabilities, supported by the stromal tissue or cytokine environment. Indeed, epidermal LCs are known to undergo self-renewal, throughout life, from endogenous progenitors (Merad et al., 2002). Similarly, some splenic DCs are thought to arise from local progenitors (Kamath et al., 2002) and 5% of splenic DCs have been shown to pass through the cell cycle (Kabashima et al., 2005), whereas most of the DCs from other tissues do not divide. It remains unclear why splenic DCs may also be renewed from resident progenitors. However, the splenic environment, which filters fast-moving antigens borne by the blood, clearly needs to maintain effector DC populations with functional properties different from those of mucosal DCs, which are involved in active tolerance and its maintenance.

Liu et al. (2007) showed in parabiotic mice that DC homeostasis is maintained by a dynamic balance between blood DC precursors, DC division, and DC death. These results are not consistent with the hypothesis that homeostasis is controlled by the presence of resident stem cell-like progenitors (Bogunovic et al., 2006; Wright et al., 2001). In the presence of inflammation, BM progenitors are mobilized and MDPs are probably rapidly recruited from the blood. However, it remains unclear how the homing and dissemination of DC progenitors is regulated.

THE BLOOD, MONOCYTES, AND DCs

CMPs in the BM develop into monoblasts, which are released into the bloodstream as monocytes—a pool of progenitor cells from which tissue DCs and macrophages develop (Geissmann et al., 2003; Fogg et al., 2006). Monocytes are defined on the basis of CD14, CD11b, and CD11c expression in humans and CD11b and F4/80 expression in mice. These cells lack all T, B, NK, and DC markers. However, the blood monocyte population is heterogeneous: 10% of the cells express CD16 (Passlick et al., 1989) and some express either low or high levels of CX3CR1, a chemokine receptor that interacts with fractalkine, a chemokine present on endothelial cells. As for other blood cells, monocyte homing is probably controlled by chemokine-chemokine interactions (Butcher and Picker, 1996). Indeed, it has been shown that a $CX3CR1^{high}$ monocyte subpopulation homes to various tissues independently of inflammatory signals in vivo, whereas the $CX3CR1^{low}$ subpopulation homes to inflamed tissues, where it acquires a DC phenotype ($CD11c^{high}$, major histocompatibility complex [MHC] class II, and priming capability) (Geissmann et al., 2003). Cells of the $CX3CR1^{low}$ subpopulation also express the CCR2 chemokine receptor for the proinflammatory mediator MCP-1 (Geissmann et al., 2003), which plays a key role in LC reconstitution in response to epidermal inflammation or damage (Palframan et al., 2001).

It has been shown that monocytes give rise to DCs in the gut, lamina propria, and lungs of mice, but that these cells do not generate splenic $CD11c^{high}$ DCs, which seem to arise from local or blood-borne precursors without the production of a monocytic intermediate (Liu et al., 2007; Varol et al., 2007). Monocytes are probably committed to the replenishment of mucosal DCs (lung and gut), whereas epidermal LCs originate from a pool of self-renewing precursors present in the skin (Merad et al., 2002) in steady-state conditions. However, in the presence of inflammatory signals inducing a loss of LCs, blood monocytes are used to replenish the LC population (Ginhoux et al., 2006). Monocytes were first shown to migrate efficiently to inflamed tissues about 20 years ago (van Furth, 1998), but our understanding of the molecular mechanisms regulating monocyte homing and differentiation remains limited.

THE SKIN, DERMAL DCs, AND LCs

The skin is the largest organ of the body and represents the first line of defense against invasion by foreign substances and organisms. The skin carries out immune surveillance through a multitude of cutaneous DCs. Paul Langerhans first reported the presence in skin of cells with "nerve ending" features in 1868 (Langerhans, 1868). However, the hematopoietic origin of these cells was not recognized until more than 100 years later (Katz et al., 1979). The DCs of normal skin comprise LCs located in the suprabasal layer of the epidermis, dermal DCs (dDCs), and interstitial DCs. LCs express the nonclassical MHC class I molecule CD1a, the intracellular adhesion molecule E-cadherin, and langerin (CD207). Like other DCs, they also express the integrin CD11c and MHC class II antigens. Langerin is a C-type lectin present in intracellular organelles known as Birbeck granules (Valladeau et al., 2000; Kissenpfennig et al., 2005a). dDCs can be distinguished from LCs on the basis of their location, their lack of CD1a, E-cadherin, and langerin expression, and their expression of CD1b and CD1c. The selective expression of these nonclassical CD1 MHC molecules allows LCs and dDCs to stimulate T cells efficiently in response to glycolipid antigens presented by CD1 molecules. LCs are also classical "professional antigen-presenting cells" that increase their expression of MHC class II and costimulatory molecules on activation and migrate to the T-cell areas of regional lymph nodes (LNs), where they initiate a systemic immune response by presenting processed antigens to naive $CD4^+$ and $CD8^+$ T cells. The origin of LCs has recently been established: studies in adult mice have shown that, under steady-state conditions, epidermal LCs are renewed locally in the skin throughout the life of the animal, from a proliferating pool of local precursors. This self-renewal property is highly unusual, having previously been demonstrated only for microglial cells in the brain (Santambrogio et al., 2001). LC homeostasis seems to be regulated differently from the homeostasis of other DCs. Indeed, after lethal irradiation and BM transplantation, host DCs other than LCs are replenished by donor DCs within 1 to 2 months of transplantation (Merad et al., 2002), whereas host LCs persist for up to 18 months after transplantation. This persistence is due to the self-renewal of resident LC progenitors. Such renewal was also demonstrated in human skin, in the first two-hand allograft study (Kanitakis et al., 2004). In contrast, in the presence of inflammatory conditions inducing LC depletion, such as exposure to a high dose of UV, LC homeostasis is maintained by the recruitment to the skin of blood monocytes, which then differentiate into LCs (Ginhoux et al., 2006). As for microglial cells, it is difficult to determine whether the functional properties of these LCs resemble those of the resident LCs.

Both LCs and dDC-like cells can be generated in vitro from lineage-negative $CD34^+$ precursors (Strobl et al., 1996) or from human peripheral blood monocytes by treatment with GM-CSF and transforming growth factor-β (TGF-β) (Geissmann et al., 1998) or GM-CSF and interleukin-15 (IL-15) (Mohamadzadeh et al., 2001). However, it remains unclear whether these pathways of differentiation in vitro are physiologically relevant. The CSF-1 receptor is required for LC development, and CSF-1 receptor-deficient mice have normal splenic DCs but no LCs in the skin (Ginhoux et al., 2006). The dependence on CSF-1 receptor expression for LC repopulation is interesting, because LC histiocytosis, a proliferative disease of unknown pathogenesis, is associated with high serum concentrations of CSF-1 (Rolland et al., 2005), consistent with the dysregulation of homeostatic control.

LCs and dDCs are believed to have different functional properties. Most of the cellular markers used to discriminate LCs from dDCs are not entirely specific, as they are expressed on both cell types, but to different extents. Kissenpfennig et al. (2005a) developed a knock-in mouse model to track LCs in vivo and distinguish them from dDCs. In this model, the *Lang-EGFP* model, the expression cassette encoding enhanced green fluorescent protein (EGFP) is inserted into the 3' untranslated region of the *langerin* gene (Kissenpfennig et al., 2005b). In steady-state conditions, most LCs are sessile, with only rare motile LCs observed, as shown by vital confocal microscopy. Mechanical trauma, induced by tape stripping, led to the production of inflammatory mediators such as tumor necrosis factor-α (TNF-α) by keratinocytes, leading to LC motility and migration away from the epidermis to the cLNs. During skin inflammation, dDC ($TRITC^+$ $EGFP^-$) migration clearly precedes the migration of LCs ($TRITC^+$ $EGFP^+$), as shown by the topical images obtained with the fluorescent dye tetramethyl rhodamine isocyanate (TRITC) in *Lang-EGFP* mice (Kissenpfennig et al., 2005b). DCs derived from LCs and from the dermis migrate into different areas of the paracortex of draining cLNs. In particular, dDCs tend to be found in the outer paracortex, just beneath the B-cell-rich follicles, whereas LCs are located in the inner paracortex. These data suggest that the microanatomy of LN areas could influence the distribution of LC- and dDC-derived DCs. In vivo, LCs do not seem to be required for the immune response to infections due to viruses, such as herpes virus, but there is growing evidence to suggest that LCs may have key regulatory functions in cutaneous immunity. Indeed, the dysregulation of cutaneous DCs has been observed in several inflammatory skin diseases, including atopic dermatitis (AD) and psoriasis. AD is a chronic inflammatory skin disease characterized by recurrent eczematous skin lesions. DC subtypes in the skin, and blood DCs, have been shown to play a key role in the generation and/or control of inflammation (Leung et al., 2004). Psoriasis vulgaris is another common inflammatory skin disease. It has many physiological features in common with AD, including the presence of marked T-cell infiltrates and modified keratinocyte differentiation. Patients with psoriasis present a Th1-mediated immune response associated with local neutrophil infiltration (Nomura et al., 2003a, 2003b). Interactions between DCs and T cells play a key role in the pathogenesis of the disease.

THE GUT AND INTESTINAL DCs

DCs play an essential role in priming the immune system against antigens. Their presence in tissues in which they may encounter bacteria soon after invasion is therefore indispensable. DCs are particularly abundant in the intestinal mucosa, which interfaces with the external environment, and in which they function as efficient sentinels.

The gut is remarkable in that exposure to food antigens or commensal bacteria leads to oral tolerance, whereas the gut retains the ability to mount a powerful immune response against pathogenic microorganisms. Furthermore, the gastrointestinal tract is covered by a simple, one-cell-thick epithelial layer, which separates the sterile environment of the body from 10^{14} microorganisms and a multitude of food and environmental antigens. Thus, in addition to its role in adsorption and defense, the mucosal tissue and associated mucosal immune system is frequently involved in tolerance induction. A failure to establish oral tolerance can lead to severe inflammatory diseases. Intestinal DCs are required to balance the roles of tolerance induction in response to commensal bacteria, food, and self-antigens, and the induction of immunity to invasive pathogens.

The mucosal surface is associated with lymphoid follicles called Peyer's patches (PPs) and mesenteric LNs. At intestinal mucosal interfaces, DCs are located either below the specialized intestinal epithelial cells—the M cells—covering the PP lymphoid follicles (Niedergang and Kraehenbuhl, 2000) or are scattered in nonlymphoid tissues, such as the lamina propria of the intestinal villi of the small intestine. The gut microenvironment is characterized by constitutive expression of cytokines such as TGF-β (Mucida et al., 2005) and IL-10 (Chirdo et al., 2005) and by the presence of intestinal DCs expressing CD103 (Smith et al., 1994). These DCs maintain gut homeostasis and promote immunoglobulin A (IgA) secretion by gut-activated B cells (Mora et al., 2006). Tolerance to commensal organisms (about 1,000 bacterial species) is rapidly induced in newborns (Lotz et al., 2006) and is clearly beneficial to the host, as it influences gut development and mucosal innate immunity (Kelly and Conway, 2005). Indeed, mice free of these organisms have a reduced mucosal immune system and produce smaller amounts of IgA (Weinstein and Cebra, 1991). Furthermore, commensal bacteria seem to modulate gut-resident T cells, including regulatory T cells (T-reg) and Th17 cells (Jung et al., 2006), probably via intestinal DCs. However, it remains unclear how intestinal DCs obtain access to the commensal flora. It was originally suggested that intestinal immunity results from exclusive antigen uptake by M cells in PPs, but this hypothesis has been called into question by the observation of oral tolerance induction in PP-deficient mice (Spahn et al., 2001). We have shown that resident lamina propria DCs can be also found as intraepithelial DCs. They extend their dendrites through the epithelial layer into the lumen, by forming tight junctions with the epithelial cells, due to the regulated expression of occludin, thereby preserving the integrity of the mucosal barrier (Rescigno et al., 2001). Transepithelial dendrite formation requires the expression on DCs of the chemokine receptor CX3CR1, in particular, in the terminal ileum, where its ligand, CX3CL1, is expressed (Niess et al., 2005). The introduction of *Salmonella enterica* serovar Typhimurium into the small intestine results in the large numbers of intra-epithelial dendrites, regardless of whether the bacterium is invasive or noninvasive (Chieppa et al., 2006). Lamina propria DCs have been shown to present antigens delivered via the oral route efficiently (Chirdo et al., 2005) and, surprisingly, most of the DCs migrating to the mesenteric LNs have been shown to be derived from the intestinal lamina propria rather than from PPs (Bimczok et al., 2005; Turnbull et al., 2005). Antigen recognition in the intestinal immune system is obligatory for oral tolerance induction and depends on CCR7-mediated cell migration (Worbs et al., 2006). Indeed, there is growing evidence to suggest that antigen recognition is restricted to the mesenteric LNs and tolerance is not induced in the spleen (Worbs et al., 2006). Mesenteric LNs act as gatekeepers, preventing commensal bacteria from obtaining access to the bloodstream. Indeed, the regional compartmentalization of lymphoid organs is well established, and is based on both anatomical and functional differences, with immunity predominantly generated by exposure to antigens in the blood and tolerance induced via mucosal routes (Kraal et al., 2006).

It has also been suggested that mucosal DCs play a role in lymphocyte homing: lamina propria DCs seem to play a very important role in imprinting T cells for the expression of $\alpha_4\beta_7$ integrin. This integrin binds to the mucosal addressin cell adhesion molecule on blood vessels in the gut (Butcher et al., 1999). The homing of lymphocytes to the gut is also mediated by the expression of CCR9, a chemokine binding to a ligand expressed on the crypts of the small intestine (Wurbel et al., 2000). Both CCR9 and $\alpha_4\beta_7$ are strongly upregulated by a vitamin A metabolite, retinoic acid (RA). Interestingly, T cells recognizing antigens on intestinal DCs, but not on other tissue DCs, return to the gut, and these DCs express a unique $CD103^+$ $CD11^+$ phenotype (Mora et al., 2003). These $CD103^+$ intestinal DCs modulate the expression of a family of retinal dehydrogenases (F. Powrie, personal communication, 2007) required for the generation of RA from vitamin A. Indeed, $\alpha_4\beta_7$ downregulation induced by inhibitors of retinal dehydrogenases (Iwata et al., 2004) leads to T-cell depletion in the lamina propria. RA deficiency not only disrupts the integrity of the mucosal barrier, it also prevents effector lymphocytes from carrying out their functions (Mora et al., 2006). RA binds to nuclear RA receptors. This binding inhibits the transcriptional activity of activating protein-1 (AP-1), a transcription factor regulating expression of the *IL2* gene in a DNA-binding complex with nuclear factor of activated T cells (NFATs). As all tissue DCs produce IL-2 in response to various stimuli, including commensal bacteria (Foti et al., 2006; Granucci et al., 2003a, 2003b) and stress signals (M. Urbano, personal communication, 2007), an autocrine autoregulatory loop may act on intestinal DCs via RA receptors. It remains unclear how this mechanism interferes with the conversion of T cells into T-reg cells.

It has been suggested that DCs may modulate innate immune signaling by commensal bacteria. The avirulent *Salmonella pullorum* has been shown to block IκB-α degradation, preventing the NF-κB-dependent production of inflammatory cytokines (Neish et al., 2000). The intestinal flora may play an important role in damping inflammatory signals and, indeed, reconstitution of the commensal flora in germ-free mice induces major beneficial changes in host gut function (Mowat, 2003; Fagarasan and Honjo, 2003), whereas the absence of commensal bacteria within the gut may lead to autoimmunity (Izcue et al., 2006; Bouma and Strober, 2003).

Lamina propria DCs probably play a very important role in controlling gut inflammation and, as recently shown, they promote the de novo generation of Foxp3 T-reg cells via RA (Sun et al., 2007; Benson et al., 2007; Coombes et al., 2007). This conversion of naive T cells into T-reg cells has been shown to depend on TGF-β and requires IL-2 in vitro (Chen et al., 2003; Zheng et al., 2004; Fantini et al., 2004). The key molecules favoring the differentiation of T cells into T-reg cells in vitro are indeed IL-2, TGF-β, and RA (Benson et al., 2007). Lamina propria DCs produce IL-2 on activation by microbial signals. One possible role for the intestinal flora would thus be to temper gut inflammation by inducing the production of IL-2 by DCs, thereby favoring the generation of T-reg cells, because a continuous supply of IL-2 is required to ensure T-reg homeostasis (Guiducci et al., 2005). IL-2-deficient mice develop a severe form of inflammatory bowel disease (Sadlack et al., 1993), this pathological process being dominated by a pathogenic T-helper type 1 (Th1) immune response (Ehrhardt et al., 1997; Strober et al., 2002). Up to 17% of the OVA-specific T cells in the gut may be converted into T-reg cells if mice are supplied with OVA in drinking water—consistent with the suggestion that T-reg cells may play a role in oral tolerance (Coombes et al., 2007; Sun et al., 2007).

Finally, the possible role of intestinal DCs in regulating IgA production in the gut has been investigated. The humoral IgA response in the gut is the most effective Ig switch pathway in individuals, resulting in the production of grams of IgA every day (Fagarasan and Honjo, 2003). In the gut, IgA controls and prevents the adhesion of bacteria to the epithelial cells, without damaging the commensal organisms, which continue to exert their beneficial functions. In mice, activated DCs trigger T-cell-independent IgA responses to commensal organisms and this process involves the activation of lamina propria $CD5^+$ B-1 cells (Macpherson and Uhr, 2004; Mora et al., 2006). Similarly, it has been shown in humans that T-cell-independent IgA responses are mediated by intestinal DCs expressing APRIL and BAFF, two CD40-related molecules also produced by mucosal epithelial cells. Genetic defects in APRIL or BAFF signaling induce a phenotype of IgA deficiency (Castigli et al., 2004, 2005). In addition to this T-cell-independent mechanism of IgA class switching, canonical T-cell-dependent class switching occurs in the germinal centers of mucosal sites, including PPs, via CD40L interaction and the production of cytokines such as TGF-β and IL-10. Differentiated IgA-secreting plasmacytoid B cells home to the lamina propria and IgA may ultimately migrate through the epithelial cell layer to reach the gut lumen (Brandtzaeg et al., 2001).

Invasive bacterial pathogens have a larger battery of mechanisms at their disposal for crossing the epithelial barrier. Invasive bacteria often use the M cells in the PPs as a portal of entry (Jones et al., 1994; Kerneis et al., 1997; Sansonetti and Phalipon, 1999). M cells have no brush border and a limited glycocalyx, and are therefore highly accessible to pathogens, facilitating transport into the underlying lymphoid tissue. In the PPs, pathogens initially encounter DCs located below the M cells, and may therefore be taken up by phagocytosis and processed. Activated DCs may then migrate to the mesenteric LNs, where they initiate an adaptive immune response.

THE LUNG AND DCs

The lung has a very large internal surface area, facilitating efficient gaseous exchange following the inhalation of large volumes of air. This property exposes this organ to various environmental stimuli. Like the gastrointestinal tract, the lung epithelium consists of a single layer of cells, maximizing gaseous exchange. However, in contrast to the gastrointestinal tract, the lower respiratory tract is sterile. DCs are found mostly in the upper conducting airways, within the epithelium and beneath the epithelial layer (the lamina propria). In the lower airways, DCs are found in the alveolar septae, interstitial space, and lamina propria, but only rarely within alveoli and bronchioles (Holt et al., 1994). Only a small number of DCs can be detected in bronchoalveolar lavage fluid. DCs are distributed at a mean density of several hundred cells per millimeter squared in the large airways, decreasing to less than a hundred DCs per millimeter squared within smaller intrapulmonary airways (Schon-Hegrad et al., 1991). However, DCs are present in larger numbers in the lung parenchyma and interalveolar septae (Schon-Hegrad et al., 1991).

In normal human lung parenchyma, myeloid DCs express CD11c and either the blood DC antigen (BDCA)-1 or BDCA-3 (Demedts et al., 2005). Even in the absence of overt inflammation, DCs and their precursors are constantly recruited from the blood to the lung in response to environmental stimuli (Holt et al., 1994). The air inhaled has a profound impact on the steady-state dynamics of DC recruitment. Inhaled antigens, such as bacteria or viral particles, induce a very rapid influx of DCs into the airways. No macrophages or lymphocytes are recruited at this early time point, suggesting that DC recruitment is an integral part of the early phase of innate immunity. Freshly isolated human pulmonary DCs specifically express CCR1 and CCR5, which bind the chemokine CCL5 (Stumbles et al., 2001), whereas CCR2 expression has been observed during the recruitment of DCs in a murine model of pulmonary *Mycobacterium tuberculosis* infection (Peters et al., 2001). Chemokines are not the only mediators of DC recruitment to the airway epithelium. Defensins and cationic peptides with bactericidal activity produced by epithelial cells (Cole and Waring, 2002) have also been shown to attract DCs: human β-defensin 2 binds to CCR6, a chemokine receptor expressed by immature DCs homing to epithelia (Harder et al., 2000; Yang et al., 1999). It remains unclear whether lung DCs are recruited from the blood as precursors or as differentiated cells (Suda et al., 1998). However, it seems unlikely that lung DCs undergo self-renewal through the local proliferation of intrapulmonary progenitors.

Pulmonary DCs, like other tissue DCs, possess a repertoire of receptors for the efficient sensing and sampling of a wide variety of airborne antigens. In humans, pulmonary myeloid DCs express several Toll-like receptors (TLRs), including TLR1, TLR2, TLR3, and TLR4. Signaling through these receptors is mediated by the myeloid differentiation primary response gene 88 (MyD88) adaptor molecule or by an independent pathway involving the TRIF molecule and the IRF3 transcription factor (Beutler, 2004). Proteins of the NF-κB family are activated in response to microbial signaling through TLRs, resulting in their translocation to the nucleus (Hofer et al., 2001), inducing the transcription of many NF-κB-dependent genes, mostly immune and inflammatory genes. Other receptors of the DC repertoire include members of the C-type family recognizing carbohydrate motifs present on the surface of several microbial organisms (Engering et al., 2002). Human pulmonary DCs express the DC-specific intercellular adhesion molecule (ICAM)-3-grabbing nonintegrin, also known as DC-SIGN—a C-type lectin receptor capable of binding various microorganisms, including viruses such as human immunodeficiency virus and pathogenic bacteria such as *M. tuberculosis* (Tailleux et al., 2003b; van Kooyk and Geijtenbeek, 2003)—and the lectin-type receptors BDCA-2 and DEC-205. DCs can also display surface molecules, such as the MARCO receptor, which we have shown to be involved in profound remodeling of the actin cytoskeleton (Granucci et al., 2003c).

Following antigen recognition, airway DCs migrate to the T-cell areas of the mediastinal LNs, this migration being mediated by CCR7 receptor expression. Expression of the CCR7 ligands, CCL19 and CCL21, has been reported in the mediastinal LNs (Itakura et al., 2001). Lipid mediators, such as leukotrienes and prostaglandins, are also important mediators of DC migration to the LNs (Hammad et al., 2003). The production of leukotriene D_4 and prostaglandin E_2 by epithelial cells has been shown to stimulate DC emigration toward the draining LNs, whereas prostaglandin D_2 has the opposite effect (Jakubzick et al., 2006). The migration of lung DCs depends less on leukotriene B_4 than the migration of cutaneous DCs does. On reaching the LNs, pulmonary DCs instruct T cells to respond to a particular antigen in the most appropriate way. For many years, it was thought that airway DCs induced a predominantly Th2-type response, preventing the classical Th1 response, which is characterized by tissue damage. Indeed, lung DCs have been shown to produce Th2-polarizing cytokines, such as IL-10 and IL-6, in some studies, whereas the production of IL-12, a Th1-inducing cytokine, is impaired (Stumbles et al., 1998; Dodge et al., 2003).

The immune response initiated by pulmonary DCs is influenced not only by the antigens inhaled, but also by numerous molecular signals originating from all of the cell types present in the airways and lung parenchyma, such as airway and alveolar epithelial cells, alveolar and interstitial macrophages, lymphocytes, endothelium, and mast cells. Macrophages and DCs are closely connected in the lung parenchyma, and alveolar macrophages have been reported to produce soluble mediators inhibiting DC functions. In particular, nitric oxide, prostaglandins, H_2O_2, TGF-β, IL-10, and decoy receptors for IL-1 and TNF-α are secreted, inhibiting MHC II upregulation and preventing T-cell stimulation (Holt et al., 1993; Lee et al., 2001). Damage to the epithelial barrier would allow the inhaled antigens to reach the pulmonary DCs, triggering DC activation. In addition, interstitial macrophages could break particulate antigens into smaller peptides, with which the pulmonary DCs could be loaded. Most allergens are immunologically inert and result in the establishment of tolerance rather than immunity. There is increasing evidence to suggest that the balance between tolerance and immunity is controlled by DCs (Steinman et al., 2003; Wakkach et al., 2003; Smits et al., 2005), particularly in the lung (Cochand et al., 1999; Steinman and Nussenzweig, 2002). Under steady-state conditions, inhaled antigens in the airways of naive mice are taken up by pulmonary DCs, which then migrate to the T-cell-rich area of the mediastinal LNs. Within the LNs, DCs induce T-cell anergy and apoptosis (Steinman and Nussenzweig, 2002). In the presence of inflammation, fully mature DCs reach the LNs, inducing the potent proliferation of antigen-specific T cells, which differentiate into effector cells.

An important mechanism in the maintenance of peripheral tolerance is the suppressor activity of T-reg cells, a heterogeneous population of T cells with immunosuppressive function. The generation of T-reg cells in vitro and in vivo is mediated by immunosuppressive cytokines, such as IL-10 and TGF-β. Various studies have shown that pulmonary DCs challenged by respiratory exposure to antigen produce IL-10, mediating inhalation tolerance through expansion of the IL-10-producing T-reg cell population (Akbari et al., 2001). Pulmonary DCs in the bronchial LNs of mice exposed to respiratory allergen induce the development of T-reg cells in a process requiring T-cell costimulation via the inducible costimulator ICOS-ICOS-ligand pathway (Akbari et al., 2002). The essential role of "tolerogenic DCs" in the lungs is clearly illustrated by the mechanism of asthma (Lewkowich et al., 2005). Indeed, humans with allergies have significantly fewer T-reg cells of more limited function than individuals without allergy (Kuipers and Lambrecht, 2004). Moreover, the T-cell-mediated tolerance induced by respiratory exposure to allergen can inhibit the development of airway hyperreactivity (Akbari et al., 2001).

In mice and humans, pulmonary DCs and macrophages control highly relevant lung pathogen infections, by triggering a specific local immune response. M. *tuberculosis* is the causal agent of tuberculosis (TB), a contagious disease of the respiratory system affecting one-third of the world's population. Like the common cold, M. *tuberculosis* is rapidly spread through the air by coughing and sneezing. Not all infected subjects contract TB disease: 90% of human carriers are asymptomatic. However, if the immune system is weakened because of other bacterial or viral infections, the chances of developing TB become very high. Activated alveolar macrophages and recruited DCs control the response to M. *tuberculosis* (Kaufmann, 2001), leading to Th1-cell responses mediated by IL-12 production (Nigou et al., 2001; Giacomini et al., 2001). M. *tuberculosis* binding to macrophages involves complement receptor (CR) 1, CR3, mannose receptors, CD14, and scavenger and TLR receptors, which are essential for the interaction of M. *tuberculosis* with phagocytic cells (Brightbill et al., 1999). Some of these receptors are present on DCs and may be involved in the binding and entry of mycobacteria. The C-type lectin DC-SIGN, which recognizes the mannose-rich molecule ManLAM, a component of the mycobacterial cell wall, is one of the major M. *tuberculosis* receptors expressed on DCs in the airway mucosa, in particular, at the submucosal and interstitial sites of the respiratory tract (Tailleux et al., 2003b). M. *tuberculosis* is captured and internalized by DC-SIGN, resulting in its presence in lysosomal compartments. After an initial reactive phase, M. *tuberculosis* infections are restricted by the cellular immune response and enter a chronic latent phase in the host. Various mechanisms of immune escape have been proposed. It has been reported that human monocyte-derived DCs, unlike macrophages, are not permissive, and that they therefore block the intracellular growth of virulent M. *tuberculosis* (Tailleux et al., 2003a). Mycobacteria-containing phagosomes may mature to late endosomes/lysosomes in DCs, resulting in degradation, whereas phagosome maturation may be arrested at an early endosomal stage in mycobacteria-containing macrophages, thereby promoting the growth of mycobacteria. It has also been demonstrated that DC-SIGN binding to ManLAM impairs DC maturation and induces production of the anti-inflammatory cytokine IL-10, leading to immunosuppression and the intracellular survival of M. *tuberculosis* (Geijtenbeek et al., 2003). However, it remains unclear how M. *tuberculosis* maintains DC infections and disseminates outside the lungs.

THE LN AND DCs

Activated antigen-loaded DCs migrate to the T-cell zone of secondary lymphoid organs to prime T-cell responses. The requirement of DCs for T-cell activation has been demonstrated in vivo for CD8$^+$ T lymphocytes. Mice temporarily deprived of CD11c$^+$ DCs are unable to mount efficient specific CD8$^+$ T-cell responses to infections with the intracellular bacterium *Listeria monocytogenes*, the parasite *Plasmodium yoelii*, lymphocytic choriomeningitis virus

(LCMV), or antigen immunization (Jung et al., 2002; Probst and van den Broek, 2005). DCs can enter the LN via the blood or lymph (Cavanagh and Von Andrian, 2002). They are initially found clustered together close to high endothelial venules. However, their distribution subsequently changes, such that, one day after reaching the LN, DCs are distributed throughout the T-cell area (Mempel et al., 2004). The interaction of DCs with $CD8^+$ T cells has been monitored in vivo over a 48-hour period following the entry of T cells into the LN. Immature DCs were injected into the footpads of mice, with or without peptide pulsing. The mice were also treated with lipopolysaccharide (LPS) to allow the migration of DCs to draining LN. Antigen-specific T cells were then injected intravenously into the mice, 18 hours after DC administration. DC-T cell interactions were analyzed over time by using intravital multiphoton microscopy (Sumen et al., 2004). In the first 8 hours after their entry into the LN, T cells rapidly scan many different DCs, establishing only short interactions of no more than a few minutes.

The T cells then become less motile, forming contacts with DCs that last longer than 1 hour. Long-lasting contacts between DCs and T cells continue for the rest of the first day after T-cell entry into the LN, until these cells begin to proliferate (Mempel et al., 2004). Stable DC-T cell interactions are necessary for the induction of T-cell priming (Hugues et al., 2004), and stable DC-T cell contacts are associated with the formation of an immunological synapse (Grakoui et al., 1999). The capacity of DCs to prime T-cell responses has not been attributed to one particular DC-specific, surface, or secreted molecule, but to a combination of multiple factors, such as the high level of expression of cell membrane costimulatory molecules, the secretion of specific cytokines, and the efficient antigen-processing machinery acquired by DCs during maturation.

In addition to their well established role in priming T lymphocytes, DCs also have the capacity to activate B-cell responses (Banchereau et al., 2000). Human tonsillar interdigitating DCs have been shown to interact with B cells in situ (Bjorck et al., 1997). In vivo, DCs have been shown to take up the antigen, retain it unprocessed, and transfer it to naive B cells to initiate an antigen-specific response (Wykes et al., 1998). The migration of DCs to the B-cell zone in the LN may be controlled by CXCR5 (Wu and Hwang, 2002). DCs induce the proliferation and IgM secretion of B cells activated via CD40 in vitro (Dubois et al., 1997).

DCs are also involved in inducing the B-cell class switch (Dubois et al., 1999). After exposure to various molecular mediators, such as IFN-γ, CD40 ligand, LPS, and IFN-α, marginal zone DCs upregulate the expression of members of the TNF family, such as B-lymphocyte stimulator protein (BlyS) and proliferation-inducing ligand (APRIL). In the presence of IL-4, IL-10, or TGF-β, activated DCs expressing BlyS and APRIL can induce B-cell class switch recombination to Cγ, Cα, and Cε in a CD40-independent manner. The synergy of BlyS and APRIL with IL-15 is required for the activation of transcription factors, including NF-κB, leading to efficient B-cell proliferation and antibody production (Litinskiy et al., 2002).

Finally, DCs in the LN may also encounter NK cells (Moretta, 2002). Activated DCs migrating to the draining LN following the specific upregulation of the CC-chemokine receptor CCR7 (Weninger and von Andrian, 2003) can also stimulate resident and newly recruited NK cells. Indeed, in the T-cell area of human LNs, the colocalization of NK cells and DCs has been reported (Ferlazzo et al., 2004). Inflammatory stimuli can induce NK cell migration to the draining LN via the expression of CXCR3. Activated DCs thus produce CXCR3 ligands (CXCR3Ls), favoring the recruitment of NK cells (Martin-Fontecha et al., 2004). Consistent with this hypothesis, we found that DCs activated with microbial stimuli (LPS and CpG) known to cause the accumulation of NK cells in the draining LN produce large amounts of two CXCR3Ls—CXCL9 and CXCL10. This suggests that the accumulation of NK cells in draining LNs may actually be due to DC-mediated CXCR3-dependent NK cell recruitment (Zanoni et al., 2005). DC-mediated NK cell activation requires cell-cell contact, and the induction of IFN-γ production by NK cells seems to occur through immunological synapses between DCs and NK cells (Grakoui et al., 1999; Borg et al., 2004). Under appropriate conditions, NK cells may also contribute to DC activation. Indeed, IL-2-activated NK cells can promote DC maturation, as shown by the upregulation of costimulatory and MHC molecules and the production of inflammatory cytokines.

THE THYMUS AND THYMIC DCs

Central tolerance is established in the thymus. Tolerance to tissue antigens is achieved through a combination of thymic and peripheral events eliminating or inactivating potentially dangerous T cells (Stockinger, 1999). The thymus is responsible for a very important initial step in the elimination of potentially dangerous self-specific T cells (Liston et al., 2003). Positive T-cell selection and negative T-cell selection are considered to be qualitatively different processes dependent on thymic compartmentalization and the cellular context of T-cell receptor (TCR)-MHC interaction (Laufer et al., 1999). Different thymic cell types can provide T cells with qualitatively different signals, such that positive and negative selection can occur sequentially following the interaction of T cells with thymic stromal cells (Laufer et al., 1999). Three different stromal cell types are present in the thymus: cortical epithelial cells, medullary epithelial cells, and bone marrow-derived cells, including DCs, macrophages, and B cells. It has been shown, in mouse models in which the expression of MHC-peptide complexes is limited to particular thymic cell types, that only the cortical epithelium can induce the positive selection of both autoreactive and nonautoreactive thymocytes (Laufer et al., 1996; Capone et al., 2001). Negative selection therefore follows positive selection, and occurs mostly in the corticomedullary junction and within the thymic medulla (Murphy et al., 1990; Surh and Sprent, 1994). In these regions, negative selection is mediated by medullary epithelial cells (Kyewski and Derbinski, 2004) and antigen-presenting cells (APCs) (Marrack et al., 1988). The capacity of medullary epithelial cells for negative selection has been attributed to their ability to express tissue-specific genes (Gotter and Kyewski, 2004). It has been suggested that the promiscuous gene expression of thymic epithelial cells is regulated by the expression of a particular transcription factor—transcriptional regulator autoimmune regulator (Aire) (Gotter and Kyewski, 2004). Aire-deficient mice display very low levels of expression for many tissue-specific genes (Anderson et al., 2002). These mice develop tissue-specific antibodies and lymphoid cell infiltration in many peripheral organs (Anderson et al., 2002). Moreover, the deletion of CD4$^+$ T lymphocytes specific for a self-antigen expressed under the rat insulin promoter is abolished in Aire-deficient mice (Liston et al., 2003).

The function of DCs in thymic central tolerance has been investigated in various experimental settings. DCs loaded with small amounts of C5, the fifth component of complement, or spontaneously presenting C5 can, together with thymic epithelial cells, delete C5-specific T lymphocytes in culture (Zal et al., 1994; Volkmann et al., 1997). Moreover, the efficient negative selection of I-E reactive Vβ5$^+$ and Vβ11$^+$ CD4$^+$ T cells has been described in an in vivo model in which MHC class II is selectively expressed by DCs under control of the CD11c promoter (Brocker et al., 1997). The role of these cells in tolerance induction has been demonstrated in the context of CD8$^+$ T cells in a similar model, in which MHC class I was expressed exclusively by DCs (Cannarile et al., 2004). It therefore seems likely that thymic DCs specialize in the induction of tolerance and cannot positively select either CD4$^+$ or CD8$^+$ T cells.

THE SPLEEN AND SPLENIC DCs

DCs have been found at different locations within the spleen. The anatomy of the spleen is complex and different from that of other lymphoid organs: its surface is smooth, with the exception of an indented region—the hilus—where the blood vessels enter and leave, and the spleen is connected solely to the bloodstream, whereas the LNs are connected to the lymph circulation. Like the LNs, the spleen is enclosed in a capsule and enmeshed in trabeculae. Its interior contains a delicate, three-dimensional network of reticulate tissue. The mesh of this network is filled with parenchyma, the free cells of the spleen. The trabecular arteries are surrounded by a lymphoid tissue, the periarteriolar lymphoid sheath (PALS), which has an inner and an outer region, differing in their cellular composition. The inner PALS contains only T lymphocytes, whereas the outer PALS contains mostly T lymphocytes but also some B cells. There are lymphoid follicles, spherical structures similar to those found in the LNs, throughout the PALS. These follicles may be primary or secondary, with secondary follicles containing germinal centers surrounded by the mantle. The PALS and its follicles form the spleen white pulp, which is clearly distinguished from the red pulp. A minor population of nonphagocytic cells, the interdigiting cells (IDCs), has been described in the T-cell zone of the PALS (Veerman and van Ewijk, 1975). Most splenic DCs (up to 75%) are derived from the edge of the T-cell zone and are called marginal DCs (Agger et al., 1992) because they originate from the so-called marginal zone, which is thickest over the lymphoid follicles. IDCs seem to be more efficient activators of T-cell responses in vitro than marginal DCs (Steinman and Swanson, 1995). However, it remains unclear which of these cells are the most efficient for antigen presentation in vivo. Marginal DCs are highly phagocytic in vivo and have a much higher turnover rate. In contrast, IDCs are not phagocytic in vivo and their turnover rate is slow (Leenen et al., 1998). The functional implications of these observations remain unclear. Circulating T cells seem to be activated in the marginal zone of the white pulp and must gain access to and migrate to T-cell areas of the follicle to initiate the adaptive response.

The splenic DCs of mice form two subsets: the CD8α$^+$ and CD8α$^-$ subsets, both of which express high levels of CD11c, MHC II, and the costimulatory molecules CD86 and CD40 (Vremec et al., 1992; Anjuere et al., 1999). CD8α$^+$ and CD8α$^-$ DCs display phenotypic differences and reside in different areas of the spleen. CD8α$^-$ cells are more abundant and are found in the marginal zone-bridging channels of the splenic lymphoid follicles. They can be induced to migrate to the PALS under the influence of proinflammatory signals, such as LPS or parasite extracts. CD8α$^+$ DCs are located in the T-cell-rich areas of the PALS, although recent studies have suggested a broader distribution in the spleen (Steinman et al., 1997; Banchereau et al., 2000; Neuenhahn et al., 2006). This CD8α$^+$ DC subset seems to be the only subset able to cross-present cell-associated antigens via the MHC class I pathway (den Haan et al., 2000; Dudziak et al., 2007). Furthermore, a substantial fraction of differentially expressed genes have been identified between splenic CD8α$^-$ and CD8α$^+$ DCs (Edwards et al., 2003; Dudziak et al., 2007). Thus, CD8α$^-$ and CD8α$^+$ DCs may be functionally different, although this hypothesis remains to be definitively demonstrated (Shortman and Liu, 2002). The development of both CD8α$^-$ and CD8α$^+$ DCs in the spleen marks this organ out as different from other organs. Late DC progenitors were recently purified from the spleen (Naik et al., 2006). These cells, accounting for 0.05% of splenocytes, included cells "precommitted" to form either CD8α$^-$ or CD8α$^+$ DCs, but not plasmacytoid DCs or cells of other lineages. The splenic precursor population could be distinguished from monocytes or blood monoblasts. Indeed, these progenitors did not respond to M-CSF, were poorly phagocytic, and differed from monocytes in surface phenotype and morphology, suggesting that the spleen contains a reservoir of immediate DC precursors, making it possible to generate substantial numbers of DCs without immediate replenishment in the form of precursors migrating from BM.

Human spleen DCs have been investigated in much less detail because of availability problems. Fresh human spleen DCs have been described as large, HLA-DR$^+$ cells, lacking the markers of all other cell lineages. CD11c^{+high} DCs are distributed in three different regions, the periarteriolar T-cell zones, the B-cell zones, and the marginal zone, in which they form a ring of cells surrounding the white pulp. This ring is located within a ring of CD14$^+$ red pulp macrophages, which appear to be more regularly organized than the marginal DC population of the mouse spleen (McIlroy et al., 2001). In humans, the T-cell zone DCs express CD11c, MHC II, and CD83. Germinal center DCs have been observed in the B-cell zones (Grouard et al., 1996). Like mouse spleen DCs, human marginal zone DCs play a sentinel role, and are able to capture and process blood-borne antigens very efficiently during their passage through the marginal zone (McIlroy et al., 2001). For this reason, the spleen is believed to control systemic immunity.

THE COMMON EFFECTOR FUNCTION OF MYELOID DCs

Microbial Uptake

DCs are activated either directly by the recognition of microbes through germ line-encoded pattern recognition receptors (PRRs) (Janeway and Medzhitov, 2002) or indirectly via microbial products generated during an infection. The PRRs recognize the molecular patterns of common bacteria and viruses (Takeda et al., 2003). Following microbe recognition, DCs undergo differentiation and produce chemokines and cytokines, such as IL-1, TNF-α, and type I IFNs (Winzler et al., 1997; Foti et al., 1999; Gallucci et al., 1999; Trevejo et al., 2001; Montoya et al., 2002). DCs may also be induced to mature by binding to Fc receptors (Jurgens et al., 1995; Regnault et al., 1999; Geissmann et al., 2001) or exposure to necrotic cells (Gallucci et al., 1999; Sauter et al., 2000). Thus, the generation of necrotic cells and/or the binding of opsonized bacteria to the Fc receptor may also stimulate DC maturation during bacterial infection. Furthermore, activated CD4$^+$ T cells induce DC differentiation via members of the TNF receptor family (Muraille et al., 2002; van Kooten and Banchereau, 2000). However, although several receptors recognize microbial structures, the TLRs are the only PRRs identified to date that directly mediate full DC maturation. TLRs recognize conserved microbe-associated molecules, such as LPS, flagellin, double-stranded RNA, and bacterial DNA containing CpG motifs (Akira, 2001). Binding to the TLRs induces nuclear translocation of the proinflammatory transcription factor NF-κB (Medzhitov and Janeway, 1997) via activation of an adaptor protein, MyD88 (Akira, 2001). However, MyD88-independent signaling has also been described (Kaisho et al., 2001; Kawai et al., 2001). PRRs other than TLRs are used by DCs for microbial uptake. These receptors include lectin receptors, such as the mannose receptor, and scavenger receptors. In addition, Fc and complement receptors are used for the internalization of bacterial products bound to antibodies and complement, respectively (Underhill and Ozinsky, 2002).

Despite several early reports on the uptake of particulate material and cells by DCs, it has long been thought that DCs have no phagocytic capacity. This view arose partly from the technical difficulty of growing immature DCs. Early procedures for DC purification and culture in vitro tended to result in DC-enriched preparations containing mostly terminally differentiated cells with a mature phenotype and function. When it became possible to grow and maintain immature mouse DCs in vitro (Winzler et al., 1997), these cells were found to have marked phagocytic activity that decreased with increasing DC differentiation. DCs have been shown to internalize latex and zymosan beads (Inaba et al., 1993; Reis e Sousa et al., 1993), apoptotic bodies (Parr et al., 1991), and a number of microbes, including many bacteria and viruses, in vitro and in vivo (Rescigno et al., 2000). In vivo, L. monocytogenes and serovar Typhimurium have been colocalized with DCs in intestinal PPs after oral infection (Hopkins et al., 2000; Pron et al., 2001) and L. monocytogenes has also been found in the DCs of the mesenteric LNs 6 to 12 h after infection (Pron et al., 2001). Splenic DCs have been found to be associated with serovar Typhimurium, Mycobacterium bovis, or bacillus Calmette-Guérin (BCG) 4 h after intravenous infection (Yrlid et al., 2001; Jiao et al., 2002). The phagocytic activity of DCs is designed to facilitate antigen processing and presentation. DCs have no bacterial scavenging function and are highly inefficient at bacterial clearance, but they do have unique mechanisms for antigen processing and for the antigen loading of MHC molecules.

Innate Responses and NK Cell Activation

DCs are part of the innate response but were first described as "natural adjuvants" capable of inducing adaptive immune responses (Steinman, 1991; Ibrahim et al., 1995; Banchereau and Steinman, 1998). Because they are found in tissues interfacing with the external environment, such as the skin, the gut, and the lungs (Sertl et al., 1986; Nestle et al., 1993; Nelson et al., 1994), they are also able to recruit and activate cells of the innate immune system (Rescigno et al., 1998; Fernandez et al., 1999; Foti et al., 1999). Indeed, before DCs can prime the adaptive immune response, they must complete a full maturation process that is initiated by direct exposure to TLR ligands or to other receptors of the innate receptor repertoire. Interaction with pathogens results in DC activation and migration to the T-cell area of LNs, where the antigen-specific cells of the adaptive immune response can be primed. Given the high plasticity of DCs, the signals determining the function of a particular DC and the type of adaptive immune response developed depend mostly on the local microenvironment and the interaction between DCs and microbial signals. These interactions are complex and differ considerably between pathogens. Genome-wide expression analyses have shown that, after interaction with microbes, DCs undergo extensive gene reprogramming. For instance, the transcriptional signature induced in mouse DCs by pathogens, such as *Schistosoma mansoni* and *Leishmania mexicana* or gram-positive bacteria, includes both common and unique regulatory networks (Trottein et al., 2004; Aebischer et al., 2005). This suggests an underlying mechanism in which the host-pathogen interaction is translated into an appropriate host inflammatory response. Indeed, we observed an inflammatory program characterized by the induction of type I IFNs followed by the upregulation of IFN-induced inflammatory molecules (Trottein et al., 2004). Moreover, the induction of a proinflammatory cytokine transcript signature, including transcripts for TNF-α, and chemokines, such as IP-10 (CXCL10), MCP-5 (CCL12), MIP-1α (CCL3), MIP-1β (CCL4), MIP-1γ (CCL9), and MIP-2 (CXCL2), is known to attract granulocytes, immature DCs, NK cells, and activated T cells. These data indicate that myeloid DCs may mediate type I IFN signaling and may be a source of IP-10 and MIP-1α production.

One of the most unanticipated findings of global gene expression analysis was the demonstration that myeloid DCs produce IL-2 transiently after encountering microbes. Such IL-2 production was first observed in DCs stimulated with gram-negative bacteria (Granucci et al., 2001b). It was later shown that many different microbial stimuli induce IL-2 production by DCs, whereas none of the inflammatory cytokines tested elicited IL-2 production by DCs (Granucci et al., 2003a). IL-2 production by DCs was transiently upregulated soon after these cells came into contact with bacteria, whereas macrophages did not produce IL-2 on bacterial stimulation. We have shown that IL-2 production by DCs controls NK cell-mediated immunity in vitro and in vivo (Granucci et al., 2004; Zanoni et al., 2005). Thus, in addition to its well-defined function in acquired immunity, IL-2 is also required for the regulation of innate immune responses, at least in response to bacterial infections. Thus, IL-2 is an additional key cytokine conferring unique NK and T-cell stimulatory properties on DCs (Granucci et al., 2001a, 2004).

Adaptive Responses and T-Cell Activation

DCs are considered to be professional APCs because they can initiate adaptive immune responses (Banchereau and Steinman, 1998). DC maturation is associated with an increase in the production of inflammatory cytokines and chemokines, a decrease in endocytic and phagocytic capacity, and the acquisition of migratory functions enabling antigen-loaded DCs to move from the marginal zones to the T-cell areas in LNs or from nonlymphoid to lymphoid tissues. After migration, mature DCs display strong cell surface MHC and costimulatory protein expression, and are able to activate $CD8^+$ and $CD4^+$ T-cell responses.

Stimuli inducing DC activation and maturation increase the efficiency of antigen processing for both the class I and class II pathways and the half-life of peptide-MHC complexes at the cell surface that would otherwise be rapidly internalized and recycled (Cella et al., 1997; Pierre et al., 1997; Rescigno et al., 1998). DCs also have a large number of the so-called multivesicular bodies (MVBs)—vesicles formed by invagination of the limiting MVB membrane and containing a pool of MHC class II molecules. In contrast, the accessory molecules H-2M and DM are physically segregated into the MVB-limiting membrane. After the activation of DCs by contact with a microbe, extensive MVB vesicle fusion occurs, resulting in the formation of long tubular compartments in which MHC class II molecules and DM molecules are closely associated and directed toward the tips of the dendrites. This reorganization of the MVBs facilitates the efficient loading of class II molecules with exogenous peptides generated in the early endosomes and the presentation of complexes at the cell surface (Kleijmeer et al., 2001).

Another interesting feature of the specific processing machinery of DCs is the ability of these cells to delay the processing of internalized antigens by antigen retention in a mildly acidic storage compartment (Lutz et al., 1997). Internalized antigens in these vesicles are not immediately degraded and fusion with the lysosomes is delayed. This mechanism seems to be coordinated with the generation of newly synthesized MHC class I molecules 12 to 18 h after DC activation (Rescigno et al., 1998). It remains unclear how degraded antigens from the exogenous pathway are loaded onto MHC class I molecules, although it has been suggested that this loading may involve endocytoplasmic reticulum-phagosome fusion (Guermonprez et al., 2003).

The induction of cytotoxic T lymphocyte (CTL) responses requires the presentation of antigen-derived peptides in complex with MHC I to specific $CD8^+$ T cells. The presented antigens are typically derived from the cytosol of the APCs. However, exogenous antigens may also gain access to the cytosol and may be loaded onto post Golgi MHC I (Bachmann et al., 1996; Rescigno et al., 1998; Reimann and Schirmbeck, 1999; Wick and Ljunggren, 1999). The importance of DCs for the induction of CTL responses in vivo was clearly demonstrated by Jung et al. (2002), who used a diphtheria toxin-based system for DC depletion. The resulting DC-depleted mice were completely devoid of CTL responses to *L. monocytogenes* and *P. yoelii* infections, which normally generate powerful CTL responses. The requirement of DCs for the activation of CTLs in vivo was also demonstrated in a study in which naive $CD8^+$ T cells were shown to cluster around vaccinia virus-infected DCs, but not infected macrophages, in a tissue-draining LN. This clustering was observed despite infected macrophages outnumbering infected DCs by approximately 10:1 (Norbury et al., 2002). These data are consistent with a role for DCs in priming naive $CD8^+$ T cells during infection. In contrast, the role of macrophages in stimulating $CD8^+$ T cells remains unclear for microbial antigens.

Adaptive Responses and T-Cell Regulation
Many tissue proteins are not expressed at sufficiently high levels in the thymus to induce clonal deletion or tolerance (Avery et al. 1995). Peripheral T cells must therefore be continually rendered tolerant, as shown by the existence of autoreactive T cells that actively recognize antigens in the periphery and undergo anergy (Schwartz, 2003), deletion (Burkly et al., 1990), or TCR downregulation (Schonrich et al., 1991). Several models have been proposed to explain the induction of tolerance in autoreactive T cells. Considerable evidence has accumulated over the past 20 years (Mueller et al., 1989) to suggest that the decision to initiate an adaptive immune response is taken by the APC rather than the antigen-specific T cell. According to the infectious nonself and noninfectious self model, APCs are maintained in a resting state until they encounter microbes or microbial cell products, which activate them and induce upregulation of the costimulatory molecules required for peripheral T-cell activation (Medzhitov et al., 1997). In the "danger" model (Matzinger, 1994), APC activation is mediated by danger signals derived from injured cells, such as those exposed to pathogens or possible stress signals. In the presence of danger, APCs are activated and express signals leading to T-cell activation. In the absence of danger, APCs are not activated and T cells (antigen-primed or naive) interacting with these resting APCs die because of a lack of costimulation (Matzinger, 2002). The danger model has been experimentally tested with antigen-presenting DCs, and it has been suggested that the activation state of DCs is relevant to the decision as to whether immune responses should be abolished or activated (Steinman et al., 2003). In particular, immature or CD40-activated DCs expressing the scavenger receptor CD205 have been targeted with antigen in vivo and the fate of $CD4^+$ and $CD8^+$ antigen-specific T cells has been monitored over time (Hawiger et al., 2001; Bonifaz et al., 2002). T lymphocytes are rendered tolerant when they encounter immature antigen-loaded DCs, whereas these cells are activated when they encounter activated antigen-presenting DCs. A model allowing inducible antigen presentation by resting or activated DCs has been designed (Probst et al., 2003). In this model, three different LCMV-derived CTL epitopes could be presented by 5% of the total DC population after induction. The presentation of LCMV-derived CTL epitopes by resting DCs resulted in antigen-specific tolerance, which was not broken down by subsequent infection with LCMV. Conversely, antigen presentation by activated DCs induced CTL activation and protective memory (Probst et al., 2003). Tolerance induction depended on the synergic effects of the PD-1 and CTLA-4 molecules (Probst et al., 2005). None of these models is entirely satisfactory, because other interpretations of the results are possible. For example, the role of antigen persistence is not assessed in these models, but may play a fundamental role in maintaining peripheral tolerance (Raimondi et al., 2006a, 2006b).

DCs have also been implicated in the functional control of T-reg cells. These lymphocytes are involved in maintaining tolerance to self-antigens by inhibiting responses mediated by effector $CD4^+$ and $CD8^+$ T cells (Sakaguchi, 2000). Some T-reg cell populations (those expressing CD4 and IL-2 receptor α chain [CD25]) originate in the thymus, whereas others (those producing IL-10) differentiate in the periphery (Roncarolo and Levings, 2000; Roncarolo et al., 2003). It has been suggested that DCs are involved in the peripheral differentiation of T-reg cells. In particular, it has been observed that repetitive stimulation of naive T cells with allogeneic immature DCs may result in the generation of IL-10-producing anergic T cells able to abolish effector

T-cell functions (Jonuleit et al., 2000). The generation of T-reg cells requires the production of IL-10 by immature DCs (Levings et al., 2005). Similar results have been obtained in vivo in humans. The injection of immature DCs pulsed with the influenza peptide induced the differentiation of peptide-specific IL-10-producing T cells and the disappearance of influenza-specific effector $CD8^+$ T cells. In contrast, a single injection of peptide-pulsed mature DCs led to the rapid expansion of specific T-lymphocyte populations (Dhodapkar et al., 2001). Moreover, semimaturation stimuli, such as TNF-α, can induce a particular population of semimature DCs, sustaining the peripheral differentiation of T-reg cells. Repetitive injections of BM-derived TNF-α-activated DCs prevent the development of experimental autoimmune encephalomyelitis (EAE), by inducing IL-10-producing T-reg cells (Menges et al., 2002). Thus, T-cell interaction with antigen-presenting immature or semimature DCs may induce the peripheral differentiation of T-reg lymphocytes.

In addition to their role in peripheral T-reg cell differentiation, DCs have been shown to affect the function of thymus-derived $CD4^+CD25^+$ T-reg cells (Steinman et al., 2003; Guiducci et al., 2005). Mature DCs are, indeed, able to induce expansion of $CD4^+CD25^+$ T-reg cell populations both in vitro and in vivo in the presence of specific antigen or IL-2 (Yamazaki et al., 2003). Following T-reg cell population expansion in response to contact with DCs, the inhibitory activity of this population is greater in vitro than ex vivo (Tarbell et al., 2004).

It has been suggested that DC-derived IL-2 may play an important role in controlling T-reg cell functions (Malek and Bayer, 2004), as IL-2 has been shown to be required for T-reg cell functionality. In this experiment, IL-2- or CD25-deficient T cells were transferred into mice harboring a monoclonal myelin basic protein-specific αβ T-cell repertoire and spontaneously developing experimental autoimmune encephalomyelitis (Lafaille et al., 1994). IL-2-deficient T cells protected against the disease, whereas CD25-negative lymphocytes did not, indicating that IL-2 is not required for the generation of T-reg cells in the thymus but is necessary for the correct functioning of T-reg cells in the periphery. Finally, DCs were recently shown to be involved in maintaining $CD4^+CD25^+$ cell homeostasis, through mechanisms involving cell-cell contact, CD40-CD40L interaction, and IL-2 production (Guiducci et al., 2005).

In conclusion, as discussed, there is growing evidence that DC plasticity is the result of both endogenous and exogenous signals; in addition, depending on the microenvironment tissue, DCs might be conditioned to acquire very different functional properties leading to immunity or tolerance.

This work was supported by fellowships and grants from the Italian Ministry of Education and Research (FIRB and COFIN projects), and M. Curie Chair (P. Ricciardi-Castagnoli).

REFERENCES

Adolfsson, J., O. J. Borge, D. Bryder, K. Theilgaard-Mönch, I. Astrand-Grundström, E. Sitnicka, Y. Sasaki, and S. E. Jacobsen. 2001. Upregulation of Flt3 expression within the bone marrow Lin(-)Sca1(+)c-kit(+) stem cell compartment is accompanied by loss of self-renewal capacity. *Immunity* 15:659–669.

Aebischer, T., C. L. Bennett, M. Pelizzola, C. Vizzardelli, N. Pavelka, M. Urbano, M. Capozzoli, A. Luchini, T. Ilg, F. Granucci, C. C. Blackburn, and P. Ricciardi-Castagnoli. 2005. A critical role for lipophosphoglycan in proinflammatory responses of dendritic cells to Leishmania mexicana. *Eur. J. Immunol.* 35:476–486.

Agger, R., M. Witmer-Pack, N. Romani, H. Stossel, W. J. Swiggard, J. P. Metlay, E. Storozynsky, P. Freimuth, and R. M. Steinman. 1992. Two populations of splenic dendritic cells detected with M342, a new monoclonal to an intracellular antigen of interdigitating dendritic cells and some B lymphocytes. *J. Leukoc. Biol.* 52:34–42.

Akbari, O., R. H. DeKruyff, and D. T. Umetsu. 2001. Pulmonary dendritic cells producing IL-10 mediate tolerance induced by respiratory exposure to antigen. *Nat. Immunol.* 2:725–731.

Akbari, O., G. J. Freeman, E. H. Meyer, E. A. Greenfield, T. T. Chang, A. H. Sharpe, G. Berry, R. H. DeKruyff, and D. T. Umetsu. 2002. Antigen-specific regulatory T cells develop via the ICOS-ICOS-ligand pathway and inhibit allergen-induced airway hyperreactivity. *Nat. Med.* 8:1024–1032.

Akira, S. 2001. Toll-like receptors and innate immunity. *Adv. Immunol.* 78:1–56.

Anderson, M. S., E. S. Venanzi, L. Klein, Z. Chen, S. P. Berzins, S. J. Turley, H. von Boehmer, R. Bronson, A. Dierich, C. Benoist, and D. Mathis. 2002. Projection of an immunological self shadow within the thymus by the aire protein. *Science* 298:1395–1401.

Anjuere, F., P. Martin, I. Ferrero, M. L. Fraga, G. M. del Hoyo, N. Wright, and C. Ardavín. 1999. Definition of dendritic cell subpopulations present in the spleen, Peyer's patches, lymph nodes, and skin of the mouse. *Blood* 93:590–598.

Avery, A. C., Z. S. Zhao, A. Rodriguez, E. K. Bikoff, M. Soheilian, C. S. Foster, and H. Cantor. 1995. Resistance to herpes stromal keratitis conferred by an IgG2a-derived peptide. *Nature* 376:431–434.

Bachmann, M. F., M. B. Lutz, G. T. Layton, S. J. Harris, T. Fehr, M. Rescigno, and P. Ricciardi-Castagnoli. 1996. Dendritic cells process exogenous viral proteins and virus-like particles for class I presentation to CD8+ cytotoxic T lymphocytes. *Eur. J. Immunol.* 26:2595–2600.

Banchereau, J., F. Briere, C. Caux, J. Davoust, S. Lebecque, Y. J. Liu, B. Pulendran, and K. Palucka. 2000. Immunobiology of dendritic cells. *Annu. Rev. Immunol.* 18:767–811.

Banchereau, J., and R. M. Steinman. 1998. Dendritic cells and the control of immunity. *Nature* 392:245–252.

Benson, M. J., K. Pino-Lagos, M. Rosemblatt, and R. J. Noelle. 2007. All-trans retinoic acid mediates enhanced T reg cell growth, differentiation, and gut homing in the face of high levels of co-stimulation. *J. Exp. Med.* 204:1765–1774.

Beutler, B. 2004. Inferences, questions and possibilities in Toll-like receptor signalling. *Nature* 430:257–263.

Bimczok, D., E. N. Sowa, H. Faber-Zuschratter, R. Pabst, and H. J. Rothkötter. 2005. Site-specific expression of CD11b and SIRPalpha (CD172a) on dendritic cells: implications for their migration patterns in the gut immune system. *Eur. J. Immunol.* 35:1418–1427.

Bjorck, P., L. Flores-Romo, and Y. J. Liu. 1997. Human interdigitating dendritic cells directly stimulate CD40-activated naive B cells. *Eur. J. Immunol.* 27:1266–1274.

Bogunovic, M., F. Ginhoux, A. Wagers, M. Loubeau, L. M. Isola, L. Lubrano, V. Najfeld, R. G. Phelps, C. Grosskreutz, E. Scigliano, P. S. Frenette, and M. Merad. 2006. Identification of a radio-resistant and cycling dermal dendritic cell population in mice and men. *J. Exp. Med.* 203:2627–2638.

Bonifaz, L., D. Bonnyay, K. Mahnke, M. Rivera, M. C. Nussenzweig, and R. M. Steinman. 2002. Efficient targeting of protein antigen to the dendritic cell receptor DEC-205 in the steady state leads to antigen presentation on major histocompatibility complex class I products and peripheral CD8+ T cell tolerance. *J. Exp. Med.* **196:**1627–1638.

Borg, C., A. Jalil, D. Laderach, K. Maruyama, H. Wakasugi, S. Charrier, B. Ryffel, A. Cambi, C. Figdor, W. Vainchenker, A. Galy, A. Caignard, and L. Zitvogel. 2004. NK cell activation by dendritic cells (DCs) requires the formation of a synapse leading to IL-12 polarization in DCs. *Blood* **104:**3267–3275.

Bouma, G., and W. Strober. 2003. The immunological and genetic basis of inflammatory bowel disease. *Nat. Rev. Immunol.* **3:**521–533.

Brandtzaeg, P., E. S. Baekkevold, and H. C. Morton. 2001. From B to A the mucosal way. *Nat Immunol* **2:**1093–1094.

Brightbill, H. D., D. H. Libraty, S. R. Krutzik, R. B. Yang, J. T. Belisle, J. R. Bleharski, M. Maitland, M. V. Norgard, S. E. Plevy, S. T. Smale, P. J. Brennan, B. R. Bloom, P. J. Godowski, and R. L. Modlin. 1999. Host defense mechanisms triggered by microbial lipoproteins through toll-like receptors. *Science* **285:**732–736.

Brocker, T., M. Riedinger, and K. Karjalainen. 1997. Targeted expression of major histocompatibility complex (MHC) class II molecules demonstrates that dendritic cells can induce negative but not positive selection of thymocytes in vivo. *J. Exp. Med.* **185:**541–550.

Burkly, L. C., D. Lo, and R. A. Flavell. 1990. Tolerance in transgenic mice expressing major histocompatibility molecules extrathymically on pancreatic cells. *Science* **248:**1364–1368.

Butcher, E. C., and L. J. Picker. 1996. Lymphocyte homing and homeostasis. *Science* **272:**60–66.

Butcher, E. C., M. Williams, K. Youngman, L. Rott, and M. Briskin. 1999. Lymphocyte trafficking and regional immunity. *Adv. Immunol.* **72:**209–253.

Cannarile, M. A., N. Decanis, J. P. van Meerwijk, and T. Brocker. 2004. The role of dendritic cells in selection of classical and nonclassical CD8+ T cells in vivo. *J. Immunol.* **173:**4799–4805.

Capone, M., P. Romagnoli, F. Beermann, H. R. MacDonald, and J. P. van Meerwijk. 2001. Dissociation of thymic positive and negative selection in transgenic mice expressing major histocompatibility complex class I molecules exclusively on thymic cortical epithelial cells. *Blood* **97:**1336–1342.

Castigli, E., S. Scott, F. Dedeoglu, P. Bryce, H. Jabara, A. K. Bhan, E. Mizoguchi, and R. S. Geha. 2004. Impaired IgA class switching in APRIL-deficient mice. *Proc. Natl. Acad. Sci. USA* **101:**3903–3908.

Castigli, E., S. A. Wilson, S. Scott, F. Dedeoglu, S. Xu, K. P. Lam, R. J. Bram, H. Jabara, and R. S. Geha. 2005. TACI and BAFF-R mediate isotype switching in B cells. *J. Exp. Med.* **201:**35–39.

Cavanagh, L. L., and U. H. Von Andrian. 2002. Travellers in many guises: the origins and destinations of dendritic cells. *Immunol. Cell Biol.* **80:**448–462.

Cella, M., A. Engering, V. Pinet, J. Pieters, and A. Lanzavecchia. 1997. Inflammatory stimuli induce accumulation of MHC class II complexes on dendritic cells. *Nature* **388:**782–787.

Chehimi, J., S. E. Starr, H. Kawashima, D. S. Miller, G. Trinchieri, B. Perussia, and S. Bandyopadhyay. 1989. Dendritic cells and IFN-alpha-producing cells are two functionally distinct non-B, non-monocytic HLA-DR+ cell subsets in human peripheral blood. *Immunology* **68:**486–490.

Chen, W., W. Jin, N. Hardegen, K. J. Lei, L. Li, N. Marinos, G. McGrady, and S. M. Wahl. 2003. Conversion of peripheral CD4+CD25- naive T cells to CD4+CD25+ regulatory T cells by TGF-beta induction of transcription factor Foxp3. *J. Exp. Med.* **198:**1875–1886.

Chieppa, M., M. Rescigno, A. Y. Huang, and R. N. Germain. 2006. Dynamic imaging of dendritic cell extension into the small bowel lumen in response to epithelial cell TLR engagement. *J. Exp. Med.* **203:**2841–2852.

Chirdo, F. G., O. R. Millington, H. Beacock-Sharp, and A. M. Mowat. 2005. Immunomodulatory dendritic cells in intestinal lamina propria. *Eur. J. Immunol.* **35:**1831–1840.

Cochand, L., P. Isler, F. Songeon, and L. P. Nicod. 1999. Human lung dendritic cells have an immature phenotype with efficient mannose receptors. *Am. J. Respir. Cell Mol. Biol.* **21:**547–554.

Cole, A. M., and A. J. Waring. 2002. The role of defensins in lung biology and therapy. *Am. J. Respir. Med.* **1:**249–259.

Coombes, J. L., K. R. Siddiqui, C. V. Arancibia-Cárcamo, J. Hall, C. M. Sun, Y. Belkaid, and F. Powrie. 2007. A functionally specialized population of mucosal CD103+ DCs induces Foxp3+ regulatory T cells via a TGF-(1)- and retinoic acid-dependent mechanism. *J. Exp. Med.* **204:**1757–1764.

D'Amico, A., and L. Wu. 2003. The early progenitors of mouse dendritic cells and plasmacytoid predendritic cells are within the bone marrow hemopoietic precursors expressing Flt3. *J. Exp. Med.* **198:**293–303.

Demedts, I. K., G. G. Brusselle, K. Y. Vermaelen, and R. A. Pauwels. 2005. Identification and characterization of human pulmonary dendritic cells. *Am. J. Respir. Cell Mol. Biol.* **32:**177–184.

den Haan, J. M., S. M. Lehar, and M. J. Bevan. 2000. CD8(+) but not CD8(-) dendritic cells cross-prime cytotoxic T cells in vivo. *J. Exp. Med.* **192:**1685–1696.

Dhodapkar, M. V., R. M. Steinman, J. Krasovsky, C. Munz, and N. Bhardwaj. 2001. Antigen-specific inhibition of effector T cell function in humans after injection of immature dendritic cells. *J. Exp. Med.* **193:**233–238.

Dodge, I. L., M. W. Carr, M. Cernadas, and M. B. Brenner. 2003. IL-6 production by pulmonary dendritic cells impedes Th1 immune responses. *J. Immunol.* **170:**4457–4464.

Dubois, B., J. M. Bridon, J. Fayette, C. Barthélémy, J. Banchereau, C. Caux, and F. Brière. 1999. Dendritic cells directly modulate B cell growth and differentiation. *J. Leukoc. Biol.* **66:**224–230.

Dubois, B., B. Vanbervliet, J. Fayette, C. Massacrier, C. Van Kooten, F. Brière, J. Banchereau, and C. Caux. 1997. Dendritic cells enhance growth and differentiation of CD40-activated B lymphocytes. *J. Exp. Med.* **185:**941–951.

Dudziak, D., A. O. Kamphorst, G. F. Heidkamp, V. R. Buchholz, C. Trumpfheller, S. Yamazaki, C. Cheong, K. Liu, H. W. Lee, C. G. Park, R. M. Steinman, and M. C. Nussenzweig. 2007. Differential antigen processing by dendritic cell subsets in vivo. *Science* **315:**107–111.

Edwards, A. D., D. Chaussabel, S. Tomlinson, O. Schulz, A. Sher, and C. Reis e Sousa. 2003. Relationships among murine CD11c(high) dendritic cell subsets as revealed by baseline gene expression patterns. *J. Immunol.* **171:**47–60.

Ehrhardt, R. O., B. R. Ludviksson, B. Gray, M. Neurath, and W. Strober. 1997. Induction and prevention of colonic inflammation in IL-2-deficient mice. *J. Immunol.* **158:**566–573.

Engering, A., T. B. Geijtenbeek, and Y. van Kooyk. 2002. Immune escape through C-type lectins on dendritic cells. *Trends Immunol.* **23:**480–485.

Fagarasan, S., and T. Honjo. 2003. Intestinal IgA synthesis: regulation of front-line body defences. *Nat. Rev. Immunol.* **3:**63–72.

Fantini, M. C., C. Becker, G. Monteleone, F. Pallone, P. R. Galle, and M. F. Neurath. 2004. Cutting edge: TGF-beta induces a regulatory phenotype in CD4+CD25- T cells through Foxp3 induction and down-regulation of Smad7. *J. Immunol.* **172:**5149–5153.

Ferlazzo, G., M. Pack, D. Thomas, C. Paludan, D. Schmid, T. Strowig, G. Bougras, W. A. Muller, L. Moretta, and C. Münz. 2004. Distinct roles of IL-12 and IL-15 in human natural killer cell activation by dendritic cells from secondary lymphoid organs. *Proc. Natl. Acad. Sci. USA* **101:**16606–16611.

Fernandez, N. C., A. Lozier, C. Flament, P. Ricciardi-Castagnoli, D. Bellet, M. Suter, M. Perricaudet, T. Tursz, E. Maraskovsky, and L. Zitvogel. 1999. Dendritic cells directly trigger NK cell functions: cross-talk relevant in innate anti-tumor immune responses in vivo. *Nat. Med.* **5:**405–411.

Fogg, D. K., C. Sibon, C. Miled, S. Jung, P. Aucouturier, D. R. Littman, A. Cumano, and F. Geissmann. 2006. A clonogenic bone marrow progenitor specific for macrophages and dendritic cells. *Science* **311:**83–87.

Foti, M., F. Granucci, D. Aggujaro, E. Liboi, W. Luini, S. Minardi, A. Mantovani, S. Sozzani, and P. Ricciardi-Castagnoli. 1999. Upon dendritic cell (DC) activation chemokines and chemokine receptor expression are rapidly regulated for recruitment and maintenance of DC at the inflammatory site. *Int. Immunol.* **11:**979–986.

Foti, M., F. Granucci, M. Pelizzola, O. Beretta, and P. Ricciardi-Castagnoli. 2006. Dendritic cells in pathogen recognition and induction of immune responses: a functional genomics approach. *J. Leukoc. Biol.* **79:**913–916.

Gallucci, S., M. Lolkema, and P. Matzinger. 1999. Natural adjuvants: endogenous activators of dendritic cells. *Nat. Med.* **5:**1249–1255.

Geijtenbeek, T. B., S. J. Van Vliet, E. A. Koppel, M. Sanchez-Hernandez, C. M. Vandenbroucke-Grauls, B. Appelmelk, and Y. Van Kooyk. 2003. Mycobacteria target DC-SIGN to suppress dendritic cell function. *J. Exp. Med.* **197:**7–17.

Geissmann, F., S. Jung, and D. R. Littman. 2003. Blood monocytes consist of two principal subsets with distinct migratory properties. *Immunity* **19:**71–82.

Geissmann, F., P. Launay, B. Pasquier, Y. Lepelletier, M. Leborgne, A. Lehuen, N. Brousse, and R. C. Monteiro. 2001. A subset of human dendritic cells expresses IgA Fc receptor (CD89), which mediates internalization and activation upon cross-linking by IgA complexes. *J. Immunol.* **166:**346–352.

Geissmann, F., C. Prost, J. P. Monnet, M. Dy, N. Brousse, and O. Hermine. 1998. Transforming growth factor beta1, in the presence of granulocyte/macrophage colony-stimulating factor and interleukin 4, induces differentiation of human peripheral blood monocytes into dendritic Langerhans cells. *J. Exp. Med.* **187:**961–966.

Giacomini, E., E. Iona, L. Ferroni, M. Miettinen, L. Fattorini, G. Orefici, I. Julkunen, and E. M. Coccia. 2001. Infection of human macrophages and dendritic cells with Mycobacterium tuberculosis induces a differential cytokine gene expression that modulates T cell response. *J. Immunol.* **166:**7033–7041.

Ginhoux, F., F. Tacke, V. Angeli, M. Bogunovic, M. Loubeau, X. M. Dai, E. R. Stanley, G. J. Randolph, and M. Merad. 2006. Langerhans cells arise from monocytes in vivo. *Nat. Immunol.* **7:**265–273.

Gotter, J., and B. Kyewski. 2004. Regulating self-tolerance by deregulating gene expression. *Curr. Opin. Immunol.* **16:**741–745.

Goud, T. J., C. Schotte, and R. van Furth. 1975. Identification and characterization of the monoblast in mononuclear phagocyte colonies grown in vitro. *J. Exp. Med.* **142:**1180–1199.

Grakoui, A., S. K. Bromley, C. Sumen, M. M. Davis, A. S. Shaw, P. M. Allen, and M. L. Dustin. 1999. The immunological synapse: a molecular machine controlling T cell activation. *Science* **285:**221–227.

Granucci, F., P. R. Castagnoli, L. Rogge, and F. Sinigaglia. 2001a. Gene expression profiling in immune cells using microarray. *Int. Arch. Allergy Immunol.* **126:**257–266.

Granucci, F., S. Feau, V. Angeli, F. Trottein, and P. Ricciardi-Castagnoli. 2003a. Early IL-2 production by mouse dendritic cells is the result of microbial-induced priming. *J. Immunol.* **170:**5075–5081.

Granucci, F., S. Feau, I. Zanoni, N. Pavelka, C. Vizzardelli, G. Raimondi, and P. Ricciardi-Castagnoli. 2003b. The immune response is initiated by dendritic cells via interaction with microorganisms and interleukin-2 production. *J. Infect. Dis.* **187**(Suppl. 2)**:**S346–S350.

Granucci, F., F. Petralia, M. Urbano, S. Citterio, F. Di Tota, L. Santambrogio, and P. Ricciardi-Castagnoli. 2003c. The scavenger receptor MARCO mediates cytoskeleton rearrangements in dendritic cells and microglia. *Blood* **102:**2940–2947.

Granucci, F., C. Vizzardelli, E. Virzi, M. Rescigno, and P. Ricciardi-Castagnoli. 2001b. Transcriptional reprogramming of dendritic cells by differentiation stimuli. *Eur. J. Immunol.* **31:**2539–2546.

Granucci, F., I. Zanoni, N. Pavelka, S. L. Van Dommelen, C. E. Andoniou, F. Belardelli, M. A. Degli Esposti, and P. Ricciardi-Castagnoli. 2004. A contribution of mouse dendritic cell-derived IL-2 for NK cell activation. *J. Exp. Med.* **200:**287–295.

Grouard, G., I. Durand, L. Filgueira, J. Banchereau, and Y. J. Liu. 1996. Dendritic cells capable of stimulating T cells in germinal centres. *Nature* **384:**364–367.

Guermonprez, P., L. Saveanu, M. Kleijmeer, J. Davoust, P. Van Endert, and S. Amigorena. 2003. ER-phagosome fusion defines an MHC class I cross-presentation compartment in dendritic cells. *Nature* **425:**397–402.

Guiducci, C., B. Valzasina, H. Dislich, and M. P. Colombo. 2005. CD40/CD40L interaction regulates CD4(+)CD25(+) T reg homeostasis through dendritic cell-produced IL-2. *Eur. J. Immunol.* **35:**557–567.

Hammad, H., H. J. de Heer, T. Soullie, H. C. Hoogsteden, F. Trottein, and B. N. Lambrecht. 2003. Prostaglandin D2 inhibits airway dendritic cell migration and function in steady state conditions by selective activation of the D prostanoid receptor 1. *J. Immunol.* **171:**3936–3940.

Harder, J., U. Meyer-Hoffert, L. M. Teran, L. Schwichtenberg, J. Bartels, S. Maune, and J. M. Schröder. 2000. Mucoid Pseudomonas aeruginosa, TNF-alpha, and IL-1beta, but not IL-6, induce human beta-defensin-2 in respiratory epithelia. *Am. J. Respir. Cell Mol. Biol.* **22:**714–721.

Hawiger, D., K. Inaba, Y. Dorsett, M. Guo, K. Mahnke, M. Rivera, J. V. Ravetch, R. M. Steinman, and M. C. Nussenzweig. 2001. Dendritic cells induce peripheral T cell unresponsiveness under steady state conditions in vivo. *J. Exp. Med.* **194:**769–779.

Hofer, S., M. Rescigno, F. Granucci, S. Citterio, M. Francolini, and P. Ricciardi-Castagnoli. 2001. Differential activation of NF-kappa B subunits in dendritic cells in response to Gram-negative bacteria and to lipopolysaccharide. *Microbes Infect.* **3:**259–265.

Holt, P. G., S. Haining, D. J. Nelson, and J. D. Sedgwick. 1994. Origin and steady-state turnover of class II MHC-bearing dendritic cells in the epithelium of the conducting airways. *J. Immunol.* **153:**256–261.

Holt, P. G., J. Oliver, N. Bilyk, C. McMenamin, P. G. McMenamin, G. Kraal, and T. Thepen. 1993. Downregulation of the antigen presenting cell function(s) of pulmonary dendritic cells in vivo by resident alveolar macrophages. *J. Exp. Med.* **177:**397–407.

Hopkins, S. A., F. Niedergang, I. E. Corthesy-Theulaz, and J. P. Kraehenbuhl. 2000. A recombinant Salmonella typhimurium vaccine strain is taken up and survives within murine Peyer's patch dendritic cells. *Cell. Microbiol.* **2:**59–68.

Hugues, S., L. Fetler, L. Bonifaz, J. Helft, F. Amblard, and S. Amigorena. 2004. Distinct T cell dynamics in lymph nodes during the induction of tolerance and immunity. *Nat. Immunol.* **5:**1235–1242.

Ibrahim, M. A., B. M. Chain, and D. R. Katz. 1995. The injured cell: the role of the dendritic cell system as a sentinel receptor pathway. *Immunol. Today* **16:**181–186.

Inaba, K., M. Inaba, M. Naito, and R. M. Steinman. 1993. Dendritic cell progenitors phagocytose particulates, including bacillus Calmette-Guerin organisms, and sensitize mice to mycobacterial antigens in vivo. *J. Exp. Med.* **178:**479–488.

Itakura, M., A. Tokuda, H. Kimura, S. Nagai, H. Yoneyama, N. Onai, S. Ishikawa, T. Kuriyama, and K. Matsushima. 2001. Blockade of secondary lymphoid tissue chemokine exacerbates Propionibacterium acnes-induced acute lung inflammation. *J. Immunol.* **166:**2071–2079.

Iwata, M., A. Hirakiyama, Y. Eshima, H. Kagechika, C. Kato, and S. Y. Song. 2004. Retinoic acid imprints gut-homing specificity on T cells. *Immunity* **21:**527–538.

Izcue, A., J. L. Coombes, and F. Powrie. 2006. Regulatory T cells suppress systemic and mucosal immune activation to control intestinal inflammation. *Immunol. Rev.* **212:**256–271.

Jakubzick, C., F. Tacke, J. Llodra, N. van Rooijen, and G. J. Randolph. 2006. Modulation of dendritic cell trafficking to and from the airways. *J. Immunol.* **176:**3578–3584.

Janeway, C. A., Jr., and R. Medzhitov. 2002. Innate immune recognition. *Annu. Rev. Immunol.* **20:**197–216.

Jiao, X., R. Lo-Man, P. Guermonprez, L. Fiette, E. Dériaud, S. Burgaud, B. Gicquel, N. Winter, and C. Leclerc. 2002. Dendritic cells are host cells for mycobacteria in vivo that trigger innate and acquired immunity. *J. Immunol.* **168:**1294–1301.

Jones, B. D., N. Ghori, and S. Falkow. 1994. Salmonella typhimurium initiates murine infection by penetrating and destroying the specialized epithelial M cells of the Peyer's patches. *J. Exp. Med.* **180:**15–23.

Jonuleit, H., E. Schmitt, G. Schuler, J. Knop, and A. H. Enk. 2000. Induction of interleukin 10-producing, nonproliferating CD4(+) T cells with regulatory properties by repetitive stimulation with allogeneic immature human dendritic cells. *J. Exp. Med.* **192:**1213–1222.

Jung, S., D. Unutmaz, P. Wong, G. Sano, K. De los Santos, T. Sparwasser, S. Wu, S. Vuthoori, K. Ko, F. Zavala, E. G. Pamer, D. R. Littman, and R. A. Lang. 2002. In vivo depletion of CD11c(+) dendritic cells abrogates priming of CD8(+) T cells by exogenous cell-associated antigens. *Immunity* **17:**211–220.

Jung, Y. W., T. R. Schoeb, C. T. Weaver, and D. D. Chaplin. 2006. Th17: an effector CD4 T cell lineage with regulatory T cell ties. *Immunity* **24:**677–688.

Jurgens, M., A. Wollenberg, D. Hanau, H. de la Salle, and T. Bieber. 1995. Activation of human epidermal Langerhans cells by engagement of the high affinity receptor for IgE, Fc epsilon RI. *J. Immunol.* **155:**5184–5189.

Kabashima, K., T. A. Banks, K. M. Ansel, T. T. Lu, C. F. Ware, and J. G. Cyster. 2005. Intrinsic lymphotoxin-beta receptor requirement for homeostasis of lymphoid tissue dendritic cells. *Immunity* **22:**439–450.

Kaisho, T., O. Takeuchi, T. Kawai, K. Hoshino, and S. Akira. 2001. Endotoxin-induced maturation of MyD88-deficient dendritic cells. *J. Immunol.* **166:**5688–5694.

Kamath, A. T., S. Henri, F. Battye, D. F. Tough, and K. Shortman. 2002. Developmental kinetics and lifespan of dendritic cells in mouse lymphoid organs. *Blood* **100:**1734–1741.

Kanitakis, J., P. Petruzzo, and J. M. Dubernard. 2004. Turnover of epidermal Langerhans' cells. *N. Engl. J. Med.* **351:**2661–2662.

Karsunky, H., M. Merad, A. Cozzio, I. L. Weissman, and M. G. Manz. 2003. Flt3 ligand regulates dendritic cell development from Flt3+ lymphoid and myeloid-committed progenitors to Flt3+ dendritic cells in vivo. *J. Exp. Med.* **198:**305–313.

Katz, S. I., K. Tamaki, and D. H. Sachs. 1979. Epidermal Langerhans cells are derived from cells originating in bone marrow. *Nature* **282:**324–326.

Kaufmann, S. H. 2001. How can immunology contribute to the control of tuberculosis? *Nat. Rev. Immunol.* **1:**20–30.

Kawai, T., O. Takeuchi, T. Fujita, J. Inoue, P. F. Mühlradt, S. Sato, K. Hoshino, and S. Akira. 2001. Lipopolysaccharide stimulates the MyD88-independent pathway and results in activation of IFN-regulatory factor 3 and the expression of a subset of lipopolysaccharide-inducible genes. *J. Immunol.* **167:**5887–5894.

Kelly, D., and S. Conway. 2005. Bacterial modulation of mucosal innate immunity. *Mol. Immunol.* **42:**895–901.

Kerneis, S., A. Bogdanova, J. P. Kraehenbuhl, and E. Pringault. 1997. Conversion by Peyer's patch lymphocytes of human enterocytes into M cells that transport bacteria. *Science* **277:**949–952.

Kissenpfennig, A., S. Ait-Yahia, V. Clair-Moninot, H. Stössel, E. Badell, Y. Bordat, J. L. Pooley, T. Lang, E. Prina, I. Coste, O. Gresser, T. Renno, N. Winter, G. Milon, K. Shortman, N. Romani, S. Lebecque, B. Malissen, S. Saeland, and P. Douillard. 2005a. Disruption of the langerin/CD207 gene abolishes Birbeck granules without a marked loss of Langerhans cell function. *Mol. Cell. Biol.* **25:**88–99.

Kissenpfennig, A., S. Henri, B. Dubois, C. Laplace-Builhé, P. Perrin, N. Romani, C. H. Tripp, P. Douillard, L. Leserman, D. Kaiserlian, S. Saeland, J. Davoust, and B. Malissen. 2005b. Dynamics and function of Langerhans cells in vivo: dermal dendritic cells colonize lymph node areas distinct from slower migrating Langerhans cells. *Immunity* **22:**643–654.

Kleijmeer, M., G. Ramm, D. Schuurhuis, J. Griffith, M. Rescigno, P. Ricciardi-Castagnoli, A. Y. Rudensky, F. Ossendorp, C. J. Melief, W. Stoorvogel, and H. J. Geuze. 2001. Reorganization of multivesicular bodies regulates MHC class II antigen presentation by dendritic cells. *J. Cell Biol.* **155:**53–63.

Kraal, G., J. N. Samsom, and R. E. Mebius. 2006. The importance of regional lymph nodes for mucosal tolerance. *Immunol. Rev.* **213:**119–130.

Kuipers, H. and B. N. Lambrecht. 2004. The interplay of dendritic cells, Th2 cells and regulatory T cells in asthma. *Curr. Opin. Immunol.* **16:**702–708.

Kyewski, B., and J. Derbinski. 2004. Self-representation in the thymus: an extended view. *Nat. Rev. Immunol.* **4:**688–698.

Lafaille, J. J., K. Nagashima, M. Katsuki, and S. Tonegawa. 1994. High incidence of spontaneous autoimmune encephalomyelitis in immunodeficient anti-myelin basic protein T cell receptor transgenic mice. *Cell* **78:**399–408.

Langerhans, P. 1868. Über die Nerven der menschlichen Haut. *Virchows Arch. Path. Anat.* **44:**325–337.

Laufer, T. M., J. DeKoning, J. S. Markowitz, D. Lo, and L. H. Glimcher. 1996. Unopposed positive selection and autoreactivity in mice expressing class II MHC only on thymic cortex. *Nature* **383:**81–85.

Laufer, T. M., L. H. Glimcher, and D. Lo. 1999. Using thymus anatomy to dissect T cell repertoire selection. *Semin. Immunol.* **11**:65–70.

Lee, P. T., P. G. Holt, and A. S. McWilliam. 2001. Ontogeny of rat pulmonary alveolar macrophage function: evidence for a selective deficiency in il-10 and nitric oxide production by newborn alveolar macrophages. *Cytokine* **15**:53–57.

Leenen, P. J., K. Radosevic, J. S. Voerman, B. Salomon, N. van Rooijen, D. Klatzmann, and W. van Ewijk. 1998. Heterogeneity of mouse spleen dendritic cells: in vivo phagocytic activity, expression of macrophage markers, and subpopulation turnover. *J. Immunol.* **160**:2166–2173.

Leung, D. Y., M. Boguniewicz, M. D. Howell, I. Nomura, and Q. A. Hamid. 2004. New insights into atopic dermatitis. *J. Clin. Invest.* **113**:651–657.

Levings, M. K., S. Gregori, E. Tresoldi, S. Cazzaniga, C. Bonini, and M. G. Roncarolo. 2005. Differentiation of Tr1 cells by immature dendritic cells requires IL-10 but not CD25+CD4+ Tr cells. *Blood* **105**:1162–1169.

Lewkowich, I. P., N. S. Herman, K. W. Schleifer, M. P. Dance, B. L. Chen, K. M. Dienger, A. A. Sproles, J. S. Shah, J. Köhl, Y. Belkaid, and M. Wills-Karp. 2005. CD4+CD25+ T cells protect against experimentally induced asthma and alter pulmonary dendritic cell phenotype and function. *J. Exp. Med.* **202**:1549–1561.

Liston, A., S. Lesage, J. Wilson, L. Peltonen, and C. C. Goodnow. 2003. Aire regulates negative selection of organ-specific T cells. *Nat. Immunol.* **4**:350–354.

Litinskiy, M. B., B. Nardelli, D. M. Hilbert, B. He, A. Schaffer, P. Casali, and A. Cerutti. 2002. DCs induce CD40-independent immunoglobulin class switching through BLyS and APRIL. *Nat. Immunol.* **3**:822–829.

Liu, K., C. Waskow, X. Liu, K. Yao, J. Hoh, and M. Nussenzweig. 2007. Origin of dendritic cells in peripheral lymphoid organs of mice. *Nat. Immunol.* **8**:578–583.

Lotz, M., D. Gutle, S. Walther, S. Ménard, C. Bogdan, and M. W. Hornef. 2006. Postnatal acquisition of endotoxin tolerance in intestinal epithelial cells. *J. Exp. Med.* **203**:973–984.

Lutz, M. B., P. Rovere, M. J. Kleijmeer, M. Rescigno, C. U. Assmann, V. M. Oorschot, H. J. Geuze, J. Trucy, D. Demandolx, J. Davoust, and P. Ricciardi-Castagnoli. 1997. Intracellular routes and selective retention of antigens in mildly acidic cathepsin D/lysosome-associated membrane protein-1/MHC class II-positive vesicles in immature dendritic cells. *J. Immunol.* **159**:3707–3716.

Macpherson, A. J., and T. Uhr. 2004. Induction of protective IgA by intestinal dendritic cells carrying commensal bacteria. *Science* **303**:1662–1665.

Malek, T. R., and A. L. Bayer. 2004. Tolerance, not immunity, crucially depends on IL-2. *Nat. Rev. Immunol.* **4**:665–674.

Marrack, P., D. Lo, R. Brinster, R. Palmiter, L. Burkly, R. H. Flavell, and J. Kappler. 1988. The effect of thymus environment on T cell development and tolerance. *Cell* **53**:627–634.

Martin-Fontecha, A., L. L. Thomsen, S. Brett, C. Gerard, M. Lipp, A. Lanzavecchia, and F. Sallusto. 2004. Induced recruitment of NK cells to lymph nodes provides IFN-gamma for T(H)1 priming. *Nat. Immunol.* **5**:1260–1265.

Matzinger, P. 1994. Tolerance, danger, and the extended family. *Annu. Rev. Immunol.* **12**:991–1045.

Matzinger, P. 2002. An innate sense of danger. *Ann. N. Y. Acad. Sci.* **961**:341–342.

McIlroy, D., C. Troadec, F. Grassi, A. Samri, B. Barrou, B. Autran, P. Debré, J. Feuillard, and A. Hosmalin. 2001. Investigation of human spleen dendritic cell phenotype and distribution reveals evidence of in vivo activation in a subset of organ donors. *Blood* **97**:470–477.

Medzhitov, R., and C. A. Janeway, Jr. 1997. Innate immunity: impact on the adaptive immune response. *Curr. Opin. Immunol.* **9**:4–9.

Medzhitov, R., P. Preston-Hurlburt, and C. A. Janeway, Jr. 1997. A human homologue of the Drosophila Toll protein signals activation of adaptive immunity. *Nature* **388**:394–397.

Mempel, T. R., S. E. Henrickson, and U. H. Von Andrian. 2004. T-cell priming by dendritic cells in lymph nodes occurs in three distinct phases. *Nature* **427**:154–159.

Menges, M., S. Rossner, C. Voigtländer, H. Schindler, N. A. Kukutsch, C. Bogdan, K. Erb, G. Schuler, and M. B. Lutz. 2002. Repetitive injections of dendritic cells matured with tumor necrosis factor alpha induce antigen-specific protection of mice from autoimmunity. *J. Exp. Med.* **195**:15–21.

Merad, M., M. G. Manz, H. Karsunky, A. Wagers, W. Peters, I. Charo, I. L. Weissman, J. G. Cyster, and E. G. Engleman. 2002. Langerhans cells renew in the skin throughout life under steady-state conditions. *Nat. Immunol.* **3**:1135–1141.

Metchinkoff, E. 1887. Uber den Kampf der Zellen gegen Erysipelkokken, ein Beitrag zur Phagocytenlehre. *Arch. Pathol. Anat.* **107**:209–249.

Mohamadzadeh, M., F. Berard, G. Essert, C. Chalouni, B. Pulendran, J. Davoust, G. Bridges, A. K. Palucka, and J. Banchereau. 2001. Interleukin 15 skews monocyte differentiation into dendritic cells with features of Langerhans cells. *J. Exp. Med.* **194**:1013–1020.

Montoya, M., G. Schiavoni, F. Mattei, I. Gresser, F. Belardelli, P. Borrow, and D. F. Tough. 2002. Type I interferons produced by dendritic cells promote their phenotypic and functional activation. *Blood* **99**:3263–3271.

Mora, J. R., M. R. Bono, N. Manjunath, W. Weninger, L. L. Cavanagh, M. Rosemblatt, and U. H. Von Andrian. 2003. Selective imprinting of gut-homing T cells by Peyer's patch dendritic cells. *Nature* **424**:88–93.

Mora, J. R., M. Iwata, B. Eksteen, S. Y. Song, T. Junt, B. Senman, K. L. Otipoby, A. Yokota, H. Takeuchi, P. Ricciardi-Castagnoli, K. Rajewsky, D. H. Adams, and U. H. von Andrian. 2006. Generation of gut-homing IgA-secreting B cells by intestinal dendritic cells. *Science* **314**:1157–1160.

Moretta, A. 2002. Natural killer cells and dendritic cells: rendezvous in abused tissues. *Nat. Rev. Immunol.* **2**:957–964.

Mowat, A. M. 2003. Anatomical basis of tolerance and immunity to intestinal antigens. *Nat. Rev. Immunol.* **3**:331–341.

Mucida, D., N. Kutchukhidze, A. Erazo, M. Russo, J. J. Lafaille, and M. A. Curotto de Lafaille. 2005. Oral tolerance in the absence of naturally occurring Tregs. *J. Clin. Invest.* **115**:1923–1933.

Mueller, D. L., M. K. Jenkins, and R. H. Schwartz. 1989. Clonal expansion versus functional clonal inactivation: a costimulatory signalling pathway determines the outcome of T cell antigen receptor occupancy. *Annu. Rev. Immunol.* **7**:445–480.

Muraille, E., C. De Trez, B. Pajak, M. Brait, J. Urbain, and O. Leo. 2002. T cell-dependent maturation of dendritic cells in response to bacterial superantigens. *J. Immunol.* **168**:4352–4360.

Murphy, K. M., A. B. Heimberger, and D. Y. Loh. 1990. Induction by antigen of intrathymic apoptosis of CD4+CD8+TCRlo thymocytes in vivo. *Science* **250**:1720–1723.

Naik, S. H., D. Metcalf, A. van Nieuwenhuijze, I. Wicks, L. Wu, M. O'Keeffe, and K. Shortman. 2006. Intrasplenic steady-state dendritic cell precursors that are distinct from monocytes. *Nat. Immunol.* **7**:663–671.

Neish, A. S., A. T. Gewirtz, H. Zeng, A. N. Young, M. E. Hobert, V. Karmali, A. S. Rao, and J. L. Madara. 2000. Prokaryotic regulation of epithelial responses by inhibition of IkappaB-alpha ubiquitination. *Science* **289:**1560–1563.

Nelson, D. J., C. McMenamin, A. S. McWilliam, M. Brenan, and P. G. Holt. 1994. Development of the airway intraepithelial dendritic cell network in the rat from class II major histocompatibility (Ia)-negative precursors: differential regulation of Ia expression at different levels of the respiratory tract. *J. Exp. Med.* **179:**203–212.

Nestle, F. O., X. G. Zheng, C. B. Thompson, L. A. Turka, and B. J. Nickoloff. 1993. Characterization of dermal dendritic cells obtained from normal human skin reveals phenotypic and functionally distinctive subsets. *J. Immunol.* **151:**6535–6545.

Neuenhahn, M., K. M. Kerksiek, M. Nauerth, M. H. Suhre, M. Schiemann, F. E. Gebhardt, C. Stemberger, K. Panthel, S. Schröder, T. Chakraborty, S. Jung, H. Hochrein, H. Rüssmann, T. Brocker, and D. H. Busch. 2006. CD8alpha+ dendritic cells are required for efficient entry of Listeria monocytogenes into the spleen. *Immunity* **25:**619–630.

Niedergang, F., and J. P. Kraehenbuhl. 2000. Much ado about M cells. *Trends Cell Biol.* **10:**137–141.

Niess, J. H., S. Brand, X. Gu, L. Landsman, S. Jung, B. A. McCormick, J. M. Vyas, M. Boes, H. L. Ploegh, J. G. Fox, D. R. Littman, and H. C. Reinecker. 2005. CX3CR1-mediated dendritic cell access to the intestinal lumen and bacterial clearance. *Science* **307:**254–258.

Nigou, J., C. Zelle-Rieser, M. Gilleron, M. Thurnher, and G. Puzo. 2001. Mannosylated lipoarabinomannans inhibit IL-12 production by human dendritic cells: evidence for a negative signal delivered through the mannose receptor. *J. Immunol.* **166:**7477–7485.

Nomura, I., B. Gao, M. Boguniewicz, M. A. Darst, J. B. Travers, and D. Y. Leung. 2003a. Distinct patterns of gene expression in the skin lesions of atopic dermatitis and psoriasis: a gene microarray analysis. *J. Allergy Clin. Immunol.* **112:**1195–1202.

Nomura, I., E. Goleva, M. D. Howell, Q. A. Hamid, P. Y. Ong, C. F. Hall, M. A. Darst, B. Gao, M. Boguniewicz, J. B. Travers, and D. Y. Leung. 2003b. Cytokine milieu of atopic dermatitis, as compared to psoriasis, skin prevents induction of innate immune response genes. *J. Immunol.* **171:**3262–3269.

Norbury, C. C., D. Malide, J. S. Gibbs, J. R. Bennink, and J. W. Yewdell. 2002. Visualizing priming of virus-specific CD8+ T cells by infected dendritic cells in vivo. *Nat. Immunol.* **3:**265–271.

Palframan, R. T., S. Jung, G. Cheng, W. Weninger, Y. Luo, M. Dorf, D. R. Littman, B. J. Rollins, H. Zweerink, A. Rot, and U. H. von Andrian. 2001. Inflammatory chemokine transport and presentation in HEV: a remote control mechanism for monocyte recruitment to lymph nodes in inflamed tissues. *J. Exp. Med.* **194:**1361–1373.

Parr, M. B., L. Kepple, and E. L. Parr. 1991. Langerhans cells phagocytose vaginal epithelial cells undergoing apoptosis during the murine estrous cycle. *Biol. Reprod.* **45:**252–260.

Passlick, B., D. Flieger, and H. W. Ziegler-Heitbrock. 1989. Identification and characterization of a novel monocyte subpopulation in human peripheral blood. *Blood* **74:**2527–2434.

Peters, W., H. M. Scott, H. F. Chambers, J. L. Flynn, I. F. Charo, and J. D. Ernst. 2001. Chemokine receptor 2 serves an early and essential role in resistance to Mycobacterium tuberculosis. *Proc. Natl. Acad. Sci. USA* **98:**7958–7963.

Pierre, P., S. J. Turley, E. Gatti, M. Hull, J. Meltzer, A. Mirza, K. Inaba, R. M. Steinman, and I. Mellman. 1997. Developmental regulation of MHC class II transport in mouse dendritic cells. *Nature* **388:**787–792.

Probst, H. C., J. Lagnel, G. Kollias, and M. van den Broek. 2003. Inducible transgenic mice reveal resting dendritic cells as potent inducers of CD8+ T cell tolerance. *Immunity* **18:**713–720.

Probst, H. C., K. McCoy, T. Okazaki, T. Honjo, and M. van den Broek. 2005. Resting dendritic cells induce peripheral CD8+ T cell tolerance through PD-1 and CTLA-4. *Nat. Immunol.* **6:**280–286.

Probst, H. C., and M. van den Broek. 2005. Priming of CTLs by lymphocytic choriomeningitis virus depends on dendritic cells. *J. Immunol.* **174:**3920–3924.

Pron, B., C. Boumaila, F. Jaubert, P. Berche, G. Milon, F. Geissmann, and J. L. Gaillard. 2001. Dendritic cells are early cellular targets of Listeria monocytogenes after intestinal delivery and are involved in bacterial spread in the host. *Cell. Microbiol.* **3:**331–340.

Raimondi, G., I. Zanoni, S. Citterio, P. Ricciardi-Castagnoli, and F. Granucci. 2006a. Induction of peripheral T cell tolerance by antigen-presenting B cells. I. Relevance of antigen presentation persistence. *J. Immunol.* **176:**4012–4020.

Raimondi, G., I. Zanoni, S. Citterio, P. Ricciardi-Castagnoli, and F. Granucci. 2006b. Induction of peripheral T cell tolerance by antigen-presenting B cells. II. Chronic antigen presentation overrules antigen-presenting B cell activation. *J. Immunol.* **176:**4021–4028.

Randolph, G. J., K. Inaba, D. F. Robbiani, R. M. Steinman, and W. A. Muller. 1999. Differentiation of phagocytic monocytes into lymph node dendritic cells in vivo. *Immunity* **11:**753–761.

Regnault, A., D. Lankar, V. Lacabanne, A. Rodriguez, C. Théry, M. Rescigno, T. Saito, S. Verbeek, C. Bonnerot, P. Ricciardi-Castagnoli, and S. Amigorena. 1999. Fcgamma receptor-mediated induction of dendritic cell maturation and major histocompatibility complex class I-restricted antigen presentation after immune complex internalization. *J. Exp. Med.* **189:**371–380.

Reimann, J., and R. Schirmbeck. 1999. Alternative pathways for processing exogenous and endogenous antigens that can generate peptides for MHC class I-restricted presentation. *Immunol. Rev.* **172:**131–152.

Reis e Sousa, C. 2006. Dendritic cells in a mature age. *Nat. Rev. Immunol.* **6:**476–483.

Reis e Sousa, C., P. D. Stahl, and J. M. Austyn. 1993. Phagocytosis of antigens by Langerhans cells in vitro. *J. Exp. Med.* **178:**509–519.

Rescigno, M., S. Citterio, C. Théry, M. Rittig, D. Medaglini, G. Pozzi, S. Amigorena, and P. Ricciardi-Castagnoli. 1998. Bacteria-induced neo-biosynthesis, stabilization, and surface expression of functional class I molecules in mouse dendritic cells. *Proc. Natl. Acad. Sci. USA* **95:**5229–5234.

Rescigno, M., F. Granucci, and P. Ricciardi-Castagnoli. 2000. Molecular events of bacterial-induced maturation of dendritic cells. *J. Clin. Immunol.* **20:**161–166.

Rescigno, M., M. Urbano, B. Valzasina, M. Francolini, G. Rotta, R. Bonasio, F. Granucci, J. P. Kraehenbuhl, and P. Ricciardi-Castagnoli. 2001. Dendritic cells express tight junction proteins and penetrate gut epithelial monolayers to sample bacteria. *Nat. Immunol.* **2:**361–367.

Rolland, A., L. Guyon, M. Gill, Y. H. Cai, J. Banchereau, K. McClain, and A. K. Palucka. 2005. Increased blood myeloid dendritic cells and dendritic cell-poietins in Langerhans cell histiocytosis. *J. Immunol.* **174:**3067–3071.

Roncarolo, M. G., S. Gregori, and M. Levings. 2003. Type 1 T regulatory cells and their relationship with CD4+CD25+ T regulatory cells. *Novartis Found. Symp.* **252:**115–131; 203–210.

Roncarolo, M. G. and M. K. Levings. 2000. The role of different subsets of T regulatory cells in controlling autoimmunity. *Curr. Opin. Immunol.* **12:**676–683.

Sadlack, B., H. Merz, H. Schorle, A. Schimpl, A. C. Feller, and I. Horak. 1993. Ulcerative colitis-like disease in mice with a disrupted interleukin-2 gene. *Cell* **75:**253–261.

Sakaguchi, S. 2000. Regulatory T cells: key controllers of immunologic self-tolerance. *Cell* **101:**455–458.

Sansonetti, P. J., and A. Phalipon. 1999. M cells as ports of entry for enteroinvasive pathogens: mechanisms of interaction, consequences for the disease process. *Semin. Immunol.* **11:**193–203.

Santambrogio, L., S. L. Belyanskaya, F. R. Fischer, B. Cipriani, C. F. Brosnan, P. Ricciardi-Castagnoli, L. J. Stern, J. L. Strominger, and R. Riese. 2001. Developmental plasticity of CNS microglia. *Proc. Natl. Acad. Sci. USA* **98:**6295–6300.

Sauter, B., M. L. Albert, L. Francisco, M. Larsson, S. Somersan, and N. Bhardwaj. 2000. Consequences of cell death: exposure to necrotic tumor cells, but not primary tissue cells or apoptotic cells, induces the maturation of immunostimulatory dendritic cells. *J. Exp. Med.* **191:**423–434.

Schon-Hegrad, M. A., J. Oliver, P. G. McMenamin, and P. G. Holt. 1991. Studies on the density, distribution, and surface phenotype of intraepithelial class II major histocompatibility complex antigen (Ia)-bearing dendritic cells (DC) in the conducting airways. *J. Exp. Med.* **173:**1345–1356.

Schonrich, G., U. Kalinke, F. Momburg, M. Malissen, A. M. Schmitt-Verhulst, B. Malissen, G. J. Hämmerling, and B. Arnold. 1991. Down-regulation of T cell receptors on self-reactive T cells as a novel mechanism for extrathymic tolerance induction. *Cell* **65:**293–304.

Schwartz, R. H. 2003. T cell anergy. *Annu. Rev. Immunol.* **21:**305–334.

Sertl, K., T. Takemura, E. Tschachler, V. J. Ferrans, M. A. Kaliner, and E. M. Shevach. 1986. Dendritic cells with antigen-presenting capability reside in airway epithelium, lung parenchyma, and visceral pleura. *J. Exp. Med.* **163:**436–451.

Shortman, K., and Y. J. Liu. 2002. Mouse and human dendritic cell subtypes. *Nat. Rev. Immunol.* **2:**151–161.

Smith, T. J., L. A. Ducharme, S. K. Shaw, C. M. Parker, M. B. Brenner, P. J. Kilshaw, and J. H. Weis. 1994. Murine M290 integrin expression modulated by mast cell activation. *Immunity* **1:**393–403.

Smits, H. H., E. C. de Jong, E. A. Wierenga, and M. L. Kapsenberg. 2005. Different faces of regulatory DCs in homeostasis and immunity. *Trends Immunol.* **26:**123–129.

Spahn, T. W., A. Fontana, A. M. Faria, A. J. Slavin, H. P. Eugster, X. Zhang, P. A. Koni, N. H. Ruddle, R. A. Flavell, P. D. Rennert, and H. L. Weiner. 2001. Induction of oral tolerance to cellular immune responses in the absence of Peyer's patches. *Eur. J. Immunol.* **31:**1278–1287.

Steinman, R. M. 1991. The dendritic cell system and its role in immunogenicity. *Annu. Rev. Immunol.* **9:**271–296.

Steinman, R. M., and Z. A. Cohn. 1973. Identification of a novel cell type in peripheral lymphoid organs of mice. I. Morphology, quantitation, tissue distribution. *J. Exp. Med.* **137:**1142–1162.

Steinman, R. M., and Z. A. Cohn. 1974. Identification of a novel cell type in peripheral lymphoid organs of mice. II. Functional properties in vitro. *J. Exp. Med.* **139:**380–397.

Steinman, R. M., D. Hawiger, and M. C. Nussenzweig. 2003. Tolerogenic dendritic cells. *Annu. Rev. Immunol.* **21:**685–711.

Steinman, R. M., and M. C. Nussenzweig. 2002. Avoiding horror autotoxicus: the importance of dendritic cells in peripheral T cell tolerance. *Proc. Natl. Acad. Sci. USA* **99:**351–358.

Steinman, R. M., M. Pack, and K. Inaba. 1997. Dendritic cells in the T-cell areas of lymphoid organs. *Immunol. Rev.* **156:**25–37.

Steinman, R. M., and J. Swanson. 1995. The endocytic activity of dendritic cells. *J. Exp. Med.* **182:**283–288.

Stockinger, B. 1999. T lymphocyte tolerance: from thymic deletion to peripheral control mechanisms. *Adv. Immunol.* **71:**229–265.

Strober, W., I. J. Fuss, and R. S. Blumberg. 2002. The immunology of mucosal models of inflammation. *Annu. Rev. Immunol.* **20:**495–549.

Strobl, H., E. Riedl, C. Scheinecker, C. Bello-Fernandez, W. F. Pickl, K. Rappersberger, O. Majdic, and W. Knapp. 1996. TGF-beta 1 promotes in vitro development of dendritic cells from CD34+ hemopoietic progenitors. *J. Immunol.* **157:**1499–1507.

Stumbles, P. A., D. H. Strickland, C. L. Pimm, S. F. Proksch, A. M. Marsh, A. S. McWilliam, A. Bosco, I. Tobagus, J. A. Thomas, S. Napoli, A. E. Proudfoot, T. N. Wells, and P. G. Holt. 2001. Regulation of dendritic cell recruitment into resting and inflamed airway epithelium: use of alternative chemokine receptors as a function of inducing stimulus. *J. Immunol.* **167:**228–234.

Stumbles, P. A., J. A. Thomas, C. L. Pimm, P. T. Lee, T. J. Venaille, S. Proksch, and P. G. Holt. 1998. Resting respiratory tract dendritic cells preferentially stimulate T helper cell type 2 (Th2) responses and require obligatory cytokine signals for induction of Th1 immunity. *J. Exp. Med.* **188:**2019–2031.

Suda, T., K. McCarthy, Q. Vu, J. McCormack, and E. E. Schneeberger. 1998. Dendritic cell precursors are enriched in the vascular compartment of the lung. *Am. J. Respir. Cell Mol. Biol.* **19:**728–737.

Sumen, C., T. R. Mempel, I. B. Mazo, and U. H. von Andrian. 2004. Intravital microscopy: visualizing immunity in context. *Immunity* **21:**315–329.

Sun, C. M., J. A. Hall, R. B. Blank, N. Bouladoux, M. Oukka, J. R. Mora, and Y. Belkaid. 2007. Small intestine lamina propria dendritic cells promote de novo generation of Foxp3 T reg cells via retinoic acid. *J. Exp. Med.* **204:**1775–1785.

Surh, C. D., and J. Sprent. 1994. T-cell apoptosis detected in situ during positive and negative selection in the thymus. *Nature* **372:**100–103.

Svensson, H., A. Johannisson, T. Nikkilä, G. V. Alm, and B. Cederblad. 1996. The cell surface phenotype of human natural interferon-alpha producing cells as determined by flow cytometry. *Scand. J. Immunol.* **44:**164–172.

Tailleux, L., O. Neyrolles, S. Honoré-Bouakline, E. Perret, F. Sanchez, J. P. Abastado, P. H. Lagrange, J. C. Gluckman, M. Rosenzwajg, and J. L. Herrmann. 2003a. Constrained intracellular survival of Mycobacterium tuberculosis in human dendritic cells. *J. Immunol.* **170:**1939–1948.

Tailleux, L., O. Schwartz, J. L. Herrmann, E. Pivert, M. Jackson, A. Amara, L. Legres, D. Dreher, L. P. Nicod, J. C. Gluckman, P. H. Lagrange, B. Gicquel, and O. Neyrolles. 2003b. DC-SIGN is the major Mycobacterium tuberculosis receptor on human dendritic cells. *J. Exp. Med.* **197:**121–127.

Takeda, K., T. Kaisho, and S. Akira. 2003. Toll-like receptors. *Annu. Rev. Immunol.* **21:**335–376.

Tarbell, K. V., S. Yamazaki, K. Olson, P. Toy, and R. M. Steinman. 2004. CD25+ CD4+ T cells, expanded with dendritic cells presenting a single autoantigenic peptide, suppress autoimmune diabetes. *J. Exp. Med.* **199:**1467–1477.

Trevejo, J. M., M. W. Marino, N. Philpott, R. Josien, E. C. Richards, K. B. Elkon, and E. Falck-Pedersen. 2001. TNF-alpha -dependent maturation of local dendritic cells is critical for activating the adaptive immune response to virus infection. *Proc. Natl. Acad. Sci. USA* **98:**12162–12167.

Trottein, F., N. Pavelka, C. Vizzardelli, V. Angeli, C. S. Zouain, M. Pelizzola, M. Capozzoli, M. Urbano, M. Capron, F. Belardelli, F. Granucci, and P. Ricciardi-Castagnoli. 2004. A type I IFN-dependent pathway induced by Schistosoma mansoni eggs in mouse myeloid dendritic cells generates an inflammatory signature. *J. Immunol.* **172:**3011–3017.

Turnbull, E. L., U. Yrlid, C. D. Jenkins, and G. G. Macpherson. 2005. Intestinal dendritic cell subsets: differential effects of systemic TLR4 stimulation on migratory fate and activation in vivo. *J. Immunol.* **174:**1374–1384.

Underhill, D. M., and A. Ozinsky. 2002. Phagocytosis of microbes: complexity in action. *Annu. Rev. Immunol.* **20:**825–852.

Valladeau, J., O. Ravel, C. Dezutter-Dambuyant, K. Moore, M. Kleijmeer, Y. Liu, V. Duvert-Frances, C. Vincent, D. Schmitt, J. Davoust, C. Caux, S. Lebecque, and S. Saeland. 2000. Langerin, a novel C-type lectin specific to Langerhans cells, is an endocytic receptor that induces the formation of Birbeck granules. *Immunity* **12:**71–81.

van Furth, R. 1998. Human monocytes and cytokines. *Res. Immunol.* **149:**719–720.

van Kooten, C., and J. Banchereau. 2000. CD40-CD40 ligand. *J. Leukoc. Biol.* **67:**2-17.

van Kooyk, Y., and T. B. Geijtenbeek. 2003. DC-SIGN: escape mechanism for pathogens. *Nat. Rev. Immunol.* **3:**697–709.

Varol, C., L. Landsman, D. K. Fogg, L. Greenshtein, B. Gildor, R. Margalit, V. Kalchenko, F. Geissmann, and S. Jung. 2007. Monocytes give rise to mucosal, but not splenic, conventional dendritic cells. *J. Exp. Med.* **204:**171–180.

Veerman, A. J., and W. van Ewijk. 1975. White pulp compartments in the spleen of rats and mice. A light and electron microscopic study of lymphoid and non-lymphoid cell types in T- and B-areas. *Cell Tissue Res.* **156:**417–441.

Volkmann, A., T. Zal, and B. Stockinger. 1997. Antigen-presenting cells in the thymus that can negatively select MHC class II-restricted T cells recognizing a circulating self antigen. *J. Immunol.* **158:**693–706.

Vremec, D., M. Zorbas, R. Scollay, D. J. Saunders, C. F. Ardavin, L. Wu, and K. Shortman. 1992. The surface phenotype of dendritic cells purified from mouse thymus and spleen: investigation of the CD8 expression by a subpopulation of dendritic cells. *J. Exp. Med.* **176:**47–58.

Wakkach, A., N. Fournier, V. Brun, J. P. Breittmayer, F. Cottrez, and H. Groux. 2003. Characterization of dendritic cells that induce tolerance and T regulatory 1 cell differentiation in vivo. *Immunity* **18:**605–617.

Weaver, C. T., L. E. Harrington, P. R. Mangan, M. Gavrieli, and K. M. Murphy. 2006. Th17: an effector CD4 T cell lineage with regulatory T cell ties. *Immunity* **24:**677–688.

Weinstein, P. D., and J. J. Cebra. 1991. The preference for switching to IgA expression by Peyer's patch germinal center B cells is likely due to the intrinsic influence of their microenvironment. *J. Immunol.* **147:**4126–4135.

Weninger, W., and U. H. von Andrian 2003. Chemokine regulation of naive T cell traffic in health and disease. *Semin. Immunol.* **15:**257–270.

Wick, M. J., and H. G. Ljunggren. 1999. Processing of bacterial antigens for peptide presentation on MHC class I molecules. *Immunol. Rev.* **172:**153–162.

Winzler, C., P. Rovere, M. Rescigno, F. Granucci, G. Penna, L. Adorini, V. S. Zimmermann, J. Davoust, and P. Ricciardi-Castagnoli. 1997. Maturation stages of mouse dendritic cells in growth factor-dependent long-term cultures. *J. Exp. Med.* **185:**317–328.

Worbs, T., U. Bode, S. Yan, M. W. Hoffmann, G. Hintzen, G. Bernhardt, R. Förster, and O. Pabst. 2006. Oral tolerance originates in the intestinal immune system and relies on antigen carriage by dendritic cells. *J. Exp. Med.* **203:**519–527.

Wright, D. E., A. J. Wagers, A. P. Gulati, F. L. Johnson, and I. L. Weissman. 2001. Physiological migration of hematopoietic stem and progenitor cells. *Science* **294:**1933–1936.

Wu, M. T., and S. T. Hwang. 2002. CXCR5-transduced bone marrow-derived dendritic cells traffic to B cell zones of lymph nodes and modify antigen-specific immune responses. *J. Immunol.* **168:**5096–5102.

Wurbel, M. A., J. M. Philippe, C. Nguyen, G. Victorero, T. Freeman, P. Wooding, A. Miazek, M. G. Mattei, M. Malissen, B. R. Jordan, B. Malissen, A. Carrier, and P. Naquet. 2000. The chemokine TECK is expressed by thymic and intestinal epithelial cells and attracts double- and single-positive thymocytes expressing the TECK receptor CCR9. *Eur. J. Immunol.* **30:**262–271.

Wykes, M., A. Pombo, C. Jenkins, and G. G. MacPherson. 1998. Dendritic cells interact directly with naive B lymphocytes to transfer antigen and initiate class switching in a primary T-dependent response. *J. Immunol.* **161:**1313–1319.

Yamazaki, S., T. Iyoda, K. Tarbell, K. Olson, K. Velinzon, K. Inaba, and R. M. Steinman. 2003. Direct expansion of functional CD25+ CD4+ regulatory T cells by antigen-processing dendritic cells. *J. Exp. Med.* **198:**235–247.

Yang, D., O. Chertov, S. N. Bykovskaia, Q. Chen, M. J. Buffo, J. Shogan, M. Anderson, J. M. Schröder, J. M. Wang, O. M. Howard, and J. J. Oppenheim. 1999. Beta-defensins: linking innate and adaptive immunity through dendritic and T cell CCR6. *Science* **286:**525–528.

Yrlid, U., M. Svensson, A. Håkansson, B. J. Chambers, H. G. Ljunggren, and M. J. Wick. 2001. In vivo activation of dendritic cells and T cells during Salmonella enterica serovar Typhimurium infection. *Infect. Immun.* **69:**5726–5735.

Zal, T., A. Volkmann, and B. Stockinger. 1994. Mechanisms of tolerance induction in major histocompatibility complex class II-restricted T cells specific for a blood-borne self-antigen. *J. Exp. Med.* **180:**2089–2099.

Zanoni, I., M. Foti, P. Ricciardi-Castagnoli, and F. Granucci. 2005. TLR-dependent activation stimuli associated with Th1 responses confer NK cell stimulatory capacity to mouse dendritic cells. *J. Immunol.* **175:**286–292.

Zheng, S. G., J. H. Wang, J. D. Gray, H. Soucier, and D. A. Horwitz. 2004. Natural and induced CD4+CD25+ cells educate CD4+CD25− cells to develop suppressive activity: the role of IL-2, TGF-beta, and IL-10. *J. Immunol.* **172:**5213–5221.

Zinkernagel, R. 2007. On observing and analyzing disease versus signals. *Nat. Immunol.* **8:**8–10.

MEMBRANE RECEPTORS: RECOGNITION, ADHESION, PHAGOCYTOSIS, CHEMOTAXIS, AND MIGRATION

II

4

Fc Receptors and Phagocytosis

STEVEN GREENBERG AND BENJAMIN M. DALE

Phagocytosis is a specialized endocytic response of eukaryotic cells to particulate stimuli, such as microbial pathogens. This response is utilized by myeloid cells of the immune system to aid in host defenses. The striking resemblance between phagocytosis mediated by simple organisms, such as *Dictyostelium*, and of higher-order eukaryotes implies that the fundamental mechanisms that govern the phagocytosis program are conserved. Broadly defined, phagocytosis can be triggered by ligands native to the phagocytic target ("nonopsonic phagocytosis") or can be facilitated by opsonins, such as immunoglobulins and complements, which generally enhance the efficiency of the phagocytic response. Among the best-characterized forms of opsonic phagocytosis is that mediated by the Fc portion of immunoglobulin G (IgG), which binds to specific receptors on leukocytes and triggers a phagocytic response.

Abbreviations: ADCC, antibody-dependent cellular cytotoxicity; $[Ca^{2+}]_i$, cytosolic free calcium concentration; CR3, complement receptor 3; FcαR, receptor for the Fc portion of IgA; FcεRI, high-affinity receptor for the Fc portion of IgE; FcγR, receptor for the Fc portion of IgG; FcRn, MHC class I-related neonatal Fc receptor; G-CSF, granulocyte colony-stimulating factor; GM-CSF, granulocyte-macrophage CSF; IL-10, interleukin-10; PLC, phospholipase C; ITAM, immunoreceptor tyrosine-based activation motif; ITIM, immunoreceptor tyrosine-based inhibitory motif; MHC, major histocompatibility complex; PLD, phospholipase D; PIP_2, phosphatidylinositol 4,5-bisphosphate; PIP_3, phosphatidylinositol 3,4,5-trisphosphate; PI-3-kinase, phosphatidylinositol-3-kinase; PKC, protein kinase C; SHIP, Src homology region 2 domain-containing inositol-5'-phosphatase; SHP, Src homology region 2 domain-containing phosphatase; Syt VII, synaptotagmin VII; TNF, tumor necrosis factor.

Steven Greenberg, Merck Research Labs, 126 East Lincoln Ave., RY34B-348, Rahway, NUJ 07065-0900. Benjamin M. Dale, Division of Infectious Diseases, Department of Medicine, Immunology Institute, Mount Sinai School of Medicine, New York, NY 10029.

STRUCTURE AND FUNCTION OF THE FAMILY OF Fc RECEPTORS

Fcγ Receptors

In human phagocytic leukocytes, there are three principal types of receptors that recognize the Fc domain of IgG (reviewed in Cohen-Solal et al., 2004; Gessner et al., 1998; Hogarth, 2002; Hulett and Hogarth, 1994; Stefanescu et al., 2004) (Fig. 1). All three types are members of the Ig superfamily and all are capable of independently triggering phagocytosis (Anderson et al., 1990a). Mice express an additional FcγR, FcγRIV, whose extracellular domain is 63% identical to that of human FcγRIII (Nimmerjahn et al., 2005). In addition, phagocytic leukocytes express FcγRn, which is a nonconvalently associated β_2-microglobulin.

Fcγ Receptor I (CD64)

In humans, FcγRI is a 72-kDa glycoprotein expressed on monocytes, macrophages, and gamma interferon (IFN-γ)-stimulated neutrophils and eosinophils (reviewed in van de Winkel et al., 1991) (Table 1). FcγRI is expressed at especially high levels in freshly harvested mouse dendritic cells from spleen, lymph node, and skin (Tan et al., 2003). Three genes for FcγRI have been identified (Ernst et al., 1992); however, only one (FcγRIA) encodes a transmembrane protein. A distinguishing feature of FcγRI is its relatively high affinity for ligand. Human and mouse FcγRI bind mouse IgG2a with affinities of 10^8 to 10^9 M^{-1} and 2×10^7 to 5×10^7 M^{-1}, respectively (Lubeck et al., 1985; Sears et al., 1990; Unkeless and Eisen, 1975). Similarly, human FcγRI binds human IgG1 and IgG3, but not IgG2 and IgG4, with an affinity of 4×10^8 M^{-1} (Okayama et al., 2000) to 1.2×10^9 M^{-1} (Canfield and Morrison, 1991). The structural basis for the high-affinity binding to Ig ligand resides, in part, in a third binding domain within the extracellular region of the receptor that is not present in Fcγ receptors II and III (Hulett et al., 1991). In addition to Ig, C-reactive protein, an acute-phase reactant, was proposed to bind directly to FcγRI; surface plasmon resonance experiments indicate that the affinity of this interaction (0.81×10^9 M^{-1}) equals or exceeds that of monomeric

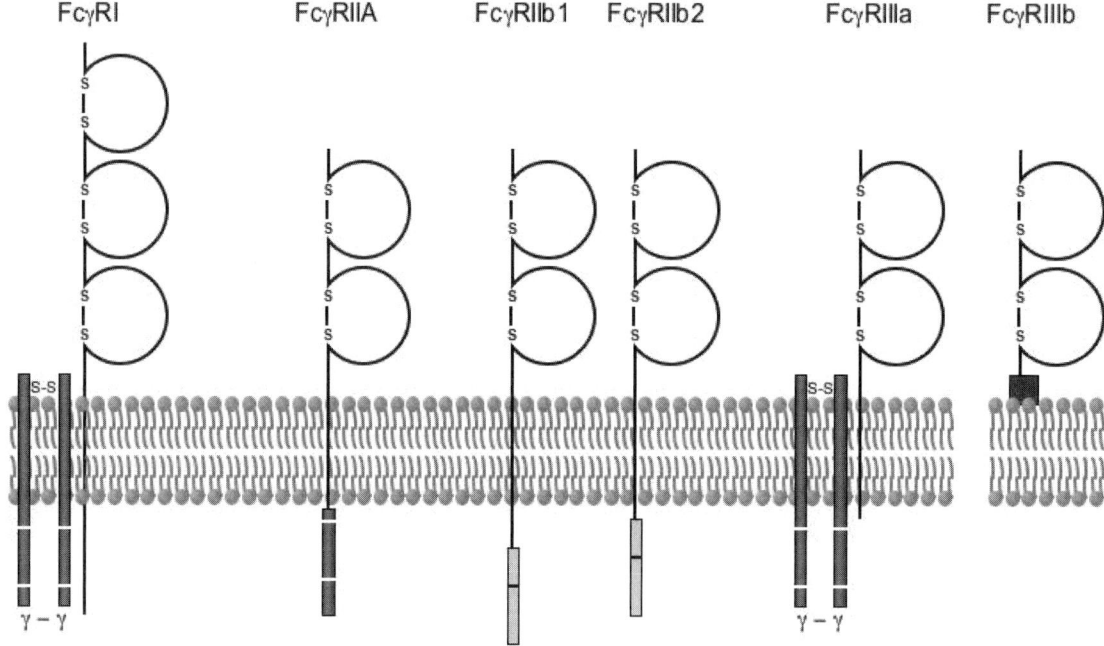

FIGURE 1 Domain structures of human FcγRs expressed in phagocytic cells. Ligand binding subunits are labeled. In dark gray are γ-subunit homodimers with white bars representing the two tyrosine residues that are contained within the consensus sequence of ITAMs. In FcγRIIb1 and FcγRIIb2 are the cytosolic domains (light gray), which contain one conserved tyrosine residue (black bar) within the consensus ITIM. FcγRIIIb is a GPI-linked protein expressed in neutrophils and activated eosinophils, but not in monocytes, macrophages, or dendritic cells.

IgG for the receptor (Bodman-Smith et al., 2002). Freshly harvested monocytes express approximately 10,000 to 40,000 FcγRI molecules on their surfaces (Perussia et al., 1987); this expression is increased by several agents, including IFN-γ (Perussia et al., 1987), G-CSF, GM-CSF (Buckle and Hogg, 1989), IL-10 (te Velde et al., 1992), glucocorticoids (Girard et al., 1987), and the complement component C5a (Yancey et al., 1985). Induction of FcγRI by IFN-γ requires the transcriptional activator, PU.1/Spi-1 (Perez et al., 1994). Not surprisingly, neutrophils isolated from patients with acute pyogenic infections express greater quantities of FcγRI (but not FcγRII or III) on their surfaces (Simms et al., 1989). IL-13 decreases expression of FcγRI on monocytes (de Waal Malefyt et al., 1993). FcγRI is associated in a noncovalent manner with a 12-kDa protein, termed "the γ-subunit," that confers enhanced surface expression and transmembrane signaling competency by the receptor. The γ-subunit contains an ITAM whose consensus is: YxxI/Lx$_{(6-12)}$YxxI/L (reviewed in Underhill and Goodridge, 2007). The γ-subunit endows FcγRI with the capacity to activate tyrosine kinases, culminating in a variety of functional signaling events (see below). FcγRI is phagocytically competent (Anderson et al., 1990) and is also capable of promoting inflammation via production of superoxide anion (Anderson et al., 1986) and TNF (Debets et al., 1988, 1990). In contrast, ligation of FcγRI in dendritic cells derived from circulating blood results in a lack of release of reactive oxygen intermediates, but enhanced antigen presentation and T-cell activation (Fanger et al., 1997). In mice in which the FcγRI gene was disrupted, there was decreased phagocytosis of IgG2a-opsonized erythrocytes and decreased lysis of polyclonal rabbit IgG-coated erythrocytes by macrophages. There was also decreased inflammation in a reverse passive Arthus reaction and, paradoxically, increased antibody production after immunization with a T-dependent antigen (Barnes et al., 2002). In another study, FcγRI knockout mice demonstrated impaired hypersensitivity responses, reduced cartilage destruction in an arthritis model, and impaired protection from infection with Bordetella pertussis (Ioan-Facsinay et al., 2002).

Although FcγRI is clearly capable of mediating inflammation, ligation of FcγRI in mouse macrophages leads to suppression of IL-12 production, due in part to production of the immunosuppressive cytokine IL-10 (Sutterwala et al., 1998). Interestingly, the activation status of the macrophage is important for this effect; in the presence of IFN-γ, IL-10-mediated responses are markedly suppressed due to FcγR- and PKC-δ-dependent decreases in IL-10 receptor expression (Ji et al., 2003). In human monocytes, ligation of all three FcγRs leads to inhibition of IL-12 production (Drechsler et al., 2002).

The concentration of IgG1 and IgG3 in the serum (approximately 5.6 g/liter, or 3.7×10^{-5} M [Schauer et al., 2003]) exceeds the K_d of IgG1 for FcγRI by 1,000- to 3,000-fold, leading to the prediction that FcγRI on blood-borne monocytes, like FcεRI on tissue mast cells, is constitutively occupied by high-affinity Ig ligand. This may also be true for FcγRI on tissue macrophages, because the concentrations of IgG in tissues (e.g., the lung) may approach that of the serum (Fahy et al., 2001). Analogously, the concentration of IgG2a in mouse serum (1.55 mg/ml, or 10^{-5} M [Gupta and Siber, 1995]) exceeds the K_d of IgG2a for mouse FcγRI by 200- to 500-fold. This implies that FcγRI would engage IgG2a-opsonized targets or IgG2a-containing immune complexes only inefficiently.

TABLE 1 Characteristics of human FcγRs[a]

Characteristic	FcγRI (CD64)	FcγRII (CD32)		FcγRIII (CD16)	
Genes	3 (A, B, C)	3 (A, B, C)		2 (A, B)	
Known coding region polymorphisms		IIA IIb1/2	131 H/R 232 I/T	IIIA IIIB	48 L/R/H 158 F/V NA-1/NA-2[b]
Transcripts	a1, a2 b1, b2, b3 c	a1, a2 b1, b2, b3 c		a1–a6 b	
Known isoforms	FcγRI	FcγRIIA1; FcγRIIA2 (S)[c] FcγRIIb1, b2		FcγRIIIa FcγRIIIb (GPI-linked)	
Associated subunits	γ-Subunit			γ-Subunit ζ-Subunit[d]; β-subunit[e]	
Affinity for monomeric IgG	10^9 M^{-1}	<10^7 M^{-1}		FcγRIIIa: 3×10^7 M^{-1} FcγRIIIb: <10^7 M^{-1}	
Isotype preference	IgG1 = 3 > 4 >>> 2	IIa IIb1, IIb2	IgG3 > 1 >> 2 > 4 IgG3 > 1 > 4 >> 2	IIIa, IIIb	IgG1 = 3 >>> 2, 4
Constitutive expression pattern	Mo, Mφ, DC	IIa: Mo, Mφ, N, Eo, Ba, DC, Plt IIb1: B IIb2: MF, DC		IIIa: Mo[f], Mφ, DC, NK, γδ T IIIb: N	
Induced expression pattern	N, Eo, Mes			IIIa: Mo, Mes IIIb: Eo	
Cytokine regulation	↑: IFN-γ, G-CSF, GM-CSF, IL-10, C5a ↓: IL-4, IL-13	IIb2	↑: IL-4, IL-10, IL-13 ↓: PGE$_2$, C5a	IIIa IIIb	↑: TGF-β ↑: IFN-γ, G-CSF, GM-CSF ↓: IL-4, IL-13
Recognized functions	Endocytosis Phagocytosis Ag presentation ADCC Mediator release[g]	IIa IIb1 IIb2	Endocytosis Phagocytosis Ag presentation ADCC Mediator release ITIM ITIM Endocytosis Ag presentation[h]	IIIa	Endocytosis Phagocytosis Ag presentation ADCC Mediator release

[a] The designations for FcγR proteins are somewhat confusing. Indicated in this table is the most commonly used nomenclature (i.e., hFcγRIIA, rather than IIa; hFcγRIIb, rather than IIB; and hFcγRIIIa and IIIb rather than IIIA and IIIB). Abbreviations: B, B cells; Ba, basophils; DC, dendritic cells; Eo, eosinophils; Mes, mesangial cells; M, monocytes; Mφ, macrophages; N, neutrophils; NK, NK cells; Plt, platelets.
[b] These alleles refer to polymorphisms in the membrane-distal Ig domain that contain two N-linked glycosylation sites.
[c] hFcγRIIA2 is a soluble protein.
[d] hFcγRIIIa in NK cells is associated with ζ- or β-subunit homo- or heterodimers; in other cells, it is associated with γ-subunit homodimers.
[e] hFcγRIIIa is associated with both the β- and γ-subunits of the high-affinity IgE receptor in mast cells.
[f] FcγRIIIa is constitutively expressed on a subpopulation of monocytes.
[g] Mediator release includes release of cytokines, reactive oxygen and nitrogen intermediates, and lipids, such as arachidonic acid and its metabolites.
[h] FcγRIIb2 mediates antigen (Ag) presentation of immune complexes to B cells (Bergtold et al., 2005).

However, the defects observed in FcγRI knockout mice suggest that FcγRI is capable of engaging clustered IgG ligand in vivo, even in the presence of a vast excess of monomeric ligand.

Fcγ Receptor II (CD32)

Isoforms of FcγRII, a 40-kDa glycoprotein, are expressed on nearly all hematopoietic cells, including platelets. In the human, three genes, termed IIA, Iib, and IIC, each encoding one or more transcripts, have been identified (Brooks et al., 1989). There is no murine homolog of FcγRIIA. Two well-described allelic variants of FcγRIIA have been identified, and termed, somewhat misleadingly, "high responder" and "low responder," corresponding to strong or weak interaction with murine IgG1. This polymorphism has been mapped to amino acid residue 131 within the second extracellular domain (Clark et al., 1991; Warmerdam et al., 1990). In contrast to results using murine IgG1, the presence of arginine at this position confers weak binding to human IgG2, whereas the presence of histidine confers strong binding to this human Ig isotype (Parren et al., 1992; Warmerdam et al., 1991). Because FcγRIIA is the only Fcγ receptor that binds hIgG2 appreciably, the expression of FcγRIIA-R131 would be expected to lead to a generalized defect in the binding of hIgG2 by phagocytic leukocytes. Expression of the R131 allele was associated with development of nephritis and the severity of immune complex disease in patients with systemic lupus erythematosus (SLE) (Duits et al., 1995; Salmon et al., 1996). Whether disease susceptibility is due to decreased clearance of IgG2-containing immune complexes remains to be established.

The surface expression of FcγRIIA varies from approximately 30,000 receptors on the human neutrophil (Selvaraj et al., 1988) to about 260,000 receptors on the human pulmonary alveolar macrophage (Rossman et al., 1989). In human neutrophils, activation with chemotactic peptide appears to increase the affinity of FcγRIIA without altering its surface expression, resulting in enhanced binding of IgG-opsonized particles and immune complexes (Nagarajan et al., 2000). The molecular basis for this phenomenon is unknown.

FcγRIIA mediates phagocytosis of IgG-coated particles by human neutrophils and mononuclear phagocytes (Anderson et al., 1990a; Odin et al., 1991). Rather than relying on a separate protein to mediate transmembrane signaling, FcγRIIA contains an ITAM in its cytosolic domain. In contrast, FcγRIIb, which is also expressed on myeloid cells and is the exclusive FcγRII isoform expressed on B cells (Brooks et al., 1989), contains an ITIM (see below). FcγRIIA also mediates ADCC (Walker et al., 1991) and superoxide anion (Huizinga et al., 1989) and TNF (Debets et al., 1990) secretion. Coligation of FcγRIIA in LPS- or IL-1β-stimulated human monocytes leads to production of the CCR8 ligand, CC chemokine ligand 1 (CCL1) (Sironi et al., 2006). The CCL1/CCR8 axis has been implicated in a Th2-polarized response and pulmonary recruitment of eosinophils after allergen challenge (Bishop and Lloyd, 2003; Chensue et al., 2001). Therefore, FcγRIIA has the potential to shape the nature of T-cell polarization during an acquired immune response.

Activating Fcγ Receptor IIA vs. Inhibitory Fcγ Receptor IIb

FcγRIIb is encoded by one gene that generates two isoforms (FcγRIIb1 and FcγRIIb2) by alternative splicing. The latter, but not the former, is localized to clathrin-coated pits (Miettinen et al., 1989), whereas both isoforms generally confer inhibitory, rather than activating ITAM-based signals. The inhibitory properties of FcγRIIb depend on an ITIM, whose consensus is: V/I/L/SxYxxL/V/I/S (Bolland and Ravetch, 1999; Sinclair, 2000; Unkeless and Jin, 1997). Tyrosine-phosphorylated ITIMs recruit SHIP, which is responsible for functional inhibition by the ITIM of FcγRIIb. The capacity of FcγRIIb to influence FcγR function has been established using mice rendered deficient for this receptor by homologous recombination (Bolland and Ravetch, 2000). Compared with wild-type mice, FcγRIIb-deficient mice were found to produce more antibodies, to exhibit exaggerated anaphylactic and Arthus reactions (Takai et al., 1996), to be more susceptible to collagen-induced arthritis (Kleinau et al., 2000; Yuasa et al., 1999), and to spontaneously develop autoimmunity (Bolland and Ravetch, 2000).

A novel function for FcγRIIb has been demonstrated recently. Perhaps because of its lack of trafficking to degradative routes in the cell, FcγRIIb has the capacity to present intact antigen to B cells. Clynes and colleagues demonstrated an FcγRIIb-mediated recycling route for immune complexes in dendritic cells that enables the recycled antigen bound in the complexes to be presented to B cells (Bergtold et al., 2005). This mechanism contrasts with more conventional modes of antigen presentation mediated by activating FcγRs in dendritic cells. When targeted to activating FcγRs, immune complexes trigger activation of Syk, which is required for endocytosis (Sedlik et al., 2003), processing of antigen within immune complexes onto MHC II molecules (and onto MHC I molecules via "cross-presentation"), dendritic cell maturation, and presentation of MHC-bound peptides to T cells (Regnault et al., 1999).

A single-nucleotide polymorphism for the FcγRIIb gene (FCGR2B c.695T>C) results in the substitution of isoleucine with threonine at position 232 in the transmembrane region. This polymorphism is associated with susceptibility to SLE in Asians (Kono et al., 2005). The functional effects of expression of FcγRIIb 232Thr is presently unclear because one study showed decreased inhibition of FcγR-dependent BCR responses compared with FcγRIIb 232Ile (Kono et al., 2005), whereas another study showed the opposite (Li et al., 2003). Another less frequent variant, 2B.4 (−386C-120A) in the FCGR2B gene promoter, was associated with the autoimmune SLE phenotype (odds ratio, 1.65) (Su et al., 2007). Although the 2B.4 variant was associated with increased expression of FcγRIIb in B cells and CD14$^+$ monocytes in healthy individuals, in individuals with active SLE it was associated with decreased expression in both memory and plasma B lymphocytes compared with naïve and memory/plasma B lymphocytes from controls. Because downregulation of FcγRIIb on myeloid-lineage cells in SLE was not observed in this study (Su et al., 2007), the level of expression of FcγRIIb in B cells, rather than myeloid cells, may be more important in the context of active autoimmune disease.

The relative expression of activating FcγRs and the inhibitory FcγRIIb is under the control of various cytokines. Many studies have documented the capacity of IFN-γ to induce the expression of FcγRI in a variety of mononuclear phagocytes. In contrast, transforming growth factor-β1 (TGF-β1) suppresses γ-subunit expression in human peripheral blood monocytes, which correlates with reduced effector function (Tridandapani et al., 2003). Expression of FcγRIIb is also regulated by cytokines; IL-4, IL-13, and IL-10, generally thought of as "inhibitory" cytokines toward myeloid cells, increase the relative expression of the inhibitory FcγRIIb (Joshi et al., 2006; Liu et al., 2005; Wijngaarden et al., 2004), whereas prostaglandin E_2 (PGE_2) and C5a have the opposite effects (Guriec et al., 2006). This led Gessner and colleagues to demonstrate a positive feedback loop, whereby FcγR-dependent and serum-independent generation of C5a by phagocytes leads to enhanced FcγR expression on C5a receptor-bearing cells (Kumar et al., 2006).

It is likely that relative changes in the ratio of activating to inhibitory FcγRs have functional consequences in various human disease states. Perhaps the emergence or clinical course of autoimmunity reflects decreased expression of FcγRIIb, similar to what is observed in mice (reviewed in Schmidt and Gessner, 2005; Stefanescu et al., 2004). Indeed, the relative expression of FcγRI and IIA was increased, relative to FcγRIIb, in monocytes from patients with rheumatoid arthritis (Wijngaarden et al, 2004). Recently, the therapeutic potential for the biasing of activating vs. inhibitory FcγR expression was highlighted by studies of immune-mediated thrombocytopenia in the mouse. Administration of IVIg, which has therapeutic efficacy in various autoimmune disease states in humans, led to reduced immune thrombocytopenia and increased expression of FcγRIIb in Mac1$^+$ splenocytes. Importantly, protection by IVIg depended on the expression and ligand binding activity of FcγRIIb (Nimmerjahn et al., 2005). This led Ravetch and colleagues to propose that the long-observed but poorly understand therapeutic efficacy of IVIg is due to induction of FcγRIIb (Samuelsson et al., 2001). At present, the effect of therapeutic IVIg on the expression of Fcγ receptors in leukocyte subpopulations in humans is unknown.

Fcγ Receptor III (CD16)

FcγRIII is a highly glycosylated protein with an apparent molecular weight of 50,000 to 80,000. It is constitutively expressed on macrophages, subpopulations of monocytes, dendritic cells, neutrophils, eosinophils, NK cells, and T cells. Two genes for FcγRIII have been identified that encode structurally distinct cell-specific proteins. FcγRIIIA is expressed as a transmembrane protein in macrophages, NK cells, and T cells, whereas FcγRIIIb is expressed as a GPI-linked protein in neutrophils and eosinophils (Hibbs et al., 1989; Huizinga et al., 1988). Like FcγRI, surface expression of the transmembrane form of FcγRIII depends on the coexpression of either the γ-subunit or the ζ-subunit of the T-cell antigen receptor/CD3 complex (Anderson et al., 1990b; Kurosaki and Ravetch, 1989; Lanier et al., 1989). Two allelic forms of FcγRIIIb have been identified (NA-1 and NA-2) that differ in four amino acids, which include two N-linked glycosylation sites. Expression of the NA-2 allele is associated with decreased phagocytic capacity of this FcγR (Salmon et al., 1990). Other polymorphisms have been detected, including one (phenylalanine to valine substitution at amino acid position 176) that is correlated with greater binding of hIgG1 and hIgG3 (Wu et al., 1997). TGF-β, but not a variety of other cytokines, increases the surface expression of FcγRIIIa on human monocytes (Welch et al., 1990), whereas IFN-γ, G-CSF, and GM-CSF increase surface expression of FcγRIIIb on human neutrophils (Buckle and Hogg, 1989). In contrast, IL-4 and IL-13 decrease expression of FcγRIIIa on monocytes (de Waal Malefyt et al., 1993; Ruppert et al., 1991). Structural studies indicate that FcγRIII binds asymmetrically to the lower hinge region of both Fc heavy chains, creating a 1:1 receptor ligand stoichiometry. Parallel FcγRIIIb dimers have been observed in the crystal lattice. Such dimerization may occur on the cell surface, increasing the avidity of the interaction and subsequently facilitating cell activation (Zhang et al., 2000).

FcγRIIIa on human macrophages mediates phagocytosis (Anderson et al., 1990a). Other functions of FcγRIII include ADCC of tumor cells by NK cells and neutrophils and clearance of immune complexes from the bloodstream (Clarkson et al., 1986). Despite its nontransmembrane topology, FcγRIIIb on human neutrophils is also capable of mediating transmembrane signaling, including increases in $[Ca^{2+}]_i$ (Kimberly et al., 1990) and F-actin content (Salmon et al., 1991). While the GPI-linked FcγRIIIb demonstrates some independent capacity to mediate phagocytosis, the function of FcγRIIIb in vivo may be to enhance binding of IgG-bound ligands and act in concert with FcγRI and/or IIA to promote phagocytosis in human neutrophils. Salmon and coworkers found that anti-FcγRIII mAb 3G8 inhibited ingestion, but not binding, of Concanavalin A-coated erythrocytes and unopsonized *Escherichia coli* by human neutrophils. They suggested that clustering of FcγRIII via its mannose-containing oligosaccharides generated a phagocytic signal (Salmon et al., 1987). Thus, FcγRIIIb, like CR3 (Altieri and Edgington, 1988; Diamond et al., 1995; Relman et al., 1990; Ross et al., 1985; Wright et al., 1988, 1989), may have more than one functional ligand binding domain.

While the exact mechanism of signaling by GPI-linked proteins is unknown, an emerging consensus is that they occupy detergent-resistant membranes (DRMs) on the cell surface, possibly physically associating with other proteins bearing enzymatic activities. Recent studies indicate that FcγRIIIb constitutively partitions with both low- and high-density DRMs; on engagement of FcγRIIIb, a significant increase in the amount of the receptor is observed in high-density DRMs (Fernandes et al., 2006). Perhaps the signaling capacity of FcγRIIIb is due to its association with other DRM-associated proteins, such as LAT, an adaptor protein linked to the PLC-dependent calcium mobilization pathway. Although mouse macrophages lack expression of FcγRIIIb, it is interesting to note that LAT-deficient bone marrow-derived macrophages display defects in phagocytosis, but not binding, of IgG-coated targets (Tridandapani et al., 2000). FcγRIIIb may cooperate with other Fcγ receptors to generate a phagocytic signal, similar to other GPI-linked proteins, which have the capacity to augment Ca^{2+} signaling by FcγRIIA when coclustered with this FcγR (Green et al., 1997). Another plausible mechanism for signaling by FcγRIIIb is through its physical association with other phagocytosis-promoting receptors, such as CR3 (Poo et al., 1995; Zhou et al., 1993).

Mice lacking FcγRIII by gene targeting displayed defects in IgG-mediated mast cell degranulation, and the mice were resistant to IgG-dependent passive cutaneous anaphylaxis and exhibited an impaired Arthus reaction. Macrophages derived from these mice were incapable of ingesting IgG1-opsonized targets (Hazenbos et al., 1996). In an aerosolized allergen challenge in OVA-sensitized mice, FcγRIII was required for early neutrophil influx, probably because of production of OVA-specific IgG1 (Taube et al., 2003). Similarly, FcγRIII was required for induction of immune complex alveolitis and production of proinflammatory cytokines and chemokines (Chouchakova et al., 2001) as well as IgG1-dependent autoimmune hemolytic anemia (Meyer et al., 1998). Thus, FcγRIII plays important roles in allergic disorders and disorders characterized by pathogenic deposition of immune complexes.

Fcα Receptor (CD89)

FcαRI is expressed on monocytes, macrophages, neutrophils, and eosinophils (reviewed in Monteiro and Van De Winkel, 2003; Wines and Hogarth, 2006). Multiple splice variants exist, although confirmation of protein expression for most of these is lacking and the functions of distinct isoforms is unclear. FcγRI binds IgA1 and IgA2 with low affinity (K_a, 10^6 M^{-1}). While monomeric IgA binding to FcαRI is transient, polymeric IgA and IgA immune complexes bind with a greater avidity (Wines et al., 2006). FcαRI binds to serum, but not secretory, IgA (van Egmond et al., 2000). The engagement of ligand by FcαRI is atypical for the Fc receptor family. Its two Ig-like domains are oriented at right angles; two FcαRI molecules can bind one IgA molecule within the EC1 domain, at a site completely different from the other FcRs (Herr et al., 2003). FcαRI associates noncovalently with a γ-subunit homodimer common to other Fc receptors (Pfefferkorn and Yeaman, 1994). However, unlike other Fc receptors, expression of FcαRI does not require the presence of this subunit for cell surface expression (Maliszewski et al., 1990). Cell surface expression of FcαRI is rapidly increased after the addition of chemoattractants, suggesting that there are latent intracellular pools of FcαRI (Hostoffer et al., 1993). Surface FcαRI expression is greater in cells obtained from bronchoalveolar lavage of patients with cystic fibrosis than from controls, and expression of FcαRI is increased following addition of TNF-α (Hostoffer et al., 1994). In contrast, TGF-β1 downregulates expression of FcαRI on human monocytes (Reterink et al., 1996). Addition of GM-CSF to human neutrophils leads to enhanced affinity of an IgA receptor and capacitates the neutrophils for IgA-dependent phagocytosis of IgA (Weisbart et al., 1988). It is not known whether this reflects activation of CD89 or another receptor recognizing IgA on these cells, although evidence suggests that CD89 is the principal IgA receptor on human neutrophils (Stewart et al., 1994). Studies using My 43, a mAb of the IgM isotype, demonstrate that FcαRI is capable of mediating phagocytosis and superoxide anion production (Shen et al., 1989).

FcαRI mediates phagocytosis of serum IgA-opsonized targets in Kupffer cells derived from transgenic mice in which human FcαRI was expressed under the control of its own promoter (van Egmond et al., 2000). This indicates that FcαRI has the capacity to function much like other phagocytosis-promoting FcγRs. However, in other contexts, signaling via FcαRI may have an entirely different outcome. Rather than leading to activation, engagement of FcαRI with monovalent anti-FcαRI Fab leads to inhibition of responses of heterologous FcγR or FcεRI. The inhibitory mechanism involves recruitment of the tyrosine phosphatase SHP-1 to the FcαRI ITAM and impairment of Syk, LAT, and ERK phosphorylation induced by FcεRI engagement. SHP-1 recruitment requires ERK activation. In contrast, sustained aggregation of FcαRI by multimeric ligands stimulates cell activation by recruiting high amounts of Syk, rather than SHP-1 (Pasquier et al., 2005). Thus, depending on ligand valency, FcαRI and its associated γ-subunit ITAM can serve either to activate cells in a receptor-autonomous fashion, or to inhibit responses induced by coligated receptors. The physiological significance of inhibition by γ-subunit ITAMs is not known, although inhibition of immune responses has been observed by other ITAMs, as well (reviewed in Hamerman and Lanier, 2006).

Other Fc Receptors Implicated in Phagocytic Signaling

Of the other known Fc receptors, the recently discovered murine FcγRIV, whose closest human ortholog is FcγRIII, mediates phagocytosis. FcγRIV is expressed on neutrophils, monocytes, macrophages, and dendritic cells. Inflammatory stimuli (e.g., LPS, IFN-γ) upregulate, whereas IL-4, IL-10, and TGF-β downregulate FcγRIV surface expression. FcγRIV binds IgG2a and IgG2b with intermediate affinity. FcγRIV is responsible for IgG2b antibody-induced thrombocytopenia in mice (Nimmerjahn et al., 2005).

FcγRn, a neonatal Fc receptor complexed to β_2-microglobulin, is a receptor on intestinal epithelial cells that mediates the transfer of maternal Ig from milk to the bloodstream of newborns (reviewed in Lencer and Blumberg, 2005). mRNA for FcγRn has been detected in a variety of rodent and human tissues and is expressed on a wide variety of myeloid cells. β_2-Microglobulin-deficient mice, which also lack expression of FcγRn, display increased clearance of IgG from the serum, corresponding to enhanced lysosomal delivery of endocytosed IgG (Israel et al., 1996; Junghans and Anderson, 1996). FcγRn thus plays a major role in the regulation of the concentration of IgG in the serum. FcγRn binds IgG with high affinity only in an acidic environment. FcγRn is also expressed within azurophilic and specific granules of neutrophils and relocates to phagolysosomes during phagocytosis of IgG-coated targets. A recent study indicated that FcγRn plays a role in phagocytosis in mouse neutrophils. Neutrophils derived from either β_2-microglobulin$^{-/-}$ or FcγRn$^{-/-}$ mice bound IgG1-opsonized targets normally, but displayed decreased ingestion. The capacity of FcγRn to enhance phagocytosis was attributed to endocytic motifs within the cytosolic domain of FcγRn (Vidarsson et al., 2006).

ACTIVATION AND DEACTIVATION SIGNALS FOLLOWING Fc RECEPTOR LIGATION

ITAM

First recognized by Reth (Reth, 1989), the ITAM consensus sequence, $YxxI/Lx_{(6-12)}YxxI/L$, is present in subunits of the T- and B-cell antigen receptor complexes, and subunits of FcRs α, γ, and ε (reviewed in Underhill and Goodridge, 2007). Using various transfected cell models, mutation of either of the tyrosine residues within the ITAMs of the γ-subunit of FcγRIIA markedly impairs receptor signaling, including phagocytosis (reviewed in Cox and Greenberg, 2001) and phagosome-lysosome fusion (Worth et al., 2001). Therefore, the ITAM is an indispensable component of the phagocytic signaling machinery of Fcγ receptors.

Tyrosine Kinases and ITAM Phosphorylation

Many aspects of FcγR signaling are inhibited by a variety of tyrosine kinase inhibitors (Allen and Aderem, 1996; Davis et al., 1995; Fallman et al., 1995; Ghazizadeh and Fleit, 1994; Greenberg et al., 1993). On oligomerization of the ligand binding subunits of Fc receptors by clustered Fc residues, the tyrosine residues within the associated γ-subunit ITAMs become phosphorylated (Duchemin et al., 1994; Greenberg et al., 1994; Paolini et al., 1991; Pfefferkorn and Yeaman, 1994). Once phosphorylated, these domains serve as high-affinity binding sites for the tyrosine kinase Syk (Agarwal et al., 1993; Benhamou et al., 1993; Kiener et al., 1993; Law et al., 1993). Syk contains tandem SH2 domains, each of which interacts with tyrosine-phosphorylated ITAMs (reviewed in Siraganian et al., 2002). The identity of the "initiating" kinase that phosphorylates tyrosine residues within the ITAM is unknown, although it is widely held that members of the Src family tyrosine kinases fulfill this role. Indeed, several workers have coprecipitated multiple Src family members and various Fc receptors (Duchemin and Anderson, 1997; Durden et al., 1995; Ghazizadeh et al., 1994; Hamada et al., 1993; Salcedo et al., 1993; Sarmay et al., 1994; Wang et al., 1994). In a study using macrophages derived from knockout mice that lack expression of Lyn, Fgr, and Hck, the Src family members normally expressed in these cells, FcγR-mediated phagocytosis occurred at a greatly reduced rate (Crowley et al., 1997). Of these three Src family members, Lyn and Hck play a positive role in phagocytosis, whereas Fgr plays a negative role (Gresham et al., 2000). Because the prior study did not address whether FcγR ligation in macrophages lacking expression of Src family members led to decreased tyrosine phosphorylation of the γ-subunit, the hierarchy of ITAM-based tyrosine kinase signaling remains an open question. Interestingly, a study of macrophages lacking Syk demonstrated that Syk was required for maximal FcγR-stimulated γ-subunit phosphorylation (Kiefer et al., 1998). This implies that Syk itself can serve as an "ITAM kinase," thus amplifying the ITAM kinase cascade that might have been triggered by other (e.g., Src family) tyrosine kinases. Alternatively, Syk may be required for maximal activation of Src family tyrosine kinases, although a role for Src family tyrosine kinases in maximal activation of Syk seems more likely (Jouvin et al., 1994), because macrophages derived from Src family knockout mice show markedly decreased activation of Syk (Crowley et al., 1997). How are Src family tyrosine kinases activated following engagement of FcγRs? While Lyn is capable of a direct association with Syk (Jouvin et al., 1994), experiments in P815 mastocytoma cells transfected with FcγRI suggest that the β-subunit of FcεR is the primary binding partner for Lyn. These data led Kinet and coworkers to suggest a model for FcεRI signaling in which the unphosphorylated β-subunit of FcεRI serves to recruit Lyn, which acts to amplify the activation of Syk that is bound to phosphorylated γ-subunits (Lin et al., 1996). However, this model is not directly applicable to FcγRs in most myeloid cells, which lack expression of the β-subunit. Other factors may influence the association with FcγRs with Src family tyrosine kinases. For example, FcγRIIA is palmitoylated at Cys208, which is required for localization to DRMs and ligand-induced enhanced tyrosine phosphorylation. Because Lyn and other Src family members are also localized to DRMs, this may be one factor that enhances Lyn-FcγR interactions (Barnes et al., 2006). However, the requirement for palmitoylation and DRM association in phagocytic signaling may not be absolute (Garcia-Garcia et al., 2007).

ITAM-Dependent Signaling and Syk

The evidence that Syk is required for ITAM-dependent signaling in macrophages and mast cells is compelling. Downregulation of Syk expression in monocytes using antisense oligonucleotides abrogated phagocytosis (Matsuda et al., 1996), and expression of a Syk construct lacking the kinase domain inhibited release of arachidonate following antigen stimulation in RBL cells (Hirasawa et al., 1995). Studies using either Syk-deficient lymphocytes (Cox et al., 1996), mast cells (Zhang et al., 1996), or macrophages (Crowley et al., 1997) indicate an essential role for Syk in mediator release, phagocytosis, and cytoskeletal assembly. In Syk-deficient mast cells, phosphorylation of the γ-subunit on FcεRI aggregation was greatly impaired, similar to what was observed in Syk-deficient macrophages upon stimulation of FcγRs, indicating that Syk plays an important role in the phosphorylation of the γ-subunit in the context of several different Fc receptors (Kiefer et al., 1998). Indeed, the constitutive association of Syk with subunits of the antigen receptor in resting B cells (Law et al., 1993) and with FcγRIIA in THP-1 cells, a human monocyte-like cell line (Ghazizadeh et al., 1995), suggest that Syk itself is a bona fide ITAM kinase. In this capacity, Syk could fulfill the role of the "initiating" kinase in ITAM-dependent phagocytosis. According to this view, a small percentage of Syk molecules promotes tyrosine phosphorylation of ITAM-bearing subunits in the absence of receptor ligation. This creates a limited number of Syk/ITAM complexes in resting cells. After engagement of Fc receptors by IgG, further recruitment of unphosphorylated ITAMs occurs; these become phosphorylated by nearby Syk/ITAM complexes, leading to further Syk recruitment from the cytosolic pool. Challenging this view is the observation that a Src family-specific tyrosine kinase inhibitor, which inhibited Lyn kinase activity in vitro, also inhibited FcγRIIA ITAM phosphorylation in THP-1 cells. Furthermore, cotransfection of Lyn, but not Syk, with FcγRIIA led to tyrosine phosphorylation of FcγRIIA in COS cells (Cooney et al., 2001). However, Syk is clearly capable of phosphorylating FcγRIIA in vitro (Greenberg et al., 1996). Collectively, these results suggest that Lyn or another Src family member phosphorylates the FcγRIIA ITAM in a DRM, which serves to recruit Syk via its SH2 domains; once recruited, Syk can phosphorylate other nearby ITAMs, thus amplifying the kinase cascade leading to further Syk recruitment.

Syk activation may result from multiple interrelated events: (i) binding to phosphorylated ITAMs and localized submembranous recruitment; (ii) activation of intrinsic Syk kinase activity by binding phosphorylated ITAMs (Kimura et al., 1996; Rowley et al., 1995; Shiue et al., 1995); (iii) Syk phosphorylation, both by neighboring Syk molecules ("autophosphorylation") and by other recruited kinases, including Src family members (Amoui et al., 1997; Kurosaki et al., 1994; Sidorenko et al., 1995; Ting et al., 1995); (iv) activation by increased cytosolic free-calcium concentration (Wang et al., 1994) and reactive oxygen species (Qin et al., 1995; Schieven et al., 1993). Interestingly, addition of N-acetylcysteine inhibited both Syk activation and effector functions following ligation of FcγRI in mast cells (Valle and Kinet, 1995), suggesting that reactive oxygen species are important activators of Syk in vivo. It is not clear whether reactive oxygen species directly enhance Syk tyrosine kinase activity. It is possible that generation of reactive oxidants promote Syk activation indirectly, perhaps via inactivation of tyrosine phosphatases by targeting the catalytic cysteine thiolate (Denu and Tanner, 1998).

ITIM

Co-ligation of FcγRIIb with the antigen receptor in B cells (BCR) leads to decreased cellular activation (Phillips and Parker, 1983). The structural basis for this observation was apparent in 1992, when Amigorena and coworkers (1992) demonstrated that a 13-amino-acid sequence in FcγRIIb1 is required for inhibition of B-cell activation. Since then, similar motifs, now termed ITIMs, have been identified in several other receptors in hematopoietic cells, including CD22 and killer cell inhibitory receptors (KIRs) (reviewed in Vivier and Daeron, 1997; Daeron and Lesourne, 2006). While the minimal sequence requirements for the ITIM are in doubt, I/VxYxxL/V has been proposed as a provisional consensus motif (Daeron, 1996). Phosphorylation of the tyrosine residue within this motif is required for delivering the inhibitory signal (Muta et al., 1994). The capacity of receptors bearing this motif to inhibit cellular activation correlates with the recruitment of either the tyrosine phosphatases SHP-1/SHP-2 (Olcese et al., 1996) or the SH2 domain-containing inositol-5′-phosphatase (SHIP) (Lioubin et al., 1996; Ono et al., 1996; Osborne et al., 1996). SHIP dephosphorylates either PIP_3 to yield phosphatidylinositol 3,4-bisphosphate, or inositol (1,3,4,5)-tetrakisphosphate to yield inositol 1,3,4-trisphosphate (Lioubin et al., 1996; Ono et al., 1996; Osborne et al., 1996). Specific receptors preferentially recruit either SHP-1/-2 (KIR) or SHIP (FcγRIIb) (Gupta et al., 1997; Ono et al., 1997; Vely et al., 1997). Inhibition of antigen receptor signaling by FcγRIIb is correlated with premature termination of BCR-induced elevations in inositol 3,4,5-trisphosphate (Bijsterbosch and Klaus, 1985) and a predominant reduction in BCR-mediated calcium influx, rather than impaired release of calcium from intracellular stores fluxes (Bolland et al., 1998). This activity requires catalytic activity of SHIP (Ono et al., 1997), which reduces recruitment of the PH domain-containing Btk to the membrane by reducing accumulation of PIP_3 (Bolland et al., 1998). By hydrolyzing PIP_3, SHIP negatively regulates phagocytosis mediated by FcγRs. Interestingly, SHIP also inhibits CR3-mediated phagocytosis, suggesting that it may serve to limit phagocytosis in general (Cox et al., 2001).

Tyrosine Phosphatases and Other Modulators of ITAM Function

Several tyrosine phosphatases have been identified that modulate ITAM-mediated responses. These include the membrane-bound tyrosine phosphatase CD45 and SHP-1/SHP-2. CD45 is required for signaling via the B-cell antigen receptors and for degranulation mediated by FcεRI in mast cells (Berger et al., 1994). The requirement for this phosphatase is most likely due to its ability to dephosphorylate a conserved tyrosine residue in the C terminus of Src family members, which is required for their activation (reviewed in Brown and Cooper, 1996; Neel, 1997). The role of CD45 in FcγR-mediated signaling is uncertain. Both SHP-1 and SHP-2 are associated with FcγRI (Kimura et al., 1997). SHP-1 directly associates with the phosphorylated amino-terminal ITAM tyrosine of FcγRIIA (Ganesan et al., 2003), in contrast to Syk, which preferentially associates with the phosphorylated carboxy-terminal ITAM tyrosine of FcγRIIA (Ganesan et al., 2003; Huang et al., 2003) and the γ-subunit (Huang et al., 2003). Overexpression of SHP-1 negatively regulates phagocytosis mediated by FcγRIIA in COS cell transfectants (Huang et al., 2003) and FcγRs in mouse macrophages (Kant et al., 2002). Other proteins have the potential to negatively regulate ITAM-mediated signaling. For example, Cbl is a 120-kDa protein that undergoes enhanced tyrosine phosphorylation after ligation of FcγRs (Marcilla et al., 1995; Matsuo et al., 1996; Tanaka et al., 1995). Its overexpression leads to inhibition of both Syk tyrosine kinase activity and association of Syk with the γ-subunit of FcγRI (Ota and Samelson, 1997).

Independence of Phagocytosis and Endocytosis

The signaling requirements for endocytosis of monomeric Ig and immune complexes and phagocytosis are distinct. This is not surprising given the relative importance of cytoskeletal assembly in phagocytosis. The larger the phagocytic particle, the more dependent the ingestion process is on cytoskeletal assembly (Koval et al., 1998). Endocytosis of FcγRIIA requires the participation of clathrin, but not AP-2, and unlike phagocytosis, does not involve participation of Src family tyrosine kinases (Huang et al., 2006; Mero et al., 2006). In addition to promoting the early signaling events required for phagocytosis, Src family tyrosine kinases seem to promote clustering of FcγRIIA and binding of IgG-coated targets (Sobota et al., 2005). This regulation of receptor clustering is reminiscent of "inside-out" signaling of integrins, and suggests a role for the regulation of FcγRIIA avidity, and therefore receptor mobility, in ligand binding. The molecular basis for this phenomenon is unknown. Interestingly, the cytosolic tail of human FcγRI, but not mouse FcγRI, binds periplakin, a protein that interacts with the cytoskeleton and acts to negatively regulate ligand binding (Beekman et al., 2004).

There is a strong requirement for protein ubiquitylation in the endocytosis of immune complexes by FcγRIIA (Booth et al., 2002). However, ubiquitylation of the receptor itself, while important for stimulating the rate of endocytosis, is not absolutely required for endocytosis (Booth et al., 2002). These results suggest that endocytosis of FcγRIIA requires ubiquitylation of multiple substrates and that there may be redundant mechanisms for endocytosis by FcγRIIA.

Despite this appealing dichotomy between endocytosis and phagocytosis, there are proteins that appear to play a role in both processes, such as dynamin 2 (Gold et al., 1999), and there is one report that clathrin itself plays a role in phagocytosis of IgG-coated particles in pulmonary alveolar macrophages (Perry et al., 1999). It is possible that components of the endocytic machinery influence phagocytosis indirectly, making it difficult to draw firm conclusions as to their precise roles in phagocytosis.

RECEPTOR COOPERATIVITY AND PHAGOCYTOSIS

While many studies of phagocytosis involve the use of artificial particles, such as erythrocytes opsonized with monospecific ligands, it is apparent that bacterial and fungal pathogens express multiple adhesins on their surfaces (reviewed in Hauschildt and Kleine, 1995; Ofek et al., 1995). For example, phagocytosis of unopsonized *Pseudomonas aeruginosa* by human macrophages is mediated by multiple receptors including the macrophage mannose receptor, CR1 and CR3, and FcγRs (Speert et al., 1988). Phagocytic receptors may function independently to promote particle ingestion, or may cooperate to generate a phagocytic signal. Cooperativity may involve activation of the phagocytic capacity of one receptor by another, including those that bind secreted cytokines and other substances (Table 2), or may reflect the physical association of two phagocytic receptors. Additionally, the production of common signaling intermediates, such as activated tyrosine kinases or Rho family GTPases (see below), could be amplified by the simultaneous activation of different classes of phagocyte surface receptors.

A novel type of receptor cooperativity involving FcγRIIA expressed on plasmacytoid dendritic cells was recently described. FcγRIIA was required for internalization of DNA-containing immune complexes obtained from sera of patients with SLE. FcγRIIA-mediated internalization led to the targeting of the DNA-containing immune complexes to a cellular compartment containing TLR9, resulting in increased production of IL-8 and IFN-α (Means et al., 2005).

Activation of One Receptor Enhances Phagocytic Activity of Another: "Inside-Out" FcγR Activation

Examples of this include activation of FcγRs on human neutrophils by the chemotactic peptide fMet-Leu-Phe and by RGD-containing proteins. Work by Brown and colleagues demonstrated that the RGD-dependent activation of both FcγR- and complement receptor-mediated phagocytosis occurs via ligation of a "leukocyte response integrin," which was later identified as the $\alpha_v\beta_3$ integrin. This integrin was found to be associated with CD47 (Brown et al., 1990; Lindberg et al., 1993). CD47-deficient mice demonstrate defects in phagocytosis and increased susceptibility to bacterial infection (Lindberg et al., 1996). Another interesting example of inside-out FcγR activation is the ability of FcγRIIIb to enhance the activity of FcγRIIA in human neutrophils. Ligation of the former induces an increase in FcγRIIA-mediated phagocytosis that is sensitive to inhibition by scavengers of reactive oxygen radicals, suggesting that the signaling pathway of GPI-linked FcγRIIIb signaling to FcγRIIA is through generation of oxidants (Salmon et al., 1995). Cooperativity between receptors during phagocytosis is likely to be important in the alveolar lining of the lung, which is the major portal for inhaled pathogenic microorganisms and noxious particles. Surfac-

TABLE 2 Secreted proteins and mediators that enhance the efficiency of FcγR-mediated phagocytosis[a]

Mediator	Reference(s)
Surfactant protein A (SP-A)	Tenner et al., 1989
Mannan binding protein (MBP)	Kuhlman et al., 1989; Tenner et al., 1995
C1q	Bobak et al., 1987, 1988
Laminin	Bohnsack et al., 1985
Fibronectin	Pommier et al., 1983; Wright et al., 1983
Serum amyloid P	Wright et al., 1983
IL-10	Capsoni et al., 1995
IL-1	Moxey-Mims et al., 1991; Simms et al., 1991
IFN-γ	Rabinovitch et al., 1980; Rollag et al., 1984
IFN-α/β	Rollag et al., 1984
GM-CSF	Capsoni et al., 1991; Collins and Bancroft, 1992
M-CSF	Sampson et al., 1991
TNF-α	Klebanoff et al., 1986
Formyl-Met-Leu-Phe	Rosales and Brown, 1991
Platelet-activating factor	Rosales and Brown, 1991
LTB_4, LTC_4	Canetti et al., 2006

[a] Enhancement of phagocytosis by these mediators occurs by different mechanisms. Cyokine-induced enhancement of phagocytosis typically requires gene induction, particularly of different FcγR isoforms, whereas lipid mediators typically enhance phagocytosis more rapidly, either by mobilization of FcγRs from latent cytoplasmic pools or by other undefined mechanisms.

tant protein A (SP-A) is the most abundant protein constituent of surfactant. SP-A is a member of a family of proteins containing collagen-like tail regions contiguous with globular lectin domains. Other members of the family include the complement protein C1q and mannose binding lectin. SP-A was shown to enhance FcγR-mediated phagocytosis (Tenner et al., 1989). Although the mechanism underlying this phenomenon is not certain, it was recently demonstrated that SP-A binds directly to the Fc portion of IgG at a site distinct from the C1q binding site (Lin and Wright, 2006).

Phagocytic Receptors Associate in Multisubunit Complexes

Biophysical evidence indicates that FcγRs and CR3 associate with each other at the cell surface (Poo et al., 1995; Zhou et al., 1993). Indeed, soluble FcγRIIIb interacts with both CD11b/CD18 and CD11c/CD18 and triggers release of IL-6 and IL-8 (Galon et al., 1996). The ability of soluble FcγRIIIb to trigger cell responses may explain how this receptor retains its signaling function when it is shed from cell surfaces. Although earlier studies attributed FcγRIIA-CR3 associations to lectin interactions between receptor ectodomains, a recent mutational analysis identified the E(253)-R(261) sequence within the $α_M$I-domain as part of the FcγRIIA binding interface within CR3 (Xiong et al., 2006). The capacity of integrins, including CR3, to interact with talin (Lim et al., 2007) and other actin binding proteins may help stabilize cytoskeletal interactions during phagocytosis. Indeed, talin accumulates in phagocytic cups of macrophages ingesting IgG-coated targets (Greenberg et al., 1990). Consistent with these observations is the finding that neutrophils that lack expression of CR3 are capable of spreading on immune complexes, but their capacity for sustained engagement is reduced (van Spriel et al., 2001).

CELLULAR EVENTS FOLLOWING ITAM PHOSPHORYLATION: CYTOSKELETAL ASSEMBLY

Phagocytosis requires the coordination of many cellular processes. As cytochalasins have long been observed to block phagocytosis by multiple receptors, actin assembly at the "barbed," or rapidly growing end of actin filaments, provides an essential driving force for particle engulfment. While the study of the regulation of actin polymerization is the focus of many laboratories around the world, here we focus on known pathways leading from activated Fcγ receptors to cytoskeletal assembly.

ITAM phosphorylation is accompanied by actin polymerization. The enzymatic steps leading to cytoskeletal assembly during phagocytosis are a matter of debate. Although Syk is required for phagocytosis, it may not be essential for all of the cytoskeletal alterations leading to phagocytosis (Crowley et al., 1997). Syk is clearly capable of autonomously triggering actin assembly (Greenberg et al., 1996) and is required for cytoskeletal assembly that accompanies complement receptor-mediated phagocytosis (Mocsai et al., 2002). However, pharmacological blockade of Syk led to decreased accumulation of F-actin-rich phagocytic "cups" beneath IgG-coated targets in mouse bone marrow-derived macrophages (Majeed et al., 2001), whereas it failed to do so in COS cells transfected with FcγRIIA (Cougoule et al., 2006). Perhaps these apparent discrepancies reflect differences in the cell types being studied. Although the precise role of Syk in cytoskeletal assembly during phagocytosis is unresolved, there is general agreement that activation of Rho family GTPases is a major driving force in the actin assembly that accompanies ITAM-based phagocytosis (reviewed in Niedergang and Chavrier, 2005).

Rho Family GTPases and Phagocytosis

Studies using dominant-negative constructs (Caron and Hall, 1998; Cox et al., 1997; Massol et al., 1998) and leukocytes from knockout mice (Hall et al., 2006; Yamauchi et al., 2004) demonstrate that Rac2 and/or Rac1 are required for FcγR-mediated phagocytosis and its underlying cytoskeletal alterations. The kinetics of recruitment of these Rac isoforms, as well as Cdc42, are distinct; Cdc42 activation is restricted to the leading margin of the cell, whereas Rac1 remains active throughout the phagocytic cup. During phagosome closure, activation of Rac1 and Rac2 increased uniformly and transiently in the actin-poor region of phagosomal membrane (Hoppe and Swanson,

2004), perhaps reflecting ongoing activation of the NADPH oxidase. Furthermore, Rac and Cdc42 appear to play distinct roles in modifying the actin-based cytoskeleton during phagocytosis. Cdc42 may control phagosomal maturation (Lerm et al., 2007). Among the downstream effectors of Cdc42 are WASP and WIP. WASP-deficient macrophages demonstrate partial phagocytic defects (Lorenzi et al., 2000).

The steps leading to activation of Rho family GTPases during phagocytosis have not been clearly delineated. Vav isoforms, leading candidates for guanine nucleotide exchange factors important in phagocytosis, turn out to be required for CR3-mediated, but not FcγR-mediated phagocytosis (Utomo et al., 2006). Their role in FcγR-mediated phagocytosis is to facilitate activation of the NADPH oxidase complex by a stoichiometric association with the NADPH oxidase component, p40^{phox} (Suh et al., 2006). Although the identity of all the relevant exchange factors for Rac and Cdc42 activation during phagocytosis is unknown, at least some of this activity has been attributed to a CrkII-DOCK180 complex (Lee et al., 2007). Regulation of Rac during phagocytosis may be even more complicated because calcium-promoted Ras inactivator (CAPRI), a Ras GTPase-activating protein, functions as an adaptor for Cdc42 and Rac1 during FcγR-mediated phagocytosis (Zhang et al., 2005).

A requirement for RhoA in FcγR-mediated phagocytosis is controversial. RhoA has been implicated in FcγR-mediated phagocytosis and early events that accompany FcγR ligation, including efficient binding of IgG ligand and FcγR capping (Hackam et al., 1997). These data are reminiscent of studies of CHO cells interacting with *Shigella*, in which addition of C3 exotoxin, an inhibitor of RhoA function, blocked early signaling events associated with phagocytosis, including enhanced protein tyrosine phosphorylation and bacterial association with the host cells (Watarai et al., 1997). Together, these studies suggest that intact RhoA function is required for maintaining the affinity and/or mobility of several classes of cell surface receptors. These findings are consistent with an earlier study showing that RhoA is required for cell adhesion via β_2 integrins (Laudanna et al., 1996). In contrast, Caron and Hall found that RhoA is required for CR3-mediated phagocytosis, but not for FcγR-mediated phagocytosis (Caron and Hall, 1998). Furthermore, two RhoA effectors have been implicated in CR3-, but not FcγR-mediated phagocytosis. Inhibition of the Rho/Rho kinase/myosin-II pathway caused a decreased accumulation of Arp2/3 complex and F-actin around bound complement-coated particles and led to a reduction in CR3-, but not FcγR-mediated phagocytosis (Olazabal et al., 2002). Another RhoA effector, mDia1, is recruited early during CR3-mediated phagocytosis and colocalizes with polymerized actin in the phagocytic cup. Interfering with mDia activity inhibits CR3-mediated phagocytosis while having no effect on FcγR-mediated phagocytosis (Colucci-Guyon et al., 2005). Thus, the bulk of the evidence supports an important primary role for Rac and Cdc42, but not RhoA, in cytoskeletal assembly mediated by FcγRs.

Actin Polymerization and Phosphatidylinositol-4-Phosphate-5-Kinase

Alternative mechanisms for activation of Rho family GTPases have been suggested, including direct association of the GTPases with phosphoinositides. PIP_2 stimulates dissociation of GDP from Cdc42 and RhoA (Zheng et al., 1996), while PIP_3 stimulates dissociation of GDP from Rac1 (Missy et al., 1998). However, the lack of sensitivity of FcγR-mediated actin polymerization to inhibitors of PI-3-kinase (Cox et al., 1999) suggests that PIP_3 is not responsible for activation of Rac1 during FcγR-mediated phagocytosis. An isoform of phosphatidylinositol-4-phosphate-5-kinase (PI5KIα) was shown to be activated during phagocytosis, and appears to be an essential early signal leading to actin assembly during FcγR-mediated phagocytosis (Coppolino et al., 2002; Scott et al., 2005). It is possible that PIP_2 synthesis contributes to Cdc42 activation during FcγR-mediated phagocytosis. In addition, many actin binding proteins bind to PIP_2 and PIP_2 may stimulate actin assembly during phagocytosis by triggering uncapping of previously capped actin filaments, thus creating new nucleation sites for actin assembly (Hartwig et al., 1995). This may underlie the requirement for the capping protein CapG in FcγR-mediated phagocytosis (Witke et al., 2001).

Actin Nucleation and Phagocytosis

Actin polymerization in vivo is catalyzed by actin-nucleating proteins. The best-characterized of these are the Arp2/3 family of proteins and formins (reviewed in Pollard, 2007). Recruitment and activation of Arp2/3 leads to formation of distinct cytoskeletal-based structures. Arp2/3 is localized to phagocytic cups and has been implicated in phagocytosis (May et al., 2000). Other cytoskeletal proteins that have been reported to localize to phagocytic cups and/or influence the efficiency of FcγR-mediated phagocytosis include coronin (Yan et al., 2005), Evl, Fyb/SLAP, VASP, WASP (Coppolino et al., 2001), galectin-3 (Sano et al., 2003), L-plastin (Jones and Brown, 1996), paxillin (Greenberg et al., 1994), and talin (Greenberg et al., 1990).

Formins comprise a family of actin binding proteins that have the interesting property of weakly capping the barbed ends of actin filaments (reviewed in Zigmond, 2004). This enables them to undergo repeated cycles of "attachment" and "detachment" from the growing actin filament, allowing for the insertion of new actin monomers into the growing actin filament. This activity may endow formins with the ability to coordinate actin assembly at the interface between the cortical actin cytoskeleton and the leading edge, thus conforming to the "Brownian ratchet" model for cell motility (Oster, 1987). Formins are likely to play a role in FcγR-mediated phagocytosis because the formin FRLα is required for efficient FcγR-mediated phagocytosis and is recruited to the phagocytic cup by Cdc42 (Seth et al., 2006).

CELLULAR EVENTS FOLLOWING ITAM PHOSPHORYLATION: MEMBRANE REMODELING

Cytoskeletal assembly is essential to the phagocytic process, but it is not the only essential step; it needs to be coordinated with a complicated array of membrane remodeling events, ultimately culminating in pseudopod extension and phagosomal maturation.

Pseudopod Extension and Phagosomal Closure

The signaling intermediates that mediate cytoskeletal assembly are different than those that signal pseudopod extension and phagosomal closure. For example, pseudopod extension and phagosomal closure, but not actin polymerization, depend on activation of PI-3-kinase(s), which represents a key nexus in signaling events downstream of FcγR engagement (Araki et al., 1996; Cox et al., 1999).

The p85α regulatory subunit (Munugalavadla et al., 2005) and the p110β (Leverrier et al., 2003) and p110α (Lee et al., 2007) catalytic subunits of type I PI-3-kinase have been implicated in FcγR-mediated phagocytosis. Type I PI-3-kinases are recruited to the phagosome by the adaptor protein Gab2 (Gu et al., 2003). Less is known about the molecular steps downstream of PI-3-kinase that contribute to phagocytosis. Several potential effectors for the lipid product of PI-3-kinase during phagocytosis have been proposed, including myosin X, an unconventional myosin with tandem PH domains (Cox et al., 2002).

Endomembrane Dynamics during Phagocytosis

Phagosome maturation is accompanied by multiple membrane fusion events. In neutrophils, in which the capacity for phagocytosis is less than that of macrophages, the immediate source of the phagosomal membrane is likely to be the plasma membrane itself (Hallett, 2006). In macrophages, the amount of plasma membrane that is internalized during large waves of sustained phagocytosis exceeds that of the surface area of unstimulated macrophages, indicating that endomembranes must replenish the internalized phagosomal membranes for phagocytosis to proceed. The source of these membranes is the subject of intense scrutiny in the field; studies have supported important roles for an endosomal pool, including a rapidly recycling, Rab11-containing pool (Cox et al., 2000), late endosomes (Braun et al., 2004), lysosomes (Czibener et al., 2006), and even endoplasmic reticulum (Gagnon et al., 2002). The potential for fusion of ER and phagosomal membranes is important from the standpoint of the acquired immune response because it provides a mechanism for "cross-presentation" of antigen by MHC class I molecules. In this form of antigen presentation, material that is ingested by phagocytosis can be further processed onto MHC class I molecules in a "phagosome to cytosol" transport of antigen. It is likely that the relative contributions of these endomembrane pools will depend on the particular leukocyte performing phagocytosis, its state of activation, and the nature of the phagosomal stimulus itself. The relative requirement for endomembranes in phagocytosis depends on the size and number of phagocytic targets that are engaged by the phagocyte. For example, Syt VII is a calcium-dependent regulator of lysosomal exocytosis. Syt VII$^{-/-}$ macrophages phagocytose normally at low particle/cell ratios but show a progressive inhibition in particle uptake under high load conditions. Complementation with Syt VII rescues this phenotype, but only when functional Ca^{2+} binding sites are retained. Reinforcing a role for Syt VII in Ca^{2+}-dependent phagocytosis, particle uptake in Syt VII$^{-/-}$ macrophages is significantly less dependent on $[Ca^{2+}]_i$ (Czibener et al., 2006).

Although a detailed discussion of the factors governing membrane dynamics during phagocytosis is beyond the scope of this chapter, many factors that regulate membrane dynamics in other contexts have been implicated in phagocytosis. These include ARF6, which, in addition to regulating membrane delivery during phagocytosis (Niedergang et al., 2003), may also stimulate cytoskeletal assembly during phagocytosis (Zhang et al., 1998), and the exocyst complex (Stuart et al., 2007).

PI-3-KINASE AND ACTIVATION OF THE NADPH OXIDASE DURING PHAGOCYTOSIS

In addition to promoting pseudopod extension, PI-3-kinase plays an important role in activation of the NADPH oxidase. Studies in neutrophils isolated from mice deficient in Vav or Rac isoforms demonstrate a critical role for Vav3 in Rac2-dependent activation of the NADPH oxidase following FcγR clustering. This was traced to a Rac- and actin polymerization-independent, but PI-3-kinase-dependent, phosphorylation of the NADPH oxidase component p40^{phox} (Utomo et al., 2006). The PI-3-kinase dependency of p40^{phox} phosphorylation is due to production of phosphatidylinositol 3-phosphate and its binding p40^{phox} by a specialized lipid-binding domain, the PX domain (Suh et al., 2006).

PHOSPHOLIPASES AND PHAGOCYTOSIS

FcγR ligation leads to the activation of one or more isoforms of phospholipase A_2 (Aderem et al., 1985; Lennartz and Brown, 1991; Suzuki et al., 1982). In human monocytes, a variety of PLA_2 inhibitors inhibit phagocytosis, which is restored in the presence of exogenous arachidonate (Lennartz and Brown, 1991). The precise role of PLA_2 in phagocytosis is not known. The requirement for PLA_2 in regulating endosomal trafficking (Mayorga et al., 1993) raises the possibility that PLA_2, like PI-3-kinase, may be required for the mobilization of a vesicular compartment necessary for phagocytosis. Ultrastructural studies support this concept and implicate PLA_2 in trafficking between this compartment and the plasma (Lennartz et al., 1997). Other phospholipases are activated following FcγR ligation, including PLC (Della Bianca et al., 1990, 1993; Dusi et al., 1994; Scholl et al., 1992; Shen et al., 1994; Ting et al., 1992) and PLD (Gewirtz and Simons, 1997; Kusner et al., 1999; Melendez et al., 1998). PLC is unlikely to be required for FcγR-mediated phagocytosis since neutrophils ingest IgG-opsonized particles under conditions that block detectable rises of inositol phosphates (Della Bianca et al., 1990), and macrophages deficient in PLC-γ2 ingest IgG-coated targets normally, despite a lack of detectable increases in $[Ca^{2+}]_i$ (Wen et al., 2002). PLC-γ may play other roles in phagocytosis, such as phagosomal remodeling (Botelho et al., 2000) and phagosome-lysosome fusion (Czibener et al., 2006) (see below). By promoting the hydrolysis of PIP_2, PLC-γ contributes to actin disassembly following particle engulfment (Scott et al., 2005).

PLD activity is required for FcγR-mediated phagocytosis (Kusner et al., 1999). Endogenous inhibition of PLD by ceramide may account for the observation that sphingomyelinase acts as a negative regulator of phagocytosis (Suchard et al., 1997). PLD1 is associated with late endosomes/lysosomes and PLD2 is associated with the cytosol and in discrete regions of the plasma membrane. Both isoforms are recruited to phagocytic cups and both are required for optimal phagocytosis (Corrotte et al., 2006). As PLD activity has been implicated in many cellular events, including synthesis of PIP_2 and vesicular fusion, PLD activity may be permissive for phagocytosis, it may be required for FcγR-stimulated signaling leading to phagocytosis, or both. Although the bulk of FcγR-stimulated PLD activation requires activation of Syk (Hitomi et al., 1999, 2001), recent work has uncovered a noncatalytic function of PLD in Syk activation; the PX domain of PLD2 associates with Syk in vitro and either catalytically active or inactive PLD2 enhances Syk-mediated phosphorylation by FcεRI ligation in mast cells (Lee et al., 2006). PLD may play other roles in phagocytic cells. For example, FcγR-stimulated PLD activation is coupled to sphingosine kinase activation, and sphingosine 1-phosphate, a product of sphingosine kinase, serves as a prosurvival factor for macrophages (Weigert et al., 2006).

PHAGOCYTOSIS AND ALTERATIONS IN $[Ca^{2+}]_i$

Many investigators have shown that FcγR ligation triggers elevations in $[Ca^{2+}]_i$ in a variety of leukocytes (reviewed in Greenberg, 1999). Peak $[Ca^{2+}]_i$ values are observed in periphagosomal regions (Sawyer et al., 1985) and correspond to a redistribution of several markers of intracellular Ca^{2+} stores (Stendahl et al., 1994). Examples of $[Ca^{2+}]_i$-dependent and -independent phagocytosis abound, but the bulk of the evidence suggests that cytosolic calcium transients are not necessary for phagocytosis to proceed. Cytosolic transients may be involved in other aspects of phagocytic function, such as chemotactic peptide enhancement of FcγR-mediated phagocytosis in neutrophils (Rosales and Brown, 1991), arachidonic acid production (Aderem et al., 1986), phagosome-lysosome fusion (Jaconi et al., 1990), and lysosome-enhanced phagocytosis (Czibener et al., 2006) (see above). An interesting study demonstrated that sequences within FcγRIIA control calcium wave propagation patterns, which correlated with phagosome-lysosome fusion in CHO cells transfected with this Fc receptor (Worth et al., 2003).

SERINE/THREONINE KINASES AND PHAGOCYTOSIS

Activation of many protein serine/threonine kinases has been reported following engagement of Fc receptors. These include PKC (Brozna et al., 1988; Zheleznyak and Brown, 1992), protein kinase A (Nitta and Suzuki, 1982; Smolen et al., 1980; Ydrenius et al., 2000), casein kinase II (Hirata and Suzuki, 1987), calcium/calmodulin-dependent protein kinase II (Liang and Huang, 1995), Akt (Tilton et al., 1997), histone H4 protein kinase (Liang and Huang, 1995), and multiple isoforms of MAP kinases (Durden et al., 1995; Liang and Huang, 1995; Park et al., 1996; Rose et al., 1997; Trotta et al., 1996). There is evidence that PKC activity is required for FcγR-mediated phagocytosis (Allen and Aderem, 1996; Zheleznyak and Brown, 1992), in particular, PKC-ε (Larsen et al., 2002). The PKC-dependent step in phagocytosis has not been clearly defined. It is likely that many more phosphorylation events will be identified during phagocytosis, some of which affect the phagocytic process itself, but many more of which influence postphagocytic events, including alterations in gene expression.

Work from this laboratory was supported by grants HL54164 and AI067502 from the National Institutes of Health and grants from the American Cancer Society, the American Heart Association, and the Stony Wold Herbert Fund.

REFERENCES

Aderem, A. A., W. A. Scott, and Z. A. Cohn. 1986. Evidence for sequential signals in the induction of the arachidonic acid cascade in macrophages. *J. Exp. Med.* **163:**139–154.

Aderem, A. A., S. D. Wright, S. C. Silverstein, and Z. A. Cohn. 1985. Ligated complement receptors do not activate the arachidonic acid cascade in resident peritoneal macrophages. *J. Exp. Med.* **161:**617–622.

Agarwal, A., P. Salem, and K. C. Robbins. 1993. Involvement of p72^{syk}, a protein-tyrosine kinase, in Fc$_γ$ receptor signaling. *J. Biol. Chem.* **268:**15900–15905.

Allen, L. A. H., and A. Aderem. 1996. Molecular definition of distinct cytoskeletal structures involved in complement- and Fc receptor-mediated phagocytosis in macrophages. *J. Exp. Med.* **184:**627–637.

Altieri, D. C., and T. S. Edgington. 1988. The saturable high affinity association of factor X to ADP-stimulated monocytes defines a novel function of the Mac-1 receptor. *J. Biol. Chem.* **263:**7007–7015.

Amigorena, S., C. Bonnerot, J. R. Drake, D. Choquet, W. Hunziker, J. G. Guillet, P. Webster, C. Sautes, I. Mellman, and W. H. Fridman. 1992. Cytoplasmic domain heterogeneity and functions of IgG Fc receptors in B lymphocytes. *Science* **256:**1808–1812.

Amoui, M., L. Draberova, P. Tolar, and P. Draber. 1997. Direct interaction of syk and lyn protein tyrosine kinases in rat basophilic leukemia cells activated via type I Fcε receptors. *Eur. J. Immunol.* **27:**321–328.

Anderson, C. L., P. M. Guyre, J. C. Whitin, D. H. Ryan, R. J. Looney, and M. W. Fanger. 1986. Soluble circulating Fcγ receptors on human mononuclear phagocytes. Antibody characterization and induction of superoxide production in a monocyte cell line. *J. Biol. Chem.* **261:**12856–12864.

Anderson, C. L., L. Shen, D. M. Eicher, M. D. Wewers, and J. K. Gill. 1990. Phagocytosis mediated by three distinct Fc$_γ$ receptor classes on human leukocytes. *J. Exp. Med.* **171:**1333–1345.

Anderson, P., M. Caliguiri, C. O'Brien, T. Manley, J. Ritz, and S. F. Schlossman. 1990. Fcγ receptor type III (CD16) is included in the ζ NK receptor complex expressed by human natural killer cells. *Proc. Natl. Acad. Sci. USA* **87:**2274–2278.

Araki, N., M. T. Johnson, and J. A. Swanson. 1996. A role for phosphoinositide 3-kinase in the completion of macropinocytosis and phagocytosis by macrophages. *J. Cell Biol.* **135:**1249–1260.

Barnes, N., A. L. Gavin, P. S. Tan, P. Mottram, F. Koentgen, and P. M. Hogarth. 2002. FcγRI-deficient mice show multiple alterations to inflammatory and immune responses. *Immunity* **16:**379–389.

Barnes, N. C., M. S. Powell, H. M. Trist, A. L. Gavin, B. D. Wines, and P. M. Hogarth. 2006. Raft localisation of FcγRIIa and efficient signaling are dependent on palmitoylation of cysteine 208. *Immunol. Lett.* **104:**118–123.

Beekman, J. M., J. E. Bakema, J. G. van de Winkel, and J. H. Leusen. 2004. Direct interaction between FcγRI (CD64) and periplakin controls receptor endocytosis and ligand binding capacity. *Proc. Natl. Acad. Sci. USA* **101:**10392–10397.

Benhamou, M., N. J. Ryba, H. Kihara, H. Nishikata, and R. P. Siraganian. 1993. Protein-tyrosine kinase p72^{syk} in high affinity IgE receptor signaling. Identification as a component of pp72 and association with the receptor γ chain after receptor aggregation. *J. Biol. Chem.* **268:**23318–23324.

Berger, S. A., T. W. Mak, and C. J. Paige. 1994. Leukocyte common antigen (CD45) is required for immunoglobulin E-mediated degranulation of mast cells. *J. Exp. Med.* **180:**471–476.

Bergtold, A., D. D. Desai, A. Gavhane, and R. Clynes. 2005. Cell surface recycling of internalized antigen permits dendritic cell priming of B cells. *Immunity* **23:**503–514.

Bijsterbosch, M. K., and G. G. Klaus. 1985. Crosslinking of surface immunoglobulin and Fc receptors on B lymphocytes inhibits stimulation of inositol phospholipid breakdown via the antigen receptors. *J. Exp. Med.* **162:**1825–1836.

Bishop, B., and C. M. Lloyd. 2003. CC chemokine ligand 1 promotes recruitment of eosinophils but not Th2 cells during the development of allergic airways disease. *J. Immunol.* **170:** 4810–4817.

Bobak, D. A., M. M. Frank, and A. J. Tenner. 1988. C1q acts synergistically with phorbol dibutyrate to activate CR1-mediated phagocytosis by human mononuclear phagocytes. *Eur. J. Immunol.* **18:**2001–2007.

Bobak, D. A., T. A. Gaither, M. M. Frank, and A. J. Tenner. 1987. Modulation of FcR function by complement: subcomponent C1q enhances the phagocytosis of IgG-opsonized targets by human monocytes and culture-derived macrophages. *J. Immunol.* **138:**1150–1156.

Bodman-Smith, K. B., A. J. Melendez, I. Campbell, P. T. Harrison, J. M. Allen, and J. G. Raynes. 2002. C-reactive protein-mediated phagocytosis and phospholipase D signalling through the high-affinity receptor for immunoglobulin G (FcγRI). *Immunology* **107:**252–260.

Bohnsack, J. F., H. K. Kleinman, T. Takahashi, J. J. O'Shea, and E. J. Brown. 1985. Connective tissue proteins and phagocytic cell function: laminin enhances complement and Fc-mediated phagocytosis by cultured human phagocytes. *J. Exp. Med.* **161:**912–923.

Bolland, S., R. N. Pearse, T. Kurosaki, and J. V. Ravetch. 1998. SHIP modulates immune receptor responses by regulating membrane association of BTK. *Immunity* **8:**509–516.

Bolland, S., and J. V. Ravetch. 1999. Inhibitory pathways triggered by ITIM-containing receptors. *Adv. Immunol.* **72:** 149–177.

Bolland, S., and J. V. Ravetch. 2000. Spontaneous autoimmune disease in FcgRIIB-deficient mice results from strain-specific epistasis. *Immunity* **13:**277–285.

Booth, J. W., M. K. Kim, A. Jankowski, A. D. Schreiber, and S. Grinstein. 2002. Contrasting requirements for ubiquitylation during Fc receptor-mediated endocytosis and phagocytosis. *EMBO J.* **21:**251–258.

Botelho, R. J., M. Teruel, R. Dierckman, R. Anderson, A. Wells, J. D. York, T. Meyer, and S. Grinstein. 2000. Localized biphasic changes in phosphatidylinositol-4,5-bisphosphate at sites of phagocytosis. *J. Cell Biol.* **151:**1353–1368.

Braun, V., V. Fraisier, G. Raposo, I. Hurbain, J. B. Sibarita, P. Chavrier, T. Galli, and F. Niedergang. 2004. TI-VAMP/VAMP7 is required for optimal phagocytosis of opsonised particles in macrophages. *EMBO J.* **23:**4166–4176.

Brooks, D. G., W. Q. Qui, A. D. Luster, and J. V. Ravetch. 1989. Structure and expression of human IgG FcRII (CD32). Functional heterogeneity is encoded by the alternatively spliced products of multiple genes. *J. Exp. Med.* **170:**1369–1385.

Brown, E., L. Hooper, T. Ho, and H. Gresham. 1990. Integrin-associated protein: a 50 kD plasma membrane antigen physically and functionally associated with integrins. *J. Cell Biol.* **111:**2785–2794.

Brown, M. T., and J. A. Cooper. 1996. Regulation, substrates and functions of src. *Biochim. Biophys. Acta* **1287:**121–149.

Brozna, J. P., N. F. Hauff, W. A. Phillips, and R. B. Johnston, Jr. 1988. Activation of the respiratory burst in macrophages. Phosphorylation specifically associated with Fc receptor-mediated stimulation. *J. Immunol.* **141:**1642–1647.

Buckle, A. M., and N. Hogg. 1989. The effect of IFN-γ and colony-stimulating factors on the expression of neutrophil cell membrane receptors. *J. Immunol.* **143:**2295–2301.

Canetti, C., D. M. Aronoff, M. Choe, N. Flamand, S. Wettlaufer, G. B. Toews, G. H. Chen, and M. Peters-Golden. 2006. Differential regulation by leukotrienes and calcium of Fc γ receptor-induced phagocytosis and Syk activation in dendritic cells versus macrophages. *J. Leukoc. Biol.* **79:**1234–1241.

Canfield, S. M., and S. L. Morrison. 1991. The binding affinity of human IgG for its high affinity Fc receptor is determined by multiple amino acids in the CH2 domain and is modulated by the hinge region. *J. Exp. Med.* **173:**1483–1491.

Capsoni, F., P. Bonara, F. Minonzio, A. M. Ongari, G. Colombo, G. P. Rizzardi, and C. Zanussi. 1991. The effect of cytokines on human neutrophil Fc receptor-mediated phagocytosis. *J. Clin. Lab. Immunol.* **34:**115–124.

Capsoni, F., F. Minonzio, A. M. Ongari, V. Carbonelli, A. Galli, and C. Zanussi. 1995. IL-10 up-regulates human monocyte phagocytosis in the presence of IL-4 and IFN-γ. *J. Leukoc. Biol.* **58:**351-358.

Caron, E., and A. Hall. 1998. Identification of two distinct mechanisms of phagocytosis controlled by different Rho GTPases. *Science* **282:**1717–1721.

Chensue, S. W., N. W. Lukacs, T. Y. Yang, X. Shang, K. A. Frait, S. L. Kunkel, T. Kung, M. T. Wiekowski, J. A. Hedrick, D. N. Cook, A. Zingoni, S. K. Narula, A. Zlotnik, F. J. Barrat, A. O'Garra, M. Napolitano, and S. A. Lira. 2001. Aberrant in vivo T helper type 2 cell response and impaired eosinophil recruitment in CC chemokine receptor 8 knockout mice. *J. Exp. Med.* **193:**573–584.

Chouchakova, N., J. Skokowa, U. Baumann, T. Tschernig, K. M. Philippens, B. Nieswandt, R. E. Schmidt, and J. E. Gessner. 2001. Fc γ RIII-mediated production of TNF-α induces immune complex alveolitis independently of CXC chemokine generation. *J. Immunol.* **166:**5193-5200.

Clark, M. R., S. G. Stuart, R. P. Kimberly, P. A. Ory, and I. M. Goldstein. 1991. A single amino acid distinguishes the high-responder from the low-responder form of Fc receptor II on human monocytes. *Eur. J. Immunol.* **21:**1911–1916.

Clarkson, S. B., R. P. Kimberly, J. E. Valinsky, M. D. Witmer, J. B. Bussel, R. L. Nachman, and J. C. Unkeless. 1986. Blockade of clearance of immune complexes by an anti-Fcγ receptor monoclonal antibody. *J. Exp. Med.* **164:** 474–489.

Cohen-Solal, J. F., L. Cassard, W. H. Fridman, and C. Sautes-Fridman. 2004. Fc gamma receptors. *Immunol. Lett.* **92:**199–205.

Collins, H. L., and G. J. Bancroft. 1992. Cytokine enhancement of complement-dependent phagocytosis by macrophages: synergy of tumor necrosis factor-alpha and granulocyte-macrophage colony-stimulating factor for phagocytosis of Cryptococcus neoformans. *Eur. J. Immunol.* **22:** 1447–1454.

Colucci-Guyon, E., F. Niedergang, B. J. Wallar, J. Peng, A. S. Alberts, and P. Chavrier. 2005. A role for mammalian diaphanous-related formins in complement receptor (CR3)-mediated phagocytosis in macrophages. *Curr. Biol.* **15:**2007–2012.

Cooney, D. S., H. Phee, A. Jacob, and K. M. Coggeshall. 2001. Signal transduction by human-restricted Fc γ RIIa involves three distinct cytoplasmic kinase families leading to phagocytosis. *J. Immunol.* **167:**844–854.

Coppolino, M. G., R. Dierckman, J. Loijens, R. F. Collins, M. Pouladi, J. Jongstra-Bilen, A. D. Schreiber, W. S. Trimble, R. Anderson, and S. Grinstein. 2002. Inhibition of phosphatidylinositol-4-phosphate 5-kinase Iα impairs localized actin remodeling and suppresses phagocytosis. *J. Biol. Chem.* **277:**43849–43857.

Coppolino, M. G., M. Krause, P. Hagendorff, D. A. Monner, W. Trimble, S. Grinstein, J. Wehland, and A. S. Sechi. 2001. Evidence for a molecular complex consisting of Fyb/SLAP, SLP-76, Nck, VASP and WASP that links the actin cytoskeleton to Fcγ receptor signalling during phagocytosis. *J. Cell Sci.* **114:**4307–4318.

Corrotte, M., S. Chasserot-Golaz, P. Huang, G. Du, N. T. Ktistakis, M. A. Frohman, N. Vitale, M. F. Bader, and N. J. Grant. 2006. Dynamics and function of phospholipase D and phosphatidic acid during phagocytosis. *Traffic* **7:**365–377.

Cougoule, C., S. Hoshino, A. Dart, J. Lim, and E. Caron. 2006. Dissociation of recruitment and activation of the small G-protein Rac during Fcγ receptor-mediated phagocytosis. *J. Biol. Chem.* **281**:8756–8764.

Cox, D., J. S. Berg, M. Cammer, J. O. Chinegwundoh, B. M. Dale, R. E. Cheney, and S. Greenberg. 2002. Myosin X is a downstream effector of PI(3)K during phagocytosis. *Nat. Cell Biol.* **4**:469-477.

Cox, D., P. Chang, T. Kurosaki, and S. Greenberg. 1996. Syk tyrosine kinase is required for immunoreceptor tyrosine activation motif-dependent actin assembly. *J. Biol. Chem.* **271**:16597–16602.

Cox, D., P. Chang, Q. Zhang, P. G. Reddy, G. M. Bokoch, and S. Greenberg. 1997. Requirements for both Rac1 and Cdc42 in membrane ruffling and phagocytosis in leukocytes. *J. Exp. Med.* **186**:1487–1494.

Cox, D., B. M. Dale, M. Kashiwada, C. D. Helgason, and S. Greenberg. 2001. A regulatory role for Src homology 2 domain-containing inositol 5'-phosphatase (SHIP) in phagocytosis mediated by Fc gamma receptors and complement receptor 3 (alpha(M)beta(2); CD11b/CD18). *J. Exp. Med.* **193**:61–71.

Cox, D., and S. Greenberg. 2001. Phagocytic signaling strategies: Fcγ receptor-mediated phagocytosis as a model system. *Semin. Immunol.* **13**:339–345.

Cox, D., D. J. Lee, B. M. Dale, J. Calafat, and S. Greenberg. 2000. A Rab11-containing rapidly recycling compartment in macrophages that promotes phagocytosis. *Proc. Natl. Acad. Sci. USA* **97**:680–685.

Cox, D., C.-C. Tseng, G. Bjekic, and S. Greenberg. 1999. A requirement for phosphatidylinositol 3-kinase in pseudopod extension. *J. Biol. Chem.* **274**:1240–1247.

Crowley, M. T., P. S. Costello, C. J. Fitzer-Attas, M. Turner, F. Meng, C. Lowell, V. L. J. Tybulewicz, and A. L. DeFranco. 1997. A critical role for Syk in signal transduction and phagocytosis mediated by Fcγ receptors on macrophages. *J. Exp. Med.* **186**:1027–1039.

Czibener, C., N. M. Sherer, S. M. Becker, M. Pypaert, E. Hui, E. R. Chapman, W. Mothes, and N. W. Andrews. 2006. Ca^{2+} and synaptotagmin VII-dependent delivery of lysosomal membrane to nascent phagosomes. *J. Cell Biol.* **174**:997–1007.

Daeron, M. 1996. Building up the family of ITIM-bearing negative coreceptors. *Immunol. Lett.* **54**:73–76.

Daeron, M., and R. Lesourne. 2006. Negative signaling in Fc receptor complexes. *Adv Immunol* **89**:39–86.

Davis, W., P. T. Harrison, M. J. Hutchinson, and J. M. Allen. 1995. Two distinct regions of Fcγ RI initiate separate signalling pathways involved in endocytosis and phagocytosis. *EMBO J.* **14**:432–441.

Debets, J. M., J. G. van de Winkel, J. L. Ceuppens, I. E. Dieteren, and W. A. Buurman. 1990. Cross-linking of both FcγRI and FcγRII induces secretion of tumor necrosis factor by human monocytes, requiring high affinity Fc-FcγR interactions. Functional activation of FcγRII by treatment with proteases or neuraminidase. *J. Immunol.* **144**:1304–1310.

Debets, J. M., C. J. Van der Linden, I. E. Dieteren, J. F. Leeuwenberg, and W. A. Buurman. 1988. Fc-receptor crosslinking induces rapid secretion of tumor necrosis factor (cachectin) by human peripheral blood monocytes. *J. Immunol.* **141**:1197–1201.

Della Bianca, V., M. Grzeskowiak, S. Dusi, and F. Rossi. 1993. Formation of inositol (1,4,5) trisphosphate and increase of cytosolic Ca^{2+} mediated by Fc receptors in human neutrophils. *Biochem. Biophys. Res. Comm.* **196**:1233–1239.

Della Bianca, V., M. Grzeskowiak, and F. Rossi. 1990. Studies on molecular regulation of phagocytosis and activation of the NADPH oxidase in neutrophils-IgG- and C3b- mediated ingestion and associated respiratory burst independent of phospholipid turnover and Ca^{2+} transients. *J. Immunol.* **144**:1411–1417.

Denu, J. M., and K. G. Tanner. 1998. Specific and reversible inactivation of protein tyrosine phosphatases by hydrogen peroxide: evidence for a sulfenic acid intermediate and implications for redox regulation. *Biochemistry* **37**:5633–5642.

de Waal Malefyt, R., C. G. Figdor, R. Huijbens, S. Mohan-Peterson, B. Bennett, J. Culpepper, W. Dang, G. Zurawski, and J. E. de Vries. 1993. Effects of IL-13 on phenotype, cytokine production, and cytotoxic function of human monocytes. Comparison with IL-4 and modulation by IFN-γ or IL-10. *J. Immunol.* **151**:6370–6381.

Diamond, M. S., R. Alon, C. A. Parkos, M. T. Quinn, and T. A. Springer. 1995. Heparin is an adhesive ligand for the leukocyte integrin Mac-1 (CD11b/CD18). *J. Cell Biol.* **130**:1473–1482.

Drechsler, Y., S. Chavan, D. Catalano, P. Mandrekar, and G. Szabo. 2002. FcγR cross-linking mediates NF-kB activation, reduced antigen presentation capacity, and decreased IL-12 production in monocytes without modulation of myeloid dendritic cell development. *J. Leukoc. Biol.* **72**:657–667.

Duchemin, A.-M., and C. L. Anderson. 1997. Association of non-receptor protein tyrosine kinases with the FcγRI/γ-chain complex in monocytic cells. *J. Immunol.* **158**:865–871.

Duchemin, A. M., L. K. Ernst, and C. L. Anderson. 1994. Clustering of the high affinity Fc receptor for immunoglobulin G (FcγRI) results in phosphorylation of its associated γ-chain. *J. Biol. Chem.* **269**:12111–12117.

Duits, A. J., H. Bootsma, R. H. Derksen, P. E. Spronk, L. Kater, C. G. Kallenberg, P. J. Capel, N. A. Westerdaal, G. T. Spierenburg, F. H. Gmelig-Meyling, et al. 1995. Skewed distribution of IgG Fc receptor IIa (CD32) polymorphism is associated with renal disease in systemic lupus erythematosus patients. *Arthritis Rheum.* **38**:1832–1836.

Durden, D. L., H. M. Kim, B. Calore, and Y. B. Liu. 1995. The FcγRI receptor signals through the activation of *hck* and MAP kinase. *J. Immunol.* **154**:4039–4047.

Dusi, S., M. Donini, B. V. Della, and F. Rossi. 1994. Tyrosine phosphorylation of phospholipase C-γ2 is involved in the activation of phosphoinositide hydrolysis by Fc receptors in human neutrophils. *Biochem. Biophys. Res. Commun.* **201**:1100–1108.

Ernst, L. K., J. G. J. van de Winkel, I.-M. Chiu, and C. L. Anderson. 1992. Three genes for the human high affinity Fc receptor for IgG (FcγRI) encode four distinct transcription products. *J. Biol. Chem.* **267**:15692–15700.

Fahy, R. J., P. T. Diaz, J. Hart, and M. D. Wewers. 2001. BAL and serum IgG levels in healthy asymptomatic HIV-infected patients. *Chest* **119**:196–203.

Fallman, M., K. Andersson, S. Hakansson, K.-E. Magnusson, O. Stendahl, and H. Wolf-Watz. 1995. *Yersinia pseudotuberculosis* inhibits Fc receptor-mediated phagocytosis in J774 cells. *Infect. Immun.* **63**:3117–3124.

Fanger, N. A., D. Voigtlaender, C. Liu, S. Swink, K. Wardwell, J. Fisher, R. F. Graziano, L. C. Pfefferkorn, and P. M. Guyre. 1997. Characterization of expression, cytokine regulation, and effector function of the high affinity IgG receptor Fc γ RI (CD64) expressed on human blood dendritic cells. *J. Immunol.* **158**:3090–3098.

Fernandes, M. J., E. Rollet-Labelle, G. Pare, S. Marois, M. L. Tremblay, J. L. Teillaud, and P. H. Naccache. 2006. CD16b associates with high-density, detergent-resistant membranes in human neutrophils. *Biochem. J.* **393**:351–359.

Gagnon, E., S. Duclos, C. Rondeau, E. Chevet, P. H. Cameron, O. Steele-Mortimer, J. Paiement, J. J. Bergeron, and M. Desjardins. 2002. Endoplasmic reticulum-mediated phagocytosis is a mechanism of entry into macrophages. *Cell* **110**:119–131.

Galon, J., J. F. Gauchat, N. Mazieres, R. Spagnoli, W. Storkus, M. Lotze, J. Y. Bonnefoy, W. H. Fridman, and C. Sautes. 1996. Soluble Fcγ receptor type III (FcγRIII, CD16) triggers cell activation through interaction with complement receptors. *J. Immunol.* **157:**1184–1192.

Ganesan, L. P., H. Fang, C. B. Marsh, and S. Tridandapani. 2003. The protein-tyrosine phosphatase SHP-1 associates with the phosphorylated immunoreceptor tyrosine-based activation motif of Fc γ RIIa to modulate signaling events in myeloid cells. *J. Biol. Chem.* **278:**35710–35717.

Garcia-Garcia, E., E. J. Brown, and C. Rosales. 2007. Transmembrane mutations to FcγRIIA alter its association with lipid rafts: implications for receptor signaling. *J. Immunol.* **178:**3048–3058.

Gessner, J. E., H. Heiken, A. Tamm, and R. E. Schmidt. 1998. The IgG Fc receptor family. *Ann. Hematol.* **76:**231–248.

Gewirtz, A. T., and E. R. Simons. 1997. Phospholipase D mediates Fcγ receptor activation of neutrophils and provides specificity between high-valency immune complexes and fMLP signaling pathways. *J. Leukoc. Biol.* **61:**522-528.

Ghazizadeh, S., J. B. Bolen, and H. B. Fleit. 1994. Physical and functional association of Src-related protein tyrosine kinases with Fcγ RII in monocytic THP-1 cells. *J. Biol. Chem.* **269:**8878–8884.

Ghazizadeh, S., J. B. Bolen, and H. B. Fleit. 1995. Tyrosine phosphorylation and association of Syk with Fcγ RII in monocytic THP-1 cells. *Biochem. J.* **305:**669–674.

Ghazizadeh, S., and H. B. Fleit. 1994. Tyrosine phosphorylation provides an obligatory early signal for Fcγ RII-mediated endocytosis in the monocytic cell line THP-1. *J. Immunol.* **152:**30–41.

Girard, M. T., S. Hjaltadottir, A. N. Fejes-Toth, and P. M. Guyre. 1987. Glucocorticoids enhance the γ-interferon augmentation of human monocyte immunoglobulin G Fc receptor expression. *J. Immunol.* **138:**3235–3241.

Gold, E. S., D. M. Underhill, N. S. Morrissette, J. Guo, M. A. McNiven, and A. Aderem. 1999. Dynamin 2 is required for phagocytosis in macrophages. *J. Exp. Med.* **190:**1849–1856.

Green, J. M., A. D. Schreiber, and E. J. Brown. 1997. Role for a glycan phosphoinositol anchor in Fcγ receptor synergy. *J. Cell Biol.* **139:**1209–1217.

Greenberg, S. 1999. Fc receptor-mediated phagocytosis, p. 149-191. *In* S. Gordon (ed.), *Phagocytosis: The Host*, vol. 5. JAI Press, Stamford, CT.

Greenberg, S., K. Burridge, and S. C. Silverstein. 1990. Colocalization of F-actin and talin during Fc receptor-mediated phagocytosis in mouse macrophages. *J. Exp. Med.* **172:**1853–1856.

Greenberg, S., P. Chang, and S. C. Silverstein. 1993. Tyrosine phosphorylation is required for Fc receptor-mediated phagocytosis in mouse macrophages. *J. Exp. Med.* **177:**529–534.

Greenberg, S., P. Chang, and S. C. Silverstein. 1994. Tyrosine phosphorylation of the γ subunit of Fc$_\gamma$ receptors, p72^{syk}, and paxillin during Fc receptor-mediated phagocytosis in macrophages. *J. Biol. Chem.* **269:**3897–3902.

Greenberg, S., P. Chang, D. Wang, R. Xavier, and B. Seed. 1996. Clustered Syk tyrosine kinase domains trigger phagocytosis. *Proc. Natl. Acad. Sci. USA* **93:**1103–1107.

Gresham, H. D., B. M. Dale, J. W. Potter, P. W. Chang, C. M. Vines, C. A. Lowell, C. F. Lagenaur, and C. L. Willman. 2000. Negative regulation of phagocytosis in murine macrophages by the Src kinase family member, Fgr. *J. Exp. Med.* **191:**515–528.

Gu, H., R. J. Botelho, M. Yu, S. Grinstein, and B. G. Neel. 2003. Critical role for scaffolding adapter Gab2 in Fc γ R-mediated phagocytosis. *J. Cell Biol.* **161:**1151-1161.

Gupta, N., A. M. Scharenberg, D. N. Burshtyn, N. Wagtmann, M. N. Lioubin, L. R. Rohrschneider, J. P. Kinet, and E. O. Long. 1997. Negative signaling pathways of the killer cell inhibitory receptor and Fc γ RIIb1 require distinct phosphatases. *J. Exp. Med.* **186:**473–478.

Gupta, R. K., and G. R. Siber. 1995. Method for quantitation of IgG subclass antibodies in mouse serum by enzyme-linked immunosorbent assay. *J. Immunol. Methods* **181:**75–81.

Guriec, N., C. Daniel, K. Le Ster, E. Hardy, and C. Berthou. 2006. Cytokine-regulated expression and inhibitory function of FcγRIIB1 and -B2 receptors in human dendritic cells. *J. Leukoc. Biol.* **79:**59–70.

Hackam, D. J., O. D. Rotstein, A. Schreiber, W. J. Zhang, and S. Grinstein. 1997. Rho is required for the initiation of calcium signaling and phagocytosis by Fcγ receptors in macrophages. *J. Exp. Med.* **186:**955–966.

Hall, A. B., M. A. Gakidis, M. Glogauer, J. L. Wilsbacher, S. Gao, W. Swat, and J. S. Brugge. 2006. Requirements for Vav guanine nucleotide exchange factors and Rho GTPases in FcγR- and complement-mediated phagocytosis. *Immunity* **24:**305–316.

Hallett, M. B. 2006. Phagocytosis of optically-trapped particles: delivery of the pure phagocytic signal. *Cell Res.* **16:**852–854.

Hamada, F., M. Aoki, T. Akiyama, and K. Toyoshima. 1993. Association of immunoglobulin G Fc receptor II with Src-like protein-tyrosine kinase Fgr in neutrophils. *Proc. Natl. Acad. Sci. USA* **90:**6305–6309.

Hamerman, J. A., and L. L. Lanier. 2006. Inhibition of immune responses by ITAM-bearing receptors. *Sci. STKE* **2006:**re1.

Hartwig, J. H., G. M. Bokoch, C. L. Carpenter, P. A. Janmey, L. A. Taylor, A. Toker, and T. P. Stossel. 1995. Thrombin receptor ligation and activated Rac uncap actin filament barbed ends through phosphoinositide synthesis in permeabilized human platelets. *Cell* **82:**643–653.

Hauschildt, S., and B. Kleine. 1995. Bacterial stimulators of macrophages. *Int. Rev. Cytol.* **161:**263–331.

Hazenbos, W. L., J. E. Gessner, F. M. Hofhuis, H. Kuipers, D. Meyer, I. A. Heijnen, R. E. Schmidt, M. Sandor, P. J. Capel, M. Daeron, J. G. van de Winkel, and J. S. Verbeek. 1996. Impaired IgG-dependent anaphylaxis and Arthus reaction in Fc γ RIII (CD16) deficient mice. *Immunity* **5:**181–188.

Herr, A. B., E. R. Ballister, and P. J. Bjorkman. 2003. Insights into IgA-mediated immune responses from the crystal structures of human FcαRI and its complex with IgA1-Fc. *Nature* **423:**614–620.

Hibbs, M. L., P. Selvaraj, O. Carpen, T. A. Springer, H. Kuster, M. E. Jouvin, and J. Kinet. 1989. Mechanisms for regulating expression of membrane isoforms of FcγRIII (CD16). *Science* **246:**1608–1611.

Hirasawa, N., A. Scharenberg, H. Yamamura, M. A. Beaven, and J.-P. Kinet. 1995. A requirement for syk in the activation of the microtubule-associated protein kinase/phospholipase A$_2$ pathway by Fc$_\varepsilon$R1 is not shared by a G protein-coupled receptor. *J. Biol. Chem.* **270:**10960–10967.

Hirata, Y., and T. Suzuki. 1987. Protein kinase activity associated with Fcγ$_{2a}$ receptor of a murine macrophage like cell line, P388D$_1$. *Biochemistry* **26:**8189–8195.

Hitomi, T., S. Yanagi, R. Inatome, J. Ding, T. Takano, and H. Yamamura. 2001. Requirement of Syk-phospholipase C-γ2 pathway for phorbol ester-induced phospholipase D activation in DT40 cells. *Genes Cells* **6:**475–485.

Hitomi, T., S. Yanagi, R. Inatome, and H. Yamamura. 1999. Cross-linking of the B cell receptor induces activation of phospholipase D through Syk, Btk and phospholipase C-γ2. *FEBS Lett.* **445:**371–374.

Hogarth, P. M. 2002. Fc receptors are major mediators of antibody based inflammation in autoimmunity. *Curr. Opin. Immunol.* **14**:798–802.

Hoppe, A. D., and J. A. Swanson. 2004. Cdc42, Rac1, and Rac2 display distinct patterns of activation during phagocytosis. *Mol. Biol. Cell* **15**:3509–3519.

Hostoffer, R. W., I. Krukovets, and M. Berger. 1994. Enhancement by tumor necrosis factor-α of Fc$_\alpha$ receptor expression and IgA-mediated superoxide generation and killing of *Pseudomonas aeruginosa* by polymorphonuclear leukocytes. *J. Infect. Dis.* **170**:82–87.

Hostoffer, R. W., I. Krukovets, and M. Berger. 1993. Increased FcαR expression and IgA-mediated function on neutrophils induced by chemoattractants. *J. Immunol.* **150**:4532–4540.

Huang, Z. Y., D. R. Barreda, R. G. Worth, Z. K. Indik, M. K. Kim, P. Chien, and A. D. Schreiber. 2006. Differential kinase requirements in human and mouse Fc-γ receptor phagocytosis and endocytosis. *J. Leukoc. Biol.* **80**:1553–1562.

Huang, Z. Y., S. Hunter, M. K. Kim, Z. K. Indik, and A. D. Schreiber. 2003. The effect of phosphatases SHP-1 and SHIP-1 on signaling by the ITIM- and ITAM-containing Fcγ receptors FcγRIIB and FcγRIIA. *J. Leukoc. Biol.* **73**:823–829.

Huizinga, T. W. J., C. E. van der Schoot, C. Jost, R. Klaassen, M. Kleijer, A. E. G. K. von dem Borne, D. Roos, and P. A. T. Tetteroo. 1988. The PI-linked receptor FcRIII is released on stimulation of neutrophils. *Nature* **333**:667–669.

Huizinga, T. W. J., F. van Kemenade, L. Koenderman, K. M. Dolman, A. E. G. Von den Borne, P. A. T. Tetteroo, and R. Roos. 1989. The 40-kDa Fc$_\gamma$ receptor (FcRII) on human neutrophils is essential for the IgG-induced respiratory burst and IgG-induced phagocytosis. *J. Immunol.* **142**:2365–2369.

Hulett, M. D., and P. M. Hogarth. 1994. Molecular basis of Fc receptor function. *Adv. Immunol.* **57**:1–127.

Hulett, M. D., N. Osman, I. F. McKenzie, and P. M. Hogarth. 1991. Chimeric Fc receptors identify functional domains of the murine high affinity receptor for IgG. *J. Immunol.* **147**:1863–1868.

Ioan-Facsinay, A., S. J. de Kimpe, S. M. Hellwig, P. L. van Lent, F. M. Hofhuis, H. H. van Ojik, C. Sedlik, S. A. da Silveira, J. Gerber, Y. F. de Jong, R. Roozendaal, L. A. Aarden, W. B. van den Berg, T. Saito, D. Mosser, S. Amigorena, S. Izui, G. J. van Ommen, M. van Vugt, J. G. van de Winkel, and J. S. Verbeek. 2002. FcγRI (CD64) contributes substantially to severity of arthritis, hypersensitivity responses, and protection from bacterial infection. *Immunity* **16**:391–402.

Israel, E. J., D. F. Wilsker, K. C. Hayes, D. Schoenfeld, and N. E. Simister. 1996. Increased clearance of IgG in mice that lack β2-microglobulin: possible protective role of FcRn. *Immunology* **89**:573–578.

Jaconi, M. E. E., D. P. Lew, J.-L. Carpentier, K. E. Magnusson, M. Sjogren, and O. Stendahl. 1990. Cytosolic free calcium elevation mediates the phagosome-lysosome fusion during phagocytosis in human neutrophils. *J. Cell Biol.* **110**:1555–1564.

Ji, J. D., I. Tassiulas, K. H. Park-Min, A. Aydin, I. Mecklenbrauker, A. Tarakhovsky, L. Pricop, J. E. Salmon, and L. B. Ivashkiv. 2003. Inhibition of interleukin 10 signaling after Fc receptor ligation and during rheumatoid arthritis. *J. Exp. Med.* **197**:1573–1583.

Jones, S. L., and E. J. Brown. 1996. FcγRII-mediated adhesion and phagocytosis induce L-plastin phosphorylation in human neutrophils. *J. Biol. Chem.* **271**:14623–14630.

Joshi, T., L. P. Ganesan, X. Cao, and S. Tridandapani. 2006. Molecular analysis of expression and function of hFcγRIIbl and b2 isoforms in myeloid cells. *Mol. Immunol.* **43**:839–850.

Jouvin, M. H., M. Adamczewski, R. Numerof, O. Letourneur, A. Valle, and J. P. Kinet. 1994. Differential control of the tyrosine kinases Lyn and Syk by the two signaling chains of the high affinity immunoglobulin E receptor. *J. Biol. Chem.* **269**:5918–5925.

Junghans, R. P., and C. L. Anderson. 1996. The protection receptor for IgG catabolism is the β2-microglobulin-containing neonatal intestinal transport receptor. *Proc. Natl. Acad. Sci. USA* **93**:5512–5516.

Kant, A. M., P. De, X. Peng, T. Yi, D. J. Rawlings, J. S. Kim, and D. L. Durden. 2002. SHP-1 regulates Fcγ receptor-mediated phagocytosis and the activation of RAC. *Blood* **100**:1852–1859.

Kiefer, F., J. Brumell, N. Al-Alawi, S. Latour, A. Cheng, A. Veillette, S. Grinstein, and T. Pawson. 1998. The Syk protein tyrosine kinase is essential for Fcγ receptor signaling in macrophages and neutrophils. *Mol. Cell. Biol.* **18**:4209–4220.

Kiener, P. A., B. M. Rankin, A. L. Burkhardt, G. L. Schieven, L. K. Gilliland, R. B. Rowley, J. B. Bolen, and J. A. Ledbetter. 1993. Cross-linking of Fcγ receptor I (FcγRI) and receptor II (FcγRII) on monocytic cells activates a signal transduction pathway common to both Fc receptors that involves the stimulation of p72^{Syk} protein tyrosine kinase. *J. Biol. Chem.* **268**:24442–24448.

Kimberly, R. P., J. W. Ahlstrom, M. E. Click, and J. C. Edberg. 1990. The glycosyl phosphatidylinositol-linked FcγRIII$_{PMN}$ mediates transmembrane signaling events distinct from FcγRII. *J. Exp. Med.* **171**:1239–1255.

Kimura, T., H. Sakamoto, E. Appella, and R. P. Siraganian. 1996. Conformational changes induced in the protein tyrosine kinase p72^{syk} by tyrosine phosphorylation or by binding of phosphorylated immunoreceptor tyrosine-based activation motif peptides. *Mol. Cell Biol.* **16**:1471–1478.

Kimura, T., J. Zhang, K. Sagawa, K. Sakaguchi, E. Appella, and R. P. Siraganian. 1997. Syk-independent tyrosine phosphorylation and association of the protein tyrosine phosphatases SHP-1 and SHP-2 with the high affinity IgE receptor. *J. Immunol.* **159**:4426–4434.

Klebanoff, S. J., M. A. Vadas, J. M. Harlan, L. H. Sparks, J. R. Gamble, J. M. Agosti, and A. M. Waltersdorph. 1986. Stimulation of neutrophils by tumor necrosis factor. *J. Immunol.* **136**:4220–4225.

Kleinau, S., P. Martinsson, and B. Heyman. 2000. Induction and suppression of collagen-induced arthritis is dependent on distinct fcγ receptors. *J. Exp. Med.* **191**:1611–1616.

Kono, H., C. Kyogoku, T. Suzuki, N. Tsuchiya, H. Honda, K. Yamamoto, K. Tokunaga, and Z. Honda. 2005. FcγRIIB Ile232Thr transmembrane polymorphism associated with human systemic lupus erythematosus decreases affinity to lipid rafts and attenuates inhibitory effects on B cell receptor signaling. *Hum. Mol. Genet.* **14**:2881–2892.

Koval, M., K. Preiter, C. Adles, P. D. Stahl, and T. H. Steinberg. 1998. Size of IgG-opsonized particles determines macrophage response during internalization. *Exp. Cell Res.* **242**:265–273.

Kuhlman, M., K. Joiner, and R. A. B. Ezekowitz. 1989. The human mannose-binding protein functions as an opsonin. *J. Exp. Med.* **169**:1733–1745.

Kumar, V., S. R. Ali, S. Konrad, J. Zwirner, J. S. Verbeek, R. E. Schmidt, and J. E. Gessner. 2006. Cell-derived anaphylatoxins as key mediators of antibody-dependent type II autoimmunity in mice. *J. Clin. Invest.* **116**:512–520.

Kurosaki, T., and J. V. Ravetch. 1989. A single amino acid in the glycosyl phosphatidyl inositol attachment domain determines the membrane topology of FcγRIII. *Nature* **342**:805–807.

Kurosaki, T., M. Takata, Y. Yamanashi, T. Inazu, T. Taniguchi, T. Yamamoto, and H. Yamamura. 1994. Syk activation by the Src-family tyrosine kinase in the B cell receptor signaling. *J. Exp. Med.* **179:**1725–1729.

Kusner, D. J., C. F. Hall, and S. Jackson. 1999. Fcγ receptor-mediated activation of phospholipase D regulates macrophage phagocytosis of IgG-opsonized particles. *J. Immunol.* **162:**2266–2274.

Lanier, L. L., S. Cwirla, G. Yu, R. Testi, and J. H. Phillips. 1989. Membrane anchoring of a human IgG Fc receptor (CD16) determined by a single amino acid. *Science* **246:**1611–1613.

Larsen, E. C., T. Ueyama, P. M. Brannock, Y. Shirai, N. Saito, C. Larsson, D. Loegering, P. B. Weber, and M. R. Lennartz. 2002. A role for PKC-ε in FcγR-mediated phagocytosis by RAW 264.7 cells. *J. Cell Biol.* **159:**939–944.

Laudanna, C., J. J. Campbell, and E. C. Butcher. 1996. Role of Rho in chemoattractant-activated leukocyte adhesion through integrins. *Science* **271:**981–983.

Law, D. A., V. W. F. Chan, S. K. Datta, and A. L. DeFranco. 1993. B-cell antigen receptor motifs have redundant signalling capabilities and bind the tyrosine kinases PTK72, Lyn and Fyn. *Curr. Biol.* **3:**645–657.

Lee, J. H., Y. M. Kim, N. W. Kim, J. W. Kim, E. Her, B. K. Kim, J. H. Kim, S. H. Ryu, J. W. Park, D. W. Seo, J. W. Han, M. A. Beaven, and W. S. Choi. 2006. Phospholipase D2 acts as an essential adaptor protein in the activation of Syk in antigen-stimulated mast cells. *Blood* **108:**956–964.

Lee, J. S., W. M. Nauseef, A. Moeenrezakhanlou, L. M. Sly, S. Noubir, K. G. Leidal, J. M. Schlomann, G. Krystal, and N. E. Reiner. 2007. Monocyte p110α phosphatidylinositol 3-kinase regulates phagocytosis, the phagocyte oxidase, and cytokine production. *J. Leukoc. Biol.* **81:**1548–1561.

Lee, W. L., G. Cosio, K. Ireton, and S. Grinstein. 2007. Role of CrkII in Fcγ receptor-mediated phagocytosis. *J. Biol. Chem.* **282:**11135–11143.

Lencer, W. I., and R. S. Blumberg. 2005. A passionate kiss, then run: exocytosis and recycling of IgG by FcRn. *Trends Cell Biol* **15:**5–9.

Lennartz, M. R., and E. J. Brown. 1991. Arachidonic acid is essential for IgG Fc receptor-mediated phagocytosis by human monocytes. *J. Immunol.* **147:**621–626.

Lennartz, M. R., A. F. C. Yuen, S. M. Masi, D. G. Russell, K. F. Buttle, and J. J. Smith. 1997. Phospholipase A_2 inhibition results in sequestration of plasma membrane into electron-lucent vesicles during IgG-mediated phagocytosis. *J. Cell Sci.* **110:**2041–2052.

Lerm, M., V. P. Brodin, I. Ruishalme, O. Stendahl, and E. Sarndahl. 2007. Inactivation of cdc42 is necessary for depolymerization of phagosomal f-actin and subsequent phagosomal maturation. *J. Immunol.* **178:**7357–7365.

Leverrier, Y., K. Okkenhaug, C. Sawyer, A. Bilancio, B. Vanhaesebroeck, and A. J. Ridley. 2003. Class I phosphoinositide 3-kinase p110β is required for apoptotic cell and Fcγ receptor-mediated phagocytosis by macrophages. *J. Biol. Chem.* **278:**38437–38442.

Li, X., J. Wu, R. H. Carter, J. C. Edberg, K. Su, G. S. Cooper, and R. P. Kimberly. 2003. A novel polymorphism in the Fcγ receptor IIB (CD32B) transmembrane region alters receptor signaling. *Arthritis Rheum.* **48:**3242–3252.

Liang, L., and C.-K. Huang. 1995. Activation of multiple protein kinase induced by cross-linking of FcγRII in human neutrophils. *J. Leukoc. Biol.* **57:**326–331.

Lim, J., A. Wiedemann, G. Tzircotis, S. J. Monkley, D. R. Critchley, and E. Caron. 2007. An essential role for talin during $\alpha_M\beta_2$-mediated phagocytosis. *Mol. Biol. Cell* **18:**976–985.

Lin, P. M., and J. R. Wright. 2006. Surfactant protein A binds to IgG and enhances phagocytosis of IgG-opsonized erythrocytes. *Am. J. Physiol. Lung Cell. Mol. Physiol.* **291:**L1199–L1206.

Lin, S. Q., C. Cicala, A. M. Scharenberg, and J. P. Kinet. 1996. The FcεRI β subunit functions as an amplifier of FcεRI γ-mediated cell activation signals. *Cell* **85:**985–995.

Lindberg, F. P., D. C. Bullard, T. E. Caver, H. D. Gresham, A. L. Beaudet, and E. J. Brown. 1996. Decreased resistance to bacterial infection and granulocyte defects in IAP-deficient mice. *Science* **274:**795–798.

Lindberg, F. P., H. D. Gresham, E. Schwarz, and E. J. Brown. 1993. Molecular cloning of integrin-associated protein: an immunoglobulin family member with multiple membrane-spanning domains implicated in $\alpha_v\beta_3$-dependent ligand binding. *J. Cell Biol.* **123:**485–496.

Lioubin, M. N., P. A. Algate, S. Tsai, K. Carlberg, A. Aebersold, and L. R. Rohrschneider. 1996. p150^{Ship}, a signal transduction molecule with inositol polyphosphate-5-phosphatase activity. *Genes Dev.* **10:**1084–1095.

Liu, Y., E. Masuda, M. C. Blank, K. A. Kirou, X. Gao, M. S. Park, and L. Pricop. 2005. Cytokine-mediated regulation of activating and inhibitory Fc γ receptors in human monocytes. *J. Leukoc. Biol.* **77:**767–776.

Lorenzi, R., P. M. Brickell, D. R. Katz, C. Kinnon, and A. J. Thrasher. 2000. Wiskott-Aldrich syndrome protein is necessary for efficient IgG-mediated phagocytosis. *Blood* **95:**2943–2946.

Lubeck, M. D., Z. Steplewski, F. Baglia, M. H. Klein, K. D. Dorrington, and H. Koprowski. 1985. The interaction of murine IgG subclass proteins with human monocyte Fc receptors. *J. Immunol.* **135:**1299–1304.

Majeed, M., E. Caveggion, C. A. Lowell, and G. Berton. 2001. Role of Src kinases and Syk in Fcγ receptor-mediated phagocytosis and phagosome-lysosome fusion. *J. Leukoc. Biol.* **70:**801–811.

Maliszewski, C. R., C. J. March, M. A. Schoenborn, S. Gimpel, and L. Shen. 1990. Expression cloning of a human Fc receptor for IgA. *J. Exp. Med.* **172:**1665–1672.

Marcilla, A., O. M. Riverolezcano, A. Agarwal, and K. C. Robbins. 1995. Identification of the major tyrosine kinase substrate in signaling complexes formed after engagement of Fcγ receptors. *J. Biol. Chem.* **270:**9115–9120.

Massol, P., P. Montcourrier, J.-C. Guillemot, and P. Chavrier. 1998. Fc receptor-mediated phagocytosis requires CDC42 and Rac1. *EMBO J.* **17:**6219–6229.

Matsuda, M., J. G. Park, D. C. Wang, S. Hunter, P. Chien, and A. D. Schreiber. 1996. Abrogation of the Fcγ receptor IIA-mediated phagocytic signal by stem-loop Syk antisense oligonucleotides. *Mol. Biol. Cell* **7:**1095–1106.

Matsuo, T., K. Hazeki, O. Hazeki, T. Katada, and M. Ui. 1996. Specific association of phosphatidylinositol 3-kinase with the protooncogene product Cbl in Fcγ receptor signaling. *FEBS Lett.* **382:**11–14.

May, R. C., E. Caron, A. Hall, and L. M. Machesky. 2000. Involvement of the Arp2/3 complex in phagocytosis mediated by FcγR or CR3. *Nat. Cell Biol.* **2:**246–248.

Mayorga, L. S., M. I. Colombo, M. Lennartz, E. J. Brown, K. H. Rahman, R. Weiss, P. J. Lennon, and P. D. Stahl. 1993. Inhibition of endosome fusion by phospholipase A_2 (PLA_2) inhibitors points to a role for PLA_2 in endocytosis. *Proc. Natl. Acad. Sci. USA* **90:**10255–10259.

Means, T. K., E. Latz, F. Hayashi, M. R. Murali, D. T. Golenbock, and A. D. Luster. 2005. Human lupus autoantibody-DNA complexes activate DCs through cooperation of CD32 and TLR9. *J. Clin. Invest.* **115:**407–417.

Melendez, A., R. A. Floto, D. J. Gilooly, M. M. Harnett, and J. M. Allen. 1998. FcγRI coupling to phospholipase D initiates sphingosine kinase-mediated calcium mobilization and vesicular trafficking. *J. Biol. Chem.* **273:**9393–9402.

Mero, P., C. Y. Zhang, Z. Y. Huang, M. K. Kim, A. D. Schreiber, S. Grinstein, and J. W. Booth. 2006.

Phosphorylation-independent ubiquitylation and endocytosis of Fc$_\gamma$RIIA. *J. Biol. Chem.* **281:**33242–33249.

Meyer, D., C. Schiller, J. Westermann, S. Izui, W. L. Hazenbos, J. S. Verbeek, R. E. Schmidt, and J. E. Gessner. 1998. FcγRIII (CD16)-deficient mice show IgG isotype-dependent protection to experimental autoimmune hemolytic anemia. *Blood* **92:**3997–4002.

Miettinen, H. M., J. K. Rose, and I. Mellman. 1989. Fc receptor isoforms exhibit distinct abilities for coated pit localization as a result of cytoplasmic domain heterogeneity. *Cell* **58:**317–327.

Missy, K., V. VanPoucke, P. Raynal, C. Viala, G. Mauco, M. Plantavid, H. Chap, and B. Payrastre. 1998. Lipid products of phosphoinositide 3-kinase interact with rad GTPase and stimulate GDP dissociation. *J. Biol. Chem.* **273:**30279–30286.

Mocsai, A., M. Zhou, F. Meng, V. L. Tybulewicz, and C. A. Lowell. 2002. Syk is required for integrin signaling in neutrophils. *Immunity* **16:**547–558.

Monteiro, R. C., and J. G. Van De Winkel. 2003. IgA Fc receptors. *Annu. Rev. Immunol.* **21:**177–204.

Moxey-Mims, M. M., H. H. Simms, M. M. Frank, E. Y. Lin, and T. A. Gaither. 1991. The effects of IL-1, IL-2, and tumor necrosis factor on polymorphonuclear leukocyte Fcγ receptor-mediated phagocytosis. *J. Immunol.* **147:**1823–1830.

Munugalavadla, V., J. Borneo, D. A. Ingram, and R. Kapur. 2005. p85α subunit of class IA PI-3 kinase is crucial for macrophage growth and migration. *Blood* **106:**103–109.

Muta, T., T. Kurosaki, Z. Misulovin, M. Sanchez, M. C. Nussenzweig, and J. V. Ravetch. 1994. A 13-amino-acid motif in the cytoplasmic domain of Fc$_\gamma$RIIB modulates B-cell receptor signalling. *Nature* **368:**70–73.

Nagarajan, S., K. Venkiteswaran, M. Anderson, U. Sayed, C. Zhu, and P. Selvaraj. 2000. Cell-specific, activation-dependent regulation of neutrophil CD32A ligand-binding function. *Blood* **95:**1069–1077.

Neel, B. G. 1997. Role of phosphatases in lymphocyte activation. *Curr. Opin. Immunol.* **9:**405–420.

Niedergang, F., and P. Chavrier. 2005. Regulation of phagocytosis by Rho GTPases. *Curr. Top. Microbiol. Immunol.* **291:**43–60.

Niedergang, F., E. Colucci-Guyon, T. Dubois, G. Raposo, and P. Chavrier. 2003. ADP ribosylation factor 6 is activated and controls membrane delivery during phagocytosis in macrophages. *J. Cell Biol.* **161:**1143–1150.

Nimmerjahn, F., P. Bruhns, K. Horiuchi, and J. V. Ravetch. 2005. FcγRIV: a novel FcR with distinct IgG subclass specificity. *Immunity* **23:**41–51.

Nitta, T., and T. Suzuki. 1982. Fc$_\gamma$2b receptor-mediated prostaglandin synthesis by a murine macrophage cell line P388D$_1$. *J. Immunol.* **128:**2527–2532.

Odin, J. A., J. C. Edberg, C. J. Painter, R. P. Kimberly, and J. C. Unkeless. 1991. Regulation of phagocytosis and [Ca^{2+}]$_i$ flux by distinct regions of an Fc receptor. *Science* **254:**1785–1788.

Ofek, I., J. Goldhar, Y. Keisari, and N. Sharon. 1995. Nonopsonic phagocytosis of microorganisms. *Annu. Rev. Microbiol.* **49:**239–276.

Okayama, Y., A. S. Kirshenbaum, and D. D. Metcalfe. 2000. Expression of a functional high-affinity IgG receptor, Fcγ RI, on human mast cells: up-regulation by IFN-γ. *J. Immunol.* **164:**4332–4339.

Olazabal, I. M., E. Caron, R. C. May, K. Schilling, D. A. Knecht, and L. M. Machesky. 2002. Rho-kinase and myosin-II control phagocytic cup formation during CR, but not FcγR, phagocytosis. *Curr. Biol.* **12:**1413–1418.

Olcese, L., P. Lang, F. Vely, A. Cambiaggi, D. Marguet, M. Blery, K. L. Hippen, R. Biassoni, A. Moretta, L. Moretta, J. C. Cambier, and E. Vivier. 1996. Human and mouse killer-cell inhibitory receptors recruit PTP1C and PTP1D protein tyrosine phosphatases. *J. Immunol.* **156:**4531–4534.

Ono, M., S. Bolland, P. Tempst, and J. V. Ravetch. 1996. Role of the inositol phosphatase SHIP in negative regulation of the immune system by the receptor FcγRIIB. *Nature* **383:**263–266.

Ono, M., H. Okada, S. Bolland, S. Yanagi, T. Kurosaki, and J. V. Ravetch. 1997. Deletion of SHIP or SHP-1 reveals two distinct pathways for inhibitory signaling. *Cell* **90:**293–301.

Osborne, M. A., G. Zenner, M. Lubinus, X. L. Zhang, Z. Songyang, L. C. Cantley, P. Majerus, P. Burn, and J. P. Kochan. 1996. The inositol 5'-phosphatase SHIP binds to immunoreceptor signaling motifs and responds to high affinity IgE receptor aggregation. *J. Biol. Chem.* **271:**29271–29278.

Oster, G. 1987. The physics of cell motility. *J. Cell Sci.* **8:**35–54.

Ota, Y., and L. E. Samelson. 1997. The product of the proto-oncogene c-cbl: a negative regulator of the Syk tyrosine kinase. *Science* **276:**418–420.

Paolini, R., M. H. Jouvin, and J. P. Kinet. 1991. Phosphorylation and dephosphorylation of the high-affinity receptor for immunoglobulin E immediately after receptor engagement and disengagement. *Nature* **353:**855–858.

Park, R. K., Y. B. Liu, and D. L. Durden. 1996. A role for Shc, Grb2, and Raf-1 in FcγRI signal relay. *J. Biol. Chem.* **271:**13342–13348.

Parren, P. W., P. A. Warmerdam, L. C. Boeije, J. Arts, N. A. Westerdaal, A. Vlug, P. J. Capel, L. A. Aarden, and J. G. van de Winkel. 1992. On the interaction of IgG subclasses with the low affinity FcγRIIa (CD32) on human monocytes, neutrophils, and platelets. Analysis of a functional polymorphism to human IgG2. *J. Clin. Invest.* **90:**1537–1546.

Pasquier, B., P. Launay, Y. Kanamaru, I. C. Moura, S. Pfirsch, C. Ruffie, D. Henin, M. Benhamou, M. Pretolani, U. Blank, and R. C. Monteiro. 2005. Identification of FcαRI as an inhibitory receptor that controls inflammation: dual role of FcRγ ITAM. *Immunity* **22:**31–42.

Perez, C., E. Coeffier, F. Moreau-Gachelin, J. Wietzerbin, and P. D. Benech. 1994. Involvement of the transcription factor PU.1/Spi-1 in myeloid cell-restricted expression of an interferon-inducible gene encoding the human high-affinity Fc gamma receptor. *Mol. Cell. Biol.* **14:**5023–5031.

Perry, D. G., G. L. Daugherty, and W. J. Martin. 1999. Clathrin-coated pit-associated proteins are required for alveolar macrophage phagocytosis. *J. Immunol.* **162:**380–386.

Perussia, B., M. Kobayashi, M. E. Rossi, I. Anegon, and G. Trinchieri. 1987. Immune interferon enhances functional properties of human granulocytes: role of Fc receptors and effect of lymphotoxin, tumor necrosis factor, and granulocyte-macrophage colony-stimulating factor. *J. Immunol.* **138:**765–774.

Perussia, B., G. Trinchieri, A. Jackson, N. L. Warner, J. Faust, H. Rumpold, D. Kraft, and L. L. Lanier. 1984. The Fc receptor for IgG on human natural killer cells: phenotypic, functional, and comparative studies with monoclonal antibodies. *J. Immunol.* **133:**180–189.

Pfefferkorn, L. C., and G. R. Yeaman. 1994. Association of IgA-Fc receptors (FcαR) with FcϵRI γ2 subunits in U937 cells: aggregation induces the tyrosine phosphorylation of γ2. *J. Immunol.* **153:**3228–3236.

Phillips, N. E., and D. C. Parker. 1983. Cross-linking of B lymphocyte Fc gamma receptors and membrane immunoglobulin inhibits anti-immunoglobulin-induced blastogenesis. *J. Immunol.* **132:**627–632.

Pollard, T. D. 2007. Regulation of actin filament assembly by arp2/3 complex and formins. *Annu. Rev. Biophys. Biomol. Struct.* **36:**451–477.

Pommier, C. G., S. Inada, L. F. Fries, T. Takahashi, M. M. Frank, and E. J. Brown. 1983. Plasma fibronectin enhances phagocytosis of opsonized particles by human peripheral blood monocytes. *J. Exp. Med.* **157:**1844–1854.

Poo, H., J. C. Krauss, L. Mayo-Bond, R. Todd, III, and H. R. Petty. 1995. Interaction of Fc$_\gamma$ receptor type IIIB with complement receptor type 3 in fibroblast transfectants: evidence from lateral diffusion and resonance energy transfer studies. *J. Mol. Biol.* **247:**597–603.

Qin, S., T. Inazu, and H. Yamamura. 1995. Activation and tyrosine phosphorylation of p72^{syk} as well as calcium mobilization after hydrogen peroxide stimulation in peripheral blood lymphocytes. *Biochem. J.* **308:**347–352.

Rabinovitch, M., S. I. Hamburg, and H. B. Fleit. 1980. Interferon-induced enhancement of Fc receptor-mediated macrophage phagocytosis. *J. Reticuloendothel. Soc.* **28:**27s–28s.

Regnault, A., D. Lankar, V. Lacabanne, A. Rodriguez, C. Thery, M. Rescigno, T. Saito, S. Verbeek, C. Bonnerot, P. Ricciardi-Castagnoli, and S. Amigorena. 1999. Fcγ receptor-mediated induction of dendritic cell maturation and major histocompatibility complex class I-restricted antigen presentation after immune complex internalization. *J. Exp. Med.* **189:**371–380.

Relman, D., T. E., S. Falkow, D. T. Golenbock, K. Saukkonen, and S. D. Wright. 1990. Recognition of a bacterial adhesion by an integrin: macrophage CR3($\alpha_M\beta_2$, CD11b/CD18) binds filamentous hemagglutinin of Bordetella pertussis. *Cell* **61:**1375–1382.

Reterink, T. J., E. W. Levarht, M. N. Klar, L. A. Van Es, and M. R. Daha. 1996. Transforming growth factor-β1 (TGF-β1) down-regulates IgA Fc-receptor (CD89) expression on human monocytes. *Clin. Exp. Immunol.* **103:**161–166.

Reth, M. 1989. Antigen receptor tail clue. *Nature* **338:**383–384.

Rollag, H., M. Degre, and G. Sonnenfeld. 1984. Effects of interferon-α/β and interferon-γ preparations on phagocytosis by mouse peritoneal macrophages. *Scand. J. Immunol.* **20:**149–155.

Rosales, C., and E. J. Brown. 1991. Two mechanisms for IgG Fc-receptor-mediated phagocytosis by human neutrophils. *J. Immunol.* **146:**3937–3944.

Rose, D. M., B. W. Winston, E. D. Chan, D. W. H. Riches, P. Gerwins, G. L. Johnson, and P. M. Henson. 1997. Fcγ receptor cross-linking activates p42, p38, and JNK/SAPK mitogen-activated protein kinases in murine macrophages: role for p42^{MAPK} in Fcγ receptor-stimulated TNF-α synthesis. *J. Immunol.* **158:**3433–3438.

Ross, G. D., J. A. Cain, and P. J. Lachmann. 1985. Membrane complement receptor type three (CR$_3$) has lectin-like properties analogous to bovine conglutinin and functions as a receptor for zymosan and rabbit erythrocytes as well as a receptor for iC3b. *J. Immunol.* **134:**3307–3315.

Rossman, M. D., E. Chen, P. Chien, M. Rotten, A. Cprek, and A. D. Schreiber. 1989. Fcγ receptor recognition of IgG ligand by human monocytes and macrophages. *Am. J. Respir. Cell Mol. Biol.* **1:**211–220.

Rowley, R. B., A. L. Burkhardt, H. G. Chao, G. R. Matsueda, and J. B. Bolen. 1995. Syk protein-tyrosine kinase is regulated by tyrosine-phosphorylated Igα/Igβ immunoreceptor tyrosine activation motif binding and autophosphorylation. *J. Biol. Chem.* **270:**11590–11594.

Ruppert, J., D. Friedrichs, H. Xu, and J. H. Peters. 1991. IL-4 decreases the expression of the monocyte differentiation marker CD14, paralleled by an increasing accessory potency. *Immunobiology* **182:**449–464.

Salcedo, T. W., T. Kurosaki, P. Kanakaraj, J. V. Ravetch, and B. Perussia. 1993. Physical and functional association of p56^{lck} with Fc$_\gamma$ RIIIA (CD16) in natural killer cells. *J. Exp. Med.* **177:**1475–1480.

Salmon, J. E., N. L. Brogle, J. C. Edberg, and R. P. Kimberly. 1991. Fc$_\gamma$ receptor III induces actin polymerization in human neutrophils and primes phagocytosis mediated by Fc$_\gamma$ receptor II. *J. Immunol.* **146:**997–1004.

Salmon, J. E., J. C. Edberg, and R. P. Kimberly. 1990. Fcγ receptor III on human neutrophils. Allelic variants have functionally distinct capacities. *J. Clin. Invest.* **85:**1287–1295.

Salmon, J. E., S. Kapur, and R. P. Kimberly. 1987. Opsonin-independent ligation of Fcγ receptors. The 3G8-bearing receptors on neutrophils mediate the phagocytosis of concanavalin A-treated erythrocytes and nonopsonized *Escherichia coli*. *J. Exp. Med.* **166:**1798–1813.

Salmon, J. E., S. Millard, L. A. Schachter, F. C. Arnett, E. M. Ginzler, M. F. Gourley, R. Ramsey Goldman, M. G. Peterson, and R. P. Kimberly. 1996. FcγRIIA alleles are heritable risk factors for lupus nephritis in African Americans. *J. Clin. Invest.* **97:**1348–1354.

Salmon, J. E., S. S. Millard, N. L. Brogle, and R. P. Kimberly. 1995. Fcγ receptor IIIb enhances Fcγ receptor IIa function in an oxidant-dependent and allele-sensitive manner. *J. Clin. Invest.* **95:**2877–2885.

Sampson, L. L., J. Heuser, and E. J. Brown. 1991. Cytokine regulation of complement receptor-mediated ingestion by mouse peritoneal macrophages. M-CSF and IL-4 activate phagocytosis by a common mechanism requiring autostimulation by IFN-β. *J. Immunol.* **146:**1005–1013.

Samuelsson, A., T. L. Towers, and J. V. Ravetch. 2001. Anti-inflammatory activity of IVIG mediated through the inhibitory Fc receptor. *Science* **291:**484–486.

Sano, H., D. K. Hsu, J. R. Apgar, L. Yu, B. B. Sharma, I. Kuwabara, S. Izui, and F. T. Liu. 2003. Critical role of galectin-3 in phagocytosis by macrophages. *J. Clin. Invest.* **112:**389–397.

Sarmay, G., I. Pecht, and J. Gergely. 1994. Protein-tyrosine kinase activity tightly associated with human type II Fcγ receptors. *Proc. Natl. Acad. Sci. USA* **91:**4140–4144.

Sawyer, D. W., J. A. Sullivan, and G. L. Mandell. 1985. Intracellular free calcium localization in neutrophils during phagocytosis. *Science* **230:**663–666.

Schauer, U., F. Stemberg, C. H. Rieger, M. Borte, S. Schubert, F. Riedel, U. Herz, M. Wick, H. D. Carr-Smith, A. R. Bradwell, and W. Herzog. 2003. IgG subclass concentrations in certified reference material 470 and reference values for children and adults determined with the binding site reagents. *Clin. Chem.* **49:**1924–1929.

Schieven, G. L., J. M. Kirihara, D. L. Burg, R. L. Geahlen, and J. A. Ledbetter. 1993. p72^{syk} tyrosine kinase is activated by oxidizing conditions that induce lymphocyte tyrosine phosphorylation and Ca^{2+} signals. *J. Biol. Chem.* **268:**16688–16692.

Schmidt, R. E., and J. E. Gessner. 2005. Fc receptors and their interaction with complement in autoimmunity. *Immunol. Lett.* **100:**56–67.

Scholl, P. R., D. Ahern, and R. S. Geha. 1992. Protein tyrosine phosphorylation induced via the IgG receptors FcγRI and FcγRII in the human monocytic cell line THP-1. *J. Immunol.* **149:**1751–1757.

Scott, C. C., W. Dobson, R. J. Botelho, N. Coady-Osberg, P. Chavrier, D. A. Knecht, C. Heath, P. Stahl, and S. Grinstein. 2005. Phosphatidylinositol-4,5-bisphosphate hydrolysis directs actin remodeling during phagocytosis. *J. Cell Biol.* **169:**139–149.

Sears, D. W., N. Osman, B. Tate, I. F. McKenzie, and P. M. Hogarth. 1990. Molecular cloning and expression of the mouse high affinity Fc receptor for IgG. *J. Immunol.* **144:**371–378.

Sedlik, C., D. Orbach, P. Veron, E. Schweighoffer, F. Colucci, R. Gamberale, A. Ioan-Facsinay, S. Verbeek, P. Ricciardi-Castagnoli, C. Bonnerot, V. L. Tybulewicz, J. Di Santo, and S. Amigorena. 2003. A critical role for Syk protein tyrosine kinase in Fc receptor-mediated antigen presentation and induction of dendritic cell maturation. *J. Immunol.* 170:846–852.

Selvaraj, P., W. F. Rosse, R. Silber, and T. A. Springer. 1988. The major Fc receptor in blood has a phosphatidylinositol anchor and is deficient in paroxysmal nocturnal haemoglobinuria. *Nature* 333:565–567.

Seth, A., C. Otomo, and M. K. Rosen. 2006. Autoinhibition regulates cellular localization and actin assembly activity of the diaphanous-related formins FRLalpha and mDia1. *J. Cell Biol.* 174:701–713.

Shen, L., R. Lasser, and M. W. Fanger. 1989. My43, a monoclonal antibody that reacts with human myeloid cells inhibits monocyte IgA binding and triggers function. *J. Immunol.* 143:4117–4122.

Shen, Z., C. T. Lin, and J. C. Unkeless. 1994. Correlations among tyrosine phosphorylation of Shc, p72^{syk}, PLCγ1, and [Ca^{2+}]$_i$ flux in Fcγ RIIA signaling. *J. Immunol.* 152:3017–3023.

Shiue, L., M. J. Zoller, and J. S. Brugge. 1995. Syk is activated by phosphotyrosine-containing peptides representing the tyrosine-based activation motifs of the high affinity receptor for IgE. *J. Biol. Chem.* 270:10498–10502.

Sidorenko, S. P., C. L. Law, K. A. Chandran, and E. A. Clark. 1995. Human spleen tyrosine kinase p72Syk associates with the Src-family kinase p53/56Lyn and a 120-kDa phosphoprotein. *Proc. Natl. Acad. Sci. USA* 92:359–363.

Simms, H. H., M. M. Frank, T. C. Quinn, S. Holland, and T. A. Gaither. 1989. Studies on phagocytosis in patients with acute bacterial infections. *J. Clin. Invest.* 83:252–260.

Simms, H. H., T. A. Gaither, L. F. Fries, and M. M. Frank. 1991. Monokines released during short-term Fcγ receptor phagocytosis up-regulate polymorphonuclear leukocytes and monocyte-phagocytic function. *J. Immunol.* 147:265–272.

Sinclair, N. R. 2000. Immunoreceptor tyrosine-based inhibitory motifs on activating molecules. *Crit. Rev. Immunol.* 20:89–102.

Siraganian, R. P., J. Zhang, K. Suzuki, and K. Sada. 2002. Protein tyrosine kinase Syk in mast cell signaling. *Mol. Immunol.* 38:1229–1233.

Sironi, M., F. O. Martinez, D. D'Ambrosio, M. Gattorno, N. Polentarutti, M. Locati, A. Gregorio, A. Iellem, M. A. Cassatella, J. Van Damme, S. Sozzani, A. Martini, F. Sinigaglia, A. Vecchi, and A. Mantovani. 2006. Differential regulation of chemokine production by Fcγ receptor engagement in human monocytes: association of CCL1 with a distinct form of M2 monocyte activation (M2b, Type 2). *J. Leukoc. Biol.* 80:342–349.

Smolen, J. E., H. M. Korchak, and G. Weissmann. 1980. Increased levels of cyclic adenosine-3', 5'-monophosphate in human polymorphonuclear leukocytes after surface stimulation. *J. Clin. Invest.* 65:1077–1085.

Sobota, A., A. Strzelecka-Kiliszek, E. Gladkowska, K. Yoshida, K. Mrozinska, and K. Kwiatkowska. 2005. Binding of IgG-opsonized particles to FcγR is an active stage of phagocytosis that involves receptor clustering and phosphorylation. *J. Immunol.* 175:4450–4457.

Speert, D. P., S. D. Wright, S. C. Silverstein, and B. Mah. 1988. Functional characterization of macrophage receptors for in vitro phagocytosis of unopsonized *Pseudomonas aeruginosa*. *J. Clin. Invest.* 82:872–879.

Stefanescu, R. N., M. Olferiev, Y. Liu, and L. Pricop. 2004. Inhibitory Fc γ receptors: from gene to disease. *J. Clin. Immunol.* 24:315–326.

Stendahl, O., K.-H. Krause, J. Krischer, P. Jerstrom, J.-M. Theler, R. A. Clark, J.-L. Carpentier, and D. P. Lew. 1994. Redistribution of intracellular Ca^{2+} stores during phagocytosis in human neutrophils. *Science* 265:1439–1441.

Stewart, W. W., R. L. Mazengera, L. Shen, and M. A. Kerr. 1994. Unaggregated serum IgA binds to neutrophil FcαR at physiological concentrations and is endocytosed but cross-linking is necessary to elicit a respiratory burst. *J. Leukoc. Biol.* 56:481–487.

Stuart, L. M., J. Boulais, G. M. Charriere, E. J. Hennessy, S. Brunet, I. Jutras, G. Goyette, C. Rondeau, S. Letarte, H. Huang, P. Ye, F. Morales, C. Kocks, J. S. Bader, M. Desjardins, and R. A. Ezekowitz. 2007. A systems biology analysis of the *Drosophila* phagosome. *Nature* 445:95–101.

Su, K., H. Yang, X. Li, X. Li, A. W. Gibson, J. M. Cafardi, T. Zhou, J. C. Edberg, and R. P. Kimberly. 2007. Expression profile of FcγRIIb on leukocytes and its dysregulation in systemic lupus erythematosus. *J. Immunol.* 178:3272–3280.

Suchard, S. J., V. Hinkovska-Galcheva, P. J. Mansfield, L. A. Boxer, and J. A. Shayman. 1997. Ceramide inhibits IgG-dependent phagocytosis in human polymorphonuclear leukocytes. *Blood* 89:2139–2147.

Suh, C. I., N. D. Stull, X. J. Li, W. Tian, M. O. Price, S. Grinstein, M. B. Yaffe, S. Atkinson, and M. C. Dinauer. 2006. The phosphoinositide-binding protein p40phox activates the NADPH oxidase during FcγIIA receptor-induced phagocytosis. *J. Exp. Med.* 203:1915–1925.

Sutterwala, F. S., G. J. Noel, P. Salgame, and D. M. Mosser. 1998. Reversal of proinflammatory responses by ligating the macrophage Fc receptor type I. *J. Exp. Med.* 188:217–222.

Suzuki, T., T. Saito-Taki, R. Sadasivan, and T. Nitta. 1982. Biochemical signal transmitted by Fcγ receptors: phospholipase A$_2$ activity of Fcγ2b receptor of murine macrophage cell line P388D$_1$. *Proc. Natl. Acad. Sci. USA* 79:591–595.

Takai, T., M. Ono, M. Hikida, H. Ohmori, and J. V. Ravetch. 1996. Augmented humoral and anaphylactic responses in Fc γ RII-deficient mice. *Nature* 379:346–349.

Tan, P. S., A. L. Gavin, N. Barnes, D. W. Sears, D. Vremec, K. Shortman, S. Amigorena, P. L. Mottram, and P. M. Hogarth. 2003. Unique monoclonal antibodies define expression of Fc γ RI on macrophages and mast cell lines and demonstrate heterogeneity among subcutaneous and other dendritic cells. *J. Immunol.* 170:2549–2556.

Tanaka, S., L. Neff, R. Baron, and J. B. Levy. 1995. Tyrosine phosphorylation and translocation of the c-Cbl protein after activation of tyrosine kinase signaling pathways. *J. Biol. Chem.* 270:14347–14351.

Taube, C., A. Dakhama, Y. H. Rha, K. Takeda, A. Joetham, J. W. Park, A. Balhorn, T. Takai, K. R. Poch, J. A. Nick, and E. W. Gelfand. 2003. Transient neutrophil infiltration after allergen challenge is dependent on specific antibodies and Fc © III receptors. *J. Immunol.* 170:4301–4309.

Tenner, A. J., S. L. Robinson, J. Borchelt, and J. R. Wright. 1989. Human pulmonary surfactant protein (SP-A), a protein structurally homologous to C1q, can enhance FcR- and CR1-mediated phagocytosis. *J. Biol. Chem.* 264:13923–13928.

Tenner, A. J., S. L. Robinson, and R. A. Ezekowitz. 1995. Mannose binding protein (MBP) enhances mononuclear phagocyte function via a receptor that contains the 126,000 M$_r$ component of the C1q receptor. *Immunity* 3:485–493.

te Velde, A. A., R. de Waal Malefijt, R. J. Huijbens, J. E. de Vries, and C. G. Figdor. 1992. IL-10 stimulates monocyte Fc γ R surface expression and cytotoxic activity. Distinct regulation of antibody-dependent cellular cytotoxicity by IFN-γ, IL-4, and IL-10. *J. Immunol.* 149:4048–4052.

Tilton, B., M. Andjelkovic, S. A. Didichenko, B. A. Hemmings, and M. Thelen. 1997. G-protein-coupled receptors and Fcγ-receptors mediate activation of Akt/protein kinase B in human phagocytes. *J. Biol. Chem.* **272:**28096–28101.

Ting, A. T., C. J. Dick, R. A. Schoon, L. R. Karnitz, R. T. Abraham, and P. J. Leibson. 1995. Interaction between lck and syk family tyrosine kinases in Fcγ receptor-initiated activation of natural killer cells. *J. Biol. Chem.* **270:**16415–16421.

Ting, A. T., L. M. Karnitz, R. A. Schoon, R. T. Abraham, and P. J. Leibson. 1992. Fcγ receptor activation induces the tyrosine phosphorylation of both phospholipase C (PLC)-γ1 and PLC-γ2 in natural killer cells. *J. Exp. Med.* **176:**1751–1755.

Tridandapani, S., T. W. Lyden, J. L. Smith, J. E. Carter, K. M. Coggeshall, and C. L. Anderson. 2000. The adapter protein LAT enhances Fc γ receptor-mediated signal transduction in myeloid cells. *J. Biol. Chem.* **275:**20480–20487.

Tridandapani, S., R. Wardrop, C. P. Baran, Y. Wang, J. M. Opalek, M. A. Caligiuri, and C. B. Marsh. 2003. TGF-β 1 suppresses myeloid Fc γ receptor function by regulating the expression and function of the common γ-subunit. *J. Immunol.* **170:**4572–4577.

Trotta, R., P. Kanakaraj, and B. Perussia. 1996. FcγR-dependent mitogen-activated protein kinase activation in leukocytes: a common signal transduction event necessary for expression of TNF-α and early activation genes. *J. Exp. Med.* **184:**1027–1035.

Underhill, D. M., and H. S. Goodridge. 2007. The many faces of ITAMs. *Trends Immunol.* **28:**66–73.

Unkeless, J. C., and H. N. Eisen. 1975. Binding of monomeric immunoglobulin G to Fc receptors of mouse macrophages. *J. Exp. Med.* **142:**1520–1533.

Unkeless, J. C., and J. Jin. 1997. Inhibitory receptors, ITIM sequences and phosphatases. *Curr. Opin. Immunol.* **9:**338–343.

Utomo, A., X. Cullere, M. Glogauer, W. Swat, and T. N. Mayadas. 2006. Vav proteins in neutrophils are required for FcγR-mediated signaling to Rac GTPases and nicotinamide adenine dinucleotide phosphate oxidase component p40^{phox}. *J. Immunol.* **177:**6388–6397.

Valle, A., and J. P. Kinet. 1995. N-Acetyl-L-cysteine inhibits antigen-mediated Syk, but not Lyn tyrosine kinase activation in mast cells. *FEBS Lett.* **357:**41–44.

van de Winkel, J. G. J., L. K. Ernst, C. L. Anderson, and I.-M. Chiu. 1991. Gene organization of the human high affinity receptor for IgG, FcγRI (CD64): characterization and evidence for a second gene. *J. Biol. Chem.* **266:**13449–13455.

van Egmond, M., E. van Garderen, A. B. van Spriel, C. A. Damen, E. S. van Amersfoort, G. van Zandbergen, J. van Hattum, J. Kuiper, and J. G. van de Winkel. 2000. FcαRI-positive liver Kupffer cells: reappraisal of the function of immunoglobulin A in immunity. *Nat. Med.* **6:**680–685.

van Spriel, A. B., J. H. Leusen, M. van Egmond, H. B. Dijkman, K. J. Assmann, T. N. Mayadas, and J. G. van de Winkel. 2001. Mac-1 (CD11b/CD18) is essential for Fc receptor-mediated neutrophil cytotoxicity and immunologic synapse formation. *Blood* **97:**2478–2486.

Vely, F., S. Olivero, L. Olcese, A. Moretta, J. E. Damen, L. Liu, G. Krystal, J. C. Cambier, M. Daeron, and E. Vivier. 1997. Differential association of phosphatases with hematopoietic co-receptors bearing immunoreceptor tyrosine-based inhibition motifs. *Eur. J. Immunol.* **27:**1994–2000.

Vidarsson, G., A. M. Stemerding, N. M. Stapleton, S. E. Spliethoff, H. Janssen, F. E. Rebers, M. de Haas, and J. G. van de Winkel. 2006. FcRn: an IgG receptor on phagocytes with a novel role in phagocytosis. *Blood* **108:**3573–3579.

Vivier, E., and M. Daeron. 1997. Immunoreceptor tyrosine-based inhibition motifs. *Immunol. Today* **18:**286–291.

Walker, B. A. M., B. E. Hagenlocker, E. B. J. Stubbs, R. R. Sandborg, B. W. Agranoff, and P. A. Ward. 1991. Signal transduction events and FcγR engagement in human neutrophils stimulated with immune complexes. *J. Immunol.* **146:**735–741.

Wang, A. V. T., P. R. Scholl, and R. S. Geha. 1994. Physical and functional association of the high affinity immunoglobulin G receptor (FcγRI) with the kinases hck and lyn. *J. Exp. Med.* **180:**1165–1170.

Wang, X. Y., K. Sada, S. Yanagi, C. Yang, K. Rezaul, and H. Yamamura. 1994. Intracellular calcium dependent activation of p72^{syk} in platelets. *J. Biochem.* **116:**858–861.

Warmerdam, P. A., J. G. van de Winkel, A. Vlug, N. A. Westerdaal, and P. J. Capel. 1991. A single amino acid in the second Ig-like domain of the human Fcγ receptor II is critical for human IgG2 binding. *J. Immunol.* **147:**1338–1343.

Warmerdam, P. A. M., J. G. J. van de Winkel, E. J. Gosselin, and P. J. A. Capel. 1990. Molecular basis for a polymorphism of human Fcγ receptor II (CD32). *J. Exp. Med.* **172:**10–25.

Watarai, M., Y. Kamata, S. Kozaki, and C. Sasakawa. 1997. rho, a small GTP-binding protein, is essential for *Shigella* invasion of epithelial cells. *J. Exp. Med.* **185:**281–292.

Weigert, A., A. M. Johann, A. von Knethen, H. Schmidt, G. Geisslinger, and B. Brune. 2006. Apoptotic cells promote macrophage survival by releasing the antiapoptotic mediator sphingosine-1-phosphate. *Blood* **108:**1635–1642.

Weisbart, R. H., A. Kacena, A. Schuh, and D. W. Golde. 1988. GM-CSF induces human neutrophil IgA-mediated phagocytosis by an IgA Fc receptor activation mechanism. *Nature* **332:**647–648.

Welch, G. R., H. L. Wong, and S. M. Wahl. 1990. Selective induction of FcγRIII on human monocytes by transforming growth factor-β. *J. Immunol.* **144:**3444–3448.

Wen, R., S. T. Jou, Y. Chen, A. Hoffmeyer, and D. Wang. 2002. Phospholipase Cγ2 is essential for specific functions of FcεR and FcγR. *J. Immunol.* **169:**6743–6752.

Wijngaarden, S., J. G. van de Winkel, K. M. Jacobs, J. W. Bijlsma, F. P. Lafeber, and J. A. van Roon. 2004. A shift in the balance of inhibitory and activating Fcγ receptors on monocytes toward the inhibitory Fcγ receptor IIb is associated with prevention of monocyte activation in rheumatoid arthritis. *Arthritis Rheum.* **50:**3878–3887.

Wines, B. D., and P. M. Hogarth. 2006. IgA receptors in health and disease. *Tissue Antigens* **68:**103–114.

Wines, B. D., H. M. Trist, P. A. Ramsland, and P. M. Hogarth. 2006. A common site of the Fc receptor γ subunit interacts with the unrelated immunoreceptors FcαRI and FcεRI. *J. Biol. Chem.* **281:**17108–17113.

Witke, K., W. Li, D. J. Kwiatkowski, and F. S. Southwick. 2001. Comparisons of CapG and gelsolin-null macrophages: demonstration of a unique role for CapG in receptor-mediated ruffling, phagocytosis, and vesicle rocketing. *J. Cell Biol.* **154:**775–784.

Worth, R. G., M. K. Kim, A. L. Kindzelskii, H. R. Petty, and A. D. Schreiber. 2003. Signal sequence within Fc γ RIIA controls calcium wave propagation patterns: apparent role in phagolysosome fusion. *Proc. Natl. Acad. Sci. USA* **100:**4533–4538.

Worth, R. G., L. Mayo-Bond, M. K. Kim, J. G. van de Winkel, R. F. Todd III, H. R. Petty, and A. D. Schreiber. 2001. The cytoplasmic domain of FcγRIIA (CD32) participates in phagolysosome formation. *Blood* **98:**3429–3434.

Wright, S. D., L. S. Craigmyle, and S. C. Silverstein. 1983. Fibronectin and serum amyloid P component stimulate C3b- and C3bi-mediated phagocytosis in cultured human monocytes. *J. Exp. Med.* **158:**1338–1343.

Wright, S. D., S. M. Levin, M. T. C. Jong, Z. Chad, and L. G. Kabbash. 1989. CR3 (CD11b/CD18) expresses one binding site for Arg-Gly-Asp-containing peptides and a second site for bacterial lipopolysaccharide. *J. Exp. Med.* **169:** 175–183.

Wright, S. D., J. I. Weitz, A. J. Huang, S. M. Levin, S. C. Silverstein, and J. D. Loike. 1988. Complement receptor type three (CD11b/CD18) of human polymorphonuclear leukocytes recognizes fibrinogen. *Proc. Natl. Acad. Sci. USA* **85:**7734–7738.

Wu, J. M., J. C. Edberg, P. B. Redecha, V. Bansal, P. M. Guyre, K. Coleman, J. E. Salmon, and R. P. Kimberly. 1997. A novel polymorphism of FcγRIIIa (CD16) alters receptor function and predisposes to autoimmune disease. *J. Clin. Invest.* **100:**1059–1070.

Xiong, Y., C. Cao, A. Makarova, B. Hyman, and L. Zhang. 2006. Mac-1 promotes FcγRIIA-dependent cell spreading and migration on immune complexes. *Biochemistry* **45:**8721–8731.

Yamauchi, A., C. Kim, S. Li, C. C. Marchal, J. Towe, S. J. Atkinson, and M. C. Dinauer. 2004. Rac2-deficient murine macrophages have selective defects in superoxide production and phagocytosis of opsonized particles. *J. Immunol.* **173:** 5971–5979.

Yan, M., R. F. Collins, S. Grinstein, and W. S. Trimble. 2005. Coronin-1 function is required for phagosome formation. *Mol. Biol. Cell* **16:**3077–3087.

Yancey, K. B., J. O'Shea, T. Chused, E. Brown, T. Takahashi, M. M. Frank, and T. J. Lawley. 1985. Human C5a modulates monocyte Fc and C3 receptor expression. *J. Immunol.* **135:**465–470.

Ydrenius, L., M. Majeed, B. J. Rasmusson, O. Stendahl, and E. Sarndahl. 2000. Activation of cAMP-dependent protein kinase is necessary for actin rearrangements in human neutrophils during phagocytosis. *J. Leukoc. Biol.* **67:**520–528.

Yuasa, T., S. Kubo, T. Yoshino, A. Ujike, K. Matsumura, M. Ono, J. V. Ravetch, and T. Takai. 1999. Deletion of fcγ receptor IIB renders H-2(b) mice susceptible to collagen-induced arthritis. *J. Exp. Med.* **189:**187–194.

Zhang, J., E. H. Berenstein, R. L. Evans, and R. P. Siraganian. 1996. Transfection of Syk protein tyrosine kinase reconstitutes high affinity IgE receptor-mediated degranulation in a Syk-negative variant of rat basophilic leukemia RBL-2H3 cells. *J. Exp. Med.* **184:**71–79.

Zhang, J., J. Guo, I. Dzhagalov, and Y. W. He. 2005. An essential function for the calcium-promoted Ras inactivator in Fcγ receptor-mediated phagocytosis. *Nat. Immunol.* **6:** 911–919.

Zhang, Q., D. Cox, C.-C. Tseng, J. G. Donaldson, and S. Greenberg. 1998. A requirement for ARF6 in Fc receptor-mediated phagocytosis in macrophages. *J. Biol. Chem.* **273:** 19977–19981.

Zhang, Y., C. C. Boesen, S. Radaev, A. G. Brooks, W. H. Fridman, C. Sautes-Fridman, and P. D. Sun. 2000. Crystal structure of the extracellular domain of a human Fc γ RIII. *Immunity* **13:**387–395.

Zheleznyak, A., and E. J. Brown. 1992. Immunoglobulin-mediated phagocytosis by human monocytes requires protein kinase C activation. Evidence for protein kinase C translocation to phagosomes. *J. Biol. Chem.* **267:**12042–12048.

Zheng, Y., J. A. Glaven, W. J. Wu, and R. A. Cerione. 1996. Phosphatidylinositol 4,5-bisphosphate provides an alternative to guanine nucleotide exchange factors by stimulating the dissociation of GDP from Cdc42Hs. *J. Biol. Chem.* **271:** 23815–23819.

Zhou, M., R. F. Todd III, J. G. van de Winkel, and H. R. Petty. 1993. Cocapping of the leukoadhesin molecules complement receptor type 3 and lymphocyte function-associated antigen-1 with Fc_γ receptor III on human neutrophils. Possible role of lectin-like interactions. *J. Immunol.* **150:**3030–3041.

Zigmond, S. H. 2004. Formin-induced nucleation of actin filaments. *Curr. Opin. Cell Biol.* **16:**99–105.

5

Chemokines and Phagocyte Trafficking

TIMOTHY J. WILLIAMS AND SARA M. RANKIN

It is essential for survival that specific types of leukocytes, in particular phagocytic cells, are induced to migrate from the bloodstream to sites of tissue infection. This process is initiated by the local production of soluble chemical signals, chemoattractants that stimulate receptors on leukocytes within venules. Leukocytes roll along the endothelium of venules by forming loose reversible adhesions via selectin molecules (Ley et al., 2007). Intracellular signaling via the chemoattractant receptors induces an upregulation of integrin adhesion molecules on the leukocytes that bind to complementary adhesion molecules of the immunoglobin superfamily on the endothelium, resulting in firm attachment. By a series of complex interactions between adhesion molecules and cytoskeletal components in the cell, initial attachment is followed by migration through the venule wall. This is followed, in turn, by migration along tissue chemotactic gradients to the source of generation of the chemoattractant. The efficiency of this system is enhanced by other soluble signals, cytokines, such as interleukin-1β (IL-1β) and tumor necrosis factor alpha (TNF-α), that upregulate the adhesion molecules on the endothelium.

The importance of leukocyte recruitment mechanisms is evidenced by rare genetic defects (LAD-1 and LAD-2) in the adhesion molecules involved; the affected individuals are unable to deal effectively with local infections and suffer severe life-threatening systemic infections. The mechanisms involved in leukocyte recruitment are an adaptation of a fundamental property of cells to detect and respond to chemotactic gradients. This property, which has been elegantly analyzed in primitive single-celled organisms, is of critical importance in the organization of multicellular organisms at every stage in their life history (Willard and Devreotes, 2006; Franca-Koh et al., 2007).

The sequence of events leading to the recruitment of phagocytes is best illustrated by reference to the complement system. Many different agents can initiate the activation of the complement system by either classical or alternative pathways, e.g., bacterial cell walls, fungi, endotoxin, and immune complexes. The complement system in tissue fluid can provide the initial recognition step for the presence of bacteria, for example, in a tissue. Complement activation consists of a cascade of enzymic reactions that can be induced in blood plasma or tissue fluid. One by-product of activation, C5a, is a cleavage fragment of one of the complement components, C5, generated extravascularly in tissue fluid. C5a acts on high-affinity receptors on neutrophils within venules to induce attachment and migration. Neutrophils then migrate along C5a gradients within the tissue toward the initiating events, e.g., bacteria that stimulated complement activation. The interaction between neutrophils and the endothelium induces plasma protein leakage (Bacon et al., 2002). This increases the supply of complement components from the blood to the tissues to regulate the local host defense process. C5a-induced mast cell activation and histamine release may also contribute to plasma protein leakage. Activation of complement can induce lysis of foreign cells (depending on the type of pathogen). Importantly, pathogens such as bacteria become opsonized by C3b, or antibodies if present, and this facilitates phagocytosis by neutrophils.

Phagocytosing neutrophils produce chemoattractants themselves to continue cell recruitment after the initial phase driven by C5a. One such chemoattractant is the arachidonate metabolite leukotriene B_4 (LTB_4) (Bray et al., 1981). Another is IL-8, one of the first members of a large family of chemokines identified. The chemokines are important in providing varying degrees of cell type specificity to leukocyte recruitment. C5a and LTB_4 are potent neutrophil chemoattractants (others are platelet-activating factor and formyl peptides derived from bacteria). However, these agents lack specificity as their receptors are on a number of leukocyte types. C5a, which can be generated very rapidly, is probably important in the earliest phase of neutrophil recruitment under certain circumstances. A first phase of C5a production and the subsequent appearance of chemokines were described in early studies using models of microbial infection and in ischemia/reperfusion injury

---

Timothy J. Williams and Sara M. Rankin, Leukocyte Biology Section, NHLI Division, Faculty of Medicine, Imperial College London, South Kensington Campus, Exhibition Road, London SW7 2AZ, United Kingdom.

(Jose et al., 1983; Beaubien et al., 1990; Collins et al., 1991; Jose et al., 1991; Ivey et al., 1995).

Chemokines were originally identified as biological activities in cell supernatants in vitro or exudates ex vivo, usually assayed by using in vitro chemotaxis assays, such as Boyden chambers. Some chemokines were purified and sequenced as proteins. Many were cloned based on common sequence characteristics, expressed, and subsequently characterized in test systems in vitro and in vivo. The specificity of different chemokines for leukocyte types provides a mechanism to explain the geographical organization of the immune system under basal conditions and the recruitment of selected leukocyte types in response to different inflammatory stimuli.

AN OVERVIEW OF CHEMOKINES

Chemokines are small peptides with a molecular mass of 8 to 17 kDa (Viola and Luster, 2008). To date, approximately 50 chemokines have been identified. Functionally, chemokines were originally defined by their ability to stimulate the selective chemotaxis of one or more types of leukocytes. The vast majority of chemokines have four conserved cysteine residues. As a family, chemokines have been subdivided into the CC and CXC chemokines based on whether the first two amino-terminal conserved cysteine residues are adjacent (CC) or separated by a single amino acid (CXC). The prototypes of these two classes of chemokines are the CC chemokine monocyte chemoattractant protein-1 (MCP-1) and the CXC chemokine IL-8. These two chemokines have differential activities on monocytes and neutrophils, MCP-1 preferentially activating monocytes, while IL-8 preferentially stimulates neutrophils.

There are two chemokines that do not fall into the original classification of chemokines into CXC and CC chemokines, namely, the CX3C chemokine, fractalkine, and the C chemokine, lymphotactin.

The cysteine residues in chemokines are important in that they form intermolecular disulfide bonds stabilizing the tertiary structure of chemokines and making them biologically active (Rajarathnam et al., 1999). Structurally, all chemokines exhibit the same protein fold, whereby three antiparallel β-sheets are overlaid by a C-terminal α-helix. Analysis of chemokines in nuclear magnetic resonance and crystallographic studies reveals that, at micromolar concentrations, chemokines exist as dimers or higher-order oligomers. However, at physiological (nanomolar) concentrations chemokines have been shown to exist as monomers.

Chemokines were originally named according to their functional activities or cellular source. Thus for example, RANTES (Regulated on Activation Normal T Expressed and Secreted) was so named because it was induced by mitogen or antigen in T lymphocytes, while the chemokine eotaxin was originally identified as an eosinophil-selective chemoattractant (Schall et al., 1990; Jose et al., 1994). However, as the number of chemokines identified increased and the complexity of chemokine biology became apparent, there was a need to simplify the nomenclature. Chemokines were therefore named systematically as CC, CXC, CX3C, and C ligands; hence, CCL, CXCL, CX3CL, and CL, followed by a number that was assigned according to their order of identification (Bacon et al., 2002). According to this nomenclature IL-8 is now referred to as CXCL8 and MCP-1 is CCL2.

CHEMOKINE RECEPTORS

Chemokine receptors are seven transmembrane G-protein-coupled receptors. To date, 19 chemokine receptors have been identified: CXCR1 to CXCR7 bind CXC chemokines, while CCR1 to CCR10 bind the CC chemokines. In addition, CX3CR1 and XCR1 selectively bind the chemokines fractalkine (CX3CL1) and lymphotactin (XCL1), respectively. Other chemokine binding proteins include D6 and Duffy that bind chemokines more promiscuously: D6 binds a number of CC chemokines and Duffy binds most CC and CXC chemokines (Rot, 2005; Graham and McKimmie, 2006).

Chemokine receptors are approximately 350 amino acids long, having an N-terminal extracellular domain followed by seven hydrophobic domains that traverse the plasma membrane leaving a cytoplasmic tail. At the cytoplasmic end of the third transmembrane helix is a well-conserved aspartate-arginine-tyrosine (DRY) motif that by reference to the rhodopsin crystal structure is likely to maintain the receptor in an inactive state, a so-called ionic lock. The exact details of receptor activation by ligand are not well defined, but molecular biological studies by several groups have led to the adoption of a two-step model for receptor activation, in which the chemokine receptor N terminus initially tethers the ligand and presents it to the remaining portions of the receptor, allowing the chemokine N terminus to project into an intrahelical pocket. This is postulated to disrupt reactions between neighboring helices and induce a receptor conformation that has high affinity for heterotrimeric G proteins. Treatment of leukocytes with pertussis toxin inhibits their chemotaxis in response to chemokines, indicating that chemokine receptors activate Gαi proteins. The binding of chemokines to their receptors stimulates the association of G proteins and facilitates the exchange of guanosine diphosphate (GDP) for guanosine triphosphate (GTP), thereby activating the G proteins. In its active state the α-subunit of the heterotrimeric G protein dissociates from the βγ-subunit, leaving the βγ-subunit free to activate phospholipase C and induce the production of diacylglycerol and inositol 1,4,5-triphosphate (IP3). While diacylglycerol is required for the activation of protein kinase C, IP3 is the second messenger responsible for mobilizing calcium from intracellular stores. Chemokine receptors are coupled to a plethora of other intracellular signaling molecules and pathways, including the small GTPases, Ras and Rho, MAP kinases, and a number of tyrosine kinases. One that has received considerable attention is the PI3 kinase pathway. Initial in vitro studies with nonselective PI3 kinase inhibitors indicated a critical role for this enzyme in neutrophil chemotaxis, and studies in mice deficient in PI3 kinase γ suggested that this isoform of PI3 kinase was particularly relevant for neutrophil responses to chemokines in vitro and in vivo (Hirsch et al., 2000; Li et al., 2000; Sasaki et al., 2000).

IL-8 and Other Neutrophil-Selective Chemokines

IL-8 is considered the prototypic neutrophil-selective CXC chemokine. It was originally purified in 1987 from the culture supernatants of stimulated monocytes by three independent groups and named neutrophil-activating protein or factor (NAP-1, NAF) (Walz et al., 1987; Yoshimura et al., 1987; Schröder et al., 1988). According to the new nomenclature IL-8 (NAP-1/NAF) is termed CXCL8. The predominant mature form of the protein consists of 72 amino acids with the CXC motif at the amino-terminal end; a longer form of the protein, 77 amino acids, is secreted by endothelial cells. Original experiments showed that IL-8 was a potent activator of neutrophils in vitro, stimulating neutrophil shape change, chemotaxis, adhesion, and respiratory burst. Importantly, it was shown that IL-8 did not stimulate the chemotaxis of naïve mononuclear cells. Furthermore, IL-8 induced edema and accumulation of massive numbers of neutrophils when injected intradermally (Beaubien et al., 1990; Colditz et al., 1990).

In addition to IL-8, several distinct neutrophil-selective chemokines were purified and sequenced from a variety of culture conditions in rapid succession between 1987 and 1993. Gro-α was identified as a gene upregulated in transformed Chinese hamster and human cells and subsequently purified by another laboratory as melanoma growth stimulatory activity (MGSA), now termed CXCL1 (Anisowicz et al., 1987; Sugano et al., 1987). In early studies, CXCL1 was sequenced as a protein, together with IL-8 in inflammatory exudate (Jose et al., 1991). Two other related genes were subsequently identified in a screen of human monocytes and named Gro-β and Gro-γ (Haskill et al., 1990). At the time, it was noted that, while the Gro gene encodes a protein with growth-related properties, its sequence was related to an expanding family of inflammatory peptides. The functional activity of Gro-α along with Gro-β and Gro-γ as neutrophil-activating chemokines was later independently recognized (Geiser et al., 1993). Coincidently, the murine homologue of this gene, KC, was discovered in a search for genes regulated by platelet-derived growth factor in fibroblasts (Oquendo et al., 1989). The chemokine NAP-2 (CXCL7) was isolated from culture supernatants of lipopolysaccharide (LPS)-stimulated monocytes and found to be a cleavage product of β-thromboglobulin (Walz and Baggiolini, 1989, 1990). Finally, ENA-78 (CXCL5) and GCP-2 (CXCL6) were purified from the supernatants of the immortalized lung epithelial cell line A549 and the MG63 osteosarcoma cell line, respectively (Walz et al., 1991; Proost et al., 1993). All of these chemokines had one unifying feature in that they stimulated chemotaxis of neutrophils, but not mononuclear cells in vitro. Further, the transgenic expression of KC in thymus or skin resulted in the selective accumulation of neutrophils in these tissues (Lira et al., 1994).

More recently, elegant intravital microscopy studies have shown that chemokines stimulate rolling neutrophils to firmly adhere to the endothelium prior to their transmigration across the endothelial wall and into tissues (Ley et al., 2007). Further, extravascular gradients of chemokines guide leukocytes to specific anatomic locations within tissues (Cara et al., 2001).

Chemokine Receptor Expression on Neutrophils

The original radioligand binding studies showed that neutrophils bound the chemokine IL-8 (CXCL8) with both low and high affinities, suggesting that neutrophils expressed two receptors for this chemokine. Subsequent chemokine competition studies supported the two-receptor hypothesis, but suggested that IL-8 bound to both of these receptors with high affinity, while other CXC chemokines, NAP-2 and GRO, bound to one of these receptors with high affinity and to the other receptor with low affinity. This was confirmed when the two receptors were cloned. Thus, Holmes et al. (1991) described the cloning of CXCR1, while simultaneously Murphy and Tiffany (1991) cloned CXCR2. Neutrophils have been shown to express equal levels of CXCR1 and CXCR2 (Holmes et al., 1991; Murphy and Tiffany, 1991).

Although these receptors were 77% homologous, they had different affinities for the CXC chemokines. Thus, CXCR2 bound IL-8, NAP-2, GRO, and ENA-78 with high affinity (K_d, 0.1 to 0.3 nM) while CXCR1 had a similar high affinity for IL-8, but bound NAP-2, GRO, and ENA-78 with low affinity (K_d, 100 to 130 nM) (Fig. 1). Using chemically synthesized analogues and scanning mutagenesis it was found that the amino-terminal Glu-Leu-Arg (ELR) sequence of these CXC chemokines was critical for receptor binding. Indeed, all CXC chemokines that stimulate neutrophils have since been shown to have the ELR motif. The importance of this motif for receptor binding was elegantly demonstrated by the fact that substitution of the first three amino acids (DLQ) of platelet factor 4 with ELR leads to the generation of a ligand for CXCR1 that competes with IL-8 for binding (Clark-Lewis et al., 1993). Conversely, Gro-α (8–73) inhibits murine neutrophil responses to KC and MIP-2 in vitro and in vivo and attenuates neutrophil recruitment into the peritoneal cavity in response to LPS or TNF-α (McColl and Clark-Lewis, 1999).

Matrix metalloproteinase 9 (MMP-9) is stored in neutrophils and released upon activation. In vitro studies have shown that purified MMP-9 truncates IL-8 (1–77) to IL-8 (7–77). This amino-terminal processing generates a truncated chemokine that binds with higher affinity to CXCR1 and is 10-fold more potent than CXCL8 in chemotaxis assays (Van Den Steen et al., 2000), although it has similar activity in tests in vivo (Nourshargh et al., 1992).

When neutrophils bind IL-8, within 10 min, 90% of CXCR1 is lost from the cell surface due to receptor internalization in clathrin-coated pits (Neel et al., 2005). While IL-8 is degraded in the lysosomes, CXCR1 is recycled to the cell surface. This process is orchestrated by the phosphorylation of the intracellular C terminus of the receptor that facilitates the docking of arrestins, thereby impeding G protein signaling, but promoting receptor endocytosis. While it was originally thought that this was solely a mechanism for desensitizing chemokine receptors, it was subsequently found that leukocytes from arrestin-2-deficient mice exhibited impaired chemotaxis in response to chemokines (McColl and Clark-Lewis, 1999; Su et al., 2005). More detailed analysis of this signaling pathway has revealed that arrestins are themselves docking proteins recruiting signaling molecules such as Jun amino-terminal kinase 3 (JNK3) and thereby activating additional signaling pathways required for chemotaxis (McDonald et al., 2000).

In vitro studies using human peripheral blood neutrophils have suggested that the functional responses induced by IL-8 acting via CXCR1 are very similar to those stimulated by activation of CXCR2. Thus, activation of either receptor stimulates neutrophil shape change, chemotaxis, and adhesion. However, there is evidence that, in contrast to IL-8, chemokines that act via CXCR2 do not activate the respiratory burst or stimulate neutrophil degranulation (Jones et al., 1996, 1997). Thus, CXCR1, but not CXCR2, is coupled to the molecular machinery required for neutrophil activation, stimulating superoxide generation and the release of granule enzymes (Jones et al., 1996, 1997).

Cellular Sources of ELR$^+$ CXC Chemokines

ELR$^+$ chemokines are generated by a wide range of cells including monocytes, T lymphocytes, fibroblasts, and en-

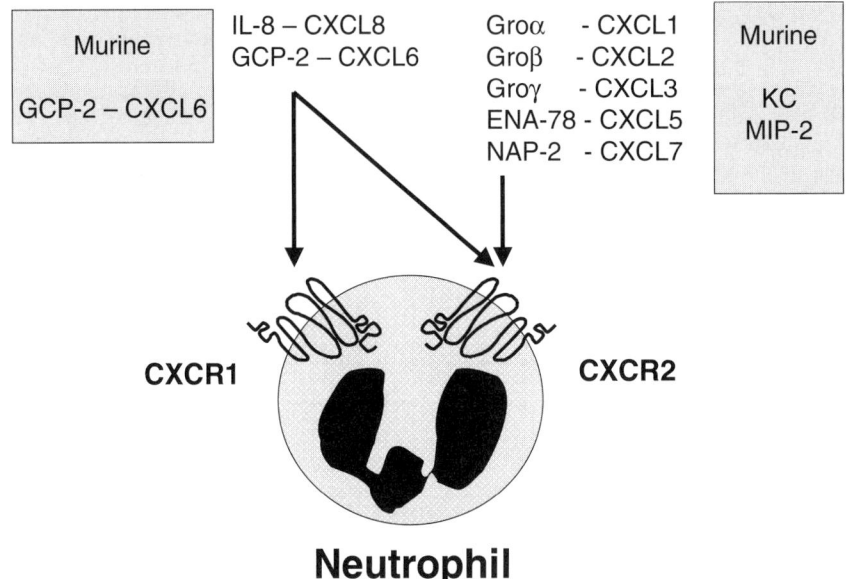

FIGURE 1 The ELR+ CXC chemokines and their binding to chemokine receptors on neutrophils.

dothelial cells when stimulated by mediators of inflammation, including TNF-α, IL-1, or bacterial products, such as LPS. Moreover, neutrophils are themselves an important source of these chemokines during an inflammatory reaction. Thus, neutrophils have the capacity to produce IL-8 (CXCL8), Gro-α (CXCL1), MIP-3β (CCL19), MIP-3α (CCL20), IP-10 (CXCL10), and MIG (CXCL9).

In mouse models, using LPS as an inflammatory stimulus, it has been shown that, while both KC and MIP-2 are generated during the inflammatory response, the cellular source of these two chemokines is distinct: KC is produced by stromal cells, while the majority of MIP-2 is derived from the neutrophils themselves (Lin et al., 2007). This differential expression of chemokines is thought to account for the differences in the kinetics and spatial distribution of these two chemokines within tissues during an inflammatory response.

ELR+ CXC Chemokines in Disease Pathology

ELR+ CXC chemokines and their receptors play a direct role in the pathogenesis of a number of inflammatory diseases, including reperfusion injury, ulcerative colitis, rheumatoid arthritis, and chronic obstructive pulmonary disease (COPD). Numerous studies in human and animal models have reported elevated expression of ELR+ chemokines in these conditions. Moreover, studies using anti-CXCL1, anti-CXCL8, and anti-CXCR2 neutralizing monoclonal antibodies have identified a role for these chemokines in both the neutrophil influx and ensuing tissue damage (Sekido et al., 1993; Boyle et al., 1998; Yagihashi et al., 1998; McColl and Clark-Lewis, 1999; Miura et al., 2001); e.g., inhibiting neutrophil recruitment into the lungs in response to *Aspergillus fumigatus* (Mehrad et al., 1999) or *Pseudomonas aeruginosa* (Tsai et al., 2000). Further, results with CXCR2-deficient mice show that CXCR2 is required for neutrophil recruitment into tissues in many inflammatory models, such as ozone exposure of the lungs (Johnston et al., 2005).

In a recent study, bone marrow transplants were performed between wild-type and $CXCR2^{-/-}$ mice to create chimeric mice selectively deficient in CXCR2 on hematopoietic or stromal cells. By using these chimeric mice it was shown that maximal neutrophil recruitment to the lung, in response to LPS, required CXCR2 expression by both the neutrophils and stromal cells (Reutershan et al., 2006). The functional response elicited in the stromal cells by these chemokines is yet to be defined.

Dynamic Expression of Chemokine Receptors on Neutrophils as They Age

The chemokine stromal-derived factor (SDF-1α, CXCL12) was originally isolated from the culture supernatants of bone marrow stromal cells (Bleul et al., 1996). It has since been shown that this chemokine is produced constitutively in the bone marrow. SDF-1α binds selectively to the chemokine receptor CXCR4, and recent data suggest that the SDF-1α/CXCR4 chemokine axis regulates both the retention of mature neutrophils in the bone marrow reserve and the homing of senescent neutrophils back to the marrow (Martin et al., 2003). Large numbers of mature neutrophils are retained in the bone marrow. These neutrophils form the bone marrow reserve, a pool of mature neutrophils that can be rapidly mobilized during inflammatory reactions. Mature neutrophils present in the bone marrow reserve express very low levels of CXCR4. However, these levels of expression appear to be functionally relevant in vivo as several lines of evidence indicate that the retention of neutrophils in the bone marrow depends on the SDF-1α/CXCR4 chemokine axis. Thus, for example, disrupting this axis with a CXCR4 antagonist induces a very rapid and dramatic mobilization of mature neutrophils (Martin et al., 2003). Under homeostatic conditions, the cytokine granulocyte colony-stimulating factor (G-CSF) regulates both granulopoiesis and the mobilization of neutrophils from the bone marrow reserve into the blood. Thus, under noninflammatory conditions G-CSF controls circulating numbers of neutrophils. Recent data have identified a molecular mechanism whereby G-CSF suppresses the expression of CXCR4 by neutrophils, and this is thought to be the mechanism whereby G-CSF regulates neutrophil mobilization (Kim et al., 2006).

Each day, 10^{11} neutrophils are released from the bone marrow under homeostatic conditions. These cells have a half-life in the blood of only 6.5 h and are cleared via the liver, spleen, and bone marrow. As neutrophils age in culture, they change their expression of chemokine receptors (Fig. 2). Specifically, they upregulate expression of CXCR4 and downregulate expression of CXCR2. This change in cell surface expression is reflected functionally by a dramatic increase in their responsiveness toward SDF-1α (CXCL12), the ligand for CXCR4, and a parallel reduction in responsiveness to chemokines such as Gro-α that act via CXCR2 (Martin et al., 2003). Examination of the kinetics of CXCR4 expression revealed that CXCR4 was rapidly upregulated on neutrophils as they aged, prior to the cells becoming apoptotic. Furthermore, while senescent CXCR4^{hi} neutrophils migrate toward a gradient of SDF-1α, this chemokine is not chemotactic for apoptotic neutrophils. Indeed, apoptotic cells are functionally unresponsive to all chemokines. As mentioned above, SDF-1α is expressed constitutively in the bone marrow, and it was shown that CXCR4^{hi} senescent neutrophils preferentially home back to the bone marrow in a CXCR4-dependent manner (Martin et al., 2003). Thus, upregulation of CXCR4 on neutrophils as they age represents a mechanism for guiding these cells back to the bone marrow for clearance (Fig. 3).

In addition to the upregulation of CXCR4 on neutrophils as they age, expression of CCR5 (normally expressed on Th1 lymphocytes and monocytes) has been reported on apoptotic neutrophils at sites of inflammation. However, apoptotic neutrophils do not respond functionally to CCL3 and CCL5, ligands of this receptor. Evidence from in vivo experiments suggests that, in this context, CCR5 functions to scavenge CCL3 and CCL5 at sites of inflammation and thereby promotes the resolution of inflammation (Ariel et al., 2006). This represents a novel function for these chemokine receptors, similar to that of DARC and D6, which are nonsignaling receptors with a role in scavenging chemokines (Rot, 2005; Graham and McKimmie, 2006).

An Unexpected Role for Chemokines in the Mobilization of Neutrophils from the Bone Marrow during Inflammation

During acute inflammatory reactions neutrophils are mobilized into the blood from the bone marrow reserve, which allows their subsequent recruitment to the site of inflammation. It has recently been shown that the ELR$^+$ CXC chemokines CXCL1 (KC) and CXCL2 (MIP-2) together with the cytokine G-CSF exhibit a combinatorial effect with respect to the rapid mobilization of neutrophils from the bone marrow in a model of acute peritonitis. It was shown that the chemokines mobilize neutrophils by stimulating their chemotaxis across the bone marrow sinusoidal endothelium into the sinuses and further are necessary for the subsequent recruitment of neutrophils from the blood into the tissue. In contrast, G-CSF had no direct effect on neutrophil migration but facilitated the chemokine-induced neutrophil mobilization by disrupting the CXCL12 (SDF-1α)/CXCR4-mediated retention of neutrophils in the bone marrow (Wengner et al., 2008).

Neutrophil Expression of Chemokine Receptors May Be Altered by the Inflammatory Environment

While it was originally believed that human peripheral blood neutrophils did not express CC chemokine receptors, a number of studies have shown that cytokines can stimulate the upregulation of these receptors; for example, GM-CSF stimulates the upregulation of CCR1 and interferon-γ upregulates CCR1 and CCR3 (Bonecchi et al., 1999;

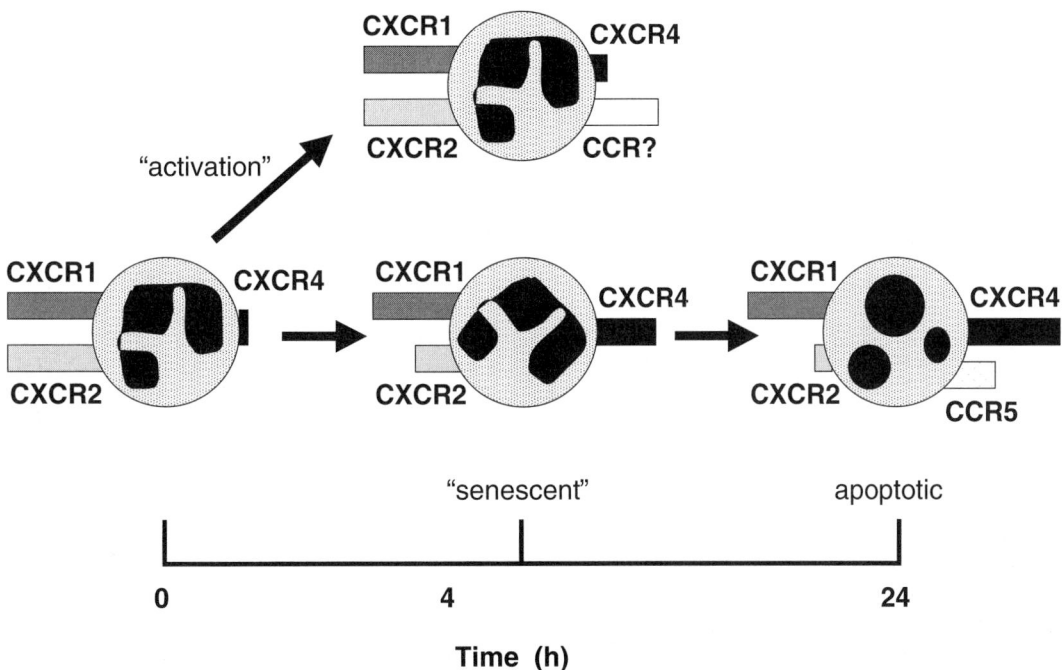

FIGURE 2 The dynamic expression of chemokine receptors on neutrophils as they age.

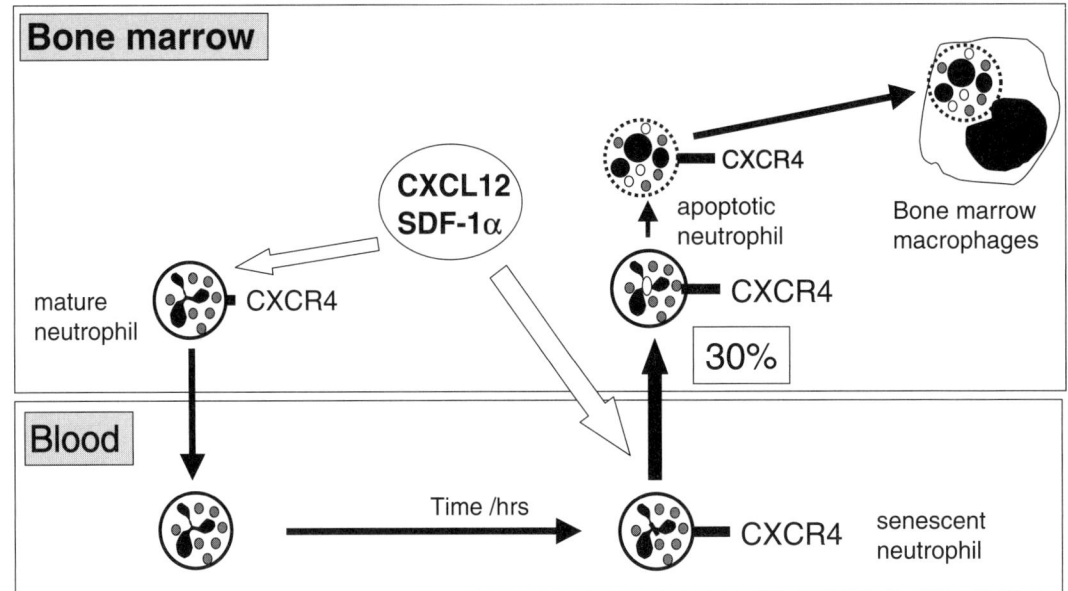

FIGURE 3 The role of the CXCL12/CXCR4 chemokine axis in neutrophil clearance via the bone marrow.

Cheng et al., 2001). In vivo, there is also evidence that CC chemokine receptors are upregulated on neutrophils during specific diseases. Thus, in a model of chronic adjuvant-induced vasculitis in rats, neutrophils were found to express CCR1 and CCR2 and responded chemotactically in vitro and in vivo to the chemokine MCP-1 (Johnston et al., 1999). Similarly, it was reported that neutrophils isolated from rodents during sepsis following cecal ligation and puncture exhibit chemotaxis to MCP-1 and MIP-1α. Further, monoclonal antibody blockade of these chemokines reduced the extent of neutrophil recruitment into the peritoneum (Speyer et al., 2004). In contrast, neutrophil recruitment into the lungs following instillation of LPS was independent of these chemokines (Speyer et al., 2004). These studies suggest that chemokine receptor expression is a dynamic process and may change during disease progression in response to the specific cytokine milieu.

Alternative Ligands for CXC Chemokine Receptors

Proteolytic fragments of the extracellular matrix have long been known to have chemotactic properties for neutrophils. It was recently shown that an acetylated tripeptide (Pro-Gly-Pro), generated from the proteolytic cleavage of collagen, stimulates neutrophil chemotaxis in vitro and accumulation in vivo. Moreover, it was shown that this acetylated tripeptide bound CXCR1 and CXCR2 (Weathington et al., 2006). Lymphatic filariasis is a chronic human parasitic disease in which the parasites repeatedly provoke both acute and chronic inflammatory reactions in the lymphatics and bloodstream. This inflammation is characterized by a neutrophilia and it has recently been shown that a worm excretory/secretory product, *Brugia malayi* asparaginyl transfer (t) RNA synthetase (*Bm*AsnRS) is chemotactic for neutrophils in vitro. Further, *Bm*AsnRS was shown to bind to the third extracellular domain of CXCR2 and stimulate chemotaxis of HEK-293 cells transfected with CXCR1 or CXCR2 (Ramirez et al., 2006).

MONOCYTES

Monocytes originate in the bone marrow, where they have a common myeloid progenitor, shared with neutrophils. Circulating monocytes released from the bone marrow comprise approximately 5 to 10% of total blood leukocytes. Under homeostatic conditions these cells infiltrate tissues and give rise to resident macrophages, while under specific inflammatory conditions they accumulate at sites of inflammation. Thus, for example, in atherosclerosis, monocytes are recruited into fatty streaks and, after uptake of modified low-density lipoprotein (LDL), become foam cells. Monocytes may also differentiate into more specialized cells, such as dendritic cells and osteoclasts.

Two Major Subsets of Monocytes Differentially Express the Chemokine Receptors CCR2 and CX3CR1

Two major subsets of circulating monocytes have been defined in humans. One subset is $CD14^+$ $CD16^+$ noninflammatory monocytes that are recruited into tissues under homeostatic conditions (Passlick et al., 1989) (Table 1). These cells and are thought to give rise to long-lived resident tissue macrophages, such as alveolar or peritoneal macrophages (Ziegler-Heitbrock et al., 1993). Conversely, a second major subset of circulating monocytes that are $CD14^{hi}$ $CD16^-$ are involved primarily in triggering an immune response and thus traffic to sites of inflammation. The heterogeneity of circulating monocytes is further revealed by their differential expression of chemokine receptors (Table 1). Thus, resident ($CD14^+$ $CD16^+$) monocytes typically express high levels of CX3CR1 and exhibit low levels of inflammatory chemokine receptors (CCR1 and CCR2) (Weber et al., 2000). In contrast, the inflammatory ($CD14^{hi}$ $CD16^-$) monocytes characteristically express high levels of CCR2 and low levels of CX3CR1. While the two subsets of monocytes were originally defined by their differential expression of CX3CR1 and CCR2, it is now rec-

TABLE 1 Monocyte subsets and their expression of chemokine receptors

Monocyte subset	"Inflammatory" monocytes	"Intermediate" subset	"Resident" monocytes
Human monocyte	CD14^{hi} CD16$^-$ CD64$^+$	CD14$^+$ CD16$^+$ CD64$^+$	CD14^{lo} CD16$^+$ CD64$^-$
Mouse monocytes Gr-1 (Ly6C/G)	Gr-1^{high}	Gr-1^{int}	Gr-1^{low}
Chemokine receptor expression (mouse/human)	CCR2^{++}	CCR2$^+$	CCR2^{lo}
	CX3CR1$^+$	CX3CR1$^+$	CX3CR1^{++}
		CCR7$^+$	
	CCR1	CCR8$^+$	CCR5$^+$
	CCR4		CXCR4$^+$
	CXCR1		
	CXCR2		

ognized that they may also express other chemokine receptors. In particular, CXCR4 and CCR5 are expressed on the resident CD14$^+$ CD16$^+$ monocytes, while multiple chemokine receptors have been reported on the inflammatory CD14^{hi} CD16$^-$ subset, including CCR1, CCR4, CCR7, CXCR1, and CXCR2.

As noted for neutrophils, chemokine receptor expression by monocytes can be modulated by their exposure to cytokines. Activation of monocytes with bacterial LPS or proinflammatory mediators including IFN-γ, TNF-α, and IL-2 downregulates their expression of CCR2 (Sica et al., 1997; Penton-Rol et al., 1998). In contrast, anti-inflammatory cytokines such as IL-10 have been shown to increase the expression of CCR2 and CCR5 (Sozzani et al., 1998).

While human monocytes have been reported to express CXCR1 and CXCR2, these receptors are normally considered nonfunctional. However, exposure of monocytes in vitro to IL-4 or IL-13 increases both the expression and coupling of CXCR1 and CXCR2 (Bonecchi et al., 2000). It has been reported that blood monocytes isolated from patients with COPD exhibit enhanced responsiveness to CXCL1 (Traves et al., 2004). Further, elevated levels of CXCL1 have been reported in the bronchoalveolar lavage fluid and sputum of patients with COPD, and it has been proposed that this chemokine/receptor axis may promote the recruitment of monocytes to the lungs in this disease (Traves et al., 2004).

Similarly in mice, two distinct subsets of monocytes have been described: one that is Gr-1^{hi}, corresponding to the inflammatory monocytes, and the other that is Gr-1^{lo} and equivalent to the resident monocytes (Geissmann et al., 2003). Like their human counterparts, inflammatory Gr-1^{hi} monocytes are CX3CR1^{lo} CCR2$^+$ and the Gr-1^{lo} monocytes are CX3CR1^{hi} CCR2^{lo}. Adoptive transfer of these two different subsets of monocytes into mice has shown that they do indeed have distinct functions, reflected by their expression of discrete chemokine receptors. Several days after intravenous injection, the CX3CR1^{hi} CCR2^{lo} monocytes were found in the blood and a number of tissues including the liver, spleen, lungs, and brain. Interestingly, in the spleen some of these monocytes had acquired the markers, CD11c and MHC class II, suggesting that they had differentiated into dendritic cells (Geissmann et al., 2003). This observation was consistent with in vitro studies of human monocyte transendothelial migration, in which migrated monocytes that reverse transmigrated, exhibited a dendritic cell phenotype (Randolph et al., 2002). In contrast, after their intravenous injection in mice, inflammatory CCR2$^+$ monocytes were short-lived in the blood and could rarely be detected in tissues, except under inflammatory conditions. A proportion of these monocytes, which were recruited to the sites of inflammation, similarly upregulated CD11c and MHC class II, taking on a dendritic cell phenotype (Geissmann et al., 2003). With the discovery that the classical subtype of monocytes that expresses CCR2 is Gr-1^{hi}, it has more recently been shown that it is indeed this Gr-1^{hi} subset of monocytes that is recruited in numerous experimental models of inflammation, including acute peritonitis and UV-induced inflammation of the skin and in response to infection with pathogens such as *Toxoplasma gondii* and *Listeria monocytogenes*.

In addition to CCR2, other chemokine receptors are involved in the specific localization of the Gr-1^{hi} monocytes during particular inflammatory reactions. For example, the CCR6/CCL20 chemokine axis has been shown to be involved in the recruitment of Gr-1^{hi} monocytes into epithelial tissues after the application of adjuvants to the skin or buccal mucosa (Merad et al., 2004; LeBorgne et al., 2006).

Monocyte-Specific Chemokines

MCP-1 (CCL2) was the first monocyte-specific chemokine to be identified. This CC chemokine was originally purified from the conditioned media generated by baboon aortic smooth muscle cells, and subsequently, the human homologue was isolated from tumor cell lines (Valente et al., 1988; Graves et al., 1989; Robinson et al., 1989). MCP-1 was originally shown to be a selective chemoattractant for mononuclear cells, as the chemokine does stimulate the chemotaxis of neutrophils. Further, the transgenic expression of MCP-1 in specific organs leads to the selective recruitment of monocytes into these tissues (Fuentes et al., 1995). The receptors for MCP-1, CCR2A and CCR2B, were identified by two independent groups (Charo et al., 1994; Yamagami et al., 1994). Subsequently, MCP-2, -3, and -4 were identified and similarly shown to stimulate the chemotaxis of monocytes, but not neutrophils. While all the MCPs activate monocytes, they also activate other leukocytes, including T lymphocytes, NK cells, eosinophils, and basophils, to a lesser or greater extent due to their different receptor specificities. Thus, while MCP-1 binds solely to CCR2, MCP-2 binds to CCR3, MCP-3 binds CCR1, CCR2, and CCR3, and MCP-4 binds to CCR2 and CCR3. The chemokine MCP-1 is generated in a number of inflammatory reactions associated with monocyte recruitment. Most significantly, MCP-1 expression has been reported in atherosclerotic lesions in human and animal models (Nelken et al., 1991; Yla-Herttuala et al., 1991). The generation of mice with a genetic deficiency in either

CCR2 or MCP-1 has highlighted the importance of the CCR2/MCP-1 chemokine axis in monocyte recruitment in numerous inflammatory models (Boring et al., 1997; Kurihara et al., 1997; Gu et al., 1998; Lu et al., 1998; Charo and Peters, 2003).

A Third Subset of Monocytes

In addition to the two major subpopulations of monocytes described above a third subset has been described that in humans is $CD14^+$ $CD16^+$ and express CD64. In mice these monocytes are identified as $CD14^+$ $CD16^+$ Ly-6C $Gr-1^{int}$, while the inflammatory macrophages are $CD14^{hi}$ $CD16^-$ Ly-6C $Gr-1^{hi}$ and the resident monocytes are $CD14^+$ $CD16^+$ Ly-6C $Gr-1^{lo}$. These $Gr-1^{int}$ populations express mRNA for CCR7 and CCR8, the chemokine receptors for CCL21 (SCL, 6Ckine, Exodus-2) and CCL1 (I-309 in humans and TCA-3 in mice), respectively. Both CCR7 and CCR8 are necessary for the accumulation of $Gr-1^{int}$ monocytes into the draining lymph nodes in both the skin and lung and acquisition of a dendritic cell phenotype (Qu et al., 2004).

Monocyte Recruitment and Retention in Atherosclerotic Plaques

Recent work highlights further layers of complexity with respect to chemokine biology and monocytes. Mice with a genetic deficiency in apoE develop atherosclerotic lesions when on a high-fat diet. $CX3CR1^{lo}$ $CCR2^+$ monocytes were preferentially recruited to the plaques, but it was shown that their accumulation unexpectedly depended on CX3CR1 in addition to CCR2 and CCR5 (Tacke et al., 2007). In contrast, $CX3CR1^{hi}$ $CCR2^{lo}$ monocytes were less frequent in plaques, but surprisingly their accumulation was independent of CX3CR1 and partially dependent on CCR5, this chemokine receptor being upregulated on this monocyte subset in $apoE^{-/-}$ mice (Tacke et al., 2007). These findings are further corroborated by studies carried out using mice with genetic deficiencies of either CCR5 or CX3CR1, suggesting a role for these receptors in atherogenesis in this model (Combadiere et al., 2003; Lesnik et al., 2003; Stein et al., 2003; Zernecke et al., 2006).

While most research has concentrated on the identification of the chemokines and receptors responsible for the initial recruitment of monocytes into atherosclerotic lesions, recent studies have focused on elucidating the potential role of chemokines and their receptors in monocyte/macrophage retention. During atherogenesis, the lipid peroxidation products, 9-HODE and 13-HODE, are generated through the oxidation of linoleic acid, the primary fatty acid in LDL (Rankin et al., 1991). These HODEs have been shown to stimulate a switch in chemokine receptor expression in monocytes, reducing expression of CCR2 and stimulating CX3CR1 expression through a peroxisome proliferator-activated receptor gamma-dependent pathway (Han et al., 2000; Barlic et al., 2006). Fractalkine (CX3CL1), the ligand for CX3CR1, is unique in that it can act either as a soluble chemokine, promoting leukocyte chemotaxis, or in its transmembrane form it can serve as an adhesion molecule. It has been reported that CX3CL1 is expressed in its transmembrane form on smooth muscle cells in the lesions of $apoE^{-/-}$ mice on a high-fat diet, and it has recently been proposed that CX3CR1 is involved in the retention of macrophages within plaques, as it mediates direct adhesion of monocytes to smooth muscle cells that express the ligand fractalkine (CX3CL1) (Barlic et al., 2007).

A Role for Chemokines in Monocyte Mobilization from the Bone Marrow

The first step in the trafficking of monocytes to sites of infection or injury is their mobilization from the bone marrow. Recently, a unique role for chemokines in this process has been identified. In a model of bacterial infection it was shown that the exit of $Gr-1^{hi}$ monocytes from the bone marrow depended on the CCR2/MCP-1 chemokine axis (Serbina, 2006). Mice with a genetic deficiency in CCR2 were shown to have fewer $Gr-1^{hi}$ monocytes in the blood, suggesting that this chemokine receptor regulates the basal release of leukocytes from the bone marrow. Further, when $CCR1^{-/-}$ mice were infected with L. monocytogenes, $Gr-1^{hi}$ monocytes accumulated in the bone marrow, but not at the site of infection, indicating a deficiency in mobilization (Serbina and Pamer, 2006). More recently, in a model of atherosclerosis, it has similarly been shown that MCP-1/CCR2 regulates the mobilization of monocytes from the bone marrow (Tsou et al., 2007). Thus, chemokines may be required, not only for orchestrating the recruitment of leukocytes from the blood to sites of inflammation, but also for their initial mobilization from the bone marrow into the blood, their first step in trafficking to sites of inflammation.

DENDRITIC CELLS

Immature dendritic cells are recruited to inflamed peripheral tissues by using inflammatory chemokine receptors such as CCR1, CCR5, and CCR6 responding to their respective chemokine ligands, which are produced at sites of inflammation (McWilliam et al., 1996). Once in tissues, dendritic cells can be activated by inflammatory cytokines such as TNF-α and IL-1α or by bacterial products such as LPS that promote cell maturation. After capturing antigen, dendritic cells are required to migrate to lymph nodes to effect antigen presentation to T lymphocytes. This is achieved by the downregulation of inflammatory chemokine receptors and an upregulation of CCR7, induced during maturation (Sallusto et al., 1998). Mature dendritic cells are attracted to the lymph nodes by the CCR7 ligand CCL21 that is produced by lymphatic endothelial cells. This chemokine is also produced by high endothelial venules attracting naïve $CCR7^+$ T lymphocytes from the blood into the lymph nodes (Gunn et al., 1998; Willimann et al., 1998). To ensure that T cells encounter antigen from dendritic cells, both cell types are positioned within the T-cell areas of the lymph node, which is mediated by the constitutive expression of both CCR7 ligands CCL21 and CCL19 by local stromal cells. CCL19 can also be produced by dendritic cells, macrophages, and endothelial cells within the T-cell areas (Ngo et al., 1998; Nagira et al., 1999; Luther et al., 2000). After priming, T cells downregulate the chemokine receptor CCR7 and upregulate other chemokine receptors, which are appropriate to localize them to their next destination.

MACROPHAGES

Tissue macrophages exhibit a high degree of heterogeneity, reflecting their diverse functional roles at specific anatomical locations. Tissue macrophages play a critical role in maintaining both blood and tissue homeostasis through the clearance of senescent cells and in driving tissue remodeling and repair following inflammation. It was originally thought that tissue macrophages were replenished solely by the recruitment and differentiation of circulating monocytes, but there is now evidence that local proliferation also plays a significant role in maintaining populations of tissue macrophages under steady state. Tissue macrophages include osteoclasts, Kupffer cells, alveolar macrophages, Langerhans cells, and splenic macrophages, macrophages that have well-defined characteristics and functions. The differentiation of human monocytes to macrophages in culture results in the loss of CCR2 expression and increased expression of CCR1 and CCR5 (Fantuzzi et al., 1999; Kaufmann et al., 2001). Thus, it has been postulated that the CCL2/CCR2 chemokine axis is required for the initial recruitment of monocytes into tissues, but CCR1 and CCR5 may be required for more precise positioning within the tissue.

Chemokine receptor expression on macrophages is also altered in hypoxic tissue microenvironments, such as the necrotic areas of tumors or in ischemic tissues. Activation of hypoxia-inducible factor-1α upregulates the expression of CXCR4 while reducing the ability of macrophages to migrate in response to CCL2 (Schioppa et al., 2003). With respect to the more specialized resident populations of macrophages, very little is known about their specific expression of chemokine receptors and the functional effects of chemokines on these cells. One notable exception is the expression of CCR7 by marginal zone macrophages in the spleen. The retention of this subset of macrophages in the marginal zone depends on the activities of CCL21 and, to a lesser extent, CCL19, the ligands for CCR7 (Ato et al., 2004).

The Effect of Alternative and Classical Activation of Macrophages on the Generation of Chemokines

Under inflammatory conditions, tissue macrophages may be polarized into M1 and M2 phenotypes by their differential activation (Mantovani et al., 2007). Thus, macrophages that are classically activated with the cytokine IFN-γ alone or in concert with microbial products, such as LPS, or cytokines, such as TNF-α, are polarized into type 1 macrophages (M1). Alternative activation of macrophages with IL-4/IL-13, activin A, and IL-21 polarizes these cells into type 2 macrophages (M2).

With respect to chemokine biology, M1 and M2 macrophages are particularly important as a source of chemokines, driving inflammation or remodeling. Activation of macrophages by the classical and alternative routes results in the generation of distinct repertoires of chemokines (Mantovani et al., 2004; Martinez et al., 2006) (Table 2). The classical activation of macrophages with inflammatory cytokines such as IFN-γ and TNF-α or with LPS leads to the production of an array of chemokines that promote the recruitment of neutrophils, Th1 cells, and NK cells, required for the host to mount a response to intracellular pathogens, for tumor resistance and delayed-type hypersensitivity reactions. These chemokines include CXCL1, -2, -3, -5, -8, -9, and -10 and CCL2, -3, -4, -5, -11, -17, and -22. In contrast, the alternative activation of macrophages leads to the production of discrete arrays of chemokines, dependent on the stimulus. The Th2 cytokines IL-4 and IL-13 stimulate the production of chemokines CCL24, CCL22, CCL17, and CCL18, involved in the recruitment of eosinophils, Th2 cells, T-reg cells, and basophils, leukocytes that are all involved in allergic responses. In contrast, activation of macrophages with LPS in conjunction with immune complexes results in the selective generation of CCL1, the ligand for CCR8, a receptor that has been reported on Th2 cells, T-reg cells, and eosinophils isolated from the allergic lung. The cytokine IL-10 stimulates macrophages to produce CCL18, CCL16, and CXCL13.

In summary, macrophages that are alternatively activated generate an array of chemokines that are involved in recruiting cells required for tissue remodeling, allergic inflammation, and tumor progression. Moreover, the cytokine and immune complexes that alternatively activate macrophages have been shown to actually suppress the generation of inflammatory chemokines stimulated by classical activation of macrophages with LPS and IFN-γ. At a molecular level this is explicable, as LPS stimulates chemokine production via activation of NF-κB and STAT-1, while IL-4, IL-10, and IL-13 reduce the activities of both NF-κB and STAT-1, thereby inhibiting the production of these proinflammatory chemokines.

OTHER FUNCTIONS OF CHEMOKINE RECEPTORS

While chemokines are primarily thought of as chemotactic factors, these proteins may also induce distinct functional responses in specific cell types. One such example explains how the influenza virus subverts the chemokine system to ensure the survival of its host cell, the macrophage. Thus, the infection of macrophages with influenza virus stimulates the production of CCL5 (Tyner et al., 2005). This chemokine acts in an autocrine manner via CCR5 to stimulate antiapoptotic pathways in the macrophage, thereby ensuring that viral clearance by macrophages does not induce macrophage death (Tyner et al., 2005).

Chemokine receptors have been shown to also serve as scavenger receptors, mopping up inflammatory chemokines to promote the resolution of inflammation. Specifically, the anti-inflammatory cytokine IL-10 has been shown to up-

TABLE 2 Differential profiles of chemokine production by classically and alternatively activated macrophages

Activation	Chemokines generated	Primary cells recruited
Classical: M1 macrophages		
LPS + IFN-γ	CXCL1, 2, 3, 5, 8, 9, 10	Neutrophils, Th1 cells, NK cells
	CCL2, 3, 4, 5, 11, 17, 22	
Alternative: M2 macrophages		
IL-4/IL-13	CCL17, 18, 22, 24	Eosinophils, Th2 cells
		T-reg cells, basophils
Immune complexes + LPS	CCL1	Th2 cells, T-reg cells, eosinophils
IL-10	CCL13, 16, 20	

regulate inflammatory chemokine receptors, CCR2 and CCR5, on the surface of macrophages. These chemokine receptors are functionally uncoupled, and serve to scavenge chemokines, augmenting the anti-inflammatory function of IL-10 (D'Amico et al., 2000).

Chemokine Receptors as Targets for Therapeutic Intervention

Chemokine receptors are G-protein-coupled receptors with a defined binding cleft for ligand, making them very good therapeutic targets suitable for small molecular antagonists. A major roadblock in the development of such chemokine receptor antagonists has been the fact that small-molecule chemokine receptor antagonists that are effective in blocking human chemokine receptors generally lack activity on rodent receptors. As such, the effectiveness of these antagonists in models of disease cannot be tested effectively in animal models. Moreover, the finding that some chemokine receptor antagonists induced cardiac complications because of their activity on the hERG channel has severely reduced the enthusiasm for antagonists (Hodgson et al., 2004). Despite this, a number of chemokine receptor antagonists are currently in development.

The apparent functional redundancy in chemokines acting via CXCR1 and CXCR2 and the fact that IL-8 binds to both of these receptors with high affinity has led to the development of broad-spectrum CXCR1/CXCR2 antagonists. A number of CXCR1/CXCR2 small peptide antagonists have been reported in the literature and shown to be effective in models such as delayed-type hypersensitivity reactions and LPS-induced pulmonary neutrophilia. Thus, specifically, the CXCR2 antagonists SB332235 and SCH-N have been reported to reduce the cigarette smoke-induced influx of neutrophils into the lungs of rats and mice, respectively (Hay and Sarau, 2001; Stevenson et al., 2005; Thatcher et al., 2005). SB332235 has also been reported to reduce neutrophil numbers in synovial fluid in acute and chronic models of arthritis (Podolin et al., 2002). The CXCR1/2 antagonist Repertaxin has been shown to reduce reperfusion injury (Bertini et al., 2004). More recently, the CXCR1/2 antagonist Sch527123 was reported to inhibit LPS-induced airway neutrophilia (Chapman et al., 2007). Some of these CXCR2 antagonists are currently undergoing phase I and II clinical trials with a view to developing them for the treatment of diseases characterized by a neutrophilic inflammation, such as COPD (Donnelly and Barnes, 2006).

SUMMARY

Mechanisms of chemoattraction are critical for the transfer of phagocytic cells from their site of production in the bone marrow to their sites of action in tissues as acute host defense cells or longer-lived resident defensive cells. Once the cells are in the tissues, depending on the context, chemokine receptor expression may change to effect subsequent transfer of the cells to other tissues, e.g., transfer to lymph nodes to present antigens to T lymphocytes. Chemoattraction is also a fundamental process involved in leukocyte disposal mechanisms. At all of these stages of the life history of the phagocyte, chemokines play an important role. We now have a comprehensive knowledge of the chemokines involved and the receptors on which they act. There has been intensive work to translate this knowledge into therapy for inflammatory diseases in the form of monoclonal antibodies to chemokines and their receptors and small-molecule receptor antagonists. The usual caveats apply as with any anti-inflammatory therapy concerning compromising host defense processes. However, the relative specificity of the mechanism mediated by chemokines offers the prospect of highly targeted therapy for different types of inflammatory diseases. This approach has so far not realized its potential, despite the elegant disease models that have been established and analyzed in detail. The drive to develop specific therapies has highlighted the fact that our understanding of the intricacies of human disease mechanisms needs considerable refinement.

REFERENCES

Anisowicz, A., L. Bardwell, and R. Sager. 1987. Constitutive overexpression of a growth-regulated gene in transformed Chinese hamster and human cells. *Proc. Natl. Acad. Sci. USA* **84:**7188–7192.

Ariel, A., G. Fredman, Y. P. Sun, A. Kantarci, T. E. Van Dyke, A. D. Luster, and C. N. Serhan. 2006. Apoptotic neutrophils and T cells sequester chemokines during immune response resolution through modulation of CCR5 expression. *Nat. Immunol.* **7:**1209–1216.

Ato, M., H. Nakano, T. Kakiuchi, and P. M. Kaye. 2004. Localization of marginal zone macrophages is regulated by C-C chemokine ligands 21/19. *J. Immunol.* **173:**4815–4820.

Bacon, K., M. Baggiolini, H. Broxmeyer, R. Horuk, I. Lindley, A. Mantovani, K. Maysushima, P. Murphy, H. Nomiyama, J. Oppenheim, A. Rot, T. Schall, M. Tsang, R. Thorpe, J. Van Damme, M. Wadhwa, O. Yoshie, A. Zlotnik, and K. Zoon. 2002. Chemokine/chemokine receptor nomenclature. *J. Interferon Cytokine Res.* **22:**1067–1068.

Barlic, J., Y. Zhang, J. F. Foley, and P. M. Murphy. 2006. Oxidized lipid-driven chemokine receptor switch, CCR2 to CX3CR1, mediates adhesion of human macrophages to coronary artery smooth muscle cells through a peroxisome proliferator-activated receptor gamma-dependent pathway. *Circulation* **114:**807–819.

Barlic, J., Y. Zhang, and P. M. Murphy. 2007. Atherogenic lipids induce adhesion of human coronary artery smooth muscle cells to macrophages by up-regulating chemokine CX3CL1 on smooth muscle cells in a TNFalpha-NFkappaB-dependent manner. *J. Biol. Chem.* **282:**19167–19176.

Beaubien, B. C., P. D. Collins, P. J. Jose, N. F. Totty, M. D. Waterfield, J. Hsuan, and T. J. Williams. 1990. A novel neutrophil chemoattractant generated during an inflammatory reaction in the rabbit peritoneal cavity in vivo. purification, partial amino acid sequence and structural relationship to interleukin 8. *Biochem. J.* **271:**797–801.

Bertini, R., M. Allegretti, C. Bizzarri, A. Moriconi, M. Locati, G. Zampella, M. N. Cervellera, V. Di Cioccio, M. C. Cesta, E. Galliera, F. O. Martinez, R. Di Bitondo, G. Troiani, V. Sabbatini, G. D'Anniballe, R. Anacardio, J. C. Cutrin, B. Cavalieri, F. Mainiero, R. Strippoli, P. Villa, M. Di Girolamo, F. Martin, M. Gentile, A. Santoni, D. Corda, G. Poli, A. Mantovani, P. Ghezzi, and F. Colotta. 2004. Noncompetitive allosteric inhibitors of the inflammatory chemokine receptors CXCR1 and CXCR2: prevention of reperfusion injury. *Proc. Natl. Acad. Sci. USA* **101:**11791–11796.

Bleul, C. C., R. C. Fuhlbrigge, J. M. Casasnovas, A. Aiuti, and T. A. Springer. 1996. A highly efficacious lymphocyte chemoattractant, stromal cell-derived factor 1 (SDF-1). *J. Exp. Med.* **184:**1101–1109.

Bonecchi, R., F. Facchetti, S. Dusi, W. Luini, D. Lissandrini, M. Simmelink, M. Locati, S. Bernasconi, P. Allavena, E. Brandt, F. Rossi, A. Mantovani, and S. Sozzani. 2000. Induction of functional IL-8 receptors by IL-4 and IL-13 in human monocytes. *J. Immunol.* **164:**3862–3869.

Bonecchi, R., N. Polentarutti, W. Luini, A. Borsatti, S. Bernasconi, M. Locati, C. Power, A. Proudfoot, T. N. Wells, C. Mackay, A. Mantovani, and S. Sozzani. 1999. Upregulation of CCR1 and CCR3 and induction of chemotaxis to CC chemokines by IFN-gamma in human neutrophils. *J. Immunol.* **162:**474–479.

Boring, L., J. Gosling, S. W. Chensue, S. L. Kunkel, R. V. Farese, Jr., H. E. Broxmeyer, and I. F. Charo. 1997. Impaired monocyte migration and reduced type 1 (Th1) cytokine responses in C-C chemokine receptor 2 knockout mice. *J. Clin. Invest.* **100:**2552–2561.

Boyle, E. M., Jr, J. C. Kovacich, C. A. Hebert, T. G. Canty, Jr, E. Chi, E. N. Morgan, T. H. Pohlman, and E. D. Verrier. 1998. Inhibition of interleukin-8 blocks myocardial ischemia-reperfusion injury. *J. Thorac. Cardiovasc. Surg.* **116:**114–121.

Bray, M. A., A. W. Ford-Hutchinson, and M. J. Smith. 1981. Leukotriene B4: an inflammatory mediator in vivo. *Prostaglandins* **22:**213–222.

Cara, D. C., J. Kaur, M. Forster, D. M. McCafferty, and P. Kubes. 2001. Role of p38 mitogen-activated protein kinase in chemokine-induced emigration and chemotaxis in vivo. *J. Immunol.* **167:**6552–6558.

Chapman, R. W., M. Minnicozzi, C. S. Celly, J. E. Phillips, T. T. Kung, R. W. Hipkin, X. Fan, D. Rindgen, G. Deno, R. Bond, W. Gonsiorek, M. M. Billah, J. S. Fine, and J. A. Hey. 2007. A novel, orally active CXCR1/2 receptor antagonist, Sch527123, inhibits neutrophil recruitment, mucus production, and goblet cell hyperplasia in animal models of pulmonary inflammation. *J. Pharmacol. Exp. Ther.* **322:**486–493.

Charo, I. F., S. J. Myers, A. Herman, C. Franci, A. J. Connolly and S. R. Coughlin. 1994. Molecular cloning and functional expression of two monocyte chemoattractant protein 1 receptors reveals alternative splicing of the carboxyl-terminal tails. *Proc. Natl. Acad. Sci. USA* **91:**2752–2756.

Charo, I. F., and W. Peters. 2003. Chemokine receptor 2 (CCR2) in atherosclerosis, infectious diseases, and regulation of T-cell polarization. *Microcirculation* **10:**259–264.

Cheng, S. S., J. J. Lai, N. W. Lukacs, and S. L. Kunkel. 2001. Granulocyte-macrophage colony stimulating factor upregulates CCR1 in human neutrophils. *J. Immunol.* **166:**1178–1184.

Clark-Lewis, I., B. Dewald, T. Geiser, B. Moser, and M. Baggiolini. 1993. Platelet factor 4 binds to interleukin 8 receptors and activates neutrophils when its N terminus is modified with Glu-Leu-Arg. *Proc. Natl. Acad. Sci. USA* **90:**3574–3577.

Colditz, I. G., R. D. Zwahlen, and M. Baggiolini. 1990. Neutrophil accumulation and plasma leakage induced in vivo by neutrophil-activating peptide-1 (NAP-1). *J. Leukoc. Biol.* **48:**129–137.

Collins, P. D., P. J. Jose, and T. J. Williams. 1991. The sequential generation of neutrophil chemoattractant proteins in acute inflammation in the rabbit in vivo: relationship between C5a and a protein with the characteristics of IL-8. *J. Immunol.* **146:**677–684.

Combadiere, C., S. Potteaux, J. L. Gao, B. Esposito, S. Casanova, E. J. Lee, P. Debre, A. Tedgui, P. M. Murphy, and Z. Mallat. 2003. Decreased atherosclerotic lesion formation in CX3CR1/apolipoprotein E double knockout mice. *Circulation* **107:**1009–1016.

D'Amico, G., G. Frascaroli, G. Bianchi, P. Transidico, A. Doni, A. Vecchi, S. Sozzani, P. Allavena, and A. Mantovani. 2000. Uncoupling of inflammatory chemokine receptors by IL-10: generation of functional decoys. *Nat. Immunol.* **1:**387–391.

Donnelly, L. E., and P. J. Barnes. 2006. Chemokine receptors as therapeutic targets in chronic obstructive pulmonary disease. *Trends Pharmacol. Sci.* **27:**546–553.

Fantuzzi, L., P. Borghi, V. Ciolli, G. Pavlakis, F. Belardelli, and S. Gessani. 1999. Loss of CCR2 expression and functional response to monocyte chemotactic protein (MCP-1) during the differentiation of human monocytes: role of secreted MCP-1 in the regulation of the chemotactic response. *Blood* **94:**875–883.

Franca-Koh, J., Y. Kamimura, and P. N. Devreotes. 2007. Leading-edge research: PtdIns(3,4,5)P3 and directed migration. *Nat. Cell Biol.* **9:**15–17.

Fuentes, M. E., S. K. Durham, M. R. Swerdel, A. C. Lewin, D. S. Barton, J. R. Megill, R. Bravo, and S. A. Lira. 1995. Controlled recruitment of monocytes and macrophages to specific organs through transgenic expression of monocyte chemoattractant protein-1. *J. Immunol.* **155:**5769–5776.

Geiser, T., B. Dewald, M. U. Ehrengruber, I. Clark-Lewis, and M. Baggiolini. 1993. The interleukin-8-related chemotactic cytokines GRO alpha, GRO beta, and GRO gamma activate human neutrophil and basophil leukocytes. *J. Biol. Chem.* **268:**15419–15424.

Geissmann, F., S. Jung, and D. R. Littman. 2003. Blood monocytes consist of two principal subsets with distinct migratory properties. *Immunity* **19:**71–82.

Graham, G. J., and C. S. McKimmie. 2006. Chemokine scavenging by D6: a movable feast? *Trends Immunol.* **27:**381–386.

Graves, D. T., Y. L. Jiang, M. J. Williamson, and A. J. Valente. 1989. Identification of monocyte chemotactic activity produced by malignant cells. *Science* **245:**1490–1493.

Gu, L., Y. Okada, S. K. Clinton, C. Gerard, G. K. Sukhova, P. Libby, and B. J. Rollins. 1998. Absence of monocyte chemoattractant protein-1 reduces atherosclerosis in low density lipoprotein receptor-deficient mice. *Mol. Cell* **2:**275–281.

Gunn, M. D., K. Tangemann, C. Tam, J. G. Cyster, S. D. Rosen, and L. T. Williams. 1998. A chemokine expressed in lymphoid high endothelial venules promotes the adhesion and chemotaxis of naive T lymphocytes. *Proc. Natl. Acad. Sci. USA* **95:**258–263.

Han, K. H., M. K. Chang, A. Boullier, S. R. Green, A. Li, C. K. Glass, and O. Quehenberger. 2000. Oxidized LDL reduces monocyte CCR2 expression through pathways involving peroxisome proliferator-activated receptor gamma. *J. Clin. Invest.* **106:**793–802.

Haskill, S., A. Peace, J. Morris, S. A. Sporn, A. Anisowicz, S. W. Lee, T. Smith, G. Martin, P. Ralph, and R. Sager. 1990. Identification of three related human GRO genes encoding cytokine functions. *Proc. Natl. Acad. Sci. USA* **87:**7732–7736.

Hay, D. W., and H. M. Sarau. 2001. Interleukin-8 receptor antagonists in pulmonary diseases. *Curr. Opin. Pharmacol.* **1:**242–247.

Hirsch, E., V. L. Katanaev, C. Garlanda, O. Azzolino, L. Pirola, L. Silengo, S. Sozzani, A. Mantovani, F. Altruda, and M. P. Wymann. 2000. Central role for G protein-coupled phosphoinositide 3-kinase gamma in inflammation [see comments]. *Science* **287:**1049–1053.

Hodgson, S., S. Charlton, and P. Warne. 2004. Chemokines and drug discovery. *Drug News Perspect.* **17:**335–338.

Holmes, W. E., J. Lee, W.-J. Kuang, G. C. Rice, and W. I. Wood. 1991. Structure and functional expression of a human interleukin-8 receptor. *Science* **253:**1278–1283.

Ivey, C. L., F. M. Williams, P. D. Collins, P. J. Jose, and T. J. Williams. 1995. Neutrophil chemoattractants generated in two phases during reperfusion of ischemic myocardium in the rabbit: evidence for a role for C5a and interleukin-8. *J. Clin. Invest.* **95:**2720–2728.

Johnston, B., A. R. Burns, M. Suematsu, T. B. Issekutz, R. C. Woodman, and P. Kubes. 1999. Chronic inflammation upregulates chemokine receptors and induces neutrophil migration to monocyte chemoattractant protein-1. *J. Clin. Invest.* **103:**1269–1276.

Johnston, R. A., J. P. Mizgerd, and S. A. Shore. 2005. CXCR2 is essential for maximal neutrophil recruitment and methacholine responsiveness after ozone exposure. *Am. J. Physiol.* **288:**L61–L67.

Jones, S. A., B. Dewald, I. Clark-Lewis, and M. Baggiolini. 1997. Chemokine antagonists that discriminate between interleukin-8 receptors. Selective blockers of CXCR2. *J. Biol. Chem.* **272:**16166–16169.

Jones, S. A., M. Wolf, S. Qin, C. R. Mackay, and M. Baggiolini. 1996. Different functions for the interleukin 8 receptors (IL-8R) of human neutrophil leukocytes: NADPH oxidase and phospholipase D are activated through IL-8R1 but not IL-8R2. *Proc. Natl. Acad. Sci. USA* **93:**6682–6686.

Jose, P. J., P. D. Collins, J. A. Perkins, B. C. Beaubien, N. F. Totty, M. D. Waterfield, J. Hsuan, and T. J. Williams. 1991. Identification of a second neutrophil chemoattractant cytokine generated during an inflammatory reaction in the rabbit peritoneal cavity in vivo: purification, partial amino acid sequence and structural relationship to melanoma growth stimulatory activity. *Biochem. J.* **278:**493–497.

Jose, P. J., M. J. Forrest, and T. J. Williams. 1983. Detection of the complement fragment C5a in inflammatory exudates from the rabbit peritoneal cavity using radioimmunoassay. *J. Exp. Med.* **158:**2177–2182.

Jose, P. J., D. A. Griffiths-Johnson, P. D. Collins, D. T. Walsh, R. Moqbel, N. F. Totty, O. Truong, J. J. Hsuan, and T. J. Williams. 1994. Eotaxin: a potent eosinophil chemoattractant cytokine detected in a guinea-pig model of allergic airways inflammation. *J. Exp. Med.* **179:**881–887.

Kaufmann, A., R. Salentin, D. Gemsa, and H. Sprenger. 2001. Increase of CCR1 and CCR5 expression and enhanced functional response to MIP-1 alpha during differentiation of human monocytes to macrophages. *J. Leukoc. Biol.* **69:**248–252.

Kim, H. K., M. De La Luz Sierra, C. K. Williams, A. V. Gulino, and G. Tosato. 2006. G-CSF down-regulation of CXCR4 expression identified as a mechanism for mobilization of myeloid cells. *Blood* **108:**812–820.

Kurihara, T., G. Warr, J. Loy, and R. Bravo. 1997. Defects in macrophage recruitment and host defense in mice lacking the CCR2 chemokine receptor. *J. Exp. Med.* **186:**1757–1762.

Le Borgne, M., N. Etchart, A. Goubier, S. A. Lira, J. C. Sirard, N. van Rooijen, C. Caux, S. Ait-Yahia, A. Vicari, D. Kaiserlian, and B. Dubois. 2006. Dendritic cells rapidly recruited into epithelial tissues via CCR6/CCL20 are responsible for CD8+ T cell crosspriming in vivo. *Immunity* **24:**191–201.

Lesnik, P., C. A. Haskell, and I. F. Charo. 2003. Decreased atherosclerosis in CX3CR1−/− mice reveals a role for fractalkine in atherogenesis. *J. Clin. Invest.* **111:**333–340.

Ley, K., C. Laudanna, M. I. Cybulsky, and S. Nourshargh. 2007. Getting to the site of inflammation: the leukocyte adhesion cascade updated. *Nat. Rev. Immunol.* **7:** 678–689.

Li, Z., H. Jiang, W. Xie, Z. Zhang, A. V. Smrcka, and D. Wu. 2000. Roles of PLC-beta2 and -beta3 and PI3Kgamma in chemoattractant-mediated signal transduction. *Science* **287:**1046–1049.

Lin, M., E. Carlson, E. Diaconu, and E. Pearlman. 2007. CXCL1/KC and CXCL5/LIX are selectively produced by corneal fibroblasts and mediate neutrophil infiltration to the corneal stroma in LPS keratitis. *J. Leukoc. Biol.* **81:**786–792.

Lira, S. A., P. Zalamea, J. N. Heinrich, M. E. Fuentes, D. Carrasco, A. C. Lewin, D. S. Barton, S. Durham, and R. Bravo. 1994. Expression of the chemokine N51/KC in the thymus and epidermis of transgenic mice results in marked infiltration of a single class of inflammatory cells. *J. Exp. Med.* **180:**2039–2048.

Lu, B., B. J. Rutledge, L. Gu, J. Fiorillo, N. W. Lukacs, S. L. Kunkel, R. North, C. Gerard, and B. J. Rollins. 1998. Abnormalities in monocyte recruitment and cytokine expression in monocyte chemoattractant protein 1-deficient mice. *J. Exp. Med.* **187:**601–608.

Luther, S. A., H. L. Tang, P. L. Hyman, A. G. Farr, and J. G. Cyster. 2000. Coexpression of the chemokines ELC and SLC by T zone stromal cells and deletion of the ELC gene in the plt/plt mouse. *Proc. Natl. Acad. Sci. USA* **97:** 12694–12699.

Mantovani, A., A. Sica, and M. Locati. 2007. New vistas on macrophage differentiation and activation. *Eur. J. Immunol.* **37:**14–16.

Mantovani, A., A. Sica, S. Sozzani, P. Allavena, A. Vecchi, and M. Locati. 2004. The chemokine system in diverse forms of macrophage activation and polarization. *Trends Immunol.* **25:**677–686.

Martin, C., P. C. E. Burdon, G. Bridger, J.-C. Gutierrez-Ramos, T. J. Williams, and S. M. Rankin. 2003. The balance between chemokines acting via CXCR4 and CXCR2 determines the release of neutrophils from the bone marrow and their return following senescence. *Immunity* **19:**583–593.

Martinez, F. O., S. Gordon, M. Locati, and A. Mantovani. 2006. Transcriptional profiling of the human monocyte-to-macrophage differentiation and polarization: new molecules and patterns of gene expression. *J. Immunol.* **177:**7303–7311.

McColl, S. R., and I. Clark-Lewis. 1999. Inhibition of murine neutrophil recruitment in vivo by CXC chemokine receptor antagonists. *J. Immunol.* **163:**2829–2835.

McDonald, P. H., C. W. Chow, W. E. Miller, S. A. Laporte, M. E. Field, F. T. Lin, R. J. Davis, and R. J. Lefkowitz. 2000. Beta-arrestin 2: a receptor-regulated MAPK scaffold for the activation of JNK3. *Science* **290:**1574–1577.

McWilliam, A. S., S. Napoli, A. M. Marsh, F. L. Pemper, D. J. Nelson, C. L. Pimm, P. A. Stumbles, T. N. Wells, and P. G. Holt. 1996. Dendritic cells are recruited into the airway epithelium during the inflammatory response to a broad spectrum of stimuli. *J. Exp. Med.* **184:**2429–2432.

Mehrad, B., R. M. Strieter, T. A. Moore, W. C. Tsai, S. A. Lira, and T. J. Standiford. 1999. CXC chemokine receptor-2 ligands are necessary components of neutrophil-mediated host defense in invasive pulmonary aspergillosis. *J. Immunol.* **163:**6086–6094.

Merad, M., P. Hoffmann, E. Ranheim, S. Slaymaker, M. G. Manz, S. A. Lira, I. Charo, D. N. Cook, I. L. Weissman, S. Strober, and E. G. Engleman. 2004. Depletion of host Langerhans cells before transplantation of donor alloreactive T cells prevents skin graft-versus-host disease. *Nat. Med.* **10:** 510–517.

Miura, M., X. Fu, Q. W. Zhang, D. G. Remick, and R. L. Fairchild. 2001. Neutralization of Gro alpha and macrophage inflammatory protein-2 attenuates renal ischemia/reperfusion injury. *Am. J. Pathol.* **159:** 2137–2145.

Murphy, P. M., and H. L. Tiffany. 1991. Cloning of complementary DNA encoding a functional human interleukin-8 receptor. *Science* **253:**1280–1283.

Nagira, M., A. Sato, S. Miki, T. Imai, and O. Yoshie. 1999. Enhanced HIV-1 replication by chemokines constitutively expressed in secondary lymphoid tissues. *Virology* **264:**422–426.

Neel, N. F., E. Schutyser, J. Sai, G. H. Fan, and A. Richmond. 2005. Chemokine receptor internalization and intracellular trafficking. *Cytokine Growth Factor Rev.* **16:**637–658.

Nelken, N. A., S. R. Coughlin, D. Gordon, and J. N. Wilcox. 1991. Monocyte chemoattractant protein-1 in human atheromatous plaques. *J. Clin. Invest.* **88:**1121–1127.

Ngo, V. N., H. L. Tang, and J. G. Cyster. 1998. Epstein-Barr virus-induced molecule 1 ligand chemokine is expressed by dendritic cells in lymphoid tissues and strongly attracts naive T cells and activated B cells. *J. Exp. Med.* **188**:181–191.

Nourshargh, S., J. A. Perkins, H. J. Showell, K. Matsushima, T. J. Williams, and P. D. Collins. 1992. A comparative study of the neutrophil stimulatory activity in vitro and pro-inflammatory properties in vivo of 72 amino acid and 77 amino acid IL-8. *J. Immunol.* **148**:106–111.

Oquendo, P., J. Alberta, D. Z. Wen, J. L. Graycar, R. Derynck, and C. D. Stiles. 1989. The platelet-derived growth factor-inducible KC gene encodes a secretory protein related to platelet alpha-granule proteins. *J. Biol. Chem.* **264**:4133–4137.

Passlick, B., D. Flieger, and H. W. Ziegler-Heitbrock. 1989. Identification and characterization of a novel monocyte subpopulation in human peripheral blood. *Blood* **74**:2527–2534.

Penton-Rol, G., N. Polentarutti, W. Luini, A. Borsatti, R. Mancinelli, A. Sica, S. Sozzani, and A. Mantovani. 1998. Selective inhibition of expression of the chemokine receptor CCR2 in human monocytes by IFN-gamma. *J. Immunol.* **160**:3869–3873.

Podolin, P. L., B. J. Bolognese, J. J. Foley, D. B. Schmidt, P. T. Buckley, K. L. Widdowson, Q. Jin, J. R. White, J. M. Lee, R. B. Goodman, T. R. Hagen, O. Kajikawa, L. A. Marshall, D. W. Hay, and H. M. Sarau. 2002. A potent and selective nonpeptide antagonist of CXCR2 inhibits acute and chronic models of arthritis in the rabbit. *J. Immunol.* **169**:6435–6444.

Proost, P., C. De Wolf-Peeters, R. Conings, G. Opdenakker, A. Billiau, and J. Van Damme. 1993. Identification of a novel granulocyte chemotactic protein (GCP-2) from human tumor cells. *J. Immunol.* **150**:1000–1010.

Qu, C., E. W. Edwards, F. Tacke, V. Angeli, J. Llodra, G. Sanchez-Schmitz, A. Garin, N. S. Haque, W. Peters, N. van Rooijen, C. Sanchez-Torres, J. Bromberg, I. F. Charo, S. Jung, S. A. Lira, and G. J. Randolph. 2004. Role of CCR8 and other chemokine pathways in the migration of monocyte-derived dendritic cells to lymph nodes. *J. Exp. Med.* **200**:1231–1241.

Rajarathnam, K., B. D. Sykes, B. Dewald, M. Baggiolini, and I. Clark-Lewis. 1999. Disulfide bridges in interleukin-8 probed using non-natural disulfide analogues: dissociation of roles in structure from function. *Biochemistry* **38**:7653–7658.

Ramirez, B. L., O. M. Howard, H. F. Dong, T. Edamatsu, P. Gao, M. Hartlein, and M. Kron. 2006. Brugia malayi asparaginyl-transfer RNA synthetase induces chemotaxis of human leukocytes and activates G-protein-coupled receptors CXCR1 and CXCR2. *J. Infect. Dis.* **193**:1164–1171.

Randolph, G. J., G. Sanchez-Schmitz, R. M. Liebman, and K. Schakel. 2002. The CD16(+) (FcgammaRIII(+)) subset of human monocytes preferentially becomes migratory dendritic cells in a model tissue setting. *J. Exp. Med.* **196**:517–527.

Rankin, S. M., S. Parthasarathy, and D. Steinberg. 1991. Evidence for a dominant role of lipoxygenase(s) in the oxidation of LDL by mouse peritoneal macrophages. *J. Lipid Res.* **32**:449–456.

Reutershan, J., M. A. Morris, T. L. Burcin, D. F. Smith, D. Chang, M. S. Saprito, and K. Ley. 2006. Critical role of endothelial CXCR2 in LPS-induced neutrophil migration into the lung. *J. Clin. Invest.* **116**:695–702.

Robinson, E. A., T. Yoshimura, E. J. Leonard, S. Tanaka, P. R. Griffin, J. Shabanowitz, D. F. Hunt, and E. Appella. 1989. Complete amino acid sequence of a human monocyte chemoattractant, a putative mediator of cellular immune reactions. *Proc. Natl. Acad. Sci. USA* **86**:1850–1854.

Rot, A. 2005. Contribution of Duffy antigen to chemokine function. *Cytokine Growth Factor Rev.* **16**:687–694.

Sallusto, F., P. Schaerli, P. Loetscher, C. Schaniel, D. Lenig, C. R. Mackay, S. Qin, and A. Lanzavecchia. 1998. Rapid and coordinated switch in chemokine receptor expression during dendritic cell maturation. *Eur. J. Immunol.* **28**:2760–2769.

Sasaki, T., J. Irie-Sasaki, R. G. Jones, A. J. Oliveira-dos-Santos, W. L. Stanford, B. Bolon, A. Wakeham, A. Itie, D. Bouchard, I. Kozieradzki, N. Joza, T. W. Mak, P. S. Ohashi, A. Suzuki, and J. M. Penninger. 2000. Function of PI3Kgamma in thymocyte development, T cell activation, and neutrophil migration [see comments]. *Science* **287**:1040–1046.

Schall, T. J., K. Bacon, K. I. Toy, and D. V. Goeddel. 1990. Selective attraction of monocytes and T lymphocytes of the memory phenotype by cytokine RANTES. *Nature* **347**:669–671.

Schioppa, T., B. Uranchimeg, A. Saccani, S. K. Biswas, A. Doni, A. Rapisarda, S. Bernasconi, S. Saccani, M. Nebuloni, L. Vago, A. Mantovani, G. Melillo, and A. Sica. 2003. Regulation of the chemokine receptor CXCR4 by hypoxia. *J. Exp. Med.* **198**:1391–1402.

Schröder, J.-M., U. Mrowietz, and E. Christophers. 1988. Purification and partial biological characterization of a human lymphocyte-derived peptide with potent neutrophil-stimulating activity. *J. Immunol.* **140**:3534–3540.

Sekido, N., N. Mukaida, A. Harada, I. Nakanish, Y. Watanade, and K. Matsushima. 1993. Prevention of lung reperfusion injury in rabbits by a monoclonal antibody against interleukin-8. *Nature* **365**:654–657.

Serbina, N. V., and E. G. Pamer. 2006. Monocyte emigration from bone marrow during bacterial infection requires signals mediated by chemokine receptor CCR2. *Nat. Immunol.* **7**:311–317.

Sica, A., A. Saccani, A. Borsatti, C. A. Power, T. N. Wells, W. Luini, N. Polentarutti, S. Sozzani, and A. Mantovani. 1997. Bacterial lipopolysaccharide rapidly inhibits expression of C-C chemokine receptors in human monocytes. *J. Exp. Med.* **185**:969–974.

Sozzani, S., S. Ghezzi, G. Iannolo, W. Luini, A. Borsatti, N. Polentarutti, A. Sica, M. Locati, C. Mackay, T. N. Wells, P. Biswas, E. Vicenzi, G. Poli, and A. Mantovani. 1998. Interleukin 10 increases CCR5 expression and HIV infection in human monocytes. *J. Exp. Med.* **187**:439–444.

Speyer, C. L., H. Gao, N. J. Rancilio, T. A. Neff, G. B. Huffnagle, J. V. Sarma, and P. A. Ward. 2004. Novel chemokine responsiveness and mobilization of neutrophils during sepsis. *Am. J. Pathol.* **165**:2187–2196.

Stein, O., Y. Dabach, M. Ben-Naim, G. Halperin, I. F. Charo, and Y. Stein. 2003. In CCR2−/− mice monocyte recruitment and egress of LDL cholesterol in vivo is impaired. *Biochem. Biophys. Res. Commun.* **300**:477–481.

Stevenson, C. S., K. Coote, R. Webster, H. Johnston, H. C. Atherton, A. Nicholls, J. Giddings, R. Sugar, A. Jackson, N. J. Press, Z. Brown, K. Butler, and H. Danahay. 2005. Characterization of cigarette smoke-induced inflammatory and mucus hypersecretory changes in rat lung and the role of CXCR2 ligands in mediating this effect. *Am. J. Physiol.* **288**:L514–L522.

Su, Y., S. K. Raghuwanshi, Y. Yu, L. B. Nanney, R. M. Richardson, and A. Richmond. 2005. Altered CXCR2 signaling in beta-arrestin-2-deficient mouse models. *J. Immunol.* **175**:5396–5402.

Sugano, S., M. Y. Stoeckle, and H. Hanafusa. 1987. Transformation by Rous sarcoma virus induces a novel gene with homology to a mitogenic platelet protein. *Cell* **49**:321–328.

Tacke, F., D. Alvarez, T. J. Kaplan, C. Jakubzick, R. Spanbroek, J. Llodra, A. Garin, J. Liu, M. Mack, N. van Rooijen, S. A. Lira, A. J. Habenicht, and G. J. Randolph. 2007. Monocyte subsets differentially employ CCR2, CCR5, and CX3CR1 to accumulate within atherosclerotic plaques. *J. Clin. Invest.* **117**:185–194.

Thatcher, T. H., N. A. McHugh, R. W. Egan, R. W. Chapman, J. A. Hey, C. K. Turner, M. R. Redonnet, K. E. Seweryniak, P. J. Sime, and R. P. Phipps. 2005. Role of CXCR2 in cigarette smoke-induced lung inflammation. *Am. J. Physiol.* **289**:L322–L328.

Traves, S. L., S. J. Smith, P. J. Barnes, and L. E. Donnelly. 2004. Specific CXC but not CC chemokines cause elevated monocyte migration in COPD: a role for CXCR2. *J. Leukoc. Biol.* **76**:441–450.

Tsai, W. C., R. M. Strieter, B. Mehrad, M. W. Newstead, X. Zeng, and T. J. Standiford. 2000. CXC chemokine receptor CXCR2 is essential for protective innate host response in murine Pseudomonas aeruginosa pneumonia. *Infect. Immun.* **68**:4289–4296.

Tsou, C. L., W. Peters, Y. Si, S. Slaymaker, A. M. Aslanian, S. P. Weisberg, M. Mack, and I. F. Charo. 2007. Critical roles for CCR2 and MCP-3 in monocyte mobilization from bone marrow and recruitment to inflammatory sites. *J. Clin. Invest.* **117**:902–909.

Tyner, J. W., O. Uchida, N. Kajiwara, E. Y. Kim, A. C. Patel, M. P. O'Sullivan, M. J. Walter, R. A. Schwendener, D. N. Cook, T. M. Danoff, and M. J. Holtzman. 2005. CCL5-CCR5 interaction provides antiapoptotic signals for macrophage survival during viral infection. *Nat. Med.* **11**:1180–1187.

Valente, A. J., D. T. Graves, C. E. Vialle-Valentin, R. Delgado, and C. J. Schwartz. 1988. Purification of a monocyte chemotactic factor secreted by nonhuman primate vascular cells in culture. *Biochemistry* **27**:4162–4168.

Van Den Steen, P. E., P. Proost, A. Wuyts, J. Van Damme, and G. Opdenakker. 2000. Neutrophil gelatinase B potentiates interleukin-8 tenfold by aminoterminal processing, whereas it degrades CTAP-III, PF-4, and GRO-alpha and leaves RANTES and MCP-2 intact. *Blood* **96**:2673–2681.

Viola, A., and A. D. Luster. 2008. Chemokines and their receptors: drug targets in immunity and inflammation. *Annu. Rev. Pharmacol. Toxicol.* **48**:171–197.

Walz, A., and M. Baggiolini. 1989. Novel cleavage product of b-thromboglobulin formed in cultures of stimulated mononuclear cells activates human neutrophils. *Biochem. Biophys. Res. Commun.* **159**:969–975.

Walz, A., and M. Baggiolini. 1990. Generation of the neutrophil-activating peptide NAP-2 from platelet basic protein or connective tissue-activating peptide III through monocyte proteases. *J. Exp. Med.* **171**:449–454.

Walz, A., R. Burgener, B. Car, M. Baggiolini, S. L. Kunkel, and R. M. Strieter. 1991. Structure and neutrophil-activating properties of a novel inflammatory peptide (ENA-78) with homology to interleukin 8. *J. Exp. Med.* **174**:1355–1362.

Walz, A., P. Peveri, H. Aschauer, and M. Baggiolini. 1987. Purification and amino acid sequencing of NAF, a novel neutrophil-activating factor produced by monocytes. *Biochem. Biophys. Res. Commun.* **149**:755–761.

Weathington, N. M., A. H. van Houwelingen, B. D. Noerager, P. L. Jackson, A. D. Kraneveld, F. S. Galin, G. Folkerts, F. P. Nijkamp, and J. E. Blalock. 2006. A novel peptide CXCR ligand derived from extracellular matrix degradation during airway inflammation. *Nat. Med.* **12**:317–323.

Weber, C., K. U. Belge, P. von Hundelshausen, G. Draude, B. Steppich, M. Mack, M. Frankenberger, K. S. Weber, and H. W. Ziegler-Heitbrock. 2000. Differential chemokine receptor expression and function in human monocyte subpopulations. *J. Leukoc. Biol.* **67**:699–704.

Wengner, A. M., S. C. Pitchford, R. C. Furze, and S. M. Rankin. 2008. The co-ordinated action of G-CSF and ELR+CXC chemokines in neutrophil mobilisation during acute inflammation. *Blood* **111**:42–49.

Willard, S. S., and P. N. Devreotes. 2006. Signaling pathways mediating chemotaxis in the social amoeba, Dictyostelium discoideum. *Eur. J. Cell Biol.* **85**:897–904.

Willimann, K., D. F. Legler, M. Loetscher, R. S. Roos, M. B. Delgado, I. Clark-Lewis, M. Baggiolini, and B. Moser. 1998. The chemokine SLC is expressed in T cell areas of lymph nodes and mucosal lymphoid tissues and attracts activated T cells via CCR7. *Eur. J. Immunol.* **28**:2025–2034.

Yagihashi, A., T. Tsuruma, K. Tarumi, T. Kameshima, T. Yajima, Y. Yanai, N. Watanabe, and K. Hirata. 1998. Prevention of small intestinal ischemia-reperfusion injury in rat by anti-cytokine-induced neutrophil chemoattractant monoclonal antibody. *J. Surg. Res.* **78**:92–96.

Yamagami, S., Y. Tokuda, K. Ishii, H. Tanaka, and N. Endo. 1994. cDNA cloning and functional expression of a human monocyte chemoattractant protein 1 receptor. *Biochem. Biophys. Res. Commun.* **202**:1156–1162.

Yla-Herttuala, S., B. A. Lipton, M. E. Rosenfeld, T. Sarkioja, T. Yoshimura, E. J. Leonard, J. L. Witztum, and D. Steinberg. 1991. Expression of monocyte chemoattractant protein 1 in macrophage-rich areas of human and rabbit atherosclerotic lesions. *Proc. Natl. Acad. Sci. USA* **88**:5252–5256.

Yoshimura, T., K. Matsushima, S. Tanaka, E. A. Robinson, E. Appella, J. J. Oppenheim, and E. J. Leonard. 1987. Purification of a human monocyte-derived neutrophil chemotactic factor that has peptide sequence similarity to other host defense cytokines. *Proc. Natl. Acad. Sci. USA* **84**:9233–9237.

Zernecke, A., E. A. Liehn, J. L. Gao, W. A. Kuziel, P. M. Murphy, and C. Weber. 2006. Deficiency in CCR5 but not CCR1 protects against neointima formation in atherosclerosis-prone mice: involvement of IL-10. *Blood* **107**:4240–4243.

Ziegler-Heitbrock, H. W., G. Fingerle, M. Strobel, W. Schraut, F. Stelter, C. Schutt, B. Passlick, and A. Pforte. 1993. The novel subset of CD14+/CD16+ blood monocytes exhibits features of tissue macrophages. *Eur. J. Immunol.* **23**:2053–2058.

6

Toll-Like Receptors

RICARDO T. GAZZINELLI, KATE FITZGERALD, AND DOUGLAS T. GOLENBOCK

The innate immune system is a complicated collection of a variety of host defenses, ranging from simple defenses such as the skin, which provides important barrier functions, to more sophisticated means of responding to pathogens, such as receptor-mediated cytokine production due to the direct recognition of microbial products. While it is difficult to discount the importance of the former types of immune defense, the latter system of host response has proved to be remarkably interesting and has captured the attention of the entire immunological community.

Perhaps the deep degree of interest in pathogen recognition began with clinical observations of the sepsis syndrome, a loosely defined event that consists of changes in physiological parameters like body temperature, heart rate, respiratory rate, a concomitant recruitment of phagocytic leukocytes into the bloodstream, and evidence of infection. Septic shock, which has become increasingly common in hospitals over the course of the past 35 years, is one of the most frustrating syndromes for the clinician to treat precisely because our understanding of its pathophysiology is highly advanced, but our therapeutic armamentarium is not. Nevertheless, investigators have always believed that more knowledge will ultimately translate into better treatments for diseases, and persevere in their efforts to better define this aspect of the innate immune system.

The sepsis syndrome suggested to investigators the importance of one of the major questions in innate immune signal transduction, i.e., how is bacterial lipopolysaccharide (LPS), the major component of the outer membrane of gram-negative bacteria, recognized by phagocytes? This question was seemingly answered in 1990, when two papers were published that described CD14, a glycosylphosphatidyl-linked receptor that had previously been thought to be an uncharacterized growth factor receptor, in LPS recognition (Ferrero et al., 1990). However, CD14 only enhanced responses to LPS, but it was not essential for LPS responses. This was predictable, of course, because CD14 had no transmembrane domain; nevertheless, alternative theories of the importance of CD14 abounded at the time and needed to be systematically disproved. Thus, by the following year, it was recognized that a CD14-associated signal transducer must exist.

The subsequent discovery of the identity of Lps^d as Toll-like receptor 4 (TLR4) (Poltorak et al., 1998), and the discovery of MD-2 (Shromm et al., 2001; Shimazu et al., 1999), rapidly closed the hole in our knowledge concerning the LPS receptor. These early reports were responsible for the enthusiasm in TLR research, and rapidly led to the recognition that each TLR has unique aspects to its ligand preference, the means in which it is activated, the downstream adapter molecules utilized, and the gene sets that are activated as a result of ligand binding. The literature is huge and confusing. The purpose of this chapter is to cut through this confusion by describing selected aspects of the existing literature and focus on examples of TLR research where questions have been particularly well addressed as potential paradigms for the entire TLR system.

TOLL RECEPTORS AND THEIR ROLE IN DEVELOPMENT OF *DROSOPHILA*

The Toll gene was first described as having a primary role in the pathway that establishes the dorsal-ventral axis of the early *Drosophila* embryos (Anderson et al., 1985a, 1985b). This gene was shown to encode a transmembrane protein that belongs to a superfamily of receptors, which contain an ectodomain formed by blocks of 24-amino-acid leucine-rich repeats (LRRs). The sequence of *Drosophila* reveals a total of nine Toll receptors (Belvin and Anderson, 1996), and null mutation genes encoding Toll or lack of associated signaling molecules suggest their primary role in embryogenesis. Toll has also been identified as a direct regulator of organogenesis in *Drosophila*, and loss of zygotic Toll induces muscle pattern, as well as motor neuron de-

---

Ricardo T. Gazzinelli, Div. of Infectious Diseases and Immunology, Dept. of Medicine, University of Massachusetts Medical School, Worcester, MA 01605; Rene Rachou Institute, Oswaldo Cruz Foundation, Institute of Biological Sciences, and Federal University of Minas Gerais, Belo Horizonte, MG, Brazil. **Kate Fitzgerald,** Division of Infectious Diseases and Immunology, Dept. of Medicine, University of Massachusetts Medical School, Worcester, MA 01605. **Douglas T. Golenbock,** Div. of Infectious Diseases and Immunology, Dept. of Medicine, University of Massachusetts Medical School, Worcester, MA 01605; Rene Rachou Institute, Oswaldo Cruz Foundation, and Institute of Biological Sciences, Belo Horizonte, MG, Brazil.

fects (Leulier and Lemaitre, 2008). All the members of the Toll family are transmembrane proteins containing a similar extracellular domain (550 to 980 amino acids), which include 18 to 31 LRRs and a cytoplasmic domain (approximately 200 amino acids) that is involved in initiating a cellular response once the extracellular domain recognizes its ligand (Anderson, 2000).

Toll is activated by an endogenous ligand, named Spätzle, which is generated by serine protease cleavage of the inactive precursor (Schneider et al., 1994; Morisato and Anderson, 1994). Activation of the Toll receptor leads to stimulation of an intracellular pathway involving the adapters Tube and MyD88 and the kinase Pelle, leading to phosphorylation and degradation of Cactus, which interacts with the NF-κB family transcription factor, named Dorsal, and retains it in the cytoplasm. Once Cactus is degraded, Dorsal is released and translocates to the nucleus, where it regulates the transcription of several genes involved in dorsoventral regionalization during Drosophila development. Most of the genes in this embryonic signaling pathway have been characterized at a molecular level and have shown striking structural similarities with the cytokine-induced activation cascade of NF-κB in mammals (Anderson, 2000).

THE ROLE OF TOLL RECEPTORS IN IMMUNE RESPONSES AND RESISTANCE OF *DROSOPHILA* TO MICROBIAL CHALLENGE

The Toll/IL-1 Receptor Identity Domain

The interleukin-1 (IL-1) receptor was the first description of a mammalian molecule with structure containing significant homology with the Drosophila Toll receptor (Gay and Keith, 1991). On the cytoplasmic signaling domain, both chains of the Toll and the mammalian type 1 IL-1 receptor (IL-1R) share a conserved motif, what is now known as the Toll/interleukin-1 receptor (TIR) domain (Gay and Keith, 1991; Gangloff et al., 2003). The IL-1R signals in response to the cytokines IL-1α and IL-1β, mediators of acute-phase response during microbial infection. Importantly, components in the signaling pathway downstream of IL-1R clearly corresponded with those in the dorsoventral patterning system; e.g., Pelle kinase/IL1R-associated kinase and Dorsal/NF-κB (Balachandran et al., 2004). These findings suggested that innate immunity might have a common and ancient evolutionary origin. This hypothesis was strongly supported by the discovery of other mammalian genes that encoded a family of proteins containing both TIR and LRR motifs and even greater homology to the Toll genes from Drosophila (Rock et al., 1998; Taguchi et al., 1996), thus being named Toll-like receptors (TLRs).

The Role of Toll Receptors in Innate Immunity and *Drosophila* Resistance to Microbial Challenge

Although the Toll signaling pathway had initially been defined as important in embryogenesis, it was later shown to be triggered during immune challenge of Drosophila, leading to Dorsal translocation and upregulation of related genes. Studies performing survival tests in mutants of the proteolytic cascade and upstream of Spätzle and Toll showed that in Drosophila the Toll pathway protects mainly against fungal and gram-positive bacterial infection (Lemaitre et al., 1996). Importantly, lack of the Toll adapter/signaling molecules, Tube and Pelle, also dramatically reduces sur-

vival after fungal challenge, further indicating the role of the Toll pathway in the immune response and Drosophila resistance to microbial infection (Lemaitre et al., 1996). However, from the nine Tolls found in Drosophila, only one seems to have a defensive function, whereas most of the other eight Tolls apparently have developmental purposes (Lemaitre et al., 1996).

THE TOLL-LIKE RECEPTORS IN MAMMALS

TLRs

A family of receptors with ectodomains containing LRRs and TIR was first identified in a human expressed sequence tag database (Rock et al., 1998; Taguchi et al., 1996), and named Toll-like receptors, based on similarity to Toll receptors from Drosophila. There are 10 and 13 genes encoding TLRs in the human and mouse genome, respectively, and most of them are expressed in immune cells. In contrast to Drosophila, in the vertebrate hosts (human and mouse) most (or all) of the TLRs have evolved to exert immunological rather than developmental functions (Janeway and Medzhitov, 2002; Akira et al., 2006). Indeed, none of the TLR knockout mice have embryonic defects. Instead, each vertebrate TLR seems to be specialized for recognizing particular pathogen-associated molecular patterns (PAMPs). Indeed, an important discovery that boosted the whole field was the finding that one member from the TLR family, namely TLR4, was responsible for recognizing LPS, the major cell wall component of gram-negative bacteria. LPS has been used for decades in immunology laboratories because of its multiple effects on innate immune cells as well as B lymphocytes and, after discovery of TLR4 as its vertebrate host counterpart receptor, has served as an important paradigm to understand the interaction and function of other TLRs with various microbial components (Beutler et al., 2006).

The Identification of TLR4 as the LPS Receptor

Molecules detected by innate immune cells began to be identified almost a century ago, and the structures of most of them were resolved several decades ago. For all those years the idea of a signaling counterpart host receptor puzzled scientists working in this field. The main prototypic microbial stimulator of innate immune cells (i.e., phagocytes) has been LPS. The failure of an infected animal to recognize LPS enhances the capability of gram-negative bacteria to invade and kill the host (Jack et al., 1997; Bernheiden et al., 2001). Conversely, an excessive response to LPS results in the uncontrolled secretion of inflammatory molecules, which in turn are responsible for a variety of symptoms observed during sepsis. Altogether, these early studies provided solid proof that sensing LPS was important to an effective innate immune response to gram-negative microbes or, conversely, that LPS was one of the best-known molecules recognized by the innate immune system.

In the 1980s and 1990s the search for the signaling element of the LPS receptor was being intensively pursued by many research groups, and the discovery of the TLRs in mammals immediately became the center of attention for those interested in identifying the LPS receptor. So in the mid- to late 1990s a large number of studies were published indicating the importance of TLRs, and more precisely TLR4, as the critical element involved in LPS recognition and signaling to the host cells. Finding the Lps^d locus assumed great practical importance because it seemed to offer the key to understanding innate immune sensing, a process of great importance in the field of immunology and infectious diseases (Beutler et al., 2006).

The dependence of LPS signaling on a single protein was revealed in the 1960s in experiments employing the spontaneously mutated C3H/HeJ, which became hyporesponsive to LPS (Heppner and Weiss, 1965). The lethal effect, the cellular effects, and the well-known adjuvant activity of LPS did not occur in these mice. The mutated gene was later traced to a single locus, termed Lps (Watson and Riblet, 1974; Watson et al., 1977). Similar results were obtained with a second strain of mice, named C57BL/10ScCr (Poltorak et al., 1998). The Lps^d mutation was widely believed to affect the LPS receptor or an essential component of the receptor, but formal proof awaited positional cloning data.

In 1997, a breakthrough study was published, showing first that activation of the cytoplasmic domain of TLR4 results in induction of proinflammatory cytokines and expression of costimulatory molecules in dendritic cells (DCs), showing results similar to what one would expect in DCs activated with LPS, and further explaining the molecular basis of LPS immunological adjuvant activity in stimulating acquired immunity (Medzhitov et al., 1997).

Soon after, in 1998, the mutation responsible for LPS unresponsiveness was found to be a missense error (P712H) altering the cytoplasmic domain of TLR4 (Poltorak et al., 1998). By early results on positional cloning of the LPS gene, and TLR4 location in the mouse genome, the gene that encoded the LPS receptor was identified. These breakthroughs, as well as the equally important discovery of MD-2, are described in more detail in "LPS Recognition by TLR4," below.

Other Members of TLR Family and Their Agonists

The term PAMP was coined to describe infectious, nonself targets of the innate immune receptors. Three main features characterize molecules containing PAMPs: they are usually expressed by microbes and not host cells, they are conserved among microorganisms of a given class, and their expression is essential for microbial survival. Whereas the first two characteristics allow recognition of microbes and not host cells, the third prevents the development of mutants that escape recognition by the host immune system. At the time TLR4 was discovered, it was known only as one of various mammalian TLR paralogs of Toll, and investigators immediately postulated that the other TLRs might sense other microbial molecules with proinflammatory and immunological adjuvant activity, e.g., double-stranded RNA (dsRNA) [poly(I:C)], unmethylated DNA CpG motifs, flagellin, and lipopeptides (Kaisho and Akira, 2002).

Bacterial PAMPs

In addition to LPS/lipid A, various other PAMPs, possess agonist activity for TLRs. Lipid A, the major phospholipid in the outer leaflet of the outer membrane of gram-negative bacteria and the active component of LPS, is the best studied of the PAMPs. Lipid A has a diglucosamine bisphosphate backbone and is covalently modified by six to eight fatty acids in *Escherichia coli* and other enteric bacteria. The fatty acid composition of lipid A determines its biological potency, and even whether the molecule will activate the TLR4/MD-2 receptor complex or function as an LPS inhibitor (Dobrovolskaia and Vogel, 2002). Soon after LPS was defined as a main agonist of TLR4, other pathogen products were assigned as agonists of different TLRs. For instance, lipoteichoic acid, peptidoglycan, and lipoproteins from bacteria have all been reported to be agonists of TLR2 (the assignment of peptidoglycan as a TLR2 agonist is controversial) (Yoshimura et al., 1999; Lien et al., 1999). TLR2 does not appear to function as a homodimer, even under experimental conditions where the receptor is cross-linked with monoclonal antibody, but rather heterodimerizes with either TLR1 or TLR6 to initiate signal transduction (Ozinsky et al., 2000). Interestingly, lipopeptides containing two and three lipid chains activate TLR2/TLR6 and TLR2/TLR1, respectively (Takeuchi et al., 2000, 2001), indicating that minor changes in fatty acid composition have major effects on TLR utilization. Other important examples of TLR ligands include flagellin, a protein purified from bacterial flagellum, which has been shown to activate TLR5 (Hayashi et al., 2001). Similarly, unmethylated CpG bacterial DNA has been identified to trigger signaling through TLR9 (Hemmi et al., 2000). Thus, a large variety of PAMPs, including glycolipids, lipoproteins, nucleic acids, and proteins, are all established TLR ligands.

Viral PAMPs

Although best characterized for their role in antibacterial responses, TLR2 and TLR4 have also been implicated in the recognition of viruses. It is now clear that TLR2 and TLR4, which are expressed on the cell surface, recognize different viral proteins. TLR2 recognizes a component of the human cytomegalovirus particle (Compton et al., 2003), hemagglutinin of measles virus (Bieback et al., 2002), and an as yet unknown component of herpes simplex virus (HSV) (Bieback et al., 2002). The fusion protein of respiratory syncytial virus (RSV) triggers proinflammatory cytokine expression via TLR4 (Kurt-Jones et al., 2000). Virions of coxsackievirus B4 (Triantafilou et al., 2002) and the envelope protein of mouse mammary tumor virus and murine leukemia virus also activate TLR4 (Rassa et al., 2002; Burzyn et al., 2004). The "endosomal" TLRs act as receptors for microbial nucleic acids. These include TLR3 (dsRNA) (Alexopoulou et al., 2001); TLR7 and TLR8 (single-stranded RNA [ssRNA]) (Diebold et al., 2004; Hemmi et al., 2002; Jurk et al., 2002); and TLR9 (unmethylated CpG DNA motifs) (Hemmi et al., 2000). There is now strong evidence to support a role for these TLRs in the detection of certain viruses. In this regard influenza virus (Diebold et al., 2004) and vesicular stomatitis virus (Lund et al., 2004) have been shown to signal via TLR7, while others like HSV (Krug et al., 2004; Lund et al., 2003) signal in a TLR9-dependent manner, particularly plasmacytoid DCs (discussed below). It has been suggested that upon receptor-mediated endocytosis of ssRNA or dsDNA viruses, the degradative environment of late endosomes and lysosomes breaks down viral particles and releases these viral genomes, which can then be detected by TLRs. Although not actually PAMPs, no discussion of TLRs would be complete without mentioning synthetic drugs that mimic PAMPs. One of the most important of these is the imidazoquinoline R848 (Jurk et al., 2002), a potent activator of TLR7 and TLR8.

Although it is certain that TLRs recognize viruses, the importance of TLRs in antiviral defenses in vivo is hard to define. The lack of a clear-cut role for TLR3 is a good example of the obscurity that characterizes our current understanding of TLRs in antiviral defenses and viral pathogenesis. TLR3 plays a clear role in promoting the cross-priming of cytotoxic T lymphocytes (CTLs). Under conditions where DCs are not directly infected by viruses, and therefore cannot present viral antigens to CTLs, TLR3 can recognize dsRNA from phagocytosed apoptotic cells that were themselves infected. As a result, these TLR3+ DCs promote CTL responses (Schulz et al., 2005). Yet, despite the additional fact that TLR3 has a clear role in interferon-β (IFN-β) gene induction, TLR3-deficient mice failed to show increased susceptibility to many viral infections, including vesicular stomatitis virus, lymphocytic choriomeningitis virus, reovirus, and murine cytomegalovirus. Immune responses to Sendai virus (SeV) and Newcastle disease virus were also unaffected in TLR3-deficient cells (reviewed in Schroder and Bowie, 2005). Although some RSV-induced responses (including induction of CXCL10 and CCL5) were affected in the absence of TLR3, CXCL8 production and viral replication were unaffected in the absence of TLR3 (Rudd et al., 2005, 2006). Further complicating our understanding is that loss of TLR3 was found to be beneficial in in vivo infection with West Nile virus (Wang et al., 2004) and influenza virus (Le Goffic et al., 2006). Similar observations of the importance of antiviral TLRs exist for HSV infection. Thus, while we conclude that TLRs are important in antiviral host defenses, it is also clear that the data suggest that there must be additional sensors for viral infection. While not discussed here, the DExD/H box RNA helicases, retinoic acid-inducible gene (RIG-I) (Yoneyama et al., 2004), and melanoma differentiation-associated antigen-5 (MDA-5), all of which are cytosolic receptors, are an important family of newly identified receptors for viruses (Gitlin et al., 2006; Andrejeva et al., 2004; Kato et al., 2006).

Protozoan-Derived PAMPs

Although the identification of protozoan-derived PAMPs is at an early stage when compared with the identification of PAMPs contained within bacterial and viral molecules, several TLR agonists derived from protozoans have been identified. Studies have shown that glycosylphosphatidylinositol (GPI) anchors (or their fragments) from *Leishmania major*, *Trypanosoma brucei*, *Trypanosoma cruzi*, *Plasmodium falciparum*, and *Toxoplasma gondii* activate cells of both lymphoid and myeloid lineages (Almeida and Gazzinelli, 2001). GPI moieties function to anchor proteins to the surface of eukaryotic cells, and they are abundantly expressed by many protozoan parasites. GPI anchors are composed of a glycan core and a lipid component. For *T. cruzi* trypomastigotes derived from mammalian cells, the proinflammatory activity of GPI anchors covalently linked to mucin-like glycoproteins (GPI mucin) expressed on the surface of the parasite depends on the GPI anchor's fine structure. Nonsaturated, fatty acyl chains and periodate-sensitive components from the GPI anchor of *T. cruzi* have been shown to be required to trigger the production of cytokines by macrophages (Camargo et al., 1997; Almeida et al., 2000; Campos et al., 2001). By using Chinese hamster ovary (CHO) cells transfected with genes encoding different TLR molecules, *T. cruzi*-derived GPI anchors were shown to trigger NF-κB activation through TLR2. In addition, recognition of the GPI anchors required CD14, a host cell surface molecule involved in the recognition of bacterial LPS by TLR4 (Dobrovolskaia and Vogel, 2002). Furthermore, the induction of proinflammatory cytokines by GPI anchors is ablated in macrophages derived from TLR2-deficient mice (Campos et al., 2001). Because macrophages that lack TLR6 expression fail to respond to *T. cruzi* GPI anchors (unpublished data), the data indicate that a complex of TLR2-TLR6 and CD14 is involved in the recognition of these parasite molecules. In addition to these *T. cruzi* TLR2-TLR6 ligands, a subset of free GPI anchors from *T. cruzi* that contains ceramide activates CHO cells transfected with TLRs. This response depends on TLR4 and CD14 but not TLR2 (Oliveira et al., 2004).

Similar to *T. cruzi*, other kinetoplastids such as *Leishmania* spp. have GPI-linked molecules that trigger TLR2 activation. At this particular stage of development, the main GPI-linked molecules are lipophosphoglycans, which contain long carbohydrate branches with repeating phosphoglycan units. Lipophosphoglycans from *L. major* have been shown to stimulate mouse macrophages and human natural killer (NK) cells through TLR2 (de Veer et al., 2003; Becker et al., 2003). Furthermore, use of RNA interference to knock down the expression of various TLRs revealed that activation of macrophages by *Leishmania donovani* also, at least in part, depends on TLR2 (Flandin et al., 2006). There is evidence that GPI-related molecules from apicomplexan parasites also trigger TLR2 and TLR4 activation. For example, GPI anchors derived from *P. falciparum* merozoites have been shown to induce tumor necrosis factor (TNF) synthesis through the interaction of the three fatty acyl chains of the GPI anchor with the TLR2-TLR1 complex, which involves a minor contribution from TLR4 (Krishnegowda et al., 2005). Native GPI anchors purified from *T. gondii* tachyzoites, as well as synthetic fragments of the proposed structure of these GPI anchors, were shown to promote NF-κB activation and stimulate TNF synthesis by a mouse macrophage cell line, and these responses also seem to be mediated through TLR2 and TLR4 (Debierre-Grockiego et al., 2007).

In addition, it is becoming clear that TLR9, well known as a receptor for unmethylated bacterial CpG DNA motifs, is important for the induction of proinflammatory cytokines during infection with protozoans (Gazzinelli and Denkers, 2006). DNA from protozoan parasites such as *T. cruzi*, *T. brucei*, and *Babesia bovis* stimulates both macrophage and DC activation (Brown and Corral, 2002; Shoda et al., 2001), probably through unmethylated CpG motifs. Indeed, the frequency of CpG motifs has been found to be quite high in the *T. cruzi* genome, approximately 5%, which is comparable to some types of bacteria. Further, the proinflammatory activity of *T. cruzi* and *T. brucei* genomic DNA has been shown to be mediated by TLR9 (Drennan et al., 2005; Bafica et al., 2006; Bartholomeu et al., 2008).

Recognition of malaria by TLRs is an extraordinary example of how complicated pathogen recognition can be. The malarial parasite is remarkably nonimmunogenic, both for the innate and the acquired immune systems. Patients with malaria are typically afebrile until parasite densities exceed 10^4 to 10^5 per ml, and parasites may reach densities exceeding 10^7 per ml without killing the host (Collins and Jeffery, 1999)! At the time of presentation, however, patients are typically quite ill, and while death is uncommon, they often have incapacitating constitutional symptoms. While there is little question that the GPI anchors of *Plasmodium* activate TLR2 (Krishnegowda et al., 2005), the low potency of GPI anchors as immunomodulators suggests that other PAMPs may be involved in innate immune activation. Hemozoin, a protoporphyrin crystalline waste product of hemoglobin metabolism, has been suggested as an innate immune activator (Jaramillo et al., 2003, 2004; Coban et al., 2005). The naked crystal itself, however, is apparently not active; instead, hemozoin is coated with malarial DNA that can activate the innate immune system if it is delivered to the appropriate endolysosomal subcellular compartment (Parroche et al., 2007). The genomic DNA from *Plasmodium* spp. has an extraordinarily high AT (70 to 80%) and low GC (20 to 30%) content. Indeed, we have described both CpG and AT-rich motifs in the malarial genome to be immunostimulatory (Parroche et al., 2007; unpublished observations). The CpG motifs are active only in the endolysosomal compartment and appear to activate TLR9. In contrast, AT-rich motifs are active only in the cell cytosol, presumably through the interaction of these motifs with an as yet unidentified cytoplasmic receptor. While the best available evidence suggests that hemozoin is not a direct ligand for TLR9, the significance of this crystal should not be underestimated. Hemozoin functions as a particulate vehicle that facilitates delivery of malaria DNA into the endosomal compartments of host cells, where it becomes available to these innate immune receptors (Parroche et al., 2007). Indeed, recent evidence suggests that crystalline structures like hemozoin are capable of damaging lysosomes, resulting in the movement of lysosomal contents (such as cathepsin B) into the cytosol (Hornung et al., 2008). We hypothesize that hemozoin sequesters malaria DNA and promotes its delivery to a lysosomal compartment. Subsequently, DNA is cleaved off of the surface of the crystals by DNases and detected by TLR9. With time, lysosomal damage occurs and the lysosome becomes "leaky," allowing AT-rich DNA to enter the cytosol. The entire sequence of events is summarized in Fig. 1.

Finally, a profilin-like protein from *T. gondii* (PFTG) has been found to activate TLR11 in mouse cells (Yarovinsky et al., 2005). PFTG is present as a relatively conserved molecule in a number of apicomplexans, indicating that these proteins might serve as another broad class of protozoan PAMPs. Although its exact function in the parasite is unknown, PFTG is predicted to bind to actin and, like flagellin, might be involved in parasite motility and invasion of the host cell. The induction of IL-12 in DCs exposed to PFTG is mediated by TLR11, as the response was abolished in DCs from TLR11-deficient mice. The TLR11 gene in humans, however, has a premature stop codon and therefore encodes a nonfunctional form of TLR11. Heat shock proteins from *T. gondii* have also been shown to activate B lymphocytes and macrophages via TLR4 and to be involved in the initial activation of innate immunity during infection with this protozoan parasite (Chen et al., 2002; Aosai et al., 2002).

STRUCTURE AND FUNCTION OF TLR COMPLEX

Cellular Localization of TLRs

TLRs can be classified into two groups based on cellular localization. The first group includes TLRs 1, 2, 4, 5, and 6, which are located on the surface of the cell. These receptors can also be found in the Golgi apparatus, a finding that is related both to their glycosyl modification there and the observation that lipid rafts (where these TLRs are concentrated) appear to shuttle between the cell surface and the Golgi. Signaling via these TLRs typically does not require internalization—and hence is cytochalasin resistant, although there is abundant evidence that signaling can occur in intracellular compartments under the correct circumstances (e.g., after phagocytosis) (Underhill et al., 1999).

In contrast to the surface TLRs, TLRs 3, 7, 8, and 9 are typically found in intracellular locations including the endoplasmic reticulum (ER) and endolysosomes (Latz et al., 2004), and hence are often referred to as the "endosomal TLRs." While these TLRs are found primarily in the ER during resting conditions, the uptake of ligand into the endolysosomal compartment is simultaneously associated with the translocation of these TLRs to the endosome. The signals that result in this apparent translocation event are unknown.

Beutler and colleagues (Tabeta et al., 2006) performed a forward genetic screen on mutagenized populations of mice and screened them for their responses to TLR ligands. Ultimately, a mouse that was unresponsive to TLR3, -7, and -9 ligands was identified (in contrast to humans, there is no known ligand for mouse TLR8, so this TLR was not tested). These mice were termed "3D" in recognition of their three deficiencies. By using positional cloning approaches, the investigators identified a mutant protein, Unc93b, as the genetic lesion in the 3D mouse. At the time, Unc93b was recognized only as a 12-membrane-spanning molecule present in the ER that appeared to have homology with known ion channels. Ploegh and colleagues (Kim et al., 2008) proved that Unc93b was necessary for the endosomal localization of TLR9. These investigators found that wild-type (but not mutant) Unc93b bound to the endosomal TLRs, and cotranslocated with the endosomal TLRs in the presence of ligand. While the exact implications of the role of Unc93b were not clear, it was evident that the molecule binds and traffics the endosomal TLRs.

Those TLRs that recognize nucleic acids constitute one of the most important classes of innate immune receptors involved in recognition of viruses. TLR9 seems to have an important role against other intracellular pathogens, including bacteria, fungi, and protozoan parasites. The second group includes TLR1, -2, -4, -5, and -6, which are all present at the plasma membrane and primarily involved in recognition of bacteria components, but are also involved in recognition of other categories of pathogens (Akira et al., 2006; Beutler et al., 2006).

LPS Recognition by TLR4

LPS is the most-studied PAMP because of its putative role as the cause of gram-negative bacterial sepsis. The active moiety of LPS is lipid A, which has all of the proinflammatory activities of endotoxin. A significant body of work has been based on the characterization of lipid A; this task was greatly aided by the correct elucidation of its structure in the late 1980s (Strain et al., 1985; Raetz, 1984; Ray et al., 1984; Takayama et al., 1983) and its successful synthesis (e.g., see Shiba et al. [1984]).

While the structure of lipid A is conserved among species of gram-negative bacteria, variation does exist. *E. coli* lipid A is highly proinflammatory due to its acyl chain composition, while the lipid A of organisms such as *Rhodobacter sphaeroides*, which differs from *E. coli* markedly in fatty acyl composition, is actually an LPS antagonist (Golenbock et al., 1991). Synthetic analogs of *Rhodobacter* spp. lipid As have inhibitory activity and have been developed by the pharmaceutical industry as therapeutics for endotoxin-mediated diseases (Christ et al., 1995). In addition, lipid A precursors have been purified from mutant strains of gram-negative bacteria; these lipid As are generally hypoacylated (4 to 5 fatty acids) and are far less

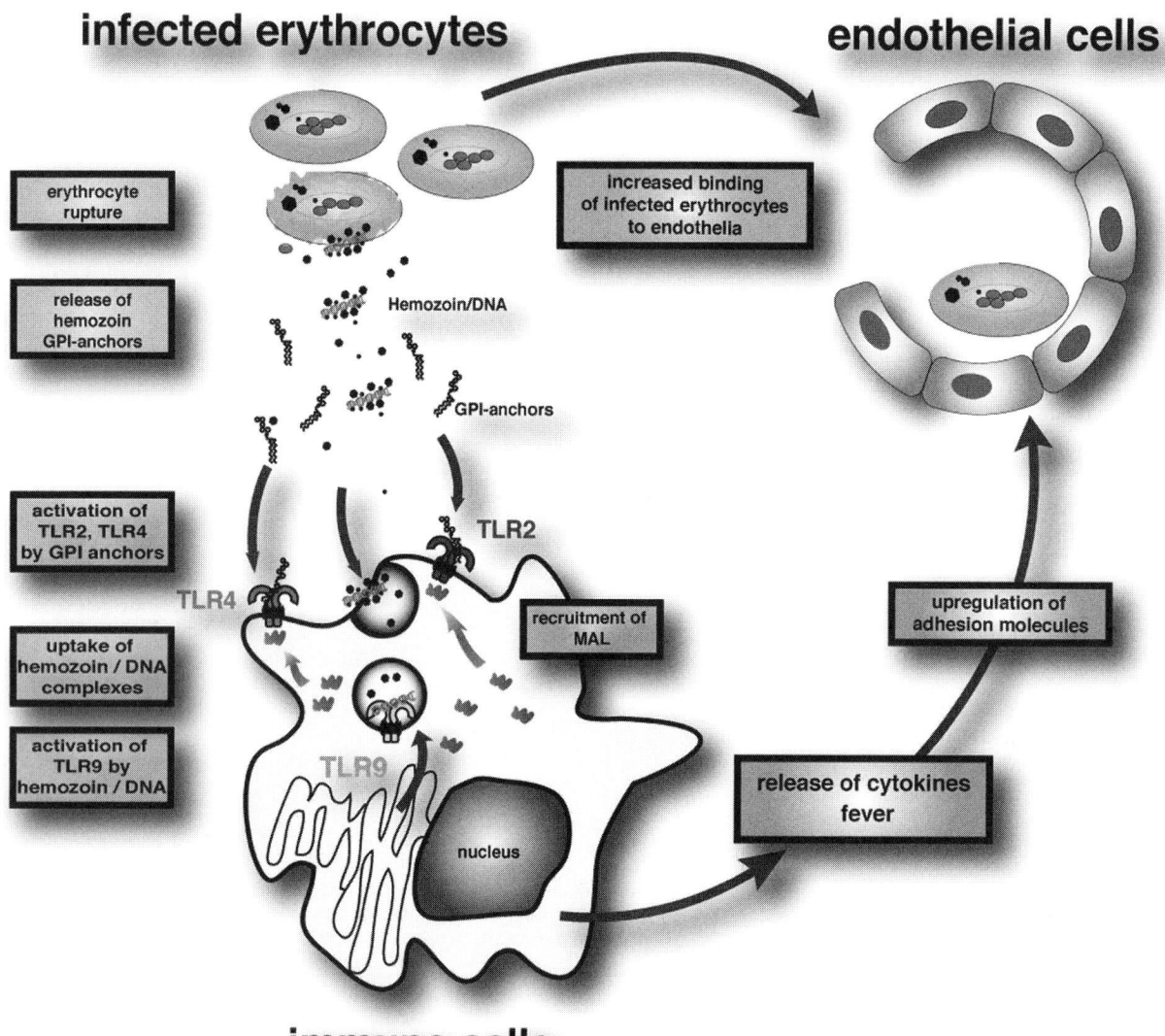

FIGURE 1 Innate immunity in malaria. During the asexual portion of the cycle, merozoites are released from erythrocytes into the systemic circulation. Parasite GPI anchors and hemozoin-bound DNA trigger cytokine production. GPI anchors bind to surface TLR2 on CD14+ monocytes, macrophages, and DCs. CpG-containing ODN is associated with hemozoin and triggers endosomal TLR9. Hemozoin-induced lysosomal destabilization releases DNA to access cytosolic sensors. AT-rich ODN triggers an as yet uncharacterized cytosolic DNA sensor and elicits IFN-β via NALP3, TBK1, and IRF1 to regulate IFN-β production. AT-rich ODN also elicits cytokine production (TNF, IL-6) and caspase-1-mediated processing of pro-IL-1β. These latter cytokines, still not entirely characterized, cause fever. Immune mediators also increase adhesion molecule expression on capillary endothelium. Infected erythrocytes, especially those with large schizonts, will bind to microcapillary beds in the brain.

potent than fully acylated *E. coli* lipid A. These hypoacylated structures are often LPS antagonists and are best characterized by lipid IVa (which corresponds to the synthetic lipid 406), a tetra-acylated molecule (Golenbock et al., 1991).

Investigators in the 1990s hypothesized that by detailing the pharmacology of lipid A and its analogs, they could make insights into the biology of the LPS receptor. This view was greatly encouraged by the observation that lipid IVa was an LPS mimetic in rodent cells, but an LPS an-

tagonist in human cells. That suggested that the identification of the true LPS receptor, the "Holy Grail" of LPS research at the time, would simply require a receptor whose rodent cDNA would encode a protein that responded to lipid IVa in a proinflammatory manner, but whose human ortholog would recognize lipid IVa as an LPS antagonist. Indeed, one piece of evidence ruling out CD14 as the true LPS signal transducer was that it did not meet these criteria (Delude et al., 1995). Strangely, while the components of the LPS receptor have now been identified by other means, the true "IVa responsive element" has been difficult to define.

In 1997, Medzhitov and Janeway identified TLR4 as an ortholog of *Drosophila* Toll, hence beginning the current era of Toll research. In their landmark paper, they described how TLR4 overexpression could drive proinflammatory events and, just as importantly, link the innate and adaptive immune systems by promoting the expression of antigen-presenting molecules on the surface of cells (Medzhitov et al., 1997).

Shortly thereafter, Rock and colleagues (1998) reported on the chromosomal location of TLRs 1 to 5. This promoted speculation in the LPS field, as TLR4 was located on an area of human chromosome 10 that was noted to be syngeneic to mouse chromosome 4 in the area of Lps^d. Lps^d was a known locus of resistance to LPS in the C3H/HeJ mouse, and many believed it to be the gene encoding the true LPS receptor. Indeed, it was only a few months later, in another landmark paper, that Poltorak and colleagues (1998) reported that Lps^d was TLR4 by using positional cloning technology. The mutant form of TLR4 expression in the C3H/HeJ mouse proved to be a single point mutation, resulting in an amino acid exchange (Pro712His) in the TIR domain of TLR4. The central role of TLR4 in LPS signal transduction was confirmed by Akira and colleagues in TLR4 knockout mice (Hoshino et al., 1999).

While many considered the LPS receptor mystery to be solved, there were problems with the concept that TLR4 was the LPS receptor. First, there was no evidence that TLR4 bound to LPS. Second, there was the odd finding that TLR4 expression in TLR4-null cells, such as HEK293, did not always result in an LPS-responsive phenotype. And, even when LPS responses were observed, they tended to be weak (Chow et al., 1999). Miyake and colleagues (Shimazu et al., 1999) greatly clarified the issue when they identified MD-2, a 20-kDa protein that lacks any transmembrane domain (or a GPI linkage site), as an important component of the LPS receptor. This discovery came from remarkable insight by these investigators that the extracellular domain of TLR4 was highly homologous with RP-105, a heterodimeric receptor that was responsible for radioprotection of cells in culture. RP-105, which, among other things, is a negative regulator of TLR4 (Divanovic et al., 2005) activity, depended on the coexpression of a 20-kDa protein known as MD-1. Hence, Miyake correctly surmised that TLR4 activity might similarly require coexpression of the MD-1 homolog, MD-2.

Indeed, it was soon clear that MD-2 was as essential for LPS responses as TLR4. Using a forward genetic screening approach, our group found that TLR4-positive transfectants that carried a mutant form of MD-2 (C95Y) were absolutely unresponsive to LPS (Schromm et al., 2001). The reason for this lack of response appeared to be that, unlike the wild-type protein, the C95Y mutant did not bind LPS. MD-2 knockout mice were generated shortly after and had a phenotype that was identical to TLR4 knockouts. Finally, at the same time, we discovered that soluble MD-2 was present in serum, and was as capable of complementing the LPS-defective phenotype of cells that lacked MD-2 expression as was the transfection of the cDNA for MD-2 into these cells (Schromm et al., 2001). This observation explained why TLR4 transfectants sometimes responded to LPS and sometimes did not: the relative abundance of soluble MD-2 in the tissue culture medium used would determine whether a response would be observed. It should be noted that MD-2 is heat labile (J.

Meng and D. T. Golenbock, unpublished results) and is probably completely destroyed when calf serum is heat treated to reduce the activity of complement.

The importance of MD-2 as a receptor component that binds LPS cannot be underestimated. To date, in contrast to MD-2 (Viriyakosol et al., 2001; Akashi et al., 2003; Visintin et al., 2005), there is no evidence of LPS binding to TLR4. Indeed, when the structure of MD-2 was reported by two different groups (Ohto et al., 2007; Kim et al., 2007), both of which resolved MD-2 in the presence of an inhibitory lipid A, it did not seem likely that lipid A made contact with TLR4, yet molecular genetic approaches have raised some doubts that this is entirely true. Expression of human TLR4 in rodent cells, for example, results in the alteration of phenotype of the response to lipid IVa, i.e., lipid IVa becomes an LPS antagonist (Poltorak et al., 2000; Lien et al., 2000). It should be noted that not all groups have reproduced this observation, and in fact have had the opposite result (Hajjar et al., 2002; Akashi et al., 2001). Transgenic mice have now been engineered that express human or mouse TLR4/MD-2 combinations under the control of their endogenous promoters (L. Hajjar, presented at Toll2008, Cascais, Portugal, September 2008). When lipid IVa responses were examined, the observed phenotype could be predicted based on the species of MD-2. In what might be the final, and most confounding word on this topic, however, *Pseudomonas* lipid A, which is pentaacylated, has been demonstrated to be an antagonist in human cells and an agonist in mouse cells. When responses in the transgenic mice were tested, *Pseudomonas* lipid A was an antagonist in the transgenic mice that express human TLR4, regardless of the species of MD-2 that was expressed. Thus, some type of interaction of lipid A with TLR4 cannot be ruled out based on the available literature.

Regardless of the fine details of lipid A/MD-2/TLR4 interactions, it does seem likely that the activating event for the TLR4/MD-2 complex is an LPS-induced conformational change in the structure of MD-2. This conformational effect might be very small, but enough to change LPS-bound MD-2 into an activating ligand. The next event in signaling appears to be the dimerization of TLR4 (Saitoh et al., 2004a, 2004b). The TLR4 homodimer is then thought to be able to recruit the adapters Mal/TIRAP and TRAM, which in turn recruit MyD88 and TRIF, respectively. Indeed, one unique feature of the response to LPS is that it involves all four of the known TIR-domain-containing adapters (reviewed in O'Neill et al., 2003). These adapters, in turn, recruit and activate IL-1R-associated kinases (IRAKs), resulting in downstream signal transduction (Fig. 2).

In many ways, the TLR4 receptor is paradigmatic of the mechanism by which TLRs are activated. Signaling is initiated by ligand-induced TLR dimerization and the recruitment of TIR-domain-containing adapter molecules. However, in several ways TLR4 is unique: it does not directly bind its major PAMP but instead binds MD-2. In contrast, TLRs 1, 2, 3, and 9 clearly bind their respective ligands (and TLRs 5, 6, 7, and 8 probably do as well) (Latz et al., 2004; Jin et al., 2007; Liu et al., 2008), and the receptor utilizes all the known adapter molecules. It is no surprise, in view of this last observation, that LPS activates the enhanced expression of nearly 1,000 genes (Bjorkbacka et al., 2004)!

DNA Recognition by TLR9

TLR9 was the first recognized "endosomal TLR." This recognition came from confocal studies in our labs (Latz et

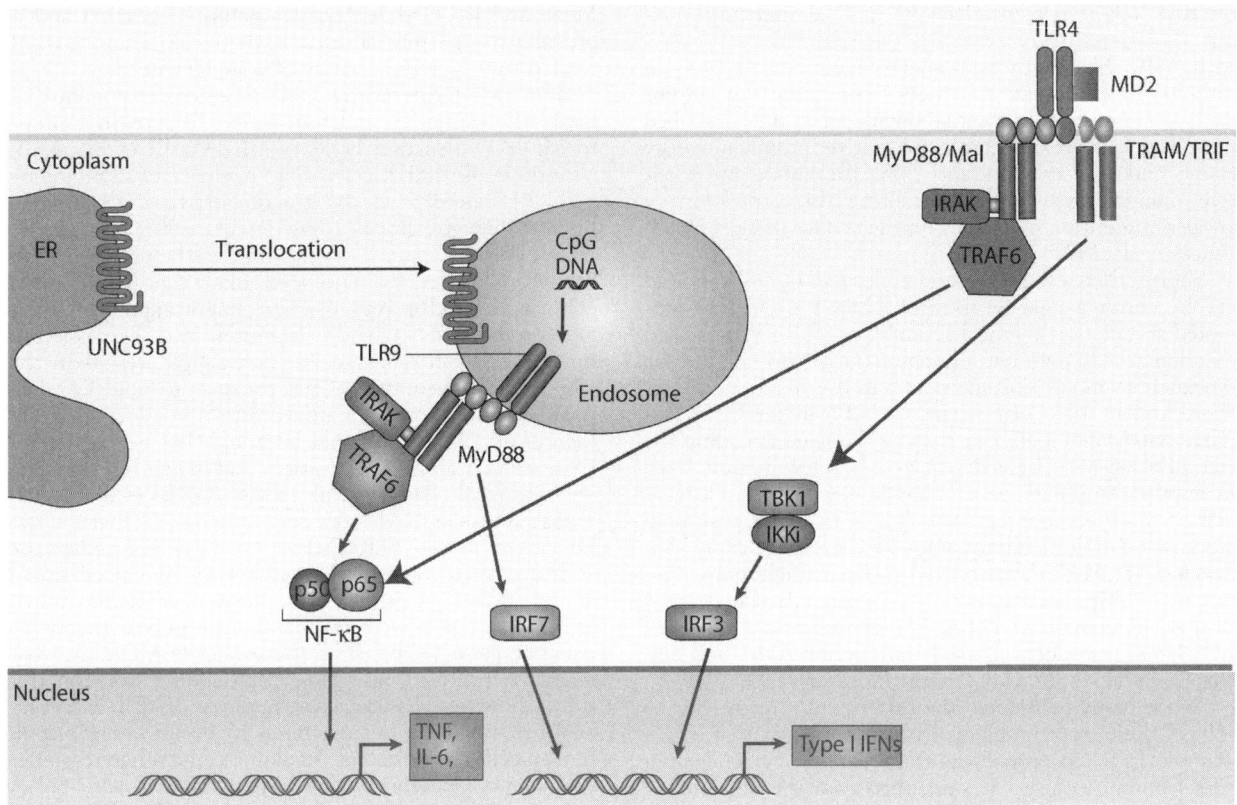

FIGURE 2 Molecular and cellular steps of TLR activation. TLR4 and TLR9 are by far the best-understood surface and endosomal TLRs, respectively. TLR4 is most often located on the cell surface and it is highly specific for gram-negative derived lipid A. The recognition and sensitivity of TLR4 signaling by bacterial lipid A highly depends on the coreceptor named CD14 and MD-2. The TLR4 intracellular signaling also depends on four adapter molecules, which work in pairs, i.e., MyD88/Mal and TRIF/TRAM, which are essential for induction of IRAK/TRAF6 and TBK1/IKKe, which are then responsible for activation of NF-κB and IRF3 responses, respectively. However, TLR9 normally resides in the ER and migrates to the endolysosomal compartment in an UNCN93Bi-dependent manner, where it is activated by its agonists. TLR9 activation also depends on the MyD88 adapter molecule and results in activation of IRAK/TRAF6, culminating in stimulation of NF-κB. During activation of TLR9, MyD88 also activates IRF7- and IFN-related responses.

al., 2004), although previous work by Wagner and colleagues (Ahmad-Nejad et al., 2002) suggested that the location of the "CpG receptor" was indeed not on the cell surface, as we originally assumed. Furthermore, it should be recognized that subcellular localization of TLRs may vary depending on the cell type that is assessed. For example, TLR9 may actually be a surface receptor on gut epithelial cells (Lee et al., 2006), in stark contrast to immune cells such as macrophages and DCs. The discussion below thus applies only to professional phagocytes, including polymorphonuclear leukocytes where TLR localizes to the ER and the endosome.

In the resting state, TLR9 can be found in great abundance in the ER. If one carefully examines early endosomes in cells that have not been exposed to a TLR9 ligand, TLR9 is occasionally observed in the endosomes. Despite a great amount of effort, TLR9 is not found on the cell surface prior to stimulation with CpG DNA, but can be found in small quantities otherwise. It may be, therefore, that TLR9 is always represented both in the endosome and the cell surface in small quantities that are nondetectable or virtually nondetectable by using standard fluorescent approaches. It is clear, however, that once CpG-rich DNA enters the endosome, there is immediate movement of TLR9 to the endosome. Furthermore, if one assesses the subcellular localization of MyD88, by cotransfecting TLR9 with a fluorescently tagged MyD88, recruitment of this key signaling adapter after DNA internalization appears to be instantaneous. Thus, DNA uptake appears to be accompanied by the immediate translocation of TLR9 from the ER to the endosome and the simultaneous initiation of signal transduction via MyD88.

Once DNA encounters TLR9, we (and others) have shown that a high-affinity binding event can occur (Latz et al., 2004). The binding of DNA to TLR9 results in a critical conformational change in both the ectodomain of the molecule and in its cytosolic domain (Latz et al., 2007). More recent evidence suggests that, in addition, TLR9 may become cleaved to an activated receptor, presumably via cathepsins (Ewald et al., 2008; Park et al., 2008). This results in the formation of stable homodimers that now have high affinity for MyD88, and hence begins the signaling process. One critical question that remains unresolved is, "what signals TLR9 to translocate from the ER to the endosome?" That question remains unanswered, although it is clear, as described earlier, that Unc93b is an important regulatory molecule in this respect. It is possible, of course, that TLR9 is expressed in low quantities on the cell surface, and TLR9 triggers its own movement to the endosome. Because TLR9 translocation was observed in a MyD88 knockout mouse (Latz et al., 2004), however, and adapter usage by TLR9 is thought to be restricted to MyD88, this seems less likely. Another possibility is that the translocation event is not a signaled event, but that endosomes are formed with TLR9 on their surface because they employ ER as a source of membrane; in this model, TLR9 is essentially trapped in the developing endosome by the presence of ligand.

Although TLR9 is only one of the endosomal TLRs, it seems likely that it is more representative of the other endosomal TLRs than TLR4 is representative of the surface TLRs. This speculation seems likely because the ligands for all of these molecules appear primarily to be nucleic acids (TLRs 3, 7, and 8 are primarily viral RNA receptors). It seems plausible that these receptors have redundant functions for one another, and that highly specific inhibitors of the endosomal TLRs will be difficult to find. Nevertheless, there is a great deal of interest in finding such inhibitors because of the role the endosomal TLRs probably play in autoimmune diseases (e.g., Means et al., 2005; Barrat et al., 2005) and the inflammation associated with a variety of serious viral infections such as influenza, HIV, and malaria.

The Adapter Molecules and Signaling through TLRs

Among the five different adapter molecules containing the TIR domain, MyD88 was the first identified and shown to be critical for TLR and IL-1R family signaling (Adachi et al., 1998; Kawai et al., 1999). MyD88 can associate with all TLRs (Medzhitov et al., 1998) with the exception of TLR3 (Oshiumi et al., 2003; Yamamoto et al., 2003a). MyD88 has an amino-terminal death domain (DD) and a carboxy-terminal TIR domain. The TIR domain is involved in the interaction with TLRs and other adapters (see below) while the death domain associates with members of the IRAK family (Martin and Wesche, 2002). IRAK-1 is recruited to MyD88 via DD-DD interactions within a complex with another protein termed Toll-interacting protein (Tollip) (Burns et al., 2000). This IRAK-1–MyD88 association triggers hyperphosphorylation of IRAK-1 by itself as well as phosphorylation by the related kinase, IRAK-4 (Li et al., 2002; Cao et al., 1996). These events lead to the dissociation of IRAK-1 from MyD88 and Tollip and its interaction with the downstream adapter TNF receptor-associated factor 6 (TRAF6) (Burns et al., 2000). TRAF6, a RING domain ubiquitin ligase, activates the TAK1 kinase through K63-linked polyubiquitination (reviewed in Chen, 2005). TAK1, in turn, activates the IKK complex, which phosphorylates IkBs and targets these NF-κB inhibitors for ubiquitination and degradation by the proteasome. NF-κB is then released and translocates to the nucleus, where it can induce several hundred target genes (reviewed in Karin and Greten, 2005). The diversity of TLR signaling pathways was revealed following the analysis of the response of MyD88-deficient macrophages to gram-negative bacteria-derived LPS (Kawai et al., 1999). LPS, which signals via TLR4 and MD-2, can still trigger the activation of NF-κB and mitogen-activated protein kinase (MAPK) in cells from MyD88 knockout mice, albeit with delayed kinetics compared with wild-type cells, whereas most other TLR ligands are completely ineffective at triggering these events in the absence of MyD88. Although MyD88-deficient mice lose the ability to induce proinflammatory cytokines in response to LPS, they are still able to upregulate costimulatory molecules and induce type I IFNs and IFN-inducible genes (ISGs) (Kaisho et al., 2001; Kawai et al., 2001). Subsequent studies from several groups identified another adapter related to TRIF that regulates these MyD88-independent pathways, which we named TRIF-related adapter molecule or TRAM because of its high degree of homology to TRIF (Fitzgerald et al., 2003b; Yamamoto et al., 2003b). TRIF knockout mice are compromised in the induction of type I IFNs and the expression of ISGs in response to LPS and the dsRNA mimetic poly(I:C), a TLR3 ligand. Both TLR4 and TLR3 signaling cascades activate the nuclear translocation and DNA binding of the transcriptional regulator IRF3, a key regulator of IFN-β and ISGs, a process mediated solely by TRIF in the case of TLR3 signaling (Yamamoto et al., 2003b; Hoebe et al., 2003a, 2003b). In the case of TLR4 signaling an additional adapter, TRAM, is also required to recruit TRIF to TLR4 (Fitzgerald et al., 2003b). TRAM is modified by N-terminal myristoylation, which is important in tethering TRAM to the plasma membrane, where it colocalizes with TLR4 (Rowe et al., 2006). This function of TRAM appears to be important in recruiting TRIF to membrane-localized TLR4. A fourth adapter molecule, Mal (also called TIRAP), also participates in TLR4 signaling. In contrast to TRIF and TRAM, however, Mal appears to be important in the recruitment of MyD88 to TLR4 to regulate inflammatory cytokine genes (Fitzgerald et al., 2001; Horng et al., 2001; Kagan and Medzhitov, 2006). TLR3-mediated NF-κB activation is also triggered by a TRIF-dependent mechanism. The C terminus of TRIF associates with the serine-threonine kinase receptor interacting protein-1 (RIP1) through a RIP homotypic interaction motif (Meylan et al., 2004). RIP1-deficient cells fail to activate NF-κB in response to poly(I:C) (Meylan et al., 2004), whereas IRF3 activation remains intact (Cusson-Hermance et al., 2005). The TRIF N-terminal region has also been shown to associate with TRAF6 in overexpression systems (Sato et al., 2003). Studies using macrophages from TRAF6-deficient mice, however, suggest that the exact requirement for TRAF6 in the TLR3 response to NF-κB is still a little unclear, probably because of functional redundancy with other TRAF proteins in certain cell types (Gohda et al., 2004). TAK1 is also involved in TLR3-mediated NF-κB and MAPK activation (Sato et al., 2005). Recent studies have also shown that TRIF and MyD88 can bind to a second TRAF family member, TRAF3, which activates IRFs to induce type I IFNs. TRAF3 does not appear to be required for the induction of proinflammatory cytokines, however (Hacker et al., 2006; Oganesyan et al., 2006). Transcriptional regulation of the IFN-β gene requires the activation of IRF3, ATF-2/c-Jun, and NF-κB. These transcription factors form a multiprotein complex, the enhanceosome or the IFN-β enhancer (Maniatis, 1986). In the resting state, IRF3 is localized to the cytoplasm. In response to a viral challenge, IRF3 is phosphorylated on multiple serine/threonine residues, which control its dimerization. In this active form IRF3 then translocates to the nucleus and associates with the coactivator CREB-binding protein (CBP)/p300 on the IFN-β enhancer. The IkB-related kinases inhibitory protein kB kinase (IKKε) (also called IKKi [Shimada et al., 1999]) and TANK binding kinase (TBK1) (also called

NAK [Tojima et al., 2000] or T2K [Bonnard et al., 2000]) phosphorylate IRF3 (Fitzgerald et al., 2003a; Sharma et al., 2003). IKKe and TBK1 are structurally related to IKKa and IKKb but, unlike IKKa or IKKb, do not appear to be involved in NF-κB activation (Sharma et al., 2003; McWhirter et al., 2004). Sharma et al. (2003) and Fitzgerald et al. (2003a) showed that blocking IKKe and TBK1 activity by using RNA interference prevented SeV-induced IRF3 phosphorylation and subsequent activation of the IFN promoter. Fitzgerald et al. also described a requirement for IKKe and TBK1 in poly(I:C)-induced IRF3 activation via TLR3 and TLR4 (Fitzgerald et al., 2003a; McWhirter et al., 2004). TBK1$^{-/-}$ embryonic fibroblasts fail to activate IRF3 and induce IFN-β, IFN-α, or ISGs in response to virus, LPS, or poly(I:C) (McWhirter et al., 2004). TBK1 is ubiquitously expressed, while IKKe expression is restricted to lymphoid cells, even if it can be inducible in several other cell types. Moreover, IKKe may be functionally redundant with TBK1 in cells where both are expressed (Hemmi et al., 2004; Perry et al., 2004). Perry et al. showed that the SeV-induced IFN response in TBK1$^{-/-}$ embryonic fibroblasts could be partially restored by reconstitution with wild-type IKKe but not with a mutant lacking the kinase activity (Perry et al., 2004).

IMMUNOLOGICAL FUNCTIONS OF TLRs

The early elimination of invasive microorganisms is a fundamental function of the innate immune system. Macrophages, DCs, and polymorphonuclear leukocytes recognize components of invasive pathogens and orchestrate an early antimicrobial defense. In mammals, the TLR family is a primary means whereby the innate immune system recognizes and rapidly responds to the presence of microbes (Akira et al., 2006; Beutler et al., 2006). To date, TLRs have been implicated in immune responses to every known category of microorganisms that cause human disease (Akira et al., 2006; Beutler et al., 2006; Gazzinelli and Denkers, 2006). TLRs are critical for all aspects of this process, including the production of chemokines and proinflammatory cytokines and consequent recruitment and activation of phagocytes to infected tissues (Akira et al., 2006). In addition, activation of TLRs can directly trigger or potentiate the generation of various effector mechanisms, such as the generation of reactive oxygen intermediates as well as reactive nitrogen intermediates involved in microbial killing (Akira et al., 2006).

These same professional phagocytic cells, especially the DCs, have important roles in initiating and shaping long-term acquired immunity. To begin with, the activation of DCs by TLRs leads to maturation of these specialized antigen-presenting cells, via both MyD88-dependent and MyD88-independent mechanisms (Kaisho et al., 2001; Kawai et al., 2001). Once they differentiate into a mature stage, the DCs migrate to the peripheral lymphoid organs and express surface costimulatory molecules, becoming allowed to initiate T-cell immune responses. In addition, the MyD88-dependent production of IL-12 is very important for the initial differentiation of CD4$^+$ T-helper lymphocytes toward the type 1 phenotype and development of cell-mediated immunity, which is highly relevant for protective immunity against viral, bacterial, and protozoan infections (Biron and Gazzinelli, 1995; Trinchieri, 2003).

Thus, TLRs are not only critical for the initial activation of the innate immune compartment, but also for directing T-cell differentiation and establishment of acquired immunity. This effect on DCs and consequently on T lymphocytes is critical for the immunological adjuvant activity of TLR agonists and their use in vaccine development. When activated in excess, however, TLRs may also mediate pathology. This is observed in the septic shock induced during infection with gram-negative bacteria or LPS intoxication, as well as during acute malaria episodes (Gazzinelli and Denkers, 2006; Lien and Ingalls, 2002). Indeed, the identification of TLR antagonists or inhibitors of TLR adapters is an intense area of research, targeting identification of new compounds to be used to prevent the clinical symptoms and lethality caused by excessive and systemic production of proinflammatory cytokines.

The most convincing data indicating the importance of the TIR signaling pathway in host resistance and pathogenesis during infectious diseases are those obtained from infections of MyD88-deficient mice (Akira et al., 2006; Beutler et al., 2006; Gazzinelli and Denkers, 2006). MyD88-deficient mice are devoid of the function of most TLRs, except for a residual TLR4 and intact TLR3 functions. Further, lack of MyD88 also results in nonfunctional IL-1, IL-18, and IL-33 receptors. These cytokines are triggered by TLRs and seem to amplify the initial proinflammatory response elicited during early stages of infection with a different category of pathogens. Therefore, MyD88-deficient mice are highly susceptible to different viral, bacterial, and protozoan infections, which are often associated with impaired production of proinflammatory cytokines and development of T-helper type 1 lymphocytes (Akira et al., 2006; Beutler et al., 2006; Gazzinelli and Denkers, 2006). In contrast, the absence of functional MyD88 often results in amelioration of clinical symptoms and signs in conditions where the pathology is mainly mediated by the immune system, as is the case with septic shock and malaria acute episodes. In these situations, the beneficial role of TLRs is often defined by a balance in the role of a TLR in control of a specific infectious agent and its involvement in eliciting an excessive and deleterious immune response.

Finally, despite the various studies indicating an important role for TLR signaling in resistance to infectious diseases, as is observed in MyD88-deficient mice, the deficiency in a single TLR only rarely causes an increase in susceptibility to infectious diseases (Trinchieri and Sher, 2007). Therefore, these results indicate that, in general, host defense against pathogens is orchestrated by multiple TLRs that, in addition to having a redundant role, might lead to distinct responses by different host cells. Further, TLR-dependent induction of cytokines that activate receptors, of which the functional activity depends on MyD88, may contribute as an important positive feedback loop, resulting in maximal MyD88-dependent activation of innate immune cells as well as T lymphocytes.

BEYOND TLRs: OTHER INNATE IMMUNE RECEPTORS

Although the TLR family is an important class of sensors for pathogens, there is rapid accumulation of information indicating the existence of other innate recognition systems that fulfill similar functions. There is increasing evidence that the combined response of various innate immune receptors is critical in recognizing different classes of pathogens and triggering effector mechanisms during infection. Two major families of non-TLRs that have received the most attention to date are the NOD-like receptor (NLR) family of receptors and the associated inflammasomes (Kanneganti et al., 2007). Recent studies indicate the existence of non-TLR pattern recognition systems that function as sensors of viral and bacterial components. For example, the peptidoglycan component of bacterial cell walls and bacterial flagellin are recognized by a family of peptidoglycan recognition proteins (PGRPs) (Lu et al., 2006). Also, NOD1 (nucleotide-binding oligomerization domain 1) and NOD2 proteins act as cytosolic sensors of peptidoglycan-derived bacterial components. Activation of NOD proteins, as with TLRs, triggers host NF-κB and MAPK pathways. In addition to the well-known RNA-binding capability of PKR (IFN-inducible dsRNA-dependent protein kinase), the RNA helicase RIG-I (also known as DDX58) recognizes dsRNA (Kawai et al., 2005; Kato et al., 2005). Likewise, the intracellular FAS-associated via death domain (FADD) molecule is involved in recognition of viral dsRNA (Balachandran et al., 2004). Whether these or other yet-to-be discovered non-self-sensing systems contribute to protozoan recognition and how they cooperate with TLR signaling pathways are exciting areas awaiting future discovery.

We thank Dr. David Russell for his patience and friendship. D.T.G., K.F., and R.T.G. are funded by the National Institutes of Health. R.G. is also funded by the Oswaldo Cruz Foundation (FIOCRUZ).

REFERENCES

Adachi, O., T. Kawai, K. Takeda, M. Matsumoto, H. Tsutsui, M. Sakagami, K. Nakanishi, and S. Akira. 1998. Targeted disruption of the MyD88 gene results in loss of IL-1- and IL-18-mediated function. *Immunity* 9:143–150.

Ahmad-Nejad, P., H. Häcker, M. Rutz, S. Bauer, R. M. Vabulas, and H. Wagner. 2002. Bacterial CpG DNA and lipopolysaccharides activate Toll-like receptors at distinct cellular compartments. *Eur. J. Immunol.* 32:1958–1968.

Akashi, S., Y. Nagai, H. Ogata, M. Oikawa, K. Fukase, S. Kusumoto, K. Kawasaki, M. Nishijima, S. Hayashi, M. Kimoto, and K. Miyake. 2001. Human MD-2 confers on mouse Toll-like receptor 4 species-specific lipopolysaccharide recognition. *Int. Immunol.* 13:1595–1599.

Akashi, S., S. Saitoh, Y. Wakabayashi, T. Kikuchi, N. Takamura, Y. Nagai, Y. Kusumoto, K. Fukase, S. Kusumoto, Y. Adachi, A. Kosugi, and K. Miyake. 2003. Lipopolysaccharide interaction with cell surface Toll-like receptor 4-MD-2: higher affinity than that with MD-2 or CD14. *J. Exp. Med.* 198:1035–1042.

Akira, S., S. Uematsu, and O. Takeuchi. 2006. Pathogen recognition and innate immunity. *Cell* 124:783–801.

Alexopoulou, L., A. C. Holt, R. Medzhitov, and R. A. Flavell. 2001. Recognition of double-stranded RNA and activation of NF-kappaB by Toll-like receptor 3. *Nature* 413:732–738.

Almeida, I. C., M. M. Camargo, D. O. Procopio, L. S. Silva, A. Mehlert, L. R. Travassos, R. T. Gazzinelli, and M. A. Ferguson. 2000. Highly purified glycosylphosphatidylinositols from *Trypanosoma cruzi* are potent proinflammatory agents. *EMBO J.* 19:1476–1485.

Almeida, I. C., and R. T. Gazzinelli. 2001. Proinflammatory activity of glycosylphosphatidylinositol anchors derived from *Trypanosoma cruzi*: structural and functional analyses. *J. Leukoc. Biol.* 70:467–477.

Anderson, K. V. 2000. Toll signaling pathways in the innate immune response. *Curr. Opin. Immunol.* 12:13–19.

Anderson, K. V., L. Bokla, and C. Nüsslein-Volhard. 1985a. Establishment of dorsal-ventral polarity in the Drosophila embryo: the induction of polarity by the Toll gene product. *Cell* 42:791-798.

Anderson, K. V., G. Jurgens, and C. Nüsslein-Volhard. 1985b. Establishment of dorsal-ventral polarity in the Drosophila embryo: genetic studies on the role of the Toll gene product. *Cell* 42:779–789.

Andrejeva, J., K. S. Childs, D. F. Young, T. S. Carlos, N. Stock, S. Goodbourn, and R. E. Randall. 2004. The V proteins of paramyxoviruses bind the IFN-inducible RNA helicase, mda-5, and inhibit its activation of the IFN-beta promoter. *Proc. Natl. Acad. Sci. USA* 101:17264–17269.

Aosai, F., M. Chen, H. K. Kang, H. S. Mun, K. Norose, L. X. Piao, M. Kobayashi, O. Takeuchi, S. Akira, and A. Yano. 2002. *Toxoplasma gondii*-derived heat shock protein HSP70 functions as a B cell mitogen. *Cell Stress Chaperones* 7:357–364.

Bafica, A., H. C. Santiago, R. Goldszmid, C. Ropert, R. T. Gazzinelli, and A. Sher. 2006. Cutting edge: TLR9 and TLR2 signaling together account for MyD88-dependent control of parasitemia in Trypanosoma cruzi infection. *J. Immunol.* 177:3515–3519.

Balachandran, S., E. Thomas, and G. N. Barber. 2004. A FADD-dependent innate immune mechanism in mammalian cells. *Nature* 432:401–405.

Barrat, F. J., T. Meeker, J. Gregorio, J. H. Chan, S. Uematsu, S. Akira, B. Chang, O. Duramad, and R. L. Coffman. 2005. Nucleic acids of mammalian origin can act as endogenous ligands for Toll-like receptors and may promote systemic lupus erythematosus. *J. Exp. Med.* 202:1131–1139.

Bartholomeu, D. C., C. Ropert, M. B. Melo, P. Parroche, C. F. Junqueira, S. M. Teixeira, C. Sirois, P. Kasperkovitz, C. F. Knetter, E. Lien, E. Latz, D. T. Golenbock, and R. T. Gazzinelli. 2008. Recruitment and endo-lysosomal activation of TLR9 in dendritic cells infected with Trypanosoma cruzi. *J. Immunol.* 181:1333–1344.

Becker, I., N. Salaiza, M. Aguirre, J. Delgado, N. Carrillo-Carrasco, L. G. Kobeh, A. Ruiz, R. Cervantes, A. P. Torres, N. Cabrera, A. Gonzalez, C. Maldonado, and A. Isibasi. 2003. Leishmania lipophosphoglycan (LPG) activates NK cells through toll-like receptor-2. *Mol. Biochem. Parasitol.* 130:65–74.

Belvin, M. P., and K. V. Anderson. 1996. A conserved signaling pathway: the Drosophila toll-dorsal pathway. *Annu. Rev. Cell. Dev. Biol.* 12:393–416.

Bernheiden, M., J. M. Heinrich, G. Minigo, C. Schutt, F. Stelter, M. Freeman, D. Golenbock, and R. S. Jack. 2001. LBP, CD14, TLR4 and the murine innate immune response to a peritoneal Salmonella infection. *J. Endotoxin Res.* 7:447–450.

Beutler, B., Z. Jiang, P. Georgel, K. Crozat, B. Croker, S. Rutschmann, X. Du, and K. Hoebe. 2006. Genetic analysis of host resistance: Toll-like receptor signaling and immunity at large. *Annu. Rev. Immunol.* 24:353–389.

Bieback, K., E. Lien, I. M. Klagge, E. Avota, J. Schneider-Schaulies, W. P. Duprex, H. Wagner, C. J. Kirschning, V. Ter Meulen, and S. Schneider-Schaulies. 2002. Hemagglutinin protein of wild-type measles virus activates Toll-like receptor 2 signaling. *J. Virol.* 76:8729–8736.

Biron, C. A., and R. T. Gazzinelli. 1995. Effects of IL-12 on immune responses to microbial infections: a key mediator in regulating disease outcome. *Curr. Opin. Immunol.* 7:485–496.

Bjorkbacka, H., K. A. Fitzgerald, F. Huet, X. Li, J. A. Gregory, M. A. Lee, C. M. Ordija, N. E. Dowley, D. T. Golenbock, and M. W. Freeman. 2004. The induction of macrophage gene expression by LPS predominantly utilizes Myd88-independent signaling cascades. *Physiol. Genomics* 19:319–330.

Bonnard, M., C. Mirtsos, S. Suzuki, K. Graham, J. Huang, M. Ng, A. Itie, A. Wakeham, A. Shahinian, W. J. Henzel, A. J. Elia, W. Shillinglaw, T. W. Mak, Z. Cao, and W. C.

Yeh. 2000. Deficiency of T2K leads to apoptotic liver degeneration and impaired NF-kappaB-dependent gene transcription. *EMBO J.* **19:**4976–4985.

Brown, W. C., and R. S. Corral. 2002. Stimulation of B lymphocytes, macrophages, and dendritic cells by protozoan DNA. *Microbes Infect.* **4:**969-974.

Burns, K., J. Clatworthy, L. Martin, F. Martinon, C. Plumpton, B. Maschera, A. Lewis, K. Ray, J. Tschopp, and F. Volpe. 2000. Tollip, a new component of the IL-1RI pathway, links IRAK to the IL-1 receptor. *Nat. Cell Biol.* **2:**346–351.

Burzyn, D., J. C. Rassa, D. Kim, I. Nepomnaschy, S. R. Ross, and I. Piazzon. 2004. Toll-like receptor 4-dependent activation of dendritic cells by a retrovirus. *J. Virol.* **78:**576–584.

Camargo, M. M., I. C. Almeida, M. E. Pereira, M. A. Ferguson, L. R. Travassos, and R. T. Gazzinelli. 1997. Glycosylphosphatidylinositol-anchored mucin-like glycoproteins isolated from Trypanosoma cruzi trypomastigotes initiate the synthesis of proinflammatory cytokines by macrophages. *J. Immunol.* **158:**5890–5901.

Campos, M. A., I. C. Almeida, O. Takeuchi, S. Akira, E. P. Valente, D. O. Procopio, L. R. Travassos, J. A. Smith, D. T. Golenbock, and R. T. Gazzinelli. 2001. Activation of Toll-like receptor-2 by glycosylphosphatidylinositol anchors from a protozoan parasite. *J. Immunol.* **167:**416–423.

Cao, Z., W. J. Henzel, and X. Gao. 1996. IRAK: a kinase associated with the interleukin-1 receptor. *Science* **271:**1128–1131.

Chen, M., F. Aosai, K. Norose, H. S. Mun, O. Takeuchi, S. Akira, and A. Yano. 2002. Involvement of MyD88 in host defense and the down-regulation of anti-heat shock protein 70 autoantibody formation by MyD88 in Toxoplasma gondii-infected mice. *J. Parasitol.* **88:**1017–1019.

Chen, Z. J. 2005. Ubiquitin signalling in the NF-kappaB pathway. *Nat. Cell Biol.* **7:**758–765.

Chow, J. C., D. W. Young, D. T. Golenbock, W. J. Christ, and F. Gusovsky. 1999. Toll-like receptor-4 mediates lipopolysaccharide-induced signal transduction. *J. Biol. Chem.* **274:**10689–10692.

Christ, W. J., O. Asano, A. L. Robidoux, M. Perez, Y. Wang, G. R. Dubuc, W. E. Gavin, L. D. Hawkins, P. D. McGuinness, M. A. Mullarkey, et al. 1995. E5531, a pure endotoxin antagonist of high potency. *Science* **268:**80–83.

Coban, C., K. J. Ishii, T. Kawai, H. Hemmi, S. Sato, S. Uematsu, M. Yamamoto, O. Takeuchi, S. Itagaki, N. Kumar, T. Horii, and S. Akira. 2005. Toll-like receptor 9 mediates innate immune activation by the malaria pigment hemozoin. *J. Exp. Med.* **201:**19–25.

Collins, W. E., and G. M. Jeffery. 1999. A retrospective examination of secondary sporozoite- and trophozoite-induced infections with Plasmodium falciparum: development of parasitologic and clinical immunity following secondary infection. *Am. J. Trop. Med. Hyg.* **61:**20–35.

Compton, T., E. A. Kurt-Jones, K. W. Boehme, J. Belko, E. Latz, D. T. Golenbock, and R. W. Finberg. 2003. Human cytomegalovirus activates inflammatory cytokine responses via CD14 and Toll-like receptor 2. *J. Virol.* **77:**4588–4596.

Cusson-Hermance, N., S. Khurana, T. H. Lee, K. A. Fitzgerald, and M. A. Kelliher. 2005. Rip1 mediates the Trif-dependent toll-like receptor 3- and 4-induced NF-{kappa}B activation but does not contribute to interferon regulatory factor 3 activation. *J. Biol. Chem.* **280:**36560–36566.

Debierre-Grockiego, F., M. A. Campos, N. Azzouz, J. Schmidt, U. Bieker, M. G. Resende, D. S. Mansur, R. Weingart, R. R. Schmidt, D. T. Golenbock, R. T. Gazzinelli, and R. T. Schwarz. 2007. Activation of TLR2 and TLR4 by glycosylphosphatidylinositols derived from Toxoplasma gondii. *J. Immunol.* **179:**1129–1137.

Delude, R. L., R. Savedra, Jr., H. Zhao, R. Thieringer, S. Yamamoto, M. J. Fenton, and D. T. Golenbock. 1995. CD14 enhances cellular responses to endotoxin without imparting ligand-specific recognition. *Proc. Natl. Acad. Sci. USA* **92:**9288–9292.

de Veer, M. J., J. M. Curtis, T. M. Baldwin, J. A. DiDonato, A. Sexton, M. J. McConville, E. Handman, and L. Schofield. 2003. MyD88 is essential for clearance of Leishmania major: possible role for lipophosphoglycan and Toll-like receptor 2 signaling. *Eur. J. Immunol.* **33:**2822–2831.

Diebold, S. S., T. Kaisho, H. Hemmi, S. Akira, and C. Reis e Sousa. 2004. Innate antiviral responses by means of TLR7-mediated recognition of single-stranded RNA. *Science* **303:**1529–1531.

Divanovic, S., A. Trompette, S. F. Atabani, R. Madan, D. T. Golenbock, A. Visintin, R. W. Finberg, A. Tarakhovsky, S. N. Vogel, Y. Belkaid, E. A. Kurt-Jones, and C. L. Karp. 2005. Negative regulation of Toll-like receptor 4 signaling by the Toll-like receptor homolog RP105. *Nat. Immunol.* **6:**571–578.

Dobrovolskaia, M. A., and S. N. Vogel. 2002. Toll receptors, CD14, and macrophage activation and deactivation by LPS. *Microbes Infect.* **4:**903–914.

Drennan, M. B., B. Stijlemans, J. Van den Abbeele, V. J. Quesniaux, M. Barkhuizen, F. Brombacher, P. De Baetselier, B. Ryffel, and S. Magez. 2005. The induction of a type 1 immune response following a *Trypanosoma brucei* infection is MyD88 dependent. *J. Immunol.* **175:**2501–2509.

Ewald, S. E., B. L. Lee, L. Lau, K. E. Wickliffe, G. P. Shi, H. A. Chapman, and G. M. Barton. 2008. The ectodomain of Toll-like receptor 9 is cleaved to generate a functional receptor. *Nature* **456:**658–662.

Ferrero, E., C. L. Hsieh, U. Francke, and S. M. Goyert. 1990. CD14 is a member of the family of leucine-rich proteins and is encoded by a gene syntenic with multiple receptor genes. *J. Immunol.* **145:**331–336.

Fitzgerald, K. A., S. M. McWhirter, K. L. Faia, D. C. Rowe, E. Latz, D. T. Golenbock, A. J. Coyle, S. M. Liao, and T. Maniatis. 2003a. IKKepsilon and TBK1 are essential components of the IRF3 signaling pathway. *Nat. Immunol.* **4:**491–496.

Fitzgerald, K. A., E. M. Palsson-McDermott, A. G. Bowie, C. A. Jefferies, A. S. Mansell, G. Brady, E. Brint, A. Dunne, P. Gray, M. T. Harte, D. McMurray, D. E. Smith, J. E. Sims, T. A. Bird, and L. A. O'Neill. 2001. Mal (MyD88-adapter-like) is required for Toll-like receptor-4 signal transduction. *Nature* **413:**78–83.

Fitzgerald, K. A., D. C. Rowe, B. J. Barnes, A. Visintin, E. Latz, B. Monks, P. M. Pitha, and D. T. Golenbock. 2003b. LPS/TLR4 signaling to IRF-3/7 and NF-kB involves the Toll adapters TRAM and TRIF. *J. Exp. Med.* **198:**1043–1055.

Flandin, J. F., F. Chano, and A. Descoteaux. 2006. RNA interference reveals a role for TLR2 and TLR3 in the recognition of Leishmania donovani promastigotes by interferon-gamma-primed macrophages. *Eur. J. Immunol.* **36:**411–420.

Gangloff, M., A. N. Weber, R. J. Gibbard, and N. J. Gay. 2003. Evolutionary relationships, but functional differences, between the Drosophila and human Toll-like receptor families. *Biochem. Soc. Trans.* **31:**659–663.

Gay, N. J., and F. J. Keith. 1991. Drosophila Toll and IL-1 receptor. *Nature* **351:**355–356.

Gazzinelli, R. T., and E. Y. Denkers. 2006. Protozoan encounters with Toll-like receptor signalling pathways: implications for host parasitism. *Nat. Rev. Immunol.* **6:**895–906.

Gitlin, L., W. Barchet, S. Gilfillan, M. Cella, B. Beutler, R. A. Flavell, M. S. Diamond, and M. Colonna. 2006. Essential role of mda-5 in type I IFN responses to polyriboinosinic:polyribocytidylic acid and encephalomyocarditis picornavirus. *Proc. Natl. Acad. Sci. USA* **103:**8459–8464.

Gohda, J., T. Matsumura, and J. Inoue. 2004. Cutting edge: TNFR-associated factor (TRAF) 6 is essential for MyD88-dependent pathway but not toll/IL-1 receptor domain-containing adaptor-inducing IFN-beta (TRIF)-dependent pathway in TLR signaling. *J. Immunol.* 173:2913–2917.

Golenbock, D. T., R. Y. Hampton, N. Qureshi, K. Takayama, and C. R. Raetz. 1991. Lipid A-like molecules that antagonize the effects of endotoxins on human monocytes. *J. Biol. Chem.* 266:19490–19498.

Hacker, H., V. Redecke, B. Blagoev, I. Kratchmarova, L. C. Hsu, G. G. Wang, M. P. Kamps, E. Raz, H. Wagner, G. Hacker, M. Mann, and M. Karin. 2006. Specificity in Toll-like receptor signalling through distinct effector functions of TRAF3 and TRAF6. *Nature* 439:204–207.

Hajjar, A. M., R. K. Ernst, J. H. Tsai, C. B. Wilson, and S. I. Miller. 2002. Human Toll-like receptor 4 recognizes host-specific LPS modifications. *Nat. Immunol.* 3:354–359.

Hayashi, F., K. D. Smith, A. Ozinsky, T. R. Hawn, E. C. Yi, D. R. Goodlett, J. K. Eng, S. Akira, D. M. Underhill, and A. Aderem. 2001. The innate immune response to bacterial flagellin is mediated by Toll-like receptor 5. *Nature* 410:1099–1103.

Hemmi, H., T. Kaisho, O. Takeuchi, S. Sato, H. Sanjo, K. Hoshino, T. Horiuchi, H. Tomizawa, K. Takeda, and S. Akira. 2002. Small anti-viral compounds activate immune cells via the TLR7 MyD88-dependent signaling pathway. *Nat. Immunol.* 3:196–200.

Hemmi, H., O. Takeuchi, T. Kawai, T. Kaisho, S. Sato, H. Sanjo, M. Matsumoto, K. Hoshino, H. Wagner, K. Takeda, and S. Akira. 2000. A Toll-like receptor recognizes bacterial DNA. *Nature* 408:740–745.

Hemmi, H., O. Takeuchi, S. Sato, M. Yamamoto, T. Kaisho, H. Sanjo, T. Kawai, K. Hoshino, K. Takeda, and S. Akira. 2004. The roles of two IkappaB kinase-related kinases in lipopolysaccharide and double stranded RNA signaling and viral infection. *J. Exp. Med.* 199:1641–1650.

Heppner, G., and D. W. Weiss. 1965. High susceptibility of strain A mice to endotoxin and endotoxin-red blood cell mixtures. *J. Bacteriol.* 90:696–703.

Hoebe, K., X. Du, P. Georgel, E. Janssen, K. Tabeta, S. O. Kim, J. Goode, P. Lin, N. Mann, S. Mudd, K. Crozat, S. Sovath, J. Han, and B. Beutler. 2003a. Identification of Lps2 as a key transducer of MyD88-independent TIR signalling. *Nature* 424:743–748.

Hoebe, K., E. M. Janssen, S. O. Kim, L. Alexopoulou, R. A. Flavell, J. Han, and B. Beutler. 2003b. Upregulation of costimulatory molecules induced by lipopolysaccharide and double-stranded RNA occurs by Trif-dependent and Trif-independent pathways. *Nat. Immunol.* 4:1223–1229.

Horng, T., G. M. Barton, and R. Medzhitov. 2001. TIRAP: an adapter molecule in the Toll signaling pathway. *Nat. Immunol.* 2:835–841.

Hornung, V., F. Bauernfeind, A. Halle, E. O. Samstad, H. Kono, K. L. Rock, K. A. Fitzgerald, and E. Latz. 2008. Silica crystals and aluminum salts activate the NALP3 inflammasome through phagosomal destabilization. *Nat. Immunol.* 9:847–856.

Hoshino, K., O. Takeuchi, T. Kawai, H. Sanjo, T. Ogawa, Y. Takeda, K. Takeda, and S. Akira. 1999. Cutting edge: Toll-like receptor 4 (TLR4)-deficient mice are hyporesponsive to lipopolysaccharide: evidence for TLR4 as the Lps gene product. *J. Immunol.* 162:3749–3752.

Jack, R. S., X. Fan, M. Bernheiden, G. Rune, M. Ehlers, A. Weber, G. Kirsch, R. Mentel, B. Furll, M. Freudenberg, G. Schmitz, F. Stelter, and C. Schutt. 1997. Lipopolysaccharide-binding protein is required to combat a murine gram-negative bacterial infection. *Nature* 389:742–745.

Janeway, C. A., Jr., and R. Medzhitov. 2002. Innate immune recognition. *Annu. Rev. Immunol.* 20:197–216.

Jaramillo, M., D. C. Gowda, D. Radzioch, and M. Olivier. 2003. Hemozoin increases IFN-gamma-inducible macrophage nitric oxide generation through extracellular signal-regulated kinase- and NF-kappa B-dependent pathways. *J. Immunol.* 171:4243–4253.

Jaramillo, M., I. Plante, N. Ouellet, K. Vandal, P. A. Tessier, and M. Olivier. 2004. Hemozoin-inducible proinflammatory events in vivo: potential role in malaria infection. *J. Immunol.* 172:3101–3110.

Jin, M. S., S. E. Kim, J. Y. Heo, M. E. Lee, H. M. Kim, S. G. Paik, H. Lee, and J. O. Lee. 2007. Crystal structure of the TLR1-TLR2 heterodimer induced by binding of a tri-acylated lipopeptide. *Cell* 130:1071–1082.

Jurk, M., F. Heil, J. Vollmer, C. Schetter, A. M. Krieg, H. Wagner, G. Lipford, and S. Bauer. 2002. Human TLR7 or TLR8 independently confer responsiveness to the antiviral compound R-848. *Nat. Immunol.* 3:499.

Kagan, J. C., and R. Medzhitov. 2006. Phosphoinositide-mediated adaptor recruitment controls Toll-like receptor signaling. *Cell* 125:943–955.

Kaisho, T., and S. Akira. 2002. Toll-like receptors as adjuvant receptors. *Biochim. Biophys. Acta* 1589:1–13.

Kaisho, T., O. Takeuchi, T. Kawai, K. Hoshino, and S. Akira. 2001. Endotoxin-induced maturation of MyD88-deficient dendritic cells. *J. Immunol.* 166:5688–5694.

Kanneganti, T. D., M. Lamkanfi, and G. Nunez. 2007. Intracellular NOD-like receptors in host defense and disease. *Immunity* 27:549–559.

Karin, M., and F. R. Greten. 2005. NF-kappaB: linking inflammation and immunity to cancer development and progression. *Nat. Rev. Immunol.* 5:749–759.

Kato, H., S. Sato, M. Yoneyama, M. Yamamoto, S. Uematsu, K. Matsui, T. Tsujimura, K. Takeda, T. Fujita, O. Takeuchi, and S. Akira. 2005. Cell type-specific involvement of RIG-I in antiviral response. *Immunity* 23:19–28.

Kato, H., O. Takeuchi, S. Sato, M. Yoneyama, M. Yamamoto, K. Matsui, S. Uematsu, A. Jung, T. Kawai, K. J. Ishii, O. Yamaguchi, K. Otsu, T. Tsujimura, C. S. Koh, C. Reis e Sousa, Y. Matsuura, T. Fujita, and S. Akira. 2006. Differential roles of MDA5 and RIG-I helicases in the recognition of RNA viruses. *Nature* 441:101–105.

Kawai, T., O. Adachi, T. Ogawa, K. Takeda, and S. Akira. 1999. Unresponsiveness of MyD88-deficient mice to endotoxin. *Immunity* 11:115–122.

Kawai, T., K. Takahashi, S. Sato, C. Coban, H. Kumar, H. Kato, K. J. Ishii, O. Takeuchi, and S. Akira. 2005. IPS-1, an adaptor triggering RIG-I- and Mda5-mediated type I interferon induction. *Nat. Immunol.* 6:981–988.

Kawai, T., O. Takeuchi, T. Fujita, J. Inoue, P. F. Muhlradt, S. Sato, K. Hoshino, and S. Akira. 2001. Lipopolysaccharide stimulates the MyD88-independent pathway and results in activation of IFN-regulatory factor 3 and the expression of a subset of lipopolysaccharide-inducible genes. *J. Immunol.* 167:5887–5894.

Kim, H., B. S. Park, J. I. Kim, S. E. Kim, J. Lee, S. C. Oh, P. Enkhbayar, N. Matsushima, H. Lee, O. J. Yoo, and J. O. Lee. 2007. Crystal structure of the TLR4-MD-2 complex with bound endotoxin antagonist Eritoran. *Cell* 130:906–917.

Kim, Y. M., M. M. Brinkmann, M. E. Paquet, and H. L. Ploegh. 2008. UNC93B1 delivers nucleotide-sensing toll-like receptors to endolysosomes. *Nature* 452:234–238.

Krishnegowda, G., A. M. Hajjar, J. Zhu, E. J. Douglass, S. Uematsu, S. Akira, A. S. Woods, and D. C. Gowda. 2005. Induction of proinflammatory responses in macrophages by the glycosylphosphatidylinositols of *Plasmodium falciparum*: cell signaling receptors, glycosylphosphatidylinositol (GPI) structural requirement, and regulation of GPI activity. *J. Biol. Chem.* **280:**8606–8616.

Krug, A., G. D. Luker, W. Barchet, D. A. Leib, S. Akira, and M. Colonna. 2004. Herpes simplex virus type 1 activates murine natural interferon-producing cells through toll-like receptor 9. *Blood* **103:**1433–1437.

Kurt-Jones, E. A., L. Popova, L. Kwinn, L. M. Haynes, L. P. Jones, R. A. Tripp, E. E. Walsh, M. W. Freeman, D. T. Golenbock, L. J. Anderson, and R. W. Finberg. 2000. Pattern recognition receptors TLR4 and CD14 mediate response to respiratory syncytial virus. *Nat. Immunol.* **1:**398–401.

Latz, E., A. Schoenemeyer, A. Visintin, K. A. Fitzgerald, B. G. Monks, C. F. Knetter, E. Lien, N. J. Nilsen, T. Espevik, and D. T. Golenbock. 2004. TLR9 signals after translocating from the ER to CpG DNA in the lysosome. *Nat. Immunol.* **5:**190–198.

Latz, E., A. Verma, A. Visintin, M. Gong, C. M. Sirois, D. C. Klein, B. G. Monks, C. J. McKnight, M. S. Lamphier, W. P. Duprex, T. Espevik, and D. T. Golenbock. 2007. Ligand-induced conformational changes allosterically activate Toll-like receptor 9. *Nat. Immunol.* **8:**772–779.

Lee, J., J. H. Mo, K. Katakura, I. Alkalay, A. N. Rucker, Y. T. Liu, H. K. Lee, C. Shen, G. Cojocaru, S. Shenouda, M. Kagnoff, L. Eckmann, Y. Ben-Neriah, and E. Raz. 2006. Maintenance of colonic homeostasis by distinctive apical TLR9 signalling in intestinal epithelial cells. *Nat. Cell. Biol.* **8:**1327–1336.

Le Goffic, R., V. Balloy, M. Lagranderie, L. Alexopoulou, N. Escriou, R. Flavell, M. Chignard, and M. Si-Tahar. 2006. Detrimental contribution of the Toll-like receptor (TLR)3 to influenza A virus-induced acute pneumonia. *PLoS Pathog.* **2:**e53.

Lemaitre, B., E. Nicolas, L. Michaut, J. M. Reichhart, and J. A. Hoffmann. 1996. The dorsoventral regulatory gene cassette spatzle/Toll/cactus controls the potent antifungal response in Drosophila adults. *Cell* **86:**973–983.

Leulier, F., and B. Lemaitre. 2008. Toll-like receptors—taking an evolutionary approach. *Nat. Rev. Genet.* **9:**165–178.

Li, S., A. Strelow, E. J. Fontana, and H. Wesche. 2002. IRAK-4: a novel member of the IRAK family with the properties of an IRAK-kinase. *Proc. Natl. Acad. Sci. USA* **99:**5567–5572.

Lien, E., and R. R. Ingalls. 2002. Toll-like receptors. *Crit. Care Med.* **30:**S1–S11.

Lien, E., T. J. Sellati, A. Yoshimura, T. H. Flo, G. Rawadi, R. W. Finberg, J. D. Carroll, T. Espevik, R. R. Ingalls, J. D. Radolf, and D. T. Golenbock. 1999. Toll-like receptor 2 functions as a pattern recognition receptor for diverse bacterial products. *J. Biol. Chem.* **274:**33419–33425.

Lien, E., T. K. Means, H. Heine, A. Yoshimura, S. Kusumoto, K. Fukase, M. J. Fenton, M. Oikawa, N. Qureshi, B. Monks, R. W. Finberg, R. R. Ingalls, and D. T. Golenbock. 2000. Toll-like receptor 4 imparts ligand-specific recognition of bacterial lipopolysaccharide. *J. Clin. Invest.* **105:**497–504.

Liu, L., I. Botos, Y. Wang, J. N. Leonard, J. Shiloach, D. M. Segal, and D. R. Davies. 2008. Structural basis of toll-like receptor 3 signaling with double-stranded RNA. *Science* **320:**379–381.

Lu, X., M. Wang, J. Qi, H. Wang, X. Li, D. Gupta, and R. Dziarski. 2006. Peptidoglycan recognition proteins are a new class of human bactericidal proteins. *J. Biol. Chem.* **281:**5895–5907.

Lund, J. M., L. Alexopoulou, A. Sato, M. Karow, N. C. Adams, N. W. Gale, A. Iwasaki, and R. A. Flavell. 2004. Recognition of single-stranded RNA viruses by Toll-like receptor 7. *Proc. Natl. Acad. Sci. USA* **101:**5598–5603.

Lund, J., A. Sato, S. Akira, R. Medzhitov, and A. Iwasaki. 2003. Toll-like receptor 9-mediated recognition of Herpes simplex virus-2 by plasmacytoid dendritic cells. *J. Exp. Med.* **198:**513–520.

Maniatis, T. 1986. Mechanisms of human beta-interferon gene regulation. *Harvey Lect.* **82:**71–104.

Martin, M. U., and H. Wesche. 2002. Summary and comparison of the signaling mechanisms of the Toll/interleukin-1 receptor family. *Biochim. Biophys. Acta* **1592:**265–280.

McWhirter, S. M., K. A. Fitzgerald, J. Rosains, D. C. Rowe, D. T. Golenbock, and T. Maniatis. 2004. IFN-regulatory factor 3-dependent gene expression is defective in Tbk1-deficient mouse embryonic fibroblasts. *Proc. Natl. Acad. Sci. USA* **101:**233–238.

Means, T. K., E. Latz, F. Hayashi, M. R. Murali, D. T. Golenbock, and A. D. Luster. 2005. Human lupus autoantibody-DNA complexes activate DCs through cooperation of CD32 and TLR9. *J. Clin. Invest.* **115:**407–417.

Medzhitov, R., P. Preston-Hurlburt, and C. A. Janeway, Jr. 1997. A human homologue of the Drosophila Toll protein signals activation of adaptive immunity. *Nature* **388:**394–397.

Medzhitov, R., P. Preston-Hurlburt, E. Kopp, A. Stadlen, C. Chen, S. Ghosh, and C. A. Janeway, Jr. 1998. MyD88 is an adaptor protein in the hToll/IL-1 receptor family signaling pathways. *Mol. Cell* **2:**253–258.

Meylan, E., K. Burns, K. Hofmann, V. Blancheteau, F. Martinon, M. Kelliher, and J. Tschopp. 2004. RIP1 is an essential mediator of Toll-like receptor 3-induced NF-kappa B activation. *Nat. Immunol.* **5:**503–507.

Morisato, D., and K. V. Anderson. 1994. The spatzle gene encodes a component of the extracellular signaling pathway establishing the dorsal-ventral pattern of the Drosophila embryo. *Cell* **76:**677–688.

Oganesyan, G., S. K. Saha, B. Guo, J. Q. He, A. Shahangian, B. Zarnegar, A. Perry, and G. Cheng. 2006. Critical role of TRAF3 in the Toll-like receptor-dependent and -independent antiviral response. *Nature* **439:**208–211.

Ohto, U., K. Fukase, K. Miyake, and Y. Satow. 2007. Crystal structures of human MD-2 and its complex with antiendotoxic lipid IVa. *Science* **316:**1632–1634.

Oliveira, A. C., J. R. Peixoto, L. B. de Arruda, M. A. Campos, R. T. Gazzinelli, D. T. Golenbock, S. Akira, J. O. Previato, L. Mendonca-Previato, A. Nobrega, and M. Bellio. 2004. Expression of functional TLR4 confers proinflammatory responsiveness to Trypanosoma cruzi glycoinositolphospholipids and higher resistance to infection with T. cruzi. *J. Immunol.* **173:**5688–5696.

O'Neill, L. A., K. A. Fitzgerald, and A. G. Bowie. 2003. The Toll-IL-1 receptor adaptor family grows to five members. *Trends Immunol.* **24:**286–290.

Oshiumi, H., M. Matsumoto, K. Funami, T. Akazawa, and T. Seya. 2003. TICAM-1, an adaptor molecule that participates in Toll-like receptor 3-mediated interferon-beta induction. *Nat. Immunol.* **4:**161–167.

Ozinsky, A., D. M. Underhill, J. D. Fontenot, A. M. Hajjar, K. D. Smith, C. B. Wilson, L. Schroeder, and A. Aderem. 2000. The repertoire for pattern recognition of pathogens by the innate immune system is defined by cooperation between toll-like receptors. *Proc. Natl. Acad. Sci. USA* **97:**13766–13771.

Park, B., M. M. Brinkmann, E. Spooner, C. C. Lee, Y. M. Kim, and H. L. Ploegh. 2008. Proteolytic cleavage in an endolysosomal compartment is required for activation of Toll-like receptor 9. *Nat. Immunol.* **9:**1407–1414.

Parroche, P., F. N. Lauw, N. Goutagny, E. Latz, B. G. Monks, A. Visintin, K. A. Halmen, M. Lamphier, M. Olivier, D. C. Bartholomeu, R. T. Gazzinelli, and D. T. Golenbock. 2007. Malaria hemozoin is immunologically inert but radically enhances innate responses by presenting malaria

DNA to Toll-like receptor 9. *Proc. Natl. Acad. Sci. USA* **104:**1919–1924.

Perry, A. K., E. K. Chow, J. B. Goodnough, W. C. Yeh, and G. Cheng. 2004. Differential requirement for TANK-binding kinase-1 in type I interferon responses to toll-like receptor activation and viral infection. *J. Exp. Med.* **199:**1651–1658.

Poltorak, A., X. He, I. Smirnova, M. Y. Liu, C. Van Huffel, X. Du, D. Birdwell, E. Alejos, M. Silva, C. Galanos, M. Freudenberg, P. Ricciardi-Castagnoli, B. Layton, and B. Beutler. 1998. Defective LPS signaling in C3H/HeJ and C57BL/10ScCr mice: mutations in Tlr4 gene. *Science* **282:**2085–2088.

Poltorak, A., P. Ricciardi-Castagnoli, S. Citterio, and B. Beutler. 2000. Physical contact between lipopolysaccharide and toll-like receptor 4 revealed by genetic complementation. *Proc. Natl. Acad. Sci. USA* **97:**2163–2167.

Raetz, C. R. 1984. The enzymatic synthesis of lipid A: molecular structure and biologic function of monosaccharide precursors. *Rev. Infect. Dis.* **6:**463–471.

Rassa, J. C., J. L. Meyers, Y. Zhang, R. Kudaravalli, and S. R. Ross. 2002. Murine retroviruses activate B cells via interaction with toll-like receptor 4. *Proc. Natl. Acad. Sci. USA* **99:**2281–2286.

Ray, B. L., G. Painter, and C. R. Raetz. 1984. The biosynthesis of gram-negative endotoxin. Formation of lipid A disaccharides from monosaccharide precursors in extracts of Escherichia coli. *J. Biol. Chem.* **259:**4852–4859.

Rock, F. L., G. Hardiman, J. C. Timans, R. A. Kastelein, and J. F. Bazan. 1998. A family of human receptors structurally related to Drosophila Toll. *Proc. Natl. Acad. Sci. USA* **95:**588–593.

Rowe, D. C., A. F. McGettrick, E. Latz, B. G. Monks, N. J. Gay, M. Yamamoto, S. Akira, L. A. O'Neill, K. A. Fitzgerald, and D. T. Golenbock. 2006. The myristoylation of TRIF-related adaptor molecule is essential for Toll-like receptor 4 signal transduction. *Proc. Natl. Acad. Sci. USA* **103:**6299–6304.

Rudd, B. D., E. Burstein, C. S. Duckett, X. Li, and N. W. Lukacs. 2005. Differential role for TLR3 in respiratory syncytial virus-induced chemokine expression. *J. Virol.* **79:**3350–3357.

Rudd, B. D., J. J. Smit, R. A. Flavell, L. Alexopoulou, M. A. Schaller, A. Gruber, A. A. Berlin, and N. W. Lukacs. 2006. Deletion of TLR3 alters the pulmonary immune environment and mucus production during respiratory syncytial virus infection. *J. Immunol.* **176:**1937–1942.

Saitoh, S., S. Akashi, T. Yamada, N. Tanimura, M. Kobayashi, K. Konno, F. Matsumoto, K. Fukase, S. Kusumoto, Y. Nagai, Y. Kusumoto, A. Kosugi, and K. Miyake. 2004a. Lipid A antagonist, lipid IVa, is distinct from lipid A in interaction with Toll-like receptor 4 (TLR4)-MD-2 and ligand-induced TLR4 oligomerization. *Int. Immunol.* **16:**961–969.

Saitoh, S., S. Akashi, T. Yamada, N. Tanimura, F. Matsumoto, K. Fukase, S. Kusumoto, A. Kosugi, and K. Miyake. 2004b. Ligand-dependent Toll-like receptor 4 (TLR4)-oligomerization is directly linked with TLR4-signaling. *J. Endotoxin Res.* **10:**257–260.

Saitoh, S., and K. Miyake. 2006. Mechanism regulating cell surface expression and activation of Toll-like receptor 4. *Chem. Rec.* **6:**311–319.

Sato, S., H. Sanjo, K. Takeda, J. Ninomiya-Tsuji, M. Yamamoto, T. Kawai, K. Matsumoto, O. Takeuchi, and S. Akira. 2005. Essential function for the kinase TAK1 in innate and adaptive immune responses. *Nat. Immunol.* **6:**1087–1095.

Sato, S., M. Sugiyama, M. Yamamoto, Y. Watanabe, T. Kawai, K. Takeda, and S. Akira. 2003. Toll/IL-1 receptor domain-containing adaptor inducing IFN-beta (TRIF) associates with TNF receptor-associated factor 6 and TANK-binding kinase 1, and activates two distinct transcription factors, NF-kappa B and IFN-regulatory factor-3, in the Toll-like receptor signaling. *J. Immunol.* **171:**4304–4310.

Schneider, D. S., Y. Jin, D. Morisato, and K. V. Anderson. 1994. A processed form of the Spatzle protein defines dorsal-ventral polarity in the Drosophila embryo. *Development* **120:**1243–1250.

Schroder, M., and A. G. Bowie. 2005. TLR3 in antiviral immunity: key player or bystander? *Trends Immunol.* **26:**462–468.

Schromm, A. B., E. Lien, P. Henneke, J. C. Chow, A. Yoshimura, H. Heine, E. Latz, B. G. Monks, D. A. Schwartz, K. Miyake, and D. T. Golenbock. 2001. Molecular genetic analysis of an endotoxin nonresponder mutant cell line: a point mutation in a conserved region of MD-2 abolishes endotoxin-induced signaling. *J. Exp. Med.* **194:**79–88.

Schulz, O., S. S. Diebold, M. Chen, T. I. Naslund, M. A. Nolte, L. Alexopoulou, Y. T. Azuma, R. A. Flavell, P. Liljestrom, and C. Reis e Sousa. 2005. Toll-like receptor 3 promotes cross-priming to virus-infected cells. *Nature* **433:**887–892.

Sharma, S., B. R. tenOever, N. Grandvaux, G. P. Zhou, R. Lin, and J. Hiscott. 2003. Triggering the interferon antiviral response through an IKK-related pathway. *Science* **300:**1148–1151.

Shiba, T., S. Kusumoto, M. Inage, M. Imoto, H. Chaki, and T. Shimamoto. 1984. Recent developments in the organic synthesis of lipid A in relation to biologic activities. *Rev. Infect. Dis.* **6:**478–482.

Shimada, T., T. Kawai, K. Takeda, M. Matsumoto, J. Inoue, Y. Tatsumi, A. Kanamaru, and S. Akira. 1999. IKK-i, a novel lipopolysaccharide-inducible kinase that is related to IkappaB kinases. *Int. Immunol.* **11:**1357–1362.

Shimazu, R., S. Akashi, H. Ogata, Y. Nagai, K. Fukudome, K. Miyake, and M. Kimoto. 1999. MD-2, a molecule that confers lipopolysaccharide responsiveness on Toll-like receptor 4. *J. Exp. Med.* **189:**1777–1782.

Shoda, L. K., K. A. Kegerreis, C. E. Suarez, I. Roditi, R. S. Corral, G. M. Bertot, J. Norimine, and W. C. Brown. 2001. DNA from protozoan parasites Babesia bovis, Trypanosoma cruzi, and T. brucei is mitogenic for B lymphocytes and stimulates macrophage expression of interleukin-12, tumor necrosis factor alpha, and nitric oxide. *Infect. Immun.* **69:**2162–2171.

Strain, S. M., I. M. Armitage, L. Anderson, K. Takayama, N. Qureshi, and C. R. Raetz. 1985. Location of polar substituents and fatty acyl chains on lipid A precursors from a 3-deoxy-D-manno-octulosonic acid-deficient mutant of Salmonella typhimurium. Studies by 1H, 13C, and 31P nuclear magnetic resonance. *J. Biol. Chem.* **260:**16089–16098.

Tabeta, K., K. Hoebe, E. M. Janssen, X. Du, P. Georgel, K. Crozat, S. Mudd, N. Mann, S. Sovath, J. Goode, L. Shamel, A. A. Herskovits, D. A. Portnoy, M. Cooke, L. M. Tarantino, T. Wiltshire, B. E. Steinberg, S. Grinstein, and B. Beutler. 2006. The Unc93b1 mutation 3d disrupts exogenous antigen presentation and signaling via Toll-like receptors 3, 7 and 9. *Nat. Immunol.* **7:**156–164.

Taguchi, T., J. L. Mitcham, S. K. Dower, J. E. Sims, and J. R. Testa. 1996. Chromosomal localization of TIL, a gene encoding a protein related to the Drosophila transmembrane receptor Toll, to human chromosome 4p14. *Genomics* **32:**486–488.

Takayama, K., N. Qureshi, P. Mascagni, M. A. Nashed, L. Anderson, and C. R. Raetz. 1983. Fatty acyl derivatives of glucosamine 1-phosphate in Escherichia coli and their relation to lipid A. Complete structure of A diacyl GlcN-1-P found in a phosphatidylglycerol-deficient mutant. *J. Biol. Chem.* **258:**7379–7385.

Takeuchi, O., A. Kaufmann, K. Grote, T. Kawai, K. Hoshino, M. Morr, P. F. Muhlradt, and S. Akira. 2000. Cutting edge: preferentially the R-stereoisomer of the mycoplasmal lipopeptide macrophage-activating lipopeptide-2 activates immune cells through a toll-like receptor 2- and MyD88-dependent signaling pathway. *J. Immunol.* **164:**554–557.

Takeuchi, O., T. Kawai, P. F. Muhlradt, M. Morr, J. D. Radolf, A. Zychlinsky, K. Takeda, and S. Akira. 2001. Discrimination of bacterial lipoproteins by Toll-like receptor 6. *Int. Immunol.* **13:**933–940.

Tojima, Y., A. Fujimoto, M. Delhase, Y. Chen, S. Hatakeyama, K. Nakayama, Y. Kaneko, Y. Nimura, N. Motoyama, K. Ikeda, M. Karin, and M. Nakanishi. 2000. NAK is an IkappaB kinase-activating kinase. *Nature* **404:**778–782.

Triantafilou, M., K. Miyake, D. T. Golenbock, and K. Triantafilou. 2002. Mediators of innate immune recognition of bacteria concentrate in lipid rafts and facilitate lipopolysaccharide-induced cell activation. *J. Cell Sci.* **115:**2603–2611.

Trinchieri, G. 2003. Interleukin-12 and the regulation of innate resistance and adaptive immunity. *Nat. Rev. Immunol.* **3:**133–146.

Trinchieri, G., and A. Sher. 2007. Cooperation of Toll-like receptor signals in innate immune defence. *Nat. Rev. Immunol.* **7:**179–190.

Underhill, D. M., A. Ozinsky, A. M. Hajjar, A. Stevens, C. B. Wilson, M. Bassetti, and A. Aderem. 1999. The Toll-like receptor 2 is recruited to macrophage phagosomes and discriminates between pathogens. *Nature* **401:**811–815.

Viriyakosol, S., P. S. Tobias, R. L. Kitchens, and T. N. Kirkland. 2001. MD-2 binds to bacterial lipopolysaccharide. *J. Biol. Chem.* **276:**38044–38051.

Visintin, A., K. A. Halmen, E. Latz, B. G. Monks, and D. T. Golenbock. 2005. Pharmacological inhibition of endotoxin responses is achieved by targeting the TLR4 coreceptor, MD-2. *J. Immunol.* **175:**6465–6472.

Wang, T., T. Town, L. Alexopoulou, J. F. Anderson, E. Fikrig, and R. A. Flavell. 2004. Toll-like receptor 3 mediates West Nile virus entry into the brain causing lethal encephalitis. *Nat. Med.* **10:**1366–1373.

Wasserman, S. A. 1993. A conserved signal transduction pathway regulating the activity of the rel-like proteins dorsal and NF-kappa B. *Mol. Biol. Cell* **4:**767–771.

Watson, J., and R. Riblet. 1974. Genetic control of responses to bacterial lipopolysaccharides in mice. I. Evidence for a single gene that influences mitogenic and immunogenic responses to lipopolysaccharides. *J. Exp. Med.* **140:**1147–1161.

Watson, J., R. Riblet, and B. A. Taylor. 1977. The response of recombinant inbred strains of mice to bacterial lipopolysaccharides. *J. Immunol.* **118:**2088–2093.

Yamamoto, M., S. Sato, H. Hemmi, K. Hoshino, T. Kaisho, H. Sanjo, O. Takeuchi, M. Sugiyama, M. Okabe, K. Takeda, and S. Akira. 2003a. Role of adaptor TRIF in the MyD88-independent toll-like receptor signaling pathway. *Science* **301:**640–643.

Yamamoto, M., S. Sato, H. Hemmi, S. Uematsu, K. Hoshino, T. Kaisho, O. Takeuchi, K. Takeda, and S. Akira. 2003b. TRAM is specifically involved in the Toll-like receptor 4-mediated MyD88-independent signaling pathway. *Nat. Immunol.* **4:**1144–1150.

Yarovinsky, F., D. Zhang, J. F. Andersen, G. L. Bannenberg, C. N. Serhan, M. S. Hayden, S. Hieny, F. S. Sutterwala, R. A. Flavell, S. Ghosh, and A. Sher. 2005. TLR11 activation of dendritic cells by a protozoan profilin-like protein. *Science* **308:**1626–1629.

Yoneyama, M., M. Kikuchi, T. Natsukawa, N. Shinobu, T. Imaizumi, M. Miyagishi, K. Taira, S. Akira, and T. Fujita. 2004. The RNA helicase RIG-I has an essential function in double-stranded RNA-induced innate antiviral responses. *Nat. Immunol.* **5:**730–737.

Yoshimura, A., E. Lien, R. R. Ingalls, E. Tuomanen, R. Dziarski, and D. Golenbock. 1999. Cutting edge: recognition of Gram-positive bacterial cell wall components by the innate immune system occurs via Toll-like receptor 2. *J. Immunol.* **163:**1–5.

7

C-Type Lectins: Multifaceted Receptors in Phagocyte Biology

ALESSANDRA CAMBI AND CARL G. FIGDOR

The term "C-type lectin" was introduced to distinguish a group of Ca^{2+}-dependent (C-type) carbohydrate-binding animal proteins (lectins) from the other, Ca^{2+}-independent types of animal lectins. Once the C-type lectin structures were biochemically analyzed and the functions of the different domains defined, it became clear that the carbohydrate-binding activity is mediated by a compact unit designated the "carbohydrate recognition domain" (CRD), only present in the Ca^{2+}-dependent lectins but not in other animal lectins (Drickamer, 1988, 1989). Crystallographic studies confirmed that the CRD has a compact globular structure, different from any known protein fold (Weis et al., 1991). This domain has been called "C-type CRD" or "C-type lectin domain." As the number of defined structures increased, it became evident that not all proteins containing C-type CRDs actually bind carbohydrates or even Ca^{2+}. Therefore, a more general term, "C-type lectin-like domains," was introduced to refer to such domains (Drickamer, 1999). In this chapter, we will focus on those members of the C-type lectin family that contain a classical CRD and are expressed and functional on phagocytic cells of the immune system, such as neutrophils, dendritic cells (DCs), and macrophages. We will refer to them as C-type lectin receptors (CLRs); discuss general aspects of CLRs concerning their carbohydrate specificity, the relevance of combining multiple CRDs, the ability of several CLRs to signal directly or to cross talk with other signaling pathways, and the interactions with other immune receptor families; and conclude by presenting potential applications of CLRs as targets in several developing immunotherapies.

STRUCTURE OF CLRs

CLRs exist either as transmembrane proteins or as soluble proteins, but all contain one or more CRDs for binding to carbohydrate structures (Fig. 1). Collectins such as mannose-binding lectin (MBL) and the lung surfactant proteins A and D (SP-A and SP-D, respectively) are examples of soluble CLRs and are important for the clearance of microorganisms (van de Wetering et al., 2004a). Transmembrane CLRs are divided into type I and II depending on the orientation of the amino terminus (Figdor et al., 2002). Examples of type I CLRs are the macrophage mannose receptor (MMR) and DEC-205, both containing several CRDs or CRD-like domains with an extracellular N terminus (Figdor et al., 2002). Type II CLRs, comprising dendritic cell-specific intercellular adhesion molecule-3 (ICAM-3)-grabbing nonintegrin (DC-SIGN), L-SIGN (liver/lymph node SIGN), langerin, dendritic cell immunoreceptor (DCIR), C-type lectin receptor 1 (CLEC-1), dendritic cell lectin (DLEC), BDCA-2, DCAL-1, DCAL-2, Dectin-1, and Dectin-2, have only one CRD and an intracellular N terminus (Figdor et al., 2002).

CRD

CLRs contain the prototype C-type lectin fold with two antiparallel β-strands and two α-helices. This common fold contains irregular loop structures, two of which are involved in the binding of monosaccharides (Kogelberg and Feizi, 2001). Ca^{2+} is required for the interaction of carbohydrates with the CRD. More specifically, carbohydrates interact with a primary ligand-binding site by coordinating bonds with the Ca^{2+} ion and further hydrogen bonding with amino acid side chains that also bind to the conserved Ca^{2+} (Weis and Drickamer, 1996). The typical CRDs can be subdivided into two broad groups: the CRDs binding mannose-type carbohydrates and those binding galactose-type carbohydrates. The mannose-type CRD group contains the triplet amino acid sequence EPN, whereas the galactose-type CRD group contains the triplet QPD (Drickamer, 1992). Despite this specific sequence, each CRD recognizes a particular panel of carbohydrates, which is further determined by the ability and flexibility of the CRD to bind multivalently to differently oriented sugars (Feinberg et al., 2005; Mitchell et al., 2001). For example, DC-SIGN and its liver homologue L-SIGN bind mannose

---

Alessandra Cambi and Carl G. Figdor, Department of Tumor Immunology, Nijmegen Centre for Molecular Life Sciences, Radboud University Nijmegen Medical Centre, Nijmegen, The Netherlands.

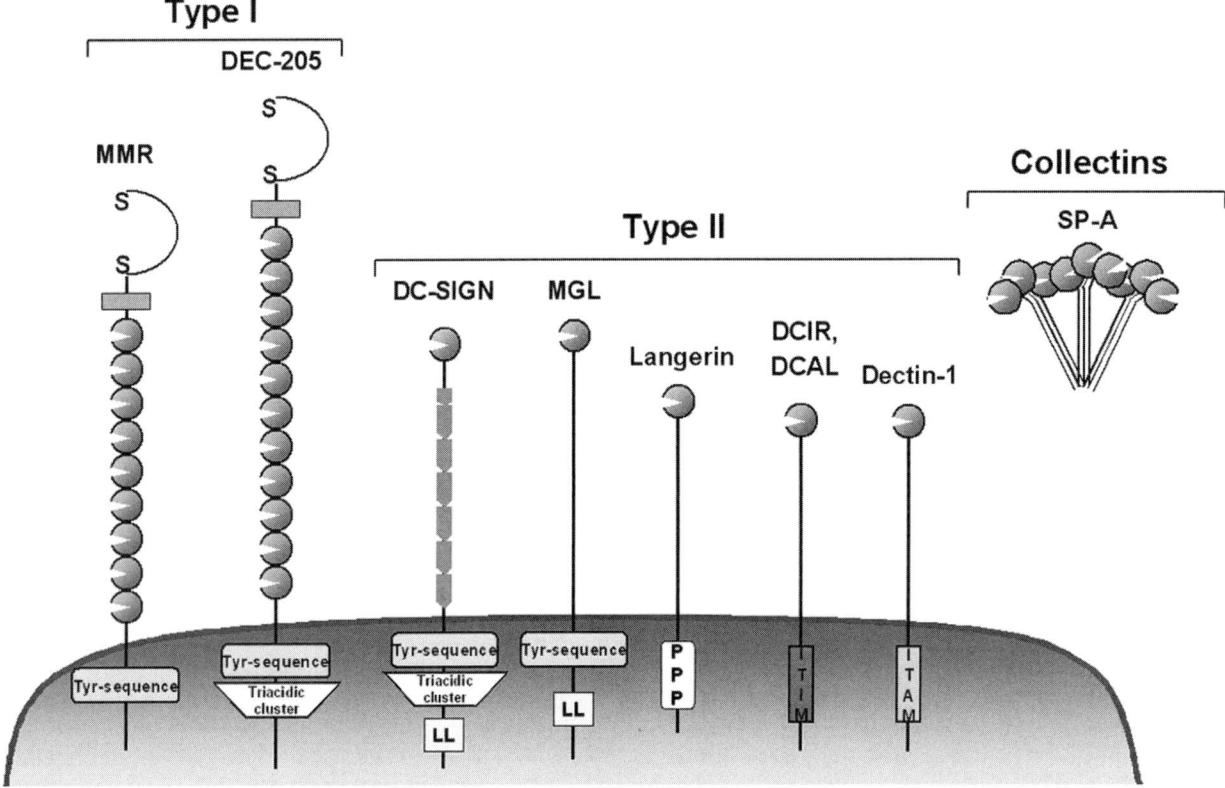

FIGURE 1 Examples of CLRs expressed at the cell surface or secreted by phagocytes. Type I CLRs (MMR and DEC-205) present an amino-terminal cysteine-rich repeat (S–S), a fibronectin type II repeat, and 8 to 10 CRDs, which bind ligand in a Ca^{2+}-dependent manner. MMR binds ligands through CRD4 and -5. Type II CLRs have only one CRD at their carboxy-terminal extracellular domain. The cytoplasmic domains of transmembrane CLRs contain several conserved motifs essential for antigen uptake such as the tyrosine-based intracellular targeting motif, a triad of acidic amino acid residues, and a dileucine motif. Other type II CLRs contain additional signaling motifs such as ITIM, ITAM, and the proline-rich region. Collectins such as MBL, SP-A, and SP-D are soluble CLRs that play important roles in innate immunity. They form large oligomers to enhance their affinity for carbohydrates exposed on microbial surfaces.

in a different manner compared with MBL. The CRD of MBL recognizes a single terminal mannose residue, and strong binding is obtained when three CRDs are combined into a trimeric complex (Mitchell et al., 2001). Within MBL trimers, the distance between individual CRDs is approximately 4.5 to 5.3 nm; therefore, MBL can only recognize mannose densely packed on large mannosylated areas of pathogens. Since the mannose residues found on endogenous proteins are usually separated by short distances (2 to 3 nm), they often cannot be bound by MBL (McGreal et al., 2004; Mitchell et al., 2001). By contrast, the tetrameric structures formed by DC-SIGN and L-SIGN can recognize other spatial arrangements and bind mannose on pathogens as well as on endogenous proteins (Mitchell et al., 2001). By homology modeling, a DC-SIGN/L-SIGN tetramer model was developed, and the surface area encompassed by each CRD was calculated to be approximately 16 nm^2 (Snyder et al., 2005b). Ligands for DC-SIGN and L-SIGN must possess a surface glycosylation level exceeding one glycan molecule per 16 nm^2 of their surface area to allow multiple interactions with one tetramer. Both pathogen-derived and endogenous proteins have this glycosylation level and can interact with DC-SIGN and L-SIGN (Snyder et al., 2005b).

Currently, glycan arrays are frequently used to further detail the carbohydrate preferences of several CLRs (Feizi and Chai, 2004; Galustian et al., 2004; Drickamer and Taylor, 2002). These arrays represent extensive panels of immobilized polysaccharides, glycoproteins, oligosaccharides, and monosaccharides (Feizi and Chai, 2004; Galustian et al., 2004; Drickamer and Taylor, 2002). For example, the binding specificity of DC-SIGN and L-SIGN was compared by using a glycan array where biotinylated oligosaccharides were immobilized onto streptavidin-coated wells (Guo et al., 2004). Fluorescently labeled extracellular domains of DC-SIGN and L-SIGN were probed to this array. Despite the 77% homology in amino acid sequence between the two CLRs, remarkable differences in carbohydrate specificity were observed. While L-SIGN only binds mannose-containing glycans, DC-SIGN is also able to bind fucose-containing glycans such as Lewis X (Guo et al., 2004). Van Liempt et al. (2004) also reported differences in the carbohydrate-binding profile of DC-SIGN and L-SIGN, but they also demonstrated that L-SIGN does bind to the fucose ligands Lewis A, B, and Y, but not to Lewis X. Both groups point to the amino acid Val-351 present in the CRD of DC-SIGN to explain the difference in fucose specificity (Guo et al., 2004; Van Liempt et al., 2004). In DC-SIGN, Val-351 creates a hydrophobic pocket that strongly interacts with the Fuc1,3/4-GlcNAc moiety of Lewis antigens. In contrast, in L-SIGN, at the corresponding position of Val-351 of DC-SIGN a serine (Ser-363) is present. This Ser-363 creates a hydrophilic pocket that excludes binding to Lewis X (Van Liempt et al., 2004).

As an additional example of subtle carbohydrate specificity, sialylation of Lewis X and ICAM-2 abrogates the binding to DC-SIGN (Guo et al., 2004; Weber et al., 2004). Platelets with heavily sialylated ICAM-2 do not bind to DC-SIGN, whereas endothelial cells with unsialylated ICAM-2 can bind to DC-SIGN (Weber et al., 2004). In this way, sialylation modulates the binding specificity of DC-SIGN, preventing the interference of DC-SIGN with the selectins, a group of CLRs with a high specificity for sialyl Lewis X (Guo et al., 2004).

Clearly, CRDs recognize subtle differences in the arrangement and branching of carbohydrates, and even a minor modification can often result in the loss of binding. On the other hand, some CLRs share overlapping ligand repertoires and may therefore contribute to redundancy.

Multimerization

Besides the density of the carbohydrates present on the cell surface, the degree of multimerization of the CRDs contributes to the regulation of the carbohydrate recognition by CLRs (Cambi et al., 2005). Multimerization of CRDs can occur at different levels.

MMR contains several CRDs within the extracellular portion of one single molecule. Each CRD has only a weak affinity for single carbohydrates, and at least three of these CRDs must cooperate to achieve a high-affinity binding to multivalent glycoconjugates (Taylor et al., 1992). Multimerization resulting in high-affinity binding can also be achieved by oligomerization of several CLRs, as observed for DC-SIGN and L-SIGN (Snyder et al., 2005a). Both CLRs form tetramers when expressed at the cell surface as well as in a recombinant soluble form (Feinberg et al., 2005; Mitchell et al., 2001). The extracellular neck domain is responsible for this tetramer formation (Mitchell et al., 2001). More specifically, the portion of the neck domain adjacent to the CRD is sufficient to mediate formation of dimers, while regions near the transmembrane part are required for stabilization of tetramers (Feinberg et al., 2005). Clearly, DC-SIGN tetramerization results in an increased binding potential for carbohydrates: a monomeric DC-SIGN CRD is still able to bind the human immunodeficiency virus type 1 (HIV-1) envelope glycoprotein gp120, but the avidity increases substantially when a tetrameric CRD is used (Snyder et al., 2005b). CLR oligomerization not only increases the avidity for oligosaccharides, but may also induce clustering of cytoplasmic internalization motifs leading to the formation of a more efficient signaling platform for internalization (Snyder et al., 2005b).

An additional level of CLR multimerization is achieved by locally increasing the receptor distribution at the cell membrane. For instance, during the differentiation of human monocyte-derived DCs, DC-SIGN organization at the cell surface is modified from a random distribution (at early differentiation stage) into nanoclusters on immature DCs. These DC-SIGN nanoclusters (150 to 200 nm in diameter) have been shown to confer to the receptor the capacity of binding and internalizing virus particles (Cambi et al., 2004). A similar clustered organization on the microvilli of human lymphocytes has been documented for selectins, which play a major role in cell adhesion but are not involved in pathogen recognition, and has been suggested to facilitate the rolling of lymphocytes on the endothelial surface (Hasslen et al., 1995).

Cytoplasmic Tail Motifs

A general characteristic of most CLRs is that they endocytose foreign ligands (antigens) that are subsequently targeted to lysosomes for degradation. Peptide fragments resulting from lysosomal degradation are eventually complexed onto major histocompatibility complex class II (MHC-II) molecules and exposed at the cell surface of antigen-presenting cells (APCs) to be inspected by T cells for self/nonself discrimination. Several conserved motifs in CLR cytoplasmic tails, such as the dileucine motif, the triacidic cluster, and tyrosine-based sequences (Fig. 1), regulate endocytosis of antigens and may also determine the specific intracellular routing of each CLR (Engering et al., 2002a). DC-SIGN contains both a dileucine motif and the triacidic cluster, but it seems that only the dileucine motif is required for internalization upon ligand binding, since mutation of this dileucine motif completely abolished internalization (Engering et al., 2002b). The triacidic cluster, expressed by DEC-205 and DC-SIGN, is a putative motif involved in routing internalized glycoconjugates toward lysosomal and MHC-II-positive late endosomes, indicating that these CLRs enable loading of peptides onto MHC-II molecules (Engering et al., 2002b; Mahnke et al., 2000). The MMR expresses only the tyrosine-based sequences (Ezekowitz et al., 1990) and recycles via early endosomes back to the cell surface, enabling endocytosis of large quantities of glycoconjugates (Sallusto et al., 1995).

Besides the presence of internalization sequences in their cytoplasmic tails, some CLRs also contain signaling motifs like the immunoreceptor tyrosine-based activation motif (ITAM) or the immunoreceptor tyrosine-based inhibitory motif (ITIM) (Fig. 1). On macrophages, the β-glucan receptor Dectin-1 contains an ITAM motif and cooperates with Toll-like receptor-2 (TLR2) in eliciting inflammatory responses against the yeast cell wall component zymosan (Underhill, 2003). Upon triggering this Dectin-1/TLR-2 pathway, the proinflammatory cytokines tumor necrosis factor alpha (TNF-α) and interleukin-12 (IL-12) are produced. Recently, it was discovered that Dectin-1 also signals via an additional TLR2 independent pathway (Rogers et al., 2005). In this pathway, Dectin-1 recruits the protein tyrosine kinase Syk in response to zymosan, resulting in the production of IL-2 and IL-10. DCs deficient for Syk are unable to produce these two cytokines, but can still produce IL-12, indicating that the Dectin-1/Syk and the Dectin-1/TLR2 pathways operate independently (Rogers et al., 2005). Other CLRs that contain a complete or partial ITAM sequence are DC-SIGN, the macrophage galactose-type C-type lectin (MGL), CLEC-1, and CLEC-2, but it is unclear whether they use a pathway similar to Dectin-1 (Engering et al., 2002a). In fact, DC-SIGN contains a YXXL motif in the cytoplasmic tail that could potentially be involved in signaling. However, Gringhuis et al. (2007) recently showed that DC-SIGN is able to modulate TLR signaling via Raf-1 kinase-dependent acetylation of NF-κB but the YXXL motif is not essential for Raf-1 activation. Furthermore, they could not detect any role for Syk in DC-SIGN signaling.

DCIR contains an ITIM sequence that inhibits B-cell receptor-mediated Ca^{2+} mobilization and tyrosine phosphorylation (Kanazawa et al., 2002). However, the exact role of this motif in modulating DC function is still unknown. Another example of an ITIM-containing type II CLR is DCAL-2 (Chen et al., 2006). The expression of DCAL-2 is restricted to immature DCs, monocytes, and CD1a+ DCs. Cross-linking of this receptor on immature DCs induces protein tyrosine phosphorylation, mitogen-activated protein kinase (MAPK) activation, and receptor internalization. Interestingly, DCAL-2 ligation upregulates CCR7 expression and IL-6 and IL-10 production on DCs. However, receptor ligation has opposite effects on CD40L versus TLR signaling: anti-DCAL-2 enhances CD40L-induced IL-12 production but suppresses TLR-induced IL-12 expression (Chen et al., 2006). Therefore, DCAL-2 triggering seems to program DCs in a different way depending on whether DCs are stimulated via T cells or via TLRs. This CLR may therefore be a potential immunotherapeutic target for regulating autoimmune diseases or for developing immune vaccines. Sialic acid-binding immunoglobulin superfamily lectins (Siglecs) also contain ITIM motifs in their cytoplasmic tail and negatively regulate the function of leukocytes upon binding sialic acids that appear on glycoproteins and glycolipids of normal cells (Crocker and Varki, 2001). For instance, the Siglec CD33 constitutively represses monocyte activation due to interactions with sialic acid residues (Lajaunias et al., 2005).

In addition to ITAM and ITIM sequences, most CLRs contain multiple serine and threonine residues that are potential phosphorylation sites (Engering et al., 2002a). MMR is reported to be involved in signaling transduction events. In one study, the anti-human MMR monoclonal antibody 19.2 was shown to inhibit the lipopolysaccharide (LPS)-induced IL-12 production by DCs by interfering with TLR signaling (Nigou et al., 2001). A different monoclonal antibody against MMR induces phenotypic and functional maturation of immature DCs that dampens inflammation and inhibits the generation of Th1-polarized immune responses (Chieppa et al., 2003). Furthermore, MMR is implicated in NF-κB activation in response to *Pneumocystis carinii* (Zhang et al., 2004). The CLR BDCA-2 does not contain any clear signaling motif, but it is able to induce Ca^{2+} mobilization (Dzionek et al., 2001) and is rapidly internalized upon cross-linking. Conceivably, CLRs lacking signaling motifs might dynamically associate with other signaling proteins to induce specific cellular responses during microbial infections.

INTERACTION WITH MEMBRANE COMPARTMENTS

A significant portion of DC-SIGN nanoclusters detected on DC plasma membrane was shown to reside in lipid rafts (Caparros et al., 2006; Cambi et al., 2004). Lipid rafts are membrane compartments with elevated cholesterol and glycosphingolipid content that have the ability to include and exclude proteins to variable extents (Simons and Toomre, 2000). The localization of DC-SIGN nanoclusters in lipid rafts may create a local nanometer-sized platform, where pathogen binding occurs and various signaling molecules are brought into contact with DC-SIGN (Cambi et al., 2005). Colocalization with lipid rafts is a potential mechanism for several CLRs, since E-selectin expressed on endothelial cells has been demonstrated to associate with lipid rafts as well. This association is important for signaling during leukocyte-endothelial interactions, since E-selectin must associate with and activate phospholipase Cγ. This will only occur when E-selectin is recruited into these lipid rafts (Kiely et al., 2003). Moreover, TLR2 and TLR4, which have been implicated in the regulation of different CLRs, were recently reported to transiently reside in lipid rafts as well upon ligand interaction (Soong et al., 2004; Triantafilou et al., 2002). This observation, together with the findings that physical interactions between TLRs and CLRs can occur (Nagaoka et al., 2005), could imply that both receptor types may reside within the same lipid platforms. Different combinational associations of CLRs and TLRs within activation clusters may possibly determine the specific responses to a variety of bacterial stimuli (Triantafilou and Triantafilou, 2003; Triantafilou et al., 2002). A different type of membrane compartment, the tetraspanin microdomain, has also been demonstrated to regulate specific CLR functions. Recently, Dectin-1 and the tetraspanin CD37 have been shown to colocalize on the surface of human APCs (Meyer-Wentrup et al., 2007). More importantly, the interaction of Dectin-1 with CD37 has been demonstrated to stabilize Dectin-1 at the cell surface and modulate Dectin-1-mediated cytokine production on macrophages, while leaving Dectin-1-mediated phagocytosis unaffected (Meyer-Wentrup et al., 2007).

The recruitment of CLRs into different membrane compartments represents an additional way to determine and broaden CLR signaling potential.

SELF AND NONSELF RECOGNITION

The main task of the immune system is to eliminate life-threatening entities, such as pathogens and tumors, while protecting the normal tissues. Therefore, the immune system developed several strategies to discriminate between self and nonself. A mechanism exploited by DCs is based on the expression of pattern recognition receptors (PRRs) (McGreal et al., 2004; Medzhitov and Janeway, 2002), each of them having a preference for binding to a certain set of molecular patterns. Molecular patterns recognized by PRRs can derive from pathogens, aberrant cells, or normal cells. PRRs such as CLRs and TLRs (reviewed in Takeda and Akira [2005]) bind to so-called pathogen-associated molecular patterns (PAMPs), which include lipids, proteins, and carbohydrates (Medzhitov and Janeway, 2002), thereby playing an important role in the immune defense against pathogens (Cambi and Figdor, 2005; McGreal et al., 2004). In addition, CLRs recognize aberrant carbohydrates on apoptotic and malignant cells, thus contributing to the clearance of abnormal cells (van Gisbergen et al., 2005a; Saevarsdottir et al., 2004). Finally, CLRs are also involved in mediating cellular interactions between hematopoietic cells and plasma glycoprotein turnover (McGreal et al., 2004).

On the surface of microbes, these carbohydrate structures are typically found in dense arrays, while on host cells the same or related carbohydrates are expressed in different configurations (Weis et al., 1998). DC-SIGN binds a broad range of ligands (see Table 1) that include not only pathogen-derived ligands like HIV-1 glycoprotein 120 (Geijtenbeek et al., 2000b), *Candida albicans* (Cambi et al., 2003), and *Mycobacterium tuberculosis* (Geijtenbeek et al., 2003), but also endogenous ligands like ICAM-2 on endothelial cells (Geijtenbeek et al., 2000a), ICAM-3 on T cells (Geijtenbeek et al., 2000c), and the $β_2$ integrin Mac-1 as well as CEACAM1 on activated neutrophils (van Gis-

TABLE 1 Overview of C-type lectins and their ligands

C-type lectins	Endogenous ligand	Pathogen
MMR	L-selectin	HIV, Pneumocystis, Mycobacterium, C. albicans
DEC-205	Unknown	Unknown
DC-SIGN	ICAM-2, ICAM-3, Mac-1, CEACAM1	HIV, HCV, dengue, H. pylori, Mycobacterium, S. mansoni, C. albicans, Aspergillus fumigatus, Leishmania, Neisseria meningitidis
DCIR	Unknown	Unknown
Dectin-1	Unknown	C. albicans, Pneumocystis
MGL	CD45	S. mansoni
SP-A	Unknown	A. fumigatus, S. aureus, influenza A virus, herpes simplex virus
DCAL	Unknown	Unknown

bergen et al., 2005b, 2005c). In addition, DC-SIGN is—based on differences in glycosylation—able to discriminate between normal and cancer cells such as the tumor-associated carcinoembryonic antigen (CEA) expressed on colon cancer cells. Interestingly, malignant transformation changes the glycosylation of CEA on colon epithelial cells, resulting in higher levels of Lewis X and de novo expression of Lewis Y on tumor-associated CEA. Immature DCs bind to colorectal cancer cells via the interaction between DC-SIGN and the Lewis X and Y moieties, whereas no binding occurs to CEA expressed on normal colon epithelium (van Gisbergen et al., 2005a). In this way, DC-SIGN may provide a way for these tumor cells to escape from immunological attack.

Another example of a CLR that recognizes both endogenous and pathogen-derived carbohydrate structures is MGL, which binds to Schistosoma mansoni soluble egg antigens (van Liempt et al., 2007) and to the endogenous ligand CD45 expressed on effector T cells (van Vliet et al., 2006). Furthermore, MGL detects glycan changes between normal and colon carcinoma-associated MUC1, preferentially binding to MUC1 on cancer cells and not to normal epithelial cells (Saeland et al., 2007). Since MGL is highly expressed on tolerogenic DCs, it has been postulated that the interaction between MGL and tumor-associated MUC1 may lead to immunosuppressive effects (Saeland et al., 2007).

Several studies suggest a tolerizing function for CLRs upon interaction with endogenous ligands. For instance, MMR contributes to tissue homeostasis and resolution of inflammation by mediating the clearance of potentially harmful endogenous products (Taylor et al., 2005; Allavena et al., 2004; McGreal et al., 2004). MMR binds soluble ligands such as lysosomal hydrolases to prevent tissue damage (Taylor et al., 2005) and agalactosyl immunoglobulins (IgGs), which are increased in autoimmune disorders (Dong et al., 1999). MGL seems to act in a similar fashion, because it shows specificity for terminal α- and β-linked GalNAc residues that naturally occur as parts of glycoproteins or glycosphingolipids (van Vliet et al., 2005). Furthermore, Bonifaz et al. (2002) showed that ovalbumin protein targeting to DEC-205 in mice results in tolerance induction when no danger signal is present.

The in vivo localization of CLRs in peripheral tissues also points toward a tolerance-inducing function for several CLRs. In the placenta, DC-SIGN and MMR are expressed on decidual immature DCs and macrophages, whereas L-SIGN is detected on endothelial cells (Dietl et al., 2006; Laskarin et al., 2005; Kammerer et al., 2003; Soilleux et al., 2000). The presence in the placenta indicates that these CLRs may maintain the balance between defense against pathogens and tolerance of the fetal allograft (Laskarin et al., 2005; Rieger et al., 2004).

The capacity of several CLRs to bind both foreign and endogenous ligands suggests that a CLR alone is not sufficient to properly instruct both the innate and the adaptive immune system for self/nonself discrimination. TLRs are thought to accomplish the role of providing the nonself-related danger signal, since they recognize pathogen-derived ligands with an exquisite specificity (Matzinger, 2002). However, an increasing number of endogenous ligands for TLRs have been recently documented (Rifkin et al., 2005; Okamura et al., 2001), suggesting that self- and nonself discrimination rather depend on the orchestrated interplay of multiple PRRs, which will eventually shape the innate and adaptive immune response.

Cell Adhesion and Trafficking

Several of the discovered CLRs are involved in cell-cell adhesion processes. Dectin-1, Dectin-2, MGL, DC-SIGN, and DCAL all mediate DC–T-cell interactions (van Vliet et al., 2006; Aragane et al., 2003; Ryan et al., 2002; Ariizumi et al., 2000; Geijtenbeek et al., 2000c). These interactions result either in increased T-cell proliferation, as shown for Dectin-1 (Ariizumi et al., 2000) and DCAL-1 (Ryan et al., 2002), or in attenuation of the immune response as observed for Dectin-2 (Aragane et al., 2003). Dectin-1 and DCAL His-tagged fusion proteins deliver costimulatory signals to T cells in the presence of anti-CD3 antibodies (Ryan et al., 2002; Ariizumi et al., 2000). MGL interacts with effector T cells via CD45 and negatively regulates T-cell-receptor-mediated signaling and T-cell-dependent cytokine responses, thereby decreasing T-cell proliferation and increasing T-cell death (van Vliet et al., 2006).

DC-SIGN is able to perform a stimulatory as well as an inhibitory role in DC–T-cell interactions depending on the strength of the T-cell stimulus (Martinez et al., 2005; Geijtenbeek et al., 2000c). Besides DC–T-cell interactions, DC-SIGN mediates DC migration across the endothelium by binding to ICAM-2 (Geijtenbeek et al., 2000a). Furthermore, DC-SIGN is involved in mediating interactions between DCs and neutrophils, by binding to the Lewis X moiety expressed on the β_2 integrin Mac-1. This is thought to occur during inflammation, when neutrophils infiltrate inflamed areas and encounter DCs, and may modulate the immune response (van Gisbergen et al., 2005c). Interestingly, only Mac-1 expressed by neutrophils and not by other leukocytes contains this Lewis X structure (van Gisbergen et al., 2005c), emphasizing the impact of specific glycan expression on CLR function.

MMR binds to sulfated carbohydrate ligands such as sulfo-Lewis A and X, and this recognition overlaps with the binding specificity of selectins, specialized in leukocyte trafficking (Leteux et al., 2000; Green et al., 1992). Through its cysteine-rich domain, MMR can bind to other macrophages in marginal zones of the spleen, and to B cells in germinal centers. This is thought to direct MMR-bearing cells toward germinal centers during an immune response. The finding that several classes of carbohydrate bind MMR provides a mechanism for regulating the trafficking and function of MMR-bearing cells (Leteux et al., 2000). Recently, MMR was identified on human lymphatic endothelium to mediate binding of lymphocytes by interacting with its counterreceptor L-selectin (Irjala et al., 2001). Interestingly, MMR is absent on high endothelial venules, indicating that L-selectin exploits distinct ligands to mediate binding at sites of lymphocyte entrance and exit within lymph nodes.

Soluble CLRs, such as collectins, have been also reported to modulate the differentiation of bone marrow-derived DCs (Brinker et al., 2003). SP-A, a representative of the innate immune system, regulates pathogen clearance and lung inflammation and modulates innate immune functions such as phagocytosis by macrophages, cytokine production, and chemotaxis. However, little is known about regulation of adaptive immunity by SP-A. Brinker et al. (2003) demonstrated that SP-A regulates the differentiation of immature DCs into potent T-cell stimulators. The incubation of immature DCs for 24 h with SP-A inhibits basal- and LPS-mediated expression of MHC-II and CD86. Stimulation of immature DCs by SP-A also inhibits the allostimulation of T cells, and alters DC chemotaxis toward RANTES and secondary lymphoid tissue chemokine.

Chemotaxis

The directed migration or chemotaxis of macrophages and neutrophils is essential to the execution of an inflammatory response. A variety of molecules stimulate chemotaxis of innate immune cells, including microbial products such as the bacterial peptide formyl-methionyl-leucyl-phenylalanine (fMLP), activated complement (C5a), cytokines such as IL-8 and macrophage inhibitory protein 2, and bioactive lipids such as leukotriene B_4. These chemoattractants bind G-protein-coupled receptors on the surface of neutrophils, thus stimulating signaling events that result in cytoskeletal rearrangements and cell migration. Schagat et al. investigated the ability of SP-A to regulate neutrophil chemotaxis and demonstrated that, although SP-A did not directly stimulate chemotaxis, it does modulate neutrophil chemotaxis toward known chemoattractants (Schagat et al., 2003). Interestingly, this regulation depended on the activation state of the neutrophil; while peripheral neutrophils showed decreased chemotaxis in the presence of SP-A, neutrophils isolated from inflamed lungs showed increased chemotaxis in the presence of SP-A (Schagat et al., 2003). This study provides evidence that SP-A regulates neutrophil migration most likely via its collagen-like domain, although the exact regulatory mechanism remains to be identified.

PATHOGEN RECOGNITION: PROTECTION OR EVASION?

The main function of CLRs after binding pathogens is internalization, which eventually leads to lysosomal degradation and subsequent loading of peptide fragments into MHC molecules. When this MHC-peptide complex is recognized by T cells, the adaptive immune system is activated (Figdor et al., 2002). However, there is increasing evidence to suggest that pathogens use several strategies to evade host immune surveillance (van Kooyk et al., 2004; van Kooyk and Geijtenbeek, 2003). For instance, at the site of infection HIV-1 binds to DCs via DC-SIGN and exploits the DC migratory capacity to gain access to T cells in the lymph node (Geijtenbeek et al., 2000b). Recently, Arrighi et al. (2004a) demonstrated that DC-SIGN is required for the formation of an infectious synapse between HIV-1-bearing DCs and resting $CD4^+$ T cells. More specifically, the use of small RNAi-expressing lentiviral vectors specifically to knock down DC-SIGN on DCs showed that DC-SIGN DCs are still able to internalize HIV-1, although the binding of virions was found to be reduced, but they cannot transfer HIV-1 infection to target cells (Arrighi et al., 2004a, 2004b). It is becoming clear that, besides HIV-1, other viruses also target DC-SIGN to evade the immune system and promote their dissemination by modulating DC function to establish chronic infections. Hepatitis C virus (HCV) has recently been shown to target DC-SIGN and L-SIGN to escape lysosomal degradation (Ludwig et al., 2004). Similarly, severe acute respiratory syndrome coronavirus binds to DC-SIGN and is transferred by DCs to susceptible target cells through a synapse-like structure (Yang et al., 2004). Besides DC-SIGN, MMR on macrophages is also exploited by HIV-1 for transmission to permissive T cells. However, the half-life of HIV-1 bound to MMR is lower than virus bound to DC-SIGN, suggesting different internalization routes for DC-SIGN and MMR (Nguyen and Hildreth, 2003). Besides targeting of CLRs by viruses for transmission to permissive cells, nonviral pathogens exploit CLRs for immune escape in different ways. M. tuberculosis targets DC-SIGN and inhibits the DC immunostimulatory function by producing the anti-inflammatory cytokine IL-10 (Geijtenbeek et al., 2003). The human gastric pathogen Helicobacter pylori also modulates the function of DCs via DC-SIGN by blocking the polarization toward a T-helper type 1 response (Bergman et al., 2004). This bacterium binds to DC-SIGN via its LPS, which exposes Lewis antigens. In particular, reversible on-off switching of specific fucosyltransferases regulates the expression of Lewis-X and -Y on the LPS, thus causing LPS phase variation (Lewis-X^+/Y^+ and Lewis-X^+/Y^-) within a strain. Interestingly, whereas Lewis-X^+/Y^- H. pylori escapes binding to DCs, the Lewis-X^+/Y^+ phase variant exploits binding to DC-SIGN to increase IL-10 levels and block the skewing of naive T cells to Th1 cells (Bergman et al., 2004).

A similar evasion strategy has also been developed by the yeast-like fungus Cryptococcus neoformans, which infects the respiratory and nervous system, to escape from aggregation by SP-D (van de Wetering et al., 2004b). Only acapsular C. neoformans is aggregated by SP-D and subsequently removed by microciliary clearance. Therefore, after deposition in the lung, C. neoformans readily starts producing a capsule, thus also releasing soluble capsular components, including glucuronoxylomannan (GXM). SP-D binds with high affinity to soluble GXM, and this interaction has been shown to inhibit SP-D aggregation of acapsular C. neoformans, thus interfering with pathogen clearance (van de Wetering et al., 2004b).

Bordetella pertussis, the causative agent of human whooping cough, uses a slightly different mechanism for resisting the bactericidal effects of SP-A (Schaeffer et al., 2004). SP-A binds to the lipid A component of LPS via its CRD. However, B. pertussis LPS has a terminal trisaccharide that apparently shields the bacteria from SP-A-mediated clearance by sterically limiting access of SP-A to the lipid A region (Schaeffer et al., 2004).

Despite these examples of immune evasion by pathogens through CLRs, we must not forget that these receptors play a fundamental role in limiting the early proliferation of infectious microorganisms. The importance of MBL in restricting the complications associated with *Staphylococcus aureus* infection is highlighted by studies involving MBL-null mice (Shi et al., 2004). This in vivo study demonstrated that 100% of the MBL-null mice died 48 h after exposure to an intravenous inoculation of *S. aureus*, compared with a 45% mortality rate in wild-type mice (Shi et al., 2004). Similarly, SP-D knockout mice displayed a delayed clearance of *P. carinii* infection, increased inflammation, and altered nitric oxide metabolism (Atochina et al., 2004).

Interestingly, although several viruses exploit DC-SIGN to escape lysosomal degradation, Moris et al. (2004) recently demonstrated that DC-SIGN does not always protect captured virions against degradation. In fact, a fraction of the incoming viral material is processed by the proteasome and leads to activation of anti-HIV-specific cytotoxic T lymphocytes by DC-SIGN-expressing cells (Moris et al., 2004). This suggests that the different routing of the incoming viral material might be directly related to the viral load. Further evidence for a protective in vivo role for a SIGN family member is provided by Lanoue et al. (2004). Mice lacking SIGN-R1 are significantly more susceptible to *Streptococcus pneumoniae* infection and fail to clear the bacteria from the circulation, showing an important role for SIGN-R1 in the protection against septicemia (Lanoue et al., 2004). Unfortunately, a complete comparison between these mouse disease models and infectious diseases in humans is difficult, since remarkable differences exist between human DC-SIGN and mouse SIGN-R1 with respect to cell-type-specific expression patterns (Koppel et al., 2005).

The Langerhans cell-specific CLR langerin has also been reported to play a crucial role in the generation of CD1a-restricted T-cell clones against a *Mycobacterium leprae* lipid antigen (Hunger et al., 2004). In addition, langerin has recently been shown to be a natural barrier to HIV-1 transmission, as HIV-1 captured by Langerhans cells via langerin is internalized into Birbeck granules and degraded, thus preventing HIV-1 infection of and transmission by Langerhans cells (de Witte et al., 2007).

CROSS TALK WITH TLRs

Besides CLRs, pathogens also interact with other receptors present on DCs, such as the TLRs. The engagement of a specific TLR with associated signaling complexes results in a specific response against the pathogen involved (Underhill, 2003). Increasing evidence suggests that TLRs and C-type lectins communicate with each other and that this cross talk is critically important for the balance between immune tolerance and immune activation. In particular, the β-glucan receptor, Dectin-1, has been shown to mediate binding and to phagocytose yeast and fungal-derived zymosan, resulting in the production of inflammatory cytokines by macrophages (Brown et al., 2003). Interestingly, TLR2 and TLR6 are also responsible for the production of zymosan-induced inflammatory cytokines (Takeda et al., 2003). Dectin-1 colocalizes with both TLR2 and TLR6 in areas of contact between zymosan particles and macrophages and its increased expression significantly enhanced TLR2-depending zymosan-induced TNF-α production (Brown et al., 2003). On the other hand, CLRs can also interfere with TLR signaling, like the collectin SP-A which has been shown to downregulate zymosan-induced signaling and TNF-α production by attenuating the binding of TLR2 to zymosan (Sato et al., 2003). Similarly, DC-SIGN is able to modulate TLR3, -4, and -5 signaling via a Raf-1 kinase-dependent acetylation of the transcription factor NF-κB (Gringhuis et al., 2007). Recently, it was reported that SIGN-R1 associates with TLR4 to capture gram-negative bacteria and to facilitate signal transduction to activate innate macrophage responses (Nagaoka et al., 2005). Apparently, unlike what is observed for the human DC-SIGN, mouse SIGN-R1 cooperates with TLR4 instead of interfering. Therefore, evidence is accumulating that microbial recognition is not the result of one microbial component interacting with one single recognition receptor, but rather a complex network of interacting receptors and ligands. Depending on the receptor type involved, and also the organization of the receptors at the cell surface, the outcome can be completely different. It can result in effective resolution of the pathogen, lead to a vigorous adaptive immune response, or lead to tolerance induction. We expect that intensive research in this area will reveal the TLR/CLR combinations and subsequent signaling pathways that result in immune evasion or elimination of pathogens and will also reveal potential targets for intervention therapy.

MODALITY OF ANTIGEN UPTAKE

The main function of C-type lectins in microbial recognition is binding and subsequent internalization for direct elimination by macrophages and/or DCs. Lysosomal degradation produces antigenic fragments that, after presentation by DCs and macrophages onto MHC molecules at the cell surface, stimulate the adaptive immune system (Figdor et al., 2002). Besides the classical MMR, DEC-205 (Mahnke et al., 2000) and DC-SIGN (Cambi et al., 2003, 2007; Ludwig et al., 2004; Engering et al., 2002b) are also known to mediate antigen uptake. While the MMR delivers antigen to early endosomes and recycles back to the cell surface, DEC-205 and DC-SIGN deliver antigens to late endosomes or lysosomes for degradation. Besides the tyrosine-based coated pit sequence uptake motif present in MMR, the cytoplasmic domains of DEC-205 and DC-SIGN contain an additional triacidic cluster important for targeting to proteolytic vacuoles (Mahnke et al., 2000). Furthermore, a dileucine motif present in the cytoplasmic domain of DC-SIGN is essential for internalization (Engering et al., 2002b), and by using antigen-coated fluorescent quantum dots, we recently demonstrated that DC-SIGN can internalize antigens via clathrin-coated pits (Cambi et al., 2007).

The DC-SIGN homologue L-SIGN is not expressed by DCs but is found on liver sinusoids (Bashirova et al., 2001). These are specialized capillary vessels characterized by the presence of resident macrophages adhering to the LSECs. The LSEC-leukocyte interactions, which require expression of adhesion molecules on the cell surfaces, appear to constitute a central mechanism of peripheral immune surveillance in the liver. MMR and now also L-SIGN are known to be expressed on LSECs and may mediate the clearance of many potentially antigenic proteins from the circulation, in a manner similar to DCs in lymphoid organs. However, whether L-SIGN is able to internalize antigen remains controversial. Ludwig et al. (2004) demonstrated that HCV particles are internalized by a monocytic cell line transfected with L-SIGN and target nonlysosomal compartments to escape degradation. In contrast, Guo et al. showed that L-SIGN expressed in fibroblasts is not able to release its ligands at low pH and does not mediate endocytosis, suggesting that L-SIGN predominantly acts as an adhesion receptor (Guo et al., 2004). Investigating the internalization capacity of L-SIGN in a more physiological context (i.e., on LSECs) is required to shed light on this controversial issue.

CLR TARGETING IN ANTITUMOR CLINICAL TRIALS

Thanks to their exquisite ability to take up, process, and present antigens (Banchereau and Steinman, 1998), DCs are key regulators of T- and B-cell immunity. A vaccination strategy aiming at induction of antigen-specific effector and memory cells involves in vivo loading of DCs with tumor- or pathogen-derived antigens via their surface receptors. Several years ago it became clear that antibodies enhance specific T-cell responses by promoting Fc receptor-mediated recognition of opsonized antigens by APCs (Chang, 1985; Celis and Chang, 1984). This led to the hypothesis that targeted delivery of antigen to cell surface molecules expressed by APCs might increase T-cell-mediated immune responses. More recently, the identification of receptors that are more or less specifically expressed by DCs stimulated the development of vaccination strategies that specifically target these professional APCs. So far, these targeting studies have revealed that the efficacy of in vivo DC vaccination depends on a multitude of factors, including the expression pattern and biological properties of the specific receptor and the maturation status of the DC. Many receptors employed for these targeting studies are CLRs. Approaches to target CLR fall into two categories: strategies based on binding of natural receptor ligands and strategies that exploit antibodies directed against a specific receptor. In this section, we will discuss several targeting studies focusing on the well-characterized CLRs, MMR, DEC-205, and DC-SIGN. Both mannose and mannan have been applied in preclinical mice studies for targeted delivery of antigens to the MMR, resulting in enhanced antigen presentation via MHC-I and MHC-II (Keler et al., 2000). In a phase I trial, patients with advanced carcinoma of the breast, colon, stomach, or rectum were treated with mannan conjugated to part of the tumor-associated antigen MUC1. This resulted in antigen-specific humoral responses in half of the patients and cytotoxic-T-cell responses in a minority of patients, although no apparent clinical responses were observed (Karanikas et al., 1997). Stage II breast cancer patients in a pilot phase III study, who were treated with oxidized mannan conjugated to MUC1 and evaluated 5 years after the last individual started treatment, revealed that all patients receiving immunotherapy were free of recurrences (Apostolopoulos et al., 2006). Although the sugar residues that were used in these vaccination studies bind to MMR, they lack receptor specificity and likely target the same antigen to multiple CLRs with overlapping binding specificities (Keler et al., 2000). The use of MMR-specific antibodies has confirmed that antigens targeted to the MMR on human DCs and macrophages augment uptake and presentation of the antigen via both the MHC-I and MHC-II pathways (He et al., 2004; Ramakrishna et al., 2004).

DEC-205 is another type I CLR that has been exploited for in vivo antigen-targeting studies. DEC-205 recycles through late endosomal/lysosomal compartments and mediates antigen presentation (Mahnke et al., 2000). In mice, DEC-205 is highly expressed by mature DCs and thymic epithelium, in low levels by B cells, and in very low levels by T cells and granulocytes (Inaba et al., 1995). Therefore, DEC-205 represents an excellent target to study DC targeting in vivo. Instead of inducing immunity, delivery of the model antigen OVA to DEC-205 without other maturation stimuli results in induction of regulatory T cells and T-cell tolerance (Mahnke et al., 2003; Bonifaz et al., 2002; Hawiger et al., 2001). On the contrary, upon coadministration of DC maturation stimuli, strong induction of OVA-specific CD4$^+$ and CD8$^+$ T-cell responses can be observed (Bonifaz et al., 2004; van Broekhoven et al., 2004; Hawiger et al., 2001). Mahnke and coworkers (2005) used a melanoma model to explore DEC-205 targeting in antitumor therapy. Targeting of two melanoma antigens to DEC-205 together with a DC maturation stimulus cured 70% of the mice from existing tumors (Mahnke et al., 2005). In humans, DEC-205 expression seems less DC-restricted than in mice, and targeting constructs could be endocytosed by several other cell types (Kato et al., 2006).

DC-SIGN is predominantly expressed on immature and at lower levels mature DCs and macrophages (Granelli-Piperno et al., 2005; Engering et al., 2004; Soilleux et al., 2002; Geijtenbeek et al., 2000c). Unfortunately, the mouse is not a suitable preclinical model to study DC-SIGN targeting, since multiple isoforms are expressed in mice that seem functionally unrelated to the human receptor (Caminschi et al., 2006; Gramberg et al., 2006). Therefore, the feasibility of targeting DC-SIGN in vivo was assessed in a monkey model, using an anti-human DC-SIGN antibody that cross-reacts with the cynomolgus monkey homologue. Upon injection of the antibody, efficient targeting of DCs in draining and mesenteric lymph nodes was observed (C. Pereira, R. Torensma, K. Hebeda, A. Kretz-Rommel, S. Faas, C. G. Figdor, et al., unpublished observations). Subsequently, by cloning the mouse hypervariable ligand binding domains into human framework regions, a humanized anti-DC-SIGN antibody was developed with a hybrid IgG2/IgG4 constant domain that prevents binding to Fc receptors (Mueller et al., 1997). Targeted delivery to human DCs of a model antigen conjugated to this humanized DC-SIGN antibody allowed presentation of the antigen via MHC-I and MHC-II, thus eliciting both naïve and recall T-cell responses in vitro (Tacken et al., 2005).

Antigen presentation upon engagement of a specific receptor is also likely to be affected by intracellular routing of the targeted receptor (Mahnke et al., 2000). Furthermore, the routing of a receptor might be influenced by the targeting moiety, as illustrated by Dectin-1, which recycles to the cell surface upon binding of laminarin, but not upon binding glucan phosphate (Herre et al., 2004).

As discussed above, some CLRs contain an ITIM in their cytoplasmic domain, while others carry an ITAM motif in their cytoplasmic domain. Also, CLRs may or may not interfere with TLR signaling. Therefore, the triggering of these various CLRs is likely to have very distinct functional consequences. In conclusion, the CLRs exploited for targeted antigen delivery are not simple portals merely shuttling antigen into the DCs. Differences in their expression patterns on distinct DC subsets, their intracellular signaling cascades, as well as their intracellular routing will have consequences for the immunological outcome of in vivo DC therapy.

REFERENCES

Allavena, P., M. Chieppa, P. Monti, and L. Piemonti. 2004. From pattern recognition receptor to regulator of homeostasis: the double-faced macrophage mannose receptor. *Crit. Rev. Immunol.* **24:**179–192.

Apostolopoulos, V., G. A. Pietersz, A. Tsibanis, A. Tsikkinis, H. Drakaki, B. E. Loveland, S. J. Piddlesden, M. Plebanski, D. S. Pouniotis, M. N. Alexis, I. F. McKenzie, and S. Vassilaros. 2006. Pilot phase III immunotherapy study in early-stage breast cancer patients using oxidized mannan-MUC1 [ISRCTN71711835]. *Breast Cancer Res.* **8:**R27.

Aragane, Y., A. Maeda, A. Schwarz, T. Tezuka, K. Ariizumi, and T. Schwarz. 2003. Involvement of dectin-2 in ultraviolet radiation-induced tolerance. *J. Immunol.* **171:**3801–3807.

Ariizumi, K., G. L. Shen, S. Shikano, S. Xu, R. Ritter III, T. Kumamoto, D. Edelbaum, A. Morita, P. R. Bergstresser, and A. Takashima. 2000. Identification of a novel, dendritic cell-associated molecule, dectin-1, by subtractive cDNA cloning. *J. Biol. Chem.* **275:**20157–20167.

Arrighi, J. F., M. Pion, E. Garcia, J. M. Escola, Y. van Kooyk, T. B. Geijtenbeek, and V. Piguet. 2004a. DC-SIGN-mediated infectious synapse formation enhances X4 HIV-1 transmission from dendritic cells to T cells. *J. Exp. Med.* **200:**1279–1288.

Arrighi, J. F., M. Pion, M. Wiznerowicz, T. B. Geijtenbeek, E. Garcia, S. Abraham, F. Leuba, V. Dutoit, O. Ducrey-Rundquist, Y. van Kooyk, D. Trono, and V. Piguet. 2004b. Lentivirus-mediated RNA interference of DC-SIGN expression inhibits human immunodeficiency virus transmission from dendritic cells to T cells. *J. Virol.* **78:**10848–10855.

Atochina, E. N., A. J. Gow, J. M. Beck, A. Haczku, A. Inch, H. Kadire, Y. Tomer, C. Davis, A. M. Preston, F. Poulain, S. Hawgood, and M. F. Beers. 2004. Delayed clearance of pneumocystis carinii infection, increased inflammation, and altered nitric oxide metabolism in lungs of surfactant protein-D knockout mice. *J. Infect. Dis.* **189:**1528–1539.

Banchereau, J., and R. M. Steinman. 1998. Dendritic cells and the control of immunity. *Nature* **392:**245–252.

Bashirova, A. A., T. B. Geijtenbeek, G. C. van Duijnhoven, S. J. van Vliet, J. B. Eilering, M. P. Martin, L. Wu, T. D. Martin, N. Viebig, P. A. Knolle, V. N. KewalRamani, Y. van Kooyk, and M. Carrington. 2001. A dendritic cell-specific intercellular adhesion molecule 3-grabbing noninte-grin (DC-SIGN)-related protein is highly expressed on human liver sinusoidal endothelial cells and promotes HIV-1 infection. *J. Exp. Med.* **193:**671–678.

Bergman, M. P., A. Engering, H. H. Smits, S. J. van Vliet, A. A. van Bodegraven, H. P. Wirth, M. L. Kapsenberg, C. M. Vandenbroucke-Grauls, Y. van Kooyk, and B. J. Appelmelk. 2004. Helicobacter pylori modulates the T helper cell 1/T helper cell 2 balance through phase-variable interaction between lipopolysaccharide and DC-SIGN. *J. Exp. Med.* **200:**979–990.

Bonifaz, L., D. Bonnyay, K. Mahnke, M. Rivera, M. C. Nussenzweig, and R. M. Steinman. 2002. Efficient targeting of protein antigen to the dendritic cell receptor DEC-205 in the steady state leads to antigen presentation on major histocompatibility complex class I products and peripheral CD8+ T cell tolerance. *J. Exp. Med.* **196:**1627–1638.

Bonifaz, L. C., D. P. Bonnyay, A. Charalambous, D. I. Darguste, S. Fujii, H. Soares, M. K. Brimnes, B. Moltedo, T. M. Moran, and R. M. Steinman. 2004. In vivo targeting of antigens to maturing dendritic cells via the DEC-205 receptor improves T cell vaccination. *J. Exp. Med.* **199:**815–824.

Brinker, K. G., H. Garner, and J. R. Wright. 2003. Surfactant protein A modulates the differentiation of murine bone marrow-derived dendritic cells. *Am. J. Physiol.* **284:**L232–L241.

Brown, G. D., J. Herre, D. L. Williams, J. A. Willment, A. S. Marshall, and S. Gordon. 2003. Dectin-1 mediates the biological effects of beta-glucans. *J. Exp. Med.* **197:**1119–1124.

Cambi, A., F. de Lange, N. M. van Maarseveen, M. Nijhuis, B. Joosten, E. M. van Dijk, B. I. de Bakker, J. A. Fransen, P. H. Bovee-Geurts, F. N. van Leeuwen, N. F. Van Hulst, and C. G. Figdor. 2004. Microdomains of the C-type lectin DC-SIGN are portals for virus entry into dendritic cells. *J. Cell Biol.* **164:**145–155.

Cambi, A., and C. G. Figdor. 2005. Levels of complexity in pathogen recognition by C-type lectins. *Curr. Opin. Immunol.* **17:**345–351.

Cambi, A., K. Gijzen, J. M. de Vries, R. Torensma, B. Joosten, G. J. Adema, M. G. Netea, B. J. Kullberg, L. Romani, and C. G. Figdor. 2003. The C-type lectin DC-SIGN (CD209) is an antigen-uptake receptor for Candida albicans on dendritic cells. *Eur. J. Immunol.* **33:**532–538.

Cambi, A., M. Koopman, and C. G. Figdor. 2005. How C-type lectins detect pathogens. *Cell Microbiol.* **7:**481–488.

Cambi, A., D. S. Lidke, D. J. Arndt-Jovin, C. G. Figdor, and T. M. Jovin. 2007. Ligand-conjugated quantum dots monitor antigen uptake and processing by dendritic cells. *Nano Lett.* **7:**970–977.

Caminschi, I., A. J. Corbett, C. Zahra, M. Lahoud, K. M. Lucas, M. Sofi, D. Vremec, T. Gramberg, S. Pöhlmann, J. Curtis, E. Handman, S. L. van Dommelen, P. Fleming, M. A. Degli-Esposti, K. Shortman, and M. D. Wright. 2006. Functional comparison of mouse CIRE/mouse DC-SIGN and human DC-SIGN. *Int. Immunol.* **18:**741–753.

Caparros, E., P. Munoz, E. Sierra-Filardi, D. Serrano-Gomez, A. Puig-Kroger, J. L. Rodriguez-Fernandez, M. Mellado, J. Sancho, M. Zubiaur, and A. L. Corbí. 2006. DC-SIGN ligation on dendritic cells results in ERK and PI3K activation and modulates cytokine production. *Blood* **107:**3950–3958.

Celis, E., and T. W. Chang. 1984. Antibodies to hepatitis B surface antigen potentiate the response of human T lymphocyte clones to the same antigen. *Science* **224:**297–299.

Chang, T. W. 1985. Regulation of immune response by antibodies: the importance of antibody and monocyte Fc receptor interaction in T-cell activation. *Immunol. Today* **6:**245–249.

Chen, C. H., H. Floyd, N. E. Olson, D. Magaletti, C. Li, K. Draves, and E. A. Clark. 2006. Dendritic-cell-associated C-type lectin 2 (DCAL-2) alters dendritic-cell maturation and cytokine production. *Blood* **107:**1459–1467.

Chieppa, M., G. Bianchi, A. Doni, A. Del Prete, M. Sironi, G. Laskarin, P. Monti, L. Piemonti, A. Biondi, A. Mantovani, M. Introna, and P. Allavena. 2003. Cross-linking of the mannose receptor on monocyte-derived dendritic cells activates an anti-inflammatory immunosuppressive program. *J. Immunol.* **171:**4552–4560.

Crocker, P. R., and A. Varki. 2001. Siglecs, sialic acids and innate immunity. *Trends Immunol.* **22:**337-342.

de Witte, L., A. Nabatov, M. Pion, D. Fluitsma, M. A. de Jong, T. de Gruijl, V. Piguet, Y. van Kooyk, and T. B. Geijtenbeek. 2007. Langerin is a natural barrier to HIV-1 transmission by Langerhans cells. *Nat. Med.* **13:**367–371.

Dietl, J., A. Honig, U. Kammerer, and L. Rieger. 2006. Natural killer cells and dendritic cells at the human fetomaternal interface: an effective cooperation? *Placenta* **27:**341–347.

Dong, X., W. J. Storkus, and R. D. Salter. 1999. Binding and uptake of agalactosyl IgG by mannose receptor on macrophages and dendritic cells. *J. Immunol.* **163:**5427–5434.

Drickamer, K. 1988. Two distinct classes of carbohydrate-recognition domains in animal lectins. *J. Biol. Chem.* **263:** 9557–9560.

Drickamer, K. 1989. Demonstration of carbohydrate-recognition activity in diverse proteins which share a common primary structure motif. *Biochem. Soc. Trans.* **17:**13–15.

Drickamer, K. 1992 Engineering galactose-binding activity into a C-type mannose-binding protein. *Nature* **360:**183–186.

Drickamer, K. 1999. C-type lectin-like domains. *Curr. Opin. Struct. Biol.* **9:**585–590.

Drickamer, K., and M. E. Taylor. 2002. Glycan arrays for functional glycomics. *Genome Biol.* **3:**REVIEWS1034.

Dzionek, A., Y. Sohma, J. Nagafune, M. Cella, M. Colonna, F. Facchetti, G. Günther, I. Johnston, A. Lanzavecchia, T. Nagasaka, T. Okada, W. Vermi, G. Winkels, T. Yamamoto, M. Zysk, Y. Yamaguchi, and J. Schmitz. 2001. BDCA-2, a novel plasmacytoid dendritic cell-specific type II C-type lectin, mediates antigen capture and is a potent inhibitor of interferon alpha/beta induction. *J. Exp. Med.* **194:**1823–1834.

Engering, A., T. B. Geijtenbeek, and Y. van Kooyk. 2002a. Immune escape through C-type lectins on dendritic cells. *Trends Immunol.* **23:** 480–485.

Engering, A., T. B. Geijtenbeek, S. J. van Vliet, M. Wijers, E. van Liempt, N. Demaurex, A. Lanzavecchia, J. Fransen, C. G. Figdor, V. Piguet, and Y. van Kooyk. 2002b. The dendritic cell-specific adhesion receptor DC-SIGN internalizes antigen for presentation to T cells. *J. Immunol.* **168:** 2118–2126.

Engering, A., S. J. van Vliet, K. Hebeda, D. G. Jackson, R. Prevo, S. K. Singh, T. B. Geijtenbeek, H. van Krieken, and Y. van Kooyk. 2004. Dynamic populations of dendritic cell-specific ICAM-3 grabbing nonintegrin-positive immature dendritic cells and liver/lymph node-specific ICAM-3 grabbing nonintegrin-positive endothelial cells in the outer zones of the paracortex of human lymph nodes. *Am. J. Pathol.* **164:** 1587–1595.

Ezekowitz, R. A., K. Sastry, P. Bailly, and A. Warner. 1990. Molecular characterization of the human macrophage mannose receptor: demonstration of multiple carbohydrate recognition-like domains and phagocytosis of yeasts in Cos-1 cells. *J. Exp. Med.* **172:**1785–1794.

Feinberg, H., Y. Guo, D. A. Mitchell, K. Drickamer, and W. I. Weis. 2005. Extended neck regions stabilize tetramers of the receptors DC-SIGN and DC-SIGNR. *J. Biol. Chem.* **280:**1327–1335.

Feizi, T., and W. Chai. 2004. Oligosaccharide microarrays to decipher the glyco code. *Nat. Rev. Mol. Cell. Biol.* **5:**582–588.

Figdor, C. G., Y. van Kooyk, and G. J. Adema. 2002. C-type lectin receptors on dendritic cells and Langerhans cells. *Nat. Rev. Immunol.* **2:**77–84.

Galustian, C., C. G. Park, W. Chai, M. Kiso, S. A. Bruening, Y. S. Kang, R. M. Steinman, and T. Feizi. 2004. High and low affinity carbohydrate ligands revealed for murine SIGN-R1 by carbohydrate array and cell binding approaches, and differing specificities for SIGN-R3 and langerin. *Int. Immunol.* **16:**853–866.

Geijtenbeek, T. B., D. J. Krooshoop, D. A. Bleijs, S. J. van Vliet, G. C. van Duijnhoven, V. Grabovsky, R. Alon, C. G. Figdor, and Y. van Kooyk. 2000a. DC-SIGN-ICAM-2 interaction mediates dendritic cell trafficking. *Nat. Immunol.* **1:**353–357.

Geijtenbeek, T. B., D. S. Kwon, R. Torensma, S. J. van Vliet, G. C. van Duijnhoven, J. Middel, I. L. Cornelissen, H. S. Nottet, V. N. KewalRamani, D. R. Littman, C. G. Figdor, and Y. van Kooyk. 2000b. DC-SIGN, a dendritic cell-specific HIV-1-binding protein that enhances trans-infection of T cells. *Cell* **100:**587–597.

Geijtenbeek, T. B., R. Torensma, S. J. van Vliet, G. C. van Duijnhoven, G. J. Adema, Y. van Kooyk, and C. G. Figdor. 2000c. Identification of DC-SIGN, a novel dendritic cell-specific ICAM-3 receptor that supports primary immune responses. *Cell* **100:**575–585.

Geijtenbeek, T. B., S. J. Van Vliet, E. A. Koppel, M. Sanchez-Hernandez, C. M. Vandenbroucke-Grauls, B. Appelmelk, and Y. Van Kooyk. 2003. Mycobacteria target DC-SIGN to suppress dendritic cell function. *J. Exp. Med.* **197:** 7–17.

Gramberg, T., I. Caminschi, A. Wegele, H. Hofmann, and S. Pohlmann. 2006. Evidence that multiple defects in murine DC-SIGN inhibit a functional interaction with pathogens. *Virology* **345:**482–491.

Granelli-Piperno, A., A. Pritsker, M. Pack, I. Shimeliovich, J. F. Arrighi, C. G. Park, C. Trumpfheller, V. Piguet, T. M. Moran, and R. M. Steinman. 2005. Dendritic cell-specific intercellular adhesion molecule 3-grabbing nonintegrin/CD209 is abundant on macrophages in the normal human lymph node and is not required for dendritic cell stimulation of the mixed leukocyte reaction. *J. Immunol.* **175:**4265–4273.

Green, P. J., T. Tamatani, T. Watanabe, M. Miyasaka, A. Hasegawa, M. Kiso, C. T. Yuen, M. S. Stoll, and T. Feizi. 1992. High affinity binding of the leucocyte adhesion molecule L-selectin to 3'-sulphated-Le(a) and -Le(x) oligosaccharides and the predominance of sulphate in this interaction demonstrated by binding studies with a series of lipid-linked oligosaccharides. *Biochem. Biophys. Res. Commun.* **188:**244–251.

Gringhuis, S. I., J. den Dunnen, M. Litjens, B. van Het Hof, Y. van Kooyk, and T. B. Geijtenbeek. 2007. C-Type lectin DC-SIGN modulates toll-like receptor signaling via Raf-1 kinase-dependent acetylation of transcription factor NF-kappaB. *Immunity* **26:**605–616.

Guo, Y., H. Feinberg, E. Conroy, D. A. Mitchell, R. Alvarez, O. Blixt, M. E. Taylor, W. I. Weis, and K. Drickamer. 2004. Structural basis for distinct ligand-binding and targeting properties of the receptors DC-SIGN and DC-SIGNR. *Nat. Struct. Mol. Biol.* **11:**591–598.

Hasslen, S. R., U. H. von Andrian, E.C. Butcher, R. D. Nelson, and S. L. Erlandsen. 1995. Spatial distribution of L-selectin (CD62L) on human lymphocytes and transfected murine L1-2 cells. *Histochem. J.* **27:** 547–554.

Hawiger, D., K. Inaba, Y. Dorsett, M. Guo, K. Mahnke, M. Rivera, J. V. Ravetch, R. M. Steinman, and M. C. Nussenzweig. 2001. Dendritic cells induce peripheral T cell unresponsiveness under steady state conditions in vivo. *J. Exp. Med.* **194:**769–779.

He, L. Z., V. Ramakrishna, J. E. Connolly, X. T. Wang, P. A. Smith, C. L. Jones, M. Valkova-Valchanova, A. Arunakumari, J. F. Treml, J. Goldstein, P. K. Wallace, T. Keler, and M. J. Endres. 2004. A novel human cancer vaccine elicits cellular responses to the tumor-associated antigen, human chorionic gonadotropin beta. *Clin. Cancer Res.* **10:** 1920–1927.

Herre, J., A. S. Marshall, E. Caron, A. D. Edwards, D. L. Williams, E., Schweighoffer, V. Tybulewicz, C. Reis e Sousa, S. Gordon, G. D. Brown. 2004. Dectin-1 uses novel mechanisms for yeast phagocytosis in macrophages. *Blood* **104:**4038–4045.

Hunger, R. E., P. A. Sieling, M. T. Ochoa, M. Sugaya, A. E. Burdick, T. H. Rea, P. J. Brennan, J. T. Belisle, A. Blauvelt, S. A. Porcelli, and R. L. Modlin. 2004. Langerhans cells utilize CD1a and langerin to efficiently present nonpeptide antigens to T cells. *J. Clin. Invest.* **113:**701–708.

Inaba, K., W. J. Swiggard, M. Inaba, J. Meltzer, A. Mirza, T. Sasagawa, M. C. Nussenzweig, and R. M. Steinman. 1995. Tissue distribution of the DEC-205 protein that is detected by the monoclonal antibody NLDC-145. I. Expression on dendritic cells and other subsets of mouse leukocytes. *Cell Immunol.* **163:**148–156.

Irjala, H., E. L. Johansson, R. Grenman, K. Alanen, M. Salmi, and S. Jalkanen. 2001. Mannose receptor is a novel ligand for L-selectin and mediates lymphocyte binding to lymphatic endothelium. *J. Exp. Med.* **194:**1033–1042.

Kammerer, U., A. O. Eggert, M. Kapp, A. D. McLellan, T. B. Geijtenbeek, J. Dietl, Y. van Kooyk, and E. Kämpgen. 2003. Unique appearance of proliferating antigen-presenting cells expressing DC-SIGN (CD209) in the decidua of early human pregnancy. *Am. J. Pathol.* **162:**887–896.

Kanazawa, N., T. Okazaki, H. Nishimura, K. Tashiro, K. Inaba, and Y. Miyachi. 2002. DCIR acts as an inhibitory receptor depending on its immunoreceptor tyrosine-based inhibitory motif. *J. Invest. Dermatol.* **118:**261–266.

Karanikas, V., L. A. Hwang, J. Pearson, C. S. Ong, V. Apostolopoulos, H. Vaughan, P. X. Xing, G. Jamieson, G. Pietersz, B. Tait, R. Broadbent, G. Thynne, and I. F. McKenzie. 1997. Antibody and T cell responses of patients with adenocarcinoma immunized with mannan-MUC1 fusion protein. *J. Clin. Invest.* **100:**2783–2792.

Kato, M., K. J. McDonald, S. Khan, I. L. Ross, S. Vuckovic, K. Chen, D. Munster, K. P. MacDonald, and D. N. Hart. 2006. Expression of human DEC-205 (CD205) multilectin receptor on leukocytes. *Int. Immunol.* **18:**857–869.

Keler, T., P. M. Guyre, L. A. Vitale, K. Sundarapandiyan, J. G. van De Winkel, Y. M. Deo, and R. F. Graziano. 2000. Targeting weak antigens to CD64 elicits potent humoral responses in human CD64 transgenic mice. *J. Immunol.* **165:**6738–6742.

Kiely, J. M., Y. Hu, G. Garcia-Cardena, and M. A. Gimbrone, Jr. 2003. Lipid raft localization of cell surface E-selectin is required for ligation-induced activation of phospholipase C gamma. *J. Immunol.* **171:**3216–3224.

Kogelberg, H., and T. Feizi. 2001. New structural insights into lectin-type proteins of the immune system. *Curr. Opin. Struct. Biol.* **11:**635–643.

Koppel, E. A., K. P. van Gisbergen, T. B. Geijtenbeek, and Y. van Kooyk. 2005. Distinct functions of DC-SIGN and its homologues L-SIGN (DC-SIGNR) and mSIGNR1 in pathogen recognition and immune regulation. *Cell Microbiol.* **7:**157–165.

Lajaunias, F., J. M. Dayer, and C. Chizzolini. 2005. Constitutive repressor activity of CD33 on human monocytes requires sialic acid recognition and phosphoinositide 3-kinase-mediated intracellular signaling. *Eur J Immunol.* **35:** 243-251.

Lanoue, A., M. R. Clatworthy, P. Smith, S. Green, M. J. Townsend, H. E. Jolin, K. G. Smith, P. G. Fallon, A. N. McKenzie. 2004. SIGN-R1 contributes to protection against lethal pneumococcal infection in mice. *J. Exp. Med.* **200:** 1383–1393.

Laskarin, G., K. Cupurdija, V. S. Tokmadzic, D. Dorcic, J. Dupor, K. Juretic, N. Strbo, T. B. Crncic, F. Marchezi, P. Allavena, A. Mantovani, L. J. Randic, and D. Rukavina. 2005. The presence of functional mannose receptor on macrophages at the maternal-fetal interface. *Hum. Reprod.* **20:** 1057–1066.

Leteux, C., W. Chai, R. W. Loveless, C. T. Yuen, L. Uhlin-Hansen, Y. Combarnous, M. Jankovic, S. C. Maric, Z. Misulovin, M. C. Nussenzweig, and T. Feizi. 2000. The cysteine-rich domain of the macrophage mannose receptor is a multispecific lectin that recognizes chondroitin sulfates A and B and sulfated oligosaccharides of blood group Lewis(a) and Lewis(x) types in addition to the sulfated N-glycans of lutropin. *J. Exp. Med.* **191:**1117–1126.

Ludwig, I. S., A. N. Lekkerkerker, E. Depla, F. Bosman, R. J. Musters, S. Depraetere, Y. van Kooyk, and T. B. Geijtenbeek. 2004. Hepatitis C virus targets DC-SIGN and L-SIGN to escape lysosomal degradation. *J. Virol.* **78:**8322–8332.

Mahnke, K., M. Guo, S. Lee, H. Sepulveda, S. L. Swain, M. Nussenzweig, and R. M. Steinman. 2000. The dendritic cell receptor for endocytosis, DEC-205, can recycle and enhance antigen presentation via major histocompatibility complex class II-positive lysosomal compartments. *J. Cell Biol.* **151:**673–684.

Mahnke, K., Y. Qian, S. Fondel, J. Brueck, C. Becker, and A. H. Enk. 2005. Targeting of antigens to activated dendritic cells in vivo cures metastatic melanoma in mice. *Cancer Res.* **65:**7007–7012.

Mahnke, K., Y. Qian, J. Knop, and A. H. Enk. 2003. Induction of CD4+/CD25+ regulatory T cells by targeting of antigens to immature dendritic cells. *Blood* **101:**4862–4869.

Martinez, O., S. Brackenridge, A. El-Idrissi Mel, and B. S. Prabhakar. 2005. DC-SIGN, but not sDC-SIGN, can modulate IL-2 production from PMA- and anti-CD3-stimulated primary human CD4 T cells. *Int. Immunol.* **17:**769–778.

Matzinger, P. 2002. The danger model: a renewed sense of self. *Science* **296:**301–305.

McGreal, E. P., L. Martinez-Pomares, and S. Gordon. 2004. Divergent roles for C-type lectins expressed by cells of the innate immune system. *Mol. Immunol.* **41:**1109–1121.

Medzhitov, R., and C. A. Janeway, Jr. 2002. Decoding the patterns of self and nonself by the innate immune system. *Science* **296:**298–300.

Meyer-Wentrup, F., C. G. Figdor, M. Ansems, P. Brossart, M. D. Wright, G. J. Adema, and A. B. van Spriel. 2007. Dectin-1 interaction with tetraspanin CD37 inhibits IL-6 production. *J. Immunol.* **178:**154–162.

Mitchell, D. A., A. J. Fadden, and K. Drickamer. 2001. A novel mechanism of carbohydrate recognition by the C-type lectins DC-SIGN and DC-SIGNR. Subunit organization and binding to multivalent ligands. *J. Biol. Chem.* **276:**28939–28945.

Moris, A., C. Nobile, F. Buseyne, F. Porrot, J. P. Abastado, and O. Schwartz. 2004. DC-SIGN promotes exogenous MHC-I-restricted HIV-1 antigen presentation. *Blood* **103:** 2648–2654.

Mueller, J. P., M. A. Giannoni, S. L. Hartman, E. A. Elliott, S. P. Squinto, L. A. Matis, and M. J. Evans. 1997. Humanized porcine VCAM-specific monoclonal antibodies with chimeric IgG2/G4 constant regions block human leukocyte binding to porcine endothelial cells. *Mol. Immunol.* **34:**441–452.

Nagaoka, K., K. Takahara, K. Tanaka, H. Yoshida, R. M. Steinman, S. Saitoh, S. Akashi-Takamura, K. Miyake, Y. S. Kang, C. G. Park, and K. Inaba. 2005. Association of SIGNR1 with TLR4-MD-2 enhances signal transduction by recognition of LPS in gram-negative bacteria. *Int. Immunol.* **17:**827–836.

Nguyen, D. G., and J. E. Hildreth. 2003. Involvement of macrophage mannose receptor in the binding and transmission of HIV by macrophages. *Eur. J. Immunol.* **33:**483–493.

Nigou, J., C. Zelle-Rieser, M. Gilleron, M. Thurnher, and G. Puzo. 2001. Mannosylated lipoarabinomannans inhibit IL-12 production by human dendritic cells: evidence for a negative signal delivered through the mannose receptor. *J. Immunol.* **166:**7477–7485.

Okamura, Y., M. Watari, E. S. Jerud, D. W. Young, S. T. Ishizaka, J. Rose, J. C. Chow, and J. F. Strauss III. 2001. The extra domain A of fibronectin activates Toll-like receptor 4. *J. Biol. Chem.* **276:**10229–10233.

Ramakrishna, V., J. F. Treml, L. Vitale, J. E. Connolly, T. O'Neill, P. A. Smith, C. L. Jones, L. Z. He, J. Goldstein,

P. K. Wallace, T. Keler, and M. J. Endres. 2004. Mannose receptor targeting of tumor antigen pmel17 to human dendritic cells directs anti-melanoma T cell responses via multiple HLA molecules. *J. Immunol.* **172:**2845–2852.

Rieger, L., A. Honig, M. Sutterlin, M. Kapp, J. Dietl, P. Ruck, and U. Kammerer. 2004. Antigen-presenting cells in human endometrium during the menstrual cycle compared to early pregnancy. *J. Soc. Gynecol. Investig.* **11:**488–493.

Rifkin, I. R., E. A. Leadbetter, L. Busconi, G. Viglianti, and A. Marshak-Rothstein. 2005. Toll-like receptors, endogenous ligands, and systemic autoimmune disease. *Immunol. Rev.* **204:**27–42.

Rogers, N. C., E. C. Slack, A. D. Edwards, M. A. Nolte, O. Schulz, E. Schweighoffer, D. L. Williams, S. Gordon, V. L. Tybulewicz, G. D. Brown, and C. Reis e Sousa. 2005. Syk-dependent cytokine induction by Dectin-1 reveals a novel pattern recognition pathway for C type lectins. *Immunity* **22:**507–517.

Ryan, E. J., A. J. Marshall, D. Magaletti, H. Floyd, K. E. Draves, N. E. Olson, and E. A. Clark. 2002. Dendritic cell-associated lectin-1: a novel dendritic cell-associated, C-type lectin-like molecule enhances T cell secretion of IL-4. *J. Immunol.* **169:**5638–5648.

Saeland, E., S. J. van Vliet, M. Backstrom, V. C. van den Berg, T. B. Geijtenbeek, G. A. Meijer, and Y. van Kooyk. 2007. The C-type lectin MGL expressed by dendritic cells detects glycan changes on MUC1 in colon carcinoma. *Cancer Immunol. Immunother.* **56:**1225–1236.

Saevarsdottir, S., T. Vikingsdottir, and H. Valdimarsson. 2004. The potential role of mannan-binding lectin in the clearance of self-components including immune complexes. *Scand. J. Immunol.* **60:**23–29.

Sallusto, F., M. Cella, C. Danieli, and A. Lanzavecchia. 1995. Dendritic cells use macropinocytosis and the mannose receptor to concentrate macromolecules in the major histocompatibility complex class II compartment: downregulation by cytokines and bacterial products. *J. Exp. Med.* **182:**389–400.

Sato, M., H. Sano, D. Iwaki, K. Kudo, M. Konishi, H. Takahashi, T. Takahashi, H. Imaizumi, Y. Asai, and Y. Kuroki. 2003. Direct binding of Toll-like receptor 2 to zymosan, and zymosan-induced NF-kappa B activation and TNF-alpha secretion are down-regulated by lung collectin surfactant protein A. *J. Immunol.* **171:**417–425.

Schaeffer, L. M., F. X. McCormack, H. Wu, and A. A. Weiss. 2004. Bordetella pertussis lipopolysaccharide resists the bactericidal effects of pulmonary surfactant protein A. *J. Immunol.* **173:**1959–1965.

Schagat, T. L., J. A. Wofford, K. E. Greene, and J. R. Wright. 2003. Surfactant protein A differentially regulates peripheral and inflammatory neutrophil chemotaxis. *Am. J. Physiol.* **284:**L140–L147.

Shi, L., K. Takahashi, J. Dundee, S. Shahroor-Karni, S. Thiel, J. C. Jensenius, F. Gad, M. R. Hamblin, K. N. Sastry, and R. A. Ezekowitz. 2004. Mannose-binding lectin-deficient mice are susceptible to infection with Staphylococcus aureus. *J. Exp. Med.* **199:**1379–1390.

Simons, K., and D. Toomre. 2000. Lipid rafts and signal transduction. *Nat. Rev. Mol. Cell. Biol.* **1:**31–39.

Snyder, G. A., M. Colonna, and P. D. Sun. 2005a. The structure of DC-SIGNR with a portion of its repeat domain lends insights to modeling of the receptor tetramer. *J. Mol. Biol.* **347:**979–989.

Snyder, G. A., J. Ford, P. Torabi-Parizi, J. A. Arthos, P. Schuck, M. Colonna, and P. D. Sun. 2005b. Characterization of DC-SIGN/R interaction with human immunodeficiency virus type 1 gp120 and ICAM molecules favors the receptor's role as an antigen-capturing rather than an adhesion receptor. *J. Virol.* **79:**4589–4598.

Soilleux, E. J., R. Barten, and J. Trowsdale. 2000. DC-SIGN; a related gene, DC-SIGNR; and CD23 form a cluster on 19p13. *J. Immunol.* **165:**2937–2942.

Soilleux, E. J., L. S. Morris, G. Leslie, J. Chehimi, Q. Luo, E. Levroney, J. Trowsdale, L. J. Montaner, R. W. Doms, D. Weissman, N. Coleman, and B. Lee. 2002. Constitutive and induced expression of DC-SIGN on dendritic cell and macrophage subpopulations in situ and in vitro. *J. Leukoc. Biol.* **71:**445–457.

Soong, G., B. Reddy, S. Sokol, R. Adamo, and A. Prince. 2004. TLR2 is mobilized into an apical lipid raft receptor complex to signal infection in airway epithelial cells. *J. Clin. Invest.* **113:**1482–1489.

Tacken, P. J., I. J. de Vries, K. Gijzen, B. Joosten, D. Wu, R. P. Rother, S. J. Faas, C. J. Punt, R. Torensma, G. J. Adema, and C. G. Figdor. 2005. Effective induction of naive and recall T-cell responses by targeting antigen to human dendritic cells via a humanized anti-DC-SIGN antibody. *Blood* **106:**1278–1285.

Takeda, K., and S. Akira. 2005. Toll-like receptors in innate immunity. *Int. Immunol.* **17:**1–14.

Takeda, K., T. Kaisho, and S. Akira. 2003. Toll-like receptors. *Annu. Rev. Immunol.* **21:**335–376.

Taylor, M. E., K. Bezouska, and K. Drickamer. 1992. Contribution to ligand binding by multiple carbohydrate-recognition domains in the macrophage mannose receptor. *J. Biol. Chem.* **267:**1719–1726.

Taylor, P. R., S. Gordon, and L. Martinez-Pomares. 2005. The mannose receptor: linking homeostasis and immunity through sugar recognition. *Trends Immunol.* **26:**104–110.

Triantafilou, M., and K. Triantafilou. 2003. Receptor cluster formation during activation by bacterial products. *J. Endotoxin Res.* **9:**331–335.

Triantafilou, M., K. Miyake, D. T. Golenbock, and K. Triantafilou. 2002. Mediators of innate immune recognition of bacteria concentrate in lipid rafts and facilitate lipopolysaccharide-induced cell activation. *J. Cell Sci.* **115:**2603–2611.

Underhill, D. M. 2003. Toll-like receptors: networking for success. *Eur. J. Immunol.* **33:**1767–1775.

van Broekhoven, C. L., C. R. Parish, C. Demangel, W. J. Britton, and J. G. Altin. 2004. Targeting dendritic cells with antigen-containing liposomes: a highly effective procedure for induction of antitumor immunity and for tumor immunotherapy. *Cancer Res.* **64:**4357–4365.

van de Wetering, J. K., L. M. van Golde, and J. J. Batenburg. 2004a. Collectins: players of the innate immune system. *Eur. J. Biochem.* **271:**1229–1249.

van de Wetering, J. K., F. E. Coenjaerts, A. B. Vaandrager, L. M. van Golde, and J. J. Batenburg. 2004b. Aggregation of Cryptococcus neoformans by surfactant protein D is inhibited by its capsular component glucuronoxylomannan. *Infect. Immun.* **72:**145–153.

van Gisbergen, K. P., C. A. Aarnoudse, G. A. Meijer, T. B. Geijtenbeek, and Y. van Kooyk. 2005a. Dendritic cells recognize tumor-specific glycosylation of carcinoembryonic antigen on colorectal cancer cells through dendritic cell-specific intercellular adhesion molecule-3-grabbing nonintegrin. *Cancer Res.* **65:**5935–5944.

van Gisbergen, K. P., I. S. Ludwig, T. B. Geijtenbeek, and Y. van Kooyk. 2005b. Interactions of DC-SIGN with Mac-1 and CEACAM1 regulate contact between dendritic cells and neutrophils. *FEBS Lett.* **579:**6159–6168.

van Gisbergen, K. P., M. Sanchez-Hernandez, T. B. Geijtenbeek, and Y. van Kooyk. 2005c. Neutrophils mediate immune modulation of dendritic cells through glycosylation-dependent interactions between Mac-1 and DC-SIGN. *J. Exp. Med.* **201:**1281–1292.

van Kooyk, Y., A. Engering, A. N. Lekkerkerker, I. S. Ludwig, and T. B. Geijtenbeek. 2004. Pathogens use carbohydrates to escape immunity induced by dendritic cells. *Curr. Opin. Immunol.* **16:**488–493.

van Kooyk, Y., and T. B. Geijtenbeek. 2003. DC-SIGN: escape mechanism for pathogens. *Nat. Rev. Immunol.* **3:**697–709.

Van Liempt, E., A. Imberty, C. M. Bank, S. J. Van Vliet, Y. Van Kooyk, T. B. Geijtenbeek, and I. Van Die. 2004. Molecular basis of the differences in binding properties of the highly related C-type lectins DC-SIGN and L-SIGN to Lewis X trisaccharide and Schistosoma mansoni egg antigens. *J. Biol. Chem.* **279:**33161–33167.

van Liempt, E., S. J. van Vliet, A. Engering, J. J. Garcia Vallejo, C. M. Bank, M. Sanchez-Hernandez, Y. van Kooyk, and I. van Die. 2007. Schistosoma mansoni soluble egg antigens are internalized by human dendritic cells through multiple C-type lectins and suppress TLR-induced dendritic cell activation. *Mol. Immunol.* **44:**2605–2615.

van Vliet, S. J., S. I. Gringhuis, T. B. Geijtenbeek, and Y. van Kooyk. 2006. Regulation of effector T cells by antigen-presenting cells via interaction of the C-type lectin MGL with CD45. *Nat. Immunol.* **7:**1200–1208.

van Vliet, S. J., E. van Liempt, E. Saeland, C. A. Aarnoudse, B. Appelmelk, T. Irimura, T. B. Geijtenbeek, O. Blixt, R. Alvarez, I. van Die, and Y. van Kooyk. 2005. Carbohydrate profiling reveals a distinctive role for the C-type lectin MGL in the recognition of helminth parasites and tumor antigens by dendritic cells. *Int. Immunol.* **17:**661–669.

Weber, K. S., R. Alon, and L. B. Klickstein. 2004. Sialylation of ICAM-2 on platelets impairs adhesion of leukocytes via LFA-1 and DC-SIGN. *Inflammation* **28:**177–188.

Weis, W. I., and K. Drickamer. 1996. Structural basis of lectin-carbohydrate recognition. *Annu. Rev. Biochem.* **65:**441–473.

Weis, W. I., R. Kahn, R. Fourme, K. Drickamer, and W. A. Hendrickson. 1991. Structure of the calcium-dependent lectin domain from a rat mannose-binding protein determined by MAD phasing. *Science* **254:**1608–1615.

Weis, W. I., M. E. Taylor, and K. Drickamer. 1998. The C-type lectin superfamily in the immune system. *Immunol. Rev.* **163:**19–34.

Yang, Z. Y., Y. Huang, L. Ganesh, K. Leung, W. P. Kong, O. Schwartz, K. Subbarao, and G. J. Nabel. 2004. pH-dependent entry of severe acute respiratory syndrome coronavirus is mediated by the spike glycoprotein and enhanced by dendritic cell transfer through DC-SIGN. *J. Virol.* **78:**5642–5650.

Zhang, J., J. Zhu, A. Imrich, M. Cushion, T. B. Kinane, and H. Koziel. 2004. Pneumocystis activates human alveolar macrophage NF-kappaB signaling through mannose receptors. *Infect. Immun.* **72:**3147–3160.

8
Integrins on Phagocytes

WOUTER L. W. HAZENBOS AND ERIC J. BROWN

All living organisms, including unicellular organisms, must respond to the extracellular environment to survive. For single-cell species, the extracellular cues are most often soluble molecules, including carbon sources, chemoattractants, and other molecules that will regulate the organism's location, metabolism, and gene regulation to optimize its growth. In metazoans, in addition to soluble molecules, a cell's environment contains spatial cues from both neighboring cells and the extracellular matrix (ECM) in addition to soluble mediators of signal transduction. Information from these cellular components and insoluble extracellular molecules is essential for normal development as well as for responses to perturbations of homeostasis, and is transmitted through plasma membrane molecules specialized to report both cell-cell interactions and the nature of the ECM. These receptors collectively are called adhesion receptors and can be structurally quite diverse. Adhesion receptors can be found among the immunoglobulin, lectin, integrin, and receptor tyrosine kinase superfamilies.

In the immune system, in addition to their roles in development and cell-cell communication within lymphoid organs, adhesion receptors are involved in leukocyte migration to sites of inflammation and infection, as well as in normal recirculation, pathogen recognition and response, phagocytosis, and activation of effector functions. For the functions involving surveillance and response to inflammatory and infectious stimuli, the integrins are the best studied of the adhesion receptors. At least two, and very often more, members of this family of heterodimeric membrane receptors are expressed on every immune cell. One subfamily of integrins that share a common β-chain, the β2 integrins, is expressed exclusively on leukocytes (Table 1). A genetic deficiency leading to absence or dramatic reduction in expression of this integrin family, known as leukocyte adhesion deficiency type I (LAD-I), causes a markedly increased propensity to infection (reviewed in Bunting et al. [2002]). A major characteristic of LAD-I is the failure to accumulate neutrophils at the site of infection. This observation led to the discovery of important roles for β2 integrins in leukocyte migration and phagocyte activation, as will be detailed below. In this chapter, we will review the expression, signaling mechanisms, and functions of integrins on phagocytes, summarizing their central roles in host defense and inflammatory processes.

INTEGRIN STRUCTURE AND EXPRESSION ON PHAGOCYTES

Integrin Structure

All integrins are heterodimers consisting of two type 1 membrane proteins intimately associated with each other (Fig. 1A). The α-chain of the heterodimer is 1 of 18 proteins; the β-chain is 1 of 8. Promiscuous pairing allows these gene products to generate 24 integrin heterodimers, with unique expression and ligand binding patterns. Expression of integrins on the plasma membrane requires interaction of the α- and β-chains in the endoplasmic reticulum; unpaired chains are presumably recognized as unfolded and ultimately degraded. There are four structural domains present in the extracellular region of all integrin α-chains (Luo et al., 2007). The most amino-terminal domain forms a β-propeller, related to the structure of the α-chain of heterotrimeric G proteins. Approximately half of all α-chains express an additional domain called I (for "inserted"), which is in close proximity to the β-propeller domain. In I domain-containing integrins, ligand recognition generally requires this domain. All integrin β-chains contain a domain homologous to the α-chain I domain, called I-like, which also is important for ligand recognition. The extracellular region of all integrin β-chains also includes four epidermal growth factor-like repeats. The cytoplasmic domains of integrin α- and β-chains are generally short, most often 50 amino acids or less; the exception to this rule is the cytoplasmic domain of integrin β4, a chain expressed only on epithelium, which is 1,000 amino acids (Suzuki and Naitoh, 1990). Despite the absence of enzymatic function, the cytoplasmic domains of the integrins are key to their function, as will be discussed in detail be-

---

Wouter L. W. Hazenbos and Eric J. Brown, Department of Microbial Pathogenesis, Genentech, Inc., 1 DNA Way, M.S. 33, South San Francisco, CA 94080.

TABLE 1 Integrins on phagocytes[a]

Phagocyte integrin	Natural ligands	Microbial ligands
αLβ2 (LFA-1; CD11a/CD18)	ICAM-1, -2, -3, -5	
αMβ2 (Mac-1; CR3; CD11b/CD18)	iC3b, fibrinogen	Lipophosphoglycan (L); FHA (BP); LPS
αDβ2	ICAM-3, VCAM-1	Unknown
αXβ2 (p150/95; CR4; CD11c/CD18)	iC3b, fibrinogen	
α1β1 (VLA-1; CD49a/CD29)	Collagen	Invasin (Y); intimin (EC)
α2β1 (VLA-2; CD49b/CD29)	Collagen	
α3β1 (VLA-3; CD49c/CD29)	Collagen, laminin	FimH (EC)
α4β1 (VLA-4; CD49d/CD29)	Fibronectin; VCAM-1	
α5β1 (VLA-5; CD49e/CD29)	Fibronectin	Phosphatidylinositol mannoside (MT); FimD (BP)
α6β1 (VLA-6; CD49f/CD29)	Laminin	
α9β1	VCAM-1, tenascin C, osteopontin, vWf, factor XIII	
αVβ3	Vitronectin, fibrinogen, other RGD-containing proteins	FHA (BP); *Mycobacterium avium*
αVβ5	Similar to αVβ3	
αVβ8	TGF-β	

[a] Abbreviations: BP, *Bordetella pertussis*; EC, *Escherichia coli*; FHA, filamentous hemagglutinin; ICAM, intercellular adhesion molecule; L, *Leishmania*; LFA-1, lymphocyte function-associated antigen; MT, *Mycobacterium tuberculosis*; TGF, transforming growth factor; VCAM, vascular cell adhesion molecule; VLA, very late antigen; Y, *Yersinia* spp.

low. These cytoplasmic domains contain sites for interactions with the cytoskeleton, and are thus are the basis for the name "integrin," which is meant to designate that these molecules integrate the ECM with the intracellular matrix, or cytoskeleton. The cytoplasmic domains also make direct or indirect connections to a variety of signaling cascades that are activated by integrin ligation, a feature discussed further later in the chapter.

Integrin Expression on Phagocytes

At least 14 different integrins can be expressed on phagocytes (Table 1). The first integrins to be recognized on phagocytic cells were in the β2 family, and were initially identified by using antibodies made against cell surface structures. The β2 family consists of four integrins (Table 1), all expressed exclusively on leukocytes. Both αLβ2 and αMβ2 are broadly expressed on phagocytic cells, including neutrophils, macrophages, and myeloid dendritic cells. Expression of αXβ2 differs somewhat between mice and humans. In mice, phagocyte expression is confined to dendritic cells and alveolar macrophages, while in humans there is much broader expression, including neutrophils and tissue macrophages, in general. The αDβ2 integrin has the narrowest expression in the family, present only on a subset of splenic macrophages, and also has the narrowest spectrum of ligands, recognizing only intercellular adhesion molecule-3 (ICAM-3) and vascular cell adhesion molecule-1 (VCAM-1), two members of the immunoglobulin superfamily (IgSF) also recognized by other integrins.

Mutations in the β2 gene that lead to severe reduction in expression of the entire family cause LAD-I, a severe immunodeficiency. LAD-I has been described in humans, cattle, and dogs, but there have been no reports of functionally significant genetic mutations of individual α-chains. This has led to the speculation that the functions of the individual α-chains may be sufficiently overlapping that there is no clinically significant defect in the absence of a single α-integrin. To test this hypothesis, mouse strains lacking αL (CD11a), αM (CD11b), αX (CD11c), and αD (CD11d) have been made. None is as severely immunodeficient as the β2 knockout mouse, but subtle defects in αL-, αM-, and αD-deficient mice have been reported, especially in response to infectious challenge (Guerau-de-Arellano et al., 2005; Hynes, 2002; Prince et al., 2001). To date, there have been no reports of abnormalities in the αX$^{-/-}$ mouse strain.

Some of the β2 integrins on neutrophils are present in secretory granules rather than at the plasma membrane. Granules containing β2 integrins have characteristics of both secondary and tertiary granules and can be rapidly mobilized to the plasma membrane when neutrophils encounter inflammatory stimuli, such as peptides of bacterial origin, activated complement components, or various products of arachidonate metabolism, such as leukotriene B4. Interestingly, while as much as 90% of neutrophil αMβ2 can be present in secretory granules, virtually all of αLβ2 is constitutively expressed at the plasma membrane. The molecular basis for the difference in intracellular trafficking of these two closely related integrins is not known.

At least seven different β1 integrins are expressed by phagocytes. Among the β1-associated α-chains, only α1 and α2 have I domains. The β1 integrins frequently recognize ECM components, including fibronectin (ligand for α4β1 and α5β1), laminin (α1β1, α2β1, and α6β1), or collagen (α1β1, α2β1, and α3β1) (reviewed in Springer [1990]). The β1 integrins with highest expression on monocytes, macrophages, and dendritic cells are α4β1 and α5β1, both of which recognize fibronectin. In addition to binding to fibronectin, α4β1 also binds to VCAM-1, an IgSF member expressed on inflamed endothelium; this interaction is necessary for monocyte migration out of the vasculature into inflamed tissue. Neutrophils express fewer β1 integrins than β2, and most of the β1 integrins are in secretory granules prior to neutrophil activation (Singer et

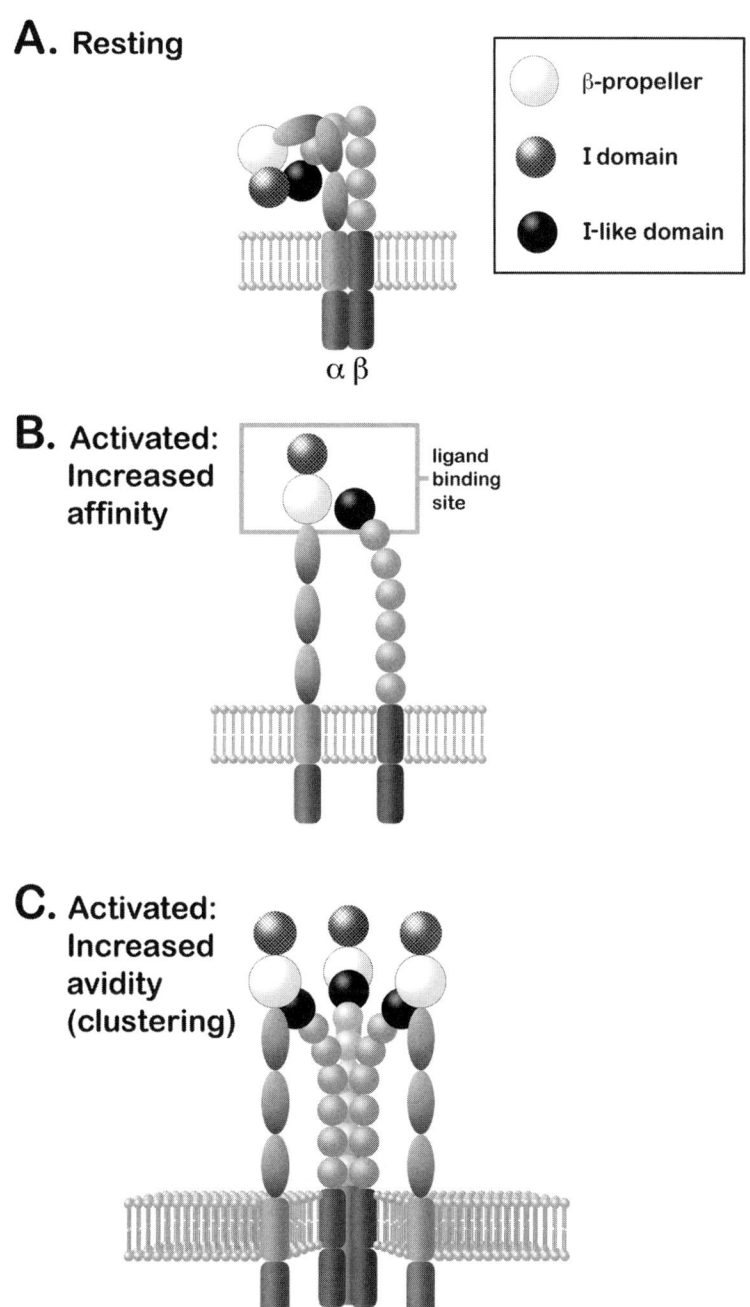

FIGURE 1 Model for integrin activation. In resting state, integrins do not efficiently bind their ligand; activation by inside-out signaling induces a conformational change that dramatically enhances integrin-ligand interactions. (A) Resting, low-affinity state. Structural analyses revealed that the low-affinity state of resting integrins correlates with a "bent" confirmation. The ligand binding site is unavailable for ligand recognition. In this state, integrins likely exist as monomers diffusely distributed on the plasma membrane. (B) Activated, enhanced affinity state. Cell activation induces inside-out signaling, leading to separation of the integrin α- and β-chain cytoplasmic tails. This separation results in an intramolecular conformational change propagated through the plasma membrane, resulting in unbending of the extracellular domains. The ligand binding site becomes exposed and undergoes additional conformational change, enhancing affinity. This conformational change is thought to occur for both I domain-containing and I domain-lacking integrins; for simplicity, only an I domain-containing molecule is shown. (C) Activated, enhanced avidity state. Inside-out signaling can also lead to clustering of multiple integrin molecules, resulting in enhanced avidity for a multivalent ligand. It has been proposed, at least for αIIbβ3, that this clustering involves oligomerization of β-chain transmembrane domains (Li et al., 2003).

al., 1989). However, α4β1 also is important in neutrophil transendothelial migration, at least under some circumstances (Bowden et al., 2002; Ibbotson et al., 2001).

There are three αV integrins expressed on phagocytes: αVβ3, αVβ5, and αVβ8. The αVβ3- and αVβ5 integrins are closely related in ligand binding specificity and in function, but ligation of αVβ3 leads to stronger constitutive association with cytoskeleton than is the case for αVβ5. Many macrophages express αVβ5 and smaller amounts of αVβ3, while αVβ3 expression is upregulated on osteoclasts and dendritic cells. The αVβ3 integrin also is expressed to a small extent on neutrophils, where it apparently can regulate both motility (Hendey et al., 1996) and phagocytosis (Blystone et al., 1999; Van Strijp et al., 1993). Both αVβ5 and αVβ3 on macrophages may act as receptors for apoptotic cells (Albert et al., 2000; Savill et al., 1990), but there is clearly great redundancy in this function, since many other nonintegrin receptors also are involved in clearance of apoptotic cells. It is interesting that granulocyte-macrophage colony-stimulating factor (GM-CSF), which favors dendritic cell maturation of bone marrow precursors, drives αVβ3 expression, while macrophage colony-stimulating factor (M-CSF), which favors differentiation into macrophages, drives αVβ5 expression (De Nichilo and Burns, 1993). The presence of αVβ8, generally considered an epithelial integrin, on dendritic cells has recently been uncovered (Travis et al., 2007). Loss of this integrin from dendritic cells leads to inflammatory bowel disease because of loss of activation of transforming growth factor-β (TGF-β). It has been known for several years that αVβ6 and αVβ8 are important TGF-β activators in vivo (Sheppard, 2005), a phenomenon dependent on the Arg-Gly-Asp (RGD) sequence present in TGF-β1 (Yang et al., 2007). The TGF-β generated during dendritic cell interaction with T cells bearing latent TGF-β on their surface, in turn, is necessary for generation of regulatory T cells that control autoimmunity. In the absence of dendritic cell αVβ8, this important brake on autoimmunity apparently does not occur, at least in the gastrointestinal tract.

Phagocyte Functions That Require Integrins

TGF-β activation by αVβ8 aside, most phagocyte integrin-dependent functions follow from the ability of integrins to act as mediators of adhesion and as transducers of signals arising from ligation during cell-cell interaction or binding to and spreading on the ECM. First, integrins are required for transendothelial migration of neutrophils and monocytes into sites of inflammation. Second, integrin ligation is required for an effective phagocyte response to inflammatory stimuli. The classic model for this phenomenon is the "adhesion-dependent respiratory burst," first described by Nathan et al. (1989). Neutrophils exposed to physiologic inflammatory stimuli, such as formylated peptides or tumor necrosis factor alpha (TNF-α), do not assemble the NADPH oxidase unless β2 integrins are engaged. While it is unclear whether integrin engagement alone induces any signals required for oxidase assembly, neutrophil adhesion markedly facilitates signal transduction through receptors recognizing inflammatory products. This role for integrins is quite parallel to their requirement in growth factor signaling in a variety of cell types, including phagocytes. For example, integrin-mediated adhesion is important for M-CSF signaling in macrophages and for osteoclast differentiation in response to receptor activator of nuclear factor κB (RANKL) (Teitelbaum, 2007). Third, some integrins can be directly involved in pathogen-phagocyte interactions, by recognizing ligands on microbes, leading to phagocytosis. Integrin ligands on pathogens can be either host derived, such as complement factor iC3b, or microbe derived. In addition, αMβ2 appears to cooperate with a variety of other phagocytic receptors, such as IgG Fc receptors (FcγR), for efficient ingestion of microorganisms, even when it is not directly involved in ligand recognition; this will be discussed in greater detail below.

INTEGRIN ACTIVATION; INSIDE-OUT SIGNALING

One key to understanding the function of integrins on phagocytes, as well as on other cells, is to understand that they can have variable affinity for their nominal ligands depending on the state of cell activation. Platelets and leukocytes circulating through the bloodstream are very poorly adhesive; only when they get to a site of inflammation or clotting is their inherent adhesive potential manifest. Teleologically, this makes sense: adhesion of circulating cells to vessel endothelia can lead to disruption of blood flow, release of inflammatory mediators, and other events potentially detrimental to the host. These functions are useful at sites where specific host defense actions are required, but they are harmful if allowed to occur in the general circulation. Mechanistically, the alteration from nonadhesive to adhesive implies a major change in the ligand binding ability of the cell's integrins. The ability of a cell to regulate its adhesive potential in response to external cues is called "integrin activation" and is the result of a signal transduction process known as "inside-out" signaling.

At the level of the integrin, three mechanisms for enhanced adhesion have been investigated. The first is a simple increase in expression of adhesive receptors at the plasma membrane. As discussed above, neutrophils contain intracellular reservoirs of integrins in granules that can be mobilized rapidly to the plasma membrane in response to secretory signals. Although increased integrin expression may contribute in part to enhanced cell-ligand interactions, early work demonstrated that release of the intracellular pool of integrins was not required for enhanced neutrophil adhesion in response to inflammatory stimuli (Vedder and Harlan, 1988). Moreover, lymphocytes demonstrate the same ability to regulate adhesion as neutrophils, but they do not have an intracellular pool of integrins in secretory granules. More recently, study of regulation of integrin-mediated adhesion has focused on two hypotheses: first, that adhesive ability is regulated by receptor clustering; and second, that differences in adhesive state represent alterations in affinity of individual receptors for ligand, due to conformational change. In the past several years, much evidence has been generated for major conformational changes in integrins that can regulate ligand affinity. This area has been reviewed recently (Luo et al., 2007). The relative contribution of integrin clustering to enhancing integrin-ligand interactions (reviewed in Hogg and Leitinger, 2001) is still uncertain, although there is increasing evidence that both mechanisms are simultaneously involved in integrin activation (Hynes, 2002).

Integrin Conformation and Affinity Changes

There are two major conformational changes in integrins that can affect ligand affinity. Within the I domain of the α-chain, there is a rearrangement that occurs upon ligand binding that stabilizes interaction with ligand. The resting conformation is called "closed" and the ligand-bound conformation is called "open" because of the relative accessibility of a divalent cation within the domain to aqueous solvent. Measurements of affinity for ICAM-1 of recombinant αL I domains that can only achieve either the open or the closed conformation show that the open conformation has 10^4 greater ligand affinity than the closed (Shimaoka et al., 2001). Although this degree of affinity change between open and closed conformation may not be as great for all integrins, or for activation in response to physiologic stimuli (Luo et al., 2007), this conformational change clearly contributes to the enhanced ligand affinity of activated integrins.

Recently, an even more remarkable conformational change involving the whole integrin for both I domain-containing and I domain-lacking integrins has been demonstrated. Detailed electron microscopic studies have provided evidence that resting integrins are in an inactive "bent" state, with the ligand binding domain in close proximity to the plasma membrane. Activation leads to an "extended" state, in which the extracellular portion of the integrin heterodimer literally straightens, so that the ligand binding sites of the α- and β-chains are now extending into the environment (Fig. 1B) (Nishida et al., 2006) (reviewed in Luo et al. [2007]). The one crystal structure obtained so far of an entire integrin heterodimer extracellular domain (αVβ3) shows the integrin in the "bent" conformation (Xiong et al., 2001). In the "bent" state, the ligand binding headpiece of the integrin is closely apposed to the juxtamembrane stalk, so that the potential ligand binding site is sterically hindered from ligand recognition (Fig. 1A). It is easy to imagine that upon straightening (which has been likened to the opening of a switchblade knife (Kim et al., 2003), the ligand binding sites are no longer inhibited from ligand interaction. How are the two conformational changes linked? While the hypothesis has not been proven, it is generally thought that the straightening of the integrin during activation puts tension on the α-helix that connects the I domain to the β-propeller in the α-chain and that this "pull" causes the I domain to assume the open conformation in a piston-like movement.

Clustering and Avidity

In addition to a conformational change of individual molecules enhancing affinity, clustering of multiple integrin molecules can contribute to integrin activation, likely by increasing the avidity for a multimeric ligand (Fig. 1C). Activation-induced integrin clustering can be detected by confocal microscopy or by immunoelectron microscopy. Unlike the affinity-related conformational changes, the molecular mechanisms of integrin clustering have not yet been well defined. Some evidence suggests that rearrangements in the intramembrane α-helices, which may occur with the switchblade-like conformational change, mediate integrin clustering (Li et al., 2003). Rearrangements of the actin cytoskeleton also are thought to play a role in activation-induced integrin clustering. This is supported by a recent study showing that neutrophils deficient in the Wiskott-Aldrich syndrome protein (WASP), which is an important intermediate in actin reorganization, are defective in activation-mediated β2 integrin clustering and β2 integrin-dependent adhesion (Zhang et al., 2006).

In addition, the Ras-related GTPase Rap1 has a role in integrin clustering. Recent studies showed that overexpression of Rap1, or of one of its effectors, regulator of adhesion and cell polarization enriched in lymphoid tissues (RAPL), leads to increased adhesion through αLβ2 (Katagiri et al., 2003; Kinashi and Katagiri, 2004; Sebzda et al., 2002). Rap1-induced αLβ2 adhesiveness correlates with both enhanced αLβ2 clustering and affinity modulation (Ghandour et al., 2007; Han et al., 2006; Katagiri et al., 2000; Sebzda et al., 2002). The mechanism by which Rap1-mediated activation of RAPL leads to integrin clustering is not known.

A number of studies have shown that on resting cells, inactive integrins are distributed diffusely at the cell surface, while on activated cells, clustering of integrins occurs as a result of increased integrin diffusion within the plane of the plasma membrane. For example, a recent study showed a correlation between chemokine-mediated lateral mobility and cluster formation of αLβ2, and increased adhesion to its ligand ICAM-1 (Constantin et al., 2000). Particle-tracking studies showed that either cell activation or the actin inhibitor cytochalasin D enhances lateral diffusion of αLβ2, leading to enhanced adhesion (Kucik et al., 1996). Thus, a current model is that in unactivated cells, monomeric integrins are homogeneously distributed across the membrane, a state actively maintained by actin constraints, and that activation-induced integrin clustering requires the release of these actin constraints. Perhaps this is an event in which Rap1 and RAPL are involved. Many integrin ligands are part of the ECM, or are expressed in restricted domains on apposing cells, such as on the tips of microvilli. Since their ligands appear in multimeric or in relatively immobile form, clustering of rapidly diffusing integrins could result from binding clustered ligand and lead to significant increases in avidity.

Extracellular Stimuli Trigger Inside-Out Integrin Activation

A variety of specific proinflammatory stimuli can trigger integrin activation through inside-out signaling and comprise proinflammatory mediators such as chemokines or cytokines including interleukin-8 and TNF-α, GM-CSF, the complement fragment C5a, lipid mediators such as platelet-activating factor and leukotriene B4, and bacterial products such as lipopolysaccharide (LPS) or formylated peptides. Apparently, diverse families of receptors can be involved in initiation of inside-out signaling, including G-protein-coupled receptors (GPCRs), cytokine receptors, and membrane tyrosine kinases. The relative physiologic importance of the various mediators is not well understood, in part because of the complexity of mediator release in inflammation and infection.

One of the more interesting and potentially physiologically relevant activators of integrin-mediated adhesion is ligation of L-selectin (Gopalan et al., 1997). Since L-selectin is involved in the initial slowing and rolling of neutrophils on endothelium overlying sites of inflammation, its ability to activate integrin-mediated adhesion would be a mechanism for integrin activation during the process of transendothelial migration. On the other hand, at least in vitro, selectin-mediated rolling is not sufficient to activate transendothelial migration, and an additional integrin activation step through GPCR ligation is required.

Another mechanism for integrin activation is via IgG Fc receptors (FcγR) (Jones et al., 1998). This has in vivo significance, since immune complexes induce αMβ2-dependent glomerular accumulation of neutrophils in a model for glomerulonephritis (Tang et al., 1997). Conversely, efficient FcγR function depends on αMβ2. This is demonstrated by reduced FcγR-mediated phagocytosis and respiratory burst in the presence of anti-αMβ2 monoclonal antibodies (Arnaout et al., 1983; Brown et al., 1988), and is consistent with reduced FcγR functions in phagocytes from LAD-I patients (Graham et al., 1993; Gresham et al., 1991). Mice deficient in αMβ2 exhibit reduced in vivo FcγR-mediated antibody-dependent cellular cytotoxicity against tumor cells (Van Spriel et al., 2001). Thus, it seems clear that there is a biologically significant mutual dependence or "cross talk" between αMβ2 and FcγR.

Ligation of other integrins on the same cell (often referred to as integrin cross talk) also can activate integrins. One example of integrin cross talk is activation of αMβ2 by ligation of the fibronectin receptor α5β1 (Hazenbos et al., 1993; Wright et al., 1984) or of the vitronectin receptor αVβ3 (Van Strijp et al., 1993). Such an αMβ2 activation mechanism might operate physiologically when migrating cells contact ECM in the tissue. *Bordetella pertussis*, a pathogen that may use αMβ2-mediated phagocytosis as a step in successful infection, has coopted this activation mechanism by expressing ligands for αMβ2, α5β1, and αVβ3 (Hazenbos et al., 1995; Ishibashi et al., 1994). Ligation of either β1 or the β3 integrins by these ligands activates αMβ2, leading to bacterial phagocytosis. A similar cascade of events is initiated by the causative agent of Q fever, *Coxiella burnetii*, which binds both αMβ2 and αVβ3; apparently the most virulent strains suppress this pathway of phagocytosis (Capo et al., 1999, 2003).

Intracellular Signals Inducing Inside-Out Integrin Activation

Much recent work has focused on understanding how "inside-out" signaling can affect the major conformational change in the integrin required for efficient ligand binding. It has been shown that upon integrin activation, the cytoplasmic tails of the α- and β-chains separate, and the current hypothesis is that this separation is propagated through the membrane to initiate the unfolding of the molecule. A number of recent studies have demonstrated that the membrane- and actin-binding cytoplasmic protein talin has a central role in this event (Vinogradova et al., 2002) (reviewed in Ginsberg et al. [2005] and Hynes [2002]). Talin is a cytoskeletal protein, a member of the 4.1 band, Ezrin, Radixin, and Moesin (FERM) family, and it is able to bind to the cytoplasmic domain of integrin β-chains. In inactive integrin molecules, the α- and β-chain cytoplasmic tails are linked by a salt bridge, and binding of talin to the β-chain may be able to disrupt this interaction (Vinogradova et al., 2002). This event likely mediates the separation of the cytoplasmic tail of the β-chain from the α-chain that is required to initiate the conformational change in the extracellular domain required for enhanced ligand affinity. Indeed, overexpression of the β-chain binding FERM domain of talin has been shown to activate integrins (Calderwood et al., 2002). A recent study further solidified the important role of talin in integrin function by showing that either RNAi knockdown of talin expression, or mutations in talin that disrupt its interactions with the cytoplasmic β-chain, inhibit integrin activation (Tadokoro et al., 2003). Two distinct interactions between talin and the integrin β-chain are required for activation. The higher-affinity interaction, between a conserved NPXY sequence in the β-chain and a phosphotyrosine binding (PTB)-like domain in talin, is required but not sufficient for integrin activation. A second, lower-affinity interaction between membrane-proximal sequences in the β-chain and talin is also required (Calderwood et al., 2002; Tadokoro et al., 2003). Perhaps, because of the necessity for dual interaction, other PTB-containing proteins that can only interact with the β-chain NPXY motif cannot activate integrins.

It is currently unclear how inside-out signaling regulates the binding of talin to integrins. Membrane association of FERM-domain-containing proteins often is regulated by phosphatidylinositol binding and by phosphorylation, both of which can induce a conformational change in the molecule. PIP2 has been shown to associate with talin and increase talin binding to the β-chain (Martel et al., 2001), and thus, it may be a factor in talin-mediated integrin activation. Talin binds an isoform of PIP kinase (PIPKIγ), which should enhance PIP2 concentration in the vicinity of talin. However, PIPKIγ binds to a similar site on talin as the integrin β-chain cytoplasmic tail (Barsukov et al., 2003), suggesting that regulation of integrin activation by PIP2 may be quite complex. Since PIP2 also stimulates the interaction of vinculin, another integrin binding protein, with Arp2/3 (DeMali et al., 2002), a complex involved in actin nucleation and polymerization, the association of PIPKIγ with talin may be important for association of activated integrins with the actin cytoskeleton. Thus, additional effects of inside-out signaling on talin, perhaps including posttranslational modification of it or its binding partners, likely are required for talin-mediated integrin activation.

As yet, the intracellular signaling pathways proximal to activation stimuli required for integrin activation are only partially understood. Two systems have been analyzed in the most detail: activation of the highest expressed platelet integrin αIIbβ3 (also known as GPIIb/IIIa) by thrombin or other platelet aggregation agonists, and activation of αLβ2 on lymphocytes following T-cell antigen receptor ligation. Thrombin activates platelet αIIbβ3 through GPCRs called protease-activated receptors (PARs), likely using $G\alpha_q$, although $G\alpha_i$ and $G\alpha_{12,13}$ can activate αIIbβ3 in the absence of $G\alpha_q$ (Stouffer and Smyth, 2003). Phosphoinositide hydrolysis by phospholipase C and generation of phosphatidylinositol triphosphate by type I phosphatidylinositol 3-kinase (PI3K) both are necessary downstream effectors of G-protein signaling for activation of αIIbβ3. Chemokines, which also bind to GPCR, are thought to induce activation of phagocyte integrins through similar pathways, although chemokine receptors generally are coupled to downstream signaling pathways through G_i rather than G_q heterotrimeric G proteins.

A number of recent studies have demonstrated that Rap1 plays a central role in β1, β2, and β3 integrin activation in leukocytes and platelets (reviewed in Rose et al. [2007]). Like other small GTPases, Rap1 also exists in either an inactive GDP-bound or an active GTP-bound form. Expression of constitutively active Rap1 in T cells or macrophages induces activation of β1 and β2 integrins (Caron et al., 2000; Reedquist et al., 2000; Sebzda et al., 2002). Various extracellular stimuli, including chemokine receptor ligands, are capable of inducing Rap1 activation leading to integrin activation. Identification of the molecular events immediately upstream and downstream of Rap1 activation for inside-out signaling remains an area of intense investigation. Recent studies have implicated the Ca^{2+} and diacylglycerol-regulated guanine nucleotide exchange factor I (CalDAG-GEFI) as an important Rap1 GEF, able to activate Rap1 and integrins. RNAi knockdown of CalDAG-GEFI impairs activation of αLβ2 in T cells (Ghandour et al., 2007). Mice deficient in CalDAG-GEFI have defects in both Rap1 activation and in activation of β1 and β2 integrins in neutrophils and of β1 and β3 integrins in platelets (Bergmeier et al., 2007). This interesting phenotype is reminiscent of a recently described human immunodeficiency, LAD-III, in which integrins on leukocytes and platelets cannot be activated efficiently by inside-out signals, despite normal integrin expression on the plasma membrane (reviewed in Alon and Etzioni [2003]). Patients with this deficiency present with increased infections, decreased neutrophil migration to infection sites, and poor leukocyte adhesion in vitro—and so look like patients with LAD-I. Very recently, two patients with LAD-III were found to have identical homozygous mutations in a splice acceptor site of the CalDAG-GEFI gene, and to have reduced amounts of CalDAG-GEFI mRNA and protein in neutrophils, platelets, and lymphocytes (Pasvolsky et al., 2007). This suggests that LAD-III, at least in these two patients, results from defective expression of the same Rap1-GEF implicated in integrin activation by in vitro studies. It remains to be determined whether this gene is the target of mutation in all LAD-III patients. However, knockdown of CalDAG-GEFI in T cells does not block α4β1 activation (Ghandour et al., 2007), hinting that there are additional complexities in Rap1 regulation of integrin activation yet to be revealed.

Two downstream Rap1 effectors recently have been proposed to link Rap1 activation to integrin activation (reviewed in Bos [2005] and Rose et al. [2007]). One is the nonkinase adapter molecule RAPL, which directly interacts with αLβ2; RAPL activates αLβ2 when overexpressed (Katagiri et al., 2003; Kinashi and Katagiri, 2004). An essential role for RAPL in inside-out signaling to integrins is suggested by the phenotype of RAPL knockout mice, in which lymphocytes and dendritic cells migrate poorly (Katagiri et al., 2004). Rap1-GTP-interacting adapter molecule (RIAM) has been proposed as another Rap1 effector, linking Rap1 with activation of at least αIIbβ3 and αLβ2. Overexpression of RIAM, or expression of a constitutively active form, induces activation of these integrins, while RNAi knockdown impairs it (Han et al., 2006; Lafuente et al., 2004). Rap1 activation was found to induce a complex of RIAM and talin in association with integrins (Han et al., 2006). It may be that RAPL and RIAM are both required for efficient inside-out signaling; alternatively, they may have roles in distinct aspects of integrin biology, with RIAM involved in integrin activation and RAPL involved in linking the activated integrin to effector pathways like migration and phagocytosis. Since Rap1 is involved in both affinity changes and clustering of integrins (see above), it is possible that these distinct events are mediated by different Rap1 effectors. Furthermore, there is some evidence that the signaling pathways for activation of different neutrophil integrins may be distinct, depending both on the GPCR ligated and the integrin activated (Heit et al., 2005), suggesting that there is still much complexity to be unraveled in this important event.

Activation of αLβ2 on T cells through T-cell-receptor (TCR) ligation is initiated by tyrosine kinase signaling rather than heterotrimeric G-protein signaling. Surprisingly, the downstream effectors, PLC, PI3K, Rap1, and talin, are required in this pathway as well, suggesting that tyrosine kinase signals and GPCR signals represent two different paths to a common final effector pathway. Several adapter molecules, including Vav1-3, ADAP, and Slp-76, have been shown to be required for TCR-mediated activation of αLβ2 to the high-affinity state. At least in part, this is because these molecules form a complex that can activate actin polymerization and consequent receptor clustering via the small GTPase Cdc42 and its effector, WASP (Griffiths and Penninger, 2002). Since the tyrosine kinase cascade activated by ligation of FcγR on phagocytes is quite similar to that activated by TCR ligation on T cells, it is thought that FcγR-mediated signaling for integrin activation also parallels what is known in T cells. Indeed, to the extent that it has been studied, this seems to be true. For example, PI3K is involved in activation of αMβ2 by ligation of FcγR with immune complexes (Jones et al., 1998; Wang and Brown, 1999). Studies in neutrophils have suggested that the Vav GEFs may be part of the final common pathway in both GPCR and tyrosine kinase integrin activation, since they are required for fMLF- and immune complex-mediated integrin activation (Gakidis et al., 2004; Hall et al., 2006).

INTEGRIN OUTSIDE-IN SIGNALING AND PHAGOCYTOSIS

Role of Tyrosine Kinases in Integrin Outside-In Signaling

In recent years there has been increasing appreciation that ligation of activated integrins not only leads to cell adhesion and spreading, but also transmits signals into cells, a phenomenon called outside-in signaling, to contrast it with inside-out signaling mediating integrin activation. Integrin outside-in signaling has been studied in many settings, both within and outside the immune system. Neutrophils have been quite informative in this regard because of the requirement for integrin-mediated adhesion in agonist-induced assembly of the NADPH oxidase. These studies have shown that both the tyrosine kinase Syk and Src family kinases (SFK) are required for adhesion-dependent signaling for oxidase assembly (Mocsai et al., 2002) (reviewed in Berton et al. [2005]).

Both SFK and Syk can associate physically with integrin β-chains, and ligation of integrins can activate both SFK and Syk (Arias-Salgado et al., 2003; Berton et al., 1994; Yan et al., 1997). Kinase activation may occur as a result of integrin clustering during adhesion. Integrin-mediated activation of Syk, the downstream focal adhesion kinase (FAK), and its relative Pyk2 all depend on activation of SFK (Arias-Salgado et al., 2003; Suen et al., 1999). Studies in mice with genetic deficiencies in SFK or Syk have supported a central role for these tyrosine kinases in phagocyte integrin signaling (reviewed in Abram and Lowell [2007]). For example, in experimental endotoxic shock and inflammation, migration of neutrophils deficient in the SFK Hck and Fgr into tissues was impaired (Lowell and Berton, 1998), and mice deficient in Fgr exhibit reduced migration of eosinophils to the lung in response to allergen challenge (Vicentini et al., 2002). Monocytes with a triple Hck, Fgr, and Lyn deficiency exhibit migratory defects in response to the inflammatory stimulus thioglycollate (Meng and Lowell, 1998). The magnitude of integrin signaling results from a balance between tyrosine kinases and tyrosine phosphatases. Neutrophils and macrophages deficient in the Ig-like receptor PIR-B, which contains a cytoplasmic immunoreceptor tyrosine-based inhibitory motif (ITIM) able to activate tyrosine phosphatases, show enhanced responses to integrin ligation (Pereira et al., 2004). Surprisingly, neutrophils deficient in either Syk or in SFK migrate normally in vitro and in vivo in response to thioglycollate (Mocsai et al., 2002), suggesting complexities in signaling for migration, possibly because of varied requirements for integrin activation prior to initiation of outside-in signaling.

No systematic study has been done on how integrin conformation and clustering affect activation of tyrosine kinase cascades and outside-in signaling. Nathan's laboratory showed that TNF-α-mediated, integrin-dependent activation of the respiratory burst requires both Syk activation and activation of a Ca^{2+}-dependent adenylyl cyclase (Han et al., 2003, 2005). Importantly, while the adenylyl cyclase is upstream of Rap1 activation, as might be expected, it is not upstream of Syk activation. Since Rap1 is required for integrin affinity change, the implication of this discovery is that some integrin outside-in signaling, including the central event of Syk activation, does not require the affinity change mediated by Rap1; thus, low-affinity integrins may contribute significantly to signaling events that arise from ligand binding. This finding adds yet another level of complexity to regulation of integrin function, i.e., the possibility that distinct signaling events arise from high- and low-affinity integrins, which has not yet been addressed.

How Syk and SFK become activated by integrins also remains an unanswered question. Recent work has used a genetic approach to argue that integrin-mediated Syk activation requires the presence of receptors containing immunoreceptor tyrosine-based activation motifs (ITAMs) of the FcR γ-chain and DAP-12 (Mocsai et al., 2006). In the absence of these ITAMs, integrin outside-in signaling was abrogated. One potential explanation is that "tonic" SFK signaling phosphorylates ITAM sequences in neutrophil receptors and that this localizes Syk to the plasma membrane, where it can interact in a phosphotyrosine-independent way with integrin β-chains after ligand binding.

Tyrosine Kinases and Integrins in Osteoclast Biology

An important physiologic context for integrin signaling though tyrosine kinases is bone resorption by osteoclasts (reviewed in Ross and Teitelbaum, 2005). Osteoclasts are myeloid in origin, differentiating from bone marrow monocytes in response to RANKL. Osteoclast-dependent bone resorption is tightly coupled to osteoblast-mediated bone deposition to achieve bone remodeling and growth, as well as homeostasis. Defective osteoclast function results in excess bone formation, abnormal bone remodeling, and loss of the marrow cavity, a syndrome called osteopetrosis. The main integrin on osteoclasts implicated in their primary function of bone resorption is αVβ3, which recognizes bone proteins containing an Arg-Gly-Asp (RGD) sequence, such as osteopontin and bone sialoprotein (Teitelbaum and Ross, 2003). The essential role of this integrin in bone homeostasis is strongly supported by β3-deficient mice, which exhibit defective bone resorption by osteoclasts leading to increased bone mass (McHugh et al., 2000). The biological function of αVβ3 in osteoclasts requires tyrosine kinase signaling. Osteoclasts deficient in the SFK c-Src or in Syk are defective in bone resorption, and mice lacking c-Src develop osteopetrosis (Mocsai et al., 2004; Soriano et al., 1991; Zou et al., 2007). Recent studies have demonstrated the existence of a constitutive direct interaction between c-Src and the β3-chain of αVβ3 required for osteoclast function (Arias-Salgado et al., 2003; Zou et al., 2007). Consistent with their essential role in polymorphonuclear leukocyte integrin signaling, deletion of both FcRγ and DAP-12 results in impaired osteoclast differentiation and osteopetrosis (Mocsai et al., 2004). In response to αVβ3 ligation on osteoclasts, these ITAM-containing molecules associate with Syk, which in turn activates both c-Src and the Rho GTPase GEF Vav3. Mice lacking Vav3 also have osteoclasts that are defective in bone resorption and therefore develop osteopetrosis (Faccio et al., 2005). Together, these studies establish an important in vivo role for integrin outside-in signaling.

Role of Rho GTPases in Integrin Outside-In Signaling

It has been difficult to disentangle the cytoskeletal effects of integrin ligation from integrin outside-in signaling. Activation of tyrosine kinases following integrin ligation is intimately linked to integrin-dependent changes in cytoskeletal organization. An excellent example of the central role of actin in integrin signaling is the demonstration that neutrophils deficient in an actin-cross-linking protein, L-plastin, have defects in β2 integrin-mediated Syk activation, oxidative burst, and bacterial killing (Chen et al., 2003). Although there is no evidence showing that integrin cytoplasmic tails directly interact with actin, they do bind various linker proteins, such as talin, vinculin, and α-actinin, which in turn interact with actin. Other molecules in the integrin adhesion complex, such as WASP and filamin, are capable of triggering actin reorganization. Among the molecules involved in integrin signaling and actin reorganization, the Rho family GTPases Rho, Rac, and Cdc42 have a major role. These enzymes are essential links between integrin ligation and actin rearrangement events, mediating downstream effector functions such as cell morphology and motility (reviewed in DeMali et al. [2003] and Etienne-Manneville and Hall [2002]). Downstream effector molecules of Rac and Cdc42 include suppressor of G-protein-coupled cyclic AMP receptor (Scar), WASP, and the p21-activated kinases (Paks) (reviewed in Thrasher [2002]). These molecules in turn can interact with the Arp2/3 complex (Scar and WASP), leading to actin nucleation and polymerization, or phosphorylate actin regulatory proteins, leading to enhanced actin polymerization (Pak). Downstream effectors of Rho include Rho-activated kinase (ROCK), a serine/threonine kinase that targets several cytoskeleton-associated proteins, and formins, which can nucleate actin polymerization by a mechanism independent of Arp2/3. In general, Rac has been implicated in actin polymerization leading to the formation of lamellipodia, membrane ruffles, and dendrites; Cdc42 has been implicated in the formation of filapodia and persistence of direction in chemotaxis; while Rho-induced myosin activation leads to stress fiber formation and contractility (reviewed in Etienne-Manneville and Hall [2002]). All of these cellular structures require integrin-mediated adhesion as well, and it is likely that integrin ligation makes a significant contribution to activation of these GTPases under physiologic conditions. Thus, integrins and the GTPases are involved in a positive feedback loop: integrin ligation activates GTPases that lead to cytoskeletal reorganization, which, in turn, reinforces integrin-mediated adhesion.

Rho Family GTPases in Phagocytosis

Rho family GTPases are also involved in at least two major pathways of phagocytosis, initiated by the major opsonins of serum, IgG and complement C3. IgG opsonization leads to recognition by FcγR; the major phagocytic receptor for C3 is the integrin αMβ2, which recognizes complement iC3b covalently deposited onto the pathogen's surface.

There are significant differences in the cell biology of αMβ2-mediated ingestion and phagocytosis through FcγR. Initial studies in the 1970s, when the molecular nature of the receptors was unknown, demonstrated that the morphologic membrane changes during complement-mediated phagocytosis were distinct from those that occurred during IgG-mediated phagocytosis. Complement-mediated phagocytosis occurred through "sinking" of the ingested particle into the phagocyte, without apparent membrane protrusions, while IgG-mediated phagocytosis was accompanied by the formation of membrane protrusions (filopodia) surrounding the particle (Kaplan, 1977). These morphologic differences correlated with differences in organization of cytoskeletal components at the phagosome and in sensitivity to the actin filament-disrupting drug cytochalasin; while actin clusters that appeared in punctate form around αMβ2 phagosomes were relatively resistant to cytochalasin, the homogeneously distributed actin around FcγR phagosomes was much more easily disrupted by the drug (Allen and Aderem, 1996b). A potential molecular basis for these morphologic and pharmacologic differences between αMβ2 and FcγR phagocytosis has been elucidated recently. While both αMβ2- and FcγR-mediated phagocytosis require the Arp2/3 complex (May et al., 2000), they apparently depend on distinct upstream Rho GTPases. In these experiments, FcγR-mediated phagocytosis requires activation of Rac and Cdc42 (Caron and Hall, 1998; Cougoule et al., 2004); this is consistent with their known roles in Arp2/3-dependent actin polymerization. In phagocytic cells, both of these GTPases are thought to be activated by the GEF Vav1 and to be able to induce polymerization of actin at the phagocytic cup. In contrast to FcγR, αMβ2 in at least some circumstances uses Rho as the GTPase required for phagocytosis (Caron and Hall, 1998; Olazabal et al., 2002). Unlike Rac and Cdc42, Rho's effects on actin are essentially contractile rather than polymerizing. While the Rac/Cdc42 enhancement of actin polymerization can explain the membrane protrusions characteristic of FcγR-mediated phagocytosis, enhanced Rho- and cytoskeleton-dependent membrane contraction surrounding an αMβ2 phagosome could explain the "sinking" of complement-opsonized particles into the ingesting phagocyte.

Unfortunately, subsequent experiments using primary phagocytes from mice genetically deficient in specific GTPases or GEFs were inconsistent with these conclusions. Instead, macrophages lacking Rac were defective in both αMβ2- and FcγR-mediated phagocytosis, and deletion of Vav rendered macrophages defective in αMβ2- but not FcγR-mediated phagocytosis (Hall et al., 2006). Several possible explanations for these conflicting conclusions have been proposed, related to technical differences between the approaches or to differences in GTPase and GEF expression in the cell types investigated (Lowell, 2006). While both approaches point to a critical role for Rho family GTPases and their GEFs in integrin-mediated phagocytosis, essential details remain controversial and unresolved, as does the basis for the striking morphologic difference in phagocytosis through αMβ2 and FcγR, which has been observed repeatedly (Allen and Aderem, 1996a, 1996b; Kaplan, 1977).

Another striking difference between complement- and IgG-mediated phagocytosis is that, at least in some circumstances, phagocytosis mediated by αMβ2 in macrophages activates less potent microbicidal activity than FcγR. Activation of the oxidative burst, release of arachidonic acid metabolites, and phagosome maturation through fusion of vesicles containing defensins and proteolytic enzymes are all less efficient during αMβ2 phagocytosis than during FcγR phagocytosis (Berton et al., 1992; Joiner et al., 1989; Wright and Silverstein, 1983). These findings are the basis for the proposal that some microorganisms benefit from αMβ2-mediated entry into host phagocytes, including *Leishmania* (Mosser and Edelson, 1985, 1987), *Histoplasma capsulatum* (Bullock and Wright, 1987; Eissenberg and Goldman, 1987), and *B. pertussis* (Relman et al., 1990; Saukkonen et al., 1991). Thus, phagocytosis of pathogenic microorganisms through αMβ2 does not necessarily lead to efficient killing, and may contribute to their intracellular survival, particularly inside macrophages, and thus to the persistence of infections.

Microbial Ligands for αMβ2

Because αMβ2 is a phagocytic receptor that is deficient in activation of inflammation, an attractive hypothesis is that certain intracellular pathogens might target this receptor for entry into macrophages. An alternative hypothesis is that αMβ2 has evolved to recognize certain bacterial structures as part of host defense, as a pattern recognition receptor similar to Toll-like receptors (TLRs) or lectins. A wide variety of bacterial products have been proposed as direct ligands for αMβ2 (reviewed in Agramonte-Hevia et al. [2002]). However, some of these studies were entirely based on the effects of anti-αMβ2 antibodies on neutrophil adhesion to the ligand under study; since αMβ2 is intimately involved in cell adhesion, and antibodies may signal through the integrin, these experiments must be regarded as inconclusive (Brown, 1991). However, there is more compelling evidence for binding of some microbial molecules to αMβ2, including filamentous hemagglutinin (FHA) of *B. pertussis* (Relman et al., 1990), lipophosphoglycan (Talamas-Rohana et al., 1990) and glycoprotein gp63 (Russell and Wright, 1988) of *Leishmania*, and LPS (Wright, 1991). While some of these ligands can mediate macrophage uptake of the presenting microbe in vitro, these interactions have yet to be shown to be important during in vivo infection.

Functional Interactions of Integrins with Other Plasma Membrane Proteins

Integrins can associate with several other proteins within the plane of the plasma membrane, which often results in enhanced integrin-mediated signaling. Tetraspanins, a family of integral membrane proteins that may create a protein web on the cell surface (Berditchevski and Odintsova, 2007), can frequently be coprecipitated with integrins, especially of the β1 family (Hemler, 2005). In phagocytes, the tetraspanin CD81 has seemed especially important, with roles in stabilizing integrin-mediated adhesion (Feigelson et al., 2003), in phagocytosis (Chang and Finnemann, 2007), and in migration (Berditchevski and Odintsova, 1999). In cells outside the immune system, physical interaction of integrins with growth factor receptors is known to enhance growth factor signaling (Damsky and Werb, 1992).

Another integrin-associated protein is the CD47 molecule (reviewed in Brown and Frazier, 2001). In addition to lateral interactions with specific integrins within the plane of the membrane, CD47 binds the ECM molecule thrombospondin (Gao et al., 1996) and the IgSF membrane protein signal-regulatory protein (SIRP) α (also known as SHPS-1) to mediate cell adhesion (Jiang et al., 1999). Ligation of an integrin-CD47 complex activates heterotrimeric G proteins, leading to signaling for particle binding, phagocytosis, and migration (Lindberg et al., 1996). This signaling pathway appears to be completely independent of the tyrosine kinase outside-in signaling pathway discussed above, and loss of CD47 does not affect tyrosine kinase activation after integrin ligation. CD47 also is required for efficient dendritic cell migration in vivo, but whether this is related to its effects on integrin signaling or is an independent result of SIRPα ligation is not known. For transendothelial migration of dendritic cells, CD47 expression seems to be required on the dendritic cells, but not on the endothelial cells (Van et al., 2006).

Several glycan phosphatidylinositol (GPI)-anchored proteins have been reported to associate physically with β2 integrins, in particular, αMβ2. These GPI-anchored proteins include the urokinase plasminogen activator receptor (uPAR) (Bohuslav et al., 1995; Xue et al., 1994), human FcγRIIIB (Zhou et al., 1993), CD14 (Pfeiffer et al., 2001), and Thy-1 (Wetzel et al., 2004). uPAR also can coprecipitate with β1 integrins, especially α3β1, and its interaction with both αMβ2 and α3β1 can be blocked with a single peptide (Simon et al., 2000). Since GPI-anchored proteins have no transmembrane domains, it is thought that they require pairing with a transmembrane protein to initiate signal transduction. Thus, interactions with integrins likely are important for the activation of signaling pathways and cellular functions upon ligation of the GPI-anchored receptor. From the perspective of the integrin, this cooperation increases the range of ligands that activate its adhesive and signaling properties and enhances access to signaling molecules clustered in membrane rafts where the GPI-anchored proteins are concentrated. Furthermore, it provides a sophisticated means of communication between cellular processes like immune recognition (in the case of FcγRIIIB) and ECM proteolysis (in the case of uPAR) that are required for integrin-dependent functions like phagocytosis and migration.

ADHERENCE AND MIGRATION

Phagocyte-Endothelium Interactions

In uninflamed tissue, there is little interaction between endothelium and leukocytes circulating in the blood. This ignorance is in part because selectins and their ligands are not expressed on uninflamed endothelium and in part because leukocyte integrins are in a low-affinity, unclustered, nonadherent state. When a local inflammatory stimulus occurs, the steady state is perturbed, and leukocytes adhere to and migrate through the overlying endothelium to the site of injury. This process of leukocyte recruitment is generally divided into four steps (for a recent review, see Vestweber [2007]); such tight regulation of adhesion and migration is important to confine inflammatory tissue damage to the region of perturbation of homeostasis.

The first step in neutrophil transendothelial migration is a decrease in the speed at which the leukocytes migrate over the endothelium. This slowing is mediated largely by repeated short-lived weak interactions between a family of Ca^{2+}-dependent lectins (called selectins) and their carbohydrate ligands. P- and E-selectin are expressed on endothelium, L-selectin is expressed on leukocytes, and each can contribute to slowing of the leukocytes. While the endothelial selectins are expressed only by endothelium overlying sites of inflammation, L-selectin expression on leukocytes is constitutive, so that they are always "poised" to receive the endothelial signal. Unusually, selectin-ligand interactions require shear forces to occur; forces equivalent to those generated by blood flowing in postcapillary venules are optimal for initiation of these bonds. Interactions requiring shear forces presumably result from molecular deformations induced by the shear itself and are called "catch bonds." Because of short individual bond lifetimes and the density of selectins and their ligands, these selectin catch bonds are transient and occur repeatedly, leading to a very characteristic rolling of leukocytes along inflamed endothelium. Some integrins also have been shown to be capable of mediating this rolling step of leukocyte-endothelial adhesion, although the mechanism is different, since integrin-mediated interactions do not require shear (Berlin et al., 1995).

The second and third steps in transendothelial migration lead to leukocyte arrest, induced by integrin activation. Integrin activation leads to ligand recognition (step 2); clustering of ligated integrins leads to cytoskeletal attachment of the integrin and firm adhesion (step 3). The relevant integrins are on leukocytes; the endothelial integrin ligands are ICAM-1 and -2, as well as VCAM-1. ICAM-1 and VCAM-1 are expressed only after NF-κB signaling in the endothelial cell, in response to a panoply of agonists at sites of inflammation. ICAM-1, an αLβ2 ligand, appears earlier than VCAM-1, an α4β1 ligand. This difference in expression may contribute to the kinetics of inflammatory cell influx: neutrophils, which rely heavily on αLβ2 at this step, appear at sites of inflammation earlier than monocytes, which are more dependent on α4β1 for transendothelial migration. As discussed above, chemokines generated at sites of inflammation can induce integrin activation, as can other cytokines produced by inflammatory cells. Ligation of L-selectin itself also may activate integrin-mediated adhesion. Activated integrins make longer-lived bonds than selectins, and integrin attachment to cytoskeleton leads to firm adhesion between leukocyte and endothelium that resists the shear forces of the passing blood.

Actual migration through the endothelial barrier is the fourth and final step of the process, and the least understood. Leukocytes can crawl through endothelial intercellular junctions (especially at vertices where three cells are in contact) or, in some cases, apparently burrow directly through an endothelial cell (Carman and Springer, 2004; Feng et al., 1998) (reviewed in Middleton et al., 2002) to leave the bloodstream. In the absence of integrin-mediated adhesion, leukocyte extravasation does not happen. After passing through the vessel wall, the leukocytes crawl through the ECM up a chemotactic gradient to the origin of the perturbation of homeostasis. For both neutrophils and monocytes, this process depends on β1 and β2 integrins. However, the relatively low expression of β1 integrins on neutrophils makes these cells extremely dependent on β2 integrins for migration from the blood to an inflammatory site; this is why LAD-I patients lacking β2 integrins fail to accumulate neutrophils at sites of infection, while monocyte accumulation is relatively spared.

Neutrophil Migration

The use of genetically deficient mice has shed more light on the relative roles of individual integrins in neutrophil migration in vivo. A comparison of mice deficient in αMβ2 and αLβ2 confirmed that both β2 integrins contribute in part to adhesion and extravasation of neutrophils, although the defects in αLβ2-deficient mice were more pronounced (Ding et al., 1999). Studies using mice deficient in αLβ2 and treated with anti-αMβ2 antibodies showed that the main role for αLβ2 is in adhesion to endothelium, while αMβ2 is more important in the transendothelial migration of inflammatory neutrophils (Henderson et al., 2001). Besides αMβ2 and αLβ2, the integrin α4β1 (VLA-4) has been proposed to play an additional, though minor, role in neutrophil migration through the endothelium and into the tissues (Bowden et al., 2002; Ibbotson et al., 2001). The α4β1 integrin is a receptor for endothelial VCAM-1 and the ECM protein fibronectin. Although α4β1 expression is rather low on circulating noninflammatory neutrophils, it may be upregulated by ligation of β2 integrins during transendothelial migration of activated inflammatory neutrophils (Ley, 2002).

Monocyte Migration

In contrast to neutrophils, monocytes extravasate constitutively from the circulation into the tissues at a low basal rate, where they differentiate into tissue macrophages (Van Furth and Cohn, 1968). This constitutive migration is important to maintain a continuous pool of tissue macrophages, which have an estimated lifetime ranging from to 4 to 15 days (Van Furth et al., 1985). It is unclear whether the low-rate constitutive monocyte extravasation requires specific stimuli, although recently a role for the chemokine CXCL14 in this process has been suggested (Kurth et al., 2001).

In inflammation, the extravasation rate of monocytes is greatly increased, and the transmigrated cells develop a phenotype different from resident tissue macrophages. These cells are called "inflammatory" because they show increased rates of phagocytosis and cytokine production. Compared with their dominant role in neutrophil migration, β2 integrins are relatively less important in inflammatory monocyte migration, which seems to depend more heavily on α4β1 (reviewed in Imhof and Aurrand-Lions, 2004). This is supported by experiments showing that monocyte extravasation, in response to the inflammatory stimulus thioglycollate, was normal in mice deficient in αLβ2. In these mice, monocyte extravasation could be blocked by anti-α4β1 antibodies, whereas anti-αMβ2 had no effect (Henderson et al., 2003).

CONCLUDING REMARKS

In the past few years, several major advances have been made in our understanding of the molecular mechanisms by which integrins become activated, and by which integrins activate phagocyte functions. Deep insights into integrin structure and conformational changes, revelation of the roles for Rho family GTPases and Rap1, and dissection of the functions of specific tyrosine kinases in phagocyte integrin signaling have led to better understanding of the central role for this family of adhesion receptors in phagocyte biology. A major area of investigation is how this knowledge, generated primarily from in vitro experiments, can be extrapolated to physiologic functions of integrins in vivo. A further question is whether and in which circumstances integrin function can be therapeutically manipulated. A major challenge will be to obtain enough understanding of the differences in integrin signaling pathways and of the specific effects of individual integrins, so that we can separate unwanted proinflammatory effects from their requirement in combating infectious diseases. When this is possible, integrins will be attractive targets for treatment of autoimmune and inflammatory diseases of many etiologies.

The authors thank J. Hiroshi Morisaki for excellent artwork.

REFERENCES

Abram, C. L., and C. A. Lowell. 2007. Convergence of immunoreceptor and integrin signaling. *Immunol. Rev.* **218:**29–44.

Agramonte-Hevia, J., A. Gonzalez-Arenas, D. Barrera, and M. Velasco-Velazquez. 2002. Gram-negative bacteria and phagocytic cell interaction mediated by complement receptor 3. *FEMS Immunol. Med. Microbiol.* **34:**255–266.

Albert, M. L., J. I. Kim, and R. B. Birge. 2000. alphavbeta5 integrin recruits the CrkII-Dock180-rac1 complex for phagocytosis of apoptotic cells. *Nat. Cell Biol.* **2:**899–905.

Allen, L. A., and A. Aderem. 1996a. Mechanisms of phagocytosis. *Curr. Opin. Immunol.* **8:**36-40.

Allen, L. A., and A. Aderem. 1996b. Molecular definition of distinct cytoskeletal structures involved in complement- and Fc receptor-mediated phagocytosis in macrophages. *J. Exp. Med.* **184:**627–637.

Alon, R., and A. Etzioni. 2003. LAD-III, a novel group of leukocyte integrin activation deficiencies. *Trends Immunol.* **24:**561–566.

Arias-Salgado, E. G., S. Lizano, S. Sarkar, J. S. Brugge, M. H. Ginsberg, and S. J. Shattil. 2003. Src kinase activation by direct interaction with the integrin beta cytoplasmic domain. *Proc. Natl. Acad. Sci. USA* **100:**13298–13302.

Arnaout, M. A., R. F. Todd III, N. Dana, J. Melamed, S. F. Schlossman, and H. R. Colten. 1983. Inhibition of phagocytosis of complement C3- or immunoglobulin G-coated particles and of C3bi binding by monoclonal antibodies to a monocyte-granulocyte membrane glycoprotein (Mol). *J. Clin. Invest.* **72:**171–179.

Barsukov, I. L., A. Prescot, N. Bate, B. Patel, D. N. Floyd, N. Bhanji, C. R. Bagshaw, K. Letinic, G. Di Paolo, P. De Camilli, G. C. Roberts, and D. R. Critchley. 2003. Phosphatidylinositol phosphate kinase type 1gamma and beta1-integrin cytoplasmic domain bind to the same region in the talin FERM domain. *J. Biol. Chem.* **278:**31202–31209.

Berditchevski, F., and E. Odintsova. 1999. Characterization of integrin-tetraspanin adhesion complexes: role of tetraspanins in integrin signaling. *J. Cell Biol.* **146:**477–492.

Berditchevski, F., and E. Odintsova. 2007. Tetraspanins as regulators of protein trafficking. *Traffic* **8:**89–96.

Bergmeier, W., T. Goerge, H. W. Wang, J. R. Crittenden, A. C. Baldwin, S. M. Cifuni, D. E. Housman, A. M. Graybiel, and D. D. Wagner. 2007. Mice lacking the signaling molecule CalDAG-GEFI represent a model for leukocyte adhesion deficiency type III. *J. Clin. Invest.* **117:**1699–1707.

Berlin, C., R. F. Bargatze, J. J. Campbell, U. H. von Andrian, M. C. Szabo, S. R. Hasslen, R. D. Nelson, E. L. Berg, S. L. Erlandsen, and E. C. Butcher. 1995. alpha 4 integrins mediate lymphocyte attachment and rolling under physiologic flow. *Cell* **80:**413–422.

Berton, G., L. Fumagalli, C. Laudanna, and C. Sorio. 1994. Beta 2 integrin-dependent protein tyrosine phosphorylation and activation of the FGR protein tyrosine kinase in human neutrophils. *J. Cell Biol.* **126:**1111–1121.

Berton, G., C. Laudanna, C. Sorio, and F. Rossi. 1992. Generation of signals activating neutrophil functions by leukocyte integrins: LFA-1 and gp150/95, but not CR3, are able to stimulate the respiratory burst of human neutrophils. *J. Cell Biol.* **116:**1007–1017.

Berton, G., A. Mocsai, and C. A. Lowell. 2005. Src and Syk kinases: key regulators of phagocytic cell activation. *Trends Immunol.* **26:**208–214.

Blystone, S. D., S. E. Slater, M. P. Williams, M. T. Crow, and E. J. Brown. 1999. A molecular mechanism of integrin crosstalk: alphavbeta3 suppression of calcium/calmodulin-dependent protein kinase II regulates alpha5beta1 function. *J. Cell Biol.* **145:**889–897.

Bohuslav, J., V. Horejsi, C. Hansmann, J. Stockl, U. H. Weidle, O. Majdic, I. Bartke, W. Knapp, and H. Stockinger. 1995. Urokinase plasminogen activator receptor, beta 2-integrins, and Src-kinases within a single receptor complex of human monocytes. *J. Exp. Med.* **181:**1381–1390.

Bos, J. L. 2005. Linking Rap to cell adhesion. *Curr. Opin. Cell Biol.* **17:**123–128.

Bowden, R. A., Z. M. Ding, E. M. Donnachie, T. K. Petersen, L. H. Michael, C. M. Ballantyne, and A. R. Burns. 2002. Role of alpha4 integrin and VCAM-1 in CD18-independent neutrophil migration across mouse cardiac endothelium. *Circ. Res.* **90:**562–569.

Brown, E. J. 1991. Complement receptors and phagocytosis. *Curr. Opin. Immunol.* **3:**76–82.

Brown, E. J., J. F. Bohnsack, and H. D. Gresham. 1988. Mechanism of inhibition of immunoglobulin G-mediated phagocytosis by monoclonal antibodies that recognize the Mac-1 antigen. *J. Clin. Invest.* **81:**365–375.

Brown, E. J., and W. A. Frazier. 2001. Integrin-associated protein (CD47) and its ligands. *Trends Cell. Biol.* **11:**130–135.

Bullock, W. E., and S. D. Wright. 1987. Role of the adherence-promoting receptors, CR3, LFA-1, and p150,95, in binding of Histoplasma capsulatum by human macrophages. *J. Exp. Med.* **165:**195–210.

Bunting, M., E. S. Harris, T. M. McIntyre, S. M. Prescott, and G. A. Zimmerman. 2002. Leukocyte adhesion deficiency syndromes: adhesion and tethering defects involving beta 2 integrins and selectin ligands. *Curr. Opin. Hematol.* **9:**30–35.

Calderwood, D. A., B. Yan, J. M. de Pereda, B. G. Alvarez, Y. Fujioka, R. C. Liddington, and M. H. Ginsberg. 2002. The phosphotyrosine binding-like domain of talin activates integrins. *J. Biol. Chem.* **277:**21749–21758.

Capo, C., F. P. Lindberg, S. Meconi, Y. Zaffran, G. Tardei, E. J. Brown, D. Raoult, and J. L. Mege. 1999. Subversion of monocyte functions by *Coxiella burnetii*: impairment of the cross-talk between alphavbeta3 integrin and CR3. *J. Immunol.* **163:**6078–6085.

Capo, C., A. Moynault, Y. Collette, D. Olive, E. J. Brown, D. Raoult, and J. L. Mege. 2003. *Coxiella burnetii* avoids macrophage phagocytosis by interfering with spatial distribution of complement receptor 3. *J. Immunol.* **170:**4217–4225.

Carman, C. V., and T. A. Springer. 2004. A transmigratory cup in leukocyte diapedesis both through individual vascular endothelial cells and between them. *J. Cell Biol.* **167:**377–388.

Caron, E., and A. Hall. 1998. Identification of two distinct mechanisms of phagocytosis controlled by different Rho GTPases. *Science* **282:**1717–1721.

Caron, E., A. J. Self, and A. Hall. 2000. The GTPase Rap1 controls functional activation of macrophage integrin alphaMbeta2 by LPS and other inflammatory mediators. *Curr. Biol.* **10:**974–978.

Chang, Y., and S. C. Finnemann. 2007. Tetraspanin CD81 is required for the {alpha}vbeta5-integrin-dependent particle-binding step of RPE phagocytosis. *J. Cell Sci.* **120:**3053–3063.

Chen, H., A. Mocsai, H. Zhang, R. X. Ding, J. H. Morisaki, M. White, J. M. Rothfork, P. Heiser, E. Colucci-Guyon, C. A. Lowell, H. D. Gresham, P. M. Allen, and E. J. Brown. 2003. Role for plastin in host defense distinguishes integrin signaling from cell adhesion and spreading. *Immunity* **19:**95–104.

Constantin, G., M. Majeed, C. Giagulli, L. Piccio, J. Y. Kim, E. C. Butcher, and C. Laudanna. 2000. Chemokines trigger immediate beta2 integrin affinity and mobility changes: differential regulation and roles in lymphocyte arrest under flow. *Immunity* **13:**759–769.

Cougoule, C., A. Wiedemann, J. Lim, and E. Caron. 2004. Phagocytosis, an alternative model system for the study of cell adhesion. *Semin. Cell Dev. Biol.* **15:**679–689.

Damsky, C. H., and Z. Werb. 1992. Signal transduction by integrin receptors for extracellular matrix: cooperative processing of extracellular information. *Curr. Opin. Cell. Biol.* **4:**772–781.

DeMali, K. A., C. A. Barlow, and K. Burridge. 2002. Recruitment of the Arp2/3 complex to vinculin: coupling membrane protrusion to matrix adhesion. *J. Cell Biol.* **159:**881–891.

DeMali, K. A., K. Wennerberg, and K. Burridge. 2003. Integrin signaling to the actin cytoskeleton. *Curr. Opin. Cell Biol.* **15:**572–582.

De Nichilo, M. O., and G. F. Burns. 1993. Granulocyte-macrophage and macrophage colony-stimulating factors differentially regulate alpha v integrin expression on cultured human macrophages. *Proc. Natl. Acad. Sci. USA* **90:**2517–2721.

Ding, Z. M., J. E. Babensee, S. I. Simon, H. Lu, J. L. Perrard, D. C. Bullard, X. Y. Dai, S. K. Bromley, M. L. Dustin, M. L. Entman, C. W. Smith, and C. M. Ballantyne. 1999. Relative contribution of LFA-1 and Mac-1 to neutrophil adhesion and migration. *J. Immunol.* **163:**5029–5038.

Eissenberg, L. G., and W. E. Goldman. 1987. Histoplasma capsulatum fails to trigger release of superoxide from macrophages. *Infect. Immun.* **55:**29–34.

Etienne-Manneville, S., and A. Hall. 2002. Rho GTPases in cell biology. *Nature* **420:**629–635.

Faccio, R., S. L. Teitelbaum, K. Fujikawa, J. Chappel, A. Zallone, V. L. Tybulewicz, F. P. Ross, and W. Swat. 2005. Vav3 regulates osteoclast function and bone mass. *Nat. Med.* **11:**284–290.

Feigelson, S. W., V. Grabovsky, R. Shamri, S. Levy, and R. Alon. 2003. The CD81 tetraspanin facilitates instantaneous leukocyte VLA-4 adhesion strengthening to vascular cell adhesion molecule 1 (VCAM-1) under shear flow. *J. Biol. Chem.* **278:**51203–51212.

Feng, D., J. A. Nagy, K. Pyne, H. F. Dvorak, and A. M. Dvorak. 1998. Neutrophils emigrate from venules by a transendothelial cell pathway in response to FMLP. *J. Exp. Med.* **187:**903–915.

Gakidis, M. A., X. Cullere, T. Olson, J. L. Wilsbacher, B. Zhang, S. L. Moores, K. Ley, W. Swat, T. Mayadas, and J. S. Brugge. 2004. Vav GEFs are required for beta2 integrin-dependent functions of neutrophils. *J. Cell Biol.* **166:**273–282.

Gao, A. G., F. P. Lindberg, M. B. Finn, S. D. Blystone, E. J. Brown, and W. A. Frazier. 1996. Integrin-associated protein is a receptor for the C-terminal domain of thrombospondin. *J. Biol. Chem.* **271:**21–24.

Ghandour, H., X. Cullere, F. W. Luscinskas, and T. N. Mayadas. 2007. Essential role for Rap1 GTPase and its guanine exchange factor CalDAG-GEFI in LFA-1 but not VLA-4 integrin-mediated human T cell adhesion. *Blood* **110:**3682–3690.

Ginsberg, M. H., A. Partridge, and S. J. Shattil. 2005. Integrin regulation. *Curr. Opin. Cell Biol.* **17:**509–516.

Gopalan, P. K., C. W. Smith, H. Lu, E. L. Berg, L. V. McIntire, and S. I. Simon. 1997. Neutrophil CD18-dependent arrest on intercellular adhesion molecule 1 (ICAM-1) in shear flow can be activated through L-selectin. *J. Immunol.* **158:**367–375.

Graham, I. L., J. B. Lefkowith, D. C. Anderson, and E. J. Brown. 1993. Immune complex-stimulated neutrophil LTB4 production is dependent on beta 2 integrins. *J. Cell Biol.* **120:** 1509–1517.

Gresham, H. D., I. L. Graham, D. C. Anderson, and E. J. Brown. 1991. Leukocyte adhesion-deficient neutrophils fail to amplify phagocytic function in response to stimulation. Evidence for CD11b/CD18-dependent and -independent mechanisms of phagocytosis. *J. Clin Invest.* **88:**588–597.

Griffiths, E. K., and J. M. Penninger. 2002. Communication between the TCR and integrins: role of the molecular adapter ADAP/Fyb/Slap. *Curr. Opin. Immunol.* **14:**317–322.

Guerau-de-Arellano, M., J. Alroy, D. Bullard, and B. T. Huber. 2005. Aggravated Lyme carditis in CD11a−/− and CD11c−/− mice. *Infect. Immun.* **73:**7637–7643.

Hall, A. B., M. A. Gakidis, M. Glogauer, J. L. Wilsbacher, S. Gao, W. Swat, and J. S. Brugge. 2006. Requirements for Vav guanine nucleotide exchange factors and Rho GTPases in FcgammaR- and complement-mediated phagocytosis. *Immunity* **24:**305–316.

Han, H., M. Fuortes, and C. Nathan. 2003. Critical role of the carboxyl terminus of proline-rich tyrosine kinase (Pyk2) in the activation of human neutrophils by tumor necrosis factor: separation of signals for the respiratory burst and degranulation. *J. Exp. Med.* **197:**63–75.

Han, H., A. Stessin, J. Roberts, K. Hess, N. Gautam, M. Kamenetsky, O. Lou, E. Hyde, N. Nathan, W. A. Muller, J. Buck, L. R. Levin, and C. Nathan. 2005. Calcium-sensing soluble adenylyl cyclase mediates TNF signal transduction in human neutrophils. *J. Exp. Med.* **202:**353–361.

Han, J., C. J. Lim, N. Watanabe, A. Soriani, B. Ratnikov, D. A. Calderwood, W. Puzon-McLaughlin, E. M. Lafuente, V. A. Boussiotis, S. J. Shattil, and M. H. Ginsberg. 2006. Reconstructing and deconstructing agonist-induced activation of integrin alphaIIbbeta3. *Curr. Biol.* **16:**1796–1806.

Hazenbos, W. L., B. M. van den Berg, C. W. Geuijen, F. R. Mooi, and R. van Furth. 1995. Binding of FimD on Bordetella pertussis to very late antigen-5 on monocytes activates complement receptor type 3 via protein tyrosine kinases. *J. Immunol.* **155:**3972–3978.

Hazenbos, W. L., B. M. van den Berg, and R. van Furth. 1993. Very late antigen-5 and complement receptor type 3 cooperatively mediate the interaction between Bordetella pertussis and human monocytes. *J. Immunol.* **151:**6274–6282.

Heit, B., P. Colarusso, and P. Kubes. 2005. Fundamentally different roles for LFA-1, Mac-1 and alpha4-integrin in neutrophil chemotaxis. *J. Cell Sci.* **118:**5205–5220.

Hemler, M. E. 2005. Tetraspanin functions and associated microdomains. *Nat. Rev. Mol. Cell Biol.* **6:**801–811.

Henderson, R. B., J. A. Hobbs, M. Mathies, and N. Hogg. 2003. Rapid recruitment of inflammatory monocytes is independent of neutrophil migration. *Blood* **102:**328–335.

Henderson, R. B., L. H. Lim, P. A. Tessier, F. N. Gavins, M. Mathies, M. Perretti, and N. Hogg. 2001. The use of lymphocyte function-associated antigen (LFA)-1-deficient mice to determine the role of LFA-1, Mac-1, and alpha4 integrin in the inflammatory response of neutrophils. *J. Exp. Med.* **194:**219–226.

Hendey, B., M. Lawson, E. E. Marcantonio, and F. R. Maxfield. 1996. Intracellular calcium and calcineurin regulate neutrophil motility on vitronectin through a receptor identified by antibodies to integrins alphav and beta3. *Blood* **87:**2038–2048.

Hogg, N., and B. Leitinger. 2001. Shape and shift changes related to the function of leukocyte integrins LFA-1 and Mac-1. *J. Leukoc. Biol.* **69:**893–898.

Hynes, R. O. 2002. Integrins: bidirectional, allosteric signaling machines. *Cell* **110:**673–687.

Ibbotson, G. C., C. Doig, J. Kaur, V. Gill, L. Ostrovsky, T. Fairhead, and P. Kubes. 2001. Functional alpha4-integrin: a newly identified pathway of neutrophil recruitment in critically ill septic patients. *Nat. Med.* **7:**465–470.

Imhof, B. A., and M. Aurrand-Lions. 2004. Adhesion mechanisms regulating the migration of monocytes. *Nat. Rev. Immunol.* **4:**432–444.

Ishibashi, Y., S. Claus, and D. A. Relman. 1994. Bordetella pertussis filamentous hemagglutinin interacts with a leukocyte signal transduction complex and stimulates bacterial adherence to monocyte CR3 (CD11b/CD18). *J. Exp. Med.* **180:**1225–1233.

Jiang, P., C. F. Lagenaur, and V. Narayanan. 1999. Integrin-associated protein is a ligand for the P84 neural adhesion molecule. *J. Biol. Chem.* **274:**559–562.

Joiner, K. A., T. Ganz, J. Albert, and D. Rotrosen. 1989. The opsonizing ligand on Salmonella typhimurium influences incorporation of specific, but not azurophil, granule constituents into neutrophil phagosomes. *J. Cell Biol.* **109:**2771–2782.

Jones, S. L., U. G. Knaus, G. M. Bokoch, and E. J. Brown. 1998. Two signaling mechanisms for activation of alphaM beta2 avidity in polymorphonuclear neutrophils. *J. Biol. Chem.* **273:**10556–10566.

Kaplan, G. 1977. Differences in the mode of phagocytosis with Fc and C3 receptors in macrophages. *Scand. J. Immunol.* **6:** 797–807.

Katagiri, K., M. Hattori, N. Minato, K. Irie, K. Takatsu, and T. Kinashi. 2000. Rap1 is a potent activation signal for leukocyte function-associated antigen 1 distinct from protein kinase C and phosphatidylinositol-3-OH kinase. *Mol. Cell Biol.* **20:**1956–1969.

Katagiri, K., A. Maeda, M. Shimonaka, and T. Kinashi. 2003. RAPL, a Rap1-binding molecule that mediates Rap1-induced adhesion through spatial regulation of LFA-1. *Nat. Immunol.* **4:**741–748.

Katagiri, K., N. Ohnishi, K. Kabashima, T. Iyoda, N. Takeda, Y. Shinkai, K. Inaba, and T. Kinashi. 2004. Crucial functions of the Rap1 effector molecule RAPL in lymphocyte and dendritic cell trafficking. *Nat. Immunol.* **5:**1045–1051.

Kim, M., C. V. Carman, and T. A. Springer. 2003. Bidirectional transmembrane signaling by cytoplasmic domain separation in integrins. *Science* **301:**1720–1725.

Kinashi, T., and K. Katagiri. 2004. Regulation of lymphocyte adhesion and migration by the small GTPase Rap1 and its effector molecule, RAPL. *Immunol. Lett.* **93:**1–5.

Kucik, D. F., M. L. Dustin, J. M. Miller, and E. J. Brown. 1996. Adhesion-activating phorbol ester increases the mobility of leukocyte integrin LFA-1 in cultured lymphocytes. *J. Clin. Invest.* **97:**2139–2144.

Kurth, I., K. Willimann, P. Schaerli, T. Hunziker, I. Clark-Lewis, and B. Moser. 2001. Monocyte selectivity and tissue localization suggests a role for breast and kidney-expressed chemokine (BRAK) in macrophage development. *J. Exp. Med.* **194:**855–861.

Lafuente, E. M., A. A. van Puijenbroek, M. Krause, C. V. Carman, G. J. Freeman, A. Berezovskaya, E. Constantine, T. A. Springer, F. B. Gertler, and V. A. Boussiotis. 2004. RIAM, an Ena/VASP and Profilin ligand, interacts with Rap1-GTP and mediates Rap1-induced adhesion. *Dev. Cell* **7:**585–595.

Ley, K. 2002. Integration of inflammatory signals by rolling neutrophils. *Immunol. Rev.* **186:**8–18.

Li, R., N. Mitra, H. Gratkowski, G. Vilaire, R. Litvinov, C. Nagasami, J. W. Weisel, J. D. Lear, W. F. DeGrado, and J. S. Bennett. 2003. Activation of integrin alphaIIbbeta3 by modulation of transmembrane helix associations. *Science* **300:**795–798.

Lindberg, F. P., D. C. Bullard, T. E. Caver, H. D. Gresham, A. L. Beaudet, and E. J. Brown. 1996. Decreased resistance to bacterial infection and granulocyte defects in IAP-deficient mice. *Science* **274**:795–798.

Lowell, C. A. 2006. Rewiring phagocytic signal transduction. *Immunity* **24**:243–245.

Lowell, C. A., and G. Berton. 1998. Resistance to endotoxic shock and reduced neutrophil migration in mice deficient for the Src-family kinases Hck and Fgr. *Proc. Natl. Acad. Sci. USA* **95**:7580–7584.

Luo, B. H., C. V. Carman, and T. A. Springer. 2007. Structural basis of integrin regulation and signaling. *Annu. Rev. Immunol.* **25**:619–647.

Martel, V., C. Racaud-Sultan, S. Dupe, C. Marie, F. Paulhe, A. Galmiche, M. R. Block, and C. Albiges-Rizo. 2001. Conformation, localization, and integrin binding of talin depend on its interaction with phosphoinositides. *J. Biol. Chem.* **276**:21217–21227.

May, R. C., E. Caron, A. Hall, and L. M. Machesky. 2000. Involvement of the Arp2/3 complex in phagocytosis mediated by FcgammaR or CR3. *Nat. Cell Biol.* **2**:246–248.

McHugh, K. P., K. Hodivala-Dilke, M. H. Zheng, N. Namba, J. Lam, D. Novack, X. Feng, F. P. Ross, R. O. Hynes, and S. L. Teitelbaum. 2000. Mice lacking beta3 integrins are osteosclerotic because of dysfunctional osteoclasts. *J. Clin. Invest.* **105**:433–440.

Meng, F., and C. A. Lowell. 1998. A beta 1 integrin signaling pathway involving Src-family kinases, Cbl and PI-3 kinase is required for macrophage spreading and migration. *EMBO J.* **17**:4391–4403.

Middleton, J., A. M. Patterson, L. Gardner, C. Schmutz, and B. A. Ashton. 2002. Leukocyte extravasation: chemokine transport and presentation by the endothelium. *Blood* **100**:3853–3860.

Mocsai, A., C. L. Abram, Z. Jakus, Y. Hu, L. L. Lanier, and C. A. Lowell. 2006. Integrin signaling in neutrophils and macrophages uses adaptors containing immunoreceptor tyrosine-based activation motifs. *Nat. Immunol.* **7**:1326–1333.

Mocsai, A., M. B. Humphrey, J. A. Van Ziffle, Y. Hu, A. Burghardt, S. C. Spusta, S. Majumdar, L. L. Lanier, C. A. Lowell, and M. C. Nakamura. 2004. The immunomodulatory adapter proteins DAP12 and Fc receptor gamma-chain (FcRgamma) regulate development of functional osteoclasts through the Syk tyrosine kinase. *Proc. Natl. Acad. Sci. USA* **101**:6158–6163.

Mocsai, A., M. Zhou, F. Meng, V. L. Tybulewicz, and C. A. Lowell. 2002. Syk is required for integrin signaling in neutrophils. *Immunity* **16**:547–558.

Mosser, D. M., and P. J. Edelson. 1985. The mouse macrophage receptor for C3bi (CR3) is a major mechanism in the phagocytosis of Leishmania promastigotes. *J. Immunol.* **135**:2785–2789.

Mosser, D. M., and P. J. Edelson. 1987. The third component of complement (C3) is responsible for the intracellular survival of Leishmania major. *Nature* **327**:329–331.

Nathan, C., S. Srimal, C. Farber, E. Sanchez, L. Kabbash, A. Asch, J. Gailit, and S. D. Wright. 1989. Cytokine-induced respiratory burst of human neutrophils: dependence on extracellular matrix proteins and CD11/CD18 integrins. *J. Cell Biol.* **109**:1341–1349.

Nishida, N., C. Xie, M. Shimaoka, Y. Cheng, T. Walz, and T. A. Springer. 2006. Activation of leukocyte beta2 integrins by conversion from bent to extended conformations. *Immunity* **25**:583–594.

Olazabal, I. M., E. Caron, R. C. May, K. Schilling, D. A. Knecht, and L. M. Machesky. 2002. Rho-kinase and myosin

II control phagocytic cup formation during CR, but not FcgammaR, phagocytosis. *Curr. Biol.* **12**:1413–1418.

Pasvolsky, R., S. W. Feigelson, S. S. Kilic, A. J. Simon, G. Tal-Lapidot, V. Grabovsky, J. R. Crittenden, N. Amariglio, M. Safran, A. M. Graybiel, G. Rechavi, S. Ben-Dor, A. Etzioni, and R. Alon. 2007. A LAD-III syndrome is associated with defective expression of the Rap-1 activator CalDAG-GEFI in lymphocytes, neutrophils, and platelets. *J. Exp. Med.* **204**:1571–1582.

Pereira, S., H. Zhang, T. Takai, and C. A. Lowell. 2004. The inhibitory receptor PIR-B negatively regulates neutrophil and macrophage integrin signaling. *J. Immunol.* **173**:5757–5765.

Pfeiffer, A., A. Bottcher, E. Orso, M. Kapinsky, P. Nagy, A. Bodnar, I. Spreitzer, G. Liebisch, W. Drobnik, K. Gempel, M. Horn, S. Holmer, T. Hartung, G. Multhoff, G. Schutz, H. Schindler, A. J. Ulmer, H. Heine, F. Stelter, C. Schutt, G. Rothe, J. Szollosi, S. Damjanovich, and G. Schmitz. 2001. Lipopolysaccharide and ceramide docking to CD14 provokes ligand-specific receptor clustering in rafts. *Eur. J. Immunol.* **31**:3153–3164.

Prince, J. E., C. F. Brayton, M. C. Fossett, J. A. Durand, S. L. Kaplan, C. W. Smith, and C. M. Ballantyne. 2001. The differential roles of LFA-1 and Mac-1 in host defense against systemic infection with Streptococcus pneumoniae. *J. Immunol.* **166**:7362–7369.

Reedquist, K. A., E. Ross, E. A. Koop, R. M. Wolthuis, F. J. Zwartkruis, Y. van Kooyk, M. Salmon, C. D. Buckley, and J. L. Bos. 2000. The small GTPase, Rap1, mediates CD31-induced integrin adhesion. *J. Cell Biol.* **148**:1151–1158.

Relman, D., E. Tuomanen, S. Falkow, D. T. Golenbock, K. Saukkonen, and S. D. Wright. 1990. Recognition of a bacterial adhesion by an integrin: macrophage CR3 (alpha M beta 2, CD11b/CD18) binds filamentous hemagglutinin of Bordetella pertussis. *Cell* **61**:1375–1382.

Rose, D. M., R. Alon, and M. H. Ginsberg. 2007. Integrin modulation and signaling in leukocyte adhesion and migration. *Immunol. Rev.* **218**:126–134.

Ross, F. P., and S. L. Teitelbaum. 2005. alphavbeta3 and macrophage colony-stimulating factor: partners in osteoclast biology. *Immunol. Rev.* **208**:88–105.

Russell, D. G., and S. D. Wright. 1988. Complement receptor type 3 (CR3) binds to an Arg-Gly-Asp-containing region of the major surface glycoprotein, gp63, of Leishmania promastigotes. *J. Exp. Med.* **168**:279–292.

Saukkonen, K., C. Cabellos, M. Burroughs, S. Prasad, and E. Tuomanen. 1991. Integrin mediated localization of Bordetella pertussis within macrophages: role in pulmonary colonization. *J. Exp. Med.* **173**:1143–1149.

Savill, J., I. Dransfield, N. Hogg, and C. Haslett. 1990. Vitronectin receptor-mediated phagocytosis of cells undergoing apoptosis. *Nature* **343**:170–173.

Sebzda, E., M. Bracke, T. Tugal, N. Hogg, and D. A. Cantrell. 2002. Rap1A positively regulates T cells via integrin activation rather than inhibiting lymphocyte signaling. *Nat. Immunol.* **3**:251–258.

Sheppard, D. 2005. Integrin-mediated activation of latent transforming growth factor beta. *Cancer Metastasis Rev.* **24**:395–402.

Shimaoka, M., C. Lu, R. T. Palframan, U. H. von Andrian, A. McCormack, J. Takagi, and T. A. Springer. 2001. Reversibly locking a protein fold in an active conformation with a disulfide bond: integrin alphaL I domains with high affinity and antagonist activity in vivo. *Proc. Natl. Acad. Sci. USA* **98**:6009–6014.

Simon, D. I., Y. Wei, L. Zhang, N. K. Rao, H. Xu, Z. Chen, Q. Liu, S. Rosenberg, and H. A. Chapman. 2000. Identification of a urokinase receptor-integrin interaction site. Promiscuous regulator of integrin function. *J. Biol. Chem.* **275:** 10228–10234.

Singer, I. I., S. Scott, D. W. Kawka, and D. M. Kazazis. 1989. Adhesomes: specific granules containing receptors for laminin, C3bi/fibrinogen, fibronectin, and vitronectin in human polymorphonuclear leukocytes and monocytes. *J. Cell Biol.* **109:**3169–3182.

Soriano, P., C. Montgomery, R. Geske, and A. Bradley. 1991. Targeted disruption of the c-src proto-oncogene leads to osteopetrosis in mice. *Cell* **64:**693–702.

Springer, T. A. 1990. Adhesion receptors of the immune system. *Nature* **346:**425–434.

Stouffer, G. A., and S. S. Smyth. 2003. Effects of thrombin on interactions between beta3-integrins and extracellular matrix in platelets and vascular cells. *Arterioscler. Thromb. Vasc. Biol.* **23:**1971–1978.

Suen, P. W., D. Ilic, E. Caveggion, G. Berton, C. H. Damsky, and C. A. Lowell. 1999. Impaired integrin-mediated signal transduction, altered cytoskeletal structure and reduced motility in Hck/Fgr deficient macrophages. *J. Cell Sci.* **112**(Pt 22)**:**4067–4078.

Suzuki, S., and Y. Naitoh. 1990. Amino acid sequence of a novel integrin beta 4 subunit and primary expression of the mRNA in epithelial cells. *EMBO J.* **9:**757–763.

Tadokoro, S., S. J. Shattil, K. Eto, V. Tai, R. C. Liddington, J. M. de Pereda, M. H. Ginsberg, and D. A. Calderwood. 2003. Talin binding to integrin beta tails: a final common step in integrin activation. *Science* **302:**103–106.

Talamas-Rohana, P., S. D. Wright, M. R. Lennartz, and D. G. Russell. 1990. Lipophosphoglycan from Leishmania mexicana promastigotes binds to members of the CR3, p150,95 and LFA-1 family of leukocyte integrins. *J. Immunol.* **144:**4817–4824.

Tang, T., A. Rosenkranz, K. J. Assmann, M. J. Goodman, J. C. Gutierrez-Ramos, M. C. Carroll, R. S. Cotran, and T. N. Mayadas. 1997. A role for Mac-1 (CDIIb/CD18) in immune complex-stimulated neutrophil function in vivo: Mac-1 deficiency abrogates sustained Fcgamma receptor-dependent neutrophil adhesion and complement-dependent proteinuria in acute glomerulonephritis. *J. Exp. Med.* **186:** 1853–1863.

Teitelbaum, S. L. 2007. Osteoclasts: what do they do and how do they do it? *Am. J. Pathol.* **170:**427–435.

Teitelbaum, S. L., and F. P. Ross. 2003. Genetic regulation of osteoclast development and function. *Nat. Rev. Genet.* **4:** 638–649.

Thrasher, A. J. 2002. WASp in immune-system organization and function. *Nat. Rev. Immunol.* **2:**635–646.

Travis, M. A., B. Reizis, A. C. Melton, E. Masteller, Q. Tang, J. M. Proctor, Y. Wang, X. Bernstein, X. Huang, L. F. Reichardt, J. A. Bluestone, and D. Sheppard. 2007. Loss of integrin alpha(v)beta(8) on dendritic cells causes autoimmunity and colitis in mice. *Nature* **449:**361–365.

Van, V. Q., S. Lesage, S. Bouguermouh, P. Gautier, M. Rubio, M. Levesque, S. Nguyen, L. Galibert, and M. Sarfati. 2006. Expression of the self-marker CD47 on dendritic cells governs their trafficking to secondary lymphoid organs. *EMBO J.* **25:**5560–5568.

Van Furth, R., and Z. A. Cohn. 1968. The origin and kinetics of mononuclear phagocytes. *J. Exp. Med.* **128:**415–435.

Van Furth, R., M. M. C. Diesselhoff-den Dulk, W. Sluiter, and J. T. Van Dissel. 1985. New perspectives on the kinetics of mononuclear phagocytes. *In* R. Van Furth (ed.), *Mononuclear Phagocytes: Characteristics, Physiology, & Function. Proceedings of the Fourth Conference on Mononuclear Phagocytes.* Martinus Nijhoff, Leiden, The Netherlands.

Van Spriel, A. B., J. H. Leusen, M. van Egmond, H. B. Dijkman, K. J. Assmann, T. N. Mayadas, and J. G. van de Winkel. 2001. Mac-1 (CD11b/CD18) is essential for Fc receptor-mediated neutrophil cytotoxicity and immunologic synapse formation. *Blood* **97:**2478–2486.

Van Strijp, J. A., D. G. Russell, E. Tuomanen, E. J. Brown, and S. D. Wright. 1993. Ligand specificity of purified complement receptor type three (CD11b/CD18, alpha m beta 2, Mac-1). Indirect effects of an Arg-Gly-Asp (RGD) sequence. *J. Immunol.* **151:**3324–3336.

Vedder, N. B., and J. M. Harlan. 1988. Increased surface expression of CD11b/CD18 (Mac-1) is not required for stimulated neutrophil adherence to cultured endothelium. *J. Clin. Invest.* **81:**676–682.

Vestweber, D. 2007. Adhesion and signaling molecules controlling the transmigration of leukocytes through endothelium. *Immunol. Rev.* **218:**178–196.

Vicentini, L., P. Mazzi, E. Caveggion, S. Continolo, L. Fumagalli, J. A. Lapinet-Vera, C. A. Lowell, and G. Berton. 2002. Fgr deficiency results in defective eosinophil recruitment to the lung during allergic airway inflammation. *J. Immunol.* **168:**6446–6454.

Vinogradova, O., A. Velyvis, A. Velyviene, B. Hu, T. Haas, E. Plow, and J. Qin. 2002. A structural mechanism of integrin alpha(IIb)beta(3) "inside-out" activation as regulated by its cytoplasmic face. *Cell* **110:**587–597.

Wang, J., and E. J. Brown. 1999. Immune complex-induced integrin activation and L-plastin phosphorylation require protein kinase A. *J. Biol. Chem.* **274:**24349–24356.

Wetzel, A., T. Chavakis, K. T. Preissner, M. Sticherling, U. F. Haustein, U. Anderegg, and A. Saalbach. 2004. Human Thy-1 (CD90) on activated endothelial cells is a counterreceptor for the leukocyte integrin Mac-1 (CD11b/CD18). *J. Immunol.* **172:**3850–3859.

Wright, S. D. 1991. Multiple receptors for endotoxin. *Curr. Opin. Immunol.* **3:**83–90.

Wright, S. D., M. R. Licht, L. S. Craigmyle, and S. C. Silverstein. 1984. Communication between receptors for different ligands on a single cell: ligation of fibronectin receptors induces a reversible alteration in the function of complement receptors on cultured human monocytes. *J. Cell Biol.* **99:** 336–339.

Wright, S. D., and S. C. Silverstein. 1983. Receptors for C3b and C3bi promote phagocytosis but not the release of toxic oxygen from human phagocytes. *J. Exp. Med.* **158:**2016–2023.

Xiong, J. P., T. Stehle, B. Diefenbach, R. Zhang, R. Dunker, D. L. Scott, A. Joachimiak, S. L. Goodman, and M. A. Arnaout. 2001. Crystal structure of the extracellular segment of integrin alpha Vbeta3. *Science* **294:**339–345.

Xue, W., A. L. Kindzelskii, R. F. Todd III, and H. R. Petty. 1994. Physical association of complement receptor type 3 and urokinase-type plasminogen activator receptor in neutrophil membranes. *J. Immunol.* **152:**4630–4640.

Yan, S. R., M. Huang, and G. Berton. 1997. Signaling by adhesion in human neutrophils: activation of the p72syk tyrosine kinase and formation of protein complexes containing p72syk and Src family kinases in neutrophils spreading over fibrinogen. *J. Immunol.* **158:**1902–1910.

Yang, Z., Z. Mu, B. Dabovic, V. Jurukovski, D. Yu, J. Sung, X. Xiong, and J. S. Munger. 2007. Absence of integrin-mediated TGFbeta1 activation in vivo recapitulates the phenotype of TGFbeta1-null mice. *J. Cell Biol.* **176:**787–793.

Zhang, H., U. Y. Schaff, C. E. Green, H. Chen, M. R. Sarantos, Y. Hu, D. Wara, S. I. Simon, and C. A. Lowell.

2006. Impaired integrin-dependent function in Wiskott-Aldrich syndrome protein-deficient murine and human neutrophils. *Immunity* **25:**285–295.

Zhou, M., R. F. Todd, 3rd, J. G. van de Winkel, and H. R. Petty. 1993. Cocapping of the leukoadhesin molecules complement receptor type 3 and lymphocyte function-associated antigen-1 with Fc gamma receptor III on human neutrophils. Possible role of lectin-like interactions. *J. Immunol.* **150:**3030–3041.

Zou, W., H. Kitaura, J. Reeve, F. Long, V. L. Tybulewicz, S. J. Shattil, M. H. Ginsberg, F. P. Ross, and S. L. Teitelbaum. 2007. Syk, c-Src, the alphavbeta3 integrin, and ITAM immunoreceptors, in concert, regulate osteoclastic bone resorption. *J. Cell Biol.* **176:**877–888.

9

Intracytosolic Sensing of Pathogens: Nucleic Acid Receptors, NLRs, and the Associated Responses during Infections and Autoinflammatory Diseases

THOMAS HENRY AND DENISE M. MONACK

Phagocytes are recruited to lesions and sites of inflammation caused by invading microbes. Typically, macrophages are inhospitable hosts for microbes because they are endowed with numerous antimicrobial effector functions. However, some pathogens have evolved mechanisms to subvert macrophage responses and can establish a persistent reservoir or a replicative niche inside these host cells. Thus, the interactions between phagocytes and microbial pathogens are crucial for the initiation of innate immune defense mechanisms.

Although nonphagocytic cells can detect pathogens, phagocytes are of prime importance in signaling in response to pathogens. The innate immune system recognizes molecules conserved within numerous microbes and pathogens which are called pathogen-associated molecular patterns, or PAMPs. PAMPs are recognized by host receptors called pattern recognition receptors (PRRs) (Janeway, 1989). PRRs can also recognize danger signals (Matzinger, 2002), such as membrane damage, or molecules present both in microbes and in the host, such as nucleic acids. However, typically they distinguish self from nonself.

The best-characterized PRRs are the Toll-like receptors (TLRs) (see chapter 10), which can sense PAMPs present extracellularly or in the endosomal system. However, viruses and some bacterial pathogens such as *Listeria monocytogenes*, *Shigella flexneri*, or *Francisella tularensis* are only transiently exposed to TLRs and quickly gain access to the phagocyte cytosol. Other bacterial pathogens such as *Salmonella enterica* serovar Typhimurium, *Legionella pneumophila*, or *Mycobacterium tuberculosis* use specialized secretion systems to secrete bacterial effectors into the cytosol and establish a vacuolar replicative niche in the phagocyte. Several PRRs present in the cytosol can detect the presence of intracellular pathogens, leading to different signals.

Most innate immune signaling pathways can be schematically presented as three-player pathways consisting of a sensor (or receptor), an adaptor, and an effector. Upon engagement of the sensor by direct or indirect recognition of a PAMP, a change in conformation is thought to reveal a domain that allows the interaction with an adaptor to occur. Adaptors are scaffolding molecules that act to bring several proteins together, which results in the activation of an effector. There are two main effector classes:

- Transcription factors such as nuclear factor-κB (NF-κB) and interferon regulatory factor-3 (IRF3) that upon activation lead to the upregulation of proinflammatory cytokines in the former case or type I interferon (IFN) secretion in the latter case
- Proteases such as caspase-1 that upon activation lead to a posttranslational response exemplified mainly by the maturation of proinflammatory cytokines and cell death

Cytosolic PRRs can be classified in three categories based on the response they trigger:

1. The PRRs RIG-I, MDA-5, and DAI, which sense nucleic acids and trigger a type I IFN response
2. The PRRs NOD1 and NOD2, which sense peptidoglycan fragments and trigger NF-κB activation
3. The PRRs, which sense various kinds of stimuli and are involved in the assembly of multiprotein complexes that leads to caspase-1 activation.

The last two classes of PRRs comprise a family of receptors called the nucleotide binding domain, leucine-rich repeat-containing family (NLR family) (Table 1).

We will review here the current knowledge concerning the cytosolic PRRs, the responses they trigger, their physiological role during infection, and their regulation. Finally, we will briefly describe the diseases associated with mutations in several of these PRRs or associated proteins.

Thomas Henry and Denise M. Monack, Stanford University, School of Medicine, Department of Microbiology and Immunology, Fairchild Building, 299 Campus Drive, Stanford, CA 94305-5124.

TABLE 1 Human NLR proteins

HGNC approved symbol[a]	HGNC approved name	Other symbols	Domain organization[b]
NOD1	Nucleotide binding oligomerization domain containing 1	NLRC1/CARD4	CARD-NBD-LRR
NOD2	Nucleotide binding oligomerization domain containing 2	NLRC2/CARD15	CARD-NBD-LRR
NLRC3	NLR family, CARD domain containing 3	NOD3	CARD-NBD-LRR
NLRC4	NLR family, CARD domain containing 4	IPAF	CARD-NBD-LRR
NLRC5	NLR family, CARD domain containing 5	NOD27	CARD-NBD-LRR
NLX1	NLR family member X1	NOD9	X-NBD-LRR
Naip	NLR family, apoptosis-inhibitory protein	NLRB1/BIRC1	BIR-NBD-LRR
NLRP1	NLR family, pyrin domain containing 1	Nalp1	PYD-NBD-LRR-CARD
NLRP2	NLR family, pyrin domain containing 2	Nalp2	PYD-NBD-LRR
NLRP3	NLR family, pyrin domain containing 3	Nalp3/Cryopyrin	PYD-NBD-LRR
NLRP4	NLR family, pyrin domain containing 4	Nalp4	PYD-NBD-LRR
NLRP5	NLR family, pyrin domain containing 5	Nalp5/Mater	PYD-NBD-LRR
NLRP6	NLR family, pyrin domain containing 6	Nalp6	PYD-NBD-LRR
NLRP7	NLR family, pyrin domain containing 7	Nalp7	PYD-NBD-LRR
NLRP8	NLR family, pyrin domain containing 8	Nalp8	PYD-NBD-LRR
NLRP9	NLR family, pyrin domain containing 9	Nalp9	PYD-NBD-LRR
NLRP10	NLR family, pyrin domain containing 10	Nalp10	PYD-NBD
NLRP11	NLR family, pyrin domain containing 11	Nalp11	PYD-NBD-LRR
NLRP12	NLR family, pyrin domain containing 12	Nalp12	PYD-NBD-LRR
NLRP13	NLR family, pyrin domain containing 13	Nalp13	PYD-NBD-LRR
NLRP14	NLR family, pyrin domain containing 14	Nalp14	PYD-NBD-LRR

[a]This table is based on the nomenclature adopted by the HUGO Gene Nomenclature Committee (HGNC) as of February 16, 2007.
[b]CARD, caspase activating and recruitment domain; LRR, leucine-rich repeat; NBD, nucleotide binding domain; BIR, baculovirus inhibitor of apoptosis repeat; PYD, pyrin domain.

SENSING OF NUCLEIC ACIDS IN THE HOST CYTOSOL TRIGGERS A TYPE I IFN RESPONSE

Several viruses such as human immunodeficiency virus (HIV) (Cassol et al., 2006) (see chapter 33), dengue virus (Blackley et al., 2007), or vaccinia virus (Tompkins et al., 1970) can target phagocytes either as a replicative niche or as a permanent reservoir. Moreover, phagocytes play a critical role in the innate immune response to viral infection, in particular, by releasing high amounts of type I IFNs including IFN-β and numerous IFNs-α, which are the main antiviral cytokines (Durbin et al., 1996). The ability of phagocytes to respond to viruses depends on their ability to detect the infection. Viruses can be detected through recognition of their nucleic acids either extracellularly or in the endosomes by TLRs (see chapter 10). They can also be detected in the cytosol by three host PRRs called retinoic acid-inducible gene-I (RIG-I/Ddx58), melanoma differentiation-associated gene-5 (MDA-5/Helicard/Ifih1), and DNA-dependent activator of IFN regulatory factors (DAI) (Takeuchi and Akira, 2008) (Fig. 1).

RIG-I and MDA-5 Detect Nonhost RNAs

RIG-I and MDA-5 are two RNA helicases responsible for the detection of numerous viruses. RIG-I binds 5′-triphosphate RNA (Hornung et al., 2006; Pichlmair et al., 2006), which is present in the host cytosol during infection with negative-strand RNA viruses (NSVs) such as influenza virus. In contrast, the posttranscriptional modifications undergone by eukaryotic RNA in the nucleus before export into the cytosol (capping of the 5′ extremity of mRNA, 5′-monophosphate of rRNA and tRNA, and nucleoside modifications) inhibit the recognition of host RNA by RIG-I, leading to the discrimination between self and viral (nonself) RNA in the cytosol. MDA-5 recognizes double-stranded RNA, a structure generated by RNA viruses during RNA-dependent RNA synthesis but absent from noninfected eukaryotic cells (Kang et al., 2002; Gitlin et al., 2006).

In addition to the helicase domain involved in ligand binding, both PRRs contain a caspase-activating and recruitment domain (CARD). After engagement, a change in conformation is thought to allow the interaction of their CARD with the CARD of the adaptor protein MAVS (mitochondrial antiviral signaling/IPS-1/VISA/CARDIF) (Kawai et al., 2005; Meylan et al., 2005; Seth et al., 2005; Xu et al., 2005). MAVS is a critical adaptor required for the induction of type I IFN in response to cytosolic nucleic acids (Seth et al., 2005). MAVS binds TRAF3 (Hacker et al., 2006; Oganesyan et al., 2006; Saha et al., 2006), an E3 ubiquitin ligase that self-ligates Lys-63 polyubiquitination (nondegradative ubiquitination) and promotes its association with TANK binding kinase-1 (TBK-1/T2K/NAK) (Xu et al., 2005; Kayagaki et al., 2007). In infected cells, TBK1 phosphorylates IRF3 (Fitzgerald et al., 2003; Sharma et al., 2003), which is normally inactive in the cytoplasm, which leads to its homodimerization, nuclear translocation, and binding to IFN-stimulated response element (ISRE) resulting in the upregulation of numerous genes including ifn-β. In some cell types, including macrophages, another kinase, inducible Ikk-B kinase (IKKi; also called IKKε) (Meylan et al., 2005), which interacts with MAVS (Meylan et al., 2005), can also phosphorylate IRF3, showing a redundant role with TBK1 (Hemmi et al., 2004; Perry et al., 2004).

Even if these signaling pathways are not restricted to phagocytes and can be mobilized in infected fibroblasts, phagocytes seem to be the main source of type I IFN during viral infection in vivo (Cella et al., 1999; Siegal et al., 1999). Depending on the virus and on the route of infec-

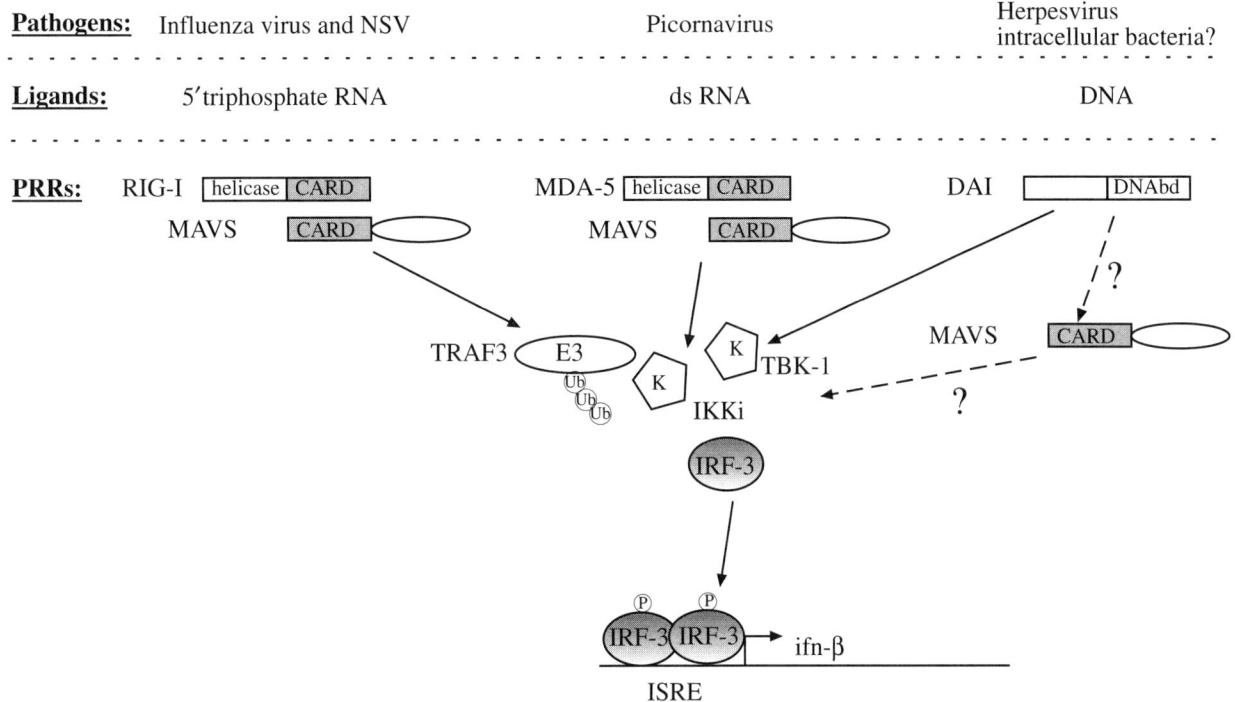

FIGURE 1 Nonself nucleic acids within the host cytosol trigger type I IFN secretion. Viruses, and possibly intracellular bacteria, release nonself nucleic acids into the host cytosol that are recognized by the PRRs RIG-I, MDA-5, and DAI. PRRs signal through the adaptor MAVS, the E3-ubiquitin ligase TRAF3, and the kinases TBK-1 and IKKi, leading to phosphorylation of the transcription factor IRF3. Phosphorylated IRF3 dimerizes and translocates into the nucleus, where it binds ISREs and induces transcription of numerous genes including ifn-β.

tion, different cells are the main producers of type I IFNs (Kumagai et al., 2007). For example, plasmacytoid dendritic cells (DCs) are characterized by their high capacity to secrete IFN-α (Cella et al., 1999) during systemic infection, while alveolar macrophages and conventional DCs play a major role in IFN-α secretion during viral infection in the lungs (Kumagai et al., 2007). Interestingly, the pathways involved in the generation of type I IFNs appear to be cell type specific. Indeed, plasmacytoid DCs produce type I IFNs mainly in a TLR-dependent manner, while other phagocytes such as conventional DCs and macrophages produce type I IFNs in a RIG-I-, MDA-5-dependent manner (Kato et al., 2005; Kumagai et al., 2007).

The critical role of RIG-I or MDA-5 in the innate immune response to viruses has been studied by using knockout mice. RIG-I$^{-/-}$ mice have been shown to be more susceptible to infection by Japanese encephalitis virus (a negative-strand RNA virus) than wild-type (WT) or MDA-5$^{-/-}$ mice, which is in agreement with the described ligand. In contrast, MDA-5$^{-/-}$ mice were more susceptible to infection with encephalomyocarditis virus, a positive-strand RNA virus (Kato et al., 2006), demonstrating that cytosolic sensing is critical to mounting an efficient antiviral response.

DAI Senses Cytosolic DNA

While it has long been known that viral RNA is recognized in the host cytosol, until recently it was still an open question whether or not viral DNA could be sensed by a PRR other than TLR9. The first direct evidence came from two studies showing that delivery of double-stranded DNA in the cytosol induces a potent type I IFN response independently of TLR, RIG-I, and MDA-5 signaling (Ishii et al., 2006; Stetson and Medzhitov, 2006). In contrast to TLR9, engagement of this cytosolic signaling pathway does not require the presence of an unmethylated CpG island since mammalian DNA in the cytosol also activates this response. The discrimination between self and nonself might be due to the difference of localization, with host DNA being sequestered in the nucleus and thus not exposed to the cytosolic DNA receptor. Similar to the RIG-I and the MDA-5 pathways, the type I IFN induction in response to cytosolic DNA is dependent on the kinases TBK1 and IKKi.

In 2007, a DNA cytosolic receptor was identified and named DAI for DNA-dependent activator of IFN regulatory factors (Takaoka et al., 2007). This receptor is capable of binding double-stranded DNA from both eukaryotic and prokaryotic origins independently of the sequence. Upon DNA binding, its association with the kinase TBK1 and the effector IRF3 increases, followed by translocation of IRF3 into the nucleus, resulting in IFN-β induction. It is likely that this PRR participates in detection of DNA viruses such as herpesvirus (Hochrein et al., 2004). Moreover, it may participate in the detection of intracellular bacteria.

Intracellular Bacteria Can Induce a TLR-Independent Type I IFN Response

While most of the knowledge concerning type I IFN secretion comes from studying viral infections, recent data indicate that some cytosolic PRRs can detect intracellular bacteria and trigger a type I IFN response.

Bacteria replicating in the host cytosol, both gram-negative bacteria, such as *S. flexneri* (Hess et al., 1987) and *F. tularensis* (Henry et al., 2007), and gram-positive bacteria, such as *L. monocytogenes* (O'Riordan et al., 2002), can trigger type I IFN secretion. In contrast, bacterial mutants that are unable to reach the cytosol do not trigger this response, suggesting that a cytosolic PRR is involved in the recognition of cytosolic bacterial products. During *Listeria* or *Francisella* infections, this response is independent of TLRs and MAVS (McCaffrey et al., 2004; Soulat et al., 2006; Henry et al., 2007).

Delivery of double-stranded DNA into the host cell cytosol has been shown to recapitulate the type I IFN secretion induced by cytosolic bacteria (Stetson and Medzhitov, 2006; Leber et al., 2008). Thus, it has been suggested that bacterial DNA could be the ligand sensed in the cytosol (Stetson and Medzhitov, 2006), leading to the hypothesis that DAI could be the PRR responsible for sensing intracytosolic bacteria. However, at this time, this remains to be proven.

In addition to cytosolic bacteria, certain intracellular bacteria residing in a vacuole/phagosome can also induce a TLR-independent IFN-β secretion. In resting or unstimulated cells, this response correlates with the presence of specialized bacterial secretion systems, such as type IV secretion systems, that are responsible for delivering bacterial molecules directly into the host cytosol. Indeed, *L. pneumophila* (see chapter 29) induces type I IFNs in a dot/Icm-type IV secretion system-dependent manner (Opitz et al., 2006b). Type IV secretion systems are conjugation-adaptive transporters allowing the export of DNA from the bacteria (Vogel et al., 1998). Thus, it has been suggested that DNA could be exported from the bacteria by the dot/Icm system into the host cytosol, where it is recognized by a cytosolic PRR that triggers type I IFN secretion (Stetson and Medzhitov, 2006). Similarly, *M. tuberculosis* resides in a vacuole during early infection and uses a specialized transport system called ESX-1 (DiGiuseppe Champion and Cox, 2007) or type VII secretion system (Abdallah et al., 2007) to translocate proteins into the host cytosol. In macrophages, *M. tuberculosis* can induce type I IFN secretion in an ESX-1-dependent manner, suggesting that bacterial molecules secreted by the ESX-1 transport apparatus are detected in the host cytosol. The PRR responsible for this detection is still unknown; however, type I IFN secretion is independent of TLRs and requires TBK1 (Stanley et al., 2007).

Finally, it was recently shown that in some cases the cytosolic recognition systems are able to sense vacuolar bacteria independently of specialized bacterial secretion systems. Indeed, in macrophages activated by IFN-γ, *L. monocytogenes* is degraded in the phagolysosome but is still capable of eliciting the cytosolic response as monitored by IFN-β induction. This response required the generation of bacterial ligands in the phagolysosome and largely depended on nucleotide binding and oligomerization domain 2, NOD2, in the cytosol. Since the NOD2-dependent response to degraded bacteria required an intact phagosomal membrane potential and the activity of lysosomal proteases, it has been proposed that the active transport of bacterial molecules can occur between the phagosome and the cytosol, leading to engagement of the cytosolic PRR (Herskovits et al., 2007).

It is still unclear if the type I IFN secretion triggered by intracellular bacteria is a virulence aspect of the pathogen or if it is a host immune defense mechanism. Indeed, while mice deficient for type I IFN signaling or IRF3 are much more susceptible to viral infection, they are more resistant to infection with *L. monocytogenes* (Auerbuch et al., 2004; Carrero et al., 2004; O'Connell et al., 2004) and *M. tuberculosis* (Stanley et al., 2007), indicating that type I IFNs contribute to the pathogenesis during these bacterial infections.

In the same way that type I IFN secretion can be triggered by engagement of TLRs or by engagement of cytosolic PRRs, NF-κB can be activated in response to bacteria either after engagement of TLRs at the plasma membrane or in the endosomes after recognition of bacterial cell wall fragment by the cytosolic PRRs NOD1 and NOD2.

SENSING OF PEPTIDOGLYCAN FRAGMENTS IN THE HOST CYTOSOL TRIGGERS NF-κB ACTIVATION

The first clue that a cytosolic PRR capable of sensing a bacterial molecule exists came from the observation that the microinjection of bacterial supernatant into the cytoplasm of epithelial cells leads to NF-κB and mitogen-activated protein kinase (MAPK) activation while there is no effect when added extracellularly (Philpott et al., 2000). Such a phenotype correlated with the capacity of invasive *S. flexneri* strains to induce NF-κB activation in epithelial cells, in contrast to noninvasive mutants (Philpott et al., 2000). Based on homology with the resistance proteins from the plant innate immune system, two cytosolic PRRs, called nucleotide binding and oligomerization domain 1 (NOD1) (Girardin et al., 2001; Inohara et al., 2001) and NOD2 (Girardin et al., 2003b), were later identified as the receptors responsible for NF-κB and MAPK activation in response to cytosolic bacteria. Both PRRs recognize fragments derived from the bacterial cell wall, which are generated during either synthesis or degradation of the peptidoglycan. NOD1 recognizes a dipeptide, γ-glutamyl-meso-diaminopimelic acid (Chamaillard et al., 2003; Girardin et al., 2003a) while NOD2 recognizes muramyl dipeptide (MDP) and muramyl tripeptide containing lysine but not diaminopimelic acid (Girardin et al., 2003b) (Fig. 2). The ability of NOD2 to sense MDP, which is the base unit of all peptidoglycans, makes it a universal sensor for peptidoglycan fragments. In contrast, NOD1 is restricted to sensing peptidoglycan fragments originating mostly from gram-negative bacteria, although some gram-positive bacteria such as *L. monocytogenes* can be sensed by NOD1 (Opitz et al., 2006a; Park et al., 2007).

NOD1 and NOD2 possess leucine-rich repeats (LRRs) involved in the recognition of bacterial ligands and a CARD required for downstream signaling events. Indeed, after recognition of a peptidoglycan fragment, the NOD proteins interact with RIP2 (RICK, Ripk2, CARDIAK) (McCarthy et al., 1998; Kobayashi et al., 2002) via homotypic CARD-CARD interaction. Interaction of the NOD proteins with RIP2 leads to its polyubiquitination, a necessary step for the recruitment of transforming growth factor β-activated kinase 1 (TAK1) complex. RIP2 constitutively interacts with NF-κB essential modifier (NEMO/IKKg/IKBKG), a member of IKK complex. The dual recruitment of TAK1 and IKK complexes by activated RIP2 is thought to lead to activation of IκB kinase complexes (IKK) by TAK1 (Abbott et al., 2007; Hasegawa et al., 2008; Yang et al., 2007). Activated IKK phosphorylates IκBs, which leads to their degradation. NF-κB is seques-

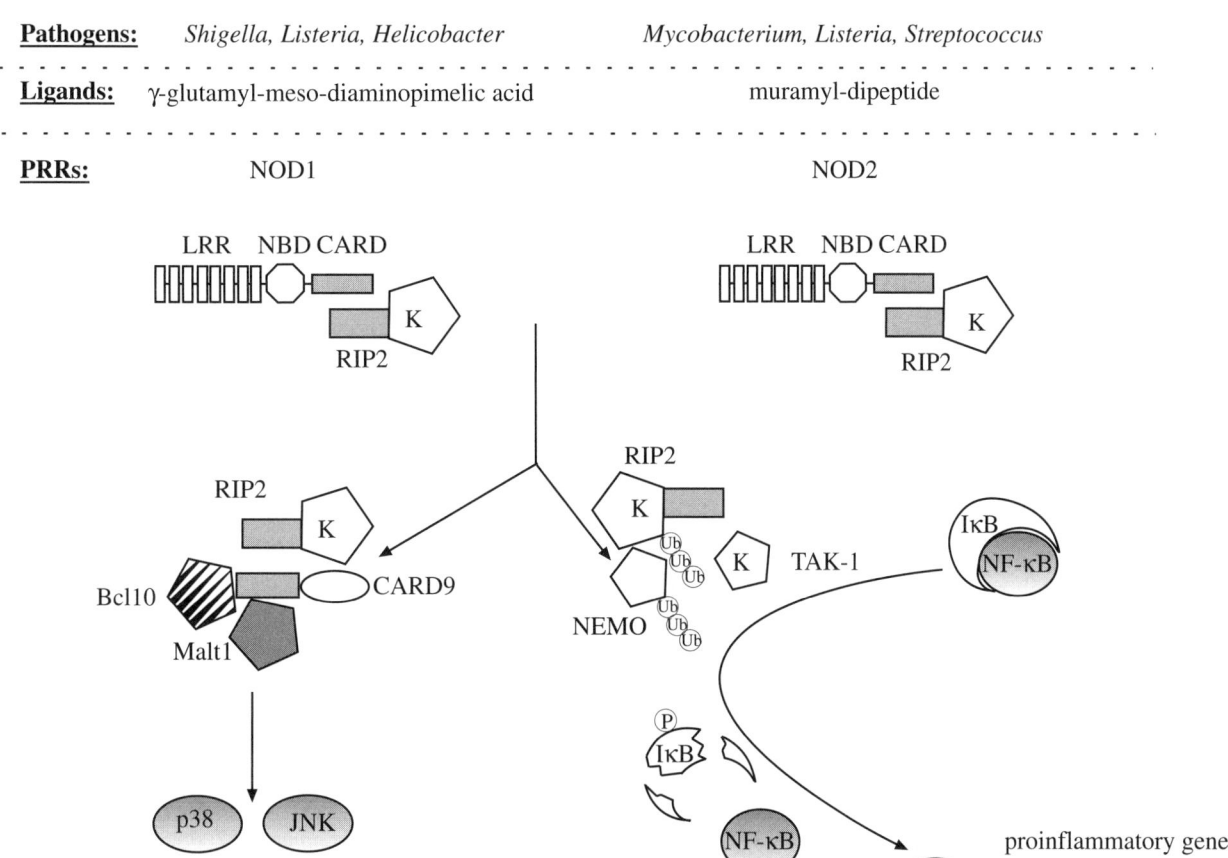

FIGURE 2 Bacterial cell wall fragments within the host cytosol trigger MAPK and NF-κB activation. Peptidoglycan fragments from various intracellular and extracellular bacteria can be delivered into the host cytosol, where they are recognized by the PRRs NOD1 or NOD2. The PRRs signal through the adaptor RIP2. RIP2 engagement leads to IκB phosphorylation in a NEMO- and TAK1-dependent manner resulting in its degradation and the translocation of NF-κB into the nucleus. Nuclear NF-κB induces the transcription of numerous proinflammatory genes including IL-8. NOD1/2 engagement also leads to MAP kinase activation through RIP2 and the tripartite complex CARD9-Bcl10-Malt1.

tered in the cytosol through its interaction with IκBs; however, the degradation of IκBs leads to NF-κB translocation to the nucleus, where it can activate the transcription of proinflammatory genes such as interleukin-8 (IL-8) (Roebuck, 1999).

NOD1 and NOD2 also activate MAPK (Girardin et al., 2001; Opitz et al., 2006a) in a RIP2-dependent manner. In contrast to the NOD-mediated NF-κB activation, MAPK activation requires CARD9 (Hsu et al., 2007). The downstream events are still unknown but may involve the tripartite complex CARD9-bcl-10-MALT1 (Hara et al., 2007).

NOD1 plays an important role in innate immune responses to *S. flexneri* by inducing the NF-κB-dependent secretion of IL-8 by intestinal epithelial cells (Philpott et al., 2000; Girardin et al., 2001), a cytokine critical for the recruitment of polymorphonuclear cells and the control of the infection (Sansonetti et al., 1999). Subsequent studies have shown a role for NOD1 in the innate immune signaling of epithelial cells exposed to a number of gram-negative pathogens such as *Campylobacter jejuni* (Zilbauer et al., 2007), *Pseudomonas aeruginosa* (Travassos et al., 2005), enteroinvasive *Escherichia coli* (Kim et al., 2004), and various *Chlamydia* subspecies (Welter-Stahl et al., 2006). Interestingly, NOD1 is involved in responses to extracellular pathogens. In particular, *Helicobacter pylori* triggers NF-κB activation in epithelial cells in a NOD1-dependent manner (Viala et al., 2004). This activation depends on the bacterial type IV secretion system encoded within the cag pathogenicity island, suggesting that peptidoglycan fragments are delivered into the host cytosol through the type IV secretion system. The relevance of the NOD1-mediated innate immune response was shown in NOD1$^{-/-}$ mice, which are more susceptible than WT mice to *H. pylori* colonization (Viala et al., 2004).

NOD2 senses gram-positive peptidoglycan fragments; indeed, NOD2 mediates recognition of *L. monocytogenes* (Park et al., 2007; Leber et al., 2008), *Streptococcus pneumoniae* (Opitz et al., 2004), and *M. tuberculosis* in macrophages (Ferwerda et al., 2005). Furthermore, NOD2-deficient mice are more susceptible than WT mice to oral administration of *L. monocytogenes* (Kobayashi et al., 2005).

While the main response triggered on engagement of NOD1 or NOD2 is transcriptional, another class of PRRs belonging to the same NLR family triggers a posttranslational response by activating caspase-1. The complex within which caspase-1 activation takes place has been called the inflammasome (Martinon et al., 2002).

THE INFLAMMASOME: A CYTOSOLIC SENSING COMPLEX LEADING TO CASPASE-1 ACTIVATION

Caspase-1: a Proinflammatory Caspase

Caspase-1 is a proinflammatory caspase that is synthesized as a proform of 55 kDa. It is activated by an autocatalytic cleavage that leads to the formation of an active heterotetramer, $(p10, p20)_2$, possessing a cysteine-protease activity (Fig. 3). The first caspase-1 substrate discovered was the proform of the proinflammatory cytokine IL-1β. IL-1β has numerous effects both locally, as a cytokine, recruiting neutrophils and monocytes, and systemically, as a pyrogen (Dinarello, 1998). It is synthesized as an inactive proform that is specifically cleaved by active caspase-1 prior to secretion. Similarly, caspase-1 matures pro-IL-18 and pro-IL-33. IL-18 is also called IFN-γ-inducing factor and synergizes with IL-12 to stimulate IFN-γ secretion. IL-33 is a cytokine inducing Th2 cytokines (IL-4, IL-5, and IL-13) (Schmitz et al., 2005). In addition to these proinflammatory roles, caspase-1 activation leads, in most cases, to cell death (Fink and Cookson, 2007).

Mechanisms of Pyroptosis: the Caspase-1-Dependent Cell Death

The mechanisms of cell death are still unclear but are distinguishable from apoptosis. The proinflammatory cell death mediated by caspase-1 has been called pyroptosis (Cookson and Brennan, 2001). Much of the original characterization of pyroptosis has been done in macrophages infected with serovar Typhimurium. Some of the hallmarks of this type of host cell death are the absence of nuclear condensation and the rapid loss of membrane integrity associated with the opening of membrane pores between 1.1 and 2.4 nm in diameter, leading to osmotic cell death (Brennan and Cookson, 2000; Fink and Cookson, 2006). Recently, some soluble caspase-1 substrates were identified by proteomic techniques (Shao et al., 2007). Numerous cytoskeleton proteins were identified as caspase-1 substrates, which is in agreement with previous studies (Kayalar et al., 1996). Interestingly, one new class of substrates belongs to the glycolysis pathway (Shao et al., 2007), a pathway that appears to be important for host cell survival (Colell et al., 2007). Caspase-1 has been localized to the plasma membrane (Singer et al., 1995). However, its putative membrane substrates, which might be responsible for the opening of the pores, are still unknown (Singer et al., 1995). Because of its dual role in inflammation and cell death, caspase-1 is classified as a proinflammatory caspase and the complex containing the NLR, the adaptor, and caspase-1 is called the inflammasome (Martinon et al., 2002).

The Inflammasome: a Multimolecular Complex Involved in Cytosolic Surveillance and Detection of Danger Signals

The mechanism of caspase-1 activation remained unknown until the isolation of a molecular complex called the inflammasome in which caspase-1 activation takes place (Martinon et al., 2002). The first inflammasome complex to be described contained the PRR called "NLR family, pyrin domain containing 1" (NLRP1/Nalp1), an adaptor protein called "apoptosis-associated speck-like protein containing a CARD" (ASC/pycard/TMS-1), and two proinflammatory caspases: caspase-1 and caspase-5. Caspase-1 is the active effector of the complex, while the role of caspase-5 and of its murine homologue caspase-11 remains unclear (Hilbi et al., 1998; Wang et al., 1998).

Numerous putative inflammasome PRRs exist (14 are encoded within the human genome and 20 within the mouse genome) (http://www.genenames.org/genefamily/nacht.html) and the inflammasome has been shown to be involved in triggering caspase-1 activation in response to numerous danger signals (Matzinger, 2002) and infections (Mariathasan and Monack, 2007), suggesting that the wide array of PRRs allows detection of diverse signals and microbial molecules. In particular, the inflammasome is involved in the innate immune response to cytosolic bacteria (*F. tularensis* [Mariathasan et al., 2005], *L. monocytogenes* [Mariathasan et al., 2006], and *S. flexneri* [Suzuki et al., 2007]), vacuolar bacteria (serovar Typhimurium [Mariathasan et al., 2004], *L. pneumophila* [Molofsky et al., 2006; Ren et al., 2006; Zamboni et al., 2006], and *M. tuberculosis* [Netea et al., 2006]), and also to toxin-expressing extracellular bacteria (*Bacillus anthracis* [Boyden and Dietrich, 2006] and *Staphylococcus aureus* [Mariathasan et al., 2006]) and type III secretion system-expressing extracellular bacteria (*P. aeruginosa* [Franchi et al., 2007; Galle et al., 2007; Sutterwala et al., 2007; Miao et al., 2008] and various *Yersinia* subspecies [Bergsbaken and Cookson, 2007; Shin and Cornelis, 2007]).

The molecular mechanisms of caspase-1 activation within the inflammasome are beginning to be understood. In particular, some of the molecular mechanisms of the activation of the NLRP1 inflammasome have recently been elucidated.

Molecular Mechanisms of Caspase-1 Activation within the Inflammasome

Human NLRP1 (Nalp1) has recently been characterized as a cytosolic receptor for MDP (Faustin et al., 2007), the universal base unit of peptidoglycan. NLRP1 is a multi-domain protein containing a pyrin domain, a nucleotide binding domain (NBD/NACHT), five LRR domains, and a CARD. Similar to other PRRs (Bell et al., 2003), the LRR domain is involved in binding the ligand, MDP (Faustin et al., 2007). Binding of MDP leads to a conformational change in NLRP1 that relieves the NBD accessibility. Binding and hydrolysis of ATP then triggers the oligomerization of NLRP1. The CARD of NLRP1 directly binds to the CARD in procaspase-1. It is believed that the oligomerization of NLRP1 brings several procaspase-1 proteins in close proximity, leading to its cross-activation, similar to the mechanism of activation of caspase-9 in the Apaf-1 apoptosome (for review, see Bao and Shi [2007]). This oligomerization does not require the adaptor protein ASC, but is facilitated by its presence (Faustin et al., 2007).

In contrast to NLRP1, most putative inflammasome PRRs do not contain a CARD domain and are thus unlikely to directly bind caspase-1. Indeed, the typical structure of these receptors is pyrin domain-NBD-LRR (Meylan et al., 2006). ASC, a small protein containing a pyrin and a CARD domain, is an adaptor/scaffolding protein that

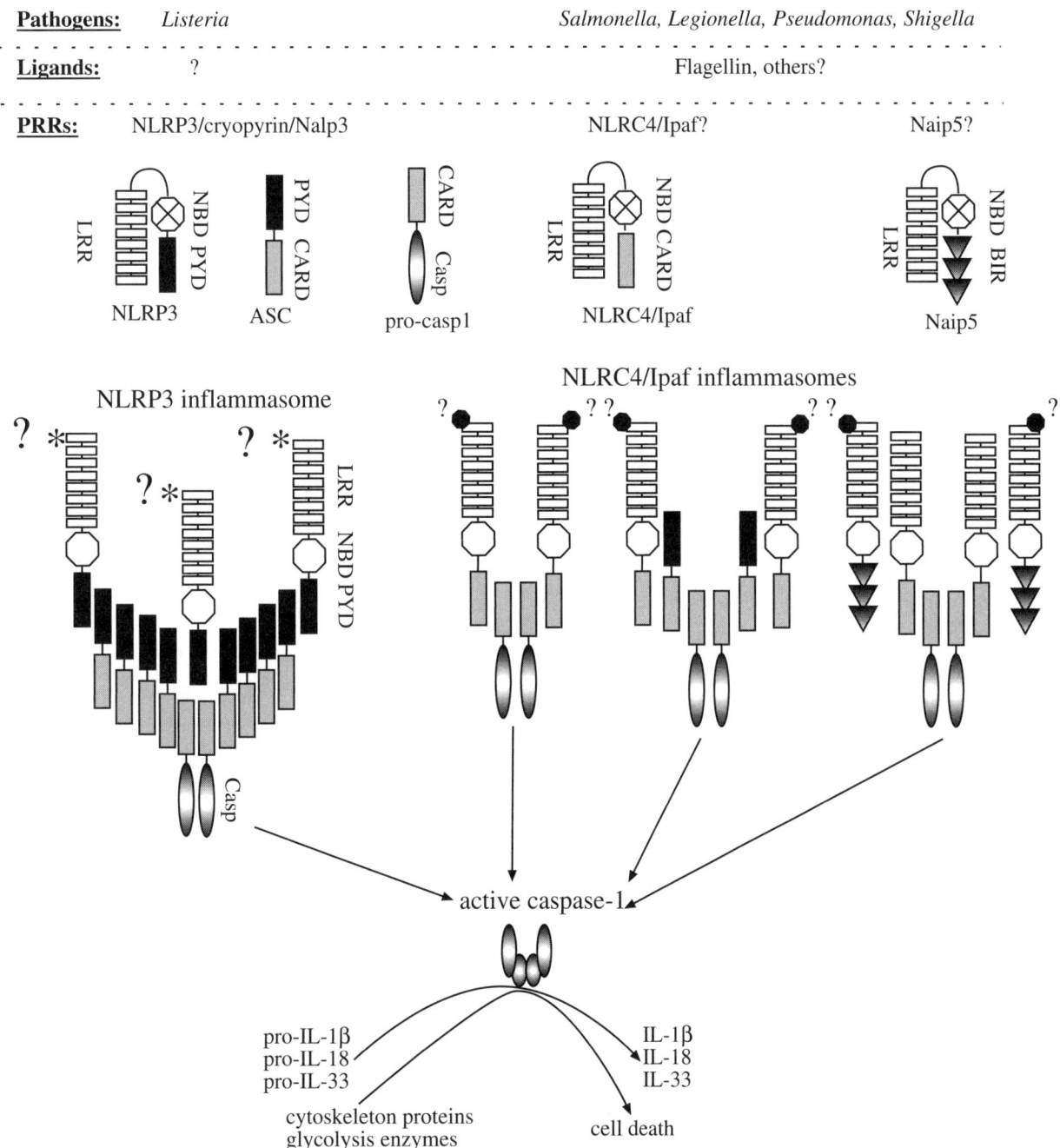

FIGURE 3 Flagellin and other unknown bacterial molecules trigger caspase-1 activation within various inflammasomes. The LRRs of the PRRs are the proposed ligand binding domains. Upon binding, a modification of conformation releases the accessibility of the NBD, allowing the PRR to interact through homotypic interactions with other partners including the adaptor ASC. ASC can oligomerize and interact both with the PRRs containing a pyrin domain and procaspase-1, leading to the formation of a multimolecular complex called the inflammasome. Dimerization of procaspase-1 within the inflammasome leads to its autocatalytic activation. Mature caspase-1 consists of a heterotetramer possessing a cysteine protease activity responsible for maturation of proinflammatory cytokines and triggering cell death of the host cell. The nature of the IPAF inflammasomes and the protein responsible for direct binding of flagellin are still unknown; three possible models are presented.

links the NLR family pyrin domain-containing proteins and caspase-1. Indeed, ASC is present as an oligomer both in cell-free systems and in cells in which the inflammasome is activated (Martinon et al., 2002). ASC is able to form oligomers called pyroptosomes in low potassium concentrations, suggesting that it can be a critical protein in the formation of the multiprotein complex in which caspase-1 activation takes place (Fernandes-Alnemri et al., 2007).

Inflammasome Activation in Response to Membrane Damage

Tissue injury can result in the release of extracellular ATP. In the presence of microbial products such as lipopolysaccharide or peptidoglycan fragments, ATP can activate the inflammasome in macrophages after signaling through the purinergic P2Z receptor, P2X7R (Ferrari et al., 1997). P2X7R signals through the membrane pore pannexin-1 (Pelegrin and Surprenant, 2006, 2007) and the PRR NLRP3 (Nalp3/cryopyrin) (Mariathasan et al., 2006), leading to activation of caspase-1 in an ASC-dependent manner (Mariathasan et al., 2004). Similarly, toxins such as maitotoxin and nigericin, which trigger membrane damage, activate the inflammasome in a pannexin-1-, NRLP3-, ASC-dependent manner (Mariathasan et al., 2004; Pelegrin and Surprenant 2007). While P2X7R could directly activate pannexin-1, the link between pannexin-1 and NLRP3 is still unclear.

Inflammasome Activation in Response to Anthrax Toxin

Anthrax lethal toxin, in contrast to the toxins previously described, does not directly cause cytoplasmic membrane damage, but is delivered into the host cytosol, where it cleaves MAPK kinases (Duesbery et al., 1998). Macrophages from different inbred mice differ in their susceptibility to anthrax lethal factor (Friedlander et al., 1993). This phenotypic difference has been linked to polymorphisms in the nlrp1b (nalp1b) gene. In macrophages from susceptible mice, anthrax lethal toxin activates the inflammasome in an NLRP1b-, caspase-1-dependent manner, which results in the secretion of IL-β and host cell death (Boyden and Dietrich, 2006). The macrophage cytotoxicity depends on the zinc-metalloprotease activity of the toxin (Klimpel et al., 1994). However, it is still unknown whether the toxin has evolved to cleave NLRP1b, which would lead to a constitutively active form of this PRR, or if NLRP1b senses another signal that is generated by the zinc-metalloprotease activity in host cells.

Inflammasome Activation in Response to Cytosolic Bacteria: *L. monocytogenes*, *S. flexneri*, and *F. tularensis*

The host cell signaling pathways that lead to caspase-1 activation in macrophages infected with three bacterial pathogens that replicate in the host cell cytosol—the gram-positive bacterium *L. monocytogenes* and the gram-negative bacteria *S. flexneri* and *F. tularensis*—have been studied so far.

L. monocytogenes activates caspase-1 in an NLRP3 (Nalp3/cryopyrin)-ASC-dependent manner in macrophages (Mariathasan et al., 2006). Bacterial mutants deficient for listeriolysin O, a pore-forming toxin that is required for escape from the phagosome and entry into the host cell cytosol, do not activate the inflammasome (Mariathasan et al., 2006). Thus, it is believed that the presence of *L. monocytogenes* in the cytosol leads to the engagement of NLRP3 and this leads to the formation of the inflammasome and caspase-1 activation. However, the bacterial ligand that is sensed by NLRP3 is not known.

Similarly, WT, cytosolic *F. tularensis* activates the inflammasome in an ASC-dependent manner, and mutants unable to escape into the macrophage cytosol fail to trigger caspase-1 activation. However, in contrast to *L. monocytogenes*, *F. tularensis*-mediated inflammasome activation is independent of NLRP3 (Mariathasan et al., 2005). This may mean that the *F. tularensis* ligand that triggers inflammasome activation is different than the *L. monocytogenes* ligand.

S. flexneri differs from these two other cytosolic bacteria because it possesses a type III secretion apparatus for injecting bacterial effectors into the host cytosol. The type III secretion system is required for escape into the cytosol and inflammasome activation. It is unclear whether inflammasome activation results from the detection of *S. flexneri* within the host cytosol (Zychlinsky et al., 1994), from the activity of IpaB, a type III secreted effector that has been reported to bind caspase-1 (Hilbi et al., 1998), or from a more direct effect of the type III secretion system such as the membrane damage associated with its insertion in the host membrane (Schroeder et al., 2007). Inflammasome activation in response to *S. flexneri* depends on "NLR family, CARD domain containing 4" (NLRC4/Ipaf) (Suzuki et al., 2007), a protein possessing LRR and a CARD domain.

Interestingly, caspase-1 activation in response to these three cytosolic bacteria involves different PRRs but it is closely associated in every case with the ability to escape from the phagosome/early vacuole into the macrophage cytosol. Caspase-1 is critical for innate immune defense against *L. monocytogenes*, *F. tularensis*, and *S. flexneri* since caspase-1-deficient mice are more susceptible to infections with any of these bacteria (Sansonetti et al., 2000; Tsuji et al., 2004; Mariathasan et al., 2005). In agreement with the inflammasome-dependent cell death of macrophages observed in vitro, caspase-1-dependent cell death is observed in mice in a lung model of shigellosis (Phalipon et al., 1995). Similarly, extensive cell death occurs in mice infected with *L. monocytogenes* and *F. tularensis* (Henry et al., 2007), although the role of caspase-1 has not been addressed in these two infections. Moreover, in agreement with the role of caspase-1 in maturing pro-IL-1β and pro-IL-18, IL-18 levels are much lower in the serum of caspase-1$^{-/-}$ mice infected with *L. monocytogenes* (Tsuji et al., 2004) and *F. tularensis* (Mariathasan et al., 2005) than in the serum of WT infected mice. Consistent with its role as an inducer of IFN-γ, the lower levels of IL-18 in *L. monocytogenes*-infected caspase-1$^{-/-}$ mice correlate with a decrease in IFN-γ (Tsuji et al., 2004). Similarly during *Shigella* or *Francisella* infections, both IL-1β and IL-18 have been shown to play a role in caspase-1-mediated immune responses (Sansonetti et al., 2000; Mariathasan et al., 2005). Importantly, caspase-1$^{-/-}$ mice harbor higher *F. tularensis* counts than IL-1β and IL-18 double-knockout mice (Henry and Monack, 2007), suggesting that in vivo both the cytokine maturation and the cell death activities of caspase-1 are important to fight the infection. *F. tularensis*, like many other intracellular bacteria, replicates in phagocytes. It is thought that macrophage cell death is beneficial for the host since it is removing the bacterial replicative niche and releasing bacteria into the extracellular environment in which neutrophils and monocytes have been recruited and activated by the proinflammatory cytokines.

Inflammasome Activation in Response to Vacuolar Pathogens

While inflammasome PRRs seem to be localized in the cytosol, some pathogens that reside in a vacuole within macrophages, such as serovar Typhimurium and *L. pneumophila*, can activate the inflammasome. However, as previously described for the type I IFN response, activation of the inflammasome depends on the presence of an intact secretion system (the SPI-1-encoded type III secretion system for serovar Typhimurium and the dot/Icm type IV secretion system for *L. pneumophila*). Interestingly, serovar Typhimurium and *L. pneumophila* flagellin mutants do not trigger caspase-1 activation, suggesting that flagellin subunits are either secreted or leaked through the secretion system into the macrophage cytosol, where it is detected by the inflammasome (Amer et al., 2006; Franchi et al., 2006; Miao et al., 2006; Molofsky et al., 2006; Ren et al., 2006). It is striking how two different innate immune pathways, TLR5 on the surface of cells (Hayashi et al., 2001) and the inflammasome in the host cell cytosol, have evolved to recognize the same PAMP, flagellin.

Both serovar Typhimurium- and *L. pneumophila*-mediated activation of the inflammasome require NLRC4/Ipaf (Mariathasan et al., 2004). In addition, NLR family, apoptosis inhibitory protein 5 (Naip5) has been identified as a PRR necessary for *L. pneumophila*-flagellin-dependent caspase-1 activation. The current model is that *L. pneumophila* flagellin is sensed in a Naip5-dependent manner and that activated Naip5 signals through NLRC4/Ipaf to activate caspase-1 (Molofsky et al., 2006).

Caspase-1$^{-/-}$ mice are more susceptible to infections with serovar Typhimurium (Lara-Tejero et al., 2006; Raupach et al., 2006) and *L. pneumophila* (Zamboni et al., 2006) compared with WT mice, illustrating that inflammasome activation is critical in vivo for innate immune responses to both cytosolic and vacuolar bacteria.

Inflammasome Activation: Beyond Cell Death and Proinflammatory Cytokine Maturation

Although the best-described activities of the inflammasome are caspase-1-mediated cell death and maturation of proinflammatory cytokines, the inflammasome and/or inflammasome components are involved in processes such as cell survival, trafficking of the *Legionella*-containing phagosome, or regulation of cathepsin B activity.

Pore-forming toxins, such as aerolysin from *Aeromonas hydrophila*, activate the inflammasome, resulting in caspase-1 activation. Although high doses of aerolysin kill target cells, it was recently shown that sublethal doses cause the activation of the inflammasome and result in cell survival in human fibroblasts (Gurcel et al., 2006). This process depends on the catalytic activity of caspase-1, which activates two enzymes, S1P and S2P, by an unknown mechanism. S1P and S2P proteolytically activate the transcription factor sterol regulatory element binding proteins (SREBPs). Active SREBPs then promote the transcription of lipogenic genes that are involved in membrane repair and cell survival (Gurcel et al., 2006). While this process was observed in human fibroblast cells, it is unknown what the balance between caspase-1-mediated cell survival and caspase-1-mediated cell death is in phagocytes.

Macrophages from different inbred mice show different susceptibility to infection with *L. pneumophila*, a phenotype that has been mapped to the Naip5 gene (Diez et al., 2003; Wright et al., 2003). While the difference in the replication rate of *L. pneumophila* in Naip5-resistant compared with Naip5-susceptible macrophages can be partially attributed to a difference in caspase-1-dependent cell death, other processes seem to be involved (Amer and Swanson, 2005; Amer et al., 2006; Molofsky et al., 2006; Fortier et al., 2007). Indeed, in Naip5-resistant macrophages, the trafficking of the *Legionella*-containing phagosome is altered, leading to more fusion with the lysosomes and less fusion with endoplasmic reticulum-derived vesicles (Amer et al., 2006; Fortier et al., 2007). This suggests that upon detection of *L. pneumophila* flagellin, the Naip5-Ipaf axis can modify the trafficking of the *Legionella*-containing phagosome to impair replication of the bacteria.

Similarly, NLRP3 (Nalp3/cryopyrin) and ASC can activate cathepsin B during *S. flexneri* infection (Willingham et al., 2007), and an ASC-dependent caspase-1-independent cell death is observed at late time points in *F. tularensis*-infected macrophages (Mariathasan et al., 2005). Overall, these results indicate that while the typical inflammasome involves a PRR, an adaptor, and caspase-1 as an effector triggering cell death and maturation of proinflammatory cytokines, numerous other complexes involving the inflammasome PRR might exist and recruit different effectors to activate different signaling pathways.

MODULATION AND ESCAPE OF CYTOSOLIC SENSING BY PATHOGENS

As described previously, the various responses triggered after engagement of cytosolic PRRs are critical to fight infections, but some pathogens have virulence factors that block the innate immune signaling. In particular, numerous strategies to block the RIG-I and MDA-5 pathways have been described for viruses (Hiscott et al., 2006). Picornaviruses escape recognition by RIG-I due to the VpG protein that binds covalently to the 5'-RNA (Hornung et al., 2006), and in cells infected with poliovirus, the prototypic picornavirus, MDA-5 is degraded in a proteasome, caspase-dependent manner (Barral et al., 2007). The downstream signaling proteins are also affected by viruses. For example, MAVS has been shown to be cleaved by the hepatitis C protein NS3/4A (Meylan et al., 2005) and an Epstein-Barr virus decoy protein, LMP1, sequesters TRAF3 (Wu et al., 2005). Interestingly, all pathogenic viruses trigger low levels of type I IFN, suggesting that pathogenicity correlates with the ability to block the cytosolic PRR signaling pathways.

Some viruses also trigger inflammasome activation (Kanneganti et al., 2006). A poxvirus protein, M13L, contains a pyrin domain and interacts with ASC, which leads to the inhibition of inflammasome activation. Importantly, a myxoma virus lacking the M13L protein is unable to inhibit host cell death and is attenuated for virulence (Johnston et al., 2005).

Much less in known about bacterial mechanisms of evasion of host cytosolic sensing pathways. *L. monocytogenes* has an atypical peptidoglycan due to the presence of *pgdA*, a peptidoglycan N-deacetylase. Modified peptidoglycan is less recognized by the NOD1 and NOD2 pathways (in addition to a decrease in lysozyme digestibility), suggesting that *pgdA* is a virulence factor promoting *L. monocytogenes* escape of the recognition by NOD1 and NOD2 (Boneca et al., 2007).

Inflammasome-mediated cell death is also targeted by bacterial virulence factors. For example, the *Yersinia* type III secreted effector YopE inhibits caspase-1-mediated release of IL-1β (Schotte et al., 2004). Similarly, hypercytotoxic *F. tularensis* mutants were recently isolated, suggesting that the corresponding proteins delay inflammasome activation (Weiss et al., 2007).

REGULATION OF INTRACYTOSOLIC PRR SIGNALING

Pathogen detection by several PRRs leads to the engagement of many signaling pathways, illustrating the host's strategy to mount quick and intense responses upon detection of pathogens. Conversely, inflammation and cell death are processes that can be deleterious for the host and thus need to be tightly regulated. Numerous processes exist to control or dampen the immune response. Here, we will review some recent examples of positive and negative regulation of cytosolic PRR signaling.

Positive Regulation of PRR Signaling

The comparison of the transcriptional responses of macrophages to cytosolic bacterial pathogens and isogenic vacuolar mutants has shown that recognition of cytosolic bacteria induces the expression of numerous genes involved in cytosolic sensing (Henry et al., 2007; Leber et al., 2008). For example, sensing of cytosolic *F. tularensis* in macrophages induces the expression of 68 genes, including the adaptor RIP2 and the cytosolic sensors PRKR (protein kinase RNA-dependent), RIG-I, NOD1, and MDA-5 (Henry et al., 2007); the last two are also induced in response to cytosolic *L. monocytogenes* (Leber et al., 2008). Similarly, TLR or NOD2 stimulation has been shown to amplify the secretion of IFN-β in response to *F. tularensis* (Cole et al., 2007; Henry et al., 2007), *L. monocytogenes*, and *M. tuberculosis* (Leber et al., 2008) infections, illustrating how different sensing events lead to an amplification of the cytosolic innate immune signaling.

While engagement of two different PRRs can modulate the intensity of the response, in certain cases the cooperation of at least two different PRRs is required for inflammasome activation. This has led to the model that activation and release of IL-1β requires two distinct signals (Dinarello, 1998). What constitutes these signals in vivo either during an infection or during autoinflammatory responses is not known definitively. However, in vitro studies indicate that the first signal can be triggered by TLR activation. Resting macrophages and DCs constitutively express the adaptor ASC and caspase-1, but express very little pro-IL-1β. Thus, the first signal leads to the engagement of a TLR, which induces NF-κB-dependent expression of pro-IL-1β. The second signal involved in caspase-1 activation and IL-1β release can be derived from a variety of intracellular stimuli that have been described above (e.g., flagellin, ATP and lipopolysaccharide, and MDP). Furthermore, inflammasome activation appears to require yet another signal in some instances. For example, caspase-1 activation in response to *L. monocytogenes* and *F. tularensis* requires type I IFN signaling (Henry et al., 2007). Thus, the first sensing event in the cytosol appears to involve type I IFN signaling, which is required for the inflammasome to be competent to respond to a second cytosolic sensing event (Henry et al., 2007). Thus, the requirement of multiple signals for IL-1β production might constitute a fail-safe mechanism that ensures activation of potent inflammatory responses occurs only in the presence of a bona fide stimulus, such as the presence of invasive pathogens (not commensal bacteria alone) and/or tissue injury. While various PRRs can synergize to amplify signaling, numerous control mechanisms also exist to inhibit or dampen the response.

Negative Regulation of PRR Signaling

At least two different mechanisms have been shown to inhibit the PRR signaling leading to type I IFN response. The first one acts at the level of the adaptor MAVS. The protein NLRX1 is a protein localized at the mitochondria (Moore et al., 2008; Tattoli et al., 2008), which interacts with MAVS and disrupts MAVS interactions with RIG-I and possibly MDA-5 (Moore et al., 2008). The second control acts at a step downstream of MAVS by controlling the association of TRAF3 and TBK1. The association of TRAF3 and TBK1 is enhanced by the nondegradative polyubiquitination of TRAF3. However, upon macrophage stimulation, deubiquitinating enzyme A (DUBA) is upregulated and deubiquitinates TRAF3, which leads to a decrease in the type I IFN secretion in response to engagement of cytosolic PRRs RIG-I and MDA-5 (Kayagaki et al., 2007).

Similarly, the NOD1/2 and inflammasome signaling pathways must be tightly regulated. In humans, several proteins such as Iceberg, Inca, and NOD2s (a small isoform of NOD2) have been described. These molecules contain only a CARD domain or only a pyrin domain and are thought to act as decoys or dominant negative molecules of caspase-1 activation and seem to function by inhibiting inflammasome-mediated oligomerization of caspase-1 by sequestering the CARD domain of either RIP2 or caspase-1 or the pyrin domain of ASC (Stehlik and Dorfleutner, 2007). However, their study has been limited due to the lack of homologues in the mouse genome.

MUTATIONS IN PRRs CAN LEAD TO AUTOINFLAMMATORY SYNDROMES

Despite the tight control of the PRR-initiated responses, an increasing number of autoinflammatory symptoms are linked to mutations in PRR signaling pathways (Table 2) (Kanneganti et al., 2007; McDermott and Tschopp, 2007).

Mutations in NOD1 and NOD2

NOD1 polymorphisms have been associated with higher immunoglobulin E levels and susceptibility to childhood asthma (Kabesch et al., 2003; Hysi et al., 2005) and atopic eczema (Weidinger et al., 2005). Since microbial exposure prevents childhood asthma, and detection of peptidoglycan by NOD1 is key to balanced adaptive immune responses (Fritz et al., 2007), it is tempting to speculate that a deficiency in sensing bacterial infection due to a defect in NOD1 signaling impairs the normal T- and B-cell responses, which leads to higher susceptibility to asthma and atopic eczema.

Similarly, polymorphisms in the LRR of NOD2 are associated with susceptibility to Crohn's disease, a chronic inflammatory bowel disease (Hugot et al., 2001; Ogura et al., 2001). It is still controversial whether the sequenced NOD2 variants are hyper- or hyporesponsive to MDP (Kanneganti et al., 2007). One hypothesis is that mutations in NOD2 trigger an inability to control intestinal bacteria, leading to recurrent inflammation in the intestinal tract (Inohara et al., 2003). Moreover, mutations in the NBD domain of NOD2 that result in a constitutively active PRR have been associated with Blau syndrome, a disease asso-

TABLE 2 Mutations in PRRs or associated proteins that can lead to autoinflammatory diseases

Disease/susceptibility	Gene mutated	References
Asthma	Nod1	Hysi et al., 2005
Atopic eczema	Nod1	Weidinger et al., 2005
Inflammatory bowel disease	Nod1	McGovern et al., 2005
Crohn's disease	Nod2	Hugot et al., 2001; Ogura et al., 2001
Blau syndrome	Nod2	Miceli-Richard et al., 2001
Vitiligo	NLRP1/Nalp1	Jin et al., 2007
Chronic infantile neurological cutaneous and articular syndrome (CINCA)	NLRP3/Nalp3/Cryopyrin	Feldmann et al., 2002
Familial cold autoinflammatory syndrome (FCAS)	NLRP3/Nalp3/Cryopyrin	Hoffman et al., 2001
Muckle-Wells syndrome (MWS)	NLRP3/Nalp3/Cryopyrin	Hoffman et al., 2001
Hydatidiform mole	NLRP7	Murdoch et al., 2006
Hereditary periodic fever syndrome	NLRP12/Nalp12	Jéru et al., 2008
Familial Mediterranean fever (FMF)	Pyrin	International FMF Consortium, 1997; French FMF Consortium, 1997
Pyogenic arthritis and syndrome of pyogenic arthritis, pyoderma gangrenosum, acne (PAPA)	PSTPIP1	Wise et al., 2002

ciated with arthritis and skin rash (Miceli-Richard et al., 2001).

Taken together, it is remarkable that different mutations in the peptidoglycan-sensing receptors can result in susceptibility to different diseases arising in very different organs/tissues (e.g., lungs, intestine, and skin).

Mutations in Inflammasome Components

Various inflammatory diseases (familial cold autoinflammatory syndrome, Muckle-Wells syndrome, chronic infantile neurological cutaneous and articular syndrome) have been associated with mutations in NLRP3/Nalp3/cryopyrin (Hoffman et al., 2001; Feldmann et al., 2002). They are all characterized by recurring fever and can be associated with other symptoms of systemic inflammation such as arthritis. Similarly, two additional autoinflammatory diseases, familial Mediterranean fever (FMF) and pyogenic arthritis and syndrome of pyogenic arthritis, pyoderma gangrenosum, and acne (PAPA) are associated with mutations in the host molecule pyrin (International FMF Consortium, 1997; French FMF Consortium, 1997) and the cytoskeleton organizing protein, proline serine threonine phosphatase-interacting protein 1 (PSTPIP1) (also known as CD2 binding protein 1, CD2BP1) (Wise et al., 2002). These two proteins interact with each other and activate the ASC inflammasome, although the physiological stimuli responsible for the activation are still unknown (Yu et al., 2007). In addition, mutations in NLRP1/Nalp1 have been associated with the autoinflammatory disease vitiligo (Jin et al., 2007), suggesting that the misregulation of peptidoglycan sensing by the NLRP1 inflammasome can contribute to this disease. NLRP1 is highly expressed in Langerhans cells, the resident skin DCs, which is consistent with its implication in a skin disorder.

Finally, mutations in NLRP7/Nalp7 have been linked to recurrent hydatidiform mole, an abnormal pregnancy (Murdoch et al., 2006). Since IL-1β plays numerous roles in normal (Simon et al., 1994) and pathological (Karhukorpi et al., 2003) pregnancies, it suggests that deregulation of the inflammasome can affect not only the immune system but also the reproductive system.

CONCLUDING REMARKS

Recent research has led to the discovery not only that host cells are capable of recognizing pathogens extracellularly via TLRs, but that several cytosolic PRRs are required for monitoring the intracellular environment. Some of these intracellular PRRs are crucial for the recognition of the presence of pathogens, such as viruses and bacteria that have escaped into the host cytosol. Interestingly, some of these cytosolic PRRs can also detect the presence of pathogenic bacteria that reside within a vacuole. The host detects bacterial molecules that are secreted via the specialized secretion system that are required by these pathogens to establish the vacuolar replicative niche. While most emphasis so far has been focused on bacteria and viruses, cytosolic PRRs might also sense cytosolic parasites such as *Trypanosoma cruzi* (Vaena de Avalos et al., 2002) or vacuolar parasites secreting proteins into the host cytosol such as *Toxoplasma gondii* (Bradley and Sibley, 2007).

The regulation and activation of the intracellular innate immune response is very sophisticated. Clearly the host has met the challenge of responding to the presence of pathogen-derived cytosolic molecules, membrane disruptions, and toxins. Most of the cytosolic PRRs and their signaling pathways have been identified very recently and many are likely undiscovered. For example, it is known that autophagy is triggered in response to numerous intracellular pathogens (Levine and Deretic, 2007), but the receptors and the pathways involved are still largely unknown. Numerous other questions remain unanswered. Indeed, the identification of the ligands remains unknown for most of the cytosolic PRRs. It is likely that as we gain more insight into the different intracellular signaling pathways, we will find that many pathogens have evolved mechanisms to modulate and evade these pathways.

Finally, while these pathways are highly regulated, some PRR mutations lead to a number of autoinflammatory diseases. A better understanding of the role of these PRRs during the normal innate immune response to intracellular pathogens is likely to give us a better understanding of the autoinflammatory diseases and will lead to the development of new therapeutics (McDermott and Tschopp, 2007).

REFERENCES

Abbott, D. W., Y. Yang, J. E. Hutti, S. Madhavarapu, M. A. Kelliher, and L. C. Cantley. 2007. Coordinated regulation of Toll-like receptor and NOD2 signaling by K63-linked polyubiquitin chains. *Mol. Cell. Biol.* 27:6012–6025.

Abdallah, A. M., N. C. Gey van Pittius, P. A. Champion, J. Cox, J. Luirink, C. M. Vandenbroucke-Grauls, B. J. Appelmelk, and W. Bitter. 2007. Type VII secretion—mycobacteria show the way. *Nat. Rev. Microbiol.* 5:883–891.

Amer, A., L. Franchi, T. D. Kanneganti, M. Body-Malapel, N. Ozoren, G. Brady, S. Meshinchi, R. Jagirdar, A. Gewirtz, S. Akira, and G. Nunez. 2006. Regulation of Legionella phagosome maturation and infection through flagellin and host Ipaf. *J. Biol. Chem.* 281:35217–35223.

Amer, A. O., and M. S. Swanson. 2005. Autophagy is an immediate macrophage response to Legionella pneumophila. *Cell. Microbiol.* 7:765–778.

Auerbuch, V., D. G. Brockstedt, N. Meyer-Morse, M. O'Riordan, and D. A. Portnoy. 2004. Mice lacking the type I interferon receptor are resistant to Listeria monocytogenes. *J. Exp. Med.* 200:527–533.

Bao, Q., and Y. Shi. 2007. Apoptosome: a platform for the activation of initiator caspases. *Cell Death Differ.* 14:56–65.

Barral, P. M., J. M. Morrison, J. Drahos, P. Gupta, D. Sarkar, P. B. Fisher, and V. R. Racaniello. 2007. MDA-5 is cleaved in poliovirus-infected cells. *J. Virol.* 81:3677–3684.

Bell, J. K., G. E. Mullen, C. A. Leifer, A. Mazzoni, D. R. Davies, and D. M. Segal. 2003. Leucine-rich repeats and pathogen recognition in Toll-like receptors. *Trends Immunol.* 24:528–533.

Bergsbaken, T., and B. T. Cookson. 2007. Macrophage activation redirects yersinia-infected host cell death from apoptosis to caspase-1-dependent pyroptosis. *PLoS Pathog.* 3:e161.

Blackley, S., Z. Kou, H. Chen, M. Quinn, R. C. Rose, J. J. Schlesinger, M. Coppage, and X. Jin. 2007. Primary human splenic macrophages, but not T or B cells, are the principal target cells for dengue virus infection in vitro. *J. Virol.* 81:13325–13334.

Boneca, I. G., O. Dussurget, D. Cabanes, M. A. Nahori, S. Sousa, M. Lecuit, E. Psylinakis, V. Bouriotis, J. P. Hugot, M. Giovannini, A. Coyle, J. Bertin, A. Namane, J. C. Rousselle, N. Cayet, M. C. Prevost, V. Balloy, M. Chignard, D. J. Philpott, P. Cossart, and S. E. Girardin. 2007. A critical role for peptidoglycan N-deacetylation in Listeria evasion from the host innate immune system. *Proc. Natl. Acad. Sci. USA* 104:997–1002.

Boyden, E. D., and W. F. Dietrich. 2006. Nalp1b controls mouse macrophage susceptibility to anthrax lethal toxin. *Nat. Genet.* 38:240–244.

Bradley, P. J., and L. D. Sibley. 2007. Rhoptries: an arsenal of secreted virulence factors. *Curr. Opin. Microbiol.* 10:582–587.

Brennan, M. A., and B. T. Cookson. 2000. Salmonella induces macrophage death by caspase-1-dependent necrosis. *Mol. Microbiol.* 38:31–40.

Carrero, J. A., B. Calderon, and E. R. Unanue. 2004. Type I interferon sensitizes lymphocytes to apoptosis and reduces resistance to Listeria infection. *J. Exp. Med.* 200:535–540.

Cassol, E., M. Alfano, P. Biswas, and G. Poli. 2006. Monocyte-derived macrophages and myeloid cell lines as targets of HIV-1 replication and persistence. *J. Leukoc. Biol.* 80:1018–1030.

Cella, M., D. Jarrossay, F. Facchetti, O. Alebardi, H. Nakajima, A. Lanzavecchia, and M. Colonna. 1999. Plasmacytoid monocytes migrate to inflamed lymph nodes and produce large amounts of type I interferon. *Nat. Med.* 5:919–923.

Chamaillard, M., M. Hashimoto, Y. Horie, J. Masumoto, S. Qiu, L. Saab, Y. Ogura, A. Kawasaki, K. Fukase, S. Kusumoto, M. A. Valvano, S. J. Foster, T. W. Mak, G. Nunez, and N. Inohara. 2003. An essential role for NOD1 in host recognition of bacterial peptidoglycan containing diaminopimelic acid. *Nat. Immunol.* 4:702–707.

Cole, L. E., K. A. Shirey, E. Barry, A. Santiago, P. Rallabhandi, K. L. Elkins, A. C. Puche, S. M. Michalek, and S. N. Vogel. 2007. Toll-like receptor 2-mediated signaling requirements for Francisella tularensis live vaccine strain infection of murine macrophages. *Infect. Immun.* 75:4127–4137.

Colell, A., J. E. Ricci, S. Tait, S. Milasta, U. Maurer, L. Bouchier-Hayes, P. Fitzgerald, A. Guio-Carrion, N. J. Waterhouse, C. W. Li, B. Mari, P. Barbry, D. D. Newmeyer, H. M. Beere, and D. R. Green. 2007. GAPDH and autophagy preserve survival after apoptotic cytochrome c release in the absence of caspase activation. *Cell* 129:983–997.

Cookson, B. T., and M. A. Brennan. 2001. Pro-inflammatory programmed cell death. *Trends Microbiol.* 9:113–114.

Diez, E., S. H. Lee, S. Gauthier, Z. Yaraghi, M. Tremblay, S. Vidal, and P. Gros. 2003. Birc1e is the gene within the Lgn1 locus associated with resistance to Legionella pneumophila. *Nat. Genet.* 33:55–60.

DiGiuseppe Champion, P. A., and J. S. Cox. 2007. Protein secretion systems in Mycobacteria. *Cell. Microbiol.* 9:1376–1384.

Dinarello, C. A. 1998. Interleukin-1 beta, interleukin-18, and the interleukin-1 beta converting enzyme. *Ann. N. Y. Acad. Sci.* 856:1–11.

Duesbery, N. S., C. P. Webb, S. H. Leppla, V. M. Gordon, K. R. Klimpel, T. D. Copeland, N. G. Ahn, M. K. Oskarsson, K. Fukasawa, K. D. Paull, and G. F. Vande Woude. 1998. Proteolytic inactivation of MAP-kinase-kinase by anthrax lethal factor. *Science* 280:734–737.

Durbin, J. E., R. Hackenmiller, M. C. Simon, and D. E. Levy. 1996. Targeted disruption of the mouse Stat1 gene results in compromised innate immunity to viral disease. *Cell* 84:443–450.

Faustin, B., L. Lartigue, J. M. Bruey, F. Luciano, E. Sergienko, B. Bailly-Maitre, N. Volkmann, D. Hanein, I. Rouiller, and J. C. Reed. 2007. Reconstituted NALP1 inflammasome reveals two-step mechanism of caspase-1 activation. *Mol. Cell* 25:713–724.

Feldmann, J., A. M. Prieur, P. Quartier, P. Berquin, S. Certain, E. Cortis, D. Teillac-Hamel, A. Fischer, and G. de Saint Basile. 2002. Chronic infantile neurological cutaneous and articular syndrome is caused by mutations in CIAS1, a gene highly expressed in polymorphonuclear cells and chondrocytes. *Am. J. Hum. Genet.* 71:198–203.

Fernandes-Alnemri, T., J. Wu, J. W. Yu, P. Datta, B. Miller, W. Jankowski, S. Rosenberg, J. Zhang, and E. S. Alnemri. 2007. The pyroptosome: a supramolecular assembly of ASC dimers mediating inflammatory cell death via caspase-1 activation. *Cell Death Differ.* 14:1590–1604.

Ferrari, D., P. Chiozzi, S. Falzoni, M. Dal Susino, L. Melchiorri, O. R. Baricordi, and F. Di Virgilio. 1997. Extracellular ATP triggers IL-1 beta release by activating the purinergic P2Z receptor of human macrophages. *J. Immunol.* 159:1451–1458.

Ferwerda, G., S. E. Girardin, B. J. Kullberg, L. Le Bourhis, D. J. de Jong, D. M. Langenberg, R. van Crevel, G. J. Adema, T. H. Ottenhoff, J. W. Van der Meer, and M. G. Netea. 2005. NOD2 and toll-like receptors are nonredundant recognition systems of Mycobacterium tuberculosis. *PLoS Pathog.* 1:279–285.

Fink, S. L., and B. T. Cookson. 2006. Caspase-1-dependent pore formation during pyroptosis leads to osmotic lysis of infected host macrophages. *Cell. Microbiol.* **8:**1812–1825.

Fink, S. L., and B. T. Cookson. 2007. Pyroptosis and host cell death responses during Salmonella infection. *Cell. Microbiol.* **9:**2562–2570.

Fitzgerald, K. A., S. M. McWhirter, K. L. Faia, D. C. Rowe, E. Latz, D. T. Golenbock, A. J. Coyle, S. M. Liao, and T. Maniatis. 2003. IKKepsilon and TBK1 are essential components of the IRF3 signaling pathway. *Nat. Immunol.* **4:**491–496.

Fortier, A., C. de Chastellier, S. Balor, and P. Gros. 2007. Birc1e/Naip5 rapidly antagonizes modulation of phagosome maturation by Legionella pneumophila. *Cell. Microbiol.* **9:** 910–923.

Franchi, L., A. Amer, M. Body-Malapel, T. D. Kanneganti, N. Ozoren, R. Jagirdar, N. Inohara, P. Vandenabeele, J. Bertin, A. Coyle, E. P. Grant, and G. Nunez. 2006. Cytosolic flagellin requires Ipaf for activation of caspase-1 and interleukin 1beta in salmonella-infected macrophages. *Nat. Immunol.* **7:**576–582.

Franchi, L., J. Stoolman, T. D. Kanneganti, A. Verma, R. Ramphal, and G. Nunez. 2007. Critical role for Ipaf in Pseudomonas aeruginosa-induced caspase-1 activation. *Eur. J. Immunol.* **37:**3030–3039.

French FMF Consortium. 1997. A candidate gene for familial Mediterranean fever. *Nat. Genet.* **17:**25–31.

Friedlander, A. M., R. Bhatnagar, S. H. Leppla, L. Johnson, and Y. Singh. 1993. Characterization of macrophage sensitivity and resistance to anthrax lethal toxin. *Infect. Immun.* **61:**245–252.

Fritz, J. H., L. Le Bourhis, G. Sellge, J. G. Magalhaes, H. Fsihi, T. A. Kufer, C. Collins, J. Viala, R. L. Ferrero, S. E. Girardin, and D. J. Philpott. 2007. Nod1-mediated innate immune recognition of peptidoglycan contributes to the onset of adaptive immunity. *Immunity* **26:**445–459.

Galle, M., P. Schotte, M. Haegman, A. Wullaert, H. J. Yang, S. Jin, and R. Beyaert. 2007. The Pseudomonas aeruginosa Type III secretion system plays a dual role in the regulation of Caspase-1 mediated IL-1beta maturation. *J. Cell. Mol. Med.* **12:**1767–1776.

Girardin, S. E., I. G. Boneca, L. A. Carneiro, A. Antignac, M. Jehanno, J. Viala, K. Tedin, M. K. Taha, A. Labigne, U. Zahringer, A. J. Coyle, P. S. DiStefano, J. Bertin, P. J. Sansonetti, and D. J. Philpott. 2003a. Nod1 detects a unique muropeptide from gram-negative bacterial peptidoglycan. *Science* **300:**1584–1587.

Girardin, S. E., I. G. Boneca, J. Viala, M. Chamaillard, A. Labigne, G. Thomas, D. J. Philpott, and P. J. Sansonetti. 2003b. Nod2 is a general sensor of peptidoglycan through muramyl dipeptide (MDP) detection. *J. Biol. Chem.* **278:** 8869–8872.

Girardin, S. E., R. Tournebize, M. Mavris, A. L. Page, X. Li, G. R. Stark, J. Bertin, P. S. DiStefano, M. Yaniv, P. J. Sansonetti, and D. J. Philpott. 2001. CARD4/Nod1 mediates NF-kappaB and JNK activation by invasive Shigella flexneri. *EMBO Rep.* **2:**736–742.

Gitlin, L., W. Barchet, S. Gilfillan, M. Cella, B. Beutler, R. A. Flavell, M. S. Diamond, and M. Colonna. 2006. Essential role of mda-5 in type I IFN responses to polyriboinosinic:polyribocytidylic acid and encephalomyocarditis picornavirus. *Proc. Natl. Acad. Sci. USA* **103:**8459–8464.

Gurcel, L., L. Abrami, S. Girardin, J. Tschopp, and F. G. van der Goot. 2006. Caspase-1 activation of lipid metabolic pathways in response to bacterial pore-forming toxins promotes cell survival. *Cell* **126:**1135–1145.

Hacker, H., V. Redecke, B. Blagoev, I. Kratchmarova, L. C. Hsu, G. G. Wang, M. P. Kamps, E. Raz, H. Wagner, G. Hacker, M. Mann, and M. Karin. 2006. Specificity in Toll-like receptor signalling through distinct effector functions of TRAF3 and TRAF6. *Nature* **439:**204–207.

Hara, H., C. Ishihara, A. Takeuchi, T. Imanishi, L. Xue, S. W. Morris, T. Inui, T. Takai, A. Shibuya, S. Saijo, Y. Iwakura, N. Ohno, H. Koseki, H. Yoshida, J. M. Penninger, and T. Saito. 2007. The adaptor protein CARD9 is essential for the activation of myeloid cells through ITAM-associated and Toll-like receptors. *Nat. Immunol.* **8:**619–629.

Hasegawa, M., Y. Fujimoto, P. C. Lucas, H. Nakano, K. Fukase, G. Nunez, and N. Inohara. 2008. A critical role of RICK/RIP2 polyubiquitination in Nod-induced NF-kappaB activation. *EMBO J.* **27:**373–383.

Hayashi, F., K. D. Smith, A. Ozinsky, T. R. Hawn, E. C. Yi, D. R. Goodlett, J. K. Eng, S. Akira, D. M. Underhill, and A. Aderem. 2001. The innate immune response to bacterial flagellin is mediated by Toll-like receptor 5. *Nature* **410:**1099–1103.

Hemmi, H., O. Takeuchi, S. Sato, M. Yamamoto, T. Kaisho, H. Sanjo, T. Kawai, K. Hoshino, K. Takeda, and S. Akira. 2004. The roles of two IkappaB kinase-related kinases in lipopolysaccharide and double stranded RNA signaling and viral infection. *J. Exp. Med.* **199:**1641–1650.

Henry, T., A. Brotcke, D. S. Weiss, L. J. Thompson, and D. M. Monack. 2007. Type I interferon signaling is required for activation of the inflammasome during Francisella infection. *J. Exp. Med.* **204:**987–994.

Henry, T., and D. M. Monack. 2007. Activation of the inflammasome upon Francisella tularensis infection: interplay of innate immune pathways and virulence factors. *Cell. Microbiol.* **9:**2543–2551.

Herskovits, A. A., V. Auerbuch, and D. A. Portnoy. 2007. Bacterial ligands generated in a phagosome are targets of the cytosolic innate immune system. *PLoS Pathog.* **3:**e51.

Hess, C. B., D. W. Niesel, Y. J. Cho, and G. R. Klimpel. 1987. Bacterial invasion of fibroblasts induces interferon production. *J. Immunol.* **138:**3949–3953.

Hilbi, H., J. E. Moss, D. Hersh, Y. Chen, J. Arondel, S. Banerjee, R. A. Flavell, J. Yuan, P. J. Sansonetti, and A. Zychlinsky. 1998. Shigella-induced apoptosis is dependent on caspase-1 which binds to IpaB. *J. Biol. Chem.* **273:**32895–32900.

Hiscott, J., T. L. Nguyen, M. Arguello, P. Nakhaei, and S. Paz. 2006. Manipulation of the nuclear factor-kappaB pathway and the innate immune response by viruses. *Oncogene* **25:**6844–6867.

Hochrein, H., B. Schlatter, M. O'Keeffe, C. Wagner, F. Schmitz, M. Schiemann, S. Bauer, M. Suter, and H. Wagner. 2004. Herpes simplex virus type-1 induces IFN-alpha production via Toll-like receptor 9-dependent and -independent pathways. *Proc. Natl. Acad. Sci. USA* **101:** 11416–11421.

Hoffman, H. M., J. L. Mueller, D. H. Broide, A. A. Wanderer, and R. D. Kolodner. 2001. Mutation of a new gene encoding a putative pyrin-like protein causes familial cold autoinflammatory syndrome and Muckle-Wells syndrome. *Nat. Genet.* **29:**301–305.

Hornung, V., J. Ellegast, S. Kim, K. Brzozka, A. Jung, H. Kato, H. Poeck, S. Akira, K. K. Conzelmann, M. Schlee, S. Endres, and G. Hartmann. 2006. 5'-Triphosphate RNA is the ligand for RIG-I. *Science* **314:**994–997.

Hsu, Y. M., Y. Zhang, Y. You, D. Wang, H. Li, O. Duramad, X. F. Qin, C. Dong, and X. Lin. 2007. The adaptor protein CARD9 is required for innate immune responses to intracellular pathogens. *Nat. Immunol.* **8:**198–205.

Hugot, J. P., M. Chamaillard, H. Zouali, S. Lesage, J. P. Cezard, J. Belaiche, S. Almer, C. Tysk, C. A. O'Morain, M. Gassull, V. Binder, Y. Finkel, A. Cortot, R. Modigliani, P. Laurent-Puig, C. Gower-Rousseau, J. Macry, J. F. Colombel, M. Sahbatou, and G. Thomas. 2001. Association of

NOD2 leucine-rich repeat variants with susceptibility to Crohn's disease. *Nature* **411**:599–603.

Hysi, P., M. Kabesch, M. F. Moffatt, M. Schedel, D. Carr, Y. Zhang, B. Boardman, E. von Mutius, S. K. Weiland, W. Leupold, C. Fritzsch, N. Klopp, A. W. Musk, A. James, G. Nunez, N. Inohara, and W. O. Cookson. 2005. NOD1 variation, immunoglobulin E and asthma. *Hum. Mol. Genet.* **14**:935–941.

Inohara, N., Y. Ogura, F. F. Chen, A. Muto, and G. Nunez. 2001. Human Nod1 confers responsiveness to bacterial lipopolysaccharides. *J. Biol. Chem.* **276**:2551–2554.

Inohara, N., Y. Ogura, A. Fontalba, O. Gutierrez, F. Pons, J. Crespo, K. Fukase, S. Inamura, S. Kusumoto, M. Hashimoto, S. J. Foster, A. P. Moran, J. L. Fernandez-Luna, and G. Nunez. 2003. Host recognition of bacterial muramyl dipeptide mediated through NOD2. Implications for Crohn's disease. *J. Biol. Chem.* **278**:5509–5512.

International FMF Consortium. 1997. Ancient missense mutations in a new member of the RoRet gene family are likely to cause familial Mediterranean fever. *Cell* 90:797–807.

Ishii, K. J., C. Coban, H. Kato, K. Takahashi, Y. Torii, F. Takeshita, H. Ludwig, G. Sutter, K. Suzuki, H. Hemmi, S. Sato, M. Yamamoto, S. Uematsu, T. Kawai, O. Takeuchi, and S. Akira. 2006. A Toll-like receptor-independent antiviral response induced by double-stranded B-form DNA. *Nat. Immunol.* **7**:40–48.

Janeway, C. A., Jr. 1989. Approaching the asymptote? Evolution and revolution in immunology. *Cold Spring Harb. Symp. Quant. Biol.* **54**(Pt 1):1–13.

Jéru, I., P. Duquesnoy, T. Fernandes-Alnemri, E. Cochet, J. W. Yu, M. Lackmy-Port-Lis, E. Grimprel, J. Landman-Parker, V. Hentgen, S. Marlin, K. McElreavey, T. Sarkisian, G. Grateau, E. S. Alnemri, and S. Amselem. 2008. Mutations in NALP12 cause hereditary periodic fever syndromes. *Proc. Natl. Acad. Sci. USA* **105**:1614–1619.

Jin, Y., C. M. Mailloux, K. Gowan, S. L. Riccardi, G. LaBerge, D. C. Bennett, P. R. Fain, and R. A. Spritz. 2007. NALP1 in vitiligo-associated multiple autoimmune disease. *N. Engl. J. Med.* **356**:1216–1225.

Johnston, J. B., J. W. Barrett, S. H. Nazarian, M. Goodwin, D. Ricciuto, G. Wang, and G. McFadden. 2005. A poxvirus-encoded pyrin domain protein interacts with ASC-1 to inhibit host inflammatory and apoptotic responses to infection. *Immunity* **23**:587–598.

Kabesch, M., W. Peters, D. Carr, W. Leupold, S. K. Weiland, and E. von Mutius. 2003. Association between polymorphisms in caspase recruitment domain containing protein 15 and allergy in two German populations. *J. Allergy Clin. Immunol.* **111**:813–817.

Kang, D. C., R. V. Gopalkrishnan, Q. Wu, E. Jankowsky, A. M. Pyle, and P. B. Fisher. 2002. mda-5: an interferon-inducible putative RNA helicase with double-stranded RNA-dependent ATPase activity and melanoma growth-suppressive properties. *Proc. Natl. Acad. Sci. USA* **99**:637–642.

Kanneganti, T. D., M. Body-Malapel, A. Amer, J. H. Park, J. Whitfield, L. Franchi, Z. F. Taraporewala, D. Miller, J. T. Patton, N. Inohara, and G. Nunez. 2006. Critical role for Cryopyrin/Nalp3 in activation of caspase-1 in response to viral infection and double-stranded RNA. *J. Biol. Chem.* **281**:36560–36568.

Kanneganti, T. D., M. Lamkanfi, and G. Nunez. 2007. Intracellular NOD-like receptors in host defense and disease. *Immunity* **27**:549–559.

Karhukorpi, J., T. Laitinen, H. Kivela, A. Tiilikainen, and M. Hurme. 2003. IL-1 receptor antagonist gene polymorphism in recurrent spontaneous abortion. *J. Reprod. Immunol.* 58:61–67.

Kato, H., S. Sato, M. Yoneyama, M. Yamamoto, S. Uematsu, K. Matsui, T. Tsujimura, K. Takeda, T. Fujita, O. Takeuchi, and S. Akira. 2005. Cell type-specific involvement of RIG-I in antiviral response. *Immunity* **23**:19–28.

Kato, H., O. Takeuchi, S. Sato, M. Yoneyama, M. Yamamoto, K. Matsui, S. Uematsu, A. Jung, T. Kawai, K. J. Ishii, O. Yamaguchi, K. Otsu, T. Tsujimura, C. S. Koh, C. Reis e Sousa, Y. Matsuura, T. Fujita, and S. Akira. 2006. Differential roles of MDA5 and RIG-I helicases in the recognition of RNA viruses. *Nature* **441**:101–105.

Kawai, T., K. Takahashi, S. Sato, C. Coban, H. Kumar, H. Kato, K. J. Ishii, O. Takeuchi, and S. Akira. 2005. IPS-1, an adaptor triggering RIG-I- and Mda5-mediated type I interferon induction. *Nat. Immunol.* **6**:981–988.

Kayagaki, N., Q. Phung, S. Chan, R. Chaudhari, C. Quan, K. M. O'Rourke, M. Eby, E. Pietras, G. Cheng, J. F. Bazan, Z. Zhang, D. Arnott, and V. M. Dixit. 2007. DUBA: a deubiquitinase that regulates type I interferon production. *Science* **318**:1628–1632.

Kayalar, C., T. Ord, M. P. Testa, L. T. Zhong, and D. E. Bredesen. 1996. Cleavage of actin by interleukin 1 beta-converting enzyme to reverse DNase I inhibition. *Proc. Natl. Acad. Sci. USA* **93**:2234–2238.

Kim, J. G., S. J. Lee, and M. F. Kagnoff. 2004. Nod1 is an essential signal transducer in intestinal epithelial cells infected with bacteria that avoid recognition by toll-like receptors. *Infect. Immun.* **72**:1487–1495.

Klimpel, K. R., N. Arora, and S. H. Leppla. 1994. Anthrax toxin lethal factor contains a zinc metalloprotease consensus sequence which is required for lethal toxin activity. *Mol. Microbiol.* **13**:1093–1100.

Kobayashi, K., N. Inohara, L. D. Hernandez, J. E. Galan, G. Nunez, C. A. Janeway, R. Medzhitov, and R. A. Flavell. 2002. RICK/Rip2/CARDIAK mediates signalling for receptors of the innate and adaptive immune systems. *Nature* **416**:194–199.

Kobayashi, K. S., M. Chamaillard, Y. Ogura, O. Henegariu, N. Inohara, G. Nunez and R. A. Flavell. 2005. Nod2-dependent regulation of innate and adaptive immunity in the intestinal tract. *Science* **307**:731–734.

Kumagai, Y., O. Takeuchi, H. Kato, H. Kumar, K. Matsui, E. Morii, K. Aozasa, T. Kawai, and S. Akira. 2007. Alveolar macrophages are the primary interferon-alpha producer in pulmonary infection with RNA viruses. *Immunity* **27**:240–252.

Lara-Tejero, M., F. S. Sutterwala, Y. Ogura, E. P. Grant, J. Bertin, A. J. Coyle, R. A. Flavell, and J. E. Galan. 2006. Role of the caspase-1 inflammasome in Salmonella typhimurium pathogenesis. *J. Exp. Med.* **203**:1407–1412.

Leber, J. H., G. T. Crimmins, S. Raghavan, N. P. Meyer-Morse, J. S. Cox, and D. A. Portnoy. 2008. Distinct TLR- and NLR-mediated transcriptional responses to an intracellular pathogen. *PLoS Pathog.* **4**:e6.

Levine, B., and V. Deretic. 2007. Unveiling the roles of autophagy in innate and adaptive immunity. *Nat. Rev. Immunol.* **7**:767–777.

Mariathasan, S., and D. M. Monack. 2007. Inflammasome adaptors and sensors: intracellular regulators of infection and inflammation. *Nat. Rev. Immunol.* **7**:1–10.

Mariathasan, S., K. Newton, D. M. Monack, D. Vucic, D. M. French, W. P. Lee, M. Roose-Girma, S. Erickson, and V. M. Dixit. 2004. Differential activation of the inflammasome by caspase-1 adaptors ASC and Ipaf. *Nature* **430**:213–218.

Mariathasan, S., D. S. Weiss, V. M. Dixit, and D. M. Monack. 2005. Innate immunity against Francisella tularensis is dependent on the ASC/caspase-1 axis. *J. Exp. Med.* **202**:1043–1049.

Mariathasan, S., D. S. Weiss, K. Newton, J. McBride, K. O'Rourke, M. Roose-Girma, W. P. Lee, Y. Weinrauch, D. M. Monack, and V. M. Dixit. 2006. Cryopyrin activates the inflammasome in response to toxins and ATP. *Nature* **440**:228–232.

Martinon, F., K. Burns, and J. Tschopp. 2002. The inflammasome: a molecular platform triggering activation of inflammatory caspases and processing of proIL-beta. *Mol. Cell* **10**:417–426.

Matzinger, P. 2002. The danger model: a renewed sense of self. *Science* **296**:301–305.

McCaffrey, R. L., P. Fawcett, M. O'Riordan, K. D. Lee, E. A. Havell, P. O. Brown, and D. A. Portnoy. 2004. A specific gene expression program triggered by Gram-positive bacteria in the cytosol. *Proc. Natl. Acad. Sci. USA* **101**:11386–11391.

McCarthy, J. V., J. Ni, and V. M. Dixit. 1998. RIP2 is a novel NF-kappaB-activating and cell death-inducing kinase. *J. Biol. Chem.* **273**:16968–16975.

McDermott, M. F., and J. Tschopp. 2007. From inflammasomes to fevers, crystals and hypertension: how basic research explains inflammatory diseases. *Trends Mol. Med.* **13**:381–388.

McGovern, D. P., P. Hysi, T. Ahmad, D. A. van Heel, M. F. Moffatt, A. Carey, W. O. Cookson and D. P. Jewell. 2005. Association between a complex insertion/deletion polymorphism in NOD1 (CARD4) and susceptibility to inflammatory bowel disease. *Hum. Mol. Genet.* **14**:1245–1250.

Meylan, E., J. Curran, K. Hofmann, D. Moradpour, M. Binder, R. Bartenschlager, and J. Tschopp. 2005. Cardif is an adaptor protein in the RIG-I antiviral pathway and is targeted by hepatitis C virus. *Nature* **437**:1167–1172.

Meylan, E., J. Tschopp, and M. Karin. 2006. Intracellular pattern recognition receptors in the host response. *Nature* **442**:39–44.

Miao, E. A., C. M. Alpuche-Aranda, M. Dors, A. E. Clark, M. W. Bader, S. I. Miller, and A. Aderem. 2006. Cytoplasmic flagellin activates caspase-1 and secretion of interleukin 1beta via Ipaf. *Nat. Immunol.* **7**:569–575.

Miao, E. A., R. K. Ernst, M. Dors, D. P. Mao, and A. Aderem. 2008. Pseudomonas aeruginosa activates caspase 1 through Ipaf. *Proc. Natl. Acad. Sci. USA* **105**:2562–2567.

Miceli-Richard, C., S. Lesage, M. Rybojad, A. M. Prieur, S. Manouvrier-Hanu, R. Hafner, M. Chamaillard, H. Zouali, G. Thomas, and J. P. Hugot. 2001. CARD15 mutations in Blau syndrome. *Nat. Genet.* **29**:19–20.

Molofsky, A. B., B. G. Byrne, N. N. Whitfield, C. A. Madigan, E. T. Fuse, K. Tateda, and M. S. Swanson. 2006. Cytosolic recognition of flagellin by mouse macrophages restricts Legionella pneumophila infection. *J. Exp. Med.* **203**:1093–1104.

Moore, C. B., D. T. Bergstralh, J. A. Duncan, Y. Lei, T. E. Morrison, A. G. Zimmermann, M. A. Accavitti-Loper, V. J. Madden, L. Sun, Z. Ye, J. D. Lich, M. T. Heise, Z. Chen, and J. P. Ting. 2008. NLRX1 is a regulator of mitochondrial antiviral immunity. *Nature* **451**:573–577.

Murdoch, S., U. Djuric, B. Mazhar, M. Seoud, R. Khan, R. Kuick, R. Bagga, R. Kircheisen, A. Ao, B. Ratti, S. Hanash, G. A. Rouleau, and R. Slim. 2006. Mutations in NALP7 cause recurrent hydatidiform moles and reproductive wastage in humans. *Nat. Genet.* **38**:300–302.

Netea, M. G., T. Azam, E. C. Lewis, L. A. Joosten, M. Wang, D. Langenberg, X. Meng, E. D. Chan, D. Y. Yoon, T. Ottenhoff, S. H. Kim, and C. A. Dinarello. 2006. Mycobacterium tuberculosis induces interleukin-32 production through a caspase-1/IL-18/interferon-gamma-dependent mechanism. *PLoS Med.* **3**:e277.

O'Connell, R. M., S. K. Saha, S. A. Vaidya, K. W. Bruhn, G. A. Miranda, B. Zarnegar, A. K. Perry, B. O. Nguyen, T. F. Lane, T. Taniguchi, J. F. Miller, and G. Cheng. 2004. Type I interferon production enhances susceptibility to Listeria monocytogenes infection. *J. Exp. Med.* **200**:437–445.

Oganesyan, G., S. K. Saha, B. Guo, J. Q. He, A. Shahangian, B. Zarnegar, A. Perry, and G. Cheng. 2006. Critical role of TRAF3 in the Toll-like receptor-dependent and -independent antiviral response. *Nature* **439**:208–211.

Ogura, Y., D. K. Bonen, N. Inohara, D. L. Nicolae, F. F. Chen, R. Ramos, H. Britton, T. Moran, R. Karaliuskas, R. H. Duerr, J. P. Achkar, S. R. Brant, T. M. Bayless, B. S. Kirschner, S. B. Hanauer, G. Nunez, and J. H. Cho. 2001. A frameshift mutation in NOD2 associated with susceptibility to Crohn's disease. *Nature* **411**:603–606.

Opitz, B., A. Puschel, W. Beermann, A. C. Hocke, S. Forster, B. Schmeck, V. van Laak, T. Chakraborty, N. Suttorp, and S. Hippenstiel. 2006a. Listeria monocytogenes activated p38 MAPK and induced IL-8 secretion in a nucleotide-binding oligomerization domain 1-dependent manner in endothelial cells. *J. Immunol.* **176**:484–490.

Opitz, B., A. Puschel, B. Schmeck, A. C. Hocke, S. Rosseau, S. Hammerschmidt, R. R. Schumann, N. Suttorp, and S. Hippenstiel. 2004. Nucleotide-binding oligomerization domain proteins are innate immune receptors for internalized Streptococcus pneumoniae. *J. Biol. Chem.* **279**:36426–36432.

Opitz, B., M. Vinzing, V. van Laak, B. Schmeck, G. Heine, S. Gunther, R. Preissner, H. Slevogt, P. D. N'Guessan, J. Eitel, T. Goldmann, A. Flieger, N. Suttorp, and S. Hippenstiel. 2006b. Legionella pneumophila induces IFNbeta in lung epithelial cells via IPS-1 and IRF3, which also control bacterial replication. *J. Biol. Chem.* **281**:36173–36179.

O'Riordan, M., C. H. Yi, R. Gonzales, K. D. Lee, and D. A. Portnoy. 2002. Innate recognition of bacteria by a macrophage cytosolic surveillance pathway. *Proc. Natl. Acad. Sci. USA* **99**:13861–13866.

Park, J. H., Y. G. Kim, C. McDonald, T. D. Kanneganti, M. Hasegawa, M. Body-Malapel, N. Inohara, and G. Nunez. 2007. RICK/RIP2 mediates innate immune responses induced through Nod1 and Nod2 but not TLRs. *J. Immunol.* **178**:2380–2386.

Pelegrin, P., and A. Surprenant. 2006. Pannexin-1 mediates large pore formation and interleukin-1beta release by the ATP-gated P2X7 receptor. *EMBO J.* **25**:5071–5082.

Pelegrin, P., and A. Surprenant. 2007. Pannexin-1 couples to maitotoxin- and nigericin-induced interleukin-1beta release through a dye uptake-independent pathway. *J. Biol. Chem.* **282**:2386–2394.

Perry, A. K., E. K. Chow, J. B. Goodnough, W. C. Yeh, and G. Cheng. 2004. Differential requirement for TANK-binding kinase-1 in type I interferon responses to toll-like receptor activation and viral infection. *J. Exp. Med.* **199**:1651–1658.

Phalipon, A., M. Kaufmann, P. Michetti, J. M. Cavaillon, M. Huerre, P. Sansonetti, and J. P. Kraehenbuhl. 1995. Monoclonal immunoglobulin A antibody directed against serotype-specific epitope of Shigella flexneri lipopolysaccharide protects against murine experimental shigellosis. *J. Exp. Med.* **182**:769–778.

Philpott, D. J., S. Yamaoka, A. Israel, and P. J. Sansonetti. 2000. Invasive Shigella flexneri activates NF-kappa B through a lipopolysaccharide-dependent innate intracellular response and leads to IL-8 expression in epithelial cells. *J. Immunol.* **165**:903–914.

Pichlmair, A., O. Schulz, C. P. Tan, T. I. Naslund, P. Liljestrom, F. Weber, and C. Reis e Sousa. 2006. RIG-I-mediated antiviral responses to single-stranded RNA bearing 5'-phosphates. *Science* **314**:997–1001.

Raupach, B., S. K. Peuschel, D. M. Monack, and A. Zychlinsky. 2006. Caspase-1-mediated activation of interleukin-1beta (IL-1beta) and IL-18 contributes to innate immune defenses against Salmonella enterica serovar Typhimurium infection. *Infect. Immun.* **74:**4922–4926.

Ren, T., D. S. Zamboni, C. R. Roy, W. F. Dietrich, and R. E. Vance. 2006. Flagellin-deficient Legionella mutants evade caspase-1- and Naip5-mediated macrophage immunity. *PLoS Pathog.* **2:**e18.

Roebuck, K. A. 1999. Regulation of interleukin-8 gene expression. *J. Interferon Cytokine Res.* **19:**429–438.

Saha, S. K., E. M. Pietras, J. Q. He, J. R. Kang, S. Y. Liu, G. Oganesyan, A. Shahangian, B. Zarnegar, T. L. Shiba, Y. Wang, and G. Cheng. 2006. Regulation of antiviral responses by a direct and specific interaction between TRAF3 and Cardif. *EMBO J.* **25:**3257–3263.

Sansonetti, P. J., J. Arondel, M. Huerre, A. Harada, and K. Matsushima. 1999. Interleukin-8 controls bacterial transepithelial translocation at the cost of epithelial destruction in experimental shigellosis. *Infect. Immun.* **67:**1471–1480.

Sansonetti, P. J., A. Phalipon, J. Arondel, K. Thirumalai, S. Banerjee, S. Akira, K. Takeda, and A. Zychlinsky. 2000. Caspase-1 activation of IL-1beta and IL-18 are essential for Shigella flexneri-induced inflammation. *Immunity* **12:**581–590.

Schmitz, J., A. Owyang, E. Oldham, Y. Song, E. Murphy, T. K. McClanahan, G. Zurawski, M. Moshrefi, J. Qin, X. Li, D. M. Gorman, J. F. Bazan, and R. A. Kastelein. 2005. IL-33, an interleukin-1-like cytokine that signals via the IL-1 receptor-related protein ST2 and induces T helper type 2-associated cytokines. *Immunity* **23:**479–490.

Schotte, P., G. Denecker, A. Van Den Broeke, P. Vandenabeele, G. R. Cornelis, and R. Beyaert. 2004. Targeting Rac1 by the Yersinia effector protein YopE inhibits caspase-1-mediated maturation and release of interleukin-1beta. *J. Biol. Chem.* **279:**25134–25142.

Schroeder, G. N., N. J. Jann, and H. Hilbi. 2007. Intracellular type III secretion by cytoplasmic Shigella flexneri promotes caspase-1-dependent macrophage cell death. *Microbiology* **153**(Pt 9):2862–2876.

Seth, R. B., L. Sun, C. K. Ea, and Z. J. Chen. 2005. Identification and characterization of MAVS, a mitochondrial antiviral signaling protein that activates NF-kappaB and IRF 3. *Cell* **122:**669–682.

Shao, W., G. Yeretssian, K. Doiron, S. N. Hussain, and M. Saleh. 2007. The caspase-1 digestome identifies the glycolysis pathway as a target during infection and septic shock. *J. Biol. Chem.* **282:**36321–36329.

Sharma, S., B. R. tenOever, N. Grandvaux, G. P. Zhou, R. Lin, and J. Hiscott. 2003. Triggering the interferon antiviral response through an IKK-related pathway. *Science* **300:**1148–1151.

Shin, H., and G. R. Cornelis. 2007. Type III secretion translocation pores of Yersinia enterocolitica trigger maturation and release of pro-inflammatory IL-1beta. *Cell. Microbiol.* **9:**2893–2902.

Siegal, F. P., N. Kadowaki, M. Shodell, P. A. Fitzgerald-Bocarsly, K. Shah, S. Ho, S. Antonenko, and Y. J. Liu. 1999. The nature of the principal type 1 interferon-producing cells in human blood. *Science* **284:**1835–1837.

Simon, C., A. Frances, G. N. Piquette, I. el Danasouri, G. Zurawski, W. Dang, and M. L. Polan. 1994. Embryonic implantation in mice is blocked by interleukin-1 receptor antagonist. *Endocrinology* **134:**521–528.

Singer, I. I., S. Scott, J. Chin, E. K. Bayne, G. Limjuco, J. Weidner, D. K. Miller, K. Chapman, and M. J. Kostura. 1995. The interleukin-1 beta-converting enzyme (ICE) is localized on the external cell surface membranes and in the cytoplasmic ground substance of human monocytes by immuno-electron microscopy. *J. Exp. Med.* **182:**1447–1459.

Soulat, D., A. Bauch, S. Stockinger, G. Superti-Furga, and T. Decker. 2006. Cytoplasmic Listeria monocytogenes stimulates IFN-beta synthesis without requiring the adapter protein MAVS. *FEBS Lett.* **580:**2341–2346.

Stanley, S. A., J. E. Johndrow, P. Manzanillo, and J. S. Cox. 2007. The type I IFN response to infection with Mycobacterium tuberculosis requires ESX-1-mediated secretion and contributes to pathogenesis. *J. Immunol.* **178:**3143–3152.

Stehlik, C., and A. Dorfleutner. 2007. COPs and POPs: modulators of inflammasome activity. *J. Immunol.* **179:**7993–7998.

Stetson, D. B., and R. Medzhitov. 2006. Recognition of cytosolic DNA activates an IRF3-dependent innate immune response. *Immunity* **24:**93–103.

Sutterwala, F. S., L. A. Mijares, L. Li, Y. Ogura, B. I. Kazmierczak, and R. A. Flavell. 2007. Immune recognition of Pseudomonas aeruginosa mediated by the IPAF/NLRC4 inflammasome. *J. Exp. Med.* **204:**3235–3245.

Suzuki, T., L. Franchi, C. Toma, H. Ashida, M. Ogawa, Y. Yoshikawa, H. Mimuro, N. Inohara, C. Sasakawa, and G. Nunez. 2007. Differential regulation of caspase-1 activation, pyroptosis, and autophagy via Ipaf and ASC in Shigella-infected macrophages. *PLoS Pathog.* **3:**e111.

Takaoka, A., Z. Wang, M. K. Choi, H. Yanai, H. Negishi, T. Ban, Y. Lu, M. Miyagishi, T. Kodama, K. Honda, Y. Ohba, and T. Taniguchi. 2007. DAI (DLM-1/ZBP1) is a cytosolic DNA sensor and an activator of innate immune response. *Nature* **448:**501–505.

Takeuchi, O., and S. Akira. 2008. MDA5/RIG-1 and virus recognition. *Curr. Opin. Immunol.* **20:**17–22.

Tattoli, I., L. A. Carneiro, M. Jehanno, J. G. Magalhaes, Y. Shu, D. J. Philpott, D. Arnoult, and S. E. Girardin. 2008. NLRX1 is a mitochondrial NOD-like receptor that amplifies NF-kappaB and JNK pathways by inducing reactive oxygen species production. *EMBO Rep.* **9:**293–300.

Tompkins, W. A., J. M. Zarling, and W. E. Rawls. 1970. In vitro assessment of cellular immunity to vaccinia virus: contribution of lymphocytes and macrophages. *Infect. Immun.* **2:**783–790.

Travassos, L. H., L. A. Carneiro, S. E. Girardin, I. G. Boneca, R. Lemos, M. T. Bozza, R. C. Domingues, A. J. Coyle, J. Bertin, D. J. Philpott, and M. C. Plotkowski. 2005. Nod1 participates in the innate immune response to Pseudomonas aeruginosa. *J. Biol. Chem.* **280:**36714–36718.

Tsuji, N. M., H. Tsutsui, E. Seki, K. Kuida, H. Okamura, K. Nakanishi, and R. A. Flavell. 2004. Roles of caspase-1 in Listeria infection in mice. *Int. Immunol.* **16:**335–343.

Vaena de Avalos, S., I. J. Blader, M. Fisher, J. C. Boothroyd, and B. A. Burleigh. 2002. Immediate/early response to Trypanosoma cruzi infection involves minimal modulation of host cell transcription. *J. Biol. Chem.* **277:**639–644.

Viala, J., C. Chaput, I. G. Boneca, A. Cardona, S. E. Girardin, A. P. Moran, R. Athman, S. Memet, M. R. Huerre, A. J. Coyle, P. S. DiStefano, P. J. Sansonetti, A. Labigne, J. Bertin, D. J. Philpott, and R. L. Ferrero. 2004. Nod1 responds to peptidoglycan delivered by the Helicobacter pylori cag pathogenicity island. *Nat. Immunol.* **5:**1166–1174.

Vogel, J. P., H. L. Andrews, S. K. Wong, and R. R. Isberg. 1998. Conjugative transfer by the virulence system of Legionella pneumophila. *Science* **279:**873–876.

Wang, S., M. Miura, Y. K. Jung, H. Zhu, E. Li, and J. Yuan. 1998. Murine caspase-11, an ICE-interacting protease, is essential for the activation of ICE. *Cell* **92:**501–509.

Weidinger, S., N. Klopp, L. Rummler, S. Wagenpfeil, N. Novak, H. J. Baurecht, W. Groer, U. Darsow, J. Heinrich, A. Gauger, T. Schafer, T. Jakob, H. Behrendt, H. E. Wichmann, J. Ring, and T. Illig. 2005. Association of NOD1 polymorphisms with atopic eczema and related phenotypes. *J. Allergy Clin. Immunol.* **116:**177–184.

Weiss, D. S., A. Brotcke, T. Henry, J. J. Margolis, K. Chan, and D. M. Monack. 2007. In vivo negative selection screen identifies genes required for Francisella virulence. *Proc. Natl. Acad. Sci. USA* **104:**6037–6042.

Welter-Stahl, L., D. M. Ojcius, J. Viala, S. Girardin, W. Liu, C. Delarbre, D. Philpott, K. A. Kelly, and T. Darville. 2006. Stimulation of the cytosolic receptor for peptidoglycan, Nod1, by infection with Chlamydia trachomatis or Chlamydia muridarum. *Cell. Microbiol.* **8:**1047–1057.

Willingham, S. B., D. T. Bergstralh, W. O'Connor, A. C. Morrison, D. J. Taxman, J. A. Duncan, S. Barnoy, M. M. Venkatesan, R. A. Flavell, M. Deshmukh, H. M. Hoffman, and J. P. Ting. 2007. Microbial pathogen-induced necrotic cell death mediated by the inflammasome components CIAS1/cryopyrin/NLRP3 and ASC. *Cell Host Microbe* **2:**147–159.

Wise, C. A., J. D. Gillum, C. E. Seidman, N. M. Lindor, R. Veile, S. Bashiardes, and M. Lovett. 2002. Mutations in CD2BP1 disrupt binding to PTP PEST and are responsible for PAPA syndrome, an autoinflammatory disorder. *Hum. Mol. Genet.* **11:**961–969.

Wright, E. K., S. A. Goodart, J. D. Growney, V. Hadinoto, M. G. Endrizzi, E. M. Long, K. Sadigh, A. L. Abney, I. Bernstein-Hanley, and W. F. Dietrich. 2003. Naip5 affects host susceptibility to the intracellular pathogen Legionella pneumophila. *Curr. Biol.* **13:**27–36.

Wu, S., P. Xie, K. Welsh, C. Li, C. Z. Ni, X. Zhu, J. C. Reed, A. C. Satterthwait, G. A. Bishop, and K. R. Ely. 2005. LMP1 protein from the Epstein-Barr virus is a structural CD40 decoy in B lymphocytes for binding to TRAF3. *J. Biol. Chem.* **280:**33620–33626.

Xu, L. G., Y. Y. Wang, K. J. Han, L. Y. Li, Z. Zhai, and H. B. Shu. 2005. VISA is an adapter protein required for virus-triggered IFN-beta signaling. *Mol. Cell* **19:**727–740.

Yang, Y., C. Yin, A. Pandey, D. Abbott, C. Sassetti, and M. A. Kelliher. 2007. NOD2 pathway activation by MDP or Mycobacterium tuberculosis infection involves the stable polyubiquitination of Rip2. *J. Biol. Chem.* **282:**36223–36229.

Yu, J. W., T. Fernandes-Alnemri, P. Datta, J. Wu, C. Juliana, L. Solorzano, M. McCormick, Z. Zhang, and E. S. Alnemri. 2007. Pyrin activates the ASC pyroptosome in response to engagement by autoinflammatory PSTPIP1 mutants. *Mol. Cell* **28:**214–227.

Zamboni, D. S., K. S. Kobayashi, T. Kohlsdorf, Y. Ogura, E. M. Long, R. E. Vance, K. Kuida, S. Mariathasan, V. M. Dixit, R. A. Flavell, W. F. Dietrich, and C. R. Roy. 2006. The Birc1e cytosolic pattern-recognition receptor contributes to the detection and control of Legionella pneumophila infection. *Nat. Immunol.* **7:**318–325.

Zilbauer, M., N. Dorrell, A. Elmi, K. J. Lindley, S. Schuller, H. E. Jones, N. J. Klein, G. Nunez, B. W. Wren, and M. Bajaj-Elliott. 2007. A major role for intestinal epithelial nucleotide oligomerization domain 1 (NOD1) in eliciting host bactericidal immune responses to Campylobacter jejuni. *Cell. Microbiol.* **9:**2404–2416.

Zychlinsky, A., B. Kenny, R. Menard, M. C. Prevost, I. B. Holland, and P. J. Sansonetti. 1994. IpaB mediates macrophage apoptosis induced by Shigella flexneri. *Mol. Microbiol.* **11:**619–627.

10

Phagocytes Are a Source of the Fluid-Phase Pattern Recognition Receptor PTX3: Interplay between Cellular and Humoral Innate Immunity

ALBERTO MANTOVANI, BARBARA BOTTAZZI, ANDREA DONI, GIOVANNI SALVATORI, PASCALE JEANNIN, AND CECILIA GARLANDA

INTRODUCTION

The innate immune system consists of a cellular and a humoral arm. Components of humoral immunity include members of the complement cascade and soluble pattern recognition receptors (PRRs), such as collectins (surfactant protein-A [SP-A] and SP-D), ficolins, and pentraxins (Bottazzi et al., 2006; Garlanda et al., 2005). Fluid-phase PRRs are therefore a heterogeneous group of molecular families, which represent functional ancestors of antibodies. They play a key role as effectors and modulators of innate resistance in animals and humans. There is evidence that this heterogeneous set of soluble PRRs interacts with cellular innate immunity, and the prototypic long pentraxin PTX3 represents a case-in-point of interplay. Here, we will review how phagocytes represent a key source of this fluid-phase PRR and how this molecule is essential for innate resistance against diverse pathogens.

Structure of Pentraxins

Pentraxins are a superfamily of proteins, phylogenetically conserved from arachnids to mammals and characterized by the presence in their carboxy terminus of a 200-amino-acid pentraxin domain, with an 8-amino-acid conserved pentraxin signature (Garlanda et al., 2005). Pentraxins recognize a wide range of exogenous pathogenic substances and altered self-molecules and, in a species-specific manner, behave as acute-phase proteins.

Based on the primary structure of the subunit, the pentraxins are divided into short and long pentraxins. C-reactive protein (CRP) and serum amyloid P component (SAP) are the prototype of the short pentraxin family: they are mainly produced in the liver in response to inflammatory signals, most prominently interleukin-6 (IL-6), and are acute-phase proteins in humans and mice, respectively. PTX3 is the prototype of the long pentraxin family, whose members were identified in the 1990s as cytokine-inducible genes or molecules expressed in specific tissues, e.g., guinea pig apexin in spermatozoa (Noland et al., 1994; Reid and Blobel, 1994), neuronal pentraxin (NP) 1 or NPTX1 (Omeis et al., 1996; Schlimgen et al., 1995), NP2, also called Narp or NPTX2 (Hsu and Perin, 1995; Tsui et al., 1996), and neuronal pentraxin receptor (NPR), a transmembrane molecule in neurons (Dodds et al., 1997; Kirkpatrick et al., 2000). PTX3 is a 45-kDa protein that assembles to form high-molecular-weight multimers linked by interchain disulfide bonds (Bottazzi et al., 1997). The C-terminal domain (203 amino acids) of PTX3 shares homology with the classic short pentraxins, whereas the N-terminal domain (178 amino acids) does not show any significant homology with other known proteins (Fig. 1). PTX3 differs from CRP and SAP also for gene organization, cellular source, and ligand-binding properties (Garlanda et al., 2005).

The analysis of the glycosylation status of PTX3 indicated that the unique glycosylation site at Asn-220 within the pentraxin domain is fully occupied and ruled out the occurrence of O-linked oligosaccharides. Oligosaccharides conjugated to Asn-220 were N-linked complex-type sugars, mainly consisting of biantennary fucosylated and variably sialylated structures. PTX3 glycans showed heterogeneity in the relative amount of bi-, tri-, and tetra-antennary structures depending on the cell type and inflammatory stimulus used to induce PTX3 (Inforzato et al., 2006).

Alberto Mantovani, Istituto Clinico Humanitas, IRCCS, via Manzoni 56, 20089, Rozzano, Milan, Italy, and Department of Translational Medicine, Faculty of Medicine, University of Milan, Milan, 20100, Italy. Barbara Bottazzi, Andrea Doni, and Cecilia Garlanda, Istituto Clinico Humanitas, IRCCS, via Manzoni 56, 20089, Rozzano, Milan, Italy. Giovanni Salvatori, Sigma-Tau R&D, via Pontina Km 30.400, 00040, Pomezia, Rome, Italy. Pascale Jeannin, INSERM, U564, Equipe Avenir, 4 rue Larrey, Angers, F-49933 France, and University Hospital of Angers, Service d'Immunologie, Angers, F-49933 France.

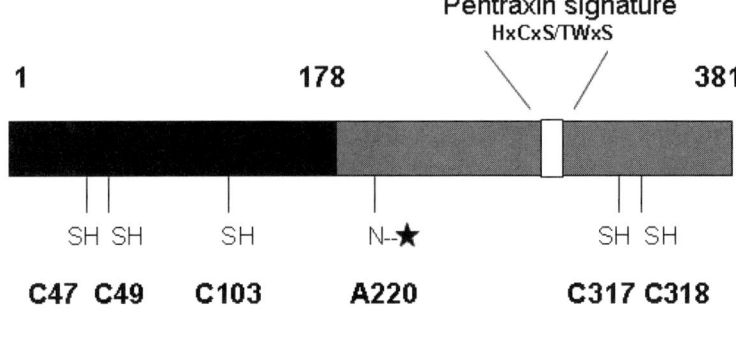

FIGURE 1 Structure of PTX3 protomer with its N- and C-terminal domains, the leader peptide, and the pentraxin signature. Cysteine residues relevant for interchain disulfide bonds and the N-linked glycosylation site at aspartic acid residue 220 are shown.

PHAGOCYTES AS A MAJOR SOURCE OF THE LONG PENTRAXIN PTX3

Gene Expression-Dependent Regulation

Mononuclear phagocytes and myeloid-derived dendritic cells (DCs) are a major source of the long pentraxin PTX3. In addition, a variety of cell types produce PTX3 in vitro upon exposure to primary inflammatory signals, such as IL-1β, tumor necrosis factor alpha (TNF-α), microbial moieties such as lipopolysaccharide (LPS), lipoarabinomannan, outer membrane protein A (OmpA), and other agonists for different members of the Toll-like receptor (TLR) family (Breviario et al., 1992; Jeannin et al., 2005). These cells include myeloid DCs, which are major producers of PTX3, endothelial cells, monocytes, adipocytes, fibroblasts, smooth muscle cells, synovial cells, and chondrocytes (Abderrahim-Ferkoune et al., 2003; Agnello et al., 2000; Breviario et al., 1992; Doni et al., 2003; Goodman et al., 2000; Introna et al., 1996; Klouche et al., 2004; Lee et al., 1993). Recently, cells of epithelial origin, for instance, renal and alveolar epithelial cells, have also been found to produce low amounts of PTX3 under stimulation (dos Santos et al., 2004; Nauta et al., 2005). IL-6, a poor inducer of PTX3 in vitro, was found to be involved in PTX3 expression in Castleman's disease (Malaguarnera et al., 2000) and in Kaposi sarcoma (Klouche et al., 2002).

Vascular endothelial and smooth muscle cells produce PTX3 in response to inflammatory signals including oxidized low-density lipoproteins (Klouche et al., 2004) and accordingly, PTX3 was observed in atherosclerosis lesions in humans (Rolph et al., 2002).

Interferon gamma (IFN-γ) and IL-10 have different effects on PTX3 production. IFN-γ, which generally has a synergistic effect with LPS (Ehrt et al., 2001), inhibits LPS-induced PTX3 expression and production in different cellular contexts (Goodman et al., 1996, 2000; Polentarutti et al., 1998), whereas IL-10 weakly induces PTX3 expression in DCs and monocytes and significantly synergizes with LPS, other TLR agonists, and IL-1β (Perrier et al., 2004). IL-10 induces a set of genes (e.g., type I collagen, fibronectin, versican, α1-antitrypsin) related to tissue remodeling (Lang et al., 2002; Perrier et al., 2004) and is involved in the chronic and resolution phase of inflammation (Moore et al., 2001). Given its role in matrix organization (Salustri et al., 2004), PTX3 expression in M2 mononuclear phagocytes and IL-10-treated DCs and fibroblasts is likely to be related to the orchestration of matrix deposition, tissue repair, and remodeling (Mantovani et al., 2004). Moreover, it is interesting that, in addition to the stimulation of B-cell differentiation and antibody production (Moore et al., 2001) (the humoral arm of adaptive antibody-mediated immunity), IL-10 also stimulates the humoral arm of innate immunity (PTX3).

A peculiar tissue is the cumulus oophorus, in which PTX3 mRNA expression is orchestrated by hormonal ovulatory stimuli (follicle-stimulating hormone or human chorionic gonadotropin), by oocyte-derived soluble factors, and in particular by a member of the transforming growth factor-β family, growth differentiation factor-9 (GDF-9) (Salustri et al., 2004; Varani et al., 2002). In this tissue, PTX3 expression is restricted to the pre-ovulatory period, showing close temporal correlation to matrix deposition by cumulus cells. Western blot and immunofluorescence analysis indicate that PTX3 is associated with the extracellular matrix of the cumulus oophorus.

Recent analysis of PTX3 gene regulation has yielded unexpected results. First, two independent studies have shown that PTX3 is a major responsive gene downstream of the FUSS-CHOP translocation involved in the pathogenesis of a subset of soft tissue sarcomas (Willeke et al., 2006). The pathophysiological significance of this finding and its value for the monitoring of disease remains to be elucidated. In addition, recent results have shown that glucocorticoid hormones (GCs) have divergent effects on PTX3 expression and production in mononuclear phagocytes and in nonhematopoietic cells (Doni et al., 2008). In myeloid DCs, GC inhibited the PTX3 production. In contrast, in fibroblasts and endothelial cells, GC alone induced and, under inflammatory conditions, enhanced and extended PTX3 production. In vivo administration of GC augmented the blood levels of PTX3 in mice and humans. Moreover, patients with Cushing's syndrome had increased levels of circulating PTX3, whereas PTX3 levels were decreased in subjects affected by iatrogenic hypocortisolism. In nonhematopoietic cells, GC receptor functioned as a ligand-dependent transcription factor (dimerization-dependent) to induce PTX3 gene expression. In contrast, in hematopoietic cells, GC receptor repressed PTX3 gene transcription in a dimerization-independent manner by interfering with the action of other signaling pathways, likely

NF-κB and AP-1. Thus, divergent effects of GC were found to be due to different GC receptor mechanisms.

The divergent effects of GC on PTX3 production are likely to reflect the different functions of this multifunctional molecule in innate immunity and in the construction of the extracellular matrix.

Neutrophils as a Source of Preformed PTX3

In an unexpected twist, we recently found that PTX3 is stored in specific granules (Color Plate 1) and undergoes release in response to microbial recognition and inflammatory signals (Jaillon et al., 2007). Upon activation, neutrophils release extracellular traps (neutrophil extracellular traps, or NETs), composed of chromatin decorated with granular proteins, which provide a high local concentration of antimicrobial molecules that bind and kill microbes (Brinkmann et al., 2004; Fuchs et al., 2007). Released PTX3 can partially localize in NETs, as shown in Color Plate 2. Eosinophils and basophils do not contain preformed PTX3. PTX3-deficient neutrophils have defective microbial recognition and phagocytosis, and PTX3 is nonredundant for neutrophil-mediated resistance against *Aspergillus fumigatus* (see below). Thus, neutrophils serve as a reservoir, ready for rapid release, of the long pentraxin PTX3, a key component of humoral innate immunity with opsonic activity. Myeloid, but not plasmacytoid, DCs and macrophages are major producers of PTX3 (Doni et al., 2003). Over a period of 24 h, DCs release approximately 50 ng of PTX3 per 10^6 cells (Doni et al., 2003). Neutrophils contain 24.9 ± 3.8 ng of this PRR per 10^6 cells ($n = 5$). Upon stimulation, they release approximately 25% of stored PTX3, a part of it remaining cell-associated, presumably with NETs. Given the abundance of neutrophils in the circulation and in the early phases of inflammatory reactions in tissues, these cells represent a major source of PTX3 covering a temporal window preceding gene expression-dependent production. Under conditions of tissue damage (e.g., myocardial infarction) or infection (e.g., sepsis), PTX3 levels increase rapidly (see below). The results reported here shed new light on PTX3 elevations in pathological conditions and on their pathophysiological implications. It is likely that rapid release of stored PTX3 by activated neutrophils plays a role in the early phases of its elevation in pathology, preceding gene expression-dependent production. PTX3 expressed by neutrophils is essential to control fungal growth in vitro and in vivo (Jaillon et al., 2007). Innate and adaptive immunity are both essential for the development of a protective antifungal immune response. Generation of a Th1-oriented *A. fumigatus*-specific immune response is associated with protection (Cenci et al., 1997; Nagai et al., 1995). Injection of PTX3 in *ptx3*-deficient mice favors the generation of a protective Th1 anti-*Aspergillus* immune response (Garlanda et al., 2002). Neutrophil-derived PTX3, in addition to DC-derived PTX3, may be involved in the orientation of the immune response toward a protective Th1 phenotype. Neutrophils, an innate cell type without professional antigen-presenting functions, may participate, via the release of this preformed soluble PRR, in the activation and orientation of adaptive immunity.

FUNCTIONAL ROLES OF PTX3

The multifunctional properties exerted by PTX3 can be, at least in part, explained by its capacity to interact with a number of different ligands, a characteristic shared with the classical short pentraxins CRP and SAP. Table 1 is a summary of the complex network of PTX3 ligands identified so far.

Unlike the classic short pentraxins CRP and SAP, whose sequence and regulation have diverged from mice to humans, PTX3 is evolutionarily conserved. Thus, results obtained by using genetic approaches in the mouse are likely to be informative for the function of PTX3 in humans.

Interaction with Complement

The first described and best characterized ligand for PTX3 is the complement component C1q (Bottazzi et al., 1997; Nauta et al., 2003a). PTX3 binds to plastic-immobilized C1q, interacting with C1q globular head (gC1q), in particular, with charged residues localized on the apex of the molecule and involving all the three C1q chains (gC1qA, gC1qB, and gC1qC) (Nauta et al., 2003a; Roumenina et al., 2006). In the same experimental conditions, PTX3 fails to interact with other components of the complement system, such as C3 and C4 (B. Bottazzi and L. Deban, unpublished observations). Because of its stable multimeric structure, PTX3 does not require a previous aggregation, unlike CRP and SAP, which show optimal interaction with C1q only after chemical cross-linking (Hicks et al., 1992).

Interaction of PTX3 with plastic-immobilized C1q, an experimental condition that could mimic the surface of microbes, results in the activation of the classical complement cascade, measured as C3 and C4 deposition. On the other hand, the presence of PTX3 resulted in a dose-dependent inhibition of C1q hemolytic activity, due to interference with C1q binding to antibody-sensitized erythrocytes. These results indicate that the binding of fluid-phase PTX3 to C1q may inhibit complement activation by competitive blocking of relevant interaction sites (Nauta et al., 2003a). These data indicate that PTX3 may exert a dual role and contrasting effects on complement activation: it supports clearance of material that is able to bind PTX3, such as microbes, whereas it may protect against unwanted complement activation in the fluid phase. Interaction with C1q and subsequent complement activation is also modulated by the extent of PTX3 glycosylation, as demonstrated by the observation that removal of sialic acid or complete deglycosylation of the protein significantly increases its binding to C1q (Inforzato et al., 2006). In accordance, PTX3 desialylation increases complement activation, as assessed by C3 and C4 deposition (Inforzato et al., 2006). Interestingly, recombinant PTX3 and PTX3 from TNF-α-stimulated fibrosarcoma cell line 8387 and from LPS-stimulated DCs exhibited N-linked sialylated complex-type sugars, heterogeneous in the relative amount of bi-, tri-, and tetra-antennary structures. These results suggest that changes in the PTX3 glycosylation pattern might be ascribed to cell type and the inflammatory context in which it is produced and might be directly implicated in tuning its biological functions (Inforzato et al., 2006).

In 1999, H. Jarva showed that CRP can modulate the alternative pathway of complement activation through interaction with Factor H, the main soluble regulator of the alternative pathway (Jarva et al., 1999). In accordance, PTX3 can also interact with Factor H (Deban et al., 2008), suggesting a more general and complex role of PTX3 in the control of complement functions.

TABLE 1 Ligands recognized by PTX3 and short pentraxins

Ligands	PTX3	Calcium requirement[a]	Short pentraxins
Complement components			
C1q	+	−	CRP and SAP
Factor H	+	+	CRP
C4b-binding protein	+[c]		SAP
Extracellular matrix proteins			
TSG-6	+	+	NT[b]
IαI	+	+	No binding with CRP
Hyaluronan	−		NT
Laminin	−		CRP and SAP
Collagen IV	−		SAP
Fibronectin	−		CRP and SAP
Growth factors			
FGF2	+	+	Low binding with CRP
FGF1 and FGF4	−		NT
Membrane moieties			
PC	−		CRP
PE	−		SAP
LPS	−		SAP
KpOmpA	+	+	NT
Pathogens			
Bacteria	+	NT	CRP and SAP
Aspergillus fumigatus	+	−	CRP
Zymosan	+	NT	CRP and SAP
Viruses	+	NT	SAP

[a]Calcium requirement for PTX3 binding.
[b]NT, not tested.
[c]L. Deban and S. Meri, unpublished observations.

The interaction of short pentraxins CRP and SAP with their ligands, in general, depends on the presence of calcium (Szalai et al., 1999). On the contrary, inductive coupled plasma/atomic emission spectroscopy shows that PTX3 does not have a specific coordination site for calcium ions (Bottazzi et al., 1997). Accordingly, calcium is not required for PTX3 interaction with C1q (Bottazzi et al., 1997). However, with the discovery of other PTX3 ligands, it appears that this is not a general rule, as summarized in Table 1.

Role in Innate Resistance

The observation that cells of the myelomonocytic lineage produce high levels of PTX3 upon stimulation with proinflammatory signals or TLR engagement supports the idea that PTX3 plays a crucial role in the defense mechanisms. Moreover, like CRP and SAP, PTX3 can interact with a number of different pathogens: fungi, virus, and bacteria. A specific binding has been observed for zymosan, *Paracoccidioides brasiliensis*, and conidia from *A. fumigatus* (Diniz et al., 2004; Garlanda et al., 2002). Interaction has been observed with some selected gram-positive and gram-negative bacteria, including *Staphylococcus aureus, Pseudomonas aeruginosa, Salmonella enterica* serovar Typhimurium, *Klebsiella pneumoniae, Streptococcus pneumoniae*, and *Neisseria meningitidis*, and with human and murine cytomegalovirus and H3N2 influenza virus (Bozza et al., 2006).

In an attempt to identify the molecular structures recognized by PTX3 on the surface of pathogens, the direct interaction of PTX3 with different bacterial moieties localized on the microbial cell wall was analyzed (Jeannin et al., 2005). PTX3 does not bind LPS, lipoteichoic acid, enterotoxin A and B, exotoxin A, and N-acetylmuramyl-L-alanyl-D-isoglutamine (MDP), whereas it binds outer membrane protein A from *K. pneumoniae* (KpOmpA), a major component of the outer membrane of gram-negative bacteria highly conserved among the *Enterobacteriaceae* family (Jeannin et al., 2005).

According with the binding properties, macrophages from PTX3-overexpressing mice show an increased phagocytic activity of zymosan and *P. brasiliensis* (Diniz et al., 2004), while macrophages and polymorphonuclear leukocytes (PMNs) from PTX3-deficient mice are characterized by a defective phagocytosis of conidia from *A. fumigatus* (Garlanda et al., 2002; Jaillon et al., 2007). These data support the idea that cells of the myelomonocytic lineage can express a receptor for PTX3 on their surface. In accordance, a specific, dose-dependent, and saturable binding of PTX3 to murine macrophages, as well as human monocytes, DCs and PMNs, has been observed (Garlanda et al., 2002).

In vivo studies with gene-modified mice indicated that PTX3 is nonredundant in selected fungal and bacterial infections (*A. fumigatus, P. brasiliensis, P. aeruginosa*, serovar Typhimurium) and irrelevant in others (*Listeria monocytogenes, S. aureus*, polymicrobic intra-abdominal sepsis) (Diniz et al., 2004; Garlanda et al., 2002), suggesting that PTX3 deficiency does not cause a generalized impairment of host resistance to microbial pathogens, and that PTX3 is involved in recognition and resistance against specific microorganisms. In particular, PTX3-deficient mice were extremely susceptible to invasive pulmonary aspergillosis and the specificity of the defect and the therapeutic potential of PTX3 could be demonstrated by the complete protective effect of treatment with recombinant PTX3 (Garlanda et al., 2002; Gaziano et al., 2004). Moreover, in this model, the defective recognition of *A. fumigatus* conidia by PTX3-deficient mice was associated with the lack of development of appropriate and protective Th1 antifungal responses and with an unbalanced cytokine profile skewed toward a Th2 response (Garlanda et al., 2002).

Recently, Bozza et al. (2006) studied the role of PTX3 in viral infections and found that PTX3 bound both human and murine cytomegalovirus (MCMV), reducing viral entry and infectivity in DCs in vitro. Consistently, PTX3-deficient mice were more susceptible to MCMV infection than PTX3 wild-type mice, and PTX3 protected susceptible BALB/c mice from MCMV primary infection and reactivation in vivo, as well as *Aspergillus* superinfection. This occurred through the activation of interferon regulatory factor-3 (IRF3) in DCs via the TLR9/MyD88-independent viral recognition sensing and the promotion of the IL-12/IFN-γ-dependent effector pathway (Bozza et al., 2006).

As PTX3 binds to C1q and modulates the activation of the classical pathway of complement cascade, an indirect complement-mediated immune response could be activated by PTX3. The in vivo relevance of complement activation by PTX3 through the classical pathway has been studied by evaluating the therapeutic potential of PTX3 in C1q-deficient mice in infections. Results suggest that PTX3 can mediate resistance, at least against *A. fumigatus* and *P. aeruginosa*, independently of C1q (Garlanda et al., 2002; F. Moalli, unpublished results).

The role played by PTX3 in innate resistance to pathogens could also be exerted in an opsonizing-independent manner: in the case of *K. pneumoniae* infection, Soares et al. (2006) could not demonstrate binding of PTX3 to *K. pneumoniae*; however, overexpression of PTX3 by transgenic mice during infection was associated with an enhanced ability to produce proinflammatory mediators, including nitric oxide and TNF-α, and, as a consequence, with protection or faster lethality, depending on the dose of inocula. Thus, according to studies with transgenic mice, PTX3 overexpression under the control of its own promoter is associated with enhanced inflammatory responses, which, depending on the model studied, can be beneficial or detrimental for the host (see below).

Interestingly, Olesen et al. (2007) recently analyzed the role of polymorphisms within PTX3 in pulmonary tuberculosis and showed that PTX3 haplotype frequencies significantly differed in patients compared with controls.

In conclusion, PTX3 is released by PMNs and produced by DCs, neighboring macrophages, and other cell types upon TLR engagement or pathogen recognition and recognizes microbial moieties; opsonizes fungi, selected gram-positive and gram-negative bacteria, and viruses; and activates complement. Opsonization results in facilitated pathogen recognition (increased phagocytosis and killing) and in innate immune cell activation (increased cytokine and nitric oxide production); moreover, opsonization by PTX3 is likely involved in the activation of an appropriate adaptive immune response (DC maturation and polarization). All these properties suggest that this long pentraxin behaves as a functional ancestor of antibodies.

Role in Inflammation

PTX3 behaves as an acute-phase response protein since its blood levels, low in normal conditions (about 25 ng/ml in the mouse, less than 2 ng/ml in humans), increase rapidly (peak at 6 to 8 h) and dramatically (200 to 800 ng/ml in humans and mice) during endotoxic shock, sepsis, and other inflammatory and infectious conditions, correlating with the severity of the disease. The in vivo role of PTX3 in inflammatory conditions has been investigated by using PTX3-overexpressing and -deficient mice. In a model of LPS toxicity and in cecal ligation and puncture, PTX3 overexpression resulted in increased resistance (Dias et al., 2001), whereas its deficiency was irrelevant (Garlanda et al., 2002). After intestinal ischemia reperfusion injury, PTX3-overexpressing mice showed exacerbated inflammatory response and reduced survival rate, due to enhanced production of proinflammatory mediators (TNF-α, in particular) (Souza et al., 2002). Limbic seizures enhance the production of proinflammatory cytokines as well as PTX3 (Ravizza et al., 2001); in a model of kainate-induced seizures, PTX3-deficient mice had more widespread and severe IL-1-induced neuronal damage. In this model, PTX3 confers resistance to neurodegeneration, possibly by binding to dying neurons and rescuing them from otherwise irreversible damage (Ravizza et al., 2001).

Among the microbial moieties recognized by PTX3 is KpOmpA (Jeannin et al., 2005). KpOmpA binds to and is internalized by DCs and macrophages (Jeannin et al., 2000), activating both these cellular types in a TLR2-dependent way. The innate immune response to KpOmpA involves recognition by the scavenger receptors LOX-1 and SREC-I (Jeannin et al., 2005). The activation program set in motion by KpOmpA involves production of PTX3, which, in turn, binds KpOmpA, amplifying the inflammatory response in vivo. The response to KpOmpA in vivo is significantly reduced in PTX3-deficient mice and is restored by exogenous administration of purified PTX3, indicating that PTX3 is part of a nonredundant amplification loop induced by KpOmpA. Experiments performed inhibiting complement activation through different approaches indicate that PTX3 amplifies the inflammatory response to KpOmpA by activating a complement-dependent humoral amplification loop of the innate response to a microbial ligand (Cotena et al., 2007).

Altogether, these results outline the delicate role exerted by PTX3 in the inflammatory context.

Role in Self/Nonself Discrimination, DC Maturation, and Responses

Similarly to CRP and SAP, PTX3 binds to apoptotic cells during late phases of apoptosis (Rovere et al., 2000). Confocal analysis shows that PTX3 binds to discrete membrane domains of late apoptotic cells but the structures recognized have not been identified so far, even if competition experiments with both CRP and SAP suggest that all three pentraxins may to some extent interact with a common site (Rovere et al., 2000). Phosphoethanolamine, phosphocoline, small nuclear ribonucleoproteins, and chromatin/nucleolar components, recognized by CRP and/or SAP (Pepys et al., 1994), redistribute to the plasma membrane during late apoptosis. PTX3 can bind histones, raising the possibility that interaction with nuclear components could actually occur, whereas it does not bind to phosphoethanolamine and phosphocoline.

While PTX3 promotes removal of selected pathogens by professional phagocytes, it inhibits removal of apoptotic cells. In fact, in the presence of PTX3, both immature DCs and macrophages failed to internalize dying cells, thus preventing inflammatory uptake of late apoptotic cells and antigen presentation by antigen-presenting cells (Rovere et al., 2000; van Rossum et al., 2004).

Interaction of PTX3 with C1q may play a crucial role in the clearance of apoptotic cells. Both C1q and PTX3 are produced by immature DCs in response to TLR engagement (Baruah et al., 2006a; Castellano, et al., 2004); moreover, both the proteins bind apoptotic cells with similar kinetics, interacting with different binding sites and remaining stably associated to the apoptotic cell membrane (Baruah, et al., 2006a). In addition, when PTX3 is incubated with apoptotic cells, it enhances the deposition of both C1q and C3 on the cell surface (Nauta et al., 2003a), suggesting a role for PTX3 in the complement-mediated removal of dying cells. On the contrary, in the fluid phase, PTX3 reduces C1q and C3 deposition on apoptotic cells as well as the C1q-mediated phagocytosis of apoptotic cells by DCs (Baruah et al., 2006a). These data suggest that PTX3 may play a dual role in the regulation of complement-mediated immune responses and further support accumulating evidence suggesting that complement components and pentraxins may participate in the handling of apoptotic cells (Nauta et al., 2003b).

PTX3 influences the maturation program of DCs induced by LPS, inhibiting TNF-α and IL-10 secretion, as well as the upregulation of membrane molecules, such as CD86, HLA-DR, and HLA-ABC (Baruah et al., 2006b). In contrast, in the presence of dying cells, PTX3 enhances cytokine production but inhibits the cross-presentation of apoptotic cell-derived epitopes of self, viral, or tumoral origin to autoreactive CD8+ T cells. Among the members of the pentraxin family, these effects are specific for PTX3 since CRP does not affect the expression of membrane molecules or the secretion of cytokines by DCs as well as the cross-presentation of apoptotic cell-associated antigens (Baruah et al., 2006b). Thus, PTX3 behaves as a flexible regulator of the function of DCs, modulating the maturation program and the secretion of soluble factors. These results have led to the speculation that PTX3 has a dual role in the protection against pathogens and in the control of autoimmunity.

Role in Angiogenesis

PTX3 binds fibroblast growth factor 2 (FGF2), but not other members of the FGF family, such as FGF1 and FGF4, or cytokines and chemokines (Rusnati et al., 2004). The FGF2 binding site has been mapped on the N-terminal domain of PTX3, as demonstrated by means of recombinant N-terminal and C-terminal PTX3 domains expressed and purified from CHO cells (Camozzi et al., 2006). The two synthetic peptides PTX3 (82 to 110) and PTX3 (97 to 110), spanning in the N-terminal portion of PTX3, were able to prevent PTX3 binding to immobilized FGF2, confirming that the FGF2-binding site is located within the N-terminal domain from amino acid 97 to 110 (Camozzi et al., 2006).

Interaction between PTX3 and FGF2 prevents FGF2 binding to endothelial cells, leading to inhibition of FGF-dependent cell proliferation in vitro. Moreover, PTX3 overexpression in FGF2-transformed endothelial cells inhibits both their proliferation in vitro and their capacity to generate vascular lesions in vivo. FGF2 plays a key role in the induction of proliferation, migration, and survival of vascular smooth muscle cells (SMCs) and excessive growth of SMCs is an important component in atherosclerosis and restenosis. Interaction between PTX3 and FGF2 inhibits SMC proliferation in vitro; in addition, PTX3 overexpression in transduced SMCs reduces intimal hyperplasia after arterial injury as a result of direct binding to FGF2 (Camozzi et al., 2005). Thus, PTX3 could act as an "FGF2 decoy" able to sequester the growth factor in an inactive form.

FGF2 plays important roles in vivo by promoting angiogenesis and neovascularization during wound healing, inflammation, atherosclerosis, and tumor growth. All these pathological conditions are also characterized by accumulation of macrophages that, together with endothelial cells, may represent a major local source of both FGF2 and PTX3. The interaction between PTX3 and FGF2 may modulate angiogenesis in various physiopathological conditions, affecting the cross-talk between inflammatory cells and endothelium. In addition, the potent inhibitory effect on FGF2-mediated activation of SMCs suggests that PTX3 may modulate SMC activation after arterial injury.

Role in Female Fertility

PTX3 deficiency is associated with a severe defect in female fertility (Garlanda et al., 2002; Salustri et al., 2004; Varani et al., 2002). Infertility of PTX3-deficient mice is associated with an abnormal cumulus oophorus characterized by an unstable extracellular matrix in which cumulus cells are uniformly dispersed instead of radiating out from a central oocyte (Salustri et al., 2004). The oocyte develops normally in the absence of PTX3 and can be fertilized in vitro, whereas the fertilization failure observed in vivo is due to the defective cumulus expansion (Salustri et al., 2004). Cumulus cells express PTX3 mRNA under ovulatory stimuli (Salustri et al., 2004; Zhang et al., 2005). PTX3 produced by cumulus cells localizes in the extracellular matrix. The major integral component of cumulus matrix is hyaluronan, a large glycosaminoglycan responsible for the viscoelastic properties of this matrix. Other proteins interact with hyaluronan and participate in the organization of cumulus matrix, such as TNF-stimulated gene-6 (TSG-6), a multifunctional protein associated with inflammation (Milner and Day, 2003; Milner et al., 2006). In vitro experiments demonstrate that PTX3 binds TSG-6 (Salustri et al., 2004), interacting with the Link module of TSG-6, the hyaluronic acid binding domain. Addition of PTX3 to PTX3-deficient cumuli restores normal cumulus morphology, demonstrating the crucial and nonredundant role exerted by PTX3 in the assembly, organization, and stabilization of the hyaluronan-rich cumulus matrix. In addition, PTX3 directly interacts with a second component of cumulus matrix, inter-alpha-trypsin inhibitor (IαI) (Scarchilli et al., 2007). The heavy chains of IαI and PTX3 colocalize in the cumulus matrix and coimmunoprecipitate from cumulus matrix extracts. The interaction is mediated through the PTX3 N-terminal region, which showed the same ability as full-length protein to enable HA organization and matrix formation by PTX3-deficient cumulus cell oocyte complexes cultured in vitro. Thereby, PTX3 may form multimolecular complexes that can cross-link HA chains through TSG-6 and IαI, playing a role as structural constituents of the cumulus oophorus extracellular matrix essential for female fertility. Other proteins participating in the organization of extracellular matrixes were investigated, but no PTX3 binding has been observed to collagen IV, fibronectin, laminin, and hyaluronic acid (Bottazzi et al., 1997; Salustri et al., 2004).

Expression of PTX3 mRNA and protein by human cumulus cells (Paffoni et al., 2006; Salustri et al., 2004; Zhang et al., 2005) suggests that this molecule might have the same role in murine and human female fertility. Studies on PTX3 mRNA in cumulus cells from fertilized oocytes compared with cumulus cells from unfertilized oocytes indicated that PTX3 might be a possible marker for oocyte quality and success in fertilization (Zhang et al., 2005). PTX3 protein is abundantly present in the follicular fluid, where its concentrations are sixfold higher than in plasma, but we could not find a correlation between its levels in follicular fluid at the time of oocyte retrieval and oocyte quality, possibly because PTX3 shedding from the cumulus matrix to the follicular fluid is not a finely regulated phenomenon (Paffoni et al., 2006).

Different gene expression studies indicated that PTX3 is one of the most upregulated genes related to inflammation and angiogenesis induced during implantation. In particular, PTX3 was induced in decidual stromal cells by the trophoblast, in in vitro systems that mimic the alteration of the local immune environment induced by the trophoblast in the process of embryo implantation (Hess et al., 2007; Popovici et al., 2006), and was upregulated in implantation sites compared to interimplantation sites in the mouse (Reese et al., 2001). Accordingly, defective decidualization and implantation rates were observed in PTX3-deficient mice (Tranguch et al., 2007).

PTX3 AS A MARKER IN HUMAN PATHOLOGY

The structural and functional similarity to the classic diagnostic CRP have given impetus to efforts aimed at assessing the usefulness of PTX3 as a marker in diverse human pathological conditions. The hypothesis driving this effort is that PTX3, unlike CRP (made in the liver and induced primarily by IL-6), may represent a rapid marker for primary local activation of innate immunity and inflammation (Fig. 2). Actually, PTX3 behaves as an acute-phase response protein since its blood levels can increase up to 1,000-fold depending on the severity of inflammation. The general characteristic emerging from studies on PTX3

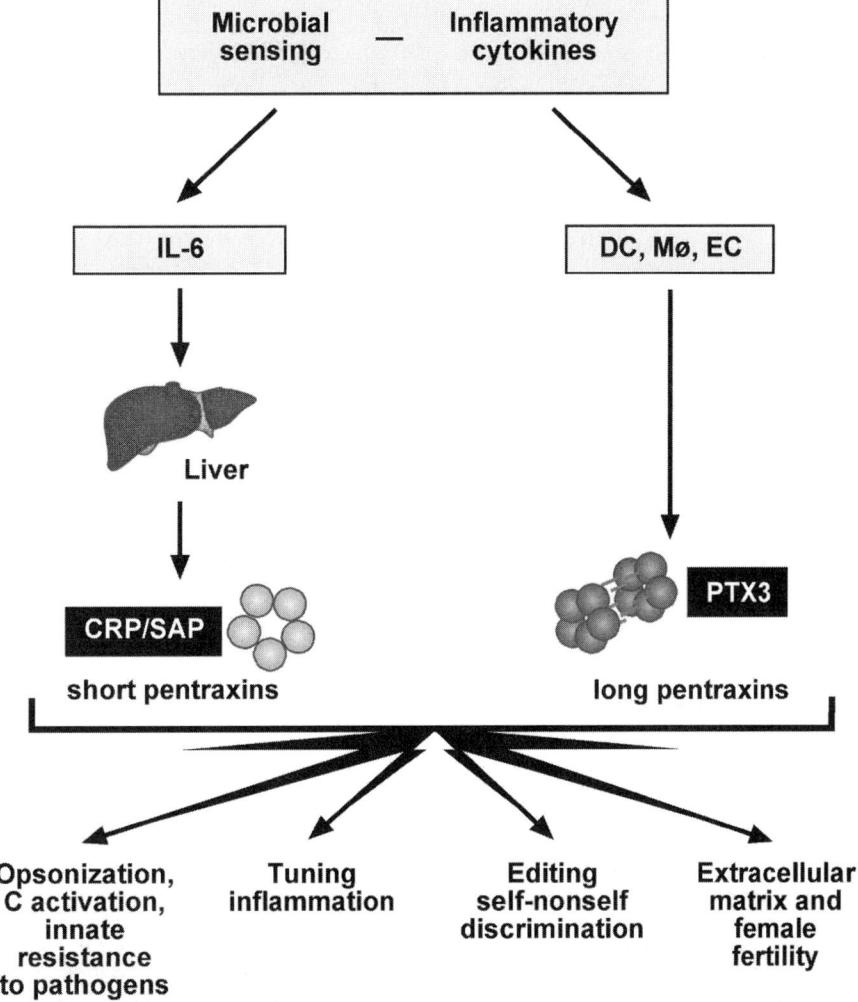

FIGURE 2 Pentraxins in innate immunity: liver-derived short pentraxins (e.g., CRP and SAP) and tissue-expressed long pentraxins (e.g., PTX3) are produced in response to microbial sensing and inflammatory cytokines and are likely to fulfill complementary functions in innate resistance to pathogens, tuning of inflammation, editing self/nonself discrimination, and participating in extracellular matrix architecture and female fertility (adapted with permission from Bottazzi et al., 2006).

blood levels in human pathology is the rapidity of its increase compared with CRP, consistent with its original identification as an immediate early gene (Peri et al., 2000), together with a lack of correlation between levels of CRP and PTX3 (Muller et al., 2001; Peri et al., 2000).

Data collected so far in different pathologies indicate a correlation between PTX3 plasma levels and severity of disease, suggesting a possible role of PTX3 as a marker of pathology. It remains to be elucidated whether the impressive correlation with outcome and severity actually reflects a role in the pathogenesis of damage, for instance, by amplifying the complement and coagulation cascades (Bottazzi et al., 1997; Napoleone et al., 2002, 2004; Nauta et al., 2003a).

Increased levels of PTX3 have been observed in diverse infectious disorders including sepsis and septic shock, A. fumigatus infection, tuberculosis, and dengue (Azzurri et al., 2005; Garlanda et al., 2002; Mairuhu et al., 2005; Muller et al., 2001). In all these conditions, PTX3 levels correlated with disease severity and had a prognostic value (Muller et al., 2001).

There is evidence linking PTX3 to ischemic heart disorders. PTX3 is induced in vascular SMCs by atherogenic modified low-density lipoprotein and is present in human atherosclerotic lesions (Klouche et al., 2004; Rolph et al., 2002). PTX3 levels increase rapidly in acute myocardial infarction (AMI), reaching a peak at approximately 7 h after the onset of symptoms (Peri et al., 2000). In a series of 748 patients with ST-elevation AMI, PTX3, measured along with established markers including CRP, emerged as the only independent predictor of mortality (Latini, et al., 2004). Patients with arterial inflammation, eligible for coronary intervention, exhibited high concentrations of plasma PTX3; in particular, patients with unstable angina pectoris exhibited PTX3 levels three times higher than the normal range (Inoue et al., 2007). Thus, PTX3 is a candidate new prognostic marker in ischemic heart disorders including AMI.

Increased levels of PTX3 have been observed in a restricted set of autoimmune disorders (e.g., in the blood in small vessel vasculitis, in the synovial fluid in rheumatoid arthritis), but not in others (e.g., systemic lupus erythematosus) (Fazzini et al., 2001; Luchetti et al., 2000). In small vessel vasculitis, PTX3 levels correlate with clinical activity of the disease and represent a candidate marker for monitoring the disease (Fazzini et al., 2001). Immunohistochemistry performed on skin sections at sites of vasculitis shows that endothelial cells are responsible for PTX3 production (van Rossum et al., 2006). Moreover, in these patients, PTX3 is abundantly present at sites of leukocytoclastic infiltration; the finding that PTX3, in contrast to the short pentraxin SAP, inhibits the uptake of apoptotic PMNs by macrophages (van Rossum et al., 2004) suggests that PTX3 is a key factor in the incomplete clearance of apoptotic and secondary necrotic PMNs observed in small-vessel vasculitis (van Rossum et al., 2006).

Recent results show that pregnancy itself, a condition associated with relevant involvement of inflammatory molecules at the implantation site (Redman et al., 1999), is associated with a slight increase in maternal circulating PTX3 levels compared with the nonpregnant condition. Higher maternal PTX3 levels were observed in pregnancies complicated by preeclampsia (Cetin et al., 2006; Rovere-Querini et al., 2006), which represents the clinical manifestation of an endothelial dysfunction as part of an excessive maternal inflammatory response to pregnancy (Benyo et al., 2001; Redman et al., 1999; Rinehart et al., 1999).

Finally, PTX3 plasma and vaginal levels were increased during pregnancy complicated by spontaneous preterm delivery and, in particular, in the cases of placenta vasculopathy (Assi et al., 2007).

CONCLUDING REMARKS

CRP was the first innate immune molecule capable of recognizing microbial moieties to be identified (Garlanda et al., 2005). Yet, despite its widespread use as a diagnostic tool in the clinic, its in vivo function has not been unequivocally defined. Indeed, the considerable differences in sequence and, most prominently, regulation (CRP is not an acute-phase protein in the mouse) have precluded the use of straightforward genetic approaches to explore its in vivo function (Hirschfield et al., 2005). By contrast, gene targeting of the prototypic, evolutionarily conserved, long pentraxin PTX3 has unequivocally defined the role of this molecule and, by inference, the role of the whole pentraxin family, as the crossroads of innate immunity, inflammation, matrix deposition, and female fertility (Garlanda et al., 2005) (Color Plate 2).

Recent progress has further defined the structure, regulation, microbial recognition, and in vivo function of PTX3. PTX3 (and presumably other members of the pentraxin superfamily) is a component of the complex and complementary network of cellular and humoral pattern recognition receptors involved in the recognition and response to microbial elements and damaged tissues. Moreover, evidence suggests that PTX3 acts as a tuner of inflammatory reactions and possibly as a component of the decoding system, which in antigen presentation discriminates between infectious nonself and apoptotic self (Baruah et al., 2006b; Diniz et al., 2004). Translational efforts suggest that PTX3 may be a new marker of innate immunity and inflammation, rapidly reflecting tissue and vascular bed involvement.

Unexpectedly, it was recently found that PTX3 is stored in a ready-made form in neutrophils localized in specific granules and is secreted in response to recognition of microbial moieties and inflammatory signals. Thus, neutrophils serve as a reservoir, ready for rapid release, of a key component of humoral innate immunity, and complement its subsequent delayed neosynthesis by macrophages and DCs.

The contribution of the European Commission (MUGEN, MUVAPRED, EMBIC), Ministero dell'Istruzione, Università e della Ricerca (MIUR) (project FIRB), Telethon (Telethon grant GGP05095), fondazione CARIPLO (project Nobel), and the Italian Association for Cancer Research (AIRC) is gratefully acknowledged.

REFERENCES

Abderrahim-Ferkoune, A., O. Bezy, C. Chiellini, M. Maffei, P. Grimaldi, F. Bonino, N. Moustaid-Moussa, F. Pasqualini, A. Mantovani, G. Ailhaud, and E. Z. Amri. 2003. Characterization of the long pentraxin PTX3 as a TNFalpha-induced secreted protein of adipose cells. *J. Lipid. Res.* **44:** 994–1000.

Agnello, D., L. Carvelli, V. Muzio, P. Villa, B. Bottazzi, N. Polentarutti, T. Mennini, A. Mantovani, and P. Ghezzi. 2000. Increased peripheral benzodiazepine binding sites and pentraxin 3 expression in the spinal cord during EAE: relation to inflammatory cytokines and modulation by dexamethasone and rolipram. *J. Neuroimmunol.* **109:**105–111.

Assi, F., R. Fruscio, C. Bonardi, A. Ghidini, P. Allavena, A. Mantovani, and A. Locatelli. 2007. Pentraxin 3 in plasma and vaginal fluid in women with preterm delivery. *BJOG* **114:**143–147.

Azzurri, A., O. Y. Sow, A. Amedei, B. Bah, S. Diallo, G. Peri, M. Benagiano, M. M. D'Elios, A. Mantovani, and G. Del Prete. 2005. IFN-gamma-inducible protein 10 and pentraxin 3 plasma levels are tools for monitoring inflammation and disease activity in Mycobacterium tuberculosis infection. *Microbes Infect.* **7:**1–8.

Baruah, P., I. E. Dumitriu, G. Peri, V. Russo, A. Mantovani, A. A. Manfredi, and P. Rovere-Querini. 2006a. The tissue pentraxin PTX3 limits C1q-mediated complement activation and phagocytosis of apoptotic cells by dendritic cells. *J. Leukoc. Biol.* **80:**87–95.

Baruah, P., A. Propato, I. E. Dumitriu, P. Rovere-Querini, V. Russo, R. Fontana, D. Accapezzato, G. Peri, A. Mantovani, V. Barnaba, and A. A. Manfredi. 2006b. The pattern recognition receptor PTX3 is recruited at the synapse between dying and dendritic cells, and edits the cross-presentation of self, viral, and tumor antigens. *Blood* **107:**151–158.

Benyo, D. F., A. Smarason, C. W. Redman, C. Sims, and K. P. Conrad. 2001. Expression of inflammatory cytokines in placentas from women with preeclampsia. *J. Clin. Endocrinol. Metab.* **86:**2505–2512.

Bottazzi, B., C. Garlanda, G. Salvatori, P. Jeannin, A. Manfredi, and A. Mantovani. 2006. Pentraxins as a key component of innate immunity. *Curr. Opin. Immunol.* **18:**10–15.

Bottazzi, B., V. Vouret-Craviari, A. Bastone, L. De Gioia, C. Matteucci, G. Peri, F. Spreafico, M. Pausa, C. D'Ettorre, E. Gianazza, A. Tagliabue, M. Salmona, F. Tedesco, M. Introna, and A. Mantovani. 1997. Multimer formation and ligand recognition by the long pentraxin PTX3. Similarities and differences with the short pentraxins C-reactive protein and serum amyloid P component. *J. Biol. Chem.* **272:**32817–32823.

Bozza, S., F. Bistoni, R. Gaziano, L. Pitzurra, T. Zelante, P. Bonifazi, K. Perruccio, S. Bellocchio, M. Neri, A. M. Iorio, G. Salvatori, R. De Santis, M. Calvitti, A. Doni, C. Garlanda, A. Mantovani, and L. Romani. 2006. Pentraxin 3 protects from MCMV infection and reactivation through TLR sensing pathways leading to IRF3 activation. *Blood* **108:**3387–3396.

Breviario, F., E. M. d'Aniello, J. Golay, G. Peri, B. Bottazzi, A. Bairoch, S. Saccone, R. Marzella, V. Predazzi, M. Rocchi, et al. 1992. Interleukin-1-inducible genes in endothelial cells. Cloning of a new gene related to C-reactive protein and serum amyloid P component. *J. Biol. Chem.* **267:**22190–22197.

Brinkmann, V., U. Reichard, C. Goosmann, B. Fauler, Y. Uhlemann, D. S. Weiss, Y. Weinrauch, and A. Zychlinsky. 2004. Neutrophil extracellular traps kill bacteria. *Science* **303:**1532–1535.

Camozzi, M., M. Rusnati, A. Bugatti, B. Bottazzi, A. Mantovani, A. Bastone, A. Inforzato, L. Vincenti, L. Bracci, D. Mastroianni, and M. Presta. 2006. Identification of an antiangiogenic FGF2-binding site in the N terminus of the soluble pattern recognition receptor PTX3. *J. Biol. Chem.* **281:**22605–22613.

Camozzi, M., S. Zacchigna, M. Rusnati, D. Coltrini, G. Ramirez-Correa, B. Bottazzi, A. Mantovani, M. Giacca, and M. Presta. 2005. Pentraxin 3 inhibits fibroblast growth factor 2-dependent activation of smooth muscle cells in vitro and neointima formation in vivo. *Arterioscler. Thromb. Vasc. Biol.* **25:**1837–1842.

Castellano, G., A. M. Woltman, A. J. Nauta, A. Roos, L. A. Trouw, M. A. Seelen, F. P. Schena, M. R. Daha, and C. van Kooten. 2004. Maturation of dendritic cells abrogates C1q production in vivo and in vitro. *Blood* **103:**3813–3820.

Cenci, E., S. Perito, K. Enssle, P. Mosci, J. Latge, L. Romani, and F. Bistoni. 1997. Th1 and Th2 cytokines in mice with invasive aspergillosis. *Infect. Immun.* **65:**564–570.

Cetin, I., V. Cozzi, F. Pasqualini, M. Nebuloni, C. Garlanda, L. Vago, G. Pardi, and A. Mantovani. 2006. Elevated maternal levels of the long pentraxin 3 (PTX3) in preeclampsia and intrauterine growth restriction. *Am. J. Obstet. Gynecol.* **194:**1347–1353.

Cotena, A., V. Maina, M. Sironi, B. Bottazzi, P. Jeannin, A. Vecchi, N. Corvaia, M. R. Daha, A. Mantovani, and C. Garlanda. 2007. Complement dependent amplification of the innate response to a cognate microbial ligand by the long pentraxin PTX3. *J. Immunol.* **179:**6311–6317.

Deban, L., H. Jarva, M. J. Lehtinen, B. Bottazzi, A. Bastone, A. Doni, T. S. Jokiranta, A. Mantovani, and S. Meri. 2008. Binding of the long pentraxin PTX3 to factor H: interacting domains and function in the regulation of complement activation. *J. Immunol.* **181:**8433–8440.

Dias, A. A., A. R. Goodman, J. L. Dos Santos, R. N. Gomes, A. Altmeyer, P. T. Bozza, M. F. Horta, J. Vilcek, and L. F. Reis. 2001. TSG-14 transgenic mice have improved survival to endotoxemia and to CLP-induced sepsis. *J. Leukoc. Biol.* **69:**928–936.

Diniz, S. N., R. Nomizo, P. S. Cisalpino, M. M. Teixeira, G. D. Brown, A. Mantovani, S. Gordon, L. F. Reis, and A. A. Dias. 2004. PTX3 function as an opsonin for the dectin-1-dependent internalization of zymosan by macrophages. *J. Leukoc. Biol.* **75:**649–656.

Dodds, D. C., I. A. Omeis, S. J. Cushman, J. A. Helms, and M. S. Perin. 1997. Neuronal pentraxin receptor, a novel putative integral membrane pentraxin that interacts with neuronal pentraxin 1 and 2 and taipoxin-associated calcium-binding protein 49. *J. Biol. Chem.* **272:**21488–21494.

Doni, A., G. Mantovani, C. Porta, J. Tuckermann, H. M. Reichardt, A. Kielman, M. Sironi, L. Rubino, F. Pasqualini, M. Nebuloni, S. Signorini, G. Peri, A. Sica, P. Beck-Peccoz, B. Bottazzi, and A. Mantovani. 2008. Cell-specific regulation of PTX3 by glucocorticoid hormones in hematopoietic and nonhematipoietic cells. *J. Biol. Chem.* **283:**29983–29992.

Doni, A., G. Peri, M. Chieppa, P. Allavena, F. Pasqualini, L. Vago, L. Romani, C. Garlanda, and A. Mantovani. 2003. Production of the soluble pattern recognition receptor PTX3 by myeloid, but not plasmacytoid, dendritic cells. *Eur. J. Immunol.* **33:**2886–2893.

dos Santos, C. C., B. Han, C. F. Andrade, X. Bai, S. Uhlig, R. Hubmayr, M. Tsang, M. Lodyga, S. Keshavjee, A. S. Slutsky, and M. Liu. 2004. DNA microarray analysis of gene expression in alveolar epithelial cells in response to TNFalpha, LPS, and cyclic stretch. *Physiol. Genomics* **19:**331–342.

Ehrt, S., D. Schnappinger, S. Bekiranov, J. Drenkow, S. Shi, T. R. Gingeras, T. Gaasterland, G. Schoolnik, and C. Nathan. 2001. Reprogramming of the macrophage transcriptome in response to interferon-gamma and Mycobacterium tuberculosis: signaling roles of nitric oxide synthase-2 and phagocyte oxidase. *J. Exp. Med.* **194:**1123–1140.

Fazzini, F., G. Peri, A. Doni, G. Dell'Antonio, E. Dal Cin, E. Bozzolo, F. D'Auria, L. Praderio, G. Ciboddo, M. G. Sabbadini, A. A. Manfredi, A. Mantovani, and P. R. Querini. 2001. PTX3 in small-vessel vasculitides: an independent indicator of disease activity produced at sites of inflammation. *Arthritis Rheum.* **44:**2841–2850.

Fuchs, T. A., U. Abed, C. Goosmann, R. Hurwitz, I. Schulze, V. Wahn, Y. Weinrauch, V. Brinkmann, and A.

Zychlinsky. 2007. Novel cell death program leads to neutrophil extracellular traps. *J. Cell Biol.* **176:**231–241.

Garlanda, C., B. Bottazzi, A. Bastone, and A. Mantovani. 2005. Pentraxins at the crossroads between innate immunity, inflammation, matrix deposition, and female fertility. *Annu. Rev. Immunol.* **23:**337–366.

Garlanda, C., E. Hirsch, S. Bozza, A. Salustri, M. De Acetis, R. Nota, A. Maccagno, F. Riva, B. Bottazzi, G. Peri, A. Doni, L. Vago, M. Botto, R. De Santis, P. Carminati, G. Siracusa, F. Altruda, A. Vecchi, L. Romani, and A. Mantovani. 2002. Non-redundant role of the long pentraxin PTX3 in anti-fungal innate immune response. *Nature* **420:**182–186.

Gaziano, R., S. Bozza, S. Bellocchio, K. Perruccio, C. Montagnoli, L. Pitzurra, G. Salvatori, R. De Santis, P. Carminati, A. Mantovani, and L. Romani. 2004. Anti-Aspergillus fumigatus efficacy of pentraxin 3 alone and in combination with antifungals. *Antimicrob. Agents Chemother.* **48:**4414–4421.

Goodman, A. R., T. Cardozo, R. Abagyan, A. Altmeyer, H. G. Wisniewski, and J. Vilcek. 1996. Long pentraxins: an emerging group of proteins with diverse functions. *Cytokine Growth Factor Rev.* **7:**191–202.

Goodman, A. R., D. E. Levy, L. F. Reis, and J. Vilcek. 2000. Differential regulation of TSG-14 expression in murine fibroblasts and peritoneal macrophages. *J. Leukoc. Biol.* **67:**387–395.

Hess, A. P., A. E. Hamilton, S. Talbi, C. Dosiou, M. Nyegaard, N. Nayak, O. Genbecev-Krtolica, P. Mavrogianis, K. Ferrer, J. Kruessel, A. T. Fazleabas, S. J. Fisher, and L. C. Giudice. 2007. Decidual stromal cell response to paracrine signals from the trophoblast: amplification of immune and angiogenic modulators. *Biol. Reprod.* **76:**102–117.

Hicks, P. S., L. Saunero-Nava, T. W. Du Clos, and C. Mold. 1992. Serum amyloid P component binds to histones and activates the classical complement pathway. *J. Immunol.* **149:**3689–694.

Hirschfield, G. M., J. R. Gallimore, M. C. Kahan, W. L. Hutchinson, C. A. Sabin, G. M. Benson, A. P. Dhillon, G. A. Tennent, and M. B. Pepys. 2005. Transgenic human C-reactive protein is not proatherogenic in apolipoprotein E-deficient mice. *Proc. Natl. Acad. Sci. USA* **102:**8309–8314.

Hsu, Y. C., and M. S. Perin. 1995. Human neuronal pentraxin II (NPTX2): conservation, genomic structure, and chromosomal localization. *Genomics* **28:**220–227.

Inforzato, A., G. Peri, A. Doni, C. Garlanda, A. Mantovani, A. Bastone, A. Carpentieri, A. Amoresano, P. Pucci, A. Roos, M. R. Daha, S. Vincenti, G. Gallo, P. Carminati, R. De Santis, and G. Salvatori. 2006. Structure and function of the long pentraxin PTX3 glycosidic moiety: fine-tuning of the interaction with C1q and complement activation. *Biochemistry* **45:**11540–11551.

Inoue, K., A. Sugiyama, P. C. Reid, Y. Ito, K. Miyauchi, S. Mukai, M. Sagara, K. Miyamoto, H. Satoh, I. Kohno, T. Kurata, H. Ota, A. Mantovani, T. Hamakubo, H. Daida, and T. Kodama. 2007. Establishment of a high sensitivity plasma assay for human pentraxin3 as a marker for unstable angina pectoris. *Arterioscler. Thromb. Vasc. Biol.* **27:**161–167.

Introna, M., V. V. Alles, M. Castellano, G. Picardi, L. De Gioia, B. Bottazzi, G. Peri, F. Breviario, M. Salmona, L. De Gregorio, T. A. Dragani, N. Srinivasan, T. L. Blundell, T. A. Hamilton, and A. Mantovani. 1996. Cloning of mouse ptx3, a new member of the pentraxin gene family expressed at extrahepatic sites. *Blood* **87:**1862–1872.

Jaillon, S., G. Peri, Y. Delneste, I. Fremaux, A. Doni, F. Moalli, C. Garlanda, L. Romani, H. Gascan, S. Bellocchio, S. Bozza, M. A. Cassatella, P. Jeannin, and A. Mantovani. 2007. The humoral pattern recognition receptor PTX3 is stored in neutrophil granules and localizes in extracellular traps. *J. Exp. Med.* **204:**793–804.

Jarva, H., T. S. Jokiranta, J. Hellwage, P. F. Zipfel, and S. Meri. 1999. Regulation of complement activation by C-reactive protein: targeting the complement inhibitory activity of factor H by an interaction with short consensus repeat domains 7 and 8-11. *J. Immunol.* **163:**3957–3962.

Jeannin, P., B. Bottazzi, M. Sironi, A. Doni, M. Rusnati, M. Presta, V. Maina, G. Magistrelli, J. F. Haeuw, G. Hoeffel, N. Thieblemont, N. Corvaia, C. Garlanda, Y. Delneste, and A. Mantovani. 2005. Complexity and complementarity of outer membrane protein A recognition by cellular and humoral innate immunity receptors. *Immunity* **22:**551–560.

Jeannin, P., T. Renno, L. Goetsch, I. Miconnet, J. P. Aubry, Y. Delneste, N. Herbault, T. Baussant, G. Magistrelli, C. Soulas, P. Romero, J. C. Cerottini, and J. Y. Bonnefoy. 2000. OmpA targets dendritic cells, induces their maturation and delivers antigen into the MHC class I presentation pathway. *Nat. Immunol.* **1:**502–509.

Kirkpatrick, L. L., M. M. Matzuk, D. C. Dodds, and M. S. Perin. 2000. Biochemical interactions of the neuronal pentraxins. Neuronal pentraxin (NP) receptor binds to taipoxin and taipoxin-associated calcium-binding protein 49 via NP1 and NP2. *J. Biol. Chem.* **275:**17786–17792.

Klouche, M., N. Brockmeyer, C. Knabbe, and S. Rose-John. 2002. Human herpesvirus 8-derived viral IL-6 induces PTX3 expression in Kaposi's sarcoma cells. *AIDS* **16:**F9–F18.

Klouche, M., G. Peri, C. Knabbe, H. H. Eckstein, F. X. Schmid, G. Schmitz, and A. Mantovani. 2004. Modified atherogenic lipoproteins induce expression of pentraxin-3 by human vascular smooth muscle cells. *Atherosclerosis* **175:**221–228.

Lang, R., D. Patel, J. J. Morris, R. L. Rutschman, and P. J. Murray. 2002. Shaping gene expression in activated and resting primary macrophages by IL-10. *J. Immunol.* **169:**2253–2263.

Latini, R., A. P. Maggioni, G. Peri, L. Gonzini, D. Lucci, P. Mocarelli, L. Vago, F. Pasqualini, S. Signorini, D. Soldateschi, L. Tarli, C. Schweiger, C. Fresco, R. Cecere, G. Tognoni, and A. Mantovani. 2004. Prognostic significance of the long pentraxin PTX3 in acute myocardial infarction. *Circulation* **110:**2349–2354.

Lee, G. W., T. H. Lee, and J. Vilcek. 1993. TSG-14, a tumor necrosis factor- and IL-1-inducible protein, is a novel member of the pentaxin family of acute phase proteins. *J. Immunol.* **150:**1804–1812.

Luchetti, M. M., G. Piccinini, A. Mantovani, G. Peri, C. Matteucci, G. Pomponio, M. Fratini, P. Fraticelli, P. Sambo, C. Di Loreto, A. Doni, M. Introna, and A. Gabrielli. 2000. Expression and production of the long pentraxin PTX3 in rheumatoid arthritis (RA). *Clin. Exp. Immunol.* **119:**196–202.

Mairuhu, A. T., G. Peri, T. E. Setiati, C. E. Hack, P. Koraka, A. Soemantri, A. D. Osterhaus, D. P. Brandjes, J. W. van der Meer, A. Mantovani, and E. C. van Gorp. 2005. Elevated plasma levels of the long pentraxin, pentraxin 3, in severe dengue virus infections. *J. Med. Virol.* **76:**547–552.

Malaguarnera, L., M. R. Pilastro, L. Vicari, R. Di Marco, M. Malaguarnera, and A. Messina. 2000. PTX3 gene expression in Castleman's disease. *Eur. J. Haematol.* **64:**132–134.

Mantovani, A., A. Sica, S. Sozzani, P. Allavena, A. Vecchi, and M. Locati. 2004. The chemokine system in diverse forms of macrophage activation and polarization. *Trends Immunol.* **25:**677–686.

Milner, C. M., and A. J. Day. 2003. TSG-6: a multifunctional protein associated with inflammation. *J. Cell Sci.* **116:**1863–1873.

Milner, C. M., V. A. Higman, and A. J. Day. 2006. TSG-6: a pluripotent inflammatory mediator? *Biochem. Soc. Trans.* **34:**446–450.

Moore, K. W., R. de Waal Malefyt, R. L. Coffman, and A. O'Garra. 2001. Interleukin-10 and the interleukin-10 receptor. *Annu. Rev. Immunol.* **19:**683–765.

Muller, B., G. Peri, A. Doni, V. Torri, R. Landmann, B. Bottazzi, and A. Mantovani. 2001. Circulating levels of the long pentraxin PTX3 correlate with severity of infection in critically ill patients. *Crit. Care Med.* **29:**1404–1407.

Nagai, H., J. Guo, H. Choi, and V. Kurup. 1995. Interferon-gamma and tumor necrosis factor-alpha protect mice from invasive aspergillosis. *J. Infect. Dis.* **172:**1554–1560.

Napoleone, E., A. Di Santo, A. Bastone, G. Peri, A. Mantovani, G. de Gaetano, M. B. Donati, and R. Lorenzet. 2002. Long pentraxin PTX3 upregulates tissue factor expression in human endothelial cells: a novel link between vascular inflammation and clotting activation. *Arterioscler. Thromb. Vasc. Biol.* **22:**782–787.

Napoleone, E., A. di Santo, G. Peri, A. Mantovani, G. de Gaetano, M. B. Donati, and R. Lorenzet. 2004. The long pentraxin PTX3 up-regulates tissue factor in activated monocytes: another link between inflammation and clotting activation. *J. Leukoc. Biol.* **76:**203–209.

Nauta, A. J., B. Bottazzi, A. Mantovani, G. Salvatori, U. Kishore, W. J. Schwaeble, A. R. Gingras, S. Tzima, F. Vivanco, J. Egido, O. Tijsma, E. C. Hack, M. R. Daha, and A. Roos. 2003a. Biochemical and functional characterization of the interaction between pentraxin 3 and C1q. *Eur. J. Immunol.* **33:**465–473.

Nauta, A. J., M. R. Daha, C. van Kooten, and A. Roos. 2003b. Recognition and clearance of apoptotic cells: a role for complement and pentraxins. *Trends Immunol.* **24:**148–154.

Nauta, A. J., S. de Haij, B. Bottazzi, A. Mantovani, M. C. Borrias, J. Aten, M. P. Rastaldi, M. R. Daha, C. van Kooten, and A. Roos. 2005. Human renal epithelial cells produce the long pentraxin PTX3. *Kidney Int.* **67:**543–553.

Noland, T. D., B. B. Friday, M. T. Maulit, and G. L. Gerton. 1994. The sperm acrosomal matrix contains a novel member of the pentaxin family of calcium-dependent binding proteins. *J. Biol. Chem.* **269:**32607–32614.

Olesen, R., C. Wejse, D. R. Velez, C. Bisseye, M. Sodemann, P. Aaby, P. Rabna, A. Worwui, H. Chapman, M. Diatta, R. A. Adegbola, P. C. Hill, L. Ostergaard, S. M. Williams, and G. Sirugo. 2007. DC-SIGN (CD209), pentraxin 3 and vitamin D receptor gene variants associate with pulmonary tuberculosis risk in West Africans. *Genes Immun.* **8:**456–467.

Omeis, I. A., Y. C. Hsu, and M. S. Perin. 1996. Mouse and human neuronal pentraxin 1 (NPTX1): conservation, genomic structure, and chromosomal localization. *Genomics* **36:**543–545.

Paffoni, A., G. Ragni, A. Doni, E. Somigliana, F. Pasqualini, L. Restelli, G. Pardi, A. Mantovani, and C. Garlanda. 2006. Follicular fluid levels of the long pentraxin PTX3. *J. Soc. Gynecol. Investig.* **13:**226–231.

Pepys, M. B., S. E. Booth, G. A. Tennent, P. J. Butler, and D. G. Williams. 1994. Binding of pentraxins to different nuclear structures: C-reactive protein binds to small nuclear ribonucleoprotein particles, serum amyloid P component binds to chromatin and nucleoli. *Clin. Exp. Immunol.* **97:**152–157.

Peri, G., M. Introna, D. Corradi, G. Iacuitti, S. Signorini, F. Avanzini, F. Pizzetti, A. P. Maggioni, T. Moccetti, M. Metra, L. D. Cas, P. Ghezzi, J. D. Sipe, G. Re, G. Olivetti, A. Mantovani, and R. Latini. 2000. PTX3, a prototypical long pentraxin, is an early indicator of acute myocardial infarction in humans. *Circulation* **102:**636–641.

Perrier, P., F. O. Martinez, M. Locati, G. Bianchi, M. Nebuloni, G. Vago, F. Bazzoni, S. Sozzani, P. Allavena, and A. Mantovani. 2004. Distinct transcriptional programs activated by interleukin-10 with or without lipopolysaccharide in dendritic cells: induction of the B cell-activating chemokine, CXC chemokine ligand 13. *J. Immunol.* **172:**7031–7042.

Polentarutti, N., G. Picardi, A. Basile, S. Cenzuales, A. Rivolta, C. Matteucci, G. Peri, A. Mantovani, and M. Introna. 1998. Interferon-gamma inhibits expression of the long pentraxin PTX3 in human monocytes. *Eur. J. Immunol.* **28:**496–501.

Popovici, R. M., N. K. Betzler, M. S. Krause, M. Luo, J. Jauckus, A. Germeyer, S. Bloethner, A. Schlotterer, R. Kumar, T. Strowitzki, and M. von Wolff. 2006. Gene expression profiling of human endometrial-trophoblast interaction in a coculture model. *Endocrinology* **147:**5662–5675.

Ravizza, T., D. Moneta, B. Bottazzi, G. Peri, C. Garlanda, E. Hirsch, G. J. Richards, A. Mantovani, and A. Vezzani. 2001. Dynamic induction of the long pentraxin PTX3 in the CNS after limbic seizures: evidence for a protective role in seizure-induced neurodegeneration. *Neuroscience* **105:**43–53.

Redman, C. W., G. P. Sacks, and I. L. Sargent. 1999. Preeclampsia: an excessive maternal inflammatory response to pregnancy. *Am. J. Obstet. Gynecol.* **180:**499–506.

Reese, J., S. K. Das, B. C. Paria, H. Lim, H. Song, H. Matsumoto, K. L. Knudtson, R. N. DuBois, and S. K. Dey. 2001. Global gene expression analysis to identify molecular markers of uterine receptivity and embryo implantation. *J. Biol. Chem.* **276:**44137–44145.

Reid, M. S., and C. P. Blobel. 1994. Apexin, an acrosomal pentaxin. *J. Biol. Chem.* **269:**32615–32620.

Rinehart, B. K., D. A. Terrone, S. Lagoo-Deenadayalan, W. H. Barber, E. A. Hale, J. N. Martin, Jr., and W. A. Bennett. 1999. Expression of the placental cytokines tumor necrosis factor alpha, interleukin 1beta, and interleukin 10 is increased in preeclampsia. *Am. J. Obstet. Gynecol.* **181:**915–920.

Rolph, M. S., S. Zimmer, B. Bottazzi, C. Garlanda, A. Mantovani, and G. K. Hansson. 2002. Production of the long pentraxin PTX3 in advanced atherosclerotic plaques. *Arterioscler. Thromb. Vasc. Biol.* **22:**e10–E14.

Roumenina, L. T., M. M. Ruseva, A. Zlatarova, R. Ghai, M. Kolev, N. Olova, M. Gadjeva, A. Agrawal, B. Bottazzi, A. Mantovani, K. B. Reid, U. Kishore, and M. S. Kojouharova. 2006. Interaction of C1q with IgG1, C-reactive protein and pentraxin 3: mutational studies using recombinant globular head modules of human C1q A, B, and C chains. *Biochemistry* **45:**4093–4104.

Rovere, P., G. Peri, F. Fazzini, B. Bottazzi, A. Doni, A. Bondanza, V. S. Zimmermann, C. Garlanda, U. Fascio, M. G. Sabbadini, C. Rugarli, A. Mantovani, and A. A. Manfredi. 2000. The long pentraxin PTX3 binds to apoptotic cells and regulates their clearance by antigen-presenting dendritic cells. *Blood* **96:**4300–4306.

Rovere-Querini, P., S. Antonacci, G. Dell'Antonio, A. Angeli, E. D. Cin, L. Valsecchi, C. Lanzani, M. G. Sabbadini, C. Doglioni, A. A. Manfredi, and M. T. Castiglioni. 2006. Plasma and tissue expression of the long pentraxin 3 during normal pregnancy and preeclampsia. *Obstet. Gynecol.* **108:**148–155.

Rusnati, M., M. Camozzi, E. Moroni, B. Bottazzi, G. Peri, S. Indraccolo, A. Amadori, A. Mantovani, and M. Presta. 2004. Selective recognition of fibroblast growth factor-2 by the long pentraxin PTX3 inhibits angiogenesis. *Blood* **104:**92–99.

Salustri, A., C. Garlanda, E. Hirsch, M. De Acetis, A. Maccagno, B. Bottazzi, A. Doni, A. Bastone, G. Mantovani, P. Beck Peccoz, G. Salvatori, D. J. Mahoney, A. J. Day, G.

Siracusa, L. Romani, and A. Mantovani. 2004. PTX3 plays a key role in the organization of the cumulus oophorus extracellular matrix and in in vivo fertilization. *Development* **131:**1577–1586.

Scarchilli, L., A. Camaioni, B. Bottazzi, V. Negri, A. Doni, L. Deban, A. Bastone, G. Salvatori, A. Mantovani, G. Siracusa, and A. Salustri. 2007. PTX3 interacts with inter-{alpha}-trypsin inhibitor: implications for hyaluronan organization and cumulus oophorus expansion. *J. Biol. Chem.* **282:**30161–30170.

Schlimgen, A. K., J. A. Helms, H. Vogel, and M. S. Perin. 1995. Neuronal pentraxin, a secreted protein with homology to acute phase proteins of the immune system. *Neuron* **14:**519–526.

Soares, A. C., D. G. Souza, V. Pinho, A. T. Vieira, J. R. Nicoli, F. Q. Cunha, A. Mantovani, L. F. Reis, A. A. Dias, and M. M. Teixeira. 2006. Dual function of the long pentraxin PTX3 in resistance against pulmonary infection with Klebsiella pneumoniae in transgenic mice. *Microbes Infect.* **8:**1321–1329.

Souza, D. G., A. C. Soares, V. Pinho, H. Torloni, L. F. Reis, M. M. Teixeira, and A. A. Dias. 2002. Increased mortality and inflammation in tumor necrosis factor-stimulated gene-14 transgenic mice after ischemia and reperfusion injury. *Am. J. Pathol.* **160:**1755–1765.

Szalai, A. J., A. Agrawal, T. J. Greenhough, and J. E. Volanakis. 1999. C-reactive protein: structural biology and host defense function. *Clin. Chem. Lab. Med.* **37:**265–270.

Tranguch, S., A. Chakrabarty, Y. Guo, H. Wang, and S. K. Dey. 2007. Maternal pentraxin 3 deficiency compromises implantation in mice. *Biol. Reprod.* **77:**425–432.

Tsui, C. C., N. G. Copeland, D. J. Gilbert, N. A. Jenkins, C. Barnes, and P. F. Worley. 1996. Narp, a novel member of the pentraxin family, promotes neurite outgrowth and is dynamically regulated by neuronal activity. *J. Neurosci.* **16:**2463–2478.

van Rossum, A. P., F. Fazzini, P. C. Limburg, A. A. Manfredi, P. Rovere-Querini, A. Mantovani, and C. G. Kallenberg. 2004. The prototypic tissue pentraxin PTX3, in contrast to the short pentraxin serum amyloid P, inhibits phagocytosis of late apoptotic neutrophils by macrophages. *Arthritis Rheum.* **50:**2667–2674.

van Rossum, A. P., H. H. Pas, F. Fazzini, M. G. Huitema, P. C. Limburg, M. F. Jonkman, and C. G. Kallenberg. 2006. Abundance of the long pentraxin PTX3 at sites of leukocytoclastic lesions in patients with small-vessel vasculitis. *Arthritis Rheum.* **54:**986–991.

Varani, S., J. A. Elvin, C. Yan, J. DeMayo, F. J. DeMayo, H. F. Horton, M. C. Byrne, and M. M. Matzuk. 2002. Knockout of pentraxin 3, a downstream target of growth differentiation factor-9, causes female subfertility. *Mol. Endocrinol.* **16:**1154–1167.

Willeke, F., A. Assad, P. Findeisen, E. Schromm, R. Grobholz, B. von Gerstenbergk, A. Mantovani, S. Peri, H. H. Friess, S. Post, M. von Knebel Doeberitz, and M. H. Schwarzbach. 2006. Overexpression of a member of the pentraxin family (PTX3) in human soft tissue liposarcoma. *Eur. J. Cancer.* **42:**2639–2646.

Zhang, X., N. Jafari, R. B. Barnes, E. Confino, M. Milad, and R. R. Kazer. 2005. Studies of gene expression in human cumulus cells indicate pentraxin 3 as a possible marker for oocyte quality. *Fertil. Steril.* **83**(Suppl 1)**:**1169–1179.

11

Leukocyte Chemotaxis

ANN P. WHEELER AND ANNE J. RIDLEY

Phagocytes, including macrophages, neutrophils, and dendritic cells, share the ability to detect and migrate toward a chemical stimulus, a process known as chemotaxis (Niggli, 2003; Kobayashi et al., 2004; Pixley and Stanley, 2004). When a tissue is damaged by wounding or by a pathogenic microorganism, the body initiates an inflammatory response. This is marked by the influx of leukocytes to the wounded tissue, often within minutes of injury. Inflammation includes vasodilation, which increases the flow of blood to the affected area, allowing more leukocytes to be brought to the wound; an increase in vascular permeability, which facilitates leukocyte movement from capillaries into the infected or damaged tissue; and chemotaxis of leukocytes to the site of infection or damage, followed by phagocytosis of microorganisms and infected cells (Albelda et al., 1994; Burke et al., 2002). The damaged tissue attracts leukocytes by secreting chemoattractants (Rot and von Andrian, 2004). Neutrophils are the first leukocytes to arrive at a wound, where they phagocytose debris and bacteria and secrete factors that promote the recruitment of macrophages and lymphocytes (Kobayashi, 2008). The sequential and regulated recruitment of leukocytes into tissues by chemoattractants is essential for effective clearance of pathogens and healing (Burke and Lewis, 2002).

MECHANICS OF MIGRATION

Chemoattractants induce neutrophils, dendritic cells, and macrophages to assume a polarized migratory morphology, with a lamellipodium at the front and a uropod or tail at the back. Both the actin and microtubule cytoskeletons become polarized: extension of lamellipodia depends on actin polymerization and filament turnover at the front, whereas myosin II accumulates at the sides and rear of migrating neutrophils, and actin cables act coordinately with myosin II to induce contractility at the rear. In neutrophils, the microtubule-organizing center (MTOC) is located behind the nucleus, whereas in macrophages and dendritic cells it is normally localized in front of the nucleus (Weiner et al., 2002; Niggli, 2003; Eng et al., 2007). Sustained migration requires dynamic actin polymerization and depolymerization, myosin II activity, microtubules, the formation of new adhesions to the extracellular matrix at the front of the cell, generation of tractional forces in the cell body, disassembly of adhesions at the back of the cell, and retraction of the cell tail or uropod.

The Rho GTPases Cdc42, Rac, and Rho are important for establishing and maintaining migratory polarity (Fig. 1). Rac is usually active at the front of cells and is required for lamellipodium extension, whereas Rho activity is restricted mostly to the rear. In neutrophils, however, there is also some active Rac in the uropod (Bokoch, 2005). Cdc42 appears to be important for stable polarization of cells and thus for sustained chemotaxis (Van Keymeulen et al., 2006) (see below). Rho GTPases cycle between an active GTP-bound form and an inactive GDP-bound form. Guanine nucleotide exchange factors (GEFs) stimulate the exchange of GDP for GTP on Rho GTPases. Signaling downstream of Rho family GTPases is switched off through hydrolysis of the γ-phosphate in GTP, a process catalyzed by GTPase-activating proteins (GAPs). Together GEFs and GAPs coordinate signaling between membrane receptors and cytoskeletal components by mediating precise spatial and temporal activation of Rho GTPases.

The migratory properties of cells, such as the speed of cell migration and the morphology of cells during migration, depends on the cell type and can be affected dramatically by the surrounding environment (Abercrombie et al., 1977; Mitchison and Cramer, 1996; Friedl, 2004; Ridley, 2004). In general, neutrophils migrate faster than macrophages and dendritic cells, reflecting their role as the front-line defense during infection and tissue damage.

CHEMOATTRACTANTS FOR PHAGOCYTES

Gradient sensing and chemotaxis does not depend on a single molecular mechanism, but consists of several signaling networks acting together to generate a polarized cell capable of persistent movement toward a chemoattractant.

Ann P. Wheeler, Imperial College London, Department of Molecular Medicine, Sir Alexander Fleming Building, London SW7 2AZ, United Kingdom. **Anne J. Ridley,** King's College London, Randall Division of Cell and Molecular Biophysics, New Hunt's House, Guy's Campus, London SE1 1UL, United Kingdom.

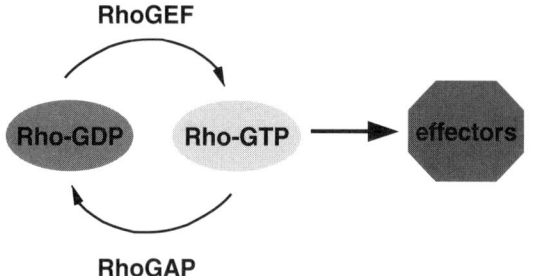
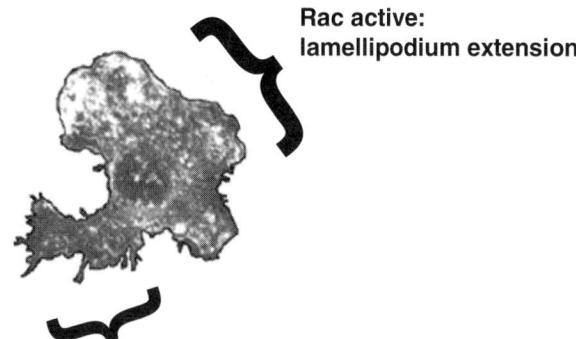

FIGURE 1 Rho GTPase signaling in migration and chemotaxis. (Top) Rho family GTPases cycle between an active GTP-bound conformation and an inactive GDP-bound conformation. This is regulated by RhoGEFs, which stimulate exchange of GDP for GTP, and RhoGAPs, which stimulate hydrolysis of GTP. When bound to GTP, Rho GTPases activate downstream effectors that mediate cellular responses. (Bottom) In migrating cells, Rac is active at the front of cells. Rac is required for lamellipodium extension. Rho activity is restricted mostly to the rear and is involved in generation of cell traction in the cell body and retraction of the tail or uropod.

Most chemoattractants for phagocytes signal either through seven transmembrane G-protein-coupled receptors (GPCRs) or tyrosine kinase receptors. Leukocytes are very sensitive to variations in concentrations of chemoattractants and can detect a difference of as little as 0.1% across a cell, in vitro (Webb et al., 1996). Cells are believed to respond to such a shallow change in gradient by intracellular amplification of the chemoattractant signal.

GPCR Signaling

Many stimuli for phagocytes, including chemokines, bacterial products, extracellular nucleotides (e.g., ATP), and complement components, activate GPCRs to induce chemotaxis (Table 1) (Lattin et al., 2007). Chemokines are a large family of secreted proteins of 8 to 10 kDa that all have two essential disulfide bonds formed by amino-terminal cysteines. Of the four families of chemokines, CXC and CC chemokines are the largest, distinguished according to the position of the first two cysteines, which are either adjacent (CC) or separated by one amino acid (CXC) (Zlotnik and Yoshie, 2000).

Neutrophils mainly respond to CXC chemokines and bacterial products (Kobayashi, 2008). Once neutrophils have been recruited, neutrophils and tissue-resident macrophages secrete CCL2 (also known as MCP-1) as well as other factors, which recruit macrophages (Lu et al., 1998; Burke and Lewis, 2002). Monocytes and macrophages express chemokine receptors of the CC type including CCR2, the receptor for CCL2 (Lattin et al., 2007). CCL2-null mice have severe impairments in the recruitment of monocytes to a wound (Zlotnik and Yoshie, 2000). This indicates that CCL2 secretion both by neutrophils and by surrounding tissues is crucial in the inflammatory response (Allen et al., 1998; Lu et al., 1998; Henderson et al., 2003).

GPCRs change conformation when a ligand binds, leading to dissociation of bound heterotrimeric G proteins into α- and $\beta\gamma$-subunits, both of which contribute to GPCR-induced signaling. Although chemoattractant receptors couple to multiple G proteins and this is cell type specific (Arai and Charo, 1996), signaling downstream of most chemokine receptors and the N-formyl-methionyl-leucyl-phenylalanine (fMLP) receptor is mainly transduced through Gi family G proteins (Myers et al., 1995; Mahadeo et al., 2007). Chemotaxis is inhibited by pertussis toxin, which specifically modifies Gαi, preventing it from binding to GPCRs and thus also inhibiting G$\beta\gamma$-dependent responses (Wettschureck and Offermanns, 2005). G$\beta\gamma$ regulates many of the known pathways activated by chemoattractants in leukocytes (Rickert et al., 2000). The $\beta\gamma$-subunits can activate adenylate cyclase and hence cylic AMP production, and phospholipase C-β (PLC-β), leading to protein kinase C (PKC) activation and an increase in cytosolic Ca^{2+}. In addition, they bind to and activate phosphatidylinositide 3-kinase γ (PI3Kγ), inducing production of phosphatidylinositol 3,4,5-trisphosphate (PIP$_3$). Of these pathways, PI3Kγ is best characterized for its role in neutrophil chemotaxis (see below).

G$\alpha_{12/13}$ subunits also play a role in chemotaxis by activating RhoGEFs and hence RhoA at the rear of neutrophils (Needham and Rozengurt, 1998; Franca-Koh et al., 2006; Van Keymeulen et al., 2006; Wong et al., 2006; Neves et al., 2002) (Fig. 2).

Signaling by GPCRs is downregulated by GPCR kinase-induced phosphorylation, leading to the recruitment of β-arrestins, which uncouple GPCRs from heterotrimeric G proteins and thus induce loss of receptor responsiveness. The binding of β-arrestin proteins to GPCRs is also important for the recruitment of the receptor to clathrin-coated vesicles and subsequent endocytosis. In addition, β-arrestins can act as docking proteins, bringing other kinases such as Src or JNK3 to the vicinity of the receptor complex (Pierce and Lefkowitz, 2001; Jimenez-Sainz et al., 2003), and thereby are believed to contribute to chemotaxis as well as downregulation of signaling. Chemotactic signaling probably continues on endosomes.

Tyrosine Kinase Receptor Signaling

Phagocytes can chemotax to cytokines that activate tyrosine kinase receptors. The best characterized is CSF-1, which stimulates chemotaxis of macrophages as well as their survival, proliferation, and differentiation (Webb et al., 1996). The CSF-1 receptor (CSF-1R, also known as c-Fms) is a tyrosine kinase receptor, and binding of CSF-1 results in dimerization of CSF-1R, resulting in autophosphorylation on several tyrosines. Signaling molecules then associate with CSF-1R through their SH2 domains, and can themselves become tyrosine phosphorylated (Yeung et al., 1998). Seven tyrosines identified in CSF-1R specifically recruit certain proteins in response to CSF-1 stimulation

TABLE 1 Chemokine receptors expressed on macrophages and neutrophils

Cell type	Chemokine	Receptor	Source of chemokine if known
Neutrophil	CXCL1/KC	CXCR1	Epithelial, endothelial, stromal, and T cells
	IL-8	CXCR2	
	IP10, MIG, I-TAC	CXCR3	
	SDF-1	CXCR4	
	MIP-1α/RANTES	CCR1	
	C5a	Complement receptor	Complement cascade
	fMLP	FPR-1	Bacterial peptide
	Leukotriene B_4	Leukotriene B_4 receptor	Phospholipid metabolite
Macrophage	GM-CSF	GM-CSFR	Many tissue types, specifically lung
	MIP-1α/RANTES	CCR1	Monocytes, neutrophils, fibroblasts
	MCP-1,2,3,4	CCR2	Epithelial, endothelial, and stromal cells, specifically endothelial calls
	MIP-1α/MIP-1β	CCR5	Neutrophils
	LARC/CCL20	CCR6	Keratinocytes, lung epithelia, and intestinal epithelia
	IL-8	CXCR2	Paracrine signal from neutrophil
Dendritic cell	MIP-1α/RANTES, MCP-3	CCR1	Epithelial, endothelial, stromal, and T cells
	MCP-1,2,3,4	CCR2	Epithelial, endothelial, and stromal cells, specifically endothelial cells
	MIP-1α/MIP-1β	CCR5	Monocytes, neutrophils, fibroblasts
	CCL21, MIP-3β	CCR7	Endothelial venules in lymph nodes and Peyer's patches
	IL-8	CXCR1	Epithelial, endothelial, and stromal cells, specifically in skin and lung
	IL-8	CXCR2	
	SDF-1	CXCR4	

(Pixley and Stanley, 2004). Of these, Y721 is probably the most important for chemotaxis because it recruits PLC-γ and the p85 subunit of class 1A PI3Ks, both of which are implicated in the initiation of chemotaxis (Bourette and Rohrschneider, 2000) (Fig. 3). CSF-1-induced activation of PI3K and its downstream target, the Akt/PKB pathway, also signal to cell survival (Kelley et al., 1999) and activation of the ERK, NF-κB, and mTOR/S6K pathways (Brach et al., 1991; Bhatt et al., 2002; Glantschnig et al., 2003). Activated CSF-1R recruits c-Cbl, which promotes receptor ubiquitination, leading to internalization of the receptor complex to endosomes and then to lysosomal deg-

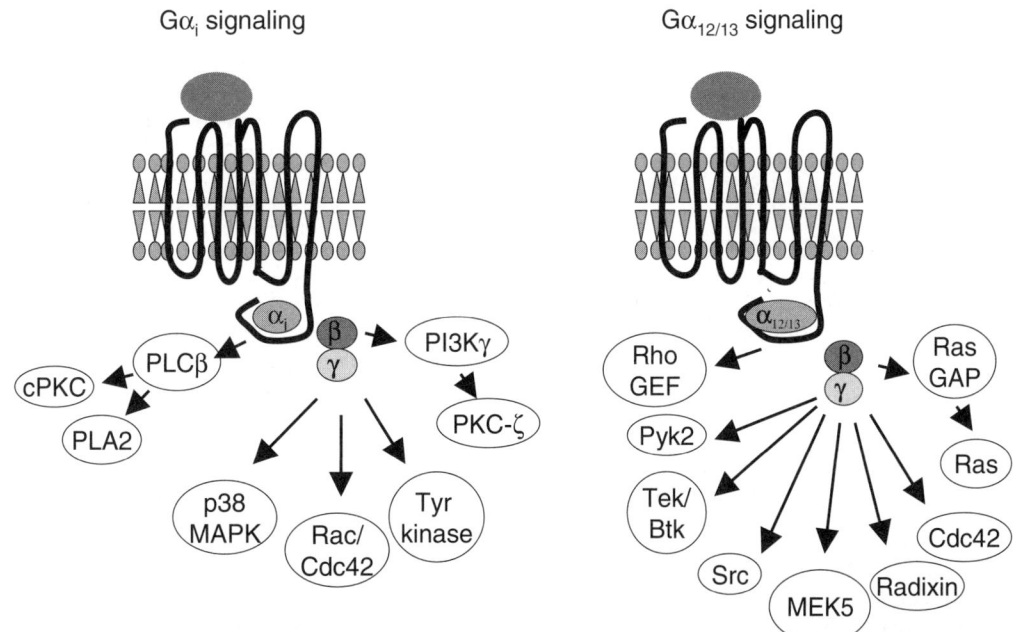

FIGURE 2 GPCR-induced signaling in chemotaxis. Ligand binding to GPCRs activates heterotrimeric G proteins, leading to dissociation of Gα from Gβγ subunits. Gα and Gβγ each activate a variety of signaling pathways that contribute to cell migration. For example, $Gα_{12/13}$ subunits can activate RhoGEFs to increase Rho activity at the rear, and Gβγ subunits activate PI3Kγ, which acts at the front of the cell to stimulate Rac.

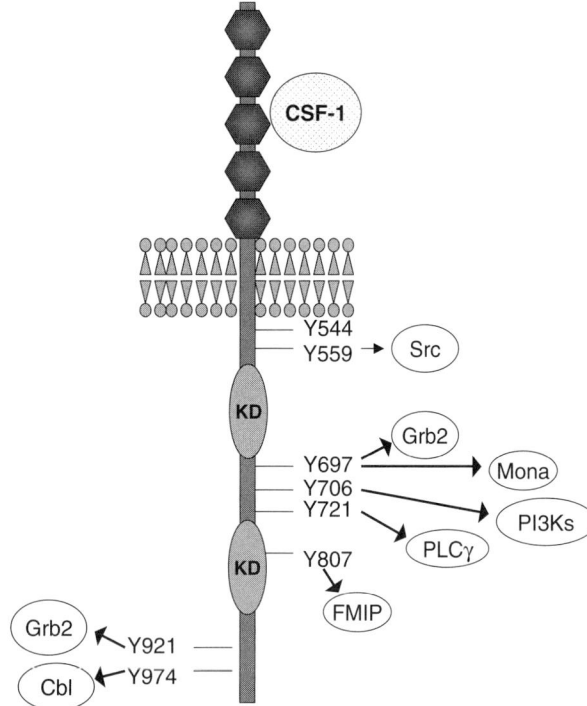

FIGURE 3 CSF-1 receptor signaling in chemotaxis. Activation of the CSF-1 receptor results in homodimerization (not depicted) and autophosphorylation on several tyrosines. Phosphorylation of Y721 is probably most important for chemotaxis because it recruits PLC-γ and the p85 subunit of class 1A PI3Ks, both of which are implicated in the initiation of chemotaxis. Several other SH2-containing proteins are also recruited to the activated CSF-1 receptor including RhoGEFS and GAPs. KD, kinase domain.

radation (Yeung and Stanley, 2003), thereby downregulating CSF-1R signaling.

INTRACELLULAR SIGNALING PATHWAYS IN CHEMOTAXIS

Several intracellular signaling complexes contribute to the polarization of phagocytes in response to chemoattractants, and they probably act together to allow optimal chemotaxis. They include Cdc42, acting via Par3/Par6/aPKC and PIX/PAK; PI3Ks; and PLCs.

Cdc42

Cdc42 was first shown to be required specifically for chemotaxis but not migration in macrophages (Allen et al., 1998). Subsequently, directed migration of many other cell types was found to depend on Cdc42, including dendritic cells (Calle et al., 2004) and neutrophils (Weiner et al., 2002; Srinivasan et al., 2003; Van Keymeulen et al., 2006). Dominant-negative Cdc42 causes formation of multiple unstable pseudopods in neutrophils, whereas in macrophages it induces the formation of one larger lamellipodium, increasing migration speed but preventing chemotaxis. Cdc42 is now implicated in multiple types of cell polarity, including axon specification, yeast mating, and epithelial polarity (Etienne-Manneville, 2004). In fibroblasts and astrocytes, directed migration into a scratch wound depends on Cdc42 interaction with Par6, which is part of a complex with Par3 and PKC-ζ, an atypical PKC. Together with cytoplasmic dynein, these proteins mediate reorientation of the MTOC toward the scratch edge (Palazzo et al., 2001; Etienne-Manneville, 2004). In neutrophil-like HL60 cells, Par6 is also essential for cell polarization (Xu et al., 2007).

A second mechanism whereby Cdc42 has been reported to induce neutrophil polarity is by interacting with the serine/threonine kinase PAK1, which binds to Gβγ and the GEF αPIX, and thereby activates Rac. By restricting Rac activation and hence lamellipodium extension to the front of the cell, this leads to stable migration in one direction (Li et al., 2003). In neutrophils, chemotaxis to the chemokine CXCL1 (Chellaiah et al., 2000) and C5a (Adamson et al., 1992) was shown to require PAK1 (Rousseau et al., 2006), whereas PAK2 contributes to macrophage chemotaxis toward the chemokine RANTES (Weiss-Haljiti et al., 2004).

Cdc42 also directly induces actin polymerization through the hematopoietic cell-specific Wiskott-Aldrich syndrome protein (WASP) and its ubiquitously expressed relative N-WASP (Miki et al., 1996; Symons et al., 1996). Cdc42 binding to WASPs activates the Arp2/3 complex, which nucleates new actin filaments that branch off the sides of existing filaments (Rohatgi et al., 1999; Welch and Mullins, 2002). WASPs play an important role in vesicle trafficking, including endocytosis and exocytosis (Takenawa and Suetsugu, 2007), but whether WASPs are required for chemotaxis is not clear. Dendritic cells from Wiskott-Aldrich syndrome patients show defective migration in vitro, and WASP is crucial for podosome formation (Burns et al., 2001, 2004; Jones et al., 2002). Podosomes are sites of cell adhesion and matrix degradation, and could thereby contribute generally to cell migration rather than specifically to chemotaxis (Linder and Aepfelbacher, 2003).

PI3Ks

In addition to Cdc42, PI3Ks contribute to cell polarity in phagocytes. There are three classes of PI3Ks, of which class 1 PI3Ks have been implicated in chemotaxis. Class 1 PI3Ks phosphorylate the 3' position of the inositol ring of phosphatidylinositol 4,5-bisphosphate (PIP_2) to create PIP_3. PIP_3 acts as a second messenger by recruiting Pleckstrin homology (PH) domain-containing proteins to the front of the cell including PKB/Akt and GEFs for Rac (Bourne and Weiner, 2002; Huang et al., 2003; Postma et al., 2004; Comer et al., 2005; Matsuoka et al., 2006), which then mediate intracellular signaling (Fig. 4).

There are two types of class 1 PI3Ks, class 1A (PI3Kα, PI3Kβ, and PI3Kδ) and class 1B (PI3Kγ). Class 1A PI3Ks consist of a regulatory subunit (p85) and catalytic subunit (p110) (Wymann and Pirola, 1998; Procko and McColl, 2005). They are activated by tyrosine kinase receptors and cytoplasmic tyrosine kinases such as Src through recruitment of p85 via its SH2 domains to phosphorylated tyrosine residues. PI3Kγ is activated by Gβγ liberated from GPCR-activated heterotrimeric G proteins and by Ras GTPases (Suire et al., 2006). It binds to adaptor molecules p101 and p85, which are thought to confer Gβγ sensitivity.

The general PI3K inhibitors LY294002 and wortmannin inhibit migration of macrophages and chemotaxis of neutrophils, but they are not specific for PI3Ks. Injection of isoform-specific antibodies to class 1A PI3Ks demonstrated that PI3Kδ and PI3Kβ but not PI3Kα were important for CSF-1-induced macrophage chemotaxis (Vanhaesebroeck

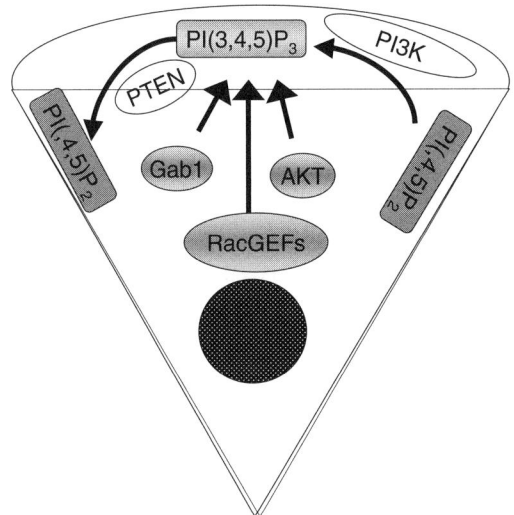

FIGURE 4 Involvement of PI3Ks and PTEN in chemotaxis. PI3Ks are selectively active at the front of polarized migrating cells, whereas PTEN is primarily localized in the cytoplasm. Production of PIP_3 by PI3Ks recruits proteins that have PH domains including the serine/threonine kinase AKT, Rho-GEFs, and adaptor proteins such as Gab1, all of which transduce signals to cell motility and chemotaxis. PIP_3 is removed by the phosphatase PTEN, which converts PIP_3 to $PI(4,5)P_2$.

et al., 1999). Clearer evidence for involvement of specific PI3K isoforms has come from studies of phagocytes derived from PI3K-null mice. Analysis of macrophages and neutrophils derived from mice lacking PI3Kγ has shown that PI3Kγ is required for chemotaxis to chemokines and fMLP (Procko and McColl, 2005). However, chemotaxis of both neutrophils and macrophages derived from PI3Kγ-null mice is not completely abrogated (Hannigan et al., 2002; Jones et al., 2003), and recent analysis in neutrophils indicates that PI3Kγ is not essential for chemotaxis, but is critical for integrin-mediated adhesion, and thereby affects the number of cells that polarize rather than their ability to polarize (Ferguson et al., 2007). Together, these results suggest that other isoforms of PI3K or different molecule pathways (e.g., Cdc42) may act together with PI3Kγ to mediate chemotaxis.

Whether or not PI3Ks are essential for chemotaxis, they are selectively active at the front of polarized migrating cells. The PH domain of Akt/PKB fused to GFP, which binds to PIP_3 and $PI(3,4)P_2$ with high affinity, is widely used as a probe to detect changes in $PIP_3/PI(3,4)P_2$ levels in living cells. This probe has demonstrated that $PIP_3/PI(3,4)P_2$ is rapidly recruited to the plasma membrane of neutrophil-like HL60 cells in response to chemoattractants, and that the intracellular gradient of PIP_3 generated by PI3K across a neutrophil exposed to a chemoattractant is far steeper than the extracellular gradient of chemoattractant (Xu et al., 2005), implying that there is a positive feedback loop acting on PI3Ks. External PIP_3 can itself induce accumulation of PH-GFP (Bourne and Weiner, 2002; Niggli, 2003). It has been proposed that PIP_3 recruits a PH-domain-containing Rac GEF, thereby activating Rac, which in turn stimulates and/or localizes PI3K activity. PIP_3-sensitive PH domains are present in several GEFs for Rac, including P-Rex1, α- and βPIX, and Vav (Li et al., 2003; Merlot and Firtel, 2003). However, it is not clear which of these Rac GEFs is critical for PIP_3-induced polarization, or whether they each contribute complementary functions. Knockout of P-Rex1 in neutrophils leads to defects in chemotaxis. P-Rex1 becomes recruited to the leading edge of neutrophils by PIP_3 and Gβγ, along with its target Rac2 (Welch et al., 2005; Zhao et al., 2007). So far, the role of P-Rex1 in macrophages has not been analyzed. αPIX has already been implicated in Cdc42-mediated signaling in chemotaxis (see above), and thus could provide a link between Cdc42 and PI3K pathways. Vav1-null neutrophils and macrophages are able to chemotax, although the migration speed of Vav1-null macrophages is slightly slower (Gakidis et al., 2004; Wells et al., 2005).

PIP_3 is removed by the phosphatase "phosphatase and tensin homolog" (PTEN), which dephosphorylates PIP_3 to $PI(4,5)P_2$ (Fig. 4). In *Dictyostelium* PTEN is recruited to the rear of chemotaxing cells, restricting accumulation of PIP_3 to the leading edge (Funamoto et al., 2002), but in leukocytes its localization is controversial. On the one hand, in primary mouse neutrophils it is recruited to the rear in a Rho-kinase-dependent pathway (Li et al., 2005), but on the other hand, in neutrophil-like HL60 cells PTEN is not specifically localized in the uropod nor is it important for gradient sensing (Lacalle et al., 2004).

Interestingly, inhibition of PI3Ks in neutrophils prevented fMLP-stimulated activation of Akt and Cdc42, but the accumulation of F-actin at the plasma membrane was not affected (Van Keymeulen et al., 2006). Conversely, reducing the amount of F-actin in cells with latrunculin A caused a large reduction in the PIP_3 accumulation at the front of the cell following chemokine stimulation. Inhibition of Rho GTPases has similar effects on buildup of both PIP_3 at the front of cells and PTEN at the rear (Matsuoka et al., 2006). Together, these results support a model where chemotaxis requires a combination of different processes that are stimulated independently but act concertedly.

PLC and Calcium

There are several PLC isoforms, of which PLCβ2 and PLCβ3 are activated by GPCR signaling in neutrophils, whereas PLCγ isoforms are activated by tyrosine kinase receptors. Neutrophils lacking PLCβ2 and PLCβ3 show no difference in chemotaxis to fMLP, but slightly enhanced chemotaxis to MIP1α (Li et al., 2000), suggesting that PLC activity could downregulate chemotaxis/migration. However, PLCγ could be important for CSF-1-induced chemotaxis (Pixley and Stanley, 2004). PLC activation leads to production of the two second messengers, diacylglycerol and inositol 3,4,5-trisphosphate (IP_3). IP_3 in turn induces the release of calcium from intracellular stores, and diacylglycerol together with calcium activates PKC isoforms. While there is little known about the roles of different PKC isoforms in phagocyte migration, calcium is believed to contribute to fMLP-induced motility of neutrophils by stimulating myosin II phosphorylation in the uropod and subsequent detachment (Eddy et al., 2000). Calcium also activates calpains, which have been shown to regulate neutrophil chemotaxis: calpain 2 localizes selectively to the front of neutrophils chemotaxing to C5a, and calpain activity is required to limit pseudopod localization to the front (Nuzzi et al., 2007). A requirement of calcium and calpain in GPCR-stimulated neutrophil chemotaxis seems to contradict the lack of involvement of PLCβ (Li et al., 2000). It is possible, however, that calcium is provided through other signaling pathways in addition to PLCβ activation.

Phospholipase A₂
Phospholipase A₂ isoforms have been shown to be required for chemotaxis of phagocytes, including monocytes to CCL2 (MCP-1) and neutrophils to fMLP, both using pharmacological inhibitors of PLA₂ and antisense-mediated knockdown (Locati et al., 1996; Carnevale and Cathcart, 2001). The mechanism whereby PLA₂ regulates chemotaxis is not clear, but it is interesting that recent results in *Dictyostelium* have revealed that PI3Ks and PLA₂ act on parallel and complementary pathways to regulate cyclic AMP-induced chemotaxis (Chen et al., 2007).

MAINTENANCE OF POLARITY: Rac AND Rho
Polarity signals act to initiate polarization of cells, but subsequent maintenance of polarity could be achieved by Rac and Rho without the requirement for additional signals. Multiple lines of evidence show that Rac acts to inhibit Rho activity, and vice versa, thereby creating a Rac zone at the front of cells and a Rho zone in the uropod. However, there is also good evidence that Rac is active in the uropod of neutrophils and contributes to uropod retraction (Gardiner et al., 2002). Cdc42 and Rac promote polarization by inhibiting Rho activity at the leading edge, augmenting Rho-dependent actomyosin contraction at the trailing edge. Rho and Rac refine each other's activity during cell polarization and migration, balancing actin polymerization, cell contraction, and adhesion essential for chemotaxis (Pestonjamasp et al., 2006). Based on these data, models have been generated that predict mutually exclusive activity of Rac and Rho (Civelekoglu-Scholey et al., 2005).

Rac
Rac stimulates formation of lamellipodia by promoting Arp2/3-mediated actin polymerization at the plasma membrane (Ridley et al., 2003). Key downstream effectors of Rac1 involved in lamellipodium extension include p140-SRA1, IRSp53, and PAKs. The roles of these proteins have largely been determined in fibroblasts, and little is known of their roles in phagocytes, although it is presumed they will act similarly. p140-SRA1 is part of a complex including the Arp2/3 complex regulator WAVE, and Rac1 binding to this WAVE complex enhances Arp2/3-driven actin polymerization (Higgs and Pollard, 1999; Millard et al., 2004). IRSp53 can also associate with the WAVE complex (Suetsugu et al., 2006) as well as with Cdc42, and appears to be a scaffold protein that, like its homolog, MIM-B, localizes proteins to actin filaments (Bompard et al., 2005). PAKs activate LIMKs, which in turn induce cofilin phosphorylation, inhibiting cofilin's ability to stimulate actin depolymerization (Bokoch, 2003). In neutrophils, cofilin phosphorylation rapidly decreased in response to fMLP (Zhan et al., 2003), reflecting transient activation. The correct spatial and temporal regulation of cofilin activity is likely to be critical both to allow lamellipodium formation at the leading edge and to prevent membrane extension at other regions.

Neutrophils and macrophages express Rac1 and Rac2 but not Rac3, and each isoform makes a different contribution to migration. Rac2-null neutrophils have severely reduced migration and defects in polarization due to a disrupted actin cytoskeleton (Li et al., 2002), whereas Rac2 deletion has only minor effects on macrophage migration (Wheeler et al., 2006). Neutrophils and macrophages derived from Rac1-null mice have defects in cell spreading and membrane ruffling but not in migration (Glogauer et al., 2003; Wells et al., 2004). Interestingly, Rac1 has been reported to play a key role in neutrophil chemotaxis (Sun et al., 2004), although it does not affect macrophage chemotaxis (Wells et al., 2004). Recent work suggests that Rac1 activation promotes RhoA-induced contractility at the rear of the cell (Pestonjamasp et al., 2006). One possible mechanism is via PAK, which regulates the activity of the Rho-specific GEF, RhoGEF-H1 (Zenke et al., 2004).

Rac1 also promotes microtubule stabilization in several cell types and could thereby set up a positive feedback loop regulating its activity at the front of migrating cells (Watanabe et al., 2005), although this has not been reported in phagocytes. IQGAPs are Rac and Cdc42 targets that regulate microtubule capture at the leading edge of cells (Briggs and Sacks, 2003; Watanabe et al., 2005), while the Rac/Cdc42 target PAK promotes microtubule stability. So far, the involvement of IQGAPs in phagocyte migration has not been tested.

Rho
It is thought that Rho is essential for generation of actomyosin-based contractility at the rear of migrating phagocytes (Ridley, 2004; Burridge and Doughman, 2006; Van Keymeulen et al., 2006). Inhibition of Rho with C3 transferase reduces the migration speed of macrophages and neutrophils (Allen et al., 1997; Niggli, 1999). There are three closely related Rho isoforms, RhoA, RhoB, and RhoC. Of these, only RhoA and RhoB are expressed in macrophages, and RhoA appears to be the dominant target for C3 transferase in migration (Wheeler and Ridley, 2007). RhoA activity is primarily restricted to the back of polarized neutrophils (Van Keymeulen et al., 2006) and is regulated by GPCR receptor signaling to RhoGEFs such as p115RhoGEF (Xu et al., 2003; Koike et al., 2006). How p115RhoGEF activation is restricted to the uropod is not clear.

An important target of Rho, involved at the rear of chemotaxing phagocytes, is ROCK, a serine/threonine kinase. ROCK enhances the phosphorylation of myosin light chain (MLC), primarily by phosphorylating and inhibiting the regulatory subunit of MLC phosphatase (Riento and Ridley, 2003). Phosphorylated MLC increases the interaction of myosin II with actin filaments and contractility. Another target of Rho proteins is mDia1, a multifunctional protein that both nucleates actin polymerization (Watanabe et al., 1997; Eisenmann et al., 2007) and is required in fibroblasts for polarized organization of microtubules (Ishizaki et al., 2001; Palazzo et al., 2004). Whether mDia proteins play a role in phagocyte migration is not yet known, although mDia1 is involved in complement-mediated phagocytosis in macrophages (Colucci-Guyon et al., 2005).

FUTURE DIRECTIONS
Our current understanding of chemotaxis indicates that several signaling pathways act in concert to induce cell polarization, including Cdc42, Par proteins, PAK/PIX, and PI3Ks. Several proteins have been best characterized as regulators of chemotaxis in cell types other than phagocytes, including the Par proteins and PLA₂, and it will be

important in the future to determine to what extent these pathways are conserved in phagocytes. Phagocytes generally move much faster than fibroblasts or astrocytes, where cell polarity pathways have been carefully dissected (Ridley et al., 2003; Etienne-Manneville, 2004; Friedl, 2004), and it is thus possible that these pathways are not so crucial in phagocytes. In addition, there are 20 different Rho GTPases in mammals, although not all of them are expressed in every cell type (Wennerberg and Der, 2004). It is not known whether other family members contribute to phagocyte migration and chemotaxis in addition to Cdc42, Rac1 and Rac2, and RhoA. It will therefore be interesting to determine the roles of each Rho GTPase in phagocytes. Inhibiting phagocyte chemotaxis could be a route to the reduction of chronic inflammatory disorders; thus, the design and testing of inhibitors of signal transduction molecules involved in migration and chemotaxis will be an important goal for the future.

REFERENCES

Abercrombie, M., G. A. Dunn, and J. P. Heath. 1977. The shape and movement of fibroblasts in culture. *Soc. Gen. Physiol. Ser.* **32:**57–70.

Adamson, P., C. J. Marshall, A. Hall, and P. A. Tilbrook. 1992. Post-translational modifications of p21rho proteins. *J. Biol. Chem.* **267:**20033–20038.

Albelda, S. M., C. W. Smith, and P. A. Ward. 1994. Adhesion molecules and inflammatory injury. *FASEB J.* **8:**504–512.

Allen, W. E., G. E. Jones, J. W. Pollard, and A. J. Ridley. 1997. Rho, Rac and Cdc42 regulate actin organization and cell adhesion in macrophages. *J. Cell Sci.* **110:**707–720.

Allen, W. E., D. Zicha, A. J. Ridley, and G. E. Jones. 1998. A role for Cdc42 in macrophage chemotaxis. *J. Cell Biol.* **141:**1147–1157.

Arai, H., and I. F. Charo. 1996. Differential regulation of G-protein-mediated signaling by chemokine receptors. *J. Biol. Chem.* **271:**21814–21819.

Bhatt, N. Y., T. W. Kelley, V. V. Khramtsov, Y. Wang, G. K. Lam, T. L. Clanton, and C. B. Marsh. 2002. Macrophage-colony-stimulating factor-induced activation of extracellular-regulated kinase involves phosphatidylinositol 3-kinase and reactive oxygen species in human monocytes. *J. Immunol.* **169:**6427–6434.

Bokoch, G. M. 2003. Biology of the P21-activated kinases. *Annu. Rev. Biochem.* **72:**743–781.

Bokoch, G. M. 2005. Regulation of innate immunity by Rho GTPases. *Trends Cell Biol.* **15:**163–171.

Bompard, G., S. J. Sharp, G. Freiss, and L. M. Machesky. 2005. Involvement of Rac in actin cytoskeleton rearrangements induced by MIM-B. *J. Cell Sci.* **118:**5393–5403.

Bourette, R. P., and L. R. Rohrschneider. 2000. Early events in M-CSF receptor signaling. *Growth Factors* **17:**155–166.

Bourne, H. R., and O. Weiner. 2002. A chemical compass. *Nature* **419:**21.

Brach, M. A., R. Henschler, R. H. Mertelsmann, and F. Herrmann. 1991. Regulation of M-CSF expression by M-CSF: role of protein kinase C and transcription factor NFκB. *Pathobiology* **59:**284–288.

Briggs, M. W., and D. B. Sacks. 2003. IQGAP1 as signal integrator: Ca2+, calmodulin, Cdc42 and the cytoskeleton. *FEBS Lett.* **542:**7–11.

Burke, B., and C. E. Lewis. 2002. *The Macrophage.* Oxford University Press, Oxford, United Kingdom.

Burke, B., S. Sumner, N. Maitland, and C. E. Lewis. 2002. Macrophages in gene therapy: cellular delivery vehicles and in vivo targets. *J. Leukoc. Biol.* **72:**417–428.

Burns, S., S. J. Hardy, J. Buddle, K. L. Yong, G. E. Jones, and A. J. Thrasher. 2004. Maturation of DC is associated with changes in motile characteristics and adherence. *Cell Motil. Cytoskeleton* **57:**118–132.

Burns, S., A. J. Thrasher, M. P. Blundell, L. Machesky, and G. E. Jones. 2001. Configuration of human dendritic cell cytoskeleton by Rho GTPases, the WAS protein, and differentiation. *Blood* **98:**1142–1149.

Burridge, K., and R. Doughman. 2006. Front and back by Rho and Rac. *Nat. Cell Biol.* **8:**781–782.

Calle, Y., H. C. Chou, A. J. Thrasher, and G. E. Jones. 2004. Wiskott-Aldrich syndrome protein and the cytoskeletal dynamics of dendritic cells. *J. Pathol.* **204:**460–469.

Carnevale, K. A., and M. K. Cathcart. 2001. Calcium-independent phospholipase A(2) is required for human monocyte chemotaxis to monocyte chemoattractant protein 1. *J. Immunol.* **167:**3414–3421.

Chellaiah, M. A., N. Soga, S. Swanson, S. McAllister, U. Alvarez, D. Wang, S. F. Dowdy, and K. A. Hruska. 2000. Rho-A is critical for osteoclast podosome organization, motility, and bone resorption. *J. Biol. Chem.* **275:**11993–12002.

Chen, L., M. Iijima, M. Tang, M. A. Landree, Y. E. Huang, Y. Xiong, P. A. Iglesias, and P. N. Devreotes. 2007. PLA2 and PI3K/PTEN pathways act in parallel to mediate chemotaxis. *Dev. Cell* **12:**603–614.

Civelekoglu-Scholey, G., A. Wayne Orr, I. Novak, J. J. Meister, M. A. Schwartz, and A. Mogilner. 2005. Model of coupled transient changes of Rac, Rho, adhesions and stress fibers alignment in endothelial cells responding to shear stress. *J. Theor. Biol.* **232:**569–585.

Colucci-Guyon, E., F. Niedergang, B. J. Wallar, J. Peng, A. S. Alberts, and P. Chavrier. 2005. A role for mammalian diaphanous-related formins in complement receptor (CR3)-mediated phagocytosis in macrophages. *Curr. Biol.* **15:**2007–2012.

Eddy, R. J., L. M. Pierini, F. Matsumura, and F. R. Maxfield. 2000. Ca2+-dependent myosin II activation is required for uropod retraction during neutrophil migration. *J. Cell Sci.* **113:**1287–1298.

Eisenmann, K. M., E. S. Harris, S. M. Kitchen, H. A. Holman, H. N. Higgs, and A. S. Alberts. 2007. Dia-interacting protein modulates formin-mediated actin assembly at the cell cortex. *Curr. Biol.* **17:**579–591.

Eng, E. W., A. Bettio, J. Ibrahim, and R. E. Harrison. 2007. MTOC reorientation occurs during FcγR-mediated phagocytosis in macrophages. *Mol. Biol. Cell* **18:**2389-2399.

Etienne-Manneville, S. 2004. Cdc42—the centre of polarity. *J. Cell Sci.* **117:**1291–1300.

Ferguson, G. J., L. Milne, S. Kulkarni, T. Sasaki, S. Walker, S. Andrews, T. Crabbe, P. Finan, G. Jones, S. Jackson, M. Camps, C. Rommel, M. Wymann, E. Hirsch, P. Hawkins, and L. Stephens. 2007. PI(3)Kγ has an important context-dependent role in neutrophil chemokinesis. *Nat. Cell Biol.* **9:**86–91.

Franca-Koh, J., Y. Kamimura, and P. Devreotes. 2006. Navigating signaling networks: chemotaxis in *Dictyostelium discoideum*. *Curr. Opin. Genet. Dev.* **16:**333–338.

Friedl, P. 2004. Prespecification and plasticity: shifting mechanisms of cell migration. *Curr. Opin. Cell Biol.* **16:**14–23.

Funamoto, S., R. Meili, S. Lee, L. Parry, and R. A. Firtel. 2002. Spatial and temporal regulation of 3-phosphoinositides by PI 3-kinase and PTEN mediates chemotaxis. *Cell* **109:**611–623.

Gakidis, M. A. M., X. Cullere, T. Olson, J. L. Wilsbacher, B. Zhang, S. L. Moores, K. Ley, W. Swat, T. Mayadas, and J. S. Brugge. 2004. Vav GEFs are required for β2 integrin-dependent functions of neutrophils. *J. Cell Biol.* **166:**273–282.

Gardiner, E. M., K. N. Pestonjamasp, B. P. Bohl, C. Chamberlain, K. M. Hahn, and G. M. Bokoch. 2002. Spatial and temporal analysis of Rac activation during live neutrophil chemotaxis. *Curr. Biol.* **12:**2029–2034.

Glantschnig, H., J. E. Fisher, G. Wesolowski, G. A. Rodan, and A. A. Reszka. 2003. M-CSF, TNFalpha and RANK ligand promote osteoclast survival by signaling through mTOR/S6 kinase. *Cell Death Differ.* **10:**1165–1177.

Glogauer, M., C. C. Marchal, F. Zhu, A. Worku, B. E. Clausen, I. Foerster, P. Marks, G. P. Downey, M. Dinauer, and D. J. Kwiatkowski. 2003. Rac1 deletion in mouse neutrophils has selective effects on neutrophil functions. *J. Immunol.* **170:**5652–5657.

Hannigan, M., L. Zhan, Z. Li, Y. Ai, D. Wu, and C. K. Huang. 2002. Neutrophils lacking phosphoinositide 3-kinase gamma show loss of directionality during N-formyl-Met-Leu-Phe-induced chemotaxis. *Proc. Natl. Acad. Sci. USA* **99:**3603–3608.

Henderson, R. B., J. A. Hobbs, M. Mathies, and N. Hogg. 2003. Rapid recruitment of inflammatory monocytes is independent of neutrophil migration. *Blood* **102:**328–335.

Higgs, H. N., and T. D. Pollard. 1999. Regulation of actin polymerization by Arp2/3 complex and WASp/Scar proteins. *J. Biol. Chem.* **274:**32531–32534.

Huang, Y. E., M. Iijima, C. A. Parent, S. Funamoto, R. A. Firtel, and P. Devreotes. 2003. Receptor-mediated regulation of PI3Ks confines PI(3,4,5)P3 to the leading edge of chemotaxing cells. *Mol. Biol. Cell* **14:**1913–1922.

Ishizaki, T., Y. Morishima, M. Okamoto, T. Furuyashiki, T. Kato, and S. Narumiya. 2001. Coordination of microtubules and the actin cytoskeleton by the Rho effector mDia1. *Nat. Cell Biol.* **3:**8–14.

Jimenez-Sainz, M. C., B. Fast, F. Mayor, Jr., and A. M. Aragay. 2003. Signaling pathways for monocyte chemoattractant protein 1-mediated extracellular signal-regulated kinase activation. *Mol. Pharmacol.* **64:**773–782.

Jones, G. E., E. Prigmore, R. Calvez, C. Hogan, G. A. Dunn, E. Hirsch, M. P. Wymann, and A. J. Ridley. 2003. Requirement for PI 3-kinase γ in macrophage migration to MCP-1 and CSF-1. *Exp. Cell Res.* **290:**120–131.

Jones, G. E., D. Zicha, G. A. Dunn, M. Blundell, and A. Thrasher. 2002. Restoration of podosomes and chemotaxis in Wiskott-Aldrich syndrome macrophages following induced expression of WASp. *Int. J. Biochem. Cell Biol.* **34:**806–815.

Kelley, T. W., M. M. Graham, A. I. Doseff, R. W. Pomerantz, S. M. Lau, M. C. Ostrowski, T. F. Franke, and C. B. Marsh. 1999. Macrophage colony-stimulating factor promotes cell survival through Akt/protein kinase B. *J. Biol. Chem.* **274:**26393–26398.

Kobayashi, H., S. Miura, H. Nagata, Y. Tsuzuki, R. Hokari, T. Ogino, C. Watanabe, T. Azuma, and H. Ishii. 2004. In situ demonstration of dendritic cell migration from rat intestine to mesenteric lymph nodes: relationships to maturation and role of chemokines. *J. Leukoc. Biol.* **75:**434–442.

Kobayashi, Y. 2008. The role of chemokines in neutrophil biology. *Front. Biosci.* **13:**2400–2407.

Koike, D., H. Obinata, A. Yamamoto, S. Takeda, H. Komori, F. Nara, T. Izumi, and T. Haga. 2006. 5-Oxo-eicosatetraenoic acid-induced chemotaxis: identification of a responsible receptor hGPCR48 and negative regulation by G protein G12/13. *J. Biochem. (Tokyo)* **139:**543–549.

Lacalle, R. A., C. Gomez-Mouton, D. F. Barber, S. Jimenez-Baranda, E. Mira, C. Martinez-A, A. C. Carrera, and S. Manes. 2004. PTEN regulates motility but not directionality during leukocyte chemotaxis. *J. Cell Sci.* **117:**6207–6215.

Lattin, J., D. A. Zidar, K. Schroder, S. Kellie, D. A. Hume, and M. J. Sweet. 2007. G-protein-coupled receptor expression, function, and signaling in macrophages. *J. Leukoc. Biol.* **82:**16–32.

Li, S., A. Yamauchi, C. C. Marchal, J. K. Molitoris, L. A. Quilliam, and M. C. Dinauer. 2002. Chemoattractant-stimulated Rac activation in wild-type and Rac2-deficient murine neutrophils: preferential activation of Rac2 and Rac2 gene dosage effect on neutrophil functions. *J. Immunol.* **169:**5043–5051.

Li, Z., X. Dong, Z. Wang, W. Liu, N. Deng, Y. Ding, L. Tang, T. Hla, R. Zeng, L. Li, and D. Wu. 2005. Regulation of PTEN by Rho small GTPases. *Nat. Cell Biol.* **7:**399–404.

Li, Z., M. Hannigan, Z. Mo, B. Liu, W. Lu, Y. Wu, A. V. Smrcka, G. Wu, L. Li, M. Liu, C. K. Huang, and D. Wu. 2003. Directional sensing requires G beta gamma-mediated PAK1 and PIX alpha-dependent activation of Cdc42. *Cell* **114:**215–227.

Li, Z., H. Jiang, W. Xie, Z. Zhang, A. V. Smrcka, and D. Wu. 2000. Roles of PLC-2 and -3 and PI3K in chemoattractant-mediated signal transduction. *Science* **287:**1046–1049.

Linder, S., and M. Aepfelbacher. 2003. Podosomes: adhesion hot-spots of invasive cells. *Trends Cell Biol.* **13:**376–385.

Locati, M., G. Lamorte, W. Luini, M. Introna, S. Bernasconi, A. Mantovani, and S. Sozzani. 1996. Inhibition of monocyte chemotaxis to C-C chemokines by antisense oligonucleotide for cytosolic phospholipase A2. *J. Biol. Chem.* **271:**6010–6016.

Lu, B., B. J. Rutledge, L. Gu, J. Fiorillo, N. W. Lukacs, S. L. Kunkel, R. North, C. Gerard, and B. J. Rollins. 1998. Abnormalities in monocyte recruitment and cytokine expression in monocyte chemoattractant protein 1-deficient mice. *J. Exp. Med.* **187:**601–608.

Mahadeo, D. C., M. Janka-Junttila, R. L. Smoot, P. Roselova, and C. A. Parent. 2007. A chemoattractant-mediated Gi-coupled pathway activates adenylyl cyclase in human neutrophils. *Mol. Biol. Cell* **18:**512–522.

Matsuoka, S., M. Iijima, T. M. Watanabe, H. Kuwayama, T. Yanagida, P. N. Devreotes, and M. Ueda. 2006. Single-molecule analysis of chemoattractant-stimulated membrane recruitment of a PH-domain-containing protein. *J. Cell Sci.* **119:**1071–1079.

Merlot, S., and R. A. Firtel. 2003. Leading the way: directional sensing through phosphatidylinositol 3-kinase and other signaling pathways. *J. Cell Sci.* **116:**3471–3478.

Miki, H., K. Miura, and T. Takenawa. 1996. N-WASP, a novel actin-depolymerizing protein, regulates the cortical cytoskeletal rearrangement in a PIP2-dependent manner downstream of tyrosine kinases. *EMBO J.* **15:**5326–5335.

Millard, T. H., S. J. Sharp, and L. M. Machesky. 2004. Signalling to actin assembly via the WASP (Wiskott-Aldrich syndrome protein)-family proteins and the Arp2/3 complex. *Biochem. J.* **380:**1–17.

Mitchison, T. J., and L. P. Cramer. 1996. Actin-based cell motility and cell locomotion. *Cell* **84:**371–379.

Myers, S. J., L. M. Wong, and I. F. Charo. 1995. Signal transduction and ligand specificity of the human monocyte chemoattractant protein-1 receptor in transfected embryonic kidney cells. *J. Biol. Chem.* **270:**5786–5792.

Needham, L. K., and E. Rozengurt 1998. Galpha12 and Galpha13 stimulate Rho-dependent tyrosine phosphorylation of focal adhesion kinase, paxillin, and p130 Crk-associated substrate. *J. Biol. Chem.* **273:**14626–14632.

Neves, S., P. Ram, and R. Iyengar. 2002. G protein pathways. *Science* **296:**1636–1639.

Niggli, V. 1999. Rho-kinase in human neutrophils: a role in signalling for myosin light chain phosphorylation and cell migration. *FEBS Lett.* **445:**69–72.

Niggli, V. 2003. Signaling to migration in neutrophils: importance of localized pathways. *Int. J. Biochem. Cell Biol.* **35:**1619–1638.

Nuzzi, P. A., M. A. Senetar, and A. Huttenlocher. 2007. Asymmetric localization of calpain 2 during neutrophil chemotaxis. *Mol. Biol. Cell* **18:**795–805.

Palazzo, A. F., C. H. Eng, D. D. Schlaepfer, E. E. Marcantonio, and G. G. Gundersen. 2004. Localized stabilization of microtubules by integrin- and FAK-facilitated Rho signaling. *Science* **303:**836–839.

Palazzo, A. F., H. L. Joseph, Y. J. Chen, D. L. Dujardin, A. S. Alberts, K. K. Pfister, R. B. Vallee, and G. G. Gundersen. 2001. Cdc42, dynein, and dynactin regulate MTOC reorientation independent of Rho-regulated microtubule stabilization. *Curr. Biol.* **11:**1536–1541.

Pestonjamasp, K. N., C. Forster, C. Sun, E. M. Gardiner, B. Bohl, O. Weiner, G. M. Bokoch, and M. Glogauer. 2006. Rac1 links leading edge and uropod events through Rho and myosin activation during chemotaxis. *Blood* **108:**2814–2820.

Pierce, K. L., and R. J. Lefkowitz. 2001. Classical and new roles of beta-arrestins in the regulation of G-protein-coupled receptors. *Nat. Rev. Neurosci.* **2:**727–733.

Pixley, F. J., and E. R. Stanley. 2004. CSF-1 regulation of the wandering macrophage: complexity in action. *Trends Cell Biol.* **14:**628–638.

Postma, M., L. Bosgraaf, H. M. Loovers, and P. J. Van Haastert. 2004. Chemotaxis: signalling modules join hands at front and tail. *EMBO Rep.* **5:**35–40.

Procko, E., and S. R. McColl. 2005. Leukocytes on the move with phosphoinositide 3-kinase and its downstream effectors. *BioEssays* **27:**153–163.

Rickert, P., O. D. Weiner, F. Wang, H. R. Bourne, and G. Servant. 2000. Leukocytes navigate by compass: roles of PI3Kgamma and its lipid products. *Trends Cell Biol.* **10:**466–473.

Ridley, A. J. 2004. Rho proteins and cancer. *Breast Cancer Res. Treat.* **84:**13–19.

Ridley, A. J., M. A. Schwartz, K. Burridge, R. A. Firtel, M. H. Ginsberg, G. Borisy, J. T. Parsons, and A. R. Horwitz. 2003. Cell migration: integrating signals from front to back. *Science* **302:**1704–1709.

Riento, K., and A. J. Ridley. 2003. Rocks: multifunctional kinases in cell behaviour. *Nat. Rev. Mol. Cell Biol.* **4:**446–456.

Rohatgi, R., L. Ma, H. Miki, M. Lopez, T. Kirchhausen, T. Takenawa, and M. W. Kirschner. 1999. The interaction between N-WASP and the Arp2/3 complex links Cdc42-dependent signals to actin assembly. *Cell* **97:**221–231.

Rot, A., and U. H. von Andrian. 2004. Chemokines in innate and adaptive host defense: basic chemokinese grammar for immune cells. *Annu. Rev. Immunol.* **22:**891–928.

Rousseau, S., I. Dolado, V. Beardmore, N. Shpiro, R. Marquez, A. R. Nebreda, J. S. Arthur, L. M. Case, M. Tessier-Lavigne, M. Gaestel, A. Cuenda, and P. Cohen. 2006. CXCL12 and C5a trigger cell migration via a PAK1/2-p38α MAPK-MAPKAP-K2-HSP27 pathway. *Cell Signal.* **18:**1897–1905.

Srinivasan, S., F. Wang, S. Glavas, A. Ott, F. Hofmann, K. Aktories, D. Kalman, and H. R. Bourne. 2003. Rac and Cdc42 play distinct roles in regulating PI(3,4,5)P3 and polarity during neutrophil chemotaxis. *J. Cell Biol.* **160:**375–385.

Suetsugu, S., K. Murayama, A. Sakamoto, K. Hanawa-Suetsugu, A. Seto, T. Oikawa, C. Mishima, M. Shirouzu, T. Takenawa, and S. Yokoyama. 2006. The RAC binding domain/IRSp53-MIM homology domain of IRSp53 induces RAC-dependent membrane deformation. *J. Biol. Chem.* **281:**35347–35358.

Suire, S., A. M. Condliffe, G. J. Ferguson, C. D. Ellson, H. Guillou, K. Davidson, H. Welch, J. Coadwell, M. Turner, E. R. Chilvers, P. T. Hawkins, and L. Stephens. 2006. Gβγ and the Ras binding domain of p110γ are both important regulators of PI3K[gamma] signalling in neutrophils. *Nat. Cell Biol.* **8:**1303–1309.

Sun, C. X., G. P. Downey, F. Zhu, A. L. Koh, H. Thang, and M. Glogauer. 2004. Rac1 is the small GTPase responsible for regulating the neutrophil chemotaxis compass. *Blood* **104:**3758–3765.

Symons, M., J. M. Derry, B. Karlak, S. Jiang, V. Lemahieu, F. McCormick, U. Francke, and A. Abo. 1996. Wiskott-Aldrich syndrome protein, a novel effector for the GTPase CDC42Hs, is implicated in actin polymerization. *Cell* **84:**723–734.

Takenawa, T., and S. Suetsugu. 2007. The WASP-WAVE protein network: connecting the membrane to the cytoskeleton. *Nat. Rev. Mol. Cell Biol.* **8:**37–48.

Vanhaesebroeck, B., G. E. Jones, W. E. Allen, D. Zicha, R. Hooshmand-Rad, C. Sawyer, C. Wells, M. D. Waterfield, and A. J. Ridley. 1999. Distinct PI(3)Ks mediate mitogenic signalling and cell migration in macrophages. *Nat. Cell Biol.* **1:**69–71.

Van Keymeulen, A., K. Wong, Z. A. Knight, C. Govaerts, K. M. Hahn, K. M. Shokat, and H. R. Bourne. 2006. To stabilize neutrophil polarity, PIP3 and Cdc42 augment RhoA activity at the back as well as signals at the front. *J. Cell Biol.* **174:**437–445.

Watanabe, N., P. Madaule, T. Reid, T. Ishizaki, G. Watanabe, A. Kakizuka, Y. Saito, K. Nakao, B. M. Jockusch, and S. Narumiya. 1997. p140mDia, a mammalian homolog of Drosophila diaphanous, is a target protein for Rho small GTPase and is a ligand for profilin. *EMBO J.* **16:**3044–3056.

Watanabe, T., J. Noritake, and K. Kaibuchi. 2005. Regulation of microtubules in cell migration. *Trends Cell Biol.* **15:**76–83.

Webb, S. E., J. W. Pollard, and G. E. Jones. 1996. Direct observation and quantification of macrophage chemoattraction to the growth factor CSF-1. *J. Cell Sci.* **109:**793–803.

Weiner, O. D., P. O. Neilsen, G. D. Prestwich, M. W. Kirschner, L. C. Cantley, and H. R. Bourne. 2002. A PtdInsP(3)- and Rho GTPase-mediated positive feedback loop regulates neutrophil polarity. *Nat. Cell Biol.* **4:**509–513.

Weiss-Haljiti, C., C. Pasquali, H. Ji, C. Gillieron, C. Chabert, M. L. Curchod, E. Hirsch, A. J. Ridley, R. H. van Huijsduijnen, M. Camps, and C. Rommel. 2004. Involvement of phosphoinositide 3-kinase gamma, Rac, and PAK signaling in chemokine-induced macrophage migration. *J. Biol. Chem.* **279:**43273–43284.

Welch, H. C., A. M. Condliffe, L. J. Milne, G. J. Ferguson, K. Hill, L. M. Webb, K. Okkenhaug, W. J. Coadwell, S. R., Andrews, M. Thelen, G. E. Jones, P. T. Hawkins, and L. R. Stephens. 2005. P-Rex1 regulates neutrophil function. *Curr. Biol.* **15:**1867–1873.

Welch, M. D., and R. D. Mullins. 2002. Cellular control of actin nucleation. *Annu. Rev. Cell Dev. Biol.* **18:**247–288.

Wells, C. M., P. J. Bhavsar, I. R. Evans, E. Vigorito, M. Turner, V. Tybulewicz, and A. J. Ridley. 2005. Vav1 and Vav2 play different roles in macrophage migration and cytoskeletal organization. *Exp. Cell Res.* **310:**303–310.

Wells, C. M., M. Walmsley, S. Ooi, V. Tybulewicz, and A. J. Ridley. 2004. Rac1-deficient macrophages exhibit defects in cell spreading and membrane ruffling but not migration. *J. Cell Sci.* **117:**1259–1268.

Wennerberg, K., and C. J. Der. 2004. Rho-family GTPases: it's not only Rac and Rho (and I like it). *J. Cell Sci.* **117:**1301–1312.

Wettschureck, N., and S. Offermanns. 2005. Mammalian G proteins and their cell type specific functions. *Physiol. Rev.* **85:**1159–1204.

Wheeler, A. P., and A. J. Ridley. 2007. RhoB affects macrophage adhesion, integrin expression and migration. *Exp. Cell Res.* **313:**3505–3516.

Wheeler, A. P., C. M. Wells, S. D. Smith, F. M. Vega, R. B. Henderson, V. L. Tybulewicz, and A. J. Ridley. 2006. Rac1 and Rac2 regulate macrophage morphology but are not essential for migration. *J. Cell Sci.* **119:**2749–2757.

Wong, K., O. Pertz, K. Hahn, and H. Bourne. 2006. Neutrophil polarization: spatiotemporal dynamics of RhoA activity support a self-organizing mechanism. *Proc. Natl. Acad. Sci. USA* **103:**3639–3644.

Wymann, M. P., and L. Pirola. 1998. Structure and function of phosphoinositide 3-kinases. *Biochim. Biophys. Acta* **1436:**127–150.

Xu, J., A. Van Keymeulen, N. M. Wakida, P. Carlton, M. W. Berns, and H. R. Bourne. 2007. Polarity reveals intrinsic cell chirality. *Proc. Natl. Acad. Sci. USA* **104:**9296–9300.

Xu, J., F. Wang, A. Van Keymeulen, P. Herzmark, A. Straight, K. Kelly, Y. Takuwa, N. Sugimoto, T. Mitchison, and H. R. Bourne. 2003. Divergent signals and cytoskeletal assemblies regulate self-organizing polarity in neutrophils. *Cell* **114:**201–214.

Xu, J., F. Wang, A. Van Keymeulen, M. Rentel, and H. R. Bourne. 2005. Neutrophil microtubules suppress polarity and enhance directional migration. *Proc. Natl. Acad. Sci. USA* **102:**6884–6889.

Yeung, Y. G., and E. R. Stanley. 2003. Proteomic approaches to the analysis of early events in colony-stimulating factor-1 signal transduction. *Mol. Cell Proteomics* **2:**1143–1155.

Yeung, Y. G., Y. Wang, D. B. Einstein, P. S. Lee, and E. R. Stanley. 1998. Colony-stimulating factor-1 stimulates the formation of multimeric cytosolic complexes of signaling proteins and cytoskeletal components in macrophages. *J. Biol. Chem.* **273:**17128–17137.

Zenke, F. T., M. Krendel, C. DerMardirossian, C. C. King, B. P. Bohl, and G. M. Bokoch. 2004. p21-activated kinase 1 phosphorylates and regulates 14-3-3 binding to GEF-H1, a microtubule-localized Rho exchange factor. *J. Biol. Chem.* **279:**18392–18400.

Zhan, Q., J. R. Bamburg, and J. A. Badwey. 2003. Products of phosphoinositide specific phospholipase C can trigger dephosphorylation of cofilin in chemoattractant stimulated neutrophils. *Cell Motil. Cytoskeleton* **54:**1-15.

Zhao, T., P. Nalbant, M. Hoshino, X. Dong, D. Wu, and G. M. Bokoch. 2007. Signaling requirements for translocation of P-Rex1, a key Rac2 exchange factor involved in chemoattractant-stimulated human neutrophil function. *J. Leukoc. Biol.* **81:**1127–1136.

Zlotnik, A., and O. Yoshie. 2000. Chemokines: a new classification system and their role in immunity. *Immunity* **12:**121–127.

PHAGOCYTOSIS: SIGNALING, CYTOSKELETON, AND THE PHAGOSOME

III

12

Signaling for Phagocytosis

JOEL A. SWANSON

Phagocytes ingest a variety of particles, cells, and microbes by use of actin-rich, contractile extensions of plasma membrane that close into intracellular membranous organelles called phagosomes. Signaling for phagocytosis begins when cell surface receptors engage particle-bound molecular ligands. Engaged receptors are modified by phosphorylation and ubiquitination, and recruit proteins into molecular signaling complexes. Recruited enzymes modify lipids and proteins that regulate actin polymerization, myosin contractility, membrane fusion, and sometimes also the synthesis of reactive oxygen species (ROS). The distinct component activities localize to subregions of forming phagosomes and may be coordinated by membrane-associated proteins and phospholipids that diffuse in the plane of the membrane bilayer. Different receptors use distinct mechanisms for regulating the actin cytoskeleton and the movements of membrane, producing a variety of morphologies for forming phagosomes. Signaling for phagocytosis is regulated at the molecular, cellular, and tissue levels by medium- and long-range feedback mechanisms that activate or suppress the activities of ligated receptors.

THE NATURE OF PHAGOCYTOSIS

Phagocytosis, the ingestion process in which a cell envelops and engulfs particles or other cells, was first observed and named by Elie Metchnikoff in his 19th-century studies of invertebrate development. He proposed that tissues and organs were shaped in part by phagocytic embryonic cells, which consumed each other in a morphogenetic competition. He later proposed that phagocytic cells, named phagocytes, were essential to host defense against microorganisms, an insight that marked the beginning of the field of immunology. The two activities of phagocytes discovered by Metchnikoff—shaping tissues and defending against infection—are still considered chief functions of phagocytosis in animals, although single-celled protists evidently use phagocytosis for sustenance. Indeed, phagocytosis is an important element of host defense and tissue maintenance.

The phagocytes emphasized here include macrophages, neutrophils, and dendritic cells. However, other cells of animal tissues, especially epithelial cells, may be induced to ingest particles or other cells, either as elements of tissue homeostasis or as targets of microbial infection.

The essential movement of phagocytosis is the wrapping of an extracellular particle with deformable extensions of the phagocyte's plasma membrane. Cup-shaped folds of the cell surface extend over the particle, then constrict at their distal margin to enclose it, such that the membrane of the inner surface of the cup becomes a plasma membrane-derived, discrete intracellular compartment containing the particle. The process is initiated by cell surface receptors that engage specific molecular ligands on the particle surface. Those ligands may be intrinsic carbohydrates, lipids, or proteins of the particle surface, or exogenous molecules—opsonins—that bind to the particle surface and render it stimulatory for phagocytosis (opsonization). Ligand binding brings receptors into clusters that then organize the movements of plasma membrane and the underlying actin cytoskeleton, as well as the fusion of intracellular vesicles with plasma membrane. Some phagocytic receptors also initiate distinct signals during phagocytosis that activate membrane protein complexes capable of delivering microbicidal ROS into the forming or nascent phagosome. Receptors may also direct the subsequent interactions of the phagosome with other membranous organelles of the phagocyte cytoplasm (Color Plate 3).

Within minutes of its formation as a new organelle inside the cell, the phagosome's content and membrane composition change by a process called maturation. The nascent phagosomal membrane looks mostly like a sample of plasma membrane, except that the phagocytic receptors are selectively concentrated in the forming phagosome. Phagocytosis recruits lipid-modifying enzymes and regulatory proteins that in turn stimulate fusion of phagosomes with other membranous organelles, as well as the selective removal of proteins and lipids. Together, these activities remodel the phagosomal membrane to resemble early endosomes and, shortly afterward, late endosomes. The luminal contents of the phagosome are altered by this differentiation; most notably, the phagosome acidifies as it matures.

Joel A. Swanson, Department of Microbiology and Immunology, University of Michigan Medical School, Ann Arbor, MI 48109-0620.

Eventually the remodeled phagosome fuses with lysosomes, acidic membranous organelles that contain hydrolytic enzymes, forming a hybrid organelle called the phagolysosome. The phagolysosome's luminal environment of pH 4.5 to 5.0 and concentrated degradative enzymes can kill and digest most organic material delivered into it by phagocytosis.

Beyond providing a simple mechanism for disposing of unnecessary or harmful debris, phagocytosis informs the phagocyte and its neighbors about the context of the encounter. If the ingested particle is foreign, that is, not of the host animal, or is a host cell that died by frank mechanical or chemical damage, then its phagocytosis initiates signals leading to differentiation of the phagocyte and the secretion of inflammatory signaling molecules. Alternatively, if the particle is a tissue cell that has died by a suicidal programmed cell death called apoptosis, then its ingestion by a phagocyte is noninflammatory and may even generate signals that suppress inflammation or specific immune responses. The phagocyte's ability to read context in this way allows it to influence subsequent differentiation responses of neighboring tissue cells, including lymphocytes and other phagocytes. Phagocytosis itself can be inhibited by diffusible molecules released from nearby phagocytes that have ingested particles. Thus, the condition of a tissue is often regulated by phagocytosis, through a network of signals emanating from phagocytes and from the cells attracted, repelled, or otherwise stimulated by those phagocytes.

The sizes of particles ingested by phagocytosis vary. It is frequently suggested that phagocytosis applies to the ingestion of particles larger than 0.2 μm, and that smaller particles such as viruses enter through some other means. Although the mechanisms of uptake for smaller particles seem to be distinct from those for phagocytosis, too little is known about the size dependence of any endocytic mechanism to define a size threshold for phagocytosis. Some phagocytes can ingest very large particles. Macrophages, which when not spread on a surface form rounded cells 12 to 14 μm in diameter, can ingest immunoglobulin G (IgG)-opsonized latex spheres as large as 20 μm in diameter. Macrophages or neutrophils plated onto IgG-coated coverslips engage the surface in a frustrated phagocytic response, as though receptor signaling is hard-wired and indifferent to the scale of the task. However, recent studies indicate that macrophages can limit their phagocytic responses based on the size, shape, and stiffness of the particles they encounter.

The signaling associated with phagocytosis involves two categories of information: constructive and interpretive. Constructive signals are those which initiate or coordinate the movements that build the phagosome, such as the signals that govern the movements of cytoplasm and membranes during Fc receptor (FcR)-mediated phagocytosis. Interpretive signals are those associated with analysis of the phagosomal contents. Phagocytes contain many receptors on their cell surfaces, on intracellular vesicles, and in their cytoplasm that cannot organize the assembly of a phagosome but instead generate signals reporting the molecular makeup of the particle. For example, Toll-like receptors (TLRs) reside on plasma membranes and intracellular endocytic compartments and, when activated by their cognate ligands, signal a number of responses related to inflammation and immunity. TLR ligands do not stimulate phagocytosis themselves; that is, particles coated with only TLR ligands are not internalized. Instead, the TLR-generated signals lead to the activation of NF-κB and other regulators of transcription. Some elements of phagocytic signaling are both constructive and interpretive. This chapter emphasizes the constructive signals.

THE CONSTRUCTION OF PHAGOSOMES

The building of a phagosome begins with binding of the phagocyte cell surface receptors to molecular ligands on the particle surface. This interaction induces conformational changes of the receptors, often leading to multimerization of receptors in the plane of the membrane, which alter the cytoplasmic domains of the receptor. These alterations increase affinities of those domains for cytoplasmic proteins, which then bind to the receptor and to each other in a sequence that quickly assembles into a molecular complex. Proteins of receptor-associated signaling complexes include enzymes that modify lipids or other proteins, as well as proteins that facilitate the docking of such enzymes. The enzymes of the receptor complexes then generate additional signals that stimulate or inhibit phagocytosis by regulating the movements of membranes and the actin cytoskeleton, or by generating other signals that modify the overall responsiveness of the cell.

Receptors and Associated Molecules

The plasma membranes of phagocytes contain a large variety of receptors, a subset of which can mediate phagocytosis. The phagocytic receptors that can bind particles and signal their internalization are transmembrane proteins whose cytoplasmic domains are chemically modified after activation, or associate with other proteins which are themselves modified by activation. FcRs, the best-characterized receptors in this category, recognize the Fc region of IgG, IgA, IgM, or IgD. Other receptors, such as the receptors CR3 and CR4, can bind particles opsonized with the complement protein C3bi, but must be activated for phagocytosis. Some receptors simply serve to link particles to phagocytes, leaving the mechanics to other regulatory molecules, while others can bind and internalize soluble ligands but not particles. Finally, some receptors stimulate cell surface movements that internalize extracellular solutes and fluid in large vesicular organelles called macropinosomes. Phagocytosis of necrotic cells and some bacteria occurs by stimulation of macropinocytosis and entry into cells via macropinosome-like spacious phagosomes.

The receptors for phagocytosis organize local cytoplasmic responses to particle surfaces. Cytoplasmic tails of the receptors, or their associated proteins, are modified by binding to particle ligands such that their affinities for cytoplasmic signaling proteins increase. The cytosolic signaling proteins bind to the receptors or associated proteins, and a complex aggregate assembles around the receptor by diffusion and trapping. This signaling complex then activates secondary signals that radiate from the receptor. Secondary signals include the enzymatic modification of phospholipids and other membrane-associated proteins that in turn organize the cytoplasmic movements of phagocytosis. Thus, the activated phagocytosis receptor generates a signal that radiates some distance from the point of interaction with the particle-bound ligand. Repeated interactions between receptors and their cognate ligands on a particle surface can generate a phagocytic cup by a kind of crawling over the surface, with each ligated receptor activating a small patch of the cell surface for the phagocytic response (Color Plate 3). This local or segmental response was identified experimentally by Samuel Silverstein and colleagues, who named it the zipper model, based on the analogous movements of a zipper. Although the zipper model cannot describe all kinds of phagocytosis, it provides a useful paradigm for analysis of the process.

Many different kinds of particle have been used to study phagocytosis, including polystyrene microspheres, bacteria, erythrocytes, lymphocytes, and yeast. One particle often used in studies of phagocytosis is zymosan, which consists of the cell walls of the yeast *Saccharomyces cerevisiae*. Some studies also use zymosan opsonized with IgG or complement. Although often revealing about the mechanism of phagocytosis, zymosan is not ideal for studies of signaling by specific receptors, because it contains on its surface several different molecular species that can signal for phagocytosis. IgG-opsonized zymosan can stimulate FcR-mediated phagocytosis in phagocytes containing FcRs, but will also elicit signals through other phagocytic receptors, such as dectin-1 and the mannose receptor. A better particle for studying receptor-specific signals for phagocytosis is one that is neither bound nor internalized in the absence of the opsonin specific for that receptor. Erythrocytes can serve well in this regard. In the absence of added ligands, they do not bind to macrophages or neutrophils. Addition of IgG molecules directed against erythrocyte antigens opsonizes them for FcR-mediated phagocytosis. Alternatively, they may be coated with IgM and complement to form C3bi-opsonized particles which, if carefully prepared to exclude IgG, should signal phagocytosis entirely through CR3.

The various receptors may all construct phagosomes differently. However, phagocytes surely encounter particles like zymosan, presenting a mixture of ligands. Many studies that use bacteria or apoptotic cells as phagocytic targets are less concerned with how individual receptors signal than with the signals downstream of receptors that mediate phagocytosis of those target cells. Little is known of how mixed signals from different receptors are coordinated to make a phagosome or how cells distinguish signals for adhesive migration from those for phagocytosis. It seems likely that receptors for phagocytosis generate some essential signals that identify particles for ingestion. These signals should be present in all kinds of phagocytosis. They may also be co-opted by microbial pathogens that enter nonphagocytic host cells by induced phagocytosis. Very often, these microbes stimulate their entry into cells by activating receptors that do not normally mediate phagocytosis but which instead stimulate growth or migration.

Actin

β-Actin is an abundant protein of the eukaryotic cytoskeleton that is capable of self-assembly into helical polymeric filaments. Actin filaments can be linked to other actin filaments by cross-linking proteins to form rigid gel-like structures near cell surfaces. They may also be linked to other kinds of filaments, such as microtubules or intermediate filaments, or to membrane proteins, anchoring them to structures and integrating the structure of cytoplasm. The anchoring of actin filaments to organelles or to each other is subject to regulation, such that actin gels may transform into more fluid structures (solation) that facilitate motility. Actin filaments also serve as tracks upon which the contractile proteins of the myosin family move, transporting organelles or other actin filaments. Actin-myosin contractility allows muscle cells to contract and nonmuscle cells, like phagocytes, to migrate through tissues and to ingest particles by phagocytosis.

In actively moving cells, actin polymers are continually assembling and disassembling near cell surfaces. Actin binds ATP, and hydrolysis of ATP by actin is coupled to actin filament assembly. This gives the actin filament a dynamic polarity, such that actin monomer addition occurs preferentially at one end, the so-called plus end (a.k.a. barbed end) of the filament. This polarity also renders the assembly and disassembly of actin polymers subject to regulation. Thus, polymerization can be regulated by proteins that bind actin monomers, proteins that bind the side of filaments, or proteins that bind preferentially to the plus or minus (a.k.a. pointed) ends of filaments. In general, regulatory processes that expose actin filament plus ends stimulate actin polymer assembly at those locations, but polymerization may also be stimulated by factors that increase actin filament turnover.

Actin filament dynamics are essential to phagocytosis. Drugs that depolymerize actin, such as latrunculin, or that inhibit actin filament growth, such as cytochalasin D, inhibit the very first movements of phagocytosis. Actin filaments can be visualized by labeling fixed cells with fluorescently labeled phalloidin, which binds to the sides of the polymer. Staining of phagocytes with fluorescent phalloidin reveals that actin filaments are present in the earliest deformations of the cell surface for phagocytosis. This suggests that the first extension of the cell around a particle during FcR-mediated phagocytosis is driven by plus-end-directed growth of actin filaments from an actin meshwork located beneath the particle-adherent plasma membrane. Actin can be visualized in living cells by fluorescence microscopy of cells injected with fluorescently labeled actin, or in cells expressing fluorescent protein chimeras of green fluorescent protein and actin. The fluorescent actin in these cells appears to advance over particles during phagocytosis, disappearing from the base of the cup even before the particle is completely enclosed into the cell. Although not yet established experimentally, this gives the impression that continued extension of a phagocytic cup requires continuous addition of actin monomers to filament plus ends at the distal margin of the cup and disassembly of actin filaments at the base of the phagocytic cup.

Proteins that regulate actin polymerization and that are also necessary for phagocytosis include Arp2/3, which binds to actin filaments and nucleates new plus ends as filament branch points; mammalian Diaphanous-related formin (mDia), which stimulates polymerization of actin filaments at plus ends; WASP and WAVE, which locally activate Arp2/3; and coronin, which stimulates actin polymerization by binding and modulating the activities of Arp2/3. The protein cofilin, which is necessary for some kinds of phagocytosis, has been shown in nonphagocytes to stimulate actin filament turnover by increasing the offrate of monomers from the minus ends of actin filaments. This indicates that actin filament disassembly is necessary for sustained extension of the phagocytic cups.

Actin filament dynamics may be different for different receptor-mediated phagocytic processes. For example, CR3-mediated phagocytosis organizes the actin cytoskeleton differently than FcR-mediated phagocytosis. Adherent, C3bi-opsonized particles sink into macrophage cytoplasm, rather than eliciting an extended phagocytic cup (Fig. 1). The underlying mechanism of phagocytosis appears to depend less on localized extension of the plasma membrane and more on controlled invagination of the cell surface by actin-myosin contractile activities. Other kinds of phagosome morphologies exist which indicate different patterns

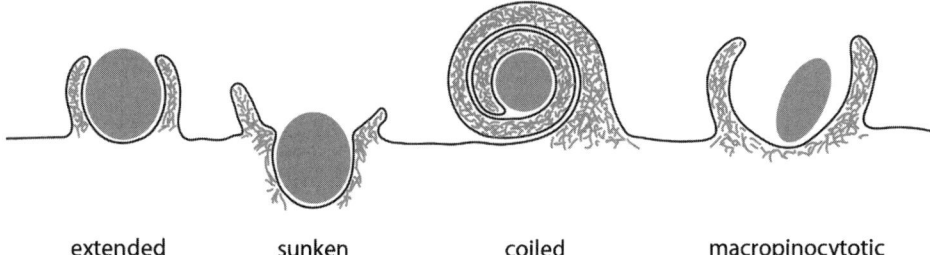

FIGURE 1 Different morphologies of phagosome formation. Phagocyte membranes (black lines) and actin filaments (gray lines) envelop particles (gray ovals) in distinct patterns. (Extended) Typical of Fc receptor-mediated phagocytosis, actin-rich pseudopodia form a closely adherent, cup-shaped extension around a particle surface. (Sunken) Particles opsonized with C3bi form cytoplasmic indentations of the cell surface as they are drawn into the cell. Actin is organized into discrete, punctate structures, possibly at the base of actin-rich cell protrusions called ruffles. (Coiled) The bacterium *Legionella pneumophila* is sometimes internalized into coil-shaped phagosomes which form by sheet-like pseudopod that rolls up around the bacterium. The mechanism by which these structures close into phagosomes is not known. (Macropinocytotic) Particles stimulate ruffle formation in nearby regions of the cell surface. They are then enclosed into loosely adherent, spacious phagosomes that resemble macropinosomes.

of actin dynamics. Some bacteria are internalized into coil-shaped phagosomes; others enter through loose-fitting, spacious phagosomes.

Myosin

The morphogenesis of phagosomes suggests that actin-myosin-like contractility is essential to phagocytosis. Particles are frequently pulled into phagocytes as they are ingested, as in the sinking movements of cytoplasm during CR3-mediated phagocytosis. Phagosomes that form by extension of a phagocytic cup close by constriction of their distal margins, suggestive of a purse-string-like contractile closure mechanism. Several experimental approaches indicate the presence of contractile activities. Direct measurements of internal pressures generated by neutrophils during phagocytosis indicated that extension of cytoplasm over particles is followed by cytoplasmic contractions in which the cytoplasm of the phagocytic cup compresses against the particle. Addition of IgG-opsonized sheep erythrocytes to cultures of macrophages crowded close together on coverslips produces frequent examples of two cells competing for the same particle. In these instances, two unrelenting macrophages each take in half of the erythrocyte by a kind of biting action that pinches it and pulls it into a dumbbell shape. These motions indicate two clear contractile activities: circumferential constriction at the distal cytoplasm of the cup and centripetal pulling of the phagocytosed fragment into the cell.

Myosin is a mechanochemical enzyme that links ATP hydrolysis with its displacement along an actin filament. The many isoforms of myosin each have distinct biophysical properties and cellular activities. The most abundant form of myosin in cells, called myosin II, is structurally similar to the myosin that powers muscle contraction. It self-assembles into short fibrous macromolecular structures with motor domains at each end. The activities of the opposite-facing motors allow the myosin molecule to power movement in opposite directions along actin filaments, consequently contracting an actin network. Other classes of myosin do not self-associate into fibers, but instead contain a motor domain at one end and an organelle-binding domain at the other, rendering them capable of transporting organelles or membranes along actin fibers.

Several classes of myosin localize to forming phagosomes, including myosins Ic, II, IX, and X. Myosins II and IX could constrict the actin meshwork of the phagocytic cup. Myosins Ic and X can associate with membranes; they may orient contractile activities for phagosome closure. Myosin II is regulated by phosphorylation of associated light chains. The enzyme myosin light chain kinase (MLCK), which is capable of activating myosin II, has been shown to participate in phagocytosis, as have several other regulatory proteins that modulate myosin contractility (RhoA, ROCK). When phagocytosis is arrested by inhibitors of myosin, macrophages make phagosomes that loosely surround the particle and that fail to constrict.

Membrane Movements

The membrane used to construct a phagosome derives partly from the plasma membrane and partly from membranes of intracellular organelles. If externally exposed molecules of macrophage plasma membranes are chemically labeled before phagocytosis, many of those labeled molecules appear in the membranes of phagosomes made by those cells, demonstrating that plasma membrane-derived molecules are incorporated into forming phagosomes. Several lines of evidence indicate that phagocytosis also entails addition of membrane from intracellular sources. First, despite the fact that phagocytosis removes plasma membrane into the cell, the total surface area of the phagocyte increases during phagocytosis. Second, membrane proteins of intracellular organelles localize to forming phagocytic cups. Third, some proteins that regulate membrane fusion localize to forming phagosomes, and interference with their function can inhibit phagocytosis.

Many different categories of intracellular compartment deliver membrane to forming phagosomes, including recycling endosomes, secretory granules, endoplasmic reticulum (ER), late endosomes, and lysosomes. Some of these are membranous compartments that typically fuse with fully formed phagosomes as part of their maturation. Their delivery to forming phagosomes may be either a rare event or one that can be induced by distinct signaling activities.

The most widely acknowledged intracellular source of phagosomal membrane in macrophages is the recycling endosome, a compartment that can be accessed (and therefore labeled) by endocytosis of soluble proteins or fluorescent probes. Recycling endosomes contain distinct markers that allow their microscopic association with forming phagosomes, including the GTPase Rab11 and the SNARE (soluble, N-ethylmalemide-sensitive factor-associated receptor) membrane fusion protein VAMP3. These proteins have been localized to forming phagosomes containing IgG-opsonized sheep erythrocytes, polystyrene particles, and zymosan. Interference with VAMP3 function inhibits phagocytosis. Secretion of tumor necrosis factor-α (TNF-α) is stimulated by phagocytosis in macrophages, and this secretion occurs by fusion of recycling endosomes with forming phagocytic cups.

Phagocytosis in neutrophils is accompanied by fusion of secretory granules with phagosomes. These granules deliver enzymes into the phagosome lumen which facilitate killing or degradation of ingested microbes. The membranes of these granules also contain proteins of the phagocyte NADPH oxidase (NOX2), which synthesizes ROS inside the phagosome. Granule fusion likely also provides membrane to the forming phagocytic cup.

The ER is an extensive, tubuloreticular organelle that can provide membrane to forming phagosomes, although the extent of its contribution is uncertain. Isolated phagosomes containing polystyrene microspheres contain many ER proteins, indicating that ER membrane comprises some of those phagosomal membranes. Electron microscopic histochemical methods show ER-specific enzyme activities in phagosomes, and other electron microscopic images suggest direct merger of ER with phagocytic cups. Particle phagocytosis can be inhibited by interference with the fusion proteins ERS24/Sec22b, which mediates fusion between ER and plasma membrane, or syntaxin-18, an ER-localized SNARE. However, studies that quantified the contributions of plasma membrane and endocytic compartments to phagosome formation during FcR-mediated phagocytosis indicated that the quantities of ER membrane in phagosomes must be negligible. The discrepant results are presently not easy to reconcile. They may be explained by differences in the extent which ER contributes membrane to different kinds of phagocytosis or different phagocytic cells. Accordingly, ER would contribute less membrane during FcR-mediated phagocytosis than during phagocytosis of microbes or polystyrene beads, whose uptake is mediated by different receptors. Also, ER fusion with phagosomes may typically occur after phagocytosis, either as part of phagosome maturation or through a distinct trafficking activity called autophagy, in which ER or ER-like membranes engulf damaged organelles inside cytoplasm. Finally, it is possible that ER membranes are called into action for large phagocytic loads after other membrane compartments have been depleted.

Although late endosomes and lysosomes typically fuse with phagosomes well after phagocytosis, they can also provide membranes during phagosome formation. Lysosomal enzymes are often secreted from cells during phagocytosis, indicating early fusion of lysosomes with phagosomes or plasma membrane. Markers of late endosomes, such as lysosome-associated membrane protein-1 (LAMP-1), can be delivered to unclosed phagocytic cups, and may appear on plasma membranes during phagocytosis. The late endosomal fusion protein VAMP7 (TI-VAMP) localizes to phagocytic cups, and interference with VAMP7 expression or activity inhibits FcR-mediated phagocytosis. Finally, phagocytosis of IgG-opsonized zymosan can be inhibited by interference with the activity of a calcium-regulated membrane protein, synaptotagmin VII, which is implicated in fusion between lysosomes and plasma membrane.

Some new membrane from intracellular compartments is inserted directly into the forming phagocytic cup. Markers of intracellular vesicles and secreted proteins localize to phagocytic cups. Plasma membrane molecules caught in forming phagosomes quickly become less concentrated in that membrane, suggesting that intracellular organelles fuse directly into the inner membranes of the phagocytic cups. However, the dense actin meshwork beneath the phagocytic cup would seem to form a barrier against fusion of intracellular organelles with plasma membrane. Perhaps localized membrane fusion requires disassembly or displacement of actin filaments from the base of the cup. It is also still possible that some compensatory membrane fusion that accompanies phagocytosis occurs at regions of plasma membrane outside of the forming phagosome.

Phagocytosis is one variant of a more general process called endocytosis, in which plasma membrane invaginates to form intracellular vesicles. Other kinds of endocytosis include macropinocytosis, which internalizes extracellular fluid and solutes into 0.2- to 5.0-μm vesicles called macropinosomes, and micropinocytosis, which makes smaller vesicles including clathrin-coated vesicles and caveolae. The mechanism of macropinosome formation appears similar in many ways to phagosome formation: cup-shaped, actin-rich extensions of plasma membrane constrict at their distal margin and pinch into vesicles in cytoplasm. However, although macropinocytosis is often triggered by receptor-mediated signaling, the macropinosome is not shaped by an opsonized surface. The ability of cells to construct large endocytic vesicles without particle surfaces to shape them suggests that macropinocytosis and some or all forms of phagocytosis require spatial coordination of signaling over large regions of cytoplasm.

Clathrin-coated vesicles typically internalize receptor-bound soluble ligands, such as transferrin, in a process called receptor-mediated endocytosis. Although clathrin-coated vesicles sometimes internalize virus particles, they are not considered phagosomes. This may be because, at 75 to 120 nm (0.12 μm) diameter, they are too small for the conventional definition of a phagosome. Clathrin is a large protein that self-associates into three-legged, propeller-shaped macromolecular complexes. Coated vesicle formation entails the assembly of clathrin and associated proteins into a multimeric, cage-like structure on membranes, which quickly curves to form a spherical invagination of clathrin-coated membrane. Clathrin-coated invaginations separate into cytoplasm by the action of dynamin-1, a large GTPase that cuts the narrow tubular stalk of membrane connecting the plasma membrane and the nascent coated vesicle. Clathrin-coated patches of membrane have been observed on forming phagosomes, prompting suggestions that clathrin contributes to phagosome formation. However, although interference with dynamin-1 or clathrin inhibits the FcR-mediated endocytosis of soluble immune complexes, it does not inhibit phagocytosis of IgG-opsonized, 0.8- to 3.0-μm-diameter microspheres. Two proteins that contribute to coated-vesicle formation, dynamin-2 and amphiphysin 2, are also necessary for phagocytosis, but their contributions remain mysterious. Dominant inhibitory versions of these proteins interfere with early stages of phagosome formation, indicating possible roles in events preceding phagosome closure. Studies in the nematode Caenorhabditis elegans implicated dynamin-1 in the phagocytic engulfment of apoptotic cells. Morphological and genetic analyses suggested that dynamin-1 mediates fusion of endocytic vesicles with forming phagosomes. Thus, the contributions of clathrin-related molecules to phagocytosis appear unrelated to their roles in coated-vesicle formation.

The mechanics of phagosome or macropinosome separation into cytoplasm from plasma membrane remain largely unexplained. The process must entail fusion of plasma membrane at a small annulus, beginning with merger of the outer leaflet of the bilayer, followed by fission of the membrane connecting the phagosome to the plasma membrane. However, none of these steps have been visualized at any resolution that reveals mechanism. Contractile activities have been associated with phagosome closure, but no molecules have been directly implicated.

REGULATORY CHEMISTRIES

What signals emanating from ligated receptors elicit the mechanical processes outlined above? The signals of phagocytosis can be organized for descriptive purposes according to their proximity to the activated receptors. Thus, short-range signals are the molecules that associate directly with activated receptors, middle-range signals are the molecules generated locally by activated receptors that are not directly bound to those receptors, and long-range signals are those which modify cellular responsiveness or the overall signaling potential of the receptors (Color Plate 4). The construction of the phagosome therefore follows from many molecular alterations on or near activated receptors, or radiating various distances from those receptors, and which may consequently regulate the information transmitted by those or other receptors. Such feedback regulation could coordinate multiple receptors for the ingestion of single particles, help to integrate signals from different classes of receptors, or sate the cell's appetite for more particles.

Before considering which features of signaling are common and essential for phagocytosis, the signals for various different receptor-mediated phagocytic processes will be reviewed.

FcR-Mediated Phagocytosis

Tyrosine kinase receptors such as FcR generally contain an extended extracellular ligand-binding domain, a transmembrane domain, and a cytoplasmic domain, the amino acids of which can be modified by phosphorylation or ubiquitinylation. They are called tyrosine kinase receptors because their cytoplasmic domains, or proteins closely associated with those domains, recruit tyrosine-specific protein kinases whose activities are essential to the signaling that follows receptor-ligand interaction. These receptors also contain tyrosine residues that are substrates of tyrosine kinases and that, as phosphotyrosines, recruit Src-homology 2 (SH2) domain-containing proteins. The amino acids that flank those tyrosines determine the nature of the signal that follows. Receptors with one sequence motif, called the immunoreceptor tyrosine-based activation motif or ITAM, generate stimulatory signals for phagocytosis.

Fc receptors provide the most completely understood examples of signaling for phagocytosis by tyrosine kinase receptors. The components of short-range signaling include the receptor and its associated membrane lipids, the cytoplasmic or membrane proteins that bind to the receptor, and the posttranslational modifications of the entire receptor complex that elicit further signaling. A variety of different FcRs are expressed on mammalian phagocytes. They share a general structure and the ability to recognize Fc domains of immunoglobulin molecules. However, different classes of FcR bind to different classes of immunoglobulin, exhibit different affinities for Fc regions of immunoglobulins, and may stimulate or inhibit phagocytosis. In unstimulated phagocytes, FcRs are uniformly distributed as single-transmembrane proteins, sometimes with associated cytoplasmic γ-chains (depending on the receptor). Signaling begins when FcRs are clustered in the plasma membrane after binding to immune complexes or to particles opsonized with immunoglobulins. The clustered receptors become targets for phosphorylation of ITAMs by Src-family kinases, which include Lyn, Hck, and Fgr in macrophages. Src kinase-mediated phosphorylation of the ITAM region increases the FcR's affinity for the tyrosine kinase Syk, which then phosphorylates the adapter protein Gab2, which then binds the p85 regulatory subunit of the lipid-modifying enzyme phosphatidylinositol 3′-kinase (PI3K) and the adapter protein phosphatase SHP-2. Gab2 association with FcR increases PI3K signaling in a positive feedback loop. Other receptor-proximal signals include modification of the receptor by ubiquitinylation, which promotes endocytosis of soluble immune complexes and the eventual degradation of FcRs in late endosomes or lysosomes.

The lipid bilayer near ligated FcRs provides a structural platform for short-range signaling. Plasma membranes contain small domains enriched in cholesterol or ceramide-1-phosphate, which locally increase the rigidity of the bilayer relative to the more abundant cholesterol-poor membrane. Ligated FcRs become enriched in these so-called lipid rafts, and early FcR signaling is inhibited by experimental depletion of cholesterol from membranes. In Fcα receptor signaling, several essential signaling molecules concentrate in lipid rafts after receptor ligation, including Lyn kinase, Bruton's tyrosine kinase (Btk), Syk, PI3K, and protein kinase C (PKC). Although some of these molecules may associate with lipid rafts indirectly by binding to receptors, their limited mobility in these microdomains could enhance short-range interactions between these enzymes or with their substrates.

The enzymes recruited to the FcRs generate signals that radiate by diffusion away from the activated receptors, effectively expanding the range of the signal. Locally produced phospholipids or water-soluble molecules recruit or activate other enzymes that organize the movements of phagocytosis. In effect, an activated receptor generates a small field of activated membrane (Color Plate 4). However, signals radiating from a single receptor may be insufficient to trigger the next stages of signaling, because diffusion should quickly reduce the local concentrations of the secondary signals. The magnitude of these diffusible signals should therefore depend on the density or arrangement of ligated receptors in a region of membrane, potentially creating thresholds for receptor signaling or context-dependent feedback controls of phagocytosis.

An example of this short- to middle-range signaling is PI3K. The regulatory subunit of type I PI3K, p85, binds to phosphorylated Gab2 and is phosphorylated by FcR-bound Syk. Phospho-p85 then activates the associated p110 catalytic subunit of PI3K, which phosphorylates the membrane phospholipid phosphatidylinositol 4,5-bisphosphate [PtdIns(4,5)P_2] at the 3' hydroxyl of the inositol head group, generating phosphatidylinositol 3,4,5-trisphosphate [PtdIns(3,4,5)P_3]. Many cytoplasmic signaling enzymes and GTP-exchange proteins contain protein domains, called pleckstrin homology or PH domains, that bind to the phosphoinositol head groups of PtdIns(3,4,5)P_3 or other phosphoinositides on the inner leaflet of the membrane. Interaction between PH domains and phosphoinositides recruits proteins to membranes or activates PH-domain-containing proteins already recruited by other proteins. Thus, local activation of PI3K on ligated FcRs generates PtdIns(3,4,5)P_3, which then recruits or activates PH-domain-containing proteins essential for phagocytosis of larger particles. If too few Fc receptors are activated, then concentrations of PtdIns(3,4,5)P_3 in membranes may not be high enough to activate signaling proteins. But, if many receptors are activated, the PtdIns(3,4,5)P_3 they generate may reach suprathreshold levels in membranes that allow 3'-phosphoinositide-dependent stages of signaling to follow.

Members of the Ras superfamily of small GTPases are critical for translating signals from activated receptors into the mechanical movements of phagocytosis. For FcR-mediated phagocytosis, these include the Rho-family proteins Rac1, Rac2, and Cdc42, which primarily regulate cytoskeletal function, and the ADP ribosylation factors ARF1 and ARF6, which regulate membrane fusion and the actin cytoskeleton. These GTPases share essential features of structure and regulation. They function primarily by regulated association with membranes and alternation between an active, GTP-bound state and an inactive, GDP-bound state. Posttranslational modifications add acyl chains to the proteins, which facilitate binding and association with membranes. Exposure of the acyl chains for association with membranes is regulated by guanosine dissociation inhibitors (GDIs), soluble proteins that bind to GTPases and repress their activities by inhibiting GTP exchange and GTPase association with membranes. Rho-family proteins contain covalently bound acyl groups at their C termini; binding to RhoGDI sequesters those groups and effectively prevents GTPase association with membranes. Once associated with membranes, the GTPases can be converted to their active, GTP-bound form by guanine nucleotide exchange factors (GEFs), which facilitate the dissociation of bound GDP and the association of the more abundant GTP. The activated GTPases bind to various effector proteins, typically enzymes, that in turn regulate changes in membrane trafficking or the cytoskeleton. The GTPases are inactivated by GTPase-activating proteins (GAPs), which stimulate hydrolysis of bound GTP and the consequent conversion of the GTPase to its inactive GDP-bound form. The small GTPases act as regulatory switches that translate signals from receptors into movements of cytoplasm. Effector activities are increased by conditions that increase levels of GTP-bound GTPases, and this can be achieved by stimulating GEF activity or inhibiting GDI or GAP activities. GEFs and GAPs can be regulated by phosphorylation and dephosphorylation, or by phospholipid binding to regulatory domains. Thus, the overall state of GTPase signaling in any subregion of the cell is governed by local concentrations of the kinases, phosphatases, or phosphoinositides that regulate GEFs and GAPs.

Several approaches have been used to implicate GTPases in signaling for phagocytosis. Some bacterial toxins inactivate particular subsets of GTPases, and they can be delivered into cells to identify processes dependent on the actions of their target GTPases. Overexpression of dominant-negative mutant GTPases can sequester GEFs and inhibit the actions of endogenous GTPases that are activated by the same GEFs. Overexpression of GDI can inactivate GTPases by sequestration from membranes. Overexpression of specific GAPs can inactivate their target GTPases. Finally, GTPases, GEFs, and GAPs can be depleted from cells by genetic deletion, by interfering with their expression by using inhibitory RNAs or by direct immunoprecipitation with microinjected antibodies. By these methods, the Rho-family GTPases Rac1 and Cdc42 have been shown to be necessary for FcR-mediated phagocytosis in macrophages. Similar studies of RhoA have been equivocal; some macrophages do not require RhoA for FcR-mediated phagocytosis, others do.

Several classes of small GTPase have been implicated in FcR-mediated phagocytosis, each controlling different effector functions. Cdc42 can activate p21-activated kinase-1 (PAK1) and WASP, which enhances Arp2/3 nucleation of actin filaments. Rac1 can activate WAVE, which similarly increases actin polymerization via Arp2/3. Other Rac1 effectors include PAK1 and LIM kinase, which activates cofilin and inhibits MLCK. Rac2 regulates activation of the NADPH oxidase NOX2. Two ARF-family proteins, ARF1 and ARF6, have been implicated in FcR-mediated phagocytosis. ARF1 may activate PKC or regulate membrane trafficking into or away from the phagosome. ARF6 affects both actin cytoskeleton dynamics and organelle trafficking near plasma membrane. Rab GTPases regulate docking and fusion of membranous organelles. Rab proteins localize to forming phagosomes, but none have been shown to be essential for FcR-mediated phagocytosis.

A number of lipases, lipid kinases, and lipid phosphatases are necessary for FcR-mediated phagocytosis. Many of these enzymes localize to forming phagosomes, although most are not directly recruited to receptor complexes. The products of these enzymes regulate other lipid kinases, protein kinases, or regulatory GTPases that in turn regulate actin filament dynamics, myosin contractile activities, or membrane trafficking. Phospholipid-hydrolyzing enzymes implicated in phagocytosis include phospholipase C-γ1 (PLCγ1), phospholipases D1 and D2 (PLD1, PLD2), and phospholipase A_2 (PLA$_2$). PLCγ1 hydrolyzes PtdIns(4,5)P_2 to diacylglycerol and inositol 1,4,5-trisphosphate (InsP$_3$). Diacylglycerol can activate protein kinase C; InsP$_3$ can stimulate release of calcium from ER. PLD1 and PLD2 hydrolyze phosphatidylcholine to phosphatidic acid, which can be further hydrolyzed to diacylglycerol. PLA$_2$ hydrolyzes phosphatidylethanolamine to arachidonic acid, which contributes to FcR-mediated phagocytosis by a mechanism that may involve regulated fusion of vesicles with phagocytic cups. Arachidonic acid also serves as a substrate for the synthesis of leukotrienes and prostaglandins, which modulate phagocytosis and inflammatory responses. Relevant phospholipid kinases are PI3K and phosphatidylinositol-4-phosphate 5-kinase (PI4P5K), which synthesizes PtdIns(4,5)P_2. Phospholipid phosphatases include phosphatase and tensin homologue (PTEN), SH2 domain-containing 5'-lipid phosphatase-1 (SHIP-1), and SHIP-2. PTEN is a 3'-phosphatidylinositol phosphatase that dephosphorylates PtdIns(3,4,5)P_3 to PtdIns(4,5)P_2 and acts as a brake on FcR signaling. Although PTEN does not localize to phagosomes, its overexpression in macrophages inhibits phagocytosis of large particles, much like inhibitors of PI3K. Its activity may therefore provide a negative feedback that governs a cell's overall capacity for phagocytosis. SHIP-1 and SHIP-2 hydrolyze PtdIns(3,4,5)P_3 to PtdIns(3,4)P_2, and inhibit phagocytosis differently than PTEN does. They localize to forming phagosomes and may provide inhibitory signals that coordinate phagosome formation. SHIP-1 is recruited to ITAMs of FcRs in the advancing membrane of the phagocytic cup, but unlike PI3K it dissociates from the membrane at the base of the cup. The brief presence of SHIP-1 may transiently suppress PtdIns(3,4,5)P_3-dependent signals necessary for later stages of phagosome formation.

The lipids generated by lipid-modifying enzymes also concentrate in forming phagosomes. Their dynamics can be observed in living cells by fluorescence microscopy of cells expressing fluorescent PH domains. For example, the PH domain of PLCδ1 (PLCPH) binds specifically to PtdIns(4,5)P_2. A chimera of green fluorescent protein fused to PLCPH expressed in macrophages binds to PtdIns(4,5)P_2 present on the cytoplasmic leaflets of membranes and displays the distributions of that phospholipid during the movements of phagocytosis. Time-lapse video microscopy of macrophages expressing these probes revealed that PtdIns(4,5)P_2 is relatively abundant on plasma membranes of unstimulated cells. It increases transiently as phagocytosis begins but then disappears from the membrane of the phagocytic cup, sometimes even before the phagosome has closed into the cell. Other chimeras of fluorescent proteins and lipid-binding domains report the distributions of PtdIns(3,4,5)P_3, PtdIns(3,4)P_2, PI3P, and diacylglycerol. Studies with those probes showed that during phagosome formation, levels of PtdIns(3,4,5)P_3 and PtdIns(3,4)P_2 increase transiently on the inner membrane of the phagocytic cup, and diacylglycerol appears and persists on the phagosome. Thus, FcR signaling generates dramatic changes in the profile of phosphoinositides on the inner membrane of the forming phagocytic cup. The steep gradients of phosphoinositides from the inner membrane of the cup to the contiguous plasma membrane outside of the cup suggest either the presence of barriers to lipid diffusion out of the phagocytic cup or restricted distributions of highly active lipid-modifying enzymes that sustain those gradients.

Membrane phosphoinositides regulate many of the proteins that control actin, myosin, endocytosis, membrane fusion, and the oxidative burst. PtdIns(4,5)P_2 binding activates a number of actin-regulatory proteins, including profilin, WASP, WAVE, and gelsolin, all of which have been implicated in regulation of FcR-mediated phagocytosis in various cell types. Binding of WASP to PtdIns(4,5)P_2 and GTP-Cdc42 induces conformational changes that allow it to recruit and activate Arp2/3 complex. Although the GEFs or GAPs necessary for FcR-mediated phagocytosis have not been identified definitively, some of those that regulate Rac1, Rac2, Cdc42, ARF1, and ARF6 contain PH domains that recognize PtdIns(3,4,5)P_3 specifically.

Diacylglycerol generated by PLCγ in phagocytic cups activates PKCα and PKCε in macrophages, and PKCζ in neutrophils, all of which localize to forming phagosomes. PKCα is activated by diacylglycerol and calcium; its substrates for phosphorylation include myristoylated, alanine-rich C-kinase substrate (MARCKS) and the related protein MacMARCKS, which also localize to phagosomes. MARCKS may help to construct the phagosome by locally regulating the sol-gel state of actin filament networks. PKCε is activated by diacylglycerol independent of calcium, and is required for phagocytosis. Its relevant substrates during FcR-mediated phagocytosis have not yet been identified. PKCζ is regulated by diacylglycerol and PtdIns(3,4,5)P_3; it regulates activation of NOX2 by phosphorylation of its subunit protein p47^{phox}.

PLCγ also generates the soluble sugar InsP$_3$, which can release calcium from intracellular stores. Increases in cytosolic free calcium ([Ca^{2+}]$_{free}$) accompany FcR phagocytosis in neutrophils, but not always in macrophages. Experimental methods that prevent increases in [Ca^{2+}]$_{free}$ inhibit FcR-mediated phagocytosis in neutrophils but not in macrophages. Therefore, calcium signaling makes variable contributions to phagosome formation. Several calcium-dependent regulatory proteins participate in phagocytosis, including PKCα and calmodulin, which also regulates MARCKS. Inhibitors of calmodulin inhibit phagocytosis, suggesting that calcium facilitates phagosome formation at some level.

Patterns of signaling by the GTPases Rac1, Rac2, Cdc42, ARF1, and ARF6 have been visualized during phagocytosis by fluorescence resonance energy transfer (FRET) microscopy of cells expressing fluorescent chimeras of GTPases and the GTPase-binding domains of their effector proteins. For example, in macrophages expressing yellow fluorescent protein (YFP)-Rac1 and a cyan fluorescent protein (CFP) chimera of the Rac-binding domain of PAK1 (CFP-PBD), FRET microscopy was used to localize activated YFP-Rac1 during phagocytosis. When GTP-bound YFP-Rac1 binds to CFP-PBD, the YFP and CFP domains of those chimeras are brought sufficiently close together that the resonance energy of excited CFP transfers to YFP and produces spectral changes that report the interaction. These quantitative FRET microscopic methods showed that although the GTPases Rac1, Rac2, Cdc42, ARF1, and ARF6 associate continuously with membranes of the phagocytic cup, they exhibit three distinct patterns of activation. Cdc42 and ARF6 are active at the leading edge of the advancing phagosome and are deactivated over the base of the phagocytic cup, Rac2 and ARF1 are active over the base of the cup, and Rac1 is active throughout the process. Both Rac1 and Rac2 exhibit pronounced increases in activity coincident with closure of the phagosome.

These patterns of GTPase activation and deactivation may correspond to distinct stages of phagosome formation (Fig. 2). Stage 1 activities are those at the leading edge that reflect the first signals triggered by FcR ligation; these include activation of Rac1, Cdc42, and ARF6 and the localized polymerization of actin for pseudopod extension. Stage 2 activities follow stage 1 and include actin-myosin contraction, membrane expansion by organelle fusion, and activation of NOX2. Corresponding GTPase activities for stage 2 would include activation of Rac2 and ARF1, deactivation of Cdc42 and ARF6, and increased activation of Rac1. This may be followed by later stages, including activation and deactivation of NOX2 and fusion of the phagosome with intracellular vesicles associated with phagosome maturation.

The patterns of GTPase activities and the corresponding patterns of movement for actin, myosin, and membranes may be coordinated by patterns of membrane lipids in the inner leaflet of the inner membrane of the phagocytic cup that appear and disappear through the course of phagocytosis. Phagocytosis of large particles can be blocked by inhibiting generation of 3'-phosphoinositides, using either pharmacological inhibitors of PI3K or overexpression

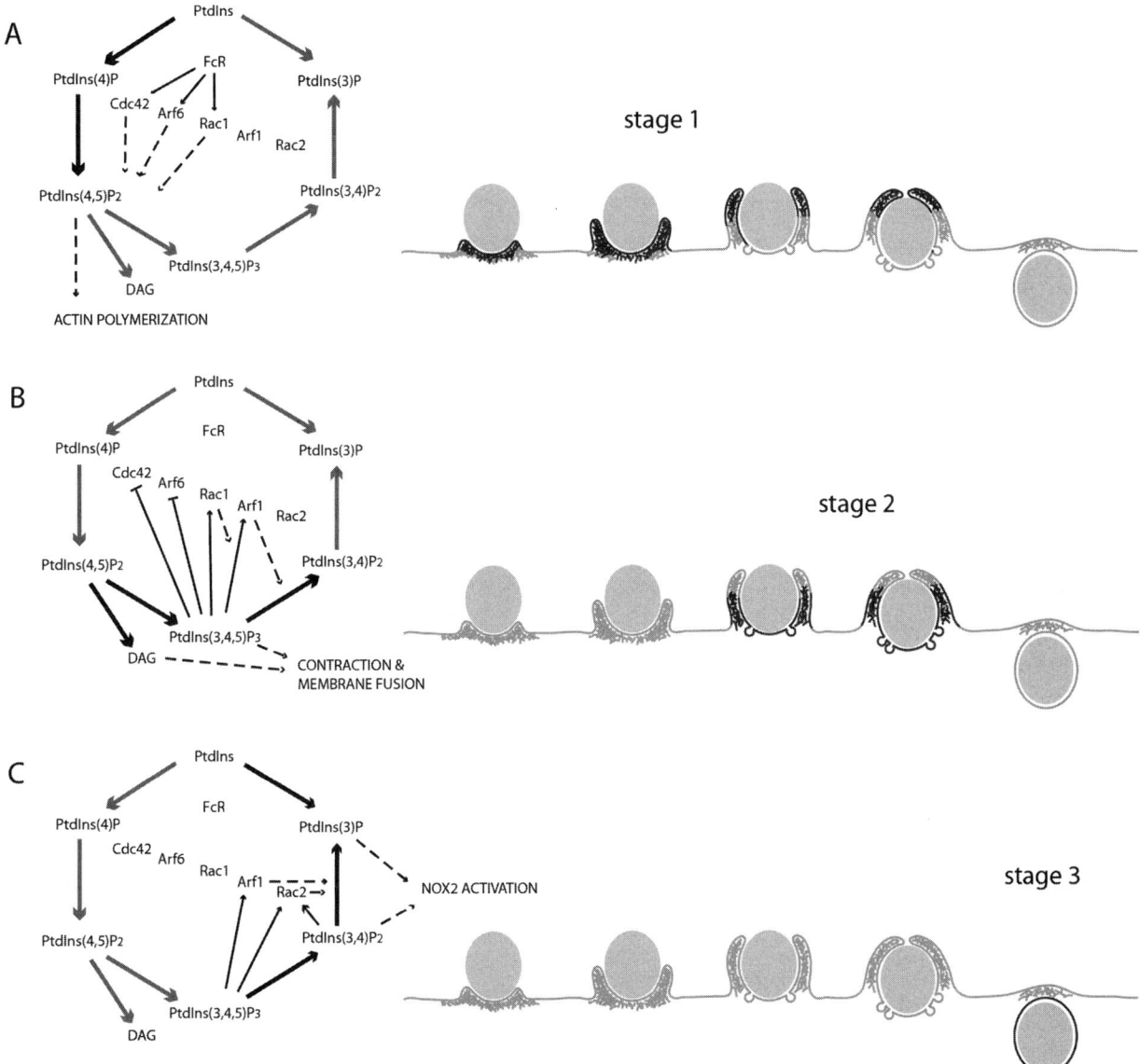

FIGURE 2 Coordination by membrane phospholipids and small GTPases of the various stages of phagosome formation. In the diagrams on the left, thick arrows mark the progression of phospholipid modifications during FcR-mediated phagocytosis, thin solid lines indicate GTPase activation (arrows) or inhibition (cross-bars) by phosphoinositides, and thin broken lines indicate general effector activities of GTPases or other proteins regulated by lipids. Diagrams on the right indicate the morphological stages of phagocytosis and the regions of the forming phagosome in which distinct stages of signaling occur. (A) In the earliest signaling (stage 1), FcRs activate Cdc42, ARF6, and Rac1, which in turn facilitate synthesis of PtdIns(4,5)P_2 (thick black arrows) and activities that promote actin polymerization. Stage 1 activities predominate early during phagocytosis and persist at the distal margin of the phagocytic cup. (B) The accumulation of PtdIns(3,4,5)P_3 and PtdIns(3,4)P_2 in the phagocytic cup initiates stage 2 activities. Cdc42 and ARF6 are inactivated, ARF1 and Rac2 are activated, and Rac1 remains active. These activities, together with those activated by diacylglycerol [DAG; generated from PtdIns(4,5)P_2], stimulate actin depolymerization, contraction, and the fusion of vesicles with the phagocytic cup. Stage 2 activities are blocked by inhibitors of PI3K. (C) In stage 3, depletion of PtdIns(4,5)P_2, accumulation of DAG, PtdIns(3,4,5)P_3, and PtdIns(3,4)P_2, and the additional generation of PtdIns(3)P lead to deactivation of Rac1 and generation of ROS by NOX2.

of PTEN. Interestingly, the inhibition arrests phagocytosis after phagocytic cups have formed; particles become caught in actin-rich cups that cannot close into phagosomes. FRET microscopic measurements of GTPase activities after inhibition of PI3K show that the unclosed phagocytic cups contain active ARF6, Cdc42, and Rac1. Moreover, Rac2 and ARF1 are not activated in these cups. These observations indicate that the signal transition from stage 1 to stage 2 that normally occurs on phagosomal membranes is coordinated by the increases in 3'-phosphoinositides in membranes of the phagocytic cup (Fig. 2). That is, PtdIns(3,4,5)P_3 or other 3'-phosphoinositides activate the GAPs that turn off Cdc42 and ARF6 or the GEFs that turn on Rac2 and ARF1. As Rac1 activity spans this signal transition, its regulation is relatively independent of these 3'-phosphoinositide-dependent signals. Continued generation of diacylglycerol and 3'-phosphoinositides may also coordinate the activation of NOX2 in the phagosome (Fig. 2).

The patterns of GTPase or other enzyme activities may be coordinated by other membrane phospholipids besides PtdIns(3,4,5)P_3. The anionic phospholipids of the inner leaflet of the plasma membrane create a net negative surface charge that attracts proteins with domains rich in basic amino acids. MARCKS contains basic domains that facilitate a calcium- and surface charge-dependent association with plasma membrane after phosphorylation by PKC. Recruitment of Rac1 to plasma membranes early during phagocytosis is facilitated by interactions between anionic phospholipids, such as PtdIns(4,5)P_2, and lysine residues of the Rac1 basic domain. Rac1 dissociation from phagosomes at the completion of phagocytosis may be driven by the decrease in net charge of membranes, through actions of PLC or PLD, which should decrease concentrations of PtdIns(4,5)P_2 in the phagocytic cup.

Signaling via Inhibitory Fc Receptors

Inhibitory Fc receptors suppress signaling for phagocytosis by recruiting to membranes activities that counteract those of the stimulatory Fc receptors. Inhibitory FcR contains an amino acid motif including tyrosine which recruits SH2 domain-containing enzymes that suppress phagocytic signaling; this motif is therefore called an immunoreceptor tyrosine-based inhibitory motif, or ITIM. Phosphorylated ITIMs of inhibitory Fc receptors bind the protein phosphatase SHP2 and the phospholipid phosphatase SHIP-1, which counteract the stimulatory signals of activating receptors. SHIP-1 inhibits PI3K signaling by dephosphorylating PtdIns(3,4,5)P_3 to PtdIns(3,4)P_2, and decreases calcium signals in response to FcR ligation.

Phagocytosis by CR3

The complement receptor CR3 recognizes the opsonin C3bi on particles. It consists of two transmembrane integrin subunits, α_M (CD11b) and β_2 (CD18), which associate together in plasma membranes. The related receptor CR4 consists of α_X (CD11c) and β_2 (CD18). In unstimulated cells, CR3 or CR4 can bind C3bi-coated particles but cannot ingest them; phagocytosis requires activation of the receptors through other signaling pathways. CR3 can be activated for phagocytosis by phorbol myristate acetate (PMA), TNF-α, lipopolysaccharide (LPS), and a variety of lipids, chemoattractants, and extracellular matrix proteins. CR3 activation in response to LPS is mediated by the small GTPase Rap1.

The zipper model may not apply to CR3-mediated phagocytosis, as receptors nucleate actin from discrete punctate sites on the particle surface. Several proteins that localize to actin-rich focal adhesions in other cell types have been localized to CR3 phagosomes, including paxillin, vinculin, α-actinin, and talin. The extracellular domain of α_M integrin binds to C3bi on particles. After activation of CR3 for phagocytosis, its binding to C3bi leads to phosphorylation of β_2 integrin, which undergoes conformational changes that permit binding of talin and the initiation of actin polymerization. Actin polymerization during CR3-mediated phagocytosis utilizes both Arp2/3 and mDia.

Several small GTPases have been indicated as necessary for CR3-mediated phagocytosis. Most phagocytes require RhoA activation, and some also require activation of Rac1. CR3 activates Rac1 via the GEF Vav. By analogy with signaling for focal adhesion assembly in nonphagocytes, Rac1 activity may regulate the early polymerization of actin via Arp2/3 and WAVE, with RhoA organizing later actin-myosin contractile activities. Two important RhoA effectors are mDia, whose actin-polymerizing activity is stimulated by binding GTP-RhoA, and ROCK (Rho kinase), which upon binding GTP-RhoA phosphorylates and inactivates myosin light chain phosphatase (MLCP). Inhibition of the phosphatase increases levels of phosphorylated myosin light chain, consequently increasing the contractile activity of myosin II.

CR3 signaling also requires elements of signaling through tyrosine kinase receptors. Despite the fact that the integrins lack ITAM domains, their signaling requires Syk and other ITAM-dependent signaling elements. It is not yet known how these signaling elements are activated by CR3 or how they contribute to CR3-mediated phagocytosis.

Phagocytosis by Other Complement Receptors

CR1 is a single-pass transmembrane protein that binds C3b-opsonized particles. Although blocking antibodies against CR1 can inhibit phagocytosis of complement-opsonized particles, the receptor may not be sufficient to mediate phagocytosis. Rather, CR1 may facilitate particle binding, with other membrane proteins mediating internalization.

A complement receptor of the immunoglobulin superfamily (CRIg) is expressed on macrophages of the liver (Kuppfer cells). It mediates phagocytosis of particles and bacteria opsonized with C3bi or C3b. The regulation of signaling for phagocytosis by CRIg is not known.

Phagocytosis by Mannose Receptors

The mannose receptor mediates ingestion of particles and soluble mannose ligands, the latter through receptor-mediated endocytosis into coated vesicles. It is important for clearance by macrophages of many kinds of bacteria and fungi. Although the cytoplasmic tail of the mannose receptor is necessary for phagocytosis, the receptor may be insufficient for phagocytosis. Full-length mannose receptors expressed in nonphagocytes can bind mannose-rich particles, but phagocytosis does not follow, despite the fact that other phagocytic receptors expressed in such cells are sufficient for phagocytosis. It is therefore likely that mannose receptors bind mannose- or fucose-rich particles to macrophages, and other membrane proteins organize phagosome formation. Signaling for mannose receptor-dependent phagocytosis by macrophages entails localized actin polymerization, recruitment of the proteins PKCα, MARCKS, and talin, activation of Cdc42, Rac, and RhoB, and phosphorylation of PAK1. Mannose receptor-dependent phagocytosis of *Pneumocystis* by human alveolar macrophages is inhibited by expression of dominant-negative Cdc42 or RhoB, or by inhibitors of ROCK. Mannose receptor-dependent phagocytosis therefore appears to be related to, but distinct from, both FcR- and CR3-mediated phagocytosis.

Phagocytosis by Dectin-1

Dectin-1 is a transmembrane receptor (type II) expressed on macrophages, neutrophils, and dendritic cells that is important for innate immune defense against several fungal pathogens. It recognizes β-glucan molecules on fungal cell walls and is the principal receptor for phagocytosis of zymosan. Dectin-1 consists of an extracellular, C-type lectin, carbohydrate-binding domain, a single transmembrane domain, and a cytoplasmic domain containing a single ITAM. It does not self-associate in membranes, but it is stabilized at the plasma membrane by association with tetraspanin molecules. Dectin-1-mediated phagocytosis is like FcR-mediated phagocytosis in that it requires actin polymerization, tyrosine phosphorylation of its ITAM, and activation of PKC and PI3K. Moreover, like FcR-mediated phagocytosis, dectin-1-mediated phagocytosis is inhibited by overexpression of dominant-negative Rac1 and Cdc42, but not by dominant-negative RhoA. Upon phosphorylation, the dectin-1 ITAM binds the tyrosine kinase Syk. Unlike FcR-mediated phagocytosis, however, recruitment of Syk to dectin-1 is not required for phagocytosis, although it is necessary for generation of reactive oxygen intermediates.

Phagocytosis of Apoptotic Cells

The ingestion and degradation of dead or dying cells are essential to development and immunity. In the normal course of animal development, many cells live briefly, die by apoptosis, and are cleared by phagocytes or by neighboring cells. In adults, phagocytosis of dead cells is essential for the normal turnover of aging cells or tissue repair after mechanical or chemical injury. Genetic studies of development in *C. elegans* identified proteins that mediate clearance of dead or dying cells by phagocytosis into neighboring cells. In animals in which those proteins are missing or mutant, cells that die during the normal course of development accumulate in tissues. In mammals, the removal and degradation of apoptotic cells by macrophages and dendritic cells occurs efficiently and without generation of inflammatory cytokines. This clearance and the associated suppression of inflammation help to maintain immune tolerance of host molecules and to prevent the development of autoimmunity. Conversely, bacterial infections are often cleared by neutrophils, which die by necrosis or apoptosis shortly after ingesting bacteria. Removal of dead neutrophils by macrophages provides an additional method for killing bacteria that might have survived the first assault. Interpretive signaling that follows phagocytosis of neutrophils can stimulate or suppress subsequent immune responses, depending on what else was internalized with the neutrophil.

A large variety of extracellular proteins and cell surface receptors participate in phagocytosis of apoptotic cells. Some molecules bridge proteins on apoptotic cells to membrane proteins on phagocytes. Some transmembrane proteins facilitate binding but are insufficient for phagocytosis and others are competent for both binding and phagocytosis. A common feature of apoptotic cells recognized by phagocytes is the exposure of the inner leaflet phospholipid phosphatidylserine (PS) or of lysophosphatidylcholine (LPC) on the outer leaflet of apoptotic cell plasma membranes. Receptors recognize either these phospholipids or proteins that bind them, such as annexin. LPC is directly recognized for phagocytic clearance by scavenger receptors, which include scavenger receptor-A (SR-A), LOX1, CD68, and CD36. LPC can also activate complement, which can opsonize apoptotic cells for clearance by CD91, CR3, or CR4. With so many receptor-ligand interactions capable of mediated clearance of apoptotic cells, it seems unlikely that one signaling pathway is used to construct all of these phagosomes.

Studies of *C. elegans* have defined two pathways for phagocytic clearance of apoptotic cells in embryos. Proteins comprising both pathways have been ordered into sequences of activities in the phagocytic cells, and homologues of those proteins in mammalian cells have been identified (indicated in parentheses). One pathway consists of a putative receptor protein Ced-1 (CD91), an ABC transport protein Ced-7, an effector molecule Ced-6, and dynamin-1. Ced-7 expression is required on both the engulfing and engulfed cell, which suggests a role for the proteins in the recognition process. Dynamin-1, which is recruited to forming phagosomes after the arrival of Ced-6, regulates fusion of endosomes into phagocytic cups and fully closed phagosomes. Ced-6 is an adapter protein that binds Ced-1 and transduces signals by unknown mechanisms. A second pathway consists of MIG-2 (RhoG), UNC-73 (the RhoG GEF TRIO), Ced-12 (ELMO), Ced-5 (Dock180), Ced-2 (CrkII), and Ced-10 (Rac). MIG-2 is activated by UNC-73 and Ced-10 is activated by Ced-12, Ced-5, and Ced-2. MIG-2 activity is required for the activation of Ced-10.

Other experimental systems have identified similar relationships among homologues of these signaling proteins in clearance of apoptotic cells. Dock180 and ELMO function as GEFs for Rac, possibly after CrkII recruitment to activated integrins, or after their activation by RhoG. In tissue culture models of apoptotic cell clearance, RhoG, Rab5, and Rac facilitate clearance, whereas RhoA is inhibitory.

Signaling for Macropinocytosis

Some bacteria and necrotic or apoptotic cells are internalized into spacious phagosomes that resemble macropinosomes. Macropinocytosis can be stimulated by growth factors and chemoattractants, and FcR-mediated phagocytosis is often accompanied by macropinosome formation near the forming phagosome. Signaling for macropinocytosis is similar to that for phagocytosis. On binding growth factors, ITAM regions of growth factor receptor cytoplasmic domains become phosphorylated and recruit kinases that modify lipids and other proteins. Actin assembles into circular ruffles which resemble particle-free phagocytic cups. Circular ruffles close into macropinosomes by PI3K-dependent contractile activities. Macropinocytosis or spacious phagosome formation can be inhibited by overexpression of dominant-negative Cdc42, Rac1, or RhoG, indicating mechanistic roles for these GTPases. The macropinocytosis-like phagocytosis described for clearance of apoptotic and necrotic cells may be regulated by different pathways.

Common Features of Phagocytosis

After actin and myosin, surprisingly few molecules can be listed as essential for phagocytosis. Small GTPases are important, but different ones matter for different receptor-mediated mechanisms. PI3K is necessary for phagocytosis of large but not small particles, and not for the phagocytosis of some bacteria. There are two likely sources for this variability. First, the construction of phagosomes does not scale uniformly with particle size. Either phagocytosis of small particles does not require the same mechanical apparatus as large particles, or small particles are already internalized before the machinery for large particle phagocytosis kicks in. Second, the variability in regulatory mechanisms indicates that the signals for phagocytosis represent more than just the instructions for building a phagosome. They must also be conveying information about context. Many regulatory molecules have overlapping functions; having three different ways to activate Arp2/3 can afford cells some flexibility in the kinds of input that trigger phagocytosis. Nonetheless, many molecules important for phagosome formation have yet to be analyzed in different receptor signaling pathways. Some of these may prove essential for the construction of all phagosomes.

FEEDBACK REGULATION

How are signals coordinated to achieve the complex movements of phagosome or macropinosome formation? Although receptors for phagocytosis are activated when they bind cognate ligands on a particle surface, the continuation of the phagocytic response requires considerable information about context at many scales. If ligand density on the particle is low, the number of activated receptors may be insufficient for a phagocytic response. Other inputs may indicate that the particle should not be ingested; it may be too big, long, soft, or flat, or it may be a healthy, neighboring tissue cell that should not be damaged. Cells might judge if an opsonized surface should be ingested or crawled over. Ligation of different receptors in the same cell may send conflicting signals for cytoskeletal movements, or signals that suppress phagocytic signaling. A phagocyte's appetite for phagocytosis may be diminished by a sense that it has already ingested all it can hold, by other cytoplasmic activities with higher priority, by signaling from other cells, or by the action of microbial toxins. Thus, phagocytosis can be regulated by short-, medium- or long-range feedback mechanisms.

As mentioned earlier, short-range regulation of signaling is that associated with ligated receptors themselves. Cell differentiation can affect the overall expression levels of stimulatory and inhibitory FcRs. A cell's ability to ingest IgG-opsonized particles may be determined by the balance of positive and negative signals generated from the mixture of receptors recruited to those particles.

Ubiquitinylation of ligated FcRs can lead to their sequestration after phagocytosis into multivesicular bodies for degradation. This could form a mechanism for satiety, as phagocytosis of IgG-coated particles and subsequent degradation of Fc receptors could leave a cell with too few receptors or associated signaling proteins for further ingestion of particles.

Medium-range regulation is that which coordinates receptors in a region of cytoplasm. For FcRs and also possibly other receptor signaling, the various stages of signaling may be coordinated by the signal molecules that radiate by diffusion from activated receptors. In membranes, these may include phosphoinositides or lipid-anchored GTPases, which expand the range of receptor signaling by diffusion in the plane of the lipid bilayer. These radiating signal molecules could organize regions of plasma membrane by creating thresholds for transitions from early to later stages of phagosome formation. Accordingly, a minimum density of activated receptors in a region of plasma membrane would be required to attain sufficient local concentrations of these intermediates for progression to those later stages of signaling. Similar thresholds could be created by water-soluble products of receptor signaling, such as $InsP_3$.

Physical properties of particles can provide another kind of medium-range feedback for phagocytosis. Although phagocytes can ingest particles considerably larger than themselves, they do show some restraint with very large or nonspherical objects. If macrophages or neutrophils are allowed to settle onto a coverslip coated with IgG, they spread differently than they do on unopsonized coverslips. Spreading is uniform and closely adherent to the planar surface, similar to the movements that build a phagocytic cup. This frustrated phagocytic response indicates that signaling for phagocytosis is independent of particle geometry. However, if macrophages that are already adhering to a coverslip are presented with variously shaped, IgG-opsonized polystyrene microparticles, they will only ingest those particles with a high radius of curvature. If a macrophage encounters the long, flat side of a prolate, ellipsoid (cigar-shaped) IgG-coated particle, it will not ingest it; but if it encounters the highly curved end of that particle, it engulfs it directly. Macrophages also sense particle rigidity. They can ingest IgG-opsonized, rigid microspheres made of cross-linked polyacrylamide, but not the softer polyacrylamide particles made with less cross-linker. This indicates that particle rigidity provides a permissive, positive feedback for phagocytosis. The sensing mechanism must be finely tuned, because macrophages routinely ingest relatively deformable cells, such as erythrocytes and apoptotic cells. The mechanisms for sensing these physical features of particles are unknown.

These phenomena indicate that receptor signaling receives feedback based on integrated properties of membrane or cytoplasm. Continued receptor signaling may require proper construction of a phagocytic cup, without which the process aborts. The radial organization of those cups, with actin polymerization at the distal margins and actin depolymerization and contraction at the central base of the cup, is reminiscent of the immunological synapse, a structured interface at contact sites between lymphocytes and antigen-presenting cells. The phagocytic cup has been called analogously the phagosomal synapse. If some structural or physical feature of this interface proves to be essential to phagosomal signaling, then the phagosomal synapse will be a useful framework for understanding the mechanisms of phagocytosis.

Long-Range Regulation of Phagocytosis: Permission To Eat

The machinery for phagocytosis is present but quiet in many metazoan cells. For example, epithelial cells are not normally phagocytic, but those that find themselves bordering a wounded epithelium quickly begin to ingest dead or dying neighbors and debris. Also, expression of receptors for phagocytosis in nonphagocytic cell lines can confer on those cells the ability to ingest opsonized particles, indicating that the machinery for phagocytosis is present but only needs the right receptor to trigger the full response.

Professional phagocytes also show discriminating tastes in targets for phagocytosis; they leave most of their healthy neighboring tissue cells unperturbed. We are only beginning to understand what controls their appetite.

Phagocytosis can be inhibited by receptor-mediated signaling. Interactions between phagocytes and apoptotic cells are influenced by positive and negative signals about what happens next ("eat-me" versus "don't-eat-me"). Living, healthy cells express surface proteins that bind receptors in macrophages that inhibit phagocytosis. CD47 on erythrocytes and other cells binds to SIRPα (CD172a) in macrophages and inhibits clearance by phagocytosis. ITIMs in cytoplasmic domains of SIRPα bind inhibitory tyrosine phosphatases SHP1 and SHP2, which counteract positive signals for phagocytosis. Similarly, FcR-mediated phagocytosis can be modulated by increased expression of inhibitory FcRs, whose ITIM domains effectively counteract the signaling via the ITAM domains of activating Fc receptors.

Suppression of phagocytosis may be limited to regulating the activities of a subset of receptors. For example, while Fc receptors in macrophages are fully capable of initiating phagocytosis, their CR3 must be activated for phagocytosis after particle binding. The inside-out signaling pathways that activate CR3 effectively permit phagocytosis only in certain circumstances (i.e., when the activating molecules are also present). The signaling pathways that activate CR3 involve cytoplasmic microtubules and Rap1. These elements may translate contextual information about the environment into permission to use CR3 for phagocytosis.

Phagocytosis can also be modulated by diffusible signals from other cells. Eicosanoids, lipids secreted from macrophages and neutrophils at sites of infection or inflammation, can augment or suppress inflammatory responses by altering macrophage abilities to ingest and kill microorganisms. The leukotriene LTB_4 stimulates phagocytosis and inflammatory responses, and the anti-inflammatory prostaglandin PGE_2 inhibits phagocytosis.

Many bacterial pathogens are capable of inhibiting phagocytosis by macrophages. Some avoid opsonization by secreting a carbohydrate capsule; others inject proteins into macrophage cytoplasm that interfere with signaling for phagosome formation and the oxidative burst. Several antiphagocytic toxins increase cytoplasmic levels of cyclic AMP. It is not known how high concentrations of cyclic AMP inhibit phagocytosis, but it appears to be essential to how prostaglandins inhibit phagocytosis, as well. Cyclic AMP has two principal receptors in macrophage cytoplasm: protein kinase A (PKA) and Epac (exchange proteins directly activated by cyclic AMP). PKA phosphorylates multiple targets in cytoplasm. Epac is a GEF for the GTPase Rap1. It is not known how either of these activities inhibits phagocytosis. They could interfere with particular elements of the mechanics of phagosome formation, or instead they could affect a more general decision about whether to eat at all. In some cells, elevation of cyclic AMP inhibits some kinds of phagocytosis but not others. This may be because cyclic AMP-dependent inhibitory mechanisms are specific for particular receptor pathways, or because their inhibitory activities cannot override very strong signals for phagocytosis.

The mechanisms governing a cell's decision to ingest a target have important implications for understanding phagocytosis itself. Experimental interventions that inhibit phagocytosis are usually interpreted as affecting the mechanism of phagosome formation, but they may instead be affecting more general decisions about whether to ingest a particle at all. Independent of receptor signaling, cells may gauge the dimensions of particles relative to their own dimensions in deciding whether to ingest them, to crawl past them, or to ignore them. If so, then one long-range regulation of phagocytosis may include the mechanisms by which cells measure or regulate their dimensions. A possible case-in-point is PTEN, whose overexpression inhibits phagocytosis despite the fact that it does not localize to the phagosome. PTEN may inhibit phagocytosis generally by controlling the phagocyte's capacity for signaling via 3′-phosphoinositides.

INTERPRETIVE SIGNALING FROM PHAGOSOMES

As pointed out earlier, the many different mechanisms for constructing phagosomes suggest that other, contextual information accompanies phagocytosis. Some kinds of phagocytosis stimulate inflammation; other kinds of phagocytosis suppress it. Some of these differences originate at the phagocytic receptors. For example, many phagocytic receptors also activate NOX2 for generation of ROS into phagosomes. Because NOX2 activity often begins before phagocytosis is completed, the signals for the two activities must be coordinated. Similar categories of regulatory molecules control phagocytosis and the oxidative burst—PKC, small GTPases, and membrane lipids—so similar principles should underlie their coordination. Other inflammatory responses are due to signaling systems that are independent of the phagocytic receptors, such as TLRs, which use distinct signal transduction pathways to report the nature of the particle as friend or foe. Until the essential activities for constructing a phagosome have been identified, the distinctions between the constructive and interpretive signaling must be considered provisional.

REFERENCES

Aderem, A., and D. M. Underhill. 1999. Mechanisms of phagocytosis in macrophages. *Annu. Rev. Immunol.* **17:**593–623.

Brown, E. J., and H. D. Gresham. 2003. Phagocytosis, p. 1105–1126. *In* W. E. Paul (ed.), *Fundamental Immunology*, 5th ed. Lippincott Williams & Wilkins, Philadelphia, PA.

Brown, G. D. 2006. Dectin-1: a signalling non-TLR pattern-recognition receptor. *Nat. Rev. Immunol.* **6:**33–43.

Greenberg, S. 1999. Modular components of phagocytosis. *J. Leukoc. Biol.* **66:**712–717.

Jutras, I., and M. Desjardins. 2005. Phagocytosis: at the crossroads of innate and adaptive immunity. *Annu. Rev. Cell Dev. Biol.* **21:**511–527.

Krysko, D. V., K. D'Herde, and P. Vandenabeele. 2006. Clearance of apoptotic and necrotic cells and its immunological consequences. *Apoptosis* **11:**1709–1726.

Lennartz, M. R. 1999. Phospholipases and phagocytosis: the role of phospholipid-derived second messengers in phagocytosis. *Int. J. Biochem. Cell Biol.* **31:**415–430.

Niedergang, F., and P. Chavrier. 2005. Regulation of phagocytosis by Rho GTPases. *Curr. Top. Microbiol. Immunol.* **291:**43–60.

Reddien, P. W., and H. R. Horvitz. 2004. The engulfment process of programmed cell death in Caenorhabditis elegans. *Annu. Rev. Cell. Dev. Biol.* **20:**193–221.

Stuart, L. M., and R. A. Ezekowitz. 2005. Phagocytosis: elegant complexity. *Immunity* **22:**539–550.

Swanson, J. A. 2008. Shaping cups into phagosomes and macropinosomes. *Nat. Rev. Mol. Cell Biol.* **9:**639–649.

Taylor, P. R., L. Martinez-Pomares, M. Stacey, H. H. Lin, G. D. Brown, and S. Gordon. 2005. Macrophage receptors and immune recognition. *Annu. Rev. Immunol.* **23:**901–944.

Underhill, D. M., and B. Gantner. 2004. Integration of Toll-like receptor and phagocytic signaling for tailored immunity. *Microbes Infect.* **6:**1368–1373.

Yeung, T., B. Ozdamar, P. Paroutis, and S. Grinstein. 2006. Lipid metabolism and dynamics during phagocytosis. *Curr. Opin. Cell Biol.* **18:**429–437.

13

Membrane Trafficking during Phagosome Formation and Maturation

GREGORY D. FAIRN, ELENA GERSHENZON,
AND SERGIO GRINSTEIN

Phagocytosis, the engulfment of particles, plays a critical role in the elimination of pathogens and in the clearance of apoptotic bodies. The process of phagocytosis can be conceptually divided into two separate stages: phagosome formation and phagosome maturation. The former refers to the entrapment of the target particle by extensions of the plasma membrane, called pseudopods, which merge and fuse, generating an intracellular vacuole or phagosome. The formation and extension of pseudopods are highly orchestrated events involving changes in cytoskeletal architecture and membrane dynamics; endomembranes are thought to fuse with the incipient phagosomal cup, contributing to pseudopod elongation. Sealing of the vacuole is followed by an extensive remodeling of its membrane and contents. The result of this remodeling, better known as maturation, is to convert the nascent phagosome into an effective microbicidal organelle. Invading microorganisms are destroyed and degraded within phagosomes and products of this degradation are delivered to major histocompatibility complex class II (MHC-II) compartments for subsequent antigen presentation. Thus, in addition to its role in the primary defense against pathogens, phagocytosis is also important to the development of acquired immunity. The primary objective of this chapter is to address the vesicular trafficking events involved in the formation and maturation of the phagosome.

Phagocytosis is initiated by the recognition of ligands on the surface of the target particles by specific receptors on the host cell membrane. Endogenous components of the particle surface, including sugars (e.g., mannose) and lipids (e.g., phosphatidylserine), can serve as binding determinants. In addition, opsonins, such as complement and immunoglobulins, are effectively recognized by receptors on the surface of phagocytes. The type and abundance of receptors vary among the different types of phagocytic cells.

With this in mind, we have attempted to highlight the specific particle, receptor, and phagocyte type employed in the individual studies cited in this chapter.

PHAGOSOME FORMATION

Increase in Surface Area

First-time observers of the ingestion process quickly realize that professional phagocytes have a seemingly unlimited appetite. In the simplest of models, phagocytosis would have the plasma membrane surround the pathogen, bud inward, and undergo fission to generate the phagocytic vacuole. However, this notion was dispelled by early studies demonstrating that macrophages could ingest the equivalent of ~100% of their surface area during phagocytosis, without any apparent reduction of their surface area (Aderem, 2002; Griffin et al., 1975; Snyderman et al., 1977). In fact, macrophages often display a net increase in total surface area during phagocytosis. This conclusion was reached by using two independent techniques. One line of evidence involved measurements of electrical capacitance. Typically, biological membranes have a capacitance of 1 $\mu F/cm^2$. Because this value is remarkably invariant among membranes, changes in capacitance detected during the course of electrophysiological measurements are indicative of changes in the surface area of the membrane. This concept has been exploited to measure exocytosis of secretory granules and synaptic vesicles in neuronal and endocrine cells. Application of this technique to the study of phagocytosis showed that the capacitance of the surface membrane of macrophages actually *increases* during the course of phagocytosis (Holevinsky and Nelson, 1998). This is consistent with the delivery of endomembranes to the cell surface, in excess over the area of membrane internalized.

A similar conclusion was reached by using a spectroscopic method. FM1-43, a fluorescent steryl dye, can efficiently and rapidly label the plasma membrane, without traversing it. If used under conditions where endocytosis is

---

Gregory D. Fairn, Elena Gershenzon, and Sergio Grinstein, Programme in Cell Biology, Hospital for Sick Children and Department of Biochemistry, University of Toronto, Toronto, Ontario, Canada.

minimized, FM1-43 staining can provide an accurate estimate of the area of the plasma membrane, which is directly proportional to the fluorescence intensity. Control cells and cells undergoing phagocytosis had their cell surface stained with FM1-43 and the fluorescence was quantified by using flow cytometry. These elegant studies demonstrated a stepwise *increase* in plasma membrane fluorescence, corresponding to the number of particles engulfed by the macrophages (Hackam et al., 1998). Macrophages that had performed phagocytosis had a >60% increase in fluorescence compared with control cells. Jointly, these two approaches clearly demonstrate that, contrary to the simplistic model of membrane invagination, the area of the cell undergoes an increase during phagocytosis, which can only be accounted for by a concomitant, exocytic delivery of endomembranes to the surface.

Sources of Additional Membrane for Phagocytosis

Several mechanisms could in principle explain the increase in surface area found to accompany phagocytosis. In principle, physical stretching of the bilayer could have occurred, but this possibility can be readily discounted. Previous work has demonstrated that simple phospholipid bilayers or more complex biological membranes can stretch or expand no more than 3% before rupturing (Evans et al., 2003; Zimmerberg, 2006). While stretching may play a role in zippering of the membrane around the particle, it cannot explain the bulk of the increase in membrane surface. Furthermore, stretching of the plasma membrane would not account for the observed large changes in electrical capacitance.

A novel and intriguing hypothesis was put forward recently to account for the ability of neutrophils to ingest large and/or multiple particles. Frequently, phagocytes are thought of as smooth spheres and their area is approximated as $4\pi R^2$. However, this model is ostensibly simplistic. Leukocyte membranes are highly convoluted, with multiple microvilli protruding from their surface. These wrinkles or microvilli could be important reservoirs of plasma membrane. Based on this fact, Hallett and Dewitt (2007) proposed that during phagocytosis the plasma membrane can be "unwrinkled," effectively increasing the area capable of participating in phagocytosis, with no net effect on cell size. At least in the case of small particles (e.g., 1.0 μm), the collapse of these microvilli might be sufficient to supply membrane for phagocytosis. This unwrinkling mechanism may be insufficient to account for the engulfment of larger particles and cannot explain the observed increases in capacitance and FM1-43 binding. Therefore, while it may be suitable to account for the behavior of neutrophils confronted with certain particles, many questions remain surrounding this hypothesis and its general applicability, which needs to be tested in other cell types.

In our view, the reported increase in surface area during phagocytosis is most readily explained by delivery of an internal pool of membranes to the plasmalemma. The evidence supporting this view and the potential mechanisms involved are discussed below.

Exocytosis during Phagocytosis

The Compartments Involved

A number of compartments of the endocytic and secretory pathways could provide membranes for pseudopod formation and extension. In macrophages, evidence has been provided that documents the fusion of early/recycling endosomes with the plasma membrane during the course of phagocytosis (Bajno et al., 2000; Collins et al., 2002). Similarly, late endosomes (Braun et al., 2004) and even lysosomes (Czibener et al., 2006) were found to fuse with the membrane and to contribute to the effective phagocytosis of large particles, as illustrated in Fig. 1. In dendritic cells, which are highly specialized for antigen presentation, the MHC-II compartment could conceivably also be mobilized at the time of phagocytosis.

The source of the endomembranes has not been studied in as much detail in neutrophils, but it appears likely that additional compartments are involved in these cells. Neutrophils possess several types of highly specialized secretory compartments that can be mobilized when receptors are engaged, including phagocytic receptors (Faurschou and Borregaard, 2003). Their primary (azurophilic) granules are akin to secretory lysosomes, but secondary (specific) and tertiary (gelatinase) granules have no obvious equivalent in macrophages or dendritic cells. In addition, neutrophils have readily mobilizable secretory vesicles that could be delivered to sites of phagosome formation. Whether related, although clearly not identical, specialized compartments also exist in macrophages or dendritic cells remains to be established.

In addition to the exocytosis of secretory granules and endocytic organelles, a proteomic study of isolated phagosomes suggested that the endoplasmic reticulum (ER) can contribute membrane during the formation of the vacuoles (Gagnon et al., 2002). The isolation of phagosomes formed by macrophages upon engulfment of latex beads and their subsequent analysis by mass spectrometry demonstrated the presence of several proteins that are normally residents of the ER. Immunocytochemical electron microscopy data further supported this contention. The contribution of the ER to the formation of phagosomes is an attractive model, because the ER is the most abundant source of membranes within the cell.

Despite these tantalizing suggestions, the actual contribution of the ER membrane to the phagosomal membrane remains unclear. Indeed, the same features that make the model attractive, i.e., the abundance and ubiquity of ER membranes, complicate the interpretation of the observations. The proximity of the ER to sites of phagocytosis could be coincidental and the presence of ER proteins in imperfectly purified phagosomal fractions may be merely an indication of contamination. Indeed, more recent quantitative investigations of the composition of the phagosomal membrane suggested that the contribution of the ER to the phagosome was minimal (Touret et al., 2005).

Mechanisms of Fusion

Vesicles form by budding and pinching off a donor compartment and travel to, dock, and fuse with a target compartment to deliver soluble cargo and membrane (Bonifacino and Glick, 2004). Fusion of vesicles with target membranes includes several steps: initial capture, docking, and compartment mixing (fusion). Once the vesicle and target membranes dock, the process of fusion is mediated by soluble, N-ethylmaleimide-sensitive factor attachment protein receptor (SNARE) proteins (Weber et al., 1998). The SNARE hypothesis postulates that a SNARE protein on a vesicle binds to a cognate SNARE on the target membrane, forming a trans-SNARE complex that bridges the two membranes. SNAREs are classified into one of two families, R- and Q-SNAREs (Fasshauer et al., 1998). The

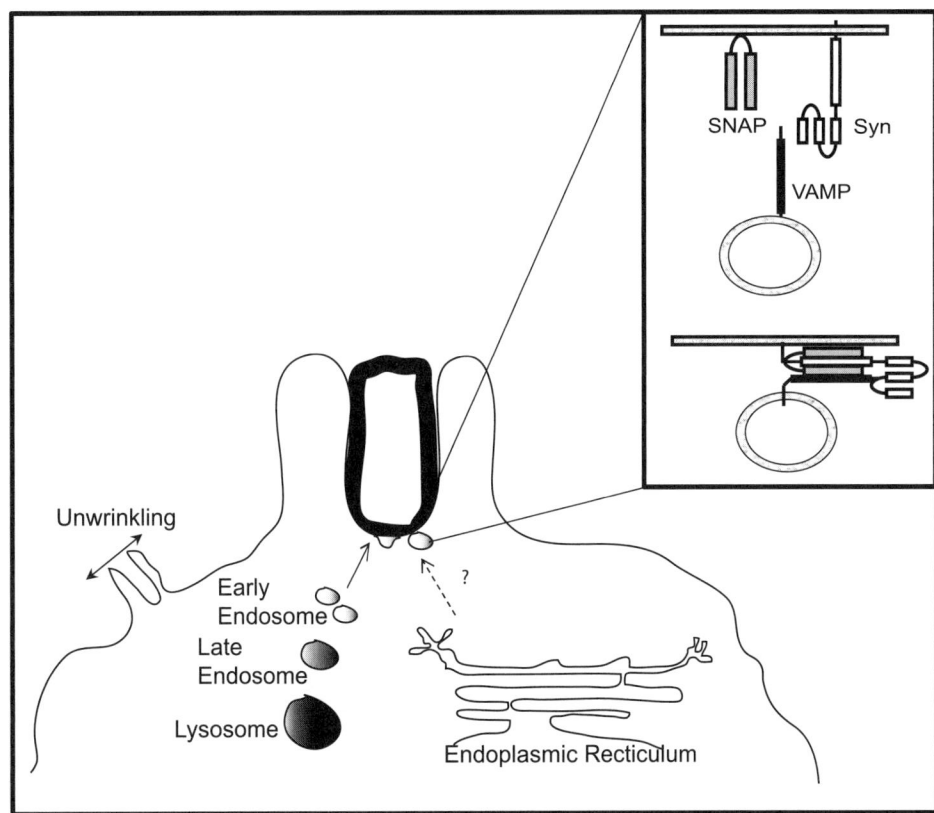

FIGURE 1 Schematic representation of the delivery of endomembranes to the site of phagocytosis. In addition to the area of the plasma membrane underlying the phagocytic target, several sources have been suggested to account for the membrane required to surround and engulf the particle. In neutrophils, which have highly convoluted plasma membranes, unwrinkling of folds was suggested to increase the effective area of the plasma membrane. In macrophages, early endosomes (white), late endosomes (gray), and lysosomes (black) have been suggested to fuse with the growing phagosome through SNARE-mediated fusion. A more detailed scheme of the SNARE components thought to be involved is shown in the inset. A contribution by the ER to phagosome formation has also been suggested, but this hypothesis has been questioned.

designation depends on the nature of the residue that occupies a key functional site, which may be an arginine (R) or a glutamine (Q). Often, but not always, R-SNAREs are present on vesicles and Q-SNAREs are found on target membranes. The selective pairing and distribution of SNAREs throughout the cell generates specificity for vesicles and target compartments. Each cell type expresses its own complement of SNARE partners, with phagocytes being no exception. The multistep maturation of phagosomes into microbicidal phagolysosomes requires several SNARE-mediated fusion events.

In macrophages, early endosomes, late endosomes, and the ER contain the appropriate SNAREs to participate in the coordinated delivery of membrane to the phagosome. The R-SNARE VAMP3, along with its likely Q-SNARE complex, consisting of syntaxin-4 (STX4) and SNAP23 (soluble N-ethylmaleimide-sensitive factor attachment protein), regulates the fusion of the early/recycling endosomes with the nascent phagosomal cup during the ingestion of both zymosan and Fcγ-opsonized particles (Bajno et al., 2000; Pagan et al., 2003). Recent experiments using bone marrow-derived macrophages from VAMP3-deficient mice demonstrated that VAMP3 is required for optimal phagocytosis of unopsonized yeast (Allen et al., 2002). However, compensatory mechanisms or an organelle that fuses via a VAMP3-independent process is likely engaged during complement and Fcγ receptor-mediated phagocytosis since these were unaffected in the VAMP3-deficient cell. According to this prediction, the R-SNARE VAMP7 was reported to mediate the delivery of late endosomes to the phagocytic cup during Fcγ receptor- or complement receptor-mediated phagocytosis. Through the use of dominant-negative SNAREs the group of Niedergang (Braun et al., 2004) demonstrated that a combination of early and late endosomes contribute membrane to the growing phagosome. It is not known whether late endosomes are delivered to forming phagosomes after or simultaneously with the fusion of early endosomes, nor is it clear whether the Q-SNARE that interacts with VAMP7 is STX2, STX3, and/or STX4 (Hackam et al., 1996; Stow et al., 2006).

SNARE proteins of the ER are known to play a role in both anterograde and retrograde transport to and from the cis-Golgi or the intervening ER to Golgi intermediate compartment. Recent work has found that manipulation of the ER/Golgi resident R-SNARE ERS24 can alter phagocytosis of immunoglobulin G (IgG)-coated particles (Becker et al., 2005). Experiments using anti-ERS24 antibodies or protein fragments of ERS24 inhibited phagocytosis of large (3.0 μm) but not small (0.8 μm) particles. These observations were interpreted to mean that the ER may fuse directly with the plasma membrane and that this process aids in the uptake of large IgG-coated particles.

Although the work described above involved primarily macrophages, a different set of membranes and therefore SNARE proteins may participate in phagocytosis by neutrophils (Stow et al., 2006), which are more efficient killers of pathogens, but present antigens poorly. Studies of the phagocytosis of IgG-opsonized zymosan by neutrophils revealed that both azurophilic and specific granules were secreted prior to phagosome sealing, as determined by the appearance at the cell surface of the markers CD63 and CD66b (Tapper and Grinstein, 1997). Recent studies demonstrated that specific and tertiary granules contain VAMP1 and VAMP2, while azurophilic granules contain VAMP1 and VAMP7, with STX4 and SNAP23 involved in the exocytosis of these distinct populations of granules (Mollinedo et al., 2006).

What is described above is the minimal model of SNARE-mediated membrane fusion. However, a growing body of evidence suggests that, at least for neuronal SNAREs, formation of SNARE complexes is not sufficient to account for all aspects of membrane fusion. An additional molecule thought to contribute to the regulation of fusion is synaptotagmin, a single-pass transmembrane protein found on synaptic vesicles. Synaptotagmin, a member of a family of proteins characterized by the presence of tandem C2 domains, can bind to phospholipids in a Ca^{2+}-dependent manner. Importantly, it is also able to bind SNARE complexes. While the precise mode of action of synaptotagmin remains unclear, current evidence suggests a role in calcium regulation of membrane fusion. Specifically, synaptotagmin I is thought to be responsible for calcium-stimulated exocytosis of synaptic vesicles in neurons (Koh and Bellen, 2003) and other isoforms of synaptotagmin are suspected to have an equivalent role in other cell types and organelles. Recently, experiments have identified synaptotagmin VII as a regulator of lysosomal exocytosis (Czibener et al., 2006). Further, both synaptotagmin VII and calcium are required for optimal phagocytosis of IgG- and complement-opsonized particles (Czibener et al., 2006). This was revealed by studies of macrophages obtained from synaptotagmin VII-null mice, which displayed impaired phagocytosis of large and/or multiple particles.

Molecules Involved in Delivery of Membranes

Early work in the genetically tractable budding yeast *Saccharomyces cerevisiae* identified 15 essential gene products (Sec proteins) that were involved in the delivery of vesicles from the Golgi to the plasma membrane (Novick et al., 1980; Schekman and Novick, 2004). Of these, six proteins (Sec3p, Sec5p, Sec6p, Sec8p, Sec10p, and Sec15p) were found to be part of an octameric complex termed the exocyst, together with two other subsequently discovered subunits, Exo70p and Exo84p (Guo et al., 1999; TerBush et al., 1996). Mammalian homologues of all eight subunits have been identified by bioinformatic means, based on their primary structure.

The function of the exocyst is to recognize secretory vesicles and promote their delivery to the plasma membrane, where they undergo fusion. In yeast cells containing one or more defective components of the exocyst, secretory vesicles accumulate within the cell (Novick et al., 1980; Schekman and Novick, 2004). Evidence suggests that the exocyst functions upstream of SNARE complex formation and that its function is regulated by members of the Rab, Ral, and Rho families.

Recently, a systems biology approach identified the exocyst as a potential regulator of phagocytosis in *Drosophila melanogaster* (Stuart et al., 2007). RNA interference-mediated decrease in Sec8, -10, and -15 resulted in a ~30% decrease in the phagocytic ability of *Drosophila* S2 cells. Additional experiments demonstrated differential requirements for Exo70 and Sec3 in the phagocytosis of two types of bacteria, *Escherichia coli* and *Staphylococcus aureus*, respectively (Stuart et al., 2007). In yeast Exo70 and Sec3 localize to target membranes and provide a platform for exocyst assembly. It is therefore possible that Exo70 and Sec3 serve an analogous function in phagocytic cells, where they may differ in their responsiveness to the particular receptor engaged.

The exocyst components Sec10 and Sec15 have been shown to interact with ARF6 and Rab11, respectively (Prigent et al., 2003; Zhang et al., 2004). In murine macrophages the expression of mutant forms of both ARF6 and Rab11 have been reported to inhibit phagocytosis (Cox et al., 2000; Niedergang et al., 2003), raising the possibility that the effects of these GTPases are mediated by the exocyst. Interestingly, ARF6 coexists with VAMP3 on endosomal membranes and an inhibitory allele of ARF6 (T27N) impairs the delivery of the SNARE to the nascent phagosomal cup and inhibits pseudopod extension (Niedergang et al., 2003). Together, these observations are consistent with a model whereby the exocyst links regulatory GTPases with SNARE-mediated fusion at sites of particle ingestion.

Two recently identified proteins, arfophilin and arfophilin-2, were described as possessing the ability to simultaneously bind to Rab11 and ARF6 (Fielding et al., 2005). It is noteworthy that immunoprecipitation of Exo70 resulted in the coprecipitation of Rab11, arfophilin, and arfophilin-2. While the role of arfophilin and arfophilin-2 in phagocytosis is currently unknown, it is tempting to speculate that these ARF-binding proteins may be part of the membrane-delivery complex.

Endocytosis during Phagocytosis

Phagosome formation is not only associated with exocytosis, but is also accompanied by an acceleration of endocytosis. At least five distinct mechanisms can mediate the entry of fluid and solutes into the cell: macropinocytosis, clathrin-dependent endocytosis, caveolae-mediated endocytosis, and clathrin- and caveolae-independent endocytosis. Some of these processes are stimulated during the course (or as a result) of phagocytosis, accounting for the retrieval of some of the membrane components delivered to the surface by exocytosis and contributing to the early stages of maturation.

As mentioned above, CD63 and CD66b are delivered to the surface of neutrophils during the phagocytosis of IgG-opsonized yeast. This same study noticed that both CD63 and CD66b were selectively removed from the plasma membrane shortly after their delivery (Tapper and Grinstein, 1997). This finding suggests that some form of endocytosis is occurring during phagocytosis. Additional experiments demonstrated that Fcγ receptor-mediated phagocytosis stimulates pinocytosis, as measured by the uptake of the fluorescent fluid-phase marker Lucifer Yellow in neutrophils. That the stimulation of pinocytosis does not rely on actin remodeling was concluded from its insensitivity to cytochalasin B. In contrast, the stimulated uptake of Lucifer Yellow required an intact microtubule network, as it was reduced by addition of colchicines (Tapper and Grinstein, 1997).

In cells treated with colchicine, vesicle secretion is also impaired. This raised the possibility that the stimulation of endocytosis may be related, possibly dependent on prior successful exocytosis. This hypothesis was tested by stimulating secretion independently of phagocytosis, using ionophores or thapsigargin to increase the cytosolic Ca^{2+} concentration. These manipulations, which were known to promote the secretion of secretory granules and vesicles in neutrophils, similarly led to a marked increase in the rate of pinocytosis (Tapper et al., 2002). The correlation between the two events is not coincidental; a comparable elevation in the Ca^{2+} concentration had no effect on pinocytosis in cytoplasts, which are degranulated neutrophils prepared by a conservative method involving density step centrifugation (Tapper et al., 2002).

Clathrin-mediated budding and fission of vesicles is important for transport from the plasma membrane to endosomes, and from the *trans*-Golgi network to and from endosomes and lysosomes. An important component of the complex machinery that enables clathrin-mediated vesicular transport is the GTPase dynamin. Dynamin is thought to aid in the scission of clathrin-coated vesicles from donor membranes. The ubiquitous isoform dynamin 2 is involved in membrane-trafficking events at the *trans*-Golgi network, endosomes, and plasma membrane, whereas the predominantly neuronal form, dynamin 1, appears restricted to the plasma membrane. Early evidence demonstrated that GTP-binding deficient (dominant negative) dynamin 2 inhibited receptor-mediated endocytosis, as well as zymosan-, IgG-, and complement-mediated phagocytosis (Gold et al., 1999). These findings seem to implicate dynamin also in the scission of phagocytic vacuoles. However, because of its broad distribution, dynamin 2 could conceivably have pleiotropic effects, affecting endomembrane compartments that may indirectly affect phagosome formation.

The use of dynamin 1, which is restricted to the plasmalemma, offered a way to ascertain whether the effects of the GTPase on phagocytosis (Aggeler and Werb, 1982) are exerted directly, i.e., at the plasma membrane, or by an indirect alteration of endomembrane traffic. Heterologous expression of dominant-negative dynamin 1 in macrophages effectively inhibited transferrin and Fcγ receptor endocytosis, two processes that normally depend on the native dynamin 2. Remarkably, under these conditions the dominant-negative form of dynamin had no discernible effect on Fcγ receptor-mediated phagocytosis (Tse et al., 2003).

This result implies that dynamin itself is not involved in the fission of the phagocytic vacuole from the membrane, but does not necessarily discount a role for clathrin. Clathrin assembly had been reported to occur under abortive phagosomes (Aggeler and Werb, 1982), perhaps suggesting that the coat protein contributes to vacuole formation. The role of clathrin in phagocytosis was assessed independently using antisense mRNA to reduce the levels of its heavy chain. The effectiveness of this procedure was validated by measuring the endocytosis of Fcγ receptors, which was reduced by 90%. However, while these conditions essentially ablated endocytosis, no effects were observed on IgG-mediated phagocytosis (Tse et al., 2003). Thus, at present, there is no convincing evidence that either dynamin or clathrin participate directly in the formation or scission of phagosomes.

Caveolae-dependent endocytosis has also been considered as a possible component of phagosome formation and remodeling. Caveolae are plasma membrane invaginations involved in cholesterol trafficking and signal transduction. They originate at microdomains that are typically enriched in cholesterol, ganglioside (GM1), and caveolin-1, which can be the main structural component. The recent generation of caveolin-1-deficient mice pointed to a role of this protein in phagocytic processes (Li et al., 2005). Characterization of the phenotype of caveolin-1-null animals revealed an increase in the number of apoptotic bodies in the spleen and thymus. This was not due to increased apoptosis, but appears instead to result from a decrease in the rate of clearance of apoptotic bodies by macrophages. This was deduced in part from in vitro experiments where caveolin-1-deficient macrophages displayed a slight, yet significant decrease in their phagocytic activity. This result is recapitulated when wild-type macrophages are treated with methyl-β-cyclodextrin, an agent widely used to remove cholesterol from the plasma membrane, which can disrupt caveolae. Whether the disruption of caveolae-mediated endocytosis in the caveolin-1-deficient macrophages is the primary cause of the decreased phagocytic ability is unclear. Alteration of signaling by receptors that reside in cholesterol-enriched microdomains may also be involved. One such receptor, CD36, is involved in the recognition of phosphatidylserine exposed by apoptotic bodies and may be affected by depletion of either cholesterol or caveolin-1.

Delivery of Microbicidal Machinery Prior to Phagosomal Cup Closure

While the full range of microbicidal tools is deployed in the mature phagolysosome, there is evidence that some of the processes are active already during phagosome formation, prior to phagosomal cup closure. This early deployment strategy is particularly important in situations where the target antigen is too large for a phagocyte to ingest, leading to the so-called frustrated phagocytosis. A key virulence factor in the pathogenesis of several fungal pathogens is the reversible morphological transition between yeast and filamentous (hyphal) forms. Neutrophils can readily ingest and kill yeast by phagocytosis, but filaments are too long to ingest. While the mechanisms by which they recognize filamentous fungi are still unclear, it is evident that neutrophils respond by producing a respiratory burst (Lavigne et al., 2006). This suggests that the NADPH oxidase can assemble in response to stimulation of phagocytic receptors even when phagosome formation is aborted, providing a means of destroying invaders that cannot be fully ingested.

Other mechanisms that can be deployed despite failure of the phagosome to close include the secretion of proteases and the delivery of vacuolar-type (V)-ATPases that pump protons and can generate a localized acidification, akin to that formed by osteoclasts in bone resorption lacunae.

PHAGOSOME MATURATION

The Importance of Maturation

The membrane of the forming phagosome is initially similar to the plasmalemma and its contents resemble the intracellular milieu. Despite their ability to initiate activation of some microbicidal processes, nascent phagosomes are not fully competent to kill and eliminate the ingested microorganisms. To overcome this limitation the phagosome undergoes a sophisticated series of rapid fission and fusion events shortly after sealing, resulting in the extensive modification of the composition of its limiting membrane and contents. This process, referred to as phagosome maturation, bestows the vacuole with a host of degradative properties that are central to its microbicidal function (Beron et al., 1995; Tjelle et al., 2000). Phagosome maturation is an organellar remodeling sequence that resembles the progression of the endocytic pathway. The similarity arises from the fact that phagosomes mature primarily as a consequence of their progressive fusion with sorting and late endosomes, and with lysosomes. This carefully coordinated sequence ultimately converts a plasmalemma-derived vacuole to a lysosome-like hybrid organelle, the phagolysosome (Beron et al., 1995; Pitt et al., 1992).

Maturation is a very dynamic process, thought to involve multiple interactions with components of the endocytic pathway. In principle, addition and removal of material can be accomplished by conventional vesicular fusion and fission events. Alternatively, the exchange of materials may occur through "kiss-and-run" interactions (Desjardins, 1995). The kiss-and-run hypothesis postulates that endosomes and lysosomes connect only transiently to phagosomes by fusion pores that allow selective exchange of some membrane and luminal components, followed by severance of the connecting link. Clearly, the two models are not mutually exclusive and can potentially coexist. Regardless of the detailed mechanism responsible for maturation, there is general agreement that it is the orderly interaction with elements of the endocytic pathway that confers lytic ability to the phagosome.

Phagosome maturation has much in common with the endocytic pathway (Fig. 2). The endocytic pathway is an organized continuum of organelles ranging from early endosomes to lysosomes. In endocytosis, after internalization of a receptor-ligand complex, the contents of an endocytic vesicle are targeted to a sorting or early endosome. Sorting endosomes organize and reroute the assortment of internalized molecules. In some instances the ligands and their receptors dissociate, following distinct intracellular routes: the receptors can return to the cell surface in recycling endosomes, while the ligand molecules continue along the degradative stages of the endocytic pathway toward lysosomes. Sorting (or early) endosomes are identified by the presence of Rab5 and/or early endosome antigen 1 (EEA1), or experimentally by applying short pulses of ligands or fluid-phase markers. They have only a mildly acidic lumen (pH 6.5 to 6.0) and limited hydrolytic capacity (Gruenberg, 2001; Mukherjee et al., 1997).

Late endosomes show gradual acidification of the lumen and are more acidic than sorting endosomes (pH 5.5). They can be identified by their multivesicular nature and by the presence of Rab7, Rab9, lysobisphosphatidic acid (LBPA), mannose-6-phosphate receptors (MPRs), and a variety of lysosomal-associated membrane proteins (LAMPs) such as LAMP-1, LAMP-2, and CD63 (Gruenberg, 2001; Mukherjee et al., 1997; Somsel Rodman and Wandinger-Ness, 2000). Lysosomes, the final stage of the endocytic sequence, are characterized by their extreme acidity (pH $\leq$ 5.0) provided by an increased number of V-ATPases, and by the presence of LAMPs and the mature form of hydrolytic enzymes such as cathepsin D (Mukherjee et al., 1997). Because lysosomes share many markers with late endosomes, such as LAMPs, the best method for lysosome identification is through "pulse-chase" labeling experiments, which consist of an extended pulse that ensures delivery of internalized probes to a late, terminal compartment, i.e., lysosome, followed by a chase of at least one hour, to allow clearance of the label from upstream compartments (Vieira et al., 2002).

As will be apparent from the following sections, much more is known about endosomal progression than about phagosomal maturation. Where information regarding specific steps in the maturation sequence is unavailable, we describe the knowledge accumulated regarding the comparable stage of endosome progression, which should provide useful guidelines for future experimentation.

The Early Phagosome

Nascent or early phagosomes (within 30 min of sealing) rapidly gain properties of sorting endosomes, as revealed by their acquisition of Rab5, EEA1, phosphatidylinositol 3-phosphate [PI(3)P], transferrin receptors, and STX-13. They also acquire small but significant amounts of vacuolar ATPases that account for the decrease in luminal pH to 6.0 to 6.5, which commences almost immediately after phagosome sealing (Alvarez-Dominguez, 1996; Collins et al., 2002; Ellson et al., 2001; Pitt et al., 1992; Vieira et al., 2001).

The earliest identified stages of phagosomal maturation are thought to be controlled by Rab5 (Duclos et al., 2000; Roberts et al., 2000). Recruitment of Rab5 to endosomal membranes was suggested to be modulated by Ras, through mediation of the guanine nucleotide exchange factor RIN1 (Barbieri et al., 1998; Hoffenberg et al., 2000; Li et al., 1997). In its GTP-bound, membrane-associated form, endosomal Rab5 can interact with several effectors, such as Rabenosyn-5, Rabaptin-5, and Rabex-5. Rabex-5 exists in association with Rabaptin-5 and is another guanine nucleotide exchange factor specific for Rab5 (Horiuchi et al., 1997; Lippe et al., 2001). It is not entirely clear at present whether RIN1, Rabex-5, or some other factors determine the recruitment and activation of Rab5 on phagosomes.

Like other small GTPases, Rab5-GTP is capable of recruiting downstream effectors. One of these is EEA1, which contains two spatially separate Rab5-binding domains, one at each terminus. The presence of dual Rab5-binding domains allows EEA1 to serve as a bridge between two early endosomes and presumably also between an endosome and an early phagosome (Mu et al., 1995). Two other factors contribute to the unique ability of EEA1 to serve as a bridge or tether between organelles bearing Rab5. First, it contains a FYVE domain that interacts with PI(3)P that is also present in early endosomes and phagosomes (Lawe et al., 2000; Patki et al., 1998; Stenmark et al., 1996). Second, EEA1 has a tendency to form antiparallel dimers (Callaghan et al., 1999), which increases the avidity of the interactions.

By recognizing and tethering two specific compartments, EEA1 facilitates their fusion, which, as discussed earlier, is mediated by the SNARE machinery. Fusion of early phagosomes and endosomes is thought to be catalyzed by STX-13 (a Q-SNARE) and VAMP3 (an R-SNARE) by a process that involves N-ethylmaleimide-sensitive factor (NSF) (Coppolino et al., 2001).

Another critical Rab5 effector is hVps34, a type III PI-3-kinase. Class I, II, and III PI 3-kinases are all capable of PI(3)P synthesis in vitro, but it is the class III kinase that is the biologically relevant source of PI(3)P in phagosome maturation (Leevers et al., 1999; Vanhaesebroeck et al.,

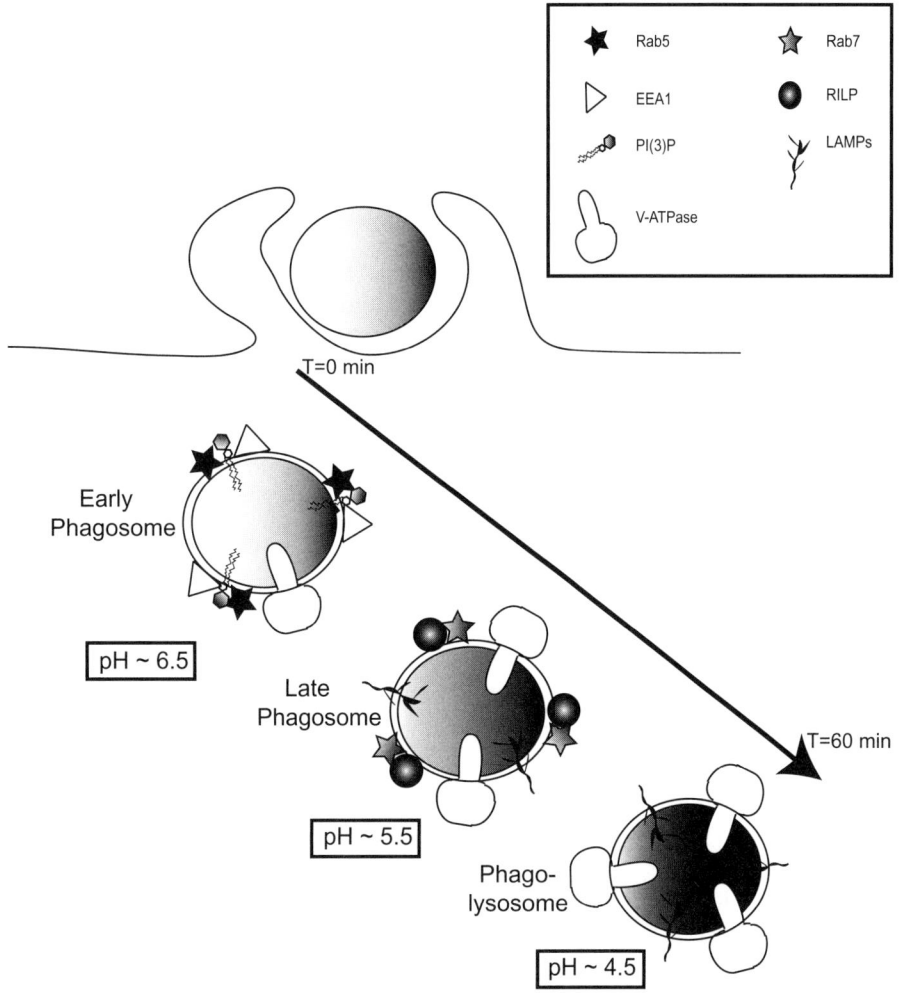

FIGURE 2 Phagosome maturation. Diagram illustrating progression from a nascent phagosome through a mature phagolysosome. In the diagram maturation proceeds downward from left to right. The presence of unique markers and contributors of the maturation process is indicated. In brief, the nascent phagosome (0 min) promptly acquires Rab5, which in turn recruits p150/Vps34 to generate PI(3)P (2 to 5 min) and attracts EEA1, which facilitates tethering and fusion of early endosomes. Rab7 is then recruited (5 to 10 min), which through its binding to RILP is instrumental in promoting fusion of lysosomes with the phagosome (30 to 60 min). The late stages of maturation are characterized by the acquisition of LAMPs. Note the decrease in phagosomal pH resulting from the V-ATPase activity during the progression of the maturation sequence. See text for more details.

1997). Vps34 exists in a complex associated with p150, a Vps15-like serine/threonine kinase subunit of the complex, which is thought to regulate the membrane association and activity of Vps34 (Murray and Backer, 2005; Murray et al., 2000, 2002). Besides EEA1, Rab5 also recruits the p150/Vps34 to generate PI(3)P in early endosomes and phagosomes. Nascent phagosomes acquire PI(3)P shortly after sealing, and this lipid is crucial in the subsequent maturation process. PI(3)P is clearly apparent on the membrane of phagosomes within 1 to 2 min of sealing and persists for approximately 10 min, disappearing subsequently. Downstream effectors of PI(3)P contain well-defined domains that recognize and bind the lipid, such as the PX domain in the p40 subunit of the NADPH oxidase complex or the FYVE domain of EEA1 (Ellson et al., 2002; Gaullier et al., 1998; Patki et al., 1998). Binding to PI(3)P through its FYVE domain is required for the firm attachment of EEA1 to the endosomal and early phagosomal membranes. In view of the critical role of EEA1 and of other ligands of PI(3)P (see below), it is not surprising that interference with the generation of this phosphoinositide using pharmacological inhibitors such as wortmannin and LY294002 causes severe impairment of phagosome maturation (Vieira et al., 2001).

Other FYVE and PX domain-containing proteins participate in the maturation sequence, although not necessarily in support of fusion. Remodeling of the phagosome involves not only addition of new components, but also the concomitant loss of membrane and luminal contents through budding, or fission of vesicles. This process is vital to ensure that maturing phagosomes maintain a constant size despite ongoing fusion events. In the endocytic pathway, internalized cell surface receptors that are destined for degradation are targeted to the inner vesicles of a subset of late endosomes. These specialized endosomes, termed multivesicular bodies (MVBs), allow the cell to physically separate proteins destined for lysosomal destruction from those that are to be conserved by the cell. Another FYVE domain-containing protein, hepatocyte growth factor-regulated tyrosine kinase substrate (Hrs), is key to MVB formation (Raiborg, 2006). In addition to a canonical FYVE domain, Hrs possesses a ubiquitin-interaction motif (UIM) and a clathrin-binding domain that enable it to interact with monoubiquitylated proteins and with clathrin. The UIM domain of Hrs is shared with a number of proteins that are known to recognize and influence the routing of ubiquitylated cargo. These include the epsins, eps15 and eps15R, that are involved in receptor internalization at the plasma membrane (Urbe et al., 2003). Hrs has been shown to sort ubiquitylated proteins to clathrin-coated microdomains on endosomal membranes, causing these proteins to be targeted to degradative compartments (Raiborg et al., 2001a, 2001b; Urbe et al., 2003). The exact role of clathrin is not clear, but it has been hypothesized that the coat protein functions as a scaffold that restricts Hrs, and therefore ubiquitylated cargo, to microdomains on the endosome membranes. MVB formation from this clathrin patch occurs through the recruitment of three protein complexes, collectively known as the endosome-associated complex required for transport (ESCRT) complexes, known as ESCRT I, II, and III (Babst, 2005, 2006; Babst et al., 2002a, 2002b). ESCRT I, comprising Vps23, Vps28, and Vps37, activates ESCRT II, which in turn recruits ESCRT III (Clague and Urbe, 2003). The Hrs-mediated recruitment of ESCRT I to endosomes is essential for proper degradation of ubiquitylated membrane proteins (Bache et al., 2003). Cargo is first recognized by Hrs in conjunction with Tsg101, and is then transferred to the Vps23 subunit of ESCRT I. The ubiquitylated cargo is then transferred to the Vps36 subunit of ESCRT II for sorting into endosomal invaginations formed through the assembly and disassembly of ESCRT III polymers. The mechanism underlying transfer of ubiquitinated cargo from one ESCRT to another is presently unknown, but deubiquitylation occurs prior to sequestration of cargo into intraluminal vesicles within the endosome (Babst, 2006; Malerod and Stenmark, 2007; Slagsvold et al., 2006). A similar sequence is likely to operate in phagosomes, inasmuch as Hrs is required for maturation (Vieira et al., 2004). The presence of all the ESCRT components and their functionality in phagosomes, however, remains to be verified experimentally.

Another fission complex involved in the endocytic pathway is the coat protein complex type I (COPI). Previous findings have indicated that normal delivery of cargo from early to late endosomes and lysosomes requires COPI (Aniento et al., 1996; Stoorvogel et al., 1996). The COPI protein complex is one of the amorphous coats distinct from ordered clathrin lattices that are detected on budding membranes and was originally shown to mediate traffic between the ER and the Golgi complex. COPI also plays a role in phagosome maturation. Although phagosome maturation proceeds to completion in the absence of functional COPI, the traffic of some components is altered. Impaired COPI function results in partial inhibition of transferrin receptor recycling (Botelho et al., 2000).

Transition between Early and Late Phagosomes

The sorting endosomal-like stage of the phagosome is transient. Its transition into the late phagosome stage is indicated by the loss of Rab5, EEA1, and PI(3)P, followed by the acquisition of the late endosomal markers Rab7, RILP, and LBPA (Vieira et al., 2002). Rab5-GTP is presumably lost from the phagosomal membrane upon its conversion to the inactive form Rab5-GDP which is sequestered in the cytosol by a binding protein (Rab-GDI). As in other systems, the transition between the active (phagosome-bound) and inactive (soluble) forms of Rab5 is felt to be regulated by GTPase-activating proteins (GAPs). Possible GAPs of Rab5 include p120RasGAP, the tumor suppressor tuberin, which is a product of the TSC2 complex, RN-tre (Lanzetti et al., 2000; Liu and Li, 1998), and the recently described RabGAP-5 (Haas et al., 2005). Overexpression of RabGAP-5 results in a loss of EEA1 from endosomes and blocks endocytic traffic, while depletion of RabGAP-5 results in increased endosome size, more endosome-associated EEA1, and altered traffic of LAMP-1 (Haas et al., 2005; Pfeffer, 2005). RabGAP-5 interacts with two endosomal proteins, Hrs and ALIX, both of which participate in the down-regulation of receptors by facilitating their incorporation into MVBs (Haas et al., 2005). In this manner, RabGAP-5 limits the amount of activated Rab5 and regulates traffic through endosomes. The presence of RabGAP-5 in phagosomes and its role in maturation is currently under investigation.

The disappearance of PI(3)P results largely from its degradation by phosphatases, primarily those of the myotubularin family. Myotubularins belong to a large subfamily of hydrolases that dephosphorylate the 3' position of PI(3)P and PI(3,5)P_2 (phosphatidylinositol 3,5-bisphosphate) (Srivastava et al., 2005). These enzymes appear to be peripheral membrane proteins that are also present in the cytosol. Structurally, myotubularins consist of a PH-GRAM (pleckstrin homology glucosyltransferases, Rab-like GTPase activators, and myotubularins) domain and a large, 370-residue PTP domain (Robinson and Dixon, 2006). Specific subclasses of myotubularins contain additional conserved protein domains, such as the FYVE domain. One such FYVE domain-containing myotubularin, MTMR3, was shown to reduce levels of both PI(3)P and PI(3,5)P_2 in yeast (Walker et al., 2001). Although the substrate specificity of myotubularins is well established, their cellular functions need to be defined more clearly. Indeed, it is worth noting that myotubularins have not been convincingly shown to localize to early endosomes, the major site of PI(3)P accumulation (Robinson and Dixon, 2006).

PIKfyve, a mammalian orthologue of the yeast PI(3)P-kinase Fab1p, provides an alternate route for elimination of PI(3)P. Unlike the myotubularins, however, PIKfyve is both an effector and downregulator of PI(3)P, which it converts into PI(3,5)P_2. In mammals, PIKfyve has been suggested to localize to early endosomes (Cabezas et al., 2006). As its name implies, PIKfyve bears a FYVE domain that it requires to recognize and convert PI(3)P to PI(3,5)P_2. The latter species is suspected to contribute to the formation of MVBs, but the mechanism involved is still obscure (Shisheva, 2001). Remarkably, there is no evidence to date that either PIKfyve or PI(3,5)P_2 is present in phagosomes or that they function in the maturation cascade.

The Late Phagosome

Late endosomes and phagosomes are more acidic (pH 5.5 to 6.0) than early phagosomes and can be identified by the presence of Rab7, RILP, Rab9, LBPA, MPR, CD63, STX-7, and LAMP isoforms. They are apparent between 15 and 45 min after phagosome sealing. Rab7 appears on phagosomes as Rab5 dissociates and functions in the transition between early and late phagosomes and/or phagolysosomes. A recent breakthrough in understanding the role of the Rab5 to Rab7 switch, referred to as Rab conversion, has given insight on the mechanism of cargo progression between early and late endosomes (Rink et al., 2005). This process occurs via the replacement of the bulk of Rab5 in an organelle by Rab7. The loss of Rab5 and the concomitant acquisition of Rab7 require the involvement of the class C Vps/HOPS complex, which is an established Rab7 guanine nucleotide exchange factor that also interacts with Rab5. Rab5 mutants defective in GTP hydrolysis produce endosomes that recruit Rab7 yet fail to displace Rab5. Therefore, it is likely that following the class C Vps/HOPS-dependent recruitment of Rab7 onto Rab5-positive endosomes, the completion of conversion involves negative feedback between Rab7 and Rab5, possibly through the action of a Rab5-GAP (Rink et al., 2005).

Three Rab7 effectors have been described, RILP, oxysterol-binding protein (OSBP)-related protein-1L (OPR1L), and Rabring7 (Cantalupo et al., 2001; Johansson et al., 2005; Jordens et al., 2001; Mizuno et al., 2003). Rabring7 is recruited by active Rab7 to late endosomes, where it controls their traffic to lysosomes (Mizuno et al., 2003). Likewise, ORP1L also localizes to late endosomes/lysosomes and interacts physically with Rab7, preferentially in its GTP-bound form (Johansson et al., 2005). Rab7 in its active GTP-bound form on late endosomes also recruits RILP, a key protein for the biogenesis of lysosomes and phagolysosomes (Jordens et al., 2001). When bound to active Rab7, RILP in turn is capable of simultaneously binding the microtubule motor dynein/dynactin. These molecular motors transport the Rab7-positive organelles toward the minus end of microtubules, i.e., from the cell periphery toward the perinuclear region (Colucci et al., 2005; Jordens et al., 2001). In addition to its interaction with motor proteins, RILP has emerged also as a key regulator of protein sorting into the MVBs (Wang and Hong, 2006). The Rab7 effector has been found to interact with Vps22 and Vps36, two subunits of the ESCRT II complex, which in turn interact with Vps28 and Vps23 (components of ESCRT I), and with Vps20 (a component of ESCRT III) (Progida et al., 2006; Wang and Hong, 2006), as illustrated in Fig. 3. The selective interaction with Vps22 and Vps36 suggests that RILP may regulate the recruitment of ESCRT II to endosomal and possibly also phagosomal membranes (Wang and Hong, 2006).

Rab9 is another small GTPase of the Rab family that is found on late phagosomes. Rab9 is required for retrograde transport from late endosomes to the *trans*-Golgi network (TGN) and for lysosome biogenesis (Pfeffer and Aivazian, 2004). Rab9 GTPase resides in late endosome microdomains together with MPRs, which carry newly synthesized lysosomal enzymes to endosomes (Ganley et al., 2004). MPRs release their ligands upon encountering the low pH within late endosomes/lysosomes and return to the TGN to reinitiate another cycle of enzyme transport (Riederer et al., 1994). Transport of MPRs from late endosomes to the TGN requires Rab9, the cargo adaptor tail-interacting protein of 47 kDa (TIP47), a Rab9 effector named p40, NSF, a SNAP, and a protein named mapmodulin (Carroll et al., 2001; Diaz and Pfeffer, 1998; Diaz et al., 1997). Interestingly, depletion of Rab9 causes a decrease in late endosome size, the loss of dense tubular and multilamellar structures, and MPR missorting to the lysosome. It has recently been shown that increased cholesterol stabilizes Rab9 on late endosomal membranes and disrupts late endosomal export of MPRs (Ganley and Pfeffer, 2006). The high cholesterol content may force and trap Rab9 into a more ordered membrane domain, which in turn could restrict the Rab9-effector interactions required to initiate vesicle budding. Likewise, cholesterol accumulation also increases the levels of Rab7 on membranes, with a concomitant decrease in the capacity of GDI to extract membrane-associated Rab7 (Lebrand et al., 2002). It is tempting to speculate that accumulation of cholesterol would affect phagosome maturation in a comparable fashion, but to our knowledge this possibility has not been examined experimentally.

The transition of the early phagosome into a more mature, late endosome-like stage is also signaled by the acquisition of LBPA. LBPA is generated by degradation of phosphatidylglycerol and cardiolipin and is abundant in the internal membranes of MVBs/late endosomes. LBPA has a unique cone-shaped structure that may induce membrane curvature and facilitate the budding that generates MVBs (Kobayashi and Hirabayashi, 2000). Recently, it has been demonstrated that the addition of LBPA to pure lipid liposomes causes them to adopt a multilayered morphology similar to MVBs, but only at an acidic pH, as is found in the phagosomal lumen (Matsuo et al., 2004). The E class Vps protein that binds LBPA, ALIX, has been identified as a regulator of MVB formation. This protein has also been identified to be an interacting partner of RabGAP-5 (Haas et al., 2005). LBPA is found in late phagosomes, where it is likely to complex ALIX and direct membrane fission and the intermediate stages of maturation.

Internal membranes of multivesicular compartments also contain significant amounts of cholesterol. Over 80% of the cholesterol in the endocytic pathway can be found in the recycling compartments and in MVBs (Mobius et al., 2003). In MVBs, most of the cholesterol is contained in the internal membrane vesicles. Despite its importance in endocytic traffic, remarkably little is known about cholesterol in phagosomes. Evidence of its presence and precise concentration at the individual stages of the maturation sequence is lacking. Preliminary results have suggested that cholesterol may regulate the fate of PI(3)P in phagosomes, influence the ability of phagosomes to bind flotillin, and even affect the microbicidal competence of phagosomes, although much regarding the role of cholesterol in phagosome maturation remains to be elucidated.

LAMPs, despite their name, can also be found on late endosomes and late phagosomes. These proteins are heavily N- and O-glycosylated and have a single transmembrane domain (Eskelinen et al., 2003). The relationship between LAMPs and Rab7 is particularly intriguing. Overexpression of dominant-negative Rab7 leads to impaired traffic from the early endosome to late endosomes and lysosomes and prevents the acquisition of LAMPs, implying that normal Rab7 function is required for LAMP recruitment to the late endosome/lysosome (Bucci et al., 2000). However, the converse relationship has also been suggested. Experiments using cells from LAMP knockout mice demonstrated that phagosomes from LAMP-deficient cells fail to acquire Rab7 and do not fuse with lysosomes (Huynh et al., 2007).

As stated previously, late phagosomes, like late endosomes, are more acidic than their earlier counterparts. The increase in intraluminal acidity is generally attributed to the recruitment of increasing numbers of proton-pumping V-ATPases. While traditionally the lower pH has been thought of as a consequence of phagosome maturation,

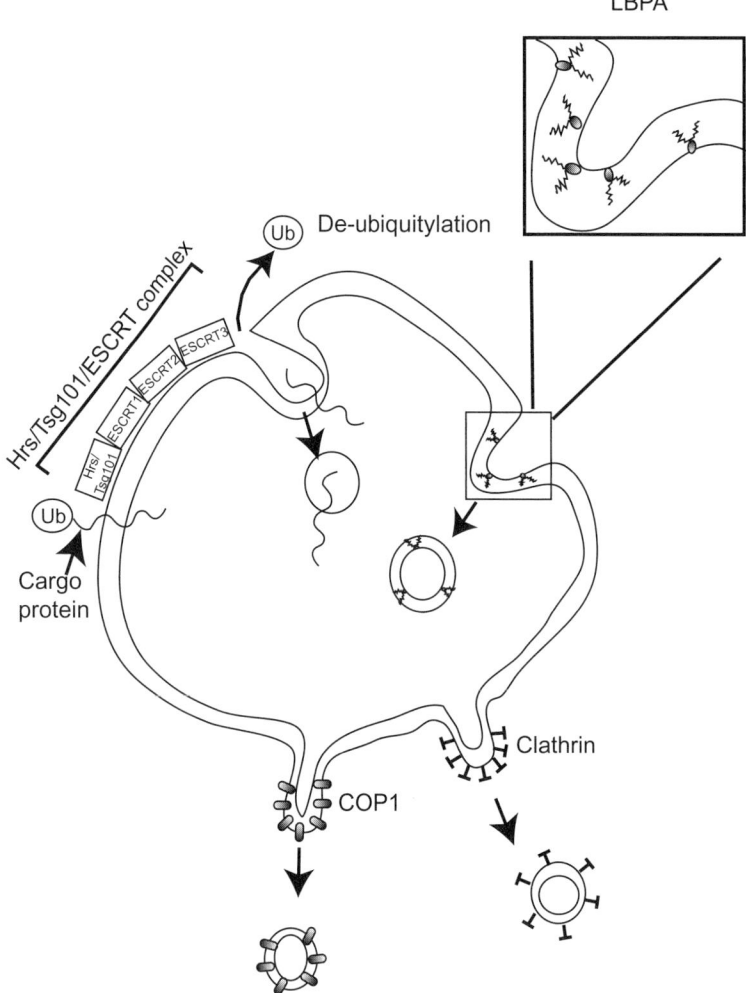

FIGURE 3 Membrane fission during phagosome maturation. Diagram illustrating the fission events that are important in the remodeling of the phagosome during maturation. MVB formation occurs through the recruitment of Hrs, Tsg101, and the ESCRT proteins to the membrane following the ubiquitylation of cargo proteins. The cargo protein is deubiquitylated as it is diverted to the luminal vesicles of MVB. LBPA is thought to play a role in MVB formation by inducing inward budding through membrane curvature (see inset). Clathrin and COPI are two other fission complexes that contribute to outward budding during the maturation process.

there is also evidence that acidification plays a more active role in directing and/or promoting maturation. When the pH of the phagosome is artificially raised by using an inhibitor of the V-ATPase, phagosome maturation is arrested (Geisow et al., 1981), and recent experiments from our laboratory revealed that Rab5 remains on the phagosome in its GTP-bound form for an inordinately long time (K. K. Huynh, W. L. Lee, A. Brech, S. Corvera, and S. Grinstein, unpublished data). The mechanism responsible for this effect is unknown, but we speculate that the low pH that develops as the phagosome matures serves as a timing device to terminate the association of Rab5, perhaps by recruiting a Rab5-GAP.

The Phagolysosomes

Phagosomes are ultimately transformed from a late-endosome-like organelle into phagolysosomes, the final destination of cargo that is intended for degradation. Phagolysosomes are identified and characterized by the presence of elevated concentrations of the mature form of hydrolytic proteases, such as cathepsin D, and by the extreme acidity of their lumen (pH 4.5 to 5.0). In general, maturation culminates with the formation of the phagolysosome approximately 40 to 60 min after the particle has been ingested. Because lysosomes and phagolysosomes share many markers with late endosomes/phagosomes, such as the presence of LAMPs and CD63, they can be best identified in pulse-chase labeling experiments where delivery of internalized probes to a terminal, lysosomal compartment is ensured.

Remarkably little is known about the molecular determinants of the transition between late phagosomes and mature phagolysosomes. Information is also lacking regarding the disposition of the membranes and fluid that are delivered to phagolysosomes on a continued basis, even after maturation is seemingly complete. On occasion entire

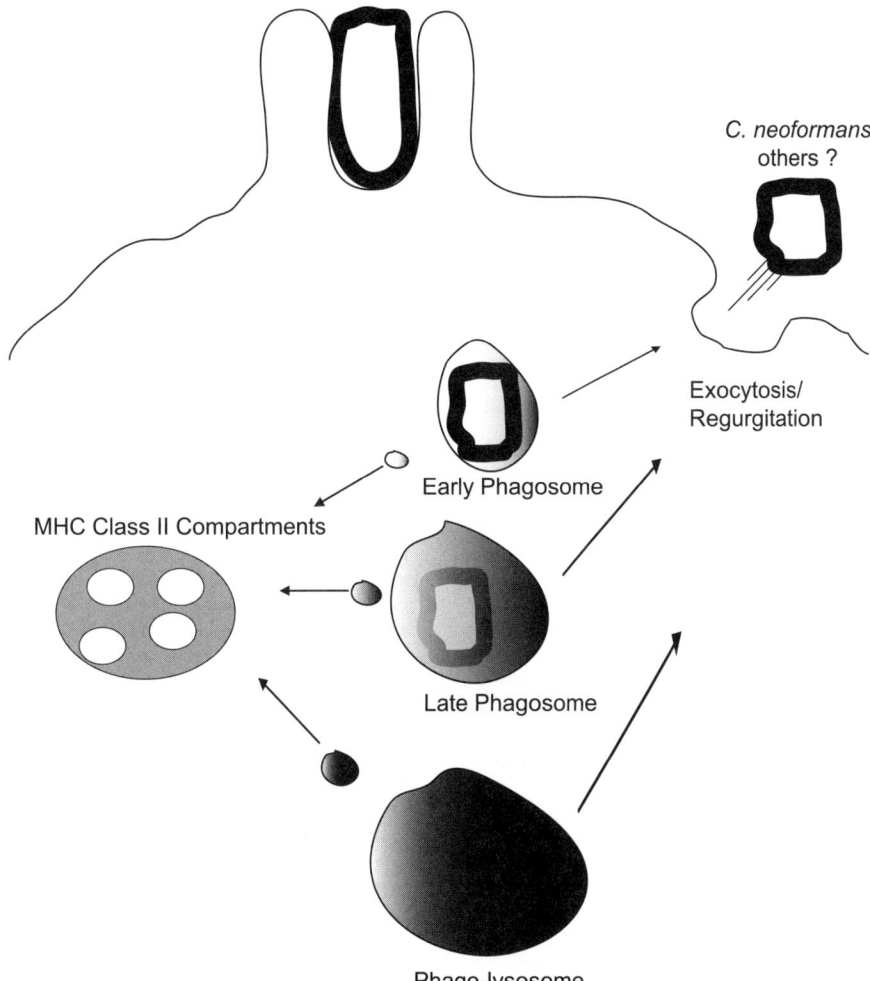

FIGURE 4 Potential fates of the phagosome. In the classical phagosomal maturation pathway the phagosome acquires microbicidal properties as it matures to become a phagolysosome. As pathogens are degraded a fraction of the resulting peptides (or other molecules) are transported by fission and fusion reactions to the MHC-II compartment, for antigen presentation. In some instances, phagosomes can be secreted. This is likely the case for *C. neoformans*, which is extruded intact from macrophages after phagocytosis. Phagocytosis followed by exocytosis of the engulfed particle may contribute to the spread of some pathogens throughout the body of the host and may facilitate crossing of the blood-brain barrier.

phagosomes are secreted, as discussed below. But this is a rare occurrence, and active processes must exist to retrieve membranes as phagosomes digest their prey or to maintain their size over extended periods when poorly digestible particles are engulfed.

The Phagosome as a Secretory Organelle

Recent work has demonstrated that phagosomes are secretory organelles (Di et al., 2002). Macrophages containing IgG-coated latex beads were reported to expel their contents into the medium when treated with a secretagogue (i.e., GTP-γS) or when exposed to heat-aggregated IgG. This study also noted that reactive oxygen species were abruptly released from the macrophage under these conditions. This ability could be especially important in situations where phagocytes are exposed to large numbers of pathogens. Under these conditions extensive digestion and antigen presentation may not be a priority for survival and the ability to partially damage multiple pathogens while releasing to the medium microbicidal molecules could be advantageous. Regardless of teleological interpretations, the work of Di et al. (2002) demonstrates that phagosomes can be secreted from macrophages in a manner akin to secretory lysosomes (Fig. 4).

Emerging evidence suggests that phagosome secretion is more than a laboratory curiosity and may be important in at least one form of pathogenesis. As described throughout this volume, pathogens have evolved several mechanisms to avoid or subvert the immune response, specifically killing by phagocytes. Recent studies demonstrate that the human pathogen *Cryptococcus neoformans* is able to escape from macrophages without damaging the macrophage. The process appears to be rapid, can occur several hours after phagocytosis, and does not require phagosome maturation to occur. The current molecular mechanisms of this "phagosomal extrusion" are unknown, but it is clear that release of the pathogen from phagocytic (invasion) vacuoles is involved (Alvarez and Casadevall, 2006; Ma et al., 2006). *C. neoformans* infections can spread rapidly throughout the body and the fungus can be isolated from circulating monocytes. Perhaps the ability to reside in phagosomes within monocytes/macrophages and to be subsequently regurgitated enables the pathogen to spread throughout the host. Of note, *Cryptococcus* reaches even the central nervous system, where it can cause encephalitis. In all likelihood the fungus traverses the blood-brain barrier while inside macrophages, managing to escape by exocytosis afterward.

CONCLUDING REMARKS

Phagosome formation and maturation are impressive microbicidal tools. For this reason pathogens have developed a remarkable variety of strategies to subvert this process. Some of the most virulent and persistent bacterial pathogens such as *Mycobacterium* and *Leishmania* in fact take advantage of the phagocytic machinery to gain access into the host cell interior, where they are able to circumvent the sophisticated killing mechanism of the maturing phagosome (Scott et al., 2003). It is clear that better management and prevention of infection will require a thorough understanding of the microbial mode of action, which in turn necessitates understanding of the microbicidal pathways deployed by the host cells. At present our knowledge of phagosome maturation is rudimentary and largely extrapolated from that garnered for the endocytic pathway. Some extrapolation is probably warranted, but many unique features of phagosomes will only be revealed by direct studies of particle, preferably microbial, ingestion.

Original work in the authors' laboratory is supported by Heart and Stroke Foundation of Canada, the Canadian Institutes of Health Research (CIHR), and the Canadian Cystic Fibrosis Foundation (CCFF). G.D.F. is the recipient of a postdoctoral fellowship from the CIHR. E.G. is the recipient of a Hospital for Sick Children Restracomp Studentship. S.G. is the current holder of the Pitblado Chair in Cell Biology. G.F. and E.G. contributed equally to this work.

REFERENCES

Aderem, A. 2002. How to eat something bigger than your head. *Cell* **110:**5–8.

Aggeler, J., and Z. Werb. 1982. Initial events during phagocytosis by macrophages viewed from outside and inside the cell: membrane-particle interactions and clathrin. *J. Cell Biol.* **94:**613–623.

Allen, L. A., C. Yang, and J. E. Pessin. 2002. Rate and extent of phagocytosis in macrophages lacking vamp3. *J. Leukoc. Biol.* **72:**217–221.

Alvarez, M., and A. Casadevall. 2006. Phagosome extrusion and host-cell survival after Cryptococcus neoformans phagocytosis by macrophages. *Curr. Biol.* **16:**2161–2165.

Alvarez-Dominguez, C. 1996. Vesicular transport of microorganisms in macrophages. *Biocell* **20:**355–365.

Aniento, F., F. Gu, R. G. Parton, and J. Gruenberg. 1996. An endosomal beta COP is involved in the pH-dependent formation of transport vesicles destined for late endosomes. *J. Cell Biol.* **133:**29–41.

Babst, M. 2005. A protein's final ESCRT. *Traffic* **6:**2–9.

Babst, M. 2006. A close-up of the ESCRTs. *Dev. Cell.* **10:**547–548.

Babst, M., D. J. Katzmann, E. J. Estepa-Sabal, T. Meerloo, and S. D. Emr. 2002a. Escrt-III: an endosome-associated heterooligomeric protein complex required for mvb sorting. *Dev. Cell.* **3:**271–282.

Babst, M., D. J. Katzmann, W. B. Snyder, B. Wendland, and S. D. Emr. 2002b. Endosome-associated complex, ESCRT-II, recruits transport machinery for protein sorting at the multivesicular body. *Dev. Cell.* **3:**283–289.

Bache, K. G., C. Raiborg, A. Mehlum, and H. Stenmark. 2003. STAM and Hrs are subunits of a multivalent ubiquitin-binding complex on early endosomes. *J. Biol. Chem.* **278:**12513–12521.

Bajno, L., X. R. Peng, A. D. Schreiber, H. P. Moore, W.S. Trimble, and S. Grinstein. 2000. Focal exocytosis of VAMP3-containing vesicles at sites of phagosome formation. *J. Cell Biol.* **149:**697–706.

Barbieri, M. A., A. D. Kohn, R. A. Roth, and P. D. Stahl. 1998. Protein kinase B/akt and rab5 mediate Ras activation of endocytosis. *J. Biol. Chem.* **273:**19367–19370.

Becker, T., A. Volchuk, and J. E. Rothman. 2005. Differential use of endoplasmic reticulum membrane for phagocytosis in J774 macrophages. *Proc. Natl. Acad. Sci. USA* **102:**4022–4026.

Beron, W., C. Alvarez-Dominguez, L. Mayorga, and P. D. Stahl. 1995. Membrane trafficking along the phagocytic pathway. *Trends Cell Biol.* **5:**100–104.

Bonifacino, J. S., and B. S. Glick. 2004. The mechanisms of vesicle budding and fusion. *Cell* **116:**153–166.

Botelho, R. J., D. J. Hackam, A. D. Schreiber, and S. Grinstein. 2000. Role of COPI in phagosome maturation. *J. Biol. Chem.* **275:**15717–15727.

Braun, V., V. Fraisier, G. Raposo, I. Hurbain, J. B. Sibarita, P. Chavrier, T. Galli, and F. Niedergang. 2004. TI-VAMP/VAMP7 is required for optimal phagocytosis of opsonised particles in macrophages. *EMBO J.* **23:**4166–76.

Bucci, C., P. Thomsen, P. Nicoziani, J. McCarthy, and B. van Deurs. 2000. Rab7: a key to lysosome biogenesis. *Mol. Biol. Cell* **11:**467–80.

Cabezas, A., K. Pattni, and H. Stenmark. 2006. Cloning and subcellular localization of a human phosphatidylinositol 3-phosphate 5-kinase, PIKfyve/Fab1. *Gene* **371:**34–41.

Callaghan, J., A. Simonsen, J. M. Gaullier, B. H. Toh, and H. Stenmark. 1999. The endosome fusion regulator early-endosomal autoantigen 1 (EEA1) is a dimer. *Biochem. J.* **338(Pt 2):**539–543.

Cantalupo, G., P. Alifano, V. Roberti, C. B. Bruni, and C. Bucci. 2001. Rab-interacting lysosomal protein (RILP): the Rab7 effector required for transport to lysosomes. *EMBO J.* **20:**683–693.

Carroll, K. S., J. Hanna, I. Simon, J. Krise, P. Barbero, and S. R. Pfeffer. 2001. Role of Rab9 GTPase in facilitating receptor recruitment by TIP47. *Science* **292:**1373–1376.

Clague, M. J., and S. Urbe. 2003. Hrs function: viruses provide the clue. *Trends Cell Biol.* **13:**603–606.

Collins, R. F., A. D. Schreiber, S. Grinstein, and W. S. Trimble. 2002. Syntaxins 13 and 7 function at distinct steps during phagocytosis. *J. Immunol.* **169:**3250–3256.

Colucci, A. M., M. C. Campana, M. Bellopede, and C. Bucci. 2005. The Rab-interacting lysosomal protein, a Rab7 and Rab34 effector, is capable of self-interaction. *Biochem. Biophys. Res. Commun.* **334:**128–133.

Coppolino, M. G., C. Kong, M. Mohtashami, A. D. Schreiber, J. H. Brumell, B. B. Finlay, S. Grinstein, and W. S. Trimble. 2001. Requirement for N-ethylmaleimide-sensitive factor activity at different stages of bacterial invasion and phagocytosis. *J. Biol. Chem.* **276:**4772–4780.

Cox, D., D. J. Lee, B. M. Dale, J. Calafat, and S. Greenberg. 2000. A Rab11-containing rapidly recycling compartment in macrophages that promotes phagocytosis. *Proc. Natl. Acad. Sci. USA* **97:**680–685.

Czibener, C., N. M. Sherer, S. M. Becker, M. Pypaert, E. Hui, E. R. Chapman, W. Mothes, and N. W. Andrews. 2006. Ca2+ and synaptotagmin VII-dependent delivery of lysosomal membrane to nascent phagosomes. *J. Cell Biol.* **174:**997–1007.

Desjardins, M. 1995. Biogenesis of phagolysosomes: the 'kiss and run' hypothesis. *Trends Cell Biol.* **5:**183–186.

Di, A., B. Krupa, V. P. Bindokas, Y. Chen, M. E. Brown, H. C. Palfrey, A. P. Naren, K. L. Kirk, and D. J. Nelson. 2002. Quantal release of free radicals during exocytosis of phagosomes. *Nat. Cell Biol.* **4:**279–285.

Diaz, E., and S. R. Pfeffer. 1998. TIP47: a cargo selection device for mannose 6-phosphate receptor trafficking. *Cell* **93:**433–443.

Diaz, E., F. Schimmoller, and S. R. Pfeffer. 1997. A novel Rab9 effector required for endosome-to-TGN transport. *J. Cell Biol.* **138:**283–290.

Duclos, S., R. Diez, J. Garin, B. Papadopoulou, A. Descoteaux, H. Stenmark, and M. Desjardins. 2000. Rab5 regulates the kiss and run fusion between phagosomes and endosomes and the acquisition of phagosome leishmanicidal properties in RAW 264.7 macrophages. *J. Cell Sci.* **113**(Pt 19)**:**3531–3541.

Ellson, C. D., K. E. Anderson, G. Morgan, E. R. Chilvers, P. Lipp, L. R. Stephens, and P. T. Hawkins. 2001. Phosphatidylinositol 3-phosphate is generated in phagosomal membranes. *Curr. Biol.* **11:**1631–1635.

Ellson, C. D., S. Andrews, L. R. Stephens, and P. T. Hawkins. 2002. The PX domain: a new phosphoinositide-binding module. *J. Cell Sci.* **115:**1099–1105.

Eskelinen, E. L., Y. Tanaka, and P. Saftig. 2003. At the acidic edge: emerging functions for lysosomal membrane proteins. *Trends Cell Biol.* **13:**137–145.

Evans, E., V. Heinrich, F. Ludwig, and W. Rawicz. 2003. Dynamic tension spectroscopy and strength of biomembranes. *Biophys. J.* **85:**2342–2350.

Fasshauer, D., R. B. Sutton, A. T. Brunger, and R. Jahn. 1998. Conserved structural features of the synaptic fusion complex: SNARE proteins reclassified as Q- and R-SNAREs. *Proc. Natl. Acad. Sci. USA* **95:**15781–15786.

Faurschou, M., and N. Borregaard. 2003. Neutrophil granules and secretory vesicles in inflammation. *Microbes Infect.* **5:**1317–1327.

Fielding, A. B., E. Schonteich, J. Matheson, G. Wilson, X. Yu, G. R. Hickson, S. Srivastava, S. A. Baldwin, R. Prekeris, and G. W. Gould. 2005. Rab11-FIP3 and FIP4 interact with Arf6 and the exocyst to control membrane traffic in cytokinesis. *EMBO J.* **24:**3389–3399.

Gagnon, E., S. Duclos, C. Rondeau, E. Chevet, P. H. Cameron, O. Steele-Mortimer, J. Paiement, J. J. Bergeron, and M. Desjardins. 2002. Endoplasmic reticulum-mediated phagocytosis is a mechanism of entry into macrophages. *Cell* **110:**119–131.

Ganley, I. G., K. Carroll, L. Bittova, and S. Pfeffer. 2004. Rab9 GTPase regulates late endosome size and requires effector interaction for its stability. *Mol. Biol. Cell* **15:**5420–5430.

Ganley, I. G., and S. R. Pfeffer. 2006. Cholesterol accumulation sequesters Rab9 and disrupts late endosome function in NPC1-deficient cells. *J. Biol. Chem.* **281:**17890–17899.

Gaullier, J. M., A. Simonsen, A. D'Arrigo, B. Bremnes, H. Stenmark, and R. Aasland. 1998. FYVE fingers bind PtdIns(3)P. *Nature* **394:**432–433.

Geisow, M. J., P. D'Arcy Hart, and M. R. Young. 1981. Temporal changes of lysosome and phagosome pH during phagolysosome formation in macrophages: studies by fluorescence spectroscopy. *J. Cell Biol.* **89:**645–652.

Gold, E. S., D. M. Underhill, N. S. Morrissette, J. Guo, M. A. McNiven, and A. Aderem. 1999. Dynamin 2 is required for phagocytosis in macrophages. *J. Exp. Med.* **190:**1849–1856.

Griffin, F. M., Jr., J. A. Griffin, J. E. Leider, and S. C. Silverstein. 1975. Studies on the mechanism of phagocytosis. I. Requirements for circumferential attachment of particle-bound ligands to specific receptors on the macrophage plasma membrane. *J. Exp. Med.* **142:**1263–1282.

Gruenberg, J. 2001. The endocytic pathway: a mosaic of domains. *Nat. Rev. Mol. Cell Biol.* **2:**721–730.

Guo, W., A. Grant, and P. Novick. 1999. Exo84p is an exocyst protein essential for secretion. *J. Biol. Chem.* **274:**23558–23564.

Haas, A. K., E. Fuchs, R. Kopajtich, and F. A. Barr. 2005. A GTPase-activating protein controls Rab5 function in endocytic trafficking. *Nat. Cell Biol.* **7:**887–893.

Hackam, D. J., O. D. Rotstein, M. K. Bennett, A. Klip, S. Grinstein, and M. F. Manolson. 1996. Characterization and subcellular localization of target membrane soluble NSF attachment protein receptors (t-SNAREs) in macrophages. Syntaxins 2, 3, and 4 are present on phagosomal membranes. *J. Immunol.* **156:**4377–4383.

Hackam, D. J., O. D. Rotstein, C. Sjolin, A. D. Schreiber, W. S. Trimble, and S. Grinstein. 1998. v-SNARE-dependent secretion is required for phagocytosis. *Proc. Natl. Acad. Sci. USA* **95:**11691–11696.

Hallett, M. B., and S. Dewitt. 2007. Ironing out the wrinkles of neutrophil phagocytosis. *Trends Cell Biol.* **17:**209–214.

Hoffenberg, S., X. Liu, L. Nikolova, H. S. Hall, W. Dai, R. E. Baughn, B. F. Dickey, M. A. Barbieri, A. Aballay, P. D. Stahl, and B. J. Knoll. 2000. A novel membrane-anchored Rab5 interacting protein required for homotypic endosome fusion. *J. Biol. Chem.* **275:**24661–24669.

Holevinsky, K. O., and D. J. Nelson. 1998. Membrane capacitance changes associated with particle uptake during phagocytosis in macrophages. *Biophys. J.* **75:**2577–2586.

Horiuchi, H., R. Lippe, H. M. McBride, M. Rubino, P. Woodman, H. Stenmark, V. Rybin, M. Wilm, K. Ashman, M. Mann, and M. Zerial. 1997. A novel Rab5 GDP/GTP exchange factor complexed to Rabaptin-5 links nucleotide exchange to effector recruitment and function. *Cell* **90:**1149–1159.

Huynh, K. K., E. L. Eskelinen, C. C. Scott, A. Malevanets, P. Saftig, and S. Grinstein. 2007. LAMP proteins are required for fusion of lysosomes with phagosomes. *EMBO J.* **26:**313–324.

Johansson, M., M. Lehto, K. Tanhuanpaa, T. L. Cover, and V. M. Olkkonen. 2005. The oxysterol-binding protein homologue ORP1L interacts with Rab7 and alters functional properties of late endocytic compartments. *Mol. Biol. Cell.* **16:**5480–5492.

Jordens, I., M. Fernandez-Borja, M. Marsman, S. Dusseljee, L. Janssen, J. Calafat, H. Janssen, R. Wubbolts, and J. Neefjes. 2001. The Rab7 effector protein RILP controls lysosomal transport by inducing the recruitment of dynein-dynactin motors. *Curr. Biol.* **11:**1680–1685.

Kobayashi, T., and Y. Hirabayashi. 2000. Lipid membrane domains in cell surface and vacuolar systems. *Glycoconj. J.* **17:**163–171.

Koh, T. W., and H. J. Bellen. 2003. Synaptotagmin I, a Ca2+ sensor for neurotransmitter release. *Trends Neurosci.* **26:**413–422.

Lanzetti, L., V. Rybin, M. G. Malabarba, S. Christoforidis, G. Scita, M. Zerial, and P. P. Di Fiore. 2000. The Eps8 protein coordinates EGF receptor signalling through Rac and trafficking through Rab5. *Nature* **408:**374–377.

Lavigne, L. M., J. E. Albina, and J. S. Reichner. 2006. Beta-glucan is a fungal determinant for adhesion-dependent human neutrophil functions. *J. Immunol.* **177:**8667–8675.

Lawe, D. C., V. Patki, R. Heller-Harrison, D. Lambright, and S. Corvera. 2000. The FYVE domain of early endosome antigen 1 is required for both phosphatidylinositol 3-

phosphate and Rab5 binding. Critical role of this dual interaction for endosomal localization. *J. Biol. Chem.* **275**:3699–3705.

Lebrand, C., M. Corti, H. Goodson, P. Cosson, V. Cavalli, N. Mayran, J. Faure, and J. Gruenberg. 2002. Late endosome motility depends on lipids via the small GTPase Rab7. *EMBO J.* **21**:1289–1300.

Leevers, S. J., B. Vanhaesebroeck, and M. D. Waterfield. 1999. Signalling through phosphoinositide 3-kinases: the lipids take centre stage. *Curr. Opin. Cell Biol.* **11**:219–225.

Li, G., C. D'Souza-Schorey, M. A. Barbieri, J. A. Cooper, and P. D. Stahl. 1997. Uncoupling of membrane ruffling and pinocytosis during Ras signal transduction. *J. Biol. Chem.* **272**:10337–10340.

Li, J., A. Scherl, F. Medina, P. G. Frank, R. N. Kitsis, H. B. Tanowitz, F. Sotgia, and M. P. Lisanti. 2005. Impaired phagocytosis in caveolin-1 deficient macrophages. *Cell Cycle* **4**:1599–1607.

Lippe, R., M. Miaczynska, V. Rybin, A. Runge, and M. Zerial. 2001. Functional synergy between Rab5 effector Rabaptin-5 and exchange factor Rabex-5 when physically associated in a complex. *Mol. Biol. Cell* **12**:2219–2228.

Liu, K., and G. Li. 1998. Catalytic domain of the p120 Ras GAP binds to RAb5 and stimulates its GTPase activity. *J. Biol. Chem.* **273**:10087–10090.

Ma, H., J. E. Croudace, D. A. Lammas, and R. C. May. 2006. Expulsion of live pathogenic yeast by macrophages. *Curr. Biol.* **16**:2156–2160.

Malerod, L., and H. Stenmark. 2007. ESCRTs. *Curr. Biol.* **17**:R42–R43.

Matsuo, H., J. Chevallier, N. Mayran, I. Le Blanc, C. Ferguson, J. Faure, N. S. Blanc, S. Matile, J. Dubochet, R. Sadoul, R. G. Parton, F. Vilbois, and J. Gruenberg. 2004. Role of LBPA and Alix in multivesicular liposome formation and endosome organization. *Science* **303**:531–534.

Mizuno, K., A. Kitamura, and T. Sasaki. 2003. Rabring7, a novel Rab7 target protein with a RING finger motif. *Mol. Biol. Cell* **14**:3741–3752.

Mobius, W., E. van Donselaar, Y. Ohno-Iwashita, Y. Shimada, H. F. Heijnen, J. W. Slot, and H. J. Geuze. 2003. Recycling compartments and the internal vesicles of multivesicular bodies harbor most of the cholesterol found in the endocytic pathway. *Traffic* **4**:222–231.

Mollinedo, F., J. Calafat, H. Janssen, B. Martin-Martin, J. Canchado, S. M. Nabokina, and C. Gajate. 2006. Combinatorial SNARE complexes modulate the secretion of cytoplasmic granules in human neutrophils. *J. Immunol.* **177**:2831–2841.

Mu, F. T., J. M. Callaghan, O. Steele-Mortimer, H. Stenmark, R. G. Parton, P. L. Campbell, J. McCluskey, J. P. Yeo, E. P. Tock, and B. H. Toh. 1995. EEA1, an early endosome-associated protein. EEA1 is a conserved alpha-helical peripheral membrane protein flanked by cysteine "fingers" and contains a calmodulin-binding IQ motif. *J. Biol. Chem.* **270**:13503–13511.

Mukherjee, S., R. N. Ghosh, and F. R. Maxfield. 1997. Endocytosis. *Physiol. Rev.* **77**:759–803.

Murray, J. T., and J. M. Backer. 2005. Analysis of hVps34/hVps15 interactions with Rab5 in vivo and in vitro. *Methods Enzymol.* **403**:789–799.

Murray, J. T., G. Craggs, L. Wilson, and S. Kellie. 2000. Mechanism of phosphatidylinositol 3-kinase-dependent increases in BAC1.2F5 macrophage-like cell density in response to M-CSF: phosphatidylinositol 3-kinase inhibitors increase the rate of apoptosis rather than inhibit DNA synthesis. *Inflamm. Res.* **49**:610–618.

Murray, J. T., C. Panaretou, H. Stenmark, M. Miaczynska, and J. M. Backer. 2002. Role of Rab5 in the recruitment of hVps34/p150 to the early endosome. *Traffic* **3**:416–427.

Niedergang, F., E. Colucci-Guyon, T. Dubois, G. Raposo, and P. Chavrier. 2003. ADP ribosylation factor 6 is activated and controls membrane delivery during phagocytosis in macrophages. *J. Cell Biol.* **161**:1143–1150.

Novick, P., C. Field, and R. Schekman. 1980. Identification of 23 complementation groups required for post-translational events in the yeast secretory pathway. *Cell* **21**:205–215.

Pagan, J. K., F. G. Wylie, S. Joseph, C. Widberg, N. J. Bryant, D. E. James, and J. L. Stow. 2003. The t-SNARE syntaxin 4 is regulated during macrophage activation to function in membrane traffic and cytokine secretion. *Curr. Biol.* **13**:156–160.

Patki, V., D. C. Lawe, S. Corvera, J. V. Virbasius, and A. Chawla. 1998. A functional PtdIns(3)P-binding motif. *Nature* **394**:433–434.

Pfeffer, S. 2005. Filling the Rab GAP. *Nat. Cell Biol.* **7**:856–857.

Pfeffer, S., and D. Aivazian. 2004. Targeting Rab GTPases to distinct membrane compartments. *Nat. Rev. Mol. Cell Biol.* **5**:886–896.

Pitt, A., L. S. Mayorga, P. D. Stahl, and A. L. Schwartz. 1992. Alterations in the protein composition of maturing phagosomes. *J. Clin. Investig.* **90**:1978–1983.

Prigent, M., T. Dubois, G. Raposo, V. Derrien, D. Tenza, C. Rosse, J. Camonis, and P. Chavrier. 2003. ARF6 controls post-endocytic recycling through its downstream exocyst complex effector. *J. Cell Biol.* **163**:1111–1121.

Progida, C., M. R. Spinosa, A. De Luca, and C. Bucci. 2006. RILP interacts with the VPS22 component of the ESCRT-II complex. *Biochem. Biophys. Res. Commun.* **347**:1074–1079.

Raiborg, C. 2006. Hrs makes receptors silent: a key to endosomal protein sorting. *Crit. Rev. Oncog.* **12**:295–296.

Raiborg, C., K. G. Bache, A. Mehlum, E. Stang, and H. Stenmark. 2001a. Hrs recruits clathrin to early endosomes. *EMBO J.* **20**:5008–5021.

Raiborg, C., K. G. Bache, A. Mehlum, and H. Stenmark. 2001b. Function of Hrs in endocytic trafficking and signalling. *Biochem. Soc. Trans.* **29**:472–475.

Riederer, M. A., T. Soldati, A. D. Shapiro, J. Lin, and S. R. Pfeffer. 1994. Lysosome biogenesis requires Rab9 function and receptor recycling from endosomes to the trans-Golgi network. *J. Cell Biol.* **125**:573–582.

Rink, J., E. Ghigo, Y. Kalaidzidis, and M. Zerial. 2005. Rab conversion as a mechanism of progression from early to late endosomes. *Cell* **122**:735–749.

Roberts, R. L., M. A. Barbieri, J. Ullrich, and P. D. Stahl. 2000. Dynamics of rab5 activation in endocytosis and phagocytosis. *J. Leukoc. Biol.* **68**:627–632.

Robinson, F. L., and J. E. Dixon. 2006. Myotubularin phosphatases: policing 3-phosphoinositides. *Trends Cell Biol.* **16**:403–412.

Schekman, R., and P. Novick. 2004. 23 genes, 23 years later. *Cell* **116**:S13–S15, 1 p following S19.

Scott, C. C., R. J. Botelho, and S. Grinstein. 2003. Phagosome maturation: a few bugs in the system. *J. Membr. Biol.* **193**:137–152.

Shisheva, A. 2001. PIKfyve: the road to PtdIns 5-P and PtdIns 3,5-P(2). *Cell Biol. Int.* **25**:1201–1206.

Slagsvold, T., K. Pattni, L. Malerod, and H. Stenmark. 2006. Endosomal and non-endosomal functions of ESCRT proteins. *Trends Cell Biol.* **16**:317–326.

Snyderman, R., M. C. Pike, D. G. Fischer, and H. S. Koren. 1977. Biologic and biochemical activities of continuous macrophage cell lines P388D1 and J774.1. *J. Immunol.* **119**:2060–2066.

Somsel Rodman, J., and A. Wandinger-Ness. 2000. Rab GTPases coordinate endocytosis. *J. Cell Sci.* **113**(Pt 2):183–192.

Srivastava, S., Z. Li, L. Lin, G. Liu, K. Ko, W. A. Coetzee, and E. Y. Skolnik. 2005. The phosphatidylinositol 3-phosphate phosphatase myotubularin-related protein 6 (MTMR6) is a negative regulator of the Ca2+-activated K+ channel KCa3.1. *Mol. Cell Biol.* **25:**3630–3638.

Stenmark, H., R. Aasland, B. H. Toh, and A. D'Arrigo. 1996. Endosomal localization of the autoantigen EEA1 is mediated by a zinc-binding FYVE finger. *J. Biol. Chem.* **271:**24048–24054.

Stoorvogel, W., V. Oorschot, and H. J. Geuze. 1996. A novel class of clathrin-coated vesicles budding from endosomes. *J. Cell Biol.* **132:**21–33.

Stow, J. L., A. P. Manderson, and R. Z. Murray. 2006. SNAREing immunity: the role of SNAREs in the immune system. *Nat. Rev. Immunol.* **6:**919–929.

Stuart, L. M., J. Boulais, G. M. Charriere, E. J. Hennessy, S. Brunet, I. Jutras, G. Goyette, C. Rondeau, S. Letarte, H. Huang, P. Ye, F. Morales, C. Kocks, J. S. Bader, M. Desjardins, and R. A. Ezekowitz. 2007. A systems biology analysis of the Drosophila phagosome. *Nature* **445:**95–101.

Tapper, H., W. Furuya, and S. Grinstein. 2002. Localized exocytosis of primary (lysosomal) granules during phagocytosis: role of Ca2+-dependent tyrosine phosphorylation and microtubules. *J. Immunol.* **168:**5287–5296.

Tapper, H., and S. Grinstein. 1997. Fc receptor-triggered insertion of secretory granules into the plasma membrane of human neutrophils: selective retrieval during phagocytosis. *J. Immunol.* **159:**409–418.

TerBush, D. R., T. Maurice, D. Roth, and P. Novick. 1996. The Exocyst is a multiprotein complex required for exocytosis in Saccharomyces cerevisiae. *EMBO J.* **15:**6483–6494.

Tjelle, T. E., T. Lovdal, and T. Berg. 2000. Phagosome dynamics and function. *BioEssays* **22:**255–263.

Touret, N., P. Paroutis, and S. Grinstein. 2005. The nature of the phagosomal membrane: endoplasmic reticulum versus plasmalemma. *J. Leukoc. Biol.* **77:**878–885.

Tse, S. M., W. Furuya, E. Gold, A. D. Schreiber, K. Sandvig, R. D. Inman, and S. Grinstein. 2003. Differential role of actin, clathrin, and dynamin in Fc gamma receptor-mediated endocytosis and phagocytosis. *J. Biol. Chem.* **278:**3331–3338.

Urbe, S., M. Sachse, P. E. Row, C. Preisinger, F. A. Barr, G. Strous, J. Klumperman, and M. J. Clague. 2003. The UIM domain of Hrs couples receptor sorting to vesicle formation. *J. Cell Sci.* **116:**4169–4179.

Vanhaesebroeck, B., S. J. Leevers, G. Panayotou, and M. D. Waterfield. 1997. Phosphoinositide 3-kinases: a conserved family of signal transducers. *Trends Biochem. Sci.* **22:**267–272.

Vieira, O. V., R. J. Botelho, and S. Grinstein. 2002. Phagosome maturation: aging gracefully. *Biochem. J.* **366:**689–704.

Vieira, O. V., R. J. Botelho, L. Rameh, S. M. Brachmann, T. Matsuo, H. W. Davidson, A. Schreiber, J. M. Backer, L. C. Cantley, and S. Grinstein. 2001. Distinct roles of class I and class III phosphatidylinositol 3-kinases in phagosome formation and maturation. *J. Cell Biol.* **155:**19–25.

Vieira, O. V., R. E. Harrison, C. C. Scott, H. Stenmark, D. Alexander, J. Liu, J. Gruenberg, A. D. Schreiber, and S. Grinstein. 2004. Acquisition of Hrs, an essential component of phagosomal maturation, is impaired by mycobacteria. *Mol. Cell Biol.* **24:**4593–4604.

Walker, D. M., S. Urbe, S. K. Dove, D. Tenza, G. Raposo, and M. J. Clague. 2001. Characterization of MTMR3. an inositol lipid 3-phosphatase with novel substrate specificity. *Curr. Biol.* **11:**1600–1605.

Wang, T., and W. Hong. 2006. RILP interacts with VPS22 and VPS36 of ESCRT-II and regulates their membrane recruitment. *Biochem. Biophys. Res. Commun.* **350:**413–423.

Weber, T., B. V. Zemelman, J. A. McNew, B. Westermann, M. Gmachl, F. Parlati, T. H. Sollner, and J. E. Rothman. 1998. SNAREpins: minimal machinery for membrane fusion. *Cell* **92:**759–772.

Zhang, X. M., S. Ellis, A. Sriratana, C. A. Mitchell, and T. Rowe. 2004. Sec15 is an effector for the Rab11 GTPase in mammalian cells. *J. Biol. Chem.* **279:**43027–43034.

Zimmerberg, J. 2006. Membrane biophysics. *Curr. Biol.* **16:**R272–R276.

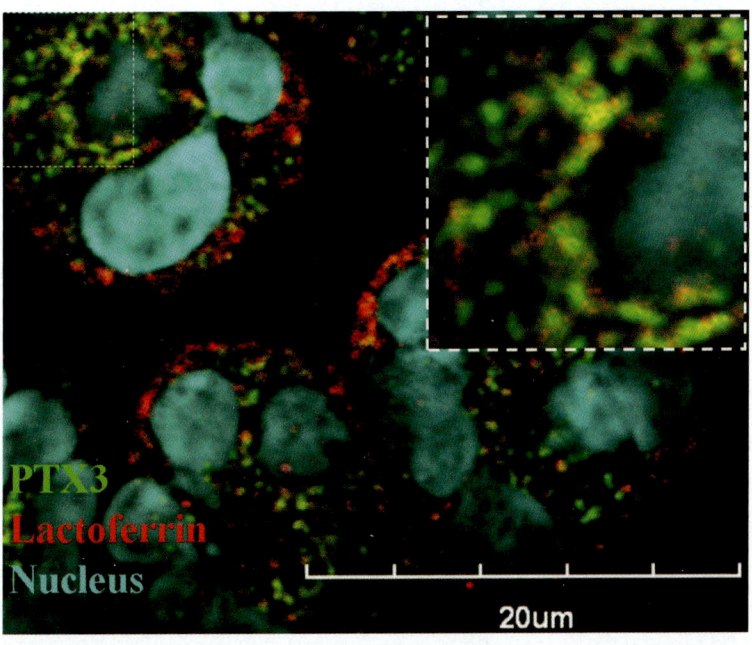

COLOR PLATE 1 (chapter 10) PTX3 is stored in neutrophil-specific granules. Human PMNs from peripheral blood were stained for PTX3 (green), lactoferrin (red), and DNA (Hoechst 33258) and analyzed by confocal microscopy. (Inset) Enlargement of the indicated area with colocalization of PTX3 with lactoferrin.

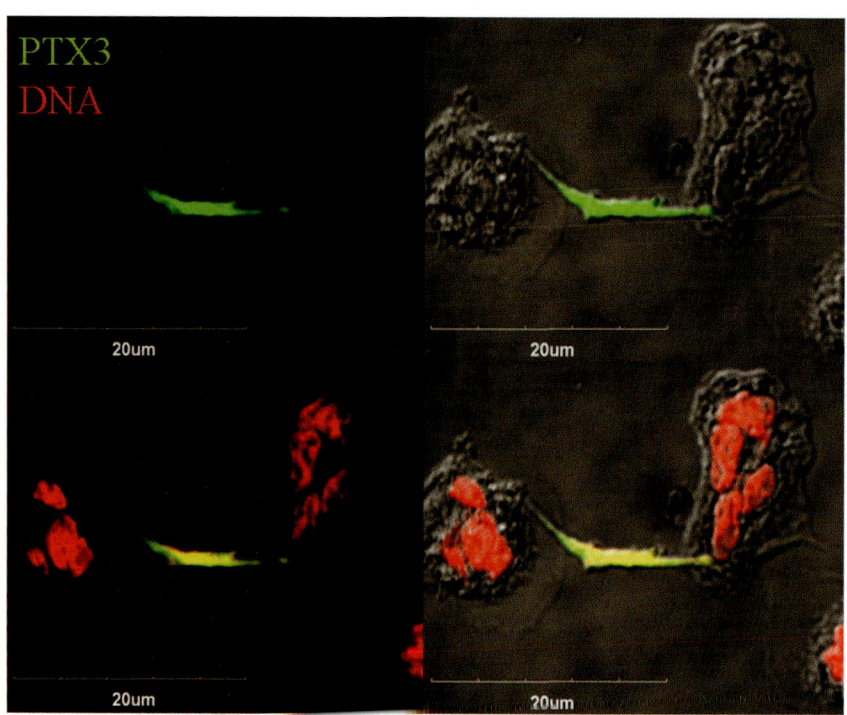

COLOR PLATE 2 (chapter 10) PTX3 is localized in NETs. Neutrophils were exposed to 100 ng/ml LPS for 40 min. PTX3 (green) and DNA (red) staining was done on nonpermeabilized neutrophils. A differential interference contrast (Nomarski technique) is shown in the right panels.

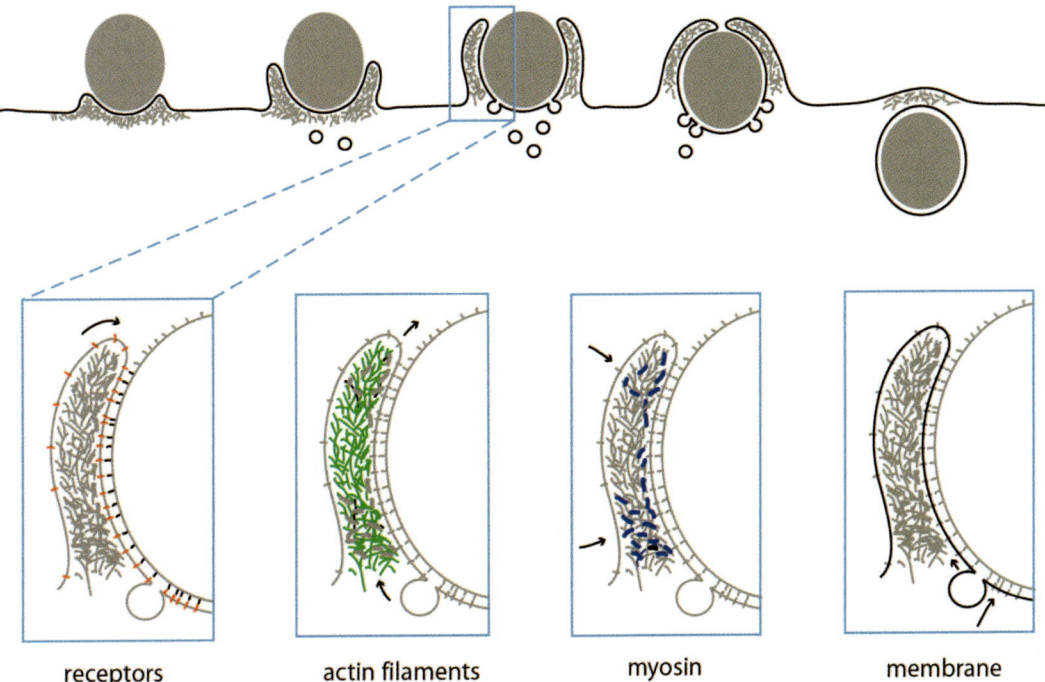

COLOR PLATE 3 (chapter 12) The movements of phagocytosis. The diagram shows a time series of the essential cellular movements during phagocytosis, viewed as a sagittal section. Binding of an opsonized particle (solid gray oval) to the phagocyte stimulates polymerization of actin (gray lines) and the extension of plasma membrane (black lines) as a cup-shaped phagocytic pseudopod. As the phagocytic cup extends outward around the particle, intracellular vesicles fuse with plasma membrane, and actin is cleared at the base of the cup. The phagosome closes at its distal margin into a discrete intracellular vacuole. Enlargements show the dynamics of essential components during pseudopod extension. As the pseudopod extends over the particle, receptors (red) diffusing in the plasma membrane engage cognate ligands (black) on the particle surface. Pseudopod extension is mediated in part by actin (green) polymerization at the distal margin of the phagocytic cup. Older actin filaments toward the base of the cup depolymerize. Several classes of myosin (blue) generate inward, circumferential contractile forces that constrict the phagosome and possibly additional contractile forces that pull the actin meshwork toward the distal margin. Phagosomal membranes derive from plasma membrane and intracellular vesicles that fuse in or near the phagocytic cup.

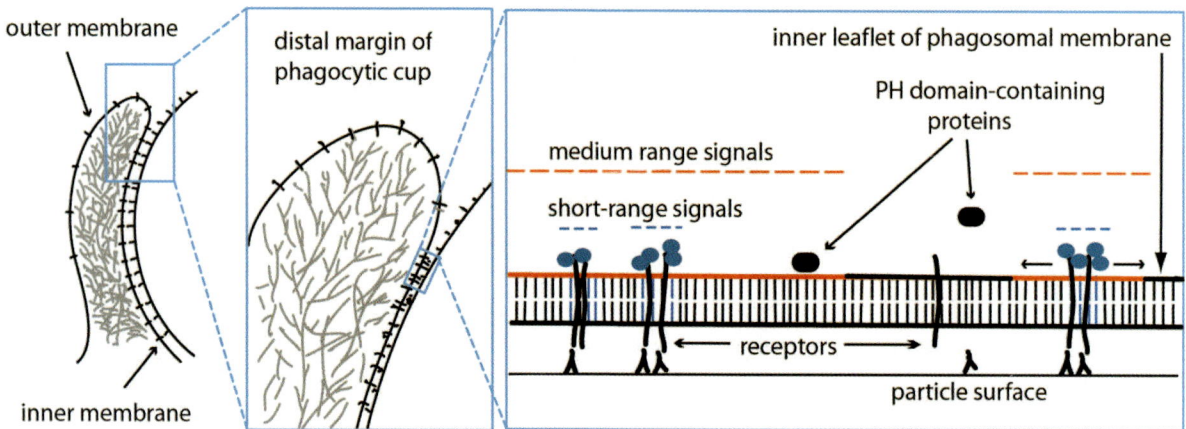

COLOR PLATE 4 (chapter 12) Short- and medium-range signaling in FcR-mediated phagocytosis. An enlarged region of the inner membrane of a phagocytic cup contains ligated and nonligated FcRs. Short-range signals that follow FcR binding to IgG include chemical modifications of the receptors, as well as the enzymes (blue circles) and lipids (blue lines) directly associated with the receptors. Lipids of the inner leaflet of the membrane are modified by receptor-associated enzymes. Diffusion of these lipids away from the receptors (red lines) creates a local area of modified membrane that can recruit or activate PH domain-containing proteins that recognize those modified lipids. Such medium range signaling can report the collective activities of many receptors, thereby coordinating or integrating the signals from multiple receptors.

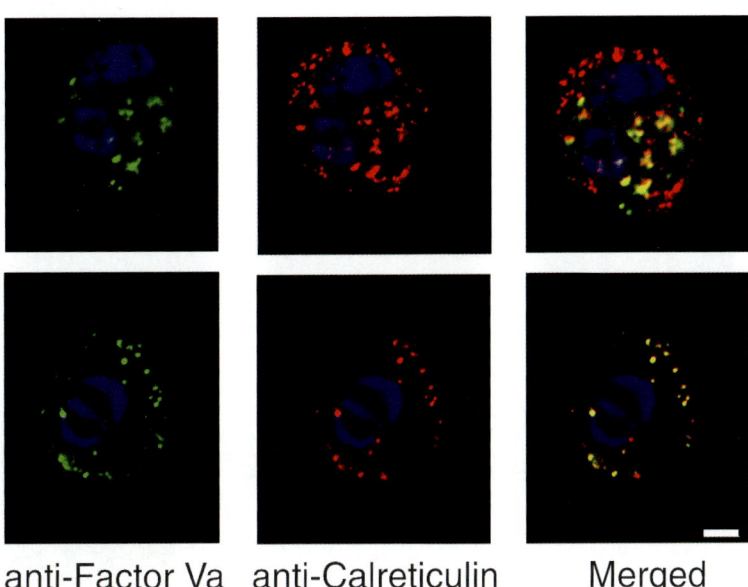

anti-Factor Va   anti-Calreticulin   Merged

COLOR PLATE 5 (chapter 21) Phosphatidylserine (PS) and calreticulin are exposed together in patches on apoptotic cells. An apoptotic neutrophil is depicted with surface PS (green) detected with fluorescent factor Va and calreticulin (red) with anticalreticulin antibody. (From Gardai et al., 2005.)

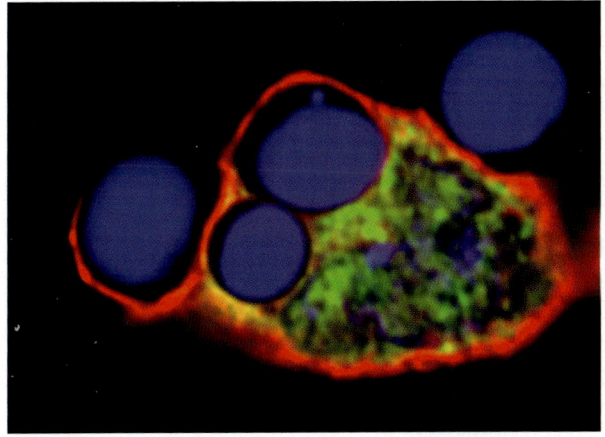

COLOR PLATE 6 (chapter 21) Phagocytosis of apoptotic cells; efferocytosis. Mouse resident peritoneal macrophage phagocytosing apoptotic (dexamethasone-treated) thymocytes. Blue = nuclei (Hoechst), green = Cell Tracker Green (macrophage cytosol only), red = anti-12/15-lipoxygenase, which in this micrograph is seen in the extended membrane ruffle. (Yury Miller, University of California, San Diego.)

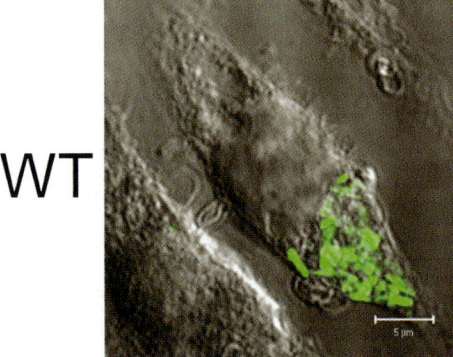

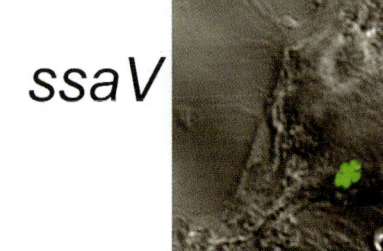

COLOR PLATE 7 (chapter 23) The SPI-2 T3SS enables *Salmonella* replication in macrophages. Mouse macrophage-like RAW cells were infected with green fluorescent protein-expressing wild-type or SPI-2 T3SS null (*ssaV*) mutant *Salmonella* for 8 h. Fluorescence and differential interference contrast images of the same cells are superimposed. Scale bar, 5 μm.

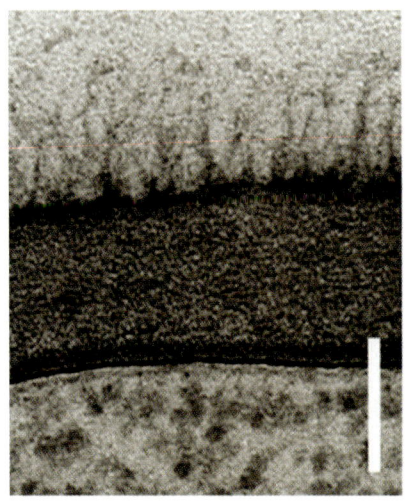
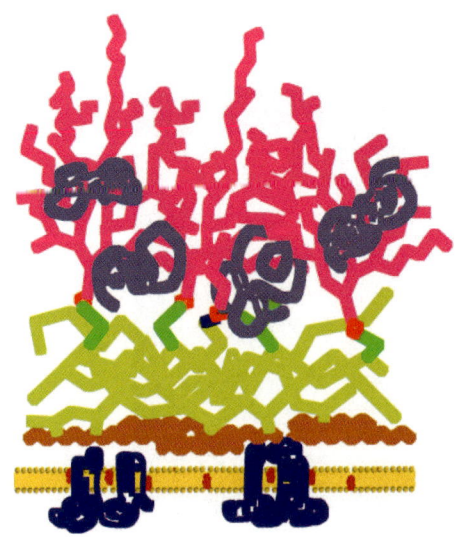

COLOR PLATE 8 (chapter 28) Structure of the cell wall of *C. albicans*. On the left is a transmission electron microscopy image clearly showing the fibrillar mannoprotein outer layer (scale bar is 100 μm); on the right is a cartoon representation of the major cell wall components. Image courtesy of Neil Gow (University of Aberdeen).

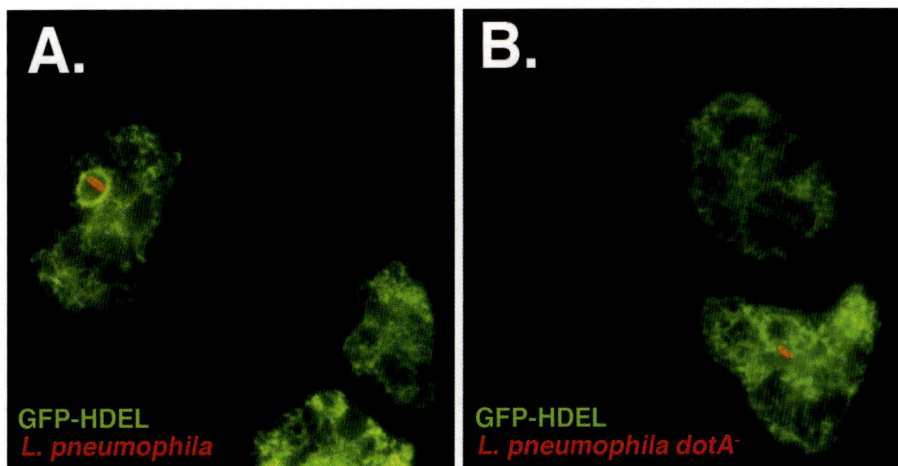

COLOR PLATE 9 (chapter 32) Recruitment of an ER-associated marker with the *L. pneumophila* replication vacuole. Fluorescent micrographs of *D. discoideum* challenged with either (A) *L. pneumophila* or (B) an *L. pneumophila dotA*[−] mutant having a nonfunctional Dot/Icm apparatus. Amoebae were challenged for 6 h with *L. pneumophila* fixed and stained with anti-*L. pneumophila* (red). The amoebae express a GFP-HDEL fusion protein (green stain) that has a 4-amino-acid, carboxyl-terminal ER-retention signal that allows labeling of the ER and intermediate compartments in the host cell. Note the large ring of GDP-HDEL about wild-type bacteria missing in the micrograph of the *dotA*[−] mutant. Micrograph is from Li et al. (2005).

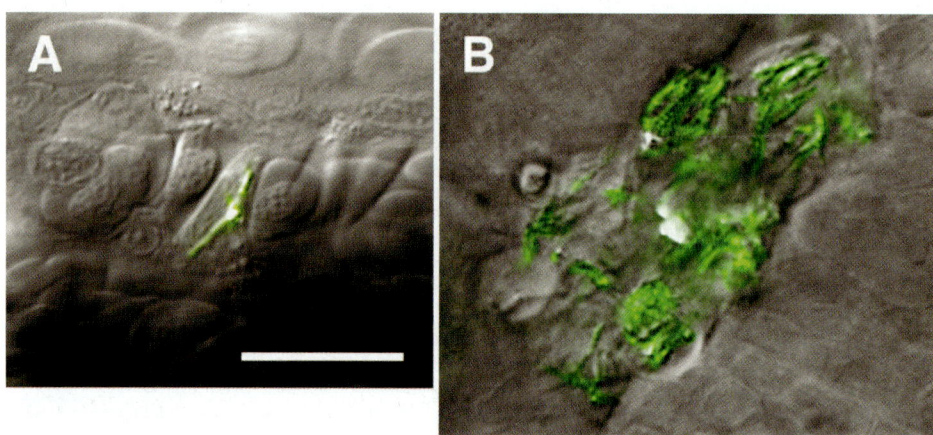

COLOR PLATE 10 (chapter 34) Larval zebrafish macrophages infected with M. *marinum* in vivo. (A) Single macrophage infected with green fluorescent M. *marinum*, surrounded by caudal hematopoietic tissue (dorsal is up). Immediately below is the caudal vein, where nucleated erythrocytes are flowing to the left (anterior). ~24 hpi. (B) Granuloma in the brain of a 5-day embryo. Green fluorescent M. *marinum* seen inside both living and dead (nonmotile) macrophages. Neuronal tissue is visible to the lower right. (Scale bar, 20 μm; both panels, same scale.)

14

Acidification of Endosomes and Phagosomes

SUSHMITA MUKHERJEE AND FREDERICK R. MAXFIELD

All nucleated mammalian cells internalize material from the extracellular milieu via one or more of the following processes: clathrin-mediated endocytosis, caveolar endocytosis, pinocytosis, macropinocytosis, phagocytosis, or various mechanisms generally grouped under uncoated membrane endocytosis. Although the details of the internalization mechanism differ in each instance, they all involve an invagination of the plasma membrane, which pinches off to form a sealed vesicle. In clathrin-mediated endocytosis, molecules internalized from the plasma membrane can be found in an endocytic organelle, the sorting endosome (SE), within 1 min of the initiation of endocytosis (Dunn et al., 1989). This organelle is acidic, with a pH of ~ 6.0 (Kielian et al., 1986; Sipe and Murphy, 1987; Yamashiro and Maxfield, 1987). In neurons expressing a synaptic vesicle protein with a pH-sensitive fluorescent chimeric protein in the luminal domain, acidification of synaptic vesicles following clathrin-mediated endocytosis was found to have a time constant of 4 to 5 s (Atluri and Ryan, 2006). Measurements with equivalent precision have not been made in nonneuronal cells, but it is likely that endosomes acidify rapidly after their severance from the plasma membrane.

In the low pH of the SE, many ligands are released from their receptors (e.g., insulin, low-density lipoproteins, α_2-macroglobulin, and asialoglycoproteins) (Mukherjee et al., 1997). When di-ferric transferrin (Fe_2Tf) bound to its receptor is internalized into the SE, the iron is released as a consequence of acidification, but the apo-Tf remains bound to its receptor at low pH (Maxfield and McGraw, 2004; Mukherjee et al., 1997). Apo-Tf is released from its receptor when it is recycled back to the cell surface.

There are several recycling routes that bring internalized membrane components back to the cell surface. A large fraction of internalized membrane lipids return directly to the plasma membrane within 1 to 2 min (Hao and Maxfield, 2000). Many receptors and other membrane-associated proteins and lipids are recycled to the plasma membrane via the endocytic recycling compartment (ERC), which has a pH somewhat more alkaline than the SE (6.0 to 6.5) (Mellman, 1992; Mukherjee et al., 1997). Some recycling molecules, such as the cation-independent mannose-6-phosphate receptor (CI-MPR), follow more complex recycling itineraries that include passage through the Golgi apparatus.

Interestingly, certain cell lines have been reported to acidify their Tf-containing SE to much lower pH values ($\sim$pH 5.4), and to not display alkalinization during recycling (Killisch et al., 1992; Sipe et al., 1991). To understand whether this difference might be an artifact of prolonged tissue culture, or whether primary cultures from mammalian cells would show such differences, early endosomal pH was measured in primary cells obtained from mouse kidney and heart explants (Rybak and Murphy, 1998). Indeed, it was found that cells from the heart explants (primarily fibroblasts and cardiac muscle cells) had a lower early endosomal pH (~ 5.5) than cells (primarily epithelial cells) from the kidney explants (~ 6.0). These different early endosomal pH values in different cell types might have functional implications.

Most of the volume and some membrane components (e.g., the transmembrane endopeptidase, furin) from the SE are delivered toward the late endosomal pathway, going from the SE to the late endosomes (LE; pH ~ 5.5) and then on to the lysosomes (LY; pH ≤ 5.0) (Mukherjee et al., 1997). Not all contents of the LE enter the LY; some, such as the CI-MPR, traffic between the LE and the *trans*-Golgi network (TGN), which maintains a mildly acidic pH (~ 5.9) (Demaurex et al., 1998). In general, the intraluminal pH of the endoplasmic reticulum is near neutral (~ 7.1) (Kim et al., 1998), and the pH is reported to decrease somewhat as the biosynthetic cargo travels through the Golgi (~ 6.5) (Kim et al., 1998) and finally reaches the TGN (~ 5.9) (Demaurex et al., 1998). The TGN thus serves as a common point between the endocytic and the biosynthetic (exocytic) pathway.

One of the major functions of the pH change between the extracellular milieu and the SE is the release of ligands from receptors, following which, the ligands traffic toward the LE/LY, whereas the empty receptors are recycled for

---

Sushmita Mukherjee and Frederick R. Maxfield, Department of Biochemistry, Weill Medical College of Cornell University, New York, NY 10065.

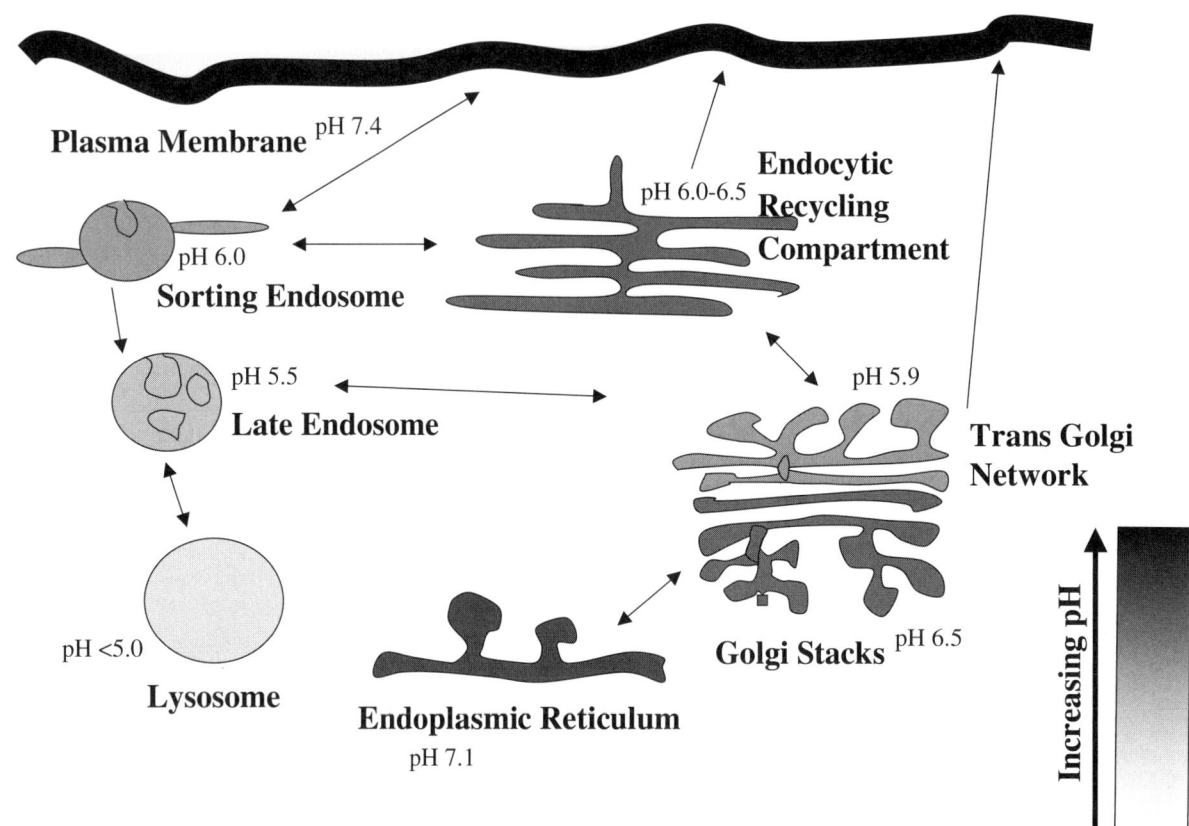

FIGURE 1 Schematic representation of various organelles that participate in phagocytic and endocytic processes. Organelles are indicated in shades of gray to represent the degree of acidity. See text for references to studies from which the pH values were taken.

further rounds of endocytosis (Mukherjee et al., 1997). The significantly lower pH of the LE/LY activates degradative enzymes such as proteases, esterases and lipases, all of which have low pH optima (Mellman, 1992; Mukherjee et al., 1997). These enzymes function in metabolic turnover of endocytosed material. Interestingly, in the case of pro-opiomelanocortin processing in the mouse pituitary AtT-20 cells, it was found that when the pH of permeabilized cells was clamped to pH values of 6.4 or 6.0 for 90 minutes, no processing of this protein took place. However, lowering the pH to 5.5 resulted in efficient processing (Schmidt and Moore, 1995). Thus, pH differences of less than 0.5 pH units can be enough to compartmentalize cellular events.

The phagocytic pathway is a specialized endocytic pathway seen in cells that are capable of phagocytosis. Certain cells, such as macrophages and neutrophils, are termed professional phagocytes since they specialize in phagocytosis as a mechanism for cells to degrade unwanted material, which includes foreign particles such as bacteria, as well as necrotic or apoptotic host cells (Vieira et al., 2002). The phagocytic pathway intersects with the generalized endocytic pathway at various stages (Scott et al., 2003; Tjelle et al., 2000; Vieira et al., 2002). Phagocytosed material is engulfed via actin-dependent mechanisms (Allison et al., 1971; Sheterline et al., 1984). Once pinched off from the plasma membrane, this structure is called a phagosome. Early phagosomes contain SE surface markers such as Rab5, EEA1, syntaxin-13, and phosphatidylinositol-3-phosphate (PI3P) (Scott et al., 2003). Later in the process, the phagosome fuses with lysosomes, creating the mature phagolysosome (Scott et al., 2003; Tjelle et al., 2000; Vieira et al., 2002).

Although the emphasis of this chapter is on the phagocytosis of pathogens, phagocytosis is in fact also essential for housekeeping functions in multicellular organisms, by way of clearing the apoptotic and necrotic cells (Krysko et al., 2006). This is an important function, since cell corpses can release cytotoxic substances that can damage neighboring cells.

The eventual delivery of the phagocytosed material to the low-pH phagolysosome serves several purposes. First, it plays a role in killing internalized pathogens. Second, the degradation allows the constituents of the phagocytosed material to be reutilized metabolically. In some cases, degradation also generates antigens, which can then be presented to immune effector cells, with the peptide antigens presented in the context of major histocompatibility complex (MHC) class I or II (Kaufmann and Schaible, 2005), and lipid and glycolipid antigens presented in the context of CD1 (Sugita et al., 2004).

For a phagocytosed pathogen to be virulent, it has to escape from the phagosomes into the cytosol; prevent the phagosomes from acidifying; or develop the ability to survive in the low-pH environment of acidified phagolysosomes. All of these strategies have been developed by one or more pathogens as an evolutionary adaptation for survival.

An example of the first kind of survival mechanism is *Listeria monocytogenes* (Pizarro-Cerda and Cossart, 2006). Once inside the phagosome of a host cell, the low pH activates a bacterial lytic enzyme, listeriolysin O, which lyses the phagocytic vacuole. Once released into the cytosol, the neutral pH inactivates the enzyme. The bacterium then utilizes another bacterial protein, ActA, which mimics the activity of the eukaryotic Wiskott-Aldrich syndrome protein (WASP) family of proteins, to recruit and polymerize actin at the bacterium surface. These actin structures propel *Listeria* in the cytosol and help it to invade neighboring cells.

The second mechanism, preventing phagosome acidification, is employed by several pathogens, including *Mycobacterium tuberculosis* (Nguyen and Pieters, 2005). *Mycobacterium* uses several strategies to prevent phagosome maturation, one of whose components is increased luminal acidification: (i) prevention of dissociation of the host coat protein coronin-1, also known as P57 or tryptophan-aspartate-containing coat protein (TACO), from bacterium-containing phagosomes; (ii) altering the biosynthesis of PI3P, whose production in temporally scheduled waves is essential for phagosome maturation since PI3P binds to several endocytic trafficking regulatory proteins containing the FYVE or phox homology domains; (iii) function of specific mycobacterial cell wall proteins either in phagosome maturation (such as mycobacterial PI3P analog glycosylated phosphatidylinositol lipoarabinomannan, or ManLAM), or in preferential fusion of the mycobacterium-containing phagosome with early endosomes (such as phosphatidylinositol mannoside, or PIM); (iv) utilization of bacterial kinases such as mycobacterial protein kinase G in preventing phagosome maturation; (v) prevention of macrophage maturation either via adaptive immune response mediated by interferon-γ, or mediated by the Toll-like receptors.

Another well-studied example of prevention of endosome acidification is the bacterium *Legionella pneumophila*. A comparative study of the properties of phagosomes containing *L. pneumophila* vs. *M. tuberculosis* shows that whereas *M. tuberculosis*-containing endosomes exhibit several characteristics of early endosomes, such as a delayed clearance of MHC-I molecules and a relatively intense staining for MHC-II and transferrin receptor, *L. pneumophila* phagosomes rapidly clear the MHC-I and do not express any of the endolysosomal markers studied (Clemens and Horwitz, 1995). Furthermore, *L. pneumophila*-containing phagosomes never acquire Rab5, whereas *M. tuberculosis*-containing phagosomes show abundant staining with Rab5 throughout their life cycle (Clemens et al., 2000a). Interestingly, however, both *L. pneumophila* and *M. tuberculosis* phagosomes acquire the LE marker Rab7, but not LAMP-1 (Clemens et al., 2000b). Thus, although both these intracellular pathogens do interact with the endolysosomal pathway to some extent, they are still able to prevent phagosome maturation, which would include acidification and the activation of lysosomal proteases and lipases. In fact, in the case of *Legionella*, the phagosomes rapidly take on the properties of the endoplasmic reticulum (ER), eventually including attached ribosomes (Roy and Tilney, 2002).

As a third survival strategy, pathogens such as *Leishmania mexicana* and *Coxiella burnetii* allow acidification of the phagolysosome, complete with the acquisition of LE/LY markers such as lysosomal hydrolases, lysosomal glycoproteins, proton ATPases, MPRs, Rab7, and microsialin (Tjelle et al., 2000). How these bacteria are able to survive in such harsh environments is not completely understood; however, *Coxiella* has been shown to produce enzymes such as acid phosphatase that eliminate or prevent the formation of toxic oxygen metabolites by the host cell (Tjelle et al., 2000).

TECHNIQUES AND SYSTEMS FOR STUDYING ORGANELLE ACIDIFICATION AND ITS REGULATION

Fluorescence Ratiometric Measurements

Fluorescence ratio measurements using pH-sensitive fluorophores have been widely used to measure the pH of endocytic organelles. One of the most frequently used indicator dyes is fluorescein, for which fluorescence excited at 495 nm decreases at acid pH while excitation at 450 nm is nearly pH independent. Thus, measuring the ratio of fluorescence excited at 495 and 450 nm excitation provides a sensitive measure of pH, with relatively little dependence on confounding variables such as fluorophore concentration, photobleaching, heterogeneity of illumination, detector sensitivity, or differences in optical pathlength (Dunn and Maxfield, 2003; Dunn et al., 1994).

These measurements can be made quantitative by calibrating the emission intensities at the two excitation wavelengths when the cells are clamped to a certain pH. This is achieved for measurement of endosomal/lysosomal pH by first creating a calibration curve in which the cells labeled with an appropriate fluorescein-labeled marker (e.g., fluorescein-dextran for LE/LY, fluorescein-transferrin for the ERC) are incubated in buffers of varying pH values in the presence of weak bases such as methylamine and/or proton ionophores such as nigericin and monensin to collapse pH gradients across membranes (Dunn and Maxfield, 2003; Dunn et al., 1994).

Ratio imaging can also be carried out effectively if the organelle marker is labeled with two different fluorophores: one pH sensitive (e.g., fluorescein) and the other pH insensitive (e.g., tetramethylrhodamine) (Majumdar et al., 2007). The advantage to this is that the fluorescein fluorescence excited at 450 nm is quite weak, whereas the fluorescence from rhodamine dyes can be equal to or greater than the fluorescein fluorescence excited at or near 495 nm. In this case fluorescence is excited sequentially at two excitation wavelengths, and the two images are collected by using two separate emission bandpass filters. Acquisition of the green and red images by laser scanning confocal microscopy can be almost simultaneous, since most commercial systems are able to acquire the images in a line-by-line scanning mode, in which the raster scanning of each line of pixels is carried out first with one and then with the other excitation wavelength. Experimental details for accurate measurements have been described (Dunn and Maxfield, 2003; Dunn et al., 1994).

Newer fluorophores have been developed that circumvent some of the difficulties of fluorescein. One of these is DM-NERF, which can be used for excitation ratio measurements (similar to fluorescein) by using a confocal microscope, since its excitation spectra at two of the major lines of the argon laser, 488 and 514 nm, are differentially sensitive to pH (Fagotto and Maxfield, 1994; Lin et al., 1999). DM-NERF has a broad emission spectrum centered at approximately 530 nm (Lin et al., 1999). The pK_a of DM-NERF is relatively low (5.4) (Lin et al., 1999), making it an effective indicator of pH in acidic organelles such as endosomes and lysosomes (Lee et al., 1996).

The pH dependence of green fluorescent proteins (GFPs) can be used to measure pH or to detect entry into acidic compartments. In particular, the pHluorins have been optimized to give a large decrease in fluorescence intensity between pH 7 and pH 5 (Miesenbock et al., 1998). The excitation spectrum of wild-type GFPs is bimodal with peaks at 395 and 475 nm, representing the protonated and the unprotonated states of a tyrosine residue, which is part of the GFP chromophore. Wild-type GFP is relatively pH insensitive at physiological pH values. By using structure-directed combinatorial mutagenesis, modified GFPs were created that interconverted between the protonated and deprotonated forms with a time constant of <20 ms and exhibited changes in fluorescence between pH 5 and 7. Two classes of pHluorins were generated: ratiometric pHluorins that exhibited a pH-independent isosbestic point and ecliptic pHluorins that lost nearly all their 475-nm excitation at pH <6.0 (eclipsed) and were barely visible at 395-nm excitation.

HOW IS THE pH IN AN ORGANELLE SO PRECISELY MAINTAINED?

V-ATPase

Vacuolar ATPases (V-ATPases) are the major proton pumps in endocytic organelles (Breton and Brown, 2007; Inoue et al., 2005). They are localized both at the plasma membranes and in the membranes' acidic organelles (endosomes, lysosomes, TGN). Depending on their location, V-ATPases pump protons from the cytosol into the lumen of acidic organelles or into the extracellular space. These pumps are electrogenic, in that they pump protons in one direction without a directly coupled transport of another ion to maintain electroneutrality. In addition to housekeeping functions, these proteins play a very important role in certain specialized cells. For example, in kidney, the plasma membrane V-ATPase in the intercalated cells of the collecting duct is critical in the regulation of systemic acid-base balance (Wagner et al., 2004). In osteoclasts, they play a very important role in bone resorption (Blair et al., 1989). In the narrow and clear cells of the epididymis and the vas deferens, a low luminal pH maintained by the V-ATPase, along with a low bicarbonate concentration, are necessary to maintain the spermatozoa in a quiescent state during maturation and storage (Au and Wong, 1980). Study of V-ATPase has been aided greatly by the discovery of specific inhibitors, such as bafilomycin A1 and concanamycin A (Breton and Brown, 2007; Inoue et al., 2005).

The structural details of the V-ATPases have been elaborated in great detail (Breton and Brown, 2007; Inoue et al., 2005). V-ATPases are composed of two subunits: V_0 and V_1. The V_1 domain is a 600- to 650-kDa peripheral complex composed of eight different subunits (subunits A to H, present in the stoichiometry $A_3B_3C_1D_1E_1F_1G_2H_2$). The V_1 domain is responsible for ATP hydrolysis, and both A and B subunits participate in nucleotide binding. The V_0 domain is a 260-kDa integral membrane complex, composed of subunits a, c, c', c'', d, and e in yeast with a stoichiometry $a_1d_1e_xc_{4-5}c'_1c''_1$. In mammalian cells, the c' subunit is absent, and instead they contain another subunit, Ac45. The V_0 domain is responsible for proton translocation. The V_1 (head) and V_0 (membrane) domains are connected by both central and peripheral stalks.

V-ATPase activity is regulated in several ways (Inoue et al., 2005). These include (i) reversible disulfide bond formation between a set of conserved cysteine residues at the catalytic site on the A subunit; (ii) changes in coupling efficiency between ATP hydrolysis and proton transport; (iii) changes in the distribution of V-ATPases between different organellar membranes; and (iv) reversible changes in the assembly of the V_1 and V_0 domains.

The organellar distribution of V-ATPase is determined by the specific isoform of subunit "a" present in the assembly (Inoue et al., 2005). In mammalian cells, subunit "a" is present in four different isoforms, a1 to a4, with a3 and a4 responsible for targeting the V-ATPase to the plasma membranes of osteoclasts and renal intercalated cells, respectively (Oka et al., 2001; Toyomura et al., 2000). Similarly, in yeast subunit Vph1p targets it to the vacuole and Stv1p targets it to a late Golgi compartment (Kawasaki-Nishi et al., 2001b). It has been shown that V-ATPases with different isoforms of subunit "a" not only differ in subcellular localization, but they also have different degrees of assembly and tightness of the coupling between proton transport and ATP hydrolysis (Kawasaki-Nishi et al., 2001a). Subunit "a" also regulates tightness of coupling (Shao et al., 2003).

Reversible dissociation of V-ATPase is controlled in several ways. In yeast glucose depletion causes a rapid dissociation that does not require new protein synthesis and does not involve signal transduction pathways known to respond to nutrient starvation (Parra and Kane, 1998). Dissociation and reassembly appear to be regulated independently of each other. Whereas dissociation requires the presence of intact microtubules (Xu and Forgac, 2001), reassembly requires a novel complex, termed RAVE, which includes a ubiquitin ligase component, Skp1p (Seol et al., 2001). The degree of dissociation is also controlled by the membrane environment in which the V-ATPases reside. Thus, V-ATPases containing Vph1p, which are targeted to the vacuole, dissociate, whereas those containing Stv1p and residing in the Golgi do not (Kawasaki-Nishi et al., 2001b). Using various strategies such as forcing Stv1p-containing V-ATPases to the vacuole, preventing Vhp1p-containing V-ATPases from getting to the vacuole, or treating cells with the weak base chloroquine to neutralize vacuolar pH, it was determined that the disassembly of V-ATPase was controlled by the luminal pH of the organelle (Kawasaki-Nishi et al., 2001b; Shao and Forgac, 2004). This might be adaptive, since organelles with a higher pH will get further alkalinized if V-ATPase components disassemble.

CLC Chloride Channels and Transporters

If the electrogenic V-ATPase were to operate by itself, acidification would be inhibited after a certain point due to the electrical potential across the organelle membrane. Chloride channels and transporters (CLCs) have been identified as one important mechanism to dissipate the electrical potential generated by the V-ATPase (Faundez and Hartzell, 2004; Jentsch, 2007). Ten different CLCs have been described, of which five (CLCs 3 to 7) reside on the membranes of various intracellular vesicles, while others are at the plasma membrane (Jentsch, 2007). Whereas CLCs 6 and 7 are true chloride channels, CLCs 4 and 5 have recently been recognized as voltage-dependent Cl^-/H^+ exchangers (Picollo and Pusch, 2005; Scheel et al., 2005). It is predicted that CLC3 might also belong to this class (Jentsch, 2007). All CLC proteins, whether ion channels or exchangers, function as homodimers, with each of the two ion translocation pathways contained within a single subunit (Faundez and Hartzell, 2004; Jentsch, 2007). Some CLC proteins have accessory subunits (Faundez and Hartzell, 2004; Jentsch, 2007).

Knockout mice for each of these CLCs are now available, and several have been implicated in human and mouse pathologies. For example, CLC5 mutations are associated with renal pathologies such as Dent's disease, with patients exhibiting urinary loss of low-molecular-weight proteins and electrolytes and a high incidence of kidney stones (Lloyd et al., 1996). Both in cell culture and in suspensions of renal cortical endosomes, disruption of CLC5 was found to reduce endosome acidification, which would compromise the reabsorption of filtered proteins from primary urine and cause proteinuria (Gunther et al., 1998).

Disruption in mice of CLC3, 6, or 7 caused degeneration of the central nervous system (Kasper et al., 2005; Poet et al., 2006; Stobrawa et al., 2001). In mice lacking CLC6 or 7, neurons displayed intracellular electron-dense deposits that stained for lysosomal markers and subunit c of ATP synthase, a protein found in a subset of human lysosomal storage disorders termed neuronal ceroid lipofuscinosis (Kasper et al., 2005; Poet et al., 2006). Immunohistochemical studies and subcellular fractionation show that CLC7 is localized in the LE/LY. Whereas CLC7 shows an almost complete overlap with the LE/LY marker LAMP-1 (Kasper et al., 2005), the overlap is partial in the case of CLC6 (Poet et al., 2006). Also, localization of epitope-tagged CLC3 and 6 place them both on LE compartments (Poet et al., 2006; Stobrawa et al., 2001). In neuronal cells, CLC3 localizes additionally at the synaptic vesicles (Stobrawa et al., 2001). These are all acidic organelles that would require the CLCs to compensate the inside positive electrochemical gradient generated by the V-ATPase.

Human patients with homozygous mutations in CLC7 exhibit infantile malignant osteopetrosis (Kornak et al., 2001), whereas heterozygotes with a missense mutation show a less severe osteopetrosis termed Albers-Schonberg disease (Cleiren et al., 2001). Similarly, the *grey lethal* mice, a spontaneous, severely osteopetrotic mouse mutant, were found to be a mutant in Ostm1, the accessory β-subunit of CLC7 (Chalhoub et al., 2003). Ostm1 is a type I membrane protein with a highly glycosylated amino terminus and a short cytoplasmic tail (Lange et al., 2006). CLC7 and Ostm1 are highly expressed in osteoclasts, and they are inserted into the ruffled border of the acidified resorption lacuna, or "extracellular lysosome," along with V-ATPase (Jentsch, 2007). This acidic compartment is essential for bone resorption. CLC7 knockout osteoclasts still attach to ivory (a bone substitute), but they do not acidify the lacuna or degrade bone. CLC7 and Ostm1 are also both localized at the LY (Lange et al., 2006). Ostm1 needs CLC7 to travel to the LY, but CLC7 can get there without Ostm1 (Lange et al., 2006). Both subunits are necessary for the stability of each other (Lange et al., 2006).

Interestingly, steady-state pH in the lysosomes of cultured neurons from CLC6 or 7 knockout mice were not altered compared with the wild type (Kasper et al., 2005; Poet et al., 2006). This observation may indicate that neuronal lysosomes have pumps or channels other than the CLCs that can shunt the electrochemical gradient generated by the V-ATPase. Alternately, the main physiological function of CLC6 and 7 may not involve neutralizing the proton gradient, but rather, their function might be maintaining a high enough Cl^- concentration in the lumen of the lysosomes, which might be essential for proper functioning of the lysosomes (Jentsch, 2007).

Another chloride channel that has been reported to be localized both at the plasma membrane and several intracellular membranes is the cystic fibrosis transmembrane conductance regulator, CFTR. Several groups, using several different cell lines expressing exogenous CFTR, have reported that acidification of endosomes or the TGN are not affected by CFTR expression (Dunn et al., 1994; Root et al., 1994; Seksek et al., 1996). However, a recent study using alveolar macrophages, which normally express functional CFTR, has shown that this protein is essential in these cells to achieve phagosome acidification and the resultant killing of pathogens (Di et al., 2006).

Na^+/K^+ ATPases

Na^+/K^+ ATPases are transmembrane pumps that utilize ATP to pump three Na^+ out of the cytosol and two K^+ into the cytosol (Yu, 2003). Consequently, they serve to increase the inside-positive membrane potential of endosomes in which they are present, minimizing the ability of V-ATPase to pump more protons into the lumen of these organelles. Na^+/K^+ ATPases have been reported to be present on the membranes of early endosomes, which maintain a more alkaline pH compared to the LE/LY (SE, ~6.0; ERC, ~6.5) (Cain et al., 1989; Fuchs et al., 1989).

Na^+/K^+ ATPases are inhibited by several cardiotonic steroids such as digitalis and ouabain, as well as by the ion vanadate (Yu, 2003). These compounds have been utilized to understand the role of Na^+/K^+ ATPases in modulating early endosomal pH. When the early endosomes of CHO cells were labeled with the pH-sensitive fluorophore fluorescein conjugated to transferrin, it was found that the pH in the SE as well as ERC could be modulated in Na^+- and K^+-containing buffers by treating the cells with ouabain or vanadate (Fuchs et al., 1989). Similar dependence of early endosome acidification on ouabain was also observed in the A549 human epidermoid carcinoma cell lines (Cain et al., 1989). The effect of ouabain was also studied in a derivative of Swiss 3T3 cell line, CHL60-64 (Cain et al., 1989), which was originally selected for its resistance to the weak bases chloroquine and ammonium chloride (Cain and Murphy, 1988). Both these weak bases accumulate in the acidic LE/LY of normal cells (including parental Swiss 3T3) and cause their vacuolation and resultant cell death. CHL60-64 cells are resistant to both chloroquine and ammonium chloride and show a markedly slowed acidification of compartments containing fluorescein isothiocyanate-labeled dextrans (LE/LY) (Cain and Murphy, 1988). Treatment of these cells with ouabain completely eliminated their resistance to chloroquine (Cain et al., 1989). These observations led the authors to hypothesize that the LE/LY acidification defect in CHL60-64 cells is because of a mislocalization of Na^+/K^+ ATPases to the LE/LY (Cain et al., 1989).

Cells of the erythroid lineage (Killisch et al., 1992; Sipe et al., 1991), as well as primary cells obtained from mouse heart explants (Rybak and Murphy, 1998), provide an interesting example of specialized pH regulation. When the pH of the Tf-containing early endosome compartment was measured in these cells, it was found to acidify to a pH below 5.5, whereas in many other cell types (e.g., CHO), the pH of the Tf-containing early endosome compartments never dips below 6.0. Interestingly also, this acidification of the Tf-containing compartment of these cells was ouabain insensitive (again, unlike CHO) (Sipe et al., 1991). It is thus possible that in these cell types, the early endosome membranes do not contain Na^+/K^+ ATPases, and this allows these endosomes to acidify to the same extent as the LE.

Na$^+$/H$^+$ Exchangers or Antiporters

Na$^+$/H$^+$ exchangers (NHEs) are another class of transmembrane transporters that are important in regulating organelle pH (Nakamura et al., 2005; Slepkov et al., 2007; Yun et al., 1995). It has been suggested that neither CLCs nor Na$^+$/K$^+$ ATPases may in fact be major players in maintaining pH in organelles with relatively alkaline pH values such as the Golgi and the TGN (Nakamura et al., 2005). The proposed reason is that the measured conductance of both Cl$^-$ and K$^+$ in Golgi and the TGN are much higher than the proton flux in these organelles. It thus seems improbable that the modulation of Cl$^-$ and/or K$^+$ conductance in these organelles would modulate intraluminal pH. Instead, it has been proposed that the steady-state pH in these organelles is controlled by a balance between proton flux mediated by V-ATPase and a "H$^+$ leak" from the organelle lumen.

The major effector for this proton leak is a class of NHEs, which are a class of integral membrane proteins that exchange one proton for one sodium ion (Nakamura et al., 2005; Slepkov et al., 2007; Yun et al., 1995). Thus, these exchangers are electroneutral. Nine isoforms of these proteins, NHE1-9, have been identified, with NHE6-9 being localized at various intracellular organelles (NHE1-5 are localized at the cell surface) (Slepkov et al., 2007). Some of these NHEs are ubiquitous, while others are tissue specific. NHE6 appears to be localized in SE and/or the ERC (Brett et al., 2002; Nakamura et al., 2005), whereas NHE9 is located at the ERC (Nakamura et al., 2005). NHE8 localizes predominantly at the TGN and NHE7 at the mid-to trans-Golgi (Nakamura et al., 2005). None of these localizations are exclusive, and there is partial overlap between the different isoforms (Nakamura et al., 2005).

By using overexpression of various epitope-tagged NHEs, it was demonstrated that overexpression of the NHEs caused an alkalinization of the compartments where they were present, confirming the ability of these proteins to affect organelle pH (Nakamura et al., 2005).

CONSEQUENCES OF ACIDIFICATION OF ENDOSOMES/PHAGOSOMES

pH-Sensitive Viral Fusion Proteins

Many enveloped viruses enter cells by endocytosis after binding their cell surface receptors. Once inside the acidic endosomes, their fusion proteins or spike glycoproteins assume an "acid conformation" that exposes a hydrophobic domain, which in turn facilitates the fusion of the viral envelope with the limiting membrane of the endosomes. This fusion releases the viral nucleocapsid into the cytosol, which can now replicate (Mellman, 1992). The pH at which this acid conformation is achieved varies for different enveloped viruses, and this determines the specific endocytic organelle from which translocation to the cytosol occurs. For example, Semliki Forest virus has a wild-type and a mutant form, which achieve the acid conformation at pH 6.3 and 5.3, respectively (Schmid et al., 1989).

pH-Dependent Cytosolic Translocation of Bacterial Toxins, and the Utilization of This Phenomenon To Isolate Mutant Cell Lines with Endocytic Acidification Defects

Several bacterial toxins, such as diphtheria toxin, *Pseudomonas* exotoxin A, and ricin are internalized from the cell surface after binding to respective receptors. Once inside the cells, the toxins translocate into the cytosol from either an endosomal or a postendosomal organelle. This is again a result of a conformational change in a subunit or a section of the toxin, which exposes hydrophobic segments, making them fusion competent with the organelle membrane (Sandvig and van Deurs, 2002, 2005). Specifically, diphtheria toxin translocates into the cytosol from the early endosomes, whereas *Pseudomonas* exotoxin and ricin enter the "retrogressive pathway," moving from the endosomes to the TGN and then down the Golgi stacks into the ER (Sandvig and van Deurs, 2002, 2005). From the ER, they translocate to the cytosol (Sandvig and van Deurs, 2002, 2005). Studies with these and other toxins (such as Shiga toxin) have provided a lot of information about postendocytic trafficking pathways, including retrograde traffic to the ER.

PHAGOSOME ACIDIFICATION AND ITS ROLE IN BACTERIAL KILLING

One question that arises when considering acidification of phagosomes is whether phagosome acidification per se is responsible for killing the pathogens. The answer to this question is ambiguous and may depend on the specific pathogen. Indeed, for pathogens such as *M. tuberculosis* and *L. pneumophila*, prevention of phagosome acidification is essential for their survival (Clemens and Horwitz, 1995; Clemens et al., 2000b; Nguyen and Pieters, 2005). In contrast, pathogens such as *L. mexicana* and *C. burnetii* allow acidification of the phagolysosome, complete with the acquisition of a complete repertoire of LE/LY markers (Tjelle et al., 2000). Finally, for *Salmonella enterica* serovar Typhimurium infecting murine macrophages, phagosome acidification was shown to be essential for survival and replication (Rathman et al., 1996).

In cases where phagosome acidification is needed for killing the pathogens, part of the function of acidification may be the activation of degradative lysosomal enzymes, which have low pH optima (Lowrie et al., 1979). In addition, both reactive oxygen intermediates (ROIs, such as superoxide, hydrogen peroxide, and hydroxyl radical) and reactive nitrogen intermediates (RNIs, such as nitric oxide, nitrate, peroxynitrite, etc.) have been reported to participate in pathogen killing inside phagosomes (Nathan and Ehrt, 2003; Nathan and Shiloh, 2000). Both ROI and RNI (nitric oxide) production requires the consumption of protons (Shu et al., 1997); thus, the low-pH environment of phagosomes could enhance cell killing by these ROI- and RNI-mediated mechanisms.

We thank Amitabha Majumdar and Lynda M. Pierini for critical reading of the manuscript.

REFERENCES

Allison, A. C., P. Davies, and S. De Petris. 1971. Role of contractile microfilaments in macrophage movement and endocytosis. *Nat. New Biol.* **232**:153–155.

Atluri, P. P., and T. A. Ryan. 2006. The kinetics of synaptic vesicle reacidification at hippocampal nerve terminals. *J. Neurosci.* **26:**2313–2320.

Au, C. L., and P. Y. Wong. 1980. Luminal acidification by the perfused rat cauda epididymidis. *J. Physiol.* **309:**419–427.

Blair, H. C., S. L. Teitelbaum, R. Ghiselli, and S. Gluck. 1989. Osteoclastic bone resorption by a polarized vacuolar proton pump. *Science* **245:**855–857.

Breton, S., and D. Brown. 2007. New insights into the regulation of V-ATPase-dependent proton secretion. *Am. J. Physiol.* **292:**F1–F10.

Brett, C. L., Y. Wei, M. Donowitz, and R. Rao. 2002. Human Na(+)/H(+) exchanger isoform 6 is found in recycling endosomes of cells, not in mitochondria. *Am. J. Physiol.* **282:**C1031–C1041.

Cain, C. C., and R. F. Murphy. 1988. A chloroquine-resistant Swiss 3T3 cell line with a defect in late endocytic acidification. *J. Cell Biol.* **106:**269–277.

Cain, C. C., D. M. Sipe, and R. F. Murphy. 1989. Regulation of endocytic pH by the Na+,K+-ATPase in living cells. *Proc. Natl. Acad. Sci. USA* **86:**544–548.

Chalhoub, N., N. Benachenhou, V. Rajapurohitam, M. Pata, M. Ferron, A. Frattini, A. Villa, and J. Vacher. 2003. Grey-lethal mutation induces severe malignant autosomal recessive osteopetrosis in mouse and human. *Nat. Med.* **9:**399–406.

Cleiren, E., O. Benichou, E. Van Hul, J. Gram, J. Bollerslev, F. R. Singer, K. Beaverson, A. Aledo, M. P. Whyte, T. Yoneyama, M. C. deVernejoul, and W. Van Hul. 2001. Albers-Schonberg disease (autosomal dominant osteopetrosis, type II) results from mutations in the ClCN7 chloride channel gene. *Human Mol. Gen.* **10:**2861–2867.

Clemens, D. L., and M. A. Horwitz. 1995. Characterization of the Mycobacterium tuberculosis phagosome and evidence that phagosomal maturation is inhibited. *J. Exp. Med.* **181:**257–270.

Clemens, D. L., B. Y. Lee, and M. A. Horwitz. 2000a. Deviant expression of Rab5 on phagosomes containing the intracellular pathogens Mycobacterium tuberculosis and Legionella pneumophila is associated with altered phagosomal fate. *Infect. Immun.* **68:**2671–2684.

Clemens, D. L., B. Y. Lee, and M. A. Horwitz. 2000b. Mycobacterium tuberculosis and Legionella pneumophila phagosomes exhibit arrested maturation despite acquisition of Rab7. *Infect. Immun.* **68:**5154–5166.

Demaurex, N., W. Furuya, S. D'Souza, J. S. Bonifacino, and S. Grinstein. 1998. Mechanism of acidification of the trans-Golgi network (TGN). In situ measurements of pH using retrieval of TGN38 and furin from the cell surface. *J. Biol. Chem.* **273:**2044–2051.

Di, A., M. E. Brown, L. V. Deriy, C. Li, F. L. Szeto, Y. Chen, P. Huang, J. Tong, A. P. Naren, V. Bindokas, H. C. Palfrey, and D. J. Nelson. 2006. CFTR regulates phagosome acidification in macrophages and alters bactericidal activity. *Nat. Cell Biol.* **8:**933–944.

Dunn, K. W., and F. R. Maxfield. 2003. Ratio imaging instrumentation. *Methods Cell Biol.* **72:**389–413.

Dunn, K. W., S. Mayor, J. N. Myers, and F. R. Maxfield. 1994. Applications of ratio fluorescence microscopy in the study of cell physiology. *FASEB J.* **8:**573–582.

Dunn, K. W., T. E. McGraw, and F. R. Maxfield. 1989. Iterative fractionation of recycling receptors from lysosomally destined ligands in an early sorting endosome. *J. Cell Biol.* **109:**3303–3314.

Dunn, K. W., J. Park, C. E. Semrad, D. L. Gelman, T. Shevell, and T. E. McGraw. 1994. Regulation of endocytic trafficking and acidification are independent of the cystic fibrosis transmembrane regulator. *J. Biol. Chem.* **269:**5336–5345.

Fagotto, F., and F. R. Maxfield. 1994. Changes in yolk platelet pH during Xenopus laevis development correlate with yolk utilization. A quantitative confocal microscopy study. *J. Cell Sci.* **107:**3325–3337.

Faundez, V., and H. C. Hartzell. 2004. Intracellular chloride channels: determinants of function in the endosomal pathway. *Science STKE* **2004:**re8.

Fuchs, R., S. Schmid, and I. Mellman. 1989. A possible role for Na+,K+-ATPase in regulating ATP-dependent endosome acidification. *Proc. Natl. Acad. Sci. USA* **86:**539–543.

Gunther, W., A. Luchow, F. Cluzeaud, A. Vandewalle, and T. J. Jentsch. 1998. ClC-5, the chloride channel mutated in Dent's disease, colocalizes with the proton pump in endocytotically active kidney cells. *Proc. Natl. Acad. Sci. USA* **95:**8075–8080.

Hao, M., and F. R. Maxfield. 2000. Characterization of rapid membrane internalization and recycling. *J. Biol. Chem.* **275:**15279–15286.

Inoue, T., Y. Wang, K. Jefferies, J. Qi, A. Hinton, and M. Forgac. 2005. Structure and regulation of the V-ATPases. *J. Bioenerg. Biomemb.* **37:**393–398.

Jentsch, T. J. 2007. Chloride and the endosomal-lysosomal pathway: emerging roles of CLC chloride transporters. *J. Physiol.* **578:**633–640.

Kasper, D., R. Planells-Cases, J. C. Fuhrmann, O. Scheel, O. Zeitz, K. Ruether, A. Schmitt, M. Poet, R. Steinfeld, M. Schweizer, U. Kornak, and T. J. Jentsch. 2005. Loss of the chloride channel ClC-7 leads to lysosomal storage disease and neurodegeneration. *EMBO J.* **24:**1079–1091.

Kaufmann, S. H., and U. E. Schaible. 2005. Antigen presentation and recognition in bacterial infections. *Curr. Opin. Immunol.* **17:**79–87.

Kawasaki-Nishi, S., K. Bowers, T. Nishi, M. Forgac, and T. H. Stevens. 2001a. The amino-terminal domain of the vacuolar proton-translocating ATPase a subunit controls targeting and in vivo dissociation, and the carboxyl-terminal domain affects coupling of proton transport and ATP hydrolysis. *J. Biol. Chem.* **276:**47411–47420.

Kawasaki-Nishi, S., T. Nishi, and M. Forgac. 2001b. Yeast V-ATPase complexes containing different isoforms of the 100-kDa a-subunit differ in coupling efficiency and in vivo dissociation. *J. Biol. Chem.* **276:**17941–17948.

Kielian, M. C., M. Marsh, and A. Helenius. 1986. Kinetics of endosome acidification detected by mutant and wild-type Semliki Forest virus. *EMBO J.* **5:**3103–3109.

Killisch, I., P. Steinlein, K. Romisch, R. Hollinshead, H. Beug, and G. Griffiths. 1992. Characterization of early and late endocytic compartments of the transferrin cycle. Transferrin receptor antibody blocks erythroid differentiation by trapping the receptor in the early endosome. *J. Cell Sci.* **103:**211–232.

Kim, J. H., L. Johannes, B. Goud, C. Antony, C. A. Lingwood, R. Daneman, and S. Grinstein. 1998. Noninvasive measurement of the pH of the endoplasmic reticulum at rest and during calcium release. *Proc. Natl. Acad. Sci. USA* **95:**2997–3002.

Kornak, U., D. Kasper, M. R. Bosl, E. Kaiser, M. Schweizer, A. Schulz, W. Friedrich, G. Delling, and T. J. Jentsch. 2001. Loss of the ClC-7 chloride channel leads to osteopetrosis in mice and man. *Cell* **104:**205–215.

Krysko, D. V., K. D'Herde, and P. Vandenabeele. 2006. Clearance of apoptotic and necrotic cells and its immunological consequences. *Apoptosis* **11:**1709–1726.

Lange, P. F., L. Wartosch, T. J. Jentsch, and J. C. Fuhrmann. 2006. ClC-7 requires Ostm1 as a beta-subunit to support bone resorption and lysosomal function. *Nature* **440:**220–223.

Lee, R. J., S. Wang, and P. S. Low. 1996. Measurement of endosome pH following folate receptor-mediated endocytosis. *Biochim. Biophys. Acta* **1312:**237–242.

Lin, H. J., H. Szmacinski, and J. R. Lakowicz. 1999. Lifetime-based pH sensors: indicators for acidic environments. *Anal. Biochem.* **269:**162–167.

Lloyd, S. E., S. H. Pearce, S. E. Fisher, K. Steinmeyer, B. Schwappach, S. J. Scheinman, B. Harding, A. Bolino, M. Devoto, P. Goodyer, S. P. Rigden, O. Wrong, T. J. Jentsch, I. W. Craig, and R. V. Thakker. 1996. A common molecular basis for three inherited kidney stone diseases. *Nature* **379:**445–449.

Lowrie, D. B., P. W. Andrew, and T. J. Peters. 1979. Analytical subcellular fractionation of alveolar macrophages from normal and BCG-vaccinated rabbits with particular reference to heterogeneity of hydrolase-containing granules. *Biochem. J.* **178:**761–767.

Majumdar, A., D. Cruz, N. Asamoah, A. Buxbaum, I. Sohar, P. Lobel, and F. R. Maxfield. 2007. Activation of microglia acidifies lysosomes and leads to degradation of Alzheimer amyloid fibrils. *Mol. Biol. Cell* **18:**1490–1496.

Maxfield, F. R., and T. E. McGraw. 2004. Endocytic recycling. *Nat. Rev. Mol. Cell Biol.* **5:**121–132.

Mellman, I. 1992. The importance of being acid: the role of acidification in intracellular membrane traffic. *J. Exp. Biol.* **172:**39–45.

Miesenbock, G., D. A. De Angelis, and J. E. Rothman. 1998. Visualizing secretion and synaptic transmission with pH-sensitive green fluorescent proteins. *Nature* **394:**192–195.

Mukherjee, S., R. N. Ghosh, and F. R. Maxfield. 1997. Endocytosis. *Physiol. Rev.* **77:**759–803.

Nakamura, N., S. Tanaka, Y. Teko, K. Mitsui, and H. Kanazawa. 2005. Four Na+/H+ exchanger isoforms are distributed to Golgi and post-Golgi compartments and are involved in organelle pH regulation. *J. Biol. Chem.* **280:**1561–1572.

Nathan, C., and S. Ehrt. 2003. *Nitric Oxide in Tuberculosis*, 2nd ed. Lippincott Williams & Wilkins, Philadelphia, PA.

Nathan, C., and M. U. Shiloh. 2000. Reactive oxygen and nitrogen intermediates in the relationship between mammalian hosts and microbial pathogens. *Proc. Natl. Acad. Sci. USA* **97:**8841–8848.

Nguyen, L., and J. Pieters. 2005. The Trojan horse: survival tactics of pathogenic mycobacteria in macrophages. *Trends Cell Biol.* **15:**269–276.

Oka, T., Y. Murata, M. Namba, T. Yoshimizu, T. Toyomura, A. Yamamoto, G. H. Sun-Wada, N. Hamasaki, Y. Wada, and M. Futai. 2001. a4, a unique kidney-specific isoform of mouse vacuolar H+-ATPase subunit a. *J. Biol. Chem.* **276:**40050–40054.

Parra, K. J., and P. M. Kane. 1998. Reversible association between the V1 and V0 domains of yeast vacuolar H+-ATPase is an unconventional glucose-induced effect. *Mol. Cell Biol.* **18:**7064–7074.

Picollo, A., and M. Pusch. 2005. Chloride/proton antiporter activity of mammalian CLC proteins ClC-4 and ClC-5. *Nature* **436:**420–423.

Pizarro-Cerda, J., and P. Cossart. 2006. Subversion of cellular functions by Listeria monocytogenes. *J. Pathol.* **208:**215–223.

Poet, M., U. Kornak, M. Schweizer, A. A. Zdebik, O. Scheel, S. Hoelter, W. Wurst, A. Schmitt, J. C. Fuhrmann, R. Planells-Cases, S. E. Mole, C. A. Hubner, and T. J. Jentsch. 2006. Lysosomal storage disease upon disruption of the neuronal chloride transport protein ClC-6. *Proc. Natl. Acad. Sci. USA* **103:**13854–13859.

Rathman, M., M. D. Sjaastad, and S. Falkow. 1996. Acidification of phagosomes containing Salmonella typhimurium in murine macrophages. *Infect. Immun.* **64:**2765–2773.

Root, K. V., J. F. Engelhardt, M. Post, J. W. Wilson, and R. W. Van Dyke. 1994. CFTR does not alter acidification of L cell endosomes. *Biochem. Biophys. Res. Commun.* **205:**396–401.

Roy, C. R., and L. G. Tilney. 2002. The road less traveled: transport of Legionella to the endoplasmic reticulum. *J. Cell Biol.* **158:**415–419.

Rybak, S. L., and R. F. Murphy. 1998. Primary cell cultures from murine kidney and heart differ in endosomal pH. *J. Cell Physiol.* **176:**216–222.

Sandvig, K., and B. van Deurs. 2005. Delivery into cells: lessons learned from plant and bacterial toxins. *Gene Ther.* **12:**865–872.

Sandvig, K., and B. van Deurs. 2002. Membrane traffic exploited by protein toxins. *Annu. Rev. Cell Dev. Biol.* **18:**1–24.

Scheel, O., A. A. Zdebik, S. Lourdel, and T. J. Jentsch. 2005. Voltage-dependent electrogenic chloride/proton exchange by endosomal CLC proteins. *Nature* **436:**424–427.

Schmid, S. R., M. Fuchs, A. Kielian, and I. Mellman. 1989. Acidification of endosome subpopulations in wild-type Chinese hamster ovary cells and temperature-sensitive acidification-defective mutants. *J. Cell Biol.* **108:**1291–1300.

Schmidt, W. K., and H. P. Moore. 1995. Ionic milieu controls the compartment-specific activation of pro-opiomelanocortin processing in AtT-20 cells. *Mol. Biol. Cell* **6:**1271–1285.

Scott, C. C., R. J. Botelho, and S. Grinstein. 2003. Phagosome maturation: a few bugs in the system. *J. Memb. Biol.* **193:**137–152.

Seksek, O., J. Biwersi, and A. S. Verkman. 1996. Evidence against defective trans-Golgi acidification in cystic fibrosis. *J. Biol. Chem.* **271:**15542–15548.

Seol, J. H., A. Shevchenko, A. Shevchenko, and R. J. Deshaies. 2001. Skp1 forms multiple protein complexes, including RAVE, a regulator of V-ATPase assembly. *Nat. Cell Biol.* **3:**384–391.

Shao, E., and M. Forgac. 2004. Involvement of the nonhomologous region of subunit A of the yeast V-ATPase in coupling and in vivo dissociation. *J. Biol. Chem.* **279:**48663–48670.

Shao, E., T. Nishi, S. Kawasaki-Nishi, and M. Forgac. 2003. Mutational analysis of the non-homologous region of subunit A of the yeast V-ATPase. *J. Biol. Chem.* **278:**12985–12991.

Sheterline, P., J. E. Rickard, and R. C. Richards. 1984. Fc receptor-directed phagocytic stimuli induce transient actin assembly at an early stage of phagocytosis in neutrophil leukocytes. *Eur. J. Cell Biol.* **34:**80–87.

Shu, Z., M. Jung, H. G. Beger, M. Marzinzig, F. Han, U. Butzer, U. B. Bruckner, and A. K. Nussler. 1997. pH-dependent changes of nitric oxide, peroxynitrite, and reactive oxygen species in hepatocellular damage. *Am. J. Physiol.* **273:**G1118–G1126.

Sipe, D. M., A. Jesurum, and R. F. Murphy. 1991. Absence of Na+,K(+)-ATPase regulation of endosomal acidification in K562 erythroleukemia cells. Analysis via inhibition of transferrin recycling by low temperatures. *J. Biol. Chem.* **266:**3469–3474.

Sipe, D. M., and R. F. Murphy. 1987. High resolution kinetics of transferrin acidification in Balb/c 3T3 cells: exposure to pH 6 followed by temperature sensitive alkalinization during recycling. *Proc. Natl. Acad. Sci. USA* **84:**7119–7123.

Slepkov, E. R., J. K. Rainey, B. D. Sykes, and L. Fliegel. 2007. Structural and functional analysis of the Na+/H+ exchanger. *Biochem. J.* **401:**623–633.

Stobrawa, S. M., T. Breiderhoff, S. Takamori, D. Engel, M. Schweizer, A. A. Zdebik, M. R. Bosl, K. Ruether, H. Jahn, A. Draguhn, R. Jahn, and T. J. Jentsch. 2001. Disruption of ClC-3, a chloride channel expressed on synaptic vesicles, leads to a loss of the hippocampus. *Neuron* **29:**185–196.

Sugita, M., M. Cernadas, and M. B. Brenner. 2004. New insights into pathways for CD1-mediated antigen presentation. *Curr. Opin. Immunol.* **16:**90–95.

Tjelle, T. E., T. Lovdal, and T. Berg. 2000. Phagosome dynamics and function. *BioEssays* **22:**255–263.

Toyomura, T., T. Oka, C. Yamaguchi, Y. Wada, and M. Futai. 2000. Three subunit a isoforms of mouse vacuolar H(+)-ATPase. Preferential expression of the a3 isoform during osteoclast differentiation. *J. Biol. Chem.* **275:**8760–8765.

Vieira, O. V., R. J. Botelho, and S. Grinstein. 2002. Phagosome maturation: aging gracefully. *Biochem. J.* **366:**689–704.

Wagner, C. A., K. E. Finberg, S. Breton, V. Marshansky, D. Brown, and J. P. Geibel. 2004. Renal vacuolar H+-ATPase. *Physiol. Rev.* **84:**1263–1314.

Xu, T., and M. Forgac. 2001. Microtubules are involved in glucose-dependent dissociation of the yeast vacuolar [H+]-ATPase in vivo. *J. Biol. Chem.* **276:**24855–24861.

Yamashiro, D. J., and F. R. Maxfield. 1987. Kinetics of endosome acidification in mutant and wild-type Chinese hamster ovary cells. *J. Cell Biol.* **105:**2713–2721.

Yu, S. P. 2003. Na(+), K(+)-ATPase: the new face of an old player in pathogenesis and apoptotic/hybrid cell death. *Biochem. Pharmacol.* **66:**1601–1609.

Yun, C. H., C. M. Tse, S. K. Nath, S. A. Levine, S. R. Brant, and M. Donowitz. 1995. Mammalian Na+/H+ exchanger gene family: structure and function studies. *Am. J. Physiol.* **269:**G1–G11.

15

Actin-Based Motility in Professional Phagocytes[†]

FREDERICK S. SOUTHWICK

Phagocytes are among the most dynamic cells in the body. Within seconds of exposure to a chemotactic peptide, they are able to convert from a rounded symmetric shape to a polarized morphology. These cells are able to sense and crawl by amoeboid movement toward chemotactic gradient. As the phagocyte crawls it extends out a broad veil of cytoplasm called the lamellipod and at the trailing end forms a narrow tail-like structure called a uropod (Fig. 1). This directional movement has been termed chemotaxis. Phagocytes are also capable of forming large arms or pseudopods that can surround foreign particles and bring them into the cell cytoplasm. The ability to ingest foreign objects is called phagocytosis. Both chemotaxis and phagocytosis require the rapid reorganization of the actin cytoskeleton, and the assembly and disassembly of actin filaments, combined with myosin contraction, provide the forces for these movements. Actin polymerization not only drives the motility of all mammalian nonmuscle cells, but also provides the propulsive force for the movement of intracellular bacteria including *Listeria*, *Shigella*, and *Rickettsia*, as well as the poxvirus vaccinia. Thus, a working knowledge of how host cells regulate their actin filament architecture is required to fully understand not only cell movement and host defense, but also bacterial and viral pathogenesis.

BASIC STRUCTURES OF THE ACTIN MONOMER AND ACTIN FILAMENT

Actin is the most abundant protein in the cytoplasm of phagocytes, representing 10 to 20% of the total cytoplasmic protein. This protein exists in two interchangeable forms, monomeric globular actin (G-actin, 43 kDa) and a two-stranded helical polymer (F-actin); G-actin polymerizes through reversible noncovalent interactions into F-actin. Atomic resolution structures of the actin monomer in both the ADP and ATP forms reveal that the actin molecule consists of four domains (Kabsch et al., 1990). Domains I and II are separated from domains III and IV by a cleft containing a high-affinity nucleotide-binding site as well as a high-affinity divalent cation-binding site, which are usually bound by ATP and magnesium, respectively, in vivo (Fig. 2). Actin is an enzyme that catalyzes the hydrolysis of the nucleoside triphosphate ATP to ADP and inorganic phosphate (P_i). G-actin has very low ATPase activity, and incorporation of the monomers into the filament significantly enhances this activity. After incorporation of actin-ATP monomers into the polymer, ATP hydrolysis occurs, resulting in an intermediate of actin-ADP-P_i, followed by the slower release of inorganic phosphate into the surrounding medium, forming an actin-ADP species (Fig. 3). The monomers retain the ADP, and it is exchanged for ATP once the monomers have dissociated from the filament. Thus, the actin-nucleotide complex "matures" within the actin filament, and an individual filament contains segments carrying ATP, ADP-P_i, and ADP. A large free-energy change is associated with ATP hydrolysis; some of that energy is stored within the filament and probably as an altered state of the monomer within the filament. The ability of actin to assume different conformational as well as phosphorylation states may regulate the strength of interactions of the actin-binding proteins along the length of the filament. For example, cofilin/ADF (a filament severing and depolymerizing protein) has a higher affinity for ADP-actin than for ATP-actin.

The crystal structure of F-actin has not been solved, and the architecture of the actin filament is presently based on modeling. The current atomic model of F-actin indicates that the large domain of the actin monomer is located near the center of the filament axis and the small domain is on the exterior (Fig. 4). The filament is stabilized by both lateral and longitudinal contacts in the form of hydrophobic interactions, hydrogen bonds, and salt bridges between the monomers. This derived model of the actin filament is consistent with previously determined radial positions of selected residues, as well as with the crystal structures of actin bound to several actin-binding proteins. The actin filament can be thought of as two right-handed intertwined helices (Aguda et al., 2005). Actin monomers assemble in a head-to-tail manner that accounts for the polar character of the filament. The two ends of the filament are not equivalent; the filaments display both structural and functional

---

[†] Some sections adapted from Southwick, 2005.

Frederick S. Southwick, Gainesville, FL 32610-0275.

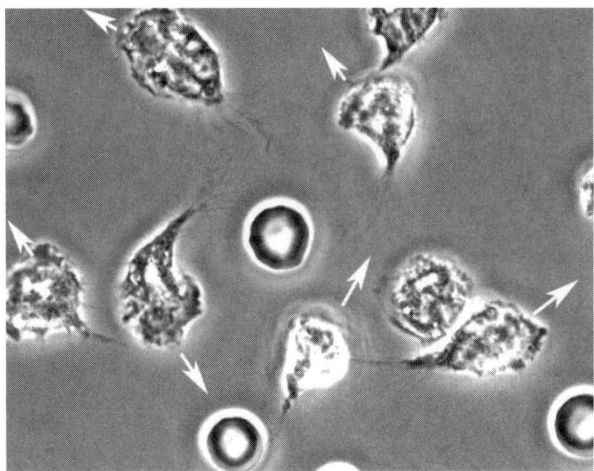

FIGURE 1 Polarized neutrophils. Arrows point to the direction the cells are moving. Note the broad lamellipodia at the front of each neutrophil and the narrow uropod at the back.

polarity. The structural polarity can be demonstrated by decorating F-actin with myosin heads that in the absence of ATP bind at a 45° angle, forming an "arrowhead"-like pattern on electron micrographs defining "barbed" (also known as plus ends) and "pointed" ends (also known as the minus ends).

ACTIN ASSEMBLY KINETICS

Actin polymerization in vitro consists of three steps: (i) nucleation (formation of actin oligomers that have a greater tendency to form filaments rather than disintegrating into monomers); (ii) elongation (growth of the filament at both the barbed and pointed ends); and (iii) treadmilling (flux of monomers along the filament).

Initiation of polymerization, also known as nucleation, is an unfavorable process and is the rate-limiting step in actin polymerization. During this first step, three to four actin monomers associate into an oligomer that serves as the template for actin filament elongation. Actin nucleation is favored by a high actin-ATP concentration, a condition that allows nuclei to persist long enough for pro-

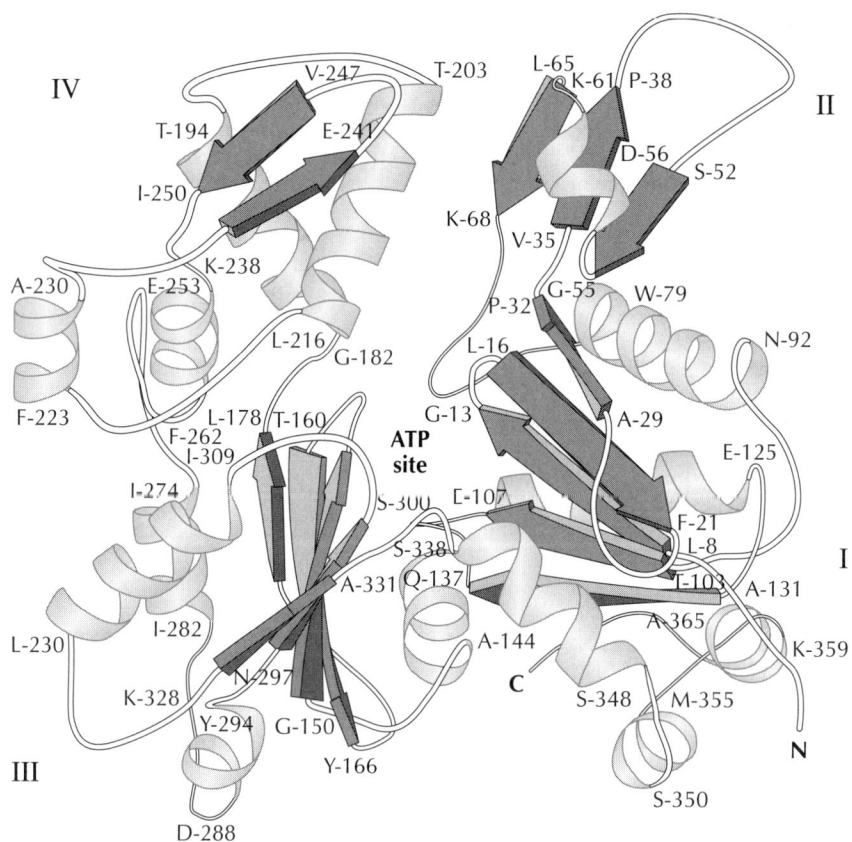

FIGURE 2 Atomic-level structure of an actin monomer showing the ATP-binding site. Based on the work of Kabsch et al., 1990. The vertical axis of the monomer (as depicted in this figure) runs parallel to the long axis of the filament. The right-hand side of the molecule is exposed to the outside of the actin filament, whereas the left-hand side is nearest the long axis of the filament. Residues 262 to 274 are thought to reach across this axis and interact with the adjacent actin monomer of the double-stranded helix. As oriented here, the polarity of the filament would correspond to the pointed end at the top and the barbed end at the bottom. (From Southwick, 2005.)

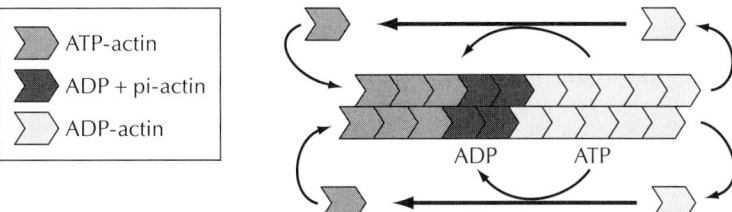

FIGURE 3 Schematic drawing of an actin filament under steady-state conditions. ATP-actin monomers add to the barbed or fast-growing end and are hydrolyzed to ADP-P_i, followed by the slower dissociation of P_i to form ADP-actin. ADP-actin monomers dissociate from the pointed or slow-growing end. Under these conditions an actin monomer added to the barbed end will eventually treadmill through the filament and dissociate from the pointed end. (From Southwick, 2005.)

ductive polymer formation. Cells contain actin monomer sequestering proteins (see below) that suppress spontaneous nucleation and prevent indiscriminate actin assembly that would result in unregulated changes in the shape and consistency of the cell.

After nucleation, elongation begins and monomers are added to the two ends of the actin filament. When the association rate is faster than the dissociation rate, net elongation of filaments occurs. The rate of elongation depends on the number of nuclei and the concentration of free actin monomers. The filament grows at both ends rapidly until a steady-state level of polymerization is reached. At steady state, the rate of monomer addition to the ends of the filament is balanced by the rate of monomer dissociation and there is no net change in the monomer or filament concentration. The concentration of monomers at the steady state is called the macroscopic critical concentration and reflects the on and off rates of both the barbed and pointed ends. The macroscopic critical concentration also represents the concentration at which actin filaments begin to assemble, and this concept is analogous to the point at which a solution becomes saturated and crystals begin to form. In a plot of polymerized actin versus actin monomer concentration, no polymer will be observed until the actin concentration exceeds the macroscopic critical concentration (Fig. 5).

The rate of association (k_{on}) and dissociation (k_{off}) of the monomers and therefore the K_Ds (k_{off}/k_{on}) or the critical concentrations of the two ends differ. The pointed end is the slow-growing end and has a higher critical concentration, while the barbed end is the fast-growing end and has a lower critical concentration. The individual ratios of the off rate over the on rate (the K_D) of the barbed end and pointed end represent the microscopic critical concentrations of the two ends (Fig. 5). Because both the on and off rates of the barbed end are faster than the pointed end, the macroscopic critical concentration primarily reflects the critical concentration of the barbed end. These differences in the microscopic critical concentration of the two ends also result in a constant flux of monomers through the filament. Under steady-state conditions, the lower critical concentration of the barbed end, compared with the pointed end, leads to net growth of the filament from the barbed end and a net loss of monomers from the pointed end. Thus, the filament maintains a constant length despite a cycling of monomers through the filament. The cycling of actin monomers through the filament is called treadmilling (Fig. 3).

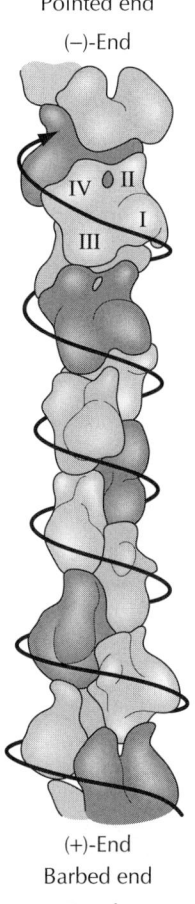

FIGURE 4 Schematic drawing of the tertiary structure of an actin filament. Roman numerals correspond to the domains shown in Fig. 2. (From Southwick, 2005.)

MODULATION OF ACTIN DYNAMICS BY ACTIN-BINDING PROTEINS

The cytoplasm of phagocytes contains high concentrations of unpolymerized actin that exceed by several orders of

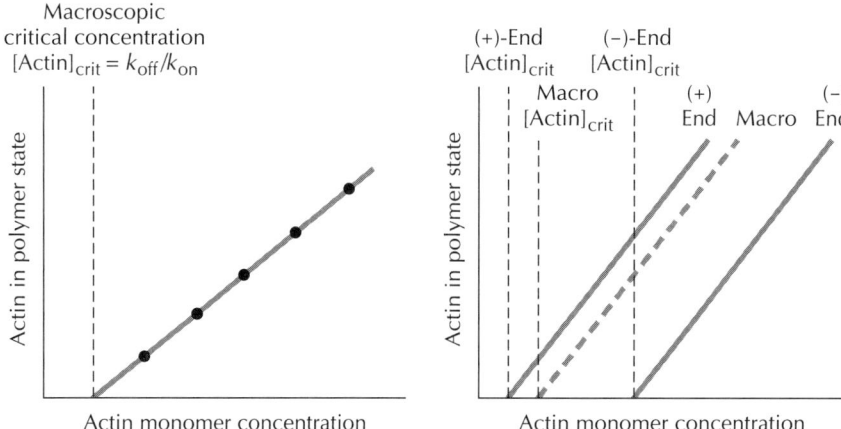

FIGURE 5 Critical-concentration behavior in actin polymerization. (Left) Plot of the steady-state actin filament concentration as a function of monomer concentration. This macroscopic behavior is often measured by the increase in fluorescence when pyrenyl-actin is incorporated into filaments. (Right) When both filament ends are uncapped, the macroscopic critical concentration lies between the microscopic critical concentrations for actin monomer interactions at the barbed end [or (+) end] and the pointed end [or (−) end]. At steady state, monomers will naturally dissociate from the pointed ends and will associate with the more stable barbed ends. This phenomenon is known as treadmilling. Because the exchange rates are higher at the barbed end, the macroscopic critical concentration is closer to the microscopic critical concentration of the barbed end. (From Southwick, 2005.)

magnitude the macroscopic critical concentration of purified actin. In the unstimulated neutrophil, 60 to 70% of the total actin exists as unpolymerized actin and only 30 to 40% is incorporated into filaments. As cells change shape and move, they are able to quickly convert this large storage pool of actin monomers into filaments. Within 30 s of exposure to a chemoattractant, the concentration of filamentous actin in neutrophils doubles. The temporal and spatial control of actin assembly and disassembly involves a myriad of actin regulatory proteins. These proteins orchestrate the dynamic changes in the actin cytoskeleton associated with cell movement. As listed in Table 1, and described in the sections below, these proteins can be classified by their mechanisms of action into six categories. It should be emphasized that no single protein can by itself explain the complex changes associated with actin-based motility. As research has advanced, it has become ever more evident that different constellations of proteins combine to generate different actin structures and account for different forms of actin-based motility.

MONOMER-BINDING PROTEINS

One mechanism for regulating actin filament assembly is by binding to and altering the ability of actin monomers to polymerize. Three actin monomer-binding proteins, thymosin β4, profilin, and heat shock protein 27, work in concert to maintain the high concentrations of unpolymerized actin and direct monomers to the sites of new actin filament assembly.

The factor primarily responsible for preventing actin monomer assembly into filaments is the 5-kDa polypeptide known as thymosin β4 (often abbreviated as Tβ4). This small protein binds to a single actin monomer to form a 1:1, or binary, complex (Safer et al., 1997). As it binds, the Tβ4 stretches along the full length of the actin monomer, blocking access to both the barbed and pointed ends and sterically hindering the ability of the actin monomer to bind to other actin monomers (Fig. 6A). The concentrations of Tβ4 in cells (100 to 400 μM range) match the concentrations of unpolymerized monomeric actin in resting nonmuscle cells. Tβ4 has a higher affinity for ATP-actin monomers (the form of actin that most readily forms actin filaments) than for ADP-actin. The affinity of Tβ4 is relatively weak (1 to 2 μM for ATP-actin monomer-Tβ4), and under the proper conditions it can quickly release actin monomers for rapid actin assembly (see below). These characteristics make Tβ4 the ideal molecule for maintaining a large storage pool of monomeric ATP-actin in phagocytes.

The second less abundant monomer-binding component is the 15-kDa protein profilin. The intracellular concentration of profilin is approximately one-fourth that of Tβ4. Like Tβ4, profilin binds actin monomers to form a 1:1 complex (termed the PA complex) and under certain conditions can form a ternary complex with Tβ4 and monomeric actin (Yarmola and Bubb, 2006). Profilin binds to the barbed end of the actin monomer and like Tβ4 prevents the spontaneous formation of actin nuclei and indiscriminate actin assembly in the cell. However, profilin primarily functions to stimulate the rapid growth of actin filaments. Profilin binds to actin monomers and markedly enhances the rate of nucleotide exchange (Fig. 6B). Low concentrations of profilin can catalyze nucleotide exchange, even in the presence of high Tβ4 concentrations. Therefore, profilin can readily convert ADP-actin to ATP-actin, the form of monomeric actin that favors actin assembly. Because the concentration of ATP-actin is unlikely to be rate limiting in unstimulated cells, this capacity to stimulate nucleotide exchange function may not be critical in the resting cell. However, in regions of the cell undergoing rapid actin filament turnover, the concentrations of ADP-actin should increase locally and the supply of ATP-actin would be rate limiting in the absence of pro-

TABLE 1 Actin regulatory proteins

Protein	Function
Monomer-binding proteins	
Tβ4	Preferentially interacts with actin-ATP; weaker binding to actin-ADP.
Profilin	Accelerates ATP-ADP nucleotide exchange; ushers actin monomers from Tβ4 to the barbed ends of actin filaments; can form a ternary complex with monomeric actin and Tβ4; binds to XP_5 sequences in VASP and other related proteins; binds PIP_2.
Hsp27	Monomer binding controlled by phosphorylation. Dephosphorylated protein binds actin monomers and concentrates in the cell periphery. Phosphorylation by the p38 MAP kinase pathway releases actin monomers and relocates Hsp27 to the cell center.
Profilin-actin complex-binding proteins	
VASP/Ena/Mena	Binds to zyxin and vinculin; has multiple GP_5 sites for binding profilin-actin; also binds to the ends of actin filaments; may deliver actin monomers from profilin to the barbed filament ends.
WASP, N-WASP	Bind Arp2/3 complexes; in the presence of GTP-bound Cdc42, Rho and Rac activate nucleation; have multiple GP_5 sites for binding profilin-actin.
Vinculin	Found in focal contacts; binds VASP, α-actinin, and actin filaments.
Zyxin	Found in focal contacts; binds VASP; multifunctional adaptor protein.
Actin-nucleating proteins	
Arp2/3 complex	Seven-protein complex including Arp2/3; nucleates actin assembly in the presence of free barbed ends and activated WASP proteins; binds to the sides of actin filaments, forms 70° angle branches; caps pointed ends.
Formins	Bind to the barbed ends and nucleate linear filament assembly.
Barbed-end capping proteins	
CapG	Calcium-sensitive binding interaction; binding of PIP_2 inhibits capping function.
Gelsolin	High-affinity binding interaction associated with actin filament severing; requires calcium; binding of PIP_2 inhibits capping function.
CapZ	Calcium-independent binding interaction; binding of PIP_2 and binding of CD2AP, CIN85, and CARMIL inhibit capping function.
Filament severing and recycling proteins	
Gelsolin	Initially binds to the sides of filaments, followed by filament severing (also see above); recycles newly formed ATP-actin filaments.
ADF/cofilin	Accelerates disassembly at the pointed end of actin filaments; thought to selectively interact with ADP-rich regions in actin filaments; severs actin filaments; recycles older actin filaments.
Cross-linking/bundling proteins	
Filamin (ABP-280)	Creates orthogonal networks responsible for gel formation.
α-Actinin	Creates colinear arrays of actin filaments; calcium-sensitive; interacts with vinculin, zyxin, and phospholipids; binds to surface adherence receptors.

filin. In addition to increasing the concentrations of ATP-actin, profilin is able to deliver actin monomers to the barbed ends of growing actin filament (Nyman et al., 2002). Profilin has higher affinity for actin monomers than Tβ4 and is able to take actin monomers from the large pool of sequestered Tβ4-actin. Profilin then ushers ATP-actin monomers onto the barbed ends of actin filaments. On binding to the actin filament, profilin quickly dissociates, leaving its actin monomer on the end of the growing filament (Kang et al., 1999). The resulting free profilin molecule can then compete with Tβ4 for another actin monomer and repeat the cycle (Fig. 6C). Calculations indicate that in nonmuscle cells over 90% of the actin monomers in filaments assemble from profilin-actin complexes rather than from free-actin monomers.

In addition to being found throughout the cytoplasm, profilin concentrates in regions where new actin filaments are assembling. Profilin's unusual affinity for poly-L-proline accounts for its localization in these regions. Profilin is the only actin regulatory protein known to bind to poly-L-proline, and a number of proteins contain oligoproline repeats that are capable of binding profilin, including vasodilator-stimulated phosphoprotein (VASP), Wiskott-Aldrich syndrome protein (WASP), and N-WASP (see below). Profilin also associates with the plasma membrane where it binds to the phosphoinositide phosphatidylinositol 4,5-bisphosphate (PIP_2). When micelles containing this phospholipid bind to a profilin-actin complex, actin is released from the complex.

The third and least abundant of the monomer-binding proteins is heat shock protein 27 (Hsp27). This protein is found in the cytoplasm of all motile cells, and its location in the cell, conformation, and ability to bind monomeric actin are all regulated by phosphorylation. Unphosphory-

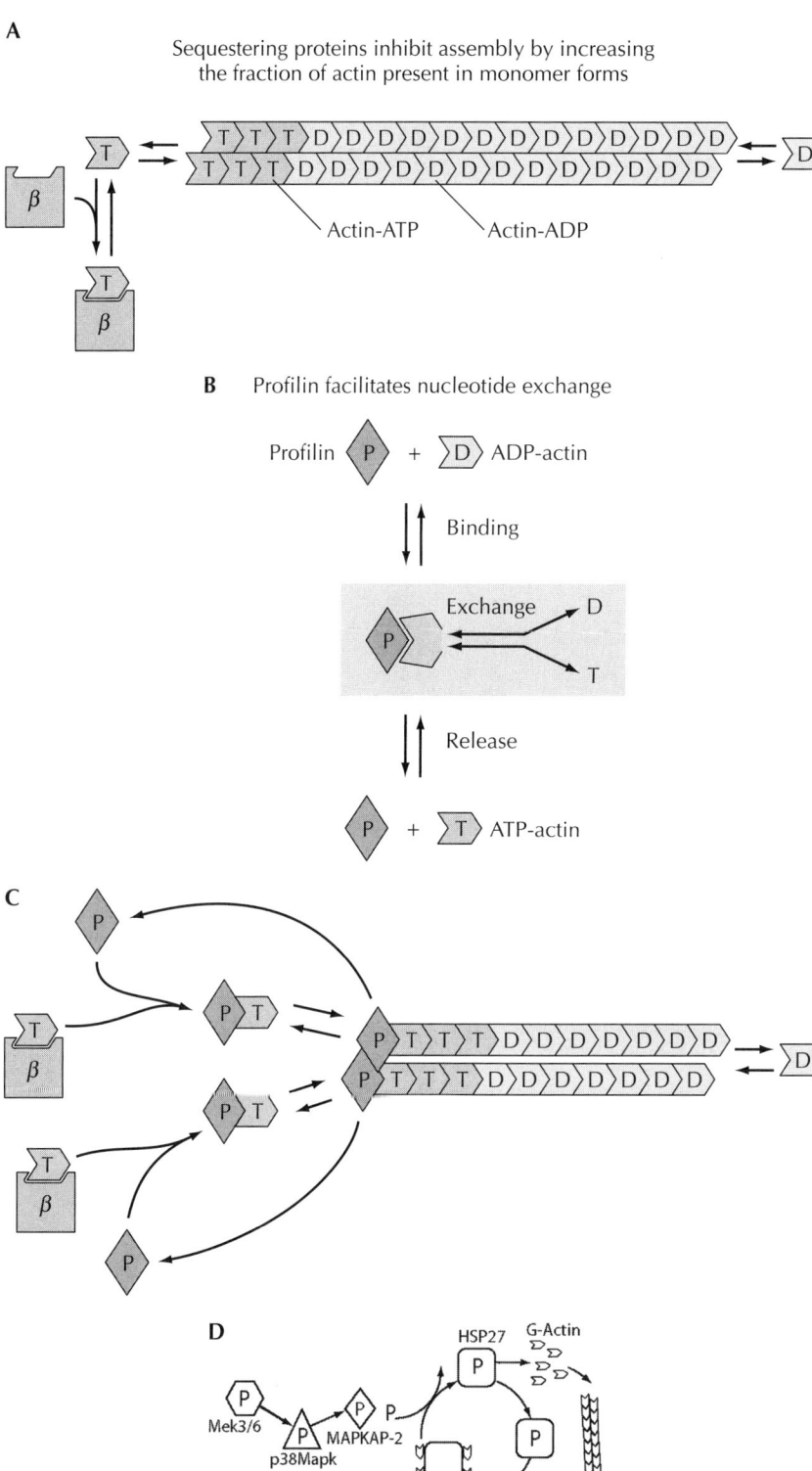

lated Hsp27 forms a homo-octomer and is able to bind and sequester monomeric actin with high affinity (K_D, 20 nM) and high capacity (20 to 30 monomers). In the unphosphorylated state, Hsp27 localizes to the leading edge of polarized neutrophils. Stimulation by chemoattractants, such as N-formyl-methionyl-leucyl-phenylalanine, activates the p38 mitogen-activated protein (MAP) kinase pathway, resulting in the phosphorylation of Hsp27. Upon phosphorylation, Hsp27 shifts from an octomer to a dimer and releases its actin monomers to allow new actin filament assembly. Phosphorylation also results in the movement of Hsp27 from the periphery to more central regions of the cell. Upon dephosphorylation, Hsp27 again binds and sequesters actin monomers and migrates to the cell periphery, where the cycle is repeated (During et al., 2007). Thus, Hsp27 serves as a shuttle to bring actin monomers to the leading edge of lamellipodia as they expand toward a chemotactic gradient during directional migration (Fig. 6D).

PROFILIN-ACTIN COMPLEX-BINDING PROTEINS

High rates of actin assembly are required to produce the rapid changes in shape observed during phagocyte amoeboid movement. To achieve these high rates of assembly, actin monomers must be concentrated in regions of active actin filament growth. At the leading edge of the lamellipod, Hsp27 provides one mechanism for accomplishing this task. In other regions of the cell, in particular in areas of the cell that adhere and form close surface contacts called focal contacts, protein complexes form that are capable of attracting profilin-actin. Adherence receptors contain high concentrations of actin filaments and are linked to the actin cytoskeleton by ezrin, radixin, and moesin (the ERM proteins). In addition, talin (a 270-kDa protein) is found in these regions and binds vinculin (120 kDa). Vinculin in turn can bind α-actinin and actin filaments, and through the amino acid sequence 840-PDFPPPPPDL-849 (called an FP_4 sequence) binds VASP. A second VASP-binding protein, zyxin, is also found in focal contacts (Krause et al., 2003). This 84-kDa protein contains two FP_4 sites and binds VASP with high affinity. VASP, a 45-kDa protein with multiple phosphorylation sites, is a central adaptor protein that binds to the FP_4-binding regions via its EVH1 domain and concentrates profilin-actin complexes in regions where actin filaments are assembling. Each VASP monomer contains four amino acid sequences of the type XPPPPP, where X is glycine, alanine, lysine, or serine, and these sequences preferentially bind profilin-actin. VASP exists as a tetramer in solution; therefore, one VASP tetramer has 16 potential profilin-binding sites. VASP is one of the founding members of the Ena/VASP protein family that includes two other mammalian counterparts, mammalian Ena (abbreviated Mena) and Ena-VASP-like protein. These proteins also contain XP_5 sequences capable of concentrating profilin. In addition to profilin-binding sequences, VASP possesses an actin monomer and an actin filament-binding site, both found in the EVH2 domain (Fig. 7A). The G-actin-binding site is adjacent to the most C-terminal profilin-binding site, and VASP can simultaneously bind both profilin and G-actin in the PA complex. Because VASP also binds to the barbed end of the actin filament, it can transfer the actin monomer from the PA complex to the growing filament end and may facilitate the release of profilin (Ferron et al., 2007). The profilin molecule is then free to usher a new actin monomer from Tβ4 to the growing filament (Fig. 7B). Loss of the Ena/VASP proteins results in a reduction in F-actin content in vascular endothelial cells and leaky vascular tight junctions (Furman et al., 2007). In addition to the Ena/VASP proteins, the WASP/WAVE protein family also possess XP_5 sequences and have sites that interact with the actin-like subunits of the Arp2/3 complex. Similarly, formins possess profilin-binding sites. Thus, in addition to delivering actin monomers to preformed filaments, profilin-binding sites are likely to deliver actin monomers to initiate the formation of new actin filaments (see below).

ACTIN NUCLEATION FACTORS

Monomer-binding proteins prevent spontaneous nucleation. Therefore, to create new actin filaments, the cell must either uncap preformed actin filaments (see below) or activate proteins that can serve as a template for the nucleation of actin assembly. The Arp2/3 complex may nucleate actin assembly, but may also amplify the growth rate of free barbed actin filament ends. First identified in *Acanthamoeba*, this complex has since been identified in mammalian cells as well as yeast and consists of seven proteins including actin-related protein 2 (Arp2) and actin-related protein 3 (Arp3). As their names imply, these two proteins have a high level of amino acid sequence homology to actin, and this characteristic is likely to explain the ability of the complex to bind actin monomers. Arp2/3 complex has two functions. First, this complex nucleates actin filament formation upon activation. Arp2/3 complex binds to the pointed end of the actin filament, stimulating filament growth at the fast-growing barbed end. The intrinsic ability of Arp2/3 complex to nucleate actin filament formation is weak, and in unstimulated cells there is minimal nucleation activity. However, upon binding of agonists to the appropriate receptors, the Rho family of proteins, Rho,

---

FIGURE 6 (A) Schematic diagram of the action of monomer-sequestering agents such as Tβ4. (B) Schematic diagram of how profilin may facilitate the exchange of ATP for ADP on an actin monomer. When profilin binds to an actin monomer, the central cleft of actin opens, making the nucleotide-binding site more accessible for release and exchange. Because the ATP concentration far exceeds the ADP concentration in living cells, ATP will readily replace ADP from the actin monomer. (C) Simplified kinetic diagram showing how free profilin can take actin monomers from the Tβ4 storage pool and usher them onto the barbed end of an actin filament. For simplicity, the reverse arrows from profilin-ATP to free profilin and to Tβ4-ATP-actin are not shown. Once profilin-ATP binds to the barbed end, profilin rapidly dissociates. (D) Model of how Hsp27 phosphorylation and dephosphorylation could serve to shuttle actin monomers to sites of new actin filament assembly and facilitate actin-based motility. Unphosphorylated Hsp27 with bound actin monomers concentrates at the leading edge of motile cells where, upon chemoattractant stimulation, the p38 MAPK signal transduction cascade induces Hsp27 phosphorylation, releasing actin monomers for new actin filament assembly. Phosphorylated Hsp27 then moves toward the center of the cell, where it is dephosphorylated, again binds actin monomers, and then shuttles back to the leading edge. (From Southwick, 2005.)

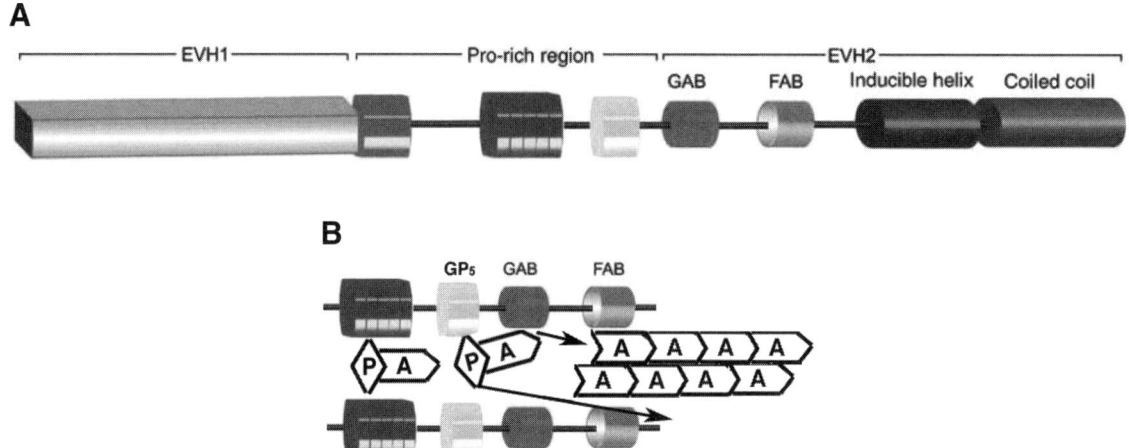

FIGURE 7 (A) Schematic diagram of the modular organization of VASP. The EVH1 domain binds to FP_4 sequences found in ActA, vinculin, and zyxin. The EVH2 domain contains both a G-actin-binding site (GAB) adjacent to the last GP_5 profilin-binding site in the proline-rich domain, and an F-actin-binding site (FAB). This region also contains a coiled-coil domain responsible for forming VASP tetramers. (B) Schematic diagram showing how VASP may usher an actin monomer from profilin to the barbed end of an actin filament tethered by the FAB. Profilin binds to the GP_5 polyproline sequences on VASP. The final GP_5 sequence is adjacent to the actin-monomer-binding site (GAB), allowing both the profilin and actin molecule to be bound simultaneously. The monomer can then be transferred to the actin filament end, and free profilin is released to bind a new ATP-actin monomer (arrows). (Images adapted from Ferron et al., 2007.)

Rac, and Cdc42, are activated and bind to and induce conformational changes in the WASP family of proteins, leading to the unmasking of binding sites for the Arp2/3 complex. The structure of inactivated Arp2/3 complex has been solved, and Arp2 and Arp3 are not in contact with each other. Binding of activated WASP protein to the Arp2/3 complex alters the conformation of the complex, bringing Arp2 and Arp3 closer together to form a template for the initiation of actin filament elongation. ActA from the pathogenic bacteria *Listeria* mimics an activated WASP protein and is able to directly bind to and activate the Arp2/3 complex. It has also been observed that Arp2/3 complex-mediated nucleation is greatly enhanced by the presence of preformed actin filaments with uncapped barbed ends. Thus, the Arp2/3 complex can work in concert with uncapping of the barbed ends to expand the actin cytoskeleton.

The second function of the Arp2/3 complex is to generate Y-shaped arrays that permit directional expansion of actin filaments, allowing the application of broad expansile forces to the peripheral cell membrane. The Arp2/3 complex is able to bind to the side of one filament and nucleate the formation of a second filament, forming a branch at a fixed angle of 70° (Fig. 8A). In fibroblasts and amoebae, Arp2/3 complex is localized near the leading edge, where expansion of peripheral cytoplasm takes place, and is not found near stable actin-myosin bundles associated with cell adhesion and contraction. It should be emphasized that the

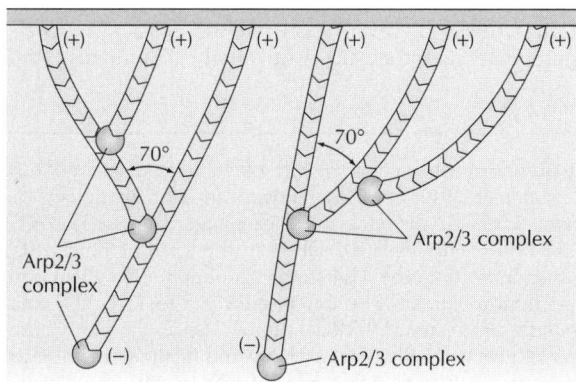
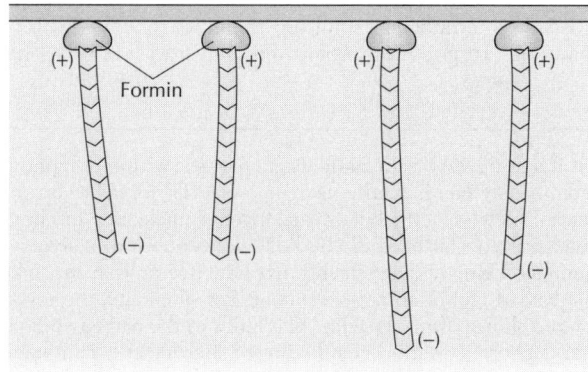

FIGURE 8 (A) Schematic view of Arp2/3 complex nucleated actin filament assembly. (B) Schematic view of formin-mediated actin filament assembly. (From Southwick, 2005.)

rheological studies prove that the Arp2/3 complex is not capable of forming a gel, but rather the branched filaments are linked into a stable network by the cross-linking protein filamin.

A second group of nucleating proteins, called the formins, are important for the formation of narrow membrane projections called filopodia. The formins are a group of multidomain proteins that are characterized by the presence of formin homology (FH) domains. The FH1 domain is rich in oligoproline sequences capable of binding profilin. The FH2 domain is required for nucleation. Curiously, the formins bind to the barbed ends of actin filaments, but do not completely cap them. In particular, in the presence of profilin, the formins nucleate the growth of actin assembly and produce linear actin filament arrays rather than a branched architecture (Fig. 8B). Formins remain at the tip of the growing filament, moving along the filament as it grows. Localization of formins at the peripheral membrane allows actin monomers to add to growing filaments at the membrane-actin cytoskeleton interface, and the resulting directional actin assembly provides the force for discrete membrane protrusions called filopodia (Pollard, 2007).

ACTIN FILAMENT-CAPPING PROTEINS

Proteins that bind to the barbed ends of actin filaments (Fig. 9A) have a profound effect on filament growth. The barbed end has a high affinity for actin monomers and in

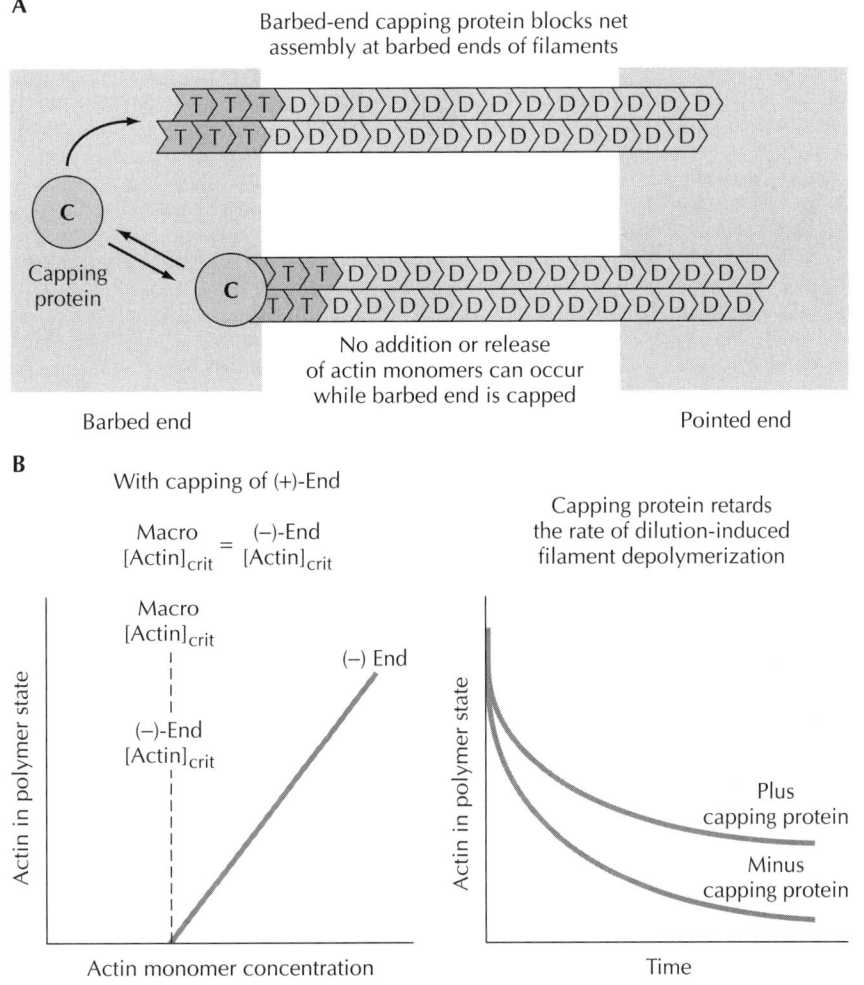

FIGURE 9 (A) Mechanism of a barbed-end capping protein binding to the barbed (or plus) end of an actin filament. Bound capping proteins prevent both association and dissociation of monomers; under such conditions, only the pointed end can interact with the actin monomer pool. (B) Effects of a barbed-end capping protein on the critical concentration and rate of depolymerization of actin. (Left) Graph of steady-state actin filament concentration as a function of actin monomer concentration. As shown in panel A, capping blocks all exchange at the barbed end. The free pointed end has a lower affinity for actin monomers, and this lower affinity is reflected as an increase in the critical concentration (see Fig. 5 for comparison). (Right) Plot of the decrease in filamentous actin versus time after diluting actin filaments to below their critical concentration in the absence and presence of a barbed-end capping protein. Capping of the barbed end retards the depolymerization because dissociation of actin monomers occurs only at the pointed end. (From Southwick, 2005.)

combination with profilin can readily compete with Tβ4 for sequestered ATP-actin monomers. When the barbed ends of actin filaments are capped, the critical concentration of the actin filament increases to that of the pointed end, 0.6 μM (Fig. 9B, left). Barbed-end capping also lowers the depolymerization rate of actin filaments because the off rate for the free pointed ends is considerably slower than that for the barbed end (Fig. 9B, right). In the presence of Tβ4 and profilin, the lower-affinity pointed end is unable to efficiently compete for actin monomers, and significant growth of actin filaments is unlikely to occur. Therefore, in the cell, the barbed filament end is the primary site for actin filament assembly. For unstimulated nonmuscle cells to maintain the high concentrations of unpolymerized actin, nearly all of the barbed ends must be blocked from competing with Tβ4 for actin monomers. Given the importance of regulating the barbed end to maintain high concentrations of monomeric actin in resting cells and to initiate new actin filament assembly during cell movement, it is not surprising that nonmuscle cells contain multiple proteins that are capable of binding the barbed end and blocking monomer exchange. These proteins are called barbed-end-capping proteins.

CapZ is a heterodimer consisting of a 36-kDa α-subunit and a 32-kDa β-subunit. The protein derives its name from the observation that it localizes to the Z line in skeletal muscle (Wear and Cooper, 2004). This protein has also been called β-actinin and capping protein. CapZ is found in all nonmuscle cells and is particularly abundant in neutrophils, where it represents 1% of the total cytoplasmic protein. This protein binds the barbed ends of actin filaments and prevents the release or addition of actin monomers. The dissociation constant for the CapZ filament end is 0.5 to 3 nM. The ability of CapZ to cap filaments is not affected by ionized calcium concentration, but is blocked by PIP_2. Several binding partners, CARMIL, CD2AP, and CIN85, also block capping. The role of these proteins in regulating actin assembly is presently under investigation. In the submicromolar Ca^{2+} concentrations found in unstimulated cells, CapZ would be expected to cap the barbed ends, preventing them from competing with Tβ4 for actin monomers and allowing the cell to maintain a high concentration of unpolymerized actin (see above). Studies of CapZ in platelets reveal that stimuli that induce platelet actin assembly result in the uncapping of the ends of actin filaments by CapZ.

Another capping protein that differs structurally from CapZ is CapG, a 38-kDa protein that is closely related to the actin filament-severing proteins known as gelsolin and villin. While most abundant in macrophages and neutrophils, CapG is also found in most other cell types, with the notable exception of platelets. The affinity of CapG (K_D = 1 nM) for the barbed end is comparable to that of CapZ. Unlike other members of the gelsolin-villin family, CapG caps the barbed ends of actin filaments but demonstrates no actin filament-severing activity. Like other members of this family, CapG is calcium sensitive and requires 1 μM Ca^{2+} for half-maximal capping of actin filaments. The phosphoinositide PIP_2 inhibits barbed-end capping by CapG, suggesting that CapG responds to two intracellular signals, PIP_2 and ionized calcium. Its calcium sensitivity may permit CapG to cap and uncap filaments in response to the brief fluctuations in ionized calcium observable in the periphery of motile phagocytes. Cycles of filament capping and uncapping may represent an essential feature of peripheral membrane ruffling, and macrophages lacking CapG demonstrate a marked decrease in receptor-mediated ruffling (Witke et al., 2001).

ACTIN FILAMENT-SEVERING AND -RECYCLING PROTEINS

One of the most abundant severing proteins in nonmuscle cells is gelsolin, an 82-kDa monomeric protein found in the cytoplasm of all nonmuscle cells. In the presence of micromolar concentrations of ionized calcium, gelsolin's tail region unfolds, exposing actin-binding sites and promoting binding to the sides of actin filaments. On binding to the filament, gelsolin interferes with monomer-monomer interactions within the filament. These actions result in the severing of actin filaments (Fig. 10A). Gelsolin is closely related to CapG; other members of the gelsolin family include villin, a 95-kDa protein found in the intestinal brush border, and adseverin (also named scinderin), found in the adrenal medulla. With the exception of CapG, all members of this family sever actin filaments. In addition, like CapG, gelsolin, as well as all other members of the gelsolin/villin family, caps the barbed ends of actin filaments. Once gelsolin caps an actin filament, the protein binds the filament end with high affinity (K_D in the subpicomolar range) and does not dissociate from the barbed end when the ionized calcium concentration is lowered. However, as observed with both CapG and CapZ, binding to actin filaments can be inhibited by PIP_2. Experiments with increasing and decreasing gelsolin levels in cells prove that, in vivo, gelsolin enhances the recycling of actin filaments and facilitates actin-based motility in nonmuscle cells (Larson et al., 2005; McGrath et al., 2000). The severing of actin filaments reduces the viscosity of the peripheral cytoplasm and allows the actin cytoskeleton to rapidly remodel. Gelsolin also plays a role in apoptosis and appears to be linked to mitochondrial porins. Reduced gelsolin levels have been noted in many cancer cells, and this condition may partly explain prolonged cancer cell survival.

Another family of proteins called actin depolymerization factor (ADF) or cofilin also plays a major role in the recycling of actin filaments (Bamburg and Wigand, 2002). These are low-molecular-mass proteins (19 to 20 kDa) that increase the dissociation rate of ADP-actin monomers from the pointed ends of actin filaments and also sever actin filaments by a distinctly different mechanism than gelsolin (Pavlov et al., 2007). Severing activity and actin filament recycling are impaired by LIM kinase phosphorylation. These proteins can bind to ADP-actin monomers with considerably higher affinity (0.1 to 0.2 μM K_D) than ATP-actin monomers (1.3 μM K_D) and enhance the dissociation rate of ADP- but not ATP-actin. They recycle actin filaments in concert with gelsolin, which is able to sever ATP-actin filaments. ADF/cofilin recycles older actin filaments, while gelsolin is capable of recycling newly formed filaments (Fig. 10B).

ACTIN FILAMENT-BUNDLING AND CROSS-LINKING PROTEINS

Electron micrographs of nonmuscle cells reveal that actin filaments are organized into a network in which many filaments appear to cross each other at right angles, forming an orthogonal mesh. The cross-linking protein ABP-280, or filamin, is responsible for organizing these networks (Nakamura et al., 2002). This spatially extended homodimeric protein consists of two 280-kDa subunits linked at a

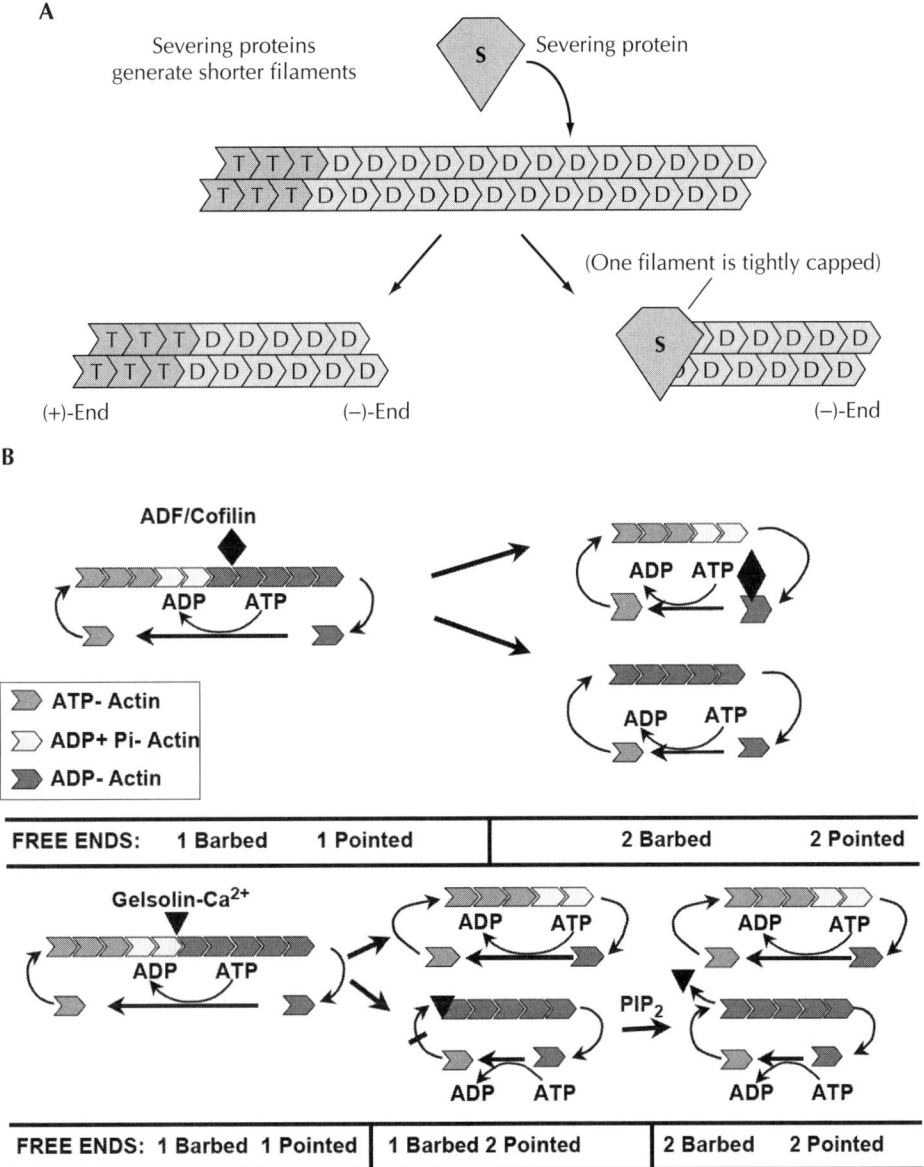

FIGURE 10 (A) Model for the actin filament-severing proteins. The severing protein first binds along the side of the actin filament, next interposes itself between neighboring actin subunits within the filament, and then remains tightly bound to the barbed end of one of the severed filaments. (From Southwick, 2005.) (B) Schematic diagram of actin filament cycling before and after addition of ADF/cofilin or gelsolin-Ca^{2+}. At steady state (left), actin monomers come on the filament at the barbed end as ATP-actin monomers. As they enter the filament, the ATP is hydrolyzed, forming an intermediate ADP + P_i and then ADP-actin. Once the ADP-actin monomer dissociates from the pointed end, ATP is exchanged for ADP, and the actin monomer can again add to the barbed end. This process is called treadmilling, and the rate of treadmilling depends on the number of free filament ends. Doubling the number of free ends of the same concentration of filamentous actin would be expected to double the rate of treadmilling. In the slow-cycling filament, significant amounts of ADP-actin exist in the filament; therefore, ADF/cofilin can bind and sever. Each time ADF severs, it doubles the filament ends. Gelsolin-Ca^{2+} has very high affinity for filaments and can bind and sever regions of the filament containing ADP- or ATP-actin. Because gelsolin also binds and caps the barbed ends, severing and capping doubles the free pointed ends but does not increase the free barbed ends. However, when chemotactic signal transduction pathways increase the concentration of phosphatidylinositol bisphosphate, gelsolin will dissociate from the barbed end and the number of free barbed ends will double. When actin filaments rapidly cycle, there is reduced time for ATP hydrolysis and the filament would be expected to have a lower content of ADP-actin. This condition would be expected to reduce the ability of ADF/cofilin to enhance treadmilling, but would not impair gelsolin.

flexible hinge region. Each subunit possesses a single actin-binding site, thereby allowing the dimer to link two actin filaments. When actin filaments are linked into a network by filamin, the solution forms a gel, giving the cell cytoplasm a firm consistency. Activation of gelsolin by calcium can abruptly shorten actin filaments and dismantle this network, causing the cytoplasm to shift from a highly viscous to a liquid consistency. Gel-sol transitions are likely to play an important role in the shape changes associated with amoeboid movement.

In addition to cross-linking proteins, nonmuscle cells possess smaller actin filament-bundling proteins, the most prominent member of this class being α-actinin (105 kDa). This protein links actin filaments into bundles of filaments in a parallel array forming stress fibers. Bundling is regulated by calcium and phosphoinositides (Broderick and Winder, 2005). α-Actinin has a number of binding partners, including vinculin and zyxin. This protein also binds to surface adherence receptors (integrins, ICAMs, and α-catenins) and is concentrated in focal contacts as well as adherens junctions. Another member of this class is plastin, a 65-kDa protein (also known as fimbrin), which occurs in the so-called T and L forms.

TROPOMYOSIN AND MYOSINS

Other proteins capable of binding to actin include tropomyosin and the myosins. There are multiple subtypes of tropomyosin, and all appear to bind in the groove of the actin filament in a fashion that prevents myosin binding and also blocks gelsolin side binding and severing (Pittenger et al., 1994). Myosins are another large family of proteins that produce force and movement by binding to actin filaments through the myosin head regions (O'Connell et al., 2007). The ATPase activity of the head region is activated by binding to the actin filament. The energy of hydrolysis is transduced into a structural change in the binding angle of the myosin head from 90° to 45°. This change in angle advances the filament toward its pointed end. There are two major classes of myosins. While myosin II is most abundant in muscle, this force-producing motor is also found in nonmuscle cells. The tail regions of myosin II self-associate and form filament bundles in which the myosin heads arrange at opposing ends with respect to each other. These heads can pull actin filaments toward the center of the bundle. Myosin I is a more recently described class of proteins possessing a shorter tail, which fails to self-associate. Myosin I tails can bind to actin filaments or membranes. Furthermore, myosin I proteins tend to localize in the leading edge of moving cells, whereas myosin II tends to localize toward the posterior region. In addition to myosin I, a large number of other unconventional myosins have been described that serve specific transport functions and result in specific morphological changes. For example, overexpression of myosin X results in the formation of filopodia in cultured tissue cells (Berg and Cheney, 2002). The specific contributions of unconventional myosins to cell motility are presently being clarified.

POLYMERIZATION ZONE MODEL

How might cells use actin polymerization to generate force during locomotion? In order to form a membrane projection such as a lamellipod, pseudopod, and filopod or to propel an intracellular bacterium, a discrete zone must be created that promotes the assembly of actin filaments. Such a polymerization zone would be expected to contain nucleating proteins, actin filaments with uncapped barbed ends, and high concentrations of profilin-ATP-actin. As new actin filaments form, they are organized by cross-linking and bundling proteins, and expansion of this actin filament network provides the thrust for expansion of the peripheral membranes during chemotaxis and phagocytosis. Myosin II is thought to retract the tail of polarized cells as they crawl toward a chemotactic gradient.

The activities within the polymerization zone can be divided into two categories: initiation of new actin filament assembly, analogous to the starter motor in a conventional gasoline engine, and delivery of ATP-actin monomers to the growing barbed ends of actin filaments, analogous to a fuel delivery system.

Generation of Free Barbed Filament Ends

In phagocytic cells, actin-based motility is initiated by the generation of new free barbed filament ends. When cells are stimulated to move and change shape, the number of free barbed filament ends markedly increases. This rapid rise in the number of free barbed ends can be accomplished in two ways. First, receptor agonists through the Rho family of proteins activate Arp2/3 complex nucleation. In their GTP form, these proteins bind to and alter the conformation of the WASP protein family, which in turn binds to and activates the Arp2/3 complex. Members of the Rho family can also activate the formins to initiate barbed-end filament growth. The second mechanism for initiating new actin assembly is by uncapping the barbed ends of preformed actin filaments. Uncapping is likely to be a more energy-efficient mechanism for initiating new actin filament assembly, and considerable evidence now points to the importance of free barbed filament ends for maximizing Arp2/3 complex nucleation. The mechanisms controlling uncapping and regulating actin filament length remain to be determined, but phosphatidylinositides, uncapping proteins such as CD2AP/CIN82 and CARMIL, as well as VASP and formins, are likely to play key roles. Combining severing and uncapping can further amplify the number of free barbed ends. In platelets as well as other nonmuscle cells, receptor agonists can produce a transient rise in ionized calcium and stimulate severing by gelsolin. Increased production of PIP_2 and/or the production of other uncapping activities subsequently dissociates gelsolin, resulting in the generation of multiple short actin filaments with free barbed ends. It is of interest that gelsolin-null platelets form long actin filaments and demonstrate a reduced number of free barbed ends following agonist stimulation. Furthermore, gelsolin-deficient platelets demonstrate aberrant localization of the Arp2/3 complex, indicating that in vivo the production of free barbed ends is critical for normal Arp2/3 complex localization and function.

Delivery of ATP-Actin Monomers to the Growing Barbed Ends of Actin Filaments

The second important activity required to support actin filament growth within the polymerization zone is the efficient delivery of assembly-competent actin monomers. Intracellular *Listeria* actin-based motility occurs at rates up to 1 μm/s and phagocyte lamellipodia can expand at similar speeds. To achieve speeds of this magnitude, monomers must be added to the filament ends at rates of 400 monomers per s and requires that 200 μM ATP-actin be immediately available for assembly. As discussed earlier, the primary pathway by which actin monomers are delivered to the barbed ends of actin filaments is via profilin-actin complexes. However, the cytoplasmic concentration of profilin-actin, even in the most motile cell, does not exceed 60 μM. To attract high concentrations of profilin-actin into the polymerization zone, the cell utilizes two consensus docking sequences, (D/E)FPPPPX(D/E) (where X is P or T), commonly called the FP_4 consensus sequence, and XPPPPP (where X is G, A, L, P, or S), sometimes termed the XP_5 consensus sequence. In *Listeria*, the bacterial surface protein ActA contains a series of four sequences of the type EFPPPPTDE. Each FP_4 sequence attracts a VASP tetramer. Because each VASP tetramer contains 16 to 20 GP_5 profilin-binding sites, one ActA molecule can potentially attract 64 to 80 profilin molecules. Because GP_5 sequences preferentially bind profilin-actin complexes over free profilin, this protein-binding amplification cascade will attract extremely high concentrations of profilin-actin to the surface of *Listeria*. Similarly, vinculin and zyxin both contain (D/E)FPPPPX(D/E) sequences that can attract VASP and profilin-actin to the leading edge of motile cells. N-WASP and WASP activators of Arp2/3 also contain multiple XP_5 sequences and can deliver profilin-actin to sites of actin nucleation. In addition to concentrating profilin-actin at sites of new actin filament assembly, VASP may transfer monomeric actin directly to the barbed ends of growing filaments through its C-terminal G-actin- and F-actin-binding sites. The role of VASP phosphorylation in mediating this exchange, as well as the role of ATP hydrolysis in releasing profilin from the barbed end and in generating the force associated with actin assembly, are subjects of intense research and speculation.

In addition to ushering actin monomers onto the barbed end, the presence of profilin within the polymerization zone ensures that the weakly polymerizing ADP-actin will undergo facilitated exchange with ATP. As the filaments extend outside the polymerization zone, ATP hydrolyzes within the filament, and these older filaments will primarily consist of ADP-actin. ADF/cofilin would be expected to bind to and depolymerize these older filaments. Thus, disassembly would be expected to occur primarily outside of the polymerization zone. Finally, it must be recognized that the binding of high concentrations of profilin-actin to a surface will not increase the profilin-actin concentration in the solution phase; therefore, the growing filament must transiently bind to the surface via proteins such as VASP to allow profilin-actin to be directly transferred to the filament end. This mechanism has the advantage of not being limited by the diffusion rate, and allows exquisite control of where new actin filaments are assembled.

Hsp27 may provide a second delivery system at the leading edge of lamellipodia. Chemoattractants stimulate the p38 MAP kinase pathway, resulting in the phosphorylation of Hsp27 and the release of actin monomers bound to Hsp27. The interplay of Hsp27 with profilin remains to be explored; however, the Hsp27 phosphorylation cycle promises to provide an additional level for the control of actin assembly in phagocytes.

CONCLUSIONS

The precise mechanisms regulating the formation of new actin filaments in the motile cell remain to be determined. Actin filament growth can be induced by formin and Arp2/3 complex nucleation as well as by the capping and uncapping of the barbed filament ends by CapZ and CapG. Arp2/3 complex activity is stimulated by GTP-bound G proteins combined with the WASP family proteins. In addition, filaments can be severed and capped by gelsolin. The combination of uncapping and severing can greatly increase the number of free barbed filament ends available for rapid actin assembly and can further stimulate the nucleating activity of the Arp2/3 complex. To explain the rapid rates of actin assembly in vivo, profilin-actin must be highly concentrated on a membrane surface through a binding amplification cascade involving FP_4 and XP_5 sequences. Within a discrete polymerization zone, surface-bound profilin-actin is directly transferred to growing actin filaments by VASP and other VASP-like proteins, and this transfer bypasses the rate limitations of diffusion. ATP-actin addition to barbed filament ends is thermodynamically favorable, and hydrolysis of filament-bound ATP-actin during filament elongation can provide the useful work needed for movement. Through the action of filamin (ABP-280) and α-actinin, newly formed filaments can be stabilized into orthogonal networks and bundles. These proteins increase the rigidity of the actin filaments, giving the peripheral cytoplasm the structure and mechanical properties required for shape change and movement. Osmotic forces and myosins may also play a role in advancing the peripheral membrane. In regions of expansion, new actin filaments are assembled, but in other areas where membranes are retracting, filaments are disassembled by gelsolin and ADF/cofilin. An increasingly large repertoire of functionally similar proteins are continually being discovered, emphasizing the redundancy of the actin regulatory system and providing a rich repertoire of pathways available for the cell to generate movement and shape change.

REFERENCES

Aguda, A. H., L. D. Burtnick, and R. C. Robinson. 2005. The state of the filament. *EMBO Rep.* **6:**220–226.

Bamburg, J. R., and O. P. Wiggan. 2002. ADF/cofilin and actin dynamics in disease. *Trends Cell Biol.* **12:**598–605.

Berg, J. S., and R. E. Cheney. 2002. Myosin-X is an unconventional myosin that undergoes intrafilopodial motility. *Nat. Cell Biol.* **4:**246–250.

Broderick, M. J., and S. J. Winder. 2005. Spectrin, alpha-actinin, and dystrophin. *Adv. Protein Chem.* **70:**203–246.

During, R. L., B. G. Gibson, W. Li, E. A. Bishai, G. S. Sidhu, J. Landry, and F. S. Southwick. 2007. Anthrax lethal toxin paralyzes actin-based motility by blocking Hsp27 phosphorylation. *EMBO J.* **26:**2240–2250.

Ferron, F., G. Rebowski, S. H. Lee, and R. Dominguez. 2007. Structural basis for the recruitment of profilin-actin complexes during filament elongation by Ena/VASP. *EMBO J.* **26:**4597–4606.

Furman, C., A. L. Sieminski, A. V. Kwiatkowski, D. A. Rubinson, E. Vasile, R. T. Bronson, R. Fassler, and F. B. Gertler. 2007. Ena/VASP is required for endothelial barrier function in vivo. *J. Cell Biol.* **179:**761–775.

Kabsch, W., H. G. Mannherz, D. Suck, E. F. Pai, and K. C. Holmes. 1990. Atomic structure of the actin:DNase I complex. *Nature* **347:**37–44.

Kang, F., D. L. Purich, and F. S. Southwick. 1999. Profilin promotes barbed-end actin filament assembly without lowering the critical concentration. *J. Biol. Chem.* **274:**36963–36972.

Krause, M., E. W. Dent, J. E. Bear, J. J. Loureiro, and F. B. Gertler. 2003. Ena/VASP proteins: regulators of the actin cytoskeleton and cell migration. *Annu. Rev. Cell Dev. Biol.* **19:**541–564.

Larson, L., S. Arnaudeau, B. Gibson, W. Li, R. Krause, B. Hao, J. R. Bamburg, D. P. Lew, N. Demaurex, and F. Southwick. 2005. Gelsolin mediates calcium-dependent disassembly of Listeria actin tails. *Proc. Natl. Acad. Sci. USA* **102:**1921–1926.

McGrath, J. L., E. A. Osborn, Y. S. Tardy, C. F. Dewey, Jr., and J. H. Hartwig. 2000. Regulation of the actin cycle in vivo by actin filament severing. *Proc. Natl. Acad. Sci. USA* **97:**6532–6537.

Nakamura, F., E. Osborn, P. A. Janmey, and T. P. Stossel. 2002. Comparison of filamin A-induced cross-linking and Arp2/3 complex-mediated branching on the mechanics of actin filaments. *J. Biol. Chem.* **277:**9148–9154.

Nyman, T., R. Page, C. E. Schutt, R. Karlsson, and U. Lindberg. 2002. A cross-linked profilin-actin heterodimer interferes with elongation at the fast-growing end of F-actin. *J. Biol. Chem.* **277:**15828–15833.

O'Connell, C. B., M. J. Tyska, and M. S. Mooseker. 2007. Myosin at work: motor adaptations for a variety of cellular functions. *Biochim. Biophys. Acta* **1773:**615–630.

Pavlov, D., A. Muhlrad, J. Cooper, M. Wear, and E. Reisler. 2007. Actin filament severing by cofilin. *J. Mol. Biol.* **365:**1350–1358.

Pittenger, M. F., J. A. Kazzaz, and D. M. Helfman. 1994. Functional properties of non-muscle tropomyosin isoforms. *Curr. Opin. Cell Biol.* **6:**96–104.

Pollard, T. D. 2007. Regulation of actin filament assembly by Arp2/3 complex and formins. *Annu. Rev. Biophys. Biomol. Struct.* **36:**451–477.

Safer, D., T. R. Sosnick, and M. Elzinga. 1997. Thymosin beta 4 binds actin in an extended conformation and contacts both the barbed and pointed ends. *Biochemistry* **36:**5806–5816.

Southwick, F. S. 2005. Actin cytoskeleton: regulation of actin filament assembly, p. 255–273. *In* P. Cossart, P. Boquet, S. Normark, and R. Rappuoli (ed.), *Cellular Microbiology*, 2nd ed. ASM Press, Washington, DC.

Wear, M. A., and J. A. Cooper. 2004. Capping protein: new insights into mechanism and regulation. *Trends Biochem. Sci.* **29:**418–428.

Witke, W., W. Li, D. J. Kwiatkowski, and F. S. Southwick. 2001. Comparisons of CapG and gelsolin-null macrophages: demonstration of a unique role for CapG in receptor-mediated ruffling, phagocytosis, and vesicle rocketing. *J. Cell Biol.* **154:**775–784.

Yarmola, E. G., and M. R. Bubb. 2006. Profilin: emerging concepts and lingering misconceptions. *Trends Biochem. Sci.* **31:**197–205.

16

Functional Analysis of the Intraphagosomal Environment of the Macrophage: Fluorogenic Reporters and the Transcriptional Responses of *Salmonella* and *Mycobacterium* spp.

ROBIN M. YATES, KYLE H. ROHDE, ROBERT B. ABRAMOVITCH, AND DAVID G. RUSSELL

Phagocytosis is the process by which particulate material is internalized by cells. In protists this process is undertaken predominantly as a means of acquiring nutrients; therefore, the phagosome is almost exclusively a degradative organelle. In multicellular organisms the phagosome has evolved to fulfill multiple roles in addition to the simple breakdown of internalized material. These roles include intracellular sensing and cell/cell communication, antigen processing and presentation, and a range of regulated microbicidal responses. The successful execution of these roles depends on, and is modulated by, the physiology of the phagosomal lumen.

The bulk of the literature documenting the process of phagosome maturation focuses on the regulatory machinery that drives and controls the membrane trafficking and fusion events in nascent phagosomes (Rupper and Cardelli, 2001; Vieira et al., 2002; Lindmo and Stenmark, 2006; Yeung et al., 2006), as discussed in chapter 13 of this volume. These events involve the recruitment of membrane to the phagosome, the interaction of the phagosome with early endosomal compartments, and its acquisition of membrane and proteins from the exocytic/secretory pathways. This process is accompanied by increased mixing of the phagosomal contents with the contents of preexisting lysosomes. Early maturation events are regulated by the GTPase Rab5, in concert with the phosphatidylinositol kinase Vps34 and the production of phosphatidylinositol 3-phosphate (PI3P), and recruitment of the early endosome antigen EEA1. At this stage the phagosome communicates with the rapid recycling pathway and is accessible to endocytosed transferrin. The phagosome progresses to show a diminished association with Rab5 and an increasing accumulation of the late endosome/lysosome-associated GTPase Rab7. This progression is accompanied by the acquisition of a wide range of lysosomal hydrolases.

Much of the data on phagosome formation and modulation have relied on immunofluorescent approaches to demonstrate altered acquisition or accumulation of phagosomal constituents of known function. These studies have been phenomenally useful in determining the basic maturation characteristics of the phagosome. However, the localization of any protein, for example the aspartic proteinase cathepsin D, implies but does not demonstrate its function. Cathepsin D is delivered to the early endosome in an inactive, proenzyme form that is activated most probably by self-cleavage or processing by cathepsin B or L (Sturgill-Koszycki et al., 1996; Ullrich et al., 1999). This activation depends on the acidification of the phagosomal compartment. Therefore, it is critical to generate a more complete picture of the physiology of this compartment if we wish to understand its functional modulation. Given the wide array of roles that the phagocyte is required to fulfill, this information is key to our appreciation of phagocyte biology.

Of all the different phagocyte lineages, the macrophage is required to assume the widest range of physiological functions. Unlike specialized phagocytes such as the dendritic cell, a dedicated antigen-presenting cell, or the osteoclast, which remodels bone, the macrophage has to fulfill a range of homeostatic, inflammatory, and microbicidal functions linked to its tissue environment. In its resting state, the macrophage is required to roam through the tissues internalizing and quietly digesting cellular debris to cleanse the body. Once confronted by a microbial insult, however, the macrophage detects and subsequently responds to the stimulus by upregulating its microbicidal be-

---

Robin M. Yates, Kyle H. Rohde, Robert B. Abramovitch, and David G. Russell, Microbiology and Immunology, College of Veterinary Medicine, Cornell University, Ithaca, NY 14853.

havior and increasing its antigen-presenting capabilities (Akira, 2006; Trinchieri and Sher, 2007). Finally, once it has stimulated T lymphocytes, it is at the receiving end of a positive feedback loop that enhances its effector activities with respect to its ability both to kill microbes and to present their antigens to lymphocytes (Mosser, 2003; Fujiwara and Kobayashi, 2005). The phagosome is at the fulcrum of all of these functions. In these cells, the simple act of degradation has been subverted to satisfy the more complex demands of both the innate and acquired immune systems. The manner in which the immune system has evolved to interface with the phagocytic process has placed operational demands on the phagosomal compartment that exceed simple degradation. However, a full appreciation of how these processes have been modified to expand their "responsibilities" is limited by the paucity of functional readouts to measure physiologically relevant shifts in phagosomal behavior.

This chapter therefore focuses on the biology of the macrophage phagosome with respect to its changing responsibilities, although the data discussed will be placed in context with other phagocytic cells.

DYNAMIC REPORTERS OF PHAGOSOME FUNCTION

Although phagosomal acidification is covered in detail in chapter 14 of this volume, its overwhelming influence over other properties of the phagosomal lumen necessitates its mention here. Until recently the only parameter of phagosomal maturation that was measured with any frequency was phagosome acidification. These observations were achieved through the exploitation of pH-sensitive fluorochromes such as fluorescein or Oregon Green (Sturgill-Koszycki et al., 1994; Christensen et al., 2002; Yates et al., 2005; Savina et al., 2006). Both of these fluors are quenched upon protonation in an acidic environment. The readout is then normalized either to a second, pH-insensitive fluor, or to the same fluor excited at a second, pH-insensitive wavelength, known as the isobestic point. When macrophages are fed immunoglobulin G (IgG)-coated beads tagged with fluorescein, one can quantify the rate of acidification following internalization (Yates et al., 2005). IgG-bead-containing phagosomes equilibrate to a pH below pH 5.0 within 10 to 12 min postinternalization (Fig. 1).

The rate of acidification appears to correlate tightly with the rate of accumulation of functional proton V-ATPase complexes (Sturgill-Koszycki et al., 1994, 1996). While other ion pumps and channels do have an impact on acidification, as discussed in greater depth in chapter 19, it is clear that the V-ATPase is responsible ultimately for this process. Treatment of cells with the V-ATPase inhibitor concanamycin A leads to a complete abrogation of phagosomal acidification (Fig. 1) (Yates et al., 2005). Moreover, treatment with other drugs known to affect phagosome maturation, such as the calmodulin inhibitor W7, also reduces phagosome acidification. This finding demonstrates the intimate interplay between pH and phagosome maturation and has led to the suggestion that an intraphagosomal pH sensor regulates this process (Hurtado-Lorenzo et al., 2006).

THE RATE OF ACQUISITION OF LYSOSOMAL CONSTITUENTS

According to several studies, the phagosome acquires its degradative capacity both through fusion with preexisting lysosomes and through delivery of vesicles containing newly synthesized lysosomal hydrolases trafficking from the *trans*-Golgi network. Routinely, the rate of phagosome/lysosome mixing is measured either through the acquisition and accumulation of "lysosomal markers" such as LAMP-1, or through mixing with lysosomal cargo, such as fluorescein isothiocyanate dextran, which has been preloaded into the lysosomes. The use of LAMP-1 has caveats because the protein is delivered into the early endosome and sorted within the endosomal/lysosomal continuum to achieve enrichment in the lysosomes (Rohrer et al., 1996). In some cell types, such as macrophages, LAMP-1 is present in appreciable amounts even in early endosomal compartments (Sturgill-Koszycki et al., 1996). The mixing with lysosomal cargo is a better readout of phagosome/lysosome fusion, but again, because this is normally assayed by fluorescence microscopy, the data are only semiquantitative.

Recently, fluorescent resonance energy transfer (FRET) was employed in the development of a quantitative assay for phagosome/lysosome fusion (Yates et al., 2005). The assay involved pulse-chasing macrophages with a hydrophilic fluorochrome, Alexa Fluor 594 hydrazide (acceptor fluor), so that the fluorochrome was localized predominantly to the lysosomes. Macrophages were then fed IgG beads that had been labeled with Alexa 488 SE (donor fluor). FRET signal was achieved when the donor fluor was excited and emitted energy that subsequently excited the acceptor fluor. This energy transfer resulted in both a quenching of the donor fluor and an excitation of the acceptor fluor. These excitation and emission profiles are demonstrated in the confocal micrograph shown in Fig. 2. Analysis of the kinetics of phagosome/lysosome mixing was measured across the macrophage population by spectrofluorometer and shown to reach equilibrium 90 min postinternalization of the IgG-coated particles (Fig. 3). Both the acidification and FRET fusion assays reach a state of equilibrium that may be treated experimentally as an end point. Terminal acidification of an IgG-bead phagosome takes approximately 15 min, while maximal concentration of fluid-phase lysosomal contents is achieved at 90 min postinternalization. This phagosome/lysosome fusion assay measures the composite product of a range of biological parameters that contribute to the concentration of fluid-phase, lysosomal cargo in the phagosome. One can, however, argue that, with respect to the hydrolytic capacity of the phagosome, the concentration of soluble hydrolases (lysosomal cargo) is one of the most important parameters, once full acidification of the phagosome has been attained.

REAL-TIME MEASUREMENTS OF HYDROLYTIC CAPACITY

The rates of acidification and phagosome/lysosome mixing have implications with respect to the rate of acquisition of hydrolytic capacity, but they do not demonstrate hydrolysis directly. To determine the degradative capacity of the phagosome following its formation, fluorogenic substrates were coupled, along with a calibration fluor, to the surface of beads coated with the phagocytic receptor ligands IgG or mannosylated BSA (Man-BSA) (Yates et al., 2005). Hydrolysis of the substrates results in dequenching of the fluorochrome. The increase in activity is then expressed a ratio of emission from both the unquenched fluor and the calibration fluor. The substrates used and the activities assayed are detailed in Table 1. These substrate-based assays

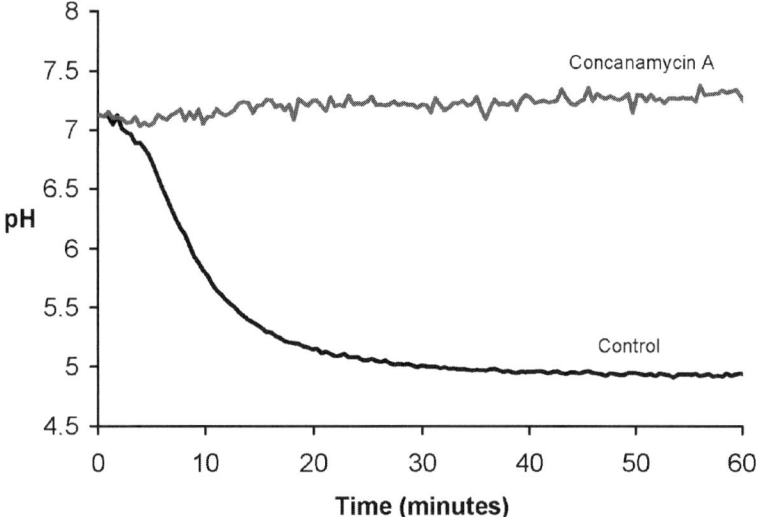

FIGURE 1 Acidification profiles of phagocytosed IgG-coupled beads labeled with fluorescein. Fluorescent emission at 520 nm was taken every 2 s, using alternating excitation wavelengths of 450 nm and 490 nm. Determination of pH of the lumen of the bead-containing phagosomes was calculated following polynomial regression of the excitation ratio with a standard curve. Inhibition of acidification was achieved with the addition of the V-ATPase specific inhibitor concanamycin A (100 nM) following binding of the beads.

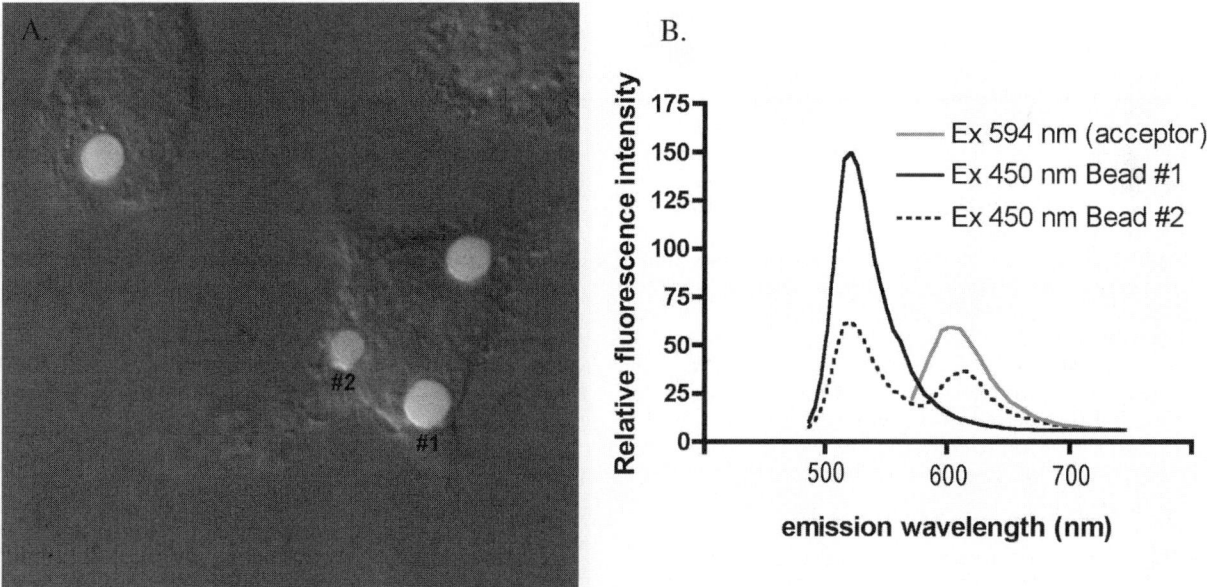

FIGURE 2 Confocal images demonstrating the FRET-based fusion phagosome/lysosome fusion assay. IgG-coupled beads labeled with Alexa Fluor 488-SE were bound to and incubated with bone marrow-derived macrophages that were preloaded with Alexa Fluor 594-hydrazide. Beads were given in two separate pulses 60 min apart and fixed 2 min after the second pulse. (A) Combined confocal images showing a bead in a phagosome that has fused with a lysosome (#2) and a bead in a phagosome that has not in fused with a lysosome (#1). (B) Spectral scan at ex/em ratios 450/520 nm of beads #1 and #2 and 594/620 of the acceptor fluor. Bead #2, in a fused phagolysosome, exhibits a biphasic emission profile, demonstrating quenching of the donor fluor signal coupled with excitation of the acceptor fluor.

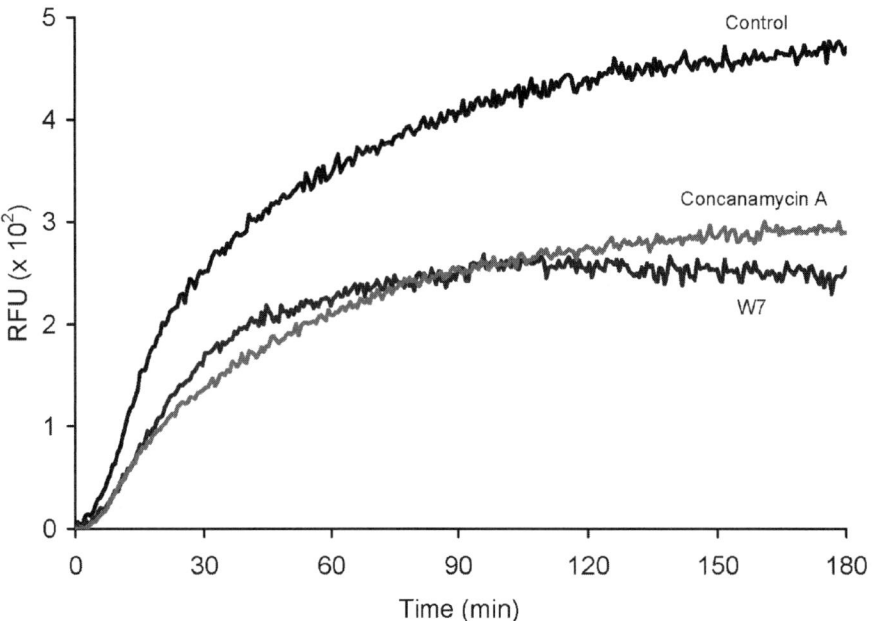

FIGURE 3 FRET-based phagosome/lysosome fusion profiles. Diminished phagosome/lysosome fusion was achieved with the calmodulin inhibitor W7 (15 μM) and the V-ATPase inhibitor concanamycin A (100 nM) in comparison with untreated (control) cells. Lysosomal fusion profiles for IgG-coupled bead-containing phagosomes were generated using the equation RFU = $F_{RT}/D_{RT} - F_{BO}/D_{BO}$ (RFU, relative fluorescent units; F_{RT}, FRET-generated fluorescent emission in real time; D_{RT}, donor emission in real time; F_{BO}, "FRET" signal contribution of the beads alone; D_{BO}, donor emission of the beads alone). Fluorescent measurements were taken every 2 s for 3 h.

reach a point of substrate limitation; therefore, the most meaningful measurements are acquired during the linear part of the readout. This rate of hydrolysis is a composite measurement that is a function of pH, as well as enzyme activity and concentration. Nonetheless, it represents a real measurement of actual hydrolytic capacity in a phagosome at a specific time point. The acidification assay and the FRET assay for phagosome/lysosome fusion both reach an end point equilibrium or maximum value. This contrasts with the hydrolysis assays that progress to substrate limitation because an enzyme, once present, will continue to hydrolyze its substrate until either the enzyme is removed or deactivated or, more likely, the substrate is exhausted.

By using this approach, reporters were generated that measure cysteine proteinase activity, bulk proteolysis, lipase activity, and β-galactosidase activity (Fig. 4) (Yates et al., 2005, 2007). Not surprisingly, the kinetics of maximal enzyme activity varied between the different substrates (Fig. 4). With the exception of the β-galactosidase assay, all the enzyme readouts reached a point of substrate limitation within a 20- to 90-min time frame. Pharmacological agents known to inhibit phagosome maturation—such as concanamycin A and W7, which affect phagosome maturation with respect to acidification and phagosome/lysosome mixing—also modified the hydrolytic activity of the phagosome.

Some of these assays exploited specific substrates processed by single, identified enzymes, whereas other substrates were more generic and were hydrolyzed by several enzymes. The caveat of assays that use generic substrates, such as DQ Green albumin, is that one does not know which enzymes are responsible for its degradation. On the flip side, however, the strengths of the assays are that they measure real activity that is the functional outcome of enzyme delivery, activation, and accumulation, as well as the pH and ionic environment within the phagosome. This approach does not require foreknowledge of the enzymes involved.

TABLE 1 Substrates used and activities assayed

Enzymatic activity	Substrate
Bulk proteinase activity	DQ Green Bodipy BSA
Cysteine proteinase activity	(Biotin-LC-Phe-Arg)$_2$-rhodamine 110
β-Galactosidase activity	5-Dodecanoylaminofluorescein di-β-D-galactopyranoside
Lipolytic activity	1-Trinitrophenyl-amino-dodecanoyl-2-pyrenedecanoyl-3-O-hexadecyl-sn-glycerol

INNATE IMMUNE SENSING AND PHAGOSOME MATURATION

Recently, these assays have been employed to probe the modulation of the macrophage phagosomes under differing biological situations such as microbial insult and cytokine-mediated activation (Yates and Russell, 2005). Toll-like receptors (TLRs) have been identified relatively recently as mediating the sensing of conserved microbial patterns and serving as an early warning of microbial invasion (Akira, 2006; Trinchieri and Sher, 2007). Macrophages express several members of this receptor family, as discussed in chapter 6 of this volume, and early studies detail the recruitment of TLR2 to the phagosome during its formation around microbial cargo (Ozinsky et al., 2000). This observation was followed up by a study that detailed accelerated maturation of phagosomes formed around particles bearing TLR agonists (Blander and Medzhitov, 2004). This idea was attractive because it suggested that naive macrophages could discriminate between microbes that were potentially infectious and benign cargo, and modulate phagosome behavior accordingly. Acceleration was detected through increased association with LAMP-1 and Lysotracker-labeled lysosomes. In a subsequent publication, the authors demonstrated a similar phenomenon in dendritic cells and reported that this accelerated maturation of the phagosome leads to enhanced antigen presentation and T-cell stimulation (Blander and Medzhitov, 2006). Data were presented indicating that this effect was mediated at the level of the individual phagosome, giving rise to the concept that the macrophage handled phagosomes as autonomous entities that were regulated by localized TLR signaling.

This concept was revisited with model particles either with or without the addition of the TLR agonists lipopolysaccharide (LPS) and Pam3Cys (Yates and Russell, 2005). Activation of TLR signaling was observed only in the presence of TLR agonists and their appropriate TLR receptor. However, despite the activation of TLRs during phagocytosis, there were no detectable effects on either the rate of acidification or the rate of phagosome/lysosome mixing, measured by FRET. Subsequently, it has been argued that the use of ManR and FcR ligands may have engaged receptors that then stimulated the signaling pathway for accelerated maturation, obviating the need for TLR stimulation to accelerate phagosome maturation (Blander and Medzhitov, 2007). However, experiments performed with phosphatidylserine-coated C18 silica particles with and without TLR agonists, to mimic uptake of apoptotic cells, also failed to reveal a TLR-dependent component regulating the kinetics of phagosome maturation (Yates and Russell, 2005). Subsequent experiments were performed on phagosomes formed around a more complex particle, *Staphylococcus aureus*, both in the presence and absence of LPS and following uptake by both wild-type and TLR2-deficient macrophages (Fig. 5). TLR signaling was detected by degradation of IκB and phosphorylation of p38 mitogen-activated protein kinase. While activation of TLR signaling was observed with both *S. aureus* and *S. aureus* with LPS in wild-type macrophages, only the *S. aureus* with LPS activated TLR signaling in the TLR2-deficient macrophages. Most significantly, however, the kinetics of phagosome/lysosome mixing were again identical irrespective of the presence or absence of TLR signaling. These data suggest strongly that TLR agonists do not modulate short-term phagosome maturation through TLR signaling. These assays all examined the immediate effect of TLR signaling on the maturation of the phagosome containing the particle carrying the TLR agonist. However, what the assays did not address was the downstream consequences of activation of the macrophage mediated by the autocrine loop induced by stimulation of TLRs.

IMMUNE ACTIVATION AND THE PHYSIOLOGY OF THE PHAGOSOME

In contrast to the debate concerning TLR signaling and short-term phagosome maturation, there is a considerable body of data documenting a wide range of altered activities in the phagosome of activated macrophages (Tsang et al., 2000; MacMicking et al., 2003; Park, 2003; Gutierrez et al., 2004; Murray et al., 2005; Prost et al., 2007a; Vidal et al., 2008). Macrophages activated for a prolonged period (>90 min) with TLR agonists and cytokines such as interferon gamma (IFN-γ) show marked changes in the physiology of their phagosomal compartments. These phagosomes show enhanced recruitment of the NADPH oxidase complex responsible for the production of oxygen radicals. In addition, pathogens such as *Mycobacterium tuberculosis* that are capable of arresting the normal maturation and acidification of their phagosomes, which equilibrate to pH 6.4, find themselves unable to regulate their intracellular compartment and are delivered to lysosomes upon activation of the phagocyte (Schaible et al., 1998; Via et al., 1998; MacMicking et al., 2003; MacMicking, 2005). It has been shown that this altered fusogenicity in the *M. tuberculosis*-containing phagosome is mediated, at least in mice, by the activity of a family of IFN-γ-inducible GTPases such as LRG-47. These GTPases appear to regulate the membrane fusion capacity of phagosomes and vesicles within the endosomal/lysosomal continuum.

These data, together with the enhanced killing capacity of activated macrophages, have led to the assumption that phagosome maturation is accelerated in activated cells and that the phagolysosome in these cells represents a more hostile, degradative environment. A recent study addressed this issue through determination of the kinetics of acidification and phagosome/lysosome mixing, and the rates of acquisition of protease, lipase, and β-galactosidase activities in phagosomes in macrophages activated by exposure to LPS, or IFN-γ, or LPS and IFN-γ (Yates et al., 2007). The rates of acidification of phagosomes containing Man-BSA-coated beads were altered subtly, with LPS-treated cells exhibiting an accelerated and enhanced pH drop. At 2 h postinternalization, the LPS-treated and LPS+IFN-γ-activated macrophage phagosomes were at pH 4.7, while untreated cells and cells treated with IFN-γ had a phagosomal pH of 4.9. The kinetics of phagosome/lysosome mixing revealed by FRET analysis also showed subtle alterations. Early on, phagosomes in cells treated with LPS + IFN-γ showed a delayed acquisition of lysosomal cargo, but from 2 h onward all of the phagosomes in activated macrophages demonstrated enhanced and sustained accumulation of lysosomal cargo up to 8 h postinternalization.

These data were consistent with an enhancement of phagosomal degradation, but macrophages activated overnight by exposure to LPS and IFN-γ exhibited a diminution in the total hydrolytic capacity of their maturing phagosomes compared with phagosomes in nonactivated cells. Moreover, the hydrolytic capacity was downregulated differentially, varying with both the activation stimulus and the enzymatic activity examined. For example, β-galactosidase activity was most strongly depressed in macrophages activated by IFN-γ and LPS, whereas proteolytic activity was most strongly downregulated by IFN-γ alone (Fig. 6). This differential modulation of the various lysosomal hydrolases implies that the cell has the capability to reprogram its degradative capabilities to process different cargoes, or to fulfill different functions.

At first glance, a reduction in the degradative capacity in the phagosomes of activated macrophages appears counterintuitive if one thinks of a macrophage purely as an immune effector cell. However, this transition is actually consistent with the changing behavior and new roles of

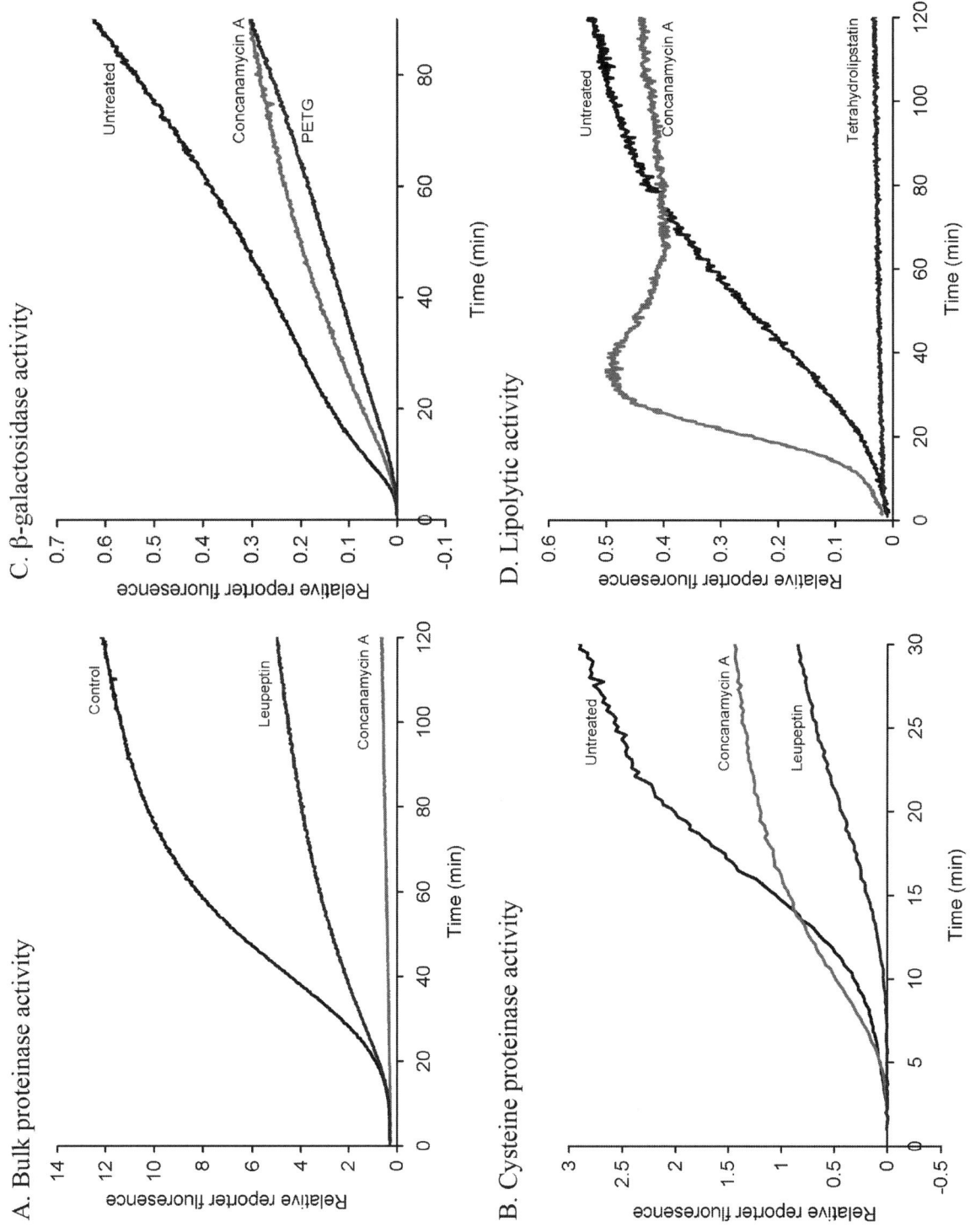

the activated phagocyte. As discussed earlier, a resting macrophage is required to clean tissues in a quiet, noninflammatory manner; this function would best be achieved by a cell that is highly degradative and immunologically silent, to avoid the induction of an autoimmune response. Such a cell would degrade nonpathogenic material without triggering a strong inflammatory response. In contrast, upon activation with either TLR agonists, or more significantly with IFN-γ, the phagocyte must now be capable of antigen processing and presentation, as well as possessing enhanced microbicidal capabilities. In general, it is accepted that macrophages are poor inducers of a primary immune response but are highly effective antigen-presenting cells in a secondary immune response. The cells that induce the primary immune response are professional antigen-presenting cells like dendritic cells, which exhibit a much-reduced proteolytic capacity compared with macrophages (Lennon-Dumenil et al., 2002; Trombetta et al., 2003; Delamarre et al., 2005; Savina et al., 2006). It has been reported that this reduced proteolysis can be achieved by alkalinization of the phagosome through the activity of the NADPH oxidase (Savina et al., 2006). The results from these studies imply that too much proteolysis hinders epitope generation and/or half-life. Recently, Delmarre and colleagues (2005) tested this hypothesis directly by immunizing mice with RNase A and RNase S, proteins that differ only in an intrachain cleavage that renders RNase S susceptible to lysosomal proteolysis. After immunization, the mice that received RNase A developed a much more robust cellular and humoral immune response. The immune response recognized both forms of the antigen, indicating that the difference was, at least predominantly, one of degree not specificity. The authors were careful to acknowledge the complexities of the in vivo experiments and the multiple benefits associated with antigen longevity. Nonetheless, the study underlines how important the proteolytic sensitivity of an antigen and the degradative capacity of the phagolysosome can be to optimizing an immune response.

How is this achieved in macrophages that do not exhibit the NADPH oxidase-dependent regulation of phagosomal pH (Savina et al., 2006), and how can the hydrolytic capacity be modulated differentially for some lysosomal enzymes (Yates et al., 2007)? Lysosomal hydrolases are routed to the phagosome from both the synthetic pathway and through fusion with pre-existing lysosomes. To test the relative contribution of these two sources, cells were treated with brefeldin A to block the trafficking of newly formed hydrolases from the trans-Golgi network (Nebenfuhr et al., 2002). This manipulation had no discernible effect on the proteolytic activity of the phagosome, indicating that de novo-synthesized hydrolases play a minimal role in the degradative capacity of the phagosome (Yates et al., 2007). To determine whether the degradative profile of the total lysosomal population was modified, macrophages were fed iron-dextran and a magnetic isolation procedure was used to isolate the lysosomes from resting versus activated macrophages (Pethe et al., 2004; Alonso et al., 2007; Yates et al., 2007). Measurement of the relative activities for lipolysis, proteolysis, and β-galactosidase activity revealed no significant differences between preparations irrespective of the activation stimulus applied to the macrophage. This result suggested that the cell is capable of generating a "mature" phagolysosome that has a hydrolase profile that differs from the bulk lysosomal population.

These data indicate that the simple idea that a full complement of acid hydrolases mediate mass hydrolysis within the phagosome following fusion with a lysosome has to be reappraised because mixing appears to be incomplete or selective. First, it is becoming increasingly clear that the term "lysosome" better represents a group of heterogeneous compartments, each containing a unique repertoire of hydrolases, rather than a single defined entity. In the "kiss-and-run" model of phagosome maturation, proposed by Desjardins and colleagues, such heterogeneity could be generated through transient, partial mixing events with multiple compartments (Desjardins, 1995; Stuart et al., 2007). It is therefore unlikely that hydrolases are delivered to the phagosome as a single complete bolus, but rather through a series of discrete, differentially regulated events. By measuring the activity of 10 acid hydrolases in purified phagosomal extracts from the macrophage-like J774 cell line, Claus and colleagues (1998) demonstrated the asynchronous delivery of these hydrolases to the maturing phagosome and their differential recruitment following pharmacological manipulation of the cell. In particular, they found that the protease cathepsin H was delivered preferentially to early phagosomes and, contrary to other hydrolases investigated, was enriched in phagosomes following treatment with bafilomycin A and chloroquine to block phagosome acidification. They went on to show that a subset of acid hydrolases were depleted in phagosomes following the induction of lysosomal secretion with acidotropic drugs, whereas others, including cathepsin H, were retained. Together these data demonstrate that certain hydrolases are heterogeneously compartmentalized in J774 cells, and this conclusion is reflected in the patterns of delivery of hydrolases to the forming phagosome. These

---

FIGURE 4 Measurement of the hydrolytic activities in macrophage phagosomes following the internalization of substrate-bearing particles. (A) Bulk proteinase activity was measured with the substrate DQ Green BSA linked to Man-BSA-coated beads. Manipulation of hydrolytic rates was achieved with inhibitors concanamycin A (100 nM) and leupeptin (100 μg/ml). (B) Cysteine proteinase activity was measured with Biotin-FR-Rhodamine 110 beads. Manipulation of hydrolytic rates was achieved with inhibitors concanamycin A (100 nM) and leupeptin (100 μg/ml). (C) β-Galactosidase activity was measured with beads bearing the substrate 5-dodecanoyl-aminofluorescein di-β-D-galactopyranoside. Manipulation of hydrolytic rates was achieved with inhibitors concanamycin A (100 nM) and PETG (phenylethyl thiogalactopyranoside) (10 μg/ml). (D) Lipolysis was measured with the quenched pyrene-containing substrate 1-trinitrophenyl-amino-dodecanoyl-2-pyrenedecanoyl-3-O-hexadecyl-sn-glycerol. Manipulation of hydrolytic rates was achieved with inhibitors concanamycin A (100 nM) and tetrahydrolipstatin (10 μg/ml). Traces were generated using the equation FU = substrate fluorescence/calibration fluor emission (where FU = arbitrary fluorescent units). Measurements were taken every second for 30 min.

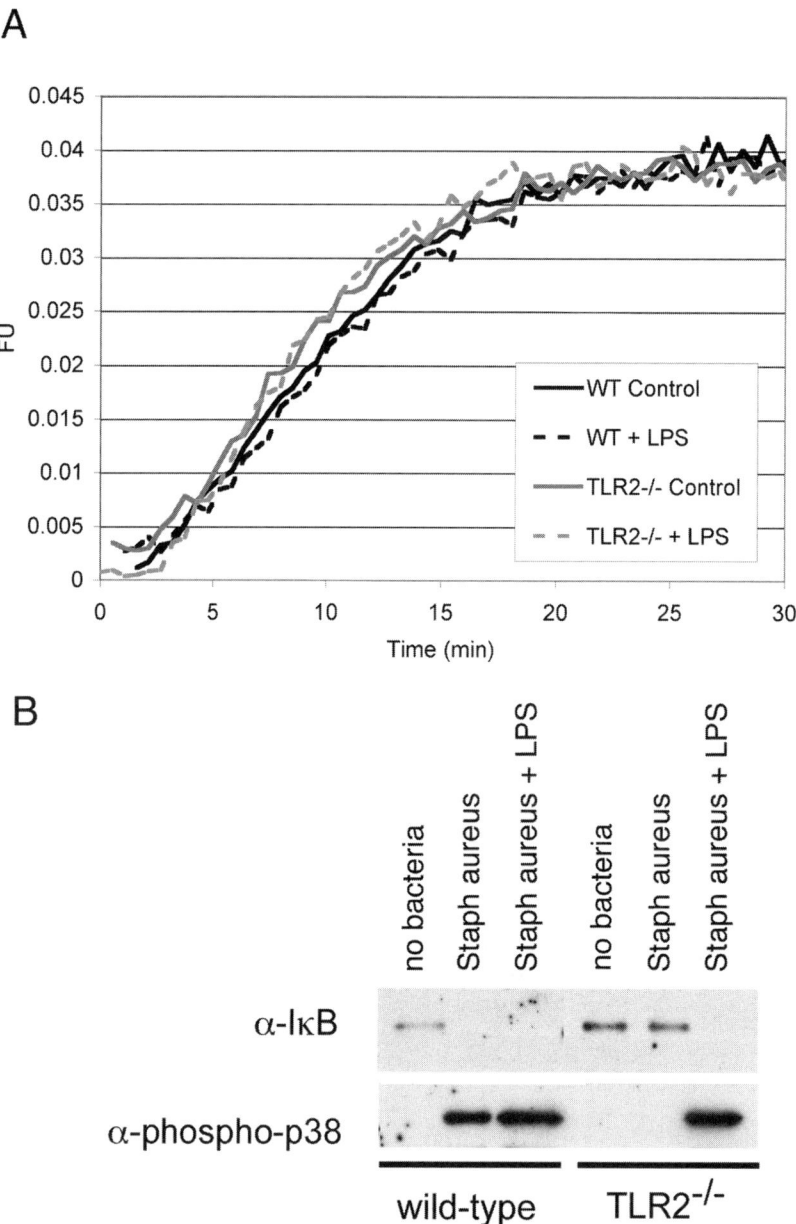

FIGURE 5 Phagosomes containing *S. aureus*, formed in the presence or absence of TLR2 or TLR4 signaling, exhibit comparable phagosome/lysosome fusion profiles, indicating that the stimulation of either TLR2 or TLR4 does not have an impact on phagosome maturation. (A) The FRET-based assay was used to quantify phagosome/lysosome mixing following uptake of formalin-fixed *S. aureus* +/− LPS in wild-type (WT) (C57BL/6) and TLR2$^{-/-}$ macrophages. Data are presented as an average over four individual sets of data. (B) Degradation of IκBα and phosphorylation of p38 mitogen-activated protein kinase in the macrophage were examined by immunoblotting following phagocytosis of *S. aureus* particles with or without incorporated LPS in WT (C57BL/6) and TLR2$^{-/-}$ macrophages. Reproduced with permission from Elsevier (Yates and Russell, 2005).

data are similar to the observation that blocking acidification of phagosomes actually enhances degradation of the lipolytic substrate 1-trinitrophenyl-amino-dodecanoyl-2-pyrenedecanoyl-3-O-hexadecyl-sn-glycerol on the surface of C18 beads following uptake by macrophages. Although the enzymes involved remain unidentified, the data suggest that neutral lipases are active in the early phagosome and are delivered to the phagosome from a nonlysosomal source (Yates et al., 2005). Heterogeneous patterns of hydrolase recruitment were also noted in a more recent study that employed stable isotope labeling by amino acids in cell culture (SILAC) to quantify the association of 382 proteins with latex bead phagosomes over time (Rogers and Foster, 2007). Beyond asynchronous recruitment of lysosomal hydrolases to the phagosome, Rogers and Foster (2007) described a biphasic pattern of recruitment of several lysosomal proteins to the maturing phagosome including LAMP-1 and the cathepsins B and S.

All these data argue persuasively that there is a level of functionally significant heterogeneity among the "lysosomes" of the phagocyte, and that the complement of hydrolases that we imagine present in a mature lysosome is actually distributed differentially among this population of vesicles. This scenario was suggested previously by Lennon-Duminel and colleagues (2002) from their analysis of proteolytic processing in phagosomes from dendritic cells.

TRANSCRIPTIONAL PROFILING FROM WITHIN THE PHAGOSOME

Unlike the synthetic reporters that we have developed to probe the physiological conditions within the phagosome, several researchers have exploited pathogenic bacteria and their responses to environmental stimuli to report directly and indirectly on the environmental changes inside the phagosome. Whole-genome microarray studies have been performed on several species of intracellular bacteria, including *Bacillus anthracis*, *Listeria*, *Staphylococcus*, *Salmonella*, and *Yersinia* (Voyich et al., 2005; Chatterjee et al., 2006; Zhou et al., 2006; Bergman et al., 2007; Fadl et al., 2007), and have started to generate some insights into the environment in which these bacteria find themselves. The two systems for which the greatest body of data is available are *Salmonella enterica* serovar Typhimurium (Faucher et al., 2006; Jansen and Yu, 2006; Martin-Orozco et al., 2006) and *M. tuberculosis* (Schnappinger et al., 2003; Talaat et al., 2004; Rohde et al., 2007a, 2007b; Waddell and Butcher, 2007). Both these pathogens are the subject of specific chapters later in this volume (see chapters 23 and 26), and therefore we are focusing specifically on the interplay between the two bacteria and the changing environments that they will experience in the phagosomes of the macrophage.

Salmonella

Salmonella gains entry into phagocytes via receptor-mediated phagocytosis or through uptake induced by effectors released into the macrophage cytoplasm through the bacterium's type III secretion system, which is encoded by the pathogenicity island SPI-1 (Schlumberger and Hardt, 2005; Mulvey et al., 2006; Ellermeier and Slauch, 2007). Although these pathways give rise to vacuoles that can be either tight or "spacious," both compartments appear to mature relatively normally. They show transient interaction with early endosomes prior to acquiring EEA1, Rab7, and other accepted constituents of the late endosome/lysosomal compartments. This compartment acidifies below pH 5.0, although the progression is slower for live bacteria (approximately 5 h) versus dead bacteria (45 min).

However, in the later stages 60 min postinfection, the bacterium initiates active remodeling of its compartment. At this time the bacterium is triggered to express components of SPI-2, a second pathogenicity island, which encodes another type III secretion system (Waterman and Holden, 2003; Guiney, 2005; Abrahams and Hensel, 2006). This apparatus inoculates the host cell cytosol with SifA and other effectors that hijack the membrane-trafficking pathways in the macrophage to generate *Salmonella*-induced filaments (Sifs), which are necessary to support the intracellular survival of the bacterium.

Transcriptional Reporters of the Intraphagosomal Environment

Researchers have been aware for many years that the two-component regulator PhoP/Q is critical for intracellular survival and regulates expression of in excess of 200 genes, including those of SPI-2 and its effector substrates. There has been debate, however, over the intraphagosomal signal that is recognized by PhoQ, the sensor kinase, and its response regulator PhoP. Previous studies had implicated [Mg^{2+}] (Chamnongpol et al., 2003), but more recent analysis indicates that the PhoP/Q system is a pH sensor that responds specifically to the acidification of the *Salmonella*-containing vacuole (Martin-Orozco et al., 2006; Prost et al., 2007a, 2007b).

Martin-Orozco and colleagues (2006) generated a reporter strain of *Salmonella* sp. in which green fluorescent protein (GFP) was expressed under regulation of the *phoP* promoter. This strain responded to both pH and [Mg^{2+}] in an additive manner, with low [Mg^{2+}] at low pH giving the greatest induction of GFP fluorescence. In macrophages, most bacteria were expressing GFP 60 min postinfection and reached maximal expression levels 120 min postinfection. The intraphagosomal [Mg^{2+}] was measured by PEBBLE (probe encapsulated by biologically localized embedding), which exploits a coumarin 343 and Texas Red loaded nanosensor to generate a ratiometric readout of [Mg^{2+}]. Despite successful manipulation of the intraphagosomal [Mg^{2+}], the researchers could not detect alterations in the transcriptional activity of the *phoP* promoter. In contrast, treatment of infected cells with the H^+-ATPase inhibitor concanamycin A abrogated the activation of the *phoP* promoter and little increase in GFP expression was observed. This study demonstrates how the bacterium can be exploited to provide real-time kinetic readouts of the intravacuolar conditions inside the phagosome and how, through pharmacological manipulation of this environment, one can demonstrate the specificity of the readout to an individual intraphagosomal stimulus, in this case pH.

Mycobacterium tuberculosis

The previous study focused on using a promoter from a characterized gene to generate a live readout of intraphagosomal conditions. The next series of studies discussed explore genome-wide approaches to examining the impact of intracellular, environmental shifts on the transcriptional activity of a bacterium.

Many labs have studied *Mycobacterium*-containing phagosomes, and a reasonable consensus has emerged that the vacuoles retain many of the characteristics common to the early endosome. *Mycobacterium*-containing phagosomes equilibrate to a luminal pH of 6.4 (Sturgill-Koszycki et al., 1994) and retain markers of the early endosome, such as transferrin receptor (TfR) and the small GTPase Rab5 (Clemens and Horwitz, 1995, 1996; Sturgill-Koszycki et al., 1996). The arrest of the normal maturation process allows the vacuole to retain access to cargo trafficking through the rapid recycling pathway, such as transferrin (Clemens and Horwitz, 1996; Sturgill-Koszycki et al., 1996), yet limits the acquisition of lysosomal constituents. The point of arrest of phagosome maturation is prior to the accumulation of the early endosomal antigen, EEA1 (Fratti et al., 2001; Vergne et al., 2003; Purdy, 2005). EEA1 is a PI3P-binding protein that associates with the cytosolic face of the phagosome once the type III phosphatidylinositol 3-

A.

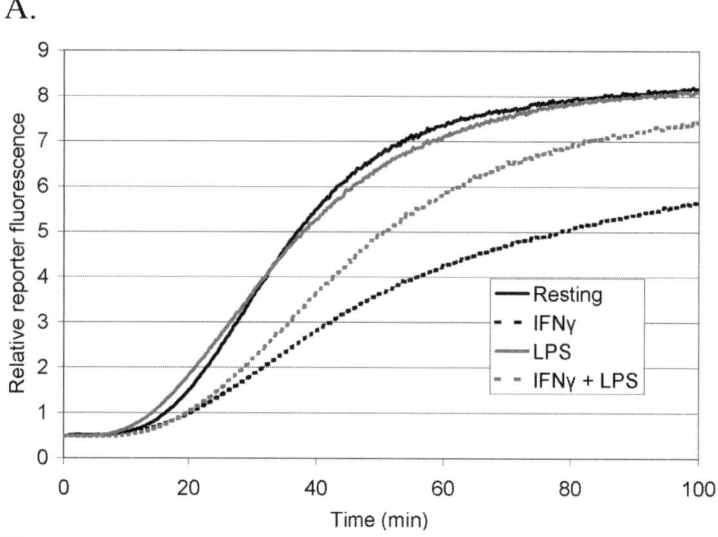

B.

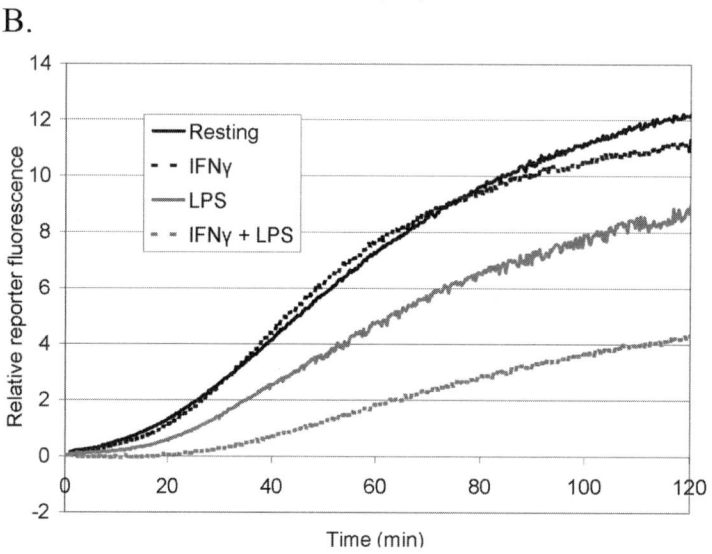

C.

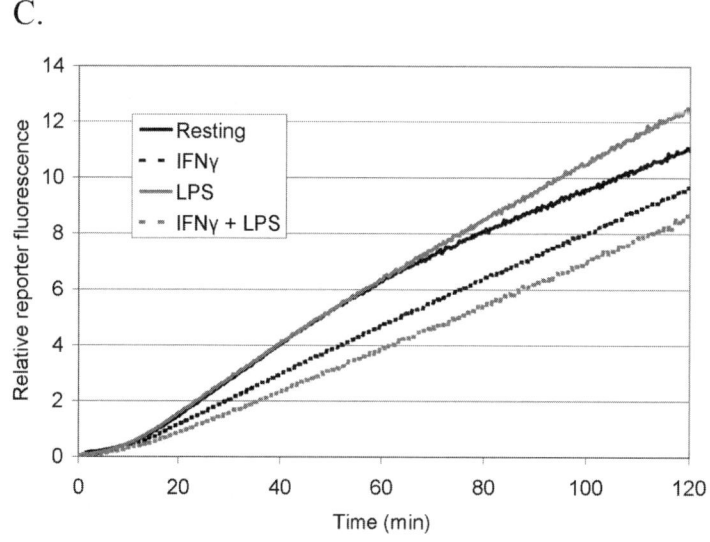

kinase Vps34 has generated sufficient density of PI3P (Fratti, 2003; Kelley and Schorey, 2004; Purdy et al., 2005). EEA1 accumulation depends on calmodulin and Ca^{2+} signaling through the recruitment and activity of sphingosine kinase. The direct manipulation of infected macrophages by calcium ionophores can promote the maturation of the Mycobacterium-containing phagosome (Malik et al., 2003; Thompson et al., 2005). All these studies tend to stress the complexity of the regulatory pathways that control intracellular fusion events associated with phagosome maturation, as discussed in chapter 13.

Despite the ability of M. tuberculosis to infect the macrophage, it is important to remember that there is an inverse correlation between the survival of intracellular Mycobacterium spp. and the activation status of the host cell. In macrophages, killing appears to depend predominantly on upregulation of the inducible nitric oxide synthase (iNOS) (Nicholson et al., 1996; MacMicking et al., 1997). However, mice defective in iNOS fare considerably better when infected with M. tuberculosis than do mice that are deficient in IFN-γ production. This difference is due to other alterations in phagosomal physiology mediated by activation (Aston et al., 1998; Gomes et al., 1999; Gutierrez et al., 2004; Hisert et al., 2004). Macrophages activated with IFN-γ prior to infection with Mycobacterium are capable of overcoming the bacterium's ability to arrest phagosome maturation and instead deliver the bacterium to an acidified compartment that fuses with lysosomes. Early studies on Mycobacterium avium calculated the pH of this compartment to be pH 5.2, which has recently been confirmed for M. tuberculosis (Schaible et al., 1998; MacMicking et al., 2003). This translocation to a more hostile environment correlates with a decline in bacterial viability. It also renders the bacterium accessible to microbicidal peptides like the ubiquitin-derived peptide found in multivesicular lysosomes (Alonso et al., 2007).

Transcriptional Profiling from within the Phagosome

McKinney and colleagues (2000) identified icl1, the gene encoding isocitrate lyase, as being transiently upregulated inside M. tuberculosis in resting macrophages and constitutively upregulated in activated macrophages. Isocitrate lyase is the gating enzyme into the glyoxylate shunt that enables organisms to retain carbon while growing on fatty acids as their single, limiting carbon source. Bacteria expressing an ICL1::GFP fusion protein under control of the icl1 promoter showed markedly increased fluorescence in the phagosome of activated macrophages, implying a link between the changing intraphagosomal environment and bacterial metabolism. Although this study demonstrated the effect of changing phagosomal environments on bacterial transcription, the exact nature of the stimulus was not elucidated.

More recently, Schnappinger and colleagues (2003, 2006) generated a comprehensive, genome-wide analysis of the transcriptional response of M. tuberculosis (H37Rv) in wild-type and NOS2-deficient macrophages in both resting and activated states. They isolated bacteria at 4, 24, and 48 h postinfection and conducted microarray analysis of the transcriptional response. They then conducted further in vitro studies exposing bacteria to comparable stresses in the test tube to link transcriptional activities to specific stimuli to which the bacterium would be exposed in the intraphagosomal environment. These stresses included low O_2, low Fe^{2+}, NO, H_2O_2, sodium dodecyl sulfate, heat shock, and palmitic acid as an alternate carbon source. These different in vitro conditions were able to induce regulation of approximately 60% of the genes that made up the differential intraphagosomal transcriptome. The researchers identified several sets of specific responses that could be attributed to individual intraphagosomal stimuli. Genes upregulated by NO in vitro and in activated, wild-type macrophages, but not in activated $NOS2^{-/-}$ macrophages, appeared to be specific to nitrosative stress. Another set of genes upregulated under conditions of low iron in vitro were also upregulated in IFN-γ-activated macrophages, suggesting that iron is scarce in the M. tuberculosis-containing phagosomes of activated phagocytes. Consistent with the observations of McKinney and colleagues (2000), the study by Schnappinger et al. also observed upregulation of genes involved in fatty acid metabolism, including icl1. This study paints a profile of an intraphagosomal environment that places nutritional stress on the bacterium and, in an activated macrophage, is accessible to NO and constitutes an iron-depleted environment. The authors compared their data with those obtained previously for Escherichia coli and S. enterica serovar Typhimurium inside macrophage phagosomes (Staudinger et al., 2002; Eriksson et al., 2003). They noted that the oxidative response in M. tuberculosis was induced by NO production, in contrast to E. coli, where this response depended on the activity of the NADPH oxidase. Whether this represents a difference at the level of the bacterial response or could be explained by reports that the NADPH oxidase complex may not be recruited to the M. tuberculosis-containing phagosome (Astarie-Dequeker et al., 1999) remains to be determined. They also noted that the apparent nutritional shift of M. tuberculosis to the utilization of fatty acids as an important carbon source differed from the case with S. enterica serovar Typhimurium, where gluconate is proposed as the primary nutrient. The authors' conclusions were that these transcriptional profiles indicated that the bacteria inhabited different intracellular niches and that the intraphagosomal

FIGURE 6 Specific hydrolase activities of phagosomes containing mannosylated beads bearing fluorogenic substrates along with a calibration fluor were measured in resting and macrophage monolayers activated by overnight incubation with LPS (10 ng/ml) and/or IFN-γ (100 U/ml). The increase in substrate fluorescence relative to the calibration fluor (relative reporter fluorescence) correlates to substrate hydrolysis and was plotted against time. (A) Phagosomal proteolysis was measured through incorporation of the generic protease substrate DQ Green Bodipy BSA. (B) Phagosomal lipolysis was measured through incorporation of the triglyceride analogue 1-trinitrophenyl-amino-dodecanoyl-2-pyrenedecanoyl-3-O-hexadecyl-sn-glycerol. (C) Phagosomal β-galactosidase activity was measured through the incorporation of the β-galactosidase substrate 5-dodecanoylaminofluorescein di-β-D-galactopyranoside. Reproduced with permission (Yates et al., 2007).

environments to which they were exposed afforded significantly different stresses and opportunities (Schnappinger et al., 2003, 2006).

Microarray analysis appears to afford novel insights into the intraphagosomal environment, with the important caveat that the physiology of these compartments is a reflection of both the host cell biology and its manipulation by the pathogen of interest. In a recent study on M. tuberculosis, Rohde and colleagues (2007b) developed a temporal picture of the changing transcriptional profile of the bacterium during the initial stages of infection (Fig. 7). They reported that, in the presence of the F-actin inhibitor cytochalasin D, bacteria bound to the surface of the macrophage, but not internalized, did not detect the presence of the phagosome. This absence of a response is significantly different from many enteric bacteria, including Salmonella, E. coli, Yersinia, etc., all of which require rapid activation of type III secretory systems to control the initial infection process to ensure a productive infection. In contrast, M. tuberculosis did not show a transcriptional response until the internalization process had been initiated. Temporal dissection of the transcriptional profile indicated that the majority of genes upregulated at 2 h showed the greatest change in transcriptional activity in the initial 5- to 40-min period of infection.

Among the groups of genes upregulated, several patterns or unifying themes emerged. Several members of the WhiB family of putative transcriptional regulators unique to actinomycetes (Hutter and Dick, 1999; Mulder et al., 1999; Soliveri et al., 2000), notably *whib3*, *whib6*, and *whib7*, were upregulated. Previous studies had shown that WhiB7 was upregulated markedly by exposure of bacteria to aminoglycosides, and mutants deficient in WhiB7 expression showed increased sensitivity to the drug; therefore, it was thought that WhiB7 influences the response of M. tuberculosis to drug pressure (Morris et al., 2005). Recently, however, it was shown that heat shock, iron starvation, growth phase, and exposure to toxic fatty acids all induce increased expression of *whib7*, implying that the gene responds to a broad range of noxious stimuli (Steyn et al., 2002; Geiman et al., 2006; Singh et al., 2007). WhiB3 binds to the sigma factor RpoV, which has been implicated both in bacterial survival and tissue pathology late in infection. Expression of WhiB3 and WhiB6 is also upregulated to some degree by aminoglycosides, ethanol, oxidants, low pH, sodium dodecyl sulfate, and heat shock, again indicating that they respond to a broad range of hostile stimuli. This family of proteins has an extremely interesting structural feature in that they all possess an iron-sulfur (FE-S) cluster. Electron paramagnetic resonance and UV-visible spectroscopy analysis of reduced WhiB3 indicates that the cluster is sensitive to O_2, which leads to the conversion of a $[4Fe-4S]^{2+}$ active form of the protein to a $[3Fe-4S]^{1+}$ inert form (Singh et al., 2007). This mechanism of cluster disassembly is also observed in the fumarate nitrate regulator (Fnr) of E. coli, which regulates the transcription of >100 genes in response to oxygen and nitric oxide (Spiro, 2006). It has been shown previously that both Fnr and WhiB3 are acutely sensitive to reactive oxygen and nitrogen intermediates such as superoxide and NO. This suggests that the WhiB family of proteins may form an extremely sensitive set of sensors for redox changes experienced by M. tuberculosis within the phagosome.

Within the same theme of stress response, there is increased expression of a number of genes that are members of the DosR regulon. The two-component signal transduction system DosSR comprises a membrane-bound histidine kinase sensor (DosS) and a cytoplasmic response regulator (DosR). Genes dependent on this regulator are upregulated in experimental models that induce growth arrest in M. tuberculosis and upon exposure of the bacterium to NO or hypoxia, which reduces the aerobic respiration of the bacterium (Kendall et al., 2004; Roberts et al., 2004; Saini et al., 2004; Voskuil et al., 2004). We also noted the increased expression of genes belonging to the regulon of another two-component signal transduction system, PhoPR (Perez et al., 2001; Gonzalo Asensio et al., 2006; Walters et al., 2006). At 2 h postinfection, 25 of the 44 genes reported

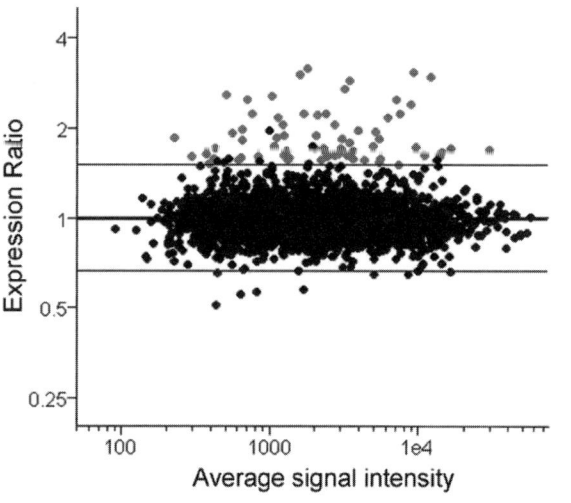 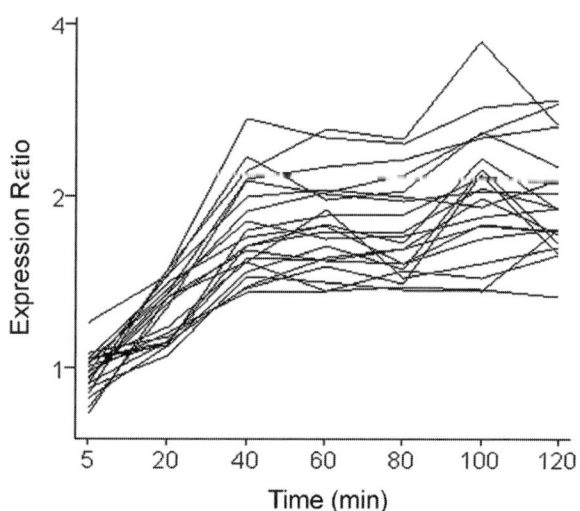

FIGURE 7 (Left) Transcriptional profile of M. *tuberculosis* 2 h postinternalization compared with bacteria in the medium control. Genes upregulated beyond the 1.6X cutoff are shown in gray. (Right) Kinetics of the transcriptional response to the transition to an intraphagosomal environment. A rapid, time-dependent induction of a gene subset is triggered within minutes of internalization in response to intracellular cues. Reproduced with permission from Elsevier (Rohde et al., 2007b).

to be controlled by PhoPR were upregulated. It has already been reported that M. *tuberculosis* mutants defective in PhoPR expression are attenuated markedly in both macrophages and mice. As mentioned earlier in this chapter, the related two-component signal transduction system in *Salmonella* functions primarily as an intravacuolar pH sensor.

Consistent with the observations of Schnappinger and coworkers, we also observed the increased expression of multiple genes linked to fatty acid metabolism, both in the β-oxidation pathway (several putative *fadD* and *fadE* genes) as well as *icl1*, the gating enzyme into the glyoxylate shunt. The big question associated with these observations is whether the fatty acids to be processed are acquired directly from the host or mobilized from triacylglycerol stores present in the bacteria. It should be noted that there is a strong degree of concordance with the themes identified in the studies of Schnappinger and Rohde, even if the identities of individual genes showed variance.

pH Shift Is a Dominant Intraphagosomal Signal to *M. tuberculosis*

The vacuole containing M. *tuberculosis* exhibits arrested maturation and fails to fuse with lysosomes; nonetheless, it does show a drop in pH from pH 7.2, in the external medium, equilibrating to a pH of 6.4 in the established vacuole (Sturgill-Koszycki et al., 1994; Russell, 2001; Pethe et al., 2004). Recent dissection of the killing processes of *Mycobacterium* by the macrophage has placed considerable emphasis on early pH-dependent, or pH-enhanced, mechanisms as mediating significant bacterial killing while the bacterium is fighting to establish an infection (Jordao et al., 2008). So, while the transcriptional response of M. *tuberculosis* as it enters the macrophage is likely triggered by multiple interdependent cues such as pH, ionic balance, and nutritional and oxidative stresses, it is pH that may be of particular significance to the bacillus. Existing studies have examined the pH response of M. *tuberculosis* by adjusting the pH of the bacterial medium in vitro, but this approach denies the contribution of other, undefined cues in the phagosome. To determine the pH-dependent response in context with the other intraphagosomal cues, Rohde and colleagues (2007b) adopted a subtractive approach by blocking acidification of M. *tuberculosis*-containing phagosomes through treatment with the V-ATPase inhibitor concanamycin A. Although the pH shift from pH 7.2 to pH 6.4 is relatively minor, considerably less than the shift that would be experienced on translocation to the lysosome, blocking this partial acidification to pH 6.4 abrogated the upregulation of approximately 40% of the genes normally upregulated on macrophage entry. The gene list included the majority of genes under regulation of PhoPR and DosRS. In support of this subtractive intracellular approach, the list of pH-sensitive genes generated within the phagosomal environment demonstrated only partial overlap (67%) with the acid-induced transcriptome generated in the test tube (Fig. 8).

CONCLUDING REMARKS

The goal of this chapter is to raise the awareness of the reader that the road to a better understanding of the functions of the phagosome lies in the development and application of readouts that actually quantify the changing physiological parameters in this compartment. Immunofluorescence has taken us about as far as it can and

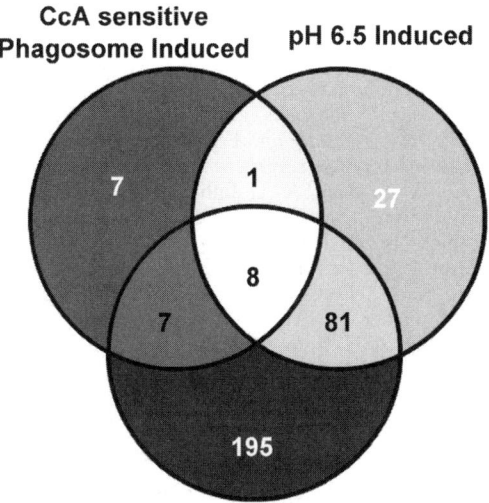

FIGURE 8 Phagosome acidification modulates M. *tuberculosis* gene expression. Prior to infection, macrophages were treated with 100 nM concanamycin A (CcA) to inhibit acidification of M. *tuberculosis*-containing phagosome from pH 7 to pH 6.4. Transcriptional profiling showed that 24 of the 64 genes that were upregulated in untreated macrophages were not upregulated in cells treated with CcA (>1.5-fold, $P < 0.05$). The Venn diagram shows the overlap of these 24 genes compared with those genes induced by in vitro acid stress at pH 6.5 and 5.5. Genes induced >1.5-fold ($P < 0.05$) in at least one condition and whose array signals passed quality filters in all three conditions were included. Reproduced with permission from Elsevier (Rohde et al., 2007b).

needs to be replaced by real-time assays of phagosomal function. From the limited number of functional assays available, it is clear that the enzymatic activities within the phagosome are modulated extensively by changes in the activation status of the macrophage, which assign a different set of goals to the phagocyte.

Much of this chapter has focused on development and application of fluorescent reporters that measure pH, phagosome/lysosome fusion, and a range of hydrolytic activities. However, the chapter also discusses emerging approaches that will allow exploitation of intracellular pathogens to report back about the environment that they are sensing. This latter approach holds great promise toward understanding not only phagosomal function, but also the intimate interface between a pathogen and its host cell. So, while we are currently struggling with the complex pathways that regulate phagosomal function, there are several avenues opening up to probe this complex and important biological organelle.

REFERENCES

Abrahams, G. L., and M. Hensel. 2006. Manipulating cellular transport and immune responses: dynamic interactions between intracellular Salmonella enterica and its host cells. *Cell. Microbiol.* **8:**728–737.

Akira, S. 2006. TLR signaling. *Curr. Top. Microbiol. Immunol.* **311:**1–16.

Alonso, S., K. Pethe, D. G. Russell, and G. E. Purdy. 2007. Lysosomal killing of Mycobacterium mediated by ubiquitin-derived peptides is enhanced by autophagy. *Proc. Natl. Acad. Sci. USA* **104:**6031–6036.

Astarie-Dequeker, C., E. N. N'Diaye, V. Le Cabec, M. G. Rittig, J. Prandi, and I. Maridonneau-Parini. 1999. The mannose receptor mediates uptake of pathogenic and nonpathogenic mycobacteria and bypasses bactericidal responses in human macrophages. *Infect. Immun.* **67:**469–477.

Aston, C., W. N. Rom, A. T. Talbot, and J. Reibman. 1998. Early inhibition of mycobacterial growth by human alveolar macrophages is not due to nitric oxide. *Am. J. Respir. Crit. Care Med.* **157**(6 Pt 1)**:**1943–1950.

Bergman, N. H., E. C. Anderson, E. E. Swenson, B. K. Janes, N. Fisher, M. M. Niemeyer, A. D. Miyoshi, and P. C. Hanna. 2007. Transcriptional profiling of Bacillus anthracis during infection of host macrophages. *Infect. Immun.* **75:**3434–3444.

Blander, J. M., and R. Medzhitov. 2004. Regulation of phagosome maturation by signals from toll-like receptors. *Science* **304:**1014–1018.

Blander, J. M., and R. Medzhitov. 2006. Toll-dependent selection of microbial antigens for presentation by dendritic cells. *Nature* **440:**808–812.

Blander, J. M., and R. Medzhitov. 2007. Reply to "Toll-like receptors and phagosome maturation." *Nat. Immunol.* **8:**217–218.

Chamnongpol, S., M. Cromie, and E. A. Groisman. 2003. Mg2+ sensing by the Mg2+ sensor PhoQ of Salmonella enterica. *J. Mol. Biol.* **325:**795–807.

Chatterjee, S. S., H. Hossain, S. Otten, C. Kuenne, K. Kuchmina, S. Machata, E. Domann, T. Chakraborty, and T. Hain. 2006. Intracellular gene expression profile of Listeria monocytogenes. *Infect. Immun.* **74:**1323–1338.

Christensen, K. A., J. T. Myers, and J. A. Swanson. 2002. pH-dependent regulation of lysosomal calcium in macrophages. *J. Cell Sci.* **115**(Pt 3)**:** 599–607.

Claus, V., A. Jahraus, T. Tjelle, T. Berg, H. Kirschke, H. Faulstich, and G. Griffiths. 1998. Lysosomal enzyme trafficking between phagosomes, endosomes, and lysosomes in J774 macrophages. Enrichment of cathepsin H in early endosomes. *J. Biol. Chem.* **273:**9842–9851.

Clemens, D. L., and M. A. Horwitz. 1995. Characterization of the Mycobacterium tuberculosis phagosome and evidence that phagosomal maturation is inhibited. *J. Exp. Med.* **181:**257–270.

Clemens, D. L., and M. A. Horwitz. 1996. The Mycobacterium tuberculosis phagosome interacts with early endosomes and is accessible to exogenously administered transferrin. *J. Exp. Med.* **184:**1349–1355.

Delamarre, L., M. Pack, H. Chang, I. Mellman, and E. S. Trombetta. 2005. Differential lysosomal proteolysis in antigen-presenting cells determines antigen fate. *Science* **307:**1630–1634.

Desjardins, M. 1995. Biogenesis of phagolysosomes: the 'kiss and run' hypothesis. *Trends Cell Biol.* **5:**183–186.

Ellermeier, J. R., and J. M. Slauch. 2007. Adaptation to the host environment: regulation of the SPI1 type III secretion system in Salmonella enterica serovar Typhimurium. *Curr. Opin. Microbiol.* **10:**24–29.

Eriksson, S., S. Lucchini, A. Thompson, M. Rhen, and J. C. Hinton. 2003. Unravelling the biology of macrophage infection by gene expression profiling of intracellular Salmonella enterica. *Mol. Microbiol.* **47:**103–118.

Fadl, A. A., C. L. Galindo, J. Sha, F. Zhang, H. R. Garner, H. Q. Wang, and A. K. Chopra. 2007. Global transcriptional responses of wild-type Aeromonas hydrophila and its virulence-deficient mutant in a murine model of infection. *Microb. Pathog.* **42:**193–203.

Faucher, S. P., S. Porwollik, C. M. Dozois, M. McClelland, and F. Daigle. 2006. Transcriptome of Salmonella enterica serovar Typhi within macrophages revealed through the selective capture of transcribed sequences. *Proc. Natl. Acad. Sci. USA* **103:**1906–1911.

Fratti, R. A., J. M. Backer, J. Gruenberg, S. Corvera, and V. Deretic. 2001. Role of phosphatidylinositol 3-kinase and Rab5 effectors in phagosomal biogenesis and mycobacterial phagosome maturation arrest. *J. Cell Biol.* **154:**631–644.

Fratti, R. A., J. Chua, and V. Deretic. 2003. Induction of p38 mitogen-activated protein kinase reduces early endosome autoantigen 1 (EEA1) recruitment to phagosomal membranes. *J. Biol. Chem.* **278:**46961–46967.

Fujiwara, N., and K. Kobayashi. 2005. Macrophages in inflammation. *Curr. Drug Targets Inflamm. Allergy* **4:**281–286.

Geiman, D. E., T. R. Raghunand, N. Agarwal, and W. R. Bishai. 2006. Differential gene expression in response to exposure to antimycobacterial agents and other stress conditions among seven Mycobacterium tuberculosis whiB-like genes. *Antimicrob. Agents Chemother.* **50:**2836–2841.

Gomes, M. S., M. Florido, T. F. Pais, and R. Appelberg. 1999. Improved clearance of Mycobacterium avium upon disruption of the inducible nitric oxide synthase gene. *J. Immunol.* **162:**6734–6739.

Gonzalo Asensio, J., C. Maia, N. L. Ferrer, N. Barilone, F. Laval, C. Y. Soto, N. Winter, M. Daffe, B. Gicquel, C. Martin, and M. Jackson. 2006. The virulence-associated two-component PhoP-PhoR system controls the biosynthesis of polyketide-derived lipids in Mycobacterium tuberculosis. *J. Biol. Chem.* **281:**1313–1316.

Guiney, D. G. 2005. The role of host cell death in Salmonella infections. *Curr. Top. Microbiol. Immunol.* **289:**131–150.

Gutierrez, M. G., S. S. Master, S. B. Singh, G. A. Taylor, M. I. Colombo, and V. Deretic. 2004. Autophagy is a defense mechanism inhibiting BCG and Mycobacterium tuberculosis survival in infected macrophages. *Cell* **119:**753–766.

Hisert, K. B., M. A. Kirksey, J. E. Gomez, A. O. Sousa, J. S. Cox, W. R. Jacobs, Jr., C. F. Nathan, and J. D. McKinney. 2004. Identification of Mycobacterium tuberculosis counterimmune (cim) mutants in immunodeficient mice by differential screening. *Infect. Immun.* **72:**5315–5321.

Hurtado-Lorenzo, A., M. Skinner, J. El Annan, M. Futai, G. H. Sun-Wada, S. Bourgoin, J. Casanova, A. Wildeman, S. Bechoua, D. A. Ausiello, D. Brown, and V. Marshansky. 2006. V-ATPase interacts with ARNO and Arf6 in early endosomes and regulates the protein degradative pathway. *Nat. Cell Biol.* **8:**124–136.

Hutter, B., and T. Dick. 1999. Molecular genetic characterisation of whiB3, a mycobacterial homologue of a Streptomyces sporulation factor. *Res. Microbiol.* **150:**295–301.

Jansen, A., and J. Yu. 2006. Differential gene expression of pathogens inside infected hosts. *Curr. Opin. Microbiol.* **9:**138–142.

Jordao, L., C. K. Bleck, L. Mayorga, G. Griffiths, and E. Anes. 2008. On the killing of mycobacteria by macrophages. *Cell. Microbiol.* **10:**529–548.

Kelley, V. A., and J. S. Schorey. 2004. Modulation of cellular phosphatidylinositol 3-phosphate levels in primary macrophages affects heat-killed but not viable Mycobacterium avium's transport through the phagosome maturation process. *Cell. Microbiol.* **6:**973–985.

Kendall, S. L., F. Movahedzadeh, S. C. Rison, L. Wernisch, T. Parish, K. Duncan, J. C. Betts, and N. G. Stoker. 2004. The Mycobacterium tuberculosis dosRS two-component system is induced by multiple stresses. *Tuberculosis (Edinb.)* **84:**247–255.

Lennon-Dumenil, A. M., A. H. Bakker, R. Maehr, E. Fiebiger, H. S. Overkleeft, M. Rosemblatt, H. L. Ploegh, and C. Lagaudriere-Gesbert. 2002. Analysis of protease activity in live antigen-presenting cells shows regulation of the phagosomal proteolytic contents during dendritic cell activation. *J. Exp. Med.* **196:**529–540.

Lindmo, K., and H. Stenmark. 2006. Regulation of membrane traffic by phosphoinositide 3-kinases. *J. Cell Sci.* 119(Pt 4):605–614.

MacMicking, J. D. 2005. Immune control of phagosomal bacteria by p47 GTPases. *Curr. Opin. Microbiol.* **8:**74–82.

MacMicking, J. D., R. J. North, R. LaCourse, J. S. Mudgett, S. K. Shah, and C. F. Nathan. 1997. Identification of nitric oxide synthase as a protective locus against tuberculosis. *Proc. Natl. Acad. Sci. USA* **94:**5243–5248.

MacMicking, J. D., G. A. Taylor, and J. D. McKinney. 2003. Immune control of tuberculosis by IFN-gamma-inducible LRG-47. *Science* **302:**654–659.

Malik, Z. A., C. R. Thompson, S. Hashimi, B. Porter, S. S. Iyer, and D. J. Kusner. 2003. Cutting edge: Mycobacterium tuberculosis blocks Ca2+ signaling and phagosome maturation in human macrophages via specific inhibition of sphingosine kinase. *J. Immunol.* **170:**2811–2815.

Martin-Orozco, N., N. Touret, M. L. Zaharik, E. Park, R. Kopelman, S. Miller, B. B. Finlay, P. Gros, and S. Grinstein. (2006). Visualization of vacuolar acidification-induced transcription of genes of pathogens inside macrophages. *Mol. Biol. Cell* **17:**498–510.

McKinney, J. D., K. Honer zu Bentrup, E. J. Munoz-Elias, A. Miczak, B. Chen, W. T. Chan, D. Swenson, J. C. Sacchettini, W. R. Jacobs, Jr., and D. G. Russell. 2000. Persistence of Mycobacterium tuberculosis in macrophages and mice requires the glyoxylate shunt enzyme isocitrate lyase. *Nature* **406:**735–738.

Morris, R. P., L. Nguyen, J. Gatfield, K. Visconti, K. Nguyen, D. Schnappinger, S. Ehrt, Y. Liu, L. Heifets, J. Pieters, G. Schoolnik, and C. J. Thompson. 2005. Ancestral antibiotic resistance in Mycobacterium tuberculosis. *Proc. Natl. Acad. Sci. USA* **102:**12200–12205.

Mosser, D. M. 2003. The many faces of macrophage activation. *J. Leukoc. Biol.* **73:**209–212.

Mulder, N. J., H. Zappe, and L. M. Steyn. 1999. Characterization of a Mycobacterium tuberculosis homologue of the Streptomyces coelicolor whiB gene. *Tuber. Lung Dis.* **79:**299–308.

Mulvey, M. R., D. A. Boyd, A. B. Olson, B. Doublet, and A. Cloeckaert. 2006. The genetics of Salmonella genomic island 1. *Microbes Infect.* **8:**1915–1922.

Murray, R. Z., J. G. Kay, D. G. Sangermani, and J. L. Stow. 2005. A role for the phagosome in cytokine secretion. *Science* **310:**1492–1495.

Nebenfuhr, A., C. Ritzenthaler, and D. G. Robinson. 2002. Brefeldin A: deciphering an enigmatic inhibitor of secretion. *Plant Physiol.* **130:**1102–1108.

Nicholson, S., G. Bonecini-Almeida Mda, J. R. Lapa e Silva, C. Nathan, Q. W. Xie, R. Mumford, J. R. Weidner, J. Calaycay, J. Geng, N. Boechat, C. Linhares, W. Rom, and J. L. Ho. 1996. Inducible nitric oxide synthase in pulmonary alveolar macrophages from patients with tuberculosis. *J. Exp. Med.* **183:**2293–2302.

Ozinsky, A., D. M. Underhill, J. D. Fontenot, A. M. Hajjar, K. D. Smith, C. B. Wilson, L. Schroeder, and A. Aderem. 2000. The repertoire for pattern recognition of pathogens by the innate immune system is defined by cooperation between toll-like receptors. *Proc. Natl. Acad. Sci. USA* **97:**13766–13771.

Park, J. B. 2003. Phagocytosis induces superoxide formation and apoptosis in macrophages. *Exp. Mol. Med.* **35:**325–335.

Perez, E., S. Samper, Y. Bordas, C. Guilhot, B. Gicquel, and C. Martin. 2001. An essential role for phoP in Mycobacterium tuberculosis virulence. *Mol. Microbiol.* **41:**179–187.

Pethe, K., D. L. Swenson, S. Alonso, J. Anderson, C. Wang, and D. G. Russell. 2004. Isolation of Mycobacterium tuberculosis mutants defective in the arrest of phagosome maturation. *Proc. Natl. Acad. Sci. USA* **101:**13642–13647.

Prost, L. R., M. E. Daley, V. Le Sage, M. W. Bader, H. Le Moual, R. E. Klevit, and S. I. Miller. 2007a. Activation of the bacterial sensor kinase PhoQ by acidic pH. *Mol. Cell* **26:**165–174.

Prost, L. R., S. Sanowar, and S. I. Miller. 2007b. Salmonella sensing of anti-microbial mechanisms to promote survival within macrophages. *Immunol. Rev.* **219:**55–65.

Purdy, G. E., R. M. Owens, L. Bennett, D. G. Russell, and B. A. Butcher. 2005. Kinetics of phosphatidylinositol-3-phosphate acquisition differ between IgG bead-containing phagosomes and Mycobacterium tuberculosis-containing phagosomes. *Cell. Microbiol.* **7:**1627–1634.

Roberts, D. M., R. P. Liao, G. Wisedchaisri, W. G. Hol, and D. R. Sherman. 2004. Two sensor kinases contribute to the hypoxic response of Mycobacterium tuberculosis. *J. Biol. Chem.* **279:**23082–23087.

Rogers, L. D., and L. J. Foster. 2007. The dynamic phagosomal proteome and the contribution of the endoplasmic reticulum. *Proc. Natl. Acad. Sci. USA* **104:**18520–18525.

Rohde, K., R. M. Yates, G. E. Purdy, and D. G. Russell. 2007a. Mycobacterium tuberculosis and the environment within the phagosome. *Immunol. Rev.* **219:**37–54.

Rohde, K. H., R. B. Abramovitch, and D. G. Russell. 2007b. Mycobacterium tuberculosis invasion of macrophages: linking bacterial gene expression to environmental cues. *Cell Host Microbe* **2:**352–364.

Rohrer, J., A. Schweizer, D. Russell, and S. Kornfeld. 1996. The targeting of Lamp1 to lysosomes is dependent on the spacing of its cytoplasmic tail tyrosine sorting motif relative to the membrane. *J. Cell. Biol.* **132:**565–576.

Rupper, A., and J. Cardelli. 2001. Regulation of phagocytosis and endo-phagosomal trafficking pathways in Dictyostelium discoideum. *Biochim. Biophys. Acta* **1525:**205–216.

Russell, D. G. 2001. Mycobacterium tuberculosis: here today, and here tomorrow. *Nat. Rev. Mol. Cell Biol.* **2:**569–577.

Saini, D. K., V. Malhotra, D. Dey, N. Pant, T. K. Das, and J. S. Tyagi. 2004. DevR-DevS is a bona fide two-component system of Mycobacterium tuberculosis that is hypoxia-responsive in the absence of the DNA-binding domain of DevR. *Microbiology* **150**(Pt 4):865–875.

Savina, A., C. Jancic, S. Hugues, P. Guermonprez, P. Vargas, I. C. Moura, A. M. Lennon-Dumenil, M. C. Seabra, G. Raposo, and S. Amigorena. 2006. NOX2 controls phagosomal pH to regulate antigen processing during crosspresentation by dendritic cells. *Cell* **126:**205–218.

Schaible, U. E., S. Sturgill-Koszycki, P. H. Schlesinger, and D. G. Russell. 1998. Cytokine activation leads to acidification and increases maturation of Mycobacterium avium-containing phagosomes in murine macrophages. *J. Immunol.* **160:**1290–1296.

Schlumberger, M. C., and W. D. Hardt. 2005. Triggered phagocytosis by Salmonella: bacterial molecular mimicry of RhoGTPase activation/deactivation. *Curr. Top. Microbiol. Immunol.* **291:**29–42.

Schnappinger, D., S. Ehrt, M. I. Voskuil, Y. Liu, J. A. Mangan, I. M. Monahan, G. Dolganov, B. Efron, P. D. Butcher, C. Nathan, and G. K. Schoolnik. 2003. Transcriptional adaptation of Mycobacterium tuberculosis within macrophages: insights into the phagosomal environment. *J. Exp. Med.* **198:**693–704.

Schnappinger, D., G. K. Schoolnik, and S. Ehrt. 2006. Expression profiling of host pathogen interactions: how Mycobacterium tuberculosis and the macrophage adapt to one another. *Microbes Infect.* **8:**1132–1140.

Singh, A., L. Guidry, J. Trombley, S. Nelson, and A. J. Steyn. 2007. Mycobacterium tuberculosis WhiB3 responds to dormancy signals through its [4Fe-4S] cluster. Keystone Symposium. Tuberculosis: From lab research to field trials, abstr. 420.

Soliveri, J. A., J. Gomez, W. R. Bishai, and K. F. Chater. 2000. Multiple paralogous genes related to the Streptomyces coelicolor developmental regulatory gene whiB are present in Streptomyces and other actinomycetes. *Microbiology* **146**(Pt 2):333–343.

Spiro, S. 2006. Nitric oxide-sensing mechanisms in Escherichia coli. *Biochem. Soc. Trans.* **34**(Pt 1):200–202.

Staudinger, B. J., M. A. Oberdoerster, P. J. Lewis, and H. Rosen. 2002. mRNA expression profiles for Escherichia coli ingested by normal and phagocyte oxidase-deficient human neutrophils. *J. Clin. Invest.* **110**:1151–1163.

Steyn, A. J., D. M. Collins, M. K. Hondalus, W. R. Jacobs, Jr., R. P. Kawakami, and B. R. Bloom. 2002. Mycobacterium tuberculosis WhiB3 interacts with RpoV to affect host survival but is dispensable for in vivo growth. *Proc. Natl. Acad. Sci. USA* **99**:3147–3152.

Stuart, L. M., J. Boulais, G. M. Charriere, E. J. Hennessy, S. Brunet, I. Jutras, G. Goyette, C. Rondeau, S. Letarte, H. Huang, P. Ye, F. Morales, C. Kocks, J. S. Bader, M. Desjardins, and R. A. Ezekowitz. 2007. A systems biology analysis of the Drosophila phagosome. *Nature* **445**:95–101.

Sturgill-Koszycki, S., U. E. Schaible, and D. G. Russell. 1996. Mycobacterium-containing phagosomes are accessible to early endosomes and reflect a transitional state in normal phagosome biogenesis. *EMBO J.* **15**:6960–6968.

Sturgill-Koszycki, S., P. H. Schlesinger, P. Chakraborty, P. L. Haddix, H. L. Collins, A. K. Fok, R. D. Allen, S. L. Gluck, J. Heuser, and D. G. Russell. 1994. Lack of acidification in Mycobacterium phagosomes produced by exclusion of the vesicular proton-ATPase. *Science* **263**:678–681.

Talaat, A. M., R. Lyons, S. T. Howard, and S. A. Johnston. 2004. The temporal expression profile of Mycobacterium tuberculosis infection in mice. *Proc. Natl. Acad. Sci. USA* **101**:4602–4607.

Thompson, C. R., S. S. Iyer, N. Melrose, R. VanOosten, K. Johnson, S. M. Pitson, L. M. Obeid, and D. J. Kusner. 2005. Sphingosine kinase 1 (SK1) is recruited to nascent phagosomes in human macrophages: inhibition of SK1 translocation by Mycobacterium tuberculosis. *J. Immunol.* **174**:3551–3561.

Trinchieri, G., and A. Sher. 2007. Cooperation of Toll-like receptor signals in innate immune defence. *Nat. Rev. Immunol.* **7**:179–190.

Trombetta, E. S., M. Ebersold, W. Garrett, M. Pypaert, and I. Mellman. 2003. Activation of lysosomal function during dendritic cell maturation. *Science* **299**:1400–1403.

Tsang, A. W., K. Oestergaard, J. T. Myers, and J. A. Swanson. 2000. Altered membrane trafficking in activated bone marrow-derived macrophages. *J. Leukoc. Biol.* **68**:487–494.

Ullrich, H. J., W. L. Beatty, and D. G. Russell. 1999. Direct delivery of procathepsin D to phagosomes: implications for phagosome biogenesis and parasitism by Mycobacterium. *Eur. J. Cell. Biol.* **78**:739–748.

Vergne, I., J. Chua, and V. Deretic. 2003. Mycobacterium tuberculosis phagosome maturation arrest: selective targeting of PI3P-dependent membrane trafficking. *Traffic* **4**:600–606.

Via, L. E., R. A. Fratti, M. McFalone, E. Pagan-Ramos, D. Deretic, and V. Deretic. 1998. Effects of cytokines on mycobacterial phagosome maturation. *J. Cell Sci.* **111** (Pt 7):897–905.

Vidal, S. M., D. Malo, J. F. Marquis, and P. Gros. 2008. Forward genetic dissection of immunity to infection in the mouse. *Annu. Rev. Immunol.* **26**:81–132.

Vieira, O. V., R. J. Botelho, and S. Grinstein. 2002. Phagosome maturation: aging gracefully. *Biochem. J.* **366**(Pt 3):689–704.

Voskuil, M. I., K. C. Visconti, and G. K. Schoolnik. 2004. Mycobacterium tuberculosis gene expression during adaptation to stationary phase and low-oxygen dormancy. *Tuberculosis (Edinb.)* **84**:218–227.

Voyich, J. M., K. R. Braughton, D. E. Sturdevant, A. R. Whitney, B. Said-Salim, S. F. Porcella, R. D. Long, D. W. Dorward, D. J. Gardner, B. N. Kreiswirth, J. M. Musser, and F. R. DeLeo. 2005. Insights into mechanisms used by Staphylococcus aureus to avoid destruction by human neutrophils. *J. Immunol.* **175**:3907–3919.

Waddell, S. J., and P. D. Butcher. 2007. Microarray analysis of whole genome expression of intracellular Mycobacterium tuberculosis. *Curr. Mol. Med.* **7**:287–296.

Walters, S. B., E. Dubnau, I. Kolesnikova, F. Laval, M. Daffe, and I. Smith. 2006. The Mycobacterium tuberculosis PhoPR two-component system regulates genes essential for virulence and complex lipid biosynthesis. *Mol. Microbiol.* **60**:312–330.

Waterman, S. R., and D. W. Holden. 2003. Functions and effectors of the Salmonella pathogenicity island 2 type III secretion system. *Cell. Microbiol.* **5**:501–511.

Yates, R. M., A. Hermetter, and D. G. Russell. 2005. The kinetics of phagosome maturation as a function of phagosome/lysosome fusion and acquisition of hydrolytic activity. *Traffic* **6**:413–4-20.

Yates, R. M., A. Hermetter, G. A. Taylor, and D. G. Russell. 2007. Macrophage activation downregulates the degradative capacity of the phagosome. *Traffic* **8**:241–250.

Yates, R. M., and D. G. Russell. 2005. Phagosome maturation proceeds independently of stimulation of toll-like receptors 2 and 4. *Immunity* **23**:409–417.

Yeung, T., B. Ozdamar, P. Paroutis, and S. Grinstein. 2006. Lipid metabolism and dynamics during phagocytosis. *Curr. Opin. Cell. Biol.* **18**:429–437.

Zhou, D., Y. Han, J. Qiu, L. Qin, Z. Guo, X. Wang, Y. Song, Y. Tan, Z. Du, and R. Yang. 2006. Genome-wide transcriptional response of Yersinia pestis to stressful conditions simulating phagolysosomal environments. *Microbes Infect.* **8**:2669–2678.

BRIDGING THE GAP BETWEEN THE INNATE AND ACQUIRED IMMUNE RESPONSE

IV

17

Novel Anti-Inflammatory and Proresolution Lipid Mediators in Induction and Modulation of Phagocyte Function

CHARLES N. SERHAN AND JULIO ALIBERTI

In general terms, a local inflammatory response is protective and ultimately rids tissues of both the cause and consequences of tissue injury that can accompany host defense (Cotran et al., 1999). Acute inflammation, defined by its cardinal signs dolor, calor, and rubor, may lead to chronic inflammation, scarring, and eventual loss of function if the tissue fails to completely resolve the inflammatory site (Cotran et al., 1999; Majno and Joris, 2004). The polymorphonuclear neutrophils (PMNs) of the first line of host defense, in this context, must also exit from the inflamed tissues (illustrated in Fig. 1) to return to homeostasis and resolve (Levy et al., 2001). It has become widely appreciated that, in addition to classic diseases associated with inflammation (for example, psoriasis, periodontal disease, and arthritis), uncontrolled inflammation governs the pathogenesis of many widely prevalent diseases including cardiovascular and cerebrovascular disease, cancer, obesity, and Alzheimer's disease (Calder, 2006a; Libby, 2002; Van Dyke and Serhan, 2006). Prostaglandins and leukotrienes are generated in the initiation of local inflammation and are key in promoting the signs of inflammation. Of interest, another class of arachidonic acid-derived mediators, the lipoxins (LXs) and aspirin-triggered lipoxins (ATLs), were the first mediators recognized to carry both endogenous anti-inflammatory and proresolving actions (Serhan et al., 2007; Serhan, 2005b), clearly indicating that not all oxidized lipids are "bad guys" in human disease (Morris et al., 2006).

Several novel enzymatic pathways were recently identified that are activated during the resolution phase and initiated from precursors eicosapentaenoic acid (EPA) and docosahexaenoic acid (DHA). Both EPA and DHA are major n-3 fatty acids, also widely known as the omega-3 PUFAs or fish oils. The new mediators are biosynthesized during the evolution of locally contained inflammatory exudates. They possess potent actions in controlling the resolution phase (Hong et al., 2003; Serhan et al., 2000, 2002). The term resolvins, resolution phase interaction products, was introduced to signify that the new structures are endogenous, local-acting mediators that carry potent anti-inflammatory and immunoregulatory signals (Serhan et al., 2002). These include novel actions that are targeted to promote resolution, namely, reducing neutrophil (PMN) infiltration and regulating the cytokine-chemokine axis and reactive oxygen species and stimulating the uptake and clearance of apoptotic PMNs, as well as lowering the magnitude of the inflammatory response and associated pain (Serhan et al., 2000, 2002; Svensson et al., 2007). The protectin family is biosynthesized in many organs, and thus the location-specific term neuro-protectin D1 was introduced when generated in neural tissue (Serhan et al., 2006). The protectin family name signifies their formation, potent anti-inflammatory (Hong et al., 2003), as well as protective actions demonstrated for the novel and potent DHA-derived 10,17-docasatriene in animal models of stroke (Marcheselli et al., 2003) and Alzheimer's disease (Lukiw et al., 2005). Both families, the resolvins and protectins, are potent local-acting agonists of endogenous anti-inflammation and promote resolution-specific processes (Serhan, 2007). The connection(s) of these new lipid mediators to the control of an acute inflammation and its timely resolution are illustrated in Fig. 1.

Omega-3 fatty acids (PUFAs) were known to possess beneficial roles in health and organ function since the early 20th century (Burr and Burr, 1929). However, they require very high concentrations in vitro, as omega-3 PUFAs decrease production of proinflammatory prostaglandins, cytokines, and reactive oxygen species held to play critical roles in inflammatory diseases (James et al., 2003). Clinically relevant anti-inflammatory properties have also been reported over the decades with high doses of omega-3 fatty

---

Charles N. Serhan, Center for Experimental Therapeutics and Reperfusion Injury, Department of Anesthesiology, Perioperative and Pain Medicine, Brigham and Women's Hospital and Harvard Medical School, Boston, MA 02115. Julio Aliberti, Divisions of Molecular Immunology and Pulmonary Medicine, Cincinnati Children's Hospital Medical Center, Cincinnati, OH 45229.

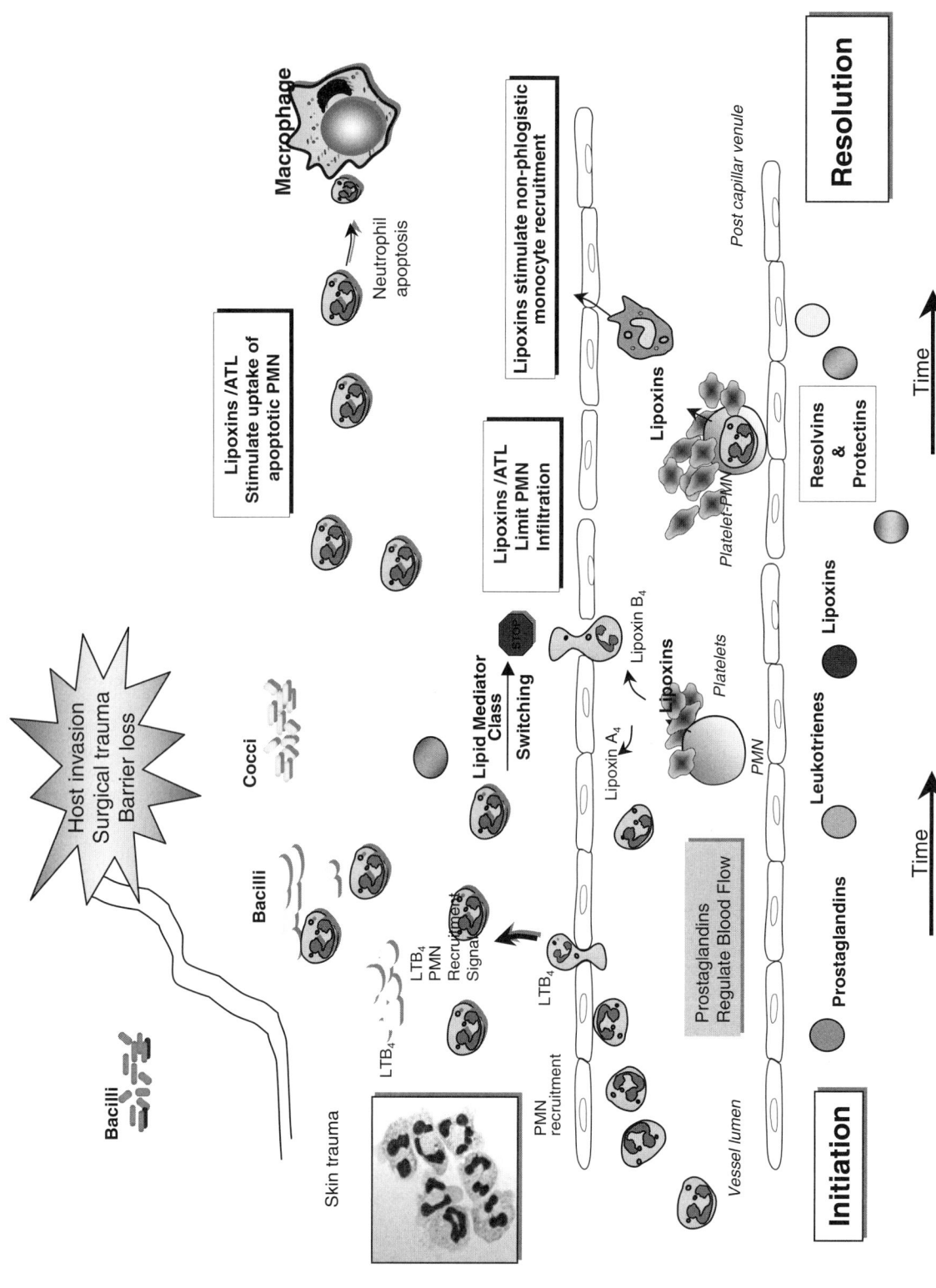

FIGURE 1 Lipid mediators during the initiation and resolution of acute inflammation.

acids, for example, in controlling rheumatoid arthritis (reviewed in Calder, 2006b) and periodontal disease (Hamazaki et al., 2006), whereas the evidence available at this time remains inconclusive for other conditions such as asthma. Also, cardiovascular disease was reduced with high-dose omega-3 in a multicenter clinical trial (GISSI-Prevenzione Investigators, 1999; Marchioli et al., 2002), and blood levels of EPA and DHA were shown to clearly reduce the risk of cardiovascular disease (Harris et al., 2007). However, it was not clear how these essential fatty acids might act; taken together, these published findings raised the question of what mechanism(s) underlies the many beneficial actions of omega-3 PUFAs.

Abbreviations. The following abbreviations and terms are used: AT, aspirin-triggered, bioactive compounds initiated via modified COX-2 when the enzyme remains active; AT-RvD1, aspirin-triggered resolvin D1 (7S,8,17R-trihydroxy-docosa-4Z,9E,11E,13Z,15E,19Z-hexaenoic acid); AT-RvD2, aspirin-triggered resolvin D2 (7S,16,17R-trihydroxy-docosa-4Z,8E,10Z,12E,14E,19Z-hexaenoic acid); AT-RvD3, aspirin-triggered resolvin D3 (4S,11,17R-trihydroxy-docosa-5,7E,9E,13Z,15E,19Z-hexaenoic acid); AT-RvD4, aspirin-triggered resolvin D4 (4S,5,17R-trihydroxy-docosa-6E,8E,10Z,13Z,15E,19Z-hexaenoic acid); eicosanoids from Greek *eicosa* ("twenty"), family of arachidonic acid-derived mediators with carbon-based structure; LC-MS-MS, liquid chromatography-tandem mass spectrometry; LC-UV-MS/MS, liquid chromatography-UV spectrometry-tandem mass spectrometry; leukotrienes, bioactive leukocyte-derived conjugated triene-containing structures biosynthesized from arachidonic acid that are potent proinflammatory mediators; lipoxins, arachidonate-derived eicosanoids possessing conjugated tetraenes, trihydroxy bioactive structures; LM, lipid mediator; LO, lipoxygenase, inserts molecular oxygen and abstract hydrogen in a stereoselective reaction with 1,4-*cis*-pentadiene units present in polyunsaturated fatty acids (the major human lipoxygenases, 5-LO, 12-LO, and 15-LO, are defined by the carbon position with arachidonic acid substrate, where molecular oxygen is enzymatically inserted to form hydroperoxy-containing intermediates, i.e., 15S-HpETE or 17S-HpDHA); LTB_4, leukotriene B_4 (5S,12R-dihydroxy-eicosa-6Z,8E,10E,14Z-tetraenoic acid); LXA_4, lipoxin A_4 (5S,6R,15S-trihydroxy-eicosa-7E,9E,11Z,13E-tetraenoic acid); LXB_4, lipoxin B_4 (5S,14R,15S-trihydroxy-eicosa-6E,8Z,10E,12E-tetraenoic acid); nonphlogistic, defined as noninflammatory, non-fever-producing; PD, protectin, the family of DHA-derived mediators possessing a conjugated triene structure as a distinguishing feature; PD1/NPD1, protectin D1/*neuro*-protectin D1 (10R,17S-dihydroxy-docosa-4Z,7Z,11E,13E,15Z,19Z-hexaenoic acid); PGE_2, 9-oxo-11a,15S-dihydroxy-prosta-5Z,13E-dien-1-oic acid; prostanoids, branch of the eicosanoid family having the general prostaglandin cyclic structures (bioactive products of cyclooxygenases; isoprostanoids can be produced via nonenzymatic oxidative mechanisms); PUFA, polyunsaturated fatty acid; Rv, resolvin, resolution-phase interaction products carrying bioactivity; RvD1, resolvin D1 (7S,8R,17R-trihydroxy-4Z,9E,11E,13Z,15E,19Z-docosahexaenoic acid); RvE1, resolvin E1 (5S,12R,18R-trihydroxy-eicosa-6Z,8E,10E,14Z,16E-pentaenoic acid).

LIPID MEDIATORS IN INFLAMMATION AND RESOLUTION

The lipid-derived local chemical mediators are well positioned to play key role(s) as signaling molecules because they are small molecules, locally acting, and rapidly biosynthesized, as well as inactivated locally. Acute inflammation is defined by a rapid infiltration of PMNs into the site, the first line of phagocytic host defense (Fig. 1). Proinflammatory prostaglandins and leukotriene B_4 control local blood flow, vascular dilation, and permeability changes needed at the site for leukocyte adhesion, diapedesis, and recruitment (Cotran et al., 1999; Levy et al., 2001; Samuelsson et al., 1987). Microbial invaders, tissue injury, or surgical trauma activate the release and formation of arachidonate-derived eicosanoids. These chemical mediators are enzymatically generated via specific cyclooxygenase (COX) and lipoxygenase (LO) pathways. As exudates form in a contained space and pustules form and are walled off (so to speak), prostaglandins initiate a number of responses relevant in inflammation (i.e., vasoconstriction, vascular permeability changes, pain, vasodilation, and edema). It is noteworthy that new results indicate that both PGE_2 and PGD_2 also signal the end by activating the transcriptional regulation of 15-LO in human neutrophils that in turn gives rise to the temporal dissociation of eicosanoids and production of lipoxins (Levy et al., 2001; Serhan and Savill, 2005). Thus, the prostaglandins and leukotrienes are rapidly generated, whereas the lipoxins, which are also produced from arachidonate, are generated later in the time course within the onset of the resolution phase (Fig. 1 and 2).

LIPOXINS IN RESOLUTION

Lipoxins are trihydroxytetraene-containing products of arachidonic acid (Fig. 2). The lipoxins were the first family of mediators identified in vivo with anti-inflammatory and proresolving actions. Their biosynthesis and actions were recently reviewed in a special journal issue (for in-depth reviews, see Serhan, 2005b) and are overviewed here in light of their role(s) in endogenous anti-inflammation and resolution. In Fig. 1, eicosanoid class switching refers to change in production within the arachidonate-derived family, for example, prostaglandins and leukotrienes, which switch to lipoxin biosynthesis, which initiates and/or is coincident with termination. Human PMNs in this case switch from leukotriene B_4 to lipoxin production (Levy et al., 2001). Lipoxins, specifically LXA_4 and LXB_4, as well as their aspirin-triggered forms (see below), stop further PMN entry into the exudates as well as counterregulate the main signs of inflammation (Table 1). As new PMNs parachute into exudates, older and apoptotic PMNs must be removed from the site in a timely fashion for inflammation to resolve (Fig. 1). Once PMNs enter an exudate, they interact with other cells (such as other leukocytes, platelets, endothelial cells, mucosal epithelial cells, and fibroblasts) in their immediate vicinity and are able to perform transcellular biosynthesis to produce LX and eventually new mediators. The process of transcellular biosynthesis is defined as the generation of new bioactive compounds that neither cell type can produce on its own. For example, human platelets do not produce LX on their own. When platelets adhere to PMNs, the platelet-PMN aggregates (Fig. 1) then become a major intravascular source of LX. The platelet-generated LX can in turn halt further PMN recruitment (reviewed in Serhan, 2005a).

When PMNs interact with mucosal epithelial cells in the lung, oral, or gastrointestinal mucosa, these PMNs generate LX from precursor 15-S-hydroxy-5, 8, 11-*cis*-13-*trans*-eicosatetraenoic acid (15-HETE) donated by interactions with mucosal epithelial cells (Bonnans et al., 2003; Gronert et al., 1998; Levy et al., 2003). These results also

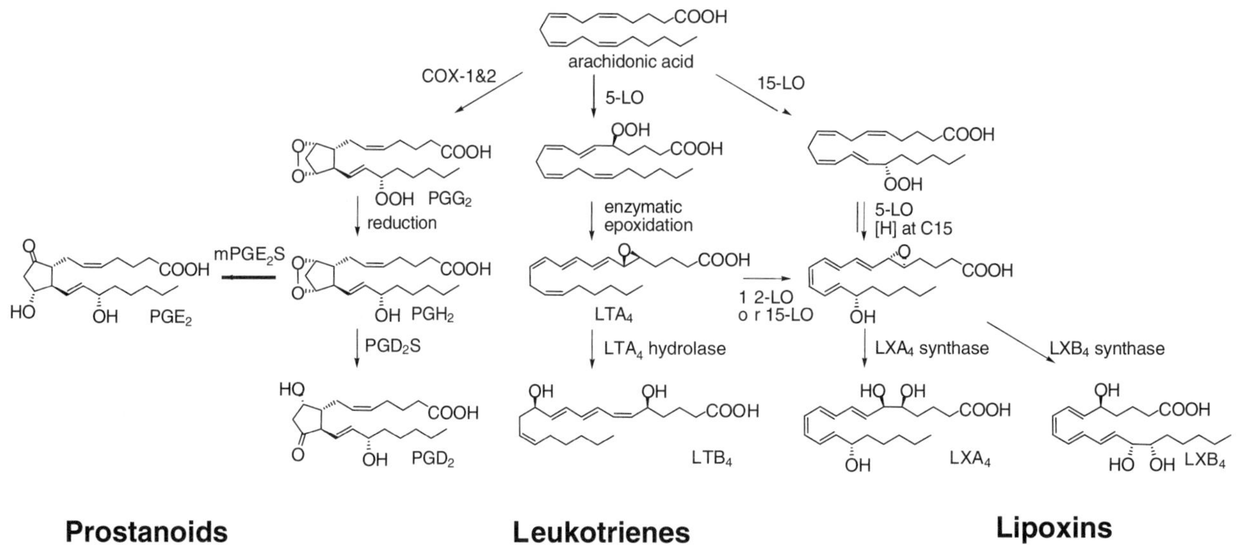

FIGURE 2 Eicosanoids and their proinflammatory and proresolution actions.

clearly demonstrated that PMNs switch their phenotype in that they change the profile of lipid mediators that they produce depending on their local environment (Levy et al., 2001; Serhan et al., 2000). Hence, they switch their lipid mediator phenotype compared with, for example, peripheral blood PMNs, which generate LTB_4 as their main product when activated (Levy et al., 2001; Serhan, 1989). During the course of inflammation and complete resolution, as discussed below, mediator switching also occurs between families of lipid mediators, namely from eicosanoids to resolvins of the E and D series as well as protectins (Bannenberg, et al., 2005; Serhan et al., 2000, 2002). The progression appears to depend on the availability of substrate within the evolving exudates and hence in the diet (Kasuga et al., 2007; Serhan et al., 2000).

BIOSYNTHESIS OF PRORESOLVING MEDIATORS VIA CYCLOOXYGENASES AND LIPOXYGENASES

Prostaglandins, for example PGE_2 and PGD_2, which are produced by both cyclooxygenase-1 (COX-1) and COX-2, are generated in the initial phase of inflammation (Fig. 1 and 2) and have a dual role in stimulating resolution (Gilroy et al., 2004). Signaling pathways leading to prostaglandins E_2 and D_2 actively switch on the transcription of enzymes (15-LO type 1) required for the generation of lipoxins (Levy et al., 2001) as well as PUFA-derived resolvins and protectins (Bannenberg et al., 2005; González-Périz et al., 2006). Selective inhibition of COX-2, blocking production of PGE_2 and PGD_2, delays the onset of resolution (Gilroy et al., 1999). Therefore, COX-2 has a role in both the initiation of acute inflammatory response and the resolution phase (Rajakariar et al., 2006). Lipoxins promote resolution by stopping entry of PMNs to sites of inflammation and decreasing reperfusion (reflow) tissue injury (for detailed reviews, see Serhan [2007]). They also reduce vascular permeability, promote nonphlogistic recruitment of monocytes, and stimulate clearance of apoptotic neutrophils via macrophages (Fig. 1) as well as reduce pain (Svensson et al., 2007). Drugs that disrupt this switch may have unwanted side effects in resolution (Gilroy et al., 2004; Gilroy and Perretti, 2005). Hence, a key event in resolution is the temporal "switch" in the lipid mediator class from pro- to anti-inflammatory eicosanoids, which has direct implications for the treatment of inflam-

TABLE 1 Lipoxins reduce inflammation and promote resolution

Lipoxin- and aspirin-triggered lipoxin actions	Reference
Regulate leukocyte traffic	Colgan et al., 1993
Stop PMN and eosinophil infiltration	Lee et al., 1989
Stimulate nonphlogistic monocyte recruitment	Maddox and Serhan, 1996
Stimulate macrophage uptake of apoptotic PMNs	Godson et al., 2000
Redirect chemokine-cytokine axis	
Block IL-8, IL-1 gene expression	Gewirtz et al., 1998
Block TNF-α actions and release	Takano et al., 1997
Stimulate transforming growth factor-β	Bannenberg et al., 2005
Reduce edema	
Regulate actions of histamine	Bandeira-Melo et al., 2000; Pouliot et al., 2000; Menezes-de-Lima et al., 2006
Block pain signals	
LX/LT regulate neuronal stem cells, proliferation, and differentiation	Serhan et al., 2001; Svensson et al., 2007; Wada et al., 2006

matory diseases. COX-2 also plays a key role in the biosynthesis of PGD_2 (Fig. 2), which is a precursor to the cyclopentenones. These include prostaglandin J_2, which is an extracellular activator of PPARγ. This property has led to the proposal that cyclopentenone prostaglandins can be anti-inflammatory and enhance the resolution of inflammation via regulation of NF-κB (reviewed in Gilroy et al. [1999, 2004]). These studies also bring to light the dual role of COX-2 in initiation versus termination of acute inflammation and the induction of proresolving lipid mediators. Continued study of the mechanism(s) underlying the actions of cyclopentenones in resolution as well as other regulators of the transcriptome could provide new leads for therapeutics in the resolution phase.

Resolution of acute inflammation in murine models involves the appearance in exudates of EPA and DHA, which follow the appearance of unesterified arachidonate (Bannenberg et al., 2005). Precursors are transformed via enzymatic mechanisms to bioactive compounds such as lipoxins, resolvins, and protectins that regulate the duration and magnitude of inflammation (shorten the period of neutrophil infiltration and initiate clearance of apoptotic PMNs). LX, Rv, and PD1 also increase the expression of CCR5 receptors on T cells and aging PMNs, which help clear local chemokine depots from the inflammatory site (Ariel et al., 2006). Apoptotic neutrophils are then phagocytized by macrophages, leading to neutrophil clearance and release of anti-inflammatory and reparative cytokines such as transforming growth factor-β1 (Bannenberg et al., 2005; Freire-de-Lima et al., 2006). A set of "resolution indices" was introduced as a quantitative means for assessing key resolution parameters and the impact of specific agents within active resolution. Temporal and differential changes in self-limited experimental murine peritonitis were investigated as a model system to identify anti-inflammatory and proresolving circuits (Bannenberg et al., 2005). With these resolution indices defined, specific lipid mediators (e.g., RvE1, ATLa, and PD1) were pinpointed to promote resolution via specific and separate mechanisms.

When grossly viewed as the same outcome or end point, namely anti-inflammation, each of these mediators could be considered anti-inflammatory (Bannenberg et al., 2005). Rather, it is now clear that active anti-inflammation and proresolution are not the same processes. These measurable indices are a useful tool applicable for evaluating the molecular basis of novel therapeutic interventions in disease models where inflammation resolution is a component and for identifying when pharmacologic agents are resolution toxic (Serhan et al., 2007).

LIPOXINS IN HUMANS AND DISEASE MODELS

LXA_4 and ATL are counterregulatory in several animal models of disease (Table 2), possess local organ-specific functions, and modulate leukotriene formation and their activities. The protective actions of LXA_4 and ATL are ligand receptor dependent, since transgenic overexpression of human ALX, the G-protein-coupled surface receptor for LXA_4, leads to decreased PMN infiltration with endogenous LXA_4 (Devchand et al., 2003). The LXA_4 receptor ALX is upregulated by glucocorticoids (Hashimoto et al., 2007); its structure-activity relationship was recently reviewed in detail (Chiang et al., 2006). Alterations in LX and ATL levels may be causally associated with the pathophysiology of several human diseases (Serhan, 2007). LXA_4 and ATL regulate tumor necrosis factor-α (TNF-α)-directed neutrophil actions and stimulate interleukin-4 (IL-4) in exudates, and thus regulate endogenous mediators in the pathogenesis of inflammatory conditions such as periodontal disease (Pouliot et al., 2000). In many human diseases, LX production appears to be deficient compared with leukotrienes (Table 2). These include cardiovascular (Brezinski et al., 1992), asthma (Levy et al., 2002), kidney inflammation (Kieran et al., 2004), cystic fibrosis (Karp et al., 2004), gastrointestinal (Mangino et al., 2006), and periodontal disease (Pouliot et al., 2000), to name a few. Designed metabolically stable analogs of LXs and ATLs are useful tools in examining the role(s) and local actions of lipoxins in vivo (Table 2) (Guilford and Parkinson, 2005; Maddox et al., 1997; Serhan et al., 1995).

Administration of LX stable analogs in animal models protects from tissue damage and inflammation (Parkinson, 2006; Serhan, 2005b) and enhances resolution (Bandeira-Melo et al., 2000). The identification of these anti-inflammatory actions of LXs and ATLs provided strong evidence for the existence of endogenous anti-inflammatory mediators derived from arachidonic acid. Along with reducing PMN influx (Takano et al., 1998), redirecting chemokines and cytokines (Hachicha et al., 1999), and reducing pain (Serhan et al., 2001; Svensson et al., 2007), LX and ATL have the ability to stimulate the removal of apoptotic PMNs by macrophages in vitro (Godson et al., 2000) and at sites of inflammation in vivo (Mitchell et al., 2002). This proresolving agonist activity is shared by annexin

TABLE 2 Lipoxin deficiencies in human disease and proresolving actions after lipoxin analog treatment in animal disease models

Human disease (reference)	Animal model (reference)
Cardiovascular disease LX/LT (Brezinski et al., 1992)	Rabbit and mouse (Jain et al., 2003; Shen et al., 1996)
Asthma	Mouse Tg hALX (Levy et al., 2002)
Aspirin-sensitive asthma (Levy et al., 2002)	
Cystic fibrosis (classic nonresolving) (Karp et al., 2004)	Mouse (Karp et al., 2004)
Rheumatoid arthritis (Thomas et al., 1995)	
Gastrointestinal disease (Mangino et al., 2006)	Mouse (Fiorucci et al., 2004; Wallace and Fiorucci, 2003)
Renal ischemia-reperfusion injury	Mouse (Kieran et al., 2003)
Glomerulonephritis (gene therapy approach) (Munger et al., 1999)	
Periodontal disease (Pouliot et al., 2000)	Rabbit TgLO (Serhan et al., 2003)
Neuropathic inflammatory pain	Mouse (Svensson et al., 2007)

1 (Gilroy and Perretti, 2005) and glucocorticoids (Rossi and Sawatzky, 2007; Ward et al., 1999) and accelerates the return of tissues to homeostasis and function.

IDENTIFICATION OF THE RESOLVINS AND PROTECTINS

Considering the role of LX in resolution and reported beneficial actions of omega-3 in humans, it was of interest to determine whether specialized mediators are involved in the resolution of self-limited inflammation. The mouse air pouch was selected for systematic analyses because acute inflammation and the contained exudates are spontaneously resolved in this dorsal skin cavity, which permitted kinetic analysis of chemical mediators and leukocyte traffic. The novel lipid mediators produced from EPA were first isolated from resolving exudates that proved to contain 18R-hydroeicosapentaenoic acid (18R-HEPE) as well as several additional related bioactive compounds (Serhan et al., 2000). The first bioactive product isolated from exudates, coined resolvin E1, reduced inflammation (Table 3) and blocked human PMN transendothelial migration (Serhan, 2007). Structural elucidation was carried out together with both GC-MS- and MS-MS-based lipidomic analysis of bioactive fractions obtained following extraction and RP-HPLC. The basic structure of this potent bioactive product in the resolving exudates was established (Serhan et al., 2000) (Fig. 3).

Recombinant COX-2 treated with aspirin generates 18R-HEPE as well as 15R-HEPE from EPA, which are blocked by selective COX-2 inhibitor (Serhan et al., 2000). Of interest, clinically used doses of acetaminophen and indomethacin permitted oxygenation of EPA to both 18R-HEPE and 15R-HEPE with isolated recombinant COX-2; the levels of these hydroxy products were significantly reduced yet nonetheless actively generated by the COX-2. These results indicate that the oxygenation of omega-3 PUFAs to generate novel bioactive mediators can also involve certain of the widely used anti-inflammatory drugs but not selective COX-2 inhibitors (Serhan et al., 2000). Next, the most likely human cell types and path-

TABLE 3 Resolvins in models of human diseases

Disease model	Mediator(s)	Actions	Reference(s)
Acute inflammation (murine peritonitis and dermal air pouch inflammation)	Resolvin E1	Reduces PMN infiltration	Serhan et al., 2000, 2002; Arita et al., 2005a; Bannenberg et al., 2005
	Resolvin E1 Protectin D1 ATL/LXA$_4$	Upregulates CCR5 expression on late apoptotic human leukocytes	Ariel et al., 2006
	Resolvin D1	Acts as "terminator" of chemokine signaling during resolution	
Colitis	Resolvin E1	Decreases PMN recruitment and proinflammatory gene expression Improves survival Reduces weight loss	Arita et al., 2005b
	Resolvin E1 Resolvin D	Genetically engineered fat-1 mice possess high levels of DHA and EPA, generate resolvins during colitis, and demonstrate reduction in gastrointestinal tissue damage	Hudert et al., 2006
Periodontitis	Resolvin E1	Reduces PMN infiltration, stops inflammation-induced tissue and bone loss	Hasturk et al., 2006
Acute inflammation (murine peritonitis)	Resolvin E2	Stops PMN infiltration	Tjonahen et al., 2006
Retinal damage	Resolvin E2	Reduces macrophages driving injury	Connor et al., 2007
Acute inflammation (murine peritonitis)	Resolvin D1	Stops PMN recruitment	Serhan et al., 2002 Sun et al., 2007
Microglial cells		Reduces microglial cell cytokine expression in vitro	Hong et al., 2003
Kidney	Resolvin D1 and protectin D1	Protects in renal ischemic injury by limiting PMN infiltration Regulates macrophages	Duffield et al., 2006
Acute inflammation (murine peritonitis and dermal air pouch inflammation)	Resolvins D2, D3, D4	Stop PMN recruitment; reduce peritonitis	Serhan et al., 2002

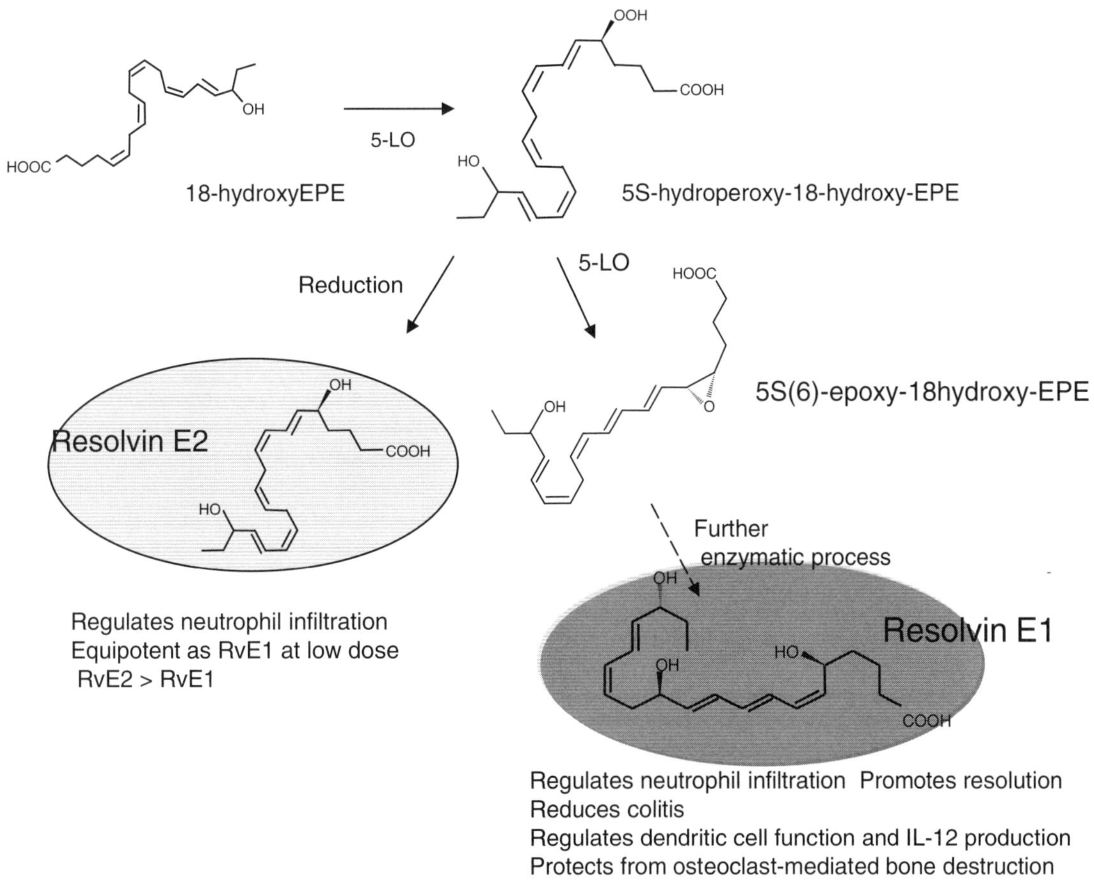

FIGURE 3 Biosynthesis of E-series resolvins.

ways that biosynthesize these bioactive mediators within the resolving exudates were reconstructed in vitro, which involved cell-cell interactions in vivo within the exudates. Isolated human cells, vascular endothelial cells treated with aspirin, convert EPA to 18R-HEPE, which is released and then rapidly converted by activated human PMNs to a 5(6)-epoxide-containing intermediate that is converted to the bioactive 5,12,18R-trihydroxyeicosapentaenoic acid, denoted resolvin E1 (RvE1). RvE1 possesses a distinct structure consisting of a conjugated triene plus a conjugated diene chromophore present within the same molecule (Fig. 3).

Both biogenic (Serhan et al., 2000) and total organic synthesis of RvE1 were achieved, and the complete stereochemical assignment was recently established (Arita et al., 2005a). RvE1 proved to be 5S,12R,18R-trihydroxy-6Z,8E,10E,14Z,16E-eicosapentaenoic acid. The synthetic RvE1 displayed potent stereoselective actions in vivo and with isolated cells, confirming the original structural assignment. To determine the receptors involved in RvE1 stereoselective actions, a screening library of G-protein-coupled receptors (GPCRs) was devised employing counter-regulation of TNF-α, a task that one might expect a resolvin to carry out in vivo (Arita et al., 2005a). The orphan receptor ChemR23 specifically bound to ³H-labeled RvE1 and attenuated TNF-activated NF-κB signaling (Arita et al., 2005a). ChemR23 is expressed on dendritic cells and monocytes. The main second messenger for RvE1 ag-

onist actions via these GPCRs appears to be activation of intracellular phosphorylation pathways. This contrasts with proinflammatory mediators that use mobilization of intracellular calcium or cyclic AMP (Arita et al., 2005a, 2005b). By use of a structure-function approach, a second high-affinity GPCR that signals in response to RvE1 was identified. RvE1 also interacts with LTB$_4$ receptor BLT1 and attenuates LTB$_4$-induced proinflammatory signals by acting as partial agonist/antagonist with the LTB$_4$ receptor, namely, BLT1 present on human PMNs (Arita et al., 2007).

Mice given aspirin plus DHA contained novel 17R-hydroxy-DHA (17R-HDHA) and two separate novel families of bioactive mediators in their resolving exudates (Fig. 4 and 5). The biosynthetic pathways were reconstructed in vitro to establish potential cellular origins for these novel compounds. Hypoxic human microvascular endothelial cells treated with aspirin release 17R-HDHA. DHA is a substrate for isolated human recombinant COX-2 producing 13-hydroxy-DHA (Serhan et al., 2002). With aspirin treatment, COX-2 switches to 17R-oxygenation with molecular oxygen to give an epimeric or aspirin-triggered form in exudates and also in blood and brain with both families of resolvins and protectins (Hong et al., 2003; Serhan et al., 2002). The aspirin-triggered forms carry a 17R alcohol group configuration instead of the carbon 17S as when biosynthesized via lipoxygenase mechanisms (Fig. 4 and 5).

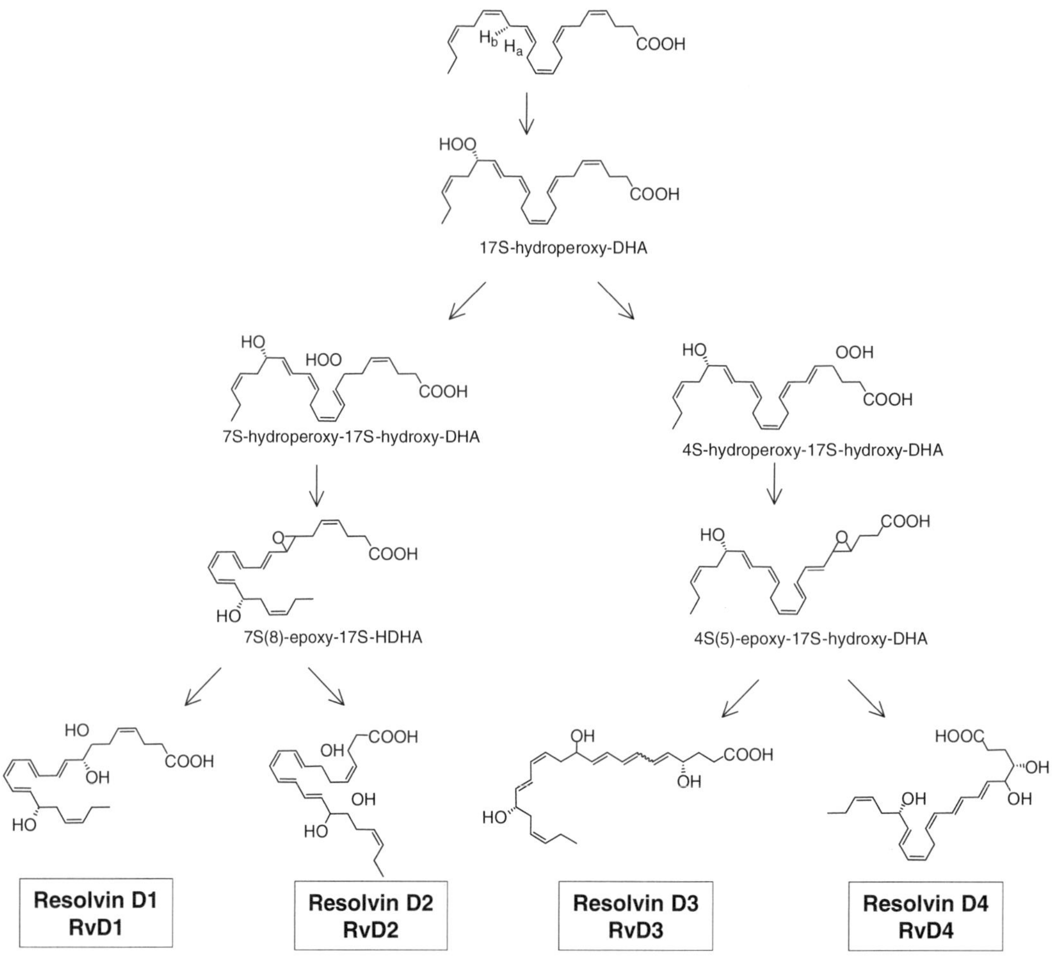

FIGURE 4 Biosynthesis of D-series resolvins.

Endogenous DHA is converted to a 17S alcohol-containing series of resolvins, even in the absence of aspirin. These are denoted RvD1 through RvD4. Also, docosa-conjugated triene-containing structures via lipoxygenase-initiated mechanisms were isolated (Hong et al., 2003; Marcheselli et al., 2003). The complete stereochemistry of protectin D1, which carries a 10,17-dihydroxydocosatriene base structure (Hong et al., 2003; Serhan et al., 2002), was determined and confirmed the original assignment (Serhan et al., 2002, 2006). Total organic synthesis of related isomers and matching studies with biologically derived materials showed that endogenous protectin D1, denoted *neuro*-protectin D1 (NPD1) when produced by neural tissues, was established in isolated human cells and murine cells in vivo as 10R,17S-dihydroxy-docosa-4Z,7Z,11E,13E,15Z,19Z-hexaenoic acid (Serhan et al., 2006). The geometry of the double bonds in PD1 and their positions during biosynthesis from intermediates in its biosynthesis, i.e., an epoxide formed at 16(17) position, indicated that PD1 biosynthesis requires enzymatic steps to produce the potent bioactive molecule carrying the necessary double bond geometry (Fig. 5). On a molar basis, PD1 proved to be log orders of magnitude more potent than its native precursor DHA (Duffield et al., 2006; Hong et al., 2003; Serhan et al., 2006).

RESOLVINS AND PROTECTINS IN ANIMAL DISEASES

RvE1, in only nanogram doses, reduces neutrophil transendothelial migration, dermal inflammation (Serhan et al., 2002), peritonitis, dendritic cell migration, and IL-12 production (Table 3). Synthetic RvE1 blocks PMN infiltration, protects from bone destruction in a rabbit model of periodontal disease (Hasturk et al., 2006), and protects against the development of colitis (Arita et al., 2005b). In several animal models of inflammatory diseases, RvE1 is a potent counterregulator that protects against leukocyte-mediated tissue injury. Resolvins of the D series also block TNF-α-induced IL-1β transcripts in microglial cells and are potent regulators of PMNs, limiting infiltration into inflamed brain, skin, and peritonitis (Hong et al., 2003; Marcheselli et al., 2003; Serhan et al., 2002) (Table 4).

Direct comparisons between resolvin E versus both resolvins of the D series (17S and 17R epimer aspirin-triggered series) at equal doses demonstrated that the 17S series RvDs generated by lipoxygenase-initiated mechanisms and the 17R series RvDs triggered by aspirin treatment, when administered intravenously at 100 ng (~3 μg/kg) in mice, display essentially similar actions, reducing PMN infiltration by ~50% in peritonitis.

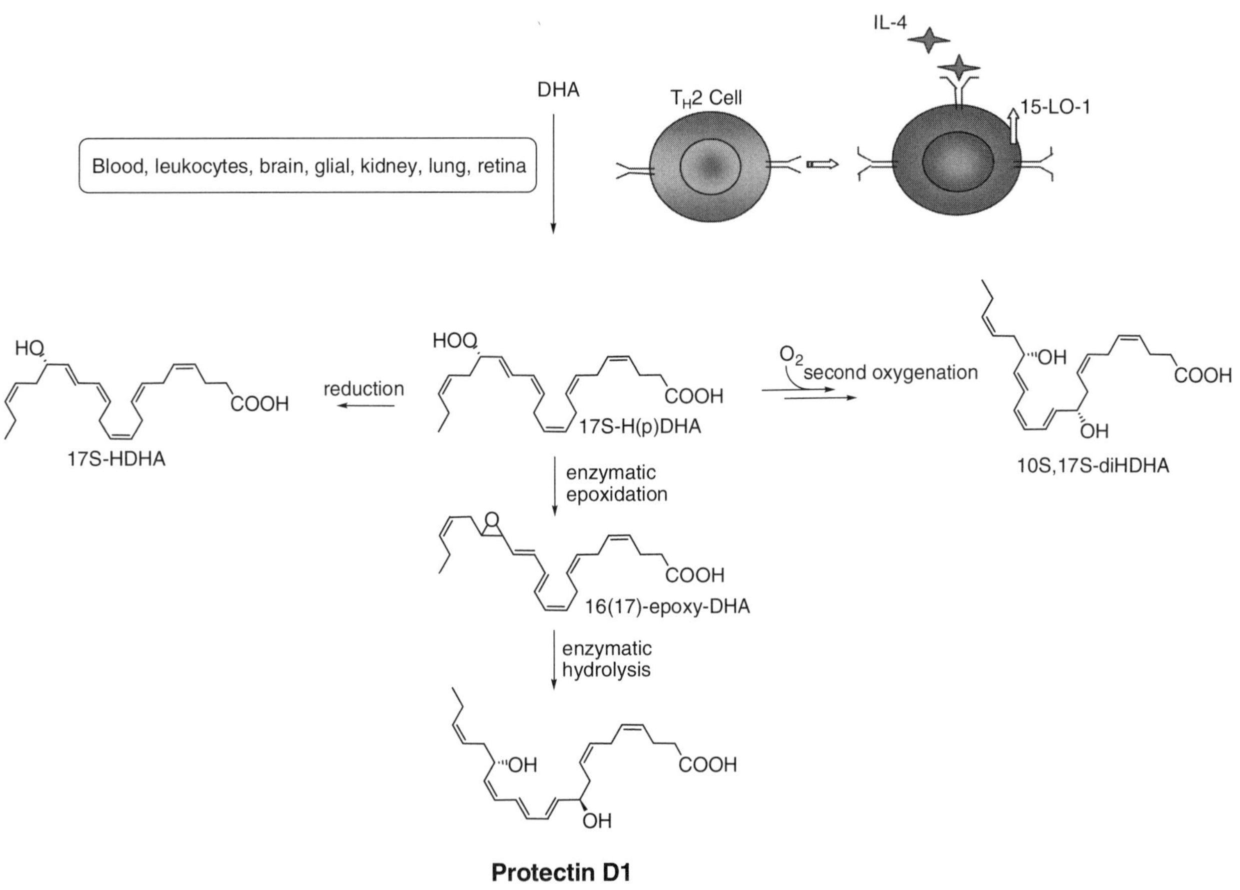

FIGURE 5 Protectin D1/*neuro*-protectin D1 biosynthesis and related dihydroxydocosatrienes.

Protectins possess the conjugated triene structure, a key feature of this family. PD1 is referred to as *neuro*-protectin D1 when generated in the nervous system (Fig. 6), and possesses potent actions in vitro and in vivo (Table 4). For example, synthetic PD1 at 10 nM attenuates human neutrophil transmigration by ~50% in vitro, whereas its Δ15-*trans*-isomer is essentially inactive. PD1 is also a potent regulator of PMNs in vivo by reducing PMN infiltration (~40% at 1 ng per mouse) in murine peritonitis. PD1 also reduced PMN infiltration when administered after the initiation of inflammation in vivo and acted in an additive fashion with RvE1 to stop PMN infiltration. PD1 is thus a potent, stereoselective anti-inflammatory molecule in vivo (Hong et al., 2003; Serhan et al., 2002, 2006). Moreover, these results and those obtained by Bazan and colleagues in neural tissues demonstrate that PD1 displays potent immunoregulatory (Table 4) (Hong et al., 2003; Serhan et al., 2002, 2006) and neuroprotective (Lukiw et al., 2005; Marcheselli et al., 2003; Mukherjee et al., 2004) actions. In addition, it promotes wound healing capacity (Gronert et al., 2005) and is antifibrotic in the kidney (Duffield et al., 2006).

PATHOGEN LIPOXYGENASE-DERIVED LIPOXINS FOR PATHOGEN EVASION

There is a growing body of evidence indicating an immunemodulatory role for lipoxins during infections. Furthermore, it is possible to speculate that pathogens may take advantage of this regulatory pathway to promote host survival, or even to allow a less toxic environment in which replication can occur.

Emerging evidence indicates that the immunomodulatory actions of several lipid mediators are exploited by pathogens, including fungi and helminths. In such cases, it seems that modulation of immunity by suppressing host proinflammatory responses is the aim. Despite the fact that dendritic cells are one of the main targets for the immunemodulatory actions of LXA_4 during *Toxoplasma gondii* infection, this cell population does not produce detectable levels of the eicosanoid (Aliberti et al., 2002b; Bannenberg et al., 2004). Instead, resident splenic macrophages upregulate 5-LO expression after in vivo stimulation with parasite extract (Aliberti et al., 2002a), indicating the participation of macrophages in the generation of lipoxins.

The action of 5-LO on arachidonic acid results in the formation of leukotriene A_4, which can be rapidly converted to LXA_4 through the actions of a second enzyme, 15-LO (Fig. 6). Although the 5-LO activity after *T. gondii* infection was known to be associated with splenic macrophages (Aliberti et al., 2002b), the 15-LO-expressing cell population was not known. In an effort to identify the sources of 15-LO activity after *T. gondii* infection, Bannenberg and colleagues identified an enzymatic activity in tachyzoite forms exposed to calcium ionophore in the presence of arachidonic acid in vitro (Bannenberg et al., 2004).

TABLE 4 Protectins in models of human diseases

Protectin	Organ system/ disease model	Action(s)	Reference
Protectin D1	Acute inflammation (peritonitis)	Reduces PMN infiltration	Hong et al., 2003
		Upregulates CCR5 expression on late apoptotic human leukocytes; "terminator" of chemokine signaling during resolution	Ariel et al., 2006
		Regulates T-cell migration	Ariel et al., 2005
	Liver	Correlates supplements with biosynthesis of PD1 and organ protection in vivo	González-Périz et al., 2006
		PD1 and 17S-HDHA attenuate peroxide-induced DNA damage and oxidative stress in hepatocytes and protect from necroinflammatory liver injury in mice	
	Lung	PD1 formation is reduced in murine models of asthma	Levy et al., 2007
		PD1 protects from lung damage in vivo	
		PD1 is generated in human asthma, protects from airway inflammation and hyperresponsiveness	
	Kidney	PD1 formed in murine kidney	Duffield et al., 2006
		Protects from ischemia-reperfusion-induced kidney damage and loss of function	
	Acute inflammation (peritonitis)	Reduces PMN infiltration	Bannenberg et al., 2005
		Shortens resolution interval (R_i)	
		Downregulates proinflammatory cytokines and chemokines	
		Stimulates anti-inflammatory cytokines and chemokines	
NPD1[a]	Retina	Protects from injury	Bazan, 2006
	Retinal damage	Reduces macrophage driving injury	Connor et al., 2007
	Alzheimer's disease	Diminished production in human Alzheimer's disease	Lukiw et al., 2005
		Promotes neural cell survival in vivo	
	Stroke	Limits ischemic damage	Marcheselli et al., 2003
		Reduces PMN entry into the brain	

[a] PD1 is termed NPD1 when biosynthesized in the neural system (see abbreviations).

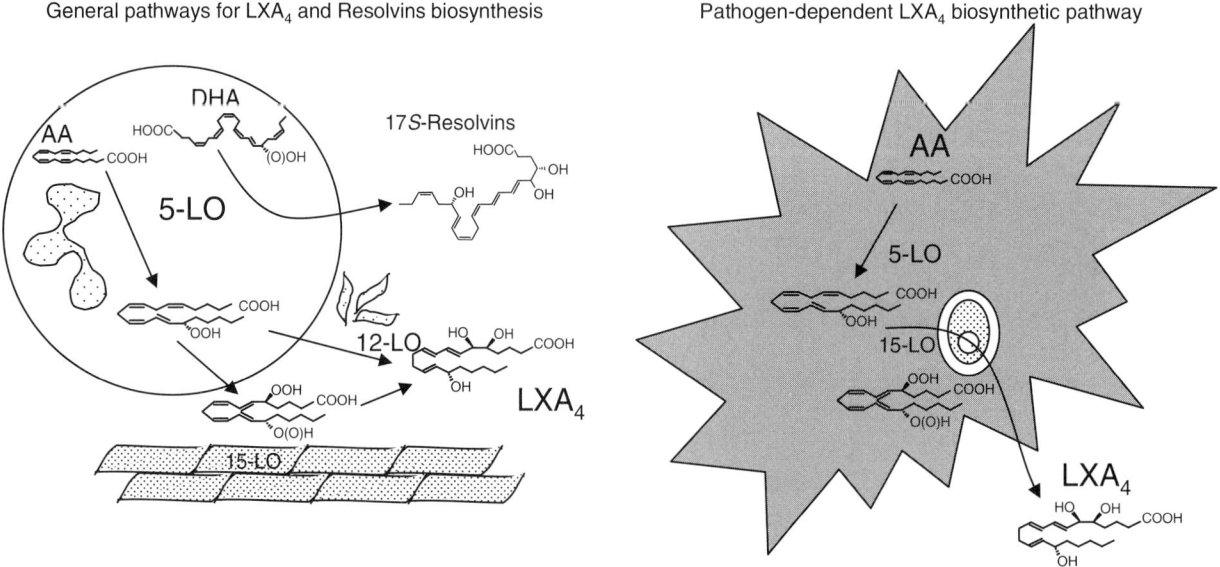

FIGURE 6 Leukocyte- and pathogen-mediated synthesis of lipoxins.

Moreover, proteomics analysis of tachyzoite-derived lysates revealed the presence of peptides homologous to plant-derived type 1 LOs (Bannenberg et al., 2004). It therefore seems probable that the induction of lipoxin biosynthesis by T. gondii has been selected through the carrying of a plant-like LO gene, which together with the actions of host-derived 5-LO results in the high-level production of lipoxins. The presence of high levels of lipoxin, in turn, dampens ongoing immune responses so that hosts can control parasite proliferation without succumbing to the damaging consequences of excessive inflammation or tissue destruction.

Although the genes responsible for the 15-LO activity in T. gondii have not been formally identified, the fact that enzymatic activity is induced after ionophore activation in vitro indicates a putative regulatory mechanism for 15-LO expression/activation (Bannenberg et al., 2004) in which it can be speculated that invasion of host cells, or even immune attack of infected cells, may trigger 15-LO activity in intracellular parasites. On the host side, the expression of 5-LO is increased after stimulation with parasite extracts or after infection (Aliberti et al., 2002a). Although the molecular basis for 15-LO induction after parasite stimulation has not been clarified, it is known that this enzyme can be induced after leukocyte exposure to a variety of stimuli, including PGE_2 (Levy et al., 2001). The interplay between these mediators, the induction of 5-LO, and the control of immune responses in vivo await further investigation.

Another intriguing point that contributes to the argument for a role of T. gondii 15-LO in immune evasion is the presence of such enzymatic activity in an organism that does not have lipids that could serve as substrates for LOs. Therefore, the substrate has to come from infected host cells. Consistent with this is the recent identification and cloning of a 15-LO-like enzyme from the bacterial pathogen Pseudomonas aeruginosa (Vance et al., 2004). Although the production of lipoxins was not formally shown in this report, injection of exogenous 15-LO into naive mice, as shown by Bannenberg et al. (2004), was sufficient to induce the production of endogenous LXA_4 and have biological actions, such as inhibition of inflammatory infiltration and IL-12 production. P. aeruginosa is most commonly associated with chronic lung infections in patients with cystic fibrosis. It is possible that the bacteria may use the 15-LO pathway leading to lipoxin biosynthesis to promote suppression of inflammation and persistence, possibly not activating innate or adaptive immune mechanisms in immune-competent individuals. However, patients with cystic fibrosis fail to generate lipoxins in the lungs, and the continuing proliferation of bacteria results in uncontrolled accumulation of activated neutrophils that ultimately lead to serious tissue damage with organ failure (Karp et al., 2004). This constitutes the major pathology for the lung form of cystic fibrosis. The relevance of pathogen-derived 15-LO, given the lack of lipoxin generation in the lungs of patients with cystic fibrosis and the severity of disease, still remains to be elucidated.

Another lung pathogen that causes a chronic disease with enormous public health relevance is Mycobacterium tuberculosis. For most individuals, infection is asymptomatic with granuloma formation preventing the spread of the bacteria, and potent cell-mediated immunity is typically found in exposed individuals (Chan and Flynn, 2004). Nevertheless, whenever immunity is lowered, bacterial proliferation increases and transmission of viable bacilli occurs, granuloma structure is disrupted, and organ function is compromised (Flynn and Chan, 2003).

M. tuberculosis invasion of the lungs is typically a latent process, with very little reaction occurring in the organ. It seems that the pathogen "slips" into the organ and establishes itself there with almost no intervention from the host inflammatory response. Several factors may be involved in the "immune silencing" phenomenon seen during M. tuberculosis infection (Flynn and Chan, 2003). Interestingly, Bafica and colleagues (2005) have shown that, in the absence of endogenously generated LXA_4, mice become more resistant to infection, with longer survival rates, lower bacterial counts, and higher type 1 cell-mediated immunity against the bacilli. Taking into account the effects of endogenously generated lipoxins during M. tuberculosis infection, it is possible to conjecture that in the previously mentioned model—P. aeruginosa infection in patients with cystic fibrosis—the resulting lung pathology due to the lack of lipoxin generation may indicate that the cystic fibrosis gene defect may affect LO activity directly, even in the presence of a pathogen-derived 15-LO. This needs to be tested directly.

Taking the two infection models studied (T. gondii and M. tuberculosis), one notices an apparent discrepancy in the outcome of infection of 5-LO-deficient animals, indicating a protective versus a host-detrimental role for endogenously produced lipoxins, respectively. From the perspective of T. gondii, which is a fast-replicating pathogen, the host must be kept alive so that transmission can occur through predation. Thus, the host needs well-balanced immunity against the parasite, the number of which is kept low but not completely eliminated. To accomplish this, lipoxins are induced locally to keep immunity present, but not intensified. By contrast, M. tuberculosis is a slow-growing, silent pathogen that requires high proliferation rates in lungs of infected hosts for transmission to occur. To achieve this, lipoxins may be generated and may lower ongoing immune responses allowing enough bacilli to expand. Both cases suggest that lipoxin-dependent inhibition of proinflammatory type 1 responses provides a favorable environment for transmission and propagation of the parasite. Hence, certain microbes have harnessed key biochemical mediators in normal, healthy resolution of acute inflammation in the host to create a local favorable microenvironment exploiting the host's own chemical mediators to escape destruction.

We thank M. Halm Small for manuscript preparation. Studies in the Serhan laboratory reviewed here were supported, in part, by National Institutes of Health grants GM38765, DK074448, and P50-DE016191 (to C.N.S.).

REFERENCES

Aliberti, J., S. Hieny, C. Reis e Sousa, C. N. Serhan, and A. Sher. 2002a. Lipoxin-mediated inhibition of IL-12 production by DCs: a mechanism for regulation of microbial immunity. *Nat. Immunol.* **3:**76–82.

Aliberti, J., C. Serhan, and A. Sher. 2002b. Parasite-induced lipoxin A(4) is an endogenous regulator of IL-12 production and immunopathology in Toxoplasma gondii infection. *J. Exp. Med.* **196:**1253–1262.

Ariel, A., G. Fredman, Y.-P. Sun, A. Kantarci, T. E. Van Dyke, A. D. Luster, and C. N. Serhan. 2006. Apoptotic neutrophils and T cells sequester chemokines during immune response resolution via modulation of CCR5 expression. *Nat. Immunol.* **7:**1209–1216.

Ariel, A., P.-L. Li, W. Wang, W.-X. Tang, G. Fredman, S. Hong, K. H. Gotlinger, and C. N. Serhan. 2005. The docosatriene protectin D1 is produced by T_H2 skewing and promotes human T cell apoptosis via lipid raft clustering. *J. Biol. Chem.* **280**:43079–43086.

Arita, M., F. Bianchini, J. Aliberti, A. Sher, N. Chiang, S. Hong, R. Yang, N. A. Petasis, and C. N. Serhan. 2005a. Stereochemical assignment, anti-inflammatory properties, and receptor for the omega-3 lipid mediator resolvin E1. *J. Exp. Med.* **201**:713–722.

Arita, M., T. Ohira, Y. P. Sun, S. Elangovan, N. Chiang, and C. N. Serhan. 2007. Resolvin E1 selectively interacts with leukotriene B_4 receptor BLT1 and ChemR23 to regulate inflammation. *J. Immunol.* **178**:3912–3917.

Arita, M., M. Yoshida, S. Hong, E. Tjonahen, J. N. Glickman, N. A. Petasis, R. S. Blumberg, and C. N. Serhan. 2005b. Resolvin E1, an endogenous lipid mediator derived from omega-3 eicosapentaenoic acid, protects against 2,4,6-trinitrobenzene sulfonic acid-induced colitis. *Proc. Natl. Acad. Sci. USA* **102**:7671–7676.

Bafica, A., C. A. Scanga, C. Serhan, F. Machado, S. White, A. Sher, and J. Aliberti. 2005. Host control of Mycobacterium tuberculosis is regulated by 5-lipoxygenase-dependent lipoxin production. *J. Clin. Invest.* **115**:1601–1606.

Bandeira-Melo, C., M. F. Serra, B. L. Diaz, R. S. B. Cordeiro, P. M. R. Silva, H. L. Lenzi, Y. S. Bakhle, C. N. Serhan, and M. A. Martins. 2000. Cyclooxygenase-2-derived prostaglandin E_2 and lipoxin A_4 accelerate resolution of allergic edema in *Angiostrongylus costaricensis*-infected rats: relationship with concurrent eosinophilia. *J. Immunol.* **164**:1029–1036.

Bannenberg, G. L., J. Aliberti, S. Hong, A. Sher, and C. N. Serhan. 2004. Exogenous pathogen and plant 15-lipoxygenase initiate endogenous lipoxin A_4 biosynthesis. *J. Exp. Med.* **199**:515–523.

Bannenberg, G. L., N. Chiang, A. Ariel, M. Arita, E. Tjonahen, K. H. Gotlinger, S. Hong, and C. N. Serhan. 2005. Molecular circuits of resolution: formation and actions of resolvins and protectins. *J. Immunol.* **174**:4345–4355.

Bazan, N. G. 2006. Survival signaling in retinal pigment epithelial cells in response to oxidative stress: significance in retinal degenerations. *Adv. Exp. Med. Biol.* **572**:531–540.

Bonnans, C., B. Mainprice, P. Chanez, J. Bousquet, and V. Urbach. 2003. Lipoxin A4 stimulates a cytosolic Ca2+ increase in human bronchial epithelium. *J. Biol. Chem.* **278**:10879–10884.

Brezinski, D. A., R. W. Nesto, and C. N. Serhan. 1992. Angioplasty triggers intracoronary leukotrienes and lipoxin A_4. Impact of aspirin therapy. *Circulation* **86**:56–63.

Burr, G. O., and M. M. Burr. 1929. A new deficiency disease produced by the rigid exclusion of fat from the diet. *J. Biol. Chem.* **82**:345–367.

Calder, P. C. 2006a. Long-chain polyunsaturated fatty acids and inflammation. *Scand. J. Food Nutr.* **50**(S2):54–61.

Calder, P. C. 2006b. n-3 polyunsaturated fatty acids, inflammation, and inflammatory diseases. *Am. J. Clin. Nutr.* **83**(Suppl.):1505S–1519S.

Chan, J., and J. Flynn. 2004. The immunological aspects of latency in tuberculosis. *Clin. Immunol.* **110**:2–12.

Chiang, N., C. N. Serhan, S.-E. Dahlén, J. M. Drazen, D. W. P. Hay, G. E. Rovati, T. Shimizu, T. Yokomizo, and C. Brink. 2006. The lipoxin receptor ALX: potent ligand-specific and stereoselective actions in vivo. *Pharmacol. Rev.* **58**:463–487.

Colgan, S. P., C. N. Serhan, C. A. Parkos, C. Delp-Archer, and J. L. Madara. 1993. Lipoxin A_4 modulates transmigration of human neutrophils across intestinal epithelial monolayers. *J. Clin. Investig.* **92**:75–82.

Connor, K. M., J. P. SanGiovanni, C. Lofqvist, C. M. Aderman, J. Chen, A. Higuchi, S. Hong, E. A. Pravda, S. Majchrzak, D. Carper, A. Hellstrom, J. X. Kang, E. Y. Chew, N. N. Salem, Jr., C. N. Serhan, and L. E. H. Smith. 2007. Increased dietary intake of omega-3 PUFA reduces pathological retinal angiogenesis. *Nat. Med.* **13**:868–873.

Cotran, R. S., V. Kumar, and T. Collins (ed.). 1999. *Robbins Pathologic Basis of Disease*, 6th ed. W.B. Saunders Co., Philadelphia, PA.

Devchand, P. R., M. Arita, S. Hong, G. Bannenberg, R.-L. Moussignac, K. Gronert, and C. N. Serhan. 2003. Human ALX receptor regulates neutrophil recruitment in transgenic mice: roles in inflammation and host-defense. *FASEB J.* **17**:652–659.

Duffield, J. S., S. Hong, V. Vaidya, Y. Lu, G. Fredman, C. N. Serhan, and J. V. Bonventre. 2006. Resolvin D series and protectin D1 mitigate acute kidney injury. *J. Immunol.* **177**:5902–5911.

Fiorucci, S., J. L. Wallace, A. Mencarelli, E. Distrutti, G. Rizzo, S. Farneti, A. Morelli, J.-L. Tseng, B. Suramanyam, W. J. Guilford, and J. F. Parkinson. 2004. A beta-oxidation-resistant lipoxin A_4 analog treats hapten-induced colitis by attenuating inflammation and immune dysfunction. *Proc. Natl. Acad. Sci. USA* **101**:15736–15741.

Flynn, J. L., and J. Chan. 2003. Immune evasion by *Mycobacterium tuberculosis*: living with the enemy. *Curr. Opin. Immunol.* **15**:450–455.

Freire-de-Lima, C. G., Y. Q. Xiao, S. J. Gardai, D. L. Bratton, W. P. Schiemann, and P. M. Henson. 2006. Apoptotic cells, through transforming growth factor-beta, coordinately induce anti-inflammatory and suppress pro-inflammatory eicosanoid and NO synthesis in murine macrophages. *J. Biol. Chem.* **281**:38376–38384.

Gewirtz, A. T., B. McCormick, A. S. Neish, N. A. Petasis, K. Gronert, C. N. Serhan, and J. L. Madara. 1998. Pathogen-induced chemokine secretion from model intestinal epithelium is inhibited by lipoxin A_4 analogs. *J. Clin. Investig.* **101**:1860–1869.

Gilroy, D. W., P. R. Colville-Nash, D. Willis, J. Chivers, M. J. Paul-Clark, and D. A. Willoughby. 1999. Inducible cyclooxygenase may have anti-inflammatory properties. *Nat. Med.* **5**:698–701.

Gilroy, D. W., T. Lawrence, M. Perretti, and A. G. Rossi. 2004. Inflammatory resolution: new opportunities for drug discovery. *Nat. Rev. Drug Discov.* **3**:401–416.

Gilroy, D. W., and M. Perretti. 2005. Aspirin and steroids: new mechanistic findings and avenues for drug discovery. *Curr. Op. Pharmacol.* **5**:405–411.

GISSI-Prevenzione Investigators. 1999. Dietary supplementation with n-3 polyunsaturated fatty acids and vitamin E after myocardial infarction: results of the GISSI-Prevenzione trial. Gruppo Italiano per lo Studio della Sopravvivenza nell'Infarto miocardico. *Lancet* **354**:447–455.

Godson, C., S. Mitchell, K. Harvey, N. A. Petasis, N. Hogg, and H. R. Brady. 2000. Cutting edge: lipoxins rapidly stimulate nonphlogistic phagocytosis of apoptotic neutrophils by monocyte-derived macrophages. *J. Immunol.* **164**:1663–1667.

González-Périz, A., A. Planagumà, K. Gronert, R. Miquel, M. López-Parra, E. Titos, R. Horrillo, N. Ferré, R. Deulofeu, V. Arroyo, J. Rodés, and J. Clària. 2006. Docosahexanenoic acid (DHA) blunts liver injury by conversion to protective lipid mediators: protectin D1 and 17S-hydroxy-DHA. *FASEB J.* **20**:2537–2539.

Gronert, K., A. Gewirtz, J. L. Madara, and C. N. Serhan. 1998. Identification of a human enterocyte lipoxin A_4 receptor that is regulated by IL-13 and IFN-γ and inhibits TNF-α-induced IL-8 release. *J. Exp. Med.* **187**:1285–1294.

Gronert, K., N. Maheshwari, N. Khan, I. R. Hassan, M. Dunn, and M. L. Schwartzman. 2005. A role for the mouse 12/15-lipoxygenase pathway in promoting epithelial wound healing and host defense. *J. Biol. Chem.* **280:**15267–15278.

Guilford, W. J. and J. F. Parkinson. 2005. Second-generation beta-oxidation resistant 3-oxa-lipoxin A_4 analogs. *Prostaglandins Leukot. Essent. Fatty Acids* **73:**245–250.

Hachicha, M., M. Pouliot, N. A. Petasis, and C. N. Serhan. 1999. Lipoxin (LX)A_4 and aspirin-triggered 15-epi-LXA$_4$ inhibit tumor necrosis factor 1α-initiated neutrophil responses and trafficking: regulators of a cytokine-chemokine axis. *J. Exp. Med.* **189:**1923–1929.

Hamazaki, K., M. Itomura, S. Sawazaki, and T. Hamazaki. 2006. Fish oil reduces tooth loss mainly through its anti-inflammatory effects? *Med. Hypotheses* **67:**868–870.

Harris, W. S., K. J. Reid, S. A. Sands, and J. A. Spertus. 2007. Blood omega-3 and trans fatty acids in middle-aged acute coronary syndrome patients. *Am. J. Cardiol.* **99:**154–158.

Hashimoto, A., Y. Murakami, H. Kitasato, I. Hayashi, and H. Endo. 2007. Glucocorticoids co-interact with lipoxin A_4 via lipoxin A_4 receptor (ALX) up-regulation. *Biomed. Pharmacother.* **61:**81–85.

Hasturk, H., A. Kantarci, T. Ohira, M. Arita, N. Ebrahimi, N. Chiang, N. A. Petasis, B. D. Levy, C. N. Serhan, and T. E. Van Dyke. 2006. RvE1 protects from local inflammation and osteoclast mediated bone destruction in periodontitis. *FASEB J.* **20:**401–403.

Hong, S., K. Gronert, P. Devchand, R.-L. Moussignac, and C. N. Serhan. 2003. Novel docosatrienes and 17S-resolvins generated from docosahexaenoic acid in murine brain, human blood and glial cells: autacoids in anti-inflammation. *J. Biol. Chem.* **278:**14677–14687.

Hudert, C. A., K. H. Weylandt, J. Wang, Y. Lu, S. Hong, A. Dignass, C. N. Serhan, and J. X. Kang. 2006. Transgenic mice rich in endogenous n-3 fatty acids are protected from colitis. *Proc. Natl. Acad. Sci. USA* **103:**11276–11281.

Jain, A., E. L. Batista, Jr., C. Serhan, G. L. Stahl, and T. E. Van Dyke. 2003. Role for periodontitis in the progression of lipid deposition in an animal model. *Infect. Immun.* **71:**6012–6018.

James, M. J., S. M. Proudman, and L. G. Cleland. 2003. Dietary n-3 fats as adjunctive therapy in a prototypic inflammatory disease: issues and obstacles for use in rheumatoid arthritis. *Prostaglandins Leukot. Essent. Fatty Acids* **68:**399–405.

Karp, C. L., L. M. Flick, K. W. Park, S. Softic, T. M. Greer, R. Keledjian, R. Yang, J. Uddin, W. B. Guggino, S. F. Atabani, Y. Belkaid, Y. Xu, J. A. Whitsett, F. J. Accurso, M. Wills-Karp, and N. A. Petasis. 2004. Defective lipoxin-mediated anti-inflammatory activity in the cystic fibrosis airway. *Nat. Immunol.* **5:**388–392.

Kasuga, K., T. Porter, and C. N. Serhan. 2007. Novel mechanisms in resolution: rapid appearance of EPA and DHA in murine exudates for resolvins and protectins (abstract). 6th Biennial Arthritis Research Conference, Stone Mountain, GA, 20 to 22 April 2007.

Kieran, N. E., P. P. Doran, S. B. Connolly, M.-C. Greenan, D. F. Higgins, M. Leonard, C. Godson, C. T. Taylor, A. Henger, M. Kretzler, M. J. Burne, H. Rabb, and H. R. Brady. 2003. Modification of the transcriptomic response to renal ischemia/reperfusion injury by lipoxin analog. *Kidney Int.* **64:**480–492.

Kieran, N. E., P. Maderna, and C. Godson. 2004. Lipoxins: potential anti-inflammatory, proresolution, and antifibrotic mediators in renal disease. *Kidney Int.* **65:**1145–1154.

Lee, T. H., C. E. Horton, U. Kyan-Aung, D. Haskard, A. E. Crea, and B. W. Spur. 1989. Lipoxin A_4 and lipoxin B_4 inhibit chemotactic responses of human neutrophils stimulated by leukotriene B_4 and N-formyl-L-methionyl-L-leucyl-L-phenylalanine. *Clin. Sci.* **77:**195–203.

Levy, B. D., C. B. Clish, B. Schmidt, K. Gronert, and C. N. Serhan. 2001. Lipid mediator class switching during acute inflammation: signals in resolution. *Nat. Immunol.* **2:**612–619.

Levy, B. D., G. T. De Sanctis, P. R. Devchand, E. Kim, K. Ackerman, B. Schmidt, W. Szczeklik, J. M. Drazen, and C. N. Serhan. 2003. Lipoxins and aspirin-triggered lipoxins in airway responses. *Adv. Exp. Med. Biol.* **525:**19–23.

Levy, B. D., G. T. De Sanctis, P. R. Devchand, E. Kim, K. Ackerman, B. A. Schmidt, W. Szczeklik, J. M. Drazen, and C. N. Serhan. 2002. Multi-pronged inhibition of airway hyper-responsiveness and inflammation by lipoxin A_4. *Nat. Med.* **8:**1018–1023.

Levy, B. D., P. Kohli, K. Gotlinger, O. Haworth, S. Hong, S. Kazani, E. Israel, K. J. Haley, and C. N. Serhan. 2007. Protectin D1 is generated in asthma and dampens airway inflammation and hyper-responsiveness. *J. Immunol.* **178:**496–502.

Libby, P. 2002. Atherosclerosis: the new view. *Sci. Am.* **286:**46–55.

Lukiw, W. J., J. G. Cui, V. L. Marcheselli, M. Bodker, A. Botkjaer, K. Gotlinger, C. N. Serhan, and N. G. Bazan. 2005. A role for docosahexaenoic acid-derived neuroprotectin D1 in neural cell survival and Alzheimer disease. *J. Clin. Investig.* **115:**2774–2783.

Maddox, J. F., M. Hachicha, T. Takano, N. A. Petasis, V. V. Fokin, and C. N. Serhan. 1997. Lipoxin A_4 stable analogs are potent mimetics that stimulate human monocytes and THP-1 cells via a G-protein linked lipoxin A_4 receptor. *J. Biol. Chem.* **272:**6972–6978.

Maddox, J. F., and C. N. Serhan. 1996. Lipoxin A_4 and B_4 are potent stimuli for human monocyte migration and adhesion: selective inactivation by dehydrogenation and reduction. *J. Exp. Med.* **183:**137–146.

Majno, G., and I. Joris. 2004. *Cells, Tissues, and Disease: Principles of General Pathology*, 2nd ed. Oxford University Press, New York, NY.

Mangino, M. J., L. Brounts, B. Harms, and C. Heise. 2006. Lipoxin biosynthesis in inflammatory bowel disease. *Prostaglandins Other Lipid Mediat.* **79:**84–92.

Marcheselli, V. L., S. Hong, W. J. Lukiw, X. Hua Tian, K. Gronert, A. Musto, M. Hardy, J. M. Gimenez, N. Chiang, C. N. Serhan, and N. G. Bazan. 2003. Novel docosanoids inhibit brain ischemia-reperfusion-mediated leukocyte infiltration and pro-inflammatory gene expression. *J. Biol. Chem.* **278:**43807–43817.

Marchioli, R., F. Barzi, E. Bomba, C. Chieffo, D. Di Gregorio, R. Di Mascio, M. G. Franzosi, E. Geraci, G. Levantesi, A. P. Maggioni, L. Mantini, R. M. Marfisi, G. Mastrogiuseppe, N. Mininni, G. L. Nicolosi, M. Santini, C. Schweiger, L. Tavazzi, G. Tognoni, C. Tucci, and F. Valagussa. 2002. Early protection against sudden death by n-3 polyunsaturated fatty acids after myocardial infarction: time-course analysis of the results of the Gruppo Italiano per lo Studio della Sopravvivenza nell'Infarto Miocardico (GISSI)-Prevenzione. *Circulation* **105:**1897–1903.

Menezes-de-Lima, O., Jr., C. A. Kassuya, A. F. Nascimento, M. G. Henriques, and J. B. Calixto. 2006. Lipoxin A_4 inhibits edema in mice: implications for the anti-edematogenic mechanism induced by aspirin. *Prostaglandins Other Lipid Mediat.* **80:**123–135.

Mitchell, S., G. Thomas, K. Harvey, D. Cottell, K. Reville, G. Berlasconi, N. A. Petasis, L. Erwig, A. J. Rees, J. Savill, H. R. Brady, and C. Godson. 2002. Lipoxins, aspirin-triggered epi-lipoxins, lipoxin stable analogues, and the resolution of inflammation: stimulation of macrophage phagocytosis of apoptotic neutrophils in vivo. *J. Am. Soc. Nephrol.* **13:**2497–2507.

Morris, T., R. Rajakariar, M. Stables, and D. W. Gilroy. 2006. Not all eicosanoids are bad. *Trends Pharmacol. Sci.* **27:**609–611.

Mukherjee, P. K., V. L. Marcheselli, C. N. Serhan, and N. G. Bazan. 2004. Neuroprotectin D1: a docosahexaenoic acid-derived docosatriene protects human retinal pigment epithelial cells from oxidative stress. *Proc. Natl. Acad. Sci. USA* **101:**8491–8496.

Munger, K. A., A. Montero, M. Fukunaga, S. Uda, T. Yura, E. Imai, Y. Kaneda, J. M. Valdivielso, and K. F. Badr. 1999. Transfection of rat kidney with human 15-lipoxygenase suppresses inflammation and preserves function in experimental glomerulonephritis. *Proc. Natl. Acad. Sci. USA* **96:**13375–13380.

Parkinson, J. F. 2006. Lipoxin and synthetic lipoxin analogs: an overview of anti-inflammatory functions and new concepts in immunomodulation. *Inflamm. Allergy Drug Targets* **5:**91–106.

Pouliot, M., C. B. Clish, N. A. Petasis, T. E. Van Dyke, and C. N. Serhan. 2000. Lipoxin A_4 analogues inhibit leukocyte recruitment to *Porphyromonas gingivalis*: a role for cyclooxygenase-2 and lipoxins in periodontal disease. *Biochemistry* **39:**4761–4768.

Rajakariar, R., M. M. Yaqoob, and D. W. Gilroy. 2006. COX-2 in inflammation and resolution. *Mol. Interv.* **6:**199–207.

Rossi, A. G., and D. A. Sawatzky (ed.). 2008. *The Resolution of Inflammation.* Birkhäuser Verlag AG, Basel, Switzerland.

Samuelsson, B., S. E. Dahlen, J. A. Lindgren, C. A. Rouzer, and C. N. Serhan. 1987. Leukotrienes and lipoxins: structures, biosynthesis, and biological effects. *Science* **237:**1171–1176.

Serhan, C. N. 1989. On the relationship between leukotriene and lipoxin production by human neutrophils: evidence for differential metabolism of 15-HETE and 5-HETE. *Biochim. Biophys. Acta* **1004:**158–168.

Serhan, C. N. 2005a. Mediator lipidomics. *Prostaglandins Other Lipid Mediat.* **77:**4–14.

Serhan, C. N. (Guest Ed.). 2005b. Special Issue on Lipoxins and Aspirin-Triggered Lipoxins. *Prostaglandins Leukot. Essent. Fatty Acids* **73(3-4):**139–321.

Serhan, C. N. 2007. Resolution phases of inflammation: novel endogenous anti-inflammatory and pro-resolving lipid mediators and pathways. *Annu. Rev. Immunol.* **25:**101–137.

Serhan, C. N., S. D. Brain, C. D. Buckley, D. W. Gilroy, C. Haslett, L. A. J. O'Neill, M. Perretti, A. G. Rossi, and J. L. Wallace. 2007. Resolution of inflammation: state of the art, definitions and terms. *FASEB J.* **21:**325–332.

Serhan, C. N., C. B. Clish, J. Brannon, S. P. Colgan, N. Chiang, and K. Gronert. 2000. Novel functional sets of lipid-derived mediators with antiinflammatory actions generated from omega-3 fatty acids via cyclooxygenase 2-nonsteroidal antiinflammatory drugs and transcellular processing. *J. Exp. Med.* **192:**1197–1204.

Serhan, C. N., I. M. Fierro, N. Chiang, and M. Pouliot. 2001. Nociceptin stimulates neutrophil chemotaxis and recruitment: inhibition by aspirin-triggered-15-epi-lipoxin A_4. *J. Immunol.* **166:**3650–3654.

Serhan, C. N., K. Gotlinger, S. Hong, Y. Lu, J. Siegelman, T. Baer, R. Yang, S. P. Colgan, and N. A. Petasis. 2006. Anti-inflammatory actions of neuroprotectin D1/protectin D1 and its natural stereoisomers: assignments of dihydroxy-containing docosatrienes. *J. Immunol.* **176:**1848–1859.

Serhan, C. N., S. Hong, K. Gronert, S. P. Colgan, P. R. Devchand, G. Mirick, and R.-L. Moussignac. 2002. Resolvins: a family of bioactive products of omega-3 fatty acid transformation circuits initiated by aspirin treatment that counter pro-inflammation signals. *J. Exp. Med.* **196:**1025–1037.

Serhan, C. N., A. Jain, S. Marleau, C. Clish, A. Kantarci, B. Behbehani, S. P. Colgan, G. L. Stahl, A. Merched, N. A. Petasis, L. Chan, and T. E. Van Dyke. 2003. Reduced inflammation and tissue damage in transgenic rabbits overexpressing 15-lipoxygenase and endogenous anti-inflammatory lipid mediators. *J. Immunol.* **171:**6856–6865.

Serhan, C. N., J. F. Maddox, N. A. Petasis, I. Akritopoulou-Zanze, A. Papayianni, H. R. Brady, S. P. Colgan, and J. L. Madara. 1995. Design of lipoxin A_4 stable analogs that block transmigration and adhesion of human neutrophils. *Biochemistry* **34:**14609–14615.

Serhan, C. N., and J. Savill. 2005. Resolution of inflammation: the beginning programs the end. *Nat. Immunol.* **6:**1191–1197.

Shen, J., E. Herderick, J. F. Cornhill, E. Zsigmond, H.-S. Kim, H. Kühn, N. V. Guevara, and L. Chan. 1996. Macrophage-mediated 15-lipoxygenase expression protects against atherosclerosis development. *J. Clin. Investig.* **98:**2201–2208.

Sun, Y.-P., S. F. Oh, J. Uddin, R. Yang, K. Gotlinger, E. Campbell, S. P. Colgan, N. A. Petasis, and C. N. Serhan. 2007. Resolvin D1 and its aspirin-triggered 17R epimer: stereochemical assignments, anti-inflammatory properties and enzymatic inactivation. *J. Biol. Chem.* **282:**9323–9334.

Svensson, C. I., M. Zattoni, and C. N. Serhan. 2007. Lipoxins and aspirin-triggered lipoxin stop inflammatory pain processing. *J. Exp. Med.* **204:**245–252.

Takano, T., C. B. Clish, K. Gronert, N. Petasis, and C. N. Serhan. 1998. Neutrophil-mediated changes in vascular permeability are inhibited by topical application of aspirin-triggered 15-epi-lipoxin A_4 and novel lipoxin B_4 stable analogues. *J. Clin. Investig.* **101:**819–826.

Takano, T., S. Fiore, J. F. Maddox, H. R. Brady, N. A. Petasis, and C. N. Serhan. 1997. Aspirin-triggered 15-epi-lipoxin A_4 and LXA_4 stable analogs are potent inhibitors of acute inflammation: evidence for anti-inflammatory receptors. *J. Exp. Med.* **185:**1693–1704.

Thomas, E., J. L. Leroux, F. Blotman, and C. Chavis. 1995. Conversion of endogenous arachidonic acid to 5,15-diHETE and lipoxins by polymorphonuclear cells from patients with rheumatoid arthritis. *Inflamm. Res.* **44:**121–124.

Tjonahen, E., S. F. Oh, J. Siegelman, S. Elangovan, K. B. Percarpio, S. Hong, M. Arita, and C. N. Serhan. 2006. Resolvin E2: identification and anti-inflammatory actions: pivotal role of human 5-lipoxygenase in resolvin E series biosynthesis. *Chem. Biol.* **13:**1193–1202.

Vance, R. E., S. Hong, K. Gronert, C. N. Serhan, and J. J. Mekalanos. 2004. The opportunistic pathogen *Pseudomonas aeruginosa* carries a novel secretable arachidonate 15-lipoxygenase. *Proc. Natl. Acad. Sci. USA* **101:**2135–2139.

Van Dyke, T. E., and C. N. Serhan. 2006. A novel approach to resolving inflammation. *Sci. Am. Oral and Whole Body Health* 42–45.

Wada, K., M. Arita, A. Nakajima, K. Katayama, C. Kudo, Y. Kamisaki, and C. N. Serhan. 2006. Leukotriene B_4 and lipoxin A_4 are regulatory signals for neural stem cell proliferation and differentiation. *FASEB J.* **20:**1785–1792.

Wallace, J. L., and S. Fiorucci. 2003. A magic bullet for mucosal protection, and aspirin is the trigger! *Trends Pharmacol. Sci.* **24:**323–326.

Ward, C., I. Dransfield, E. R. Chilvers, C. Haslett, and A. G. Rossi. 1999. Pharmacological manipulation of granulocyte apoptosis: potential therapeutic targets. *Trends Pharmacol. Sci.* **20:**503–509.

18

Cytokines and Macrophages and Dendritic Cells: Key Modulators of Immune Responses

FRANK KAISER AND ANNE O'GARRA

GENERAL INTRODUCTION ON CYTOKINES AND THEIR ROLE IN REGULATING IMMUNE RESPONSES

This chapter will discuss the fundamental role of phagocytes, with specific emphasis on macrophages and dendritic cells (DCs), as potent cytokine producers and responders, establishing them as primary effector cells as well as regulators of innate and adaptive immune responses.

Infection of a host with a pathogen first results in the activation of cells of the innate immune response, especially phagocytes such as macrophages and DCs (Banchereau et al., 2000; Gordon, 1998; Gordon and Taylor, 2005). These phagocytes act as sentinels: upon recognition of pathogens or pathogen-derived products (Janeway and Medzhitov, 2002; Taylor et al., 2005) they immediately induce a multitude of direct antimicrobial activities, including phagocytosis and the subsequent production of intracellular antimicrobial effector molecules (e.g., reactive oxygen intermediates, inducible nitric oxide synthase, GTPases) (Bogdan et al., 2000; Kaufmann, 1993). This "innate" activation of macrophages and DCs is accompanied by the secretion of a series of soluble mediators, most importantly cytokines (Trinchieri, 1997).

Cytokines are defined as small soluble proteins (approximately 8 to 80 kDa) that principally can deliver signals to cells in an autocrine or paracrine manner. Once released, they exert their functions on target cells at picomolar and sometimes even femtomolar concentrations and have the potency to control the communication between immune cells. Cytokines generally act in a network: the response of an individual cell will depend on the pattern of cytokines it is subjected to, and the set of cytokine receptors it expresses.

The initiation and regulation of microbicidal functions of innate cells at early stages of primary infection is crucial to protect the host from pathogens, either by completely eliminating or at least by restricting the expansion of a wide variety of pathogens before the adaptive immune response is initiated. In this context, cytokines produced by macrophages and DCs provide signals which can activate (by the action of proinflammatory cytokines, e.g., interferons [IFNs], tumor necrosis factor [TNF], interleukin-6 [IL-6]) or suppress (by the action of immunosuppressive cytokines, e.g., IL-10, transforming growth factor-β [TGF-β]) the antimicrobial effector functions of macrophages and DCs themselves via autocrine mechanisms (Gordon et al., 1995). Macrophage- and DC-derived cytokines additionally deliver paracrine signals to direct the activation or suppression of other effector cell populations of the innate immune system, e.g., neutrophils and natural killer (NK) cells (Trinchieri, 1997).

DCs are also major activators of adaptive immune responses by their ability to process and present antigens to B and T cells (Banchereau and Steinman, 1998). Furthermore, many cytokines released by macrophages and DCs regulate the migration, expansion, and differentiation of T and B cells (Medzhitov and Janeway, 1997). Therefore, the innate immune system functions not only to initially protect the host directly from infection while slower adaptive immune responses are developing, but also to direct the qualitative and quantitative nature of the adaptive arm of the immune response (O'Garra, 1998). This will determine whether a host is able to mount a protective T-cell response with minimum damage to itself (O'Garra and Vieira, 2004).

During the past two decades there has been growing appreciation of the immunoregulatory role of phagocyte-derived cytokines, both in modulating cellular constituents of innate immunity and in shaping downstream adaptive immune responses. Furthermore, macrophages and DCs themselves respond to cytokines released by other cells of the innate and adaptive immune response (Gordon, 2003). Such cytokines provide feedback loops for the function of macrophages and DCs such that innate and adaptive immune responses are not simply sequential and complementary mechanisms of resistance to pathogens, but additionally regulate each other and act in concert (Sher and

---

Frank Kaiser and Anne O'Garra, National Institute for Medical Research, Division of Immunoregulation, The Ridgeway, Mill Hill, London NW7 1AA, United Kingdom.

Coffman, 1992). This intricate interplay between innate and adaptive immune responses results in a complex and tightly regulated cytokine network. A fine balance between pro- and anti-inflammatory cytokines is essential for effective eradication of pathogens with minimum damage to the host (O'Garra and Vieira, 2004).

In this context, our chapter focuses first on the underlying cellular mechanisms and molecular determinants for cytokine production by macrophages and DCs in response to pathogens or their products. Subsequently, we discuss the contribution of macrophage- and DC-derived proinflammatory (e.g., type I IFNs, the IL-12 family, TNF, IL-6) and anti-inflammatory cytokines (e.g., IL-10, TGF-β) to the balance of cellular components of the innate immune response, including their immunoregulatory roles and in determining the type of adaptive immune response being elicited in response to initial challenge with a pathogen or its products. Finally, we will draw attention to the role of T-cell-derived cytokines produced during adaptive Th1 and Th2 responses, IFN-γ and IL-4/IL-13, respectively, and their fundamental role in modulating macrophage and DC function.

CAPACITY OF MACROPHAGES AND DCs TO SECRETE CYTOKINES

The tightly regulated expression of cytokines by phagocytes of the innate immune system, including macrophages and DCs, results in the establishment of a finely balanced cytokine microenvironment. This section discusses the cellular and molecular mechanisms that enable macrophages and DCs to express complex cytokine patterns in response to pathogens, which are key to mounting effective innate and downstream adaptive immune responses.

Innate Immune Recognition by Pattern Recognition Receptors in Macrophages and DCs

The innate immune response is not triggered unspecifically, as was originally thought, but rather is able to discriminate between self and a variety of pathogens and microbial-derived products. Macrophages and DCs recognize highly conserved microbial structures via a limited number of germ line-encoded pattern recognition receptors (PRRs) (Reis e Sousa, 2004; Taylor et al., 2005). Macrophages and DCs use a variety of PRRs that can be expressed on the cell surface and/or in intracellular compartments (Gordon, 2002; Janeway and Medzhitov, 2002). The engagement of PRRs quickly activates proinflammatory signaling pathways in macrophages and DCs and most importantly leads to the expression and secretion of cytokines, and furthermore can induce opsonization, phagocytosis, and sometimes apoptosis. The cellular response induced by the engagement of PRRs is often referred to as the "innate" activation of phagocytes, in contrast to the "classical" and "alternative" activation of macrophages by cytokines (Gordon, 1998, 2003) (as discussed in "The Regulation of Innate and Adaptive Immune Responses by Macrophages and DCs: Cytokine Producers and Responders," below).

Distinct types of PRRs have the potential to induce differential cytokine production in macrophages and DCs. While the scavenger receptor expressed on macrophages is thought to primarily exhibit ligand-binding properties for intact microbes and microbial surface constituents (Pearson, 1996), the ligation of the mannose receptor expressed on macrophages and DCs by microbes or microbial products not only mediates their phagocytosis and lysosomal degradation, but also leads to cytokine expression (Fraser et al., 1998). However, the molecular mechanisms mediating cytokine expression upon ligation of the macrophage mannose receptor are relatively undefined. Binding of ligands to PRRs leads to induction of cytokines and includes the induction of intracellular signaling pathways such as NF-κB and the mitogen-activated protein (MAP) kinases (Williams, 1999). It has recently been shown that classical DCs will produce IFN-α upon stimulation with double-stranded RNA (dsRNA) in a dsRNA-dependent protein kinase R-dependent fashion (Diebold et al., 2003). Some members of the growing family of cytosolic NOD proteins have also been implicated in the innate recognition of bacteria. At least some NOD proteins function as PRRs and can induce inflammatory cytokine responses via activation of the NF-κB signaling pathway (Inohara and Nunez, 2003). The induction of the immunosuppressive cytokine IL-10 in macrophages and DCs was initially thought to be preferentially mediated by the recognition of pathogen-derived lectins that signal through the membrane-bound PRRs dectin-1 (Brown, 2006; Rogers et al., 2005) and the DC-specific DC-SIGN (Geijtenbeek et al., 2003; van Kooyk and Geijtenbeek, 2003). However, recent studies have demonstrated that IL-10, together with proinflammatory cytokines, can also be induced via the family of Toll-like receptors (TLRs) (Boonstra et al., 2006; Dillon et al., 2004).

Signaling in Macrophages Resulting from TLR Ligation Leads to Proinflammatory and Anti-Inflammatory Cytokine Production

Among the various membrane-bound and cytosolic PRRs in macrophages and DCs, TLRs are well characterized with respect to their induction of cytokine expression (Akira et al., 2006; Beutler, 2005; Medzhitov, 2001). The founding member of this family was initially identified in *Drosophila* and later shown to play a critical role in the antifungal response of flies (Lemaitre et al., 1996). To date, 12 members of the TLR family have been identified in mammals. TLRs detect multiple pathogen-associated molecular patterns, including lipopolysaccharide (detected by TLR4), bacterial lipoproteins and lipoteichoic acids (detected by TLR2), flagellin (detected by TLR5), unmethylated CpG DNA motifs of bacteria and viruses (detected by TLR9), dsRNA (detected by TLR3), and single-stranded viral RNA (detected by TLR7). It has been proposed that the membrane surface-bound TLRs 2, 4, and 5 may specialize in the recognition of bacterial compounds. On the other hand, TLRs 3, 7, and 9, which are localized to intracellular compartments, have been suggested to specialize in the detection of viral nucleic acids. Distinct TLRs differ from each other not only in their ligand specificities, but also in their expression patterns, use of downstream adaptor molecules, and signaling pathways, and thereby presumably in the target genes they can induce, including cytokines.

Upon ligation, TLRs on macrophages and DCs induce a complex antimicrobial host defense response, which includes the production of antimicrobial effector molecules and, in particular, the expression and secretion of proinflammatory cytokines such as type I IFNs, TNF, IL-6, and IL-12, as well as the immunosuppressive cytokine IL-10 (Akira et al., 2006; Boonstra et al., 2006; Dillon et al., 2004).

The diverse responsiveness of macrophages and DC subsets to pathogen-derived products to produce distinct cytokines (Bauer et al., 2001; Hochrein et al., 2001) can be explained, in part, by their differential expression of TLRs (Boonstra et al., 2003; Kadowaki et al., 2001; Shortman and Liu, 2002). For example, while macrophages, monocytes, and myeloid DCs express TLR2 and TLR4 and therefore can sense gram-positive and -negative bacteria, these receptors are not expressed on plasmacytoid DCs, which instead express only TLR7 and TLR9 (Boonstra et al., 2003). This suggests that plasmacytoid DCs specialize in sensing different pathogens such as viruses (Jarrossay et al., 2001; Kadowaki et al., 2001). However, although macrophages and myeloid DCs express TLR7 in mice and humans, TLR9 is expressed only in plasmacytoid DCs and activated B cells in humans, but more widely expressed in macrophages, myeloid DCs, and B cells in addition to plasmacytoid DCs in mice, suggesting greater complexity for the immune response to pathogens (Boonstra et al., 2003; Shortman and Liu, 2002). Therefore, the differential TLR expression on phagocytes may in part facilitate a predominant pathogen-specific activation of precisely those innate immune cell populations that provide the most effective antimicrobial activity and the pattern of cytokines required for such a response.

The cytoplasmic region of TLRs share a protein motif known as the Toll/IL-1R (TIR) domain (Takeda and Akira, 2005). Upon ligand binding, this domain mediates homo- and heteromeric associations between TLRs and TIR-domain-containing adaptor molecules. Four adaptor molecules have been identified, namely MyD88, TIR-associated protein (TIRAP)/MyD88-adaptor-like (MAL), TIR-domain-containing adaptor protein inducing IFN-β (TRIF)/TIR-domain-containing molecule 1 (TICAM-1), and TRIF-related adaptor molecule (TRAM) (reviewed in Medzhitov [2001] and Yamamoto and Akira [2005]). TLR4 utilizes all four adaptors, whereas TLR7 and TLR9 use only MyD88. While MyD88 is a universal TLR adaptor and is utilized by all TLRs with the exception of TLR3, TRIF is involved only in TLR3 and TLR4 signaling and activates an alternative pathway, which is particularly important for induction of type I IFNs. The differential cytokine responses that are induced upon ligation of distinct TLRs in one and the same cell can be explained in part as discussed before by differential expression of TLR and potentially in part by the selective usage of TLR adaptor molecules (Boonstra et al., 2006; Fitzgerald et al., 2001; Oshiumi et al., 2003; Yamamoto et al., 2002).

In addition to the differential expression and signaling specificities of TLRs, macrophages and DC subsets appear to have different intrinsic capacities to produce particular patterns of cytokines in response to the same TLR ligand (Boonstra et al., 2006; Shortman and Liu, 2002). For example, macrophage stimulation through TLR4 and TLR9 induces high levels of IL-10, some TNF and IFN-β, but low levels of IL-12 (Boonstra et al., 2006). In contrast, upon stimulation of the same TLRs, myeloid DCs produce less IL-10, similar levels of TNF and IFN-β, but most notably higher levels of IL-12. Strikingly, in response to TLR9 ligation, plasmacytoid DCs do not produce IL-10, but very high amounts of IL-12 (Boonstra et al., 2003) and type I IFNs (Liu, 2005). It has also been suggested that tissue-specific macrophages (Gordon and Taylor, 2005) have different intrinsic capacities to express differential cytokine patterns (Smythies et al., 2005), but the underlying mechanisms for the different cytokine profiles induced in macrophage and DC subsets are still poorly characterized.

In addition to differential TLR expression and different intrinsic capacities of macrophages and DCs to produce cytokines in response to TLR ligation, other factors may influence their capacity to produce cytokines. For example, T cells have been demonstrated to enhance the production of proinflammatory cytokines by phagocytes (see "The Regulation of Innate and Adaptive Immune Responses by Macrophages and DCs: Cytokine Producers and Responders," below) through cell-cell contact involving the interaction of CD40L on activated T cells with CD40 on phagocytes (Cella et al., 1996; Koch et al., 1996; Schulz et al., 2000) or the action of T-cell-derived cytokines such as IFN-γ. Furthermore, the finding that different TLRs can not only synergize with other TLR family members (Napolitani et al., 2005), but also cooperate with other non-TLR receptors (Creagh and O'Neill, 2006; Trinchieri and Sher, 2007), emphasizes the complexity of receptor-induced cytokine expression by macrophages and DCs that will ensue upon challenge of a host with a pathogen or microorganisms included in the gut flora.

Signaling Pathways Controlling Cytokine Expression in Macrophages and DCs

The stimulation of TLRs on macrophages and DCs induces the activation of several intracellular signal transduction pathways, including those regulating the MAP kinases (MAPK) (Chen et al., 2001; Dong et al., 2002; Widmann et al., 1999), phosphoinositide 3-kinases (PI3K) (Fukao et al., 2002; Katso et al., 2001), NF-κB transcription factors (Li and Verma, 2002), and interferon regulatory factor (IRF)-dependent signaling pathways (Honda and Taniguchi, 2006).

MAPK Signaling in Macrophages and DCs Results in the Production of Proinflammatory and Anti-Inflammatory Cytokines

MAPKs are particularly important in regulating cytokine gene expression (Dong et al., 2002; Johnson and Lapadat, 2002). The MAPK signaling cascade is one of the most ancient and evolutionarily conserved signaling pathways (Widmann et al., 1999), and is widely used in all aspects of immune responses, including TLR signaling (Symons et al., 2006). Activation of MAPKs results in the phosphorylation of numerous substrate proteins, which include other protein kinases, phospholipases, cytoskeletal proteins, and most importantly transcription factors specific for cytokine gene expression in a number of immune cells including macrophages and DCs (Chen et al., 2001). MAPK signaling pathways comprise a series of three protein kinases (MAP3K, MAP2K, and MAPK), which phosphorylate and activate one another (Pearson et al., 2001; Roux and Blenis, 2004). The further regulation of MAPK specificities by scaffolding proteins (Morrison and Davis, 2003) and MAPK activities by dual-specificity MAPK phosphatases (DUSPs) (Lang et al., 2006; Liu et al., 2007) adds to the complexity of these signal transduction pathways in controlling cytokine gene expression during inflammatory and immune responses.

Mammalian cells have at least three subfamilies of MAPKs: the extracellular signal-related kinases (ERK)-1/2, Jun amino-terminal kinases (JNK), and p38. TLR stimulation induces the activation of each of the major MAPK subtypes (Symons et al., 2006) in macrophages and DCs. However, the precise contribution of each of the MAPK families to the expression of a particular pro- or anti-inflammatory cytokine in macrophages and DCs is incompletely defined to date. The activation of the ERK MAPK signaling pathway via the MAPK3K TPL2 (Salmeron et al., 1996) has been shown to regulate the production of TNF in macrophages upon TLR stimulation, in part, by regulating its biosynthesis at the posttranscriptional level via AU-rich elements in the 3′-untranslated region of TNF transcripts (Dumitru et al., 2000). Phosphorylation of ERK has also been shown to negatively regulate the production of the proinflammatory cytokine IL-12 in macrophages (Hacker et al., 1999; Tomczak et al., 2006) and DCs (Agrawal et al., 2003), and it is suggested that this occurs via the stabilization of the transcription factor c-Fos. It has been shown that the inhibition of IL-12 production in macrophages and DCs can be explained to large extent by the action of the immunosuppressive cytokine IL-10, which is induced by ERK activation upon TLR ligation (Dillon et al., 2004; Yi et al., 2002). P38 substrates such as MAPKAP kinase 2 have also been demonstrated to regulate TNF production at the posttranscriptional level through AU-rich element regions in TNF transcripts (Kotlyarov et al., 1999). Studies using macrophages and DCs with deficiencies in the p38 signaling pathway have suggested p38 as a positive regulator for IL-12 production (Lu et al., 1999). However, this is controversial since studies could demonstrate increased IL-12 production by macrophages and DCs upon inhibition of p38 (Marriott et al., 2001; Salmon et al., 2001).

With their central role in regulating cytokine production by macrophages and DCs, MAPKs and their antagonist DUSPs have recently gained much attention as therapeutic targets for inflammatory diseases, and intense efforts are under way to develop highly specific pharmacological inhibitors for these signaling molecules (English and Cobb, 2002; Jeffrey et al., 2007; Kumar et al., 2003). However, the concern remains that blocking these signaling pathways may result in different effects not only on proinflammatory cytokine production, but also on the production of suppressive cytokines, and may thus lead to adverse rather than ameliorative effects.

NF-κB Signaling in Macrophages and DCs Results in the Production of Proinflammatory and Anti-Inflammatory Cytokines

The extensively studied NF-κB/Rel transcription factor family has a central role in controlling macrophage, DC, and lymphocyte function in both innate and adaptive immune responses, respectively (Li and Verma, 2002; Silverman and Maniatis, 2001). As prototypical transcription factors, NF-κBs are one of the pivotal regulators of proinflammatory gene expression in immune and inflammatory responses (Perkins, 2000). In macrophages and DCs, the activation of NF-κBs induces a plethora of microbicidal effector mechanisms, growth factors, cell surface receptors, adhesion molecules, and most importantly the expression of proinflammatory cytokines (Kulms and Schwarz, 2006).

The NF-κB signaling pathway in macrophages and DCs is activated rapidly in response to a wide range of stimuli, including stress signals, proinflammatory cytokines such as TNF and IL-1, and upon engagement of TLRs by pathogens or microbial-derived products (Hayden and Ghosh, 2004; Zhang and Ghosh, 2001). The mechanisms by which NF-κB integrates multiple stimuli to generate a specific outcome suitable for specific situations are incompletely understood to date.

All TLRs universally activate NF-κB transcription factors (Akira et al., 2006; Medzhitov, 2001) via downstream activation of the IκB kinases (IKKs) (Wang et al., 2001). The IKKs directly phosphorylate the NF-κB precursor molecules NF-κB p105 and NF-κB p100, as well as the members of the inhibitory IκB protein family (IκBs), which associate with NF-κB to retain it in the cytosol in its inactive form. Once phosphorylated, NF-κB precursors and NF-κB/IκB complexes become ubiquitinated and subsequently degraded by the proteasome, and this enables the translocation of NF-κB to the nucleus to induce transcription of target genes such as cytokines. How these different signal transduction pathways selectively activate different NF-κB complexes in a coordinated manner to potentially exert possible distinct functions in innate and adaptive immunity, including effects on macrophage and DC cytokine gene expression, is not clearly defined and is a very active area of investigation (Beinke and Ley, 2004; Bonizzi and Karin, 2004; Pomerantz and Baltimore, 2002). The members of the NF-κB family have been demonstrated to be involved in proinflammatory responses and to be essential for the production of IL-6 (Grigoriadis et al., 1996) and IL-12p40 (Mason et al., 2002) in macrophages. Importantly, it has been shown that, upon stimulation with lipopolysaccharide, production of TNF was reduced in IKK2-deleted macrophages, but a major reduction in the anti-inflammatory cytokine IL-10 was also reported (Kanters et al., 2003). This study showed that the inhibition of the NF-κB pathway in macrophages resulted in more severe atherosclerosis in mice, possibly because of its effect on the pro- and anti-inflammatory cytokine balance. The requirement of NF-κB for optimal gene expression of IL-10 in macrophages was supported by a study demonstrating NF-κB-specific sites in the IL-10 locus (Saraiva et al., 2005). TNF and IL-1 also strongly activate NF-κB, which plays an important role in amplifying the innate immune response of macrophages and DCs (O'Neill and Dinarello, 2000; Wallach et al., 1999).

IRFs: Major Inducers of Type I IFNs by Macrophages and DCs in Response to Viruses and TLR Ligands

The members of the IRF family of transcription factors, which comprises nine members in mammalian cells (Taniguchi et al., 2001), were initially described as being involved in the induction of genes that encode type I IFNs (Taniguchi and Takaoka, 2002). More recently they have been demonstrated to have distinct roles in the development and function of immune cells, including macrophages and DCs. There is now growing evidence that the IRF family represents an intricate gene-regulatory network of the immune system, which has the capacity to rapidly alter the expression of cytokine genes, a critical requirement for efficient immune responses against pathogens (Honda and Taniguchi, 2006). IRFs have been extensively studied in the context of host defense against viral infection.

Two highly homologous IRFs, IRF3 and IRF7, dimerize and translocate to the nucleus, where they activate transcription of IFN type I genes (Taniguchi et al., 2001) and are regarded as key regulators of type I IFN gene expression elicited by viruses. The family of IRFs has recently gained more attention with the discovery that upon stimulation of TLRs distinct IRF family members are activated by MyD88-dependent and MyD88-independent pathways in macrophages and DCs (Moynagh, 2005). TLR3 and TLR4 ligands have been demonstrated to activate IRF3 via activation of TBK1 (Doyle et al., 2002; Fitzgerald et al., 2003). Signaling through TLR7 and TLR9 activates IRF7, which directly interacts with MyD88 and its adaptor molecule TRAF6 (Honda et al., 2004; Kawai et al., 2004). The TLR-induced IRF7 signaling pathway probably explains the large amounts of type I IFNs that are produced by plasmacytoid DCs in response to TLR7 and TLR9 ligands (Boonstra et al., 2003; Colonna et al., 2004; Hoshino et al., 2006). Plasmacytoid DCs detect viruses by using endosomal TLR7 and TLR9 instead of cytosolic PRRs such as RIG-I and MDA-5 (Diebold et al., 2004; Hemmi et al., 2003). In addition to IRF7, IRF5 also interacts directly with the TLR adaptor molecules MyD88 and TRAF6 (Takaoka et al., 2005). It is possible that IRF5 may be involved selectively in the induction of proinflammatory cytokines in macrophages and DCs such as TNF, IL-6, and IL12p40 in response to TLR signaling. IRF4 is suggested to compete with IRF5, but not IRF7, for binding MyD88 and negatively regulates cytokine expression in macrophages (Negishi et al., 2005). IRF8 has been suggested to promote the production of proinflammatory cytokines in macrophages and DCs by increasing TLR4- and TLR9-induced IRF signaling by binding and activating TRAF6 (Tsujimura et al., 2004; Zhao et al., 2006). In summary, there is considerable evidence that the IRF family of transcription factors are not only important regulators of cytokine expression in antiviral responses, but also play a crucial role in TLR-dependent signaling in macrophages and DCs. However, their exact role in the rapidly increasing complexity of TLR signaling pathways and their precise contribution to cell-specific cytokine gene expression are still not completely understood.

THE REGULATION OF INNATE AND ADAPTIVE IMMUNE RESPONSES BY MACROPHAGES AND DCs: CYTOKINE PRODUCERS AND RESPONDERS

Upon recognition of pathogens or microbial-derived products, macrophages and DCs are induced to produce particular cytokine profiles that direct both the innate as well as the subsequent adaptive immune response, which in return feeds back to regulate the innate response by releasing cytokines. As a result, a complex network of multiple cytokines is induced that will determine whether the pathogen is cleared successfully with minimum damage to the host (Fig. 1).

Macrophages and DCs are the primary source of many key cytokines in early innate immune responses. These include type I IFNs, proinflammatory cytokines with pleiotropic functions such as IL-1β, TNF, IL-12, IL-6, and many others, as well as immunosuppressive cytokines like IL-10 and TGF-β. These cytokines are critical modulators, in an autocrine fashion, of macrophage and DC function and physiology, and in addition, they affect many other cells such as endothelial cells, epithelial cells, and keratinocytes.

Although innate resistance mediated by macrophages and DCs can be efficient at either preventing an infection or at least reducing the pathogen burden, sterile cure or ultimate control of an infection is usually achieved when adaptive immunity is induced.

The type of immune response elicited in response to pathogens determines whether the pathogen is eradicated

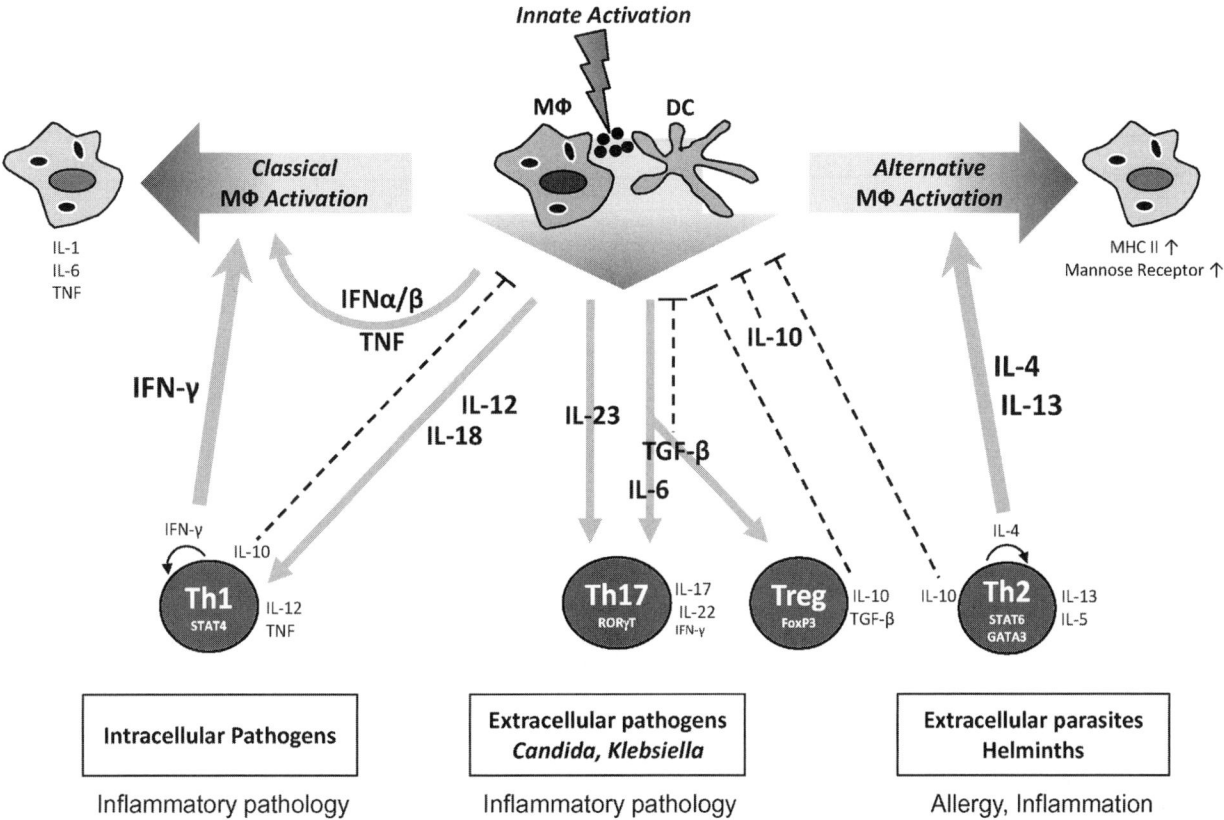

FIGURE 1 Key cytokines, produced by or exerting their effects on macrophages and DCs, regulate, direct, and balance the innate and adaptive immune response against pathogens. Gray arrows, stimulatory signals; broken lines, inhibitory signals.

with minimum pathology or whether chronic infection sets in (O'Garra and Robinson, 2004). The pattern of cytokines that are present during the clonal expansion of antigen-specific T cells early at the initiation of an immune response critically determines the subsequent induction of different effector CD4$^+$ T cells, including Th1, Th2, or Th17 subsets (Mosmann and Coffman, 1989; Murphy and Reiner, 2002; O'Garra and Robinson, 2004; Sher and Coffman, 1992; Weaver et al., 2006) and regulatory T cells (Sakaguchi et al., 2006; Zhang and Ghosh, 2001). Macrophage- and DC-derived cytokines are therefore essential in directing the deviation of T cells toward either type of adaptive effector response (discussed in more detail later). For example, members of the IL-12 family have been recognized to have a central role in directing Th1 responses (O'Garra and Robinson, 2004; Trinchieri, 2003), and in contrast, IL-6 together with TGF-β and IL-23 have been found to be essential for the development of the Th17 lineage (Stockinger and Veldhoen, 2007; Weaver et al., 2006).

The hallmark cytokines of distinct CD4$^+$ T-cell responses, i.e., IFN-γ produced in Th1 or IL-4/IL-13 produced in Th2 responses, respectively, are themselves potent factors to direct macrophage and DC function. For example, IFN-γ results in the cellular responses against intracellular pathogens, while IL-4 and IL-13 alternatively activate macrophages and promote Th2-type responses to extracellular pathogens (Goodbourn et al., 2000; Gordon, 2003). Accordingly, T-cell-derived cytokines, whose production has initially been regulated by the cytokine milieu secreted by macrophages and DCs, provide important regulatory feedback signals for macrophage and DC effector function and are fundamentally important for the effective eradication of different pathogens, with minimum damage to the host.

Cytokines Produced by Macrophages and DCs and Their Effects on the Innate and Adaptive Immune Responses

Type I IFNs: More than Just Antiviral

One of the earliest cytokine responses of macrophages and DCs upon infection is the secretion of type I IFNs, IFN-α or IFN-β, and in some cases both, as seen in plasmacytoid DCs, which produce vast quantities of these cytokines (Colonna et al., 2004). Also monocytes and macrophages (Eloranta and Alm, 1999; Fleit and Rabinovitch, 1981; Toshchakov et al., 2002), myeloid and lymphoid DCs (Cella et al., 1999; Hochrein et al., 2001), as well as neutrophils (Shirafuji et al., 1990), NK cells (Peter et al., 1980), and T cells (Conta et al., 1983), have been described to release the type I IFN, IFN-β, in response to viral, mitogenic, and/or microbial stimuli, although less efficiently than plasmacytoid DCs. Some of these cells show a selective expression of IFN-β or of some other IFN subtypes in response to certain stimuli (Hoss-Homfeld et al., 1989; Toshchakov et al., 2002).

Type I IFNs are considered as the first line of defense of the immune system against viruses and have been best characterized for their essential function for restricting viral spread (Goodbourn et al., 2000; Stark et al., 1998). They elicit antiviral activity of the target cells to inhibit viral replication and can selectively induce apoptosis in virally infected cells (Guidotti and Chisari, 2001; Tanaka et al., 1998).

Two distinct IFN-inducing pathways are known to exist. One is the ubiquitous cytosolic viral detection system that enables probably any nucleated cell to produce type I IFNs upon an appropriate viral stimulus. The other pathway is restricted to endosomes of macrophages and DCs and mediated by TLRs (Stetson and Medzhitov, 2006). Based on their induction by nonviral TLR ligands (Honda et al., 2005), type I IFNs are now considered not only to coordinate immunity to viruses, but also to be relevant for the pathogenesis or control of certain bacterial and protozoan infections (Bogdan et al., 2004). In fact, type I IFNs have recently been demonstrated to exert important indirect antimicrobial effects during the innate immune response to certain nonviral pathogens that were previously thought to be exclusively controlled by IFN-γ (Decker et al., 2005; Hertzog et al., 2003). However, detrimental effects of IFN-β on the innate immune response to bacterial infections have been reported (Auerbuch et al., 2004; Carrero et al., 2004; O'Connell et al., 2004). This attenuation of innate immunity was shown to be due, in part, to the production of the anti-inflammatory cytokine IL-10 by phagocytic cells (Carrero et al., 2006).

Macrophage- and DC-derived type I IFNs are important immunomodulators with pleiotropic effects that link the innate and adaptive immune system (Biron, 2001). During the innate response, type I IFNs are crucial for the activation of macrophages and NK cells (Biron et al., 1999; Bogdan, 2000). Type I IFNs can promote different DC subpopulations and inhibit the differentiation of monocytes into myeloid DCs (Kadowaki et al., 2000). Secretion of type I IFNs together with IL-12 increases NK-cell-mediated cytotoxicity, IFN-γ production, and early antiviral resistance (Dalod et al., 2003; Krug et al., 2004). On the other hand, type I IFNs also exert negative regulatory functions, e.g., they can inhibit the production of IL-12, in particular, the p40 subunit (Dalod et al., 2002).

In addition to their regulation of innate immune responses, macrophage- and DC-derived type I IFNs have also been described to directly regulate T- and B-cell responses. For example, type I IFNs can induce Th1 differentiation and long-term T-cell survival (Agnello et al., 2003; Tough et al., 1997), demonstrating that type I IFNs are positively linked with the activation and expansion of lymphocytes that are important for the control of intracellular infections. In humans type I IFNs can induce the differentiation of Th1 cells (Rogge et al., 1998); however, whether this effect of type I IFNs is operating in the mouse has been controversial (Biron, 2001; Farrar and Murphy, 2000). Furthermore, type I IFNs together with IL-6 can induce human B cells to differentiate into plasma cells and produce immunoglobulin (Ig) (Jego et al., 2003).

The IL-12 Cytokine Family: Key Cytokines Influencing the Development of Effector T Cells

Differentiation of CD4$^+$ T cells into Th1 cells requires signaling via STAT1 and STAT4, and the expression of the Th1-specific transcription factor T-bet upon encounter with certain microbes or their products (Glimcher and Murphy, 2000), in part, through the action of macrophage- and DC-derived IL-12 (Hsieh et al., 1993; Macatonia et al., 1995). Th1 differentiation independently of IL-12 has also been reported (Jankovic et al., 2002; Skokos and Nussenzweig, 2007). The Th1 response, usually accompanied by the activation of NK cells and CD8$^+$ T cells producing IFN-γ, is critical for the eradication of intracellular pathogens (O'Garra and Vieira, 2004).

IL-12 is the prototype of a small family of heterodimeric cytokines which includes IL-23 (formed by p40 and p19 chains) and IL-27 (p40 and p28 chains) and which is structurally related to IL-6 (Trinchieri, 2003). The bioactive heterodimer IL-12 [IL-12(p70)] is formed by a p40 heavy chain and a p35 light chain (Kobayashi et al., 1989). IL-12p40 is produced in large excess over the IL-12(p70) heterodimer (D'Andrea et al., 1992). Coordinated expression is needed to produce biologically active heterodimers; in the absence of p35, p40 is secreted as a monomer or a homodimer (Ma and Trinchieri, 2001; Snijders et al., 1996).

Macrophages and DCs produce the heterodimeric cytokine IL-12 in response to pathogens during infection (Hsieh et al., 1993). It has been shown in vivo that CD8$^+$ DCs in the spleens of mice and not macrophages (Hsieh et al., 1993; Macatonia et al., 1995) are the first cells to produce IL-12 upon challenge with microbial products (Reis e Sousa et al., 1997). In response to viral infections in the mouse, IL-12 (together with IFN-α) is mainly produced by plasmacytoid DCs and in smaller amounts by myeloid DCs (Dalod et al., 2002). In macrophages and DCs, TLR ligands often induce only low levels of the IL-12 heterodimer in phagocytes, but the key cytokine IFN-γ of the Th1 response increases the ability to produce IL-12 in a positive feedback loop (Ma et al., 1996). Interestingly, the two hallmark cytokines of Th2 responses, IL-4 and IL-13, can be potent enhancers of IL-12 production (D'Andrea et al., 1995; Hochrein et al., 2000). In addition, IL-12 production can be greatly enhanced by direct cell-cell interactions, mostly through ligands of the TNF family. For example, the interaction of CD40L on T cells with CD40 on DCs or macrophages leads to markedly enhanced IL-12p70 production (Cella et al., 1996; Koch et al., 1996; Schulz et al., 2000).

The production of IL-12 can be efficiently inhibited by IL-10, which blocks the transcription of both IL-12p40 and IL12p35 encoding genes (Aste-Amezaga et al., 1998; D'Andrea et al., 1993). Another potent inhibitor of IL-12 production is TGF-β, which has also been shown to reduce the stability of IL-12p40 mRNA (Du and Sriram, 1998). Furthermore, type I IFNs (Cousens et al., 1999) and TNF (Ma, 2001) in certain contexts can suppress IL-12 production.

IL-12 induces the production of IFN-γ in NK cells and T cells, which leads to the polarization of naive T cells into Th1 effector cells (Moser and Murphy, 2000). Th cell differentiation most probably is determined early after infection by the balance between cytokines such as IL-12 and IL-4, which favor Th1- and Th2-cell development, respectively. Low levels of IFN-γ produced by T or NK cells are sufficient to induce T-bet, a transcription factor critical for Th1-cell development, which is amplified by macrophage- and DC-derived IL-12 (Glimcher and Murphy, 2000). IL-12 synergizes with IL-18, also released by macrophages and DCs, to induce the high levels of IFN-γ production by Th1 and NK cells often required for eradication of intracellular pathogens (Okamura et al., 1998).

Many of the proinflammatory activities of IL-12 are mediated by the induction of IFN-γ in Th1 cells (Chan et al., 1991; Kubin et al., 1994). Furthermore, IL-12 has a direct proliferative effect on preactivated T and NK cells (Perussia et al., 1992) and enhances the generation of cytotoxic T lymphocytes (Trinchieri, 1998).

In summary, the ability of macrophage- and DC-derived IL-12 to favor Th1 responses exemplifies its function as an important immunoregulatory cytokine that links innate resistance with the development of adaptive Th1 immune responses, which are dominated by the production of IFN-γ and are essential for resistance to many infections, in particular, with bacteria and intracellular parasites.

IL-23 and IL-27: Master Regulators of Innate and Adaptive Immunity

In addition to IL-12(p70), macrophages and DCs produce IL-23 and IL-27, two IL-12-related heterodimeric cytokines that have been identified recently (Hunter, 2005; Kastelein et al., 2007; Langrish et al., 2004). IL-23 shares the common p40 subunit with IL-12 and has a unique p19 chain (Oppmann et al., 2000). Instead, IL-27 is formed by a p40 and p28 chain (Pflanz et al., 2002).

IL-23 enhances the survival and growth of IL-17-producing Th cells, subsequent to their differentiation (Weaver et al., 2006). IL-6 together with TGF-β are required for the differentiation of naive T-cell precursors toward the Th17 subset (Veldhoen et al., 2006a). This CD4$^+$ Th-cell subset has recently been identified to produce high levels of proinflammatory IL-17 (Stockinger and Veldhoen, 2007), along with other cytokines such as TNF and IL-22, and is supposed to have a dominant role in the development and maintenance of autoimmune inflammation and the clearance of extracellular pathogens (Veldhoen et al., 2006b). In addition, IL-23 has a small effect on the induction of IFN-γ, in particular, in the absence of IL-12(p70), and can partially compensate for IL-12p35 deficiency to stimulate a Th1 response (Khader et al., 2005; Wozniak et al., 2006).

While the function of IL-12(p70) and IL-23 is linked with the Th1 or Th17 effector cell subset, macrophage- and DC-derived IL-27 seems to have broad pro- as well as anti-inflammatory effects on T-cell function (Villarino et al., 2003). IL-27 has been characterized initially as a proinflammatory cytokine based on its role in the development of Th1-cell responses. In the presence of IL-12 and IL-18, IL-27 enhances IFN-γ production and proliferation of naive T cells (Pflanz et al., 2002). There is also compelling evidence that IL-27 can antagonize both adaptive T-cell responses and has been suggested to have immunosuppressive properties (Villarino et al., 2004). For example, IL-27 seems to have a direct inhibitory effect on the generation of Th2-cell responses that is independent of its ability to enhance IFN-γ production, e.g., by suppressing the production of IL-2 by CD4$^+$ T cells (Owaki et al., 2006). Furthermore, IL-27 inhibits IL-17 production, independent of the IFN-γ-mediated suppression of IL-17 (Stumhofer et al., 2006). In summary, macrophages and DCs seem to play an important role in limiting the intensity and duration of adaptive immune responses by the secretion of IL-27.

IL-1, IL-6, and TNF: Pleiotropic Proinflammatory Cytokines in the Innate Immune Response

Upon the innate recognition of pathogens or their products, macrophages and DCs produce a collection of potent pleiotropic proinflammatory cytokines. These include the cytokines IL-1, IL-6, and TNF, which all have a broad range of activities. They play important roles in immunity and inflammation, to a lesser extent in the control of cell proliferation and differentiation (and apoptosis in the case of TNF), and these cytokines may also contribute to host damage. Principle producers of IL-1β, IL-6, and TNF are macrophages and DCs following stimulation via TLRs, but these proinflammatory cytokines are also secreted by many other cell types, including monocytes, lymphocytes, NK cells, and fibroblasts.

IL-1, a Prototypic Proinflammatory Cytokine

IL-1 is a major highly proinflammatory cytokine secreted by macrophages and DCs early upon infection and inflammation (Dinarello, 2004). IL-1 is composed of two distinct proteins, IL-1α and IL-1β, that are expressed together with a third member of the IL-1 superfamily, the naturally occurring IL-1 receptor antagonist IL-1RA, which competes with the IL-1 agonists, thereby tightly controlling the production and activity exerted by IL-1 (Dinarello, 1997).

Originally termed pyrogen-inducing factor, IL-1 activates the hypothalamus thermoregulatory center during infection, leading to fever (Netea et al., 2000). IL-1 stimulates the production of other proinflammatory cytokines, in particular, IL-6 and the entire chemokine family, which explains the rather promiscuous and apparently nonspecific nature of its many properties. In fact, IL-1 affects nearly every tissue and organ system, often in concert with TNF and IL-6, which links IL-1 to the release of hepatic acute-phase proteins, peripheral neutrophilia, and thrombocytosis (Dinarello, 1996). IL-1 augments lymphocyte responses and increases the expression of adhesion factors on endothelial cells to enable transmigration of leukocytes. IL-4, IL-13, IL-10, TGF-β, and surprisingly IL-6 can suppress the IL-1 gene expression and synthesis, suggesting that an antiinflammatory cytokine network regulates the disease outcome by affecting the production or activity of IL-1 (Tilg et al., 1994; Vannier et al., 1992). Although the highly proinflammatory cytokine IL-1 contributes to the pathogenesis of many diseases (Dinarello, 2005; Pascual et al., 2005), a small and tightly controlled amount appears to be required to combat infection successfully.

IL-6: Functions in Innate and Adaptive Immune Responses

The proinflammatory cytokine IL-6 is secreted by macrophages and DCs following stimulation via TLRs, but can also be produced by T and B cells and nonlymphoid cells such as fibroblasts, keratinocytes, endothelial cells, and several tumor cells (Heinrich et al., 2003; Kishimoto, 2005). IL-6 was discovered as a B-cell differentiation factor stimulating Ig secretion by activated B cells and causing their terminal differentiation into plasma cells (Kishimoto, 2005; Muraguchi et al., 1988). IL-6 is a typical example of a pleiotropic and pluripotent cytokine. Unlike TNF and IL-1β, IL-6 does not upregulate major inflammatory mediators such as prostaglandins, nitric oxide, or matrix metalloproteinases, but rather induces the expression of various acute-phase genes (Castell et al., 1988) and shows a broad range of activities, affecting and regulating immune functions. Furthermore, IL-6 has a crucial role in mediating hematopoiesis, liver and neuronal regeneration, embryonal development, and fertility. Hence, deregulated levels of IL-6 are associated with various disease conditions, such as inflammatory, autoimmune, and malignant diseases (Naka et al., 2002; Nishimoto and Kishimoto, 2006).

In addition to its broad functions as a proinflammatory cytokine, IL-6 has recently been identified as an essential factor for the Th17 lineage development (Veldhoen and Stockinger, 2006). In macrophages, IL-6 can induce terminal differentiation (Shabo et al., 1988). IL-6 induces T-cell proliferation and cytotoxic-T-cell differentiation by augmenting the expression of the IL-2 receptor and the production of IL-2, while maintaining the cytolytic function of T cells (Le et al., 1988; Lotz et al., 1988). The production of IL-6 by DCs has been found to be critical to overcome the suppressive effects of $CD4^+CD25^+$ Treg cells (Pasare and Medzhitov, 2003). Also, the cytolytic capability of NK cells can be enhanced by IL-6 (Luger et al., 1989). However, IL-6 has also been reported to exert negative immunoregulatory functions. For example, IL-6 can inhibit the synthesis of TLR-induced IL-1 and TNF (Aderka et al., 1989, 1993).

TNF Is a Double-Edged Sword: Controller of Infection and Promoter of Host Damage

The principal producers of proinflammatory TNF are macrophages and DCs. B and T lymphocytes, eosinophils, and NK cells can also secrete TNF (Tracey and Cerami, 1994). In macrophages, the NF-κB (Kanters et al., 2003) and the MAPK TPL2/ERK (Dumitru et al., 2000) signaling pathways have been shown to be critical for the production of TNF. In vivo, TNF is produced within minutes after encountering an inflammatory stimulus as part of the immediate early-response genes (Barnes and Karin, 1997).

Binding of TNF to its receptor causes the activation of two major transcription factors, AP-1 and NF-κB, which in turn induce genes generally involved in chronic and acute inflammatory responses (Baud and Karin, 2001). Hence, TNF elicits a particular broad spectrum of cellular and organismal responses (Pfeffer, 2003). One principle physiological function is to stimulate the recruitment of monocytes and neutrophils to sites of infection and to activate these cells to eradicate microbes. TNF mediates these effects by inducing the expression of adhesion molecules and by stimulating the expression of cytokines and chemokines in macrophages, DCs, and endothelial cells. In a broad range of cell types, TNF is a key mediator of cell activation, regulates MHC-I and -II expression, and induces secondary inflammatory mediators. TNF synergizes with IFN-γ for the optimal activation of mycobacterial killing by macrophages (Britton et al., 1998). TNF alone is suggested to be insufficient to activate microbial killing by macrophages, but stimulates phagocytosis in macrophages in synergy with IFN-γ to stimulate maximal induction of microbicidal effector molecules.

In addition to its proinflammatory and costimulatory effects, exposure of cells to TNF can result in activation of a caspase cascade leading to apoptosis (Tartaglia et al., 1991). Surprisingly, TNF has been found to suppress the production of IL-10 independently of other cytokines (Ma et al., 2000) and can, in some contexts, play a regulatory role against an uncontrolled Th1 response (Zganiacz et al., 2004). These findings may suggest a context-specific inhibitory role for TNF, in addition to its well accepted proinflammatory properties (Ma, 2001). Furthermore, TNF has recently be shown to enhance Th17 lineage development (Veldhoen et al., 2006a).

TNF is a pleiotropic cytokine produced by many cells, including macrophages and DCs, that is critical for the eradication of intracellular pathogens (Barnes and Karin, 1997; Tracey and Cerami, 1994) by its proinflammatory effects. However, as with many cytokines, findings with TNF or its antagonists demonstrate the double-edged sword of the immune response. For example, blockade of TNF activity has been shown to ameliorate the symptoms of rheumatoid arthritis and Crohn's disease (Feldmann et al., 2005) and yet reactivates tuberculosis in patients previously exposed to the pathogen *Mycobacterium tuberculosis* (Keane, 2005).

Cytokines Produced by Cells of the Adaptive Immune Response Influence the Function of Macrophages and DCs

Many of the cytokines produced by macrophages and DCs operate in an autocrine fashion, as previously discussed. Macrophage and DC function, including cytokine expression, is also critically regulated by cytokines that are produced by other cells of the innate and adaptive immune response, most importantly by NK cells and T cells. Effector $CD4^+$ T cells, initially polarized by macrophage- and DC-derived cytokines, produce hallmark cytokines such as IFN-γ or IL-4 and IL-13, which themselves directly feedback to regulate macrophage and DC function. In addition, the T-cell-derived immunosuppressive cytokine IL-10 inhibits the function of macrophages and DCs, including their production of cytokines. These positive and negative feedback loops from the adaptive to the innate immune system are critical to orchestrate effective immune responses with minimal damage to the host.

IFN-γ, the Hallmark Th1 Response Cytokine: Microbicidal and Regulatory Functions

IFN-γ, also referred to as immune or type II IFN, is one of the most important stimuli that a macrophage responds to (Boehm et al., 1997; Schroder et al., 2004). The historical name "macrophage activation factor" emphasizes the role of IFN-γ as the principle cytokine to activate microbicidal effector functions in macrophages (Nathan et al., 1984). In addition to its role in macrophage activation, IFN-γ coordinates a diverse array of cellular programs involved in many aspects of innate and adaptive immune responses (Boehm et al., 1997). Therefore, IFN-γ is an essential cytokine for the resistance to many intracellular pathogens.

During the innate immune response, IFN-γ is produced mainly by NK cells in response to IL-12 and other proinflammatory cytokines (Pien et al., 2000; Trinchieri, 1997). In the adaptive immune response, IFN-γ as the hallmark cytokine of Th1 responses is predominantly produced by Th1 cells, but also by $CD8^+$ cytotoxic T cells (Frucht et al., 2001). The production of IFN-γ in T cells is initiated and controlled by macrophage- and DC-derived cytokines such as IL-12 and IL-18, which serve as a bridge to link infection with IFN-γ production and the induction of specific immunity (Trinchieri, 1997, 2003). IL-12-induced IFN-γ production by T cells participates in positive feedback by further promoting IL-12 production in macrophages. This positive feedback loop to amplify IFN-γ is probably important for the stabilization of Th1 responses (Trinchieri, 1997; Boehm et al., 1997). T-cell-derived IFN-γ thus further skews the adaptive immune response toward a Th1 phenotype and promotes characteristic Th1 effector functions, a main part of which is the activation of macrophages and DCs to enhance the response against intracellular pathogens (O'Garra and Vieira, 2004). The IFN-γ response is itself regulated by interaction with responses to other cytokines: proinflammatory macrophage- and DC-derived cytokines such as type I IFNs and TNF synergize with the effects of IFN-γ (Ohmori et al., 1997), while negative regulators such as IL-10 and TGF-β, as well as Th2-cell-derived IL-4, can antagonize the functions of IFN-γ (Schindler et al., 2001). On the other hand, IL-4 can enhance IL-12 production by DCs under different conditions (Hochrein et al., 2001) via a mechanism involving the downregulation of IL-10 (Yao et al., 2005).

In addition to directing the differentiation of naive $CD4^+$ T cells to the Th1 subset while inhibiting Th2 cells, the many diverse immunoregulatory properties of IFN-γ include orchestrating leukocyte-endothelium interactions (Ebnet et al., 1996), regulation of B-cell functions such as Ig production and isotype switching to IgG2a (Finkelman et al., 1988), and activation of neutrophils and stimulation of cytotoxic activity of NK cells (Carnaud et al., 1999). However, the most important function of IFN-γ for cell-mediated immunity to infection with intracellular pathogens is probably the activation of microbicidal effector functions on macrophages, also referred to as "classical" macrophage activation (Boehm et al., 1997; Schroder et al., 2004).

IFN-γ-stimulated macrophages display increased pinocytosis and receptor-mediated phagocytosis and exert enhanced microbial killing ability through induction of several antimicrobial effector mechanisms (Kaufmann, 1993). These include the production of reactive oxygen and nitrogen intermediates via induction of the NADPH-dependent phagocyte oxidase ("respiratory burst") and inducible nitric oxide synthase 2 (Bogdan, 2000; MacMicking et al., 1997; Nathan, 1997), tryptophan depletion to starve microbes and parasites (Pfefferkorn, 1984), and upregulation of lysosomal enzymes to promote microbe destruction (Lah et al., 1995). Macrophage activation can also lead to downregulation of the degradative capacity of the phagosome (Yates et al., 2007).

Similar to type I IFNs, IFN-γ upregulates multiple functions within the class I antigen presentation pathway to increase the quantity, quality, and repertoire of peptides presented to $CD8^+$ T cells in the context of MHC-I (Stark et al., 1998). This serves to increase immune surveillance and to promote the induction of cell-mediated immunity. However, of the IFNs, only IFN-γ can efficiently upregulate the class II antigen-processing and presentation pathway in macrophages, DCs, and B cells, thus promoting antigen-specific activation of $CD4^+$ T cells (Mach et al., 1996).

Effects of the Th2 Cytokines IL-4 and IL-13 on Macrophage and DC Function

IL-4 is the key cytokine to induce the differentiation of naive $CD4^+$ T cells into Th2 cells (Nelms et al., 1999), which are important for the eradication of extracellular parasites, such as helminths. Th2 responses have however also been implicated in atopic and allergic manifestations (Romagnani, 1994; Sher and Coffman, 1992). IL-4 is produced by many different cells, including T cells, eosinophils, and mast cells. The main source that initially produces IL-4 in the body to direct an immune response is determined by the nature of the challenge to the host, which can range from parasitic infection to allergic challenge (Nelms et al., 1999). Upon activation by IL-4, Th2 cells subsequently produce additional IL-4, which induces STAT6 signaling and the expression of the Th2-specific transcription factor GATA-3 critical for the differentiation of Th2 cells (O'Garra and Arai, 2000). IL-4 stimulates B-cell proliferation and class switching to IgE and thus is a key regulator in humoral immunity and allergic responses (Romagnani, 1994; Sher and Coffman, 1992; Tangye et al., 2002). IL-13, produced by Th2 cells but also by other cell types, shares several structural characteristics with IL-4, whereby its functions overlap considerably but not completely with those of IL-4 (Wynn, 2003).

Key cytokines of Th2 immune responses also exert their function on macrophages and thus provide an important link between adaptive and innate immunity. The effects of IL-4 and IL-13, produced by Th2 cells, on macrophages include a moderate inhibition of a range of antigen-presenting and effector functions (Doyle et al., 1994). On account of this, Th2 cytokines were initially grouped together with other immunosuppressive cytokines such as IL-10. However, IL-4 and IL-13 have been demonstrated to induce cell surface and other phenotypic changes in macrophages that are distinct from those induced by other proinflammatory or anti-inflammatory cytokines (Montaner et al., 1999). Hence, in analogy to the well-characterized "classical" activation of macrophages by the hallmark Th1 cytokine IFN-γ, the term "alternative" activation refers to those modifications of macrophage function induced by the Th2 cytokines IL-4 and IL-13 (Goerdt and Orfanos, 1999; Gordon, 2003).

IL-4 and IL-13 have broadly similar effects on macrophages and other target cells because they share a common receptor chain (Chomarat and Banchereau, 1998). Both cytokines antagonize the actions of IFN-γ and inhibit the production of proinflammatory cytokines such as TNF (de Waal Malefyt et al., 1993). IL-4 and IL-13 moderately suppress intracellular microbicidal effector functions of macrophages such as the respiratory burst. However, IL-4 and IL-13 do not act solely as immunosuppressive cytokines for macrophage function. Rather, their apparent inhibitory function is the result of reprogramming the response of macrophages that most importantly comprises the induction of selective chemokines and the upregulation of the mannose receptor and MHC-II molecules (Stein et al., 1992). IL-4 and IL-13 also induce intracellular enzymes that are implicated in cell recruitment and granuloma formation (Cheever et al., 1994). Therefore, the distinctive macrophage phenotype induced by the Th2-type cytokines IL-4 and IL-13, also referred to as alternatively activated macrophages, plays an important role in dampening classical cell-mediated immune responses while reprogramming the response toward an appropriate type that promotes an efficient eradication of extracellular pathogens. Alternatively activated macrophages have been shown to be induced by nematode infection and to inhibit immune cell proliferation via cell-to-cell contact, suggesting a role in immunoevasion (Loke et al., 2000). However, this type of macrophage activation is particularly important for humoral immunity and repair, and may also be implicated in inflammatory and allergic conditions such as asthma, atherosclerosis, and probably tumor immunity (Gordon, 2003).

Immunosuppressive Cytokines Have Major Effects on Macrophages and DCs

So far, cytokines have been described that directly stimulate innate immunity and/or positively induce and regulate the development of adaptive immune responses. In contrast, the immunosuppressive cytokines IL-10 and TGF-β are potent inhibitors of immune responses, in particular, responses involving macrophages. IL-10 and TGF-β balance immune responses to inhibit various types of inflammatory pathologies by limiting inflammatory responses.

Suppressive Effects of IL-10 on Macrophages and DCs

IL-10 is a potent immunosuppressive cytokine that limits immune responses to prevent damage to the host (Gazzinelli et al., 1996; Moore et al., 2001). Mice deficient in IL-10 develop enterocolitis, but this depends on the microbial colonization of their gut (Kuhn et al., 1993; Moore et al., 2001). Although infections with intracellular pathogens may be resolved more efficiently in IL-10-deficient mice, this can also lead to severe damage to the host (Gazzinelli et al., 1996; Moore et al., 2001; O'Garra and Vieira, 2004). The broad anti-inflammatory properties of IL-10 result to a great extent from its ability to inhibit the function of macrophages and DCs, including their production of proinflammatory cytokines and expression of MHC-II and costimulatory molecules (Fiorentino et al., 1991; Ding et al., 1993). In contrast to its major immunosuppressive effects on phagocytes and most other hemopoietic cells, IL-10 can paradoxically also positively regulate certain immune responses, which includes the induction of growth and differentiation of $CD8^+$ T cells, stimulation of mast cells, and enhancement of IgA responses by activating B cells, suggesting a potential role in mucosal immunity as well as immunoregulation (Moore et al., 2001).

The secretion of proinflammatory cytokines such as IL-1, IL-6, TNF, and IL-12 by macrophages and DCs upon stimulation by pathogens or pathogen-derived products is accompanied by the release of immunosuppressive IL-10, which potently limits the production of proinflammatory cytokines in an autocrine fashion (Boonstra et al., 2006; de Waal Malefyt et al., 1991; Dillon et al., 2004; Fiorentino et al., 1991; Rogers et al., 2005). It is noteworthy that macrophages and DC subsets differ profoundly in their capacity to produce IL-10 (Boonstra et al., 2006). Additional producers of IL-10 are B cells (O'Garra et al., 1992), mast cells and eosinophils (Moore et al., 2001), and most importantly many T-cell subsets, including Th1, Th2, Treg ($Foxp3^+$), and $CD8^+$ T cells (O'Garra and Robinson, 2004). It is likely that IL-10 from different cellular sources may play a role to control immune responses to different pathogens at different stages of infection and at different anatomical locations. In this context, Th1-cell-derived IL-10 may be an important feedback regulator, inhibiting macrophages and DCs from further production of proinflammatory cytokines and continued effector T-cell responses, thus preventing pathologies associated with exaggerated inflammation (O'Garra and Vieira, 2007). The molecular mechanisms that regulate the expression of IL-10 in these diverse cell types remain poorly understood.

The major immunoregulatory function of IL-10 is the suppression of the production of proinflammatory cytokines by macrophages and DCs, including TNF, IL-1, IL-6, IL-12, IL-18, as well as most inducible chemokines involved in inflammation. Furthermore, IL-10 can enhance the expression of natural cytokine antagonists such as IL-1RA (Jenkins et al., 1994). The inhibition of TNF production by IL-10 is particularly important to prevent the development of colitis as a result of deregulated TNF levels, and establishes IL-10 as an important regulator of mucosal immunity (Kontoyiannis et al., 2001; Powrie et al., 1994). By inhibiting the production of IL-12 in macrophages and DCs, IL-10 subsequently suppresses Th1 responses and reduces IFN-γ levels, which in synergy with TNF are critical for effective cell-mediated immune reactions against intracellular microbes (Fiorentino et al., 1991; Flesch and Kaufmann, 1990). The inhibition of MHC-II expression (de Waal Malefyt et al., 1991; Koppelman et al., 1997) and costimulatory molecules (Ding and Shevach, 1992) on macrophages and DCs by IL-10 generally results in an inhibition of T-cell activation and serves to ultimately regulate T-cell responses. In keeping with this, IL-10 not only keeps in check Th1 responses critical for the eradication of intracellular pathogens (O'Garra and Vieira, 2004) but has also been shown to suppress Th2-cell responses (Hawrylowicz and O'Garra, 2005).

TGF-β Is a Context-Dependent Inflammatory or Suppressive Cytokine

TGF-β is another immunosuppressive cytokine mainly secreted by macrophages and DCs, but also by T cells, in which its expression is often accompanied by the production of immunosuppressive IL-10 (Li et al., 2006; Wahl, 1994).

TGF-β is a pleiotropic, multifunctional cytokine and acts on a wide variety of cell types. TGF-β functions to inhibit immune and inflammatory responses, largely by counteracting the effects of proinflammatory cytokines (Letterio and Roberts, 1998). Furthermore, it controls the differentiation, proliferation, and state of activation in all immune cells, including macrophages, DCs, and T cells. Interestingly, the action of TGF-β is context dependent, i.e., it is not only dependent on the cell type and its state of differentiation, but also on the total milieu of cytokines present (Sporn and Roberts, 1992) and the anatomical location (Veldhoen and Stockinger, 2006). This is in keeping with previous observations that, for example, TGF-β generally inhibits T-cell development, differentiation, and proliferation (Kehrl et al., 1986), but paradoxically can also enhance the growth of naive T cells (Cerwenka et al., 1994) and promote T-cell expansion by inhibiting T-cell apoptosis (Zhang et al., 1995). Depending on whether macrophages or DCs are induced to produce IL-6 in the context of TGF-β will determine whether a Th17 or regulatory T-cell response (Foxp3$^+$ Treg) develops. This may result from distinct signals presumably initiated by microbial products (Bettelli et al., 2006; Veldhoen et al., 2006a, 2006b; Weaver et al., 2006).

In general, the actions of TGF-β on tissue macrophages are suppressive and contribute to the resolution of inflammatory responses. Similar to the immunosuppressive effects of IL-4 and IL-10 on macrophages, TGF-β limits the production of proinflammatory cytokines and increases the expression of the natural antagonist for IL-1, IL-1RA (Turner et al., 1991). Possibly the most important deactivating effect of TGF-β on macrophages is its ability to decrease the production of cytotoxic reactive oxygen and nitrogen intermediates and to suppress the oxidative respiratory burst capacity (Assoian et al., 1987).

Similar to the previously described pleiotropic effects of TGF-β on lymphocytes, TGF-β can also act as a stimulatory modulator on macrophages depending on their differentiation status and tissue origin (Fan et al., 1992). In this respect, the stimulation of IL-1 and TNF production via TGF-β and the enhancement of the expression of adhesion receptors in peripheral blood monocytes, promoting their subsequent transmigration into tissues, can serve as a paradigm for the pleiotropic nature of TGF-β (Wahl et al., 1987, 1993). However, the immunoregulatory capacity of TGF-β to initially enhance the induction of an immune response is often accompanied by the increased expression of TGF-β itself, and its context-dependent activity then often serves to dampen the response or inhibit the cell populations once they have become activated.

CONCLUDING REMARKS

In this chapter we describe macrophages and DCs as primary effector cells as well as regulators of innate and adaptive immune responses, largely dictated by the cytokines that they produce. Recognition of microorganisms or their products initiates complex signaling pathways in these cells that determine the pattern of cytokines produced and the subsequent effector response, critical for the eradication of pathogens with minimum damage to the host.

REFERENCES

Aderka, D., J. M. Le, and J. Vilcek. 1989. IL-6 inhibits lipopolysaccharide-induced tumor necrosis factor production in cultured human monocytes, U937 cells, and in mice. *J. Immunol.* **143:**3517–3523.

Aderka, D., Y. Maor, D. Novick, H. Engelmann, Y. Kahn, Y. Levo, D. Wallach, and M. Revel. 1993. Interleukin-6 inhibits the proliferation of B-chronic lymphocytic leukemia cells that is induced by tumor necrosis factor-alpha or -beta. *Blood* **81:**2076–2084.

Agnello, D., C. S. Lankford, J. Bream, A. Morinobu, M. Gadina, J. J. O'Shea, and D. M. Frucht. 2003. Cytokines and transcription factors that regulate T helper cell differentiation: new players and new insights. *J. Clin. Immunol.* **23:**147–161.

Agrawal, S., A. Agrawal, B. Doughty, A. Gerwitz, J. Blenis, T. Van Dyke, and B. Pulendran. 2003. Cutting edge: different Toll-like receptor agonists instruct dendritic cells to induce distinct Th responses via differential modulation of extracellular signal-regulated kinase-mitogen-activated protein kinase and c-Fos. *J. Immunol.* **171:**4984–4989.

Akira, S., S. Uematsu, and O. Takeuchi. 2006. Pathogen recognition and innate immunity. *Cell* **124:**783–801.

Assoian, R. K., B. E. Fleurdelys, H. C. Stevenson, P. J. Miller, D. K. Madtes, E. W. Raines, R. Ross, and M. B. Sporn. 1987. Expression and secretion of type beta transforming growth factor by activated human macrophages. *Proc. Natl. Acad. Sci. USA* **84:**6020–6024.

Aste-Amezaga, M., X. Ma, A. Sartori, and G. Trinchieri. 1998. Molecular mechanisms of the induction of IL-12 and its inhibition by IL-10. *J. Immunol.* **160:**5936–5944.

Auerbuch, V., D. G. Brockstedt, N. Meyer-Morse, M. O'Riordan, and D. A. Portnoy. 2004. Mice lacking the type I interferon receptor are resistant to Listeria monocytogenes. *J. Exp. Med.* **200:**527–533.

Banchereau, J., F. Briere, C. Caux, J. Davoust, S. Lebecque, Y. J. Liu, B. Pulendran, and K. Palucka. 2000. Immunobiology of dendritic cells. *Annu. Rev. Immunol.* **18:**767–811.

Banchereau, J., and R. M. Steinman. 1998. Dendritic cells and the control of immunity. *Nature* **392:**245–252.

Barnes, P. J., and M. Karin. 1997. Nuclear factor-kappaB: a pivotal transcription factor in chronic inflammatory diseases. *N. Engl. J. Med.* **336:**1066–1071.

Baud, V., and M. Karin. 2001. Signal transduction by tumor necrosis factor and its relatives. *Trends Cell Biol.* **11:**372–377.

Bauer, M., V. Redecke, J. W. Ellwart, B. Scherer, J. P. Kremer, H. Wagner, and G. B. Lipford. 2001. Bacterial CpG-DNA triggers activation and maturation of human CD11c-, CD123+ dendritic cells. *J. Immunol.* **166:**5000–5007.

Beinke, S., and S. C. Ley. 2004. Functions of NF-kappaB1 and NF-kappaB2 in immune cell biology. *Biochem. J.* **382:**393–409.

Bettelli, E., Y. Carrier, W. Gao, T. Korn, T. B. Strom, M. Oukka, H. L. Weiner, and V. K. Kuchroo. 2006. Reciprocal developmental pathways for the generation of pathogenic effector TH17 and regulatory T cells. *Nature* **441:**235–238.

Beutler, B. 2005. The Toll-like receptors: analysis by forward genetic methods. *Immunogenetics* **57:**385–392.

Biron, C. A. 2001. Interferons alpha and beta as immune regulators—a new look. *Immunity* **14:**661–664.

Biron, C. A., K. B. Nguyen, G. C. Pien, L. P. Cousens, and T. P. Salazar-Mather. 1999. Natural killer cells in antiviral defense: function and regulation by innate cytokines. *Annu. Rev. Immunol.* **17:**189–220.

Boehm, U., T. Klamp, M. Groot, and J. C. Howard. 1997. Cellular responses to interferon-gamma. *Annu. Rev. Immunol.* **15:**749–795.

Bogdan, C. 2000. The function of type I interferons in antimicrobial immunity. *Curr. Opin. Immunol.* **12:**419–424.

Bogdan, C., J. Mattner, and U. Schleicher. 2004. The role of type I interferons in non-viral infections. *Immunol. Rev.* **202:**33–48.

Bogdan, C., M. Rollinghoff, and A. Diefenbach. 2000. Reactive oxygen and reactive nitrogen intermediates in innate and specific immunity. *Curr. Opin. Immunol.* **12:**64–76.

Bonizzi, G., and M. Karin. 2004. The two NF-kappaB activation pathways and their role in innate and adaptive immunity. *Trends Immunol.* **25:**280–288.

Boonstra, A., C. Asselin-Paturel, M. Gilliet, C. Crain, G. Trinchieri, Y. J. Liu, and A. O'Garra. 2003. Flexibility of mouse classical and plasmacytoid-derived dendritic cells in directing T helper type 1 and 2 cell development: dependency on antigen dose and differential toll-like receptor ligation. *J. Exp. Med.* **197:**101–109.

Boonstra, A., R. Rajsbaum, M. Holman, R. Marques, C. Asselin-Paturel, J. P. Pereira, E. E. Bates, S. Akira, P. Vieira, Y. J. Liu, G. Trinchieri, and A. O'Garra. 2006. Macrophages and myeloid dendritic cells, but not plasmacytoid dendritic cells, produce IL-10 in response to MyD88- and TRIF-dependent TLR signals, and TLR-independent signals. *J. Immunol.* **177:**7551–7558.

Britton, W. J., N. Meadows, D. A. Rathjen, D. R. Roach, and H. Briscoe. 1998. A tumor necrosis factor mimetic peptide activates a murine macrophage cell line to inhibit mycobacterial growth in a nitric oxide-dependent fashion. *Infect. Immun.* **66:**2122–2127.

Brown, G. D. 2006. Dectin-1: a signalling non-TLR pattern-recognition receptor. *Nat. Rev. Immunol.* **6:**33–43.

Carnaud, C., D. Lee, O. Donnars, S. H. Park, A. Beavis, Y. Koezuka, and A. Bendelac. 1999. Cutting edge: cross-talk between cells of the innate immune system: NKT cells rapidly activate NK cells. *J. Immunol.* **163:**4647–4650.

Carrero, J. A., B. Calderon, and E. R. Unanue. 2004. Type I interferon sensitizes lymphocytes to apoptosis and reduces resistance to Listeria infection. *J. Exp. Med.* **200:**535–540.

Carrero, J. A., B. Calderon, and E. R. Unanue. 2006. Lymphocytes are detrimental during the early innate immune response against Listeria monocytogenes. *J. Exp. Med.* **203:**933–940.

Castell, J. V., M. J. Gomez-Lechon, M. David, T. Hirano, T. Kishimoto, and P. C. Heinrich. 1988. Recombinant human interleukin-6 (IL-6/BSF-2/HSF) regulates the synthesis of acute phase proteins in human hepatocytes. *FEBS Lett.* **232:**347–350.

Cella, M., M. Salio, Y. Sakakibara, H. Langen, I. Julkunen, and A. Lanzavecchia. 1999. Maturation, activation, and protection of dendritic cells induced by double-stranded RNA. *J. Exp. Med.* **189:**821–829.

Cella, M., D. Scheidegger, K. Palmer-Lehmann, P. Lane, A. Lanzavecchia, and G. Alber. 1996. Ligation of CD40 on dendritic cells triggers production of high levels of interleukin-12 and enhances T cell stimulatory capacity: T-T help via APC activation. *J. Exp. Med.* **184:**747–752.

Cerwenka, A., D. Bevec, O. Majdic, W. Knapp, and W. Holter. 1994. TGF-beta 1 is a potent inducer of human effector T cells. *J. Immunol.* **153:**4367–4377.

Chan, S. H., B. Perussia, J. W. Gupta, M. Kobayashi, M. Pospisil, H. A. Young, S. F. Wolf, D. Young, S. C. Clark, and G. Trinchieri. 1991. Induction of interferon gamma production by natural killer cell stimulatory factor: characterization of the responder cells and synergy with other inducers. *J. Exp. Med.* **173:**869–879.

Cheever, A. W., M. E. Williams, T. A. Wynn, F. D. Finkelman, R. A. Seder, T. M. Cox, S. Hieny, P. Caspar, and A. Sher. 1994. Anti-IL-4 treatment of Schistosoma mansoni-infected mice inhibits development of T cells and non-B, non-T cells expressing Th2 cytokines while decreasing egg-induced hepatic fibrosis. *J. Immunol.* **153:**753–759.

Chen, Z., T. B. Gibson, F. Robinson, L. Silvestro, G. Pearson, B. Xu, A. Wright, C. Vanderbilt, and M. H. Cobb. 2001. MAP kinases. *Chem. Rev.* **101:**2449–2476.

Chomarat, P., and J. Banchereau. 1998. Interleukin-4 and interleukin-13: their similarities and discrepancies. *Int. Rev. Immunol.* **17:**1–52.

Colonna, M., G. Trinchieri, and Y. J. Liu. 2004. Plasmacytoid dendritic cells in immunity. *Nat. Immunol.* **5:**1219–1226.

Conta, B. S., M. B. Powell, and N. H. Ruddle. 1983. Production of lymphotoxin, IFN-gamma and IFN-alpha, beta by murine T cell lines and clones. *J. Immunol.* **130:**2231–2235.

Cousens, L. P., R. Peterson, S. Hsu, A. Dorner, J. D. Altman, R. Ahmed, and C. A. Biron. 1999. Two roads diverged: interferon alpha/beta- and interleukin 12-mediated pathways in promoting T cell interferon gamma responses during viral infection. *J. Exp. Med.* **189:**1315–1328.

Creagh, E. M., and L. A. O'Neill. 2006. TLRs, NLRs and RLRs: a trinity of pathogen sensors that co-operate in innate immunity. *Trends Immunol.* **27:**352–357.

Dalod, M., T. Hamilton, R. Salomon, T. P. Salazar-Mather, S. C. Henry, J. D. Hamilton, and C. A. Biron. 2003. Dendritic cell responses to early murine cytomegalovirus infection: subset functional specialization and differential regulation by interferon alpha/beta. *J. Exp. Med.* **197:**885–898.

Dalod, M., T. P. Salazar-Mather, L. Malmgaard, C. Lewis, C. Asselin-Paturel, F. Briere, G. Trinchieri, and C. A. Biron. 2002. Interferon alpha/beta and interleukin 12 responses to viral infections: pathways regulating dendritic cell cytokine expression in vivo. *J. Exp. Med.* **195:**517–528.

D'Andrea, A., M. Aste-Amezaga, N. M. Valiante, X. Ma, M. Kubin, and G. Trinchieri. 1993. Interleukin 10 (IL-10) inhibits human lymphocyte interferon gamma-production by suppressing natural killer cell stimulatory factor/IL-12 synthesis in accessory cells. *J. Exp. Med.* **178:**1041–1048.

D'Andrea, A., X. Ma, M. Aste-Amezaga, C. Paganin, and G. Trinchieri. 1995. Stimulatory and inhibitory effects of interleukin (IL)-4 and IL-13 on the production of cytokines by human peripheral blood mononuclear cells: priming for IL-12 and tumor necrosis factor alpha production. *J. Exp. Med.* **181:**537–546.

D'Andrea, A., M. Rengaraju, N. M. Valiante, J. Chehimi, M. Kubin, M. Aste, S. H. Chan, M. Kobayashi, D. Young, E. Nickbarg, et al. 1992. Production of natural killer cell stimulatory factor (interleukin 12) by peripheral blood mononuclear cells. *J. Exp. Med.* **176:**1387–1398.

Decker, T., M. Muller, and S. Stockinger. 2005. The yin and yang of type I interferon activity in bacterial infection. *Nat. Rev. Immunol.* **5:**675–687.

de Waal Malefyt, R., J. Abrams, B. Bennett, C. G. Figdor, and J. E. de Vries. 1991. Interleukin 10(IL-10) inhibits cytokine synthesis by human monocytes: an autoregulatory role of IL-10 produced by monocytes. *J. Exp. Med.* **174:**1209–1220.

de Waal Malefyt, R., C. G. Figdor, R. Huijbens, S. Mohan-Peterson, B. Bennett, J. Culpepper, W. Dang, G. Zurawski, and J. E. de Vries. 1993. Effects of IL-13 on phenotype, cytokine production, and cytotoxic function of human monocytes. Comparison with IL-4 and modulation by IFN-gamma or IL-10. *J. Immunol.* **151:**6370–6381.

de Waal Malefyt, R., J. Haanen, H. Spits, M. G. Roncarolo, A. te Velde, C. Figdor, K. Johnson, R. Kastelein, H. Yssel, and J. E. de Vries. 1991. Interleukin 10 (IL-10) and viral IL-10 strongly reduce antigen-specific human T cell proliferation by diminishing the antigen-presenting capacity of monocytes via downregulation of class II major histocompatibility complex expression. *J. Exp. Med.* **174:**915–924.

Diebold, S. S., T. Kaisho, H. Hemmi, S. Akira, and C. Reis e Sousa. 2004. Innate antiviral responses by means of TLR7-mediated recognition of single-stranded RNA. *Science* **303:**1529–1531.

Diebold, S. S., M. Montoya, H. Unger, L. Alexopoulou, P. Roy, L. E. Haswell, A. Al-Shamkhani, R. Flavell, P. Borrow, and C. Reis e Sousa. 2003. Viral infection switches non-plasmacytoid dendritic cells into high interferon producers. *Nature* **424:**324–328.

Dillon, S., A. Agrawal, T. Van Dyke, G. Landreth, L. McCauley, A. Koh, C. Maliszewski, S. Akira, and B. Pulendran. 2004. A Toll-like receptor 2 ligand stimulates Th2 responses in vivo, via induction of extracellular signal-regulated kinase mitogen-activated protein kinase and c-Fos in dendritic cells. *J. Immunol.* **172:**4733–4743.

Dinarello, C. A. 1996. Biologic basis for interleukin-1 in disease. *Blood* **87:**2095–2147.

Dinarello, C. A. 1997. Interleukin-1. *Cytokine Growth Factor Rev.* **8:**253–265.

Dinarello, C. A. 2004. Infection, fever, and exogenous and endogenous pyrogens: some concepts have changed. *J. Endotoxin Res.* **10:**201–222.

Dinarello, C. A. 2005. Blocking IL-1 in systemic inflammation. *J. Exp. Med.* **201:**1355–1359.

Ding, L., P. S. Linsley, L. Y. Huang, R. N. Germain, and E. M. Shevach. 1993. IL-10 inhibits macrophage costimulatory activity by selectively inhibiting the up-regulation of B7 expression. *J. Immunol.* **151:**1224–1234.

Ding, L., and E. M. Shevach. 1992. IL-10 inhibits mitogen-induced T cell proliferation by selectively inhibiting macrophage costimulatory function. *J. Immunol.* **148:**3133–3139.

Dong, C., R. J. Davis, and R. A. Flavell. 2002. MAP kinases in the immune response. *Annu. Rev. Immunol.* **20:**55–72.

Doyle, A. G., G. Herbein, L. J. Montaner, A. J. Minty, D. Caput, P. Ferrara, and S. Gordon. 1994. Interleukin-13 alters the activation state of murine macrophages in vitro: comparison with interleukin-4 and interferon-gamma. *Eur. J. Immunol.* **24:**1441–1445.

Doyle, S., S. Vaidya, R. O'Connell, H. Dadgostar, P. Dempsey, T. Wu, G. Rao, R. Sun, M. Haberland, R. Modlin, and G. Cheng. 2002. IRF3 mediates a TLR3/TLR4-specific antiviral gene program. *Immunity* **17:**251–263.

Du, C., and S. Sriram. 1998. Mechanism of inhibition of LPS-induced IL-12p40 production by IL-10 and TGF-beta in ANA-1 cells. *J. Leukoc. Biol.* **64:**92–97.

Dumitru, C. D., J. D. Ceci, C. Tsatsanis, D. Kontoyiannis, K. Stamatakis, J. H. Lin, C. Patriotis, N. A. Jenkins, N. G. Copeland, G. Kollias, and P. N. Tsichlis. 2000. TNF-alpha induction by LPS is regulated posttranscriptionally via a Tpl2/ERK-dependent pathway. *Cell* **103:**1071–1083.

Ebnet, K., E. P. Kaldjian, A. O. Anderson, and S. Shaw. 1996. Orchestrated information transfer underlying leukocyte endothelial interactions. *Annu. Rev. Immunol.* **14:**155–177.

Eloranta, M. L., and G. V. Alm. 1999. Splenic marginal metallophilic macrophages and marginal zone macrophages are the major interferon-alpha/beta producers in mice upon intravenous challenge with herpes simplex virus. *Scand. J. Immunol.* **49:**391–394.

English, J. M., and M. H. Cobb. 2002. Pharmacological inhibitors of MAPK pathways. *Trends Pharmacol. Sci.* **23:**40–45.

Fan, K., Q. Ruan, L. Sensenbrenner, and B. Chen. 1992. Transforming growth factor-beta 1 bifunctionally regulates murine macrophage proliferation. *Blood* **79:**1679–1685.

Farrar, J. D., and K. M. Murphy. 2000. Type I interferons and T helper development. *Immunol. Today* **21:**484–489.

Feldmann, M., F. M. Brennan, B. M. Foxwell, P. C. Taylor, R. O. Williams, and R. N. Maini. 2005. Anti-TNF therapy: where have we got to in 2005? *J. Autoimmun.* **25**(Suppl.):26–28.

Finkelman, F. D., I. M. Katona, T. R. Mosmann, and R. L. Coffman. 1988. IFN-gamma regulates the isotypes of Ig secreted during in vivo humoral immune responses. *J. Immunol.* **140:**1022–1027.

Fiorentino, D. F., A. Zlotnik, T. R. Mosmann, M. Howard, and A. O'Garra. 1991. IL-10 inhibits cytokine production by activated macrophages. *J. Immunol.* **147:**3815–3822.

Fitzgerald, K. A., S. M. McWhirter, K. L. Faia, D. C. Rowe, E. Latz, D. T. Golenbock, A. J. Coyle, S. M. Liao, and T. Maniatis. 2003. IKKepsilon and TBK1 are essential components of the IRF3 signaling pathway. *Nat. Immunol.* **4:**491–496.

Fitzgerald, K. A., E. M. Palsson-McDermott, A. G. Bowie, C. A. Jefferies, A. S. Mansell, G. Brady, E. Brint, A. Dunne, P. Gray, M. T. Harte, D. McMurray, D. E. Smith, J. E. Sims, T. A. Bird, and L. A. O'Neill. 2001. Mal (MyD88-adapter-like) is required for Toll-like receptor-4 signal transduction. *Nature* **413:**78–83.

Fleit, H. B., and M. Rabinovitch. 1981. Production of interferon by in vitro derived bone marrow macrophages. *Cell. Immunol.* **57:**495–504.

Flesch, I. E., and S. H. Kaufmann. 1990. Activation of tuberculostatic macrophage functions by gamma interferon, interleukin-4, and tumor necrosis factor. *Infect. Immun.* **58:**2675–2677.

Fraser, I. P., H. Koziel, and R. A. Ezekowitz. 1998. The serum mannose-binding protein and the macrophage mannose receptor are pattern recognition molecules that link innate and adaptive immunity. *Semin. Immunol.* **10:**363–372.

Frucht, D. M., T. Fukao, C. Bogdan, H. Schindler, J. J. O'Shea, and S. Koyasu. 2001. IFN-gamma production by antigen-presenting cells: mechanisms emerge. *Trends Immunol.* **22:**556–560.

Fukao, T., M. Tanabe, Y. Terauchi, T. Ota, S. Matsuda, T. Asano, T. Kadowaki, T. Takeuchi, and S. Koyasu. 2002. PI3K-mediated negative feedback regulation of IL-12 production in DCs. *Nat. Immunol.* **3:**875–881.

Gazzinelli, R. T., M. Wysocka, S. Hieny, T. Scharton-Kersten, A. Cheever, R. Kuhn, W. Muller, G. Trinchieri, and A. Sher. 1996. In the absence of endogenous IL-10, mice acutely infected with Toxoplasma gondii succumb to a lethal immune response dependent on CD4+ T cells and accompanied by overproduction of IL-12, IFN-gamma and TNF-alpha. *J. Immunol.* **157:**798–805.

Geijtenbeek, T. B., S. J. Van Vliet, E. A. Koppel, M. Sanchez-Hernandez, C. M. Vandenbroucke-Grauls, B. Appelmelk, and Y. Van Kooyk. 2003. Mycobacteria target DC-SIGN to suppress dendritic cell function. *J. Exp. Med.* **197:**7–17.

Glimcher, L. H., and K. M. Murphy. 2000. Lineage commitment in the immune system: the T helper lymphocyte grows up. *Genes Dev.* **14:**1693–1711.

Goerdt, S., and C. E. Orfanos. 1999. Other functions, other genes: alternative activation of antigen-presenting cells. *Immunity* **10:**137–142.

Goodbourn, S., L. Didcock, and R. E. Randall. 2000. Interferons: cell signalling, immune modulation, antiviral response and virus countermeasures. *J. Gen. Virol.* **81:**2341–2364.

Gordon, S. 1998. The role of the macrophage in immune regulation. *Res. Immunol.* **149:**685–688.

Gordon, S. 2002. Pattern recognition receptors: doubling up for the innate immune response. *Cell* **111:**927–930.

Gordon, S. 2003. Alternative activation of macrophages. *Nat. Rev. Immunol.* **3:**23–35.

Gordon, S., S. Clarke, D. Greaves, and A. Doyle. 1995. Molecular immunobiology of macrophages: recent progress. *Curr. Opin. Immunol.* **7:**24–33.

Gordon, S., and P. R. Taylor. 2005. Monocyte and macrophage heterogeneity. *Nat. Rev. Immunol.* **5:**953–964.

Grigoriadis, G., Y. Zhan, R. J. Grumont, D. Metcalf, E. Handman, C. Cheers, and S. Gerondakis. 1996. The Rel subunit of NF-kappaB-like transcription factors is a positive and negative regulator of macrophage gene expression: distinct roles for Rel in different macrophage populations. *EMBO J.* **15:**7099–7107.

Guidotti, L. G., and F. V. Chisari. 2001. Noncytolytic control of viral infections by the innate and adaptive immune response. *Annu. Rev. Immunol.* **19:**65–91.

Hacker, H., H. Mischak, G. Hacker, S. Eser, N. Prenzel, A. Ullrich, and H. Wagner. 1999. Cell type-specific activation of mitogen-activated protein kinases by CpG-DNA controls interleukin-12 release from antigen-presenting cells. *EMBO J.* **18:**6973–6982.

Hawrylowicz, C. M., and A. O'Garra. 2005. Potential role of interleukin-10-secreting regulatory T cells in allergy and asthma. *Nat. Rev. Immunol.* **5:**271–283.

Hayden, M. S., and S. Ghosh. 2004. Signaling to NF-kappaB. *Genes Dev.* **18:**2195–2224.

Heinrich, P. C., I. Behrmann, S. Haan, H. M. Hermanns, G. Muller-Newen, and F. Schaper. 2003. Principles of interleukin (IL)-6-type cytokine signalling and its regulation. *Biochem. J.* **374:**1–20.

Hemmi, H., T. Kaisho, K. Takeda, and S. Akira. 2003. The roles of Toll-like receptor 9, MyD88, and DNA-dependent protein kinase catalytic subunit in the effects of two distinct CpG DNAs on dendritic cell subsets. *J. Immunol.* **170:**3059–3064.

Hertzog, P. J., L. A. O'Neill, and J. A. Hamilton. 2003. The interferon in TLR signaling: more than just antiviral. *Trends Immunol.* **24:**534–539.

Hochrein, H., M. O'Keeffe, T. Luft, S. Vandenabeele, R. J. Grumont, E. Maraskovsky, and K. Shortman. 2000. Interleukin (IL)-4 is a major regulatory cytokine governing bioactive IL-12 production by mouse and human dendritic cells. *J. Exp. Med.* **192:**823–833.

Hochrein, H., K. Shortman, D. Vremec, B. Scott, P. Hertzog, and M. O'Keeffe. 2001. Differential production of IL-12, IFN-alpha, and IFN-gamma by mouse dendritic cell subsets. *J. Immunol.* **166:**5448–5455.

Honda, K., and T. Taniguchi. 2006. IRFs: master regulators of signalling by Toll-like receptors and cytosolic pattern-recognition receptors. *Nat. Rev. Immunol.* **6:**644–658.

Honda, K., H. Yanai, T. Mizutani, H. Negishi, N. Shimada, N. Suzuki, Y. Ohba, A. Takaoka, W. C. Yeh, and T. Taniguchi. 2004. Role of a transductional-transcriptional processor complex involving MyD88 and IRF-7 in Toll-like receptor signaling. *Proc. Natl. Acad. Sci. USA* **101:**15416–15421.

Honda, K., H. Yanai, A. Takaoka, and T. Taniguchi. 2005. Regulation of the type I IFN induction: a current view. *Int. Immunol.* **17:**1367–1378.

Hoshino, K., T. Sugiyama, M. Matsumoto, T. Tanaka, M. Saito, H. Hemmi, O. Ohara, S. Akira, and T. Kaisho. 2006. IkappaB kinase-alpha is critical for interferon-alpha production induced by Toll-like receptors 7 and 9. *Nature* **440:**949–953.

Hoss-Homfeld, A., E. C. Zwarthoff, and R. Zawatzky. 1989. Cell type specific expression and regulation of murine interferon alpha and beta genes. *Virology* **173:**539–550.

Hsieh, C. S., S. E. Macatonia, C. S. Tripp, S. F. Wolf, A. O'Garra, and K. M. Murphy. 1993. Development of TH1 CD4+ T cells through IL-12 produced by Listeria-induced macrophages. *Science* **260:**547–549.

Hunter, C. A. 2005. New IL-12-family members: IL-23 and IL-27, cytokines with divergent functions. *Nat. Rev. Immunol.* **5:**521–531.

Inohara, N., and G. Nunez. 2003. NODs: intracellular proteins involved in inflammation and apoptosis. *Nat. Rev. Immunol.* **3:**371–382.

Janeway, C. A., Jr., and R. Medzhitov. 2002. Innate immune recognition. *Annu. Rev. Immunol.* **20:**197–216.

Jankovic, D., M. C. Kullberg, S. Hieny, P. Caspar, C. M. Collazo, and A. Sher. 2002. In the absence of IL-12, CD4(+) T cell responses to intracellular pathogens fail to default to a Th2 pattern and are host protective in an IL-10(−/−) setting. *Immunity* **16:**429–439.

Jarrossay, D., G. Napolitani, M. Colonna, F. Sallusto, and A. Lanzavecchia. 2001. Specialization and complementarity in microbial molecule recognition by human myeloid and plasmacytoid dendritic cells. *Eur. J. Immunol.* **31:**3388–3393.

Jeffrey, K. L., M. Camps, C. Rommel, and C. R. Mackay. 2007. Targeting dual-specificity phosphatases: manipulating MAP kinase signalling and immune responses. *Nat. Rev. Drug Discov.* **6:**391–403.

Jego, G., A. K. Palucka, J. P. Blanck, C. Chalouni, V. Pascual, and J. Banchereau. 2003. Plasmacytoid dendritic cells induce plasma cell differentiation through type I interferon and interleukin 6. *Immunity* **19:**225–234.

Jenkins, J. K., M. Malyak, and W. P. Arend. 1994. The effects of interleukin-10 on interleukin-1 receptor antagonist and interleukin-1 beta production in human monocytes and neutrophils. *Lymphokine Cytokine Res.* **13:**47–54.

Johnson, G. L., and R. Lapadat. 2002. Mitogen-activated protein kinase pathways mediated by ERK, JNK, and p38 protein kinases. *Science* **298:**1911–1912.

Kadowaki, N., S. Antonenko, J. Y. Lau, and Y. J. Liu. 2000. Natural interferon alpha/beta-producing cells link innate and adaptive immunity. *J. Exp. Med.* **192:**219–226.

Kadowaki, N., S. Ho, S. Antonenko, R. W. Malefyt, R. A. Kastelein, F. Bazan, and Y. J. Liu. 2001. Subsets of human dendritic cell precursors express different toll-like receptors and respond to different microbial antigens. *J. Exp. Med.* **194:**863–869.

Kanters, E., M. Pasparakis, M. J. Gijbels, M. N. Vergouwe, I. Partouns-Hendriks, R. J. Fijneman, B. E. Clausen, I. Forster, M. M. Kockx, K. Rajewsky, G. Kraal, M. H. Hofker, and M. P. de Winther. 2003. Inhibition of NF-kappaB activation in macrophages increases atherosclerosis in LDL receptor deficient mice. *J. Clin. Investig.* **112:**1176–1185.

Kastelein, R. A., C. A. Hunter, and D. J. Cua. 2007. Discovery and biology of IL-23 and IL-27: related but functionally distinct regulators of inflammation. *Annu. Rev. Immunol.* **25:**221–242.

Katso, R., K. Okkenhaug, K. Ahmadi, S. White, J. Timms, and M. D. Waterfield. 2001. Cellular function of phosphoinositide 3-kinases: implications for development, homeostasis, and cancer. *Annu. Rev. Cell Dev. Biol.* **17:**615–675.

Kaufmann, S. H. 1993. Immunity to intracellular bacteria. *Annu. Rev. Immunol.* **11:**129–163.

Kawai, T., S. Sato, K. J. Ishii, C. Coban, H. Hemmi, M. Yamamoto, K. Terai, M. Matsuda, J. Inoue, S. Uematsu, O. Takeuchi, and S. Akira. 2004. Interferon-alpha induction through Toll-like receptors involves a direct interaction of IRF7 with MyD88 and TRAF6. *Nat. Immunol.* **5:**1061–1068.

Keane, J. 2005. TNF-blocking agents and tuberculosis: new drugs illuminate an old topic. *Rheumatology* (Oxford) **44:**714–720.

Kehrl, J. H., L. M. Wakefield, A. B. Roberts, S. Jakowlew, M. Alvarez-Mon, R. Derynck, M. B. Sporn, and A. S. Fauci. 1986. Production of transforming growth factor beta by human T lymphocytes and its potential role in the regulation of T cell growth. *J. Exp. Med.* **163:**1037–1050.

Khader, S. A., J. E. Pearl, K. Sakamoto, L. Gilmartin, G. K. Bell, D. M. Jelley-Gibbs, N. Ghilardi, F. deSauvage, and A. M. Cooper. 2005. IL-23 compensates for the absence of IL-12p70 and is essential for the IL-17 response during tuberculosis but is dispensable for protection and antigen-specific IFN-gamma responses if IL-12p70 is available. *J. Immunol.* **175:**788–795.

Kishimoto, T. 2005. Interleukin-6: from basic science to medicine—40 years in immunology. *Annu. Rev. Immunol.* **23:**1–21.

Kobayashi, M., L. Fitz, M. Ryan, R. M. Hewick, S. C. Clark, S. Chan, R. Loudon, F. Sherman, B. Perussia, and G. Trinchieri. 1989. Identification and purification of natural killer cell stimulatory factor (NKSF), a cytokine with multiple biologic effects on human lymphocytes. *J. Exp. Med.* **170:**827–845.

Koch, F., U. Stanzl, P. Jennewein, K. Janke, C. Heufler, E. Kampgen, N. Romani, and G. Schuler. 1996. High level IL-12 production by murine dendritic cells: upregulation via MHC class II and CD40 molecules and downregulation by IL-4 and IL-10. *J. Exp. Med.* **184:**741–746.

Kontoyiannis, D., A. Kotlyarov, E. Carballo, L. Alexopoulou, P. J. Blackshear, M. Gaestel, R. Davis, R. Flavell, and G. Kollias. 2001. Interleukin-10 targets p38 MAPK to modulate ARE-dependent TNF mRNA translation and limit intestinal pathology. *EMBO J.* **20:**3760–3770.

Koppelman, B., J. J. Neefjes, J. E. de Vries, and R. de Waal Malefyt. 1997. Interleukin-10 down-regulates MHC class II alphabeta peptide complexes at the plasma membrane of monocytes by affecting arrival and recycling. *Immunity* **7:**861–871.

Kotlyarov, A., A. Neininger, C. Schubert, R. Eckert, C. Birchmeier, H. D. Volk, and M. Gaestel. 1999. MAPKAP kinase 2 is essential for LPS-induced TNF-alpha biosynthesis. *Nat. Cell. Biol.* **1:**94–97.

Krug, A., A. R. French, W. Barchet, J. A. Fischer, A. Dzionek, J. T. Pingel, M. M. Orihuela, S. Akira, W. M. Yokoyama, and M. Colonna. 2004. TLR9-dependent recognition of MCMV by IPC and DC generates coordinated cytokine responses that activate antiviral NK cell function. *Immunity* **21:**107–119.

Kubin, M., M. Kamoun, and G. Trinchieri. 1994. Interleukin 12 synergizes with B7/CD28 interaction in inducing efficient proliferation and cytokine production of human T cells. *J. Exp. Med.* **180:**211–222.

Kuhn, R., J. Lohler, D. Rennick, K. Rajewsky, and W. Muller. 1993. Interleukin-10-deficient mice develop chronic enterocolitis. *Cell* **75:**263–274.

Kulms, D., and T. Schwarz. 2006. NF-kappaB and cytokines. *Vitam. Horm.* **74:**283–300.

Kumar, S., J. Boehm, and J. C. Lee. 2003. p38 MAP kinases: key signalling molecules as therapeutic targets for inflammatory diseases. *Nat. Rev. Drug Discov.* **2:**717–726.

Lah, T. T., M. Hawley, K. L. Rock, and A. L. Goldberg. 1995. Gamma-interferon causes a selective induction of the lysosomal proteases, cathepsins B and L, in macrophages. *FEBS Lett.* **363:**85–89.

Lang, R., M. Hammer, and J. Mages. 2006. DUSP meet immunology: dual specificity MAPK phosphatases in control of the inflammatory response. *J. Immunol.* **177:**7497–7504.

Langrish, C. L., B. S. McKenzie, N. J. Wilson, R. de Waal Malefyt, R. A. Kastelein, and D. J. Cua. 2004. IL-12 and IL-23: master regulators of innate and adaptive immunity. *Immunol. Rev.* **202:**96–105.

Le, J. M., G. Fredrickson, L. F. Reis, T. Diamantstein, T. Hirano, T. Kishimoto, and J. Vilcek. 1988. Interleukin 2-dependent and interleukin 2-independent pathways of regulation of thymocyte function by interleukin 6. *Proc. Natl. Acad. Sci. USA* **85:**8643–8647.

Lemaitre, B., E. Nicolas, L. Michaut, J. M. Reichhart, and J. A. Hoffmann. 1996. The dorsoventral regulatory gene cassette spatzle/Toll/cactus controls the potent antifungal response in Drosophila adults. *Cell* **86:**973–983.

Letterio, J. J., and A. B. Roberts. 1998. Regulation of immune responses by TGF-beta. *Annu. Rev. Immunol.* **16:**137–161.

Li, M. O., Y. Y. Wan, S. Sanjabi, A. K. Robertson, and R. A. Flavell. 2006. Transforming growth factor-beta regulation of immune responses. *Annu. Rev. Immunol.* **24:**99–146.

Li, Q., and I. M. Verma. 2002. NF-kappaB regulation in the immune system. *Nat. Rev. Immunol.* **2:**725–734.

Liu, Y., E. G. Shepherd, and L. D. Nelin. 2007. MAPK phosphatases—regulating the immune response. *Nat. Rev. Immunol.* **7:**202–212.

Liu, Y. J. 2005. IPC: professional type 1 interferon-producing cells and plasmacytoid dendritic cell precursors. *Annu. Rev. Immunol.* **23:**275–306.

Loke, P., A. S. MacDonald, A. Robb, R. M. Maizels, and J. E. Allen. 2000. Alternatively activated macrophages induced by nematode infection inhibit proliferation via cell-to-cell contact. *Eur. J. Immunol.* **30:**2669–2678.

Lotz, M., F. Jirik, P. Kabouridis, C. Tsoukas, T. Hirano, T. Kishimoto, and D. A. Carson. 1988. B cell stimulating factor 2/interleukin 6 is a costimulant for human thymocytes and T lymphocytes. *J. Exp. Med.* **167:**1253–1258.

Lu, H. T., D. D. Yang, M. Wysk, E. Gatti, I. Mellman, R. J. Davis, and R. A. Flavell. 1999. Defective IL-12 production in mitogen-activated protein (MAP) kinase kinase 3 (Mkk3)-deficient mice. *EMBO J.* **18:**1845–1857.

Luger, T. A., J. Krutmann, R. Kirnbauer, A. Urbanski, T. Schwarz, G. Klappacher, A. Kock, M. Micksche, J. Malejczyk, E. Schauer, et al. 1989. IFN-beta 2/IL-6 augments the activity of human natural killer cells. *J. Immunol.* **143:**1206–1209.

Ma, X. 2001. TNF-alpha and IL-12: a balancing act in macrophage functioning. *Microbes Infect.* **3:**121–129.

Ma, X., J. M. Chow, G. Gri, G. Carra, F. Gerosa, S. F. Wolf, R. Dzialo, and G. Trinchieri. 1996. The interleukin 12 p40 gene promoter is primed by interferon gamma in monocytic cells. *J. Exp. Med.* **183:**147–157.

Ma, X., J. Sun, E. Papasavvas, H. Riemann, S. Robertson, J. Marshall, R. T. Bailer, A. Moore, R. P. Donnelly, G. Trinchieri, and L. J. Montaner. 2000. Inhibition of IL-12 production in human monocyte-derived macrophages by TNF. *J. Immunol.* **164:**1722–1729.

Ma, X., and G. Trinchieri. 2001. Regulation of interleukin-12 production in antigen-presenting cells. *Adv. Immunol.* **79:**55–92.

Macatonia, S. E., N. A. Hosken, M. Litton, P. Vieira, C. S. Hsieh, J. A. Culpepper, M. Wysocka, G. Trinchieri, K. M. Murphy, and A. O'Garra. 1995. Dendritic cells produce IL-12 and direct the development of Th1 cells from naive CD4+ T cells. *J. Immunol.* **154:**5071–5079.

Mach, B., V. Steimle, E. Martinez-Soria, and W. Reith. 1996. Regulation of MHC class II genes: lessons from a disease. *Annu. Rev. Immunol.* **14:**301–331.

MacMicking, J., Q. W. Xie, and C. Nathan. 1997. Nitric oxide and macrophage function. *Annu. Rev. Immunol.* **15:**323–350.

Marriott, J. B., I. A. Clarke, and A. G. Dalgleish. 2001. Inhibition of p38 MAP kinase during cellular activation results in IFN-gamma-dependent augmentation of IL-12 production by human monocytes/macrophages. *Clin. Exp. Immunol.* **125:**64–70.

Mason, N., J. Aliberti, J. C. Caamano, H. C. Liou, and C. A. Hunter. 2002. Cutting edge: identification of c-Rel-dependent and -independent pathways of IL-12 production during infectious and inflammatory stimuli. *J. Immunol.* **168:**2590–2594.

Medzhitov, R. 2001. Toll-like receptors and innate immunity. *Nat. Rev. Immunol.* **1:**135–145.

Medzhitov, R., and C. A. Janeway, Jr. 1997. Innate immunity: impact on the adaptive immune response. *Curr. Opin. Immunol.* **9:**4–9.

Montaner, L. J., R. P. da Silva, J. Sun, S. Sutterwala, M. Hollinshead, D. Vaux, and S. Gordon. 1999. Type 1 and type 2 cytokine regulation of macrophage endocytosis: differential activation by IL-4/IL-13 as opposed to IFN-gamma or IL-10. *J. Immunol.* **162:**4606–4613.

Moore, K. W., R. de Waal Malefyt, R. L. Coffman, and A. O'Garra. 2001. Interleukin-10 and the interleukin-10 receptor. *Annu. Rev. Immunol.* **19:**683–765.

Morrison, D. K., and R. J. Davis. 2003. Regulation of MAP kinase signaling modules by scaffold proteins in mammals. *Annu. Rev. Cell Dev. Biol.* **19:**91–118.

Moser, M., and K. M. Murphy. 2000. Dendritic cell regulation of TH1-TH2 development. *Nat. Immunol.* **1:**199–205.

Mosmann, T. R., and R. L. Coffman. 1989. TH1 and TH2 cells: different patterns of lymphokine secretion lead to different functional properties. *Annu. Rev. Immunol.* **7:**145–173.

Moynagh, P. N. 2005. TLR signalling and activation of IRFs: revisiting old friends from the NF-kappaB pathway. *Trends Immunol.* **26:**469–476.

Muraguchi, A., T. Hirano, B. Tang, T. Matsuda, Y. Horii, K. Nakajima, and T. Kishimoto. 1988. The essential role of B cell stimulatory factor 2 (BSF-2/IL-6) for the terminal differentiation of B cells. *J. Exp. Med.* **167:**332–344.

Murphy, K. M., and S. L. Reiner. 2002. The lineage decisions of helper T cells. *Nat. Rev. Immunol.* **2:**933–944.

Naka, T., N. Nishimoto, and T. Kishimoto. 2002. The paradigm of IL-6: from basic science to medicine. *Arthritis Res.* **4(Suppl 3):**S233–S242.

Napolitani, G., A. Rinaldi, F. Bertoni, F. Sallusto, and A. Lanzavecchia. 2005. Selected Toll-like receptor agonist combinations synergistically trigger a T helper type 1-polarizing program in dendritic cells. *Nat. Immunol.* **6:**769–776.

Nathan, C. 1997. Inducible nitric oxide synthase: what difference does it make? *J. Clin. Investig.* **100:**2417–2423.

Nathan, C. F., T. J. Prendergast, M. E. Wiebe, E. R. Stanley, E. Platzer, H. G. Remold, K. Welte, B. Y. Rubin, and H. W. Murray. 1984. Activation of human macrophages. Comparison of other cytokines with interferon-gamma. *J. Exp. Med.* **160:**600–605.

Negishi, H., Y. Ohba, H. Yanai, A. Takaoka, K. Honma, K. Yui, T. Matsuyama, T. Taniguchi, and K. Honda. 2005. Negative regulation of Toll-like-receptor signaling by IRF-4. *Proc. Natl. Acad. Sci. USA* **102:**15989–15994.

Nelms, K., A. D. Keegan, J. Zamorano, J. J. Ryan, and W. E. Paul. 1999. The IL-4 receptor: signaling mechanisms and biologic functions. *Annu. Rev. Immunol.* **17:**701–738.

Netea, M. G., B. J. Kullberg, and J. W. Van der Meer. 2000. Circulating cytokines as mediators of fever. *Clin. Infect. Dis.* **31(Suppl 5):**S178–S184.

Nishimoto, N., and T. Kishimoto. 2006. Interleukin 6: from bench to bedside. *Nat. Clin. Pract. Rheumatol.* **2:**619–626.

O'Connell, R. M., S. K. Saha, S. A. Vaidya, K. W. Bruhn, G. A. Miranda, B. Zarnegar, A. K. Perry, B. O. Nguyen, T. F. Lane, T. Taniguchi, J. F. Miller, and G. Cheng. 2004. Type I interferon production enhances susceptibility to *Listeria monocytogenes* infection. *J. Exp. Med.* **200:**437–445.

O'Garra, A. 1998. Cytokines induce the development of functionally heterogeneous T helper cell subsets. *Immunity* **8:**275–283.

O'Garra, A., and N. Arai. 2000. The molecular basis of T helper 1 and T helper 2 cell differentiation. *Trends Cell Biol.* **10:**542–550.

O'Garra, A., R. Chang, N. Go, R. Hastings, G. Haughton, and M. Howard. 1992. Ly-1 B (B-1) cells are the main source of B cell-derived interleukin 10. *Eur. J. Immunol.* **22:**711–717.

O'Garra, A., and D. Robinson. 2004. Development and function of T helper 1 cells. *Adv. Immunol.* **83:**133–162.

O'Garra, A., and P. Vieira. 2004. Regulatory T cells and mechanisms of immune system control. *Nat. Med.* **10:**801–805.

O'Garra, A., and P. Vieira. 2007. T(H)1 cells control themselves by producing interleukin-10. *Nat. Rev. Immunol.* **7:**425–428.

Ohmori, Y., R. D. Schreiber, and T. A. Hamilton. 1997. Synergy between interferon-gamma and tumor necrosis factor-alpha in transcriptional activation is mediated by cooperation between signal transducer and activator of transcription 1 and nuclear factor kappaB. *J. Biol. Chem.* **272:**14899–14907.

Okamura, H., H. Tsutsui, S. Kashiwamura, T. Yoshimoto, and K. Nakanishi. 1998. Interleukin-18: a novel cytokine that augments both innate and acquired immunity. *Adv. Immunol.* **70:**281–312.

O'Neill, L. A., and C. A. Dinarello. 2000. The IL-1 receptor/toll-like receptor superfamily: crucial receptors for inflammation and host defense. *Immunol. Today* **21:**206–209.

Oppmann, B., R. Lesley, B. Blom, J. C. Timans, Y. Xu, B. Hunte, F. Vega, N. Yu, J. Wang, K. Singh, F. Zonin, E. Vaisberg, T. Churakova, M. Liu, D. Gorman, J. Wagner, S. Zurawski, Y. Liu, J. S. Abrams, K. W. Moore, D. Rennick, R. de Waal-Malefyt, C. Hannum, J. F. Bazan, and R. A. Kastelein. 2000. Novel p19 protein engages IL-12p40 to form a cytokine, IL-23, with biological activities similar as well as distinct from IL-12. *Immunity* **13:**715–725.

Oshiumi, H., M. Matsumoto, K. Funami, T. Akazawa, and T. Seya. 2003. TICAM-1, an adaptor molecule that participates in Toll-like receptor 3-mediated interferon-beta induction. *Nat. Immunol.* **4:**161–167.

Owaki, T., M. Asakawa, S. Kamiya, K. Takeda, F. Fukai, J. Mizuguchi, and T. Yoshimoto. 2006. IL-27 suppresses CD28-mediated IL-2 production through suppressor of cytokine signaling 3. *J. Immunol.* **176:**2773–2780.

Pasare, C., and R. Medzhitov. 2003. Toll pathway-dependent blockade of CD4+CD25+ T cell-mediated suppression by dendritic cells. *Science* **299:**1033–1036.

Pascual, V., F. Allantaz, E. Arce, M. Punaro, and J. Banchereau. 2005. Role of interleukin-1 (IL-1) in the pathogenesis of systemic onset juvenile idiopathic arthritis and clinical response to IL-1 blockade. *J. Exp. Med.* **201:**1479–1486.

Pearson, A. M. 1996. Scavenger receptors in innate immunity. *Curr. Opin. Immunol.* **8:**20–28.

Pearson, G., F. Robinson, T. Beers Gibson, B. E. Xu, M. Karandikar, K. Berman, and M. H. Cobb. 2001. Mitogen-activated protein (MAP) kinase pathways: regulation and physiological functions. *Endocr. Rev.* **22:**153–183.

Perkins, N. D. 2000. The Rel/NF-kappa B family: friend and foe. *Trends Biochem. Sci.* **25:**434–440.

Perussia, B., S. H. Chan, A. D'Andrea, K. Tsuji, D. Santoli, M. Pospisil, D. Young, S. F. Wolf, and G. Trinchieri. 1992. Natural killer (NK) cell stimulatory factor or IL-12 has differential effects on the proliferation of TCR-alpha beta+, TCR-gamma delta+ T lymphocytes, and NK cells. *J. Immunol.* **149:**3495–3502.

Peter, H. H., H. Dallugge, R. Zawatzky, S. Euler, W. Leibold, and H. Kirchner. 1980. Human peripheral null lymphocytes. II. Producers of type-1 interferon upon stimulation with tumor cells, Herpes simplex virus and Corynebacterium parvum. *Eur. J. Immunol.* **10:**547–555.

Pfeffer, K. 2003. Biological functions of tumor necrosis factor cytokines and their receptors. *Cytokine Growth Factor Rev.* **14:**185–191.

Pfefferkorn, E. R. 1984. Interferon gamma blocks the growth of Toxoplasma gondii in human fibroblasts by inducing the host cells to degrade tryptophan. *Proc. Natl. Acad. Sci. USA* **81:**908–912.

Pflanz, S., J. C. Timans, J. Cheung, R. Rosales, H. Kanzler, J. Gilbert, L. Hibbert, T. Churakova, M. Travis, E. Vaisberg, W. M. Blumenschein, J. D. Mattson, J. L. Wagner, W. To, S. Zurawski, T. K. McClanahan, D. M. Gorman, J. F. Bazan, R. de Waal Malefyt, D. Rennick, and R. A. Kastelein. 2002. IL-27, a heterodimeric cytokine composed of EBI3 and p28 protein, induces proliferation of naive CD4(+) T cells. *Immunity* **16:**779–790.

Pien, G. C., A. R. Satoskar, K. Takeda, S. Akira, and C. A. Biron. 2000. Cutting edge: selective IL-18 requirements for induction of compartmental IFN-gamma responses during viral infection. *J. Immunol.* **165:**4787–4791.

Pomerantz, J. L., and D. Baltimore. 2002. Two pathways to NF-kappaB. *Mol. Cell* **10:**693–695.

Powrie, F., M. W. Leach, S. Mauze, S. Menon, L. B. Caddle, and R. L. Coffman. 1994. Inhibition of Th1 responses prevents inflammatory bowel disease in scid mice reconstituted with CD45RBhi CD4+ T cells. *Immunity* **1:**553–562.

Reis e Sousa, C. 2004. Activation of dendritic cells: translating innate into adaptive immunity. *Curr. Opin. Immunol.* **16:**21–25.

Reis e Sousa, C., S. Hieny, T. Scharton-Kersten, D. Jankovic, H. Charest, R. N. Germain, and A. Sher. 1997. In vivo microbial stimulation induces rapid CD40 ligand-independent production of interleukin 12 by dendritic cells and their redistribution to T cell areas. *J. Exp. Med.* **186:**1819–1829.

Rogers, N. C., E. C. Slack, A. D. Edwards, M. A. Nolte, O. Schulz, E. Schweighoffer, D. L. Williams, S. Gordon, V. L. Tybulewicz, G. D. Brown, and C. Reis e Sousa. 2005. Syk-dependent cytokine induction by Dectin-1 reveals a novel pattern recognition pathway for C type lectins. *Immunity* **22:**507–517.

Rogge, L., D. D'Ambrosio, M. Biffi, G. Penna, L. J. Minetti, D. H. Presky, L. Adorini, and F. Sinigaglia. 1998. The role of Stat4 in species-specific regulation of Th cell development by type I IFNs. *J. Immunol.* **161:**6567–6574.

Romagnani, S. 1994. Regulation of the development of type 2 T-helper cells in allergy. *Curr. Opin. Immunol.* **6:**838–846.

Roux, P. P., and J. Blenis. 2004. ERK and p38 MAPK-activated protein kinases: a family of protein kinases with diverse biological functions. *Microbiol. Mol. Biol. Rev.* **68:**320–344.

Sakaguchi, S., M. Ono, R. Setoguchi, H. Yagi, S. Hori, Z. Fehervari, J. Shimizu, T. Takahashi, and T. Nomura. 2006. Foxp3+ CD25+ CD4+ natural regulatory T cells in dominant self-tolerance and autoimmune disease. *Immunol. Rev.* **212:**8–27.

Salmeron, A., T. B. Ahmad, G. W. Carlile, D. Pappin, R. P. Narsimhan, and S. C. Ley. 1996. Activation of MEK-1 and SEK-1 by Tpl-2 proto-oncoprotein, a novel MAP kinase kinase kinase. *EMBO J.* **15:**817–826.

Salmon, R. A., X. Guo, H. S. Teh, and J. W. Schrader. 2001. The p38 mitogen-activated protein kinases can have opposing roles in the antigen-dependent or endotoxin-stimulated production of IL-12 and IFN-gamma. *Eur. J. Immunol.* **31:**3218–3227.

Saraiva, M., J. R. Christensen, A. V. Tsytsykova, A. E. Goldfeld, S. C. Ley, D. Kioussis, and A. O'Garra. 2005. Identification of a macrophage-specific chromatin signature in the IL-10 locus. *J. Immunol.* **175:**1041–1046.

Schindler, H., M. B. Lutz, M. Rollinghoff, and C. Bogdan. 2001. The production of IFN-gamma by IL-12/IL-18-activated macrophages requires STAT4 signaling and is inhibited by IL-4. *J. Immunol.* **166:**3075–3082.

Schroder, K., P. J. Hertzog, T. Ravasi, and D. A. Hume. 2004. Interferon-gamma: an overview of signals, mechanisms and functions. *J. Leukoc. Biol.* **75:**163–189.

Schulz, O., A. D. Edwards, M. Schito, J. Aliberti, S. Manickasingham, A. Sher, and C. Reis e Sousa. 2000. CD40 triggering of heterodimeric IL-12 p70 production by dendritic cells in vivo requires a microbial priming signal. *Immunity* **13:**453–462.

Shabo, Y., J. Lotem, M. Rubinstein, M. Revel, S. C. Clark, S. F. Wolf, R. Kamen, and L. Sachs. 1988. The myeloid blood cell differentiation-inducing protein MGI-2A is interleukin-6. *Blood* **72:**2070–2073.

Sher, A., and R. L. Coffman. 1992. Regulation of immunity to parasites by T cells and T cell-derived cytokines. *Annu. Rev. Immunol.* **10:**385–409.

Shirafuji, N., S. Matsuda, H. Ogura, K. Tani, H. Kodo, K. Ozawa, S. Nagata, S. Asano, and F. Takaku. 1990. Granulocyte colony-stimulating factor stimulates human mature neutrophilic granulocytes to produce interferon-alpha. *Blood* **75:**17–19.

Shortman, K., and Y. J. Liu. 2002. Mouse and human dendritic cell subtypes. *Nat. Rev. Immunol.* **2:**151–161.

Silverman, N., and T. Maniatis. 2001. NF-kappaB signaling pathways in mammalian and insect innate immunity. *Genes Dev.* **15:**2321–2342.

Skokos, D., and M. C. Nussenzweig. 2007. CD8- DCs induce IL-12-independent Th1 differentiation through Delta 4 Notch-like ligand in response to bacterial LPS. *J. Exp. Med.* **204:**1525–1531.

Smythies, L. E., M. Sellers, R. H. Clements, M. Mosteller-Barnum, G. Meng, W. H. Benjamin, J. M. Orenstein, and P. D. Smith. 2005. Human intestinal macrophages display profound inflammatory anergy despite avid phagocytic and bacteriocidal activity. *J. Clin. Investig.* **115:**66–75.

Snijders, A., C. M. Hilkens, T. C. van der Pouw Kraan, M. Engel, L. A. Aarden, and M. L. Kapsenberg. 1996. Regulation of bioactive IL-12 production in lipopolysaccharide-stimulated human monocytes is determined by the expression of the p35 subunit. *J. Immunol.* **156:**1207–1212.

Sporn, M. B., and A. B. Roberts. 1992. Autocrine secretion—10 years later. *Ann. Intern. Med.* **117:**408–414.

Stark, G. R., I. M. Kerr, B. R. Williams, R. H. Silverman, and R. D. Schreiber. 1998. How cells respond to interferons. *Annu. Rev. Biochem.* **67:**227–264.

Stein, M., S. Keshav, N. Harris, and S. Gordon. 1992. Interleukin 4 potently enhances murine macrophage mannose receptor activity: a marker of alternative immunologic macrophage activation. *J. Exp. Med.* **176:**287–292.

Stetson, D. B., and R. Medzhitov. 2006. Type I interferons in host defense. *Immunity* **25:**373–381.

Stockinger, B., and M. Veldhoen. 2007. Differentiation and function of Th17 T cells. *Curr. Opin. Immunol.* **19:**281–286.

Stumhofer, J. S., A. Laurence, E. H. Wilson, E. Huang, C. M. Tato, L. M. Johnson, A. V. Villarino, Q. Huang, A. Yoshimura, D. Sehy, C. J. Saris, J. J. O'Shea, L. Hennighausen, M. Ernst, and C. A. Hunter. 2006. Interleukin 27 negatively regulates the development of interleukin 17-producing T helper cells during chronic inflammation of the central nervous system. *Nat. Immunol.* **7:**937–945.

Symons, A., S. Beinke, and S. C. Ley. 2006. MAP kinase kinase kinases and innate immunity. *Trends Immunol.* **27:**40–48.

Takaoka, A., H. Yanai, S. Kondo, G. Duncan, H. Negishi, T. Mizutani, S. Kano, K. Honda, Y. Ohba, T. W. Mak, and T. Taniguchi. 2005. Integral role of IRF-5 in the gene induction programme activated by Toll-like receptors. *Nature* **434**:243–249.

Takeda, K., and S. Akira. 2005. Toll-like receptors in innate immunity. *Int. Immunol.* **17**:1–14.

Tanaka, N., M. Sato, M. S. Lamphier, H. Nozawa, E. Oda, S. Noguchi, R. D. Schreiber, Y. Tsujimoto, and T. Taniguchi. 1998. Type I interferons are essential mediators of apoptotic death in virally infected cells. *Genes Cells* **3**:29–37.

Tangye, S. G., A. Ferguson, D. T. Avery, C. S. Ma, and P. D. Hodgkin. 2002. Isotype switching by human B cells is division-associated and regulated by cytokines. *J. Immunol.* **169**:4298–4306.

Taniguchi, T., K. Ogasawara, A. Takaoka, and N. Tanaka. 2001. IRF family of transcription factors as regulators of host defense. *Annu. Rev. Immunol.* **19**:623–655.

Taniguchi, T., and A. Takaoka. 2002. The interferon-alpha/beta system in antiviral responses: a multimodal machinery of gene regulation by the IRF family of transcription factors. *Curr. Opin. Immunol.* **14**:111–116.

Tartaglia, L. A., R. F. Weber, I. S. Figari, C. Reynolds, M. A. Palladino, Jr., and D. V. Goeddel. 1991. The two different receptors for tumor necrosis factor mediate distinct cellular responses. *Proc. Natl. Acad. Sci. USA* **88**:9292–9296.

Taylor, P. R., L. Martinez-Pomares, M. Stacey, H. H. Lin, G. D. Brown, and S. Gordon. 2005. Macrophage receptors and immune recognition. *Annu. Rev. Immunol.* **23**:901–944.

Tilg, H., E. Trehu, M. B. Atkins, C. A. Dinarello, and J. W. Mier. 1994. Interleukin-6 (IL-6) as an anti-inflammatory cytokine: induction of circulating IL-1 receptor antagonist and soluble tumor necrosis factor receptor p55. *Blood* **83**:113–118.

Tomczak, M. F., M. Gadjeva, Y. Y. Wang, K. Brown, I. Maroulakou, P. N. Tsichlis, S. E. Erdman, J. G. Fox, and B. H. Horwitz. 2006. Defective activation of ERK in macrophages lacking the p50/p105 subunit of NF-kappaB is responsible for elevated expression of IL-12 p40 observed after challenge with Helicobacter hepaticus. *J. Immunol.* **176**:1244–1251.

Toshchakov, V., B. W. Jones, P. Y. Perera, K. Thomas, M. J. Cody, S. Zhang, B. R. Williams, J. Major, T. A. Hamilton, M. J. Fenton, and S. N. Vogel. 2002. TLR4, but not TLR2, mediates IFN-beta-induced STAT1alpha/beta-dependent gene expression in macrophages. *Nat. Immunol.* **3**:392–398.

Tough, D. F., S. Sun, and J. Sprent. 1997. T cell stimulation in vivo by lipopolysaccharide (LPS). *J. Exp. Med.* **185**:2089–2094.

Tracey, K. J., and A. Cerami. 1994. Tumor necrosis factor: a pleiotropic cytokine and therapeutic target. *Annu. Rev. Med.* **45**:491–503.

Trinchieri, G. 1997. Cytokines acting on or secreted by macrophages during intracellular infection (IL-10, IL-12, IFN-gamma). *Curr. Opin. Immunol.* **9**:17–23.

Trinchieri, G. 1998. Interleukin-12: a cytokine at the interface of inflammation and immunity. *Adv. Immunol.* **70**:83–243.

Trinchieri, G. 2003. Interleukin-12 and the regulation of innate resistance and adaptive immunity. *Nat. Rev. Immunol.* **3**:133–146.

Trinchieri, G., and A. Sher. 2007. Cooperation of Toll-like receptor signals in innate immune defence. *Nat. Rev. Immunol.* **7**:179–190.

Tsujimura, H., T. Tamura, H. J. Kong, A. Nishiyama, K. J. Ishii, D. M. Klinman, and K. Ozato. 2004. Toll-like receptor 9 signaling activates NF-kappaB through IFN regulatory factor-8/IFN consensus sequence binding protein in dendritic cells. *J. Immunol.* **172**:6820–6827.

Turner, M., D. Chantry, P. Katsikis, A. Berger, F. M. Brennan, and M. Feldmann. 1991. Induction of the interleukin 1 receptor antagonist protein by transforming growth factor-beta. *Eur. J. Immunol.* **21**:1635–1639.

van Kooyk, Y., and T. B. Geijtenbeek. 2003. DC-SIGN: escape mechanism for pathogens. *Nat. Rev. Immunol.* **3**:697–709.

Vannier, E., L. C. Miller, and C. A. Dinarello. 1992. Coordinated antiinflammatory effects of interleukin 4: interleukin 4 suppresses interleukin 1 production but up-regulates gene expression and synthesis of interleukin 1 receptor antagonist. *Proc. Natl. Acad. Sci. USA* **89**:4076–4080.

Veldhoen, M., R. J. Hocking, C. J. Atkins, R. M. Locksley, and B. Stockinger. 2006a. TGFbeta in the context of an inflammatory cytokine milieu supports de novo differentiation of IL-17-producing T cells. *Immunity* **24**:179–189.

Veldhoen, M., R. J. Hocking, R. A. Flavell, and B. Stockinger. 2006b. Signals mediated by transforming growth factor-beta initiate autoimmune encephalomyelitis, but chronic inflammation is needed to sustain disease. *Nat. Immunol.* **7**:1151–1156.

Veldhoen, M., and B. Stockinger. 2006. TGFbeta1, a "Jack of all trades": the link with pro-inflammatory IL-17-producing T cells. *Trends Immunol.* **27**:358–361.

Villarino, A., L. Hibbert, L. Lieberman, E. Wilson, T. Mak, H. Yoshida, R. A. Kastelein, C. Saris, and C. A. Hunter. 2003. The IL-27R (WSX-1) is required to suppress T cell hyperactivity during infection. *Immunity* **19**:645–655.

Villarino, A. V., E. Huang, and C. A. Hunter. 2004. Understanding the pro- and anti-inflammatory properties of IL-27. *J. Immunol.* **173**:715–720.

Wahl, S. M. 1994. Transforming growth factor beta: the good, the bad, and the ugly. *J. Exp. Med.* **180**:1587–1590.

Wahl, S. M., J. B. Allen, B. S. Weeks, H. L. Wong, and P. E. Klotman. 1993. Transforming growth factor beta enhances integrin expression and type IV collagenase secretion in human monocytes. *Proc. Natl. Acad. Sci. USA* **90**:4577–4581.

Wahl, S. M., D. A. Hunt, L. M. Wakefield, N. McCartney-Francis, L. M. Wahl, A. B. Roberts, and M. B. Sporn. 1987. Transforming growth factor type beta induces monocyte chemotaxis and growth factor production. *Proc. Natl. Acad. Sci. USA* **84**:5788–5792.

Wallach, D., E. E. Varfolomeev, N. L. Malinin, Y. V. Goltsev, A. V. Kovalenko, and M. P. Boldin. 1999. Tumor necrosis factor receptor and Fas signaling mechanisms. *Annu. Rev. Immunol.* **17**:331–367.

Wang, C., L. Deng, M. Hong, G. R. Akkaraju, J. Inoue, and Z. J. Chen. 2001. TAK1 is a ubiquitin-dependent kinase of MKK and IKK. *Nature* **412**:346–351.

Weaver, C. T., L. E. Harrington, P. R. Mangan, M. Gavrieli, and K. M. Murphy. 2006. Th17: an effector CD4 T cell lineage with regulatory T cell ties. *Immunity* **24**:677–688.

Widmann, C., S. Gibson, M. B. Jarpe, and G. L. Johnson. 1999. Mitogen-activated protein kinase: conservation of a three-kinase module from yeast to human. *Physiol. Rev.* **79**:143–180.

Williams, B. R. 1999. PKR; a sentinel kinase for cellular stress. *Oncogene* **18**:6112–6120.

Wozniak, T. M., A. A. Ryan, and W. J. Britton. 2006. Interleukin-23 restores immunity to Mycobacterium tuberculosis infection in IL-12p40-deficient mice and is not required for the development of IL-17-secreting T cell responses. *J. Immunol.* **177**:8684–8892.

Wynn, T. A. 2003. IL-13 effector functions. *Annu. Rev. Immunol.* **21**:425–456.

Yamamoto, M., and S. Akira. 2005. TIR domain-containing adaptors regulate TLR signaling pathways. *Adv. Exp. Med. Biol.* **560**:1–9.

Yamamoto, M., S. Sato, K. Mori, K. Hoshino, O. Takeuchi, K. Takeda, and S. Akira. 2002. Cutting edge: a novel Toll/IL-1 receptor domain-containing adapter that preferentially activates the IFN-beta promoter in the Toll-like receptor signaling. *J. Immunol.* **169:**6668–6672.

Yao, Y., W. Li, M. H. Kaplan, and C. H. Chang. 2005. Interleukin (IL)-4 inhibits IL-10 to promote IL-12 production by dendritic cells. *J. Exp. Med.* **201:**1899–1903.

Yates, R. M., A. Hermetter, G. A. Taylor, and D. G. Russell. 2007. Macrophage activation downregulates the degradative capacity of the phagosome. *Traffic* **8:**241–250.

Yi, A. K., J. G. Yoon, S. J. Yeo, S. C. Hong, B. K. English, and A. M. Krieg. 2002. Role of mitogen-activated protein kinases in CpG DNA-mediated IL-10 and IL-12 production: central role of extracellular signal-regulated kinase in the negative feedback loop of the CpG DNA-mediated Th1 response. *J. Immunol.* **168:**4711–4720.

Zganiacz, A., M. Santosuosso, J. Wang, T. Yang, L. Chen, M. Anzulovic, S. Alexander, B. Gicquel, Y. Wan, J. Bramson, M. Inman, and Z. Xing. 2004. TNF-alpha is a critical negative regulator of type 1 immune activation during intracellular bacterial infection. *J. Clin. Investig.* **113:**401–413.

Zhang, G., and S. Ghosh. 2001. Toll-like receptor-mediated NF-kappaB activation: a phylogenetically conserved paradigm in innate immunity. *J. Clin. Investig.* **107:**13–19.

Zhang, X., L. Giangreco, H. E. Broome, C. M. Dargan, and S. L. Swain. 1995. Control of CD4 effector fate: transforming growth factor beta 1 and interleukin 2 synergize to prevent apoptosis and promote effector expansion. *J. Exp. Med.* **182:**699–709.

Zhao, J., H. J. Kong, H. Li, B. Huang, M. Yang, C. Zhu, M. Bogunovic, F. Zheng, L. Mayer, K. Ozato, J. Unkeless, and H. Xiong. 2006. IRF-8/interferon (IFN) consensus sequence-binding protein is involved in Toll-like receptor (TLR) signaling and contributes to the cross-talk between TLR and IFN-gamma signaling pathways. *J. Biol. Chem.* **281:**10073–10080.

19

Macrophage Classical Activation

DONALD C. VINH AND STEVEN M. HOLLAND

INTRODUCTION

Macrophages are bone marrow-derived, peripheral blood monocytes that have entered tissues in response to various chemotactic and/or inflammatory stimuli. In the process they differentiate into morphologically, histochemically, and functionally distinct tissue macrophage subpopulations with critical roles in host defense. These cells are key components of innate immunity, and by virtue of their ability to recruit and activate other inflammatory cells, modulate the subsequent adaptive immune response as well. In order to affect such functions, though, the macrophage must first be activated, which can occur either by a classical pathway, discussed in this chapter, or by an alternative process, reviewed in chapter 20. We will focus on the critical steps of classical macrophage activation that have been emphasized by rare human diseases in which such key components are naturally deficient.

Physiology

The concept of macrophage activation is ascribed to Elie Metchnikoff, who noted that the mononuclear phagocytes of animals resistant to infection with a particular microbe possessed the ability to effectively ingest and kill the challenge organism. Mackaness and colleagues characterized macrophages that were appropriately activated to respond to intracellular pathogens as being large and angry. These large and "angry" (i.e., activated) macrophages were subsequently demonstrated to also possess antineoplastic capacity (Mackaness, 1970). Thus, the activated macrophage was defined as one that was capable of mediating antimicrobial or antitumor activity (Adams and Hamilton, 1984).

The classical pathway of macrophage activation requires two signals: a priming signal, followed by a triggering signal. This sequential stimulation results in a proinflammatory macrophage, whereby cytokines and reactive metabolite intermediates are generated for effective killing of pathogens, in particular, intracellular ones. Because these phagocytes develop within a T-helper lymphocyte type 1 (Th1)-type cytokine milieu, they have also been referred to as type 1 macrophages or macrophage-1.

Priming and Activation

"Priming" is the process by which, in response to an external stimulus (e.g., a ligand binding to its cognate receptor on the cell surface), the macrophage is not rendered fully cytolytic, but rather develops enhanced responsiveness to a secondary (triggering) signal. It is the delivery of this latter signal that "activates" the macrophage and precipitates the expression of killing. In the absence of the priming signal, normal macrophages may still be able to respond to activating stimuli, albeit in a quantitatively weaker capacity. For example, larger amounts of triggering stimuli may be required to elicit the same response, or the corollary, lower levels of cytokines and reactive metabolites may be produced or produced more slowly. Thus, "priming" lowers the macrophage's threshold to unleash its microbicidal arsenal. The prototypical priming signal for classical macrophage activation is interferon-γ (IFN-γ) acting on its receptor, IFN-γR.

Traditionally, the triggering signal for classical macrophage activation has been lipopolysaccharide (LPS), a glycolipid component derived from the outer membrane of gram-negative bacteria, resulting in the production of tumor necrosis factor-α (TNF-α). Thus, classical macrophage activation is historically defined as the IFN-γ-primed, LPS-triggered macrophage production of a proinflammatory response, characterized particularly by the production of TNF-α. However, priming can also be achieved in vitro with other substances, such as IFN-α, IFN-β, and neutrophil-derived antimicrobial peptides (Speer et al., 1984) binding to their respective receptors, as well as in vivo with total body irradiation (Hill et al., 1997). However, IFN-γ is approximately 850 to 1,000 times more effective than other interferons (Pace et al., 1983). Other biological agents are also capable of activating the macrophage's cytotoxicity capacity, including bacterial peptidoglycan (Gupta et al., 1995), lipoteichoic acid from gram-positive bacteria (Grunfeld et al., 1999), mycobacterial cell wall derivatives (e.g., muramyl dipeptide) (Fevrier et al., 1978; Wahl et al., 1979), lipoarabinomannan (Roach et al., 1993; van Crevel et al., 2002), and heat

---

Donald C. Vinh and Steven M. Holland, Immunopathogenesis Section, Laboratory of Clinical Infectious Diseases, National Institute of Allergy and Infectious Diseases, National Institutes of Health, Bethesda, MD 20892-1684.

shock proteins (Kol et al., 2000). IFN-γ priming does not necessarily have to be delivered prior to the activating signal; metachronous delivery of IFN-γ relative to the triggering signal can still lead to effective priming (Schroder et al., 2006). The constituents of the macrophage proinflammatory response have also expanded to include the production of a respiratory burst with the generation of reactive oxygen species (ROS), the induction of nitric oxide synthase (NOS) with the generation of reactive nitrogen species (RNS), efficient peptide processing with enhanced antigen presentation via HLA/major histocompatibility complex (MHC) molecules, increased surface expression of various ligands and receptors, and the elaboration of multiple proinflammatory cytokines and chemokines.

Therefore, in a broad sense, classical macrophage activation is an IFN-γ-mediated generation of a proinflammatory response. Pathogen moieties will usually activate the macrophage and initiate phagocytosis, while pathogen-stimulated IFN-γ will prime the macrophage. Subsequently, the macrophage will attempt to kill the intracellular microbe, while increasing surface expression of key molecules and producing cytokines. TNF-α will be secreted and further stimulates macrophages. Macrophage production of IL-12 will augment IFN-γ production by natural killer (NK) cells and T cells, which will amplify the microbicidal response. In conjunction with other cells, macrophages will form granulomas. Numerous molecules are definitely or putatively involved in this process, and a few have been identified by natural mutations. We will review these through an arbitrary and artificial division of the proinflammatory response into premacrophage, intramacrophage, and postmacrophage phases.

PREMACROPHAGE PHASE

IFN-γ Pathway

Three classes of IFNs are defined, based on differences in structure, cognate receptor specificity, and effector functions. Type I IFNs are α, β, τ, and ω, with the first two being the most extensively characterized; all are encoded on human chromosome 9. The sole type II IFN is IFN-γ, encoded on human chromosome 12. The type III IFNs (IFN-λ1/interleukin-29 [IL-29], IFN-λ2/IL-28A, IFN-λ3/IL 28B) are encoded on human chromosome 19. IFN-γ has the most well-defined role in classical macrophage activation.

IFN-γ is produced chiefly by NK cells and T cells in a biphasic manner. Encounter of the pathogen results in early production of IFN-γ by NK cells, as well as by antigen-presenting cells, i.e., tissue macrophages and monocyte-derived dendritic cells (DCs) (Frucht et al., 2001; Gessani et al., 1998). Recruitment and activation of T cells provides IFN-γ for the later (adaptive) phase of the immune response.

IFN-γ works as a homodimer binding to its receptor, IFN-γR, present on essentially all cells, except mature erythrocytes (Stark et al., 1998). The IFN-γR consists of two subunits: IFN-γR1 and IFN-γR2. IFN-γR1 (α-chain or CD119w) is a 90-kDa polypeptide encoded by the IFNGR1 gene on chromosome 6; it mediates ligand binding as well as signal transduction. IFN-γR2 (β-chain or accessory factor-1) is a 62-kDa polypeptide encoded by the IFNGR2 gene on chromosome 21; it serves a minor role in ligand binding but is required for intracellular signaling.

In unstimulated cells, the IFN-γR subunits are not associated with each other. However, each subunit is preassociated with a tyrosine kinase of the Janus kinase (JAK) family: IFN-γR1 has bound JAK-1, while JAK-2 binds IFN-γR2. The functionally active IFN-γR consists of two IFN-γR1 subunits assembled with two IFN-γR2 subunits, which occurs in the presence of IFN-γ. This process brings into proximity the corresponding JAK-1 and JAK-2 molecules on the intracytoplasmic surface of the cell membrane. Through auto- and transphosphorylation, the JAKs self-activate, forming phosphorylated tyrosine residues that serve as docking sites for signal transducers and activators of transcription-1 (STAT-1). Two STAT-1 molecules are phosphorylated by the JAKs, dissociate from the receptor complex, and form a reciprocal homodimer that translocates to the nucleus. Within the nucleus, STAT-1 homodimers bind to IFN-γ-activated site elements and induce transcription of the relevant genes.

Defects in this IFN-γ pathway cause enhanced susceptibility to severe and/or recurrent disease with nontuberculous mycobacteria (NTM) or with bacillus Calmette-Guérin (BCG). These deficiencies are grouped as Mendelian susceptibility to mycobacterial infections (MIM209950), although their phenotype extends beyond NTM infection.

IFN-γ

Human IFN-γ, at chromosome 12q15, spans ~5.4 kb and contains four exons encoding a protein of 146 amino acids (with a 20-amino-acid signal peptide) (Henri et al., 2002). Mutations in the IFN-γ gene itself have not been published. IFN-γ knockout mice have demonstrated increased susceptibility and/or augmented severity of disease to a wide range of pathogens. The level of IFN-γ production is modulated by IL-12, by a variable CA repeat in intron 2 of the IFN-γ gene (Rossouw et al., 2003), and can be post-translationally affected by autoantibodies to IFN-γ (Höflich et al., 2004).

IFN-γR1

IFN-γR1, encoded by IFNGR1, consists of seven exons, providing an extracellular ligand-binding domain (exons 1 to 5 and part of exon 6), a transmembrane domain (exon 6), and a short intracytoplasmic tail involved in signal transduction (exon 7) (Bach et al., 1997). Mutations involving IFN-γR1 occur in recessive and dominant forms and have been well characterized as having a tight correlation between IFNGR1 genotype, cellular phenotype, and clinical disease (Dorman et al., 2004).

The complete recessive forms of IFN-γR1 deficiency typically result from null mutations occurring in the extracellular domain of IFN-γR1 which either abrogate protein or interfere with its capacity for ligand (IFN-γ) binding (Dorman et al., 2004; Newport et al., 1996; Jouanguy et al., 1996). Partial IFN-γR1 deficiencies can be dominant or recessive. Dominant forms are due to mutations resulting in premature stops in exon 6, producing truncated forms of IFN-γR1. These truncated receptors are lacking the intracellular domain needed for signal transduction and also missing the critical receptor recycling domains needed for removal of receptor from the cell surface (Jouanguy et al., 1999). Therefore, mutant receptors accumulate on the cell surface, thus exerting a dominant effect over the normal alleles. Recessive partial IFN-γR1 deficiencies are due to missense mutations attenuating IFN-γ recognition of the receptor (Jouanguy et al., 1997). Therefore, three variants of IFN-γR1 deficiency exist: recessive complete, partial dominant, and partial recessive.

Recessive complete IFN-γR1 deficiency is clinically the most severe form: mean age of onset of first NTM disease is 3.1 years; disease is usually disseminated; mortality rates can be high (Dorman et al., 2004). Hematopoietic stem cell transplantation is curative and recommended, but only once optimal control of mycobacterial disease is established (Roesler et al., 2004).

Mutations producing dominant partial IFN-γR1 deficiency congregate in "hot spots," such as the 4-bp deletion at positions 816, 817, 818, and 819; collectively, these mutations are termed 818del4. Deletion of the thymidine nucleotide at position 818 or 819 is designated 818delT. These two categories of mutations, 818del4 and 818delT, have the same consequences (Jouanguy et al., 1999). Another hot spot, also characterized by a 4-bp deletion in or around nucleotide position 561 (designated 561del4), was subsequently identified (Rosenzweig et al., 2002). Clinically, patients with partial dominant IFN-γR1 deficiency manifest milder disease. *Mycobacterium avium* complex is the most common pathogen (Dorman et al., 2004). The average age of onset of first NTM disease is 13.4 years, and NTM osteomyelitis involving multiple bones is common (Dorman et al., 2004). The lytic lesions have led to the incorrect diagnosis of Langerhans' cell histiocytosis (Edgar et al., 2001). Osteomyelitis with *Histoplasma capsulatum* has also been described in a patient with partial dominant IFN-γR1 deficiency (Zerbe and Holland, 2005). Mycobacterial disease in partial dominant deficiency is usually curable with antimycobacterial treatment and is frequently followed by periods without obvious NTM disease. Survival rates for patients with this disorder are excellent, with most living beyond 60 years of age.

IFN-γR2

IFNGR2, located on chromosome 21q22.1, consists of seven exons, similar to its subunit counterpart. Exons 1 through 5 encode the extracellular domain, exon 6 encodes the transmembrane region and the early portion of the intracellular domain, and exon 7 produces the intracytoplasmic domain (Bach et al., 1997). Here again there are complete recessive (Dorman and Holland, 1998), partial recessive (Doffinger et al., 2000), and partial dominant IFN-γR2 deficiencies (Rosenzweig et al., 2004). In addition, there is an unusual class of mutations due to gains of glycosylation on IFN-γR2 (Vogt et al., 2005). Although the numbers of patients identified to date have been lower than those for IFN-γR1 deficiency, IFN-γR2 deficiency appears to demonstrate a similar correlation between genotype with cellular and clinical phenotype.

JAK-1 and JAK-2

Two Janus kinases, JAK-1 and JAK-2, are intimately involved in IFN-γR signaling. These two kinases, along with JAK-3 and Tyk2, form the JAK family (Yamaoka et al., 2004). JAK-3 and Tyk2 are both located on human chromosome 19, and the absence of either is associated with a primary immunodeficiency. JAK-3 deficiency causes autosomal recessive severe combined immunodeficiency (Yamaoka et al., 2004), while Tyk2 deficiency causes a condition with immunoglobulin E (IgE) elevation, viral skin infections, and bacterial and mycobacterial susceptibility (Minegishi et al., 2006). In contrast to JAK-3 and Tyk2, no human cases of germ line JAK-1 or JAK-2 deficiency have been reported. JAK-1, encoded on chromosome 1, is ubiquitously expressed and is crucial for effective functioning of a variety of cytokines and growth factors (Yamaoka et al., 2004). JAK-2, encoded on chromosome 9, is also ubiquitously expressed; it too is crucial for signal transduction of many hormones, hematopoietic growth factors, and cytokines. JAK-2 deficiency in the knockout mouse results in embryonic lethality due to failure of definitive erythropoiesis.

STAT-1

STAT-1 phosphorylation is a defining feature of the early IFN-γ response pathway. Once STAT-1 is phosphorylated in association with the ligand-activated IFN-γR complex, there is homodimerization of STAT-1, producing the transcription factor gamma-activating factor (GAF). GAF translocates to the nucleus and binds to gamma-activating sequences (GASs) in the promoter regions of genes that constitute the effector response to IFN-γ.

STAT-1 is activated by various IFNs binding to their respective cell surface receptors, leading to hetero-oligomerization with itself, activated STAT-2, and other signal-transducing molecules including p48/IRF-9/ISGF-3G, which forms the complex, interferon-stimulated gene factor-3 (ISGF-3) (Kim and Lee, 2007). Thus, STAT-1 deficiency has the potential to impair the IFN-γ and IFN-α/β pathways. Two forms of STAT-1 deficiency in humans have been identified thus far: recessive complete and dominant partial, each with unique mutations, modes of defect, and clinical phenotype.

STAT-1, encoded on chromosome 2, consists of five domains (Kim and Lee, 2007). Dominant partial STAT-1 deficiency results in a heterozygous loss-of-function allele that cannot be phosphorylated properly (Dupuis et al., 2001). This single null allele impairs STAT-1 activation to allow formation of only about 25% of normal GAF. Similar heterozygous mutations have been described with mild phenotypes (Chapgier et al., 2006b). Recessive complete STAT-1 deficiency is more dramatic, with severe susceptibility to bacterial and viral infections (Dupuis et al., 2003; Chapgier et al., 2006a). These cases highlight the crucial role that STAT-1 plays in both IFN-γ and IFN-αβ signaling, and provide strong credence for their involvement in human mycobacterial and viral immunity.

Activation by LPS

LPS is historically the most commonly used ligand for classical macrophage activation, with the capacity to exert a spectrum of effects in macrophages: activate macrophages to produce an appropriate proinflammatory response, excessively activate macrophages to produce a potentially deleterious systemic inflammatory response syndrome, or induce tolerance (i.e., deactivate) macrophages. In classical activation there is a cognate interaction between LPS and the "LPS receptor complex" at the cell surface of the macrophage. This complex has three components: Toll-like receptor 4 (TLR-4), CD14, and myeloid differentiation protein-2 (MD-2).

TLRs

The mammalian TLRs, type I transmembrane innate immune receptors, are characterized by three domains: a leucine-rich region and cysteine-rich domain in the ligand-binding extracellular part; a transmembrane region; and a Toll/IL-1 receptor (TIR) homology domain responsible for initiating the signaling cascade intracellularly. Ten TLRs have been identified in humans. They are thought to activate innate immunity by recognizing conserved pathogen-associated molecular patterns (PAMPs). Different TLRs recognize different ligands (Fig. 1), either independently or through heterodimerization (with other TLRs and non-TLR proteins). Phylogenetic analysis reveals that the human TLRs can be grouped into three evolutionarily distinct clades: clade A (TLRs 2, 10, 1, 6), clade B (TLR4), and clade C (TLRs 3, 5, 7, 8, 9) (Zhou et al., 2007). TLRs are expressed by a wide variety of cell types, but monocyte derivatives (i.e., macrophages, DCs) express almost the full repertoire. Some of the TLRs are predominantly expressed at the cell surface (TLRs 1, 2, 4, 5, 6), while others are localized intracellularly (TLRs 7, 8, 9); the specific localization of TLR-3 is unresolved (Ulevitch, 2004). Activation of the TLRs results in the recruitment of one or more of the myeloid differentiation primary response protein-88 (MyD88)-adaptor family of proteins; the members of this family include MyD88, MyD88-adaptor-like (Mal, also known as TIRAP, the term used in this chapter), TIR-domain-containing adaptor protein inducing IFN (TRIF), TRIF-related adaptor molecule (TRAM), and sterile α- and armadillo-motif-containing protein (SARM). TLR-mediated classical macrophage activation is fundamentally based on pathways centered on TLR-4.

LPS induces classical macrophage activation through TLR-4. Engagement of TLR-4 by LPS also requires CD14 and MD-2. CD14 is constitutively expressed primarily on the surface of monocytes, macrophages, and neutrophils as membrane-bound CD14 (mCD14). Soluble CD14 (sCD14) is abundant in serum and derived from both secretion of CD14 and enzymatically cleaved glycosylphosphatidylinositol (GPI)-anchored mCD14 (Haziot et al., 1988). The transfer of LPS monomers to CD14 is mediated, at least in some tissues, by LPS-binding protein (LBP) (Fenton and Golenbock, 1998). In addition to facilitating transfer to immune cells, LBP can also shuttle LPS into high-density lipoprotein (HDL) particles, which neutralizes LPS (Fenton and Golenbock, 1998). The LPS/CD14 complex is then recognized by TLR-4 associated with MD-2. The heterotrimer, activated by LPS, then mediates signaling via the intracellular domain of TLR-4 (Akira and Takeda, 2004).

TLR-4 stimulation results in the activation of two intracellular signaling pathways, one that is MyD88 dependent and one that is MyD88 independent (Ulevitch, 2004; Akira and Takeda, 2004; Watters et al., 2007). MyD88 is recruited via TIRAP to the TIR domain on the intracytoplasmic region of TLR-4 (Fig. 1). MyD88 then binds IL-1 receptor-associated kinase 1 (IRAK-1) and IRAK-4. IRAK-4 phosphorylates IRAK-1, which then autophosphorylates, permitting its dissociation from the receptor complex. Activated IRAK-1 then associates with TNFR-associated factor-6 (TRAF-6), resulting in the ubiquitination of TRAF-6. The ubiquitinated TRAF-6 subsequently activates TGF-β-activating kinase 1 (TAK-1) via interaction with TAK-1 binding protein 2 (TAB-2). The activated TAK-1 then activates the IκB kinase (IKK) complex; this complex is composed of two catalytic subunits, IKKα and IKKβ, and the regulatory subunit IKKγ (also known as NF-κB essential modulator, or NEMO, the term used in this chapter). Activated IKK complex leads to phosphorylation of inhibitor of NF-κB (IκB), allowing NF-

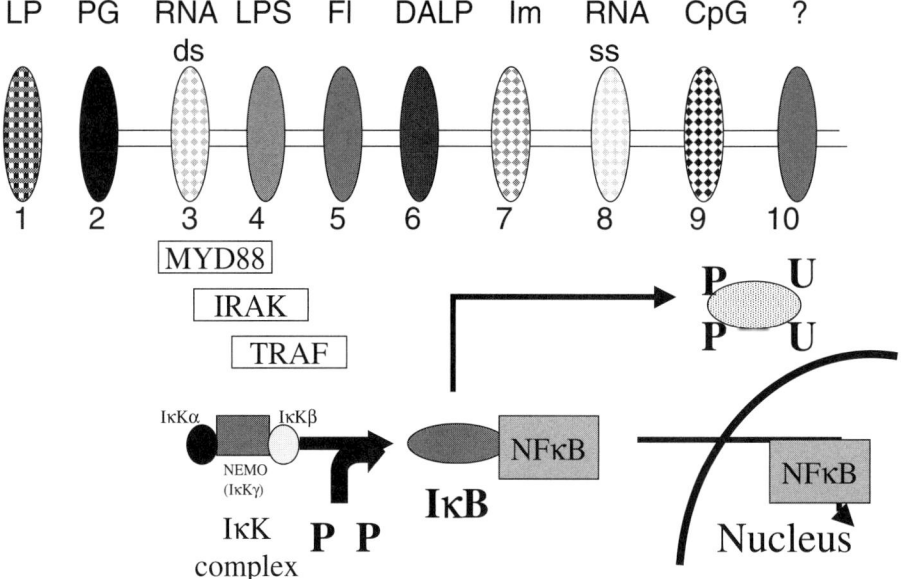

FIGURE 1 TLRs are pattern-recognition receptors that distinguish among PAMPs. Structurally, TLRs contain extracellular leucine-rich repeats responsible for recognition of their respective PAMPs, a transmembrane domain that determines their cellular localization and an intracellular domain that mediates signaling pathways. Ligand binding induces TLR oligomerization or heterodimerization, responsible for PAMP specificity; subsequent recruitment of a particular network of adaptor proteins allows for specific intracellular signaling. Here, TLR4-mediated signaling is depicted (see text for details). Selected ligands for the 10 human TLRs are shown here as follows: TLR1, triacyl lipopeptides (LP); TLR2, peptidoglycan (PG); TLR3, double-stranded RNA (dsRNA); TLR4, lipopolysaccharide (LPS); TLR5, flagellin (Fl); TLR6, diacyl lipopeptides (DALP); TLR7, imiquimod (Im); TLR8, single-stranded RNA (ssRNA); TLR9, unmethylated CpG dinucleotides (CpG); TLR10, undefined.

κB to translocate to the nucleus and induce the transcription of proinflammatory cytokine genes (e.g. TNF-α, IL-1β, IL-6, IL-12). In addition to stimulating the IKK/NF-κB cascade, TRAF-6 also initiates the mitogen-activated protein kinase (MAPK) pathway, via activation of specific MAP-3 kinases (MAP-3K), which then activate specific MAP-2 kinases (MAP-2K), and eventually activating MAPK (i.e., the ERK kinases, the JNK kinases, or the p38 kinases), leading to the production of proinflammatory cytokines (Watters et al., 2007).

LPS can also stimulate macrophages via a MyD88-independent pathway (also known as the TRIF-dependent pathway). Agonist binding of TLR-4 leads to the recruitment of TRIF via binding of the bridging molecule, TRAM, to the TIR domain. Activated TRIF stimulates a kinase, TBK-1, to phosphorylate interferon-regulatory factor 3 (IRF-3), which then translocates to the nucleus and induces the transcription of interferon-related genes, e.g., glucocorticoid-attenuated response gene 16 (GARG-16)/IFN-induced 56-kDa protein (IFI 56K), CXC-chemokine ligand 10 (CXCL10)/IFN-γ-inducible protein 10 kDa (IP-10) (Akira and Takeda, 2004; Niikura et al., 1997). IP-10 is a potent chemoattractant for activated T lymphocytes and NK cells (Liu et al., 2000). The MyD88-independent process is not mutually exclusive to the MyD88-dependent pathway, since TRIF can also activate receptor-interacting protein 1 (RIP-1), which then activates TRAF-6 and the MyD88-dependent process. The ability of TLR-4 to use two distinct pathways to generate an inflammatory response is unique among the TLRs.

Defects of the TLR-4 Signaling Pathway

Three major targets of mutation in this pathway are IRAK-4, NEMO, and IκBα. However, TLR-4 at the population level is highly polymorphic, in particular, in the extracellular ligand-binding domain, some of which are genetic risks for infections and rheumatologic and cardiovascular diseases (Smirnova et al., 2003).

IRAK-4

IRAK-4 is essential in mediating cellular activation following TLR engagement. As well, IRAK-4 mediates response to the IL-1/IL-18 superfamily. IRAK-4 deficiency is autosomal recessive, leading to invasive and life-threatening infections caused mostly by *Streptococcus pneumoniae*, *Staphylococcus aureus*, and *Neisseria meningitidis* (Ku et al., 2007). Infectious episodes are typically characterized by a blunted systemic inflammatory response (i.e., no or low-grade fever, minimal leukocytosis, low C-reactive protein). All infections have manifested before 14 years of age, and typically before 2 years of age. Commonly, in particular, for invasive disease, these infections are life-threatening, with 12 of the known 28 patients dying and all doing so prior to 8 years of age. All surviving patients demonstrated immunological compensation, such that their frequency of invasive infections decreased with age. In fact, no evidence of such infections was documented among the 6 patients over the age of 14 years, even in the absence of prophylaxis, suggesting that IRAK-4 deficiency is an immune defect primarily manifest in infancy and childhood. Cells from patients with IRAK-4 deficiency demonstrate profoundly impaired proinflammatory cytokine production when stimulated to TLRs 1, 2, 4, 5, 6, 7, 8, and 9. Why patients with IRAK-4 deficiency manifest susceptibility to only a limited range of pathogens and only in childhood remains unknown. IRAK-4 may be nonredundant for anti-*S. pneumoniae* immunity early in childhood, but may be redundant for other pathogens.

NEMO

In humans, NF-κB is a family of proteins consisting of reticuloendotheliosis (Rel), RelA (p65), RelB, p50, and p52. NF-κB is maintained in the cytosol in an inactive state by its association with inhibitors of NF-κB (IκB). The IκB family of proteins includes IκBα, IκBβ, IκBγ, IκBε, Bcl-3, p100, and p105 (Orange et al., 2005). The canonical pathway of NF-κB activation, as exemplified by TLR ligation, induces the assembly of the IKK complex that mediates the phosphorylation, ubiquitination, and degradation of IκB, thereby liberating NF-κB dimers, which then translocate to the nucleus, bind to NF-κB sites, and induce transcription. Gene mutations interfering with the NF-κB canonical pathway are clearly linked to susceptibility to infection.

Proteins of the NF-κB family are essential nuclear transcription factors expressed in a variety of tissues and have pleiotropic effects, particularly in development and immunity. Thus, while defects in cell surface-expressed receptors usually affect development or immunological functions, dysfunction of the NF-κB molecules or their regulatory elements typically produces defects in both development and immunity. This process is exemplified by mutations in NEMO and in IκBα.

NEMO is encoded by *IKBKG*, located on chromosome Xq28. NEMO self-assembles into trimers (Vinolo et al., 2006), allowing it to function as the scaffold for the IKK complex, bringing into proximity IKKα and IKKβ, which then permits cross-phosphorylation; this allows the subsequent phosphorylation and degradation of IκB and activation of NF-κB. NEMO also interacts with other proteins that regulate NF-κB activation, such as RIP and A20 (Caromody and Chen, 2007). NEMO permits the interaction of various modulators of NF-κB that ultimately result in proper NF-κB function.

IKBKG has 10 exons and encodes a 419-amino-acid protein with an N-terminal coiled-coil domain (CC1), a central coiled-coil domain (CC2), a leucine zipper, and a C-terminal zinc finger (Orange et al., 2005). The N-terminal CC1 region is responsible for interacting with IKKα and IKKβ, promoting the assembly of the IKK complex (May et al., 2000). There is an *IKBKG* pseudogene (termed ΔNEMO), which contains only exons 3 to 10. Although this pseudogene has neither transcript nor function, its presence has implications for the occurrence of mutations and the difficulty of genetic testing (Smahi et al., 2000).

The sex-linked aspect of *IKBKG* leads to some critical consequences. Amorphic NEMO mutations result in the complete lack of NF-κB activation via the canonical pathway. Hemizygous males with amorphic NEMO mutations usually die in utero before the second trimester. In contrast, females with heterozygous amorphic NEMO mutations often have incontinentia pigmenti (IP; Bloch-Sulzberger syndrome). The incidence of IP is estimated at between 1:10,000 and 1:100,000 (Courtois, 2005). The clinical presentation of IP in females consists of skin defects, in conjunction with neurological, ophthalmologic, and/or odontologic abnormalities. The skin abnormalities progress through four major stages: stage I (vesicular stage), characterized by vesicles/blisters within 2 weeks of birth and associated with massive eosinophilic granulocyte infiltration into the epidermis; stage II (verrucous stage), in which hyperkeratotic warty lesions develop, then wax and wane over time; stage III (hyperpigmented stage), with the typical hyperpigmented linear cutaneous streaks developing along the lines of Blaschko, representing melanin accumulation along paths of ectodermal cell migration during embryologic development of the skin; and stage IV, where there is atrophy, scarring, and hypopigmentation of the skin. In addition to skin, other ectoderm-derived structures are affected, including eccrine (sweat) glands, hair, and nails. Females with IP may demonstrate extensive X inactivation skewing (i.e., early selection against cells with the mutated X chromosome); this process during early life eventually results in the majority of surviving cells carrying the mutation on the inactive condensed X chromosome (Courtois, 2005). Studies of cells from spontaneously aborted fetuses have found that IP can be a complex rearrangement with excision of the region between exons 4 and 10 (Smahi et al., 2000). This large deletion produces a truncated protein, containing only the first three exons, that can still interact with IKKs but prevents the degradation of IκB and subsequent NF-κB activation. Females with IP do not demonstrate any consistent immune deficiency, nor any specific susceptibility to infections, presumably because extensive X inactivation skewing allows for normal immunity.

It is apparent from the IP phenotype that NF-κB is critical for ectoderm development. Ectodermal dysplasia (ED) is a clinically heterogeneous condition due to disordered development of the ectoderm. A clinical variant of ED manifests with anhidrosis (or hypohidrosis) and is termed ectodermal dysplasia with anhidrosis (EDA); it is characterized by the presence of sparse hair, abnormal (e.g., conical) or missing teeth, and inability to sweat. Ectodermal dysplasia causing mutations can occur in the gene for the ligand, ectodysplasin (*EDA1*), its receptor (EDA-R), or an adaptor molecule (EDARADD) (Orange et al., 2005). This signaling pathway is NEMO dependent and activates NF-κB; thus, NEMO mutations can also produce EDA. Patients with EDA associated with immunodeficiency (ID) have EDA-ID. EDA-ID can be X-linked (XL-EDA-ID) or autosomal dominant (AD-EDA-ID).

Whereas IP is often due to amorphic mutations and absent NEMO function in mosaic females, XL-EDA-ID results from hypomorphic NEMO mutations, which produce impaired, but not abolished, NF-κB signaling. XL-EDA-ID affects males in early childhood, usually <1 year of age (Orange et al., 2004). The estimated incidence of XL-EDA-ID is ~1:250,000 live male births (Orange et al., 2005). XL-EDA-ID manifests with unusually severe bacterial infections involving the lower respiratory tract, gastrointestinal tract, skin and soft tissues, and/or bones, and may be accompanied by meningitis or septicemia. *S. pneumoniae*, *S. aureus*, *Haemophilus influenzae*, and *Pseudomonas aeruginosa* can cause severe infections. Patients may have frontal bossing, sparse hair/eyebrows/eyelashes, depressed nasal bridge, abnormal dentition (especially conical incisors in either the decidual or the permanent teeth), and dry skin with heat intolerance (Orange et al., 2004). In addition to bacterial infections, approximately one-third of patients have also developed severe mycobacterial infection, mostly with M. avium complex. *Pneumocystis jiroveci*, candidiasis, cytomegalovirus colitis, herpes simplex virus oropharyngitis, chronic molluscum contagiosum, and diffuse cutaneous human papillomavirus can also be problematic in these patients (Orange et al., 2004). The most consistent immunological abnormality has been poor serum-antibody response to polysaccharide antigens; the levels of the other immunoglobulins (IgM, IgA, IgE) are variable. A hyper-IgM-like phenotype occurs in some cases (Jain et al., 2001). Impaired TNF-α responsiveness may explain the predisposition to mycobacteria. Location of the mutation within NEMO likely influences the quality and degree of hypomorphism, which may account for some of the clinical heterogeneity.

Multiple cases of immunodeficiency without EDA (ID sine EDA) have been reported, indicating that the immune and ectodermal functions are separable, even within the same family (Niehues et al., 2004; Puel et al., 2006; Filipe-Santos et al., 2006).

IκBα

A child with EDA-ID with associated severe T-cell immunodeficiency (Courtois et al., 2003) had a history of multiple pyogenic infections with both gram-positive and gram-negative organisms from the age of 2 months, requiring bone marrow transplantation at age 1 year. He had an autosomal dominant form of EDA-ID (AD-EDA-ID) due to a gain-of-function mutation that impaired phosphorylation and degradation of IκBα; thus, the inhibitory function of IκBα onto NF-κB was enhanced. Because the patient was heterozygous, both forms of IκBα (i.e., wild-type and mutant) were expressed, yet the mutant protein exerted a dominant negative effect, preventing NF-κB activation. Since that original description of AD-EDA-ID, an additional case has been reported (Janssen et al., 2004). This young boy suffered from recurrent group A streptococcal infection (meningitis, pneumonia/bacteremia) from the age of 2 months. He also developed *P. jiroveci* (*carinii*) pneumonia, recurrent mucosal candidiasis, and had a "hyper-IgM-like" phenotype. Features of EDA developed after the first months of life (i.e., conical teeth, periorbital wrinkling). Interestingly, the father had a *forme fruste* variant, with persistent *Salmonella enterica* serovar Typhimurium infection, initially presenting with intestinal manifestations as a young adult, but complicated by disseminated extraintestinal disease over the span of 1.5 years. The father did not have features of EDA, nor did he have a history of other recurrent or severe infections. He did not have T lymphocytosis and possessed both naive and memory subsets in his circulation. Functional studies of monocyte function in vitro demonstrated that the severity of the father's immune deficiencies was intermediate between that of his son and those of healthy controls. His T-cell function was relatively normal. Sequence analysis revealed that the father had the exact same mutation in IκBα. However, the father displayed complex mosaicism: CD45 RA (naive) T cells were heterozygous for the *IKBA* mutation, whereas CD45 RO (memory) T cells were wild-type, suggesting in vivo reversion and selection for wild-type T cells in the father.

EDA-ID results from defective signaling through NF-κB, either due to hypomorphic NEMO mutations or, less commonly, from dominant negative mutations in IκBα. The range of mutations is broad and expanding. Amorphic NEMO mutations result in incontinentia pigmenti in females and hypomorphic mutations cause XL-EDA-ID (or its variants) in males.

MACROPHAGE PHASE

Phagocytosis

Phagocytosis is an essential mechanism used by macrophages for both immunity and scavenging. It is a complex yet coordinated process, commencing with a cognate ligand/receptor interaction that initiates signaling cascades, resulting in formation of the phagocytic cup, membrane remodeling, vacuole formation, particle internalization, phagosome maturation, and ultimately digestion of the engulfed substance.

Ingestion

Phagocytosis is initiated when ligands interact with their receptors on the cell surface of the macrophage. The best characterized are opsonin dependent, i.e., the Fcγ receptor (FcγR) and the complement receptor 3 (CR3). The mannose receptor (MR), or the more novel receptors (dectin-1 [Brown et al., 2002], DC-SIGN [Zhang et al., 2006], CD44 [Vachon et al., 2006], macrophage C-type galactose/N-acetylgalactosamine-specific lectin [MGL]/CD301, and DEC-205-associated C-type lectin-1 [DCL-1]/CD302 [Kato et al., 2007]) have less well defined mechanisms of phagocytosis induction.

FcγR-Mediated Phagocytosis

The FcγRs recognize the Fc domain of IgG. Human macrophages mediate phagocytosis by FcγR-I, FcγR-IIA, and FcγR-IIIA. FcγR-IIA is a complete molecule, consisting of an extracellular Fc binding domain, a transmembrane domain, and an intracytoplasmic tail containing an immunoglobulin-gene-family tyrosine activation motif (ITAM) that functions in intracellular signaling. FcγR-I and FcγR-IIIA have only extracellular Fc binding domains, transmembrane domains, and short tails without ITAM motifs. Therefore, they must associate with a dimer of a small transmembrane protein containing ITAM motifs, the γ-chain, to signal. Clustering of FcγRs at the cell surface by IgG results in tyrosine phosphorylation of the ITAMs by Src, a family of membrane-associated tyrosine kinases (Fig. 2). The phosphorylated tyrosine residues on the ITAMs then serve as docking sites for SH2 domains of Syk kinases. Syk is critical in FcγR-mediated signaling: inhibition of Syk, either through knockout (Crowley et al., 1997) or knockdown (Matsuda et al., 1996), significantly reduces phagocytosis. Syk is particularly important for actin polymerization and formation of the phagocytic cup, the concave depression composed of actin filaments that assemble beneath the opsonized particle during FcγR-mediated phagocytosis.

Many enzymes produce secondary messenger molecules, including protein kinase C (PKC), various phospholipases (A2, Cγ, and D), phosphatidylinositol 3-kinase (PI3K), extracellular signal-regulated kinase (ERK), and the Rho family of GTPases. PI3K, a lipid kinase, has three major functions during phagocytosis (Garcia-Garcia and Rosales, 2002): (i) local recruitment of molecules containing pleckstrin homology (PH) domains (e.g., myosin X; Vav proteins); (ii) regulation of membrane events at the phagocytic cup, resulting in pseudopod extension necessary for particle engulfment; and (iii) activation of ERK via intermediary molecules. ERK has the ability to activate phospholipase A2 and produce arachidonic acid, as well as modulating actin dynamics.

The actin cytoskeleton provides a scaffold on which myosins interact to move intracellular organelles. Activation of myosin is by myosin light-chain kinase (MLCK), whose activation is ERK dependent. The Rho family of small GTPases, including Rho, Rac, and Cdc42, are crucial for the actin cytoskeleton rearrangement that occurs during phagocytosis, diapedesis, chemotaxis, etc. Activation of Cdc42 results in its accumulation on the intracytoplasmic aspect of the cell membrane, beneath the phagocytic cup, where it triggers actin assembly and formation of pseudopods. Rac activation also results in localized accumulation at the nascent phagosome, but it is involved in particle internalization. Activation of Rac and Cdc42 is mediated by guanine-nucleotide exchange factors (GEFs), a family of molecules that activate the Rho family by conjugating them to GTP. One group of these GEFs, the Vav proteins, is important not only in the phagocytic process, but also in the generation of the oxidative burst (Miletic et al., 2007).

The activated (GTP-bound) forms of Rac and Cdc42 accumulate at the site of phagosome formation, where they exert their action through the Wiskott-Aldrich syndrome protein (WASP). WASP is a cytoplasmic protein of 502 amino acids, encoded at chromosome Xp11.22-Xp11.23 (Ochs, 2001). It is found only in hematopoietic cells and consists of six functional domains (Ochs and Notarangelo, 2005). It binds constitutively to WASP-interacting protein (WIP), resulting in inhibition of WASP protein in T lymphocytes until engagement of the T-cell receptor leads to phosphorylation of WIP and liberation/activation of WASP at the immune synapse (Sasahara et al., 2002). The assembled WASP/WIP complex is also required for phagocytic cup formation (Tsuboi and Meerloo, 2007), as well as podosome assembly (Tsuboi, 2007). Interaction with a lipid product of PI3K, phosphatidylinositol 4,5-biphosphate [PI(4,5)P2, or PIP2], enhances WASP effector activity. Rac and Cdc42 can also activate p21-activated kinase (PAK1), which phosphorylates/activates LIM kinase (LIMK) (Edwards et al., 1999). Activated LIMK phosphorylates the actin-depolymerizing factor, cofilin, and in doing so, inhibits it; this permits stabilization of actin filaments. Rac activates the ADP-ribosylating factor 6 (ARF-6) via POR-1 (D'Souza-Schorey et al., 1997). ARF-6 stimulates actin polymerization, and it induces focal exocytosis of internal vesicles at or near the site of phagocytosis, thereby providing enough cellular membrane for phagocytosis to occur (Zhang et al., 1998; Niedergang et al., 2003). Through pathways involving PI3K, ERK, and activated (GTP-bound) Rac and Cdc42, the macrophage undergoes cytoskeletal actin polymerization and rearrangement, in conjunction with membrane recycling, to produce pseudopods at focal areas of the cell membrane where particles have bound to surface receptors.

CR3-Mediated Phagocytosis

Phagocytosis can also be mediated by complement receptor 3 (CR3, CD11b/CD18, or Mac1), an $\alpha_M\beta_2$ integrin heterodimer that recognizes inactivated complement 3b (iC3b). This process utilizes RhoA, rather than Cdc42/Rac (Wiedemann et al., 2006). Subsequently, RhoA activates Rho-kinase, which interacts with myosin II to regulate the Arp2/3 complex allowing for actin accumulation and formation of the phagocytic cup (Olazabal et al., 2002).

The Wiskott-Aldrich Syndrome: a Defect in Phagocytosis Too

The Wiskott-Aldrich syndrome is an X-linked recessive primary immunodeficiency with recurrent otitis media, eczema, thrombocytopenia, and bloody diarrhea. Classic Wiskott-Aldrich syndrome (WAS) has recurrent infections, eczema, and microthrombocytopenia. However, this classic constellation was seen in only 27% of patients subsequently confirmed to have WAS (Sullivan et al., 1994). A clinically milder variant, X-linked thrombocytopenia, does not have the associated immunodeficiency/increased

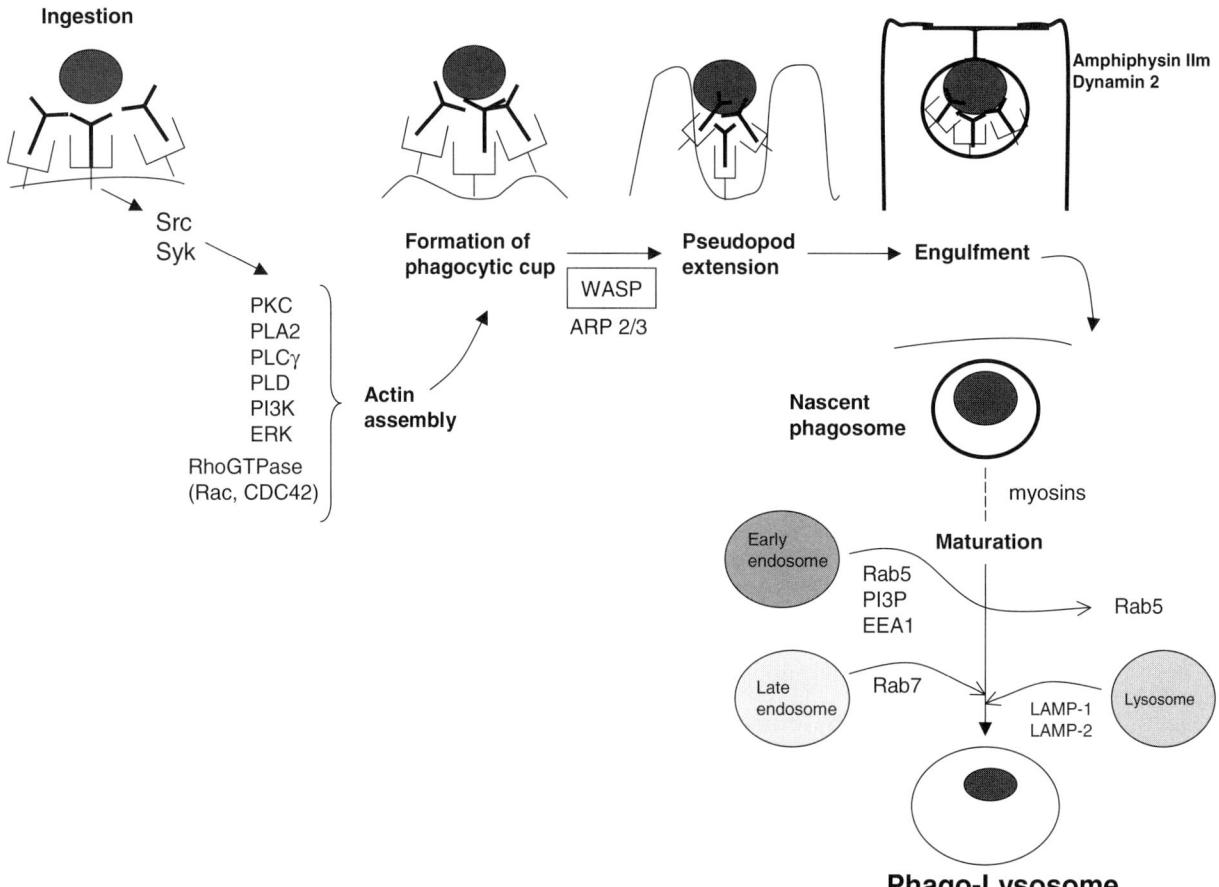

FIGURE 2 Critical components for macrophage phagocytosis and phagosomal maturation to form the phagolysosome. Key molecules are listed (see text for details). Opsonization with IgG allows for FcγR-mediated internalization of pathogens or particles (ingestion). Subsequently, various kinases and secondary messenger molecules allow for reorganization of the macrophage cytoskeleton, specifically, actin assembly at the plasma membrane area engaged with the opsonized pathogen and the formation of cell membrane protrusions (phagocytic cup). Continuous growth of these protrusions allows for pseudopods to surround the pathogen, allowing for subsequent engulfment. The nascent phagosome, containing the pathogen, moves centripetally via myosins and other intracellular machinery. Maturation of the phagosome requires coordinated fusion with various vesicles, including the early endosome, the late endosome, and the lysosome, during which the protein content of both the phagosome membrane and the phagosome lumen are modified.

susceptibility to infections or dermatologic involvement. The most common infections involved the upper and lower respiratory tracts and were presumed bacterial (Sullivan et al., 1994). Otitis media was typical during the first 6 months of life (Ochs, 2001). Presumed infectious diarrhea was also common. Life-threatening infections (i.e., meningitis, sepsis) were occasional initial manifestations. Viral infections came after bacterial infections and were typically with herpesviruses, molluscum contagiosum, and human papillomaviruses. P. jiroveci (carinii) pneumonia, superficial candidiasis, as well as increased risks for autoimmunity and malignancy are common (Ochs, 2001).

Although WAS is classically viewed as a combined B- and T-cell deficiency, monocytes/macrophages are also significantly impaired. WASP-deficient macrophages demonstrate grossly abnormal morphology. They have impaired podosome formation, which is critical for cell migration and macrophage chemotaxis (Tsuboi, 2006). Peripheral blood monocytes and primary macrophage cultures from patients with classical WAS also demonstrate impaired FcγR-mediated phagocytosis due to defective phagocytic cup formation (Tsuboi and Meerloo, 2007).

In summary, WASP plays a critical role both in macrophage chemotaxis and phagocytic cup formation, as well as in T/B cell interactions. Dysfunctional WASP leads to defective cytoskeletal mobilization, impairs phagocytosis, affects the earliest moment when the host recognizes pathogen, and interferes with the immunological synapse.

Internalization

For a particle to be properly engulfed, membrane remodeling has to occur so that enough of the cellular lipid bilayer is available at the site of phagocytosis. Directionality is conferred by proteins on the vesicle membrane, v-SNARES, binding to cognate proteins on the targeted budding phagosome, t-SNAREs. This specificity is further ensured by Rab proteins, small GTPases of the Ras superfamily involved in vesicle trafficking/docking (Cai et al., 2007). Rabs regulate discrete steps in the phagocytic/endocytic as well as the exocytic pathways. More than 60 Rab proteins are identified in mammalian cells (Chavrier and Goud, 1999; Gissen and Maher, 2007). Rab proteins are located on the cytoplasmic surface of the vesicle. In the resting state, Rab proteins are inactively bound to GDP; this arrangement is maintained by the GDP-dissociation inhibitor (GDI). This association prevents indiscriminate membrane-membrane interactions, thus providing a degree of specificity during the process of membrane recruitment. Rab-GDP binds to vesicle membrane, releasing GDI. GDP is then exchanged for GTP by GEF, thus activating Rab. The Rab-GTP complex interacts with various effector molecules, including myosin, which allows propagation along the actin cytoskeleton, and specific molecules on the acceptor membrane, permitting accurate localized delivery of the vesicle. After fusion of the membranes, the GTP is hydrolyzed, and the Rab-GDP complex can reinitiate the cycle (Cai et al., 2007).

Numerous and diverse Rab proteins exist. Human diseases associated with defects in Rab proteins have predominantly neurological and/or neurodevelopmental abnormalities, such as Griscelli syndrome type II, which is due to mutation of Rab27 (Seabra et al., 2002).

After the substrate has been bound by macrophage surface receptors and engulfed by its pseudopods, the nascent phagosome has to be internalized. Particle internalization is regulated by amphiphysin IIm and dynamin 2, which cleave the budding phagosome from the intracytoplasmic surface of the cell membrane (Ren et al., 2006). Myosins, by interacting with the actin cytoskeleton, subsequently provide the intracellular "muscle" to propel the phagosome along the maturation process (Swanson and Hoppe, 2004).

Phagosome Maturation

Once internalized, the phagosome matures through a series of interactions with endosomal membranes, eventually fusing with lysosomes. The resultant phagolysosome is enriched in hydrolytic enzymes, ROS, and antimicrobial peptides that promote the killing and degradation of internalized microorganism. Fusion with the lysosome also permits the phagosomal membrane to acquire proteins critical for function, such as the vacuolar-ATPase (V-ATPase) that renders the lumen of the phagolysosome acidic. Phagosome maturation is key in determining the fate of certain engulfed pathogens, and some organisms have evolved mechanisms that interfere with specific steps in this process, providing them with the capacity to survive intracellularly.

Phagosome maturation begins with fusion to early endosomes, acquiring Rab5 (also known as Rab5a), which in turn recruits a specific type III PI3K that generates PI-3-phosphate (PI3P) (Henry et al., 2004). The concomitant presence of Rab5 and PI3P is necessary for recruitment of early endosomal antigen 1 (EEA1), which mediates vesicle docking and fusion (Henry et al., 2004). PI3P can also facilitate assembly of the NADPH oxidase complex via recruitment of p40^{phox} and p47^{phox} (Kanai et al., 2001; Zhan et al., 2002). The maturing phagosome subsequently accumulates lysosome-associated membrane proteins 1 and 2 (LAMP-1 and LAMP-2) from lysosomes, has Rab5 removed, and incorporates Rab7 from late endosomes and lysosomes. This synchronous Rab5-Rab7 replacement is termed Rab conversion (Rink et al., 2005). The LAMPs promote docking of phagosomes to the cytoskeleton and permit centripetal movement via interactions with dynein/dynactin motor units (Huynh et al., 2007). Rab7, via its effector protein Rab7-interacting lysosomal protein (RILP), also permits locomotion using the dynein/dynactin complex and directs fusion of phagosomes to lysosomes (Sun et al., 2007). The actual fusion of membranes is mediated by SNARE proteins. Thus, phagosome maturation requires a coordinated acquisition of Rab5, Rab7, and LAMPs, which permit secondary interactions that culminate in fusion with the lysosome, resulting in a microbicidal organelle.

No human immunodeficiencies due to mutations in Rab5 have been identified. Rab7 mutations cause Charcot-Marie-Tooth disease, axonal, type 2B, an autosomal dominant peripheral neuropathy (Verhoeven et al., 2003). LAMP-2 deficiency causes Danon's disease, an X-linked dominant neuromuscular condition characterized by hypertrophic cardiomyopathy, cardiac conduction abnormalities, skeletal vacuolar myopathy, accumulation of autophagic material intracellularly, variable mental retardation, and peripheral pigmentary retinopathy (Eskelinen, 2006). However, none of these intracellular trafficking protein defects appear to confer significant susceptibility to infections.

Certain pathogens have developed methods to interfere with the process of phagosome maturation, especially mycobacteria. The phagosomes containing *Mycobacterium tuberculosis*, *Mycobacterium bovis* BCG, and *M. avium* display significantly reduced acidification compared with those containing killed organisms or latex beads. This effect is associated with a paucity of V-ATPase pumps, suggesting that mycobacteria either exclude the incorporation of such pumps into early phagosomes or induce their degradation (Sturgill-Koszycki et al., 1994). Mycobacterial phagosomes also demonstrate persistence of early endosomal markers (Rab5) and limited or no acquisition of the late endosomal markers Rab7 (Via et al., 1997) and LAMP-1 (Clemens et al., 2000b). The mechanism of impaired Rab5-Rab7 conversion is probably multifactorial, including impaired recruitment of the early endosomal tethering factor, EEA1 (which can be reproduced by the mycobacterial lipid, lipoarabinomannan, LAM) (Fratti et al., 2001). Other mycobacterial impediments to phagolysosomal maturation and function include interference with PI3P activity (Chua and Deretic, 2004; Vergne et al., 2005), preferential recruitment of alternative proteins such as Rab14 (Kyei et al., 2006) and Rab22a (Roberts et al., 2006) that preclude the acquisition of Rab7, and degradation of SNARE proteins (Fratti et al., 2002). *S. enterica* serovar Typhimurium (Smith et al., 2007), *Legionella pneumophila* (Clemens et al., 2000a, 2000b), *Chlamydia* spp. (Rzomp et al., 2003), and *Leishmania donovani* (Scianimanico et al., 1999) all actively interfere with and subvert phagolysosomal fusion, acquisition of proteins, and maturation.

The Respiratory Burst

Like neutrophils, macrophages use a respiratory burst, in which oxygen is consumed in response to a stimulus. This process depends on the assembly and activation of a multicomponent complex, the reduced NADPH oxidase. The NADPH oxidase catalyzes the generation of superoxide (O_2^-) from molecular oxygen ($2 O_2 + NADPH \rightarrow 2 O_2^- + NADP^+ + H^+$). At rest, the components of the NADPH oxidase are in a membrane-bound form, the flavocytochrome b_{558}, composed of gp91^{phox} and p22^{phox} subunits; and a soluble cytoplasmic form consisting of p47^{phox}, p67^{phox}, p40^{phox}, and Rac1/Rac2 (Fig. 3). Flavocytochrome b$_{558}$ is a dimer of gp91^{phox} and p22^{phox} that binds NADPH, heme, and FAD, the latter two of which are necessary for electron transport. p67^{phox} and p47^{phox} are structural proteins necessary for superoxide generation, while p40^{phox} functions as a regulator and increases the affinity of p47^{phox} for flavocytochrome b$_{558}$ (Cross, 2000). Once the cell is activated by engagement of specific cell surface receptors, phospholipase C mediates release of diacylglycerol and inositol triphosphate (IP$_3$), which causes release of Ca^{2+} from the endoplasmic reticulum (ER), in conjunction with increased calcium influx from the extracellular environment (Iles and Forman, 2002). Protein kinase Cα (PKCα) is activated by the increased intracellular Ca^{2+} concentration, which in turn activates an isoform of phospholipase A$_2$ sensitive to calcium (cPLA$_2$), allowing for the production of arachidonic acid. Another isoform of PKC, PKCδ, regulates both the phosphorylation of the cytosolic components, as well as their translocation to the membrane (Zhao et al., 2005), a task facilitated by arachidonic acid (Cathcart et al., 2004). The activated p47^{phox}, p67^{phox}, and p40^{phox} proteins then translocate *en bloc* to associate with the flavocytochrome b$_{558}$. Rac1/Rac2 translocates independently to the assembling oxidase, after separating from the GDP-dissociation inhibitor (RhoGDI) and acquiring GTP (DeLeo et al., 1999). In the macrophage, Rac1 is the predominant Rac isoform, while neutrophils preferentially express Rac2 (Zhao et al., 2003).

The fully assembled activated NADPH oxidase then generates superoxide anion by transferring electrons from NADPH in the cytoplasmic space to O$_2$ in the vacuolar or extracellular space. This process is supported by redirecting glucose metabolism through the hexose monophosphate (HMP) shunt, which allows glutathione reductase to regenerate NADPH. Superoxide anion is highly reactive and either combines with other molecules or is converted by superoxide dismutase (SOD) to hydrogen peroxide (H$_2$O$_2$). In the neutrophil, H$_2$O$_2$ is converted to hypochlorous acid (HOCl) by myeloperoxidase (MPO), and to hydroxyl anion (OH$^-$), hydroxyl radical (OH$^•$), nitryl chloride (NO$_2$Cl), and peroxynitrite anion (ONOO$^-$). The lattermost product occurs through the interaction of O$_2^-$ with nitric oxide (NO), produced through the oxidation of L-arginine by inducible NOS (iNOS). Collectively, these compounds are ROS or RNS, respectively. Murine experimental systems have shown definitive roles for RNS in host defense, but human data for the clinical relevance of NO/RNS have been less voluminous (Nathan, 2006).

Differences in the NADPH oxidase characteristics exist between neutrophils and macrophages. In neutrophils, the resting NADPH oxidase components are both in the specific granules and in the cell membrane; after stimulation, most of the activity is intracellular, from the specific granules. In contrast, macrophages display no or little oxidase in granule/lysosomal membranes (Johansson et al., 1995). Thus, macrophages must rely on cell membrane-bound oxidase components to be included in the phagosomal

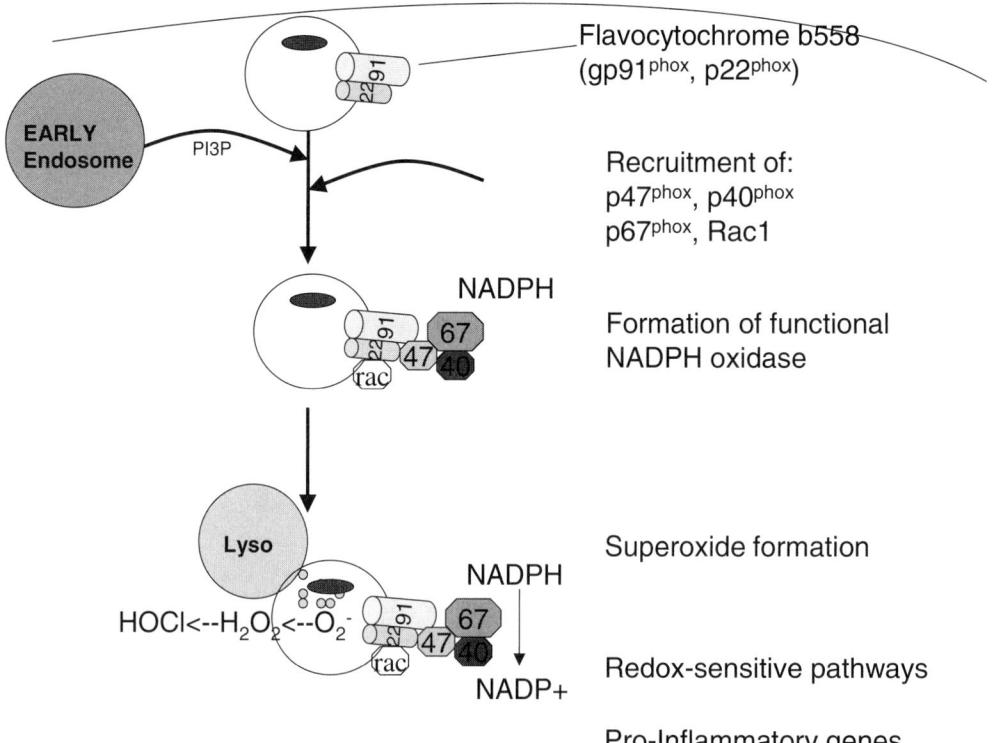

FIGURE 3 The macrophage respiratory burst. Early in the process following phagocytosis, the NADPH oxidase assembles on the phagosome. This multisubunit enzyme complex allows for the respiratory (or oxidative) burst, whereby ROS (i.e., superoxide anion [O$_2^-$] and hydrogen peroxide [H$_2$O$_2$]) are generated within the phagosome. The ROS contribute to pathogen killing, primarily by activating pathways culminating in the expression of proinflammatory genes.

membrane during phagocytosis for intracellular ROS production. Generation of ROS into the extracellular space by plasma membrane NADPH oxidase may be important when macrophages encounter pathogens too large to be phagocytosed. Whereas neutrophils possess MPO in their azurophilic (primary) granules, the MPO content of monocytes is ~50% of that in neutrophils; most (>99%) of it is lost during differentiation of monocytes into macrophages (Johansson et al., 1995). Thus, conversion of H_2O_2 to HOCl is low in macrophages.

The kinetics of ROS production differs between neutrophils and macrophages. Although monocytes/macrophages possess approximately the same amount of gp91^{phox} as neutrophils (Johansson et al., 1995), the amount of O_2^- and H_2O_2 produced by the respiratory burst is much less in macrophages (Iles and Forman, 2002). The time-to-peak NADPH oxidase activity is shorter in neutrophils, with most ROS produced within 2 to 10 min of the stimulus (Cathcart, 2004). In contrast, the macrophage response is protracted, with peak production occurring approximately 1 h after stimulation, but waning over several hours. Macrophages also demonstrate repetitive excitability, being able to mount repeat responses after sufficient recovery (Cathcart, 2004).

Historically, ROS were considered antimicrobial molecules whose primary purpose was to kill phagocytosed microorganisms. Extensive early work demonstrated that O_2^- and H_2O_2 were not the molecules primarily used for killing, because O_2^- was relatively innocuous and H_2O_2 was only weakly microbicidal (Babior, 1984). Instead, these molecules were thought to generate potent antimicrobial agents, including oxidized halogens (i.e., HOCl, chloramines/R-NH_2Cl) and oxidizing radicals (i.e., OH·). In vitro, these molecules can be shown to inflict damage on bacterial proteins and nucleic acids and to induce genetic mutations. It had even been proposed that phagocytes, by virtue of their ability to produce reactive compounds and mutagens, may be mediators of carcinogenesis in response to specific insults (Babior, 1984).

However attractive the reactive oxidant hypotheses, they were challenged by creation of mice deleted for neutrophil elastase and cathepsin G (Tkalcevic et al., 2000). These animals showed susceptibility to staphylococci and *Aspergillus*, characteristic chronic granulomatous disease (CGD) pathogens, despite normal superoxide generation. These gene knockout mice suggested that superoxide per se was not sufficient for bactericidal or fungicidal activity. This led to investigation of the molecular mechanisms governing susceptibility in CGD. The resulting discovery of a potassium channel in neutrophil vacuoles has refocused attention from superoxide as a central cidal molecule to a crucial signal molecule that governs the magnitude and efficacy of killing (Reeves et al., 2002). In this hypothesis, the NADPH oxidase generates O_2^-, thereby depolarizing the phagosomal membrane of the neutrophil, while O_2^- consumes intraphagosomal protons. This results in a compensatory K^+ influx into the phagosome. Azurophilic (primary) granule contents within the phagosome, specifically cathepsin and elastase, which are bound to a glycocalyx at rest, are liberated by the movement of K^+ into the phagosome, permitting activation of the antimicrobial peptides that mediate pathogen killing (Segal, 2005).

Although this new perspective on the function of the NADPH oxidase has altered our perception of the role of ROS in intracellular activity in the neutrophil, macrophages do not possess the same arsenal of granular antimicrobial peptides characteristic of neutrophils. Specifically, macrophages lack primary (azurophilic) granules, secondary granules, and secretory granules, although they do possess various glycosidases and proteases (Savina and Amigorena, 2007). Therefore, ROS in macrophages do not perform the exact same functions as they do in neutrophils. However, in the macrophage, NADPH oxidase generates ROS that also function as second messengers, thus activating signaling pathways within the macrophage.

Forman and Torres suggest four criteria for candidate second messengers: (i) they are either enzymatically generated or regulated by channels/pumps; (ii) they are enzymatically degraded; (iii) their concentration can be increased or decreased within a short period; and (iv) they are specific in action (Forman and Torres, 2001). The ROS generated by NADPH oxidase fulfill these criteria: O_2^- and H_2O_2 are produced by NADPH oxidase and SOD, respectively, while they are degraded by SOD and catalase or glutathione peroxidase, respectively. Their levels are tightly regulated within discrete periods of time, and both demonstrate specificity, although by seemingly contradictory means. Because O_2^- is unstable, its site of action is within a very short radius of its site of production. This spatial restriction for O_2^- reactivity translates into a restricted location of action, conferring specificity (Forman and Torres, 2002). H_2O_2, on the other hand, is relatively stable, reacting rapidly only with thiolates (S^-) and transition metals (Forman and Torres, 2001).

Signaling pathways activated by ROS are termed redox sensitive. In particular, those resulting in the stimulation of certain transcription factors (e.g., NF-κB, AP-1) or some members of the MAPK signaling pathways (e.g., ERK, p38^{mapk}, JNK) (Iles and Forman, 2002; Forman and Torres, 2001, 2002; Tephly and Carter, 2007a, 2007b), are redox sensitive; many of these are critical in the expression of proinflammatory genes. The exact mechanisms of ROS-induced activation remain incompletely defined. Furthermore, the precise pathway activated by ROS varies with the inciting stimulus, as well as the specific monocyte-derived cell type (e.g., alveolar macrophage versus Kupffer cell versus macrophage cell lines). Nonetheless, it is clear that the role of ROS in macrophages extends beyond simple direct bacterial killing to signaling pathways responsible for classical macrophage activation and for the subsequent coordination of the inflammatory response.

Defects in the Respiratory Burst: CGD

CGD is an inherited disorder characterized by defects in the ability of phagocytes to generate ROS. CGD is caused by mutations in one of four genes encoding the structural components of the NADPH oxidase: gp91^{phox} (encoded by CYBB), p47^{phox} (NCF-1), p22^{phox} (CYBA), and p67^{phox} (NCF-2). Cases involving gp91^{phox} are the most common (~70%) and are inherited in an X-linked recessive manner. Mutations involving the other genes are inherited in an autosomal recessive manner, with p47^{phox} mutations being second most common (~25%), followed by p22^{phox} and p67^{phox} (<5% each) (Segal et al., 2000; Winkelstein et al., 2000). The minimum estimate for CGD is between 1/200,000 and 1/250,000 live births. A single patient with a dominant-negative mutation in the Rac2 gene has been described; this patient was a 5-week-old infant male who presented with severe bacterial infections, poor wound healing, and absence of pus in the wounds (Ambruso et al., 2000; Williams et al., 2000). No human cases of Rac1 deficiency have been reported. Since the Rac1 knockout mouse dies early in development, Rac1 deficiency may not be compatible with life (Wells et al., 2004). To date, no human case of p40^{phox} deficiency has been reported, but this is unlikely to have a classic CGD phenotype, since p40^{phox} is thought to have a regulatory role in the NADPH oxidase, and not a structural one.

CGD is characterized by recurrent life-threatening infections, typically with select catalase-positive bacteria and filamentous fungi. Pneumonia is the most common serious infection, caused by both *Aspergillus* spp. and bacteria. The presentation and outcome of *Aspergillus* infection in CGD is distinct from that encountered in the more common settings of neutropenia or organ transplantation. On histopathology, aspergillosis in CGD is not angio-invasive (Almyroudis et al., 2005; Stergiopoulou et al., 2007). The corollary to this histopathologic difference is that the radiographic manifestations of pulmonary aspergillosis in CGD typically do not include the "classical" features from neutropenia of the halo sign, pulmonary infarction, or the air-crescent sign. Despite frank angio-invasion, aspergillosis in CGD patients is tissue invasive and disseminated disease may occur. However, the most common progression of aspergillosis is local.

After pneumonia, the most common infections are suppurative adenitis (*S. aureus*), subcutaneous abscess (*S. aureus*), liver abscess (*S. aureus*), osteomyelitis (*Serratia marcescens*), and sepsis (*Burkholderia cepacia* complex, *Salmonella* spp.) (Winkelstein et al., 2000). Infections account for most CGD deaths. Although *Aspergillus* spp. reports dominate the published literature, the advent of azole antifungals for prophylaxis and therapy has dramatically altered the spectrum of infections and causes of mortality in CGD (Gallin et al., 2003; Segal et al., 2005). X-linked CGD is generally more severe than the autosomal recessive forms, as shown by earlier age of onset/diagnosis, more frequent granulomatous complications, and higher mortality. Patients with CGD often have exuberant inflammation and granuloma formation evinced by inflammatory bowel disease, urinary tract obstruction, and wound dehiscence, as well as autoimmune phenomena like juvenile idiopathic arthritis, idiopathic thrombocytopenic purpura, and discoid lupus erythematosus.

The diagnosis of CGD can be made on the basis of functional assays of O_2^- production by neutrophils, demonstration of protein loss, or by genetic determination. The nitroblue tetrazolium (NBT) assay relies on the ability of neutrophils to reduce clear yellow NBT to blue-black formazan within the neutrophils. The NBT assay has been used for decades, is simple, and uses a microscope to read the assay. However, it is relatively insensitive to hypomorphic mutations and is potentially compromised by observer bias. The dihydrorhodamine oxidation assay relies on flow cytometry and is readily able to distinguish X-linked, hypomorphic, and autosomal forms of CGD (Vowells et al., 1996).

Previous explanations for the selective infection susceptibility in CGD relied on microbial catalase as a critical virulence factor. The argument was that, whereas CGD phagocytes lacked superoxide and hydrogen peroxide production, hydrogen peroxide-producing microbes (the vast majority) would complement the defect in the CGD cell unless they also produced catalase to degrade their own hydrogen peroxide. Therefore, catalase-negative organisms would inundate the phagosome with their own H_2O_2, thereby providing the substrate that was lacking and facilitating their own demise. However, laboratory-generated catalase-negative strains of high-grade CGD pathogens S. aureus and *Aspergillus nidulans* showed no difference in virulence between catalase-positive and -negative strains, indicating that catalase is not a crucial virulence factor in CGD (Chang et al., 1998; Messina et al., 2002). Therefore, the mechanism of selective susceptibility in CGD is still unknown, but is clearly not due to microbial catalase.

In summary, CGD neutrophils and macrophages are defective in O_2^- production. Superoxide and its metabolites are critical in intra- and intercellular signal molecules whose roles in the regulation of inflammation and microbial killing are more complex than previously assumed.

The Nitrogen Burst

IFN-γ is central for the generation of macrophages with antimicrobial activity against intracellular pathogens through both oxygen-dependent and oxygen-independent pathways. In mouse macrophages, RNS are important mediators of antimicrobial activity; indeed, production of RNS is a hallmark of classically activated murine macrophages, produced concomitantly with ROS (Fang, 2004; Nathan and Shiloh, 2000). ·NO was initially recognized as a vasodilatator (Furchgott and Zawadzki, 1980), but as its biology was studied, it became clear that it also played a role in infection. Although macrophages produce less ROS than neutrophils, they produce considerably more RNS than their polymorphonuclear counterparts (Fang, 2004). The crucial role of RNS in mouse macrophage immunity is supported by the fact that inhibition of RNS production blocks the antimicrobial activity of IFN-γ-stimulated mouse macrophages against a variety of pathogens (Fang, 2004; Nathan and Shiloh, 2000).

On macrophage activation, iNOS is upregulated and catalyzes the conversion of L-arginine to nitric oxide (·NO) according to the following stoichiometric formula: 3 NADPH + 2 L-arginine + 3 O_2 + H^+ → 3 $NADP^+$ + 2 ·NO + 2 OH^- + 2 citrulline (Forman and Torres, 2001). Under conditions of reduced substrate (like L-arginine) or cofactor availability, iNOS can also produce O_2^-, emphasizing the interaction between ·NO and ROS. Furthermore, ·NO can react with O_2^- to form peroxynitrite anion ($ONOO^-$), a radical with potential for membrane and DNA damage.

The role of ·NO or of iNOS in human susceptibility to infections is unclear. To date, human iNOS deficiency has not been reported. However, polymorphisms in the iNOS promoter have been associated with human susceptibility to infection or severe disease with *M. tuberculosis* (Gomez et al., 2007), *Plasmodium falciparum* (Mombo et al., 2003), and brucellosis (Orozco et al., 2003). However, such associations were not seen in other studies (Levesque et al., 1999). Given its rapid metabolism and rapid mutually annihilating interactions with superoxide, ·NO is likely to play an important role in the generation or modulation of human inflammation, as it does in mice (van der Veen et al., 2000).

Intracellular Killing by Macrophages

How are phagocytosed microbes actually inactivated and destroyed by macrophages? The phagosome gradually acquires molecules necessary for lysosomal fusion to produce the phagolysosome. In doing so, it also acquires the acidification machinery and the lysosomal degradative enzymes necessary to digest microbes.

In macrophages, acidification of the phagosome is accomplished primarily by the V-ATPases obtained during phagolysosomal fusion (Lukacs et al., 1990). The V-ATPase is a multisubunit enzyme with a membrane-bound component (V_0 domain) and an intracellular catalytic component (V_1 domain) that functions to transport H^+ into luminal/extracellular spaces. The V_0 domain, consisting of eight components, is responsible for proton translocation, while the V_1 domain, constructed by seven components, is responsible for ATP hydrolysis (Forgac, 2007). Several isoforms of V-ATPase have been identified, demonstrating tissue-specific distribution and characteristics (e.g., phagocytes, renal cells, osteoclasts, spermatid-producing cells) (Sun-Wada et al., 2003). In macrophages, these proton-pump channels are recruited to the phagosome within minutes of ingestion of bacteria, decreasing the phagosomal pH from 7.4 (i.e., extracellular milieu) to <6 by 20 min after engulfment; the estimated rate of pH decline is 0.2 to 0.4 pH units per minute (Savina and Amigorena, 2007). The centrality of the V-ATPase to acidification of the phagosome is demonstrated by the abrogation of this effect with inhibitors like bafilomycin A and concanamycin B (Drose and Altendorf, 1997).

Acidification of the phagosome may be assisted by other molecules. Some, like natural resistance-associated macrophage protein gene 1 *(Nramp1)* (Hackam et al., 1997), are of unclear clinical significance in humans, while others, such as Na/H exchange or the cystic fibrosis transmembrane receptor (CFTR), have been shown to have only modest relevance to phagosomal acidification (Haggie and Verkman, 2007). Establishing an acidic phagosomal compartment during phagocytosis assists in denaturation of proteins, increasing their accessibility and susceptibility to proteolysis, and activation of the lysosomal hydrolases.

Lysosomal hydrolases degrade intracellular material. Although macrophages lack some of the classic antimicrobial peptides characteristic of neutrophils, such as defensins, they possess their own arsenal of enzymes, including the glycosidases hexosaminidase, galactosidase, glucuronidase, mannosidase; and the proteases cathepsins -B, -D, -H, -L, and -S; furin; and dipeptidyl peptidase II (Ramachandra et al., 1999). These peptides are carried in endosomes and lysosomes during the resting state and are transferred to phagosomes upon activation. These enzymes are recruited to the phagosomes at varying stages of phagosomal maturation (Claus et al., 1998) and have different pH optima (Savina and Amigorena, 2007), facilitating the macrophage's capacity to respond to a broad range of pathogens over a broad range of conditions, especially against pathogens like *M. tuberculosis*, *M. avium*, *Toxoplasma gondii*, and *H. capsulatum* that alter phagosome acidification (Newman et al., 2006).

Phagosome-lysosome fusion can occur directly, or indirectly by autophagy. Autophagy regulates cytoplasm homeostasis, whereby discrete portions of the cytoplasm are sequestered into a specialized double-membrane vacuole, the autophagosome, and delivered to lysosomes for degradation (Shintani and Klionsky, 2004). The autophagosome provides the cell with the ability to overcome pathogen-induced phagolysosomal maturation arrest (Deretic et al., 2006).

Defects in Lysosomal Pathways

Defects in lysosomal degradation typically result in intracellular accumulation of undegraded substrates, characterized by neurodegeneration without associated immunodeficiency. In contrast, conditions related to defective lysosome secretion include immunodeficiencies with hypopigmentation/albinism (i.e., Chediak-Higashi syndrome, Griscelli syndrome type 2, Hermansky-Pudlak syndrome type 2, p14 deficiency).

Antigen Processing and Presentation

In addition to their central role in innate immunity, macrophages also coordinate the adaptive immune response. Once a pathogen is phagocytosed and degraded, its peptides are presented on the macrophage cell surface to T and B lymphocytes in the context of MHC molecules, thereby initiating an adaptive immune response. There are two classes of MHC molecules, classes I and II, termed in humans human leukocyte antigens (HLA) I and II. HLA I antigens are composed of -A, -B, and -C, all of which are expressed as heterodimers formed as distinct transmembrane heavy chains but possessing a common β_2-microglobulin chain. Similarly, HLA II is composed of HLA-DR, -DQ, and -DP. These transmembrane heterodimers are composed of distinct α- and β-chains. The core genes encoding HLA I proteins are clustered on the short arm of chromosome 6, although associated genes have been reported at other loci (Radosavljevic et al., 2002). The genes encoding HLA II proteins are also clustered on the short arm of chromosome 6, where two nonclassical HLA molecules, HLA-DM and HLA-DO, are colocated. Nearby, in the MHC-II locus, are two genes, designated the transporter associated with antigen presentation-1 *(TAP-1)* and *TAP-2*, which encode their respective subunits that form the functional TAP complex. An additional nonclassical HLA II, the invariant chain (Ii), is encoded on chromosome 5 (Genuardi and Saunders, 1988). All nucleated cells express HLA I, while HLA II expression is usually limited to "professional" antigen-presenting cells like macrophages. The expression of HLA II is constitutive on B cells, macrophages, DCs, and thymic epithelial cells and transcriptionally upregulated by IFN-γ on other cells (Paulnock-King et al., 1985; Celada et al., 1989).

All classical (-DR, -DQ, -DP) and nonclassical (-DM, -DO, Ii) HLA II genes possess a unique regulatory element composed of four subsequences, referred to as S, X, X2, and Y boxes in their promoters, and collectively referred to as the S-Y module (Krawczyk and Reith, 2006). The nucleotide sequences of these boxes, as well as their positional organization on the chromosomes, are conserved. Functionally, the S-Y module is a binding site for the MHC-II enhanceosome complex, which consists of nuclear factor Y (NF-Y), which binds the Y box; cyclic AMP response element-binding protein (CREB), which binds the X2 box; regulatory factor X (RFX), which binds the X box; and an unidentified component that binds the S box. The RFX protein is a trimer, consisting of RFX-associated protein containing ankyrin repeats (RFXANK), the fifth member of RFX (RFX5), and the RFX-associated protein (RFXAP). Occupation of the S-Y module by all components of the enhanceosome is necessary for HLA II gene expression, but further requires recruitment of the class II transactivator (CIITA). The *CIITA* gene is induced by IFN-γ, and it mediates, in part, induction of HLA II by IFN-γ (Krawczyk and Reith, 2006). The enhanceosome leads to HLA II expression by recruiting the general transcription machinery, as well as by promoting acetylation/methylation of histones.

In the ER, the newly synthesized α- and β-chains dimerize. The Ii chain associates with the HLA II molecule and targets the HLA II:Ii complex to the endocytic pathway. The Ii chain also inhibits the binding of endogenous self-peptides to HLA II but eventually undergoes proteolytic degradation. HLA II associated with a fragment of Ii is called the class II-associated Ii peptide (CLIP). Endocytic vesicles containing HLA II:CLIP complexes encounter antigenic peptides within the phagolysosome in specialized endosomes called MHC-II compartments (MIIC). Facilitated by the acidic pH environment, HLA-DM binds to the HLA II:CLIP complex, catalyzes the dissociation of CLIP from this complex, and allows binding of the antigenic peptides (Karlsson, 2005). The function of HLA-DM is regulated by HLA-DO in DCs and in B cells. HLA-DO is not expressed in monocytes/macrophages, although the DOα chain can be upregulated, raising the possibility of an HLA-DO-like process regulating HLA-DM during antigen loading within macrophages (Hornell et al., 2003). The antigenic peptide is loaded into the HLA II cleft and expressed at the cell surface.

Costimulatory molecule-ligand pairs important in the generation of a robust macrophage:T-cell interaction are CD28/B7 (Carreno and Collins, 2002; Sharpe and Freeman, 2002) and CD40/CD40-ligand (CD40L, CD154) (Quezada et al., 2004). Engagement of CD40 on macrophages by CD40L on T cells results in its trimerization, permitting recruitment of TNF-receptor associated factors (TRAFs) to the cytoplasmic tail of CD40. This signaling pathway activates NF-κB in a MyD88-independent process, through the activation of NF-κB-inducing kinase (NIK)- and receptor-interacting protein (RIP)-mediated activation of MAPK. Stimulation of CD40 on macrophages results in prolonged presentation of the HLA:peptide complex, as well as upregulation of HLA II, CD80, and CD86 expression and production of cytokines, including IL-12 (Quezada et al., 2004).

HLA I-Based Antigen Presentation in Macrophages

Peptides are generated by the proteolytic activity of the multisubunit complex immunoproteasome, a slightly modified version of the constitutive proteasome. Peptide fragments generated by the immunoproteasome are transported into the lumen of the ER via TAP, where they are further modified and then loaded onto newly synthesized HLA I. Expression of the TAP transporter complex and of the molecular activators of the proteasome is enhanced by IFN-γ (Kloetzel and Ossendorp, 2004). The native HLA I molecules are restrained in the ER, in association with various chaperones (e.g., calnexin, calreticulin, ERp57, tapasin). In the absence of peptide loading, the HLA I dimer is unstable and dissociates. Binding of peptide stabilizes HLA I assembly. The HLA I:peptide complex then exits the ER via secretory vesicles and is transported to the cell surface, where they are exposed to the TcR of $CD8^+$ T lymphocytes.

Defects in MHC-I Regulation: the Bare Lymphocyte Syndrome Type 1

The most common cause of the bare lymphocyte syndrome type 1 is defective TAP complex function, resulting in reduced HLA I surface expression (Gadola et al., 2000). Patients with bare lymphocyte syndrome type 1 typically suffer from recurrent bacterial upper and lower respiratory tract infections, as well as necrotizing granulomatous skin lesions. Infections usually manifest within the first 6 years of life, first involving the upper respiratory tract. In the second decade of life, recurrent episodes of bronchitis/pneumonia occur, eventually leading to bronchiectasis. Severe viral infections are not characteristic, and patients usually demonstrate serologic evidence of prior infections, suggesting effective humoral immunity (Gadola et al., 2000). The necrotizing granulomatous skin lesions are typically distributed on the extremities and on the midline of the face, sometimes including saddle nose deformity. Immunoglobulin levels and subclasses are typically normal. IFN-γ upregulates TAP1 and TAP2 in primary human macrophages (Schiffer et al., 2002).

Cross-Presentation: Emerging Concepts

The presentation of phagocytosed rather than cytosol-derived antigens by HLA I molecules is called cross-presentation (Randolph et al., 2008). Early phagosomes retain ER proteins until they fuse with lysosomes, thus acquiring lysosomal proteins. However, these phagolysosomal vesicles subsequently receive additional waves of ER membrane. During this fusion process, the phagosome acquires the tools necessary for HLA I presentation, including the proteasome, TAP, and Sec61, a heterotrimeric membrane-protein channel complex that transports proteins across the ER (Rapoport, 2007). In the cross-presenting phagosome, antigens are translocated by Sec61 from the lumen of the phagosome into the cytosol, where they undergo proteasomal degradation. The peptides are then transported back into the lumen of the phagosome by TAP, where they are loaded onto the newly acquired HLA I molecules. The HLA I:peptide complex in the phagosome is then redirected for cell surface expression. The direction of phagocytosed antigens toward HLA I or HLA II presentation is not random, but is influenced in part by the cell surface receptor that recognized the antigen and triggered phagocytosis (Burgdorf et al., 2006). Cross-presentation has been predominantly demonstrated in DCs, as well as on some murine macrophage cell lines (Ackerman and Cresswell, 2004; Brode and Macary, 2004), but its clinical relevance in humans remains to be defined.

Production of Proinflammatory Cytokines

The classically activated macrophage produces proinflammatory cytokines (e.g., IL-1β, IL-15, IL-18, TNF-α, IL-12) and chemokines (e.g., CCL15, CCL20, CXCL13) that act as early response and secondary response genes (Saccani et al., 2001). TNF-α is a prototypical early response gene, while IL-12 is secondary response one (Fig. 4).

TNF-α

TNF-α is a mediator of both innate and adaptive immunity. The TNF superfamily consists of several proteins, including TNF-α, TNF-β (lymphotoxin/LT-α), LT-β, fibroblast-associated surface ligand (FasL, CD95L), and CD40L. TNF-α is synthesized primarily by activated monocytes/macrophages, but also by activated T, B, and NK lymphocytes, as well as neutrophils and nonhematopoietic cells. The gene encoding TNF-α, *TNFA*, is located within the TNF locus in the MHC-III region on chromosome 6, upstream of the MHC-I locus and downstream of the MHC-II locus (Hehlgans and Männel, 2002). TNF-α is produced as either membrane-bound (26 kDa) or soluble (17 kDa), both of which are active. It is first produced as a type II membrane protein, anchored to the cell membrane by a hydrophobic presequence. The soluble form

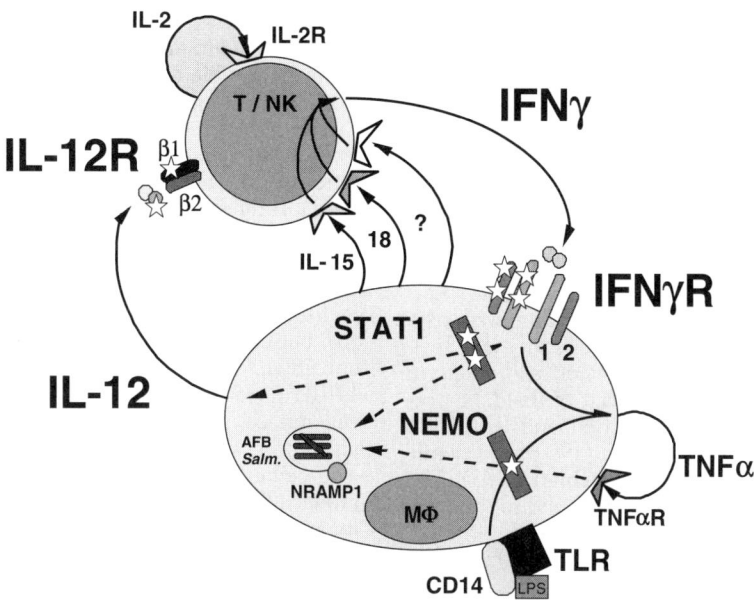

FIGURE 4 Cytokine networks critical for classical macrophage activation. On interaction with select intracellular pathogens, activated macrophages produce proinflammatory cytokines that act in a paracrine or autocrine manner to augment intracellular killing mechanisms.

is produced by cleavage of the membrane-bound form by TNF-α-converting enzyme (TACE), a member of the adamalysin class of zinc-binding metalloproteases or ADAM (a disintegrin and metalloprotease) family (Ceretti et al., 1999). Cleaved TNF-α exists in solution as a homotrimer.

The biological effects of TNF-α are mediated by two TNF-α receptors, TNFR-1 and TNFR-2. TNFR-1 (55 kDa, p55) is expressed almost ubiquitously by hematopoietic and nonhematopoietic cells including fibroblasts and endothelial cells, whereas TNFR-2 (75 kDa, p75) is confined mainly to lymphoid tissues (Hehlgans and Männel, 2002). TNFR-1 and TNFR-2 have similar extracellular structures but differ in their intracellular domains. TNFR-1 contains a cytoplasmic death domain involved in apoptosis signal transduction. TNFR-1 is preferentially bound by soluble TNF-α (sTNF-α), which is then internalized. The death domain of the activated TNFR-1 complex recruits TNF-α receptor-associated death domain (TRADD), which in turn recruits death domain-binding proteins (e.g., Fas-associated death domain [FADD]), leading to the activation of the proteolytic caspase cascade leading to DNA degradation and cell death. TRADD also recruits TNFR-associated factor 2 (TRAF-2), which can activate kinases, leading to activation of NF-κB and AP-1 and induction of both antiapoptotic and proinflammatory genes. Thus, the TNF receptors activate divergent pathways which are modulated by regulation of receptor/ligand expression, soluble decoy receptor expression, and antiapoptotic ligand induction (Hehlgans and Pfeffer, 2005).

The regulation of TNF-α transcription is tissue specific, with monocytes/macrophages and activated T cells the predominant, but not exclusive producers. Myeloid cells are especially sensitive to engagement of TLRs, IFN-γ, IL-1, GM-CSF, and specific lipid mediators. Transcription is initiated within minutes and protein production occurs within hours (Sullivan et al., 2007). Priming by IFN-γ influences TNF-α by increasing the amount and stability of the transcript, but additional regulation occurs through splicing, polyadenylation of message, cis-acting AUUUA repeats, and message transport (Anderson et al., 2004; Kontoyiannis et al., 1999). Downregulation of TNF-α transcription is one mechanism for the anti-inflammatory properties of glucocorticoids (Joyce et al., 2001).

TNF-α and IFN-γ activate macrophages, rendering them better able to control intracellular infections. Murine mycobacterial infections are controlled in part by enhanced iNOS activity. In murine *Salmonella* infections, TNF-α may maximize the concentration of NADPH oxidase incorporated into organism-bearing phagosomes (Vazquez-Torres et al., 2001), while in *Listeria* infections, TNF-α-mediated killing may be through enhancement of both ROS and RNS production (Muller et al., 1999). However, TNF-α-induced apoptosis accounts for only 5 to 10% killing of macrophages chronically infected with M. *tuberculosis*, so other factors must be involved (Marino et al., 2007).

TNF-α is essential for granuloma formation (Roach et al., 2002) and maintenance. In mice, the granuloma-specific properties of TNF-α are mediated by TNFR-1 (Florido and Appelberg, 2007). Mycobacteria-infected mice deficient in TNF-α have granuloma disintegration during CD4$^+$ and CD8$^+$ T-cell activation, enhanced IFN-γ production, and surface expression of a death-inducing ligand, TRAIL, in T cells and macrophages (Florido and Appelberg, 2007). TNF-α stimulates production of indoleamine 2,3-dioxygenase (IDO), a key enzyme in tryptophan catabolism, by human DCs and macrophages. In granulomas induced by *Listeria monocytogenes*, expression of IDO was concurrent with the absence of T cells, suggesting that IDO may inhibit T-cell-mediated destruction of *Listeria*-induced granulomas (Popov et al., 2006).

Tuberculosis is epidemiologically significantly associated with iatrogenic TNF-α blockade, particularly by infliximab (Wallis et al., 2004). However, other frequently reported infections include histoplasmosis, listeriosis, candidiasis, aspergillosis, coccidiomycosis, and nontuberculous mycobacteria (Gardam et al., 2003; Filler et al., 2005; Winthrop et al., 2008).

IL-12

Classically activated macrophages produce IL-12p70, a heterodimer composed of a 35-kDa light chain (IL-12p35 or IL-12α; referred to as p35) and a 40-kDa heavy chain (IL-12p40 or IL-12β; referred to as p40). p35 is encoded by *IL12A* on chromosome 3, while p40 is encoded by *IL12B* on chromosome 5. p40 is produced and secreted in excess of p35 and can circulate in monomeric form, while p35 is retained in cells and secreted only in association with p40. Thus, coexpression of both chains is required for IL-12p70 (Trinchieri, 2003). However, homodimeric p40 forms in mice, where it may antagonize the effects of IL-12. The IL-12 receptor is a heterodimer of IL-12Rβ1 and IL-12Rβ2. *IL12RB1* is located at chromosome 19 p13.1, while *IL12RB2* is at chromosome 1p31.2 (Trinchieri, 2003). The intracytoplasmic domain of IL-12Rβ1 is bound to the nonreceptor protein tyrosine kinase (Tyk2) while the intracytoplasmic domain IL-12Rβ2 is bound to (JAK2). Expression of IL-12R is mainly on activated CD4$^+$ and CD8$^+$ T lymphocytes and activated NK cells, but macrophages also upregulate surface expression of IL-12R, allowing autocrine stimulation, which is potentiated by IFN-γ (Grohmann et al., 2001). IL-12 binding to IL-12R leads to Tyk2 and JAK-2 transactivation, resulting in phosphorylation of STAT-4, which homodimerizes and translocates to the nucleus to induce gene expression predominantly in NK cells and in T lymphocytes (Trinchieri, 2003). IL-12 synergizes with IL-15 and IL-18 to induce IFN-γ production, which in turn acts on monocytes/macrophages to augment IL-12 secretion. IL-12 enhances the cytolytic capacity of CD8$^+$ T lymphocytes by inducing their proliferation, differentiation, and production of cytotoxic granules (i.e., granzyme B, perforin) thereby contributing to the acute adaptive immune response.

IL-12 is the prototype of a family of heterodimeric cytokines, including IL-23 and IL-27. IL-23 is composed of the IL-12 p40 subunit combined with a homologue of p35, p19. Likewise, the IL-23 receptor is composed of the common IL-12Rβ1 and IL-23R. IL-27 consists of a homologue of the IL-12 p40 subunit, the Epstein-Barr virus-induced gene 3 (EBI3), complexed with another p35 homologue, p28. The IL-27 receptor is formed by the common cytokine receptor molecule, gp130, and WSX1. IL-27 has proinflammatory and anti-inflammatory components (Kastelein et al., 2007).

IL-12 and IL-23 are produced by macrophages and DCs and stimulate IFN-γ production by T and NK cells (Kastelein et al., 2007). IL-12 primarily maintains the newly differentiated Th1 repertoire, while IL-23 acts primarily to induce the proliferation of memory T cells. IL-12 stimulates the production of Th1 cytokines (e.g., IFN-γ), while IL-23 promotes the production of Th17 cells, which produce IL-6, IL-17, IL-22, and IL-25. Although both IL-12R and IL-23R use Tyk2 and JAK-2 and activate STAT-1, STAT-3, STAT-4, and STAT-5, IL-12 preferentially activates STAT-4, while IL-23 preferentially activates STAT-3 (Parham et al., 2002).

Defects in the IL-12 Pathway

Genetic abnormalities in the IL-12 pathway include recessive mutations in IL-12p40 (*IL12B*) and IL-12Rβ1. Mutations in IL-12p35 (*IL12A*), p19, p38, IL-12Rβ2, and STAT-4 have not yet been identified, suggesting that the correct disease phenotype has not been investigated. Patients with mutations in IL-12p40 or IL-12Rβ1 are especially susceptible to salmonellae and mycobacteria (van de Vosse and Ottenhoff, 2006). Less common infections in these patients include *Nocardia* and *Paracoccidioides braziliensis*. Complete deficiency of IL-12 is due to recessive mutations in the *IL12B* gene on chromosome 5 (Picard et al., 2002). Interestingly, only three mutations have been identified, and all cases show a founder effect in diverse ethnic groups. Clinically, patients with p40 deficiency develop BCG infection as well as other mycobacteria and *Nocardia*.

Mutations in *IL12RB1* can be either complete or partial, but complete deficiency is much more common (Fieschi et al., 2003). In most cases, complete recessive IL-12Rβ1 deficiency leads to no IL-12Rβ1 expression on the cell surface. However, complete IL-12Rβ1 deficiency can also occur with surface-expressed nonfunctional receptor. The cellular phenotype is failure of the patients' cells to produce IFN-γ in response to exogenous IL-12, distinguishing it from IL-12p40 deficiency. IL-12Rβ1 deficiency is variable in penetrance and somewhat redundant in protection against primary BCG infection. However, following significant mycobacterial infection of any sort, patients are protected from subsequent mycobacterial disease, suggesting that IL-12 is redundant for adaptive immunity to mycobacteria. Interestingly, T cells in vitro can produce IFN-γ in an IL-12Rβ1-/STAT-4-independent, MAPK-dependent process.

IL-12 pathway defects are especially susceptible to *Salmonella* spp. Almost half of all IL-12 pathway-defective patients have developed extraintestinal salmonellosis, while they have developed lower levels of nontuberculous mycobacterial disease. The situation in IFN-γ receptor defects is the opposite, with comparatively more mycobacterial and fewer *Salmonella* infections (MacLennan et al., 2004). Surprisingly, defects in these pathways do not appear to predispose to significantly higher rates of allergic disease, contrary to conventional expectation (Wood et al., 2005).

Autosomal recessive Tyk-2 deficiency also impairs IL-12 signaling. One patient is reported with a history of disseminated BCG and an episode of nontyphoidal *Salmonella* gastroenteritis who also had infections with molluscum contagiosum, herpes simplex, *S. aureus*, and mild mucocutaneous candidiasis (Minegishi et al., 2006). He had impaired IL-12 and IFN-α signaling, with normal STAT-4 expression. Because Tyk-2 is an adaptor protein in several pathways, including IL-12, IL-23, type I IFNs, IL-6, and IL-10, this patient had defects in multiple axes, explaining his broad susceptibility.

CONCLUSIONS

The classically activated macrophage is an IFN-γ-primed cell that recognizes, via specific cell surface receptors, microbes and their products. Overlapping and convergent signal pathways enable the cell to phagocytose intracellular pathogens via endogenous microbicidal mechanisms. Concurrently, the macrophage elaborates a variety of proinflammatory mediators that recruit additional leukocytes, whose coordinated efforts both augment the macrophage's killing capacity and form the adaptive immune response. Although experimental animal studies have broadened our understanding of the molecular means of such pathways, natural human deficiencies have provided profound insight as to the clinical relevance of these pathways.

This work was supported by the Division of Intramural Research, National Institute of Allergy and Infectious Diseases, National Institutes of Health.

REFERENCES

Ackerman, A. L., and P. Cresswell. 2004. Cellular mechanisms governing cross-presentation of exogenous antigens. *Nat. Immunol.* **5:**678–684.

Adams, D. O., and T. A. Hamilton. 1984. The cell biology of macrophage activation. *Annu. Rev. Immunol.* **2:**283–318.

Akira, S., and K. Takeda. 2004. Toll-like receptor signalling. *Nat. Rev. Immunol.* **4:**499–511.

Almyroudis, N. G., S. M. Holland, and B. H. Segal. 2005. Invasive aspergillosis in primary immunodeficiencies. *Med. Mycol.* **43:**247–259.

Ambruso, D. R., C. Knall, A. N. Abell, J. Panepinto, A. Kurkchubasche, G. Thurman, G. Gonzalez-Aller, A. Hiester, M. deBoer, R. J. Harbeck, R. Oyer, G. L. Johnson, and D. Roos. 2000. Human neutrophil immunodeficiency syndrome is associated with an inhibitory Rac2 mutation. *Proc. Natl. Acad. Sci. USA* **97:**4654–4659.

Anderson, P., K. Phillips, G. Stoecklin, and N. Kedersha. 2004. Post-transcriptional regulation of proinflammatory proteins. *J. Leukoc. Biol.* **76:**42–47.

Babior, B. M. 1984. The respiratory burst of phagocytes. *J. Clin. Investig.* **73:**599–601.

Bach, E. A., M. Aguet, and R. D. Schreiber. 1997. The IFN[gamma] receptor: a paradigm for cytokine receptor signaling. *Annu. Rev. Immunol.* **15:**563–591.

Brode, S., and P. A. Macary. 2004. Cross-presentation: dendritic cells and macrophages bite off more than they can chew! *Immunology* **112:**345–351.

Brown, G. D., P. R. Taylor, D. M. Reid, J. A. Willment, D. L. Williams, L. Martinez-Pomares, S. Y. Wong, and S. Gordon. 2002. Dectin-1 is a major [beta]-glucan receptor on macrophages. *J. Exp. Med.* **196:**407–12.

Burgdorf, S., V. Lukacs-Kornek, and C. Kurts. 2006. The mannose receptor mediates uptake of soluble but not of cell-associated antigen for cross-presentation. *J. Immunol.* **176:**6770–6776.

Cai, H., K. Reinisch, and S. Ferro-Novick. 2007. Coats, tethers, rabs, and SNAREs work together to mediate the intracellular destination of a transport vesicle. *Dev. Cell* **12:**671–682.

Carmody, R. J., and Y. H. Chen. 2007. Nuclear factor-kappaB: activation and regulation during toll-like receptor signaling. *Cell. Mol. Immunol.* **4:**31–41.

Carreno, B. M., and M. Collins. 2002. The B7 family of ligands and its receptors: new pathways for costimulation and inhibition of immune responses. *Annu. Rev. Immunol.* **20:**29–53.

Cathcart, M. K. 2004. Regulation of superoxide anion production by NADPH oxidase in monocytes/macrophages: contributions to atherosclerosis. *Arterioscler. Thromb. Vasc. Biol.* **24:**23–28.

Celada, A., M. J. Klemsz, and R. A. Maki. 1989. Interferon-gamma activates multiple pathways to regulate the expression of the genes for major histocompatibility class II I-A beta, tumor necrosis factor and complement component C3 in mouse macrophages. *Eur. J. Immunol.* **19:**1103–1109.

Cerretti, D. P., K. Poindexter, B. J. Castner, G. Means, N. G. Copeland, D. J. Gilbert, N. A. Jenkins, R. A. Black, and N. Nelson. 1999. Characterization of the cDNA and gene for mouse tumour necrosis factor [alpha] converting enzyme (TACE/ADAM17) and its location to mouse chromosome 12 and human chromosome 2p25. *Cytokine* **11:**541–551.

Chang, Y. C., B. H. Segal, S. M. Holland, G. F. Miller, and K. J. Kwon-Chung. 1998. Virulence of catalase-deficient aspergillus nidulans in p47(phox)−/− mice. Implications for fungal pathogenicity and host defense in chronic granulomatous disease. *J. Clin. Investig.* **101:**1843–1845.

Chapgier, A., S. Boisson-Dupuis, E. Jouanguy, G. Vogt, J. Feinberg, A. Prochnicka-Chalufour, A. Casrouge, K. Yang, C. Soudais, C. Fieschi, O. F. Santos, J. Bustamante, C. Picard, L. de Beaucoudrey, J. F. Emile, P. D. Arkwright, R. D. Schreiber, C. Rolinck-Werninghaus, A. Rösen-Wolff, K. Magdorf, J. Roesler, and J. L. Casanova. 2006. Novel STAT1 alleles in otherwise healthy patients with mycobacterial disease. *PLoS Genet.* **2:**e131.

Chapgier, A., R. F. Wynn, E. Jouanguy, O. Filipe-Santos, S. Zhang, J. Feinberg, K. Hawkins, J. L. Casanova, and P. D. Arkwright. 2006. Human complete Stat-1 deficiency is associated with defective type I and II IFN responses in vitro but immunity to some low virulence viruses in vivo. *J. Immunol.* **176:**5078–5083.

Chavrier, P., and B. Goud. 1999. The role of ARF and Rab GTPases in membrane transport. *Curr. Opin. Cell. Biol.* **11:**466–475.

Chua, J., and V. Deretic. 2004. Mycobacterium tuberculosis reprograms waves of phosphatidylinositol 3-phosphate on phagosomal organelles. *J. Biol. Chem.* **279:**36982–36992.

Claus, V., A. Jahraus, T. Tjelle, T. Berg, H. Kirschke, H. Faulstich, and G. Griffiths. 1998. Lysosomal enzyme trafficking between phagosomes, endosomes, and lysosomes in J774 macrophages. Enrichment of cathepsin H in early endosomes. *J. Biol. Chem.* **273:**9842–9851.

Clemens, D. L., B. Y. Lee, and M. A. Horwitz. 2000a. Deviant expression of Rab5 on phagosomes containing the intracellular pathogens Mycobacterium tuberculosis and Legionella pneumophila is associated with altered phagosomal fate. *Infect. Immun.* **68:**2671–2684.

Clemens, D. L., B.-Y. Lee, and M. A. Horwitz. 2000b. Mycobacterium tuberculosis and Legionella pneumophila phagosomes exhibit arrested maturation despite acquisition of Rab7. *Infect. Immun.* **68:**5154–5166.

Courtois, G. 2005. The NF-kappaB signaling pathway in human genetic diseases. *Cell. Mol. Life Sci.* **62:**1682–1691.

Courtois, G., A. Smahi, J. Reichenbach, R. Döffinger, C. Cancrini, M. Bonnet, A. Puel, C. Chable-Bessia, S. Yamaoka, J. Feinberg, S. Dupuis-Girod, C. Bodemer, S. Livadiotti, F. Novelli, P. Rossi, A. Fischer, A. Israël, A. Munnich, F. Le Deist, and J. L. Casanova. 2003. A hypermorphic I[kappa]B[alpha] mutation is associated with autosomal dominant anhidrotic ectodermal dysplasia and T cell immunodeficiency. *J. Clin. Investig.* **112:**1108–1115.

Cross, A. R. 2000. p40(phox) participates in the activation of NADPH oxidase by increasing the affinity of p47(phox) for flavocytochrome b(558). *Biochem. J.* **349:**113–117.

Crowley, M. T., P. S. Costello, C. J. Fitzer-Attas, M. Turner, F. Meng, C. Lowell, V. L. Tybulewicz, and A. L. DeFranco. 1997. A critical role for Syk in signal transduction and phagocytosis mediated by fcgamma receptors on macrophages. *J. Exp. Med.* **186:**1027–1039.

DeLeo, F. R., L. A. Allen, M. Apicella, and W. M. Nauseef. 1999. NADPH oxidase activation and assembly during phagocytosis. *J. Immunol.* **163:**6732–6740.

Deretic, V., S. Singh, S. Master, J. Harris, E. Roberts, G. Kyei, A. Davis, S. de Haro, J. Naylor, H. H. Lee, and I. Vergne. 2006. Mycobacterium tuberculosis inhibition of phagolysosome biogenesis and autophagy as a host defence mechanism. *Cell. Microbiol.* **8:**719–727.

Doffinger, R., E. Jouanguy, S. Dupuis, M. C. Fondanèche, J. L. Stephan, J. F. Emile, S. Lamhamedi-Cherradi, F. Altare, A. Pallier, G. Barcenas-Morales, E. Meinl, C. Krause, S. Pestka, R. D. Schreiber, F. Novelli, and J. L. Casanova. 2000. Partial interferon gamma receptor signaling chain deficiency in a patient with Bacille Calmette Guérin and *Mycobacterium abscessus* infection. *J. Infect. Dis.* **181**:379–384.

Dorman, S. E., and S. M. Holland. 1998. Mutation in the signal-transducing chain of the interferon-gamma receptor and susceptibility to mycobacterial infection. *J. Clin. Investig.* **101**:2364–2369.

Dorman, S. E., C. Picard, D. Lammas, K. Heyne, J. T. van Dissel, R. Baretto, S. D. Rosenzweig, M. Newport, M. Levin, J. Roesler, D. Kumararatne, J. L. Casanova, and S. M. Holland. 2004. Clinical features of dominant and recessive interferon [gamma] receptor 1 deficiencies. *Lancet* **364**:2113–2121.

Drose, S., and K. Altendorf. 1997. Bafilomycins and concanamycins as inhibitors of V-ATPases and P-ATPases. *J. Exp. Biol.* **200**:1–8.

D'Souza-Schorey, C., R. L. Boshans, M. McDonough, P. D. Stahl, and L. Van Aelst. 1997. A role for POR1, a Rac1-interacting protein, in ARF6-mediated cytoskeletal rearrangements. *EMBO J.* **16**:5445–5454.

Dupuis, S., C. Dargemont, C. Fieschi, N. Thomassin, S. Rosenzweig, J. Harris, S. M. Holland, R. D. Schreiber, and J. L. Casanova. 2001. Impairment of mycobacterial but not viral immunity by a germline human STAT1 mutation. *Science* **293**:300–303.

Dupuis, S., E. Jouanguy, S. Al-Hajjar, C. Fieschi, I. Z. Al-Mohsen, S. Al-Jumaah, K. Yang, A. Chapgier, C. Eidenschenk, P. Eid, A. Al Ghonaium, H. Tufenkeji, H. Frayha, S. Al-Gazlan, H. Al-Rayes, R. D. Schreiber, I. Gresser, and J. L. Casanova. 2003. Impaired response to interferon-[alpha]/[beta] and lethal viral disease in human STAT1 deficiency. *Nat. Genet.* **33**:388–391.

Edgar, J. D. M., A. E. Smyth, J. Pritchard, D. Lammas, E. Jouanguy, R. Hague, J. Novelli, S. Dempsey, L. Sweeney, A. J. Taggart, D. O'Hara, J. L. Casanova, and D. S. Kumararatne. 2001. Interferon-[gamma] receptor deficiency mimicking Langerhans' cell histiocytosis. *J. Pediatr.* **139**:600–603.

Edwards, D. C., L. C. Sanders, G. M. Bokoch, and G. N. Gill. 1999. Activation of LIM-kinase by Pak1 couples Rac/Cdc42 GTPase signalling to actin cytoskeletal dynamics. *Nat. Cell Biol.* **1**:253–259.

Eskelinen, E. L. 2006. Roles of LAMP-1 and LAMP-2 in lysosome biogenesis and autophagy. *Mol. Asp. Med.* **27**:495–502.

Fang, F. C. 2004. Antimicrobial reactive oxygen and nitrogen species: concepts and controversies. *Nat. Rev. Microbiol.* **2**:820–832.

Fenton, M. J., and D. T. Golenbock. 1998. LPS-binding proteins and receptors. *J. Leukoc. Biol.* **64**:25–32.

Fevrier, M., J. L. Birrien, C. Leclerc, L. Chedid, and P. Liacopoulos. 1978. The macrophage, target cell of the synthetic adjuvant muramyl dipeptide. *Eur. J. Immunol.* **8**:558–562.

Fieschi, C., S. Dupuis, E. Catherinot, J. Feinberg, J. Bustamante, A. Breiman, F. Altare, R. Baretto, F. Le Deist, S. Kayal, H. Koch, D. Richter, M. Brezina, G. Aksu, P. Wood, S. Al-Jumaah, M. Raspall, A. J. Da Silva Duarte, D. Tuerlinckx, J. L. Virelizier, A. Fischer, A. Enright, J. Bernhöft, A. M. Cleary, C. Vermylen, C. Rodriguez-Gallego, G. Davies, R. Blütters-Sawatzki, C. A. Siegrist, M. S. Ehlayel, V. Novelli, W. H. Haas, J. Levy, J. Freihorst, S. Al-Hajjar, D. Nadal, D. De Moraes Vasconcelos, O. Jeppsson, N. Kutukculer, K. Frecerova, D. Lammas, D. S. Kumararatne, L. Abel, and J. L. Casanova. 2003. Low penetrance, broad resistance, and favorable outcome of interleukin 12 receptor beta1 deficiency: medical and immunological implications. *J. Exp. Med.* **197**:527–535.

Filipe-Santos, O., J. Bustamante, M. H. Haverkamp, E. Vinolo, C. L. Ku, A. Puel, D. M. Frucht, K. Christel, H. von Bernuth, E. Jouanguy, J. Feinberg, A. Durandy, B. Senechal, A. Chapgier, G. Vogt, L. de Beaucoudrey, C. Fieschi, C. Picard, M. Garfa, J. Chemli, M. Bejaoui, M. N. Tsolia, N. Kutukculer, A. Plebani, L. Notarangelo, C. Bodemer, F. Geissmann, A. Israël, M. Véron, M. Knackstedt, R. Barbouche, L. Abel, K. Magdorf, D. Gendrel, F. Agou, S. M. Holland, and J. L. Casanova. 2006. X-linked susceptibility to mycobacteria is caused by mutations in NEMO impairing CD40-dependent IL-12 production. *J. Exp. Med.* **203**:1745–1745.

Filler, S., M. Yeaman, and D. Sheppard. 2005. Tumor necrosis factor inhibition and invasive fungal infections. *Clin. Infect. Dis.* **41**:S208–S212.

Florido, M., and R. Appelberg. 2007. Characterization of the deregulated immune activation occurring at late stages of mycobacterial infection in TNF-deficient mice. *J. Immunol.* **179**:7702–7708.

Forgac, M. 2007. Vacuolar ATPases: rotary proton pumps in physiology and pathophysiology. *Nat. Rev. Mol. Cell Biol.* **8**:917–929.

Forman, H. J., and M. Torres. 2001. Redox signaling in macrophages. *Mol. Asp Med.* **22**:189–216.

Forman, H. J., and M. Torres. 2002. Reactive oxygen species and cell signaling: respiratory burst in macrophage signaling. *Am. J. Respir. Crit. Care Med.* **166**:4S–8S.

Fratti, R. A., J. M. Backer, J. Gruenberg, S. Corvera, and V. Deretic. 2001. Role of phosphatidylinositol 3-kinase and Rab5 effectors in phagosomal biogenesis and mycobacterial phagosome maturation arrest. *J. Cell Biol.* **154**:631–644.

Fratti, R. A., J. Chua, and V. Deretic. 2002. Cellubrevin alterations and *Mycobacterium tuberculosis* phagosome maturation arrest. *J. Biol. Chem.* **277**:17320–17326.

Furchgott, R. F., and J. V. Zawadzki. 1980. The obligatory role of endothelial cells in the relaxation of arterial smooth muscle by acetylcholine. *Nature* **288**:373–376.

Gadola, S. D., H. T. Moins-Teisserenc, J. Trowsdale, W. L. Gross, and V. Cerundolo. 2000. TAP deficiency syndrome. *Clin. Exp. Immunol.* **121**:173–178.

Gallin, J. I., D. W. Alling, H. L. Malech, R. Wesley, D. Koziol, B. Marciano, E. M. Eisenstein, M. L. Turner, E. S. DeCarlo, J. M. Starling, and S. M. Holland. 2003. Itraconazole to prevent fungal infections in chronic granulomatous disease. *N. Engl. J. Med.* **348**:2416–2422.

Garcia-Garcia, E., and C. Rosales. 2002. Signal transduction during Fc receptor-mediated phagocytosis. *J. Leukoc. Biol.* **72**:1092–1108.

Gardam, M. A., E. C. Keystone, R. Menzies, S. Manners, E. Skamene, R. Long, and D. C. Vinh. 2003. Anti-tumour necrosis factor agents and tuberculosis risk: mechanisms of action and clinical management. *Lancet Infect. Dis.* **3**:148–155.

Genuardi, M., and G. F. Saunders. 1988. Localization of the HLA class II-associated invariant chain gene to human chromosome band 5q32. *Immunogenetics* **28**:53–56.

Gissen, P., and E. R. Maher. 2007. Cargos and genes: insights into vesicular transport from inherited human disease. *J. Med. Genet.* **44**:545–555.

Gomez, L. M., J. M. Anaya, J. R. Vilchez, J. Cadena, R. Hinojosa, L. Vélez, M. A. Lopez-Nevot, and J. Martín. 2007. A polymorphism in the inducible nitric oxide synthase gene is associated with tuberculosis. *Tuberculosis* **87**:288–294.

Grohmann, U., M. L. Belladonna, C. Vacca, R. Bianchi, F. Fallarino, C. Orabona, M. C. Fioretti, and P. Puccetti. 2001. Positive regulatory role of IL-12 in macrophages and modulation by IFN-[gamma]. *J. Immunol.* **167**:221–227.

Grunfeld, C., M. Marshall, J. K Shigenaga, A. H. Moser, P. Tobias, and K. R. Feingold. 1999. Lipoproteins inhibit macrophage activation by lipoteichoic acid. *J. Lipid Res.* **40**:245–252.

Gupta, D., Y. P. Jin, and R. Dziarski. 1995. Peptidoglycan induces transcription and secretion of TNF-alpha and activation of lyn, extracellular signal-regulated kinase, and rsk signal transduction proteins in mouse macrophages. *J. Immunol.* **155:**2620–2630.

Hackam, D. J., O. D. Rotstein, W. J. Zhang, N. Demaurex, M. Woodside, O. Tsai, and S. Grinstein. 1997. Regulation of phagosomal acidification. Differential targeting of Na+/H+ exchangers, Na+/K+-ATPases, and vacuolar-type H+-ATPases. *J. Biol. Chem.* **272:**29810–29820.

Haggie, P. M., and A. S. Verkman. 2007. Cystic fibrosis transmembrane conductance regulator-independent phagosomal acidification in macrophages. *J. Biol. Chem.* **282:**31422–31428.

Haziot, A., S. Chen, E. Ferrero, M. G. Low, R. Silber, and S. M. Goyert. 1988. The monocyte differentiation antigen, CD14, is anchored to the cell membrane by a phosphatidylinositol linkage. *J. Immunol.* **141:**547–552.

Hehlgans, T., and D. N. Männel. 2002. The TNF-TNF receptor system. *Biol. Chem.* **383:**1581–1585.

Hehlgans, T., and K. Pfeffer. 2005. The intriguing biology of the tumour necrosis factor/tumour necrosis factor receptor superfamily: players, rules and the games. *Immunology* **115:**1–20.

Henri, S., F. Stefani, D. Parzy, C. Eboumbou, A. Dessein, and C. Chevillard. 2002. Description of three new polymorphisms in the intronic and 3'UTR regions of the human interferon gamma gene. *Genes Immun.* **3:**1–4.

Henry, R. M., A. D. Hoppe, N. Joshi, and J. A. Swanson. 2004. The uniformity of phagosome maturation in macrophages. *J. Cell Biol.* **164:**185–194.

Hill, G. R., J. M. Crawford, K. R. Cooke, Y. S. Brinson, L. Pan, and J. L. Ferrara. 1997. Total body irradiation and acute graft-versus-host disease: the role of gastrointestinal damage and inflammatory cytokines. *Blood* **90:**3204–3213.

Höflich, C., R. Sabat, S. Rosseau, B. Temmesfeld, H. Slevogt, W. D. Döcke, G. Grütz, C. Meisel, E. Halle, U. B. Göbel, H. D. Volk, and N. Suttorp. 2004. Naturally occurring anti-IFN-gamma autoantibody and severe infections with *Mycobacterium chelonae* and *Burkholderia cocovenenans*. *Blood* **103:**673–675.

Hornell, T. M. C., G. W. Beresford, A. Bushey, J. M. Boss, and E. D. Mellins. 2003. Regulation of the class II MHC pathway in primary human monocytes by granulocyte-macrophage colony-stimulating factor. *J. Immunol.* **171:**2374–2383.

Huynh, K. K., E. L. Eskelinen, C. C. Scott, A. Malevanets, P. Saftig, and S. Grinstein. 2007. LAMP proteins are required for fusion of lysosomes with phagosomes. *EMBO J.* **26:**313–324.

Iles, K. E., and H. J. Forman. 2002. Macrophage signaling and respiratory burst. *Immunol. Res.* **26:**95–105.

Jain, A., C. A. Ma, S. Liu, M. Brown, J. Cohen, and W. Strober. 2001. Specific missense mutations in NEMO result in hyper-IgM syndrome with hypohidrotic ectodermal dysplasia. *Nat. Immunol.* **2:**223–228.

Janssen, R., A. van Wengen, M. A. Hoeve, M. ten Dam, M. van der Burg, J. van Dongen, E. van de Vosse, M. van Tol, R. Bredius, T. H. Ottenhoff, C. Weemaes, J. T. van Dissel, and A. Lankester. 2004. The same I[kappa]B[alpha] mutation in two related individuals leads to completely different clinical syndromes. *J. Exp. Med.* **200:**559–568.

Johansson, A., A. J. Jesaitis, H. Lundqvist, K. E. Magnusson, C. Sjölin, A. Karlsson, and C. Dahlgren. 1995. Different subcellular localization of cytochrome b and the dormant NADPH-oxidase in neutrophils and macrophages: effect on the production of reactive oxygen species during phagocytosis. *Cell. Immunol.* **161:**61–71.

Jouanguy, E., F. Altare, S. Lamhamedi, P. Revy, J. F. Emile, M. Newport, M. Levin, S. Blanche, E. Seboun, A. Fischer, and J. L. Casanova. 1996. Interferon-gamma-receptor deficiency in an infant with fatal bacille Calmette-Guérin infection. *N. Engl. J. Med.* **335:**1956–1961.

Jouanguy, E., S. Lamhamedi-Cherradi, D. Lammas, S. E. Dorman, M. C. Fondanèche, S. Dupuis, R. Döffinger, F. Altare, J. Girdlestone, J. F. Emile, H. Ducoulombier, D. Edgar, J. Clarke, V. A. Oxelius, M. Brai, V. Novelli, K. Heyne, A. Fischer, S. M. Holland, D. S. Kumararatne, R. D. Schreiber, and J. L. Casanova. 1997. Partial interferon-gamma receptor 1 deficiency in a child with tuberculoid Bacillus Calmette-Guerin infection and a sibling with clinical tuberculosis. *J. Clin. Investig.* **100:**2658–2664.

Jouanguy, E., S. Lamhamedi-Cherradi, D. Lammas, S. E. Dorman, M. C. Fondanèche, S. Dupuis, R. Döffinger, F. Altare, J. Girdlestone, J. F. Emile, H. Ducoulombier, D. Edgar, J. Clarke, V. A. Oxelius, M. Brai, V. Novelli, K. Heyne, A. Fischer, S. M. Holland, D. S. Kumararatne, R. D. Schreiber, and J. L. Casanova. 1999. A human IFNGR1 small deletion hotspot associated with dominant susceptibility to mycobacterial infection. *Nat. Genet.* **21:**370–378.

Joyce, D. A., G. Gimblett, and J. H. Steer. 2001. Targets of glucocorticoid action on TNF-a release by macrophages. *Inflamm. Res.* **50:**337–340.

Kanai, F., H. Liu, S. J. Field, H. Akbary, T. Matsuo, G. E. Brown, L. C. Cantley, and M. B. Yaffe. 2001. The PX domains of p47phox and p40phox bind to lipid products of PI(3)K. *Nat. Cell Biol.* **3:**675–678.

Karlsson, L. 2005. DM and DO shape the repertoire of peptide-MHC-class-II complexes. *Curr. Opin. Immunol.* **17:**65–70.

Kastelein, R. A., C. A. Hunter, and D. J. Cua. 2007. Discovery and biology of IL-23 and IL-27: related but functionally distinct regulators of inflammation. *Annu. Rev. Immunol.* **25:**221–242.

Kato, M., S. Khan, E. d'Aniello, K. J. McDonald, and D. N. Hart. 2007. The novel endocytic and phagocytic C-type lectin receptor DCL-1/CD302 on macrophages is colocalized with F-actin, suggesting a role in cell adhesion and migration. *J. Immunol.* **179:**6052–6063.

Kim, H. S., and M.-S. Lee. 2007. STAT1 as a key modulator of cell death. *Cell Signal.* **19:**454–465.

Kloetzel, P. M., and F. Ossendorp. 2004. Proteasome and peptidase function in MHC-class-I-mediated antigen presentation. *Curr. Opin. Immunol.* **16:**76–81.

Kol, A., A. H. Lichtman, R. W. Finberg, P. Libby, and E. A. Kurt-Jones. 2000. Cutting edge: heat shock protein (HSP) 60 activates the innate immune response: CD14 is an essential receptor for HSP60 activation of mononuclear cells. *J. Immunol.* **164:**13–17.

Kontoyiannis, D., M. Pasparakis, T. T. Pizarro, F. Cominelli, and G. Kollias. 1999. Impaired on/off regulation of TNF biosynthesis in mice lacking TNF AU-rich elements: implications for joint and gut-associated immunopathologies. *Immunity* **10:**387–398.

Krawczyk, M., and W. Reith. 2006. Regulation of MHC class II expression, a unique regulatory system identified by the study of a primary immunodeficiency disease. *Tissue Antigens* **67:**183–197.

Ku, C. L., H. von Bernuth, C. Picard, S. Y. Zhang, H. H. Chang, K. Yang, M. Chrabieh, A. C. Issekutz, C. K. Cunningham, J. Gallin, S. M. Holland, C. Roifman, S. Ehl, J. Smart, M. Tang, F. J. Barrat, O. Levy, D. McDonald, N. K. Day-Good, R. Miller, H. Takada, T. Hara, S. Al-Hajjar, A. Al-Ghonaium, D. Speert, D. Sanlaville, X. Li, F. Geissmann, E. Vivier, L. Maródi, B. Z. Garty, H. Chapel, C. Rodriguez-Gallego, X. Bossuyt, L. Abel, A. Puel, and J. L.

Casanova. 2007. Selective predisposition to bacterial infections in IRAK-4-deficient children: IRAK-4-dependent TLRs are otherwise redundant in protective immunity. *J. Exp. Med.* **204:**2407–2422.

Kyei, G. B., I. Vergne, J. Chua, E. Roberts, J. Harris, J. R. Junutula, and V. Deretic. 2006. Rab14 is critical for maintenance of *Mycobacterium tuberculosis* phagosome maturation arrest. *EMBO J.* **25:**5250–5259.

Levesque, M., M. R. Hobbs, N. M. Anstey, T. N. Vaughn, J. A. Chancellor, A. Pole, D. J. Perkins, M. A. Misukonis, S. J. Chanock, D. L. Granger, and J. B. Weinberg. 1999. Nitric oxide synthase type 2 promoter polymorphisms, nitric oxide production, and disease severity in Tanzanian children with malaria. *J. Infect. Dis.* **180:**1994–2002.

Liu, M. T., B. P. Chen, P. Oertel, M. J. Buchmeier, D. Armstrong, T. A. Hamilton, and T. E. Lane. 2000. Cutting edge: the t cell chemoattractant IFN-inducible protein 10 is essential in host defense against viral-induced neurologic disease. *J. Immunol.* **165:**2327–2330.

Lukacs, G. L., O. D. Rotstein, and S. Grinstein. 1990. Phagosomal acidification is mediated by a vacuolar-type H(+)-ATPase in murine macrophages. *J. Biol. Chem.* **265:**21099–21107.

MacLennan, C., C. Fieschi, D. A. Lammas, C. Picard, S. E. Dorman, O. Sanal, J. M. MacLennan, S. M. Holland, T. H. Ottenhoff, J. L. Casanova, and D. S. Kumararatne. 2004. Interleukin (IL)-12 and IL-23 are key cytokines for immunity against *Salmonella* in humans. *J. Infect. Dis.* **190:**1755–1757.

Marino, S., D. Sud, H. Plessner, P. L. Lin, J. Chan, J. L. Flynn, and D. E. Kirschner. 2007. Differences in reactivation of tuberculosis induced from anti-TNF treatments are based on bioavailability in granulomatous tissue. *PLoS Comput. Biol.* **3:**1909–1924.

Matsuda, M., J. G. Park, D. C. Wang, S. Hunter, P. Chien, and A. D. Schreiber. 1996. Abrogation of the Fc gamma receptor IIA-mediated phagocytic signal by stem-loop Syk antisense oligonucleotides. *Mol. Biol. Cell* **7:**1095–1106.

May, M. J., F. D'Acquisto, L. A. Madge, J. Glöckner, J. S. Pober, and S. Ghosh. 2000. Selective inhibition of NF-kappa B activation by a peptide that blocks the interaction of NEMO with the Ikappa B kinase complex. *Science* **289:**1550–1554.

Messina, C. G., E. P. Reeves, J. Roes, and A. W. Segal. 2002. Catalase negative Staphylococcus aureus retain virulence in mouse model of chronic granulomatous disease. *FEBS Lett.* **518:**107–110.

Miletic, A. V., D. B. Graham, V. Montgrain, K. Fujikawa, T. Kloeppel, K. Brim, B. Weaver, R. Schreiber, R. Xavier, and W. Swat. 2007. Vav proteins control MyD88-dependent oxidative burst. *Blood* **109:**3360–3368.

Minegishi, Y., M. Saito, T. Morio, K. Watanabe, K. Agematsu, S. Tsuchiya, H. Takada, T. Hara, N. Kawamura, T. Ariga, H. Kaneko, N. Kondo, I. Tsuge, A. Yachie, Y. Sakiyama, T. Iwata, F. Bessho, T. Ohishi, K. Joh, K. Imai, K. Kogawa, M. Shinohara, M. Fujieda, H. Wakiguchi, S. Pasic, M. Abinun, H. D. Ochs, E. D. Renner, A. Jansson, B. H. Belohradsky, A. Metin, M. Shimizu, S. Mizutani, T. Miyawaki, S. Nonoyama, and H. Karasuyama. 2006. Human tyrosine kinase 2 deficiency reveals its requisite roles in multiple cytokine signals involved in innate and acquired immunity. *Immunity* **25:**745–755.

Mombo, L.-E., F. Ntoumi, C. Bisseye, S. Ossari, C. Y. Lu, R. L. Nagel, and R. Krishnamoorthy. 2003. Human genetic polymorphisms and asymptomatic *Plasmodium falciparum* malaria in Gabonese schoolchildren. *Am. J. Trop. Med. Hyg.* **68:**186–190.

Müller, M., R. Althaus, D. Fröhlich, K. Frei, and H. P. Eugster. 1999. Reduced antilisterial activity of TNF-deficient bone marrow-derived macrophages is due to impaired superoxide production. *Eur. J. Immunol.* **29:**3089–3097.

Nathan, C. 2006. Role of iNOS in human host defense. *Science* **312:**1874–1875

Nathan, C., and M. U. Shiloh. 2000. Reactive oxygen and nitrogen intermediates in the relationship between mammalian hosts and microbial pathogens. *Proc. Natl. Acad. Sci. USA* **97:**8841–8848.

Naylor, S. L., A. Y. Sakaguchi, T. B. Shows, M. L. Law, D. V. Goeddel, and P. W. Gray. 1983. Human immune interferon gene is located on chromosome 12. *J. Exp. Med.* **157:**1020–1027.

Newman, S. L., L. Gootee, J. Hilty, and R. E. Morris. 2006. Human macrophages do not require phagosome acidification to mediate fungistatic/fungicidal activity against *Histoplasma capsulatum*. *J. Immunol.* **176:**1806–1813.

Newport, M. J., C. M. Huxley, S. Huston, C. M. Hawrylowicz, B. A. Oostra, R. Williamson, and M. Levin. 1996. A mutation in the interferon-gamma-receptor gene and susceptibility to mycobacterial infection. *N. Engl. J. Med.* **335:**1941–1949.

Niedergang, F., E. Colucci-Guyon, T. Dubois, G. Raposo, and P. Chavrier. 2003. ADP ribosylation factor 6 is activated and controls membrane delivery during phagocytosis in macrophages. *J. Cell Biol.* **161:**1143–1150.

Niehues, T., J. Reichenbach, J. Neubert, S. Gudowius, A. Puel, G. Horneff, E. Lainka, U. Dirksen, H. Schroten, R. Döffinger, J. L. Casanova, and V. Wahn. 2004. Nuclear factor [kappa]B essential modulator-deficient child with immunodeficiency yet without anhidrotic ectodermal dysplasia. *J. Allergy Clin. Immunol.* **114:**1456–1462.

Niikura, T., R. Hirata, and S. C. Weil. 1997. A novel interferon-inducible gene expressed during myeloid differentiation. *Blood Cells Mol. Dis.* **23:**337–349.

Ochs, H. D. 2001. The Wiskott-Aldrich syndrome. *Clin. Rev. Allergy Immunol.* **20:**61–86.

Ochs, H. D., and L. D. Notarangelo. 2005. Structure and function of the Wiskott-Aldrich syndrome protein. *Curr. Opin. Hematol.* **12:**284–291.

Olazabal, I. M., E. Caron, R. C. May, K. Schilling, D. A. Knecht, and L. M. Machesky. 2002. Rho-kinase and myosin-II control phagocytic cup formation during CR, but not Fc[gamma]R, phagocytosis. *Curr. Biol.* **12:**1413–1418.

Orange, J. S., A. Jain, Z. K. Ballas, L. C. Schneider, R. S. Geha, and F. A. Bonilla. 2004. The presentation and natural history of immunodeficiency caused by nuclear factor [kappa]B essential modulator mutation. *J. Allergy Clin. Immunol.* **113:**725–733.

Orange, J. S., O. Levy, and R. S. Geha. 2005. Human disease resulting from gene mutations that interfere with appropriate nuclear factor-kappaB activation. *Immunol. Rev.* **203:**21–37.

Orozco, G., E. Sánchez, M. A. López-Nevot, A. Caballero, M. J. Bravo, P. Morata, J. de Dios Colmenero, A. Alonso, and J. Martín. 2003. Inducible nitric oxide synthase promoter polymorphism in human brucellosis. *Microbes Infect.* **5:**1165–1169.

Pace, J. L., S. W. Russell, R. D. Schreiber, A. Altman, and D. H. Katz. 1983. Macrophage activation: priming activity from a T-cell hybridoma is attributable to interferon-gamma. *Proc. Natl. Acad. Sci. USA* **80:**3782–3786.

Parham, C., M. Chirica, J. Timans, E. Vaisberg, M. Travis, J. Cheung, S. Pflanz, R. Zhang, K. P. Singh, F. Vega, W. To, J. Wagner, A. M. O'Farrell, T. McClanahan, S. Zurawski, C. Hannum, D. Gorman, D. M. Rennick, R. A. Kastelein, R. de Waal Malefyt, and K. W. Moore. 2002. A receptor for the heterodimeric cytokine IL-23 is composed of IL-12R[beta]1 and a novel cytokine receptor subunit, IL-23R. *J. Immunol.* **168:**5699–5708.

Paulnock-King, D., K. C. Sizer, Y. R. Freund, P. P. Jones, and J. R. Parnes. 1985. Coordinate induction of Ia alpha, beta, and Ii mRNA in a macrophage cell line. *J. Immunol.* **135**:632–636.

Picard, C., C. Fieschi, F. Altare, S. Al-Jumaah, S. Al-Hajjar, J. Feinberg, S. Dupuis, C. Soudais, I. Z. Al-Mohsen, E. Génin, D. Lammas, D. S. Kumararatne, T. Leclerc, A. Rafii, H. Frayha, B. Murugasu, L. B. Wah, R. Sinniah, M. Loubser, E. Okamoto, A. Al-Ghonaium, H. Tufenkeji, L. Abel, and J. L. Casanova. 2002. Inherited interleukin-12 deficiency: IL12B genotype and clinical phenotype of 13 patients from six kindreds. *Am. J. Hum. Genet.* **70**:336–348.

Popov, A., Z. Abdullah, C. Wickenhauser, T. Saric, J. Driesen, F. G. Hanisch, E. Domann, E. L. Raven, O. Dehus, C. Hermann, D. Eggle, S. Debey, T. Chakraborty, M. Krönke, O. Utermöhlen, and J. L. Schultze. 2006. Indoleamine 2,3-dioxygenase-expressing dendritic cells form suppurative granulomas following Listeria monocytogenes infection. *J. Clin. Invest.* **116**:3160–3170.

Puel, A., J. Reichenbach, J. Bustamante, C. L. Ku, J. Feinberg, R. Döffinger, M. Bonnet, O. Filipe-Santos, L. de Beaucoudrey, A. Durandy, G. Horneff, F. Novelli, V. Wahn, A. Smahi, A. Israel, T. Niehues, and J. L. Casanova. 2006. The NEMO mutation creating the most-upstream premature stop codon is hypomorphic because of a reinitiation of translation. *Am. J. Hum. Genet.* **78**:691–701.

Quezada, S. A., L. Z. Jarvinen, E. F. Lind, and R. J. Noelle. 2004. CD40/CD154 interactions at the interface of tolerance and immunity. *Annu. Rev. Immunol.* **22**:307–328.

Radosavljevic, M., B. Cuillerier, M. J. Wilson, O. Clément, S. Wicker, S. Gilfillan, S. Beck, J. Trowsdale, and S. Bahram. 2002. A cluster of ten novel MHC class I related genes on human chromosome 6q24. 2-q25. 3. *Genomics* **79**:114–123.

Ramachandra, L., E. Noss, W. H. Boom, and C. V. Harding. 1999. Phagocytic processing of antigens for presentation by class II major histocompatibility complex molecules. *Cell. Microbiol.* **1**:205–214.

Randolph, G. J., C. Jakubzick, and C. Qu. 2008. Antigen presentation by monocytes and monocyte-derived cells. *Curr. Opin. Immunol.* **20**:52–60.

Rapoport, T. A. 2007. Protein translocation across the eukaryotic endoplasmic reticulum and bacterial plasma membranes. *Nature* **450**:663–669.

Reeves, E. P., H. Lu, H. L. Jacobs, C. G. Messina, S. Bolsover, G. Gabella, E. O. Potma, A. Warley, J. Roes, and A. W. Segal. 2002. Killing activity of neutrophils is mediated through activation of proteases by K+ flux. *Nature* **416**:291–297.

Ren, G., P. Vajjhala, J. S. Lee, B. Winsor, and A. L. Munn. 2006. The BAR domain proteins: molding membranes in fission, fusion, and phagy. *Microbiol. Mol. Biol. Rev.* **70**:37–120.

Rink, J., E. Ghigo, Y. Kalaidzidis, and M. Zerial. 2005. Rab conversion as a mechanism of progression from early to late endosomes. *Cell* **122**:735–749.

Roach, T. I., C. H. Barton, D. Chatterjee, and J. M. Blackwell. 1993. Macrophage activation: lipoarabinomannan from avirulent and virulent strains of Mycobacterium tuberculosis differentially induces the early genes c-fos, KC, JE, and tumor necrosis factor-alpha. *J. Immunol.* **150**:1886–1896.

Roach, D. R., A. G. Bean, C. Demangel, M. P. France, H. Briscoe, and W. J. Britton. 2002. TNF regulates chemokine induction essential for cell recruitment, granuloma formation, and clearance of mycobacterial infection. *J. Immunol.* **168**:4620–4627.

Roberts, E. A., J. Chua, G. B. Kyei, and V. Deretic. 2006. Higher order Rab programming in phagolysosome biogenesis. *J. Cell Biol.* **174**:923–929.

Roesler, J., M. E. Horwitz, C. Picard, P. Bordigoni, G. Davies, E. Koscielniak, M. Levin, P. Veys, U. Reuter, A. Schulz, C. Thiede, T. Klingebiel, A. Fischer, S. M. Holland, J. L. Casanova, and W. Friedrich. 2004. Hematopoietic stem cell transplantation for complete IFN-[gamma] receptor 1 deficiency: a multi-institutional survey. *J. Pediatr.* **145**:806–812.

Rosenzweig, S., S. E. Dorman, J. Roesler, J. Palacios, M. Zelazko, and S. M. Holland. 2002. 561del4 defines a novel small deletion hotspot in the interferon-[gamma] receptor 1 chain. *Clin. Immunol.* **102**:25–27.

Rosenzweig, S. D., S. E. Dorman, G. Uzel, S. Shaw, A. Scurlock, M. R. Brown, R. H. Buckley, and S. M. Holland. 2004. A novel mutation in IFN-[gamma] receptor 2 with dominant negative activity: biological consequences of homozygous and heterozygous states. *J. Immunol.* **173**:4000–4008.

Rossouw, M., H. J. Nel, G. S. Cooke, P. D. van Helden, and E. G. Hoal. 2003. Association between tuberculosis and a polymorphic NFkappaB binding site in the interferon gamma gene. *Lancet* **361**:1871–1872.

Rzomp, K. A., A. R. Moorhead, and M. A. Scidmore. 2003. Rab GTPases are recruited to chlamydial inclusions in both a species-dependent and species-independent manner. *Infect. Immun.* **71**:5855–5870.

Saccani, S., S. Pantano, and G. Natoli. 2001. Two waves of nuclear factor [kappa]B recruitment to target promoters. *J. Exp. Med.* **193**:1351–1360.

Sasahara, Y., R. Rachid, M. J. Byrne, M. A. de la Fuente, R. T. Abraham, N. Ramesh, and R. S. Geha. 2002. Mechanism of recruitment of WASP to the immunological synapse and of its activation following TCR ligation. *Mol. Cell* **10**:1269–1281.

Savina, A., and S. Amigorena. 2007. Phagocytosis and antigen presentation in dendritic cells. *Immunol. Rev.* **219**:143–156.

Schiffer, R., J. Baron, G. Dagtekin, W. Jahnen-Dechent, and G. Zwadlo-Klarwasser. 2002. Differential regulation of the expression of transporters associated with antigen processing, TAP1 and TAP2, by cytokines and lipopolysaccharide in primary human macrophages. *Inflamm. Res.* **51**:403–408.

Schroder, K., M. J. Sweet, and D. A. Hume. 2006. Signal integration between IFN[gamma] and TLR signalling pathways in macrophages. *Immunobiology* **211**:511–524.

Scianimanico, S., M. Desrosiers, J. F. Dermine, S. Méresse, A. Descoteaux, and M. Desjardins. 1999. Impaired recruitment of the small GTPase rab7 correlates with the inhibition of phagosome maturation by Leishmania donovani promastigotes. *Cell. Microbiol.* **1**:19–32.

Seabra, M. C., E. H. Mules, and A. N. Hume. 2002. Rab GTPases, intracellular traffic and disease. *Trends Mol. Med.* **8**:23–30.

Segal, A. W. 2005. How neutrophils kill microbes. *Annu. Rev. Immunol.* **23**:197–223.

Segal, B., L. A. Barnhart, V. L. Anderson, T. J. Walsh, H. L. Malech, and S. M. Holland. 2005. Posaconazole as salvage therapy in patients with chronic granulomatous disease and invasive filamentous fungal infection. *Clin. Infect. Dis.* **40**:1684–1688.

Segal, B. H., T. L. Leto, J. I. Gallin, H. L. Malech, and S. M. Holland. 2000. Genetic, biochemical, and clinical features of chronic granulomatous disease. *Medicine* (Baltimore) **79**:170–200.

Sharpe, A. H., and G. J. Freeman. 2002. The B7-CD28 superfamily. *Nat. Rev. Immunol.* **2**:116–126.

Shintani, T., and D. J. Klionsky. 2004. Autophagy in health and disease: a double-edged sword. *Science* **306**:990–995.

Smahi, A., G. Courtois, P. Vabres, S. Yamaoka, S. Heuertz, A. Munnich, A. Israël, N. S. Heiss, S. M. Klauck, P. Kioschis, S. Wiemann, A. Poustka, T. Esposito, T. Bardaro, E. Gianfrancesco, A. Ciccodicola, M. D'Urso, H. Woffendin, T. Jakins, D. Donnai, H. Stewart, S. J. Kenwrick, S. Aradhya, T. Yamagata, M. Levy, R. A. Lewis, and D. L. Nelson. 2000. Genomic rearrangement in NEMO impairs NF-[kappa]B activation and is a cause of incontinentia pigmenti. *Nature* **405:**466–472.

Smirnova, I., N. Mann, A. Dols, H. H. Derkx, M. L. Hibberd, M. Levin, and B. Beutler. 2003. Assay of locus-specific genetic load implicates rare Toll-like receptor 4 mutations in meningococcal susceptibility. *Proc. Natl. Acad. Sci. USA* **100:**6075–6080.

Smith, A. C., W. D. Heo, V. Braun, X. Jiang, C. Macrae, J. E. Casanova, M. A. Scidmore, S. Grinstein, T. Meyer, and J. H. Brumell. 2007. A network of Rab GTPases controls phagosome maturation and is modulated by *Salmonella enterica* serovar Typhimurium. *J. Cell Biol.* **176:**263–268.

Speer, C. P., M. J. Pabst, H. B. Hedegaard, R. F. Rest, and R. B. Johnston, Jr. 1984. Enhanced release of oxygen metabolites by monocyte-derived macrophages exposed to proteolytic enzymes: activity of neutrophil elastase and cathepsin G. *J. Immunol.* **133:**2151–2156.

Stark, G. R., I. M. Kerr, B. R. Williams, R. H. Silverman, and R. D. Schreiber. 1998. How cells respond to interferons. *Annu. Rev. Biochem.* **67:**227–264.

Stergiopoulou, T., J. Meletiadis, E. Roilides, D. E. Kleiner, R. Schaufele, M. Roden, S. Harrington, L. Dad, B. Segal, and T. J. Walsh. 2007. Host-dependent patterns of tissue injury in invasive pulmonary aspergillosis. *Am. J. Clin. Pathol.* **127:**349–355.

Sturgill-Koszycki, S., P. H. Schlesinger, P. Chakraborty, P. L. Haddix, H. L. Collins, A. K. Fok, R. D. Allen, S. L. Gluck, J. Heuser, and D. G. Russell. 1994. Lack of acidification in Mycobacterium phagosomes produced by exclusion of the vesicular proton-ATPase. *Science* **263:**678–681.

Sullivan, K. E., C. A. Mullen, R. M. Blaese, and J. A. Winkelstein. 1994. A multiinstitutional survey of the Wiskott-Aldrich syndrome. *J. Pediatr.* **125:**876–885.

Sullivan, K. E., A. B. Reddy, K. Dietzmann, A. R. Suriano, V. P. Kocieda, M. Stewart, and M. Bhatia. 2007. Epigenetic regulation of tumor necrosis factor alpha. *Mol. Cell Biol.* **27:**5147–5160.

Sun, J., A. E. Deghmane, H. Soualhine, T. Hong, C. Bucci, A. Solodkin, and Z. Hmama. 2007. Mycobacterium bovis BCG disrupts the interaction of Rab7 with RILP contributing to inhibition of phagosome maturation. *J. Leukoc. Biol.* **82:**1437–1445.

Sun-Wada, G. H., Y. Wada, and M. Futai. 2003. Lysosome and lysosome-related organelles responsible for specialized functions in higher organisms, with special emphasis on vacuolar-type proton ATPase. *Cell. Struct. Funct.* **28:**455–463.

Swanson, J. A., and A. D. Hoppe. 2004. The coordination of signaling during Fc receptor-mediated phagocytosis. *J. Leukoc. Biol.* **76:**1093–2103.

Tephly, L. A., and A. B. Carter. 2007a. Constitutive NADPH oxidase and increased mitochondrial respiratory chain activity regulate chemokine gene expression. *Am. J. Physiol.* **293:**L1143–L1155.

Tephly, L. A., and A. B. Carter. 2007b. Differential expression and oxidation of MKP-1 modulates TNF-alpha gene expression. *Am. J. Respir. Cell. Mol. Biol.* **37:**366–367.

Tkalcevic, J., M. Novelli, M. Phylactides, J. P. Iredale, A. W. Segal, and J. Roes. 2000. Impaired immunity and enhanced resistance to endotoxin in the absence of neutrophil elastase and cathepsin G. *Immunity* **12:**201–210.

Trinchieri, G. 2003. Interleukin-12 and the regulation of innate resistance and adaptive immunity. *Nat. Rev. Immunol.* **3:**133–146.

Tsuboi, S. 2006. A complex of Wiskott-Aldrich syndrome protein with mammalian verprolins plays an important role in monocyte chemotaxis. *J. Immunol.* **176:**6576–6585.

Tsuboi, S. 2007. Requirement for a complex of Wiskott-Aldrich syndrome protein (WASP) with WASP interacting protein in podosome formation in macrophages. *J. Immunol.* **178:**2987–2995.

Tsuboi, S., and J. Meerloo. 2007. Wiskott-Aldrich syndrome protein is a key regulator of the phagocytic cup formation in macrophages. *J. Biol. Chem.* **282:**34194–34203.

Ulevitch, R. J. 2004. Therapeutics targeting the innate immune system. *Nat. Rev. Immunol.* **4:**512–520.

Vachon, E., R. Martin, J. Plumb, V. Kwok, R. W. Vandivier, M. Glogauer, A. Kapus, X. Wang, C. W. Chow, S. Grinstein, and G. P. Downey. 2006. CD44 is a phagocytic receptor. *Blood* **107:**4149–4158.

van Crevel, R., T. H. M. Ottenhoff, and J. W. M. van der Meer. 2002. Innate immunity to Mycobacterium tuberculosis. *Clin. Microbiol. Rev.* **15:**294–309.

van der Veen, R. C., T. A. Dietlin, F. M. Hofman, L. Pen, B. H. Segal, and S. M. Holland. 2000. Superoxide prevents nitric oxide-mediated suppression of helper T lymphocytes: decreased autoimmune encephalomyelitis in nicotinamide adenine dinucleotide phosphate oxidase knockout mice. *J. Immunol.* **164:**5177–5783.

van de Vosse, E., and T. H. M. Ottenhoff. 2006. Human host genetic factors in mycobacterial and Salmonella infection: lessons from single gene disorders in IL-12/IL-23-dependent signaling that affect innate and adaptive immunity. *Microbes Infect.* **8:**1167–1173.

Vazquez-Torres, A., G. Fantuzzi, C. K. Edwards III, C. A. Dinarello, and F. C. Fang. 2001. Defective localization of the NADPH phagocyte oxidase to *Salmonella*-containing phagosomes in tumor necrosis factor p55 receptor-deficient macrophages. *Proc. Natl. Acad. Sci. USA* **98:**2561–2565.

Vergne, I., J. Chua, H. H. Lee, M. Lucas, J. Belisle, and V. Deretic. 2005. Mechanism of phagolysosome biogenesis block by viable *Mycobacterium tuberculosis*. *Proc. Natl. Acad. Sci. USA* **102:**4033–4038.

Verhoeven, K., P. De Jonghe, K. Coen, N. Verpoorten, M. Auer-Grumbach, J. M. Kwon, D. FitzPatrick, E. Schmedding, E. De Vriendt, A. Jacobs, V. Van Gerwen, K. Wagner, H. P. Hartung, and V. Timmerman. 2003. Mutations in the small GTP-ase late endosomal protein RAB7 cause Charcot-Marie-Tooth type 2B neuropathy. *Am. J. Hum. Genet.* **72:**722–727.

Via, L. E., D. Deretic, R. J. Ulmer, N. S. Hibler, L. A. Huber, and V. Deretic. 1997. Arrest of mycobacterial phagosome maturation is caused by a block in vesicle fusion between stages controlled by rab5 and rab7. *J. Biol. Chem.* **272:**13326–13331.

Vinolo, E., H. Sebban, A. Chaffotte, A. Israël, G. Courtois, M. Véron, and F. Agou. 2006. A point mutation in NEMO associated with anhidrotic ectodermal dysplasia with immunodeficiency pathology results in destabilization of the oligomer and reduces lipopolysaccharide- and tumor necrosis factor-mediated NF-[kappa]B activation. *J. Biol. Chem.* **281:**6334–6348.

Vogt, G., A. Chapgier, K. Yang, N. Chuzhanova, J. Feinberg, C. Fieschi, S. Boisson-Dupuis, A. Alcais, O. Filipe-Santos, J. Bustamante, L. de Beaucoudrey, I. Al-Mohsen, S. Al-Hajjar, A. Al-Ghonaium, P. Adimi, M. Mirsaeidi, S. Khalilzadeh, S. Rosenzweig, O. de la Calle Martin, T. R. Bauer, J. M. Puck, H. D. Ochs, D. Furthner, C. Engelhorn, B. Belohradsky, D. Mansouri, S. M. Holland, R. D. Schreiber, L. Abel, D. N. Cooper, C. Soudais, and J. L. Casanova. 2005. Gains of glycosylation comprise an unexpectedly large group of pathogenic mutations. *Nat. Genet.* **37:**692–700.

Vowells, S. J., T. A. Fleisher, S. Sekhsaria, D. W. Alling, T. E. Maguire, and H. L. Malech. 1996. Genotype-dependent variability in flow cytometric evaluation of reduced nicotinamide adenine dinucleotide phosphate oxidase function in patients with chronic granulomatous disease. *J. Pediatr.* **128:**104–107.

Wahl, S. M., L. M. Wahl, J. B. McCarthy, L. Chedid, and S. E. Mergenhagen. 1979. Macrophage activation by mycobacterial water soluble compounds and synthetic muramyl dipeptide. *J. Immunol.* **122:**2226–2231.

Wallis, R. S., M. S. Broder, J. Y. Wong, M. E. Hanson, and D. O. Beenhouwer. 2004. Granulomatous infectious diseases associated with tumor necrosis factor antagonists. *Clin. Infect. Dis.* **39:**1255–1256.

Watters, T. M., E. F. Kenny, and L. A. O'Neill. 2007. Structure, function and regulation of the Toll/IL-1 receptor adaptor proteins. *Immunol. Cell Biol.* **85:**411–419.

Wells, C. M., M. Walmsley, S. Ooi, V. Tybulewicz, and A. J. Ridley. 2004. Rac1-deficient macrophages exhibit defects in cell spreading and membrane ruffling but not migration. *J. Cell Sci.* **117:**1259–1268.

Wiedemann, A., J. C. Patel, J. Lim, A. Tsun, Y. van Kooyk, and E. Caron. 2006. Two distinct cytoplasmic regions of the [beta]2 integrin chain regulate RhoA function during phagocytosis. *J. Cell Biol.* **172:**1069–1079.

Williams, D. A., W. Tao, F. Yang, C. Kim, Y. Gu, P. Mansfield, J. E. Levine, B. Petryniak, C. W. Derrow, C. Harris, B. Jia, Y. Zheng, D. R. Ambruso, J. B. Lowe, S. J. Atkinson, M. C. Dinauer, and L. Boxer. 2000. Dominant negative mutation of the hematopoietic-specific Rho GTPase, Rac2, is associated with a human phagocyte immunodeficiency. *Blood* **96:**1646–1654.

Winkelstein, J. A., M. C. Marino, R. B. Johnston, Jr., J. Boyle, J. Curnutte, J. I. Gallin, H. L. Malech, S. M. Holland, H. Ochs, P. Quie, R. H. Buckley, C. B. Foster, S. J. Chanock, and H. Dickler. 2000. Chronic granulomatous disease. Report on a national registry of 368 patients. *Medicine* (Baltimore) **79:**155–169.

Winthrop, K. L., S. Yamashita, S. E. Beekmann, P. M. Polgreen; Infectious Diseases Society of America Emerging Infections Network. 2008. Mycobacterial and other serious infections in patients receiving anti-tumor necrosis factor and other newly approved biologic therapies: case finding through the emerging infections network. *Clin. Infect. Dis.* **46:**1738–1740.

Wood, P. M., C. Fieschi, C. Picard, T. H. Ottenhoff, J. L. Casanova, and D. S. Kumararatne. 2005. Inherited defects in the interferon-gamma receptor or interleukin-12 signalling pathways are not sufficient to cause allergic disease in children. *Eur. J. Pediatr.* **164:**741–747.

Yamaoka, K., P. Saharinen, M. Pesu, V. E. Holt III, O. Silvennoinen, and J. J. O'Shea. 2004. The Janus kinases (Jaks). *Genome Biol.* **5:**253.

Zerbe, C., and S. Holland. 2005. Disseminated histoplasmosis in persons with interferon gamma receptor 1 deficiency. *Clin. Infect. Dis.* **41:**e38–e41.

Zhan, Y., J. V. Virbasius, X. Song, D. P. Pomerleau, and G. W. Zhou. 2002. The p40phox and p47phox PX domains of NADPH oxidase target cell membranes via direct and indirect recruitment by phosphoinositides. *J. Biol. Chem.* **277:**4512–4518.

Zhang, P., S. Snyder, P. Feng, P. Azadi, S. Zhang, S. Bulgheresi, K. E. Sanderson, J. He, J. Klena, and T. Chen. 2006. Role of N-acetylglucosamine within core lipopolysaccharide of several species of gram-negative bacteria in targeting the DC-SIGN (CD209). *J. Immunol.* **177:**4002–4011.

Zhang, Q., D. Cox, C. C. Tseng, J. G. Donaldson, and S. Greenberg. 1998. A requirement for ARF6 in Fcgamma receptor-mediated phagocytosis in macrophages. *J. Biol. Chem.* **273:**19977–19981.

Zhao, X., K. A. Carnevale, and M. K. Cathcart. 2003. Human monocytes use Rac1, not Rac2, in the NADPH oxidase complex. *J. Biol. Chem.* **278:**40788–40792.

Zhao, X., B. Xu, A. Bhattacharjee, C. M. Oldfield, F. B. Wientjes, G. M. Feldman, and M. K. Cathcart. 2005. Protein kinase Cdelta regulates p67phox phosphorylation in human monocytes. *J. Leukoc. Biol.* **77:**414–420.

Zhou, H., J. Gu, S. J. Lamont, and X. Gu. 2007. Evolutionary analysis for functional divergence of the toll-like receptor gene family and altered functional constraints. *J. Mol. Evol.* **65:**119–123.

20

The Functional Heterogeneity of Activated Macrophages

XIA ZHANG AND DAVID M. MOSSER

A process that is central to our understanding of cell-mediated immunity (CMI) is the activation of tissue macrophages into immune effector cells. As our understanding of CMI advances, so does our appreciation of the importance of the activated macrophage (see chapter 19). Over the past decade or so, we have come to realize that not all activated macrophages are the same, and not all activated macrophages develop in response to cell-mediated immune responses. Different populations of macrophages with distinct physiologies can develop in response to different stimuli. In fact, it is likely that the number of different macrophage populations that can arise may be as diverse as the activating stimuli that induce them. Some of these stimuli can instruct macrophages to kill microbes (see section below), or they can instruct these cells to secrete anti-inflammatory cytokines to terminate inflammation, or to lay down extracellular matrix components to promote wound healing.

New and improved ways to phenotype cells in tissue have led to a better understanding of the heterogeneity of activated macrophages. As our understanding of the various macrophage populations increases, so does the potential for therapeutic intervention. For example, the targeting of specific macrophage subpopulations will likely evolve from basic studies to identify biomarkers that are subpopulation specific. Furthermore, as our understanding of the processes that give rise to macrophage subpopulations with immunostimulatory or immunoinhibitory properties increases, so will the potential to generate these cells for therapeutic purposes. For these reasons, several different in vitro-generated macrophage subpopulations have been described by various groups. In the present chapter, we will describe some of these populations and compare and contrast their functional properties.

MONOCYTES AND MONOCYTE-DERIVED MACROPHAGES

Macrophages are derived from progenitor cells in the bone marrow (Akashi et al., 2000; Orkin, 2000). The differentiative process that results in the development of a mature tissue macrophage is quite complex, involving many intermediates. In simplified terms, monocytes originate from progenitor cells in the bone marrow which differentiate through the monoblast, promonocyte, and monocyte stages before their release into the circulation (van Furth and Cohn, 1968). The average half-life of circulating monocytes in the bloodstream is approximately 22 h in mice, 42 h in rats, and 71 h in humans (van Furth and Cohn, 1968; Volkman and Collins, 1974; Whitelaw, 1972). These cells differentiate further into macrophages as they leave the blood and enter tissue (van Furth and Cohn, 1968). Macrophages in tissue can respond to activation stimuli and dramatically change their physiology. Each of the intermediate stages of macrophage development can be identified by distinct markers and by their unique physiologies (Gordon and Taylor, 2005; Mantovani et al., 2007). We will briefly describe the monocyte differentiative process, and then focus our attention on the terminal development of activated macrophages.

Monocyte/dendritic cell (DC) precursors in the bone marrow express both c-kit ($CD117^+$) and the fractalkine receptor (CX_3CR1^+) and are lacking markers of lineage-committed precursors (Lin^-) (Fogg et al., 2006; Kumar and Jack, 2006). They belong to $CD117^+Sca1^-IL-7R\alpha^-$ myeloid progenitors (Akashi et al., 2000). Monocyte precursors in the bone marrow begin to express the monocyte markers, such as CD11b and, in the murine system, Gr-1 (Ly6C). After their release into the bloodstream, the nondividing precursor monocytes soon differentiate into monocytes. Monocytes are generally defined as mononuclear cells with surface markers of CD11b and F4/80 in mice and CD11b, CD11c, and CD14 in humans, but without T, B, NK, and DC markers (Fogg et al., 2006; Hume, 2006). Circulating monocytes have been further subdivided by the differential expression of certain surface markers such as

Xia Zhang and David M. Mosser, Department of Cell Biology and Molecular Genetics and the Maryland Pathogen Research Institute, University of Maryland, College Park, MD 20742.

CD16 (FcγRIII) or the chemokine receptors CCR2 and CX$_3$CR1 (Geissmann et al., 2003; Tacke and Randolph, 2006; Ziegler-Heitbrock, 2007).

In murine blood, two predominant populations of monocytes have been identified and characterized (Geissmann et al., 2003; Tacke and Randolph, 2006). The Gr-1^{low} CCR2$^-$CX$_3$CR1^{high}CD62L$^-$ population appears functionally to be the resident subset that has a relatively extended tenure in the bloodstream. The Gr-1^{high}CCR2$^+$CX$_3$CR1^{low}CD62L$^+$ population may represent a less mature subset of cells that are newly released from bone marrow and can still undergo a further maturation process in blood. The human monocyte subset CD14^{low}CD16$^+$CD64$^-$CCR2$^-$CD62L$^-$ shares morphological characteristics with the murine Gr-1$^{-/low}$CCR2$^-$CX$_3$CR1^{high}CD62L$^-$ subpopulation, whereas CD14$^+$CD16$^-$CD64$^+$CCR2$^+$CD62L$^+$ monocytes correspond morphologically to the murine Gr-1^{high}CCR2$^+$CX$_3$CR1^{low}CD62L$^+$ subpopulation. A third population of monocytes has been identified in the mouse which are Gr-1^{int}CCR2$^+$CX$_3$CR1^{low}CD62L$^+$. These cells also express CCR7 and CCR8. This subset may be analogous to the human CD14$^+$CD16$^+$CD64$^+$ cells identified by Randolph and colleagues (Tacke and Randolph, 2006). It is not yet clear whether this population represents a transitional form from Gr-1^{high} monocytes to Gr-1^{low} monocytes (Tacke et al., 2006; Tacke and Randolph, 2006), or whether they are independently derived (Geissmann et al., 2003). Although morphological similarities are shared between the two major monocyte populations in mice and humans, the two subpopulations may have markedly different functional properties in vivo. Furthermore, in mice, the two subpopulations of monocytes are almost equally represented in blood (Geissmann et al., 2003), but the CD14^{low}CD16$^+$ population accounts for only 10 to 15% of total human peripheral blood monocytes (Weber et al., 2000). Therefore, one must be careful to stipulate the criteria by which these subpopulations are being compared. Clearly, the biology of monocyte heterogeneity is an area that warrants additional study.

Once monocytes migrate from the bloodstream into tissue, they further differentiate into macrophages. They acquire distinct morphological and functional properties dictated largely by the tissue microenvironment in which they reside. These cells remain quite plastic and can further respond to inflammatory or immunological stimuli present in tissue. For this reason, the term "macrophage" actually refers to a heterogeneous collection of cells with distinct morphologies, physiologies, and functionalities, depending on the tissue in which they reside and the immune/inflammatory events occurring in this tissue.

Tissue-specific macrophage populations include liver Kupffer cells, lung alveolar macrophages, a variety of splenic macrophages, peritoneal macrophages, skin dermal macrophages, and even bone osteoclasts and brain microglia. It appears that under steady-state conditions each of these cell populations can be maintained in the periphery by localized tissue-resident colony-forming cells that have enough proliferative capacity to maintain the local tissue macrophage population (Gordon and Taylor, 2005). Another source of tissue macrophages under steady-state conditions may be from the Gr-1^{low}CCR2$^-$CX$_3$CR1^{high}CD62L$^-$ subpopulation of monocytes in mice or the CD14^{low}CD16$^+$CCR2$^-$CD62L$^-$ subpopulation in humans (Geissmann et al., 2003). These Gr-1^{low} monocytes can function as warning sensors to detect inflammation signals because, under noninflammatory conditions, they patrol the tissue by crawling on the resting endothelium of local blood vessels (Auffray et al., 2007). High expression of CX$_3$CR1 and LFA-1 on this subset of monocytes is essential for their patrolling behavior. These patrolling monocytes appear to have enhanced capacity for trans-endothelial migration, and upon the onset of inflammation, they quickly enter into the tissue to differentiate into macrophages for initiation of an early immune response.

During inflammation, the murine Gr-1^{high}CCR2$^+$CX$_3$CR1^{low}CD62L$^+$ (or human CD14$^+$CD16$^-$CD64$^+$CCR2$^+$CD62L$^+$) subpopulation of monocytes appears to contribute as a second wave of monocytes that infiltrate into the inflammatory site. These newly migrating immature blood monocytes rapidly differentiate into macrophages and expand the local macrophage populations. Of note, in the absence of inflammation, these immature Gr-1^{high} blood monocytes can migrate back to the bone marrow and convert into Gr-1^{low} monocytes (Varol et al., 2007). The anatomical location can also be a factor in determining which monocyte population will differentiate into macrophages. In the lung under inflammatory or noninflammatory conditions, it appears that only the Gr-1^{low}CCR2$^-$CX$_3$CR1^{high}CD62L$^-$ monocyte subpopulation can differentiate into macrophages, although both Gr-1^{high} and Gr-1^{low} subsets can give rise to DCs (Landsman et al., 2007).

At the onset of inflammation, the murine Gr-1^{high}CCR2$^+$CX$_3$CR1^{low}CD62L$^+$ subpopulation or human CD14$^+$CD16$^-$CD64$^+$CCR2$^+$CD62L$^+$ monocytes appear to be able to differentiate not only into macrophages but also inflammatory DCs such as Tip DC (tumor necrosis factor-α [TNF-α] and inducible nitric oxide- [iNOS]-producing DC) (Leon et al., 2007; Serbina et al., 2003; Shortman and Naik, 2007; Villadangos, 2007). Therefore, both monocyte populations have the potential to present antigen and propagate adaptive immune responses in tissue (Shortman and Naik, 2007). It should be pointed out that the differentiation of monocytes into DCs could be blocked if bacteria were present at the site of infection (Rotta et al., 2003). The detailed discussion and speculation of how monocytes differentiate into DCs at the site of infection and the lineage origin of steady-state DCs are beyond the scope of this review.

Overall, the scientific community has made dramatic recent progress in understanding the differentiation of monocytes into macrophages. The identification of two major subsets of circulating monocytes with distinct phenotypes and perhaps different potential to become a unique population of macrophages with different functions in the peripheral tissues has paved a path for us to better understand the activation of macrophages and their corresponding functions.

MACROPHAGE ACTIVATION

Once in tissue, macrophages can undergo profound physiological changes in response to the combination of cytokines and inflammatory stimuli they encounter there. This response is collectively referred to as macrophage activation (Cohn, 1978; MacMicking et al., 1997). That term was initially applied to these cells because activated macrophages were larger, more spread out, and capable of enhanced killing of microbes, especially intracellular microbes. For many years it was assumed that all activated macrophages shared similar capabilities, pertaining to this enhanced ability to kill microbes. We now know that this is not true. Macrophages are remarkably plastic cells that can respond to a variety of different stimuli and undergo distinct physiological changes in response to environmental cues.

A number of different macrophage populations have been described in association with different disease states. Attempts to assign biochemical and functional markers to these cell populations are under way, and several groups have made substantial progress in characterizing specific activated macrophage populations (Gordon and Taylor, 2005; Mantovani et al., 2007; Mills et al., 2000; Van Ginderachter et al., 2006). In some ways, however, this task is akin to describing the color of a chameleon. As the environment in which a macrophage resides changes, so too will the functional and physiological properties of these cells change. This plasticity has led to controversy and consternation regarding the various macrophage subpopulations and their physiological significance.

In the present chapter we will review some of the different activated macrophage populations that have been described. In many cases, these cells were developed under defined in vitro conditions, and therefore it is not always readily apparent how these cells correspond to macrophages in situ. Nevertheless, these studies reveal the differentiative potential of these cells and provide hints about how we can exploit the plasticity of this cell to manipulate the adaptive immune response to influence disease progression.

CLASSICAL MACROPHAGE ACTIVATION

Classically activated macrophages are the prototypical immune effector cells that exhibit enhanced microbicidal capabilities and secrete prodigious amounts of inflammatory cytokines and mediators (see chapter 19, this volume). These cells represent a key component of host defense, but they are also important mediators of autoimmune/inflammatory pathologies. These cells were described in detail in chapter 19, but some points bear repetition to illustrate how these cells are distinct from the alternative forms of activated macrophages that will be detailed below. Classically activated macrophages arise in response to priming with gamma interferon (IFN-γ) and stimulation with any number of inflammatory stimuli. These stimuli include any of the Toll-like receptor (TLR) agonists and many endogenous "danger signals." The combination of priming plus stimulation induces macrophage activation (Van Ginderachter et al., 2006).

The cytokines most closely associated with classically activated macrophages are the interleukin-12 (IL-12) family of cytokines (Trinchieri and Sher, 2007; Verreck et al., 2004). The production of this family of cytokines from macrophages, including IL-12p40/p35 and IL-23p40/p19, particularly depends on IFN-γ priming (Liu et al., 2003; Verreck et al., 2004). The effect of IFN-γ priming on enhancing macrophage activities is due to the activation of transcription factors, including signal transducer and activator of transcription (STAT) proteins and interferon regulatory factors (IRFs). It is believed that IL-12 is primarily controlled at the level of transcription of the p40 gene via several putative sequence motifs in the 5' regulatory region of the p40 promoter (Trinchieri and Sher, 2007). These putative sequence motifs are highly conserved between both humans and mice, and include an element for interferon consensus sequence-binding protein (ICSBP)/IRF8 that responds to priming with IFN-γ. IFN-γ triggers responses through a Janus kinase (Jak) and Stat signaling pathway. Binding of IFN-γ to its receptor activates Jak1 and Jak2, leading to the phosphorylation and dimerization of Stat1 (Levy and Darnell, 2002). Activated Stat1 migrates into the nucleus and binds to Stat1-binding elements (GAS elements) in the regulatory region of the ICSBP/IRF8 gene to induce ICSBP/IRF8 expression (Wang et al., 2000). Thus, IFN-γ priming and its synergistic activation effect with lipopolysaccharide (LPS) on IL-12p40 gene expression require de novo protein synthesis. Of note, p40 is also a subunit of IL-23 and therefore the regulation of p40 gene expression will affect IL-23 as well. IL-27 is the newest member of this IL-12 family of heterodimer cytokines (Pflanz et al., 2002). IL-27 is composed of a p28 subunit and the Epstein-Barr virus-induced gene 3, which is synergistically induced by LPS and IFN-γ. IRF-1 is implicated as a major transcriptional regulator that mediates the effect of IFN-γ on p28 gene expression (Liu et al., 2007).

Several other inflammatory cytokines are also produced by classically activated macrophages. TNF-α and IL-1 are the prototypical inflammatory cytokines associated with classical macrophage activation (Gordon and Taylor, 2005). Importantly, these cytokines are produced in response to macrophage stimulation alone and generally do not require IFN-γ priming. This priming-independent production of inflammatory cytokines by macrophages gave rise to some early confusion about the requirement for two signals for macrophage activation. In some settings, high levels of stimulation alone can sometimes overcome the requirement for IFN-γ priming, but the combination of IFN-γ priming and stimulation is clearly the most efficient way to induce the fully microbicidal state of macrophage activation, characterized by the robust production of nitric oxide (NO) and reactive oxygen intermediates.

The most abundant source of IFN-γ for macrophage priming comes from activated antigen-specific Th1 T cells (Young, 2006). Thus, while macrophage activation is not considered an antigen-specific event, the sustained activation of macrophages in lesions is generally an antigen-specific event at the level of the T cell. Innate immune cells can also be important early sources of IFN-γ, which can contribute to early macrophage activation. NK cells and NK-T cells are particularly rich sources of early IFN-γ (Young, 2006).

Because classically activated macrophages increase their expression of the costimulatory molecules B7.1 and B7.2 (CD80 and CD86, respectively) along with major histocompatibility class II (MHC-II), these cells are capable of presenting antigen to T cells (Greenwald et al., 2005; Harding et al., 2003; Menendez-Benito and Neefjes, 2007). It is likely that this presentation occurs in granulomatous lesions where activated macrophages and T cells are in close apposition. Naive T cells initially encounter antigen in the lymph node in association with DCs, then migrate to tissue, where they can be restimulated by classically activated macrophages. These activated T cells, in turn, can sustain macrophage activation through contact-dependent mechanisms. Activated T cells express CD40L, which binds to CD40 on macrophages to induce macrophage IL-12 production. This cytokine promotes Th1 differentiation and the production of IFN-γ (van Kooten and Banchereau, 2000), thereby propagating cell-mediated immune responses.

Several mechanisms can work to diminish antigen presentation by classically activated macrophages. Activated macrophages can produce high levels of NO, which can inhibit T-cell proliferation and expansion (Wood and Sawitzki, 2006). Activated macrophages can also express FasL, which can contribute to T-cell apoptosis (Brown and Savill, 1999). TNF produced by activated macrophages also contributes to terminate the response by virtue of its ability to bind to the death domain-containing TNFR-1, thereby inducing cellular apoptosis (Shen and Pervaiz, 2006). Apoptotic cells are efficiently eliminated by macrophages through phagocytosis. This process can result in the production of transforming growth factor-β (TGF-β) by macrophages (Huynh et al. 2002), a cytokine that is well known as a strong suppressor of T-cell activation and antibody secretion by B cells. Thus, macrophages can both propagate and terminate cellular adaptive immune responses.

THE ALTERNATIVELY ACTIVATED MACROPHAGE

A population of cells with a phenotype that was distinct from the classically activated macrophages was identified when macrophages were exposed to the Th2-derived cytokines IL-4 or IL-13. These "alternatively activated" macrophages (AA-Mφ) were originally described by Gordon and colleagues (Gordon, 2003). In contrast to classically activated macrophages, which produce high levels of IL-12 or IL-23, AA-Mφ produced little or no IL-12 or IL-23, but exhibited a significant increase in the production of IL-10. AA-Mφ generally show an increase in surface expression of innate recognition receptors such as the scavenger, mannose, and galactose-type receptors. Expression of a β-CC chemokine, CCL18 (DC-CK-1: dendritic cell-derived CC chemokine-1/AMAC-1: alternative macrophage activation-associated CC-chemokine-1), has been associated with AA-Mφ (Goerdt and Orfanos, 1999). In addition to increased receptor expression, two markers have been developed that appear to be specifically expressed on AA-Mφ. They are FIZZ1 (found in inflammatory zone-1) and YM1 (chitinase 3-like 3) (Gordon, 2003). These markers are not expressed on resident nor on classically activated macrophages, and therefore they represent a powerful way to identify AA-Mφ in tissue. Neither FIZZ1 nor YM-1 is expressed on the surface of cells, making immunohistochemistry somewhat more difficult. This has unfortunately diminished the frequency with which investigators use these markers to definitively identify AA-Mφ in tissue.

It is not yet clear how these molecules contribute to the AA-Mφ phenotype. FIZZ1 has been reported to exert an antiapoptotic effect on myofibroblasts (Liu et al., 2004) and thereby contribute to the fibrosis that has been associated with the accumulation of AA-Mφ (Gordon, 2003). YM-1 shares approximately 30% homology with microbial chitinases. Recently, Locksley and colleagues showed that chitin was a potent inducer of IL-4 from innate immune cells such as eosinophils and basophils (Reese et al., 2007). Thus, it seemed plausible that chitin would induce the AA-Mφ phenotype, which would, in turn, result in the production of a chitinase-like molecule that would be directly toxic to the helminths and fungi that express it. However, YM-1 does not appear to express chitinase activity. Rather, YM-1 appears to be a novel lectin that can bind to heparin sulfate and in doing so may participate in tissue remodeling.

The most dramatic departure from classically activated macrophages by AA-Mφ was revealed in an elegant series of studies by Wynn and colleagues, who demonstrated that in these macrophages the metabolism of arginine is altered and shifted away from the production of NO (Hesse et al., 2001). This is due primarily to the overexpression of arginase I, and to a lesser extent to arginase II, by AA-Mφ. The induction appears to occur through a responsive element containing STAT6 and C/EBPβ sites, located about 3 kb upstream of the transcription start site of the arginase-1 gene (Gray et al., 2005; Pauleau et al., 2004), and a functional liver X receptor (LXR)-response element that mediates promoter induction by LXR/retinoid X receptor (RXR) in the 5′-flanking region of the arginase II gene (Marathe et al., 2006). The enzyme allows macrophages to metabolize arginine to ornithine, a precursor of polyamines and collagen. Consistent with this alteration in arginine metabolism, AA-Mφ produce minimal amounts of NO and therefore are not efficient at killing microbes. It has been suggested that the ability to synthesize ornithine, a precursor of collagen, contributes to the role of this cell in wound healing (Mills et al., 2000; Mills, 2001). SHIP (Src homology 2-containing inositol-5′-phosphatase), a potent negative regulator of the phosphatidylinositol 3-kinase (PI3K) pathway in hematopoietic cells, has been shown to play a critical role in repressing the generation of AA-Mφ. Krystal and colleagues reported that both peritoneal and alveolar macrophages from SHIP$^{-/-}$ mice produced reduced NO relative to control littermates due to the constitutively enhanced expression of arginase I (Rauh et al., 2005). Other molecules, such as members of the peroxisome proliferator-activated receptors (PPARs) family, have also been identified as potential regulators which induce the production of AA-Mφ (Henson, 2003).

Importantly, while there is consensus that AA-Mφ produce markedly reduced amounts of NO in response to stimulation, there remains controversy about the utility of arginase as a marker for the AA-Mφ. This enzyme appears to be produced in variable amounts by other macrophage subpopulations, including macrophages that have previously produced high amounts of NO. Therefore, the sole reliance on this activity to identify AA-Mφ has proven to be problematic.

Because Th2 cytokines are responsible for inducing this alternative form of macrophage activation, it is not surprising that these cells have been associated with Th2-dominated nematode and helminthic infections. Several groups have shown that these cells accumulate in mice during the course of worm infections (Maizels et al., 2004; Taylor et al., 2006; Wynn, 2003). Several parasite-derived substances have been shown to induce AA-Mφ, including carbohydrate-rich antigens from schistosome eggs and even cruzipain, an antigen from African trypanosomes. There is little debate that AA-Mφ can contribute to the pathology of schistosomiasis, and these cells have also been shown to contribute to allergic pulmonary inflammation. What is lacking is definitive information about how these cells provide protection against helminths and what molecules secreted by these cells are the mediators of this protection. Several studies have suggested that chitinase is directly toxic to worms and fungi (Nair et al., 2006; Roberts et al., 1988). Chitin is the second most abundant naturally occurring biopolymer. It has a polysaccharide structure of N-acetyl-β-D-glucosamine and constitutes the semitransparent component of the cell walls of worms and fungi. Chitin can be broken down by enzymes called chitinase, which include AMCase (acidic mammalian chitinase). AMCase was shown to be produced by AA-Mφ in a mouse model of Th2-associated allergic inflammation (Zhu et al., 2004). Recent studies by Locksley and colleagues indicate that chitin from the parasite *Nippostrongylus brasiliensis* can initiate the massive migration of eosinophils and basophils associated with allergic responses (Reese et al., 2007). These eosinophils and basophils produce large amounts of IL-4 that, in turn, trigger alternative activation of tissue-resident macrophages with expression of their signature gene, arginase I. These induced mammalian chitinases can degrade worm-derived chitin to limit allergic responses. These data emphasize a critical role of mammalian chitinase to control allergic responses to a commonly encountered allergen, and further indicate that exposure to chitin results in the recruitment of cells producing Th2 cytokines, resulting in alternative activation of macrophages.

Another criterion by which AA-Mϕ are defined is the overproduction of IL-10 (Gordon, 2003). Several groups have shown that these cells produce increased IL-10 relative to classically activated macrophages. We have measured IL-10 production by macrophages primed in vitro with IL-4 and then stimulated in vitro with LPS. These cells do produce more IL-10 than classically activated macrophages (Fig. 1). However, the amount of IL-10 produced under these circumstances is greatly reduced relative to the amount produced by a different form of activated macrophage, the so-called type II macrophage (Mϕ-II), discussed below. Thus, simply relying on IL-10 as a measure of alternative activation of macrophages can be misleading.

The time of IL-4 exposure relative to stimulation can have a dramatic influence on cytokine production. In 1995, D'Andrea et al. (1995) reported that human monocytes/macrophages primed with IL-4 or IL-13 for over 20 h prior to LPS or *Staphylococcus aureus* stimulation produced significantly more IL-12 (both p40 subunit and p70 heterodimer) and TNF-α. This was in contrast to the reduced production of these cytokines observed when cells were simultaneously stimulated in the presence of IL-4 or IL-13. IL-6 production in response to IL-4 priming was also increased in a similar fashion to TNF-α (Kambayashi et al., 1996). IL-10 was inhibited in macrophages that were primed with IL-4, but the simultaneous administration of IL-4 and LPS resulted in an increase in IL-10 production (Kambayashi et al. 1996). Thus, IL-4 can synergistically or antagonistically cast its impact on macrophages, depending on the sequence of treatment and the timing of stimulus.

A considerable amount of controversy remains regarding the ability of the AA-Mϕ to present antigen to T cells. It appears as though the different conclusions may arise from differences in the source of the cells used to present antigen and the experimental model systems. When these cells were originally described, they were reported to be capable of antigen presentation. Similarly, AA-Mϕ taken from mice infected with African trypanosomes were shown to be capable of mitogen or superantigen presentation (Namangala et al., 2001). Working in the *Taenia* model, Rodriguez-Sosa et al. (2002) demonstrated that macrophages from chronic infections could not only present antigen to T cells, but also bias T-cell responses toward a Th2-type response. Importantly, these cells were never biochemically analyzed to confirm that they were in fact AA-Mϕ. However, this type of antigen presentation and T-cell biasing by AA-Mϕ would be consistent with the sustained immunopathology observed in schistosomiasis, where ongoing Th2 responses are clearly involved in the pathological fibrotic responses during chronic infections. In contrast to these findings, other groups have found scant evidence for antigen presentation by AA-Mϕ (Boven et al., 2004; Desnues et al., 2005; Garn et al., 2003; Mantovani et al., 2007; Mosser, 2003). We have generated AA-Mϕ in vitro by exposure to IL-4. These cells failed to upregulate costimulatory molecules, and they failed to support T-cell proliferation (Fig. 2). Others have added AA-Mϕ into lesions of nematode infections and shown that these cells actively inhibited T-cell proliferation (Loke et al., 2000). Furthermore, in a bone marrow transplantation model, an inhibition of B7/CD28 costimulation in humans prevents the appearance of graft-versus-host disease, due to the generation of AA-Mϕ with decreased ability to present antigen (Tzachanis et al., 2002).

Currently, the scientific community lacks a consensus to fully accept the definition of alternative macrophage activation put forward by Gordon. As initially defined (Stein et al., 1992), AA-Mϕ should have the phenotype of cells "activated" by IL-4 or IL-13 to express high levels of macrophage mannose receptor (MMR). They were specifically distinguished from the cells "deactivated" in the presence of IL-10 (Gordon, 2003; Stein et al., 1992). However, the concept of AA-Mϕ has subsequently been extended to include the Th1/Th2 dichotomy. Mantovani and colleagues (2002) proposed a simplified system to classify different subsets of activated macrophages. In this system, they grouped classically activated macrophages into an M1

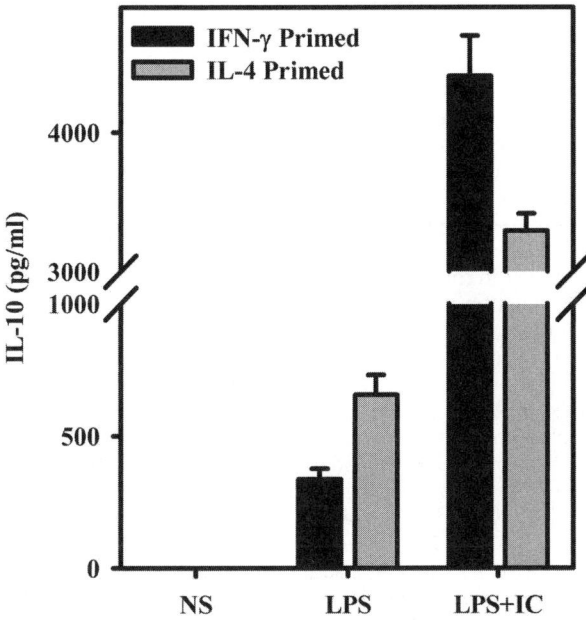

FIGURE 1 IL-10 cytokine production from macrophages following differential priming and stimulation. Macrophages were primed overnight with either 100 U/ml IFN-γ (black bars) or 10 U/ml IL-4 (gray bars). The next morning, cells were stimulated with either 10 ng/ml of lipopolysaccharide alone (LPS) or LPS plus immune complexes consisting of OVA:anti-OVA (LPS+IC). Some cells received no stimulation (NS). After 18 h, IL-10 in supernatants was measured by ELISA. No detectable IL-10 was found in unstimulated cells regardless of the priming. Macrophages primed with IL-4 made more IL-10 in response to LPS than did classically activated (IFN-γ-primed) macrophages. Cells stimulated with LPS in the presence of IC made substantially more IL-10, regardless of how they were primed.

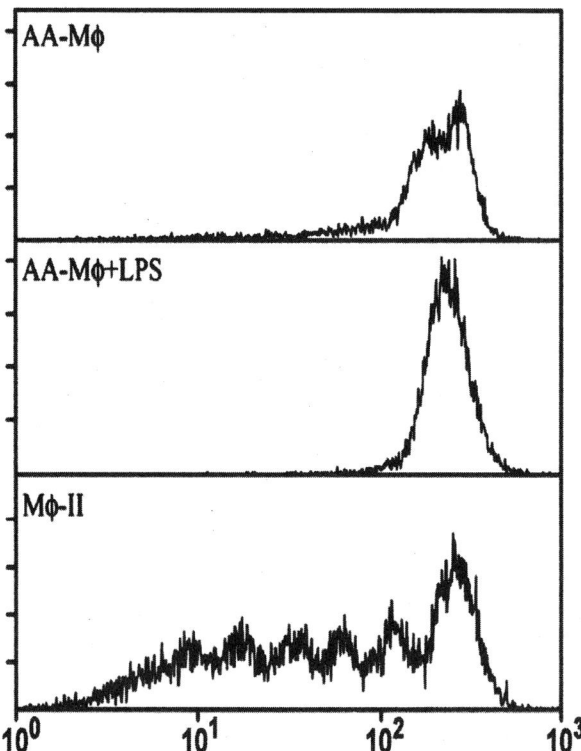

FIGURE 2 T-cell proliferation in response to antigen presented by different populations of macrophages. Macrophages were primed overnight with IL-4 (top two panels) or IFN-γ (bottom panel). The following morning, 150 μg/ml OVA was added to each population. Macrophages in the top panel received OVA alone. Macrophages in the middle and bottom panels received OVA along with 10 ng/ml LPS. After 96 h of coculture with carboxyfluorescein diacetate, the proliferation of succinimidyl ester (CFSE)-labeled CD4+ T cells from DO11.10 mice was measured by the dilution of CFSE. IL-4-primed macrophages (top panel) support only modest T-cell proliferation, and the addition of LPS to stimulate IL-4-primed macrophages (middle panel) does not enhance their ability to present antigen. Mφ-II (bottom panel) readily induce T-cell proliferation as evidenced by the high degree of CFSE dilution.

group, whereas other forms of macrophage activation that were distinct from M1 were classified within the M2 group. Regardless of the different nomenclature used, it is becoming increasingly clear that each of the different macrophage populations arises in response to its residential microenvironmental influences, and each can cast its own impact on the immune response. It may be that the diversity of macrophage populations, with different transcriptome profiles and functional plasticity, may actually reflect a continuum of phenotypes rather than distinct and different populations (Martinez et al., 2006; Stout et al., 2005; Stout and Suttles, 2004).

ALTERNATIVE FORMS OF AA-Mφ

There are many examples in the literature where the term "alternatively activated" macrophage has been used casually and sometimes erroneously. The tendency has been to group all non-classically activated macrophages into this category, despite the fact that there is little evidence to link these cells together except for the lack of classical activation. This section will focus on several examples of Mφ that were generated using stimuli other than IL-4 and/or IL-13 alone. These macrophages include: (i) Mφ that have engulfed apoptotic cells, (ii) Mφ exposed to adenosine, (iii) Mφ that have been activated through the stem cell-derived tyrosine kinase (STK)/recepteur d'origine nantais (RON) receptor, and (iv) Mφ that have been stimulated with a combination of IL-4 and glucocorticoids.

Mφ That Have Engulfed Apoptotic Cells

Many infectious diseases are associated with an induction of cellular apoptosis. The phagocytosis of apoptotic cells by macrophages, via phosphatidylserine on apoptotic cells and the phosphatidylserine receptor on macrophages (Fadok et al., 2000), results in an altered phenotype that in some ways resembles AA-Mφ (Gordon, 2003). Arginase activity appears to be increased in macrophages following the phagocytosis of apoptotic cells, and NO production is decreased (Johann et al., 2007). The production of TGF-β has been implicated in the anti-inflammatory effect of apoptotic cells (Fadok et al., 1998), although it is important to stress that in vitro the addition of apoptotic cells alone to macrophages typically does not induce TGF-β production. Rather, apoptotic cells "reprogram" macrophages to produce TGF-β in response to stimuli, such as LPS or even necrotic cells. The production of IL-10 following contact with apoptotic cells and LPS may be different between macrophages and monocytes. Lucas et al. (2003) showed that murine bone marrow-derived macrophages stimulated by a combination of LPS and apoptotic human neutrophils produced reduced amounts of TNF and IL-10 relative to LPS alone. Byrne and Reen (2002) demonstrated that human monocytes stimulated in this way produced slightly more IL-10, and this induction was subsequently mapped to an ACRE (apoptotic cell response element) in the IL-10 promoter (Chung et al., 2007). TGF-β production was increased in both human monocytes and macrophages that were stimulated with LPS in the presence of apoptotic cells compared with LPS alone. IL-12 production was also altered by the presence of apoptotic cells. IL-12p40 production was virtually abolished in LPS-treated macrophages upon contact with apoptotic cells (Lucas et al., 2003), and neither TGF-β nor IL-10 was needed for this downregulation of IL-12 production. Kim et al. (2004) showed that a novel zinc finger nuclear protein, GC-BP (GC binding protein), was essential in inhibiting IL-12p35 gene transcription by apoptotic cells. Both exposure of phosphatidylserine on the surface of apoptotic cells and phagocytosis of apoptotic cells may be critical for the induction of GC-BP.

The capability of antigen presentation by the macrophages following their uptake of apoptotic or necrotic cells was investigated. Barker et al. (1999) showed that macrophages that had taken up necrotic cell debris presented antigens to T lymphocytes with greater efficiency, whereas those that had engulfed apoptotic cells were ineffective at presenting antigen, primarily due to TGF-β production. Both necrotic and apoptotic cell death can be found in sepsis. In a murine sepsis model, Hotchkiss et al. (2003) found that mice injected with apoptotic cells had much higher mortality than those injected with necrotic cells. Thus, the type of cell death can be instrumental in regulating the balance between immune activation and immune tolerance, due to alterations in antigen-presenting cell (APC) physiology (Henson et al., 2001).

Mϕ Exposure to Adenosine

Adenosine is a purine nucleoside that can accumulate in the extracellular space following stress or inflammation. Macrophages have four adenosine receptors, A1, A2A, A2B, and A3, all of which are seven-transmembrane, G-protein-coupled receptors. The A2 and A3 receptors, in particular, have been shown to broadly inactivate inflammatory responses of macrophages. Cells treated with agonists for the receptors show reduced IL-12 and in some cases increased IL-10 production in response to stimuli. In human monocyte/macrophages, adenosine at a concentration of 100 μM can induce IL-10 production by twofold (Le Moine et al., 1996), but completely inhibits IL-12 production in murine macrophages (Hasko et al., 2000a). In early studies, it was shown that macrophage activation by either an A2 or A3 agonist resulted in inhibition of IL-12 as well as other proinflammatory cytokines and chemokines (Szabo et al., 1998). This inhibition was independent of IL-10 production, and the MAP kinase pathways were not involved (Hasko et al., 2000a).

Another purine nucleoside, inosine, was associated with an inhibitory macrophage phenotype. Inosine is a metabolite of adenosine that protected mice from septic shock (Hasko et al., 2000b). Like adenosine, inosine inhibited IL-12 production, but unlike adenosine it had no effect on IL-10 production. The increased IL-10 associated with the occupancy of adenosine receptors may occur via different molecular mechanisms. Activation of the A2B receptor was shown to relieve the translational repressive effect of IL-10 mRNA 3'-UTR (untranslational region) via increased protein binding to this region (Nemeth et al., 2005). Activation of the A2A receptor, in contrast, may play a more direct role in activating IL-10 transcription (Csoka et al., 2007). A C/EBP binding cis-element in the IL-10 promoter region appears to be critical for IL-10 induction by adenosine, because C/EBPβ-deficient macrophages fail to response to adenosine by increasing IL-10 production.

How adenosine affects antigen presentation by macrophages is not well documented. The presence of high concentrations of adenosine (250 μM) appears to affect the differentiation of monocytes into macrophages. These cells have high accessory function, more closely resembling inflammatory DCs than macrophages (Najar et al., 1990; Shortman and Naik, 2007). These DCs can present antigen and induce Th2-mediated immune responses, as shown by reduced production of IFN-γ and upregulated release of IL-5 by T cells primed with LPS-matured DCs in the presence of adenosine (Panther et al. 2003).

Macrophage Activation by Macrophage-Stimulating Protein

Macrophage-stimulating protein (MSP), also known as hepatocyte growth factor-like (HGFL), is a plasma protein belonging to the plasminogen-related growth factor (PRGF) family (Leonard and Skeel, 1978; Skeel et al., 1991). MSP is secreted by the liver into circulation as an inactive single-chain pro-MSP. The conserved triple disulfide loops (kringles) of pro-MSP place it in a group of structurally related serine proteases of the coagulation and fibrolytic systems. Similar to plasminogen, MSP is activated by macrophage membrane-associated proteases that cleave pro-MSP at a single site to yield a biologically active disulfide-linked heterodimeric MSP (Wang et al., 1996). MSP binds to its receptor, the STK/RON receptor, a receptor tyrosine kinase in the MET proto-oncogene family (Gaudino et al., 1994; Wang et al., 1994). RON activates multiple signaling pathways by recruiting several positive regulatory molecules to its docking sites, including PLC-γ, Shc, Grb2/Soc complex, and PI3 kinase, as well as the Src and FAK tyrosine kinases (Danilkovitch et al., 1999; Wang et al., 1994). Stimulation of MSP can also recruit a negative regulator, the c-Cbl ubiquitin ligase, to the docking site (Penengo et al., 2003; Wang et al., 1994) to promote the ubiquitination and proteosomal degradation of RON.

MSP activation of RON appears to inhibit inflammatory responses by macrophages, making these cells less responsive to inflammation (Chen et al., 1998; Morrison and Correll, 2002). Mice that are hemizygous for RON (Ron$^{+/-}$) (Muraoka et al., 1999), mice with a mutation in the first exon of RON (Correll et al., 1997), or mice with a germ line ablation of the tyrosine kinase domain (Waltz et al., 2001) are all susceptible to LPS-induced septic shock. All display increased delayed-type hypersensitivity reactions, suggesting that MSP/RON represent another way to modulate macrophage responses to activating stimuli. In mice lacking RON, NO production induced by LPS and IFN-γ was increased, and arginase activity was decreased. Conversely, MSP activation increases arginase activity (Morrison and Correll, 2002) and inhibits iNOS expression and NO production in the macrophages stimulated with endotoxin and/or IFN-γ (Liu et al., 1999). The expression of other marker genes associated with alternative activation, such as an increase in scavenger receptor A and IL-1R antagonist, are also increased in MSP-treated macrophages. The pretreatment of macrophages with MSP significantly diminished IL-12p40 production in an IL-10-independent fashion (Morrison et al., 2004). Thus, MSP-treated macrophages appear to have a phenotype similar to AA-Mϕ, characterized by a reduction of NO and an increase in arginase activity. These cells also exhibit increased expression of scavenger receptor A and IL-1R antagonist and a reduction in IL-12 production. However, the increased production of IL-10 has not been definitively shown to occur in MSP-stimulated macrophages. Further studies to comprehensively compare the transcriptomes of MSP-stimulated and IL-4-primed Mϕ are certainly warranted.

Macrophages Stimulated with Glucocorticoids and IL-4

Treatment of macrophages with glucocorticoids (GCs) has a profound inhibitory effect on the production of proinflammatory cytokines, including TNF and IL-12. This inhibition can skew immune responses toward a Th2 response and prevent macrophage-mediated inflammation. The concentration of GCs is a critical determinant because different concentrations of GCs can have markedly different effects on macrophage cytokine production. In general, relatively high-dose GCs (10^{-6} M) readily inhibit macrophage cytokine production, whereas lower doses can have variable effects and sometimes even enhance macrophage cytokine production.

Stimulation of macrophages with a combination of IL-4 and low doses of dexamethasone (10^{-7} M) (Kzhyshkowska et al., 2006a) elicits a macrophage profile that is remarkably similar to that of AA-Mϕ (IL-4 or IL-13 stimulation alone) (Goerdt and Orfanos, 1999). These macrophages highly express stabilin-1, a surface marker that is now routinely used to identify AA-Mϕ. Stabilin-1 is involved in endocytosis/recycling and in trafficking between early/sorting endosomes and the *trans*-Golgi network to sort newly synthesized cargo (Kzhyshkowska et al., 2006a). The ligands or interacting partners for stabilin-1 include acetylated LDL, SI-CLP (stabilin-1 interacting chitinase-like protein), and SPARC (secreted protein acidic and rich in cysteine) (Kzhyshkowska et al., 2006a, 2006b).

SI-CLP is particularly interesting because this molecule is a chitinase-like secreted protein. These chitinase-like proteins can be involved in cell differentiation, proliferation, migration, and adhesion, and it has been suggested that chitinase-like proteins should be defined as a novel class of cytokines (Kzhyshkowska et al., 2006a). SPARC is a matricellular protein that can function in extracellular matrix synthesis and remodeling. Thus, macrophages expressing high levels of stabilin-1 may regulate extracellular SPARC concentrations and play an important role in resolution of chronic inflammation.

It has been known for more than two decades that corticosteroids have profound negative effects on antigen presentation to T cells by macrophages (Snyder and Unanue, 1982). It is likely that the negative effects of GCs will predominate over any positive effects that IL-4 priming may have on APC function, and that the combination of GCs and IL-4 will result in a population of macrophages that is largely unable to present antigen to T cells.

The Type II Activated Macrophage (Mϕ-II)

Several years ago, we demonstrated that the in vitro activation of macrophages in the presence of immune complexes had a profound effect on their ability to synthesize and secrete cytokines (Anderson et al., 2002; Anderson and Mosser, 2002b). Cells activated in the presence of immune complexes turned off IL-12 and produced high levels of IL-10. This reciprocal alteration in the production of these two cytokines depended on signaling through the macrophage Fcγ receptors by immune complexes. Because these cells turned off IL-12 and induced IL-10, they appeared to be similar to the AA-Mϕ described above, at least with regard to the inverted IL-12:IL-10 ratio. However, a detailed analysis of these cells revealed that they were distinct from AA-Mϕ by many other criteria, and in fact shared many features with classically activated macrophages. Like classically activated macrophages, these cells produced high amounts of NO and they rapidly upregulated the expression of costimulatory molecules B7.1 and B7.2 (CD80, CD86). These macrophages failed to express FIZZ1 or YM-1 (Edwards et al., 2006), markers associated with AA-Mϕ.

We therefore temporarily gave them a different name, based on functional activity. Because these cells expressed high levels of MHC-II and CD80/86, we reasoned that they would efficiently present antigen to T cells. These cells not only were efficient antigen presenters, but they were quite adept at skewing antigen-specific cytokine production from CD4+ T cells. These macrophages induced T cells to produce high levels of IL-4 and decreased amounts of IFN-γ. This in vitro T-cell biasing of cytokine production was preserved even when the primed T cells were subsequently stimulated under nonbiased conditions (Anderson and Mosser, 2002b). The ability of these macrophages to influence an antibody response was also examined. Mice were injected with ovalbumin (OVA) along with the various activated macrophage populations as the adjuvant. Mice vaccinated with OVA in the presence of these (immune complex-activated) macrophages produced significantly more immunoglobulin G (IgG) with the IgG1 isotype as the dominant subclass, whereas mice vaccinated in the presence of classically activated macrophages yielded only modest levels of IgG1 in response to OVA (Fig. 3). Thus, these macrophages induced T cells to produce IL-4, which in turn acted on B cells to make IgG1 in response to the antigen. Therefore, we named these cells type II activated macrophages (Mϕ-II) (Anderson and Mosser, 2002a; Mosser, 2003).

We have begun to identify potential markers for Mϕ-II. Our initial studies examined the transcriptome of these cells and identified several transcripts that were elevated in Mϕ-II relative to classically activated macrophages and AA-Mϕ (Edwards et al., 2006). One of these transcripts encodes SPHK-1 (sphingosine kinase-1), an enzyme that catalyzes sphingosine to sphingosine 1-phosphate (S1P). A second transcript encodes LIGHT/TNFSF14 (homologous to lymphotoxins, shows inducible expression, and competes with herpes simplex virus glycoprotein D for herpes virus entry mediator [HVEM]/TNF-related 2), a molecule that can costimulate T-cell responses (Granger and Rickert, 2003; Mauri et al., 1998). Because these transcripts have not been comprehensively examined to show protein expression over the lifetime of these cells, they cannot yet be considered definitive markers for Mϕ-II. However, they illustrate the potential of this approach for assigning biomarkers for specific macrophage subpopulations. Furthermore, these molecules may contribute to the physiology of the Mϕ-II.

S1P, the product of SPHK-1 (sphingosine kinase-1), is an important signaling molecule with both intracellular and extracellular functions (Oskouian and Saba, 2007; Spiegel and Milstien, 2007). Intracellularly, S1P acts like

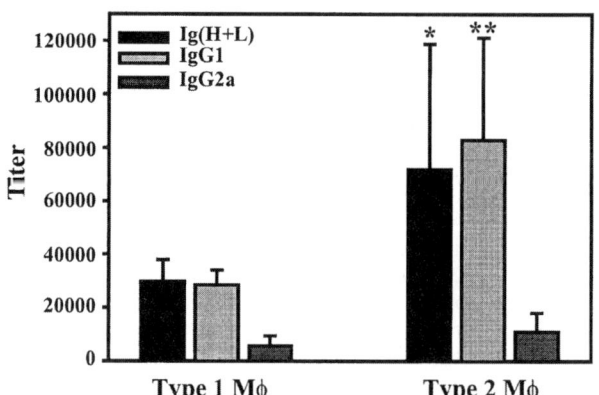

FIGURE 3 Antibody production in DO11.10 mice after immunization with OVA along with different macrophage populations. Macrophages from BALB/c mice were activated in vitro with LPS alone (Type 1 Mϕ) or LPS plus an irrelevant immune complex, E-IgG (Type 2 Mϕ). One hour after activation, 2×10^6 Mϕ were injected intraperitoneally into mice along with 50 μg of OVA in the absence of adjuvant. This procedure was repeated 10 days later. Nine days after the second immunization, mice were bled, and OVA-specific antibody (Ig) (black bars), IgG1 (open bars), and IgG2a (dark gray bars) were measured by ELISA (*$P < 0.05$; **$P < 0.01$).

ceramide in regulating calcium mobilization from intracellular pools. Altered calcium levels may exert an influence on cell differentiation and/or apoptosis. Extracellularly, S1P acts through its cell surface receptor, S1PR (S1P cell surface receptor), which belongs to the EDG family of G-protein-coupled receptors (GPCRs). Interactions of S1P with S1PRs modulate a wide range of cellular events such as cell mobility, survival, migration, and cell-cell communication. S1P has been reported to play important roles in the regulation of immune cell trafficking, vascular maturation, and cardiogenesis.

LIGHT/TNFSF14 was originally identified as a ligand for HVEM (Wang and Fu, 2004; Ware, 2005). Its expression has been found on T cells, NK cells, monocytes, and DCs, whereas HVEM is expressed on T cells, monocytes, and DCs. LIGHT can also bind to LTβR (lymphotoxin β receptor), while HVEM can interact with LTα (lymphotoxin α). The interaction of LIGHT with HVEM on T cells can trigger cell proliferation and cytokine production. This may help to explain why Mφ-II are such good APCs. LIGHT$^{-/-}$ mice and HVEM$^{-/-}$ mice have no obvious abnormalities in the development of lymphoid organs and lymphocytes. However, LIGHT$^{-/-}$ mice have a defect in CD8$^+$ T cell response to antigen. In vitro mixed leukocyte reactions are also defective in LIGHT$^{-/-}$ mice. It is not clear whether the expression of LIGHT by Mφ-II plays a role in the ability of these cells to bias T-cell responses toward IL-4 production. In other experimental models, LIGHT overexpression has been linked to the development of severe colitis, a Th1 disease (Wang et al., 2005; Wang and Fu, 2005). Furthermore, in LIGHT transgenic mice, dramatic inflammation and infiltration were found in mucosal sites, and these mice eventually developed colitis. Thus, the biology of LIGHT expression on Mφ-II and their ability to function as APCs remains, at this time, only a correlation.

The biological activity of Mφ-II largely depends on the high amounts of IL-10 that are produced by these cells. This remains the defining characteristic of these cells, and it is the reason that Mφ-II are potent anti-inflammatory cells. This activity was best illustrated in a murine lethal endotoxemia model (Gerber and Mosser, 2001). The administration of a small number (1×10^6) of Mφ-II into mice completely protected them from a lethal dose of endotoxemia. Classically activated macrophages had no effect in this model. The exploitation of the anti-inflammatory activity of these cells may lead to clinical interventions to protect patients from autoimmune or inflammatory diseases.

The overwhelming production of IL-10 by these cells can also be detrimental to the host during infectious or neoplastic disease. Mφ-II are generated during visceral leishmaniasis. In patients with this disease, high levels of IL-10 are predictive of disease severity, and there is a direct correlation between circulating antibody levels and IL-10 production in both patients and experimental animal models of this disease (Miles et al., 2005). A similar scenario may exist in neoplastic disease. Tumor-associated macrophages produce high levels of IL-10 and low levels of IL-12 (Mantovani et al., 2007; Sica et al., 2000; Sica and Bronte, 2007). Although these cells have not been definitively characterized as Mφ-II, high IgG titers are frequently associated with tumor progression, and many studies have demonstrated that the depletion of B cells can actually diminish tumor growth and progression (de Visser et al., 2005, 2006; Siegel et al., 2000; Tan and Shi, 2003).

Thus, tumor-associated macrophages bear some similarities to Mφ-II. It is not clear whether products produced by the tumor itself induce this altered macrophage phenotype, or whether the humoral immune response to the tumor may actually work to the detriment of the host by inducing IL-10 production from tumor-associated macrophages.

THE MECHANISM OF IL-10 REGULATION IN Mφ-II

At least two signals are required for the overproduction of IL-10 by Mφ-II (Sutterwala et al., 1998). Signal one is an inflammatory stimulus, such as LPS or any other TLR agonist. Exogenous danger signals, such as cleaved hyaluronic acid or heat shock proteins, can also provide this first signal. The second signal is provided by the cross-linking of the FcγRs by immune complexes (ICs). Other second signals can include prostaglandin E_2, cyclic AMP inducers, and others. The combination of the two signals results in a dramatic overproduction of IL-10 relative to signal one alone (i.e., classical activation). The two signals (stimulus + ICs) must be administered in close temporal association to maximally induce IL-10.

Coincident with the increase in IL-10 is a dramatic decrease in IL-12. The downmodulation of IL-12 does not depend on the production of IL-10, and similarly the increase in IL-10 occurs independent of IL-12. Furthermore, whereas ICs appear to be uniquely capable of inducing IL-10, the ligation of many different macrophage receptors results in an ablation of IL-12 synthesis. Both of these alterations in cytokine synthesis occur at the level of transcription, and cytokine protein production generally correlates well with mRNA levels (Lucas et al., 2005; Yang et al., 2007; Zhang et al., 2006). Importantly, the decrease in IL-12 appears to occur via a molecular mechanism that is distinct from and unrelated to the induction of IL-10. Many different groups have noticed this reciprocal regulation of these two cytokines (Trinchieri and Sher, 2007), and despite the unrelated molecular mechanisms of regulation, there are now a fairly large number of circumstances where IL-10 rises and IL-12 decreases, analogous to a playground seesaw. The same cannot be stated for the converse, however, because there are many examples where IL-12 is reduced without a parallel induction of IL-10.

The ligation of macrophage FcγRs in the absence of LPS or other related stimuli induces a transient activation of ERK (Lucas et al., 2005; Yang et al., 2007). This ERK1/2 activation is mediated through Syk (spleen tyrosine kinase), which associates with the immunoreceptor tyrosine-based activation motifs (ITAMs) of FcR-associated γ-chain. Importantly, this transient activation of ERK does not trigger any cytokine release from the cells, demonstrating that ERK activation alone is not sufficient to induce IL-10 production from macrophages. Stimulation with LPS in the presence of ICs results in the rapid and prolonged phosphorylation of ERK. The magnitude of ERK activation is also increased following stimulation in the presence of ICs. Under these conditions, IL-10 is superinduced. Inhibiting ERK activation, either with pharmacological inhibitors or with siRNA, prevents the superinduction of IL-10. Thus, activation of ERK is required for IL-10 production, but it is not sufficient. The presence of a stimulus, such as a TLR ligand along with ICs, is required for the superinduction of IL-10 by Mφ-II. It should be noted that activation of the p38 MAP kinase is also required for IL-10, and blocking this activation abrogates IL-10 production.

The induction of *il-10* gene expression following activation of macrophages in the presence of immune complexes is remarkably rapid (Lucas et al., 2005; Zhang et al., 2006). This initiation of transcription is as rapid as any other cytokine produced by macrophages under any experimental conditions. Activation in the presence of ICs does not affect mRNA stability and, therefore, the production of IL-10 correlates well with increased transcription.

ERK activation results in a dynamic and transient phosphorylation of histones associated with the IL-10 promoter (Lucas et al., 2005; Zhang et al., 2006). This phosphorylation changes the chromatin conformation, making the DNA more accessible to the transcription factors that bind to the IL-10 promoter. There are several aspects of this gene regulation that are quite remarkable. The first is the speed with which the histones are phosphorylated. Temporally, this rapid peak of phosphorylation corresponds to the peak of *il-10* transcription. The second remarkable aspect of this process is the transient nature of histone phosphorylation. Thus, this appears to represent a unique form of chromatin modification that does not lead to long-term epigenetic modifications of the gene. For example, there is little evidence that inducible histone acetylation is involved in initiating the upregulation of *il-10* transcription. The modest amounts of histone acetylation at the *il-10* promoter are not observed until after transcription has peaked, and treatment of macrophages with histone deacetylases (HDACs) does little to change IL-10 mRNA levels. The final surprising aspect of this regulation is the exquisite spatial specificity of histone phosphorylation at the regulatory regions of *il-10* gene, corresponding to the binding sites for transcription factors such as Sp1 and Stat3 (Lucas et al., 2005; Zhang et al., 2006). The mechanism for establishing this specificity has not yet been determined.

There is ample evidence that histone phosphorylation due to ERK activation results in increased DNA accessibility. The DNA comprising the proximal *il-10* promoter shows an increased susceptibility to cleavage by DNases and endonucleases following histone phosphorylation (Zhang et al., 2006). This susceptibility to cleavage corresponds temporally with histone phosphorylation, and it can be prevented by inhibiting ERK activation.

In summary, in eukaryotic cells double-stranded DNA is packaged by histone proteins into chromatin and this packaging can limit the access of transcription factors and RNA polymerase to the promoter region of some genes. During gene expression, nucleosomes can either be removed or their positions can be changed by remodeling factors to allow access of transcription factors and RNA polymerase. In resting cells, the *il-10* gene is particularly inaccessible to transcription factors. ERK activation signal results in the phosphorylation of serine 10 on histone H3 of the nucleosomes associated with the promoter region of *il-10* gene. This leads to increased accessibility of transcription factors and RNA polymerase, presumably due to removal and/or positional changes of nucleosomes (Lucas et al., 2005; Zhang et al., 2006). Thus, histone phosphorylation may play a central role in the control of IL-10 expression in Mϕ-II.

It should be emphasized that IL-10 is not only expressed in myeloid cells such as monocytes and macrophages, but also by lymphoid cells. Recently, IL-10 production by Tregs and by Th1 cells has captured some of the attention that was previously reserved for Th2 cells, which produce high levels of IL-10 when activated. In each of these cells the stimuli that induce *il-10* gene expression are different and, therefore, it is likely that some of the mechanisms to induce IL-10 production in each may be distinct. Furthermore, IL-10 expression in these cells is preserved in "daughter cells," unlike the scenario in nondividing macrophages. Therefore, comparisons between the mechanisms responsible for IL-10 expression in each of these cells is clearly warranted.

MACROPHAGE ACTIVATION DURING DISEASE

There is no question that some autoimmune diseases can be precipitated by the presence of ICs. The best studies of these diseases include experimental autoimmune encephalitis in the mouse, and rheumatoid arthritis and systemic lupus erythematosus in humans. In all of these cases, disease severity correlates directly with antibody titers, and ICs are associated with disease pathology. However, there are now several examples of diseases, both infectious and neoplastic, where ICs can actually prevent inflammation and/or immunity. In these diseases this inhibition appears to work at the level of macrophages, which produce IL-10 in response to ICs. A few of these diseases, in which atypical activation responses of macrophages may contribute to disease progression, and a disease where induction of Mϕ-II may have therapeutic effects, are highlighted below.

Septic Shock

Septic shock is an often fatal condition caused by bacterial infections that have reached the bloodstream (bacteremia). Vasodilation and damage to the vascular endothelium by endotoxin causes a precipitous drop in blood pressure and eventually septic shock. The decreased tissue perfusion leads to dysfunction of the heart, brain, liver, and kidney. Bacterial endotoxins, consisting primarily of LPS, are released from bacteria into the bloodstream. LPS binds to an LPS binding protein (LBP), which then transfers LPS to CD14 on the macrophage. The CD14-associated LPS is then likely transferred to MD-2 and TLR4 to transduce activating signals into the macrophage. These activating signals result in a dramatic transcriptional reprogramming by LPS-stimulated cells, resulting in the production of myriad inflammatory cytokines and mediators. The TNF that is produced by stimulated macrophages is a major contributor to septic shock, and antibodies to the cytokine or its receptor have been shown to ameliorate symptoms in this model (Aggarwal, 2003). Some individuals have relatively high levels of so-called "naturally occurring" antibodies to LPS (Nys et al., 1996). We would predict that some of these antibodies have the potential to form ICs with LPS and signal through the macrophage FcγR to induce the production of IL-10. We have previously demonstrated that an irrelevant IC could behave similarly and induce high levels of IL-10 in response to LPS (Gerber and Mosser, 2001). The overproduction of IL-10 by these macrophages prevented septic shock. Thus, even though the ligation of macrophage FcγR does essentially nothing to directly affect TNF production, the reprogramming of macrophages to overproduce IL-10 transforms them into potent anti-inflammatory cells capable of mediating protection from lethal endotoxemia.

Leishmaniasis

Leishmaniasis is caused by protozoan parasites of *Leishmania* spp. This organism is an intracellular pathogen that lives within phagolysosomes of tissue-resident macrophages. The amastigote form of this organism is able to replicate within macrophages and spread from cell to cell. Leishmaniasis can be thought of as a spectral disease ranging from a self-healing cutaneous form of the disease to a nonhealing, often fatal, visceral form. Because this organism is an intracellular pathogen, the development of cell-mediated immunity and classical macrophage activation is critical for disease resolution.

In both human leishmaniasis and murine models of leishmaniasis, IL-10 has been identified as an important susceptibility factor. In humans, IL-10 levels correlate with disease severity in visceral leishmaniasis (Karp et al., 1993). Using an experimental model of visceral leishmaniasis, we demonstrated that IL-10 production from macrophages was due to IgG on the surface of amastigotes (Miles et al., 2005). This IgG ligated macrophage FcγR upon amastigote phagocytosis and induced IL-10 production from infected macrophages. Macrophages that were exposed to IL-10 were refractory to priming by the classical macrophage activator IFN-γ. Thus, in this murine model of infection, IgG was actually detrimental to the host because of its ability to induce IL-10. The depletion of IgG resulted in smaller lesions with less IL-10. In humans, there was a similarly strong positive correlation between high IgG levels and disease severity. There was an inverse correlation between IgG levels and skin test positivity, an indicator of CMI. Thus, one should exercise caution in designing antibody-based immunotherapy to treat leishmaniasis and perhaps other intracellular organisms.

Tumor-Associated Macrophages

In both primary and secondary tumors, macrophages represent a major component of the inflammatory infiltrate. These cells display a unique phenotype that is clearly distinct from classically activated macrophages. These cells have been termed tumor-associated macrophages (TAMs). TAMs mainly consist of a polarized macrophage population that typically stains positively for F4/80 and CD206. These cells have a higher expression of IL-10 and lower levels of proinflammatory cytokines (Mantovani et al., 2007; Sica et al., 2000; Sica and Bronte, 2007). TAMs are generally better at scavenging debris from apoptotic cells, promoting angiogenesis, and repairing and remodeling wounded/damaged tissues. TAMs are usually poor at presenting antigen to T cells, and they can sometimes produce factors that suppress T-cell functions. An increase in the accumulation of TAMs in some tumors is often considered an indicator of a poor prognosis (Lin and Pollard, 2007; Mantovani et al., 2006).

Despite the wealth of studies on these cells, and the general agreement that these cells may actually contribute to tumor progression, there is no consensus as to the identity of TAMs. In fact, it is quite likely that different tumors can give rise to macrophages with distinct properties. Thus, the term TAM may actually describe a large collection of macrophage subtypes, each with its own markers and functions. There is also a lack of consensus as to the role of antibody in cancer. While it is clear that in some experimental systems antibodies to tumor-associated antigens can be effective, there are also a number of instances where antibody depletion actually leads to reduced tumor growth (de Visser et al., 2006).

Cervical carcinoma may be an example where antibody may be detrimental to the host (de Visser et al., 2005). Cervical carcinoma is caused by infection with human papillomavirus (HPV). HPV-16 mice are transgenic mice expressing HPV genes on epithelial cells in the skin. As cancer progresses, macrophages move into the site and a chronic inflammation progresses. This inflammation promotes the proliferation of cancerous epithelial cells, and it induces angiogenesis and eventually invasive carcinogenesis. In the HPV-16 mice that are deficient of both T and B cells, angiogenesis and tissue remodeling are blocked and carcinogenesis is arrested at the stage of overproliferation. When antibodies from HPV-16 mice are transferred back to the HPV-16 mice deficient in lymphocytes, angiogenesis and cancer progression are restored. The antibodies are actually found in the underlying dermis layer, not next to the overproliferating cells, suggesting that ICs might be acting on innate immune cells such as macrophages to promote tumor progression.

It appears that IL-10 production by TAMs is one of the most important factors to subvert tumor-specific immunity (Sica et al., 2000). IL-10 not only subverts the development of classical macrophage activation responses, but it also stimulates macrophages to express B7-H4, a novel member of the B7 family of T-cell costimulator molecules (Kryczek et al., 2006). B7-H4 inhibits T-cell proliferation, cell cycle progression, and cytokine production. More than 70% of macrophages isolated from the ascites of patients with ovarian cancers express cell surface B7-H4 antigens. B7-H4$^+$ macrophages exert an immunosuppression independent of B7-H1, arginase, and iNOS. Gene expression profiling of TAMs isolated from a murine fibrosarcoma reveals characteristic alterations in gene expression which are consistent with some alternative type of macrophage activation. A lower expression of IL-12, and a higher expression of IL-10 and the IFN-γ inducible chemokines CXCL9, CXCL10, and CXCL16, suggest that TAMs exhibit a unique transcriptional program (Biswas et al., 2006; Saccani et al., 2006).

Inflammatory Bowel Disease

Inflammatory bowel disease (IBD) comprises the two chronic disorders that cause inflammation of the intestines: ulcerative colitis and Crohn's disease (Baumgart and Carding, 2007). The exact cause of IBD is not known, but genetic factors are probably involved. About 15% to 30% of people with IBD have a relative with the disease. A vigorous local mucosal immune response appears to be responsible for initiation and aggravation of the inflammatory processes. The onset of IBD could be caused by an increased number of activated macrophages that express upregulated costimulatory molecules (e.g., CD80/CD86) with a cytokine profile favoring a type I proinflammatory response (Rugtveit et al., 1997). Depletion of intestine macrophages by poly-D,L-lactic acid microspheres containing dichloromethylene diphosphonate significantly suppresses the development of chronic colitis in an animal model of human IBD. IL-23 (heterodimer of IL-12p40 and IL-23p19), a type I proinflammatory cytokine, is found overexpressed in inflamed tissues from IBD mouse models (Langrish et al., 2005; Yen et al., 2006). Deletion of IL-23 significantly reduced inflammation of the bowel with little impact on systemic T-cell immunity, both in terms of cells and proinflammatory cytokine production. In the mice that lack B and T lymphocytes, IL-23 can still induce intestinal inflammation, confirming the important role of the innate immune system, including macrophages, in IBD. Thus, an unfavorable phenotype of local macrophages could be one cause of IBD etiology. It has been shown that mice deficient in IL-10 after bacterial infections manifest pathological changes and symptoms that are very similar to human IBD (Ebert et al., 2005; Iwakura and Ishigame, 2006). Administration of IL-10 has provided therapeutic benefit not only to IL-10-deficient mice but also to other murine models of human IBD (Baumgart and Sandborn, 2007; Iwakura and Ishigame, 2006; Rogy et al., 2000). IL-10 exerts its compensatory activity through downregulation of type I proinflammatory cytokines and/or through possible modulation of Th17 cells (Yen et al., 2006). Th17 T cells belong to a novel group of CD4$^+$ T-helper cells that are found highly upregulated in bowels of diseased mice and highly implicated in IBD (Iwakura and Ishigame, 2006). Thus, if the local intestine macrophages of IBD could be remodulated to become Mφ-II, the local production of IL-10 could prevent disease progression.

CONCLUSIONS

Macrophages are an essential component of both the innate and adaptive immune systems. They work both at the initiation of adaptive immune responses and as effector cells during CMI. Because of their remarkable plasticity, these cells can often function as control switches, securing the balance between pro- and anti-inflammatory reactions. A lack of balance in the immune response can have catastrophic consequences. Too much inflammation can lead to autoimmunity, which is characterized by the dysregulated overproduction of a variety of inflammatory mediators from classically activated macrophages. Conversely, the overproduction of anti-inflammatory mediators can prevent an effective immune response. There are many examples of both viruses and bacteria that preferentially induce IL-10 production from macrophages to prevent productive immune responses. TAMs may be similarly immunosuppressive. There are likewise many examples of pathogens that are highly inflammatory, and in some cases the innate immune responses that they trigger can lead to immune-mediated pathology. Such appears to be the case with the pandemic influenza virus of 1918 (Kash et al., 2004; Palese, 2004) and with severe acute respiratory syndrome (SARS) (Gu and Korteweg, 2007; Perlman and Dandekar, 2005; Xu and Gao, 2004).

As we begin to understand the molecular nature of macrophage responses to stimuli, we get closer to being able to exploit this understanding to experimentally tip the balance of immunity in either direction. The TLR pathway represents an activating avenue that has the potential to lead to new adjuvants and better vaccines. Conversely, the uncovering of the signaling pathway leading to IL-10 overproduction has the potential to lead to a novel class of anti-inflammatory compounds that induce IL-10 production from macrophages. Studies to assign biochemical signatures to each macrophage population are under way. These studies have the potential to lead to the depletion of pathogenic macrophage subsets to either enhance immunity or prevent autoimmunity. Thus, therapeutic interventions at the level of the macrophage have the potential to restore balance to a dysregulated immune response.

REFERENCES

Aggarwal, B. B. 2003. Signalling pathways of the TNF superfamily: a double-edged sword. *Nat. Rev. Immunol.* **3:**745–756.

Akashi, K., D. Traver, T. Miyamoto, and I. L. Weissman. 2000. A clonogenic common myeloid progenitor that gives rise to all myeloid lineages. *Nature* **404:**193–197.

Anderson, C. F., J. S. Gerber, and D. M. Mosser. 2002. Modulating macrophage function with IgG immune complexes. *J. Endotoxin. Res.* **8:**477–481.

Anderson, C. F., and D. M. Mosser. 2002a. A novel phenotype for an activated macrophage: the type 2 activated macrophage. *J. Leukoc. Biol.* **72:**101–106.

Anderson, C. F., and D. M. Mosser. 2002b. Cutting edge: biasing immune responses by directing antigen to macrophage Fc gamma receptors. *J. Immunol.* **168:**3697–3701.

Auffray, C., D. Fogg, M. Garfa, G. Elain, O. Join-Lambert, S. Kayal, S. Sarnacki, A. Cumano, G. Lauvau, and F. Geissmann. 2007. Monitoring of blood vessels and tissues by a population of monocytes with patrolling behavior. *Science* **317:**666–670.

Barker, R. N., L. Erwig, W. P. Pearce, A. Devine, and A. J. Rees. 1999. Differential effects of necrotic or apoptotic cell uptake on antigen presentation by macrophages. *Pathobiology* **67:**302–305.

Baumgart, D. C., and S. R. Carding. 2007. Inflammatory bowel disease: cause and immunobiology. *Lancet* **369:**1627–1640.

Baumgart, D. C., and W. J. Sandborn. 2007. Inflammatory bowel disease: clinical aspects and established and evolving therapies. *Lancet* **369:**1641–1657.

Biswas, S. K., L. Gangi, S. Paul, T. Schioppa, A. Saccani, M. Sironi, B. Bottazzi, A. Doni, B. Vincenzo, F. Pasqualini, L. Vago, M. Nebuloni, A. Mantovani, and A. Sica. 2006. A distinct and unique transcriptional program expressed by tumor-associated macrophages (defective NF-kappaB and enhanced IRF-3/STAT1 activation). *Blood* **107:**2112–2122.

Boven, L. A., M. van Meurs, R. G. Boot, A. Mehta, L. Boon, J. M. Aerts, and J. D. Laman. 2004. Gaucher cells demonstrate a distinct macrophage phenotype and resemble alternatively activated macrophages. *Am. J. Clin. Pathol.* **122:**359–369.

Brown, S. B., and J. Savill. 1999. Phagocytosis triggers macrophage release of Fas ligand and induces apoptosis of bystander leukocytes. *J. Immunol.* **162:**480–485.

Byrne, A., and D. J. Reen. 2002. Lipopolysaccharide induces rapid production of IL-10 by monocytes in the presence of apoptotic neutrophils. *J. Immunol.* **168:**1968–1977.

Chen, Y. Q., J. H. Fisher, and M. H. Wang. 1998. Activation of the RON receptor tyrosine kinase inhibits inducible nitric oxide synthase (iNOS) expression by murine peritoneal exudate macrophages: phosphatidylinositol-3 kinase is required for RON-mediated inhibition of iNOS expression. *J. Immunol.* **161:**4950–4959.

Chung, E. Y., J. Liu, Y. Homma, Y. Zhang, A. Brendolan, M. Saggese, J. Han, R. Silverstein, L. Selleri, and X. Ma. 2007. Interleukin-10 expression in macrophages during phagocytosis of apoptotic cells is mediated by homodomain proteins Pbx1 and Prep-1. *Immunity* **27:**952–964.

Cohn, Z. A. 1978. Activation of mononuclear phagocytes: fact, fancy, and future. *J. Immunol.* **121:**813–816.

Correll, P. H., A. Iwama, S. Tondat, G. Mayrhofer, T. Suda, and A. Bernstein. 1997. Deregulated inflammatory response in mice lacking the STK/RON receptor tyrosine kinase. *Genes Funct.* **1:**69–83.

Csoka, B., Z. H. Nemeth, L. Virag, P. Gergely, S. J. Leibovich, P. Pacher, C. X. Sun, M. R. Blackburn, E. S. Vizi, E. A. Deitch, and G. Hasko. 2007. A2A adenosine receptors and C/EBP[beta] are crucially required for IL-10 production by macrophages exposed to E. coli. *Blood* **110:**2685–2695.

D'Andrea, A., X. Ma, M. Aste-Amezaga, C. Paganin, and G. Trinchieri. 1995. Stimulatory and inhibitory effects of interleukin (IL)-4 and IL-13 on the production of cytokines by human peripheral blood mononuclear cells: priming for IL-12 and tumor necrosis factor alpha production. *J. Exp. Med.* **181:**537–546.

Danilkovitch, A., A. Skeel, and E. J. Leonard. 1999. Macrophage stimulating protein-induced epithelial cell adhesion is mediated by a PI3-K-dependent, but FAK-independent mechanism. *Exp. Cell Res.* **248:**575–582.

Desnues, B., H. Lepidi, D. Raoult, and J. L. Mege. 2005. Whipple disease: intestinal infiltrating cells exhibit a transcriptional pattern of M2/alternatively activated macrophages. *J. Infect. Dis.* **192:**1642–1646.

de Visser, K. E., A. Eichten, and L. M. Coussens. 2006. Paradoxical roles of the immune system during cancer development. *Nat. Rev. Cancer* **6:**24–37.

de Visser, K. E., L. V. Korets, and L. M. Coussens. 2005. De novo carcinogenesis promoted by chronic inflammation is B lymphocyte dependent. *Cancer Cell.* **7:**411–423.

Ebert, E. C., V. Mehta, and K. M. Das. 2005. Activation antigens on colonic T cells in inflammatory bowel disease: effects of IL-10. *Clin. Exp. Immunol.* **140:**157–165.

Edwards, J. P., X. Zhang, K. A. Frauwirth, and D. M. Mosser. 2006. Biochemical and functional characterization of three activated macrophage populations. *J. Leukoc. Biol.* **80:**1298–1307.

Fadok, V. A., D. L. Bratton, A. Konowal, P. W. Freed, J. Y. Westcott, and P. M. Henson. 1998. Macrophages that have ingested apoptotic cells in vitro inhibit proinflammatory cytokine production through autocrine/paracrine mechanisms involving TGF-beta, PGE2, and PAF. *J. Clin. Investig.* **101:**890–898.

Fadok, V. A., D. L. Bratton, D. M. Rose, A. Pearson, R. A. Ezekewitz, and P. M. Henson. 2000. A receptor for phosphatidylserine-specific clearance of apoptotic cells. *Nature* **405:**85–90.

Fogg, D. K., C. Sibon, C. Miled, S. Jung, P. Aucouturier, D. R. Littman, A. Cumano, and F. Geissmann. 2006. A clonogenic bone marrow progenitor specific for macrophages and dendritic cells. *Science* **311:**83–87.

Garn, H., A. Siese, S. Stumpf, P. J. Barth, B. Muller, and D. Gemsa. 2003. Shift toward an alternatively activated macrophage response in lungs of NO2-exposed rats. *Am. J. Respir. Cell Mol. Biol.* **28:**386–396.

Gaudino, G., A. Follenzi, L. Naldini, C. Collesi, M. Santoro, K. A. Gallo, P. J. Godowski, and P. M. Comoglio. 1994. RON is a heterodimeric tyrosine kinase receptor activated by the HGF homologue MSP. *EMBO J.* **13:**3524–3532.

Geissmann, F., S. Jung, and D. R. Littman. 2003. Blood monocytes consist of two principal subsets with distinct migratory properties. *Immunity* **19:**71–82.

Gerber, J. S., and D. M. Mosser. 2001. Reversing lipopolysaccharide toxicity by ligating the macrophage Fc gamma receptors. *J. Immunol.* **166:**6861–6868.

Goerdt, S., and C. E. Orfanos. 1999. Other functions, other genes: alternative activation of antigen-presenting cells. *Immunity* **10:**137–142.

Gordon, S. 2003. Alternative activation of macrophages. *Nat. Rev. Immunol.* **3:**23–35.

Gordon, S., and P. R. Taylor. 2005. Monocyte and macrophage heterogeneity. *Nat. Rev. Immunol.* **5:**953–964.

Granger, S. W., and S. Rickert. 2003. LIGHT-HVEM signaling and the regulation of T cell-mediated immunity. *Cytokine Growth Factor Rev.* **14:**289–296.

Gray, M. J., M. Poljakovic, D. Kepka-Lenhart, and S. M. Morris, Jr. 2005. Induction of arginase I transcription by IL-4 requires a composite DNA response element for STAT6 and C/EBPbeta. *Gene* **353:**98–106.

Greenwald, R. J., G. J. Freeman, and A. H. Sharpe. 2005. The B7 family revisited. *Annu. Rev. Immunol.* **23:**515–548.

Gu, J., and C. Korteweg. 2007. Pathology and pathogenesis of severe acute respiratory syndrome. *Am. J. Pathol.* **170:**1136–1147.

Harding, C. V., L. Ramachandra, and M. J. Wick. 2003. Interaction of bacteria with antigen presenting cells: influences on antigen presentation and antibacterial immunity. *Curr. Opin. Immunol.* **15:**112–119.

Hasko, G., D. G. Kuhel, J. F. Chen, M. A. Schwarzschild, E. A. Deitch, J. G. Mabley, A. Marton, and C. Szabo. 2000a. Adenosine inhibits IL-12 and TNF-[alpha] production via adenosine A2a receptor-dependent and independent mechanisms. *FASEB J.* **14:**2065–2074.

Hasko, G., D. G. Kuhel, Z. H. Nemeth, J. G. Mabley, R. F. Stachlewitz, L. Virag, Z. Lohinai, G. J. Southan, A. L. Salzman, and C. Szabo. 2000b. Inosine inhibits inflammatory cytokine production by a posttranscriptional mechanism and protects against endotoxin-induced shock. *J. Immunol.* **164:**1013–1019.

Henson, P. 2003. Suppression of macrophage inflammatory responses by PPARs. *Proc. Natl. Acad. Sci. USA* **100:**6295–6296.

Henson, P. M., D. L. Bratton, and V. A. Fadok. 2001. The phosphatidylserine receptor: a crucial molecular switch? *Nat. Rev. Mol. Cell Biol.* **2:**627–633.

Hesse, M., M. Modolell, A. C. La Flamme, M. Schito, J. M. Fuentes, A. W. Cheever, E. J. Pearce, and T. A. Wynn. 2001. Differential regulation of nitric oxide synthase-2 and arginase-1 by type 1/type 2 cytokines in vivo: granulomatous pathology is shaped by the pattern of L-arginine metabolism. *J. Immunol.* **167:**6533–6544.

Hotchkiss, R. S., K. C. Chang, M. H. Grayson, K. W. Tinsley, B. S. Dunne, C. G. Davis, D. F. Osborne, and I. E. Karl. 2003. Adoptive transfer of apoptotic splenocytes worsens survival, whereas adoptive transfer of necrotic splenocytes improves survival in sepsis. *Proc. Natl. Acad. Sci. USA* **100:**6724–6729.

Hume, D. A. 2006. The mononuclear phagocyte system. *Curr. Opin. Immunol.* **18:**49–53.

Huynh, M. L., V. A. Fadok, and P. M. Henson. 2002. Phosphatidylserine-dependent ingestion of apoptotic cells promotes TGF-beta1 secretion and the resolution of inflammation. *J. Clin. Investig.* **109:**41–50.

Iwakura, Y., and H. Ishigame. 2006. The IL-23/IL-17 axis in inflammation. *J. Clin. Investig.* **116:**1218–1222.

Johann, A. M., V. Barra, A. M. Kuhn, A. Weigert, A. von Knethen, and B. Brune. 2007. Apoptotic cells induce arginase II in macrophages, thereby attenuating NO production. *FASEB J.* **21:**2704–2712.

Kambayashi, T., C. O. Jacob, and G. Strassmann. 1996. IL-4 and IL-13 modulate IL-10 release in endotoxin-stimulated murine peritoneal mononuclear phagocytes. *Cell Immunol.* **171:**153–158.

Karp, C. L., S. H. el-Safi, T. A. Wynn, M. M. Satti, A. M. Kordofani, F. A. Hashim, M. Hag-Ali, F. A. Neva, T. B. Nutman, and D. L. Sacks. 1993. In vivo cytokine profiles in patients with kala-azar. Marked elevation of both interleukin-10 and interferon-gamma. *J. Clin. Investig.* **91:**1644–1648.

Kash, J. C., C. F. Basler, A. Garcia-Sastre, V. Carter, R. Billharz, D. E. Swayne, R. M. Przygodzki, J. K. Taubenberger, M. G. Katze, and T. M. Tumpey. 2004. Global host immune response: pathogenesis and transcriptional profiling of type A influenza viruses expressing the hemagglutinin and neuraminidase genes from the 1918 pandemic virus. *J. Virol.* **78:**9499–9511.

Kim, S., K. B. Elkon, and X. Ma. 2004. Transcriptional suppression of interleukin-12 gene expression following phagocytosis of apoptotic cells. *Immunity* **21:**643–653.

Kryczek, I., L. Zou, P. Rodriguez, G. Zhu, S. Wei, P. Mottram, M. Brumlik, P. Cheng, T. Curiel, L. Myers, A. Lackner, X. Alvarez, A. Ochoa, L. Chen, and W. Zou. 2006. B7-H4 expression identifies a novel suppressive macrophage population in human ovarian carcinoma. *J. Exp. Med.* **203**: 871–881.

Kumar, S., and R. Jack. 2006. Origin of monocytes and their differentiation to macrophages and dendritic cells. *J. Endotoxin. Res.* **12**:278–284.

Kzhyshkowska, J., S. Mamidi, A. Gratchev, E. Kremmer, C. Schmuttermaier, L. Krusell, G. Haus, J. Utikal, K. Schledzewski, J. Scholtze, and S. Goerdt. 2006a. Novel stabilin-1 interacting chitinase-like protein (SI-CLP) is up-regulated in alternatively activated macrophages and secreted via lysosomal pathway. *Blood* **107**:3221–3228.

Kzhyshkowska, J., G. Workman, M. Cardo-Vila, W. Arap, R. Pasqualini, A. Gratchev, L. Krusell, S. Goerdt, and E. H. Sage. 2006b. Novel function of alternatively activated macrophages: stabilin-1-mediated clearance of SPARC. *J. Immunol.* **176**:5825–5832.

Landsman, L., C. Varol, and S. Jung. 2007. Distinct differentiation potential of blood monocyte subsets in the lung. *J. Immunol.* **178**:2000–2007.

Langrish, C. L., Y. Chen, W. M. Blumenschein, J. Mattson, B. Basham, J. D. Sedgwick, T. McClanahan, R. A. Kastelein, and D. J. Cua. 2005. IL-23 drives a pathogenic T cell population that induces autoimmune inflammation. *J. Exp. Med.* **201**:233–240.

Le Moine, O., P. Stordeur, L. Schandené, A. Marchant, D. de Groote, M. Goldman, and J. Devière. 1996. Adenosine enhances IL-10 secretion by human monocytes. *J. Immunol.* **156**:4408–4414.

Leon, B., M. Lopez-Bravo, and C. Ardavin. 2007. Monocyte-derived dendritic cells formed at the infection site control the induction of protective T helper 1 responses against *Leishmania. Immunity* **26**:519–531.

Leonard, E. J., and A. H. Skeel. 1978. Isolation of macrophage stimulating protein (MSP) from human serum. *Exp. Cell Res.* **114**:117–126.

Levy, D. E., and J. E. Darnell, Jr. 2002. Stats: transcriptional control and biological impact. *Nat. Rev. Mol. Cell Biol.* **3**: 651–662.

Lin, E. Y., and J. W. Pollard. 2007. Tumor-associated macrophages press the angiogenic switch in breast cancer. *Cancer Res.* **67**:5064–5066.

Liu, J., S. Cao, L. M. Herman, and X. Ma. 2003. Differential regulation of interleukin (IL)-12 p35 and p40 gene expression and interferon (IFN)-gamma-primed IL-12 production by IFN regulatory factor 1. *J. Exp. Med.* **198**:1265–1276.

Liu, J., X. Guan, and X. Ma. 2007. Regulation of IL-27 p28 gene expression in macrophages through MyD88- and interferon-gamma-mediated pathways. *J. Exp. Med.* **204**: 141–152.

Liu, Q. P., K. Fruit, J. Ward, and P. H. Correll. 1999. Negative regulation of macrophage activation in response to IFN-gamma and lipopolysaccharide by the STK/RON receptor tyrosine kinase. *J. Immunol.* **163**:6606–6613.

Liu, T., S. M. Dhanasekaran, H. Jin, B. Hu, S. A. Tomlins, A. M. Chinnaiyan, and S. H. Phan. 2004. FIZZ1 stimulation of myofibroblast differentiation. *Am. J. Pathol.* **164**:1315–1326.

Loke, P., A. S. MacDonald, A. Robb, R. M. Maizels, and J. E. Allen. 2000. Alternatively activated macrophages induced by nematode infection inhibit proliferation via cell-to-cell contact. *Eur. J. Immunol.* **30**:2669–2678.

Lucas, M., L. M. Stuart, J. Savill, and A. Lacy-Hulbert. 2003. Apoptotic cells and innate immune stimuli combine to regulate macrophage cytokine secretion. *J. Immunol.* **171**: 2610–2615.

Lucas, M., X. Zhang, V. Prasanna, and D. M. Mosser. 2005. ERK activation following macrophage FcgammaR ligation leads to chromatin modifications at the IL-10 locus. *J. Immunol.* **175**:469–477.

MacMicking, J., Q. W. Xie, and C. Nathan. 1997. Nitric oxide and macrophage function. *Annu. Rev. Immunol.* **15**: 323–350.

Maizels, R. M., A. Balic, N. Gomez-Escobar, M. Nair, M. D. Taylor, and J. E. Allen. 2004. Helminth parasites—masters of regulation. *Immunol. Rev.* **201**:89–116.

Mantovani, A., T. Schioppa, C. Porta, P. Allavena, and A. Sica. 2006. Role of tumor-associated macrophages in tumor progression and invasion. *Cancer Metastasis Rev.* **25**:315–322.

Mantovani, A., A. Sica, and M. Locati. 2007. New vistas on macrophage differentiation and activation. *Eur. J. Immunol.* **37**:14–16.

Mantovani, A., S. Sozzani, M. Locati, P. Allavena, and A. Sica. 2002. Macrophage polarization: tumor-associated macrophages as a paradigm for polarized M2 mononuclear phagocytes. *Trends Immunol.* **23**:549–555.

Marathe, C., M. N. Bradley, C. Hong, F. Lopez, C. M. Ruiz de Galarreta, P. Tontonoz, and A. Castrillo. 2006. The arginase II gene is an anti-inflammatory target of liver X receptor in macrophages. *J. Biol. Chem.* **281**:32197–32206.

Martinez, F. O., S. Gordon, M. Locati, and A. Mantovani. 2006. Transcriptional profiling of the human monocyte-to-macrophage differentiation and polarization: new molecules and patterns of gene expression. *J. Immunol.* **177**:7303–7311.

Mauri, D. N., R. Ebner, R. I. Montgomery, K. D. Kochel, T. C. Cheung, G. L. Yu, S. Ruben, M. Murphy, R. J. Eisenberg, G. H. Cohen, P. G. Spear, and C. F. Ware. 1998. LIGHT, a new member of the TNF superfamily, and lymphotoxin alpha are ligands for herpesvirus entry mediator. *Immunity* **8**:21–30.

Menendez-Benito, V., and J. Neefjes. 2007. Autophagy in MHC class II presentation: sampling from within. *Immunity* **26**:1–3.

Miles, S. A., S. M. Conrad, R. G. Alves, S. M. Jeronimo, and D. M. Mosser. 2005. A role for IgG immune complexes during infection with the intracellular pathogen Leishmania. *J. Exp. Med.* **201**:747–754.

Mills, C. D. 2001. Macrophage arginine metabolism to ornithine/urea or nitric oxide/citrulline: a life or death issue. *Crit. Rev. Immunol.* **21**:399–425.

Mills, C. D., K. Kincaid, J. M. Alt, M. J. Heilman, and A. M. Hill. 2000. M-1/M-2 macrophages and the Th1/Th2 paradigm. *J. Immunol.* **164**:6166–6173.

Morrison, A. C., and P. H. Correll. 2002. Activation of the stem cell-derived tyrosine kinase/RON receptor tyrosine kinase by macrophage-stimulating protein results in the induction of arginase activity in murine peritoneal macrophages. *J. Immunol.* **168**:853–860.

Morrison, A. C., C. B. Wilson, M. Ray, and P. H. Correll. 2004. Macrophage-stimulating protein, the ligand for the stem cell-derived tyrosine kinase/RON receptor tyrosine kinase, inhibits IL-12 production by primary peritoneal macrophages stimulated with IFN-gamma and lipopolysaccharide. *J. Immunol.* **172**:1825–1832.

Mosser, D. M. 2003. The many faces of macrophage activation. *J. Leukoc. Biol.* **73**:209–212.

Muroaka, R. S., W. Y. Sun, M. C. Colbert, S. E. Waltze, D. P. Witte, J. L. Degen, and S. J. Friezner-Degen. 1999. The Ron/STK receptor tyrosine kinase is essential for peri-implantation development in the mouse. *J. Clin. Invest.* **103**: 1277–1285.

Nair, M. G., K. J. Guild, and D. Artis. 2006. Novel effector molecules in type 2 inflammation: lessons drawn from helminth infection and allergy. *J. Immunol.* **177**:1393–1399.

Najar, H. M., S. Ruhl, A. C. Bru-Capdeville, and J. H. Peters. 1990. Adenosine and its derivatives control human monocyte differentiation into highly accessory cells versus macrophages. *J. Leukoc. Biol.* **47**:429–439.

Namangala, B., B. P. De, W. Noel, L. Brys, and A. Beschin. 2001. Alternative versus classical macrophage activation during experimental African trypanosomosis. *J. Leukoc. Biol.* **69:** 387–396.

Nemeth, Z. H., C. S. Lutz, B. Csoka, E. A. Deitch, S. J. Leibovich, W. C. Gause, M. Tone, P. Pacher, E. S. Vizi, and G. Hasko. 2005. Adenosine augments IL-10 production by macrophages through an A2B receptor-mediated posttranscriptional mechanism. *J. Immunol.* **175:**8260–8270.

Nys, M., R. Laub, P. Damas, D. Sondag, G. Goethals, D. Jamaer, L. Joassin, and M. Lamy. 1996. Screening and characterization of specific anti-lipopolysaccharide antibodies in Belgian blood donors by enzyme-linked immunosorbent assays. *Eur. J. Clin. Invest.* **26:**1134–1142.

Orkin, S. H. 2000. Diversification of haematopoietic stem cells to specific lineages. *Nat. Rev. Genet.* **1:**57–64.

Oskouian, B., and J. Saba. 2007. Sphingosine-1-phosphate metabolism and intestinal tumorigenesis: lipid signaling strikes again. *Cell Cycle* **6:**522–527.

Palese, P. 2004. Influenza: old and new threats. *Nat. Med.* **10:** S82–S87.

Panther, E., S. Corinti, M. Idzko, Y. Herouy, M. Napp, A. la Sala, G. Girolomoni, and J. Norgauer. 2003. Adenosine affects expression of membrane molecules, cytokine and chemokine release, and the T-cell stimulatory capacity of human dendritic cells. *Blood* **101:**3985–3990.

Pauleau, A. L., R. Rutschman, R. Lang, A. Pernis, S. S. Watowich, and P. J. Murray. 2004. Enhancer-mediated control of macrophage-specific arginase I expression. *J. Immunol.* **172:**7565–7573.

Penengo, L., C. Rubin, Y. Yarden, and G. Gaudino. 2003. c-Cbl is a critical modulator of the Ron tyrosine kinase receptor. *Oncogene* **22:**3669–3679.

Perlman, S., and A. A. Dandekar. 2005. Immunopathogenesis of coronavirus infections: implications for SARS. *Nat. Rev. Immunol.* **5:**917–927.

Pflanz, S., J. C. Timans, J. Cheung, R. Rosales, H. Kanzler, J. Gilbert, L. Hibbert, T. Churakova, M. Travis, E. Vaisberg, W. M. Blumenschein, J. D. Mattson, J. L. Wagner, W. To, S. Zurawski, T. K. McClanahan, D. M. Gorman, J. F. Bazan, M. R. de Waal, D. Rennick, and R. A. Kastelein. 2002. IL-27, a heterodimeric cytokine composed of EBI3 and p28 protein, induces proliferation of naive CD4(+) T cells. *Immunity* **16:**779–790.

Rauh, M. J., V. Ho, C. Pereira, A. Sham, L. M. Sly, V. Lam, L. Huxham, A. I. Minchinton, A. Mui, and G. Krystal. 2005. SHIP represses the generation of alternatively activated macrophages. *Immunity* **23:**361–374.

Reese, T. A., H. E. Liang, A. M. Tager, A. D. Luster, N. van Rooijen, D. Voehringer, and R. M. Locksley. 2007. Chitin induces accumulation in tissue of innate immune cells associated with allergy. *Nature* **447:**92–96.

Roberts, W. K., B. E. Laue, and C. P. Selitrennikoff. 1988. Antifungal proteins from plants. *Ann. N. Y. Acad. Sci.* **544:** 141–151.

Rodriguez-Sosa, M., A. R. Satoskar, R. Calderon, L. Gomez-Garcia, R. Saavedra, R. Bojalil, and L. I. Terrazas. 2002. Chronic helminth infection induces alternatively activated macrophages expressing high levels of CCR5 with low interleukin-12 production and Th2-biasing ability. *Infect. Immun.* **70:**3656–3664.

Rogy, M. A., B. G. Beinhauer, W. Reinisch, L. Huang, and P. Pokieser. 2000. Transfer of interleukin-4 and interleukin-10 in patients with severe inflammatory bowel disease of the rectum. *Hum. Gene Ther.* **11:**1731–1741.

Rotta, G., E. W. Edwards, S. Sangaletti, C. Bennett, S. Ronzoni, M. P. Colombo, R. M. Steinman, G. J. Randolph, and M. Rescigno. 2003. Lipopolysaccharide or whole bacteria block the conversion of inflammatory monocytes into dendritic cells in vivo. *J. Exp. Med.* **198:**1253–1263.

Rugtveit, J., A. Bakka, and P. Brandtzaeg. 1997. Differential distribution of B7.1 (CD80) and B7.2 (CD86) costimulatory molecules on mucosal macrophage subsets in human inflammatory bowel disease (IBD). *Clin. Exp. Immunol.* **110:**104–113.

Saccani, A., T. Schioppa, C. Porta, S. K. Biswas, M. Nebuloni, L. Vago, B. Bottazzi, M. P. Colombo, A. Mantovani, and A. Sica. 2006. p50 nuclear factor-kappaB overexpression in tumor-associated macrophages inhibits M1 inflammatory responses and antitumor resistance. *Cancer Res.* **66:**11432–11440.

Serbina, N. V., T. P. Salazar-Mather, C. A. Biron, W. A. Kuziel, and E. G. Pamer. 2003. TNF/iNOS-producing dendritic cells mediate innate immune defense against bacterial infection. *Immunity* **19:**59–70.

Shen, H. M., and S. Pervaiz. 2006. TNF receptor superfamily-induced cell death: redox-dependent execution. *FASEB J.* **20:**1589–1598.

Shortman, K., and S. H. Naik. 2007. Steady-state and inflammatory dendritic-cell development. *Nat. Rev. Immunol.* **7:**19–30.

Sica, A., and V. Bronte. 2007. Altered macrophage differentiation and immune dysfunction in tumor development. *J. Clin. Investig.* **117:**1155–1166.

Sica, A., A. Saccani, B. Bottazzi, N. Polentarutti, A. Vecchi, J. van Damme, and A. Mantovani. 2000. Autocrine production of IL-10 mediates defective IL-12 production and NF-kappa B activation in tumor-associated macrophages. *J. Immunol.* **164:**762–767.

Siegel, C. T., K. Schreiber, S. C. Meredith, G. B. Beck-Engeser, D. W. Lancki, C. A. Lazarski, Y. X. Fu, D. A. Rowley, and H. Schreiber. 2000. Enhanced growth of primary tumors in cancer-prone mice after immunization against the mutant region of an inherited oncoprotein. *J. Exp. Med.* **191:**1945–1956.

Skeel, A., T. Yoshimura, S. D. Showalter, S. Tanaka, E. Appella, and E. J. Leonard. 1991. Macrophage stimulating protein: purification, partial amino acid sequence, and cellular activity. *J. Exp. Med.* **173:**1227–1234.

Snyder, D. S., and E. R. Unanue. 1982. Corticosteroids inhibit murine macrophage Ia expression and interleukin 1 production. *J. Immunol.* **129:**1803–1805.

Spiegel, S., and S. Milstien. 2007. Functions of the multifaceted family of sphingosine kinases and some close relatives. *J. Biol. Chem.* **282:**2125–2129.

Stein, M., S. Keshav, N. Harris, and S. Gordon. 1992. Interleukin 4 potently enhances murine macrophage mannose receptor activity: a marker of alternative immunologic macrophage activation. *J. Exp. Med.* **176:**287–292.

Stout, R. D., C. Jiang, B. Matta, I. Tietzel, S. K. Watkins, and J. Suttles. 2005. Macrophages sequentially change their functional phenotype in response to changes in microenvironmental influences. *J. Immunol.* **175:**342–349.

Stout, R. D., and J. Suttles. 2004. Functional plasticity of macrophages: reversible adaptation to changing microenvironments. *J. Leukoc. Biol.* **76:**509–513.

Sutterwala, F. S., G. J. Noel, P. Salgame, and D. M. Mosser. 1998. Reversal of proinflammatory responses by ligating the macrophage Fcgamma receptor type I. *J. Exp. Med.* **188:**217–222.

Szabo, C., G. S. Scott, L. Virag, G. Egnaczyk, A. L. Salzman, T. P. Shanley, and G. Hasko. 1998. Suppression of macrophage inflammatory protein (MIP)-1alpha production and collagen-induced arthritis by adenosine receptor agonists. *Br. J. Pharmacol.* **125:**379–387.

Tacke, F., F. Ginhoux, C. Jakubzick, N. van Rooijen, M. Merad, and G. J. Randolph. 2006. Immature monocytes acquire antigens from other cells in the bone marrow and present them to T cells after maturing in the periphery. *J. Exp. Med.* **203:**583–597.

Tacke, F., and G. J. Randolph. 2006. Migratory fate and differentiation of blood monocyte subsets. *Immunobiology* **211:**609–618.

Tan, E. M., and F. D. Shi. 2003. Relative paradigms between autoantibodies in lupus and autoantibodies in cancer. *Clin. Exp. Immunol.* **134:**169–177.

Taylor, M. D., A. Harris, M. G. Nair, R. M. Maizels, and J. E. Allen. 2006. F4/80+ alternatively activated macrophages control CD4+ T cell hyporesponsiveness at sites peripheral to filarial infection. *J. Immunol.* **176:**6918–6927.

Trinchieri, G., and A. Sher. 2007. Cooperation of Toll-like receptor signals in innate immune defence. *Nat. Rev. Immunol.* **7:**179–190.

Tzachanis, D., A. Berezovskaya, L. M. Nadler, and V. A. Boussiotis. 2002. Blockade of B7/CD28 in mixed lymphocyte reaction cultures results in the generation of alternatively activated macrophages, which suppress T-cell responses. *Blood* **99:**1465–1473.

van Furth, R., and Z. A. Cohn. 1968. The origin and kinetics of mononuclear phagocytes. *J. Exp. Med.* **128:**415–435.

Van Ginderachter, J. A., K. Movahedi, G. G. Hassanzadeh, S. Meerschaut, A. Beschin, G. Raes, and P. De Baetseller. 2006. Classical and alternative activation of mononuclear phagocytes: picking the best of both worlds for tumor promotion. *Immunobiology* **211:**487–501.

van Kooten, C., and J. Banchereau. 2000. CD40-CD40 ligand. *J. Leukoc. Biol.* **67:**2–17.

Varol, C., L. Landsman, D. K. Fogg, L. Greenshtein, B. Gildor, R. Margalit, V. Kalchenko, F. Geissmann, and S. Jung. 2007. Monocytes give rise to mucosal, but not splenic, conventional dendritic cells. *J. Exp. Med.* **204:**171–180.

Verreck, F. A., T. de Boer, D. M. Langenberg, M. A. Hoeve, M. Kramer, E. Vaisberg, R. Kastelein, A. Kolk, R. de Waal-Malefyt, and T. H. Ottenhoff. 2004. Human IL-23-producing type 1 macrophages promote but IL-10-producing type 2 macrophages subvert immunity to (myco)bacteria. *Proc. Natl. Acad. Sci. USA* **101:**4560–4565.

Villadangos, J. A. 2007. Hold on, the monocytes are coming! *Immunity* **26:**390–392.

Volkman, A., and F. M. Collins. 1974. The cytokinetics of monocytosis in acute salmonella infection in the rat. *J. Exp. Med.* **139:**264–277.

Waltz, S. E., L. Eaton, K. Toney-Earley, K. A. Hess, B. E. Peace, J. R. Ihlendorf, M. H. Wang, K. H. Kaestner, and S. J. Degen. 2001. Ron-mediated cytoplasmic signaling is dispensable for viability but is required to limit inflammatory responses. *J. Clin. Investig.* **108:**567–576.

Wang, I. M., C. Contursi, A. Masumi, X. Ma, G. Trinchieri, and K. Ozato. 2000. An IFN-gamma-inducible transcription factor, IFN consensus sequence binding protein (ICSBP), stimulates IL-12 p40 expression in macrophages. *J. Immunol.* **165:**271–279.

Wang, J., R. A. Anders, Y. Wang, J. R. Turner, C. Abraham, K. Pfeffer, and Y. X. Fu. 2005. The critical role of LIGHT in promoting intestinal inflammation and Crohn's disease. *J. Immunol.* **174:**8173–8182.

Wang, J., and Y. X. Fu. 2004. The role of LIGHT in T cell-mediated immunity. *Immunol. Res.* **30:**201–214.

Wang, J., and Y. X. Fu. 2005. Tumor necrosis factor family members and inflammatory bowel disease. *Immunol. Rev.* **204:**144–155.

Wang, M. H., C. Ronsin, M. C. Gesnel, L. Coupey, A. Skeel, E. J. Leonard, and R. Breathnach. 1994. Identification of the ron gene product as the receptor for the human macrophage stimulating protein. *Science* **266:**117–119.

Wang, M. H., A. Skeel, and E. J. Leonard. 1996. Proteolytic cleavage and activation of pro-macrophage-stimulating protein by resident peritoneal macrophage membrane proteases. *J. Clin. Invest.* **97:**720–727.

Ware, C. F. 2005. Network communications: lymphotoxins, LIGHT, and TNF. *Annu. Rev. Immunol.* **23:**787–819.

Weber, C., K. U. Belge, P. von Hundelshausen, G. Draude, B. Steppich, M. Mack, M. Frankenberger, K. S. Weber, and H. W. Ziegler-Heitbrock. 2000. Differential chemokine receptor expression and function in human monocyte subpopulations. *J. Leukoc. Biol.* **67:**699–704.

Whitelaw, D. M. 1972. Observations on human monocyte kinetics after pulse labeling. *Cell Tissue Kinet.* **5:**311–317.

Wood, K. J., and B. Sawitzki. 2006. Interferon gamma: a crucial role in the function of induced regulatory T cells in vivo. *Trends Immunol.* **27:**183–187.

Wynn, T. A. 2003. IL-13 effector functions. *Annu. Rev. Immunol.* **21:**425–456.

Xu, X., and X. Gao. 2004. Immunological responses against SARS-coronavirus infection in humans. *Cell Mol. Immunol.* **1:**119–122.

Yang, Z., D. M. Mosser, and X. Zhang. 2007. Activation of the MAPK, ERK, following *Leishmania amazonensis* infection of macrophages. *J. Immunol.* **178:**1077–1085.

Yen, D., J. Cheung, H. Scheerens, F. Poulet, T. McClanahan, B. McKenzie, M. A. Kleinschek, A. Owyang, J. Mattson, W. Blumenschein, E. Murphy, M. Sathe, D. J. Cua, R. A. Kastelein, and D. Rennick. 2006. IL-23 is essential for T cell-mediated colitis and promotes inflammation via IL-17 and IL-6. *J. Clin. Investig.* **116:**1310–1316.

Young, H. A. 2006. Unraveling the pros and cons of interferon-gamma gene regulation. *Immunity* **24:**506–507.

Zhang, X., J. P. Edwards, and D. M. Mosser. 2006. Dynamic and transient remodeling of the macrophage IL-10 promoter during transcription. *J. Immunol.* **177:**1282–1288.

Zhu, Z., T. Zheng, R. J. Homer, Y. K. Kim, N. Y. Chen, L. Cohn, Q. Hamid, and J. A. Elias. 2004. Acidic mammalian chitinase in asthmatic Th2 inflammation and IL-13 pathway activation. *Science* **304:**1678–1682.

Ziegler-Heitbrock, L. 2007. The CD14+ CD16+ blood monocytes: their role in infection and inflammation. *J. Leukoc. Biol.* **81:**584–592.

21

Recognition and Removal of Apoptotic Cells

PETER M. HENSON AND DONNA L. BRATTON

Many cells have the ability to recognize and internalize dead and dying cells via phagocytic mechanisms. Targets include cells undergoing various forms of programmed cell death (PCD) (especially apoptosis), as well as those that have become necrotic, either directly or subsequent to apoptosis. Most studies of these mechanisms for cell deletion in mammals involve uptake into so-called professional phagocytes, i.e., macrophages and dendritic cells (DCs). However, apoptotic cell removal is seen in all metazoa, including the myriad organisms that do not possess defined populations of macrophages. In mammals too, uptake can occur into fibroblasts, endothelial, epithelial, smooth muscle, and stromal cells and likely many more cell types; in essence, removal of dying cells by near neighbors, often of the same cell type as the cell that is dying. PCD and subsequent phagocytic removal is physiologically essential for tissue modeling and remodeling during development as well as for normal cell turnover in the adult animal. In mammals this may be most clearly exemplified by the constant production and removal of blood cells, some of which have life spans in the circulation of only a few hours. In inflammation and the normal functioning of the immune system, excess cells must constantly be removed—again a function of PCD and uptake into phagocytic cells. Additionally, the same general processes for cell removal appear to be operative following cell injury and death induced by foreign organisms or toxic conditions, including intracellular forms of cellular stress imposed on nucleus, mitochondria, endoplasmic reticulum (ER), or general redox and the metabolic state of the cell.

Apoptosis is one form of PCD and has received the most attention. It is often contrasted with forms of cell death that have been loosely lumped together under the term "necrosis." In simple terms, apoptosis results in cells or fragmented cells (apoptotic bodies) that maintain cell membrane integrity, whereas "necrotic" cells to varying degrees exhibit permeable cell membranes. This has further led to the concept that recognition and removal of apoptotic cells is anti-inflammatory and anti-immunogenic, whereas necrotic cells induce inflammation and immunologic responses. However, these two concepts are vastly oversimplified. Cell death by various means cannot be so simply categorized. For example, if they are not phagocytosed, apoptotic cells lose their energy supply and their membrane integrity at rates that depend on cell type, stimulus, and intracellular signaling pathways. Thus, neutrophils generally maintain membrane integrity for a significant time after apoptosis induction and the characteristic nuclear alterations of apoptosis. However, T cells undergoing one form of activation-induced cell death appear to show caspase-dependent nuclear changes simultaneously with non-caspase-dependent membrane disruption (Hildeman et al., 1999). Examples of such disparity are so common that it would seem appropriate to characterize each cell type and system separately.

Similarly, as will be discussed below, recognition of the dying cells by macrophages or any other cell type cannot so easily be characterized as pro- or anti-inflammatory. More realistically, it seems that the response represents a balance between various recognition ligands, bridge molecules, and receptors leading to different patterns of pro- or anti-inflammatory responses and mediator production. So-called necrotic cells expose many of the same ligands as intact apoptotic cells do, and the actual phagocytic clearance mechanisms for each may be essentially identical. The differences in inflammatory or immunologic responses therefore can be seen as reflecting the above-mentioned balance rather than a black-and-white difference between the two forms of death.

The critical point is that throughout life, from the earliest embryologic development, cells need to be removed. This can be because they have fulfilled their function and are no longer needed or because they have become damaged or senescent. Various forms of cell "death" followed by uptake into phagocytes is the means whereby this removal is achieved. In the whole animal, this process is remarkably efficient and usually goes unnoticed. In fact, following studies such as those by Scott et al. (2001) outlined below, we have suggested that detection of apoptotic cells in tissues should lead one to at least question the presence of a local apoptotic cell clearance defect. Increas-

---

Peter M. Henson and Donna L. Bratton, Cell Biology and Allergy Divisions, Department of Pediatrics, National Jewish Medical and Research Center, 1400 Jackson Street, Denver, CO 80206.

ingly we recognize that abnormalities in the death and removal processes can lead to widely differing forms of pathology, from neoplasia to persistent inflammation to autoimmunity. However, a striking observation in the study of apoptosis, and also in apoptotic cell removal, is the substantial redundancy that can be seen in receptors, signaling pathways, and cell processes that result in this cell deletion. One cannot help but think that this redundancy is a reflection of how important the processes are to normal development and tissue homeostasis.

The ingestion of whole cells, apoptotic bodies, or cell fragments appears to involve unique and highly conserved recognition and signaling pathways and a physical uptake mechanism that is dependent on the low-molecular-weight GTPase Rac1 and has features in common with the process of macropinocytosis. These unique features have led us to coin the term "efferocytosis" (from "to carry to the grave, to bury") to describe the uptake cells dying by apoptosis and other forms of PCD (deCathelineau and Henson, 2003; Gardai et al., 2005, 2006). Its evolutionary conservation in all multicellular animals suggests longstanding opportunities for parasite exploitation as a means to gain access to the interior of host cells and thence to escape the effects of extracellular innate or adaptive immune systems. However, as an added element of complexity, parasites may make use of the responses of phagocytes to apoptotic cells, including their altered phenotype and/or generation of anti-inflammatory mediators, to establish a foothold in the host. Removal requires the dead and dying cells to be recognized as efficiently as recognition of foreign organisms by the innate immune system. In fact, there appear to be many similarities between recognition processes of the innate immune system and those for apoptotic cells. A notable exception seems to be the Toll-like receptors (TLRs) and NOD proteins, which, at least so far, have not been shown to play a major role in deletion of apoptotic cells (but see Shiratsuchi et al., 2004).

In this chapter we will first discuss the recognition of apoptotic cells by phagocytes, the mechanisms of apoptotic cell phagocytosis (efferocytosis), and some of the consequences of this recognition. This has become a huge subject of significant general interest. Accordingly, it has been necessary to be less than all-inclusive in our choice of references and to rely in a number of areas on reviews rather than the primary papers. We apologize in advance to investigators whose work may have received less than optimal discussion as a consequence. We have also chosen to emphasize the mechanisms of apoptotic cell recognition with less detailed focus on uptake signaling, the consequences of recognition, or the relative selectivity between recognition of different forms of dying cell by different types of phagocyte. This last is itself a rapidly growing area, in particular with regard to subsets of DCs. In the last portion of the chapter we will illustrate, with a few nonexhaustive examples, some of the possibilities for contribution of apoptotic cell recognition and/or its mechanisms to host-parasite interrelationships.

RECOGNITION OF APOPTOTIC OR NECROTIC CELLS: LIGANDS ON THE CELL SURFACE

In general, one can consider PCD as serving two functions: first, an active destruction of the cell's replicative potential that will prevent abnormal growth, and second, the induction of recognition signals that promote deletion of the cell, i.e., its removal by phagocytes. Colloquially, these have been called "eat me" signals. Surprisingly, less experimental attention has been given to these surface changes than to the recognition receptors or the soluble bridge molecules that then secondarily act as ligands for phagocyte receptors and mediate the clearance. Just as there appears to be great redundancy in these ligands, so we would expect, and are beginning to find, a significant and almost certainly underrepresented number of surface changes on apoptotic cells that contribute to their recognition. We have suggested that some ligand combinations primarily promote tethering of the apoptotic cell to the phagocyte, whereas others drive uptake and still others serve both functions together. The interaction and uptake has thus been seen as a two-step process, a "tether and tickle" mechanism (Hoffmann et al., 2001) (Fig. 1). Listed in Table 1 are some of the surface changes on apoptotic cells that have been suggested to contribute to their recognition and removal. Distinctions have often been made between "early" and "late" apoptotic cells and then cells undergoing "secondary" necrosis or postapoptotic cytolysis (for example, see Ren et al. [2001]). In some ways these stages may be seen as a continuum, and depend in large measure on how the cells being examined are defined and characterized. Thus, in many studies, early and late apoptotic cells are distinguished by membrane permeability to propidium iodide in the latter. Others might call these necrotic, but both early and late apoptotic cells (or their apoptotic bodies) remain intact and can be ingested in this form. Necrotic cells in most in vitro studies are generated by heating or freeze thawing the cells—generating various degrees of cell disruption, and neither occurring very often in real life. Moreover, "necrotic cells induced by heat versus mechanical stress lead to quite different responses in terms of uptake and its consequences" (Munoz et al., 2007a). Many of the ligands in Table 1 are also present in and on necrotic cell corpses. Thus, ligands that are normally intracellular but become exposed during apoptosis may, by definition, become available to receptors and bridge molecules if the cell membrane becomes permeable. In addition, there are many forms of PCD that have features in common with classical, extrinsically and intrinsically driven, caspase-dependent apoptosis. They include various cellular stress effects such as the unfolded protein response, autophagy, mitotic catastrophe, and induced necrosis and senescence (Okada and Mak, 2004). For the most part, analysis of surface changes during these different forms of cell death has not been carried out. However, it is highly probable that many of the alterations resemble those seen in classical apoptosis.

Phosphatidylserines

A characteristic alteration of the plasma membrane in apoptosis is the exposure of phosphatidylserine (PS) (Fadok et al., 1992; Martin et al., 1995; van den Eijnde et al., 1998) often appearing in patches and on membrane blebs that may constitute recognition platforms (Gardai et al., 2005). In viable cells, PS is generally located on the inner leaflet of the plasma membrane, which is also enriched for phosphatidylethanolamine, whereas sphingomyelin is localized to the outer leaflet along with most of the phosphatidylcholine (PC). The asymmetric distribution of PS is thought to be maintained by the balance between an ongoing low level of phospholipid flip-flop (flip for inward movement, and flop for outward) between the two leaflets and activity of an aminophospholipid translocase (APLT)

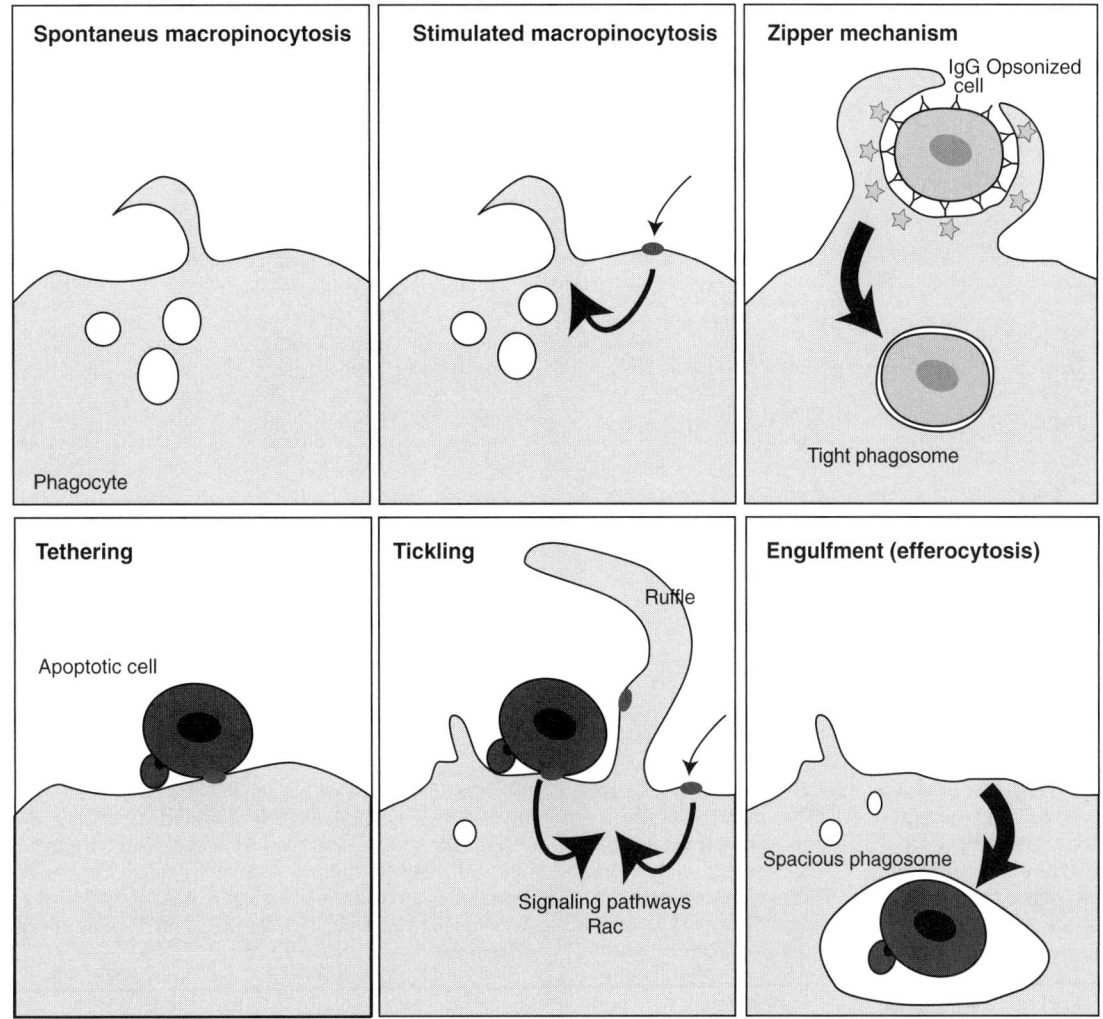

FIGURE 1 Macropinocytosis, phagocytosis, and uptake of apoptotic cells (efferocytosis). The proposed tether-and-tickle mechanism (Hoffmann et al., 2001) for apoptotic cell uptake is depicted with ingestion into spacious phagosomes. (With thanks to Aimee deCathelineau.)

that maintains the normal asymmetric state by actively transporting PS back to the inner leaflet. A novel family of P-type ATPases (Tang et al., 1996) with yeast analogs, Dnf1p, Dnf2p, and Drs2p (Pomorski et al., 2003), have been proposed as APLT candidates.

PS exposure during apoptosis is envisioned as a two-part process: (i) inhibition of the APLT by oxidants, nitrosative stress, ATP depletion, or high levels of intracellular calcium (Bratton et al., 1997; Daleke, 2003; Tyurina et al., 2007), and (ii) enhanced PS outward movement. Less clarity exists for candidates that flop PS from the plasma membrane inner leaflet to the outer leaflet. Outwardly directed floppases, members of the ABC transporter family, have been described for various sphingolipids and glycerophospholipids including PS exposure (Hamon et al., 2000; Mitra et al., 2006; van Helvoort et al., 1996). Given the near-universal exposure of PS on most apoptosing cells, however, no single ABC transporter has been definitively assigned this role, nor is there clarity regarding activation of such transporters or substrate specificity during apoptosis. Finally, there is little evidence for their participation in inward phospholipid flip, part of the bidirectional flip-flop

that accompanies apoptosis. This latter appears to be independent of head group resulting in generalized "scrambling" of phospholipids and loss of plasma membrane phospholipid asymmetry. The phospholipid scramblase family (Wang et al., 2007; Zhao et al., 1998) has been proposed to serve this role, but without support, as yet, from murine knockout models (Zhou et al., 2002). Finally, alteration of plasma membrane lipids (e.g., oxidized PS, ceramides, and cholesterol content) or of liposome composition (in the presence of transmembrane peptides) has been shown to induce phospholipid flip-flop, raising the possibility that alteration of the lipid bilayer composition during apoptosis is itself responsible for flip-flop (Boon et al., 2003; Elliott et al., 2006; Tyurina et al., 2004; Lange et al., 2007). One caveat of most investigations of flip-flop is that transbilayer movement is often explored by using phospholipids with hydrophilic substituents such as fluorescently tagged acyl groups (e.g., the 7-nitro-2-1,3-benzoxadiazol-4-yl [NBD]-labeled phospholipid analogs), and it is unclear as to how well the movements of these probes represent those of native phospholipids during apoptosis.

Under certain circumstances, activation of cells leads to increased flip-flop and, as a consequence, exposure of PS

TABLE 1 Ligands, bridge molecules, and receptors involved in clearance of apoptotic cells

Ligand on apoptotic cells	Bridge molecule	"Receptor" on phagocyte
PSs		PS recognition structures (BAI1, TIM4 and TIM1, stabilin-2)
		Scavenger receptors (SR-A, SR-B, etc.)
	Gas6, protein S	Mer family of receptor tyrosine kinases
	MFG-E8, Del-1	$\alpha_v\beta_3/\beta_5$ integrins
	β_2-Glycoprotein 1	
	Annexins 1 and 2	Homophilic interaction
		PS recognition structures
Oxidized PSs		CD36
		Other scavenger receptors
		Antibodies
Oxidized PCs	Pentraxins (e.g., CRP)	Antibodies
Lyso-PC		Antibodies
Calreticulin		LRP (CD91)
CD43		Nucleolin
CD14		ICAM-3
DNA	Collectins and ficolins	LRP
		Unknown
Unknown	Thrombospondin	CD36
		LRP
	Collectins	LRP
C3 (after fixation to apoptotic cell)		β_2 integrin
	Adiponectin	Calreticulin/LRP

(Dillon et al., 2001; Frasch et al., 2004). However, in the normal viable cell, the active APLT soon rectifies the increased PS symmetrization. This raises important questions as to discrimination by phagocytes between viable and apoptotic cells (see below). Conversely, activated PS-exposing platelets demonstrate little residual APLT activity, and similarly, apoptosis is accompanied both by an increase in flip-flop and by inactivation of the APLT (Bratton et al., 1997) so that PS exposure is permanent, perhaps providing adequate opportunity for its recognition.

PS exposure is usually detected by use of the PS-binding protein annexin V. Indeed, the exposure of PS has also led to a standard assay for apoptosis. When accompanied by simultaneous exclusion of the membrane-impermeable propidium iodide, it serves to detect an apoptotic cell that has not yet become membrane permeable. It is clear that the inner leaflet location of PS in cell membranes means that plasma membrane permeabilization (depending on the degree) may allow PS-recognizing bridge molecules (see below) access to the internal PS. Late apoptotic cells with permeable membranes or necrotic cells would therefore be expected to bind annexin and also the bridge molecules that can contribute to cell removal.

Modified and Oxidized Phospholipids

A number of investigators have shown the presence of oxidized PS and PC species on apoptotic cells and have suggested that these are recognized by receptors (such as scavenger receptors), by bridge molecules, or in the context of innate or adaptive immunity (e.g., Bayir et al., 2006; Chang et al., 2004; Greenberg et al., 2006; Kagan et al., 2002). Kagan et al. have proposed that PS, localized to the plasma membrane inner leaflet, is selectively oxidized by cytochrome c released from mitochondria during apoptosis (Tyurina et al., 2004), and that oxidative stress is required for efficient clearance of apoptotic cells by macrophages (Kagan et al., 2002). Further, PS oxidation may enhance its own flip-flop (Tyurina et al., 2004). An intriguing variant on this is the suggestion that the APLT may not be able to flip oxidized forms of PS back into the inner leaflet (above), leaving them selectively exposed for recognition. Signaling of oxidized PS species has recently been shown to signal through CD36 for the engulfment of apoptotic cells (Greenberg et al., 2006).

Several oxidized PC species have been identified on apoptotic cells with the unsaturation of the sn-2 fatty acid determining biological effects (Chang et al., 2004; Subbanagounder et al., 2000). These oxidized species resemble those found in oxidized low-density lipoprotein (oxLDL) and are reportedly recognized by various receptors (including the scavenger receptors CD36 and SR-A, discussed below, and LOX-1, CD14, TLR4, and the platelet-activating factor receptor), and such lipids/oxLDL have been shown to inhibit uptake of apoptotic cells (Miller et al., 2003; Oka et al., 1998; Subbanagounder et al., 2000). Oxidized species of PC are recognized by natural antibodies or in some instances are immunogenic and lead to proinflammatory consequences (Chang et al., 2004) (see below). In addition, lyso-PC has been shown to be produced during apoptosis as a result of calcium-independent PLA_2 ($iPLA_2$) activity and plays a role in the attraction of macrophages (Lauber et al., 2003). Lyso-PC, like several oxidized species, is also recognized by natural antibodies and appears to play a role in recognition of apoptotic cells (Kim et al., 2002; Lauber et al., 2003).

Calreticulin

Recent studies have implicated calreticulin as a ligand on apoptotic cells that can be recognized and contributes to their removal (Gardai et al., 2005; Kuraishi et al., 2007; Obeid et al., 2007; Taylor et al., 2007). Calreticulin is a highly conserved protein primarily located within the ER of cells, where it serves as a calcium store and protein chaperone. However, it is also found, to a variable extent, on the surface of most cells examined. Since it contains a KDEL, ER-retention signal, it is at present unclear how it gains access to the cell surface, but its exposure on apoptotic cells is dependent on caspases (Taylor et al., 2007). The absence of a transmembrane domain or other sites for membrane association suggests interaction with one or more membrane proteins. One of these seems to be LDL receptor-related protein, LRP-1 (and probably also megalin, LRP-2) (Sim et al., 1998). As noted below, this association of calreticulin and LRP constitutes a receptor complex for collectin family bridge molecules in their contribution to apoptotic cell ingestion. However, calreticulin on apoptotic cells also appears capable of directly binding LRP-1 on phagocytes (Gardai et al., 2005), thereby inducing a *trans* activation of the LRP ingestion receptor. Studies of calreticulin expression suggest that most apoptotic cells examined showed increased amounts on the surface, but all apoptotic cells exhibited redistribution into patches. These often coincided with the patched exposure of PS (Gardai et al., 2005, 2006) (Color Plate 5). Once again, the fact that viable cells can also expose calreticulin needs to be considered in the context of phagocyte recognition of apoptotic, but not viable, cells.

Other Surface Ligands

Early studies of apoptotic cells showed alterations in surface charge (Savill et al., 1989) and likely increased levels of exposed amino sugars. The mechanisms for, and nature of, these changes have not been elucidated. Apoptotic hepatocytes exhibit altered surface sugars, including mannose that appeared to be recognized by so far unidentified lectins on the phagocytes (Dini, 2000; Dini et al., 1995). The polygalactosamine side chains of CD43 have been identified as an early candidate ligand on apoptotic Jurkats (Eda et al., 2004; Hirano et al., 2005; Yamanaka et al., 2005), apparently reacting with nucleolin on the macrophage. CD43, too, is seen as patches or caps on the apoptotic cell. A role for CD14 on the phagocyte has been identified by blockade of apoptotic cell removal in vitro and in vivo by antibody and genetic deletion of CD14 (Devitt et al., 1998, 2004). A putative ligand for this molecule on the apoptotic cell was identified as ICAM3 (Gregory et al., 1998). Another set of possible ligands that have recently been identified are components of the cell nucleus, i.e., DNA itself (Palaniyar et al., 2003, 2004) and possibly other components from the nucleus (Gebska et al., 2002). How common this is and how much it occurs in vivo in normal (nonautoimmune) states still needs to be determined.

Immune Recognition of Apoptotic Cells

Incubation of apoptotic cells with serum results in fixation of C3 (Mevorach et al., 1998; Verbovetski et al., 2002) and a number of studies have suggested that complement can be involved in the uptake of apoptotic cells (e.g., Botto et al., 1998; Gershov et al., 2000; Ghebrehiwet and Peerschke, 2004; Roos et al., 2004; Taylor et al., 2000). Complement activation could arise from direct recognition of apoptotic cells by the classical pathway, complement-activating, collectin family proteins, C1q and mannose-binding lectin (MBL), although the ligands on apoptotic cells that are recognized by these have not been clearly defined. Exposed mannose (see above) might provide the ligand for MBL, but it should be noted that collectin family molecules are promiscuous (they serve as classic pattern recognition molecules) in their recognition functions. Incubation of apoptotic cells with normal serum also leads to binding of so-called "natural" antibodies. The full spectrum of antigens potentially recognized by such immunoglobulins is not yet clear but may include lyso-PC, oxidized cardiolipin (Tuominen et al., 2006), and other oxidized phospholipid species (Chang et al., 2004). Studies by Elkon and his associates showed that PCs, especially lyso-PC, presumably on the outer leaflet of the apoptotic cell, constituted one set of candidate antigens (Kim et al., 2002, 2003). Lyso-PC in apoptotic cells has been shown to result from the action of caspase-activated iPLA$_2$ (Kim et al., 2002; Lauber et al., 2003), or could be derived from oxidation of unsaturated PCs during the apoptotic process. Oxidation of PC during apoptosis has also been shown to be immunogenic (Chang et al., 2004), perhaps the source of the above-mentioned natural antibodies. At this point, it is not clear what roles antibodies play in apoptotic cell removal, either directly through Fcγ receptors or following activation of complement. It is also relevant here to note that in autoimmune situations antibodies can be detected against many components that are expressed on apoptotic cells, including PSs, calreticulin, and some of the bridge molecules such as C1q itself.

In this context it should be noted that most studies of apoptotic cell uptake in vitro are performed in serum-free conditions or in relatively low concentrations of, often heat-inactivated, fetal calf serum. The true roles for antibody or complement contributions to apoptotic cell clearance in vivo are not clear. C1q deficiency is associated with defective clearance in humans and mice in vitro and in vivo (Botto and Walport, 2002; Cortes-Hernandez et al., 2004; Pittoni and Valesini, 2002; Roos et al., 2004), but C3 deficiency so far has not been identified with comparable abnormalities. Since C1q has a direct effect on apoptotic cell uptake by phagocytes acting through calreticulin and LRP (see below), the knockout data raise questions that need to be addressed experimentally as to how important complement activation pathways beyond C1 are in the usual clearance process.

Differences between Apoptotic Cells

The discussion of apoptotic cell ligands has been treated as if all apoptotic cells (or, for that matter, cells undergoing different forms of PCD) expose the same structures. While this seems highly unlikely, it is certainly true that, when examined, the commonality is striking, differing more in amount than kind. The overall importance and conservation of apoptotic cell removal may have contributed to similar recognition processes exhibited by a wide variety of apoptotic cells. However, an important note of caution here is that, for reasons of experimental simplicity, most studies have been carried out with apoptotic cells of hematological origin or cell lines cultured in vitro and may not reflect accurately the ligands present on tissue cells becoming apoptotic in situ induced by different apoptotic signaling pathways. It seems likely that "executioner" caspases induce a common pattern of surface changes in apoptotic cells, including exposure of PS and patching of such structures in the plane of the membrane following known caspase effects on cytoskeleton and membranes. Likewise, many apoptotic processes are accompanied by changes in redox state (more oxidative), and these too may contribute to common responses, including, for example, exposure of oxidized phospholipids. However, it would seem profitable to examine noncaspase and/or non-oxidant-dependent death pathways for similar or different surface alterations that can also contribute to cell removal.

BRIDGE MOLECULES (OPSONINS) FOR APOPTOTIC CELLS AND THEIR RECEPTORS

PS-Binding Proteins

A large number of proteins within the cell or in the extracellular environment are known to bind PS. These extend from intracellular protein kinase Cs to plasma coagulation components. The large annexin family may be intracellular, membrane bound, or extracellular, and annexin I has indeed been suggested to contribute to apoptotic cell uptake in a complex model along with exposed PS in the apoptotic cell (Arur et al., 2003). The annexin V that is commonly used for detecting surface PS is also found in vivo and may play a role in modifying uptake (Munoz et al., 2007a). A number of serum proteins have been clearly implicated in recognizing PS on apoptotic cells and then mediating uptake of such cells by engaging phagocyte "receptors."

Gas6, Protein S, and Mer

Gas6 (growth arrest specific 6) and protein S of the coagulation system are vitamin K-dependent proteins that bind PS (and oxidized PS) on apoptotic cells (Hafizi and Dahlback, 2006; Ishimoto et al., 2000). In turn they interact with members of the Mer, Axl, Sky family of receptor tyrosine kinases (Cohen et al., 2002; Sather et al., 2007; Scott et al., 2001; Seitz et al., 2007; Wu et al., 2005) on the phagocyte. Mer, or triple-knockout, animals show decreased ability to clear apoptotic cells in vivo (Scott et al., 2001; Seitz et al., 2007) and exhibit autoimmunity (Lu and Lemke, 2001). Isolated macrophages from these animals are likewise inefficient at such ingestion. The original studies on the Mer kinase-dead mouse (which is in fact a knockout for Mer) is additionally interesting because it provides one of the best examples for the efficiency of apoptotic cell removal in vivo. Induction of essentially complete thymocyte apoptosis in vivo by dexamethasone was not, in wild-type animals, accompanied by substantial evidence of apoptosis within the thymus, even though the cells were all dying. In the Mer$^{-/-}$ mice, however, large numbers of apoptotic cells now became visible (Scott et al., 2001). The clear implication is that the apoptotic cells were normally removed in such an efficient manner that, at any one point in time, almost no apoptotic cells were visible and that only when the clearance process was defective could the cells be seen. Importantly, as with a number of mouse models with defective apoptotic cell clearance, the Mer$^{-/-}$ animals develop spontaneous autoimmunity and also defects in the eye (see below).

MFG-E8 and Del-1

Milk fat globule protein E8 (MFG-E8) was first described as a protein in milk but later shown to be secreted by macrophages. It too binds PS, oxidized PS, and apoptotic cells and can serve as a bridge molecule to integrins on the phagocyte by virtue of an RGD domain (Akakura et al., 2004; Borisenko et al., 2004; Hanayama et al., 2004b). Early studies of apoptotic cell uptake had implicated $\alpha_v\beta_3$ and $\alpha_v\beta_5$ integrins on the basis of antibody or peptide inhibition studies (Hall et al., 1994; Savill et al., 1990), although the bridge molecule implicated in these early studies was thrombospondin. These two integrins serve as major signaling targets for the MFG-E8. MFG-E8$^{-/-}$ mice show defective apoptotic cell clearance in lymphoid organs and in the involuting mammary gland and also develop autoimmunity (Hanayama et al., 2004b, 2006). Mutation of the MFG-E8 RGD sequence to non-integrin-binding RGE has allowed development of a PS binding and blocking molecule that can be used in vitro and in vivo to implicate apoptotic cell PS in removal processes. This was used very effectively to show that clearance of extruded nuclei from developing erythrocytes depended on PS and probably was mediated by mechanisms similar to those used for whole apoptotic cells or apoptotic bodies (Yoshida et al., 2005). Del-1, developmental endothelial locus-1, is structurally and functionally related to MFG-E8 and has also been implicated as a bridge molecule for apoptotic cell clearance (Hanayama et al., 2004a, 2006).

Anti-PS Antibodies

As noted, anti-PS antibodies are common in autoimmunity and would be expected to bind surface PS on apoptotic cells with possible consequences for their removal (Manfredi et al., 1998). However, apoptotic cell uptake in systemic lupus erythematosus has been reported as being inefficient (see, for example, Gaipl et al. [2007]), so a precise role for such antibodies may only serve in the context of an overall inadequacy of the clearance processes.

β_2 Glycoprotein 1

The serum protein β_2 glycoprotein 1 (β2GP1) has been shown to bind PS liposomes and PS-exposing apoptotic cells and enhance their uptake into macrophages (Balasubramanian et al., 1997; Balasubramanian and Schroit, 1998) or DCs (Rovere et al., 1998). The issue is made more complex by the observation that antiphospholipid antibodies appear to utilize β2GP1 in some way to recognize and bind to apoptotic cells (Manfredi et al., 1998; Price et al., 1996). The nature of the receptor with which the β2GP1 interacts on the phagocyte is not at this point clear. However, the ability of the protein to bind phospholipids and "present" them to the immune system (Subang et al., 2000) raises the possibility that its effect in apoptotic cell clearance is likewise to present PS more optimally to PS receptors or bridge molecules. Of some interest, the normally anti-inflammatory consequences of apoptotic cell removal may be reversed in the presence of antiphospholipid antibodies and B2GP1 (Manfredi et al., 1998).

Thrombospondin

Early studies by Savill, Haslett, and colleagues implicated thrombospondin (TSP) in the recognition of apoptotic cells (Haslett et al., 1994; Ren and Savill, 1995; Ren et al., 2001). They proposed a trimolecular complex that involved TSP binding to the apoptotic cell surface followed by interaction of the bound TSP with both the vitronectin receptor ($\alpha_v\beta_3$ and $\alpha_v\beta_5$ integrin) and the scavenger receptor CD36 on the phagocyte. The involvement of this complex has not yet been formally proven, but it may well serve as part of the recognition system. In subsequent studies, each of the three components has also been implicated in other pathways and receptor systems that also participate in apoptotic cell uptake, making the whole system much more complex. Thus, TSP itself binds to the ingestion receptor LRP (Greenaway et al., 2007; Orr et al., 2003) (see below), CD36 can interact with anionic phospholipids, and the α_v integrins are candidate receptors for stimulation by MFG-E8 (see above). The possibility that TSP can act as a glue to hold everything together is intriguing. At this point, evidence for defects in apoptotic cell clearance in the double TSP1 and TSP2 knockout (redundancy between the two isoforms would require the double knockout) has not to our knowledge been reported. Invocation of TSP in the recognition complex and in the uptake may have additional importance in some of the subsequent consequences of apoptotic cell recognition, as it is a major activator of the anti-inflammatory mediator transforming growth factor-β (TGF-β; see below).

Collectin Family

An honorary member of the collectin family, C1q, was shown to bind to blebs on apoptotic cells (Korb and Ahearn, 1997; Navratil et al., 2001). Subsequently, the true collectins, MBL and surfactant proteins A and D (SP-A and SP-D), as well as C1q were all shown to interact directly with apoptotic cells and to mediate their subsequent uptake into macrophages (Ogden et al., 2001; Reidy and Wright, 2003; Schagat et al., 2001; Vandivier et al., 2002). SP-D$^{-/-}$ mice showed defective uptake in the lungs (Palaniyar et al., 2003; Vandivier et al., 2002) and C1q$^{-/-}$ animals similarly reduced uptake in kidney, peritoneum, and atheromatous lesions (Botto et al., 1998; Taylor et al., 2000; Bhatia et al., 2007). MBL$^{-/-}$ mice (there are two isoforms in mice and both had to be deleted) also showed reduced apoptotic cell phagocytosis (Stuart et al., 2005). Intriguingly, as with a number of the apoptotic cells that are known to act systemically in recognition systems, deletion of C1q resulted in developing autoimmunity (Botto et al., 1998; Botto and Walport, 2002) as well as increased detectable apoptotic cells in the kidney. This is in keeping with the known association between C1q deficiency and systemic lupus erythematosus in humans and the demonstration of impaired C1q-calreticulin-LRP apoptotic cell uptake in this disease (Donnelly et al., 2006). The SP-D-deficient mice develop spontaneous lung emphysematous changes and exhibit increased numbers of detectable apoptotic cells within the lungs (Clark et al., 2002; Wert et al., 2000). The ficolin family of molecules is related structurally to the collectins, and recent reports suggest that they too may contribute to recognize apoptotic cells (Garlatti et al., 2007; Jensen et al., 2007) and, in fact, activate complement as a consequence (Kuraya et al., 2005).

The ligands on the apoptotic cell that are recognized by the ficolins, collectins, and C1q are not fully characterized. SP-A and SP-D are known to interact with phospholipids via their globular head groups and may bind the exposed PS or oxidized PS. They, ficolins, and MBL are also lectins (hence "collectin") and could certainly recognize surface carbohydrates such as newly exposed mannose or other glycosyl moieties. Recently, it has been suggested that SP-D (Palaniyar et al., 2003, 2004) and ficolins (Jensen et al., 2007) also directly bind to DNA that has become exposed on the surface of apoptotic cells. In other words, the collectins are pattern recognition molecules and their head groups may bind many different molecules on the apoptotic cell surface. C1q is known to bind components of the mitochondrial membrane, including cardiolipin (Kovacsovics et al., 1985). This raises the intriguing possibility that during PCD, especially after some degree of membrane disruption, mitochondrial constituents might gain access to the cell surface and be available for recognition by C1q and other bridge molecules. If true, a relationship between recognition of apoptotic cells and some microorganisms, for example by collectins, might be considered.

Over the years, a variety of receptors on phagocytes for C1q and collectins have been described (Kishore et al., 2006). One of the key issues here is that these may bind either the globular head groups or the collagenous tails of these complex molecules. The heads contain the pattern recognition and lectin domains and are presumed in many systems, including for apoptotic cells, to recognize and bind the "foreign" structure. This implies interaction of the tails with receptors on the phagocyte that would then mediate internalization. One candidate here (Ogden et al., 2001; Vandivier et al., 2002) is the cC1q receptor (c for collagenous) (Ghebrehiwet, 1989; Peerschke et al., 1994) that was later shown to be identical with calreticulin (Sim et al., 1998) and to be a receptor for not only C1q, but also for the collagenous tails of the other family members as well. As noted above, calreticulin is found on the surface of cells, including macrophages and DCs, but because it lacks a transmembrane domain it must interact with a companion molecule to mediate internalization. A variety of studies implicate LRP as such a partner for calreticulin-mediated internalization and therefore as the effector receptor for collectin and C1q recognition of apoptotic cells.

Pentraxins

In general, pentraxins seem to bind to late apoptotic cells, i.e., while still intact but yet with permeable cell membranes. C-reactive protein (CRP) has been shown to recognize apoptotic cells (Bijl et al., 2003; Mold et al., 2002) and to enhance their uptake. On the apoptotic cell surface CRP may bind to and modulate C1q effects, or to oxidized PC (Chang et al., 2002) or perhaps to surface DNA, and may also induce activation of complement (Gershov et al., 2000). Serum amyloid P component also binds to apoptotic cells and can promote their uptake (Bijl et al., 2003; Mold et al., 2002). Both it and CRP may act in part through binding to Fcγ receptors on the macrophage (Mold et al., 2002). By contrast, while the long pentraxin PTX3 also recognizes apoptotic cells, it may act in a negative role by inhibiting uptake of apoptotic cells by DCs (Rovere et al., 2000). The authors suggest this may serve to divert the apoptotic cells away from DCs, presumably toward their clearance by macrophages (although this was not demonstrated).

Other Bridge Molecules

An intriguing participation of annexins was suggested by Fan et al. (2004) following their demonstration that PS is exposed at the site of uptake on the macrophage as well as on the apoptotic cell. Bivalent annexins I and II were implicated in a cross-linking of the PS on the two cell surfaces, presumably serving to help tether the target. We would presume that PS exposure on the macrophage at the site of activation reflects the activation-induced flip-flop noted above. The adipocyte-derived hormone adiponectin was found to bind early apoptotic cells, and its absence led to defective in vivo clearance (Takemura et al., 2007). It appears to interact with calreticulin on the phagocyte, presumably mediating uptake via LRP.

RECEPTORS FOR APOPTOTIC CELL SURFACE LIGANDS

PS Receptors

There are also receptors on the phagocyte that can directly recognize PS on the apoptotic cell without the intervention of a bridge molecule. The best characterized of these are various members of the scavenger receptor families. A key element for apoptotic cell recognition is their general ability to bind anionic phospholipids and, therefore, exposed PSs, including oxidized forms of these. Thus, a recent study by Hazen and coworkers has implicated direct interaction of surface-exposed oxidized PS with CD36 on the phagocyte (Greenberg et al., 2006). In fact, this scavenger receptor has long been known to participate in apoptotic cell removal (see below) and may interact with a number of additional ligands, bridge molecules, and receptors.

A candidate direct PS receptor (PSR) was proposed in 2000 (Fadok et al., 2000) that appeared to be expressed on many different cell types and which was downregulated by exposure to PS. It was identified by a monoclonal antibody of immunoglobulin M (IgM) isotype (Mab217) raised against macrophages that were able to ingest apoptotic cells in a PS-inhibitable fashion. Binding of the antibody to cells was also blocked with PS. However, subsequent studies have shown that this is not a PS receptor (see, for example, Bose et al., 2004; Cikala et al., 2004; Mitchell et al., 2006; Williamson and Schlegel, 2004). Rather, it is primarily located in the cell nucleus and its exact functions are currently unknown. We currently believe that the original observation arose from a binding of Mab217 in the phage display used to identify the "PSR" that was nonspecific—probably a result of the known "stickiness" of IgM. Knockouts were developed in mice (Bose et al., 2004; Kunisaki et al., 2004; Li et al., 2003), Drosophila (Krieser et al., 2007; Li and Baker, 2007), Caenorhabditis elegans (Wang et al., 2003), and zebrafish (Hong et al., 2004). In mice its deletion was lethal in the perinatal time period, and in zebrafish larval alterations were apparent. Less dramatic phenotypic alterations were seen in the worm and fly. The question of a possible participation of this molecule in apoptotic cell clearance is still unanswered; for example, defective uptake was reported in two of the knockout mice (Kunisaki et al., 2004; Li et al., 2003); not in the third (Bose et al., 2004), in zebrafish, or in the nematode; but with variable results in Drosophila. However, at this point in time one must conclude that if this molecule is indeed a participant it will function indirectly, i.e., not as a receptor itself. In the light of this conclusion, it would seem time to change the name from "PSR," and since the molecule exhibits a Jumonji domain and has been implicated as a histone arginine demethylase, it may now be termed JMJD6 (Chang et al., 2007).

On the other hand, there remain a number of observations to suggest that true membrane-associated PS receptors do exist. For one, blockade of apoptotic cell uptake into macrophages was achieved by PS, glycerophosphoserine, and even to some degree phosphoserine itself and shown to be stereospecific (Fadok et al., 1992). In the past year a bevy of new candidate PS receptors have been recognized. Brain-specific angiogenesis inhibitor (BAI1) was identified (Park et al., 2007) as an upstream PS-recognizing receptor in mammals for the ELMO/DOCK180 intracellular signaling pathway already shown to be involved in apoptotic cell uptake (see below). Tim4 and Tim1 (T-cell immunoglobulin- and mucin-domain-containing molecule) were shown by three groups to recognize PS and participate in apoptotic cell uptake (Kobayashi et al., 2007; Miyanishi et al., 2007). However, the mechanism by which these molecules signal for uptake remains unclear—they may utilize as yet uncharacterized signaling partners. Stabilin-2 is a multifunctional receptor that binds a large assortment of ligands and is known to serve scavenger, endocytic, and adhesion functions. This too has now been identified as a candidate PS receptor (Park et al., 2008). Intriguingly, each of these receptors seems to use different sequence structures to recognize PS (IgV domain in TIM4, TSP repeats in BAI1, and unknown sequences in stabilin-2, which exhibits fasiculin, EGF, and hyaluronan binding domains), and these may be different again from such recognition domains in the bridge molecules Gas6 and MFG-E8, or the scavenger receptors, or for that matter those of other known PS-binding proteins such as protein kinase Cs, annexins, or other coagulation proteins. A crystal structure for TIM4 bound to PS has been reported (Santiago et al., 2007), but until crystal structures are available for each of these, much remains unknown as to how PS is recognized by each of these diverse molecular structures (e.g., whether as a monomer or patched or oxidized) or how they signal. Much work will also be required to sort out the different contributions of all these PS-recognizing "receptors," both for apoptotic cell removal and for the other functional responses to apoptotic cells noted below.

LRP (LDL Receptor-Related Protein)

LRP-1 and LRP-2 (megalin) are internalization receptors (Fisher and Howie, 2006; Gonias et al., 2004). They bind a very wide variety of ligands (over 50 different molecules, including proteins, lipids, and carbohydrates) and, although not classed in the scavenger receptor family, do in fact subserve similar activity for general internalization functions. As noted, through attached calreticulin, LRP mediates uptake of particles, including apoptotic and necrotic cells, that have bound collectins—a cis-acting role for the calreticulin (Bartl et al., 2001; Gardai et al., 2005; Ogden et al., 2001; Vandivier et al., 2002). We have also provided evidence for calreticulin acting in a trans-activation mode when upregulated on the apoptotic cell surface and directly stimulating LRP on the phagocyte (Gardai et al., 2005). Among its many ligands, LRP also recognizes a variety of heat shock proteins, and since these are upregulated during cellular stress that can lead to apoptosis, one wonders if these and other LRP ligands may act in similar ways. This could also extend to the lipidic changes on the apoptotic cell. Since LRP knockouts are embryonic lethal, its relative importance in apoptotic cell removal in vivo has yet to be determined.

A candidate Drosophila ortholog, Draper, has been shown to participate in apoptotic cell clearance (Awasaki et al., 2006; Li and Baker, 2007; Manaka et al., 2004). Moreover, the intracellular domain of LRP is homologous with CED1, a C. elegans molecule known to participate in apoptotic cell clearance. Extracellular CED1 resembles the mammalian scavenger receptor expressed by endothelial cells, SREC (Zhou et al., 2001). This is a type F scavenger receptor, also found on phagocytes (Tamura et al., 2004), that, intriguingly, is also known to bind calreticulin (Berwin et al., 2004). Its role in apoptotic cell clearance has not been addressed, but the possibility that the SREC-like domains on CED1 serve to recognize apoptotic cell calreticulin in C. elegans, as does the extracellular domain of LRP in mammals, would make an interesting parallel between the different animal groups.

Scavenger Receptors

As noted, scavenger receptors have been identified as candidates for direct and indirect recognition of apoptotic cells. Early studies with CD36 suggested its participation in vitro (Savill et al., 1991) and it has also been shown to be effective in some systems in vivo (see, for example, Greenberg et al., 2006). In particular, the daily removal of discarded retinal outer segments by the retinal epithelium represents a huge and constant phagocytic capacity of the latter and has been shown to involve CD36 (Ryeom et al., 1996) as well as Mer and α_v integrins (Finnemann and Silverstein, 2001; Sun et al., 2006). In fact, mice defective in the Mer family of receptor tyrosine kinases or α_v integrins exhibit retinal dystrophy (Duncan et al., 2003; Nandrot et al., 2004). This type of observation is consistent with the ongoing importance of the removal at this site, but also with the probability that for any apoptotic cell removal process in vivo, many different pathways for recognition and phagocyte stimulation are acting simultaneously—another reflection of the extensive redundancy. A *Drosophila* homologue of CD36, croquemort, has been shown to participate in apoptotic cell clearance in this organism (Franc et al., 1996). As noted, there are a number of candidate ligands or bridge molecules for CD36, including oxidized and anionic phospholipids and TSP (Savill et al., 1992), although a recent study suggested that CD36 only recognizes oxidized forms of PS (Greenberg et al., 2006).

SR-A has been shown to be capable of contributing to apoptotic cell uptake into macrophages (Platt et al., 1996), although clear evidence of a critical role in vivo based on defective uptake in knockout animals has not been as evident (Platt et al., 2000). In addition, a possible role for removal of pieces of target cell membrane by SR-A-expressing macrophages has been noted (Harshyne et al., 2003), although a similar role for apoptotic cells was not explored. LOX-1 (lectin-like oxidized low-density lipoprotein receptor 1) has also been reported to recognize PS and apoptotic cells (Murphy et al., 2006). The degree to which other members of the scavenger receptor families are involved in apoptotic cell removal, or in fact removal of cells undergoing different forms of PCD and at different stages of dissolution, is yet to be determined. An important point to make here, which applies to all of these separate receptors and bridge molecules, is that the increasing evidence for overall redundancy means that in vitro or in vivo, removal or inhibition of any one system does not result in complete absence of clearance and may not be easily visible at all. This may be the case with the type A scavenger receptors. On the other hand, in specific sites and circumstances, any one of these systems for apoptotic cell recognition may gain significant biological prominence.

Integrins

The α_v integrins have for some time been candidates for apoptotic cell uptake, although, for the most part, in response to bridge molecules to recognize the dying cells—first, TSP (Ren et al., 2001; Savill et al., 1992), and more recently, MFG-E8 or Del-1 (see above). A role for $\alpha_v\beta_5$ was reported for uptake into DCs (Albert et al., 1998) and for the retinal epithelium (Finnemann et al., 1997; Lin and Clegg, 1998). At this point, however, there is little evidence that mice deficient in either or both of $\alpha_v\beta_3$ or $\alpha_v\beta_5$ exhibit general defects in apoptotic cell clearance, except for the clearance of retinal outer segments in the eye. Redundancy may be operative here. Moreover, MFG-E8 interacts with the integrins via its RGD sequence, whose recognition is also shared by other integrins, so the effect of this important bridge molecule may involve numerous potential signaling partners. A cooperative effect of α_v integrins and Mer has also been reported (Wu et al., 2005). In fact, we suggest that most of these receptors work cooperatively to mediate optimal control of recognition and uptake.

A second group of integrins that may contribute are the receptors for C3 and perhaps C4 (members of the β_2 integrin group) (Dalgaard et al., 2005; Mevorach et al., 1998; Skoberne et al., 2006). Again, C3R-deficient animals (or macrophages in vitro [Ren et al., 2001]) are not known to show defective uptake, but one never knows how well this has been looked for. Intriguingly, unlike C1q deficiency, they also do not show spontaneous autoimmunity.

Other Receptors

We anticipate that, as time progresses, more receptors will be identified. For example, Beppu and colleagues have identified CD43 as a potential ligand on apoptotic Jurkats and, more recently, that nucleolin may be its cognate receptor on the macrophage (Eda et al., 2004; Hirano et al., 2005; Yamanaka et al., 2005). TREM2 (triggering receptor expressed on myeloid cells-2) may also serve as a receptor for apoptotic cell clearance, for example, on microglia (Neumann and Takahashi, 2007; Takahashi et al., 2005). Its ligand on apoptotic cells is not known. It associates with an immunoreceptor tyrosine-based activation motif-containing adaptor protein, DAP12, and seems associated with pro- rather than anti-inflammatory effects.

DISTINGUISHING BETWEEN APOPTOTIC AND VIABLE CELLS

The key feature of apoptotic cell removal is the ability of phagocytes to distinguish between living versus apoptosing, or dying, cells. As noted above, although ligands can be distinguished on apoptotic (and necrotic) cells that can serve to identify them to phagocytes, in most cases these same molecules and changes can also be seen on viable cells. Where then is the distinction? We suggest that there are two main ways in which this is achieved.

Ligand Intensity and Distribution

A key concept here is that of low-affinity–high-avidity recognition. That is, we suggest that the amount and distribution of the ligands on the apoptotic cells are key features for the effective recognition and response of the phagocyte. Thus, PS and calreticulin exposure on apoptotic cells is more intense than on stimulated viable cells and is also seen as patches on the membrane surface (Gardai et al., 2005, 2006) (Color Plate 5). In fact, the two ligands may be patched together (Gardai et al., 2005), thereby potentially enhancing the dual stimulation through their combined receptors and/or bridge molecules. The collectins and C1q bind to many viable cell types in culture, but weakly and with a relatively uniform distribution. However, on apoptotic cells they bind more intensely and also in patches (Ogden et al., 2001; Vandivier et al., 2002). What could be considered relative inefficiency at the individual ligand/receptor interaction might then avoid casual induction of the removal sequence when transient membrane alterations occur on viable cells.

"Don't Eat Me" Signals

A second concept is that of direct inhibition of removal processes by stimuli from viable and not apoptotic cells. When CD31 (platelet endothelial cell adhesion molecule, or PECAM) is functionally present on both target cell and phagocyte, homologous interactions can lead to active liberation of the viable (potential) target cell from the phagocyte surface (Brown et al., 2002). Apoptotic target cells do not respond to the CD31 stimulation, are not released, and therefore may be ingested. In another system, many cells express CD47 on their surface. This can interact with inhibitory receptors such as signal regulatory protein SIRPα on the phagocytes and directly block the signal pathways that initiate uptake, i.e., providing "don't eat me" signals (Fig. 2) (Gardai et al., 2005; see also Elward and Gasque, 2003, and J. Savill [personal communication] for use of this term). As predicted from this model, absence of CD47 can result in uptake of viable cells. Apoptotic cells appear to either lose their surface CD47 or to redistribute it into patches that are distinct from the patches of pro-ingestion ligands such as PS and calreticulin. This may reduce the inhibitory CD47-SIRPα signaling at local sites on the phagocyte membrane and thereby permit ingestion of the apoptotic cell. These are but two mechanisms for prevention of viable cell removal. The pentraxin PTX3 may be another—in this case acting as an inhibitory bridge molecule (see above). We fully anticipate many additional "don't eat me" signal systems, i.e., a probable plethora of redundant prevention processes to match the abundance of ligands, receptors, and signaling pathways seen for the actual removal of the cells when they do become apoptotic. In fact, we have even wondered whether cell removal is a default process in multicellular organisms that has to be actively kept at bay to maintain viable tissues and cells (Gardai et al., 2005).

Conclusions

The general conclusion from this discussion of receptors and bridge molecules is that recognition of apoptotic cells is a highly redundant process. This probably explains the usual experimental results wherein blockade of any one of these leads to only partial prevention of clearance, and the related observation that the knockouts are seldom lethal. A further implication is that there are almost certainly more recognition structures out there that have not yet been described. The next stages of investigation will undoubtedly start sorting out which "receptors" act together, which transduce signals for uptake, and how the individual molecules participate in uptake of different target cells undergoing different forms of PCD and upon which types of phagocyte.

UPTAKE MECHANISMS

In this section we briefly discuss the uptake mechanisms for apoptotic cells and outline some of the signal pathways that have been ascribed to the process.

Uptake

It has been suggested that ingestion of apoptotic cells involves substantial ruffling of the phagocyte membrane, an apparent palpation of the target cell, and an engulfment that resembles macropinocytosis (Hoffmann et al., 2005; Ogden et al., 2001). Mechanisms and signaling pathways can be distinguished from uptake via Fc or C3 receptors (see Caron and Hall, 1998), and many in vitro studies of apoptotic cell uptake have directly contrasted phagocytosis of immune-opsonized cells. In particular, apoptotic cells do not seem to be ingested through the zipper-like process originally described by Griffin and Silverstein (Griffin et al., 1975). This form of uptake involves sequential engagement of receptors on the phagocyte membrane and results, by definition, in close apposition of the phagosome membrane with the particle or cell being ingested. Little or no concurrent uptake of extracellular fluid accompanies the process. By contrast, uptake of apoptotic cells appears to result in spacious phagosomes and, like macropinocytosis, significant concurrent uptake of fluid (Hoffmann et al., 2005; Ogden et al., 2001) (Fig. 1 and Color Plate 6). It is suggested to involve extension of large membrane ruffles that can surround the apoptotic cell and, by implication, only limited points of phagosome contact with the target. This likely reflects the patching of ligands on the

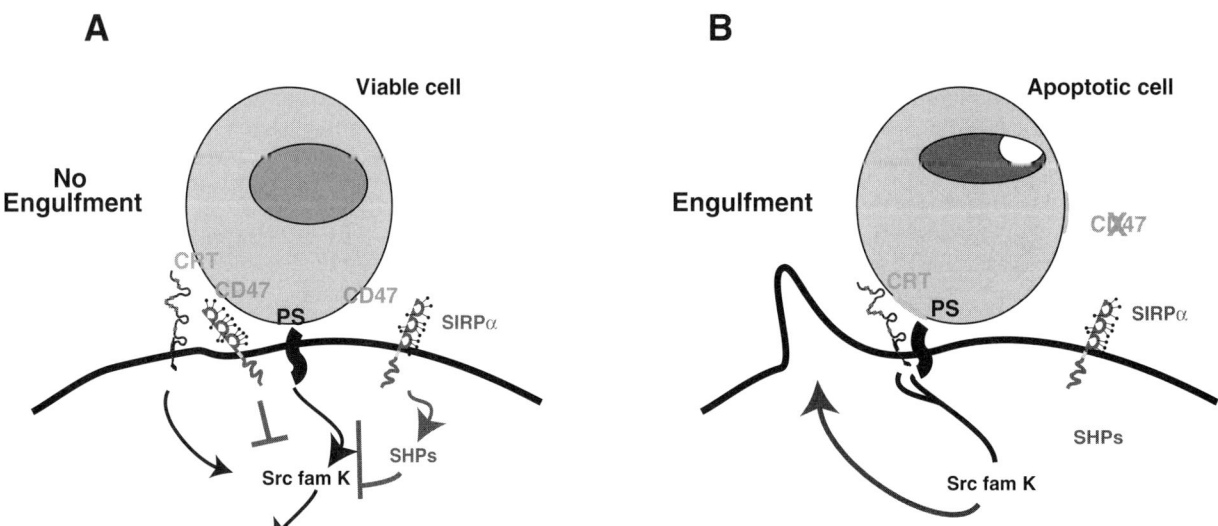

FIGURE 2 Proposed mechanisms by which CD47 serves as a "don't eat me" signal for viable cells, and which are inactivated when the cells become apoptotic. Ligation of SIRPα by CD47 on the viable cell activates inhibitory tyrosine phosphatases (SHPs) that block uptake signal pathways. Downregulation and/or redistribution of CD47 on apoptotic cells releases this inhibition and allows the uptake signaling to proceed. (From Gardai et al., 2006.)

apoptotic cell surface that was noted above. Of note, the process can lead to ingestion of intact apoptotic cells of more than 20 μm in diameter, much larger than usually attributed to macropinocytosis (Conner and Schmid, 2003). On the other hand, Krysko has provided evidence of apoptotic cell ingestion by mechanisms that do appear to involve zipper-like processes and tight phagosomes (Krysko et al., 2006b), whereas necrotic cells in these studies followed a pathway more akin to macropinocytosis. It may be that the mechanisms employed vary among different phagocytes, although the membrane ruffling and spacious phagocyte type has been observed in many different macrophages, DCs, and nonprofessional phagocytes both in vitro and in vivo. More likely it varies with the recognition and signaling molecules employed, which may in turn depend on milieu and the dying cell itself. Clearly, apoptotic cells opsonized with IgG antibody or C3bi will have a tendency to be ingested by pathways and mechanisms initiated by these stimuli. However, other ligands, receptors, and signal pathways will also be present, leaving the question of which competing process is dominant in any given system.

Another manifestation of an activated macropinocytosis type of uptake to explain ingestion of apoptotic cells is that of bystander uptake. The system is exemplified in studies by Galan and colleagues (Galan, 2001) with penetration of *Salmonella* sp. into epithelial cells. Attachment of the bacteria to the cell membrane is normally accompanied by injection via type III secretory mechanisms of virulence factors that lead to activation of low-molecular-weight GTPases, membrane ruffling, and internalization of the attached bacteria. Organisms lacking the virulence factors still bound to the cells but were not internalized. However, internalization in this system could be achieved by separate stimulation of the epithelial cells to undergo macropinocytosis, leading to inadvertent uptake of any particles (bacteria) that were attached. Extending this to the apoptotic cell uptake scenario, attachment of nonapoptotic cells (for example, erythrocytes) to potential phagocytes normally does not result in engulfment. However, uptake of such tethered cells could be initiated by apoptotic cell uptake stimuli added in soluble form to the phagocytes, i.e., a bystander uptake process following stimulation at sites different on the phagocyte membrane from the sites of attachment of the target cells (Hoffmann et al., 2001). These same stimuli also induce membrane ruffling and, where examined, uptake of extracellular fluid (Gardai et al., 2005; Ogden et al., 2001). Bystander uptake may also explain the studies by Finnemann and Silverstein (2001) showing that stimulation with potentially activating anti-CD36 was able to induce uptake of previously tethered retinal outer segments. The observations raise the more general question of external factors enhancing uptake of apoptotic cells that themselves may be able to stimulate ingestion only weakly at their contact points when tethered to the phagocyte. Growth factors such as macrophage colony-stimulating factor for macrophages are good inducers of membrane ruffling and could play such an auxiliary role.

In directly stimulating the different phagocyte receptors, one may reasonably expect a significant molecular complex to develop at the site of apoptotic cell-phagocyte contact. Different ligands, bridge molecules, and receptors may all become associated at these sites, not to mention accumulation of intracellular adaptors and signaling proteins. These would involve structures whose primary role is to tether the apoptotic cell onto the phagocyte surface as well as ligand-receptor complexes that lead to transduction of signals to the inside of the phagocyte. This concept is the basis for the two-step tether-and-tickle mechanism for the ingestion of apoptotic cells (Hoffmann et al., 2001) (Fig. 1). The many receptors that appear to initiate the uptake and the apparent redundancy of ligands on the apoptotic cell suggest that under most circumstances, efferocytosis is not driven by a single ligand, receptor, and signal pathway but by combinations of these acting together. A key issue for the future will be to determine how they all work together. As noted below, different balances between such pathways may contribute to some of the other consequences of apoptotic cell recognition, even if they all act together to mediate the uptake.

Signal Pathways

Much work has been carried out on the signal pathways involved in uptake of apoptotic cells. The subject has been extensively discussed in a recent review (Wu et al., 2006) and will be addressed in only a brief fashion here. Much of the early information on the signal pathways came from genetic studies in *C. elegans*. A number of gene knockdowns were shown to be accompanied by defective apoptotic cell clearance that fell into two complementary pathways (Fig. 3). Mammalian homologues for most of these were soon identified. Interestingly, most of these nematode genes turned out to be for signaling molecules; only CED1 seemed to code for an apoptotic cell receptor (see discussion above). This may reflect the redundancy in receptors but relative conservation of signaling. In mammals, uptake seems to depend on the low-molecular-weight GTPase Rac (Santy et al., 2005), presumably Rac1, although the relative contribution of Rac2 has not been clearly delineated. The studies in *C. elegans* had Rac (CED10) in the CED 5,2,12 (DOCK180, CrkII, ELMO in mammals) pathway that was suggested to be independent of a CED1,7,6 path. A later report suggested that CED10 might also be involved in the latter (Kinchen et al., 2005), i.e., representing a merge point that presumably led to coordination of the cytoskeletal elements needed for uptake. This would certainly be in keeping with the probable common participation of Rac in the mammalian systems. However, the association of CED10 with the CED1,7,6 path was fairly weak, and more studies in the nematode are probably needed to sort this out. Most of the PS-dependent systems seem to act via the DOCK180 path, and it is suggested that calreticulin and the collectins, acting through LRP, initiate the mammalian equivalent of the CED1,7,6 sequence (see above). However, while there is some progress (see Wu et al., 2006), the precise connections between the candidate receptors and the downstream pathways need significantly more investigation. Bypass of proximal signals may occur in spontaneously macropinocytosing DCs, which seem to exhibit a high level of intrinsically activated Rac. Also it is unreasonable to assume that these pathways are really linear, and complex interactions between them, as well as from other ongoing processes, are to be expected. We are also making a broad assumption here that the signaling pathways involved in apoptotic cell uptake are similar among different phagocytes. Their evolutionary conservation would support this contention, and where investigated, cell-to-cell conservation is also seen, even to nonprofessional phagocytes such as fibroblasts or epithelial cells.

As noted above, the different receptors and pathways likely work together, e.g., Mer and α_v integrins (Wu et al.,

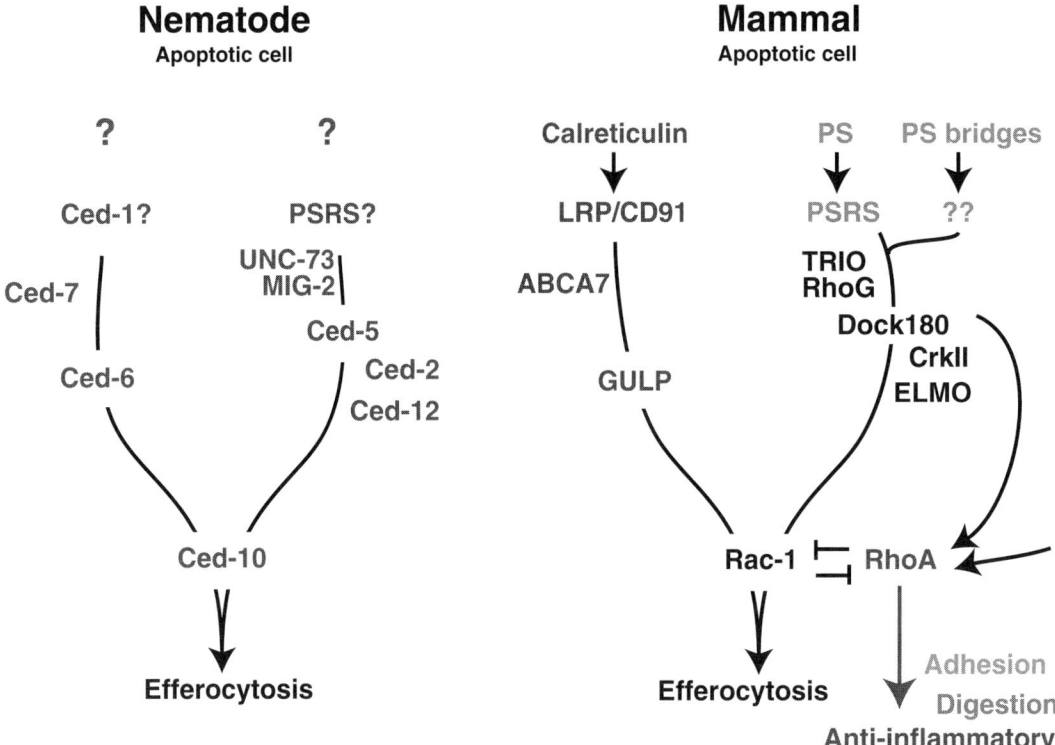

FIGURE 3 Synopsis of possible signaling pathways associated with engulfment of apoptotic cells in *C. elegans* and in mammals. PSRS, phosphatidylserine recognition structure, i.e., as yet unidentified phosphatidylserine receptors. (From Gardai et al., 2006.)

2005), and are almost certainly highly regulated. Another feature that has not received as much attention is the time dependence of the signaling processes. For example, while interaction of intact apoptotic cells with macrophages consistently leads to an early activation of Rac, within a short period of time (20 min or so) increased levels of activated Rho can be detected. The relevance here is that a number of studies have shown that Rho is inhibitory for apoptotic cell uptake (Hoffmann et al., 2001; Leverrier and Ridley, 2001; Tosello-Trampont et al., 2003), incidentally showing another difference from FcR-mediated phagocytosis. On the other hand, Rho activation may participate in later stages of the apoptotic phagosome maturation (Erwig et al., 2006) as well as in some of the other processes associated with apoptotic cell-phagocyte interaction, including the production of anti-inflammatory mediators such as TGF-β (see below). In fact, very little attention has so far been placed on the disposition of the ingested apoptotic cell within its phagosome and the likely unique features associated with this stage of the clearance process (see Erwig et al., 2006; Shiratsuchi et al., 2004).

THE PHAGOCYTES INVOLVED IN APOPTOTIC CELL REMOVAL

The proportional contribution of the mononuclear phagocyte system (MPS), such as macrophages and DCs, to the removal of apoptotic cells in situ, compared with the removal by neighboring tissue cells, is not at all clear. As noted above, organisms that do not have such clearly defined professional phagocytes (e.g., *C. elegans*) still remove apoptotic cells very efficiently. When PU-1$^{-/-}$ mice, lacking most of their MPS, were examined, removal of the interdigital webs during formation of the feet was indeed delayed, but only by 12 to 24 h, and the structures eventually developed normally (Wood et al., 2000). Most studies of apoptotic cell clearance to date have focused on MPS cells. However, this may give a somewhat misleading emphasis of their overall importance, driven in part by the interests of the investigators and the ease with which MPS cells can be studied. Nevertheless, the focus of this chapter is on MPS cells, and the contribution of fixed tissue cells—many of which are quite capable of ingesting apoptotic cells in vitro and are increasingly known to do so in vivo—will be left to other fora for consideration.

Macrophages

Studies in vitro and in vivo have examined many different sources and types of macrophages for their ability to ingest apoptotic cells, and most have been shown to carry out this function, though with differing degrees of efficiency (reviewed in Xu et al., 2006a). As usual, an important point here is to be cautious about differences between macrophage abilities manifested in vitro being able to extend to differences in vivo. A potential example is the contribution of the adherence state of macrophages or macrophage cell lines in vitro to their ability to ingest apoptotic cells. Highly adherent cells (e.g., the RAW264 cell line) ingest poorly, unless grown on surfaces to which they do not adhere as strongly (unpublished data). Relating phenomena such as this back to the environmental exposures of macrophages in vivo is extremely difficult. Nevertheless, even with such caveats in mind, macrophage "phenotypes" do seem to vary in their ability to remove apoptotic cells, at least in vitro. Whether one adheres to the concept of defined, stable macrophage phenotypes or suggests that mononuclear phagocytes are highly adaptable cells that respond to their environment of the moment, the experimental observations are the same. Certain stimuli and maturation signals lead to an enhanced, and others to a reduced ability of macrophages to ingest apoptotic cells. As one example of the variation, resident macrophages from the murine or human lung are inefficient at apoptotic cell uptake in vitro, whereas those from the mouse peritoneum are much more active. As a consequence, macrophages from inflammatory lesions in lung are more active in this regard than the pre-existing resident cells, whereas those in the peritoneum may be less effective than their resident counterparts. Despite these differences in vitro, in real life, apoptotic cells are effectively cleared from both naïve and inflamed lungs and peritonea.

Alternatively activated (M2) macrophages (see chapter 20, this volume) have been suggested to be more efficient at the uptake of apoptotic cells when compared with classically activated macrophages (Xu et al., 2006b). In this study, human monocytes were matured toward the M1 form by granulocyte-macrophage colony-stimulating factor and toward M2 cells by macrophage colony-stimulating factor. The latter were more efficient at ingesting apoptotic cells. While it can be questioned how closely these two cell types represent the classical or alternatively activated macrophages described by others, for example in the mouse, the study makes some relevant points about macrophage heterogeneity for this apoptotic cell uptake function. The M2 macrophages were suggested to exhibit higher levels of spontaneous macropinocytosis—a known effect of growth factors. They also showed enhanced CD14-induced tethering of the apoptotic cells—in keeping with the tether-and-tickle mechanism. Finally, they secreted increased levels of interleukin-10 (IL-10), a cytokine that is reported to enhance apoptotic cell uptake (Ogden et al., 2005). These observations provide potential mechanisms for heterogeneity in macrophage ability to ingest apoptotic cells. In other circumstances, e.g., the lung macrophage, inhibitory processes are probably involved in their inefficiency. Finally, uptake of apoptotic cells itself (Erwig et al., 1999) induces changes in the macrophage (and DCs, see below) that might be considered as changing its phenotype.

From these comments one can raise a number of key points in the consideration of apoptotic cell removal efficiency. (i) The mode of maturation of the macrophages is clearly important—processes that can be redefined as the effects of different growth and maturation stimuli. (ii) Differences in apoptotic cell ligands and bridge molecules will also make a huge difference. Thus, macrophage efficiency may often be determined by the character of the target. In the study discussed above, the two macrophage types handled late apoptotic cells differently from early apoptotic cells (Xu et al., 2006b). We are approaching the point where definition of the mechanisms may be necessary in consideration of differences in efficiency. (iii) Few in vitro experiments are ever carried out in 100% plasma, let alone in interstitial fluid or exudates. This obvious point may be particularly important in these systems since the availability of different ligands (resulting, for example, from complement fixation) or different bridge molecules in the in vivo situation are likely to substantially change the efficacy of the uptake. (iv) We would suggest that the physical environment in which the macrophage contacts the apoptotic cell may also be critical, especially the difference between two- and three-dimensional systems and the intercurrent signals to the macrophage coming from its environment, matrix, and intercellular contacts. The other mechanistic point to make here is to reiterate the comments about the intracellular balance between Rho and Rac. Macrophages that ingest apoptotic cells better tend to have, and develop, higher levels of active Rac and as a consequence show more active membrane ruffling. Those that ingest poorly tend to have higher levels of active Rho and less ruffling. While other such balanced intracellular signaling pathways may contribute to the heterogeneity, it is noteworthy that immature DCs are very active at ingesting apoptotic cells and also exhibit high levels of active Rac and spontaneous membrane ruffling.

DCs

It is not our intention here to probe in detail the complexity of different types of DCs. In fact, the subject of apoptotic cell uptake by DCs, and its consequences for antigen presentation and immune stimulation or suppression, is large enough to warrant a completely separate chapter (see reviews: Liu et al., 2006; Skoberne et al., 2005; Viorritto et al., 2007; Xu et al., 2006a, and also chapter 18 of this volume). Certainly immature DCs are capable of ingesting apoptotic cells. In fact, their propensity to undergo constitutive macropinocytosis may particularly promote this function if suitable receptors are present on their surface to tether the apoptotic cells. It may be that specific stimulating receptors and pathways are not always necessary for uptake and that mere binding of the apoptotic cells can lead to uptake by the ongoing macropinocytotic processes. This would put more emphasis on tethering ligands for providing the ability of the apoptotic cell to be recognized and then ingested by DCs. However, additional stimulation may be important, perhaps to enhance uptake by leading to the development of large enough ruffles to achieve ingestion of large, intact apoptotic cells. Thus apoptotic cells have been shown to utilize $\alpha_v\beta_5$ integrins and CD36 (Akakura et al., 2004; Albert et al., 1998; Dalgaard et al., 2005; but see Schulz et al., 2002), probably following opsonization with bridge molecules such as MFG-E8 (Akakura et al., 2004). A comparison between integrins suggested that α_v integrins were more efficient for ingestion, but β_2 integrins were more effective at mediating tolerogenic DCs (Skoberne et al., 2006). PS does block DC maturation, suggesting that PS-recognizing receptors or bridge molecules are available (Chen et al., 2004), although the study did not demonstrate a role for PS in the uptake of apoptotic cells. A subset of DCs has also been shown to express LRP (Hart et al., 2004). However, in general, the ligands, bridge molecules, receptors, and signaling pathways have not been studied for DCs to the degree that they have for macrophages.

As with macrophages, there have been studies that suggest that particular DC phenotypes are more or less efficient at ingesting apoptotic cells (see reviews by Viorritto et al., 2007, and Xu et al., 2006a, and also Ferguson and Kazama, 2005; Huang et al., 2000; Iyoda et al., 2002; Schulz et al., 2002). Thus CD8+ murine DCs may be particularly effective in this regard. However, the comments made above, about differing macrophage uptake efficiencies, extend equally to DCs, and once again little is known about the specific mechanisms underlying these differences. In particular, it would seem important to address the efficiency with which apoptotic cells become tethered to different DC types as well as to explore the requirement, if any, for specific activating receptors. As the DCs mature, their altered Rho/Rac balance and decreased ability to undergo membrane ruffling and macropinocytosis would also be expected to have a profound effect on their ability to ingest and process apoptotic cells.

STIMULI THAT MODULATE APOPTOTIC CELL INGESTION

As noted, in addition to the receptors and bridge molecules that promote uptake of apoptotic cells, the phagocytic capability (for apoptotic cell ingestion) of the phagocytes themselves can be altered by external environmental factors and stimuli.

Enhancing Stimuli

A variety of molecules have been reported to enhance the ability of macrophages to ingest apoptotic cells without interacting with the apoptotic cells themselves. These include ligands or antibodies against CD44 (Hart et al., 1997; Vivers et al., 2004), corticosteroids (Heasman et al., 2003; Liu et al., 1999), annexin I (Maderna et al., 2002; Scannell et al., 2007), lipoxins (Mitchell et al., 2002), the nucleotide UDP (Koizumi et al., 2007), interferon gamma (unpublished data), and as noted, cytokines associated with the alternative activated macrophage. The mechanisms underlying these effects have not always been delineated and, in some cases, may be related to each other, for example, the possible release of annexin I fragments in response to corticosteroids (Maderna et al., 2002; Scannell et al., 2007). However, in a number of these cases, these mechanisms seem to reflect the balance between activated Rho family GTPases (see above), i.e., leading to increased Rac and cytoskeletal assembly response to the apoptotic cells (e.g., Giles et al., 2001; Maderna et al., 2002; Reville et al., 2006).

Suppressive Stimuli

On the other side, a number of stimuli and environmental factors may suppress uptake of apoptotic cells—again probably acting in a general fashion on the phagocyte and not directly on the target. These include TNF-α and oxidants (McPhillips et al., 2007) and, as noted, adhesion. There are likely many more. However, these three can all be reversed by blockade of Rho or the downstream Rho kinase and, therefore, may also reflect altered balance at the level of the Rho family GTPases.

It also worth questioning the possibility that in vivo some of the ligands on apoptotic cells may under some circumstances be blocked and thereby lead to reduced uptake of altered balance between the different effects of different ligands. One possibility for blockade is the annexins, including annexin V, that are found within cells. If released following membrane permeability—late apoptosis or necrosis—these could cover up exposed PS and limit the roles for this both in uptake and in suppression of inflammation (see below), perhaps leading to uptake pathways skewed more toward the potentially proinflammatory and proimmunogenic LRP effects. Certainly exogenous administration of annexin V or RGE-containing MFG-E8 has been used to similarly block the effects of PS exposure (Asano et al., 2004; Bondanza et al., 2004; Stach et al., 2000; Yoshida et al., 2005).

ATTRACTION OF PHAGOCYTES TO APOPTOTIC CELLS

An important question for understanding the role of MPS cells in apoptotic cell removal is how they find the apoptotic cells in the first place. What are the "come and get me" signals? A study by Lauber et al. (2003) addressed this question by examination of molecules released from apoptotic cells that were chemotactic for THP-1 cells in vitro. They showed generation of lyso-PC in response to caspase activation of an intracellular, iPLA$_2$ (which cleaves the sn-2 acyl bond of PC). The study is intriguing and provides a candidate for attraction of monocytes or macrophages to sites of apoptotic cell generation. However, at the moment there is little evidence for its role in vivo, and the apoptotic cells used in the study were probably in a state of late apoptosis or necrosis, begging the question of such signals generated during early apoptosis. It should be noted that PCD resulting from autophagic processes also exhibited increased levels of lyso-PC (Qu et al., 2007), and indeed, autoantibodies that are directed toward apoptotic cells in normal and autoimmune circumstances may also recognize this molecule (see above). An alternative possibility to direct release of "come and get me" molecules is a secondary generation of such signals from interaction of the apoptotic cells with near-neighbor tissue cells, even if the latter do not themselves indulge in their phagocytosis. An example here is involution of the corpus luteum, which involves attraction of macrophages to the sites of apoptosis, apparently in response to MCP-1 (monocyte chemoattractant protein-1, CCL2) generated by nearby viable cells of the corpus luteum (Nagaosa et al., 2003). One might wonder here if chemokines such as MCP-1 might also in some cases be directly generated and released during the process of apoptosis. An intriguing recent study by Koizumi et al. (2007) suggests that different nucleotides released from dying cells in the brain may act as attractants for microglia (ATP) and also to enhance microglial phagocytic activity (UDP).

Once again, it may be suggested that these observations represent just the iceberg's tip and that there are likely many other candidates for MPS cell attraction to cells undergoing PCD still to be identified. Other important questions also remain. For example, in the standard model of a usually noninflammatory removal of local apoptotic cells, one would anticipate an attraction of tissue macrophages to the apoptotic cells but not a substantial attraction of monocytes, which would normally be considered a manifestation of inflammation. A possible selective attraction of tissue macrophages adds additional constraints on candidate molecules. On the other hand, if blood monocytes are attracted, their interaction with the apoptotic cells would presumably direct a noninflammatory state (see below). Early studies of dying cells (apoptosis) in the brain did suggest mononuclear phagocyte accumulation with eventual maturation of the cells toward microglial phenotypes (Perry et al., 1985; see also Koizumi et al., 2007).

CONSEQUENCES OF APOPTOTIC CELL INTERACTION WITH PHAGOCYTES FOR INFLAMMATION AND IMMUNITY

The first and probably most important consequence is the efficient removal of the dying cell and/or cell debris. In addition, as noted earlier and as discussed in a number of reviews (e.g., Aderem and Underhill, 1999; Gaipl et al.,

2007; Gaipl et al., 2006; Gardai et al., 2006; Krysko et al., 2006a; Munoz et al., 2007a, 2007b; Roos et al., 2004; Savill et al., 2002, 2003; Serhan and Savill, 2005; Skoberne et al., 2005), recognition of apoptotic cells has been shown to be both noninflammatory and actively anti-inflammatory. In oversimplified mechanistic terms, this means no direct induction of proinflammatory mediators and the active generation of anti-inflammatory molecules then acting in autocrine or paracrine mode. The process is also anti-immunogenic. The logical concept is that normal cell clearance occurs in situ and is quiet and nonreactive. Given the huge ongoing removal of cells, this makes sense. The original observation was that apoptotic cell uptake did not induce proinflammatory mediators such as thromboxane (Meagher et al., 1992; Stern et al., 1996). Subsequently, macrophages that were ingesting apoptotic cells were shown not to respond to external stimuli such as TLR engagement with the normal production of proinflammatory mediators (Fadok et al., 1998; Voll et al., 1997)—an observation later extended in vivo (e.g., Huynh et al., 2002). Some of the effect is mediated by the production of anti-inflammatory mediators such as TGF-β (Fadok et al., 1998; Huynh et al., 2002), IL-10 (Voll et al., 1997), and anti-inflammatory eicosanoids (Fadok et al., 1998; Freire-de-Lima et al., 2006). Molecules such as these may have effects beyond the suppression of inflammation, extending to blockade of adaptive immunity or induction of fibrosis. In addition, more direct inhibitory effects of apoptotic cell interaction on NF-κB activation have been reported (Cvetanovic et al., 2006), perhaps mediated through Mer (Sen et al., 2007) and not requiring TGF-β. An alternative way in which apoptotic cell clearance may mitigate inflammation is by the removal of membranes (or cells) expressing proinflammatory oxidized lipids (Dahl et al., 2007).

The data to date support the concept that the apoptotic cell ligand most involved in the anti-inflammatory effects is the PS, presumably acting through one or more of its bridge molecules and receptors. Which of these is the major source of the anti-inflammatory signals still remains to be determined. On the other hand, C3bi has also been reported not to initiate inflammatory consequences (Edwards et al., 2001; Marth and Kelsall, 1997; Morelli et al., 2003) and to suppress maturation of DCs (Skoberne et al., 2006). Collectin tails and calreticulin/LRP stimulation are more likely to be proinflammatory (Gardai et al., 2003, 2005). Much work still needs to be carried out on these processes, but a current conclusion would be that there is a constant balance between pro- and anti-inflammatory stimulation of the responding phagocyte depending on relative amounts and presentation of the different ligands and bridge molecules acting in turn on the different receptors. This would explain the apparently discrepant observations that apoptotic cells can be proinflammatory (Chang et al., 2004; Kawagishi et al., 2001; Uchimura et al., 1997) and the reports of varying degrees of anti- or pro-inflammatory effects of late-apoptotic and necrotic cells. Similarly, combining lipopolysaccharide with apoptotic cells can lead to early stimulation of proinflammatory mediators with later suppression (Lucas et al., 2003), the latter perhaps due to the delayed production of suppressors such as TGF-β. It has long been emphasized that generation of anti-inflammatory mediators does not require actual uptake of apoptotic cells; i.e., engagement of the appropriate bridge molecules/receptors is sufficient, even if the downstream phagocytic processes are inactive (as, for example, in the RAW264 murine macrophage cell line, which is remarkably inefficient at apoptotic cell ingestion but fully responsive for anti-inflammatory mediator production). One must also note that each pro- or anti-inflammatory mediator has its own induction signaling pathway and is often regulated at many different stages of production, and it is somewhat dangerous to try and pigeonhole responses at this level as generally pro- or anti-inflammatory. As yet, cells dying by other forms of PCD have not been examined in as much detail for these secondary effects following their recognition and removal, and we should also note that different mononuclear phagocytes are differently preprogrammed with regard to their ability to synthesize and secrete pro- or anti-inflammatory mediators.

There are clearly some potential parallels between the anti-inflammatory effects generated by interaction between apoptotic cells and macrophages and the anti-immunogenic effects of apoptotic cells mentioned above for DCs. It seems reasonable to suppose that PS on the apoptotic cell may contribute to the suppression of the DC (Chen et al., 2004). On the other hand, Stuart et al. (2002) at the single-cell level showed that only the DCs that had actually taken up apoptotic cells exhibited reduction of maturation markers, suggesting that release of soluble factors such as TGF-β or IL-10 was not involved in their in vitro system. Nevertheless, DCs are highly responsive to TGF-β, and in a real site of immunization in vivo generation of TGF-β from apoptotic cell interaction with macrophages (or DCs, if it occurs under other circumstances) would be expected to have an effect. Furthermore, apoptotic cell suppression of inflammation at a local site could substantially limit the available inflammatory cytokines that normally drive maturation of any DCs that were present. The point again is that, while it is important to tease out mechanisms and individual pathways in vitro, in the animal they are all acting at once in varying degrees of balance with each other.

Ingestion of apoptotic cells by DCs has been shown to lead to productive antigen presentation and adaptive immunity, but also to suppression and immune tolerance. Of particular interest is the propensity of apoptotic-cell-associated antigens to be cross-presented on MHC1. Much emphasis is currently being placed on utilizing some of these effects in enhancing immune responses to cancer, on the one hand (Obeid et al., 2007), or suppressing immune responses, for example in transplantation, on the other (Wang et al., 2006). One issue is whether, when apoptotic cells are ingested, they, without additional stimuli, induce maturation of the DCs. In general, this is thought not to happen with apoptotic cells, possibly, we suggest, because of inhibitory signals derived from surface PS (Chen et al., 2004), perhaps via Mer signaling (Sen et al., 2007) or oxidized phospholipids (Bluml et al., 2005) from C3bi (Verbovetski et al., 2002) or CD36 (Urban et al., 2001). DC suppression induced by malaria-infected erythrocytes (Urban et al., 1999) may also be a consequence of the PS that is known to be exposed on such erythrocytes. On the other hand, C1q has been shown to induce DC maturation (Csomor et al., 2007) and might play a similar role when attached to apoptotic cells. Thus, as with macrophages, we suggest that the final outcome may reflect a balance between different stimuli supplied from the apoptotic cell and its bridge molecules. It is furthermore suggested that DCs with ingested apoptotic cells and associated (auto)antigens can act in a suppressive manner. At issue here might be whether antigens can be processed and presented on MHC without DC maturation and expression of costimulatory molecules, and whether such persistently immature apoptotic-cell-containing DCs can migrate to the lymph nodes. External activation of such DCs through TLRs or cytokines would presumably lead to appropriate migration and antigen presentation with the result of active immunity. On the other hand, uptake of necrotic cells by DCs can lead to their maturation and antigen presentation, in part due to release from the necrotic cells of stimulating materials such as heat shock proteins (Basu et al., 2000; Feng et al., 2002). Presumably, these molecules act through external receptors on the DCs; otherwise they should also act when liberated from the apoptotic cells within the phagosome. Other structures on the surface of the necrotic cell may also stimulate DC maturation. An additional question may be when, during the progression from early to late apoptosis and then to secondary necrosis, does the cell become able to directly stimulate DC maturation. While these concepts are almost certainly oversimplistic, the questions they raise are critical in the context of suppression, or induction, of autoimmunity (see, for example, Gaipl et al., 2006, 2007; Liu et al., 2006; Pittoni and Valesini, 2002; Viorritto et al., 2007).

One cannot help but speculate that parasitic organisms might have learned to exploit these anti-inflammatory and anti-immunogenic effects in their constant attempts to adapt to the host. Thus, recognition of apoptotic cells itself changes the phenotype of the macrophage, perhaps differently depending on the balance of different stimuli. Autocrine/paracrine effects of the TGF-β that are generally produced in response to apoptotic cells lead to a TGF-β macrophage. An example of this effect is the suppression of inducible nitric oxide (NO) synthase (iNOS) induction (Freire-de-Lima et al., 2006), and, thereby, it would be expected to alter some of the potential mechanisms for dealing with infection.

EXPLOITATION OF APOPTOTIC CELL RECOGNITION BY PARASITES

Here, we present a few examples to illustrate the potential for, and to raise questions about, other possible interactions between parasites and apoptotic cell recognition, ingestion, and consequences. In each of these cases, the observations are intriguing but require much more investigation to clearly demonstrate the extent and consequences of the exploitation and, in turn, to generate ways that we may in turn exploit the exploitation to help control the parasite.

Leishmania
The intracellular parasite *Leishmania* sp. appears to use varied means to gain entry into its preferred location in the macrophage phagosome, including uptake via Fc receptors. However, interesting reports (de Freitas Balanco et al., 2001; de Souza et al., 2003; Wanderley et al., 2006) suggest that *Leishmania* parasites express PS on their surface that may also contribute to uptake, and by inference would likely also alter the macrophage phenotype toward a less proinflammatory state, perhaps contributing to a more advantageous intracellular habitat. A related, but more complex, Trojan horse mechanism has been described by van Zandbergen (Laskay et al., 2003; van Zandbergen et al., 2004), in which the *Leishmania* cells are first taken up into neutrophils and then secondarily into macrophages when the neutrophils (still containing the living *Leishmania*) become apoptotic. The secondary uptake of the apoptotic neutrophils was, as usual, associated with the production of TGF-β with its potentially important anti-inflammatory and anti-immunogenic consequences.

Trypanosoma cruzi
Trypanosoma cruzi is related to *Leishmania* and is capable of inducing substantial apoptosis in the host. Evidence has been presented to support the notion that the local response to the apoptosis (Freire-de-Lima et al., 2000) is necessary for optimal infection, in part, due to production of prostaglandin E_2 and TGF-β. Suppression of NO production by TGF-β and IL-10 produced in response to the apoptotic cells may serve to minimize killing of the parasites and allow the infection to gain a foothold (Silva et al., 2003). Apoptosis of the infected macrophage has also been suggested to allow persistence and spread of the parasite without triggering enhanced inflammation (de Souza et al., 2003).

Listeria
Listeria is a facultative intracellular pathogen of macrophages. It is internalized into phagosomes from which it then escapes to enter the cytoplasm. It can subsequently pass from macrophage to macrophage by presentation in pseudopods from donor to recipient. Two possible connections may be noted between apoptotic cell effects and *Listeria* infection. Initial infection is often accompanied by lymphocyte apoptosis, which appears to play an enhancing role in establishing the infection by downregulation of innate immune responses (Carrero et al., 2006; Carrero and Unanue, 2006). One mechanism for this may be the generation of the inhibitory cytokine IL-10 in response to the apoptotic cell recognition. As noted, the cytokine could also enhance the apoptotic cell uptake, thereby providing a positive-feedback loop for the process. From a completely different direction, it is also intriguing to wonder whether the *Listeria*-containing pseudopods expose apoptotic-cell-type ligands on their surface that allow removal of such structures by the recipient macrophages, thereby promoting cell-cell transfer. For example, SR-A has been shown to participate in DC "nibbling" of membrane from live cells (Harshyne et al., 2003).

Salmonella
As noted above, *Salmonella* sp. penetrates epithelial cells by mechanisms that are similar to apoptotic cell uptake processes, even if the proximal receptors and signaling pathways are not the same. *Salmonella enterica* serovar Typhimurium can function as a facultative intracellular pathogen in macrophage phagosomes. Induction of apoptosis of infected macrophages by the bacteria has been reported to result in subsequent uptake of the apoptotic macrophages by DCs that were then able to present *Salmonella* antigens to the immune system (Yrlid and Wick, 2000).

Mycobacterium
In another example, *Mycobacterium tuberculosis* induces neutrophil apoptosis, and apoptotic neutrophil uptake into infected macrophages may provide better killing of intracellular organisms because of secondary actions by ingested neutrophil granules (Tan et al., 2006). However, this could possibly be offset by potentially decreased levels of iNOS and NO after interaction with apoptotic cells. Winau et al. (2005) have suggested that *M. tuberculosis*-induced apoptosis in macrophages can lead to uptake into DCs with subsequent cross-presentation of bacterial antigens to po-

tentially protective CD8 T cells. This theme may be repeated in many different types of infection.

Viruses

A number of examples may be derived from the virus world. For example, and in a similar fashion to that mentioned for *T. cruzi*, human immunodeficiency virus (HIV) replication has been reported to be enhanced by interaction of apoptotic cells with macrophages, apparently related to the production of molecules such as prostaglandin E_2 (Lima et al., 2006). There is also great likelihood in the virus world for DCs to pick up and present viral antigen from virus-induced apoptotic cells (e.g., Fleeton et al., 2004; Subklewe et al., 2001; Van Zanten et al., 2002). In another direction, removal of virus-infected apoptotic cells may also serve as an important mechanism for removal of the virus and minimization of the infection (Hashimoto et al., 2007; Watanabe et al., 2005). Envelope viruses derive their membrane from the host cell plasma membrane. Maintenance of membrane phospholipid asymmetry in the cell is ATP dependent. Is such asymmetry maintained on virus particles, and if so, how? Does the possible exposure of PS lead to interaction with the PS-binding bridge molecules and receptors associated with apoptotic cell recognition? At least one example may be a candidate role for exposed PS in HIV infection of macrophages (Callahan et al., 2003). An interesting extension of the viral concept are papers suggesting that functional DNA itself can be transferred by uptake of apoptotic cells (Bergsmedh et al., 2001; Holmgren et al., 1999).

This work was supported by National Institutes of Health grants GM61031, HL81151, AI058228, HL68864, and HL34303.

REFERENCES

Aderem, A., and D. M. Underhill. 1999. Mechanisms of phagocytosis in macrophages. *Annu. Rev. Immunol.* **17:**593–623.

Akakura, S., S. Singh, M. Spataro, R. Akakura, J. I. Kim, M. L. Albert, and R. B. Birge. 2004. The opsonin MFG-E8 is a ligand for the alphavbeta5 integrin and triggers DOCK180-dependent Rac1 activation for the phagocytosis of apoptotic cells. *Exp. Cell Res.* **292:**403–416.

Albert, M. L., S. F. Pearce, L. M. Francisco, B. Sauter, P. Roy, R. L. Silverstein, and N. Bhardwaj. 1998. Immature dendritic cells phagocytose apoptotic cells via alphavbeta5 and CD36, and cross-present antigens to cytotoxic T lymphocytes. *J. Exp. Med.* **188:**1359–1368.

Arur, S., U. E. Uche, K. Rezaul, M. Fong, V. Scranton, A. E. Cowan, W. Mohler, and D. K. Han. 2003. Annexin I is an endogenous ligand that mediates apoptotic cell engulfment. *Dev. Cell* **4:**587–598.

Asano, K., M. Miwa, K. Miwa, R. Hanayama, H. Nagase, S. Nagata, and M. Tanaka. 2004. Masking of phosphatidylserine inhibits apoptotic cell engulfment and induces autoantibody production in mice. *J. Exp. Med.* **200:**459–467.

Awasaki, T., R. Tatsumi, K. Takahashi, K. Arai, Y. Nakanishi, R. Ueda, and K. Ito. 2006. Essential role of the apoptotic cell engulfment genes draper and ced-6 in programmed axon pruning during Drosophila metamorphosis. *Neuron* **50:**855–867.

Balasubramanian, K., J. Chandra, and A. J. Schroit. 1997. Immune clearance of phosphatidylserine-expressing cells by phagocytes. The role of beta2-glycoprotein I in macrophage recognition. *J. Biol. Chem.* **272:**31113–31117.

Balasubramanian, K., and A. J. Schroit. 1998. Characterization of phosphatidylserine-dependent beta2-glycoprotein I macrophage interactions. Implications for apoptotic cell clearance by phagocytes. *J. Biol. Chem.* **273:**29272–29277.

Bartl, M. M., T. Luckenbach, O. Bergner, O. Ullrich, and C. Koch-Brandt. 2001. Multiple receptors mediate apoJ-dependent clearance of cellular debris into nonprofessional phagocytes. *Exp. Cell Res.* **271:**130–141.

Basu, S., R. J. Binder, R. Suto, K. M. Anderson, and P. K. Srivastava. 2000. Necrotic but not apoptotic cell death releases heat shock proteins, which deliver a partial maturation signal to dendritic cells and activate the NF-kappa B pathway. *Int. Immunol.* **12:**1539–1546.

Bayir, H., B. Fadeel, M. J. Palladino, E. Witasp, I. V. Kurnikov, Y. Y. Tyurina, V. A. Tyurin, A. A. Amoscato, J. Jiang, P. M. Kochanek, S. T. DeKosky, J. S. Greenberger, A. A. Shvedova, and V. E. Kagan. 2006. Apoptotic interactions of cytochrome c: redox flirting with anionic phospholipids within and outside of mitochondria. *Biochim. Biophys. Acta* **1757:**648–659.

Bergsmedh, A., A. Szeles, M. Henriksson, A. Bratt, M. J. Folkman, A. L. Spetz, and L. Holmgren. 2001. Horizontal transfer of oncogenes by uptake of apoptotic bodies. *Proc. Natl. Acad. Sci. USA* **98:**6407–6411.

Berwin, B., Y. Delneste, R. V. Lovingood, S. R. Post, and S. V. Pizzo. 2004. SREC-I, a type F scavenger receptor, is an endocytic receptor for calreticulin. *J. Biol. Chem.* **279:**51250–51257.

Bhatia, V. K., S. Yun, V. Leung, D. C. Grimsditch, G. M. Benson, M. B. Botto, J. J. Boyle, and D. O. Haskard. 2007. Complement C1q reduces early atherosclerosis in low-density lipoprotein receptor-deficient mice. *Am. J. Pathol.* **170:**416–426.

Bijl, M., G. Horst, J. Bijzet, H. Bootsma, P. C. Limburg, and C. G. Kallenberg. 2003. Serum amyloid P component binds to late apoptotic cells and mediates their uptake by monocyte-derived macrophages. *Arthritis Rheum.* **48:**248–254.

Bluml, S., S. Kirchberger, V. N. Bochkov, G. Kronke, K. Stuhlmeier, O. Majdic, G. J. Zlabinger, W. Knapp, B. R. Binder, J. Stockl, and N. Leitinger. 2005. Oxidized phospholipids negatively regulate dendritic cell maturation induced by TLRs and CD40. *J. Immunol.* **175:**501–508.

Bondanza, A., V. S. Zimmermann, P. Rovere-Querini, J. Turnay, I. E. Dumitriu, C. M. Stach, R. E. Voll, U. S. Gaipl, W. Bertling, E. Poschl, J. R. Kalden, A. A. Manfredi, and M. Herrmann. 2004. Inhibition of phosphatidylserine recognition heightens the immunogenicity of irradiated lymphoma cells in vivo. *J. Exp. Med.* **200:**1157–1165.

Boon, J. M., T. N. Lambert, A. L. Sisson, A. P. Davis, and B. D. Smith. 2003. Facilitated phosphatidylserine (PS) flip-flop and thrombin activation using a synthetic PS scramblase. *J. Am. Chem. Soc.* **125:**8195–8201.

Borisenko, G. G., S. L. Iverson, S. Ahlberg, V. E. Kagan, and B. Fadeel. 2004. Milk fat globule epidermal growth factor 8 (MFG-E8) binds to oxidized phosphatidylserine: implications for macrophage clearance of apoptotic cells. *Cell Death Differ.* **11:**943–945.

Bose, J., A. D. Gruber, L. Helming, S. Schiebe, I. Wegener, M. Hafner, M. Beales, F. Kontgen, and A. Lengeling. 2004. The phosphatidylserine receptor has essential functions during embryogenesis but not in apoptotic cell removal. *J. Biol.* **3:**15.

Botto, M., C. Dell'Agnola, A. E. Bygrave, E. M. Thompson, H. T. Cook, F. Petry, M. Loos, P. P. Pandolfi, and M. J. Walport. 1998. Homozygous C1q deficiency causes glomerulonephritis associated with multiple apoptotic bodies. *Nat. Genet.* **19:**56–59.

Botto, M., and M. J. Walport. 2002. C1q, autoimmunity and apoptosis. *Immunobiology* **205:**395–406.

Bratton, D. L., V. A. Fadok, D. A. Richter, J. M. Kailey, L. A. Guthrie, and P. M. Henson. 1997. Appearance of phosphatidylserine on apoptotic cells requires calcium-mediated nonspecific flip-flop and is enhanced by loss of the aminophospholipid translocase. *J. Biol. Chem.* 272:26159–26165.

Brown, S., I. Heinisch, E. Ross, K. Shaw, C. D. Buckley, and J. Savill. 2002. Apoptosis disables CD31-mediated cell detachment from phagocytes promoting binding and engulfment. *Nature* 418:200–203.

Callahan, M. K., P. M. Popernack, S. Tsutsui, L. Truong, R. A. Schlegel, and A. J. Henderson. 2003. Phosphatidylserine on HIV envelope is a cofactor for infection of monocytic cells. *J. Immunol.* 170:4840–4845.

Caron, E., and A. Hall. 1998. Identification of two distinct mechanisms of phagocytosis controlled by different Rho GTPases. *Science* 282:1717–1721.

Carrero, J. A., B. Calderon, and E. R. Unanue. 2006. Lymphocytes are detrimental during the early innate immune response against Listeria monocytogenes. *J. Exp. Med.* 203:933–940.

Carrero, J. A., and E. R. Unanue. 2006. Lymphocyte apoptosis as an immune subversion strategy of microbial pathogens. *Trends Immunol.* 27:497–503.

Chang, B., Y. Chen, Y. Zhao, and R. K. Bruick. 2007. JMJD6 is a histone arginine demethylase. *Science* 318:444–447.

Chang, M. K., C. J. Binder, Y. I. Miller, G. Subbanagounder, G. J. Silverman, J. A. Berliner, and J. L. Witztum. 2004. Apoptotic cells with oxidation-specific epitopes are immunogenic and proinflammatory. *J. Exp. Med.* 200:1359–1370.

Chang, M. K., C. J. Binder, M. Torzewski, and J. L. Witztum. 2002. C-reactive protein binds to both oxidized LDL and apoptotic cells through recognition of a common ligand: phosphorylcholine of oxidized phospholipids. *Proc. Natl. Acad. Sci. USA* 99:13043–13048.

Chen, X., K. Doffek, S. L. Sugg, and J. Shilyansky. 2004. Phosphatidylserine regulates the maturation of human dendritic cells. *J. Immunol.* 173:2985–2994.

Cikala, M., O. Alexandrova, C. N. David, M. Proschel, B. Stiening, P. Cramer, and A. Bottger. 2004. The phosphatidylserine receptor from Hydra is a nuclear protein with potential Fe(II) dependent oxygenase activity. *BMC Cell Biol.* 5:26.

Clark, H., N. Palaniyar, P. Strong, J. Edmondson, S. Hawgood, and K. B. Reid. 2002. Surfactant protein D reduces alveolar macrophage apoptosis in vivo. *J. Immunol.* 169:2892–2899.

Cohen, P. L., R. Caricchio, V. Abraham, T. D. Camenisch, J. C. Jennette, R. A. Roubey, H. S. Earp, G. Matsushima, and E. A. Reap. 2002. Delayed apoptotic cell clearance and lupus-like autoimmunity in mice lacking the c-mer membrane tyrosine kinase. *J. Exp. Med.* 196:135–140.

Conner, S. D., and S. L. Schmid. 2003. Regulated portals of entry into the cell. *Nature* 422:37–44.

Cortes-Hernandez, J., L. Fossati-Jimack, F. Petry, M. Loos, S. Izui, M. J. Walport, H. T. Cook, and M. Botto. 2004. Restoration of C1q levels by bone marrow transplantation attenuates autoimmune disease associated with C1q deficiency in mice. *Eur. J. Immunol.* 34:3713–3722.

Csomor, E., Z. Bajtay, N. Sandor, K. Kristof, G. J. Arlaud, S. Thiel, and A. Erdei. 2007. Complement protein C1q induces maturation of human dendritic cells. *Mol. Immunol.* 44:3389–3397.

Cvetanovic, M., J. E. Mitchell, V. Patel, B. S. Avner, Y. Su, P. T. van der Saag, P. L. Witte, S. Fiore, J. S. Levine, and D. S. Ucker. 2006. Specific recognition of apoptotic cells reveals a ubiquitous and unconventional innate immunity. *J. Biol. Chem.* 281:20055–20067.

Dahl, M., A. K. Bauer, M. Arredouani, R. Soininen, K. Tryggvason, S. R. Kleeberger, and L. Kobzik. 2007. Protection against inhaled oxidants through scavenging of oxidized lipids by macrophage receptors MARCO and SR-AI/II. *J. Clin. Investig.* 117:757–764.

Daleke, D. L. 2003. Regulation of transbilayer plasma membrane phospholipid asymmetry. *J. Lipid Res.* 44:233–242.

Dalgaard, J., K. J. Beckstrom, F. L. Jahnsen, and J. E. Brinchmann. 2005. Differential capability for phagocytosis of apoptotic and necrotic leukemia cells by human peripheral blood dendritic cell subsets. *J. Leukoc. Biol.* 77:689–698.

deCathelineau, A. M., and P. M. Henson. 2003. The final step in programmed cell death: phagocytes carry apoptotic cells to the grave. *Essays Biochem.* 39:105–117.

de Freitas Balanco, J. M., M. E. Moreira, A. Bonomo, P. T. Bozza, G. Amarante-Mendes, C. Pirmez, and M. A. Barcinski. 2001. Apoptotic mimicry by an obligate intracellular parasite downregulates macrophage microbicidal activity. *Curr. Biol.* 11:1870–1873.

de Souza, E. M., T. C. Araujo-Jorge, C. Bailly, A. Lansiaux, M. M. Batista, G. M. Oliveira, and M. N. Soeiro. 2003. Host and parasite apoptosis following Trypanosoma cruzi infection in in vitro and in vivo models. *Cell Tissue Res.* 314:223–235.

Devitt, A., O. D. Moffatt, C. Raykundalia, J. D. Capra, D. L. Simmons, and C. D. Gregory. 1998. Human CD14 mediates recognition and phagocytosis of apoptotic cells. *Nature* 392:505–509.

Devitt, A., K. G. Parker, C. A. Ogden, C. Oldreive, M. F. Clay, L. A. Melville, C. O. Bellamy, A. Lacy-Hulbert, S. C. Gangloff, S. M. Goyert, and C. D. Gregory. 2004. Persistence of apoptotic cells without autoimmune disease or inflammation in CD14−/− mice. *J. Cell Biol.* 167:1161–1170.

Dillon, S. R., A. Constantinescu, and M. S. Schlissel. 2001. Annexin V binds to positively selected B cells. *J. Immunol.* 166:58–71.

Dini, L. 2000. Recognizing death: liver phagocytosis of apoptotic cells. *Eur. J. Histochem.* 44:217–227.

Dini, L., A. Lentini, G. D. Diez, M. Rocha, L. Falasca, L. Serafino, and F. Vidal-Vanaclocha. 1995. Phagocytosis of apoptotic bodies by liver endothelial cells. *J. Cell Sci.* 108(Pt 3):967–973.

Donnelly, S., W. Roake, S. Brown, P. Young, H. Naik, P. Wordsworth, D. A. Isenberg, K. B. Reid, and P. Eggleton. 2006. Impaired recognition of apoptotic neutrophils by the C1q/calreticulin and CD91 pathway in systemic lupus erythematosus. *Arthritis Rheum.* 54:1543–1556.

Duncan, J. L., M. M. LaVail, D. Yasumura, M. T. Matthes, H. Yang, N. Trautmann, A. V. Chappelow, W. Feng, H. S. Earp, G. K. Matsushima, and D. Vollrath. 2003. An RCS-like retinal dystrophy phenotype in mer knockout mice. *Invest. Ophthalmol. Vis. Sci.* 44:826–838.

Eda, S., M. Yamanaka, and M. Beppu. 2004. Carbohydrate-mediated phagocytic recognition of early apoptotic cells undergoing transient capping of CD43 glycoprotein. *J. Biol. Chem.* 279:5967–5974.

Edwards, J. L., E. J. Brown, K. A. Ault, and M. A. Apicella. 2001. The role of complement receptor 3 (CR3) in Neisseria gonorrhoeae infection of human cervical epithelia. *Cell. Microbiol.* 3:611–622.

Elliott, J. I., A. Sardini, J. C. Cooper, D. R. Alexander, S. Davanture, G. Chimini, and C. F. Higgins. 2006. Phosphatidylserine exposure in B lymphocytes: a role for lipid packing. *Blood* 108:1611–1617.

Elward, K., and P. Gasque. 2003. "Eat me" and "don't eat me" signals govern the innate immune response and tissue repair in the CNS: emphasis on the critical role of the complement system. *Mol. Immunol.* 40:85–94.

Erwig, L. P., S. Gordon, G. M. Walsh, and A. J. Rees. 1999. Previous uptake of apoptotic neutrophils or ligation of integrin receptors downmodulates the ability of macrophages to ingest apoptotic neutrophils. *Blood* **93**:1406–1412.

Erwig, L. P., K. A. McPhilips, M. W. Wynes, A. Ivetic, A. J. Ridley, and P. M. Henson. 2006. Differential regulation of phagosome maturation in macrophages and dendritic cells mediated by Rho GTPases and ezrin-radixin-moesin (ERM) proteins. *Proc. Natl. Acad. Sci. USA* **103**:12825–12830.

Fadok, V. A., D. L. Bratton, A. Konowal, P. W. Freed, J. Y. Westcott, and P. M. Henson. 1998. Macrophages that have ingested apoptotic cells in vitro inhibit proinflammatory cytokine production through autocrine/paracrine mechanisms involving TGF-beta, PGE2, and PAF. *J. Clin. Investig.* **101**:890–898.

Fadok, V. A., D. L. Bratton, D. M. Rose, A. Pearson, R. A. Ezekewitz, and P. M. Henson. 2000. A receptor for phosphatidylserine-specific clearance of apoptotic cells. *Nature* **405**:85–90.

Fadok, V. A., D. R. Voelker, P. A. Campbell, J. J. Cohen, D. L. Bratton, and P. M. Henson. 1992. Exposure of phosphatidylserine on the surface of apoptotic lymphocytes triggers specific recognition and removal by macrophages. *J. Immunol.* **148**:2207–2216.

Fan, X., S. Krahling, D. Smith, P. Williamson, and R. A. Schlegel. 2004. Macrophage surface expression of annexins I and II in the phagocytosis of apoptotic lymphocytes. *Mol. Biol. Cell* **15**:2863–2872.

Feng, H., Y. Zeng, M. W. Graner, and E. Katsanis. 2002. Stressed apoptotic tumor cells stimulate dendritic cells and induce specific cytotoxic T cells. *Blood* **100**:4108–4115.

Ferguson, T. A., and H. Kazama. 2005. Signals from dying cells: tolerance induction by the dendritic cell. *Immunol. Res.* **32**:99–108.

Finnemann, S. C., V. L. Bonilha, A. D. Marmorstein, and E. Rodriguez-Boulan. 1997. Phagocytosis of rod outer segments by retinal pigment epithelial cells requires alpha(v)beta5 integrin for binding but not for internalization. *Proc. Natl. Acad. Sci. USA* **94**:12932–12937.

Finnemann, S. C., and R. L. Silverstein. 2001. Differential roles of CD36 and alphavbeta5 integrin in photoreceptor phagocytosis by the retinal pigment epithelium. *J. Exp. Med.* **194**:1289–1298.

Fisher, C. E., and S. E. Howie. 2006. The role of megalin (LRP-2/Gp330) during development. *Dev. Biol.* **296**:279–297.

Fleeton, M. N., N. Contractor, F. Leon, J. D. Wetzel, T. S. Dermody, and B. L. Kelsall. 2004. Peyer's patch dendritic cells process viral antigen from apoptotic epithelial cells in the intestine of reovirus-infected mice. *J. Exp. Med.* **200**:235–245.

Franc, N. C., J. L. Dimarcq, M. Lagueux, J. Hoffmann, and R. A. Ezekowitz. 1996. Croquemort, a novel *Drosophila* hemocyte/macrophage receptor that recognizes apoptotic cells. *Immunity* **4**:431–443.

Frasch, S. C., P. M. Henson, K. Nagaosa, M. B. Fessler, N. Borregaard, and D. L. Bratton. 2004. Phospholipid flip-flop and phospholipid scramblase 1 (PLSCR1) co-localize to uropod rafts in formylated Met-Leu-Phe-stimulated neutrophils. *J. Biol. Chem.* **279**:17625–17633.

Freire-de-Lima, C. G., D. O. Nascimento, M. B. Soares, P. T. Bozza, H. C. Castro-Faria-Neto, F. G. de Mello, G. A. DosReis, and M. F. Lopes. 2000. Uptake of apoptotic cells drives the growth of a pathogenic trypanosome in macrophages. *Nature* **403**:199–203.

Freire-de-Lima, C. G., Y. Q. Xiao, S. J. Gardai, D. L. Bratton, W. P. Schiemann, and P. M. Henson. 2006. Apoptotic cells, through transforming growth factor-beta, coordinately induce anti-inflammatory and suppress pro-inflammatory eicosanoid and NO synthesis in murine macrophages. *J. Biol. Chem.* **281**:38376–38384.

Gaipl, U. S., L. E. Munoz, G. Grossmayer, K. Lauber, S. Franz, K. Sarter, R. E. Voll, T. Winkler, A. Kuhn, J. Kalden, P. Kern, and M. Herrmann. 2007. Clearance deficiency and systemic lupus erythematosus (SLE). *J. Autoimmun.* **28**:114–121.

Gaipl, U. S., A. Sheriff, S. Franz, L. E. Munoz, R. E. Voll, J. R. Kalden, and M. Herrmann. 2006. Inefficient clearance of dying cells and autoreactivity. *Curr. Top. Microbiol. Immunol.* **305**:161–176.

Galan, J. E. 2001. Salmonella interactions with host cells: type III secretion at work. *Annu. Rev. Cell Dev. Biol.* **17**:53–86.

Gardai, S. J., D. L. Bratton, C. A. Ogden, and P. M. Henson. 2006. Recognition ligands on apoptotic cells: a perspective. *J. Leukoc. Biol.* **79**:896–903.

Gardai, S. J., K. A. McPhillips, S. C. Frasch, W. J. Janssen, A. Starefeldt, J. E. Murphy-Ullrich, D. L. Bratton, P. A. Oldenborg, M. Michalak, and P. M. Henson. 2005. Cell-surface calreticulin initiates clearance of viable or apoptotic cells through trans-activation of LRP on the phagocyte. *Cell* **123**:321–334.

Gardai, S. J., Y. Q. Xiao, M. Dickinson, J. A. Nick, D. R. Voelker, K. E. Greene, and P. M. Henson. 2003. By binding SIRPalpha or calreticulin/CD91, lung collectins act as dual function surveillance molecules to suppress or enhance inflammation. *Cell* **115**:13–23.

Garlatti, V., N. Belloy, L. Martin, M. Lacroix, M. Matsushita, Y. Endo, T. Fujita, J. C. Fontecilla-Camps, G. J. Arlaud, N. M. Thielens, and C. Gaboriaud. 2007. Structural insights into the innate immune recognition specificities of L- and H-ficolins. *EMBO J.* **26**:623–633.

Gebska, M. A., I. Titley, H. F. Paterson, R. M. Morilla, D. C. Davies, A. M. Gruszka-Westwood, V. V. Kakkar, S. Eccles, and M. F. Scully. 2002. High-affinity binding sites for heparin generated on leukocytes during apoptosis arise from nuclear structures segregated during cell death. *Blood* **99**:2221–2227.

Gershov, D., S. Kim, N. Brot, and K. B. Elkon. 2000. C-Reactive protein binds to apoptotic cells, protects the cells from assembly of the terminal complement components, and sustains an antiinflammatory innate immune response: implications for systemic autoimmunity. *J. Exp. Med.* **192**:1353–1364.

Ghebrehiwet, B. 1989. Functions associated with the C1q receptor. *Behring Inst. Mitt.* July:204–215.

Ghebrehiwet, B., and E. I. Peerschke. 2004. Role of C1q and C1q receptors in the pathogenesis of systemic lupus erythematosus. *Curr. Dir. Autoimmun.* **7**:87–97.

Giles, K. M., K. Ross, A. G. Rossi, N. A. Hotchin, C. Haslett, and I. Dransfield. 2001. Glucocorticoid augmentation of macrophage capacity for phagocytosis of apoptotic cells is associated with reduced p130Cas expression, loss of paxillin/pyk2 phosphorylation, and high levels of active Rac. *J. Immunol.* **167**:976–986.

Gonias, S. L., L. Wu, and A. M. Salicioni. 2004. Low density lipoprotein receptor-related protein: regulation of the plasma membrane proteome. *Thromb. Haemost.* **91**:1056–1064.

Greenaway, J., J. Lawler, R. Moorehead, P. Bornstein, J. Lamarre, and J. Petrik. 2007. Thrombospondin-1 inhibits VEGF levels in the ovary directly by binding and internalization via the low density lipoprotein receptor-related protein-1 (LRP-1). *J. Cell. Physiol.* **210**:807–818.

Greenberg, M. E., M. Sun, R. Zhang, M. Febbraio, R. Silverstein, and S. L. Hazen. 2006. Oxidized phosphatidylserine-CD36 interactions play an essential role in macrophage-dependent phagocytosis of apoptotic cells. *J. Exp. Med.* **203**:2613–25.

Gregory, C. D., A. Devitt, and O. Moffatt. 1998. Roles of ICAM-3 and CD14 in the recognition and phagocytosis of apoptotic cells by macrophages. *Biochem. Soc. Trans.* **26:**644–649.

Griffin, F. M., Jr., J. A. Griffin, J. E. Leider, and S. C. Silverstein. 1975. Studies on the mechanism of phagocytosis. I. Requirements for circumferential attachment of particle-bound ligands to specific receptors on the macrophage plasma membrane. *J. Exp. Med.* **142:**1263–1282.

Hafizi, S., and B. Dahlback. 2006. Gas6 and protein S. Vitamin K-dependent ligands for the Axl receptor tyrosine kinase subfamily. *FEBS J.* **273:**5231–5344.

Hall, S. E., J. S. Savill, P. M. Henson, and C. Haslett. 1994. Apoptotic neutrophils are phagocytosed by fibroblasts with participation of the fibroblast vitronectin receptor and involvement of a mannose/fucose-specific lectin. *J. Immunol.* **153:**3218–3227.

Hamon, Y., C. Broccardo, O. Chambenoit, M. F. Luciani, F. Toti, S. Chaslin, J. M. Freyssinet, P. F. Devaux, J. McNeish, D. Marguet, and G. Chimini. 2000. ABC1 promotes engulfment of apoptotic cells and transbilayer redistribution of phosphatidylserine. *Nat. Cell Biol.* **2:**399–406.

Hanayama, R., K. Miyasaka, M. Nakaya, and S. Nagata. 2006. MFG-E8-dependent clearance of apoptotic cells, and autoimmunity caused by its failure. *Curr. Dir. Autoimmun.* **9:**162–172.

Hanayama, R., M. Tanaka, K. Miwa, and S. Nagata. 2004a. Expression of developmental endothelial locus-1 in a subset of macrophages for engulfment of apoptotic cells. *J. Immunol.* **172:**3876–3882.

Hanayama, R., M. Tanaka, K. Miyasaka, K. Aozasa, M. Koike, Y. Uchiyama, and S. Nagata. 2004b. Autoimmune disease and impaired uptake of apoptotic cells in MFG-E8-deficient mice. *Science* **304:**1147–1150.

Harshyne, L. A., M. I. Zimmer, S. C. Watkins, and S. M. Barratt-Boyes. 2003. A role for class A scavenger receptor in dendritic cell nibbling from live cells. *J. Immunol.* **170:**2302–2309.

Hart, J. P., M. D. Gunn, and S. V. Pizzo. 2004. A CD91-positive subset of CD11c+ blood dendritic cells: characterization of the APC that functions to enhance adaptive immune responses against CD91-targeted antigens. *J. Immunol.* **172:**70–78.

Hart, S. P., G. J. Dougherty, C. Haslett, and I. Dransfield. 1997. CD44 regulates phagocytosis of apoptotic neutrophil granulocytes, but not apoptotic lymphocytes, by human macrophages. *J. Immunol.* **159:**919–925.

Hashimoto, Y., T. Moki, T. Takizawa, A. Shiratsuchi, and Y. Nakanishi. 2007. Evidence for phagocytosis of influenza virus-infected, apoptotic cells by neutrophils and macrophages in mice. *J. Immunol.* **178:**2448–2457.

Haslett, C., J. S. Savill, M. K. Whyte, M. Stern, I. Dransfield, and L. C. Meagher. 1994. Granulocyte apoptosis and the control of inflammation. *Philos. Trans. R. Soc. Lond. B Biol. Sci.* **345:**327–333.

Heasman, S. J., K. M. Giles, C. Ward, A. G. Rossi, C. Haslett, and I. Dransfield. 2003. Glucocorticoid-mediated regulation of granulocyte apoptosis and macrophage phagocytosis of apoptotic cells: implications for the resolution of inflammation. *J. Endocrinol.* **178:**29–36.

Hildeman, D. A., T. Mitchell, T. K. Teague, P. Henson, B. J. Day, J. Kappler, and P. C. Marrack. 1999. Reactive oxygen species regulate activation-induced T cell apoptosis. *Immunity* **10:**735–744.

Hirano, K., Y. Miki, Y. Hirai, R. Sato, T. Itoh, A. Hayashi, M. Yamanaka, S. Eda, and M. Beppu. 2005. A multifunctional shuttling protein nucleolin is a macrophage receptor for apoptotic cells. *J. Biol. Chem.* **280:**39284–39293.

Hoffmann, P. R., A. M. deCathelineau, C. A. Ogden, Y. Leverrier, D. L. Bratton, D. L. Daleke, A. J. Ridley, V. A. Fadok, and P. M. Henson. 2001. Phosphatidylserine (PS) induces PS receptor-mediated macropinocytosis and promotes clearance of apoptotic cells. *J. Cell Biol.* **155:**649–659.

Hoffmann, P. R., J. A. Kench, A. Vondracek, E. Kruk, D. L. Daleke, M. Jordan, P. Marrack, P. M. Henson, and V. A. Fadok. 2005. Interaction between phosphatidylserine and the phosphatidylserine receptor inhibits immune responses in vivo. *J. Immunol.* **174:**1393–1404.

Holmgren, L., A. Szeles, E. Rajnavolgyi, J. Folkman, G. Klein, I. Ernberg, and K. I. Falk. 1999. Horizontal transfer of DNA by the uptake of apoptotic bodies. *Blood* **93:**3956–3963.

Hong, J. R., G. H. Lin, C. J. Lin, W. P. Wang, C. C. Lee, T. L. Lin, and J. L. Wu. 2004. Phosphatidylserine receptor is required for the engulfment of dead apoptotic cells and for normal embryonic development in zebrafish. *Development* **131:**5417–5427.

Huang, Y. M., J. S. Yang, L. Y. Xu, H. Link, and B. G. Xiao. 2000. Autoantigen-pulsed dendritic cells induce tolerance to experimental allergic encephalomyelitis (EAE) in Lewis rats. *Clin. Exp. Immunol.* **122:**437–444.

Huynh, M. L., V. A. Fadok, and P. M. Henson. 2002. Phosphatidylserine-dependent ingestion of apoptotic cells promotes TGF-beta1 secretion and the resolution of inflammation. *J. Clin. Investig.* **109:**41–50.

Ishimoto, Y., K. Ohashi, K. Mizuno, and T. Nakano. 2000. Promotion of the uptake of PS liposomes and apoptotic cells by a product of growth arrest-specific gene, gas6. *J. Biochem.* **127:**411–417.

Iyoda, T., S. Shimoyama, K. Liu, Y. Omatsu, Y. Akiyama, Y. Maeda, K. Takahara, R. M. Steinman, and K. Inaba. 2002. The CD8+ dendritic cell subset selectively endocytoses dying cells in culture and in vivo. *J. Exp. Med.* **195:**1289–1302.

Jensen, M. L., C. Honore, T. Hummelshoj, B. E. Hansen, H. O. Madsen, and P. Garred. 2007. Ficolin-2 recognizes DNA and participates in the clearance of dying host cells. *Mol. Immunol.* **44:**856–865.

Kagan, V. E., B. Gleiss, Y. Y. Tyurina, V. A. Tyurin, C. Elenstrom-Magnusson, S. X. Liu, F. B. Serinkan, A. Arroyo, J. Chandra, S. Orrenius, and B. Fadeel. 2002. A role for oxidative stress in apoptosis: oxidation and externalization of phosphatidylserine is required for macrophage clearance of cells undergoing Fas-mediated apoptosis. *J. Immunol.* **169:**487–499.

Kawagishi, C., K. Kurosaka, N. Watanabe, and Y. Kobayashi. 2001. Cytokine production by macrophages in association with phagocytosis of etoposide-treated P388 cells in vitro and in vivo. *Biochim. Biophys. Acta* **1541:**221–230.

Kim, S. J., D. Gershov, X. Ma, N. Brot, and K. B. Elkon. 2002. I-PLA(2) activation during apoptosis promotes the exposure of membrane lysophosphatidylcholine leading to binding by natural immunoglobulin M antibodies and complement activation. *J. Exp. Med.* **196:**655–665.

Kim, S. J., D. Gershov, X. Ma, N. Brot, and K. B. Elkon. 2003. Opsonization of apoptotic cells and its effect on macrophage and T cell immune responses. *Ann. N. Y. Acad. Sci.* **987:**68–78.

Kinchen, J. M., J. Cabello, D. Klingele, K. Wong, R. Feichtinger, H. Schnabel, R. Schnabel, and M. O. Hengartner. 2005. Two pathways converge at CED-10 to mediate actin rearrangement and corpse removal in C. elegans. *Nature* **434:**93–99.

Kishore, U., T. J. Greenhough, P. Waters, A. K. Shrive, R. Ghai, M. F. Kamran, A. L. Bernal, K. B. Reid, T. Madan, and T. Chakraborty. 2006. Surfactant proteins SP-A and

SP-D: structure, function and receptors. *Mol. Immunol.* **43:** 1293–1315.

Kobayashi, N., P. Karisola, V. Pena-Cruz, D. M. Dorfman, M. Jinushi, S. E. Umetsu, M. J. Butte, H. Nagumo, I. Chernova, B. Zhu, A. H. Sharpe, S. Ito, G. Dranoff, G. G. Kaplan, J. M. Casasnovas, D. T. Umetsu, R. H. Dekruyff, and G. J. Freeman. 2007. TIM-1 and TIM-4 glycoproteins bind phosphatidylserine and mediate uptake of apoptotic cells. *Immunity* **27:**927–940.

Koizumi, S., Y. Shigemoto-Mogami, K. Nasu-Tada, Y. Shinozaki, K. Ohsawa, M. Tsuda, B. V. Joshi, K. A. Jacobson, S. Kohsaka, and K. Inoue. 2007. UDP acting at P2Y6 receptors is a mediator of microglial phagocytosis. *Nature* **446:** 1091–1095.

Korb, L. C., and J. M. Ahearn. 1997. C1q binds directly and specifically to surface blebs of apoptotic human keratinocytes: complement deficiency and systemic lupus erythematosus revisited. *J. Immunol.* **158:**4525–4528.

Kovacsovics, T., J. Tschopp, A. Kress, and H. Isliker. 1985. Antibody-independent activation of C1, the first component of complement, by cardiolipin. *J. Immunol.* **135:**2695–2700.

Krieser, R. J., F. E. Moore, D. Dresnek, B. J. Pellock, R. Patel, A. Huang, C. Brachmann, and K. White. 2007. The *Drosophila* homolog of the putative phosphatidylserine receptor functions to inhibit apoptosis. *Development* **134:**2407–2414.

Krysko, D. V., K. D'Herde, and P. Vandenabeele. 2006a. Clearance of apoptotic and necrotic cells and its immunological consequences. *Apoptosis* **11:**1709–1726.

Krysko, D. V., G. Denecker, N. Festjens, S. Gabriels, E. Parthoens, K. D'Herde, and P. Vandenabeele. 2006b. Macrophages use different internalization mechanisms to clear apoptotic and necrotic cells. *Cell Death Differ.* **13:**2011–2022.

Kunisaki, Y., S. Masuko, M. Noda, A. Inayoshi, T. Sanui, M. Harada, T. Sasazuki, and Y. Fukui. 2004. Defective fetal liver erythropoiesis and T lymphopoiesis in mice lacking the phosphatidylserine receptor. *Blood* **103:**3362–3364.

Kuraishi, T., J. Manaka, M. Kono, H. Ishii, N. Yamamoto, K. Koizumi, A. Shiratsuchi, B. L. Lee, H. Higashida, and Y. Nakanishi. 2007. Identification of calreticulin as a marker for phagocytosis of apoptotic cells in Drosophila. *Exp. Cell Res.* **313:**500–510.

Kuraya, M., Z. Ming, X. Liu, M. Matsushita, and T. Fujita. 2005. Specific binding of L-ficolin and H-ficolin to apoptotic cells leads to complement activation. *Immunobiology* **209:** 689–697.

Lange, Y., J. Ye, and T. L. Steck. 2007. Scrambling of phospholipids activates red cell membrane cholesterol. *Biochemistry* **46:**2233–2238.

Laskay, T., G. van Zandbergen, and W. Solbach. 2003. Neutrophil granulocytes—Trojan horses for *Leishmania major* and other intracellular microbes? *Trends Microbiol.* **11:**210–214.

Lauber, K., E. Bohn, S. M. Krober, Y. J. Xiao, S. G. Blumenthal, R. K. Lindemann, P. Marini, C. Wiedig, A. Zobywalski, S. Baksh, Y. Xu, I. B. Autenrieth, K. Schulze-Osthoff, C. Belka, G. Stuhler, and S. Wesselborg. 2003. Apoptotic cells induce migration of phagocytes via caspase-3-mediated release of a lipid attraction signal. *Cell* **113:**717–730.

Leverrier, Y., and A. J. Ridley. 2001. Requirement for Rho GTPases and PI 3-kinases during apoptotic cell phagocytosis by macrophages. *Curr. Biol.* **11:**195–199.

Li, M. O., M. R. Sarkisian, W. Z. Mehal, P. Rakic, and R. A. Flavell. 2003. Phosphatidylserine receptor is required for clearance of apoptotic cells. *Science* **302:**1560–1563.

Li, W., and N. E. Baker. 2007. Engulfment is required for cell competition. *Cell* **129:**1215–1225.

Lima, R. G., L. Moreira, J. Paes-Leme, V. Barreto-de-Souza, H. C. Castro-Faria-Neto, P. T. Bozza, and D. C. Bou-Habib. 2006. Interaction of macrophages with apoptotic cells enhances HIV Type 1 replication through PGE2, PAF, and vitronectin receptor. *AIDS Res. Hum. Retroviruses* **22:**763–769.

Lin, H., and D. O. Clegg. 1998. Integrin alphavbeta5 participates in the binding of photoreceptor rod outer segments during phagocytosis by cultured human retinal pigment epithelium. *Investig. Ophthalmol. Vis. Sci.* **39:**1703–1712.

Liu, G., C. Wu, Y. Wu, and Y. Zhao. 2006. Phagocytosis of apoptotic cells and immune regulation. *Scand. J. Immunol.* **64:**1–9.

Liu, Y., J. M. Cousin, J. Hughes, J. Van Damme, J. R. Seckl, C. Haslett, I. Dransfield, J. Savill, and A. G. Rossi. 1999. Glucocorticoids promote nonphlogistic phagocytosis of apoptotic leukocytes. *J. Immunol.* **162:**3639–3646.

Lu, Q., and G. Lemke. 2001. Homeostatic regulation of the immune system by receptor tyrosine kinases of the Tyro 3 family. *Science* **293:**306–311.

Lucas, M., L. M. Stuart, J. Savill, and A. Lacy-Hulbert. 2003. Apoptotic cells and innate immune stimuli combine to regulate macrophage cytokine secretion. *J. Immunol.* **171:** 2610–2615.

Maderna, P., D. C. Cottell, G. Berlasconi, N. A. Petasis, H. R. Brady, and C. Godson. 2002. Lipoxins induce actin reorganization in monocytes and macrophages but not in neutrophils: differential involvement of rho GTPases. *Am. J. Pathol.* **160:**2275–2283.

Manaka, J., T. Kuraishi, A. Shiratsuchi, Y. Nakai, H. Higashida, P. Henson, and Y. Nakanishi. 2004. Draper-mediated and phosphatidylserine-independent phagocytosis of apoptotic cells by *Drosophila* hemocytes/macrophages. *J. Biol. Chem.* **279:**48466–48476.

Manfredi, A. A., P. Rovere, S. Heltai, G. Galati, G. Nebbia, A. Tincani, G. Balestrieri, and M. G. Sabbadini. 1998. Apoptotic cell clearance in systemic lupus erythematosus. II. Role of beta2-glycoprotein I. *Arthritis Rheum.* **41:**215–223.

Marth, T., and B. L. Kelsall. 1997. Regulation of interleukin-12 by complement receptor 3 signaling. *J. Exp. Med.* **185:** 1987–1995.

Martin, S. J., C. P. Reutelingsperger, A. J. McGahon, J. A. Rader, R. C. van Schie, D. M. LaFace, and D. R. Green. 1995. Early redistribution of plasma membrane phosphatidylserine is a general feature of apoptosis regardless of the initiating stimulus: inhibition by overexpression of Bcl-2 and Abl. *J. Exp. Med.* **182:**1545–1556.

McPhillips, K., W. J. Janssen, M. Ghosh, A. Byrne, S. Gardai, L. Remigio, D. L. Bratton, J. L. Kang, and P. Henson. 2007. TNF-alpha inhibits macrophage clearance of apoptotic cells via cytosolic phospholipase A2 and oxidant-dependent mechanisms. *J. Immunol.* **178:**8117–8126.

Meagher, L. C., J. S. Savill, A. Baker, R. W. Fuller, and C. Haslett. 1992. Phagocytosis of apoptotic neutrophils does not induce macrophage release of thromboxane B2. *J. Leukoc. Biol.* **52:**269-2-73.

Mevorach, D., J. O. Mascarenhas, D. Gershov, and K. B. Elkon. 1998. Complement-dependent clearance of apoptotic cells by human macrophages. *J. Exp. Med.* **188:**2313–2320.

Miller, Y. I., S. Viriyakosol, C. J. Binder, J. R. Feramisco, T. N. Kirkland, and J. L. Witztum. 2003. Minimally modified LDL binds to CD14, induces macrophage spreading via TLR4/MD-2, and inhibits phagocytosis of apoptotic cells. *J. Biol. Chem.* **278:**1561–1568.

Mitchell, J. E., M. Cvetanovic, N. Tibrewal, V. Patel, O. R. Colamonici, M. O. Li, R. A. Flavell, J. S. Levine, R. B. Birge, and D. S. Ucker. 2006. The presumptive phosphatidylserine receptor is dispensable for innate anti-inflammatory recognition and clearance of apoptotic cells. *J. Biol. Chem.* **281:**5718–5725.

Mitchell, S., G. Thomas, K. Harvey, D. Cottell, K. Reville, G. Berlasconi, N. A. Petasis, L. Erwig, A. J. Rees, J. Savill, H. R. Brady, and C. Godson. 2002. Lipoxins, aspirin-triggered epi-lipoxins, lipoxin stable analogues, and the resolution of inflammation: stimulation of macrophage phagocytosis of apoptotic neutrophils in vivo. *J. Am. Soc. Nephrol.* **13:**2497–2507.

Mitra, P., C. A. Oskeritzian, S. G. Payne, M. A. Beaven, S. Milstien, and S. Spiegel. 2006. Role of ABCC1 in export of sphingosine-1-phosphate from mast cells. *Proc. Natl. Acad. Sci. USA* **103:**16394–16399.

Miyanishi, M., K. Tada, M. Koike, Y. Uchiyama, T. Kitamura, and S. Nagata. 2007. Identification of Tim4 as a phosphatidylserine receptor. *Nature* **450:**435–439.

Mold, C., R. Baca, and T. W. Du Clos. 2002. Serum amyloid P component and C-reactive protein opsonize apoptotic cells for phagocytosis through Fcgamma receptors. *J. Autoimmun.* **19:**147–154.

Morelli, A. E., A. T. Larregina, W. J. Shufesky, A. F. Zahorchak, A. J. Logar, G. D. Papworth, Z. Wang, S. C. Watkins, L. D. Falo, Jr., and A. W. Thomson. 2003. Internalization of circulating apoptotic cells by splenic marginal zone dendritic cells: dependence on complement receptors and effect on cytokine production. *Blood* **101:**611–620.

Munoz, L. E., S. Franz, F. Pausch, B. Furnrohr, A. Sheriff, B. Vogt, P. M. Kern, W. Baum, C. Stach, D. von Laer, B. Brachvogel, E. Poschl, M. Herrmann, and U. S. Gaipl. 2007a. The influence on the immunomodulatory effects of dying and dead cells of Annexin V. *J. Leukoc. Biol.* **81:**6–14.

Munoz, L. E., B. Frey, F. Pausch, W. Baum, R. B. Mueller, B. Brachvogel, E. Poschl, F. Rodel, K. von der Mark, M. Herrmann, and U. S. Gaipl. 2007b. The role of annexin A5 in the modulation of the immune response against dying and dead cells. *Curr. Med. Chem.* **14:**271–277.

Murphy, J. E., D. Tacon, P. R. Tedbury, J. M. Hadden, S. Knowling, T. Sawamura, M. Peckham, S. E. Phillips, J. H. Walker, and S. Ponnambalam. 2006. LOX-1 scavenger receptor mediates calcium-dependent recognition of phosphatidylserine and apoptotic cells. *Biochem. J.* **393:**107–115.

Nagaosa, K., A. Shiratsuchi, and Y. Nakanishi. 2003. Concomitant induction of apoptosis and expression of monocyte chemoattractant protein-1 in cultured rat luteal cells by nuclear factor-kappaB and oxidative stress. *Dev. Growth Differ.* **45:**351–359.

Nandrot, E. F., Y. Kim, S. E. Brodie, X. Huang, D. Sheppard, and S. C. Finnemann. 2004. Loss of synchronized retinal phagocytosis and age-related blindness in mice lacking alphavbeta5 integrin. *J. Exp. Med.* **200:**1539–1545.

Navratil, J. S., S. C. Watkins, J. J. Wisnieski, and J. M. Ahearn. 2001. The globular heads of C1q specifically recognize surface blebs of apoptotic vascular endothelial cells. *J. Immunol.* **166:**3231–3239.

Neumann, H., and K. Takahashi. 2007. Essential role of the microglial triggering receptor expressed on myeloid cells-2 (TREM2) for central nervous tissue immune homeostasis. *J. Neuroimmunol.* **184:**92–99.

Obeid, M., A. Tesniere, F. Ghiringhelli, G. M. Fimia, L. Apetoh, J. L. Perfettini, M. Castedo, G. Mignot, T. Panaretakis, N. Casares, D. Metivier, N. Larochette, P. van Endert, F. Ciccosanti, M. Piacentini, L. Zitvogel, and G. Kroemer. 2007. Calreticulin exposure dictates the immunogenicity of cancer cell death. *Nat. Med.* **13:**54–61.

Ogden, C. A., A. deCathelineau, P. R. Hoffmann, D. Bratton, B. Ghebrehiwet, V. A. Fadok, and P. M. Henson. 2001. C1q and mannose binding lectin engagement of cell surface calreticulin and CD91 initiates macropinocytosis and uptake of apoptotic cells. *J. Exp. Med.* **194:**781–795.

Ogden, C. A., J. D. Pound, B. K. Batth, S. Owens, I. Johannessen, K. Wood, and C. D. Gregory. 2005. Enhanced apoptotic cell clearance capacity and B cell survival factor production by IL-10-activated macrophages: implications for Burkitt's lymphoma. *J. Immunol.* **174:**3015–3023.

Oka, K., T. Sawamura, K. Kikuta, S. Itokawa, N. Kume, T. Kita, and T. Masaki. 1998. Lectin-like oxidized low-density lipoprotein receptor 1 mediates phagocytosis of aged/apoptotic cells in endothelial cells. *Proc. Natl. Acad. Sci. USA* **95:**9535–9540.

Okada, H., and T. W. Mak. 2004. Pathways of apoptotic and non-apoptotic death in tumour cells. *Nat. Rev. Cancer* **4:**592–603.

Orr, A. W., C. E. Pedraza, M. A. Pallero, C. A. Elzie, S. Goicoechea, D. K. Strickland, and J. E. Murphy-Ullrich. 2003. Low density lipoprotein receptor-related protein is a calreticulin coreceptor that signals focal adhesion disassembly. *J. Cell Biol.* **161:**1179–1189.

Palaniyar, N., H. Clark, J. Nadesalingam, S. Hawgood, and K. B. Reid. 2003. Surfactant protein D binds genomic DNA and apoptotic cells, and enhances their clearance, in vivo. *Ann. N. Y. Acad. Sci.* **1010:**471–475.

Palaniyar, N., J. Nadesalingam, H. Clark, M. J. Shih, A. W. Dodds, and K. B. Reid. 2004. Nucleic acid is a novel ligand for innate, immune pattern recognition collectins surfactant proteins A and D and mannose-binding lectin. *J. Biol. Chem.* **279:**32728–32736.

Park, D., A. C. Tosello-Trampont, M. R. Elliott, M. Lu, L. B. Haney, Z. Ma, A. L. Klibanov, J. W. Mandell, and K. S. Ravichandran. 2007. BAI1 is an engulfment receptor for apoptotic cells upstream of the ELMO/Dock180/Rac module. *Nature* **450:**430–434.

Park, S. Y., M. Y. Jung, H. J. Kim, S. J. Lee, S. Y. Kim, B. H. Lee, T. H. Kwon, R. W. Park, and I. S. Kim. 2008. Rapid cell corpse clearance by stabilin-2, a membrane phosphatidylserine receptor. *Cell Death Differ.* **15:**192–201.

Peerschke, E. I., K. B. Reid, and B. Ghebrehiwet. 1994. Identification of a novel 33-kDa C1q-binding site on human blood platelets. *J. Immunol.* **152:**5896–5901.

Perry, V. H., D. A. Hume, and S. Gordon. 1985. Immunohistochemical localization of macrophages and microglia in the adult and developing mouse brain. *Neuroscience* **15:**313–326.

Pittoni, V., and G. Valesini. 2002. The clearance of apoptotic cells: implications for autoimmunity. *Autoimmun. Rev.* **1:**154–161.

Platt, N., H. Suzuki, T. Kodama, and S. Gordon. 2000. Apoptotic thymocyte clearance in scavenger receptor class A-deficient mice is apparently normal. *J. Immunol.* **164:**4861–4867.

Platt, N., H. Suzuki, Y. Kurihara, T. Kodama, and S. Gordon. 1996. Role for the class A macrophage scavenger receptor in the phagocytosis of apoptotic thymocytes in vitro. *Proc. Natl. Acad. Sci. USA* **93:**12456–12460.

Pomorski, T., R. Lombardi, H. Riezman, P. F. Devaux, G. van Meer, and J. C. Holthuis. 2003. Drs2p-related P-type ATPases Dnf1p and Dnf2p are required for phospholipid translocation across the yeast plasma membrane and serve a role in endocytosis. *Mol. Biol. Cell* **14:**1240–1254.

Price, B. E., J. Rauch, M. A. Shia, M. T. Walsh, W. Lieberthal, H. M. Gilligan, T. O'Laughlin, J. S. Koh, and J. S. Levine. 1996. Anti-phospholipid autoantibodies bind to apoptotic, but not viable, thymocytes in a beta 2-glycoprotein I-dependent manner. *J. Immunol.* **157:**2201–2208.

Qu, X., Z. Zou, Q. Sun, K. Luby-Phelps, P. Cheng, R. N. Hogan, C. Gilpin, and B. Levine. 2007. Autophagy gene-dependent clearance of apoptotic cells during embryonic development. *Cell* **128**:931–946.

Reidy, M. F., and J. R. Wright. 2003. Surfactant protein A enhances apoptotic cell uptake and TGF-beta1 release by inflammatory alveolar macrophages. *Am. J. Physiol.* **285**:L854–L861.

Ren, Y., and J. Savill. 1995. Proinflammatory cytokines potentiate thrombospondin-mediated phagocytosis of neutrophils undergoing apoptosis. *J. Immunol.* **154**:2366–2374.

Ren, Y., L. Stuart, F. P. Lindberg, A. R. Rosenkranz, Y. Chen, T. N. Mayadas, and J. Savill. 2001. Nonphlogistic clearance of late apoptotic neutrophils by macrophages: efficient phagocytosis independent of beta 2 integrins. *J. Immunol.* **166**:4743–4750.

Reville, K., J. K. Crean, S. Vivers, I. Dransfield, and C. Godson. 2006. Lipoxin A4 redistributes myosin IIA and Cdc42 in macrophages: implications for phagocytosis of apoptotic leukocytes. *J. Immunol.* **176**:1878–1888.

Roos, A., W. Xu, G. Castellano, A. J. Nauta, P. Garred, M. R. Daha, and C. van Kooten. 2004. Mini-review: a pivotal role for innate immunity in the clearance of apoptotic cells. *Eur. J. Immunol.* **34**:921–929.

Rovere, P., A. A. Manfredi, C. Vallinoto, V. S. Zimmermann, U. Fascio, G. Balestrieri, P. Ricciardi-Castagnoli, C. Rugarli, A. Tincani, and M. G. Sabbadini. 1998. Dendritic cells preferentially internalize apoptotic cells opsonized by anti-beta2-glycoprotein I antibodies. *J. Autoimmun.* **11**:403–411.

Ryeom, S. W., J. R. Sparrow, and R. L. Silverstein. 1996. CD36 participates in the phagocytosis of rod outer segments by retinal pigment epithelium. *J. Cell Sci.* **109**(Pt 2):387–395.

Santiago, C., A. Ballesteros, L. Martinez-Munoz, M. Mellado, G. G. Kaplan, G. J. Freeman, and J. M. Casasnovas. 2007. Structures of T cell immunoglobulin mucin protein 4 show a metal-ion-dependent ligand binding site where phosphatidylserine binds. *Immunity* **27**:941–951.

Santy, L. C., K. S. Ravichandran, and J. E. Casanova. 2005. The DOCK180/Elmo complex couples ARNO-mediated Arf6 activation to the downstream activation of Rac1. *Curr. Biol.* **15**:1749–1754.

Sather, S., K. D. Kenyon, J. B. Lefkowitz, X. Liang, B. C. Varnum, P. M. Henson, and D. K. Graham. 2007. A soluble form of the Mer receptor tyrosine kinase inhibits macrophage clearance of apoptotic cells and platelet aggregation. *Blood* **109**:1026–1233.

Savill, J., I. Dransfield, C. Gregory, and C. Haslett. 2002. A blast from the past: clearance of apoptotic cells regulates immune responses. *Nat. Rev. Immunol.* **2**:965–975.

Savill, J., I. Dransfield, N. Hogg, and C. Haslett. 1990. Vitronectin receptor-mediated phagocytosis of cells undergoing apoptosis. *Nature* **343**:170–173.

Savill, J., C. Gregory, and C. Haslett. 2003. Cell biology. Eat me or die. *Science* **302**:1516–1517.

Savill, J., N. Hogg, and C. Haslett. 1991. Macrophage vitronectin receptor, CD36, and thrombospondin cooperate in recognition of neutrophils undergoing programmed cell death. *Chest* **99**:6S–7S.

Savill, J., N. Hogg, Y. Ren, and C. Haslett. 1992. Thrombospondin cooperates with CD36 and the vitronectin receptor in macrophage recognition of neutrophils undergoing apoptosis. *J. Clin. Investig.* **90**:1513–1522.

Savill, J. S., P. M. Henson, and C. Haslett. 1989. Phagocytosis of aged human neutrophils by macrophages is mediated by a novel "charge-sensitive" recognition mechanism. *J. Clin. Investig.* **84**:1518–1527.

Scannell, M., M. B. Flanagan, A. deStefani, K. J. Wynne, G. Cagney, C. Godson, and P. Maderna. 2007. Annexin-1 and peptide derivatives are released by apoptotic cells and stimulate phagocytosis of apoptotic neutrophils by macrophages. *J. Immunol.* **178**:4595–4605.

Schagat, T. L., J. A. Wofford, and J. R. Wright. 2001. Surfactant protein A enhances alveolar macrophage phagocytosis of apoptotic neutrophils. *J. Immunol.* **166**:2727–2733.

Schulz, O., D. J. Pennington, K. Hodivala-Dilke, M. Febbraio, and C. Reis e Sousa. 2002. CD36 or alphavbeta3 and alphavbeta5 integrins are not essential for MHC class I cross-presentation of cell-associated antigen by CD8 alpha+ murine dendritic cells. *J. Immunol.* **168**:6057–6065.

Scott, R. S., E. J. McMahon, S. M. Pop, E. A. Reap, R. Caricchio, P. L. Cohen, H. S. Earp, and G. K. Matsushima. 2001. Phagocytosis and clearance of apoptotic cells is mediated by MER. *Nature* **411**:207–211.

Seitz, H. M., T. D. Camenisch, G. Lemke, H. S. Earp, and G. K. Matsushima. 2007. Macrophages and dendritic cells use different Axl/Mertk/Tyro3 receptors in clearance of apoptotic cells. *J. Immunol.* **178**:5635–5642.

Sen, P., M. A. Wallet, Z. Yi, Y. Huang, M. Henderson, C. E. Mathews, H. S. Earp, G. Matsushima, A. S. Baldwin, Jr., and R. M. Tisch. 2007. Apoptotic cells induce Mer tyrosine kinase-dependent blockade of NF-kappaB activation in dendritic cells. *Blood* **109**:653–660.

Serhan, C. N., and J. Savill. 2005. Resolution of inflammation: the beginning programs the end. *Nat. Immunol.* **6**:1191–1197.

Shiratsuchi, A., I. Watanabe, O. Takeuchi, S. Akira, and Y. Nakanishi. 2004. Inhibitory effect of Toll-like receptor 4 on fusion between phagosomes and endosomes/lysosomes in macrophages. *J. Immunol.* **172**:2039–2047.

Silva, J. S., F. S. Machado, and G. A. Martins. 2003. The role of nitric oxide in the pathogenesis of Chagas disease. *Front. Biosci.* **8**:s314–s325.

Sim, R. B., S. K. Moestrup, G. R. Stuart, N. J. Lynch, J. Lu, W. J. Schwaeble, and R. Malhotra. 1998. Interaction of C1q and the collectins with the potential receptors calreticulin (cC1qR/collectin receptor) and megalin. *Immunobiology* **199**:208–224.

Skoberne, M., A. S. Beignon, M. Larsson, and N. Bhardwaj. 2005. Apoptotic cells at the crossroads of tolerance and immunity. *Curr. Top. Microbiol. Immunol.* **289**:259–292.

Skoberne, M., S. Somersan, W. Almodovar, T. Truong, K. Petrova, P. M. Henson, and N. Bhardwaj. 2006. The apoptotic-cell receptor CR3, but not alphavbeta5, is a regulator of human dendritic-cell immunostimulatory function. *Blood* **108**:947–955.

Stach, C. M., X. Turnay, R. E. Voll, P. M. Kern, W. Kolowos, T. D. Beyer, J. R. Kalden, and M. Herrmann. 2000. Treatment with annexin V increases immunogenicity of apoptotic human T-cells in Balb/c mice. *Cell Death Differ.* **7**:911-5.

Stern, M., J. Savill, and C. Haslett. 1996. Human monocyte-derived macrophage phagocytosis of senescent eosinophils undergoing apoptosis. Mediation by alpha v beta 3/CD36/thrombospondin recognition mechanism and lack of phlogistic response. *Am. J. Pathol.* **149**:911–921.

Stuart, L. M., M. Lucas, C. Simpson, J. Lamb, J. Savill, and A. Lacy-Hulbert. 2002. Inhibitory effects of apoptotic cell ingestion upon endotoxin-driven myeloid dendritic cell maturation. *J. Immunol.* **168**:1627–1635.

Stuart, L. M., K. Takahashi, L. Shi, J. Savill, and R. A. Ezekowitz. 2005. Mannose-binding lectin-deficient mice display defective apoptotic cell clearance but no autoimmune phenotype. *J. Immunol.* **174**:3220–3226.

Subang, R., J. S. Levine, A. S. Janoff, S. M. Davidson, T. F. Taraschi, T. Koike, S. R. Minchey, M. Whiteside, M. Tannenbaum, and J. Rauch. 2000. Phospholipid-bound beta 2-glycoprotein I induces the production of anti-phospholipid antibodies. *J. Autoimmun.* **15**:21–32.

Subbanagounder, G., N. Leitinger, D. C. Schwenke, J. W. Wong, H. Lee, C. Rizza, A. D. Watson, K. F. Faull, A. M. Fogelman, and J. A. Berliner. 2000. Determinants of bioactivity of oxidized phospholipids. Specific oxidized fatty acyl groups at the sn-2 position. *Arterioscler. Thromb. Vasc. Biol.* **20:**2248–2254.

Subklewe, M., C. Paludan, M. L. Tsang, K. Mahnke, R. M. Steinman, and C. Munz. 2001. Dendritic cells cross-present latency gene products from Epstein-Barr virus-transformed B cells and expand tumor-reactive CD8(+) killer T cells. *J. Exp. Med.* **193:**405–411.

Sun, M., S. C. Finnemann, M. Febbraio, L. Shan, S. P. Annangudi, E. A. Podrez, G. Hoppe, R. Darrow, D. T. Organisciak, R. G. Salomon, R. L. Silverstein, and S. L. Hazen. 2006. Light-induced oxidation of photoreceptor outer segment phospholipids generates ligands for CD36-mediated phagocytosis by retinal pigment epithelium: a potential mechanism for modulating outer segment phagocytosis under oxidant stress conditions. *J. Biol. Chem.* **281:**4222–4230.

Takahashi, K., C. D. Rochford, and H. Neumann. 2005. Clearance of apoptotic neurons without inflammation by microglial triggering receptor expressed on myeloid cells-2. *J. Exp. Med.* **201:**647–657.

Takemura, Y., N. Ouchi, R. Shibata, T. Aprahamian, M. T. Kirber, R. S. Summer, S. Kihara, and K. Walsh. 2007. Adiponectin modulates inflammatory reactions via calreticulin receptor-dependent clearance of early apoptotic bodies. *J. Clin. Investig.* **117:**375–386.

Tamura, Y., J. Osuga, H. Adachi, R. Tozawa, Y. Takanezawa, K. Ohashi, N. Yahagi, M. Sekiya, S. Okazaki, S. Tomita, Y. Iizuka, H. Koizumi, T. Inaba, H. Yagyu, N. Kamada, H. Suzuki, H. Shimano, T. Kadowaki, M. Tsujimoto, H. Arai, N. Yamada, and S. Ishibashi. 2004. Scavenger receptor expressed by endothelial cells I (SREC-I) mediates the uptake of acetylated low density lipoproteins by macrophages stimulated with lipopolysaccharide. *J. Biol. Chem.* **279:**30938–30944.

Tan, B. H., C. Meinken, M. Bastian, H. Bruns, A. Legaspi, M. T. Ochoa, S. R. Krutzik, B. R. Bloom, T. Ganz, R. L. Modlin, and S. Stenger. 2006. Macrophages acquire neutrophil granules for antimicrobial activity against intracellular pathogens. *J. Immunol.* **177:**1864–1871.

Tang, X., M. S. Halleck, R. A. Schlegel, and P. Williamson. 1996. A subfamily of P-type ATPases with aminophospholipid transporting activity. *Science* **272:**1495–1497.

Taylor, P. R., A. Carugati, V. A. Fadok, H. T. Cook, M. Andrews, M. C. Carroll, J. S. Savill, P. M. Henson, M. Botto, and M. J. Walport. 2000. A hierarchical role for classical pathway complement proteins in the clearance of apoptotic cells in vivo. *J. Exp. Med.* **192:**359–366.

Taylor, R. C., G. Brumatti, S. Ito, M. O. Hengartner, W. B. Derry, and S. J. Martin. 2007. Establishing a blueprint for CED-3-dependent killing through identification of multiple substrates for this protease. *J. Biol. Chem.* **282:**15011–15021.

Tosello-Trampont, A. C., K. Nakada-Tsukui, and K. S. Ravichandran. 2003. Engulfment of apoptotic cells is negatively regulated by Rho-mediated signaling. *J. Biol. Chem.* **278:**49911–49919.

Tuominen, A., Y. I. Miller, L. F. Hansen, Y. A. Kesaniemi, J. L. Witztum, and S. Horkko. 2006. A natural antibody to oxidized cardiolipin binds to oxidized low-density lipoprotein, apoptotic cells, and atherosclerotic lesions. *Arterioscler. Thromb. Vasc. Biol.* **26:**2096–2102.

Tyurina, Y. Y., L. V. Basova, N. V. Konduru, V. A. Tyurin, A. I. Potapovich, P. Cai, H. Bayir, D. Stoyanovsky, B. R. Pitt, A. A. Shvedova, B. Fadeel, and V. E. Kagan. 2007. Nitrosative stress inhibits the aminophospholipid translocase resulting in phosphatidylserine externalization and macrophage engulfment: implications for the resolution of inflammation. *J. Biol. Chem.* **282:**8498–8509.

Tyurina, Y. Y., K. Kawai, V. A. Tyurin, S. X. Liu, V. E. Kagan, and J. P. Fabisiak. 2004. The plasma membrane is the site of selective phosphatidylserine oxidation during apoptosis: role of cytochrome C. *Antioxid. Redox Signal.* **6:**209–225.

Uchimura, E., T. Kodaira, K. Kurosaka, D. Yang, N. Watanabe, and Y. Kobayashi. 1997. Interaction of phagocytes with apoptotic cells leads to production of pro-inflammatory cytokines. *Biochem. Biophys. Res. Commun.* **239:**799–803.

Urban, B. C., D. J. Ferguson, A. Pain, N. Willcox, M. Plebanski, J. M. Austyn, and D. J. Roberts. 1999. *Plasmodium falciparum*-infected erythrocytes modulate the maturation of dendritic cells. *Nature* **400:**73–77.

Urban, B. C., N. Willcox, and D. J. Roberts. 2001. A role for CD36 in the regulation of dendritic cell function. *Proc. Natl. Acad. Sci. USA* **98:**8750–8755.

van den Eijnde, S. M., L. Boshart, E. H. Baehrecke, C. I. De Zeeuw, C. P. Reutelingsperger, and C. Vermeij-Keers. 1998. Cell surface exposure of phosphatidylserine during apoptosis is phylogenetically conserved. *Apoptosis* **3:**9–16.

Vandivier, R. W., C. A. Ogden, V. A. Fadok, P. R. Hoffmann, K. K. Brown, M. Botto, M. J. Walport, J. H. Fisher, P. M. Henson, and K. E. Greene. 2002. Role of surfactant proteins A, D, and C1q in the clearance of apoptotic cells in vivo and in vitro: calreticulin and CD91 as a common collectin receptor complex. *J. Immunol.* **169:**3978–3986.

van Helvoort, A., A. J. Smith, H. Sprong, I. Fritzsche, A. H. Schinkel, P. Borst, and G. van Meer. 1996. MDR1 P-glycoprotein is a lipid translocase of broad specificity, while MDR3 P-glycoprotein specifically translocates phosphatidylcholine. *Cell* **87:**507–517.

van Zandbergen, G., M. Klinger, A. Mueller, S. Dannenberg, A. Gebert, W. Solbach, and T. Laskay. 2004. Cutting edge: neutrophil granulocyte serves as a vector for *Leishmania* entry into macrophages. *J. Immunol.* **173:**6521–6525.

Van Zanten, J., G. A. Hospers, M. C. Harmsen, T. H. The, N. H. Mulder, and L. F. De Leij. 2002. Dendritic cells present an intracellular viral antigen derived from apoptotic cells and induce a T-cell response. *Scand. J. Immunol.* **56:**254–259.

Verbovetski, I., H. Bychkov, U. Trahtemberg, I. Shapira, M. Hareuveni, O. Ben-Tal, I. Kutikov, O. Gill, and D. Mevorach. 2002. Opsonization of apoptotic cells by autologous iC3b facilitates clearance by immature dendritic cells, down-regulates DR and CD86, and up-regulates CC chemokine receptor 7. *J. Exp. Med.* **196:**1553–1561.

Viorritto, I. C., N. P. Nikolov, and R. M. Siegel. 2007. Autoimmunity versus tolerance: can dying cells tip the balance? *Clin. Immunol.* **122:**125–134.

Vivers, S., S. J. Heasman, S. P. Hart, and I. Dransfield. 2004. Divalent cation-dependent and -independent augmentation of macrophage phagocytosis of apoptotic neutrophils by CD44 antibody. *Clin. Exp. Immunol.* **138:**447–452.

Voll, R. E., M. Herrmann, E. A. Roth, C. Stach, J. R. Kalden, and I. Girkontaite. 1997. Immunosuppressive effects of apoptotic cells. *Nature* **390:**350–351.

Wanderley, J. L., M. E. Moreira, A. Benjamin, A. C. Bonomo, and M. A. Barcinski. 2006. Mimicry of apoptotic cells by exposing phosphatidylserine participates in the establishment of amastigotes of *Leishmania* (L) *amazonensis* in mammalian hosts. *J. Immunol.* **176:**1834–1839.

Wang, X., J. Wang, K. Gengyo-Ando, L. Gu, C. L. Sun, C. Yang, Y. Shi, T. Kobayashi, S. Mitani, X. S. Xie, and D. Xue. 2007. C. elegans mitochondrial factor WAH-1 promotes phosphatidylserine externalization in apoptotic cells through phospholipid scramblase SCRM-1. *Nat. Cell Biol.* **9:**541–549.

Wang, X., Y. C. Wu, V. A. Fadok, M. C. Lee, K. Gengyo-Ando, L. C. Cheng, D. Ledwich, P. K. Hsu, J. Y. Chen, B. K. Chou, P. Henson, S. Mitani, and D. Xue. 2003. Cell corpse engulfment mediated by C. elegans phosphatidylserine receptor through CED-5 and CED-12. *Science* 302:1563–1566.

Wang, Z., A. T. Larregina, W. J. Shufesky, M. J. Perone, A. Montecalvo, A. F. Zahorchak, A. W. Thomson, and A. E. Morelli. 2006. Use of the inhibitory effect of apoptotic cells on dendritic cells for graft survival via T-cell deletion and regulatory T cells. *Am. J. Transplant.* 6:1297–1311.

Watanabe, Y., Y. Hashimoto, A. Shiratsuchi, T. Takizawa, and Y. Nakanishi. 2005. Augmentation of fatality of influenza in mice by inhibition of phagocytosis. *Biochem. Biophys. Res. Commun.* 337:881–886.

Wert, S. E., M. Yoshida, A. M. LeVine, M. Ikegami, T. Jones, G. F. Ross, J. H. Fisher, T. R. Korfhagen, and J. A. Whitsett. 2000. Increased metalloproteinase activity, oxidant production, and emphysema in surfactant protein D gene-inactivated mice. *Proc. Natl. Acad. Sci. USA* 97:5972–5977.

Williamson, P., and R. A. Schlegel. 2004. Hide and seek: the secret identity of the phosphatidylserine receptor. *J. Biol.* 3:14.

Winau, F., G. Hegasy, S. H. Kaufmann, and U. E. Schaible. 2005. No life without death—apoptosis as prerequisite for T cell activation. *Apoptosis* 10:707–715.

Wood, W., M. Turmaine, R. Weber, V. Camp, R. A. Maki, S. R. McKercher, and P. Martin. 2000. Mesenchymal cells engulf and clear apoptotic footplate cells in macrophageless PU.1 null mouse embryos. *Development* 127:5245–5252.

Wu, Y., S. Singh, M. M. Georgescu, and R. B. Birge. 2005. A role for Mer tyrosine kinase in alphavbeta5 integrin-mediated phagocytosis of apoptotic cells. *J. Cell Sci.* 118:539–553.

Wu, Y., N. Tibrewal, and R. B. Birge. 2006. Phosphatidylserine recognition by phagocytes: a view to a kill. *Trends Cell Biol.* 16:189–197.

Xu, W., A. Roos, M. R. Daha, and C. van Kooten. 2006a. Dendritic cell and macrophage subsets in the handling of dying cells. *Immunobiology* 211:567–575.

Xu, W., A. Roos, N. Schlagwein, A. M. Woltman, M. R. Daha, and C. van Kooten. 2006b. IL-10-producing macrophages preferentially clear early apoptotic cells. *Blood* 107:4930–4937.

Yamanaka, M., S. Eda, and M. Beppu. 2005. Carbohydrate chains and phosphatidylserine successively work as signals for apoptotic cell removal. *Biochem. Biophys. Res. Commun.* 328:273–280.

Yoshida, H., K. Kawane, M. Koike, Y. Mori, Y. Uchiyama, and S. Nagata. 2005. Phosphatidylserine-dependent engulfment by macrophages of nuclei from erythroid precursor cells. *Nature* 437:754–758.

Yrlid, U., and M. J. Wick. 2000. Salmonella-induced apoptosis of infected macrophages results in presentation of a bacteria-encoded antigen after uptake by bystander dendritic cells. *J. Exp. Med.* 191:613–624.

Zhao, J., Q. Zhou, T. Wiedmer, and P. J. Sims. 1998. Level of expression of phospholipid scramblase regulates induced movement of phosphatidylserine to the cell surface. *J. Biol. Chem.* 273:6603–6606.

Zhou, Q., J. Zhao, T. Wiedmer, and P. J. Sims. 2002. Normal hemostasis but defective hematopoietic response to growth factors in mice deficient in phospholipid scramblase 1. *Blood* 99:4030–4038.

Zhou, Z., E. Hartwieg, and H. R. Horvitz. 2001. CED-1 is a transmembrane receptor that mediates cell corpse engulfment in C. elegans. *Cell* 104:43–56.

22

Regulation and Antimicrobial Function of Inducible Nitric Oxide Synthase in Phagocytes

CHRISTIAN BOGDAN

Phagocytes are essential components of the innate and adaptive immune response to infectious pathogens. They produce a spectrum of either constitutive or inducible antimicrobial effector molecules that are generated by oxygen-independent or oxygen-dependent pathways and help to contain and kill viruses, bacteria, protozoa, fungi, and helminths (Table 1). Most of these mechanisms have been reported for both neutrophils and macrophages. However, there are also notable differences. Whereas (human) neutrophils are prepacked with a load of antimicrobial peptides, proteins, enzymes, and chelators in their primary (azurophilic) (e.g., α-defensins, bactericidal/permeability-increasing protein [BPI], serine proteases with microbicidal activity ["serprocidins," i.e., cathepsin G, elastase, protease 3, azuricidin/CAP37]), secondary (specific), or tertiary (gelatinase) granules (e.g., lysozyme, lactoferrin, lipocalin, cathelicidin CAP18/LL37, matrix metalloproteases MMP8, MMP9, and MMP25), only a very limited set of antimicrobial proteins and peptides is found in monocytes and macrophages of different origin (Levy, 2004). For example, a family of arginine- and lysine-rich murine microbicidal proteins with high homology to histones has been described in mouse-resident peritoneal macrophages (Hiemstra et al., 1993), and α-defensins have been detected in human blood monocytes (Agerbeth et al., 2000). Recent data suggest that human macrophages might become secondary carriers of antimicrobial proteins and peptides (e.g., α-defensin HNP-1) via phagocytosis of apoptotic neutrophils (Tan et al., 2006) or acquire enhanced microbicidal activity when exposed to proteolytic enzymes of neutrophils (Ribeiro-Gomes et al., 2007; Speer et al., 1984). High amounts of myeloperoxidase are usually only found in neutrophils and blood monocytes, but not in mature tissue macrophages (Kaneda et al., 1980; Klebanoff et al., 1983). The inducible or type 2 nitric oxide synthase (iNOS, NOS2) was originally discovered in macrophages (Stuehr and Marletta, 1985, 1987) and later also described in neutrophils, but, at least in vitro, macrophages are a much more potent source of nitric oxide (NO) than granulocytes (Padgett and Pruett, 1995; Wheeler et al., 1997). The expression of iNOS is not restricted to cells of the immune system (macrophages, neutrophils, dendritic cells, NK cells, and possibly also T cells), but can be also induced in many other cell types such as keratinocytes, epithelial cells, endothelial cells, fibroblasts, hepatocytes, and chondrocytes (Bogdan, 2000, 2001a; Vig et al., 2004).

In this chapter I will discuss the regulation and function of iNOS in phagocytes. The focus will be on more recent aspects of the expression of iNOS in macrophages, the molecular mechanisms of iNOS-dependent control of infectious pathogens, and the cross-regulation of iNOS and arginase. For detailed summaries of the role of iNOS/NO in the immune system and during infectious diseases, the reader is referred to earlier in-depth reviews (Bogdan, 2000, 2001a, 2004; Fang, 2004; Nathan and Shiloh, 2000).

INDUCTION AND REGULATION OF iNOS

iNOS is a homodimeric enzyme that in the presence of molecular oxygen converts the amino acid L-arginine into N^{ω}-hydroxy-L-arginine and further into citrulline and ·NO radical (in the following abbreviated as "NO"). Dimerization of the enzyme requires the binding of calmodulin and the incorporation of iron protoporphyrin IX (heme), is supported by the presence of L-arginine and (6R)-tetrahydrobiopterin (BH_4), and is stabilized by divalent zinc cations (Kolodziejski et al., 2003; Stuehr et al., 2004). The complex oxidoreductase reaction involves multiple cofactors and cosubstrates (BH_4, NADPH, flavin adenine mononucleotide [FMN], flavin adenine dinucleotide [FAD], heme, thiol) (Stuehr et al., 2004) (Fig. 1).

In contrast to the constitutively expressed neuronal NO synthase (nNOS or NOS1) and endothelial NO synthase (eNOS or NOS3), which primarily fulfill homeostatic functions in the central nervous system or cardiovascular system, respectively, iNOS mRNA and protein are not readily detectable in strictly resting, i.e., nonstimulated macrophages. Expression of iNOS in macrophages is strongly and

---

Christian Bogdan, University Clinic of Erlangen, Erlangen, Germany.

TABLE 1 Overview of antimicrobial effector mechanisms of phagocytes

Mechanism	Constitutive/rapidly available	Cytokine/TLR-inducible
Oxygen-independent	Enzymes (e.g., lysozyme, esterase, gelatinase)	Acidification
	Antimicrobial peptides (defensins, bactericidal/permeability-increasing protein [BPI], serprocidins) and proteins (histones)	Tryptophan depletion (indolamine 2,3-dioxygenase)
	Iron chelators (lactoferrin)	Arginine depletion (arginase)
	Mn^{2+} chelators (calprotectin)	Iron chelators (lipocalin-2)
	DNA webs ("neutrophil extracellular traps," NETs)	Small GTPases
		Tumor necrosis factor
Oxygen-dependent	NADPH-oxidase (phox)	iNOS, NOS2
	Myeloperoxidase (MPO)	NADPH oxidase (phox)
	Haber-Weiss reaction	
	Catalytic antibodies	

rapidly induced by a proinflammatory cytokine stimulus (prototypically interferon-γ [IFN-γ]) in combination with a viral or microbial compound that functions as a Toll-like receptor (TLR) ligand (e.g., lipopolysaccharide [LPS]; flagellin; mycobacterial proteins; *Helicobacter pylori* urease; hemozoin; *Trypanosoma cruzi*; viral, bacterial, or protozoan DNA; unmethylated CpG desoxyoligonucleotides; double-stranded RNA) or with the ligand or cross-linker of a costimulatory cell surface receptor (e.g., CD40 ligand, anti-CD8, anti-Fcε receptor IIb [anti-CD23]) (Adachi et al., 2006; Bergeron and Olivier, 2006; Bogdan, 2000, 2004; Gobert et al., 2001; Jaramillo et al., 2003; Pindado et al., 2007; Punturieri et al., 2004; Sable et al., 2007). Additional factors that have been reported to coinduce iNOS in macrophages include (i) environmental factors such as hypoxia (Albina et al., 1995; Melillo et al., 1995; Mi et al., 2008), acidic pH (Bellocq et al., 1998;), hyperthermia (Pritchard et al., 2005), ionizing radiation (McKinney et al., 2000), or monosodium urate crystals (Jaramillo et al.,

2004) (the latter presumably activate macrophages via CD14 and/or the NALP3 inflammasome [Martinon, 2008; Scott et al., 2006]); (ii) estrogens (Karpuzoglu and Ahmed, 2006; Lezama-Davila et al., 2007; Osorio et al., 2008); and (iii) potassium channels (Kaushal et al., 2007; Lowry et al., 1998).

iNOS induction and suppression in macrophages are regulated at multiple levels: gene transcription, mRNA stability, mRNA translation, protein stability, and substrate availability.

iNOS Gene Transcription

There are three major signaling pathways for the activation of the iNOS promoter in macrophages, as follows.

- The first pathway is triggered by IFN-γ (type II IFN) or IFN-α/β (type I IFN), which cause the phosphorylation, dimerization, and nuclear translocation of the signal transducer and activator of transcription 1 (STAT1) or STAT1 and STAT2, respectively (Platanias, 2005). STAT1 homodimers or STAT1/STAT2 heterodimers will activate the transcription of iNOS either directly via binding to γ-activated sites (GAS) or IFN-stimulated response elements (ISREs) within the iNOS promoter, or indirectly via induction of the interferon-regulatory factor-1 (IRF-1), which will bind to ISREs of the iNOS enhancer. In addition, IFN-γ also triggers the release of endogenous tumor necrosis factor (TNF), which in an autocrine manner will then activate NF-κB binding to the iNOS promoter (see below, second pathway) (Fig. 2). The importance of these signaling cascades is underscored by the observation that macrophages with a targeted disruption of the STAT1, IRF-1, or TNF gene show hardly any production of NO in response to stimulation with IFN-γ or IFN-γ plus LPS (Durbin et al., 1996; Gao et al., 1997, 1998; Kamijo et al., 1994; Martin et al., 1994; Meraz et al., 1996; Rodig et al., 1998; Vila-del Sol et al., 2007; Wilhelm et al., 2001).

- The second pathway is triggered by ligands of the Toll-like receptors TLR2, TLR4, TLR5, or TLR9 that use the myeloid differentiation primary-response protein 88 (MyD88) as a signaling adaptor and results in the nuclear translocation of NF-κB and/or the activation of mitogen-activated protein kinases (MAPKs) and the formation of the transcription factor-activated protein-1 (AP-1) (Kleinert et al., 2004; Uematsu and Akira, 2008; Xie et al., 1994) (Fig. 2). NF-κB and AP-1 will directly bind to the respective sites in the iNOS pro-

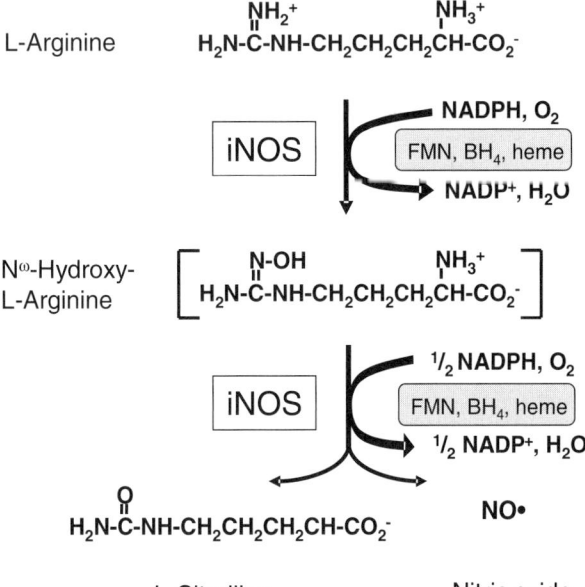

FIGURE 1 Reaction catalyzed by iNOS. FAD, flavin adenine dinucleotide; FMN, flavin adenine mononucleotide; NADP, nicotinamide adenine dinucleotide phosphate; H_4B, tetrahydrobiopterin.

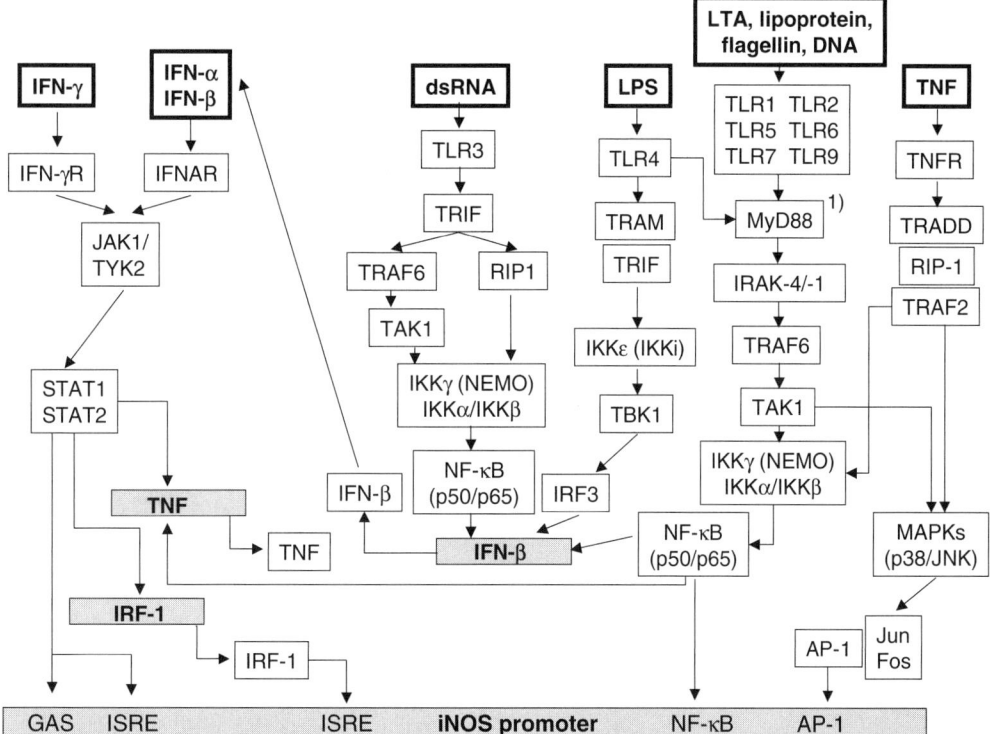

FIGURE 2 Overview of the major signaling pathways leading to the induction of iNOS in macrophages. Please note that not all signaling molecules involved are depicted. iNOS-inducing stimuli are surrounded by bold boxes; promoters are shown in gray. AP-1, activated protein-1; GAS, gamma-activated site; IFN, interferon; IKK, inhibitor of nuclear factor κB (IκB) kinase (IKKε is also termed inducible IKK or IKKi); IRAK, IL-1 receptor-associated kinase; IRF, interferon regulatory factor; ISRE, interferon-stimulated response element; JAK, Janus kinase; JNK, c-Jun NH_2-terminal kinase; NF-κB, nuclear factor κB; MAPK, mitogen-activated protein kinase; MyD88, myeloid differentiation primary response protein 88; NEMO, NF-κB essential modulator (also termed IKKγ); RIP1, receptor-interacting protein 1; STAT, signal transducer and activator of transcription; TAB1, TAK1-binding protein 1; TAK1, transforming-growth-factor-β-activated kinase; TBK1, TRAF-family-member-associated NF-κB activator (TANK)-binding kinase 1; TLR, toll-like receptor; TNF, tumor necrosis factor; TRADD, TNF receptor-associated death domain; TRAF, tumor necrosis factor receptor-associated factor; TRAM, TRIF-related adaptor molecule, also known as TIR-domain-containing molecule 2 (TICAM2); TRIF, Toll/interleukin 1 receptor (TIR)-domain-containing adaptor protein inducing IFN-β (also known as TIR-domain-containing molecule 1, TICAM1); TYK2, tyrosine kinase 2. In the case of TLR2 and TLR4, a second adaptor protein termed TIRAP (TIR-domain-containing adaptor protein) or MAL (MyD88-adaptor-like protein) is required, which is not shown.

moter or act indirectly by stimulating the production of TNF (see above) or IFN-α/β (see below, third pathway).

- The third pathway is elicited by TLR3 or TLR4 ligands (e.g., viral double-stranded RNA and LPS), is independent of MyD88 but dependent on the Toll/interleukin-1 (IL-1)-receptor (TIR) domain containing adaptor protein inducing IFN-β (TRIF), and leads to the activation of both NF-κB and IRF-3, which then switch on the IFN-β gene and the IFN-β feedback loop (Gao et al., 1998; Uematsu and Akira, 2006) (Fig. 2). There is also evidence that TLR3 ligands (viral double-stranded RNA) can induce iNOS via activation of group IVA cytosolic phospholipase A_2 (cPLA$_2$α) (not depicted in Fig. 2). cPLA$_2$α promotes the release of free arachidonic acid, which is converted by cyclooxygenase-2 into pros-

taglandin E_2 that is capable of upregulating iNOS expression (Pindado et al., 2007).

In macrophages as well as other cell types, numerous additional transcription factors (e.g., octamer factor, peroxisome proliferator-activated receptors, hypoxia-inducible factor-1, nuclear factor IL-6, zinc finger Kruppel-like factor 4) have been described that might contribute to the activation of the iNOS promoter in vitro and in vivo, as reviewed elsewhere (Feinberg et al., 2005; Kleinert et al., 2004). The induction of iNOS can be positively or negatively modulated by the activity of various signaling enzymes and their products that intersect with the above-mentioned pathways (e.g., protein kinase C, phosphoinositide-3-kinase, protein tyrosine phosphatases, phosphatase, and tensin homologue deleted on chromosome 10 [Pten], and MAPKs such as extracellular signal-regulated kinase 1 and 2 [ERK1/2]) (Blanchette et al., 2007; Bogdan, 2000, 2001a; Jaramillo et al., 2003; Kristof et al., 2006; Kuroda et al., 2008; Pindado et al., 2007).

iNOS mRNA Stability

Several studies have demonstrated that the expression of iNOS is also regulated posttranscriptionally on the level of mRNA stability. Since the first demonstration of regulation of iNOS mRNA stability by transforming growth factor-β1 (TGF-β1) (Vodovotz et al., 1993), several proteins have been described that bind to AU-rich elements within in the 3'-untranslated region of the iNOS mRNA and either stabilize (e.g., tristetraproline [TTP], embryonic lethal abnormal vision-like protein HuR, human polypyrimidine tract binding protein [PTB], also termed heterogenous nuclear ribonucleoprotein I, hnRNP I]) or destabilize the iNOS mRNA (e.g., KH-type splicing regulatory protein [KSRP], mouse PTB) (Korhonen et al., 2007; Soderberg et al., 2007; and references therein). iNOS mRNA stability can also be modulated by infectious pathogens (Bergeron and Olivier, 2006).

iNOS Protein Synthesis, Stability, and Activity

Several mechanisms have been reported that allow for modulation of the expression of iNOS protein and NO production even if the level of iNOS mRNA remains constant. TGF-β inhibited the synthesis and the stability of iNOS protein in inflammatory macrophages and RAW264.7 macrophage-like cells, with the latter effect most likely resulting from enhanced ubiquitination and proteasomal degradation (Kolodziejski et al., 2002; Mitani et al., 2005; Takaki et al., 2006; Vodovotz et al., 1993). Analogues of cyclic AMP (cAMP) or activators of cAMP synthesis inhibited the ubiquitination of iNOS in macrophages and thereby enhanced the stability of iNOS protein (Won et al., 2004). Spermine, a product of the ornithine metabolism (see below), impeded the de novo synthesis of iNOS protein in RAW264.7 cells (Bussiere et al., 2005). Cultures of IFN-γ/LPS-stimulated macrophages, which were severely deficient for extracellular arginine (concentration, <100 μM), not only generated strongly reduced amounts of NO, but also showed a strikingly reduced synthesis of iNOS protein, whereas the level of iNOS mRNA and the overall protein synthesis remained unaltered (Chaturvedi et al., 2007; El-Gayar et al., 2003). Thus, arginine is required for iNOS activity (substrate function) as well as for iNOS translation (regulatory function). The molecular mechanism underlying the arginine-dependent regulation of iNOS is a matter of current investigations. An intermediate concentration of arginine (e.g., 100 to 400 μM) or the targeted disruption of the cationic amino acid transporter CAT2 (encoded by Scl7A2) impeded the production of NO without strong alteration of the expression of iNOS protein (El-Gayar et al., 2003; Nicholson et al., 2001; Rutschman et al., 2001; Yeramian et al., 2006).

LOCALIZATION OF iNOS IN MACROPHAGES

The intracellular localization of iNOS is critical for its function and has been best studied in various populations of mouse macrophages (bone marrow-derived macrophages, peritoneal exudate macrophages, and macrophage cell lines RAW264.7 and J774). Analyses by Western blots, electron microscopy, and confocal laser scanning microscopy uniformly revealed that, in uninfected IFN-γ/LPS-stimulated macrophages, iNOS protein was equally distributed in the cytosol and a membranous (particulate) compartment that consisted of 50- to 80-nm vesicles. The vesicles are distinct from lysosomes and peroxisomes and have been provisionally termed "nitroxosomes" (Miller et al., 2004; Vodovotz et al., 1995; Webb et al., 2001). In macrophages, iNOS also associates with the cortical actin cytoskeleton immediately below the plasma membrane (Webb et al., 2001).

Upon infection of macrophages with wild-type (virulent) strains of mycobacteria (Mycobacterium avium, M. bovis BCG, M. smegmatis, and M. tuberculosis), Salmonella enterica serovar Typhimurium, or Leishmania mexicana, either IFN-γ/LPS-induced iNOS did not colocalize with endosomal, lysosomal, or phagolysosomal compartments, or it was targeted toward pathogen-containing vacuoles only after a delay of 1 to 2 days. In contrast, after phagocytosis of immunoglobulin G- or complement-coated latex beads, Escherichia coli, M. smegmatis lacking the antioxidant repair enzyme methionine sulfoxide reductase A, and serovar Typhimurium mutants that were deficient of the Salmonella pathogenicity island 2 (SPI2)-activated macrophages, all expressed iNOS protein in the close vicinity of the phagosomes (Chakravortty et al., 2002; Douglas et al., 2004; Miller et al., 2004; Vodovotz et al., 1993; Webb et al., 2001). iNOS recruitment required a functional actin cytoskeleton and the interaction of iNOS with the scaffolding adaptor proteins α-actinin-4 and ezrin/radizin/moesin-binding phosphoprotein 50 (EBP50) (Daniliuc et al., 2003; Davis et al., 2007; Miller et al., 2004; Webb et al., 2001). EBP50 binds to iNOS and ezrin, a linker between the plasma membrane and the actin cytoskeleton. Phagosomes harboring live M. bovis BCG or M. tuberculosis H37Rv showed a reduced capacity to retain EBP50 and a lower recruitment of iNOS to the phagosomes (Davis et al., 2007). Together, these data illustrate that the (re-)localization of iNOS is highly relevant for its antimicrobial activity or, conversely, for the survival of intracellular pathogens.

ANTIMICROBIAL FUNCTION OF iNOS

One of the key functions of iNOS, its product NO, and its oxidation products (collectively termed reactive nitrogen intermediates [RNIs]) is the antimicrobial activity, which, from an operational point of view, can be divided into four major categories, as follows (Fig. 3).

1. Direct ("toxic") effects of NO on structural components, the replication machinery, nucleic acids, virulence factors, metabolic enzymes, and pathways of infectious pathogens
2. iNOS-dependent antimicrobial effects that are independent of its product NO
3. NO-mediated inhibition of microbial evasion and resistance mechanisms
4. Immunostimulatory function of NO.

Direct Antimicrobial and Cytotoxic Effects of NO

Major targets of NO are DNA molecules, where NO can cause mutations and oligonucleosomal fragmentation as well as inhibition of synthesis and repair; structural proteins, which are altered by S-nitrosylation, ADP-ribosylation, or tyrosine nitration; enzymes, which can be inactivated by disruption of Fe-S clusters, zinc fingers, or heme groups; and membrane lipids that can be modified by peroxidation. The multiplex chemical reactivities of RNIs

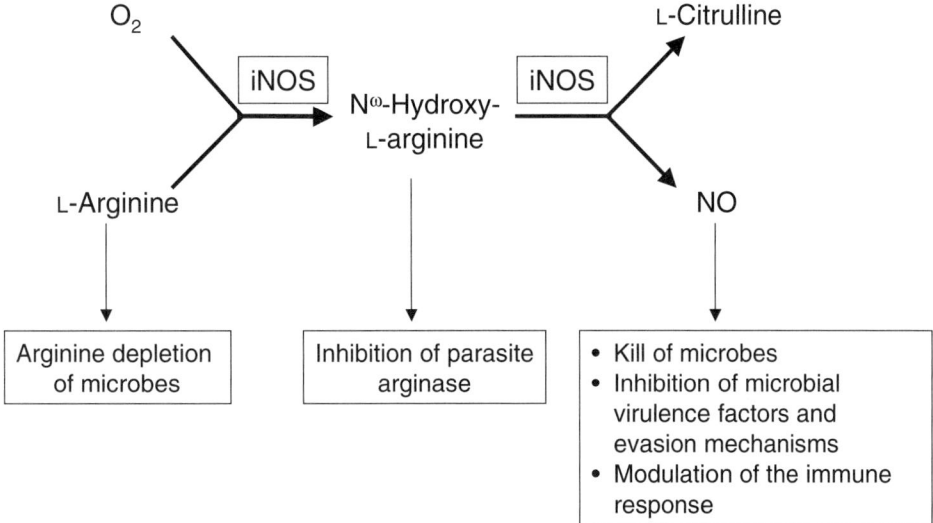

FIGURE 3 Mechanisms of the antimicrobial activity of iNOS/NO. For details see text.

and their relevance for the direct antimicrobial activity of NO has been discussed in detail elsewhere (Bogdan, 2001b; Fang, 2004; Holzmuller et al., 2002). NO is capable to react with superoxide anions (O_2^-) generated by the phagocyte NADPH oxidase or by the pathogen itself and thereby lead to the formation of peroxynitrite ($ONOO^-$), an antimicrobial effector molecule in its own right. Importantly, the cytotoxic function of NO and peroxynitrite is not restricted to pathogens, but may also extend to host cells. This explains why a strong expression of iNOS in infected organs is frequently associated with significant tissue destruction (Bogdan, 2000; Zaki et al., 2005).

iNOS-Dependent, but NO-Independent Antimicrobial Effects

iNOS activity not only leads to the production of NO, but also to the depletion of L-arginine and the intermittent generation of N^ω-hydroxy-L-arginine. Deficiency of L-arginine was reported to impede the growth of a number of diverse pathogens such as *T. cruzi*, African trypanosomes, *Giardia lamblia*, or *Schistosoma mansoni* (Eckmann et al., 2000; Gobert et al., 2000; Olds et al., 1980; Piacenza et al., 2001). The presence of an arginine biosynthetic pathway in obligatively intracellular bacteria such as *Ehrlichia chaffensis* might rescue these bacteria after arginine depletion by iNOS (Hotopp et al., 2006). N^ω-Hydroxy-L-arginine is a potent inhibitor of arginases, including arginases of eukaryotic pathogens (Kropf et al., 2005), which has led to the suggestion that this intermediate of the iNOS reaction might exert antimicrobial activity (Iniesta et al., 2001). To date, it has not yet been formally demonstrated that iNOS-derived N^ω-hydroxy-L-arginine will be able to block arginases of pathogens in vivo and thereby contributes to their control.

Inhibition of Microbial Evasion and Resistance

The antimicrobial activity of NO can also result from the invalidation of microbial mechanisms that allow pathogens to withstand the innate and adaptive immune system. In this case, NO does not cause the death, but rather initiates signaling processes that modulate the phenotype of the microbes. For example, NO was found to antagonize the acid tolerance response that is under control of the PhoPQ regulon and protects *Salmonella* against the acidity of the stomach (Bourret et al., 2008). Similarly, NO caused dispersal of the biofilms formed by *Pseudomonas aeruginosa*, which otherwise protect these bacteria from the action of phagocytes, antimicrobial peptides, and antibiotics (Barraud et al., 2006). iNOS-derived NO was also reported to overcome the inhibition of phagosomal maturation seen in macrophages infected with *Leishmania donovani*: in the presence of iNOS and NO the *Leishmania*-induced breakdown of periphagosomal F-actin was inhibited so that phagosome-lysosome fusion was no longer blocked (Winberg et al., 2007). There is also evidence that NO inhibits the escape of *Listeria monocytogenes* from the phago(lyso)some and thereby promotes the bacterial degradation within this hostile environment (Myers et al., 2003). Another example for NO-mediated modification of intracellular compartments was seen in *Coxiella burnetii*-infected macrophages, where iNOS/NO blocked the formation of mature (large) acidic phagolysosome-like vacuoles that represent the preferred site of replication for *C. burnetii* (Howe et al., 2002; Zamboni and Rabinovitch, 2004). Finally, iNOS/NO was reported to inhibit the gene expression, protein synthesis, or function of virulence factors of a variety of pathogens (e.g., pathogenicity island-2-encoded effector molecules of *S. enterica* serovar Typhimurium, Shiga toxin of enterohemorrhagic *E. coli*, and virulence-associated thiolactone autoinducer of *Staphylococcus aureus*) (McCollister et al., 2005; Rothfork et al., 2004; Vareille et al., 2007), which virtually disarms them and/or makes them susceptible to rapid control and kill.

The results discussed above might generate the impression that iNOS/NO blocks replication and survival of pathogens in the host organism without eliciting a counterregulatory response. This is not the case. In fact, numerous studies have demonstrated that in bacteria, protozoa, and fungi nitrosative stress will induce potent mechanisms of resistance to NO (Bogdan, 2004; de Jesus-Berrios et al., 2003; Fang, 2004; Poole, 2005; Seib et al., 2006). Thus, it is the balance between iNOS/NO-dependent direct or indirect antimicrobial effector pathways and NO-induced antioxidant molecules and processes that determine the fate of an NO-sensitive pathogen.

Immunostimulatory Function of NO

Various if not all types of cells of the immune system are susceptible to the activating or inhibitory effects of NO (Bogdan, 2000, 2001a, 2001b; Niedbala et al., 2002, 2007). It was therefore not surprising to see that iNOS-derived NO can also pave or enhance certain functions of phagocytes and antigen-presenting cells, which secondarily helps to control an infectious pathogen. Microarray analyses of wild-type and iNOS-deficient IFN-γ-activated macrophages revealed that iNOS-derived NO controls the transcription of a considerable number of genes (Ehrt et al., 2001). Processes in neutrophils, macrophages, or dendritic cells that were recently discovered to be subject to positive regulation by iNOS-derived NO include (i) the processing and presentation of bacterial carbohydrate antigens by the major histocompatibility complex class II (MHC-II) pathway (Duan et al., 2008); (ii) the induction of MHC-II and of the costimulatory molecules CD80 and CD86 on the surface of immature dendritic cells (Huang et al., 2008); (iii) the inhibition of premature degradation of the MHC-II-associated invariant chain (CD74) in maturing dendritic cells, which allows for correct MHC-II folding and targeted transport to the late endosomes (Huang et al., 2008); and (iv) the induction of apoptosis of macrophages infected with *Streptococcus pneumoniae*, which occurs in parallel to the killing of the bacteria and contributes to the shutdown of inflammation (Marriott and Dockrell, 2006). With respect to this proapoptotic function of NO, it is important to mention that equally convincing data exist that demonstrate an antiapoptotic function of NO in infected host cells. NO-mediated protection of macrophages or dendritic cells from apoptosis might result from the induction of catalase, which shields these cells against the cytotoxic activity of hydrogen peroxide (Yoshioka et al., 2006). These antiapoptotic effects of NO help to prevent rapid death of immune cells and uncontrolled tissue damage during infections (Alam et al., 2008).

iNOS VERSUS ARGINASE IN INFECTIOUS DISEASES

In mammals, L-arginine is a substrate not only for iNOS and the other NO synthases, but also for arginases and arginine decarboxylase (ADC). Whereas the ADC-catalyzed degradation of arginine to agmatine appears to have a minor impact in vivo, the two isoforms of arginase, arginase I and arginase II, are truely important for the metabolism of arginine (Mori, 2007). Arginase I is primarily expressed by hepatocytes (therefore, sometimes also termed hepatic arginase) and by myeloid cells (neutrophils, macrophages, and dendritic cells in the mouse; [meta]myelocytes and neutrophils only in humans) (Munder et al., 1999). The enzyme is localized in the cytosol with the exception of human neutrophils, which store arginase I in azurophilic or gelatinase granules (Jacobsen et al., 2007). Arginase II (extrahepatic arginase) is a mitochondrial enzyme found in the kidney, the small intestine, and the brain, but can be also expressed by macrophages (Gobert et al., 2002; Gotoh and Mori, 1999; Johann et al., 2007). Arginases are homotrimeric enzymes that use manganese cations as cofactors and catalyze the conversion of L-arginine into ornithine and urea (Fig. 4). In macrophages, arginase I is induced by cytokines (Th2 cytokines [IL-4, IL-10, IL-13], TGF-β, and macrophage colony-stimulating factor), microbial products and TLR ligands (e.g., LPS, *Trypanosoma brucei brucei*, and *H. pylori*), hypoxia, prostaglandins, cAMP, 1α,25-dihydroxyvitamin D_3, and antibiotics (e.g., azithromycin) (Duleu et al., 2004; Ehrchen et al., 2007; Gobert et al., 2002; Morris, 2002; Murphy et al., 2008; Rodriguez et al., 2005). In macrophages the protein levels of induced arginase I are generally much higher than the constitutive expression of arginase II (Munder et al., 1999).

Arginases and their products fulfill four major metabolic functions. First, both arginase I and II can deprive cells of their (exogenous) arginine supply. Some of the possible

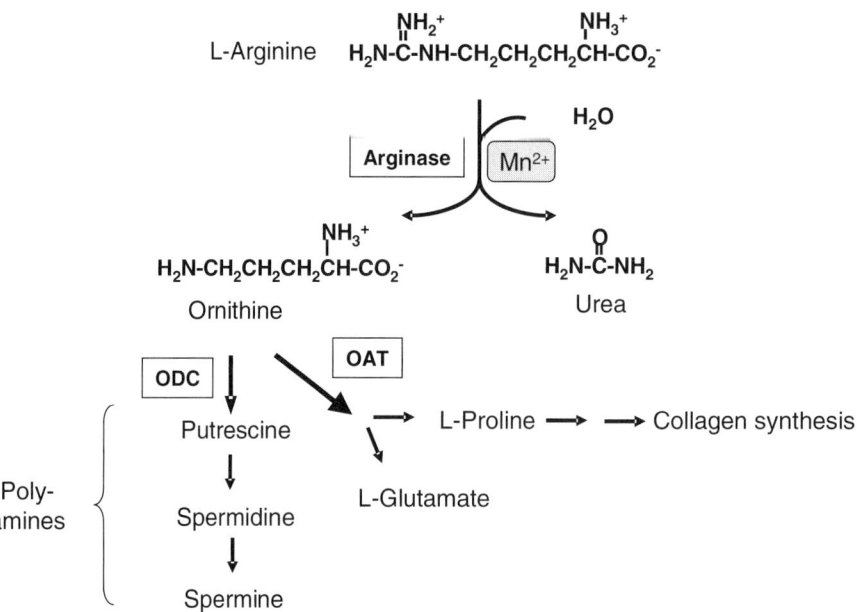

FIGURE 4 Overview of the arginase pathway and subsequent reactions. For details see text. Mn^{2+}, manganese cations; OAT, ornithine aminotransferase; ODC, ornithine decarboxylase.

consequences are discussed below. Second, arginase I is one of the enzymes of the urea cycle in the liver, which serves to generate excretable urea. Third, ornithine is a substrate for the ornithine aminotransferase, which takes part in the generation of proline, a precursor of collagen synthesis by fibroblasts. This process is critical for tissue regeneration and wound healing. Fourth, ornithine can be metabolized by ornithine decarboxylase to produce the polyamine putrescine that is further converted to the polyamines spermidine and spermine. Polyamines are important regulators of the proliferation, migration, and apoptosis of cells (Morris, 2002). These pathways help to explain the diverse functions of arginase during immune responses to infectious pathogens as well as the cross-regulation between arginase and iNOS as discussed below.

Arginase Counteracts iNOS-Dependent Pathogen Killing

Although arginase has an approximately 1,000-fold lower affinity for arginine compared with iNOS, it exhibits an at least 1,000-fold higher substrate conversion rate (V_{max}) (Wu and Morris, 1998). Therefore, expression of arginase by host macrophages (or by the pathogen itself) can very efficiently deplete arginine. Depending on the residual arginine concentration in the extracellular milieu the NO-generating activity or even the protein expression of iNOS was drastically reduced in the presence of arginase (Chaturvedi et al., 2007; El-Gayar et al., 2003; Rutschman et al., 2001). Accordingly, the iNOS-dependent control of NO-sensitive pathogens such as *Leishmania major*, *T. brucei brucei*, *T. cruzi*, or *H. pylori* was significantly improved in vivo when the host cell arginase I was blocked (Gobert et al., 2000; Iniesta et al., 2005; Kropf et al., 2005; Stempin et al., 2004) or the pathogen arginase was deleted (Chaturvedi et al., 2007; Gobert et al., 2001). Because *Leishmania* and *Trypanosoma* require polyamines for growth (Gaur et al., 2007; Piacenza et al., 2001), it is conceivable that host cell arginase supports parasite survival not only by impeding NO production, but also indirectly by supplying polyamines (*Leishmania* otherwise express their own arginase). In addition, polyamines such as spermine inhibit the production of proinflammatory cytokines (Zhang et al., 1997) and the translation of iNOS protein (Bussiere et al., 2005). Arginine depletion by macrophage or neutrophil arginase I or by *H. pylori* arginase was reported to downregulate the protein expression of the CD3ζ chain and to suppress the proliferation of T cells, which might further abrogate pathogen control (Munder et al., 2006; Zea et al., 2006).

Arginase Promotes Pathogen Control

On theoretical grounds, hyperexpression of arginase by myeloid cells of the host organism might deprive protozoan or metazoan pathogens of the essential amino acid arginine. In this respect, arginase could act synergistically with iNOS. Indeed, macrophage-derived arginase was shown to mediate killing of *S. mansoni* schistosomula in vitro (89). Likewise, arginine depletion promoted apoptotic death of *T. cruzi*, although there was no formal demonstration that arginase might act in this manner against trypanosomes in vivo (Piacenza et al., 2001). There is also recent evidence that arginase I of human granulocytes translocates toward the phago(lyso)some after phagocytosis of yeasts and contributes to the early antifungal activity of neutrophils, presumably through arginine depletion (Munder et al., 2005).

In a *Citrobacter rodentium* model of murine colitis, arginine supplementation and arginase I activity were strictly required to prevent the iNOS-dependent tissue damage triggered by this enteropathogenic bacterium. Treatment of the mice with an inhibitor of arginase or an inhibitor of ornithine decarboxylase (which suppresses the generation of polyamines) drastically exacerbated the colonic lesions and promoted the death of the mice (Gobert et al., 2004). These data indicate that arginase-mediated antagonism of iNOS can be host protective when tissue-repair processes (such the formation of collagen) and the control of inflammation (e.g., by polyamines) are more relevant for the resolution of the disease than the elimination of the eliciting pathogen.

Arginase Promotes Inflammation and Tissue Fibrosis

Murine experimental schistosomiasis is an excellent example to highlight the third category of arginase function during infectious diseases. Immunization and infection with *S. mansoni* eggs leads to a progressive granulomatous response in the lung or liver, depending on the exact infection model. The fibrotic granulomata of the lung or liver formed around the eggs are driven by Th2 cells and require the activity of arginase and ornithine aminotransferase. Th1 cells cause cytotoxic tissue damage in these models, but at the same time limit collagen deposition and tissue fibrosis in an iNOS-dependent manner. Thus, a reciprocal antagonism between arginase and iNOS is required to prevent both excess cytotoxicity and uncontrolled granuloma formation (Hesse et al., 2000, 2001; Thompson et al., 2008).

CONCLUSION

Since the discovery of NO production by LPS-stimulated macrophages in 1985 (111), both the regulation and the function of iNOS in macrophages have become far more complex than anticipated. Until now, there is a still growing spectrum of stimuli and intracellular signaling pathways that lead to the transcriptional or posttranscriptional induction of iNOS. Likewise, the number of cell-autonomous and intercellular functions that have been assigned to macrophage-derived NO goes far beyond the initially described direct killing of microbes and tumor cells. Today, we know that even the antimicrobial activity of iNOS in macrophages can originate from various direct or indirect processes that do not necessarily require high-output generation of NO. Furthermore, the interplay between iNOS and arginase has led to additional layers of regulation and function. The long-desired therapeutic application of NO donors and arginase inhibitors only begins to emerge (Iniesta et al., 2005; Kropf et al., 2005; Valdez et al., 2008).

REFERENCES

Adachi, Y., A. L. Kindzelskii, A. R. Petty, J. B. Huang, N. Maeda, S. Yotsumoto, Y. Aratani, N. Ohno, and H. R. Petty. 2006. IFN-gamma primes RAW264 macrophages and human monocytes for enhanced oxidant production in response to CpG DNA via metabolic signaling: roles of TLR9 and myeloperoxidase trafficking. *J. Immunol.* 176:5033–5040.

Agerbeth, B., J. Charo, J. Werr, B. Olsson, F. Idali, L. Lindbom, R. Kiessling, H. Jornvall, H. Wigzell, and G. H. Gudmundsson. 2000. The human antimicrobial and chemotactic peptides LL-37 and α-defensins are expressed by specific lymphocyte and monocyte populations. *Blood* **96:**3086–3093.

Alam, M. S., M. H. Zaki, T. Sawa, S. Islam, K. A. Ahmed, S. Fujii, T. Okamoto, and T. Akaike. 2008. Nitric oxide produced in Peyer's patches exhibits antiapoptotic activity contributing to an antimicrobial effect in murine salmonellosis. *Microbiol. Immunol.* **52:**197-208.

Albina, J. E., W. L. Henry, B. Mastrofrancesco, B.-A. Martin, and J. S. Reichner. 1995. Macrophage activation by culture in an anoxic environment. *J. Immunol.* **155:**4391–4396.

Barraud, N., D. J. Hassett, S. H. Hwang, S. A. Rice, S. Kjelleberg, and J. S. Webb. 2006. Involvement of nitric oxide in biofilm dispersal of *Pseudomonas aeruginosa*. *J. Bacteriol.* **188:**7344–7353.

Bellocq, A., S. Suberville, C. Philippe, F. Bertrand, J. Perez, B. Fouqueray, G. Cherqui, and L. Baud. 1998. Low environmental pH is responsible for the induction of nitric oxide synthase in macrophages. *J. Biol. Chem.* **273:**5086–5092.

Bergeron, M., and M. Olivier. 2006. Trypanosoma cruzi-mediated IFN-gamma-inducible nitric oxide output in macrophages is regulated by iNOS mRNA stability. *J. Immunol.* **177:**6271–6280.

Blanchette, J., P. Pouliot, and M. Olivier. 2007. Role of protein tyrosine phosphatases in the regulation of interferon-[gamma]-induced macrophage nitric oxide generation: implication of ERK pathway and AP-1 activation. *J. Leukoc. Biol.* **81:**835–844.

Bogdan, C. 2000. The function of nitric oxide in the immune system, p. 443–492. *In* B. Mayer (ed.), *Handbook of Experimental Pharmacology*. Volume: *Nitric Oxide*. Springer, Heidelberg, Germany.

Bogdan, C. 2001a. Nitric oxide and the immune response. *Nat. Immunol.* **2:**907–916.

Bogdan, C. 2001b. Nitric oxide and the regulation of gene expression. *Trends Cell Biol.* **11:**66–75.

Bogdan, C. 2004. Reactive oxygen and reactive nitrogen metabolites as effector molecules against infectious pathogens, p. 357–396. *In* S. H. E. Kaufmann, R. Medzhitov, and S. Gordon (ed.), *The Innate Immune Response to Infection*. ASM Press, Washington, DC.

Bourret, T. J., S. Porwollik, M. McClelland, R. Zhao, T. Greco, H. Ischiropoulos, and A. Vazquez-Torres. 2008. Nitric oxide antagonizes the acid tolerance response that protects *Salmonella* against innate gastric defenses. *PLoS ONE* **3:**e1833.

Bussiere, F. I., R. Chaturvedi, Y. Cheng, A. P. Gobert, M. Asim, D. R. Blumberg, H. Xu, P. Y. Kim, A. Hacker, R. A. Casero, Jr., and K. T. Wilson. 2005. Spermine causes loss of innate immune response to *Helicobacter pylori* by inhibition of inducible nitric-oxide synthase translation. *J. Biol. Chem.* **280:**2409–2412.

Chakravortty, D., I. Hansen-Wester, and M. Hensel. 2002. *Salmonella* pathogenicity island 2 mediates protection of intracellular *Salmonella* from reactive nitrogen intermediates. *J. Exp. Med.* **195:**1155–1166.

Chaturvedi, R., M. Asim, N. D. Lewis, H. M. Algood, T. L. Cover, P. Y. Kim, and K. T. Wilson. 2007. L-arginine availability regulates inducible nitric oxide synthase-dependent host defense against *Helicobacter pylori*. *Infect. Immun.* **75:**4305–4315.

Daniliuc, S., H. Bitterman, M. A. Rahat, A. Kinarty, D. Rosenzweig, and N. Lahat. 2003. Hypoxia inactivates inducible nitric oxide synthase in mouse macrophages by disrupting its interaction with alpha-actinin 4. *J. Immunol.* **171:**3225–3232.

Davis, A. S., I. Vergne, S. S. Master, G. B. Kyei, J. Chua, and V. Deretic. 2007. Mechanism of inducible nitric oxide synthase exclusion from mycobacterial phagosomes. *PLOS Pathog.* **3:**1887–1894.

de Jesus-Berrios, M., L. Liu, J. C. Nussbaum, G. M. Cox, J. S. Stamler, and J. Heitman. 2003. Enzymes that counteract nitrosative stress promote fungal virulence. *Curr. Biol.* **13:**1963–1968.

Douglas, T., D. S. Daniel, B. K. Parida, C. Jagannath, and S. Dhandayuthapani. 2004. Methionine sulfoxide reductase A (MsrA) deficiency affects the survival of Mycobacterium smegmatis within macrophages. *J. Bacteriol.* **186:**3590–3598.

Duan, J., F. Y. Avci, and D. L. Kasper. 2008. Microbial carbohydrate depolymerization by antigen-presenting cells: deamination prior to presentation by the MHCII pathway. *Proc. Natl. Acad. Sci. USA* **105:**5183–5188.

Duleu, S., P. Vincendeau, P. Courtois, S. Semballa, I. Lagroye, S. Daulouede, J. L. Boucher, K. T. Wilson, B. Veyret, and A. P. Gobert. 2004. Mouse strain susceptibility to trypanosome infection: an arginase-dependent effect. *J. Immunol.* **172:**6298–6303.

Durbin, J. E., R. Hackenmiller, M. C. Simon, and D. E. Levy. 1996. Targeted disruption of the mouse Stat1 gene results in compromised innate immunity to viral disease. *Cell* **84:**443–450.

Eckmann, L., F. Laurent, T. D. Langford, M. L. Hetsko, J. R. Smith, M. F. Kagnoff, and F. D. Gillin. 2000. Nitric oxide production by human intestinal epithelial cells and competition for arginine as potential determinants of host defense against the lumen-dwelling pathogen *Giardia lamblia*. *J. Immunol.* **164:**1478–1487.

Ehrchen, J., L. Helming, G. Varga, B. Pasche, K. Loser, M. Gunzer, C. Sunderkotter, C. Sorg, J. Roth, and A. Lengeling. 2007. Vitamin D receptor signaling contributes to susceptibility to infection with *Leishmania major*. *FASEB J.* **21:**3208–3218.

Ehrt, S., D. Schnappinger, S. Bekiranov, J. Drenkow, S. Shi, T. R. Gingeras, T. Gaasterland, G. Schoolnik, and C. Nathan. 2001. Reprogramming of the macrophage transcriptome in response to interferon-γ and *Mycobacterium tuberculosis*: signaling roles of nitric oxide synthase-2 and phagocyte oxidase. *J. Exp. Med.* **194:**1123–1139.

El-Gayar, S., H. Thüring-Nahler, J. Pfeilschifter, M. Röllinghoff, and C. Bogdan. 2003. Translational control of inducible nitric oxide synthase by IL-13 and arginine availability in inflammatory macrophages. *J. Immunol.* **171:**4561–4568.

Fang, F. C. 2004. Antimicrobial reactive oxygen and nitrogen species: concepts and controversies. *Nat. Rev. Immunol.* **2:**820–832.

Feinberg, M. W., Z. Cao, A. K. Wara, M. A. Lebedeva, S. Senbanerjee, and M. K. Jain. 2005. Kruppel-like factor 4 is a mediator of proinflammatory signaling in macrophages. *J. Biol. Chem.* **280:**38247–38258.

Gao, J., D. C. Morrison, T. J. Parmely, S. W. Russell, and W. J. Murphy. 1997. An interferon-γ-activated site (GAS) is necessary for full expression of the mouse iNOS gene in response to interferon-γ and lipopolysaccharide. *J. Biol. Chem.* **272:**1226–1230.

Gao, J. J., M. B. Filla, M. J. Fultz, S. N. Vogel, S. W. Russell, and W. J. Murphy. 1998. Autocrine/paracrine IFN-α/β mediates the lipopolysaccharide-induced activation of transcription factor Stat1α in mouse macrophages: pivotal role of Stat1α in induction of the inducible nitric oxide synthase gene. *J. Immunol.* **161:**4803–4810.

Gaur, U., S. C. Roberts, R. P. Dalvi, I. Corraliza, B. Ullman, and M. E. Wilson. 2007. An effect of parasite-encoded arginase on the outcome of murine cutaneous leishmaniasis. *J. Immunol.* **179:**8446–8453.

Gobert, A. P., Y. Cheng, M. Akhtar, B. D. Mersey, D. R. Blumberg, R. K. Cross, R. Chaturvedi, C. B. Drachenberg, J. L. Boucher, A. Hacker, R. A. Casero, Jr., and K. T.

Wilson. 2004. Protective role of arginase in a mouse model of colitis. *J. Immunol.* **173:**2109–2117.

Gobert, A. P., Y. Cheng, J. Y. Wang, J. L. Boucher, R. K. Iyer, S. D. Cederbaum, R. A. Casero, Jr., J. C. Newton, and K. T. Wilson. 2002. Helicobacter pylori induces macrophage apoptosis by activation of arginase II. *J. Immunol.* **168:**4692–4700.

Gobert, A. P., S. Daulouede, M. Lepoivre, J. L. Boucher, B. Bouteille, A. Buguet, R. Cespuglio, B. Veyret, and P. Vincendeau. 2000. L-Arginine availability modulates local nitric oxide production and parasite killing in experimental trypanosomiasis. *Infect. Immun.* **68:**4653–4657.

Gobert, A. P., D. J. McGee, M. Akhtar, G. L. Mendz, J. C. Newton, Y. Cheng, H. L. T. Mobley, and K. T. Wilson. 2001. Helicobacter pylori arginase inhibits nitric oxide production by eukaryotic cells: a strategy for bacterial survival. *Proc. Natl. Acad. Sci. USA* **98:**13844–13849.

Gobert, A. P., B. D. Mersey, Y. Cheng, D. R. Blumberg, J. C. Newton, and K. T. Wilson. 2002. Urease release by Helicobacter pylori stimulates macrophage inducible nitric oxide synthase. *J. Immunol.* **168:**6002–6006.

Gotoh, T., and M. Mori. 1999. Arginase II downregulates nitric oxide (NO) production and prevents NO-mediated apoptosis in murine macrophage-derived RAW264.7 cells. *J. Cell Biol.* **144:**427–434.

Hesse, M., A. W. Cheever, D. Jankovic, and T. A. Wynn. 2000. NOS-2 mediates the protective anti-inflammatory and anti-fibrotic effects of the Th1-inducing adjuvant, IL-12, in a Th2 model of granulomatous disease. *Am. J. Pathol.* **157:**945–955.

Hesse, M., M. Modolell, A. C. La Flamme, M. Schito, J. M. Fuentes, A. W. Cheever, E. J. Pearce, and T. A. Wynn. 2001. Differential regulation of nitric oxide synthase-2 and arginase-1 by type 1/type 2 cytokines in vivo: granulomatous pathology is shaped by the pattern of L-arginine metabolism. *J. Immunol.* **167:**6533–6544.

Hiemstra, P. S., P. B. Eisenhauer, S. S. L. Harwig, M. T. van den Barselaar, R. van Furth, and R. I. Lehrer. 1993. Antimicrobial peptides of murine macrophages. *Infect. Immun.* **61:**3038–3046.

Holzmuller, P., D. Sereno, M. Cavaleyra, I. Mangot, S. Daulouede, P. Vincendeau, and J.-L. Lemesre. 2002. Nitric oxide-mediated proteasome-dependent oligonucleosomal DNA fragmentation in Leishmania amazonensis amastigotes. *Infect. Immun.* **70:**3727–3735.

Hotopp, J. C., M. Lin, R. Madupu, J. Crabtree, S. V. Angiuoli, J. Eisen, R. Seshadri, Q. Ren, M. Wu, T. R. Utterback, S. Smith, M. Lewis, H. Khouri, C. Zhang, H. Niu, Q. Lin, N. Ohashi, N. Zhi, W. Nelson, L. M. Brinkac, R. J. Dodson, M. J. Rosovitz, J. Sundaram, S. C. Daugherty, T. Davidsen, A. S. Durkin, M. Gwinn, D. H. Haft, J. D. Selengut, S. A. Sullivan, N. Zafar, L. Zhou, F. Benahmed, H. Forberger, R. Halpin, S. Mulligan, J. Robinson, O. White, Y. Rikihisa, and H. Tettelin. 2006. Comparative genomics of emerging human ehrlichiosis agents. *PLoS Genet.* **2:**e21.

Howe, D., L. F. Barrows, N. M. Lindstrom, and R. A. Heinzen. 2002. Nitric oxide inhibits Coxiella burnetii replication and parasitophorous vacuole maturation. *Infect. Immun.* **70:**5140–5147.

Huang, C. J., I. U. Haque, P. N. Slovin, R. B. Nielsen, X. Fang, and J. W. Skimming. 2002. Environmental pH regulates LPS-induced nitric oxide formation in murine macrophages. *Nitric Oxide* **6:**73–78.

Huang, D., D. T. Cai, R. Y. Chua, D. M. Kemeny, and S. H. Wong. 2008. Nitric-oxide synthase 2 interacts with CD74 and inhibits its cleavage by caspase during dendritic cell development. *J. Biol. Chem.* **283:**1713–1722.

Iniesta, V., J. Carcelen, I. Molano, P. M. V. Peixoto, E. Redondo, P. Parra, M. Mangas, I. Monroy, M. L. Campo, C. G. Nieto, and I. M. Corraliza. 2005. Arginase I induction during Leishmania major infection mediates the development of disease. *Infect. Immun.* **73:**6085–6090.

Iniesta, V., L. C. Gomez-Nieto, and I. Corraliza. 2001. The inhibition of arginase by Nw-hydroxy-L-arginine controls the growth of Leishmania inside macrophages. *J. Exp. Med.* **193:**777–783.

Jacobsen, L. C., K. Theilgaard-Mönch, E. I. Christensen, and N. Borregaard. 2007. Arginase I is expressed in myelocytes/metamyelocytes and localized in gelatinase granules of human neutrophils. *Blood* **109:**3084–3087.

Jaramillo, M., D. C. Gowda, D. Radzioch, and M. Olivier. 2003. Hemozoin increases IFN-gamma-inducible macrophage nitric oxide generation through extracellular signal-regulated kinase- and NF-kappa B-dependent pathways. *J. Immunol.* **171:**4243–4253.

Jaramillo, M., P. H. Naccache, and M. Olivier. 2004. Monosodium urate crystals synergize with IFN-gamma to generate macrophage nitric oxide: involvement of extracellular signal-regulated kinase 1/2 and NF-kappa B. *J. Immunol.* **172:**5734–5742.

Johann, A. M., V. Barra, A. M. Kuhn, A. Weigert, A. von Knethen, and B. Brune. 2007. Apoptotic cells induce arginase II in macrophages, thereby attenuating NO production. *FASEB J.* **21:**2704–2712.

Kamijo, R., H. Harada, T. Matsuyama, M. Bosland, J. Gerecitano, D. Shapiro, J. Le, S. I. Koh, T. Kimura, S. J. Green, T. W. Mak, T. Taniguchi, and J. Vilcek. 1994. Requirement for transcription factor IRF-1 in NO synthase induction in macrophages. *Science* **263:**1612–1615.

Kaneda, M., K. Kakinuma, T. Yamaguchi, and K. Shimada. 1980. Comparative studies on alveolar macrophages and polymorphonuclear leukocytes. II. The ability of guinea pig alveolar macrophages to produce H2O2. *J. Biochem.* **88:**1159–1165.

Karpuzoglu, E., and S. A. Ahmed. 2006. Estrogen regulation of nitric oxide and inducible nitric oxide synthase (iNOS) in immune cells: implications for immunity, autoimmune diseases, and apoptosis. *Nitric Oxide* **15:**177–186.

Kaushal, V., P. D. Koeberle, Y. Wang, and L. C. Schlichter. 2007. The Ca2+-activated K+ channel KCNN4/KCa3.1 contributes to microglia activation and nitric oxide-dependent neurodegeneration. *J. Neurosci.* **27:**234–244.

Klebanoff, S. J., R. M. Locksley, E. C. Jong, and H. Rosen. 1983. Oxidative response of phagocytes to parasite invasion. *Ciba Found. Symp.* **99:**92–112.

Kleinert, H., A. Pautz, K. Linker, and P. M. Schwarz. 2004. Regulation of the expression of inducible nitric oxide synthase. *Eur. J. Pharmacol.* **500:**255–266.

Kolodziejski, P., A. Musial, J.-S. Koo, and N. T. Eissa. 2002. Ubiquitination of inducible nitric oxide synthase is required for its degradation. *Proc. Natl. Acad. Sci. USA* **99:**12315–12320.

Kolodziejski, P. J., M. B. Rashid, and N. T. Eissa. 2003. Intracellular formation of "undisruptable" dimers of inducible nitric oxide synthase. *Proc. Natl. Acad. Sci. USA* **100:**14263–14268.

Korhonen, R., K. Linker, A. Pautz, U. Forstermann, E. Moilanen, and H. Kleinert. 2007. Post-transcriptional regulation of human inducible nitric-oxide synthase expression by the Jun N-terminal kinase. *Mol. Pharmacol.* **71:**1427–1434.

Kristof, A. S., J. Fielhaber, A. Triantafillopoulos, S. Nemoto, and J. Moss. 2006. Phosphatidylinositol 3-kinase-dependent suppression of the human inducible nitric-oxide synthase promoter is mediated by FKHRL1. *J. Biol. Chem.* **281:**23958–23968.

Kropf, P., J. M. Fuentes, E. Fahnrich, L. Arpa, S. Herath, V. Weber, G. Soler, A. Celada, M. Modolell, and I. Müller. 2005. Arginase and polyamine synthesis are key factors in the regulation of experimental leishmaniasis in vivo. *FASEB J.* **19:**1000–1002.

Kuroda, S., M. Nishio, T. Sasaki, Y. Horie, K. Kawahara, M. Sasaki, M. Natsui, T. Matozaki, H. Tezuka, T. Ohteki, I. Forster, T. W. Mak, T. Nakano, and A. Suzuki. 2008. Effective clearance of intracellular *Leishmania major* in vivo requires Pten in macrophages. *Eur. J. Immunol.* **38:**1331–1340.

Levy, O. 2004. Antimicrobial proteins and peptides: anti-infective molecules of mammalian leukocytes. *J. Leukoc. Biol.* **76:**909–925.

Lezama-Davila, C. M., A. P. Isaac-Marquez, J. Barbi, S. Oghumu, and A. R. Satoskar. 2007. 17Beta-estradiol increases *Leishmania mexicana* killing in macrophages from DBA/2 mice by enhancing production of nitric oxide but not pro-inflammatory cytokines. *Am. J. Trop. Med. Hyg.* **76:**1125–1127.

Lowry, M. A., J. I. Goldberg, and M. Belosevic. 1998. Induction of nitric oxide (NO) synthesis in murine macrophages requires potassium channel activity. *Clin. Exp. Immunol.* **111:**597–603.

Marriott, H. M., and D. H. Dockrell. 2006. *Streptococcus pneumoniae*: the role of apoptosis in host defense and pathogenesis. *Int. J. Biochem. Cell Biol.* **38:**1848–1854.

Martin, E., C. Nathan, and Q.-W. Xie. 1994. Role of interferon regulatory factor-1 in induction of nitric oxide synthase. *J. Exp. Med.* **180:**977–984.

Martinon, F. 2008. Detection of immune danger signals by NALP3. *J. Leukoc. Biol.* **83:**507–511.

McCollister, B. D., T. J. Bourret, R. Gill, J. Jones-Carson, and A. Vazquez-Torres. 2005. Repression of SPI2 transcription by nitric oxide-producing, IFNgamma-activated macrophages promotes maturation of *Salmonella* phagosomes. *J. Exp. Med.* **202:**625–635.

McKinney, L. C., E. M. Aquilla, D. Coffin, D. A. Wink, and Y. Vodovotz. 2000. Ionizing radiation potentiates the induction of nitric oxide synthase by interferon-gamma and/or lipopolysaccharide in murine macrophage cell lines. Role of tumor necrosis factor-alpha. *Ann. N. Y. Acad. Sci.* **899:**61–68.

Melillo, G., T. Musso, A. Sica, L. S. Taylor, G. W. Cox, and L. Varesio. 1995. A hypoxia-responsive element mediates a novel pathway of activation of the inducible nitric oxide synthase promoter. *J. Exp. Med.* **182:**1683–1693.

Meraz, M. A., J. M. White, K. C. F. Sheehan, E. A. Bach, S. J. Rodig, A. S. Dighe, D. H. Kaplan, J. K. Riley, A. C. Greenlund, D. Campbell, K. Carver-Moore, R. N. DuBois, R. Clark, M. Aguet, and R. D. Schreiber. 1996. Targeted disruption of the Stat1 gene in mice reveals unexpected physiologic specificity in the Jak-STAT signaling pathway. *Cell* **84:**431–442.

Mi, Z., A. Rapisarda, L. Taylor, A. Brooks, M. Creighton-Gutteridge, G. Melillo, and L. Varesio. 2008. Synergistic induction of HIF-1alpha transcriptional activity by hypoxia and lipopolysaccharide in macrophages. *Cell Cycle* **7:**232–241.

Miller, B. H., R. A. Fratti, J. F. Poschet, G. S. Timmins, S. S. Master, M. Burgos, and M. A. Marletta. 2004. Mycobacteria inhibit nitric oxide synthase recruitment to phagosomes during macrophage infection. *Infect. Immun.* **2004:**2872–2878.

Mitani, T., M. Terashima, H. Yoshimura, Y. Nariai, and Y. Tanigawa. 2005. TGF-beta1 enhances degradation of IFN-gamma-induced iNOS protein via proteasomes in RAW 264.7 cells. *Nitric Oxide* **13:**78–87.

Mori, M. 2007. Regulation of nitric oxide synthesis and apoptosis by arginase and arginine recycling. *J. Nutr.* **137:**1616S–1620S.

Morris, S. M., Jr. 2002. Regulation of enzymes of the urea cycle and arginine metabolism. *Annu. Rev. Nutr.* **22:**87–105.

Munder, M., M. Eichmann, J. M. Moran, F. Centeno, G. Soler, and M. Modolell. 1999. Th1/Th2-regulated expression of arginase isoforms in murine macrophages and dendritic cells. *J. Immunol.* **163:**3771–3777.

Munder, M., F. Mollinedo, J. Calafat, J. Canchado, C. Gil-Lamaignere, J. M. Fuentes, C. Luckner, G. Doschko, G. Soler, K. Eichmann, F. M. Muller, A. D. Ho, M. Goerner, and M. Modolell. 2005. Arginase I is constitutively expressed in human granulocytes and participates in fungicidal activity. *Blood* **105:**2549–2556.

Munder, M., H. Schneider, C. Luckner, T. Giese, C.-D. Langhans, J. M. Fuentes, P. Kropf, I. Mueller, A. Kolb, M. Modolell, and A. D. Ho. 2006. Suppression of T cell functions by human granulocyte arginase. *Blood* **108:**1627–1634.

Murphy, B. S., V. Sundareshan, T. J. Cory, D. Hayes, Jr., M. I. Anstead, and D. J. Feola. 2008. Azithromycin alters macrophage phenotype. *J. Antimicrob. Chemother.* **61:**554–560.

Myers, J. T., A. W. Tsang, and J. A. Swanson. 2003. Localized reactive oxygen and nitrogen intermediates inhibit escape of *Listeria monocytogenes* from vacuoles in activated macrophages. *J. Immunol.* **171:**5447–5453.

Nathan, C., and M. U. Shiloh. 2000. Reactive oxygen and nitrogen intermediates in the relationship between mammalian hosts and microbial pathogens. *Proc. Natl. Acad. Sci. USA* **97:**8841–8848.

Nicholson, B., C. K. Manner, J. Kleeman, and C. L. MacLeod. 2001. Sustained nitric oxide production in macrophages requires the arginine transporter CAT2. *J. Biol. Chem.* **276:**15881–15885.

Niedbala, W., B. Cai, H. Liu, N. Pitman, L. Chang, and F. Y. Liew. 2007. Nitric oxide induces CD4+CD25+ Foxp3 regulatory T cells from CD4+CD25 T cells via p53, IL-2, and OX40. *Proc. Natl. Acad. Sci. USA* **104:**15478–15483.

Niedbala, W., X.-Q. Wei, C. Campbell, D. Thomson, M. Komai-Koma, and F. Y. Liew. 2002. Nitric oxide preferentially induces type 1 T cell differentiation by selectively up-regulating IL-12 receptor β2 expression via cGMP. *Proc. Natl. Acad. Sci. USA* **99:**16186–16191.

Olds, G. R., J. J. Ellner, L. A. Kearse, J. W. Kazura, and A. A. F. Mahmoud. 1980. Role of arginase in killing of schistosomula of *Schistosoma mansoni. J. Exp. Med.* **151:**1557–1562.

Osorio, Y., D. L. Bonilla, A. G. Peniche, P. C. Melby, and B. L. Travi. 2008. Pregnancy enhances the innate immune response in experimental cutaneous leishmaniasis through hormone-modulated nitric oxide production. *J. Leukoc. Biol.* **83:**1413–1422.

Padgett, E. L., and S. B. Pruett. 1995. Rat, mouse and human neutrophils stimulated by a variety of activating agents produce much less nitrite than rodent macrophages. *Immunology* **84:**135–141.

Piacenza, L., G. Peluffo, and R. Radi. 2001. L-Arginine-dependent suppression of apoptosis in *Trypanosoma cruzi*: contribution of the nitric oxide and polyamine pathways. *Proc. Natl. Acad. Sci. USA* **98:**7301–7306.

Pindado, J., J. Balsinde, and M. A. Balboa. 2007. TLR3-dependent induction of nitric oxide synthase in RAW 264.7 macrophage-like cells via a cytosolic phospholipase A2/cyclooxygenase-2 pathway. *J. Immunol.* **179:**4821–4828.

Platanias, L. C. 2005. Mechanisms of type-I- and type-II-interferon-mediated signalling. *Nat. Rev. Immunol.* **5:**375–386.

Poole, R. K. 2005. Nitric oxide and nitrosative stress tolerance in bacteria. *Biochem. Soc. Trans.* **33:**176–180.

Pritchard, M. T., Z. Li, and E. A. Repasky. 2005. Nitric oxide production is regulated by fever-range thermal stimulation of murine macrophages. *J. Leukoc. Biol.* **78:**630–638.

Punturieri, A., R. S. Alviani, T. Polak, P. Copper, J. Sonstein, and J. L. Curtis. 2004. Specific engagement of TLR4 or TLR3 does not lead to IFN-beta-mediated innate signal amplification and STAT1 phosphorylation in resident murine alveolar macrophages. *J. Immunol.* **173:**1033–1042.

Ribeiro-Gomes, F. L., M. C. Moniz-de-Souza, M. S. Alexandre-Moreira, W. B. Dias, M. F. Lopes, M. P. Nunes, G. Lungarella, and G. A. DosReis. 2007. Neutrophils activate macrophages for intracellular killing of *Leishmania major* through recruitment of TLR4 by neutrophil elastase. *J. Immunol.* **179:**398–3994.

Rodig, S. J., M. A. Meraz, J. M. White, P. A. Lampe, J. K. Riley, C. D. Arthur, K. L. King, K. C. F. Sheehan, L. Yin, D. Pennica, E. M. Johnson, and R. D. Schreiber. 1998. Disruption of the Jak1 gene demonstrates obligatory and nonredundant roles of the Jaks in cytokine-induced biologic responses. *Cell* **93:**373–383.

Rodriguez, P. C., C. P. Hernandez, D. Quiceno, S. M. Dubinett, J. Zabaleta, J. B. Ochoa, J. Gilbert, and A. C. Ochoa. 2005. Arginase I in myeloid suppressor cells is induced by COX-2 in lung carcinoma. *J. Exp. Med.* **202:**931–939.

Rodriguez, P. C., A. H. Zea, J. DeSalvo, K. S. Culotta, J. Zabaleta, D. G. Quiceno, J. B. Ochoa, and A. C. Ochoa. 2003. L-arginine consumption by macrophages modulates the expression of CD3ζ chain in T lymphocytes. *J. Immunol.* **17:**1232–1239.

Rothfork, J. M., G. S. Timmins, M. N. Harris, X. Chen, A. J. Lusis, M. Otto, A. L. Cheung, and H. D. Gresham. 2004. Inactivation of a bacterial virulence pheromone by phagocyte-derived oxidants: new role for the NADPH oxidase in host defense. *Proc. Natl. Acad. Sci. USA* **101:**13867–13872.

Rutschman, R., R. Lang, M. Hesse, J. N. Ihle, T. A. Wynn, and P. J. Murray. 2001. Stat6-dependent substrate depletion regulates nitric oxide production. *J. Immunol.* **166:**2173–2177.

Sable, S. B., D. Goyal, I. Verma, D. Behera, and G. K. Khuller. 2007. Lung and blood mononuclear cell responses of tuberculosis patients to mycobacterial proteins. *Eur. Respir. J.* **29:**337–346.

Scott, P., H. Ma, S. Viriyakosol, R. Terkeltaub, and R. Liu-Bryan. 2006. Engagement of CD14 mediates the inflammatory potential of monosodium urate crystals. *J. Immunol.* **177:**6370–6378.

Seib, K. L., H. J. Wu, S. P. Kidd, M. A. Apicella, M. P. Jennings, and A. G. McEwan. 2006. Defenses against oxidative stress in *Neisseria gonorrhoeae*: a system tailored for a challenging environment. *Microbiol. Mol. Biol. Rev.* **70:**344–361.

Soderberg, M., F. Raffalli-Mathieu, and M. A. Lang. 2007. Identification of a regulatory cis-element within the 3'-untranslated region of the murine inducible nitric oxide synthase (iNOS) mRNA; interaction with heterogeneous nuclear ribonucleoproteins I and L and role in the iNOS gene expression. *Mol. Immunol.* **44:**434–442.

Speer, C. P., M. J. Pabst, H. B. Hedegaard, R. F. Rest, and R. B. Johnston, Jr. 1984. Enhanced release of oxygen metabolites by monocyte-derived macrophages exposed to proteolytic enzymes: activity of neutrophil elastase and cathepsin G. *J. Immunol.* **133:**2151–2156.

Stempin, C. C., T. B. Tanos, O. A. Coso, and F. M. Cerban. 2004. Arginase induction promotes *Trypanosoma cruzi* intracellular replication in Cruzipain-treated J774 cells through the activation of multiple signaling pathways. *Eur. J. Immunol.* **34:**200–209.

Stuehr, D. J., and M. A. Marletta. 1985. Mammalian nitrite biosynthesis: mouse macrophages produce nitrite and nitrate in response to *Escherichia coli* lipopolysaccharide. *Proc. Natl. Acad. Sci. USA* **82:**7738–7742.

Stuehr, D. J., and M. A. Marletta. 1987. Induction of nitrite/nitrate synthesis in murine macrophages by BCG infection, lymphokines or interferon-γ. *J. Immunol.* **139:**518–525.

Stuehr, D. J., J. Santolini, Z.-Q. Wang, C.-C. Wei, and S. Adak. 2004. Update on mechanism and catalytic regulation in the NO synthases. *J. Biol. Chem.* **279:**36167–36170.

Takaki, H., Y. Minoda, K. Koga, G. Takaesu, A. Yoshimura, and T. Kobayashi. 2006. TGF-beta1 suppresses IFN-gamma-induced NO production in macrophages by suppressing STAT1 activation and accelerating iNOS protein degradation. *Genes Cells* **11:**871–882.

Tan, B. H., C. Meinken, M. Bastian, H. Bruns, A. Legaspi, M. T. Ochoa, S. R. Krutzik, B. R. Bloom, T. Ganz, R. L. Modlin, and S. Stenger. 2006. Macrophages acquire neutrophil granules for antimicrobial activity against intracellular pathogens. *J. Immunol.* **177:**1864–1871.

Thompson, R. W., J. T. Pesce, T. Ramalingam, M. S. Wilson, S. White, A. W. Cheever, S. M. Ricklefs, S. F. Porcella, L. Li, L. G. Ellies, and T. A. Wynn. 2008. Cationic amino acid transporter-2 regulates immunity by modulating arginase activity. *PLoS Pathog.* **4:**e1000023.

Uematsu, S., and S. Akira. 2006. Toll-like receptors and innate immunity. *J. Mol. Med.* **84:**712–725.

Uematsu, S., and S. Akira. 2008. Toll-Like receptors (TLRs) and their ligands. *Handb. Exp. Pharmacol.* 1–20.

Valdez, C. A., J. E. Saavedra, B. M. Showalter, K. M. Davies, T. C. Wilde, M. L. Citro, J. J. Barchi, Jr., J. R. Deschamps, D. Parrish, S. El-Gayar, U. Schleicher, C. Bogdan, and L. K. Keefer. 2008. Hydrolytic reactivity trends among potential prodrugs of the O(2)-glycosylated diazeniumdiolate family. Targeting nitric oxide to macrophages for antileishmanial activity. *J Med Chem.* **51:**3961–3970.

Vareille, M., T. de Sablet, T. Hindre, C. Martin, and A. P. Gobert. 2007. Nitric oxide inhibits Shiga-toxin synthesis by enterohemorrhagic *Escherichia coli*. *Proc. Natl. Acad. Sci. USA* **104:**10199–101204.

Vig, M., S. Srivastava, U. Kandpal, H. Sade, V. Lewis, A. Sarin, A. George, V. Bal, J. M. Durdik, and S. Rath. 2004. Inducible nitric oxide synthase in T cells regulates T cell death and immune memory. *J. Clin. Invest.* **113:**1734–1742.

Vila-del Sol, V., M. D. Diaz-Munoz, and M. Fresno. 2007. Requirement of tumor necrosis factor alpha and nuclear factor-kappaB in the induction by IFN-gamma of inducible nitric oxide synthase in macrophages. *J. Leukoc. Biol.* **81:**272–283.

Vodovotz, Y., C. Bogdan, J. Paik, Q.-W. Xie, and C. Nathan. 1993. Mechanisms of suppression of macrophage nitric oxide release by transforming growth factor-β. *J. Exp. Med.* **178:**605–613.

Vodovotz, Y., D. Russell, Q.-W. Xie, C. Bogdan, and C. Nathan. 1995. Vesicle membrane association of nitric oxide synthase in primary mouse macrophages. *J. Immunol.* **154:**2914–2925.

Webb, J. L., M. W. Harvey, D. W. Holden, and T. J. Evans. 2001. Macrophage nitric oxide synthase associates with cortical actin but is not recruited to phagosomes. *Infect. Immun.* **69:**6391–6400.

Wheeler, M. A., S. D. Smith, G. Garcia-Cardena, C. F. Nathan, R. M. Weiss, and W. C. Sessa. 1997. Bacterial infection induces nitric oxide synthase in human neutrophils. *J. Clin. Invest.* **99:**110–116.

Wilhelm, P., U. Ritter, S. Labbow, N. Donhauser, M. Röllinghoff, C. Bogdan, and H. Körner. 2001. Rapidly fatal leishmaniasis in resistant C57BL/6 mice lacking tumor necrosis factor. *J. Immunol.* **166:**4012–4019.

Winberg, M. E., B. Rasmusson, and T. Sundqvist. 2007. *Leishmania donovani*: inhibition of phagosomal maturation is rescued by nitric oxide in macrophages. *Exp. Parasitol.* **117:** 165–170.

Won, J. S., Y. B. Im, A. K. Singh, and I. Singh. 2004. Dual role of cAMP in iNOS expression in glial cells and macrophages is mediated by differential regulation of p38-MAPK/ATF-2 activation and iNOS stability. *Free Radic. Biol. Med.* **37:**1834–1844.

Wu, G., and S. M. Morris. 1998. Arginine metabolism: nitric oxide and beyond. *Biochem. J.* **336:**1–17.

Xie, Q.-W., Y. Kasshiwabara, and C. Nathan. 1994. Role of transcription factor NF-κB/Rel in induction of nitric oxide synthase. *J. Biol. Chem.* **269:**4705–4708.

Yeramian, A., L. Martin, N. Serrat, L. Arpa, C. Soler, J. Bertran, C. McLeod, M. Palacin, M. Modolell, J. Lloberas, and A. Celada. 2006. Arginine transport via cationic amino acid transporter 2 plays a critical regulatory role in classical or alternative activation of macrophages. *J. Immunol.* **176:** 5918–5924.

Yoshioka, Y., T. Kitao, T. Kishino, A. Yamamuro, and S. Maeda. 2006. Nitric oxide protects macrophages from hydrogen peroxide-induced apoptosis by inducing the formation of catalase. *J. Immunol.* **176:**4675–4681.

Zabaleta, J., D. J. McGee, A. H. Zea, C. P. Hernandez, P. C. Rodriguez, R. A. Sierra, P. Correa, and A. C. Ochoa. 2004. Helicobacter pylori arginase inhibits T cell proliferation and reduces the expression of the TCR zeta-chain (CD3zeta). *J. Immunol.* **173:**586–593.

Zaki, M. H., T. Akuta, and T. Akaike. 2005. Nitric oxide-induced nitrative stress involved in microbial pathogenesis. *J. Pharmacol. Sci.* **98:**117–129.

Zamboni, D. S., and M. Rabinovitch. 2004. Phagocytosis of apoptotic cells increases the susceptibility of macrophages to infection with *Coxiella burnetii* phase II through downmodulation of nitric oxide production. *Infect. Immun.* **72:** 2075–2080.

Zea, A. H., K. S. Culotta, J. Ali, C. Mason, H. J. Park, J. Zabaleta, L. F. Garcia, and A. C. Ochoa. 2006. Decreased expression of CD3zeta and nuclear transcription factor kappa B in patients with pulmonary tuberculosis: potential mechanisms and reversibility with treatment. *J. Infect. Dis.* **194:** 1385–1393.

Zhang, M., T. Caragine, H. Wang, P. S. Cohen, G. Botchkina, K. Soda, M. Bianchi, P. Ulrich, A. Cerami, B. Sherry, and K. J. Tracey. 1997. Spermine inhibits proinflammatory cytokine synthesis in human mononuclear cells: a counterregulatory mechanism that restrains the immune response. *J. Exp. Med.* **185:**1759–1768.

PATHOGENS OF THE PROFESSIONAL PHAGOCYTE

V

23

The Multiple Interactions between *Salmonella* and Phagocytes

JESSICA A. THOMPSON AND DAVID W. HOLDEN

INTRODUCTION

Salmonella spp. comprise a large group of enteric pathogens that cause disease in a wide range of animals. In humans, these infections are caused by serovars of *Salmonella enterica*. *S. enterica* serovar Typhi (serovar Typhi) is strictly host-adapted to primates and causes typhoid fever, whereas *S. enterica* serovar Typhimurium (serovar Typhimurium) produces a self-limiting gastroenteritis in humans. However, this serovar is also able to infect other mammals with varying degrees of severity, and infection of certain mouse strains by serovar Typhimurium produces a typhoid-like disease that is a valuable model for the study of typhoid fever. Indeed, most of our understanding of *Salmonella* virulence has come from studies using serovar Typhimurium. In this chapter, "*Salmonella*" will refer to serovar Typhimurium unless otherwise stated.

Throughout infection, *S. enterica* encounters phagocytic cells, and the interactions that take place between them are central in shaping the progression and outcome of disease. While being the main mediators of resistance to *Salmonella* infection, phagocytes also provide a means for dissemination of the pathogen and a niche for bacterial survival and replication. This paradox between the protective function of phagocytes and their ability to act as a conduit to establishment of systemic infection results from the opposing action of the antibacterial responses of macrophages, neutrophils, and dendritic cells (DCs), and the repertoire of virulence proteins produced by the bacteria that enables it to evade, exploit, or combat these responses. It is important to bear in mind that if left untreated, even typhoid fever has a relatively low mortality rate. This reminds us that final resolution of the encounters between the pathogen and phagocytes is usually clearance and resistance. However, this resolution follows bacterial growth in the face of an early innate immune response and depends on the later development of robust acquired immunity. A series of challenges are presented to *Salmonella* by phagocytic cells, some of which curtail infection within the individual cell, while others have a broader effect and lead to the generation of specific immunity. In turn, this can provide signals for the pathogen to upregulate its own responses and prolong bacterial presence within the tissues. This chapter will outline current understanding of these host-pathogen interactions, illustrating the close evolutionary relationship between *Salmonella* and phagocytic cells.

The Route to Systemic Infection: Where and When Do *Salmonella* Meet Phagocytes?

After oral ingestion, a proportion of *Salmonella* is able to withstand the acidic pH of the stomach and colonize the small intestine. Here, bacteria translocate across the follicular epithelium using at least three distinct mechanisms: invasion of antigen-sampling M cells at Peyer's patches (Clark et al., 1994, 1996), invasion of the adjacent enterocytes (Jones et al., 1994), and phagocytosis by CD18[+] phagocytes (Vazquez-Torres et al., 1999), presumed to be DCs (Fig. 1). The bacteria access and colonize the underlying lamina propria and mesenteric lymph nodes (MLNs) (Carter and Collins, 1974; Cirillo et al., 1998), where large populations of phagocytic cells are present that engulf the bacteria. Although DCs and neutrophils frequent the gut lumen, it is in the subepithelial compartment that *Salmonella* has its first major encounter with large populations of phagocytes. The ability of *Salmonella* to reside within DCs and macrophages provides its means of transport via the lymphatics and blood (where a transient bacteremia occurs) to the major sites of systemic colonization, the spleen and liver.

During the initial phases of hepatosplenic colonization, large numbers of neutrophils infiltrate both organs, and *Salmonella* is associated with neutrophil-rich lesions that develop into distinct, macrophage-rich granulomas. These increase in number rather than size as disease progresses (Sheppard et al., 2003). The influx of further innate immune effector cells to the spleen and liver, and resulting inflammation, give rise to the hepatosplenomegaly that is characteristic of systemic *Salmonella* infection. Throughout this phase of infection, the bacteria are intimately associ-

---

Jessica A. Thompson and David W. Holden, Department of Microbiology, Imperial College London, London, United Kingdom.

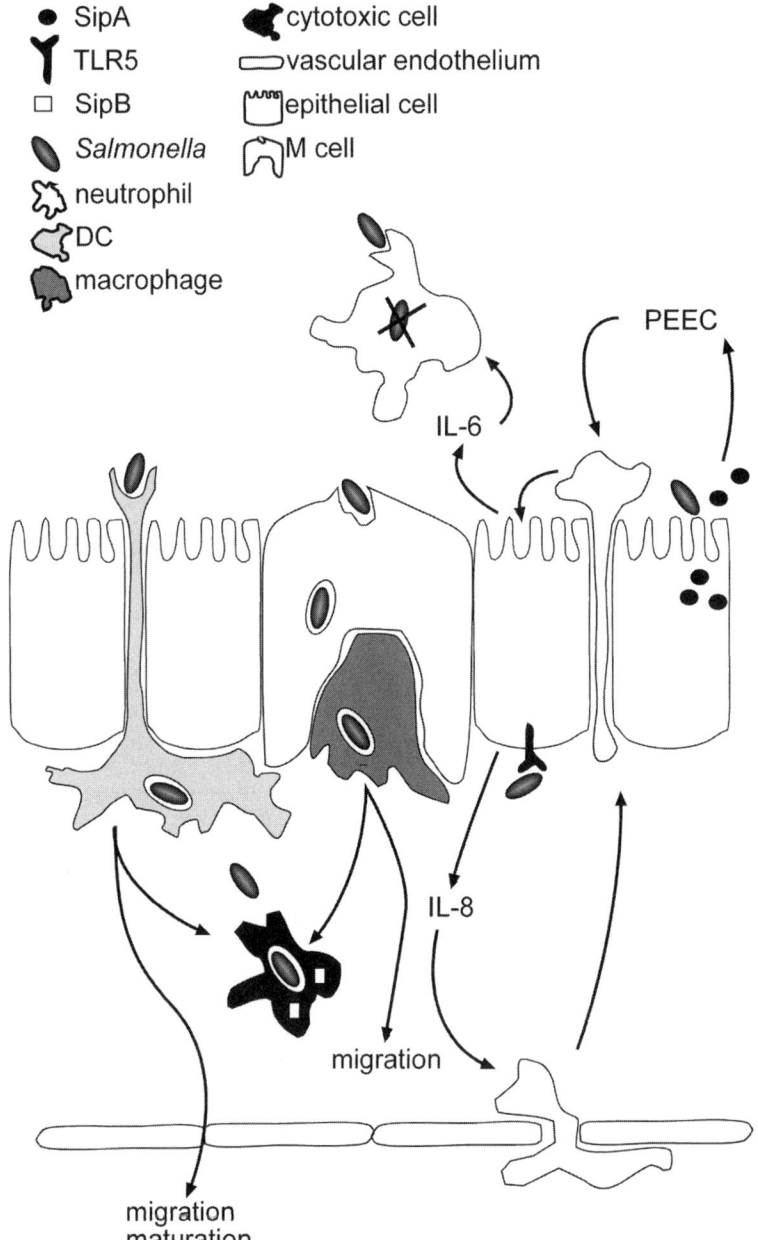

FIGURE 1 Interactions between *Salmonella* and phagocytes at the intestinal epithelium. *Salmonella* access the SED through invasion of M cells, enterocytes, and phagocytosis by CD18[+] DCs. Flagellin binding to TLR5 elicits IL-8 signaling that recruits neutrophils from the circulation. Interaction between SipA and epithelial cells leads to secretion of PEEC and neutrophil migration to the gut lumen. Epithelial cells secrete IL-6, which activates the antibacterial activities of neutrophils. Phagocytosis of *Salmonella* by DCs and macrophages leads to SipB-dependent cytotoxicity or migration and maturation of infected cells.

ated with macrophages, in a predominantly intracellular location within *Salmonella*-containing vacuoles (SCVs), where replication occurs. If growth within hepatic or splenic cells is not controlled, a second bacteremia ultimately ensues and the host succumbs to infection, whether through reinfection and perforation of the gut as in typhoid fever, or septicemia and multiorgan failure as in the typhoid-like disease experienced by mice.

Bacterial adherence factors and their cognate receptors on the intestinal epithelium are likely to play a role in determining host specificity of the different serovars, and a major contribution is made by the ability of phagocytic cells to contain infection. During serovar Typhimurium infection of humans, the pathogen is unable to replicate in macrophages, and infection is contained until adaptive immunity is raised; *Salmonella* is eliminated without hepatosplenic colonization. In contrast, when mice are infected with serovar Typhimurium, or humans with serovar Typhi, the phagocytes are less able to control or kill intracellular bacteria and the spleen and liver become extensively colonized. Although this is a crucial aspect in *Salmonella* infection, the molecular mechanisms that underlie host cell specificity are yet to be characterized.

Anti-*Salmonella* Immune Responses Depend on Phagocytes at Early and Late Stages of Infection

In the immunocompetent host, resistance depends on containment by the innate immune response and subsequent clearance by adaptive immunity. Ligation of pattern recognition receptors (PRRs) on macrophages, DCs, and neutrophils by pathogen-associated molecular patterns (PAMPs) leads to intracellular signaling followed by elicitation of both oxidative and nonoxidative bactericidal responses, as well as cytokine release. Although neutrophils contribute toward early protection, macrophages and DCs play the more important role overall in elimination of *Salmonella*: as antigen-presenting cells they are able to bridge the gap between nonspecific and specific immunity. Maturing antigen-presenting cells migrate to the lymphoid follicles after encounters with the pathogen, and on reaching this location they present bacterial antigens on major histocompatibility complex class II (MHC-II) to naive $CD4^+$ T cells. Costimulation is required before these T-cell populations start to mature and expand, with cross-talk between T and B cells providing additional signals for this response. A protective force is generated that is predominantly Th1 in nature. The associated cytokines, including interleukin 12 (IL-12), IL-15, IL-18, and interferon gamma (IFN-γ), increase the activation of macrophages, which in turn augments the antibacterial activity of these cells: expression of proteins involved in the respiratory burst is increased, as are the number of lysosomes and the hydrolytic enzymes that accumulate within them. Tumor necrosis factor alpha (TNF-α) also contributes toward the stimulation of oxidative responses by phagocytes and is crucial to the development of organized granulomas that contain bacteria within discrete foci in infected tissues.

Cross-presentation of antigens on MHC-I by DCs raises cytotoxic $CD8^+$ T cells that strengthen the cellular response required against intracellular *Salmonella*. The immune response is not strictly cell mediated, however, as B cells generate antibodies against flagellin, lipopolysaccharide (LPS), V_i antigen of serovar Typhi, outer membrane proteins, and other bacterial surface molecules. Effective immunological memory involves both these antibody responses and the activity of Th1 $CD4^+$ cells (Mastroeni, 2002).

Salmonella Virulence Factors Enable Bacterial Replication

Salmonella expresses a large number of virulence factors that combat host defenses and enable bacterial replication. Most of these are encoded by numerous pathogenicity islands, phage elements, and plasmids. In serovar Typhi, these comprise at least 7% of the bacterial genome (Parkhill et al., 2001); in general, they are the result of horizontal gene transfer, identifiable by their altered GC content, characteristic transposon insertion sequences, or other mobile DNA elements. So far, 12 *Salmonella* pathogenicity islands (SPIs) have been identified, and when smaller pathogenicity islets are included, this number rises to approximately 60. Among the genes carried by these elements are those encoding fimbriae, pili, and other surface antigens that mediate adherence and immune evasion, secretion systems, toxins, and other factors important for cell invasion and intracellular survival. The combination and structure of SPIs carried by *S. enterica* varies from serovar to serovar. This heterogeneity contributes to the ability of these serovars to cause diseases of varying severity in a range of host species.

The environments to which *Salmonella* is exposed change as infection progresses; these conditions frequently provide signals for the controlled temporal and spatial expression of bacterial virulence factors. Tight regulation of gene expression prevents the unnecessary outlay of bacterial energy resources and also limits the availability of potential antigens for host recognition. Throughout infection, the bacterium adapts to its surroundings, initially responding to the low oxygen and high osmolarity in the gut lumen through expression of one of its type III secretion systems (T3SSs) encoded by SPI-1. T3SSs are needle-like complexes that assemble at the bacterial surface and translocate protein effectors across three membranes into the eukaryotic cell cytosol (Hueck, 1998). SPI-1 T3SS-mediated delivery of proteins into host cells enables invasion of nonphagocytic cells and translocation of the epithelial cell layer, and it induces rapid host cell cytotoxicity in macrophages during the gut phase of infection (Hersh et al., 1999), both of which promote dissemination of *Salmonella* away from the small intestine. In addition, electrolyte loss and fluid secretion is stimulated (Norris et al., 1998), giving rise to diarrheal symptoms that aid dissemination through bacterial shedding.

The signals present in the intravacuolar environment then activate a second T3SS encoded by the SPI-2 chromosomal region of *S. enterica* (Waterman and Holden, 2003). Proteins are delivered from bacteria within SCVs, across the vacuolar membrane into the eukaryotic cytosol or onto the vacuolar membrane itself. After translocation, these proteins maintain the integrity of the SCV membrane and create conditions conducive to bacterial replication, enabling *Salmonella* to establish itself within macrophages. The importance of this secretion system (and therefore of intramacrophage replication) in systemic disease is evident by the severe attenuation in virulence of SPI-2 null mutant bacteria on infection of mice by *Salmonella*. Orally inoculated bacteria colonize the Peyer's patches but do not reach the MLN (Cirillo et al., 1998), while those injected intraperitoneally fail to proliferate in the spleen and liver before being cleared from these organs (Shea et al., 1999).

The acidification of the SCV provides one of the signals for activation of another set of approximately 40 virulence factors controlled by the PhoP/Q two-component regulatory system (Alpuche Aranda et al., 1992; Groisman et al., 1989). This represses expression of the SPI-1 T3SS and alters the bacterial cell envelope with protein and lipid modifications in ways that enable bacteria to withstand antimicrobial peptides and ion-limiting conditions. Not surprisingly, *phoP/Q* mutant strains are also profoundly attenuated in virulence (Fields et al., 1989; Miller et al., 1989).

Several serovars of *S. enterica* also contain the five-gene *spv* operon. This operon is usually carried on the *Salmonella* virulence plasmid, but is occasionally present in the chromosome, and contributes to bacterial replication (Matsui et al., 2000), cytotoxicity (Kurita et al., 2003), and blocking of mitogen-activated protein kinase (MAPK) signaling (Mazurkiewicz et al., 2008).

The coordinated expression of virulence factors enables *Salmonella* to first reach its target cells within the host, then enter and survive within them, before disseminating to new targets. Extracellular and intracellular conditions that normally contribute to elimination of pathogens thus facilitate the adaptation of *Salmonella* to a physiological state that enables it to colonize macrophages and cause systemic disease. Our current understanding of the interactions between *Salmonella* and phagocytes will be reviewed below, but it should be emphasized that many of the mechanisms through which *Salmonella* responds to attack by neutrophils, macrophages, and DCs are yet to be characterized.

INTERACTIONS BETWEEN *SALMONELLA* AND NEUTROPHILS

Neutrophils contribute to the early innate immune response against *Salmonella*, providing a means for control of infection until macrophages and more specific immune effector cells and molecules can eliminate the pathogen. These phagocytes are the first to respond and are rapidly recruited to sites of infection, where they employ both oxidative and nonoxidative bactericidal mechanisms, primarily against extracellular bacteria. Neutrophil recruitment and responses vary with the different diseases caused by serovars of *S. enterica*; the implications for disease progression will also be outlined in the following sections.

Neutrophils Degranulate and Kill *Salmonella*

Neutrophil-mediated killing of *Salmonella* involves primary and secondary granules. Bacteria are phagocytosed by neutrophils, whereupon fusion between granules and SCVs occurs, delivering bactericidal molecules into the same compartment as *Salmonella*. In addition, the release of granule contents into the extracellular environment targets those bacteria that are in the vicinity of neutrophils but remain extracellular.

Secondary granules contain the p22phox component of NADPH oxidase; formation of NADPH oxidase following degranulation leads to the synthesis of superoxide. This reactive oxygen species (ROS) produces several other oxidative compounds, including hydrogen peroxide, which kill *Salmonella* through disruption of bacterial membranes and DNA damage. Myeloperoxidase, released from primary granules, uses hydrogen peroxide as a substrate for the synthesis of hypochlorous acid that, in conjunction with N-chloroamines, augments the antimicrobial respiratory burst.

Oxygen-independent killing of *Salmonella* also depends on the degranulation of secondary granules. The iron chelator, lactoferrin, is released, as are antimicrobial peptides that exert their toxic effects by inserting into and disrupting the integrity of bacterial membranes.

Serovar Typhimurium Infection Leads to Infiltration of the Gut by Neutrophils

Salmonella and neutrophils need to encounter each other for the bacterial killing mechanisms mentioned above to come into action. In nontyphoidal salmonellosis, the gastrointestinal phase of infection is characterized by a massive influx of neutrophils into the terminal ileum and colon at the sites of *Salmonella* colonization; this increases the likelihood of interactions between *Salmonella* and neutrophils. The recruitment of immune effector cells depends on cytokine signaling elicited after stimulation of the intestinal epithelium by bacterial molecules.

SPI-1 T3SS-mediated translocation of epithelial cells provides access for both *Salmonella* and their PAMPs to the basolateral surface of this layer, which is also the site of Toll-like receptor 5 (TLR5) expression. TLR5 ligation by flagellin leads to the secretion of IL-8 from the basal side of intestinal epithelial cells, which acts in conjunction with that released by macrophages on recognition of flagellin (by TLR5) and LPS (by TLR4). IL-8 is a chemotactic factor for neutrophils and stimulates these cells to extravasate and migrate from the microvasculature to the subepithelial area.

The SPI-1 T3SS also stimulates IL-8-independent recruitment of neutrophils, through the translocated effector SipA. Whether this protein binds a cell surface receptor that induces intracellular signaling, or enters the cell via receptor-mediated endocytosis (which can be bypassed by bacterial translocation of the protein), the outcome of its interaction with the apical surface of the intestinal epithelium is the release of pathogen-elicited epithelial chemoattractant (PEEC) (Lee et al., 2000). Downstream signaling, presumably via MAPK pathways, leads to release of PEEC from the apical surface of epithelial cells; neutrophils respond by transmigrating from the subepithelial area to the gut lumen.

The process of transmigration exposes neutrophils to activatory signals, augmenting the antibacterial responses of these cells as they near the site of recruitment. During transepithelial migration, neutrophils release $5'$-AMP which is converted to adenosine after interacting with CD73 on epithelial cells. The receptor for adenosine is found on the apical surface of the epithelial layer, ligation of which leads to release of another neutrophil chemoattractant, IL-6. This chemokine also primes cells for granule release by increasing the intracellular stores of calcium ions (Nadeau et al., 2002). Therefore recruitment of neutrophils not only raises the chances of *Salmonella* encountering these cells in the gut, but also heightens the antibacterial activity of those cells that reach this location and thus the probability of effective bacterial killing (summarized in Fig. 1).

The Contribution of Neutrophils to Resistance in Different *S. enterica* Infections

The recruitment of neutrophils to the gastrointestinal phase of infection is not present in all diseases caused by serovars of *S. enterica*. Serovar Typhi infection clearly suppresses the early inflammatory response in the human gut; the neutrophil-rich infiltrate is absent, and diarrheal symptoms are rarely present in typhoid pathogenesis. This serovar expresses several antigens on its surface that are associated with greater likelihood of clinical disease: the O antigen component of LPS and the V_i antigen (specific to serovar Typhi).

The V_i antigen has been implicated in inhibition of proinflammatory responses that are normally elicited by *Salmonella* interaction with the intestinal epithelium. This capsular structure is a linear polymer of α-1,4(2-deoxy)-2-N-acetylgalacturonic acid, which can be variably O-acetylated at the C3 position. Flagellin and LPS purified from strains expressing V_i antigen remain immunogenic; however, these strains stimulate a reduced IL-8 response from epithelial cells and macrophages. The V_i antigen somehow masks these PAMPs and inhibits signaling via TLR4 and TLR5. The consequential lack of IL-8 secretion decreases the chemotactic signals for neutrophils, and therefore their recruitment to the gastrointestinal tract (Raffatellu et al., 2005) and chances of encountering serovar Typhi in this location.

When neutrophils do encounter serovar Typhi during infection, their responses are also altered. The presence of V_i antigen does not affect phagocytosis or bacterial killing, but oxygen metabolism of infected cells is reduced, which presumably leads to a decrease in the oxidative responses of these cells (Kossack et al., 1981). It is thought that the differential ligation of PRRs by different *Salmonella* PAMPs is responsible for reduced neutrophil responses to serovar Typhi.

The large neutrophil presence in the gut during gastrointestinal infections undoubtedly provides some protection against systemic infection, perhaps by acting as a protective barrier and limiting bacterial access to intracellular sites. However, the inability of nontyphoidal *S. enterica* to replicate in human macrophages must also restrict the ability of these serovars to disseminate into deeper tissues.

Salmonella associate with neutrophil-rich lesions in the early stages of spleen and liver colonization (Mastroeni and Sheppard, 2004), and mice rendered neutropenic through antibody depletion of the neutrophil cell population are more susceptible to infection and exhibit increased colonization of systemic sites than immunocompetent mice (Conlan, 1997). Although it has been reported that neutrophils can act as an intracellular niche for *Salmonella* in the murine spleen (Dunlap et al., 1992), it is now clear that in this organ the majority of *Salmonella* reside within macrophages (Salcedo et al., 2001). How neutrophils contribute to resistance during a phase of infection in which most bacteria are intracellular and therefore hidden from these cells is not yet known.

Despite the different contributions of neutrophils to protection in typhoid fever and gastroenteritis, in neither disease does there seem to be the same exploitation of neutrophils by *Salmonella* as a replicative niche or vehicle for dissemination as occurs with macrophages and DCs. Therefore our current understanding indicates that the role of neutrophils during the course of infection is solely protective.

INTERACTIONS BETWEEN *SALMONELLA* AND DCs

DCs are antigen-presenting cells that sample, process, and present antigen, and their presence in the periphery and in secondary lymphoid tissue means that they encounter *Salmonella* during both intestinal and systemic phases of infection. Bacteria are phagocytosed by DCs and the stimulation of DC migration upon maturation provides a means of transport for *Salmonella* away from peripheral sites.

Salmonella Infect DCs in the Intestinal and Systemic Phases of Infection

The subepithelial dome (SED) contains large numbers of phagocytes, many of which are $CD11c^+$ DCs (Kelsall and Strober, 1996); these are likely to be the first phagocytic cell type that *Salmonella* encounters following its translocation across the epithelial cell layer. Indeed bacteria have been visualized within $CD11c^+$ DCs in the MLN as well as the SED by immunofluorescence confocal microscopy (Hopkins et al., 2000). These interactions are promoted by the further *Salmonella*-induced recruitment of DCs to the SED, as illustrated in the serovar Typhimurium-infected mouse model. Bacterial flagellin ligates to TLR5 on the follicle-associated epithelium (FAE) enterocytes; these cells then secrete CCL20 (MIP-3α) and CCL9, the receptor for which, CCR6, is found on immature DCs. The DCs subsequently respond by chemotaxis toward the increasing concentration of chemokines (Biedzka-Sarek and Skurnik, 2006; Iwasaki and Kelsall, 2000).

Once at this site of bacterial colonization, DCs phagocytose *Salmonella* and antigen in the SED and from M cells. The chemokine CX_3CL1 stimulates these cells underlying the FAE to extend dendrites between epithelial cells, enabling the uptake of material directly from the gut lumen (Niess et al., 2005). Expression of tight-junction proteins on DC dendrites is upregulated after exposure to *Salmonella*; dendrite extension then occurs without disrupting the integrity of the epithelial cell barrier (Rescigno et al., 2001). Uptake of material from the gut lumen not only provides antigen for processing but also an SPI-1-independent route for bacteria to access the liver and spleen. $CD18^+$ cells of monocyte lineage containing serovar Typhimurium transmigrate across the intestinal epithelium and enable the dissemination of bacteria from the intestinal lumen via the bloodstream (Vazquez-Torres et al., 1999). This cell population is thought to consist of DCs due to their presence in the periphery and subsequent migration back to the secondary lymphoid tissue.

After dissemination from the gut, *Salmonella* have also been observed within DCs in both the murine spleen and liver. There is no evidence of a bacterial tropism for particular DC subsets in either organ: studies show that all are capable of internalizing bacteria (Johansson and Wick, 2004; Yrlid and Wick, 2002). The distribution of the pathogen between different DC types might therefore be a reflection of which cells it comes into contact with most frequently.

Salmonella Survive in and Modify DCs

There are several possible outcomes of *Salmonella* phagocytosis by DCs. When the SPI-1 T3SS is expressed, cytotoxicity can be triggered in DCs; alternatively, in some cells, a quiescent state is reached where the bacterium survives intracellularly, independently of both the SPI-2 T3SS (despite its expression) and the PhoP/Q regulon, with no apparent replication or killing of bacteria (Jantsch et al., 2003; Niedergang et al., 2000). Maturation and migration of the infected DCs to the secondary lymphoid tissue follow, but *Salmonella* is able to inhibit the development of specific immunity that is normally the consequence of this process.

Induction of DC cytotoxicity occurs in the gastrointestinal phase of infection and shares some features with early induced macrophage cell death: in both cases the SPI-1 T3SS protein SipB activates caspase-1, leading to death of the host cell (van der Velden et al., 2003). Caspase-1-mediated apoptosis antagonizes clearance of the pathogen from the SED, and is important for colonization of the Peyer's patches and bacterial dissemination to the lymph nodes and spleen (see Fig. 1). It is not without immunogenic consequences, however, as this enzyme cleaves and activates the proinflammatory cytokines IL-1β and IL-18 (Dreher et al., 2002), which are released from the cell. IL-18 stimulates the recruitment and activation of neutrophils and DCs, increasing the number of immune effector cells in the locality of *Salmonella* infection. Apoptotic bodies are sampled by bystander DCs, providing a source of antigenic material for processing and presentation to naive T cells; however, this process could also provide a means for dissemination as live bacteria can also be phagocytosed from apoptotic bodies.

Those bacteria not expressing SipB that are phagocytosed by DCs remain in a nonreplicative state within

SCVs. The vacuole matures similarly to those in macrophages (see below): late endosomal proteins such as LAMP-1 are delivered to the vacuolar membrane but fusion with lysosomes is generally avoided. Activation of PhoP/Q and SPI-2 T3SS enables avoidance of cathepsin D acquisition by the SCV: when strains of *Salmonella* mutated in *phoP* and *ssaV* infect DCs, a larger number of SCVs acquire this lysosomal protease. Nevertheless, the less degradative nature of DC lysosomes (compared with those in macrophages, for example) means that the survival of this mutant strain is similar to that of wild-type bacteria that avoid lysosomes.

After interaction with *Salmonella* (or LPS), DCs downregulate the chemokine receptor involved in DC recruitment to the SED, CCR6, and upregulate CCR7. The heightened levels of CCR7 expression enables cells to respond to and migrate toward the sources of CCL19 and CCL20: T-zone stromal cells in secondary lymphoid tissue (Cheminay et al., 2002; Zhao et al., 2006). DCs secrete CCL22, which further promotes interaction with the Th2 subset of $CD4^+$ T cells that express the corresponding receptor for this chemokine, CCR4. As cells move toward the lymphoid follicle, expression of MHC- I and -II, CD40, CD54, CD80, and CD86 increases; cells lose their phagocytic potential and switch to a phenotype more adapted to antigen presentation and stimulation of specific responses. Both infected and uninfected DCs undergo migration and maturation during infection, with uninfected cells requiring *Salmonella*-induced TNF-α-dependent stimulation (Sundquist and Wick, 2005) before they respond.

DCs secrete IL-1β, IL-18, and IL-6, which contribute to anti-*Salmonella* immunity (Marriot et al.,1999): the latter two are involved in neutrophil recruitment to the SED. Small amounts of IL-12 are released by hepatic and splenic DCs (Johansson and Wick, 2004; Yrlid and Wick, 2002), which leads to IFN-γ secretion and amplification of the initial IL-12 signal, macrophage activation, and Th cell differentiation to a Th1 phenotype. Some TNF-α is secreted by DCs, but the contribution to the total pool is likely to be minimal (Yrlid and Wick, 2002); the mouse model of infection shows the vast majority of splenic *Salmonella* within macrophages (Salcedo et al., 2001), not DCs, and it is macrophages that contribute most to secretion of TNF-α.

The migratory and maturation processes triggered by infection increase the proximity of DCs to T cells and enable them to prime T cells against bacteria through interactions between MHC- I, MHC-II, costimulatory molecules CD40 and CD86, and their counterparts on naive T cells (Yrlid et al., 2001). This, in conjunction with the effects of cytokine signaling, is likely to be detrimental to *Salmonella* persistence within the tissues. However, bacteria can benefit from DC migration as they are delivered to those systemic sites in which they can replicate due to the presence of a significant number of resident macrophages. Once there, *Salmonella*-induced inhibition of specific immunity to some extent counteracts the outcomes of interactions with DCs that are adverse to bacterial pathogenesis.

Splenic and hepatic subsets of DCs are able to present *Salmonella* antigens on MHC-I to $CD8^+$ T cells and MHC-II to $CD4^+$ T cells. However, serovar Typhimurium-infected DCs exhibit reduced stimulation of $CD4^+$ T-cell proliferation in vitro (Cheminay et al., 2005). In vivo, *Salmonella* can inhibit T-cell responses such as IL-2 (Tobar et al., 2006). Although bacterial infection of DCs induces the production of nitric oxide (NO), the presence of this immunosuppressive molecule alone is not responsible for the lack of T-cell responsiveness. Live, protein-synthesizing bacteria are required; it appears that both the SPI-2 T3SS and NO act independently to bring about inhibition of T-cell proliferation by serovar Typhimurium (Cheminay et al. 2005).

Growing evidence for SPI-2 T3SS-mediated inhibition of antigen presentation may also explain the lack of T-cell responsiveness during *Salmonella* infections. Although the biosynthesis and loading of MHC-II is unaffected, translocation of effectors by bacteria through the SPI-2 T3SS into host cells somehow downregulates the presence of antigen in the context of MHC-II on the cell surface of both DCs and macrophages (Mitchell et al., 2004; Cheminay et al., 2005). *Salmonella* also inhibits the presentation of antigen on MHC-I, although opsonization of the pathogen by immunoglobulin G and uptake via FcγR stimulates antigen processing and presentation and, to some extent, overrides this inhibition (Tobar et al., 2004).

Whether this inhibition is induced indirectly as a result of *Salmonella* interfering with normal cellular trafficking, reflects direct targeting of antigen presentation, or is a combination of the two has yet to be determined. Nevertheless, inhibition of antigen presentation, NO-mediated T-cell suppression, and avoidance of lysosomes all provide a means to the same end: delaying the expansion and maturation of T-cell populations specific to *Salmonella*.

INTERACTIONS BETWEEN *SALMONELLA* AND MACROPHAGES

Salmonella Replicates within Macrophages Both In Vitro and In Vivo

Despite the many bactericidal mechanisms that macrophages possess, many serovars of *S. enterica* are able to survive within them. This adaptation is essential for bacterial pathogenesis, since mutant strains of serovar Typhimurium that are unable to survive within macrophages are attenuated upon infection of mice (Fields, 1986). Furthermore, the pathogen is not limited simply to surviving in the face of macrophage-killing activities, but is able to replicate intracellularly (Color Plate 7).

Replication of serovar Typhimurium occurs in both murine macrophage-like cell lines, such as RAW and J774 cells, as well as primary bone marrow-derived and elicited peritoneal macrophages from susceptible mouse strains (e.g., $Nramp^-$). However, macrophage activation by IFN-γ, microbial products, and TNF-α promotes their control of *Salmonella* replication; the macrophages become larger, more pinocytic, and more vacuolated, with increased bactericidal activity.

It is likely that, in vivo, *Salmonella* first encounters macrophages in the SED, but because the majority of phagocytic cells in this location are immature DCs, the main sites of *Salmonella*-macrophage interactions are the spleen and liver. Serovar Typhimurium has been visualized within macrophages in these organs in mice. In the spleen, the large majority of infected cells are macrophages, and these are extensively colonized, with approximately 35% of the splenic macrophage population containing *Salmonella* at late stages of infection (Salcedo et al., 2001). Of these macrophages, it is the red pulp and marginal zone macrophages that are predominantly associated with bacteria, rather than the less phagocytic marginal metallophilic subset. This distribution of the pathogen might be the result of the differing phagocytic potential of these cell types, or it may be that the location of marginal metallophilic macrophages closer to the white pulp simply means that they encounter and therefore engulf *Salmonella* less often than the other cells. In the liver, bacteria have been observed within murine CD18+ cells, presumed to be macrophages. *Salmonella* are predominantly associated with an infiltrating population rather than resident Kupffer cells as determined by confocal microscopy and labeling for these resident macrophages (Richter-Dahlfors et al., 1997).

Macrophages Target Intracellular *Salmonella* Using Oxidative and Nonoxidative Mechanisms

Although the specific receptors and ligands that mediate *Salmonella* phagocytosis by macrophages are unknown, it is likely that a combination of Fcγ, complement (CR), scavenger receptors, and TLRs is involved; adherence and uptake sets in motion several antibacterial responses that target the contents of the maturing phagosome. The most rapid of these is the oxidative burst, triggered by assembly of the multiprotein NADPH oxidase complex on the nascent phagosome. This delivers ROS into the lumen of the phagosome, where they exert a bactericidal effect. Subsequent fusion of vacuoles containing inducible NO synthase (iNOS) with phagosomes releases NO and other reactive nitrogen intermediates (RNIs) into the vicinity of bacteria. The extreme sensitivity of knockout mice lacking the intact NADPH oxidase to *Salmonella* infection indicates that ROS are essential to successful early host defense against this pathogen (Mastroeni et al., 2000). Although iNOS$^{-/-}$ mice also show increased sensitivity to *Salmonella*, it is not until approximately 3 weeks after inoculation, after a period of increased bacterial burden in the liver and spleen. Both neutrophils and macrophages can synthesize ROS and RNIs, so it is difficult to determine the individual contributions of each of these cells at the different stages of infection. However, it is clear that RNIs act at later stages in infection than ROS and are involved in limiting the growth of *Salmonella* in the spleen and liver.

The divalent metal cation transporter Nramp1 (natural resistance-associated macrophage protein) is also rapidly delivered to vacuoles containing bacteria, where it augments the toxic activity of ROS and RNIs within. Nramp1 transports transition metal ions such as Fe^{2+}, Mn^{2+}, and Zn^{2+} across the phagosome membrane, but opinion differs as to the direction in which this traffic occurs: pH-dependent influx or efflux of metal ions in exchange for protons. The direction of ion flux has implications for the bactericidal effects of this protein. Transported ions are cofactors for some of the enzymes that metabolize ROS and RNIs; therefore, limiting ion availability by efflux also inhibits the ability of the bacterium to cope with oxidative metabolites. On the other hand, increasing the concentration of iron in the vacuole would promote the formation of ferryl intermediates from ROS that cause DNA damage. Two alleles of the murine gene encoding Nramp1 exist that are linked to resistance and susceptibility in the mouse model of infection; however, no link has been established between the human homologue, NRAMP1, and susceptibility to typhoid fever.

Aside from the delivery of oxidative molecules to SCVs, *Salmonella* can be killed by the activity of oxygen-independent antimicrobial peptides (AMPs) synthesized by macrophages and neutrophils. For the most part, AMPs are cationic molecules between 15 and 50 amino acids in size (Peschel, 2002). Their positive charge means that on contacting *Salmonella*, AMPs interact electrostatically with negative charges on LPS, leading to disruption of the outer membrane. The peptides then bind to and subsequently disrupt the inner membrane, permeabilizing the bacteria. Antimicrobial activity is found in the cytosol of macrophages and the granules of neutrophils; delivery of these to SCVs and the extracellular environment targets both intra- and extracellular *Salmonella*. Nevertheless, *S. enterica* is resistant to several structurally unrelated AMPs—polymyxin, protegrin-1, magainin, and cathelicidin-related antimicrobial peptide (CRAMP)—and this resistance depends on the function of PhoP/Q and another bacterial two-component regulatory system, PmrA/B (see below).

The SPI-1 T3SS Induces Macrophage Cytotoxicity in the Gut Phase of Infection

Salmonella can induce cytotoxicity in macrophages via the SPI-1 T3SS in the gastrointestinal tissue during early stages of host colonization. Translocation of SipB into macrophages by the SPI-1 T3SS activates two separate pathways to host cell death; clues to the existence of these different pathways were given by the conflicting descriptions of cell death. High multiplicity of infection (MOI) and SPI-1 expression are likely to increase the delivery of SipB into the host cytosol, and cell death can be induced as rapidly as 45 min after bacterial uptake. These conditions give rise to necrotic cell death, including loss of organelle and membrane integrity in the absence of caspase-3 activation. In contrast, low MOI and SPI-1 expression lead at later time points to chromatin fragmentation, caspase-3 activation leading to poly(ADP ribose) polymerase (PARP) cleavage, and nucleosomes visible in the cytoplasm of infected cells, features that are more characteristic of conventional apoptosis.

Neither of these two SipB-induced pathways use the classical routes to cell death via either caspase-8 or caspase-9. The first, more rapid pathway involves SipB-mediated activation of caspase-1 (Hersh et al., 1999), either through direct interaction or by acting as a PAMP and binding Nod1. Nod1 is a receptor for microbial products that on ligand binding either acts on caspase-1 directly or via the serine/threonine kinase, RIP2. Caspase-1 activation leads to the degradation of Raf1, a signaling protein generally involved in cell survival pathways; its removal enables the induction of cell death. The more delayed apoptotic cell death is independent of caspase-1 activation but is linked to the ability of SipB to localize to mitochondria after translocation. It is thought that SipB induces autophagy by damaging mitochondria, leading to type II programmed cell death (Hernandez et al., 2003).

The ability of bacteria to induce cytotoxicity in those macrophages that they encounter enables the pathogen to evade killing and is likely to promote dissemination to and colonization of systemic sites within the host. Infection of caspase-1-deficient mice by wild-type bacteria leads to poor colonization of the spleen and liver. Interestingly, observations of human and murine macrophages in vitro reveals reduced cytotoxicity of serovar Typhi compared with serovar Typhimurium. This might limit the release of inflammatory mediators from infected cells and, coupled with the serovar Typhi-mediated inhibition of IL-8 signaling, contribute to the decrease in inflammation and neutrophil recruitment in the gut that is characteristic of typhoid fever in humans.

Intramacrophage SCVs Interact Selectively with the Endocytic Pathway

Those *Salmonella* cells that do not induce early apoptosis in host cells remain within SCVs; however, these compartments are not conventional phagosomes. The normal trafficking of a phagocytosed particle in macrophages leads to acidification of the phagosome followed by delivery of lysosomal hydrolases, which degrade the phagocytosed particle and provide material for antigen presentation. A major component of *S. enterica* virulence is its ability to modify this process.

Early vacuole maturation is characterized by delivery of vacuolar ATPase to the SCV and its subsequent acidification (Rathman et al., 1996). Indeed, vacuole acidification is important for *Salmonella* virulence because it is a signal for induction of the SPI-2 T3SS. Rab GTPases are likely to influence the biogenesis of the SCV, which usually develops into a specialized compartment, acquiring lysosomal glycoproteins (such as LAMP-1 and LAMP-2) but avoiding those proteins delivered to late endosomes via the mannose-6-phosphate receptor (MPR) such as the lysosomal cysteine proteases, cathepsin D and L (Rathman et al., 1997) (illustrated in Fig. 2). However, it should be borne in mind that this does not occur in all macrophages and that, even in the same macrophage, not all SCVs escape phagolysosomal fusion.

Rapid Activation of PhoP/Q Enables Bacterial Survival in Macrophages

During the first hour after *Salmonella* phagocytosis by macrophages, the increasing acidity of the SCV activates the bacterial two-component regulatory system, PhoP/Q (Martin-Orozco et al., 2006; Prost et al., 2007). The drop in pH is detected by the sensor kinase, PhoQ, found in the inner membrane as a homodimer. This protein then undergoes a conformational change that stimulates its kinase activity; autophosphorylation occurs, followed by transfer of this phosphate to cytoplasmic PhoP. Phospho-PhoP has increased affinity for DNA binding and is thus able to regulate the expression of approximately 40 bacterial genes,

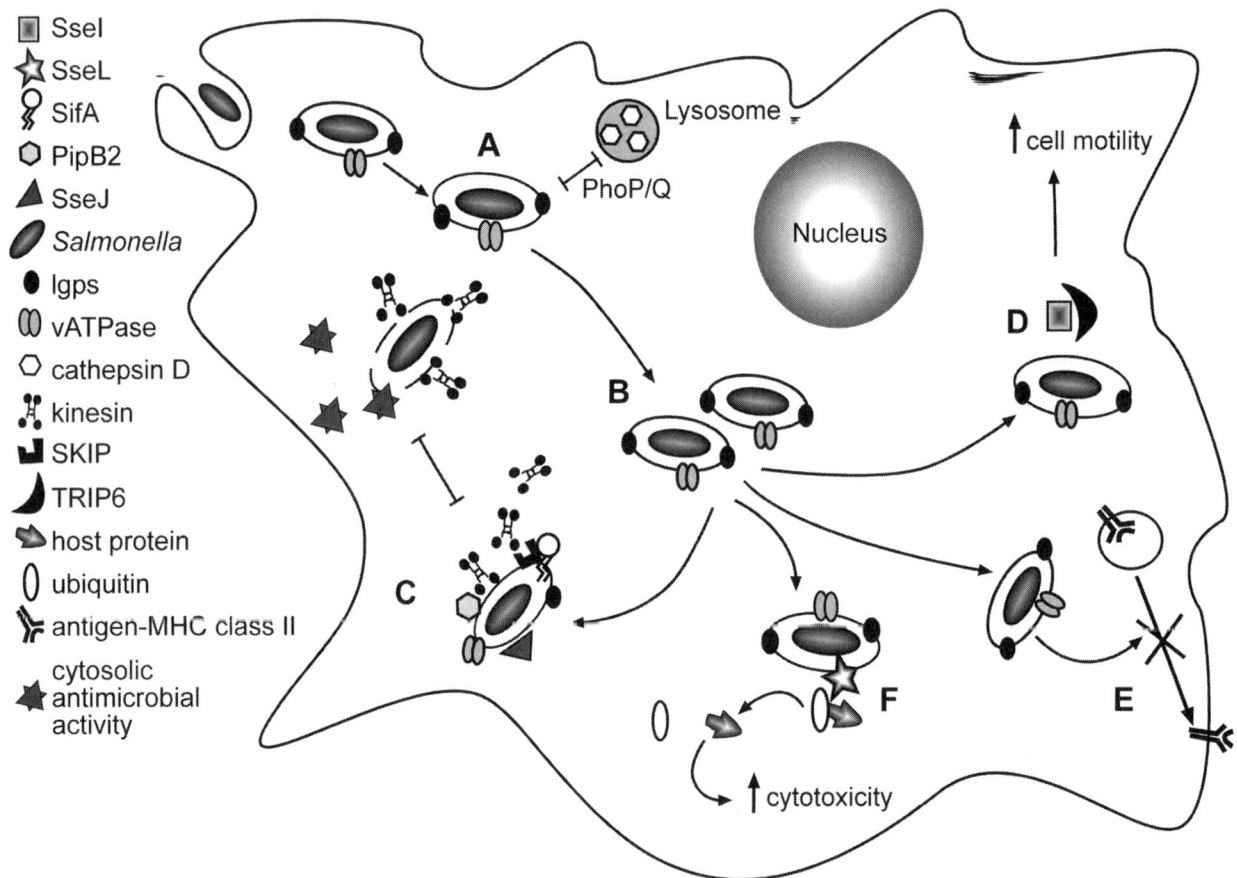

FIGURE 2 *Salmonella* virulence proteins modify macrophage behavior. (A) *Salmonella* is phagocytosed by macrophages and resides within the SCV, which selectively interacts with the endocytic pathway. *pag* gene products of the PhoP/Q regulon modify LPS and enable resistance to AMPs. PhoP/Q activity also blocks fusion between the maturing SCV and lysosomes. (B) The SPI-2 T3SS is activated and bacteria begin to replicate. (C) PipB2 binds kinesin on the SCV while SifA interaction with SKIP prevents excessive kinesin recruitment, loss of the vacuole membrane, and cytosolic killing of bacteria. SseJ contributes to membrane stability. (D) SseI binds TRIP6 and increases cell motility. (E) Unknown SPI-2 effector(s) inhibit MHC-II antigen presentation, and (F) at late stages of infection SseL deubiquitinates host protein(s) and induces host cell death.

which contribute to intramacrophage survival of the bacterium (Groisman et al., 1989; Miller et al., 1989).

Although this system can be regulated in vitro by divalent cations such as Mg^{2+}, the intravacuolar conditions are such that PhoP activation appears to be independent of their concentration. However, if sublethal concentrations of AMPs are found in the periplasm, binding of these peptides to PhoQ also stimulates its kinase activity (Bader et al., 2005). This adds to the activation of PhoP brought about by the decreased pH of the SCV (Prost et al., 2007).

Gene expression regulated by PhoP equips bacteria to withstand those pressures that provided the initial signals for activation. Adaptation includes downregulation of PhoP-repressed genes (*prgs*), including those encoding the SPI-1 T3SS, concurrent with upregulation of PhoP-activated genes (*pags*). These include genes encoding proteins that modify the cell envelope, particularly LPS, increasing its resistance to AMPs. PagP catalyzes the addition of palmitate to the lipid A component of LPS (Bishop et al., 2000), decreasing the fluidity of this molecule and increasing bacterial resistance to the AMP protegrin. Lipid A is also modified by PagQ through its oxygen-dependent formation of 2-hydroxymyristate; PagL removes 3-hydroxymyristate (Trent et al., 2001). UgtL and PcgL contribute to resistance against magainin-2 and polymyxin B, possibly through the remodeling of LPS or peptidoglycan, while VirK and Mig-14 inhibit antimicrobial peptide binding to the bacterial surface and increase resistance to CRAMP by unknown mechanisms (Brodsky et al., 2005; Detweiler et al., 2003).

PhoP also activates the response regulator of another two-component system, PmrA/B, through a small basic protein, PmrD. This regulates additional genes encoding LPS-modifying enzymes (Gunn and Miller, 1996, Gunn et al., 1998, 2000). For example, PmrE (also called PagA or Ugd) is a UDP-glucose dehydrogenase that synthesizes a 4-aminoarabinose precursor; the *pmrF* operon encodes seven genes involved in 4-aminoarabinose biosynthesis and its addition to LPS; phosphoethanolamine is incorporated into lipid A by PmrC and CptA. The modification of lipid A with 4-aminoarabinose and phosphoethanolamine further decreases the fluidity of LPS, reducing the ability of AMPs to insert into the bacterial outer membrane. It also increases the positivity of LPS, which inhibits the initial binding of cationic AMPs such as polymyxin and α-helical peptides. In addition to the increased protection from AMPs that these modifications provide, antibacterial responses stimulated by TLR4 recognition of LPS are dampened (Kawasaki et al., 2004).

The function of the PhoP/Q regulon in not limited to increasing resistance against AMPs. Its activation appears to prevent fusion of SCVs with lysosomes at later stages of vacuolar trafficking in macrophages (Garvis et al., 2001). However, the mechanisms underlying PhoP/Q-mediated avoidance of SCV-lysosome fusion have yet to be characterized (Fig. 2).

The Multiple Functions of the SPI-2 T3SS Contribute to *Salmonella* Replication

The second major *Salmonella* virulence system that operates inside macrophages is the SPI-2 T3SS. This translocates approximately 20 effectors onto the cytoplasmic face of the SCV membrane and into the host cell cytosol. These effectors modify the SCV and help create an environment that is conducive to bacterial replication and thereby enable the pathogen to colonize the spleen and liver (Color Plate 7).

Through studies of the intracellular phenotypes of SPI-2 mutant strains, as well as biochemical analysis of individual effectors, a few functions of the SPI-2 T3SS have become clear (summarized in Fig. 2). The effector about which most is known is SifA: this is translocated onto the vacuolar membrane, where it is prenylated at its C terminus by the host enzyme protein geranyl geranyl transferase I. Subsequent S-acylation occurs; these modifications to the protein help to anchor SifA in the SCV membrane (Reinicke et al., 2005). In this location, the bacterial effector recruits and interacts with its eukaryotic partner, SifA, and kinesin-interacting protein (SKIP) (Boucrot et al., 2005). This inhibits recruitment of the plus-end microtubule motor kinesin to the vacuolar membrane. When SifA or SKIP are lacking, excessive amounts of kinesin are active at the vacuolar membrane and provide a driving force for removal of this membrane from around bacteria (Beuzon et al., 2000). The pathogen is exposed to and killed by a potent bactericidal activity present in the cytosol of activated macrophages (Beuzon et al., 2002). Therefore, through binding by SKIP and maintaining the integrity of the SCV membrane, SifA protects *Salmonella* from macrophage killing and facilitates bacterial replication.

Interestingly, recruitment of kinesin to the SCV in the absence of SifA is mediated by another SPI-2 effector, PipB2, which is capable of binding kinesin directly (Henry et al., 2006). Therefore, by interacting with SKIP, SifA appears to inhibit PipB2-dependent recruitment of kinesin. However, SifA and PipB2 do not simply oppose each other's function, because the vacuole enclosing a *sifA*, *pipB2* double mutant is as unstable as that of the *sifA* mutant.

Another effector that affects SCV stability is the acyl-transferase SseJ (Nawabi et al., 2008). This enzyme contributes to the breakdown of the vacuolar membrane around a *sifA* mutant (Ruiz-Albert et al., 2002) and probably esterifies cholesterol in the vacuolar membrane.

Between 12 and 24 h after uptake by cultured macrophages, *Salmonella* induces an SPI-2 T3SS-dependent cytotoxicity. This is mediated in part by SseL, an effector with deubiquitinating activity (Rytkönen et al., 2007). The ubiquitinated protein(s) that are targeted by SseL and their relationships to the cytotoxic process are important questions that remain to be addressed. However, this phenomenon is likely to be important in vivo, because necrotic and apoptotic macrophages are frequently observed in spleens and livers of infected mice (Richter-Dahlfors et al., 1997), and *sseL* mutant strains are attenuated in virulence.

Dissemination of intramacrophage *Salmonella* also takes place independently of the induction of host cell death. After SPI-2 T3SS-dependent translocation into host cells, SseI (SrfH) interacts with thyroid receptor interacting protein 6 (TRIP6), an adaptor that binds to proteins in the Rac family. Signaling pathways downstream of Rac are often involved in cell motility and NF-κB responses: the mouse model of infection has been used to show that increased migration of *Salmonella*-containing $CD18^+$ cells away from the gut lumen via the bloodstream is stimulated by SseI (Worley et al., 2006).

It has been reported that the SPI-2-encoded protein SpiC is a translocated effector that has an important role in trafficking of the SCV (Uchiya et al., 1999). However, more recent data implicate this protein in regulating the release of effector proteins, acting from within the bacterial cell (Freeman et al., 2002; Yu et al., 2004). Several other characteristics of *Salmonella* intramacrophage life have been attributed to the function of the SPI-2 T3SS: the inhibition of delivery of vesicles containing NADPH oxidase and iNOS to the SCV (Vazquez-Torres et al., 2000; Chakravortty et al., 2002), accumulation of cholesterol and GPI-anchored CD55 to the vacuolar membrane (Catron et al., 2002), and formation of a cage-like actin meshwork around SCVs (Meresse et al., 2001). The effectors that contribute to these processes have yet to be identified. Furthermore, additional SPI-2 effectors (such as SseF and SseG) have been identified that contribute to intramacrophage replication but for which biochemical functions have not been elucidated.

CONCLUSIONS

Salmonella encounters neutrophils, DCs, and macrophages during systemic infection and exploits macrophages, in particular, as an intracellular niche for its replication. Not surprisingly, therefore, the activities of many *Salmonella* virulence factors counteract mechanisms for bacterial elimination by macrophages. Furthermore, it inhibits antigen presentation in macrophages and DCs, which facilitates its persistence within the host as a whole, and can induce host cell death, which enables its spread through host tissues. Despite the fact that many *Salmonella* virulence genes are known to contribute to these processes, our understanding of the functions of many of them is quite rudimentary. A great deal remains to be learned about the mechanisms that enable bacterial replication in cells whose primary function is the destruction of ingested material. Of course, the hope is that this knowledge can be exploited in the rational design of vaccines against *Salmonella*, and its use as a vector against other pathogens. Progress continues to be made with attenuated strains of *S. enterica* as vaccine candidates, including those lacking a functional SPI-2 T3SS. Finally, increasing our understanding of how *Salmonella* disrupts the functions of phagocytes is likely to shed light on normal cell biology and the regulation of antimicrobial mechanisms in these cells. Developments in this area are likely to prove extremely useful, not only in shedding light upon host immune responses, but also in providing new avenues for investigation with other pathogens that are also able to infect phagocytes.

We apologize to those colleagues whose work it was not possible to cite due to space constraints. We also thank Piotr Mazurkiewicz for his help and comments during the development of this chapter.

REFERENCES

Alpuche Aranda, C. M., J. A. Swanson, W. P. Loomis, and S. I. Miller. 1992. *Salmonella* Typhimurium activates virulence gene transcription within acidified macrophage phagosomes. *Proc. Natl. Acad. Sci. USA* **89:**10079–10083.

Bader, M. W., S. Sanowar, M. E. Daley, A. R. Schneider, U. Cho, W. Xu, R. E. Klevit, H. Le Moual, and S. I. Miller. 2005. Recognition of antimicrobial peptides by a bacterial sensor kinase. *Cell* **122:**461–472.

Beuzon, C. R., S. Meresse, K. E. Unsworth, J. Ruiz-Albert, S. Garvis, S. R. Waterman, T. A. Ryder, E. Boucrot, and D. W. Holden. 2000. *Salmonella* maintains the integrity of its intracellular vacuole through the action of SifA. *EMBO J.* **19:**3235–3249.

Beuzon, C. R., S. P. Salcedo, and D. W. Holden. 2002. Growth and killing of a *Salmonella enterica* serovar Typhimurium *sifA* mutant strain in the cytosol of different host cell lines. *Microbiology* **148:**2705–2715.

Biedzka-Sarek, M., and M. Skurnik. 2006. How to outwit the enemy: dendritic cells face *Salmonella*. *APMIS* **114:**589–600.

Bishop, R. E., H. S. Gibbons, T. Guina, M. S. Trent, S. I. Miller, and C. R. Raetz. 2000. Transfer of palmitate from phospholipids to lipid A in outer membranes of Gram-negative bacteria. *EMBO J.* **19:**5071–5080.

Boucrot, E., T. Henry, J. P. Borg, J. P. Gorvel, and S. Meresse. 2005. The intracellular fate of *Salmonella* depends on the recruitment of kinesin. *Science* **308:**1174–1178.

Brodsky, I. E., N. Ghori, S. Falkow, and D. Monack. 2005. Mig-14 is an inner membrane-associated protein that promotes *Salmonella typhimurium* resistance to CRAMP, survival within activated macrophages and persistent infection. *Mol. Microbiol.* **55:**954–972.

Carter, P. B., and F. M. Collins. 1974. The route of enteric infection in normal mice. *J. Exp. Med.* **139:**1189–1203.

Catron, D. M., M. D. Sylvester, Y. Lange, M. Kadekoppala, B. D. Jones, D. M. Monack, S. Falkow, and K. Haldar. 2002. The *Salmonella*-containing vacuole is a major site of intracellular cholesterol accumulation and recruits the GPI-anchored protein CD55. *Cell. Microbiol.* **4:**315–328.

Chakravortty, D., I. Hansen-Wester, and M. Hensel. 2002. *Salmonella* pathogenicity island 2 mediates protection of intracellular *Salmonella* from reactive nitrogen intermediates. *J. Exp. Med.* **195:**1155–1166.

Cheminay, C., A. Mohlenbrink, and M. Hensel. 2005. Intracellular *Salmonella* inhibit antigen presentation by dendritic cells. *J. Immunol.* **174:**2892–2899.

Cheminay, C., M. Schoen, M. Hensel, A. Wandersee-Steinhäuser, U. Ritter, H. Körner, M. Röllinghoff, and J. Hein. 2002. Migration of *Salmonella* Typhimurium-harbouring bone marrow-derived dendritic cells towards the chemokines CCL-19 and CCL-21. *Microb. Pathog.* **32:**207–218.

Cirillo, D. M., R. H. Valdivia, D. M. Monack, and S. Falkow. 1998. Macrophage-dependent induction of the *Salmonella* pathogenicity island 2 type III secretion system and its role in intracellular survival. *Mol. Microbiol.* **30:**175–188.

Clark, M. A., M. A. Jepson, N. L. Simmons, and B. H. Hirst. 1994. Preferential interaction of *Salmonella* Typhimurium with mouse Peyer's patch M cells. *Res. Microbiol.* **145:**543–552.

Clark, M. A., K. A. Reed, J. Lodge, J. Stephen, B. H. Hirst, and M. A. Jepson. 1996. Invasion of murine intestinal M cells by *Salmonella* Typhimurium *inv* mutants severely deficient for invasion of cultured cells. *Infect. Immun.* **64:**4363–4368.

Conlan, J. W. 1997. Critical roles of neutrophils in host defense against experimental systemic infections of mice by *Listeria monocytogenes*, *Salmonella* Typhimurium, and *Yersinia enterocolitica*. *Infect. Immun.* **65:**630–635.

Detweiler, C. S., D. M. Monack, I. E. Brodsky, H. Mathew, and S. Falkow. 2003. *virK*, *somA* and *rcsC* are important for systemic *Salmonella enterica* serovar Typhimurium infection and cationic peptide resistance. *Mol. Microbiol.* **48:**385–400.

Dreher, D., M. Kok, C. Obregon, S. G. Kiama, P. Gehr, and L. P. Nicod. 2002. *Salmonella* virulence factor SipB induces activation and release of IL-18 in human dendritic cells. *J. Leukoc. Biol.* **72:**743–751.

Dunlap, N. E., W. H. Benjamin, A. K. Berry, J. H. Eldridge, and D. E. Briles. 1992. A "safe-site" for *Salmonella* Typhimurium is within splenic polymorphonuclear cells. *Microb. Pathog.* **13:**181–190.

Fields, P. I., E. A. Groisman, and F. Heffron. 1989. A *Salmonella* locus that controls resistance to microbicidal proteins from phagocytic cells. *Science* 243:1059–1062.

Fields, P. I., R. V. Swanson, C. G. Haidaris, and F. Heffron. 1986. Mutants of *Salmonella* Typhimurium that cannot survive within the macrophage are avirulent. *Proc. Natl. Acad. Sci. USA* 83:5189–5193.

Freeman, J. A., C. Rappl, V. Kuhle, M. Hensel, and S. I. Miller. 2002. SpiC is required for translocation of *Salmonella* pathogenicity island 2 effectors and secretion of translocon proteins SseB and SseC. *J. Bacteriol.* 184:4971–4980.

Garvis, S. G., C. R. Beuzon, and D. W. Holden. 2001. A role for the PhoP/Q regulon in inhibition of fusion between lysosomes and *Salmonella*-containing vacuoles in macrophages. *Cell. Microbiol.* 3:731–744.

Groisman, E. A., E. Chiao, C. J. Lipps, and F. Heffron. 1989. *Salmonella* Typhimurium *phoP* virulence gene is a transcriptional regulator. *Proc. Natl. Acad. Sci. USA* 86:7077–7781.

Gunn, J. S., K. B. Lim, J. Krueger, K. Kim, L. Guo, M. Hackett, and S. I. Miller. 1998. PmrA-PmrB-regulated genes necessary for 4-aminoarabinose lipid A modification and polymyxin resistance. *Mol. Microbiol.* 27:1171–1182.

Gunn, J. S., and S. I. Miller. 1996. PhoP-PhoQ activates transcription of *pmrAB*, encoding a two-component regulatory system involved in *Salmonella* Typhimurium antimicrobial peptide resistance. *J. Bacteriol.* 178:6857–6864.

Gunn, J. S., S. S. Ryan, J. C. Van Velkingurgh, R. K. Ernst, and S. I. Miller. 2000. Genetic and functional analysis of a PmrA-PmrB-regulated locus necessary for lipopolysaccharide modification, antimicrobial peptide resistance, and oral virulence of *Salmonella enterica* serovar Typhimurium. *Infect. Immun.* 69:6139–6146.

Henry, T., C. Couillault, P. Rockenfeller, E. Boucrot, A. Dumont, N. Schroeder, A. Hermant, L. A. Knodler, P. Lecine, O. Steele-Mortimer, J. P. Borg, J. P. Gorvel, and S. Meresse. 2005. The *Salmonella* effector protein PipB2 is a linker for kinesin-1. *Proc. Natl. Acad. Sci. USA* 103:13497–13502.

Hernandez, L. D., M. Pypaert, R. A. Flavell, and J. E. Galan. 2003. A *Salmonella* protein causes macrophage cell death by inducing autophagy. *J. Cell Biol.* 163:1123–1131.

Hersh, D., D. M. Monack, M. R. Smith, N. Ghori, S. Falkow, and A. Zychlinsky. 1999. The *Salmonella* invasin SipB induces macrophage apoptosis by binding to caspase-1. *Proc. Natl. Acad. Sci.* 96:2396–2401.

Hopkins, S. A., F. Niedergang, I. E. Corthesy-Theulaz, and J. P. Kraehenbuhl. 2000. A recombinant *Salmonella* Typhimurium vaccine strain is taken up and survives within murine Peyer's patch dendritic cells. *Cell. Microbiol.* 2:59–68.

Hueck, C. J. 1998. Type III protein secretion systems in bacterial pathogens of animals and plants. *Microbiol. Mol. Biol. Rev.* 62:379–433.

Iwasaki, A., and B. L. Kelsall. 2000. Localization of distinct Peyer's patch dendritic cell subsets and their recruitment by chemokines macrophage inflammatory protein (MIP)-3α, MIP-3β, and secondary lymphoid organ chemokine. *J. Exp. Med.* 191:1381–1394.

Jantsch, J., C. Cheminay, D. Chakravortty, T. Lindig, J. Hein, and M. Hensel. 2003. Intracellular activities of *Salmonella enterica* in murine dendritic cells. *Cell. Microbiol.* 5:933–945.

Johansson, C., and M. J. Wick. 2004. Liver dendritic cells present bacterial antigens and produce cytokines upon *Salmonella* encounter. *J. Immunol.* 172:2496–2503.

Jones, B. D., N. Ghori, and S. Falkow. 1994. *Salmonella* Typhimurium initiates murine infection by penetrating and destroying the specialized epithelial M cells of the Peyer's patches. *J. Exp. Med.* 180:15–23.

Kawasaki, K., R. K. Ernst, and S. I. Miller. 2004. Deacylation and palmitoylation of lipid A by *Salmonellae* outer membrane enzymes modulate host signaling through Toll-like receptor 4. *J. Endotoxin Res.* 10:439–444.

Kelsall, B. L., and W. Strober. 1996. Distinct populations of dendritic cells are present in the subepithelial dome and T cell regions of the murine Peyer's patch. *J. Exp. Med.* 183:237–247.

Kossack, R. E., R. L. Guerrant, P. Densen, J. Schadelin, and G. L. Mandell. 1981. Diminished neutrophil oxidative metabolism after phagocytosis of virulent *Salmonella* Typhi. *Infect. Immun.* 31:674–678.

Kurita, A., H. Gotoh, M. Eguchi, N. Okada, S. Matsuura, H. Matsui, H. Danbara, and Y. Kikuchi. 2003. Intracellular expression of the *Salmonella* plasmid virulence protein, SpvB, causes apoptotic cell death in eukaryotic cells. *Microb. Pathog.* 35:43–48.

Lee, C. A., M. Silva, A. M. Siber, A. J. Kelly, E. Galyov, and B. A. McCormick. 2000. A secreted *Salmonella* protein induces a proinflammatory response in epithelial cells, which promotes neutrophil migration. *Proc. Natl. Acad. Sci. USA* 97:12283–12288.

Marriot, I., T. G. Hammond, E. K. Thomas, and K. L. Bost. 1999. *Salmonella* efficiently enter and survive within cultured CD11c+ dendritic cells initiating cytokine secretion. *Eur. J. Immunol.* 29: 1107–1115.

Martin-Orozco, N., N. Touret, M. L. Zaharik, E. Park, R. Kopelman, S. Miller, B. B. Finlay, P. Gros, and S. Grinstein. 2006. Visualization of vacuolar acidification-induced transcription of genes of pathogens inside macrophages. *Mol. Biol. Cell.* 17:498–510.

Mastroeni, P. 2002. Immunity to systemic *Salmonella* infections. *Curr. Mol. Med.* 2:393–406.

Mastroeni, P., and M. Sheppard. 2004. *Salmonella* infections in the mouse model: host resistance factors and in vivo dynamics of bacterial spread and distribution in the tissues. *Microbes Infect.* 6:398–405.

Mastroeni, P., A. Vazquez-Torres, F. C. Fang, Y. Xu, S. Khan, C. E. Hormaeche, and G. Dougan. 2000. Antimicrobial actions of the NADPH phagocyte oxidase and inducible nitric oxide synthase in experimental salmonellosis. II. Effects on microbial proliferation and host survival *in vivo*. *J. Exp. Med.* 192:237–248.

Matsui, H., M. Eguchi, and Y. Kikuchi. 2000. Use of confocal microscopy to detect *Salmonella* Typhimurium within host cells associated with Spv-mediated intracellular proliferation. *Microb. Pathog.* 29:53–59

Mazurkiewicz, P., J. Thomas, J. A. Thompson, M. Liu, L. Arbibe, P. Sansonetti, and D. W. Holden. 2008. SpvC is a *Salmonella* effector with phosphothreonine lyase activity on host mitogen-activated protein kinases. *Mol. Microbiol.* 67:1371–1383.

Meresse, S., K. E. Unsworth, A. Habermann, G Griffiths, F. Fang, M. J. Martinez-Lorenzo, S. R. Waterman, J. P. Gorvel, and D. W. Holden. 2001. Remodelling of the actin cytoskeleton is essential for replication of intravacuolar *Salmonella*. *Cell. Microbiol.* 3:567–577.

Miller, S. I., A. M. Kukral, and J. J. Mekalanos. 1989. A two-component regulatory system (phoP phoQ) controls *Salmonella* Typhimurium virulence. *Proc. Natl. Acad. Sci. USA* 86:5054–5058.

Mitchell, E. K., P. Mastroeni, A. P. Kelly, and J. Trowsdale. 2004. Inhibition of cell surface MHC class II expression by *Salmonella*. *Eur. J. Immunol.* 34:2559–2567.

Nadeau, W. J., T. G. Pistole, and B. A. McCormick. 2002. Polymorphonuclear leukocyte migration across model intestinal epithelia enhances *Salmonella* Typhimurium killing via the epithelial derived cytokine, IL-6. *Microb. Infect.* 4:1379–1387.

Nawabi, P., D. M. Catron, and K. Haldar. 2008. Esterification of cholesterol by a type III secretion effector during intracellular *Salmonella* infection. *Mol. Microbiol.* **68**:173–185.

Niedergang, F., J. Sirard, C. T. Blanc, and J. Kraehenbuhl. 2000. Entry and survival of *Salmonella* Typhimurium in dendritic cells and presentation of recombinant antigens do not require macrophage-specific virulence factors. *Proc. Natl. Acad. Sci. USA* **97**:14650–14655.

Niess, J. H., S. Brand, X. Gu, L. Landsman, S. Jung, B. A. McCormick, J. M. Vyas, M. Boes, H. L. Ploegh, F. G. Fox, D. R. Littman, and H. Reinecker. 2005. CX_3CR1-mediated dendritic cell access to the intestinal lumen and bacterial clearance. *Science* **307**:254–258.

Norris, F. A., M. P. Wilson, T. S. Wallis, E. E. Galyov, and P. W. Majerus. 1998. SopB, a protein required for virulence of *Salmonella* Dublin, is an inositol phosphate phosphatase. *Proc. Natl. Acad. Sci. USA* **95**:14057–14059.

Parkhill, J., G. Dougan, K. D. James, N. R. Thomson, D. Pickard, J. Wain, C. Churcher, K. L. Mungall, S. D. Bentley, M. T. Holden, M. Sebaihia, S. Baker, D. Basham, K. Brooks, T. Chillingworth, P. Connerton, A. Cronin, P. Davis, R. M. Davies, L. Dowd, N. White, J. Farrar, T. Feltwell, N. Hamlin, A. Haque, T. T. Hien, S. Holroyd, K. Jagels, A. Krogh, T. S. Larsen, S. Leather, S. Moule, P. O'Gaora, C. Parry, M. Quail, K. Rutherford, M. Simmonds, J. Skelton, K. Stevens, S. Whitehead, and B. G. Barrell. 2001. Complete genome sequence of a multiple drug resistant *Salmonella enterica* serovar Typhi CT18. *Nature* **413**:848–852.

Peschel, A. 2002. How do bacteria resist human antimicrobial peptides? *Trends Microbiol.* **10**:179–186.

Prost, L. R., M. E. Daley, V. Le Sage, M. W. Bader, H. Le Moual, R. E. Klevit, and S. I. Miller. 2007. Activation of the bacterial sensor kinase PhoQ by acidic pH. *Mol. Cell* **26**:165–174.

Raffatellu, M., D. Chessa, R. P. Wilson, R. Dusold, S. Rubino, and A. J. Baumler. 2005. The Vi capsular antigen of *Salmonella enterica* serotype Typhi reduces Toll-like receptor-dependent interleukin-8 expression in the intestinal mucosa. *Infect Immun.* **73**:3367–3374.

Rathman, M., L. P. Barker, and S. Falkow. 1997. The unique trafficking pattern of *Salmonella* Typhimurium-containing phagosomes in murine macrophages is independent of the mechanism of bacterial entry. *Infect. Immun.* **65**:1475–1485.

Rathman, M., M. D. Sjaastad, and S. Falkow. 1996. Acidification of phagosomes containing *Salmonella* Typhimurium in murine macrophages. *Infect. Immun.* **64**:2765–2773.

Reinicke, A. T., J. L. Hutchinson, A. I. Magee, P. Mastroeni, J. Trowsdale, and A. P. Kelly. 2005. A *Salmonella* Typhimurium effector protein SifA is modified by host cell prenylation and S-acylation machinery. *J. Biol. Chem.* **280**:14620–14627.

Rescigno, M., M. Urbano, B. Valzasina, M. Francolini, G. Rotta, R. Bonasio, F. Granucci, J. Kraehenbuhl, and P. Ricciardi-Castagnoli. 2001. Dendritic cells express tight junction proteins and penetrate gut epithelial monolayers to sample bacteria. *Nat. Immunol.* **2**:361–367.

Richter-Dahlfors, A., A. M. Buchan, and B. B. Finlay. 1997. Murine salmonellosis studied by confocal microscopy: *Salmonella* Typhimurium resides intracellularly inside macrophages and exerts a cytotoxic effect on phagocytes in vivo. *J. Exp. Med.* **186**:569–580.

Ruiz-Albert, J., X. J. Yu, C. R. Beuzon, A. N. Blakey, E. E. Galyov, and D. W. Holden. 2002. Complementary activities of SseJ and SifA regulate dynamics of the *Salmonella* Typhimurium vacuolar membrane. *Mol. Microbiol.* **44**:645–661.

Rytkönen, A., J. Poh, J. Garmendia, C. Boyle, A. Thompson, M. Liu, P. Freemont, J. C. Hinton, and D. W. Holden. 2007. SseL, a *Salmonella* deubiquitinase required for macrophage killing and virulence. *Proc. Natl. Acad. Sci. USA* **104**:3502–3507.

Salcedo, S. P., M. Noursadeghi, J. Cohen, and D. W. Holden. 2001. Intracellular replication of *Salmonella* Typhimurium strains in specific subsets of splenic macrophages in vivo. *Cell Microbiol.* **3**:587–597.

Shea, J. E., C. R. Beuzon, C. Gleeson, R. Mundy, and D. W. Holden. 1999. Influence of the *Salmonella* Typhimurium pathogenicity island 2 type III secretion system on bacterial growth in the mouse. *Infect. Immun.* **67**:213–219.

Sheppard, M., C. Webb, F. Heath, V. Mallows, R. Emilianus, D. Maskell, and P. Mastroeni. 2003. Dynamics of bacterial growth and distribution within the liver during *Salmonella* infection. *Cell Microbiol.* **5**:593–600.

Sundquist, M., and M. J. Wick. 2005. TNF-alpha-dependent and -independent maturation of dendritic cells and recruited CD11c(int)CD11b+ Cells during oral *Salmonella* infection. *J. Immunol.* **175**:3287–3298.

Tobar, J. A., L. J. Carreno, S. M. Bueno, P. A. Gonzalez, J. E. Mora, S. A. Quezada, and A. M. Kalergis. 2006. Virulent *Salmonella enterica* serovar Typhimurium evades adaptive immunity by preventing dendritic cells from activating T cells. *Infect. Immun.* **74**:6438–6448.

Tobar, J. A., P. A. Gonzalez, and A. M. Kalergis. 2004. *Salmonella* escape from antigen presentation can be overcome by targeting bacteria to Fc gamma receptors on dendritic cells. *J. Immunol.* **173**:4058–4065.

Trent, M. S., W. Pabich, C. R. Raetz, and S. I. Miller. 2001. A PhoP/PhoQ-induced lipase (PagL) that catalyzes 3-O-deacylation of lipid A precursors in membranes of *Salmonella* Typhimurium. *J. Biol. Chem.* **276**:9083–9092.

Uchiya, K., M. A. Barbieri, K. Funato, A. H. Shah, P. D. Stahl, and E. A. Groisman. 1999. A *Salmonella* virulence protein that inhibits cellular trafficking. *EMBO J.* **18**:3924–3933.

Van der Velden, A. W., M. K. Copass, and M. N. Starnbach. 2005. *Salmonella* inhibit T cell proliferation by a direct, contact-dependent immunosuppressive effect. *Proc. Natl. Acad. Sci. USA* **102**:17769–17774.

Vazquez-Torres, A., J. Jones-Carson, A. J. Bäumler, S. Falkow, R. Valdivia, W. Brown, M. Le, R. Berggren, W. T. Parks, and F. C. Fang. 1999. Extraintestinal dissemination of *Salmonella* by CD18-expressing phagocytes. *Nature* **401**:804–808.

Vazquez-Torres, A., Y. Xu, J. Jones-Carson, D. W. Holden, S. M. Lucia, M.C. Dinauer, P. Mastroeni, and F. C. Fang. 2000. *Salmonella* pathogenicity island 2-dependent evasion of the phagocyte NADPH oxidase. *Science* **287**:1655–1658.

Waterman, S. R., and D. W. Holden. 2003. Functions and effectors of the *Salmonella* pathogenicity island 2 type III secretion system. *Cell. Microbiol.* **5**:501–511.

Worley, M. J., G. S. Nieman, K. Geddes, and F. Heffron. 2006. *Salmonella* Typhimurium disseminates within its host by manipulating the motility of infected cells. *Proc. Natl. Acad. Sci. USA* **103**:17915–17920.

Yrlid, U., M. Svensson, A. Kirby, and M. J. Wick. 2001. Antigen-presenting cells and anti-*Salmonella* immunity. *Microb. Infect.* **3**:1239–1248.

Yrlid, U., and M. J. Wick. 2002. Antigen presentation capacity and cytokine production by murine splenic dendritic cell subsets upon *Salmonella* encounter. *J. Immunol.* **169**:108–116.

Yu, X. J., M. Liu, and D. W. Holden. 2004. SsaM and SpiC interact and regulate secretion of *Salmonella* pathogenicity island 2 type III secretion system effectors and translocators. *Mol. Microbiol.* **54**:604–819.

Zhao, C., M. W. Wood, E. E. Galyov, U. E. Hopken, M. Lipp, H. C. Bodmer, D. F. Tough, and R. W. Carter. 2006. *Salmonella* Typhimurium infection triggers dendritic cells and macrophages to adopt distinct migration patterns in vivo. *Eur. J. Immunol.* **36**:2939–2950.

24

Legionella pneumophila, a Pathogen of Amoebae and Macrophages

MICHELE S. SWANSON AND ANDREW BRYAN

AN ENVIRONMENTAL PATHOGEN

Legionella pneumophila is a gram-negative bacterium that has evolved strategies to parasitize professional phagocytes. Most *Legionella* species, including *L. pneumophila*, are aquatic microbes; other species thrive in soil. There, the microbe naturally infects amoebae and protozoa, but, when given the opportunity, *L. pneumophila* can infect human alveolar macrophages and cause the severe pneumonia Legionnaires' disease (Fig. 1).

Aerosols of contaminated water or soil transmit the disease agent to humans. Because the bacterium is ubiquitous in nature, water sources of all kinds can act as points for transmission. The point source is most often identified in epidemic outbreaks, where exposures can be evaluated by using case controls. Epidemiologists usually implicate manmade equipment, including water-cooling towers, whirlpools, and showers. The origin of sporadic disease is more difficult to assess, as is the contribution by natural water sources to disease burden.

A significant portion of Legionnaires' disease occurs in the health care setting, and these cases account for the highest fatality rates. Contributing factors include the vulnerability of the patient population and the institutional setting, where virulent isolates can be efficiently transmitted to the large number of susceptible individuals housed there.

Significantly, during the 30 years the Centers for Disease Control and Prevention has tracked the disease, there have been no reports of human-to-human transmission. Evidently, factors that promote spread in freshwater microbial communities are not sufficient for *L. pneumophila* to pass safely and efficiently from the lungs of one susceptible human to another. It is also possible that amoebae are required for the biogenesis of infectious particles that can withstand transmission in aerosols.

BURDEN OF DISEASE

Considered along with *Chlamydia pneumoniae* and *Mycoplasma pneumoniae* to be an "atypical" etiological agent of pneumonia, *L. pneumophila* actually causes considerable morbidity and mortality (Edelstein and Cianciotto, 2005). Although estimates vary widely, between 18,000 and 88,000 cases are thought to occur annually in the United States. Of these, the Centers for Disease Control and Prevention calculates that between 8,000 and 18,000 require hospitalization. Mainly afflicting immunocompromised individuals, the disease carries a case-fatality rate of 10 to 46%, with the outcome primarily dependent on the immunological state of the person and the choice of therapy. The high mortality rate is in contrast to pneumococcal pneumonia, which, although more common, is less often fatal. While interpreting the trends in overall incidence of Legionnaires' disease is complicated by increased diagnosis, it is clear that fatality rates are decreasing—a pattern likely attributable to more convenient diagnostic tools and the rising use of empiric antimicrobial therapy with activity against *L. pneumophila*. *L. pneumophila* is also associated with Pontiac fever, a milder, febrile, nonpneumonic illness whose etiology is not clear. Since Pontiac fever does not appear to require viable bacteria, it is beyond the scope of this chapter.

OUTBREAKS

Although epidemics of Legionnaires' disease comprise the minority of cases, they are a rich source of information about the illness (Edelstein and Cianciotto, 2005). The namesake 1976 outbreak, which occurred at a Philadelphia convention of American Legion members, highlights both the epidemic potential of the institutional setting and the predilection of disease in susceptible individuals. The convention was attended by approximately 4,000 members of the Pennsylvania State American Legion, a group predisposed primarily by age, but also by preexisting chronic lung disease and tobacco use. After exposure to contaminated aerosols from the convention hotel's air conditioning sys-

Michele S. Swanson and Andrew Bryan, Department of Microbiology and Immunology, University of Michigan Medical School, Ann Arbor, MI 48109-0620.

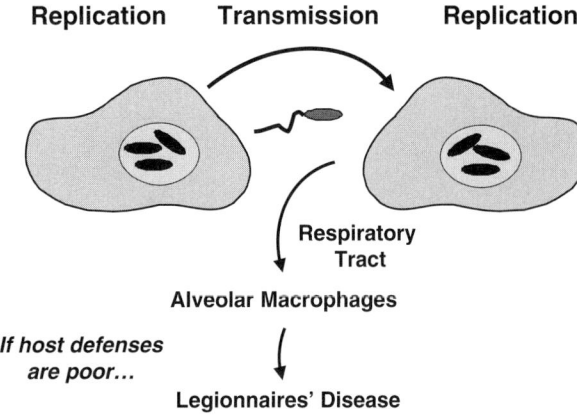

FIGURE 1 L. pneumophila is an environmental bacterium that causes opportunistic infections of humans. L. pneumophila is transmitted to humans by the respiratory route from water or soil that harbors its natural hosts, amoebae and protozoa. In the absence of a robust cell-mediated immune response, the bacteria can replicate in alveolar macrophages and cause the severe pneumonia, Legionnaires' disease.

tem, 149 Legionnaires and 33 non-Legionnaires were eventually diagnosed with the disease, and 29 died. Epidemic Legionnaires' disease continues to be a problem. For example, in October 2005, 127 residents of a home for the elderly in Toronto, Ontario, became ill with Legionnaires' disease, and 20 of them died. The source of the outbreak was eventually determined to be a cooling tower on the roof that spread contaminated aerosols into the building's air-handling system. Typically, the attack rate in epidemics is ~5%, and those who become ill have preexisting damage to their airways or weakened cell-mediated immune systems. Healthy individuals can clear the infection. Once activated by interferon-γ, human macrophages can restrict L. pneumophila replication by limiting the microbe's iron supply.

CLINICAL FEATURES, DIAGNOSIS, AND TREATMENT

The clinical presentation of Legionnaires' disease can range from the average febrile pneumococcal-appearing pneumonia to symptoms less commonly associated with pneumonia, such as abdominal pain, diarrhea, memory loss, severe headache, and low serum sodium (Edelstein and Cianciotto, 2005). Cough may be unimpressive, and can range from nonproductive to bloody. Chest radiography yields pathological changes, frequently confluent lobular and bilateral, but nothing diagnostic for Legionnaires' disease. Because of the inconsistent nature of symptoms, diagnostic formulas with acceptable positive predictive values have been elusive. Nevertheless, experienced clinicians with an appropriate index of suspicion are good judges of disease and, by applying an appropriate diagnostic test, can identify the illness with reasonable frequency.

Lung pathology is usually consistent with acute bronchiolitis and alveolitis. Histology of human, rodent, and primate animal models generally reveals fibrinopurulent infiltrates, with macrophages usually comprising a significant portion of the total leukocytes. Neutrophils make up most of the remainder of the infiltrate. Erythrocytes are also commonly present, indicative of vascular damage that can accompany the disease. Lysed white blood cells are also evident. Interestingly, alveolar macrophages appear to phagocytose L. pneumophila to a greater extent than neutrophils. Phagosomes that contain multiple intact bacteria are often observed, suggestive of intracellular replication. The in situ observations of macrophage infection, intracellular replication, and phagocyte lysis are mirrored by current in vitro experimental models, including interactions with primary macrophages derived from the bone marrow of A/J mice.

Legionella retains Gram's stain poorly, which can lead to false-negative sputum smears and pathology sections. The difficulty of detecting L. pneumophila by Gram stain is not only a current diagnostic consideration but is also of historical interest, because it contributed to the agonizing delay before experts identified the bacterium as the etiological agent of Legionnaires' disease in 1976. Culture is still the gold standard of diagnosis, but the advent of the urine antigen test has made Legionnaires' disease possibly the easiest respiratory illness to diagnose with speed and certainty (Edelstein and Cianciotto, 2005). The test's specificity approaches 100%, and it avoids the wait needed for cultures. However, because the urine antigen test only detects L. pneumophila serogroup 1, an unintended consequence is that other species and serogroups are underrecognized. The focus on serogroup 1 also complicates interpretations of epidemiological data that were not collected with an unbiased diagnostic tool. Alternatives now available include direct fluorescent antibody staining, serology, and PCR assays.

Rapid therapeutic intervention is critical for resolution of disease with a low mortality rate, and the appropriate chemotherapeutic agent is necessary to achieve clearance (Edelstein and Cianciotto, 2005). Because of its insensitivity to β-lactam antibiotics, Legionella has contributed to a shift in empiric antimicrobial therapy for pneumonia. Rather than administer narrow-range antipneumococcal agents, clinicians now rely first on drugs that cover a broad range of microorganisms. Macrolides and quinolones are the most common classes of antimicrobials with good activity against Legionella. Acquired resistance to antimicrobials has not been an issue, likely due to lack of human-to-human transmission and the acute nature of the illness. Interestingly, the susceptibility pattern of Legionella highlights the intracellular nature of the pathogen. The effective and commonly used drug gentamycin is efficacious in vitro and useful in the research laboratory, but its inability to penetrate host cells limits its clinical value.

EPIDEMIOLOGY

The genus of Legionella is quite diverse, comprising 50 distinct species. Nevertheless, the species L. pneumophila contributes to more than 90% of the human disease burden in the United States and Europe, with smaller percentages contributed by L. micdadei, L. longbeachae, and several others. Within L. pneumophila, there are at least 15 defined serogroups. Even though serogroup 1 strains account for less than 40% of environmental isolates, they cause the majority of Legionnaires' disease. Since predisposed patients are more readily infected with less pathogenic isolates, the serogroup distribution in hospital-acquired cases mirrors their prevalence in the environment: ~53% are due to serogroup 1. In contrast, healthy individuals in the community typically only become ill when exposed to the more virulent serogroup 1 isolates, which occurs in ~76% of these cases.

All Legionella species are adept at infecting environmental phagocytes, amoebae, and protozoa. But what attributes equip serogroup 1 isolates to more readily infect the alveolar macrophage? The different features of each serogroup have begun to be addressed by molecular epidemiology and comparative genomics. Only three genes appear to distinguish the serogroup 1 isolates, and all of the corresponding proteins are predicted to affect lipopolysaccharide (LPS). A role for *L. pneumophila* LPS in disease would not be entirely surprising, considering its contribution to the pathogenesis of many other bacterial agents, including intracellular pathogens such as *Brucella abortus* and *Coxiella burnetii*. It is unclear, however, whether the differences in LPS contribute to increased bacterial fitness in the environment, survival during aerosolization, or entry, survival, or replication in macrophages of the lung.

A DYNAMIC GENOME RIDDLED WITH EUKARYOTIC DNA

L. pneumophila is equipped to internalize foreign DNA and to recombine the genetic material onto its chromosome. Although the ~3.5-Mb chromosome is predicted to encode ~3,000 proteins, only ~2,400 of these are common to all four of the *L. pneumophila* strains that have been sequenced. Moreover, there is an extraordinarily high frequency of eukaryotic motifs embedded in the *L. pneumophila* genome. Nearly 3.5% of the predicted proteins have eukaryotic motifs (Bruggemann et al., 2006a). These include ankyrin domains, which mediate protein-protein interactions; F-box and U-box domains, a feature of components of the ubiquitination pathway; a homologue of sphingosine-1-phosphate lyase, an enzyme that modulates whether cells induce autophagy or cell death; two apyrases, secreted proteins that degrade extracellular nucleotide di- or triphosphates; and three serine-threonine kinases, signal transduction proteins. Presumably, its environmental hosts supply ample eukaryotic DNA which *L. pneumophila* efficiently incorporates, ultimately retaining those DNA sequences that increase fitness. It is tempting to speculate that eukaryotic motifs that are broadly conserved in the species equip *L. pneumophila* to modulate the biology of its host phagocyte. However, a priori, it is not possible to predict whether functions gained via horizontal transmission of DNA act at the host-microbe interface or instead within the bacterial cell.

THE *L. PNEUMOPHILA* LIFE CYCLE

To endure fluctuating environments, microbes must alter their physiology. Bacterial adaptation may be accompanied by a drastic remodeling of the cell, as occurs during sporulation or germination. Cellular differentiation is also crucial for pathogens, which must replicate in one host and then spread to the next. Because *L. pneumophila* is not transmitted from person to person, selective pressures in aquatic ecosystems must shape its life cycle. Unfortunately, many of the traits that contribute to its fitness in the environment also equip *L. pneumophila* to replicate in alveolar macrophages.

L. pneumophila can differentiate into multiple cell types (Molofsky and Swanson, 2004). In his pioneering studies, Rowbotham infected amoebae and then watched as *L. pneumophila* switched between two forms readily distinguished by their motility, shape, surface, and stores of energy-rich polymers. If subjected to prolonged starvation, the bacteria differentiate into a mature intracellular form, a thick-walled, pleomorphic cell type with a low basal respiration rate that is not only infectious but also extraordinarily resistant to biocides and antibiotics. In nature, the microbes likely persist for extended periods within biofilms, hardy adherent microbial communities that divide labor among specialized cell types. When phagocytic amoebae or protozoa graze on biofilms, the ingested *L. pneumophila* can establish a replication niche within a vacuole that does not merge with lysosomes. As conditions deteriorate, the progeny stop multiplying and instead express multiple factors likely to promote transmission to a new niche; these traits include cytotoxicity, motility, and the ability to inhibit phagosome-lysosome fusion (Fig. 2).

Many aspects of the pathogen's life cycle can also be observed in broth culture, a discovery that has expedited identification of the circuitry that governs *L. pneumophila* differentiation. Characterization of each cell type has also been aided by the development of several genetic tools, including transformation via natural competence, conjugation, or electroporation; transposon mutagenesis; recombinant green fluorescent protein (*gfp*) genes; genome sequences of four *L. pneumophila* strains; and genomic microarrays.

Transmission traits are coordinately induced as replicating cells exit the logarithmic growth phase (Molofsky and Swanson, 2004). Among the traits specifically expressed in the stationary phase are most of the activities identified classically as "virulence" factors. For example, entry into host cells is promoted by FlaA, DotO, and DotH, proteins that are expressed throughout the escape and invasion periods, but not during the replication phase in macrophages. Likewise, the pathogen absolutely requires type IV secretion to establish its protective vacuolar niche; however, during bacterial replication, type IV secretion is dispensable. That replication and transmission are reciprocal and genetically separable phases can be illustrated by manipulating expression of CsrA, a master repressor of transmission traits. Bacteria that lack the CsrA repressor efficiently infect and survive within macrophages, but they fail to replicate. Conversely, when engineered to constitutively express this repressor of transmission, *L. pneumophila* infect macrophages poorly, as most are rapidly digested in lysosomes; yet, those bacteria that survive phagocytosis replicate as well as the wild type.

To convert between replication and transmission, *L. pneumophila* alters the expression of a substantial share of its genome (Bruggemann et al., 2006b). As it multiplies in either broth or amoebae, the microbe's transcriptional profile includes many genes dedicated to nutrient acquisition. These include proteins predicted to transport or degrade peptides or amino acids. After the replication period in either broth or amoebae, *L. pneumophila* induces expression of a distinct set of ~400 genes, many of which encode factors that target pathways of host cells. Examples include RalF, LidA, SdeA, SdcA, and SidC, proteins that are delivered by type IV secretion to the host cytoplasm, where they impede particular cell pathways. Also strongly induced in the transmission period is the flagellar regulon, which equips the progeny to disperse while also making *L. pneumophila* vulnerable to detection by the innate immune system of mammalian hosts.

METABOLIC CUES GOVERN THE LIFE CYCLE

From genetic and molecular studies of the microbe's life cycle in broth and macrophages, a paradigm has emerged. By gauging its metabolic capacity, *L. pneumophila* commits

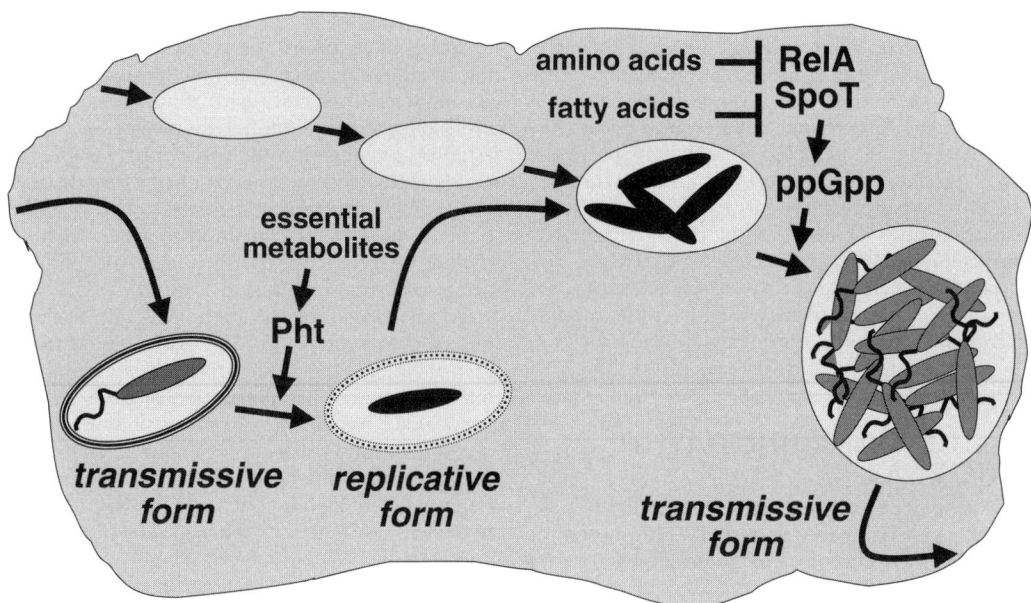

FIGURE 2 Metabolic cues govern the *L. pneumophila* life cycle. In macrophages and broth, the bacterium responds to metabolic cues by alternating between a replicative form and a cell type equipped for transmission to a new host. After phagocytosis, the bacteria rely on Pht proteins to acquire essential metabolites. When amino acids become scarce or fatty acid biosynthesis is strained, the enzymes RelA or SpoT, respectively, generate (p)ppGpp, a second messenger that coordinates the bacterium's exit from logarithmic phase with the expression of multiple transmission traits, including cytotoxicity, motility, osmotic resistance, motility, and the capacity to inhibit phagosome-lysosome fusion.

to either replication or transmission (Fig. 2; Molofsky and Swanson, 2004). To obtain nutrients from its host, *L. pneumophila* can employ a dozen proteins of the major facilitator superfamily. In particular, the phagosomal transporter protein A (PhtA) enables intracellular bacteria to acquire threonine from macrophages and amoebae, whereas PhtJ behaves as a valine transporter, and PhtC and PhtD are critical for thymidine assimilation. Bacteria that lack PhtA infect macrophages efficiently and can persist for days, but the mutants fail to reenter the replicative phase until PhtA expression is restored or excess threonine is provided. Thus, differentiation of transmissive bacteria to the replicative state occurs if and when nutrients are ample.

Metabolic cues also trigger replicating bacteria to differentiate to the transmissive state (Fig. 2). In particular, when amino acids are scarce or fatty acid biosynthesis is strained, *L. pneumophila* accumulate (p)ppGpp, a second messenger of the stringent response pathway. By altering the set of promoters activated by RNA polymerase, this alarmone then coordinates exit from logarithmic-phase growth with expression of traits that promote survival and transmission to a new host, including cytotoxicity, motility, osmotic resistance, and lysosome evasion. Depending on the bacterium's metabolic status, the bifunctional enzyme SpoT either generates or degrades (p)ppGpp. *L. pneumophila* that lack SpoT replicate efficiently in macrophages, but the mutants are killed during the transmission period. Thus, the stringent response pathway coordinates intracellular progeny escape from macrophages when conditions deteriorate.

TYPE II SECRETION SYSTEM PROMOTES INFECTION OF AMOEBAE, MACROPHAGES, AND MOUSE LUNGS

One mechanism that equips *L. pneumophila* to exploit host cells as a replication niche is type II secretion (DebRoy et al., 2006). Derived from a type 4 pilus apparatus, this machinery translocates proteins from the bacterial periplasm across a pore in the outer membrane. Before it can do so, cargo is first translocated from the bacterial cytoplasm into the periplasm by either the Sec or the Tat machinery, two broadly conserved systems that operate at the cytoplasmic membrane. At least 25 *L. pneumophila* proteins are substrates of type II secretion, as judged by comparing the culture supernatants of wild-type and secretion mutant bacteria. Many of these proteins are digestive enzymes, namely proteases, lipases, chitinases, phosphatases, and an RNase. Others have eukaryotic features, e.g., a collagen-like protein, an acid phosphatase, and a glucoamylase. In some manner, the secreted proteins promote growth of *L. pneumophila* in amoebae, macrophages derived from human peripheral blood, and the lungs of A/J mice, as judged by comparing the yield of type II secretion-competent *L. pneumophila* with corresponding mutants.

SUBVERTING MACROPHAGE MEMBRANE TRAFFIC PATHWAYS

When ingested by amoebae or macrophages, *L. pneumophila* alters the course of its phagosome to establish a replication niche (Fig. 3) (Horwitz, 1983). Within minutes, the phago-

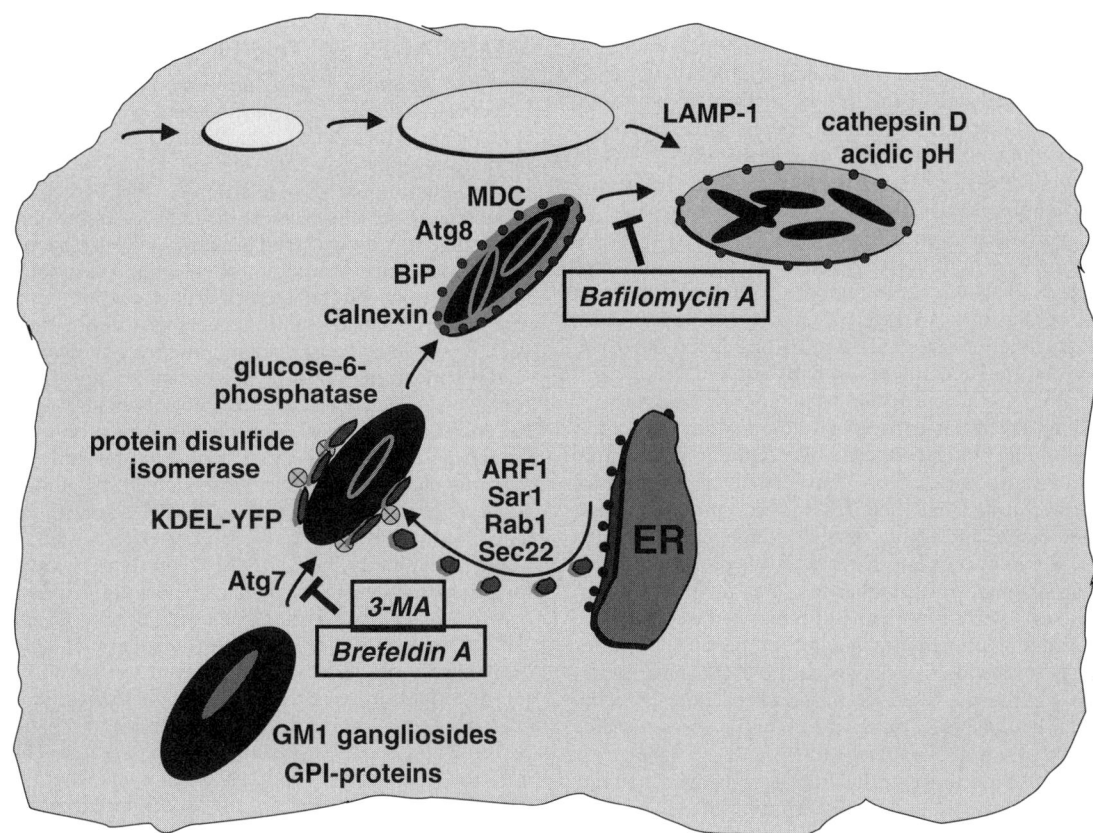

FIGURE 3 Trafficking of *L. pneumophila* in macrophages derived from bone marrow of A/J *naip5* mice. After phagocytosis, *L. pneumophila* evades immediate delivery to lysosomes. Instead, its vacuole acquires membranes from the secretory pathway and also several features typical of autophagosomes. Once the bacterium differentiates to the replicative form, its vacuole slowly merges with the lysosomal compartment, where replication continues. How quickly and extensively the *L. pneumophila* replication vacuole interacts with the endosomal compartment depends on the host cell. For example, macrophages from resistant C57Bl/6 mice rapidly deliver *L. pneumophila* to lysosomes, whereas permissive human macrophages derived from peripheral blood monocytes harbor replicating *L. pneumophila* within a compartment derived from ER. See text for details.

somal membrane thins, suggesting rapid changes to its lipid composition. At the same time, small vesicles from the early secretory pathway attach via hair-like projections to the phagosome's cytoplasmic face (Tilney et al., 2001). Within hours, these vesicles flatten, then fuse, encasing the phagosome within a double membrane rich in markers of the endoplasmic reticulum (ER), including BiP, calnexin, glucose-6-phosphatase, protein disulfide isomerase, or the recombinant protein KDEL-YFP (O'Conner et al., 2008). After several hours, the barrier between the bacteria and the ER lumen dissipates, as judged by the appearance of glucose-6-phosphatase within the *L. pneumophila* vacuole. Thus, rather than following the canonical phagosome-lysosome pathway, *L. pneumophila* evades immediate delivery to lysosomes and instead begins to replicate within a double-membraned vacuole that resembles rough ER. Since Horwitz first described the extraordinary morphology of the *L. pneumophila* replication vacuole, a major focus of the field has been to determine what interactions between host and bacterial proteins reroute the *L. pneumophila* vacuole away from the microbicidal lysosomes to associate stably with membranes from the secretory pathway.

L. PNEUMOPHILA LPS CORRELATES WITH VIRULENCE

A variety of studies indicate that glycoconjugates affixed to or shed from the surfaces of microbes perturb membrane traffic. For example, release of lipophosphoglycan by *Leishmania donovani* or lipoarabinomannan by *Mycobacterium tuberculosis* promotes survival of each microbe in phagosomes that do not merge with lysosomes. The surface composition of synthetic beads also dictates how efficiently these inert particles are delivered to lysosomes. Likewise, a number of observations link virulence of *L. pneumophila* to its surface glycolipid, LPS.

The major component of the outer membrane of gram-negative bacteria, LPS comprises three domains: lipid A, core polysaccharide, and O antigen. The chemical properties of LPS are determined by several factors, since the number and types of sugars in the core polysaccharide and O antigen, their linkages, and the number and type of nonstoichiometric substitutions to the O polysaccharide can vary. The O chain of *L. pneumophila* LPS is unusually hydrophobic, a trait also characteristic of synthetic particles that macrophages do not deliver to lysosomes efficiently. The L. pneumophila O antigen is a homopolymer of legionaminic acid, an unusual sugar that resembles sialic acid. Whether the similarity of the LPS O antigen to sialic acid affects the interaction between L. pneumophila and macrophages has not been addressed experimentally.

Because LPS is the dominant component of the bacterial outer membrane, a single LPS biosynthetic enzyme can convert a pathogen's serotype. For example, bacteriophages that encode O-acetyltransferases and glucosyltransferases are sufficient to switch the serotypes of Salmonella enterica serovar Typhimurium, Shigella flexneri, and Pseudomonas aeruginosa. Genetic phase variation of one L. pneumophila strain correlates LPS composition with virulence in an animal model. In this particular isolate, a 30-kb locus of phage origin alternates between an episomal and chromosomal state. The status of the mobile element determines not only LPS structure and the capacity of the bacteria to replicate in cultured macrophages and in guinea pigs, but also other traits, including surface charge and motility. However, the molecular mechanism of this phase variation remains an enigma, since neither the large mobile element nor its site of insertion contains LPS biosynthetic genes.

Several developmentally regulated properties of the L. pneumophila surface do correlate with the pathogen's ability to inhibit phagosome-lysosome fusion (Fernandez-Moreira et al., 2006). Compared with replicating L. pneumophila, transmissive phase cells express LPS species of higher molecular weight, bind more strongly to the sialic acid-specific lectin Limulus polyphemus hemagglutinin, and adhere less avidly to the hydrocarbon hexadecane. Biochemical studies indicate that the predominant modification to L. pneumophila LPS is acetylation of its O antigen. The L. pneumophila genome encodes several acetylases and deacetylases, along with other enzymes predicted to assemble or modify LPS. Whether the activity of particular LPS biosynthetic or modifying enzymes contributes to the developmental changes to the L. pneumophila surface has not been analyzed.

Like all gram-negative bacteria, L. pneumophila sheds LPS-rich vesicles from its outer membrane, and the composition of these small vesicles is developmentally regulated. Moreover, membrane vesicles shed by L. pneumophila are sufficient to inhibit phagosome-lysosome fusion. When coated with membrane vesicles purified from transmissive phase wild-type L. pneumophila, only ~25% of beads traffic to lysosomes, whereas ~75% of control bead preparations do so (Fernandez-Moreira et al., 2006). Together these observations suggest that, as it enters the transmissive phase, L. pneumophila may modify its LPS and shed outer membrane vesicles that, by some mechanism, inhibit fusion of phagosomes with the degradative lysosomal compartment.

Given the extraordinary plasticity of the L. pneumophila genome, the preponderance of serogroup 1 LPS among Legionnaires' disease isolates is especially striking. However, to date, the few direct tests of the role of LPS in L. pneumophila pathogenesis have yielded negative results. For example, virulence, as judged by serum resistance and by growth in cultured monocytes or amoebae, is not affected by null mutations in five different LPS biosynthetic genes. Therefore, additional genetic and biochemical studies are needed to understand the mechanism(s) that account for the prevalence of L. pneumophila serogroup 1 strains among clinical isolates.

ESTABLISHING RESIDENCE WITHIN MEMBRANES OF THE ER

Multiple redundant host pathways likely contribute to biogenesis of the L. pneumophila replication vacuole (O'Conner et al., 2008). For example, in an experimental model that exploits a Dictyostelium discoideum cell line as a host, depletion by double-stranded RNA technology of particular components of the secretory traffic machinery has little effect on bacterial growth. However, bacterial replication is clearly inhibited when components of two different portions of the secretory pathway are depleted simultaneously. Therefore, it appears that the L. pneumophila vacuole can receive host membrane from multiple sources.

The prevailing concept holds that, to replicate in macrophages, L. pneumophila intercepts vesicular traffic from the secretory pathway. A number of host proteins that are known to regulate vesicular traffic in the early secretory pathway either decorate or contribute to biogenesis of the L. pneumophila replication vacuole (O'Conner et al., 2008). These include Rab proteins, which tether vesicles to their target membranes, and SNARE proteins, which trigger membrane fusion. In particular, the small GTPase proteins Rab1, Rab5c, ARF1, and Sar1 and the SNARE protein Sec22b are host proteins known to regulate vesicular traffic between the ER and the Golgi apparatus that also contribute to formation of the L. pneumophila replication vacuole (Fig. 3).

Rab1 and Sec22b associate with the L. pneumophila vacuole for a few hours. Rab1 does so even when traffic of COP1-coated secretory vesicles is inhibited by treatment with brefeldin A. Rapid Sec22b association also requires secretory traffic, as it occurs more slowly in cells expressing dominant negative Rab1 or Sar1, and it is inhibited by treatment with brefeldin A or dominant negative ARF1. However, the cognate SNARE proteins, namely Membrin, Syntaxin 5, and Bet1, cannot be detected on the vacuole, and their depletion by double-stranded RNA has no effect on L. pneumophila replication in a Drosophila cell model system. Perhaps the bacterial vacuole associates with monomeric Sec22b, a form that is not sufficient to mediate the fusion reactions that drive progress along the secretory pathway. By some mechanism, host membrane traffic in the early secretory pathway promotes growth of intracellular L. pneumophila, since cells that express dominant negative Rab1, Sar1, or ARF1 mutant proteins yield fewer progeny.

TYPE IV SECRETION SYSTEM, A CONDUIT TO HOST MACHINERY

To establish a replication vacuole within a professional phagocyte, L. pneumophila relies on a specialized secretion system that can deliver bacterial proteins directly to the host cytosol (Vogel et al., 1998). Derived from machinery that transfers plasmids during conjugation with another bacterium, type IV secretion systems can transport DNA or protein across both the bacterial and the phagosomal membranes (Fig. 4) (Vincent et al., 2006). Such adapted conjugation systems are utilized by a variety of microbes, including the plant pathogen Agrobacterium tumefaciens and the mammalian pathogens Helicobacter pylori, Bordetella pertussis, Bartonella henselae, and C. burnetii.

Encoded on two genomic islands of at least 26 genes, known as dot (defective organelle trafficking) or icm (intracellular multiplication), the L. pneumophila Dot/Icm type IV secretion system resembles the conjugation system

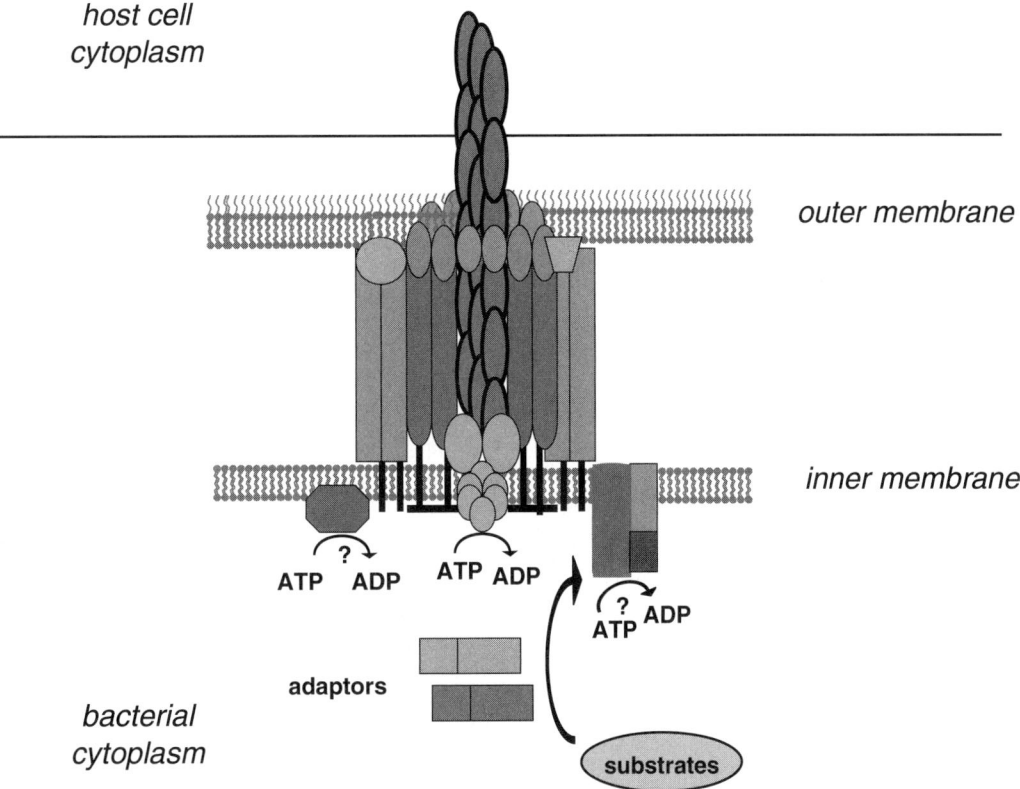

FIGURE 4 The type IV secretion system provides a conduit to the host cytoplasm. Composed of dozens of proteins encoded by the *dot* and *icm* genes, the secretion apparatus spans the *L. pneumophila* inner and outer membranes. A pilus may contact the host plasma membrane. Three Dot proteins have been shown or are predicted to hydrolyze ATP; these may drive the assembly of the complex or regulate its activity. Substrates of the apparatus are thought to rely on particular adaptor proteins for their delivery. For details, see the text and Sexton and Vogel (2002).

of IncI plasmids (Vincent et al., 2006). Mutants that lack a *dot* or *icm* gene fail to associate with the ER or to replicate in host cells. Although type IV secretion substrates have not been detected in *L. pneumophila* broth cultures, a number of clever genetic screens have identified bacterial proteins that gain access to the host cytoplasm by a mechanism that requires the *dot/icm* genes (Luo and Isberg, 2004; Shohdy et al., 2005). From this approach, the current picture is of an extraordinarily promiscuous apparatus that can transfer more than 200 proteins across the bacterial and phagosomal membranes. How the cargo delivered by type IV secretion manipulates the biology of the host phagocyte remains a primary focus of research in the *Legionella* field.

Genetic and cell biological studies have established that the type IV secretion system must act during macrophage phagocytosis for *L. pneumophila* to establish a protective niche within membranes derived from the ER (O'Conner et al., 2008). Although the concept that *L. pneumophila* secretes virulence proteins to inhibit the host endocytic fusion machinery holds great appeal, a toxin with such biochemical activity has not yet been identified for this or any other intracellular pathogen. Furthermore, when ingested by mouse macrophages, *L. pneumophila* that lack a functional type IV secretion system remain viable for many days. The mutants persist in vacuoles that are rich in the late endosomal and lysosomal protein LAMP-1, but lack several lysosomal features, such as the protease cathepsin D and accessibility to endosomal tracers. Even when formalin-killed, wild-type *L. pneumophila* remain competent to inhibit phagosome-lysosome fusion. Therefore, during its coevolution with predatory amoebae, *L. pneumophila* apparently acquired multiple mechanisms to avoid lysosomal killing, some but not all of which rely on type IV secretion.

BACTERIAL EFFECTORS DELIVERED BY TYPE IV SECRETION

Particular components of *L. pneumophila*'s type IV secretion system payload are thought to alter the fate of its vacuole, based on the predicted or measured biochemical activities of certain substrates (Table 1) (O'Conner et al., 2008). To date, the best-characterized secretion substrate is RalF, a bacterial protein identified on the basis of its homology to the eukaryotic guanine nucleotide exchange factor (GEF) Sec7. As expected for a GEF, RalF stimulates binding of GTP-γ-S to ARF1 and ARF3 in vitro and also inhibits secretory traffic when expressed ectopically by CHO cells, as judged by quantifying secreted alkaline phosphatase. However, neither of these RalF activities is abrogated by a mutation to its conserved catalytic site, indicating that this bacterial GEF acts by a noncanonical mechanism. In contrast, unlike wild type, *L. pneumophila* that encode the

TABLE 1 Effectors delivered by type IV secretion that target host pathways

Protein	Features	Host targets
RalF	Guanine nucleotide exchange factor	Arf1, Arf3
SidM/DrrA	Guanine nucleotide exchange factor	Rab1
LidA	Acts with SidM/DrrA	Rab1, Rab6, Rab8
SdhA	Prolongs macrophage viability	Apoptosis
SidH	Prolongs macrophage viability	BNIP3, Bcl-rambo

catalytic site mutant RalF protein reside in phagosomes that are not decorated by ARF1. Accordingly, RalF mediates ARF1 persistence on *L. pneumophila* phagosomes by a mechanism that is distinct from its ability to inhibit host secretory traffic. Genetic or pharmacological treatments that block tethering of secretory vesicles to the nascent phagosome and subsequent bacterial replication do not diminish the decoration of *L. pneumophila* phagosomes by ARF1. Furthermore, bacterial mutants that fail to recruit ARF1 replicate in macrophages as well as wild-type *L. pneumophila* do. Therefore, the functional significance of ARF1 association with the *L. pneumophila* vacuole remains to be established.

Another GEF that *L. pneumophila* translocates across the phagosomal membrane is SidM (also known as DrrA; Table 1) (O'Conner et al., 2008). This type IV secretion substrate can bind to the inactive GDP-bound form of Rab1 and stimulate nucleotide exchange in vitro. In macrophages, SidM mediates the stable association of Rab1 with *L. pneumophila* phagosomes by cooperating with LidA, another secreted bacterial protein that binds to Rab1, Rab6, and Rab8. When either SidM or LidA are ectopically expressed in COS1 cells, the integrity of the Golgi structure is lost, consistent with the idea that these bacterial proteins target host proteins critical for secretory traffic. Perhaps by stabilizing activated Rab1 on the *L. pneumophila* phagosome, LidA and SidM prolong its interaction with early secretory vesicles, thereby precluding interactions with the bactericidal lysosomal pathway.

Determining whether RalF, SidM, or other substrates of the type IV secretion system contribute to *L. pneumophila* fitness has been complicated by the negative results obtained when the respective mutants are challenged to grow in macrophages. With the exception of SdhA, *L. pneumophila* mutants engineered to lack a putative effector protein replicate as efficiently as the wild-type bacteria. For example, in the absence of RalF or SidM, *L. pneumophila* nevertheless replicates efficiently in amoebae and macrophages. Accordingly, the activity of each effector is postulated to be redundant with another factor in the virulence payload. Efforts to ascribe function to particular substrates is further complicated by the extraordinary promiscuity of the type IV secretion system, as it is thought to translocate ~200 different proteins. Another unsettled question is the source of selective pressure needed to maintain functions that are each dispensable for *L. pneumophila* to grow in any of the experimental systems applied. It is necessary to invoke encounters with a large variety of host phagocytes to account for the acquisition and maintenance by *L. pneumophila* of multiple redundant factors that each override the formidable quality control mechanisms of the secretory or endosomal pathways of professional phagocytes.

AUTOPHAGY, A REGULATED INTERSECTION OF SECRETORY AND ENDOSOMAL TRAFFIC

An alternative model holds that sequestration of the *L. pneumophila* phagosome within ER membranes is a host innate immune response to infection (Swanson, 2006). All eukaryotic cells can efficiently capture organelles within secretory membranes for delivery to digestive lysosomes by the broadly conserved pathway known as autophagy. A wide variety of bacteria, viruses, and protozoa encounter autophagy within host cells, and biogenesis of the *L. pneumophila* vacuole shares many features with the autophagy pathway.

In macrophages from A/J *naip5* mutant mice, the *L. pneumophila* vacuole becomes decorated with Atg7 in the same period that small secretory vesicles attach to its cytoplasmic face (Fig. 3). Since in yeast Atg7 mediates homotypic fusion of membranes to generate autophagosomes, it is tempting to speculate that mouse Atg7 contributes to biogenesis of the ER-derived replication vacuole. Engulfment of the *L. pneumophila* phagosome within ER requires Sar1, a host protein that regulates vesicular traffic between the ER and the Golgi apparatus. Likewise, membrane flow through the early secretory pathway is required for biogenesis of autophagosomes in yeast. Several hours after ingestion by A/J mouse macrophages, vacuoles harboring *L. pneumophila* acquire several features typical of autophagolysosomes. These include an acidic pH, the autophagy enzyme Atg8, the late endosomal and lysosomal protein LAMP-1, the lysosomal acid hydrolase cathepsin D, and retention of monodansylcadaverine, a fluorescent dye that preferentially accumulates in autophagosomes. Moreover, when compared with restrictive C57Bl/6 mouse macrophages, permissive A/J *naip5* macrophages that are subjected to amino acid starvation exhibit markedly slower maturation of autophagosomes. Thus, in A/J *naip5* mouse macrophages, *L. pneumophila* appears to replicate in autophagosomal vacuoles that merge with lysosomes slowly (Swanson, 2006).

Remarkably, *L. pneumophila* continues to replicate for hours within acidic autophagolysosomal vacuoles of A/J mouse macrophages. Whereas stimulation of autophagy increases the yield of bacteria, *L. pneumophila* replicates poorly either when autophagosome formation is inhibited by 3-MA, recruitment of Atg7 is disrupted by brefeldin A, or vacuole acidification and fusion with lysosomes is blocked by bafilomycin A (Fig. 4). Therefore, flux through both the secretory and the endosomal pathways promotes replication of intracellular *L. pneumophila*.

Perhaps modification of the *L. pneumophila* phagosome by either LPS-rich vesicles or the type IV secretion apparatus marks the vacuole as a target for the autophagy machinery of mouse macrophages. To avoid rapid delivery to lysosomes, *L. pneumophila* may rely on type IV secretion substrates such as RalF, SidM, and LidA to stabilize its association with the host secretory machinery, thereby prolonging its residence within an immature autophagosome. Indeed, when *L. pneumophila* lacks IcmS, a cytoplasmic protein that interacts with certain type IV secretion substrates, its phagosome first associates transiently with secretory vesicles before fusing with lysosomes, which degrade the mutant bacterium. Also, like the nascent *L. pneumophila* phagosome, Tilney found that the plasma membrane of uninfected macrophages is connected by hairlike projections to small vesicles (Tilney et al., 2001). Therefore, *L. pneumophila* may prolong a membrane-trafficking process of macrophages that normally is too transient to detect readily. By this model, over the course of its coevolution with amoebae, *L. pneumophila* has acquired by horizontal gene transfer multiple dominant negative functions that each interfere with the dynamic membrane traffic machinery of its host.

Interactions between *L. pneumophila* and the autophagy pathway have also been analyzed genetically by using the soil amoeba *D. discoideum* as an experimental host model. The ability of *L. pneumophila* to replicate in *D. discoideum* is not affected by mutations in each of several components of the autophagy pathway. However, the expression of multiple autophagy genes is sensitive to infection. Atg8 and Atg12 are each downregulated 24 h after ingestion of type IV secretion-competent *L. pneumophila* but not avirulent *Legionella hackeliae*. At the same time, the host protein Atg9 is upregulated in response to infection by wild-type *L. pneumophila* but not by type IV secretion mutants. Whether during its coevolution with amoebae *L. pneumophila* acquired mechanisms to modulate the autophagy pathway is a question that merits direct testing.

AVERTING PROGRAMMED CELL DEATH OF MACROPHAGES

While delivering its virulence payload to the cytoplasm, the type IV secretion system inadvertently triggers host defenses. Although infection activates programmed cell death of macrophages, *L. pneumophila* is equipped to prolong the life of its host (O'Conner et al., 2008). For example, macrophages that contain replicating *L. pneumophila* are resistant to agents known to trigger apoptotic death. Although human and A/J mouse macrophages that are infected by type IV secretion-competent *L. pneumophila* accumulate activated caspase-3, an executioner of apoptosis, the host cells do not succumb for several hours.

One host pathway that protects macrophages from type IV secretion-competent *L. pneumophila* is the NF-κB signaling cascade (O'Conner et al., 2008). Within hours of infection, NF-κB translocates into the nucleus, where it persists and induces expression of multiple genes; some of these encode proinflammatory cytokines and others act either as pro- or antiapoptotic factors. When NF-κB signaling is inhibited genetically or pharmacologically during infection, macrophage death increases and bacterial yield decreases. Therefore, to replicate in macrophages, secretion-competent *L. pneumophila* prevent host cell death by manipulating regulation of apoptosis that is mediated by NF-κB.

One bacterial factor that prolongs the life of infected mouse macrophages is SdhA, a protein that is translocated by type IV secretion across the phagosomal membrane throughout the replication period (Table 1) (O'Conner et al., 2008). In the absence of SdhA, *L. pneumophila* induces mitochondrial damage, caspase activation, and cell death of A/J mouse macrophages, and the intracellular mutant bacteria cannot replicate. The target of SdhA appears to be cell-type dependent, since *sdhA* is dispensable for *L. pneumophila* to replicate in a human monocyte cell line or in *D. discoideum*. Presumably programmed cell death is regulated somewhat differently by each of these host cells.

Another type IV secretion substrate that targets the mouse macrophage cell death machinery is SidH (Table 1) (O'Conner et al., 2008). By physically interacting with two members of the Bcl2 family of proapoptotic proteins, BNIP3 and Bcl-rambo, SidH makes infected macrophages more refractory to programmed cell death. Like SdhA, SidH contributes to *L. pneumophila* growth in mouse macrophages, but not human monocytic cells or *D. discoideum*. Indeed, the compartment where *L. pneumophila* resides appears to depend on the host cell. For example, the bacterial replication vacuole slowly merges with the lysosomal compartment in A/J mouse macrophages, but not in macrophages derived from human peripheral blood. Thus, during its coevolution with a variety of environmental phagocytes, *L. pneumophila* likely acquired a number of factors whose potencies differ in different host cells.

L. PNEUMOPHILA AS A PROBE OF INNATE IMMUNE MECHANISMS OF MOUSE MACROPHAGES

L. pneumophila causes pneumonia by replicating in alveolar macrophages, mimicking its natural lifestyle within phagocytic protozoa and amoebae that inhabit fresh water. Despite the microbe's versatility, mice are naturally resistant to *L. pneumophila*. An exception is the A/J mouse, in which polymorphisms in the NOD-like receptor (NLR) protein Naip5 make both the animal and its isolated macrophages susceptible to infection. Much of the cell biological research designed to understand how *L. pneumophila* alters macrophage membrane traffic has relied on the A/J *naip5* mutant mouse. From bone marrow cells, large numbers of relatively homogenous primary macrophages can be obtained that are well suited for either fluorescence microscopy or quantitative growth studies.

In contrast to A/J mice, C57Bl/6 Naip5$^+$ macrophages efficiently restrict *L. pneumophila* replication. When mouse macrophages encounter flagellated *L. pneumophila*, the host cells induce a caspase-1-dependent proinflammatory cell death known as pyroptosis (Fig. 5) (Delbridge and O'Riordan, 2007). As the type IV secretion machinery delivers its payload, flagellin apparently escapes into the cytoplasm. Flagellin, a well-known microbe-associated molecular pattern, is then detected directly or indirectly by the NLR proteins Naip5 or Ipaf. As a result, NLR proteins are predicted to assemble inflammasomes, the protein complexes in which caspase-1 is activated. Consequently, the plasma membrane becomes permeable, the macrophage dies by osmotic lysis, and the inflammatory cytokines interleukin-1β (IL-1β) and IL-18 are released. In this manner, infected mouse macrophages not only deny the microbe a safe haven, but also recruit leukocytes to the site of infection. Remarkably, when introduced into the lungs of C57Bl/6 mice, *L. pneumophila* mutants that do not encode flagellin escape innate immune defenses. Instead, the flagellin mutants replicate to the levels observed for wild-type *L. pneumophila* in A/J *naip5* mutant mouse lungs. Accordingly, recognition of *L. pneumophila* flagellin by the NLR proteins Naip5 and Ipaf triggers an effective innate immune response in the mouse lung.

While some C57Bl/6 macrophages that ingest *L. pneumophila* commit pyroptosis, others in the culture withstand the infection. In these cells, *L. pneumophila* establish replication vacuoles, and the bacterial yield slowly increases.

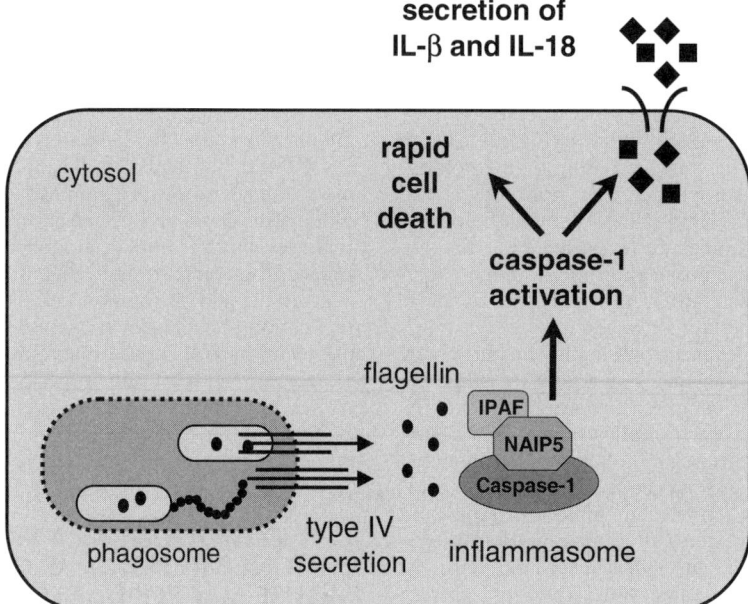

FIGURE 5 Mouse macrophages recognize cytosolic flagellin to restrict *L. pneumophila* replication. Macrophages that encode the NOD-like receptor proteins Naip5 and Ipaf respond to flagellin of type IV secretion-competent bacteria by activating a cell death program mediated by caspase-1. As a consequence, the macrophages deny *L. pneumophila* its replication niche while releasing the proinflammatory cytokines IL-1β and IL-18.

But, with time, the bacteria become dispersed throughout the cytoplasm, individually wrapped in membranes marked by the late endosomal and lysosomal protein LAMP-1. Degraded *L. pneumophila* become apparent, and bacterial viability steadily declines. These viable infected macrophages also accumulate numerous vacuoles decorated with the autophagy enzyme LC3, indicating a strong autophagic response. In the absence of flagellin, *L. pneumophila* replicate in C57Bl/6 Naip5$^+$ macrophages, and LC3-labeled vacuoles do not accumulate. Also, the amount of the NLR protein Naip5 correlates with both the speed of delivery of the microbes to lysosomes and the speed of autophagosome maturation (Swanson, 2006). Thus, it is tempting to speculate that particular NLR proteins equip macrophages to respond to infection by inducing autophagy as a first defense or pyroptosis as a fail-safe emergency response.

CONCLUDING REMARKS

After ingestion by macrophages, *L. pneumophila* encounters multiple arms of the innate immune system. The pathogen evades immediate delivery to degradative lysosomes, stimulates autophagy, induces NF-κB signaling, and protects its host from apoptotic death. By dissecting the cell biology of infected macrophages using bacterial and mouse genetics, transcriptional profiling of intracellular *L. pneumophila* and infected macrophages, the microbe's genome sequence, and specific fluorescence probes of bacterial and host proteins, a number of key players in *L. pneumophila* pathogenesis have been identified, and a number of testable conceptual frameworks have been developed. Workers in the field are now in an excellent position to exploit this fascinating microbe as a powerful probe of the many mechanisms that equip macrophages to fight infection.

REFERENCES

Bruggemann, H., C. Cazalet, and C. Buchrieser. 2006a. Adaptation of *Legionella pneumophila* to the host environment, role of protein secretion, effectors and eukaryotic-like proteins. *Curr. Opin. Microbiol.* **9:**86–94.

Bruggemann, H., A. Hagman, M. Jules, O. Sismeiro, M. A. Dillies, C. Gouyette, F. Kunst, M. Steinert, K. Heuner, J. Y. Coppee, and C. Buchrieser. 2006b. Virulence strategies for infecting phagocytes deduced from the *in vivo* transcriptional program of *Legionella pneumophila*. *Cell. Microbiol.* **8:**1228–1240.

DebRoy, S., J. Dao, M. Soderberg, O. Rossier, and N. P. Cianciotto. 2006. *Legionella pneumophila* type II secretome reveals unique exoproteins and a chitinase that promotes bacterial persistence in the lung. *Proc. Natl. Acad. Sci. USA* **103:**19146–19151.

Delbridge, L. M., and M. X. O'Riordan. 2007. Innate recognition of intracellular bacteria. *Curr. Opin. Immunol.* **19:**10–16.

Edelstein, P. H., and N. P. Cianciotto. 2005. *Legionella*, p. 2711–2724. *In* G. L. Mandell, E. Bennett, and R. Dolin (ed.), *Principles and Practice of Infectious Diseases*, 6th ed. Elsevier, Philadelphia, PA.

Fernandez-Moreira, E., J. H. Helbig, and M. S. Swanson. 2006. Membrane vesicles shed by *Legionella pneumophila* inhibit fusion of phagosomes with lysosomes. *Infect. Immun.* **74:**3285–3295.

Horwitz, M. A. 1983. Formation of a novel phagosome by the Legionnaires' disease bacterium (*Legionella pneumophila*) in human monocytes. *J. Exp. Med.* **158:**1319–1331.

Luo, Z. Q., and R. R. Isberg. 2004. Multiple substrates of the *Legionella pneumophila* Dot/Icm system identified by interbacterial protein transfer. *Proc. Natl. Acad. Sci. USA* **101:**841–846.

Molofsky, A. B., and M. S. Swanson. 2004. Differentiate to thrive: lessons from the *Legionella pneumophila* life cycle. *Mol. Microbiol.* **53:**29–40.

O'Conner, T. O., M. Heidtman, and R. R. Isberg. 2008. Mechanisms of intracellular survival and replication of *Legionella pneumophila*, p. 181–211. *In* K. Heuner and M. Swanson (ed.), Legionella *Molecular Microbiology*. Horizon Scientific Press, Norwich, United Kingdom.

Sexton, J. A., and J. P. Vogel. 2002. Type IVB secretion by intracellular pathogens. *Traffic* **3:**178–185.

Shohdy, N., J. A. Efe, S. D. Emr, and H. A. Shuman. 2005. Pathogen effector protein screening in yeast identifies *Legionella* factors that interfere with membrane trafficking. *Proc. Natl. Acad. Sci. USA* **102:**4866–4871.

Swanson, M. S. 2006. Autophagy: eating for good health. *J. Immunol.* **177:**4945–4951.

Tilney, L. G., O. S. Harb, P. S. Connelly, C. G. Robinson, and C. R. Roy. 2001. How the parasitic bacterium *Legionella pneumophila* modifies its phagosome and transforms it into rough ER: implications for conversion of plasma membrane to the ER membrane. *J. Cell Sci.* **114:**4637–4650.

Vincent, C. D., J. R. Friedman, K. C. Jeong, E. C. Buford, J. L. Miller, and J. P. Vogel. 2006. Identification of the core transmembrane complex of the Legionella Dot/Icm type IV secretion system. *Mol. Microbiol.* **62:**1278–1291.

Vogel, J. P., H. L. Andrews, S. K. Wong, and R. R. Isberg. 1998. Conjugative transfer by the virulence system of *Legionella pneumophila*. *Science* **279:**873–876.

25

The Role of Phagocytic Cells during *Shigella* Invasion of the Colonic Mucosa

GUY TRAN VAN NHIEU AND PHILIPPE SANSONETTI

The capacity of professional phagocytes to internalize particles is a key property that allows them to perform several functions. It allows the removal of self-apoptotic cells or destruction of nonself, such as bacterial pathogens, by virtue of their microbicidal effects. It also allows the presentation of major histocompatibility complex class II (MHC-II) bound antigen to the adaptive immune system. The etymology referring to the process of cell feeding evokes the notion of immune cells acting as predators preying on foreign organisms and stresses the paradox in the widely used expression "bacterial-induced phagocytosis" when referring to invasive microorganisms. Some pathogenic bacteria have the property to induce their own internalization into normally nonphagocytic cells, such as epithelial cells. This property is used by pathogens to breach an epithelial barrier and to reach a specific niche. In this case, since the bacteria are the active player that tricks the host cell into a process that it does not normally fulfill, the role of predator is reversed. Studying the various facets of the strategies employed by pathogens to invade cells has revealed the wide diversity of molecular tools that bacteria have engineered to divert host cell functions and taught us about fundamental aspects of the cell biological processes that govern the dynamics of the cell cytoskeleton (Cossart and Sansonetti, 2004). The interaction with epithelial cells, however, often only represents the primary stage at the onset of the infectious process, and bacteria need to successfully cope with various cell types and tissues. In the heart of these multiple encounters, dealing with professional phagocytes is a major issue for the invading pathogen, because of their direct role in bacterial clearance and their key function in shaping the innate and adaptive immune responses. Also, because of their capacity to release proinflammatory mediators, macrophages or neutrophils play a major role in the mounting of the inflammation that leads to tissue destruction. In acute shigellosis, bacterial invasion of the colonic mucosa is characterized by the development of an intense inflammatory reaction that is responsible for tissue destruction. The severity of the disease is determined by the capacity of the bacteria to invade and to disseminate in the colonic epithelium and by the bacterial ability to counter host innate immune defenses to elicit and survive inflammation. In this chapter, we will review the role of *Shigella* determinants with an emphasis on the role of phagocytic cells during the establishment of infection.

Shigella is the causative agent of the bacillary dysentery responsible for a large number of deaths related to diarrheal diseases, in particular among children under the age of 5 years in developing countries. *Shigella* spp. are enteroinvasive bacteria that are transmitted by the fecal-oral route directly from individual to individual, or through the ingestion of contaminated foodstuff. There are four described *Shigella* spp.: *S. sonnei*, *S. boydii*, *S. flexneri*, and *S. dysenteriae*. While *S. sonnei* is usually associated with mild watery diarrhea in developed areas, *S. flexneri* and *S. dysenteriae* are responsible for the dysenteric syndrome associated with destruction of the colonic mucosa and the emission of bloody and mucopurulent stools. *Shigella* does not invade polarized intestinal epithelial cells efficiently at their apical side in vitro, while invasion from the basolateral side of these cells is highly efficient (Mounier et al., 1992). Discrete invasion events occur from the apical side, but whether these events occur during infection in vivo is difficult to assess (Fig. 1). There may be factors in vivo that limit *Shigella* invasion of enterocytes from the intestinal lumen. These include the presence of the mucus and glycocalyx layer that overlays the epithelium and limits bacterial access to plasma membranes. The role of the endogenous colonic flora in *Shigella* infection is also by and large uncharacterized, but it may contribute to priming or to fencing the innate immune system. For example, defensins that are released by Paneth cells at the levels of intestinal crypts in the small intestine, and by intestinal epithelial cells in general, presumably play an important role in con-

---

Guy Tran Van Nhieu, Unité de Communication Intercellulaire et Infections Microbiennes, Inserm U971, Collège de France, 11, Place Marcelin Berthelot, 75005 Paris, France. **Philippe Sansonetti,** Unité de Pathogénie Microbienne Moléculaire, Institut Pasteur, 28, rue du Dr. Roux, 75724 Paris Cedex 15, France.

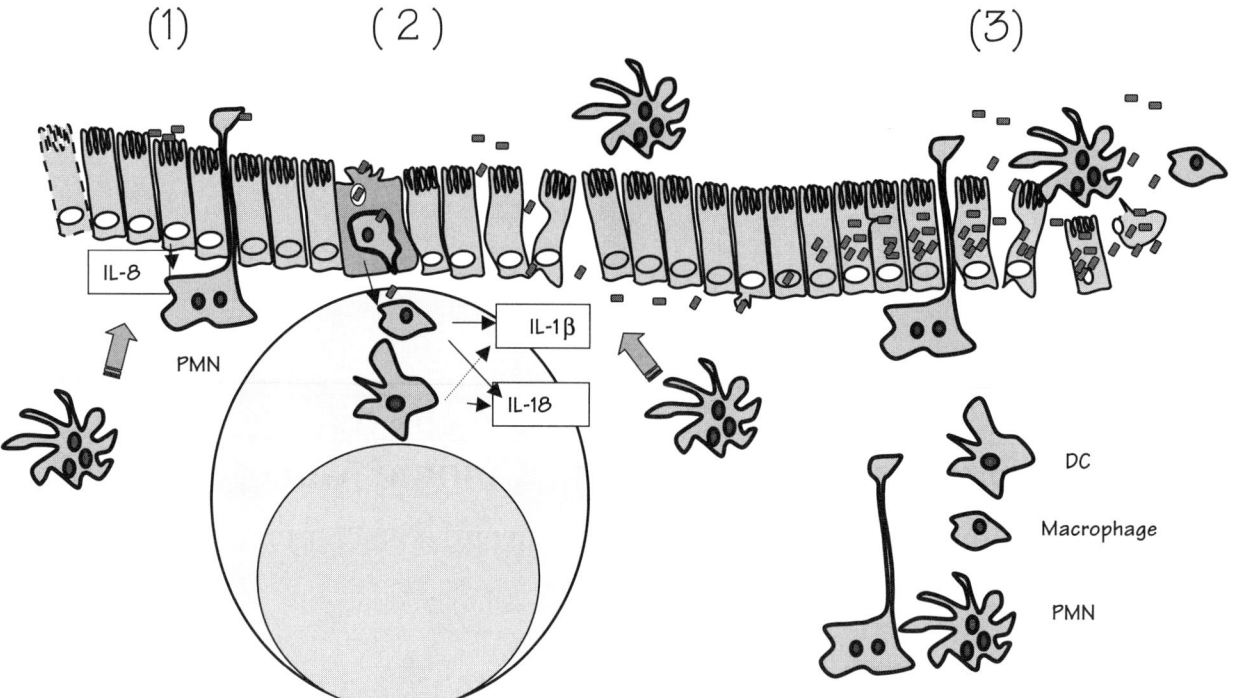

FIGURE 1 Phagocytic cells and *Shigella* invasion of the colonic epithelium. (1) *Shigella* invasion is not efficient at the apical side of enterocytes but may occur occasionally, leading to the release of IL-8, which acts as a chemoattractant for PMNs. These discrete events account for successful invasion or bacterial crossing of the epithelium. (2) *Shigella* invasions preferentially occur at the levels of M cells overlying solitary lymphoid follicles. Bacterial phagocytosis by macrophages results in cell death and the release of IL-1β and IL-18 which, in turn, trigger the recruitment of PMNs at the site of infection. The influx of incoming PMNs destabilizes the epithelium, favoring access to the enterocyte's basal side, where *Shigella* invasion readily occurs. (3) Destruction of the epithelium following bacterial invasion, replication, and dissemination in enterocytes is determined by the proinflammatory action of PMNs and macrophages. DC, dendritic cell.

trolling bacterial infection, since butyrate treatment that increases their expression has a protective role against *Shigella* (Raqib et al., 2006). The role of defensins is further highlighted by the observation that successful infections by *Shigella* correlate with a decrease in the LL-37 defensin expression, while remission is accompanied by increased levels of LL-37 (Zasloff, 2006). As will be discussed further, local invasion events may be controlled by neutrophils that guard the epithelium integrity but, in some instances, could also increase tissue injury. Experimental evidence indicates that, in macaque monkeys, infection with an *icsA* mutant strain that is unable to spread from cell to cell, shows that the primary infection sites lay at the levels of M cells overlying solitary lymphoid follicles (Sansonetti and Phalipon, 1999; Sansonetti et al., 1999). This has led to the generally admitted notion that *Shigella* invasion occurs mostly through M cells. Following internalization by M cells, bacteria face a mixed population of phagocytes, including resident macrophages and dendritic cells present in the lamina propria at the level of the lymphoid domes, as well as monocytes recruited from the peripheral blood that become activated at the site of infection (Fig. 1). Massive killing of phagocytes is usually observed during *Shigella* infection at the level of the infected mucosa. This observation, which is a hallmark of the inflammatory response induced by *Shigella* in humans, also stands out from immunological staining of histopathological sections of infected rabbit ileal loops, where massive death of phagocytes is observed at the level of the lamina propria of intestinal villi (Zychlinsky et al., 1996). The discovery of a mechanism allowing *Shigella* to specifically kill macrophages has represented a major breakthrough in the understanding of the manipulation of the host inflammatory responses by this pathogen.

MODELS FOR SHIGELLOSIS

Shigella is specific for humans, and as few as 100 bacteria are sufficient to cause the disease. There are no animal models that faithfully reproduce the disease, and consequently the sequence of events leading to bacterial invasion of the colonic mucosa and to the mounting of the inflammatory reaction in its natural host is not precisely known. Mice are resistant to *Shigella* infection by the oral route, but pulmonary infections that lead to severe bronchopneumonia provide a model to study the acute inflammation occurring during shigellosis (Mallett et al., 1993). Although the infected tissue is not related to the colonic mucosa, this model has proven very useful to decipher cytokine profiles and immune responses induced during infection and benefits from the availability of mice genetically impaired for specific pathways (Le-Barillec et al., 2005). The Sereny test, which consists of corneal inoculation of bacterial strains in guinea pigs, has also been used to test the capacity of *Shigella* strains to invade and to disseminate from cell to cell (Sereny et al., 1971). The use of xenotransplants of human intestinal tissue in SCID mice offers interesting perspectives to study the effects of *Shigella* infection on host gene profile expression (Zhang et al., 2001). Because of its complex setup, however, this model will probably remain restricted to specific applications. Finally, the rabbit ligated ileal loop is presumably the most amenable model that reflects the bacterial ability to destroy the intestinal mucosa, with abscess formation and a massive infiltration of polynuclear cells in the intestinal lumen (Sansonetti et al., 1983). Obvious limitations of this model, however, reside in the fact that (i) infection takes place in the small and not the large intestine; (ii) large numbers of bacteria are required to induce clear histological symptoms (about 10^9 bacteria per ligated loop); and (iii) the availability of immunological reagents for rabbits is limited. Perhaps the experiments most referred to that are key to the model of initial M-cell entry, followed by subsequent induction of the inflammation and destabilization of the epithelium, are infections of macaque monkeys with an *icsA* mutant of *Shigella* that is unable to spread from cell to cell (Sansonetti et al., 1999). In these experiments, the sites of infection that were clearly identified corresponded to isolated lymphoid follicles, suggesting that these represented the primary infectious sites. However, since IcsA is required for dissemination in enterocytes, it is possible that the use of such a mutant prevents the detection of discrete invasion events in epithelial cells, while introducing a bias toward inherently phagocytic sites (i.e., M cells). In addition, IcsA has also been involved in the signaling that leads to the recruitment of polymorphonuclear leukocytes (PMNs) at sites of infection, so that the use of an *icsA* mutant may not trigger the levels of inflammation required for the detection of invasion sites in the epithelium (Fernandez et al., 2001). Although imperfect, animal models for shigellosis proved to be key in allowing the validation or emergence of new concepts that arise from cellular or molecular findings. The use of in vitro cultured cells has proven extremely useful to identify the function of T3S effectors involved in the invasion and dissemination process.

THE *SHIGELLA* ARSENAL

The genomes of several *S. flexneri* serotype 2a strains and of a serotype 5a strain have been sequenced and show high levels of similarity both among themselves and with the genome of *Escherichia coli* K-12 (Jin et al., 2002; Nie et al., 2006; Wei et al., 2003). With the exception of insertion sequences, the majority of the housekeeping genes are conserved between *E. coli* K-12 and *Shigella*. Of note, *S. flexneri* serotype 2a also contains the two pathogenicity islands SHI-1 and SHI-2, whereas SHI-1 appears to have been deleted from *S. flexneri* serotype 5a. In addition, there are about 900 genes that are absent or inactivated in *S. flexneri* serotype 2a compared to *E. coli* K-12. As will be discussed further, some of these genes were probably eliminated or inactivated during evolution because they encode antivirulence factors when their *E. coli* orthologues are expressed in *Shigella*. Conversely, about 195 genes are specific for *Shigella*. These genes are typically clustered in islets across the *Shigella* genome and encode factors that are important for virulence, such as iron uptake systems, toxins, and enzymes involved in the modification of the O antigen. The majority of the virulence determinants are encoded by the large virulence plasmid (approximately 220 kb), also present in strains of enteroinvasive *E. coli* (Buchrieser et al., 2000). The large virulence plasmid comprises about 100 genes, including those involved in the synthesis of TT3S, a macromolecular device devoted to the injection of bacterial effectors in the host cell cytoplasm upon cell contact. Remarkably, genes responsible for bacterial invasion and dissemination from cell to cell, for induction of apoptosis of the macrophage, are encoded by a 30-kb region on the virulence plasmid. Consistent with their specific role during infection, these genes are downregulated at 30°C and upregulated at 37°C through the plasmid-encoded transcriptional activators VirB and VirF, or by high-osmolarity conditions through the chromosomally encoded two-component system EnvZ/OmpR (Dorman and Porter, 1998). In addition, the activity of the type III secretion system (T3SS) also regulates the expression of a subset of T3S effectors at transcriptional levels (Parsot, 2005). Besides genes for T3S effectors, the plasmid-borne IcsA gene encodes a bacterial surface protein that drives actin polymerization required for intracellular motility and dissemination from cell to cell.

The *Shigella* Type III Secretory Apparatus

Various T3SS identified to date in gram-negative bacteria share significant similarities in their genetic organization and structures (Cornelis, 2006; Galan and Wolf-Watz, 2006). The scientific community has largely benefited from important advances in the structural organization of T3SS by the assignment of roles of orthologues in other species, such as *Yersinia* or *Salmonella*. Several examples of these flagella-related structures have been resolved by electron microscopy from purified preparations (Blocker et al., 1999; Kubori et al., 2000). Even if they present some minor variations, T3SS share common features and can be divided in two distinct parts: (i) a basal part that resembles the flagella basal body spanning the inner and outer membranes, and (ii) a needle complex, corresponding to a needle whose length and diameter vary according to the species, that connects the part of the basal body associated with the inner and outer membranes and protrudes at the bacterial cell surface from the lipopolysaccharide (LPS) layer. In the case of *Shigella*, the T3SS needle was estimated to be about 50 nm long (Blocker et al., 2001). Single-particle reconstruction based on the averaging of electron microscopy images and X-ray fiber diffraction has provided a detailed architecture of the T3SS. Consistent with a microinjection device, the T3SS needle appears hollow, with an estimated inner diameter of 2 nm and an outer diameter of 10 nm. In the case of *Shigella*, a cytoplasmic bulb-like structure has been visualized in association with the basal part, which is not always visualized in the different structures that have been resolved, presumably because its association with the rest of the T3SS is more sensitive to disruption during the purification procedures. There are T3SS substrates that are dispensable for the assembly of this tripartite structure but which are essential for its translocating function. In particular, two orthologues, in the case of *Shigella* the IpaB and IpaC proteins, insert into host cell membrane and are required for injection of T3S effectors in the cell cytosol. IpaB and IpaC are the main proteins recovered from host cell membranes following bacterial contact, and their insertion in erythrocytes' membranes is associated with the formation of a pore with an estimated size of 2 to 3 nm (Blocker et al., 1999). Since the estimated size of this IpaB-IpaC-dependent pore corresponds roughly to the size of the inner part of the T3S needle, it is thought that these membrane-inserted components act as adaptors at the tip of the needle through which T3S effectors are channeled to get access to the cell cytosol. Studies on T3S regulation have indicated that IpaB and IpaD also regulate secretion, leading to the model that these proteins act as a clog of the T3SS (Parsot et al., 1995). This model was supported by electron microscopy and fractionation data indicating that the IpaB and IpaD are exposed at the tip of the T3S needle under nonsecretion conditions and therefore may act as a sensor responsible for the triggering of T3S upon cell contact (Sani et al., 2007; Veenendaal et al., 2007). The role of IpaB as part of a sensor is also consistent with evidence that IpaB binds to the hyaluronic receptor CD44 or to cholesterol, allowing the targeting of T3S to microdomains (Skoudy et al., 2000; Lafont et al., 2002). IpaB pre-positioning at the needle tip also ensures that this component, which is required for the injection, is inserted prior to other T3S effectors. IpaD is required for the association of IpaB with the needle tip before secretion is induced, and may be involved in a transient association of IpaC with the needle after insertion of IpaB in host cell membranes (Veenendaal et al., 2007). Structural studies indicate that the IpaD N terminus presents similarities with T3S chaperones, suggesting that IpaD prevents premature oligomerization of IpaB in or at the tip of the needle and assists in the proper insertion of IpaB and IpaC in host cell membranes (Johnson et al., 2007). IpaC is not required for IpaB insertion into host cell membranes and for membrane destabilization, but for the formation of a 2- to 3-nm pore that is detected upon challenge of red blood cells with invasive *Shigella* (Blocker et al., 1999). The emerging picture is that IpaB at the tip of the T3SS needle acts as a sensor for cell receptors such as CD44 or β_1 integrins, or lipids. On cell contact, IpaB is released from the T3SS and inserts into host cell membranes. This would be followed by the secretion of IpaC, which also inserts into host cell membranes and organizes the oligomerization of the translocator components to form a channel connecting the T3SS to the host cell cytosol (Fig. 2). The continuum between the needle and the translocator component has not yet been definitely proven, but needs to be envisioned since, in epithelial cells, T3S injection is not associated with cytotoxicity and pore formation cannot be detected. This continuum, however, has to be transient since translocator components are detected away from the initial bacteria-cell interaction site immediately after secretion has been triggered (Enninga et al., 2005). Obviously, much is yet to be learned about the precise mechanism of T3S-mediated injection of effectors. A major limitation in the characterization of this process is the transient nature of the interactions at the tip of the T3S needle and the rapid kinetics of secretion.

There are 26 identified substrates for the T3SS (Parsot, 2005). Although most T3SS substrates are encoded by the large virulence plasmid, copies of genes corresponding to the IpaH family proteins are located on the chromosome. These substrates are classified according to the regulation mode of their expression: (i) 13 are subjected to the VirB/VirF regulation; (ii) 4 are controlled by VirB/VirF and upregulated following host cell contact and secretion; and (iii) 9 are independent of the VirB/VirF regulation and are only expressed after secretion has been induced (Fig. 2). Control of the expression levels of T3S substrates is assured through the MxiE transcriptional activator from the AraC family, whose activity depends on the intracellular concentration of IpgC, the molecular chaperone of IpaB and IpaC (Mavris et al., 2002). Since bacterial uptake occurs after host cell contact, T3S effectors that induce bacterial invasion obviously need to belong to the first category (Fig. 2). Constitutive expression and storage in the bacterial cytosol of these T3S effectors allows the release of their total pool within a few minutes after cell contact. This first category of T3S substrates is the best characterized, even if the function for some of them remained undetermined. For those substrates with an ill-defined function, inactivation of the corresponding genes results in bacterial mutant strains that show no apparent defect in invasion or dissemination in epithelial cells. Most likely, this apparent lack of function rather underscores the need to develop relevant systems or models that reflect bacterial functions that are needed during in vivo infection. The function of the T3S effectors, whose expression is upregulated on T3S, points to their role in the control of the inflammatory reaction (Fig. 2).

The "Early" *Shigella* T3S Effectors

Current knowledge on the function of T3S effectors is mostly derived from their action in epithelial cells. Whether these functions are transposable to phagocytes or nonepithelial cells is unknown. In light of the importance in downregulating inflammation, and given the importance of phagocytic cells in the control of inflammation, it is possible that the function of some T3S effectors, in particular, the second-line subset of effectors, target-specific cell types. Also, some T3S effectors may have different effects according to the cell type. For example, in addition to its role in T3S translocation in epithelial cells, IpaB has been shown to induce apoptosis in macrophage (Chen et al., 1996).

Early T3S Effectors Triggering Bacterial Invasion

Many T3S effectors that reorganize the cytoskeleton have been implicated in *Shigella* invasion of epithelial cells (Cossart and Sansonetti, 2004; Galan and Wolf-Watz, 2006). In addition to its role in translocation, IpaC has been shown to induce actin polymerization that is essential for the formation of membrane ruffles that surround the bacterium (Tran Van Nhieu et al., 2005). The precise mechanism of IpaC-mediated actin polymerization is still not completely resolved and could rely on several modes. In cell systems, IpaC-mediated effects need the Cdc42 and Rac GTPases, whereas in vitro, purified IpaC has the capacity to directly nucleate actin polymerization (Tran Van Nhieu et al., 1999; Kueltzo et al., 2003). It is possible that these two activities reflect functions of IpaC that intervene at different stages of the invasion process. The role of all three Cdc42, Rac, and Rho GTPases during *Shigella* invasion has been demonstrated by the use of inhibitory dominant-negative forms (Mounier et al., 1999). The implication of Cdc42 in *Shigella* or IpaC-mediated actin polymerization has become questionable because the use of dominant-negative forms may not allow differentiation between the requirements for Cdc42 or Rac. Rather, Rac could represent the only Rho GTPase involved in bacterial-induced actin polymerization, since, in the case of *Salmonella* invasion, a process that also implicates T3SS-dependent membrane ruffling, anti-Cdc42 small interfering RNAs did not affect bacterial internalization (Patel and Galan, 2006). As opposed to Rac, Rho is not involved in *Shigella*-induced actin polymerization but, rather, is required for the recruitment in the cell extensions of other components, such as ezrin or the Src tyrosine kinase, that participate in the completion of the internalization process

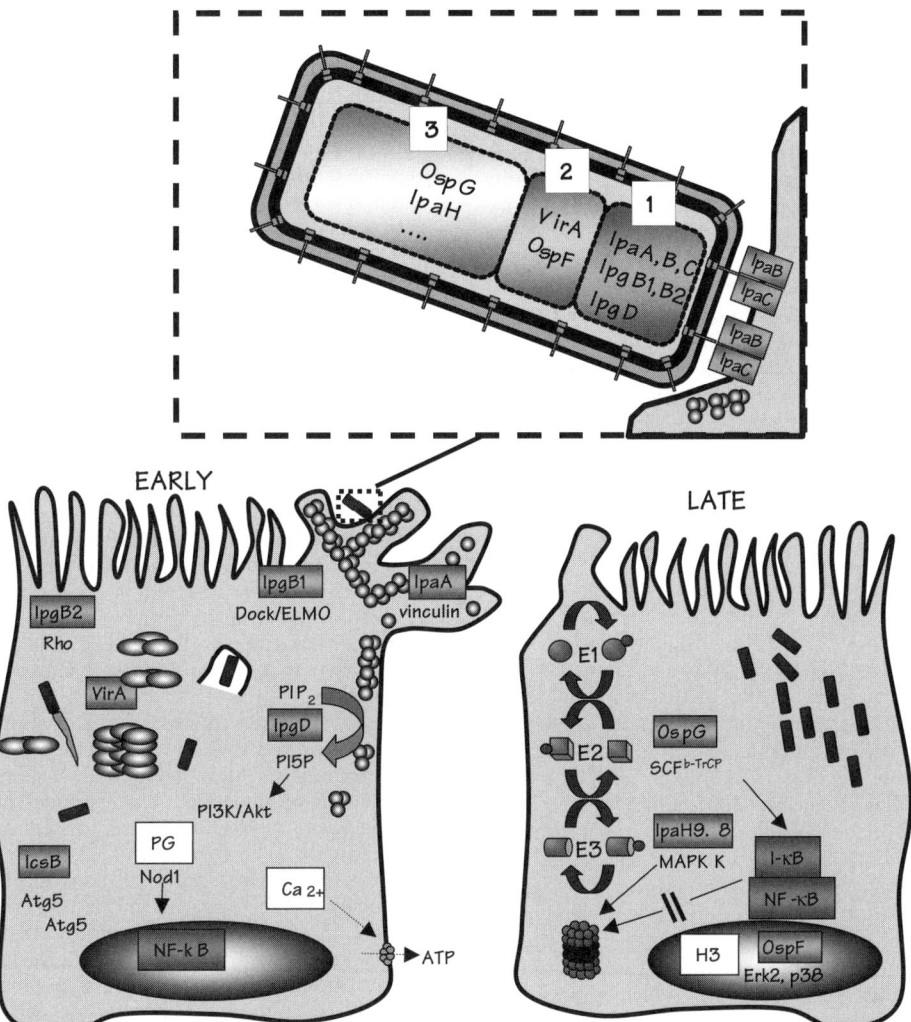

FIGURE 2 The *Shigella* arsenal of T3S effectors. (Inset) *Shigella* T3S effectors can be divided into (1) constitutively expressed early effectors that are injected following cell contact—among these, the IpaB and IpaC proteins insert into host cell membranes to form the "translocon"; (2) effectors that are upregulated after secretion; and (3) late effectors that are only expressed after T3S has been induced. (EARLY) T3S effectors that promote bacterial invasion include IpaC, which induces actin polymerization; IpgB1, which activates the Dock/ELMO pathway; and IpaA, which binds to vinculin and induces actin depolymerization. IpgB2 is a mimic of the GTPase Rho. IpgD hydrolyzes PI(4,5)P2 and activates the PI3-K/Akt pathway. VirA inhibits microtubule polymerization and favors intracellular actin-based motility. IcsB binds to Atg5 and prevents autophagy. The early stages of invasion are associated with proinflammatory signals since intracellular PG stimulates Nod1-dependent activation of NF-κB. *Shigella* invasion also induces Ca^{2+} signaling and the release of extracellular ATP through hemichannels. (LATE) Late T3S effectors downregulate inflammatory signals. OspG binds to UbcH5, a E2 ubiquitin-conjugating enzyme, and prevents degradation of I-κB by the proteasome. OspF inhibits the activation of the MAPKs Erk2 and p38. OspF also alters the histone code by inhibiting the phosphorylation of histone H3. IpaH9.8 is an E3 ubiquitin ligase that may target the degradation of a MAPK by the proteasome.

(Dumenil et al., 2000). In addition to its role during the completion of the *Shigella* internalization process, Src is also important for the initial events of actin polymerization that lead to membrane ruffling at entry sites (Dumenil et al., 1998). Promptly after bacterial-host cell contact, Src allows the tyrosyl phosphorylation of cortactin, a cytoskeletal protein that binds to filamentous actin (Dehio et al., 1995).

Cortactin affects actin dynamics and filament organization by performing various functions. Cortactin binds to and activates the Arp2/3 complex, but as opposed to other actin nucleators such as the WASP family proteins, activation of the Arp2/3 complex is highly favored by cortactin binding to actin filaments (Weaver et al., 2003). Alternatively, tyrosylphosphorylated cortactin enhances N-WASP-mediated actin nucleation by allowing binding to the Nck adaptor and the WASP-interacting protein WIP (Tehrani et al., 2007). Because cortactin also stabilizes the branching of actin filaments, the combination of both nucleation and branching activities leads to the formation of a dendritic network at the plasma membrane (Egile et al., 2005; Weaver et al., 2001). *Shigella*-induced Src activation upon invasion triggers the recruitment of tyrosyl-phosphorylated cortactin at the plasma membrane through interaction with the Crk adaptor protein (Bougnères et al., 2004). Because mutations in the cortactin Arp2/3 binding domain impair actin polymerization leading to ruffling formation at *Shigella* entry sites, the cortactin actin nucleation activity appears to be involved in this process. Clearly, however, cortactin activation is not sufficient for *Shigella*-induced ruffling, which implicates a combined activity of the Rac GTPase and Src tyrosine kinase. Other studies also point at a role for Crk in *Shigella* invasion, but acting upstream of the Rac GTPase and downstream of Abl/Arg tyrosine kinases (Burton et al., 2003). Whether Crk acts upstream of Rac or together with cortactin, presumably downstream of this GTPase, is unclear. It is possible that these Crk functions are not exclusive and that they participate at different steps of bacterial entry foci formation.

Finally, IpgB1, another *Shigella* T3S effector, also plays a role in bacterial-induced ruffling at entry sites by activating the ELMO/Dock pathway. Similar to what is observed upon inhibition of the Src tyrosine kinase pathway, an *ipgB1* mutant induces smaller actin foci that do not promote efficient bacterial internalization. IpgB1 was shown to act as a RhoG mimic that binds to the ELMO adaptor protein and recruits Dock180, an exchange factor (GEF) for Rac (Handa et al., 2007; Ohya et al., 2005). Together, *Shigella*-induced ruffling appears to be a combined action of IpaC, which is indispensable for bacterial uptake, and IpgB1, which activates Rac through the ELMO/Dock pathway in combination with Src-dependent signaling to propagate actin polymerization at a distance from the initial bacterial-cell contact site. For productive bacterial internalization, bacterial-induced actin polymerization also needs to be controlled, and in particular, actin needs to depolymerize for the completion of the entry process. During *Shigella* invasion of epithelial cells, this is achieved by the injection of the IpaA T3S effector. IpaA binds to the focal adhesion protein vinculin, a focal adhesion protein that participates in anchoring of the cytoskeleton to the membrane (Bourdet-Sicard et al., 1999). Interestingly, the vinculin binding domain on IpaA corresponds to a carboxy-terminal region containing two individual vinculin binding sites (VBSs) that act as a "super mimic" of talin VBSs, which bind with high affinity to vinculin and promote the helical-bundle conversion that characterizes its activated state (Izard et al., 2006). In vitro, the purified IpaA vinculin binding domain shows partial capping activity of the barbed ends of actin filaments (Ramarao et al., 2007). This is in contrast to what is observed with binding of full-length IpaA to vinculin, which leads to actin depolymerization (Bourdet-Sicard et al., 1999). From experiments implicating the transfection of truncated IpaA derivatives, the first two-thirds of IpaA were shown to convey the actin depolymerization activity that appears to rely on a simultaneous activation of the Rho GTPase combined with the ability of IpaA to compete with talin for binding to the cytoplasmic tail of β_1 integrins (Demali et al., 2006). It is unclear why IpaA targets vinculin to perform such functions, but the mode of action of this T3S effector is likely to reflect mechanistic processes controlling the dynamics of cytoskeletal anchorage during cell adhesion.

In support for their role in *Shigella* invasion, inactivation of the *ipaC*, *ipaA*, and *ipgB1* genes leads to bacterial mutants with invasive defects. It is clear, however, that these effectors act in concert to organize bacterial invasion and that they do not play redundant functions. This is illustrated by their relative importance in invasion assessed in in vitro cultured cells. In these assays, IpaC is absolutely required for actin polymerization and cytoskeletal reorganization leading to bacterial invasion, and an *ipaC* mutant is noninvasive. In contrast, while IpgB1 amplifies actin polymerization to allow the formation of large membrane leaflets at bacterial entry sites, an *ipgB1* mutant is about twofold defective for bacterial invasion. Finally, in line with its role in modulating invasion, an *ipaA* mutant shows a tenfold decrease in the efficiency of invasion. Besides the translocator component IpaB and IpaC, none of the "injected" T3S effectors are essential for bacterial uptake.

Early Effectors That May Be Involved in Processes Other Than Invasion

Other front-line T3S effectors have no detectable phenotype in invasion when tested in in vitro cultured cells. Although this does not exclude a potential role for these effectors in regulating invasion in tissue, there is mounting evidence that some of these effectors play a role in pathogenesis that is not connected to bacterial invasion of epithelial cells.

IpgB2 is related to IpgB1 but was shown to mimic the activated form of the RhoA GTPase and to activate stress fiber formation (Alto et al., 2006). The precise role of IpgB2 remains undefined, since it does not appear to be involved in invasion or bacterial dissemination in in vitro cultured cells. The IpgD and VirA T3S effectors were initially assigned a role in *Shigella* invasion, but this role is challenged by the fact that mutants in the corresponding genes show no defect in invasion and by findings pointing to their role in cell-to-cell spread, or in the control of host transcriptional responses. IpgD is a phosphatidylinositol (PI) (4,5) bisphosphate-phosphatase, whose hydrolase activity was reported to loosen cortical actin connection to the plasma membrane, thereby favoring de novo actin polymerization at bacterial entry sites (Niebuhr et al., 2002). Its unusual phosphatase at position 4 of PI(4,5)P2 leads to the production of PI(5)P, which has been involved in the activation of the PI-3 kinase/Akt pathway, implicating IpgD in modulating cellular transcriptional responses involved in cell survival (Pendaries et al., 2006). VirA was shown to bind to hetero-oligomers of tubulin and to inhibit microtubule polymerization (Yoshida et al., 2002). Because it is generally admitted that microtubule depolymerization leads to activation of a GEF for Rho GTPases and actin polymerization, VirA was initially proposed to act as a T3S effector favoring bacterial invasion (Yoshida et al., 2002). The transcriptional control of VirA, however, indicates upregulation after T3S is activated, suggesting that VirA plays a role intracellularly. More recently, the microtubule depolymerization activity of VirA has been shown to favor bacterial actin-based motility (Yoshida et al., 2006). Consistently, a *virA* mutant invades epithelial cells but is unable to disseminate from cell to cell. Finally, IcsB favors bacterial cell-to-cell spreading in epithelial cells by inhibiting autophagic responses following vacuolar lysis and release of the bacterium in the cell cytoplasm (Ogawa et al., 2005). In a counterintuitive way, *Shigella* appears to be targeted to autophagy because of the recognition of IcsA/VirG by the Atg5/LC3 autophagic proteins. The T3S effector IcsB, however, was shown to bind to IcsA/VirG and to prevent its recognition by Atg5 (Ogawa et al., 2005).

INNATE PROINFLAMMATORY SIGNALS LINKED TO *SHIGELLA* INVASION OF EPITHELIAL CELLS

Recognition of LPS and PG by Toll-Like and Nod-Like Receptors

The role of Toll-like receptors (TLRs) in innate immunity, through the recognition of microbial products such as LPS or peptidoglycan (PG) at the cell surface, has been well established (Uematsu and Akira, 2006). In particular, in the macrophage, the concomitant engagement of TLRs with receptors involved in phagocytosis during bacterial internalization has been shown to determine the proinflammatory responses that are not triggered during phagocytosis of apoptotic cells or the activation of antimicrobial phagocytic cell properties (Ulevitch et al., 2004; Blander and Medzhitov, 2006). More recent years have highlighted the importance of Nod-like receptors (NLRs) in the recognition of intracellular bacterial products or stimuli in the triggering of TLR-independent responses that may combine with TLR signaling to induce the inflammatory response (Fritz et al., 2006). NLR signaling may be particularly important in the surveillance of the integrity of the epithelium, since intestinal epithelial cells express low levels of TLRs or other pattern recognition receptors (PRRs) such as CD14. These NLRs have various ligand specificities that may allow them to diagnose the type of microbes involved in the infectious process. For example, Nod1 and Nod2, respectively, are more specifically activated by the PG products meso-diaminopimelate (mainly produced by gram-negative bacteria) and muramyl dipeptide (which is found both in gram-negative and gram-positive bacteria) (Girardin and Philpott, 2004). Such NLRs are particularly relevant in the case of intracellular pathogens, such as *Shigella*, that multiply freely within the cell cytosol rupture of the phagocytic vacuole. Compelling evidence that NLRs act by directly binding to intracellular pathogen-associated molecular patterns (PAMPs) is lacking. There are some observations that some NLRs, like Nod2, are associated with the plasma membrane, suggesting that pathogen recognition by these molecules occurs upon alteration of the host cell plasma membrane integrity (Lecine et al., 2007). Major hindrances in the precise assignment in the function of NLRs are their low expression levels, sometimes at the verge of detection limits, and their propensity to aggregate when slightly overexpressed. Cytosolic NLR protein and adaptor complexes leading to the oligomerization and activation of caspase 1 have been described and named "inflammasomes" (Sutterwala et al., 2007). For example, the NALP3 NLR inflammasome shows a pattern of activation that is reminiscent of that induced by *Shigella*. NALP3 is activated by bacterial RNA in the case of *Listeria*, or by bacterial toxins such as aerolysin that promote a change in free-ion intracellular concentration (Mariathasan and Monack, 2007). In combination with ATP stimulation, NALP3 is part of a so-called inflammasome that leads to the activation of caspase 1 and caspase 5 and to proinflammatory signals linked to the release of interleukin-1 (IL-1) and IL-18. Based on the study of deficient macrophages, however, NALP3 does not appear to be involved in the activation of proinflammatory signals linked to *Shigella* infection. Instead, the IPAF inflammasome that senses other stimuli was shown to play a role in the activation of caspase 1 mediated by *Shigella*, since macrophages that are deficient for ASC, an adaptor component of the IPAF inflammasome, show reduced caspase-1 activation upon *Shigella* infection (Mariathasan and Monack, 2007). The picture is probably more complex, since bacterial-mediated killing appears similar to wild-type in *asc*-deficient macrophage, suggesting that *Shigella*, through IpaB, activates caspase-1-independent signals that lead to cell death.

Calcium Signaling during *Shigella* Invasion

Shigella invasion of epithelial cells induces calcium (Ca^{2+}) signaling. This signaling is dependent on the T3SS and corresponds to atypical oscillatory Ca^{2+} responses, which are reminiscent of those induced by pore-forming toxins (Tran Van Nhieu et al., 2003). Because Ca^{2+} signaling is associated with the activation of NF-κB, *Shigella* may thus induce proinflammatory signals during the initial phases of bacterial invasion (Gewirtz et al., 2000). *Shigella* was also shown to induce paracrine responses linked to the opening of hemichannels and the release of extracellular ATP. Paradoxically, extracellular release of ATP was shown to favor *Shigella* invasion and dissemination in neighboring epithelial cells (Tran Van Nhieu et al., 2003). Since ATP is an agonist of Ca^{2+} responses, it could also act as a paracrine messenger that amplifies inflammatory responses during *Shigella* invasion of epithelial cells.

ROLE OF "LATE" T3S EFFECTORS

The function of three "late" T3S effectors has been identified and points at their role in controlling or modulating inflammatory and immune responses (see Fig. 2). Whether these effectors act in redundant ways to limit the inflammation, or whether they act in concert to more precisely shape a specific response, is yet unclear.

OspG is a kinase that binds in vitro to UbcH5, an E2 ubiquitin-conjugating enzyme that is part of the SCF$^{\beta-TrCP}$ complex (Kim et al., 2005). This complex ubiquitinates phosphorylated IκB, the inhibitor of the transcriptional activator NF-κB, and targets it for degradation to the proteasome. In the presence of OspG, however, phosphorylation of IκB still occurs, but ubiquitination and degradation is blocked. As a result, OspG prevents the activation of NF-κB and the transcription of proinflammatory signals. Consistently, in the rabbit ileal loop, a *Shigella ospG* mutant induces extensive tissue destruction linked to an inflammatory response that is remarkably more intense than that observed for wild-type *Shigella*. OspF is also involved in the downregulation of proinflammatory signals. OspF either dephosphorylates (Arbibe et al., 2007) or prevents the phosphorylation of the Erk2 and p38 mitogen-activated protein kinases (MAPKs) by using an atypical mechanism that involves the cleavage of phosphothreonine residue in the MAPK activation loop at the level of the C-OP bond (Li et al., 2007). As opposed to a phosphatase activity, this cleavage leads to an irreversible inactivation of the MAPK, since phosphorylation cannot occur on the cleaved substrate. Although OspF targets all Erk2, p38, and JNK in vitro, only Erk2 and p38 are affected by OspF during infection, suggesting that perhaps other bacterial effectors control MAPK activation, or that the spatial localization of OspF restricts target accessibility (Arbibe et al., 2007; Li et al., 2007). Along those lines, OspF-dependent MAPK inactivation in the nucleus also prevents the phosphorylation of histone H3. Because phosphorylated H3 favors chromatin access to NF-κB, OspF prevents the transcription of MAPK, but also NF-κB-dependent genes, such as IL-8 (Arbibe et al., 2007). As for the *ospG* mutant, rabbit ileal loops infected by an *ospF* mutant show massive recruitment of PMNs in the lamina propria and in the lumen that is more intense than what is observed with wild-type *Shigella*. OspF therefore acts at least as a regulator of PMN transmigration across the intestinal epithelium. Finally, the IpaH9.8 T3S effector belongs to the IpaH family, which consists of related proteins with a highly conserved carboxy-terminal half and a variable amino-terminal leucine-rich repeat-containing moiety. Five *ipaH* genes are carried by the large virulence plasmid and seven by the chromosome. Individual deletions of *ipaH* family members do not lead to a significant difference in the mice pulmonary model of infection, but infection with a *Shigella* mutant strain in which all the chromosomal copies of *ipaH* genes have been deleted induces higher levels of inflammation than does wild-type *Shigella*, indicating that, like OspG and OspF, the IpaH effectors participate in the downregulation of inflammatory responses during *Shigella* invasion (Ashida et al., 2007). Consistently, IpaH9.8, like its *Salmonella* orthologue SspH1, has a ubiquitin ligase activity in vitro and belongs to the E3 ubiquitin ligase family (Rohde et al., 2007). In a yeast system, IpaH9.8 ubiquitinates ubiquitin and the MAPKK Ste7, leading to its degradation by the proteasome, but targets of IpaH9.8 in human cells remain unidentified (Rohde et al., 2007).

SHIGELLA-INDUCED MACROPHAGE DEATH

A key feature of *Shigella* virulence is its ability to rapidly kill macrophages and dendritic cells. Macrophage death is an immediate consequence of *Shigella* internalization (Fig. 3). As for epithelial cells, macrophage treatment with the cholesterol-depleting drug methylcyclodextrin prevents bacterial phagocytosis and subsequent bacterial-induced cell death. This agrees with a previous demonstration that translocator components are involved in apoptotic death, since IpaB and IpaC binding to cholesterol was shown to enhance T3S and injection of bacterial effectors (Lafont et al., 2002). However, when cholesterol depletion was performed with a timing that did not prevent phagocytosis and escape from the vacuole, IpaB-mediated apoptosis was still inhibited, indicating that cholesterol is also required for intracellular signaling leading to cell death (Schroeder and Hilbi, 2007). Following internalization by macrophage, *Shigella* escapes from the phagocytic vacuole and induces mac-

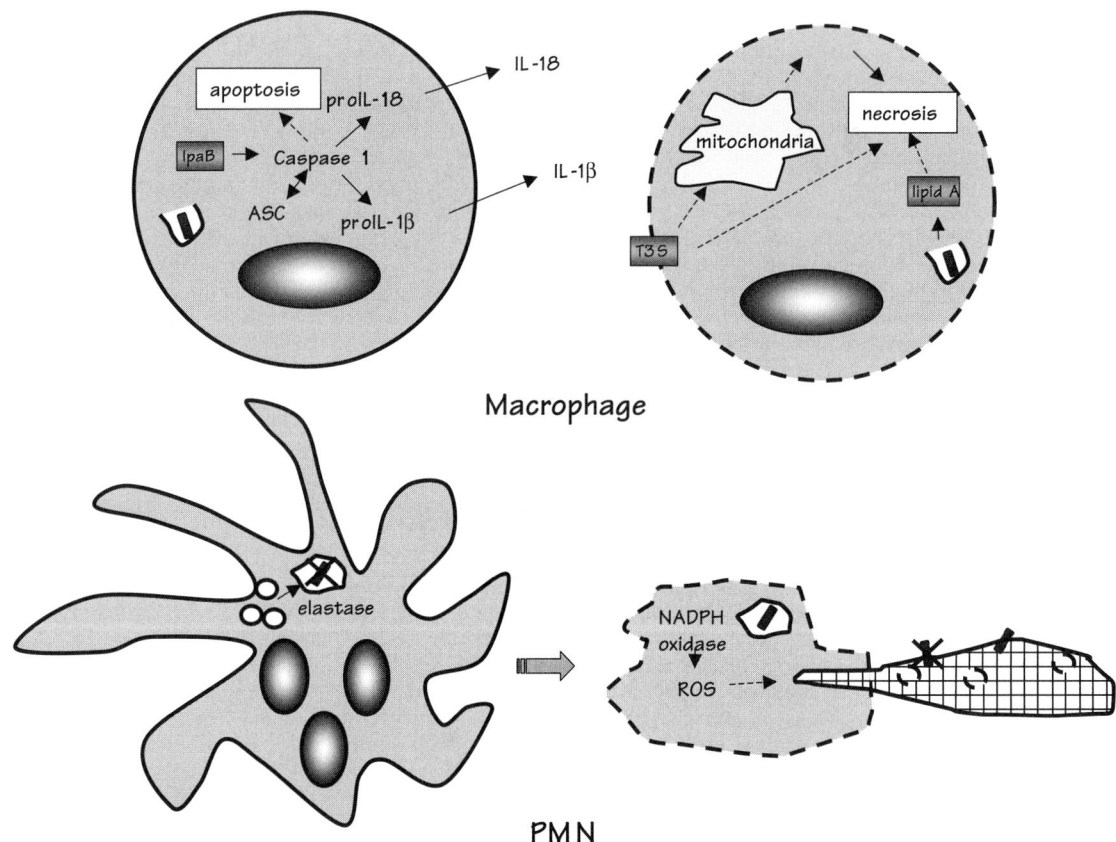

FIGURE 3 *Shigella* and phagocytes: cell or bacterial killing. *Shigella* internalization by the macrophage leads to cell death. Bacterial-induced apoptosis dependent on caspase 1/IpaB leads to the release of IL-1β and IL-18. The ASC-containing IPAF inflammasome may be involved in this process. Alternatively, macrophages may die of caspase-1-independent necrosis linked to T3S-dependent plasma membrane permeabilization, potentially mediated by the targeting of mitochondria by invasive bacteria. Necrosis was also reported to result from the release of lipid A by intracellular bacteria. Neutrophils (PMNs) kill internalized *Shigella* through the fusion of elastase-containing granules with the phagocytic vacuole. The activation of the NADPH oxidase and the production of reactive oxygen species (ROS) trigger the formation of neutrophil extracellular traps with PMN postmortem bactericidal activity.

rophage death. Based on the fact that cytotoxicity does not occur upon cytochalasin treatment which inhibits bacterial uptake by the macrophage, it is thought that *Shigella* internalization by the macrophage is required for bacterial-induced cell death. Subsequent studies provided a comprehensive model of *Shigella*'s ability to kill the macrophage. First, it was shown that escape from the vacuole also appears to be a requisite for *Shigella*-induced cell death, suggesting that effectors of the T3SS were involved in cell killing (Zychlinsky et al., 1994). By expressing the *E. coli* hemolysin to force the escape of *Shigella* strains deficient for Ipa proteins, it was shown that, as opposed to *ipaC* or *ipaD* mutants, an *ipaB* strain was unable to cause macrophage death, pointing at a major role for IpaB in this process (Zychlinsky et al., 1994). IpaB was then shown to bind to caspase 1 (Chen et al., 1996). IpaB binding to caspase 1 leads to two outcomes: it triggers the apoptotic process downstream of caspase1, leading to the macrophage programmed cell death, and it stimulates the cleavage by caspase 1 of pro-IL-1β and pro-IL-18 to allow the release of mature IL-1β and IL-18 in the extracellular milieu, where they act as potent proinflammatory mediators (Chen et al., 1996). These findings are further supported by experiments using the pulmonary infection model and mice genetically deficient for caspase 1. Casp $1^{-/-}$ mice show significant delayed inflammation upon infection of the lung, combined with a deficit in bacterial clearance. In contrast, casp $3^{-/-}$ or casp $11^{-/-}$ mice developed an inflammatory response that was similar to that of wild-type mice. Interestingly, caspase 1 deficiency appears to reflect the deficiencies in IL-1β and IL-18 that play seemingly complementary roles in controlling infection (Sansonetti et al., 2000). In IL-1β-deficient mice, a significant delay in the inflammatory response to *Shigella* infection is observed, although bacterial clearance appears unaffected. This argues for a role of IL-1β in boosting a rapid inflammation upon *Shigella* infection, but with little implications on bacterial killing. In contrast, IL-18-deficient mice show an even stronger inflammatory response to *Shigella* infection but bacteria persist longer, indicating that IL-18 is necessary for the control of the appropriate inflammatory response required for bacterial killing (Sansonetti et al., 2000). These findings provide an explanation for the massive inflammation in correlation with the massive macrophage cell death observed during acute shigellosis. They also illustrate the ambiguous facets of bacterial-induced inflammation: inflammation is required for efficient bacterial colonization of the colonic mucosa, presumably at the early step of the infectious process. However, as the infection proceeds and the response is aggravated, the intensity and probably the nature of the inflammatory response will also determine whether bacteria will eventually be cleared.

Although this scheme provides a straightforward explanation for the mechanism of IpaB-induced apoptosis, many issues need to be clarified. Strikingly, the mutagenesis of IpaB indicated that IpaB binding to caspase 1 is not sufficient for cytotoxicity, since a small deletion in the large hydrophobic region of IpaB resulted in a construction that could still bind to caspase 1, but could not induce cytotoxicity after microinjection of the purified protein (Guichon et al., 2001). Since this construction did not complement an *ipaB* mutant for invasion, it is possible that, in addition to IpaB binding to caspase 1, signals that are triggered during invasion also contribute to *Shigella*-induced apoptosis. Alternatively, it is possible that purified recombinant derivatives of IpaB do not reproduce IpaB function because they do not adopt the proper configuration. This is becoming particularly critical in light of recent developments suggesting that IpaB, as part of the T3SS-translocating machinery, may form oligomers that insert into host cell membranes to participate in the formation of a pore (Veenendaal et al., 2007). Also, it was observed that, upon macrophage infection by *Shigella*, secreted IpaB could be detected throughout the cell cytosol, a staining that is never observed in epithelial cells (Chen et al., 1996). Whether the macrophage cytosolic staining represents a soluble or a vesicular-associated form of IpaB bound to caspase 1 is unclear. These findings suggest that IpaB is subjected to different mechanistic constraints and follows a different fate according to the cell type. In addition to its role at the plasma membrane in translocation in epithelial cells, IpaB is injected in the cytosol of macrophages to induce cell death.

Other controversies concern the type of cell death induced by *Shigella* upon cell invasion. For example, other teams have found that *Shigella* infection led to necrosis of J774 cells or U937 cells that were subjected to all-*trans*-retinoic acid to allow their differentiation into macrophages, with a loss of membrane integrity and the nuclear fragmentation without condensation (Nonaka et al., 2003). In this case, necrosis was not accompanied with the activation of caspase 1, but osmoprotectants that prevented pore formation mediated by the T3SS could also prevent cell death. Caspase-1-independent necrosis linked to increased membrane permeability during macrophage infection by *Shigella* was speculated to occur through the targeting of mitochondria and a drop in mitochondria membrane potential (Koterski et al., 2005). These studies suggest that *Shigella* can promote the macrophage death by various pathways. Cell death could result from necrosis following pore formation mediated by the T3SS at the plasma membrane or, alternatively, from apoptosis following IpaB binding to caspase 1. Another caspase-1/IpaB-independent pathway leading to cell death by necrosis has been linked to the intracellular release of lipid A in the cytosol of *Shigella*-infected macrophages (Suzuki et al., 2005). Cell lysis in a process called oncosis or "accidental cell death" was also observed following *Shigella* infection of human macrophages derived from monocytes from peripheral blood (Fernandez-Prada et al., 1997). These studies were performed because it is generally admitted that microbial killing in mouse macrophages occurs primarily through the production of reactive oxygen species, while human macrophages are thought to generate a much weaker reactive oxygen response. Exposure to oxidative stress appears to be a relevant issue in macrophage since comparative transcriptional analysis indicates that *Shigella* upregulates genes involved in the thioredoxin pathway involved in the repair of components damaged by oxidation (Lucchini et al., 2005).

Since *Shigella*-induced apoptosis of human macrophages was also observed in other independent studies, the reasons for these paradoxical findings are unclear (Nonaka et al., 2003). It is possible that these inconsistencies are due to the type of macrophage used, or to variations in the conditions used for their preparation. These various types of *Shigella*-induced killing of macrophages may also reflect processes occurring at different stages of the inflammatory process, or in the midst of the inflammatory response, where macrophages exist under different stages of activation.

The mode of bacterial internalization could also account for the differences observed in the various studies, since *Shigella* internalization could result from the macrophage's phagocytic property or from the bacterial invasion process mediated by the T3SS. In work performed on J774 and RAW264.7 mouse macrophages, it was observed that invasive *Shigella* enters up to 10 times more efficiently, suggesting that invasion mediated by the T3SS is also an important feature to enhance cell death (Kuwae et al., 2001). *Shigella* also induces cell death upon internalization by dendritic cells, accompanied by a loss of membrane integrity. *Shigella*-induced death of dendritic cells appears to differ from that described in macrophages in that it results from caspase-1-dependent as well as caspase-1-independent processes. These conclusions were reached after observation that YVAD, an inhibitor of caspase 1, only partially inhibits *Shigella*-induced cytotoxicity of dendritic cells, under conditions where full inhibition of IL-1β release was obtained. Conversely, glycine, which was previously shown to inhibit cytotoxicity induced by the *Salmonella* orthologue SipB, partially inhibited cell death without affecting IL-1β release. The glycine-dependent pathway is sensitive to the pan-caspase inhibitor Z-VAD, suggesting that it implicates a yet unidentified cysteine protease (Edgeworth et al., 2002). The combination of the two inhibitors, however, led to full inhibition of dendritic cell death. Interestingly, as opposed to macrophages, *Shigella*-induced cytotoxicity in dendritic cells does not lead to significant amounts of IL-18 release. Since death of dendritic cells induced by *Shigella* occurs extremely rapidly in vitro (ca. 30 min), it is possible that this efficient removal of dendritic cells by *Shigella* plays an important role in shaping the immune response to the infection.

SHIGELLA MEETS THE NEUTROPHILS: WHOSE CATCH?

Neutrophils are key players in the control of infections of the intestinal mucosa (see Fig. 3). These front-line defense cells are equipped with an arsenal of bactericidal functions, including their ability to release in the extracellular milieu cationic antimicrobial peptides, proteases, and lactoferrin, which sequester iron and prevent bacterial growth (Mayer-Scholl et al., 2004). Upon phagocytosis, their abilities to produce high levels of reactive oxygen species and to fuse granular contents with the bacteria-containing phagosomes further contribute to bacterial killing. In addition to classical lysosomal proteins, granules also contain specific enzymes, such as elastase, which has been implicated in the control of bacterial infections. Release of granular contents by neutrophils can also be induced by activated PMNs during bacterial phagocytosis, or through activation by soluble products. Neutrophils were also shown to release granules in complex with extracellular chromatin. These extracellular structures, termed "NETs" for neutrophil extracellular traps, are presumed to allow the trapping of bacteria in places where antimicrobial substances such as granules or histones are concentrated, thereby optimizing their bactericidal function (Brinkmann et al., 2004). Altered peripheral neutrophil functions have been associated with inflammatory bowel diseases, such as Crohn's disease, because of their propensity to mediate or to amplify tissue injury upon activation. As for other bacterial infections, neutrophils play a major role in the development of the disease and in the severity of the symptoms linked to the infectious process. Acute mucosal inflammatory response to *Shigella* infection is linked to a massive infiltration of neutrophils in the colonic mucosa. Fatal complications linked to *Shigella* infections are often associated with vascular endothelial damage. These complications include the hemolytic-uremic syndrome, which consists of a combination of thrombocytopenia, hemolytic anemia, and renal failure, and intestinal perforations following toxic megacolon. During these complications, the high counts of neutrophils recruited at the sites of infection are largely responsible for tissue injury. This is mostly due to the release of toxic granular proteases by dying neutrophils. Under physiological conditions, neutrophils have a short life span and are removed from the organism through an apoptotic process. During acute inflammation such as that occurring during shigellosis, the extent of neutrophil lysis is likely to determine the extent of tissue injury.

Neutrophil Recruitment: the IL-1 and IL-18 Pathways

Following M-cell invasion, the release of IL-1 and IL-18 from *Shigella*-infected macrophages undergoing apoptosis triggers the recruitment of neutrophils at the site of infection (Sansonetti et al., 2000; McCormick, 2003). The key role of IL-1 in neutrophil recruitment and neutrophil-induced tissue damage is supported by experiments in the rabbit ileal loop model, where it was shown that treatment with IL-1Ra, which antagonizes the IL-1 receptor function, prevents neutrophil recruitment and destruction of the mucosa associated with *Shigella* infection in this model. The massive influx of activated neutrophils at the site of infection is held to be responsible for the destabilization of the colonic epithelium that allows *Shigella* access to the basolateral sides of enterocytes where bacteria can trigger invasion (Sansonetti et al., 1999). If neutrophils are the key in mounting the acute inflammatory reaction during shigellosis, other innate immune cells also contribute to bacterial clearance. These include NK cells, which produce interferon-γ as indicated by studies using the rag2γc-deficient mouse pulmonary model of shigellosis (Le-Barillec et al., 2005). Also, mucosal mast cells and eosinophils may contribute to efficient bacterial clearance, since these cells show delayed recruitment in pediatric patients who are particularly sensitive to *Shigella* infections (Raqib et al., 2003). Besides IL-1, bacterial invasion of enterocytes leading to the release of IL-8, bacterial LPS contributes to neutrophil recruitment and transmigration across the infected intestinal epithelium (Kohler et al., 2002).

Specific features indicate that *Shigella* may also upregulate signals that have been described to favor neutrophil transmigration. The IpaA protein that has been implicated in *Shigella* invasion was shown to act as a chemotactic agent favoring transmigration (Fernandez et al., 2001). These results are interesting because they raise the possibility that the T3S apparatus plays a role that is different from a microinjection device, since in the case of IpaA the T3S effector is thought to modulate cell function after its release in the extra milieu. There is in vitro and in vivo evidence that T3S effectors are released extracellularly. Many compounds in vitro, including serum albumin, were shown to induce T3S, indicating that the regulation of secretion is not restricted to contact of host cell membranes (Bahrani et al., 1997). The relevance of these findings is unclear for T3S for which intracellular targets have been identified, because effectors such as the translocator component IpaB and IpaC may not fold properly and may lose their function upon extracellular secretion. However, the fact that significant amounts of T3S effectors are found extracellularly in association with the bacterial surface in histological sections of infected rabbit ileal loops suggests that some of these effectors could play a role from an extracellular location (Wenneras et al., 2000). Since compounds that activate secretion, such as serum albumin, are normally not found in the intestinal lumen, it is possible that extracellular secretion of T3S effectors, and hence their putative extracellular function, occurs at later stages of the infectious process, after significant cell lysis and tissue destruction has occurred.

Besides the presence of bacterial virulence determinants, *Shigella* during the course of its evolution appears to have been engineered to facilitate the recruitment of neutrophils through deletions of so-called "antivirulence" genes. In contrast to most nonpathogenic *E. coli*, to which *Shigella* is related, *Shigella* lacks *cadA*, a gene encoding a lysine decarboxylase responsible for the production of cadaverine (Maurelli et al., 1998). Consistent with *cadA* interfering with virulence, introduction of the *E. coli cadA* gene in *Shigella* prevents its ability to cause lesions in rabbit ileal loops (McCormick et al., 1999). Cadaverine was further shown to prevent vacuolar escape of *Shigella* following invasion of intestinal epithelial cells, presumably because its polybasic nature may stabilize the phagosomal membrane or may interfere with the pH acidification required for IpaC-mediated vacuolar lysis. The inability of *Shigella* to access the epithelial cell cytosol leads to a decreased ability to induce IL-8 signaling and to recruit neutrophils. This signaling deficiency may be related to a decrease in ability to activate the Nod1 pathway through free cytoplasmic PG derivatives.

Following internalization by neutrophils, it was observed that *Shigella* do not lyse the phagocytic vacuole and are eventually killed (Weinrauch et al., 2002). Because *Shigella* escape the phagocytic vacuole of neutrophils from elastase-deficient mice, elastase stands out as a key factor controlling bactericidal escape from the vacuole. *Shigella*-soluble IpaA, B, and C proteins are sensitive to very low concentrations of neutrophil elastase, suggesting that these proteins are rapidly degraded following T3S (Weinrauch et al., 2002). Since IpaB and IpaC are required for escape from the vacuole, their rapid proteolysis by elastase provides an plausible explanation of the absence of vacuolar escape in neutrophils. In other studies, however, the inability of *Shigella* to escape from the vacuole did not appear to be required for its ability to induce the death of neutrophils by necrosis 2 h after bacterial infection (Francois et al., 2000). Since in these studies bacterial-induced necrosis required a functional T3S apparatus and, in particular, the IpaB and IpaC proteins, elastase may not fully restrict T3S in neutrophils. Necrosis leading to the release of proinflammatory mediators in the milieu would further amplify inflammation and participate in tissue destruction. Whether bacterial clearance by neutrophils or bacterial-induced neutrophil necrosis reflects outcomes that occur simultaneously during the infectious process is unclear. It is possible that, depending on their state of activation or on the surrounding environment (cytokines released, bacterial load, etc.), neutrophils may or may not succeed in overcoming *Shigella* infection. These situations may also reflect outcomes that are preferentially occurring during the early or later stages of the bacterial infectious process.

CONCLUDING REMARKS: UP- OR DOWNREGULATION OF IMMUNE RESPONSES

Understanding the sequence of events that lead to successful colonization of the colonic epithelium by *Shigella* represents a formidable task, for which bits of the puzzle are just being gathered. This task is rendered even more complex because many of these effectors are bound to act in concert to modulate the response, and because this response, as discussed above, may vary according to the cell type or to the inflammatory context. These issues, however, will probably be the major challenge in coming years, and should lead us to consider the relevance of in vitro-identified signaling during in vivo infections. In order to better define the role of the various tissues during infection, a legitimate concern would be to identify the cells that are being targeted in vivo by T3S effectors, and the subsequent signaling being induced. This will necessitate tracking the invading bacteria and their multiplication in tissues, while following the recruitment of incoming innate phagocytic cells and their impact on the mounting of inflammatory responses. There are obviously paradoxical findings concerning the bacterial strategy to induce inflammation to better colonize tissue, versus more recent discoveries pointing at the role of *Shigella* second-line T3S effectors as antiinflammatory factors. Because a hallmark of *Shigella* acute infection is the extensive inflammation that leads to the destruction of the colonic mucosa, researchers were prompted to look for virulence factors that favor proinflammatory signals. However, findings that some T3S effector *Shigella* mutants induce tissue destruction and inflammation that is even more acute than caused by the wild type raise the need to reconsider or refine our initial view of this bacterium's strategy. Obviously, the function of these second-line T3S effectors stresses the need for the bacterium to control the inflammatory responses during the infectious process. That, from in vitro experiments, *Shigella* shows a poor ability to breach the epithelial barrier from the intestinal luminal side, is consistent with other experimental data indicating that *Shigella* uses M cells as the main initial route of entry. The high populations of macrophages and dendritic cells present following crossing of the epithelium at the levels of M cells will lead to a very different inflammatory context than if crossing occurred elsewhere, after invasion of polarized intestinal epithelial cells.

It is certainly difficult to rule out completely the possibility that *Shigella* invasion occurs at low levels in enterocytes. If such discrete invasion events were to occur in vivo, it might be of interest for the invading bacteria to tone down inflammatory responses during the early stages of infection, to allow multiplication before being cleared by innate immune cells.

We thank past and present members of the PMM laboratory for their continuous help and support. P.S. is a Howard Hughes investigator.

REFERENCES

Alto, N. M., F. Shao, C. S. Lazar, R. L. Brost, G. Chua, S. Mattoo, S. A. McMahon, P. Ghosh, T. R. Hughes, C. Boone, and J.-E. Dixon. 2006. Identification of a bacterial type III effector family with G protein mimicry functions. *Cell* **124**:133–145.

Arbibe, L., D. W. Kim, E. Batsche, T. Pedron, B. Mateescu, C. Muchardt, C. Parsot, and P. J. Sansonetti. 2007. An injected bacterial effector targets chromatin access for transcription factor NF-kappaB to alter transcription of host genes involved in immune responses. *Nat. Immunol.* **8**:47–56.

Ashida, H., T. Toyotome, T. Nagai, and C. Sasakawa. 2007. *Shigella* chromosomal IpaH proteins are secreted via the type III secretion system and act as effectors. *Mol. Microbiol.* **63**:680–693.

Bahrani, F. K., P. J. Sansonetti, and C. Parsot. 1997. Secretion of Ipa proteins by *Shigella flexneri*: inducer molecules and kinetics of activation. *Infect. Immun.* **65:**4005–4010.

Blander, J. M., and R. Medzhitov. 2006. On regulation of phagosome maturation and antigen presentation. *Nat. Immunol.* **7:**1029–1035.

Blocker, A., P. Gounon, E. Larquet, K. Niebuhr, V. Cabiaux, C. Parsot, and P. Sansonetti. 1999. The tripartite type III secretion of *Shigella flexneri* inserts IpaB and IpaC into host membranes. *J. Cell Biol.* **147:**683–693.

Blocker, A., N. Jouihri, E. Larquet, P. Gounon, F. Ebel, C. Parsot, P. Sansonetti, and A. Allaoui. 2001. Structure and composition of the *Shigella flexneri* "needle complex," a part of its type III secretion. *Mol. Microbiol.* **39:**652–663.

Bougnères, L., S. E. Girardin, S. A. Weed, A. V. Karginov, J. C. Olivo-Marin, J. T. Parsons, P. J. Sansonetti, and G. Tran Van Nhieu. 2004. Cortactin and Crk cooperate to trigger actin polymerization during *Shigella* invasion of epithelial cells. *J. Cell Biol.* **166:**225–235.

Bourdet-Sicard, R., M. Rudiger, B. M. Jockusch, P. Gounon, P. J. Sansonett G. Tran Van Nhieu. 1999. Binding of the *Shigella* protein IpaA to vinculin induces F-actin depolymerization. *EMBO J.* **18:**5853–5862.

Brinkmann, V., U. Reichard, C. Goosmann, B. Fauler, Y. Uhlemann, D. S. Weiss, Y. Weinrauch, and A. Zychlinsky. 2004. Neutrophil extracellular traps kill bacteria. *Science* **303:**1532–1535.

Buchrieser, C., P. Glaser, C. Rusniok, H. Nedjari, H. D'Hauteville, F. Kunst, P. Sansonetti, and C. Parsot. 2000. The virulence plasmid pWR100 and the repertoire of proteins secreted by the type III secretion apparatus of *Shigella flexneri*. *Mol. Microbiol.* **38:**760–771.

Burton, E. A., R. Plattner, and A. M. Pendergast. 2003. Abl tyrosine kinases are required for infection by *Shigella flexneri*. *EMBO J.* **22:**5471–5479.

Chen, Y., M. R. Smith, K. Thirumalai, and A. Zychlinsky. 1996. A bacterial invasin induces macrophage apoptosis by binding directly to ICE. *EMBO J.* **15:**3853–3860.

Cornelis, G. R. 2006. The type III secretion injectisome. *Nat. Rev. Microbiol.* **4:**811–825.

Cossart, P., and P. J. Sansonetti. 2004. Bacterial invasion: the paradigms of enteroinvasive pathogens. *Science* **304:**242–248.

Dehio, C., M. C. Prevost, and P. J. Sansonetti. 1995. Invasion of epithelial cells by *Shigella flexneri* induces tyrosine phosphorylation of cortactin by a pp60c-src-mediated signalling pathway. *EMBO J.* **14:**2471–2482.

Demali, K. A., A. L. Jue, and K. Burridge. 2006. IpaA targets beta1 integrins and rho to promote actin cytoskeleton rearrangements necessary for *Shigella* entry. *J. Biol. Chem.* **281:**39534–39541.

Dorman, C. J., and M. E. Porter. 1998. The *Shigella* virulence gene regulatory cascade: a paradigm of bacterial gene control mechanisms. *Mol. Microbiol.* **29:**677–684.

Dumenil, G., J. C. Olivo, S. Pellegrini, M. Fellous, P. J. Sansonetti, and G. Tran Van Nhieu. 1998. Interferon alpha inhibits a Src-mediated pathway necessary for *Shigella*-induced cytoskeletal rearrangements in epithelial cells. *J. Cell Biol.* **143:**1003–1012.

Dumenil, G., P. Sansonetti, and G. Tran Van Nhieu. 2000. Src tyrosine kinase activity down-regulates Rho-dependent responses during *Shigella* entry into epithelial cells and stress fibre formation. *J. Cell Sci.* **113**(Pt 1)**:**71–80.

Edgeworth, J. D., J. Spencer, A. Phalipon, G. E. Griffin, and P. J. Sansonetti. 2002. Cytotoxicity and interleukin-1beta processing following *Shigella flexneri* infection of human monocyte-derived dendritic cells. *Eur. J. Immunol.* **32:**1464–1471.

Egile, C., I. Rouiller, X. P. Xu, N. Volkmann, R. Li, and D. Hanein. 2005. Mechanism of filament nucleation and branch stability revealed by the structure of the Arp2/3 complex at actin branch junctions. *PLoS Biol.* **3:**e383.

Enninga, J., J. Mounier, P. Sansonetti, and G. Tran Van Nhieu. 2005. Secretion of type III effectors into host cells in real time. *Nat. Methods* **2:**959–965.

Fernandez, I. M., M. Silva, R. Schuch, W. A. Walker, A. M. Siber, A. T. Maurelli, and B. A. McCormick. 2001. Cadaverine prevents the escape of *Shigella flexneri* from the phagolysosome: a connection between bacterial dissemination and neutrophil transepithelial signaling. *J. Infect. Dis.* **184:**743–753.

Fernandez-Prada, C. M., D. L. Hoover, B. D. Tall, and M. M. Venkatesan. 1997. Human monocyte-derived macrophages infected with virulent *Shigella flexneri* in vitro undergo a rapid cytolytic event similar to oncosis but not apoptosis. *Infect. Immun.* **65:**1486–1496.

Francois, M., V. Le Cabec, M. A. Dupont, P. J. Sansonetti, and I. Maridonneau-Parini. 2000. Induction of necrosis in human neutrophils by *Shigella flexneri* requires type III secretion, IpaB and IpaC invasions, and actin polymerization. *Infect. Immun.* **68:**1289–1296.

Fritz, J. H., R. L. Ferrero, D. J. Philpott, and S. E. Girardin. 2006. Nod-like proteins in immunity, inflammation and disease. *Nat. Immunol.* **7:**1250–127.

Galan, J. E., and H. Wolf-Watz. 2006. Protein delivery into eukaryotic cells by type III secretion machines. *Nature* **444:**567–573.

Gewirtz, A. T., A. S. Rao, P. O. Simon, Jr., D. Merlin, D. Carnes, J. L. Madara, and A. S. Neish. 2000. *Salmonella typhimurium* induces epithelial IL-8 expression via Ca(2+)-mediated activation of the NF-kappaB pathway. *J. Clin. Invest.* **105:**79–92.

Girardin, S. E., and D. J. Philpott. 2004. Mini-review: the role of peptidoglycan recognition in innate immunity. *Eur. J. Immunol.* **34:**1777–1782.

Guichon, A., D. Hersh, M. R. Smith, and A. Zychlinsky. 2001. Structure-function analysis of the *Shigella* virulence factor IpaB. *J. Bacteriol.* **183:**1269–1276.

Handa, Y., M. Suzuki, K. Ohya, H. Iwai, N. Ishijima, A. J. Koleske, Y. Fukui, and C. Sasakawa. 2007. *Shigella* IpgB1 promotes bacterial entry through the ELMO-Dock180 machinery. *Nat. Cell Biol.* **9:**121–128.

Izard, T., G. Tran Van Nhieu, and P. R. Bois. 2006. *Shigella* applies molecular mimicry to subvert vinculin and invade host cells. *J. Cell Biol.* **175:**465–475.

Jin, Q., Z. Yuan, J. Xu, Y. Wang, Y. Shen, W. Lu, J. Wang, H. Liu, J. Yang, F. Yang, X. Zhang, J. Zhang, G. Yang, H. Wu, D. Qu, J. Dong, L. Sun, Y. Xue, A. Zhao, Y. Gao, J. Zhu, B. Kan, K. Ding, S. Chen, H. Cheng, Z. Yao, B. He, R. Chen, D. Ma, B. Qiang, Y. Wen, Y. Hou, and J. Yu. 2002. Genome sequence of *Shigella flexneri* 2a: insights into pathogenicity through comparison with genomes of Escherichia coli K-12 and O157. *Nucleic Acids Res.* **30:**4432–4441.

Johnson, S., P. Roversi, M. Espina, A. Olive, J. E. Deane, S. Birket, T. Field, W. D. Picking, A. J. Blocker, E. E. Galyov, W. L. Picking, and S. M. Lea. 2007. Self-chaperoning of the type III secretion system needle tip proteins IpaD and BipD. *J. Biol. Chem.* **282:**4035–4044.

Kim, D. W., G. Lenzen, A. L. Page, P. Legrain, P. J. Sansonetti, and C. Parsot. 2005. The *Shigella flexneri* effector OspG interferes with innate immune responses by targeting ubiquitin-conjugating enzymes. *Proc. Natl. Acad. Sci. USA* **102:**14046–14051.

Kohler, H., S. P. Rodrigues, and B. A. McCormick. 2002. *Shigella flexneri* interactions with the basolateral membrane domain of polarized model intestinal epithelium: role of lipopolysaccharide in cell invasion and in activation of the mitogen-activated protein kinase ERK. *Infect. Immun.* **70:** 1150–1158.

Koterski, J. F., M. Nahvi, M. M. Venkatesan, and B. Haimovich. 2005. Virulent *Shigella flexneri* causes damage to mitochondria and triggers necrosis in infected human monocyte-derived macrophages. *Infect. Immun.* **73:**504–513.

Kubori, T., A. Sukhan, S. I. Aizawa, and J. E. Galan. 2000. Molecular characterization and assembly of the needle complex of the *Salmonella typhimurium* type III protein secretion system. *Proc. Natl. Acad. Sci. USA* **97:**10225–10230.

Kueltzo, L. A., J. Osiecki, J. Barker, W. L. Picking, B. Ersoy, W. D. Picking, and C. R. Middaugh. 2003. Structure-function analysis of invasion plasmid antigen C (IpaC) from *Shigella flexneri*. *J. Biol. Chem.* **278:**2792–2798.

Kuwae, A., S. Yoshida, K. Tamano, H. Mimuro, T. Suzuki, and C. Sasakawa. 2001. *Shigella* invasion of macrophage requires the insertion of IpaC into the host plasma membrane. Functional analysis of IpaC. *J. Biol. Chem.* **276:**32230–32239.

Lafont, F., G. Tran Van Nhieu, K. Hanada, P. Sansonetti, and F. G. van der Goot. 2002. Initial steps of *Shigella* infection depend on the cholesterol/sphingolipid raft-mediated CD44-IpaB interaction. *EMBO J.* **21:**4449–4457.

Le-Barillec, K., J. G. Magalhaes, E. Corcuff, A. Thuizat, P. J. Sansonetti, A. Phalipon, and J. P. Di Santo. 2005. Roles for T and NK cells in the innate immune response to *Shigella flexneri*. *J. Immunol.* **175:**1735–1740.

Lecine, P., S. Esmiol, J. Y. Metais, C. Nicoletti, C. Nourry, C. McDonald, G. Nunez, J. P. Hugot, J. P. Borg, and V. Ollendorff. 2007. The NOD2-RICK complex signals from the plasma membrane. *J. Biol. Chem.* **282:**15197–15207.

Li, H., H. Xu, Y. Zhou, J. Zhang, C. Long, S. Li, S. Chen, J. M. Zhou, and F. Shao. 2007. The phosphothreonine lyase activity of a bacterial type III effector family. *Science* **315:** 1000–1003.

Lucchini, S., H. Liu, Q. Jin, J. C. Hinton, and J. Yu. 2005. Transcriptional adaptation of *Shigella flexneri* during infection of macrophages and epithelial cells: insights into the strategies of a cytosolic bacterial pathogen. *Infect. Immun.* **73:**88–102.

Mallett, C. P., L. VanDeVerg, H. H. Collins, and T. L. Hale. 1993. Evaluation of *Shigella* vaccine safety and efficacy in an intranasally challenged mouse model. *Vaccine* **11:**190–196.

Mariathasan, S., and D. M. Monack. 2007. Inflammasome adaptors and sensors: intracellular regulators of infection and inflammation. *Nat. Rev. Immunol.* **7:**31–40.

Maurelli, A. T., R. E. Fernandez, C. A. Bloch, C. K. Rode, and A. Fasano. 1998. "Black holes" and bacterial pathogenicity: a large genomic deletion that enhances the virulence of *Shigella* spp. and enteroinvasive *Escherichia coli*. *Proc. Natl. Acad. Sci. USA* **95:**3943–3948.

Mavris, M., A. L. Page, R. Tournebize, B. Demers, P. Sansonetti, and C. Parsot. 2002. Regulation of transcription by the activity of the *Shigella flexneri* type III secretion apparatus. *Mol. Microbiol.* **43:**1543–1553.

Mayer-Scholl, A., P. Averhoff, and A. Zychlinsky. 2004. How do neutrophils and pathogens interact? *Curr. Opin. Microbiol.* **7:**62–66.

McCormick, B. A. 2003. The use of transepithelial models to examine host-pathogen interactions. *Curr. Opin. Microbiol.* **6:**77–81.

McCormick, B. A., M. I. Fernandez, A. M. Siber, and A. T. Maurelli. 1999. Inhibition of *Shigella flexneri*-induced transepithelial migration of polymorphonuclear leucocytes by cadaverine. *Cell Microbiol.* **1:**143–155.

Mounier, J., V. Laurent, A. Hall, P. Fort, M. F. Carlier, P. J. Sansonetti, and C. Egile. 1999. Rho family GTPases control entry of *Shigella flexneri* into epithelial cells but not intracellular motility. *J. Cell Sci.* **112**(Pt 13)**:**2069–2080.

Mounier, J., T. Vasselon, R. Hellio, M. Lesourd, and P. J. Sansonetti. 1992. *Shigella flexneri* enters human colonic Caco-2 epithelial cells through the basolateral pole. *Infect. Immun.* **60:**237–248.

Nie, H., F. Yang, X. Zhang, J. Yang, L. Chen, J. Wang, Z. Xiong, J. Peng, L. Sun, J. Dong, Y. Xue, X. Xu, S. Chen, Z. Yao, Y. Shen, and Q. Jin. 2006. Complete genome sequence of *Shigella flexneri* 5b and comparison with *Shigella flexneri* 2a. *BMC Genomics* **7:**173.

Niebuhr, K., S. Giuriato, T. Pedron, D. J. Philpott, F. Gaits, J. Sable, M. P. Sheetz, C. Parsot, P. J. Sansonetti, and B. Payrastre. 2002. Conversion of PtdIns(4,5)P(2) into PtdIns(5)P by the *S. flexneri* effector IpgD reorganizes host cell morphology. *EMBO J.* **21:**5069–5078.

Nonaka, T., T. Kuwabara, H. Mimuro, A. Kuwae, and S. Imajoh-Ohmi. 2003. *Shigella*-induced necrosis and apoptosis of U937 cells and J774 macrophages. *Microbiology* **149:**2513–2527.

Ogawa, M., T. Yoshimori, T. Suzuki, H. Sagara, N. Mizushima, and C. Sasakawa. 2005. Escape of intracellular *Shigella* from autophagy. *Science* **307:**727–731.

Ohya, K., Y. Handa, M. Ogawa, M. Suzuki, and C. Sasakawa. 2005. IpgB1 is a novel *Shigella* effector protein involved in bacterial invasion of host cells. Its activity to promote membrane ruffling via Rac1 and Cdc42 activation. *J. Biol. Chem.* **280:**24022–24034.

Parsot, C. 2005. *Shigella* spp. and enteroinvasive *Escherichia coli* pathogenicity factors. *FEMS Microbiol. Lett.* **252:**11–18.

Parsot, C., R. Menard, P. Gounon, and P. J. Sansonetti. 1995. Enhanced secretion through the *Shigella flexneri* Mxi-Spa translocon leads to assembly of extracellular proteins into macromolecular structures. *Mol. Microbiol.* **16:**291–300.

Patel, J. C., and J. E. Galan. 2006. Differential activation and function of Rho GTPases during *Salmonella*-host cell interactions. *J. Cell Biol.* **175:**453–463.

Pendaries, C., H. Tronchere, L. Arbibe, J. Mounier, O. Gozani, L. Cantley, M. J. Fry, F. Gaits-Iacovoni, P. J. Sansonetti, and B. Payrastre. 2006. PtdIns5P activates the host cell PI3-kinase/Akt pathway during *Shigella flexneri* infection. *EMBO J.* **25:**1024–1034.

Ramarao, N., C. Le Clainche, T. Izard, R. Bourdet-Sicard, E. Ageron, P. J. Sansonetti, M. F. Carlier, and G. Tran Van Nhieu. 2007. Capping of actin filaments by vinculin activated by the *Shigella* IpaA carboxyl-terminal domain. *FEBS Lett.* **581:**853–857.

Raqib, R., P. K. Moly, P. Sarker, F. Qadri, N. H. Alam, M. Mathan, and J. Andersson. 2003. Persistence of mucosal mast cells and eosinophils in *Shigella*-infected children. *Infect. Immun.* **71:**2684–2692.

Raqib, R., P. Sarker, P. Bergman, G. Ara, M. Lindh, D. A. Sack, K. M. Nasirul Islam, G. H. Gudmundsson, J. Andersson, and B. Agerberth. 2006. Improved outcome in shigellosis associated with butyrate induction of an endogenous peptide antibiotic. *Proc. Natl. Acad. Sci. USA* **103:**9178–9183.

Rohde, R., A. Breikreutz, A. Chenal, P. Sansonetti, and C. Parsot. 2007. Type III secretion effectors of the IpaH family are E3 ubiquitin ligases. *Cell Host Microbes* **1:**77–83.

Sani, M., A. Botteaux, C. Parsot, P. Sansonetti, E. J. Boekema, and A. Allaoui. 2007. IpaD is localized at the tip of the *Shigella flexneri* type III secretion apparatus. *Biochim. Biophys. Acta* **1770:**307–311.

Sansonetti, P. J., T. L. Hale, G. J. Dammin, C. Kapfer, H. H. Collins, Jr., and S. B. Formal. 1983. Alterations in the pathogenicity of *Escherichia coli* K-12 after transfer of plasmid and chromosomal genes from *Shigella flexneri*. *Infect. Immun.* **39:** 1392–1402.

Sansonetti, P. J., and A. Phalipon. 1999. M cells as ports of entry for enteroinvasive pathogens: mechanisms of interaction, consequences for the disease process. *Semin. Immunol.* **11:**193–203.

Sansonetti, P. J., A. Phalipon, J. Arondel, K. Thirumalai, S. Banerjee, S. Akira, K. Takeda, and A. Zychlinsky. 2000. Caspase-1 activation of IL-1beta and IL-18 are essential for *Shigella flexneri*-induced inflammation. *Immunity* **12:**581–590.

Sansonetti, P. J., G. Tran Van Nhieu, and C. Egile. 1999. Rupture of the intestinal epithelial barrier and mucosal invasion by *Shigella flexneri*. *Clin. Infect. Dis.* **28:**466–475.

Schroeder, G. N., and H. Hilbi. 2007. Cholesterol is required to trigger caspase-1 activation and macrophage apoptosis after phagosomal escape of *Shigella*. *Cell. Microbiol.* **9:**265–278.

Sereny, B., C. Tenner, and P. Racz. 1971. Immunogenicity of living attenuated shigellae. *Acta Microbiol. Acad. Sci. Hung.* **18:**239–245.

Skoudy, A., J. Mounier, A. Aruffo, H. Ohayon, P. Gounon, P. Sansonetti, and G. Tran Van Nhieu. 2000. CD44 binds to the *Shigella* IpaB protein and participates in bacterial invasion of epithelial cells. *Cell. Microbiol.* **2:**19–33.

Sutterwala, F. S., Y. Ogura, and R. A. Flavell. 2007. The inflammasome in pathogen recognition and inflammation. *J. Leukoc. Biol.* **82:**259–264.

Suzuki, T., K. Nakanishi, H. Tsutsui, H. Iwai, S. Akira, N. Inohara, M. Chamaillard, G. Nunez, and C. Sasakawa. 2005. A novel caspase-1/toll-like receptor 4-independent pathway of cell death induced by cytosolic *Shigella* in infected macrophages. *J. Biol. Chem.* **280:**14042–14050.

Tehrani, S., N. Tomasevic, S. Weed, R. Sakowicz, and J. A. Cooper. 2007. Src phosphorylation of cortactin enhances actin assembly. *Proc. Natl. Acad. Sci. USA* **104:**11933–11938.

Tran Van Nhieu, G., E. Caron, A. Hall, and P. J. Sansonetti. 1999. IpaC induces actin polymerization and filopodia formation during *Shigella* entry into epithelial cells. *EMBO J.* **18:**3249–3262.

Tran Van Nhieu, G., C. Clair, R. Bruzzone, M. Mesnil, P. Sansonetti, and L. Combettes. 2003. Connexin-dependent inter-cellular communication increases invasion and dissemination of *Shigella* in epithelial cells. *Nat. Cell Biol.* **5:**720–726.

Tran Van Nhieu, G., J. Enninga, P. Sansonetti, and G. Grompone. 2005. Tyrosine kinase signaling and type III effectors orchestrating *Shigella* invasion. *Curr. Opin. Microbiol.* **8:**16–20.

Uematsu, S., and S. Akira. 2006. Innate immunity and toll-like receptor. *Nippon Naika Gakkai Zasshi* **95:**1115–1121.

Ulevitch, R. J., J. C. Mathison, and J. da Silva Correia. 2004. Innate immune responses during infection. *Vaccine* **22**(Suppl 1)**:**S25–S30.

Veenendaal, A. K., J. L. Hodgkinson, L. Schwarzer, D. Stabat, S. F. Zenk, and A. J. Blocker. 2007. The type III secretion system needle tip complex mediates host cell sensing and translocon insertion. *Mol. Microbiol.* **63:**1719–1730.

Weaver, A. M., A. V. Karginov, A. W. Kinley, S. A. Weed, Y. Li, J. T. Parsons, and J. A. Cooper. 2001. Cortactin promotes and stabilizes Arp2/3-induced actin filament network formation. *Curr. Biol.* **11:**370–374.

Weaver, A. M., M. E. Young, W. L. Lee, and J. A. Cooper. 2003. Integration of signals to the Arp2/3 complex. *Curr. Opin. Cell Biol.* **15:**23–30.

Wei, J., M. B. Goldberg, V. Burland, M. M. Venkatesan, W. Deng, G. Fournier, G. F. Mayhew, G. Plunkett, III, D. J. Rose, A. Darling, B. Mau, N. T. Perna, S. M. Payne, L. J. Runyen-Janecky, S. Zhou, D. C. Schwartz, and F. R. Blattner. 2003. Complete genome sequence and comparative genomics of *Shigella flexneri* serotype 2a strain 2457T. *Infect. Immun.* **71:**2775–2786.

Weinrauch, Y., D. Drujan, S. D. Shapiro, J. Weiss, and A. Zychlinsky. 2002. Neutrophil elastase targets virulence factors of enterobacteria. *Nature* **417:**91–94.

Wenneras, C., P. Ave, M. Huerre, J. Arondel, R. J. Ulevitch, J. C. Mathison, and P. Sansonetti. 2000. Blockade of CD14 increases *Shigella*-mediated invasion and tissue destruction. *J. Immunol.* **164:**3214–3221.

Yoshida, S., Y. Handa, T. Suzuki, M. Ogawa, M. Suzuki, A. Tamai, E. Abe, E. Katayama, and C. Sasakawa. 2006. Microtubule-severing activity of *Shigella* is pivotal for intercellular spreading. *Science* **314:**985–989.

Yoshida, S., E. Katayama, A. Kuwae, H. Mimuro, T. Suzuki, and C. Sasakawa. 2002. *Shigella* delivers an effector protein to trigger host microtubule destabilization, which promotes Rac1 activity and efficient bacterial internalization. *EMBO J.* **21:**2923–2935.

Zasloff, M. 2006. Inducing endogenous antimicrobial peptides to battle infections. *Proc. Natl. Acad. Sci. USA* **103:**8913–8914.

Zhang, Z., L. Jin, G. Champion, K. B. Seydel, and S. L. Stanley, Jr. 2001. *Shigella* infection in a SCID mouse-human intestinal xenograft model: role for neutrophils in containing bacterial dissemination in human intestine. *Infect. Immun.* **69:**3240–3247.

Zychlinsky, A., B. Kenny, R. Menard, M. C. Prevost, I. B. Holland, and P. J. Sansonetti. 1994. IpaB mediates macrophage apoptosis induced by *Shigella flexneri*. *Mol. Microbiol.* **11:**619–627.

Zychlinsky, A., K. Thirumalai, J. Arondel, J. R. Cantey, A. O. Aliprantis, and P. J. Sansonetti. 1996. In vivo apoptosis in *Shigella flexneri* infections. *Infect. Immun.* **64:**5357–5365.

26

Autophagy: a Fundamental Cytoplasmic Sanitation Process Operational in All Cell Types Including Macrophages

VOJO DERETIC

Autophagic degradation is a cell-autonomous mechanism for direct elimination of intracellular microbes, but also plays a broader role in innate and adaptive immunity and in general cytoplasmic homeostasis. Autophagy is a highly conserved, specialized intracellular degradation pathway operational in all eukaryotic cells. Autophagy was initially recognized as a ubiquitous biological process of cleaning the cell's interior, whereby portions of the cytoplasm, including large objects such as organelles, get sequestered by autophagic membranes for delivery to lysosomes and degradation of the captured contents. The process of autophagy plays many functions, including feeding the cells under starvation conditions or upon withdrawal of growth factors. Furthermore, damaged or surplus organelles are removed by autophagy. Autophagy has been implicated in cancer, neurodegeneration, development, aging, and innate and adaptive immunity. Autophagy plays both housekeeping and immune functions in macrophages and dendritic cells. In one model, phagocytosis, a property usually reserved for phagocytic cells such as macrophages, can be considered as a special case of autophagy. In this model unifying phagosomes and autophagosomes, a cell's exterior is topologically equivalent to an intracellular organelle, and the phagosome can be considered as a special autophagosome of the exterior space delimited by the plasma membrane. Unlike phagocytosis, which particularly endows cells of the reticuloendothelial system, autophagy is more egalitarian and allows all cells, including neurons, to clear up their interiors, be it by removing a damaged organelle such as a leaky mitochondrion lest cells undergo unscheduled apoptosis, a toxic protein aggregate, or a pathogen.

TYPES OF AUTOPHAGY

The word "autophagy" ("self-ingestion") is a catchall term used to describe several related but distinct intracellular membrane-trafficking and protein-sorting processes. These include, among others, macroautophagy (bulk cytoplasmic autophagy), microautophagy (a specialized process often seen in yeast [*Saccharomyces cerevisiae*] that can target various organelles, including its manifestation as piecemeal microautophagy of the nucleus), pexophagy (autophagy of peroxisomes), mitophagy (autophagy of mitochondria), and chaperone-mediated autophagy (CMA), which imports individual cytoplasmic proteins directly into the lysosomes for degradation. In addition to these forms of autophagy, the recent progress in uncovering new roles of autophagy has introduced additional terms, such as xenophagy (removal of foreign microbial intruders into host cell cytoplasm), immunophagy (autophagy supporting innate and adaptive immunity processes), and virophagy (autophagy related to viral infections). Unless otherwise specified, autophagy is most commonly used as a term describing specifically the process of macroautophagy.

EXECUTION STAGES OF AUTOPHAGY

Following induction, autophagy sequesters a damaged organelle or a portion of cell's cytosol into a nascent autophagosomal structure called the isolation membrane or phagophore (Fig. 1). The phagophore is enlarged by the addition of a new membrane that is of undefined origin but is suspected to come from the endoplasmic reticulum or a combination of sources including Golgi and endosomes. The isolation membrane eventually seals to form an autophagosome. An autophagosome is distinguished from the conventional phagosome by the presence of a double delimiting membrane (two lipid bilayers) and intraluminal cytosolic content including membranes originating from the captured organelles. A subsequent fusion of autophagosomes with lysosomes, referred to as flux or maturation, results in a degradative compartment of the autolysosome.

Most cells undergo baseline autophagy, removing protein aggregates and spuriously damaged mitochondria or other organelles, or adjusting the cellular biomass. Auto-

---

Vojo Deretic, Department of Molecular Genetics and Microbiology, University of New Mexico School of Medicine, Albuquerque, NM 87131.

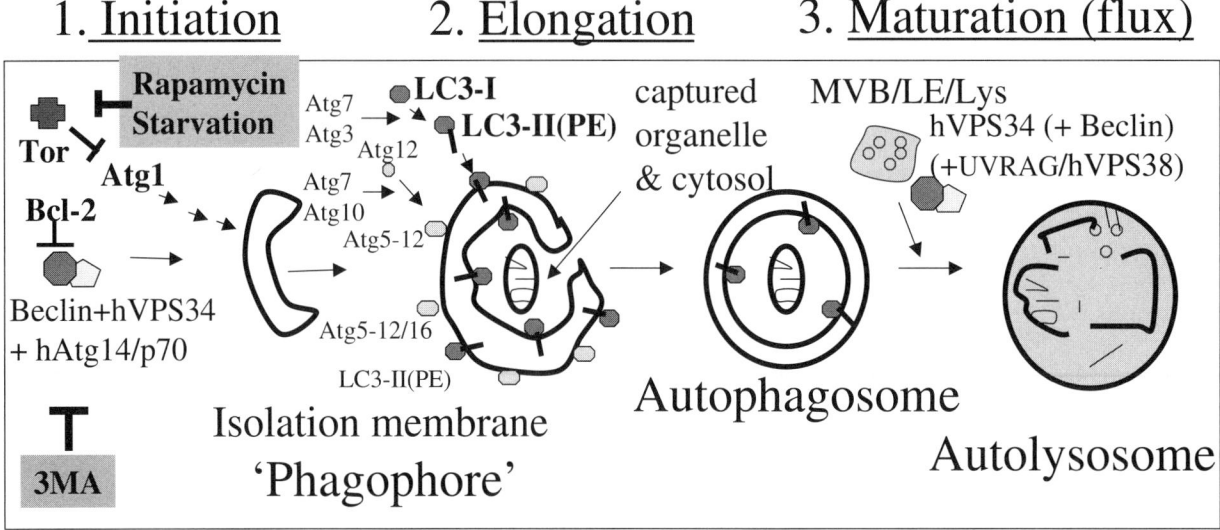

FIGURE 1 Execution stages of the autophagic pathway. See text for explanations. Reproduced with permission from Deretic, 2005.

phagy is regulated by a core signaling pathway (Fig. 2; activators in light gray, inhibitors in dark gray), with two marquee players: (i) Tor (target of rapamycin), a cellular biomass sentinel that switches between biomass increase and decrease, with the latter involving autophagy; and (ii) phosphatidylinositol 3-kinase (PI3K), both type I and type III, acting at multiple stages along the signaling pathway. The classical physiological inducers of autophagy are amino acid starvation or absence of growth factors (Lum et al., 2005). A pharmacological induction of autophagy can be achieved by using rapamycin, a drug that inhibits the Ser/Thr kinase Tor (Wullschleger et al., 2006). In contrast to type I PI3K, which generates phosphatidylinositol(3,4,5)P3 (PIP3) on plasma membrane and inhibits autophagy through Tor activation, hVPS34 generates phosphatidylinositol-3-phosphate (PI3P) on endomembranes and acts at several steps along the autophagy signaling and execution cascade. This is in contrast to its multiple positive effects during the execution stages of autophagy, when hVPS34 acts in complex with Beclin 1 (yeast Atg6), an autophagy-inducing tumor suppressor required specifically for the role of hVPS34 in autophagy (Pattingre et al., 2005). Beclin 1 is not needed for signaling processes controlled by hVPS34, i.e., the upstream Tor signaling cascade, and in the regular endosomal/lysosomal pathway (Furuya et al., 2005; Zeng et al., 2006). Recent studies have identified Bcl-2 (Pattingre et al., 2005) and UVRAG (Liang et al., 2006), two Beclin 1-interacting partners, as an inhibitor and an activator of autophagy, respectively. Lower levels of Beclin 1 are associated with lower levels of autophagy, compatible with cell survival, while higher levels of Beclin 1 induce too much autophagy, which leads to cell death (Lavieu et al., 2006; Scarlatti et al., 2004). In contrast to PI3K inhibitory effects on signaling (Fig. 2) leading to autophagy, the execution stages of autophagy (Fig. 1) depend on PI3P production on endomembranes, and thus require the type III PI3K hVPS34. The hVPS34 lipid kinase is the target for the well-known pharmacological inhibitor of autophagy, 3-methyl adenine (3-MA).

The most detailed picture on autophagy protein (Atg) function is available in the yeast system. However, only about a dozen yeast Atg factors have their orthologs in mammals, emphasizing both the conservation of the basic apparatus and a divergence in its use by metazoans (Fig. 1). Inhibition of Tor activity leads to the initiation of autophagy via Atg1. Overexpression of *Drosophila* Atg1 is sufficient to induce autophagy (Scott et al., 2007), thus confirming its pivotal role. Recent studies examining effects on the redistribution of a sole integral membrane Atg protein, Atg9, have implicated ULK1 as the putative mammalian Atg1 ortholog (Young et al., 2006). Autophagosome enlargement and wrapping around its target in the elongation stage is facilitated by the two specialized protein conjugation systems (Fig. 1). The Atg5-12/16 complex stimulates a second conjugation system whereby Atg8 (also known as LC3) undergoes conversion from its free C-terminus state (LC3-I) to its C-terminally lipidated form (LC3-II) covalently modified by phosphatidylethanolamine (Fig. 1). The lipidated LC3-II localizes to the membrane on both sides of a growing phagophore (Fig. 1). On autophagosome closure, sealing the typical double-membrane organelle, Atg5-12/16 (and LC3, now delipidated by Atg4) dissociate from the outer autophagosomal membrane and get recycled. The LC3 associated with the luminal membrane remains trapped in the autophagosome and is degraded during maturation, which involves fusion with late endosomal and lysosomal organelles (Fig. 1). LC3 lipidation and LC3-II association with autophagosomal membranes provide the basis for two frequently employed assays of autophagy: (i) autophagosome formation can be scored by immunofluorescence microscopy as a transition of LC3 from its diffuse cytosolic appearance to a membrane-associated, punctate intracellular distribution; and (ii) the LC3-I to L3-II conversion can be monitored by immunoblotting. The yeast Atg8 has multiple paralogs in mammals termed LC3A (with two splice isoforms, a and b, in humans), LC3B, GABARAP, GABARAPL1, and GABARAPL2 (GATE-16), all encoded on different chromosomes, with the exception of LC3B and GABARAPL2, which are linked on chromosome 16 in humans and chromosome 8 in the mouse. The LC3 referred to in publications often corresponds to LC3B.

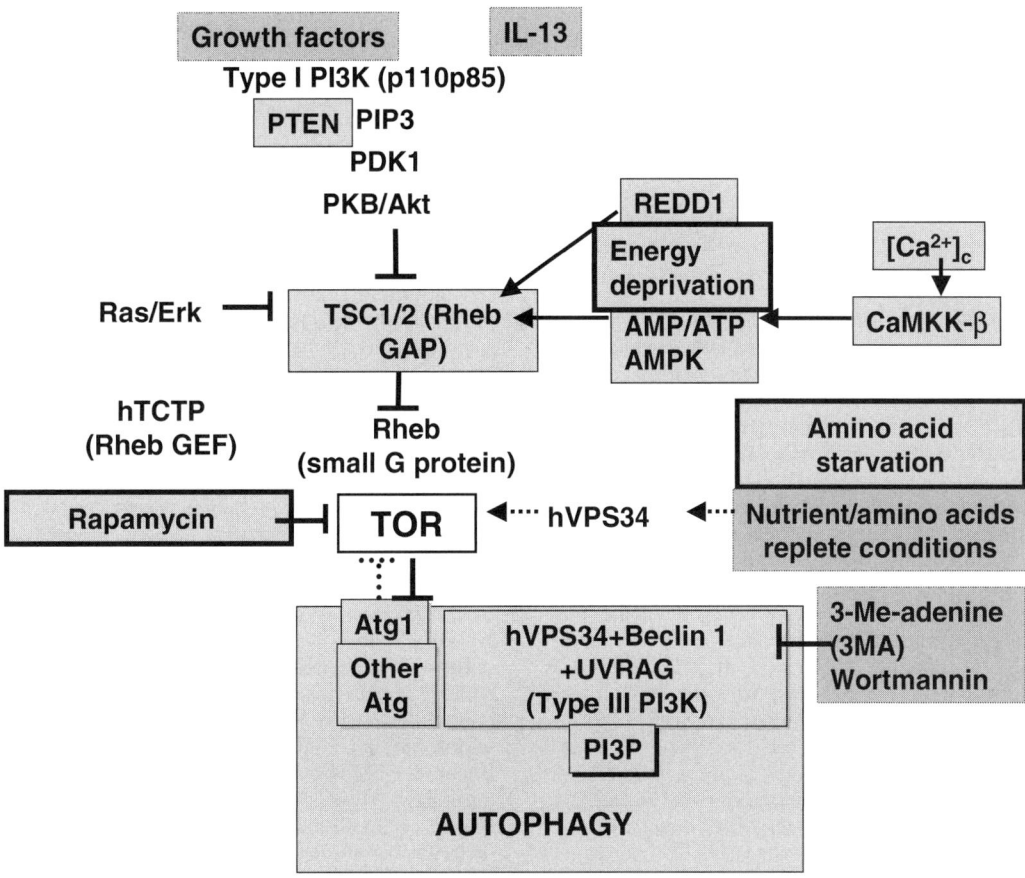

FIGURE 2 Core signaling pathways regulating autophagy. Light gray boxes, agonists of autophagy (positive regulators); dark gray boxes, antagonists of autophagy (negative regulators). See text for description and explanations.

Recent work has connected LC3 and another protein, p62 (sequestosome 1), which recognizes polyubiquitinated protein aggregates too big to be disposed of by proteasome and delivers them for degradation to autophagosomes (Bjorkoy et al., 2005). Figure 2 depicts one possible model of how polyubiquinated cargo may serve as a nidus for nascent autophagosomal assembly, thus providing the only presently available glimpse at how targets get tagged and recognized for capture by the autophagic machinery. It remains to be settled whether additional signals or molecular tags can earmark organelles, cytosolic components, and other unwanted objects, including invading pathogens, for autophagy. Note that in the model by Bjorkoy (Bjorkoy et al., 2005), p62 serves as an adapter between the ubiquinated target and LC3, and that membrane assembly or its delivery may be linked or coordinated with the oligomerization of p62 and LC3.

The origin of membrane for nascent and growing autophagosomes is not clear, although endoplasmic reticulum (Dunn, 1990), Golgi (Yamamoto et al., 1990), and late endosomes (Young et al., 2006) have all been implicated as contributors. Atg9, an evolutionarily conserved and the sole Atg integral membrane protein, has been utilized as a marker to trace the source of membrane. It has been placed with or in the vicinity of mitochondrial membranes in yeast, while the mammalian mAtg9 has been localized as shuttling between *trans*-Golgi network and late endosomes, where it colocalizes with TGN46, the mannose-6-phosphate receptor Rab7, and Rab9 (Young et al., 2006). The yeast Atg9 traffics between mitochondria (Reggiori et al., 2005) and the specialized organelle PAS (preautophagosomal structure), which is believed to be a specialized staging point for autophagosome formation in yeast but has not been identified in mammalian cells. On starvation, the mammalian Atg9 redistributes between its resting location to LC3-positive endomembranes, in an ULK1 (Atg1)- and PI3K-dependent manner (Young et al., 2006). Atg9 in yeast does not share the subsequent fate of the autophagic membrane, and instead is recycled for reuse. The retrograde transport (recycling) of Atg9 in yeast depends on Atg18 (Reggiori et al., 2004), a conserved PI3P-binding protein also present in mammals (known as WIPI-1α) (Proikas-Cezanne et al., 2004), which binds PI3P on lipid dot blots but displays $PI3,5P_2$ specificity when lipids are presented within liposomes in binding assays (Dove et al., 2004). Although $PI3,5P_2$ has been reportedly excluded as an autophagy regulator (Krick et al., 2006), the enzyme PIKfyve that converts PI3P into $PI3,5P_2$ is conspicuously under regulation by PKB/Akt (Berwick et al., 2004), one of the key regulators of Tor and the autophagic signaling pathway, discussed below.

At its final, maturation stage, a completed, double-membrane autophagosome fuses with late endosomes (e.g., multivesicular bodies) to form an intermediate termed amphistome and lysosomal organelles. The double membrane loses its inner bilayer, transforming into a degradative autolysosome where the digestion of the captured cytoplasmic material takes place. The type III PI3K VPS34, most likely without its autophagy-specific interacting partners described below, plays a role in the formation of late endosomal multivesicular bodies and lysosomal organelles, thus providing an essential function for the terminal maturation stages of autophagy.

SIGNALING PROCESSES REGULATING AUTOPHAGY

Under certain physiological conditions, autophagy can be upregulated or downregulated by a core signaling pathway, with multiple factors either inducing or inhibiting autophagy (Fig. 2; activators in light gray, inhibitors in dark gray). This pathway includes two kinases, a Ser-Thr protein kinase (Tor) and a lipid kinase (hVPS34): (i) Tor (target of rapamycin) controls initiation of autophagy, whereas (ii) hVPS34 (also known as type III PI3K) controls both the initiation and maturation stages of autophagy.

A classical physiological inducer of autophagy is amino acid starvation, in particular, withdrawal of leucine. Another signal is absence of growth factors, which normally authorize cells to take up nutrients (Lum et al., 2005). A pharmacological induction of autophagy can be achieved by using rapamycin, a drug that targets the Ser/Thr kinase Tor (Wullschleger et al., 2006). Tor is a gatekeeper of cellular response to nutritional conditions and integrates various inputs including growth hormone signals (Fig. 2). Activation of Tor suppresses autophagy, while inactivation of Tor promotes autophagy (Fig. 2). In yeast, Tor negatively regulates Atg1, a factor that sets off the autophagy execution cascade. Hence, when Tor is inhibited by rapamycin, autophagy is induced. Tor is normally regulated by a small GTPase, Rheb (Fig. 2), which binds to the N-terminal portion of the Tor kinase domain. As with other small signaling GTPases that act as molecular switches, the GDP-bound form of Rheb is in the OFF position, while the GTP-bound form is in the ON position. Rheb activity is regulated by the GTPase-activating protein (GAP) TSC1/2, which receives and integrates various upstream inputs (Wullschleger et al., 2006) (Fig. 2) that come from (i) growth factor receptor signaling via the Akt/PKB pathway, inhibiting autophagy; (ii) energy status via AMPK, a kinase that responds to the AMP/ATP ratio, with active AMPK stimulating autophagy (Meley et al., 2006), and the recently recognized contributor REDD1, which acts independently of AMPK (Sofer et al., 2005); and (iii) Ca^{2+} effects on CaMKK-β and phosphorylation and activation of AMPK and thus induction of autophagy (Hoyer-Hansen et al., 2007).

The best-understood pathway controlling TSC1/2 is the one stimulated by growth factors (Fig. 2). Binding of insulin growth factors to receptors recruits insulin receptor substrates (IRSs) and, via IRS, type I PI3K p110/p85, which generates PIP3. PIP3 recruits PDK1 and Akt/PKB to the plasma membrane, where PDK1 phosphorylates and activates Akt/PKB. Active PKB phosphorylates and inactivates TSC1/2. By inactivating the TSC1/2 GAP, this cascade enhances Rheb-GTP-dependent activation of Tor and phosphorylation of Tor targets. This in turn inhibits autophagy. If growth factors or amino acids are withheld, autophagy is augmented. Recent work (Hsu et al., 2007) has identified a long-missing nucleotide exchange factor for Rheb, which loads Rheb with GTP, thus providing another arm of Rheb regulation, and hence Tor activation (which should result in autophagy inhibition), which yet remains to be explored.

MULTIPLE ROLES OF TYPE III PI3K hVPS34 IN AUTOPHAGY

The type III PI3K Vps34 in yeast, or its hVPS34 equivalent in humans, is an evolutionarily old lipid kinase antedating the type I PI3K, which represents a more recent addition in higher metazoans transducing growth factor inputs. Recognized only recently, but probably representing an ancient signaling mechanism, hVPS34 appears to transduce the amino acid replete conditions to Tor and in this capacity plays a negative role in signaling upstream of autophagy initiation (Byfield et al., 2005; Nobukuni et al., 2005) (Fig. 2). This contrasts with the requirement for hVPS34 during the execution stages of autophagy, when hVPS34 is complexed with Beclin 1 (Pattingre et al., 2005). In yeast, Vps34 plays a role in the formation of a specialized preautophagosomal structure and thus positively controls autophagy execution at its initiation stage (Klionsky, 2005). In S. cerevisiae, two separate protein complexes containing Vps34 control autophagic and endosomal degradative pathways: complex I, consisting of Atg6 (Beclin 1 ortholog), Atg14, and Vps34, specializes in autophagy, while complex II, with Vps38 instead of Atg14, specializes in endosomal-vacuolar sorting (Obara et al., 2006). Although Beclin 1 (Atg6) was recognized early on as a regulator of autophagy (Liang et al., 1999), putative equivalents of Vps38 or Atg14 have been missing, and their potential orthologs or paralogs are only now beginning to be identified, with several candidates found (UVRAG, p70/hAtg14, Bif1, p130, and Ambra1). The mechanism of UVRAG action is not known at present, but it could play a role of Vps38 in the lysosomal pathway, with indirect or partially direct roles in the autophagic pathway.

The newly identified Beclin 1-interacting partners, Bcl-2 (Pattingre et al., 2005) and UVRAG (Liang et al., 2006), act as an inhibitor and an activator of autophagy, respectively. Bcl-2 binding to Beclin 1 may be akin to the Bcl-2 inhibitory role in apoptosis (Pattingre et al., 2005), i.e., Bcl-2 binding possibly alters the Beclin 1 state and disables it from driving autophagy (Maiuri et al., 2007). This may be a way of regulating the extent of autophagic degradation, since levels of Beclin 1 upon stimulation with certain bioactive lipids have been correlated with levels of resulting autophagy (Lavieu et al., 2006; Scarlatti et al., 2004). Lower levels of Beclin 1 are associated with lower levels of autophagy, compatible with cell survival, while higher levels of Beclin 1 induce too much autophagy, which leads to cell death (Lavieu et al., 2006; Scarlatti et al., 2004).

The initiation and maturation stages of autophagy depend on PI3P production on endomembranes, although the exact location of PI3P in these processes remains elusive. The hVPS34 PI3K is the enzyme inhibited by the classical inhibitor of autophagy, 3MA. Note that 3MA (and another PI3K inhibitor, wortmannin) inhibits both the upstream (type I) PI3K and the downstream (type III) PI3K, but the downstream kinase (hVPS34) dominates the net effect. The same dominant effect of the terminal positive role of hVPS34 explains why the positive regulation of Tor by hVPS34, under amino acid-replete conditions (Fig. 3) (Byfield et al., 2005; Nobukuni et al., 2005), does not contradict the net effect of 3MA or wortmannin as inhibitors of autophagy. Thus, 3MA is invariably used to pharmacologically test whether a given process involves autophagy.

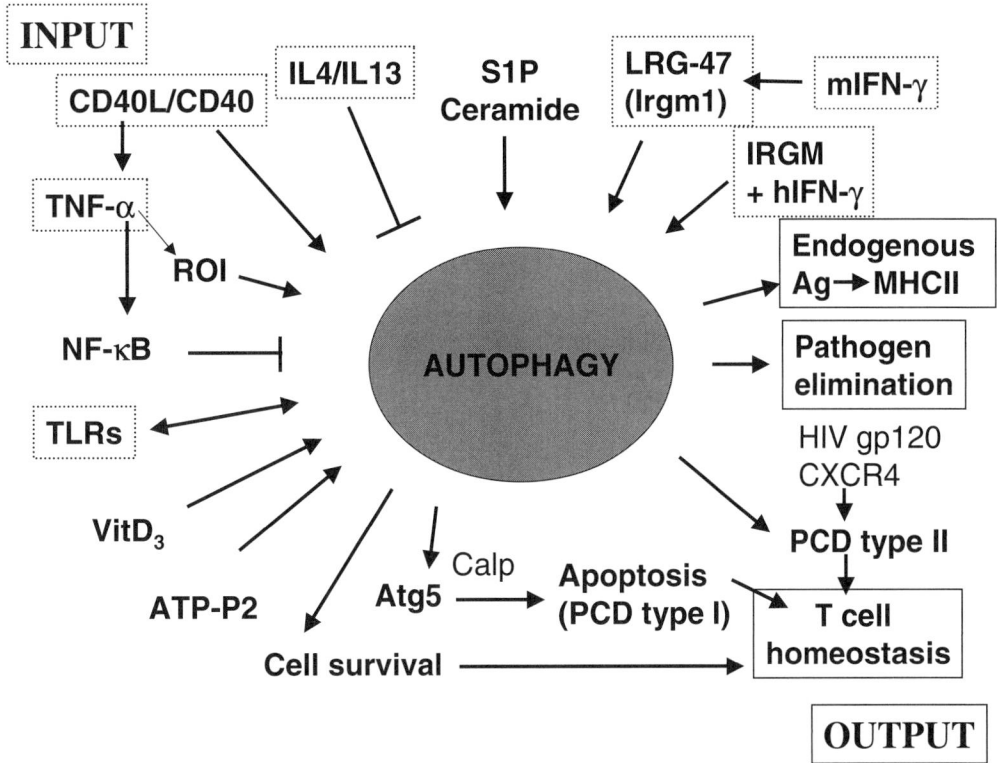

FIGURE 3 Immunologically relevant inputs (dotted boxes) and outputs (solid boxes) of autophagy. See text for details and references. Adapted from Deretic, 2006.

AUTOPHAGY AS AN IMMUNITY MECHANISM: IMMUNOPHAGY

The various connections (inputs and outputs) of autophagy with immunological processes are tallied in Fig. 3. Given the cytoplasmic housekeeping role of autophagy and its physical capacity to sequester and degrade large objects such as organelles (ranging in size from mitochondria to the entire nucleus), it should not be surprising that it plays a role in eliminating intracellular pathogens, including bacteria, protozoa, and viruses, as recently shown in a stream of publications (Andrade et al., 2006; Birmingham et al., 2006; Checroun et al., 2006; Gutierrez et al., 2004; Ling et al., 2006; Nakagawa et al., 2004; Ogawa et al., 2005; Orvedahl et al., 2007; Py et al., 2007). This function of autophagy represents a cell-autonomous capability of fighting microbes that invade a cell's interior, and is evolutionarily very old, hailing back to the initial interactions between eukaryotic cells and bacteria or viruses.

In vertebrates, the intrinsic immunity/defense role of autophagy, in addition to preserving the ancient regulation by Tor, has become connected (Ling et al., 2006; Singh et al., 2006) to a control by immunity-related GTPases, known as p47 GTPases (Taylor et al., 2004) or IRG (Bekpen et al., 2005), shown to control intracellular pathogens but without a clear mechanism of action (MacMicking et al., 2003) until the connection with autophagy was established (Gutierrez et al., 2004; Ling et al., 2006; Singh et al., 2006).

In mammalian cells, the role of autophagy has been linked to immunological signals. In mice and humans it is under the control of cytokines and agonists regulating innate and adaptive immunity, such as interferon gamma (IFN-γ) (Inbal et al., 2002; Pyo et al., 2005), tumor necrosis factor alpha (TNF-α) (Djavaheri-Mergny et al., 2006), and interleukin 13 (IL-13) (Arico et al., 2001; Petiot et al., 2000), and CD40L-CD40 (Andrade et al., 2006). Additional small bioactive molecules in our bodies, e.g., 1,25-dihydroxyvitamin D_3 (Hoyer-Hansen et al., 2005, 2007), sphingosine 1-phosphate (Lavieu et al., 2006), and ceramide (Scarlatti et al., 2004), can act as agonists or antagonists of autophagy in the context of immune regulation: 1,25-dihydroxyvitamin D_3 has been implicated in resistance to certain infectious diseases (Liu et al., 2006), and sphingosine 1-phosphate (Thompson et al., 2005) and ceramide (Anes et al., 2003) have been shown to activate macrophages for transfer of pathogens into lysosomal compartments.

Not only is autophagy a direct affront to the pathogens, but it also plays an active role in antigen processing (Paludan et al., 2005; Schmid et al., 2007), delivery of pathogen-associated pattern ligands to innate immunity receptors such as Toll-like receptors (TLRs) (Lee et al., 2007), T-cell homeostasis (Li et al., 2006; Pua et al., 2007), and survival of T cells following thymic selection (Pua et al., 2007). Furthermore, a genetic predisposition to Crohn's disease has now been linked to a polymorphism in one of the basal autophagic machinery genes, Atg16 (Hampe et al., 2007). All this attests to the fact that evolution has led to multiple layers of immunological autophagy (immunophagy) (Deretic, 2005) and that immunity and autophagy in mammalian cells have established a two-way communication to coordinate defenses against infection, autoimmune diseases, or chronic inflammation.

REFERENCES

Andrade, R. M., M. Wessendarp, M. J. Gubbels, B. Striepen, and C. S. Subauste. 2006. CD40 induces macrophage anti-

Toxoplasma gondii activity by triggering autophagy-dependent fusion of pathogen-containing vacuoles and lysosomes. *J. Clin. Invest.* **116:**2366–2377.

Anes, E., M. P. Kuhnel, E. Bos, J. Moniz-Pereira, A. Habermann, and G. Griffiths. 2003. Selected lipids activate phagosome actin assembly and maturation resulting in killing of pathogenic mycobacteria. *Nat. Cell. Biol.* **5:**793–802.

Arico, S., A. Petiot, C. Bauvy, P. F. Dubbelhuis, A. J. Meijer, P. Codogno, and E. Ogier-Denis. 2001. The tumor suppressor PTEN positively regulates macroautophagy by inhibiting the phosphatidylinositol 3-kinase/protein kinase B pathway. *J. Biol. Chem.* **276:**35243–35246.

Bekpen, C., J. P. Hunn, C. Rohde, I. Parvanova, L. Guethlein, D. M. Dunn, E. Glowalla, M. Leptin, and J. C. Howard. 2005. The interferon-inducible p47 (IRG) GTPases in vertebrates: loss of the cell autonomous resistance mechanism in the human lineage. *Genome Biol.* **6:**R92.

Berwick, D. C., G. C. Dell, G. I. Welsh, K. J. Heesom, I. Hers, L. M. Fletcher, F. T. Cooke, and J. M. Tavare. 2004. Protein kinase B phosphorylation of PIKfyve regulates the trafficking of GLUT4 vesicles. *J. Cell Sci.* **117:**5985–5993.

Birmingham, C. L., A. C. Smith, M. A. Bakowski, T. Yoshimori, and J. H. Brumell. 2006. Autophagy controls *Salmonella* infection in response to damage to the *Salmonella*-containing vacuole. *J. Biol. Chem.* **281:**11374–11383.

Bjorkoy, G., T. Lamark, A. Brech, H. Outzen, M. Perander, A. Overvatn, H. Stenmark, and T. Johansen. 2005. p62/SQSTM1 forms protein aggregates degraded by autophagy and has a protective effect on huntingtin-induced cell death. *J. Cell Biol.* **171:**603–614.

Byfield, M. P., J. T. Murray, and J. M. Backer. 2005. hVps34 is a nutrient-regulated lipid kinase required for activation of p70 S6 kinase. *J. Biol. Chem.* **280:**33076–33082.

Checroun, C., T. D. Wehrly, E. R. Fischer, S. F. Hayes, and J. Celli. 2006. Autophagy-mediated reentry of *Francisella tularensis* into the endocytic compartment after cytoplasmic replication. *Proc. Natl. Acad. Sci. USA* **103:**14578–14583.

Deretic, V. 2005. Autophagy in innate and adaptive immunity. *Trends Immunol.* **26:**523–528.

Deretic, V. 2006. Autophagy as an immune defense mechanism. *Curr. Opin. Immunol.* **18:**375–382.

Djavaheri-Mergny, M., M. Amelotti, J. Mathieu, F. Besancon, C. Bauvy, S. Souquere, G. Pierron, and P. Codogno. 2006. NF-kappaB activation represses tumor necrosis factor-alpha-induced autophagy. *J. Biol. Chem.* **281:**30373–30382.

Dove, S. K., R. C. Piper, R. K. McEwen, J. W. Yu, M. C. King, D. C. Hughes, J. Thuring, A. B. Holmes, F. T. Cooke, R. H. Michell, P. J. Parker, and M. A. Lemmon. 2004. Svp1p defines a family of phosphatidylinositol 3,5-bisphosphate effectors. *EMBO J.* **23:**1922–1933.

Dunn, W. A., Jr. 1990. Studies on the mechanisms of autophagy: formation of the autophagic vacuole. *J. Cell Biol.* **110:**1923–1933.

Furuya, T., J. Yu, M. Byfield, S. Pattingre, and B. Levine. 2005. The evolutionarily conserved domain of Beclin 1 is required for Vps34 binding, autophagy and tumor suppressor function. *Autophagy* **1:**46–52.

Gutierrez, M. G., S. S. Master, S. B. Singh, G. A. Taylor, M. I. Colombo, and V. Deretic. 2004. Autophagy is a defense mechanism inhibiting BCG and *Mycobacterium tuberculosis* survival in infected macrophages. *Cell* **119:**753–766.

Hampe, J., A. Franke, P. Rosenstiel, A. Till, M. Teuber, K. Huse, M. Albrecht, G. Mayr, F. M. De La Vega, J. Briggs, S. Günther, N. J. Prescott, C. M. Onnie, R. Häsler, B. Sipos, U. R. Fölsch, T. Lengauer, M. Platzer, C. G. Mathew, M. Krawczak, and S. Schreiber. 2007. A genome-wide association scan of nonsynonymous SNPs identifies a susceptibility variant for Crohn disease in ATG16L1. *Nat. Genet.* **39:**207–211.

Høyer-Hansen, M., L. Bastholm, I. S. Mathiasen, F. Elling, and M. Jäättelä. 2005. Vitamin D analog EB1089 triggers dramatic lysosomal changes and Beclin 1-mediated autophagic cell death. *Cell Death Differ.* **12:**1297–1309.

Høyer-Hansen, M., L. Bastholm, P. Szyniarowski, M. Campanella, G. Szabadkai, T. Farkas, K. Bianchi, N. Fehrenbacher, F. Elling, R. Rizzuto, I. S. Mathiasen, and M. Jäättelä. 2007. Control of macroautophagy by calcium, calmodulin-dependent kinase kinase-beta, and Bcl-2. *Mol. Cell* **25:**193–205.

Hsu, Y. C., J. J. Chern, Y. Cai, M. Liu, and K. W. Choi. 2007. *Drosophila* TCTP is essential for growth and proliferation through regulation of dRheb GTPase. *Nature* **445:**785–788.

Inbal, B., S. Bialik, I. Sabanay, G. Shani, and A. Kimchi. 2002. DAP kinase and DRP-1 mediate membrane blebbing and the formation of autophagic vesicles during programmed cell death. *J. Cell Biol.* **157:**455–468.

Klionsky, D. J. 2005. The molecular machinery of autophagy: unanswered questions. *J. Cell. Sci.* **118:**7–18.

Krick, R., J. Tolstrup, A. Appelles, S. Henke, and M. Thumm. 2006. The relevance of the phosphatidylinositolphosphate-binding motif FRRGT of Atg18 and Atg21 for the Cvt pathway and autophagy. *FEBS Lett.* **580:**4632–4638.

Lavieu, G., F. Scarlatti, G. Sala, S. Carpentier, T. Levade, R. Ghidoni, J. Botti, and P. Codogno. 2006. Regulation of autophagy by sphingosine kinase 1 and its role in cell survival during nutrient starvation. *J. Biol. Chem.* **281:**8518–8527.

Lee, H. K., J. M. Lund, B. Ramanathan, N. Mizushima, and A. Iwasaki. 2007. Autophagy-dependent viral recognition by plasmacytoid dendritic cells. *Science* **315:**1398–1401.

Li, C., E. Capan, Y. Zhao, J. Zhao, D. Stolz, S. C. Watkins, S. Jin, and B. Lu. 2006. Autophagy is induced in CD4+ T cells and important for the growth factor-withdrawal cell death. *J. Immunol.* **177:**5163–5168.

Liang, C., P. Feng, B. Ku, I. Dotan, D. Canaani, B. H. Oh, and J. U. Jung. 2006. Autophagic and tumour suppressor activity of a novel Beclin1-binding protein UVRAG. *Nat. Cell Biol.* **8:**688–699.

Liang, X. H., S. Jackson, M. Seaman, K. Brown, B. Kempkes, H. Hibshoosh, and B. Levine. 1999. Induction of autophagy and inhibition of tumorigenesis by beclin 1. *Nature* **402:**672–676.

Ling, Y. M., M. H. Shaw, C. Ayala, I. Coppens, G. A. Taylor, D. J. Ferguson, and G. S. Yap. 2006. Vacuolar and plasma membrane stripping and autophagic elimination of *Toxoplasma gondii* in primed effector macrophages. *J. Exp. Med.* **203:**2063–2071.

Liu, P. T., S. Stenger, H. Li, L. Wenzel, B. H. Tan, S. R. Krutzik, M. T. Ochoa, J. Schauber, K. Wu, C. Meinken, D. L. Kamen, M. Wagner, R. Bals, A. Steinmeyer, U. Zügel, R. L. Gallo, D. Eisenberg, M. Hewison, B. W. Hollis, J. S. Adams, B. R. Bloom, and R. L. Modlin. 2006. Toll-like receptor triggering of a vitamin D-mediated human antimicrobial response. *Science* **311:**1770–1773.

Lum, J. J., D. E. Bauer, M. Kong, M. H. Harris, C. Li, T. Lindsten, and C. B. Thompson. 2005. Growth factor regulation of autophagy and cell survival in the absence of apoptosis. *Cell* **120:**237–248.

MacMicking, J. D., G. A. Taylor, and J. D. McKinney. 2003. Immune control of tuberculosis by IFN-gamma-inducible LRG-47. *Science* **302:**654–659.

Maiuri, M. C., G. Le Toumelin, A. Criollo, J. C. Rain, F. Gautier, P. Juin, E. Tasdemir, G. Pierron, K. Troulinaki, N. Tavernarakis, J. A. Hickman, O. Geneste, and G. Kroemer. 2007. Functional and physical interaction between Bcl-X(L) and a BH3-like domain in Beclin-1. *EMBO J.* **26:**2527–2539.

Meley, D., C. Bauvy, J. H. Houben-Weerts, P. F. Dubbelhuis, M. T. Helmond, P. Codogno, and A. J. Meijer. 2006. AMP-activated protein kinase and the regulation of autophagic proteolysis. *J. Biol. Chem.* **281**:34870–34879.

Nakagawa, I., A. Amano, N. Mizushima, A. Yamamoto, H. Yamaguchi, T. Kamimoto, A. Nara, J. Funao, M. Nakata, K. Tsuda, S. Hamada, and T. Yoshimori. 2004. Autophagy defends cells against invading group A *Streptococcus*. *Science* **306**:1037–1040.

Nobukuni, T., M. Joaquin, M. Roccio, S. G. Dann, S. Y. Kim, P. Gulati, M. P. Byfield, J. M. Backer, F. Natt, J. L. Bos, F. J. Zwartkruis, and G. Thomas. 2005. Amino acids mediate mTOR/raptor signaling through activation of class 3 phosphatidylinositol 3OH-kinase. *Proc. Natl. Acad. Sci. USA* **102**:14238–14243.

Obara, K., T. Sekito, and Y. Ohsumi. 2006. Assortment of phosphatidylinositol 3-kinase complexes—Atg14p directs association of complex I to the preautophagosomal structure in *Saccharomyces cerevisiae*. *Mol. Biol. Cell.* **17**:1527–1539.

Ogawa, M., T. Yoshimori, T. Suzuki, H. Sagara, N. Mizushima, and C. Sasakawa. 2005. Escape of intracellular *Shigella* from autophagy. *Science* **307**:727–731.

Orvedahl, A., D. Alexander, Z. Tallóczy, Q. Sun, Y. Wei, W. Zhang, D. Burns, D. Leib, and B. Levine. 2007. HSV-1 ICP34.5 confers neurovirulence by targeting the Beclin 1 autophagy protein. *Cell Host Microbe* **1**:23–35.

Paludan, C., D. Schmid, M. Landthaler, M. Vockerodt, D. Kube, T. Tuschl, and C. Munz. 2005. Endogenous MHC class II processing of a viral nuclear antigen after autophagy. *Science* **307**:593–596.

Pattingre, S., A. Tassa, X. Qu, R. Garuti, X. H. Liang, N. Mizushima, M. Packer, M. D. Schneider, and B. Levine. 2005. Bcl-2 antiapoptotic proteins inhibit Beclin 1-dependent autophagy. *Cell* **122**:927–939.

Petiot, A., E. Ogier-Denis, E. F. Blommaart, A. J. Meijer, and P. Codogno. 2000. Distinct classes of phosphatidylinositol 3'-kinases are involved in signaling pathways that control macroautophagy in HT-29 cells. *J. Biol. Chem.* **275**:992–998.

Proikas-Cezanne, T., S. Waddell, A. Gaugel, T. Frickey, A. Lupas, and A. Nordheim. 2004. WIPI-1alpha (WIPI49), a member of the novel 7-bladed WIPI protein family, is aberrantly expressed in human cancer and is linked to starvation-induced autophagy. *Oncogene* **23**:9314–9325.

Pua, H. H., I. Dzhagalov, M. Chuck, N. Mizushima, and Y. W. He. 2007. A critical role for the autophagy gene Atg5 in T cell survival and proliferation. *J. Exp. Med.* **204**:25–31.

Py, B. F., M. M. Lipinski, and J. Yuan. 2007. Autophagy limits *Listeria monocytogenes* intracellular growth in the early phase of primary infection. *Autophagy* **3**:117–125.

Pyo, J. O., M. H. Jang, Y. K. Kwon, H. J. Lee, J. I. Jun, H. N. Woo, D. H. Cho, B. Choi, H. Lee, J. H. Kim, N. Mizushima, Y. Oshumi, and Y. K. Jung. 2005. Essential roles of Atg5 and FADD in autophagic cell death: dissection of autophagic cell death into vacuole formation and cell death. *J. Biol. Chem.* **280**:20722–20729.

Reggiori, F., T. Shintani, U. Nair, and D. J. Klionsky. 2005. Atg9 cycles between mitochondria and the pre-autophagosomal structure in yeasts. *Autophagy* **1**:101–109.

Reggiori, F., K. A. Tucker, P. E. Stromhaug, and D. J. Klionsky. 2004. The Atg1-Atg13 complex regulates Atg9 and Atg23 retrieval transport from the pre-autophagosomal structure. *Dev. Cell.* **6**:79–90.

Scarlatti, F., C. Bauvy, A. Ventruti, G. Sala, F. Cluzeaud, A. Vandewalle, R. Ghidoni, and P. Codogno. 2004. Ceramide-mediated macroautophagy involves inhibition of protein kinase B and up-regulation of beclin 1. *J. Biol. Chem.* **279**:18384–18391.

Schmid, D., M. Pypaert, and C. Munz. 2007. Antigen-loading compartments for major histocompatibility complex class II molecules continuously receive input from autophagosomes. *Immunity* **26**:79–92.

Scott, R. C., G. Juhasz, and T. P. Neufeld. 2007. Direct induction of autophagy by Atg1 inhibits cell growth and induces apoptotic cell death. *Curr. Biol.* **17**:1–11.

Singh, S. B., A. S. Davis, G. A. Taylor, and V. Deretic. 2006. Human IRGM induces autophagy to eliminate intracellular mycobacteria. *Science* **313**:1438–1441.

Sofer, A., K. Lei, C. M. Johannessen, and L. W. Ellisen. 2005. Regulation of mTOR and cell growth in response to energy stress by REDD1. *Mol. Cell. Biol.* **25**:5834–5845.

Taylor, G. A., C. G. Feng, and A. Sher. 2004. p47 GTPases: regulators of immunity to intracellular pathogens. *Nat. Rev. Immunol.* **4**:100–109.

Thompson, C. R., S. S. Iyer, N. Melrose, R. VanOosten, K. Johnson, S. M. Pitson, L. M. Obeid, and D. J. Kusner. 2005. Sphingosine kinase 1 (SK1) is recruited to nascent phagosomes in human macrophages: inhibition of SK1 translocation by *Mycobacterium tuberculosis*. *J. Immunol.* **174**:3551–3561.

Wullschleger, S., R. Loewith, and M. N. Hall. 2006. TOR signaling in growth and metabolism. *Cell* **124**:471–484.

Yamamoto, A., R. Masaki, and Y. Tashiro. 1990. Characterization of the isolation membranes and the limiting membranes of autophagosomes in rat hepatocytes by lectin cytochemistry. *J. Histochem. Cytochem.* **38**:573–580.

Young, A. R., E. Y. Chan, X. W. Hu, R. Kochl, S. G. Crawshaw, S. High, D. W. Hailey, J. Lippincott-Schwartz, and S. A. Tooze. 2006. Starvation and ULK1-dependent cycling of mammalian Atg9 between the TGN and endosomes. *J. Cell Sci.* **119**:3888–3900.

Zeng, X., J. H. Overmeyer, and W. A. Maltese. 2006. Functional specificity of the mammalian Beclin-Vps34 PI 3-kinase complex in macroautophagy versus endocytosis and lysosomal enzyme trafficking. *J. Cell Sci.* **119**:259–270.

27

Brucella, a Perfect Trojan Horse in Phagocytes

SUZANA P. SALCEDO AND JEAN-PIERRE GORVEL

BRUCELLA SPP. AND DISEASE

Brucella was first isolated in 1887 in Malta by a British Army physician named David Bruce from the spleens of several soldiers with fatal cases of the disease, commonly known as undulant fever or Malta fever. At the present time, brucellosis is the most common zoonotic disease worldwide (Pappas et al., 2006) and affects a number of mammals, with a particular economic impact on livestock in endemic regions. The genus *Brucella*, characterized by small, gram-negative bacteria, belongs to the α-2 subdivision of Proteobacteria. This group of bacteria have coevolved with animal or plant hosts, some in a beneficial symbiotic manner such as *Sinorhizobium melitoti* (a plant symbiont), while others have evolved as pathogens such as *Agrobacterium tumefaciens* (a plant pathogen) and *Rickettsia* spp. and *Bartonella* spp. (both animal pathogens).

Several species of the genus *Brucella* are responsible for brucellosis: *Brucella melitensis* (goats, sheep, camels), *B. abortus* (cattle, camels), *B. suis* (pigs), *B. canis* (dogs), *B. ovis* (goats, sheep), and *B. neotomae* (desert wood rat). *Brucella* has also been isolated from a variety of terrestrial wildlife mammal species such as elk, buffalo, reindeer, and bison as well as from marine mammals, illustrating its broad host range. Currently, of the six main species of *Brucella*, four can cause disease in humans, with *B. melitensis* having the most severe pathogenicity, followed by *B. suis*, which is more rare but still presents severe pathogenicity, and then *B. abortus* and *B. canis*, which are less pathogenic in humans.

In animals, brucellosis mainly affects the reproductive system, often leading to sterility in male animals and abortion in pregnant female animals after bacterial colonization of the placenta. High numbers of bacteria are shed in milk, via an aborted fetus or other reproductive tract discharges. In addition, animals can become carriers and remain infected for many years. The chronic nature of brucellosis is a key characteristic of this disease that develops not only in the natural host but also in humans, a secondary host. Human brucellosis is mainly transmitted by ingestion of contaminated food products (such as unpasteurized milk) or by occupational exposure to infected animals. *Brucella* is also highly infectious through inhalation of contaminated aerosols; data from laboratory-acquired infection indicate that inhalation of 10 to 100 bacteria is sufficient to cause disease. For this reason, *Brucella* is classified as a potential agent of biological warfare. In fact, it became the first biological weapon developed by the United States in 1954, which produced *B. suis*-containing cluster bombs for the Air Force. Although *Brucella* spread via aerosols would cause significant morbidity and economic loss, brucellosis does not spread from person to person and could therefore be easily contained. Nonetheless, due to the chronic nature of the disease, its tendency to relapse, and its propensity to affect joints, human brucellosis can result in relatively long-term disability. Despite the nonspecific symptoms such as recurrent fever and malaise that characterize the onset of the disease, patients can present with severe complications such as liver abscess formation, neurobrucellosis, and endocarditis. Mortality from brucellosis is rare when appropriate and prompt treatment with multiple antibiotics is administered for a prolonged time. Although several vaccines are being used in animals, no efficient or safe vaccine is yet available for humans.

An essential aspect of *Brucella* pathogenicity lies in its ability to survive and replicate within host cells. In animals, abortion is associated with rapid proliferation of *Brucella* within placental trophoblasts (Meador and Deyoe, 1989; Anderson and Cheville, 1986), which are key cellular components of the epithelial layers of the placenta. The presence of high bacterial loads within these cells in the end results in disruption of the placenta and infection of the fetus. Trophoblasts are therefore important cellular targets for *Brucella* in the natural host, but little is known about the infectious process in these cells.

The establishment of a chronic infection by *Brucella* in either animals or humans is also linked to its ability to persist for prolonged periods within host cells, predominantly macrophages. Intracellular survival within phagocytic cells is an essential aspect of *Brucella* virulence.

Suzana P. Salcedo and Jean-Pierre Gorvel, Centre d'Immunologie de Marseille-Luminy, Aix Marseille Université, Faculté de Sciences de Luminy, Case 906, Marseille, 13288 Cedex 9, France; INSERM, U631, Marseille, 13288, France; and CNRS, UMR6102, Marseille, 13288, France.

BRUCELLA PATHOGENESIS IN MACROPHAGES (MODEL SYSTEM)

Significant progress has been made in our understanding of the virulence factors that enable *Brucella* to reside within phagocytic cells and escape their killing mechanisms. The vast majority of in vitro studies have relied on cultured macrophages as a model system to analyze the intracellular survival of this pathogen. In essence, *Brucella* is able to proliferate extensively within both macrophages and non-phagocytic epithelial cells without affecting their basic cellular functions or inducing cell death. It efficiently controls its own intracellular trafficking in order to avoid degradation within lysosomes and reach an intracellular compartment suited for replication, which we commonly designate as the *Brucella*-containing vacuole (BCV).

Early Stages of Intracellular Trafficking

The early interactions of *Brucella* with the cell are key to determining its intracellular survival. Both opsonized and nonopsonized bacteria are efficiently taken up by macrophages, suggesting that complement or antibody-mediated phagocytosis as well as direct *Brucella*-host cell contact contribute to bacterial entry. Indeed, nonphagocytic cells such as trophoblasts are easily infected by *Brucella*. In the case of nonopsonized bacteria, entry into macrophages is mediated by lipid rafts (Naroeni and Porte, 2002; Watarai et al., 2002). Two receptors have been proposed to mediate entry of nonopsonized *Brucella*: (i) the class A scavenger receptor, by interaction with the lipid A part of the *Brucella* lipopolysaccharide (LPS) (Kim et al., 2004), and (ii) the cellular prion protein PrPc, which seems to bind the *Brucella* heat shock protein Hsp60 (Watarai et al., 2003). However, the specific role of either of these receptors in *Brucella* infection remains elusive and controversial (Fontes et al., 2005). In addition, the O-polysaccharide chain of the *Brucella* LPS is necessary for efficient entry into host cells. Bacterial mutants that lack this O chain (rough LPS) do not enter through lipid rafts and are quickly directed to lysosomes where they are degraded (Porte et al., 2003). It is possible that the O chain of the LPS is modifying the fusogenic properties of the early BCV since heat-killed *Brucella* that contain an intact LPS (smooth LPS) have delayed fusion with lysosomes compared with the rough mutants.

It is these early stages of BCV maturation that therefore help determine whether *Brucella* will survive within macrophages. In fact, in vitro studies have shown that the vast majority of bacteria taken up by these phagocytic cells are efficiently eliminated and only a few are able to escape and replicate intracellularly. These successful early BCVs undergo transient interactions with specific compartments of the endocytic pathway. It is well documented that in cultured cells BCVs acquire markers of early endosomes immediately after entry, including the early endosomal antigen EEA1 and the small GTP-binding protein Rab5 (Celli et al., 2003; Chaves-Olarte et al., 2002; Pizarro-Cerda et al., 1998). These interactions are fast and occur during the first minutes of BCV maturation; for example, in the case of bone marrow-derived murine macrophages, most of these early markers are already excluded 30 min following uptake (Celli et al., 2003). Unlike latex beads and heat-killed bacteria, BCVs do not acquire most of the markers of late endosomes and lysosomes, suggesting that they efficiently segregate from the endocytic pathway. Interestingly, as BCVs lose early endosomal markers they acquire the late endosomal/lysosomal membrane protein LAMP1, suggesting some interaction with late endocytic-derived compartments (Celli et al., 2003; Pizarro-Cerda et al., 1998). Alternatively, LAMP1 could be acquired directly from the Golgi network, but further work is required to understand how the BCV membrane becomes enriched with this protein and whether it plays a role in *Brucella* virulence.

Recent work has identified a new *Brucella* virulence determinant involved in the control of BCV trafficking, the cyclic β-1,2-glucan (Arellano-Reynoso et al., 2005). This molecule is able to extract cholesterol from eukaryotic membranes and has therefore been proposed to modify cholesterol-rich lipid rafts present on the BCV membrane to prevent fusion with lysosomes. BCVs are also enriched in flotillin-1, a protein involved in lipid raft signaling that has been associated with phagosome maturation along the endocytic pathway.

The process of BCV maturation proceeds for several hours as BCVs progressively exclude LAMP1. In bone marrow-derived murine macrophages, LAMP1 on BCVs diminishes significantly by 12 h, and only very few BCVs still retain LAMP1 by 24 h, when intracellular replication is well under way. An important outcome from phagosome maturation is the acidification of BCVs, which is essential for *Brucella* survival within host cells. In cultured macrophages infected with *B. suis*, BCVs acidify to a pH of about 4.0 to 4.5 within 60 min (Porte et al., 1999). The lowering of the vacuolar pH provides an important signal for the expression of *Brucella* genes required for virulence. In fact, inhibition of BCV acidification at early stages of the phagosome maturation (within 1 h) completely abolishes intracellular replication. BCV acidification induces the expression of the *virB* operon approximately 3 h after infection (Boschiroli et al., 2002). This operon encodes for a type IV secretion system essential for the establishment of the *Brucella* replication niche. The VirB type IV secretion system is necessary at later stages of phagosome maturation, when BCVs directly interact with the endoplasmic reticulum (ER) to set up a compartment in which *Brucella* begins to replicate.

Replication and Interaction with the ER

Following the initial exchanges with the endocytic pathway BCVs undergo extensive interactions with the ER. In cultured macrophages and epithelial cell lines, BCVs acquire numerous proteins of the ER such as the lectin chaperones calnexin and calreticulin, as well as the ER protein translocator Sec61 (Celli et al., 2003; Pizarro-Cerda et al., 1998). Through sustained exchanges between BCVs and the ER, the vacuolar membrane becomes decorated with ribosomes visible by electron microscopy, and finally BCVs mature into an ER-derived vacuole (Celli et al., 2003). In addition, glucose-6-phosphatase normally present in the lumen of the ER can be found within BCVs, confirming that fusion events have occurred. At this stage, *Brucella* has reached a safe niche for intracellular replication.

Formation of these specialized *Brucella* compartments, which coincides with onset of bacterial replication, requires extensive membrane recruitment since each dividing bacterium is found within an individual vacuole. Therefore, the ER is an important source of membrane for replicating bacteria but also creates an environment suited for *Brucella* replication, perhaps by providing the necessary nutrients. *Brucella* multiplies efficiently within these ER-derived vacuoles, reaching such high bacterial numbers that it affects the normal structure of the ER in macrophages, which becomes reorganized toward the replicating vacuoles. It is curious that induction of vacuolation of the ER by treating cells with the toxin aerolysin from *Aeromonas hydrophila* also results in vacuolation of BCVs (Celli et al., 2003), highlighting the nature of these *Brucella* compartments which acquire ER properties and behave as "extensions" of the ER. Interestingly, *Brucella* can also be found in ER-associated compartments in placental trophoblasts by analyzing tissue obtained from *B. abortus*-infected cattle and goats (Anderson and Cheville, 1986). This highlights the importance of an ER-derived intracellular niche for infection.

The molecular mechanisms that enable *Brucella* to initiate fusion with the ER are still unclear. It is known that the type IV secretion system encoded by the *virB* operon is necessary for sustained interactions with the ER (Celli et al., 2003; Comerci et al., 2001). In bone marrow-derived macrophages, *Brucella* mutant strains lacking specific genes of the *virB* operon survive during the early stages of the infection and can be found in close contact with the ER by electron microscopy. Yet, vacuoles containing *virB* mutant bacteria are then unable to fuse with the ER and eventually acquire lysosomal markers such as cathepsin D, indicating that bacteria are in lysosomes where they are finally degraded. Based on other type IV secretion systems it is likely that the *Brucella* VirB system secretes effector molecules that mediate the fusion events between the BCV membrane and the ER. However, these molecules remain to be identified.

Recent data have begun to shed some light on the cellular machinery involved in BCV-ER interaction. As the initial contact with the ER is established, BCVs intercept the early secretory pathway that is initiated at ER exit sites. In these subdomains of the ER highly dynamic fusion and fission events are taking place under the control of the small GTPase Sar1 and the assembly of the COPII-coated complex on the ER membrane. Molecules to be exported to the Golgi network or back to the ER are afterward packaged into vesicles by the Arf1 GTPase and the coat complex COPI. Unlike *Legionella*, which intercepts COPI-coated vesicles, BCVs interact with the Sar1/COPII complex at ER exit sites (Celli et al., 2005). Inhibition of the Sar1 activity, which results in disruption of ER exit sites, prevents BCVs from fusing with the ER. When the integrity and function of ER exit sites is compromised *Brucella* is no longer able to establish a niche permissive for intracellular replication.

Virulence Factors Required for Intracellular Survival

Several virulence factors have been implicated in *Brucella* resistance to cellular defensive mechanisms and interaction with cellular pathways to create the environment suited for its intracellular survival.

LPS

Brucella LPS has particular characteristics that make it a very important virulence factor that enhances both extracellular and intracellular survival of *Brucella* (Lapaque et al., 2005). In comparison with classical LPS from enterobacteria, *Brucella* LPS has several significant structural differences. Its lipid A contains longer acyl groups, linked to the core by amide bonds, not ester bonds, and has a diaminoglucose backbone instead of a glucosamine. Together these features make it several hundred times less toxic than classical LPSs and able to avoid the deposition of complement at the bacterial surface, thereby enhancing *Brucella* extracellular survival. *Brucella* LPS is also highly resistant to many antimicrobial cationic peptides including lysozyme and the defensin NP-2 (Martinez de Tejada et al., 1995). This resistance is probably due to a low number of charged phosphate groups in the core lipid A. With fewer of these anionic groups *Brucella* LPS forms aggregates that hamper penetration of cationic bactericidal peptides. Another important consequence of the particular chemical structure of the *Brucella* LPS is the altered pathogen-associated molecular pattern (PAMP). Although still recognized by the Toll-like receptor-4 (TLR4) pathway (Lapaque et al., 2006a), it fails to induce proinflammatory cytokine secretion, inducible nitric oxide synthase activation, or expression of antimicrobial proteins such as p47 GTPase (Lapaque et al., 2006a; Jimenez de Bagues et al., 2004; Rasool et al., 1992; Riley and Robertson, 1984a).

As mentioned earlier, the *Brucella* LPS also has an important role in the early stages of intracellular maturation of *Brucella* compartments in macrophages. The lipid A portion of the LPS mediates entry into macrophages by interacting with the class A scavenger receptor in lipid rafts. However, this lipid A-receptor interaction is not solely capable of mediating a "safe" entry for *Brucella*, as mutants lacking the O chain of the LPS are not able to enter macrophages via lipid rafts and instead fuse rapidly with lysosomes (Porte et al., 2003). The mechanism by which the O chain is mediating entry and affecting the early maturation stages of BCVs remains uncharacterized, although it is possible that an additional receptor is involved in specific recognition of the O chain.

Finally, the *Brucella* LPS has also been shown to have immunomodulatory properties. Once within host cells, LPS molecules resist degradation and are directed toward major histocompatibility complex class II (MHC-II)-enriched compartments (Forestier et al., 1999). The LPS is then recycled to the cell surface, where it clusters with MHC-II molecules in structures designated macrodomains (Lapaque et al., 2006b; Forestier et al., 1999, 2000). These structures contain certain components of detergent-resistant membranes but are dense and have minimal membrane fluidity (Lapaque et al., 2006b). Although antigen processing is not affected by the presence of LPS, this clustering at the cell surface interferes with efficient presentation of antigenic peptides loaded on MHC-II molecules to CD4-positive T cells (Lapaque et al., 2006b; Forestier et al., 2000). Because these macrodomains are very stable, maintenance of LPS on the cell surface contributes to the long-term persistence of *Brucella* within the macrophages.

Cyclic β-1,2-Glucan

Bacterial cyclic β-1,2-glucans belong to the family of periplasmic glucans that are involved in osmoregulation and structure stabilization of the bacterial envelope (Bohin, 2000). However, in the case of the plant symbiont *Sinorhizobium*, the plant pathogen *Agrobacterium*, and *Brucella*, all members of the α-2 subdivision of *Proteobacteria*, cyclic β-1,2-glucans have additional functions related to survival and virulence. For example, in the case of the plant symbiont *S. melitoti*, the cyclic β-1,2-glucans have been implicated in suppression of phytoalexin synthesis (a plant defense mechanism) in order to survive intracellularly. In *Brucella*, the cyclic β-1,2-glucans are required for intracellular replication and virulence in the mouse model of brucellosis (Arellano-Reynoso et al., 2005; Briones et al., 2001; Inon de Iannino et al., 1998). Purified *Brucella* cyclic β-1,2-glucan has been shown to extract cholesterol from eukaryotic membranes, leading to the hypothesis that it may be altering cholesterol-rich lipid rafts present on the BCV membrane. This would in turn alter the fusogenic properties of BCVs and prevent fusion with lysosomes. Consistent with this observation, Brucella mutant strains lacking a functional cyclic β-1,2-glucan synthase are not able to segregate from the endocytic pathway, and instead BCVs acquire lysosomal markers such as cathepsin D and bacteria are eventually killed. Furthermore, the cyclic β-1,2-glucan enhanced recruitment of flotillin-1, an important signaling molecule characteristic of lipid rafts previously implicated in control of phagosome maturation. Addition of purified cyclic β-1,2-glucan to mutants lacking a functional synthase was able to restore virulence and bacteria were found replicating within ER-derived BCVs. It still remains unclear how this high-molecular-weight compound, which is normally periplasmic, is modifying the membrane composition of BCVs to prevent fusion with lysosomes, but there is no doubt of the importance of the cyclic β-1,2-glucan in Brucella virulence.

Type IV Secretion System

The Brucella virB operon was discovered by using insertional mutagenesis-based approaches to identify bacterial factors required for virulence in B. suis, B. melitensis, and B. abortus (Sieira et al., 2000; O'Callaghan et al., 1999). This chromosomal region is homologous to the A. tumefaciens virB operon, which contains 12 genes encoding a type IV secretion system able to translocate bacterial products from the bacterial cytoplasm into plant cells. Type IV secretion systems can secrete a wide range of molecules ranging from single-stranded DNA–protein complexes to multicomponent toxins. They form a complex of up to 11 proteins that span the bacterial membrane and in certain cases the vacuolar membrane that surrounds intracellular pathogens such as Brucella. Examples of other pathogens containing type IV secretion systems include Bordetella pertussis, Helicobacter pylori, and Legionella pneumophila. In the case of Brucella, the VirB system is essential for virulence in the murine model of chronic brucellosis and for intracellular replication in vitro (O'Callaghan et al., 1999; Celli et al., 2003; Comerci et al., 2001; Delrue et al., 2001; Foulongne et al., 2000; Hong et al., 2000; Sieira et al., 2000). Experiments in bacteria grown in vitro have shown that the expression of the virB operon is dependent on the growth phase, being highest in the early exponential phase. Acid shock to a pH of 4 was also a strong inducer of transcription of the VirB operon. This was further confirmed in infected macrophages, where expression of the virB operon is induced 3 h after infection upon acidification of BCVs (Boschiroli et al., 2002).

As mentioned earlier, the VirB apparatus is essential for the intracellular trafficking of Brucella inside macrophages. VirB-deficient strains are unable to sustain the acquisition of ER markers and eventually fuse with lysosomes, where they are degraded (Celli et al., 2003). This demonstrates that the VirB type IV secretion system is required for the biogenesis of the Brucella replicative compartment in cultured macrophages. Identification of the molecules secreted by Brucella via its type IV system will be essential to unravel the mechanism by which BCVs fuse with the ER at the molecular level.

In the mouse model of brucellosis, Brucella can be found in many tissues such as the spleen and liver. Curiously, Brucella was found to replicate in salivary glands, which could be of significance in relation to human infection, where inoculation occurs through ingestion of contaminated food (Rajashekara et al., 2005). In addition, mice presented chronic infection of tail joints with Brucella resembling osteoarticular brucellosis in humans. In this model, virB mutants were efficient in spreading to different tissues but failed to replicate and persist within the host, highlighting the function of the type IV secretion system in long-term survival and proliferation of Brucella within host cells.

Other recent studies have also implicated the VirB type IV secretion system in control of the host immune response. Mutant mice lacking different components of adaptive immunity were used to analyze the ability to eliminate the virB mutant from the spleen during a mixed infection with the wild type (Rolan and Tsolis, 2007). The idea was to identify mouse mutations that can rescue the virB mutant but have no effect on wild type and, in this way, distinguish immune mechanisms circumvented by Brucella through its type IV secretion system. Although there was no effect on IFN-$\gamma^{-/-}$ mice, in Rag1$^{-/-}$ mice there was partial rescue of the virB mutant. Rag1$^{-/-}$ mice cannot generate functional B and T cells, thus completely inactivating adaptive immunity. In addition, Igh6$^{-/-}$ mice that lack B cells are also able to rescue the virB mutant. However, there was no effect when macrophages obtained from these knockout mice were infected in vitro, suggesting that the decreased capacity of Rag1$^{-/-}$ and Igh6$^{-/-}$ mice to control virB infection is not related to the ability of macrophages to control intracellular replication. The results suggest that, besides its clear role in ensuring the survival and replication of Brucella within host cells, the type IV secretion system may contribute to evasion of adaptive immune mechanisms by Brucella, but further work is needed to validate this hypothesis.

The BvrR/BvrS Regulatory System

The Brucella BvrR-BvrS two-component regulatory system controls the expression of genes involved in maintenance of the bacterial cell envelope. In addition, the BvrR-BvrS regulatory system controls the expression of genes that modify the composition of the lipid A portion of the Brucella LPS. Mutants were originally identified for their sensitivity to bactericidal polycations, suggesting that their outer membrane was modified. Importantly, mutants in the BvrR-BvrS regulatory system are defective for invasion and subsequent intracellular replication in both macrophages and epithelial cultured cells (Sola-Landa et al., 1998). Consequently, they are also extremely attenuated in the mouse model of infection. Within host cells, they fail to establish an ER-derived compartment and instead are degraded after fusion with lysosomes. It is possible that modification of the bacterial envelope renders them more susceptible to neutrophil killing in the case of the mouse infection. This would also reduce their resistance to phagosome acidification during the early stages of intracellular trafficking. In summary, the BvrR-BvrS regulatory system controls several genes that strongly influence Brucella virulence.

The VjbR Quorum-Sensing Regulator

VjbR, for vacuolar hijacking Brucella regulator, was recently described as a quorum-sensing-related transcriptional regulator. vjbR encodes a protein of 260 amino acids with significant homology to transcriptional regulators of the LuxR family that respond to the presence of the pheromone N-acylhomoserine lactone. Interestingly, Brucella *vjbR* mutants are unable to replicate intracellularly, but instead their vacuoles fuse with lysosomes following the same kinetics as a *virB* mutant strain; it is also highly attenuated in mice (Delrue et al., 2005). Indeed, the expression of the *virB* operon during infection is downregulated in the *vjbR* mutant. Besides inducing the expression of the *virB* operon, VjbR also upregulates expression of flagellar genes. Curiously, the homoserine lactone produced by *Brucella* has been proposed to negatively regulate VjbR, which in turn would repress the *virB* operon and flagellar genes. It is not yet known how this regulation is established during the infection nor if VjbR acts directly by binding the *virB* promoter. Nonetheless, it is clear that the VjbR quorum-sensing molecule is a key regulator of *Brucella* virulence genes.

Other Virulence Factors

Several virulence proteins have recently been described for which functions have not yet been ascribed. One such factor is the *Brucella* phosphatidylcholine. This is typically a eukaryotic phospholipid absent from most prokaryotes but that is present in the *Brucella* outer membrane. Mutation of a key enzyme of the *Brucella* phosphatidylcholine biosynthetic pathway leads to reduced ability to avoid BCV-lysosome fusion. In the mouse model, these mutants are attenuated, suggesting that this phospholipid contributes to *Brucella* virulence (Comerci et al., 2006; Conde-Alvarez et al., 2006).

Another virulence protein recently identified is BvfA, a small protein of 11 kDa that is unique to the genus *Brucella* (Lavigne et al., 2005). It was identified in a random screen using the *Yersinia* YopP as a reporter system, which induces apoptosis when secreted into the host cytosol. As YopP alone cannot be secreted by *Brucella*, this *Yersinia* protein can be used to identify proteins that have a secretory signal. BvfA is necessary for intracellular survival within both human and murine macrophages, and *bvfA* mutants are highly attenuated in mice. Although its promoter is induced intracellularly upon acidification, it does not seem to be an effector of the VirB type IV secretion system. The function of BvfA is still unknown, and it is unclear how this small protein predicted to be periplasmic is secreted into host cells.

Several other virulence proteins are required for *Brucella* to survive the harsh conditions in the macrophage. The *hfq* gene product called host factor I (HF-I) is an RNA-binding protein that modulates gene expression in stationary phase. A *Brucella* mutant in the *hfq* gene is able neither to maintain chronic infection in mice nor to survive within cultured macrophages. This mutant is highly susceptible to oxidative killing and low pH (Robertson and Roop, 1999). One set of genes whose expression is modulated by HF-I is the *cydAB* operon that encodes a cytochrome *bd* oxidase, known to protect bacteria in microaerobic conditions, as well as to increase resistance to reactive oxygen intermediates (Endley et al., 2001). Therefore, the CydAB cytochrome oxidase probably protects *Brucella* during infection of macrophages. Another stress encountered by *Brucella* during the early stages of intracellular trafficking is nutrient deprivation. It is therefore not surprising that *Brucella* auxotrophic mutants in essential metabolic pathways are avirulent. This is the case for *Brucella* purine biosynthesis mutants that are highly attenuated in both cultured macrophages and in mice. In response to nutrient starvation, *Brucella* expresses *rsh*, which encodes a ppGpp synthase. Rsh is necessary for virulence, and curiously, it modulates expression of the *virB* operon, highlighting the multitude of factors involved in *Brucella* intracellular survival (Dozot et al., 2006). It also underlines the complexity of virulence regulation and the interaction of various operons within *Brucella* during infection.

Other virulence factors recently identified are necessary for virulence in vivo but are not specifically associated with intracellular survival in cultured macrophages. These include PrpA, which is required for chronic infection in mice and is directly involved in the immune modulation of the host (Spera et al., 2006). PrpA belongs to the proline-racemase family and was shown to elicit B-lymphocyte polyclonal activation and interleukin-10 (IL-10) secretion. Another example is the flagellar apparatus of *Brucella*. Genes encoding for distinct parts of the apparatus including the basal ring, the motor protein, the secretion apparatus, the hook, and the filament were all necessary for the establishment of chronic infection in mice (Fretin et al., 2005). This raises an interesting question regarding the biological role of the flagella apparatus in *Brucella*, which is a nonmotile pathogen. It is possible that components of the apparatus may interact directly with host components or perhaps it is acting as a secretion system for other virulence proteins.

Evading the Immune Response

An innate capacity to avoid cellular killing mechanisms is an essential feature of *Brucella* virulence, which thrives on its ability to remain unnoticed by the host. Although there is increased knowledge about the virulence factors that enable *Brucella* to survive inside host cells, it is still unclear how *Brucella* is able to persist within the host for long periods of time, hidden from the immune system, and cause chronic disease. The escape from the lysosomal pathway and fusion with the ER may interfere with the ability of host cells to mount an efficient immune response against *Brucella* since these constitute important sites for antigen processing, loading, and transport by MHC-II and -I, respectively.

As mentioned previously, one of the main components of the outer bacterial membrane, the LPS, is of very low endotoxicity. Unlike classical LPSs like that of *Escherichia coli*, the atypical *Brucella* LPS does not efficiently induce proinflammatory cytokines (such as tumor necrosis factor-α [TNF-α]) nor antimicrobial proteins (such as the interferon-γ [IFN-γ]-inducible p47 GTPases IIGP and IGTP). It is interesting that *Brucella* LPS is still being recognized via the TLR4 pathway like classical LPSs, without significant induction of proinflammatory cytokines (Lapaque et al., 2006a). In vitro studies using purified *Brucella* LPS have found that induction of TNF-α depended on the adaptor MyD88, whereas induction of the p47 GTPases was MyD88-independent. This highlights how the *Brucella* LPS can trigger different signaling pathways through the same TLR. Further work is now necessary to understand how these differences arise.

Despite the fact that *Brucella* LPS is up to 100 times less toxic than other LPSs, host cells can still detect *Brucella* PAMPs other than LPS and induce cytokine secretion at later stages of the infection. A recent study has shown that a lumazine synthase from *Brucella* can activate dendritic cells via TLR4 (Berguer et al., 2006). In addition, heat-killed *Brucella* was shown to induce TLR9 and promote Th1-mediated responses in both dendritic cells and macrophages, probably after bacterial degradation in lysosomes. Because the majority of internalized *Brucella* are efficiently eliminated by macrophages, they would provide numerous signals to activate these cells. A separate study has implicated *Brucella* lipoproteins in the induction of proinflammatory cytokines by heat-killed bacteria (Giambartolomei et al., 2004). Therefore, it is possible that *Brucella* has also developed strategies to efficiently reduce the host immune response by directly manipulating components of the immune system. One example is the ability of *Brucella* LPS to cluster MHC-II molecules at the cell surface of macrophages, preventing efficient antigen presentation, mentioned earlier. In this way, *Brucella* renders the infected macrophage less visible from an immune standpoint. Another strategy is the ability of *Brucella* to prevent apoptosis of infected macrophages. It is striking that heavily infected cells are still able to perform several cycles of nuclear division (Barquero-Calvo et al., 2007), showing that *Brucella* replication has very reduced toxic effects on host cells. In one study, analysis of *Brucella* infection rendered human monocytes resistant to Fas ligand- or IFN-γ-induced apoptosis, which suggests that *Brucella* infection protects host cells from cytotoxic processes related to the immune response (Gross et al., 2000). *Brucella* inhibition of apoptosis in infected human monocytes requires intracellular replication. Separate studies have shown that *Brucella* infection triggers downregulation of several apoptosis-related genes in mouse macrophages (He et al., 2006; Eskra et al., 2003). It has been proposed that *Brucella* is inhibiting activation of caspase cascades by blocking release of cytochrome c and production of reactive oxygen species in the mitochondria (He et al., 2006). Further experimental work is necessary to determine the bacterial factors contributing to inhibition of apoptosis and which of the host's cellular pathways are being blocked.

Immunity against *Brucella* requires cell-mediated mechanisms that result in the production of cytokines such as IL-12 and IFN-γ. However, in cultured murine macrophages, virulent *Brucella* induces only low levels of cytokines such as IL-12 and TNF-α in comparison with *Salmonella* or heat-killed *Brucella*. Infection of murine macrophages with the vaccine S19 strain, which efficiently replicates intracellularly, showed that both cytokine secretion and TLR activation were delayed in comparison to *Salmonella*-infected cells (Weiss et al., 2005). At early stages of the infection (within the first 6 h) no cytokines were detected in culture supernatants of macrophages infected with *Brucella*, but low levels were detected at 24 h after infection. Secretion of IL-12 and TNF-α was dependent on both TLR4 and TLR2 pathways, consistent with the hypothesis that molecules other than the LPS are being recognized by the host macrophages. Cytokine secretion was also dependent on MyD88. Infection of a series of knockout mice revealed that MyD88$^{-/-}$ mice do not clear the infection as efficiently as the wild-type or TLR2/TLR4$^{-/-}$. Therefore, MyD88 activation is delayed during *Brucella* infection, but it is still important for host resistance against this pathogen. Studies of the global immune response elicited during infection with virulent *Brucella* strains are lacking, so our understanding of *Brucella* pathogenesis in vivo and all the immune effectors involved remains scarce. A recent study analyzed the innate immune response to *Brucella* during the early stages of the infection in mice (Barquero-Calvo et al., 2007). There was no significant proinflammatory response for the first 48 h of infection, and although *Brucella* hampered neutrophil function, the absence of these cells had no immediate effect on *Brucella* survival in the host. In addition, they also showed that the absence of proinflammatory response was due to low induction of the innate immune system by *Brucella* PAMPs rather than active inhibition by specific *Brucella* virulence factors. In this study, neither TLR2 nor TLR4 had any influence on bacterial replication in vivo. Interestingly, they found that activation of macrophages after establishment of an ER-derived replication niche had no effect on bacterial survival. Therefore, *Brucella* virulence relies on its ability to remain unnoticed by the host cell long enough to reach the ER, where it can replicate in a safe haven, protected from macrophage killing.

It is possible that impairment of macrophage and dendritic cell immune functions favors infection and/or promotes the establishment of the chronic phase of the disease. *Brucella* may be using the tolerogenic properties of immune cells such as dendritic cells to subvert the immune response. Consistent with this hypothesis, reduction of IL-10 levels in mice has been shown to improve host resistance to *Brucella* infection, suggesting that this anti-inflammatory cytokine plays a role in brucellosis (Fernandes et al., 1996). In addition, a new *Brucella* virulence protein (PrpA) was recently identified as a potent B-cell mitogen and IL-10 inducer (Spera et al., 2006). Indeed, this protein is necessary for the early immunosuppression observed in *Brucella*-infected mice and the establishment of chronic disease.

OTHER PHAGOCYTIC CELLULAR TARGETS BESIDES MACROPHAGES

Besides the well-characterized murine macrophage model described above, *Brucella* has been shown to survive within other phagocytic cells, namely human monocytes. In these cells, trafficking of opsonized *Brucella* seems to differ from macrophages and epithelial cells (Bellaire et al., 2005; Rittig et al., 2001). Soon after uptake, *Brucella* can be found within LAMP1-positive vacuoles that are negative for ER markers throughout the course of the infection (Bellaire et al., 2005). In this case, *Brucella* seems to survive and replicate within a compartment derived from the endocytic pathway.

Brucella has also been shown to survive within neutrophils. Very little information is available regarding the intracellular behavior of *Brucella* within these cells at present. Some studies suggest that *Brucella* inhibits degranulation in bovine neutrophils, preventing release of antimicrobial hydrogen peroxide derivatives that would kill the bacteria (Iyankan and Singh, 2002; Orduna et al., 1991; Riley and Robertson, 1984b). *Brucella* has also been described to inhibit the respiratory burst in bovine neutrophils. The LPS also has an important role in these cells and it only induces very low level of superoxide and lysosyme production, which contributes to *Brucella* survival. Neutrophils may therefore constitute an important target for *Brucella* infection in the natural host, but further work is required to characterize this cell model.

A final important cellular target recently identified is the dendritic cell. *Brucella* was found to efficiently proliferate within human dendritic cells (Billard et al., 2005). As in macrophages, *Brucella* entry depends on lipid rafts and intracellular replication on the VirB type IV secretion system. In our laboratory, we have also obtained similar results in murine bone marrow-derived dendritic cells. In these cells *Brucella* replicates within the ER as previously described for macrophages, but the cyclic β-1,2-glucan is not required for avoiding fusion with lysosomes. Dendritic cells may constitute an important niche for *Brucella* during infection, which could take advantage of their migratory properties. In addition, *Brucella* may use dendritic cells to control the host immune response and establish chronic disease.

CONCLUSION AND PERSPECTIVES

The ability of *Brucella* to survive and replicate within phagocytic cells is essential for its virulence. In recent years great advances have been made in our understanding of its pathogenesis, due in particular to the availability of the genome sequences. Numerous virulence genes have been identified and analyzed in the context of infection, principally in vitro. It will be important to perform more comparative studies between different cellular models as the cellular specificity of some virulence factors is becoming apparent. In addition, it is now crucial to integrate all the data in the context of in vivo infections where complex cellular and immune factors interplay. This will be essential to grasp how chronic or relapsing brucellosis develops in the host. Not only is *Brucella* able to infect host cells without significant recognition, it is also actively modulating the host immune response to enhance its survival. *Brucella* is a true stealth pathogen.

REFERENCES

Anderson, T. D., and N. F. Cheville. 1986. Ultrastructural morphometric analysis of *Brucella abortus*-infected trophoblasts in experimental placentitis. Bacterial replication occurs in rough endoplasmic reticulum. *Am. J. Pathol.* **124:**226–237.

Arellano-Reynoso, B., N. Lapaque, S. Salcedo, G. Briones, A. E. Ciocchini, R. Ugalde, E. Moreno, I. Moriyon, and J. P. Gorvel. 2005. Cyclic beta-1,2-glucan is a *Brucella* virulence factor required for intracellular survival. *Nat. Immunol.* **6:**618–625.

Barquero-Calvo, E., E. Chaves-Olarte, D. S. Weiss, C. Guzman-Verri, C. Chacon-Diaz, A. Rucavado, I. Moriyon, and E. Moreno. 2007. *Brucella abortus* uses a stealthy strategy to avoid activation of the innate immune system during the onset of infection. *PLoS ONE* **2:**e631.

Bellaire, B. H., R. M. Roop, II, and J. A. Cardelli. 2005. Opsonized virulent *Brucella abortus* replicates within nonacidic, endoplasmic reticulum-negative, LAMP-1-positive phagosomes in human monocytes. *Infect. Immun.* **73:**3702–3713.

Berguer, P. M., J. Mundinano, I. Piazzon, and F. A. Goldbaum. 2006. A polymeric bacterial protein activates dendritic cells via TLR4. *J. Immunol.* **176:**2366–2372.

Billard, E., C. Cazevieille, J. Dornand, and A. Gross. 2005. High susceptibility of human dendritic cells to invasion by the intracellular pathogens *Brucella suis*, *B. abortus*, and *B. melitensis*. *Infect. Immun.* **73:**8418–8424.

Bohin, J. P. 2000. Osmoregulated periplasmic glucans in Proteobacteria. *FEMS Microbiol. Lett.* **186:**11–19.

Boschiroli, M. L., S. Ouahrani-Bettache, V. Foulongne, S. Michaux-Charachon, G. Bourg, A. Allardet-Servent, C. Cazevieille, J. P. Liautard, M. Ramuz, and D. O'Callaghan. 2002. The *Brucella suis* virB operon is induced intracellularly in macrophages. *Proc. Natl. Acad. Sci. USA* **99:**1544–1549.

Briones, G., N. Inon de Iannino, M. Roset, A. Vigliocco, P. S. Paulo, and R. A. Ugalde. 2001. *Brucella abortus* cyclic beta-1,2-glucan mutants have reduced virulence in mice and are defective in intracellular replication in HeLa cells. *Infect. Immun.* **69:**4528–4535.

Celli, J., C. de Chastellier, D. M. Franchini, J. Pizarro-Cerda, E. Moreno, and J. P. Gorvel. 2003. *Brucella* evades macrophage killing via VirB-dependent sustained interactions with the endoplasmic reticulum. *J. Exp. Med.* **198:**545–556.

Celli, J., S. P. Salcedo, and J. P. Gorvel. 2005. *Brucella* coopts the small GTPase Sar1 for intracellular replication. *Proc. Natl. Acad. Sci. USA* **102:**1673–1678.

Chaves-Olarte, E., C. Guzman-Verri, S. Meresse, M. Desjardins, J. Pizarro-Cerda, J. Badilla, J. P. Gorvel, and E. Moreno. 2002. Activation of Rho and Rab GTPases dissociates *Brucella abortus* internalization from intracellular trafficking. *Cell. Microbiol.* **4:**663–676.

Comerci, D. J., S. Altabe, D. de Mendoza, and R. A. Ugalde. 2006. *Brucella abortus* synthesizes phosphatidylcholine from choline provided by the host. *J. Bacteriol.* **188:**1929–1934.

Comerci, D. J., M. J. Martinez-Lorenzo, R. Sieira, J. P. Gorvel, and R. A. Ugalde. 2001. Essential role of the VirB machinery in the maturation of the *Brucella abortus*-containing vacuole. *Cell. Microbiol.* **3:**159–168.

Conde-Alvarez, R., M. J. Grillo, S. P. Salcedo, M. J. de Miguel, E. Fugier, J. P. Gorvel, I. Moriyon, and M. Iriarte. 2006. Synthesis of phosphatidylcholine, a typical eukaryotic phospholipid, is necessary for full virulence of the intracellular bacterial parasite *Brucella abortus*. *Cell. Microbiol.* **8:**1322–1335.

Delrue, R. M., C. Deschamps, S. Leonard, C. Nijskens, I. Danese, J. M. Schaus, S. Bonnot, J. Ferooz, A. Tibor, X. De Bolle, and J. J. Letesson. 2005. A quorum-sensing regulator controls expression of both the type IV secretion system and the flagellar apparatus of *Brucella melitensis*. *Cell. Microbiol.* **7:**1151–1161.

Delrue, R. M., M. Martinez-Lorenzo, P. Lestrate, I. Danese, V. Bielarz, P. Mertens, X. De Bolle, A. Tibor, J. P. Gorvel, and J. J. Letesson. 2001. Identification of *Brucella* spp. genes involved in intracellular trafficking. *Cell. Microbiol.* **3:**487–497.

Dozot, M., R. A. Boigegrain, R. M. Delrue, R. Hallez, S. Ouahrani-Bettache, I. Danese, J. J. Letesson, X. De Bolle, and S. Kohler. 2006. The stringent response mediator Rsh is required for *Brucella melitensis* and *Brucella suis* virulence, and for expression of the type IV secretion system virB. *Cell. Microbiol.* **8:**1791–1802.

Endley, S., D. McMurray, and T. A. Ficht. 2001. Interruption of the cydB locus in *Brucella abortus* attenuates intracellular survival and virulence in the mouse model of infection. *J. Bacteriol.* **183:**2454–2462.

Eskra, L., A. Mathison, and G. Splitter. 2003. Microarray analysis of mRNA levels from RAW264.7 macrophages infected with *Brucella abortus*. *Infect. Immun.* **71:**1125–1133.

Fernandes, D. M., and C. L. Baldwin. 1995. Interleukin-10 downregulates protective immunity to *Brucella abortus*. *Infect. Immun.* **63:**1130–1133.

Fontes, P., M. T. Alvarez-Martinez, A. Gross, C. Carnaud, S. Kohler, and J. P. Liautard. 2005. Absence of evidence for the participation of the macrophage cellular prion protein in infection with *Brucella suis*. *Infect. Immun.* **73:**6229–6236.

Forestier, C., F. Deleuil, N. Lapaque, E. Moreno, and J. P. Gorvel. 2000. *Brucella abortus* lipopolysaccharide in murine peritoneal macrophages acts as a down-regulator of T cell activation. *J. Immunol.* **165:**5202–5210.

Forestier, C., E. Moreno, J. Pizarro-Cerda, and J. P. Gorvel. 1999. Lysosomal accumulation and recycling of lipopolysaccharide to the cell surface of murine macrophages, an in vitro and in vivo study. *J. Immunol.* **162:**6784–6791.

Foulongne, V., G. Bourg, C. Cazevieille, S. Michaux-Charachon, and D. O'Callaghan. 2000. Identification of *Brucella suis* genes affecting intracellular survival in an in vitro human macrophage infection model by signature-tagged transposon mutagenesis. *Infect. Immun.* **68:**1297–1303.

Fretin, D., A. Fauconnier, S. Kohler, S. Halling, S. Leonard, C. Nijskens, J. Ferooz, P. Lestrate, R. M. Delrue, I. Danese, J. Vandenhaute, A. Tibor, X. DeBolle, and J. J. Letesson. 2005. The sheathed flagellum of *Brucella melitensis* is involved in persistence in a murine model of infection. *Cell. Microbiol.* **7:**687–698.

Giambartolomei, G. H., A. Zwerdling, J. Cassataro, L. Bruno, C. A. Fossati, and M. T. Philipp. 2004. Lipoproteins, not lipopolysaccharide, are the key mediators of the proinflammatory response elicited by heat-killed *Brucella abortus*. *J. Immunol.* **173:**4635–4642.

Gross, A., A. Terraza, S. Ouahrani-Bettache, J. P. Liautard, and J. Dornand. 2000. In vitro *Brucella suis* infection prevents the programmed cell death of human monocytic cells. *Infect. Immun.* **68:**342–351.

He, Y., S. Reichow, S. Ramamoorthy, X. Ding, R. Lathigra, J. C. Craig, B. W. Sobral, G. G. Schurig, N. Sriranganathan, and S. M. Boyle. 2006. *Brucella melitensis* triggers time-dependent modulation of apoptosis and down-regulation of mitochondrion-associated gene expression in mouse macrophages. *Infect. Immun.* **74:**5035–5046.

Hong, P. C., R. M. Tsolis, and T. A. Ficht. 2000. Identification of genes required for chronic persistence of *Brucella abortus* in mice. *Infect. Immun.* **68:**4102–4107.

Inon de Iannino, N., G. Briones, M. Tolmasky, and R. A. Ugalde. 1998. Molecular cloning and characterization of cgs, the *Brucella abortus* cyclic beta(1-2) glucan synthetase gene: genetic complementation of *Rhizobium meliloti* ndvB and *Agrobacterium tumefaciens* chvB mutants. *J. Bacteriol.* **180:**4392–4400.

Iyankan, L., and D. K. Singh. 2002. The effect of *Brucella abortus* on hydrogen peroxide and nitric oxide production by bovine polymorphonuclear cells. *Vet. Res. Commun.* **26:**93–102.

Jimenez de Bagues, M. P., A. Terraza, A. Gross, and J. Dornand. 2004. Different responses of macrophages to smooth and rough *Brucella* spp.: relationship to virulence. *Infect. Immun.* **72:**2429–2433.

Kim, S., M. Watarai, H. Suzuki, S. Makino, T. Kodama, and T. Shirahata. 2004. Lipid raft microdomains mediate class A scavenger receptor-dependent infection of *Brucella abortus*. *Microb. Pathog.* **37:**11–19.

Lapaque, N., F. Forquet, C. de Chastellier, Z. Mishal, G. Jolly, E. Moreno, I. Moriyon, J. E. Heuser, H. T. He, and J. P. Gorvel. 2006a. Characterization of *Brucella abortus* lipopolysaccharide macrodomains as mega rafts. *Cell. Microbiol.* **8:**197–206.

Lapaque, N., I. Moriyon, E. Moreno, and J. P. Gorvel. 2005. *Brucella* lipopolysaccharide acts as a virulence factor. *Curr. Opin. Microbiol.* **8:**60–66.

Lapaque, N., O. Takeuchi, F. Corrales, S. Akira, I. Moriyon, J. C. Howard, and J. P. Gorvel. 2006b. Differential inductions of TNF-alpha and IGTP, IIGP by structurally diverse classic and non-classic lipopolysaccharides. *Cell. Microbiol.* **8:**401–413.

Lavigne, J. P., G. Patey, F. J. Sangari, G. Bourg, M. Ramuz, D. O'Callaghan, and S. Michaux-Charachon. 2005. Identification of a new virulence factor, BvfA, in *Brucella suis*. *Infect. Immun.* **73:**5524–5529.

Martinez de Tejada, G., J. Pizarro-Cerda, E. Moreno, and I. Moriyon. 1995. The outer membranes of *Brucella* spp. are resistant to bactericidal cationic peptides. *Infect. Immun.* **63:**3054–3061.

Meador, V. P., and B. L. Deyoe. 1989. Intracellular localization of *Brucella abortus* in bovine placenta. *Vet. Pathol.* **26:**513–515.

Naroeni, A., and F. Porte. 2002. Role of cholesterol and the ganglioside GM(1) in entry and short-term survival of *Brucella suis* in murine macrophages. *Infect. Immun.* **70:**1640–1644.

O'Callaghan, D., C. Cazevieille, A. Allardet-Servent, M. L. Boschiroli, G. Bourg, V. Foulongne, P. Frutos, Y. Kulakov, and M. Ramuz. 1999. A homologue of the *Agrobacterium tumefaciens* VirB and *Bordetella pertussis* Ptl type IV secretion systems is essential for intracellular survival of *Brucella suis*. *Mol. Microbiol.* **33:**1210–1220.

Orduna, A., C. Orduna, J. M. Eiros, M. A. Bratos, P. Gutierrez, P. Alonso, and A. Rodriguez Torres. 1991. Inhibition of the degranulation and myeloperoxidase activity of human polymorphonuclear neutrophils by *Brucella melitensis*. *Microbiologia* **7:**113–119.

Pappas, G., P. Papadimitriou, N. Akritidis, L. Christou, and E. V. Tsianos. 2006. The new global map of human brucellosis. *Lancet Infect. Dis.* **6:**91–99.

Pizarro-Cerda, J., S. Meresse, R. G. Parton, G. van der Goot, A. Sola-Landa, I. Lopez-Goni, E. Moreno, and J. P. Gorvel. 1998. *Brucella abortus* transits through the autophagic pathway and replicates in the endoplasmic reticulum of nonprofessional phagocytes. *Infect. Immun.* **66:**5711–5724.

Porte, F., J. P. Liautard, and S. Kohler. 1999. Early acidification of phagosomes containing *Brucella suis* is essential for intracellular survival in murine macrophages. *Infect. Immun.* **67:**4041–4047.

Porte, F., A. Naroeni, S. Ouahrani-Bettache, and J. P. Liautard. 2003. Role of the *Brucella suis* lipopolysaccharide O antigen in phagosomal genesis and in inhibition of phagosome-lysosome fusion in murine macrophages. *Infect. Immun.* **71:**1481–1490.

Rajashekara, G., D. A. Glover, M. Krepps, and G. A. Splitter. 2005. Temporal analysis of pathogenic events in virulent and avirulent *Brucella melitensis* infections. *Cell. Microbiol.* **7:**1459–1473.

Rasool, O., E. Freer, E. Moreno, and C. Jarstrand. 1992. Effect of *Brucella abortus* lipopolysaccharide on oxidative metabolism and lysozyme release by human neutrophils. *Infect. Immun.* **60:**1699–1702.

Riley, L. K., and D. C. Robertson. 1984a. Brucellacidal activity of human and bovine polymorphonuclear leukocyte granule extracts against smooth and rough strains of *Brucella abortus*. *Infect. Immun.* **46:**231–236.

Riley, L. K., and D. C. Robertson. 1984b. Ingestion and intracellular survival of *Brucella abortus* in human and bovine polymorphonuclear leukocytes. *Infect. Immun.* **46:**224–230.

Rittig, M. G., M. T. Alvarez-Martinez, F. Porte, J. P. Liautard, and B. Rouot. 2001. Intracellular survival of *Brucella* spp. in human monocytes involves conventional uptake but special phagosomes. *Infect. Immun.* **69:**3995–4006.

Robertson, G. T., and R. M. Roop, Jr. 1999. The *Brucella abortus* host factor I (HF-I) protein contributes to stress resistance during stationary phase and is a major determinant of virulence in mice. *Mol. Microbiol.* **34:**690–700.

Rolan, H. G., and R. M. Tsolis. 2007. Mice lacking components of adaptive immunity show increased *Brucella abortus* virB mutant colonization. *Infect. Immun.* **75:**2965–2973.

Sieira, R., D. J. Comerci, D. O. Sanchez, and R. A. Ugalde. 2000. A homologue of an operon required for DNA transfer in *Agrobacterium* is required in *Brucella abortus* for virulence and intracellular multiplication. *J. Bacteriol.* **182:**4849–4855.

Sola-Landa, A., J. Pizarro-Cerda, M. J. Grillo, E. Moreno, I. Moriyon, J. M. Blasco, J. P. Gorvel, and I. Lopez-Goni. 1998. A two-component regulatory system playing a critical role in plant pathogens and endosymbionts is present in *Brucella abortus* and controls cell invasion and virulence. *Mol. Microbiol.* **29:**125–138.

Spera, J. M., J. E. Ugalde, J. Mucci, D. J. Comerci, and R. A. Ugalde. 2006. A B lymphocyte mitogen is a *Brucella abortus* virulence factor required for persistent infection. *Proc. Natl. Acad. Sci. USA* **103:**16514–16519.

Watarai, M., S. Kim, J. Erdenebaatar, S. Makino, M. Horiuchi, T. Shirahata, S. Sakaguchi, and S. Katamine. 2003. Cellular prion protein promotes *Brucella* infection into macrophages. *J. Exp. Med.* **198:**5–17.

Watarai, M., S. Makino, Y. Fujii, K. Okamoto, and T. Shirahata. 2002. Modulation of *Brucella*-induced macropinocytosis by lipid rafts mediates intracellular replication. *Cell. Microbiol.* **4:**341–355.

Weiss, D. S., K. Takeda, S. Akira, A. Zychlinsky, and E. Moreno. 2005. MyD88, but not toll-like receptors 4 and 2, is required for efficient clearance of *Brucella abortus*. *Infect. Immun.* **73:**5137–5143.

28

Interaction of *Candida albicans* with Phagocytes

INÊS FARO-TRINDADE AND GORDON D. BROWN

The immune system of healthy individuals is sufficient to control infection with most fungi, except for a few species such as *Coccidioides*. However, immunosuppressed hosts are much more susceptible, and the past few decades have seen an increase in the number of these individuals, as a result of modern medical interventions and AIDS. Consequently, fungi are now the fourth most common cause of nosocomial bloodstream infections (Kauffman, 2006), and despite the availability of antifungal agents, systemic infections with these organisms have poor prognosis and high rates of mortality. Much effort is now being placed on understanding the mechanisms of immunity to these pathogens.

One of the central players in antifungal immunity is phagocytic cells. Recognition by these cells leads to fungal uptake and killing and the induction of an inflammatory response. This ultimately initiates protective T_H1-type adaptive immunity, which in turn enhances the fungicidal mechanisms of phagocytes, through the actions of cytokines and other factors, such as gamma interferon (IFN-γ) (Romani, 2004). The increased risk of fungal infection resulting from perturbations in the levels of the number of phagocytes emphasizes the importance of these cells in the control of mycoses (Ashman et al., 2004).

Here we will discuss the role of phagocytes in immunity to *Candida albicans*. Although this fungus is a normal commensal in most healthy individuals, it is an opportunistic pathogen that can cause a variety of diseases ranging from mucosal forms, such as vulvovaginal candidiasis, to systemic infections (candidemia). In immunocompromised individuals, infection with *C. albicans* is a prevalent cause of invasive fungal disease, and candidemia has an attributable mortality rate of nearly 50% (Gudlaugsson et al., 2003; Wisplinghoff et al., 2004). In this chapter, we will cover the mechanisms that phagocytic cells use to recognize, ingest, and kill this pathogen, as well as those mechanisms involved in the induction of inflammatory and adaptive response to these organisms. Along the way, we will highlight some of the strategies utilized by this pathogen to subvert the phagocyte and escape these responses.

Inês Faro-Trindade and Gordon D. Brown, Institute of Infectious Disease and Molecular Medicine, Division of Immunology, University of Cape Town, Observatory, 7925, Cape Town, South Africa.

RECOGNITION OF *CANDIDA*

The innate detection of *Candida* is achieved through the sensing of pathogen-associated molecular patterns (PAMPs) found primarily in the cell wall, although fungal DNA may also be detected (Bellocchio et al., 2004). The cell wall is a rigid yet dynamic structure, giving the organism its shape and providing protection from harmful external factors such as turgor pressure. Consisting predominantly of carbohydrates, the cell wall is thought to be layered and made up of an inner meshwork of β-glucans and chitin, covered by an outer layer of mannosylated proteins (mannan) (Color Plate 8) (Klis et al., 2001). Some of the inner meshwork, however, is exposed on the cell surface in specific areas, such as bud scars, allowing immune recognition of these structures (Gantner et al., 2005).

All of the major fungal cell wall macromolecules act as PAMPs and are recognized by a number of germ line-encoded pattern recognition receptors (PRRs), although most of these receptors appear to recognize the various mannan-based structures present in the outer cell wall layer (Table 1). Recognition occurs both nonopsonically, through membrane-bound PRRs, and opsonically, where soluble PRRs, such as complement, opsonize the organism and allow recognition through receptors such as complement receptor 3 (CR3). Opsonic recognition is also enhanced following the initiation of the adaptive response, by the production of specific anti-*Candida* antibodies that are present in the serum of most individuals. Although involving multiple interactions with multiple receptors, different phagocytes appear to utilize different combinations of receptors to recognize *Candida*. The recognition of mannan, for example, is mediated primarily by the mannose receptor (MR) on macrophages (Netea et al., 2006), whereas both the MR and DC-SIGN contribute to the recognition of these structures in dendritic cells (DCs) (Cambi et al., 2003).

Recognition of *Candida* also depends on the morphological form of the pathogen. *C. albicans* is polymorphic and can reversibly switch between yeast and filamentous forms (pseudohyphae and hyphae; Fig. 1). All of these forms are present in infected tissues, and the ability of *Candida* to undergo morphogenic switching is thought to con-

TABLE 1 Selected nonopsonic PRRs that recognize *C. albicans*

Location	PRR	PAMP(s)	Reference
Soluble	SP-A	Mannan	Kishore et al., 2006
	SP-D	Mannan	Kishore et al., 2006
	Galectin-3	β-1,2-Mannan	Kohatsu et al., 2006
	Mannose-binding lectin	Mannan	Kilpatrick, 2002
Membrane	TLR2	Phospholipomannan	Jouault et al., 2003
	TLR4	O-Mannan	Netea et al., 2006
	Dectin-1	β-Glucan	Brown, 2006
	Dectin-2	Hyphal mannan	Sato et al., 2006
	CR3	β-Glucan, mannan	Brown and Gordon, 2005
	DC-SIGN	Mannan	Cambi et al., 2003
	MR	N-Mannan	Taylor et al., 2005
	CD14	Mannan	Tada et al., 2002
	?	Chitin	Reese et al., 2007
Intracellular	TLR9	DNA	Bellocchio et al., 2004

tribute to virulence (Lo et al., 1997). Recognition of these morphological forms gives rise to different immune responses (see below), which are likely to be mediated by recognition through different combinations of receptors (Romani et al., 2004). Although this is an emerging area of investigation, the evidence suggests that hyphal forms of *Candida*, for example, are preferentially recognized by selected receptors, such as Dectin-2 (Sato et al., 2006). Furthermore, filamentous forms of *Candida* do not display certain components on their surface, such as β-glucan, which prevents recognition by receptors such as Dectin-1 (Gantner et al., 2005). Indeed, the masking of these components may be an immune evasion strategy employed by fungi, and recent evidence suggests that inducing their "unmasking" may provide a novel therapeutic approach (Rappleye et al., 2007; Wheeler and Fink, 2006).

PHAGOCYTOSIS OF *CANDIDA*

Recognition of *Candida* leads to ingestion of the fungus through the actin-dependent process of phagocytosis. The uptake of *Candida* involves several different phagocytic mechanisms which depend on the cell type, presence of opsonins, and the morphological form of the fungus (Marodi et al., 1991a, 1991b; Romani, et al., 2004). In DCs, for example, the ingestion of yeasts occurs through coiling phagocytosis, whereas hyphae are ingested by a classical "zipper-type" mechanism (d'Ostiani et al., 2000) (Fig. 2). Several opsonic (complement and Fc receptors) and nonopsonic receptors (the MR, DC-SIGN, and Dectin-1) implicated in this process have mostly been studied with the yeast form of the pathogen. Phagocytosis may also be influenced by the anatomical location in which the phagocyte-fungal interaction occurs, as recent data suggest that the uptake of *Candida* is less efficient in two-dimensional environments, such as mucosal surfaces, than in three dimensions, such as within tissues (Behnsen et al., 2007). Here, we will focus on fungal phagocytosis mediated through the nonopsonic receptors, and the reader is referred to earlier chapters that detail the mechanisms involved in phagocytosis mediated by the opsonic (Fc and complement) receptors.

MR-Mediated Phagocytosis

MR (CD206) was one of the first fungal PRRs to be identified and has been the focus of considerable interest, particularly given the exposure of mannose-based structures on the outermost surface of fungal cell walls (Klis et al., 2001). This receptor is widely expressed on tissue macrophages, but also by some DCs and other cells in tissues, such as lymphatic and hepatic endothelia, although the majority of the MR is located intracellularly, within the endocytic pathway (McKenzie et al., 2007; Taylor et al., 2005a). The MR is a transmembrane protein possessing an extracellular cysteine-rich NH_2-terminal domain, a fibronectin type II repeat domain, and eight carbohydrate recognition domains (CRDs) linked to a short cytoplasmic tail. A soluble form of the receptor, generated through proteolytic cleavage and containing the entire extracellular portion of the MR, can be shed into the serum (Martinez-Pomares et al., 1998). In macrophages, fungi are able to enhance MR shedding, which inhibits the nonopsonic recognition of these organisms, and may therefore be an immune evasion mechanism (Fraser et al., 2000).

The MR is an endocytic receptor that recognizes a number of endogenous and exogenous ligands, including several pathogens, through its various extracellular domains. CRDs 4 to 8, in particular, recognize carbohydrates, such as the branched N-mannans of *C. albicans*, in a calcium-dependent fashion (Netea et al., 2006; Taylor et al., 2005a). The MR can mediate the phagocytosis of yeast in transfected Chinese hamster ovary cells, an activity that depends on the cytoplasmic tail of the receptor (Ezekowitz et al., 1990). Further, characterization in macrophages sug-

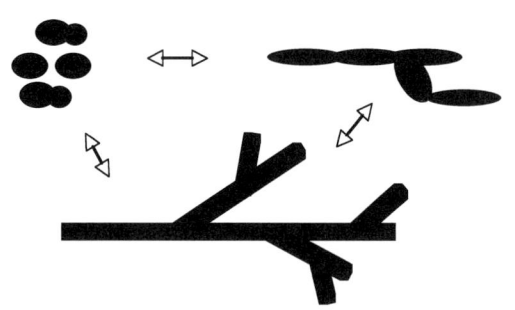

FIGURE 1 Cartoon representation of the various morphologies (yeast, hyphae, and pseudohyphae) of *C. albicans*. Reprinted with permission from Heinsbroek et al. (2005).

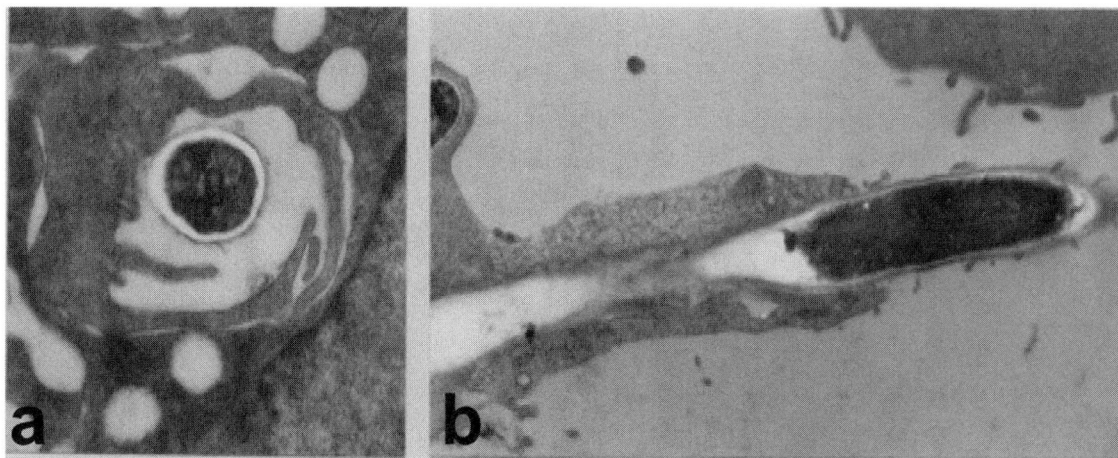

FIGURE 2 DC phagocytosis of yeast and hyphae of C. *albicans*. (a) Phagocytosis of yeasts occurs through coiling phagocytosis, whereas (b) phagocytosis of hyphae occurs through a "zipper-like" mechanism. Reprinted with permission from d'Ostiani et al. (2000).

gested that yeast uptake was different from that mediated by the Fc or complement receptors, and involved Cdc42 and RhoB guanosine triphosphatases (GTPases), as well as utilization of cytoskeletal proteins, such as F-actin, talin, and MARCKS, but not vinculin and paxillin (Allen and Aderem, 1995, 1996; Zhang et al., 2005). However, the MR lacks defined phagocytic motifs in its cytoplasmic tail and the exact mechanism behind this activity is unclear. More recently, the phagocytic ability of this receptor has been questioned, as expression of this receptor in a number of other cell lines did not confer an ability to ingest yeast particles (Le Cabec et al., 2005). Furthermore, phagocytosis of *Candida* was unaffected in leukocytes lacking the MR (Lee et al., 2003). It has been proposed that the MR may primarily be involved in mediating fungal binding, and that other protein(s) are required to induce uptake (Le Cabec et al., 2005).

DC-SIGN-Mediated Phagocytosis

Human DC-specific ICAM-3-grabbing nonintegrin (DC-SIGN; CD209) is a C-type lectin receptor (CLR) that possesses a single extracellular CRD, a neck region composed of seven tandem repeat sequences which enables tetramerization of the receptor, and a cytoplasmic tail containing internalization motifs (Koppel et al., 2005). DC-SIGN is primarily expressed by immature DCs, but has also been described on selected macrophage subsets and endothelium (Lai et al., 2006; Tailleux et al., 2005). Through the calcium-dependent recognition of carbohydrates, such as mannan, DC-SIGN recognizes a number of microbial pathogens, including C. *albicans* and several endogenous ligands (Cambi et al., 2003). However, only two of the eight murine orthologues (SIGNR1 and SIGNR3) have been shown to recognize fungi (Powlesland et al., 2006; Takahara et al., 2004; Taylor et al., 2004). DC-SIGN becomes enriched on *Candida* phagosomes in immature DCs, and has been proposed to mediate the uptake of these organisms (Cambi et al., 2003; Serrano-Gomez et al., 2004). Although DC-SIGN can internalize soluble antigens through endocytosis, as yet there is no direct evidence that supports an active role for DC-SIGN in phagocytosis (Engering et al., 2002). Moreover, one murine orthologue (SIGNR1) was shown to be poorly phagocytic (Taylor et al., 2004).

Dectin-1-Mediated Phagocytosis

Dectin-1 is a nonclassical CLR that possesses a single extracellular CRD, a stalk region, and a cytoplasmic tail containing an immunoreceptor tyrosine-based activation-like (ITAM-like) motif (Brown, 2006). Another isoform of Dectin-1, which lacks the stalk region, is generated by alternative splicing and has slightly different characteristics (Brown, 2006; Heinsbroek et al., 2006). The receptor is expressed by a number of phagocytes, including macrophages, DCs, and neutrophils, but also on certain lymphocytes (Brown, 2006). Dectin-1 specifically recognizes β-1,3-glucans in a calcium-independent fashion, through an unknown mechanism, and is the major leukocyte receptor for these carbohydrates (Brown, 2006; Palma et al., 2006).

Through recognition of these carbohydrates, Dectin-1 can bind a variety of fungi, including the yeast form of C. *albicans*, and can mediate their uptake by phagocytosis (Brown, 2006; Gantner et al., 2005). This activity depends on the cytoplasmic ITAM-like motif of the receptor, but involves only the membrane proximal tyrosine (Brown, 2006; Herre et al., 2004). With a variety of inhibitors, Dectin-1-mediated phagocytosis was shown to require tyrosine phosphorylation and involve phosphatidylinositol 3-kinase, protein kinase C, and the Rho GTPases Cdc42 and Rac-1 (Herre et al., 2004). In macrophages, Dectin-1-mediated phagocytosis is independent of p72Syk, a protein kinase essential for phagocytosis mediated by the ITAM motif of Fc receptors (Herre et al., 2004). However, the requirement for Syk may vary in different leukocytes, as yeast uptake in DCs, for example, depends in part on this kinase (Rogers et al., 2005). Dectin-1 may also associate with other receptors, such as CD63 and PTX3, during fungal phagocytosis, but the significance of these interactions is unclear (Diniz et al., 2004; Mantegazza et al., 2004).

PHAGOSOME MATURATION AND *CANDIDA*

After internalization, the phagosome matures through a number of sequential steps involving extensive vesicle budding and fusion, generating an increasingly antimicrobial environment (the phagolysosome) in which the ingested microbe is ultimately killed and digested. The mechanisms and processes involved in phagosome maturation are described in detail in earlier chapters. The fate of the receptors involved in mediating microbial uptake is determined by their cytoplasmic domains, and they either traffic with the microbe and are degraded in phagolysosomes, such as occurs with Dectin-1, or they are recycled back to the cell surface, such as occurs with the MR (Herre et al., 2004; Taylor et al., 2005a).

Interestingly, the intracellular fate and rate of surface recovery of Dectin-1 depends on the nature of the β-glucan ligand. With intact fungal particles, Dectin-1 traffics down to phagolysosomes in macrophages and the recovery of the receptor at the cell surface depends on de novo protein synthesis. In contrast, little de novo synthesis has been obtained with soluble β-glucan polymers, and Dectin-1 has been observed to recycle back to the cell surface with short polymers, but is retained intracellularly, in an undefined intracellular vesicle, with longer β-glucan polymers, leading to prolonged reduction in the surface levels of this receptor (Herre et al., 2004; Ozment-Skelton et al., 2006). Patients with systemic fungal infections, including candidemia, have been described to possess high levels of circulating β-glucans in their plasma, and this has been proposed to be an indicator of mycotic infection (Obayashi et al., 1995). It is therefore likely that the ability of these carbohydrates to modulate the levels of Dectin-1 on the leukocyte surface influences the immune response to these pathogens.

In macrophages, the phagocytosis of Candida is accompanied by the rapid recruitment of endosomes and lysosomes (Kaposzta et al., 1999). It has been suggested that the rate of internalization and phagosome maturation can be regulated by signals from the Toll-like receptors (TLRs) (Blander and Medzhitov, 2004). The TLRs are PRRs that play an essential role in the initiation of inflammatory responses (see below), and consist of a family of proteins that possess extracellular leucine-rich repeat ligand-binding domains and a conserved intracellular Toll/interleukin-1 (IL-1) receptor (TIR) signaling domain. Ligand recognition leads to the initiation of specific signaling cascades mediated through intracellular TIR-containing adaptors, including MyD88 (Akira and Takeda, 2004). While the absence of MyD88 does slightly impair Candida internalization, the absence of individual TLRs (TLR2, TLR4, and TLR9) had no effect on uptake (Bellocchio et al., 2004; Marr et al., 2003). These findings are in line with a more recent study demonstrating that phagosome maturation occurs independently of the TLRs, but can be affected by MyD88 deficiency (Yates and Russell, 2005).

Candida can escape from the leukocyte following uptake. In both DCs and macrophages, the fungus can generate hyphae within the phagolysosome, an activity dependent on pH (Fig. 3). The formation of hyphae leads to the vesicle membrane becoming distended and rupturing, ultimately leading to destruction of the host cell (d'Ostiani et al., 2000; Kaposzta et al., 1999). This activity depends on a number of Candida-derived secreted aspartyl proteases that are known virulence factors of this pathogen (Borg-von Zepelin et al., 1998). In contrast, neutrophils are able to block hyphal development and thus prevent escape of Candida from the phagosome (Fradin et al., 2005). Candida may also be able to escape through the induction of host cell apoptosis, and although the mechanisms of this process are unclear, it requires living fungal cells and cell wall phospholipomannan (Ibata-Ombetta et al., 2003).

CANDIDA KILLING BY PHAGOCYTES

Phagocytes play essential roles in killing extracellular and internalized Candida, and defects in their antimicrobial functions lead to an increased risk of fungal infection (Romani, 2004). However, the antimicrobial ability of these cells depends on their state of activation, which can be modulated by cytokines and other agents. Nonactivated macrophages, for example, have a poor candidacidal activity, which can be greatly enhanced by activation with IFN-γ (Vazquez-Torres and Balish, 1997). The various phagocytes also have different abilities to kill Candida, with neutrophils being the most potent effector cells, followed

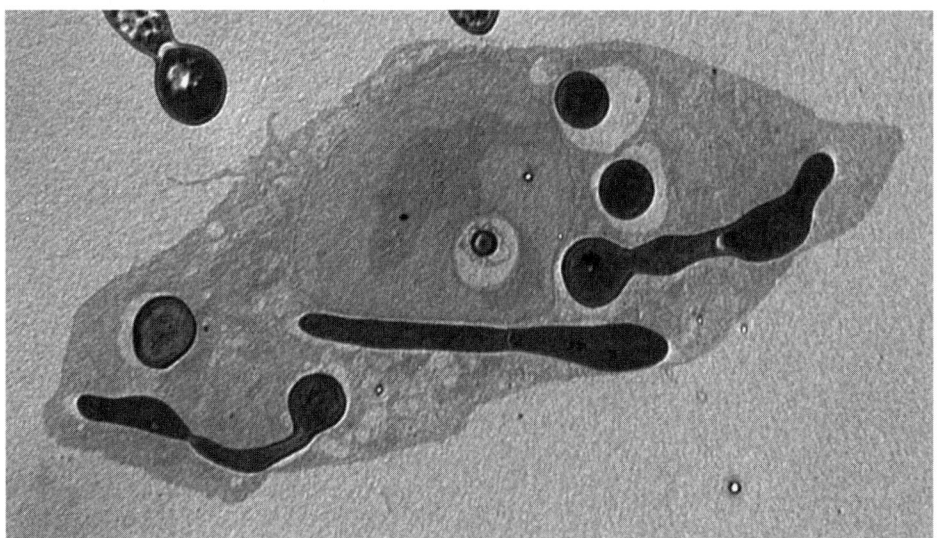

FIGURE 3 Transmission electron microscopy image demonstrating C. albicans undergoing yeast-to-hyphal transition following phagocytosis in macrophages. Reprinted with permission from Heinsbroek et al. (2005).

by monocytes, macrophages, and then DCs (Fradin et al., 2005; Netea et al., 2004a; Vonk et al., 2002). The killing of Candida is achieved through the synergistic actions of a number of oxidative and nonoxidative antimicrobial mechanisms.

Oxidative Candidacidal Mechanisms

One of the key oxidative candidacidal activities of phagocytes is the respiratory burst, which culminates in the production of reactive oxygen intermediates (ROIs) and is mediated by the phagocyte NADPH oxidase (Phox), a membrane-associated protein complex that generates superoxide through the transfer of electrons from NADPH to O_2. Activation of the NADPH oxidase complex involves numerous proteins, including phosphatidylinositol 3-kinase, phospholipase C, protein kinase C, and GTP-binding proteins, in signal transduction pathways leading from the cellular receptor involved in microbial recognition to assembly of the complex around flavocytochrome b at the membrane (Clark, 1999; Karlsson and Dahlgren, 2002; Quinn and Gauss, 2004). The importance of this complex is highlighted by chronic granulomatous disease (CGD), a disease caused by genetic mutations that result mostly in the absence of a functional oxidase complex (Heyworth et al., 2003). Patients with CGD suffer from recurrent, and often fatal, bacterial and fungal infections, including those caused by Candida (Antachopoulos et al., 2007). This disease, and the associated increased susceptibility to candidiasis, has been recapitulated in murine models (Aratani et al., 2002).

The superoxide generated by the NADPH complex has limited, if any, toxicity, but can be converted into a variety of ROIs, such as H_2O_2 and hydroxyl radicals, which have candidacidal activity (Vazquez-Torres and Balish, 1997). Hydrogen peroxide can further be converted into hypochlorous acid, an extremely toxic and effective candidacidal oxidant (Diamond et al., 1980; Marodi et al., 1991a, 1991b). This reaction is catalyzed by myeloperoxidase (MPO), an enzyme located in azurophilic granules of neutrophils and in lysosomes of monocytes. Although macrophages are deficient in MPO, they can scavenge this enzyme, through their MRs, and traffic the enzyme to lysosomes where it can contribute to the killing of ingested Candida (Lefkowitz et al., 1996; Shepherd and Hoidal, 1990). In contrast to CGD, most people with MPO deficiency are asymptomatic, although this deficiency can predispose individuals to infections with Candida (Kitahara et al., 1981). Deficiency of MPO in mice, however, increased the susceptibility to infections with this pathogen (Aratani et al., 1999).

A number of PRRs involved in fungal recognition have been implicated in triggering the respiratory burst, including Galectin-3, TLRs, Dectin-1, and the Fcγ receptors. Galectin-3 is a member of the β-galactosidase-binding family lectins, consisting of one CRD fused to tandem repeats of short amino acid sequences, which mediates oligomerization on binding of multivalent ligands. The receptor has many functions and is produced by a variety of cells including monocytes, macrophages, and epithelial cells. It is located both intracellularly and extracellularly, being secreted through nonclassical pathways, and can recognize β-1,2-linked mannans present on the Candida surface (Kohatsu et al., 2006; Rabinovich and Gruppi, 2005). Galectin-3 also has fungicidal activity, being able to kill Candida on binding (Kohatsu et al., 2006). In macrophages, Galectin-3 contributes to phagocytosis mediated by the Fcγ and other receptors, but it does not appear to be required for Candida uptake (Jouault et al., 2006; Sano et al., 2003). Galectin-3 induces the respiratory burst in activated neutrophils, although the mechanism is undefined (Karlsson et al., 1998).

There is evidence to suggest that the TLRs can induce and/or prime the respiratory burst in leukocytes. In neutrophils, for example, stimulation with purified TLR2 and TLR4 agonists directly induces the respiratory burst and primes for an enhanced response, when the cells are subsequently stimulated with other microbial agonists (Sabroe et al., 2003). TLR stimulation can also prime macrophages for an enhanced Dectin-1-mediated respiratory burst to fungal particles (Gantner et al., 2003).

Dectin-1 is capable of directly triggering the respiratory burst in macrophages and neutrophils in response to fungi (Gantner et al., 2003; Kennedy et al., 2007; Taylor et al., 2007). For both Dectin-1 and the Fcγ receptors, this response depends on the cytoplasmic ITAM sequences and signaling via Syk kinase (Gantner et al., 2003; Swanson and Hoppe, 2004; Underhill et al., 2005). However, in contrast to the Fcγ receptors (discussed in detail in a previous chapter), the association of Syk with Dectin-1 occurs through a novel mechanism involving only one cytoplasmic tyrosine (Rogers et al., 2005). In addition, the ability of Dectin-1 to trigger the respiratory burst appears to be restricted to specific subset(s) of macrophages, which are capable of activating Syk (Underhill et al., 2005).

Another oxidative system involved in the antimicrobial activity of phagocytes is the production of reactive nitrogen intermediates (RNIs) by the inducible nitric oxide synthase (iNOS or NOS2). Induction of iNOS occurs either individually or synergistically by TLR agonists, such as lipopolysaccharide, and cytokines, such as tumor necrosis factor (TNF) and IFN-γ. Induction of iNOS leads to the production of nitric oxide (NO), through the oxidative deamination of L-arginine (MacMicking et al., 1997). While NO itself has poor candidacidal activity, it can react with superoxide, generated by the respiratory burst, to produce peroxynitrite, which is very effective at killing Candida (Vazquez-Torres et al., 1996). Although Candida infections have not been studied in iNOS-deficient mice, animals treated with iNOS inhibitors show more severe forms of disease (Vazquez-Torres et al., 1995). In a relatively recent study, mice deficient in both iNOS and Phox were found to display severe forms of candidiasis; however, when examined in vitro, cells from wild-type or singly or doubly deficient mice displayed an equal capacity to kill Candida (Balish et al., 2005). While these results support a role for ROIs and RNIs in the control of fungal infections in vivo, they also demonstrate the importance of the other systems utilized by phagocytes to kill this fungus.

Nonoxidative Candidacidal Mechanisms

Phagocytes, particularly neutrophils, possess a number of nonoxidative candidacidal mechanisms, including antimicrobial peptides, enzymes, and proteins. The α-defensins are a group of four small cationic peptides (HNP1 to HNP4) found in the azurophilic granules of human, but not murine, neutrophils (Eisenhauer and Lehrer, 1992). Defensins target Candida membranes, inducing nonlytic permeabilization and release of cellular ATP (Edgerton et al., 2000). Although the antimicrobial activity of these peptides can be inhibited by physiological salt concentrations and serum components, casting some doubt on their role in vivo, these inhibitory effects are thought to be overcome by the high concentrations of these peptides that are found in the granules (Selsted and Ouellette, 2005). In addition to their antimicrobial activities, defensins can also act as chemoattractants for monocytes, DCs, and selected lymphocytes (Yang et al., 2000). It is notable that β-defensins, which are distinguishable from the α-defensins based on their arrangement of disulfide linkages, also have significant antifungal activity. These peptides, which are primarily expressed by epithelial cells, can be induced by a variety of agents, including TLR agonists, as well as monocyte/macrophage-derived factors, such as IL-1 (Selsted and Ouellette, 2005).

The cathelicidins are defined by the presence of an N-terminal cathelin domain that is proteolytically cleaved from the C-terminal cationic peptide domain, which has antimicrobial activity. In contrast to other mammals, humans and mice possess only one cathelicidin, termed hCAP-18 and mCRAMP, respectively (Zanetti, 2004). hCAP-18 is expressed in the specific granules of neutrophils, but is also produced by other cells, including monocytes, NK cells, lymphocytes, and a variety of epithelia. hCAP-18 is found in a number of secretions, such as sweat and seminal fluid, and in the plasma, where it is bound to lipoproteins (Sorensen et al., 1999). Proteolytic cleavage of hCAP-18 produces LL-37, an antimicrobial peptide that kills *Candida* by inducing pores in the cellular membrane (Lopez-Garcia et al., 2005). hCAP-18 can also be differentially processed, on secretion onto the skin, for example, generating multiple peptides that have enhanced candidacidal activity (Murakami et al., 2004). Similarly to the defensins, LL-37 can also act as a chemoattractant for neutrophils, monocytes, and lymphocytes. In addition, LL-37 can induce histamine release from mast cells, alter the transcriptional responses in macrophages, and play a role in wound repair (Zanetti, 2004). In mice, however, deficiency of mCRAMP did not appear to affect the in vitro or in vivo responses to *Candida* (Lopez-Garcia et al., 2005).

Lysozyme is an antimicrobial enzyme expressed by a variety of phagocytes, including granulocytes, monocytes, and macrophages, but is also found at high levels in various tissues and secretions, such as saliva. Although traditionally associated with antibacterial activity, through its ability to cleave peptidoglycan, lysozyme also has candidacidal activity, although only at high concentrations (Levy, 2004). The fungicidal activity of lysozyme is thought to occur through enzymatic hydrolysis of N-glycosidic bonds within the fungal cell wall and injury to the cell membrane (Marquis et al., 1982).

The serprocidins are a family of cationic serine proteases stored within neutrophil granules, including protease-3, cathepsin G, and elastase. These proteases are involved in many cellular processes including, for example, the cleavage of cathelicidins, cellular activation, and chemotaxis, but they also possess antimicrobial activity (Owen and Campbell, 1999). In a landmark article, Segal and colleagues demonstrated that in vivo and in vitro resistance to *Candida* infection required elastase, but not cathepsin G (Reeves et al., 2002). Furthermore, they showed that this required the respiratory burst, which induces a potassium influx and rise in pH, allowing release of the proteases from the granule proteoglycan matrix and providing optimal conditions for enzyme activity. From these data they have proposed, somewhat controversially, that activation of these proteases is a primary mechanism of microbial killing. However, a serprocidin homologue, azurocidin, which lacks protease activity, is also candidacidal, suggesting that the antimicrobial activities of these proteins are not solely related to their enzymatic activities (McCabe et al., 2002).

Another effective candidacidal mechanism is the limitation of nutrients that are essential for fungal growth. This is achieved both through the containment of the microbe within the phagosome, as well as systems designed to further restrict the availability of certain nutrients. Intracellular iron, for example, is targeted by downregulation of transferrin receptors, which reduces delivery of iron to the phagosome via the endosomal recycling pathway, and natural resistance-associated macrophage protein-1 (Nramp-1), which removes iron and other divalent cations from the phagosome after microbial uptake (Puliti et al., 1995). Another iron-binding protein, lactoferrin, which is found in a variety of secretions and neutrophil granules, possesses candidacidal activity, not only through its ability to sequester iron, but also directly by altering membrane permeability (Kuipers et al., 1999). Secreted lactoferrin can also protect against infection with *Candida* by enhancing neutrophil and macrophage functions (Tanida et al., 2001).

The sequestration of zinc is another mechanism of nutrient limitation with potent antifungal activity (Lulloff et al., 2004). Zinc sequestration is largely mediated by calprotectin, a dimeric molecule composed of two calcium-binding proteins (MRP8 and MRP14), and which is a major constituent of the neutrophil cytoplasm. Although calprotectin is not actively secreted by neutrophils, it is released following neutrophil death induced by live *Candida* and at sites of inflammation (Sohnle et al., 2001; Voganatsi et al., 2001).

Finally, neutrophils can undergo a novel ROS-dependent death pathway, on activation, which results in the formation of antimicrobial extracellular structures that have been termed neutrophil extracellular traps, or NETs (Brinkmann et al., 2004; Fuchs et al., 2007). These NETs consist of DNA, histones, and a number of granule proteins, including elastase and MPO. *Candida* can induce the formation of NETs, which trap and kill both the yeast and hyphal forms of this pathogen (Urban et al., 2006).

Candida Antimicrobial Evasion Mechanisms

Candida has a number of mechanisms to defend against the antimicrobial activities of phagocytes. We have already discussed morphogenic switching, which is important for escape from the phagosome, as well as the masking of certain cell wall structures, such as β-glucan, for example, which prevents recognition by Dectin-1, and the initiation of protective downstream responses, such as the respiratory burst (Donini et al., 2007) (Fig. 4). *Candida* may even specifically target receptors, such as CR3, which do not induce antimicrobial responses, such as the respiratory burst (Romani et al., 2004).

Candida can utilize a number of mechanisms to resist the antimicrobial activities of phagocytes, several of which have been demonstrated to contribute to the virulence of the pathogen. To counter oxidative stress, for example, *Candida* upregulates antioxidant enzymes, such as catalase and superoxide dismutase, and deals with oxidative damage to nucleic acids and proteins by upregulating DNA damage repair systems and heat shock proteins (Chauhan et al., 2006; Lorenz et al., 2004). *Candida* can also suppress iNOS function and inhibit the expression of defensins, although the mechanisms behind these activities are unclear (Lu et al., 2006; Schroppel et al., 2001; Shin et al., 2005).

The acquisition of limiting nutrients, such as iron and zinc, is achieved through the upregulation of metal ion

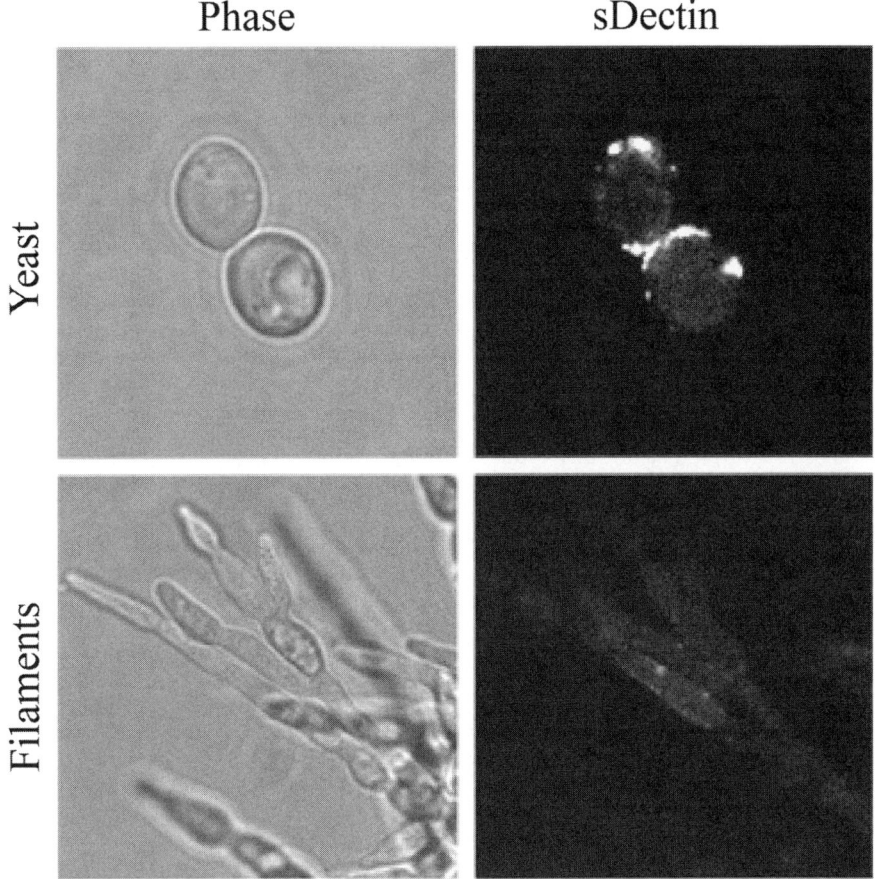

FIGURE 4 Exposure of β-glucans, detected by soluble recombinant Dectin-1 (sDectin), on selected regions of *Candida* yeast, but not hyphae. Reprinted with permission from Gantner et al. (2005).

transporters, such as the high-affinity iron permease CaFTR1 (Ramanan and Wang, 2000). *Candida* also upregulates the amino acid biosynthesis and nitrogen assimilation machinery on ingestion by phagocytes, and undergoes a significant metabolic change, shifting from the glycolytic pathway to the glyoxylate cycle and gluconeogenesis, presumably to allow utilization of alternative carbon sources available in the phagosome (Barelle et al., 2006; Fradin et al.; 2005; Lorenz and Fink, 2001). Finally, a number of *Candida* proteins without known function are upregulated on contact with phagocytes, suggesting that *Candida* is likely to possess other, as yet undefined, mechanisms for resisting host attack (Lorenz et al., 2004).

CANDIDA AND PHAGOCYTE-DERIVED SOLUBLE MEDIATORS

In addition to uptake and killing, recognition of *Candida* by phagocyte PRRs leads to the induction of soluble mediators, such as cytokines, chemokines, and lipids, which initiate and modulate the inflammatory response and ultimately the type of adaptive immunity mounted toward the pathogen. These factors include eicosanoids, TNF, IL-1, macrophage inflammatory protein 1 (MIP-1), IL-6, keratinocyte chemoattractant (KC; CXCL1), MIP-2, IL-8, and IL-12, which is important for inducing a T_H1 response and the production of IFN-γ (see below). Understanding of the mechanisms of intracellular signal transduction following pathogen sensing, leading to the production of these mediators, began primarily after the discovery of the TLRs. The contribution of other non-TLR PRRs was realized only more recently, particularly from studies of the receptors involved in antifungal immunity, and it is now appreciated that these receptors collaborate to induce optimal immune responses to pathogens. For *Candida*, the TLRs, the MR, Dectin-2, and Dectin-1 have been implicated in intracellular signaling leading to the induction of soluble mediators.

TLRs

The TLRs were first identified based on their ability to control fungal infection in *Drosophila*, and are now recognized as one of the major transducers of intracellular signals upon pathogen recognition (Lemaitre et al., 1996). The TLRs are expressed either at the cell surface or in intracellular compartments, including endosomes, and recognize a diverse, but receptor-specific range of microbial structures. The intracellular signaling triggered by these receptors results in the activation of several transcription factors, such as NF-κB and interferon regulatory factor 3, inducing the production of cytokines and chemokines as well as TLR-specific patterns of gene expression (Akira et al., 2006).

In mammals, three TLRs have been implicated in the recognition of *Candida*: TLR4, TLR2, and TLR9. Although mice deficient in MyD88 are extremely susceptible to candidiasis, and macrophages from these animals display defects in cytokine production, phagocytosis, and intracellular killing, the role of the individual TLRs in the control of *Candida* is still controversial (Bellocchio et al., 2004; Marr et al., 2003; Villamon et al., 2004a). Furthermore, TLR recognition of *Candida* is influenced by the morphological form of the fungus as well as costimulatory signals from other receptors such as Dectin-1 (Brown, 2006; van der Graaf et al., 2005).

TLR4 is thought to recognize O-linked mannan, and there is evidence both for and against a role for this receptor in anti-*Candida* immunity (Netea et al., 2006). Evidence in support of a role for TLR4 includes data showing that TLR4$^{-/-}$ mice had higher fungal loads (although this depended on the route of infection and mouse strain), a defective T_H1 response (increased IL-4 and reduced IFN-γ producing T cells), and reduced proinflammatory cytokine production (TNF) in vivo. In addition, leukocytes from these animals were shown to produce reduced levels of chemokines (KC and MIP-2), but not proinflammatory cytokines (TNF and IL-1), in response to *Candida* in vitro (Bellocchio et al., 2004; Netea et al., 2002). Further supporting evidence from studies in humans includes data showing that TLR4 antibodies are able to block cytokine responses (TNF, IFN-γ) to *Candida* in vitro and the identification of receptor polymorphisms that may be linked to susceptibility to candidiasis (Netea et al., 2006; Van der Graaf et al., 2005, 2006).

In contrast, it has been suggested that TLR4 plays no real role in anti-*Candida* immunity (Gil and Gozalbo, 2006). In all cases studied so far, TLR4 deficiency does not affect the ability of TLR4 mice to survive infection with *Candida*, and even protects against infection with hyphae (Bellocchio et al., 2004; Murciano et al., 2006). Furthermore, there are data which suggest that proinflammatory cytokine production and T_H1 responses are normal in TLR4-deficient animals (Murciano et al., 2006).

Controversy also surrounds the role of TLR2, which can recognize *Candida* phospholipomannan and collaborates with TLR6 to detect fungal particles (Jouault et al., 2003; Ozinsky et al., 2000; Underhill et al., 1999). Although there is no doubt that TLR2 is involved in immunity to *Candida*, it is unclear whether it has a protective or detrimental role. Evidence from both human (blocking monoclonal antibodies) and murine (TLR2$^{-/-}$) leukocytes clearly demonstrates that TLR2 is involved in proinflammatory cytokine production (TNF, IL-1, MIP-2) in vitro (Netea et al., 2002; Villamon et al., 2004a). However, in vivo, one group has shown that TLR2 deficiency leads to decreased survival on *Candida* infection, which was proposed to be due to reduced proinflammatory cytokine and chemokine production (TNF and MIP-2) and a reduced recruitment of neutrophils (Villamon et al., 2004b). In contrast, two other groups have shown that TLR2 deficiency is protective, enhancing survival and reducing fungal loads and proinflammatory cytokines in infected organs (Bellocchio et al., 2004; Netea et al., 2004b). This protective effect was linked to decreased IL-10 production, reduced immunosuppressive T cells, and enhanced T_H1 (IL-12, IFN-γ) responses (see below).

One study has implicated TLR9, a lysosomal sensor of unmethylated DNA, in the immune response to *Candida* (Bellocchio et al., 2004; Latz et al., 2004). Although TLR9 deficiency did not affect resistance to infection with yeast forms of *Candida*, it was protective against infection with hyphae. However, fungal loads were reduced in mice infected with both morphological forms, although proinflammatory cytokine production (TNF) was unaltered. Paradoxically, hyphal infection was associated with a T_H2 profile (increased IL-4 and reduced IFN-γ producing T cells) in these animals. In vitro, TLR9$^{-/-}$ neutrophils displayed an enhanced capacity to kill hyphae.

MR

The MR has been implicated in the production of a number of cytokines, including IL-12, IL-1β, IL-6, TNF, and granulocyte-macrophage colony-stimulating factor (GM-CSF), in response to *Candida* in vitro (Netea et al., 2006; Romani et al., 2004; Yamamoto et al., 1997). How the MR mediates this activity is unclear, because the receptor lacks classical signaling motifs in its cytoplasmic tail. In mice, deficiency of the MR did not alter the susceptibility to infection with *Candida*, although slightly higher fungal burdens were noted in selected tissues (Lee et al., 2003).

Dectin-2

Dectin-2 is expressed on Langerhans cells, DCs, and tissue macrophages and is structurally similar, although unrelated, to Dectin-1 (Ariizumi et al., 2000; Taylor et al., 2005b). Dectin-2 is a cation-dependent mannose/fucose lectin with specificity for high-mannose structures (McGreal et al., 2006). The receptor preferentially recognizes *C. albicans* hyphae but is also able to weakly recognize yeast (McGreal et al., 2006; Sato et al., 2006). Although not possessing any signaling motifs of its own, Dectin-2 associates with the signaling adaptor, Fcγ, and can induce IL-1Ra and TNF in response to *Candida* (Sato et al., 2006). The role of Dectin-2 in antifungal immunity in vivo is unknown.

Dectin-1

In response to fungal particles in vitro, Dectin-1 has been shown to induce the production of numerous cytokines and chemokines, including GM-CSF, IL-1, TNF, MIP-2, IL-12, IL-2, IL-10, IL-6, and IL-23, as well as arachidonic acid release and prostaglandin production, through the activation and regulation of PLA$_2$ and COX2 (Brown, 2006; Leibundgut-Landmann et al., 2007; Suram et al., 2006). Induction of soluble mediators is dependent on signaling via Syk kinase, which leads to the activation of NF-κB, and involves downstream components including CARD9, MALT1, Bcl-10, and mitogen-activated protein kinases (Dillon et al., 2006; Gross et al., 2006; Slack et al., 2007). Although signaling from Dectin-1 is sufficient for many of these responses, others, particularly the production of proinflammatory cytokines and chemokines, requires collaborative signaling from the TLRs (Brown, 2006) (see below).

In vivo, the role of Dectin-1 is less clear. Dectin-1 knockout mice made by one group demonstrated enhanced susceptibility to the fungus, which correlated with defects in inflammatory cytokine and chemokine production (TNF, IL-6, MCP-1, MIP-1, C-CSF, and GM-CSF), enhanced fungal dissemination and growth in tissues (Fig. 5), a modest defect in fungal killing, and reduced neutrophil recruitment (Taylor et al., 2007). Mice deficient in the downstream signaling component, CARD9, were similarly susceptible to *C. albicans* infection (Gross et al., 2006). In contrast, no defect in cytokine production or resistance to infection was observed with Dectin-1$^{-/-}$ mice generated by another group (Saijo et al., 2007).

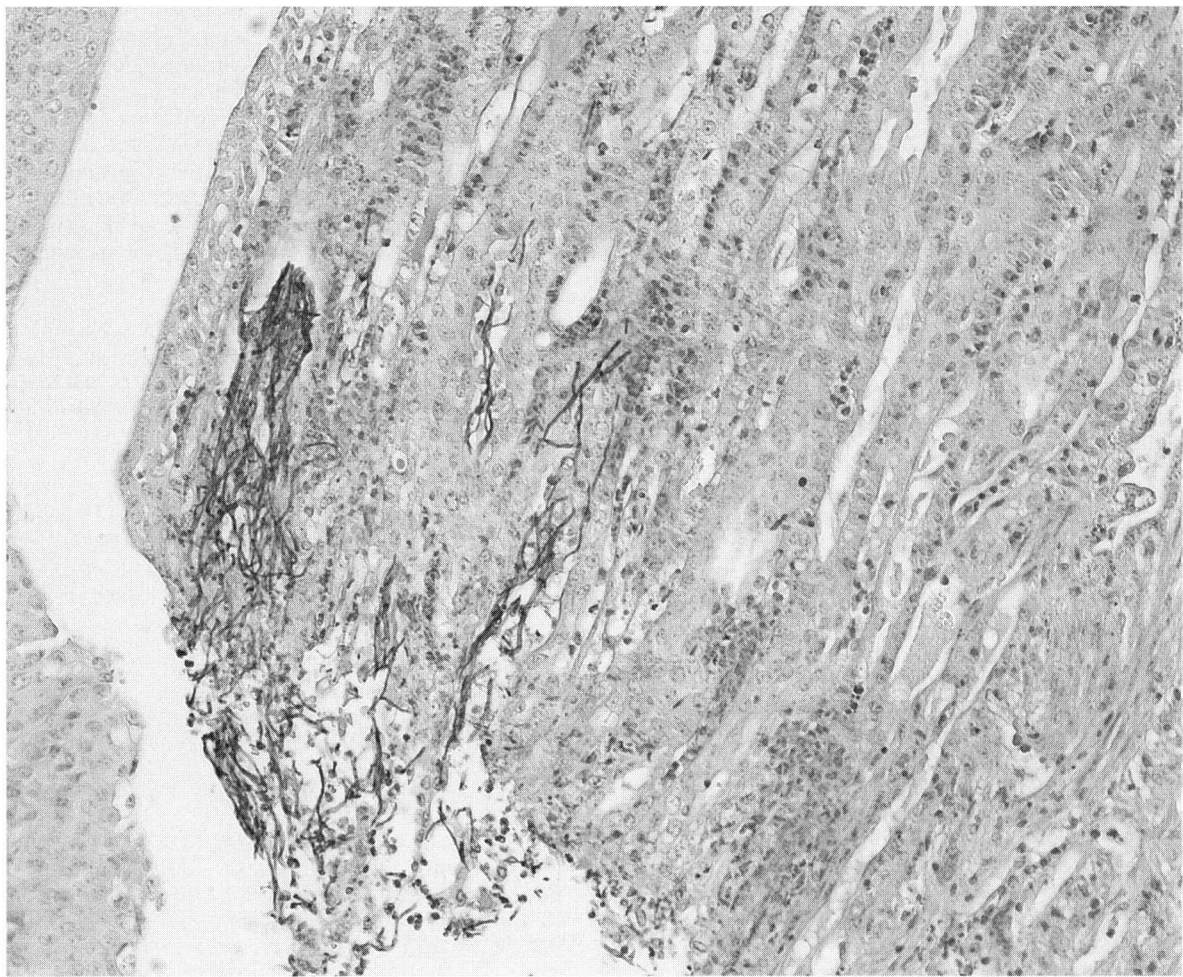

FIGURE 5 Photomicrograph showing the presence of invasive candidiasis in the kidney of a Dectin-1-deficient mouse. Reprinted with permission from Taylor et al. (2007).

Collaborative PRR Responses to *Candida*

The recognition of intact microbes involves many receptors, and it is now appreciated that cooperation of these receptors is needed for induction of an optimal immune response required to resist infection (Trinchieri and Sher, 2007). This is particularly evident in fungal immunology, where cooperative signaling between Dectin-1 and TLR2 is required for inflammatory responses to fungi (Brown et al., 2003; Gantner et al., 2003). More recently, the optimal immune response to *Candida* was shown to involve cooperative recognition of mannans and glucans by several receptors, including TLR2, TLR4, Dectin-1, and the MR (Netea et al., 2006). Other demonstrations of receptor cooperation in leukocyte response to *Candida* involve interactions between CD14 and TLR4, Galectin-3 and TLR2, and Dectin-1 and SIGNR1 (Jouault et al., 2006; Tada et al., 2002; Taylor et al., 2004). Given that several receptors are involved in the recognition and response to *Candida*, it is likely that some of the confusion/controversy regarding the role of the various receptors has arisen from the study of these molecules in isolation, as well as the variation in surface PAMPs of different *Candida* strains and morphotypes (van der Graaf et al., 2005).

Suppression of Inflammatory Responses

The interaction of *Candida* mannan with certain receptors can also be suppressive. Examples include the interaction of *Candida* with TLR2, already described, which can result in the induction of IL-10 and regulatory T cells (Netea et al., 2004b). CR3 and the Fcγ receptors, in addition, can suppress IL-12 production and enhance the production of IL-10 and IL-4 in response to *Candida*, presumably favoring nonprotective $T_H 2$ responses (Romani et al., 2004). Consistent with this, CR3- and FcγR-deficient mice show enhanced resistance to systemic candidiasis (Romani et al., 2004). Activation of the Raf kinase pathway by DC-SIGN, another "suppressive" receptor thought to be targeted by many pathogens, including *Candida*, results in the modulation of TLR-mediated responses and the induction of high levels of IL-10 by DCs (Gringhuis et al., 2007). While the suppressive responses induced by these receptors could be important for limiting inflammatory pathology and/or maintaining commensalism, they may be exploited by the fungus to circumvent or modify protective immune responses. Induction of these "suppressive" responses by *Candida* may also be aided by morphogenic switching from yeast to hyphal forms, whereby selected PAMPs are displayed or masked, enabling or preventing recognition by distinct PRRs, as described earlier.

CANDIDA, PHAGOCYTES, AND ADAPTIVE IMMUNITY

Following uptake and killing, microbial antigens are processed and presented, which, along with selected cytokines (IL-12, IL-4, etc.), leads to the induction and shaping of adaptive immunity. The presentation of antigen to naïve T cells is mediated by DCs, a process that is enabled following maturation of these cells, as occurs after Candida uptake. The mechanisms underlying these processes are described in detail in previous chapters. As discussed earlier, resistance to Candida is generally thought to require a T_H1 response (characterized by IFN-γ-producing CD4$^+$ T cells), which enhances the candidacidal activities of phagocytes, whereas T_H2 responses (characterized by IL-4-producing CD4$^+$ T cells) induce susceptibility. However, a balance between these two types of responses is actually required for optimal immunity to Candida (Romani, 1999).

The importance of T_H1 responses has been clearly demonstrated in the control of mucosal infections, but there is controversy regarding the importance of adaptive immunity in systemic infections (Ashman et al., 2004). The ability of mice lacking T and B cells, IFN-γ, or IL-12 to resist systemic candidiasis, for example, suggests that the innate components are sufficient for the control of this type of infection (Jones-Carson et al., 2000; Mahanty et al., 1988; Mencacci et al., 1998; Qian and Cutler, 1997). However, following primary systemic infection, T_H1 deficiency does result in susceptibility to reinfection with the pathogen (Romani, 1999).

The induction of a correct adaptive response to Candida can be influenced by the morphological form of the fungus and the phagocyte receptors with which it interacts. Although murine DCs can process and present antigens from both morphological forms of Candida, the interaction with yeasts has been shown to induce T_H1 responses, whereas the interaction with hyphae induces T_H2 responses (d'Ostiani et al., 2000). In humans, the morphological form of the pathogen can influence the differentiation of monocytes into DCs, but there is no influence on the ability of DCs to induce T_H1 responses (Romagnoli et al., 2004; Torosantucci et al., 2004). In terms of receptors, the interaction with TLR4 promotes T_H1 responses, for example, whereas the interaction with FcγR or TLR2 promotes T_H2 responses (Netea et al., 2004b, 2004c; Romani et al., 2004). TLR2 can also induce regulatory T cells that suppress immunity to Candida (Dillon et al., 2004; Netea et al., 2004b). Recently, the interaction of Candida with Dectin-1 has been shown to induce T_H17 responses (characterized by IL-17-producing CD4$^+$ T cells). In humans, this response was linked to the hyphal form of the fungus (Acosta-Rodriguez et al., 2007). Although the significance of T_H17 responses to Candida is not yet understood, these responses appear to be required for resistance to infection, and have also been linked to fungal-induced autoimmune disease (Huang et al., 2004; Yoshitomi et al., 2005).

CONCLUSIONS

As we have seen, phagocytes are the central players involved in the control of Candida infections, particularly as effector cells in both the innate and adaptive arms of the immune response. Although we are starting to understand some of the mechanisms underlying the induction of protective or detrimental immune responses to this organism, many questions remain. For example, we require a better understanding of the differences in yeast and hyphal recognition, and how this influences immune response and outcome of infection. We also know little of the systems in place regulating commensalism with Candida, and how perturbations in these systems can lead to infection. Ultimately, it is hoped that by gaining more insight into the factors governing the immune response to this organism, alternative therapeutic approaches, such as immunomodulation, can be developed.

We thank the Wellcome Trust, CANSA South Africa, University of Cape Town, National Research Foundation, Medical Research Council (South Africa) for financial support. G.D.B. is a Wellcome Trust Senior Research Fellow in Biomedical Science in South Africa.

REFERENCES

Acosta-Rodriguez, E. V., L. Rivino, J. Geginat, D. Jarrossay, M. Gattorno, A. Lanzavecchia, F. Sallusto, and G. Napolitani. 2007. Surface phenotype and antigenic specificity of human interleukin 17-producing T helper memory cells. *Nat. Immunol.* **8:**639–646.

Akira, S., and K. Takeda. 2004. Toll-like receptor signalling. *Nat. Rev. Immunol.* **4:**499–511.

Akira, S., S. Uematsu, and O. Takeuchi. 2006. Pathogen recognition and innate immunity. *Cell* **124:**783–801.

Allen, L. H., and A. Aderem. 1995. A role for MARCKS, the alpha isozyme of protein kinase C and myosin I in zymosan phagocytosis by macrophages. *J. Exp. Med.* **182:**829–840.

Allen, L. A., and A. Aderem. 1996. Molecular definition of distinct cytoskeletal structures involved in complement- and Fc receptor-mediated phagocytosis in macrophages. *J. Exp. Med.* **184:**627–637.

Antachopoulos, C., T. J. Walsh, and E. Roilides. 2007. Fungal infections in primary immunodeficiencies. *Eur. J. Pediatr.* **166:**1099–1117.

Aratani, Y., H. Koyama, S. Nyui, K. Suzuki, F. Kura, and N. Maeda. 1999. Severe impairment in early host defense against Candida albicans in mice deficient in myeloperoxidase. *Infect. Immun.* **67:**1828–1836.

Aratani, Y., F. Kura, H. Watanabe, H. Akagawa, Y. Takano, K. Suzuki, M. C. Dinauer, N. Maeda, and H. Koyama. 2002. Critical role of myeloperoxidase and nicotinamide adenine dinucleotide phosphate-oxidase in high-burden systemic infection of mice with Candida albicans. *J. Infect. Dis.* **185:**1833–1837.

Ariizumi, K., G. L. Shen, S. Shikano, R. Ritter, 3rd, P. Zukas, D. Edelbaum, A. Morita, and A. Takashima. 2000. Cloning of a second dendritic cell-associated C-type lectin (dectin-2) and its alternatively spliced isoforms. *J. Biol. Chem.* **275:**11957–11963.

Ashman, R. B., C. S. Farah, S. Wanasaengsakul, Y. Hu, G. Pang, and R. L. Clancy. 2004. Innate versus adaptive immunity in Candida albicans infection. *Immunol. Cell Biol.* **82:**196–204.

Balish, E., T. F. Warner, P. J. Nicholas, E. E. Paulling, C. Westwater, and D. A. Schofield. 2005. Susceptibility of germfree phagocyte oxidase- and nitric oxide synthase 2-deficient mice, defective in the production of reactive metabolites of both oxygen and nitrogen, to mucosal and systemic candidiasis of endogenous origin. *Infect. Immun.* **73:**1313–1320.

Barelle, C. J., C. L. Priest, D. M. Maccallum, N. A. Gow, F. C. Odds, and A. J. Brown. 2006. Niche-specific regulation of central metabolic pathways in a fungal pathogen. *Cell. Microbiol.* **8:**961–971.

Behnsen, J., P. Narang, M. Hasenberg, F. Gunzer, U. Bilitewski, N. Klippel, M. Rohde, M. Brock, A. A. Brakhage, and M. Gunzer. 2007. Environmental dimensionality controls the interaction of phagocytes with the pathogenic fungi *Aspergillus fumigatus* and *Candida albicans*. *PLoS Pathog.* **3:** e13.

Bellocchio, S., C. Montagnoli, S. Bozza, R. Gaziano, G. Rossi, S. S. Mambula, A. Vecchi, A. Mantovani, S. M. Levitz, and L. Romani. 2004. The contribution of the toll-like/IL-1 receptor superfamily to innate and adaptive immunity to fungal pathogens in vivo. *J. Immunol.* **172:**3059–3069.

Blander, J. M., and R. Medzhitov. 2004. Regulation of phagosome maturation by signals from toll-like receptors. *Science* **304:**1014–1018.

Borg-von Zepelin, M., S. Beggah, K. Boggian, D. Sanglard, and M. Monod. 1998. The expression of the secreted aspartyl proteinases Sap4 to Sap6 from *Candida albicans* in murine macrophages. *Mol. Microbiol.* **28:**543–554.

Brinkmann, V., U. Reichard, C. Goosmann, B. Fauler, Y. Uhlemann, D. S. Weiss, Y. Weinrauch, and A. Zychlinsky. 2004. Neutrophil extracellular traps kill bacteria. *Science* **303:**1532–1535.

Brown, G. D. 2006. Dectin-1: a signalling non-TLR pattern-recognition receptor. *Nat. Rev. Immunol.* **6:**33–43.

Brown, G. D., and S. Gordon. 2005. Immune recognition of fungal beta-glucans. *Cell. Microbiol.* **7:**471–479.

Brown, G. D., J. Herre, D. L. Williams, J. A. Willment, A. S. J. Marshall, and S. Gordon. 2003. Dectin-1 mediates the biological effects of beta-glucan. *J. Exp. Med.* **197:**1119–1124.

Cambi, A., K. Gijzen, J. M. de Vries, R. Torensma, B. Joosten, G. J. Adema, M. G. Netea, B. J. Kullberg, L. Romani, and C. G. Figdor. 2003. The C-type lectin DC-SIGN (CD209) is an antigen-uptake receptor for *Candida albicans* on dendritic cells. *Eur. J. Immunol.* **33:**532–538.

Chauhan, N., J. P. Latge, and R. Calderone. 2006. Signalling and oxidant adaptation in *Candida albicans* and *Aspergillus fumigatus*. *Nat. Rev. Microbiol.* **4:**435–444.

Clark, R. A. 1999. Activation of the neutrophil respiratory burst oxidase. *J. Infect. Dis.* **179**(Suppl. 2)**:**S309–S317.

Diamond, R. D., R. A. Clark, and C. C. Haudenschild. 1980. Damage to *Candida albicans* hyphae and pseudohyphae by the myeloperoxidase system and oxidative products of neutrophil metabolism in vitro. *J. Clin. Investig.* **66:**908–917.

Dillon, S., A. Agrawal, T. Van Dyke, G. Landreth, L. McCauley, A. Koh, C. Maliszewski, S. Akira, and B. Pulendran. 2004. A Toll-like receptor 2 ligand stimulates Th2 responses in vivo, via induction of extracellular signal-regulated kinase mitogen-activated protein kinase and c-Fos in dendritic cells. *J. Immunol.* **172:**4733–4743.

Dillon, S., S. Agrawal, K. Banerjee, J. Letterio, T. L. Denning, K. Oswald-Richter, D. J. Kasprowicz, J. Kellar, J. Pare, T. van Dyke, S. Ziegler, D. Unutmaz, and B. Pulendran. 2006. Yeast zymosan, a stimulus for TLR2 and dectin-1, induces regulatory antigen-presenting cells and immunological tolerance. *J. Clin. Investig.* **116:**916–928.

Diniz, S. N., R. Nomizo, P. S. Cisalpino, M. M. Teixeira, G. D. Brown, A. Mantovani, S. Gordon, L. F. Reis, and A. A. Dias. 2004. PTX3 function as an opsonin for the dectin-1-dependent internalization of zymosan by macrophages. *J. Leukoc. Biol.* **75:**649–656.

Donini, M., E. Zenaro, N. Tamassia, and S. Dusi. 2007. NADPH oxidase of human dendritic cells: role in *Candida albicans* killing and regulation by interferons, dectin-1 and CD206. *Eur. J. Immunol.* **37:**1194–1203.

d'Ostiani, C. F., G. Del Sero, A. Bacci, C. Montagnoli, A. Spreca, A. Mencacci, P. Ricciardi-Castagnoli, and L. Romani. 2000. Dendritic cells discriminate between yeasts and hyphae of the fungus *Candida albicans*. Implications for initiation of T helper cell immunity in vitro and in vivo. *J. Exp. Med.* **191:**1661–1674.

Edgerton, M., S. E. Koshlukova, M. W. Araujo, R. C. Patel, J. Dong, and J. A. Bruenn. 2000. Salivary histatin 5 and human neutrophil defensin 1 kill *Candida albicans* via shared pathways. *Antimicrob. Agents Chemother.* **44:**3310–3316.

Eisenhauer, P. B., and R. I. Lehrer. 1992. Mouse neutrophils lack defensins. *Infect. Immun.* **60:**3446–3447.

Engering, A., T. B. Geijtenbeek, S. J. van Vliet, M. Wijers, E. van Liempt, N. Demaurex, A. Lanzavecchia, J. Fransen, C. G. Figdor, V. Piguet, and Y. van Kooyk. 2002. The dendritic cell-specific adhesion receptor DC-SIGN internalizes antigen for presentation to T cells. *J. Immunol.* **168:**2118–2126.

Ezekowitz, R. A., K. Sastry, P. Bailly, and A. Warner. 1990. Molecular characterization of the human macrophage mannose receptor: demonstration of multiple carbohydrate recognition-like domains and phagocytosis of yeasts in Cos-1 cells. *J. Exp. Med.* **172:**1785–1794.

Fradin, C., P. De Groot, D. MacCallum, M. Schaller, F. Klis, F. C. Odds, and B. Hube. 2005. Granulocytes govern the transcriptional response, morphology and proliferation of *Candida albicans* in human blood. *Mol. Microbiol.* **56:**397–415.

Fraser, I. P., K. Takahashi, H. Koziel, B. Fardin, A. Harmsen, and R. A. Ezekowitz. 2000. *Pneumocystis carinii* enhances soluble mannose receptor production by macrophages. *Microbes Infect.* **2:**1305–1310.

Fuchs, T. A., U. Abed, C. Goosmann, R. Hurwitz, I. Schulze, V. Wahn, Y. Weinrauch, V. Brinkmann, and A. Zychlinsky. 2007. Novel cell death program leads to neutrophil extracellular traps. *J. Cell Biol.* **176:**231–241.

Gantner, B. N., R. M. Simmons, S. J. Canavera, S. Akira, and D. M. Underhill. 2003. Collaborative induction of inflammatory responses by Dectin-1 and Toll-like receptor 2. *J. Exp. Med.* **197:**1107–1117.

Gantner, B. N., R. M. Simmons, and D. M. Underhill. 2005. Dectin-1 mediates macrophage recognition of *Candida albicans* yeast but not filaments. *EMBO J.* **24:**1277–1286.

Gil, M. L., and D. Gozalbo. 2006. TLR2, but not TLR4, triggers cytokine production by murine cells in response to *Candida albicans* yeasts and hyphae. *Microbes Infect.* **8:**2299–2304.

Gringhuis, S. I., J. den Dunnen, M. Litjens, B. van Het Hof, Y. van Kooyk, and T. B. Geijtenbeek. 2007. C-Type lectin DC-SIGN modulates Toll-like receptor signaling via Raf-1 kinase-dependent acetylation of transcription factor NF-kappaB. *Immunity* **26:**605–616.

Gross, O., A. Gewies, K. Finger, M. Schafer, T. Sparwasser, C. Peschel, I. Forster, and J. Ruland. 2006. Card9 controls a non-TLR signalling pathway for innate anti-fungal immunity. *Nature* **442:**651–656.

Gudlaugsson, O., S. Gillespie, K. Lee, J. Vande Berg, J. Hu, S. Messer, L. Herwaldt, M. Pfaller, and D. Diekema. 2003. Attributable mortality of nosocomial candidemia, revisited. *Clin. Infect. Dis.* **37:**1172–1177.

Heinsbroek, S. E., G. D. Brown, and S. Gordon. 2005. Dectin-1 escape by fungal dimorphism. *Trends Immunol.* **26:**352–354.

Heinsbroek, S. E., P. R. Taylor, M. Rosas, J. A. Willment, D. L. Williams, S. Gordon, and G. D. Brown. 2006. Expression of functionally different Dectin-1 isoforms by murine macrophages. *J. Immunol.* **176:**5513–5518.

Herre, J., A. J. Marshall, E. Caron, A. D. Edwards, D. L. Williams, E. Schweighoffer, V. L. Tybulewicz, C. Reis e Sousa, S. Gordon, and G. D. Brown. 2004. Dectin-1 utilizes novel mechanisms for yeast phagocytosis in macrophages. *Blood* **104:**4038–4045.

Heyworth, P. G., A. R. Cross, and J. T. Curnutte. 2003. Chronic granulomatous disease. *Curr. Opin. Immunol.* **15:** 578–584.

Huang, W., L. Na, P. L. Fidel, and P. Schwarzenberger. 2004. Requirement of interleukin-17A for systemic anti-*Candida albicans* host defense in mice. *J. Infect. Dis.* **190:**624–631.

Ibata-Ombetta, S., T. Idziorek, P. A. Trinel, D. Poulain, and T. Jouault. 2003. *Candida albicans* phospholipomannan promotes survival of phagocytozed yeasts through modulation of Bad phosphorylation and macrophage apoptosis. *J. Biol. Chem.* **278:**13086–13093.

Jones-Carson, J., A. Vazquez-Torres, T. Warner, and E. Balish. 2000. Disparate requirement for T cells in resistance to mucosal and acute systemic candidiasis. *Infect. Immun.* **68:** 2363–2365.

Jouault, T., M. El Abed-El Behi, M. Martinez-Esparza, L. Breuilh, P. A. Trinel, M. Chamaillard, F. Trottein, and D. Poulain. 2006. Specific recognition of *Candida albicans* by macrophages requires galectin-3 to discriminate *Saccharomyces cerevisiae* and needs association with TLR2 for signaling. *J. Immunol.* **177:**4679–4687.

Jouault, T., S. Ibata-Ombetta, O. Takeuchi, P. A. Trinel, P. Sacchetti, P. Lefebvre, S. Akira, and D. Poulain. 2003. *Candida albicans* phospholipomannan is sensed through toll-like receptors. *J. Infect. Dis.* **188:**165–172.

Kaposzta, R., L. Marodi, M. Hollinshead, S. Gordon, and R. P. da Silva. 1999. Rapid recruitment of late endosomes and lysosomes in mouse macrophages ingesting *Candida albicans*. *J. Cell Sci.* **112**(Pt 19)**:**3237–3248.

Karlsson, A., and C. Dahlgren. 2002. Assembly and activation of the neutrophil NADPH oxidase in granule membranes. *Antioxid. Redox Signal.* **4:**49–60.

Karlsson, A., P. Follin, H. Leffler, and C. Dahlgren. 1998. Galectin-3 activates the NADPH-oxidase in exudated but not peripheral blood neutrophils. *Blood* **91:**3430–3438.

Kauffman, C. A. 2006. Fungal infections. *Proc. Am. Thorac. Soc.* **3:**35–40.

Kennedy, A. D., J. A. Willment, D. W. Dorward, D. L. Williams, G. D. Brown, and F. R. DeLeo. 2007. Dectin-1 promotes fungicidal activity of human neutrophils. *Eur. J. Immunol.* **37:**467–478.

Kilpatrick, D. 2002. Mannan-binding lectin: clinical significance and applications. *Biochim. Biophys. Acta* **1572:**401.

Kishore, U., T. J. Greenhough, P. Waters, A. K. Shrive, R. Ghai, M. F. Kamran, A. L. Bernal, K. B. Reid, T. Madan, and T. Chakraborty. 2006. Surfactant proteins SP-A and SP-D: structure, function and receptors. *Mol. Immunol.* **43:** 1293–1315.

Kitahara, M., H. J. Eyre, Y. Simonian, C. L. Atkin, and S. J. Hasstedt. 1981. Hereditary myeloperoxidase deficiency. *Blood* **57:**888–893.

Klis, F. M., P. de Groot, and K. Hellingwerf. 2001. Molecular organization of the cell wall of *Candida albicans*. *Med. Mycol.* **39**(Suppl 1)**:**1–8.

Kohatsu, L., D. K. Hsu, A. G. Jegalian, F. T. Liu, and L. G. Baum. 2006. Galectin-3 induces death of *Candida* species expressing specific beta-1,2-linked mannans. *J. Immunol.* **177:**4718–4726.

Koppel, E. A., K. P. van Gisbergen, T. B. Geijtenbeek, and Y. van Kooyk. 2005. Distinct functions of DC-SIGN and its homologues L-SIGN (DC-SIGNR) and mSIGNR1 in pathogen recognition and immune regulation. *Cell. Microbiol.* **7:** 157–165.

Kuipers, M. E., H. G. de Vries, M. C. Eikelboom, D. K. Meijer, and P. J. Swart. 1999. Synergistic fungistatic effects of lactoferrin in combination with antifungal drugs against clinical *Candida* isolates. *Antimicrob. Agents Chemother.* **43:** 2635–2641.

Lai, W. K., P. J. Sun, J. Zhang, A. Jennings, P. F. Lalor, S. Hubscher, J. A. McKeating, and D. H. Adams. 2006. Expression of DC-SIGN and DC-SIGNR on human sinusoidal endothelium: a role for capturing hepatitis C virus particles. *Am. J. Pathol.* **169:**200–208.

Latz, E., A. Schoenemeyer, A. Visintin, K. A. Fitzgerald, B. G. Monks, C. F. Knetter, E. Lien, N. J. Nilsen, T. Espevik, and D. T. Golenbock. 2004. TLR9 signals after translocating from the ER to CpG DNA in the lysosome. *Nat. Immunol.* **5:**190–198.

Le Cabec, V., L. J. Emorine, I. Toesca, C. Cougoule, and I. Maridonneau-Parini. 2005. The human macrophage mannose receptor is not a professional phagocytic receptor. *J. Leukoc. Biol.* **77:**934–943.

Lee, S. J., N. Y. Zheng, M. Clavijo, and M. C. Nussenzweig. 2003. Normal host defense during systemic candidiasis in mannose receptor-deficient mice. *Infect. Immun.* **71:**437–445.

Lefkowitz, S. S., M. P. Gelderman, D. L. Lefkowitz, N. Moguilevsky, and A. Bollen. 1996. Phagocytosis and intracellular killing of *Candida albicans* by macrophages exposed to myeloperoxidase. *J. Infect. Dis.* **173:**1202–1207.

Leibundgut-Landmann, S., O. Gross, M. J. Robinson, F. Osorio, E. C. Slack, S. V. Tsoni, E. Schweighoffer, V. Tybulewicz, G. D. Brown, J. Ruland, and C. Reis e Sousa. 2007. Syk- and CARD9-dependent coupling of innate immunity to the induction of T helper cells that produce interleukin 17. *Nat. Immunol.* **8:**630–638.

Lemaitre, B., E. Nicolas, L. Michaut, J. M. Reichhart, and J. A. Hoffmann. 1996. The dorsoventral regulatory gene cassette spatzle/Toll/cactus controls the potent antifungal response in *Drosophila* adults. *Cell* **86:**973–983.

Levy, O. 2004. Antimicrobial proteins and peptides: anti-infective molecules of mammalian leukocytes. *J. Leukoc. Biol.* **76:**909–925.

Lo, H. J., J. R. Kohler, B. DiDomenico, D. Loebenberg, A. Cacciapuoti, and G. R. Fink. 1997. Nonfilamentous *C. albicans* mutants are avirulent. *Cell* **90:**939–949.

Lopez-Garcia, B., P. H. Lee, K. Yamasaki, and R. L. Gallo. 2005. Anti-fungal activity of cathelicidins and their potential role in *Candida albicans* skin infection. *J. Invest. Dermatol.* **125:**108–115.

Lorenz, M. C., J. A. Bender, and G. R. Fink. 2004. Transcriptional response of *Candida albicans* upon internalization by macrophages. *Eukaryot. Cell.* **3:**1076–1087.

Lorenz, M. C., and G. R. Fink. 2001. The glyoxylate cycle is required for fungal virulence. *Nature* **412:**83–86.

Lu, Q., J. A. Jayatilake, L. P. Samaranayake, and L. Jin. 2006. Hyphal invasion of *Candida albicans* inhibits the expression of human beta-defensins in experimental oral candidiasis. *J. Invest. Dermatol.* **126:**2049–2056.

Lulloff, S. J., B. L. Hahn, and P. G. Sohnle. 2004. Fungal susceptibility to zinc deprivation. *J. Lab. Clin. Med.* **144:** 208–214.

MacMicking, J., Q. W. Xie, and C. Nathan. 1997. Nitric oxide and macrophage function. *Annu. Rev. Immunol.* **15:** 323–350.

Mahanty, S., R. A. Greenfield, W. A. Joyce, and P. W. Kincade. 1988. Inoculation candidiasis in a murine model of severe combined immunodeficiency syndrome. *Infect. Immun.* **56:**3162–3166.

Mantegazza, A. R., M. M. Barrio, S. Moutel, L. Bover, M. Weck, P. Brossart, J. L. Teillaud, and J. Mordoh. 2004. CD63 Tetraspanin slows down cell migration and translocates to the endosomal/lysosomal/MIICs route after extracellular stimuli in human immature dendritic cells. *Blood* **104:**1183–1190.

Marodi, L., J. R. Forehand, and R. B. Johnston, Jr. 1991a. Mechanisms of host defense against *Candida* species. II. Biochemical basis for the killing of *Candida* by mononuclear phagocytes. *J. Immunol.* **146:**2790–2794.

Marodi, L., H. M. Korchak, and R. B. Johnston, Jr. 1991b. Mechanisms of host defense against *Candida* species. I. Phagocytosis by monocytes and monocyte-derived macrophages. *J. Immunol.* **146:**2783–2789.

Marquis, G., S. Montplaisir, S. Garzon, H. Strykowski, and P. Auger. 1982. Fungitoxicity of muramidase. Ultrastructural damage to *Candida albicans*. *Lab. Invest.* **46:**627–636.

Marr, K. A., S. A. Balajee, T. R. Hawn, A. Ozinsky, U. Pham, S. Akira, A. Aderem, and W. C. Liles. 2003. Differential role of MyD88 in macrophage-mediated responses to opportunistic fungal pathogens. *Infect. Immun.* **71:**5280–5286.

Martinez-Pomares, L., J. A. Mahoney, R. Kaposzta, S. A. Linehan, P. D. Stahl, and S. Gordon. 1998. A functional soluble form of the murine mannose receptor is produced by macrophages in vitro and is present in mouse serum. *J. Biol. Chem.* **273:**23376–23380.

McCabe, D., T. Cukierman, and J. E. Gabay. 2002. Basic residues in azurocidin/HBP contribute to both heparin binding and antimicrobial activity. *J. Biol. Chem.* **277:**27477–27488.

McGreal, E. P., M. Rosas, G. D. Brown, S. Zamze, S. Y. Wong, S. Gordon, L. Martinez-Pomares, and P. R. Taylor. 2006. The carbohydrate recognition domain of Dectin-2 is a C-type lectin with specificity for high-mannose. *Glycobiology* **16:**422–430.

McKenzie, E. J., P. R. Taylor, R. J. Stillion, A. D. Lucas, J. Harris, S. Gordon, and L. Martinez-Pomares. 2007. Mannose receptor expression and function define a new population of murine dendritic cells. *J. Immunol.* **178:**4975–4983.

Mencacci, A., E. Cenci, F. Bistoni, A. Bacci, G. Del Sero, C. Montagnoli, C. Fe d'Ostiani, and L. Romani. 1998. Specific and non-specific immunity to *Candida albicans*: a lesson from genetically modified animals. *Res. Immunol.* **149:**352–361; discussion 517–519.

Murakami, M., B. Lopez-Garcia, M. Braff, R. A. Dorschner, and R. L. Gallo. 2004. Postsecretory processing generates multiple cathelicidins for enhanced topical antimicrobial defense. *J. Immunol.* **172:**3070–3077.

Murciano, C., E. Villamon, D. Gozalbo, P. Roig, J. E. O'Connor, and M. L. Gil. 2006. Toll-like receptor 4 defective mice carrying point or null mutations do not show increased susceptibility to *Candida albicans* in a model of hematogenously disseminated infection. *Med. Mycol.* **44:**149–157.

Netea, M. G., K. Gijzen, N. Coolen, I. Verschueren, C. Figdor, J. W. Van der Meer, R. Torensma, and B. J. Kullberg. 2004a. Human dendritic cells are less potent at killing *Candida albicans* than both monocytes and macrophages. *Microbes Infect.* **6:**985–989.

Netea, M. G., N. A. Gow, C. A. Munro, S. Bates, C. Collins, G. Ferwerda, R. P. Hobson, G. Bertram, H. B. Hughes, T. Jansen, L. Jacobs, E. T. Buurman, K. Gijzen, D. L. Williams, R. Torensma, A. McKinnon, D. M. MacCallum, F. C. Odds, J. W. Van der Meer, A. J. Brown, and B. J. Kullberg. 2006. Immune sensing of *Candida albicans* requires cooperative recognition of mannans and glucans by lectin and Toll-like receptors. *J. Clin. Investig.* **116:**1642–1650.

Netea, M. G., R. Sutmuller, C. Hermann, C. A. Van der Graaf, J. W. Van der Meer, J. H. van Krieken, T. Hartung, G. Adema, and B. J. Kullberg. 2004b. Toll-like receptor 2 suppresses immunity against *Candida albicans* through induction of IL-10 and regulatory T cells. *J. Immunol.* **172:**3712–3718.

Netea, M. G., C. Van der Graaf, J. W. Van der Meer, and B. J. Kullberg. 2004c. Recognition of fungal pathogens by Toll-like receptors. *Eur. J. Clin. Microbiol. Infect. Dis.* **23:**672–676.

Netea, M. G., C. A. Van Der Graaf, A. G. Vonk, I. Verschueren, J. W. Van Der Meer, and B. J. Kullberg. 2002. The role of toll-like receptor (TLR) 2 and TLR4 in the host defense against disseminated candidiasis. *J. Infect. Dis.* **185:**1483–1489.

Obayashi, T., M. Yoshida, T. Mori, H. Goto, A. Yasuoka, H. Iwasaki, H. Teshima, S. Kohno, A. Horiuchi, A. Ito, et al. 1995. Plasma (1→3)-beta-D-glucan measurement in diagnosis of invasive deep mycosis and fungal febrile episodes. *Lancet* **345:**17–20.

Owen, C. A., and E. J. Campbell. 1999. The cell biology of leukocyte-mediated proteolysis. *J. Leukoc. Biol.* **65:**137–150.

Ozinsky, A., D. M. Underhill, J. D. Fontenot, A. M. Hajjar, K. D. Smith, C. B. Wilson, L. Schroeder, and A. Aderem. 2000. The repertoire for pattern recognition of pathogens by the innate immune system is defined by cooperation between toll-like receptors. *Proc. Natl. Acad. Sci. USA* **97:**13766–13771.

Ozment-Skelton, T. R., M. P. Goldman, S. Gordon, G. D. Brown, and D. L. Williams. 2006. Prolonged reduction of leukocyte membrane-associated Dectin-1 levels following beta-glucan administration. *J. Pharmacol. Exp. Ther.* **318:**540–546.

Palma, A. S., T. Feizi, Y. Zhang, M. S. Stoll, A. M. Lawson, E. Diaz-Rodriguez, M. A. Campanero-Rhodes, J. Costa, S. Gordon, G. D. Brown, and W. Chai. 2006. Ligands for the beta-glucan receptor, Dectin-1, assigned using "designer" microarrays of oligosaccharide probes (neoglycolipids) generated from glucan polysaccharides. *J. Biol. Chem.* **281:**5771–5779.

Powlesland, A. S., E. M. Ward, S. K. Sadhu, Y. Guo, M. E. Taylor, and K. Drickamer. 2006. Widely divergent biochemical properties of the complete set of mouse DC-SIGN-related proteins. *J. Biol. Chem.* **281:**20440–20449.

Puliti, M., D. Radzioch, R. Mazzolla, R. Barluzzi, F. Bistoni, and E. Blasi. 1995. Influence of the Bcg locus on macrophage response to the dimorphic fungus *Candida albicans*. *Infect. Immun.* **63:**4170–4173.

Qian, Q., and J. E. Cutler. 1997. Gamma interferon is not essential in host defense against disseminated candidiasis in mice. *Infect. Immun.* **65:**1748–1753.

Quinn, M. T., and K. A. Gauss. 2004. Structure and regulation of the neutrophil respiratory burst oxidase: comparison with nonphagocyte oxidases. *J. Leukoc. Biol.* **76:**760–781.

Rabinovich, G. A., and A. Gruppi. 2005. Galectins as immunoregulators during infectious processes: from microbial invasion to the resolution of the disease. *Parasite Immunol.* **27:**103–114.

Ramanan, N., and Y. Wang. 2000. A high-affinity iron permease essential for *Candida albicans* virulence. *Science* **288:**1062–1064.

Rappleye, C. A., L. G. Eissenberg, and W. E. Goldman. 2007. *Histoplasma capsulatum* alpha-(1,3)-glucan blocks innate immune recognition by the beta-glucan receptor. *Proc. Natl. Acad. Sci. USA* **104:**1366–1370.

Reese, T. A., H. E. Liang, A. M. Tager, A. D. Luster, N. Van Rooijen, D. Voehringer, and R. M. Locksley. 2007. Chitin induces accumulation in tissue of innate immune cells associated with allergy. *Nature* **447:**92–96.

Reeves, E. P., H. Lu, H. L. Jacobs, C. G. Messina, S. Bolsover, G. Gabella, E. O. Potma, A. Warley, J. Roes, and A. W. Segal. 2002. Killing activity of neutrophils is mediated through activation of proteases by K+ flux. *Nature* **416:**291–297.

Rogers, N. C., E. C. Slack, A. D. Edwards, M. A. Nolte, O. Schulz, E. Schweighoffer, D. L. Williams, S. Gordon, V. L. Tybulewicz, G. D. Brown, and C. Reis e Sousa. 2005. Syk-dependent cytokine induction by dectin-1 reveals a novel pattern recognition pathway for C-type lectins. *Immunity* **22:**507–517.

Romagnoli, G., R. Nisini, P. Chiani, S. Mariotti, R. Teloni, A. Cassone, and A. Torosantucci. 2004. The interaction of human dendritic cells with yeast and germ-tube forms of *Candida albicans* leads to efficient fungal processing, dendritic cell maturation, and acquisition of a Th1 response-promoting function. *J. Leukoc. Biol.* **75:**117–126.

Romani, L. 1999. Immunity to *Candida albicans*: Th1, Th2 cells and beyond. *Curr. Opin. Microbiol.* **2:**363–367.

Romani, L. 2004. Immunity to fungal infections. *Nat. Rev. Immunol.* **4:**1–23.

Romani, L., C. Montagnoli, S. Bozza, K. Perruccio, A. Spreca, P. Allavena, S. Verbeek, R. A. Calderone, F. Bistoni, and P. Puccetti. 2004. The exploitation of distinct recognition receptors in dendritic cells determines the full range of host immune relationships with *Candida albicans*. *Int. Immunol.* **16:**149–161.

Sabroe, I., L. R. Prince, E. C. Jones, M. J. Horsburgh, S. J. Foster, S. N. Vogel, S. K. Dower, and M. K. Whyte. 2003. Selective roles for Toll-like receptor (TLR)2 and TLR4 in the regulation of neutrophil activation and life span. *J. Immunol.* **170:**5268–5275.

Saijo, S., N. Fujikado, T. Furuta, S. H. Chung, H. Kotaki, K. Seki, K. Sudo, S. Akira, Y. Adachi, N. Ohno, T. Kinjo, K. Nakamura, K. Kawakami, and Y. Iwakura. 2007. Dectin-1 is required for host defense against *Pneumocystis carinii* but not against *Candida albicans*. *Nat. Immunol.* **8:**39–46.

Sano, H., D. K. Hsu, J. R. Apgar, L. Yu, B. B. Sharma, I. Kuwabara, S. Izui, and F. T. Liu. 2003. Critical role of galectin-3 in phagocytosis by macrophages. *J. Clin. Investig.* **112:**389–397.

Sato, K., X. L. Yang, T. Yudate, J. S. Chung, J. Wu, K. Luby-Phelps, R. P. Kimberly, D. Underhill, P. D. Cruz, Jr., and K. Ariizumi. 2006. Dectin-2 is a pattern recognition receptor for fungi that couples with the Fc receptor gamma chain to induce innate immune responses. *J. Biol. Chem.* **281:**38854–38866.

Schroppel, K., M. Kryk, M. Herrmann, E. Leberer, M. Rollinghoff, and C. Bogdan. 2001. Suppression of type 2 NO-synthase activity in macrophages by *Candida albicans*. *Int. J. Med. Microbiol.* **290:**659–668.

Selsted, M. E., and A. J. Ouellette. 2005. Mammalian defensins in the antimicrobial immune response. *Nat. Immunol.* **6:**551–557.

Serrano-Gomez, D., A. Dominguez-Soto, J. Ancochea, J. A. Jimenez-Heffernan, J. A. Leal, and A. L. Corbi. 2004. Dendritic cell-specific intercellular adhesion molecule 3-grabbing nonintegrin mediates binding and internalization of *Aspergillus fumigatus* conidia by dendritic cells and macrophages. *J. Immunol.* **173:**5635–5643.

Shepherd, V. L., and J. R. Hoidal. 1990. Clearance of neutrophil-derived myeloperoxidase by the macrophage mannose receptor. *Am. J. Respir. Cell. Mol. Biol.* **2:**335–340.

Shin, Y. K., K. Y. Kim, and Y. K. Paik. 2005. Alterations of protein expression in macrophages in response to *Candida albicans* infection. *Mol. Cells* **20:**271–279.

Slack, E. C., M. J. Robinson, P. Hernanz-Falcon, G. D. Brown, D. L. Williams, E. Schweighoffer, V. L. Tybulewicz, and C. Reis e Sousa. 2007. Syk-dependent ERK activation regulates IL-2 and IL-10 production by DC stimulated with zymosan. *Eur. J. Immunol.* **37:**1600–1612.

Sohnle, P. G., B. L. Hahn, and R. Karmarkar. 2001. Effect of metals on *Candida albicans* growth in the presence of chemical chelators and human abscess fluid. *J. Lab. Clin. Med.* **137:**284–289.

Sorensen, O., T. Bratt, A. H. Johnsen, M. T. Madsen, and N. Borregaard. 1999. The human antibacterial cathelicidin, hCAP-18, is bound to lipoproteins in plasma. *J. Biol. Chem.* **274:**22445–22451.

Suram, S., G. D. Brown, M. Ghosh, S. Gordon, R. Loper, P. R. Taylor, S. Akira, S. Uematsu, D. L. Williams, and C. C. Leslie. 2006. Regulation of cytosolic phospholipase A2 activation and cyclooxygenase 2 expression in macrophages by the beta-glucan receptor. *J. Biol. Chem.* **281:**5506–5514.

Swanson, J. A., and A. D. Hoppe. 2004. The coordination of signaling during Fc receptor-mediated phagocytosis. *J. Leukoc. Biol.* **76:**1093–1103.

Tada, H., E. Nemoto, H. Shimauchi, T. Watanabe, T. Mikami, T. Matsumoto, N. Ohno, H. Tamura, K. Shibata, S. Akashi, K. Miyake, S. Sugawara, and H. Takada. 2002. *Saccharomyces cerevisiae-* and *Candida albicans*-derived mannan induced production of tumor necrosis factor alpha by human monocytes in a CD14- and Toll-like receptor 4-dependent manner. *Microbiol. Immunol.* **46:**503–512.

Tailleux, L., N. Pham-Thi, A. Bergeron-Lafaurie, J. L. Herrmann, P. Charles, O. Schwartz, P. Scheinmann, P. H. Lagrange, J. de Blic, A. Tazi, B. Gicquel, and O. Neyrolles. 2005. DC-SIGN induction in alveolar macrophages defines privileged target host cells for mycobacteria in patients with tuberculosis. *PLoS Med.* **2:**e381.

Takahara, K., Y. Yashima, Y. Omatsu, H. Yoshida, Y. Kimura, Y. S. Kang, R. M. Steinman, C. G. Park, and K. Inaba. 2004. Functional comparison of the mouse DC-SIGN, SIGNR1, SIGNR3 and Langerin, C-type lectins. *Int. Immunol.* **16:**819–829.

Tanida, T., F. Rao, T. Hamada, E. Ueta, and T. Osaki. 2001. Lactoferrin peptide increases the survival of *Candida albicans*-inoculated mice by upregulating neutrophil and macrophage functions, especially in combination with amphotericin B and granulocyte-macrophage colony-stimulating factor. *Infect. Immun.* **69:**3883–3890.

Taylor, P. R., G. D. Brown, J. Herre, D. L. Williams, J. A. Willment, and S. Gordon. 2004. The role of SIGNR1 and the beta-glucan receptor (Dectin-1) in the nonopsonic recognition of yeast by specific macrophages. *J. Immunol.* **172:**1157–1162.

Taylor, P. R., S. Gordon, and L. Martinez-Pomares. 2005a. The mannose receptor: linking homeostasis and immunity through sugar recognition. *Trends Immunol.* **26:**104–110.

Taylor, P. R., D. M. Reid, S. E. Heinsbroek, G. D. Brown, S. Gordon, and S. Y. Wong. 2005b. Dectin-2 is predominantly myeloid restricted and exhibits unique activation-dependent expression on maturing inflammatory monocytes elicited in vivo. *Eur. J. Immunol.* **35:**2163–2174.

Taylor, P. R., S. V. Tsoni, J. A. Willment, K. M. Dennehy, M. Rosas, H. Findon, K. Haynes, C. Steele, M. Botto, S. Gordon, and G. D. Brown. 2007. Dectin-1 is required for beta-glucan recognition and control of fungal infection. *Nat. Immunol.* **8:**31–38.

Torosantucci, A., G. Romagnoli, P. Chiani, A. Stringaro, P. Crateri, S. Mariotti, R. Teloni, G. Arancia, A. Cassone, and R. Nisini. 2004. *Candida albicans* yeast and germ tube forms interfere differently with human monocyte differentiation into dendritic cells: a novel dimorphism-dependent mechanism to escape the host's immune response. *Infect. Immun.* **72:**833–843.

Trinchieri, G., and A. Sher. 2007. Cooperation of Toll-like receptor signals in innate immune defence. *Nat. Rev. Immunol.* **7:**179–190.

Underhill, D. M., A. Ozinsky, A. M. Hajjar, A. Stevens, C. B. Wilson, M. Bassetti, and A. Aderem. 1999. The Toll-like receptor 2 is recruited to macrophage phagosomes and discriminates between pathogens. *Nature* **401:**811–815.

Underhill, D. M., E. Rossnagle, C. A. Lowell, and R. M. Simmons. 2005. Dectin-1 activates Syk tyrosine kinase in a dynamic subset of macrophages for reactive oxygen production. *Blood* **106:**2543–2550.

Urban, C. F., U. Reichard, V. Brinkmann, and A. Zychlinsky. 2006. Neutrophil extracellular traps capture and kill *Candida albicans* yeast and hyphal forms. *Cell. Microbiol.* **8:**668–676.

Van der Graaf, C. A., M. G. Netea, S. A. Morre, M. Den Heijer, P. E. Verweij, J. W. Van der Meer, and B. J. Kullberg. 2006. Toll-like receptor 4 Asp299Gly/Thr399Ile polymorphisms are a risk factor for *Candida* bloodstream infection. *Eur. Cytokine Netw.* **17:**29–34.

van der Graaf, C. A., M. G. Netea, I. Verschueren, J. W. van der Meer, and B. J. Kullberg. 2005. Differential cytokine production and Toll-like receptor signaling pathways by *Candida albicans* blastoconidia and hyphae. *Infect. Immun.* **73:**7458–7464.

Vazquez-Torres, A., and E. Balish. 1997. Macrophages in resistance to candidiasis. *Microbiol. Mol. Biol. Rev.* **61:**170–192.

Vazquez-Torres, A., J. Jones-Carson, and E. Balish. 1996. Peroxynitrite contributes to the candidacidal activity of nitric oxide-producing macrophages. *Infect. Immun.* **64:**3127–3133.

Vazquez-Torres, A., J. Jones-Carson, T. Warner, and E. Balish. 1995. Nitric oxide enhances resistance of SCID mice to mucosal candidiasis. *J. Infect. Dis.* **172:**192–198.

Villamon, E., D. Gozalbo, P. Roig, C. Murciano, J. E. O'Connor, D. Fradelizi, and M. L. Gil. 2004a. Myeloid differentiation factor 88 (MyD88) is required for murine resistance to *Candida albicans* and is critically involved in *Candida*-induced production of cytokines. *Eur. Cytokine Netw.* **15:**263–271.

Villamon, E., D. Gozalbo, P. Roig, J. E. O'Connor, D. Fradelizi, and M. L. Gil. 2004b. Toll-like receptor-2 is essential in murine defenses against *Candida albicans* infections. *Microbes Infect.* **6:**1–7.

Voganatsi, A., A. Panyutich, K. T. Miyasaki, and R. K. Murthy. 2001. Mechanism of extracellular release of human neutrophil calprotectin complex. *J. Leukoc. Biol.* **70:**130–134.

Vonk, A. G., C. W. Wieland, M. G. Netea, and B. J. Kullberg. 2002. Phagocytosis and intracellular killing of *Candida albicans* blastoconidia by neutrophils and macrophages: a comparison of different microbiological test systems. *J. Microbiol. Methods* **49:**55–62.

Wheeler, R. T., and G. R. Fink. 2006. A drug-sensitive genetic network masks fungi from the immune system. *PLoS Pathog.* **2:**e35.

Wisplinghoff, H., T. Bischoff, S. M. Tallent, H. Seifert, R. P. Wenzel, and M. B. Edmond. 2004. Nosocomial bloodstream infections in US hospitals: analysis of 24,179 cases from a prospective nationwide surveillance study. *Clin. Infect. Dis.* **39:**309–317.

Yamamoto, Y., T. W. Klein, and H. Friedman. 1997. Involvement of mannose receptor in cytokine interleukin-1beta (IL-1beta), IL-6, and granulocyte-macrophage colony-stimulating factor responses, but not in chemokine macrophage inflammatory protein 1beta (MIP-1beta), MIP-2, and KC responses, caused by attachment of *Candida albicans* to macrophages. *Infect. Immun.* **65:**1077–1082.

Yang, D., Q. Chen, O. Chertov, and J. J. Oppenheim. 2000. Human neutrophil defensins selectively chemoattract naive T and immature dendritic cells. *J. Leukoc. Biol.* **68:**9–14.

Yates, R. M., and D. G. Russell. 2005. Phagosome maturation proceeds independently of stimulation of toll-like receptors 2 and 4. *Immunity* **23:**409–417.

Yoshitomi, H., N. Sakaguchi, K. Kobayashi, G. D. Brown, T. Tagami, T. Sakihama, K. Hirota, S. Tanaka, T. Nomura, I. Miki, S. Gordon, S. Akira, T. Nakamura, and S. Sakaguchi. 2005. A role for fungal β-glucans and their receptor Dectin-1 in the induction of autoimmune arthritis in genetically susceptible mice. *J. Exp. Med.* **201:**949–960.

Zanetti, M. 2004. Cathelicidins, multifunctional peptides of the innate immunity. *J. Leukoc. Biol.* **75:**39–48.

Zhang, J., J. Zhu, X. Bu, M. Cushion, T. B. Kinane, H. Avraham, and H. Koziel. 2005. Cdc42 and RhoB activation are required for mannose receptor-mediated phagocytosis by human alveolar macrophages. *Mol. Biol. Cell* **16:**824–834.

29

The Parasite Point of View: Insights and Questions on the Cell Biology of *Trypanosoma* and *Leishmania* Parasite-Phagocyte Interactions

KEITH GULL

In this essay I will deal mainly with the cell biology of three kinetoplastid parasites: *Trypanosoma cruzi*, the *Leishmania* species, and *Trypanosoma brucei*. Each of these parasites causes a devastating disease (Chagas' disease, leishmaniasis, and African sleeping sickness, respectively) affecting tens to hundreds of thousands of individuals each year in the developing world, with many millions at risk (Maudlin, 2006; Piscopo and Mallia, 2006; Teixeira et al., 2006). In each case there is a great need for new drugs and vaccines (Nwaka and Hudson, 2006). *T. cruzi* can potentially infect most cells in the mammalian body including macrophages, whereas *Leishmania* species specifically target the professional phagocytes of the body. *T. brucei* is a very similar parasite in many aspects of its cell biology but remains extracellular using antigenic variation of a surface coat to avoid the immune response (Mansfield and Paulnock, 2005; Pays, 2005; Taylor and Rudenko, 2006). The form of *T. brucei* that infects humans in East Africa also has specific and novel mechanisms for eluding the innate immune system (Pays, 2006; Pays et al., 2006). Therefore, in many ways *T. brucei* is a useful comparative parasite in terms of the integrative cell biology designed to facilitate intracellular growth of *T. cruzi* and *Leishmania* spp.

In this review I have made my observations from the standpoint of the cell biology of the parasite, not of the mammalian cell. I have been rather selective, and what follows is not intended to be a review of this extensive area of host-pathogen interactions. Instead, I have chosen to discuss some well-studied and some less well-studied aspects of parasite cell biology important for interaction with the host mammalian cell. In addition I have presented some arguments whereby I suggest that there are lacunae in our knowledge of the intrinsic life cycle and parasite cell biology underpinning the pathology of these diseases. Again, I have been selective and I have often focused on aspects of motility, cell cycle control, and parasite cell differentiation in mammalian host cells and tissues. I have sometimes set these descriptions in the context of how early 20th century textbooks viewed these parasites, given that many were discovered and intensively studied during the first two decades of the last century. By this device I wish to acknowledge the vast progress that we have made, but want to point out that some aspects of the parasite cell biology of the infection process are still rather cryptic. At a time when we are celebrating the achievements of parasite genome data and new postgenomic opportunities, it might be wise to remind ourselves that we may still have work to do in describing the basic parasite attributes and phenotypes involved in disease pathology.

T. CRUZI MOTILITY AND INVASION

The etiological agent of Chagas' disease, *T. cruzi*, is transmitted by various members of the *Reduviidae* family of insects. A sylvatic cycle of transmission from infected insect to small vertebrates to insect is widespread in South America. Infections in humans arise mainly by cohabitation of infected vectors, domestic animals and humans, in close living conditions. Epimastigote and metacyclic trypomastigote forms of the parasite are released by defecation, and the parasite enters through the skin wounds made by the "kissing bugs" or via a mucosal membrane infection. These actively mobile trypomastigote forms are able to invade many different cell types including the macrophage. The trypomastigote parasite initially enters a lysosome-derived parasitophorous vacuole. Proliferation occurs after escape of the parasite from the vacuole to the cytoplasm. This replication cycle then involves another form of the parasite—the amastigote (Fig. 1).

There is evidence that distinct clinical forms of Chagas' disease may involve selection of different *T. cruzi* clonal populations in the body. At what level such clonal selection occurs is not clear. Persistence of parasites during infection mostly involves their location in muscle tissues.

Keith Gull, Sir William Dunn School of Pathology, University of Oxford, South Parks Road, Oxford OX1 3RE, United Kingdom.

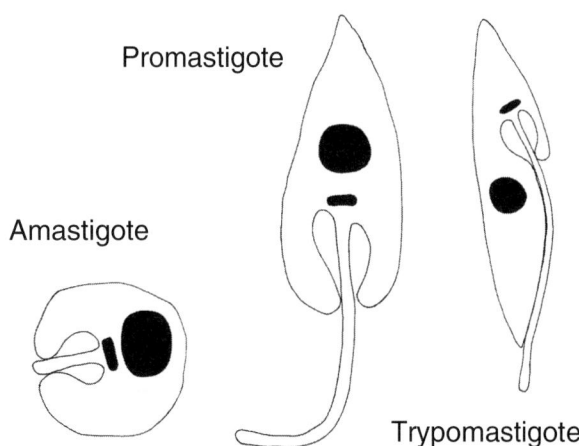

FIGURE 1 Illustration of the structure of the three main forms of the kinetoplastid parasites. In each case, the nucleus is represented by the black circle, the kinetoplast is represented by the black oblong, and the flagellum emerges into the flagellar pocket.

Hence, much work has focused on the mechanisms parasites use to infect such nonphagocytic cells. Moreover, a theme emerges here (to be repeated later in my discussion of *Leishmania*) as to how an actively motile parasite restrains itself (or is restrained) within the infected cell and then differentiates into a nonmotile amastigote form. *T. cruzi* trypomastigotes are highly motile cells and motility appears to continue for some minutes after invasion. Could it be that the orchestration of the cortical actin cytoskeleton and lysosomal fusion events are critical to ensure retention of these actively motile parasites inside the host cells (Andrade and Andrews, 2005)?

In respect to *T. cruzi* invasion, it turns out that two mechanisms have been described for the process and that both differ from phagocytosis in being actin-polymerization independent. The trypomastigote *T. cruzi* parasite develops a tight association with the host cell plasma membrane during the entry event. Once inside, the trypomastigote is enclosed in the host-derived vacuole before finally escaping to the cytoplasm. This invasion process is active, it is not inhibited by cytochalasin D (an anti-actin drug), and there is an active recruitment of host cell lysosomes to the entry site (Tardieux et al., 1992). Lysosome fusions with the plasma membrane create the entry via a calcium-signaling pathway (Rodriguez et al., 1999). However, another mechanism also exists—host cells can be invaded by invaginated plasma membrane, which, after forming a vacuole, then fuses with lysosomes (Andrade and Andrews, 2005; Schenkman and Mortara, 1992). This invasion mechanism is also independent of actin polymerization, and the two mechanisms differ in their involvement of phosphatidylinositol-3-kinase activity as defined by inhibitors such as wortmannin, which preferentially ablates the lysosome-mediated pathway. By using inhibitor studies it has been demonstrated that lysosomal fusion is essential for the retention of *T. cruzi* inside host cells (Andrade and Andrews, 2004). In its absence, the invading trypomastigotes gradually exit from the host cell. The nonmotile amastigote forms of the parasite can invade host cells but do so by a mechanism that depends on actin polymerization. The entry and retention mediated by lysosomal fusion appear to be critical to the invasion by the motile trypomastigote forms. Experiments involving cytochalasin D treatment and dominant negative RhoA suggest that similar conditions facilitate not only on entry, but also exit. An intact cytoskeleton appears to be important for parasite retention, and *T. cruzi* replicates for nine generations in the cytoplasm before the immotile amastigotes differentiate into actively motile trypomastigotes (Andrade and Andrews, 2005). A general point might be made here: flagellar motility in a eukaryotic intracellular pathogen may be just too deleterious for the host cell. Hence, the formation of the amastigote cell type. However, this raises the issue of what part motility plays in the release of the newly flagellated *T. cruzi* trypomastigote progeny.

THE INTRACELLULAR AMASTIGOTE FORM— THE IMPORTANCE OF THE FLAGELLUM

In considering the interaction of phagocytic cells and kinetoplastid parasites such as *Leishmania*, it is intriguing that the earliest accounts rather mixed them up! The first observer to see one of the parasites, which are now regarded as belonging to the genus *Leishmania*, was Cunningham in 1885 in India, who described "Peculiar Parasite Organisms in the Tissue of a Specimen of Delhi Boil." In fact, the parasite organisms referred to were the large macrophages, which were assumed to be amoebae while the *Leishmania* inside them were regarded as spores (Wenyon, 1926). It was Leishman's 1903 paper on "The Possibility of the Occurrence of Trypanosomiasis in India" that clearly described the parasites which he had found 3 years before in cases of dum-dum fever.

The intracellular forms of parasites such as *Leishmania* and *T. cruzi* are often referred to as the "aflagellate amastigote form." While not wishing to be a pedant over this issue, I observe that this form of terminology has led to views that the flagellum is unimportant or not present in the amastigote form of these parasites. I believe this is a mistake and I will argue that this is an area of critical importance. Moreover, it has likely masked from attention some cryptic cell biology pertaining to the operational success of these parasites within the parasitophorous vacuole (*Leishmania* sp.) or cytoplasm (*T. cruzi*).

Wenyon in his 1926 textbook described the morphology of the *Leishmania* parasite inside the macrophage. He commented that in stained material a line (first described by Christophers in 1904) can be traced from the blepharoplast (basal body) to the surface of the parasite. He clearly indicates that this line leading from the kinetoplast to the plasma membrane is the axoneme of the amastigote forms that do not possess a flagellum. Thus, in a textbook of the early 20th century three things were recognized as being distinct—the blepharoplast (basal body), the axoneme, and the flagellum. In essence, these early reports viewed the flagellum as the motile part of the organelle that protruded from the cell. This distinction was, and still is, important since in parasites such as *Giardia* the axonemes traverse a long distance through the cytoplasm before finally emerging as motile flagella (Briggs et al., 2004; Elmendorf et al., 2003). With the advent of the electron microscope it became clear that in the case of *Leishmania* and *T. cruzi* the axoneme in the amastigote form is not enclosed by cytoplasm. It is in fact a small, apparently nonmotile flagellum that emerges from the plasma membrane and protrudes through a cup-shaped flagellar pocket that is internal to the cell body (Alexander et al., 1999; Pan and Pan, 1986). Hence, although it is true that there is no external flagellum in one sense (i.e., an elongated motile structure lying outside the general cell profile), there is an apparently nonmotile, short flagellum that traverses the flagella pocket and may protrude a short way out of the pocket. Recognition of the latter is important for a number of reasons. However, before discussing these let us recognize that the evolutionary cell biology of flagella allows for organisms to construct fully flagellated and completely nonflagellated cell types and to switch between them at will. Good examples of this come in the form of the amoeboflagellates such as *Physarum* and *Naegleria*, where the amoeboid forms of the organisms exhibit no flagellum and no axoneme; they retain merely the basal bodies as centrioles or even do away with these (Glyn and Gull, 1990; Kim et al., 2005). Thus, retention of the short flagellum in amastigote parasites appears to be an important phenomenon in their life cycle.

The first clue to the importance of the short flagellum in the intracellular amastigote forms of *Leishmania* and *T. cruzi* parasites comes when we consider the flagellar pocket itself. The flagellar pocket is a key feature of both free-living and pathogenic kinetoplastid protozoa. The plasma membrane of the organisms is characterized by a subpellicular corset of microtubules that defines the shape and form of the parasite. However, a consequence of this subpellicular corset covering the inner face of the plasma membrane is that it hinders vesicle traffic to and from the cell surface. This is solved by the flagellar pocket. This is a microtubule-free, internal area of plasma membrane, and it is to this area that all vesicle traffic, both uptake and secretory, is directed (De Souza, 2006; Field et al., 2007; Gull, 2003; Landfear and Ignatushchenko, 2001; Morgan et al., 2002a, 2002b). The flagellar pocket also allows these kinetoplastid parasites to locate particular receptors (such as the transferrin receptors in *T. brucei*) within the pocket. Thus, the flagellar pocket is a critical feature of these parasites and, given the earlier point about evolutionary possibilities, presumably must be maintained even in the absence of the need for an external motile flagellum.

The flagellum pocket is defined by two boundaries. The first is the basal boundary where the basal body is connected to the flagellar pocket membrane. However, critically, the neck region of the flask structure of the pocket is defined by an electron-dense cytoplasmic flagellar pocket collar (Sherwin and Gull, 1989). This collar region is also intimately associated with the plasma membrane and forms the pocket neck. The membrane of the emerging flagellum exits through this neck and the membranes are in close proximity. Thus, it is likely that an axoneme of a certain length is actually a requirement in the amastigote cell to allow morphogenesis of the flagellar pocket neck. Without a flagellar pocket the amastigote would be unlikely to be able to take up material from the luminal material of the parasitophorous vacuole.

Second, these parasites are defined by the possession of the kinetoplast, a mass of over 1,000 circular, catenated mitochondrial DNA molecules that constitute the mitochondrial genome (Shlomai, 2004). This mass of mitochondrial DNA is connected through the mitochondrial membranes to the flagellum basal body, and duplication and segregation of the flagellar basal bodies segregates the mitochondrial genome (Ogbadoyi et al., 2003). It is likely that maintenance of a flagellum and not just a basal body is critical for efficient segregation of the kinetoplast.

Finally, however, it is important to realize that flagella are not only organelles of motility, but they also operate as sensory organelles. Mammalian cells use these cilia in a number of important sensory situations (Dawe et al., 2007; Pan et al., 2005; Praetorius and Spring, 2005). Such sensory cilia (flagella) present specific receptors/signaling channels on their membranes and are often immotile. Thus, given that there are specific flagella membrane-located proteins, then I suggest here that the amastigote uses this short flagellum as a sensory organelle. In this respect, I believe that it is the trypanosome equivalent of the primary and sensory cilia of mammalian cells (Dawe et al., 2007; Pan et al., 2005; Praetorius and Spring, 2005).

Thus, although the original light microscopic descriptions of the amastigote form of kinetoplastid parasites emphasized the existence of a short axoneme, this was revealed by electron microscopic studies to be a short flagellum that is present within or just protruding from a flagellar pocket. A closer research focus on this aspect of amastigote cell biology is likely to be informative in terms of its importance for cell morphogenesis and interaction with the host macrophage.

THE TRANSFORMATION TO AND FROM THE FLAGELLATED CELL

It is interesting to return to Wenyon's 1926 textbook to emphasize another area of our ignorance of the cell biology of parasite development in the host cell. The original infection-founding, invading cell in the mammalian stage of the life cycle of both *Leishmania* and *T. cruzi* is flagellated. In *T. cruzi* this is the trypomastigote and in *Leishmania* species it is the metacyclic promastigote. There is very little modern cell biological evidence of exactly how these cells transform after invasion into the amastigote. Normally in the cell cycle of *Leishmania* and *Trypanosoma* species the old flagellum is retained and a new one develops from the probasal body that formed in the last cell cycle (Gull, 1999). So, in this transformation from a long, external, motile flagellum to a short, immotile flagellum in an amastigote, does the flagellum shorten or is there an asymmetric first division to produce the first amastigote cell? On page 486, writing about *T. cruzi*, Wenyon (1926) states that "multiplication appears to commence after the invasion of a cell by a single trypanosome which, losing its membrane and flagellum, becomes a *Leishmania* form." There is little modern cell biology to confirm that this is indeed the process and, if so, how it is achieved.

In a similar manner, one can ask whether or not, in the transformation back to a flagellated cell, the short flagellum of the individual amastigote extends or whether again there is an asymmetric division? Again, Wenyon (1926) gives an indication of what might occur in stating that in *T. cruzi* "at a certain stage each of the leishmanial forms develops a flagellum, and by gradual changes in the arrangements of its parts becomes transformed into a crithidial form and finally into a trypanosome of the blood type." No time scale is given; however, in his Fig. 207 (page 491), Wenyon shows the different dividing forms and rather suggests that the model is flagellum extension and then division. It is difficult, however, to discriminate between the various possible scenarios without further evidence. One further early observation again has importance for understanding parasite cell cycle control in the host cell. Wenyon states that, "In any group of organisms the changes affect all the individuals at the same time and at approximately the same rate, so that they all arrive at maturity together." This suggests an important cell cycle coordination and differentiation communication between individual parasites within an infected host cell. Again, it is likely that a modern cell biological view of these cryptic events would be both useful and interesting.

LEISHMANIASIS: VISCERAL VERSUS CUTANEOUS

Although all species of Leishmania invade phagocytic cells, they produce individually characteristic disease symptoms falling within two main groups: the visceral and the subcutaneous leishmaniases. The former is a chronic, disseminated visceral disease of the liver and spleen characteristic of L. infantum and L. donovani. The latter is characterized by self-limiting cutaneous lesions caused by species such as L. mexicana, where self-healing can result in lifelong immunity. Leishmania species are spread by sand flies and the parasite develops as a flagellated promastigote form in the sand fly gut. Metacyclic promastigotes are introduced into the skin when the sand fly bites. After engulfment by a phagocytic cell, the parasite transforms into an amastigote form that resides in a parasitophorous vacuole and undergoes active division (Handman and Bullen, 2002; Kima, 2007; McConville and Handman, 2007; Murray et al., 2005; Piscopo and Mallia, 2006; Vannier-Santos et al., 2002).

Visceral leishmaniasis caused by L. donovani and L. infantum reveals itself in approximately 550,000 new cases per annum in India, Sudan, and Brazil. It differs from cutaneous leishmaniasis in that most clinical cases lead to death if untreated. Interestingly, the vast majority of infections do not progress to the clinical state because the parasites remain contained in a localized granulomatous tissue response (Murray, 2001). It is unclear whether specific parasite stages lead to the visceral infection or whether it originates from a direct bloodstream infection or via infected phagocyte migration. One of the major issues for molecular parasitology is what pathogen factors drive L. donovani and L. infantum to exhibit this viscerotropic tendency compared with the cutaneous Leishmania species.

There are various phases in the experimental L. donovani infection, and there are specific and different outcomes in the liver and spleen (Engwerda et al., 2004; Kaye, 2006). In the liver there is a fast invasion and expansion by amastigote parasites, but this is controlled by the granulomatous response (involving Kupffer cells, monocytes, and $CD4^+$ and $CD8^+$ T cells) by day 28. There is extensive cytokine involvement (interferon-γ, interleukin-12, interleukin-4, and tumor necrosis factor). As the hepatic infection is resolved, the progressive infection in the spleen is manifest and a splenomegaly results with loss of spleen microarchitecture. In this case, the tissue damage and immunological dysfunction appear orchestrated by an excess of tumor necrosis factor (Engwerda et al., 2004; Kaye, 2006; Murray, 2001). Are these differences in target tissue outcomes a reflection of uptake of parasites by different types of mononuclear phagocytes, perhaps with concomitant variation in the receptors involved in recognition of these visceral leishmaniasis parasites? Electron microscopy of human visceral infections has shown amastigote forms of L. donovani within Kupffer cells and hepatocytes, leading to the suggestion that such sites may be possible reservoirs for recrudescence of the disease (Duarte et al., 1989; Rittig and Bogdan, 2000).

BY WHICH ROUTE AND BY WHICH CELL?

When a sand fly bites, it ingests Leishmania amastigote parasites along with its blood meal, and their subsequent differentiation back into promastigote-flagellated parasites then initiates another cycle of vector infection. Examination of these steps from the parasite point of view reveals some intriguing gaps in our knowledge of important stages in the interaction of Leishmania parasites with phagocytic cells. In general, it is accepted that metacyclic promastigotes invade a phagocytic cell in a tissue local to the infection site. However, it could be that it is more complex and that parasites could move via lymph traffic to a lymph node and the first infection occurs there; or, because the fly may inflict local damage in the skin, there could be capillary damage and direct entry followed by monocyte uptake. As discussed above, it is possible that invasion routes and different subtypes of phagocytic cells play a role in the infections of different target tissues by different Leishmania species.

Over the past 7 years, there has been an explosion of data relating to the organization and sequence of the genomes of many kinetoplastid parasites, including a number of Leishmania species (Berriman et al., 2005; El-Sayed et al., 2005; Ivens et al., 2005). A major challenge now exists to turn these data into knowledge. One such area receiving attention is the use of comparative genomics coupled with reverse genetics to shed light on the genetic basis for important pathogenic features such as visceralization, tissue tropism, and specific targeting of subsets of phagocytes. The published sequence of the L. major genome (Ivens et al., 2005) and the near-complete L. infantum genome (http://www.sanger.ac.uk/Projects/Protozoa/) offer the start of such comparative genomic surveys.

While comparative genomic surveys may shed light on differing parasite pathologies, other postgenomic approaches seek to understand the expression of specific gene sets in specific niches within the mammalian host and vector. The trypanosomatid parasites exhibit the common feature of polycistronic transcription and trans-splicing (Palenchar and Bellofatto, 2006). Thus, the differential expression of proteins is essentially controlled at a posttranscriptional level. This leaves open the question of how insightful the use of transcriptomic analysis by DNA microarray will be in these parasites. Indeed, the results from initial applications of this technology have revealed limited stage-specific expression of genes, with virtually all genes expressed in all stages of the life cycle studied (Cohen-Fruee et al., 2007). Here, the experiments have focused on studying comparisons of the selectively expressed genes of the infecting metacyclic promastigotes and the amastigote forms isolated from lesions or grown axenically. Thus, these initial surveys have not revealed specific pathways or factors likely to be significant in defining the interactions with host phagocytic cells discussed above.

However, there are parasite-specific differences (McConville and Handman, 2007), and hope that we will see insights emerge from these postgenomic studies of parasites has come from studies of the interesting A2 gene locus. This gene family was first identified as a locus in L. donovani that is specifically expressed in the amastigote stage, and A2 remains as one of the few identified amastigote-specific proteins. Reverse genetics on the A2 locus identify it as a pathogenicity factor given that A2-deficient parasites are attenuated in terms of visceralization (Zhang and Matlashewski, 1997, 2001). Recently, the L. donovani (visceral leishmaniasis) A2 locus was compared with that in L. major (cutaneous leishmaniasis). L. donovani contains multiple tandemly repeated copies of the A2 gene interspersed with copies of an A2rel gene family. The A2 proteins have different numbers of a repetitive domain. However, the L. major genome lacks this multiple gene organization, and the remaining sole gene appears different in organization, particularly in terms of the repetitive domain region of the putative protein (Garin et al., 2005; Zhang et al., 2003). Ectopic L. donovani A2 amastigote-specific expression in L. major was engineered, and interestingly, such A2-expressing parasites could not establish the normal cutaneous infection in mice, reminiscent of the natural situation in L. donovani. This suggests that the differences in the A2 protein and its expression contribute to the different patterns of tissue phagocyte infection seen in L. donovani and L. major.

RECOGNITION ZONES ON THE PARASITE SURFACE?

The early interaction of both types of parasite—the motile metacyclic promastigote and the immotile amastigote—with the engulfing macrophage is of some interest. Are there particular surface zones on the parasite for recognition by the macrophage and does parasite motility influence the means of phagocytosis? Rittig and Bogdan (2000) have argued that the flagellum appears to be the predominant site for early interaction of promastigotes of Leishmania species with the phagocyte and that adhesion and phagocytosis are almost exclusively unipolar, with the anterior (flagellum tip) end providing the initial contact point. They commented that perhaps L. donovani is an exception since this species has been observed to attach via its flagellum and the nonflagellar end in roughly equal proportions (Pearson et al., 1983; Rittig and Bogdan, 2000). However, Courret et al. (2002) studied the phagocytosis of both promastigotes and amastigotes of L. amazonensis and L. major by macrophages derived from mouse bone marrow, preexposed or not to interferon-γ. They believed that the first interactions between promastigotes and the macrophage were through either the flagellum, the cell body, or the entire parasite. They also found that binding of the flagellum could be followed by phagocytosis starting with the cell body. In fact, internalized promastigotes were often orientated with the flagella pointing toward the macrophage periphery. A transient accumulation of host cell actin was seen around phagocytosed parasites within 10 min of the phagocytic process. The accumulation of the late endosome/lysosomal markers characteristic of the phagosomes was similar regardless of whether metacyclic promastigotes or amastigote parasites were internalized. However, the development of the large parasitophorous vacuoles characteristic of L. amazonensis (see below) was faster when amastigotes were internalized (Courret et al., 2002).

Thus, some debate remains as to whether there is a specific zone on the parasite that is important for macrophage interaction. One can easily argue for a structural or biochemical differentiation of the flagellum from the rest of the cell body. There is no doubt that the flagellar surface of these kinetoplastid parasites is different from the cell body surface. For instance, in organisms such as Leishmania and T. brucei the differentiated parasites form highly structured and intimate associations via hemidesmosomal-like plaques with insect vector cell surfaces. Structural differences leading to "flagellum-first" phagocytosis may include simple issues such as this being the first "collision" point, since kinetoplastid parasites move with their flagellum pointing forward. From the earliest description of these organisms there are many reports of Leishmania flagellated forms tending to aggregate by their flagella in culture or attaching to surfaces via their flagella tips. Hence, there is little doubt that phenotypically the flagellum membrane and tip are differentiated in some way. Moreover, there are reports of Leishmania flagellar antigens being located at the tip, and more recently, sets of Leishmania proteins have been shown to locate to the flagellar membrane rather that the cell body membrane (Fridberg et al., 2007; Tull et al., 2004). Determining the nature of putative phagocyte recognition ligands at flagellar surfaces is of obvious importance in providing insights to these critical early events of Leishmania infection. A differentiation of the other pole of the cell is also possible, since the subpellicular microtubule cytoskeleton of these parasites is highly polarized, giving ample cell biological opportunities for membrane differentiation at the nonflagellum pole of the cell.

The phagocytic internalization process itself proceeds via two mechanisms. The first is a classical "zipper" type of phagocytosis, where cytoskeletal rearrangements in the phagocyte produce a pseudopodial extension that advances along the flagellum/cell body and sequentially engulfs the parasite into a phagosome. It has been argued that this is the main mechanism (Courret et al., 2002). An additional mechanism, termed coiling phagocytosis, has been described, in which asymmetric, multilayered stacks of membrane surround the parasite (Courret et al., 2002; Rittig et al., 1998). There are often unexplained delays between engulfment of the flagellum and the cell body itself.

The group of African trypanosomes that include T. brucei are spread by the bite of an infected tsetse fly. The metacyclic parasite that establishes the infection is a pre-differentiated cell that already possesses a variable surface glycoprotein coat and is in G_0 of the cell cycle (Matthews, 2005). On entering the mammalian host, it differentiates into the slender, proliferative bloodstream form of the parasite. However, this parasite is designed to infect via a parasitemia in the blood rather than to invade host cells. In this regard one imagines that these types of kinetoplastid parasites may possess cryptic strategies for avoiding macrophage attack. However, the rising parasitemia is recognized and lysed by the immune response and the parasite population is then efficiently cleared by macrophages. This clearance strategy results in the parasitemia wave being limited. New antigen switch variants establish forthcoming waves of parasitemia, thus ensuring vertical transmission of the parasite population in the same host (Pays et al., 2004; Taylor and Rudenko, 2006).

The African trypanosomes such as T. brucei have a variable surface glycoprotein (VSG) antigenic coat and appear able to evade phagocytosis by macrophages. However, antibody response to the VSG coat changes this, and both immunoglobulin M and G antibodies to the VSG were observed to facilitate macrophage phagocytosis of T. congolense (Shi et al., 2004). Only a small phagocytic clearance of trypanosomes by Kupffer cells was apparent at 5 days after infection, but there was a dramatic increase in this activity only 1 day later. Earlier work in T. gambiense had defined the requirement for antibody and that attachment did not occur in the absence of antibody that recognized the trypanosome cell surface. Attached parasites, however, were phagocytosed within about 90 min, and there was some indication that trypanosomes adhered to the macrophage by their anterior ends (Takayana et al., 1974a, 1974b). Thus, antibody-mediated clearance of parasites is an intrinsic part of the cycle of parasitemia, associated with antigenic switching and subsequent "escape" of new serovars, that maintains the infection. Unfortunately, we have little information on whether these parasites possess particular attributes that facilitate their avoidance of phagocytosis in the absence of antibody in situations such as the site of the tsetse bite or in the first days as the parasitemia builds.

Motility could be an important aspect of an "avoidance" mechanism, and we recently sought to test this by engineering motility deficient-inducible RNAi mutants in *T. brucei* (Broadhead et al., 2006). However, this produced the surprising result that in the bloodstream form of the parasite such interference with the flagellar function led to a failure of cytokinesis, with a resulting monstrous cell phenotype. These monsters were the result of multiple attempts at cytokinesis with continued nuclear divisions and formation of new flagella. Ultimately, such cells were unviable. Experiments in vivo where the RNAi phenotype was induced in such mutants during an infection showed that the turbulence of the bloodstream did not facilitate cytokinesis and the parasitemia was quickly cleared (Griffiths et al., 2007). Although one of the original aims of the work now becomes rather academic, these results revealed some very interesting aspects of trypanosome biology and established the flagellum as a clear drug target for future intervention opportunities.

THE COMPLEX GLYCOBIOLOGY OF THE PARASITE SURFACE

Leishmania parasites exhibit an intriguing array of glycoconjugate surface molecules, including membrane-bound lipophosphoglycan (LPG) and proteophosphoglycan as well as secreted phosphoglycan, proteophosphoglycan and acid phosphatase. The surface is also populated by many types of glycosylphosphatidylinositol-anchored glycoconjugates including glycoproteins and glycoinositolphospholipids (Field et al., 2007; Naderer et al., 2004; Turco et al., 2001).

Surface molecules on *Leishmania* such as gp63 and the LPGs are important parasite molecules implicated in acting as attachment ligands for phagocyte receptors (among others, these include the complement receptors type I [CR1] and type III [CR3]). The gp63 protein is a surface-located metalloprotease abundant on promastigotes but absent from amastigotes. This gp63 metalloprotease cleaves host complement component C3b into iC3b, which prevents complement-mediated lysis and facilitates phagocytosis via the above-mentioned receptors. Thus, some of the most important ligands for parasite uptake are not parasite derived.

LPG has been proposed to be important for delaying phagosome maturation and so facilitating the development of the amastigote parasite form that is resistant to the acid environment of the phagolysosome (see below). LPG incorporation into the phagosome membrane may impair its fusigenic capability, thus delaying maturation by inhibiting fusion with other endomembrane components. Moreover, there is evidence of LPG's role in subverting signal transduction and cytokine production in the phagocyte. The continuing development of excellent genetic tools for modification of *Leishmania* has led to direct tests of the function of LPG by creating mutants unable to synthesize this material. Stephen Beverley and colleagues created a null mutant of *L. major* lacking the *LPG1* gene encoding a galactofuranosyltransferase required to build the LPG glycan core. Promastigote forms of this mutant entered macrophages as normal (given the opsonization process described earlier) but were eliminated within a few days (Spath et al., 2000). As expected (since wild-type macrophages express very little LPG on their surface), amastigote forms infected macrophages and proliferated as normal. In terms of animal infectivity, the LPG null mutant had significantly delayed lesion formation in mice. This careful and insightful study revealed how the surface molecules of *Leishmania* are of critical importance as pathogenicity factors. However, it appears that the arsenal of *Leishmania* cell surface molecules is orchestrated in different tactical strategies during infection by different species of *Leishmania*. In a similar series of experiments in *L. mexicana*, Thomas Ilg (Ilg, 2000) found that LPG null mutants showed no significant impairment in infectivity to macrophages or animals. Thus, a well-studied and highly expressed molecule appears to be a virulence factor in the mouse model of one pathogen but not the other.

The development of mutant *Leishmania* parasites deficient in putative virulence factors, and their use in infection studies in concert with mouse mutants deficient in host defense systems, has been one of the most important developments in molecular parasitology over the past 10 years. These studies have started to clarify the importance and role of the different surface molecules (Spath et al, 2003; Turco et al., 2001; Zufferey et al., 2003). For instance, the LPG null strains of *L. major* together with mouse mutants have proven useful in dissecting the events of macrophage invasion and survival by metacyclic promastigote stages of the parasite (Spath et al., 2003). The LPG null *L. major* was highly susceptible to human complement, was oxidant sensitive, and exhibited an inability to inhibit phagolysosomal fusion transiently. These mutant parasites bound C3b and resisted low pH and proteases normally, suggesting that *Leishmania* cells have an abundance of alternative C3 acceptors on their cell surface. They entered macrophages efficiently and continued to inhibit the host cell signaling pathways. However, combination experiments with the LPG mutants and mouse knockouts ($STAT1^-$, $phox^-$) produced some important conclusions in relation to the view that promastigote *L. major* parasites transiently inhibit phagolysosomal fusion in an LPG manner (as in *L. donovani* [Desjardins and Descoteaux, 1997]) to allow differentiation to the amastigote form in the first 2 days. Although the transient LPG-mediated phagolysosomal fusion inhibition occurs, it does not appear to play a major role in allowing the establishment of macrophage infections. Hence, the application of modern molecular approaches is revealing the *Leishmania* parasites to be a complex group whose members orchestrate a rich arsenal of virulence molecules to meet individual needs with resulting differences in pathology.

THE PARASITOPHOROUS VACUOLE

Life inside the cell involves a series of differentiation events whereby the phagosome is remodeled into a parasitophorous vacuole and the parasite differentiates into the amastigote form. This obviously represents a new cellular compartment and varies in structure and properties from parasite to parasite (de Souza, 2005).

An intriguing difference exists between different species of *Leishmania* in their lifestyle within the macrophage after their establishment in the phagolysosomal parasitophorous vacuole. All parasitophorous vacuoles exhibit membrane markers, such as the lysosomal-associated membrane proteins (LAMPs), plus an acidic environment and luminal hydrolases. Modification of the parasitophorous vacuole is via Nramp1, a lysosomal membrane protein that is recruited to the membrane of the parasitophorous vacuole. Here, it acts as a proton-dependent efflux pump that removes iron (Fe^{2+}) ions from the vacuole containing the amastigote parasite, thus acting as part of a host defense system depriving the parasite of iron. This battle over the requirement for rare iron is faced by every pathogen invading the human body. Recent studies reveal that *Leishmania* counters the debilitating effects of Nramp1 by placing LIT1 proteins of the ZIP family of Fe^{2+} transporters in the plasma membrane of the intracellular amastigote form of the parasite (Huynh et al., 2006; Marquis and Gros, 2007). LIT1 therefore turns out to be an important virulence determinant.

However, there are two distinct types of parasitophorous vacuole, and each is specific to a particular group of *Leishmania* pathogens. Old World species of *Leishmania* (e.g., *L. donovani* and *L. tropica*) form small parasitophorous vacuoles (type I) enclosing individual parasites, whereas New World species (e.g., *L. mexicana*) form a large, single parasitophorous vacuole (type II) enclosing many amastigotes (Antoine et al., 1998; Castro et al., 2006). A recent study of the parasitophorous vacuoles formed by *L. major* in macrophages in the lymph node of infected mice provides some of the best published images of type I vacuoles and surveys the patchy literature relating to the ultrastructure of infected macrophages (Castro et al., 2006). These authors point out that the implication of a type I pattern is that, as the amastigotes of *L. major*, *L. tropica*, and *L. donovani* divide and proliferate, so the vacuole must be divided and segregated with each daughter amastigote. There is, however, little evidence for how this might occur or for the implication that there is some intimate association between the amastigote's surface and the surrounding parasitophorous vacuole membrane. Some evidence exists that some amastigotes present in type II multi-amastigote large vacuoles connect to the surrounding parasitophorous membrane at specific sites on their cells (Castro et al., 2006; de Souza, 2005; Rittig and Bogdan, 2000). The molecular basis for cell-to-vacuole interactions is unclear and deserves attention. Different occurrences and distributions of parasitophorous vacuole-membrane-interacting parasite surface molecules might shed light on this important aspect of pathology. However, this distinct difference in the cell biology of the New and Old World *Leishmania* species is perhaps relevant to the distinct species differences seen between phenotypes of the surface molecule null mutants described above.

MODE OF RELEASE FROM CELLS AND THE NATURE OF THE RELEASED PARASITE

Many diagrams of the *Leishmania* life cycle show the end of the macrophage cycle as a "burst" event, with sudden lysis of the cell and release of the enclosed amastigotes. In an evidence-poor context, and with the general observation that other intracellular pathogens have evolved rather sophisticated ways of "leaving" infected cells, it has been argued that an alternative view is possible. This view is that amastigotes might be released in a controlled manner over many hours from peripheral sites in the macrophage (Rittig and Bogdan, 2000). If this is the case in natural infections, does it indicate that the parasite is able to subvert the exocytic machinery of the macrophage? The ultimate fate of the macrophage in this process is unclear.

However, this also raises the question of the nature of the parasite that is released from the infected phagocyte (or possibly the parasite that remains in the phagocyte?). There is little discussion of this in the literature. The issue is that the parasite has two potential alternative routes to ensure its survival and progression. The first is vertical transmission, where, to ensure establishment of the infection, amastigote parasites need to infect new macrophages and continue to proliferate. However, the parasite also needs to prepare for the potential of horizontal transmission to the vector. This imposes a major and sudden change in environment that the parasite needs to react to in a very short period of time—from mammalian body to fly vector gut! A comparison with the well-studied parts of other parasite life cycles shows that these two routes of transmission, horizontal and vertical, are often accomplished by two different cell types. Regardless of the parasite, the cell type engaged in horizontal transmission to the vector displays certain characteristics. Such cells are often predifferentiated and have exited the cell cycle and are in a G_1/G_0 state. Two examples will illustrate this. In *T. brucei* this horizontal transmission from bloodstream to tsetse fly midgut is achieved by the stumpy-form parasites that have already differentiated in the activation of metabolic functions required in the tsetse procyclic form and have exited the proliferative cell cycle (Matthews, 2005). In malaria, the transmission from mammalian bloodstream to the mosquito gut is achieved by the gametocytes, which again exhibit cell cycle status and differentiations very different from the cells (merozoites) that will continue the vertical transmission within the erythrocytes of the infected individual (Barillas-Mury and Kumar, 2005; Khan and Waters, 2004; Whitten et al., 2006). Given the very different and harsh environment of the vector gut (which is designed to digest the normal cell and plasma contents of the blood meal), a fast reentry to the proliferative cell cycle by a preprepared differentiated cell type would appear to be a sensible evolutionary strategy. The question thus arises as to whether all amastigotes produced in the *Leishmania* life cycle are equivalent. There is symmetry in the life cycle of protozoan parasites, in that a predifferentiated and nondividing cell type is also used to effect the transfer of the parasite from the vector back to the mammalian host. In the case of *T. brucei*, this is the metacyclic form, and in the case of malaria, it is the sporozoite. *Leishmania* has developed such a predifferentiated, nonproliferative form of parasite—the metacyclic promastigote—which is the cell type that initiates the initial phagocyte infection. Two issues therefore emerge that may merit future research. The first is the examination of the precise subtype(s) of host phagocytes infected by this metacyclic promastigote stage in the natural infection caused by the bite of a sand fly. The second would be a closer questioning as to whether all amastigotes produced during the infection are the same, or whether a particular type (perhaps associated with a particular mode of exit) is associated with horizontal transmission to the vector. These areas might reveal particular insights to the interplay between *Leishmania* and the host phagocytic cell and form part of the cryptic cell biology of the kinetoplastid pathogens that has been the theme of this essay.

Work in my laboratory is funded by the Wellcome Trust, BBSRC, HFSP, and the EP Abraham Trust. I hold a Wellcome Trust Principal

Research Fellowship. I thank past and present members of my group and other researchers for useful discussions.

REFERENCES

Alexander, J., A. R. Satoskar, and D. G. Russell. 1999. Leishmania species: models of intracellular parasitism. *J. Cell Sci.* **112:**2993–3002.

Andrade, L. O., and N. W. Andrews. 2004. Lysosomal fusion is essential for the retention of *Trypanosoma cruzi* inside host cells. *J. Exp. Med.* **200:**1135–1143.

Andrade, L. O., and N. W. Andrews. 2005. The *Trypanosoma cruzi*-host-cell interplay: location, invasion, retention. *Nat. Rev. Microbiol.* **3:**819–823.

Antoine, J. C., E. Prina, T. Lang, and N. Courret. 1998. The biogenesis and properties of the parasitophorous vacuoles that harbour *Leishmania* in murine macrophages. *Trends Microbiol.* **6:**392–401.

Barillas-Mury, C., and S. Kumar. 2005. *Plasmodium*-mosquito interactions: a tale of dangerous liaisons. *Cell. Microbiol.* **7:**1539–1545.

Berriman, M., E. Ghedin, C. Hertz-Fowler, G. Blandin, H. Renauld, D. C. Bartholomeu, N. J. Lennard, E. Caler, N. E. Hamlin, B. Haas, U. Böhme, L. Hannick, M. A. Aslett, J. Shallom, L. Marcello, L. Hou, B. Wickstead, U. C. Alsmark, C. Arrowsmith, R. J. Atkin, A. J. Barron, F. Bringaud, K. Brooks, M. Carrington, I. Cherevach, T. J. Chillingworth, C. Churcher, L. N. Clark, C. H. Corton, A. Cronin, R. M. Davies, J. Doggett, A. Djikeng, T. Feldblyum, M. C. Field, A. Fraser, I. Goodhead, Z. Hance, D. Harper, B. R. Harris, H. Hauser, J. Hostetler, A. Ivens, K. Jagels, D. Johnson, J. Johnson, K. Jones, A. X. Kerhornou, H. Koo, N. Larke, S. Landfear, C. Larkin, V. Leech, A. Line, A. Lord, A. Macleod, P. J. Mooney, S. Moule, D. M. Martin, G. W. Morgan, K. Mungall, H. Norbertczak, D. Ormond, G. Pai, C. S. Peacock, J. Peterson, M. A. Quail, E. Rabbinowitsch, M. A. Rajandream, C. Reitter, S. L. Salzberg, M. Sanders, S. Schobel, S. Sharp, M. Simmonds, A. J. Simpson, L. Tallon, C. M. Turner, A. Tait, A. R. Tivey, S. Van Aken, D. Walker, D. Wanless, S. Wang, B. White, O. White, S. Whitehead, J. Woodward, J. Wortman, M. D. Adams, T. M. Embley, K. Gull, E. Ullu, J. D. Barry, A. H. Fairlamb, F. Opperdoes, B. G. Barrell, J. E. Donelson, N. Hall, C. M. Fraser, S. E. Melville, and N. M. El-Sayed. 2005. The genome of the African trypanosome *Trypanosoma brucei*. *Science* **309:**416–422.

Briggs, L. J., J. A. Davidge, B. Wickstead, M. L. Ginger, and K. Gull. 2004. More than one way to build a flagellum: comparative genomics of parasitic protozoa. *Curr. Biol.* **14:**R611–R612.

Broadhead, R., H. R. Dawe, H. Farr, S. Griffiths, S. R. Hart, N. Portman, M. K. Shaw, M. L. Ginger, S. J. Gaskell, P. G. McKean, and K. Gull. 2006. Flagellar motility is required for the viability of the bloodstream trypanosome. *Nature* **440:**224–227.

Castro, R., K. Scott, T. Jordan, B. Evans, J. Craig, E. L. Peters, and K. Swier. 2006. The ultrastructure of the parasitophorous vacuole formed by *Leishmania major*. *J. Parasitol.* **92:**1162–1170.

Christophers, S. R. 1904. A preliminary report on a parasite found in persons suffering from enlargement of the spleen in India. *Sci. Mem. Off. Med. San. Dep. Gov. India* N.S. **VIII.**

Cohen-Fruee, G., T. R. Holzer, J. D. Forney, and W. R. McMaster. 2007. Global gene expression in *Leishmania*. *Int. J. Parasitol.* **37:**1077–1086.

Courret, N., C. Frehel, N. Gouhier, M. Pouchelet, E. Pina, P. Roux, and J. C. Antoine. 2002. Biogenesis of *Leishmania*-harbouring parasitophorous vacuoles following phagocytosis of the metacyclic promastigote or amastigote stages of the parasites. *J. Cell Sci.* **115:**2303–2316.

Cunningham, D. D. 1885. On the presence of peculiar parasite organisms in the tissue of a specimen of Delhi boil. *Sci. Mem. Med. Off. Army India* **I:**21.

Dawe, H. R., H. Farr, and K. Gull. 2007. Centriole/basal body morphogenesis and migration during ciliogenesis in animal cells. *J. Cell Sci.* **120:**7–15.

Desjardins, M., and A. Descoteaux. 1997. Inhibition of phagolysosomal biogenesis by the *Leishmania* lipophosphoglycan. *J. Exp. Med.* **185:**2061–2068.

de Souza, W. 2005. Microscopy and cytochemistry of the biogenesis of the parasitophorous vacuole. *Histochem. Cell Biol.* **123:**1–18.

De Souza, W. 2006. Secretory organelles of pathogenic protozoa. *An. Acad. Bras. Cienc.* **78:**271–291.

Duarte, M. I. S., O. N. Mariano, and C. E. P. Corbett. 1989. Liver parenchymal-cell parasitism in human visceral leishmaniasis. *Virchows Arch. A Pathol. Anat. Histopathol.* **415:**1–6.

Elmendorf, H. G., S. C. Dawson, and M. McCaffery. 2003. The cytoskeleton of *Giardia lamblia*. *Int. J. Parasitol.* **33:**3–28.

El-Sayed, N. M., P. J. Myler, D. C. Bartholomeu, D. Nilsson, G. Aggarwal, A. N. Tran, E. Ghedin, E. A. Worthey, A. L. Delcher, G. Blandin, S. J. Westenberger, E. Caler, G. C. Cerqueira, C. Branche, B. Haas, A. Anupama, E. Arner, L. Aslund, P. Attipoe, E. Bontempi, F. Bringaud, P. Burton, E. Cadag, D. A. Campbell, M. Carrington, J. Crabtree, H. Darban, J. F. da Silveira, P. de Jong, K. Edwards, P. T. Englund, G. Fazelina, T. Feldblyum, M. Ferella, A. C. Frasch, K. Gull, D. Horn, L. Hou, Y. Huang, E. Kindlund, M. Klingbeil, S. Kluge, H. Koo, D. Lacerda, M. J. Levin, H. Lorenzi, T. Louie, C. R. Machado, R. McCulloch, A. McKenna, Y. Mizuno, J. C. Mottram, S. Nelson, S. Ochaya, K. Osoegawa, G. Pai, M. Parsons, M. Pentony, U. Pettersson, M. Pop, J. L. Ramirez, J. Rinta, L. Robertson, S. L. Salzberg, D. O. Sanchez, A. Seyler, R. Sharma, J. Shetty, A. J. Simpson, E. Sisk, M. T. Tammi, R. Tarleton, S. Teixeira, S. Van Aken, C. Vogt, P. N. Ward, B. Wickstead, J. Wortman, O. White, C. M. Fraser, K. D. Stuart, and B. Andersson. 2005. The genome sequence of *Trypanosoma cruzi*, etiologic agent of Chagas disease. *Science* **309:**409–415.

Engwerda, C. R., M. Ato, and P. M. Kaye. 2004. Macrophages, pathology and parasite persistence in experimental visceral leishmaniasis. *Trends Parasitol.* **20:**524–530.

Field, M. C., S. K. A. Natesan, C. Gabernet-Castello, and V. L. Koumandou. 2007. Intracellular trafficking in the trypanosomatids. *Traffic* **8:**629–639.

Fridberg, A., K. T. Buchanan, and D. M. Engman. 2007. Flagellar membrane trafficking in kinetoplastids. *Parasitol. Res.* **100:**205–212.

Garin, Y. J. F., P. Meneceur, F. Pratlong, J. P. Dedet, F. Derouin, and F. Lorenzo. 2005. A2 gene of Old World cutaneous *Leishmania* is a single highly conserved functional gene. *BMC Infect. Dis.* **5:**18.

Glyn, M., and K. Gull. 1990. Flagellum retraction and axoneme depolymerization during the transformation of flagellates to amebas in physarum. *Protoplasma* **158:**130–141.

Griffiths, S., N. Portman, P. R. Tatlor, S. Gordon, M. L. Ginger, and K. Gull. 2007. RNA interference mutant induction in vivo demonstrates the essential nature of trypanosome flagellar function during mammalian infection. *Eukaryot. Cell* **6:**1248–1250.

Gull, K. 1999. The cytoskeleton of trypanosomatid parasites. *Annu. Rev. Microbiol.* **53:**629–655.

Gull, K. 2003. Host-parasite interactions and trypanosome morphogenesis: a flagellar pocketful of goodies. *Curr. Opin. Microbiol.* **6:**365–370.

Handman, E., and D. V. R. Bullen. 2002. Interaction of *Leishmania* with the host macrophage. *Trends Parasitol.* **18:**332–334.

Huynh, C., D. L. Sacks, and N. W. Andrews. 2006. A *Leishmania amazonensis* ZIP family iron transporter is essential for parasite replication within macrophage phagolysosomes. *J. Exp. Med.* **203:**2363–2375.

Ilg, T. 2000. Lipophosphoglycan is not required for infection of macrophages or mice by *Leishmania mexicana*. *EMBO J.* **19:**1953–1962.

Ivens, A. C., C. S. Peacock, E. A. Worthey, L. Murphy, G. Aggarwal, M. Berriman, E. Sisk, M. A. Rajandream, E. Adlem, R. Aert, A. Anupama, Z. Apostolou, P. Attipoe, N. Bason, C. Bauser, A. Beck, S. M. Beverley, G. Bianchettin, K. Borzym, G. Bothe, C. V. Bruschi, M. Collins, E. Cadag, L. Ciarloni, C. Clayton, R. M. Coulson, A. Cronin, A. K. Cruz, R. M. Davies, J. De Gaudenzi, D. E. Dobson, A. Duesterhoeft, G. Fazelina, N. Fosker, A. C. Frasch, A. Fraser, M. Fuchs, C. Gabel, A. Goble, A. Goffeau, D. Harris, C. Hertz-Fowler, H. Hilbert, D. Horn, Y. Huang, S. Klages, A. Knights, M. Kube, N. Larke, L. Litvin, A. Lord, T. Louie, M. Marra, D. Masuy, K. Matthews, S. Michaeli, J. C. Mottram, S. Müller-Auer, H. Munden, S. Nelson, H. Norbertczak, K. Oliver, S. O'Neil, M. Pentony, T. M. Pohl, C. Price, B. Purnelle, M. A. Quail, E. Rabbinowitsch, R. Reinhardt, M. Rieger, J. Rinta, J. Robben, L. Robertson, J. C. Ruiz, S. Rutter, D. Saunders, M. Schäfer, J. Schein, D. C. Schwartz, K. Seeger, A. Seyler, S. Sharp, H. Shin, D. Sivam, R. Squares, S. Squares, V. Tosato, C. Vogt, G. Volckaert, R. Wambutt, T. Warren, H. Wedler, J. Woodward, S. Zhou, W. Zimmermann, D. F. Smith, J. M. Blackwell, K. D. Stuart, B. Barrell, and P. J. Myler. 2005. The genome of the kinetoplastid parasite, *Leishmania major*. *Science* **309:**436–442.

Kaye, P. M. 2006. Macrophage-*Leishmania* interaction: complexities and uncertainties from the study of leishmaniasis in vivo, p. 11. *In* E. Y. Denkers and R. T. Gazzinelli (ed.), *Protozoans in Macrophages*. Landes Biosciences, Austin, TX.

Khan, S. M., and A. P. Waters. 2004. Malaria parasite transmission stages: an update. *Trends Parasitol.* **20:**575–580.

Kim, H. K., J. G. Kang, S. Yumura, C. J. Walsh, J. W. Cho, and J. Lee. 2005. De novo formation of basal bodies in *Naegleria gruberi*: regulation by phosphorylation. *J. Cell Biol.* **169:**719–724.

Kima, P. E. 2007. The amastigote forms of *Leishmania* are experts at exploiting host cell processes to establish infection and persist. *Int. J. Parasitol.* **37:**1087–1096.

Landfear, S. M., and M. Ignatushchenko. 2001. The flagellum and flagellar pocket of trypanosomatids. *Mol. Biochem. Parasitol.* **115:**1–17.

Leishman, W. B. 1903. On the possibility of trypanosomiasis in India. *Br. Med. J.* **I:**1252.

Mansfield, J. M., and D. M. Paulnock. 2005. Regulation of innate and acquired immunity in African trypanosomiasis. *Parasite Immunol.* **27:**361–371.

Marquis, J. F., and P. Gros. 2007. Intracellular *Leishmania*: your iron or mine? *Trends Microbiol.* **15:**93–95.

Matthews, K. R. 2005. The developmental cell biology of *Trypanosoma brucei*. *J. Cell Sci.* **118:**283–290.

Maudlin, I. 2006. African trypanosomiasis. *Ann. Trop. Med. Parasitol.* **100:**679–701.

McConville, M. J., and E. Handman. 2007. The molecular basis of *Leishmania* pathogenesis. *Int. J. Parasitol.* **37:**1047–1051.

Morgan, G. W., B. S. Hall, P. W. Denny, M. Carrington, and M. C. Field. 2002a. The kinetoplastida endocytic apparatus. Part I: A dynamic system for nutrition and evasion of host defences. *Trends Parasitol.* **18:**491–496.

Morgan, G. W., B. S. Hall, P. W. Denny, M. C. Field, and M. Carrington. 2002b. The endocytic apparatus of the kinetoplastida. Part II: Machinery and components of the system. *Trends Parasitol.* **18:**540–546.

Murray, H. W. 2001. Tissue granuloma structure-function in experimental visceral leishmaniasis. *Int. J. Exp. Pathol.* **82:**249–267.

Murray, H. W., J. D. Berman, C. R. Davies, and N. G. Saravia. 2005. Advances in leishmaniasis. *Lancet* **366:**1561–1577.

Naderer, T., J. E. Vince, and M. J. McConville. 2004. Surface determinants of *Leishmania* parasites and their role in infectivity in the mammalian host. *Curr. Mol. Med.* **4:**649–665.

Nwaka, S., and A. Hudson. 2006. Innovative lead discovery strategies for tropical diseases. *Nat. Rev. Drug Discov.* **5:**941–955.

Ogbadoyi, E. O., D. R. Robinson, and K. Gull. 2003. A high-order trans-membrane structural linkage is responsible for mitochondrial genome positioning and segregation by flagellar basal bodies in trypanosomes. *Mol. Biol. Cell* **14:**1769–1779.

Palenchar, J. B., and V. Bellofatto. 2006. Gene transcription in trypanosomes. *Mol. Biochem. Parasitol.* **146:**135–141.

Pan, A. A., and S. C. Pan. 1986. *Leishmania mexicana*: comparative fine structure of amastigotes and promastigotes in vitro and in vivo. *Exp. Parasitol.* **62:**254–265.

Pan, J. M., Q. Wang, and W. J. Snell. 2005. Cilium-generated signaling and cilia-related disorders. *Lab. Investig.* **85:**452–463.

Pays, E. 2005. Regulation of antigen gene expression in *Trypanosoma brucei*. *Trends Parasitol.* **21:**517–520.

Pays, E. 2006. The variant surface glycoprotein as a tool for adaptation in African trypanosomes. *Microbes Infect.* **8:**930–937.

Pays, E., L. Vanhamme, and D. Perez-Morga. 2004. Antigenic variation in *Trypanosoma brucei*: facts, challenges and mysteries. *Curr. Opin. Microbiol.* **7:**369–374.

Pays, E., B. Vanhollebeke, L. Vanhamme, F. Paturiaux-Hanocq, D. P. Nolan, and D. Perez-Morga. 2006. The trypanolytic factor of human serum. *Nat. Rev. Microbiol.* **4:**477–486.

Pearson, R. D., J. A. Sullivan, D. Roberts, R. Romito, and G. L. Mandell. 1983. Interaction of *Leishmania donovani* promastigotes with human phagocytes. *Infect. Immun.* **40:**411–416.

Piscopo, T. V., and A. C. Mallia. 2006. Leishmaniasis. *Postgrad. Med. J.* **82:**649–657.

Praetorius, H. A., and K. R. Spring. 2005. A physiological view of the primary cilium. *Annu. Rev. Physiol.* **67:**515–529.

Rittig, M. G., and C. Bogdan. 2000. *Leishmania*-host-cell interaction: complexities and alternative views. *Parasitol. Today* **16:**292–297.

Rittig, M. G., K. Schroppel, K. H. Seack, U. Sander, E. N. N'Diaye, I. Maridonneau-Parini, W. Solbach, and C. Bogdan. 1998. Coiling phagocytosis of trypanosomatids and fungal cells. *Infect. Immun.* **66:**4331–4339.

Rodriguez, A., I. Martinez, A. Chung, C. H. Berlot, and N. W. Andrews. 1999. cAMP regulates Ca2+-dependent exocytosis of lysosomes and lysosome-mediated cell invasion by trypanosomes. *J. Biol. Chem.* **274:**16754–16759.

Schenkman, S., and R. A. Mortara. 1992. Hela cells extend and internalize pseudopodia during active invasion by *Trypanosoma cruzi* trypomastigotes. *J. Cell Sci.* **101:**895–905.

Sherwin, T., and K. Gull. 1989. The cell-division cycle of *Trypanosoma brucei brucei*: timing of event markers and cytoskeletal modulations. *Philos. Trans. R. Soc. Lond. B* **323:**573–588.

Shi, M. Q., G. J. Wei, W. L. Pan, and H. Tabel. 2004. *Trypanosoma congolense* infections: antibody-mediated phagocytosis by Kupffer cells. *J. Leukoc. Biol.* **76:**399–405.

Shlomai, J. 2004. The structure and replication of kinetoplast DNA. *Curr. Mol. Med.* **4:**623–647.

Spath, G. F., L. Epstein, B. Leader, S. M. Singer, H. A. Avila, S. J. Turco, and S. M. Beverley. 2000. Lipophosphoglycan is a virulence factor distinct from related glycoconjugates in the protozoan parasite *Leishmania major*. *Proc. Natl. Acad. Sci. USA* **97:**9258–9263.

Spath, G. F., L. A. Garraway, S. J. Turco, and S. M. Beverley. 2003. The role(s) of lipophosphoglycan (LPG) in the establishment of *Leishmania major* infections in mammalian hosts. *Proc. Natl. Acad. Sci. USA* **100:**9536–9541.

Takayana, T., Y. Nakatake, and G. L. Enriquez. 1974a. *Trypanosoma gambiense*—phagocytosis in vitro. *Exp. Parasitol.* **36:**106–113.

Takayana, T., Y. Nakatake, and G. L. Enriquez. 1974b. Attachment and ingestion of *Trypanosoma gambiense* to rat macrophage by specific antiserum. *J. Parasitol.* **60:**336–339.

Tardieux, I., P. Webster, J. Ravesloot, W. Boron, J. A. Lunn, J. E. Heuser, and N. W. Andrews. 1992. Lysosome recruitment and fusion are early events required for trypanosome invasion of mammalian cells. *Cell* **71:**1117–1130.

Taylor, J. E., and G. Rudenko. 2006. Switching trypanosome coats: what's in the wardrobe? *Trends Genet.* **22:**614–620.

Teixeira, A. R. L., N. Nitz, M. C. Guimaro, C. Gomes, and C. A. Santos-Buch. 2006. Chagas disease. *Postgrad. Med. J.* **82:**788–798.

Tull, D., J. E. Vince, J. M. Callaghan, T. Naderer, T. Spurck, G. I. McFadden, G. Currie, K. Ferguson, A. Bacic, and M. J. McConville. 2004. SMP-1, a member of a new family of small myristoylated proteins in kinetoplastid parasites, is targeted to the flagellum membrane in *Leishmania*. *Mol. Biol. Cell* **15:**4775–4786.

Turco, S. J., G. F. Spath, and S. M. Beverley. 2001. Is lipophosphoglycan a virulence factor? A surprising diversity between *Leishmania* species. *Trends Parasitol.* **17:**223–226.

Vannier-Santos, M. A., A. Martiny, and W. de Souza. 2002. Cell biology of *Leishmania* spp.: invading and evading. *Curr. Pharm. Design* **8:**297–318.

Wenyon, C. M. 1926. *Protozoölogy*. William Wood and Co., New York, NY.

Whitten, M. M. A., S. H. Shiao, and E. A. Levashina. 2006. Mosquito midguts and malaria: cell biology, compartmentalization and immunology. *Parasite Immunol.* **28:**121–130.

Zhang, W. W., and G. Matlashewski. 1997. Loss of virulence in *Leishmania donovani* deficient in an amastigote-specific protein, A2. *Proc. Natl. Acad. Sci. USA* **94:**8807–8811.

Zhang, W. W., and G. Matlashewski. 2001. Characterization of the A2-A2rel gene cluster in *Leishmania donovani*: involvement of A2 in visceralization during infection. *Mol. Microbiol.* **39:**935–948.

Zhang, W. W., S. Mendez, A. Ghosh, P. Myler, A. Ivens, J. Clos, D. L. Sacks, and G. Matlashewski. 2003. Comparison of the A2 gene locus in *Leishmania donovani* and *Leishmania major* and its control over cutaneous infection. *J. Biol. Chem.* **278:**35508–35515.

Zufferey, R., S. Allen, T. Barron, D. R. Sullivan, P. W. Denny, I. C. Almeida, D. F. Smith, S. J. Turco, M. A. J. Ferguson, and S. M. Beverley. 2003. Ether phospholipids and glycosylinositolphospholipids are not required for amastigote virulence or for inhibition of macrophage activation by *Leishmania major*. *J. Biol. Chem.* **278:**44708–44718.

30

Phagocyte Interactions with the Intracellular Protozoan *Toxoplasma gondii*

ERIC Y. DENKERS

The intracellular protozoan *Toxoplasma gondii* is a pathogen of humans and an important zoonotic infection in domestic animals (Dubey, 2007). The parasite is capable of living within virtually all nucleated cell types, but macrophages, dendritic cells, and neutrophils are important targets of early infection (Denkers et al., 2004). The outcome of parasite encounter with cells of the phagocyte lineage is key to determining whether the host survives or succumbs during infection.

T. gondii is globally distributed, and infection normally results in long-lasting asymptomatic disease. It is estimated that 10 to 30% of the human population is latently infected with *Toxoplasma* (Dubey, 1998). Cats serve as the definitive host for *Toxoplasma* (Dubey, 2007). Ingestion of tissue cysts or sporulated oocysts in felids initiates an enteroepithelial infection that culminates in formation of gamonts. After fusion to form oocysts, shedding into the environment followed by sporulation generates infectious sporozoites contained within the oocyst. In other animals (including cats), ingestion of oocysts and tissue cysts results in generation of tachyzoites. This parasite stage displays rapid dissemination throughout host tissues. Coincident with development of adaptive immunity, tachyzoites differentiate into slow-growing bradyzoites that form cysts in tissues of the skeletal muscle and central nervous system. Tissue cysts can be stable for the lifetime of the host, and transmission of the parasite in this form relies on carnivorism.

Toxoplasma is a quintessential opportunistic pathogen. While chronic infection is almost invariably asymptomatic, host immunodeficiency results in cyst reactivation associated with emergence of tachyzoites that may cause tissue destruction and proinflammatory pathology. The resultant toxoplasmic encephalitis, most dramatically highlighted in chronically infected AIDS patients, is lethal if not appropriately controlled by drug therapy (Liesenfeld et al., 1999;

Luft et al., 1984; Luft and Remington, 1992). Toxoplasmosis is also an important congenital infection (Remington et al., 2001). Maternal transmission of the parasite during primary infection can cause many lesions including hydrocephalus, intracerebral calcification, and chorioretinitis. Congenital infection may ultimately lead to blindness, mental retardation, and death in the newborn, and the parasite can emerge later in life with many sequelae of infection.

In recent years it has become clear that *T. gondii* strain type exerts an important influence on disease pathogenesis. The parasite possesses a unique population structure that is dominated by three clonal lineages in Europe and North America (Howe and Sibley, 1995; Sibley and Boothroyd, 1992; Sibley et al., 2002). Strain type I is highly lethal in mice, and death is associated with overproduction of proinflammatory cytokines (Gavrilescu and Denkers, 2001; Mordue et al., 2001). Strain types II and III are less virulent, and the immune response is correspondingly less extreme. There is evidence to suggest that type I *Toxoplasma* strains also cause more severe disease in humans (Boothroyd and Grigg, 2002; Khan et al., 2005).

The ability of *T. gondii* to elicit strong cell-mediated immunity is key to long-term survival and establishment of a stable relationship with the immunocompetent host (Alexander and Hunter, 1998; Denkers and Gazzinelli, 1998). Numerous studies in mouse infection models, as well as the AIDS experience in humans, have demonstrated the importance of an intact T-cell compartment in control of infection. $CD4^+$ and $CD8^+$ T-cell-derived gamma interferon (IFN-γ) is necessary to control acute infection and to prevent recrudescence of parasites during chronic infection (Gazzinelli et al., 1991, 1992; Saavedra et al., 1996; Scharton-Kersten et al., 1996; Suzuki et al., 1988, 1989; Suzuki and Remington, 1988). From the perspective of the parasite, a tightly regulated, cell-mediated immune response that permits host survival is necessary to ensure establishment of latent infection and long-term persistence.

Phagocytes are major players in igniting protective cell-mediated immunity. *Toxoplasma* triggers dendritic cell, macrophage, and neutrophil signaling pathways leading to

Eric Y. Denkers, Department of Microbiology and Immunology, College of Veterinary Medicine, Cornell University, Ithaca, NY 14853-6401.

interleukin-12 (IL-12) production that is key to T-helper (Th) type 1 response initiation (Alberti et al., 2004; Denkers et al., 2004). At the same time, T. gondii actively disassembles certain proinflammatory signaling pathways during intracellular infection (Denkers et al., 2003, 2004). In other situations, cytokine-activated phagocytes are potent microbicidal effectors that kill intracellular tachyzoites through mechanisms that are only now becoming clear (Taylor et al., 2004). This chapter describes the multiple facets of early encounters between T. gondii and the professional phagocytes, considering how these interactions lead to resistance or disease as the ultimate outcome of infection.

TOXOPLASMA INTRACELLULAR NICHE

T. gondii actively invades the host cell and creates a parasitophorous vacuole that is segregated from the phagocyte endosomal-lysosomal machinery (Carruthers and Boothroyd, 2007; Sibley et al., 2007). In this specialized niche, the parasite undergoes successive rounds of division culminating in egress of tachyzoites and concomitant host cell destruction. This lifestyle is unlike other intracellular protozoa such as Leishmania that reside within an acidified endocytic vacuole, or Trypanosoma cruzi that escapes into the cytoplasm after phagocytic uptake (Bogdan and Rollinghoff, 1999).

Invasion of the host cell is a complex process requiring translocation via a parasite actin-myosin-based motor (Dobrowolski and Sibley, 1996; Morisaki et al., 1995). Entry into the cell is accomplished within 20 to 30 s. In this way, tachyzoites outmaneuver the endocytic machinery of the phagocyte that operates at a much slower rate. Invasion is initiated by tachyzoite apical attachment to the host cell plasma membrane. This is mediated by release of adhesin molecules contained within parasite apical organelles called micronemes (Carruthers et al., 1999; Carruthers and Sibley, 1997; Mital et al., 2005). During entry, the host cell plasma membrane invaginates to form the nascent parasitophorous vacuole (Charron and Sibley, 2004; Suss-Toby et al., 1996). The close apposition of the tachyzoite and target cell forms a tight junction from which host plasma membrane proteins, with the exception of those tethered by a glycosylphosphatidylinositol (GPI) anchor, are excluded (Mordue et al., 1999). Thus, the parasitophorous vacuole membrane lacks transmembrane proteins that direct endocytic trafficking, and the vacuole fails to acquire endosomal markers such as LAMP-1 or the transferrin receptor, as well as Rab molecules that direct endosomal-lysosomal trafficking (Mordue and Sibley, 1997; Sibley et al., 1985).

During early invasion, a second set of apical organelles—the rhoptries—are discharged. Some molecules contained within rhoptries insert into the parasitophorous vacuole membrane. This is the fate of ROP2, which appears to play a role in the tethering of host mitochondria to the parasitophorous vacuole membrane (Sinai and Joiner, 2001). Other rhoptry proteins are injected into the host cell cytoplasm contained within vacuoles ("evacuoles") (Hakansson et al., 2001). Recent evidence (discussed below) suggests that rhoptry proteins inserted into the host cytoplasm target signaling pathways involved in early immune response initiation (Gilbert et al., 2007; Saeij et al., 2007).

The final phase of invasion that completes formation of the mature parasitophorous vacuole is release of proteins contained within dense granules (Mercier et al., 2005). Dense-granule proteins such as GRA1 locate to the lumen of the vacuole. Others, such as GRA3 and GRA5, insert into the parasitophorous vacuole membrane. GRA protein function is poorly understood but may involve nutrient acquisition from the host cell. The parasitophorous vacuole membrane itself is porous to molecules under approximately 2,000 Da, an adaptation that is likely to facilitate acquisition of low-molecular-mass nutrients (Schwab et al., 1994). Within this intracellular niche, tachyzoites divide by a process known as endodyogeny that involves formation of daughter cells from within the tachyzoite. In the parasite-laden cell, tachyzoite egress occurs in a Ca^{2+}-dependent process that is associated with dissolution of membrane of both the parasitophorous vacuole and the host cell (Moudy et al., 2001).

Although Toxoplasma has been viewed as sequestered from the host cell by virtue of enclosure within the parasitophorous vacuole, recent evidence suggests that this is not entirely the case. In addition to rhoptry protein injection into the host cytoplasm, transgenic parasites expressing ovalbumin release this model antigen into the cytoplasm from within the parasitophorous vacuole, resulting in peptide presentation by major histocompatibility complex class I (Gubbels et al., 2005). Similarly, Cre protein secreted by transgenic parasites has been shown capable of entering the host cell nucleus to mediate DNA recombination (Gubbels et al., 2005). The molecular basis for how Toxoplasma accomplishes these remarkable feats is the subject of ongoing investigation.

PHAGOCYTES AS EARLY TARGETS OF INFECTION

The ability of T. gondii to invade virtually any cell type suggests that the parasite establishes itself in the host by nonspecific dissemination after invasion in the gut. However, earlier work found that mouse secondary lymphoid organs become heavily infected after oral administration of T. gondii cysts, arguing that the parasite exploits the host lymphatic system for dissemination (Sumyuen et al., 1995). Despite their promiscuous invasive behavior, tachyzoites display distinct cell tropisms. Added to human peripheral blood leukocytes in culture, Toxoplasma preferentially infects monocytes and dendritic cells rather than neutrophils or lymphocytes (Channon et al., 2000). Several studies suggest a strategy of exploiting dendritic cells and monocyte/macrophage lineage cells for establishment in the host (Courret et al., 2006; Lambert et al., 2006; Mordue and Sibley, 2003).

In an infection model employing intraperitoneal tachyzoite injection, a novel population of $CD68^+$ monocytes was identified that was preferentially targeted during early infection (Mordue and Sibley, 2003). Interestingly, parasite strain type influences recruitment of this population. Thus, $Gr-1^+CD68^+$ monocytes were recruited to the peritoneal cavity in a CCR2-dependent manner during low-virulence ME49/PTG infection (Robben et al., 2005). In contrast, intraperitoneal infection with the high-virulence RH strain induces a rapid neutrophil influx that depends on chemokine receptor CXCR2 (Bliss et al., 2000; Del Rio et al., 2001; Ma et al., 2004; Mordue and Sibley, 2003). The cells express the Gr-1 determinant in both cases, complicating efforts to decipher the relative importance of each (Egan et al., 2008).

Other studies suggest that dendritic cells are important in early dissemination of *Toxoplasma*. Infection of bone marrow-derived dendritic cells triggers a hypermotility response in vitro, and mice inoculated with infected dendritic cells display an accelerated course of parasite dissemination compared with those animals infected with extracellular tachyzoites (Lambert et al., 2006). After intragastric inoculation of *T. gondii*, parasites infect CD11c$^+$ dendritic cells and CD11b$^+$ monocytes/macrophages in the lamina propria, Peyer's patches, and mesenteric lymph nodes (Courret et al., 2006). Nevertheless, infected cells in the peripheral blood are restricted to those expressing the macrophage/monocyte marker CD11b.

The ability of *T. gondii* to target dendritic cells and monocytes/macrophages during in vivo infection may seem paradoxical insofar as these cells, famously, are central mediators of proinflammatory cytokine responses, and in addition can function as microbial effectors. However, as described further below, several lines of evidence suggest that *Toxoplasma* blocks the ability of macrophages and dendritic cells to produce proinflammatory molecules during intracellular infection. These proinflammatory factors include inducible nitric oxide synthase (iNOS), IL-12p40, and tumor necrosis factor-α (TNF-α), mediators that are central to host control of infection (Lee et al., 2006; Luder et al., 2003; McKee et al., 2004). By downregulating the ability of macrophages and dendritic cells to produce such dangerous molecules, *T. gondii* may transform these hematopoietic cell populations into safe havens for early growth and dissemination.

During the natural route of infection, intestinal epithelial cells are likely the first to encounter *Toxoplasma*. An emerging "Trojan horse" model for the pathogenesis of *T. gondii* infection is that enterocytes lining the intestinal villi produce chemokines during early oral infection (Kasper et al., 2004). Chemokine-dependent recruitment of phagocytes would assemble a population of infection-susceptible cells. There is evidence that infected neutrophils themselves produce chemokines that recruit dendritic cells (Bennouna et al., 2003; van Gisbergen et al., 2005a). Infected dendritic cells and/or monocytes would then disseminate to secondary lymph nodes, and similarly infected cells would eventually transport parasites to brain and skeletal muscle tissue for differentiation into bradyzoites and establishment of chronic infection.

IMPORTANCE OF PHAGOCYTES AS TRIGGERS OF CELL-MEDIATED IMMUNITY

IL-12 Induction

The cytokine IL-12 is well known as the key in igniting cell-mediated immunity, and indeed, *Toxoplasma* was one of the initial model organisms used to establish this paradigm (Gazzinelli et al., 1993, 1994). These early studies showed that bone marrow-derived macrophages produce IL-12 in response to soluble tachyzoite extracts, and at the same time it was found that IL-12 production was essential for host IFN-γ production and survival in murine models of infection (Gazzinelli et al., 1993, 1994; Hunter et al., 1995; Khan et al., 1994). More recently, it has been found that bone marrow-derived macrophages respond to *Toxoplasma* in a strain-specific manner, in that high-virulence parasites induce less IL-12 relative to low-virulence strain types (Kim et al., 2006; Robben et al., 2004).

Later, with the emergence of dendritic cells as important immune initiators, their importance in the response to *Toxoplasma* was recognized. For example, within a few hours of intravenous injection of parasite antigen, CD11c$^+$CD8α$^+$ dendritic cells become positive for IL-12 while simultaneously relocalizing to T-cell regions of the spleen (Reis e Sousa et al., 1997). Nevertheless, only recently was the importance of dendritic cells during in vivo resistance to *T. gondii* infection formally established (Liu et al., 2006). Thus, transgenic mice expressing diphtheria toxin receptor under the control of the CD11c promoter undergo transient depletion of dendritic cells after toxin injection. Animals depleted of dendritic cells in this manner develop high susceptibility to *T. gondii* accompanied by defective IL-12 and IFN-γ production. In humans, peripheral blood monocyte-derived dendritic cells respond to live, but not heat-inactivated, tachyzoites by upregulating costimulatory molecules (Subauste and Wessendarp, 2000). In this situation, human dendritic cells also secrete IL-12 in a process dependent on CD40-CD40L interaction. A similar CD40-CD40L interaction regulates IL-12 production in cocultures of infected human monocytes and T lymphocytes (Subauste et al., 1999).

It is also recognized that neutrophils produce IL-12 and other proinflammatory cytokines during infection with microbial pathogens such as *Toxoplasma* (Cassatella, 1999; Denkers et al., 2003). IL-12-positive neutrophils are rapidly recruited to the site of infection during intraperitoneal infection with high-virulence tachyzoites (Bliss et al., 2000). These cells may influence adaptive immunity directly through IL-12 production (Bliss et al., 1999, 2001). An additional possibility is that parasite-triggered neutrophils, by producing chemokines such as CCL3, -4, and -5, and through production of TNF-α, act to recruit and activate dendritic cells during infection (Bennouna et al., 2003; van Gisbergen et al., 2005b). In this manner, neutrophils may program dendritic cells to assume their role as orchestrators of T-cell immune responses (Megiovanni et al., 2006). Support for an immunoregulatory role for neutrophils during in vivo *T. gondii* infection comes from studies showing that genetic inactivation of CXCR2, the receptor for neutrophil chemoattractant chemokines, results in an increase in susceptibility accompanied by a decreased Th1 response profile (Del Rio et al., 2001). In addition, antibody-mediated depletion of cells positive for Gr-1, a marker expressed at high level on neutrophils, results in an acute susceptibility phenotype and lack of proinflammatory cytokine responses (Bliss et al., 2001).

Studies using Gr-1-reactive antibody to deplete neutrophils must be interpreted with caution insofar as this marker is now recognized to be expressed by other myeloid cell populations (Egan et al., 2008). A mouse CX3CR1^{low}CCR2$^+$Gr-1$^+$ monocyte subset displays CCR2-dependent recruitment to inflamed tissues and is capable of differentiating into CD11c$^+$ dendritic cells (Geissmann et al., 2003). A similar population of iNOS-producing dendritic cells (TipDCs) is recruited to the spleen during *Listeria monocytogenes* infection (Serbina et al., 2003). Here, TipDCs are thought to mediate bacterial containment through nitric oxide and TNF-α production. A low-virulence *Toxoplasma* strain also recruits in a CCR2-dependent manner a population of Gr-1$^+$CD68$^+$ myeloid cells during infection in the peritoneal cavity (Mordue and Sibley, 2003). The cells produce IL-12, but their major role may be control of the parasite (Robben et al., 2005). This is because impaired recruitment of Gr-1$^+$CD68$^+$ cells in

CCR2$^{-/-}$ animals results in lack of control of *T. gondii* infection despite unimpaired production of IL-12. From these studies, it can be concluded that several distinct phagocyte subpopulations can contribute to IL-12 production during infection. This may vary depending on parameters such as location, time, and parasite and host genetic background.

Molecular Basis of Recognition

The Toll-like receptor (TLR) family members are well established as major host sensors of pathogen-associated molecular patterns (Akira et al., 2006; O'Neill and Bowie, 2007). Twelve distinct TLRs have been identified in mammals (Roach et al., 2005). Most signal through the common adaptor molecule MyD88. Nevertheless, TLR3, a receptor for viral double-stranded RNA, uses an alternate adaptor molecule called Toll/IL-1 receptor (TIR) homology domain-containing adaptor protein-inducing IFN-β (TRIF). The receptor for bacterial lipopolysaccharide (LPS), TLR4, mediates signaling through both MyD88 and TRIF. Originally defined for their role in detection of bacterial, fungal, and viral molecules, it is now known that TLRs also possess protozoan recognition properties (Gazzinelli and Denkers, 2006).

The first indication for TLR-based recognition of *Toxoplasma* (and for protozoa in general) came from the observation that MyD88-deficient mice were hypersusceptible to infection, and death was associated with defective IFN-γ and IL-12 responses (Scanga et al., 2002). Nevertheless, signaling through receptors for both IL-1 and IL-18 also proceeds through MyD88. Therefore, it is significant that mice deficient in IL-1β-coverting enzyme (ICE$^{-/-}$), which cannot produce functional IL-1 or IL-18, are nevertheless able to control *T. gondii* infection (Hitziger et al., 2005). This important finding provides strong evidence that the dramatic effects of MyD88 inactivation are the result of disrupted TLR signaling rather than the inability to respond to IL-1/IL-18.

A biochemical approach subsequently identified tachyzoite profilin as a *T. gondii* protein that was a potent MyD88-dependent inducer of IL-12 from mouse splenic dendritic cells (Yarovinsky et al., 2005). Interestingly, profilin triggers these cells through mouse TLR11. This little-understood receptor was previously implicated in recognition of uropathogenic bacteria and is not expressed as a functional molecule in humans (Lauw et al., 2005). The consequence of profilin-TLR11 interaction extends beyond immediate triggering of dendritic cell proinflammatory cytokine production. This is because profilin appears to be an immunodominant antigen in the subsequent CD4$^+$ T-cell response to total *Toxoplasma* lysate antigens (Yarovinsky and Sher, 2006). Immunodominance requires TLR11 recognition, MyD88 signaling, and antigen presentation within the same cell population, providing evidence that TLR-MyD88 signaling directs instruction of subsequent T-cell responses.

Yet, despite the potent IL-12-inducing activity of *Toxoplasma* profilin, TLR11$^{-/-}$ mice survive infection, although the animals display increased susceptibility in terms of brain cyst burden (Yarovinsky et al., 2005). Because MyD88-deficient mice uniformly succumb to *T. gondii* during acute infection, the implication is that additional TLRs are likely important in *Toxoplasma* recognition. In this regard, in vitro studies suggest that macrophage TLR2 recognizes components derived from soluble tachyzoites, and, reinforcing these data, TLR2$^{-/-}$ animals are susceptible to high-dose parasite infection (Del Rio et al., 2004; Mun et al., 2003). Furthermore, GPI anchors derived from *T. gondii* trigger both TLR2 and TLR4, and TLR2/4 double-knockout mice display an increase in susceptibility to the parasite, although still not to the same hypersusceptibility phenotype as MyD88$^{-/-}$ animals (Debierre-Grockiego et al., 2007). GPI anchors derived from *T. cruzi* and *Plasmodium falciparum* are also known to activate TLR2 and TLR4. So, protozoan GPI may serve as a general class of pathogen-associated molecular pattern for recognition of this class of pathogen (Gazzinelli and Denkers, 2006).

It has been suggested that TLR9, a receptor for bacterial and viral DNA, plays a role in the response to oral *T. gondii* infection, based on decreased gut pathology in infected TLR9-deficient mice (Minns et al., 2006). However, in this case it is possible that TLR9-dependent effects are the result of bacteria that breach the intestinal wall during oral *T. gondii* infection, rather than direct recognition of the parasite itself (Heimesaat et al., 2007). Regardless, these combined results suggest that TLR recognition of *Toxoplasma* and other complex pathogens is likely to be an interaction involving multiple TLRs that may be expressed on different cell types, possibly at distinct times during the immune response (Bafica et al., 2006) (Fig. 1).

Phagocytes also employ non-TLR-based detection systems for *Toxoplasma* recognition, although these are less well defined. An 18-kDa cyclophilin protein (C-18) was identified that stimulates dendritic cell IL-12 production (Aliberti et al., 2000, 2003). Interestingly, C-18 appears to act as a chemokine mimic, triggering dendritic cells through interaction with the chemokine-binding protein CCR5.

There is additional evidence for other recognition pathways. The high-virulence *Toxoplasma* strain RH induces IL-12 that is independent of both MyD88 and CCR5 in macrophages (Kim et al., 2006). Interestingly, infection of the same cells with the low-virulence *Toxoplasma* ME49/PTG strain triggers IL-12 by both MyD88-dependent and MyD88-independent signaling pathways (Kim et al., 2006). Both pathways involve p38 mitogen-activated protein kinase (MAPK), and, at least for high-virulence parasites, a role for host Ca^{2+} and protein kinase C has been implicated (Kim et al., 2005, 2006; Masek et al., 2006). Also related to these studies, c-Rel, the NF-κB family member responsible for LPS induced IL-12 production, is not required for *T. gondii*-triggered IL-12 production or resistance to infection (Mason et al., 2002). The discovery that *Toxoplasma* strain type exerts an effect on intracellular signaling pathways that are activated during infection may account for the ability of low-virulence parasites such as ME49/PTG to induce high levels of IL-12 relative to strains such as RH (Robben et al., 2004) (Fig. 1).

Phagocytes also respond to *T. gondii* through G-protein-coupled receptor (GPCR)-mediated interactions. The GPCR family, including CCR5 and other CC and CXC chemokine receptors, are seven-transmembrane-spanning receptors that are associated with heterotrimeric G proteins (Pierce et al., 2002). In macrophages, *Toxoplasma* infection leads to G-protein-mediated activation of phosphatidyl-inositol-3-kinase (PI3K), that in turn leads to activation of protein kinase B/Akt and extracellular signal-related kinase 1/2 (ERK1/2) MAPK (Kim and Denkers, 2006). This pathway is intact in CCR5-deficient macrophages, implicating non-CCR5 receptors in recognition of the parasite. Pertussis toxin and wortmannin, respective inhibitors of

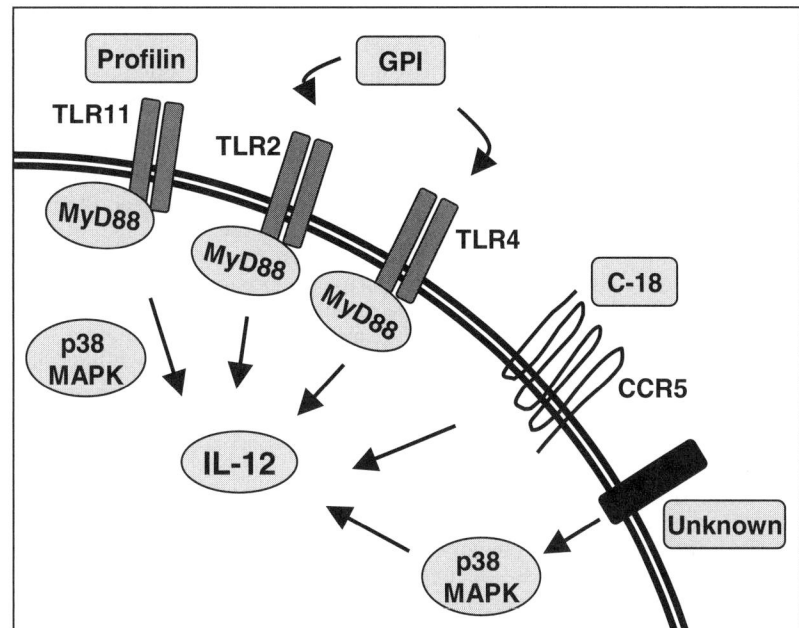

FIGURE 1 Multiple pathways lead to IL-12 production during *T. gondii* infection. A parasite profilin molecule triggers proinflammatory cytokine production through TLR11/MyD88 signaling. In addition, GPI molecules from the parasite surface activate both TLR2 and TLR4 for cytokine production. The TLR signaling pathways are likely to also involve p38 MAPK. The cyclophilin-18 molecule expressed by tachyzoites appears to act through the seven-transmembrane chemokine receptor CCR5 to induce dendritic cell IL-12. Finally, another pathway that does not involve MyD88, CCR5, or G-protein-mediated signaling also leads to IL-12 induction. Little is known about this last pathway except that p38 MAPK is an essential component.

GPCR and PI3K signaling, do not affect macrophage IL-12 production.

If the above GPCR pathway is not involved in *T. gondii*-induced IL-12, what is its importance? Infection with *T. gondii* is known to render host cells resistant to apoptosis, possibly a survival mechanism that permits host cell persistence in the overwhelmingly proinflammatory cytokine environment characterizing in vivo infection (Luder and Gross, 2005). Chemical inhibition studies suggest that the parasite triggers the GPCR-PI3K-PKB/Akt pathway as a mechanism to induce the antiapoptotic state because infected cells treated with wortmannin or pertussis toxin lose their resistance to inducers of programmed cell death (Kim and Denkers, 2006). These collective studies demonstrate that encounter of *T. gondii* with phagocytes results in MyD88- and GPCR-dependent signaling cascades that instruct host cell responses, in turn influencing host resistance and parasite survival during infection.

PHAGOCYTE TOXOPLASMACIDAL EFFECTOR MECHANISMS

Host resistance to *Toxoplasma* highly depends on IFN-γ (Suzuki et al., 1988). Lack of IFN-γ, or defects in IFN-γ signaling, results in uncontrolled tachyzoite proliferation, rapidly culminating in host death (Gavrilescu et al., 2004; Lieberman et al., 2004; Scharton-Kersten et al., 1996). The importance of IFN-γ in protective immunity is further reinforced by mouse vaccination studies with attenuated *T. gondii* strains. In these cases, vaccination-induced protective immunity depends on high-level IFN-γ production by primed T cells (Fox and Bzik, 2002; Gazzinelli et al., 1991). The host is, therefore, equipped with potent IFN-γ-dependent mechanisms for controlling *T. gondii*. Phagocytes, in particular macrophages and macrophage-lineage cells, are likely to be a major killer effector population that responds to the effects of IFN-γ. Indeed, it has long been recognized that human peripheral blood monocytes and neutrophils possess the ability to destroy intracellular tachyzoites (Wilson and Remington, 1979). Mouse macrophages stimulated with IFN-γ and TNF-α exert antitoxoplasmacidal activity, an effect that is reversed by the antiinflammatory cytokine IL-10 (Gazzinelli et al, 1992a, 1992b; Langermans et al., 1992). Several distinct IFN-γ-dependent, as well as IFN-γ-independent, *T. gondii*-killing mechanisms can readily be demonstrated in vitro, but for some the in vivo importance is uncertain.

Immunity-Related GTPases (IRG)/p47 GTPases

The immunity-related GTPase (IRG) proteins are part of a larger family of IFN-γ-inducible GTPase molecules that play dominant roles in resistance to infection (Taylor, 2007; Taylor et al., 2004). They are expressed in macrophages and other cell types in association with subcellular membranes including endoplasmic reticulum and Golgi. It is estimated that there are at least 21 distinct functional IRG genes in mice. Several have been shown to mediate resistance to bacterial and protozoan infection. In humans, only two IRG genes have been identified, and these do not appear to be regulated by IFN-γ. However, one human IRG molecule, IRGM, has recently been implicated in resistance to mycobacteria (Singh et al., 2006).

Four immunity-related GTPase (IRG) proteins have been shown to be important in control of Toxoplasma during infection, namely, Irgm1/LRG-47, Irgm3/IGTP, Irga6/IIGP1, and Irgd/IRG-47. Mice deficient in expression of Irgm1 and Irgm3 display normal inflammatory cytokine responses, but the animals are unable to control the parasite, resulting in death during acute infection (Collazo et al., 2001; Taylor et al., 2000). Mice that lack expression of Irgd also display increased susceptibility to T. gondii, but for reasons that are unclear the animals do not succumb until the chronic stage of infection (Collazo et al., 2001).

Increased susceptibility of IRG-deficient mice is recapitulated in vitro during infection of macrophages and astrocytes (Butcher et al., 2005a; Collazo et al., 2001; Martens et al., 2005). Unlike normal cells, IFN-γ-activated macrophages that are defective in Irgm1 and Irgm3 expression fail to control intracellular tachyzoite replication. Irgm3 has recently been reported to localize to the parasitophorous vacuole membrane (Ling et al., 2006). Although the exact target has not been identified, Irgm3 participates in PI3K-dependent vacuolar membrane disruption, followed by parasite plasma membrane denudation, enclosure within an autophagic-like structure, and delivery to lysosomes. Related to these studies, IFN-γ-activated astrocytes display antitoxoplasmacidal activity, and this is associated with accumulation of IRG proteins Irgm1 and Irgm6 at the parasitophorous vacuole membrane (Martens et al., 2005). This results in disintegration of vacuolar and parasite membrane, but here there was no evidence for autophagic parasite elimination. Interestingly, although Irga6 also controls parasite survival, it appears to do so without colocalizing to the parasitophorous vacuole membrane. The IRG proteins are clearly important in resistance to intracellular pathogens such as T. gondii. Nonetheless, their precise role in promoting autophagic or nonautophagic parasite elimination is presently unclear.

Human and mouse macrophages also display autophagic elimination of tachyzoites after stimulation through the CD40 molecule (Andrade et al., 2006). Remarkably, CD40-driven rerouting of the parasite to lysosomal compartments occurs even after establishment of the mature parasitophorous vacuole. In mouse macrophages, CD40-stimulated antimicrobial function occurs independently of IFN-γ, arguing that the killing mechanism does not involve IRG activity (Andrade et al., 2003, 2005; Subauste and Wessendarp, 2006). Thus, distinct pathways appear to converge on common, or highly related, mechanisms of parasite eradication (Fig. 2).

iNOS/Nitric Oxide Synthase-2 (NOS-2)

Depending on their source, some activated mouse macrophages can be shown to mediate anti-Toxoplasma activity through induction of iNOS and subsequent generation of nitric oxide (NO) (Adams et al., 1990; Scharton-Kersten et al., 1997). NO, and reactive nitrogen intermediates, are short-lived but highly toxic molecules that affect many metabolic processes of the cell (MacMicking et al., 1997). Nevertheless, mice lacking iNOS gene expression survive early infection, although they ultimately succumb to toxoplasmic encephalitis (Scharton-Kersten et al., 1997). In-

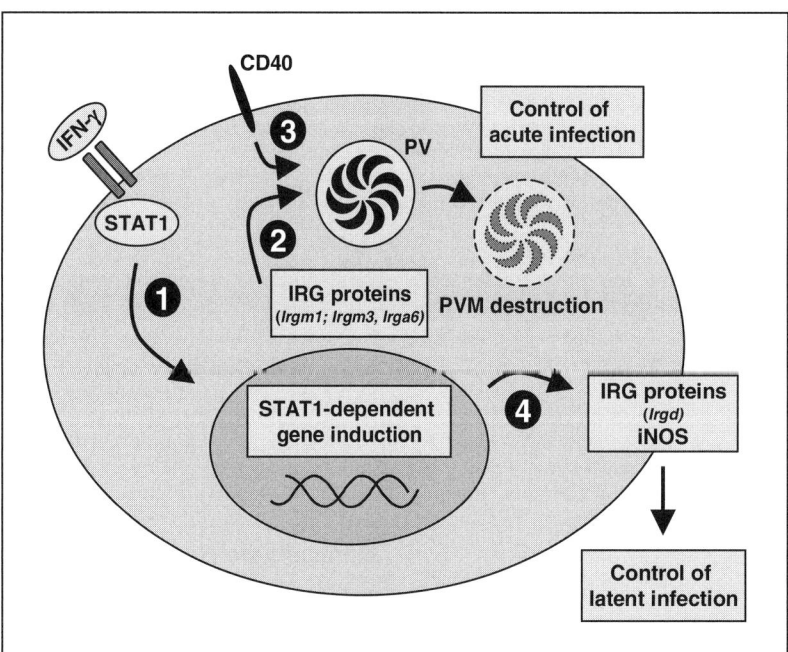

FIGURE 2 Major phagocytic killing mechanisms operating against Toxoplasma. (1) IFN-γ signaling through STAT1 is a major mechanism for inducing microbicidal activity. Among STAT1-inducible proteins are the IRG family members. (2) Some IRG proteins (Irgm1, Irgm3, Irga6) act to induce parasitophorous vacuole membrane breakdown and parasite elimination involving, in some cases, autophagy. (3) Independently of the IFN-γ pathway, CD40 ligation transmits signals for parasite killing that also involve autophagy. (4) Other IRG proteins (Irgd) as well as iNOS are induced by STAT1 signaling, but are only required for survival during chronic infection. How these molecules function is enigmatic.

duction of iNOS in the brain at this stage of infection is mediated through TNF receptor I (Deckert-Schluter et al., 1998).

Inasmuch as arginine serves as the substrate for iNOS-mediated NO generation, it may be of immunological significance that *Toxoplasma* is an arginine auxotroph (Fox et al., 2004). NO, and arginine starvation itself, induces tachyzoite-to-bradyzoite conversion in vitro (Bohne et al., 1994; Fox et al., 2004). Therefore, it is possible that iNOS-mediated arginine limitation deviates parasite differentiation away from cyst formation and toward the tachyzoite stage, resulting in fulminating toxoplasmic encephalitis. For the parasite, the importance of preventing iNOS-mediated NO production during acute infection is suggested by the observation that tachyzoites potently downregulate macrophage production of this inflammatory mediator in infected macrophages (Luder et al., 2003).

Reactive Oxygen Intermediates

Macrophages and neutrophils are well known for their ability to produce reactive oxygen intermediates (ROIs) that exert antimicrobial activity. Nevertheless, whether ROIs contribute to intracellular killing of *Toxoplasma* remains uncertain and may depend on the phagocyte population under study. Peritoneal macrophages from *T. gondii*-exposed mice were found to block parasite growth in a process inhibitable by oxygen intermediate scavengers (Murray et al., 1979). Human monocytes also display partially ROI-dependent inhibition of parasite replication, and this activity is less in cells from patients with chronic granulomatous disease (Murray et al., 1985).

Nevertheless, it has also been reported that *T. gondii* fails to trigger the respiratory burst during human macrophage infection (Wilson et al., 1980). In addition, $p47^{phox-/-}$ mice that lack inducible respiratory burst activity display normal resistance to infection (Scharton-Kersten et al., 1997). *Toxoplasma* also appears to be resistant to oxidative metabolites, and indeed, the parasite possesses its own antioxidant system (Chang and Pechere, 1989; Ding et al., 2004). By producing antioxidant enzymes, *Toxoplasma* may render itself resistant to this potent killing system.

Phagocytosis

There is little direct evidence to support a role for phagocytosis, a major function of macrophages, dendritic cells, and neutrophils, in immune control of *T. gondii*. This may be a consequence of the *Toxoplasma* invasion process, which can outpace phagocytic engulfment. Nevertheless, in vitro studies demonstrate that antibody-coated parasites can be taken up by Fc receptor-positive cells, and this method of entry directs the tachyzoite into the phagolysosomal pathway of degradation (Joiner, 1993; Joiner et al., 1990). Antibody isotypes associated with strong opsonizing activity are favored by Th1 responses that characterize *T. gondii* infection. Furthermore, B-cell-deficient (μMT) mice are susceptible to *T. gondii*, and mortality and pathology can be reversed by administration of anti-*Toxoplasma* immunoglobulin G (Johnson and Sayles, 2002; Kang et al., 2000; Sayles et al., 2000). Thus, antibody may act to opsonize parasites for phagocytic uptake. This would be expected to promote presentation of antigens for parasite-specific $CD4^+$ T-lymphocyte activation, as well as to enhance killing of *Toxoplasma*. Nevertheless, it is also possible that immunoglobulins induce tachyzoite lysis through antibody-dependent complement activation at the parasite surface in its brief extracellular phases (Sayles et al., 2000).

SUBVERSION OF PHAGOCYTE INTRACELLULAR SIGNALING

T. gondii plays an active role in downmodulating proinflammatory signaling cascades in cells such as macrophages (Denkers et al., 2003). The parasite also downregulates dendritic cell activation and downmodulates neutrophil responses to TLR ligands (Bennouna et al., 2006; McKee et al., 2004). These activities may be driven by selective pressure to avoid proinflammatory immune elimination during infection. Alternatively, it is possible that targeting proinflammatory signaling is necessary to minimize immunopathology that can be lethal to the host. The details of how *Toxoplasma* targets host cell signaling networks are lacking. Nevertheless, as related below, it is becoming clear that tachyzoite rhoptries contain kinase and phosphatase molecules that impinge on host signaling pathways (El Hajj et al., 2007; Gilbert et al., 2007; Saeij et al., 2006; Taylor et al., 2006). At least some of these molecules are targeted to the host cell nucleus during infection.

Exploitation of STAT3 Signaling

Macrophages infected with the virulent RH strain of *Toxoplasma* do not initiate TNF-α synthesis, and low levels of IL-12 are produced only at later time points of infection (Butcher et al., 2001; Kim et al., 2005). Lack of these cytokines is the result of active inhibition by the parasite because infected cells that are stimulated with TLR ligands such as LPS fail to produce these cytokines (Butcher and Denkers, 2002; Lee et al., 2006). Inhibition does not occur if cells are treated with heat-inactivated tachyzoites or soluble tachyzoite extracts. The blocking effects of parasite infection are restricted to infected cells, arguing against macrophage release of soluble downregulatory mediators that would be expected to target infected and noninfected cells alike. While the inhibitory effects of *Toxoplasma* on induction of LPS-responsive genes extend beyond IL-12 and TNF-α, they are not global because a subset of cytokine genes continues to respond to LPS stimulation (Lee et al., 2006). In particular, LPS-induced IL-10 synthesis escapes inhibition by *T. gondii*.

Recognizing the well-known property of IL-10 in downmodulating proinflammatory cytokines, including IL-12 and TNF-α, Butcher and colleagues (2001) directly addressed the function of this cytokine by using macrophages derived from $IL-10^{-/-}$ mice. Even in the absence of IL-10, infection results in inability to produce IL-12 and TNF-α in response to LPS. Nevertheless, *T. gondii* infection triggers rapid (within minutes), long-lasting (up to 22 h), and potent activation of STAT3, a molecule that transduces IL-10-mediated signaling from its receptor to the host cell nucleus (Butcher et al., 2005b; Williams et al., 2004). Importantly, experiments using $STAT3^{-/-}$ macrophages show that *T. gondii* requires this host molecule to suppress responses to LPS (Butcher et al., 2005b).

The molecular basis for how the parasite triggers STAT3 activation is becoming clear, but many questions remain. The molecule is normally phosphorylated on Tyr705 by Jak1/Tyk2 activation following receptor ligation, most prominently by the cytokine IL-10 (Moore et al., 2001). After dimerization and nuclear translocation, STAT3 undergoes Ser727 phosphorylation to achieve optimal transcriptional activity. Tyrosine phosphorylation of STAT3 and STAT6 is induced in a *Toxoplasma* strain-specific manner, inasmuch as while each of the three strain types trigger early STAT3/6 phosphorylation, type II strains fail to maintain persistent activation. A genetic cross between a type II and type III strain was performed followed by a genome-wide scan on F_1 progeny for association of genetic markers with STAT3/6 activation (Saeij et al., 2007). In this way, ROP16, a putative serine-threonine kinase molecule localized to the rhoptries, was identified as a parasite molecule mediating STAT3/6 phosphorylation.

The ROP16 molecule localizes to the nucleus early during host cell invasion, although nuclear repositioning per se does not appear necessary for STAT3/6 activation. Because parasite-induced STAT3 and STAT6 activation involves tyrosine residue phosphorylation, it is unlikely that ROP16 kinase activity is directly responsible for phosphorylating these STAT molecules. Apparently, there are yet-to-be-discovered host or parasite molecules that lie in the pathway between ROP16 and STAT3/6. Regardless, ROP16 from type I parasites appears to mediate STAT3-mediated downregulation of IL-12 production. Thus, type II tachyzoites engineered to express ROP16 from a type I strain trigger long-term STAT3 activation, and IL-12 production is downmodulated to levels that approach those seen during infection with type I and III tachyzoites (Saeij et al., 2007). These combined studies therefore suggest that *Toxoplasma* injects a kinase during invasion that hijacks STAT3 signaling to downmodulate proinflammatory signaling.

In addition to ROP16, several other rhoptry proteins appear to be either active or defective kinases. One in particular, ROP18, is also linked to strain-specific virulence in mice and confers growth advantage during in vitro infection (El Hajj et al., 2007; Saeij et al., 2006; Taylor et al., 2006). The emerging picture is that rhoptry proteins are dominant players at the interface between the parasite and signaling networks of the host cell, and that one of the major pathways targeted is the STAT3 signaling cascade.

Downregulation of NF-κB and MAPK Signaling Cascades

Both Rel/NF-κB and MAPK pathways lie directly downstream of TLR/MyD88 signaling, and they are major transduction cascades from the cell surface to the nucleus. The Rel/NF-κBs are a heterodimeric family of molecules composed of RelA/p65, RelB, c-Rel, NF-κB1 (p50/p105), and NF-κB2 (p52/p100) (Karin and Ben-Neriah, 2000). Among these, only c-Rel and RelA are known to contain transcriptional activating domains. NF-κB is normally retained in the cytoplasm through masking of a nuclear localization sequence (NLS) by the inhibitory molecule IκBα. Classical activation of NF-κB involves phosphorylation of IκBα by a trimeric IκB kinase complex. The IκBα molecule subsequently undergoes ubiquitination followed by proteasomal degradation, revealing the NF-κB NLS and enabling nuclear translocation of the latter molecule. The NF-κB transcription factor binds to κB sites in promoters of several genes encoding inflammatory and antiapoptotic mediators. Expression of IκB is also controlled by NF-κB, and in this manner an autoregulatory loop is established (Chiao et al., 1994).

Although infection by *T. gondii* induces IκBα phosphorylation-dependent degradation in macrophages, NF-κB fails to accumulate in the nucleus (Butcher et al., 2001; Shapira et al., 2002). The block in nuclear accumulation is not permanent. This is because while macrophages infected for 60 to 120 min fail to display NF-κB nuclear accumulation after LPS stimulation, cells infected for longer time periods undergo normal NF-κB activation upon subsequent LPS stimulation (Kim et al., 2004). Recent evidence indicates that lack of NF-κB nuclear accumulation may be a consequence of increased export from the nucleus, rather than decreased import (Shapira et al., 2005). Lack of macrophage NF-κB activation by *Toxoplasma* may also be a strain-specific phenomenon, in that it may be a general property of type I but not type II strains (Robben et al., 2004). Nevertheless, in contrast to these studies, it has also been reported that type I strains trigger NF-κB activation in fibroblasts (Molestina et al., 2003). This discrepancy is so far unresolved, but possibly is a consequence of using different cell types as targets of infection.

There are three major MAPK family members, consisting of p38 MAPK, stress-activated protein kinase (SAPK)/c-Jun N-terminal kinase (JNK)1/2, and ERK1/2 (Dong et al., 2002). These serine-threonine kinases activate molecules that are either transcription factors themselves or kinases that activate transcription factors. MAPK molecules are activated through dual phosphorylation of threonine-X-tyrosine motifs by MAPK kinases (MKKs). The MKK molecules are similarly activated by a large number of MKK kinases (MKKK/M3K) as a result of TLR/MyD88 and other signaling initiators. Deactivation of MAPK signaling is achieved through the activity of dual-specificity MAPK phosphatases (Camps et al., 2000).

Infection of bone marrow-derived macrophages by *T. gondii* induces rapid p38, SAPK/JNK1/2, and ERK1/2 phosphorylation and nuclear translocation of downstream transcription factors that include c-Jun, activating transcription factor-2 (ATF-2), and MAPK-activated protein kinase 2 (MAPKAP2) (Kim et al., 2004; Valere et al., 2003). Each MAPK subsequently undergoes dephosphorylation. When *T. gondii*-infected cells are subjected to LPS stimulation, reactivation of MAPK fails to occur (Kim et al., 2004). Because infected cells possess high levels of activated MKK3, a major kinase involved in p38 MAPK activation, the parasite may induce host cell MAPK phosphatase activity, or may itself produce phosphatases that are directed to the host cell nucleus (Gilbert et al., 2007). Deactivation of both MAPK and NF-κB pathways that mediate TLR/MyD88 signaling may reflect the need to nullify the activity of the parasite's own TLR ligands. Alternatively, breaches in the intestinal epithelial integrity during oral infection may require that the parasite renders infected host tissues resistant to proinflammatory effects of TLR ligands expressed by endogenous gut flora (Heimesaat et al., 2006, 2007).

Negative Regulation of IFN-γ/STAT1-Mediated Signaling

The cytokine IFN-γ is a major determinant of host and parasite survival during *T. gondii* infection (Suzuki et al., 1988). After cytokine binding at the cell surface, receptor-associated Jak1 and Jak2 molecules transphosphorylate each other, resulting in STAT1 recruitment, activation, and translocation to the nucleus in a manner similar to STAT3 (Platanias, 2005). After dimerization, STAT1 translocates to the nucleus and undergoes Ser727 phosphorylation for full transcriptional activity. *Toxoplasma* is reported to downmodulate IFN-γ-induced STAT1 activity in several cell types including primary macrophages, macrophage cell lines, fibroblasts, as well as astrocytes and microglia isolated from rat brain (Kim et al., 2007; Luder et al., 2001, 2003; Zimmermann et al., 2006). In a genome-wide microarray analysis in human fibroblasts, virtually all of >100 IFN-γ-responsive genes were refractory after infection (Kim et al., 2007). Each of the three *T. gondii* strain types displayed this suppressive activity, as measured by STAT1-dependent induction of IFN regulatory factor 1 (IRF-1).

There is not yet a clear molecular picture of how *T. gondii* downregulates expression of STAT1-responsive genes. Despite the profound and widespread suppression of IFN-γ/STAT1-dependent genes by *Toxoplasma*, phosphorylation of upstream Jak1 and Jak2, as well as STAT1 Tyr701 and Ser727 phosphorylation, proceeds normally in infected cells (Kim et al., 2007). Initial reports indicated defective STAT1 nuclear translocation, despite normal phosphorylation, in infected cells (Luder et al., 2001). However, this now appears uncertain (Kim et al., 2007; Lang et al., 2006). Instead, the parasite may modulate the phosphorylation state of the STAT1 Tyr701 residue (Kim et al., 2007). In this regard, it is of interest that a *Toxoplasma* phosphatase has recently been reported to be targeted to the host cell nucleus (Gilbert et al., 2007). It has also been found that the parasite induces suppressor of cytokine signaling-1 (SOCS-1) (Zimmermann et al., 2006). This negative regulator of STAT1 signaling affects Jak activity and may also target signal transducers for proteasomal degradation (Alexander and Hilton, 2004). Nevertheless, others have not observed induction of this modulatory protein (Kim et al., 2007), discrepancies that may be the result of the fact that different cell types were used in these studies. Regardless of the mechanism, downregulation of STAT1-dependent genes such as iNOS appears to facilitate parasite replication in vitro and may therefore promote early dissemination in vivo (Luder et al., 2003).

The contemporary picture is that *Toxoplasma* possesses multiple mechanisms to downmodulate proinflammatory signaling initiated through IFN-γR/STAT1 and TLR/MyD88 (Fig. 3). The details of how this occurs are far from understood. It seems possible, even likely, that host cell type, as well as parasite strain type, determines which signaling pathways are targeted for deactivation. Elucidating the underlying mechanisms is a major challenge for the future.

CONCLUSIONS AND FUTURE DIRECTIONS

The encounter of phagocytes with *Toxoplasma* occurs in a dynamic arena in which interactions between the host and parasite are dramatically played out. For the host, it is essential to control the parasite to prevent death from infection. The demands are more subtle from the perspective of the parasite. *Toxoplasma* must avoid or actively suppress potent host immunity to avoid elimination. Nevertheless, there is selective pressure on the *Toxoplasma* to be recog-

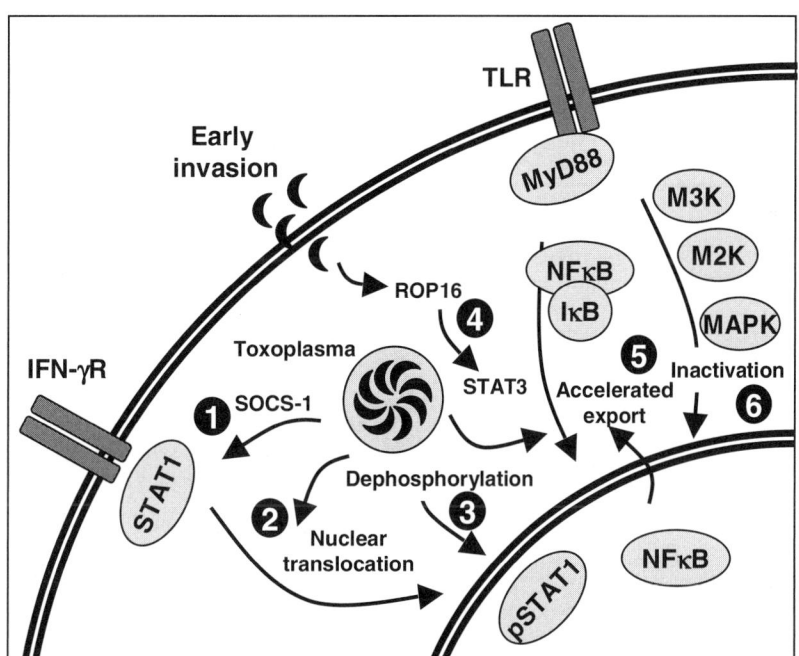

FIGURE 3 *T. gondii* mechanisms of subverting phagocyte immune function. Once inside the cell, *Toxoplasma* renders cells nonresponsive to IFN-γ. This appears to occur at several levels. (1) The parasite may induce the negative regulator SOCS-1 that blocks STAT1 phosphorylation and induces its degradation. (2) There is evidence that the parasite also prevents nuclear translocation of phosphorylated STAT1, although this is less certain. (3) Some studies also suggest that *Toxoplasma* may dephosphorylate STAT1 that has translocated into the nucleus in response to IFN-γ. (4) Rhoptry protein discharge during invasion; in particular, the putative serine-threonine kinase ROP16 induces STAT3 activation that downregulates proinflammatory signaling. (5) The parasite also interferes with TLR-mediated NF-κB activity in a process that may involve accelerated export from the nucleus. (6) *Toxoplasma* also deactivates proinflammatory MAPK signaling cascades initiated through TLR, in a process that may involve either parasite or host phosphatase activity.

nized by the host immune system, for if not, the consequence is infection that is lethal to the host and, in turn, the parasite. In the future, we can expect a major drive to determine at the molecular level how this complex balance is achieved, and indeed, this research effort is already well under way. It is hoped that such an understanding will allow development of new immunotherapeutic tools, in turn facilitating better treatment and prevention of diseases caused by Toxoplasma and other microbial pathogens.

I thank B. Butcher for critical review of this manuscript, and past and present members of my lab for useful discussion.

REFERENCES

Adams, L. B., J. B. Hibbs, Jr., R. R. Taintor, and J. L. Krahenbuhl. 1990. Microbiostatic effect of murine-activated macrophages for Toxoplasma gondii: role for synthesis of inorganic nitrogen oxides from L-arginine. *J. Immunol.* **144:** 2725–2729.

Akira, S., S. Uematsu, and O. Takeuchi. 2006. Pathogen recognition and innate immunity. *Cell* **124:**783–801.

Alexander, J., and C. A. Hunter. 1998. Immunoregulation during toxoplasmosis. *Chem. Immunol.* **70:**81–102.

Alexander, W. S., and D. J. Hilton. 2004. The role of suppressors of cytokine signaling (SOCS) proteins in regulation of the immune response. *Annu. Rev. Immunol.* **22:**503–529.

Aliberti, J., D. Jankovic, and A. Sher. 2004. Turning it on and off: regulation of dendritic cell function in Toxoplasma gondii infection. *Immunol. Rev.* **201:**26–34.

Aliberti, J., C. Reis e Sousa, M. Schito, S. Hieny, T. Wells, G. B. Huffnage, and A. Sher. 2000. CCR5 provides a signal for microbial induced production of IL-12 by CD8α+ dendritic cells. *Nat. Immunol.* **1:**83–87.

Aliberti, J., J. G. Valenzuela, V. B. Carruthers, S. Hieny, J. Andersen, H. Charest, C. Reis e Sousa, A. Fairlamb, J. M. Ribeiro, and A. Sher. 2003. Molecular mimicry of a CCR5 binding-domain in the microbial activation of dendritic cells. *Nat. Immunol.* **4:**485–490.

Andrade, R. M., J.-M. C. Portillo, M. Wessendarp, and C. S. Subauste. 2005. CD40 signaling in macrophages induces activity against an intracellular pathogen independently of gamma interferon and reactive oxygen intermediates. *Infect. Immun.* **73:**3115–3125.

Andrade, R. M., M. Wessendarp, M. J. Gubbels, B. Striepen, and C. S. Subauste. 2006. CD40 induces macrophage anti-Toxoplasma gondii activity by triggering autophagy-dependent fusion of pathogen-containing vacuoles and lysosomes. *J. Clin. Investig.* **116:**2366–2377.

Andrade, R. M., M. Wessendarp, and C. S. Subauste. 2003. CD154 activates macrophage antimicrobial activity in the absence of IFN-γ through a TNF-α-dependent mechanism. *J. Immunol.* **171:**6750–6756.

Bafica, A., H. C. Santiago, R. Goldszmid, C. Ropert, R. T. Gazzinelli, and A. Sher. 2006. Cutting edge: TLR9 and TLR2 signaling together account for MyD88-dependent control of parasitemia in Trypanosoma cruzi infection. *J. Immunol.* **177:**3515–3519.

Bennouna, S., S. K. Bliss, T. J. Curiel, and E. Y. Denkers. 2003. Cross-talk in the innate immune system: neutrophils instruct early recruitment and activation of dendritic cells during microbial infection. *J. Immunol.* **171:**6052–6058.

Bennouna, S., W. Sukhumavasi, and E. Y. Denkers. 2006. Toxoplasma gondii inhibits Toll-like receptor 4 ligand-induced mobilization of intracellular tumor necrosis factor alpha to the surface of mouse peritoneal neutrophils. *Infect. Immun.* **74:**4274–4281.

Bliss, S. K., B. A. Butcher, and E. Y. Denkers. 2000. Rapid recruitment of neutrophils with prestored IL-12 during microbial infection. *J. Immunol.* **165:**4515–4521.

Bliss, S. K., L. C. Gavrilescu, A. Alcaraz, and E. Y. Denkers. 2001. Neutrophil depletion during Toxoplasma gondii infection leads to impaired immunity and lethal systemic pathology. *Infect. Immun.* **69:**4898–4905.

Bliss, S. K., Y. Zhang, and E. Y. Denkers. 1999. Murine neutrophil stimulation by Toxoplasma gondii antigen drives high level production of IFN-γ-independent IL-12. *J. Immunol.* **163:**2081–2088.

Bogdan, C., and M. Rollinghoff. 1999. How do protozoan parasites survive inside macrophages? *Parasitol. Today* **15:**22–28.

Bohne, W., J. Heesemann, and U. Gross. 1994. Reduced replication of Toxoplasma gondii is necessary for induction of bradyzoite-specific antigens: a possible role for nitric oxide in triggering stage conversion. *Infect. Immun.* **62:**1761–1767.

Boothroyd, J. C., and M. E. Grigg. 2002. Population biology of Toxoplasma gondii and its relevance to human infection: do different strains cause different disease? *Curr. Opin. Microbiol.* **5:**438–442.

Butcher, B. A., and E. Y. Denkers. 2002. Mechanism of entry determines ability of Toxoplasma gondii to inhibit macrophage proinflammatory cytokine production. *Infect. Immun.* **70:** 5216–5224.

Butcher, B. A., R. I. Greene, S. C. Henry, K. L. Annecharico, J. B. Weinberg, E. Y. Denkers, A. Sher, and G. A. Taylor. 2005a. p47 GTPases regulate Toxoplasma gondii survival in activated macrophages. *Infect. Immun.* **73:**3278–3286.

Butcher, B. A., L. Kim, P. F. Johnson, and E. Y. Denkers. 2001. Toxoplasma gondii tachyzoites inhibit proinflammatory cytokine induction in infected macrophages by preventing nuclear translocation of the transcription factor NFκB. *J. Immunol.* **167:**2193–2201.

Butcher, B. A., L. Kim, A. Panopoulos, S. S. Watowich, P. J. Murray, and E. Y. Denkers. 2005b. Cutting edge: IL-10-independent STAT3 activation by Toxoplasma gondii mediates suppression of IL-12 and TNF-α in host macrophages. *J. Immunol.* **174:**3148–3152.

Camps, M., A. Nichols, and S. Arkinstall. 2000. Dual specificity phosphatases: a gene family for control of MAP kinase function. *FASEB J.* **14:**6–16.

Carruthers, V., and J. C. Boothroyd. 2007. Pulling together: an integrated model of Toxoplasma cell invasion. *Curr. Opin. Microbiol.* **10:**83–89.

Carruthers, V. B., O. K. Giddings, and L. D. Sibley. 1999. Secretion of micronemal proteins is associated with toxoplasma invasion of host cells. *Cell. Microbiol.* **1:**225–235.

Carruthers, V. B., and L. D. Sibley. 1997. Sequential protein secretion from three distinct organelles of Toxoplasma gondii accompanies invasion of human fibroblasts. *Eur. J. Cell Biol.* **73:**114–123.

Cassatella, M. A. 1999. Neutrophil-derived proteins: selling cytokines by the pound. *Adv. Immunol.* **73:**369–509.

Chang, H. R., and J. C. Pechere. 1989. Macrophage oxidative metabolism and intracellular Toxoplasma gondii. *Microb. Pathog.* **7:**37–44.

Channon, J. Y., R. M. Seguin, and L. H. Kasper. 2000. Differential infectivity and division of Toxoplasma gondii in human peripheral blood leukocytes. *Infect. Immun.* **68:**4822–4826.

Charron, A. J., and L. D. Sibley. 2004. Molecular partitioning during host cell penetration by Toxoplasma gondii. *Traffic* **5:**855–867.

Chiao, P. J., S. Miyamato, and I. M. Verma. 1994. Autoregulation of IκBα activity. *Proc. Natl. Acad. Sci. USA* **91:** 28–32.

Collazo, C. M., G. S. Yap, G. D. Sempowski, K. C. Lusby, L. Tessarollo, G. F. Vande Woude, A. Sher, and G. A. Taylor. 2001. Inactivation of LRG-47 and IRG-47 reveals a family of interferon-γ-inducible genes with essential, pathogen-specific roles in resistance to infection. *J. Exp. Med.* **194:** 181–187.

Courret, N., S. Darche, P. Sonigo, G. Milon, D. Buzoni-Gatel, and I. Tardieux. 2006. CD11c and CD11b expressing mouse leukocytes transport single *Toxoplasma gondii* tachyzoites to the brain. *Blood* **107:**309–316.

Debierre-Grockiego, F., M. A. Campos, N. Azzouz, J. Schmidt, U. Bieker, M. G. Resende, D. S. Mansur, R. Weingart, R. R. Schmidt, D. T. Golenbock, R. T. Gazzinelli, and R. T. Schwarz. 2007. Activation of TLR2 and TLR4 by glycosylphosphatidylinositols derived from *Toxoplasma gondii*. *J. Immunol.* **179:**1129–1137.

Deckert-Schluter, M., H. Bluethmann, A. Rang, H. Hof, and D. Schluter. 1998. Crucial role for TNF receptor Type 1 (p55), but not TNF receptor Type 2 (p75), in murine toxoplasmosis. *J. Immunol.* **160:**3427–3436.

Del Rio, L., S. Bennouna, J. Salinas, and E. Y. Denkers. 2001. CXCR2 deficiency confers impaired neutrophil recruitment and increased susceptibility during *Toxoplasma gondii* infection. *J. Immunol.* **167:**6503–6509.

Del Rio, L., B. A. Butcher, S. Bennouna, S. Hieny, A. Sher, and E. Y. Denkers. 2004. *Toxoplasma gondii* triggers MyD88-dependent and CCL2(MCP-1) responses using distinct parasite molecules and host receptors. *J. Immunol.* **172:**6954–6960.

Denkers, E. Y., B. A. Butcher, L. Del Rio, and S. Bennouna. 2004. Neutrophils, dendritic cells and *Toxoplasma*. *Int. J. Parasitol.* **34:**411–421.

Denkers, E. Y., B. A. Butcher, L. Del Rio, and L. Kim. 2004. Manipulation of mitogen-activated protein kinase/nuclear factor-κB-signaling cascades during intracellular *Toxoplasma gondii* infection. *Immunol. Rev.* **201:**191–205.

Denkers, E. Y., L. D. Del Rio, and S. Bennouna. 2003. Neutrophil production of IL-12 and other cytokines during microbial infection. *Chem. Immunol. Allergy* **83:**95–114.

Denkers, E. Y., and R. T. Gazzinelli. 1998. Regulation and function of T cell-mediated immunity during *Toxoplasma gondii* infection. *Clin. Microbiol. Rev.* **11:**569–588.

Denkers, E. Y., L. Kim, and B. A. Butcher. 2003. In the belly of the beast: subversion of macrophage proinflammatory signaling cascades during *Toxoplasma gondii* infection. *Cell. Microbiol.* **5:**75–83.

Ding, M., L. Y. Kwok, D. Schluter, C. Clayton, and D. Soldati. 2004. The antioxidant systems in *Toxoplasma gondii* and the role of cytosolic catalase in defence against oxidative injury. *Mol. Microbiol.* **51:**47–61.

Dobrowolski, J. M., and L. D. Sibley. 1996. *Toxoplasma* invasion of mammalian cells is powered by the actin cytoskeleton of the parasite. *Cell* **84:**933–939.

Dong, C., R. J. Davis, and R. A. Flavell. 2002. MAP kinases in the immune response. *Annu. Rev. Immunol.* **20:**55–72.

Dubey, J. P. 1998. Advances in the life cycle of *Toxoplasma gondii*. *Int. J. Parasitol.* **28:**1019–1024.

Dubey, J. P. 2007. The history and life-cycle of *Toxoplasma gondii*, p. 1–17. In L. M. Weiss and K. Kim (ed.), *Toxoplasma gondii. The Model Apicomplexan: Perspective and Methods.* Academic Press, San Diego, CA.

Egan, C. E., W. Sukhumavasi, A. L. Bierly, and E. Y. Denkers. 2008. Understanding the multiple functions of Gr-1+ cell subpopulations during microbial infection. *Immunol. Res.* **40:**35–48.

El Hajj, H., M. Lebrun, S. T. Arold, H. Vial, G. Labesse, and J. F. Dubremetz. 2007. ROP18 is a rhoptry kinase controlling the intracellular proliferation of *Toxoplasma gondii*. *PLoS Pathog.* **3:**e14.

Fox, B. A., and D. J. Bzik. 2002. De novo pyrimidine biosynthesis is required for virulence of *Toxoplasma gondii*. *Nature* **415:**926–929.

Fox, B. A., J. P. Gigley, and D. J. Bzik. 2004. *Toxoplasma gondii* lacks the enzymes required for de novo arginine biosynthesis and arginine starvation triggers cyst formation. *Int. J. Parasitol.* **34:**323–331.

Gavrilescu, L. C., B. A. Butcher, L. Del Rio, G. A. Taylor, and E. Y. Denkers. 2004. STAT1 is essential for antimicrobial function but dispensable for gamma interferon production during *Toxoplasma gondii* infection. *Infect. Immun.* **72:** 1257–1264.

Gavrilescu, L. C., and E. Y. Denkers. 2001. IFN-γ overproduction and high level apoptosis are associated with high but not low virulence *Toxoplasma gondii* infection. *J. Immunol.* **167:**902–909.

Gazzinelli, R., Y. Xu, S. Hieny, A. Cheever, and A. Sher. 1992. Simultaneous depletion of $CD4^+$ and $CD8^+$ T lymphocytes is required to reactivate chronic infection with *Toxoplasma gondii*. *J. Immunol.* **149:**175–180.

Gazzinelli, R. T., and E. Y. Denkers. 2006. Protozoan encounters with Toll-like receptor signalling pathways: implications for host parasitism. *Nat. Rev. Immunol.* **6:**895–906.

Gazzinelli, R. T., F. T. Hakim, S. Hieny, G. M. Shearer, and A. Sher. 1991. Synergistic role of $CD4^+$ and $CD8^+$ T lymphocytes in IFN-γ production and protective immunity induced by an attenuated *T. gondii* vaccine. *J. Immunol.* **146:** 286–292.

Gazzinelli, R. T., S. Hieny, T. Wynn, S. Wolf, and A. Sher. 1993. IL-12 is required for the T-cell independent induction of IFN-γ by an intracellular parasite and induces resistance in T-cell-deficient hosts. *Proc. Natl. Acad. Sci. USA* **90:** 6115–6119.

Gazzinelli, R. T., I. P. Oswald, S. Hieny, S. James, and A. Sher. 1992a. The microbicidal activity of interferon-γ-treated macrophages against *Trypanosoma cruzi* involves an L-arginine-dependent, nitrogen oxide-mediated mechanism inhibitable by interleukin-10 and transforming growth factor-β. *Eur. J. Immunol.* **22:**2501–2506.

Gazzinelli, R. T., I. P. Oswald, S. James, and A. Sher. 1992b. IL-10 inhibits parasite killing and nitrogen oxide production by IFN-γ activated macrophages. *J. Immunol.* **148:**1792–1796.

Gazzinelli, R. T., M. Wysocka, S. Hayashi, E. Y. Denkers, S. Hieny, P. Caspar, G. Trinchieri, and A. Sher. 1994. Parasite-induced IL-12 stimulates early IFN-γ synthesis and resistance during acute infection with *Toxoplasma gondii*. *J. Immunol.* **153:**2533–2543.

Geissmann, F., S. Jung, and D. R. Littman. 2003. Blood monocytes consist of two principal subsets with distinct migratory properties. *Immunity* **19:**71–82.

Gilbert, L. A., S. Ravindran, J. M. Turetzky, J. C. Boothroyd, and P. J. Bradley. 2007. *Toxoplasma gondii* targets a protein phosphatase 2C to the nuclei of infected host cells. *Eukaryot. Cell* **6:**73–83.

Gubbels, M.-J., B. Striepen, N. Shastri, M. Turkoz, and E. A. Robey. 2005. Class I major histocompatibility complex presentation of antigens that escape from the parasitophorous vacuole of *Toxoplasma gondii*. *Infect. Immun.* **73:**703–711.

Hakansson, S., A. J. Charron, and L. D. Sibley. 2001. *Toxoplasma* evacuoles: a two-step process of secretion and fusion forms the parasitophorous vacuole. *EMBO J.* **20:**3132–3144.

Heimesaat, M. M., S. Bereswill, A. Fischer, D. Fuchs, D. Struck, J. Niebergall, H. K. Jahn, I. R. Dunay, A. Moter, D. M. Gescher, R. R. Schumann, U. B. Gobel, and O. Liesenfeld. 2006. Gram-negative bacteria aggravate murine small intestinal Th1-Type immunopathology following oral infection with *Toxoplasma gondii*. *J. Immunol.* **177:**8785–8795.

Heimesaat, M. M., A. Fischer, H. K. Jahn, J. Niebergall, M. Freudenberg, M. Blaut, O. Liesenfeld, R. R. Schumann, U. B. Gobel, and S. Bereswill. 2007. Exacerbation of murine ileitis by toll-like receptor 4 meditated sensing of lipopolysaccharide from commensal *Escherichia coli*. *Gut* **56:**941–948.

Hitziger, N., I. Dellacasa, B. Albiger, and A. Barragan. 2005. Dissemination of *Toxoplasma gondii* to immunoprivileged organs and role of Toll/interleukin-1 receptor signalling for host resistance assessed by in vivo bioluminescence imaging. *Cell. Microbiol.* **6:**837–848.

Howe, D. K., and L. D. Sibley. 1995. *Toxoplasma gondii* comprises three clonal lineages: correlation of parasite genotype with human diseases. *J. Infect. Dis.* **172:**1561–1566.

Hunter, C. A., E. Candolfi, C. Subauste, V. Van Cleave, and J. S. Remington. 1995. Studies on the role of IL-12 in murine toxoplasmosis. *Immunology* **84:**16–21.

Johnson, L. L., and P. C. Sayles. 2002. Deficient humoral responses underlie susceptibility to *Toxoplasma gondii* in CD4-deficient mice. *Infect. Immun.* **70:**185–191.

Joiner, K. A. 1993. Cell entry by *Toxoplasma gondii*: all paths do not lead to success. *Res. Immunol.* **144:**34–38.

Joiner, K. A., S. A. Fuhrman, H. M. Miettinen, L. H. Kasper, and I. Mellman. 1990. *Toxoplasma gondii*: fusion competence of parasitophorous vacuoles in Fc receptor-transfected fibroblasts. *Science* **249:**641–646.

Kang, H., J. S. Remington, and Y. Suzuki. 2000. Decreased resistance of B cell-deficient mice to infection with *Toxoplasma gondii* despite unimpaired expression of IFN-γ, TNF-α, and inducible nitric oxide synthase. *J. Immunol.* **164:**2629–2634.

Karin, M., and Y. Ben-Neriah. 2000. Phosphorylation meets ubiquitination: the control of NF-κB activity. *Annu. Rev. Immunol.* **18:**621–663.

Kasper, L., N. Courret, S. Darche, S. Luangsay, F. Mennechet, L. Minns, N. Rachinel, C. Ronet, and D. Buzoni-Gatel. 2004. *Toxoplasma gondii* and mucosal immunity. *Int. J. Parasitol.* **34:**401–409.

Khan, A., C. Su, M. German, G. A. Storch, D. B. Clifford, and L. D. Sibley. 2005. Genotyping of *Toxoplasma gondii* strains from immunocompromised patients reveals high prevalence of type I strains. *J. Clin. Microbiol.* **43:**5881–5887.

Khan, I. A., T. Matsuura, and L. H. Kasper. 1994. Interleukin-12 enhances murine survival against acute toxoplasmosis. *Infect. Immun.* **62:**1639–1645.

Kim, L., B. A. Butcher, and E. Y. Denkers. 2004. *Toxoplasma gondii* interferes with lipopolysaccharide-induced mitogen-activated protein kinase activation by mechanisms distinct from endotoxin tolerance. *J. Immunol.* **172:**3003–3010.

Kim, L., B. A. Butcher, C. W. Lee, S. Uematsu, S. Akira, and E. Y. Denkers. 2006. *Toxoplasma gondii* genotype determines MyD88-dependent signaling in infected macrophages. *J. Immunol.* **177:**2584–2591.

Kim, L., L. Del Rio, B. A. Butcher, T. H. Mogensen, S. Paludan, R. A. Flavell, and E. Y. Denkers. 2005. p38 MAPK autophosphorylation drives macrophage IL-12 production during intracellular infection. *J. Immunol.* **174:**4178–4184.

Kim, L., and E. Y. Denkers. 2006. *Toxoplasma gondii* triggers Gi-dependent phosphatidylinositol 3-kinase signaling required for inhibition of host cell apoptosis. *J. Cell Sci.* **119:**2119–2126.

Kim, S. K., A. E. Fouts, and J. C. Boothroyd. 2007. *Toxoplasma gondii* dysregulates IFN-gamma-inducible gene expression in human fibroblasts: insights from a genome-wide transcriptional profiling. *J. Immunol.* **178:**5154–5165.

Lambert, H., N. Hitziger, I. Dellacasa, M. Svensson, and A. Barragan. 2006. Induction of dendritic cell migration upon *Toxoplasma gondii* infection potentiates parasite dissemination. *Cell. Microbiol.* **8:**1611–1623.

Lang, C., M. Algner, N. Beinert, U. Gross, and C. G. Luder. 2006. Diverse mechanisms employed by *Toxoplasma gondii* to inhibit IFN-gamma-induced major histocompatibility complex class II gene expression. *Microbes Infect.* **8:**1994–2005.

Langermans, J. A. M., M. E. B. Van der Hulst, P. H. Nibbering, P. S. Hiemstra, L. Fransen, and R. Van Furth. 1992. IFN-γ-induced L-arginine-dependent toxoplasmastatic activity in murine peritoneal macrophages is mediated by endogenous tumor necrosis factor-α. *J. Immunol.* **148:**568–578.

Lauw, F. N., D. R. Caffrey, and D. T. Golenbock. 2005. Of mice and man: TLR11 (finally) finds profilin. *Trends Immunol.* **26:**509–511.

Lee, C. W., S. Bennouna, and E. Y. Denkers. 2006. Screening for *Toxoplasma gondii* regulated transcriptional responses in LPS-activated macrophages. *Infect. Immun.* **74:**1916–1923.

Lieberman, L. A., M. Banica, S. L. Reiner, and C. A. Hunter. 2004. STAT1 plays a critical role in the regulation of antimicrobial effector mechanisms, but not in the development of Th1-type responses during toxoplasmosis. *J. Immunol.* **172:**457–463.

Liesenfeld, O., S. Y. Wong, and J. S. Remington. 1999. Toxoplasmosis in the setting of AIDS, p. 225–259. *In* J. G. Bartlett, T. C. Merigan, and D. Bolognesi (ed.), *Textbook of AIDS Medicine*. Williams and Wilkins, Baltimore, MD.

Ling, Y. M., M. H. Shaw, C. Ayala, I. Coppens, G. A. Taylor, D. J. Ferguson, and G. S. Yap. 2006. Vacuolar and plasma membrane stripping and autophagic elimination of *Toxoplasma gondii* in primed effector macrophages. *J. Exp. Med.* **203:**2063–2071.

Liu, C. H., Y. T. Fan, A. Dias, L. Esper, R. A. Corn, A. Bafica, F. S. Machado, and J. Aliberti. 2006. Cutting edge: dendritic cells are essential for in vivo IL-12 production and development of resistance against *Toxoplasma gondii* infection in mice. *J. Immunol.* **177:**31–35.

Luder, C. G. K., M. Algner, C. Lang, N. Bleicher, and U. Gross. 2003. Reduced expression of the inducible nitric oxide synthase after infection with *Toxoplasma gondii* facilitates parasite replication in activated murine macrophages. *Int. J. Parasitol.* **33:**833–844.

Luder, C. G. K., and U. Gross. 2005. Apoptosis and its modulation during infection with *Toxoplasma gondii*: molecular mechanisms and role in pathogenesis. *Curr. Top. Microbiol. Immunol.* **289:**219–238.

Luder, C. G. K., C. Lang, M. Giraldo-Velasquez, M. Algner, J. Gerdes, and U. Gross. 2003. *Toxoplasma gondii* inhibits MHC class II expression in neural antigen-presenting cells by down-regulating the class II transactivator CIITA. *J. Neuroimmunol.* **134:**12–24.

Luder, C. G. K., W. Walter, B. Beuerle, M. J. Maeurer, and U. Gross. 2001. *Toxoplasma gondii* down-regulates MHC class II gene expression and antigen presentation by murine macrophages via interference with nuclear translocation of STAT1α. *Eur. J. Immunol.* **31:**1475–1484.

Luft, B. J., R. G. Brooks, F. K. Conley, R. E. McCabe, and J. S. Remington. 1984. Toxoplasmic encephalitis in patients with acquired immune response deficiency syndrome. *JAMA* **252:**913–917.

Luft, B. J., and J. S. Remington. 1992. Toxoplasmic encephalitis in AIDS. *Clin. Infect. Dis.* **15:**211–222.

Ma, W., K. Gee, W. Lim, K. Chambers, J. B. Angel, M. Kozlowski, and A. Kumar. 2004. Dexamethasone inhibits IL-12p40 production in lipopolysaccharide-stimulated human monocytic cells by down-regulating the activity of c-Jun N-terminal kinase, the activation protein-1, and NF-kappa B transcription factors. *J. Immunol.* **172:**318–330.

MacMicking, J., Q. Xie, and C. Nathan. 1997. Nitric oxide and macrophage function. *Annu. Rev. Immunol.* **15:**323–350.

Martens, S., I. Parvanova, J. Zerrahn, G. Griffiths, G. Schell, G. Reichmann, and J. C. Howard. 2005. Disruption of *Toxoplasma gondii* parasitophorous vacuoles by the mouse p47-resistance GTPases. *PLoS Pathog.* **1:**e24.

Masek, K. S., J. Fiore, M. Leitges, S. F. Yan, B. D. Freedman, and C. A. Hunter. 2006. Host cell Ca2+ and protein kinase C regulate innate recognition of *Toxoplasma gondii*. *J. Cell Sci.* **119:**4565–4573.

Mason, N., J. Aliberti, J. C. Caamano, H. C. Liou, and C. A. Hunter. 2002. Identification of c-Rel-dependent and -independent pathways of Il-12 production during infectious and inflammatory stimuli. *J. Immunol.* **168:**2590–2594.

McKee, A. S., F. Dzierszinski, M. Boes, D. S. Roos, and E. J. Pearce. 2004. Functional inactivation of immature dendritic cells by the intracellular parasite *Toxoplasma gondii*. *J. Immunol.* **173:**2632–2640.

Megiovanni, A. M., F. Sanchez, M. Robledo-Sarmiento, C. Morel, J. C. Gluckman, and S. Boudaly. 2006. Polymorphonuclear neutrophils deliver activation signals and antigenic molecules to dendritic cells: a new link between leukocytes upstream of T lymphocytes. *J. Leukoc. Biol.* **79:**977–988.

Mercier, C., K. D. Adjogble, W. Daubener, and M. F. Delauw. 2005. Dense granules: are they key organelles to help understand the parasitophorous vacuole of all apicomplexa parasites? *Int. J. Parasitol.* **35:**829–849.

Minns, L. A., L. C. Menard, D. M. Foureau, S. Darche, C. Ronet, D. W. Mielcarz, D. Buzoni-Gatel, and L. H. M. Kasper. 2006. TLR9 is required for the gut-associated lymphoid tissue response following oral infection of *Toxoplasma gondii*. *J. Immunol.* **176:**7589–7597.

Mital, J., M. Meissner, D. Soldati, and G. E. Ward. 2005. Conditional expression of *Toxoplasma gondii* apical membrane antigen-1 (TgAMA1) demonstrates that TgAMA1 plays a critical role in host cell invasion. *Mol. Biol. Cell* **16:**4341–4349.

Molestina, R. E., T. M. Payne, I. Coppens, and A. P. Sinai. 2003. Activation of NF-κB by *Toxoplasma gondii* correlates with increased expression of antiapoptotic genes and localization of phosphorylated IκB to the parasitophorous vacuole membrane. *J. Cell Sci.* **116:**4359–4371.

Moore, K. A., R. de Waal Malefyt, R. L. Coffman, and A. O'Garra. 2001. Interleukin-10 and the interleukin-10 receptor. *Annu. Rev. Immunol.* **19:**683–765.

Mordue, D. G., N. Dessai, M. Dustin, and L. D. Sibley. 1999. Invasion by *Toxoplasma gondii* establishes a moving junction that selectively excludes host cell plasma membrane proteins on the basis of their membrane anchoring. *J. Exp. Med.* **190:**1783–1792.

Mordue, D. G., F. Monroy, M. La Regina, C. A. Dinarello, and L. D. Sibley. 2001. Acute toxoplasmosis leads to lethal overproduction of Th1 cytokines. *J. Immunol.* **167:**4574–4584.

Mordue, D. G., and L. D. Sibley. 1997. Intracellular fate of vacuoles containing *Toxoplasma gondii* is determined at the time of formation and depends upon the mechanism of entry. *J. Immunol.* **159:**4452–4459.

Mordue, D. G., and L. D. Sibley. 2003. A novel population of Gr-1+-activated macrophages induced during acute toxoplasmosis. *J. Leukoc. Biol.* **74:**1–11.

Morisaki, J. H., J. E. Heuser, and L. D. Sibley. 1995. Invasion of *Toxoplasma gondii* occurs by active penetration of the host cell. *J. Cell Sci.* **108:**2457–2464.

Moudy, R., T. J. Manning, and C. J. Beckers. 2001. The loss of cytoplasmic potassium upon host cell breakdown triggers egress of *Toxoplasma gondii*. *J. Biol. Chem.* **276:**41492–41501.

Mun, H.-S., F. Aosai, K. Norose, M. Chen, L.-X. Piao, O. Takeuchi, S. Akira, H. Ishikura, and A. Yano. 2003. TLR2 as an essential molecule for protective immunity against *Toxoplasma gondii* infection. *Int. Parasitol.* **15:**1081–1087.

Murray, H. W., C. W. Juangbhanich, C. F. Nathan, and Z. A. Cohn. 1979. Macrophage oxygen-dependent antimicrobial activity. II. The role of oxygen intermediates. *J. Exp. Med.* **150:**950–964.

Murray, H. W., B. Y. Rubin, S. M. Carriero, A. M. Harris, and E. A. Jaffee. 1985. Human mononuclear phagocyte antiprotozoal mechanisms: oxygen-dependent vs oxygen-independent activity against intracellular *Toxoplasma gondii*. *J. Immunol.* **134:**1982–1988.

O'Neill L, A., and A. G. Bowie. 2007. The family of five: TIR-domain-containing adaptors in Toll-like receptor signalling. *Nat. Rev. Immunol.* **7:**353–364.

Pierce, K. L., R. T. Premont, and R. J. Lefkowitz. 2002. Seven-transmembrane receptors. *Nat. Rev. Mol. Cell Biol.* **3:**639–650.

Platanias, L. C. 2005. Mechanisms of type-I- and type-II-interferon-mediated signalling. *Nat. Rev. Immunol.* **5:**375–386.

Reis e Sousa, C., S. Hieny, T. Scharton-Kersten, D. Jankovic, H. Charest, R. N. Germain, and A. Sher. 1997. In vivo microbial stimulation induces rapid CD40L-independent production of IL-12 by dendritic cells and their re-distribution to T cell areas. *J. Exp. Med.* **186:**1819–1829.

Remington, J. S., R. McLeod, P. Thuliez, and G. Desmonts. 2001. Toxoplasmosis, p. 205–346. *In* J. S. Remington and J. O. Klein (ed.), *Infectious Diseases of the Fetus and Newborn Infant*, 3rd ed. W. B. Saunders Co., Philadelphia, PA.

Roach, J. C., G. Glusman, L. Rowen, A. Kaur, M. K. Purcell, K. D. Smith, L. E. Hood, and A. Aderem. 2005. The evolution of vertebrate Toll-like receptors. *Proc. Natl. Acad. Sci. USA* **102:**9577–9582.

Robben, P. M., M. LaRegina, W. A. Kuziel, and L. D. Sibley. 2005. Recruitment of Gr-1+ monocytes is essential for control of acute toxoplasmosis. *J. Exp. Med.* **201:**1761–1769.

Robben, P. M., D. G. Mordue, S. M. Truscott, K. Takeda, S. Akira, and L. D. Sibley. 2004. Production of IL-12 by macrophages infected with *Toxoplasma gondii* depends on the parasite genotype. *J. Immunol.* **172:**3686–3694.

Saavedra, R., M. A. Becerril, C. Dubeaux, R. Lippens, M. J. De Vos, P. Herion, and A. Bollen. 1996. Epitopes recognized by human T lymphocytes in the ROP2 protein antigen of *Toxoplasma gondii*. *Infect. Immun.* **64:**3858–3862.

Saeij, J. P., J. P. Boyle, S. Coller, S. Taylor, L. D. Sibley, E. T. Brooke-Powell, J. W. Ajioka, and J. C. Boothroyd. 2006. Polymorphic secreted kinases are key virulence factors in toxoplasmosis. *Science* **314:**1780–1783.

Saeij, J. P., S. Coller, J. P. Boyle, M. E. Jerome, M. W. White, and J. C. Boothroyd. 2007. *Toxoplasma* co-opts host gene expression by injection of a polymorphic kinase homologue. *Nature* **445:**324–327.

Sayles, P. C., G. W. Gibson, and L. L. Johnson. 2000. B cells are essential for vaccination-induced resistance to virulent *Toxoplasma gondii*. *Infect. Immun.* **68:**1026–1033.

Scanga, C. A., J. Aliberti, D. Jankovic, F. Tilloy, S. Bennouna, E. Y. Denkers, R. Medzhitov, and A. Sher. 2002. Cutting edge: MyD88 is required for resistance to *Toxoplasma gondii* infection and regulates parasite-induced IL-12 production by dendritic cells. *J. Immunol.* **168:**5997–6001.

Scharton-Kersten, T., G. Yap, J. Magram, and A. Sher. 1997. Inducible nitric oxide is essential for host control of persistent but not acute infection with the intracellular pathogen *Toxoplasma gondii*. *J. Exp. Med.* **185:**1–13.

Scharton-Kersten, T. M., T. A. Wynn, E. Y. Denkers, S. Bala, L. Showe, E. Grunvald, S. Hieny, R. T. Gazzinelli, and A. Sher. 1996. In the absence of endogenous IFN-γ mice develop unimpaired IL-12 responses to *Toxoplasma gondii* while failing to control acute infection. *J. Immunol.* **157:**4045–4054.

Schwab, J. C., C. J. M. Beckers, and K. A. Joiner. 1994. The parasitophorous vacuole membrane surrounding intracellular *Toxoplasma gondii* functions as a molecular sieve. *Proc. Natl. Acad. Sci. USA* **91:**509–513.

Serbina, N. V., T. P. Salazar-Mathar, C. A. Biron, W. A. Kuziel, and E. A. Pamer. 2003. TNF/iNOS-producing dendritic cells mediate innate immune defense against bacterial infection. *Immunity* **19:**59–70.

Shapira, S., O. Harb, J. Margarit, M. Matrajt, J. Han, A. Hoffmann, B. Freedman, M. J. May, D. S. Roos, and C. A. Hunter. 2005. Initiation and termination of NFκB signaling by the intracellular protozoan parasite *Toxoplasma gondii*. *J. Cell Sci.* **118:**3501–3508.

Shapira, S. S., K. Speirs, A. Gerstein, J. Caamano, and C. A. Hunter. 2002. Suppression of NF-κB activation by infection with *Toxoplasma gondii*. *J. Infect. Dis.* **185:**S66–S72.

Sibley, L. D., and J. C. Boothroyd. 1992. Virulent strains of *Toxoplasma gondii* comprise a single clonal lineage. *Nature* **359:**82–85.

Sibley, L. D., A. Charron, S. Hakansson, and D. Mordue. 2007. Invasion and intracellular survival by Toxoplasma, p. 16–24. *In* E. Y. Denkers and R. T. Gazzinelli (ed.), *Protozoans in Macrophages*. Landes Bioscience, Austen, TX.

Sibley, L. D., R. Lawson, and J. L. Krahenbuhl. 1985. Phagosome acidification blocked by intracellular *Toxoplasma gondii*. *Nature* **315:**416–419.

Sibley, L. D., D. G. Mordue, C. Su, P. M. Robben, and D. K. Howe. 2002. Genetic approaches to studying virulence and pathogenesis in *Toxoplasma gondii*. *Philos. Trans. R. Soc. Lond.* **357:**81–88.

Sinai, A. P., and K. A. Joiner. 2001. The *Toxoplasma gondii* protein ROP2 mediates host organelle association with the parasitophorous vacuole membrane. *J. Cell Biol.* **154:**95–108.

Singh, S. B., A. S. Davis, G. A. Taylor, and V. Deretic. 2006. Human IRGM induces autophagy to eliminate intracellular mycobacteria. *Science* **313:**1438–1441.

Subauste, C. S., and M. Wessendarp. 2000. Human dendritic cells discriminate between viable and killed *Toxoplasma gondii* tachyzoites: dendritic cell activation after infection with viable parasites results in CD28 and CD40 ligand signaling that controls IL-12-dependent and -independent T cell production of IFN-γ. *J. Immunol.* **165:**1498–1505.

Subauste, C. S., and M. Wessendarp. 2006. CD40 restrains in vivo growth of Toxoplasma gondii independently of gamma interferon. *Infect. Immun.* **74:**1573–1579.

Subauste, C. S., M. Wessendarp, R. U. Sorensen, and L. E. Leiva. 1999. CD40-CD40 ligand interaction is central to cell-mediated immunity against *Toxoplasma gondii*: patients with hyper IgM syndrome have a defective Type I immune response that can be restored by soluble CD40 ligand trimer. *J. Immunol.* **162:**6690–6700.

Sumyuen, M. H., Y. J. Garin, and F. Derouin. 1995. Early kinetics of *Toxoplasma gondii* infection in mice infected orally with cysts of an avirulent strain. *J. Parasitol.* **81:**327–329.

Suss-Toby, E., E. J. Zimmerberg, and G. E. Ward. 1996. Toxoplasma invasion: the parasitophorous vacuole is formed from host cell plasma membrane and pinches off via a fusion pore. *Proc. Natl. Acad. Sci. USA* **93:**8413–8418.

Suzuki, Y., F. K. Conley, and J. S. Remington. 1989. Importance of endogenous IFN-γ for the prevention of toxoplasmic encephalitis in mice. *J. Immunol.* **143:**2045–2050.

Suzuki, Y., M. A. Orellana, R. D. Schreiber, and J. S. Remington. 1988. Interferon-γ: the major mediator of resistance against *Toxoplasma gondii*. *Science* **240:**516–518.

Suzuki, Y., and J. S. Remington. 1988. Dual regulation of resistance against *Toxoplasma gondii* infection by Lyt-2$^+$ and Lyt1$^+$,L3T4$^+$ T cells in mice. *J. Immunol.* **140:**3943–3946.

Taylor, G. A. 2007. IRG proteins: key mediators of interferon-regulated host resistance to intracellular pathogens. *Cell. Microbiol.* **9:**1099–1107.

Taylor, G. A., C. M. Collazo, G. S. Yap, K. Nguyen, T. A. Gregorio, L. S. Taylor, B. Eagleson, L. Secret, E. A. Southon, S. W. Reid, L. Tessarollo, M. Bray, D. W. McVicar, K. L. Komschlies, H. A. Young, C. A. Biron, A. Sher, and G. F. Vande Woude. 2000. Pathogen-specific loss of host resistance in mice lacking the IFN-γ-inducible gene IGTP. *Proc. Natl. Acad. Sci. USA* **97:**751–755.

Taylor, G. A., C. G. Feng, and A. Sher. 2004. p47 GTPases: regulators of immunity to intracellular pathogens. *Nat. Rev. Immunol.* **4:**100–109.

Taylor, S., A. Barragan, C. Su, B. Fux, S. J. Fentress, K. Tang, W. L. Beatty, H. E. Hajj, M. Jerome, M. S. Behnke, M. White, J. C. Wootton, and L. D. Sibley. 2006. A secreted serine-threonine kinase determines virulence in the eukaryotic pathogen *Toxoplasma gondii*. *Science* **314:**1776–1780.

Valere, A., R. Garnotel, I. Villena, M. Guenounou, J. M. Pinon, and D. Aubert. 2003. Activation of the cellular mitogen-activated protein kinase pathways ERK. p38 and JNK during *Toxoplasma gondii* invasion. *Parasite* **10:**59–64.

van Gisbergen, K. P., T. B. Geijtenbeek, and Y. van Kooyk. 2005a. Close encounters of neutrophils and DCs. *Trends Immunol.* **26:**626–631.

van Gisbergen, K. P., M. Sanchez-Hernandez, T. B. Geijtenbeek, and Y. van Kooyk. 2005b. Neutrophils mediate immune modulation of dendritic cells through glycosylation-dependent interactions between Mac-1 and DC-SIGN. *J. Exp. Med.* **201:**1281–1292.

Williams, L., L. Bradley, A. Smith, and B. Foxwell. 2004. Signal transducer and activator of transcription 3 is the dominant mediator of the anti-inflammatory effects of IL-10 in human macrophages. *J. Immunol.* **172:**567–576.

Wilson, C. B., and J. S. Remington. 1979. Activity of human blood leukocytes against *Toxoplasma gondii*. *J. Infect. Dis.* **140:**890–895.

Wilson, C. B., V. Tsai, and J. S. Remington. 1980. Failure to trigger the oxidative metabolic burst by normal macrophages: possible mechanism for survival of intracellular pathogens. *J. Exp. Med.* **151:**328–346.

Yarovinsky, F., and A. Sher. 2006. Toll-like receptor recognition of *Toxoplasma gondii*. *Int. J. Parasitol.* **36:**255–259.

Yarovinsky, F., D. Zhang, J. F. Anderson, G. L. Bannenberg, C. N. Serhan, M. S. Hayden, S. Hieny, F. S. Sutterwala, R. A. Flavell, S. Ghosh, and A. Sher. 2005. TLR11 activation of dendritic cells by a protozoan profilin-like protein. *Science* **308:**1626–1629.

Zimmermann, S., P. J. Murray, K. Heeg, and A. H. Dalpke. 2006. Induction of suppressor of cytokine signaling-1 by Toxoplasma gondii contributes to immune evasion in macrophages by blocking IFN-[gamma] signaling. *J. Immunol.* **176:**1840–1847.

31

Macrophages in Helminth Infection: Effectors, Regulators, and Wound Healers

JUDITH E. ALLEN AND THOMAS A. WYNN

INTRODUCTION

Helminths of Key Public Health Importance

Parasitic worms that infect people and animals cover an enormous phylogenetic spectrum within the animal kingdom. Nematodes (round worms) and platyhelminths (flukes and tapeworms) come from two separate phyla, far more distantly related in terms of phylogeny than, for example, a human is to a sea urchin. However, because of their common ability to infect mammals, all of these parasites are "classified" as helminths. It is quite remarkable that parasitism as a strategy for survival has evolved so many times independently within the different branches of the animal kingdom, most particularly among the nematodes (Blaxter, 2003). This diversity among parasitic species manifests itself not only in fundamental differences in biochemistry and structure, but life histories including places of residence or migration in the mammalian host. There is not an organ in the body that cannot be infected by one of these metazoan invaders, and each parasite uses a diverse range of mechanisms to subvert or utilize the host response to successfully complete their objective: reproductive success. With genome sizes over half the size of our own, the complexity of the potential interactions between host and parasite can seem beyond comprehension.

Helminths of key public health importance illustrate this complexity. The *filarial nematodes* that cause elephantiasis are transmitted during a blood meal taken by a mosquito. The infective larvae migrate into the nearby lymphatics and take up residence, where they mature, mate, and produce microfilaria that circulate in the bloodstream. Many infected individuals appear to be in a state of immunological tolerance and remain relatively asymptomatic. A proportion, however, make vigorous immunological responses to the nematode, resulting in damage to lymphatic vessels. Infection by *schistosome parasites* involves exposure to water containing infected snails. The infectious cercariae literally burrow through the skin, enter the bloodstream, and migrate to the lungs. After a maturation stage in the lung, parasites take up residence in the blood vessels of the gut or bladder, depending on species. Each adult pair of parasites produces hundreds to thousands of eggs per day. Eggs that lodge in the tissues, particularly the liver, gut, and bladder wall, are the main cause of pathology (see "Macrophages as Modulators of the Immune System during Helminth Infection," below). The *gastrointestinal nematodes* represent the most abundant parasites, infecting over a third of the world's human population and a much higher percentage of animals, both wild and domestic. Some of these species develop entirely within the gut, while many such as *Ascaris* and the hookworms have lung-migrating stages. Finally, the *tapeworms*, although the most poorly studied, can lead to some of the most severe diseases caused by helminths, with species that form cysts in the tissues, including the brain and eyes. With a few exceptions, these infections are chronic, with individual adult parasites living for as long as 30 years.

Helminths Activate the Th2 Arm of Immunity

Despite their diversity, metazoan parasites share a common feature, the propensity to induce the Th2 arm of the immune system (Thomas and Harn, 2004; Jankovic et al., 2006). Historically, before the knowledge that CD4$^+$ T-helper cells could be polarized into lineages with distinct cytokine secretion profiles, helminth infection was associated with elevated eosinophils, mast cells, and immunoglobulin E (IgE). We now know that all of these features rely on cytokines produced by Th2 cells, which include interleukin 4 (IL-4), IL-5, IL-9, IL-10, IL-13, IL-21, and IL-25, to name just a few. For example, eosinophils require IL-5 for their development and maturation, while mast cell development is regulated by IL-3 and IL-9. B-cell isotype switching toward IgE production requires IL-4 and IL-13 and is antagonized by IL-21 (Pesce et al., 2006). Perhaps because circulating numbers or physical appearance did not

---

Judith E. Allen, Institutes of Evolution, Immunology and Infection Research, University of Edinburgh, Edinburgh EH9 3JT, United Kingdom. Thomas A. Wynn, Immunopathogenesis Section, National Institute of Allergy and Infectious Diseases, National Institutes of Health, Bethesda, MD 20892-8003.

change significantly on helminth infection, monocytes/macrophages were not typically considered when immune responses to helminths were assessed. Although the role of Th2 immunity in helminth infection has not always been apparent, the evidence from animal models strongly suggests that Th2 responses mediate worm killing or expulsion (Urban et al., 1992). It has recently been appreciated that, to avoid destruction, helminths are particularly adept at stimulating the regulatory arm of the immune system (Maizels et al., 2004). Unrestricted Th2 immunity can be physically damaging and is associated with conditions such as asthma and fibrosis (see below) as well as acute helminth infections in an inappropriate host (such as Katayama fever induced by schistosomes). Thus, it is to both the parasite's and the host's benefit to modify this response, and indeed most chronic helminth infections are characterized by what has been termed a "modified Th2 response," in which Th2 immune responses are dampened by a host of regulatory mechanisms, including regulatory T cells (Tregs), decoy receptors, and cross-regulatory Th1 responses (Maizels and Yazdanbakhsh, 2003). This ability of helminths to hijack regulatory pathways is being considered for the treatment of autoimmune diseases and asthma where an overactive immune system is to blame for disease. Thus, understanding the mechanistic details of how worms modulate immunity is of enormous therapeutic potential. However, the regulatory mechanisms and pathways induced by helminths are vast and likely to differ with each host-parasite combination.

Macrophage Activation by Type 2 Cytokines

See chapter 20 for more in-depth coverage.

Macrophages activated by Th1 cytokines (gamma interferon [IFN-γ], tumor necrosis factor alpha [TNF-α]) and other proinflammatory mediators are able to effectively destroy microbial pathogens. This process is central to our survival, and thus this "classical" activation pathway is well studied and reasonably well understood. In contrast, until recently the importance and role of macrophages in type 2 immunity have been largely ignored. Most textbooks, even today, will show the effector arm of Th2 immunity as mediated by IgE, eosinophils, mast cells, and goblet cells as described above, with little mention, if any, of macrophages. In 1992, Siamon Gordon and colleagues noted that treatment of macrophages in vitro with IL-4 led to an alternative activation state. Subsequently, IL-13 was shown to have the same effects as IL-4 (Stein et al., 1992; Doyle et al., 1994). This in vitro phenotype was characterized by elevated class II levels, the mannose receptor (CD206), and arginase activity (Stein et al., 1992; Doyle et al., 1994; Modolell et al., 1995; Schebesch et al., 1997; Goerdt and Orfanos, 1999; Gordon 1999, 2003).

In the early 2000s, two groups looked at gene expression profiles of macrophages recruited to Th2-dominated infection sites to assess the phenotype of these cells in vivo. These studies supported the in vitro work by noting significant upregulation of arginase but also identified two new highly abundant gene products that were expressed by macrophages in an IL-4-dependent manner (Raes et al., 2001; Loke et al., 2002). It appeared that in vivo alternatively activated macrophages (AAMφ) represented a truly novel phenotype characterized by the production of two secreted proteins: a chitinase-like molecule, Ym1, and a small cysteine-rich molecule now known as RELMα, or FIZZ1. The closely related proteins Ym2, acidic mammalian chitinase (AMCase) (Boot et al., 2001), and RELMβ/FIZZ2 (Artis et al., 2004) were subsequently identified, and these families of chitinase-like molecules and fizz proteins (ChaFFs) were found to be a typical feature of nematode infection (Nair et al., 2005), although not all were produced by macrophages. It soon became evident that, along with arginase, Ym1 and RELMα were reliable markers for alternative activation in a variety of noninfection settings (Webb et al., 2001; Welch et al., 2002; Sandler et al., 2003; Zimmermann et al., 2003; Liu et al., 2004b; Misson et al., 2004) as well as virtually all helminth infections studied to date (Reyes and Terrazas, 2007). Although first identified in vivo, Ym1 and the closely related Ym2, as well as RELMα, have been shown to be directly induced by IL-4 and IL-13 in vitro (Raes et al., 2001; Nair et al., 2003), and several studies have defined their dependence on the IL-4 receptor and STAT6 (Webb et al., 2001; Welch et al., 2002; Liu et al., 2004b) in vivo (Rutschman et al., 2001). The actual functions of Ym1/2 and RELMα have yet to be fully elucidated, but many properties have been attributed to them, which is helping to unravel the roles macrophages play during helminth infection and other Th2 settings (discussed below).

In addition to the expression of Ym1/2, RELMα, and arginase, AAMφ express a range of other proteins (Noel et al., 2003) and exhibit several IL-4/IL-13-dependent features that are providing valuable insight into their potential activities in vivo. These include the IL-4/IL-13-dependent ability to block the proliferation of cells in coculture and IL-4/IL-13-dependent reduction in both proinflammatory chemokines (Loke et al., 2002) and prostaglandins such as prostaglandin E_2 (Mosca et al., 2007), as well as the regulation of antitumor immunity (Van Ginderachter et al., 2006b), nutrient homeostasis (Odegaard et al., 2007), and nitric oxide (NO) production (Rutschman et al., 2001). In the context of helminth infection, macrophages will be exposed to many signals beyond IL-4 and IL-13 that may modulate or enhance the alternatively activated phenotype. Consistent with this, the distinctive rounded shape of macrophages during nematode infection is not lost in IL-4/IL-13-deficient animals even in the face of a switch to a more classically activated macrophage (CAMφ) phenotype (Nair et al., 2005). IL-10 (Modolell et al., 1995), glucocorticoids (Heasman et al., 2004), IL-21 (Pesce et al., 2006), and immune complexes (Mosser, 2003), among others, are likely to be important additional signals to the macrophage during chronic Th2-mediated inflammation. These ligands may act synergistically with IL-4 to mediate anti-inflammatory functions of AAMφ, consistent with the ability of IL-4 to upregulate PPARγ and the glucocorticoid receptor (GR) (Henson, 2003). IL-10, which typically "deactivates" macrophages (Gordon, 2003), can synergize with IL-4 for enhanced levels of arginase production (Kropf et al., 2005). Although macrophages have been usefully classified as AAMφ and CAMφ (also called M1 and M2 macrophages in some publications), the range of actual phenotypes is likely to be much broader, with these designations representing two ends of a wide spectrum.

The human in vivo equivalents of these divergent phenotypes discovered in mice have yet to be fully elucidated. The picture is complicated by an absence of human Ym1/2 genes, but several proteins of unknown function have the potential to be the functional homologues of murine chilectins (Elias et al., 2005; Bargagli et al., 2007). Importantly, most of the human molecules have yet to be investigated in the context of Th2 cytokine regulation. Recently

Kzhyshkowska identified a human gene expressed in macrophages, SI-CLP, which has sequence similarity with Ym1 and is upregulated by IL-4 (Kzhyshkowska et al., 2006). The controversy about the human AAMϕ equivalent is further complicated by the apparent lack of arginase in human macrophages, although expression in neutrophils has been identified (Munder et al., 2005) and arginase expression in macrophages can be induced by IL-4 under the right conditions (Erdely et al., 2006). This suggests that the failure to detect arginase may be due to technical limitations on the sources of human cells, as seen with the controversy over inducible NO synthase (Fang and Nathan, 2007).

Macrophages Are Abundant in Helminth Infection Sites

In the context of helminth infection, macrophages get little coverage in most immunology or parasitology textbooks. However, they frequently represent the most abundant cell type in many if not most immune responses to metazoan parasites (MacDonald et al., 2003; Ramesh et al., 2007). A recent review by Reyes and Terrazas (2007) extensively covers the many different helminth animal models in which AAMϕ have been described. Although direct parallels in terms of specific gene products have yet to be made between mice and humans, there is little question that human macrophages exhibit distinct properties on exposure to Th2-type cytokines (Gordon, 2003) and, as in the mouse models, macrophages are abundant at the site of helminth infection (Edgeworth et al., 1993). We can presume that, in these similarly polarized contexts, macrophages activated by Th2 cytokines are playing similar roles. Yet, despite their abundance, we remain remarkably ignorant about the contribution of AAMϕ to the outcome of most helminth infections. The remainder of this review will consider the potential roles of macrophages during helminth infection and ask the question: Are they effectors, regulators, healers, or all of the above?

MACROPHAGES AS ANTIHELMINTHIC EFFECTORS

Evidence that Macrophages Can Kill

Prior to the knowledge that macrophages activated by Th2 cytokines represented a novel phenotype, it was generally assumed that the large numbers of macrophages associated with dying worms were killing via the release of oxygen or nitrogen radicals (James et al., 1982; Mackenzie et al., 1985; Oswald et al., 1992; Taylor et al., 1996; Thomas et al., 1997; Allen and Loke, 2001). Indeed, it has been demonstrated in vitro that thioglycollate-elicited murine peritoneal macrophages kill schistosomula of *Schistosoma mansoni* in vitro by an L-arginine-dependent mechanism that involves the production of reactive nitrogen oxides (Oswald et al., 1992). Similarly, J774 cells classically activated by lipopolysaccharide and IFN-γ were shown to kill larval forms of the filarial nematode *Brugia malayi* (Thomas et al., 1997). However, with our new understanding that AAMϕ are the predominant cell type associated with helminth infection, we have to ask whether AAMϕ (or CAMϕ) and/or their associated products have the capacity to kill large extracellular parasites in vivo (Coulson et al., 1998; James et al., 1998) or whether they play more indirect functions in this regard, for example, parasite disposal rather than destruction.

The most direct evidence to date that AAMϕ have antiworm effector function comes from the study of Anthony et al. (2006), who demonstrated that the depletion of AAMϕ from the intestines of mice abrogated protective immunity seen during secondary infection with *Heligmosomoides polygyrus*. Macrophage depletion studies in a wider range of helminth infection will inform us as to whether this is the principal role for macrophages in a wider range of metazoan infections. Arginase was implicated in the *H. polygyrus* study (see below), and indeed a role for arginase activity and arginine metabolism in the survival of schistosomes was proposed over 40 years ago (Senft, 1966). However, the actual mechanisms by which macrophages activated by type 2 cytokines target helminth parasites remain unclear.

Th2-Inducible Macrophage-Associated Proteins Implicated in Antiworm Effector Function

One way to address the possible function of AAMϕ during helminth infection is to consider the properties of the most abundant proteins produced by these cells. We thus need to ask whether these have cytotoxic activity against helminths.

The most abundant protein produced by AAMϕ during helminth infection is the lectin Ym1 (Loke et al., 2002). Ym1 is a member of a family of chitinases, some of which have chitin-cleaving ability, while some do not. Ym1 lacks this enzymatic activity but presumably can still bind chitin. Some but not all helminths contain chitin and not necessarily at every stage of development (Foster et al., 2005). Whether this potential to bind the parasite translates into antiparasite effector function has yet to be demonstrated. Ym1 has a remarkable capacity to form crystals under appropriate pH conditions (Guo et al., 2000; Harbord et al., 2001), and chitin can be found in the gut lining of some nematodes. One possible mechanism by which Ym1 may act is to form damaging crystals within the parasite. Consistent with this, the ability of Ym1 to induce cellular damage has been observed during fungal infections of the lung (Huffnagle et al., 1998). The ability of Ym1 to interact with a variety of host sugars (Chang et al., 2001) raises the possibility that Ym1 may promote the deposition of host matrix components interfering with parasite function or "trapping" the parasite in a matrix that limits mobility or normal function.

Ym1 is not the only lectin produced in abundance by AAMϕ, and indeed, lectin production appears to be a characteristic feature of alternative macrophage activation. The mannose receptor was the first identified marker of alternative activation (Stein et al., 1992) and is typically upregulated on macrophages during helminth infection (Noel et al., 2003). Raes and colleagues (2003) have identified IL-4/IL-13-dependent upregulation of two members of the mouse macrophage galactose-type C-type lectin gene family, mMGL1 and mMGL2, during *Taenia crassiceps* infection. With the exception of a study providing evidence that the mannose receptor can bind schistosome egg components (Linehan et al., 2003), studies to identify whether these lectins bind specific helminth carbohydrates have not been performed. Even if these interactions are identified, they may not reveal their function. Lectin binding could promote cell recruitment and matrix deposition on the worm, or bind to parasite regulatory molecules and thus prevent downregulation of host responses. Considerably more research is needed on both the identification of helminth carbohydrates and the function of the mammalian lectins before these questions will be answered.

RELMα is also a highly abundant macrophage-derived protein produced during helminth infection (Loke et al., 2002; Nair et al., 2005; Sandler et al., 2003). Although there is no direct evidence that RELMα has antiparasite effector properties, there is evidence that the very closely related RELMβ molecule, which is produced by epithelial cells of the intestine and is similarly IL-4 inducible, can interfere with the chemosensory function of the nematode, *Trichuris muris* (Artis et al., 2004).

The most compelling data thus far come from arginase inhibition studies in which inactivation of this AAMϕ-derived enzyme reduces expulsion of a gastrointestinal nematode (Anthony et al., 2006). The mechanism by which arginase is acting against the parasite is currently unclear. One possibility might be that arginase is depriving the worm of a needed resource, arginine. Perhaps more likely, interference with the arginine metabolism pathway may have many effects on both immune and nonimmune cells. In particular, inhibition of arginase will lead to increases in NO production that may modulate other antiworm effector pathways. Of direct relevance are data demonstrating that macrophage-derived arginase mediates killing of schistosomula (Olds et al., 1980). Future research needs to focus on the actual mechanism of arginase activity against helminths, as this has broad implications.

Despite direct or indirect evidence for macrophage-derived products in parasite killing, it is apparent that these proteins alone are not sufficient. In a mouse model of filarial infection, depletion of Treg cells leads to parasite killing but does not alter macrophage activation status (Taylor et al., 2005). Further, in this model system and others, there is evidence of large numbers of AAMϕ and their products such as arginase in the presence of healthy living worms (Hesse et al., 2001). So although Ym1/2 and RELMα/β and arginase may contribute to parasite killing or expulsion, they are not alone sufficient for killing. Additionally, despite the overwhelming evidence that Th2 responses are involved in antihelminth immunity, there are several examples in which the Th2 response itself appears insufficient and that Th1 immunity, potentially with the involvement of CAMϕ, may be needed or contribute to worm killing (Hoffmann et al., 1999; Saeftel et al., 2001).

When Alternative Activation Is Not Enough

Despite abundant evidence that Th2 immunity is acting to limit helminth infection and, by association, AAMϕ, there are exceptions to the rule. The *Litomosoides sigmodontis* mouse model has allowed a detailed investigation of the parameters required for control of filarial nematodes, parasites that live in the tissue rather than the gut (Allen et al., 2008). These studies found that killing of the larval stages requires Th2 immunity, although the role of macrophages has yet to be addressed. In contrast, destruction of the adult stage is enhanced by Th1 immunity, which can act synergistically with type 2 cytokines in this setting (Saeftel et al., 2003). In IL-4Rα-deficient animals, macrophages at the site of infection produce high levels of NO and no arginase (J. E. Allen, unpublished work). Adult worm killing is accelerated in their presence, strongly suggesting a role for CAMϕ in nematode destruction. Importantly, in these same mice there are extraordinarily high levels of blood-circulating first-stage larvae. Thus, even within a single infection, different arms of the immune system are acting at differences stages of infection.

Another exception to the Th2 worm-killing paradigm is the tapeworm, *T. crassiceps*, a model for the life-threatening helminth infection neurocysticercosis. These cestode parasites induce Th2 immunity, but this appears to confer susceptibility rather than resistance (Rodriguez-Sosa et al., 2004). Using inhibitors of NO, these investigators have shown that NO is the critical macrophage-derived factor needed to kill these parasites (Alonso-Trujillo et al., 2007). Finally, in the case of schistosomes, the parasites appear to be resistant to both arms of the immune response and can successfully establish infections in animals that develop highly polarized Th1- and Th2-type responses (Hoffmann et al., 2000). These exceptions raise important questions regarding the role of macrophages in helminth infection, as there are clear cases where classical activation is required for parasite control. However, this does not appear to be a typical pathway induced in response to most helminth infections, perhaps due to the potential for host damage. What are the factors that determine whether classical activation or alternative activation is the most appropriate response to a metazoan parasite?

Granulomas—Encapsulation Rather than Direct Killing?

One of the remarkable aspects of studying helminth immunity is that despite detailed knowledge of the cytokine requirements, we have only limited understanding of how worms that reside in the tissues are killed by the host immune system. In many cases, it may be that killing the worm proves too great a task and the "sensible" approach of the immune system is to simply wall it off. This approach is certainly not unique to large multicellular parasites since mycobacterial and other microbial infections can be contained by a granuloma rather than by actual destruction of the pathogen. In granuloma formation, it is clear that macrophages are playing a central role.

Macrophage-containing granulomas are a feature of many diverse helminth infections, ranging across parasite phyla and anatomical location. These include nodules in the skin of people and animals with onchocerciasis (river blindness) (Henson et al., 1979; Mackenzie et al., 1985), the granuloma around the egg lodged in the tissues of individuals with schistosomiasis (Sandor et al., 2003), and the cyst encapsulating the larval stages of the tapeworm *Taenia solium* that leads to irreversible brain damage (Restrepo et al., 2001). In all these cases, granuloma formation is, at least in part, driven by Th2 cytokines, particularly IL-4 and IL-13 (Chiaramonte et al., 1999a, 1999b). An intriguing feature of macrophages involved in granuloma formation is their propensity for cell fusion leading to multinucleate cells. This fusion is an IL-4-dependent process (Helming and Gordon, 2007) and is characteristic of all macrophages that are deposited on large foreign bodies. This suggests that IL-4-dependent immunity and the associated macrophage activation process is an ancient one akin to that described by Metchnikoff himself when he stuck a thorn into the starfish larvae and made the first discovery of macrophages.

AAMϕ not only are thought to represent a major component of helminth-associated granulomas, but also are key cells in orchestrating the process. For example, recent data suggest that AAMϕ are responsible for the recruitment of eosinophils, a predominant feature of Th2-induced granulomas (Voehringer et al., 2007). A more detailed discussion of the role of AAMϕ in granuloma formation is discussed below with their role in pathology. When encapsulation results in death, it is difficult to determine whether death occurs before or after granuloma formation (Wertheim et al., 1987). The granulomatous structure may simply be a clearance mechanism for a parasite too big to be phagocytosed, or the encapsulation process may deprive the parasite of needed resources and lead to its death. However, most data would suggest that the worm has at least been weakened prior to encapsulation, suggesting that other effector mechanisms are contributing, including antibodies and eosinophils (James and Colley, 1978; Chandrashekar et al., 1985; Hoffmann et al., 1999). Whether macrophage-derived products—for example, resistin family members that interfere with chemosensory function (Artis et al., 2004)—also contribute has yet to be elucidated.

AAMϕ IN THE PATHOLOGY OF HELMINTH INFECTION

Schistosomiasis as a Model for Granulomatous Pathology

The role of macrophages in granuloma formation and control not only is important in the consideration of parasite containment, but also is one of the most important processes associated with the pathology of helminth infection. This is best illustrated with schistosomiasis, a disease in which granuloma formation leads to severe morbidity in several million people worldwide. Similar pathological processes may occur during other helminth infections, such as the nodules in people with river blindness or the fibrotic tissue associated with lymphatic filariasis and elephantiasis. However, only in schistosomiasis has this process been extensively studied, due in large part to the availability of good animal models of pathology. In the murine model of *S. mansoni*, parasites reside in mesenteric veins where they lay hundreds of eggs per day 4 to 5 weeks postinfection. Some eggs are trapped in the microvasculature of the liver and gut, where they induce a vigorous granulomatous response (Cheever and Andrade, 1967). Subsequently, fibrosis and portal hypertension develop. Consequently, much of the symptomatology of schistosomiasis is attributed to the egg-induced granulomatous response.

The Pathology-Reducing Effects of NOS-2 and NO

Studies aimed at dissecting the respective roles of Th1 and Th2 CD4$^+$ T-cell-associated cytokines in granuloma formation showed that the granulomatous response evolves from an early Th1 to a sustained and dominant Th2-type cytokine response (Grzych et al., 1991; Pearce et al., 1991; Vella and Pearce, 1992; Wynn et al., 1993), with 20 to 30% of the granulomatous lesions composed of AAMϕ. The importance of Th2 cells to the pathogenesis of schistosomiasis was confirmed in experiments in which mice vaccinated with egg antigen plus IL-12 to induce an egg-specific CD4$^+$ Th1 response upon subsequent infection developed smaller lesions and less severe fibrosis than did nonsensitized Th2-polarized controls (Wynn et al., 1994, 1995; Hoffmann et al., 1998; Boros and Whitfield, 1999). The decreased pathology was associated with a reduced Th2 response, markedly fewer AAMϕ, and increased inducible NO synthase (NOS-2) expression.

Numerous studies showed that NOS-2 is an important regulator of Th1 responses. In one study, NOS-2 triggered immune suppression and significantly reduced the adjuvant effects of IL-12. NO has also been shown to suppress Th1 cell development in vitro. Because IL-12 promotes the differentiation of Th1 cells, and IFN-γ and TNF-α upregulate NOS-2 expression, it was thought that NOS-2 might antagonize the Th2-suppressing activity of IL-12 and, consequently, its full antipathology effect in schistosomiasis. Indeed, it was hypothesized that the efficacy of IL-12 as a Th2-reducing agent might be improved if NOS-2 activity was neutralized, because the antiproliferative effects of NO on Th1 cells would be eliminated as well as its potentially tissue-destructive and proinflammatory activities (Hesse et al., 2000). Experiments with NOS-2-deficient mice showed that while relatively normal CD4$^+$ Th1 cell responses were established in egg/IL-12-sensitized NOS-2$^{-/-}$ mice, they were incapable of controlling the egg-induced inflammatory response. Indeed, they displayed more than an eightfold increase in granuloma size and completely failed to inhibit the development of hepatic fibrosis as compared with egg/IL-12-sensitized wild-type mice. These data demonstrated that while normal or possibly improved Th2-to-Th1 immune deviation occurred in egg/IL-12-sensitized NOS-2$^{-/-}$ mice, the downstream anti-inflammatory and antifibrotic effects of the egg-specific Th1 response were completely lost in the absence of NOS-2. As such, these data revealed a critical host-protective function for NOS-2. Immune deviation strategies have been proposed for other Th2-mediated diseases including allergy and asthma. The results with NOS-2$^{-/-}$ mice in the schistosomiasis model suggest that the ultimate success of these strategies may rely not only on the successful establishment of a Th1-dominant response, but also on the simultaneous and efficient activation of NOS-2 expression in CAMϕ and other important effector cells. These studies were the first to point to macrophages and other NOS-2-expressing cells as key regulators of pathology in schistosomiasis.

Contrasting Roles for NOS-2 and Arginase-1 in Granuloma Formation

The failure to inhibit granuloma formation and hepatic fibrosis in egg/IL-12-sensitized NOS-2-deficient mice revealed a previously unrecognized protective function for NO and CAMϕ. This unexpected finding suggested that disturbances in the urea/L-arginine biosynthetic pathway might directly or indirectly regulate the pathogenesis of schistosomiasis and other Th2-associated diseases. Numerous studies have suggested that macrophages and fibroblasts are key effector cells in the pathogenesis of fibrosis. Thus, although the phenotype of CD4$^+$ T cells is clearly important (Wynn et al., 1995; Chiaramonte et al., 1999a), their primary function may be to control the activation and recruitment of macrophages, fibroblasts, and other NOS-2-expressing cells (Hesse 2000; Hesse 2001). Th1 cytokines activate NOS-2 expression in "classically activated" macrophages, whereas the Th2 cytokines IL-4, IL-13, and IL-21 preferentially stimulate arginase-1 (Arg-1) activity in AAMϕ (Munder et al., 1998; Gordon, 2003). Interestingly, dendritic cells (DCs) (Munder et al., 1999) and fibroblasts (Witte, 2002) show a similar pattern of NOS-2 and Arg-1 expression when stimulated with Th1 and Th2 cytokines. L-Arginine serves as the substrate for both enzymes, with NOS-2 generating L-hydroxyarginine, L-citrulline, and NO, and arginase-1 promoting urea and L-ornithine production. L-Ornithine serves as the substrate for two additional enzymes, ornithine decarboxylase (ODC) and ornithine aminotransferase (OAT), which generate polyamines and L-proline, respectively. Because polyamines are critical for cell growth and proline serves as a substrate for collagen synthesis, ODC and OAT are both believed to be important regulators of tissue repair (Hesse et al., 2001). Indeed, early studies in the schistosomiasis model showed that the development of hepatic fibrosis is highly dependent on the conversion of L-ornithine to proline (the basic building block of collagen), a process regulated by arginase. Therefore, the preferential activation of Arg-1 versus NOS-2 in macrophages and fibroblasts has been proposed as a possible explanation for the potent profibrotic activity of IL-13 (Chiaramonte et al., 1999a) and antifibrotic activity of IFN-γ (Wynn et al., 1995; Hesse et al., 2000). Because the NOS-2$^{-/-}$ mice treated with IL-12 were deficient in the arginase antagonist L-hydroxyarginine, the arginase biosynthetic pathway may have been inefficiently blocked, which could explain the failure to reduce collagen deposition in these mice. These observations suggest that macrophages and other NOS-2/arginase-expressing cells play critical roles in the pathogenesis of schistosomiasis.

The Arg-1/NOS-2 pathway has been investigated in detail in the schistosomiasis model of fibrosis. In these studies, schistosome eggs were initially compared with *Mycobacterium avium* (a Th1-inducing pathogen) for their ability to stimulate Arg-1 and NOS-2 activity in vivo (Hesse et al., 2001). In addition, several cytokine-deficient mice were used to determine whether Th1 and Th2 cytokines differentially regulate NOS-2 and Arg-1 expression in vivo. In agreement with previous in vitro studies, schistosome eggs preferentially stimulated Arg-1 expression, while *M. avium* triggered a dominant NOS-2 response in the granulomatous tissues, which was consistent with their ability to promote Th2- and Th1-polarized CD4$^+$ T-cell responses, respectively. Studies conducted with both knockout mice (Hesse et al., 2001) and DNA microarrays (Sandler et al., 2003; Zimmermann et al., 2003) confirmed that Arg-1 expression was highly associated with IL-4, IL-13, and STAT6 activity. More importantly, however, these studies established a strong link between arginase activity and the development of fibrosis, since the schistosome granulomas were associated with extensive fibrosis while the mycobacterium lesions were much less fibrotic. Nevertheless, to directly test the function of Arg-1, a variety of approaches were used to inhibit the Arg-1/NOS-2 pathways in vivo, including NO and ODC inhibitors, as well as NOS-2 knockout mice (Numaguchi et al., 1995; Hogaboam et al., 1998; Hesse et al., 2000, 2001). The combined results from these studies suggested that NOS-2 inhibits fibrosis while Arg-1 activity exacerbates the development of the disease. Th2 cytokine-stimulated macrophages also produced large quantities of proline through an Arg-1-dependent mechanism (Hesse et al., 2001), suggesting a possible paracrine role for AAMφ in the promotion of collagen synthesis by fibroblasts. Because fibroblasts also express high levels of Arg-1 when activated (Witte et al., 2002), direct activation of the Arg-1 pathway in fibroblasts may serve as an additional mechanism to augment collagen deposition in tissues. Regardless of the exact mechanisms involved, these findings provide strong support for pharmacologically targeting arginine metabolism in the treatment of fibrosis and other chronic Th2-associated diseases. They also provide a likely explanation for the anti- and profibrotic activities of Th1 and Th2 cytokine responses, respectively.

Arginine Transporters in the Regulation of Macrophage Effector Function

Since extracellular L-arginine is required for sustained NO and L-ornithine production (Closs et al., 2000), mechanisms controlling L-arginine transport are believed to critically regulate NOS-2 and Arg-1 activity. Among the transport systems that facilitate L-arginine uptake, system y$^+$ is considered to be the major L-arginine transporter in most cells and tissues (MacLeod et al., 1994). Encoded by the solute carrier 7a1-3 (*Slc7a1-3*) family of genes, y$^+$ is an Na$^+$-independent high-affinity amino acid transport system. CAT2 is the most dynamically regulated of the three transporters, with CAT1 operating as the product of a constitutively expressed "housekeeping" gene and CAT3 expressed primarily in the brain (MacLeod, 1996; Hosokawa et al., 1997). Several proinflammatory mediators including lipopolysaccharide regulate the expression of CAT2; thus, it is thought to function as the key L-arginine transporter during inflammatory responses. Studies with CAT2-deficient mice showed that sustained NO production in macrophages is dependent on CAT2 (Nicholson et al., 2001). Thus, it appears to be the essential L-arginine transporter in macrophages, at least in classically activated cells. Surprisingly, however, while CAT2 was studied extensively in the context of NOS-2 regulation (Nicholson et al., 2001), few studies have addressed its role in regulating Arg-1 activity.

To dissect the function of the Slc7a2 gene in vivo, CAT2$^{-/-}$ mice were infected with either *S. mansoni* or *Toxoplasma gondii*, pathogens that induce highly polarized Th2 and Th1 responses, respectively (Wynn, 2004; Gazzinelli and Denkers, 2006). When infected with the Th1-inducing pathogen *T. gondii*, the CAT2$^{-/-}$ mice were significantly more susceptible, succumbing to the infection at a rate similar to NOS-2$^{-/-}$ mice (Wynn, 2007). The increased susceptibility was attributed to an attenuated NO response, which led to uncontrolled parasite replication. After infection with the Th2-inducing pathogen *S. mansoni*, the CAT2$^{-/-}$ mice developed granulomas that were three to four times larger than in the wild type, and hepatic fibrosis (a feature of severe disease) was significantly exacerbated (Kaplan et al., 1996; Chiaramonte et al., 1999a; Jankovic et al., 1999; Fallon et al., 2000), indicating a general worsening of Th2-associated pathology in the absence of CAT2. The pathological changes in the CAT2$^{-/-}$ mice were also associated with increased arginase activity in fibroblasts and AAMφ, suggesting that CAT2 functions as a negative regulator of Arg-1 activity. Thus, CAT2 appears to promote NOS-2 function while antagonizing arginase activity. As such, these studies identify CAT2 as a powerful regulator of Th1 and Th2 effector responses, which may have major implications for a variety of infectious and inflammatory diseases.

TISSUE DAMAGE REPAIR BY MACROPHAGES DURING HELMINTH INFECTION

AAMφ as Central Players in Tissue Repair

Granuloma formation, whether as a host-protective response or a host-damaging response, is closely related to the process of wound healing. Indeed, fibrosis, the major cause of pathology during schistosome infection, is an inevitable consequence of tissue injury if the repair process is not fully or appropriately controlled. Macrophages have a well-documented and essential role in all stages of tissue repair (Wilson et al., 2004; Duffield et al., 2005; Eming et al., 2007a), and AAMφ are described in the literature as having wound-healing properties (Goerdt and Orfanos, 1999; Gordon, 2003). However, there has been little direct in vivo evidence for AAMφ in tissue repair except for their role in fibrosis described above. Certainly, the mainstream wound-healing literature makes little if any reference to type 2 cytokines or AAMɸ. Nonetheless, the evidence coming from helminth models is that macrophage activation by IL-4 and IL-13 may be a critical part of the tissue repair response or possibly involved in the resolution of the wound-healing response (Wynn, 2007). This was recently illustrated when the AAMɸ markers arginase, Ym1, and RELMα were induced in an IL-4/IL-13-dependent manner following a sterile incision in the peritoneal wall (Loke et al., 2007). Alternative activation was short-lived in the absence of further Th2 stimulus, and the presence of a helminth parasite was required for a sustained response. Evidence that AAMɸ have tissue repair as a primary role is supported by the growing data that the three most abundant proteins produced by AAMɸ (Loke et al., 2002) are associated with injury (Hung et al., 2002; Witte and Barbul, 2003; Eming et al., 2007b; Loke et al., 2007) and repair (Sandler et al., 2003). As described above, arginase has well-documented roles in wound healing, in part through promoting collagen deposition and thus wound strength (Witte and Barbul, 2003). RELMα has angiogenic and vasoconstrictive properties and can promote the differentiation of myofibroblasts (Teng et al., 2003; Liu et al., 2004a), while Ym1 may assist extracellular matrix deposition and tissue remodeling (Chang et al., 2001). This raises fundamental questions with regard to the specific functions of AAMɸ during helminth infection and tissue repair responses in general.

Repair of Gut and Lung Damage during Helminth Infection

The association of helminths with Th2 responses and AAMɸ may in part be explained by the requirement for rapid tissue repair in many infection settings. The propensity of helminth parasites to induce potentially lethal tissue damage may have provided sufficient evolutionary pressure for the development of a worm-specific tissue repair process (Graham et al., 2005). Hookworm parasites penetrate the gut wall to feed, while schistosomes use proteolytic enzymes to enter the skin and pass through the gut wall, and larval forms of both parasites traverse the vasculature of the lungs. Lesions in the skin or gut could lead to sepsis unless repair is rapid and effective. Consistent with this hypothesis, AAMɸ-deficient mice die of fatal sepsis after *S. mansoni* egg deposition (Herbert et al., 2004). Similarly, while lung-migratory phases of many different helminth parasites can lead to hemorrhage and loss of tissue integrity, repair is rapid and long-term sequelae are rare in animals or humans.

A dramatic example of repair is seen during murine *Nippostrongylus brasiliensis* infection. Infective larvae migrate from the skin via the vasculature to the lung, and most larvae pass through the lungs in 48 to 72 h. Larval migration causes severe pulmonary damage with leakage of serum proteins into the lavage fluid. This is associated with a profound inflammatory response that begins to resolve by 8 days, and by 2 weeks the lung has undergone a remarkable level of repair (Keir et al., 2004). Maximal Th2 responses occur well after the parasite has left the lung (Harris et al., 1999; Tomita et al., 2000; Voehringer et al., 2004), and Ym1/RELMα levels in the lungs of *N. brasiliensis*-infected mice continue to increase for over 2 weeks (Nair et al., 2005; Reece et al., 2006; Reese et al., 2007). This suggests that Th2 immune responses in the lung have as much to do with repair or "cleaning up" as with antiparasitic effector function. Despite elegant studies of Th2-mediated inflammation in the early lung-migratory phase (Shinkai et al., 2002; Voehringer et al., 2004), few studies have focused on aspects of tissue repair, a process that typically requires macrophages (Duffield et al., 2005). Although *N. brasiliensis* expulsion from the gut requires IL-13, it does not depend on AAMɸ (Herbert et al., 2004), suggesting these cells have other roles and raising the possibility that AAMɸ are required for effective lung repair following *N. brasiliensis* migration.

The study of macrophages in wound healing has implications far beyond helminth infection, including tissue remodeling associated with asthma and pulmonary fibrosis. Additionally, several recent studies have demonstrated that AAMɸ are associated with tumor growth and metastasis (Liu et al., 2003; Biswas et al., 2006; Rauh et al., 2005; Sinha et al., 2005). Indeed, Ym1, FIZZ1, and arginase were the most prominent markers in tumor-infiltrating F4/80[+] macrophages (Van Ginderachter et al., 2006a), reflecting a phenotype nearly identical to that observed during nematode infection (Loke et al., 2002). This may in part reflect the ability of AAMɸ to act competitively against tumoricidal CAMɸ (Liu et al., 2003), but it also suggests that tumors exploit the evolutionary response to metazoan parasites as a means to build the extracellular matrix and angiogenesis necessary for tumor growth and survival. Understanding the regulation and function of AAMɸ during helminth infection may provide valuable insight into tumorigenesis.

MACROPHAGES AS MODULATORS OF THE IMMUNE SYSTEM DURING HELMINTH INFECTION

T-Cell Hyporesponsiveness: a Cardinal Feature of Helminth Infection

Immune downregulation is a cardinal feature of helminth infection and has been extensively documented in both human disease and animal models (Maizels and Yazdanbakhsh, 2003). A remarkable finding has been the loss of antigen-specific in vitro T-cell responses in infection, a major interest in the field (Harnett and Harnett, 2006) because it correlates with both reduced pathology and a failure of the immune system to kill the parasite. T-cell hyporesponsiveness has been extensively documented during helminth infection (Yazdanbakhsh, 1999; Maizels and Yazdanbakhsh, 2003; Maizels et al., 2004), and although intrinsic T-cell defects and Treg cells are at play (Doetze et al., 2000; Taylor et al., 2005), failure of T cells to respond to antigenic or mitogenic stimuli can often be attributed to suppressive antigen-presenting cells including macrophages. The ability of macrophages from helminth-infected mice and people to block the proliferation of T cells has been demonstrated across the enormous spectrum of helminth diseases including filariasis (Piessens et al., 1980; Allen et al., 1996; Allen and MacDonald, 1998), gastrointestinal nematodes (Price and Holt, 1986), schistosomes (Ottesen, 1979; Todd et al., 1979; Olds et al., 1983; Terrazas et al., 2001), and tapeworms (Dai and Gottstein, 1999) and is a feature of both animal models and human studies. Perhaps not surprisingly, considering the diversity of infection scenarios in which suppressive macrophages are observed, many different mechanisms have been described. These include transforming growth factor β (TGF-β) production (Taylor et al., 2006), programmed death ligand interactions (Smith et al., 2004; Terrazas et al., 2005), lipid mediator release (Brys et al., 2005), and IL-10 production (Osborne and Devaney, 1999; Hesse et al., 2004). Although proliferative suppression is typically mediated by IL-4/IL-13-dependent AAMϕ (Loke et al., 2000b; Terrazas et al., 2005; Brys et al., 2005; Taylor et al., 2006), *Echinococcus* tapeworm infection suppression is mediated via NO-producing CAMϕ. In this setting, NO promotes chronic infection rather than limiting parasite growth (Dai et al., 2003).

AAMϕ as Antigen-Presenting Cells during Helminth Infection

Although macrophage-mediated suppression is often localized to site of helminth infection (Villa and Kuhn, 1996), in the *L. sigmodontis* mouse model of filariasis, suppressive AAMϕ, initially found only at the infection site, become detectable in the draining lymph nodes as the infection becomes more chronic (Taylor et al., 2006). It is difficult to ascertain whether this is the result of increased exposure to parasite products or trafficking of macrophages to the draining lymph nodes. Regardless, several lines of evidence suggest that AAMϕ can influence T-cell differentiation. Studies of AAMϕ ex vivo have shown they can prime naive T cells to differentiate into Th2 cells (Loke et al., 2000a; Rodriguez-Sosa et al., 2002). This can occur even in the context of proliferative suppression. Suppression only occurs during cell contact, and target cells can readily divide once disassociated from the AAMϕ (Loke et al., 2000a, 2000b). In addition, DCs, macrophages, and B cells, but not T cells, express Ym1 and RELMα (Nair et al., 2005), suggesting that the expression of these alternative activation markers is associated with antigen presentation. Furthermore, Ym1 has been shown to be an important DC factor involved in Th2 differentiation (Arora et al., 2006). Together with the knowledge that inflammatory macrophages can traffic to draining lymph nodes and prime T-cell responses (Randolph et al., 1999), these data suggest that AAMϕ are involved in promoting Th2 responses. However, the order of events, the location of cells, or the ability of AAMϕ to differentiate into DCs has not been addressed.

AAMϕ Are "Anti-Inflammatory"

AAMϕ are often described as anti-inflammatory despite their presence in patently "inflammatory" situations such as the schistosome granuloma (Sandor et al., 2003), body cavities containing filarial parasites (MacDonald et al., 2003), or following nematode migration in the lung (Ramaswamy et al., 1991). Despite this apparent contradiction, AAMϕ do exhibit many anti-inflammatory characteristics. For example, synthesis by macrophages of inflammatory prostaglandins such as prostaglandin E_2 is inhibited by IL-4 and IL-13 (Mosca et al., 2007), as are proinflammatory chemokines (Loke et al., 2002). Overall, in terms of cytokine secretion, chemokine profiles, and prostaglandin synthesis, macrophages activated in a Th2 environment exhibit what could be described as a noninflammatory profile. How does one reconcile this with the role of AAMϕ as potential effector cells recruited to the site of infection and their role as mediators of repair and fibrosis, a process that has many inflammatory components and itself needs to be regulated? It is in the analysis of the wound-healing response that an answer can be found. For effective tissue repair to take place, "classical" inflammatory processes must end (Ashcroft et al., 2003; Eming et al., 2007a). Thus, although macrophages during helminth infection are actively recruited in an inflammatory context, the nature of this inflammation differs from that seen during classical macrophage activation by microbial products and IFN-γ. Indeed, Th1 and Th2 responses are not only cross-regulated at the T-cell differentiation level, but also at the macrophage activation level. This is best illustrated with the arginase metabolism pathway described above in which macrophages must compete for arginine. The induction of arginase by IL-4/IL-13 will reduce NO levels (Modolell et al., 1995; Hesse et al., 2001), thus acting to reduce one of the main mediators of "classical inflammation."

Consistent with their roles as immune regulators, macrophages recruited to helminth infection produce high levels of the downregulatory cytokines IL-10 and TGF-β. This has been seen in both animal models and analysis of macrophages from helminth-infected humans. Thus, in combination with their ability to block cellular proliferation (Loke et al., 2000b), helminth-induced macrophages fulfill many of the criteria known to induce Treg cells. For example, TGF-β in combination with a block in T-cell proliferation can promote the development of Foxp3[+] Treg cells from naive T cells (Kretschmer et al., 2005). Furthermore, helminth-induced macrophages are consistently high in F4/80 (Taylor et al., 2006), a molecule that has been implicated in the generation of antigen-specific Treg cells (Lin et al., 2005). However, the link between alternative macrophage activation and preferential Treg cell expansion has yet to be tested.

Parasite Products Can Directly Alter Macrophage Activation State

Infection with helminths induces high levels of IL-4 and IL-13, thus leading to the alternative activation of macrophages. However, a direct role for parasite products in macrophage activation status is likely, considering the increasing evidence that helminths can produce bioactive molecules, including homologues of mammalian cytokines such as TGF-β (Maizels et al., 2004). The best characterized of these parasite products is ES-62, a phosphorylcholine-containing molecule of filarial nematodes that has potent anti-inflammatory properties and inhibits IL-12, TNF-α, and IL-6 production by macrophages exposed to Toll-like receptor ligands (Goodridge et al., 2001, 2005).

At least two different products, thioredoxin peroxidase from *Fasciola hepatica* (Donnelly et al., 2005) and macrophage migration inhibitory factor from *B. malayi* (Falcone et al., 2001), have been shown to induce macrophages to produce alternative activation markers such as Ym1. However, the published data do not address the possibility that these parasite products act indirectly by inducing cytokines, which are then responsible for the phenotype observed. Because of the potential for parasite products to be developed as anti-inflammatory therapies (McInnes et al., 2003), this is an area of research that is likely to be actively pursued in the future.

FUTURE DIRECTIONS

In the past several years there has been a significant paradigm shift with regard to macrophages in helminth infection. Previously, they were ignored, thought of as "suppressor cells," or considered as no different from cells that kill microbes through reactive intermediates. With our new understanding that macrophages recruited to helminth infection sites exhibit a distinct phenotype characterized by the production of a range of novel proteins, our thinking has begun to change. As illustrated in this review, their roles as "suppressors" and effectors are being clarified and the details are becoming elucidated. More radically, we have begun to appreciate that macrophages in helminth infection have major roles in wound healing, with implications for our understanding of the evolution of Th2 immunity. Despite these advances, this area of research is in relative infancy, with nearly every study raising more questions than it answers. Below is a list of just a few of the many questions that need to be answered over the next several years.

- Where and when are macrophages in general, and AAMϕ in particular, essential for control of helminth infection, in terms of either parasite destruction or protection from pathology?
- What is the relative importance of the AAMϕ versus the eosinophil in helminth infection?
- What are the requirements for AAMϕ induction beyond the need for IL-4 and/or IL-13? For example, is TNF required?
- What other mediators such as IL-21 influence AAMϕ during helminth infection? How do parasite products influence the AAMϕ phenotype?
- What is the primary role of macrophage-associated arginase?
- How does arginase mediate its effects against *H. polygyrus*, and to which helminth parasites does this mechanism apply?
- What are the roles of the chitinase family of molecules, and which functions require active enzymatic function?
- What are the distinct roles for RELMα and RELMβ? Do they have regulatory, effector, and/or tissue repair functions? Are their functions context dependent with regard to tissue localization and specific pathogen?
- Evidence suggests that, unlike Th1/Th2 cells, AAMϕ and CAMϕ can rapidly alter their phenotype in response to environmental or pathogen cues (Stout et al., 2005). How plastic are these cells? Does this offer opportunities for therapeutic intervention?
- Can we identify better diagnostic tools, specifically cell surface molecules, to differentiate CAMϕ and AAMϕ in vivo?

The rapid expansion in the availability of reporter mice (Reese et al., 2007) and mice with cell-type-specific gene deletion (Herbert et al., 2004; Siewe et al., 2006) will provide detailed answers to the questions above, as will new technologies for cell-specific depletion and replacement (Cailhier et al., 2006). However, the greatest challenge lies with translating this work to humans infected with helminths. To apply our new understanding of alternative macrophage activation to people or animals infected with helminths, we urgently need to characterize and identify these cells and the relevant homologues of the key AAMϕ gene products in humans (Elias et al., 2005) or animals of veterinary importance (Knight et al., 2007). For example, identifying whether chitinase-like molecules (or active chitinases) play protective or regulatory roles may offer opportunities for intervention or therapy as specific inhibitors exist or are being developed (Rao et al., 2005). The findings that AAMϕ can positively affect fat regulation (Odegaard et al., 2007) and negatively affect cancer outcome (Van Ginderachter et al., 2006a) highlight further the major implications of work on helminth infections that induce these cells in such dramatic numbers. Finally, the evolutionary raison d'être for Th2 response and thus AAMϕ must certainly be to contain or control the damage caused by helminth parasites. By increasing our understanding of the "normal" functions of AAMϕ during helminth infection, we will also better understand the circumstances in which they act inappropriately, such as allergic asthma, fibrosis, and some forms of cancer. It is hoped that this will open doors to new avenues of therapeutic intervention.

REFERENCES

Allen, J. E., O. Adjei, O. Bain, A. Hoerauf, W. Hoffmann, B. Makepeace, H. Schulz-Key, V. Tanya, A. J. Trees, S. Wanji, and D. W. Taylor. 2008. Of mice, cattle, and humans: the immunology and treatment of river blindness. *PLoS Negl. Trop. Dis.* **2:**e217.

Allen, J. E., R. A. Lawrence, and R. M. Maizels. 1996. APC from mice harboring the filarial nematode, *Brugia malayi*, prevent cellular proliferation but not cytokine production. *Int. Immunol.* **8:**143–151.

Allen, J. E., and P. Loke. 2001. Divergent roles for macrophages in lymphatic filariasis. *Parasite Immunol.* **23:**345–352.

Allen, J. E., and A. S. MacDonald. 1998. Profound suppression of cellular proliferation mediated by the secretions of nematodes. *Parasite Immunol.* **20:**241–247.

Alonso-Trujillo, J., I. Rivera-Montoya, M. Rodriguez-Sosa, and L. I. Terrazas. 2007. Nitric oxide contributes to host resistance against experimental *Taenia crassiceps* cysticercosis. *Parasitol. Res.* **100:**1341–1350.

Anthony, R. M., J. F. Urban, Jr., F. Alem, H. A. Hamed, C. T. Rozo, J. L. Boucher, N. Van Rooijen, and W. C. Gause. 2006. Memory T(H)2 cells induce alternatively activated macrophages to mediate protection against nematode parasites. *Nat. Med.* **12:**955–960.

Arora, M., L. Chen, M. Paglia, I. Gallagher, J. E. Allen, Y. M. Vyas, A. Ray, and P. Ray. 2006. Simvastatin promotes Th2-type responses through the induction of the chitinase family member Ym1 in dendritic cells. *Proc. Natl. Acad. Sci. USA* **103:**7777–7782.

Artis, D., M. L. Wang, S. A. Keilbaugh, W. He, M. Brenes, G. P. Swain, P. A. Knight, D. D. Donaldson, M. A. Lazar, H. R. Miller, G. A. Schad, P. Scott, and G. D. Wu. 2004. RELMbeta/FIZZ2 is a goblet cell-specific immune-effector molecule in the gastrointestinal tract. *Proc. Natl. Acad. Sci. USA* **101:**13596–13600.

Ashcroft, G. S., S. J. Mills, K. Lei, L. Gibbons, M. J. Jeong, M. Taniguchi, M. Burow, M. A. Horan, S. M. Wahl, and T. Nakayama. 2003. Estrogen modulates cutaneous wound healing by downregulating macrophage migration inhibitory factor. *J. Clin. Investig.* **111:**1309–1318.

Bargagli, E., M. Margollicci, N. Nikiforakis, A. Luddi, A. Perrone, S. Grosso, and P. Rottoli. 2007. Chitotriosidase activity in the serum of patients with sarcoidosis and pulmonary tuberculosis. *Respiration* **74:**548–552.

Biswas, S. K., L. Gangi, S. Paul, T. Schioppa, A. Saccani, M. Sironi, B. Bottazzi, A. Doni, V. Bronte, F. Pasqualini, L. Vago, M. Nebuloni, A. Mantovani, and A. Sica. 2006. A distinct and unique transcriptional programme expressed by tumor-associated macrophages: defective NF-[kappa]B and enhanced IRF-3/STAT1 activation. *Blood* **107:**2112–2122.

Blaxter, M. L. 2003. *Nematoda*: genes, genomes and the evolution of parasitism. *Adv. Parasitol.* **54:**101–195.

Boot, R. G., E. F. Blommaart, E. Swart, K. Ghauharali-van der Vlugt, N. Bijl, C. Moe, A. Place, and J. M. Aerts. 2001. Identification of a novel acidic mammalian chitinase distinct from chitotriosidase. *J. Biol. Chem.* **276:**6770–6778.

Boros, D. L., and J. R. Whitfield. 1999. Enhanced Th1 and dampened Th2 responses synergize to inhibit acute granulomatous and fibrotic responses in murine schistosomiasis mansoni. *Infect. Immun.* **67:**1187–1193.

Brys, L., A. Beschin, G. Raes, G. H. Ghassabeh, W. Noel, J. Brandt, F. Brombacher, and P. De Baetselier. 2005. Reactive oxygen species and 12/15-lipoxygenase contribute to the antiproliferative capacity of alternatively activated myeloid cells elicited during helminth infection. *J. Immunol.* **174:**6095–6104.

Cailhier, J. F., D. A. Sawatzky, T. Kipari, K. Houlberg, D. Walbaum, S. Watson, R. A. Lang, S. Clay, D. Kluth, J. Savill, and J. Hughes. 2006. Resident pleural macrophages are key orchestrators of neutrophil recruitment in pleural inflammation. *Am. J. Respir. Crit. Care Med.* **173:**540–547.

Chandrashekar, R., U. R. Rao, and D. Subrahmanyam. 1985. Serum dependent cell-mediated immune reactions to *Brugia pahangi* infective larvae. *Parasite Immunol.* **7:**633–641.

Chang, N. C., S. I. Hung, K. Y. Hwa, I. Kato, J. E. Chen, C. H. Liu, and A. C. Chang. 2001. A macrophage protein, Ym1, transiently expressed during inflammation is a novel mammalian lectin. *J. Biol. Chem.* **276:**17497–17506.

Cheever, A. W., and Z. A. Andrade. 1967. Pathological lesions associated with *Schistosoma mansoni* infection in man. *Trans. R. Soc. Trop. Med. Hyg.* **61:**626–639.

Chiaramonte, M. G., D. D. Donaldson, A. W. Cheever and T. A. Wynn. 1999a. An IL-13 inhibitor blocks the development of hepatic fibrosis during a T-helper type 2-dominated inflammatory response. *J. Clin. Investig.* **104:**777–785.

Chiaramonte, M. G., L. R. Schopf, T. Y. Neben, A. W. Cheever, D. D. Donaldson, and T. A. Wynn. 1999b. IL-13 is a key regulatory cytokine for Th2 cell-mediated pulmonary granuloma formation and IgE responses induced by *Schistosoma mansoni* eggs. *J. Immunol.* **162:**920–930.

Closs, E. I., J. S. Scheld, M. Sharafi, and U. Forstermann. 2000. Substrate supply for nitric-oxide synthase in macrophages and endothelial cells: role of cationic amino acid transporters. *Mol. Pharmacol.* **57:**68–74.

Coulson, P. S., L. E. Smythies, C. Betts, N. A. Mabbott, J. M. Sternberg, X. G. Wei, F. Y. Liew, and R. A. Wilson. 1998. Nitric oxide produced in the lungs of mice immunized with the radiation-attenuated schistosome vaccine is not the major agent causing challenge parasite elimination. *Immunology* **93:**55–63.

Dai, W. J., and B. Gottstein. 1999. Nitric oxide-mediated immunosuppression following murine *Echinococcus multilocularis* infection. *Immunology* **97:**107–116.

Dai, W. J., A. Waldvogel, T. Jungi, M. Stettler, and B. Gottstein. 2003. Inducible nitric oxide synthase deficiency in mice increases resistance to chronic infection with *Echinococcus multilocularis*. *Immunology* **108:**238–244.

Doetze, A., J. Satoguina, G. Burchard, T. Rau, C. Löliger, B. Fleischer, and A. Hoerauf. 2000. Antigen-specific cellular hyporesponsiveness in a chronic human helminth infection is mediated by Th3/Tr1-type cytokines IL-10 and transforming growth factor-β but not by a Th1 to Th2 shift. *Int. Immunol.* **12:**623–630.

Donnelly, S., S. M. O'Neill, M. Sekiya, G. Mulcahy, and J. P. Dalton. 2005. Thioredoxin peroxidase secreted by *Fasciola hepatica* induces the alternative activation of macrophages. *Infect. Immun.* **73:**166–173.

Doyle, A. G., G. Herbein, L. J. Montaner, A. J. Minty, D. Caput, P. Ferrara, and S. Gordon. 1994. Interleukin-13 alters the activation state of murine macrophages in vitro: comparison with interleukin-4 and interferon-gamma. *Eur. J. Immunol.* **24:**1441–1445.

Duffield, J. S., S. J. Forbes, C. M. Constandinou, S. Clay, M. Partolina, S. Vuthoori, S. Wu, R. Lang, and J. P. Iredale. 2005. Selective depletion of macrophages reveals distinct, opposing roles during liver injury and repair. *J. Clin. Investig.* **115:**56–65.

Edgeworth, J. D., A. Abiose, and B. R. Jones. 1993. An immunohistochemical analysis of onchocercal nodules: evidence for an interaction between macrophage MRP8/MRP14 and adult *Onchocerca volvulus*. *Clin. Exp. Immunol.* **92:**84–92.

Elias, J. A., R. J. Homer, Q. Hamid, and C. G. Lee. 2005. Chitinases and chitinase-like proteins in T(H)2 inflammation and asthma. *J. Allergy Clin. Immunol.* **116:**497–500.

Eming, S. A., T. Krieg, and J. M. Davidson. 2007a. Inflammation in wound repair: molecular and cellular mechanisms. *J. Invest. Dermatol.* **127:**514–525.

Eming, S. A., S. Werner, P. Bugnon, C. Wickenhauser, L. Siewe, O. Utermohlen, J. M. Davidson, T. Krieg, and A. Roers. 2007b. Accelerated wound closure in mice deficient for interleukin-10. *Am. J. Pathol.* **170:**188–202.

Erdely, A., D. Kepka-Lenhart, M. Clark, P. Zeidler-Erdely, M. Poljakovic, W. J. Calhoun, and S. M. Morris, Jr. 2006. Inhibition of phosphodiesterase 4 amplifies cytokine-dependent induction of arginase in macrophages. *Am. J. Physiol.* **290:**L534–L539.

Falcone, F. H., P. Loke, X. Zang, A. S. MacDonald, R. M. Maizels, and J. E. Allen. 2001. A *Brugia malayi* homologue of mammalian MIF reveals an important link between macrophages and eosinophil recruitment during nematode infection. *J. Immunol.* **167:**5348–5354.

Fallon, P. G., E. J. Richardson, G. J. McKenzie, and A. N. McKenzie. 2000. Schistosome infection of transgenic mice defines distinct and contrasting pathogenic roles for IL-4 and IL-13: IL-13 is a profibrotic agent. *J. Immunol.* **164:**2585–2591.

Fang, F. C., and C. F. Nathan. 2007. Man is not a mouse: reply. *J. Leukoc. Biol.* **81:**580.

Foster, J. M., Y. Zhang, S. Kumar, and C. K. Carlow. 2005. Parasitic nematodes have two distinct chitin synthases. *Mol. Biochem. Parasitol.* **142:**126–132.

Gazzinelli, R. T., and E. Y. Denkers. 2006. Protozoan encounters with Toll-like receptor signalling pathways: implications for host parasitism. *Nat. Rev. Immunol.* **6:**895–906.

Goerdt, S., and C. E. Orfanos. 1999. Other functions, other genes: alternative activation of antigen-presenting cells. *Immunity* **10:**137–142.

Goodridge, H. S., F. A. Marshall, K. J. Else, K. M. Houston, C. Egan, L. Al-Riyami, F. Y. Liew, W. Harnett, and M. M. Harnett. 2005. Immunomodulation via novel use of TLR4 by the filarial nematode phosphorylcholine-containing secreted product, ES-62. *J. Immunol.* **174:**284–293.

Goodridge, H. S., E. H. Wilson, W. Harnett, C. C. Campbell, M. M. Harnett, and F. Y. Liew. 2001. Modulation of macrophage cytokine production by ES-62, a secreted product of the filarial nematode *Acanthocheilonema viteae*. *J. Immunol.* **167:**940–945.

Gordon, S. 1999. Macrophages and the immune response, p. 533–544. *In* W. E. Paul (ed.), *Fundamental Immunology*. Lippincott-Raven Publishers, Philadelphia, PA.

Gordon, S. 2003. Alternative activation of macrophages. *Nat. Rev. Immunol.* **3:**23–35.

Graham, A. L., J. E. Allen, and A. F. Read. 2005. Evolutionary causes and consequences of immunopathology. *Annu. Rev. Ecol. Evol. Syst.* **36:**373–398.

Grzych, J. M., E. J. Pearce, A. Cheever, Z. A. Caulada, P. Caspar, S. Heiny, F. Lewis, and A. Sher. 1991. Egg deposition is the major stimulus for the production of Th2 cytokines in murine schistosomiasis mansoni. *J. Immunol.* **146:**1322–1327.

Guo, L., R. S. Johnson, and J. C. Schuh. 2000. Biochemical characterization of endogenously formed eosinophilic crystals in the lungs of mice. *J. Biol. Chem.* **275:**8032–8037.

Harbord, M., M. Novelli, B. Canas, D. Power, C. Davis, J. Godovac-Zimmermann, J. Roes, and A. W. Segal. 2001. Ym1 is a neutrophil granule protein that crystallizes in p47phox deficient mice. *J. Biol. Chem.* **277:**5468–5475.

Harnett, W., and M. M. Harnett. 2006. What causes lymphocyte hyporesponsiveness during filarial nematode infection? *Trends Parasitol.* **22:**105–110.

Harris, N. L., R. J. Peach, and F. Ronchese. 1999. CTLA4-Ig inhibits optimal T helper 2 cell development but not protective immunity or memory response to *Nippostrongylus brasiliensis*. *Eur. J. Immunol.* **29:**311–316.

Heasman, S. J., K. M. Giles, A. G. Rossi, J. E. Allen, C. Haslett, and I. Dransfield. 2004. Interferon gamma suppresses glucocorticoid augmentation of macrophage clearance of apoptotic cells. *Eur. J. Immunol.* **34:**1752–1761.

Helming, L., and S. Gordon. 2007. Macrophage fusion induced by IL-4 alternative activation is a multistage process involving multiple target molecules. *Eur. J. Immunol.* **37:**33–42.

Henson, P. 2003. Suppression of macrophage inflammatory responses by PPARs. *Proc. Natl. Acad. Sci. USA* **100:**6295–6296.

Henson, P. M., C. D. Mackenzie, and W. G. Spector. 1979. Inflammatory reactions in onchocerciasis: a report on current knowledge and recommendations for further study. *Bull. W. H. O.* **57:**667–682.

Herbert, D. R., C. Holscher, M. Mohrs, B. Arendse, A. Schwegmann, M. Radwanska, M. Leeto, R. Kirsch, P. Hall, H. Mossmann, B. Claussen, I. Forster, and F. Brombacher. 2004. Alternative macrophage activation is essential for survival during schistosomiasis and downmodulates T helper 1 responses and immunopathology. *Immunity* **20:**623–635.

Hesse, M., A. W. Cheever, D. Jankovic, and T. A. Wynn. 2000. NOS-2 mediates the protective anti-inflammatory and antifibrotic effects of the Th1-inducing adjuvant, IL-12, in a Th2 model of granulomatous disease. *Am. J. Pathol.* **157:**945–955.

Hesse, M., M. Modolell, A. C. La Flamme, M. Schito, J. M. Fuentes, A. W. Cheever, E. J. Pearce, and T. A. Wynn. 2001. Differential regulation of nitric oxide synthase-2 and arginase-1 by type 1/type 2 cytokines in vivo: granulomatous pathology is shaped by the pattern of L-arginine metabolism. *J. Immunol.* **167:**6533–6544.

Hesse, M., C. A. Piccirillo, Y. Belkaid, J. Prufer, M. Mentink-Kane, M. Leusink, A. W. Cheever, E. M. Shevach, and T. A. Wynn. 2004. The pathogenesis of schistosomiasis is controlled by cooperating IL-10-producing innate effector and regulatory T cells. *J. Immunol.* **172:**3157–3166.

Hoffmann, K. F., P. Caspar, A. W. Cheever, and T. A. Wynn. 1998. IFN-γ, IL-12, and TNF-α are required to maintain reduced liver pathology in mice vaccinated with *Schistosoma mansoni* eggs and IL-12. *J. Immunol.* **161:**4201–4210.

Hoffmann, K. F., A. W. Cheever, and T. A. Wynn. 2000. IL-10 and the dangers of immune polarization: excessive type 1 and type 2 cytokine responses induce distinct forms of lethal immunopathology in murine schistosomiasis. *J. Immunol.* **164:**6406–6416.

Hoffmann, K. F., S. L. James, A. W. Cheever, and T. A. Wynn. 1999. Studies with double cytokine-deficient mice reveal that highly polarized Th1- and Th2-type cytokine and antibody responses contribute equally to vaccine-induced immunity to *Schistosoma mansoni*. *J. Immunol.* **163:**927–938.

Hogaboam, C. M., C. S. Gallinat, C. Bone-Larson, S. W. Chensue, N. W. Lukacs, R. M. Strieter, and S. L. Kunkel. 1998. Collagen deposition in a non-fibrotic lung granuloma model after nitric oxide inhibition. *Am. J. Pathol.* **153:**1861–1872.

Hosokawa, H., T. Sawamura, S. Kobayashi, H. Ninomiya, S. Miwa, and T. Masaki. 1997. Cloning and characterization of a brain-specific cationic amino acid transporter. *J. Biol. Chem.* **272:**8717–8722.

Huffnagle, G. B., M. B. Boyd, N. E. Street, and M. F. Lipscomb. 1998. IL-5 is required for eosinophil recruitment, crystal deposition, and mononuclear cell recruitment during a pulmonary *Cryptococcus neoformans* infection in genetically susceptible mice (C57BL/6). *J. Immunol.* **160:**2393–2400.

Hung, S. I., A. C. Chang, I. Kato, and N. C. Chang. 2002. Transient expression of Ym1, a heparin-binding lectin, during developmental hematopoiesis and inflammation. *J. Leukoc. Biol.* **72:**72–82.

James, S. L., A. W. Cheever, P. Caspar, and T. A. Wynn. 1998. Inducible nitric oxide synthase-deficient mice develop enhanced type 1 cytokine-associated cellular and humoral immune responses after vaccination with attenuated *Schistosoma mansoni* cercariae but display partially reduced resistance. *Infect. Immun.* **66:**3510–3518.

James, S. L., and D. G. Colley. 1978. Eosinophil-mediated destruction of *Schistosoma mansoni* eggs in vitro. II. The role of cytophilic antibody. *Cell. Immunol.* **38:**35–47.

James, S. L., J. K. Lazdins, M. S. Meltzer, and A. Sher. 1982. Macrophages as effector cells of protective immunity in murine schistosomiasis. *Cell. Immunol.* **67:**255–266.

Jankovic, D., M. C. Kullberg, N. Noben-Trauth, P. Caspar, J. M. Ward, A. W. Cheever, W. E. Paul, and A. Sher. 1999. Schistosome-infected IL-4 receptor knockout (KO) mice, in contrast to IL-4 KO mice, fail to develop granulomatous pathology while maintaining the same lymphokine expression profile. *J. Immunol.* **163:**337–342.

Jankovic, D., S. Steinfelder, M. C. Kullberg, and A. Sher. 2006. Mechanisms underlying helminth-induced Th2 polarization: default, negative or positive pathways? *Chem. Immunol. Allergy* **90:**65–81.

Kaplan, M. H., U. Schindler, S. T. Smiley, and M. J. Grusby. 1996. Stat6 is required for mediating responses to IL-4 and for the development of th2 cells. *Immunity* **4:**313–319.

Keir, P. A., D. M. Brown, A. Clouter-Baker, Y. M. Harcus, and L. Proudfoot. 2004. Inhibition of neutrophil recruitment by ES of *Nippostrongylus brasiliensis*. *Parasite Immunol.* **26:**137–139.

Knight, P. A., J. Pate, W. D. Smith, and H. R. Miller. 2007. Veterinary immunology and immunopathology An ovine chitinase-like molecule, chitinase-3 like-1 (YKL-40), is upregulated in the abomasum in response to challenge with the gastrointestinal nematode, *Teladorsagia circumcincta*. *Vet. Immunol. Immunopathol.* **120:**55–60.

Kretschmer, K., I. Apostolou, D. Hawiger, K. Khazaie, M. C. Nussenzweig, and H. von Boehmer. 2005. Inducing and expanding regulatory T cell populations by foreign antigen. *Nat. Immunol.* **6:**1219–1227.

Kropf, P., J. M. Fuentes, E. Fahnrich, L. Arpa, S. Herath, V. Weber, G. Soler, A. Celada, M. Modolell, and I. Muller. 2005. Arginase and polyamine synthesis are key factors in the regulation of experimental leishmaniasis in vivo. *FASEB J.* **19:**1000–1002.

Kzhyshkowska, J., S. Mamidi, A. Gratchev, E. Kremmer, C. Schmuttermaier, L. Krusell, G. Haus, J. Utikal, K. Schledzewski, J. Scholtze, and S. Goerdt. 2006. Novel stabilin-1 interacting chitinase-like protein (SI-CLP) is upregulated in alternatively activated macrophages and secreted via lysosomal pathway. *Blood* **107:**3221–3228.

Lin, H. H., D. E. Faunce, M. Stacey, A. Terajewicz, T. Nakamura, J. Zhang-Hoover, M. Kerley, M. L. Mucenski, S. Gordon, and J. Stein-Streilein. 2005. The macrophage F4/80 receptor is required for the induction of antigen-specific efferent regulatory T cells in peripheral tolerance. *J. Exp. Med.* **201:**1615–1625.

Linehan, S. A., P. S. Coulson, R. A. Wilson, A. P. Mountford, F. Brombacher, L. Martinez-Pomares, and S. Gordon.

2003. IL-4 receptor signaling is required for mannose receptor expression by macrophages recruited to granulomata but not resident cells in mice infected with *Schistosoma mansoni*. *Lab. Invest.* **83:**1223–1231.

Liu, T., S. M. Dhanasekaran, H. Jin, B. Hu, S. A. Tomlins, A. M. Chinnaiyan, and S. H. Phan. 2004a. FIZZ1 stimulation of myofibroblast differentiation. *Am. J. Pathol.* **164:**1315–1326.

Liu, T., H. Jin, M. Ullenbruch, B. Hu, N. Hashimoto, B. Moore, A. McKenzie, N. W. Lukacs, and S. H. Phan. 2004b. Regulation of found in inflammatory zone 1 expression in bleomycin-induced lung fibrosis: role of IL-4/IL-13 and mediation via STAT-6. *J. Immunol.* **173:**3425–3431.

Liu, Y., J. A. Van Ginderachter, L. Brys, P. De Baetselier, G. Raes, and A. B. Geldhof. 2003. Nitric oxide-independent CTL suppression during tumor progression: association with arginase-producing (M2) myeloid cells. *J. Immunol.* **170:**5064–5074.

Loke, P., I. Gallagher, M. G. Nair, X. Zang, F. Brombacher, M. Mohrs, J. P. Allison, and J. E. Allen. 2007. Alternative activation is an innate response to injury that requires CD4+ T cells to be sustained during chronic infection. *J. Immunol.* **179:**3926–3936.

Loke, P., A. S. MacDonald, and J. E. Allen. 2000a. Antigen presenting cells recruited by *Brugia malayi* induce Th2 differentiation of naive CD4+ T cells. *Eur. J. Immunol.* **30:**1127–1135.

Loke, P., A. S. MacDonald, A. O. Robb, R. M. Maizels, and J. E. Allen. 2000b. Alternatively activated macrophages induced by nematode infection inhibit proliferation via cell to cell contact. *Eur. J. Immunol.* **30:**2669–2678.

Loke, P., M. G. Nair, D. Guiliano, J. Parkinson, M. L. Blaxter, and J. E. Allen. 2002. IL-4 dependent alternatively-activated macrophages have a distinctive in vivo gene expression phenotype. *Biomed. Central* **3:**7.

MacDonald, A. S., P. Loke, R. A. Martynoga, I. Dransfield, and J. E. Allen. 2003. Cytokine-dependent inflammatory cell recruitment patterns in the peritoneal cavity of mice exposed to the parasitic nematode, *Brugia malayi*. *Med. Microbiol. Immunol.* **192:**33–40.

Mackenzie, C. D., S. L. Oxenham, D. A. Liron, D. Grennan, and D. A. Denham. 1985. The induction of functional mononuclear and multinuclear macrophages in murine Brugian filariasis: morphological and immunological properties. *Trop. Med. Parasitol.* **36:**163–170.

MacLeod, C. L. 1996. Regulation of cationic amino acid transporter (CAT) gene expression. *Biochem. Soc. Trans.* **24:**846–852.

MacLeod, C. L., K. D. Finley, and D. K. Kakuda. 1994. y(+)-Type cationic amino acid transport: expression and regulation of the mCAT genes. *J. Exp. Biol.* **196:**109–121.

Maizels, R. M., A. Balic, N. Gomez-Escobar, M. Nair, M. D. Taylor, and J. E. Allen. 2004. Helminth parasites—masters of regulation. *Immunol. Rev.* **201:**89–116.

Maizels, R. M., and M. Yazdanbakhsh. 2003. Immune regulation by helminth parasites: cellular and molecular mechanisms. *Nat. Rev. Immunol.* **3:**733–744.

McInnes, I. B., B. P. Leung, M. Harnett, J. A. Gracie, F. Y. Liew, and W. Harnett. 2003. A novel therapeutic approach targeting articular inflammation using the filarial nematode-derived phosphorylcholine-containing glycoprotein ES-62. *J. Immunol.* **171:**2127–2133.

Misson, P., S. van den Brule, V. Barbarin, D. Lison, and F. Huaux. 2004. Markers of macrophage differentiation in experimental silicosis. *J. Leukoc. Biol.* **76:**926–932.

Modolell, M., I. M. Corraliza, F. Link, G. Soler, and K. Eichmann. 1995. Reciprocal regulation of the nitric oxide synthase/arginase balance in mouse bone marrow-derived macrophages by TH1 and TH2 cytokines. *Eur. J. Immunol.* **25:**1101–1104.

Mosca, M., N. Polentarutti, G. Mangano, C. Apicella, A. Doni, F. Mancini, M. De Bortoli, I. Coletta, L. Polenzani, G. Santoni, M. Sironi, A. Vecchi, and A. Mantovani. 2007. Regulation of the microsomal prostaglandin E synthase-1 in polarized mononuclear phagocytes and its constitutive expression in neutrophils. *J. Leukoc. Biol.* **82:**320–326.

Mosser, D. M. 2003. The many faces of macrophage activation. *J. Leukoc. Biol.* **73:**209–212.

Munder, M., K. Eichmann, and M. Modolell. 1998. Alternative metabolic states in murine macrophages reflected by the nitric oxide synthase/arginase balance: competitive regulation by CD4+ T cells correlates with Th1/Th2 phenotype. *J. Immunol.* **160:**5347–5354.

Munder, M., K. Eichmann, J. M. Moran, F. Centeno, G. Soler, and M. Modolell. 1999. Th1/Th2-regulated expression of arginase isoforms in murine macrophages and dendritic cells. *J. Immunol.* **163:**3771–3777.

Munder, M., F. Mollinedo, J. Calafat, J. Canchado, C. Gil-Lamaignere, J. M. Fuentes, C. Luckner, G. Doschko, G. Soler, K. Eichmann, F. M. Muller, A. D. Ho, M. Goerner, and M. Modolell. 2005. Arginase I is constitutively expressed in human granulocytes and participates in fungicidal activity. *Blood* **105:**2549–2556.

Nair, M. G., D. W. Cochrane, and J. E. Allen. 2003. Macrophages in chronic type 2 inflammation have a novel phenotype characterized by the abundant expression of Ym1 and Fizz1 that can be partly replicated in vitro. *Immunol. Lett.* **85:**173–180.

Nair, M. G., I. Gallagher, M. Taylor, P. Loke, P. S. Coulson, R. A. Wilson, R. M. Maizels, and J. E. Allen. 2005. Chitinase and Fizz family members are a generalized feature of nematode infection with selective upregulation of Ym1 and Fizz1 by antigen-presenting cells. *Infect. Immun.* **73:**385–394.

Nicholson, B., C. K. Manner, J. Kleeman, and C. L. MacLeod. 2001. Sustained nitric oxide production in macrophages requires the arginine transporter CAT2. *J. Biol. Chem.* **276:**15881–15885.

Noel, W., G. Raes, G. Hassanzadeh Ghassabeh, P. De Baetselier, and A. Beschin. 2003. Alternatively activated macrophages during parasite infections. *Trends Parasitol.* **20:**126–133.

Numaguchi, K., K. Egashira, M. Takemoto, T. Kadokami, H. Shimokawa, K. Sueishi, and A. Takeshita. 1995. Chronic inhibition of nitric oxide synthesis causes coronary microvascular remodeling in rats. *Hypertension* **26**(6 Pt 1):957–962.

Odegaard, J. I., R. R. Ricardo-Gonzalez, M. H. Goforth, C. R. Morel, V. Subramanian, L. Mukundan, A. Red Eagle, D. Vats, F. Brombacher, A. W. Ferrante, and A. Chawla. 2007. Macrophage-specific PPARgamma controls alternative activation and improves insulin resistance. *Nature* **447:**1116–1120.

Olds, G. R., J. J. Ellner, L. A. Kearse, Jr., J. W. Kazura, and A. A. Mahmoud. 1980. Role of arginase in killing of schistosomula of *Schistosoma mansoni*. *J. Exp. Med.* **151:**1557–1562.

Olds, G. R., A. E. Kholy, and J. J. Ellner. 1983. Two distinctive patterns of monocyte immunoregulatory and effector functions in heavy human infections with *Schistosoma mansoni*. *J. Immunol.* **131:**954–958.

Osborne, J., and E. Devaney. 1999. Interleukin-10 and antigen-presenting cells actively suppress Th1 cells in BALB/c mice infected with the filarial parasite *Brugia pahangi*. *Infect. Immun.* **67:**1599–1605.

Oswald, I. P., R. T. Gazzinelli, A. Sher, and S. L. James. 1992. IL-10 synergizes with IL-4 and transforming growth factor-beta to inhibit macrophage cytotoxic activity. *J. Immunol.* **148:**3578–3582.

Ottesen, E. A. 1979. Modulation of the host response in human schistosomiasis. I. Adherent suppressor cells that inhibit lymphocyte proliferative responses to parasite antigens. *J. Immunol.* **123:**1639–1644.

Pearce, E. J., P. Caspar, J.-M. Grzych, F. A. Lewis, and A. Sher. 1991. Downregulation of Th1 cytokine production accompanies induction of Th2 responses by a parasitic helminth, *Schistosma mansoni*. *J. Exp. Med.* **173:**159–166.

Pesce, J., M. Kaviratne, T. R. Ramalingam, R. W. Thompson, J. F. Urban Jr., A. W. Cheever, D. A. Young, M. Collins, M. J. Grusby, and T. A. Wynn. 2006. The IL-21 receptor augments Th2 effector function and alternative macrophage activation. *J. Clin. Investig.* **116:**2044–2055.

Piessens, W. F., S. Ratiwayanto, S. Tuti, J. H. Palmieri, P. W. Piessens, I. Koiman, and D. T. Dennis. 1980. Antigen-specific suppressor cells and suppressor factors in human filariasis with *Brugia malayi*. *N. Engl. J. Med.* **302:**833–837.

Price, P., and P. G. Holt. 1986. Immunological consequences of intestinal helminth infections: antigen presentation and immunosuppression by peritoneal cells. *Aust. J. Exp. Biol. Med. Sci.* **64**(Pt 5)**:**399–413.

Raes, G., P. D. Baetselier, W. Noel, A. Beschin, F. Brombacher, and G. Hassanzadeh Gh. 2001. Differential expression of FIZZ1 and Ym1 in alternatively versus classically activated macrophages. *J. Leukoc. Biol.* **71:**597–602.

Raes, G., L. Brys, B. K. Dahal, J. Brandt, J. Grooten, W. Noel, T. Boonefaes, A. Kindt, P. De Baetselier, and G. Hassanzadeh Gh. 2003. Macrophage galactose-type C-type lectins 1 and 2 as novel markers for alternatively activated macrophages elicited during parasite infections and allergic asthma. *J. Leukoc. Biol.* **77:**321–327.

Ramaswamy, K., G. T. De Sanctis, F. Green, and D. Befus. 1991. Pathology of pulmonary parasitic migration: morphological and bronchoalveolar cellular responses following *Nippostrongylus brasiliensis* infection in rats. *J. Parasitol.* **77:**302–312.

Ramesh, M., N. Paciorkowski, Y. Dash, L. Shultz, and T. V. Rajan. 2007. Acute but not chronic macrophage recruitment in filarial infections in mice is dependent on C-C chemokine ligand 2. *Parasite Immunol.* **29:**395–404.

Randolph, G. J., K. Inaba, D. F. Robbiani, R. M. Steinman, and W. A. Muller. 1999. Differentiation of phagocytic monocytes into lymph node dendritic cells in vivo. *Immunity* **11:**753–761.

Rao, F. V., D. R. Houston, R. G. Boot, J. M. Aerts, M. Hodkinson, D. J. Adams, K. Shiomi, S. Omura, and D. M. van Aalten. 2005. Specificity and affinity of natural product cyclopentapeptide inhibitors against *A. fumigatus*, human, and bacterial chitinases. *Chem. Biol.* **12:**65–76.

Rauh, M. J., V. Ho, C. Pereira, A. Sham, L. M. Sly, V. Lam, L. Huxham, A. I. Minchinton, A. Mui, and G. Krystal. 2005. SHIP represses the generation of alternatively activated macrophages. *Immunity* **23:**361–374.

Reece, J. J., M. C. Siracusa, and A. L. Scott. 2006. Innate immune responses to lung-stage helminth infection induce alternatively activated alveolar macrophages. *Infect. Immun.* **74:**4970–4981.

Reese, T. A., H. E. Liang, A. M. Tager, A. D. Luster, N. Van Rooijen, D. Voehringer, and R. M. Locksley. 2007. Chitin induces accumulation in tissue of innate immune cells associated with allergy. *Nature* **447:**92–96.

Restrepo, B. I., J. I. Alvarez, J. A. Castano, L. F. Arias, M. Restrepo, J. Trujillo, C. H. Colegial, and J. M. Teale. 2001. Brain granulomas in neurocysticercosis patients are associated with a Th1 and Th2 profile. *Infect. Immun.* **69:**4554–4560.

Reyes, J. L., and L. I. Terrazas. 2007. The divergent role of activated macrophages in helminthic infections. *Parasite Immunol.* **29:**609–619.

Rodriguez-Sosa, M., R. Calderon, A. R. Satoskar, R. Saavedra, L. Terrazas, and R. Bojalil. 2002. Chronic helminth infection induces alternatively activated macrophages expressing high levels of CCR5 with low IL-12 production and Th2-biasing ability. *Infect. Immun.* **70:**3656–3664.

Rodriguez-Sosa, M., R. Saavedra, E. P. Tenorio, L. E. Rosas, A. R. Satoskar, and L. I. Terrazas. 2004. A STAT4-dependent Th1 response is required for resistance to the helminth parasite Taenia crassiceps. *Infect. Immun.* **72:**4552–4560.

Rutschman, R., R. Lang, M. Hesse, J. N. Ihle, T. A. Wynn, and P. J. Murray. 2001. Cutting edge: Stat6-dependent substrate depletion regulates nitric oxide production. *J. Immunol.* **166:**2173–2177.

Saeftel, M., M. Arndt, S. Specht, L. Volkmann, and A. Hoerauf. 2003. Synergism of gamma interferon and interleukin-5 in the control of murine filariasis. *Infect. Immun.* **71:**6978–6985.

Saeftel, M., L. Volkmann, S. Korten, N. Brattig, K. M. Al-Qaoud, B. Fleischer, and A. Hoerauf. 2001. Lack of interferon-γ confers impaired neutrophil granulocyte function and imparts prolonged survival of adult filarial worms in murine filariasis. *Microbes Infect.* **3:**203–213.

Sandler, N. G., M. M. Mentink-Kane, A. W. Cheever, and T. A. Wynn. 2003. Global gene expression profiles during acute pathogen-induced pulmonary inflammation reveal divergent roles for Th1 and Th2 responses in tissue repair. *J. Immunol.* **171:**3655–3667.

Sandor, M., J. V. Weinstock, and T. A. Wynn. 2003. Granulomas in schistosome and mycobacterial infections: a model of local immune responses. *Trends Immunol.* **24:**44–52.

Schebesch, C., V. Kodelja, C. Muller, N. Hakij, S. Bisson, C. E. Orfanos, and S. Goerdt. 1997. Alternatively activated macrophages actively inhibit proliferation of peripheral blood lymphocytes and CD4+ T cells in vitro. *Immunology* **92:**478–486.

Senft, A. W. 1966. Studies in arginine metabolism by schistosomes. I. Arginine uptake and lysis by *Schistosoma mansoni*. *Comp. Biochem. Physiol.* **18:**209–216.

Shinkai, K., M. Mohrs, and R. M. Locksley. 2002. Helper T cells regulate type-2 innate immunity in vivo. *Nature* **420:**825–829.

Siewe, L., M. Bollati-Fogolin, C. Wickenhauser, T. Krieg, W. Muller, and A. Roers. 2006. Interleukin-10 derived from macrophages and/or neutrophils regulates the inflammatory response to LPS but not the response to CpG DNA. *Eur. J. Immunol.* **36:**3248–3255.

Sinha, P., V. K. Clements, and S. Ostrand-Rosenberg. 2005. Interleukin-13-regulated M2 macrophages in combination with myeloid suppressor cells block immune surveillance against metastasis. *Cancer Res.* **65:**11743–11751.

Smith, P., C. M. Walsh, N. E. Mangan, R. E. Fallon, J. R. Sayers, A. N. McKenzie, and P. G. Fallon. 2004. *Schistosoma mansoni* worms induce anergy of T cells via selective upregulation of programmed death ligand 1 on macrophages. *J. Immunol.* **173:**1240–1248.

Stein, M., S. Keshav, N. Harris, and S. Gordon. 1992. Interleukin 4 potently enhances murine macrophage mannose receptor activity: a marker of alternative immunologic macrophage activation. *J. Exp. Med.* **176:**287–292.

Stout, R. D., C. Jiang, B. Matta, I. Tietzel, S. K. Watkins, and J. Suttles. 2005. Macrophages sequentially change their functional phenotype in response to changes in microenvironmental influences. *J. Immunol.* **175:**342–349.

Taylor, M. D., A. Harris, M. G. Nair, R. M. Maizels, and J. E. Allen. 2006. F4/80+ alternatively activated macrophages control CD4+ T cell hypo-responsiveness at sites peripheral to filarial infection. *J. Immunol.* **176:**6918–6927.

Taylor, M. D., L. Le Goff, A. Harris, J. E. Allen, and R. M. Maizels. 2005. Removal of regulatory T cell activity reverses hyporesponsiveness and leads to filarial parasite clearance in vivo. *J. Immunol.* **174:**4924–4933.

Taylor, M. J., H. F. Cross, A. A. Mohammed, A. J. Trees, and A. E. Bianco. 1996. Susceptibility of *Brugia malayi* and *Onchocerca lienalis* microfilariae to nitric oxide and hydrogen peroxide in cell-free culture and from IFN γ-activated macrophages. *Parasitology* **112:**315–322.

Teng, X., D. Li, H. C. Champion, and R. A. Johns. 2003. FIZZ1/RELMalpha, a novel hypoxia-induced mitogenic factor in lung with vasoconstrictive and angiogenic properties. *Circ. Res.* **92:**1065–1067.

Terrazas, L. I., D. Montero, C. A. Terrazas, J. L. Reyes, and M. Rodriguez-Sosa. 2005. Role of the programmed Death-1 pathway in the suppressive activity of alternatively activated macrophages in experimental cysticercosis. *Int. J. Parasitol.* **35:**1349–1358.

Terrazas, L. I., K. L. Walsh, D. Piskorska, E. McGuire, and D. A. Harn, Jr. 2001. The schistosome oligosaccharide lacto-n-neotetraose expands Gr1(+) cells that secrete anti-inflammatory cytokines and inhibit proliferation of naive CD4(+) cells: a potential mechanism for immune polarization in helminth infections. *J. Immunol.* **167:**5294–5303.

Thomas, G. R., M. McCrossan, and M. E. Selkirk. 1997. Cytostatic and cytotoxic effects of activated macrophages and nitric oxide donors on *Brugia malayi*. *Infect. Immun.* **65:**2732–2739.

Thomas, P. G., and D. A. Harn, Jr. 2004. Immune biasing by helminth glycans. *Cell. Microbiol.* **6:**13–22.

Todd, C. W., R. W. Goodgame, and D. G. Colley. 1979. Immune responses during human schistosomiasis mansoni. V. Suppression of schistosome antigen-specific lymphocyte blastogenesis by adherent/phagocytic cells. *J. Immunol.* **122:**1440–1446.

Tomita, M., T. Kobayashi, H. Itoh, T. Onitsuka, and Y. Nawa. 2000. Goblet cell hyperplasia in the airway of *Nippostrongylus brasiliensis*-infected rats. *Respiration* **67:**565–569.

Urban, J. F., K. B. Madden, A. Sveti'c, A. Cheever, P. P. Trotta, W. C. Gause, I. M. Katona, and F. D. Finkelman. 1992. The importance of Th2 cytokines in protective immunity to nematodes. *Immunol. Rev.* **127:**205–220.

Van Ginderachter, J. A., S. Meerschaut, Y. Liu, L. Brys, K. De Groeve, G. Hassanzadeh Ghassabeh, G. Raes, and P. De Baetselier. 2006a. Peroxisome proliferator-activated receptor [gamma] (PPAR[gamma]) ligands reverse CTL suppression by alternatively activated (M2) macrophages in cancer. *Blood* **108:**525–535.

Van Ginderachter, J. A., K. Movahedi, G. Hassanzadeh Ghassabeh, S. Meerschaut, A. Beschin, G. Raes, and P. De Baetselier. 2006b. Classical and alternative activation of mononuclear phagocytes: picking the best of both worlds for tumor promotion. *Immunobiology* **211:**487–501.

Vella, A. T., and E. J. Pearce. 1992. CD4+ Th2 response induced by *Schistosoma mansoni* eggs develops rapidly, through an early, transient, Th0-like stage. *J. Immunol.* **148:**2283–2288.

Villa, O. F., and R. E. Kuhn. 1996. Mice infected with the larvae of *Taenia crassiceps* exhibit a Th2-like immune response with concomitant anergy and downregulation of Th1-associated phenomena. *Parasitology* **112(Pt 6):**561–570.

Voehringer, D., K. Shinkai, and R. M. Locksley. 2004. Type 2 immunity reflects orchestrated recruitment of cells committed to IL-4 production. *Immunity* **20:**267–277.

Voehringer, D., N. van Rooijen, and R. M. Locksley. 2007. Eosinophils develop in distinct stages and are recruited to peripheral sites by alternatively activated macrophages. *J. Leukoc. Biol.* **81:**1434–1444.

Webb, D. C., A. N. McKenzie, and P. S. Foster. 2001. Expression of the Ym2 lectin-binding protein is dependent on interleukin (IL)-4 and IL-13 signal transduction: identification of a novel allergy-associated protein. *J. Biol. Chem.* **276:**41969–41976.

Welch, J. S., L. Escoubet-Lozach, D. B. Sykes, K. Liddiard, D. R. Greaves, and C. K. Glass. 2002. TH2 cytokines and allergic challenge induce Ym1 expression in macrophages by a STAT6-dependent mechanism. *J. Biol. Chem.* **277:**42821–42829.

Wertheim, G., H. Zylberman, and G. S. Hamada. 1987. Macrophage-nematode interaction in vivo: *Nippostrongylus brasiliensis* infective larvae in the peritoneum of unsensitized rats. *Ann. Parasitol. Hum. Comp.* **62:**47–57.

Wilson, H. M., D. Walbaum, and A. J. Rees. 2004. Macrophages and the kidney. *Curr. Opin. Nephrol. Hypertens.* **13:**285–290.

Witte, M. B., and A. Barbul. 2003. Arginine physiology and its implication for wound healing. *Wound Repair Regen.* **11:**419–423.

Witte, M. B., A. Barbul, M. A. Schick, N. Vogt, and H. D. Becker. 2002. Upregulation of arginase expression in wound-derived fibroblasts. *J. Surg. Res.* **105:**35–42.

Wynn, T. A. 2004. Fibrotic disease and the T(H)1/T(H)2 paradigm. *Nat. Rev. Immunol.* **4:**583–594.

Wynn, T. A. 2007. Common and unique mechanisms regulate fibrosis in various fibroproliferative diseases. *J. Clin. Investig.* **117:**524–529.

Wynn, T. A., A. W. Cheever, D. Jankovic, R. W. Poindexter, P. Caspar, F. A. Lewis, and A. Sher. 1995. An IL-12-based vaccination method for preventing fibrosis induced by schistosome infection. *Nature* **376:**594–596.

Wynn, T. A., I. Eltoum, A. W. Cheever, F. A. Lewis, W. C. Gause, and A. Sher. 1993. Analysis of cytokine mRNA expression during primary granuloma formation induced by eggs of *Schistosoma mansoni*. *J. Immunol.* **151:**1430–1440.

Wynn, T. A., I. Eltoum, I. P. Oswald, A. W. Cheever, and A. Sher. 1994. Endogenous interleukin 12 (Il-12) regulates granuloma formation induced by eggs of *Schistosoma mansoni* and exogenous IL-12 both inhibits and prophylactically immunizes against egg pathology. *J. Exp. Med.* **179:**1551–1561.

Yazdanbakhsh, M. 1999. Common features of T cell reactivity in persistent helminth infections: lymphatic filariasis and schistosomiasis. *Immunol. Lett.* **65:**109–115.

Zimmermann, N., N. E. King, J. Laporte, M. Yang, A. Mishra, S. M. Pope, E. E. Muntel, D. P. Witte, A. A. Pegg, P. S. Foster, Q. Hamid, and M. E. Rothenberg. 2003. Dissection of experimental asthma with DNA microarray analysis identifies arginase in asthma pathogenesis. *J. Clin. Investig.* **111:**1863–1874.

MODELS OF HOST-PATHOGEN INTERACTIONS

VI

32

Dictyostelium discoideum: a Model Phagocyte and a Model for Host-Pathogen Interactions

ZHIRU LI AND RALPH R. ISBERG

WHY USE HARMLESS ENVIRONMENTAL AMOEBAE TO STUDY PATHOGEN-MACROPHAGE INTERACTIONS?

Researchers focus on the interplay between pathogens and various arms of host defense to dissect how a microorganism causes disease (Finlay and Falkow, 1997). For each individual pathogen, this approach has led to the development of specific criteria that define the factors necessary for the microorganism either to cause damage or to establish a replicative niche within the host. In the process, workers attempt to identify all the strategies used by the microorganism that circumvent host antimicrobial tactics, and then determine how the host responds to these strategies, analyzing a dance that either supports or restricts the disease process. There are several levels at which this analysis can take place. At the broadest level, the site that the pathogen enters and spreads within the host is revealed, followed by determining whether there exists any tissue tropism of the microbe or whether particular tissue sites exist that display disease symptoms. At a finer level, the worker wishes to define the host cell subsets that modulate the disease process as well as the accompanying cell biological responses. The molecular components involved in these cellular interactions and the mechanistic details of how the pathogen manipulates the host are then uncovered. Finally, the strategies that the host uses to detect, prevent, or respond to the microorganism must be worked out. In the end, a complete understanding of this process requires a meticulous teasing apart of the details of the host-microbe interplay.

With so much emphasis on how the pathogen combats its multicellular host, it is not obvious why anyone would be interested in the interaction between microorganisms and simple unicellular eukaryotes such as amoebae, which supposedly show none of the complexity of the immune response seen in higher organisms. After all, one could argue, simple unicellular eukaryotes do not challenge the pathogen with the diversity of cell types that are encountered by the pathogen during the disease process, and nothing similar to the innate or adaptive immune systems seems to be involved in modulating the growth and colonization patterns of potential pathogens within amoebae. Furthermore, simple differences such as the fact that many virulence-associated proteins are temperature regulated (Klein and Tebbets, 2007; Marceau, 2005), and that unicellular eukaryotics cannot control temperature in the fashion of warm-blooded animals, emphasize the differences between these types of hosts.

These objections ignore the pressures that may have led to the microorganism acquiring proteins necessary to cause disease. For instance, it is striking that pathogens that have evolved to colonize plant tissues may have the same virulence proteins as those that cause disease in humans (Rahme et al., 1995, 2000). In fact, mutations affecting the expression of single proteins in plant-pathogenic *Pseudomonas* species result in defects in virulence in both plant and mammalian animal models (Rahme et al., 1995). This suggests that, although the pathogen defense systems of plants and animals may be quite different (Nurnberger et al., 2004), a protein that was selected to overcome plant resistance may also allow a bacterium to bypass mammalian host defenses. Just because a protein is known to be important for disease in mammals does not demonstrate that the selective pressures that led to the acquisition of this factor occurred within mammals.

It now appears that we can take this argument to hosts that are quite phylogenetically unrelated to mammals, since many human pathogens reside in environmental sites that have no clear association with multicellular organisms. Furthermore, in some cases such as in *Mycobacterium smegmatis*, nonpathogenic soil bacteria that are closely related to the causative agent of tuberculosis possess gene clusters known to be important for intracellular survival of the pathogen (Berthet et al., 1998; Gey Van Pittius et al., 2001). It is unclear why the soil microorganism must main-

Zhiru Li, Department of Molecular Biology and Microbiology, Tufts University School of Medicine, 150 Harrison Avenue, Boston, MA 02111. **Ralph R. Isberg,** Howard Hughes Medical Institute and Department of Molecular Biology and Microbiology, Tufts University School of Medicine, 150 Harrison Avenue, Boston, MA 02111.

tain these clusters, but these genes may express proteins that are important for protecting the nonpathogen from predation by other soil microorganisms. Amoebal species are prevalent in a number of environmental sources, so it is reasonable to suppose that the ability to resist killing by amoebae exerts strong selective pressure on a number of microbial species that are not known to be pathogenic for higher eukaryotes. Therefore, from this point of view, rather than being harmless unicellular eukaryotes, amoebae possess tools that limit the replication of other microorganisms. The ability to battle back against amoebae may result in the acquisition of virulence factors that allow the potential pathogen to survive and grow within mammalian hosts.

Amoebae possess a number of characteristics that make them similar to macrophages. As a result, they represent excellent environmental incubators that force adaptation of microorganisms to overcome killing mechanisms that are shared by both amoebae and macrophages. The concept that survival in macrophages may be a consequence of adaptation to overcoming amoebal killing mechanisms has been an undercurrent in the literature since the original discovery that *Legionella pneumophila* replicates within amoebae (Rowbotham, 1986). An extensive review of the wide variety of microorganisms that appear to be pathogenic for both amoebae and humans argues convincingly for the importance of amoebae as selective agents in this regard (Greub and Raoult, 2004). The main argument in favor of this idea is that bacteria provide the primary food source for amoebae, so there must be an arsenal of antimicrobial strategies similar to that observed in phagocytic cells of higher eukaryotes. Similar cytoskeletal elements are used to internalize bacteria in both amoebae and phagocytes (Clarke and Maddera, 2006), and killing of the prey is an important property of both cell types (Maselli et al., 2002). The killing mechanisms in both phagocytes and amoebae may also be similar. There exists a fully functional NADPH oxidase, generating toxic free radicals that are potentially antimicrobial (Bloomfield and Pears, 2003; Lardy et al., 2005), although there is no indication of whether this complex plays a role in killing bacteria (Benghezal et al., 2006). In addition, similar to phagocytic cells, amoebae make antimicrobial peptides called amoebapores that appear to be central players in the killing of bacteria. Interestingly, some of these cytolytic proteins show sequence similarity to peptides released from granules within mammalian natural killer (NK) cells, further emphasizing the similarity between the antimicrobial response of amoebae and mammalian innate immunity strategies (Andra et al., 2003; Kumar et al., 2001; Leippe and Herbst, 2004; Zhai and Saier, 2000).

There is significant evidence that some microorganisms pathogenic for mammals may have evolved to withstand the hostile environment of macrophages as a result of acquiring genes allowing survival within amoebal species. The strongest support for this model comes from *L. pneumophila*, the causative agent of Legionnaires' disease (Pedro-Botet and Sabria, 2005). Disease by this gram-negative bacterium is initiated after inhalation of aerosols from contaminated water supplies (Muder et al., 1986; Woo et al., 1992). Although the microorganism is readily found in a number of environmental sources such as streams, plumbing systems, and even distilled water supplies (George et al., 1980; Pine et al., 1979), it has complex nutrient requirements for growth in vitro (Fliermans et al., 1979, 1981; Tobin et al., 1980). The apparent explanation for this paradox is that the bacterium readily infects and grows within the amoebae that inhabit these water sources, as demonstrated originally by Rowbotham (Rowbotham, 1980, 1986; Skinner et al., 1983). In fact, the number of *Legionella* species known to cause disease in humans is just a small subset of the spectrum of the legionellae known to be either pathogenic for, or to grow within, amoebae (Atlas, 1999). Included among these are related organisms, such as the *Legionella*-like amoebal pathogens (Adeleke et al., 2001; Greub et al., 2004; Hookey et al., 1995), for which there is scant evidence for any role in disease of higher eukaryotes. The close association of *L. pneumophila* with amoebae originally suggested to workers that the true infectious bolus that initiates disease consists of heavily infected amoebae that dump their bacteria-laden payload in the lungs of target hosts (Rowbotham, 1980). In fact, there is experimental evidence that the presence of amoebae during lung infection of mice enhances the pathogenicity of the bacterium (Brieland et al., 1996).

The growth of *L. pneumophila* in amoebae and macrophages appears extremely similar (Abu Kwaik, 1996; Gao and Kwaik, 2000). For both amoebae and macrophages, the bacterium is internalized in a vacuole that bypasses fusion with the host cell endocytic pathway, avoiding localization with lysosomal material (Bozue and Johnson, 1996; Horwitz, 1983). Once inside, membranes from the endoplasmic reticulum (ER) sequester around this compartment, and the bacterium proceeds to replicate (Abu Kwaik, 1996; Dietrich et al., 1995; Kagan and Roy, 2002; Robinson and Roy; 2006; Tilney et al., 2001). Required for formation and replication within this compartment in both amoebae and macrophages is the bacterial Dot/Icm system (Segal and Shuman, 1999; Solomon et al., 2000), a protein translocation complex that deposits proteins across the vacuolar membrane. The close association of *Legionella* species with amoebae in water supplies, the presence of organisms very similar to *L. pneumophila* that appear to have absolute tropism for amoebae, the very similar growth cycles of *L. pneumophila* within amoebae and macrophages, and similar bacterial determinants required for growth within both types of cells argue strongly that interaction with amoebae selected for the tools that allow the bacterium to grow intracellularly.

L. pneumophila is not the only pathogen that appears to have developed strategies that allow survival in both amoebae and in multicellular organisms, although with these other microorganisms, the connections between amoebal growth and pathogenesis in higher eukaryotes are not as compelling. In some cases, the importance of being able to kill or grow within amoebae is predicted based on the ecological niche in which certain classes of pathogens are expected to reside. For instance, soil amoebae, such as the dictyostelids, are more prevalent in areas where there is a high density of plant life (Cavender et al., 2005; Landolt et al., 2006; Vadell and Cavender, 2007). The expectation, therefore, is that for a plant-pathogenic microorganism to be successful, it first has to be able to combat attack by soil amoebae, which may effectively be an arm of innate immune defense for plants susceptible to the pathogen. Consistent with this observation, human-pathogenic *Pseudomonas* species, which have been shown to have the potential to cause disease in plants (Rahme et al., 1995), have the ability to kill *Dictyostelium discoideum* amoebae, suggesting common strategies for survival in the soil and virulence in humans (Pukatzki et al., 2002). Furthermore, the *Mycobacterium tuberculosis* ESAT-6-related secretion apparatus, which is required for survival in macrophages and pathogenesis in mammals, is found in nonpathogenic soil mycobacterial species (Berthet et al., 1998; Converse and Cox, 2005; Gey Van Pittius et al., 2001, 2006). Presumably, the primary selective pressure for acquisition and retention of this apparatus by soil mycobacteria is to ensure their survival in the face of predators, with the chief predator being amoebae. Consistent with this proposition is the observation that both pathogenic and nonpathogenic mycobacterial species are capable of survival within a number of amoebal species, presumably because they have common antiamoebal determinants (Adekambi et al., 2006; Cirillo et al., 1997; Miltner and Bermudez, 2000; Skriwan et al., 2002; Solomon et al., 2003).

In addition to the fact that some pathogens may have acquired unique virulence determinants in response to selective pressures exerted by amoebae, there are also dedicated amoebal pathogens that show strong taxonomic similarities to human pathogens. Perhaps most interesting is members of the *Parachlamydiaceae* which are dedicated pathogens of amoebae that are highly similar to human-pathogenic *Chlamydiaceae* (Greub and Raoult, 2002). As might be guessed by the names of the two members of this family, *Parachlamydia acanthamoeba* and *Neochlamydia hartmannellae*, the organisms were originally isolated from amoebae, where they grow within membrane-bound vacuoles as replicative forms that are similar to the reticulate bodies described for *Chlamydia* species (Greub et al., 2003). As is true for the human-pathogenic *Chlamydia*, these organisms undergo a developmental cycle during intracellular growth and are liberated by the amoebae in the form of metabolically inert, but rather tough, elementary bodies (Greub and Raoult, 2002). Not only do these organisms appear very similar to human pathogens, evidence is accumulating that they may be involved in respiratory diseases, with both zoonotic and human cases reported (Corsaro and Venditti, 2004). Therefore, as appears to be the case with *L. pneumophila*, amoebae disseminate the infectious agent as well as provide the environmental niches that cause the selective pressure for bacteria to acquire genes necessary for intracellular growth.

WHY USE *D. DISCOIDEUM* TO STUDY PATHOGEN-MACROPHAGE INTERACTIONS?

Given the recognized importance of the amoeba as both a model for the macrophage and a potential incubator for the acquisition of virulence factors that allow survival within cells of higher eukaryotes (Greub and Raoult, 2004), there must be an ideal host organism that could be used to analyze the interaction of amoebae with microbes. For this purpose, workers have turned to *D. discoideum*, one of the few amoebal species in which there has been extensive genetic analysis. (Table 1 lists pathogens studied with *D. discoideum*.) As is true of most amoebae, this single-celled organism has both a vegetative stage and a dormant stage (Kessin, 2006). Like most of the amoebae from which pathogens have been isolated, *D. discoideum* is a free-living organism, although it is normally found in the soil, rather than in water supplies, as is often the case for amoebae harboring pathogens such as *L. pneumophila*. Also, different from most of the known amoebal hosts of microbial pathogens, the dormant stage of *D. discoideum* does not exist as a simple cyst, but results from a process of a developmentally defined series of steps ending in the formation of a multicellular fruiting body that suspends a package of spores (Weijer, 2004). This process of multicellularization, which occurs in response to nutrient deprivation, involves the cessation of cell division, initiation of aggregation, differentiation into precursor cell types, and eventual formation of two distinct cell types in the dormant state: spore and stalk cells. The resulting fruiting body consists of a pocket of spores sitting on top of dead stalk cells. Although this process is considerably more elaborate than is observed with natural amoebal hosts of pathogens such as *Acanthamoeba*, the group migration and social interactions of the differentiating *D. discoideum* is reminiscent of a variety of chemotactic events involved in both phagocyte migration and multicellular development in higher eukaryotes (Williams et al., 2006).

D. discoideum is also attractive because of the ease of genetic analysis relative to other amoebal species. The microorganism is haploid, and this has greatly facilitated performing functional genetics over the years (Eichinger et al., 2005; Kuspa and Loomis, 1992). A large collection of developmental mutants have been isolated that have allowed analysis of the differentiation process, and the ability to isolate directed knockout mutants has resulted in a more detailed biological analysis than can be pursued in most other amoebae (Weijer, 2004). The complete genome sequence of *D. discoideum*, predicted to encode greater than 12,000 proteins, is now a critical tool that allows quick disruption of any nonessential gene (Eichinger et al., 2005). In addition, because the components of the cytoskeletal machinery appear to be very similar to those found in higher organisms, directed knockout mutations affecting

TABLE 1 Microorganisms that are pathogenic for *D. discoideum*

Microorganism	Nature of virulence	Reference(s)
Legionella pneumophila	Intracellular growth cytotoxicity	Hagele et al., 2000; Solomon et al., 2000
Legionella-like amoebal pathogens	Intracellular growth	Skriwan et al., 2002
Mycobacterium avium	Intracellular growth	Skriwan et al., 2002
Mycobacterium marinum	Intracellular growth	Hagedorn and Soldati, 2007; Solomon et al., 2003
Acinetobacter baumannii	Cytotoxicity	Smith et al., 2007
Pseudomonas aeruginosa	Cytotoxicity	Pukatzki et al., 2002
Vibrio cholerae	Cytotoxicity	Pukatzki et al., 2006
Aeromonas hydrophila	Cytotoxicity	Froquet et al., 2007
Aeromonas salmonicida	Cytotoxicity	Froquet et al., 2007
Cryptococcus neoformans	Intracellular growth	Steenbergen et al., 2001
Klebsiella pneumoniae	Cytotoxicity	Benghezal et al., 2006
Neochlamydia hartmannellae	Intracellular growth	Horn et al., 2000

the cytoskeleton have allowed workers to push ahead in analyzing basic events involving cell motility, chemotaxis, and phagocytosis (De Lozanne and Spudich, 1987; Kimmel and Parent, 2003; Parent and Devreotes, 1999; Van Haastert and Devreotes, 2004). Furthermore, most of the basic components of the ERK mitogen-activated protein (MAP) kinase pathway are present in *D. discoideum*, allowing detailed analysis of global signaling networks in *D. discoideum* to be applied to both higher eukaryotes and other amoebal species (Ma et al., 1997; Mendoza et al., 2005). Finally, random marked insertion mutagenesis strategies are available that allow rapid recovery and mapping of gene disruptions using restriction enzyme-mediated integration (REMI [Kuspa and Loomis, 1992; Mann et al., 1998]). Together, all these properties make *D. discoideum* the obvious host when searching for amoebal species that allow detailed molecular analyses.

Even with all the above desirable qualities of the organism, *D. discoideum* would not be useful if it did not have biological properties that mimicked phagocytes. In the amoebal form, it grows by feeding on bacteria, in a phagocytic process that involves internalization of prey in a membrane-bound compartment followed by digestion. This process is reminiscent of the most effective phagocytic cells in multicellular organisms. Equally important is the use of axenic mutants of *D. discoideum* that are able to grow in liquid medium in the absence of bacteria (Kessin, 2006). This allows the maintenance of cells in monolayer cultures that can be challenged by either particles or pathogens for the purpose of analyzing host-target relationships. Many of the proteins thought to play roles in phagocytic processes in higher eukaryotes have been analyzed in *D. discoideum* through the use of directed deletion mutations (Bretschneider et al., 2004; Cardelli, 2001; De Lozanne and Spudich, 1987; Insall et al., 2001; Rupper et al., 2001; Seastone et al., 2001). In many of these cases, the isolation of deletion mutations in this organism allowed workers to obtain the first clear evidence for the importance of these proteins in any phagocytic process. Finally, genetic analysis of phagocytosis has allowed the discovery of proteins that had not previously been implicated in phagocytosis. For instance, the first demonstration that coronin was involved in phagocytosis was through the analysis of knockout mutations in a very clear series of studies (Maniak et al, 1995; Rauchenberger et al., 1997), while it has taken about a decade to identify the precise role of this protein in macrophage phagocytosis (Yan et al., 2005, 2007). Similarly, the first demonstration that unconventional myosins play a role in phagocytosis was through the use of *D. discoideum* knockout mutations (Titus, 1999, 2000), and this was later confirmed during analysis of phagocytic cells from higher eukaryotes (Cox et al., 2002).

Perhaps the most surprising parallel to higher eukaryotes is the recent discovery that *D. discoideum* possesses something akin to innate immunity (Chen et al., 2007). If this connection can be solidified, then amoebae may respond to pathogens by either enhancing their antimicrobial activity, signaling to other cells, or initiating a death cascade that terminates pathogen infection by inducing amoebal suicide. The first hint that there may be innate immune functions within the amoeba came from the *D. discoideum* genome sequence (Eichinger et al., 2005), which raised the possibility that there are ways in which this unicellular organism can recognize potential pathogens and mount an effective response that terminates replication of the pathogen.

In both plants and animals, families of proteins exist that recognize common patterns found in a broad spectrum of microorganisms (Hoffmann, 2003; Imler and Hoffmann, 2003). These pattern recognition molecules are found localized on membrane compartments and in the cytoplasm of sensor cells. The best characterized of the membrane-associated proteins are the Toll-like receptors (TLRs), originally identified in *Drosophila* but also found in a large swath of multicellular organisms (Iwasaki and Medzhitov, 2004). In the cytoplasm, at least two families of recognition molecules exist: (i) those involved in initiating the type I interferon response (Gack et al., 2007; Ishii et al., 2006) and (ii) NOD-like receptors (NLRs) (Bourhis and Werts, 2007; Inohara et al., 2005; Rairdan and Moffett, 2007; Wilmanski et al., 2008). The latter group is found throughout the animal and plant kingdoms (Leipe et al., 2004). The result of recognition by these sensors is that downstream signaling molecules are activated that cause transcriptional upregulation of antimicrobial responses (Girardin et al., 2003; Iwasaki and Medzhitov, 2004; Kufer et al., 2006). In *Drosophila*, this response can involve the production of antimicrobial peptides, while in mammals, phagocytic activation and recruitment of immune cells occur. In both plants and animals, host cell death pathways are manipulated to limit the infectious disease (Werts et al., 2006). The importance of these recognition systems in simple multicellular organisms is emphasized by the results of the genome sequence of the sea urchin, which encodes more than 400 members of the TLR and NLR families (Rast et al., 2006).

The genome sequence of *D. discoideum* shows at least eight open reading frames having leucine-rich repeats (Eichinger et al., 2005), which are often found in plant resistance or involved in pattern recognition of microbes (Burch-Smith and Dinesh-Kumar, 2007; Meyers et al., 2003; Takeda et al., 2003; Ting and Davis, 2005). In fact, one of these proteins is annotated as an ortholog of the NLR protein NOD3. In addition, two genes encode proteins with TIR domains (TirA and TirB), which are involved in transferring information from recognition proteins to downstream signaling cascades. These intriguing similarities open up the possibility that *D. discoideum* may have a form of pattern recognition that leads to an innate immune response, either in the form of induction of host cell death or enhanced microbial killing. An unexpected connection to this process came out of studies on detoxification of fluorescent dyes during development. After aggregates form in the presence of nutrient-limiting conditions, amoebae assume a slug-like association that results in group motility and chemotaxis. After addition of fluorescent dyes to the slugs, a single cell type called the "sentinel" or S cell appears to accumulate the toxic compound. When two of the hypothesized proteins involved in pattern recognition were analyzed, one having a leucine-rich repeat and the other a TIR domain, both were found to have higher expression levels in S cells than in other cells in the slug (Chen et al., 2007). Most interesting was the fact that the S cells appeared to be the primary phagocytic cells in the slug, and that when challenged with *L. pneumophila*, which is pathogenic for *D. discoideum*, the S cells sequester the bacteria and became separated from the rest of the slug. The presence of one of the TIR domain-containing proteins (TirA) was required for survival in the presence of both pathogenic and normally nonpathogenic bacteria, as exposure of a mutant lacking the protein resulted in death to vegetative cells exposed to bacteria. In this way, the protein is similar to other proteins that are involved in pattern recognition pathways, as signaling in response to microorganisms does not distinguish between pathogens and nonpathogens, but rather involves recognition of common motifs found on a variety of microorganisms.

Interaction of *D. discoideum* with Nonpathogenic Microorganisms

During phagocytosis of either nonpathogens or particles, internalization of the target by *D. discoideum* proceeds via a process that is very reminiscent of phagocytic uptake by macrophages. Using *D. discoideum* marked with a fluorescent protein fusion to LimE, a protein that binds only to actin in the filamentous form (Fischer et al., 2004), Clarke and Maddera (2006) found that there was rapid polymerization of actin about nascent phagosomes surrounding the *Escherichia coli* prey. Shortly after closure is complete, the actin structure starts to dissipate, but this is followed up by new polymerization that takes place at one pole of the phagosome, in a process reminiscent of actin comet formation during *Listeria* or *Shigella* actin-based motility (Carlsson and Brown, 2006). The Arp2/3 complex appears to be heavily involved in this second wave of polymerization in *D. discoideum*, which is also reminiscent of actin-based motility. Interestingly, similar actin tails have been observed forming about phagosomes in mammalian cells after internalization, supporting the idea that this event is a typical step during phagocytosis in all cells (Yam and Theriot, 2004). The reason for this second wave of polymerization is apparently to propel the phagosome along the endocytic pathway, as this precedes phagosome acidification, facilitating maturation of the phagosome and ensuring degradation of the internalized bacteria (Clarke and Maddera, 2006).

Bacterial entry into degradative compartments can be followed by tracking the recruitment of the vacuolar H^+-ATPase (V-ATPase) to the phagosome after internalization (Clarke et al., 2002), a protein that appears to be similar to the identically named complex found in multicellular organisms (Inoue et al., 2003). The V-ATPase is associated with the amoebal contractile vacuole (which maintains cytoplasmic osmotic balance) as well as the amoebal endocytic network, where the protein is responsible for acidification of the luminal contents in a fashion similar to that observed in higher eukaryotes (Clarke et al., 2002; Gerisch et al., 2002). It is clear from videomicroscopy that delivery of new membrane containing the V-ATPase to internalized yeast particles is the consequence of recruitment of small endosomes to the phagosome, followed by fusion (Clarke et al., 2002). These small vesicles may be identical to the rapidly moving endocytic compartments carrying fluid-phase markers that were identified in a study using total internal reflectance fluorescence microscopy (TIRF) (Neuhaus et al., 2002), or to the "ring of dots" of the lysosomal protease cathepsin D observed surrounding latex beads or yeast internalized by *D. discoideum* (Gotthardt et al., 2002). The next event that occurs, however, is distinctly amoebal: indigestible portions of the prey are spit out by the amoebae in an exocytotic process, with the V-ATPase cycling off prior to completion of this process (Clarke et al., 2002). Exocytosis of incompletely digested organisms is a well-characterized phenomenon in amoebae, and the sequential steps involved in this process have been studied in beautiful detail by Maniak and coworkers (Rauchenberger et al., 1997). Furthermore, unlike the situation in higher eukaryotes, where the lysosome is a terminal acidic compartment, the degradative compartment in *D. discoideum* neutralizes (hence, the loss of the V-ATPase) before exocytosis of the material remaining in the compartment (Rauchenberger et al., 1997). This also raises an important point regarding strategies that intracellular pathogens must use to ensure that they establish a successful replicative niche within amoebae. Not only must the pathogen take steps to ensure that it can avoid being killed by the host cell, once it has avoided being digested it must ensure that the unsuccessful amoebal phagocyte does not expel it.

Other than the postlysosomal step, much of the endocytic pathway appears very similar morphologically in amoebae and higher eukaryotes. For instance, there appears to be tubulation of early endosomes that leads to the lysosomal network (Neuhaus et al., 2002), as was seen in animal cells (Hollenbeck and Swanson, 1990; Kreitzer et al., 2000). In addition, there is evidence for a late endosomal compartment that is morphologically similar to multivesicular bodies in other cells (Williams and Urbe, 2007). Also similar to higher eukaryotes, fusion between vesicles after their contact appears to be a rare event (Neuhaus et al., 2002), as most contact between vesicles results in transient docking without complete fusion, similar to what Swanson and coworkers had observed during macropinocytosis in macrophage cultures (Racoosin and Swanson, 1993). Analysis of the cycling of endocytic components onto phagosomes harboring either latex beads or yeast cells also indicates that trafficking is very reminiscent of that observed in macrophages (Gotthardt et al., 2002, 2006). Early endocytic markers, such as coronin, and plasma membrane material cycle off latex bead-containing phagosomes shortly after uptake, to be replaced by proteins that are found associated with late endosomal and lysosomal organelles in the cell, such as a subset of CD36 family members and lysosome-associated proteases. Intermediate in time between proteins associated with these two compartments was recruitment of proteins, such as Rab7, which are important for trafficking from the early to the late endosome (Gotthardt et al., 2006). In summary, very little regarding both the kinetics and constitution of maturing phagosomes seems different in *D. discoideum* when compared with mammalian cells.

One surprising result from the analysis of purified latex-bead-containing phagosomes from *D. discoideum* was evidence for the presence of ER-associated proteins and some components of the translation machinery (Gotthardt et al., 2006). This again is reminiscent of studies on latex-bead-containing phagosomes isolated from mammalian cells by Desjardins and coworkers (Gagnon et al., 2002; Garin et al., 2001). Consistent with a role for ER in uptake of particles, Gerisch and coworkers found that *D. discoideum* deleted for two ER-localized chaperones, calreticulin and calnexin, was defective for phagocytosis of yeast particles (Muller-Taubenberger et al., 2001). Furthermore, using green fluorescent protein (GFP)-labeled derivatives of these proteins, the authors observed early colocalization of these proteins with nascent *D. discoideum* phagosomes bearing yeast, arguing that recruitment of ER-derived material plays a central role in the early stages of phagocytosis. This model is disputed in another work, however, that shows no evidence for colocalization of calnexin-GFP with internalized yeast during internalization by *D. discoideum* (Lu and Clarke, 2005). Even this discordance is reminiscent of arguments regarding phagocytosis of latex beads in macrophages, in which the importance of ER in phagocytosis and its colocalization with nascent phagosomes is questioned (Touret et al., 2005).

GENETIC ANALYSIS OF PHAGOCYTOSIS IN D. DISCOIDEUM

One of the advantages of the D. discoideum system is that it allows the development of selection systems for the identification of novel proteins that resist identification in mammalian systems. Even so, there has not been an overabundance of attempts to identify random mutants that alter rates of phagocytosis in D. discoideum. For the most part, cytoskeletal and regulatory proteins that are thought to affect actin dynamics in mammalian systems have been targeted for directed mutations using homologous recombination. With use of this approach, unconventional myosins, phosphatidylinositol-3-kinases (PI3Ks), a variety of Rho family members, and trimeric G proteins have all been implicated in playing roles in phagocytic uptake in amoebae (Cardelli, 2001; Peracino et al., 1998; Rupper and Cardelli, 2001; Titus, 1999; Wu et al., 1995). The original attempt to isolate D. discoideum phagocytosis-defective mutants that employed a selection strategy was one that was published more than 25 years ago, although the nature of the mutations isolated in this study has never been worked out (Duffy and Vogel, 1984; Vogel et al., 1980). The procedure used for isolation of these mutants is still worth considering, however, because it could allow a rather broad swath of strains to be recovered. Chemically mutagenized cultures of D. discoideum were allowed to grow at 20°C prior to shifting to 27°C and challenging with tungsten particles. Amoebae defective for phagocytosis at the higher (nonpermissive) temperature were enriched on the basis of their lighter density compared with wild-type cells. The mutants that were isolated in this fashion were temperature sensitive both for phagocytosis and for growth, so the procedure allowed for the isolation of mutations in genes that were potentially essential for viability. The problems with this approach include the fact that identification of the gene that is altered in the mutants is not simple and it was difficult to absolutely demonstrate that the temperature-sensitive phenotype and phagocytosis defect were linked. In fact, revertant studies indicated that the temperature-sensitive phenotype and the phagocytosis defect were unlinked (Vogel et al., 1980). However, with new technology, such as the construction of complete gene banks containing D. discoideum cDNA clones of mRNAs, it is now possible to try to map mutations of this sort by complementation cloning. A strategy like this one might be worth repeating.

Analysis of the above mutants indicated that the defects in the mutants isolated were in particle recognition, so to uncover mutations that alter functions involved in the mechanics of uptake, other approaches may be necessary. A second attempt to isolate mutants by using chemical mutagenesis was performed by Mellman and coworkers (Cohen et al., 1994), employing fluorescence-activated cell sorter (FACS) sorting to isolate amoebae that failed to internalize fluorescently labeled E. coli. The mutants isolated all showed normal growth under axenic conditions, but were growth defective on lawns of bacteria, and were also defective for uptake of latex beads. Unlike the above set of mutants, adhesion to particles appeared to be normal, although there was a common defect in the ability of the mutants to bind and spread on solid substrates such as culture dishes. There was no attempt to clone the affected genes, so the causes of the defects were never determined. Therefore, although classic techniques demonstrated that mutants could be isolated that altered various adhesion events involved in phagocytosis, they were unable to define proteins that mediate this process.

The use of REMI mutagenesis solves the problem of mapping random mutants (Kuspa and Loomis, 1992), although this strategy requires that the isolated knockout mutations retain viability in axenic growth conditions. Furthermore, even if REMI is used, isolating diverse sets of mutants affecting many processes requires hunts that do not select against certain classes of mutants. Cosson and coworkers (Cornillon et al., 2000; Gebbie et al., 2004) performed a mutant hunt that was similar to one performed previously (Cohen et al., 1994), taking a pool of D. discoideum REMI mutants and challenging them with fluorescent hydrophilic latex beads prior to FACS isolation of amoebae, isolating pools that have lowered levels of internalization of the fluorescent marker. As was true of the above mutant hunts, this procedure did not distinguish between mutants that failed to adhere to target particles and those that failed to internalize bound particles. Even so, each of the three genes described in this hunt was quite different in nature. Two (called phg1 and phg2) showed defects in recognition of particles, while the third was in the gene for the unconventional myosin VII, which had been previously shown to be defective for phagocytosis and appeared to be normal for particle adhesion (Titus, 1999). The phg1 mutation was in a gene encoding a polytopic membrane protein that was found in purified latex bead phagosomes (Cornillon et al., 2000). Mutants lacking this protein showed specific defects in internalization of hydrophilic latex beads and E. coli as well as some other bacteria, but were competent for uptake of Klebsiella and showed adhesion defects similar to those described in other hunts for phagocytosis mutants (Vogel et al., 1980). The protein itself is unlikely to be a surface receptor for particles, because the phg1 mutant causes pleiotropic effects on the cell, including showing greatly altered plasma membrane protein composition in the mutant compared to wild-type cells (Benghezal et al., 2003).

The second novel mutant isolated by this FACS enrichment strategy, in the phg2 gene, is quite interesting and appears to cause a defect in linking cytoskeletal processes to the cell surface (Gebbie et al., 2004). The REMI insertion in this strain is located in a gene encoding a serine/threonine phosphotransferase that has a Ras and phosphoinositol 4,5-bisphosphate (PIP2) binding region (Blanc et al., 2005). As was true of phg1, mutations in phg2 caused a general defect in adhesion to solid matrices, but this did not appear to be due to loss of a receptor. Instead, there appeared to be a defect in promoting actin polymerization as a response to supporting the formation of adhesive structures at the cell surface. It appears that the Phg2 kinase is located at a very high point in the hierarchy of factors necessary to promote phagocytosis, with Phg2 important for stabilizing adhesion events by recruiting cytoskeletal elements to sites of interaction of the amoeba with target surfaces. That the protein probably works at a very early step in phagocytosis is supported by the fact that PIP2, which is responsible for localization of Phg2, only accumulates during the formation of the nascent phagosome, and then dissipates during closure about particles (Blanc et al., 2005). This indicates that accumulation of Phg2 at the site of particle uptake only occurs prior to membrane closure.

The connection between cytoskeletal processes involved in phagocytosis and vesicle-trafficking events is poorly understood, with very few proteins being identified that could link the two processes together. Therefore, the identification of a tripartite complex between Rab21 and the two LIM domain proteins LimF and ChLim was an important observation (Khurana et al., 2005). Rab 21 is a poorly characterized member of the Rab family of small ras-like GTPases that regulate a variety of vesicle trafficking, tethering, and organelle movement events (Jordens et al., 2005). Evidence for direct modulation of actin polymerization events by Rab proteins, such as those involved in phagocytosis, is rare. D. discoideum mutants that are deleted for limF or express a dominant interfering form of Rab21 are defective for phagocytosis of yeast particles, whereas loss of expression of ChLim resulted in an enhancement of phagocytosis (Khurana et al., 2005). Unlike the previously described phg1, phg2, and myoVII mutants, there is no defect in cellular adhesion in mutants lacking any single component of the Rab21 complex, indicating a specific defect for phagocytosis. Despite this fact, there is no evidence that this complex signals directly from the site of particle adhesion to cytoskeleton, as there is no specific localization of Rab21 about the phagocytic cup (Khurana et al., 2005). Overproduction of constitutively active Rab21 results in large-scale actin-mediated ruffling of the cell surface, so it appears that the complex may be associated with establishing a state of high phagocytosis competence within D. discoideum that allows the cell to effectively internalize a particle after contact with the plasma membrane.

ANALYSIS OF PATHOGEN INTERACTION WITH D. DISCOIDEUM

Klebsiella pneumoniae

Further analysis of the phg1 mutation indicated that the lesion results in a more complex defect than the simple inability to internalize selected bacterial species. In contrast to the defect in phagocytosis of E. coli and hydrophilic latex beads observed in this mutant, uptake of the pathogen K. pneumoniae is unaffected, but the mutant cannot degrade the bacterium after internalization (Benghezal et al., 2006). In fact, D. discoideum phg1 mutants succumb to K. pneumoniae, even though wild-type amoebae efficiently kill the pathogen. Furthermore, the bacterium proliferated in these mutants. It is not known whether proliferation of the bacteria is a reflection of the fact that phagosomes harboring it disintegrate and allow the pathogen to replicate freely in the amoebal cytoplasm, or whether there is a defect in trafficking of the phagosome into a lysosomal compartment. It is known that phg1 mutants show inefficient retention of a major endosomal protein, which seems to leach out onto the amoebal cell surface as a result of improper trafficking (Benghezal et al., 2003), but this result is consistent with either model, as endocytic compartments may be unstable in the mutant. Therefore, either phagosomes from the phg1 have reduced integrity relative to wild-type strains, or they do not interact properly with the endocytic pathway to allow trafficking of internalized bacteria into a degradative compartment.

As the ability to clear pathogens is a central feature of phagocytic cells, particularly in higher eukaryotes, more D. discoideum mutants were sought that had similar defects in resistance to attack by K. pneumoniae. One other gene, called kil1, was identified by searching for genes that when overexpressed would be able to suppress the phg1$^-$ defect in digestion of K. pneumoniae. When the chromosomal KIL1 gene was inactivated in D. discoideum, it was found to result in reduced survival of the amoebae in the presence of K. pneumoniae. Little is known about the role that the Kil1 protein plays in supporting clearing of K. pneumoniae other than the fact that the protein is a sulfotransferase (Benghezal et al., 2006). It was also of interest to determine whether mutations that caused a defect in clearing pathogens by D. discoideum would also exert these effects when crossed into multicellular eukaryotic species. This allowed a direct test of the model that a protein involved in supporting phagocytosis in D. discoideum plays a central role in innate immune defense in multicellular eukaryotes. In fact, when the ortholog of phg1 in Drosophila was knocked out, the animals showed a specific defect in clearing of K. pneumoniae but not other pathogens (Benghezal et al., 2006). Therefore, orthologs of the Phg1 protein appear to play an important role in host defense against a subset of bacterial pathogens.

Legionella pneumophila

The details of the interaction of phagocytic cells with K. pneumoniae are poorly understood, and the relationship of this pathogen with amoebae is unclear. There are number of microbial pathogens discussed in this chapter, however, that are documented to interact with amoebae in the environment, making them attractive microorganisms to study with respect to D. discoideum. The pathogen most closely associated with amoebae is L. pneumophila.

L. pneumophila traffics identically in D. discoideum, macrophages, and environmental amoebae, using the same bacterial determinants for intracellular growth. As is seen in all cell types, intracellular growth of L. pneumophila in D. discoideum requires the Dot/Icm protein translocation system, which deposits protein substrates into host cells that are necessary for formation of the bacterial replication vacuole (Hagele et al., 2000; Solomon et al., 2000). The reported yields of bacteria and the amount of time it takes to establish replication within D. discoideum vary between investigators, but if media are used that limit replication of D. discoideum, then intracellular growth is robust (Li et al., 2005; Solomon et al., 2000). In fact, although amoebal culture requires ambient temperatures, initiation of intracellular growth is only slightly delayed in D. discoideum relative to macrophages, and the terminal yield after 3 to 4 days of culture is difficult to distinguish between the two cell types (Li et al., 2005). In response to high-multiplicity incubation (MOI) of L. pneumophila with D. discoideum, the amoebae die and fail to replicate the pathogen, similar to what is observed in macrophages (Solomon et al., 2000). Interaction of L. pneumophila with amoebae incubated in suspension is poor, because D. discoideum under these conditions is resistant to high MOI challenge and unable to support L. pneumophila intracellular replication at low MOI (Solomon et al., 2000). Presumably, part of the reason for this defect is because efficient contact between the bacterium and D. discoideum requires immobilization of amoebae on solid substrates, but in addition, intracellular growth may require specific uncharacterized cell cycle conditions that respond to surface adherence (Z. Li and R. R. Isberg, unpublished observations).

Similar to what has been observed in macrophages, shortly after uptake into D. discoideum, the bacterium is found in a membrane-bound compartment that recruits material from the ER (Fajardo et al., 2004; Li et al., 2005;

Lu and Clarke, 2005). This recruitment has been detected by localizing fluorescent protein tags such as HDEL-GFP, a protein harboring an ER-retention signal (Li et al., 2005; Lu and Clarke, 2005), or calnexin- and calreticulin-GFP, which are ER-associated chaperones. There is some disagreement regarding how rapidly ER is recruited to the *Legionella*-containing vacuole, with one group reporting almost immediate sequestration of calnexin- and calreticulin-GFP about phagocytic cups (Fajardo et al., 2004), indicating that ER may be involved in the phagocytic event, with another seeing association of HDEL-GFP no sooner than 15 min after uptake (Lu and Clarke, 2005). Similar differences in time of detection have been observed in mammalian phagocytes (Derre and Isberg, 2004; Kagan and Roy, 2002; Kagan et al., 2004), but these variations in detection are of little consequence with regard to interpretation of data. The important point is that ER is observed associated with this compartment, and it appears to be essential for intracellular replication. Mutants lacking either calreticulin or calnexin are defective for supporting intracellular replication of *L. pneumophila* (Fajardo et al., 2004). This indicates that a dysfunctional ER interferes with intracellular replication (Color Plate 9).

Within 2 h of uptake, the ER-associated vacuole begins to expand in volume, and by 6 h, the majority of replication vacuoles are quite spacious (Li et al., 2005; Lu and Clarke, 2005). The importance of this expansion is unclear, but it should be pointed out that a *D. discoideum* mutant defective for production of phosphoinositol 3-phosphate lipids supports more efficient replication of *L. pneumophila* than wild-type strains despite the fact that the replication compartment fails to expand in this mutant strain background (Weber et al., 2006).

L. pneumophila Dot/Icm mutants also appear to traffic within *D. discoideum* in a fashion that is reminiscent of macrophages. In the absence of various components of the secretion apparatus, the replication-deficient mutants are found localized in a compartment that is rich in lysosomal proteins (166), the lysosome-associate CD36 family member LIMP (76), and the V-ATPase (Chen et al., 2004; Lu and Clarke, 2005), with no evidence of ER association. Wild-type strains of *L. pneumophila*, on the other hand, are generally devoid of these markers (Hagele et al., 2000; Lu and Clarke, 2005; Solomon and Isberg, 2000). Even at later time points, in replication vacuoles that have large numbers of bacteria, there is little evidence of V-ATPase association with the replication compartment (Lu and Clarke, 2005). These observations are all consistent with a range of studies indicating that there is general avoidance of the macrophage endocytic pathway by *L. pneumophila* throughout the replication cycle (Horwitz, 1983; Kagan and Roy, 2002; Swanson and Isberg, 1995; Tilney et al., 2001). There is a report that the *L. pneumophila* replication vacuole associates with late endosomal compartments as a critical determinant of replication within mouse bone marrow-derived macrophages (Sturgill-Koszycki and Swanson, 2000), but this does not appear to be the case with *D. discoideum* as a host (Lu and Clarke, 2005).

A number of *D. discoideum* mutants defective in a variety of cytoskeletal and membrane-trafficking functions have been analyzed. As pointed out previously, the common thread of each of these mutants is that the amoebae are all viable in axenic medium, indicating that only nonessential genes have been analyzed. One of the most striking results is that elimination of several different proteins involved in phagocytosis results in enhanced *L. pneumophila* growth. For instance, elimination of myosin I, coronin, and paxillin in *D. discoideum* all cause increased yields of bacteria relative to wild type after several days of incubation (Hagele et al., 2000; Solomon and Isberg, 2000). The strongest enhancement of intracellular replication is a consequence of total elimination of class I PI3K in the cell (Weber et al., 2006). It had been previously reported that uptake of *L. pneumophila* by macrophages differed from most other reported phagocytic events in that it was not reduced by pharmacological inhibition of PI3K (Khelef et al., 2001; Laguna et al., 2006), so it was not surprising that the absence of phosphoinositol 3-phosphate lipids would not reduce uptake. The total lack of restriction in these mutants, however, was unexpected. The most likely explanation for this result is that these mutants reduce trafficking of the *Legionella*-containing vacuole into the endocytic network. This predicts that interaction of the bacterium with host cells is the sum of two competing processes: one that facilitates replication, with the other terminating it. In mutants lacking components of the competing pathway, the relative efficiency of replication vacuole formation is increased. Coronin, paxillin, and PI3K are all known to be involved in phagocytic events leading to the lysosome, and presumably are dispensable for *L. pneumophila* uptake, so they can be considered proteins that promote an antimicrobial pathway leading to termination of intracellular replication. In many ways, this parallels results in mammalian cells, in which interference with the endocytic pathway by the use of dominant inhibitory Rab5 and Rab7 proteins results in enhanced intracellular replication of *L. pneumophila* (Kagan and Roy, 2002; Kagan et al., 2004).

D. discoideum mutants that show defects in phagocytosis are not uniformly more permissive for intracellular replication of *L. pneumophila*, indicating that there may be a unique cytoskeletal pathway for uptake and growth of the microorganism in host cells that shares some components with the default phagocytic system. For instance, the heterotrimeric G-protein regulatory subunit Gβ is required for efficient latex bead phagocytosis by *D. discoideum* (Gotthardt et al., 2006; Peracino et al., 2006; Wu et al., 1995), while both the efficiency of uptake and the rate of intracellular growth of *L. pneumophila* in Gβ$^-$ *D. discoideum* are reduced. Similar results were obtained with an α-actinin mutant and a mutant eliminating multiple Lim-domain-containing proteins of unknown functions (Hagele et al., 2000; Solomon and Isberg, 2000).

One of the mysteries of the *Legionella* intracellular growth is the source of the ER-derived material found associated with the replication vacuole. As host calreticulin and calnexin are required for intracellular replication, it seems likely that functional ER is a prerequisite for intracellular growth (Fajardo et al., 2004). When the ribosome-coated membranes recruited to the vacuole were first identified as ER derived, the similarity of the replication vacuole to autophagic compartments was noted (Swanson and Isberg, 1995). Autophagy is a process in which soluble and organellar material from the cytoplasm of cells is packaged in bulk in a double-membrane-bound compartment for eventual delivery to lysosomes (Levine and Klionsky, 2004). The process is initiated by a PI3K complex and the Atg7-conjugating enzyme which, among other things, transfers the Atg8 protein onto the membrane of the nascent autophagosome (Levine and Klionsky, 2004). Although the *L. pneumophila* replication compartment avoids trafficking to the lysosome, one model holds that biogenesis of the vacuole involves initiating the formation of an autophagic compartment, which is prematurely terminated by *L. pneumophila* virulence determinants to allow efficient replication vacuole formation. Consistent with this hypothesis, the Atg7 and Atg8 proteins associated with nascent phagosomes localize to the *L. pneumophila* replication vacuole (Amer and Swanson, 2005). By this model, either proteins involved in autophagy should enhance intracellular replication, or the termination of autophagy is a necessary step in ensuring replication vacuole formation. *D. discoideum* mutants exist that have lesions in several steps of the autophagy process, including the initiator serine/threonine kinase, the PI3K, Atg7, and Atg8 (Otto et al., 2003, 2004). Mutations in these proteins have no profound effect on intracellular growth in *D. discoideum* (Otto et al., 2004). Although this means it is unlikely that autophagy plays a positive role in *L. pneumophila* replication, it does not argue against the idea that in mammalian cells there may be a battle between the autophagy machinery and *L. pneumophila*, with the pathogen trying to limit an antimicrobial response exerted by the Atg proteins. The fact that autophagy proteins found associated with the replication vacuole in mammalian cells are not observed on the replication vacuole within *D. discoideum* (Otto et al., 2004) is consistent with the proposition that autophagy is a unique antimicrobial response of higher eukaryotic cells that *L. pneumophila* is able to block to allow formation of the replication vacuole.

Although it is not clear whether the *L. pneumophila* relationship with autophagy is more intimate in mammalian cells than it is in *D. discoideum*, there appears to be at least one host protein involved in limiting bacterial replication that may behave differently in amoebae than it does in mammalian cells. One of the best-characterized variants in inbred mouse strains that alters replication of intracellular pathogens is located in the macrophage Nramp1 protein (Fortier et al., 2005b). Mutations that disrupt function of this phagosomal ion transporter result in enhanced replication within macrophages of *Salmonella*, *Leishmania*, and *Mycobacterium bovis* BCG (Fortier et al., 2005b). The interaction of *L. pneumophila* with mouse macrophage is well characterized, and macrophages from different inbred mouse strains show varying capacities to support intracellular replication of the pathogen (Beckers et al., 1995; Dietrich et al., 1995; Fortier et al., 2005a; Wright et al., 2003). In no case has there ever been a demonstration that altered levels of susceptibility are associated with polymorphisms in the Nramp1 gene (Beckers et al., 1995; Yoshida et al., 1991), and it appears that there is no role for restriction of *L. pneumophila* by active alleles of the protein. The situation in *D. discoideum* is remarkably different (Peracino et al., 2006). *L. pneumophila* has enhanced replication in amoebae having a disrupted Nramp1 gene. Overproduction of Nramp1 causes a severe depression in intracellular growth of this pathogen (Peracino et al., 2006), indicating that the channel acts as an active antimicrobial protein. One explanation for the different results in mouse macrophages and *D. discoideum* is that the localization properties of Nramp1 are different in the two species. In macrophages, Nramp1 is associated with late endosomal and lysosomal network (Searle et al., 1998), whereas in *D. discoideum* it appears to localize in nonacidic compartments such as the *trans*-Golgi network and the postlysosomal compartment (Peracino et al., 2006). Since there is no clear corollary of the postlysosomal compartment in macrophages, the nature of the interaction of this protein with intracellular *L. pneumophila* may be quite different in macrophages and amoebae.

In addition to cytoskeletal and degradative processes, *L. pneumophila* manipulates host cell vesicle trafficking pathways (Derre and Isberg, 2004; Dorer and Isberg, 2006). Therefore it would be desirable to directly target genes that encode proteins associated with the docking and fusion of vesicles in a variety of cellular compartments within *D. discoideum*. Although there has been extensive work on these factors in *D. discoideum* (Harris et al., 2001; Khurana et al., 2005; Powell and Temesvari, 2004; Rupper et al., 2001), most of these proteins are essential for viability in species ranging from yeast to humans (Duhon and Cardelli, 2002; Lee et al., 2004). The use of directed deletions or REMI strategies is only useful for nonessential vesicle-trafficking genes, which constitutes a rather limited subset. As a result, studies in *D. discoideum* that focus on vesicle-trafficking proteins use dominant inhibitory or constitutively active mutations located in Rab proteins, which are small GTPases that regulate specific trafficking events in host cells (Rupper et al., 2001). A few mutations, however, do exist that affect expression of proteins involved in vesicle trafficking without causing loss of viability in *D. discoideum*, and one of these is the RtoA protein (Brazill et al., 2001; Wood et al., 1996).

Mutations in *rtoA* were isolated as being defective in development, with lowered efficiency of fruiting body formation relative to wild type, with the small fraction of fruiting bodies formed in the mutant showing aberrant morphologies (Wood et al., 1996). In amoebae, the *rtoA* mutant shows reduced exocytotic rates, decreased pH of endocytic vesicles, a general accumulation of small vesicles, and apparent lack of a postlysosomal compartment (Brazill et al., 2001). All these phenotypes point to a role in processes associated with vesicle trafficking, although there are no clear orthologs in other organisms. Mutations in *rtoA* result in delayed intracellular growth of *L. pneumophila*, and once growth is initiated, the rate of replication appears slower relative to that observed in wild-type *D. discoideum* (Li et al., 2005). In addition, there is a mild defect in uptake of the pathogen, which may reflect the reduced exocytosis rates of the mutant, since this process appears to be important for sequestering membrane material about nascent phagosomes (Bajno et al., 2000; Lee et al., 2004). The less efficient contact of *L. pneumophila* with the mutant may also explain why *rtoA* mutants are more resistant than wild-type strains to high MOI cytotoxicity (Li et al., 2005). When recruitment of ER to the replication vacuole was analyzed, it was clear that this process was disrupted in the *rtoA* mutant. Although the GFP-HDEL marker for ER appeared to be recruited to the *Legionella*-containing compartment within 2 h after uptake, this association appeared unstable. By 8 h postinfection, when there was some evidence of replication in wild-type amoebae, the mutant showed significant loss of GFP-HDEL marker, indicating that the RtoA protein may be required to maintain contact of the replication vacuole with the ER (Li et al., 2005). It was also noted that unlike the spacious wild-type vacuoles supporting replicating *L. pneumophila*, the vacuolar membrane was tightly apposed to the bacterium. As mentioned regarding the PI3K mutant, the presence of a spacious vacuole does not appear to be a requirement for replication of *L. pneumophila* in *D. discoideum* (Weber et al., 2006), but its absence in the *rtoA* mutant could reflect the fact that the compartment has been aberrantly formed, with the simplest readout for this dysfunctional compartment being its unusual morphology (Fig. 1).

The transcriptional response of D. discoideum to L. pneumophila has been analyzed using microarray technology (Farbrother et al., 2006). Interestingly, there did not appear to be a large overlap with the data obtained for the transcriptional response of a macrophage cell line to the pathogen, although in the case of macrophages, there is a significant enhancement of expression of genes under NF-κB control, a regulator that is not present in amoebae (Losick and Isberg, 2006). If these genes are ignored, then a common theme in the two cell types is that various classes of stress response genes are induced, with proteins associated with ubiquitin-dependent degradation processes in both cell types induced in response to the L. pneumophila. One important aspect of the transcriptional response in D. discoideum is an across-the-board induction of tRNA synthetases as well as a pair of translation factors (Farbrother et al., 2006). It is unclear whether the D. discoideum transcripts regulated by L. pneumophila represent an attempt by the amoebae to limit bacterial replication or whether the transcripts are modulated by L. pneumophila to support survival and intracellular growth of the bacteria. The best guess is that both processes are taking place, and what is being observed is a dynamic interplay between events that support and restrict bacterial replication. In this vein, the fact that there is wholesale induction of tRNA synthetase and modifying enzyme can be interpreted as a kind of stress response of the amoebae, with the goal of stimulating progression through the cell cycle. Although it cannot be said for certain that this allows D. discoideum to limit growth of L. pneumophila, culture density and variations in amoebal culture conditions do have profound effects on the replication efficiency of the pathogen (Solomon et al., 2000), so manipulation of proteins that drive the amoeba into different stages in the cell cycle could be one way that growth of L. pneumophila is controlled.

Interaction of Mycobacterial Species with *D. discoideum*

As described in the discussion of the interaction of Mycobacteria with amoebae, the resemblance of pathogenic mycobacterial species with common environmental isolates suggests that selection for virulence could be partially explained by acquiring resistance to killing by soil amoebae. Therefore, it should come as no surprise that both Mycobacterium avium and Mycobacterium marinum are able to grow within D. discoideum (Skriwan et al., 2002; Solomon et al., 2003). Because the difficulty of working with mycobacteria is greater than with many other pathogens, the ability to work with a simple host system is an important technical advance. In the case of M. marinum, which is a pathogen of cold-blooded animals, the ambient temperature culture conditions of D. discoideum are particularly favorable for this pathogen.

There are some technical difficulties in analyzing growth of Mycobacterium spp. that are not faced by L. pneumophila. The most glaring difference is that there is some uncertainty whether the trafficking patterns of Mycobacterium spp. are identical in amoebae and in macrophages, although the work by Hagedorn and Soldati (2007) described below indicates that this may not be a concern. Secondly, unlike L. pneumophila, mycobacteria have the ability to grow in standard D. discoideum culture medium (Ramakrishnan and Falkow, 1994). Thus, all studies on intracellular growth require the addition of antibiotics to the infection medium, so analyses of spread and reinfection cannot be pursued. Given these caveats, the analysis of M. marinum is relatively facile, and studies on its interaction with D. discoideum have been amenable to relatively straightforward strategies.

M. marinum grows robustly within D. discoideum to yields of 20 or more bacteria per amoeba (Solomon et al., 2003). This growth seems to mimic the situation observed

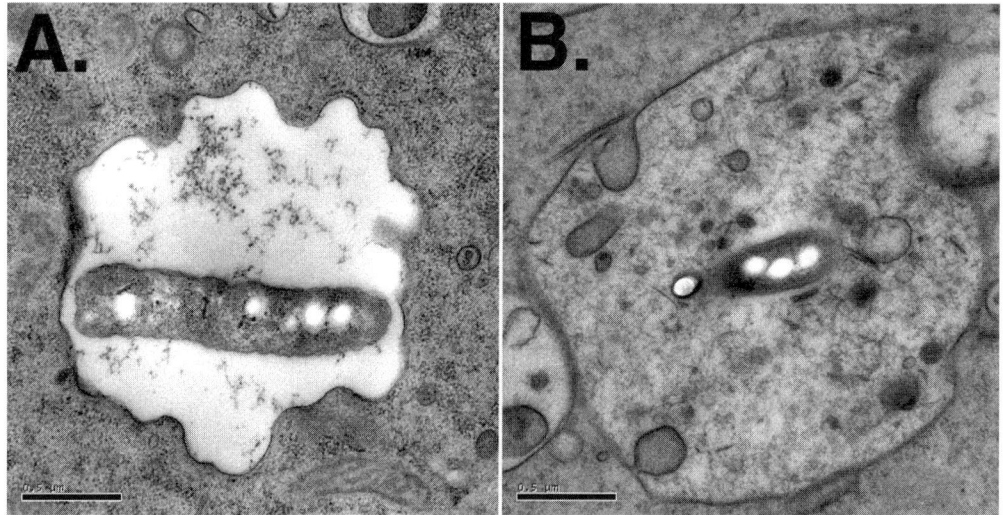

FIGURE 1 A D. discoideum rtoA⁻ mutant is inefficient at maintaining an L. pneumophila replication compartment. L. pneumophila was incubated with either (A) the wild-type AX4 strain of D. discoideum or (B) an AX4 rtoA⁻ strain for 6 h, fixed, and prepared for electron microscopy. The wild-type strain is found within a spacious vacuole, whereas the rtoA⁻ mutant appears to be in a vacuole that has fused with other compartments in the amoebae. Micrograph is from Li et al. (2005).

either in mammalian cells or in phagocytic cells from cold-blooded animals such as frogs and zebrafish (Bouley et al., 2001; Davis et al., 2002; Ramakrishnan et al., 1997, 2000). A well-characterized mutation in the M. marinum glycine-rich PE protein *mag24-1*, which results in defective growth within macrophages and lowered virulence in a variety of animal models (Ramakrishnan et al., 2000), is defective for growth within D. discoideum (Hagedorn and Soldati, 2007; Solomon et al., 2003). Furthermore, the promoter for this gene is activated by entry into phagocytic cells and is similarly activated after uptake by D. discoideum (Hagedorn and Soldati, 2007; Ramakrishnan et al., 2000), further arguing that the intracellular environment encountered by the bacterium is similar in both host cells.

A convincing argument has been made that the growth of M. marinum within D. discoideum takes place in two phases, based on the localization of membrane-associated markers (Hagedorn and Soldati, 2007). This behavior is very reminiscent of trafficking studies within macrophages seen with other related species, such as M. tuberculosis (Fratti et al., 2001; Sturgill-Koszycki et al., 1994, 1996). After uptake into D. discoideum, the pathogen is localized in a compartment that accumulates the V-ATPase marker of late endosomes, and there is some loss of bacterial viability (Hagedorn and Soldati, 2007). This marker, however, is eventually shed and, as the bacteria begin to replicate, any further progress along the endocytic path is arrested. This is in contrast to the avirulent *mag24-1* mutant, which appears to proceed through a degradative pathway that captures the lysosomal protein cathepsin, with eventual transit to the postlysosomal compartment and exocytotic exit. During the course of replication of the wild-type M. marinum, the bacterial compartment accumulates vacuolin about its surface, which is normally a marker of the postlysosomal compartment, but there is little evidence that the bacterium-containing vacuole has any other features in common with this compartment (Hagedorn and Soldati, 2007).

Vacuolin is an ortholog of the mammalian protein flotillin, which is a somewhat mysterious membrane-associated protein that shows some structural similarity to caveolin and appears to be a scaffolding protein that can promote clathrin-independent endocytosis (Glebov et al., 2006). Showing conservation of this localization, M. marinum within peripheral blood monocytes is also localized around the replication vacuole. This association is biologically significant, because gene knockouts of one of the two vacuolin isoforms, VacB, blocked replication of M. marinum and resulted in the formation of a bloated bacteria-containing compartment that had features of a late endosome. One of the remarkable features of M. marinum is that, despite the complexity of the development of the membrane compartment harboring the bacterium, it appears that the membrane eventually gets degraded, releasing bacteria into the amoebal cytoplasm, where much of the replication takes place (Hagedorn and Soldati, 2007). Entry of bacteria via a degraded phagosome is quite similar to what was observed with a population of M. marinum growing within mammalian macrophages (Gao et al., 2004; Stamm et al., 2003) (Fig. 2).

A number of D. discoideum mutants have been described that support enhanced replication of Mycobacterium species relative to wild-type strains. A model has been proposed that the mammalian ortholog of the D. discoideum coronin protein is a critical determinant for supporting intracellular growth of M. bovis BCG within mammalian cells (Ferrari et al., 1999; Jayachandran et al., 2007). To test this model, a D. discoideum knockout mutation in coronin was tested for resistance to M. marinum growth (Solomon et al., 2003). Surprisingly, the knockout mutant internalized M. marinum more efficiently than wild-type D. discoideum strains, and the yields of bacteria were significantly higher in the mutant as well. In fact, replication was so robust in the absence of coronin that the growth cycle terminated and bacteria were released hours prior to what was observed in wild-type strains (Solomon et al., 2003). Either D. dis-

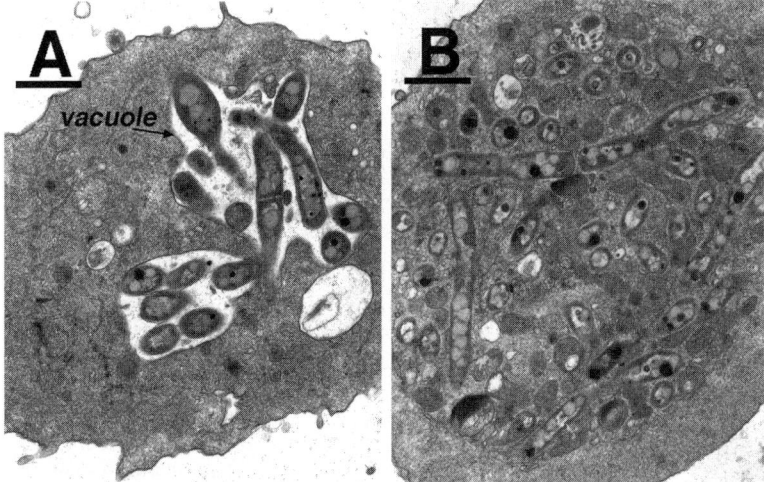

FIGURE 2 M. marinum resides in two different locales within D. discoideum. Micrographs from (A) Skriwan et al. (2002) and (B) Solomon et al. (2003). Amoebae were incubated with M. marinum for 24 h, fixed, and processed for transmission electron microscopy. (A) Groups of bacteria found in a single vacuole. (B) Bacteria apparently found in the amoebal cytoplasm. Bacteria free in cytoplasm were originally described as being in individual phagosomes (Skriwan et al., 2002; Solomon et al., 2003), but more recent work suggests that the M. marinum may not be in membrane-bound compartments in panel B (Stamm et al., 2003; Hagedorn and Soldati, 2007).

coideum behaves very differently from mammalian macrophages, or M. *bovis* BCG and M. *marinum* have very different determinants of intracellular growth. Less surprising were results with M. *avium* during challenge of *D. discoideum* Nramp-1 knockout cells. As mentioned previously, a set of inbred mouse strains show enhanced resistance to a variety of mycobacterial species, and an important determinant of this resistance maps to the mouse *nramp-1* gene encoding a metal transporter (Fortier et al., 2005b). Macrophages from mice that have a nonfunctional Nramp-1 protein show less restriction of mycobacteria, relative to macrophages from mice that have an intact transporter. When M. *avium* was used to challenge a *D. discoideum* mutant having a knockout in *nramp-1*, the bacteria showed greatly enhanced proliferation relative to that observed in wild-type amoebae (Peracino et al., 2006).

Finally, a mutation that eliminates expression of the RacH protein (Somesh et al., 2006), which controls vacuolin localization in *D. discoideum*, was analyzed (Hagedorn and Soldati, 2007). The RacH protein is a member of the Rho family GTPases that does not clearly fit into any specific subfamily of these proteins, so its specific mammalian ortholog is unclear (Okuwa et al., 2001; Rivero et al., 2001). However, its absence has interesting effects on vesicular traffic within the cell, such as causing disruption of endocytosis, dysfunction of endocytic compartments, and dispersion of vacuolin localization away from the postlysosomal compartment (Somesh et al., 2006). Conveniently, unlike many other mutations altering vesicle trafficking, amoebae harboring RacH knockout mutations are still viable. Because the absence of VacB protein profoundly interferes with M. *marinum* intracellular growth, it was rather surprising that the absence of RacH would result in a *D. discoideum* strain that was less restrictive for the pathogen (Hagedorn and Soldati, 2007). There are several explanations for this result. First, as *D. discoideum* racH$^-$ amoebae have less-acidic endosomes and a disrupted endolysosomal network, a population of bacteria that would normally be targeted for lysosomal degradation in wild-type amoebae could be rescued by the mutation. Second, dispersion of vacuolin away from the postlysosomal compartment in the mutant could free up this protein to participate in formation of the M. *marinum*-containing compartment, allowing more efficient recruitment of vacuolin than is found in wild-type cells. Finally, as transit out of the replication compartment into the amoebal cytoplasm appears to be a necessary step in efficient growth within *D. discoideum* (Hagedorn and Soldati, 2007), the absence of RacH could affect the integrity of the compartment, facilitating release of the bacteria into the cytosol at the optimal time during the replication cycle.

Intracellular Survival of *C. neoformans*

The soil fungus *Cryptococcus neoformans* is an opportunistic pathogen of humans that mainly causes disease in immunocompromised patients, such as those with a genetically defective immune system, organ transplant patients on immunosuppressive therapy, and individuals with human immunodeficiency virus infection (Zhou and Murphy, 2006). Disease, which in its most dramatic form is meningitis, is associated with growth of the encapsulated variants of the fungus within macrophages, with shedding of capsular material in a process that may be important for intracellular survival. As a common inhabitant of soil, like L. *pneumophila*, C. *neoformans* initiates disease via aerosol inoculation and is not associated with person-to-person spread. Therefore, this fungus likely represents an example of a pathogen that has been selected for virulence by its ability to survive in the presence of amoebae in the soil (Greub and Raoult, 2004).

C. *neoformans* can grow within common *Acanthamoeba* in a process that requires fungal products known to be involved in intracellular survival in macrophages (Steenbergen et al., 2001). The capsular component of the fungus, hypothesized to be important for altering trafficking properties within host cells, is also important for intracellular growth within amoebae, since an unencapsulated derivative fails to grow although it is internalized more efficiently than wild-type strains. The capsule itself appears to be dispersed throughout vesicles in the amoebae, as is seen in phagocytes.

Similar results were observed in the interaction of C. *neoformans* with *D. discoideum* (Steenbergen et al., 2003). In this case, both the acapsular mutant and a pseudohyphal derivative that had lowered virulence were incapable of growth within *D. discoideum*. In response to the virulent strains of C. *neoformans*, the amoebae die after continued replication of the pathogen, while the amoebae survive in the presence of the avirulent mutants. The interaction of C. *neoformans* with a *D. discoideum* myoVII mutant was striking. Although the mutant showed a defect in phagocytosing the fungus, there was more abundant growth in amoebae lacking myosin VII (Steenbergen et al., 2003). This is reminiscent of results with L. *pneumophila*, in which a variety of *D. discoideum* mutants that were defective for phagocytosis were found to be less restrictive of intracellular growth than wild-type strains (Fajardo et al., 2004; Solomon et al., 2000). This result was most pronounced with acapsular C. *neoformans*. Surprisingly, the acapsular mutant showed quite efficient growth in the myoVII knockout strain, indicating that the capsule may be important for interfering with a myosin VII-dependent phagocytic pathway that restricts C. *neoformans* growth (Steenbergen et al., 2003). When an rtoA$^-$ mutant was analyzed, there was a defect in phagocytosis similar to that observed for uptake of L. *pneumophila* (Solomon et al., 2000); however, unlike what was observed for this bacterium, C. *neoformans* grew similarly in both wild-type and rtoA$^-$ amoebal strains. Also similar to results observed previously with L. *pneumophila* and amoebae (Brieland et al., 1996), coculture of C. *neoformans* and *D. discoideum* enhanced the virulence of the fungus after intraperitoneal injection of mice. This is consistent with the intriguing proposition that during natural infection of humans, pathogenesis is facilitated by inhalation of aerosols containing amoebae that are laden with C. *neoformans*, just as L. *pneumophila* disease may occur.

USE OF *D. DISCOIDEUM* TO IDENTIFY MICROBIAL VIRULENCE FACTORS

Most of the studies on pathogen-*D. discoideum* interactions have used the amoebae as probes for host cell factors that support or restrict intracellular replication of the microorganism. *D. discoideum* is also a useful host for identifying microbial factors that target host cells during the disease process. There are a number of surrogate hosts that have been used to simplify analysis of virulence factors of pathogens that attack mammalian hosts, most notably *Caenorhabditis elegans*, *Drosophila melanogaster*, and *Arabidopsis thaliana* (Hilbi et al., 2007). The attraction of these systems is that intact organisms can be used to test virulence properties of pathogens, but unlike with mammalian hosts, large-scale infections can be performed and saturating mutant hunts are feasible. Although incubation of *D. discoideum* amoebae with pathogens is only a single-cell infection model, one major advantage over other systems of host cell culture analysis is that amoebae can be propagated on lawns of bacteria growing on solid medium, allowing the virulence properties of microorganisms to be analyzed by using standard bacteriological techniques. Productive growth of the amoebae on a lawn of bacteria results in the formation of plaques containing viable *D. discoideum*. By using this approach, entire gene banks of bacterial mutants can be analyzed using a reasonable number of culture dishes, so that it is feasible to identify every protein necessary for a productive interaction with *D. discoideum*.

The most obvious virulence property that can be assayed is inhibition of growth of amoebae by the pathogen growing on solid medium, with the pathogen preventing plaque formation by the amoebae. Pathogen mutants are then identified as strains that cannot interfere with plaque formation, so that instead of the pathogen disrupting *D. discoideum* growth, it ends up being a food source for the amoebae. The significance of the mutant identified can then be verified in an appropriate multicellular animal or plant infection model system.

Identification of Virulence-Associated Proteins Encoded by *Pseudomonas aeruginosa*

A number of simple eukaryotic host systems exist for the analysis of human-pathogenic *P. aeruginosa*, including pathogenesis of the nematode *C. elegans* and the plant *Arabidopsis* (Rahme et al., 2000). *P. aeruginosa* mutants defective for causing disease in these model organisms are also deranged for virulence in more complex mammalian hosts such as mice (Rahme et al., 2000). Therefore, this gram-negative bacterium was deemed a likely candidate to cause disruption of *D. discoideum* growth and to use this strategy to identify virulence factors active in mammals. As is true of many of the microorganisms studied as partners for *D. discoideum*, human disease by *P. aeruginosa* is largely opportunistic, especially targeting the lung tissues of patients with cystic fibrosis and causing localized and systemic diseases in burn patients (Gomez et al., 2006; Pruitt et al., 1998). The bacterium is also highly resistant to a number of antibiotics.

In many ways, the interaction of *P. aeruginosa* with *D. discoideum* mirrors its relationship with other model hosts (Pukatzki et al., 2002). The pathogen can be demonstrated to cause lethality to the target host with high efficiency, and virulence-associated proteins that are known to be required for disease in mammals are required for causing death to *D. discoideum*. When the amoebae are plated on lawns of *P. aeruginosa* growing on bacteriological medium, plaque formation is inhibited. Plaque formation can occur and *P. aeruginosa* can be used as an amoebal food source if lawns are made from a bacterial mutant lacking the LasR transcriptional regulator that responds to density-dependent signals (Pukatzki et al., 2002). This inability to kill *D. discoideum* mirrors the lowered virulence of this mutant in animal models (Rumbaugh et al., 1999; Tang et al., 1996). Similar to many other gram-negative pathogens, several of the well-characterized virulence-associated proteins encoded by *P. aeruginosa* are substrates of the type III secretion apparatus that injects proteins into host cells via a multiprotein needle complex (Baldwin and Barbieri, 2005). There are at least four substrates of this apparatus, and they induce misregulation of the host cell cytoplasm, dysfunction of host cell vesicle trafficking, and degradation of specific phospholipids (Baldwin and Barbieri, 2005; Cuzick et al., 2006; Phillips et al., 2003; Sato et al., 2003). Mutational loss of the type III secretion system prevents killing of *D. discoideum*, and the amoebae are able to form plaques on lawns of a mutant defective in assembly of this translocation apparatus (Pukatzki et al., 2002). Furthermore, mutational loss of just one of these substrates, the ExoU phospholipase, abolishes the ability to kill *D. discoideum* and allows plaque formation to occur. Therefore, some of the best-characterized and most important virulence proteins associated with *P. aeruginosa* pathogenesis in mammalian models are required to inhibit *D. discoideum* plaque formation.

Identification of a New Protein Translocation System Encoded by Multiple Organisms Using *V. cholerae*-*D. discoideum* Interaction

As indicated with the work on *P. aeruginosa*, interaction of pathogens with *D. discoideum* provides the opportunity to identify novel virulence proteins because of the ease of performing genetic screens for defects in interaction with amoebae. This system provided an excellent opportunity to analyze virulence factors encoded by a novel isolate of *Vibrio cholerae*. Most epidemic strains of *V. cholerae* promote diarrheal diseases as a result of the secretion of cholera toxin and colonization mediated by toxin-coregulated pili (Klose, 2001). Sporadic outbreaks of the diarrheal diseases by unusual strains of *V. cholerae*, however, result from the presence of other sets of proteins that are poorly characterized (Dziejman et al., 2002). These isolates, called non-O1, non-O139 strains, are of great interest because some of them cause diseases that are cholera-like even though the organisms are very different from the heavily studied epidemic strains (Dziejman et al., 2002). *D. discoideum* allowed workers to make significant inroads into the molecular basis of disease caused by these isolates.

Unlike strains of *V. cholerae* associated with large-scale epidemics, a non-O1, non-O139 isolate from an outbreak in Sudan was able to kill *D. discoideum* (Pukatzki et al., 2006). On lawns of more common *V. cholerae* isolates, *D. discoideum* forms impressive plaques on their vibrio food source, whereas no such plaque formation was found on the unusual isolate. This allowed a rather simple genetic screen for bacterial mutants that are defective for killing amoebae, but instead are potential prey for *D. discoideum*. *V. cholera* transposon mutants were incubated on a plate that was seeded with a large number of amoebae, and colonies were identified that appeared nibbled by the *D. discoideum* (Pukatzki et al., 2006). Mutants isolated in this fashion were extremely defective for killing of amoebae, and they were used as a food source as efficiently as standard strains used to feed *D. discoideum*.

Using this approach and a large-scale screen, seven genetically linked insertion mutations were identified (Pukatzki et al., 2006). Within this cluster of genes, called the *vas* locus, are two orthologs of components of the *L. pneumophila* Dot/Icm apparatus, IcmF and IcmH(DotU) (Van Rheenen et al., 2004; Zusman et al., 2004), indicating that the genes in this cluster encode components of a protein translocation apparatus. Because there was no evidence for any other orthologs of Dot/Icm, and orthologs of IcmF and IcmH(DotU) are much more widely distributed than other components of the Dot/Icm apparatus, it seemed apparent that a novel protein translocation system had been identified, which is now called a type VI secretion system (Pukatzki et al., 2006). Two other *vas* genes were similar to genes required for *Rhizobium* association with plants, and also may be members of the secretion apparatus (Bladergroen et al., 2003). In addition, the region includes a sigma-54-dependent activator that may control expression of virulence-associated loci, and also includes two paralogs of the previously characterized VgrG1 protein, which had been shown to promote actin cross-linking in target eukaryotic cells (Sheahan et al., 2004). Therefore, there are two to four potential components of a secretion system, and three open reading frames encoding potential substrates that are translocated to host cells. In fact, there was good evidence that such a secretion system is functional, because in the absence of the putative secretion system components, both a VgrG paralog and a major *V. cholerae*-secreted protein called Hcp (encoded by two similar genes) are no longer found in culture supernatants (Pukatzki et al., 2006).

The newly identified secretion system is active on mammalian phagocytic cells and is likely to be involved in virulence at some level. The non-O1, non-O139 strain used in these studies was cytotoxic for the mouse macrophage cell line J774 (Pukatzki et al., 2006). This cytotoxicity was likely due to several secreted toxins, because if monolayers were challenged with a bacterial strain that still harbored the *vas* locus, but which contained deletions for previously characterized toxins, there was still disruption of J774 monolayers. If a mutation in *vasK* was placed in this strain background, however, then the monolayers remained intact, indicating that the secretion system was responsible for cytotoxicity in this multiply deleted strain background. In fact, it now appears clear that at least VgrG1 is translocated into J774 cells in a *vas*-dependent fashion, based on monitoring of its ability to promote actin cross-linking after contact of bacteria with host cells (Pukatzki et al., 2007).

Identification of Virulence-Associated Proteins Using *A. baumannii*-*D. discoideum* Interaction

Acinetobacter baumannii is an emerging opportunistic bacterial pathogen associated with respiratory diseases in patients residing in hospitals, particularly those undergoing respiratory therapy (Joly-Guillou, 2005). Very few virulence models exist for the microorganism, but similar to *P. aeruginosa*, the bacterium is virulent for *C. elegans* and causes growth inhibition of *D. discoideum* (Smith et al., 2007). Using a scheme that mimics the one used to identify the *V. cholerae vas* loci, *A. baumannii* mutants that allow plaque formation of *D. discoideum* were isolated (Smith et al., 2007). Thirty mutants were identified and mapped that affected both *D. discoideum* growth and *C. elegans* pathogenesis. Unfortunately, the authors withheld information on the sites of insertion of these mutants, making it impossible to judge the power of this approach (Smith et al., 2007).

PROBLEMS AND OPPORTUNITIES

From the discussion of the work performed on *D. discoideum*, it is clear that this host has great potential as a model for interaction with pathogens. It is also obvious that the full power of the system has not been totally exploited. A comprehensive analysis of the total complement of amoebal genes required for modulating interaction with an organism has yet to be performed, and it may be that technical advances are necessary to allow this goal to be achieved. Unlike yeast, there is no well-defined sexual system to allow the construction of complicated strains, although parasexual genetics appears to work quite well (King and Insall, 2006; Zaki et al., 2007). In addition, strategies for analyzing conditional-lethal mutations have not been exploited to their fullest. Finally, short interfering RNA (siRNA) techniques are not as well developed as they are in other systems.

Most of these problems can probably be solved. However, there are strategies that can be used with this organism that would be difficult with other systems. The identification of the type VI secretion system in *V. cholerae* was undoubtedly accelerated by the use of *D. discoideum*, as simple bacteriological techniques were used to solve a problem in phagocyte-microorganism interaction (Pukatzki et al., 2006). Furthermore, the analysis of host cell mutants is more straightforward than in other systems. Although siRNA is a powerful strategy, knockout mutations are more straightforward to analyze, and there are very few techniques that allow knockouts to be analyzed in phagocytic cells other than by isolating mouse mutants.

Over and above the fact that *D. discoideum* is a useful model system, it is now appreciated that many pathogens have been host-adapted to amoebae and that this adaptation allows them to confront immune strategies used by multicellular eukaryotes (Greub and Raoult, 2004). At this point, with the years of intensive study of *D. discoideum* and the numerous genetic techniques developed for its analysis, it is certain that this is the amoebal species of choice with regard to both investigation of strategies used by amoebae for dealing with pathogens and identification of the arsenal of virulence-associated proteins acquired by pathogens to attack these free-living organisms. There are, of course, differences between phagocytes in higher eukaryotes and amoebae. Some of the killing mechanisms of microorganisms are probably quite different in amoebae and macrophages. For instance, oxygen-dependent killing is an important strategy used by mammalian phagocytes that is apparently not employed by *D. discoideum*. In addition, the organelle system is not identical in the two cell types. There is no organelle similar to the amoebal contractile vacuole in higher eukaryotes. The lysosome is not a terminal compartment in amoebae, as a postlysosomal compartment coordinates exit of partially degraded prey. Despite this fact, adaptation of the pathogen to some of these unique features of amoebae appears to have allowed intracellular organisms to use these strategies in higher eukaryotes lacking specialized organelles. One of the most interesting examples is from the work on *M. marinum*, in which a flotillin family member was shown to be a critical determinant of intracellular survival within *D. discoideum* (Hagedorn and Soldati, 2007). This protein is associated with the postlysosomal compartment in *D. discoideum* and then is hijacked for its use by the bacterial replication vacuole, allowing the microorganism to be situated in a niche that bypasses the endolysosomal system. Even though there is no comparable compartment in mammalian macrophages, the replication vacuole in higher eukaryotes still recruits flotillin. This is an exciting result because it indicates that even in the absence of direct overlap between two different cell types, the bacterium targets similar sets of proteins in the two hosts, although the normal functions of these proteins may be different in the two cell types.

The future holds considerable promise for using *D. discoideum* as a model phagocyte. As fresh techniques are developed, and there is more appreciation for the intimate association of amoebae with pathogens in the environment, then this simple eukaryote will certainly grow in importance for the analysis of pathogens in cell culture.

REFERENCES

Abu Kwaik, Y. 1996. The phagosome containing *Legionella pneumophila* within the protozoan *Hartmannella vermiformis* is surrounded by the rough endoplasmic reticulum. *Appl. Environ. Microbiol.* **62:**2022–2028.

Adekambi, T., S. Ben Salah, M. Khlif, D. Raoult, and M. Drancourt. 2006. Survival of environmental mycobacteria in *Acanthamoeba polyphaga. Appl. Environ. Microbiol.* **72:**5974–59781.

Adeleke, A. A., B. S. Fields, R. F. Benson, M. I. Daneshvar, J. M. Pruckler, R. M. Ratcliff, T. G. Harrison, R. S. Weyant, R. J. Birtles, D. Raoult, and M. A. Halablab. 2001. *Legionella drozanskii* sp. nov., *Legionella rowbothamii* sp. nov. and *Legionella fallonii* sp. nov.: three unusual new *Legionella* species. *Int. J. Syst. Evol. Microbiol.* **51:**1151–1160.

Amer, A. O., and M. S. Swanson. 2005. Autophagy is an immediate macrophage response to *Legionella pneumophila. Cell. Microbiol.* **7:**765–778.

Andra, J., R. Herbst, and M. Leippe. 2003. Amoebapores, archaic effector peptides of protozoan origin, are discharged into phagosomes and kill bacteria by permeabilizing their membranes. *Dev. Comp. Immunol.* **27:**291–304.

Atlas, R. M. 1999. *Legionella*: from environmental habitats to disease pathology, detection and control. *Environ. Microbiol.* **1:**283–293.

Bajno, L., X. R. Peng, A. D. Schreiber, H. P. Moore, W. S. Trimble, and S. Grinstein. 2000. Focal exocytosis of VAMP3-containing vesicles at sites of phagosome formation. *J. Cell Biol.* **149:**697–706.

Baldwin, M. R., and J. T. Barbieri. 2005. The type III cytotoxins of *Yersinia* and *Pseudomonas aeruginosa* that modulate the actin cytoskeleton. *Curr. Top. Microbiol. Immunol.* **291:**147–166.

Beckers, M. C., S. Yoshida, K. Morgan, E. Skamene, and P. Gros. 1995. Natural resistance to infection with *Legionella pneumophila*: chromosomal localization of the Lgn1 susceptibility gene. *Mamm. Genome* **6:**540–545.

Benghezal, M., S. Cornillon, L. Gebbie, L. Alibaud, F. Bruckert, F. Letourneur, and P. Cosson. 2003. Synergistic control of cellular adhesion by transmembrane 9 proteins. *Mol. Biol. Cell* **14:**2890–2899.

Benghezal, M., M. O. Fauvarque, R. Tournebize, R. Froquet, A. Marchetti, E. Bergeret, B. Lardy, G. Klein, P. Sansonetti, S. J. Charette, and P. Cosson. 2006. Specific host genes required for the killing of *Klebsiella* bacteria by phagocytes. *Cell. Microbiol.* **8:**139–148.

Berthet, F. X., P. B. Rasmussen, I. Rosenkrands, P. Andersen, and B. Gicquel. 1998. A *Mycobacterium tuberculosis* operon encoding ESAT-6 and a novel low-molecular-mass culture filtrate protein (CFP-10). *Microbiology* **144**(Pt. 11):3195–3203.

Bladergroen, M. R., K. Badelt, and H. P. Spaink. 2003. Infection-blocking genes of a symbiotic *Rhizobium leguminosarum* strain that are involved in temperature-dependent protein secretion. *Mol. Plant Microbe Interact.* **16:**53–64.

Blanc, C., S. Charette, N. Cherix, Y. Lefkir, P. Cosson, and F. Letourneur. 2005. A novel phosphatidylinositol 4,5-bisphosphate-binding domain targeting the Phg2 kinase to the membrane in *Dictyostelium* cells. *Eur. J. Cell Biol.* **84:**951–960.

Bloomfield, G., and C. Pears. 2003. Superoxide signalling required for multicellular development of *Dictyostelium. J. Cell Sci.* **116:**3387–3397.

Bouley, D. M., N. Ghori, K. L. Mercer, S. Falkow, and L. Ramakrishnan. 2001. Dynamic nature of host-pathogen interactions in *Mycobacterium marinum* granulomas. *Infect. Immun.* **69:**7820–7831.

Bourhis, L. L., and C. Werts. 2007. Role of Nods in bacterial infection. *Microbes Infect.* **9:**629–636.

Bozue, J. A., and W. Johnson. 1996. Interaction of *Legionella pneumophila* with *Acanthamoeba castellanii*: uptake by coiling phagocytosis and inhibition of phagosome-lysosome fusion. *Infect. Immun.* **64:**668–673.

Brazill, D. T., L. R. Meyer, R. D. Hatton, D. A. Brock, and R. H. Gomer. 2001. ABC transporters required for endocytosis and endosomal pH regulation in *Dictyostelium. J. Cell Sci.* **114:**3923–3932.

Bretschneider, T., S. Diez, K. Anderson, J. Heuser, M. Clarke, A. Muller-Taubenberger, J. Kohler, and G. Gerisch. 2004. Dynamic actin patterns and Arp2/3 assembly at the substrate-attached surface of motile cells. *Curr. Biol.* **14:**1–10.

Brieland, J., M. McClain, L. Heath, C. Chrisp, G. Huffnagle, M. LeGendre, M. Hurley, J. Fantone, and C. Engleberg. 1996. Coinoculation with *Hartmannella vermiformis* enhances replicative *Legionella pneumophila* lung infection in a murine model of Legionnaires' disease. *Infect. Immun.* **64:**2449–2456.

Burch-Smith, T. M., and S. P. Dinesh-Kumar. 2007. The functions of plant TIR domains. *Sci. STKE* **2007:**pe46.

Cardelli, J. 2001. Phagocytosis and macropinocytosis in *Dictyostelium*: phosphoinositide-based processes, biochemically distinct. *Traffic* **2:**311–320.

Carlsson, F., and E. J. Brown. 2006. Actin-based motility of intracellular bacteria, and polarized surface distribution of the bacterial effector molecules. *J. Cell. Physiol.* **209:**288–296.

Cavender, J. C., E. Vadell, J. C. Landolt, and S. L. Stephenson. 2005. New species of small dictyostelids from the Great Smoky Mountains National Park. *Mycologia* **97:**493–512.

Chen, G., O. Zhuchenko, and A. Kuspa. 2007. Immune-like phagocyte activity in the social amoeba. *Science* **317:**678–681.

Chen, J., K. S. de Felipe, M. Clarke, H. Lu, O. R. Anderson, G. Segal, and H. A. Shuman. 2004. *Legionella* effectors that promote nonlytic release from protozoa. *Science* **303:**1358–1361.

Cirillo, J. D., S. Falkow, L. S. Tompkins, and L. E. Bermudez. 1997. Interaction of *Mycobacterium avium* with environmental amoebae enhances virulence. *Infect. Immun.* **65:**3759–3767.

Clarke, M., J. Kohler, Q. Arana, T. Liu, J. Heuser, and G. Gerisch. 2002. Dynamics of the vacuolar H(+)-ATPase in the contractile vacuole complex and the endosomal pathway of *Dictyostelium* cells. *J. Cell Sci.* **115:**2893–2905.

Clarke, M., and L. Maddera. 2006. Phagocyte meets prey: uptake, internalization, and killing of bacteria by *Dictyostelium* amoebae. *Eur. J. Cell Biol.* **85:**1001–1010.

Cohen, C. J., R. Bacon, M. Clarke, K. Joiner, and I. Mellman. 1994. *Dictyostelium discoideum* mutants with conditional defects in phagocytosis. *J. Cell Biol.* **126:**955–966.

Converse, S. E., and J. S. Cox. 2005. A protein secretion pathway critical for *Mycobacterium tuberculosis* virulence is conserved and functional in *Mycobacterium smegmatis. J. Bacteriol.* **187:**1238–1245.

Cornillon, S., E. Pech, M. Benghezal, K. Ravanel, E. Gaynor, F. Letourneur, F. Bruckert, and P. Cosson. 2000. Phg1p is a nine-transmembrane protein superfamily member involved in *Dictyostelium* adhesion and phagocytosis. *J. Biol. Chem.* **275:**34287–34292.

Corsaro, D., and D. Venditti. 2004. Emerging chlamydial infections. *Crit. Rev. Microbiol.* **30:**75–106.

Cox, D., J. S. Berg, M. Cammer, J. O. Chinegwundoh, B. M. Dale, R. E. Cheney, and S. Greenberg. 2002. Myosin X is a downstream effector of PI(3)K during phagocytosis. *Nat. Cell Biol.* **4:**469–477.

Cuzick, A., F. R. Stirling, S. L. Lindsay, and T. J. Evans. 2006. The type III pseudomonal exotoxin U activates the c-Jun NH2-terminal kinase pathway and increases human epithelial interleukin-8 production. *Infect. Immun.* **74:**4104–4113.

Davis, J. M., H. Clay, J. L. Lewis, N. Ghori, P. Herbomel, and L. Ramakrishnan. 2002. Real-time visualization of *Mycobacterium*-macrophage interactions leading to initiation of granuloma formation in zebrafish embryos. *Immunity* **17:**693–702.

De Lozanne, A., and J. A. Spudich. 1987. Disruption of the *Dictyostelium* myosin heavy chain gene by homologous recombination. *Science* **236:**1086–1091.

Derre, I., and R. R. Isberg. 2004. *Legionella pneumophila* replication vacuole formation involves rapid recruitment of proteins of the early secretory system. *Infect. Immun.* **72:**3048–3053.

Dietrich, W. F., D. M. Damron, R. R. Isberg, E. S. Lander, and M. S. Swanson. 1995. Lgn1, a gene that determines susceptibility to *Legionella pneumophila*, maps to mouse chromosome 13. *Genomics* **26:**443–450.

Dorer, M. S., and R. R. Isberg. 2006. Non-vertebrate hosts in the analysis of host-pathogen interactions. *Microbes Infect.* **8:**1637–1646.

Duffy, K. T., and G. Vogel. 1984. Linkage analysis of two phagocytosis receptor loci in *Dictyostelium discoideum*. *J. Gen. Microbiol.* **130:**2071–2077.

Duhon, D., and J. Cardelli. 2002. The regulation of phagosome maturation in *Dictyostelium*. *J. Muscle Res. Cell Motil.* **23:**803–808.

Dziejman, M., E. Balon, D. Boyd, C. M. Fraser, J. F. Heidelberg, and J. J. Mekalanos. 2002. Comparative genomic analysis of *Vibrio cholerae*: genes that correlate with cholera endemic and pandemic disease. *Proc. Natl. Acad. Sci. USA* **99:**1556–1561.

Eichinger, L., J. A. Pachebat, G. Glockner, M. A. Rajandream, R. Sucgang, M. Berriman, J. Song, R. Olsen, K. Szafranski, Q. Xu, B. Tunggal, S. Kummerfeld, M. Madera, B. A. Konfortov, F. Rivero, A. T. Bankier, R. Lehmann, N. Hamlin, R. Davies, P. Gaudet, P. Fey, K. Pilcher, G. Chen, D. Saunders, E. Sodergren, P. Davis, A. Kerhornou, X. Nie, N. Hall, C. Anjard, L. Hemphill, N. Bason, P. Farbrother, B. Desany, E. Just, T. Morio, R. Rost, C. Churcher, J. Cooper, S. Haydock, N. van Driessche, A. Cronin, I. Goodhead, D. Muzny, T. Mourier, A. Pain, M. Lu, D. Harper, R. Lindsay, H. Hauser, K. James, K. Quiles, M. Madan Babu, T. Saito, C. Buchrieser, A. Wardroper, M. Felder, M. Thangavelu, D. Johnson, A. Knights, H. Loulseged, K. Mungall, K. Oliver, C. Price, M. A. Quail, H. Urushihara, J. Hernandez, E. Rabbinowitsch, D. Steffen, M. Sanders, J. Ma, Y. Kohara, S. Sharp, M. Simmonds, S. Spiegler, A. Tivey, S. Sugano, B. White, D. Walker, J. Woodward, T. Winckler, Y. Tanaka, G. Shaulsky, M. Schleicher, G. Weinstock, A. Rosenthal, E. C. Cox, R. L. Chisholm, R. Gibbs, W. F. Loomis, M. Platzer, R. R. Kay, J. Williams, P. H. Dear, A. A. Noegel, B. Barrell, and A. Kuspa. 2005. The genome of the social amoeba *Dictyostelium discoideum*. *Nature* **435:**43–57.

Fajardo, M., M. Schleicher, A. Noegel, S. Bozzaro, S. Killinger, K. Heuner, J. Hacker, and M. Steinert. 2004. Calnexin, calreticulin and cytoskeleton-associated proteins modulate uptake and growth of *Legionella pneumophila* in *Dictyostelium discoideum*. *Microbiology* **150:**2825–2835.

Farbrother, P., C. Wagner, J. Na, B. Tunggal, T. Morio, H. Urushihara, Y. Tanaka, M. Schleicher, M. Steinert, and L. Eichinger. 2006. *Dictyostelium* transcriptional host cell response upon infection with *Legionella*. *Cell. Microbiol.* **8:**438–456.

Ferrari, G., H. Langen, M. Naito, and J. Pieters. 1999. A coat protein on phagosomes involved in the intracellular survival of mycobacteria. *Cell* **97:**435–447.

Finlay, B. B., and S. Falkow. 1997. Common themes in microbial pathogenicity revisited. *Microbiol. Mol. Biol. Rev.* **61:**136–169.

Fischer, M., I. Haase, E. Simmeth, G. Gerisch, and A. Muller-Taubenberger. 2004. A brilliant monomeric red fluorescent protein to visualize cytoskeleton dynamics in *Dictyostelium*. *FEBS Lett.* **577:**227–232.

Fliermans, C. B., W. B. Cherry, L. H. Orrison, S. J. Smith, D. L. Tison, and D. H. Pope. 1981. Ecological distribution of *Legionella pneumophila*. *Appl. Environ. Microbiol.* **41:**9–16.

Fliermans, C. B., W. B. Cherry, L. H. Orrison, and L. Thacker. 1979. Isolation of *Legionella pneumophila* from nonepidemic-related aquatic habitats. *Appl. Environ. Microbiol.* **37:**1239–1242.

Fortier, A., E. Diez, and P. Gros. 2005. Naip5/Birc1e and susceptibility to *Legionella pneumophila*. *Trends Microbiol.* **13:**328–335.

Fortier, A., G. Min-Oo, J. Forbes, S. Lam-Yuk-Tseung, and P. Gros. 2005. Single gene effects in mouse models of host: pathogen interactions. *J. Leukoc. Biol.* **77:**868–877.

Fratti, R. A., J. M. Backer, J. Gruenberg, S. Corvera, and V. Deretic. 2001. Role of phosphatidylinositol 3-kinase and Rab5 effectors in phagosomal biogenesis and mycobacterial phagosome maturation arrest. *J. Cell Biol.* **154:**631–644.

Froquet, R., N. Cherix, S. E. Burr, J. Frey, S. Vilches, J. M. Tomas, and P. Cosson. 2007. Alternative host model to evaluate *Aeromonas* virulence. *Appl. Environ. Microbiol.* **73:**5657–5659.

Gack, M. U., Y. C. Shin, C. H. Joo, T. Urano, C. Liang, L. Sun, O. Takeuchi, S. Akira, Z. Chen, S. Inoue, and J. U. Jung. 2007. TRIM25 RING-finger E3 ubiquitin ligase is essential for RIG-I-mediated antiviral activity. *Nature* **446:**916–920.

Gagnon, E., S. Duclos, C. Rondeau, E. Chevet, P. H. Cameron, O. Steele-Mortimer, J. Paiement, J. J. Bergeron, and M. Desjardins. 2002. Endoplasmic reticulum-mediated phagocytosis is a mechanism of entry into macrophages. *Cell* **110:**119–131.

Gao, L. Y., S. Guo, B. McLaughlin, H. Morisaki, J. N. Engel, and E. J. Brown. 2004. A mycobacterial virulence gene cluster extending RD1 is required for cytolysis, bacterial spreading and ESAT-6 secretion. *Mol. Microbiol.* **53:**1677–1693.

Gao, L. Y., and Y. A. Kwaik. 2000. The mechanism of killing and exiting the protozoan host *Acanthamoeba polyphaga* by *Legionella pneumophila*. *Environ. Microbiol.* **2:**79–90.

Garin, J., R. Diez, S. Kieffer, J. F. Dermine, S. Duclos, E. Gagnon, R. Sadoul, C. Rondeau, and M. Desjardins. 2001. The phagosome proteome: insight into phagosome functions. *J. Cell Biol.* **152:**165–180.

Gebbie, L., M. Benghezal, S. Cornillon, R. Froquet, N. Cherix, M. Malbouyres, Y. Lefkir, C. Grangeasse, S. Fache, J. Dalous, F. Bruckert, F. Letourneur, and P. Cosson. 2004. Phg2, a kinase involved in adhesion and focal site modeling in *Dictyostelium*. *Mol. Biol. Cell* **15:**3915–3925.

George, J. R., L. Pine, M. W. Reeves, and W. K. Harrell. 1980. Amino acid requirements of *Legionella pneumophila*. *J. Clin. Microbiol.* **11:**286–291.

Gerisch, G., J. Heuser, and M. Clarke. 2002. Tubular-vesicular transformation in the contractile vacuole system of *Dictyostelium*. *Cell. Biol. Int.* **26:**845–852.

Gey Van Pittius, N. C., J. Gamieldien, W. Hide, G. D. Brown, R. J. Siezen, and A. D. Beyers. 2001. The ESAT-6 gene cluster of *Mycobacterium tuberculosis* and other high

G+C Gram-positive bacteria. *Genome Biol.* **2:**RESEARCH0044.

Gey van Pittius, N. C., S. L. Sampson, H. Lee, Y. Kim, P. D. van Helden, and R. M. Warren. 2006. Evolution and expansion of the *Mycobacterium tuberculosis* PE and PPE multigene families and their association with the duplication of the ESAT-6 *(esx)* gene cluster regions. *BMC Evol. Biol.* **6:**95.

Girardin, S. E., I. G. Boneca, L. A. Carneiro, A. Antignac, M. Jehanno, J. Viala, K. Tedin, M. K. Taha, A. Labigne, U. Zahringer, A. J. Coyle, P. S. DiStefano, J. Bertin, P. J. Sansonetti, and D. J. Philpott. 2003. Nod1 detects a unique muropeptide from gram-negative bacterial peptidoglycan. *Science* **300:**1584–1587.

Glebov, O. O., N. A. Bright, and B. J. Nichols. 2006. Flotillin-1 defines a clathrin-independent endocytic pathway in mammalian cells. *Nat. Cell Biol.* **8:**46–54.

Gomez, M. I., M. O'Seaghdha, M. Magargee, T. J. Foster, and A. S. Prince. 2006. *Staphylococcus aureus* protein A activates TNFR1 signaling through conserved IgG binding domains. *J. Biol. Chem.* **281:**20190–20196.

Gotthardt, D., R. Dieckmann, V. Blancheteau, C. Kistler, F. Reichardt, and T. Soldati. 2006. Preparation of intact, highly purified phagosomes from *Dictyostelium*. *Methods Mol. Biol.* **346:**439–448.

Gotthardt, D., H. J. Warnatz, O. Henschel, F. Bruckert, M. Schleicher, and T. Soldati. 2002. High-resolution dissection of phagosome maturation reveals distinct membrane trafficking phases. *Mol. Biol. Cell* **13:**3508–3520.

Greub, G., B. La Scola, and D. Raoult. 2004. Amoebae-resisting bacteria isolated from human nasal swabs by amoebal coculture. *Emerg. Infect. Dis.* **10:**470–477.

Greub, G., B. La Scola, and D. Raoult. 2003. *Parachlamydia acanthamoeba* is endosymbiotic or lytic for *Acanthamoeba polyphaga* depending on the incubation temperature. *Ann. N. Y. Acad. Sci.* **990:**628–634.

Greub, G., and D. Raoult. 2004. Microorganisms resistant to free-living amoebae. *Clin. Microbiol. Rev.* **17:**413–433.

Greub, G., and D. Raoult. 2002. Parachlamydiaceae: potential emerging pathogens. *Emerg. Infect. Dis.* **8:**625–630.

Hagedorn, M., and T. Soldati. 2007. Flotillin and RacH modulate the intracellular immunity of *Dictyostelium* to *Mycobacterium marinum* infection. *Cell. Microbiol.* **9:**2716–2733.

Hagele, S., R. Kohler, H. Merkert, M. Schleicher, J. Hacker, and M. Steinert. 2000. *Dictyostelium discoideum*: a new host model system for intracellular pathogens of the genus *Legionella*. *Cell. Microbiol.* **2:**165–171.

Harris, E., K. Yoshida, J. Cardelli, and J. Bush. 2001. Rab11-like GTPase associates with and regulates the structure and function of the contractile vacuole system in *Dictyostelium*. *J. Cell Sci.* **114:**3035–3045.

Hilbi, H., S. S. Weber, C. Ragaz, Y. Nyfeler, and S. Urwyler. 2007. Environmental predators as models for bacterial pathogenesis. *Environ. Microbiol.* **9:**563–575.

Hoffmann, J. A. 2003. The immune response of *Drosophila*. *Nature* **426:**33–38.

Hollenbeck, P. J., and J. A. Swanson. 1990. Radial extension of macrophage tubular lysosomes supported by kinesin. *Nature* **346:**864–866.

Hookey, J. V., R. J. Birtles, and N. A. Saunders. 1995. Intergenic 16S rRNA gene (rDNA)-23S rDNA sequence length polymorphisms in members of the family Legionellaceae. *J. Clin. Microbiol.* **33:**2377–2381.

Horn, M., M. Wagner, K. D. Müller, E. N. Schmid, T. R. Fritsche, K. H. Schleifer, and R. Michel. 2000. *Neochlamydia hartmannellae* gen. nov., sp. nov. (Parachlamydiaceae), an endoparasite of the amoeba *Hartmannella vermiformis*. *Microbiology* **146**(Pt. 5):1231–1239.

Horwitz, M. A. 1983. The Legionnaires' disease bacterium (*Legionella pneumophila*) inhibits phagosome-lysosome fusion in human monocytes. *J. Exp. Med.* **158:**2108–2126.

Imler, J. L., and J. A. Hoffmann. 2003. Toll signaling: the TIReless quest for specificity. *Nat. Immunol.* **4:**105–106.

Inohara, N., M. Chamaillard, C. McDonald, and G. Nunez. 2005. NOD-LRR proteins: role in host-microbial interactions and inflammatory disease. *Annu. Rev. Biochem.* **74:**355–383.

Inoue, T., S. Wilkens, and M. Forgac. 2003. Subunit structure, function, and arrangement in the yeast and coated vesicle V-ATPases. *J. Bioenerg. Biomembr.* **35:**291–299.

Insall, R., A. Muller-Taubenberger, L. Machesky, J. Kohler, E. Simmeth, S. J. Atkinson, I. Weber, and G. Gerisch. 2001. Dynamics of the *Dictyostelium* Arp2/3 complex in endocytosis, cytokinesis, and chemotaxis. *Cell Motil. Cytoskelet.* **50:**115–128.

Ishii, K. J., C. Coban, H. Kato, K. Takahashi, Y. Torii, F. Takeshita, H. Ludwig, G. Sutter, K. Suzuki, H. Hemmi, S. Sato, M. Yamamoto, S. Uematsu, T. Kawai, O. Takeuchi, and S. Akira. 2006. A Toll-like receptor-independent antiviral response induced by double-stranded B-form DNA. *Nat. Immunol.* **7:**40–48.

Iwasaki, A., and R. Medzhitov. 2004. Toll-like receptor control of the adaptive immune responses. *Nat. Immunol.* **5:**987–995.

Jayachandran, R., V. Sundaramurthy, B. Combaluzier, P. Mueller, H. Korf, K. Huygen, T. Miyazaki, I. Albrecht, J. Massner, and J. Pieters. 2007. Survival of mycobacteria in macrophages is mediated by coronin 1-dependent activation of calcineurin. *Cell* **130:**37–50.

Joly-Guillou, M. L. 2005. Clinical impact and pathogenicity of *Acinetobacter*. *Clin. Microbiol. Infect.* **11:**868–873.

Jordens, I., M. Marsman, C. Kuijl, and J. Neefjes. 2005. Rab proteins, connecting transport and vesicle fusion. *Traffic* **6:**1070–1077.

Kagan, J. C., and C. R. Roy. 2002. *Legionella* phagosomes intercept vesicular traffic from endoplasmic reticulum exit sites. *Nat. Cell Biol.* **4:**945–954.

Kagan, J. C., M. P. Stein, M. Pypaert, and C. R. Roy. 2004. *Legionella* subvert the functions of Rab1 and Sec22b to create a replicative organelle. *J. Exp. Med.* **199:**1201–1211.

Kessin, R. H. 2006. The secret lives of *Dictyostelium*. *Methods Mol. Biol.* **346:**3–14.

Khelef, N., H. A. Shuman, and F. R. Maxfield. 2001. Phagocytosis of wild-type *Legionella pneumophila* occurs through a wortmannin-insensitive pathway. *Infect. Immun.* **69:**5157–5161.

Khurana, T., J. A. Brzostowski, and A. R. Kimmel. 2005. A Rab21/LIM-only/CH-LIM complex regulates phagocytosis via both activating and inhibitory mechanisms. *EMBO J.* **24:**2254–2264.

Kimmel, A. R., and C. A. Parent. 2003. The signal to move: *D. discoideum* go orienteering. *Science* **300:**1525–1527.

King, J., and R. Insall. 2006. Parasexual genetics using axenic cells. *Methods Mol. Biol.* **346:**125–135.

Klein, B. S., and B. Tebbets. 2007. Dimorphism and virulence in fungi. *Curr. Opin. Microbiol.* **10:**314–319.

Klose, K. E. 2001. Regulation of virulence in *Vibrio cholerae*. *Int. J. Med. Microbiol.* **291:**81–88.

Kreitzer, G., A. Marmorstein, P. Okamoto, R. Vallee, and E. Rodriguez-Boulan. 2000. Kinesin and dynamin are required for post-Golgi transport of a plasma-membrane protein. *Nat. Cell Biol.* **2:**125–127.

Kufer, T. A., D. J. Banks, and D. J. Philpott. 2006. Innate immune sensing of microbes by Nod proteins. *Ann. N. Y. Acad. Sci.* **1072:**19–27.

Kumar, J., S. Okada, C. Clayberger, and A. M. Krensky. 2001. Granulysin: a novel antimicrobial. *Expert Opin. Investig. Drugs* **10:**321–329.

Kuspa, A., and W. F. Loomis. 1992. Tagging developmental genes in *Dictyostelium* by restriction enzyme-mediated integration of plasmid DNA. *Proc. Natl. Acad. Sci. USA* **89:** 8803–8807.

Laguna, R. K., E. A. Creasey, Z. Li, N. Valtz, and R. R. Isberg. 2006. A *Legionella pneumophila*-translocated substrate that is required for growth within macrophages and protection from host cell death. *Proc. Natl. Acad. Sci. USA* **103:** 18745–18750.

Landolt, J. C., S. L. Stephenson, and J. C. Cavender. 2006. Distribution and ecology of dictyostelid cellular slime molds in Great Smoky Mountains National Park. *Mycologia* **98:** 541–549.

Lardy, B., M. Bof, L. Aubry, M. H. Paclet, F. Morel, M. Satre, and G. Klein. 2005. NADPH oxidase homologs are required for normal cell differentiation and morphogenesis in *Dictyostelium discoideum*. *Biochim. Biophys. Acta* **1744:**199–212.

Lee, M. C., E. A. Miller, J. Goldberg, L. Orci, and R. Schekman. 2004. Bi-directional protein transport between the ER and Golgi. *Annu. Rev. Cell Dev. Biol.* **20:**87–123.

Lee, W. L., D. Mason, A. D. Schreiber, and S. Grinstein. 2007. Quantitative analysis of membrane remodeling at the phagocytic cup. *Mol. Biol. Cell* **18:**2883–2892.

Leipe, D. D., E. V. Koonin, and L. Aravind. 2004. STAND, a class of P-loop NTPases including animal and plant regulators of programmed cell death: multiple, complex domain architectures, unusual phyletic patterns, and evolution by horizontal gene transfer. *J. Mol. Biol.* **343:**1–28.

Leippe, M., and R. Herbst. 2004. Ancient weapons for attack and defense: the pore-forming polypeptides of pathogenic enteric and free-living amoeboid protozoa. *J. Eukaryot. Microbiol.* **51:**516–521.

Levine, B., and D. J. Klionsky. 2004. Development by self-digestion: molecular mechanisms and biological functions of autophagy. *Dev. Cell* **6:**463–477.

Li, Z., J. M. Solomon, and R. R. Isberg. 2005. *Dictyostelium discoideum* strains lacking the RtoA protein are defective for maturation of the *Legionella pneumophila* replication vacuole. *Cell. Microbiol.* **7:**431–442.

Losick, V. P., and R. R. Isberg. 2006. NF-kappaB translocation prevents host cell death after low-dose challenge by *Legionella pneumophila*. *J. Exp. Med.* **203:**2177–2189.

Lu, H., and M. Clarke. 2005. Dynamic properties of *Legionella*-containing phagosomes in *Dictyostelium* amoebae. *Cell. Microbiol.* **7:**995–1007.

Ma, H., M. Gamper, C. Parent, and R. A. Firtel. 1997. The *Dictyostelium* MAP kinase kinase DdMEK1 regulates chemotaxis and is essential for chemoattractant-mediated activation of guanylyl cyclase. *EMBO J.* **16:**4317–4332.

Maniak, M., R. Rauchenberger, R. Albrecht, J. Murphy, and G. Gerisch. 1995. Coronin involved in phagocytosis: dynamics of particle-induced relocalization visualized by a green fluorescent protein Tag. *Cell* **83:**915–924.

Mann, S. K. O., P. N. Devreotes, S. Eliott, K. Jermyn, A. Kuspa, M. Fechheimer, R. Furukawa, C. A. Parent, J. Segall, G. Shaulsky, P. H. Vardy, J. Williams, K. L. Williams, and R. A. Firtel. 1998. Cell biological, molecular genetic, and biochemical methods to examine Dictyostelium, p. 431–465. *In* J. E. Celis (ed.), *Cell Biology—A Laboratory Handbook*, 2nd ed. Academic Press, San Diego, CA.

Marceau, M. 2005. Transcriptional regulation in *Yersinia*: an update. *Curr. Issues Mol. Biol.* **7:**151–177.

Maselli, A., G. Laevsky, and D. A. Knecht. 2002. Kinetics of binding, uptake and degradation of live fluorescent (DsRed) bacteria by *Dictyostelium discoideum*. *Microbiology* **148:**413–420.

Mendoza, M. C., F. Du, N. Iranfar, N. Tang, H. Ma, W. F. Loomis, and R. A. Firtel. 2005. Loss of SMEK, a novel, conserved protein, suppresses MEK1 null cell polarity, chemotaxis, and gene expression defects. *Mol. Cell Biol.* **25:** 7839–7853.

Meyers, B. C., A. Kozik, A. Griego, H. Kuang, and R. W. Michelmore. 2003. Genome-wide analysis of NBS-LRR-encoding genes in *Arabidopsis*. *Plant Cell* **15:**809–834.

Miltner, E. C., and L. E. Bermudez. 2000. *Mycobacterium avium* grown in *Acanthamoeba castellanii* is protected from the effects of antimicrobials. *Antimicrob. Agents Chemother.* **44:** 1990–1994.

Muder, R. R., V. L. Yu, and A. H. Woo. 1986. Mode of transmission of *Legionella pneumophila*. A critical review. *Arch. Intern. Med.* **146:**1607–1612.

Muller-Taubenberger, A., A. N. Lupas, H. Li, M. Ecke, E. Simmeth, and G. Gerisch. 2001. Calreticulin and calnexin in the endoplasmic reticulum are important for phagocytosis. *EMBO J.* **20:**6772–6782.

Neuhaus, E. M., W. Almers, and T. Soldati. 2002. Morphology and dynamics of the endocytic pathway in *Dictyostelium discoideum*. *Mol. Biol. Cell* **13:**1390–1407.

Nurnberger, T., F. Brunner, B. Kemmerling, and L. Piater. 2004. Innate immunity in plants and animals: striking similarities and obvious differences. *Immunol. Rev.* **198:**249–266.

Okuwa, T., T. Morlo, T. Saito, Y. Masamune, and H. Yasukawa. 2001. Complete sequences and expression kinetics of *racG*, *racH*, *racI* and *racJ* genes in *Dictyostelium discoideum*. *Biol. Pharm. Bull.* **24:**84–87.

Otto, G. P., M. Y. Wu, M. Clarke, H. Lu, O. R. Anderson, H. Hilbi, H. A. Shuman, and R. H. Kessin. 2004. Macroautophagy is dispensable for intracellular replication of *Legionella pneumophila* in *Dictyostelium discoideum*. *Mol. Microbiol.* **51:**63–72.

Otto, G. P., M. Y. Wu, N. Kazgan, O. R. Anderson, and R. H. Kessin. 2003. Macroautophagy is required for multicellular development of the social amoeba *Dictyostelium discoideum*. *J. Biol. Chem.* **278:**17636–17645.

Parent, C. A., and P. N. Devreotes. 1999. A cell's sense of direction. *Science* **284:**765–770.

Pedro-Botet, M. L., and M. Sabria. 2005. Legionellosis. *Semin. Respir. Crit. Care Med.* **26:**625–634.

Peracino, B., J. Borleis, T. Jin, M. Westphal, J. M. Schwartz, L. Wu, E. Bracco, G. Gerisch, P. Devreotes, and S. Bozzaro. 1998. G protein beta subunit-null mutants are impaired in phagocytosis and chemotaxis due to inappropriate regulation of the actin cytoskeleton. *J. Cell Biol.* **141:**1529–1537.

Peracino, B., C. Wagner, A. Balest, A. Balbo, B. Pergolizzi, A. A. Noegel, M. Steinert, and S. Bozzaro. 2006. Function and mechanism of action of *Dictyostelium* Nramp1 (Slc11a1) in bacterial infection. *Traffic* **7:**22–38.

Phillips, R. M., D. A. Six, E. A. Dennis, and P. Ghosh. 2003. *In vivo* phospholipase activity of the *Pseudomonas aeruginosa* cytotoxin ExoU and protection of mammalian cells with phospholipase A2 inhibitors. *J. Biol. Chem.* **278:**41326–41332.

Pine, L., J. R. George, M. W. Reeves, and W. K. Harrell. 1979. Development of a chemically defined liquid medium for growth of *Legionella pneumophila*. *J. Clin. Microbiol.* **9:** 615–626.

Powell, R. R., and L. A. Temesvari. 2004. Involvement of a Rab8-like protein of *Dictyostelium discoideum*, Sas1, in the formation of membrane extensions, secretion and adhesion during development. *Microbiology* **150:**2513–2525.

Pruitt, B. A., Jr., A. T. McManus, S. H. Kim, and C. W. Goodwin. 1998. Burn wound infections: current status. *World J. Surg.* **22:**135–145.

Pukatzki, S., R. H. Kessin, and J. J. Mekalanos. 2002. The human pathogen *Pseudomonas aeruginosa* utilizes conserved virulence pathways to infect the social amoeba *Dictyostelium discoideum*. *Proc. Natl. Acad. Sci. USA* **99:**3159–3164.

Pukatzki, S., A. T. Ma, A. T. Revel, D. Sturtevant, and J. J. Mekalanos. 2007. Type VI secretion system translocates a phage tail spike-like protein into target cells where it cross-links actin. *Proc. Natl. Acad. Sci. USA* **104:**15508–15513.

Pukatzki, S., A. T. Ma, D. Sturtevant, B. Krastins, D. Sarracino, W. C. Nelson, J. F. Heidelberg, and J. J. Mekalanos. 2006. Identification of a conserved bacterial protein secretion system in *Vibrio cholerae* using the *Dictyostelium* host model system. *Proc. Natl. Acad. Sci. USA* **103:**1528–1533.

Racoosin, E. L., and J. A. Swanson. 1993. Macropinosome maturation and fusion with tubular lysosomes in macrophages. *J. Cell Biol.* **121:**1011–1020.

Rahme, L. G., F. M. Ausubel, H. Cao, E. Drenkard, B. C. Goumnerov, G. W. Lau, S. Mahajan-Miklos, J. Plotnikova, M. W. Tan, J. Tsongalis, C. L. Walendziewicz, and R. G. Tompkins. 2000. Plants and animals share functionally common bacterial virulence factors. *Proc. Natl. Acad. Sci. USA* **97:**8815–8821.

Rahme, L. G., E. J. Stevens, S. F. Wolfort, J. Shao, R. G. Tompkins, and F. M. Ausubel. 1995. Common virulence factors for bacterial pathogenicity in plants and animals. *Science* **268:**1899–1902.

Rairdan, G., and P. Moffett. 2007. Brothers in arms? Common and contrasting themes in pathogen perception by plant NB-LRR and animal NACHT-LRR proteins. *Microbes Infect.* **9:**677–686.

Ramakrishnan, L., and S. Falkow. 1994. *Mycobacterium marinum* persists in cultured mammalian cells in a temperature-restricted fashion. *Infect. Immun.* **62:**3222–3229.

Ramakrishnan, L., N. A. Federspiel, and S. Falkow. 2000. Granuloma-specific expression of *Mycobacterium* virulence proteins from the glycine-rich PE-PGRS family. *Science* **288:**1436–1439.

Ramakrishnan, L., R. H. Valdivia, J. H. McKerrow, and S. Falkow. 1997. *Mycobacterium marinum* causes both long-term subclinical infection and acute disease in the leopard frog (*Rana pipiens*). *Infect. Immun.* **65:**767–773.

Rast, J. P., L. C. Smith, M. Loza-Coll, T. Hibino, and G. W. Litman. 2006. Genomic insights into the immune system of the sea urchin. *Science* **314:**952–956.

Rauchenberger, R., U. Hacker, J. Murphy, J. Niewohner, and M. Maniak. 1997. Coronin and vacuolin identify consecutive stages of a late, actin-coated endocytic compartment in *Dictyostelium*. *Curr. Biol.* **7:**215–218.

Rivero, F., H. Dislich, G. Glockner, and A. A. Noegel. 2001. The *Dictyostelium discoideum* family of Rho-related proteins. *Nucleic Acids Res.* **29:**1068–1079.

Robinson, C. G., and C. R. Roy. 2006. Attachment and fusion of endoplasmic reticulum with vacuoles containing *Legionella pneumophila*. *Cell. Microbiol.* **8:**793–805.

Rowbotham, T. J. 1986. Current views on the relationships between amoebae, *Legionellae* and man. *Isr. J. Med. Sci.* **22:**678–689.

Rowbotham, T. J. 1980. Preliminary report on the pathogenicity of *Legionella pneumophila* for freshwater and soil amoebae. *J. Clin. Pathol.* **33:**1179–1183.

Rumbaugh, K. P., J. A. Griswold, and A. N. Hamood. 1999. Contribution of the regulatory gene *lasR* to the pathogenesis of *Pseudomonas aeruginosa* infection of burned mice. *J. Burn Care Rehabil.* **20:**42–49.

Rupper, A., and J. Cardelli. 2001. Regulation of phagocytosis and endo-phagosomal trafficking pathways in *Dictyostelium discoideum*. *Biochim. Biophys. Acta* **1525:**205–216.

Rupper, A., B. Grove, and J. Cardelli. 2001. Rab7 regulates phagosome maturation in *Dictyostelium*. *J. Cell Sci.* **114:**2449–2460.

Sato, H., D. W. Frank, C. J. Hillard, J. B. Feix, R. R. Pankhaniya, K. Moriyama, V. Finck-Barbancon, A. Buchaklian, M. Lei, R. M. Long, J. Wiener-Kronish, and T. Sawa. 2003. The mechanism of action of the *Pseudomonas aeruginosa*-encoded type III cytotoxin, ExoU. *EMBO J.* **22:**2959–2969.

Searle, S., N. A. Bright, T. I. Roach, P. G. Atkinson, C. H. Barton, R. H. Meloen, and J. M. Blackwell. 1998. Localisation of Nramp1 in macrophages: modulation with activation and infection. *J. Cell Sci.* **111**(Pt 19)**:**2855–2866.

Seastone, D. J., E. Harris, L. A. Temesvari, J. E. Bear, C. L. Saxe, and J. Cardelli. 2001. The WASp-like protein scar regulates macropinocytosis, phagocytosis and endosomal membrane flow in *Dictyostelium*. *J. Cell Sci.* **114:**2673–2683.

Segal, G., and H. A. Shuman. 1999. *Legionella pneumophila* utilizes the same genes to multiply within *Acanthamoeba castellanii* and human macrophages. *Infect. Immun.* **67:**2117–2124.

Sheahan, K. L., C. L. Cordero, and K. J. Satchell. 2004. Identification of a domain within the multifunctional *Vibrio cholerae* RTX toxin that covalently cross-links actin. *Proc. Natl. Acad. Sci. USA* **101:**9798–9803.

Skinner, A. R., C. M. Anand, A. Malic, and J. B. Kurtz. 1983. Acanthamoebae and environmental spread of *Legionella pneumophila*. *Lancet* **2:**289–290.

Skriwan, C., M. Fajardo, S. Hagele, M. Horn, M. Wagner, R. Michel, G. Krohne, M. Schleicher, J. Hacker, and M. Steinert. 2002. Various bacterial pathogens and symbionts infect the amoeba *Dictyostelium discoideum*. *Int. J. Med. Microbiol.* **291:**615–624.

Smith, M. G., T. A. Gianoulis, S. Pukatzki, J. J. Mekalanos, L. N. Ornston, M. Gerstein, and M. Snyder. 2007. New insights into *Acinetobacter baumannii* pathogenesis revealed by high-density pyrosequencing and transposon mutagenesis. *Genes Dev.* **21:**601–614.

Solomon, J. M., and R. R. Isberg. 2000. Growth of *Legionella pneumophila* in *Dictyostelium discoideum*: a novel system for genetic analysis of host-pathogen interactions. *Trends Microbiol.* **8:**478–480.

Solomon, J. M., G. S. Leung, and R. R. Isberg. 2003. Intracellular replication of *Mycobacterium marinum* within *Dictyostelium discoideum*: efficient replication in the absence of host coronin. *Infect. Immun.* **71:**3578–3586.

Solomon, J. M., A. Rupper, J. A. Cardelli, and R. R. Isberg. 2000. Intracellular growth of *Legionella pneumophila* in *Dictyostelium discoideum*, a system for genetic analysis of host-pathogen interactions. *Infect. Immun.* **68:**2939–2947.

Somesh, B. P., C. Neffgen, M. Iijima, P. Devreotes, and F. Rivero. 2006. *Dictyostelium* RacH regulates endocytic vesicular trafficking and is required for localization of vacuolin. *Traffic* **7:**1194–1212.

Stamm, L. M., J. H. Morisaki, L. Y. Gao, R. L. Jeng, K. L. McDonald, R. Roth, S. Takeshita, J. Heuser, M. D. Welch, and E. J. Brown. 2003. *Mycobacterium marinum* escapes from phagosomes and is propelled by actin-based motility. *J. Exp. Med.* **198:**1361–1368.

Steenbergen, J. N., J. D. Nosanchuk, S. D. Malliaris, and A. Casadevall. 2003. *Cryptococcus neoformans* virulence is enhanced after growth in the genetically malleable host *Dictyostelium discoideum*. *Infect. Immun.* **71:**4862–4872.

Steenbergen, J. N., H. A. Shuman, and A. Casadevall. 2001. *Cryptococcus neoformans* interactions with amoebae suggest an explanation for its virulence and intracellular pathogenic strategy in macrophages. *Proc. Natl. Acad. Sci. USA* **98:**15245–15250.

Sturgill-Koszycki, S., U. E. Schaible, and D. G. Russell. 1996. *Mycobacterium*-containing phagosomes are accessible to early endosomes and reflect a transitional state in normal phagosome biogenesis. *EMBO J.* **15:**6960–6968.

Sturgill-Koszycki, S., P. H. Schlesinger, P. Chakraborty, P. L. Haddix, H. L. Collins, A. K. Fok, R. D. Allen, S. L. Gluck, J. Heuser, and D. G. Russell. 1994. Lack of acidification in *Mycobacterium* phagosomes produced by exclusion of the vesicular proton-ATPase. *Science* **263**:678–681.

Sturgill-Koszycki, S., and M. S. Swanson. 2000. *Legionella pneumophila* replication vacuoles mature into acidic, endocytic organelles. *J. Exp. Med.* **192**:1261–1272.

Swanson, M. S., and R. R. Isberg. 1995. Association of *Legionella pneumophila* with the macrophage endoplasmic reticulum. *Infect. Immun.* **63**:3609–3620.

Takeda, K., T. Kaisho, and S. Akira. 2003. Toll-like receptors. *Annu. Rev. Immunol.* **21**:335–376.

Tang, H. B., E. DiMango, R. Bryan, M. Gambello, B. H. Iglewski, J. B. Goldberg, and A. Prince. 1996. Contribution of specific *Pseudomonas aeruginosa* virulence factors to pathogenesis of pneumonia in a neonatal mouse model of infection. *Infect. Immun.* **64**:37–43.

Tilney, L. G., O. S. Harb, P. S. Connelly, C. G. Robinson, and C. R. Roy. 2001. How the parasitic bacterium *Legionella pneumophila* modifies its phagosome and transforms it into rough ER: implications for conversion of plasma membrane to the ER membrane. *J. Cell Sci.* **114**:4637–4650.

Ting, J. P., and B. K. Davis. 2005. CATERPILLER: a novel gene family important in immunity, cell death, and diseases. *Annu. Rev. Immunol.* **23**:387–414.

Titus, M. A. 1999. A class VII unconventional myosin is required for phagocytosis. *Curr. Biol.* **9**:1297–1303.

Titus, M. A. 2000. The role of unconventional myosins in *Dictyostelium* endocytosis. *J. Eukaryot. Microbiol.* **47**:191–196.

Tobin, J. O., J. Beare, M. S. Dunnill, S. Fisher-Hoch, M. French, R. G. Mitchell, P. J. Morris, and M. F. Muers. 1980. Legionnaires' disease in a transplant unit: isolation of the causative agent from shower baths. *Lancet* **2**:118–121.

Touret, N., P. Paroutis, M. Terebiznik, R. E. Harrison, S. Trombetta, M. Pypaert, A. Chow, A. Jiang, J. Shaw, C. Yip, H. P. Moore, N. van der Wel, D. Houben, P. J. Peters, C. de Chastellier, I. Mellman, and S. Grinstein. 2005. Quantitative and dynamic assessment of the contribution of the ER to phagosome formation. *Cell* **123**:157–170.

Vadell, E. M., and J. C. Cavender. 2007. Dictyostelids living in the soils of the Atlantic forest, Iguazu region, Misiones, Argentina: description of new species. *Mycologia* **99**:112–124.

Van Haastert, P. J., and P. N. Devreotes. 2004. Chemotaxis: signalling the way forward. *Nat. Rev. Mol. Cell Biol.* **5**:626–634.

VanRheenen, S. M., G. Dumenil, and R. R. Isberg. 2004. IcmF and DotU are required for optimal effector translocation and trafficking of the *Legionella pneumophila* vacuole. *Infect. Immun.* **72**:5972–5982.

Vogel, G., L. Thilo, H. Schwarz, and R. Steinhart. 1980. Mechanism of phagocytosis in *Dictyostelium discoideum*: phagocytosis is mediated by different recognition sites as disclosed by mutants with altered phagocytic properties. *J. Cell Biol.* **86**:456–465.

Weber, S. S., C. Ragaz, K. Reus, Y. Nyfeler, and H. Hilbi. 2006. *Legionella pneumophila* exploits PI(4)P to anchor secreted effector proteins to the replicative vacuole. *PLoS Pathog.* **2**:e46.

Weijer, C. J. 2004. *Dictyostelium* morphogenesis. *Curr. Opin. Genet. Dev.* **14**:392–398.

Werts, C., S. E. Girardin, and D. J. Philpott. 2006. TIR, CARD and PYRIN: three domains for an antimicrobial triad. *Cell Death Differ.* **13**:798–815.

Williams, R. L., and S. Urbe. 2007. The emerging shape of the ESCRT machinery. *Nat. Rev. Mol. Cell Biol.* **8**:355–368.

Williams, R. S., K. Boeckeler, R. Graf, A. Muller-Taubenberger, Z. Li, R. R. Isberg, D. Wessels, D. R. Soll, H. Alexander, and S. Alexander. 2006. Towards a molecular understanding of human diseases using *Dictyostelium discoideum*. *Trends Mol. Med.* **12**:415–424.

Wilmanski, J. M., T. Petnicki-Ocwieja, and K. S. Kobayashi. 2008. NLR proteins: integral members of innate immunity and mediators of inflammatory diseases. *J. Leukoc. Biol.* **83**:13–30.

Woo, A. H., A. Goetz, and V. L. Yu. 1992. Transmission of *Legionella* by respiratory equipment and aerosol generating devices. *Chest* **102**:1586–1590.

Wood, S. A., R. R. Ammann, D. A. Brock, L. Li, T. Spann, and R. H. Gomer. 1996. RtoA links initial cell type choice to the cell cycle in *Dictyostelium*. *Development* **122**:3677–3685.

Wright, E. K., S. A. Goodart, J. D. Growney, V. Hadinoto, M. G. Endrizzi, E. M. Long, K. Sadigh, A. L. Abney, I. Bernstein-Hanley, and W. F. Dietrich. 2003. Naip5 affects host susceptibility to the intracellular pathogen *Legionella pneumophila*. *Curr. Biol.* **13**:27–36.

Wu, L., R. Valkema, P. J. Van Haastert, and P. N. Devreotes. 1995. The G protein beta subunit is essential for multiple responses to chemoattractants in *Dictyostelium*. *J. Cell Biol.* **129**:1667–1675.

Yam, P. T., and J. A. Theriot. 2004. Repeated cycles of rapid actin assembly and disassembly on epithelial cell phagosomes. *Mol. Biol. Cell* **15**:5647–5658.

Yan, M., R. F. Collins, S. Grinstein, and W. S. Trimble. 2005. Coronin-1 function is required for phagosome formation. *Mol. Biol. Cell* **16**:3077–3087.

Yan, M., C. Di Ciano-Oliveira, S. Grinstein, and W. S. Trimble. 2007. Coronin function is required for chemotaxis and phagocytosis in human neutrophils. *J. Immunol.* **178**:5769–5778.

Yoshida, S., Y. Goto, Y. Mizuguchi, K. Nomoto, and E. Skamene. 1991. Genetic control of natural resistance in mouse macrophages regulating intracellular *Legionella pneumophila* multiplication in vitro. *Infect. Immun.* **59**:428–432.

Zaki, M., J. King, K. Futterer, and R. H. Insall. 2007. Replacement of the essential *Dictyostelium* Arp2 gene by its *Entamoeba* homologue using parasexual genetics. *BMC Genet.* **8**:28.

Zhai, Y., and M. H. Saier, Jr. 2000. The amoebapore superfamily. *Biochim. Biophys. Acta* **1469**:87–99.

Zhou, Q., and W. J. Murphy. 2006. Immune response and immunotherapy to *Cryptococcus* infections. *Immunol. Res.* **35**:191–208.

Zusman, T., M. Feldman, E. Halperin, and G. Segal. 2004. Characterization of the *icmH* and *icmF* genes required for *Legionella pneumophila* intracellular growth, genes that are present in many bacteria associated with eukaryotic cells. *Infect. Immun.* **72**:3398–3409.

33

Phagocytosis in *Drosophila melanogaster* Immune Response

VINCENT LECLERC, ISABELLE CALDELARI,
NATALIA VERESCEAGHINA, AND JEAN-MARC REICHHART

Drosophila is a genetically tractable model organism that was a system of choice to study and identify most of the major genes and processes that govern embryonic and postembryonic development. For little more than a decade it has gained new interest as a valuable model to study physiological or pathological processes such as cancer and the immune response. The *Drosophila* immune system relies on innate defenses with two major arms (Cherry and Silverman, 2006; Wang and Ligoxygakis, 2006; Lemaitre and Hoffman, 2007). The first arm consists of a humoral response. After infection, the recognition of bacteria or fungi by pattern recognition molecules activates two signaling pathways (namely, the Toll and IMD pathways) in the fat body cells (an equivalent of the mammalian liver), leading to the secretion of antimicrobial peptides (AMPs) into the fly's open circulatory system. The second arm is a cellular response involving circulating blood cells or hemocytes.

Three types of hemocytes have been identified in *Drosophila* (Meister, 2004; Williams, 2007). Crystal cells secrete phenoloxydase, an enzyme responsible for melanization at the site of injuries or around foreign particles trapped in the body cavity, participating in wound healing. Melanization could potentially be toxic for microorganisms but is not essential for survival after an infection (Leclerc et al., 2006). Lamellocytes are specialized blood cells that encapsulate intruders such as parasitic wasp eggs that are too big to be phagocytosed. They sequester parasites from the body cavity, contributing to their killing by isolation. Plasmatocytes, the third class of hemocytes, are the most abundant and represent almost 95% of circulating blood cells. These macrophage-like cells are responsible for the phagocytosis of both microorganisms and apoptotic bodies (Wood and Jacinto, 2007). They play a crucial role during metamorphosis as they engulf and recycle doomed cells. Plasmatocytes also produce and secrete a number of peptides and proteins like extracellular matrix proteins, AMPs, and possibly signals that inform distant tissues of an infection. There is no equivalent of the lymphoid lineage in insects and no obvious mammalian counterparts for lamellocytes or crystal cells.

In *Drosophila* all adult hemocytes are produced during the third larval stage because hematopoiesis is restricted to embryonic and larval stages (Crozatier and Meister, 2007). Healthy larvae contain plasmatocytes and crystal cells. Lamellocytes differentiate from plasmatocytes only after parasite sensing.

During an infection, phagocytosis of fluorescent bacteria can be observed in vivo. The process can be easily blocked by prior injection of polystyrene or latex beads that saturate the macrophage-like plasmatocytes. However, analysis of phagocytosis in *Drosophila* has been greatly facilitated by the characterization of the phagocytic properties of S2 cells, an embryonic hemocyte-derived cell line that is readily amenable to RNA-mediated interference (RNAi), opening the field to systematic studies.

Drosophila is now used as a powerful model organism for the identification of conserved molecular mechanisms required for phagocytosis. In addition, analysis of the function of phagocytosis during the immune response and a possible interaction with the humoral defense has recently gained renewed interest.

GENOMIC AND PROTEOMIC ANALYSIS OF PHAGOCYTOSIS IN *DROSOPHILA*

In *Drosophila*, genetic screens are successfully applied to discover new genes involved in a specific process. However, tests that are available today to measure phagocytosis of microorganisms are not simple and easy enough to screen thousands of putative mutants. Instead, identification of new genes involved in phagocytosis relies either on genome-wide RNAi screens in S2 cell culture, proteomic analysis of the plasmatocyte phagosome, or in silico homology screening.

Vincent Leclerc, Isabelle Caldelari, Natalia Veresceaghina, and Jean-Marc Reichhart, UPR 9022 Institut de Biologie Moléculaire et Cellulaire, 67084 Strasbourg, France.

Because of low redundancy between genes of the same family, RNAi screens are quite powerful in *Drosophila*, resulting in a comprehensive view of the phagocytic machinery. Several RNAi screens have been conducted in S2 cells testing for the inability to phagocytose yeast or bacteria (Ramet et al., 2002; Pearson et al., 2003; Agaisse et al., 2005; Philips et al., 2005; Stroschein-Stevenson et al., 2006). These screens allowed the identification of a large number of proteins involved in the general mechanism of phagocytosis and conserved in other animal models. They include actin cytoskeleton and vesicle-trafficking regulators, DNA binding proteins, signaling molecules, and proteins involved in meta- and catabolism of polypeptides. Comparison of data from several screens that used different microbial targets allowed for the discrimination between genes involved in the basic phagocytic machinery and those involved in uptake of specific microorganisms. The latter were mostly pattern recognition molecules, either receptors or opsonins (see below).

RNAi screens allow only the identification of genes whose inactivation are not lethal to cells and are phenotypically visible. Using the method devised by Desjardin (Garin et al., 2001), Stuart and collaborators (2007) conducted a proteomic analysis of *Drosophila* S2 cell phagosomes containing latex beads. Phagosomes are the subcellular organelles that normally internalize microorganisms, but they can be loaded with latex beads and purified by flotation. This analysis identified 617 proteins potentially associated with the *Drosophila* phagosome, 70% of which had mammalian orthologues. Of the 140 proteins previously identified in the mammalian phagosome, 70% had orthologues within the *Drosophila* phagosome (Garin et al., 2001; Desjardins and Griffiths, 2003). Proteomic data were further organized by using protein-protein interaction databases to generate the phagosome interactome. Contribution of the candidate genes to bacterial phagocytosis was tested by RNAi inactivation. Using a combination of several approaches, Stuart and colleagues generated a detailed model of the phagosome, which was extended to the mammalian phagosome. Some of the newly identified molecules could be tested as putative targets for pathogens evading host defense by hiding within the phagosome. In one outcome of this research, the authors proposed a model explaining the role of the evolutionarily conserved exocyst complex during phagocytosis.

Phagocytosis contributes to the fight against invading microorganisms. Identification of molecules specifically required for pathogen uptake and analysis of their inactivation phenotypes either in S2 cells or in live animals is of great interest. Several opsonins and receptors have been identified either by genome-wide studies or by homology screening, but few have been validated by in vivo experiments.

RECEPTOR DIVERSITY

Several receptors required for phagocytosis of microorganisms have been identified in *Drosophila*. They belong to four families of evolutionarily conserved molecules: class B and C Scavenger Receptors (SRs), Nim Repeat-Containing Receptors, Peptidoglycan Recognition Proteins (PGRPs), and immunoglobulin repeat containing receptors. Table 1 provides a list of identified receptors and a description of their in vitro or in vivo specificities.

Croquemort and Peste, CD36-Related Class B Scavenger Receptors

SRs represent a large family of structurally unrelated distinct gene products, expressed by myeloid or selected endothelial cells and able to recognize modified low-density lipoproteins. Receptors of this class also bind and internalize a variety of microbial pathogens, as well as modified or endogenous molecules derived from the host, and contribute to a range of physiological or pathological processes (Mukhopadhyay and Gordon, 2004). Three families (SR-A, SR-B, and SR-C) have been identified, with mammals encoding only class A and B members, whereas the *Drosophila* genome encodes class B and C receptors. In *Drosophila* embryos, Croquemort, a member of the SR-B family, was identified as a receptor required for clearance of apoptotic corpses but not for phagocytosis of bacteria (Franc et al., 1999). However, more recently, a high-throughput RNAi screen in S2 cells for *Drosophila* genes involved in phagocytosis revealed that *croquemort* silencing resulted in a specific inhibition of *Staphylococcus aureus* phagocytosis compared with that of *Escherichia coli* (Stuart et al., 2005). Furthermore, CD36, the mammalian-encoded paralogue of Croquemort, also directly phagocytoses bacteria and plays a role in the host immune response to *S. aureus* both by clearing bacteria and as a cofactor in Toll-like receptor (TLR) signaling (Stuart et al., 2005).

The gene *peste* encodes another member of the SR-B family. It was discovered in a RNAi screen for defects in internalization of the intracellular bacteria *Mycobacterium fortuitum* (Philips et al., 2005) and is required in S2 cells for uptake of intracellular bacteria of the *Mycobacterium* and *Listeria* genera but not for the phagocytosis of *E. coli* and *S. aureus*. This work also demonstrated that several mammalian class B SRs were able to induce internalization of mycobacteria in human cells.

Drosophila Scavenger Receptor CI

Drosophila SR-CI (dSR-CI) is a bacterial pattern recognition receptor that accounts for 20% to 30% of total bacterial binding activity in *Drosophila* S2 cells. It binds both gram-positive (*S. aureus*) and gram-negative (*E. coli*) bacteria, but it is not required for phagocytosis of the yeast *Candida silvativa* (Ramet et al., 2001). Sequencing of genomic DNA encoding dSR-CI revealed an extensive polymorphism for *dSR-CI* compared with the cDNA sequence derived from S2 cells. This high-frequency polymorphism of *dSR-CI* may promote individual variations in pathogen recognition capacity.

Nimrod and Eater, Phagocytic Receptors with EGF-Like Repeats

Another *Drosophila* receptor required for phagocytosis is Nimrod (NimC1; Kurucz et al., 2007). In vivo *NimC1* suppression showed an important decrease in *S. aureus* phagocytosis index, but phagocytosis of *E. coli* was not significantly affected. However, NimC1 may also play a redundant role in *E. coli* phagocytosis, as overexpression of *NimC1* in S2 cells (which do not express *NimC1*) stimulated the uptake of both *S. aureus* and *E. coli*.

NimC1 is a 90- to 100-kDa single-pass transmembrane protein with 10 characteristic epidermal growth factor (EGF)-like repeats (Nim repeats). The *NimC1* gene is a part of a cluster of 10 related *nimrod* genes located at 34E on the *Drosophila* second chromosome. Proteins with NimC1 CCxGY motifs followed by one or several Nim repeats can also be found in humans.

Related to NimC1, the *Drosophila* Eater protein functions as a phagocytic receptor for both gram-positive and gram-negative bacteria (Kocks et al., 2005). *eater* is expressed in S2 cells and in larval or adult macrophages. The extracellular domain of the Eater protein carries EGF-like repeats. If direct binding of NimC1 to bacteria is unclear, Eater has been proven to be involved in direct recognition of bacteria, with the N-terminal 199 amino acids of Eater participating in ligand binding.

An *eater* loss-of-function mutation has been generated, showing that Eater is required for phagocytosis of several types of bacteria in vivo. Furthermore, it is the first receptor whose inactivation was shown to lead to increased susceptibility to bacterial infections. Transcriptional silencing of *eater* abolished the major part of phagocytic activity in S2 cells compared with *SR-CI* and *PGRP-LC* knockdown, but additional silencing of *SR-CI* and *PGRP-LC* further decreased phagocytosis and cell surface binding, indicating that multiple receptors cooperate in recognition and internalization of microbes. Similarly to other receptors, Eater is able to recognize multiple ligands and, like mammalian class A SRs, contains only a short intracellular tail.

draper, the *Drosophila* homologue of *Caenorabitis elegans CED-1*, is a third member of the family that appears to be required only for phagocytosis of apoptotic cells (Manaka et al., 2004).

Down Syndrome Cell Adhesion Molecule (Dscam), a Receptor with Large Isoform Diversity

Dscam is a striking example of the molecular complexity that can found in invertebrates. Originally identified as a guidance receptor in the *Drosophila* nervous system, this member of the immunoglobulin superfamily is also expressed by phagocytic hemocytes (Schmucker et al., 2000; Watson et al., 2005). *Drosophila* immune-competent cells can potentially express more than 18,000 Dscam isoforms as a result of alternative splicing. A series of experiments performed by Schmucker and colleagues revealed that *Drosophila* hemocytes, in which Dscam isoforms had been specifically knocked down, showed a substantially reduced rate of phagocytosis, with less than 60% of the cells containing bacteria (Watson et al., 2005). Two of three tested Dscam isoforms were shown by flow cytometry to be able to bind directly to live *E. coli* bacteria. The lack of binding of the third isoform may reflect different binding capabilities by diverse isoforms. Whether Dscam functions as a phagocytic receptor or signaling coreceptor in hemocytes is not known today. Dscam could also be involved in opsonization of invading pathogens, as some isoforms have been shown to be secreted into the hemolymph.

PGRPs: Phagocytic Receptors or Opsonins?

PGRP-SA and PGRP-SD have been shown to recognize gram-positive bacteria leading to Toll pathway activation, whereas PGRP-LC and PGRP-LE recognize gram-negative bacteria and activate the IMD pathway (Royet and Dziarski, 2007). In parallel to its function as pattern recognition receptor upstream of the IMD pathway, PGRP-LC has also been described as a phagocytic receptor in S2 cells, but no defect in phagocytosis could be demonstrated in vivo in *PGRP-LC* mutant flies (Ramet et al., 2002; Garver et al., 2006).

Drosophila mutant for *PGRP-SA* and another potential receptor, *PGRP-SC1*, showed two independent phenotypes: a reduced phagocytosis of *S. aureus* in vivo (Garver et al., 2006) and a lack of activation of the Toll pathway. Both proteins are secreted and could function as opsonins. However, unlike PGRP-SA, PGRP-SC1 is catalytically active and able to cleave peptidoglycan. Garver and colleagues demonstrated that this activity is required for *S. aureus* phagocytosis and that the addition of free peptidoglycan could substitute for the catalytic activity, suggesting that generation of peptidoglycan fragments could be a signal for phagocytosis. However the molecular mechanism by which both PGRP molecules currently function is still unknown.

Drosophila Teps as Opsonins?

The α_2-macroglobulins protein superfamily includes three subfamilies (Dodds and Law, 1998; Armstrong, 2006). The α_2-macroglobulins themselves are protease inhibitors that trap the target protease after cleavage, leading to its endocytosis. The complement factors C3, C4, and C5 are also activated by proteolytic cleavage via the host convertase complex. The resulting products then act as opsonins, covalently binding to the pathogen and promoting phagocytosis (Law and Reid, 1995). Teps, belonging to the third subfamily, had originally been discovered in insects (Levashina et al., 2001) and shown to be able to bind microorganisms. The *Drosophila* genome includes four *Tep* genes, *TepI* to *-IV*, and their close relative *Mcr* (Lagueux et al., 2000).

Drosophila Teps have a hypervariable or bait-like region near the center of the coding sequence (Jiggins and Kim, 2006). Alternative splicing of the *TepII* transcript results in isoforms with five different bait-like regions (Lagueux et al., 2000), suggesting that sequence variation in this region is functionally important. Jiggins and Kim (2006) believe that this bait-like region is a candidate target for host-parasite coevolution. One of their arguments relies on the fact that this particular region of the *Tep* gene evolves rapidly under positive selection in the crustacean *Daphnia* (Little et al., 2004). The Tep proteins show specificity for certain classes of pathogens in vitro: Mcr is necessary for *Candida albicans* phagocytosis, TepII is required for efficient *E. coli* phagocytosis, and TepIII is required for *S. aureus* phagocytosis (Stroschein-Stevenson et al., 2006).

Receptor Diversity: the Questions

Drosophila research on phagocytosis gave a wealth of highly informative data. The tractability of *Drosophila* S2 cells in testing phagosome function and the complex protein interactions during phagosome maturation in both *Drosophila* and mammalian systems was a key to success. Identification of genes involved in various aspects of *Drosophila* phagocytosis has already led to the characterization of their human counterparts and will undoubtedly provide new data in the future.

The second important conclusion from this brief overview of *Drosophila* phagocytic receptors concerns the high redundancy in the receptors. As in mammalian macrophages, several receptors recognize the same pathogen. Probably reflecting the lack of strong mutant phenotype, none of these receptors was identified by genetic inactivation in vivo, underlining the complexity of phagocytosis in *Drosophila*. As in mammalian cells, the most probable interpretation is that, in general, conserved ligands on the surface of microorganisms can undergo subtle alterations that redefine the binding affinities of the receptor for this particular pathogen. To deal with the evasive strategy adopted by infectious agents, phagocytes do not rely on a single receptor for pathogen recognition but use cooperative binding, as shown for Eater, PGRP-LC, and dSR-CI.

TABLE 1 Phagocytic receptors and opsonizing molecules[a]

Name	Family	Phagocytic receptor or opsonin	Mode of identification	Specificity tested in vitro	Specificity tested in vivo	Phagocytosis of apoptotic cells	Reference(s)
Croquemort	CD36-related (SR-B)	Receptor	Mutant; RNAi screen in S2 cells	S. aureus; not E. coli	No effect detected (Franc et al., 1999)	Yes (in vivo)	Franc et al., 1999; Stuart et al., 2005
Peste	CD36-related (SR-B)	Receptor	In vitro; RNAi screen in S2 cells	Intracellular bacteria: M. fortuitum, Mycobacterium smegmatis, Listeria monocytogenes; not E. coli or S. aureus	ND	ND	Philips et al., 2005
dSR-CI	SR-C	Receptor	In vitro inhibitory test and homology with mammalian receptor	S. aureus and E. coli; not C. silvatica (yeast)	ND	ND, but expression during metamorphosis	Ramet et al., 2001
Nimrod (NimC1)	Nim (EGF-like) repeat containing	Receptor	Plasmatocyte-specific antigen	Ectopic expression stimulates the uptake of both S. aureus and E. coli	S. aureus; not E. coli	ND	Kurucz et al., 2007
Eater	Nim (EGF-like) repeat containing	Receptor	RNAi screen in S2 cells	S. aureus, E. coli, Serratia marcescens, C. silvatica (yeast)	S. aureus, E. coli, and S. marcescens; susceptibility to S. marcescens in gut infection	ND	Kocks et al., 2005

				Not required for phagocytosis of zymosan		Yes (in vitro and in vivo)	
Draper	Nim (EGF-like) repeat containing	Receptor	Homology with *C. elegans* Ced1		ND	Yes (in vitro and in vivo)	Manaka et al., 2004
PGRP-LC	PGRP	Receptor	RNAi screen in S2 cells	*E. coli*; not *S. aureus*	No effect in the mutant	ND	Ramet et al., 2002; Garver et al., 2006
PGRP-SC1a	PGRP	Opsonin? catalytic activity	Mutant from a genetic screen	ND	*S. aureus*; not *E. coli*	ND	Garver et al., 2006
PGRP-SA	PGRP	Opsonin?	Genetic screen	ND	*S. aureus*; not *E. coli*	ND	Garver et al., 2006
Dscam	Immunoglobulin domains, multiple alternative splicing	Receptor or opsonin (transmembrane and secreted isoforms)	Expression profiling	*E. coli* for some isoforms	*E. coli*	ND	Watson et al., 2005
Macroglobulin complement related (Mcr)	α_2-Macroglobulin	Opsonin	RNAi screen in S2 cells	*C. albicans*; not *E. coli* nor *S. aureus*	ND	ND	Stroschein-Stevenson et al., 2006
TepII	α_2-Macroglobulin	Opsonin	Same family as Mcr	*E. coli*; not *C. albicans* or *S. aureus*	ND	ND	Stroschein-Stevenson et al., 2006
TepIII	α_2-Macroglobulin	Opsonin	Same family as Mcr	*S. aureus*; not *C. albicans* or *E. coli*	ND	ND	Stroschein-Stevenson et al., 2006

[a]ND, not tested.

It is still unknown whether some of the above-described molecules act as receptors or opsonins. Comparison of different receptor activities is also difficult because they have not been tested in the same conditions. Clearly, functional in vivo studies are required to reveal the precise role of each receptor during the process of infection.

ROLE OF PHAGOCYTOSIS IN INNATE IMMUNITY

In whole flies, resistance to infection relies on both cellular and humoral reactions. Phagocytosis contributes to the natural defenses of insects against all microorganisms, including bacteria, yeast, and parasites, but has mainly been studied in the case of bacteria. Invading pathogens must first be recognized by specific surface receptors before being taken up by hemocytes. Intracellular pathways are then activated, leading to microbial clearance and/or to activation of other defense mechanisms. Although the repertoire of known surface receptors is increasing, very few studies have focused on these other roles of phagocytosis, and the interplay between cell-based and humoral immune defense mechanisms is not well understood. In the next section, we will review what is known about in vivo models of phagocytosis in Drosophila and try to shed some light on the interactions that underlie immune response activation.

In Vivo Models Reveal the Importance of Phagocytosis in Fighting against Infections

The finding that many bacteria proliferate in the circulatory cavity of mutant flies lacking hemocytes demonstrated the role of phagocytosis. Larvae mutant for *domino* (Braun et al., 1998) or for both Rel transcription factors Dif and Dorsal (Matova and Anderson, 2006) were devoid of hemocytes and did not develop to adulthood. The lethality was explained by the inability to control natural infections, as larvae fed on a bacteria-free medium developed until the pupal stage. Because of the crucial function of hemocytes in clearing apoptotic corpses during metamorphosis, the development of pupae was compromised.

Few studies have investigated the role of phagocytosis during bacterial disease in *Drosophila*. The outcomes of such diseases are diverse and depend on route of infection and life stage of the fly. By using green fluorescent protein-tagged bacteria, several microorganisms were shown to localize and proliferate within isolated hemocytes or macrophage-like S2 cells. Blocking phagocytosis with polystyrene or latex beads enhanced fly lethality when infected with *E. coli* (Elrod-Erickson et al., 2000) or the fly pathogen *Serratia marcescens* (Nehme et al., 2007). In the latter model, flies were infected by feeding, and bacteria persisting in the gut induced a local immune response. However, some bacteria were able to escape through the gut wall into the hemolymph, where they were recognized by the hemocyte-bound Eater and eliminated by phagocytosis. *eater* null mutant flies were more sensitive to bacterial infection than wild-type flies and either *eater* mutant or flies injected with latex beads had a higher bacterial load in the hemolymph (Kocks et al., 2005; Nehme et al., 2007). These are the only studies that demonstrated a crucial role for phagocytosis in fighting infection and clearing bacteria.

Mechanisms of bacterial clearance in hemocytes are not well characterized. Recently, Psidin was identified as a lysosomal protein, required for degradation of ingested bacteria in *Drosophila* blood cells (Brennan et al., 2007). Despite strong homologies with *Saccharomyces cerevisiae* NatB N-acetyltransferase, the molecular function of Psidin is not yet understood. The importance of phagocytosis has also been stressed by recent experiments with the opportunistic pathogen *Pseudomonas aeruginosa*. *P. aeruginosa* uses the type III secretion system to inject a toxin that suppresses phagocytosis in hemocytes, thus affecting fly resistance to infection by *P. aeruginosa* but also *S. aureus* (Avet-Rochex et al., 2005). In other cases, hemocytes fail to eliminate pathogens. For instance, intracellular pathogens from the genera *Mycobacterium* and *Listeria* grow well within hemocytes during early stages of infection and kill the flies (Dionne et al., 2003; Mansfield et al., 2003). Here, bacteria have developed mechanisms to escape hemocyte killing, as when they infect mammalian cells.

Contribution of phagocytosis to the fight against microorganisms is well established, but the interaction between cellular and humoral immune responses is still controversial. Is phagocytosis of bacteria a first step, required for induction of the humoral response? Is bacterial digestion required in the phagosome for bacterial cell wall determinants to be sensed by pattern recognition receptors? Or do hemocytes send a cytokine-like signal to the fat body cells? In the next section we will discuss what is known about signaling between hemocytes and fat body cells.

Signaling between Hemocytes and the Fat Body

In *Drosophila*, several studies have investigated the interactions between hemocytes and fat body by using mutant flies. The precursor article by Braun et al. (1998) showed that *domino* mutant larvae, which are devoid of hemocytes, were able to activate a normal systemic antimicrobial response after injection of bacteria or fungi, and survived septic injury. Using *Dif/dorsal* double-mutant larvae, another study confirmed these results. The mutant larvae were able to activate expression of AMPs in response to injected *E. coli* (Matova and Anderson, 2006). Specific expression of *Dif* or *dorsal* in the blood cells was sufficient to restore all the defects of the double mutants following an infection with gram-negative bacteria. Basset et al. (2000) confirmed that direct injection of the phytopathogen *Erwinia carotovora* also led to expression of AMPs in the *domino* mutant larvae. These experiments suggested that the humoral response is not impaired in the absence of hemocytes. However, when flies were infected by natural routes, i.e., oral feeding with infection through the gut epithelium instead of by using a needle, the results were different. Orally infected larvae with *E. carotova* were able to produce AMPs in the fat body and survived to infection (Basset, 2000). However, in *domino* mutant larvae, the humoral response is no more activated in this food-borne infection model. This suggested that the induction of AMPs in the fat body cells required a dual-activation system, with direct recognition of bacteria by pattern recognition receptors and the subsequent stimulation of the Toll and IMD pathways as a first signal and a second signal emanating either from hemocytes in case of gut infection or from the injury (Fig. 1).

Two studies with opposite conclusions were in contradiction with this model. The first study showed that, apart from its role in the phagocytic degradation of microorganisms, Psidin was also required in the phagocytes to activate the fat body after injection of either gram-negative or -positive bacteria (Brennan et al., 2007). Furthermore, in *domino* mutant larvae, the expression of Defensin, an AMP that had not been observed in previous studies, was not

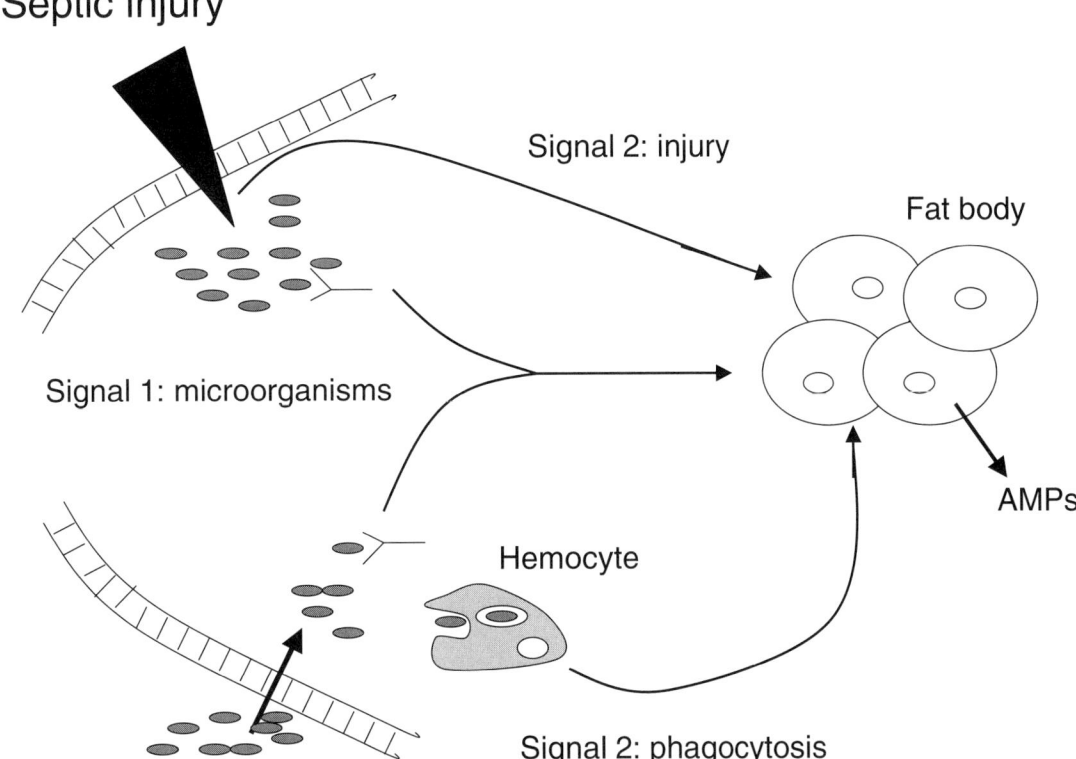

FIGURE 1 Model for the interactions between hemocytes and fat body. This model fits with most available data, but is still controversial on some aspects (see text for details). Two complementary signals would be required to activate AMP synthesis in the fat body. The first signal consists of the activation of the Toll and IMD pathways through direct recognition of microorganisms. After a septic injury, the second signal would be produced by the injury itself. The existence and the nature of this signal have not been demonstrated, but observations suggest that a clean injury results in a weak response by the fat body. After natural infections, the second signal would be generated by the hemocytes, activated by phagocytosis of microorganisms (or latex beads in experimental conditions). The nature of this signal is also unknown, but Upd3 or Spaetzle could be candidates.

induced after a septic injury with *E. coli*. This indicated that bacterial digestion products generated in the phagosome, for example, free peptidoglycan, would be required for the synthesis of AMPs by the fat body. However, because only Defensin induction is impaired in *domino* or *psidin* mutant larvae, the involvement of phagocytosis in the activation of the fat body response needs further clarification.

Using a model of food-borne infection, the second study demonstrated that the systemic response is not activated in adult flies infected with *S. marcescens*, the flies dying 6 days after the start of bacterial feeding (Nehme et al., 2007). Blocking phagocytosis with latex beads restored production of AMPs, proving that the humoral pathway was still functional. The explanation for the lack of activation of the humoral response in this infection model could be that there is not enough free peptidoglycan (produced during bacterial cell division) in the hemolymph to trigger the humoral response, unless phagocytosis is blocked and the number of bacteria is increasing. This indicates clearly that bacterial phagocytosis is not required to activate the humoral pathway. However, the hemocytes, which are still present (unlike in *domino* or *Dif/dorsal* mutant larvae) and actively phagocytose latex beads, would still be able to send a signal to the fat body. The discrepancy in the results obtained with natural infections—*Erwinia* but not *Serratia* activating the AMP-based humoral response—may be explained by differences in the life stage of the host. *Erwinia* infections were performed in *Drosophila* larvae and *Serratia* infections in adult flies. Insects have very different body structure, physiology, and environmental conditions during their life. Larvae crawl in rotted fruits and are totally surrounded by microorganisms, whereas adults only feed on the surface of fruits. Thus, larvae and adults probably have different immune responses, adapted to their particular lifestyle and the pathogens they might encounter. We expect that additional studies will resolve this discrepancy.

If a functional link exists between hemocytes and fat body, Upd3 and Spaetzle are obvious candidates for the signaling molecules. Upd3 is produced by adult plasmatocytes in response to bacterial challenge and activates the JAK/STAT pathway in the fat body. The exact role of the

JAK/STAT pathway in bacterial infections is not known yet (Agaisse et al., 2003). According to a genome-wide analysis of larval hemocytes, the gene encoding the cytokine Spaetzle is exclusively expressed in hemocytes and strongly upregulated after a bacterial challenge (Irving et al., 2005). Knowing that the major role of cleaved Spaetzle is to activate the Toll pathway at the surface of the fat body cells gives evident credit to the existence of a crosstalk between cellular and humoral immune response.

CONCLUSION

In this review, we intended to illustrate that in *Drosophila* phagocytosis plays a major role in the removal of microorganisms following any type of infections. We hope that we made clear that *Drosophila* can be used as a model organism to detail the phagocytic process as it is used to study developmental genetics and cell biology. The work that has been done up to now and that we have summarized here has paved the way for future in-depth analyses of phagocytosis during the immune response in *Drosophila* and hopefully in mammals. Many questions remain to be addressed. A large number of receptors or opsonins have now been discovered, but their exact function in bacterial recognition needs further investigation. The analysis of RNAi genome-wide screen data, with the help of *Drosophila* genetics, will shed light on the mechanisms of bacterial clearance in the phagosome. In particular, understanding the resistance of intracellular bacteria to phagocytosis and identification of the receptors for their uptake will be of special interest.

After a decade of intensive work in dissection of the signaling pathways leading to secretion of AMPs into the hemolymph, the analysis of host-pathogen interactions, including phagocytosis, is now attracting renewed interest. The major challenge is to decipher the relationships between cellular and humoral responses and the relative contribution of phagocytosis and AMP bacterial killing in various infection models.

The work in our laboratory is supported by the Centre National de la Recherche Scientifique, the Agence Nationale pour la Recherche, and National Institutes of Health Grant AI44220.

REFERENCES

Agaisse, H., L. S. Burrack, J. A. Philips, E. J. Rubin, N. Perrimon, and D. E. Higgins. 2005. Genome-wide RNAi screen for host factors required for intracellular bacterial infection. *Science* **309**:1248–1251.

Agaisse, H., U. M. Petersen, M. Boutros, B. Mathey-Prevot, and N. Perrimon. 2003. Signaling role of hemocytes in *Drosophila* JAK/STAT-dependent response to septic injury. *Dev. Cell* **5**:441–450.

Armstrong, P. B. 2006. Proteases and protease inhibitors: a balance of activities in host-pathogen interaction. *Immunobiology* **211**:263–281.

Avet-Rochex, A., E. Bergeret, I. Attree, M. Meister, and M. O. Fauvarque. 2005. Suppression of *Drosophila* cellular immunity by directed expression of the ExoS toxin GAP domain of *Pseudomonas aeruginosa*. *Cell. Microbiol.* **7**:799–810.

Basset, A., R. S. Khush, A. Braun, L. Gardan, F. Boccard, J. A. Hoffmann, and B. Lemaitre. 2000. The phytopathogenic bacteria *Erwinia carotovora* infects *Drosophila* and activates an immune response. *Proc. Natl. Acad. Sci. USA* **97**:3376–3381.

Braun, A., J. A. Hoffmann, and M. Meister. 1998. Analysis of the *Drosophila* host defense in domino mutant larvae, which are devoid of hemocytes. *Proc. Natl. Acad. Sci. USA* **95**:14337–14342.

Brennan, C. A., J. R. Delaney, D. S. Schneider, and K. V. Anderson. 2007. Psidin is required in *Drosophila* blood cells for both phagocytic degradation and immune activation of the fat body. *Curr. Biol.* **17**:67–72.

Cherry, S., and N. Silverman. 2006. Host-pathogen interactions in drosophila: new tricks from an old friend. *Nat. Immunol.* **7**:911–917.

Crozatier, M., and M. Meister. 2007. *Drosophila* haematopoiesis. *Cell. Microbiol.* **9**:1117–1126.

Desjardins, M., and G. Griffiths. 2003. Phagocytosis: latex leads the way. *Curr. Opin. Cell. Biol.* **15**:498-503.

Dionne, M. S., N. Ghori, and D. S. Schneider. 2003. *Drosophila melanogaster* is a genetically tractable model host for *Mycobacterium marinum*. *Infect. Immun.* **71**:3540–3550.

Dodds, A. W., and S. K. Law. 1998. The phylogeny and evolution of the thioester bond-containing proteins C3, C4 and alpha 2-macroglobulin. *Immunol. Rev.* **166**:15–26.

Elrod-Erickson, M., S. Mishra, and D. Schneider. 2000. Interactions between the cellular and humoral immune responses in *Drosophila*. *Curr. Biol.* **10**:781–784.

Franc, N. C., P. Heitzler, R. A. Ezekowitz, and K. White. 1999. Requirement for *croquemort* in phagocytosis of apoptotic cells in *Drosophila*. *Science* **284**:1991–1994.

Garin, J., R. Diez, S. Kieffer, J. F. Dermine, S. Duclos, E. Gagnon, R. Sadoul, C. Rondeau, and M. Desjardins. 2001. The phagosome proteome: insight into phagosome functions. *J. Cell Biol.* **152**:165–180.

Garver, L. S., J. Wu, and L. P. Wu. 2006. The peptidoglycan recognition protein PGRP-SC1a is essential for Toll signaling and phagocytosis of *Staphylococcus aureus* in *Drosophila*. *Proc. Natl. Acad. Sci. USA* **103**:660–665.

Irving, P., J. M. Ubeda, D. Doucet, L. Troxler, M. Lagueux, D. Zachary, J. A. Hoffmann, C. Hetru, and M. Meister. 2005. New insights into *Drosophila* larval haemocyte functions through genome-wide analysis. *Cell. Microbiol.* **7**:335–350.

Jiggins, F. M., and K. W. Kim. 2006. Contrasting evolutionary patterns in *Drosophila* immune receptors. *J. Mol. Evol.* **63**:769–780.

Kocks, C., J. H. Cho, N. Nehme, J. Ulvila, A. M. Pearson, M. Meister, C. Strom, S. L. Conto, C. Hetru, L. M. Stuart, T. Stehle, J. A. Hoffmann, J. M. Reichhart, D. Ferrandon, M. Ramet, and R. A. Ezekowitz. 2005. Eater, a transmembrane protein mediating phagocytosis of bacterial pathogens in *Drosophila*. *Cell* **123**:335–346.

Kurucz, E., R. Markus, J. Zsamboki, K. Folkl-Medzihradszky, Z. Darula, P. Vilmos, A. Udvardy, I. Krausz, T. Lukacsovich, E. Gateff, C. J. Zettervall, D. Hultmark, and I. Ando. 2007. Nimrod, a putative phagocytosis receptor with EGF repeats in *Drosophila* plasmatocytes. *Curr. Biol.* **17**:649–654.

Lagueux, M., E. Perrodou, E. A. Levashina, M. Capovilla, and J. A. Hoffmann. 2000. Constitutive expression of a complement-like protein in toll and JAK gain-of-function mutants of *Drosophila*. *Proc. Natl. Acad. Sci. USA* **97**:11427–11432.

Law, S. K., and K. B. M. Reid. 1995. *Complement*, 2nd ed. IRL Press, Oxford, United Kingdom.

Leclerc, V., N. Pelte, L. El Chamy, C. Martinelli, P. Ligoxygakis, J. A. Hoffmann, and J. M. Reichhart. 2006. Prophenoloxidase activation is not required for survival to microbial infections in *Drosophila*. *EMBO Rep.* **7**:231–235.

Lemaitre, B., and J. Hoffmann. 2007. The host defense of *Drosophila melanogaster*. *Annu. Rev. Immunol.* **25**:697–743.

Levashina, E. A., L. F. Moita, S. Blandin, G. Vriend, M. Lagueux, and F. C. Kafatos. 2001. Conserved role of a complement-like protein in phagocytosis revealed by dsRNA knockout in cultured cells of the mosquito, *Anopheles gambiae*. *Cell* **104:**709–718.

Little, T. J., J. K. Colbourne, and T. J. Crease. 2004. Molecular evolution of daphnia immunity genes: polymorphism in a gram-negative binding protein gene and an alpha-2-macroglobulin gene. *J. Mol. Evol.* **59:**498–506.

Manaka, J., T. Kuraishi, A. Shiratsuchi, Y. Nakai, H. Higashida, P. Henson, and Y. Nakanishi. 2004. Draper-mediated and phosphatidylserine-independent phagocytosis of apoptotic cells by *Drosophila* hemocytes/macrophages. *J. Biol. Chem.* **279:**48466–48476.

Mansfield, B. E., M. S. Dionne, D. S. Schneider, and N. E. Freitag. 2003. Exploration of host-pathogen interactions using *Listeria monocytogenes* and *Drosophila melanogaster*. *Cell. Microbiol.* **5:**901–911.

Matova, N., and K. V. Anderson. 2006. Rel/NF-kappaB double mutants reveal that cellular immunity is central to *Drosophila* host defense. *Proc. Natl. Acad. Sci. USA* **103:**16424–16429.

Meister, M. 2004. Blood cells of *Drosophila*: cell lineages and role in host defence. *Curr. Opin. Immunol.* **16:**10–15.

Mukhopadhyay, S., and S. Gordon. 2004. The role of scavenger receptors in pathogen recognition and innate immunity. *Immunobiology* **209:**39–49.

Nehme, N., S. Liégeois, B. Kele, P. Giammarinaro, E. Pradel, J. A. Hoffmann, J. J. Ewbank, and D. Ferrandon. 2007. A model of bacterial intestinal infections in *Drosophila melanogaster*. *PLOS Pathog.* **3:**e173.

Pearson, A. M., K. Baksa, M. Ramet, M. Protas, M. McKee, D. Brown, and R. A. Ezekowitz. 2003. Identification of cytoskeletal regulatory proteins required for efficient phagocytosis in *Drosophila*. *Microbes Infect.* **5:**815–824.

Philips, J. A., E. J. Rubin, and N. Perrimon. 2005. *Drosophila* RNAi screen reveals CD36 family member required for mycobacterial infection. *Science* **309:**1251–1253.

Ramet, M., P. Manfruelli, A. Pearson, B. Mathey-Prevot, and R. A. Ezekowitz. 2002. Functional genomic analysis of phagocytosis and identification of a *Drosophila* receptor for *E. coli*. *Nature* **416:**644–648.

Ramet, M., A. Pearson, P. Manfruelli, X. Li, H. Koziel, V. Gobel, E. Chung, M. Krieger, and R. A. Ezekowitz. 2001. *Drosophila* scavenger receptor CI is a pattern recognition receptor for bacteria. *Immunity* **15:**1027–1038.

Royet, J., and R. Dziarski. 2007. Peptidoglycan recognition proteins: pleiotropic sensors and effectors of antimicrobial defences. *Nat. Rev. Microbiol.* **5:**264–277.

Schmucker, D., J. C. Clemens, H. Shu, C. A. Worby, J. Xiao, M. Muda, J. E. Dixon, and S. L. Zipursky. 2000. *Drosophila* Dscam is an axon guidance receptor exhibiting extraordinary molecular diversity. *Cell* **101:**671–684.

Stroschein-Stevenson, S. L., E. Foley, P. H. O'Farrell, and A. D. Johnson. 2006. Identification of *Drosophila* gene products required for phagocytosis of *Candida albicans*. *PLoS Biol.* **4:**e4.

Stuart, L. M., J. Boulais, G. M. Charriere, E. J. Hennessy, S. Brunet, I. Jutras, G. Goyette, C. Rondeau, S. Letarte, H. Huang, P. Ye, F. Morales, C. Kocks, J. S. Bader, M. Desjardins, and R. A. Ezekowitz. 2007. A systems biology analysis of the *Drosophila* phagosome. *Nature* **445:**95–101.

Stuart, L. M., J. Deng, J. M. Silver, K. Takahashi, A. A. Tseng, E. J. Hennessy, R. A. Ezekowitz, and K. J. Moore. 2005. Response to *Staphylococcus aureus* requires CD36-mediated phagocytosis triggered by the COOH-terminal cytoplasmic domain. *J. Cell Biol.* **170:**477–485.

Wang, L., and P. Ligoxygakis. 2006. Pathogen recognition and signalling in the *Drosophila* innate immune response. *Immunobiology* **211:**251–261.

Watson, F. L., R. Puttmann-Holgado, F. Thomas, D. L. Lamar, M. Hughes, M. Kondo, V. I. Rebel, and D. Schmucker. 2005. Extensive diversity of Ig-superfamily proteins in the immune system of insects. *Science* **309:**1874–1878.

Williams, M. J. 2007. *Drosophila* hemopoiesis and cellular immunity. *J. Immunol.* **178:**4711–4716.

Wood, W., and A. Jacinto. 2007. *Drosophila melanogaster* embryonic haemocytes: masters of multitasking. *Nat. Rev. Mol. Cell Biol.* **8:**542–551.

34

The Zebrafish as a Model of Host-Pathogen Interactions

J. MUSE DAVIS AND LALITA RAMAKRISHNAN

STRENGTHS AND WEAKNESSES OF THE ZEBRAFISH AS A MODEL OF HOST-PATHOGEN INTERACTIONS

With any new model system, sheer novelty is initially enough to excite great enthusiasm (or disdain), often out of proportion to what the system has to offer. The zebrafish as a model for host-pathogen interactions has now matured to the point that we can reflect on what it truly has to offer, where it is helpful, and how it complements other models. As a model for the study of early development, the zebrafish came to prominence because of the easy access to early events (hundreds of developing embryos developing outside the mother, under simple laboratory conditions) and the speed with which they take place (from fertilization to a semimotile organism in 36 h, to a feeding larva in 5 days) (Westerfield, 2000). Seizing on these advantages, zebrafish researchers have made huge strides to propel the zebrafish model forward through the development of imaging and fixation techniques and forward and reverse genetics. Only after these vast amounts of work had been done by developmental biologists did the study of infection in the zebrafish become so attractive. As they stand now, the chief advantages afforded by this model range from the practical, to the genetic, to the observational.

Practical advantages mainly relate to the relatively low cost and ease of husbandry and breeding. Thousands of fish can be kept in a room that would hold only 100 or 200 mice. Importantly, many different strains can be housed efficiently in these small spaces by using small tanks that can fit alongside large tanks in the same racks. A female fish can lay several hundred eggs at once, most of which develop to maturity. Several breeding pairs can be set up in a small space so that progeny of different fish mutant or reporter lines can be obtained or crossed at one time. Ethical and regulatory issues can be less constraining than for mouse research, especially when fish embryos and larvae are involved. The zebrafish embryo is not considered to be a vertebrate until after hatching or even beyond; the exact window of time for this dispensation varies among institutions.

The genetic advantages of zebrafish are often considered the strongest. Forward genetic screens are relatively easy, again because of the large number of progeny derived and the speed with which they can be assessed. Chemical mutagenesis with N-ethyl-N-nitrosourea is the most popular method, but some laboratories have developed techniques for retroviral insertion mutagenesis (Patton and Zon, 2001). Additional screening techniques, such as the creation of gynecogenetic diploids (the production of offspring from only maternal DNA), allow for the screening of recessive mutations in the second generation, cutting down on one generation from the classical Mendelian screens. The more recent availability of transgenic technologies and resultant lines with fluorescently marked cells of different lineages has allowed for the full use of visualization techniques for more rapid and increasingly sophisticated screens, such as for cell migration phenotypes. The zebrafish genome is in the final stages of annotation (www.sanger.ac.uk/Projects/D_rerio), so that mutant mapping is progressively easier. Excelling so in forward genetics, the zebrafish had thus far complemented the mouse model, which offers reverse genetics by homologous recombination into and implantation of embryonic stem cells. However, new tricks are being employed to correct this deficiency in the zebrafish: mutants in desired genes are produced by a variety of methods (Wienholds et al., 2003; Skromne and Prince, 2008). In addition, transient gene knockdowns are readily created by injecting modified antisense oligonucleotides (morpholinos) into very early stage embryos (Nasevicius and Ecker, 2000). Morpholino technology makes for a rapid and relatively inexpensive tool for the study of gene function for up to 7 to 10 days of life.

With chemical biology and genetics coming to the forefront as ways to probe biological processes including host-pathogen interactions (Hung et al., 2005), we note that the zebrafish is eminently suited for such small-molecule screens. Their aquatic habitat allows for the small molecule

J. Muse Davis, Immunology and Molecular Pathogenesis Program, Emory University, 954 Gatewood Drive, Atlanta, GA 30322. Lalita Ramakrishnan, Department of Microbiology, University of Washington, 1959 NE Pacific Street, Seattle, WA 98195.

to be administered simply dissolved in the water. This advantage is highlighted when using larvae that can be placed in 96-well plates for a high-throughput screen, and a variety of reagents with known molecular actions are in use (Peterson et al., 2004). Such an approach has been elegantly used to find and characterize a small molecule that reverses a genetic defect leading to coarctation of aorta, even when administered for only 6 h during early development (Peterson et al., 2004). The zebrafish model as a means to find relevant drugs for both developmental abnormalities and pathogenic processes has extraordinary potential.

The zebrafish embryo is naturally pigmentless for the first 24 h of life, and this see-through state can be extended for 7 to 10 days by treatment with phenylthiourea (Westerfield, 2000), which has few other discernible consequences. It is this transparency that has allowed the direct viewing of both normal and abnormal development (Kimmel et al., 1995) and has allowed the detailed studies of hematopoietic and immune cell development, as will be detailed in the following sections (Traver et al., 2003). This transparency alone allowed the excellent early studies of the zebrafish embryonic macrophage (Herbomel et al., 1999, 2001), but when combined with fluorescent reagents or with transgenic fish strains, it allows an unprecedented view into the pathogenesis of a variety of diseases (Langenau et al., 2003; Mathias et al., 2006; Meijer et al., 2008). Because both the overall organism and also small areas can be imaged repeatedly with minimal harm to the embryo, this model goes far beyond the endpoint bacterial counts and postmortem morphology available in more conventional animal models. In addition, some vital dyes are capable of penetration by simple soaking (Cooper et al., 2005; Santos et al., 2006) or intravenous injection, allowing in vivo high-resolution visualization of structures previously seen only in cultured cells. The observational advantages come mostly with the use of the transparent early stages, although transparent adults are being developed, as they have been for the medaka (White et al., 2008; Broussard and Ennis, 2007; Wakamatsu et al., 2001).

The zebrafish model presents notable difficulties as well. One issue plaguing the use of the zebrafish to study immunology is the relative paucity of immunological reagents. While the available genetic tools, as mentioned above, are impressive, fish-specific antibodies have been slow in coming. Some existing mouse antibodies do identify homologous fish proteins, but the zebrafish is a long way from enjoying the number of reagents available for mouse work. Publication of successful monoclonal production (Crosnier et al., 2005) and the efforts of private companies (e.g., Phylonics) show that this deficiency may not be permanent. Owing to the small size of zebrafish and the lack of inbred lines, cell transfer and reconstitution experiments that have been the backbone of cellular immunology in the mouse model are not yet facile, although some techniques have been developed (Langenau et al., 2004). Finally, it is thought that the remnants of at least one whole-genome duplication exist in many fishes including zebrafish. The large number of hox genes present in zebrafish is a famous example of this (Amores et al., 1998; Jozefowicz et al., 2003), as is the duplication of granulin and interferon-γ genes discussed below. When taking advantage of the genetic accessibility of this species, therefore, it is important to beware of duplicates.

OVERVIEW OF THE IMMUNE SYSTEM OF THE ZEBRAFISH

Zebrafish immunology as a field is only about 15 years old, but researchers can also draw on functional immunology studies in other teleost fish such as trout, catfish, and sea bass, which appear to have very similar immune systems and have been under study for decades (see the Appendix for a discussion of fish taxonomy). The zebrafish immune system has been reviewed in detail (Traver et al., 2003; Trede et al., 2004). The major features of mammalian innate and adaptive immunity are represented in fish, including the complement system, chemokines, cytokines, and adhesion molecules. The different types of phagocytes and B and T lymphocytes are found, as are major histocompatibility complex class I and II (MHC-I and MHC-II) molecules. Some components, such as T-lymphocyte subset markers, have as yet only been identified in silico. Innate immunity develops earlier in ontogeny than adaptive immunity, similar to the case with mammals. Because the early innate-only stages of the zebrafish are accessible to experimentation, researchers can use this aspect of the developmental program to isolate the contribution of innate and adaptive immunity to pathogenesis (Davis et al., 2002). There may well be differences in the relative contribution of some immune determinants to immune function in mammals and fish. For instance, the complement system is more polymorphic in fish than in mammals, and components such as C3 and B are present as multiple copies in the zebrafish, leading to the speculation that complement may allow the recognition of a broader range of pathogens by innate immunity in the fish and compensate for the relatively minor (compared with mammals) antibody response that fish generate. In this context it is interesting to note that there is no evidence of isotype switching per se in fish, although the zebrafish AID protein is capable of mediating this process in human cells (Wakae et al., 2006). Fish B cells produce immunoglobulin M (IgM), IgD, and the fish-specific IgZ classes of heavy-chain immunoglobulin components, but the details of expression are still being worked out (Danilova et al., 2005). There is also evidence that the B lymphocytes have phagocytic potential (Li et al., 2006), a feature that may be unique to the fish or may simply have not yet been identified in mammals. Where antigen presentation occurs has posed another intriguing question, as no lymph nodes are obvious and the spleen consists of mixed pulp without the discrete zones of mammalian spleens. Lymphatics have been identified, reducing the intrigue surrounding this matter (Kuchler et al., 2006; Yaniv et al., 2006).

In summary, most immune mechanisms are common to fish and mammals, and the study of phagocytes and their pathogens appears to be on particularly solid ground in its relevance to mammalian systems, as will be discussed in the sections below.

PHAGOCYTES OF THE ZEBRAFISH

Macrophages

Macrophages are the first phagocytes, and indeed the first immune cells, to appear in zebrafish development (although the precise timing of neutrophil development is not certain—see below). The origins and capabilities of these earliest phagocytic cells have been reviewed extensively (Traver et al., 2003), and here we will attempt only an overview. Both developmental and inflammatory responses of early macrophages can be studied in exquisite detail in the zebrafish. The first macrophage precursors are identified in the anterior-most lateral mesoderm at approximately the seven-somite stage (about 12 h postfertilization [hpf]), from where they migrate laterally toward the anterior surface of the yolk sac (Fig. 1A) and express CSF-1 receptor and L-plastin (Herbomel et al., 1999, 2001). Once they arrive at the yolk sac, they change shape to become small, active "premacrophages," with scant cytoplasm, which further differentiate into macrophages by approximately 22 hpf. This location on the surface of the yolk sac is below the outer periderm and the dermis of the embryo, but within the blood flow, and will become the yolk circulation valley, also known as the duct of Cuvier (Fig. 1C). As development continues, this space becomes enclosed by vascular epithelium and remains part of the circulatory pathway. It is this detail of yolk sac-associated circulation (specific only to a few teleosts) that makes the viewing of early macrophages at high resolution so convenient in zebrafish (Herbomel et al., 1999) (Fig. 2A through C).

Although the heart has been beating since 22 hpf, it is not until after the proerythroblasts have made their way from the posterior blood island to the yolk sac at 24 to 26 h that systemic circulation begins, allowing the macrophages to circulate into a variety of tissues. Some of the macrophages find other tissues without traveling in the blood, however. Starting by 26 hpf, some embryonic macrophages migrate into the head mesenchyme, where no vasculature is yet formed (Fig. 1B). By 36 hpf the vasculature of the head is still scarce (Isogai et al., 2001), but this area is populated with roughly 100 macrophages (Herbomel et al., 1999). By this time in development the embryonic macrophages are capable of eradicating nonpathogenic bacteria injected into the bloodstream, and the introduction of bacteria into the hindbrain ventricle (Fig. 1C) accelerates their appearance in this space (Herbomel et al., 1999). Once the embryonic macrophages move into various tissues, they begin to modify their gene expression and phagocytic capacities (Traver et al., 2003). The best-studied example of this is the upregulation of apolipoprotein E by macrophages colonizing the brain and other neural tissues. It is possible that these cells become or give rise to microglia in the adult (Herbomel et al., 2001).

How these embryonic macrophages (or those of any species) compare functionally with the later adult macrophages, and how long they remain active in the organism, are still unknown.

Monocytes and Macrophages

Hematopoiesis

Although the kidney is the site of hematopoiesis in the adult fish (Fig. 1E), the predecessors of the stem cells responsible for definitive hematopoiesis first appear in a region of the trunk called the aorta-gonad-mesonephros (AGM) at about 24 hpf, whence they move to another temporary hematopoietic site in the ventral tail called the caudal hematopoietic tissue (CHT) (Fig. 1C). The origin and migrations of these cells have been shown by an elegant series of experiments from the Herbomel group (Murayama et al., 2006). By 4 days postfertilization, hematopoietic stem cells begin to populate the kidney and thymus (Fig. 1D), but new cells are still produced in the CHT until about day 14. Thus, by 5 to 7 days postfertilization, there may be macrophages of three different origins sharing the same tissues: true embryonic macrophages that are derived from anterior mesoderm, macrophages originating from the CHT, and full adult macrophages from the kidney. The monocyte stage of development of the latter two has not yet been described in zebrafish, although monocytes have been identified in the adult (Trede et al., 2004). More study of the monocyte/macrophage distinction has been done in the goldfish *Carassius auratus*, where Belosevic and colleagues have shown that the adult goldfish kidney contains three subsets of macrophage-like cells (Barreda et al., 2000; Neumann et al., 2000). The first, called R1, is a hematopoietic precursor cell, which is capable of giving rise to the other two types. R2 is phenotypically most like an adult macrophage, and the development of R1 cells to R2 cells suggests a pathway for macrophage maturation with no monocyte intermediate. The R3 subset is monocyte-like and can give rise to R2 cells, approximating the usual mammalian system of monocyte-derived macrophages (Neumann et al., 2000). Surprisingly, the most differentiated R2 cells are also capable of replication (Barreda et al., 2000). The appearance of replicating macrophages has also been noted in some cases in mammals (Alliot et al., 1991; Sorokin and Hoyt, 1987). The unexpected appearance of dual differentiation pathways in the primary hematopoietic tissue of goldfish may occur in the zebrafish as well.

Another unexpected finding in goldfish myelopoiesis is the role of granulin as a growth factor (Hanington et al., 2006), a role also suggested by studies in carp (Belcourt et al., 1993). The granulin in question was found to be expressed only in the kidney and spleen, in contrast to the more ubiquitous expression seen in mammals. The confusion has been somewhat relieved by the discovery of four granulins in zebrafish, with only one (granulin a) expressed in a way that suggests a hematopoietic role (Cadieux et al., 2005). Of the other three, one (granulin b), an ortholog of the mammalian gene, and two paralogs (granulins 1 and 2) showed more ubiquitous expression. Granulin a, which seems to be a hematopoietic growth factor, does not appear necessary for embryonic development, while the others do. Thus, it is thought that partial genome duplication (perhaps more than once) is responsible for multiple copies of what is in the mammalian line a single gene.

Molecular Markers

CSF1R remains the major molecular marker for macrophages in zebrafish and other teleosts. A line of zebrafish mutant in the *fms* gene that encodes CSF1R was originally discovered as a stripeless variety in a pet store (Parichy and Turner, 2003). Other strains of fish with loss-of-function mutations in this gene have been called *panther* (Parichy and Turner, 2003). The functional status of the gene in this and related lines remains uncertain, as the main distinguishing points of the *panther* mutant are altered pigment patterns and an altered macrophage migration pattern in early development (Herbomel et al., 2001). More recently, a deficiency in resistance to *Mycobacterium marinum* has also been found (L. Swaim and L. Ramakrishnan, unpublished data). Whether there is a duplicate gene that is preventing the severe phenotype seen in the *csf1r* knock-out mouse (Dai et al., 2002), or whether the macrophage colony-stimulating factor (M-CSF) pathway plays a role more restricted to hematopoiesis in the fish, is not clear. A

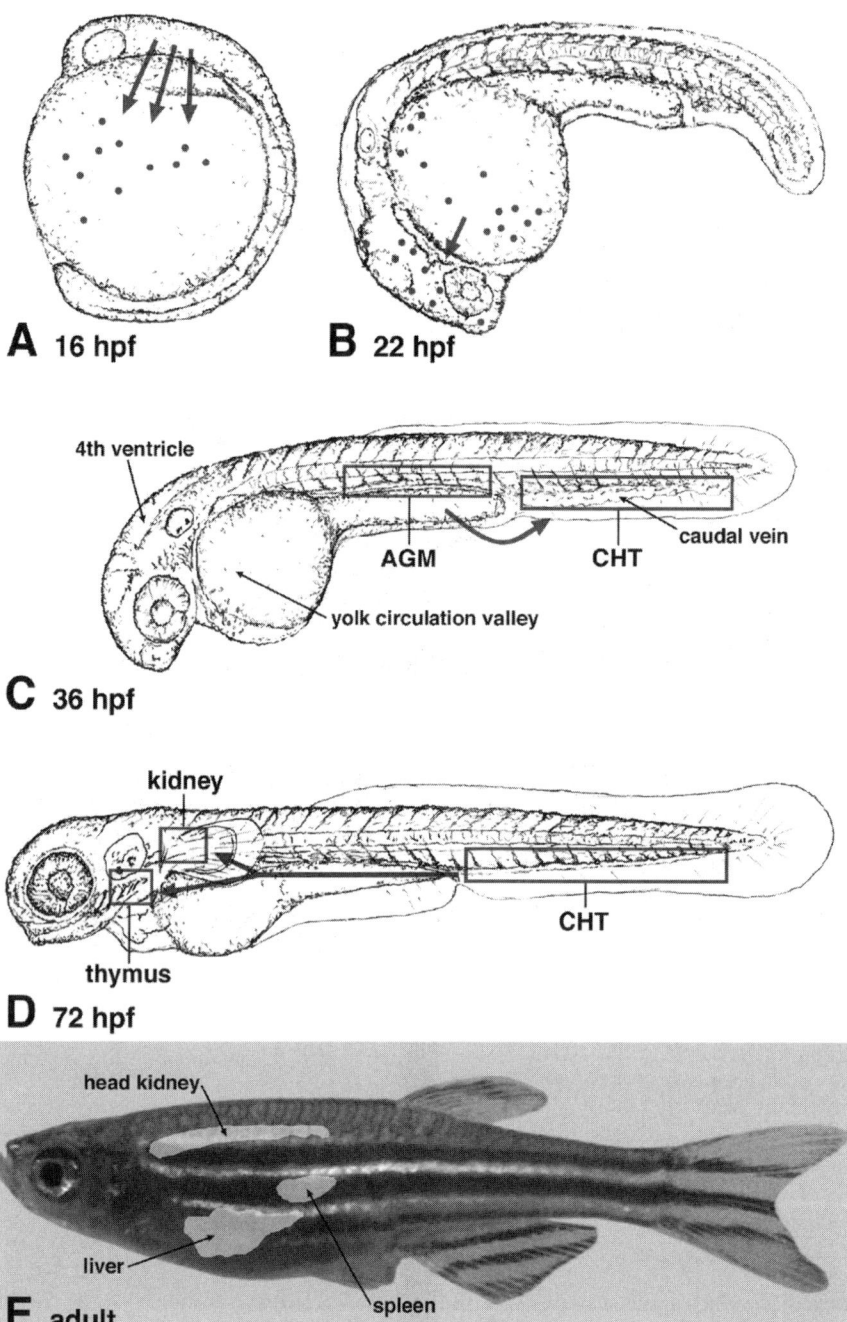

FIGURE 1 General anatomy and location of myelopoiesis during progressive stages of development. (A–D) Line drawings based on sketches from Kimmel et al. (1995). Path of hematopoietic cells as reported in Herbomel et al. (1999) and Murayama et al. (2006). (A) Path of embryonic macrophages from lateral mesoderm to anterior yolk. (B) Before the onset of circulation, embryonic macrophages have spread over the yolk and begun to infiltrate the brain. (C) Definitive hematopoiesis begins in the aorta-gonad-mesonephros (AGM), but hematopoietic precursors soon migrate to the caudal hematopoietic tissue (CHT). (D) By 4 days postfertilization, hematopoiesis is taking place in the CHT, but hematopoietic cells are also transferring to the thymus and kidney. (E) Location of organs important to hematopoiesis and infection in the adult.

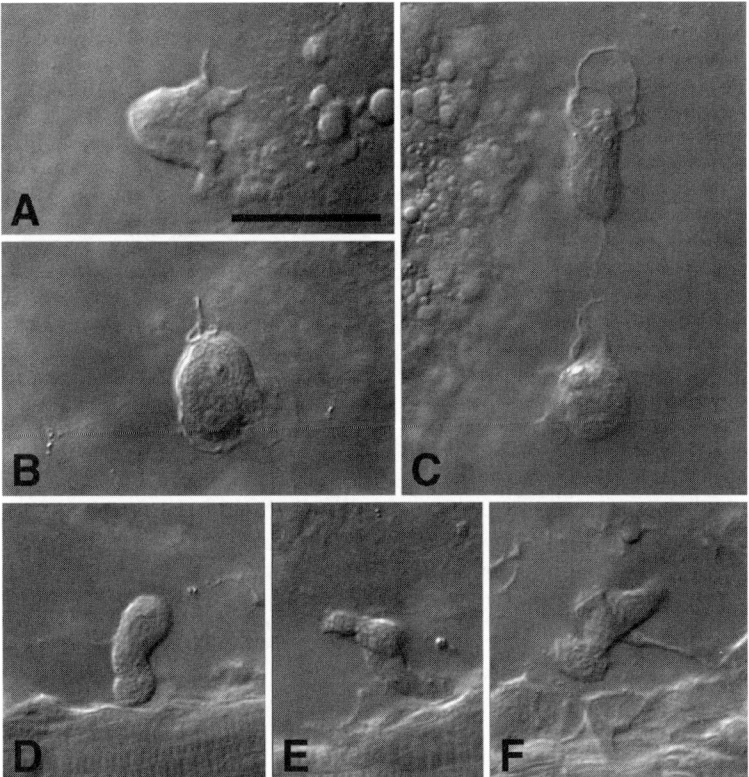

FIGURE 2 Examples of macrophages and neutrophils visible with DIC microscopy during embryonic and larval development. (A–C) Embryonic macrophages located near caudal vein (ventral is up), with muscle tissue nearby. (A) Early embryonic macrophage at yolk surface, ~30 hpf, just before the onset of circulation. (Scale bar, 20 μm; all panels, same scale.) (B) More mature embryonic macrophage in yolk circulation valley at ~48 hpf. (C) Two embryonic macrophages in yolk circulation valley, with many cellular processes and connected by a "tether." (D–F) Neutrophils located just superficial to caudal hematopoietic tissue. Note the more slender proportions and plentiful cytoplasmic granules.

distinct possibility is that the lack of bone marrow in teleost fish may simply make osteoclast deficiency, and the ensuing metabolic and bony structural defects that dominate in M-CSF pathway defects in mammals, less important in the fish.

A major impetus for finding a molecular marker of zebrafish macrophages has been the hope of producing a transgenic line with fluorescent macrophages, thereby taking advantage of the excellent optical characteristics of the embryo in studying infection. Some studies of embryonic macrophages have taken advantage of neutral red accumulation in macrophages (Davis et al., 2002; Herbomel et al., 2001), but this dye is toxic within hours of administration. Pilot studies with injected PKH26 (Sigma) have also shown bright but incomplete fluorescent marking of phagocytic cells (R. Lesley and L. R. Ramakrishnan, unpublished data). As for genetic approaches to this problem, whole-mount in situ hybridization shows CSF1R to be a reliable marker of macrophages at least in the embryo, but efforts by multiple laboratories to produce a transgenic line by using the promoter have been fruitless thus far. Existing transgenic zebrafish lines with some macrophage expression include *Tg(fli1::EGFP)^{y1}* (in which all vasculature expresses EGFP, as do macrophages for the first 24 to 36 h) and *Tg(zpu.1::EGFP)* (Hsu et al., 2004) (in which all myeloid cells express EGFP in the embryo). Due to transient expression, expression in other tissues, or a combination of both, neither of these has proven broadly useful. More recently, transgenic lines based on the lysozyme C promoter (*lysC::EGFP* and *lysC::dsRED2*) have been introduced (Hall et al., 2007). As published, this line has shown great promise, but with caveats: expression patterns show incomplete overlap with *csf1r* expression (still the gold standard), suggesting incomplete coverage of all macrophages, and overlap with *mpo* expression, suggesting that some neutrophils are also marked (Hall et al., 2007). Our own recent studies with these transgenic lines suggest that lysozyme C expression is restricted to granulocytes, and any overlap with macrophage-like gene expression likely occurs only in early primitive hematopoiesis (J. M. Davis and L. Ramakrishnan, unpublished data). For this reason, we do not consider lysozyme C to be a viable macrophage marker (see "Neutrophils," below, for further discussion). Another proposed marker molecule is L-plastin, which appears to be expressed by all macrophages and at least some neutrophils, but expression dies out in the first 2 to 4 days of life (Herbomel et al., 1999) (a more rapid downregulation in neutrophils may account for the apparent incomplete coverage of this group).

Chemotaxis

In the zebrafish embryo, bacteria such as *Escherichia coli* and *M. marinum* can be injected into the hindbrain ventricle at 21 to 24 hpf, and macrophages will be attracted to the site to phagocytose the foreign matter (Clay et al., 2007; Davis et al., 2002; Herbomel et al., 1999; Volkman et al., 2004). At this point in normal development, the hindbrain ventricle is devoid of macrophages, so the arriving cells have come from some distance. Injections of phosphate-buffered saline or untreated latex beads do not attract this response, suggesting that it is not due to the trauma of wounding (Clay et al., 2007). The exact nature of the signal is not known, although CSF1R is not absolutely required (Davis et al., 2002). Genes encoding a number of chemokines and chemokine receptors have been detected and cloned in the zebrafish, including those for SDF-1a and CXCR4b (Knaut et al., 2003; Miyasaka et al., 2007), SDF-1b and CXCR4a (Chong et al., 2007), and CXCR7 (Dambly-Chaudiere et al., 2007), and these have roles in germ cell migration and neural development. A large number of CC chemokines have been detected in the zebrafish and catfish genomes (Peatman et al., 2005, 2006; Peatman and Liu, 2006), but functional studies of any chemokines in the service of immunity have yet to be published. The effect of pertussis toxin, which blocks all G_i proteins including all chemokine receptors, on zebrafish development has been studied (Hammerschmidt and McMahon, 1998), and pilot studies in our laboratory are ongoing to learn its effect on infectious processes. The importance of microtubules in zebrafish macrophage motility was shown by visualization studies after laser-induced wounding (Redd et al., 2006). Work with zebrafish CXCR4 and SDF-1 has led to other experiments concerning subcellular mechanics of cell motility (Blaser et al., 2006). Overall, the zebrafish, presenting the unique view of cell movements that it does, is only beginning to be exploited for the study of the chemokine networks that direct these movements.

Phagocytosis

Functionally, thus far, much more is known about macrophages of the zebrafish embryo than those of the adult. As noted above, macrophages derived from "definitive" myelopoiesis are present in the stages used for embryonic observations of infection, but it is not certain how similar these are to later adult macrophages, or how they can be distinguished (if at all) from the still numerous true embryonic macrophages. Nevertheless, the cells present by 48 hpf, whatever their origin, are readily capable of phagocytosis of both living and heat-killed organisms, including gram-positive and gram-negative bacteria (Herbomel et al., 1999), spirochetes (J. M. Davis, D. Haake, and L. Ramakrishnan, in preparation), and yeast (Traver et al., 2003). Studies with adult goldfish macrophages also show uptake of *Leishmania major* (Stafford et al., 1999), so the general classes of pathogens are all documented as being taken in by fish macrophages. Given that there is no functional adaptive immune system in the zebrafish embryos of the former studies, the phagocytosis is not antibody mediated. A study of trout head kidney macrophages exposed to mycobacteria suggested that some form of opsonization is needed for efficient uptake of bacteria (Chen et al., 1998) and that trout serum and trout antibody both perform well. Opsonization by complement is quite possible in the 48-hpf zebrafish, as multiple complement components are induced by *M. marinum* infection at this time (T. Pozos, H. E. Volkman, J. F. Rawls, and L. Ramakrishnan, unpublished data).

Receptors

The presence of complement receptor and mannose receptor has been identified in other teleosts (Rodriguez et al., 2003; Schraml et al., 2006; Sorensen et al., 2001), although they have yet to be definitively found in zebrafish. Many Toll-like receptor (TLR)-related genes have been found in the zebrafish genome, including TLRs 1 to 5 and 7 to 9, MyD88, MAL, TRIF, TRAF6, and IRAK-4 (Jault et al., 2004; Meijer et al., 2004; Phelan et al., 2005a). In both embryonic and adult infections, TLR3 and TRAF6 were shown to be upregulated in response to snakehead rhabdovirus infection, and TLR3, TRAF6, and IRAK-4 were upregulated on infection with *Edwardsiella tarda* (Phelan et al., 2005a). The TLR genes are not only upregulated but also functioning, as shown by the reduced ability of zebrafish embryos to clear *Salmonella enterica* when MyD88 function was blocked (van der Sar et al., 2006). Work on intracellular receptors is also beginning, and NOD2 has been identified in the fugu and zebrafish genomes (Ogura et al., 2003).

Effector Pathways

Microbial killing by reactive oxygen or nitrogen species is well documented in fish macrophages (reviewed by Neumann et al., 2001) and uses a familiar set of enzymes. One interesting feature of fish macrophages shown in trout and goldfish is their relative autonomy in initiating antimicrobial free radicals. In vitro studies with cultured or explanted macrophages showed that lipopolysaccharide alone can induce an oxidative burst and/or nitric oxide species within 6 h (Chen et al., 1998; Neumann et al., 2000). This is in contrast to the general need for interferon-γ in mammals, although two interferon-γ homologs are present in the zebrafish genome (Igawa et al., 2006). Another contrast to mammalian systems is the use of cleaved transferrin as an activator of macrophages to produce nitric oxide (Stafford and Belosevic, 2003; Stafford et al., 2001).

Melanomacrophages

Melanomacrophages are a subset of macrophages found in fish, amphibians, and reptiles. They are most commonly found as a part of melanomacrophage centers (MMCs) in the spleen, liver, and sometimes kidney (as reviewed by Agius and Roberts, 2003), but are also seen singly. Specific studies of zebrafish melanomacrophages have not been published, but there is plentiful literature on those of other teleost species.

It is very likely that macrophages and melanomacrophages are in fact the same cell type, and the latter has happened to ingest pigmented material (predominantly lipofuscin, melanin, and hemosiderin [Agius and Roberts, 2003]) and are henceforth associated with MMCs. However, it has been asserted (Sichel et al., 1997; Zuasti et al., 1989) that these cells actually produce melanin by a unique pathway. The majority view, however, is that melanomacrophages, and MMCs, are storage and destruction sites for hazardous materials, resistant bacteria, and excess iron from dead erythrocytes. This view is supported by evidence that indigestible materials such as India ink and thorium hydroxide traffic to MMCs after intraperitoneal injection (Ellis et al., 1976; Herraez and Zapata, 1991). Also, older fish, or fish enduring toxic environments, have more and larger MMCs.

Vaccine antigens also traffic to MMCs, suggesting an antigen storage/presentation role for these structures. The assertion that MMCs were sites of antigen presentation was first made by Ellis and De Sousa in 1974 (Ellis and De Sousa, 1974), and since then other evidence has accrued to suggest that these structures represent functional homologs of mammalian germinal centers. The recent finding that melanomacrophages, both single and within MMCs, stain with CNA-42 (an antibody specific for human follicular dendritic cells) supports this relationship (Vigliano et al., 2006). A homolog of CD83, a dendritic cell marker in humans, has been found expressed on the SHK-1 line of Atlantic salmon macrophages; this same cell line was found to possess the machinery for melanin production (Haugarvoll et al., 2006). Thus, further research will be required to determine the specific identity of melanomacrophages and their possible relationship to dendritic cells and other macrophages.

Neutrophils

In earlier publications on the subject, the neutrophil-like granulocytes of some fish, including zebrafish, were termed "heterophils," based on the fact that their cytoplasmic granules are not uniform but of two distinct shapes. This term has gradually given way to "neutrophils," and we will refer to them as such here.

Hematopoiesis

Neutrophils in the zebrafish embryo probably appear only slightly later in development than do macrophages. By electron microscopy, they were first detected at 48 hpf (Willett et al., 1999), but studies relying on gene expression show neutrophils as early as 18 hpf (Bennett et al., 2001). Myeloperoxidase (MPO)-expressing cells in the transgenic model (see below) are reported by others to be on the yolk surface among the embryonic macrophages as early as 22 hpf (Mathias et al., 2006). Evidence from the Herbomel group (Le Guyader et al., 2008) and our own observations suggest that at least some of the earliest phagocytes seen in the yolk circulation valley at 25 to 30 hpf possess granules and will become neutrophils. It is intriguing that 32 hpf is the only time point published at which the macrophage markers Fms and lysozyme C (which we find expressed only by granulocytes after 40 hpf) are seen expressed in the same cell. Also, we find all of the yolk surface macrophages to express the *mpo* transgene (see below) at this time. Based on these observations, we suspect that the macrophage or neutrophil identity of these cells is not fully determined until after 32 hpf. By this logic, we cannot be certain whether macrophages or neutrophils appear first, or both are preceded by an uncommitted precursor.

Molecular Markers

The primary molecular marker for neutrophils in the zebrafish is MPO, also called myeloid-specific peroxidase. Two different lines of MPO transgenic zebrafish have been derived (Mathias et al., 2006; Renshaw et al., 2006), with apparently equivalent results in neutrophil specificity (although the line from Renshaw et al. appears brighter in our hands). As mentioned above, L-plastin and lysozyme C have been considered imperfect candidates for marking neutrophils, although our own observations suggest that lysozyme C is a very good neutrophil marker indeed after 40 hpf. A recent report of a promoter trap transgenic in which a subset of neutrophils express YFP (Meijer et al., 2008) offers another as yet unidentified candidate, but once again expression is transient and is best at 3 to 4 days postfertilization. Transient expression may again explain the apparent expression by only a subset of MPO+ neutrophils.

Chemotaxis

Expression of specific chemotactic receptors by neutrophils has not yet been addressed. Thanks to the existence of transgenic embryos, compelling studies have been published showing that neutrophilic inflammation may resolve either by apoptosis of neutrophils (Renshaw et al., 2006) or, surprisingly, their return to the vasculature (Mathias et al., 2006). Molecular mechanisms for this or other neutrophil movements have not been well studied. One group has suggested that the zebrafish homolog of CD31, acting through the membrane channel ERG (Ether-a-go-go like gene), is responsible for the attraction of neutrophils to fin wounds (Brown et al., 2007).

Phagocytosis

It has been proposed that the zebrafish neutrophil is highly attracted to infection and inflammation, but is very rarely phagocytic (Le Guyader et al., 2008). Our own observations show that embryonic neutrophils are capable of phagocytosing M. *marinum*, although this is a relatively rare event (J. M. Davis and L. Ramakrishnan, unpublished data). *Streptococcus iniae* (Ramakrishnan, unpublished data) and *Pseudomonas aeruginosa* (Brannon, et al. 2009) are predominantly taken up by neutrophils, but specific studies on the molecular features and mechanisms of neutrophil phagocytosis, as distinct from macrophage phagocytosis, have yet to be done.

Effector Pathways

There is evidence for a strong oxidative burst (Kemenade et al., 1994) and degranulation (Neumann et al., 2001) by fish neutrophils, and detailed accounts of neutrophil extracellular trap release and degranulation in zebrafish (Palic et al., 2007a) and fathead minnows (Palic et al., 2007b) have appeared recently. The visual access provided by zebrafish embryos offers an excellent opportunity for studying such phenomena.

Dendritic Cells

As stated above (see "Melanomacrophages"), the two main dendritic cell markers for which homologs have been identified in fish are CD83, in Atlantic salmon (Haugarvoll et al., 2006) and trout (Ohta et al., 2004), and three C-type lectins similar to DC-SIGN in Atlantic salmon (Soanes et al., 2004). The zebrafish embryo, in which macrophages and neutrophils have been so well observed, is not likely to contain dendritic cells or their equivalent, because mammalian species do not develop this cell type until late in development (Dakic et al., 2004). Therefore, although genomic evidence in other fish suggests their presence, more study with adult zebrafish blood cells will be required to determine whether dendritic cells function in this organism as they do in mammals.

ZEBRAFISH MODELS OF INFECTION

Despite some of the immunological tools and reagents lagging behind the mouse, the power of real-time visualization in the zebrafish has opened up the black box of host-pathogen interactions leading to critical end-stage pathologies such as granuloma formation. Our laboratory first started to use the zebrafish to study M. *marinum* and *Salmonella* serovar Arizonae infection in 2000 (Davis et al., 2002), and the model has since been used to study the pathogenesis of other bacteria as well as bacterial toxins (Davis et al., 2002; Hamm et al., 2006; Neely et al., 2002;

van der Sar et al., 2003; Voth et al., 2005). The pathogenesis of the diseases caused by these organisms in the zebrafish is remarkably similar to the ones caused by the related mammalian pathogens in their hosts, highlighting the idea that an organism might have first evolved into a pathogen in the context of a particular host species and then adapted to other hosts by further evolution leading to speciation. In the following we will highlight the work done thus far with the various pathogen species.

Mycobacteria

The zebrafish has long been known to be a natural host to M. marinum infection. Fish tuberculosis, as it is commonly referred to, is a disease dreaded by fish fanciers and zebrafish facility managers alike (Westerfield, 2000). A detailed analysis of the infection and pathology demonstrates that M. marinum infection in the zebrafish produces a caseating granulomatous disease with pathology very similar to that of active human tuberculosis (Swaim et al., 2006). Furthermore, control of infection depends on adaptive immunity because rag1 mutant zebrafish are hypersusceptible to infection, similar to the case with mammals. The close genetic relationship of M. marinum and Mycobacterium tuberculosis and their similar pathology and pathogenesis in their respective natural hosts (ectotherms and humans) have made the M. marinum-zebrafish model an ideal one in which to gain a clearer understanding of the host-pathogen interactions that lead to the ultimate pathology and outcomes of tuberculosis.

Tuberculosis in humans is a systemic disease that most frequently affects the lungs but can involve virtually any other organ or tissue. It is postulated to occur in distinct steps, with the infecting bacteria being phagocytosed by macrophages and gaining entry into tissues via these phagocytic cells (Dannenberg, 1993). Pathogenic mycobacteria have devised strategies for host survival that may range from avoiding phagosomal acidification to a subset of the bacteria entering the cytoplasm and being propelled within the cell and into others by host actin (Honer zu Bentrup and Russell, 2001; Stamm et al., 2003; van der Wel et al., 2007). Indeed, mycobacteria have been thought to be such well-adapted macrophage pathogens that they grow optimally within these cells. Additional macrophages are next recruited to the site of infection to form a granuloma, a complex immune structure that ultimately consists of a variety of immune cells and highly differentiated macrophages that have undergone epithelioid transformation. The granuloma was previously thought to require adaptive immunity for its formation. Functionally, infection is often eradicated within a granuloma but can also persist indefinitely therein. Therefore, the granuloma has been considered solely a host-beneficial structure even though it is not fully effective in this capacity. Indeed, immune defects associated with poor granuloma formation such as T-lymphocyte deficiencies are associated with a worse outcome from M. tuberculosis infection (Cooper et al., 1993; Flynn et al., 1995). One paradox of tuberculosis in humans has been the growing recognition that reinfection occurs despite a seemingly effective adaptive immune response with good granuloma formation (Cosma et al., 2004; Chang and Riley, 2005). This was attributed to the reinfecting mycobacteria surviving by avoiding preexisting granulomas where adaptive immunity was concentrated (Cosma et al., 2003).

The zebrafish model has proved most telling in matters pertaining to the cellular movements and cell-cell interactions that abound in the pathogenesis of tuberculosis.

Macrophages respond very quickly to infection by arriving at infection sites and phagocytosing bacteria within 1 h (Cosma et al., 2003; Davis et al., 2002). Cytokines are induced and the macrophages carry the bacteria back into the deeper tissues (Color Plate 10A). Macrophage depletion by using a morpholino directed against the myeloid transcription factor PU.1 showed that macrophages curb bacterial growth from the very earliest stages of infection so that the macrophage is not a preferred mycobacterial niche if growth alone is considered (Clay et al., 2007). However, the bacteria require macrophages to transport them across epithelial barriers and blood vessels [detailed by the use of the $Tg(fli1::EGFP)^{y1}$ line] (Clay et al., 2007), so that they appear to tolerate a suboptimal growth niche to establish systemic infection. This work exemplifies one type of discovery that is feasible to make in the zebrafish (versus in the mouse) using morpholino and transgenic technologies in combination with real-time visualization.

Once infected macrophages reach the tissues, other macrophages arrive and undergo the characteristic epithelioid transformation to form granulomas despite the lack of adaptive immunity at this stage (Davis et al., 2002) (Color Plate 10B). This discovery was possible only because the zebrafish embryo can be readily infected at a stage when adaptive immunity has not yet come into play. The visual clarity of the embryo also allowed unprecedented microscopic access to the pathogenic events.

The model has proved useful in defining the exact role played by individual bacterial determinants as well. For instance, the M. tuberculosis Erp virulence determinant, a secreted protein, was the first one to be identified (Berthet et al., 1998), and by examining the M. marinum Erp mutant in the zebrafish in the presence and absence of macrophages, it has been possible to show that this determinant exerts its effects on virulence by specifically endowing the bacteria with the capacity to grow within macrophages (Clay et al., 2007; Cosma et al., 2006). In contrast, the RD1 virulence locus, famously the cause of attenuation of the vaccine strain BCG and also encoding a secretion system, does not appear to affect bacterial growth within individual macrophages (Volkman et al., 2004). Rather, it enhances granuloma formation and by so doing increases the number of infected macrophages and the total number of bacteria. Thus, the use of the zebrafish has shown that the granuloma, heretofore considered a host-protective structure, is coopted by the bacteria to enhance infection at least at its early stages (Volkman et al., 2004; Davis and Ramakrishnan, 2009). Furthermore, even mature granulomas developing in the context of adaptive immunity in adult animals appear to benefit mycobacteria (Cosma et al., 2004). Superinfecting bacteria home rapidly to preexisting granulomas and survive within them to the same extent as the bacteria that were present right from the inception of the granuloma. These discoveries made in a relatively young model give hope that a yet deeper understanding of tuberculosis will come from the zebrafish with the use of forward and reverse genetics and more sophisticated imaging tools.

Salmonella spp.

Salmonella comprises another genus of intramacrophage pathogens, which occupy a phagosomal compartment distinct from Mycobacterium (Linehan and Holden, 2003) and can produce either acute or chronic infection. Salmonella serovar Typhi is the cause of typhoid fever, a systemic disease of the reticuloendothelial system in humans. Salmonella serovar Typhimurium is commonly used to model human typhoid infection in the mouse. *Salmonella* serovar Arizonae, a pathogen of ectotherms, is lethal to zebrafish embryos even at very small inocula (5 to 20 bacteria per embryo). In the embryo, it infects macrophages (Fig. 3A), where it produces the typical spacious phagosomes seen in other models (Fig. 3B) and soon overwhelms and causes lysis of macrophages very early after infection (Fig. 3C), thus reproducing the features of *Salmonella* encounters with mammalian cells (Davis et al., 2002). *S. enterica* serovar Typhimurium also causes a lethal infection of zebrafish embryos that depends on key virulence determinants of mammalian infection (van der Sar et al., 2003). Therefore, it is possible that the zebrafish will similarly provide new insights about *Salmonella* pathogenesis.

Streptococcus spp.

Streptococci are pyogenic organisms and therefore would be expected to interact with neutrophils during infection. Our laboratory has observed in the zebrafish embryo the readiness with which neutrophils engulf *Streptococcus iniae*. This is in contrast to our findings with M. *marinum*, which is taken in far more often by macrophages than by neutrophils (J. M. Davis, O. Humbert, and L. Ramakrishnan, unpublished data; Clay et al., 2007). Published studies to date have focused on adult zebrafish infection with *S. iniae*, a fish pathogen, and *Streptococcus pyogenes*, the notorious human pathogen causing pyogenic pharyngitis ("strep throat"), necrotizing fasciitis, and toxic shock syndrome (Neely et al., 2002). Inoculation of adult fish intramuscularly at the dorsal muscle results in distinct courses of infection with these two pathogens. Although both species are lethal within a few days, *S. iniae* spreads systemically by around 26 h postinfection (hpi) and is fully disseminated by 40 hpi. At the site of injection, inflammatory cells (of uncertain identity) and necrosis are plentiful; systemically, bacteria are copious in the bloodstream and notably appear intracellularly in Kupffer cells of the liver and in endothelial cells of the central nervous system (Miller and Neely, 2004; Neely et al., 2002). *S. pyogenes*, on the other hand, induces very little inflammatory cell recruitment to the injection site, but induces death within 12 to 24 h, possibly as a result of the systemic introduction of toxins from the rapidly growing initial lesion (Miller and Neely, 2004; Neely et al., 2002). Because of these variations in pathogenesis, *S. iniae* infection of adult zebrafish has been proposed as a model for systemic human pathogens such as *S. pneumoniae*, while *S. pyogenes* in the zebrafish mimics human necrotizing fasciitis caused by the same organism (Miller and Neely, 2004). In *S. iniae*, a large-scale mutant screen for bacterial virulence factors (Lowe et al., 2007; Miller and Neely, 2005) has found a preponderance of factors directly related to capsule formation and has suggested that while growth in the injection site was less dependent on this structure, the transition to systemic infection required it. These findings demonstrate that the capsule is important during zebrafish infection, as it has long been held to be for avoidance of phagocytosis and ensuing bloodborne dissemination, a completely different strategy from that used by mycobacteria (Clay et al., 2007).

Lending credence to the *S. pyogenes*-zebrafish model is the finding that several bacterial genes suspected to play a role in virulence in humans have been confirmed as important in pathogenesis in the adult zebrafish (Bates et al., 2005; Brenot et al., 2004; Montanez et al., 2005). Germane to the pathogen-phagocyte relationship are the findings that *S. pyogenes* actually survives within macrophages and that this ability is a major factor in its virulence. In recent work, Phelps and Neely (2007) have described an attenuated *S. pyogenes* mutant (SalY) that induces far more inflammatory infiltration than does the wild type and is far less resistant to killing by macrophages. Depletion of macrophages in zebrafish before infection induced a major recovery in virulence in the mutant, compared with a mild one in the wild type (Phelps and Neely, 2007). How this relates to the relative paucity of macrophages in wild-type *S. pyogenes* lesions awaits further study and exploitation of the zebrafish as a genetically tractable host.

Edwardsiella sp.

Edwardsiella tarda is a relatively common pathogen of fish and, more rarely, of humans. Pressley et al. (2005) established baseline data for infection of embryos and adults through a variety of exposure routes, and more recently the same group has examined the effect of low levels of arsenic on the immune response of zebrafish embryos to this pathogen (Nayak et al., 2007). Control of bacteria was significantly reduced concurrent with a loss of oxidative burst by macrophages and neutrophils. Chemokine levels were also

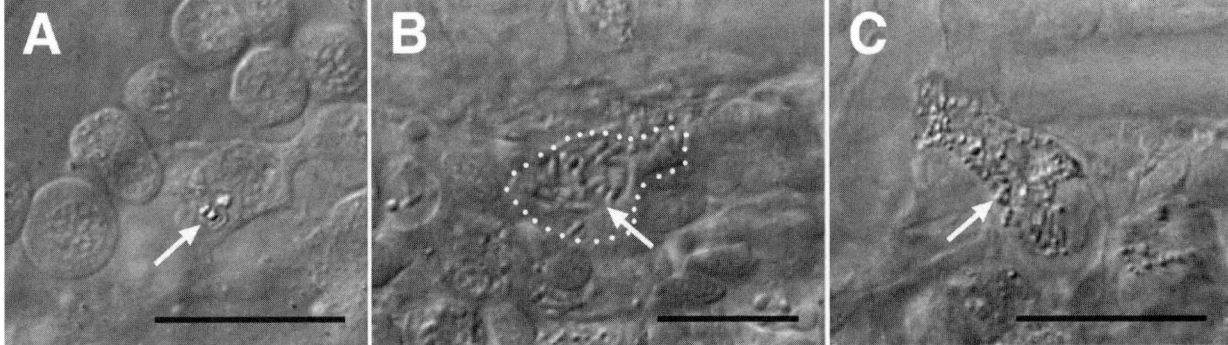

FIGURE 3 Larval zebrafish macrophages infected with *Salmonella* serovar Arizonae in vivo. (A) A macrophage in the yolk circulation valley contains three bacteria (white arrow) in a phagosome, ca. 1 h post-intravenous infection. (B) At ca. 18 hpi, a macrophage (outlined with white dots) near the CHT contains many bacteria, most within a large phagosome (white arrow). (C) At ca. 24 hpi, a macrophage crawling in a venule in the tail is overrun with intracellular bacteria. (All scale bars, 20 μm.)

affected. This reduction in effector function appears to be due to the combination of blunted tumor necrosis factor and interleukin-1β production along with a direct effect on the respiratory burst itself. Use of the embryo in these experiments allowed isolation of the innate immune response as affected by arsenic. Another study demonstrated the upregulation of TLR3 and parts of the TLR signaling pathway during *E. tarda* infection (Phelan et al., 2005a) (see "Receptors," above).

Other Microbial Pathogens

Several other pathogens, bacterial and viral, have been subject to studies in zebrafish infection. Lin and colleagues (2007) studied the acute-phase reaction of adult zebrafish to *Staphylococcus aureus* and *Aeromonas salmonicida*, gram-positive and -negative pathogens, respectively, in a gene expression study. Several viruses, including snakehead rhabdovirus (Phelan et al., 2005a, 2005b; Pressley et al., 2005), spring viremia of carp virus (Sanders et al., 2003), and viral hemorrhagic septicemia virus (Novoa et al., 2006), have also been featured in studies using zebrafish. These studies have focused mainly on establishing susceptibility and survival baselines and on the use of TLRs in the recognition of pathogens. Thus far, none of these models has expanded to address the role of phagocytes in host-pathogen interactions.

SUMMARY

The zebrafish offers several unique advantages for the study of host-pathogen interactions. The optical transparency of the larvae allows easy whole-embryo examinations of specific steps in pathogenesis. We have found that the zebrafish embryo is a magnificent host for studying the innate immune system in virtual isolation, since a functional adaptive immune system does not arise for at least 2 weeks after fertilization. The isolation of innate immunity, and the close visual access allowed by the larvae, make for an excellent opportunity to witness pathogen-phagocyte interactions first hand. Working with embryos also allows use of morpholino technology, the effects of which only last for the first 10 days of life. The genetic tractability of the organism and its fully ramified adaptive immune system also make the adult fish a promising model host. However, given the relative lack of immunological reagents and inbred lines, it is important to choose appropriate questions for study to minimize the disadvantages of the model while capitalizing on its strengths. In this way one can take advantage of the best attributes of the zebrafish—optical transparency, fecundity, and genetic tools—to make discoveries in phagocyte biology that complement those being made in other models.

The authors thank David Tobin for critical review of the manuscript. This work was supported by a National Defense Science and Engineering Fellowship (to J.M.D.), by grants from the National Institutes of Health (to L.R.), and by a Burroughs Wellcome award (to L.R.).

APPENDIX

FISH TAXONOMY

The variety of fish species is phenomenal, and their taxonomy can be difficult to grasp. Since the terminology is often lightly used in the discussion of fish immunology, a brief explanation is appropriate here. There are three main groups of fish—jawless, cartilaginous, and bony—all of which share common external features such as a mouth, gill openings, fins, nostrils, and a muscular trunk and tail. The jawless fish (Agnatha), such as the lamprey, have a sucker mouth and median fin folds. The cartilaginous group (Chondroichthyes) includes sharks and rays, which have immune systems somewhat different from the third group, Osteichthyes. Osteichthyes, the bony fish group, is incredibly diverse, and contains fish with both bony and cartilaginous features. The major subgroup of bony fish are the ray-finned fishes (Actinopterygii—with fins structured by rays of bony material). The teleosts are the largest group of ray-finned fish and also the largest group of bony fish. Teleosts are strictly bony, and the defining feature is fully moveable upper and lower jaws. The vast majority of familiar fish are teleosts. Here are the major species listed in immune and genomics literature, by their subgroup:

- **Cyprinids:** goldfish (*Carrasius auratus*), common carp (*Cyprinus carpio*), zebrafish (*Danio rerio*)
- **Salmonids:** rainbow trout (*Oncorhynchus mykiss*), Atlantic salmon (*Salmo salar*)
- **Perciforms:** sea bream (*Sparus aurata*), sea bass (*Dicentrarchus labrax*)
- **Siluriforms:** channel catfish (*Ictalurus punctatus*)
- **Tetraodons:** pufferfish, "fugu" (*Takifugu rubripes*)

REFERENCES

Agius, C., and R. J. Roberts. 2003. Melano-macrophage centres and their role in fish pathology. *J. Fish Dis.* **26:**499–509.

Alliot, F., E. Lecain, B. Grima, and B. Pessac. 1991. Microglial progenitors with a high proliferative potential in the embryonic and adult mouse brain. *Proc. Natl. Acad. Sci. USA* **88:**1541–1545.

Amores, A., A. Force, Y. L. Yan, L. Joly, C. Amemiya, A. Fritz, R. K. Ho, J. Langeland, V. Prince, Y. L. Wang, M. Westerfield, M. Ekker, and J. H. Postlethwait. 1998. Zebrafish hox clusters and vertebrate genome evolution. *Science* **282:**1711–1714.

Barreda, D. R., N. F. Neumann, and M. Belosevic. 2000. Flow cytometric analysis of PKH26-labeled goldfish kidney-derived macrophages. *Dev. Comp. Immunol.* **24:**395–406.

Bates, C. S., C. Toukoki, M. N. Neely, and Z. Eichenbaum. 2005. Characterization of MtsR, a new metal regulator in group A streptococcus, involved in iron acquisition and virulence. *Infect. Immun.* **73:**5743–5753.

Belcourt, D. R., C. Lazure, and H. P. Bennett. 1993. Isolation and primary structure of the three major forms of granulin-like peptides from hematopoietic tissues of a teleost fish (*Cyprinus carpio*). *J. Biol. Chem.* **268:**9230–9237.

Bennett, C. M., J. P. Kanki, J. Rhodes, T. X. Liu, B. H. Paw, M. W. Kieran, D. M. Langenau, A. Delahaye-Brown, L. I. Zon, M. D. Fleming, and A. T. Look. 2001. Myelopoiesis in the zebrafish, *Danio rerio*. *Blood* **98:**643–651.

Berthet, F. X., M. Lagranderie, P. Gounon, C. Laurent-Winter, D. Ensergueix, P. Chavarot, F. Thouron, E. Maranghi, V. Pelicic, D. Portnoi, G. Marchal, and B. Gicquel. 1998. Attenuation of virulence by disruption of the *Mycobacterium tuberculosis* erp gene. *Science* **282:**759–762.

Blaser, H., M. Reichman-Fried, I. Castanon, K. Dumstrei, F. L. Marlow, K. Kawakami, L. Solnica-Krezel, C. P. Heisenberg, and E. Raz. 2006. Migration of zebrafish primordial germ cells: a role for myosin contraction and cytoplasmic flow. *Dev. Cell* **11:**613–627.

Brannon, M. K., J. M. Davis, J. R. Mathias, C. J. Hall, J. C. Emerson, P. S. Crosier, A. Huttenlocher, L. Ramakrishnan, and S. M. Moskowitz. 2009. *Pseudomonas aeruginosa* type III secretion system interacts with phagocytes to modulate systemic infection of zebrafish embryos. *Cell. Microbiol.* **11:**755–768.

Brenot, A., K. Y. King, B. Janowiak, O. Griffith, and M. G. Caparon. 2004. Contribution of glutathione peroxidase to the virulence of *Streptococcus pyogenes*. *Infect. Immun.* **72:**408–413.

Broussard, G. W., and D. G. Ennis. 2007. *Mycobacterium marinum* produces long-term chronic infections in medaka: a new animal model for studying human tuberculosis. *Comp. Biochem. Physiol. C Toxicol. Pharmacol.* **145:**45–54.

Brown, S. B., C. S. Tucker, C. Ford, Y. Lee, D. R. Dunbar, and J. J. Mullins. 2007. Class III antiarrhythmic methanesulfonanilides inhibit leukocyte recruitment in zebrafish. *J. Leukoc. Biol.* **82:**79–84.

Cadieux, B., B. P. Chitramuthu, D. Baranowski, and H. P. Bennett. 2005. The zebrafish progranulin gene family and antisense transcripts. *BMC Genomics* **6:**156.

Chen, S. C., A. Adams, K. D. Thompson, and R. H. Richards. 1998. Electron microscope studies of the in vitro phagocytosis of *Mycobacterium* spp. by rainbow trout *Oncorhynchus mykiss* head kidney macrophages. *Dis. Aquat. Organ.* **32:**99–110.

Chiang, C. Y., and L. W. Riley. 2005. Exogenous reinfection in tuberculosis. *Lancet Infect. Dis.* **5:**629–636.

Chong, S. W., L. M. Nguyet, Y. J. Jiang, and V. Korzh. 2007. The chemokine, Sdf-1, and its receptor, Cxcr4, are required for formation of muscle in zebrafish. *BMC Dev. Biol.* **7:**54.

Clay, H., J. M. Davis, D. Beery, A. Huttenlocher, S. E. Lyons, and L. Ramakrishnan. 2007. Dichotomous role of the macrophage in early *Mycobacterium marinum* infection of the zebrafish. *Cell Host Microbe* **2:**29–39.

Cooper, A. M., D. K. Dalton, T. A. Stewart, J. P. Griffin, D. G. Russell, and I. M. Orme. 1993. Disseminated tuberculosis in interferon gamma gene-disrupted mice. *J. Exp. Med.* **178:**2243–2247.

Cooper, M. S., D. P. Szeto, G. Sommers-Herivel, J. Topczewski, L. Solnica-Krezel, H. C. Kang, I. Johnson, and D. Kimelman. 2005. Visualizing morphogenesis in transgenic zebrafish embryos using BODIPY TR methyl ester dye as a vital counterstain for GFP. *Dev. Dyn.* **232:**359–368.

Cosma, C. L., O. Humbert, and L. Ramakrishnan. 2004. Superinfecting mycobacteria home to established tuberculous granulomas. *Nat. Immunol.* **5:**828–835.

Cosma, C. L., K. Klein, R. Kim, D. Beery, and L. Ramakrishnan. 2006. *Mycobacterium marinum* Erp is a virulence determinant required for cell wall integrity and intracellular survival. *Infect. Immun.* **74:**3125–3133.

Cosma, C. L., D. R. Sherman, and L. Ramakrishnan. 2003. The secret lives of the pathogenic mycobacteria. *Annu. Rev. Microbiol.* **57:**641–676.

Crosnier, C., N. Vargesson, S. Gschmeissner, L. Ariza-McNaughton, A. Morrison, and J. Lewis. 2005. Delta-Notch signalling controls commitment to a secretory fate in the zebrafish intestine. *Development* **132:**1093–1104.

Dai, X. M., G. R. Ryan, A. J. Hapel, M. G. Dominguez, R. G. Russell, S. Kapp, V. Sylvestre, and E. R. Stanley. 2002. Targeted disruption of the mouse colony-stimulating factor 1 receptor gene results in osteopetrosis, mononuclear phagocyte deficiency, increased primitive progenitor cell frequencies, and reproductive defects. *Blood* **99:**111–120.

Dakic, A., Q. X. Shao, A. D'Amico, M. O'Keeffe, W. F. Chen, K. Shortman, and L. Wu. 2004. Development of the dendritic cell system during mouse ontogeny. *J. Immunol.* **172:**1018–1027.

Dambly-Chaudiere, C., N. Cubedo, and A. Ghysen. 2007. Control of cell migration in the development of the posterior lateral line: antagonistic interactions between the chemokine receptors CXCR4 and CXCR7/RDC1. *BMC Dev. Biol.* **7:**23.

Danilova, N., J. Bussmann, K. Jekosch, and L. A. Steiner. 2005. The immunoglobulin heavy-chain locus in zebrafish: identification and expression of a previously unknown isotype, immunoglobulin Z. *Nat. Immunol.* **6:**295–302.

Dannenberg, A. M., Jr. 1993. Immunopathogenesis of pulmonary tuberculosis. *Hosp. Pract.* **28:**51–58.

Davis, J. M., H. Clay, J. L. Lewis, N. Ghori, P. Herbomel, and L. Ramakrishnan. 2002. Real-time visualization of *Mycobacterium*-macrophage interactions leading to initiation of granuloma formation in zebrafish embryos. *Immunity* **17:**693–702.

Davis, J. M., and L. Ramakrishnan. 2009. The role of the granuloma in expansion and dissemination of early tuberculous infection. *Cell* **136:**37–49.

Ellis, A. E., and M. De Sousa. 1974. Phylogeny of the lymphoid system. I. A study of the fate of circulating lymphocytes in plaice. *Eur. J. Immunol.* **4:**338–343.

Ellis, A. E., A. L. S. Munro, and R. J. Roberts. 1976. Defence mechanisms in fish: fate of intraperitoneally introduced carbon in the plaice (*Pleuronectes platessa*). *J. Fish Biol.* **8:**67–78.

Flynn, J. L., M. M. Goldstein, J. Chan, K. J. Triebold, K. Pfeffer, C. J. Lowenstein, R. Schreiber, T. W. Mak, and B. R. Bloom. 1995. Tumor necrosis factor-alpha is required in the protective immune response against *Mycobacterium tuberculosis* in mice. *Immunity* **2:**561–572.

Hall, C., M. V. Flores, T. Storm, K. Crosier, and P. Crosier. 2007. The zebrafish lysozyme C promoter drives myeloid-specific expression in transgenic fish. *BMC Dev. Biol.* **7:**42.

Hamm, E. E., D. E. Voth, and J. D. Ballard. 2006. Identification of *Clostridium difficile* toxin B cardiotoxicity using a zebrafish embryo model of intoxication. *Proc. Natl. Acad. Sci. USA* **103:**14176–14181.

Hammerschmidt, M., and A. P. McMahon. 1998. The effect of pertussis toxin on zebrafish development: a possible role for inhibitory G-proteins in hedgehog signaling. *Dev. Biol.* **194:**166–171.

Hanington, P. C., D. R. Barreda, and M. Belosevic. 2006. A novel hematopoietic granulin induces proliferation of goldfish (*Carassius auratus* L.) macrophages. *J. Biol. Chem.* **281:**9963–9970.

Haugarvoll, E., J. Thorsen, M. Laane, Q. Huang, and E. O. Koppang. 2006. Melanogenesis and evidence for melanosome transport to the plasma membrane in a CD83 teleost leukocyte cell line. *Pigment Cell Res.* **19:**214–225.

Herbomel, P., B. Thisse, and C. Thisse. 1999. Ontogeny and behaviour of early macrophages in the zebrafish embryo. *Development* **126:**3735–3745.

Herbomel, P., B. Thisse, and C. Thisse. 2001. Zebrafish early macrophages colonize cephalic mesenchyme and developing brain, retina, and epidermis through a M-CSF receptor-dependent invasive process. *Dev. Biol.* **238:**274–288.

Herraez, M. P., and A. G. Zapata. 1991. Structural characterization of the melano-macrophage centres (MMC) of goldfish *Carassius auratus*. *Eur. J. Morphol.* **29:**89–102.

Honer zu Bentrup, K., and D. G. Russell. 2001. Mycobacterial persistence: adaptation to a changing environment. *Trends Microbiol.* **9:**597–605.

Hsu, K., D. Traver, J. L. Kutok, A. Hagen, T. X. Liu, B. H. Paw, J. Rhodes, J. N. Berman, L. I. Zon, J. P. Kanki, and A. T. Look. 2004. The pu.1 promoter drives myeloid gene expression in zebrafish. *Blood* **104:**1291–1297.

Hung, D. T., E. A. Shakhnovich, E. Pierson, and J. J. Mekalanos. 2005. Small-molecule inhibitor of *Vibrio cholerae* virulence and intestinal colonization. *Science* **310:**670–674.

Igawa, D., M. Sakai, and R. Savan. 2006. An unexpected discovery of two interferon gamma-like genes along with interleukin (IL)-22 and -26 from teleost: IL-22 and -26 genes have been described for the first time outside mammals. *Mol. Immunol.* **43:**999–1009.

Isogai, S., M. Horiguchi, and B. M. Weinstein. 2001. The vascular anatomy of the developing zebrafish: an atlas of embryonic and early larval development. *Dev. Biol.* **230:**278–301.

Jault, C., L. Pichon, and J. Chluba. 2004. Toll-like receptor gene family and TIR-domain adapters in *Danio rerio*. *Mol. Immunol.* **40:**759–771.

Jozefowicz, C., J. McClintock, and V. Prince. 2003. The fates of zebrafish Hox gene duplicates. *J. Struct. Funct. Genomics* **3:**185–194.

Kemenade, B., A. Groeneveld, B. Rens, and J. Rombout. 1994. Characterization of macrophages and neutrophilic granulocytes from the pronephros of carp (*Cyprinus carpio*). *J. Exp. Biol.* **187:**143–158.

Kimmel, C. B., W. W. Ballard, S. R. Kimmel, B. Ullmann, and T. F. Schilling. 1995. Stages of embryonic development of the zebrafish. *Dev. Dyn.* **203:**253–310.

Knaut, H., C. Werz, R. Geisler, and C. Nusslein-Volhard. 2003. A zebrafish homologue of the chemokine receptor Cxcr4 is a germ-cell guidance receptor. *Nature* **421:**279–282.

Kuchler, A. M., E. Gjini, J. Peterson-Maduro, B. Cancilla, H. Wolburg, and S. Schulte-Merker. 2006. Development of the zebrafish lymphatic system requires VEGFC signaling. *Curr. Biol.* **16:**1244–1248.

Langenau, D. M., A. A. Ferrando, D. Traver, J. L. Kutok, J. P. Hezel, J. P. Kanki, L. I. Zon, A. T. Look, and N. S. Trede. 2004. In vivo tracking of T cell development, ablation, and engraftment in transgenic zebrafish. *Proc. Natl. Acad. Sci. USA* **101:**7369–7374.

Langenau, D. M., D. Traver, A. A. Ferrando, J. L. Kutok, J. C. Aster, J. P. Kanki, S. Lin, E. Prochownik, N. S. Trede, L. I. Zon, and A. T. Look. 2003. Myc-induced T cell leukemia in transgenic zebrafish. *Science* **299:**887–890.

Lawson, N. D., and B. M. Weinstein. 2002. In vivo imaging of embryonic vascular development using transgenic zebrafish. *Dev. Biol.* **248:**307–318.

Le Guyader, D., M. J. Redd, E. Colucci-Guyon, E. Murayama, K. Kissa, V. Briolat, E. Mordelet, A. Zapata, H. Shinomiya, and P. Herbomel. 2008. Origins and unconventional behavior of neutrophils in developing zebrafish. *Blood* **111:**132–141.

Li, J., D. R. Barreda, Y. A. Zhang, H. Boshra, A. E. Gelman, S. Lapatra, L. Tort, and J. O. Sunyer. 2006. B lymphocytes from early vertebrates have potent phagocytic and microbicidal abilities. *Nat. Immunol.* **7:**1116–1124.

Lin, B., S. Chen, Z. Cao, Y. Lin, D. Mo, H. Zhang, J. Gu, M. Dong, Z. Liu, and A. Xu. 2007. Acute phase response in zebrafish upon *Aeromonas salmonicida* and *Staphylococcus aureus* infection: striking similarities and obvious differences with mammals. *Mol. Immunol.* **44:**295–301.

Linehan, S. A., and D. W. Holden. 2003. The interplay between *Salmonella typhimurium* and its macrophage host—what can it teach us about innate immunity? *Immunol. Lett.* **85:**183–192.

Lowe, B. A., J. D. Miller, and M. N. Neely. 2007. Analysis of the polysaccharide capsule of the systemic pathogen *Streptococcus iniae* and its implications in virulence. *Infect. Immun.* **75:**1255–1264.

Mathias, J. R., M. E. Dodd, K. B. Walters, J. Rhodes, J. P. Kanki, A. T. Look, and A. Huttenlocher. 2007. Live imaging of chronic inflammation caused by mutation of zebrafish Hai1. *J. Cell Sci.* **120:**3372–3383.

Mathias, J. R., B. J. Perrin, T. X. Liu, J. Kanki, A. T. Look, and A. Huttenlocher. 2006. Resolution of inflammation by retrograde chemotaxis of neutrophils in transgenic zebrafish. *J. Leukoc. Biol.* **80:**1281–1288.

Meijer, A. H., S. F. Gabby Krens, I. A. Medina Rodriguez, S. He, W. Bitter, B. Ewa Snaar-Jagalska, and H. P. Spaink. 2004. Expression analysis of the Toll-like receptor and TIR domain adaptor families of zebrafish. *Mol. Immunol.* **40:**773–783.

Meijer, A. H., A. M. van der Sar, C. Cunha, G. E. Lamers, M. A. Laplante, H. Kikuta, W. Bitter, T. S. Becker, and H. P. Spaink. 2008. Identification and real-time imaging of a myc-expressing neutrophil population involved in inflammation and mycobacterial granuloma formation in zebrafish. *Dev. Comp. Immunol.* **32:**36–49.

Miller, J. D., and M. N. Neely. 2005. Large-scale screen highlights the importance of capsule for virulence in the zoonotic pathogen *Streptococcus iniae*. *Infect. Immun.* **73:**921–934.

Miller, J. D., and M. N. Neely. 2004. Zebrafish as a model host for streptococcal pathogenesis. *Acta Trop.* **91:**53–68.

Miyasaka, N., H. Knaut, and Y. Yoshihara. 2007. Cxcl12/Cxcr4 chemokine signaling is required for placode assembly and sensory axon pathfinding in the zebrafish olfactory system. *Development* **134:**2459–2468.

Montanez, G. E., M. N. Neely, and Z. Eichenbaum. 2005. The streptococcal iron uptake (Siu) transporter is required for iron uptake and virulence in a zebrafish infection model. *Microbiology* **151:**3749–3757.

Murayama, E., K. Kissa, A. Zapata, E. Mordelet, V. Briolat, H. F. Lin, R. I. Handin, and P. Herbomel. 2006. Tracing hematopoietic precursor migration to successive hematopoietic organs during zebrafish development. *Immunity* **25:**963–975.

Nasevicius, A., and S. C. Ekker. 2000. Effective targeted gene "knockdown" in zebrafish. *Nat. Genet.* **26:**216–220.

Nayak, A. S., C. R. Lage, and C. H. Kim. 2007. Effects of low concentrations of arsenic on the innate immune system of the zebrafish (*Danio rerio*). *Toxicol. Sci.* **98:**118–124.

Neely, M. N., J. D. Pfeifer, and M. Caparon. 2002. *Streptococcus*-zebrafish model of bacterial pathogenesis. *Infect. Immun.* **70:**3904–3914.

Neumann, N. F., D. R. Barreda, and M. Belosevic. 2000. Generation and functional analysis of distinct macrophage sub-populations from goldfish (*Carassius auratus* L.) kidney leukocyte cultures. *Fish Shellfish Immunol.* **10:**1–20.

Neumann, N. F., J. L. Stafford, D. Barreda, A. J. Ainsworth, and M. Belosevic. 2001. Antimicrobial mechanisms of fish phagocytes and their role in host defense. *Dev. Comp. Immunol.* **25:**807–825.

Novoa, B., A. Romero, V. Mulero, I. Rodriguez, I. Fernandez, and A. Figueras. 2006. Zebrafish (*Danio rerio*) as a model for the study of vaccination against viral haemorrhagic septicemia virus (VHSV). *Vaccine* **24:**5806–5816.

Ogura, Y., L. Saab, F. F. Chen, A. Benito, N. Inohara, and G. Nunez. 2003. Genetic variation and activity of mouse Nod2, a susceptibility gene for Crohn's disease. *Genomics* **81:**369–377.

Ohta, Y., E. Landis, T. Boulay, R. B. Phillips, B. Collet, C. J. Secombes, M. F. Flajnik, and J. D. Hansen. 2004. Homologs of CD83 from elasmobranch and teleost fish. *J. Immunol.* **173:**4553–4560.

Palic, D., C. B. Andreasen, J. Ostojic, R. M. Tell, and J. A. Roth. 2007. Zebrafish (*Danio rerio*) whole kidney assays to measure neutrophil extracellular trap release and degranulation of primary granules. *J. Immunol. Methods* **319:**87–97.

Palic, D., J. Ostojic, C. B. Andreasen, and J. A. Roth. 2007. Fish cast NETs: neutrophil extracellular traps are released from fish neutrophils. *Dev. Comp. Immunol.* **31:**805–816.

Parichy, D. M., and J. M. Turner. 2003. Temporal and cellular requirements for Fms signaling during zebrafish adult pigment pattern development. *Development* **130:**817–833.

Patton, E. E., and L. I. Zon. 2001. The art and design of genetic screens: zebrafish. *Nat. Rev. Genet.* **2:**956–966.

Peatman, E., B. Bao, P. Baoprasertkul, and Z. Liu. 2005. In silico identification and expression analysis of 12 novel CC chemokines in catfish. *Immunogenetics* **57:**409–419.

Peatman, E., B. Bao, X. Peng, P. Baoprasertkul, Y. Brady, and Z. Liu. 2006. Catfish CC chemokines: genomic clustering, duplications, and expression after bacterial infection with *Edwardsiella ictaluri*. *Mol. Genet. Genomics* **275:**297–309.

Peatman, E., and Z. Liu. 2006. CC chemokines in zebrafish: evidence for extensive intrachromosomal gene duplications. *Genomics* **88:**381–385.

Peterson, R. T., S. Y. Shaw, T. A. Peterson, D. J. Milan, T. P. Zhong, S. L. Schreiber, C. A. MacRae, and M. C. Fishman. 2004. Chemical suppression of a genetic mutation in a zebrafish model of aortic coarctation. *Nat. Biotechnol.* **22:**595–599.

Phelan, P. E., M. T. Mellon, and C. H. Kim. 2005. Functional characterization of full-length TLR3, IRAK-4, and TRAF6 in zebrafish (*Danio rerio*). *Mol. Immunol.* **42:**1057–1071.

Phelan, P. E., M. E. Pressley, P. E. Witten, M. T. Mellon, S. Blake, and C. H. Kim. 2005. Characterization of snakehead rhabdovirus infection in zebrafish (*Danio rerio*). *J. Virol.* **79:**1842–1852.

Phelps, H. A., and M. N. Neely. 2007. SalY of *Streptococcus pyogenes* lantibiotic locus is required for full virulence and intracellular survival in macrophages. *Infect. Immun.* **75:**4541–4551.

Pressley, M. E., P. E. Phelan III, P. E. Witten, M. T. Mellon, and C. H. Kim. 2005. Pathogenesis and inflammatory response to *Edwardsiella tarda* infection in the zebrafish. *Dev. Comp. Immunol.* **29:**501–513.

Redd, M. J., G. Kelly, G. Dunn, M. Way, and P. Martin. 2006. Imaging macrophage chemotaxis in vivo: studies of microtubule function in zebrafish wound inflammation. *Cell. Motil. Cytoskelet.* **63:**415–422.

Renshaw, S. A., C. A. Loynes, D. M. Trushell, S. Elworthy, P. W. Ingham, and M. K. Whyte. 2006. A transgenic zebrafish model of neutrophilic inflammation. *Blood* **108:**3976–3978.

Rodriguez, A., M. A. Esteban, and J. Meseguer. 2003. A mannose-receptor is possibly involved in the phagocytosis of *Saccharomyces cerevisiae* by seabream (*Sparus aurata* L.) leucocytes. *Fish Shellfish Immunol.* **14:**375–388.

Sanders, G. E., W. N. Batts, and J. R. Winton. 2003. Susceptibility of zebrafish (*Danio rerio*) to a model pathogen, spring viremia of carp virus. *Comp. Med.* **53:**514–521.

Santos, F., G. MacDonald, E. W. Rubel, and D. W. Raible. 2006. Lateral line hair cell maturation is a determinant of aminoglycoside susceptibility in zebrafish (*Danio rerio*). *Hear. Res.* **213:**25–33.

Schraml, B., M. A. Baker, and B. D. Reilly. 2006. A complement receptor for opsonized immune complexes on erythrocytes from *Oncorhynchus mykiss* but not *Ictalarus punctatus*. *Mol. Immunol.* **43:**1595–1603.

Sichel, G., M. Scalia, F. Mondio, and C. Corsaro. 1997. The amphibian Kupffer cells build and demolish melanosomes: an ultrastructural point of view. *Pigment Cell Res.* **10:**271–287.

Skromne, I., and V. E. Prince. 2008. Current perspectives in zebrafish reverse genetics: moving forward. *Dev. Dyn.* **237:**861–882.

Soanes, K. H., K. Figuereido, R. C. Richards, N. R. Mattatall, and K. V. Ewart. 2004. Sequence and expression of C-type lectin receptors in Atlantic salmon (*Salmo salar*). *Immunogenetics* **56:**572–584.

Sorensen, K. K., O. K. Tollersrud, G. Evjen, and B. Smedsrod. 2001. Mannose-receptor-mediated clearance of lysosomal alpha-mannosidase in scavenger endothelium of cod endocardium. *Comp. Biochem. Physiol. A Mol. Integr. Physiol.* **129:**615–630.

Sorokin, S. P., and R. F. Hoyt, Jr. 1987. Pure population of nonmonocyte derived macrophages arising in organ cultures of embryonic rat lungs. *Anat. Rec.* **217:**35–52.

Stafford, J., N. F. Neumann, and M. Belosevic. 1999. Inhibition of macrophage activity by mitogen-induced goldfish leukocyte deactivating factor. *Dev. Comp. Immunol.* **23:**585–596.

Stafford, J. L., and M. Belosevic. 2003. Transferrin and the innate immune response of fish: identification of a novel mechanism of macrophage activation. *Dev. Comp. Immunol.* **27:**539–554.

Stafford, J. L., N. F. Neumann, and M. Belosevic. 2001. Products of proteolytic cleavage of transferrin induce nitric oxide response of goldfish macrophages. *Dev. Comp. Immunol.* **25:**101–115.

Stamm, L. M., J. H. Morisaki, L. Y. Gao, R. L. Jeng, K. L. McDonald, R. Roth, S. Takeshita, J. Heuser, M. D. Welch, and E. J. Brown. 2003. *Mycobacterium marinum* escapes from phagosomes and is propelled by actin-based motility. *J. Exp. Med.* **198:**1361–1368.

Swaim, L. E., L. E. Connolly, H. E. Volkman, O. Humbert, D. E. Born, and L. Ramakrishnan. 2006. *Mycobacterium marinum* infection of adult zebrafish causes caseating granulomatous tuberculosis and is moderated by adaptive immunity. *Infect. Immun.* **74:**6108–6117.

Traver, D., P. Herbomel, E. E. Patton, R. D. Murphey, J. A. Yoder, G. W. Litman, A. Catic, C. T. Amemiya, L. I. Zon, and N. S. Trede. 2003. The zebrafish as a model organism to study development of the immune system. *Adv. Immunol.* **81:**253–330.

Trede, N. S., D. M. Langenau, D. Traver, A. T. Look, and L. I. Zon. 2004. The use of zebrafish to understand immunity. *Immunity* **20:**367–379.

van der Sar, A. M., R. J. Musters, F. J. van Eeden, B. J. Appelmelk, C. M. Vandenbroucke-Grauls, and W. Bitter. 2003. Zebrafish embryos as a model host for the real time analysis of *Salmonella typhimurium* infections. *Cell. Microbiol.* **5:**601–611.

van der Sar, A. M., O. W. Stockhammer, C. van der Laan, H. P. Spaink, W. Bitter, and A. H. Meijer. 2006. MyD88 innate immune function in a zebrafish embryo infection model. *Infect. Immun.* **74:**2436–2441.

van der Wel, N., D. Hava, D. Houben, D. Fluitsma, M. van Zon, J. Pierson, M. Brenner, and P. J. Peters. 2007. *M. tuberculosis* and *M. leprae* translocate from the phagolysosome to the cytosol in myeloid cells. *Cell* **129:**1287–1298.

Vigliano, F. A., R. Bermudez, M. I. Quiroga, and J. M. Nieto. 2006. Evidence for melano-macrophage centres of teleost as evolutionary precursors of germinal centres of higher vertebrates: an immunohistochemical study. *Fish Shellfish Immunol.* **21:**467–471.

Volkman, H. E., H. Clay, D. Beery, J. C. Chang, D. R. Sherman, and L. Ramakrishnan. 2004. Tuberculous granuloma formation is enhanced by a mycobacterium virulence determinant. *PLoS Biol.* **2:**1946–1956.

Voth, D. E., E. E. Hamm, L. G. Nguyen, A. E. Tucker, I. I. Salles, W. Ortiz-Leduc, and J. D. Ballard. 2005. *Bacillus anthracis* oedema toxin as a cause of tissue necrosis and cell type-specific cytotoxicity. *Cell. Microbiol.* **7:**1139–1149.

Wakae, K., B. G. Magor, H. Saunders, H. Nagaoka, A. Kawamura, K. Kinoshita, T. Honjo, and M. Muramatsu. 2006. Evolution of class switch recombination function in fish activation-induced cytidine deaminase, AID. *Int. Immunol.* **18:**41–47.

Wakamatsu, Y., S. Pristyazhnyuk, M. Kinoshita, M. Tanaka, and K. Ozato. 2001. The see-through medaka: a fish model that is transparent throughout life. *Proc. Natl. Acad. Sci. USA* **98:**10046–10050.

Weinholds, E., F. van Eeden, M. Kosters, J. Mudde, R. H. A. Plasterk, and E. Cuppen. 2003. Efficient target-selected mutagenesis in zebrafish. *Genome Res.* **13:**2700–2707.

Westerfield, M. 2000. *The Zebrafish Book. A Guide for the Use of Zebrafish (Danio rerio).* University of Oregon Press, Eugene, OR.

White, R. M., A. Sessa, C. Burke, T. Bowman, J. LeBlanc, C. Ceol, C. Bourque, M. Dovey, W. Goessling, C. E. Burns, and L. I. Zon. 2008. Transparent adult zebrafish as a tool for in vivo transplantation analysis. *Cell Stem Cell* **2:**183–189.

Willett, C. E., A. Cortes, A. Zuasti, and A. G. Zapata. 1999. Early hematopoiesis and developing lymphoid organs in the zebrafish. *Dev. Dyn.* **214:**323–336.

Yaniv, K., S. Isogai, D. Castranova, L. Dye, J. Hitomi, and B. M. Weinstein. 2006. Live imaging of lymphatic development in the zebrafish. *Nat. Med.* **12:**711–716.

Zuasti, A., J. R. Jara, C. Ferrer, and F. Solano. 1989. Occurrence of melanin granules and melanosynthesis in the kidney of *Sparus auratus*. *Pigment Cell Res.* **2:**93–99.

35

Whole Genome Screens in Macrophages

BABAK JAVID AND ERIC J. RUBIN

Pathogens and hosts both make contributions to their interactions. Since pathogens that inhabit the host niche evolved to destroy invading microbes, the macrophage, the ensuing complexities are befitting of any long-term relationship. Probing this relationship from the perspective of both the pathogen (see part V of this volume) and the host can lead to insights into the basic biology of both organisms.

Several approaches have been used to investigate the host side of this interaction. Individual candidate genes, targeted because of their presumed functions, can be interrogated and examined for their effect on pathogen control, or unbiased screens can be performed in which the gene(s) involved are not known a priori. Classical forward genetic screens, usually performed in the mouse as a model organism, have identified many critical host genes responsible for macrophage function (Dietrich, 2001; Kramnik and Boyartchuk, 2002). However, these approaches are dependent on identifying naturally occurring alleles. Thus, the amount of possible genetic diversity in these screens is limited to natural variation. "Whole genome screens" that do not rely on allelic variation provide much more potential diversity and are, therefore, very likely to result in previously unidentified genes and pathways essential for macrophage function.

Although such assessments are not genetic screens, strictly speaking, measuring whole genome transcriptional responses to both pathogenic and nonpathogenic bacteria (Nau et al., 2002; Apidianakis et al., 2005; Ichikawa et al., 2005) has shed light on the host response to infection or to particular virulence factors. Transcriptional profiling can yield clues regarding potentially important pathways. However, transcriptional studies in macrophages rely on key elements of macrophage function being controlled at a transcriptional level. Vital genes that are constitutively transcribed, with no up- or downregulation in response to a stimulus, would not be identified in such screens. However, there are several cases where transcriptional profiling has yielded important insights into macrophage function. For example, a study comparing the profiles of phagocytic S2 cells with those of S2 cells "reprogrammed" to be nonphagocytic identified a type I transmembrane protein, Eater, that was necessary for the phagocytosis of *Escherichia coli* and *Staphylococcus aureus* (Kocks et al., 2005), as well as the efficient uptake of double-stranded RNA (dsRNA) (Ulvila et al., 2006)—a property of S2 cells that makes them attractive to RNA interference (RNAi) screens (see below).

The development of whole genome RNAi by using dsRNA in *Drosophila* cells opened up the potential to screen the entire genome of macrophage-like cells for host proteins that modulate bacterial infection. This chapter will focus on how these screens have been used in in vitro studies with several different pathogens, both intra- and extracellular.

ALTERNATIVE MODELS FOR STUDYING THE ROLE OF HOST GENES

Although most studies examining macrophage function have been performed on human and mouse macrophages, alternative models of macrophage function using nonmammalian cells have proven useful. The amoeba *Dictyostelium discoideum* is a professional phagocyte and has been used extensively as a model for phagocytosis (de Hostos et al., 1993; Maniak et al., 1995) as well as in disease models (Li et al., 2005), although no whole genome screens for genes important in interactions with specific hosts have been performed. While many such screens have been done evaluating pathogens in organisms such as *Arabidopsis* and *Caenorhabditis elegans*, both of these hosts lack macrophages. Thus, many investigators have turned to the fruit fly *Drosophila melanogaster* as a model organism. It is genetically extremely tractable and has the advantage over protozoa of having a developed innate immune system that is highly similar to that of mammals. As well as the archetypal Toll/IMD pathogen-sensing signaling pathway and antimicrobial peptides, fruit flies (unlike *C. elegans*) have hemocytes called plasmatocytes that exhibit macrophage-like behavior (Rizki and Rizki, 1990; Franc et al., 1999). The S2 cell line is thought to be derived from embryonic plasmatocytes and behaves similarly to primary *Drosophila* macrophages (Ramet et al., 2001). Coupled with the fact that *Drosophila*

Babak Javid and Eric J. Rubin, Harvard School of Public Health, 200 Longwood Avenue, Boston, MA 02115.

lack the interferon response, allowing use of (relatively) inexpensive dsRNAs (unlike mammalian cells), and that S2 cells take up dsRNAs passively through the receptor Eater (see above), it is reasonable that whole genome screens for macrophage function using RNAi first became viable in *Drosophila* cell lines.

DEVELOPMENT OF HIGH-THROUGHPUT RNAi SCREENS

Genome-wide RNAi screens in *Drosophila* cell lines were made possible by the publication of the fruit fly genome (Adams, et al., 2000) and the discovery that long dsRNAs added to *Drosophila* cells elicit potent and specific RNAi (Clemens et al., 2000; Hammond et al., 2000) and gene expression knockdown. Several different dsRNA libraries have been constructed (Echeverri and Perrimon, 2006) with varying coverage of the genome. In experiments on macrophage function, two libraries in particular have been exploited. One, with 7,216 dsRNAs of 300 to 600 bp in length (Foley and O'Farrell, 2004), covers most *Drosophila* genes that have been phylogenetically conserved with mammalian genes. Another, with 21,396 dsRNAs with an average length of 400 bp (Boutros et al., 2004), covers almost the entire (>95%) *Drosophila* genome. Despite the additional cost and complexity, the latter has the advantage that no a priori filtering on genes of potential interest has been performed, enabling truly universal coverage.

Screen Design

Any high-throughput assay using RNAi requires significant laboratory investment, with robotics for dispensing dsRNA and cells in media onto 96- or 384-well plates, computing infrastructure, and data management and data acquisition systems (Echeverri and Perrimon, 2006). Some screens into macrophage function have used automated, high-throughput microscopes (Agaisse et al., 2005; Philips et al., 2005), whereas others have required painstaking manual microscopy to verify phenotypes (see below). Because of the large number of samples that need to be processed, automated methods for dsRNA delivery are required. S2 cells have the distinct advantage that they take up dsRNA efficiently merely when bathed in serum-free media (Clemens et al., 2000). This efficient transfection, presumably due to expression of *eater* (Ulvila et al., 2006), obviates any requirement for transfection reagents. After a specified period of time to allow efficient mRNA knockdown and decreased protein expression (usually 3 to 5 days), the target cells are infected by the pathogen of interest. Following infection, the cells are fixed and stained before being visualized. For genome-wide screens to be feasible, a robust, reproducible readout is required for reliable assessment of thousands of samples. Studies examining macrophage function in S2 cells have used fluorescent microbes, either labeled with fluorescein isothiocyanate or directly expressing green fluorescent protein (GFP). GFP-expressing pathogens avoid different levels of fluorescence per microbe, which could lead to over- or underestimates of microbial uptake and/or replication within the macrophage. The specific readout would depend on the phenotype being screened (Fig. 1). For example, phagocytosis can be assayed by monitoring decreased fluorescence after nonphagocytosed bacteria have been removed by washing.

Other studies have screened for vacuolar escape of *Listeria monocytogenes* (Cheng et al., 2005) or other specific phenotypes (e.g., the "spot" phenotype in *Listeria* [Agaisse

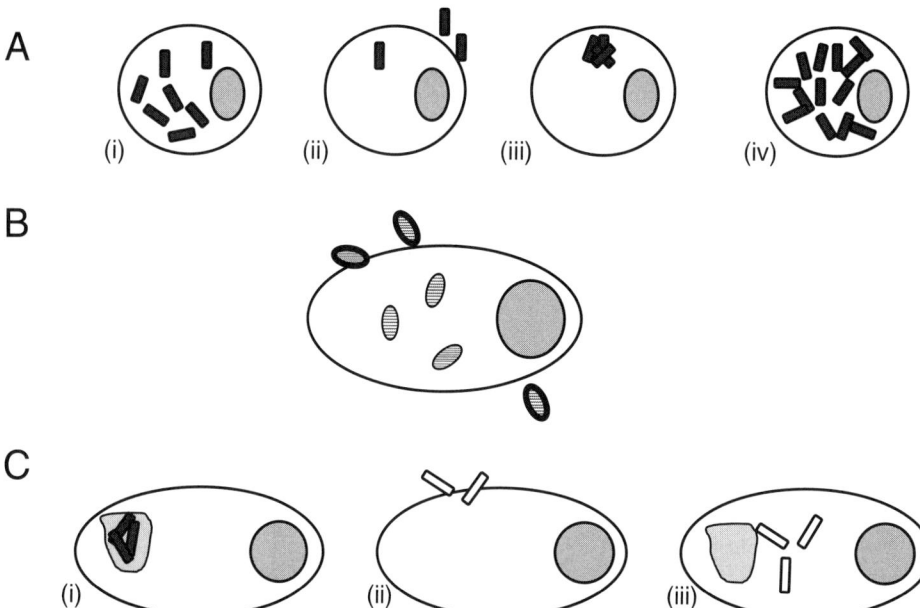

FIGURE 1 Schematic outline of different phenotypes in fluorescent-based macrophage-pathogen screens. (A) *L. monocytogenes* (Agaisse et al., 2005): (i) normal phagocytosis by S2 cells; (ii) decreased phagocytosis after RNAi; (iii) "spot" phenotype (see text); (iv) increased bacterial viability within S2 cell. (B) *C. albicans* (Stroschein-Stevenson et al., 2006). The screen utilized GFP-expressing *Candida*. Nonphagocytosed cells were distinguished from phagocytosed yeast by secondary staining (thick outline) with an anti-*Candida* antibody. (C) *M. fortuitum* (Philips et al., 2005). This screen utilized GFP under a macrophage-activated promoter in the bacteria (i). Nonfluorescence could signify nonphagocytosed cells (ii) or phagosome-vacuolar escape (iii).

et al., 2005]—see below and Fig. 1). Philips et al. (2005) used a GFP-expressing construct, but rather than being driven by a constitutive promoter, it was tied to a promoter specific for macrophage-activated genes (Ramakrishnan et al., 2000). Thus GFP^{lo}-expressing cells could represent lack of phagocytosis by S2 cells, escape from the phagolysosome, or alteration of the phagolysosome environment, leading to altered activity of the promoter (Fig. 1).

Overview of Screen Results

To date, five genome-wide RNAi screens have been performed in *Drosophila* S2 cells to investigate host-pathogen interactions in *Mycobacterium fortuitum*, *L. monocytogenes*, *E. coli* (and *S. aureus*), and *Candida albicans* (Ramet et al., 2001; Agaisse et al., 2005; Cheng et al., 2005; Philips et al. 2005; Stroschein-Stevenson et al., 2006). As discussed above, these screens have utilized different RNAi libraries of varying complexity and genome coverage.

GENES REQUIRED FOR PHAGOCYTOSIS OF PATHOGENS

The first large-scale screen for host factors important in infection used a random library of 1,000 dsRNAs (Ramet et al., 2001). In their study, Ramet and colleagues examined genes required for phagocytosis of *S. aureus* and *E. coli* by S2 cells. They incubated S2 cells with heat-killed, fluorescein isothiocyanate-labeled bacteria after treatment with dsRNA and analyzed the cells by flow cytometry. Trypan blue treatment of cells was used to quench fluorescence of bound, but not internalized bacteria. They discovered 34 genes that caused at least 20% reduction in phagocytosis without causing significant cell death. Inhibition of most of the identified genes resulted in decreased phagocytosis of both microbes. One gene, however, encoding peptidoglycan recognition protein (PRGP-LC), was involved in the phagocytosis of *E. coli* only. Subsequent studies have shown that PRGP-LC is a peptidoglycan recognition receptor involved in the recognition of gram-negative bacteria (Kaneko et al., 2006) and is involved in the *Drosophila* innate signaling IMD pathway (Choe et al., 2005). The related gene product, PRGP-Sc1a, appears to be involved in the recognition of *S. aureus* (Garver et al., 2006).

Among the dsRNAs identified were those that targeted vesicle-trafficking, chaperone, signaling, and actin-interacting genes (Ramet et al., 2001). Notably, a small subset of genes identified were involved in hemocyte development, such as *serpent* and *nejire*, suggesting that these genes had roles not only as differentiation factors, but also in constitutive macrophage functions. Subsequent RNAi studies also examined genes required for the phagocytosis of the other pathogens investigated (Agaisse et al., 2005; Cheng et al., 2005; Philips et al., 2005; Stroschein-Stevenson et al., 2006). These studies used different dsRNA libraries and methodologies. From the subset of common genes identified by these studies, a core set of presumably "generic" genes required for phagocytic functions can be deduced (see Table 1).

A remarkably small group of genes appear to be required for phagocytosis of all the pathogens studied so far, and most of these are concerned with actin remodeling and vesicle trafficking. There could be several reasons for this. As has already been discussed, the dsRNA libraries used in these studies differed in complexity and coverage of the genome. Some genes identified in one screen but not another might not have been represented in the latter library. However, this does not explain all of the variation. For example, in the *C. albicans* screen, some, but not all, Arp2/3 complex genes—involved in trafficking and previously implicated in phagocytosis—were identified, despite all of the Arp2/3 complex being represented in the library (Stroschein-Stevenson et al., 2006). This might have been due to other factors such as the quality of the dsRNA and knockdown, protein and mRNA stability, and half-life and timing of the experiment or, in some cases, differing biological roles for different components.

Pathogen-Independent Genes for Phagocytosis

As can be seen in Table 1, across at least three independent screens, only 15 genes appear to be required for phagocy-

TABLE 1 Genes implicated in phagocytosis by at least three independent screens

Gene function	Gene	Alternative gene name	Function	Pathogens[a]
Involved in vesicle trafficking	CG7961	α-COP	COPI-coated vesicles	Mf, Lm, Ca
	CG1528	γ-COP	COPI-coated vesicles	Mf, Lm, Ca
	CG14813	δ-COP	COPI-coated vesicles	Ec, Sa, Mf, Lm, Ca
	CG3948	ζ-COP	COPI-coated vesicles	Mf, Lm, Ca
	CG4214	Syntaxin-5a	t-SNARE	Mf, Lm, Ca
	CG6625	SNAP	NSF attachment protein	Mf, Lm, Ca
Chaperones and transporters	CG1404	Ran	Nuclear transporter, Ras GTPase	Ec, Sa, Mf, Lm
Signaling molecules	CG9749	Abi	Abl-interacting, signal transducer	Ec, Sa, Mf, Ca
	CG11624	Ubi-p63E	Ubiquitination	Ec, Sa, Mf, Lm, Ca
Actin interacting	CG4027	Actin 5C	Actin	Ec, Sa, Mf, Lm, Ca
	CG12530	Cdc42	Ras small GTPase	
	CG5972	Arc-p20	Arp2/3 complex, actin binding	Mf, Lm, Ca
	CG7558	Arp66B	Arp2/3 complex	Mf, Lm, Ca
	CG9881	p16-Arc	Arp2/3 complex	Mf, Lm, Ca
Lipid metabolism	CG3523	Fatty acid synthase	Lipid metabolism	Ec, Sa, Mf, Lm

[a] Mf, *M. fortuitum*; Lm, *L. monocytogenes*; Ca, *C. albicans*; Ec, *E. coli*; Sa, *S. aureus*.

tosis in general. However, patterns can be discerned from the classes of genes that, when knocked down, caused a decrease in phagocytosis, regardless of the pathogen involved (Table 2). All of the studies identified several components of actin rearrangement and vesicle-trafficking genes as being necessary for phagocytosis. Several signaling pathways, such as the Abl-interacting Abi gene, appear necessary. In addition, ubiquitination plays a key role, and it is not surprising, therefore, that several pathogens have evolved mechanisms to interfere with the ubiquitination pathway (Rytkonen and Holden, 2007).

Pathogen-Specific Genes for Phagocytosis

All but one of the screens (Cheng et al., 2005) utilized two pathogens side by side. This comparison allowed identification of "pathogen-independent" factors required for phagocytosis (see above), but also pathogen-specific factors. As already noted, Ramet and colleagues identified PRGP-LC as an *E. coli* receptor (later confirmed as being a gram-negative receptor). Perrimon's group identified *peste*, a member of the CD36 family of receptors, as necessary for the engulfment of *M. fortuitum* and *L. monocytogenes*, but not *E. coli* or *S. aureus*. However, in mammalian cells, CD36 was sufficient for the uptake of both *M. fortuitum* and *E. coli* (Philips et al., 2005). The function of Peste in vivo has yet to be verified, but it is noteworthy that another CD36 family member in *Drosophila*, Croquemort, is involved in the phagocytosis of apoptotic cell bodies (Franc et al., 1999).

A member of the Tep (thioester-containing protein) family, TepVI (also known as Mcr, macroglobulin complement-related), was identified as a receptor for the pathogenic yeast *C. albicans* but not the nonpathogen *Saccharomyces cerevisiae*, nor bacteria (Stroschein-Stevenson et al., 2006). Tep proteins are related to human α_2-macroglobulin, which is a secreted protein that binds to proteases. Johnson's group also tested the other Tep family members for effects on phagocytosis and found that TepII had a specific effect on *E. coli* uptake, whereas TepIII had a similar effect on *S. aureus* (Stroschein-Stevenson et al., 2006) (Fig. 2).

Vacuolar Escape by *Listeria* and Requirements for Intracellular Growth

Portnoy's group assayed three separate phenotypes in *Legionella* in an elegant study: (i) entry into the cell, (ii) escape from the phagocytic vacuole, and (iii) intracellular growth. A wild-type strain of the bacterium was used in the first screen, much as in the other screens discussed. In keeping with the other studies, a large proportion of the identified genes (42%) were involved with endocytosis and vesicle trafficking. For the second screen, they utilized a *Legionella*-like organism (LLO)-minus strain of *Listeria*. LLO is normally required for the escape of *Listeria* from the phagolysosome into the cytosol. By use of this mutant, the screen identified those host factors that, when knocked down, allowed the LLO requirement to be bypassed. Intriguingly, the group identified many components of the "endosomal sorting complex required for transport" (ESCRT) pathway that were required for vacuolar escape, as well as genes involved in trafficking to and from the multivesicular body (Cheng et al., 2005). This implied that the blocking of the late endosomal pathway bypassed the requirements for LLO.

Finally, Cheng and colleagues used a mutant strain of *Listeria* that lacked a PEST-like sequence normally required for strict compartmentalization of LLO. This mutant is toxic to cells because its action results in host cell permeabilization. By using this mutant, they identified that the host proteasome is involved in the degradation of LLO, and that host cell sphingolipid biosynthesis is also involved in the host inactivation of LLO (Cheng et al., 2005).

FALSE POSITIVES AND FALSE NEGATIVES IN GENOME-WIDE RNAi SCREENS

The power and usefulness of genome-wide screens utilizing RNAi technology depends on the critical assumption that gene silencing is a very specific process. That is, the dsRNA targeting one gene only knocks down that gene and has no other effects. Despite RNAi technology arriving later

TABLE 2 Summary of genome-wide phagocytosis screens

Pathogen(s)	No. of dsRNAs in library	No. of genes necessary for phagocytosis	Comments	Reference
E. coli and *S. aureus*	1,000	34	First attempt at broad coverage; PRGP-LC identified as *E. coli* receptor	Ramet et al., 2001
M. fortuitum and *E. coli*	21,300	86 for *M. fortuitum*; 54 for both bacteria	Vesicle-trafficking and actin rearrangement genes implicated; CD36 family receptor (Peste) required for *M. fortuitum* but not *E. coli* uptake	Philips et al., 2005
L. monocytogenes and *M. fortuitum*	21,300	160 for *Listeria*; 91 for both pathogens	Vesicle-trafficking, signal transduction, and actin cytoskeletal genes; Peste also for *Listeria* uptake	Agaisse et al., 2005
L. monocytogenes	7,216	89	Also examined for escape from intracellular vacuole (further 29 genes, including ESCRT components)	Cheng et al., 2005
C. albicans (and *E. coli*)	7,216	184	Only ~30 genes in common with other screens: ? due to *Candida* as extracellular pathogen; Mcr identified as *Candida* receptor.	Stroschein-Stevenson et al., 2006

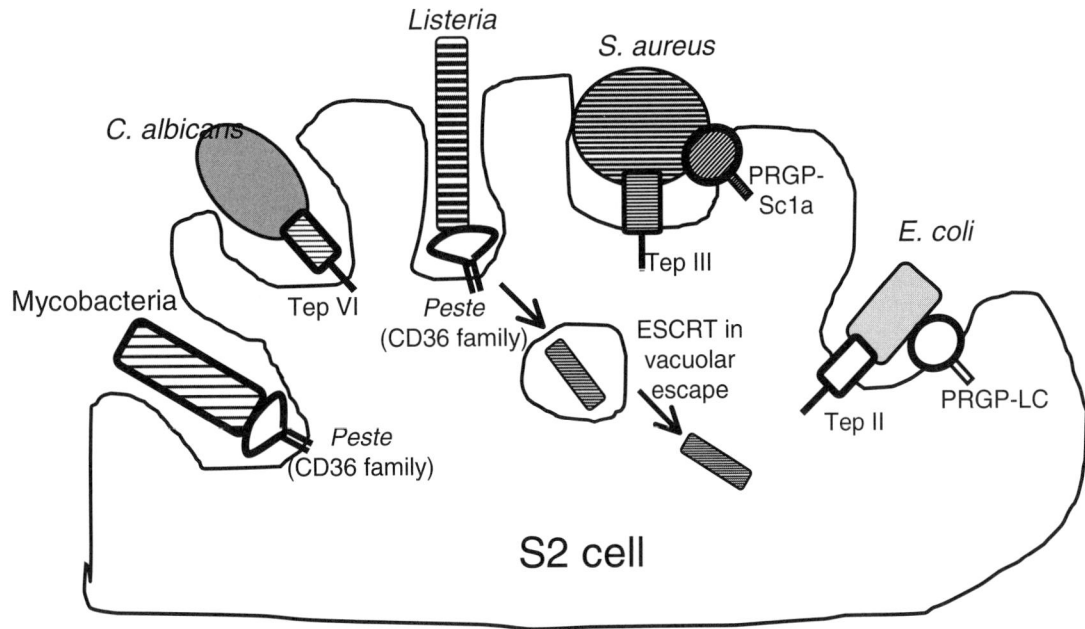

FIGURE 2 Schematic showing pathogen-specific factors in S2 cells identified by RNAi and other screens.

(through short interfering RNA, siRNA) in the mammalian field, it was here that off-target effects (OTEs) were first recognized. Nontargeted mRNAs can be degraded if there is a cross-hybridizing region to the siRNA (Jackson et al., 2003). There can also be nonspecific, sequence-independent effects, such as the interferon response (Echeverri and Perrimon, 2006). Although RNAi had been used more extensively in nonvertebrates, such as C. elegans and D. melanogaster, OTEs were thought to be less of an issue because of the lack of an interferon response and the use of longer dsRNAs rather than siRNAs. In a study of 30 genome-wide screens, however, there was significant evidence of OTEs when there were ≥19 nucleotides that matched unintended targets (Kulkarni et al., 2006). Trinucleotide repeats such as $(CAN)_n$, which are widespread in many fly genes, can also cause false positives (Perrimon and Mathey-Prevot, 2007). One way of minimizing these false positives in genome-wide screens is through the use of redundant dsRNA libraries; an effect is more likely to be real if two or more dsRNAs targeting the gene result in the same phenotype (Echeverri et al., 2006).

Another problem in any large genetic screen is that of false negatives. Of the five genome-wide screens, three used phagocytosis of E. coli either as a primary (Ramet et al., 2001) or secondary screen (Philips et al., 2005; Stroschein-Stevenson et al., 2006). However, the results of the three screens, although broadly confirmatory, identified different sets of genes required for phagocytosis in each case. These were frequently related genes in the same class (e.g., endocytosis-related), validating the approach. But the lack of even near-identical overlap suggests that such broad screens are highly dependent on the experimental conditions (RNAi library utilized, degree of knockdown, length of incubation, etc.) and that only legitimate positive results can be interpreted. Even when exactly the same organism was used, with the same dsRNA library but under slightly different protocols (such as M. fortuitum [Agaisse et al., 2005; Philips et al., 2005]), subtly different results were obtained. Other effects, such as general cellular toxicity and nonviability when essential genes are knocked down, are other possible explanations. In one study of Legionella infection, which targeted the early endosomal pathway of S2 cells (Dorer et al., 2006), the authors found a high number of false negatives if only one dsRNA targeted a particular pathway. If, however, two or more dsRNAs were directed at the same pathway, the false-negative rate dropped significantly.

It should therefore be noted that lack of identification of a gene in such a screen does not imply that it has no role in the phenotype being studied. For example, in the M. fortuitum infection screen of S2 cells, 86 dsRNAs were found to affect infection (Philips et al., 2005). These dsRNAs were utilized in a subsequent screen involving the nonpathogenic Mycobacterium smegmatis. Four were identified to promote intracellular growth of M. smegmatis, and of these, two were involved in the ESCRT machinery. Further analysis and knockdown of ESCRT machinery components confirmed the original result, but the greatest effect was seen when an ESCRT component, TSG101, was knocked down, which was not identified in the original screen (Philips et al., 2008).

THE WAY FORWARD

This chapter has shown how the relatively recent possibilities of whole genome RNAi screens in Drosophila macrophage-like cell lines have provided new insights into macrophage function, in particular the basic cellular processes involved in phagocytosis. Many questions remain, however. For example, it is not yet clear how consistently the results obtained in insect cells will predict mammalian macrophage function. The precise functions of many of the genes identified to date are far from clear in many cases. And although some studies have assessed the nonphagocytic functions of macrophages (Cheng et al. 2005), further questions on the host-pathogen interaction remain to be resolved at the genome-wide level to determine how pathogens subvert and utilize macrophage functions to serve their own needs. New types of RNAi screens, involving gain of function, may also prove promising. One example of this would be to use nonpathogenic strains of bacteria such as M. smegmatis. These bacteria are normally killed by macrophages, and RNAi screens now looking at survival would determine which host genes are required for the killing of these organisms. In turn, it may be that related pathogenic bacteria (M. tuberculosis in this case) have presumably evolved mechanisms to evade such host responses, thereby allowing insights into pathogen as well as host biology.

Further advances in technology are close at hand, with the possibility of examining macrophage function in higher animals and whole organisms. So far, we have only been able to determine which host genes are required for macrophage function in insect cells. Although whole organism RNAi is possible in C. elegans, its lack of phagocytic hemocytes precludes study of macrophage-like function. New developments in Drosophila genetics, however, have raised the possibility of performing whole organism knockdown of genes in intact flies, and there are currently attempts at cultivating fly collections covering the entire genome (http://www.hhmi.org/janelia/). Furthermore, collaborative efforts such as the RNAi Consortium and other groups (http://www.broad.mit.edu/genome_bio/trc) have developed methods for high-throughput siRNA screens in mammalian cells by using lentiviral delivery of short hairpin RNAi (Bernards et al., 2006). Although many thousands of mammalian genes are included in these collections, cost and other considerations have prevented true "genome-wide" coverage.

Therefore, in this as in most branches of biology, the "postgenome" era brings with it more questions than answers. In the case of the macrophage and its interaction with pathogens, addressing these questions will take, as in all good relationships, its own proper time.

REFERENCES

Adams, M. D., S. E. Celniker, R. A. Holt, C. A. Evans, J. D. Gocayne, P. G. Amanatides, S. E. Scherer, P. W. Li, R. A. Hoskins, R. F. Galle, R. A. George, S. E. Lewis, S. Richards, M. Ashburner, S. N. Henderson, G. G. Sutton, J. R. Wortman, M. D. Yandell, Q. Zhang, L. X. Chen, R. C. Brandon, Y. H. Rogers, R. G. Blazej, M. Champe, B. D. Pfeiffer, K. H. Wan, C. Doyle, E. G. Baxter, G. Helt, C. R. Nelson, G. L. Gabor, J. F. Abril, A. Agbayani, H. J. An, C. Andrews-Pfannkoch, D. Baldwin, R. M. Ballew, A. Basu, J. Baxendale, L. Bayraktaroglu, E. M. Beasley, K. Y. Beeson, P. V. Benos, B. P. Berman, D. Bhandari, S. Bolshakov, D. Borkova, M. R. Botchan, J. Bouck, P. Brokstein, P. Brottier, K. C. Burtis, D. A. Busam, H. Butler, E. Cadieu, A. Center, I. Chandra, J. M. Cherry, S. Cawley, C. Dahlke, L. B. Davenport, P. Davies, B. de Pablos, A. Delcher, Z. Deng, A. D. Mays, I. Dew, S. M. Dietz, K. Dodson, L. E. Doup, M. Downes, S. Dugan-Rocha, B. C. Dunkov, P. Dunn, K. J. Durbin, C. C. Evangelista, C. Ferraz, S. Ferriera, W. Fleischmann, C. Fosler, A. E. Gabrielian, N. S. Garg, W. M. Gelbart, K. Glasser, A. Glodek, F. Gong, J. H. Gorrell, Z. Gu, P. Guan, M. Harris, N. L. Harris, D. Harvey, T. J. Heiman, J. R. Hernandez, J. Houck, D. Hostin, K. A. Houston, T. J. Howland, M. H. Wei, C. Ibegwam, M. Jalali, F. Kalush, G. H. Karpen, Z. Ke, J. A. Kennison, K. A. Ketchum, B. E. Kimmel, C. D. Kodira, C. Kraft, S. Kravitz, D. Kulp, Z. Lai, P. Lasko, Y. Lei, A. A. Levitsky, J. Li, Z. Li, Y. Liang, X. Lin, X. Liu, B. Mattei, T. C. McIntosh, M. P. McLeod, D. McPherson, G. Merkulov, N. V. Milshina, C. Mobarry, J. Morris, A. Moshrefi, S. M. Mount, M. Moy, B. Murphy, L. Murphy, D. M. Muzny, D. L. Nelson, D. R. Nelson, K. A. Nelson, K. Nixon, D. R. Nusskern, J. M. Pacleb, M. Palazzolo, G. S. Pittman, S. Pan, J. Pollard, V. Puri, M. G. Reese, K. Reinert, K. Remington, R. D. Saunders, F. Scheeler, H. Shen, B. C. Shue, I. Siden-Kiamos, M. Simpson, M. P. Skupski, T. Smith, E. Spier, A. C. Spradling, M. Stapleton, R. Strong, E. Sun, R. Svirskas, C. Tector, R. Turner, E. Venter, A. H. Wang, X. Wang, Z. Y. Wang, D. A. Wassarman, G. M. Weinstock, J. Weissenbach, S. M. Williams, T. Woodage, K. C. Worley, D. Wu, S. Yang, Q. A. Yao, J. Ye, R. F. Yeh, J. S. Zaveri, M. Zhan, G. Zhang, Q. Zhao, L. Zheng, X. H. Zheng, F. N. Zhong, W. Zhong, X. Zhou, S. Zhu, X. Zhu, H. O. Smith, R. A. Gibbs, E. W. Myers, G. M. Rubin, and J. C. Venter. 2000. The genome sequence of Drosophila melanogaster. Science 287:2185–2195.

Agaisse, H., L. S. Burrack, J. A. Philips, E. J. Rubin, N. Perrimon, and D. E. Higgins. 2005. Genome-wide RNAi screen for host factors required for intracellular bacterial infection. Science 309:1248–1251.

Apidianakis, Y., M. N. Mindrinos, W. Xiao, G. W. Lau, R. L. Baldini, R. W. Davis, and L. G. Rahme. 2005. Profiling early infection responses: Pseudomonas aeruginosa eludes host defenses by suppressing antimicrobial peptide gene expression. Proc. Natl. Acad. Sci. USA 102:2573–2578.

Bernards, R., T. R. Brummelkamp, and R. L. Beijersbergen. 2006. shRNA libraries and their use in cancer genetics. Nat. Methods 3:701–706.

Boutros, M., A. A. Kiger, S. Armknecht, K. Kerr, M. Hild, B. Koch, S. A. Haas, R. Paro, and N. Perrimon. 2004. Genome-wide RNAi analysis of growth and viability in Drosophila cells. Science 303:832–835.

Cheng, L. W., J. P. Viala, N. Stuurman, U. Wiedemann, R. D. Vale, and D. A. Portnoy. 2005. Use of RNA interference in Drosophila S2 cells to identify host pathways controlling compartmentalization of an intracellular pathogen. Proc. Natl. Acad. Sci. USA 102:13646–03651.

Choe, K. M., H. Lee, and K. V. Anderson. 2005. Drosophila peptidoglycan recognition protein LC (PGRP-LC) acts as a signal-transducing innate immune receptor. Proc. Natl. Acad. Sci. USA 102:1122–1126.

Clemens, J. C., C. A. Worby, N. Simonson-Leff, M. Muda, T. Maehama, B. A. Hemmings, and J. E. Dixon. 2000. Use of double-stranded RNA interference in Drosophila cell lines to dissect signal transduction pathways. Proc. Natl. Acad. Sci. USA 97:6499–6503.

de Hostos, E. L., C. Rehfuess, B. Bradtke, D. R. Waddell, R. Albrecht, J. Murphy, and G. Gerisch. 1993. Dictyostelium mutants lacking the cytoskeletal protein coronin are defective in cytokinesis and cell motility. J. Cell Biol. 120:163–173.

Dietrich, W. F. 2001. Using mouse genetics to understand infectious disease pathogenesis. Genome Res. 11:325–331.

Dorer, M. S., D. Kirton, J. S. Bader, and R. R. Isberg. 2006. RNA interference analysis of Legionella in Drosophila cells: exploitation of early secretory apparatus dynamics. PLoS Pathog. 2:e34.

Echeverri, C. J., P. A. Beachy, B. Baum, M. Boutros, F. Buchholz, S. K. Chanda, J. Downward, J. Ellenberg, A. G. Fraser, N. Hacohen, W. C. Hahn, A. L. Jackson, A. Kiger, P. S. Linsley, L. Lum, Y. Ma, B. Mathey-Prevot, D. E. Root, D. M. Sabatini, J. Taipale, N. Perrimon, and R. Bernards. 2006. Minimizing the risk of reporting false positives in large-scale RNAi screens. Nat. Methods 3:777–779.

Echeverri, C. J., and N. Perrimon. 2006. High-throughput RNAi screening in cultured cells: a user's guide. *Nat. Rev. Genet.* **7:**373–384.

Foley, E., and P. H. O'Farrell. 2004. Functional dissection of an innate immune response by a genome-wide RNAi screen. *PLoS Biol.* **2:**E203.

Franc, N. C., P. Heitzler, R. A. Ezekowitz, and K. White. 1999. Requirement for croquemort in phagocytosis of apoptotic cells in Drosophila. *Science* **284:**1991–1994.

Garver, L. S., J. Wu, and L. P. Wu. 2006. The peptidoglycan recognition protein PGRP-SC1a is essential for Toll signaling and phagocytosis of Staphylococcus aureus in Drosophila. *Proc. Natl. Acad. Sci. USA* **103:**660–665.

Hammond, S. M., E. Bernstein, D. Beach, and G. J. Hannon. 2000. An RNA-directed nuclease mediates post-transcriptional gene silencing in Drosophila cells. *Nature* **404:**293–296.

Ichikawa, J. K., S. B. English, M. C. Wolfgang, R. Jackson, A. J. Butte, and S. Lory. 2005. Genome-wide analysis of host responses to the Pseudomonas aeruginosa type III secretion system yields synergistic effects. *Cell. Microbiol.* **7:**1635–1646.

Jackson, A. L., S. R. Bartz, J. Schelter, S. V. Kobayashi, J. Burchard, M. Mao, B. Li, G. Cavet, and P. S. Linsley. 2003. Expression profiling reveals off-target gene regulation by RNAi. *Nat. Biotechnol.* **21:**635–637.

Kaneko, T., T. Yano, K. Aggarwal, J. H. Lim, K. Ueda, Y. Oshima, C. Peach, D. Erturk-Hasdemir, W. E. Goldman, B. H. Oh, S. Kurata, and N. Silverman. 2006. PGRP-LC and PGRP-LE have essential yet distinct functions in the drosophila immune response to monomeric DAP-type peptidoglycan. *Nat. Immunol.* **7:**715–723.

Kocks, C., J. H. Cho, N. Nehme, J. Ulvila, A. M. Pearson, M. Meister, C. Strom, S. L. Conto, C. Hetru, L. M. Stuart, T. Stehle, J. A. Hoffmann, J. M. Reichhart, D. Ferrandon, M. Ramet, and R. A. Ezekowitz. 2005. Eater, a transmembrane protein mediating phagocytosis of bacterial pathogens in Drosophila. *Cell* **123:**335–346.

Kramnik, I., and V. Boyartchuk. 2002. Immunity to intracellular pathogens as a complex genetic trait. *Curr. Opin. Microbiol.* **5:**111–117.

Kulkarni, M. M., M. Booker, S. J. Silver, A. Friedman, P. Hong, N. Perrimon, and B. Mathey-Prevot. 2006. Evidence of off-target effects associated with long dsRNAs in Drosophila melanogaster cell-based assays. *Nat. Methods* **3:**833–838.

Li, Z., J. M. Solomon, and R. R. Isberg. 2005. Dictyostelium discoideum strains lacking the RtoA protein are defective for maturation of the Legionella pneumophila replication vacuole. *Cell. Microbiol.* **7:**431–442.

Maniak, M., R. Rauchenberger, R. Albrecht, J. Murphy, and G. Gerisch. 1995. Coronin involved in phagocytosis: dynamics of particle-induced relocalization visualized by a green fluorescent protein Tag. *Cell* **83:**915–924.

Nau, G. J., J. F. Richmond, A. Schlesinger, E. G. Jennings, E. S. Lander, and R. A. Young. 2002. Human macrophage activation programs induced by bacterial pathogens. *Proc. Natl. Acad. Sci. USA* **99:**1503–1508.

Perrimon, N., and B. Mathey-Prevot. 2007. Applications of high-throughput RNA interference screens to problems in cell and developmental biology. *Genetics* **175:**7–16.

Philips, J. A., M. C. Porto, H. Wang, E. J. Rubin, and N. Perrimon. 2008. ESCRT factors restrict mycobacterial growth. *Proc. Natl. Acad. Sci. USA* **105:**3070–3075.

Philips, J. A., E. J. Rubin, and N. Perrimon. 2005. Drosophila RNAi screen reveals CD36 family member required for mycobacterial infection. *Science* **309:**1251–1253.

Ramakrishnan, L., N. A. Federspiel, and S. Falkow. 2000. Granuloma-specific expression of Mycobacterium virulence proteins from the glycine-rich PE-PGRS family. *Science* **288:**1436–1439.

Ramet, M., A. Pearson, P. Manfruelli, X. Li, H. Koziel, V. Gobel, E. Chung, M. Krieger, and R. A. Ezekowitz. 2001. Drosophila scavenger receptor CI is a pattern recognition receptor for bacteria. *Immunity* **15:**1027–1038.

Rizki, R. M., and T. M. Rizki. 1990. Parasitoid virus-like particles destroy Drosophila cellular immunity. *Proc. Natl. Acad. Sci. USA* **87:**8388–8392.

Rytkonen, A., and D. W. Holden. 2007. Bacterial interference of ubiquitination and deubiquitination. *Cell Host Microbe* **1:**13–22.

Stroschein-Stevenson, S. L., E. Foley, P. H. O'Farrell, and A. D. Johnson. 2006. Identification of Drosophila gene products required for phagocytosis of Candida albicans. *PLoS Biol.* **4:**e4.

Ulvila, J., M. Parikka, A. Kleino, R. Sormunen, R. A. Ezekowitz, C. Kocks, and M. Ramet. 2006. Double-stranded RNA is internalized by scavenger receptor-mediated endocytosis in Drosophila S2 cells. *J. Biol. Chem.* **281:**14370–14375.

Index

A

Acid phosphatase, in *Leishmania* infections, 458
Acidification, of endosomes and phagosomes, 225–233, 313
  in bacterial killing, 230
  consequences of, 230
  dynamic reporters of, 250
  pH maintenance in, 228–230
  principles of, 225–227
  study systems for, 227–228
Acinetobacter baumanni, *Dictyostelium discoideum* interactions with, 505–506
ActA, *Listeria monocytogenes*, 227
Actin, 235–248
  barbed end of, 237, 243–247
  depolymerization of, 244, 410
  dynamics of, 197–198
  filament-bundling proteins for, 244
  filament-severing proteins for, 244–246
  monomeric globular (G-actin), 235
  myosin binding to, 246
  polymerization of, 79–80, 145, 197
    actin nucleation factors in, 241–243
    in chemotaxis, 186
    filament-capping proteins in, 243–244
    kinetics of, 236–237
    in migration, 183
    modulation of, 237–238
    monomer-binding proteins in, 238–241
    profilin-actin complex binding proteins in, 241
    zone model for, 246–247
  recycling proteins for, 244
  reorganization of, in integrin dysfunction, 141, 144
  structure of, 235–236
  tropomyosin binding to, 246
  as two-stranded helical polymer (F-actin), 235–236
Actin depolymerization factor, 246
Actinin, in actin polymerization, 239, 246
Actin-related proteins (Arp2/3 complex), 80, 144–145, 197, 241–243
Activating protein-1, of dendritic cells, 54
Acute monoblastic leukemia, in colony-stimulating factor defects, 7

Acute-phase proteins, of neutrophils, 5
Adapter molecules, 115–116, 153
Adaptive immunity, *Candida* and, 446
Adaptive strategies, of pathogens; see *specific pathogens*, survival mechanisms of
Adenosine, in macrophage activation, 331
Adherence, of phagocytes, integrins in, 146
Adhesion
  of macrophages, 30
  of neutrophils, 9–11
Adhesion molecules, in neutrophil chemotaxis, 8–11
Adhesion receptors, 137, 140
"Adhesion-dependent respiratory burst," 140
Adrenal gland, macrophages in, 28
Adseverin, in actin polymerization, 244
Aerolysin, in inflammasome activation, 161
Aeromonas hydrophila, in inflammasome activation, 161
Aeromonas salmonicida infections, in zebrafish, 532
Affinity changes, in integrins, 140–141
African sleeping sickness, see *Trypanosoma brucei*
Airways, dendritic cells in, 55–56
Albers-Schonberg disease, in chloride channel defects, 229
ALIX protein, in phagosome maturation, 217
Alternatively activated macrophages, 328–333
Alveolar macrophages, 28, 394–395
Alzheimer's disease
  protectins in, 276
  scavenger receptors and, 29, 32
Amastigotes, 454–455
Aminophospholipid translocase, in apoptosis, 342–345
Amoebae
  Legionella pneumophila in, 395–396
  as model for macrophages, see *Dictyostelium discoideum*
AMP, regulatory effect of, 206
Amphiphysin, in endocytosis, 199–200
Angiogenesis, pentraxin in, 176
Animal infections

brucellosis, 427
helminth, see Helminths
protectins in, 274–276
salmonellosis, 381
toxoplasmosis, 463
Annexins, in apoptosis, 342–345, 347, 354
Anthrax lethal toxin, in inflammasome activation, 160
Antigen(s), uptake of, C-type lectin receptors in, 129–130
Antigen presentation
  after apoptosis, 330–331
  in antitumor vaccination, 130–131
  dendritic cells in, 53–55, 60–61
  Fcγ receptor II in, 74
  in helminth infections, 484
  macrophages in, 38, 313–314, 327–328
Anti-inflammatory agents
  activated macrophages as, 484
  in apoptosis, 354–356
  lipid mediators as, see Lipid mediators
Antimicrobial peptides, in *Salmonella* infections, 387
Antioxidants
  in *Candida* infections, 442
  in *Toxoplasma gondii* infections, 469
α1-Antitrypsin, of neutrophils, 5
AP3 protein, defects of, in neutropenia, 6
Apoptotic cells, 341–365
  bridge molecules for, 346–347
  differences between, 345
  "don't eat me" signals on, 349–350
  early, 342
  immune recognition of, 345
  ingestion of
    noninflammatory, 196
    stimuli for, 354
  late, 342
  pathogen interactions with, 356–357
  phagocyte attraction to, 354
  phagocytosis of, 205
  receptors on, 347–349
  recognition of, 342–345
  removal of (apoptosis)
    dendritic cells in, 353–354
    inflammation and immune response suppression during, 354–356
    Legionella pneumophila aversion of, 401

Apoptotic cells (continued)
  macrophages in, 30, 32, 330–331, 352–353
  neutrophils in, 20–21
  in Shigella infections, 413–414
  signal pathways for, 351–352
  surface ligands on, 342–345
  uptake mechanisms for, 350–352
  versus viable cells, 349–350
ARF proteins, function of, 201–204
Arfophilins, in exocytosis, 212
Arginase
  in helminth infections, 478, 480
  versus inducible nitric oxide synthase, 372–373
  in macrophage activation, 328
  in Schistosoma infections, 481–482
Arginine
  arginine synthesis from, 372
  deficiency of, bacterial growth inhibition in, 371
  in Toxoplasma gondii infections, 469
Arp2/3 complex
  in actin nucleation, 80
  in actin polymerization, 197, 241–243, 246
  in integrin signaling, 144–145
  in Shigella infections, 409
Aspergillus, infections due to, in chronic granulomatous disease, 312
Aspergillus fumigatus, pentraxin and, 173–175
Aspirin-triggered lipid mediators, 269, 271–272
Asthma
  in lipoxin deficiency, 271
  pattern recognition receptor defects in, 162
  protectins in, 276
Atg (autophagy) proteins, 420–421
Atherosclerosis
  macrophage receptors and, 31
  monocyte recruitment and retention in, 100
  pentraxin-3 in, 178
  scavenger receptors and, 29
Atopic dermatitis, dendritic cells in, 53
ATP, hydrolysis of, 235
Autoimmune disorders
  pattern recognition receptor defects in, 162–163
  pentraxin-3 in, 178
Autophagosomes, 419–421
Autophagy, 419–425
  definition of, 419
  execution stages of, 419–422
  as immunity mechanism, 423–424, 429
  importance of, 419
  Legionella pneumophila, 400–401
  in macrophages, 38, 313
  types of, 419
Autophagy proteins, 420–421
Avidity, of integrins, 141
Axl proteins, in apoptosis, 346
Azurophil granules, of neutrophils, 5, 19

B
B cells
  activation of, Fc receptors in, 77–78
  dendritic cell interactions with, 57
Bacillary dysentery, Shigella; see Shigella, infections due to

Bacillus subtilis, macrophage scavenger receptors for, 31
Bactericidal permeability-increasing protein, of neutrophils, 19
BAI1 (brain-specific angiogenesis inhibitor), in apoptosis, 348
Barbed end, of actin, 237, 243–247
Bare lymphocyte syndrome type I, 314
Barth syndrome, 7
Bcl-2 proteins
  in autophagy, 422
  in neutrophil apoptosis, 20–21
BDCA antigen, in dendritic cells, 55
Beclins, in autophagy, 420, 422
"Bent state," of integrins, 141
Biofilms, Legionella pneumophila in, 395
Biomarkers, pentraxin-3 as, 177–178
Blau syndrome, 162–163
Bone, remodeling of
  integrins in, 144
  macrophages in, 28
Bone marrow
  dendritic cell progenitors in, 52
  neutrophil mobilization from, 97
Bordetella pertussis, survival mechanisms of, 128–129, 142, 145
Brain-specific angiogenesis inhibitor (BAI1), in apoptosis, 348
Bridge molecules, in apoptosis, 346–347
Brucella, 427–435
  animal diseases due to, 427
  in dendritic cells, 432–433
  discovery of, 427
  endoplasmic reticulum interactions with, 428–429
  infections due to, 427
  intracellular trafficking of, 428
  in monocytes, 432
  in neutrophils, 432
  pathogenesis of, 428–433
  species of, 427
  survival mechanisms of
    immune response evasion, 431–432
    virulence factors, 429–431
Brucella abortus, 427, 429, 430
Brucella canis, 427
Brucella melitensis, 427, 430
Brucella neotomae, 427
Brucella ovis, 427
Brucella suis, 427, 430
Brucella-containing vacuole, 428
Brugia malayi, 479, 484
Bruton's tyrosine kinase, 200
BvfA protein, in Brucella infections, 431
BvR/BvS regulatory system, in Brucella infections, 430
Bystander uptake, in apoptosis, 351

C
Cadherins, in dendritic cells, 53
Calcium
  in chemotaxis, 187–188
  Fc receptor effects on, 82
  lectins dependent on, see C-type lectin receptors
  in Shigella infections, 411
Calcium and diacylglycerol-regulated guanine nucleotide exchange factor I (CalDAG-GEFI), in integrin activation, 142–143
Calnexin, in Brucella infections, 428
Calpains, in chemotaxis, 187–188

Calprotectin, in Candida infections, 442
Calreticulin
  in apoptosis, 344–345, 348
  in Brucella infections, 428
Campylobacter jejuni, peptidoglycan fragments of, 157
Cancer
  C-type lectin receptor-targeted therapy for, 130–131
  macrophages associated with, 335
Candida albicans, 437–451
  genome-wide screen for, 539, 540
  in macrophages, 40, 41
  morphology of, 437–438
  phagocytosis of
    adaptive immunity and, 446
    DC-SIGN in, 439
    dectin-1 in, 439
    killing in, 440–443
    mannose receptors in, 438–439
    phagosome maturation in, 439–440
    soluble mediators derived from, 443–446
  recognition of, 437–438
  survival mechanisms of, 442–443
Cap proteins, in actin polymerization, 239, 244
Carbohydrate recognition domain, of C-type lectin receptors, 123–125
Carcinoembryonic antigen, C-type lectin recognition of, 127
CARD (caspase recruitment domain) proteins
  in macrophage function, 33–34
  in neutrophil function, 14–17
  of pattern recognition receptors, 154
Cardiovascular disease, in lipoxin deficiency, 271
Cartilage-hair hypoplasia, 6
Caspase(s)
  activation of, 158–161, 387
  cell death due to, 158
  function of, 158
  in neutrophils, 14–17
  in Salmonella infections, 387
  in Shigella infections, 413–414
  structure of, 158
Caspase recruitment domain proteins, see CARD (caspase recruitment domain) proteins
CAT proteins, in helminth infections, 482
Catalase, in Candida infections, 442
CATERPILLER proteins, see NOD-like receptors
Cathelicidins, in Candida infections, 442
Cathepsin G
  in Candida infections, 442
  in neutrophil development, 3–4
Cation-independent mannose-6-phosphate receptor, 225
Caveolin, in endocytosis, 213
Cbl protein, in Fc receptor function, 78
CD11b, in Toxoplasma gondii infections, 465
CD11c
  in dendritic cells, 55
  in Salmonella infections, 385
CD14
  in apoptosis, 345
  integrin interaction with, 146
  in macrophage activation, 304
CD16, see Fcγ receptor III (CD16)
CD28/B7, in antigen presentation, 314

CD31, in apoptosis, 349–350
CD32, see Fcγ receptor II (CD32)
CD36
  in apoptosis, 348–349
  function of, 540
  as macrophage receptor, 31–32
CD40
  in antigen presentation, 314
  in *Toxoplasma gondii* infections, 468
CD43, in apoptosis, 345, 349
CD44
  antibodies to, in apoptosis, 354
  in neutrophil action, 9
CD45, in Fc receptor function, 78
CD47 (integrin-associated protein)
  in apoptosis, 350
  integrin interaction with, 145–146
  in neutrophil action, 11
CD62 (selectins), in neutrophil action, 9
CD63, in endocytosis, 212–213
CD64 (Fcγ receptor I), 71–73
CD66, in neutrophil action, 9
CD66b, in endocytosis, 212–213
CD68, in *Toxoplasma gondii* infections, 464
CD89 (Fcα receptor), 75–76
CD200R, in neutrophil function, 17
CD206 (mannose receptor), of macrophages, 33
CD207 (langerin), 53
CD209 (DC-SIGN), see DC-SIGN (CD209)
Cdc42
  in chemotaxis, 186
  function of, 201, 203, 307
  in integrin signaling, 144–145
  in migration, 183
  in *Shigella* infections, 408–409
CED proteins
  in apoptosis, 348, 351
  functions of, 205
Cell death
  caspase-induced, 158, 160, 161
  programmed, 341; see also Apoptotic cells
Cervical cancer, macrophage activation in, 335
Chagas' disease, see *Trypanosoma cruzi*
Chaperone(s), function of, 539
Chaperone-mediated autophagy, 419
Charcot-Marie-Tooth disease, 309
Chediak-Higashi syndrome, 6–7
Chemoattractants, in chemotaxis, 183–186
Chemokine(s), 93–106
  in chemotaxis, 184
  classification of, 94
  of dendritic cells, 52–53, 55–57, 100
  importance of, 93–94
  in integrin activation, 141, 142
  in macrophage activation, 328
  of macrophages, 38–39, 100–101
  of monocytes, 98–100
  of neutrophils
    in apoptosis, 21
    in chemotaxis, 8–9
    in development, 3–4
    overview of, 94
    receptors for, see Chemokine receptors
  in *Salmonella* infections, 385
  structures of, 94
  in zebrafish model, 528
Chemokine receptors, 94–98
  aging of, 96–97

alternative ligands for, 98
ELR+, 95–96
functions of, 101–102
in inflammation, 97–98
of neutrophils, 94–98
neutrophil-selective, 94–95
in *Salmonella* infections, 386
structures of, 94
as therapeutic targets, 102
in *Toxoplasma gondii* infections, 465–466
Chemotaxis, 183–192; see also Migration
  chemoattractants in, 183–186
  C-type lectin receptors in, 128
  of neutrophils, 8–9
  polarity maintenance in, 188
  signaling in, 186–188
  in zebrafish model, 527–529
Chitinase, production of, 328–329
Chlamydia
  peptidoglycan fragments of, 157
  survival mechanisms of, 309
Chloride channels and transporters, in pH maintenance, 228–229
Cholesterol
  macrophage receptors for, 32
  in phagosome maturation, 217
Chronic granulomatous disease, 20, 311–312
Chronic infantile neurological cutaneous and articular syndrome, 163
Citrobacter rodentium, survival mechanisms of, 373
c-Jun N-terminal kinase, in macrophage receptor signaling, 35–37
CLA-1 (SR-B1), as macrophage receptor, 31–32
Class II transactivator, in antigen presentation, 313
Class II-associated Ii peptide (CLIP), in antigen presentation, 314
Clathrin, in endocytosis, 19–200, 78, 213
CLIP (class II-associated Ii peptide), in antigen presentation, 314
"Closed conformation," of integrins, 140–141
Clostridium perfringens, macrophage receptors for, 33
CLRs, see C-type lectin receptors
Clustering, of integrins, 141
Coat protein complex type I (COPI), 216
Cofilin, in actin polymerization, 197, 244
Colitis, *Shigella*; see *Shigella*, infections due to
Collagenase, of neutrophils, 19
Collagenous macrophage receptors, 31
Collectins
  in apoptosis, 347
  in cell adhesion and trafficking, 128
  of macrophages, 34
  structures of, 123
  Toll-like receptor interactions with, 129
Complement, see also Complement receptors
  activation of, 93
  in apoptosis, 345, 347
  function of, 145
  in integrin activation, 141
  pentraxin interaction with, 173–174
Complement receptors
  in apoptosis, 349
  in *Candida* infections, 437–438
  functions of, 204

of neutrophils, 17–19
in opsonization, 34–35
in phagocytosis, 204, 207, 307
in *Salmonella* infections, 387
in zebrafish model, 528
Conformational changes, in integrins, 140–141
Constructive signaling, 196
Cooperativity, in phagocytosis, 78–79
COPI (coat protein complex type I), 216, 539
Coronin
  in actin polymerization, 197
  in *Legionella pneumophila* infections, 500
  in *Mycobacterium* infections, 504
Cortactin, in *Shigella* infections, 409–410
Corticosteroids
  antibodies to, in apoptosis, 354
  in macrophage activation, 331–332
  pentraxin-3 and, 172–173
Coxiella burnetii, survival mechanisms of
  acid resistance, 227, 230
  integrin resistance, 142
  nitric oxide inhibition, 371
Coxsackie virus, Toll-like receptor interactions with, 109
C-reactive protein
  in apoptosis, 347
  as biomarker, 177–178
  complement interaction with, 173–174
  function of, 173
  in innate resistance, 174
  in self-discrimination, 175
  structure of, 171
Crk protein, in *Shigella* infections, 410
Crohn's disease
  macrophage activation in, 335–336
  pattern recognition receptor defects in, 162
Croquemort and croquemort receptor, 349, 514, 516
Cross talk, integrin, 142
Cross-presentation, by macrophages, 38, 314
Cryptococcus neoformans
  Dictyostelium discoideum interactions with, 504
  survival mechanisms of, 128, 219–220
CSF-1 receptor
  in chemotaxis, 184–186
  in zebrafish model, 525, 527
C-type lectin receptors, 123–135
  in antigen uptake, 129–130
  as antitumor therapy targets, 130–131
  carbohydrate recognition domain of, 123–125
  in cell adhesion, 127–128
  in chemotaxis, 128
  cytoplasmic tail motifs of, 125–126
  galactose-binding, 123–125
  as macrophage receptors, 32–33
  mannose-binding, 123–124
  membrane compartment interactions with, 126
  multimerization of, 125
  in pathogen recognition, 128–129
  recognition function of, 126–128
  in self-recognition, 126–128
  structures of, 123–126
  tolerizing function of, 126–128
  Toll-like receptor interactions with, 129
Cumulus oophorus, pentraxin-3 in, 172, 176–177

548 ■ INDEX

Cutaneous leishmaniasis, 456; see also
  Leishmania major
Cyclic β-1,2-glucans, in Brucella
  infections, 429–430
Cyclic neutropenia, 6
Cyclooxygenase-2, function of, 270–271
Cyclophilins, in Toxoplasma gondii
  infections, 466
Cystatins, of neutrophils, 5
Cystic fibrosis, in lipoxin deficiency, 271
Cystic fibrosis transmembrane
  conductance regulator, defects of,
  229
Cytokine(s), see also individual cytokines
  anti-inflammatory, 38–39
  definition of, 281
  of dendritic cells, 281–299
    in immune response regulation, 285–291
    secretion of, 282–285
  in Fcγ receptor II activation, 74
  in helminth infections, 478–479
  immunosuppressive, 290–291
  in integrin activation, 141
  in macrophage activation, 28, 327–328
  in macrophage deactivation, 28
  in macrophage development, 27
  in macrophage receptor signaling, 37
  in macrophage regulation, 289–291
  of macrophages, 38–39, 281–299
    in immune response regulation, 285–291
    secretion of, 282–285
  proinflammatory, 38–39, 287–288, 314–316
  in Salmonella infections, 383
Cytokine signaling-1 (SOCS-1), in
  Toxoplasma gondii infections, 471
Cytomegalovirus
  pentraxin interaction with, 174–175
  Toll-like receptor interactions with, 109
Cytoplasmic tail motifs, of C-type lectin
  receptors, 125–126
Cytoskeleton, assembly of, Fc receptors in,
  79–80

D

DAI (DNA-dependent activator of IFN-
  regulatory factors), in nucleic acid
  sensing, 155
Danon's disease, 309
DCAL proteins
  in cell adhesion and trafficking, 127
  structures of, 126
DCIR (dendritic cell immunoreceptor),
  126
DC-SIGN (CD209)
  in antigen uptake, 129–130
  as antitumor therapy target, 130–131
  in Candida infections, 439
  in cell adhesion and trafficking, 127
  cytoplasmic tail motif of, 125
  function of, 123–125
  of macrophages, 32–33
  membrane compartment interactions
    with, 126
  multimerization of, 125
  pathogen evasion of, 128
  in self and nonself recognition, 126–127
  Toll-like receptor interactions with, 129
Deactivation, of macrophages, 28

DEC-205
  in antigen uptake, 129
  as antitumor therapy target, 130–131
  cytoplasmic tail motif of, 125
Dectin(s)
  in Candida infections, 439–441, 444
  in cell adhesion and trafficking, 127
  function of, 125, 205
  as macrophage receptor, 33
  Toll-like receptor interactions with, 129
Defensins
  in Candida infections, 441–442
  of dendritic cells, 55
  in Shigella infections, 405–406
Degranulation, of neutrophils, 19–20, 384
Del-1 (developmental endothelial locus
  1), in apoptosis, 346
Dendritic cell(s)
  activation of, Toll-like receptors in, 116
  in antitumor vaccine, 130–131
  in apoptosis, 353–356
  Brucella interactions with, 432
  chemokines of, 100
  chemotaxis by, see Chemotaxis;
    Migration
  cytokines of, 281–299
    effects of, 289–291
    in immune response regulation, 285–291
    secretion of, 282–285
  maturation of, 176
  myeloid, 51–68
    in adaptive response, 59–61
    dermal, 53
    development of, 52
    discovery of, 51
    distribution of, 51
    function of, 58–61
    in innate response, 59
    intestinal, 53–55
    in lymph nodes, 56–57
    in microbial uptake, 58–59
    monocyte development into, 52–53
    in natural killer cell activation, 59
    as pentraxin-3 source, 172–173
    pulmonary, 55–56
    of skin, 53
    splenic, 58
    in T-cell activation and regulation, 59–61
    thymic, 57–58
    types of, 51–52
  pattern recognition receptors of, 282
  phagocytic activity of, 59
  plasmacytoid
    interferon production in, 155
    Toll-like receptors of, 283
  Salmonella enterica serovar Typhimurium
    interactions with, 385–386
  Toxoplasma gondii interactions with,
    464–465
  in zebrafish model, 529
Dendritic cell immunoreceptor (DCIR),
  structure of, 126
Dendritic cell-specific intercellular
  adhesion molecule-3-grabbing
  nonintegrin, see DC-SIGN
  (CD209)
Dengue virus, sensing of, 154
Dermal dendritic cells, 53
Dermal macrophages, 28
Dermatitis
  atopic, dendritic cells in, 53

pattern recognition receptor defects in,
  162
Developmental endothelial locus 1, in
  apoptosis, 346
Dexamethasone, in macrophage
  activation, 331–332
Diacylglycerol, in phagocytosis, 202–203
Diarrhea, Shigella; see Shigella, infections
  due to
Dictyostelium discoideum, as macrophage
  model, 493–512, 537
  in Acinetobacter baumanni infections,
    506
  in Cryptococcus neoformans infections,
    504
  genetic analysis of, 498–499
  in Klebsiella pneumoniae infections, 499
  in Legionella pneumophila infections,
    499–502
  in Mycobacterium infections, 502–504
  nonpathogenic organism interactions
    with, 497–498
  in Pseudomonas aeruginosa infections,
    505
  rationale for, 493–498
  in Vibrio cholerae infections, 505–506
  virulence factors and, 504–506
Diphtheria toxin, translocation of, 230
DNA
  nitric oxide action on, 370–371
  sensing of, 155
  Toll-like receptor recognition by, 113–115
DNA-dependent activator of IFN-
  regulatory factors (DAI), in
  nucleic acid sensing, 155
DOCK proteins
  in apoptosis, 351
  in neutrophil chemotaxis, 8
  in Shigella infections, 410
Docosahexaenoic acid, lipid mediators
  derived from; see Lipid mediators
"Don't eat me" signals, in apoptosis, 349–350
Down syndrome cell adhesion molecule,
  Drosophila, 515
Draper, in apoptosis, 348
Draper receptor, Drosophila, 517
Drosophila, phagocytosis in, 513–521, 537–539
  innate immunity and, 518–520
  proteomic analysis of, 513–514
  receptors in, 514–518
  Toll-like receptors in, 107–108
DrrA (SidM) protein, in Legionella
  pneumophila infections, 400
Dscam receptor, Drosophila, 515, 517
Dynamins, in endocytosis, 78, 199–200,
  213
Dysentery, Shigella; see Shigella, infections
  due to
Dyskeratosis congenita, 6

E

Early endosomal antigen 1 (EEA1), in
  phagosome maturation, 214–216,
  257, 259
Eat, permission to, 206–207
Eater receptor, Drosophila, 514–516
Ectodermal dysplasia, 306
EDA gene, mutations of, 306

Edwardsiella infections, in zebrafish, 531–532
EEA1 (early endosomal antigen) protein-1, in phagosome maturation, 214–216, 257, 259
Efferocytosis, 342
Ehrlichia chaffeensis, growth retardation of, in arginine deficiency, 371
Eicosapentaenoic acid, lipid mediators derived from, *see* Lipid mediators
ELA2 gene mutations, in neutropenia, 6
Elastase(s)
   in *Candida* infections, 442
   defects of, in neutropenia, 6
   in neutrophil development, 3–4
Elongation, in actin polymerization, 237
ELR chemokines, 95–96
Ena proteins
   in actin polymerization, 241
   function of, 95
Endocytosis, 199, 212–213
   phagocytosis interactions with, 226
   Salmonella-containing vacuole and, 388
   signaling requirements for, 78
Endomembrane dynamics, 81
Endoplasmic reticulum
   Brucella interactions with, 428–429
   Legionella pneumophila interactions with, 398, 500
   in phagosome formation, 199, 210
Endosomal TLR (TLR9), 113–115
Endosome(s)
   acidification of, 225–233
   in phagosome formation, 198–199
Endosome-associated complex required for transport (ESCRT), in phagosome maturation, 216–217
Endothelial cell(s)
   as pentraxin-3 source, 172
   phagocyte adherence to, 146
Endothelial cell-selective adhesion molecule, in neutrophil action, 10
Endothelial nitric oxide synthase, 367–368
Environmental pathogens, *Legionella*, 393
Epac (exchange proteins directly activated by cyclic AMP), 206
Epithelial cells, *Shigella* invasion of, 411
ERK (extracellular-related kinase), *see* Extracellular-related kinase (ERK)
Erythrocytes, for phagocytosis studies, 197
Escherichia coli
   chemotaxis of, in zebrafish model, 527–528
   genome-wide screen for, 539, 540
   macrophage scavenger receptors for, 31, 32
   peptidoglycan fragments of, 157
   Shigella similarity to, 407
   survival mechanisms of, 371
   Toll-like receptor recognition of, 111–112
ESCRT (endosome-associated complex required for transport), in phagosome maturation, 216–217
Evasive characteristics, of pathogens; *see specific pathogens*, survival mechanisms of
Exchange proteins directly activated by cyclic AMP (Epac), 206
Exocyst, 212
Exocytosis
   compartments involved in, 210
   fusion mechanisms for, 210–212
   molecules involved in, 212
   during phagocytosis, 212–213
"Extended state," of integrins, 141
Extracellular-related kinase (ERK)
   function of, 307
   in macrophage activation, 333–334
   in macrophage receptor signaling, 35–37
   in *Toxoplasma gondii* infections, 470
Extravasation, of neutrophils, 9–11

F
Familial cold autoinflammatory syndrome, 163
Familial Mediterranean fever, 163
Fanconi anemia, 6
Fasciola hepatica, macrophage activation and, 484
Fat body, signaling to, 518–520
Fatty acids, macrophage receptors for, 32
Fc receptors, 71–92, 196; *see also* Fcγ receptors
   function of, 200–204
   inhibitory, 204
   in *Toxoplasma gondii* infections, 469
Fcα receptor (CD89), 75–76
Fcγ receptor(s)
   in activation and deactivation, 76–78
   calcium levels and, 82
   in *Candida* infections, 441
   complexes of, 79
   in cytoskeletal assembly, 79–80
   genes of, 17–19
   in integrin activation, 141–142, 145
   in membrane remodeling, 80–81
   in NADPH oxidase activation, 81
   neutrophil interaction with, 17–19
   p38 pathway, 35–37
   in phagocytosis enhancement, 78–79, 307
   in phospholipase activation, 81–82
   in *Salmonella* infections, 387
   in serine/threonine kinase activation, 82
   structures of, 71–75
   types of, 17
Fcγ receptor I (CD64)
   function of, 72–73
   structure of, 71–72
Fcγ receptor II (CD32)
   in activation and deactivation, 77–78
   activation of, 74
   cells expressing, 17–18
   function of, 73–74
   genes of, 17–19
   isoforms of, 73–74
   in phagocytosis enhancement, 78–79
   polymorphisms of, 18–19
   structure of, 73–74
Fcγ receptor III (CD16)
   cells expressing, 17–18
   function of, 75
   genes of, 17–19
   in phagocytosis enhancement, 79
   polymorphisms of, 18–19
   structure of, 74–75
Fcγ receptor IV, 76
Fcγ receptor n (neonatal), 76
Fertility, female, pentraxin in, 176–177
Fgr protein, function of, 200

Fibroblast growth factor 2, pentraxin-3 interactions with, 176
Fibronectin, integrin interaction with, 138, 140, 142
Fibrosis, in *Schistosoma* infections, 481–482
Ficolins
   in apoptosis, 347
   of neutrophils, 5
Filamin
   in actin polymerization, 239, 246
   in integrin signaling, 144
Filariasis, 98, 477
Filopodia, 145
Fish, as model; *see* Zebrafish model
Fish oils, lipid mediators derived from; *see* Lipid mediators
FIZZ proteins
   in helminth infections, 478
   in macrophage activation, 328
Flagella
   Leishmania, 454–455
   Shigella, 407
   Trypanosoma cruzi, 454–455
Flagellin
   in caspase activation, 159
   as Toll-like receptor agonist, 109
Flavocytochromes
   in macrophage function, 310
   in neutrophil function, 19–20
Flip-flop mechanism, in apoptosis, 342–345
Floppases, in apoptosis, 343
Flotilin, in *Brucella* infections, 428
Fluid-phase pattern recognition receptors, *see* Pentraxin-3
Fluorescence radiometry, in acidification measurement, 227–228
Fluorescent resonance energy transfer, in phagosome studies, 250–251
Focal adhesion kinase, in integrin signaling, 143
Formins
   in actin nucleation, 80
   in actin polymerization, 243
Fractalkine (neurogactin), 8–9, 94
Francisella tularensis
   in inflammasome activation, 160
   inflammasome interaction with, 158
   interferon response from, 156
   in macrophages, 40, 41
   pattern recognition signaling in, 162
   survival mechanisms of, 153, 162
Frustrated phagocytosis, 213
Fungi, *see also Candida albicans*
   pathogen-associated molecular patterns of, 11
Fusion, in exocytosis, 210–212

G
Galectins, in *Candida* infections, 441
Gamma-activating factor, 303
Gas6 (growth arrest specific 6) protein, in apoptosis, 346
Gastrointestinal tract, dendritic cells in, 53–55
GCP-2 protein, function of, 95
Gelatinase, of neutrophils, 19
Gelsolin, in actin polymerization, 239, 244–246
Genome-wide screens, in macrophages, 537–543

Genome-wide screens, in macrophages (*continued*)
  alternative models for, 537–538
  false positives and false negatives in, 540–541
  genes required for phagocytosis, 539–540
  high-throughput methods for, 538–539
  intracellular growth and, 540
Gfi1 defects, in neutropenia, 6
Giardia lamblia
  growth retardation of, in arginine deficiency, 371
  intracellular amastigote form of, 454
Glomerulonephritis, in integrin dysfunction, 141–142
Glucocorticoids, *see* Corticosteroids
Glycan phosphatidylinositol-anchored proteins, integrin interaction with, 146
Glycogen storage disease, type 1b, neutropenia in, 7
Glycolipids, as Toll-like receptor agonists, 109
β_2-Glycoprotein, in apoptosis, 346
gp63, in *Leishmania* infections, 458
G-protein-coupled receptors
  in chemotaxis, 184
  in integrin activation, 141
  in *Toxoplasma gondii* infections, 466–467
GRA proteins, in *Toxoplasma gondii* infections, 464
Granules
  neutrophilic, 5, 19, 384
  in phagosome formation, 199
Granulins, in zebrafish model, 525
Granulocyte(s), types of, 3
Granulocyte colony-stimulating factor
  function of, 7, 96–97
  in neutrophil development, 3–4
Granulocyte-monocyte colony-stimulating factor
  in chemokine stimulation, 97–98
  function of, 7
  integrin interactions with, 140, 141
  in neutrophil development, 3
Granulomas
  formation of, tumor necrosis factor-α in, 315
  in helminth infections, 480–482
  in zebrafish, 530
Green fluorescent protein, 227–228
Gro proteins, function of, 95
Growth arrest specific 6 protein, in apoptosis, 346
Growth factors, in autophagy, 422
GTPase(s)
  in integrin signaling, 144–145
  in neutrophil chemotaxis, 8–9
  in phagocytosis, 201, 203
GTPase-activating proteins, in phagosome maturation, 216
Guanine-nucleotide exchange factors, 201, 307

H

Haemophilus influenzae, macrophage scavenger receptors for, 31
HAX1 gene mutations, neutropenia in, 6
hCAP, in *Candida* infections, 442
Hck protein
  in Fc receptor function, 76
  function of, 200
Heart disease, pentraxin-3 in, 178
Heat shock protein 27, in actin polymerization, 238–241, 247
Helicobacter pylori
  peptidoglycan fragments of, 157
  survival mechanisms of, 128, 373
Helminths, 477–490
  immune response to, 477–478
  infections due to
    effector mechanisms in, 479–481
    granuloma formation in, 480–482
    reactive nitrogen intermediates in, 481–482
    tissue repair in, 482–483
  macrophage interactions with, 328, 478–484
  public health significance of, 477
  types of, 477
Hematopoiesis, in zebrafish model, 525, 529
Hemocytes, in *Drosophila*, 513, 518–520
Hepatitis C virus
  in macrophages, 42
  survival mechanisms of, 161
Hepatocyte growth factor (macrophage-stimulating protein), 331
Hepatocyte growth-factor-regulated tyrosine kinase substrate, in phagosome maturation, 216
Hepcidin, in macrophage function, 38
Hermansky-Pudlak syndrome, type 2, 6
Herpes simplex virus, Toll-like receptor interactions with, 109
Heterodimer structure, of integrins, 137–139
Histone phosphorylation, in macrophage activation, 334
Histoplasma capsulatum
  infections due to, in interferon-γ defects, 303
  survival mechanisms of, 145
HODE lipid peroxidation products, in atherosclerosis, 100
Host factor I protein, in *Brucella* infections, 431
Human immunodeficiency virus
  sensing of, 154
  survival mechanisms of, 128–129, 357
Human leukocyte antigens, in antigen presentation, 313–314
Human papillomavirus, macrophage activation due to, 335
Humoral receptors, of macrophages, 34
hVPS34 protein
  in autophagy, 420, 422–423
  in phagosome maturation, 214–215
Hydatidiform mole, pattern recognition receptor defects in, 163
Hypoxia, macrophage recruitment in, 101

I

I (inserted) domain, of integrins, 137, 139
ICEBERG protein, in neutrophil function, 14–15
ICE-protease activating factor (IPAF), in neutrophil function, 14
Icm proteins, in *Vibrio cholerae* infections, 505–506
IFN-regulatory factor family, 284–285
IκB proteins
  activation of, 156–157
  function of, 305–306
  in macrophage activation, 306
  in Toll-like receptor signaling, 14
  in *Toxoplasma gondii* infections, 470
Immune response, suppression of, in apoptosis, 354–356
Immunity-related GTPases, in *Toxoplasma gondii* infections, 467–468
Immunodeficiency
  in IκBα defects, 306
  in interferon-γ defects, 302–303
  in interleukin-12 defects, 316
  in IRAK defects, 305
  in Janus kinase defects, 303
  in lysosomal defects, 313
  in major histocompatibility complex defects, 314
  in NEMO defects, 305–306
  in respiratory burst defects, 311–312
  in STAT defects, 303
  in Toll-like receptor defects, 305
  in Tyk2 deficiency, 316
  in Wiskott-Aldrich syndrome, 307–308
Immunoglobulin A, production of, 55
Immunoglobulin G
  Fc receptors of, *see* Fc receptors; Fcγ receptors
  in opsonization, 34–35
  receptors for, neutrophil interaction with, 17–19
Immunophagy, 423–424, 429
Immunoreceptor tyrosine activation motif (ITAM), *see* ITAM (immunoreceptor tyrosine activation motif)
Immunoreceptor tyrosine inhibitory motif (ITIM)
  in C-type lectin receptors, 125–126
  in Fcγ receptors, 77–78, 204
  in permission to eat, 207
INCA protein, in neutrophil function, 14–15
Incontinentia pigmenti, 305–306
Inducible Ikk-B kinase, in nucleic acid sensing, 154
Inducible nitric oxide synthase
  versus arginase, 372–373
  dimerization of, 367
  gene transcription of, 368–370
  induction of, 367–370
  localization of, 370
  nitric oxide produced by, 370–372
  regulation of, 367–370
  in *Schistosoma* infections, 481–482
  stability of, 370
  in *Toxoplasma gondii* infections, 468–469
Inflammasomes
  activation of, 158–161
  mutations involving, 163
Inflammation
  arginase promotion of, 373
  chemokines causing, 96–98
  C-type lectin receptor action in, 127–129
  cytokine secretion in, 38–39
  dendritic cell recruitment to, *see* Dendritic cell(s)
  lipid mediators in, *see* Lipid mediators
  macrophage function in, 100–101
  pathways for, 267–268
  pentraxin in, 175

phagocyte adherence in, integrins in, 146
sensing of, *see* Sensing, of pathogens
in sepsis syndrome, 107, 111–113
suppression of
in apoptosis, 354–356
in *Candida* infections, 445–446
Inflammatory bowel disease, macrophage activation in, 335–336
Influenza virus
chemokine interactions with, 101
pentraxin interactions with, 174
sensing of, 154
Toll-like receptor interactions with, 109–110
Ingestion
of apoptotic cells
"don't eat me" signals in, 349–350
noninflammatory, 196
permission to eat in, 206–207
stimuli for, 354
by macrophages, 307
self-, *see* Autophagy
Innate immunity, pentraxin in, 174–175
Inosine, in macrophage activation, 331
Inside-out signaling, in integrin activation, 140–143
Integrin(s), 137–152
activation of, inside-out signaling in, 140–143
in adherence, 146
in apoptosis, 349
avidity of, 141
clustering of, 141
conformational changes in, 140–141
in dendritic cells, 53
expression of, 138–140
extracellular stimuli for, 141–142
in Fc receptor function, 78–79
intracellular stimuli for, 142–143
in migration, 146–147
in neutrophil chemotaxis, 8–9
in osteoclast biology, 144
plasma membrane protein interactions with, 145–146
receptors for, in neutrophil action, 9–10
structures of, 137–140
types of, 137
"Integrin activation," 140
Integrin-associated protein, in neutrophil action, 11
Intercellular adhesion molecule-1
integrin affinity for, 141, 146
in neutrophil action, 9, 10
Interferon(s)
in *Brucella* infections, 432
classes of, 302
functions of, 286
in inducible nitric oxide synthase gene transcription, 368
as macrophage activation primers, 301–302
in macrophage function, 38
of macrophages, 282–283
in pathogen sensing, 154–156
in pentraxin regulation, 172
in phagosomes, 253, 255, 258
production of, 286
in Toll-like receptor signaling, 115–116
Interferon-γ
in apoptosis, 354
function of, 289, 302–303
in helminth infections, 478

in macrophage activation, 302–303, 327
production of, 289
receptors for, 302–303
structure of, 302
in *Toxoplasma gondii* infections, 467, 470–471
Interferon-regulatory factor family, 153, 157, 284–285
Interleukin(s)
in *Brucella* infections, 432
chemoattractant function of, 93–94
in helminth infections, 477–479
in macrophage activation, 28, 327–328
in macrophage development, 27
of macrophages, 38–39, 282–283
in *Salmonella* infections, 386
in *Shigella* infections, 413
Interleukin-1, 287–288
Interleukin-1 receptor, in Toll-like receptor signaling, 13
Interleukin-1β, in *Shigella* infections, 414–415
Interleukin-2, of dendritic cells, 59
Interleukin-4
function of, 289–290
in macrophage activation, 328–329, 331–332
production of, 289–290
Interleukin-6, 287–288
Interleukin-8
function of, 94–95
in integrin activation, 141
of neutrophils, 95
structure of, 94–95
Interleukin-10
function of, 290–291
in leishmaniasis, 335
in macrophage regulation, 333–334
of macrophages, 282
overproduction of, 329, 332
pentraxin interactions with, 172
in *Toxoplasma gondii* infections, 469
Interleukin-12
defects of, 316
downmodulation of, in macrophages, 333
function of, 286–287, 316
in macrophage activation, 327
production of, 286–287
in *Schistosoma* infections, 481
structure of, 316
in *Toxoplasma gondii* infections, 465–466, 469
Interleukin-13
function of, 289–290
in macrophage activation, 328–329
production of, 289–290
Interleukin-18, in *Shigella* infections, 414–415
Interleukin-23
function of, 287
in inflammatory bowel disease, 335–336
production of, 287
Interleukin-27, 287
Internalization, of particles, 308–309
Interpretive signaling, 196, 207
Intestines, dendritic cells in, 53–55
Ipa proteins, in *Shigella* infections, 407–408, 410–413
IpgB proteins, in *Shigella* infections, 410
IRAK proteins
defects of, 305

in Toll-like receptor signaling, 13–14, 115, 305
IRF3, in Toll-like receptor signaling, 115–116
IRG proteins (immunity-related GTPases), in *Toxoplasma gondii* infections, 467–468
Ischemic heart disorders, pentraxin-3 in, 178
Isocitrate lyase, of *Mycobacterium tuberculosis*, 259
ITAM (immunoreceptor tyrosine activation motif)
in C-type lectin receptors, 125
in Fcγ receptors, 76–81
in tyrosine kinases, 143–144
ITIM (immunoreceptor tyrosine inhibitory motif)
in C-type lectin receptors, 125–126
in Fcγ receptors, 77–78, 204
in permission to eat, 207

J
Jak proteins
in interferon signaling, 302–303
in *Toxoplasma gondii* infections, 470–471
JAMs (junctional adhesion molecules), in neutrophil action, 10–11
Janus kinases
in interferon signaling, 302–303
in macrophage receptor signaling, 37
Junctional adhesion molecules (JAMs), in neutrophil action, 10–11

K
Kidney disorders
in lipoxin deficiency, 271
protectins in, 276
Kil proteins, function of, 499
Kinetoplasts, 455
Kiss-and-run hypothesis, for phagosome maturation, 214
Klebsiella pneumoniae
Dictyostelium discoideum interactions with, 499
macrophage receptors for, 33
pentraxin interaction with, 174–175
Kostmann syndrome, 6
KpOmpA, pentraxin interaction with, 175
Kupffer cells, 28

L
Lactoferrin
in *Candida* infections, 442
of neutrophils, 19
Lamellipodia, 235
in chemotaxis, 188
in migration, 183
Lamina propria
dendritic cells in, 53–55
macrophages in, 28
LAMPs (lysosomal-associated membrane proteins), 199, 214, 217, 250, 309
Langerhans cells, 28, 53
Langerin, 53, 129
Late endosomes, 214
LBPA protein, in phagosome maturation, 217
LC3 proteins, in autophagy, 420–421

LDL receptor-related proteins (LRPs), in apoptosis, 345, 348
Lectin(s), see also C-type lectin receptors
 in helminth infections, 479
 of neutrophils, 5
Lectin-like oxidized low-density lipoprotein receptor 1 (LOX-1), in apoptosis, 349
Legionella longbeachae, 394
Legionella micdadei, 394
Legionella pneumophila, 393–403
 in amoebae, 494–495
 autophagy and, 400–401, 500–501
 Dictyostelium discoideum interactions with, 499–502
 in environment, 393–394
 epidemiology of, 394–395
 genome of, 395
 infections due to
  burden of, 393
  clinical features of, 394
  diagnosis of, 394
  mouse model of, 401–402
  outbreaks of, 393–394
  treatment of, 394
 inflammasome interaction with, 158, 161
 interferon response from, 156
 intracellular growth of, 500–501
 life cycle of, 395–396
 macrophage scavenger receptors for, 31
 in macrophages, 39
 survival mechanisms of
  acid resistance, 227, 230
  apoptosis resistance, 401
  DNA modulation, 395
  effector secretion, 153
  endoplasmic reticulum residence, 398
  lipopolysaccharide, 397–398
  macrophage membrane disruption, 396–397
  phagosome maturation interference, 309
  type II secretion system, 396
  type IV secretion system, 398–400
  virulence factors, 395
 trafficking of, 499–500
Legionnaire's disease, see *Legionella pneumophila*, infections due to
Leishmania
 flagellated form of, 455–456
 intracellular amastigote form of, 454–455
 macrophage activation due to, 334–335
 in parisitophorous vacuole, 458–459
 phagocyte interactions with, 456
 release of, 459
 surface glycobiology of, 458
 survival mechanisms of
  apoptosis avoidance, 356
  integrin resistance, 145
 Toll-like receptor interactions with, 110
Leishmania amazonensis, 457
Leishmania donovani
 infections due to, 456
 in parisitophorous vacuole, 459
 phagocyte interactions with, 456–457
 recognition of, 457
 surface glycobiology of, 458
 survival mechanisms of, 309, 371
Leishmania infantum, 456
Leishmania major
 phagocyte interactions with, 456–457
 surface glycobiology of, 458
 survival mechanisms of, 373
Leishmania mexicana
 infections due to, 456
 in macrophages, 40, 41
 in parisitophorous vacuole, 459
 surface glycobiology of, 458
 survival mechanisms of, 227, 230
Leishmania tropica, 459
Lethal toxin, anthrax, in inflammasome activation, 160
Leucine-rich repeats, in Toll receptors, 107–108
Leukemia, acute monoblastic, 7
Leukocyte adhesion deficiency
 chemokine defects in, 93
 integrin dysfunction in, 137, 138, 143
 neutrophil dysfunction in, 8, 9
Leukotrienes
 definition of, 269
 function of, 93–94, 269
 in integrin activation, 141
 in neutrophil chemotaxis, 8–9
 switching to lipoxin synthesis, 269
LidA protein, in *Legionella pneumophila* infections, 400
LIGHT/TNFSF14, 331–332
LIM kinase, function of, 307
Lipid A, as Toll-like receptor agonist, 109, 111–113
Lipid bilayer, in phagocytosis, 200
Lipid mediators, 267–280
 biosynthesis of, 270–272
 lipoxins, 269–270, 275, 277
 protectins, 272–276
 resolvins, 272–275
 types of, 269
Lipid rafts, of DC-SIGN, membrane compartment interactions with, 126
Lipopeptides, as Toll-like receptor agonists, 109
Lipophosphoglycans, in *Leishmania* infections, 458
Lipopolysaccharide
 Brucella, 428, 429, 431–432
 in inflammasome activation, 160
 in integrin activation, 141
 Legionella pneumophila, 397–398
 in macrophage activation, 301–306
 pentraxin interactions with, 172
 in septic shock, 334
 Shigella, 411
 Toll-like receptors and, 108–109, 111–113, 303–305
Lipoprotein(s)
 low density, scavenger receptors for, 29
 macrophage receptors for, 32
 as Toll-like receptor agonists, 109
Lipoteichoic acid, in macrophage function, 30
Lipoxins
 in apoptosis, 354
 deficiencies of, 271–272
 function of, 269–270
 in pathogen evasion, 275, 277
 in resolution, 269–270
 types of, 269
Lipoxygenases, definition of, 269
Listeria
 infections due to, in *Drosophila*, 518
 survival mechanisms of, 356
 vacuolar escape by, 540
Listeria monocytogenes
 genome-wide screen for, 538–540
 granuloma formation in, 315
 inflammasome interaction with, 158, 160
 interferon response from, 156
 in macrophages, 39–41
 pattern recognition signaling in, 162
 peptidoglycan fragments of, 157
 survival mechanisms of
  acid resistance, 226–227
  cytosolic sensing escape, 161
  effector secretion, 153
  nitric oxide inhibition of, 371
Listeriolysin O, 227
Liver disorders, protectins in, 276
LOX-1 (lectin-like oxidized low-density lipoprotein receptor 1), in apoptosis, 349
L-plastin, in integrin signaling, 144
LPS receptor complex, 303
LRP (LDL receptor-related) proteins, in apoptosis, 345, 348
L-SIGN
 in antigen uptake, 129–130
 function of, 123–125
 multimerization of, 125
Lung, dendritic cells in, 55–56
Lymph nodes
 dendritic cells of, 56–57
 macrophages of, 28
Lymphotactin, 8–9, 94
Lyn protein
 in Fc receptor function, 76–77
 function of, 200
Lyso-phosphatidylcholine, in apoptosis, 345
Lysosomal hydrolases, function of, 313
Lysosomal-associated membrane proteins (LAMPs), 199, 214, 217, 250, 309
Lysosome(s)
 acidification of, 225–226
 constituents of, 250
 defects of, 313
 in phagosome formation, 199
 phagosome fusion with, 313
 in *Trypanosoma cruzi* infections, 454
Lysozyme
 in *Candida* infections, 442
 of neutrophils, 19

M

M cells, dendritic cells in, 54, 55
Mac-1 integrin receptor, in neutrophil chemotaxis, 9–10
Macroglobulin complement related receptor, *Drosophila*, 517
Macrophage(s), 27–50
 activation of, 28, 101, 326–328
  adenosine in, 331
  alternate pathways for, 328–333
  in apoptosis, 330–331
  classical, 301–323, 327–328
  in disease, 334–336
  glucocorticoid stimulation of, 331–332
  ingestion in, 307
  interferon-γ pathway in, 302–303
  interleukin-4 in, 331–332
  lipopolysaccharide in, 303–306
  macrophage phase of, 306–316

macrophage-stimulating protein in, 331
parasite product effects on, 484
physiology of, 301
pre-macrophage stage of, 302–306
priming and, 301–302
type II activated macrophage in, 332–334
adhesion of, 30
adrenal, 28
alveolar, 28, 394–395
amoebae as models for, see *Dictyostelium discoideum*
antigen presentation to, 38, 313–314
antimicrobial mechanisms of, 37–38
in apoptosis, 30, 32, 352–353
in bone (osteoclasts), 28, 144
chemokines of, 100–101
chemotaxis by, see Chemotaxis; Migration
complement receptor interactions with, 307
cross-presentation to, 38, 314
cytokines of, 281–299
effects of, 289–291
in immune response regulation, 285–291
production of, 315–316
proinflammatory, 315–316
secretion of, 38–39, 282–285
deactivation of, 28
defects of, 307–308
definition of, 27, 301
derivation of, 325–326
dermal, 28
development of, 27–28
discovery of, 27
endomembrane dynamics in, 81
in epithelium (Langerhans cells), 28, 53
Fc receptor effects on, 307
function of, 306–316
genome-wide screens in, 537–543
helminth interactions with, 478–484
heterogeneity of, 27–29
inducible nitric oxide synthase localization in, 370
in infections, 39–42
Brucella, 428–433
Candida albicans, 40, 41
Francisella tularensis, 40, 41
Legionella pneumophila, 39, 40
Leishmania mexicana, 40, 41
Listeria monocytogenes, 39–41
Mycobacterium tuberculosis, 39, 40
Salmonella enterica serovar Typhimurium, 386–390
Shigella, 412–414
viral, 40–42
ingestion by, 307
interleukin-10 regulation by, 333–334
intracellular killing by, 312–313
in lamina propria, 28
in liver (Kupffer cells), 28
in lymph nodes, 28
lysosomal defects in, 313
microglial cells as, 28
nitrogen burst in, 312
particle internalization in, 308–309
pattern recognition receptors of, 282
peritoneal, 28
phagocytic action of, 34–35
phagosomes of, 199, 249–264
pleural, 28
precursors of, 325–326
receptors for, see Macrophage receptors
recruitment of, 27–29
in reproductive tissue, 28
respiratory burst in, 309–312
signal transduction by, 35–37
splenic, 28
thymic, 28
types of, 27–28, 326
in zebrafish model, 524–528
Macrophage colony-stimulating factor, integrin interactions with, 140
Macrophage galactose-type C-type lectin, 127
Macrophage mannose receptor
in antigen uptake, 129
as antitumor therapy target, 130–131
in cell adhesion and trafficking, 127–128
cytoplasmic tail motif of, 125, 126
multimerization of, 125
pathogen evasion of, 128
in self and nonself recognition, 127
Macrophage receptors, 29–34
with a collagenous structure (MARCO), 31
C-type lectins as, 32–33
cytokine interactions with, 37
humoral, 34
interaction of, 34
scavenger, 29–32
sensing, 33–34, 37
in signaling, 35–37
Macrophage-stimulating protein, 331
Macrophagy, 419
Macropinocytosis, 199, 205, 350–351
Maitotoxin, in inflammasome activation, 160
Major histocompatibility complex
in antigen presentation, 38, 313–314
defects of, 314
in T-cell activation, 60
Mal protein, in Toll-like receptor signaling, 13
Malta fever, 427
Mammalian diaphanous-related formin, in actin polymerization, 197
Mannose receptors
in *Candida* infections, 438–439, 444
C-type lectins as, 123–124
functions of, 204–205
in helminth infections, 479
of macrophages, 33
in phagosome maturation, 214
in zebrafish model, 528
Mannose-binding lectin
in apoptosis, 345, 347
pathogen evasion of, 129
MARCKS (myristoylated, alanine-rich C-kinase substrate), 202
MARCO (macrophage receptor with a collagenous structure), 31
Matrix metalloproteinases, 3–4, 95
Maturation
of dendritic cells, 176
of phagosomes, 195, 213–220
MAVS (mitochondrial antiviral signaling) protein
in nucleic acid sensing, 154
in pattern recognition receptor regulation, 162
mCRAMP (cathelicidin), in *Candida* infections, 442
MD-1 and MD-2 proteins, Toll-like receptor interactions with, 113
MDA (melanoma differentiation-associated gene) product, in nucleic acid sensing, 154–155
Measles virus, Toll-like receptor interactions with, 109
Mechnikov, Ilya, on phagocytosis, 27, 195
Melanoma, C-type lectin receptor-targeted therapy for, 130
Melanoma differentiation-associated gene (MDA) product, in nucleic acid sensing, 154–155
Melanomacrophages, in zebrafish model, 528–529
Membrane, in phagosome formation, 198–200, 210–212
Mer proteins, in apoptosis, 346, 349
Metchnikoff, Elie, on phagocytosis, 27, 195
MFG-E8 (milk fat globule protein E8), in apoptosis, 346
Microbicidal machinery, for phagosome, 213
Microglial cells, 28
Microphagy, 419
Micropinocytosis, 199
Microtubule-organizing center, in migration, 183
Microvilli, 210
Migration
integrins in, 146–147
mechanics of, 183
prostaglandins in, 56
Milk fat globule protein E8, in apoptosis, 346
Mitochondrial antiviral signaling (MAVS) protein
in nucleic acid sensing, 154
in pattern recognition receptor regulation, 162
Mitogen-activated protein kinases
in cytokine regulation, 283–284
in macrophage activation, 305
in macrophage receptor signaling, 35–37
in pathogen detection, 156–157
in *Shigella* infections, 411–412
subfamilies of, 283
in *Toxoplasma gondii* infections, 466, 470
Mitophagy, 419
Monoblasts, dendritic cells derived from, 52–53
Monocyte(s)
in atherosclerosis, 100
chemokines of, 98–100
dendritic cells derived from, 52–53
macrophages derived from, 325–326
migration of, 147
physiology of, 325–326
subsets of, 99–100
Toxoplasma gondii interactions with, 464–465
in zebrafish model, 525–528
Monocyte chemoattractant proteins, function of, 99–100
Monocyte colony-stimulating factor, in macrophage development, 27
Motility, actin; see Actin, polymerization of
MRP proteins, in *Candida* infections, 442

MUC1 protein, C-type lectin recognition of, 127
Muckle-Wells syndrome, pattern recognition receptor defects in, 163
Multivesicular bodies, in phagosome maturation, 216
Muramyl dipeptide, sensing of, 156
Murine leukemia virus, Toll-like receptor interactions with, 109
Mycobacterium, infections due to
  in *Drosophila*, 518
  in zebrafish, 530
Mycobacterium avium
  Dictyostelium discoideum interactions with, 502–504
  infections due to, in interferon-γ defects, 303
  survival mechanisms of, 309
Mycobacterium bovis, survival mechanisms of, 309
Mycobacterium fortuitum, genome-wide screen for, 53–54
Mycobacterium leprae, survival mechanisms of, 129
Mycobacterium marinum
  chemotaxis of, in zebrafish model, 527–528
  Dictyostelium discoideum interactions with, 502–504
  infections due to, in zebrafish, 530
Mycobacterium smegmatis, genome-wide screen for, 541
Mycobacterium tuberculosis
  in amoebae, 494–496
  granuloma formation related to, 315
  immune response to, 56
  infections due to, in zebrafish, 530
  inflammasome interaction with, 158
  interferon response from, 156
  macrophage receptors for, 33
  in macrophages, 39
  peptidoglycan fragments of, 157
  in phagosomes, 253, 257, 259–261
  survival mechanisms of
    acid resistance, 227, 230
    apoptosis avoidance, 356–357
    lipid mediator defenses, 277
    pattern recognition, 128
    phagosome maturation interference, 309
MyD88
  in *Brucella* infections, 432
  in inducible nitric oxide synthase gene transcription, 368–369
  in macrophage activation, 304–305
  in Toll-like receptor signaling, 13–14, 115, 116, 283
  in *Toxoplasma gondii* infections, 466
Myelodysplastic syndrome, 7
Myeloid dendritic cells, *see under* Dendritic cells
Myeloperoxidase
  in *Candida* infections, 441
  of neutrophils, 19
  in zebrafish model, 529
Myocardial infarction, pentraxin-3 in, 178
Myosin
  actin binding to, 246
  activation of, 307
  in *Cryptococcus neoformans* infections, 504
  in phagosome formation, 198

Myosin light chain kinase, 198
Myotubularins, in phagosome maturation, 216
Myristoylated, alanine-rich C-kinase substrate (MARCKS), 202
Myxoma virus, survival mechanisms of, 161

N
NADPH oxidase complex
  activation of, 81
  in *Candida* infections, 441
  Fc receptors and, 80
  in macrophage function, 37–38
  in neutrophil function, 19–20
  in respiratory burst, 309–312, 387; *see also* Respiratory burst
  in *Salmonella* infections, 387
  structure of, 309–310
NAIFs (neuronal apoptosis inhibitor proteins), in neutrophil function, 14
NALPs (NACHT-LRR and pyrin-domain-containing proteins)
  defects in, 162–163
  in inflammasomes, 158–160
  in macrophage function, 33–34, 37
  in neutrophil function, 14–15
  in *Shigella* infections, 411
NAP-2 protein, function of, 95
Natural killer cells, dendritic cell interactions with, 57, 59
Natural resistance-associated macrophage protein (Nramp1), 313
  in *Candida* infections, 442
  in *Legionella pneumophila* infections, 501
  in *Leishmania* infections, 459
  in *Salmonella* infections, 387
Necrotic cells, recognition of, 342–345
Neisseria, macrophage scavenger receptors for, 31
Neisseria meningitidis, pentraxin interaction with, 174
Nematodes, *see also* Helminths
  definition of, 477
  macrophage activation due to, 328
NEMO protein, 305–306
Neonatal Fcγ receptor n, 76
Neovascularization, pentraxin-3 in, 176
Neuronal apoptosis inhibitor proteins (NAIFs), in neutrophil function, 14
Neuronal nitric oxide synthase, 367–368
Neutropenia, 6–7
Neutrophil(s), 3–26
  activation of, 8–9
  adhesion of, 9–11
  apoptosis of, 20–21
  Brucella interactions with, 432
  chemokine receptors of, 94–98
  in chemotaxis, 8–9; *see also* Chemotaxis; Migration
  deficiency of (neutropenia), 6–7
  development of, 3–5
  elastase defects and, 6
  endomembrane dynamics in, 81
  extravasation of, 9–11
  granules in, 5, 19, 384
  immunoglobulin G receptors of, 17–19
  integrins of, 138–140
  life span of, 4
  microbicidal activity of, 19–20

  migration of, 146–147
  nomenclature of, 3
  pattern recognition by, 11–12
  as pentraxin-3 source, 173
  phagocytic action of, 19–20
  phagosomes of, 199
  respiratory burst in, 310–311
  rolling and streaming action of, 9–11
  Salmonella enterica serovar Typhimurium interactions with, 384–385
  Shigella interactions with, 414–415
  Toll-like receptors and, 5, 11–14
  in zebrafish model, 529
Neutrophil elastase
  defects of, in neutropenia, 6
  in neutrophil development, 3–4
Neutrophil extracellular traps (NETs)
  in *Candida* infections, 442
  in *Shigella* infections, 414
Nigericin, in inflammasome activation, 160
Nimrod receptor, *Drosophila*, 514–516
Nitric oxide, antimicrobial function of, 370–372
Nitric oxide synthase, types of, 367–368
Nitrogen burst, of macrophages, 312
Nitrogen compounds, reactive; *see* Reactive nitrogen intermediates
NLRPs, *see* NALPs (NACHT-LRR and pyrin-domain-containing proteins)
Nocardia brasiliensis, infections due to, 483
NOD proteins, *see* Nucleotide-binding oligomerization domain (NOD) proteins
NOD-like receptors, 14–17
  immune functions of, 116–117
  of macrophages, 33–34
  in *Shigella* infections, 411
Nodules, in helminth infections, 480–482
Nramp 1 (natural resistance-associated macrophage protein), 313
  in *Candida* infections, 442
  in *Legionella pneumophila* infections, 501
  in *Leishmania* infections, 459
  in *Salmonella* infections, 387
Nuclear factor-κB (NF-κB)
  activation of, 156–158, 305–306
  in cytokine regulation, 284
  function of, 153
  in inducible nitric oxide synthase gene transcription, 368–369
  in pathogen detection, 156–158
  in *Shigella* infections, 411–412
  in Toll-like receptor signaling, 115–116
  in *Toxoplasma gondii* infections, 470
Nucleation, in actin polymerization, 236–237, 242–243
Nucleic acids, sensing of, 154–156
Nucleotide binding domain, leucine-rich repeat containing (NLR) proteins, 153–156
Nucleotide-binding oligomerization domain (NOD) proteins, 14
  mutations in, 162–163
  in pathogen sensing, 161–162
  in pattern recognition receptor regulation, 162
  in peptidoglycan sensing, 156–158
Nutritional limitation, in *Candida* infections, 442

O
Omega-3 fatty acids, lipid mediators derived from; *see* Lipid mediators

Oocyte, pentraxin-3 in, 172, 176–177
"Open conformation," of integrins, 140–141
OPRIL (oxysterol-binding protein-related protein 1L), in phagosome maturation, 217
Opsonin(s)
  in apoptosis, 346–347
  in phagocytosis, 34–35
  receptors for, neutrophil interaction with, 17–19
Oral tolerance, dendritic cells and, 53–55
Organelles
  acidification of, 225–233
  formation of, 195–196, 199
  secretory, phagosomes as, 199–220
Ornithine, in *Schistosoma* infections, 481–482
OspG protein, in *Shigella* infections, 411–412
Osteoclasts, 28, 144
Osteopetrosis, in chloride channel defects, 229
Outside-in signaling, in integrin activation, 141–146
Oxidative burst, *see* Respiratory burst
Oxygen compounds, reactive, *see* Reactive oxygen intermediates
Oxysterol-binding protein-related protein-1L (OPRIL), in phagosome maturation, 217

P

P57, *Mycobacterium tuberculosis*, 227
p38 pathway, in macrophage receptor signaling, 35–37
PAK1 protein, in chemotaxis, 186
PAMPs, *see* Pathogen-associated molecular patterns (PAMPs)
Pannexins, in inflammasome activation, 160
Paracoccidioides brasiliensis, pentraxin interaction with, 174
Parasites, pathogen-associated molecular patterns of, 11
Parasitophorous vacuoles, 458–459, 464
Particle size, in phagocytosis, 196
Pathogen-associated molecular patterns (PAMPs), 153
  bacterial, 109
  in *Candida* infections, 437–438
  of macrophages, 29
  of microorganisms, 11–12
  protozoan-derived, 110–111
  in *Shigella* infections, 411
  Toll-like receptors as, 109–111
  viral, 109–110
Pathogenicity islands, *Salmonella*, 383
Pattern recognition receptors, 11–12; *see also* Toll-like receptors
  in *Candida* infections, 437–438, 445
  C-type lectin receptors as, 128–129
  cytosolic, 153
  defects of, autoimmune disorders in, 162–163
  of dendritic cells, 282
  fluid-phase, *see* Pentraxin-3
  of macrophages, 29–34, 282
  regulation of, 162
  in *Shigella* infections, 411
  types of, 282

Paxillin, in *Legionella pneumophila* infections, 500
PECAMs (platelet endothelial cell adhesion molecules)
  in apoptosis, 349–350
  in neutrophil action, 10
Pentraxin(s)
  in apoptosis, 347
  of macrophages, 34
  of neutrophils, 5
  structures of, 171
Pentraxin-3, 171–182
  in angiogenesis, 176
  as biomarker, 177–178
  complement interactions with, 173–174
  in dendritic cell maturation, 176
  in female fertility, 176–177
  function of, 173–177
  genes of, 172–173
  in inflammation, 175
  in innate resistance, 174–175
  preformed, 173
  regulation of, 172–173
  sources of, 172–173
  structures of, 171
  in tolerance, 175–176
Peptidoglycan(s), sensing of, 156–158
Peptidoglycan recognition proteins, 117, 515, 517
Periodontal disease, in lipoxin deficiency, 271
Peritoneal macrophages, 28
Peritonitis, protectins in, 276
Permission to eat, 206–207
Peroxyphagy, 419
Peste receptor, *Drosophila*, 514, 516
Peyer's patches, dendritic cells in, 54
pH, in organelles; *see also* Acidification
  maintenance of, 228–230
Phagocytosis
  definition of, 306–307
  overview of, 195–196, 306–316
Phagolysosomes, 196, 218–219, 226
Phagophores, 419–420
Phagosomal transporter proteins, in *Legionella pneumophila* infections, 396
Phagosomes, *see also* Phagolysosomes
  acidification of, *see* Acidification, of endosomes and phagosomes
  actin in, 197–198
  additional membrane for, 210–212
  construction of, 196–200
  definition of, 195
  early, 214–216
  endocytosis and, 212–213
  exocytosis and, 210–212
  formation of, 209–213
  functions of, 249–250
  hydrolysis in, 250, 252
  immune activation and, 254, 255–257
  late, 216–218
  Legionella pneumophila in, 394, 396–397
  lysosomal constituents of, 196, 218–219, 226, 250–251, 313
  lysosome fusion with, 313
  macropinocytosis in, 205
  maturation of, 195, 213–220, 249, 309
    in *Candida* infections, 439–440
    innate immune sensing and, 252–253
    interference with, 309
  membrane movements in, 198–200
  microbial machinery addition to, 213

molecules associated with, 196–197
Mycobacterium tuberculosis in, 257, 259–261
myosin in, 198
physiology of, 253, 255–257
receptors of, 196–197
regulation of, 200–207
Salmonella in, 257
as secretory organelle, 219–220
surface area increase of, 209–210
transcriptional profiling in, 257, 259–261
Phg proteins, function of, 498–499
PhoP/Q
  function of, 257
  in *Salmonella* infections, 388–389
Phosphatase and tensin homolog, function of, 187, 201–202
Phosphatidylcholine, in *Brucella* infections, 431
Phosphatidylinositol (3,4,5)-triphosphate, function of, 201–204
Phosphatidylinositol-3-kinase
  in autophagy, 420, 422–424
  in chemotaxis, 185–187
  function of, 200–201, 307
  inhibition of, 227
  in *Legionella pneumophila* infections, 500
  in NADPH oxidase activation, 81
  in phagosome maturation, 215, 216, 309
  types of, 186–187
Phosphatidylinositol-4-phosphate-5-kinase, in actin polymerization, 80
Phosphatidylserine(s), in apoptosis, 342–344
  antibodies, 346
  receptors for, 347–348
Phosphatidylserine-binding proteins, in apoptosis, 346
Phosphoglycans, in *Leishmania* infections, 458
Phosphoinositol 4,5-bisphosphonate binding region, 498
Phospholipase(s)
  activation of, 81–82
  function of, 201, 203, 307
  in respiratory burst, 310
Phospholipase A2, in chemotaxis, 188
Phospholipase C, in chemotaxis, 187–188
Phospholipids, modified, in apoptosis, 345
phox proteins, in neutrophil function, 19–20
Picornaviruses, survival mechanisms of, 161
PIKfyve, in phagosome maturation, 216
PIP kinase, in integrin activation, 142
PIR-B protein, in integrin signaling, 143
Placenta
  C-type lectin receptors in, 127
  macrophages in, 28
Plasmodium falciparum
  macrophage scavenger receptors for, 32
  Toll-like receptor interactions with, 110–111
Plastins
  in actin polymerization, 244
  in zebrafish model, 527
Platelet endothelial cell adhesion molecules (PECAMs)
  in apoptosis, 349–350
  in neutrophil action, 10

Platelet-activating factor
 in integrin activation, 141
 in neutrophil chemotaxis, 8–9
Platyhelminths, 477
Pleural macrophages, 28
Pneumocystis carinii
 macrophage receptors for, 33
 survival mechanisms of, 129
Pneumonia, *Legionella pneumophila; see*
   Legionella pneumophila, infections
   due to
Polarity, in chemotaxis, 188
Polymorphonuclear leukocytes, *see*
   Neutrophil(s)
Pontiac fever, 393
Poxvirus, survival mechanisms of, 161
(p)ppGpp, in *Legionella pneumophila*
   infections, 396
Pregnancy, pentraxin-3 in, 178
Premacrophages, in zebrafish model, 525
Priming, for macrophage activation, 301–302
Profilin
 in actin polymerization, 238–241, 247
 in *Toxoplasma gondii* infections, 466
Promastigotes, 455
Propeller structure, of integrins, 137, 139, 141
Prostaglandin(s)
 in dendritic cell migration, 56
 function of, 269
 switching to lipoxin synthesis, 269
 synthesis of, 271
 types of, 269
Protease(s), in *Candida* infections, 442
Protease-activated receptors, in integrin
   activation, 142
Protectins
 definition of, 267
 identification of, 272–274
 types of, 269
Protein kinase A, regulatory effect of, 206
Protein kinase C
 function of, 307
 in respiratory burst, 310
Protein S, in apoptosis, 346
Proteophosphoglycans, in *Leishmania*
   infections, 458
Protozoa
 Legionella pneumophila in, 395
 Toll-like receptor interactions with, 110–111
PrpA protein, in *Brucella* infections, 431
Pseudo-ICE protein, in neutrophil
   function, 14–15
Pseudomonas aeruginosa
 Dictyostelium discoideum interactions
   with, 505
 infections due to, in *Drosophila*, 518
 inflammasome interaction with, 158
 pentraxin interaction with, 174–175
 peptidoglycan fragments of, 157
 phagocytosis of, 78
 survival mechanisms of, 277, 371
Pseudomonas exotoxin, translocation of, 230
Pseudopods, 19, 80–81, 209, 235
Psoriasis, dendritic cells in, 53
Pyk2 protein, in integrin signaling, 143
Pyogenic arthritis and syndrome of
   pyogenic arthritis, pyoderma
   gangrenosum and acne (PAPA), 163

PYPAF proteins, *see* NOD-like receptors
Pyrin(s), in neutrophils, 14–17
PYRIN domain, of NOD-like receptors, 14–17
Pyroptosis, 158

R
Rab proteins
 function of, 309, 499
 in *Legionella pneumophila* infections, 398
 in phagosome maturation, 214–218, 309
Rabaptin, 214
Rabenosyn, 214
Rabex, 214
Rab7-interacting lysosomal protein
   (RILP), in phagosome maturation, 217
Rabring, in phagosome maturation, 217
Rac proteins
 in apoptosis, 342, 351
 in chemotaxis, 188
 defects of, 311–312
 Fc receptor interactions with, 79–80
 function of, 201, 202, 204, 307
 GTPases of, in neutrophil chemotaxis, 8–9
 in integrin signaling, 144–145
 in migration, 183
 in *Mycobacterium* infections, 504
 in neutrophil function, 19–20
 in *Shigella* infections, 408
Rag proteins, in *Brucella* infections, 430
RalF protein, in *Legionella pneumophila*
   infections, 399–400
RANTES (regulated on activation normal
   T expressed and secreted), 94
Rap1 protein
 in integrin activation, 142–143
 in integrin clustering, 141
 in neutrophil chemotaxis, 8
Rap-1-GTP-interacting adapter molecule
   (RIAM), in integrin activation, 143
RAPL (regulator of adhesion and cell
   polarization enriched in lymphoid
   tissues), in integrin clustering, 141, 143
Ras proteins, function of, 201
Reactive nitrogen intermediates
 in *Candida* infections, 441
 inducible nitric oxide synthase and, 367–378
 killing properties of, 230
 in macrophage function, 38, 312, 331
 macrophages producing, 328
 in *Salmonella* infections, 387
 in *Schistosoma* infections, 481–482
 in *Toxoplasma gondii* infections, 468–469
 in zebrafish model, 528
Reactive oxygen intermediates
 in *Candida* infections, 441
 in macrophage function, 37–38
 in neutrophil chemotaxis, 8–9
 of neutrophils, 19–20
 in respiratory burst, 309–312
 in *Salmonella* infections, 387
 in *Toxoplasma gondii* infections, 469
 in zebrafish model, 528
Receptors, for phagocytosis, 196–197; *see*
   also specific types

Recycling, of actin, 244
Recycling endosomes, 198–199, 225
Regulation
 of pattern recognition receptor, 162
 of pentraxin-3, 172–173
 of phagosomes, 200–207
 T-cell, 59–61
Regulator of adhesion and cell
   polarization enriched in lymphoid
   tissues (RAPL), in integrin
   clustering, 141, 143
Rel proteins
 in nuclear factor-κB activation, 305–306
 in *Toxoplasma gondii* infections, 470
RELM proteins, in helminth infections, 478–480
Repertaxin, 102
Resolution, of inflammation, lipid
   mediators in; *see* Lipid mediators
Resolvins (resolution phase interaction
   products)
 in animal disease, 274–275
 definition of, 267
 identification of, 272–274
 types of, 269
Respiratory burst
 adhesion-dependent, 140
 in *Candida* infections, 441
 of macrophages, 309–312
 in *Salmonella* infections, 387
 in *Toxoplasma gondii* infections, 469
 in zebrafish model, 528, 529
Retinal dehydrogenases, in dendritic cell
   homing, 54
Retinal disorders, protectins in, 276
Retinoic acid, in dendritic cell homing, 54
Retinoic acid-inducible gene (RIG)
   product, in nucleic acid sensing, 154–155
RFX protein, in antigen presentation, 313
Rheb protein, in autophagy, 422
Rho proteins
 in actin polymerization, 241–242, 246
 in apoptosis, 352
 in chemotaxis, 183, 188
 Fc receptor interactions with, 79–80
 GTPases of, in neutrophil chemotaxis, 8–9
 in integrin signaling, 144–145
 in migration, 183
 in phagocytosis, 307
 in *Shigella* infections, 410
Rho-activated kinase (ROCK)
 in chemotaxis, 188
 in integrin signaling, 144–145
Rhodobacter sphaeroides, 111
Rhoptry proteins, in *Toxoplasma gondii*
   infections, 464, 470
RIAM (Rap-1-GTP-interacting adapter
   molecule), in integrin activation, 143
Ricin, translocation of, 230
RIG (retinoic acid-inducible gene)
   product, in nucleic acid sensing, 154–155
RILP (Rab7-interacting lysosomal
   protein), in phagosome maturation, 217
Rip proteins
 in neutrophil function, 14–16
 in peptidoglycan sensing, 156

RNA
  genome-wide screens, 537–543
  sensing of, 154–155
Rolling and streaming, of neutrophils, 9–11
ROP proteins, in *Toxoplasma gondii* infections, 464, 470
Rous sarcoma virus, Toll-like receptor interactions with, 109
Rto proteins, in *Legionella pneumophila* infections, 501

S

Salmonella
  infections due to
    in interleukin-12 defects, 316
    in zebrafish, 530–531
  macrophage scavenger receptors for, 32
  in phagosomes, 257
  survival mechanisms of, 356, 371
Salmonella enterica serovar Arizonae infections, in zebrafish, 531
Salmonella enterica serovar Typhi, 381
  infections due to, in zebrafish, 530–531
  survival mechanisms of
    dendritic cell modification, 385–386
    macrophage cytotoxicity, 387
    neutrophil inhibition, 384–385
    type III secretion system, 382–383, 386–390
    virulence factors, 383
Salmonella enterica serovar Typhimurium, 381–392
  cell death due to, 158
  dendritic cell interactions with, 385–386
  immune response to, 383
  infections due to
    pathogenesis of, 381–383
    in zebrafish, 530–531
  in inflammasome activation, 161
  macrophage interactions with, 386–390
  neutrophil interactions with, 384–385
  pentraxin interaction with, 174
  survival mechanisms of
    acid resistance, 230
    apoptosis avoidance, 356–357
    effector secretion, 153
    macrophage cytotoxicity, 387
    nitric oxide inhibition of, 371
    phagosome maturation interference, 309
    virulence factors, 383–384
Salmonella pullorum, dendritic cell interactions with, 54
Salmonella-containing vacuoles (SCVs), 382–383, 386–390
SAPK (stress-activated protein kinase), in *Toxoplasma gondii* infections, 470
Sarcomas, pentraxin-3 in, 172
Scavenger receptors
  in apoptosis, 348–349
  Brucella, 428
  Drosophila, 514–516
  macrophage, 29–32
  Salmonella, 387
Schistosoma mansoni
  C-type lectin recognition of, 127
  growth retardation of, in arginine deficiency, 371
  macrophage interactions with, 479
  survival mechanisms of, 373

Schistosomiasis, 328, 477, 481–482
Scramblases, in apoptosis, 343
SdhA protein, in *Legionella pneumophila* infections, 400, 401
Sec proteins
  in *Brucella* infections, 428
  in exocytosis, 212
  in *Legionella pneumophila* infections, 398
Secreted protein acidic and rich in cysteine (SPARC), in macrophage activation, 332
Secretory leukocyte protease inhibitor, of neutrophils, 5
Selectin(s)
  in adherence, 146
  in integrin activation, 141–142
  in neutrophil action, 9
Selectin ligands, in neutrophil action, 9
Self-tolerance
  C-type lectin receptors in, 126–128
  pentraxin-3 in, 175–176
Semliki Forest virus, fusion protein of, 230
Sensing, of pathogens
  intracytosolic, 153–169
    autoinflammatory syndromes and, 162–163
    escape mechanisms in, 161–162
    by inflammasomes, 158–161
    nucleic acids, 154–156
    peptidoglycan fragments, 156–158
    regulation of, 162
  by macrophage receptors, 33–34, 37
Sepsis syndrome
  macrophage activation in, 334
  pentraxin-3 in, 178
  Toll-like receptors and, 107, 111–113
Sequestosome, in autophagy, 421
Serprocidins, in *Candida* infections, 442
Serratia marcescens infections, in *Drosophila*, 518
Severe acute respiratory syndrome coronavirus, survival mechanisms of, 128
SH2-containing inositol phosphatase (SHIP) protein
  in Fc receptor function, 77–78
  function of, 201–202
  in neutrophil function, 17
SHI pathogenicity islands, *Shigella*, 407
Shigella, 405–418
  infections due to
    epidemiology of, 405–406
    fatal complications of, 414
    models for, 406–407
    pathogenesis of, 405–406, 411
  neutrophil interactions with, 414–415
  survival mechanisms of
    macrophage death, 412–414
    type III secretion system, 407–412, 414–415
Shigella boydii, 405
Shigella dysenteriae, 405
Shigella flexneri
  in inflammasome activation, 160, 161
  interferon response from, 156
  macrophage scavenger receptors for, 31
  peptidoglycan fragments of, 157
  survival mechanisms of, 153
Shigella sonnei, 405
SHIP (SH2-containing inositol phosphatase)
  in Fc receptor function, 77–78

function of, 201–202
  in neutrophil function, 17
Shock, septic, *see* Sepsis syndrome
SHP proteins, in Fc receptor function, 78
Shwachman-Diamond syndrome, 6
SI-CLP (stabilin-1 interacting chitinase-like protein), 332
SidH protein, in *Legionella pneumophila* infections, 400, 401
SidM (DrrA) protein, in *Legionella pneumophila* infections, 400
SifA protein, in *Salmonella* infections, 389
SIGN family, *see also* DC-SIGN (CD209); L-SIGN
  pathogen evasion of, 129
  Toll-like receptor interactions with, 129
Signal transducer and activator of transcription, *see* STAT (signal transducer and activator of transcription)
Signaling
  in apoptosis, 205, 351–352
  in autophagy, 422
  in chemotaxis, 184–188
  complement receptor in, 204
  constructive, 196
  for cytokine regulation, 283–285
  dectin-1 in, 205
  in endocytosis, 78
  Fc receptors in, 200–204
  feedback regulation of, 206–207
  in inducible nitric oxide synthase synthesis, 368–370
  integrin
    inside-out, 140–143
    outside-in, 141–146
  interpretive, 196, 207
  macrophage receptor, 35–37
  in macropinocytosis, 205
  mannose receptors in, 204–205
  pattern recognition patterns, 162
  in phagosome construction, 196–200
  in *Shigella* infections, 411
  Toll-like receptor, 115–116, 282–283
  in *Toxoplasma gondii* infections, 469–471
Sip proteins, in *Salmonella* infections, 385–386
Skin
  abnormalities of, in NEMO defects, 305–306
  dendritic cells in, 53
  macrophages in, 28
SKIP protein, in *Salmonella* infections, 389
Sky proteins, in apoptosis, 346
Sleeping sickness, African; *see* *Trypanosoma brucei*
SLPI (secretory leukocyte protease inhibitor), of neutrophils, 5
SNAP proteins, in exocytosis, 211
SNAREs (soluble N-ethylmaleimide-sensitive factor accessory protein receptors)
  in exocytosis, 210–212
  function of, 308–309
  in *Legionella pneumophila* infections, 398
  of neutrophils, 5
  in phagosome formation, 199
  in phagosome maturation, 214
SOCS-1 (cytokine signaling-1), in *Toxoplasma gondii* infections, 471
Sodium/hydrogen ion exchangers or antiporters, in pH maintenance, 230

Sodium/potassium ATPases, in pH maintenance, 229–230
Soluble β-1,2-glucans, in *Candida* infections, 440
Sorting endosomes, 214, 225–233
SPARC (secreted protein acidic and rich in cysteine), in macrophage activation, 332
Sphingomonas capsulata, macrophage scavenger receptors for, 31
Sphingosine kinase-1, production of, 331–332
SPI proteins, *Salmonella*, 383–390
Spleen
  dendritic cells of, 58
  macrophages in, 28
SPs, see Surfactant proteins
SR-B1 (CLA-1), as macrophage receptor, 31–32
Src tyrosine kinases
  in Fc receptor function, 76–77
  function of, 200
  in integrin signaling, 143–144
SREBPs (sterol regulatory element binding proteins), in inflammasome activation, 161
Sse protein, in *Salmonella* infections, 389
Stabilin(s)
  in apoptosis, 348
  as macrophage activation marker, 332
Stabilin-1 interacting chitinase-like protein (SI-CLP), 332
Staphylococcus aureus
  genome-wide screen for, 539, 540
  infections due to
    in chronic granulomatous disease, 312
    in zebrafish, 532
  inflammasome interaction with, 158
  macrophage scavenger receptors for, 32
  pentraxin interaction with, 174
  survival mechanisms of, 129, 371
STAT (signal transducer and activator of transcription)
  in inducible nitric oxide synthase gene transcription, 368
  in interferon signaling, 303
  in macrophage activation, 327
  in *Toxoplasma gondii* infections, 469–471
Sterol regulatory element binding proteins, in inflammasome activation, 161
Streptococcus
  infections due to, in zebrafish, 531
  macrophage scavenger receptors for, 31
Streptococcus iniae infections, in zebrafish, 531
Streptococcus pneumoniae
  infections due to, in zebrafish, 531
  macrophage receptors for, 31, 33
  pentraxin interaction with, 174
  peptidoglycan fragments of, 157
  survival mechanisms of, 129
Streptococcus pyogenes infections, in zebrafish, 531
Stress-activated protein kinase, in *Toxoplasma gondii* infections, 470
Stroke, protectins in, 276
Stromal-derived factor, function of, 96
Superoxide anion, in respiratory burst, 310

Surfactant proteins
  in apoptosis, 347, 349
  in cell adhesion and trafficking, 128
  in chemotaxis, 128
  in Fc receptor function, 78–79
  of macrophages, 34
  pathogen evasion of, 128–129
  structures of, 123
  Toll-like receptor interactions with, 129
Survival mechanisms, of pathogens; see *specific pathogens*, survival mechanisms of
Syk protein
  in actin polymerization, 79
  in Fc receptor function, 76–77
  in integrin signaling, 143–144
Synaptotagmin, in exocytosis, 212
Syntaxin, in exocytosis, 211
Syt VII protein, 81

T
T cell(s)
  activation of, dendritic cells in, 59–60
  cytotoxic, 60
  helminths interactions with, 477–478
  hyporesponsiveness of, in helminth infections, 483–484
  regulation of, dendritic cells in, 60–61
T-cell receptors, integrin interaction with, 143
Tachyzoites, *Toxoplasma gondii*, 464
TACO (tryptophan-aspartate-containing coat protein), *Mycobacterium tuberculosis*, 227
TAK1 kinase, in Toll-like receptor signaling, 115
Talin, in integrin activation, 142
TAMs (tumor-associated macrophages), 335
Tank binding kinase, in nucleic acid sensing, 154
Tapeworms, 477
Target of rapamycin (Tor), in autophagy, 420, 422–423
Tep proteins
  Drosophila, 515, 517
  function of, 540
Testis, macrophages in, 28
"Tether and tickle" mechanism, of apoptosis, 342, 351
Tetraspanins, 126, 145
Thrombospondin, in apoptosis, 346–347
Thymosin, in actin polymerization, 238, 240
Thymus
  dendritic cells of, 57–58
  macrophages of, 28
Tim proteins, in apoptosis, 348
TIR domain, of Toll-like receptors, 12–13, 108, 115, 116, 283
TIRAP (TIR-associated protein), 283, 304
Tissue repair, in helminth infections, 482–483
TLRs, see Toll-like receptors
TNF receptor-associated factor (TRAF)
  in pattern recognition receptor regulation, 162
  in Toll-like receptor signaling, 115
Tolerance, 56
  central, 57–58
  of commensal organisms, in gastrointestinal tract, 53–55

  C-type lectin receptors in, 126–128
  pentraxin in, 175–176
Toll receptors, in *Drosophila* development and immune response, 107–108
Toll-interacting protein (Tollip), 115
Toll/interleukin-1 receptor (TIR) domain, 12–13, 108, 115, 116, 283
Toll-like receptors, 107–122, 196
  activation of, 108, 304
  adapter molecules of, 115–116
  agonists of, 109–111
  in *Brucella* infections, 431
  in *Candida* infections, 440, 441, 443–444
  cellular localization of, 111
  C-type lectin receptor interactions with, 126
  of dendritic cells, 55
  discovery of, 107
  in DNA recognition, 113–115
  family of, 108
  functions of, 111–116
  immunologic functions of, 116
  in inducible nitric oxide synthase gene transcription, 368–369
  lipopolysaccharide and, 108–109, 111–113, 303–305
  in macrophage activation, 28, 33, 36, 37
  molecules detected by, 282–283
  of neutrophils, 5, 11–14
  in phagosome maturation, 252–253
  in *Salmonella* infections, 384, 387
  in *Shigella* infections, 411
  signaling through, 115–116, 282–283
  structures of, 111–116, 303–304
  TLR4
    function of, 108–109
    lipopolysaccharide recognition by, 111–113
  TLR9 (endosomal TLR), 113–115
  in *Toxoplasma gondii* infections, 466
  in zebrafish model, 528
Tor (target of rapamycin), in autophagy, 420, 422–423
Toxoplasma gondii, 463–476
  arginine transporters in, 482
  distribution of, 463
  infections due to, 463
  intracellular niche of, 464
  microbiology of, 463
  phagocyte interactions with, 464–471
    cell-mediated immunity in, 465–467
    downmodulating signaling in, 469–471
    early, 464–465
    effector mechanisms in, 467–469
    survival mechanisms of, 275, 277, 469–471
    Toll-like receptor interactions with, 110–111
TRAF proteins
  in cytokine regulation, 285
  in macrophage activation, 304–305
TRAM (TRIF-related adaptor molecule), 283
  in macrophage activation, 304
  in Toll-like receptor signaling, 13, 115
Transferrin, internalization of, 225
Transforming growth factor-β
  apoptosis and, 330–331, 355
  function of, 291

in inducible nitric oxide synthase regulation, 370
integrin regulation of, 140
Transforming growth factor-β-activated kinase (TAK), in peptidoglycan sensing, 156
trans-Golgi network, in acidification, 225
Transporter associated with antigen presentation-1, 313
Treadmilling, in actin polymerization, 237
T-reg cells, generation of, 60–61
in gastrointestinal tract, 54–55
in lung, 56
TREM proteins
in apoptosis, 349
in neutrophil function, 17
TRIF protein, in Toll-like receptor signaling, 13
TRIF-related adapter molecule (TRAM), 115, 283
Triggering receptor expressed on myeloid cells (TREM), 17, 349
Tropomyosin, actin binding to, 246
Trypanosoma, Toll-like receptor interactions with, 110
Trypanosoma brucei
recognition of, 457–458
release of, 459
survival mechanisms of, 373
Trypanosoma cruzi
flagellated form of, 455–456
growth retardation of, in arginine deficiency, 371
intracellular amastigote form of, 454–455
invasion of, 453–454
motility of, 453–454
survival mechanisms of, 356, 373
Trypanosoma gambiense, 457
Trypomastigotes, 454, 455
Tryptophan-aspartate-containing coat protein (TACO), Mycobacterium tuberculosis, 227
Tumor(s)
C-type lectin receptor-targeted therapy for, 130–131
macrophages associated with, 335
Tumor necrosis factor-α
in apoptosis, 354
in Brucella infections, 432
family members of, 314–315
function of, 287–288, 314–316
in helminth infections, 478
in inducible nitric oxide synthase gene transcription, 368–369
in integrin activation, 141
in macrophage activation, 327–328
of macrophages, 282
production of, 287–289
receptors for, 315
regulation of, 315
Tyk2 protein
deficiency of, 316
in interferon signaling, 303
Type II secretion system, Legionella pneumophila, 396
Type III secretion system
Pseudomonas aeruginosa, 505
Salmonella, 383–390
Shigella, 407–412, 414–415
Type IV secretion system
Brucella, 430

Legionella pneumophila, 398–400
Type VI secretion system, Vibrio cholerae, 506
Tyrosine kinases, see also specific kinases
in chemotaxis, 184–186
in integrin signaling, 143–144

U

Ubiquitin-interaction motif, in phagosome maturation, 216
Ubiquitylation, in endocytosis, 78
Ulcerative colitis, macrophage activation in, 335–336
Undulant fever, 427
Urokinase plasminogen activator receptor, integrin interaction with, 146
Uropods, 235
Uterus, macrophages in, 28

V

Vaccination, antitumor, C-type lectin receptor targeted in, 130–131
Vaccinia virus, sensing of, 154
Vacuolar ATPase, in pH maintenance, 228, 313
Vacuoles
Brucella-containing, 428
parasitophorous, 458–459, 464
Salmonella-containing, 382–383, 386–390
Vacuolin, in Mycobacterium infections, 503
VAMPs (vesicle-associated membrane proteins)
in exocytosis, 211–212
of neutrophils, 5
in phagosome formation, 199
Variable surface glycoprotein coating, Trypanosoma brucei, 457
Vascular cell adhesion molecule-1
integrin affinity for, 146
in neutrophil development, 3–4
Vasculitis, pentraxin-3 in, 178
Vasodilator-stimulated phosphoprotein, in actin polymerization, 239, 241, 247
Vav proteins, in neutrophil chemotaxis, 8
Vesicle-associated membrane proteins (VAMPs), of neutrophils, 5
Vibrio cholerae, Dictyostelium discoideum interactions with, 505–506
Villin, in actin polymerization, 244
Vinculin, in actin polymerization, 239, 241, 247
Vir proteins
in Brucella infections, 429–430
in Shigella infections, 408, 410
Virophagy, 419
Virulence, adaptation for; see specific pathogens, survival mechanisms of
Virulence factors
Acinetobacter baumanni, 506
Brucella, 429–431
identification of, Dictyostelium discoideum model for, 505
Legionella pneumophila, 395
Pseudomonas aeruginosa, 505
Salmonella enterica serovar Typhi, 383
Salmonella enterica serovar Typhimurium, 383–384
Vibrio cholerae, 505–506

Viruses
fusion proteins of, 230
infections due to, in zebrafish, 532
in macrophages, 40–42
pathogen-associated molecular patterns of, 11
survival mechanisms of, 356
Toll-like receptor interactions with, 109–110
Visceral leishmaniasis, 456
Vitamin B_{12}-binding protein, of neutrophils, 19
Vitiligo, 163
Vitronectin receptor, integrin interaction with, 142
VjbR quorum-sensing regulator, in Brucella infections, 430–431

W

WASP-interacting protein (WIP), 307
Water supplies, Legionella in, 393–394
WAVE
in actin polymerization, 197
function of, 201
Wiskott-Aldrich syndrome protein (WASP)
in actin polymerization, 197, 239, 246
in chemotaxis, 186
deficiency of, 141, 307–308
function of, 201, 307
in integrin signaling, 144
in Shigella infections, 409
Worms, see Helminths

X

Xenophagy, 419
X-linked disorders
chronic granulomatous disease, 311–312
NEMO defects, 306
Wiskott-Aldrich syndrome, 307–308

Y

Yersinia, survival mechanisms of, 161–162
Ym proteins, in helminth infections, 478–479

Z

Zebrafish model, 523–536
dendritic cells in, 529
immune system and, 524
of infections, 529–532
macrophages and monocytes in, 524–528
melanomacrophages in, 528–529
neutrophils in, 529
strengths of, 523–524
weaknesses of, 523–524
Zinc sequestration, in Candida infections, 442
"Zipper mechanism"
in Candida infections, 438
in parasitic infections, 457
Zoonotic infections, see Animal infections
Zymosan
pentraxin interaction with, 174
for phagocytosis studies, 197
Zyxin, in actin polymerization, 239, 241, 247